# 42nd EPS Conference on Plasma Physics (EPS 2015)



Europhysics Conference Abstracts Volume 39E

Lisbon, Portugal
22-26 June 2015

Part 1 of 2

**Editors:**

**R. Bingham**
**S. Atzeni**
**K. McClements**
**C. Silva**
**W. Suttrop**
**R. Foest**
**B. Goncalves**
**R. Coelho**

ISBN: 978-1-5108-2946-6

**Printed from e-media with permission by:**

Curran Associates, Inc.
57 Morehouse Lane
Red Hook, NY 12571

**Some format issues inherent in the e-media version may also appear in this print version.**



Printed by Curran Associates, Inc. (2016)

For permission requests, please contact European Physical Society (EPS)
at the address below.

European Physical Society (EPS)
6 Rue des Freres Lumiere
68200 Mulhouse
France

Phone: 33 389 32 94 40
Fax: 33 389 32 94 49

www.eps.org

**Additional copies of this publication are available from:**

Curran Associates, Inc.
57 Morehouse Lane
Red Hook, NY 12571 USA
Phone: 845-758-0400
Fax: 845-758-2633
Email: curran@proceedings.com
Web: www.proceedings.com

# TABLE OF CONTENTS

## PART 1

## PART 2

# Characterization of positive electrode biasing experiments in the tokamak ISTTOK with the Heavy Ion Beam Diagnostic

R.B. Henriques, A. Malaquias, C. Silva, H. Figueiredo, I.S. Nedzelskiy, H. Fernandes
*Instituto de Plasma e Fusão Nuclear, Instituto Superior Técnico, Universidade de Lisboa, 1049-001 Lisboa, Portugal*

## 1. Introduction

Edge plasma biasing experiments have been performed on several tokamaks demonstrating a clear correlation between the $E\times B$ flow shear velocity and turbulence control [1]. The control of the $E\times B$ flow shear layer is therefore an important tool to modify the transport in tokamaks and increase the confinement. Electrode biasing (EB) experiments have been extensively studied in ISTTOK [2]; however, the analysis was mainly focused on the plasma edge using Langmuir probes (LPs) data. The Heavy Ion Beam Diagnostic (HIBD) at ISTTOK was upgraded [3, 4] to perform local fluctuation measurements of the $n_e\sigma_{eff}(T_e)$ product on the core region along ~70% of the minor radius of the tokamak. The $n_e\sigma_{eff}(T_e)$ quantity is the product of the local plasma density ($n_e$) with the effective ionization cross-section of the double ionization process ($\sigma_{eff}(T_e)$) which depends on the local electron temperature ($T_e$). In this way, plasma fluctuations can be measured in the whole minor radius using LPs (edge) and the HIBD (core and inner edge) diagnostics together. This paper presents the evolution of the profiles and fluctuations during positive EB experiments in ISTTOK using the HIBD data.

## 2. Experimental set-up

ISTTOK is an iron core, large aspect ratio and circular cross-section tokamak ($R$=46 cm, $a$=8.5 cm, $B_T$=0.5 T). For the experimental results reported in this work, the typical values of the ISTTOK discharge parameters were: plasma current $I_p \approx 4$ kA and line averaged density $<n_e> \approx 4\times10^{18}$ m$^{-3}$.

A movable graphite electrode has been used to polarize the plasma edge and consequently control the radial profile of the plasma potential. Both electrode position and biasing voltage were optimized to maximize the improvement in confinement. Typically, the electrode is located 2.5 cm inside the last closed flux surface with a bias voltage of 80 V with respect to the vessel.

Two sets of Langmuir probes were used for this experiment: (i) a radial array (rake probe) to measure the radial profile of the floating potential consisting of a boron nitride head carrying nine tungsten tips and covering an extension of 27 mm; (ii) a poloidal array with three pins used mainly to estimate the turbulent particle flux at the plasma edge.

The HIBD performs measurements of $n_e\sigma_{eff}(T_e)$ at $-0.7a < r < 0.7a$ with a spatial resolution of ~1 cm in 12 sample volumes along the primary beam trajectory. The detection time difference between the first and last sample volume is ~ 0.5μs. Since the sampling period of ISTTOK data acquisition system is 0.5μs, the measurement of a whole $n_e\sigma_{eff}(T_e)$ profile can be considered instantaneous. In order to remove the background noise from the HIBD signals, the ion beam can be modulated up to 150 kHz. Fast beam modulation ($f_{mod}$ > 5 kHz) is mainly used for $n_e\sigma_{eff}(T_e)$ profile measurements, while slow beam modulation ($f_{mod}$ < 1 kHz) is used for $n_e\sigma_{eff}(T_e)$ fluctuations measurements. When fast beam modulation is used, the HIBD signals are down-sampled to the modulation frequency ($f_{mod}$) after the background removing.

**3. Improved confinement (ISTTOK)**

Typical improved confinement regimes in ISTTOK are characterized by an averaged plasma density (<$n_e$>) increase, $H_\alpha$ radiation decrease, edge turbulent transport ($\Gamma_{E\times B}$) decrease and an averaged plasma temperature (<$T_e$>) decrease. Figure 1 presents the evolution of a proxy for the particle confinement (<$n_e$>/$H_\alpha$ ratio), the EB current, <$n_e$>, <$T_e$> and $H_\alpha$.

Positive EB leads to the reversal of the edge electric field and to an increase in the $E\times B$ shear flow near the LCFS. This is consistent with (i) the large reduction (~90%) of the turbulent transport presented in figure 3 and (ii) the improvement of particle confinement (~65%) illustrated in figure 1.

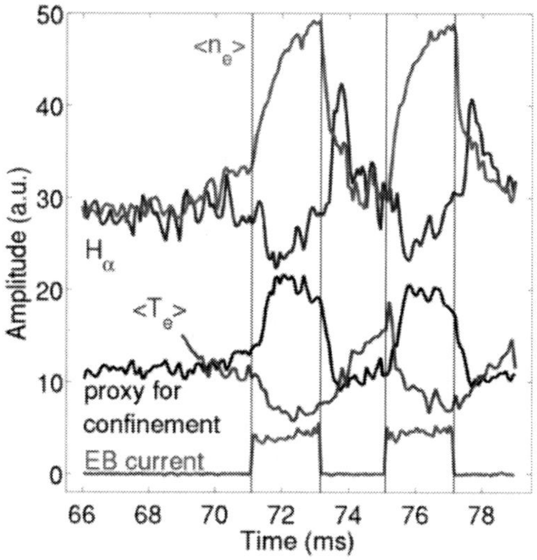


**Figure 1:** Time evolution of: electrode biasing (EB) current; $H_\alpha$ radiation; line integrated density <$n_e$> from the interferometer; line integrated temperature <$T_e$> retrieved from $n_e\sigma_{eff}(T_e)$; and, the proxy for particle confinement: <$n_e$>/$H_\alpha$ ratio.

**4. Profiles and time evolution**

Figure 2 presents the $n_e\sigma_{eff}(T_e)$ and $I_{sat}$ (ion saturation current measured by LP: $I_{sat} \propto n_e(T_e)^{1/2}$) profiles during normal and improved confinement plasma regimes. During EB, $n_e\sigma_{eff}(T_e)$ and $I_{sat}$ profiles change in a different way: the $n_e\sigma_{eff}(T_e)$ profile increases around 15% while the $I_{sat}$ profile decreases significantly. The result is a peaking of the $n_e\sigma_{eff}(T_e)$ profile and a flattening of the $I_{sat}$ profile, as observed in figure 2. These profile changes are in agreement with the transport barrier location, which is expected to be in the region nearby the electrode radial position.

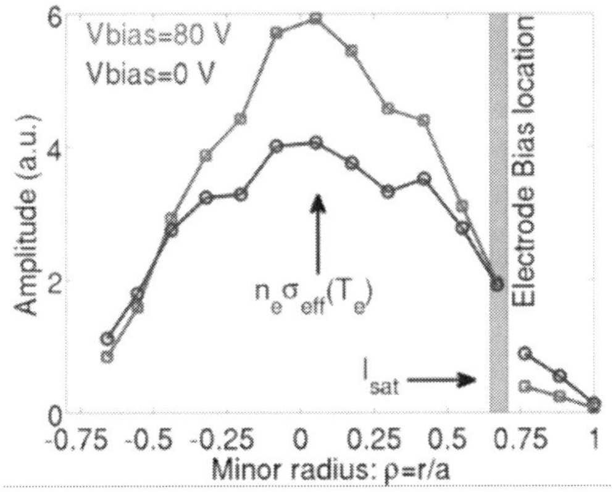


**Figure 2:** $n_e\sigma_{eff}(T_e)$ and $I_{sat}$ profiles averaged during 0.5ms on biased and non-biased plasma regimes. The electrode position is represented by the grey rectangle.

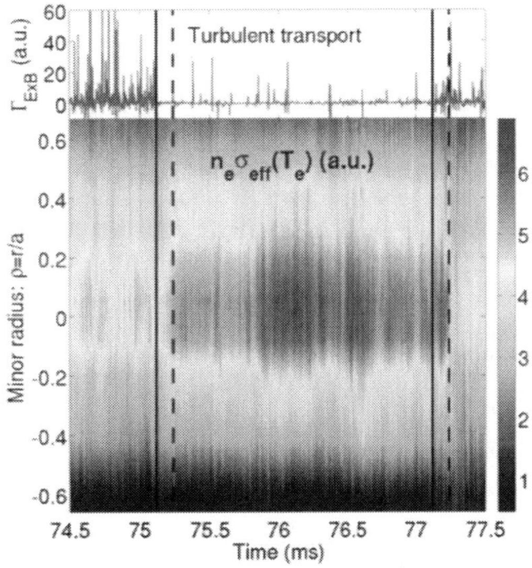


**Figure 3:** Time evolution of the edge particle flux transport (obtained by LPs at $r=a$) and $n_e\sigma_{eff}(T_e)$ profile during a whole electrode biasing cycle. The solid vertical lines indicate the start (left) and stop (right) timings of the electrode biasing.

The evolution of the turbulent transport and the $n_e\sigma_{eff}(T_e)$ profile during a full EB cycle is presented in figure 3. Just after EB transitions (solid vertical lines) the transport is modified within the ~10μs time scale (turbulence characteristic time). The $n_e\sigma_{eff}(T_e)$ profile major changes, indicated by the dashed lines in figure 3, only happen after ~150μs (particle confinement time). It is also observed that the $n_e\sigma_{eff}(T_e)$ profile reaches its maximum about 1ms after the EB is applied.

The line integrated temperature ($<T_e>$), presented in figure 1, was retrieved from the $n_e\sigma_{eff}(T_e)$ profile using simulation codes and assuming a profile for $n_e$: (i) normalized to $<n_e>$ (from the interferometer) and (ii) similarly shaped to the $n_e\sigma_{eff}(T_e)$ profile. The retrieved $<T_e>$ decreases during EB indicating that the improved particle confinement may not necessarily reflect an improvement on the total energy confinement.

**5. Fluctuations**

Figure 4 shows that the amplitude of fluctuations increases in the plasma core and decreases in the plasma periphery during EB. A similar behavior is observed for the signal averaged quantities (ratio of the profiles in figure 2). It has been observed that peripheral HIBD channels have a lower fluctuation level than LPs. This observation may be explained by the fact that HIBD is mainly sensitive to large fluctuations, ~1 cm, that are sheared apart during biasing, while LPs have a finer spatial resolution.

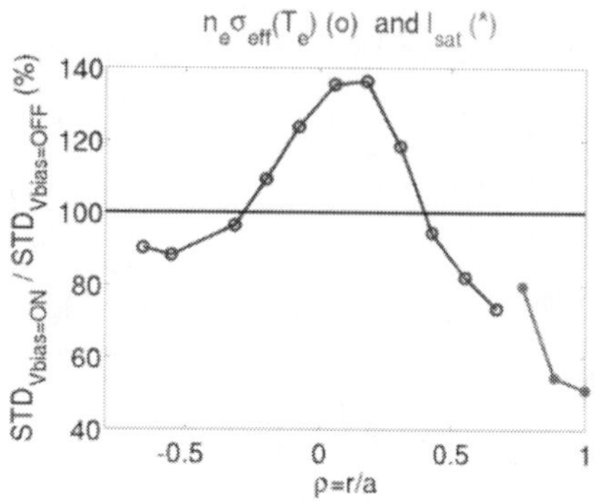


**Figure 4:** Profiles of the standard deviation (STD) ratio between the two EB regimes, ON and OFF, for both $n_e\sigma_{eff}(T_e)$ (blue) and $I_{sat}$ (red) signals.

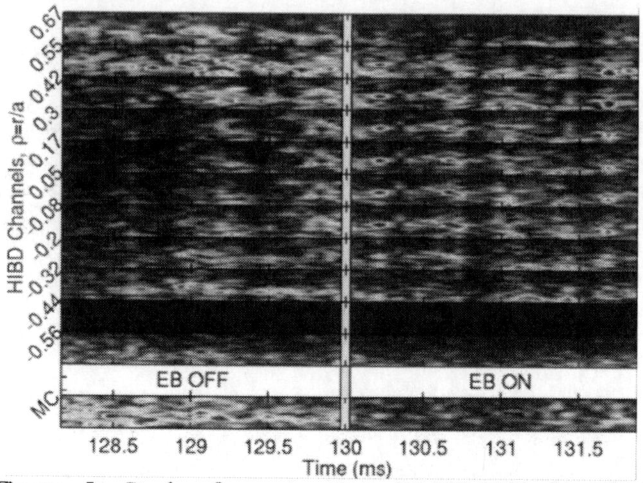


**Figure 5:** Stack of spectrograms (30-150 kHz) for HIBD channels (ch. at ρ=-0.44 not working) and Mirnov coil no. 9 (MC). The left half of the spectrograms is without biasing and the right half is with biasing.

During non EB, HIBD peripheral signals are usually dominated by MHD activity fluctuations (measured by Mirnov coils). When EB is applied, the MHD activity strongly decreases on both peripheral HIBD channels and Mirnov coils as illustrated in figure 5. It is also observed an increase of fluctuations in the central HIBD channels.

## Summary and conclusions

A full coverage of the plasma minor radius was obtained with the Heavy Ion Beam Diagnostic and the Langmuir probes in positive electrode biasing experiments. It has been observed that profiles change significantly during EB, by peaking in the center and flattening at the periphery, being consistent with the location of the edge transport barrier. The particle turbulent transport at the periphery is modified in a faster time scale (~10μs) than the $n_e\sigma_{eff}(T_e)$ core profile (~150μs). The amplitude of fluctuations increases in the core and decreases at the periphery during EB.

**Acknowledgements:** IST activities also received financial support from "Fundação para a Ciência e Tecnologia" through project UID/FIS/50010/2013. The views and opinions expressed herein do not necessarily reflect those of the European Commission.

**References**

[1] – G. Van Oost, et al, Plasma Phys. Control. Fusion, **45** (2003) 621

[2] – C. Silva, et al, Plasma Phys. Control. Fusion, **48** (2006) 727-744

[3] – R. B. Henriques, et al, Rev. Sci. Instrum., **83**, 10D705 (2012)

[4] – R. B. Henriques, et al, Rev. Sci. Instrum., **85**, 11D848 (2014)

# Influence of magnetic well on electromagnetic turbulence in the TJ-II stellarator

F. Castejón, A. M. Aguilera, E. Ascasíbar, T. Estrada, C. Hidalgo, A. López-Fraguas, M. A. Ochando, S. Yamamoto[a], A.V. Melnikov[b,c], L.G. Eliseev[b], S.V. Perfilov[b], and the TJ-II Team
Laboratorio Nacional de fusión. CIEMAT. Av Complutense 22. 28040 Madrid. Spain
[a]Institute of Advanced Energy, Kyoto University, Uji, Japan
[b]National Research Centre, NRC 'Kurchatov Institute', 123182, Moscow, Russia
[c]National Research Nuclear University MEPhI, Moscow, Russia

## 1.- Introduction.

Stellarator optimization takes into account several criteria, including neoclassical transport and stability (see e. g. [1]). Mercier criterion provides stability against interchange modes and is usually considered among the optimization criteria[2]. The magnetic well, related to the plasma shape, is a key stabilizing ingredient and, hence, the optimized configurations are usually complex and must be created by complicated coils, difficult to build and to maintain, which is especially relevant for a stellarator-based reactor, where the operations of maintenance must be performed using remote handling techniques. Therefore the relaxing of such a criterion can help to design simpler and cheaper stellarators.

The flexible heliac TJ-II allows one to vary the magnetic well, keeping constant the rotational transform profile, although the volume values of the magnetic configurations are reduced as the magnetic well decreases. Figure 1 shows the magnetic well and the rotational transform profiles and the volume values for the family of configurations used for these experiments.

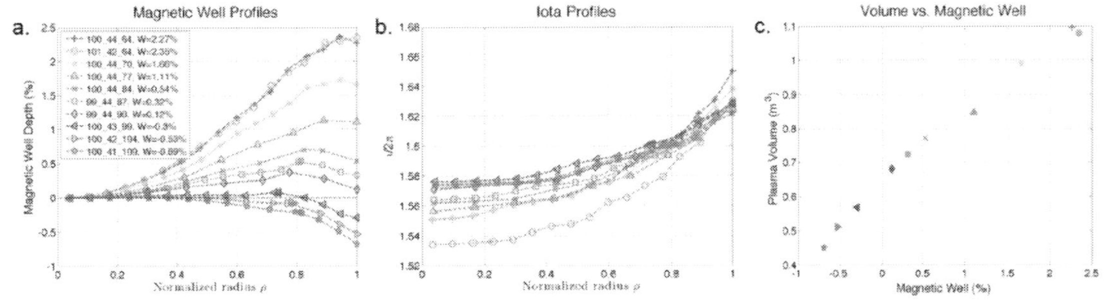


Figure 1: a) Profiles of magnetic well of the family of TJ-II configurations used for these experiments; b) rotational transform profiles and c) volumes of these configurations.

The previous magnetic well scan experiments on TJ-II have shown the existence of stable plasmas in Mercier-unstable magnetic configurations[3]. The position of LCFS is the same as the predicted by the theory and the confinement time presents a weak degradation for configurations with magnetic hill, as well as the electrostatic turbulence shows a moderate increase for those configurations[4]. In this work, we explore the electromagnetic properties of the turbulence during the magnetic well scan.

**2.- General Experimental Results**

The experimental set-up consists of Langmuir Probes, Mirnov coils, HIBP, arrays of bolometers and a Doppler reflectometer. In the configuration scan, Mirnov coils detect three groups of modes of different nature. First of all, it is detected a family of high frequency (above 100 kHz) modes together with others with middle frequencies (tens of kHz), depending on the NBI co- or counter-injection and the magnetic field structure of the configuration. The population of high frequency modes, being of Alfvénic nature, is much richer for counter injection. Regarding the intermediate frequencies, in some cases, a coherent mode appears with decreasing frequency as the magnetic well decreases. Finally, the family of configurations in this magnetic well scan is characterised for having the resonant value 8/5 of the rotational transform at $\rho\approx0.8$ in vacuum. The onset of an oscillation at $f\approx10$ - 20 KHz happens in several cases, showing a destabilization of the mode in configurations with low magnetic well or magnetic hill, although this behavior is not systematic.

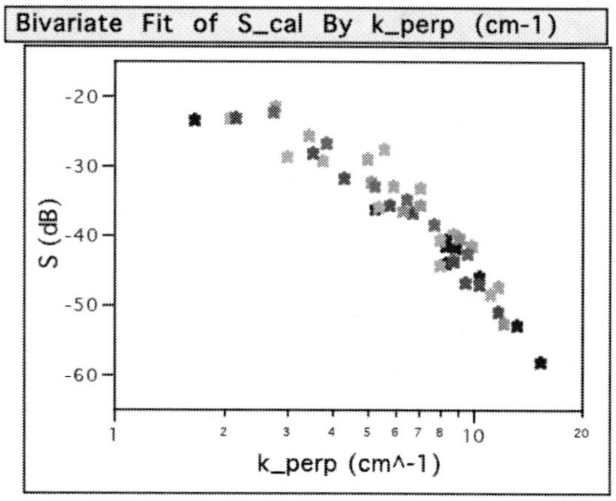


Figure 2. Wave vector spectrum of the explored family of configurations. The colours correspiond to the different configurations (see figure 3).

The broadband turbulence is measured by Doppler reflectometer on this configuration scan. For the positions detected by the reflectometer (0.7 < ρ < 0.9), it can be seen that the wave number spectrum is the same, for the same values of densities, in all the cases, as

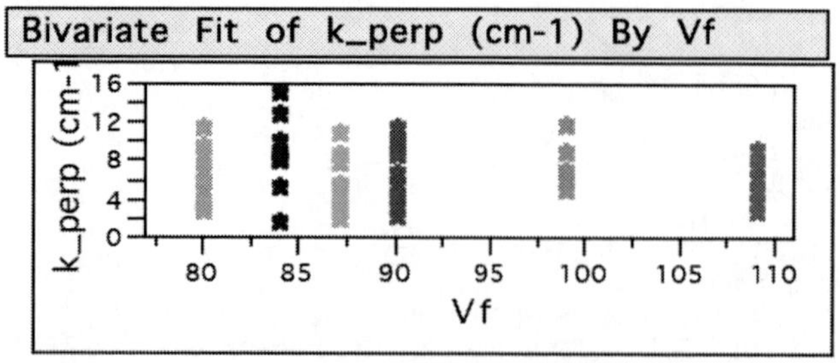


Figure 3. Explored values of the wave vector for the different configurations (Vf is the value of the current of the vertical field coil, the last number of the configuration name in Fig. 1).

can be seen in Figure 2. This fact shows that even in the configurations with magnetic hill, the structure of the turbulence is similar. The values of the wave vector explored with the reflectometer can be seen in Figure 3, showing that similar values of this vector are explored in all the cases.

**3.- Low Frequencies**

The family of configurations in this magnetic well scan is characterised for having the resonance 8/5 at $\rho\approx0.8$ in vacuum. The onset of an oscillation at $f\approx10$ - 20 KHz happens in several cases, showing that the mode creates a rotating island mainly in configurations with low magnetic well or magnetic hill, although this behaviour is not systematic. The behaviour of the island depends strongly on plasma current, which is able to move the resonance outwards in the plasma, when it takes negative values, or to inner positions for Ip>0. The onset of the island could have influence on the Alfvénic spectrum[5], since the structure of magnetic field is modified.

**4.- High and intermediate frequencies**

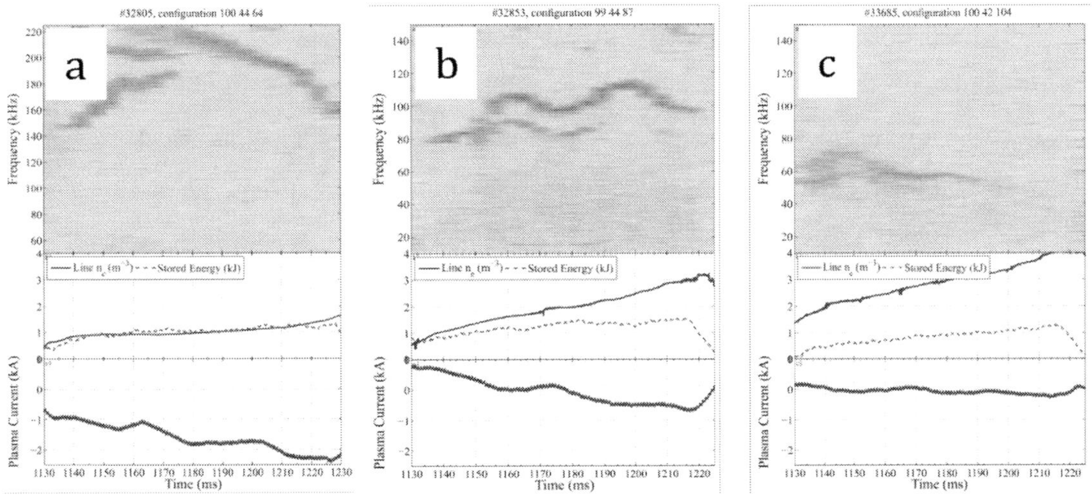


Figure 4. Spectrum of the turbulence measured by Langmuir probes, together with the evolution of plasma current, density and storage energy, in three configurations: a) 100_44_64 (W=2.27); b) 99_44_87 (W=0.32%); c) 100_42_104 (W=-0.69%).

Despite of the fact that the vacuum rotational transform is very similar in all the cases, the mode structure changes drastically when decreasing the magnetic well, showing a non-monotonic behaviour of the amplitude, and a decrease of the typical frequencies. Figure 4 a,b,c) shows the onset of the Alfvén mode for three configurations with co-NBI, the first one with deep magnetic well, the second with an intermediate value and the third one presenting magnetic hill. The mode is of Alfvénic nature in the three cases (The mode of spectrum 4a that appears between 1135 and 1170 ms at freq. between 145 and 180 is an artefact due to aliasing), since the frequency variation due to the changes in the density and the plasma current is given by[6]:

$$2\pi f = k_{\parallel} v_A \propto \frac{1}{\sqrt{n}} \left| m\iota - n \right| \approx \frac{1}{\sqrt{n}} \left| m(C(\rho) I_p + \iota_{vac}) - n \right| \sim \pm \frac{\left| I_p \right|}{\sqrt{n}}$$

Where Ip is the plasma current and C is a function with weak dependence of ρ, extracted from the experimental results. The changes observed from one configuration to another one must be attributed to the changes in the magnetic configuration, as it is shown by calculations performed with the code STELLGAP[7], which can justify the strong frequency variation from one configuration to another one, as well as the spatial location. Regarding the intermediate frequencies, a coherent mode appears for the cases with intermediate magnetic well with non-monotonic frequency evolution and mode location as the magnetic well decreases. The role of plasma current on this behaviour is under study. This mode has been observed previously and identified possibly as a GAM, which is driven by fast particle driven Alfvén modes[8].

**5.- Discussion and Conclusions**

The former studies of the effect of reducing magnetic well have shown a modest influence, if any, of magnetic well on global confinement. An increase of the level of electrostatic turbulence, detected by the Langmuir probes, happens for decreasing magnetic well. The level of fluctuations detected by Langmuir probes strongly depends on the magnetic configuration. The spectrum of wav e vector does not depend on magnetic well, showing the same rtole of the scales in the magnetic well scan. The typical frequencies of the Alfvén modes that appear in the co-NBI heating tend to decrease with the magnetic well, as does their radial postion. These findings show that the effect of magnetic well is very limited in the general confinement properties, although the changes in magnetic turbulence could affect the fast ion transport.

**Acknowledgements**

This work has been carried out within the framework of the EUROfusion Consortium and has received funding from the Euratom research and training programme 2014-2018 under grant agreement No 633053. The views and opinions expressed herein do not necessarily reflect those of the European Commission.

---

1 A. A. Subbotin et al. Nucl. Fusion 46 (2006 ) 921
2 F. Castejón et al. Plasma Phys. Control. Fusion 55 (2013) 014003
3 F. Castejón et al. Proc. of the 25th IAEA-FEC. Paper EX/P4-45. St. Petersburg, Russia, 2014.
4 A. M. Aguilera et al. "Magnetic well scan and confinement in the TJ-II stellarator". Submitted to Nucl. Fusion
5 C. R. Cook and C. C. Hegna. Physics of Plasmas 22, 042517 (2015)
6 A. Melnikov et al. Nucl. Fusion 54 (2014) 123002 (11pp)
7 D. A. Spong, R. Sanchez, A. Weller. Phys. Plasmas 10 (2003) 3217
8 T. Estrada et al. Nucl. Fusion xx (2015) xx

# Influence of resonant magnetic perturbations on transient heat load deposition – a short review

M. Rack[1], B. Sieglin[2], T. Eich[2], Y. Liang[1], I. Balboa[3], P. Denner[1], J. Pearson[1], A. Wingen[4] and JET Contributors*

*EUROfusion Consortium, JET, Culham Science Centre, Abingdon, OX14 3DB, UK*

[1]*Forschungszentrum Jülich GmbH, Institut für Energie- und Klimaforschung – Plasmaphysik, 52425 Jülich, Germany*

[2]*Max-Planck-Institute für Plasmaphysik, 85748 Garching, Germany*

[3]*CCFE Fusion Association, Culham Science Centre, Abingdon, OX14 3DB, UK*

[4]*Oak Ridge National Laboratory, PO Box 2008, Oak Ridge, Tennessee 37831-6169, USA*

## Introduction

Over the last decade, the macroscopic effects of resonant magnetic perturbations (RMPs) on edge localized modes (ELMs) have been studied in detail [1–4] and theories to explain their effects have been developed [5–8]. However, a full understanding has not yet been achieved. Especially, the study of the actual influence on the target energy deposition was restricted due to mechanical issues in the past. After a significant improvement of the infra-red viewing systems [9, 10] of the world's largest tokamak Joint European Torus (JET), this important aspect can now be studied.

## Dynamic structures prior to an edge localized mode crash

Applying RMP fields above a certain strength (at JET: $I_{\mathrm{EFCC}} > 2.5\,\mathrm{kA}$ ($\times 16$ turns), in $n = 2$ configuration) leads to a significant modification of the heat deposition on the outer divertor target: several ms before the major energy deposition of an ELM appears, heat flux structures propagating radially outward are seen. These structures form either near the original strike-line or at a distance of up to several cm away from it, depending on the plasma configuration. The distance depends on the magnetic topology, experimentally controlled by the edge safety-factor. Additionally, the propagation speed of the structures is found to be altered at different edge safety-factors. A relatively slow propagation between $7\,\mathrm{m\,s^{-1}}$ to $20\,\mathrm{m\,s^{-1}}$ is observed.

These structures appear as several parallel lines in the heat flux profile and propagate until the major energy deposition of the ELM reaches the target, see fig. 1 (a). The ELM energy

*See the Appendix of F. Romanelli et al., Proceedings of the 25th IAEA Fusion Energy Conference 2014, Saint Petersburg, Russia

deposition profile shows increased heat fluxes at the locations of the structures, which indicate that these pre-ELM structures directly affect the final ELM heat deposition. In some cases, a few of the structures even seem to continue to propagate during the major ELM energy deposition on the target, causing large heat fluxes.

A statistical analysis of the creation time of the pre-ELM structures with respect to the major energy deposition of the ELM gives a time delay of about 3 ms. This is much longer than any ELM timescale known before. Due to this long delay it appears impossible to understand the pre-ELM structures as rotating ELM filaments, which are reported to have a fast radial motion from $0.5\,\mathrm{km\,s^{-1}}$ to $2\,\mathrm{km\,s^{-1}}$ [12] and therefore much shorter lifetimes.

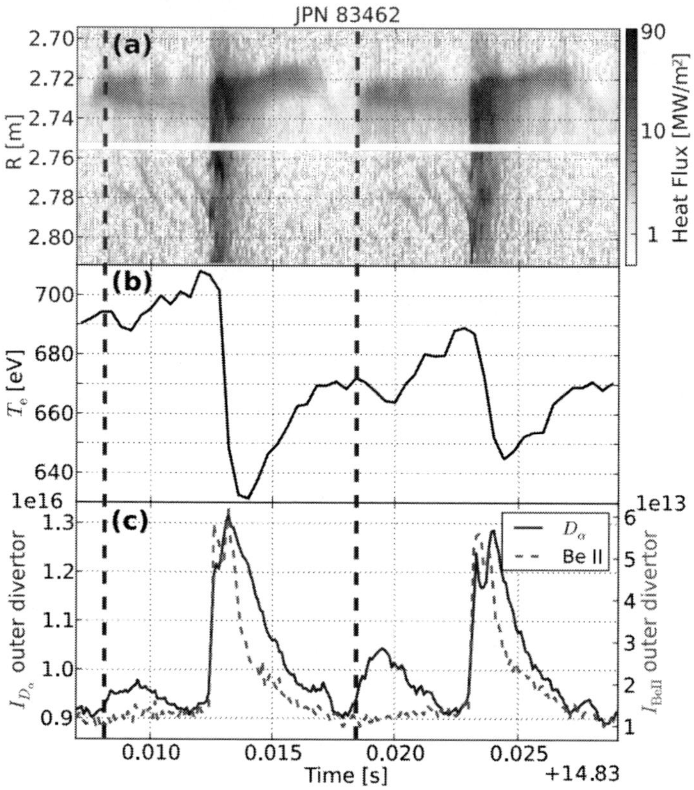


**Figure 1:** Two ELMs during the application of RMPs are shown: time-traces of the heat flux profile on the outer target (a), electron temperature at the plasma edge (b), and intensity of the $D_\alpha$ and Be II emission (c). Reprinted with permission from [11]. Copyright 2014, Euratom.

The pre-ELM structures are also accompanied by additional effects shown through different plasma parameters. A probable explanation for the observations is a reconnection process which triggers the formation of the observed structures. The stochastization of the plasma edge, caused by a reconnection process, would lead to the loss of fast electrons, seen by the drop in electron temperature when the pre-ELM structures are created (fig. 1 (b)) and the increased ion influx (seen via the $D_\alpha$ signal) with no effect on the Be II light emission (fig. 1 (c)). Furthermore, the loss of electrons can start a self-amplification process of thermoelectric currents as suggested by Evans [13], which finally results in the ELM crash. For testing this hypothesis, the thermoelectric current model [14,15] has been applied and compared to the experimental data. A good qualitative agreement was found, which further supports the hypothesis. In the next step, measurements of the thermoelectric currents are required; but those are beyond the capabilities of JET and need to be performed elsewhere, such as at

ASDEX Upgrade (AUG). The reader is referred to [11] for further details on the experimental measurements, its analysis and modelling.

**Heat load splitting during the major energy deposition**

The latest experiments with RMPs on JET enabled the further study of the previously found heat load splitting, during the ELM crash, on the DIII-D tokamak [16]. It is observed that theoretical predictions made, based on the DIII-D results, do not hold for the findings on JET and need further refinement.

Uncontrolled type-I ELMs show a random heat deposition on the divertor target during their crash [18]. Taking an average over a large number of these ELMs results in a smooth radial decay of the heat load along the target. In contrast, ELMs in the presence of RMPs appear to have radially predefined locations for the heat deposition. If RMPs are applied, the averaged heat deposition profiles, under the same conditions, show preferred heat deposition at distinct radial positions. This distinct heat deposition can be described as splitting during the ELM crash in reference to strike-line splitting during low-confinement mode (L-mode) operation. The structure of the splitting changes when the magnetic topology varies, for instance by a difference in the applied perturbation field or a changed edge safety-factor of the plasma.

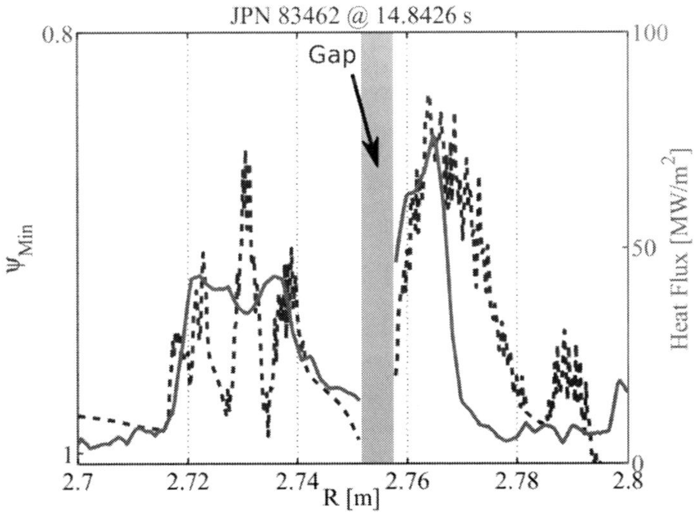


**Figure 2:** Comparison of the experimental heat flux profile (solid) with the predicted field line penetration depth, $1-\psi_{\mathrm{Min}}$, of the thermoelectric current model (dashed) during the peak heat deposition of one ELM. Reprinted with permission from [17]. Copyright 2014, Euratom.

For a better understanding, the magnetic topology for two ELM crashes at different perturbation strengths has been modelled based on the vacuum approach and compared to experimental observations. The predictions indicate the correct trend, but a precise comparison at the measurement location shows a strong discrepancy between measurement and prediction: a much stronger splitting is measured than predicted. Previous publications [14, 15] have demonstrated that consideration of additional thermoelectric currents explains the heat deposition of standard type-I ELM. This provides motivation for also applying the thermoelectric current model to controlled ELMs in the presence of RMP fields. Application of this extended model results in a good qualitative agreement between the experiment and predictions, see fig. 2. The reader is referred to [17] for further details on the experimental results and its modelling.

**Conclusions**

The new finding of pre-ELM structures extends the understanding of ELM control by RMPs: although RMPs lead to an increased ELM frequency, dynamic heat flux structures occur which are very slow compared to typical ELM time-scales. It appears that due to the RMPs, this dynamic process decouples from the major energy ejection. This may contributes to the further development of theories for the understanding of RMP ELM control.

Studying the strike-line splitting during ELM crashes lead to the discovery of an important aspect: due to the fact that a much larger splitting is observed than initially predicted by the simple vacuum approach, severe damage can occur if regions which are not prepared for such heat loads come into contact with the plasma. Therefore, it is crucial to consider an improved model – including thermoelectric edge currents – to predict the expected strike-line splitting.

**Acknowledgments**

M.R. is thankful for the help of Evgenij Bleile and Götz Lehmann. This work has been carried out within the framework of the EUROfusion Consortium and has received funding from the Euratom research and training programme 2014-2018 under grant agreement No 633053. The views and opinions expressed herein do not necessarily reflect those of the European Commission. Additional support from the Helmholtz Association in frame of the Helmholtz-University Young Investigators Group VH-NG-410 is gratefully acknowledged.

**References**

[1] T. E. Evans *et al. Physical Review Letters*, **92** (23), 235003 (2004).
[2] Y. Liang *et al. Plasma Physics and Controlled Fusion*, **49** (12B), B581 (2007).
[3] A. Kirk *et al. Plasma Physics and Controlled Fusion*, **53** (6), 065011 (2011).
[4] W. Suttrop *et al. Physical Review Letters*, **106**, 225004 (2011).
[5] J. Pearson *et al. Nuclear Fusion*, **52** (7), 074011 (2012).
[6] P. B. Snyder *et al. Physics of Plasmas*, **19** (5), 056115 (2012).
[7] I. T. Chapman *et al. Physics of Plasmas*, **20** (5), 056101 (2013).
[8] A. Kirk *et al. Plasma Physics and Controlled Fusion*, **55** (12), 124003 (2013).
[9] T. Eich *et al. Journal of Nuclear Materials*, **415** (1, Supplement), S856 (2011).
[10] I. Balboa *et al. Review of Scientific Instruments*, **83** (10), 10D530 (2012).
[11] M. Rack *et al. Nuclear Fusion*, **54** (7), 072004 (2014).
[12] A. Kirk *et al. Journal of Nuclear Materials*, **390-391** (0), 727 (2009).
[13] T. E. Evans *et al. Journal of Nuclear Materials*, **390-391**, 789 (2009).
[14] A. Wingen *et al. Physical Review Letters*, **104** (17), 175001 (2010).
[15] M. Rack *et al. Nuclear Fusion*, **52** (7), 074012 (2012).
[16] M. Jakubowski *et al. Nuclear Fusion*, **49** (9), 095013 (2009).
[17] M. Rack *et al. Nuclear Fusion*, **54** (6), 064012 (2014).
[18] T. Eich *et al. Plasma Physics and Controlled Fusion*, **47** (6), 815 (2005).

# Multi-tokamak scaling of main SOL heat flux width in limiter plasmas

J. Horacek[1], R.A. Pitts[2], J. Gunn[3], C. Silva[4], D. Rudakov[5], G. Arnoux[6], S. Marsen[7], P. Vondracek[1,8], G. Maddaluno[9], V. Pericoli[9], B. Viola[9], G. S. Xu[10], H. Wang[10], L. Nie[11], B. LaBombard[12], S. Brezinsek[13], Y. Xu[13], M. Shimada[2], J. Adamek[1], J.-G. Bak[14], Tsv. Popov[15], M. Dimitrova[1], J. Seidl[1], F. Janky[1], J. Havlicek[1,8] , R. Panek[1], R.J. Goldston[16], P.C. Stangeby[17]

[1] *Institute of Plasma Physics ASCR, Prague, Czech Republic*

[2] *ITER Organization, Route de Vinon-sur-Verdon, CS90046-13067, St Paul Lez Durance Cedex, France*

[3] *CEA, IRFM, F-13108 Saint-Paul-lez-Durance, France*

[4] *Instituto de Plasmas e Fusao Nuclear, IST, Av. Rovisco Pais, 1049-001 Lisbon, Portugal*

[5] *University of California, San Diego, Center for Energy Research, Fusion Division, 9500 Gilman Dr, Mail code 0417, EBU II, Rm 468, La Jolla, CA 92093-0417, USA*

[6] *JET-EFDA, Culham Science Centre, Abingdon, OX14 3DB, UK*

[7] *Max-PLanck-Institut fur Plasmaphysik, Teilinstitut Greifswald, 17491 Greifswald, Germany*

[8] *Faculty of Mathematics and Physics, Charles University, Czech Republic*

[9] *ENEA-UT Fusione, Centro Ricerche Frascati, Italy*

[10] *Institute of Plasma Physics, Chinese Academy of Sciences, Hefei 230031, China*

[11] *Southwestern Institute of Physics, Chengdu, China*

[12] *Massachusetts Institute of Technology, Plasma Science and Fusion Center, 175 Albany St., Cambridge, MA 02139, USA*

[13] *Juelich Forschungszentrum, Germany*

[14] *National Fusion Research Institute, Yusung, Daejeon, Korea*

[15] *Faculty of Physics, St. Kliment Ohridski University of Sofia, J. Bourchier Blvd., 1164 Sofia, Bulgaria*

[16] *Princeton, Plasma Physics Laboratory, Princeton NJ 08543, USA*

[17] *University of Toronto, Institute for Aerospace Studies, Toronto, M3H 5T6, Canada*

## Introduction

Discharge on the ITER tokamak will preferably start and end up with plasma touching the inboard (HFS) limiter before shaping into the divertor configuration. In order to avoid exceeding the designed limit of the Beryllium enhanced heat flux limiters ($4.7MW/m^2$), its shape has to be optimized in order to receive the power everywhere equally. The optimum shape has been derived based on observation that the SOL radial profile of the power flux decays mostly exponentially, $q_{||} = q_{0||}exp(-r/\lambda_q^{\mathrm{omp}})$. Therefore, a logarithmic shape

$$y = -\lambda_q^{\mathrm{omp}} \cdot \ln(1 - \frac{C \cdot x}{\lambda_q^{\mathrm{omp}}}), \tag{1}$$

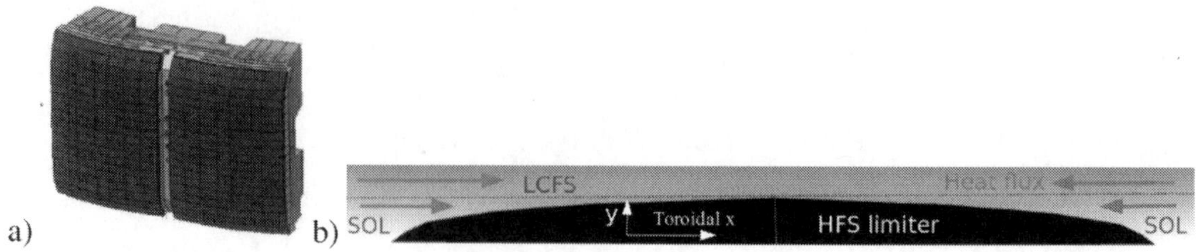


Figure 1: *a) 3D drawing of the ITER inner wall limiter [1]. b) Limiter optimized for constant power everywhere [1] for SOL plasma characterised by exponential decay length* $\lambda_q^{\text{omp}}$.

| | R [m] | a [m] | $I_p$[kA] | B [T] | data | heating | isotope | Meas. of $T_e$ | processed by |
|---|---|---|---|---|---|---|---|---|---|
| **ITER** | 6.0 | 2.0 | 3500-7500 | 5.3 | 3 | NBI | H,D,He | - | Shimada |
| **JET** | 2.8 | 0.98 | 1500,2500 | 2.8 | 33 | none | D | swept Langmuir | Silva, Horacek |
| **Tore Supra** | 2.2 | 0.65 | 500-1200 | 2.6-4.1 | 121 | none | D | tunnel + RFA | Gunn |
| **DIII-D** | 1.7 | 0.6 | 600-1200 | 1.9 | 23 | none | D | harmonic technique | Tsui, Rudakov |
| **C-Mod** | 0.68 | 0.22 | 400-1100 | 4-7 | 19 | none | D | scanning Mirror [6] | LaBombard |
| **KSTAR** | 1.78 | 0.47 | 400 | 1.99 | 1 | NBI | D | swept Langmuir | J.-G. Bak |
| **TEXTOR** | 1.73 | 0.46 | ± 200-400 | ± 1.3-2.6 | 55 | NBI | D | triple probe | Horacek |
| **EAST** | 1.85 | 0.46 | 300 | 1.96 | 2 | none | D | triple probe | G.S. Xu |
| **HL-2A** | 1.67 | 0.36 | 100-220 | 1.36 | 39 | ECH+NBI | D | triple probe | L. Nie, Wang |
| **FTU** | 0.94 | 0.28 | 250-500 | 2.7-7.5 | 3x9 | none | D | swept Langmuir | Maddaluno, Pericoli |
| **COMPASS** | 0.55 | 0.2 | 80-180 | 1.15 | 91 | none | H, D | swept Langmuir, BPP | Horacek, Seidl |
| **CASTOR** | 0.4 | 0.08 | 9 | 1.3 | 3 | none | H | swept Langmuir | ref. [2] |

Table 1: *Overview of parameters of the used tokamak plasmas. TEXTOR includes both directions of B and* $I_p$. *The data being processed by different persons and temperature measured by different techniques yield probably to systematic errors in* $\lambda_q^{\text{omp}}$.

shown in Figure 1, yields theoretically an almost constant power flux across the entire surface of the limiter. Scaling of $\lambda_q^{\text{omp}}$ is, however, available only for diverted plasmas, specified in the ITER Physics basis [4]. Experiments on Tore Supra demonstrated [5] that this scaling is not valid for circular limiter plasmas and that a single parameter determines $\lambda_q^{\text{omp}}$: the ohmic power.

Recently, a theory-based model capable of a credible prediction of $\lambda_q^{\text{omp}}$ has been developed [3], concluding that $\lambda_q^{\text{omp}}$ should scale only with $q_{95}$, $\rho^*$, $\nu_{gbs}$. Those parameters rely, however, on probe measurements of $T_e, n_e$ at LCFS which is subject of large error on current tokamaks (especially $T_e$ and exact localization of LCFS) as well as $n_e^{LCFS}$ in ITER is subject to factor of 2 uncertainty. Alternative (perhaps better) choice is thus a scaling based on engineering or those global plasma parameters well-known for ITER.

Scaling based on a single tokamak [5] is not sufficient for extrapolation to a larger tokamak, ITER. Therefore, we collected data from 450 probe strokes on 11 tokamaks worldwide, varying from the smallest plasmas (CASTOR with plasma volume 0.06 m$^3$, $I_p = 9$kA) up to JET with 70 m$^3$, $I_p = 2500$kA. Overview of the used tokamaks is compiled in Table 1.

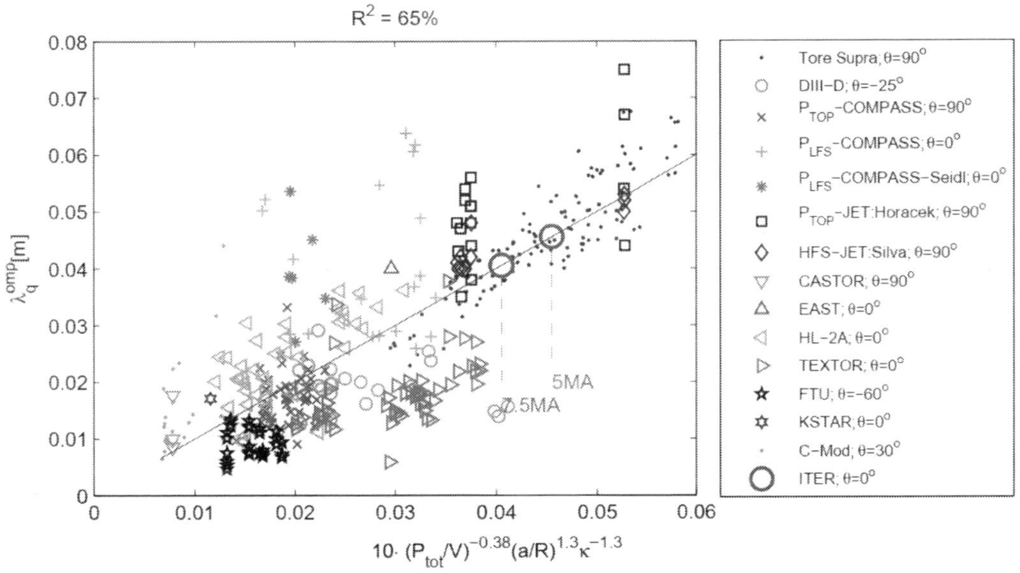


Figure 2: *An engineering scaling. The corresponding ITER predictions lie on the line.*

**Experimental results**

The experiments were performed with various reciprocating probes. Each measured radial profile of the ion saturation current density $j_{sat}$ and electron temperature $T_e$. The parallel heat flux is then $q \propto j_{sat}T_e$ and the desired $\lambda_q^{omp}$ is then the decay length of the radial profile fit.

16 global plasma parameters were measured at the same time, thus constructing a valuable matrix of $16 \times 450$ data-points. We use the statistical package Gretl to perform multi-parameter least-squares fit with robust standard errors, weighting each value of $\lambda_q^{omp}$ by the radial-profile fit quality ($R^2$ combined with $\lambda_q^{omp}/\lambda_q^{error}$ ratio).

We find that the dominant parameter of the scaling is either the total input power (consistent with [5]), $P_{tot}$, $P_{SOL}$, $I_p$, or LCFS plasma pressure $\beta$, which vary from one discharge to another within each tokamak. Due to natural high mutual correlation of $P_{tot}$, $P_{SOL}$, $I_p$, $\beta$ only one of those can be used in a particular scaling. All the other parameters then only slightly refine $\lambda_q^{omp}$ for a particular tokamak. Combinations with high mutual correlations must be excluded.

**Prediction for ITER**

A dozen of reasonable scalings are shown in Table 2 using various combinations of global plasma parameters. **Fortunately, we observe that those scalings yield very similar prediction for ITER, even though using very different parameter combinations. Mapped to inboard midplane (with flux expansion of 1.6), the worst case (7.5 MA) corresponds to optimum toroidal shaping of the FW panel with $\lambda_q^{omp} = 7 \pm 2$cm, thus consolidating the initial design**

| Scalings (dimensionless or in SI units) | $\lambda_q^{omp}$[mm] ITER prediction 2.5MA | 5.0MA | 7.5MA | $R^2$ |
|---|---|---|---|---|
| $72.1 \cdot q_{95}^{-0.81}(a/R)^{3.64}R[m]^{0.54}\nu_{gbs}^{-0.18}\rho^{*-1.13}I_p[A]^{-1}$ | $70 \pm 17$ | $49 \pm 12$ | $43 \pm 10$ | 77% |
| $q_{95}^{-0.72}(a/R)^{2.96}\nu_{gbs}^{-0.21}\rho^{*-1.24}I_p[A]^{-0.81}$ | $59 \pm 13$ | $44 \pm 10$ | $40 \pm 9$ | 73% |
| $0.00158 \cdot \beta^{-0.33}(a/R)^{0.86}\kappa^{0.59}$ | $57 \pm 13$ | $53 \pm 14$ | $43 \pm 11$ | 73% |
| $4.07 \cdot (I_p/A[A/m^2])^{-0.4}(a/R)^{1.54}\kappa^{-1.34}\rho^{*-0.29}$ | $76 \pm 21$ | $49 \pm 15$ | $44 \pm 12$ | 70% |
| $10 \cdot (P_{tot}/V[W/m^3])^{-0.38}(a/R)^{1.3}\kappa^{-1.3}$ | $67 \pm 19$ | $47 \pm 15$ | $42 \pm 13$ | 65% |
| $501 \cdot (P_{tot}/V[W/m^3])^{-0.28}(a/R)^{1.46}\kappa^{-1.11} < n_e > [m^{-3}]^{-0.11}$ | $76 \pm 22$ | $50 \pm 16$ | $44 \pm 13$ | 64% |
| $20.6 \cdot (a/R)^{2.04}R[m]^{0.94}\kappa^{-0.82}P_{tot}[W]^{-0.34}$ | $67 \pm 19$ | $47 \pm 15$ | $41 \pm 12$ | 63% |
| $7.08 \times 10^6 \cdot (I_p/A[A/m^2])^{-0.77}(a/R)^{1.61}\kappa^{-1.8} < n_e > [m^{-3}]^{-0.14}B_{pol}[T]^{0.48}$ | $93 \pm 30$ | $53 \pm 18$ | $48 \pm 15$ | 62% |
| $6.76 \times 10^4 \cdot (I_p/A[A/m^2])^{-0.89}(a/R)^{1.5}\kappa^{-1.89}B_{pol}[T]^{0.52}$ | $91 \pm 29$ | $55 \pm 18$ | $50 \pm 16$ | 61% |
| $0.00108 \cdot q_{95}^{0.29}\nu_{gbs}^{-0.1}\rho^{*-0.29}$ | $86 \pm 24$ | $70 \pm 19$ | $57 \pm 16$ | 60% |
| $1.72 \cdot (P_{tot}/V[W/m^3])^{-0.28}q_{95}^{0.17}(a/R)^{0.96}$ | $88 \pm 26$ | $73 \pm 21$ | $56 \pm 16$ | 58% |
| **Average** | $\mathbf{76 \pm 21}$ | $\mathbf{54 \pm 16}$ | $\mathbf{46 \pm 13}$ | |

Table 2: *The final scalings and consequent ITER predictions (the 95%-confidence intervals) of the SOL width at outboard midplane for the three start-up scenarios.*

**choice of 5 cm.** Note that much steeper gradients have been found in the region near separatrix ($\lambda_q^{omp}$ shorter by order of magnitude) at high-field side of tokamaks, described in [7]. Both predictions of the near and the here-scaled main SOL decay lengths are required for best ITER HFS limiter, designed finally in [8]. Details of this work can be found in [9].


## Acknowledgement

This work was supported by projects GA CR P205/12/2327 and MSMT LM2011021, the US DOE DE-FG02-07ER54917 and DE-AC02-09CH11466, Eurofusion, Euratom 2014-2018 under grant agreement No 633053.


## References


[1] A.R. Raffray et al *Nucl. Fusion* **55** (2014) 033004

[2] J. Stockel et al . J. Physics: Conference Series **63** (2007) 012001

[3] F.D. Halpern et al . *Plasma Phys. Contr. Fus.* (2015). This Special Issue

[4] ITER physics basis *Nuclear Fusion* **39** (1999) 2423

[5] J. Gunn et al , *J. Nucl. Mat.* **438**, Supplement (2013) S184

[6] LaBombard, B. et al *Rev. Sci. Instrum.* **78** (2007) 073501

[7] J. Horacek et al . *Journal of Nuclear Materials* **463** (2015) 385

[8] M. Kocan et al . *Nucl. Fusion* **55** (2015) 033019

[9] J. Horacek et al . To be published in Special Issue of *Plasma Phys. Contr. Fus.* (2015)

# Towards a general scaling of the Scrape-off Layer density width

D. Carralero[1], H. J. Sun[1], S. A. Artene[1,2], P. Manz[3], H. W. Müller[1,9], M. Groth[4,5], M. Komm[6], J. Adamek[6], L. Aho-Mantila[7], G. Birkenmeier[1,3], M. Brix[4], U. Stroth[1], N. Vianello[8,10], E. Viezzer[1], M. Wischmeier[1], E. Wolfrum[1], ASDEX Upgrade Team[1], COMPASS Team[6] and JET Contributors[5,*]

*[1]Max-Planck-Institut für Plasmaphysik, Garching, Germany. [2]Universidad Complutense de Madrid, Madrid, Spain. [3]Physik-Department E28, Technische Universität München, Garching, Germany. [4]Aalto University, Espoo, Finland. [5]EUROfusion Consortium, JET, Culham Science Centre, Abingdon, UK. [6]Institute of Plasma Physics AS CR Association, Praha, Czech Republic. [7]VTT Technical Research Center of Finland, Helsinki, Finland. [8] Consorzio RFX, Padova, Italy. [9]Institute of Materials Chemistry and Research, University of Vienna, Vienna, Austria. [10]Ecole Polytechnique Fédérale de Lausanne, Centre de Recherches en Physique des Plasmas, Lausanne, Switzerland. [*]See the Appendix of F. Romanelli et al., 25th IAEA Fusion Energy Conference 2014, Saint Petersburg, Russia.*

The width of the Scrape-off Layer –defined here as the typical radial distance travelled by particles ejected at the separatrix before hitting some PFC- is a key feature for next generation tokamaks, as it will determine erosion levels and heat loads at many critical plasma facing components, and most critically at the main chamber first wall. Recent L-mode experiments carried out on the ITER stepladder (including COMPASS, ASDEX Upgrade and JET tokamaks) [1] have proven the link between an increase in filament size at the outer midplane and the onset of the density profile flattening known as "shoulder" [2-4]. This study indicates that the increased effective collisionality, $\Lambda_{div}$, [5,6] in the divertor region at the onset of detachment leads to a change in the regime of filamentary propagation, greatly enhancing radial particle and heat transport. This has been explained as the effect of filament disconnection from the wall: when collisionality along the field line increases over a certain value, the parallel term of the charge conservation equation becomes negligible and the filament goes from Sheath Limited (SL) regime [7] to Inertial (IN) regime [8]. In order to refine the findings in [1] and confirm the validity of this model, a series of dedicated experiments have been carried out in AUG.

First, the dependence of shoulder formation on $\Lambda_{div}$ has been investigated. For this, several L-mode density ramps were carried out at different heating powers (ohmic heating plus 0, 300 and 600 kW of ECH power) in conditions equivalent to those described in [4]. By these means, the density threshold of the onset of divertor detachment is changed without

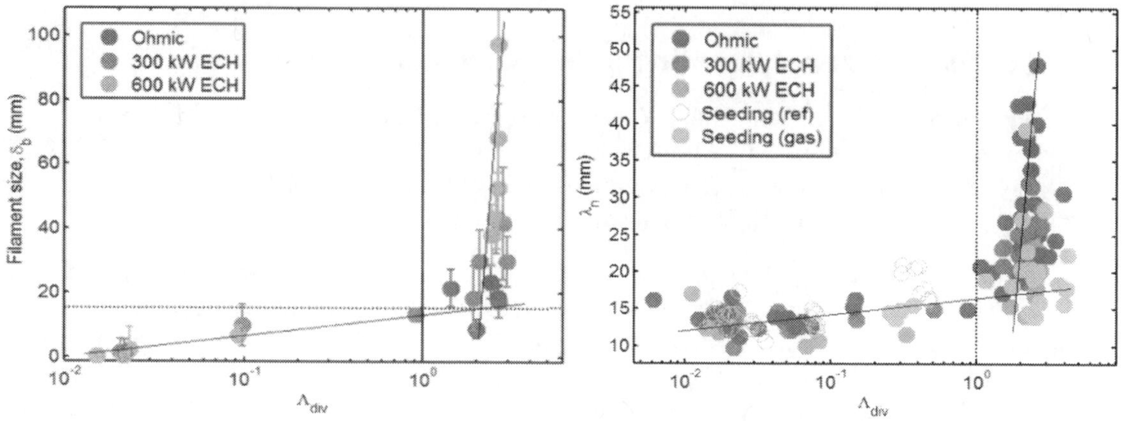


*Figure 1: Filament size (left) and density e-folding length (right) as a function of divertor collisionality, $\Lambda_{div}$. Two different regimes can be seen on both sides of $\Lambda_{div}$ ~ 1.*

affecting greatly the midplane conditions, thus varying $\Lambda_{div}$ over several orders of magnitude ($\Lambda_{div}$ ϵ [$10^{-2}$-5]), while keeping the midplane collisionality, $\Lambda_{mid}$, fixed ($\Lambda_{mid}$ ~ 0.6). $\Lambda_{div}$ and $\Lambda_{mid}$ are calculated as in [1]. Meanwhile, filament size at the far SOL (around 20 mm outside the separatrix) was measured with a multipin probe in the midplane manipulator, and edge density profile was measured using the Li Beam [4]. As a result, it has been observed that both filament size and density e-folding length, $\lambda_n$, scale with $\Lambda_{div}$ (Fig. 1). No dependence on heating power was found, indicating that neither Greenwald density fraction, $f_{GW}$, nor $\Lambda_{mid}$ are directly responsible for the transition. The same scaling is obtained when the detachment is induced by Nitrogen seeding in equivalent discharges, implying that the mechanism causing the high divertor collisionality is not important. Finally, the same results were obtained when an equivalent analysis was carried out in JET, studying the formation of the shoulder in L-mode [1]. In agreement with predictions in the literature [6], two different regimes are observed, roughly separated by the $\Lambda_{div}$ ~ 1 threshold. For $\Lambda_{div} < 1$, the filament is still connected to the wall and remains in SL regime. For $\Lambda_{div} > 1$, collisionality dominates and filaments are in the IN regime. This has been confirmed by observing the dependence of perpendicular velocity, $v_\perp$, with filament size, $\delta_b$ (Fig. 2). In agreement with the models, scaling of filament velocity transits from a second power law (SL, $v_\perp \alpha \delta_b^{-2}$) to a square root one (IN, $v_\perp \alpha \delta_b^{1/2}$). Interestingly, filaments in the IN regime are closer to the theoretical prediction for cold ions (in the SL the difference is not so pronounced). Second, a Reversed Field Analyser (RFA) installed in the midplane manipulator [9] was used to characterize the ion energies of filaments and background before and after the shoulder formation. This study was carried out in a series of L-mode

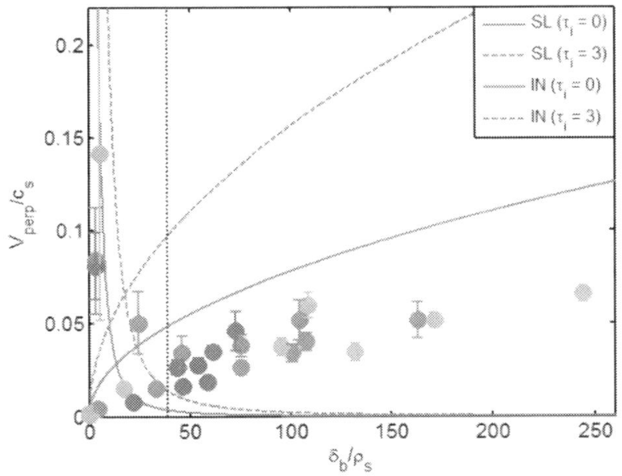


*Figure 2: Filament perpendicular velocity as a function of size (normalized with sound speed and Larmor radius, respect.). SL and IN scaling are displayed for hot/cold ions ($\tau_i = T_i/T_e = 3/ \tau_i = 0$).*

discharges equivalent to those already described [4]. In this case, low/high constant density values were kept to stay below/over the shoulder formation threshold, while the plasma was displaced horizontally with respect to the midplane manipulator (in order to observe the dependence on radial distance to the separatrix). The results are displayed in Fig. 3, with temperatures normalized to values measured in the separatrix by CXRS, to account for the fact that, since heating is kept constant at 400 kW of ECH, higher density plasmas have lower temperatures. Below the shoulder formation, filaments retain a large fraction of the ion energy at the separatrix, and also remain substantially hotter than the background. Besides, $T_i$ drops clearly with the radial distance to the separatrix. This is all in good agreement with previous measurements carried out in equivalent conditions [9]. However, after the onset of the shoulder, filaments cool down to a small fraction of the $T_{i,sep}$, losing the radial dependence and coming close to the background values. These results are consistent with the cold ion scaling observed for the IN regime in Fig 2. However, it remains to be explained which mechanism cools down the plasma in the $\rho \in$ [1.00-1.02] region. A plausible explanation for this would be an increase of recycling in the main wall as a result of the shoulder formation [10].

Last, L-mode density and electron temperature profiles in the near SOL ($\rho \in$ [1.00-1.01]) have also been observed by Thomson scattering. Results are similar to those obtained in the far SOL: both density and temperature e-folding lengths ($\lambda_n$ and $\lambda_T$) increase substantially after a certain density threshold. Again, they scale with divertor collisionality, displaying a threshold for a value around $\Lambda_{div} \sim 1$. Similar observations have been carried out during H-mode (see Fig. 3). In this case, the density threshold after which $\lambda_n$ and $\lambda_T$ show a steep increase is substantially higher (in the range of $f_{GW} \sim 0.8$). However, the scaling with $\Lambda_{div}$ remains the same as in L-mode. These results are consistent with further work carried out in ASDEX Upgrade (AUG) [11], in which the formation of an inter-ELM density shoulder in ITER-relevant H-modes has also been reported. This shoulder occurred as well at higher

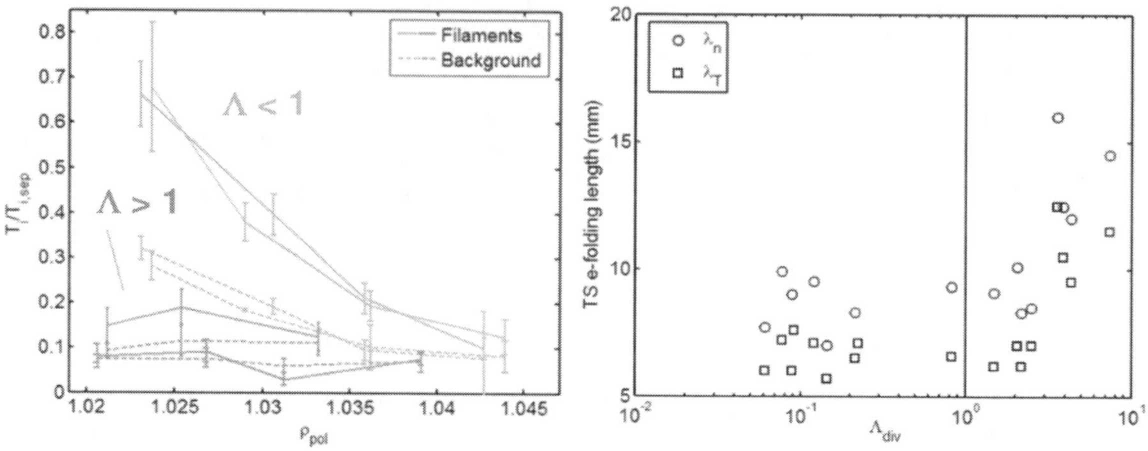


*Figure 3: Left, $T_i$ measurements carried out by RFA on filaments and background (solid/dashed lines) normalized with the separatrix $T_i$. Red/blue curves are measured before/after the formation of the shoulder. Right, Thomson scattering H-mode measurements of $\lambda_n$ and $\lambda_T$ in the near SOL ($\rho \in [1\text{-}1.01]$) as a function of $\Lambda_{div}$.*

$f_{GW}$ than in L-mode, and was not preceded by the onset of divertor detachment, but $\Lambda_{div} \sim 1$ was measured at the transition (the higher densities of H-mode yield collisional conditions even still in the high recycling phase). These results provide a valid framework to understand the physics of the shoulder in L-mode, including a measurable parameter, $\Lambda_{div}$, which determines the threshold for its formation. Besides, the basic mechanism behind filament transition is equally valid in H-mode, and several experimental observations seem to indicate that a similar dependence on $\Lambda_{div}$, holds in this regime. This paves the way for a general scaling of the SOL width, capable of improving current predictions for ITER & DEMO first wall particle and heat loads.

*This work has been carried out within the framework of the EUROfusion Consortium and has received funding from the Euratom research and training programme 2014-2018 under grant agreement No 633053. The views and opinions expressed herein do not necessarily reflect those of the European Commission.*

[1] D. Carralero, H. W. Müller et al., Journal of Nucl. Mat. (2014).

[2] B. LaBombard, R. L. Boivin, M. Greenwald et al. Phys. Plasmas 8 (2001) 2107.

[3] O.E. Garcia, J. Horacek, R. A. Pitts et al., Plasma Phys. Control. Fusion (2006) 48 L1–10

[4] D. Carralero, G. Birkenmeier, H. W. Müller et al., Nucl. Fusion, 54 (2014) 123005.

[5] J.R. Myra, D.A. Russell and D.A. D'Ippolito Phys. Plasmas, 13 (2006 ) 112502.

[6] P. Manz, D. Carralero, G. Birkenmeier et al., Phys. of Plasmas, 20 (2013), 102307.

[7] S. I. Krasheninnikov, D. A. D'Ippolito and J. R. Myra, J. Plasma Phys., 74 (2008), 679717.

[8] O.E. Garcia, N.H. Bian and W. Fundamenski, Phys.Plasmas 13 (2006), 082309.

[9] M. Kocan, F. P. Gennrich, A. Kendl et al., Plasma Phys. Control. Fusion 54 (2012) 085009

[10] B. Lipschultz, D. Whyte and B. LaBombard, Plasma Phys. Control. Fusion 47 (2005), 1559.

[11] H.W. Müller, M. Bernert, D. Carralero, Journal of Nucl. Mat. (2014).

# Electron kinetics inferred from observations of microwave bursts during edge localised modes in the Mega-Amp Spherical Tokamak

S.J. Freethy[1], K.G. McClements[2], S.C. Chapman[3], R.O. Dendy[2,3],
W.N. Lai[3], S.J.P. Pamela[2], V.F. Shevchenko[2], R.G.L. Vann[4]
[1] *Max Plank Institute for Plasma Physics, Boltzmann Str., Garching, Germany*
[2] *CCFE, Culham Science Centre, Abingdon, OX14 3DB, UK*
[3] *Centre for Fusion, Space and Astrophysics, Warwick University, Coventry, CV4 7AL, UK*
[4] *York Plasma Institute, University of York, York, YO10 5DQ , UKS*

The instability that initiates ELMs is relatively well understood at the fluid level of description (see e.g. [1]). However the observations reported here of electromagnetic radiation from plasmas with ELM activity indicate that the physics of this process incorporates electron kinetic effects that are not included in fluid descriptions, and occur on relatively short lengthscales and timescales. In the Mega Amp Spherical Tokamak (MAST) [2], ELMs are accompanied by intense bursts of microwave emission (BMEs) in the electron cyclotron (EC) frequency range. These can reach brightness temperatures 3-4 orders of magnitude higher than the thermal background and detailed experimental accounts of similar emission in other devices are given in [3, 4, 5, 6]. Recently, it has been shown that a plausible explantation for these bursts of emission is through the mechanism of the Anomalous Doppler Instability (ADI) [7].

An electron with velocity parallel to the magnetic field $v_{\|}$ is in resonance with a wave of frequency $\omega$ and parallel wavevector $k_{\|}$ if

$$\omega = \ell\Omega_e + k_{\|}v_{\|}, \tag{1}$$

where $\ell$ is an integer and $\Omega_e$ is the EC frequency. The ADI occurs when waves are excited via this resonance condition with $\ell < 0$; in applications of the ADI to tokamak plasmas, the case $\ell = -1$ is generally found to be relevant [8]. As the ADI proceeds, the suprathermal electrons acquire perpendicular momentum comparable to their parallel momentum on timescales of a few hundred EC periods, $\tau_c$. While predominantly electrostatic, the waves can be converted to electromagnetic waves, and thus propagate freely to antennas outside the plasma [9].

BME intensities relative to the average thermal background level peak at about 35dB and extend up to 40 dB. Measurements show the peak of the microwave burst activity occurs around 20$\mu$s before the peak in midplane deuterium-$\alpha$ ($D_\alpha$) line emission. In many cases a burst of X-ray emission coincides with the microwave burst, with the peak emission again preceding the $D_\alpha$ emission by around 20$\mu$s in the crash phase [10].

Simulations of ELMs in MAST performed with the nonlinear resistive MHD code JOREK [11] reveal that parallel electric fields of up to $2\mathrm{kVm}^{-1}$ can be expected in localised regions (Fig. 1). Mirnov coil measurements during ELMs in MAST imply that the modes associated with these fields have periods of the order of a few microseconds; individual electrons would be accelerated to tens of keV if they encountered $2\mathrm{kVm}^{-1}$ parallel electric fields for only a small fraction of this time.

To investigate the stability of field-aligned superthermal electrons, particle-in-cell (PIC) simulations have been carried out in one space dimension and three velocity dimensions, the magnetic field being tilted at an angle of 45° with respect to the space axis. Thus, any excited waves have wavevectors with components parallel ($k_{\|}$) and perpendicular ($k_{\perp}$) to **B** of equal magnitude, i.e. $k_{\|} = k_{\perp}$. The location in the MAST plasma of the instability producing the microwave bursts is not known precisely. PIC simulations have been carried out with a range of values of initial bulk electron density $n_e$, from $2 \times 10^{18}\mathrm{m}^{-3}$ to $4 \times 10^{19}\mathrm{m}^{-3}$, covering the MAST edge conditions. In all cases the initial electron distribution comprises a Maxwellian bulk with a temperature $T_e$ of either 100eV or 20eV and a flat magnetic field-aligned tail, containing 5% or 10% of the total electron population and extending to 20 or 40 times $v_B$ where $v_B = (2T_e/m_e)^{1/2}$ is the initial bulk electron thermal speed, $m_e$ being the electron mass. It is known that the growth-rate of the ADI is linearly dependant on the fractional density of fast electrons $n_f/n_e$, so we may extrapolate the measured growth rates to smaller fractional densities if necessary. In all cases the equilibrium magnetic field $B_0$ is 0.4T, a typical value in the outer midplane edge of MAST plasmas. Analysis of the PIC data reveals propagating waves close the the upper-hybrid-frequency [7]. These waves may convert to electromagnetic waves in this region and be detected by antennas outside the plasma.

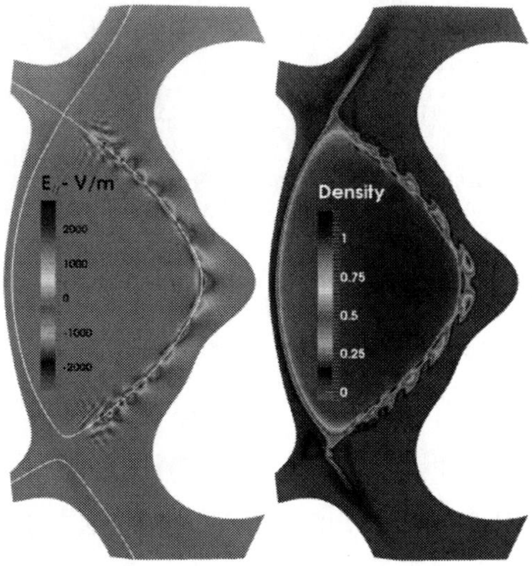

**Figure 1:** Poloidal snapshots of parallel electric field (left plot) and plasma density normalised to the magnetic axis value (right plot) in a JOREK simulation of an ELM in MAST [11]. The white curve in the left plot shows the separatrix.

Fig. 2 shows (a) the initial electron velocity distribution in this simulation and (b) the distribution after $379.4\tau_c \simeq 30\mathrm{ns}$. It is apparent that, in this short time, the electrons in the tail acquire perpendicular momenta comparable to their initial parallel momenta. We conclude that strongly field-aligned energetic electron distributions in the edge plasma of MAST are very rapidly isotropised via the ADI. We note that the forward-propagating waves in the upper hybrid frequency range have wavevectors $k \sim 0.1\omega_{pe}/v_B$ and hence $k_{\parallel} \sim 0.1\omega_{pe}/(\sqrt{2}v_B)$. Using this value of $k_{\parallel}$ in Eq. (1) and setting $\omega = 1.2\omega_{pe}$, $\ell = -1$ we find that the anomalous Doppler resonance condition is satisfied by electrons with $v_{\parallel} \sim 30v_B$; we note from Fig. 2 that this is close to the parallel velocity at which electrons have acquired the largest boost in $v_{\perp}$, and conclude from this that the ADI explains the excitation of the high amplitude waves observed in this simulation.

Similar phenomena are observed in PIC simulations with higher $n_e$ ($\geq 10^{19}\text{m}^{-3}$) and $T_e$ (100eV), more characteristic of the top of the H-mode pedestal region in MAST plasmas. However, as in the low density case discussed above, high amplitude waves are only excited in these simulations at frequencies either below $\Omega_e$ or above $\omega_{pe}$. Neither of these frequency ranges is compatible with the observed frequency peak, if $B_0 = 0.4$ and $n_e \geq 10^{19}\text{m}^{-3}$. Only in the low density simulation do we see wave excitation at frequencies that are consistent with the BME spectra, suggesting that the energetic electrons causing the emission undergo acceleration very close to the MAST plasma edge. This conclusion is bolstered by the predictions from JOREK shown in Fig. 1. A corollary of this last conclusion is that the energetic electrons are likely to have a very short confinement time, particularly in view of the fact that the magnetic field in the plasma edge region is strongly distorted by the ELM instability. In these circumstances we expect the electrons to be subject to rapid cross-field transport of the type discussed by Rechester and Rosenbluth [12], an idea explored more quantitatively in [7]. Analysis of the soft X-ray data reveals that a rather modest fraction of super thermal electrons is required to produce the observed rise, of the order of $10^{-3}$ or less.

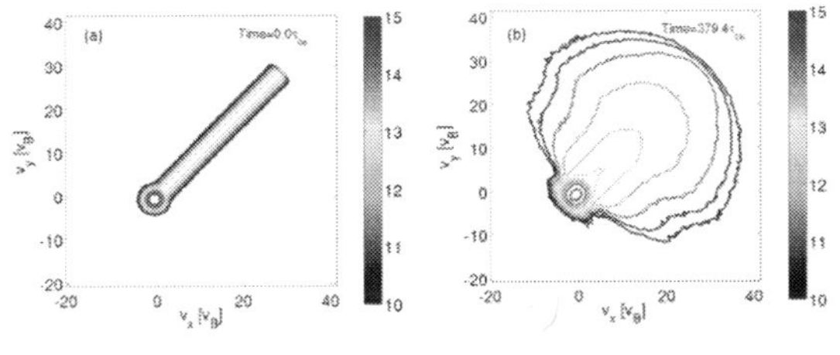


**Figure 2:** Contours of (a) initial and (b) final electron velocity distribution in PIC simulation with initial electron density $2 \times 10^{18}\text{m}^{-3}$ and nonthermal electron tail fraction of 5%. The colour bar has a logarithmic scale.

In conclusion, measurements of microwave and soft X-ray emission during ELMs in spherical tokamak plasmas provide strong evidence for the transient presence in the edge plasma of highly suprathermal electrons. The paradoxical nature of the intensity of the emission, given the absence of any RF wave heating is resolved by considering the response of electrons accelerated by a parallel electric field and invoking the collective anomalous Doppler instability. Particle-in-cell simulations show that magnetic field-aligned energetic electron distributions, of the kind inferred to result from parallel electric fields generated by ELMs, excite electromagnetic waves in the electron cyclotron range. Further, the PIC simulations predict emission frequencies between $\omega_{pe}$ and $\omega_{UH}$, which is consistent with the predictions from JOREK regarding the radial location of the accelerating field. While soft X-ray and Thomson scattering data indicate that the fraction of accelerated electrons is small, their active role suggests that purely fluid models of ELMs are incomplete. For example the radial current associated with the rapid radial transport of these electrons could have an effect on ELM dynamics, and, if sufficiently energetic and present in sufficiently large numbers, they could cause damage to plasma-facing components. For these reasons it is hoped that the present study will prompt further investigations of energetic electron production during ELMs.

This project has received funding from the RCUK Energy Programme (grant number EP/I501045) and from Euratom. Assistance from Luca Garzotti and Rory Scannell on the interpretation of soft X-ray and Thomson scattering data is gratefully acknowledged. To obtain further information on the data and models underlying this paper please contact PublicationsManager@ccfe.ac.uk. The views and opinions expressed herein do not necessarily reflect those of the European Commission.

## References

[1] G.T.A. Huysmans *et al.*, Plasma Phys. Controlled Fusion **51**, 124012 (2009).
[2] B. Lloyd *et al.*, Nucl. Fusion **51**, 094013 (2011).
[3] C.H. Fuchs, M.E. Austin, Phys. Plasmas **8**, 5 (2001).
[4] G. Taylor *et al.*, Nucl. Fusion **32**, 1867 (1992).
[5] M.E. Austin *et al.*, *Proc. 14th Joint Workshop on ECE and ECRH, Santorini* (2006).
[6] D.V. Bartlett *et al.*, *Proc. 8th Joint Workshop on ECE and ECRH, Gut Ising* (1993).
[7] S.J. Freethy *et al.*, Phys. Rev. Let. **114**, 125004 (2015).
[8] W.N. Lai, S.C. Chapman, and R.O. Dendy, Phys. Plasmas **20**, 102122 (2013).
[9] A.K. Ram, A. Bers, and C.N. Lashmore-Davies, Phys. Plasmas **9**, 409 (2002).
[10] A. Kirk *et al*, Phys. Rev. Lett., **96**, 185001 (2006)
[11] S.J.P. Pamela *et al*, Plasma Phys. Control. Fusion **55**, 095001 (2013)
[12] A.B. Rechester, M.N. Rosenbluth, Phys. Rev. Lett. **40**, 38 (1978).
[13] H.A. Kramers, Phil. Mag. Series 6 **46**, 836 (1923).

# Stochastization and pump-out in edge plasma caused by ELMs

V Rozhansky[1], E Kaveeva[1] and M Tendler[2]

[1] *Peter the Great St.Petersburg Polytechnic University, St.Petersburg, Russia*

[2] *Alfven Laboratory Royal Institute of Technology, Stockholm, Sweden*

## 1. Introduction

Filaments have been observed in the H-mode during onset of edge localized modes (ELMs) [1-2]. There is experimental evidence that rather strong bipolar [3-4] or monopolar [5-6] currents of the order of 200-400A flow in the filaments. In [7-8] an idea was put forward that such large currents could create a stochastic layer in the pedestal region leading to the pedestal density and temperature reduction [1-2]. We propose a model for pump out during onset of ELMs type I. The key element of the model is increase of the radial outward particle and heat convective fluxes due to the deviation of radial electric field from the neoclassical value in the stochastic layer during life time of current carrying filaments.

## 2. Currents in the filaments

In the SOL the filament is getting polarized by vertical $\nabla B$ and curvature current. The radial acceleration of filament measured in the experiments is not big enough to provide the closure of major part of those currents by the polarization current inside the filament. Therefore the external currents perpendicular to the magnetic field should arise. The bipolar parallel currents in the filament connecting these external currents and filament can be estimated as:

$$j_{\parallel}^{f} \approx \frac{2n_f(T_{ef}+T_{if})}{BR}\frac{l_{\parallel f}}{l_{\perp f}}, \tag{1}$$

where $n_f, T_{ef}, T_{if}$ are the filament density and temperatures and $l_{\parallel f}, l_{\perp f}$ are the filament sizes along and across magnetic field. For the typical parameters of filaments during ELMs in MAST ($n_f = 2\cdot 10^{19} m^{-3}$, $T_{ef} = T_{if} = 80eV$, $l_{\parallel f} = 8m$, $l_{\perp f} = 4cm$, $B = 0.3T$, $R = 1.5m$ [2,4]) $j_{\parallel}^{f} \approx 400 kA/m^2$. Partially the parallel currents can be short-circuited through the divertor plates, still in many discharges these currents are bigger than ion saturation current at the plates and therefore are closed through the ambient plasma. In [4] the distortion of the magnetic flux tube near the *X*-point leading to the big polarization currents in this region is discussed as a mechanism of filament current closure. The polarization current in the divertor plasma can also contribute to the closing of bipolar parallel currents.

Current density at the plasma edge inside the separatrix could be estimated as $1000 kA/m^2$ [9]. This current can be inherited by the filament when it detaches. Such monopolar current density is too large to be short-circuited through the divertor. The life times of monopolar current can be estimated for typical MAST parameters as $\tau_I^{vis} \approx 10^{-8} \div 10^{-9} s$ if it is closed by the current associated with perpendicular viscosity and $\tau_I^{n} \approx 10^{-6} \div 10^{-7} s$ if it is closed by the current associated with ion-neutral collisions in the divertor region. They are much smaller than the life-time of the filament, therefore the evolution of the filament current can be described by MHD equations assuming only perpendicular polarization currents. The solution is an Alfven standing wave localized inside the filament. The magnetic signal at the Alfven frequency was indeed detected in [2]. The steady-state currents exceeding the ion saturation current to the divertor plates observed in experiments most probably are the part of the system of dipole currents. At the same time the monopolar current in the filament contributes to the plasma edge ergodisation while it is still attached to the core plasma for the time of about $50\mu s$ [2].

### 3. Penetration of magnetic perturbations

Magnetic perturbations produced by the filaments currents start penetrating into the ETB. The estimate for the characteristic time of the linear growth of the separate island in the theory of forced reconnection [10] is $\tau_{B1} = \tau_A^{2/5}\tau_R^{3/5}$ for the magnetic perturbation with poloidal wave vector $\vec{k}$, $\tau_A = (\mu_0 m_i n_e / B_\theta^2)^{1/2} q/q'$, $\tau_R = \mu_0 \sigma_\parallel / k^2$. Further penetration is determined by the characteristic time for nonlinear island growth $\tau_{B2} = \tau_R k\delta$, where $\delta$ is the saturated island width. For the typical parameters of MAST plasma boundary ($n_e = 2\cdot 10^{19} m^{-3}$, $T_e = 200eV$, $q/q' \approx 0.1m$ [11], $\vec{k} \approx 40m^{-1}$ for ELM filaments, $\delta \approx 0.5cm$) $\tau_{B1} \approx 10 \div 100\mu s$ and $\tau_{B2} \approx 100\mu s \div 1ms$. From the experiments with RMPs it is known that stochastisation starts at the level of the RMPs about $b_{y\vec{k}}^{stoch} = B_{y\vec{k}}^{stoch} / B \approx 10^{-4}$ [11]. The magnetic field perturbation which is produced by the current 1 kA flowing inside the filament at the distance of few centimeters from resonant layer is of the order of $b_{y\vec{k}}^{sat} \approx 10^{-3}$. Therefore a stochastic layer should arise in the ETB region before the islands reach full saturation, the corresponding time could be estimated as $\tau_{stoch} \approx \tau_{B2} b_{y\vec{k}}^{stoch} / b_{y\vec{k}}^{sat} \approx 10 \div 100\mu s$.

After the stochastisation the characteristic time for the further resonant magnetic field penetration can be estimated as $\tau_s = (q/nq')(\mu_0 / k_x)\sigma_{pen}$ with the effective conductivity $\sigma_{pen} = i_\delta \sqrt{\pi} R n_e e^2 / \sqrt{2m_e T_e}$. This estimate is based on the approach on integration of electron kinetic equation [12] and the evaluation of toroidal currents [13]. For MAST $\tau_s \approx 10 \div 100 \mu s$ and the perturbation can fully penetrate during the life time of a filament $\tau_f \approx 200 \mu s$ [2].

## 4. Dynamics of radial electric field and pump-out

Radial electric field in a steady state $E_r^{Eq} = (\sigma_{St} E_r^{St} + \sigma_{NEO} E_r^{NEO})/(\sigma_{St} + \sigma_{NEO})$ is determined by the condition of ambipolarity $j_r^e = -j_r^i$ where the ion radial current $j_r^i = \sigma_{NEO}(E_r - E_r^{NEO})$ is a neoclassical current caused by the deviation of the radial electric field from the neoclassical value and the radial current of electrons $j_r^e = \sigma_{St}(E_r - E_r^{St})$ is a radial projection of a parallel current in the stochastic magnetic field [13-14]. Here

$$E_r^{St} = -(T_e / e) d \ln n_e / dr - \alpha (T_e / e) d \ln T_e / dr. \qquad (2)$$

In the collisionless limit ($\lambda_e / \tilde{L}_k >> 1$, where $\lambda_e$ is the electron mean free path, $\tilde{L}_k$ is a correlation length, of the order of Kolmogorov length; for the tokamak $\tilde{L}_k \sim qR$), $\sigma_{St} = i_\sigma e^2 n_e D_{St} \sqrt{2} / \sqrt{\pi m_e T_e}$, with $i_\sigma$ being a numerical factor of the order of unity, $\alpha = 0.5$.

$$E_r^{NEO} = T_i / e(d \ln n_e / dr + k_T d \ln T_i / dr) + B_p U_T, \qquad (3)$$

where $U_T$ is a toroidal velocity, $\sigma_{NEO} = 3\mu_{i1} / (2R^2 B^2)$. The viscosity coefficient [15] in the low collisionality regime is $\mu_{i1} = \varepsilon^{3/2} v_*^2 n_e T_i / v_i$, where $v_* = qR v_i \varepsilon^{-3/2} / \sqrt{2T_i / m_i}$.

For time-dependent radial electric field the additional radial current of ions exists associated with the toroidal inertia, see e.g. [16]:

$$j_r^i = \sigma_{NEO}(E_r - E_r^{NEO}) + n_e m_i (1 + 2q^2) / B^2 \partial E_r / \partial t. \qquad (4)$$

The equation describing the relaxation of the radial electric field to the stationary value is

$$\partial E_r / \partial t = -(E_r - E_r^{Eq}) / \tau_E, \qquad (5)$$

where $\tau_E = n_e m_i (1 + 2q^2) B^{-2} / (\sigma_{St} + \sigma_{NEO})$. When $\sigma_{St} = 0$, $\tau_E = \tau_E^{NEO}$ [16]. In the low collisionality regime $\tau_E^{NEO} \approx v_i^{-1}$. The stochastic conductivity reduces the relaxation time.

Let us consider two limits of slow $\tau_s >> \tau_E$ and fast $\tau_s << \tau_E$ time of the magnetic field penetration. In the first case the radial electric field is always close to the equilibrium one for the given level of perturbed magnetic field. The outward radial flux arises:

$$\Gamma = \sigma_{NEO}(E_r^{Eq} - E_r^{NEO}) / e = \sigma_{St}(E_r^{St} - E_r^{Eq}) / e. \qquad (6)$$

In the opposite case $\tau_s << \tau_E$ the period of the order of $\tau_E$ exists when the electric field is not an equilibrium one and the dynamic pump-out effect can be important. The loss of particles with steady-state electric field is $\delta N^{static} \approx \tau_f \Gamma$. The loss of particles during $\tau_E$ while the electric field is changing and stays of the order of $E_r^{NEO}$ is $\delta N^{dynamic} \approx \tau_E \sigma_{St}(E_r^{St} - E_r^{NEO})/e$, if this time is smaller than $\tau_f$. For low collisionality this mechanism can be the main one.

Let us make estimates for parameters of the MAST pedestal. The neoclassical relaxation time $\tau_E^{NEO} \approx \nu_i^{-1} \approx 300\mu s$. It is of the order of the filament life-time and therefore the dynamic and static losses of particles are of the same order. The neoclassical conductivity estimate is $\sigma_{NEO} \approx (5 \div 7) \cdot 10^{-3} S \cdot m^{-1}$, $E_r^{St} - E_r^{NEO} \approx 40kV/m$. The particle loss per ELMs event is $\delta N \approx S\tau_f \sigma_{NEO}(E_r^{St} - E_r^{NEO})/e \approx 5 \cdot 10^{18}$ particles, where $S$ – the flux surface area. According to experiments [11] the particle loss in the ELM event is 2-3% of the particles in the discharge while full amount of the particles is $(1 \div 2) \cdot 10^{20}$. The estimate shows that the pump-out particle loss is of the same order as experimental one.

## 5. Conclusions

The ETB region is getting stochastized during ELMs events by the magnetic field perturbations produced by currents flowing in the filaments. The radial electric field in the edge is getting positive or less negative and the convective particle flux arises for the period of filaments life time. This leads to the reduction of the pedestal density and temperature.

**Acknowledgements**

The work was supported by grant of Russian Ministry of Education and Science "Filaments and stochastization"

**References**

[1] Kirk A *et al* 2006 *Phys. Rev. Lett.* **96** 185001

[2] Kirk A *et al* 2006 *Plasma Phys. Control. Fusion* **48** B433

[3] Rozhansky V and Kirk A 2008 *Plasma Phys. Control. Fusion* **50** 025008

[4] Furno I *et al* 2011 *Phys. Rev. Lett.* **106** 245001

[5] Vianello N *et al* 2011 *Phys. Rev. Lett.* **106** 125002

[6] Muller H W *et al* 2011 *Nucl. Fusion.* **51** 073023

[7] Aladio F *et al* 2008 *Plasma Phys. Control. Fusion* **50** 124019

[8] Evans T *et al* 2009 *J. Nucl. Mater.* **390–391** 789

[9] Koh S, Chang C S, Ku S, Menard J E, Weitzner H and Choe W 2012 *Phys. Plasmas* **19** 072505

[10] Hahm T S and Kulsrud R M 1985 *Phys. Fluids* **23**(8) 2412

[11] Kirk A. *et al* 2010 *Nucl. Fusion* **50** 034008

[12] Kaganovich I and Rozhansky V 1998 *Phys. Plasmas* **5** 3901

[13] Kaveeva E and Rozhansky V 2012 *Nucl. Fusion* **52** 054011

[14] Rozhansky V *et al* 2011 *Nucl. Fusion* **51** 083009

[15] Hirshman S P and Sigmar D J 1981 *Nucl. Fusion* **21** 1079

[16] Rozhansky V and Tendler M 1996 *Reviews of Plasma Phys.* vol 19, ed. B.B. Kadomtsev, (New-York – London: Consultants Bureau) p 147

# Impact of nitrogen seeding on the AUG pedestal: experiments and modelling

M. G. Dunne[1], L. Frassinetti[2], M. Beurskens[3], M. Cavedon[1,4], R. Fischer[1], G.T.A. Huijsmans[5], F. M. Laggner[6], R. M. McDermott[1], G. Tardini[1], E. Viezzer[1], E. Wolfrum[1], the EUROfusion MST1 Team*and the ASDEX-Upgrade Team[1]

[1]*Max-Planck-Institut für Plasmaphysik, D-85748 Garching,Germany*

[2]*Division of Fusion Plasma Physics, Association EURATOM-VR, KTH, SE-10044 Stockholm, Sweden*

[3]*EURATOM/CCFE Fusion Association, Culham Science Centre, Abingdon, OX14 3DB,UK*

[4]*Physik-Department E28, Technische universität München, Garching Germany*

[5]*ITER Organization, Route de Vinon sur Verdon, F-13115 St Paul Lez Durance, France*

[6]*Institute of Applied Physics, TU Wien, Fusion@ÖAW, 1040 Vienna, Austria*

## Introduction

Improved confinement with nitrogen seeding has been widely reported on AUG and JET in recent years [1, 2]. However, despite intensive experimental and theoretical investigation on both devices, an explanation for this phenomenology has thus far remained elusive. One currently working hypothesis [3] for this effect starts with the assumption of a peeling-ballooning limited pedestal. When nitrogen is seeded, the edge current density is transiently reduced by the increased $Z_{eff}$, allowing access to a higher pedestal pressure gradient. This then propagates into a higher thermal energy in the core plasma, which in turn acts to further stabilise the pedestal via an increased Shafranov shift.

A set of experiments were carried out during the last AUG experimental campaign with the aim of experimentally testing this hypothesis and providing data for modelling. Testing of such a hypothesis involves both interpretive and predictive modelling components to verify the above mechanism.

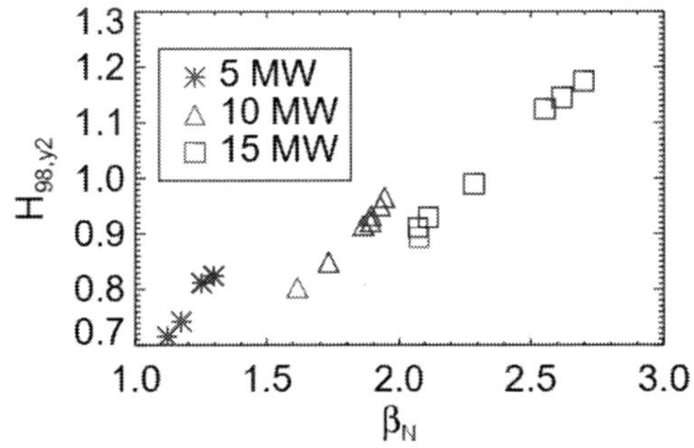


**Figure 1:** Normalised confinement time plotted against normalised plasma beta. These two measures of the plasma stored energy increase with increasing heating power (different symbols) and impurity seeding (blue symbols).

## Experiments

The experiments were carried out in a similar configuration to the scenario where the highest confinement improvement (40%) [1] was observed on AUG; 1 MA plasma current, −2.5 T toroidal magnetic field, medium triangularity ($\delta_{av}$~0.25), with a deuterium gas puff of $2.7 - 3.0 \times 10^{22}$electronss$^{-1}$. Three power levels

*see http://www.euro-fusionscipub.org/mst1

were used to investigate any beta dependence and also to investigate the behaviour of unseeded pedestals in response to a higher core beta. The additional neutral beam injection (NBI) power was varied from 5-10-12.5 MW. At the highest level of input NBI power, an additional 2 MW of ICRH power were also applied. A constant ECRH power of 1 MW was applied in the core of all discharges to prevent impurity accumulation. An overview of the confinement in these discharges is shown in a $H_{98,y2}$-$\beta_N$ plot in figure 1.

Both measures of the plasma stored energy increase with increasing heating power and also with increasing impurity seeding level. The highest levels of confinement improvement obtained were 8, 15 and 25 % in the low, medium, and high input power discharges, respectively.

Only low levels of nitrogen seeding could be applied at the lowest heating power to avoid outer divertor detachment; this causes a large increase of the plasma density, changing the confinement regime significantly. Althought the highest confinement is reached with the highest input power and the highest level of nitrogen seeding, it is difficult to separate these two effects as less nitrogen seeding could be applied at the lower power levels.

Figure 2 shows the normalised plasma beta as a function of pedestal top temperature. A strong dependence of the stored energy on the temperature is observed, while the density pedestal (not shown) has only a minor role in confinement changes, consistent with previous results [1, 4, 5]. From the different behaviour of the temperature and density pedestals, we can creates a picture of two constraints on the AUG pedestals; one on the density pedestal, which recovers first after the ELM crash in most AUG scenarios [6], and, ultimately, a peeling-ballooning limited pressure pedestal.

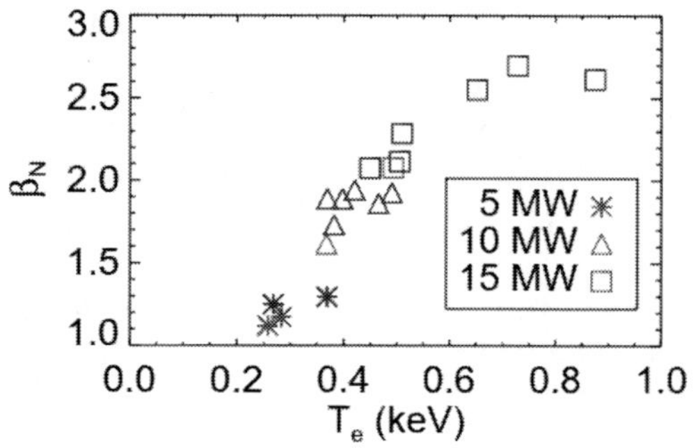

**Figure 2:** Normalised plasma beta plotted as a function of pedestal top electron temperature, showing a dominant role of the electron temperature in the stored energy increases in this databae.

In order to further vary the range of pedestal top parameters, a triangularity scan was also performed as part of this discharge set. An overview of the pedestal parameters for both shapes, with and without impurity seeding, is shown in figure 3. Circles denote the low triangularity ($\delta_{av} = 0.25$) portions of these discharges, with triangles showing data from the high triangularity ($\delta_{av} = 0.35$) phases. Red indicates an unseeded reference discharge, blue shows the seeded counterparts. The increasing levels of fill indicate increasing power levels. Only one seeding level was performed for each heating step in these discharges. Also shown is a comparison discharge with methane ($CD_4$) puffing with both low and high triangularity phases (see [7] for more information).

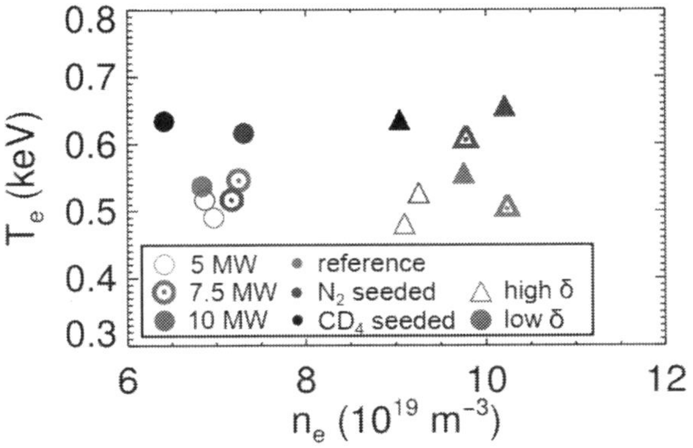


**Figure 3:** Pedestal top temperature as a function of pedestal top density for low (circles) and high (triangles) plasma shaping, with three heating levels (different fill levels) and with (blue) and without (red) impurity seeding). A discharge with $CD_4$ puffing is also shown in black.

Three key pieces of information regarding these AUG scenarios are shown in this figure: 1) increasing triangularity always increases the pedestal top in contrast with JET baseline discharges; 2) $CD_4$ has a comparable effect to $N_2$[7] and; 3) impurity seeding acts *only* on the temperature pedestal while plasma shaping acts *only* on the density pedestal. While prediction of the pedestal top density requires significant theory development to understand, predicting the combined pressure pedestal height should be within the capabilities of present day peeling-ballooning (PB) theory.

**Modelling**

Within the framework of PB, a predictive model was used to vary the plasma beta, shape, and impurity content in a controlled fashion not accessible in the experiment. To perform these variations, sample profiles were created in the same manner as in the EPED model [8], i.e. the pedestal top density is specified and the temperature height is varied to provide the different test pressure pedestal heights. The pedestal width was determined using the $\sqrt{\beta_{\mathrm{pol,ped}}}$ scaling [8]. The current density was then calculated using the Sauter model [9] for the bootstrap current and neoclassical resistivity. The toroidal electric field was assumed to be constant over the plasma radius and used to scale the Ohmic current density until the desired plasma current was reached.

The stability of each of these pressure-current density profile pairs was then tested for a sample of mode numbers ($1 < n < 70$) using the MISHKA-fast code, which is a variant of the MISHKA-1 code [10] which analyses a radially dependent array of poloidal mode numbers depending on the local q value. The results of this modelling are shown in figure 4.

The red solid line indicates the predicted total pressure ($p_i + p_e$) pedestal top as a function of plasma beta at low triangularity and a $Z_{eff}$ of 1.3, while the blue solid line shows the same for the high triangularity shape used in this work. The dashed lines correspond to simulations with a $Z_{eff}$ of 2.

Several key results are presented in this figure. First, the code predicts that, for this parameter set, increasing the plasma shaping will always result in an increase of the pedestal top. This was observed in the experiments, where the pedestal top increased even at the lowest input power. Secondly, at high beta, the increase of the pedestal top as a function of global beta saturates; this is due to the pedestal moving from a ballooning limited to a peeling limited regime. Different experimental conditions with, for example, a low beta operational point at low current density and pressure gradient, as is the case for JET BL plasmas, may not see this saturation over the available range of global beta. Thirdly, the model does predict an increased pedestal top with an increased $Z_{eff}$.

However, the magnitude of this increase (~5%) is significantly lower than in the experiment, and in the low triangularity shape and at low beta, the model predicts a *decrease* of confinement, which is not observed in experiments on either AUG or JET before the plasma transitions to a Type-III ELMing regime. This indicates that there is an (are) additional physics mechanism(s) at play other than the impact of $Z_{eff}$ on the edge current density alone.

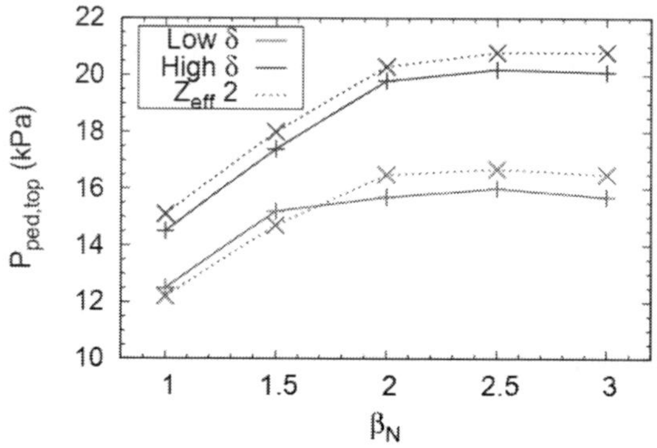


**Figure 4:** Predicted pedestal top as a function of plasma beta for low (red) and high (blue) triangularity with $Z_{eff}$ of 1.3 (solid) and 2.0 (dashed).

## Conclusions

A series of experiments on AUG has revealed that impurity seeding increases the plasma stored energy by increasing the temperature pedestal. Experiments at high triangularity have shown a separation of the impact of shaping and seeding on the density and temperature pedestals, respectively. Predictive modelling of these effects shows that the effect of triangularity on pedestal pressure is clear. However, the same modelling demonstrates that the change in $Z_{eff}$ in the pedestal and its corresponding impact on the bootstrap current does not give rise to the observed increases; further analysis taking the effects of divertor radiation into account in some manner (such as a change of the separatrix temperature) is likely to be required to explain these effects on stability.

## References

[1] A. Kallenbach, et al. *Plasma Physics and Controlled Fusion*, 55(12), 2013.

[2] C. Giroud, et al. *Nuclear Fusion*, 53(11), 2013.

[3] M G Dunne, et al. In *41st EPS Conference on Plasma Physics*, Berlin, 2014.

[4] J. Schweinzer, et al. *Nuclear Fusion*, 51(11), 2011.

[5] G Tardini, et al. *Plasma Physics and Controlled Fusion*, 55(1), 2013.

[6] A. Burckhart, et al. *Plasma Physics and Controlled Fusion*, 52(10), 2010.

[7] M.N.A. Beurskens, et al. In *42nd EPS Conference on Plasma Physics*, Lisbon, 2015.

[8] P.B. Snyder, et al. *Nuclear Fusion*, 49(8), 2009.

[9] O. Sauter, et al. *Physics of Plasmas*, 6(7), 1999.

[10] A. B. Mikhailovskii, et al. *Plasma Physics Reports*, 23, 1997.

This work has been carried out within the framework of the EUROfusion Consortium and has received funding from the Euratom research and training programme 2014-2018 under grant agreement No 633053. The views and opinions expressed herein do not necessarily reflect those of the European Commission.

# Poloidal asymmetries in edge pedestals on Alcator C-Mod

C. Theiler[1,2], R.M. Churchill[1,3], I.H. Hutchinson[1], B. Lipschultz[1,4], P.J. Catto[1], C.S. Chang[3], E. Edlund[3], P. Ennever[1], D.R. Ernst[1], R. Hager[3], A.E. Hubbard[1], J.W. Hughes[1], M. Landreman[5], E.S. Marmar[1], F.I.. Parra[6], M.L. Reinke[1,4], J.L. Terry[1], J.R. Walk[1], D. Whyte[1], and the Alcator C-Mod team

[1]*PSFC-MIT, Cambridge, USA* [2]*CRPP-EPFL, Lausanne, CH* [3]*PPPL, Princeton, USA* [4]*Department of Physics, University of York, UK* [5]*Institute for Research in Electronics and Applied Physics, University of Maryland, USA* [6]*Rudolf Peierls Centre for Theoretical Physics, University of Oxford, UK*

Due to the free streaming of charged particles along magnetic field lines, plasma parameters usually vary much more strongly across the field than parallel to it. In toroidally confined plasmas, quantities such as density, temperature, and plasma potential are therefore often considered to be flux functions. Different mechanisms such as fast rotation or steep radial gradients can, however, drive poloidal asymmetries. According to neoclassical theory, as the ratio of poloidal ion Larmor radius $\rho_i^\theta$ and radial scale length $L_\perp$ increases, the quantity which usually first develops a poloidal asymmetry is the density of heavy impurities [1]. Asymmetries in other quantities are expected as well, particularly in the pedestal where $\rho_i^\theta/L_\perp$ can approach unity [1, 2]. Besides being of basic physics interest, poloidal asymmetries can have important practical consequences. The radial transport of impurities, for example, can be strongly modified by poloidal asymmetries (see e.g. [3]). In the following, we summarize recent evidence of substantial poloidal asymmetries of both the impurity and main plasma species in edge pedestals of the Alcator C-Mod tokamak.

On Alcator C-Mod, simultaneous measurements of density, velocity, and temperature of the boron minority ions are achieved across the pedestal region at both the inboard (HFS) and outboard (LFS) edge using gas puff charge exchange recombination spectroscopy (GP-CXRS) [4]. Instead of using a high-energy neutral beam as in standard CXRS, GP-CXRS uses a simple neutral gas puff to locally induce charge exchange reactions. As gas puffs can be installed almost everywhere around the periphery of the plasma, this technique opens up new possibilities to explore the poloidal structure of the edge/pedestal. On C-Mod, two systems with a complete set of toroidal and poloidal optics are operational. Their location is shown at the left of Fig. 1 together with a magnetic reconstruction of a typical C-Mod plasma. In the following, we focus on measurements obtained with these systems in I-mode and EDA H-mode plasmas, two high-confinement regimes which typically do not feature ELMs. At the right of Fig. 1, we show some characteristic profiles for these two regimes. I-mode is a low collisionality regime ($0.1 \leq \nu^\star \leq 1$ at the pedestal top). It features L-mode like density profiles together with a clear pedestal in temperature, and the radial temperature scale length $L_T$ becomes comparable to $\rho_i^\theta$, Fig. 1 (b). EDA H-mode is a high collisionality regime ($\nu^\star \geq 1$) and features a pedestal both in density and temperature, such that $L_T$ and $L_{n_e}$ both approach $\rho_i^\theta$ in the pedestal. These pedestals are thus well in the regime $\rho_i^\theta / L_\perp \approx 1$ where poloidal asymmetries are expected.

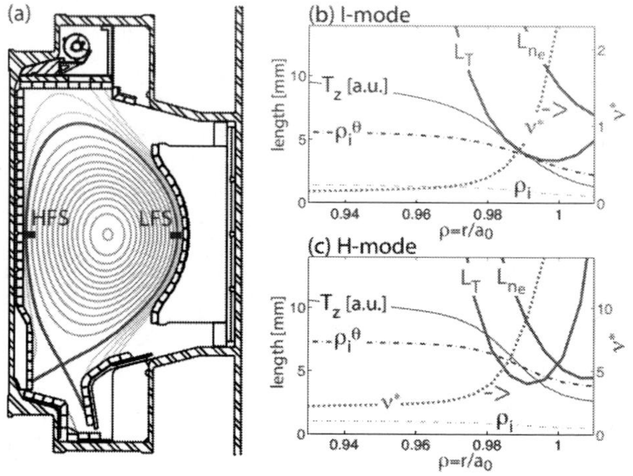


Figure 1: *(a): C-Mod cross-section with GP-CXRS measurement locations indicated. (b-c): Radial edge profiles of typical I-mode and H-mode plasmas [5].*

GP-CXRS measurements indeed reveal that isosurfaces of $T_z$ and plasma potential $\Phi$ do not coincide in these pedestals and these quantities can therefore not simultaneously be flux functions [5]. In Fig. 2, we show radial profiles of impurity temperature $T_z$ and radial electric field $E_r$ at both the HFS and the LFS midplane for I-mode and EDA H-mode. In both regimes, clear temperature pedestals and $E_r$ wells are found at the HFS and the LFS. Despite considerable freedom in the alignment of HFS profiles relative to the LFS ones (mainly due to uncertainties in the equilibrium reconstruction), it is not possible to simultaneously align all the profiles. If we align HFS and LFS $T_z$ profiles, the HFS $E_r$ well is shifted out with respect to the one at the LFS by $\Delta\rho \approx 0.012$ (I-mode) and $\Delta\rho \approx 0.015$ (H-mode). If we instead align the $E_r$ wells, the LFS temperature exceeds the temperature at the HFS by up to 70% [5].

Progress in finding a physics based alignment of HFS and LFS profiles is achieved as follows. An analysis of parallel and perpendicular heat transport time scales suggests that electron temperature is a flux function in these pedestals, while the radial heat flux estimated from global power balance is strong enough to drive poloidal asymmetries in $T_i$ [5]. This analysis furthermore suggests that at least in H-mode, the boron population and the main ions are thermally well coupled, such that we expect that $T_i \approx T_z$. From the constancy of $T_e$ on a flux surface and assuming a low enough friction force for the electrons, the electrons can be shown to follow the Boltzmann relation. Introducing this condition into the total (ion and electron) parallel momentum equation and neglecting contributions from the impurities results in the following relation between LFS and HFS profiles [5, 6].

$$\Phi^{LFS} = \Phi^{HFS} - \frac{T_e}{e}\log\left(\frac{T_e + T_i^{LFS}}{T_e + T_i^{HFS}}\right) + \frac{T_e}{e}\log\left(\frac{p_{tot}^{LFS}}{p_{tot}^{HFS}}\right). \tag{1}$$

Here, $p_{tot} = p_e + p_i$ is the total pressure. Its poloidal variation is mainly dependent on main ion inertia and viscosity. If we assume, without complete justification, that these terms are weak, $p_{tot}$ is a flux function and drops out of Eq. (1). In this case, Eq. (1) becomes a simple condition relating HFS and LFS profiles of $T_i$ and $\Phi$, or, after taking radial derivatives, profiles of $T_i$ and $E_r$. We use this constraint, combined with the assumption that $T_i \approx T_z$, to align HFS and LFS profiles and refer to it as "total pressure"-alignment in the following [6]. This alignment is applied in Fig. 2. The HFS profiles are positioned such that the HFS $E_r$ profile estimated from the above procedure (red curve) best agrees with the actually measured HFS $E_r$ profile. This alignment suggests that the ion temperature at the LFS substantially exceeds the HFS one in the region of steep gradients. As the HFS $E_r$ well is slightly shifted in with respect to the one at the LFS, the plasma potential is also not constant on a flux surface. With the Boltzmann relation for electrons, this results in an electron density about twice as large in the HFS pedestal compared to the LFS [6]. Recently, a C-Mod H-mode plasma has been modeled with XGCa, a total-f gyrokinetic neoclassical PIC code, an axisymmetric version of XGC1 [7]. These simulations show asymmetries qualitatively and quantitatively similar to the measurements using the "total pressure"-alignment approach [6].

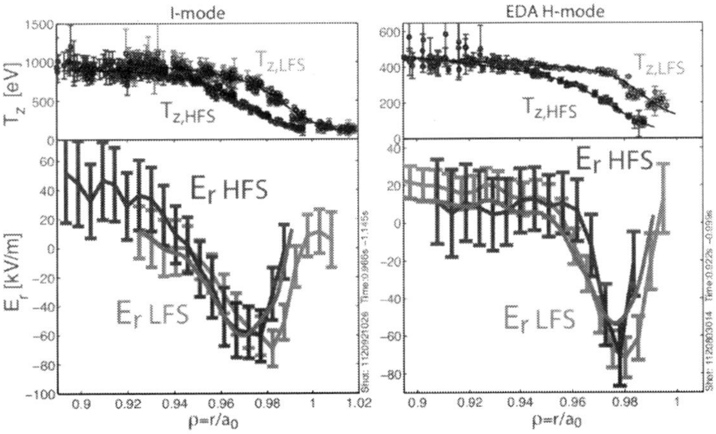


Figure 2: *HFS and LFS $T_z$ and $E_r$ profiles in I-mode (left) and H-mode (right). The "total pressure"-alignment [6] has been used to align HFS and LFS profiles. The red curve shows the HFS $E_r$ profile expected from this procedure.*

In light of these substantial $T_i$, $n_e$, and $\Phi$ poloidal asymmetries on C-Mod, it does not come as a surprise that impurity density $n_z$ also strongly varies poloidally [8]. First indications of a poloidal $n_z$ variation in C-Mod pedestals, in this case top-out, have been obtained from passive soft x-ray emission measurements [9], with the caveat that the reported emissivity asymmetries were dependent on the accuracy of the magnetic reconstruction. More recently, an in-out $n_z$

asymmetry was inferred from the discrepancy of the measured in-out impurity flow variations compared to those expected in the case of a poloidally uniform $n_z$ and negligible radial flows [10]. With the capability of directly measuring the HFS $n_z$, strong in-out asymmetries, with HFS $n_z$ values exceeding the LFS values by up to a factor six, were demonstrated in H-mode plasmas, while asymmetries in L- and I-mode plasmas were found to be negligible [8, 6].

Comparing these findings to other machines, we note that GP-CXRS measurements on ASDEX-Upgrade reveal in-out asymmetries of impurity density in H-mode pedestals of about a factor three [11, 12], considerably weaker than on C-Mod. Furthermore, ion temperature and plasma potential were found, within error bars, to be flux functions [13]. These weaker poloidal variations than on C-Mod could be related to the lower values of $\rho_i^\theta/L_\perp$ on ASDEX-Upgrade. More recently, strong poloidal potential variations were identified in the edge of the TJ-II stellarator using Langmuir probe measurements [14].

In conclusion, in steep-gradient pedestals the impurities and the bulk plasma densities and ion temperatures can have substantial in-out asymmetries on a flux surface, with potentially important consequences for pedestal transport properties. The difference in the amplitude of these variations between C-Mod and ASDEX-Upgrade could be due to the higher values of $\rho_i^\theta/L_\perp$ in C-Mod pedestals. However, a better understanding is clearly needed to predict the importance of edge poloidal asymmetries in future devices.

This work was supported by the U.S. DOE Cooperative Agreement No. DE-FC02-99ER54512. The support from the EUROfusion Researcher Fellowship programme under grant number WP14-FRF-EPFL/Theiler is gratefully acknowledged.

**References**


[1] P. Helander, Phys. Plasmas **5**, 3999 (1998)

[2] M. Landreman and D. Ernst, Plasma Phys. Control. Fusion **54**, 115006 (2012)

[3] C. Angioni and P. Helander, Plasma Phys. Control. Fusion **56**, 124001 (2014)

[4] R.M. Churchill, C. Theiler, B. Lipschultz, et al., Rev. Sci. Instrum. **84**, 093505 (2013)

[5] C. Theiler, R.M. Churchill, B. Lipschultz, et al., Nucl. Fusion **54**, 083017 (2014)

[6] R.M. Churchill, C. Theiler, B. Lipschultz, et al., Phys. Plasmas **22**, 056104 (2015)

[7] C.S. Chang, S. Ku, P.H. Diamond, et al., Phys. Plasmas **16**, 056108 (2009)

[8] R.M. Churchill, B. Lipschultz, and C. Theiler, Nucl. Fusion **53**, 122002 (2013)

[9] T.S. Pedersen, R.S. Granetz, E.S. Marmar, et al., Phys. Plasmas **9**, 4188 (2002)

[10] K.D. Marr, B. Lipschultz, P.J. Catto, et al., Plasma Phys. Control. Fusion **52**, 055010 (2010)

[11] T. Pütterich, E. Viezzer, R. Dux, and R.M. McDermott, Nucl. Fusion **52**, 083013 (2012)

[12] E. Viezzer, T. Pütterich, E. Fable, et al., Plasma Phys. Control. Fusion **55**, 124037 (2013)

[13] E. Viezzer, E. Fable, T. Pütterich, et al., Nucl. Fusion, submitted

[14] M.A. Pedrosa, J.A. Alonso, J.M. García-Regaña, et al., Nucl. Fusion **55**, 052001 (2015)

# Generation of High-Flux Neutron Beam from Radiation Pressure driven Deuterium Ion Bunches

A. Alejo[1], S. Kar[1*], H. Ahmed[1], A.G. Krygier[2], R. Clarke[3], R.R. Freeman[2], J. Fuchs[4], A. Green[1], J.S. Green[3], D. Jung[1], A. Kleinschmidt[5], J.T. Morrison[6], Z. Najmudin[7], H. Nakamura[7], P. Norreys[3,8], M. Notley[3], M. Oliver[8], M. Roth[5], L. Vassura[4], M. Zepf[1], M. Borghesi[1]

[1]*Centre for Plasma Physics, Queen's University Belfast, BT7 1NN, UK*

[2]*Department of Physics, The Ohio State University, Columbus, Ohio 43210, USA*

[3]*Central Laser Facility, Rutherford Appleton Laboratory, Didcot, Oxfordshire, OX11 0QX, UK*

[4]*LULI, Ecole Polytechnique, CNRS, Route de Saclay, 91128 Palaiseau Cedex, France*

[5]*Institut für Kernphysik, Technische Universität Darmstadt, D-64289 Darmstadt, Germany*

[6]*Propulsion Systems Directorate, Wright Patterson Air Force Base, Ohio 45433, USA*

[7]*Blackett Laboratory, Department of Physics, Imperial College, London SW7 2AZ, UK*

[8]*Department of Physics, University of Oxford, Oxford, OX1 3PU, UK*

An ultra-short burst of fast neutrons has a wide range of applications in science, industry, healthcare and security. However, the limitation in accessing large facilities has led to an increasing interest in finding a table-top sources. Laser-driven ion accelerators have been investigated as a possible solution[1], mainly by using the ions accelerated via the Target-Normal Sheath Acceleration (TNSA)[2] mechanism impinging on suitable neutron converter targets. Despite of being a robust mechanism, TNSA exhibits a slow ion energy scaling with respect to the incident laser intensity ($E_{ion} \propto \sqrt{I_L}$), with a beam predominantly formed by protons, reducing the range of possible nuclear reactions involved in the neutron generation.

Here we report on the neutron generation using radiation-pressure driven ion acceleration (RPA) [3]. This mechanism holds the advantage of generating higher energy ion beams, including heavy ions, with high laser-ion conversion efficiency and low divergence. Unlike typical experiments, where a secondary converter target is needed, a high-flux, directional neutron beam was produced from the primary target itself when RPA mechanism is involved.

## Experimental Setup

The experiment was carried out at the Rutherford Appleton Laboratory (RAL), STFC, UK by employing the petawatt arm of the VULCAN laser (wavelength ($\lambda$)=1053nm) system. A

*s.kar@qub.ac.uk

schematic of the experimental setup is shown in Fig. 1, where the laser is focussed on the target foil at normal incidence by an $f/3$ parabola down to a $\sim 5$ μm spot, delivering a total energy of $100-250$ J on the target in $850 \pm 150$ fs pulses and leading to a peak intensity ($I_0$) on the target in the range $(1-3) \times 10^{20}$ W cm$^{-2}$. The targets were made of deuterated plastic ($(C_2D_4)_n$) with thickness in the range $l = 90-900$ nm and 10μm. The ions accelerated during the interaction were diagnosed using five Thomson Parabola Spectrometers (TPS). Due to the overlapping traces of other ion species with similar charge-to-mass ratios ($C^{6+}$, $O^{8+}$), differential filtering technique [4] was implemented to obtain the full spectrum of deuterons. The neutrons generated from the laser irradiated target were diagnosed along the laser axis and $\sim 135°$ off axis by using absolutely calibrated fast plastic scintillator detectors[5] in time-of-flight (nToF) configuration, shielded against Bremsstrahlung and $\gamma$ radiations by $\sim 5$ cm of lead.

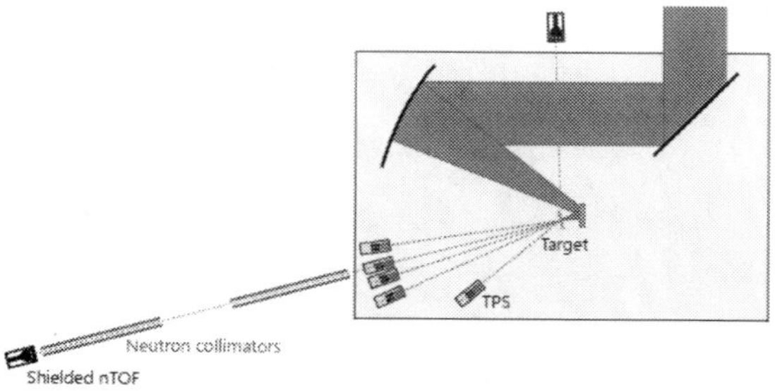


Figure 1: *Experimental setup*

**Neutron generation**

Most of the efforts regarding the generation of a bright, directional neutron beam have been focussed on the use of a pitcher-catcher configuration, in which the laser-driven ions impinge in a secondary target, where the nuclear reactions take place. The need of such double-target setup arises from the fact that the laser irradiated targets produce low flux neutrons isotropically. In order to compare the neutron generation from the laser-target interaction itself depending on the underlying mechanism, both the neutron and ion spectra where recorded for a series of shots, where CD-foil targets varied between 90nm and 10μm thick while keeping the laser conditions as constant as possible.

As expected from the TNSA mechanism, quasi-exponential ion spectra were obtained from the irradiation of thick targets (10μm) as shown in Fig. 2(a). Since the TNSA mechanism favours acceleration of highest charge-to-mass ratio ion species (protons supplied by the hydrocarbon contamination layer on the target surface), a non-efficient acceleration of the deuterium ions (upto a maximum energy of ~5 MeV/nucleon) was observed in this case. However, when the target thickness was reduced to hundreds of nanometres, narrow-bandwidth spectral features appeared towards the high energy end of the deuteron spectra, with significant increase in cut-off energy. Such spectral behaviour for heavy ions from ultra-thin targets has been previously reported by Kar *et al.* [3], attributing to the onset of RPA-LS mechanism in a TNSA hybrid regime. Finally, a further reduction in the target thickness to 90nm led to disappearance of the bunched structure, despite of the increase in deuteron cut-off energy.

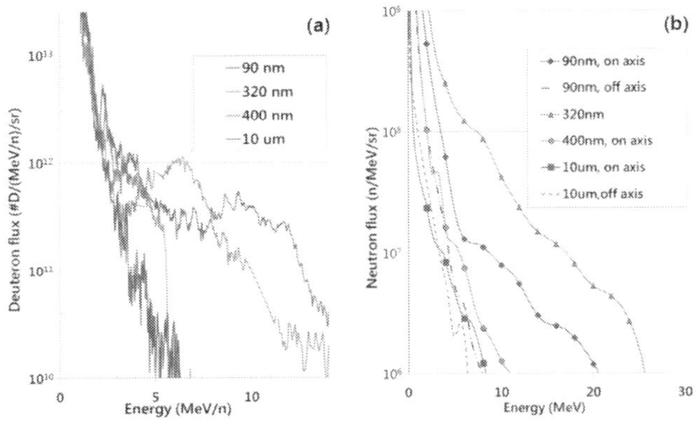


Figure 2: ***(a)** Appearance of narrow-bandwidth features in the deuterium spectra. **(b)** Variation of the neutron spectrum depending on the target thickness.*

A comparison between the neutron spectra obtained for different target thickness is shown in Fig. 2(a). For clarity, the off-axis spectra were only included for the two extreme thicknesses, showing how the emission at that angle was similar for all the targets. As expected for thick targets (10μm), where the energetic ions are driven by TNSA with an exponential energy spectra and large divergence (Fig. 2a), an isotropic, low-flux emission of neutrons was measured. In this case the neutrons are produced most likely at the target front surface due to the ions driven by hole-boring mechanism impinging the bulk of the target. However, as the thickness of the target was reduced to the RPA range, a substantial improvement in on-axis neutron flux was observed, which led to an increase in neutron energy and beam anisotropicity, as the off-axis emission remained constant. Finally, a large drop in the neutron flux was measured when the thickness was further reduced to 90nm. It is interesting to note here that the increase in neutron flux doesnot correlate to the increase in ion energy. Therefore the enhancement in neutron flux must relate to the conditions during the ion acceleration.

In order to explain the increase in flux when the RPA range is approached, as well as the eventual drop, a plasma cylinder model can be considered, in which neutron yield is given by

$$Y_n \simeq \frac{\tau_{burn}}{2} \int n_d^2 \langle \sigma v \rangle \, dV \tag{1}$$

where, $n_d$ is the deuteron number density interacting for an interval $\tau_{burn}$, called fusion burn time, and $\langle \sigma v \rangle$ is the velocity averaged fusion reactivity. With such model, the high neutron flux generated in the RPA case can be explained in terms of the high deuterium density ($n_d$) within the simultaneously-accelerated bunch, explaining the increase with respect to the TNSA/HB case. Nevertheless, a further reduction of the thickness can lead to the so-called Self-Induced Transparency, either due to relativistically induced transparency or a drop in the density following the target expansion. In such case, the bunch structure will promptly disappear, which explains the ion spectra for 90nm in Fig.2a. Also, the large drop in the interacting time of the ions reflects in the plasma cylinder model as a reduction of the burning time, which in turn explains the low efficiency in neutron generation in that case.

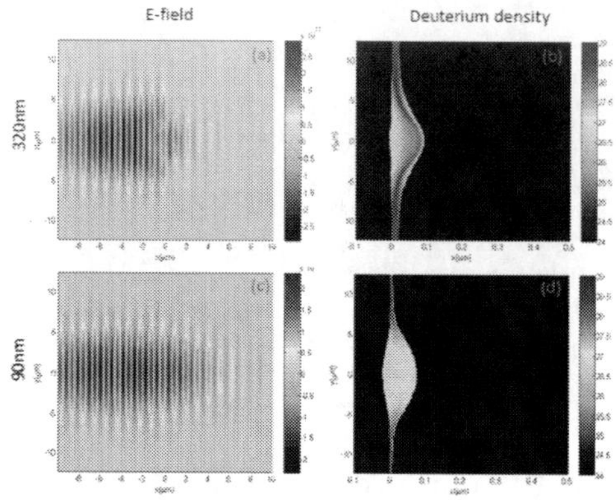


Figure 3: *Simulation results*

To substantiate the observed drop in efficiency for ultrathin targets, 2D PIC simulations were performed using the EPOCH code. Due to the limited computational resources, the parameters of the laser-plasma interaction were scaled down maintaining $I_0/c^3$ and $a_0^2\tau_p/\chi$ constant. When the 320nm foil is used (Fig. 3a,b), the laser beam is still reflected on the target surface, allowing for an optimum LS acceleration, with only a minor part of it travelling through as a consequence of an evanescent wave being able to reach the back of the thin target. However, when the thickness is further reduced to 90nm, the target becomes fully transparent rapidly before the laser reaches its peak intensity, as it can be seen from the electric field travelling through the plasma (Fig. 3c), leading to the heating of the whole plasma and removing the bunched-structure (Fig. 3d).

## Conclusion

The cohesive acceleration of ion bunches during the LS phase is underpinned implicitly by the observed increase in neutron flux and anisotropy, so as by the observation of abrupt drop in neutron generation with sufficiently thin targets, most likely becoming relativistically transparent. The neutron spectroscopy of the interaction was not only an useful diagnostic to understand and optimise the ion acceleration mechanism, but neutron generation from the LS driven ion bunches has been found to be an efficient route to produce a compact laser-driven neutron source, alternative to traditionally used neutron converters in front of the laser irradiated target.

## References

[1] K. Lancaster et al., Phys. Plasmas, **11**, 3404 (2004)

[2] S. Wilks et al., Phys. Plasmas, **8**, 542 (2001).

[3] S. Kar et al., Physical Review Letters **109**(18) (2012)

[4] A. Alejo et al., Review of Scientific Instruments **85**(9) (2014)

[5] S.R. Mirfayzi et al., Review of Scientific Instruments **86**(7) (2015)

# Three-dimensional magnetized and rotating hot plasma equilibrium and stability in a gravitational field

Peter J. Catto[1], Sergei I. Krasheninnikov[2], Istvan Pusztai[3]

[1] *Plasma Science and Fusion Center, MIT, Cambridge, MA 02139, USA*

[2] *University of California at San Diego, La Jolla, CA 92093, USA*

[3] *Department of Applied Physics, Chalmers U. of Technology, 41296 Gothenburg, Sweden*

**Abstract**

We present analytic solutions for three-dimensional magnetized axisymmetric equilibria confining rotating hot plasma in a gravitational field. Our solution to the full Grad-Shafranov equation can exhibit strong equatorial plane localization of the plasma density and current, resulting in disk equilibria for the plasma density. Unlike in [1], we find a toriodal magnetic field is necessary to find a equilibrium in the presence of gravity for most cases of interest. We expect our results to provide impetus to re-investigate magneto-rotational stability [2,3].

**Introduction:** Fully self-consistent three dimensional global equilibrium of hot, rotating plasma confined by gravity and magnetic field are of interest for astrophysical and space plasma applications, but have proven difficult to find. Previous magnetohydrodynamic (MHD) models assume strict incompressibility with constant density [4], assume the density is a flux function [5], ignore the frozen in constraint [6], and/or require poloidal flow [5-7] or an adiabatic equation of state [8,9], which are not allowed kinetically. To satisfy constraints imposed on a drifting Maxwellian by the Fokker-Planck equation, only toroidal flow is allowed and the temperature must be a flux function [1,10-12]. Moreover, in the presence of rotation and gravity the density must be allowed to vary poloidally as well as radially since strict Keplerian motion is only possible at the equatorial plane [1]. It is anticipated that the equilibria we find will be useful in setting up global simulations to investigate magneto-rotational stability [2,3] in accretion disks and help shed light on the detailed mechanisms by which mass accretion occurs at a black hole as momentum is transported outward.

**Grad-Shafranov equation:** The flux surfaces for an axisymmetric equilibrium must satisfy a Grad-Shafranov equation that we find from Ampere's law and pressure balance by taking the magnetic field to be given by

$$\vec{B} = I\nabla\zeta + \nabla\psi \times \nabla\zeta , \qquad (1)$$

where $\zeta$ is the toroidal angle, $\psi$ is the poloidal flux function, $I = RB_T$ with $B_T$ the toroidal magnetic field and R the cylindrical radius from the axis of symmetry. To satisfy the kinetic constraints [10-12] the velocity $\vec{V}$ must be toroidal

$$\vec{V} = \Omega R^2 \nabla\zeta , \tag{2}$$

with $\Omega = c\, d\Phi/d\psi$ the toroidal rotation frequency. The magnetic field is frozen into the flow so that $c\nabla\Phi = \vec{V}\times\vec{B}$ gives $\vec{B}\cdot\nabla\Phi = 0$, requiring the electrostatic potential $\Phi$ to be a flux function to lowest order. In addition, we write the gravitational potential G as

$$G = -G_O M_O / r \tag{3}$$

with $G_O$ the gravitational constant, r the spherical radius, and $M_O$ the mass of the astrophysical body that is assumed to be a compact source centered at r = 0, such that R = r sin$\vartheta$ with $\vartheta$ the angle from the axis of symmetry. Our spherical and cylindrical coordinates are defined to satisfy $r\nabla\vartheta = R\nabla\zeta\times\nabla r$ and $\nabla z = R\nabla R\times\nabla\zeta$.

Conservation of momentum requires

$$c^{-1}\vec{J}\times\vec{B} = Mn(\nabla G - \Omega^2 R\nabla R) + \nabla p . \tag{4}$$

The parallel component requires that the ion density n depend on poloidal angle

$$n = n(\psi,\vartheta) = \eta(\psi)e^{\kappa(\psi,\vartheta)} = \eta(\psi)e^{\kappa(\psi,\vartheta)} , \tag{5}$$

where the density and normalizing or "pseudo" density $\eta$ are related via the generalized Maxwell-Boltzmann exponential factor $\kappa$ that retains the poloidal dependence due to the centrifugal and gravitational potential,

$$4T\kappa/M = \Omega^2 R^2 - 2G , \tag{6}$$

with $T = T(\psi)$ and $p = n(T_i+ZT_e) = 2nT$ in a quasi-neutral plasma ($Zn = n_e$ = electron density) of ion charge number Z and ion and electron temperatures $T_i$ and $T_e$. The toroidal momentum balance requires the current density, $\vec{J}$, across a flux surface vanish, $\vec{J}\cdot\nabla\psi = 0$, and then Ampere's law gives $I = I(\psi)$. Ampere's law also gives $4\pi c\ ^{1}\vec{J}\cdot\nabla\vartheta = \vec{B}\cdot\nabla\vartheta dI/d\psi$. Then the current density can be conveniently decomposed into toroidal and parallel components by writing it as $\vec{J} = RJ_*\nabla\zeta + (c/4\pi)dI/d\psi\vec{B}$, allowing us to obtain the useful relation $\vec{J}\cdot\vec{B} = I\vec{J}\cdot\nabla\zeta + (cB_P^2/4\pi R^2)dI/d\psi$, with $\vec{B}_P = \nabla\psi\times\nabla\zeta$ the poloidal magnetic field and $B_P^2 = R^{-2}|\nabla\psi|^2$. Combining this result with the $\nabla\psi$ component of force balance, $R^2B^2\vec{J}\cdot\nabla\zeta = I\vec{J}\cdot\vec{B} + c\nabla\psi\cdot[\nabla p + Mn(\nabla G - \Omega^2 R\nabla R)]$, and the toroidal component of Ampere's law yields the Grad-Shafranov equation [13]

$$\nabla\cdot(R^{-2}\nabla\psi) = -IR^{-2}dI/d\psi - 4\pi R^{-2}B_P^2\nabla\psi\cdot[\nabla p + Mn(\nabla G - \Omega^2 R\nabla R)] . \tag{7}$$

**Separable form:** To find a separable form for the Grad-Shafranov equation we assume

$$\psi = \psi_O H(\mu)(R_O/r)^\alpha , \tag{8}$$

where $\mu = \cos\vartheta$ and our normalization is $H(\mu=0) = 1$. The vacuum limit $H = 1 - \mu^2$ recovers a homogeneous magnetic field for $\alpha = -2$ and a point dipole solution if $\alpha = 1$, and $\psi_O$ is a constant reference value at the reference location $R_O$ (reference values are defined at the equatorial plane $\mu = 0$ and denoted by a subscript "o"). We obtain an ordinary differential equation for H by assuming

$$\Omega^2 = \Omega_O^2(\psi/\psi_O)^{3/\alpha} \propto 1/r^3 , \tag{9}$$

$$T = T_O(\psi/\psi_O)^{1/\alpha} , \tag{10}$$

$$I = I_O(\psi/\psi_O)^{1+1/\alpha} , \tag{11}$$

$$n = n_O(\psi/\psi_O)^{2+3/\alpha} e^{\kappa(\mu)} , \tag{12}$$

with $n_O = n(\psi = \psi_O, \mu = 0)$. Defining the positive constants

$$g = \frac{8\pi G_O M_O M n_O}{B_{PO}^2 R_O} = \left.\frac{8\pi G_O M_O M n}{B_P^2 r}\right|_{\mu=0} , \tag{13}$$

$$\omega^2 = \frac{4\pi M n_O \Omega_O^2 R_O^2}{B_{PO}^2} = \left.\frac{4\pi M n \Omega^2 r^2}{B_P^2}\right|_{\mu=0} , \tag{14}$$

$$\beta = \frac{16\pi n_O T_O}{B_{PO}^2} = \left.\frac{8\pi p}{B_P^2}\right|_{\mu=0} , \tag{15}$$

and

$$b^2 = \frac{I_O^2}{R_O^2 B_{PO}^2} = \left.\frac{I^2}{R^2 B_P^2}\right|_{\mu=0} = \left.\frac{B_T^2}{B_P^2}\right|_{\mu=0} , \tag{16}$$

an ordinary second order nonlinear differential equation for H is obtained from (7) [13]:

$$\frac{d^2H}{d\mu^2} + \frac{\alpha(\alpha+1)}{1-\mu^2}H = \alpha\left[\frac{g}{2}H^{-1/\alpha} - \omega^2(1-\mu^2)H^{2/\alpha} - (\alpha+2)\beta - \frac{(\alpha+1)b^2}{(1-\mu^2)H^{2/\alpha}}\right]H^{1+4/\alpha}e^{\kappa} , \tag{17}$$

where

$$\kappa = -(g/\beta)(1 - H^{-1/\alpha}) - (\omega^2/\beta)[1 - (1-\mu^2)H^{2/\alpha}] , \tag{18}$$

**Solution technique:** Solutions to this Grad-Shafanov equation can be found by the techniques illustrated in [1] and references therein, however, some care is needed. The solution H must be up-down symmetric and monotonically decreasing from unity until the poloidal magnetic field vanishes at the axis of symmetry ($1 \geq H \geq 0$). This behavior seems to require $d^2H/d\mu^2 < 0$ for all $\mu$. Consequently, we desire $g - (\alpha+1)(2b^2+1) > 2\omega^2 + 2(\alpha+2)\beta$ at $\mu = 0$. Moreover, gravity forces us to assume $\alpha < 0$ to avoid singular behavior as $H \to 0$, with rotation requiring $H \leq (1-\mu^2)^{-\alpha/2}$ to avoid exponential growth at the poles. An exact solution,

$H=(1-\mu^2)^{-\alpha/2}$, exists if G = 0 [1] when $\alpha=-[2(\beta+1)+\omega^2+b^2]/(\beta+1+b^2)$. The poloidal magnetic field decreases with radius if $\alpha > -2$ and the toroidal magnetic field remains finite at the poles if $\alpha < -1$ so we expect solutions with $-1 > \alpha \geq -2$ are of most interest in the presence of gravity. The g = 0 solution suggests maintaining $d^2H/d\mu^2 < 0$ as $\mu^2 \to 1$ requires $[-(\alpha+1)]b^2 > \omega^2 + (\alpha+2)\beta$, a condition that is needed even when g is retained since $H^{-1/\alpha} \to 0$. This restriction implies a toroidal magnetic field is needed to obtain a fully self-consistent gravitational equilibrium. However, the procedures of [1] remain valid for obtaining solutions for $-1 > \alpha \geq -2$ with $b^2$ retained. Consequently, it is possible to find plasma disk equilibrium solutions to the Grad-Shafranov equation (17) and (18) that predict the rotation frequency bound, and the conditions for a plasma disk as well as its thickness.

**Magneto-rotational instability (MRI) without gravity:** In the absence of gravity the toroidal magnetic field can be neglected. In this case $\alpha < -2$, but the MRI calculation is more tractable [2]. The equilibrium associated with our exact solution for H is unstable if $\beta >> 1$ and $\Omega^2R^2 < 10T/M$ in the incompressible limit (roughly in agreement with [2]), and when $\beta < 3/2$ and $\Omega^2R^2 >> 2T/M$ in the compressible limit not considered in [2]. In the presence of gravity [3] the toroidal magnetic field must be retained to investigate MRI instability.

**Work supported by the US Department of Energy grants DE-FG02-91ER-54109 at MIT and DE-FG02-04ER54739 at UCSD & by an International Career Grant from Vetenskapsrådet.**

[1] CATTO, P. J. & KRASHENINNIKOV S. I. 2015 A rotating and magnetized three-dimensional hot plasma equilibrium in a gravitating field. *J. Plasma Physics* **81**, 105810301.

[2] VELIKHOV, E. P. 1959 Stability of an ideally conducting liquid flowing between cylinders rotating in a magnetic field. *Sov. Phys. JETP* **36**, 995-998 (1959).

[3] BALBUS, S. A. & HAWLEY, J. F. 1991 A powerful local shear instability in weakly magnetized disks. I. Linear analysis. *Astrophys. J.* **376**, 214-222.

[4] CHANDRASEKHAR, S. 1956 Axisymmetric magnetic fields and fluid motions. *ApJ* **124**, 232-243.

[5] THROUMOULOPOULOS, G. N. & TASSO, H. 2001 Axisymmetric equilibria of a gravitating plasma with incompressible flows. *Geophys. Astroph. Fluid Dynamics* **94**, 249-262.

[6] PRASANNA, A. R., TRIPATHY, S. C. & DAS, A. C. 1989 Equilibrium structure for a plasma magnetosphere around compact objects. *J. Astrophys. Astr.* **10**, 21-34.

[7] MCCLEMENTS, K. C. & THYAGARAYA, A. 2001 Azimuthally symmetric magnetohydrodynamic and two-fluid equilibria with arbitrary flows. *Mon. Not. Roy. Astr. Soc.* **323**, 733-742.

[8] LOVELACE, R. V. E., MEHANIAN, C., MOBARRY, C. M. & SULKANEN, M. E. 1986 Theory of axisymmetric magnetohydrodynamic flows: disks. *ApJS* **62**, 1-37.

[9] OGILVIE, G. I. 1997 The equilibrium of a differentially rotating disc containing a poloidal magnetic field *Mon. Not. R. Astron. Soc.* **288**, 63-77.

[10] HINTON, F. L. & WONG, S. K. 1985 Neoclassical ion transport in rotating axisymmetric plasmas. *Phys. Fluids* **28**, 3082-3098.

[11] CATTO, P. J., BERNSTEIN, I. B. & TESSAROTTO, M. 1987 Ion transport in toroidally rotating tokamak plasmas. *Phys. Fluids* **30**, 2784-2795.

[12] HELANDER, P. 2014 Theory of plasma confinement in non-axisymmetric magnetic fields. *Rep. Prog. Phys.* **77**, 087001-35.

[13] KRASHENINNIKOV, S. I., CATTO, P. J. & HAZELTINE, R. D. 2000 Plasma equilibria in dipolar magnetic configurations. *Phys. Plasmas* **7**, 1831-1838.

# Cosmic-ray acceleration at perpendicular shocks

J.G. Kirk[1], M. Takamoto[2]

[1]*Max-Planck-Institut für Kernphysik, Heidelberg, Germany*

[2]*Department of Earth and Planetary Science, The University of Tokyo, 7-3-1 Hongo, Bunkyo-ku, Tokyo 113-0033, Japan*

Diffusive shock acceleration (DSA) in supernova remnants is the dominant paradigm for the acceleration of cosmic rays in the Galaxy. However, the maximum possible energy to which parallel shocks can accelerate protons in these objects is only roughly 1 PeV [1], in tension with the observed cosmic ray spectrum. Perpendicular shocks present a possible alternative [4], and are expected to predominate if the supernova progenitor is a rotating, massive star with a powerful wind [2]. However, stochastic acceleration in this situation is complicated, because the underlying transport process is not necessarily diffusive, in which case the standard results concerning particle spectra and acceleration rates lose their validity [6]. This paper presents a new approximate, analytical approach to the test particle problem at perpendicular shocks, using an expansion in eigenfunctions of the pitch-angle and phase dependent scattering operator. An explicit expression for the spectral index of accelerated particles is found and compared with the results of Monte-Carlo simulations. Full details can be found in [7].

The goal is to solve for the stationary cosmic-ray distribution $f$ at a perpendicular shock as a function of momentum $\vec{p}$ and distance from the shock, assuming that, in addition to their gyro-motion, particles are continually deflected by magnetic fluctuations whose effect can be modeled as isotropic diffusion in the direction of motion, i.e., as diffusion on the sphere of the end-point of the unit vector $\vec{p}/p$. In both the downstream and upstream plasmas, $f$ then obeys a Fokker-Planck equation. When $\vec{x}$ and $t$ are measured in the shock rest frame and $\vec{p}$ in the local fluid frame this takes the form:

$$\frac{\partial f}{\partial t}+(v_z-u)\frac{\partial f}{\partial z}=\omega_g\left\{-\frac{\partial f}{\partial\phi}+\frac{1}{2\eta}\left[\frac{\partial}{\partial\mu}\left(1-\mu^2\right)\frac{\partial f}{\partial\mu}+\frac{1}{1-\mu^2}\frac{\partial^2 f}{\partial\phi^2}\right]\right\} \qquad (1)$$

Here, $u$ (assumed $\ll 1$) is the shock speed and $v$ the particle speed (both in units of $c$), $\omega_g$ the gyro-frequency, $\mu$ the pitch-angle cosine and $\phi$ the gyro-phase. The uniform magnetic field lies along the $y$-axis and the shock is in the $x$–$y$ plane, so that the component of the particle velocity along the shock normal is $v_z = v\sin\theta\sin\phi$. The ratio of $\omega_g$ to the "collision frequency" is denoted by $\eta$. Solutions are found by separating the variables, leading to the eigenvalue problem

$$\Lambda_i\,(v_z - u)\,Q_i = \left\{ -\frac{\partial}{\partial \phi} + \frac{1}{2\eta}\left[\frac{\partial}{\partial \mu}\left(1-\mu^2\right)\frac{\partial}{\partial \mu} + \frac{1}{1-\mu^2}\frac{\partial^2}{\partial \phi^2}\right]\right\} Q_i \qquad (2)$$

together with boundary conditions on $Q_i(\mu,\phi)$ that ensure regularity and single-valuedness on the sphere. Equation (2) has both positive and negative eigenvalues (labelled with $i > 0$ and $i < 0$ respectively), in addition to the eigenvalue $\Lambda_0 = 0$ with the isotropic eigenfunction $Q_0 =$ constant. These govern the spatial dependence of the solution: the isotropic part is independent of $z$, whereas the eigenfunctions with $i < 0$ decay exponentially upstream ($z > 0$), and grow exponentially downstream ($z < 0$) on a length scale that decreases as $|i|$ increases; eigenfunctions with $i > 0$ have the opposite behaviour. Following the procedure used for relativistic shocks [5], boundary conditions upstream are applied by limiting the expansion upstream to terms with $i < 0$. Matching conditions at the shock (essentially Liouville's theorem) introduce a dependence on $p$ into the problem and enable one to compute the corresponding downstream distribution. This is then subjected to the boundary conditions far downstream by requiring it to be orthogonal to the eigenfunctions with $i < 0$. In the absence of a source term, $f$ is a scale-free power law in momentum: $f \propto p^{-s}$ and $s$, as well as the spatial and angular dependence of $f$, is determined by the above procedure.

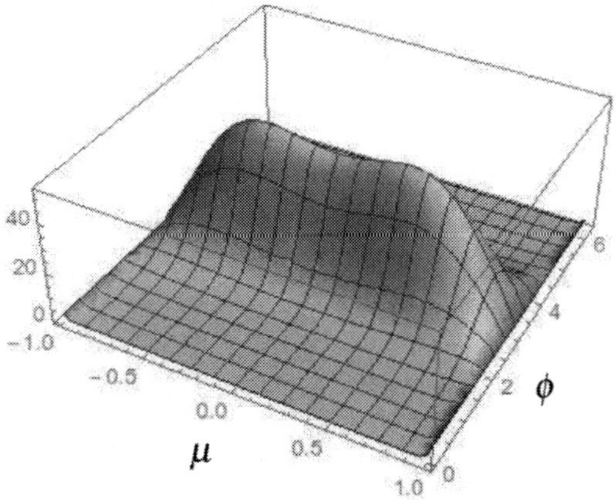


Figure 1: The leading eigenfunction (3) as a function of $\mu$ and $\phi$, for $\eta u = 2$ and $v = 1$ (for which $\Lambda_{-1} = -5.2$).

The eigenfunction with $i = -1$ plays a special role here, since it corresponds to the only term in the expansion that survives at large distance upstream, and, therefore, may not change sign in the physically relevant range ($0 < \phi \leq 2\pi$, $-1 \leq \mu \leq 1$). Using this single term to represent the upstream distribution was found to give a reasonably good approximation in the one-dimensional case studied in [5]. This is the strategy followed here, motivated partly by the relatively large numerical effort involved in computing the higher-order functions in the current, two-dimensional problem, compared to the straightforward and efficient solution available via Monte-Carlo methods.

In the nonrelativistic limit $u \to 0$, this eigenfunction has a small ($\sim u$) anisotropy proportional to $\sqrt{1-\mu^2}\,(\eta \cos\phi - \sin\phi)$, and an eigenvalue $\Lambda_{-1} \approx -3u\eta$. Its anisotropic part gives the two-dimensional equivalent of the singular "diffusion" solution discussed by Fisch & Kulsrud [3], who were concerned with the incompleteness of the eigenfunctions in the case $u = 0$. In this limit, it is straightforward to show that the standard result of DSA follows: $s = 3r/(r-1)$, where the shock compression ratio $r = u_s/u_d$ is ratio of the shock speed in the upstream medium $u_s$ to that in the downstream medium $u_d$. However, this result is valid only for $u_{s,d} \sim \varepsilon \ll 1/\eta$. In supernova remnants, $u_{s,d} \sim 10^{-1}$–$10^{-2}$, and may be comparable or larger than $1/\eta$. Therefore, an approximation scheme that sets $u \sim 1/\eta \sim \varepsilon \ll 1$ is more appropriate. The resulting eigenvalue is then $\sim \varepsilon^0$, and the eigenfunction, shown in Fig 1, is [7]

$$Q_{-1}(\mu,\phi) \approx e^{\Lambda v\sqrt{1-\mu^2}\cos\phi}\mathrm{Ps}_0^0\left(\mu, -\Lambda^2/2\right) \tag{3}$$

where $\mathrm{Ps}_n^m$ is the angular, oblate, spheroidal wave function. This eigenfunction describes a strongly anisotropic fan-beam concentrated in the plane of the shock and drifting across the magnetic field in the same direction as the drift of unperturbed trajectories. For large $\eta u_s$, the width of this beam in phase is approximately $(\eta u_s)^{-1/2}$, which is just the diffusive spread caused when the collision operator in (1) operates on a phase-collimated beam for a time $1/(\omega_g u_s)$ roughly equal to that needed for an unperturbed trajectory to traverse the shock.

In general, evaluation of the resulting power-law index requires a numerical integration. However, using a Taylor expansion of the eigenfunctions in the variable $\eta u_s$ one finds [7]

$$s = \frac{3r}{r-1} + \frac{9(r+1)}{20r(r-1)}\eta^2 {u_s}^2 + \mathrm{O}\left(\eta^4 {u_s}^4\right) . \tag{4}$$

In Fig 2, this result is compared with those of fully relativistic Monte-Carlo simulations [7], for $r = 4$, ten values of $\eta$ between 1 and 100, and thirty values of $u_s$ between 0.01 and 0.2. Qualitatively, the agreement is good. In particular, the simulations confirm that, to a good approximation, $s$ depends on $\eta$ and $mu_s$ only in $\eta u_s$, since, for a fixed value of this parameter, Fig 2 shows only a weak variation of $s$ with the colour coding (which denotes the $\eta$ value). Also, the softening of the spectrum for $\eta u_s > 1$ is confirmed. However, there is a quantitative discrepancy, which may result either from

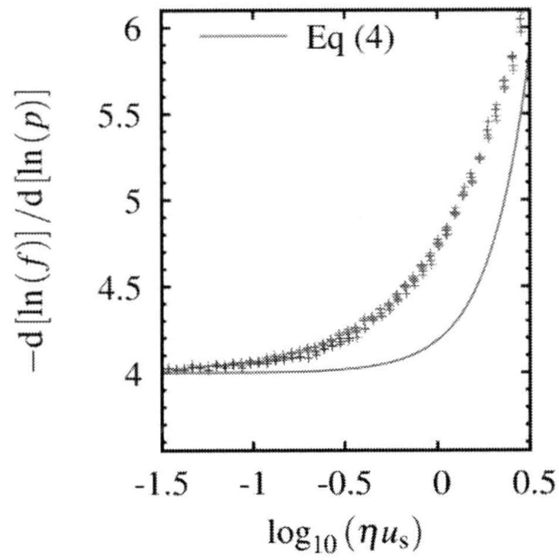


Figure 2: Comparison of the spectral index calculated using the leading eigenfunction, (Eq. (4)) with the results of fully relativistic Monte-Carlo simulations [7], for a matrix of parameters, colour-coded according to the value of $\eta$ ranging from 1 (blue) to 100 (red).

the asymptotic expression used for the eigenfunction (which is valid only in the limit $u_s \to 0$ and $\eta \to \infty$), or from the use of only the leading eigenfunction to represent the distribution, or, of course, from both of these approximations.

Nevertheless, we conclude that non-relativistic, perpendicular shocks are effective accelerators provided $\eta u_s \leq 1$. Combining this with the conjecture advanced in [4] and confirmed in [7], that the acceleration timescale at these shocks decreases monotonically with increasing $\eta$, leads to the conclusion that the optimal configuration for acceleration at a nonrelativistic shock front is one where the shock is perpendicular and the turbulence is relatively weak, such that $\eta u_s \approx 1$. In such a configuration, acceleration to energies significantly above 1 PeV appears possible for parameters thought appropriate for supernova that explode in the winds of massive stars [8].

**References**

[1] A.R. Bell, K.M. Schure, B. Reville, G. Giacinti, MNRAS 431, 415 (2013)

[2] P.L. Biermann, The ASTRA Proceedings, 1, 29–31 (2014)
www.astra-proceedings.net/1/29/2014/

[3] N.J. Fisch, M.D. Kruskal, J. Math. Phys. 21, 740 (1980)

[4] J.R Jokipii, ApJ 313, 842 (1987)

[5] J.G. Kirk, A.W. Guthmann, Y.A. Gallant, A. Achterberg, ApJ 542, 235 (2000)

[6] M. Takamoto, J.G. Kirk, Proceedings of the EPS Conference on Plasma Physics, (2014)

[7] M. Takamoto, J.G. Kirk, arXiv:1506.04354, submitted to ApJ (2015)

[8] H.J. Völk, E.G. Berezhko, L.T. Ksenofontov, A&A 433, 229 (2005)

# Collisionless Magnetic Annihilation in Relativistic Laser-Plasma Interactions

Yanjun Gu[1], Ondrej Klimo[1,2], Deepak Kumar[1], Yue Liu[1], Sushil Singh[1], Sergei V. Bulanov[3], Timur Zh. Esirkepov[3], Stefan Weber[1], and Georg Korn[1]

[1]Institute of Physics of the ASCR, ELI-Beamlines project, Na Slovance 2, Prague 18221, Czech Republic;

[2]FNSPE, Czech Technical University in Prague, 11519 Prague, Czech Republic;

[3]Advanced Photon Research Center, Japan Atomic Energy Agency, 8-1-7 Umemidai, Kizugawa-shi, Kyoto 619-0215, Japan

## ABSTRACT

Magnetic reconnection is one of the fundamental phenomena in space and astrophysics. Through the rearrangement of magnetic field topology, it converts the magnetic energy to kinetic energy. Recently, ultra-high intensity laser pulses provide a new regime to generate and investigate magnetic reconnection with laser-plasma interactions. In this paper, we report a regime to induce collisionless magnetic annihilation by ultra-short laser pulses and plasmas. Two magnetic dipoles with amplitudes as high as giga-Gauss are propagating and expanding in the target . The magnetic field lines with opposite direction annihilate each other in the central plane and form a current sheet. The inductive longitudinal electric field grows in the current sheet and accelerates the electrons backwards. This regime may be beneficial for the potential experiments to be carried out on large laser facilities such as ELI.

**Keywords:** magnetic reconnection, magnetic vortex, laser-plasma interaction

## 1. INTRODUCTION

The process of magnetic reconnection is always accompanied with the formation of a current sheet, where the magnetic field lines with opposite directions annihilate each other. Therefore, the research on magnetic annihilation and the corresponding energy dissipation becomes important. Early works, such as famous Sweet-Parker model,[1] based on resistive magnetohydrodynamics (MHD) theory regard Ohmic dissipation as the main energy conversion mechanism. However, the predicted reconnection rates are too slow to explain the observations in universe. This suggests that many magnetic reconnections actually occur in collisionless plasmas such as solar corona. The corresponding theoretical and experimental investigations have been reviewed in many papers.[2–5]

Recently, with the development of laser facilities, TW ($10^{12}$W) and PW ($10^{15}$W) pulses are available which allows to investigate magnetic reconnection in laboratory by the new method of laser-plasma interactions. Pioneering experiments were made by Nilson *et al.*[6] and Li *et al.*[7] using long pulses (1ns) and solid targets. J. Y. Zhong *et al.*[8] obtained X-ray source by reconnection outflows with nanosecond pulses and solid Aluminum targets. In all the above experiments, the magnetic fields are created by the hot electrons, which are accelerated by the laser ponderomotive force. The magnetic field lines flow and expand with the electrons and the time scales for reconnection are from ps ($10^{-12}$s) to ns $10^{-9}$s. Except the experiments, PIC (Particle-in-cell) kinetic simulations and MHD simulations provide researches in explaining the mechanisms of magnetic reconnections.[9–11]

In this proceeding, we report a fast collisionless magnetic annihilation in relativistic laser-plasma interactions by 2.5-dimensional PIC simulations. The relativistic electromagnetic code 'EPOCH' is used in the simulations.[12] The schematic of the simulation is shown in Figure 1. A fully ionized hydrogen plasma locates in $20\ (\lambda) < x < 122\ (\lambda)$ and $-20\ (\lambda) < y < 20\ (\lambda)$. In transverse direction, the density distribution of the target is uniform. In longitudinal direction, the density increases linearly from 0 to 0.1 $n_c$ in $20\ (\lambda) < x < 22\ (\lambda)$ and remains constant for $22\ (\lambda) < x < 62\ (\lambda)$, where $n_c = \frac{m_e \omega^2}{4\pi e^2}$ is the plasma critical density, $m_e$, $\omega$ and e represent the electron mass, laser angular frequency and electric charge, respectively. The long downramp region is in $62\ (\lambda) < x < 122\ (\lambda)$, where the density linearly decreases to 0. Two s-polarized Gaussian pulses with the peak

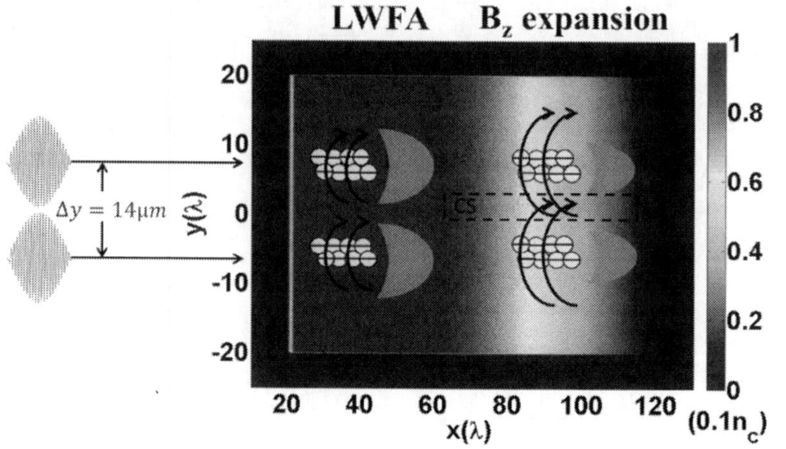


Figure 1. (color online) Schematic of the simulation model.

intensity of $10^{21}\mathrm{W/cm^2}$ propagate along longitudinal direction are focused on the target left boundary. The separation between two pulses are 14 $\lambda$ as shown in Fig. 1, i.e. the optical axes of two pulses are $\pm 7\ \lambda$. Different from the experiments and simulations described above, the pulses in our simulations are ultra-short with the pulse duration $\tau = 15\ fs$. The lasers spot sizes is 3 $\lambda$ and the laser wavelength is $\lambda = 1\ \mu m$. Free boundary conditions are employed and the initial plasma temperature is 0K.

The whole process can be divided in two stages. In the first stage, two electron beams are formed by the LWFA (laser wakefield acceleration) effect. The corresponding magnetic dipole structures are produced by the electron beams according to Ampere's law. Both the energy of electron beam and the strength of the magnetic field are growing in the density plateau region in this stage. In the second stage, the magnetic fields expand in transverse direction after they enter the downramp region. The magnetic fields with opposite direction annihilate in the current sheet. During the annihilation, it is the displacement current which balances the variation of the magnetic field. A strong inductive electric field is generated in the current sheet which intensively accelerates charged particles. In our simulations, we find the high energy electrons accelerated in backward by this field in the current sheet.

## 2. ELECTRON BEAM ACCELERATION AND MAGNETIC FIELD GENERATION

In this part, we discuss the electron acceleration and the corresponding magnetic fields generation in the first stage. It is well known that short laser pulse generate strong wakefield when it is propagating in a plasma. The strength of this longitudinal electric field is given by[13]

$$E_{max} = 0.38(2\gamma)(1+\gamma)^{(-1/2)}\sqrt{(n_0)} \approx 34\ \mathrm{GV/cm}. \tag{1}$$

Here $\gamma$ is the relativistic factor of the electrons, which is approximately equal to the dimensionless amplitude of laser pulse $a_0 = eE_0/m_e\omega c$. The electron density distribution at 51 $T_0$ is plotted in Fig. 2(a). Two bubble structures have formed and two electron beams are trapped by each bubble. The electrons are accelerated by the longitudinal wakefield to several tens MeV inside the bubble. Fig. 2(b) presents the electron momentum distribution along y-direction (i.e. $y - p_x$). The electron highest energy reaches about 50 $MeV$ at this moment. The return electrons with maximum momentum locate along the laser pulse. However, between the two pulses (i.e. $-\lambda < y < \lambda$), the electrons momentum is rather small. The distribution of backward accelerated electrons is an important point to distinguish the outflow in magnetic annihilation which will be presented below.

According to the Ampere circuital law, electron current generates magnetic loop surrounding it. In Fig. 2(c), two symmetric magnetic dipoles are produced by the LWFA electron beams. The amplitude of the magnetic fields can be estimated by Ampere-Maxwell law as

$$c\nabla \times B = 4\pi J + \partial E/\partial t. \tag{2}$$

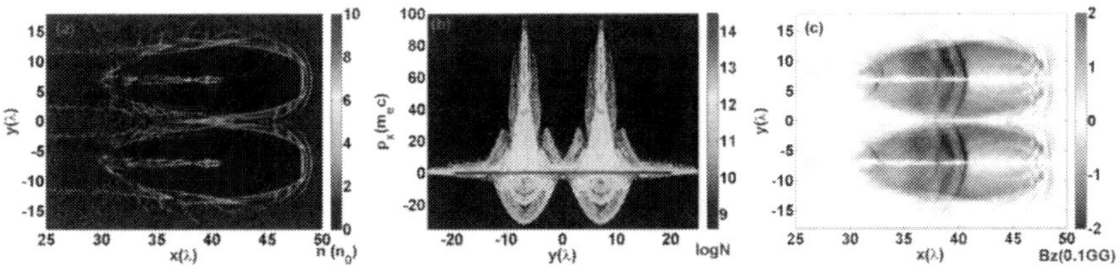


Figure 2. (color online) At 51 $T_0$, (a) The electron density distribution. (b) Electron momentum $y - p_x$ distribution. (c) The $B_z$ distribution.

Assuming the quasi electrostatic condition as $\partial_t E = 0$ and the maximum radius of the dipole is $R \approx c\sqrt{a_0}/\omega_p$. It gives the amplitude of magnetic field $B_z = 4\pi n_0 ec\sqrt{\gamma}/\omega_p \approx 0.2GG$.[14] At this stage, the two magnetic dipoles are separate. One can find the amplitude of $B_z$ in the simulation is well agree with the theory estimation.

## 3. MAGNETIC EXPANSION, ANNIHILATION AND PARTICLE ACCELERATION

After entering the density downramp, the magnetic dipoles begin expanding in the transverse direction. It is because the force acting on the bubbles which is proportional to $\nabla n \times \Omega$, where $\Omega$ is the potential vorticity.[15] In the center region, there is no enough space for the dipoles to expand freely. The magnetic fields with opposite directions attach each other and start the process of annihilation. Fig. 3(a) is the magnetic field distribution at 141 $T_0$, the dipole structures are asymmetric with large size of the outer part and small size of the inner part. In 105 $\lambda < x <$ 115 $\lambda$, the inner magnetic fields have been completely annihilated. At the same time, the annihilation of magnetic fields accompany with the growth of longitudinal electric field in the annihilation region, which is one of the signatures of magnetic annihilation. The layer where magnetic fields annihilate and electric fields grow is also called current sheet. In Fig. 3(b), we compare the profile of $E_x$ along laser axis and current sheet at 141 $T_0$. One can find in the region where magnetic fields cancelled each other, (i.e. 105 $\lambda < x <$ 115 $\lambda$), there is a strong longitudinal electric field in the current sheet, which is even higher than the field along laser axis. The growth of electric field indicates energy conversion between magnetic field and electric field.

To explain the generation of the inductive electric field, we plot the contributions of each term in Ampere-Maxwell law (Eq. (2)) in Fig. 3(c). All the curves are spatial averaged over the current sheet ($-\lambda < y < \lambda$). In the region of annihilation, the electron density is very low due to the downramp distribution, therefore the corresponding electron current is also small. In this case, the variation of the magnetic field, $< c\nabla \times B/4\pi >_z$, cannot be fully compensated by the electron current. The growth of the displacement current, $J_D = \partial E/\partial t$, is necessary to balance the equation and the corresponding electric field is induced. A strong outflow particles is another signature in magnetic annihilation. Due to the growth of the inductive electric field, a bunch of high energy electrons ejects backwards in the current sheet is expected. In Fig. 3(d), the distribution of momentum shows the acceleration effect in the current sheet and the maximum momentum reaches 100 $m_e c$. Compared with the distribution in Fig. 2(b), we find the particles in current sheet get more energy than the particles along laser axis after magnetic annihilation occurs. This proves the magnetic field release its energy to particles in annihilation process via the displacement current and inductive electric field.

## 4. CONCLUSIONS

In this proceeding, a fast magnetic annihilation is obtained with two ultra-short PW laser pulses and non-uniform underdense plasma. Strong inductive electric field and high energy outflow particles are detected. The displacement current plays an important role in magnetic annihilation, which balances the variation of magnetic fields and induces the inductive electric field. The results are beneficial for the potential experiments in the large laser facilities such as ELI project.[16]

## ACKNOWLEDGMENTS

This work was supported by the ELI (Project No. CZ.1.05/1.1.00/02.0061). Computational resources were provided by the MetaCentrum under the program LM2010005 and by IT4Innovations Centre of Excellence

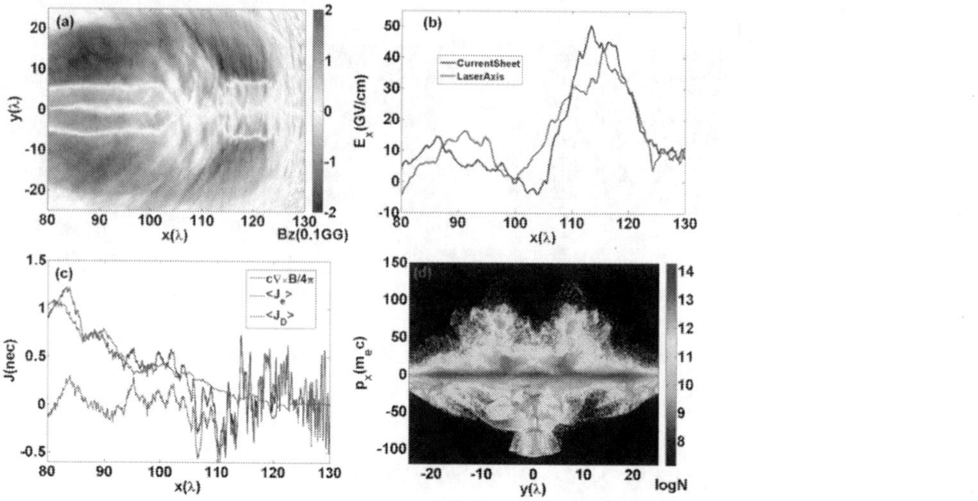


Figure 3. (color online) At 141 $T_0$, (a) The magnetic field distribution. (b) The longitudinal electric field profile along current ($y = 0$) and laser axis ($y = 7\ \lambda$). (c) The contributions of terms in Maxwell-Ampere equation. (d) Electron momentum $y - p_x$ distribution.

under projects CZ.1.05/1.1.00/02.0070 and LM2011033. The EPOCH code was developed as part of the UK EPSRC funded projects EP/G054940/1.

## REFERENCES


1. E. N. Parker *J. Geophys. Res.* **62**, pp. 509–520, 1957.
2. D. Biskamp *Phys. Fluids* **29**, pp. 1520–1531, 1986.
3. J. F. Drake, R. G. Kleva, and M. E. Mandt *Phys. Rev. Lett.* **73**, pp. 1251–1254, 1994.
4. S. V. Bulanov, T. Z. Esirkepov, D. Habs, F. Pegoraro, and T. Tajima *Eur. Phys. J. D* **55**, pp. 483–507, 2009.
5. S. V. Bulanov, T. Z. Esirkepov, M. Kando, J. Koga, K. Kondo, and G. Korn *Plasma Phys. Rep.* **41**, pp. 1–51, 2015.
6. P. M. Nilson, L. Willingale, M. C. Kaluza, C. Kamperidis, S. Minardi, M. S. Wei, P. Fernandes, M. Notley, S. Bandyopadhyay, M. Sherlock, R. J. Kingham, M. Tatarakis, Z. Najmudin, W. Rozmus, R. G. Evans, M. G. Haines, A. E. Dangor, and K. Krushelnick *Phys. Rev. Lett.* **97**, p. 255001, 2006.
7. C. K. Li, F. H. Sguin, J. A. Frenje, J. R. Rygg, R. D. Petrasso, R. P. J. Town, O. L. Landen, J. P. Knauer, and V. A. Smalyuk *Phys. Rev. Lett.* **99**, p. 055001, 2007.
8. J. Zhong, Y. Li, X. Wang, J. Wang, Q. Dong, C. Xiao, S. Wang, X. Liu, L. Zhang, L. An, F. Wang, J. Zhu, Y. Gu, X. He, G. Zhao, and J. Zhang *Nat. Phys.* **6**, pp. 984–987, 2010.
9. M. Hoshino, T. Mukai, T. Terasawa, and I. Shinohara *J. Geophys. Res. Space Physics* **106**, pp. 25979–25997, 2001.
10. Y. L. Ping, J. Y. Zhong, Z. M. Sheng, X. G. Wang, B. Liu, Y. T. Li, X. Q. Yan, X. T. He, J. Zhang, and G. Zhao *Phys. Rev. E* **89**, p. 031101, 2014.
11. J. F. Drake, D. Biskamp, and A. Zeiler *Geophys. Res. Lett.* **24**, pp. 2921–2924, 1997.
12. C. P. Ridgers, J. G. Kirk, R. Duclous, T. G. Blackburn, C. S. Brady, K. Bennett, T. D. Arber, and A. R. Bell *J. Comput. Phys.* **260**, pp. 273–285, 2014.
13. P. Sprangle, E. Esarey, A. Ting, and G. Joyce *Appl. Phys. Lett.* **53**, pp. 2146–2148, 1988.
14. G. A. Askaran, S. V. Bulanov, F. Pegoraro, and A. M. Pukhov *JETP Lett.* **60**, p. 251, 1994.
15. J. Nycander and M. B. Isichenko *Phys. Fluids B* **2**, pp. 2042–2047, 1990.
16. G. Mourou, G. Korn, W. Sandner, and J. Collier *CNRS, Paris* , 2011.

# Measurement and resonant control of modulated diocotron modes in RF-excited trapped plasmas

G. Maero[1,2], S. Chen[3,2], B. Paroli[1,2], R. Pozzoli[1,2], M. Romé[1,2]

[1] *Dipartimento di Fisica, Università degli Studi di Milano, Milano, Italy*

[2] *INFN Sezione di Milano, Milano, Italy*

[3] *Institute of Fluid Physics, China Academy of Engineering Physics, Mianyang, Sichuan 621900, China*

Radio-frequency (RF) electric fields are routinely used to create plasmas through the generation of discharges in neutral gases, where usual features include the injection of high RF power and pressures as high as $10^{-3} - 10$ mbar in order to sustain the discharge. On the contrary, Penning-Malmberg [1] traps confine non-neutral, magnetized plasmas under radically different conditions, namely in the Ultra-High Vacuum (UHV) regime and with a charged particle sample under study coming in general from an external source, like a thermo-, photo- or field-emission cathode in the case of an electron plasma. The reproducible sample injection combined with the low collision rate in UHV and the high degree of symmetry of magnetic and electric fields allows long-time confinement and precise control and manipulation of the trapped particles.

Experiments performed in recent years in the ELTRAP device [2, 3, 4] demonstrated that as an alternative to common trap sources an unconventional technique can be exploited that borrows from principles used in gas discharges (albeit in the very different context represented by Penning traps), i.e. the continuous application of a weak periodic excitation to any inner electrode of the trap when the endcaps are simultaneously set to a suitable confining potential. Experiments show in particular that a few-volt RF drive may result in significant electron heating and ionization of the residual gas when the drive frequency is of the order or larger than the typical axial bounce frequencies of few-eV electrons, eventually leading to the build-up of a confined electron plasma. A qualitative explanation on the basis of a simple one-dimensional stochastic heating model was given elsewhere [2, 5]. While often RF-generated plasmas exhibit low densities ($10^5 - 10^6$ cm$^{-3}$) and cover a large part of the trap cross-section, empirical tuning of the control parameters (geometry, magnetic field strength, RF parameters) results in the formation of narrow-radius columns that can reach densities of the order of $10^7$ cm$^{-3}$, i.e. comparable to those of conventional sources. It was experimentally found that the narrow columns produced with this technique generally exhibit a radial displacement with respect to the trap axis, although they show interesting stability properties against the well-known growth of the $l = 1$ mode and can have an indefinite equilibrium at a constant offset as a result of the balance

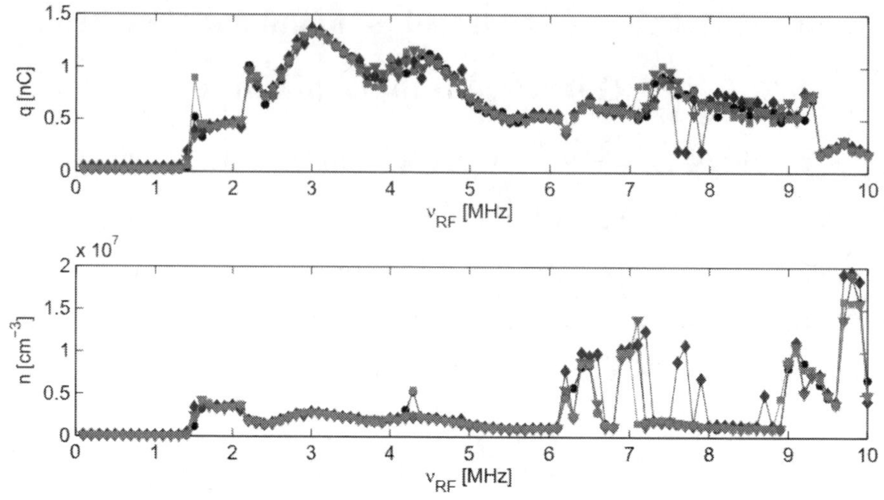


Figure 1: *Repeatability test. The four frequency scan measurements were recorded with the following set of parameters: Trapping length* 570 *mm, magnetic field* $B = 0.1$ *T, excitation amplitude* 1.5 *V. Top diagram: Total confined charge. Bottom diagram: Mean plasma core density.*

between ionization and losses in the presence of a continuous RF application.

The main disadvantage lies in the fact that the generation mechanism is extremely sensitive to experimental conditions and as a consequence the control of the properties of the resulting plasma is not straightforward. Therefore an assessment of the reproducibility of the RF-produced plasmas under constant experimental parameters was performed. In the standard experimental sequence, the frequency $\nu_{RF}$ of the drive was continuously swept at a 20 kHz/s rate in the $0.1-10$ MHz range while one of the endcap potentials was turned off every 5 s to release the plasma and record the image formed on a phosphor screen with a charge-coupled device (CCD) camera. Figure 1 shows the main results of four runs with identical experimental conditions, namely a 1.5 V excitation on the electrode adjacent to one of the endcaps in a 570 mm trapping length, endcap potentials $-80$ V, axial magnetic field 0.1 T. The top and bottom diagrams indicate the total confined electron charge and mean core density, respectively. Both curves are fairly well repeated over different sweep sequences. In particular, the main patterns present in the charge diagram show vertical (total charge) fluctuations within 10% while horizontal (frequency) fluctuations amount to one data point ($\pm 50$ kHz) at most. This guarantees an acceptable statistical reproducibility of the results obtained with RF-generated plasmas. The largest deviations are represented by few sharp spikes (see the range $7-8$ MHz in the picture), which correspond to phenomena like the transition from a diffuse plasma to a narrow and dense column or the modulation of the $l = 1$ diocotron mode of off-centered plasmas. In

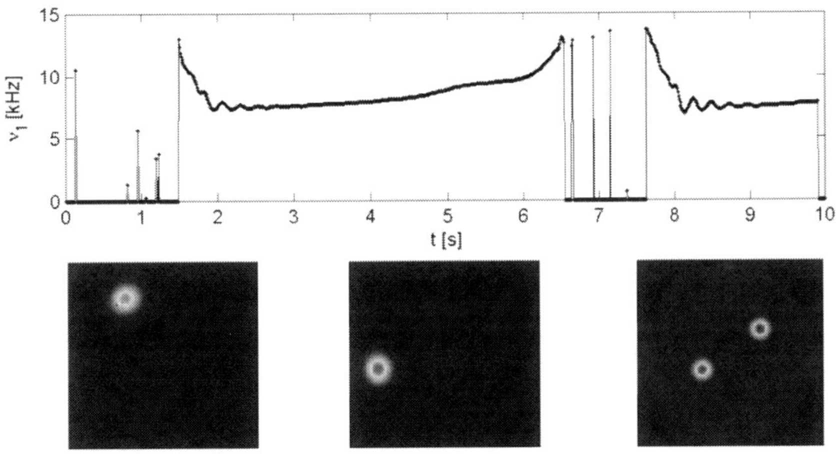


Figure 2: *Evolution of a RF-generated plasma with formation of two stable vortices. Top row: Electrostatic signal of the $l = 1$ diocotron mode showing the cyclic formation, evolution and disappearance of a single plasma column. Bottom row: CCD images at plasma dump. The plasma consists of a narrow column rotating at large offset; the vortex is initially background-free (left), but a diffuse background develops during the evolution (center). After the disappearance of the column, a stable double vortex may be produced (right). CCD images are cropped to the trap cross-section. All figures are normalized to the peak density of* $1.5 \cdot 10^7$ $cm^{-3}$.

the latter case, the observed phenomenon is a steady modulation of both frequency and amplitude of the $l = 1$ mode, which is associated to a periodic variation of the confined charge [3] (possibly due to the damping of diocotron modes mediated by radial losses or other transport phenomena [7]) as well as to an oscillation of the radial displacement. The application of a dipole RF excitation at the modulation frequency (typically in the Hertz range) results in the amplification of the modulation amplitude. Preliminary studies have been performed showing that autoresonant excitation [8] of this slow modulation is possible, leading to new possibilities of manipulation of the plasma column (e.g., radial displacement control).

We have observed that besides diffuse plasmas and narrow columns, under particular conditions a system of two vortices can be generated. Some characteristic features of this event are summarized in Fig. 2. The plasma is created by means of a 9.9 MHz, 0.9 V sinusoidal excitation on two opposite patches of a fourfold-split electrode in a 570 mm long trap at $B = 0.15$ T. As shown by the $l = 1$ mode recorded from another sectored electrode and depicted in the top diagram of Fig. 2, a cyclic evolution takes place where a plasma column is produced, lasts few seconds ($t \simeq 1.5$ to 6.5 s in the diagram) and disappears. The whole sequence can be repeated several times. This dynamics is consistently reproduced over many experimental cycles, allowing us to record CCD images at plasma dump and reconstruct the plasma evolution by associating the image to a corresponding instant in the vortex dynamics. The bottom row of Fig. 2 displays some of such pictures. The image on the left shows the plasma about 1.5 s after the appearance of the $l = 1$ signal, i.e. right after the fast oscillations disappear: A single, narrow column at large distance from the trap axis. The center image is taken at a time corresponding to an evolution of about 4.5 s, and highlights the formation of a background opposite to the position of the dense clump. In some cases, after the disappearance of the single column, a different system is formed where two well-separated vortices can be found (rightmost image). These vortices have approximately the same charge and density and rotate around the axis at a constant separation distance that is larger than their mean radius. Once they are formed, their configuration appears to be stable within our observation times, i.e. at least up to 10 s, and thus orders of magnitude longer than typical collisional time scales. An important feature of this system is the presence of a strong background plasma with a density up to $10^6$ cm$^{-3}$, i.e. only one order of magnitude less than the vortex peak density. Although a thorough explanation for the occurrence of this phenomenon is still missing, the experimental evidence shows that the background plays an essential role both in the formation and in the equilibrium of the system under the continuous external forcing and the ensuing balance between ionization and losses.

**References**

[1] J. H. Malmberg and J. S. deGrassie, Phys. Rev. Lett. **35**, 577 (1975)

[2] B. Paroli, F. De Luca, G. Maero R. Pozzoli and M. Romé, Plasma Sources Sci. Technol. **19**, 045013 (2010)

[3] B. Paroli, G. Maero R. Pozzoli and M. Romé, Phys. Plasmas **21**, 122102 (2014)

[4] G. Maero, S. Chen, R. Pozzoli and M. Romé, accepted for publication in J. Plasma Phys. (2015)

[5] G. Maero, B. Paroli, R. Pozzoli and M. Romé, "Modelling of electron heating in a Penning-Malmberg trap by means of a chaotic map", in *Proceedings of the 41st EPS Conference on Plasma Physics*, Europhysics Conference Abstracts **38F**, P1.127 (2014)

[6] E. D. Leonel and P. V. E. McClintock, Phys. Rev. E **70**, 016214 (2004)

[7] A. A. Kabantsev, C. Y. Chim, T. M. O'Neil and C. F. Driscoll, Phys. Rev. Lett. **112**, 115003 (2014)

[8] J. Fajans, E. Gilson and L. Friedland, Phys. Plasmas **6**, 4497 (1999)

# The relation between divertor conditions and the L-H threshold on JET

E. Delabie[1], A.V. Chankin[2], C.F. Maggi[3], H. Meyer[3], T.M. Biewer[1], C. Bourdelle[4], M. Brix[3], I. Carvalho[5], P. Drewelow[6], C. Guillemaut[5], N.C. Hawkes[3], J. Hillesheim[3], D. Keeling[3],C. Klepper[1], A. Meigs[3], L. Meneses[5], F. Rimini[3], G. Sips[3], E. Solano[7], M. Stamp[3], J. Svensson[6] and JET contributors*

*EUROfusion Consortium JET, Culham Science Centre, Abingdon, OX14 3DB, UK*
*[1]Oak Ridge National Laboratory, Oak Ridge, Tennessee, USA*
*[2]Max-Planck-institut für Plasmaphysik, Garching, Germangy*
*[3]CCFE, Culham Science Centre, Abingdon, UK*
*[4]CEA, IRFM, F-13108 Saint-Paul-lez-Durance, France*
*[5]Associacao EURATOM-IST, Lisboa, Portugal*
*[6]MPI für Plasmaphysik, Greifswald, Germany*
*[7]Asociación EURATOM-CIEMAT, Madrid, Spain*
*See the Appendix of F. Romanelli et al., Proc. 25th IAEA Fusion Energy Conf. 2014, Saint Petersburg, Russia

## INTRODUCTION

H-mode access will be required from the early stages of ITER operation onwards to develop plasma scenarios and test control techniques (ELM mitigation, H-mode exit). Knowledge of the L-H threshold power ($P_{L-H}$) is therefore inherent to the planning of the experimental timeline and availability of systems. The anticipated threshold power on ITER (52MW at a density of 5 $10^{19}m^{-3}$ in deuterium) is extrapolated from a multi-machine scaling [1], based on density, magnetic field and machine size. Many other 'hidden parameters' are known to affect the threshold. This induces considerable uncertainty in the extrapolation. Because the available heating power on ITER (73MW max.) is only marginally above the expected threshold, there is an ongoing effort to understand the physics behind these additional dependencies and to evaluate if any could be used to lower the power threshold on ITER.

## EXPERIMENTS

Experiments on several machines have reported a strong effect of the divertor configuration and conditions on $P_{L-H}$ (e.g. JET-C [2], C-Mod [3], DIII-D [4]). Dedicated experiments using slow power ramps (1MW/s) have been carried out on JET-ILW in 2013/14 to investigate this effect with the Be/W wall [5]. A strong increase of $P_{L-H}$ in the high density branch is found for discharges with the outer strike point (OSP) on the vertical target (VT) compared to discharges in the horizontal target configuration (HT) [6,7]. The threshold decreases further in the HT configuration if the lower triangularity is increased (HT-HLT), moving the inner strike point higher onto the vertical target [5]. Fig. 1a shows $P_{L-H}$ for the different divertor configurations at 2.4T. No minimum in $P_{L-H}$ as function of density is observed at 2.4T for the VT configuration. At increased toroidal field (3T) the minimum that is observed in $P_{L-H}$ for the HT configuration shifts to higher density (see fig. 1b) [5] and is now also observed for the VT configuration, but at lower density than for the HT configuration.

## INTERPRETATION

The currently accepted view is that the L-H transition is caused by the suppression of turbulence by $E_r$xB shear and that the radial electric field well is sustained by the ion diamagnetic term in the force balance equation [8]. Therefore we have investigated if the change in $P_{L-H}$ could be explained by changes in the edge density and temperature profiles when the divertor configuration is changed [6]. Edge density profiles (Li-beam, reflectometry, Thomson scattering) and temperature profiles (ECE, Thomson scattering, edge

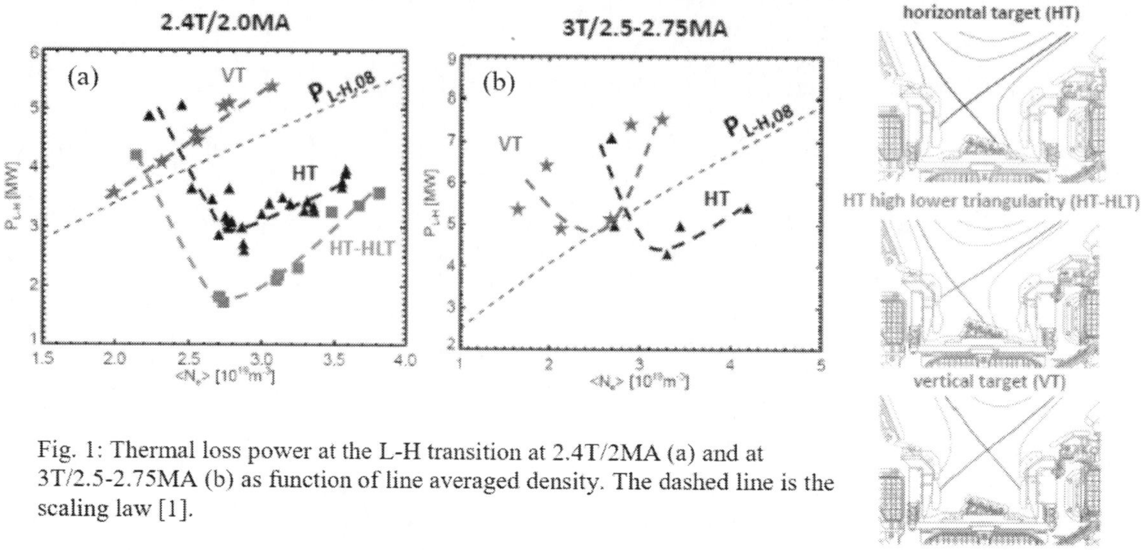


Fig. 1: Thermal loss power at the L-H transition at 2.4T/2MA (a) and at 3T/2.5-2.75MA (b) as function of line averaged density. The dashed line is the scaling law [1].

CXRS) measured in the L-mode phase before the transition have been fitted. For all discharges with edge CXRS data, $T_e$ was observed to be equal to $T_i$ in both the high and low density branch. The diamagnetic radial electric field has been derived from the fitted profiles. The density dependence of the threshold shows reasonable agreement with the existence of a critical depth of the diamagnetic $E_r$ well before the transition for each of the divertor configurations but cannot explain the differences between them as the discharges with higher threshold also exhibit higher edge temperatures and a concomitant deeper $E_r$ well before the transition [6,7].

The strong variation in threshold power due to relatively small changes in the divertor configuration hints towards changes in the scrape off layer (SOL) directly affecting the transition, rather than indirectly through changes in the upstream kinetic profiles. The shear in the outer part of the $E_r$ well, where the pedestal forms, is determined not only by the $E_r$ structure in the plasma edge but also by the radial electric field in the near SOL. We have no direct measurements of the radial electric field structure in the SOL between divertor configurations for these discharges, but we present indications that (1) $E_r$ in the SOL depends on the divertor configuration and that (2) the SOL $E_r$ plays a role in the transition.

## SOL $E_r$ IN HORIZONTAL AND VERTICAL TARGET CONFIGURATION

The potential in the SOL relates to the target $T_e$ profiles through the potential difference over the Debye and magnetic pre-sheath, leading to $E_r \approx -3\mathrm{grad}(T_e)/e$ in the SOL [9]. This neglects contributions from radial gradients in the parallel current and pressure gradients, which is a better approximation at the outer than inner strike point. Discharges with the outer strike point (OSP) on the horizontal target have a strongly peaked target $T_e$ profile, as derived from Langmuir probe data (fig. 2a). This corresponds to a strong positive $E_r$ in the near SOL. Discharges with the OSP on the vertical target have a much broader target $T_e$ profile (fig. 2b), leading to a flatter $E_r$. To corroborate this interpretation, EDGE2D-EIRENE [10,11] simulations including drifts and parallel currents have been carried out on two pulses, one in HT and one in VT configuration, using identical transport coefficients ($D_\perp = 1 m^2 s^{-1}$, $\chi_{e,i} = 2 m^2 s^{-1}$), input power (2.6MW) and similar separatrix density ($N_{sep} = 1.0\ 10^{19}\ m^{-3}$ (VT) and $1.2\ 10^{19}\ m^{-3}$ (HT)). The code results qualitatively reproduce the observed target $T_e$ profiles (fig. 2ab) and indeed predict a very strong positive $E_r$ close to the separatrix for the HT configuration, whereas for the VT configuration only a mildly positive $E_r$ is reached in the outer SOL (see fig. 2c). The strong difference between the two configurations in the code is attributed to the difference in neutral recycling. In VT configuration, the neutrals recycling of the outer divertor target plate are scattered towards the outer divertor leg, broadening the target $T_e$ profiles, whereas the neutrals in the HT configuration are recycled away from the

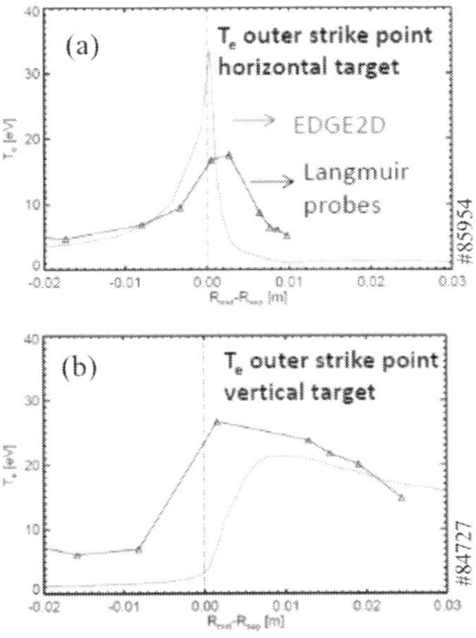


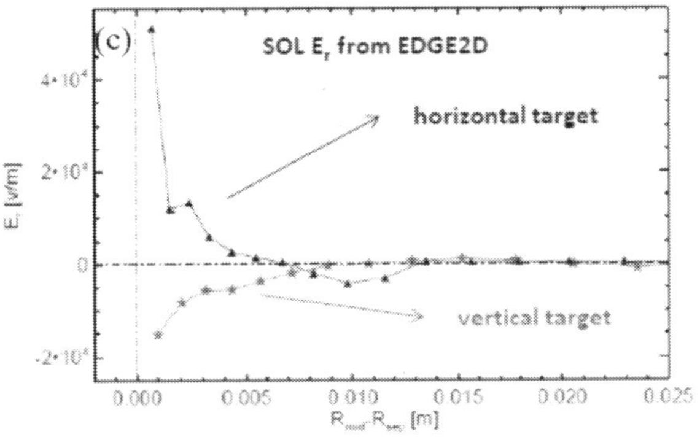


Fig. 2: (a and b) Target $T_e$ at the outer strike point from Langmuir probe data and from EDGE2D-EIRENE for a horizontal and vertical target discharge at identical power and similar density. (c) Radial electric field in the scrape off layer from EDGE2D-EIRENE for VT and HT configuration.

outer divertor leg. These results were proven to be robust against small variations in assumed input power and separatix density in the code.

The difference in observed target temperature profiles indicates a more positive $E_r$ in the SOL for the HT configuration, creating a higher shear in the outer part of the $E_r$ well, and the EDGE2D-EIRENE modelling supports these observations due to a difference in the recycling pattern of neutrals between HT and VT configuration.

## CORRELATION BETWEEN DIVERTOR RECYCLING ASYMMETRY AND $P_{L-H}$

The presence of a strong $E_r$ in the SOL is also consistent with the effects the E×B drifts have on distributing the particle and heat fluxes between the inner and outer divertor targets. For B×grad(B) directed towards the divertor, the E×B drift in the main SOL is directed from the inner towards the outer divertor and from the outer to the inner in the private flux region and along the divertor legs. The combined effect of these drifts is an increase in the inner divertor density compared to a situation with no drifts [12]. This asymmetry between inner and outer targets is already observed at power levels well below the H-mode threshold and is most pronounced in the divertor $D_\alpha$ emission. Fig. 3a shows the evolution of the divertor emission, target temperature and ion fluxes as the power is ramped up in L-mode for a HT configuration discharge. A gradually increasing inner-outer asymmetric recycling pattern is observed, consistent with an increasing $E_r$×B drift effect in the SOL and divertor. This is observed for both HT and VT discharges. For HT discharges, for which we expect a stronger SOL $E_r$, a sudden drop of the inner strike point temperature and ion saturation current and a large increase of the $D_\alpha$ emission in front of the entire inner target is observed. The ratio $D_\delta/D_\alpha$ increases by a factor 2-3, indicating inner target detachment. This sudden transition is labelled ‘inner detachment’ in fig. 3. These sudden changes in the divertor conditions are usually modulated by the heat fluxes from sawtooth crashes and are therefore referred to as ‘divertor oscillations’ [13]. No change to the upstream profiles is observed during the divertor oscillations in JET-ILW. Remarkably, a sudden increase in the target temperature peaking at the outer strike point is observed, which implies a sudden increase in $E_r$ in the SOL, as the inner divertor detaches. The critical power at which this occurs always precedes the L-H transition in the high density branch, as illustrated in fig. 3b. At the highest densities in the L-H threshold density scans, the detachment of the inner divertor will immediately trigger an L-H transition, which demonstrates that this state, associated with a higher SOL $E_r$, is beneficial for H-mode access.

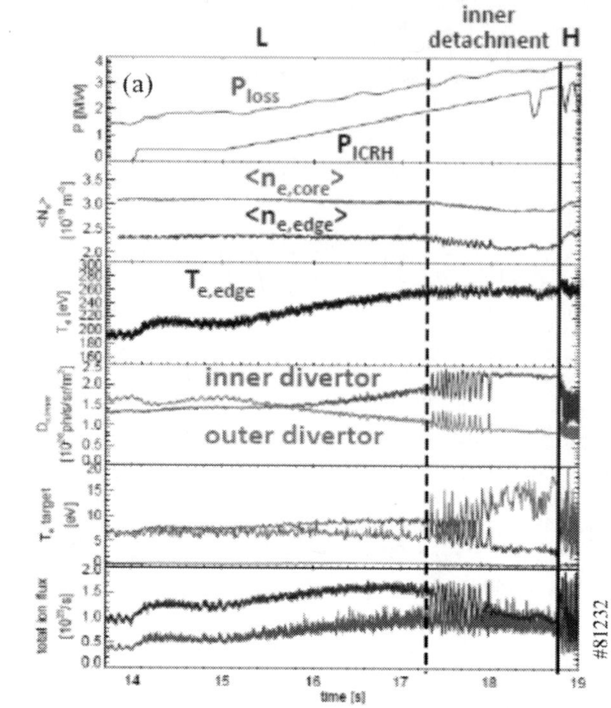


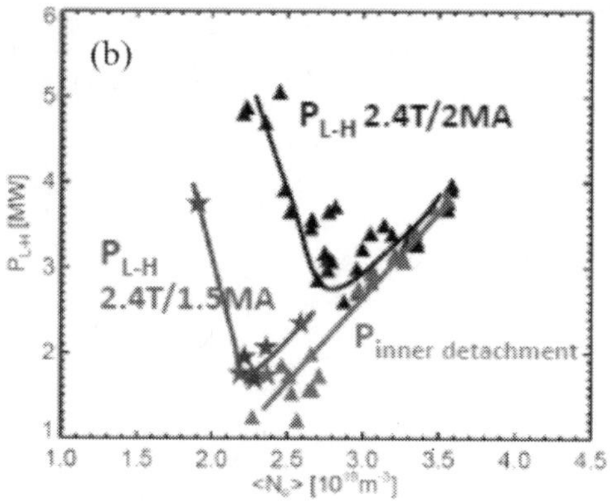


Fig. 3: (a) Evolution of deuterium emission and target temperature and ion flux as the power is ramped up at constant density in L-mode. (b) L-H threshold power and power threshold for detachment of the inner divertor leg.

Dithering L-H transitions (~130Hz), which are only observed in the high density branch, could also be understood through dynamics involving the divertor and SOL conditions. Dithering L-H transitions show very similar oscillations in the deuterium emission pattern as divertor oscillations, except for a short H-mode phase that is triggered from the state with the detached inner divertor leg, associated with the higher SOL $E_r$. The H-mode transition itself triggers the re-attachment of the inner divertor and the more symmetric recycling pattern associated with the lower SOL $E_r$ leading to the H-L back transition.

## OUTLOOK

An increase in edge $E_r \times B$ shear through the SOL $E_r$ is proposed as a mechanism to explain the divertor configuration effect on the L-H threshold. Observations on JET-ILW indicate that configurations with a strong inner/outer asymmetry in the divertor recycling pattern are beneficial for H-mode access at reduced power. A better understanding of the physics behind this mechanism could open up prospects of lowering the L-H threshold on ITER through influencing the divertor and SOL conditions.

## ACKNOWLEDGEMENT

This work has been carried out within the framework of the EUROfusion Consortium and has received funding from the Euratom research and training programme 2014-2018 under grant agreement No 633053. The views and opinions expressed herein do not necessarily reflect those of the European Commission.

## REFERENCES

[1] Y.R. Martin et al., J. Phys.: Conf. Ser. **123** (2008) 012033
[2] L.D. Horton et al., Proceedings of the 26th EPS (1999)
[3] Y.Ma et al., Plasma Phys. Control. Fusion 54 (2012) 082002
[4] P. Gohil et al., Nucl. Fusion 51 (2011) 103020
[5] C.F. Maggi et al., Nucl. Fus. **54** (2014) 023007
[6] H. Meyer et al., Proceedings of the 41st EPS conference, Berlin (2014)
[7] E. Delabie et al., proceeding of the 24th IAEA conference, St. Petersburg (2014)
[8] F. Ryter et al., Nucl. Fus. **54** (8) (2014) 083003
[9] A.V. Chankin et al., Nucl. Fus. **47** (2007) 479-489
[10] R. Simoni et al., Contrib. Plasma Phys. 34, 368 (1994)
[11] S. Wiesen et al., EDGE2D-EIRENE coupling at JET (http://www.eirene.de/e2deir_report_30jun06.pdf)
[12] A.V. Chankin et al.,Journal of Nucl. Materials **241-243** (1997) 199-213
[13] A. Loarte et al., Phys. Rev. Letters **83** (18) (1999) 3657-3660

# Experimental investigation of geodesic acoustic modes on JET using Doppler reflectometry

C. Silva[1], J. C. Hillesheim[2], C. Hidalgo[3], C. Maggi[2], L. Meneses[1], E. Belonohy[4], D. Testa[5], M. Tsalas[6], and JET Contributors*

*EUROfusion Consortium, JET, Culham Science Centre, Abingdon, OX14 3DB, UK*

[1]*Instituto de Plasmas e Fusão Nuclear, Instituto Superior Técnico, Universidade Lisboa, PT*

[2]*CCFE, Culham Science Centre, Abingdon, OX14 3DB, UK*

[3]*Laboratorio Nacional de Fusión, CIEMAT, 28040 Madrid, Spain*

[4]*Max-Planck-Institut fur Plasmaphysik, Boltzmannstr. 2, D-85748 Garching, Germany*

[5]*EPFL, Centre de Recherches en Physique des Plasmas, 1015 Lausanne, Switzerland*

[5]*FOM Inst. DIFFER, Nieuwegein, Netherlands*

## Introduction

Geodesic acoustic modes (GAMs) are characterized by a symmetric potential structure in the poloidal and toroidal directions and a rapidly varying radial structure with finite wavelength. Because of their small radial structure, the E×B shearing rate due to GAMs can be important before and during the L-H transition when the mean shear flow is modest [1]. Recently, the influence of isotope mass on the GAM amplitude has been studied [2] showing a systematic increase in the GAM amplitude during the transition from H to D dominated plasmas. Consequently, understanding GAMs may yield important implications for the dynamics of the L-H transition.

This contribution focuses on the characterization of GAMs in the edge plasma of JET Ohmic discharges using mainly Doppler reflectometry. Coherent oscillations consistent with the GAM theoretical predictions are identified and characterized. In addition, the dependence of GAM amplitude on the driving and damping mechanisms are investigated.

## Measurement technique

Doppler reflectometry is a microwave diagnostic for density fluctuation measurements that measures the radially localized propagation velocity and fluctuation level of intermediate wave number turbulent structures. This diagnostic has contributed extensively to the characterization of coherent oscillations such as the GAM [e.g. 3]. Motion of the density turbulence at the cutoff layer will induce a Doppler frequency shift ($f_D$) in the reflected signal given by $f_D = u_\perp k_\perp/2\pi$, where $u_\perp = v_{E\times B} + v_{phase}$ is the perpendicular velocity of the turbulence moving in the plasma [e.g. 4]. For the vertical target (VT) configuration (plasmas with both

* See the Appendix of F. Romanelli et al., Proceedings of the 25th IAEA Fusion Energy Conference 2014, Saint Petersburg, Russia

divertor strike points on the vertical targets) the JET correlation reflectometer [5] works for Doppler backscattering as a deliberate oblique angle between the launched beam and the normal to the plasma cutoff layer is created. The scattering wavenumber of the density fluctuations is determined via ray tracing [6]. For the data presented here the typical probed $k_\perp$ is ~ 3 $cm^{-1}$. The Doppler frequency is obtained from the complex amplitude spectrum of the reflectometer in-phase ($I = A\cos\varphi$) and quadrature ($Q = A\sin\varphi$) signals. The instantaneous Doppler shift is estimated directly from the rate of change of the reflectometer phase signal $d\varphi/dt$, where the phase is obtained from $\varphi(t) = \tan^{-1}[Q(t)/I(t)]$ and the instantaneous signal amplitude from $A(t) = [I^2(t)+Q^2(t)]^{1/2}$. The JET correlation reflectometer consists of two X-mode fast frequency hopping channels launched from the LFS midplane designed for normal incidence. Each channel can be pre-programmed with a specified launch frequency pattern, which is repeated continuously throughout the discharge, allowing a radial scan of the measurement location. For the data presented here, channel 1 (master) was set to a 11 point frequency sweep (from 74.6 to 92.6 GHz), while channel 2 (slave) had a 15 point frequency sweep of 2 ms duration around each master frequency.

**GAM identification and location**

Experiments were performed in Ohmic vertically-shifted plasmas for different values of plasma current and density. Figure 1 shows the temporal evolution of the line-averaged density and master probing frequency, together with the spectrogram of the master signal phase derivative. As illustrated, the frequency spectrum of the phase derivative is sharply peaked at the GAM frequency ($f_{GAM}$), which follows the local plasma temperature (the solid line shown in the spectrogram indicates the calculated $f_{GAM}=c_s/(2\pi R)$, where $c_s=\sqrt{[(T_e+T_i)/m_i]}$ assuming $T_e=T_i$. The local $T_e$ given by the ECE diagnostic is used). The GAM amplitude has been estimated as: (i) amplitude of the $d\varphi/dt$ spectrum at $f_{GAM}$; and (ii) rms of the $d\varphi/dt$ bandpass filtered around $f_{GAM}$. A good agreement is found between the two approaches. Figure 2a shows the electron density radial profile, obtained with the profile reflectometry system, for two periods of the discharge #87808 (t = 12 and 15 s) with different densities. The location of the master probing frequencies is also shown

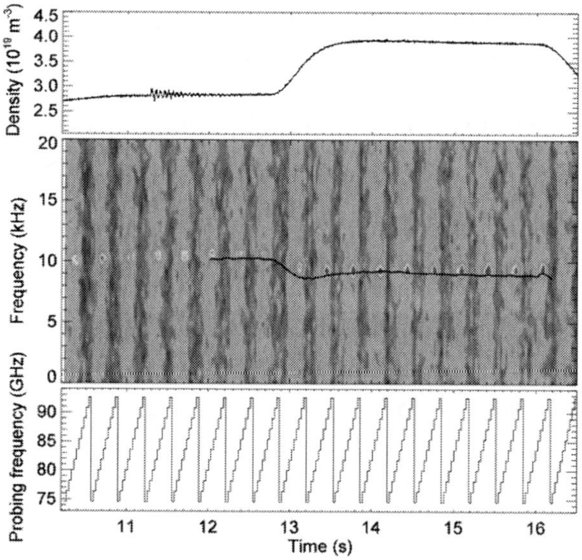

**Figure 1**: Temporal evolution of the line-averaged density, spectrogram of the master signal phase derivative and master probing frequency for discharge #87808 ($I_p$ = 2.75 MA, $B_T$ = 3 T).

(symbols). Figures 2b,c present the GAM amplitude and frequency respectively, while figure 2d shows the radial profile of the mean perpendicular velocity. As illustrated, GAMs are generally most intense in the edge density gradient region with a radial extension of about 2 cm, coinciding with the location of the $E_r$ well. The GAM has a constant frequency with radius, not varying with the local temperature (the calculated $f_{GAM}$ using the local $T_e$ is indicated by the solid line in figure 2c). The GAM rms $E_r{\times}B$ flow velocity is up to 1.5 km/s, corresponding to about 50% of the local mean value.

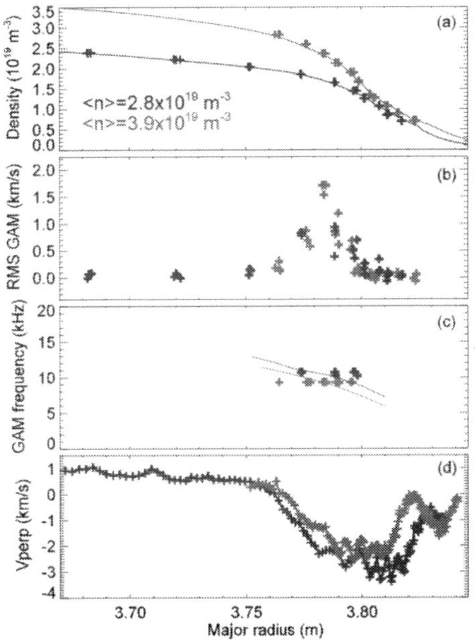


**Figure 2:** Radial profiles of density (a), GAM amplitude (b) and frequency (c) and mean perpendicular velocity (d) for two discharge periods with different densities (#87808).

**GAM damping and drive**

There are two factors determining the magnitude of the GAM: its drive and damping. Collisionless damping strongly depends on the safety factor ($\gamma \propto \exp[-q^2]$), while the collisional damping has a weaker q dependence ($\gamma \propto \nu_{ii}/q$). A series of six Ohmic discharges where performed in hydrogen for different values of plasma current ($1.5 < I_p < 2.75$ MA) and line-averaged density ($1.5 < n < 4{\times}10^{19}$ m$^{-3}$). Figure 3 shows the GAM amplitude (maximum of the GAM rms value across the radial regions scanned by the correlation reflectometer) as a function of the expected GAM damping rate. The GAM amplitude is observed to decrease with q, in contradiction with the anticipated for the collisionless damping. Furthermore, no decrease of the GAM amplitude with $\nu_{ii}/q$ is observed. GAMs on JET Ohmic plasmas appear to be regulated by the turbulence drive. Figure 4 shows the density and temperature inverse scale length ($1/L_{n,T}$) for different discharge densities (the inverse gradient scale lengths are the drive terms for drift-wave type turbulence). For the lowest discharge density a smaller $1/L_{n,T}$ is observed and no GAMs are detected. In general, the density fluctuation level, as well as the edge plasma density and temperature inverse scale lengths, increase with plasma current and line-averaged density, concurrent with the enhancement of the GAM amplitude.

**Isotope effect**

The impact of isotope mass on GAM amplitude has also been investigated on JET. As illustrated in figure 5, the GAM amplitude is apparently larger for deuterium plasmas in agreement with previous findings in TEXTOR [2] and FT-2 [7] suggesting the importance of multi-scale physics for unravelling the physics of the isotope effect in fusion plasmas.

However, the limited dataset available and the complex dependence of the GAM amplitude on plasma parameters prevent a more definitive conclusion.

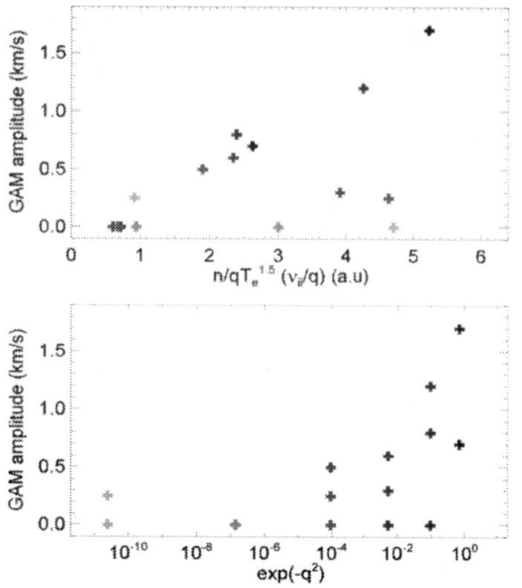


**Figure 3:** Dependence of the GAM amplitude on the collisionless and collisional GAM damping (colour scheme follows plasma current value).

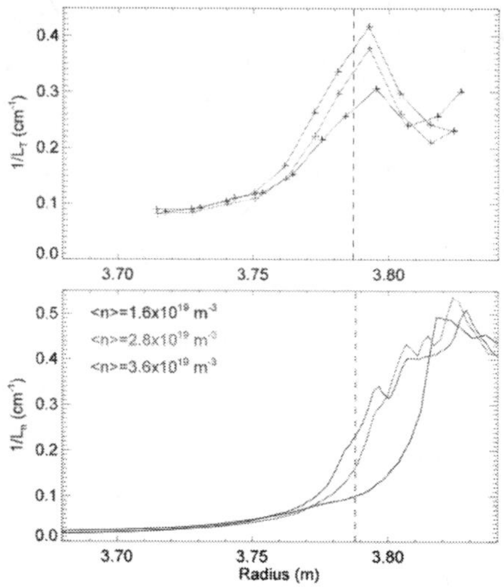


**Figure 4:** Density and temperature inverse scale-length for different discharge densities. The dashed vertical line indicates the radial location of the GAM maximum amplitude.

**Summary**

GAMs have been investigated in the JET edge plasma for Ohmic discharges using mainly Doppler reflectometry. GAMs were found to be located in a narrow layer at the edge density gradient region with amplitude corresponding to about 50% of the mean local perpendicular velocity value. GAMs on JET appear to be regulated by the turbulence drive rather than by their damping rate. Finally, it was shown that the GAM amplitude is slightly larger in deuterium than in hydrogen plasmas. The present work was limited to Ohmic plasmas; however, the importance of GAMs when approaching L-H transition will be investigated in the near future.

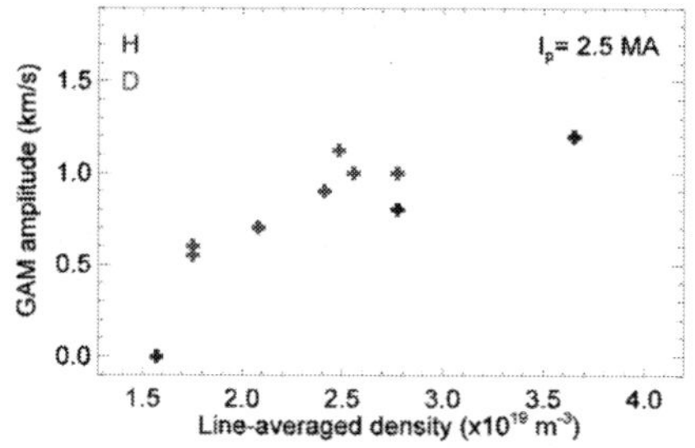


**Figure 5:** Dependence of the GAM amplitude on the plasma density for hydrogen and deuterium plasmas.

**References**

[1] G. D. Conway et al., Phys. Rev. Lett. 106 (2011) 065001; [2] Y. Xu et al., Phys. Rev. Lett. 110 (2013) 265005; [3] J. C. Hillesheim et al., Physics of plasmas 19 (2012) 02230; [4] M. Hirsch et al, Plasma Phys. Control. Fusion 43 (2001) 1641; [5] A. Sirinelli, L. Meneses, et al, 11th International Reflectometry Workshop, 2013; [6] E. Poli et al, Computer Physics Communications 136, 90 (2001); [7] A. Gurchenko et al., 25th IAEA Int. Conf. on Fusion Energy St Petersburg 2014, EX/11-2Ra.

**Acknowledgements**

This work has been carried out within the framework of the EUROfusion Consortium and has received funding from the Euratom research and training programme 2014-2018 under grant agreement No 633053. IST activities also received financial support from "Fundação para a Ciência e Tecnologia" through project UID/FIS/50010/2013. The views and opinions expressed herein do not necessarily reflect those of the European Commission.

# Modeling few-cycle shadowgraphy of laser-wakefield accelerators

E. Siminos[1], S. Skupin[2], A. Sävert[3], J. M. Cole[4], S. P. D. Mangles[4], M. C. Kaluza[3,5]

[1] *Max Planck Institute for the Physics of Complex Systems, 01187 Dresden, Germany*

[2]*Univ. Bordeaux - CNRS- CEA, CELIA, UMR 5107, 33405 Talence, France*

[3] *Institut für Optik und Quantenelektronik, Abbe-Center of Photonics, Friedrich-Schiller-Universität, 07743 Jena, Germany*

[4]*The John Adams Institute for Accelerator Science, The Blackett Laboratory, Imperial College London, London SW7 2AZ, United Kingdom*

[5]*Helmholtz-Institut Jena, Friedrich-Schiller-Universität, 07743 Jena, Germany*

**Abstract**

Ultrafast shadowgraphy utilizes few cycle probe pulses in order to image density gradients in a plasma allowing to probe structures, such as laser-driven wakes, moving close to the speed of light. Here we study the process of shadowgraphic image formation in the interaction of a few cycle probe pulse with a laser-driven wake using particle-in-cell (PIC) simulations. The output of the PIC code is then post-processed by means of Fourier optics in order to take into account the effect of a typical imaging setup. This allows to construct synthetic shadowgrams which can be compared with experimentally recorded ones. This facilitates the correct interpretation of such involved measurements.

The interaction of intense laser pulses with plasmas can lead to the excitation of nonlinear structures in the plasma, such as wakes, solitons, shocks and filaments. The time evolution of such structures is highly non-trivial and a lot of effort has been invested in developing theoretical tools and experimental diagnostics to investigate it. In a typical Laser-wakefield accelerator (LWFA) setup a relativistically intense laser pulse (the 'pump' pulse) excites an electron plasma wave (the 'wake') which propagates with a phase velocity $v_p$ equal to the pump pulse group velocity $v_g \simeq c$ through the plasma. Ambient electrons can be trapped and accelerated in the field of the wake, producing quasi-monoenergetic electron beams [1]. Stable operation of such accelerators requires a thorough understanding of nonlinear laser-plasma interaction processes motivating the development of novel diagnostics. The characteristic length scale of the wake is the plasma wavelength $\lambda_p$. Therefore, optical probing of such a wake with a transversely propagating probe pulse requires a probe pulse duration: $c\,\tau_{\mathrm{pr}} \ll \lambda_p$. Typically this corresponds to a duration of few laser cycles $\tau_L = 2\pi/\omega_L$, where $\omega_L$ the frequency of the probe pulse. Recently, pump-probe experiments utilized probe pulses of few femtosecond duration to obtain the first direct shadowgraphic images of a wake propagating in an underdense plasma [2, 3].

Conventional shadowgraphy can be analyzed using geometrical optics and formulated as an inverse problem: for a given shadowgraphic image local gradients in index of refraction can be determined, see, e.g., Ref. [4]. In ultrafast shadowgraphy of laser induced wakes this is not in general straightforward. The effect of longitudinal motion of the wake as the probe pulse traverses, the presence of strong magnetic fields as well as relativistic effects are not negligible and need to be taken into account. At the same time diffraction effects can be significant when the probe pulse wavelength $\lambda_{\rm pr}$ is comparable to the plasma skin depth $\lambda_s$ which is the characteristic length scale of density gradients in the plasma. For parameters used in recent experiments [3] this is often the case and the effect of diffraction has to be accounted for.

In a previous publication [3] synthetic shadowgrams generated using this methodology turned out to be crucial for the correct interpretation of experimental measurements. In this paper the shadowgraphic image formation process is analyzed through the use of three-dimensional particle-in-cell (PIC) simulations of the full pump-probe setup. Propagation of the pump pulse through the plasma is simulated and at different delays a probe pulse propagating transversely to the pump is launched. After the probe traverses the wake, post-processing in Fourier space allows to take into account the effect of a typical imaging setup in shadowgraphic image formation.

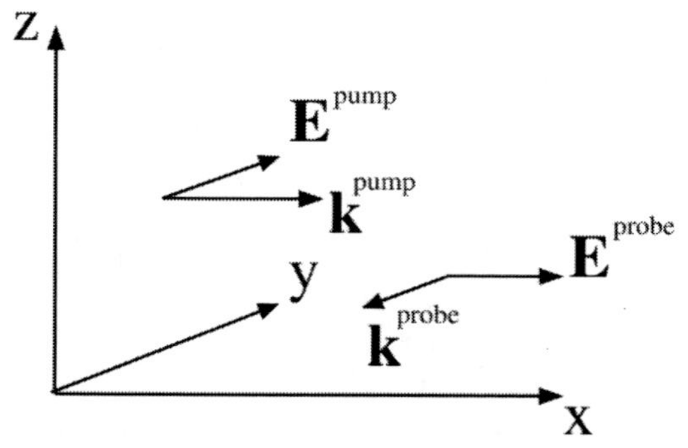


Figure 1: PIC simulation setup: The pump propagates along the $x$ direction and is polarized along $y$, while the probe propagates along the (negative) $y$ direction and is polarized along $x$.

We performed simulations of LWFA with the PIC code EPOCH with parameters similar to the ones used the experiments of Ref. [3]. A pump laser pulse propagating along the $x$ direction, linearly polarized along the $y$ direction, with an intensity FWHM duration of 36 fs and central wavelength $\lambda_L = 810\,\mathrm{nm}$ was focused to a spot size of $18.84\,\mu\mathrm{m}$ (intensity FWHM) at $x_f = 300\,\mu\mathrm{m}$ into a plasma of initial electron density $n_0 = 1.7\times10^{19}\,\mathrm{cm}^{-3}$. The pump pulse maximum intensity (in vacuum) was $I_0 = 2.5\times10^{18}\,\mathrm{W/cm}^2$. The computational domain was a 'sliding window' of size $150\,\mu\mathrm{m}\times70\,\mu\mathrm{m}\times70\,\mu\mathrm{m}$ moving at $c$. Probe propagation has been also fully simulated in 3D with EPOCH. For the probe simulations the moving window was stopped and a probe pulse was injected from the side of the box, propagating along the negative $y$ direction, i.e. perpendicularly to the pump propagation direction, see Fig. 1. Although we varied the parameters of the probe pulse, in most runs the pulse parameters were similar to the ones in recent experiments [3]: central wavelength $\lambda_{\text{pr}} = 0.75\,\mu$m, bandwidth limited duration of $\tau_{BL} = 4.4$ fs (intensity FWHM) and a negative linear chirp that increased duration to $\tau_{\text{pr}} = 12$ fs and maximum intensity $I_{\text{pr}} = 8.6 \times 10^{14}\,\text{W/cm}^2$.

We allowed the probe pulse to propagate past the wake structure, until its center reached approximately $y_B = -20\mu$m. In Fig. 2(a-c) we track the evolution of the envelope of the probe electric field as it crosses the wake. Figure 2(b-c) indicates that modulations in the intensity of the probe pulse occur as the latter interacts with density gradients in the wake.

Once the probe pulse reaches $y_B$ there are no substantial density perturbations in the plasma and therefore any local phase differences induced by the wake have been imprinted to the probe pulse. However, if we simply try to reconstruct a shadowgram by recording the time-integrated intensity passing through the plane $y_B$, as shown Fig. 2(d), there are two problems with the image we obtain. In the front of the wake there is strong scattering of pump light which is of much higher intensity than the probe intensity. This scattered light is not present in the experimental shadowgrams [2, 3]. As we will see the reason for this is that the aperture of the imaging system eliminates most of the side-scattered light. Moreover, the wake structure appears blurred in Fig. 2(d). The reason for this is strong diffraction, since the length scale for density gradients (the plasma skin depth, $\lambda_s = 1.3\,\mu$m here) is comparable to the wavelength of the probe pulse ($\lambda_{\text{pr}} = 0.75\mu$m). Thus, in order to be able to compare PIC simulation results to experimental shadowgrams, we have to take into account the influence of a typical imaging system in shadowgram

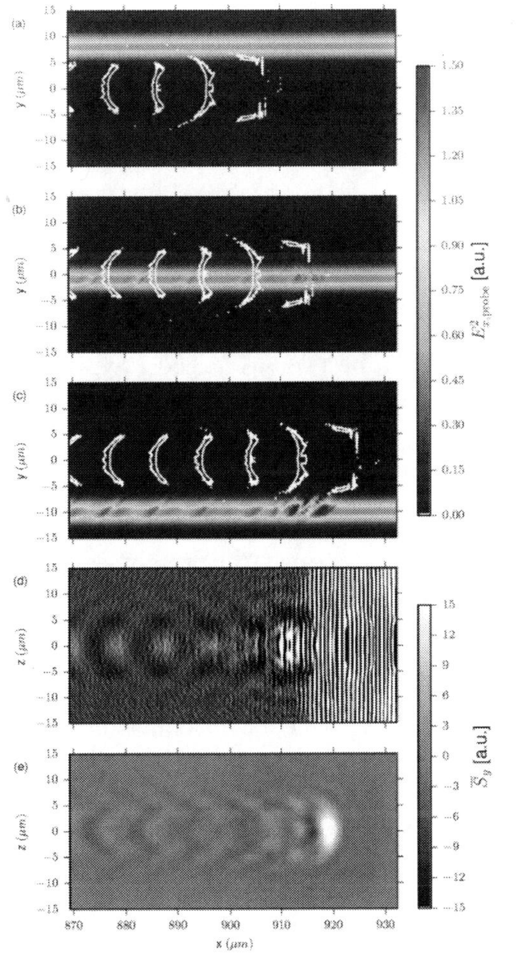


Figure 2: Probe pulse propagation past the wake. Panels (a)-(c): Three snapshots from the PIC simulation showing the probe electric field amplitude $E_x^{\text{probe}}$ and contours of the plasma density corresponding to $n = 2n_0$. (d) Image obtained by recording the time-integrated Poynting flux $\langle S_y \rangle$ passing trough the plane $y = -20\,\mu$m (e) Time-integrated Poynting flux $\langle S_y \rangle$ after adjusting for focusing optics, assuming the object plane is at $y_o = 0$.

formation.

In order to reconstruct shadowgrams from our PIC simulation results, we will use two major simplifications. Firstly, the transverse size of the simulation box in our PIC simulations is limited to a few tens of microns, and it is not possible to directly account for the optical imaging. However, once the probe pulse has propagated through the plasma wake, all information related to the laser-plasma interaction process are already imprinted in its wavefronts. Thus, we will assume that the probe pulse propagates further on in vacuum. Secondly, the probe pulse, after it has passed the plasma wake, propagates mainly in one direction, and the imaging system has a certain numerical aperture (NA), i.e., acceptance angle. Here, we assume that the NA of our imaging system is small enough to justify a paraxial description of the imaging process. The imaging system was modeled in Fourier space, allowing to take into account its aperture and image plane position. To adjust for the latter, we propagated the probe pulse backwards and recorded the time-averaged Poynting flux through the object plane, which in the images presented here was taken to be at $y = 0$. This resulted in the simulated shadowgram of Fig. 2(e), which resembles the experimental shadowgrams [3].

The methodology developed here thus allows direct comparison with experimental shadowgrams and facilitates their interpretation. In particular, our ability to track probe propagation indicates that the strongest modulations in the shadowgrams originate from the front of the 'bubble' (the first period of the wakefield), see Fig. 2, and this allowed the determination of bubble length in recent experiments [3]. Shadowgrams obtained at different delays allow to track the evolution of the size of the bubble as it propagates through the plasma. It is shown that, as suggested by recent models [5], the process of expansion of the bubble induced by self-focusing and self-compression of the pump laser pulse leads to self-injection of electrons into the bubble and their subsequent acceleration.

*Acknowledgments:* EPOCH was developed under UK EPSRC grants EP/G054940/1, EP/G055165/1 and EP/G056803/1.

## References

[1] E. Esarey, et al. *Rev. Mod. Phys.* **81**, 1229, 2009.

[2] M. B. Schwab, et al. *Applied Physics Letters* **103**, 191118, 2013.

[3] A. Sävert, et al. Direct observation of the injection dynamics of a laser wakefield accelerator using few-femtosecond shadowgraphy, 2015. Submitted.

[4] P. K. Panigrahi and K. Muralidhar. *Schlieren and Shadowgraph Methods in Heat and Mass Transfer*. SpringerBriefs in Applied Sciences and Technology. Springer New York, 2012.

[5] S. Kalmykov, et al. *Phys. Rev. Lett.* **103**, 135004, 2009.

# Study of plasma wakefield acceleration mechanism for emittance dominated regimes via hybrid and pic simulations

A. Marocchino[1], F. Massimo[1], S. Sinigardi[2], A. Mostacci[1], E. Chiadroni[3]

[1] *Dipartimento SBAI, Università di Roma "La Sapienza" e INFN, Italy*

[2] *Dipartimento di Fisica e Astronomia - Università di Bologna e INFN, Bologna, Italy*

[3] *INFN - Istituto Nazionale di Fisica Nucleare, Frascati, Italy*

**Abstract**

Electron plasma wakefield acceleration (PWFA) mechanism is a promising non conventional acceleration scheme. Nonetheless further investigation is still needed to fully uncover the instability mechanisms so to mitigate them and make PWFA an effective tool. This work focuses in this direction, we discuss the necessity to use well matched driver bunches to further mitigate witness instabilities. Specifically we propose to inject driver bunches with larger emittance than the matched one (overcompressed bunch) so to let the system reach the matching condition by itself. This preliminary results lead us to the following consideration: while a limited number of cases can be studied with a particle-in-cell code, we understand the necessity for fast systematic analysis: we briefly introduce the hybrid code *Architect*.

## Introduction

Plasma wakefield acceleration is a very promising particle accelerating mechanism, however, in order to bring such a technique to even more promising applications it is necessary to investigate the underlying plasma instabilities thus to mitigate them.

For this study we assume a standard two bunches configuration: the first bunch - the driver - is characterised with a charge of about 200 pC, a second bunch - witness - follows at about half a plasma wavelength ($1/2\ \lambda_p$, with a background number density $n_0 = 10^{16}\ \mathrm{cm}^{-3}$) with a much smaller charge: 10 pC. The driver induces the wakefield wave while the witness experiencing the field is accelerated.

We have observed that any driver betatron oscillations causes bubble oscillations, that eventually seed witness instabilities. Consequently, in order to further mitigate witness instabilities the driver bunch needs to be well matched to the background plasma. The matching condition [1] can be simply retrieved by the envelope formulation,

$$\sigma_{x,\mathrm{mtch}} = \sqrt[4]{\frac{d}{\gamma}}\sqrt{\frac{\varepsilon_x}{k_p}} \tag{1}$$

where $\sigma_{x,\mathrm{mtch}}$ is the transverse-rms matched dimension, $\gamma$ is the relativistic gamma, $\varepsilon_x$ is the emittance and $k_p$ the plasma wavenumber. $d$ is the bubble geometry parameter, $d = 2$ assumes a cylindrical bubble, $d = 3$ is used for spherical bubbles. The shape depends upon the bunch charge and dimensions. If we assume d=2, $\gamma = 200$ and $\varepsilon_x = 3$ mm-mrad, e.g. plausible values for the Sparc_Lab facility, we observe the driver being matched with a transverse size of 4.0 $\mu$m. We might consider a follower witness characterised by an emittance of $\varepsilon_x = 1$ mm-mrad -higher quality than the driver-; for such emittance value the matching transverse dimension is 2.5 $\mu$m. We can also make a physical consideration on the accelerator lattice: it is reasonable to assume that the linac lattice can be optimised for only one of the two bunches, and consequently while parameters can be accurately tailored for the witness, less control will be given to the driver.

**The over-compressed driver-bunch**

In the scenario described in the introduction: one driver plus a witness, we have introduced the necessity to control the driver matching conditions. We have also observed how controlling emittance and energy spread simultaneously for driver and witness is rather difficult nor whether impossibile. A possible solution may be achieved using an overcompressed driver bunch. The driver bunch is injected within the plasma channel with a transverse-rms

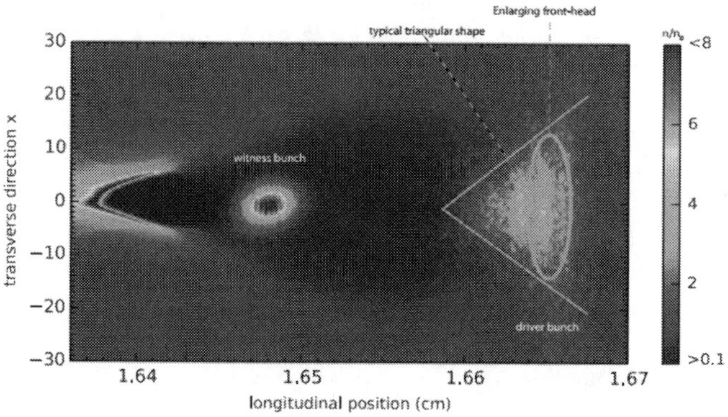


Figure 1: Two driver setup: a driver in front with a follower, reduced charge, witness. The driver characterised by a large emittance reach the characteristic cone profile.

Table 1: Equilibrium transverse size dimension, calculated by Eq.(2) and verified with the particle-in-cell code ALaDyn

| emittance $\varepsilon_x$ (mm-mrad) | $\sigma_{x,\mathrm{eps}}$ ($\mu$m) | $\sigma_x$-ALaDyn sim ($\mu$m) |
|---|---|---|
| 1 | 2.0 | 2.4 |
| 2 | 4.1 | 3.9 |
| 3 | 6.3 | 6.0 |
| 4 | 8.4 | 8.8 |
| 5 | 10.4 | 10.0 |

dimension, $\sigma_x$, smaller than the matching one: $\sigma_x(t=0) < \sigma_{x,\mathrm{mtch}}$. This configuration is fairly easy to achieve by increasing the emittance. Physically: the emittance, that represents the internal bunch pressure, pushes to enlarge the bunch while the self generated electric field works to compress it. If the emittance is rather small (a few mm-mard) the bunch expansion is rather smooth and then tends to auto-stabilize around a matched configuration. The matched configuration is found by balancing the emittance pressure with the bubble electrostatic pressure, that reads:

$$\sigma_{x,\mathrm{eps}} = \sqrt{\frac{d^2(2\pi)^{\frac{3}{2}}\varepsilon_0}{2en_0^2 m_e}\frac{\sigma_z}{Q}} \qquad (2)$$

with $\sigma_z$ driver-rms size, $Q$ total bunch charge, $e$ electron charge, $m_e$ electron mass and $\varepsilon_0$ the electric permeability in vacuum. Details of the calculation will be given elsewhere.

Eq.(2) theoretical calculations have been compared against the ALaDyn code[2, 3], results are summarised in table 1. ALaDyn is a particle-in-cell code recently adapted for electron plasma wakefield acceleration problems [4]. Table 1 shows that using $d = 2.5$, a bubble between a cylindrical shape and a spherical shape, the theoretical approximation well estimates the numerical calculations.

**Systematic scans with an Hybrid approach: Architect**

The study of the dynamics of both matched and expanding drivers, as the ones discussed in previous sections, need numerical systematic analysis; systematic scans are also necessary to fully uncover the sensitivity of the possible rising instabilities due to both plasma and bunch characterisation. Despite PIC codes are, today, heavily parallel on dedicated clusters-computers the amount of computational resources and the amount of computational operations restrict the use of such codes to proof of principles numerical investigations. While PIC codes do not seem a suitable numerical technique for heavy systematic scans, reduce models appears as the desired tool able to deliver a solution in a reduced time (minutes) without loss of accuracy. Architect, Fig.2, is a hybrid cylindrical symmetric code, it treats the bunches with

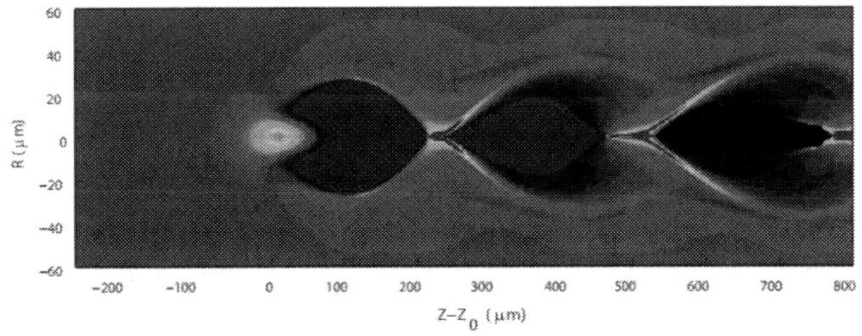


Figure 2: Electron number density of both driver bunch and background plasma in weakly nonlinear regime of PWFA, simulated with *Architect*

a kinetic approximation while the background plasma is model as a fluid.

Since the background plasma, up to the quasi non linear regimes ($\widetilde{\mathbf{Q}} < 0.5$), exhibits an ensemble behaviour, it can be treated as a single large particle: a fluid cell. On the contrary, to catch the kinetic nature of the bunches a PIC-like approach is used. The critical, and somehow interesting part, is the interaction between the kinetic bunch scale, i.e. kinetic set of equations, and the fluid background electron plasma, i.e. the fluid equations. The interaction between the two scales is given by superposing the current induced independently by the bunches and the background plasma, inducing the total electromagnetic fields: $\mathbf{E}_{total}$ and $\mathbf{B}_{total}$. The overall evolution loop is depicted in Fig.3.Both electromagnetic and fluid quantities are evolved in a computational moving window shifted accordingly to the driver center of mass. Architect represents an effective and quick way to tackle the problem, nonetheless it appears a key tool for the experimental team since it can be used during experimental run to help the experimental team to take online strategic decisions.

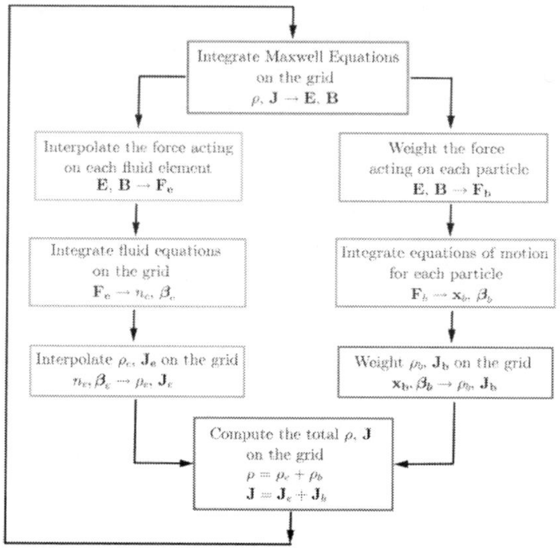


Figure 3: Architect loop.

This work has been partially funded by the Italian Minister of Research in the framework of FIRB - Fondo per gli Investimenti della Ricerca di Base, Project no. RBFR12NK5K and by Sapienza, university of Rome, project n. 2014-C26A14K3WK.

## References

[1] M Ferrario Laser-plasma Acceleration : Proceedings of the International School of Physics: Enrico Fermi **179** 41-63 (2012)

[2] C Benedetti, A Sgattoni, G Turchetti and P Londrillo Plasma Science, IEEE Transactions **36** 1790-1798 (2008)

[3] P Londrillo, C Benedetti, A Sgattoni and G Turchetti, Nuclear Instruments and Methods in Physics Research Section A: Accelerators, Spectrometers, Detectors and Associated Equipment **1** 28-35 (2010)

[4] P Londrillo, C Gatti and M Ferrario Nuclear Inst. and Methods in Physics Research, A **740** 236-241 (2014)

[5] C K Birdsall, A B Langdon Plasma Physics Via Computer Simulation CRC Press, Boca Raton (2004).

[6] S T Zalesak J Comput Phys 1-28 (1979)

# Non-monotonic features in the runaway electron tail

I. Pusztai[1], E. Hirvijoki[1], J. Decker[2], O. Embréus[1], A. Stahl[1] and T. Fülöp[1]

[1] *Applied Physics, Chalmers University of Technology, Göteborg, Sweden*

[2] *Ecole Polytechnique Fédérale de Lausanne, Centre de Recherches en Physique des Plasmas, Lausanne, Switzerland*

Radiation losses can significantly affect the momentum space dynamics of energetic runaway electrons. We find that synchrotron radiation reaction can lead to the appearance of a non-monotonic feature – a "bump" – in the runaway tail. It can be a potential source for kinetic instabilities limiting the formation of runaway beams. We derive threshold conditions for the bump to appear and to be physically relevant, and provide a lower bound on its location in momentum space. Numerical calculations with CODE [1] support the analytical results.

**Kinetic equation** Radiation reaction increases with perpendicular momentum. The range of perpendicular momenta where the accelerating field can overcome the friction plus the radiation reaction then becomes limited. Pitch-angle scattering of electrons out of this "runaway region" leads to an exponential decay of the electron distribution in the far-tail. For moderate (but relativistic) momenta, the return fluxes of electrons into the runaway region due to collisional friction and radiation reaction can overcome the outflow due to pitch-angle scattering, which can lead to the appearance of a local maximum in the distribution function. In the following we assume that such a non-monotonic feature of the steady state distribution exists and we study its properties.

The kinetic equation used here is derived from the guiding center kinetic equation for the gyro-angle averaged distribution function $F$, discussed in [2]. We consider a homogeneous plasma and a straight magnetic field line geometry, and assume the momentum range of interest to be much higher than the thermal momentum of electrons, which leads to

$$\frac{\partial F}{\partial \tau}+\frac{1}{s^2}\frac{\partial}{\partial s}\left[s^2\left(\hat{E}\xi-\sigma\gamma s(1-\xi^2)-\frac{\gamma^2}{s^2}\right)F\right]+\frac{\partial}{\partial \xi}\left[(1-\xi^2)\left(\hat{E}\frac{F}{s}+\frac{\sigma\xi}{\gamma}F-\frac{\gamma}{s}\frac{1+Z_{\rm eff}}{2s^2}\frac{\partial F}{\partial \xi}\right)\right]=0,$$

where $s=p/(m_e c)$ is the dimensionless momentum, $\xi=s_\parallel/s$ is the pitch-angle cosine, $\tau=eE_c t/(m_e c)$ is the normalized time with $E_c=n_e e^3\ln\Lambda/(4\pi\varepsilon_0^2 m_e c^2)$ the critical field for runaway generation, $n_e$ the electron density, $m_e$ the electron rest mass, $c$ the speed of light, $\epsilon_0$ the vacuum permittivity, $e$ the elementary charge, and $\ln\Lambda$ the Coulomb logarithm. Furthermore $\hat{E}=-E_\parallel/E_c$ is the normalized parallel electric field, $\gamma=(1+s^2)^{1/2}$ is the relativistic factor, and $Z_{\rm eff}$ is the effective ion charge, $\sigma=2\Omega_e^2/(3\ln\Lambda\,\omega_{pe}^2)$ quantifies the strength of the radiation reaction (relative to collisional friction), where $\Omega_e=eB/m_e$ and $\omega_{pe}^2=e^2 n_e/(\epsilon_0 m_e)$ are the gyro-, and plasma frequencies.

It is convenient to work in parallel and perpendicular momenta, $s_\parallel$ and $s_\perp$, since we will focus on the distribution function for small $s_\perp$. At an extremum of the distribution function located along the $s_\parallel$ axis (satisfying $(\partial_{s_\parallel}F)|_{(s_\parallel,0)} = 0$) the kinetic equation simplifies to

$$L(s_\parallel) \equiv 2\left(\sigma s_\parallel + \sqrt{1+s_\parallel^2}\right) - (1+s_\parallel^2)\sigma\kappa/\bar{E} = 0 \tag{1}$$

where $\bar{E} = (\hat{E}-1)/[2(1+Z_{\rm eff})]$ and $\kappa(s_\parallel) = -(\hat{E}-1)/(2\sigma F)(\partial_{s_\perp s_\perp}F)|_{s_\perp=0}$. The order unity parameter $\kappa$ is assumed to be slowly varying with $s_\parallel$, and it is defined so that it approaches 1 asymptotically in the far tail, where the distribution should approach Eq. (12) of Ref. [3].

**Thresholds and bump location** To find the threshold for the bump formation we use that the distribution should have an inflection point when the bump is about to appear. In that case $L(s_\parallel)$ of Eq. (1) and its derivative $L'$ vanish simultaneously. It can be shown that the inflection point will not appear at high $s_\parallel$. Thus we may use $(1+s_\parallel^2)^{1/2} \approx 1+s_\parallel^2/2$ to obtain simpler expressions (for more details consult [2]). Neglecting $\kappa'$ corrections we find that the bump in the distribution appears if $\bar{E} > \kappa \sim 1$ or $\sigma$ is below the threshold given by

$$\sigma_0 = \frac{3\kappa/\bar{E} + \sqrt{8+\kappa^2/\bar{E}^2}}{2\left(\kappa^2/\bar{E}^2 - 1\right)}. \tag{2}$$

Right at the threshold the inflection point is located at $s_{\parallel 0} = \sigma[(1+\sqrt{1+4\sigma^2})/2]^{-1} < 1$.

We can find an approximate expression for the bump location when it appears at high $s_\parallel$, in which case we may use $(1+s_\parallel^2)^{1/2} \approx s_\parallel$. In this limit, by assuming the characteristic perpendicular width of the distribution ($\propto \kappa^{-1/2}$) to be approximately constant, and looking for separable solutions of the form $F = h(s_\parallel)g(s_\perp)$, we find from the kinetic equation that

$$(\hat{E}-1)s_\parallel h' - 2(1+\sigma)h = -(1+Z_{\rm eff})W^2 s_\parallel h, \tag{3}$$

with $W^2 = 2\sigma\kappa/(\hat{E}-1)$, for which $h(s_\parallel) \sim s_\parallel^{2(1+\sigma)/(\hat{E}-1)}\exp[-W^2(1+Z_{\rm eff})/(\hat{E}-1)s_\parallel]$ is a solution. The function $h(s_\parallel)$ has a maximum at

$$s_\parallel = \frac{2\bar{E}}{\kappa}\frac{1+\sigma}{\sigma}. \tag{4}$$

Numerical investigation suggests that the order unity parameter $\kappa$ is usually smaller than 1 at the bump. Thus Eq. (4) for $\kappa = 1$ is a lower bound for the bump location in momentum space.

For small values of $\sigma$, the bump would appear at high parallel momenta. Defining some upper limit of physical interest, $s_{\parallel,\rm L}$, Eq. (4) may be used to find an estimate for a lower "practical limit" in $\sigma$ for the appearance of the bump. Namely, if $\sigma$ is smaller than

$$\sigma_{\rm L} = \left[(s_{\parallel,\rm L}\kappa)/(2\bar{E}) - 1\right]^{-1}, \tag{5}$$

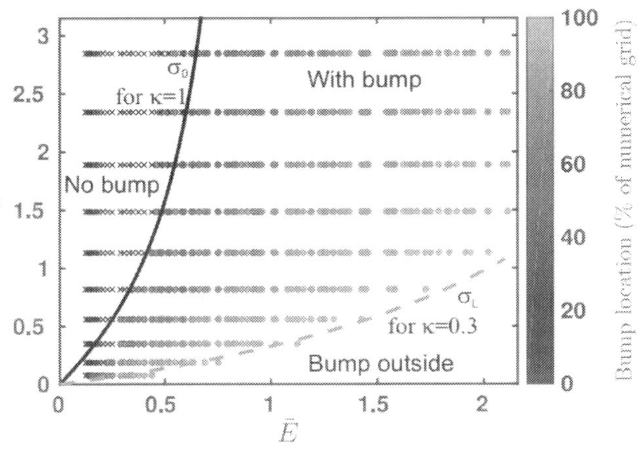


Figure 1: *CODE simulations yielding steady state solutions with (circles) or without (crosses) a bump in the runaway tail. Bumps above $s_{\|} = 27$ are excluded ("Bump outside"). Solid line: bump formation threshold Eq. (2) for $\kappa = 1$. Dashed line: practical threshold Eq. (5) for $s_{\|,\mathrm{L}} = 27$ and $\kappa = 0.3$. The color coding shows the location of the bump relative to $s_{\|} = 34$.*

for $\kappa = 1$, then a bump would only appear at some large parallel momentum above $s_{\|,\mathrm{L}}$, which is then deemed physically irrelevant. Note that if the bump is in the far tail, $\kappa$ can be significantly less than unity, as will be shown in the next section using numerical simulations. Letting $\kappa < 1$ increases the practical limit in $\sigma$. For a normalized electric field higher than $\bar{E} = s_{\|,\mathrm{L}}/2$, the bump *always* appears above $s_{\|,\mathrm{L}}$ for any value of $\sigma$.

**Numerical results** The numerical results were obtained using the continuum simulation tool CODE (COllisional Distribution of Electrons), used in its time-independent mode. CODE solves the two dimensional momentum space kinetic equation in a homogeneous plasma, using a linearized Fokker-Planck operator valid for arbitrary energies, and including the radiation reaction force [4].

To investigate the validity of our analytical calculations, we performed a scan in the parameter space with CODE. The electron temperature and density were held constant at $T_e = 1\,\mathrm{keV}$ and $n_e = 5 \cdot 10^{18}\,\mathrm{m}^{-3}$, while the magnetic field, the electric field, and the effective ion charge were varied over the ranges $B \in [1,6]\,\mathrm{T}$, $\hat{E} \in [2,14]$ and $Z_{\mathrm{eff}} \in [1,3]$. The simulations used 950 momentum grid points, 130 Legendre modes in $\xi$, and a highest resolved momentum of $s = 34$.

The results of the scan are presented in Fig. 1, where circle (cross) symbols correspond to distributions with (without) a bump. The color coding reflects the location of the bump, with 100% in the color bar corresponding to $s_{\|} = 34$. Simulations with a bump above $s_{\|} = 27$ are excluded from the figure. Increasing the normalized electric field $\bar{E}$ or decreasing the radiation reaction strength $\sigma$ moves the bump towards larger momenta, as expected from Eq. (4).

Good agreement is found between the numerical calculations and the bump formation threshold $\sigma_0$ (solid curve), Eq. (2) for $\kappa = 1$. The "no bump" solutions mostly obey the analytical threshold and fall to the left of the threshold curve. There are some solutions with bump in this region as well, since $\kappa$ at the bump can be less than unity, which would move the threshold to the left. While the qualitative behavior of the threshold is still captured, the expression for the threshold begins to fail quantitatively for $\sigma < 0.5$, showing that the slowly varying $\kappa$ approximation breaks down. The lower right corner of the plot is not populated, because the simulations where the bump would have appeared above $s_\parallel > 27$ were excluded, since they would be affected by the proximity to the end of the momentum grid. With $s_{\parallel,\mathrm{L}} = 27$, $\kappa = 0.3$ is needed for the practical threshold, $\sigma_L$ of Eq. (5) (dashed line), to correspond well to the boundary of the region of excluded points. Thus, $\kappa$ can be significantly lower than unity for a bump at large momentum.

**Conclusions** The time-asymptotic steady-state runaway distribution can become non-monotonic due to synchrotron radiation reaction. This non-monotonic feature presents a potential source for bump-on-tail instabilities, which can play a role in limiting the formation of large runaway beams. A threshold condition for the appearance of the bump and a lower limit for its location in momentum space were derived. Our analytical results show good agreement with numerical simulations obtained using the CODE solver.

For a normalized electric field $\bar{E} > 1$, the steady state electron distribution always exhibits a bump, independently of the strength of the radiation reaction (for $\sigma > 0$). For $\bar{E} < 1$, the appearance of the bump is well correlated with the $\sigma$ threshold, Eq. (2). However, if the bump would appear at a very high parallel momentum it can take too long time for it to develop, compared to other time scales of the physical system. (The temporal evolution of the distribution function is discussed in [5].) If $\sigma$ is lower than a "practical" threshold, Eq. (5), the bump will appear at a momentum above some pre-specified limit, $s_{\parallel,\mathrm{L}}$. For $\bar{E} > s_{\parallel,\mathrm{L}}/2$, this is satisfied for any $\sigma$.

**Acknowledgments** The authors are grateful to M. Landreman and P. Helander for fruitful discussions. The corresponding author is supported by the International Career Grant of Vetenskapsrådet.

## References

[1] M. Landreman, A. Stahl and T. Fülöp 2014 *Comp. Phys. Comm.* **185** 847.

[2] E. Hirvijoki, I. Pusztai, J. Decker, O. Embréus, A. Stahl and T. Fülöp 2015 "Radiation reaction induced non-monotonic features in runaway electron distributions". To appear in *J. Plasma Phys.* arXiv: 1502.03333

[3] F. Andersson, P. Helander and L.-G. Eriksson 2001 *Phys. Plasmas* **8** 5221

[4] A. Stahl, E. Hirvijoki, J. Decker, O. Embréus and T. Fülöp 2015 *Phys. Rev. Lett.* **114** 115002

[5] J. Decker, E. Hirvijoki, O. Embréus, Y. Peysson, A. Stahl, I. Pusztai and T. Fülöp 2015 "Bump formation in the runaway electron tail". Submitted to *Plasma Phys. Control. Fusion* arXiv: 1502.03333

# Absorption dynamics of nitric oxide in gas mixtures excited by pulsed e-beam sustained discharge

S.P. Derevyashkin[1,2], A.A. Ionin[1], Yu.M. Klimachev[1], I.O. Kinyaevskiy[1], A.A. Kotkov[1], A.Yu. Kozlov[1], A.K. Kurnosov[3]

[1] *P.N. Lebedev Physical Institute of the Russian Academy of Sciences, Moscow, Russia*

[2] *National Nuclear Research University «MEPHI», Moscow, Russia*

[3] *Troitsk Institute for Innovation and Fusion Research, Moscow, Russia*

Vibrational relaxation of NO molecules excited by a pulsed e-beam sustained discharge (EBSD) was studied with the usage of CO laser probing. We developed numerical model of vibrational kinetics in ensemble of NO molecules and compared the experimental and calculated data on absorption dynamics of vibrational excited NO molecules.

Relaxation properties of NO molecules were investigated mainly on lower vibrational levels ($v = 1 \div 3$) [1-5]. Analysis of experimental data showed that the vibrational relaxation of NO molecules is a very fast process [2, 3, 6] in comparison with the vibrational relaxation of CO molecules. Vibrational relaxation is a complex process that includes vibration-to-translation (VT) relaxation and vibration-to-vibration (VV) exchange. To explain fast vibrational relaxation of NO molecules, it was assumed that the rate of VT relaxation of NO molecules is five orders of magnitude higher than the rate of VT relaxation of CO molecules [3]. Although it was believed that the rate of VV exchange of NO molecules is close to the rate of VV exchange of CO molecules [2-6].

We suggest a different explanation for the fast vibrational relaxation of NO molecules. We assume that the rate of VV exchange of NO molecules is much greater than the rate of VV exchange of CO molecules. It is possible, considering that the NO molecule is a doublet $^2\Pi$ in the ground electronic state, and has two systems of rotational-vibrational (briefly ro-vibrational) levels separated by the energetic gap of 120 cm$^{-1}$. In our numerical model of vibrational kinetics of NO molecules we proceeded following this assumption. The purpose of this paper is to study in detail the vibrational relaxation of NO molecules. In our experiments, we applied a powerful-pulsed pump of lower vibrational levels of NO molecules and a CO laser probing population dynamics on higher vibrational levels.

## THEORY

The theoretical model of vibrational kinetics in an ensemble of NO molecules and buffer gases was developed by comparing the experimental and calculated data on absorption dynamics of vibrational excited NO molecules. The theoretical model includes a self-consistent numerical solution of the equations of vibrational kinetics in mixtures of NO:He,

NO:Ar и NO:$N_2$ and Boltzmann equation for the distribution function of electron energy of pulsed EBSD. The rate constants for $N_2$-$N_2$ VV-exchange processes were taken from the work [7]. As to processes NO($u$)+NO($v$)→NO($u$+1)+NO($v$-1), the corresponding rate coefficients were calculated with an approximate analytical expression taking into account the dominant role of dipole-dipole interaction between NO molecules:

$$K_{v,v-1}^{u,u+1} = b \cdot \frac{Z}{T} \cdot \frac{v}{(1-v\cdot\delta)} \cdot \frac{(u+1)}{(1-(u+1)\cdot\delta)} \cdot \exp\left(\frac{\Delta E}{2\cdot T}\right) \cdot \exp\left(-\frac{\Delta E^2}{C\cdot T}\right). \qquad (1)$$

The structure of this formula is the same as was proposed in [8]. Here $\omega_e$ is the fundamental frequency, $x_e\omega_e$ is an anharmonicity constant, $\delta = x_e\omega_e / \omega_e$, $Z=\pi\sigma^2 V_M$, $V_M$ is the average relative translational velocity, $\pi\sigma^2$ is the gas kinetic collision cross-section in the hard spheres model [9]; $\Delta E$ is energy mismatch of VV process in K, $b$ and $C$ are the semi-empirical parameters found in comparison between the theory and experiment. For gas temperature $T$=300 K the expression (1) with coefficients $b$=1.92 K and $C$=31.46 K enables a good agreement with experimentally measured VV rate constant [4] for process: NO(1)+NO(1)→NO(2)+ NO(0).

The VV′ processes between NO and $N_2$ molecules are nowadays also poorly known. On the first step of our study we have calculated the rate constants for these processes with analytic expression used early for calculations of rate constants for VV′ processes between $N_2$ and CO molecules [7]. Such approach is based on the qualitative similarity of the intermolecular potentials for two pairs of molecules: CO-$N_2$ and NO-$N_2$. Thus the rate constants for VV′ processes NO($v$)+$N_2$($u$)→NO($v$-1)+$N_2$($u$+1) were determined with an analytic expression:

$$\begin{aligned} K_{v,v-1}^{u,u+1} &= a \cdot Z \cdot T \cdot \frac{v}{(1-v\cdot\delta)} \cdot \frac{(u+1)}{(1-(u+1)\cdot\delta)} \cdot \exp\left(\frac{\Delta E}{2\cdot T}\right) \cdot f(y) \cdot F + \\ &+ b \cdot \frac{Z}{T} \cdot \frac{v}{(1-v\cdot\delta)} \cdot (u+1) \cdot \exp\left(\frac{\Delta E}{2\cdot T}\right) \cdot \exp\left(-\frac{\Delta E^2}{C\cdot T}\right). \end{aligned} \qquad (2)$$

The first additive term in expression (2) corresponds to the joint action of the short range repulsion and van der Waals attraction, the second term corresponds to dipole-quadrupole interaction. Here $f(y)$ is the adiabatic function calculated in accordance with [8], $y = l_{ST} \cdot |\Delta E| \cdot \pi^2 \cdot \sqrt{\frac{2\cdot\mu\cdot k}{h^2 \cdot T}}$, $l_{ST}$ - is the semi-empirical parameter introduced in Schwartz-Slawsky-Herzfeld (SSH) theory for the case of 1D model and characterizing the specific size of the short-range repulsion, $\mu$ is the reduced mass of the colliding molecules, while $h$ and $k$

are the Planck and Boltzmann constants respectively, $F$ is the Shin's factor for approximate description of van der Waals attraction [8]. The values of the semi-empiric parameters $a$, $b$, $C$ and $l_{ST}$ were taken the same as for $N_2$-CO VV′ rate constants [7]. Data on the rate constants for the VT relaxation of $NO(1) + NO(0) \rightarrow NO(0) + NO(0)$ were taken from paper [10]. The calculations of rate constant for the VT relaxation of molecules for higher vibrational levels $v$ were made with the use of the SSH theory [8].

## EXPERIMENT

In our experiments, the NO molecules excited in mixtures NO:Ar and NO:$N_2$ by the pulsed EBSD with duration up to ~100 μs. The experiments were performed on a pulsed EBSD laser system with active medium length $L$ = 1.4 m, described in detail in [11], with the possibility to excite a large volume (up to 2.4 liters) of gas mixtures. For probing the dynamics of vibrational excited states of NO molecules the radiation of frequency-selective cryogenic low pressure CO laser [12] was used. This probe CO laser operated in a band of fundamental vibrational transitions on more than 200 spectral lines.

The optical scheme of experiments and methodic of measuring of the temporal dynamics of absorption of NO molecules is described in detail in [13].

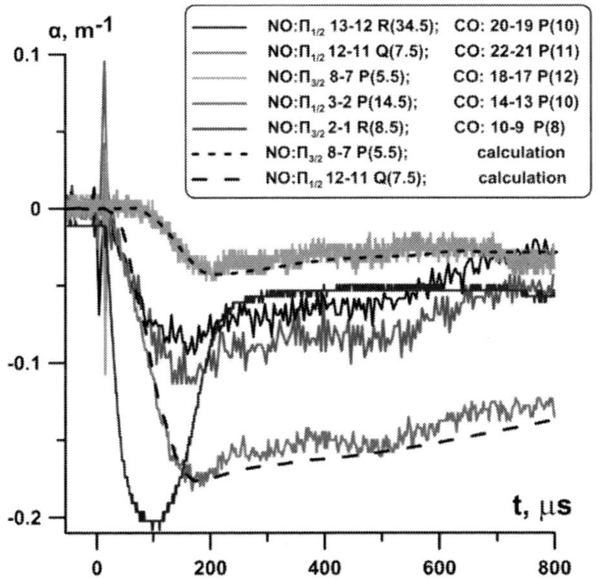


Fig 1. Temporal behavior of absorption coefficients in gas mixture NO:Ar=1:5, pressure 0.24 atm, $T$=293 K. Inset shows NO transitions and CO laser lines.

An absorption dynamics for different probe CO laser lines in the mixture NO:Ar=1:5 at specific input energy (SIE) $Q_{in}$=50 J/(l atm) is presented in Fig. 1. In all gas mixtures at initial gas temperature $T$=293 K the absorption at the lowest NO transition $\Pi_{1/2}$ 2-1 R(8.5) appears almost immediately with the start of the EBSD ($t$=0). We observed well-defined maximum of absorption at $t$=100 μs for the NO transition $\Pi_{1/2}$ 2-1 R(8.5) with time duration ~180 μs on the level 0.5 and then slowly falling tail up to 3 ms. The maximum absorption coefficient at the NO transition $\Pi_{1/2}$ 2-1 R(8.5) was 5 times higher in the nitrogen-reach gas mixture at SIE 250 J/(l atm) than in the argon-reach mixture at SIE 50 J/(l atm). As it is presented in Fig. 1

absorption at higher vibrational transitions of NO from 3-2 to 13-12 band peaked almost simultaneously at $t$=160-200 μs. Then absorption at these transitions decreased slowly during a few milliseconds.

We calculated rate constants for VV exchange between NO molecules and VV′ exchange between $N_2$ and NO for our experimental conditions. The calculated temporal behavior of the absorption coefficients on NO transitions $\Pi_{1/2}$ 12-11 Q(7.5) and $\Pi_{3/2}$ 8-7 P(5.5) is also presented in Fig. 1 and is in good agreement with the measured absorption.

Another behavior was observed at probing the absorption in gas mixtures NO:Ar = 1:6 and NO:$N_2$=1:20 cooled down to $T$=122±2 K (see Fig 2). Dynamics of absorption slowed down with decreasing gas temperature. Absorption on NO transition $\Pi_{1/2}$ 2-1 R(8.5) and higher transitions was observed for about 3 s at the level 0.1 of maximum.

This fact suggests that under these conditions the gas mixture accumulates a sufficiently large amount of vibrational energy and the energy relaxes very slowly to heat due to VV′ exchange between nitrogen and NO molecules and VT-relaxation of the NO molecules.

This research was supported by the Russian Foundation for Basic Research (Project 13-02-01135) and P.N. Lebedev Physical Institute Educational-Scientific Complex.

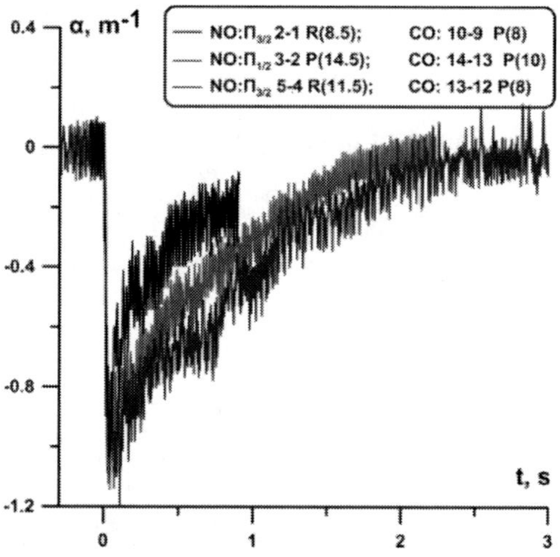


Fig 2. Temporal behavior of absorption coefficients in gas mixture NO:$N_2$=1:20, at T=122±2 K. Inset shows NO transitions and CO laser lines. $Q_{in}$=180 J/(l*Amagat).

REFERENCES

[1] Y. Nachshon, P.D. Coleman, J. Chem. Phys. 61 (1974) 2520-2524.
[2] J.C. Stephenson, J. Chem. Phys. 59 (1973) 1523-1527.
[3] J.C. Stephenson, J. Chem. Phys. 60 (1974) 4289-4294.
[4] I.J. Wysong, J. Chem. Phys. 101 (1994) 2800-2810.
[5] I.J. Wysong, Chem. Phys. Lett. 227 (1994) 69-73.
[6] H. Horiguchi, S. Tsuchiya, Japanese J. of Appl. Phys. 18 (1974) 1207-1212.
[7] A.K. Kurnosov, A.P. Napartovich, S.L. Shnyrev and M. Cacciatore, Plasma Sources Sci. Technol. 19 (2010) 045015-1-045015-13.
[8] N.S. Smith, H.A. Hassan, AIAA Journal, 14 (1976) 374-381.
[9] J.O. Hirschfelder, C.F. Curtiss, and R.B. Bird. Molecular theory of gases and liquids. John Wiley & Sons, Inc., New York (1967).
[10] L. Doyennette, M. Margottin-Maclow, J. Chem. Phys. 84 (1986) 6668–6678.
[11] A.A. Ionin, Yu.M. Klimachev, Yu.B. Konev, et al. Quantum Electronics, 30 (2000) 573–579.
[12] S. Vetoshkin, A. Ionin, Yu. Klimachev, et al. J. of Russian Laser Research, 27 (2006) 33-69.
[13] S.P. Derevyashkin, A.A. Ionin, I.O. Kinyaevskiy, et al. Physics Procedia (2015) (in print).

# Fast non-Maxwellian atoms in a linear magnetized plasma

C. Brandt, O. Marchuk, A. Pospieszczyk, S. Brezinsek, M. Reinhart, B. Unterberg

*Forschungszentrum Jülich GmbH, Institut für Energie- und Klimaforschung - Plasmaphysik, 52425 Jülich, Germany*

The sheaths between the unperturbed plasma and plasma boundary, such as surfaces or diagnostic probes, play a fundamental role in plasma research [1, 2]. The potential difference $\phi_{\text{sheath}} = \phi_{\text{wall}} - \phi_{\text{plasma}}$ results in acceleration of either positive or negative ions towards the wall, depending on the boundary conditions (e.g., the potential at the wall). After being neutralized at the wall surface, reflected neutral atoms have a kinetic energy in the order of magnitude $E_{\text{kin}} \sim Ze_0\phi_{\text{sheath}}$, with $Ze_0$ being the ion charge. The kinetic energy of this population of fast non-Maxwellian neutral atoms is higher than the average thermal energy. Observations of such *fast atoms* have been reported in a variety of plasmas, e.g., rf plasmas [3], glow discharge plasmas [4] or atmospheric plasmas [5, 6, 7]. Even though fast atoms can often be present in the plasma, they do not necessarily radiate and so can not be characterized at all. The different atomic, molecular or chemical processes can contribute to the population of excited states of hydrogen atoms. In fusion plasmas, for example, temperatures are high enough to efficiently excite neutral hydrogen by charge exchange or electron collisions [8, 9, 10, 11]. In low temperature plasmas these processes are less efficient, but emission of fast hydrogen increases with the presence of argon [3]. The source of emission of fast atoms is, however, controversially discussed in literature [3, 4].

This work reports on the observations of fast non-Maxwellian hydrogen atoms in hydrogen-argon plasmas as well as first observations of fast atoms in hydrogen mix plasmas with other noble gases, i.e., D/He, D/Ne, D/Kr and D/Xe [12]. The experiments are conducted in the linear magnetized plasma of the device PSI-2 [13, 14]. In the present work the electron densities are in the range of $n_e \sim 10^{17} - 10^{18}\,\text{m}^{-3}$ and the electron temperatures between $T_e \sim 8 - 20\,\text{eV}$. A tungsten target (surface area of $1\,\text{cm}^2$ with target normal facing towards the source) is introduced into the plasma column at $r \approx 2.4\,\text{cm}$ where the hollow plasma profile peaks. An additional negative potential can be applied to the target to further accelerate positive ions and produce reflected fast atoms. The plasma in front of the target is simultaneously observed using two lines of sight, shown schematically in Fig. 1. The imaging ($\approx 0.1$ Å/ch) and Echelle ($\approx 0.01$ Å/ch) spectrometer observe the plasma at the

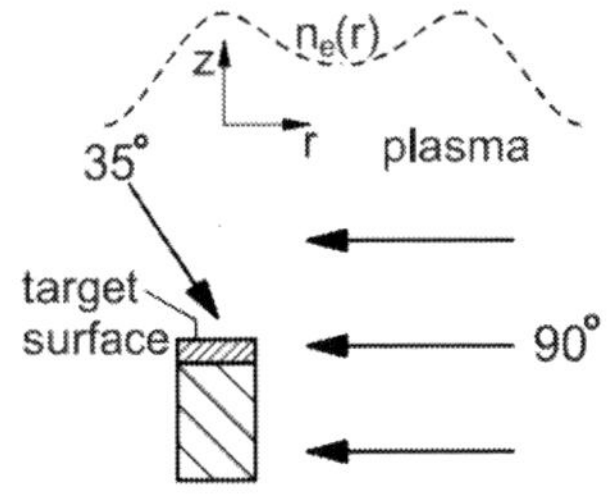

Figure 1: *Spectroscopic observation directions.*

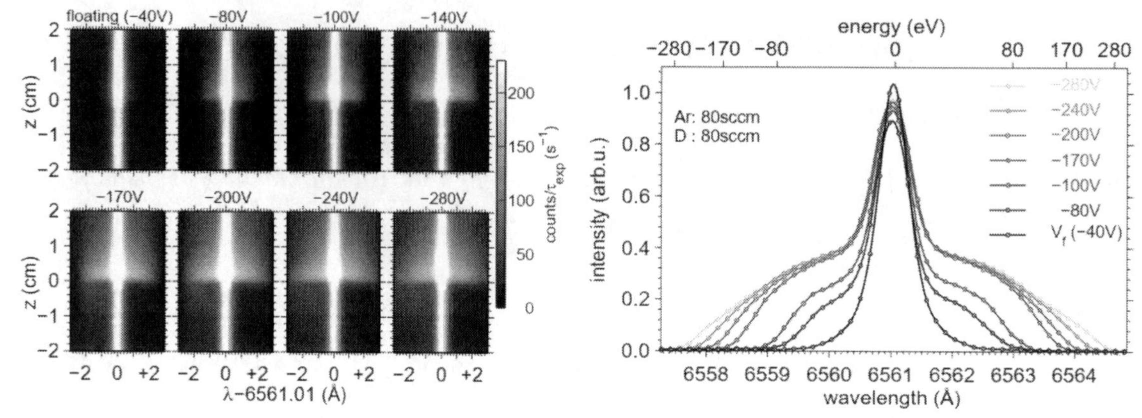


Figure 2: *(left) Measurement of $D_\alpha$ for a scan of the acceleration potential at the target in an Ar/D mix plasma (target surface at $z = 0$cm). (right) Spectra obtained by averaging between $z = 1-6$mm.*

angles of 90° and 35° towards the z axis.

In pure deuterium plasma, the emission of fast hydrogen is spectroscopically not detected above the signal-to-noise level, independent of the applied potential (0 to $-280$V) at any used gas pressures ($p = 0.01 - 0.3$Pa). However, the emission line profile of $D_\alpha$ drastically changes as soon as argon gas is added. Figure 2 illustrates the emission spectra measured with the imaging spectrometer. The emission of fast atoms, namely the blue- and red-shifted components around the $D_\alpha$ line, appear by increasing the negative potential. If the potential is more negative, the measured emission is more intense and the Doppler shift becomes larger. The blue- and red-shifted components are equal in intensity resulting in a symmetric emission profile of $D_\alpha$. According to the Doppler shift $v/c = \Delta\lambda/\lambda_0$ the kinetic energy $E = (1/2)m_D v^2$ is calculated, with $c$ being the speed of light in vacuum and $m_D$ the mass of the deuterium atom. The emission of the Doppler-shifted components can be ascribed to the emission from fast neutral atoms, since the kinetic energy of these neutral D atoms is proportional to the applied acceleration potential. Fast atoms are also observed by means of the lines shift for $D_\beta$ and $D_\gamma$.

Figure 3 depicts measurements of the $D_\alpha$ line at 35° observation. The spectrum in Fig. 3(a) shows a large passive component and blue- and red-shifted "wings". Compared to the intensity of the wings, the cold component is larger than for the 90° observation. The reason is the longer line-of-sight through the plasma from this steep angle of view. However, as plot (b) in Fig. 3 reveals, the main difference is the asymmetry between the blue- and red-shifted wings. The intensity of the red-shifted wing is smaller by a factor of $\sim 0.5 - 0.6$ than the intensity of the blue-shifted wing. The schematic in Fig. 3 illustrates the interpretation of this effect by means of reflected emission on the target surface [3]. The photon $\gamma$ emitted from a fast excited atom

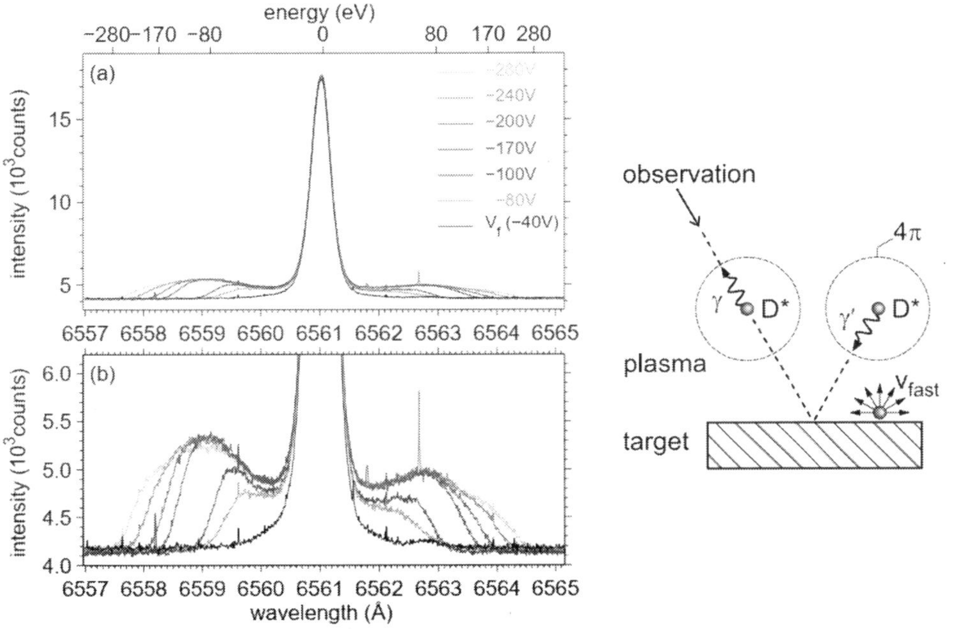


Figure 3: *(left) Scan of acceleration potential: Measurement of $D_\alpha$ from the 35° observation. (right) With 35° observation direct light is seen with a blue Doppler shift, reflected emission is seen with a red Doppler shift.*

D* moving away from the target towards the plasma is detected with a Doppler shift towards smaller wavelengths, whereas the backwards radiated photon $\gamma'$, being reflected at the target surface, is detected with a Doppler shift towards larger wavelengths.

The observation of emission of fast atoms in hydrogen mix plasmas with the other noble gases Kr, Xe, Ne and He leads to the following results. First, the emission of fast atoms in all cases is significantly lower than in the case of Ar/D plasmas. While $D_\alpha$ emission of fast atoms could be still distinctly observed in Kr/D plasma, in hydrogen mix plasmas with helium, neon and xenon this emission could hardly be detected above the signal-to-noise level. Second, when the gas flow ratio between argon and deuterium is scanned at constant pressure from pure deuterium to pure argon, the maximum of fast atom emission is observed at the gas flow ratio of approximately Ar/D=1:1. The emission is zero when either argon or deuterium flow is zero. Thus, the source driving excitation of the fast atoms is the binary collision process between argon and deuterium. Two processes could be of importance. The first process (i) is the excitation of hydrogen atoms by collisions with argon ground state atoms: $D_{fast} + Ar \rightarrow D^*_{fast}+Ar$. The second process (ii) is the so-called *excitation transfer* between excited states of argon and hydrogen atoms: $D_{fast} + Ar^* \rightarrow D^*_{fast}+Ar$ [16]. The cross sections for process (i) for Balmer-line emission for H impact on Ar, He, Ne, Kr and Xe are given in Refs. [17, 18, 19, 20]. Krypton and argon exhibit the largest cross sections ($\sim 5\times10^{-17}\,\mathrm{cm}^2$), followed by Xe ($\sim 1\times10^{-17}\,\mathrm{cm}^2$), He and Ne ($\sim 3\times10^{-18}\,\mathrm{cm}^2$). The expected excitation of fast atoms, according to these cross sections, does not really match the experimental observations. Even though the emission of fast atoms is the highest observed for Ar/D and Kr/D plasmas, the discrepancy of emission in Kr/D being smaller by a factor of $\approx 0.25\times$ than in Ar/D plasma seems too large. The comparison of the energy levels of hydrogen, argon and krypton shows that the two lowest metastable states of Ar and Kr are considerably closer to D($n=3$) (Ar: $3p^54s$, $\Delta E = 0.36\,\mathrm{eV}$ and $\Delta E = 0.53\,\mathrm{eV}$; Kr: $4p^55s$, $\Delta E\sim 1.53\,\mathrm{eV}$ and $\Delta E \sim 2.17\,\mathrm{eV}$). It is, therefore, very probable that the quasi-resonant excitation transfer between the metastable states of argon (or krypton) and hydrogen (n=3, ...) is the source for the pronounced emission of fast atoms via the Balmer series ($D_{\alpha,\beta,\gamma}$).

We are not aware about any theoretical cross sections for the excitation transfer leading to the emission of $D_{\alpha,\beta,\gamma}$ lines. We hope that this work stimulates such calculations. By making visible the emission of fast atoms, one obtains not only new insights in the theory of plasma-wall interaction, but also a completely new diagnostic tool for testing theoretical models and kinetic codes.

## References

[1] I. H. Hutchinson, *Principles of Plasma Diagnostics*, Cambridge University Press (1987)
[2] P. C. Stangeby, *The Plasma Boundary of Magnetic Fusion Devices*, IOP Publishing London (2000)
[3] T. Babkina *et al.*, Europhys. Lett. **72**, 235 (2005)
[4] N.M. Šišović *et al.*, Eur. Phys. J. D **32**, 347 (2005)
[5] J.T. Clarke *et al.*, Geophys. Res. Lett. **16**, 587 (1989)
[6] J.M. Ajello *et al.*, J. Geophys. Res. **100**, 26,411 (1995)
[7] C. Oliveira *et al.*, Appl. Phys. Lett. **93**, 041503 (2008)
[8] U. Samm *et al.*, Plasma Phys. Controlled Fusion **29**, 1231 (1987)
[9] D. Reiter *et al.*, J. Nucl. Mat. **196–198**, 1059 (1992)
[10] P. Bogen *et al.*, J. Nucl. Mater. **162–164**, 545 (1989)
[11] A. Pospieszczyk, "Diagnostics of Edge Plasmas by Optical Methods", Amsterdam: Elsevier, 213, (1993)
[12] O. Marchuk, to be submitted
[13] H. Kastelewicz and G. Fussmann, Contrib. Plasma Phys. **44**, 352 (2004)
[14] A. Kreter *et al.*, Fusion Sci. Technol. **68**, (2015)
[15] refractiveindex.info
[16] M. A. Clyne *et al.*, Chemical Physics **47**, 179 (1980)
[17] B. Van Zyl *et al.*, Phys. Rev. A **21**, 716 (1980)
[18] B. Van Zyl *et al.*, Phys. Rev. A **28**, 176 (1983)
[19] B. Van Zyl *et al.*, Phys. Rev. A **31**, 2922 (1985)
[20] B. Van Zyl *et al.*, Phys. Rev. A **33**, 2093 (1986)

# Microwave micro-plasmas in air: simulations and experiment

P. Coche[1], V. Guerra[1], K. Gadonna[2], G.D. Stancu[3], O. Leroy[2], T. Minea[2] and L.L. Alves[1]

[1]*Instituto de Plasmas e Fusão Nuclear, Instituto Superior Técnico, Universidade de Lisboa, Portugal*
[2]*LPGP, UMR 8578: CNRS-Université Paris-Sud, 91405 Orsay, France*
[3]*CentraleSupélec, EM2C, CNRS UPR 288, Grande Voie des Vignes, 92295 Châtenay-Malabry, France*

## 1. Introduction

Micro-plasmas generated in capillaries have recently aroused research interest in several fields like photonics, medicine and plasma-surface processing. Here, we report on very recent experimental and modelling results, obtained from the study of micro-plasmas generated in dry air ($N_2$ 80% - $O_2$ 20%) using microwaves (2.45 GHz excitation frequency and ~100 W coupled power), confined within a small radius capillary ($R$=345 μm) at low pressure ($p$=300 Pa).

## 2. Modelling

The simulations are performed using the in-house IST-LoKI (LisbOn KInetics) numerical code, which keeps the same algorithmic structure and calculation blocks used in previous publications [1,2]. The 0D (volume averaged) model couples the homogeneous two-term electron Boltzmann equation (EBE) (for the gas mixture considered, including first and second kind collisions) to a system of rate balance equations for molecules $N_2$(X, v=0-59), $N_2$(A, B, C, a, a', w), $O_2$(X, a, b), NO(X, A, B), $NO_2$(X, A), $O_3$, $O_3$*; atoms N($^4$S, $^2$D, $^2$P) and O($^3$P, $^1$D); positive ions $N^+$, $N_2^+$(X, B), $N_3^+$, $N_4^+$, $O^+$, $O_2^+$ and $NO^+$; and negative ion $O^-$. All cross sections used are or will be available in the near future, at the IST-LISBON database of LXCat [3].

Working at low pressure (< 1000 Pa) and low radius (< 500 μm) results in a high reduced electric field and a non-classical ambipolar transport for charged particles. To meet these conditions, the kinetic-energy working-grid in the EBE was extended up to 1000 eV, to describe the effect of the high-energy mechanisms impacting significantly on the tail of the electron energy distribution function, and a transport theory was developed for $pR$ values beyond the validity limit of the classical ambipolar theory. Under our typical operating conditions (electron density $n_e \simeq 4\times10^{12}$ cm$^{-3}$, gas temperature $T_g \simeq 1000$ K and electron temperature $T_e \simeq 4$-7 eV), we have $\Lambda/\lambda_D \sim 15$ ($\Lambda = R/2.405$ is the diffusion length for infinite cylindrical geometry and $\lambda_D$ is the Debye length) corresponding to a dense plasma in "ambipolar regime", and $\Lambda/\lambda_i$~0.5 ($\lambda_i$ is the ion mean-free-path) corresponding to low-pressure. A transport theory embedding the transition from low to high pressures in dense plasmas was derived by Self and Ewald [4] and later revisited by Ferreira and Ricard [5], for a single type of positive ion. Here, we have generalized the theory of Self and Ewald for $N_2$-$O_2$ mixtures at low pressure, accounting for several positive ions and one (low-density) negative ion species (listed at the beginning of this section), recovering a system of equations formally identical to the one of [4], whose numerical solution can be avoided using the results of [5] as an abacus.

Figure 1 shows the self-consistent calculated values of the reduced electric field, *E/N*, and of the electron kinetic temperature, $T_e$, as a function of *R*, using three different descriptions for the charged-particle transport: the current formulation for several positive ions and one negative ion; the current formulation neglecting the effect of negative ions; and the classical ambipolar diffusion.

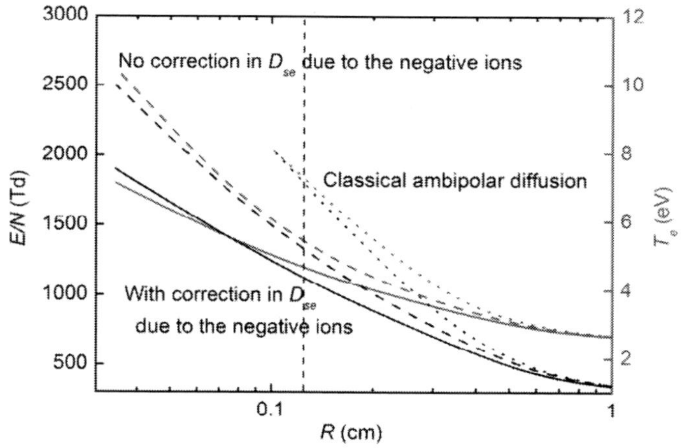


**Figure 1:** Reduced electric field (left axis, black) and electron temperature (right axis, red) for $p$=300 Pa, $T_g$=1100 K and $n_e$=3.84x10$^{12}$ cm$^{-3}$, as a function of the tube radius, calculated using: the present theory (full line); the present theory neglecting the effect of negative ions (dashed line); the classical ambipolar theory (dotted line). The short-dashed vertical line limits the *R*-region of low-pressure conditions.

Our model calculations in the presence of $O^-$ ions should be restricted to low-pressures ($\Lambda/\lambda_i \leq 1$, corresponding to $R \leq 0.12$ cm), i.e., to the region on the left of the short-dashed vertical line. However, as *R* increases the influence of negative ions becomes negligible and the present formulation can still be used, albeit with caution. Moreover, the different transport descriptions considered yield similar results at high *R*, thus confirming the classical ambipolar diffusion limit at $\Lambda/\lambda_i \gg 1$.

### 3. Experiment

The experimental setup, operating at LPGP, is represented in figure 2 (more details in [6]).

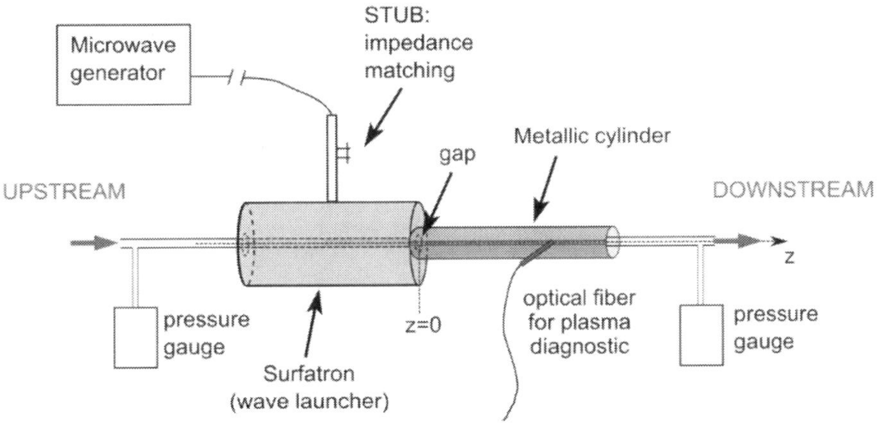


**Figure 2:** Schematic representation of the experimental setup.

The capillary is inserted into a surfatron especially adapted to the system geometrical dimensions, ensuring a maximum electric field at the gap. Dry air is introduced from the left end of the capillary, at negligible gas flow, yielding a constant pressure along the plasma column. Non-intrusive axially-resolved optical emission spectroscopy (OES) diagnostics are used to obtain: (i) the electron density, measured from the contribution of the Stark effect to the broadening of the Balmer line $H_\beta$ (486.1 nm), which provides an upper-limit estimation $n_e < 1.4\text{x}10^{13}$ cm$^{-3}$. The latter agrees with a semi-empirical analysis of the power coupled to the plasma, based on the solution to Maxwell's equations for the excitation structure; (ii) the absolute densities of $N_2$(C), $N_2$(B), $N_2^+$(B), N(3p$^4$S) and O(3p$^5$P), and the relative intensities of the second positive system $N_2$(C,v)$\rightarrow$$N_2$(B,v') (whose spectrum is used to determine also the rotational/gas temperature), the first negative system $N_2^+$(B,v)$\rightarrow$$N_2^+$(X,v'), and the atomic transitions N(3p$^4$S)$\rightarrow$N(3s$^4$P) and O(3p$^5$P)$\rightarrow$O(3s$^5$S). The absolute calibration of the spectra was performed at EM2C, using a standard tungsten lamp for which the spectral radiance is known. The same detection system (including the spectrometer and the optical collection system) is employed for recording both the discharge and the lamp spectra, the former being converted in radiance units by using the sensitivity response function. The gas temperature and the absolute density of a particular electronic excited state are obtained by comparing measured and Specair-simulated [7] spectra.

### 4. Results and discussion

Figure 2 compares measurements and simulations of (i) the relative intensity with the atomic transition of triplets N(3p$^4$S)$\rightarrow$N(3s$^4$P) and O(3p$^5$P)$\rightarrow$O(3s$^5$S) and (ii) the absolute densities of $N_2(C)$, $N_2(B)$ and $N_2^+(B)$, as a function of the axial position in the plasma column. Model results are in qualitative agreement with measurements, predicting the correct trend for the axial evolution of the atomic line transitions and under- / over- estimating the densities of $N_2(B,C)$ / $N_2^+(B)$ by factors below 10 times.

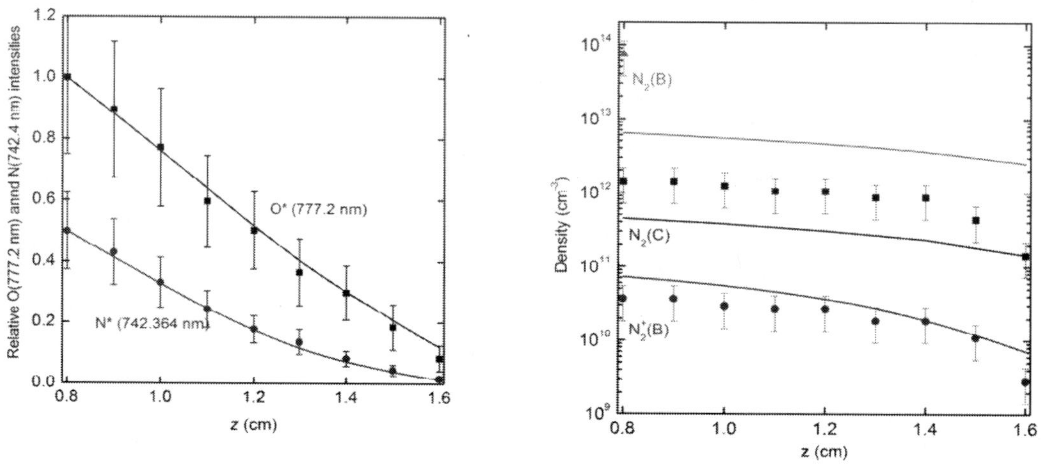


**Figure 2.** Axial profiles, calculated (lines) and measured (points) at 129 W coupled power, for: left, the relative intensities with atomic line transitions $N(3p^4S) \rightarrow N(3s^4P)$ (blue) and $O(3p^5P) \rightarrow O(3s^5S)$ (black); right, the absolute densities of $N_2(C)$ (black), $N_2(B)$ (red) and $N_2^+(B)$ (blue).

## 5. Final remarks

A low pressure, small radius microwave micro-plasma discharge has been analysed, using both simulations and measurements. Under these work conditions, the charge-particle transport description, including the effect of one negative ion species, has been successfully developed. OES diagnostics were employed to measure the absolute densities of several species and the relative intensities of some radiative transitions. Simulation results evidence good qualitative agreement with measurements.

**6. Acknowledgements**

This work was partially supported by FCT - Fundação para a Ciência e a Tecnologia, under Project UID/FIS/50010/2013 and Project PLASMA-ADIST, and by the Labex LaSIPS, under Project MicroCap.

**7. References**

[1] V. Guerra and J. Loureiro, Plasma Sources Sci. Technol. **6** (1997) 373

[2] V. Guerra and J. Loureiro, Plasma Sources Sci. Technol. **8** (1999) 110

[3] L.L. Alves, J. Phys.: Conf. Ser. **565** (2014) 012007; www.lxcat.net

[4] S.A. Self and H.N. Ewald, Phys. Fluids **9** (1966) 2486

[5] C.M. Ferreira and A. Ricard, J. Appl. Phys. **54** (1983) 2261

[6] S. Dap et al, Plasma Sources Sci. Technol. (2015) submitted

[7] Specair www.specair-radiation.net

# Optimization of the snowflake divertor by means of EMC3-Eirene simulations and experiments

T. Lunt[1], G.P. Canal[2], B.P. Duval[2], Y. Feng[1], B. Labit[2], P. McCarthy[3], H. Reimerdes[2], W.A.J. Vijvers[2], M. Wischmeier[1] and the EUROfusion MST1 Team*

[1] *Max Planck Institute for Plasma Physics, Garching/Greifswald, Germany*

[2] *Centre de Recherches en Physique des Plasmas, EPFL, Lausanne, Switzerland*

[3] *Department of Physics, University College Cork, Cork, Ireland*

## Introduction

Power exhaust is considered as one of the biggest challenges for the realization of a fusion reactor based on the divertor tokamak design [1]. In order to mitigate the heat flux density occurring near the divertor strike points (SP) the 'snowflake' (SF) configuration was proposed by Ryutov [2] and for the first time realized experimentally on TCV [3]. The SF configuration is characterized by two additional SPs and a second-order null point, where the poloidal magnetic field vanishes together with its spatial derivatives. Since it would require an infinitely accurate control of the toroidal currents, an *exact* SF cannot be achieved in the experiment, but only approximated by approaching a secondary X-point to the 'primary' one (that determines the boundary of the confinement region). Following the convention of Ryutov [4], we will speak of a 'snowflake plus' ($SF^+$), when the secondary X-point is located in the private flux region (PFR) of the primary separatrix and of a 'snowflake minus', when it is located in the common flux region (CFR). In Ref. [5] we reported on the first EMC3-Eirene simulations on $SF^+$ configurations compared to experiments. While the simulations predicted power fluxes to the secondary SPs of only about 1% of the input power (assuming constant perpendicular particle- and power diffusivities $D_\perp$ and $\chi_\perp$), about 10% were found experimentally, indicating an enhanced transport across the separatrix in the SF configuration.

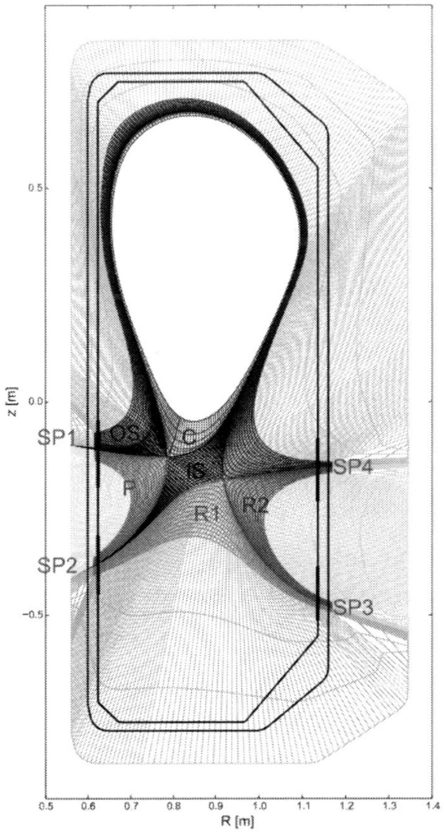


Figure 1: *Poloidal projection of the flux surface aligned computational grid for a LFS $SF^-$ configuration with $\rho_{x2} = 1.0075$.*

*See http://www.euro-fusionscipub.org/mst1

If this transport is driven e.g. by $\beta_{pol}$ instabilities as predicted by Ryutov [6] or if other mechanisms like drifts [7] are responsible needs to be investigated. As the power diffusing into the CFR is much larger than that to the PFR, a much larger power redistribution effect (combined with a reduction of the peak heat flux density at the primary targets) can be expected for a SF$^-$ without relying on this additional transport channel. Here we report on simulations including such configurations with the secondary X-point on the low-field side (LFS) of the primary one (LFS SF$^-$) with the goal of finding the optimum configuration. More details on these results can be found in Ref. [8]. Experimentally the SF$^-$ was investigated in Ref. [9].

Due to an in/out asymmetry observed experimentally [10] (for the favorable forward field conditions) in attached conditions and an asymmetric detachment [11], the outer target is loaded significantly stronger than the inner one for a single-null (SN) configuration and presently the limiting factor for the exhaustable power of the device. This asymmetry is often attributed to the drifts occurring in the SOL [12]. Another reason is the difference between connection lengths $L_{c,inner}$ and $L_{c,outer}$ from the outboard mid-plane (OMP) to the inner and outer targets respectively, as shown in Fig. 2 for the SN (green curves) in ASDEX Upgrade (AUG). Figure 2 also shows that $L_{c,outer}$ is significantly increased in the near-SOL region for a LFS SF$^-$ configuration computed for the realistic coil geometry but with currents that are presently beyond the tolerable force limits for AUG. If we assume that the detachment depends on $L_c$ we can expect a detachment of the outer targets at significantly lower densities for the LFS SF$^-$.

**Simulations**

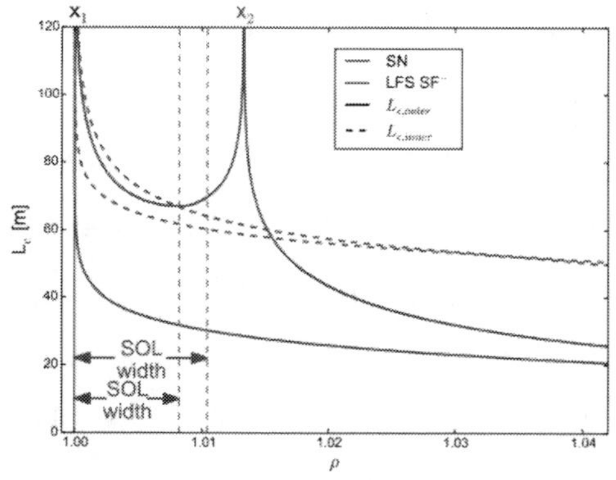


Figure 2: *Outboard mid-plane to inner ($L_{c,inner}$) and outer ($L_{c,outer}$) target connection length in AUG for a SN and a LFS SF$^-$configuration.*

Plasma transport is simulated by the Edge Monte-Carlo 3D (EMC3) code that solves Braginskii's *fluid* equations, while EIRENE [13] solves the kinetic transport equation for the neutral particles. Both codes apply a Monte-Carlo technique to solve the equations. These equations and further details on the physics and the code coupling of the two codes can be found in Ref. [14]. Note that neither drifts, nor volumetric recombination are implemented in the code so far. An example of a computational grid used for the simulations is shown in Fig. 1. It comprises six zones, the confinement region (C), the inner (IS) and outer SOL (OS), the PFR (P) and the remote areas R1 and R2. Fig. 1 also shows the four SPs enumerated in counter-clockwise direction starting from the inner target. In Ref. [8] parameters $\rho_{x2}, \vartheta_{x2}$ were intro-

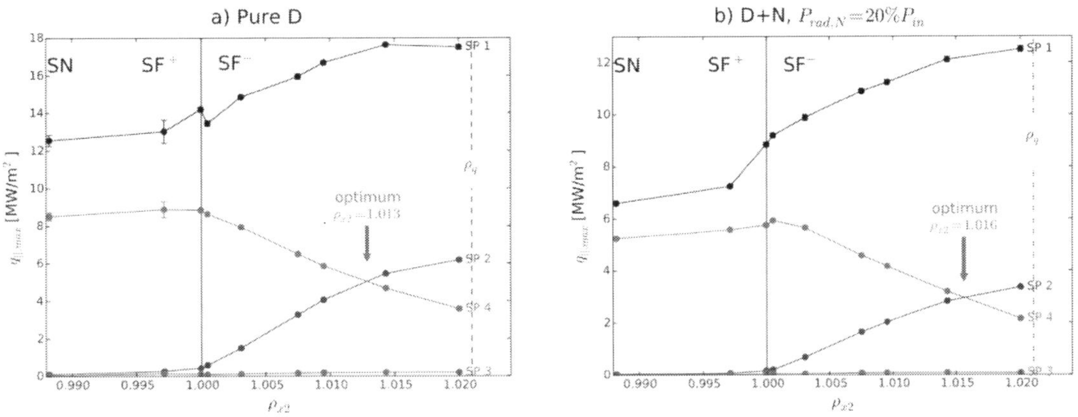


Figure 3: *Peak parallel power flux density $q_{||,max}$ near the different strike points SP 1...SP 4 for a series of nine simulations with different radial positions $\rho_{x2}$ of the secondary X-point. The transport parameters were $D_\perp = 0.5$ m²/s and $\chi_\perp = 1.5$ m²/s. a) pure deuterium, b) with nitrogen radiating 20% of $P_{in}$.*

duced to classify the configuration, as an alternative to the parameters $\sigma, \theta$ used in Refs. [3, 5]. $\rho_{x2} = ((\Psi_{x2} - \Psi_o)/(\Psi_{x1} - \Psi_o))^{0.5}$ is the radial coordinate, where $\Psi_o$ is the poloidal magnetic flux on the magnetic axis and $\Psi_{x1}$ and $\Psi_{x2}$ those at the primary and secondary X-point, respectively. For reference, $\rho_{x2} < 1$ for the SF$^+$, $\rho_{x2} = 1$ for the exact SF and $\rho_{x2} > 1$ for the SF$^-$. $\vartheta_{x2}$ specifies the poloidal position of $x_2$ and is defined in Ref. [8]. All simulations were performed with deuterium as the main ion species, an input power of $P_{in} = 300$ kW equally distributed between the electrons and ions and a separatrix density at the OMP of $n_{OMP} = 1.5 \cdot 10^{19}$ m$^{-3}$, typical for medium density TCV discharges. The transport coefficients were chosen to be $D_\perp = 0.5$ m²/s and $\chi_\perp = 1.5$ m²/s and spatially constant to match typical target profiles.

**Results**

A series of nine configurations were simulated with $\rho_{x2}$ covering the range of 0.988...1.020. For $\rho_{x2} = 0.988$ the secondary X-point is far from the primary one and we refer to this case as the single-null (SN) reference. The upstream $n_e$ and $T_e$ profiles vary only by a few percent with $\rho_{x2}$ and thus does the position $\rho_q = 1.021$ of the flux surface on which the power has fallen to 1/e of the separatrix value (cf. Fig. 7 in Ref. [8]). Fig. 3 shows the maximum parallel power flux density $q_{||,max}$ near the four strike points SP 1...SP 4 as a function of $\rho_{x2}$. Note that in contrast to the experimental observation $q_{||,max,SP_1}$ (black curve) is significantly larger than $q_{||,max,SP_4}$ (red). We attribute this discrepancy to the lack of drifts in the code and focus only on the outer SP 4 and SP 2 in the following[1]. From Fig. 3 a), $q_{||,max,SP_4}$ decreases monotonically with increasing $\rho_{x2}$ for $\rho_{x2} > 1$ (the SF$^-$ case), while $q_{||,max,SP_2}$ increases. The configuration with $\rho_{x2} = 1.013$ where $q_{||,max,SP_2} = q_{||,max,SP_4}$ constitutes the one with the minimum load on the target and is thus

[1] Although SP 2 is located on the HFS, here we refer to it as an 'outer' SP for topological reasons.

considered optimal under the conditions considered so far. It was shown in Ref. [8] (Fig. 10) that (at least in attached conditions) this result depends only little on $\vartheta_{x2}$.
Experimentally a similar effect was observed for the inner SP in a HFS SF$^-$ configuration in TCV, where a reduction of about 50% was found [15].
Figure 3 b) shows the same simulations as a) but with nitrogen impurities radiating a fixed amount of 20% of the input power. While the impurities radiate predominantly on the HFS for a SN and SF$^+$and therefore reduce $q_{||,max,SP_1}$, the fluxes to the outer SPs $q_{||,max,SP_2}$ and $q_{||,max,SP_4}$ are more affected for the LFS SF$^-$ where the impurities accumulate around the secondary X-point (cf. Fig. 12 in Ref. [8]). The radiation from carbon impurities was observed to be shifted to the LFS in a LFS SF$^-$ in the experiment.

## Summary

A series of SN, SF$^+$ and LFS SF$^-$ configurations characterized by $\rho_{x2}$ ranging from 0.988 to 1.020 was analyzed and simulated successfully by EMC3-Eirene. It was found that the maximum power flux density to the outer target can be reduced by a factor of two or presumably significantly more in a LFS SF$^-$ configuration due to three effects 1) a power splitting between SP 2 and SP 4 combined with steeper gradients and enhanced diffusion 2) an asymmetric irradiation of power removing predominantly power from the outer targets and 3) a longer connection length possibly facilitating the access to detachment at the outer targets.

## Acknowledgement

This work has been carried out within the framework of the EUROfusion Consortium and has received funding from the Euratom research and training programme 2014-2018 under grant agreement No 633053. The views and opinions expressed herein do not necessarily reflect those of the European Commission.

## References

[1] Zohm H. *et al.* 2013 *Nucl. Fusion* **53** 073019
[2] Ryutov D.D. *et al.* 2007 *Phys. of Plasmas* **14** 064502
[3] Piras F. *et al.* 2009 *Plasma Phys. Contr. Fusion* **51** 055009
[4] Ryutov D.D. *et al.* 2010 *Plasma Phys. Contr. Fusion* **52** 105001
[5] Lunt T. *et al.* 2014 *Plasma Phys. Control. Fusion* **56** 035009
[6] Ryutov D.D. *et al.* 2012 *Plasma Phys. Control. Fusion* **54** 124050
[7] Canal G. *et al.* 2015 *submitted to Nucl. Fusion*
[8] Lunt T. *et al.* 2015 *Submitted to Plasma Phys. Control. Fusion*
[9] Vijvers W.A.J. *et al.* 2015 *this conference*
[10] Eich T. *et al.* 2007 *Plasma Phys. Control. Fusion* **49** 573–604
[11] Potzel S. *et al.* 2014 *Nucl. Fusion* **54** 013001
[12] Aho-Mantila L. *et al.* 2012 *Nucl. Fusion* **52** 103006
[13] Reiter D. *et al.* 2005 *Fus. Sci. Tec.* **47** 172–186
[14] Feng Y. *et al.* 2004 *Contrib. Plasma Phys.* **44**, No. 1–3, 57–69
[15] Reimerdes H. *et al.* 2013 *Plasma Phys. Contr. Fusion* **55** 124027

# TECXY modelling studies of alternative EAST magnetic configurations

B. VIOLA[1], V. Pericoli Ridolfini[1], G. Calabró[1], F. Crisanti[1], Y. Duan[3], Y. Guo[3], L. Wang[3], Z. P. Luo[3], B. J. Xiao[3,4], J. C. Xu[3], R. Zargóski[2], B. Zhang[3]

[1] *ENEA for EUROfusion, via E. Fermi 45, 00044 Frascati (Roma), Italy.* [2]*Institute of Plasma Physics and Laser Microfusion, 01-497 Warsaw, Poland.* [3]*Institute of Plasma Physics, Chinese Academy of Sciences, Hefei, 230031, China.* [4] *School of Nuclear Science and Technology, University of Science and Technology of China, Hefei, 230026, China.*

## Introduction

Several alternative magnetic divertor configurations have been proposed to handle high exhaust power. The majority of them relies on the presence of a second null, more or less close to the first one. The main physics difference being whether the flux lines converge or flare in front of the divertor plates. Several authors refer to experimental scenarios dedicated to the understanding of the role of the two nulls as quasi Snow Flakes (QSF) scenarios. The first experiments [1, 2] on the Experimental Advanced Superconducting Tokamak (EAST) were mainly devoted to study the feasibility of QSF configurations for different distances of the two nulls. This contribution reports on predictive and interpretative simulations focusing on the comparison of two QSF configurations versus a standard single null configurations (SN). Consistency is found between the experimental and calculated mitigation of the power load onto the divertor targets for a QSF at low scrape-off layer (SOL) density and low additional power. Predictions on how further mitigation can be achieved at high density will be presented together with preliminary studies on the reciprocal distance between the two produced nulls.

## TECXY interpretative simulation results

For its very high flexibility and possibility to easily perform parameters scan, the TECXY code [3] is the modelling tool chosen to study the EAST edge plasma. It takes into account all the main physics processes occurring in the SOL, but simplifies the neutral dynamics using an analytical model instead of the more rigorous Monte Carlo method. In addition, the divertor plates are always assumed to be perpendicular to the flux surfaces and the private flux region is neglected. The large saving of computational time implied by the above simplifications motivates this choice when general trends and/or global comparisons of different scenarios are searched for and/or when a preliminary optimization/exploration of the operating parameter space is attempted. This is especially true when alternative complex configurations have to be analyzed, and several parameters quickly varied to have the basic relative idea about the role of each parameter. In our case the most important point being the exploration of the reciprocal distance of the nulls, compared to density and power variations. Moreover, TECXY agrees with the results of more sophisticated codes within its application limits described just above, as discussed in [6, 7]. In this section we use TECXY to explain the zero order differences experimentally found between the QSF and SN features, discharges #48971 and #47038 respectively. Their shapes reconstructed by a equilibrium code are shown in Fig. 1. For QSF #48971 it is: t=4.5s, $\beta_p = 0.76$ and $l_i = 1.28$, the distance of the secondary X-point from the primary one is 70 cm. For SN it is t=4.5s, with $\beta_p = 0.58$ and $l_i = 1.56$. The main quantities used as input for the TECXY calculations are experimentally determined as follows. 1) input power to the SOL: $P_{SOL}$ = 431.7 kW for SN and $P_{SOL}$ = 414.5 kW for QSF case, which has neutral beam auxiliary power $P_{NBI} = 513.8kW$ but a slightly higher radiation; 2) electron density at the outer mid-plane (OMP) separatrix $n_{e,LCMS} = 6 \times 10^{18} m^{-3}$ for both discharges. No dominant impurity has been considered since, as mentioned above, here we are only looking for the zero order effects and since we are not concerned with absolute predictions, but with comparing the relative behaviour of two similar discharges. The fraction of $P_{SOL}$ deposited on the plates is quite similar at this density, namely 51% for the SN and 46% for the QSF. Almost all the mitigation for QSF has to be attributed to the flux expansion (FE). For these rather low densities, a very similar trend was found in previous studies [8] with EDGE2D/EIRENE run for comparison purposes with TECXY. In Fig. 2 the power density measured by the infrared (IR) camera diagnostic on the lower outer target is shown versus the distance along the target and compared to the simulated heat loads. Corrections for the actual target tilting have been applied. From the qualitative point

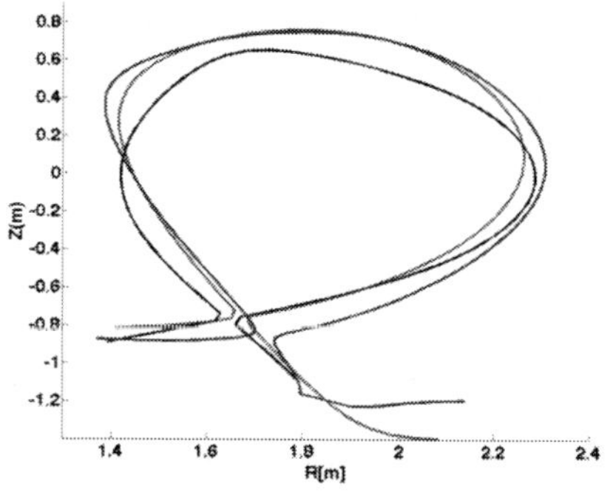


Fig. 1: Plasma boundary of the analyzed equilibria the experimental QSF (red line), reference SN equilibria (black line) and the QSF with close nulls (blue line)

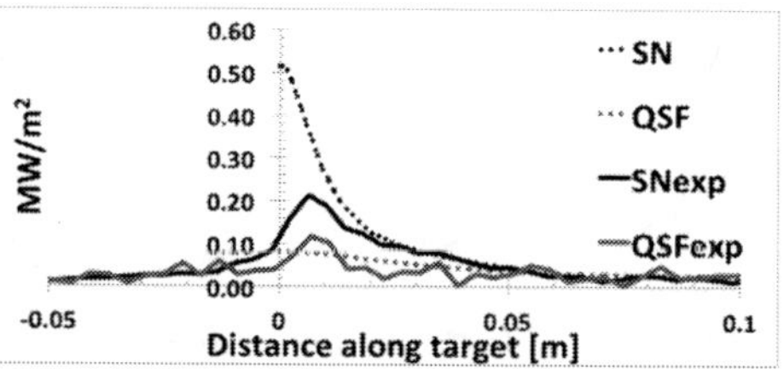


Fig. 2: Head load flux on the outer target: comparison between IR thermography data (solid lines) and TECXY simulation (dashed lines).

of view the code well reproduces the profile shapes and measured power mitigation in QSF. The largest quantitative discrepancy refers to the peak values in the case of the SN. The most likely reason for this discrepancy is the diffusion into the private region, which is not included in the calculations. Indeed the differences of the integrals over this region are consistent with the integral of the experimental curve on the left side of the graph. This effect should be much less pronounced for smaller gradients that develop in the QSF. Therefore, this could also account for the discrepancy in the resultant mitigation. Conversely, the QSF discrepancies could be attributable to experimental inaccuracies. This matter is presently under further investigation. A recent upgrade of TECXY would also allow to consider the private region, too.

**TECXY predictive simulation results**

A power and density scan have also been carried out to study the behaviour of the actually realized configurations in different scenarios, as far as target load and volume losses are concerned. Three different values of $P_{SOL}$ and $n_{e,LCMS}$ have been considered: $400kW$, $1.5MW$ and $3MW$ for the SOL input power and 3, 6 and $8.8 \times 10^{18} m^{-3}$ for the SOL outer mid-plane density. In Fig. 3 the mitigation factor (ratio of the SN to QSF peak load) is plotted versus $n_{e,LCMS}$ for the three $P_{SOL}$ values. The experimental datum is also shown. As previously found for the proposed tokamak FAST [8], high SOL density can exalt the load mitigating properties of the QSF, as evidenced by the 400 kW data (black squares). This is due to the higher volume losses that occur when the connection lengths are strongly increased as in the QSF. Clearly high volume losses develop as a macroscopic effect if a non-negligible loss already exists with SN, namely if the SOL conditions do allow any volume losses to occur. This is prevented when the power flowing into the SOL is increased so that the temperature grows too high where the largest difference in the connection lengths is found in the investigated configurations. This occurs already for 1.5 MW: in this experimental situation, by further increasing the plasma current and density, a similar behaviour would be reproduced. A further step in the investigation of the

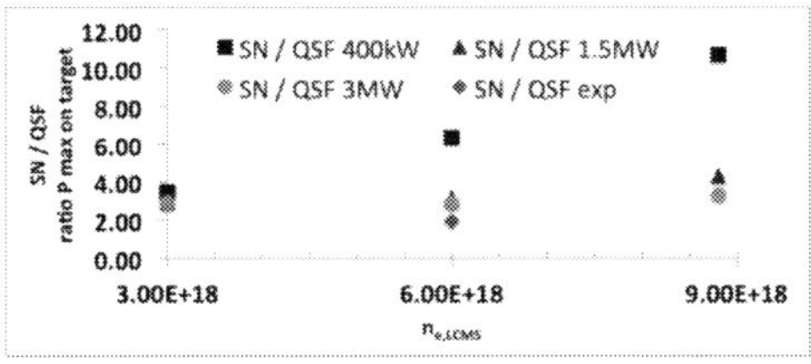


Fig. 3: Predictive power and density scan based on the experimental SN and QSF equilibria

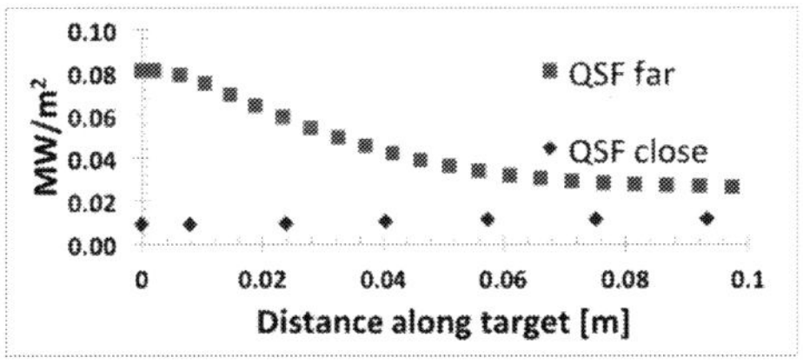


Fig. 4: Comparison of the power density calculated by TECXY at the outer target for the two QSF equilibria, $n_{e,LCMS} = 3 \times 10^{18} m^{-3}$

alternative configurations that are realizable by the PF coils system of EAST has been to study a QSF configuration with close nulls (QSFcn), also shown in Fig. 1 . The purpose of this configuration is to further prolong the connection lengths and also to shift the second null towards the primary X-point. In this configuration the proximity of the secondary null, only 43 cm far from the primary one, increases the connection length, evaluated at the outer target plate, from 33 m of the SD to 110 m (75.6 m in the experimental QSF, QSFexp). Also a significantly larger FE is generated. If we quantify the FE as the target to OMP ratio of the poloidal distance of the flux surfaces from separatrix, we obtain for SD, QSFexp and QSFcn respectively the values of 3, 10 and 33. Furthermore by comparing the poloidal flux we see that for the QSFexp we obtain a monotonic decrease (contracting), (SF [4] like, when a second null is present) whereas for QSFcn a flaring (XD like [5]) of the flux tubes. The predicted power load by TECXY on the outer target for this case is compared in Fig. 4 with that of QSFexp for $n_{e,LCMS} = 3 \times 10^{18} m^{-3}$ and $P_{SOL}$ =414.5kW. The peak of the profile is reduced by more than 4 times, i.e. more than implied by the expanded flux ($\sim$ 3 times).

## Conclusions

The first experimental QSF configuration has been obtained in the EAST tokamak by creating a far secondary X-point, allowing to increase the connection length by $\sim$ 30% and the FE in the outer divertor region by a factor $\sim$ 4. IR camera measurements have been compared with those obtained in a similar conventional SN configuration and have been interpreted with the TECXY edge code. The general trends are reproduced for the different magnetic configurations. The profiles are well reproduced except in the close proximity of the strike point, but the discrepancies can be explained by having neglected the private region. The heat load mitigation is almost entirely due to the FE since the low working density is not suitable for enhancing significantly the volume losses, achievable by increasing the connection lengths. Numeric studies predict that a reduction in the total load should be observed at higher densities, the exact value of which however depends on the power flowing into the SOL. Higher power should increase the SOL temperature and hinder the dissipative processes. In addition a QSF configuration with close nulls that can be realized with the PF coils system has been modelled with TECXY. The potentiality of this configuration in affecting the power exhaust more efficiently than the QSF just realized is evidenced.

## Acknowledgments

This work has been carried out within the framework of the EUROfusion Consortium and has received funding from the Euratom research and training programme 2014-2018 under grant agreement No 633053. The views and opinions expressed herein do not necessarily reflect those of the European Commission.

## References

[1] G. Calabró et al., Proc. of 2014 IAEA Conference, EX/P3-4.

[2] G. Calabró et al., submitted to Nuclear Fusion (2014) .

[3] R. Zagórski, H. Gerhauser, Phys. Scr. 70 (Part 2/3) (2004) 173.

[4] Ryutov et. al., Phys. of Plasmas 14, 064502 2007.

[5] Kotschenreuther M., et al., Phys. Plasmas 20 (2013) 102507.

[6] B. Viola et al., Contrib. Plasma Phys., Vol. 54, Issue 4-6, June 2014, Pages 459-463.

[7] B. Viola et al., doi:10.1016/j.jnucmat.2014.11.027.

[8] Pericoli Ridolfini V., et al., Fus. Eng. Des. 88 (2013) 1677.

# Toroidally Symmetric Plasma Vortex at Tokamak Divertor Null Point

M.V. Umansky and D.D. Ryutov

*Lawrence Livermore National Laboratory, Livermore, CA 94550, USA*

## Introduction

In next generation tokamak experiments and fusion reactors the divertor heat flux will challenge conventional means of peak heat flux reduction which makes it urgent to find innovative approaches to solving the divertor heat exhaust problem. One recent idea for tokamak divertor is using a higher order null (dubbed "snowflake") instead of the standard x-point [1]. Snowflake divertor configuration has a characteristic hexagonal separatrix structure, and it has a number of geometric properties that may affect edge plasma and may be helpful for alleviating the divertor heat flux problem: stronger fanning of the poloidal flux, stronger magnetic shear in the edge region, larger radiating volume, and larger connection length in the scrape-off layer [2].

Theoretical considerations [3, 4] and recent experimental observations from snowflake divertor experiments on several tokamaks indicate that presence of a near-second-order null of poloidal field may give rise to strong plasma mixing near the magnetic divertor null point, thereby providing sharing of heat and particle flux between multiple (three to four) strike points [5, 2].

Further, a semi-quantitative analytic model [6] was proposed to describe plasma dynamics driven by the pressure gradient and magnetic curvature near the null-point where the poloidal magnetic field is small. The analytic model predicts oscillatory twisting motion of plasma (dubbed "the churning mode") localized at the null point, and the size of the convective zone is much larger for snowflake than for a regular x-point configuration, for realistic tokamak parameters. Although the basic physics of the churning mode has been identified in the analytic model [6], the assumptions of the analytic model may be too restrictive, in particular, the motion was postulated to be in a particular form of differential rotation, the role of the resistive dissipation of the twisted poloidal field was not included, and there was no account for the heat conduction in the convectively mixed plasma. In the present study the formation of the convective cell is investigated by direct numerical simulation, solving plasma fluid equations describing time-evolution of plasma thermal energy and electric and magnetic fields.

## Reduced MHD model

The standard ideal MHD equations include equations for mass continuity, momentum balance, adiabatic gas law, and magnetic induction [7]. The specific interest here is toroidally symmetric plasma motion in the vicinity of the null point $(R_0, Z_0)$, where $(R, \theta, Z)$ are the usual cylindrical coordinates; so we'll further assume (i) toroidal symmetry, $\partial_\theta$=0, (ii) large aspect ratio, $l_\perp/R \ll 1$, (iii) strong toroidal guide field, $B_P/B_t \ll 1$, and (iv) uniform density $\rho$=const. Instead of the cylindrical coordinates we'll be using more convenient here Cartesian coordinates $(x,y)$ in the poloidal plane with the origin at the null-point, see Fig. (1), $x = R - R_0, y = Z - Z_0$

The poloidal magnetic field $B_p$ is described by the poloidal flux function $\psi(x,y)$, $B_p = -(1/R_0)e_\theta \times \nabla\psi$, where $\nabla = (\partial_x, \partial_y)$, and the toroidal magnetic field is described by the reference value $B_{t0}$ given at the reference point $(R_0, Z_0)$, $B_t = e_\theta B_{t0}/(1 + x/R_0)$. The electric field is described by axisymmetric potential $\phi(x,y)$, and the fluid velocity $V$ is poloidal and given by $V = -(c/B_{t0})(e_\theta \times \nabla\phi)$. From the magnetic induction equation one can obtain the evolution equation for the poloidal magnetic flux function, and expressing the poloidal current from the equation of motion and imposing a constraint, $\nabla \cdot J_p$=0 one finds an evolution equation for the vorticity which is defined as $\varpi = (c/B_{t0})\rho\nabla^2\phi$. For normalization of the equations, several reference parameters are chosen at the null point (subscripted 0) and at the outer mid-plane: $R_0$, $B_{t0}$, $P_0$, $a_{mid}$, $B_{pmid}$. These are combined into a reference speed and time unit: $C_{s0} = \sqrt{P_0/\rho}$, $t_0 = a_{mid}/C_{s0}$. Length units are normalized by $a_{mid}$ and time units are normalized by $t_0$. Next, the pressure is normalized by $P_0$ so at the null-point the normalized pressure is 1.0. The poloidal flux is normalized by $\psi_0 = B_{pmid}R_0 a_{mid}$ so the normalized flux is on the order of unity for normalized distance $\sqrt{x^2 + y^2} \sim 1$. Normalization of the electric potential $\phi$ is such that $V_E = [e_\theta \times \nabla\phi]$, so $\varpi = \nabla^2\phi$. The dissipation coefficients are normalized by $a_{mid}C_{s0}$.

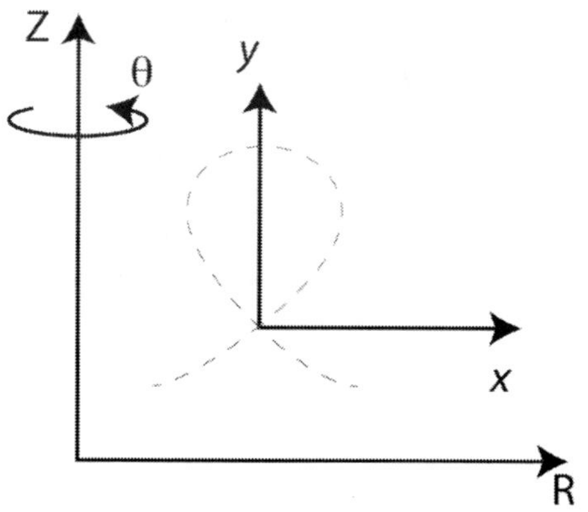


Figure 1: *Coordinates used for the model*

Finally, with dissipative terms added, the equations in normalized form are summarized as follows

$$\begin{cases} \dfrac{d}{dt}\varpi = 2\varepsilon\dfrac{dP}{dy} + \dfrac{2}{\beta_p}\{\psi, \nabla^2\psi\} + \mu\nabla^2\varpi \\ \dfrac{d}{dt}P = \chi\nabla^2 P \\ \dfrac{d}{dt}\psi = D_m\nabla^2\psi, \end{cases} \tag{1}$$

where $d/dt = \partial_t + V_E \cdot \nabla$, $V_E = [e_\theta \times \nabla\phi]$, and $\varpi = \nabla^2\phi$.

The non-dimensional parameters in the normalized equations are the aspect ratio, $\varepsilon = a_{mid}/R_0$, and the poloidal plasma beta, $\beta_p = 8\pi P_0/B^2_{pmid}$, $\mu$ is the kinematic viscosity, $\chi$ is the thermal diffusivity, $D_m$ is the magnetic diffusivity. In the equations one can identify the driving term arising from the pressure gradient and curvature, the stabilizing forcing term from magnetic field deformation, and dissipative terms. Note that the toroidal magnetic field does not enter the normalized equations, and the toroidal geometry effects show up only in the curvature drive term.

**Numerical simulations**

For numerical solution, the equations are discretized by finite-difference on a computational grid and then time-integrated using standard numerical software. The simulations are carried out in a square domain $L \times L$ centered at the null point location. On the domain boundaries the values of all evolved variables are maintained constant, which is consistent with the expectation that the studied physical phenomena are localized near the null point.

The initial vorticity is $\varpi$=0, and the initial magnetic flux $\psi_0$ is chosen as either a first order null x-point configuration (XPT), $\psi_{(1)} = x^2 - y^2$, or as an exact second order null snowflake configuration (SNF), $\psi_{(2)} = y^3 - 3x^2y$, or as an inexact second order null configuration, snowflake-minus (SNF-) or snowflake-plus (SNF+), $\psi_{(2\mp)} = y^3 - 3x^2y \pm 3\sigma^2 y$, where for SNF- two first order nulls aligned horizontally and located at $(\pm\sigma, 0)$, and for SNF+ two first order nulls aligned vertically and located at $(0, \pm\sigma)$.

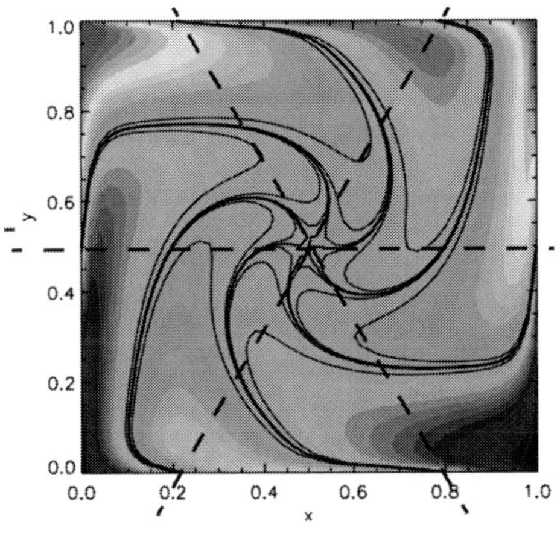


Figure 2: *Plasma vortex forming in SNF configuration*

Time-integration of Eqs. (1) shows formation of a vortex-like structure localized at the null point. Depending on the model parameters, there are various scenarios of evolution: (i) relaxation to a rotated quasi-steady state where the vertical pressure gradient $dP/dy$ is near zero, or (ii) rotation of the central kernel with occasional relaxation events, or (iii) turbulent motion in the central zone. One can see in the simulations that for a SNF configuration there is always a convective zone at the null point, and the size of it increases with $\beta_p$. For the XPT configuration, there is a threshold value of $\beta_p$ when convection starts. In general, the closer the configuration is to SNF the larger is the scale of convective motion. Furthermore, using parameters of a generic mid-size tokamak, one can conclude that in SNF configuration, during ELM the scale of convection should be significant for redistribution of thermal energy across the null-point. On the other hand, for same parameters, in XPT configuration there is virtually no perturbation of plasma pressure and magnetic field. This is consistent with the TCV experiments where redistribution of thermal energy between primary and secondary strike points was seen to be particularly large during ELM strikes in near snowflake configurations [5]. The churning mode can be a significant player in transport of thermal energy near the null point, in particular during ELMs when $\beta_p$ is large at the null point, and in particular in near-snowflake configurations. Aside from the direct action of thermal convection, there may be other effects such as inducing magnetic stochasticity in the null point region due to perturbation of the poloidal magnetic field.

Note that the churning mode physics is very similar to free thermal convection of neutral fluid, in a vertical gravity field, driven by a horizontal temperature gradient, as described in, e.g., [8, 9, 10]. The hydrodynamics equations that are used for the the thermal convection problem can be cast in a form very similar to Eqs. (1), with the magnetic curvature playing the role of effective gravity. However one important difference is that the plasma equations have the restoring term due to magnetic field perturbations which plays the role of "elasticity" resisting plasma twisting.

## Acknowledgments

This work was performed under the auspices of the US Department of Energy by Lawrence Livermore National Security, LLC, Lawrence Livermore National Laboratory, under Contract DE-AC52-07NA27344.

## References

[1] D.D. Ryutov et al., Physics of Plasmas 14, 064502 (2007).
[2] V. A. Soukhanovskii et al., Journal of Nuclear Materials 438, S96-S101 (2013).
[3] D.D. Ryutov et al., Plasma Physics and Controlled Fusion 54 (2012).
[4] D.D. Ryutov et al., Contributions to Plasma Physics 52, 5-6, 539-543 (2012).
[5] W A J Vijvers et al., Nuclear Fusion 54, 2, 023009, (2014).
[6] D.D. Ryutov et al., Physica Scripta 89, 8, 088002 (2014).
[7] J. P. Freidberg, "Ideal Magnetohydrodynamics" (Plenum, New York, 1987).
[8] D. A. Bratsun et al., International Journal of Heat and Fluid Flow, 24, 6, 835 (2003).
[9] K. Buhler, Heat and Mass Transfer, 39, 631, (2003).
[10] G. Z. Gershuni and E. M. Zhukhovitskii, "Convective Stability of Incompressible Fluids" IPST (1976).

# The X-point radiation regime in detached H-Modes in full-tungsten ASDEX Upgrade

F. Reimold[1], M. Bernert[2], A. Kallenbach[2], B. Lipschultz[3], G. Meisl[2], S. Potzel[2], M.L. Reinke[3], M. Wischmeier[2], D. Wünderlich[2], the EUROfusion MST1 team* and the ASDEX Upgrade team

[1]*Forschungszentrum Jülich, IEK-4, 52425 Jülich, Germany*
[2]*Max Planck Institute for Plasma Physics, 85748 Garching, Germany*
[3]*University of York, York Plasma Institute, Heslington, York, United Kingdom YO1 9DD*

## Introduction

Operation with a detached divertor is mandatory in future fusion devices, such as ITER and DEMO, in order to reconcile the large power fluxes into the divertor with the heat load and erosion limits of the divertor target. In ASDEX Upgrade (AUG) H-Mode discharges additional seeding of impurities has been necessary to dissipate the power in the Scrape-Off Layer (SOL) by radiation losses to achieve complete detachment with larger heating power ($P_H$). In standard operation of AUG at plasma currents above $I_P = 0.8$ MA, the vertical inner divertor target is usually detached. Partial detachment at the outer strike-point with a strong drop of plasma pressure along field lines and an ion flux reduction can be obtained with impurity seeding. In order to detach the outer target completely [1], a transition into an X-point radiation regime, which is characterized by intense, localized radiation at the X-point inside the confined plasma, seems to be required at AUG. Similar transitions have also been observed in JET [2], DIII-D [4] and JT-60U [6]. With a carbon first-wall such a transition was generally avoided as the radiation at the X-point was an unstable operational scenario. Recent observations [1, 2] and SOLPS modeling [3] have shown that in an all-metal machine such as AUG and JET the X-point radiation regime can be stable, controllable and reversible. In JET the X-point radiation regime has been observed for a number of impurities such as N, Ne and Ar [2]. In AUG high radiation scenarios with detachment of various degree have been developed for N, Ar and Kr [1, 5, 7]. With nitrogen seeding a transition into an X-point radiation regime occurs that is beneficial for power exhaust.

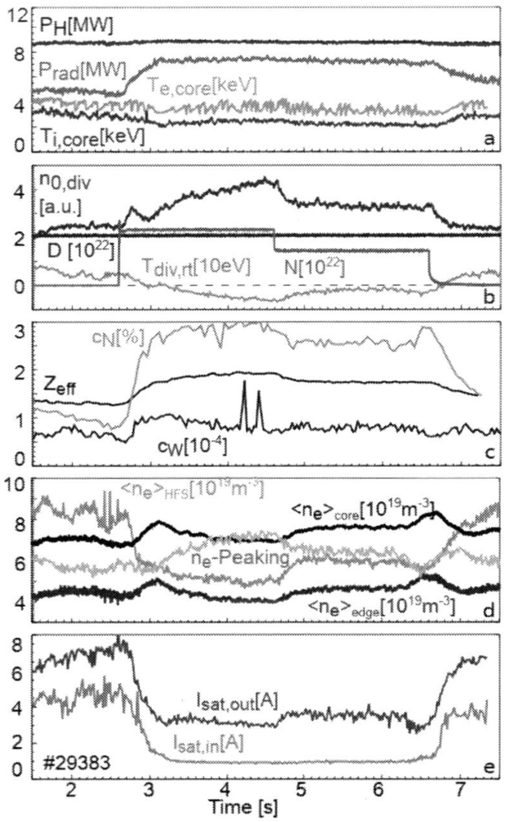


Figure 1: *Discharge overview for #29383.*

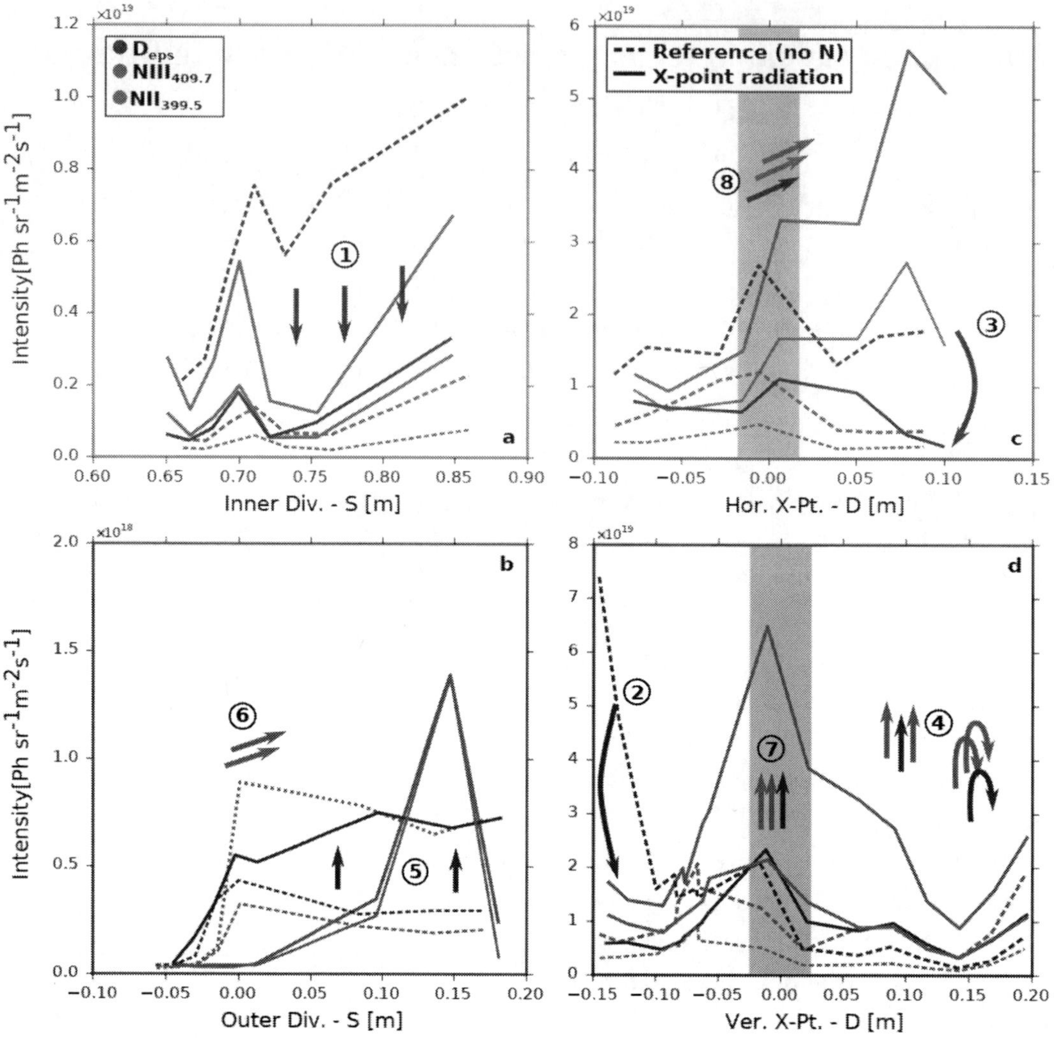


Figure 2: *Line integrated line intensities for the spectroscopic lines of sight (Fig. 3) along the inner (a) and outer (b) divertor target as well as the horizontal (c) and vertical (d) views across the X-point are shown for Balmer, NII, NIII as dashed lines for a reference before N-seeding and as solid lines during the X-point radiation regime. The grey shaded regions indicate the uncertainty of the X-point coordinates. The arrows and numbers indicate the sequence of the intensity profiles evolution.*

## Observation of the X-point radiation regime

Stable H-mode operation at power levels of $P_H/R = 3$–$12$ MW/m at $H_{98} > 0.85$ with $I_P =$ 0.8–1.2 MA and with both targets detached has been demonstrated in the all-tungsten tokamak AUG [1]. Empirically, complete detachment coincides with the observation of stable, strongly localized and intense radiation inside the confined region at the X-point. This X-point radiation regime has been observed with the application of intense nitrogen seeding in AUG H-Modes ($\Gamma_N \sim 10^{22}$ e$^-$/s). With the transition into the X-point radiation regime the divertor plasma changes and complete detachment in H-Mode discharges can be achieved even with high heating power (> 10 MW). SOLPS5.0 simulations indicate that the power dispersal within the confined plasma by the intense X-point radiation starves the divertor of power [8]. In conjunction with an increased radial transport in the divertor, the simulations show that the experimental

*See http://www.euro-fusionscipub.org/mst1.

reduction of the power loads as well as the level of recycling fluxes that can be sustained at the divertor plate are reproduced [3].

Timetraces of a discharge with medium heating power ($P_H = 8.5$ MW) that exhibits the X-point radiation regime for $t > 3.2$ s and with completely detached targets for $4.0\text{ s} < t < 4.5$ s are shown in Fig. 1. In order to analyze the effect of the X-point radiation on the radiation pattern in the divertor and at the X-point the, line intensities of a Balmer ($D_\varepsilon$ - 396.9 nm), a NIII ($2s^2 2p(3s-3p)$ - 409.7 nm) and a NII ($2s^2(3s-3p)$ - 399.5 nm) transition are plotted in Fig. 2. The intensity profiles for the inner (a) and outer (b) divertor as well as for the horizontal (c) and vertical (d) lines of sight across the X-point as shown in Fig. 3 are plotted for a reference time before nitrogen seeding without X-point radiation (2.6 s) and during the X-point radiation phase with intense nitrogen seeding (4.5 s).

A detailed description of the sequence of detachment that is triggered by the impurity radiation can be found in Ref. [1]. A high-field side high density (HFSHD) in the inner divertor as described in Ref. [9] is reduced within 200 ms of the application of the nitrogen seeding (Fig. 1.d). This coincides with an initial density increase by about 15 % [1, 7]. The collapse of the HFSHD is reflected in a strong decrease of the Balmer radiation in far SOL at and close to the inner target by more than an order of magnitude ((1-3) in Fig. 2). It can also be observed in the electron densities derived from Stark broadening. The formation of the X-point radiation takes place in about 500 ms after the application of nitrogen seeding. The radiated power fraction ($f_{rad}$) increases from about 55 to 80 %. In Fig. 4, the time evolution of the Balmer intensity seen by the horizontal views above the X-point is shown. The increased power losses lead to an initial drop of the Balmer intensity (black arrow) due to the collapse of the HFSHD in the inner divertor, that is followed by a slower evolution of the X-point radiation over the next 1.5 s, which leads to a further increase of $f_{rad}$ to 85 %. During this evolution the nitrogen line intensities at the outer divertor first increase, then roll-over and move out from the strikepoint into the far SOL (4+6 in Fig. 2). The reduction in the initially dominant NIII line is stronger than in the NII line. Consistent with the drop of the realtime divertor target temperature signal $T_{div,rt}$ in Fig. 1.b, both are indicators of a temperature drop in the outer divertor that shifts the ionization balance towards lower ionization stages. In line with Langmuir probes and Stark broadening measurements of an increased density in the outer divertor, the Balmer line intensities increase along the whole outer target (4+5 in Fig. 2) with seeding. At the same time, the outer divertor volume shows a strong increase of the nitrogen and Balmer line radiation that is most pronounced in the X-point region (7-8 in Fig. 2). The stronger increase of radiation in the horizontal views across the X-point also suggest that the radiation region is poloidally elongated along the fluxsurface.

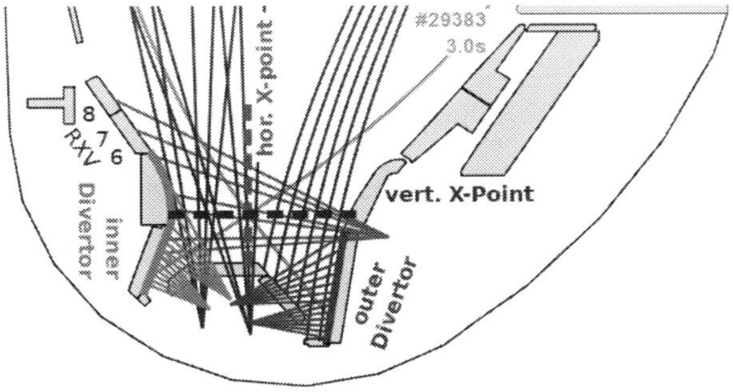

Figure 3: *Line of sight geometry and reference planes are shown.*

As shown in Fig. 4, Balmer radiation also emerges in the confined plasma above the X-point after 3.1 s and keeps increasing by factor of 2–10 during the further evolution. At the same time, the line integrated core density decreases again and the previous increase is almost reversed. At about 4.0 s (dotted line in Fig. 4), the coincident increase of Balmer radiation in RXV7 and the drop of nitrogen radiation in RXV6 indicates a slow upward motion of the radiation layers. A radial extent of a region with temperatures below 5 eV of about 5 cm can be estimated from the separation of the lines of sight and the change in the relative intensities of Balmer and nitrogen radiation.

The existence of low temperatures of the order of 1 eV close to the X-point is also suggested by measurements of the Balmer line intensity ratio $D_\varepsilon/D_\delta$. Fig. 4.b shows that the line ratio close to the X-point attains values that imply a dominantly recombining plasma before the seeding is reduced (blue arrow). However, it has to be noted that the interpretation of line intensity ratios from line-integrated measurements in non-uniform plasmas is challenging and verification of this first indications are necessary.

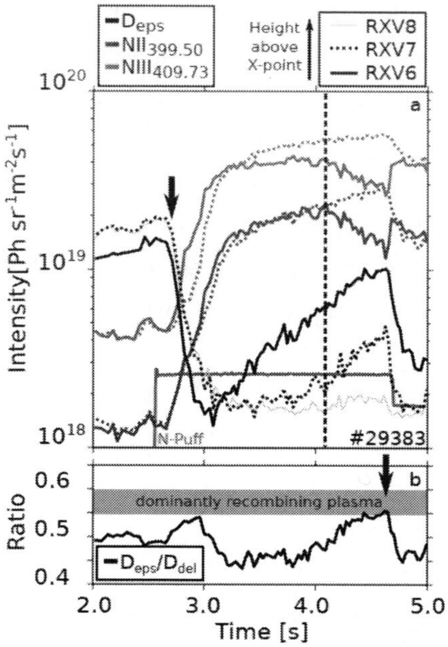


Figure 4: *Time evolution of line intensities (a) and Balmer (b) line intensity ratios for horizontal lines of sight above the X-point (see Fig. 3).*

## Summary & Outlook

In summary, it has been shown that the X-point radiation regime is a stable operational scenario with completely detached targets that features intense, localized radiation at the X-point inside the confined plasma. In the presented feed-forward seeding scheme, the X-point radiation shows a slow radial inward motion that leads to the formation of a dense, low temperature region at the X-point with a possibly recombining plasma.

Similar analysis including data for higher charge states of impurity line radiation need further development of the analysis tools and will be addressed in the future. Forward calculations using validated plasma simulations from codes such as SOLPS combined with collisional radiative models will be used to analyze the line ratios in order to prove or falsify the suggestion of a low temperature, recombining plasma at the X-point. The X-point radiation regime with Ne as a seed impurity as well as the stability and stationarity of the regime will be further investigated.


## Acknowledgements

This work has been carried out within the framework of the EUROfusion Consortium and has received funding from the Euratom research and training programme 2014-2018 under grant agreement No 633053. The views and opinions expressed herein do not necessarily reflect those of the European Commission.


## References


[1] F. Reimold, et al., Nucl. Fusion 55, (2015) 033004
[2] M. Wischmeier, et al., 25th FEC (IAEA), St. Petersburg (2014)
[3] F. Reimold, et al., Journal of Nuc. Mat. (2015) 0022-3115
[4] T. W. Petrie, et al., Journal of Nuc. Mat. 241-243:639-644 (1997)
[5] M. Bernert, et al., this conference
[6] T. Ishijima, et al., Plasma Phys. Control. Fusion 41 (1999) 1155-1666
[7] A.Kallenbach, et al., Nucl. Fusion 55 (2015) 053026
[8] B. Lipschultz, et al., Journal of Nuc. Mat. 266-269 (1999) 370-375
[9] S. Potzel, et al., Journal of Nuc. Mat. (2015)

# Pumping in vertical and horizontal target configurations on JET in L-mode; an interpretive study using EDGE2D-EIRENE

D. Moulton[1,2], M. Groth[2], D. Harting[1], S. Wiesen[3], S. Varoutis[4], C. Gleason-González[4] and JET Contributors*

*EUROfusion Consortium, JET, Culham Science Centre, Abingdon, OX14 3DB, UK; [1]CCFE, Culham Science Centre, Abingdon, Oxon, OX14 3DB, UK; [2]Aalto University, Association EURATOM-Tekes, Espoo, Finland; [3]Forschungszentrum Jülich GmbH, Institut für Energie- und Klimaforschung - Plasmaphysik, 52425 Jülich, Germany; [4]Karlsruhe Institute of Technology, Institute for Technical Physics, 76344 Karlsruhe, Germany; *See the Appendix of F. Romanelli et al., Proceedings of the 25th IAEA Fusion Energy Conference 2014, Saint Petersburg, Russia*

## Introduction

Multifluid edge codes such as SOLPS have been used to predict the pumped flux on ITER [1]. However, calculated pumped fluxes have rarely been validated against current experiments. On JET, this has been due to the lack of a sub-divertor model in the in-house multifluid edge code EDGE2D-EIRENE [2, 3, 4]. In this paper we present such a model, and benchmark it against experiments in vertical target (VT) and horizontal target (HT) configurations. The experiment modelled here has been described previously in [5]. L-mode density scans were carried out in VT and HT configurations, with $I_p = 2.5$ MA, $B_T = 2.5$ T, $P_{in} = 3$ MW. The simulations presented here are based on those discussed in [5] but with a sub-divertor model included.

## The EDGE2D-EIRENE sub-divertor module

Figure 1a shows the geometry used in the simulations. Outlines of the EDGE2D (plasma-solving) grids are shown for the HT (red) and VT (blue) configurations. With the new sub-divertor module included in the code, the EIRENE (neutral-solving) grid now spans the entire sub-divertor domain. The location of the baratron pressure gauge, used to measure the experimental sub-divertor pressure $p_{subdiv}$, lies inside the simulation domain, as shown by the red asterisk. The simulation puff was placed at the inner divertor base, as was predominantly the case in experiment. At this stage we have assumed the sub-divertor geometry to be toroidally symmetric. The simulated sub-divertor walls were set to 300°C everywhere except for the radiation baffles, poloidal field coils and baratron pipe, which were set to 20°C. Neutral-neutral collisions were not simulated. We assume here that deuterium retention and/or outgassing by the ITER-like wall in L-mode is negligible compared to the $D_2$ puff rate $\Phi_{puff}$ [6]. In the steady state, it follows that the pumped $D_2$ flux is equal to the puff: $\Phi_{pump} = \Phi_{puff}$.

Figure 1b shows an example of the simulated $D_2$ density near the cryopump, with the radiation-blocking baffles included. In the absence of neutral-neutral collisions, the density at any point is set by the number of straight-line neutral trajectories passing through that point. Thus, the density drops beyond the divertor throat baffle (which blocks neutrals coming from their source in the divertor), and also towards the upper portion of the cryopump surface, where neutral trajectories are no longer received directly from the throat baffle.

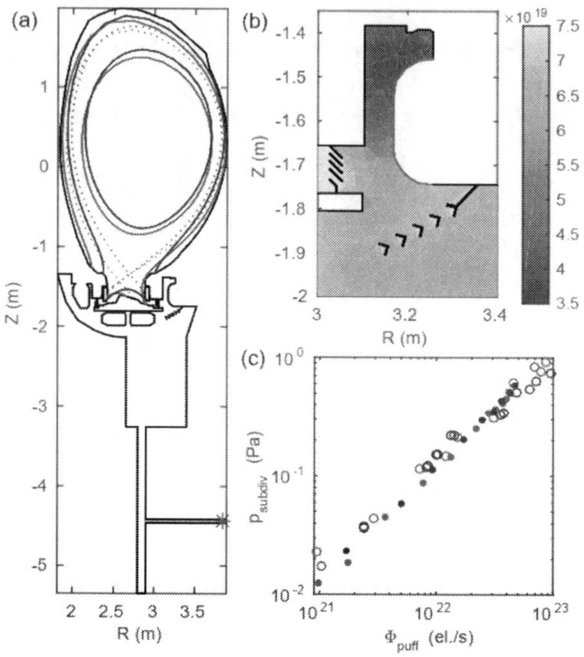


Figure 1: *(a) The simulation geometry. (b) The $D_2$ density ($m^{-3}$) near the cryopump surface (magenta line). (c) $p_{subdiv}$ as a function of $\Phi_{puff}$ in experiment (open circles) and simulation (solid circles) and in HT (red) and VT (blue) configurations.*

Consider the experimentally measured $p_{subdiv}$ as a function of $\Phi_{puff}$ (or equivalently $\Phi_{pump}$), shown in figure 1c for both HT (red open circles) and VT (blue open circles) configurations. In the absence of neutral-neutral collisions, the pumped flux is the fraction of the surface-integrated one-way Maxwellian flux which sticks: $\Phi_{pump} = (1-\alpha_{pump})\int n_{D2pump}.18.1\sqrt{T_{D2pump}}.dA_{pump}$, where $n_{D2pump}$ and $T_{D2pump}$ are the $D_2$ density ($m^{-3}$) and temperature (K) in front of the pump surface, and $\alpha_{pump}$ is the pump albedo. Our simulations show that the particle sources into the sub-divertor through the inner and outer throats remain equally balanced (within 20%) in VT for all values of $\Phi_{puff}$, while in HT the flux through the outer throat always dominates (by at least a factor four). If this is also the case in experiment, then the $D_2$ density at the baratron will be a linear function of the density at the pump and $\Phi_{puff}/p_{subdiv}$ = constant, as seen to a good approximation in figure 1c (we assume that $T_{D2pump}$ is set by the wall temperature so is independent of $\Phi_{puff}$).

The action of the cryopump in the simulations is modelled by placing a pumping surface around the pump inlet (magenta line in figure 1b). Ideally, the albedo of this surface would be set by the cryopump speed measured in a gas of constant density. The ratio $\Phi_{puff}/p_{subdiv}$ would then be an output from the code. Unfortunately such a measurement was not available; only *in-situ*, effective pumping speeds have been published for the JET cryopump (e.g. [7]). Instead, a simulated albedo of $\alpha_{pump} = 0.92$ was set such that the experimental ratio of $\Phi_{puff}/p_{subdiv}$ was recovered, as shown by the solid circles in figure 1c. Although this methodology means that $\Phi_{puff}/p_{subdiv}$ is not an output from the code, the absolute magnitude of $\Phi_{puff}$ for a given upstream density is. This is in contrast to versions of the code without the sub-divertor module, in which the EIRENE grid was cut off at the inner and outer divertor corners and pumping was modelled by placing surfaces with a prescribed albedo $\alpha_{pump}$ in those corners (e.g. [8]). The value used for $\alpha_{pump}$ was typically chosen such that the simulated puff required to reach a given upstream separatrix density was similar to that required in experiment; $\Phi_{puff}$ cannot then be considered an output from the code.

**Simulation results**

Plasma quantities calculated by EDGE2D-EIRENE simulations without the subdivertor module have been compared to this experiment previously in [5]. Plasma quantities are minimally affected by the inclusion of the subdivertor module because the recycling target flux dominates over $\Phi_{puff}$ [9]. We therefore focus on a comparison of $\Phi_{puff}$. Figure 2 plots $\Phi_{puff}$ as a function of the outer mid-plane separatrix electron density $n_{e,sep,OMP}$, in HT (red) and VT (blue) configurations, and in experiment (open circles) and simulation (solid circles). It is assumed here the experimentally measured line-averaged edge density is twice $n_{e,sep,OMP}$ [10]. The puff required to achieve a given density was a factor 2-3 larger in VT than in HT. Simulation results are within a factor ~2 of experiment, with a tendency to overestimate $\Phi_{puff}$ at low density and underestimate $\Phi_{puff}$ at high density. Furthermore, the simulations successfully recover the higher $\Phi_{puff}$ in VT.

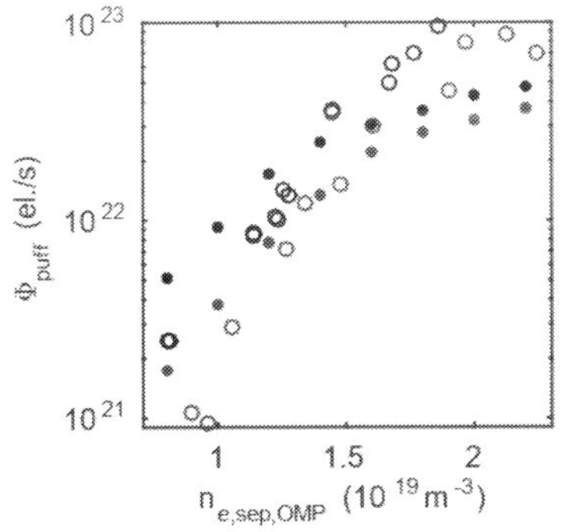


Figure 2: $\Phi_{puff}$ *as a function of* $n_{e,sep,OMP}$ *in experiment (open circles) and simulation (solid circles) and in HT (red) and VT (blue) configurations.*

To understand why a higher $\Phi_{puff}$ is required in VT than in HT, we focus on a low density simulation with $n_{e,sep,OMP} = 8.5 \times 10^{18}$ m$^{-3}$, for which a good agreement between experimental and simulation target data was found. At higher densities the argeement with experiment worsens [5], so that the reason for higher pumping in VT can be less trusted. Figure 3 shows the core penetration efficiency, defined as the percentage of a target's recycling flux which ionises inside the separatrix, from the inner and outer targets and for HT and VT configurations. From both targets, the fraction of neutrals penetrating the separatrix and contributing to $n_{e,sep,OMP}$ is significantly higher in HT than in VT. Thus, an increased target recycling flux (two times higher in our simulations) is required to achieve $n_{e,sep,OMP} = 8.5 \times 10^{18}$ m$^{-3}$ in VT, which in turn requires a higher $\Phi_{puff}$. Note that this effect is self reinforcing since a higher target recycling flux leads to a higher divertor density which leads to a decreased core penetration.

We hypothesise the increased penetration in HT to be caused by a combination of two factors: the x-point is 1.9 times closer to the inner target in HT than in VT, and neutrals preferentially recycle near the separatrix in VT compared to HT. Disentanglement of the relative role of these two factors is future work, however the fact that the difference between HT and VT penetration efficiencies is largest at the inner target suggests that the location of the x-point plays an important role in setting $\Phi_{puff}$ at this density.

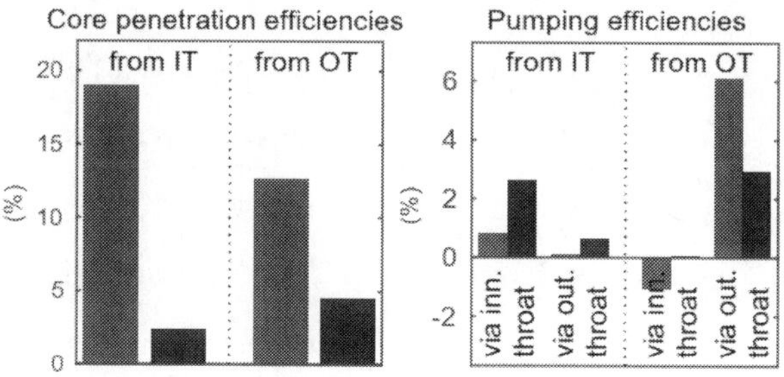


Figure 3: *Simulated core penetration and pumping efficiencies HT (red) and VT (blue) for a low density case, from outer and inner targets. Pumping efficiencies are further divided into fluxes via the inner and outer divertor throats.*

Also shown in figure 3 are the simulated pumping efficiencies, defined as the percentage of a particular target's recycling flux which is pumped. These have been further divided by the throat via which the neutrals flow before being pumped. The pumping in VT is seen to be more balanced compared to HT, in which most of the pumped flux is from the outer target, via the outer throat into which the outer target is angled. However, the *total* pumping efficiencies, i.e. the total recycling flux from both targets divided by the total pumped flux, are very similar in HT and VT (3.3% in HT compared to 3.2% in VT). The higher $\Phi_{puff}$ in VT at this density cannot therefore be attributed to an improved pumping efficiency. Furthermore, at this $n_{e,sep,OMP}$ (unusually for this dataset), both VT and HT configurations were actually puffed entirely from the top of the machine, in experiment and simulation. Direct pumping of the puff can therefore be ruled out as a reason for the higher $\Phi_{puff}$ in VT, as previously speculated [5]. Rather, at this low density it is the increased core penetration in HT that is the dominant cause of an increased $\Phi_{puff}$ in VT.

*This work has been carried out within the framework of the EUROfusion Consortium and has received funding from the Euratom research and training programme 2014-2018 under grant agreement No 633053. The views and opinions expressed herein do not necessarily reflect those of the European Commission.*

## References


[1] A. S. Kukushkin *et al.*, Fusion Eng. Des. **86**, 2865 (2011)
[2] R. Simonini *et al.*, Contrib. Plasma Phys. **34**, 368 (1994)
[3] D. Reiter *et al.*, J. Nucl. Mater. **196-198**, 80 (1992)
[4] S. Wiesen *et al.*, ITC project report 2005/6, http://www.eirene.de/e2deir_report_30jun06.pdf
[5] M. Groth *et al.*, J. Nucl. Mater., doi:10.1016/j.jnucmat.2014.12.030 (2014)
[6] S. Brezinsek *et al.*, Nucl. Fusion **53**, 083023 (2013)
[7] V. Philipps *et al.*, J. Nucl. Mater. **390-391**, 478 (2009)
[8] D. Moulton *et al.*, J. Nucl. Mater. **415**, S509 (2011)
[9] A. V. Chankin *et al.*, Plasma Phys. Contr. F. **48**, 839 (2006)
[10] M. Groth *et al.*, J. Nucl. Mater. **438**, S175 (2013)

# The Plasma E × B Staircase: Turbulence Self-Regulation through Spontaneous Flow Patterning

G. Dif-Pradalier[1], G. Hornung[2], Ph. Ghendrih[1], F. Clairet[1], P. H. Diamond[4,5], Y. Sarazin[1], L. Vermare[3], J. Abiteboul[6], T. Cartier-Michaud[1], C. Ehrlacher[1], D. Estève[1], X. Garbet[1], V. Grandgirard[1], Ö.D. Gürcan[3], P. Hennequin[3], Y. Kosuga[7], G. Latu[1], P. Morel[3], C. Norscini[1] and R. Sabot[1]

[1]*CEA, IRFM, F-13108 St. Paul-lez-Durance cedex, France*

[2]*Department of Applied Physics, Ghent University, 9000 Gent, Belgium*

[3]*Laboratoire de Physique des Plasmas, Ecole Polytechnique, Palaiseau, France*

[4]*CASS and CMTFO, University of California at San Diego, CA, USA*

[5]*WCI Center for Fusion Theory, NFRI, Daejeon, Korea*

[6]*Max-Planck-Institut für Plasmaphysik, Garching, Germany*

[7]*Institute for Advanced Study, Kyushu University, Fukuoka 816-8580, Japan*

The **E** × **B** staircase [1, 2] is a spontaneously formed, turbulence-driven self-organising pattern of quasi-regular, long-lived and localised shear flow [3] and stress layers. These layers coincide with long-lived pressure corrugations and are interspersed between regions of turbulent avalanching. The typical spacing between these layers is mesoscale, noted $\Delta \sim 25 - 30\rho_i$ [1, 4] —in-between the turbulence auto-correlation length $\ell_c \sim 5\rho_i$ at micro scales and the profile macroscale $L \gtrsim 100\rho_i$, see e.g. Fig.2 in [1]— and sets the outer scale of the turbulent avalanching. Here $\rho_i$ is ion Larmor radius. Whilst arresting, statistically, to mesoscales the detrimental avalanching these layers, located at the "steps of the staircase" are beneficial to confinement. The **E**×**B** staircase is thus best understood as a self-organising and dynamical set of weak or permeable transport barriers. Strong mean zonal flows are generated and endure at the steps of the staircase, resulting in localised deviations of the poloidal flow from its oft-assumed neoclassical prediction [5].

Through the existence of the staircase, at least two typical scale lengths for the turbulence are thus predicted [1]: $\ell_c$ and $\Delta$, the latter being a signature of the avalanching between the above staircase flow layers. This is shown in Fig.1 where the presence of the fat tail in the autocorrelation function directly points toward the nonlocal, avalanche-mediated transport regulated by the inter-step spacing of the staircase. Recent experimental findings tend to accredit this interpretation, showing typical micro- and mesoscale turbulence lengths [6].

The most visible manifestation of the plasma $\mathbf{E}\times\mathbf{B}$ staircase is the characteristic pattern of flows and stresses that long-lastingly define "valleys" of hindered transport where a strong and localised mean flow shear radially organises the heat and momentum fluxes. This is especially visible on the strong dipolar structure of the flux-surface averaged $\mathbf{E}\times\mathbf{B}$ shear $\gamma_{\mathbf{E}\times\mathbf{B}} = r\partial_r(E_r/rB)$ shown on the top and third panels of Fig.2. This figure is based on GYSELA computations mimicking the experimental parameters of the ToreSupra shot #45511 [9] and adapted from [2].

At the mean shear location are found lasting corrugations of the mean plasma profiles, clearly visible on the second panel of Fig.2 through localised and large values of the mean temperature gradient. These corrugations of the mean profiles appear with the quasi-constant mesoscale spacing $\Delta$ that corresponds to the steps of the $\mathbf{E}\times\mathbf{B}$ staircase. This is visible on the third panel of Fig.2: the profile corrugations are co-located with strong and localised turbulence-driven poloidal flows that depart from the oft-invoked neoclassical prediction Ref.[5].

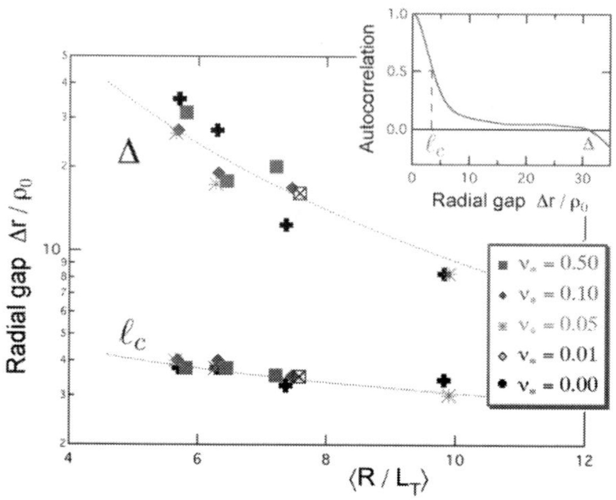


FIG. 1: Autocorrelation lengths of the electric potential from GYSELA [7] as a function of both the collisionality $\nu_\star$ [8] and the normalised temperature gradient $\langle R/L_T\rangle$ at $\rho_\star = 1/128$.

In addition, long-lasting strong and localised Reynolds stresses $\langle v_r v_\theta\rangle$ (third panel of Fig.2) organise the fluctuations on flux surfaces and provide the elasticity required for the flows to survive the impinging bath of heat and momentum avalanches. These flows define "valleys" of minimum turbulent transport (bottom panel of Fig.2) and thus act as weak or permeable transport barriers.

In-between these flow layers, nonlocal avalanching takes place, discussed in [1]. The essential conclusion is that the heat transport computed in flux-driven gyrokinetics — i.e. embedding both local/diffusive and nonlocal/non-diffusive transport with no assumption *a priori* on their respective existence nor on their respective weights— is irreconcilable with a local and diffusive approach: in-between the corrugations the typical turbulence correlation length is $\Delta \gg \ell_c$. For an experimental characterisation of nonlocality, see the review [10]. An interesting proposition as to how mesoscales could emerge as a heat flux "jamming" process may be found in [4, 11];.this approach really emphasises the plasma as a heat engine. Interestingly, the heat avalanching also goes hand-in-hand with a momentum avalanching [12, 13].

The staircase pattern is observed in computations in near-marginal turbulence drive conditions [2] and is thus reminiscent of situations described in self-organised criticality (SOC) [10, 14–19]. In the avalanching regions especially [20, 21] the plasma profiles are stiff. The staircase pattern results from a global and dynamic organisation of the turbulence at all scales: small-scale turbulence organises into avalanches and fluctuating zonal flows [3]; avalanches saturate at mesoscales through the organisation of the fluctuating zonal flows into mean zonal flows. Both avalanches and mean flows separate in space and the pattern repeats itself at macroscales.

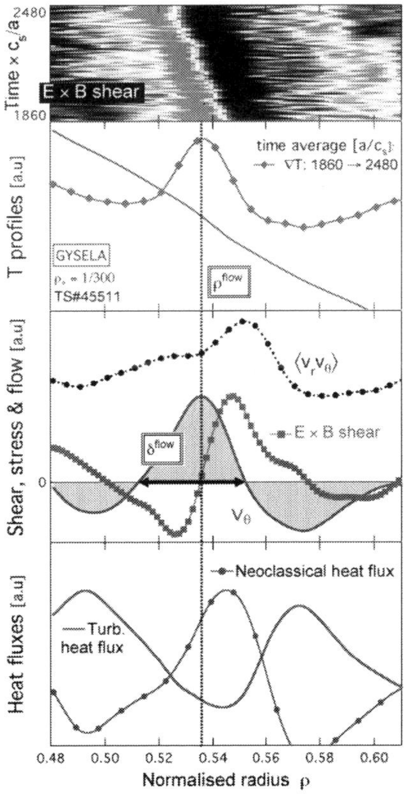


FIG. 2: Details next to a staircase step [a corrugation]. Adapted from [2].

The plasma staircase also displays a dynamics of its own: it tends to remain at constant drive $R/L_T$ of the turbulence, it meanders within the plasma volume and when destroyed by the occasional large heat front reforms in its wake, not necessarily at its former location. The overall transport within the plasma is really that of a self-organised state in which the staircase is a key dynamical player.

Based upon the predictions from GYSELA regarding its domain of existence and dependence with key plasma parameters we went on hunting for its existence in actual experiments. The spatially resolved measurement of poloidal flows or of mean plasma profile gradients across an extended radial domain of the plasma is notoriously difficult in tokamaks. From a transport perspective corrugations act indeed so as to partially decouple at regular radial intervals the upstream from the downstream plasma. The experimental characterisation of the staircase could thus be possible provided quality measurements of turbulent fluctuations with a sufficient radial and temporal precision could be made. Fast acquisitions ($\sim$ ms), due to the potential dynamics of the structure) of radially-resolved full-radius profiles of turbulent fluctuations on the other hand are now possible on Tore Supra using fast-sweeping reflectometry [22]. A routine construction of radial profiles of turbulence correlation lengths [23] can thus be inferred from the experiment. The staircase pattern is thus identified through radially localised, abrupt and quasi-regularly spaced minima of these correlation profiles. Details may be found in [2]; 170 occurrences of this structure have so far been observed in the ToreSupra database.

The staircase is a turbulence-borne structure and shows no clear connection with low-order safety factor $q$ rationals. This proves important to disentangling experimentally the plasma staircase from magnetohydrodynamic (MHD) activity. Ongoing work is concerned with a thorough experimental characterisation of the experimental parameter space in which the $\mathbf{E} \times \mathbf{B}$ staircase is observed [24].

---


[1] G. Dif-Pradalier et al. *Phys. Rev. E*, 82:025401(R), 2010.
[2] G. Dif-Pradalier et al. *Phys. Rev. Letters*, 114(085004), 2015.
[3] P. H. Diamond et al. *Plasma Phys. Control. Fusion*, 47:R35–R161, 2005.
[4] Y. Kosuga et al. *Physics of Plasmas (1994-present)*, 21:055701, 2014.
[5] G. Dif-Pradalier et al. *Physical Review Letters*, 103(6):065002, 2009.
[6] P. Hennequin et al. *42nd EPS Conference, Lisbon, Portugal*, 2015.
[7] Y. Sarazin et al. *Nuclear Fusion*, 51(10):103023, 2011.
[8] G. Dif-Pradalier et al. *Physics of Plasmas*, 18(6):062309, 2011.
[9] L. Vermare et al. *Physics of Plasmas (1994-present)*, 18(1):012306, 2011.
[10] K. Ida et al. *Nuclear Fusion*, 55(1):013022, 2015.
[11] Y. Kosuga et al. *Phys. Rev. Lett.*, 110:105002, Mar 2013.
[12] S. Ku et al. *Nuclear Fusion*, 52(6):063013, 2012.
[13] J. Abiteboul et al. *Plasma Physics and Controlled Fusion*, 55(7):074001, 2013.
[14] P. H. Diamond and T. S. Hahm. *Physics of Plasmas*, 2(10):3640–3649, 1995.
[15] D. E. Newman et al. *Physics of Plasmas*, 3(5):1858–1866, 1996.
[16] B. A. Carreras et al. *Physics of Plasmas*, 3(8):2903–2911, 1996.
[17] X. Garbet and R. E. Waltz. *Physics of Plasmas*, 5(8):2836–2845, 1998.
[18] R. Sanchez et al. *Physics of Plasmas*, 12(5):056105, 2005.
[19] D. del-Castillo-Negrete. *Physics of Plasmas*, 13(8):082308, 2006.
[20] P.A. Politzer. *Physical Review Letters*, 84(6):1192–1195, February 2000.
[21] P. Beyer et al. *Physical Review Letters*, 85(23):4892–4895, December 2000.
[22] F. Clairet et al. *Review of Scientific Instruments*, 82(8):083502, 2011.
[23] G. Hornung et al. *Plasma Physics and Controlled Fusion*, 55(12):125013, 2013.
[24] G. Hornung et al. *in preparation*, 2015.

# Two - plasmon decay as a reason for anomalous absorption and backscattering in second harmonic ECRH experiments

A.Yu. Popov, E.Z. Gusakov
*Ioffe Institute, St. Petersburg, Russia*

**1. Introduction.** Electron cyclotron resonance heating (ECRH) at power level of up to 1 MW in a single microwave beam is used nowadays in tokamak and stellarator experiments and considered for application in ITER for neoclassical tearing mode control. The analysis of parametric decay instabilities (PDIs) potentially dangerous for mm-waves at ECR frequencies was performed more than a couple of decades ago [1] under the assumption of a monotonous background density profile. It revealed the typical microwave power thresholds in the range much higher than 1 MW and predicted the nonlinear effects to be not important in the ECRH experiments utilizing gyrotrons. However, during the last decade many experiments, which were carried out in the presence of a non-monotonous density profile, originated due to the magnetic island or the density pump-out effect, have demonstrated excitation of anomalous phenomena in the ECRH experiments at Textor, TCV, TJ-II, ASDEX-U and FTU. The clearest evidence of nonlinear effect has been obtained at the Textor tokamak [2, 3] where strong backscattering signal down–shifted in frequency by approximately 1 GHz and synchronized to the MHD mode rotation was observed in the 200 – 600 kW power level 2nd harmonic X mode ECRH experiment.

In the present paper we introduce a model taking into account, as distinct from the standard theory [1], the presence of a non-monotonous density profile in the magnetic island [4] and interpret the generation of the backscattering signal as a secondary nonlinear process, which accompanies a primary low – threshold two–upper hybrid (UH) – plasmon PDI of the X wave [5]. The power threshold of the primary absolute PDI appears to be drastically smaller than predicted by [1] due to the UH waves trapping in a vicinity of the O-point of the magnetic island where the UH frequency is close to one-half the pump frequency. The PDI is excited during each passage of the island across the ECRH beam.

The primary PDI enhancing the UH wave fluctuations from the thermal noise level is saturated in our model due to a secondary low-threshold decay of the daughter UH wave leading to excitation of the secondary UH wave down-shifted in frequency and the low hybrid (LH) wave. The threshold of this cascading process can be easily overcome when the secondary UH wave is trapped in the radial direction due to a local maximum of the density profile. We derived and then solved numerically a set of equations describing the cascading process. The results of numerical modelling are compared to the analytical estimations of the growth rates of the primary and secondary parametric decays and the saturation level. It is demonstrated that for the Textor conditions [2, 3] approximately 20% of the pump wave power can be anomalously absorbed by the decay waves.

The generation of the backscattering signal is explained in terms of the nonlinear coupling of different daughter UH waves. This mechanism appears capable of reproducing the fine details of both the frequency spectrum and the absolute value of the observed backscattering signal [2, 3]. The saturation of exponential growth of the daughter waves amplitudes and the transition to a quasi-stationary regime of the PDI are shown below to

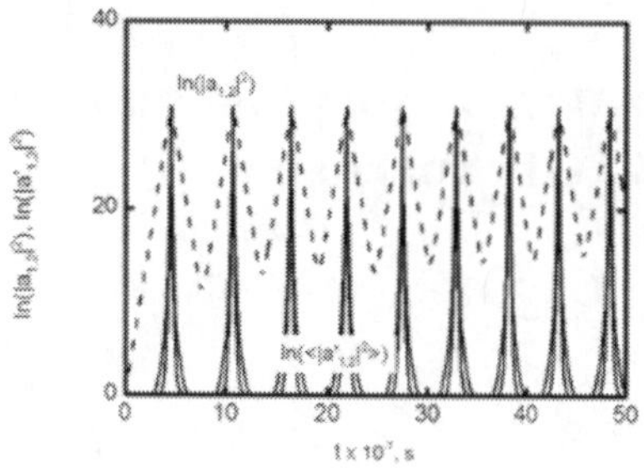


Fig. 1: Evolution of the number of the primary (dashed curve) and secondary (solid curve) plasmons in the presence of a homogeneous pump $a_0 = const$.

occur just after the one-stage cascade. Taking into account the multiple-stage cascade does not change the estimation of the saturation level derived within the one-stage cascade model.

**2. Equations describing the X-mode pump PDI cascade.** To elucidate the physics of the X-mode pump PDI, we consider the most simple but nevertheless relevant to the experiment three-wave resonance interaction model and analyse the behaviour of the pump X wave propagating inwards plasma in the tokamak mid-plane almost perpendicular to the magnetic field $\mathbf{H} = H\mathbf{e}_z$ in the density inhomogeneity direction along a unit vector $\mathbf{e}_x$. The basic set of the equations, describing the pump X wave decay into a couple of the electrostatic UH plasmons trapped in the radial direction and propagating in opposite directions, as well as the decay of the second primary daughter plasmon to the radially trapped secondary UH plasmon and the low hybrid (LH) plasmon, possessing frequencies satisfying decay conditions $\omega_0 = \omega_1 + \omega_2$, $\omega_2 = \omega_1' + \omega_2'$ reads as [6]

$$\frac{\partial a_1}{\partial t} + u_{1y}\frac{\partial a_1}{\partial y} - i\Lambda_{1z}\frac{\partial^2 a_1}{\partial z^2} = \nu_0 a_0(y,z)a_2, \quad \frac{\partial a_2}{\partial t} - u_{2y}\frac{\partial a_2}{\partial y} + i\Lambda_{2z}\frac{\partial^2 a_2}{\partial z^2} = \nu_0^* a_0^*(y,z)a_1$$

$$\frac{\partial a_1'}{\partial t} + u_{1y}'\frac{\partial a_1'}{\partial y} + i\Lambda_{1y}'\frac{\partial^2 a_1'}{\partial y^2} - u_{1z}'\frac{\partial a_1'}{\partial z} = \nu_1^* a_2 a_2', \quad \frac{\partial a_2'}{\partial t} - i\Lambda_{2y}'\frac{\partial^2 a_2'}{\partial y^2} + u_{2z}'\frac{\partial a_2'}{\partial z} = \nu_1 a_2 a_1'^* \qquad (1)$$

where $u_{1y}$, $u_{2y}$ $u_{1y}'$ are the poloidal group velocities averaged over the UH plasmons radial localization areas and describing their convective energy loss from the decay region in the poloidal direction, $\Lambda_{1z}$, $\Lambda_{2z}$, $\Lambda_{1y}'$ are the coefficients averaged over the UH plasmons radial localization areas and describing their diffraction energy loss along the magnetic field, $u_{2z}'$ and $\Lambda_{2y}'$ describe the toroidal convective and poloidal diffractive energy loss of the secondary LH wave, $\nu_0$ and $\nu_1$ describe the parametric coupling of the two primary UH plasmons and the primary and secondary UH plasmons, respectively, $|a_0|^2$, $|a_{1,2}|^2$, $|a_{1,2}'|^2$ are the number of photons, primary and secondary plasmons. We solve (1) numerically in a 2D box $2y_B \times 2z_B$ and impose the boundary conditions in a way to take full account of the2D translational symmetry over the poloidal and toroidal coordinates which allows improving stability of the numerical scheme. As initial conditions for decay wave amplitudes we take the thermal noise spectral power density integrated over the wave number space region for plasmons participating in the PDI

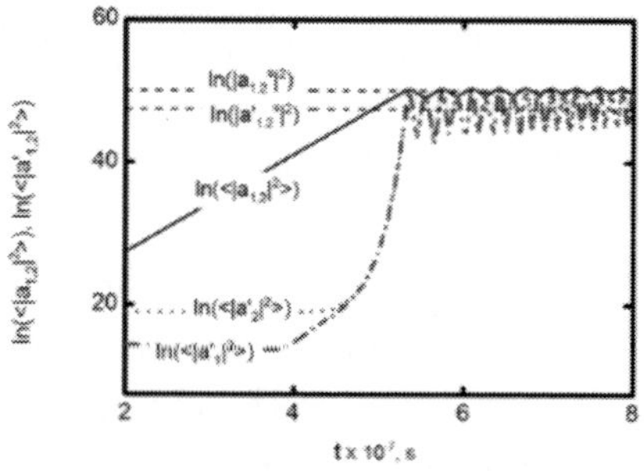


Fig. 2: The number of the primary plasmons (solid blue curves) and secondary UH (dash-dotted red curve) and LH (shot-dashed green curve) plasmons. The horizontal dashed lines correspond to the saturation level (2) of the primary (upper line) and secondary daughter plasmons (lower line).

$|a_{1,2}|^2(0,\mathbf{r}) = T_e/(\omega_{1,2}\pi w^2)$, $|a'_{1,2}|^2(0,\mathbf{r}) = T_e/(\omega'_{1,2}\pi w^2)$ ($w = 1cm$ is the pump beam radius). The coefficients of the system (1) are constant.

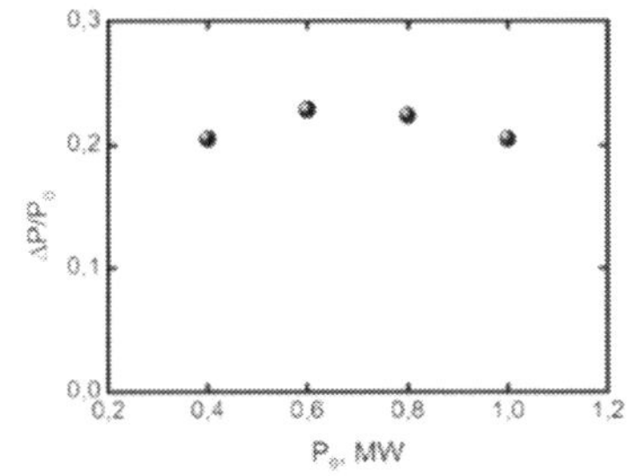


Fig. 3: Dependence $\Delta P/P_0$ ($\Delta P$ is the power which the daughter waves gain due to the PDI (3)) on the pump power $P_0$ for the typical Textor conditions.

We start with analysis of a homogeneous pump $a_0 = const$. In this case numerical solution of system (1) shown in fig. 1, demonstrates oscillatory regime of the daughter waves interaction.

In the case of the realistic finite-size pump beam ($P_0 = 600kW$) the results of numerical modeling for the typical Textor experimental parameters are shown in fig. 2 where the temporal dependencies $\ln[<|a_{1,2}(t)|^2>]$ (solid blue curves) and $\ln[<|a'_{1,2}(t)|^2>]$ (dash-dotted red and shot-dashed green curves) are depicted (The averaging is defined as $<|a_{1,2}(t)|^2> = \int_{-y_B}^{y_B}\int_{-z_B}^{z_B} dydz\,|a_{1,2}(y,z,t)|^2/(2y_B 2z_B)$). The exponential growth of the number of the primary plasmons is demonstrated at the beginning of computing. As soon as the number of the primary plasmons turns to be large enough to exceed the threshold of the secondary decay, the number of the secondary plasmons starts to grow as well. When the secondary PDI power threshold is exceeded significantly, the temporal gain coefficient can be estimated as $2\Gamma_s(t) \approx 2|v_1|\sqrt{<|a_2(t)|^2>}/(|v_0||a_0|)$. Keeping in mind the last estimation of the temporal gain coefficient along with the steady balance between the fractions of power gained by the primary UH wave from the pump X wave and by the secondary UH wave from the primary one, we get, eventually, the estimate of the saturation level of the primary and secondary plasmons

$$|a_2^s|^2 = \left[\frac{|v_0|}{|v_1|}|a_0|\ln\left(\frac{|v_0|}{|v_1|}\frac{|a_0|}{\sqrt{2}a'_1(0)}\right)\right]^2, \quad |a_1'^s|^2 = |a_0|^2\frac{|v_0|^2}{|v_1|^2}\ln\left(\frac{1}{\sqrt{2}}\frac{|v_0|}{|v_1|}\frac{|a_0|}{a'_1(0)}\right) \tag{2}$$

These estimations (2) shown in fig. 2 are in a reasonable agreement with the results of numerical modeling. As we can see, in the quasi-stationary regime weak oscillations of the number of the daughter plasmons around the corresponding saturation level occur. Nevertheless, these oscillations, especially in the case of the primary plasmons, are much weaker than those in the case of the homogeneous pump (compare to fig. 1). Then, we get numerically the power which the daughter waves gain

$$\Delta P = \frac{\partial}{\partial t}\int_S dydz\left[\omega_1|a_1(y,z,t)|^2 + \omega_2|a_2(y,z,t)|^2 + \omega'_1|a'_1(y,z,t)|^2 + \omega'_2|a'_2(y,z,t)|^2\right] \tag{3}$$

In fig. 3 the dependence of the relative power loss of the pump X wave on the pump power for the typical Textor tokamak conditions is shown. It demonstrates that for the Textor conditions [2,3] up to 22% of the pump wave power can be anomalously absorbed by the decay waves.

**3. Comparison to the experimental observations [3].** The experimental observations of strong anomalous backscattering of mm-waves in Textor plasmas were achieved with the use of the collective Thomson scattering (CTS) technique [3] a scheme of which is sketched out in fig. 4. The backscattering signal collected by the receiving CTS antenna located around 20 cm above the gyrotron launcher can be explained in terms of the nonlinear coupling of the primary ($\omega_1/2\pi = 69.82GHz$) and secondary ($\omega_1'/2\pi = 69.22GHz$) UH waves leading to excitation of the X wave propagating outwards. The frequency of the latter, $\omega_a/2\pi = (\omega_1' + \omega_1)/2\pi = 139.04GHz$, corresponds to the first down-shifted spectral line observed experimentally (fig. 9 in [3]).

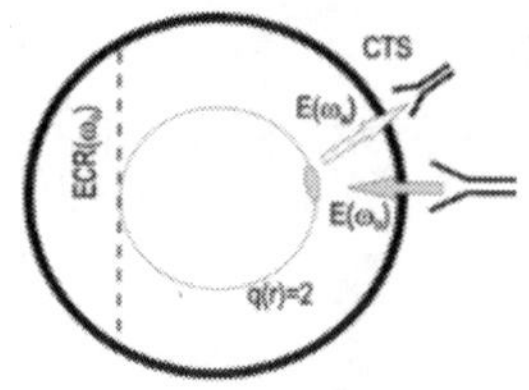

Fig. 4: Tokamak cross-section illustrating the Textor experimental setup [3].

Utilizing the approach based on the reciprocity theorem [7], we get the power of a signal collected by the microwave antenna

$$p_s(\omega_a) = \frac{\alpha}{c\pi w_a^2}\frac{W_1 W_1'}{w_H} \quad (4)$$

where $\alpha$ is the dimensionless geometrical factor, $w_a \simeq 1.5cm$ is a radius of the horn of the receiving antenna, $W_1$, $W_1'$ are energies of the radially trapped primary and secondary UH plasmons calculated numerically, and $w_H = H^2/4\pi$ is energy density of the external magnetic field. For the parameters of the ECRH experiments at Textor (see captions for fig. 1) in which the strong anomalous backscattering of the X-mode pump was observed the estimate of the anomalously backscattered X wave spectral line intensity (4) is $p_s(\omega_a)/\Delta\nu = 0.602MeV$ with $\Delta\nu \approx 0.2GHz$ being experimentally measured spectrum line-width. This calculated value is of the order of the spectral power density value provided by the CTS measurements $p_s(\omega_a)/\Delta\nu \approx 1MeV$ (see figs. 9, 11 in [3]).

**4. Conclusions.** Summarizing, we can conclude that the proposed mechanism of the backscattering signal generation based on the low threshold excitation in the magnetic island of the two-plasmon PDI and its cascading saturation appears capable of reproducing both the fine details of the frequency spectrum and absolute level of the observed anomalous backscattering signal [3]. It should be stressed that this mechanism predicts substantial anomalous absorption of the pump. In particular, up to 22% of the pump power is anomalously absorbed by the decay waves for the TEXTOR experimental parameters.

*The computations were performed at facilities of USD "Globus-M". Partial financial support of RFBR grant 13-02-00683, BRFBR-RFBR grant 14-02-90003 and RF MES (Agreement 14.619.21.0001, id RFMEFI61914X0001) is acknowledged.*

[1] M. Porkolab, B.I. Cohen 1988 *Nucl. Fusion* **28** 239
[2] E. Westerhof, S. Nielsen, J. W. Oosterbeek et al. 2009 *Phys. Rev. Lett.* **103** 125001
[3] S.K. Nielsen, M. Salewski, E. Westerhof et al. 2013 *Plasma Phys. Control.Fusion* **55** 115003
[4] M.Yu. Kantor 2009 *Plasma Phys. Control. Fusion* **51** 055002
[5] A.Yu. Popov, E.Z. Gusakov 2015 *Plasma Phys. Control. Fusion* **57** 025022
[6] A.Yu. Popov, E.Z. Gusakov 2015 JETP **121** to be published in vol 1
[7] A D Piliya, A Yu Popov 2002 *Plasma Phys. Control. Fusion* **44** 467

# Electron Cyclotron current drive studies on KSTAR using imaging MSE

C.A. Michael[1], A. Thorman[1], J. Chung[2], YS Bae[2],

J. H. Jeong[2], J. Decker[3], J. Howard[1]

[1] *Australian National University, Canberra, A.C.T.,2601, Australia*

[2] *National Fusion Research Institute, Daejeon, Republic of Korea*

[3] *CRPP, Lausanne, Switzerland*

The production of non-inductive current drive through electron cyclotron resonance is one candidate method for long pulse plasma sustainment. Electron cyclotron current drive systems have been deployed on tokamaks worldwide, including DIII-D, ASDEX and TCV. Several codes for modelling the current drive efficiency have been developed, and are continuously being improved. There are to date, however, only a few cases where these codes have been benchmarked against measurements. The net current produced by electron cyclotron current drive (ECCD) can be inferred from external magnetics, as well as temperature measurements (to model the resistivity). This technique was used in DIII-D [1], and also in in demonstrating full non-inductive scenarios produced by ECCD in TCV [2].

As demonstrated on both DIII-D [3], motional Stark effect (MSE) measurements provide a far superior method for distinguishing the non-inductive ECCD currents. In this paper, the ECCD induced changes in core current are measured using an imaging MSE System (IMSE) [4-6] installed on the KSTAR tokamak. Imaging MSE allows a 2D map of the vertical magnetic field distribution to be obtained, up to the extent of the height of the neutral beam. Experiments were undertaken at total constant current using the 170GHz (700kW) gyrotron [7], modulated at 2-2.5Hz, in various combinations and launch directions (co and counter), and at various field strengths including deposition near the magnetic axis in plasmas with sawtooth instabilities.

Electron cyclotron resonance heating and current drive ECH/CD is also known to have an effect on sawteeth, which redistribute heat and current to regions outside the sawtooth inversion radius. ECH/CD-induced changes in sawteething may also cause a net average reduction in the current inside the sawtooth inversion radius, and reduce the non-inductive ECCD current.

The evolution of current and loop-voltage is governed by the current diffusion equation. The heating promptly decreases the resistivity $\eta$ and reduces the loop voltage and flux consumption, in addition to the effects of the driven non-inductive current, which appears only after the resistive timescale $\tau=\mu_0/\eta l^2$, where l is the length scale of the current perturbation. For a typical value of l=5cm with $T_e$ =1.5keV, $Z_{eff}$=2, we get $\tau$=150ms. As the total current is held constant in these experiments, any current driven in the core will be compensated by a "back-reaction current" elsewhere in the plasma.

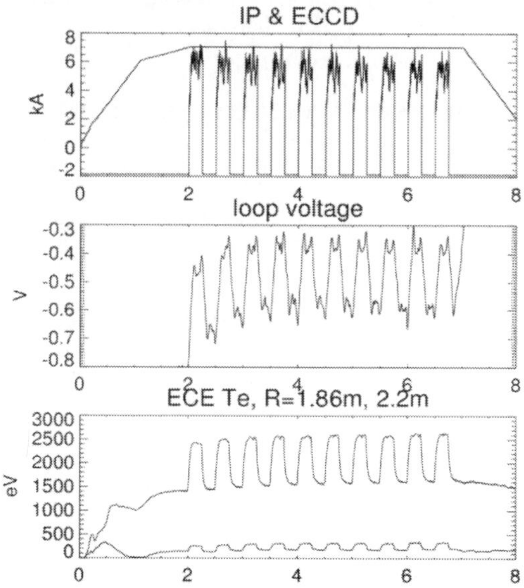


Figure 1: Waveforms of current, ECCD power, loop voltage, core and edge electron temperature for co-injected shot #9323

## ECCD Modulation experiments

To detect the small change in pitch angle due to ECCD, the power was modulated at 2-2.5Hz (slower than the current diffusion time) for 6-8 cycles. Waveforms of the current, ECCD power, and loop voltage and edge/core $T_e$ are plotted in Fig. 1, for a shot with co ECCD at 3.0T, which has a resonance position on axis (R=1.8m). It is clear that the loop voltage drops upon ECCD, however, as the edge temperature increases, the loop voltage change may be attributable to a change in resistivity alone. Thus it is difficult to conclude about the direction or magnitude of the driven current. An image of the time/space history of the MSE polarization, angle on the midplane, for the same shot, is shown in Fig 2. The ECCD waveform modulation has a clear signature in the MSE data.

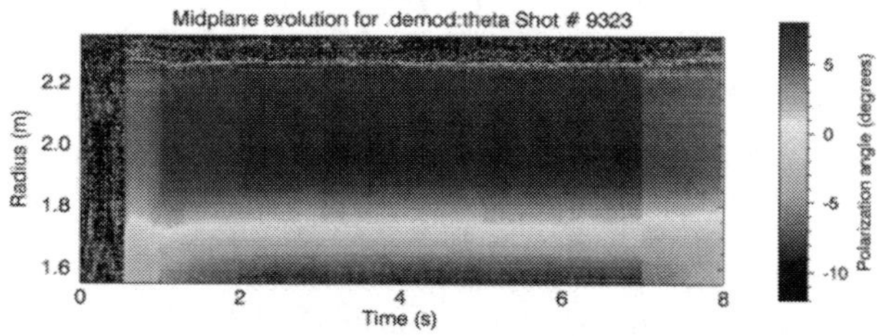


Figure 2: Evolution of IMSE mid-plane pitch angle as a function of time for discharge shown

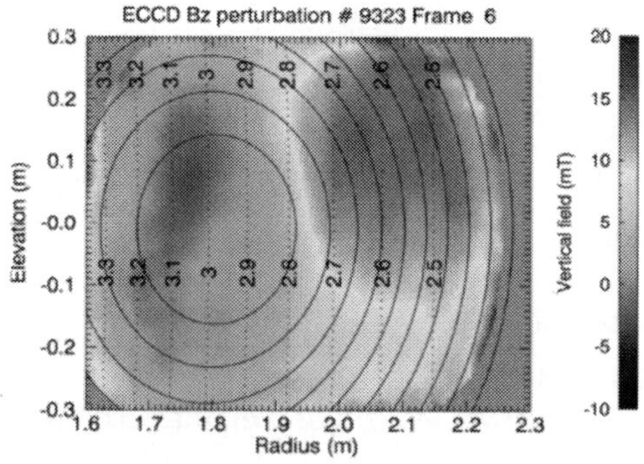


Figure 3: Cycle-averaged perturbation in Bz from IMSE, 150ms after ECCD switch on.

A 2D image of the change in $B_Z$, inferred from MSE, averaged over 6 modulation cycles, compared with the magnetics-constrained EFIT flux surfaces and |B| contours, is shown in Fig. 3. The vertical field exhibits large changes around R=1.9m and 1.6m. A component of the current density can be obtained from the radial derivative in $B_Z$, and is shown in Fig. 4, as a function of time and space. The current density is seen to respond over a timescale of ~100ms, and has a major peak around R=1.9m, which is 10cm away from the cold resonance. This may be due to the resonance of the "hot electrons" giving rise to a Doppler shift. Furthermore, there is a "negative" current peak around R=1.6m. This unexpected peak indicates that there may be a shift in the equilibrium. Such a shift would not be unexpected, due to a change in the Shafranov shift, as the stored energy increases ~30% upon ECCD application. In the above-mentioned discharge, the Shafranov shift is being opposed by the vertical field control system to attempt to maintain constant plasma position, however, there would still be a relative displacement of different surfaces, due to a change in the pressure profile. Thus, while the (R,Z) coordinate system is appropriate for comparison with the resonance condition, a flux coordinate system (ρ) is more appropriate under this strong pressure change.

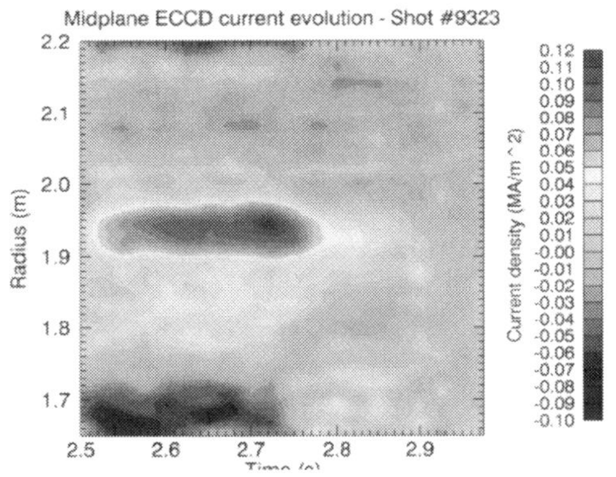

Figure 4:Spatio-temporal evolution of current density on midplane, formed from cycle average of 6 ECCD pulses.

**Enclosed current perturbation in co and counter-ECCD discharges**

Using the integral form of Ampere's law, and using a fixed EFIT reconstruction for estimation of the mapping ρ(R), a perturbation analysis reveals that the change in enclosed current as a function of flux label can be well approximated from the in-out asymmetry of the field perturbation: $\delta I(\rho) = 2\pi r(\rho)(\delta B_{out}(\rho) - \delta B_{in}(\rho))$. However, this is only possible for inner radii down to the beam tangency radius and IMSE field of view (effectively limiting this analysis to ρ=0.3-0.4 as seen in Fig. 3). The profile of the amplitude and phase of the fundamental component of the perturbed current is plotted in Fig. 4. The net current is in the co direction, and lags in phase by approx. 90 degrees in the region ρ>0.2. This long delay is characteristic of the resistive timescale, suggesting this is truly non-inductive current. Also, the current may yet not have reached equilibrium. The magnitude of the enclosed current is

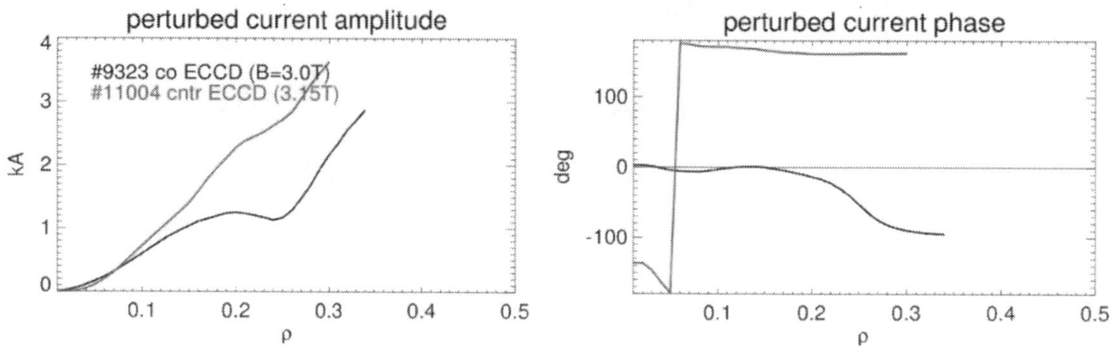


Figure 4: Amplitude and phase of fundamental component of the fundamental enclosed current perturbation in discharges with co (black) and counter (red) injection

reaches a maximum of ~3kA at ρ=0.3, possibly increasing at larger radii. Theoretical estimates based on the pre-computed current drive efficiency tabulated at ρ=0 [7] give values around 8kA total driven current. Whilst sawteeth instabilities may broaden the induced current profile, sawteeth were stabilized by ECCD in this discharge. Also in Fig. 4 is plotted the derived perturbed current profile for a shot with counter ECCD injection at 2Hz. There is a 180-160 deg phase lead, indicating a counter current, as expected.

An additional case with co-injection into an H-mode discharge at 2.0T has a magnitude of ~10kA at ρ=0.3, with phase lag of ~160 deg everywhere, indicating a dominantly counter injection behaviour. This unexpected behaviour might be caused by a change in the sawtooth behaviour during the application of ECCD in this case.

The complete response of the plasma will be modelled with the ray tracing and Fokker-Planck codes CPO/LUKE, including the effects of sawtooth and current diffusion. Future experiments will focus on changing the sawtoothing characteristics, and deposition location with respect to the sawtooth inversion radius, to better understand this effect.

1. James, R.A., G. Giruzzi, B. Degentile, L. Rodriguez, et al., Physical Review A, 1992. **45**(12): p. 8783-8786.
2. Sauter, O., C. Angioni, S. Coda, P. Gomez, et al., Physics of Plasmas (1994-present), 2001. **8**(5): p. 2199-2207.
3. Petty, C.C., R. Prater, J. Lohr, T.C. Luce, W.R. Fox, R.W. Harvey, J.E. Kinsey, L.L. Lao, and M.A. Makowski, Nuclear Fusion, 2002. **42**(12): p. 1366-1375.
4. Howard, J., Plasma Physics and Controlled Fusion, 2008. **50**(12): p. 125003.
5. Chung, J., J. Ko, J. Howard, C. Michael, G.v. Nessi, A. Thorman, and M.D. Bock, Journal of the Korean Physical Society, 2014. **65**(8): p. 1257-1260.
6. Thorman, A., C. Michael, and J. Howard, Review of Scientific Instruments 2013. **84**: p. 063507.
7. Bae, Y.S., M. Joung, H.L. Yang, W. Namkung, M.H. Cho, H. Park, R. Prater, R.A. Ellis, and J. Hosea, Fusion Science and Technology, 2011. **59**(4): p. 640-646.

# Real-time plasma profile state reconstruction on ASDEX-Upgrade

F. Felici[1], C.J. Rapson[2], W. Treutterer[2], L. Giannone[2], E. Maljaars[1], H. van den Brand[1,3], M. Reich[2], O. Sauter[4], A. Teplukhina[4], D. Kim[4], P. Piovesan[5], C. Piron[5], L. Barrera[2], M. Willensdorfer[2], A. Bock[2], E. Fable[2], B. Geiger[2], G. Tardini[2]
the ASDEX-Upgrade team[2] and the EUROfusion MST1 team[6]

[1] *Eindhoven University of Technology, P.O. Box 513, 5600 MB Eindhoven, The Netherlands*
[2] *Max-Planck-Institut für Plasmaphysik, 85748 Garching, Germany.*
[3] *FOM institute DIFFER, Eindhoven, The Netherlands.*
[4] *École Polytechnique Fédérale de Lausanne, CRPP-EPFL, Lausanne, Switzerland*
[5] *Consorzio RFX, Padova, Italy,* [6] *See http://www.euro-fusionscipub.org/mst1*

## Integrated control and the need for centralized state reconstruction

The focus of tokamak control is slowly shifting from support of physics experiments to increasingly integrated operation of the tokamak to increase performance and reliability in view of large tokamaks such as ITER. Consequently, increasingly sophisticated plasma monitoring and supervision systems are being developed for plasma control systems worldwide. This evolution also calls for a centralized approach to estimating the state of the plasma in real-time: one algorithm where all the available real-time diagnostic data is collected, analyzed, and merged into a consistent estimate of the plasma state [1].

Existing approaches such as equilibrium reconstruction, spline fitting of profiles, learning-based regime identification approaches are essentially static methods which do not take the dynamics of the process into account. In previous contributions [2] we have proposed to use a dynamic state observer, more specifically an Extended Kalman Filter (or EKF, an established tool in the signal processing/control engineering community), to approach this *data fusion* problem for tokamak fusion reactors. This method has the great advantage that physics model knowledge is combined in real-time with diagnostic measurements, which hence can be checked and validated in real-time.

This work describes new results and analysis from the implementation of an EKF for plasma profile reconstruction on ASDEX-Upgrade (AUG), undertaken as part of the MST1 2014 experimental campaign. The EKF algorithm is first described and it is shown how the RAPTOR real-time 1D plasma transport code is included. Then, some technical details of the implementation on the ASDEX-Upgrade Discharge Control System are discussed. Finally we show a comparison to off-line analysis using TRANSP for a shot with NBCD, as well as a shot illustrating the identification of unexpected diagnostic signals in real-time.

## Extended Kalman Filter for estimating plasma state and disturbances

The Extended Kalman Filter (EKF) is a recursive algorithm to compute an estimate $\hat{x}_k$ of the state $x_k$ of a system based on a time series of diagnostic measurements $y_k$ for all time indices $k$. It employs a (nonlinear) dynamic model of the system and measurements which is represented in discrete-time form as $x_{k+1} = f(x_k, u_k)$, $y_k = h(x_k)$. The state estimation is done in two steps: first, the physics model of the system is used to predict the next state of the system based on the previous state estimate $\hat{x}_{k|k-1} = f(\hat{x}_{k-1}, u_{k-1})$. Then, the predicted diagnostic measurements for this new state are computed using a forward model of the diagnostics $\hat{y}_k = h(\hat{x}_{k|k-1})$. The model-based state estimate is then corrected using the difference between measured and predicted diagnostic signals: $\hat{x}_{k|k} = \hat{x}_{k|k-1} + K(y_k - \hat{y}_k)$. The observer gain matrix $K$ can be recursively computed using the knowledge of the model functions $f, h$ and statistical properties of the noise acting on the system, which is assumed to be Gaussian. Details of the algorithm can be found in [3] and its application to plasma profile estimation is discussed in [4]. To handle systematic differences between the true plasma evolution and the diagnostic measurements, the model is extended by assuming that a constant disturbance $d_k$ acts on the system: $x_{k+1} = f(x_k, u_k) + d_k$. By defining an augmented state vector $z_k = [x_k^T, d_k^T]^T$, the EKF can easily be reformulated to estimate the state and disturbance simultaneously [4]. Effectively, the disturbance absorbs any systematic discrepancy between the modeled and measured plasma response. Its signature can be used to classify the nature of the discrepancy.

## Real-time profile evolution model RAPTOR and its implementation on ASDEX-Upgrade

The key component of the EKF is the forward model $f$. This model should accurately represent the dynamic evolution of the plasma profiles and be sufficiently fast to run in real-time. For the purpose of estimating the plasma profiles, the 1D profile evolution equations for magnetic flux and electron temperature are used [5], which are simulated with the the RAPTOR code [6], [7].

Details of the implementation of the EKF state observer on ASDEX-Upgrade and the RAPTOR code were presented earlier [2] and will be briefly mentioned here. The state observer takes several AUG real-time diagnostic signals, checks their validity using the provided status flags, and feeds these to the RAPTOR profile state observer. Some quantities, such as plasma current, density profile, and actuator power, are directly used as inputs to the simulator. Others, in particular the ECE and boundary flux, are used as measurement inputs to the EKF. The system runs routinely in real-time every 10ms. Note that no real-time internal measurements of the current density profile are used, hence the reconstruction of the $q$ profile relies entirely on the physics model.

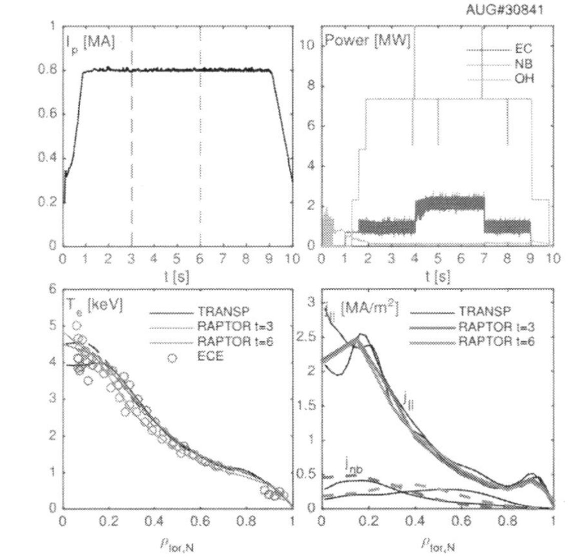


**Figure 1:** Comparison of RAPTOR- and TRANSP-reconstructed $T_e$, $j_{nb}$ (– –) and $j_{\|}$ (-) profiles for a shot with Neutral beam current drive.

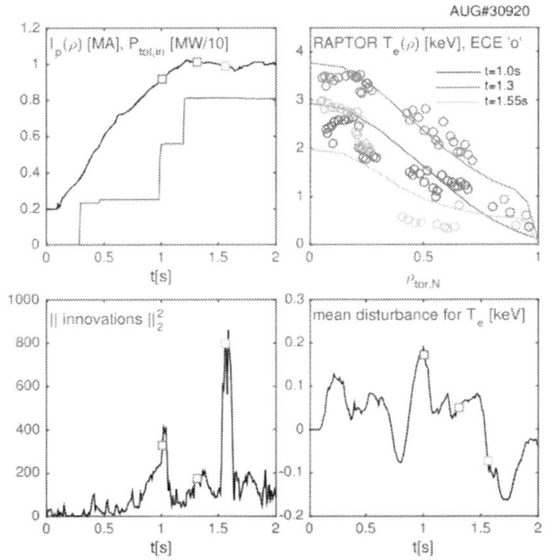


**Figure 2:** Plasma parameters, norm of innovation sequence and mean $T_e$ disturbance for a shot with a large MHD mode at $t = 1.5$s.

## Comparison to off-line profile reconstruction using TRANSP

To validate the results from the RAPTOR state observer, the reconstructed profiles are compared, in Figure 1, to more detailed TRANSP interpretative simulation analyses for a particular shot featuring on and off-axis neutral beam current drive (NBCD) [8]. The temperature profiles are similar, with the differences explained by the fact that RAPTOR uses the ECE diagnostic while for the TRANSP run several diagnostics are used and the Grad-Shafranov equilibrium is re-computed as part of the data fitting process. As a result the total parallel current is also very similar, since this is mostly composed of ohmic and bootstrap current, which are calculated using the same equations in RAPTOR and TRANSP. The neutral beam driven current in TRANSP is computed using the Monte-Carlo code NUBEAM to simulate the beam-driven fast ions. The neutral beam current density profiles in RAPTOR are approximated using Gaussian profiles, which were scaled to match TRANSP predictions for similar representative plasmas.

Indeed, the RAPTOR-based state observer is expected at best to approach more sophisticated physics modelling codes to an accuracy sufficient for real-time control purposes. Analyses using more complete physics-based codes remain essential to compute quantities that are difficult or impossible to measure or calculate in real-time and for detailed physics studies. Parametrizations of results from more complete codes continue to play a key role to obtain models that are tractable in real-time.

## Real-time detection of unexpected events

Finally, we present an example of using the innovation sequence $(\hat{y}_k - y_k)$ and disturbance estimate $d_k$ to detect and diagnose unexpected events in the plasma (Figure 2). The norm of the innovation sequence, a scalar measure for the discrepancy between measured and predicted diagnostic signals, is shown. The two spikes in this signal are caused by sudden discrepancies between the modelled and measured plasma behaviour. In the first case, the plasma has transitioned into H-mode, as is visible on the ECE (blue, edge channels), while RAPTOR-EKF has not. As soon as the model transitions into H-mode the discrepancy disappears. The second spike is caused by an MHD mode: its appearance around $t = 1.5$s causes a strongly degraded confinement and inconsistent ECE measurements, caused by cutoff (yellow). The disturbance estimate (bottom right) shows how the disturbance term adapts to the model-measurement mismatch and gives information about the nature of the discrepancy: the H-mode case shows that a positive correction in the temperature is needed to match the measurements, while the confinement-degrading MHD mode leads to a negative temperature correction as expected. At $t = 1.3$s (red), the model and measurements can be combined with good agreement.

## Conclusions and outlook

Diagnostic measurements and physics model predictions are ‘fused’ using the EKF algorithm to estimate key plasma profiles in real-time on ASDEX-Upgrade. The quality of the reconstruction can approach that of more complex post-shot analysis, though manual tuning of some parameters based on more complete physics code is sometimes required. Estimates of an additive model disturbance can be used to diagnose mismatch between model and measurements. Future work will focus on extending this classification to be able to provide information to a higher-level plasma supervision system. The reconstructed plasma profiles will also be fed to the real-time Grad-Shafranov code to yield more accurate real-time equilibria.

## References

[1] Humphreys, D. *et al.* 2015 *Physics of Plasmas (1994-present)* **22** 021806

[2] Felici, F. *et al.* 2014 in *41st EPS Conference on Plasma Physics, Berlin, Germany*

[3] Anderson, B. *et al.* 1979 *Optimal filtering* Dover Publications, Mineola, NY (USA)

[4] Felici, F. *et al.* 2014 in *American Control Conference (ACC), 2014* 4816–4823

[5] Hinton, F.L. *et al.* 1976 *Rev. Mod. Phys.* **48** 239

[6] Felici, F. *et al.* 2011 *Nuclear Fusion* **51** 083052

[7] Felici, F. *et al.* 2012 *Plasma Physics and Controlled Fusion* **54** 025002

[8] Geiger, B. *et al.* 2015 *Accepted for publication in Nuclear Fusion*

This project is sponsored in part by the Netherlands Organisation for Scientific Research and the Swiss National Science foundation. This work has been carried out within the framework of the EUROfusion Consortium and has received funding from the Euratom research and training programme 2014-2018 under grant agreement No 633053. The views and opinions expressed herein do not necessarily reflect those of the European Commission.

# Helium transport investigations at ASDEX Upgrade

A. Kappatou[1,2], R.M. McDermott[1], C. Angioni[1], T. Pütterich[1], R. Dux[1], R.J.E. Jaspers[3], E. Viezzer[1], M. Cavedon[1,4], R. Fischer[1], M. Willensdorfer[1], G. Tardini[1], the EUROfusion MST1 Team* and the ASDEX Upgrade Team

[1]*Max-Planck-Institut für Plasmaphysik, Garching, Germany*
[2]*FOM Institute DIFFER, Eindhoven, The Netherlands*
[3]*Technische Universiteit Eindhoven, Eindhoven, The Netherlands*
[4]*Technische Universität München, Garching, Germany*

## Introduction

Understanding low-Z impurity transport in fusion plasmas, particularly helium transport, is critical for the successful operation of future fusion reactors. Experimentally validated theoretical tools for the description of the transport mechanisms which define the shape of the helium density profile are necessary. In this way, predictions can be made and the extent of the accumulation of helium "ash" in the plasma core, leading to dilution of the fusion fuel, can be evaluated. To this end, helium transport investigations have been undertaken at ASDEX Upgrade (AUG) by means of charge exchange spectroscopy and quasi-linear gyrokinetic simulations.

## Forward modelling of the helium plume emission

To evaluate the helium CX spectra correctly and obtain accurate helium density profiles, forward modelling of the helium plume emission is necessary [1]. The existence of the helium plume emission in the CX spectra measured at AUG has been identified through an apparent discrepancy between the $T_i$ and $v_\phi$ profiles obtained from $He^{2+}$ and $B^{5+}$ and through an apparent discrepancy between the $He^{2+}$ density profiles measured with different diagnostic geometries.

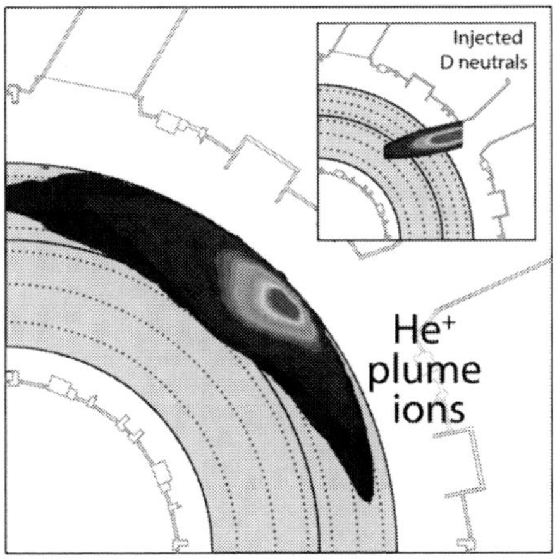


Figure 1: *Spatial distribution of helium plume ions.*

A model to describe the plume emission has been developed at AUG [2]. First, the density of the helium plume ions ($He^+$) that are born in the beam volume from charge exchange reactions between fully stripped helium ions and the beam neutrals is calculated. Then, the spatial distribution of these ions along the magnetic field lines (see Fig. 1) is evaluated in a full 3D geometry, taking into account the magnetic equilibrium and the geometries of the neutral beam and the diagnostic. Transport perpendicular to the magnetic field lines is negligible within the lifetime of a $He^+$ ion and is, therefore, neglected. Subsequently, some of the $He^+$ ions undergo electron impact excitation in the lines-of-sight of the diagnostic, and the associated plume photon flux is derived using the appropriate photon emission coefficients.

The accurate description of the plume ion distribution in real and velocity space requires non-Maxwellian effects to be taken into account. The plume particles are not quickly equilibrated

* See http://www.euro-fusionscipub.org/mst1.

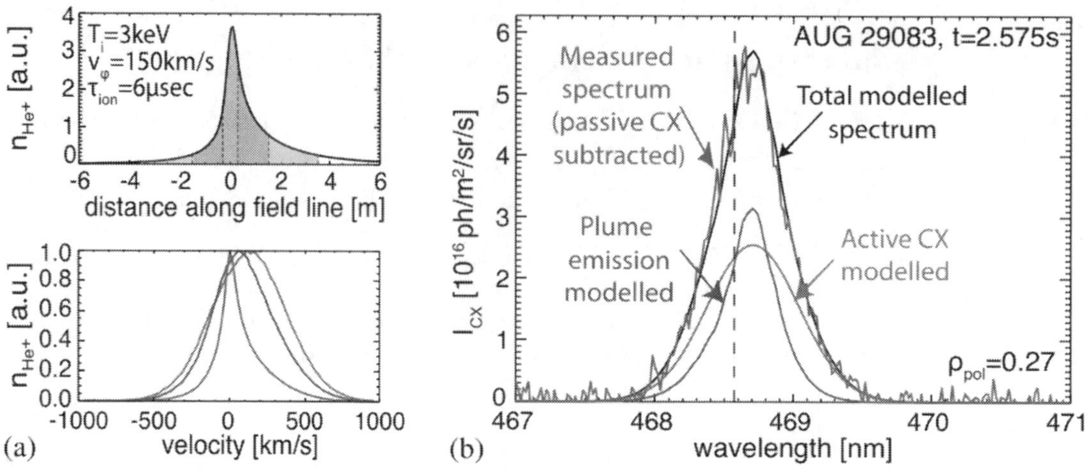


Figure 2: *a) Distribution of plume ions along a field line (which intersects the neutral beam at 0 m) and the corresponding velocity distributions, summed over the ranges indicated. b) Reconstruction of a helium CX spectrum (the passive emission is subtracted applying neutral beam modulation). The sum of the modelled plume and prompt emissions agrees well with the measured spectra.*

with the background ions, as the momentum exchange time is much larger than the ionisation time. As a result, the faster particles of the initial Maxwellian distribution leave the observation volume, while the slower ones remain, leading to apparent lower ion temperatures and rotation values in the observed plume emission.

This effect is illustrated in Fig. 2a, where the distribution of $He^+$ ions along a magnetic field line is shown for typical core plasma parameters. Assuming lines-of-sight tangential to the field lines and no magnetic field line curvature, all of the particles will be observed. For instance, if all particles along the field line are summed together (~7 m, red shaded region), then the velocity distribution is almost Gaussian (bottom plot, red line). In reality, however, the sum is over a much narrower range (~60 cm, green shaded region), resulting in a much narrower velocity distribution and thus a “colder” and “slower” emission line (green line). In order to deal with this situation, a Monte Carlo approach is used to describe the distribution of the plume ions along the magnetic field lines. The Monte Carlo modelling provides the spread of the plume ions and their velocities, thus enabling the non-Maxwellian features to be accurately described.

The developed plume emission model reproduces the experimental CX spectra, as shown in Fig. 2b. The sum of the reconstructed active CX and plume emission lines are shown to agree well with the measured spectrum, from which the passive emission line is subtracted by means of neutral beam modulation. It is, therefore, concluded that the underlying physics mechanisms are adequately described. Furthermore, the $T_i$ and $v_\phi$ profiles obtained from the measured spectra, once the plume emission line is subtracted, agree within errorbars with those obtained from $B^{5+}$ measurements.

**Experimental database for the investigation of helium transport**

The high quality spectroscopic data and the plume correction enable detailed helium transport studies to be performed at AUG [2]. A database of 131 helium and boron density profiles was assembled from dedicated experiments ($I_p = 600\,kA, B_t = -2.5\,T$). Different levels of central electron cyclotron resonance heating (ECRH) were applied ($P_{ECRH} = 0-4\,MW$) to vary the normalised gradient of the electron density ($R/L_{n_e} = -(R/n_e)(dn_e/dr) = 0.03-3.6$) and of the toroidal rotation profiles ($u' = -0.15-1.1$). Strong peaking of the electron density profile is known to occur with the application of central ECRH if the plasma turbulence regime transitions from ITG towards TEM [3]. At the same time, flat or even hollow rotation profiles have been observed [4]. Different levels of neutral beam heating ($P_{NBI} = 2.5-7.5\,MW$) as well as on- and off-axis beams were used to scan the ion temperature gradient ($R/L_{T_i} = 3.0-7.0$) and Mach number. The effective collisionality is $\nu_{eff} = \nu_{ei}/(c_s/R) = 0.3-1.9$ (all values for $\rho_{tor} = 0.5$).

Both helium and boron, the dominant low-Z impurities after boronisation at AUG ($c_B = 0.3-0.8\%$, $c_{He} = 0.5-1.5\%$), were measured simultaneously, allowing a multi-species study. A wide variety of experimental helium and boron density profiles were obtained, with helium density profiles varying from almost flat to peaked ($R/L_{n_{He}} = -0.1-4$) and boron density profiles varying from slightly hollow to peaked ($R/L_{n_B} = -0.4-2.3$). The peaking of the helium density profile is comparable to that of the electron density at mid-radius, consistent with results obtained in DIII-D [5].

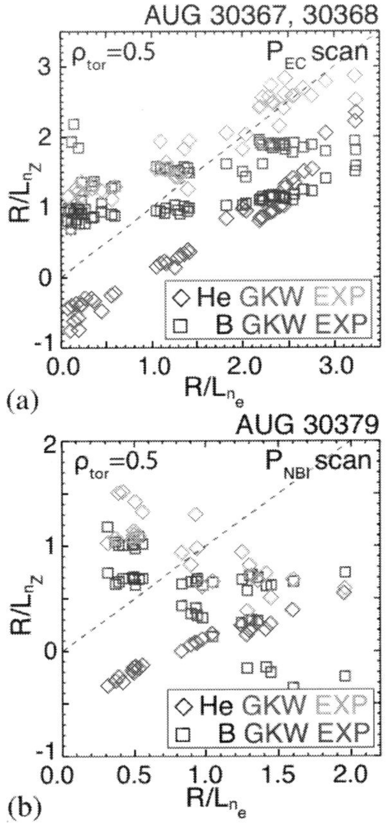


Figure 3: *$R/L_{n_{He}}$ and $R/L_{n_B}$ from gyrokinetic modelling and experimental measurements, plotted against $R/L_{n_e}$.*

**Comparison of experimental results to the predictions of gyrokinetic modelling**

The assembled database enables comparison of the experiment with gyrokinetic simulations over a wide parameter range. Neoclassical transport, calculated with NEOART [6], was confirmed to be negligible around mid-radius in comparison to the turbulent transport. Local quasi-linear gyrokinetic simulations of the turbulent transport have been performed with GKW [7]. The simulations are performed over a spectrum of $k_y\rho = 0.23-0.75$, providing the diagonal and off-diagonal contributions of impurity transport: the predicted impurity density gradients are the sum of thermodiffusive, rotodiffusive and pure convection terms. Previous work at AUG showed that the logarithmic boron density gradients can be qualitatively [3] and even quantitatively [8] reproduced, provided that "roto-diffusion" is included.

The experimentally deduced $R/L_{n_{He}}$ and $R/L_{n_B}$ are compared with the gyrokinetic modelling predictions in Fig. 3a for a subset of the database, corresponding to discharges 30367 and 30368, plotted for illustration against $R/L_{n_e}$. The centrally applied ECRH power was scanned up to a maximum of 4 MW on a background of 2.5 MW of NBI heating. Clear differences are observed between helium and boron, a behaviour predicted theoretically, as thermodiffusion is more important for lower-Z impurities, due to the inverse charge dependence of the thermodiffusive term. Only a qualitative agreement for both impurities is reproduced, as the theory underpredicts the experimental gradients, in contrast to previous results.

However, cases also exist in the database in which opposite trends are observed. In discharge 30379, a scan of NBI power was performed on a background of 0.5 MW of ECRH power. The experimental gradients of both helium and boron are found to decrease with increasing $R/L_{n_e}$ (shown in Fig. 3b), as well as with increasing $R/L_{T_i}$ and $u'$ (not shown). This behaviour is not reproduced by the gyrokinetic simulations.

An explicit explanation for the observed features is not straightforward. In the discharges in the database, strong sawteeth or MHD modes were not identified. The effect of ELMs on the profiles is minimised by selecting inter-ELM profiles. The sensitivity of the gyrokinetic modelling to the used magnetic equilibrium [8] was tested, but the uncertainties in the magnetic equilibrium are not big enough to explain the observations. The effect of a helium source at the tokamak wall on the helium density profile gradient was assessed with STRAHL [9] and was found to be too small to explain the discrepancies between the experiment and theory. And examining each turbulent transport term that contributes to the impurity density gradients, obtained from the gyrokinetic modelling, does not lead to a direct conclusion on which term is responsible for the discrepancy. Disentangling the dependencies of the impurity density gradients on the different plasma parameters in order to address these discrepancies is the aim of further experimental and modelling work.

## Conclusions

Investigations of light impurity transport have been performed at ASDEX Upgrade including, for the first time, accurate helium measurements, supported by the plume emission modelling. The shape of the helium density profile is found experimentally to follow closely the shape of the electron density profile. The current theoretical understanding of light impurity transport, as derived from quasi-linear gyrokinetic simulations, is compared against the experiment, revealing some troubling discrepancies. As the impurity density profiles are defined by the interplay of many plasma parameters, further work from both the theoretical and the experimental side are needed in order to decorrelate the parameters that define low-Z impurity transport.

## References

[1] R.J. Fonck *et al*, Phys. Rev. A **29**, 3288 (1984)

[2] A. Kappatou, PhD thesis, Eindhoven University of Technology, the Netherlands (2014)

[3] C. Angioni *et al*, Nucl. Fusion **51**, 023006 (2011)

[4] R.M. McDermott *et al*, Plasma Phys. Control. Fusion **53**, 035007 (2011)

[5] D.F. Finkenthal, PhD thesis, University of California, Berkeley (1994)

[6] R. Dux and A.G. Peeters, Nucl. Fusion **40**, 1722 (2000)

[7] A.G. Peeters *et al*, Comput. Phys. Commun. **180**, 2650 (2009)

[8] F.J. Casson *et al*, Nucl. Fusion **53**, 063026 (2013)

[9] R. Dux, STRAHL, Annual IPP report **10/30** (2006)

This work has been carried out within the framework of the EUROfusion Consortium and has received funding from the Euratom research and training programme 2014-2018 under grant agreement No 633053. The views and opinions expressed herein do not necessarily reflect those of the European Commission.

# Interplay between central ECRH and MHD in mitigating tungsten accumulation in ASDEX Upgrade

M. Sertoli[1], J. M. Garcia-Regaña[1,2], T. Odstrčil[1], C. Angioni[1], H. Smith[3], Y. Turkin[3], W. A. Cooper[4], A. de Bustos[5], D. Vezinet[1], the ASDEX Upgrade team and the EUROfusion MST1 Team*

[1] *Max-Planck-Institut für Plasmaphysik, D-85748 Garching, Germany,* [2] *EUROfusion PMU Garching, Boltzmannstrasse 1, D-85748, Garching, Germany,* [3] *Max-Planck-Institut für Plasmaphysik, 17491 Greifswald, Germany,* [4] *École Polytechnique Fédérale de Lausanne (EPFL, CRPP), 1015 Lausanne, Switzerland,* [5] *Departamento de Física, Universidad Carlos III, Leganés 28911, Madrid, Spain,* * *See http://www.euro-fusionscipub.org/mst1*

In tokamaks with reactor relevant high-Z plasma facing components (PFCs), the control of impurity accumulation in the plasma centre is of major importance to avoid performance degradation and the possible disruptive termination of the discharges due to excessive radiative power losses. At ASDEX Upgrade (AUG), whose first wall is fully covered with tungsten (W), the application of small amounts of electron cyclotron resonance heating (ECRH) has been found to be effective in mitigating central W accumulation which otherwise occurred in discharges where neutral beam (NBI) was used as sole auxiliary heating. The mitigation occurs only if the ECRH is deposited sufficiently central (typically inside the $q = 1$ surface) and usually in the presence of saturated $(m,n) = (1,1)$ MHD activity in-between sawtooth crashes. The interplay between ECRH and saturated $(1,1)$ modes in the mitigation of W accumulation is still not understood. Due to the helical equilibrium arising in the presence of the $(1,1)$ mode, typical experimental and theoretical methods employing 1D flux-surface-averaged quantities are not suited for the investigation of the processes in action. New experimental methods have been developed enabling a mode-resolved 2D measurement of the intrinsic W density by combining SXR tomographic reconstructions [1] with 2D rotation-tomography of the 1D electron cyclotron emission (ECE) electron temperature measurements [2]. These methods are here applied to a typical AUG H-mode discharge and the results compared with theoretical estimates evaluated with the coupled neoclassical and gyro-kinetic codes GKW [3] and NEO [4] (see [5] for details on the coupling method).

The plasma emission detected by the soft X-ray (SXR) diagnostic in most AUG discharges is dominated by W, typically accounting for 80% of the total emission in the plasma centre. In such cases, the tungsten density $n_W$ can be evaluated from the experimental SXR emissivity $\varepsilon_{exp}^{SXR}$ through the following equation:

$$n_W^{SXR}(\mathbf{r};t) \propto \varepsilon_{exp}^{SXR}(\mathbf{r};t) \,/\, [n_e(\mathbf{r};t)\, L_W^{SXR}(\mathbf{r},t)] \quad (1)$$

where $n_e$ is the electron density and $L_W^{SXR}$ the radiation loss parameter (also called *cooling factor*) filtered according to the SXR diagnostic filter function (photon energy range $\sim 1-20$ keV at AUG) calculated assuming local ionization equilibrium. All quantities in equation (1) are defined in the three-dimensional space $\mathbf{r}$ of the tokamak and are a function of time. In the presence of a 3D helical equilibrium due to macroscopic MHD activity, equation (1) has to be solved at least on a 2D poloidal cross-section to resolve the various quantities on the non-axisymmetric equilibrium. This requires mode-resolved 2D maps of the SXR emissivity (tomographic reconstructions) and of the electron density for direct introduction in equation (1), as well as of the electron temperature for the calculation of the radiation loss parameter $L_W^{SXR}$. In this contribution, the 2D mode-resolved $T_e$ maps have been evaluated following [2], while the $n_e$ perturbation has been neglected and only the mode-averaged profiles have been used. The latter assumption is reasonable since the $n_e$ profiles in the analysed experiment are relatively flat (see figure 2). Refer to [6, 7] for an in-depth explanation of the combined data analysis of SXR tomography and the 2D mode-resolved $T_e$ maps for the correct application of equation (1).

The example reported is that of a standard AUG lower-single-null H-mode discharge (# 31114) with $I_p = 1$ MA, $B_t = -2.5$ T, $P_{NBI} = 7.5$ MW and $P_{ECRH} = 1.4$ MW. The ECRH power is equally distributed over two gyrotrons, the first one depositing at $\rho^H \sim 0.2$, the second one adjusted for co-ECCD for sawtooth control (deposition position scanned in range $\rho^{CD} \sim 0.2 \rightarrow 0.6$, leading to a total driven current $I_{CD} \sim 11 \rightarrow 5$ kA). In the analysed time range, the CD gyrotron sits close to the inversion radius $\rho_{inv} \sim 0.4$ (see figure 2). The radial coordinate $\rho = \sqrt{(\psi - \psi_a)/(\psi_s - \psi_a)}$ is the normalised poloidal flux radius, where $\psi$ is the poloidal flux and the indices $s$ and $a$ refer to the separatrix and the magnetic axis.

The flux-surface-averaged W density has been first of all evaluated applying equation (1) with $\mathbf{r} = \rho$ and averaging all quantities over 10 ms (corresponding to $\sim$ 100 mode rotations) to obtain mode-averaged profiles. The SXR emissivity has been evaluated through 1D Abel inversion of cameras $I$ and $J$ . As performed in [6] the obtained $n_W^{SXR}$ has been cross-calibrated using the independent W concentration measurement of the grazing incidence spectrometer (GIW). The time evolution of the post-calibration W density at three radial positions is shown in figure 1a. The evolution of the simulated GIW measurement $c_{W,l}^{SXR}$ (post-calibration, dashed black line with red error bars in figure 1b) matches very well that of the spectrometer measurement (thick continuous black line). The sawtooth crashes, clearly visible in the time evolution of the W density and GIW measurement, are *inverted*, i.e. the profiles are hollow before the crash occurs, flattening inside $\rho \leq 0.4$ at the crash (already shown in the past for AUG in [8] ). Strong MHD activity lasting more than half of the sawtooth period is visible in the spectrogram of a magnetic coil positioned at the LFS mid-plane (figure 1c). Both modes at frequencies $\sim$ 10 kHz and $\sim$ 13 kHz are $(1,1)$ modes (mode numbers evaluated from SXR and magnetic coils).

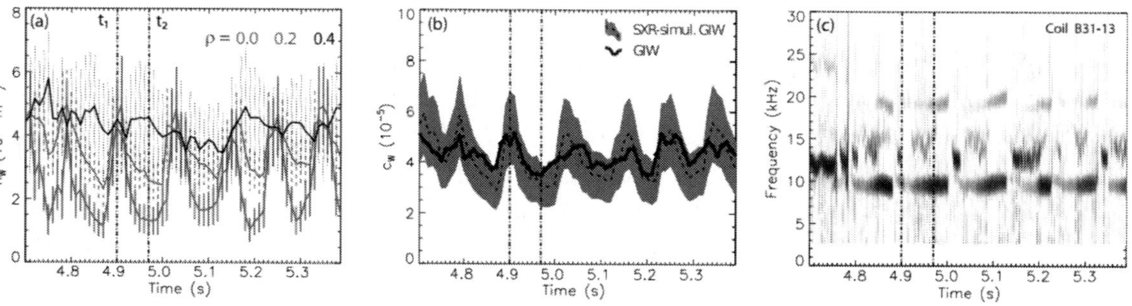


Figure 1: Time evolution of: (a) W density from SXR (all at three different radii); (b) comparison of simulated and experimental most central GIW measurement, (c) spectrogram of magnetic coil $B31-13$.

The evolution of the profiles during one sawtooth cycle are analysed more in detail for two time-points: $t_1 = 4.90$ - shortly after the sawtooth crash; $t_2 = 4.97$ - at mode saturation, shortly before the next crash occurs (shown as dashed black vertical lines in figure 1). The profiles of $T_e$, $n_e$, $T_i$ and $v_{rot}$ show minor changes (figure 2a-d), while major excursions are visible in the SXR emissivity (e), reflected in the W density profiles (figure 2f). Just after the crash, $n_W$ is flat inside $\rho_{inv} \sim 0.4$ and develops a deep and wide hollowness before the next crash occurs. This behaviour is detected by the GIW spectrometer as well, despite its averaging over large radial domains (full circles and triangles in figure 2f). Modelling the flux-surface-averaged W density profiles with the coupled codes GKW and NEO including both neoclassical and turbulent effects as well as the effects of toroidal rotation, neither the peaking for $\rho > 0.4$ nor the hollowness inside this radius are reproduced (red diamonds in figure 2f, rescaled to the experimental value at $\rho = 0.7$). Nonetheless the order of magnitude is compatible with experiment, while if turbulent effects are excluded and only the neoclassical contributions accounted for, the W density in the centre increases by two orders of magnitude. Since the sawtooth crash is not modelled, these results are shown for $t_2$ only.

For best comparison with the 1D profiles, the 2D mode-resolved maps have been evaluated close to the same time-points. Figure 3a shows the time evolution of the 2D $T_e$ maps (and its isothermal contours) at the start of the sawtooth cycle (labelled $At \sim t_1$) and at four time-points

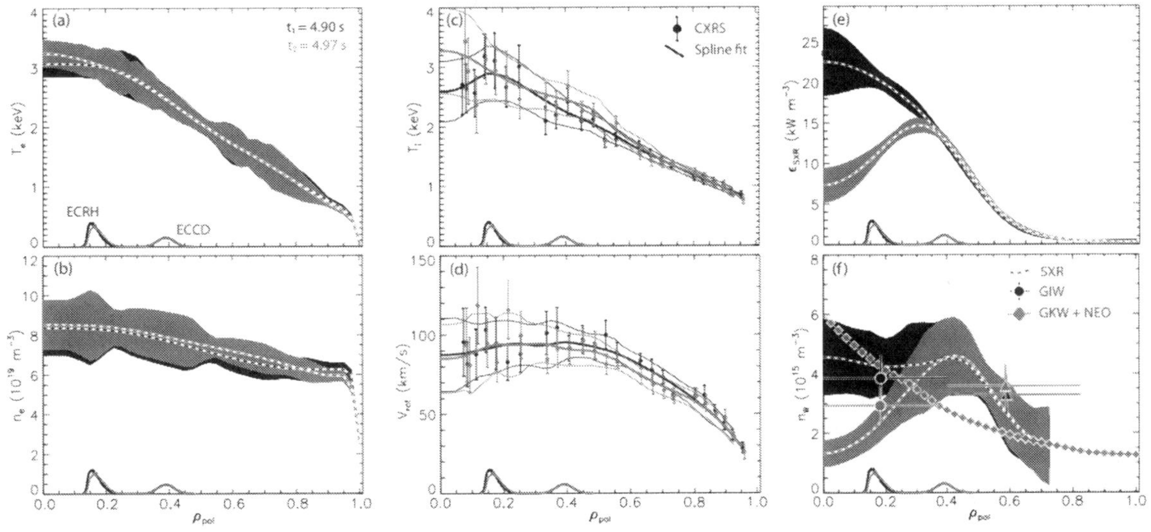


Figure 2: Profiles of (a) $T_e$, (b) $n_e$, (c) $T_i$, (d) $v_{rot}$, (e) $\varepsilon_{exp}^{SXR}$ emissivity and (f) $n_W^{SXR}$ compared to the measurements of the GIW spectrometer (full circles and triangles) and theoretical estimate using GKW + NEO (diamonds, rescaled to the experiment at $\rho = 0.7$) for two time-points (vertical lines in figure 1).

close to $t_2$ (labelled *Close to* $t_2$), the latter four occurring within a 43 $\mu s$ time window (i.e. approximately half mode rotation). Also shown are the deposition position and the inversion radius (see labels) as well as the magnetic axis from the equilibrium reconstruction (black diamond).

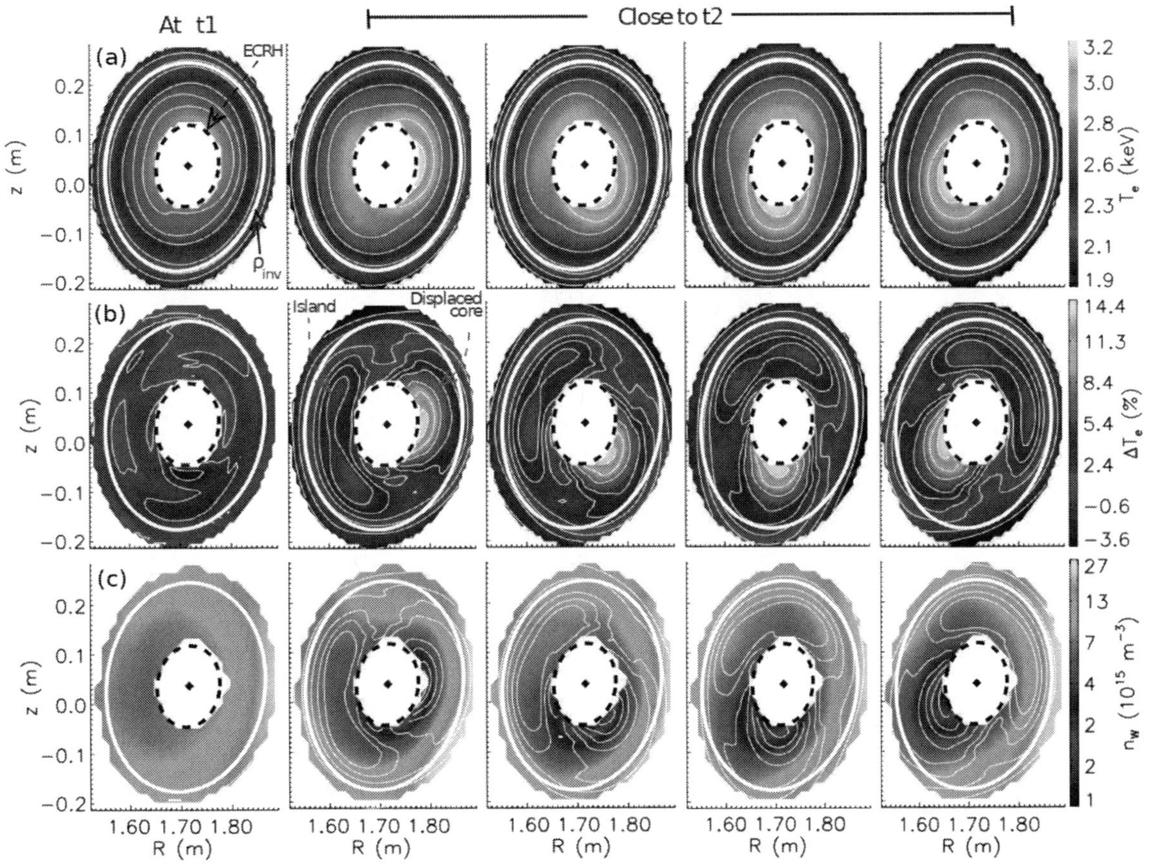


Figure 3: 2D maps: (a) $T_e$ evolution, (b) $T_e$ evolution with respect to the start of the sawtooth cycle ($\sim t_1$); (c) $n_W^{SXR}$ (shaded surface) with the contours of (b) over-plotted for correlation with the mode structure.

The central white area is a region of plasma not covered by the ECE diagnostic and therefore not accounted for in the 2D analysis. Figure 3b shows the percentage change in $T_e$ with respect to the start of the sawtooth cycle ($\Delta T_e(t) = (T_e(t) - \langle T_e(t_1) \rangle) / \langle T_e(t_1) \rangle$). From the isothermal contours of both figures 3a and 3b the rotating displaced core is clearly distinguishable, figure 3b clearly shows the mode structure, with the displaced core and the magnetic island (see labels) rotating in the laboratory frame in the clockwise direction. The 2D W density (figure 3c) inside the inversion radius is relatively flat at the start of the cycle, exhibiting a slight poloidal asymmetry consistent with LFS accumulation due to centrifugal effects. This asymmetry is visible during mode saturation close to the inversion radius, while inside $\rho_{inv}$ the W density is hollow. This *impurity hole*, already visible in the 1D profiles (figure 2), is located inside the displaced core and rotates with it. The magnetic island exhibits instead a flat W density in all mode phases. Comparing the last of these 2D W density plots (figure 4c) with the modelling from GKW+NEO (figure 4a), while, as already clear from the 1D profiles (figure 2f), the hollowness inside the inversion radius is not reproduced, the poloidal asymmetry around and outside of $\rho_{inv}$ is very well matched.

It is thus clear that modelling in 3D is required to account for the helically displaced core of the $(1,1)$ mode as well as the presence of the magnetic island. Neoclassical, turbulent and MHD simulations investigating the transport of W in a non-axisymmetric AUG geometry in the presence of an ideal $(1,1)$ mode are ongoing using the code EUTERPE [9, 10], GENE [11] and XTOR-2F [12, 13] respectively. Due to the stellarator-like equilibrium arising in the presence of the saturated $(1,1)$ mode (effective magnetic field ripple $\varepsilon_{eff}$ of the order of what encountered in LHD and higher than what expected for W7X, figure 4c), the effects of the non-axisymmetry are expected to be present also outside the $q = 1$ surface where complete axisymmetry is still not recovered ($\varepsilon_{eff} > 0$) [10].

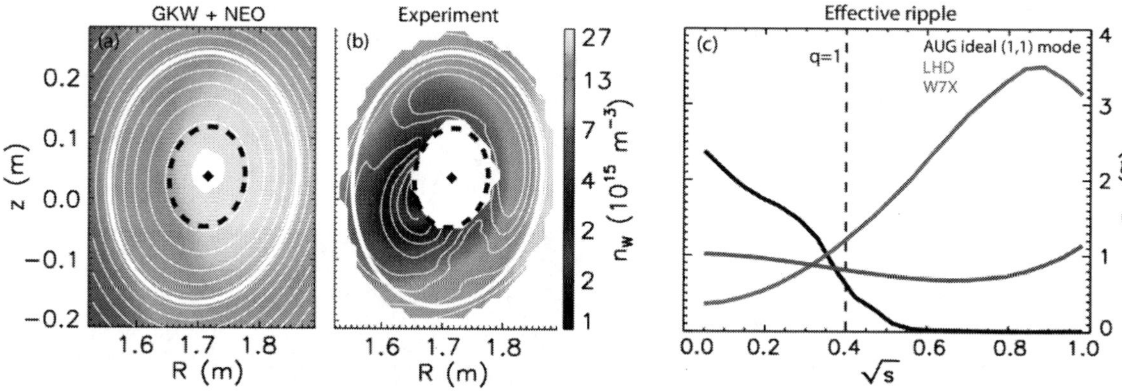


Figure 4: 2D W density maps from GKW+NEO (a) and experiment (b). Effective ripple of AUG discharge #31114 with ideal internal kink ($\xi \sim 10$ cm) compared to standard LHD and W7X configurations (c) [10]. The coordinate $\sqrt{s}$ corresponds to the normalised toroidal flux.

**Acknowledgments**: This work has been carried out within the framework of the EUROfusion Consortium and has received funding from the Euratom research and training programme 2014-2018 under grant agreement No 633053. The views and opinions expressed herein do not necessarily reflect those of the European Commission.

**References**: **[1]** M. Odstrcil et al., Nucl. Instr. and Methods in Phys. Research A **686**, 156 (2012), **[2]** M. Sertoli et al., Nucl. Fusion **53**, 053015 (2013), **[3]** A. Peeters et al., Comp. Phys. Comm. **180**(12), 2650 (2009), **[4]** E. A. Belli et al., Plasma Phys. Control. Fusion **54**(1), 015015 (2012), **[5]** C. Angioni et al., Nucl. Fusion **54**(8), 083028 (2014), **[6]** M. Sertoli et al., Plasma Phys. Control. Fusion **57**(7), 075004 (2015), **[7]** M. Sertoli et al., Submitted to Nucl. Fusion, **[8]** A. Gude et al., 37th EPS Conference, P4.124 (2010), **[9]** G. Jost et al., Phys. Plasmas **8**(7) (2001), **[10]** J. Garcia-Regaña, This conference, P2.170, **[11]** F. Jenko et al., Phys. Plasmas **7** (2000) 1904, **[12]** H. Lütjens et al., J. Comput. Phys. **229**(21), 8130 (2010), **[13]** J. Ahn, This conference, P1.103

# Modelling of core tungsten transport and its control in ITER H-mode scenarios

A. Loarte[1], M. Hosokawa[1], A. Polevoi[1], F. Köchl[2], V. Parail[3], M.L. Reinke[4] E. Belli[5], J. Candy[5], R. Dumont[6], D. Zarzoso[7]

[1]*ITER Organization, Route de Vinon-sur-Verdon, 13067 St Paul Lez Durance, France*
[2]*Atominstitut, Technische Universität Wien, Stadionallee 2, 1020 Vienna, Austria*
[3]*CCFE, Culham Science Centre, Abingdon, OX14 3DB, United Kingdom*
[4]*Oak Ridge National Laboratory, Oak Ridge, Tennessee 37831-8072, USA*
[5]*General Atomics, P.O. Box 85608, San Diego, CA 92186-5608, USA*
[6]*CEA, IRFM, F-13108 Saint-Paul-lez-Durance, France*
[7]*PIIM-UMR 7345-Universit´e d'Aix-Marseille-CNRS, 13397 Marseille Cedex 20, France*

## 1. Introduction

The decision to start ITER operation in the non-active phase (H/He plasmas) with a W divertor has brought increased attention to physics issues related to high Z impurity transport and control. Lack of impurity control could lead to the radiative collapse of plasmas by W accumulation and increased disruptivity, which is detrimental to ITER operation. Control of W in H-mode plasmas requires, as a first step, the control of W production and its transport into the core plasma through the SOL and edge transport barrier. In addition, even when the concentration of W at the pedestal is kept at low levels, unfavourable core W transport can lead to its uncontrolled accumulation and to loss of the H-mode due increased radiation in present experiments (e.g. JET and ASDEX-Upgrade) [1, 2], where schemes have been developed to avoid W accumulation by the application of additional heating to provide stable H-mode operation with W plasma facing components.

An extensive modelling study has been carried out to evaluate the transport of W in the core plasmas and the demonstration of its control by the available heating and fuelling schemes to provide stable H-mode operation in ITER. Stationary H-mode phases have been modelled with the ASTRA and NEO codes and confinement transients (i.e. H-L transitions) have been modelled with the JINTRAC code suite. In these simulations transport is modelled by neoclassical transport plus anomalous transport from the GLF23 model. The study encompasses a range of ITER DT H-mode plasmas from 7.5 to 15 MA. A range of assumptions regarding electron heat transport and DT particle transport in the central region of the plasma (where anomalous transport is found to be negligible) have been considered.

## 2. Modelling of W transport in stationary phases of ITER DT scenarios

The ITER simulations show that W transport is anomalous for most of the plasma cross section except in the central region (typically $r/a \leq 0.2$-$0.3$) and is characterized by a large value of the diffusion coefficient with a negligible pinch velocity, as shown in Fig. 1. In the central region W transport is neoclassical and thus determined by the local ion temperature and density gradients, which are found to be modified by the ITER heating and current drive systems even for plasmas dominated by alpha heating; the ITER RF-based heating systems provide peaked power deposition profiles so that the electron (ECRH) and ion (ICRH-$He^3$ minority) power densities exceed those of alpha heating in the central plasma region (see Fig. 2).

In general, for 15 MA Q ~ 10 plasmas with $P_{NBI}$ = 33 MW and $P_{RF}$ = 20 MW (ECRH or ICRH, with a range of deposition locations), it is found that the level of core W peaking in

ITER is very modest (see Fig. 3 and Fig.4), which is in agreement with ITER-like experimental results in Alcator C-Mod [3]. This is caused by the low central density gradients in ITER due to the very small particle density source rate provided by the NBI system ($E_{NBI}$ = 1 MeV). The ITER particle density source rate is more than $10^2$ times lower than that of JET for the same levels of $P_{NBI}$ so that the central density gradient is solely determined by neoclassical transport physics for the DT ions, when ion-scale turbulent processes are considered for anomalous transport. Recently it has been pointed out that electron-scale turbulent processes may have an effect on energy and particle transport [4]. When this effect is considered in the modelling in a simplified way (by assuming an anomalous DT particle transport coefficient proportional to the electron thermal diffusivity from ETG modes) then W density profiles in ITER can become hollow for cases with central ECRH or ICRH heating (see Fig. 4).

Modelling for 7.5 MA/2.65T plasmas shows similar results regarding W transport and W density profiles to those for 15 MA Q = 10 plasmas. In this case the differences between the various auxiliary heating combinations explored is larger ($P_{NBI}$ = 33 MW and $P_{RF}$ = 0 and 20 MW (ECRH or ICRH)) with somewhat more peaked W profiles for the configurations with lower central power densities. This is expected because of the low level of alpha heating in these 7.5 MA plasma conditions. On the other hand, as $\langle n_e \rangle$ is lower in 7.5 MA than in 15 MA plasmas, the electron and ion temperatures are less coupled leading to a larger $T_e/T_i$ ratio at 7.5 MA than at 15 MA in ITER. Thus, when a residual effect of ETG turbulence on DT particle transport is assumed in the simulations, the DT profiles become very flat in the central region and correspondingly the W density profiles are very hollow. In summary we can conclude that, on the basis of our study, W transport is expected to be similar for 7.5 MA and 15 MA plasmas in ITER so that the results of the development of W control strategies at the beginning of operations at low plasma currents will be applicable to 15 MA high Q plasmas.

**3. Modelling of effects of toroidal rotation and fast particles on neoclassical W transport**

The calculations carried out with ASTRA and JINTRAC contain a simplified model for neoclassical W transport (NCLASS) that does not describe effects which have been identified to be of importance for high Z impurity transport [5]. One of them is the centrifugal force effect due to the toroidal rotation of impurities with the main DT ions (which can become supersonic for high Z impurities). The other is the effect of high energy particles which are ubiquitous in ITER plasmas (3.5 MeV alpha particles, 1 MeV NBI ions and ICRH accelerated minority ions). These two effects have been modelled with the NEO code [6] using plasma parameters (including toroidal rotation profiles) and fast particle energies from ASTRA as well as impurity minority energy distributions from EVE/AQL [7]. NEO simulations show that both the diffusion and pinch velocity for W increase in the same proportion with the plasma toroidal rotation velocity so that the stationary W core density peaking remains constant for a wide range or Mach numbers (up to M = 0.4); for larger Mach numbers the pinch velocity decreases and can reverse sign, which is beneficial for W accumulation control. The toroidal rotation Mach number predicted by ASTRA is very low for ITER (M = 0.02) due to the momentum diffusivity assumed ($\chi_\phi = \chi_i$) and the low momentum input provided by the 1MeV NBI. Thus no major effect of toroidal rotation on W transport is expected (i.e. the simulations carried out are appropriate) unless intrinsic toroidal rotation is very large in ITER, which seems unlikely.

Analysis of the effects of fast particles on W transport shows that these are negligible for alpha particles and NBI ions due to their large energy (low collisionality with W ions), low density and moderate density/energy radial gradients. On the contrary, the effects associated with $He^3$ ICRH accelerated ions are significant because of their moderate density ($n_{He}{}^3/<n_e> \sim 2\%$) and their very steep radial energy gradients due to the localized ICRH power absorption in the resonant layer. The screening of W provided by these steep energy radial gradients acts positively on core W transport and can even reverse the sign of the inwards pinch (i.e. leading to hollow W density profiles) for the highest levels of $P_{ICRH}$ considered (20 MW), as shown in Fig. 5. However, this positive effect requires fine-tuning of the ICRH scenario ($P_{ICRH}$, $n_{He}{}^3/<n_e>$, $f_{ICRH}$) for each plasma condition considered in order to ensure that the collisionality of the $He^3$ minority ions with W and their radial energy gradients in the area of interest (plasma centre) are optimized for their beneficial effect on W transport.

**4. W transport in the exit phase of ITER Q = 10 H-modes**

While the situation regarding core W control during the stationary phases of ITER H-modes is rather positive, core W control is found to be more complex in our first studies of the termination phases of ITER H-modes. For optimum radial position control of the plasma in ITER, it is desirable that the plasma energy decreases with timescales of $t_{decay\text{-}min} \sim 5$ - 10 s in the termination phase of high Q H-modes. The $t_{decay\text{-}min}$ timescale is determined by the penetration of the vertical field through the ITER vacuum vessel and the time constants of the ITER Poloidal Field Coils. The decrease of the plasma energy in such timescale can be achieved by a slow ramp-down of the auxiliary heating power, which together with alpha heating, extends the H-mode phase and thus the timescale of plasma energy decay over $t_{decay\text{-}min}$ [8]. Modelling of W transport with the JINTRAC code has been carried out for two such Q = 10 H-mode terminations: one in which the auxiliary heating ($P_{NBI}$ = 33 MW and $P_{ECRH}$ = 20 MW) is abruptly switched-off and one in which it is ramped-down gradually in 10 s. The results of the simulations for two core DT particle transport assumptions (see Fig. 6) show that strong W accumulation (with timescales of ~ 5 s) can occur when the auxiliary heating power is gradually decreased, which is optimum for radial position control. Further high Q H-mode termination studies will be carried out to optimize the ramp-down of the auxiliary heating power (as well as the type of heating scheme applied in this phase) and of the pellet fuelling so that the requirements regarding radial position and W control can be fulfilled simultaneously.

It is important to note that the quantitative features of the ITER simulations in this paper depend on physics assumptions on: a) the residual level of ion and electron thermal transport in the core region, where anomalous transport is low and b) the particle transport physics in the core region, in particular on the magnitude of the neoclassical DT pinch and on whether there is a residual low level of particle transport ($D >> D_{neo} \sim$ few $10^{-2}$ $m^2/s$) in this region or not. It is thus crucial to improve the central plasma transport physics basis and its experimental validation to refine the predictions for core W transport and its control in ITER.

*Disclaimer: The views and opinions expressed herein do not necessarily reflect those of the ITER Organization.*

**6. References**

*[1] C Angioni, et al, Nucl. Fusion **54**(2014)083028.*
*[2] R Dux, et al, Pl. Ph. Cont. Fus. **45**(2003)1815.*
*[5] FJ Casson, et al, Pl. Ph. Cont. Fus. **57**(2015)014031.*
*[6] E Belli, et al, Pl. Ph. Cont. Fus. **54**(2012)015015.*

*[3] A Loarte, et al, Phys. Plas. 22(2015)056117.*
*[4] N T Howard, et al, Phys. Plas. **21**(2014)112510.*
*[7] RJ Dumont, et al, Nucl. Fusion **53**(2013)013002.*
*[8] A Loarte, et al, Nuclear Fusion **54**(2014)123014.*

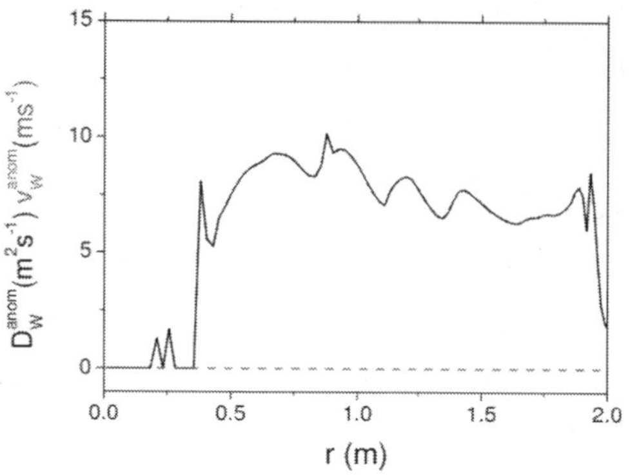


Figure 1. Anomalous diffusion coefficient and pinch for W transport in typical ITER Q = 10 plasma conditions.

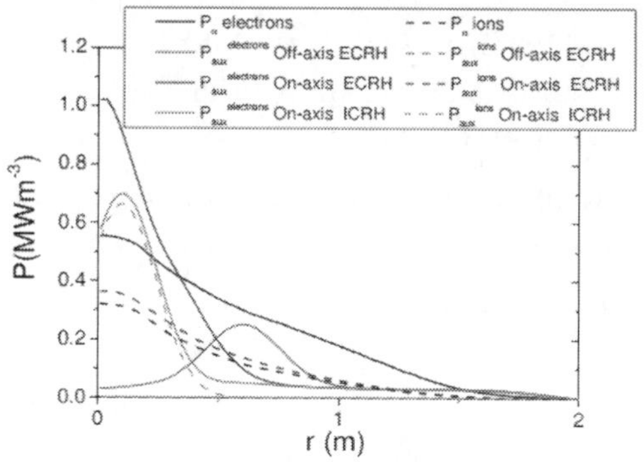


Figure 2. Ion and electron power density profiles for alpha heating and NBI+ECRH or NBI+ICRH ($He^3$ minority 50:50 electron-ion) heating for ITER Q = 10 plasma conditions.

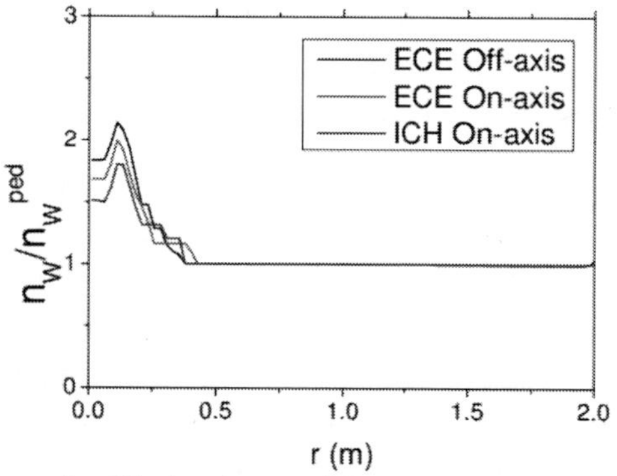


Figure 3. W density profiles, normalized to the pedestal value, for ITER Q = 10 plasmas with 33 MW of NBI and a range of RF heating schemes. Core particle and energy transport assumptions are those of neoclassical transport for all species (ions and electrons).

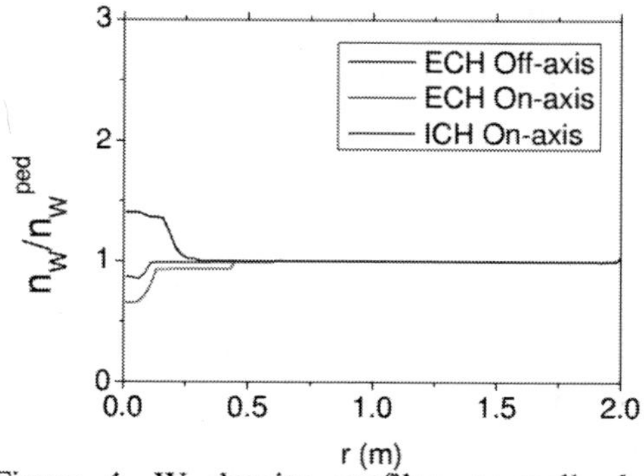


Figure 4. W density profiles, normalized to the pedestal value, for ITER Q = 10 plasmas with 33 MW of NBI and a range of RF heating schemes. Energy transport assumptions are: $\chi_{i,e} = \chi_{i,e}^{neo} + \chi_{i,e}^{GLF}$. Particle transport assumptions are: $D_{DT} = D_{DT}^{neo} + D_{DT}^{GLF} + 0.1\ \chi_e^{ETG}$, where the last term accounts for particle transport driven by ETG turbulence.

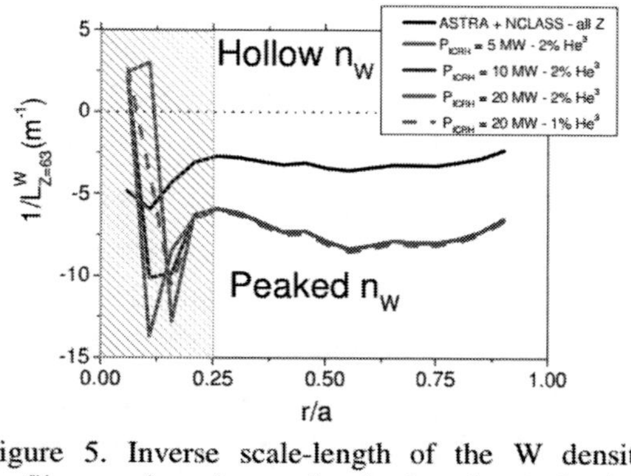


Figure 5. Inverse scale-length of the W density profiles evaluated on the basis of neoclassical transport with NCLASS (thermal plasma) and NEO including the effect of high energy $He^3$ minority ions at 2% level plus one case with 1% (not self-consistently evaluated). The dashed region corresponds to the plasma core where neoclassical transport alone is found to describe W transport.

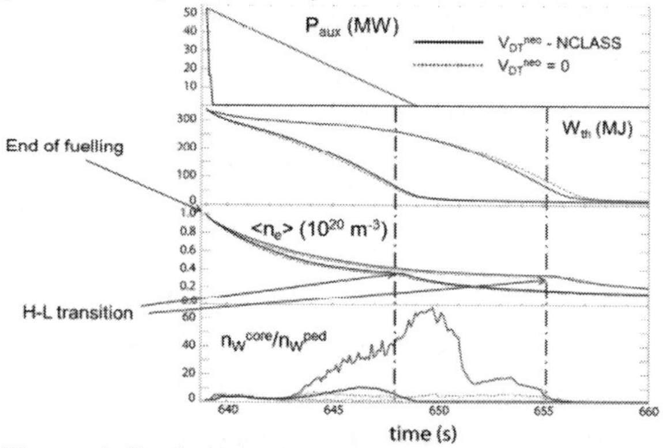


Figure 6. Evolution of auxiliary heating power and plasma parameters (energy, density and W peaking) in the termination phase of ITER Q=10 plasmas with a fast and a slow ramp of the auxiliary heating power. Full lines correspond to simulations which include a neoclassical pinch for the DT ions and dashed lines to those that do not include it.

# Study on sawtooth collapses using two-fluid equations

Q. Yu, S. Günter and K. Lackner
*Max-Planck-Institut für Plasmaphysik, 85748 Garching, Germany*

## 1. Introduction

As a typical phenomenon observed in tokamak discharges, sawtooth collapse has attracted much research interest [e.g. 1-7]. Using the large aspect-ratio tokamak approximation, sawtooth collapses are studied here numerically based on two fluid equations, including the electron continuity equation, generalized Ohm's law, the equation of motion in the parallel and the perpendicular direction (after taking $\mathbf{e}_t \cdot \nabla \times$), and the electron energy transport equation [8],

$$\frac{dn_e}{dt} = d_1 \nabla_{||} j - \nabla_{||}(n_e v_{||}) + \nabla \cdot (D_\perp \nabla n_e) + S_n \quad (1)$$

$$\frac{d\psi}{dt} = E_0 - \eta j - \frac{\eta}{\nu_{ei}} \frac{dj}{dt} + \eta \frac{\mu_e}{\nu_{ei}} \nabla_\perp^2 j + \Omega(\nabla_{||} n_e + \nabla_{||} T_e) \quad (2)$$

$$\frac{dv_{||}}{dt} = -C_s^2 \nabla_{||} p / n_e + \mu \nabla_\perp^2 v_{||} \quad (3)$$

$$\frac{dU}{dt} = S^2 \nabla_{||} j + \mu \nabla_\perp^2 U \quad (4)$$

$$\frac{3}{2} n_e \frac{dT_e}{dt} = d_1 T_e \nabla_{||} j - T_e n_e \nabla_{||} v_{||} + n_e \nabla \cdot (\chi_{||} \nabla_{||} T_e) + n_e \nabla \cdot (\chi_\perp \nabla_\perp T_e) + S_p \quad (5)$$

where the magnetic field $\mathbf{B}=B_{0t}(\mathbf{e}_t-\mathbf{e}_\theta k_t/k_\theta)+\nabla\psi\times\mathbf{e}_t$, $\psi$ is the helical flux function, $k_\theta=m/r$ and $k_t=n/R$, m and n are the poloidal and toroidal mode numbers, r and R the minor and major radius, and the subscript 0 denotes an equilibrium quantity. The ion velocity $v=v_{||}\mathbf{e}_{||}+\mathbf{v}_\perp$, $\mathbf{v}_\perp=\nabla\phi\times\mathbf{e}_t$, $\phi$ is the stream function, the subscripts $||$ and $\perp$ denote the parallel and perpendicular components, $d/dt=\partial/\partial t+\mathbf{v}_\perp\cdot\nabla$, $j$ is the plasma current density along the $\mathbf{e}_t$ direction, $U=-\nabla_\perp^2\phi$ the plasma vorticity, $\mu$ the ion viscosity, $\chi$ the heat conductivity, and D the particle diffusivity. $S_n$ and $S_p$ are the particle and heat source, $E_0$ is the equilibrium electric field, $S=\tau_R/\tau_A$, $p=p_e=n_eT_e$, $d_1=\omega_{ce}/\nu_{ei}$, $\Omega=\beta_e d_1$, $C_s=[T_e/m_i]^{1/2}/(a/\tau_R)$, $\beta_e=4\pi n_e T_e/B_{0t}^2$, $\omega_{ce}$ is the electron cyclotron frequency, $\mu_e$ the perpendicular electron viscosity, $\upsilon_{ei}$ the electron-ion collisional frequency, $\tau_A=a/V_A$, and $V_A$ is defined using $B_{0t}$.

## 2. Numerical results

Typical ASDEX Upgrade plasma parameters, $B_{0t}$=2T, a=0.5m, R=1.7m, $T_e$=2keV and $n_e=3\times10^{19}m^{-3}$ at q=1 surface, are utilized, leading to $S=2.6\times10^8$, $\Omega=9.4\times10^4$, $C_s=2.0\times10^7(a/\tau_R)$, $d_1=3.1\times10^7$, and $\nu_{ei}=2.2\times10^4$/s. Furthermore, $\mu_e/\upsilon_{ei}=10^{-4}a^2$, $\chi_\perp=\mu_\perp=0.2/m^2/s=19(a^2/\tau_R)$,

$D_\perp=\mu/5$, $\chi_\parallel/\chi_\perp=8.0\times10^8$, and a monotonic profile for the safety factor $q$ is assumed, with $q_0=0.91$ and $r_{q=1}= 0.3a$, where $q_0$ is the original safety factor at r=0, and $r_{q=1}$ the minor radius of the q=1 surface. The original equilibrium electron temperature profile has the form $T_e=T_{e0}$ $[1-(r/a)^2]^\kappa$, and the equilibrium electron density profile is taken to be a constant.

In figure 1 the time evolution of the helical flux during the sawtooth collapse at the x- and the o-point of the island as well as at the original magnetic axis are shown for $\kappa=2$. The helical flux at the x-point varies from that at the island's o-point to that at the axis, indicating a full magnetic reconnection. The reconnection process takes a time period less than $100\mu s$ ($1\,\tau_A$ $=8.7\times10^{-8}$ s) when neglecting the small flux change in the early phase, in agreement with ASDEX Upgrade experimental results [2].

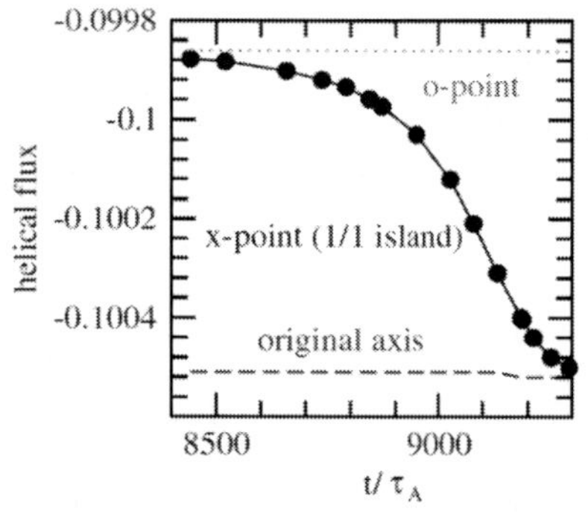


Figure 1 Time evolution of the helical flux at the x- and the o-point of the island as well as at the original magnetic axis for $\kappa=2$.

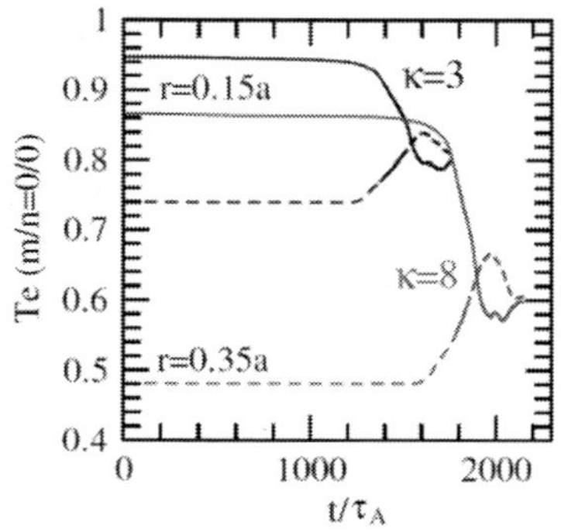


Figure 2 Time evolution of the electron temperature at the minor radius both inside and outside the $q=1$ surface for $\kappa=3$ and $8$.

The time evolution of the electron temperature during the sawtooth collapses at the minor radius both inside and outside the $q=1$ surface is shown in figure 2 for $\kappa=3$ and $8$. When the island width is still small, the island growth is faster for a smaller value of $\kappa$ (or electron diamagnetic drift frequency $\omega_{*e0}$) due to a weaker stabilizing effect of the diamagnetic drift. However, a fast decrease in the electron temperature is found for both cases once the island width is sufficiently large. This can be explained by the flattening of the electron temperature profile inside a large island.

Slow sawtooth collapses are found only for a sufficiently small $(1-q_0)$ and a large local $\omega_{*e0}$ at the $q=1$ surface. Figure 3 shows the time evolution of the electron temperature both inside and outside the $q=1$ surface for $q_0=0.95$ and $r_{q=1}=0.21a$. In this case, a much longer collapse time, ~ $10^4\,\tau_A \sim 10^{-3}s$ , is found for $\kappa=4$ and $6$. In the hybrid operational mode with a small $(1-q_0)$ and weak magnetic shear in the central region, slow sawtooth collapses or even continuous $1/1$ mode activity were observed in experiments [3,4].

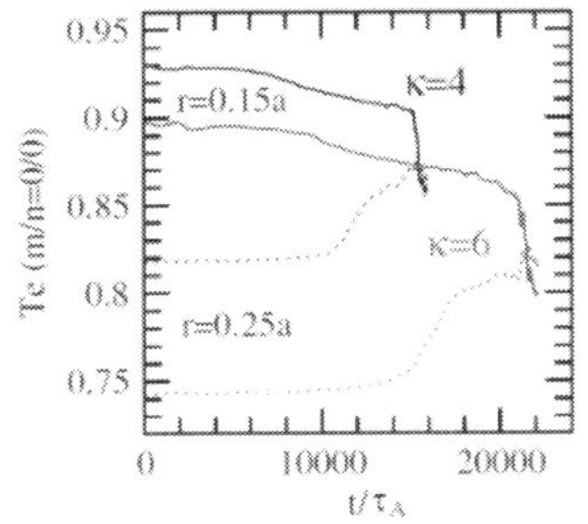


Figure 3 Time evolution of the electron temperature both inside and outside the $q=1$ surface for $q_0=0.95$, $r_{q=1}=0.21a$ and $\kappa=4$ and $6$.

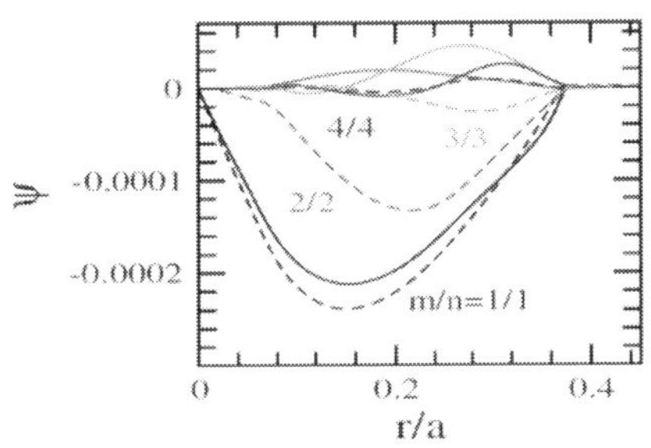


Figure 4 Radial profiles of the normalized (to $aB_{0t}$) helical flux functions $\psi_{1/1}$, $\psi_{2/2}$, $\psi_{3/3}$ and $\psi_{4/4}$ for the simulation of figure 1 at the time $t=8919\tau_A$

The amplitudes of high harmonic perturbations are usually found to be comparable to that of the 1/1 component during the collapse, as seen in experiments [2]. Fig. 4 shows the radial profiles of the normalized (to $aB_{0t}$) flux functions $\psi_{1/1}$, $\psi_{2/2}$, $\psi_{3/3}$ and $\psi_{4/4}$ for the simulation of figure 1 at the time $t=8919\tau_A$. The solid (dashed) curves represent the real (imaginary) part.

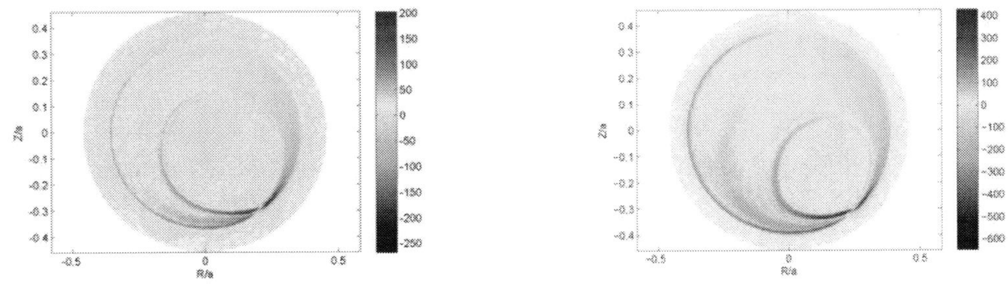

Figure 5 Contours of negative parallel electric field, $-E_\parallel=d\psi/dt$ (in the unit $aB_{0t}/\tau_R$), at $t=8851$ (left) and $8956\tau_A$ (right). R=0 corresponds to the major radius of the original magnetic axis, and Z is along the vertical direction. The parallel electric field is increased during the collapse by $1000$ times in the island's x-point region.

During the sawtooth collapse, strong parallel electric field is induced. Corresponding to Fig. 1, contours of the negative parallel electric field, $-E_\parallel=d\psi/dt$ (in the unit $aB_{0t}/\tau_R$), are shown at $t=8851$ and $8956\tau_A$ in figure 5. The original equilibrium electric field is $-0.55(aB_{0t}/\tau_R)$. The parallel electric field is increased during the sawtooth collapse compared to the equilibrium field by about $1000$ times in the island's x-point region. The maximum electron velocity produced by the reconnection is found to be $1.05*10^8 m/s$, corresponding to an electron energy of $31.6$ $keV$ and a collision time of $1.3$ $ms$. This indicates that supra-thermal electrons can be generated during the sawtooth collapse in existing tokamak plasmas as observed in experiments [5,6]. As the generated fast electron energy is proportional to $(aB_{0t})^2$, relativistic runaway electrons could readily be generated during sawtooth collapses in ITER.

In figure 6 the linear mode growth rate and mode frequency are shown as a function of the local $\omega_{*e0}$, obtained for $C_s=\mu_e=0$ and zero electron inertia. The mode is stable for a sufficiently large $\omega_{*e0}$ due to the stabilization by diamagnetic drift. The mode frequency does not linearly increase with $\omega_{*e0}$. The difference between these two frequencies leads to a plasma current density perturbation being out of phase with $\psi$, resulting in an $m/n=0/0$ component electromagnetic torque $r\cdot\boldsymbol{j}\times\boldsymbol{B}$, localized around the $q=1$ surface as shown in figure 7 in the linear phase for $\omega_{*e0}=5.0\times10^4\,\tau_R$. The driven plasma rotation is in the counter (co-) current direction inside (outside) the $q=1$ surface in the linear phase, propagates towards the magnetic axis during the nonlinear phase and reverses the rotation direction after the sawtooth collapse, in agreement with TCV experimental observations [7]. The amplitude of the driven rotation velocity reaches the level of the electron diamagnetic drift velocity.

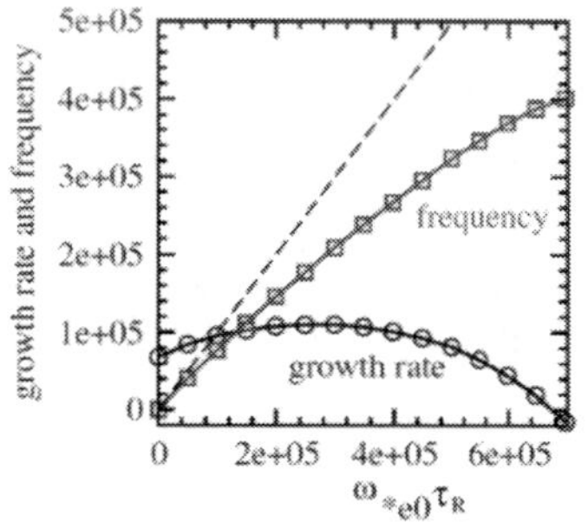


Figure 6 Linear mode growth rate and frequency versus the local $\omega_{*e0}\tau_R$ at the $q=1$ surface.

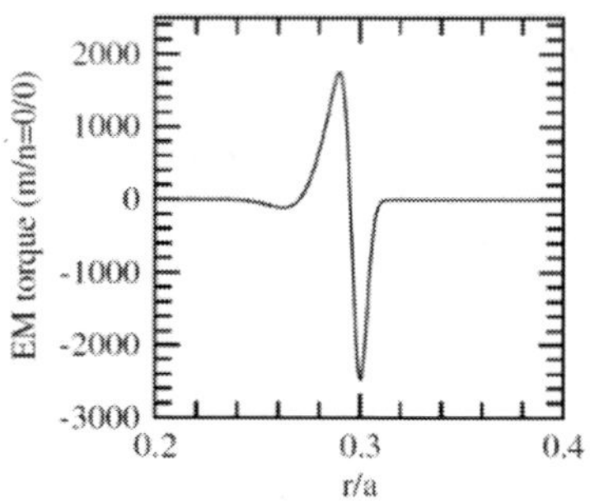


Figure 7 Radial profile of the m/n=0/0 component electromagnetic torque $r\cdot\boldsymbol{j}\times\boldsymbol{B}$ in linear phase.

## 3. Summary

(a) Based on two-fluid equations, fast sawtooth collapse, ~100μs or shorter, is obtained for typical ASDEX Upgrade parameters, in agreement with experimental observations. Slow collapses are also found for small values of $(1\text{-}q_0)$ and large local diamagnetic drift frequencies.
(b) During the sawtooth collapse, strong parallel electric field is formed in the x-point region of the magnetic island, which could generate super-thermal electrons in existing tokamaks and runaway electrons in ITER.

[1] S. Von Goeler *et al.*, Phys. Rev. Lett. **33** (1974) 1201.
[2] V. Igochine, J. Boom, I. Classen *et al.*, Plasmas **17**, 122506 (2010).
[3] P. Buratti, B. Alper, S.V. Annibaldi, *et al.*, Plasma Phys. Control. Fusion, **48**, 1005 (2006).
[4] J. Stober, A.C.C. Sips, C. Angioni, *et al.*, Nucl. Fusion **47**, 728 (2007).
[5] P. V. Savrukhin, Phys. Rev. Lett. 86 (2001) 3036.
[6] I. Klimanov, *et al.*, Plasma Phys. Control. Fusion, **49**, L1-L6 (2007).
[7] B. P. Duval, *et al.*, Phys.Plasmas **15**, 056113 (2008); 23rd IAEA FEC (2010), IAEA-CN-180.
[8] Q. Yu, Nucl. Fusion **50**, 025014 (2010).

# Advanced MHD models of anisotropy, flow and chaotic fields

M. J. Hole[1], M. Fitzgerald[1], G. R. Dennis[1], S. R Hudson[2], R. L. Dewar[1], G. T. von Nessi[1]

[1]*Res. School of Physics and Eng., Australian National University, ACT, Australia*

[2] *Princeton Plasma Physics Laboratory, PO Box 451, Princeton, NJ 08543, USA*

e-mail address of submitting author: matthew.hole@anu.edu.au

Large scale neutral beam heating, the growing importance of either deliberate or spontaneous equilibrium 3D structure, and the increased diversity, accuracy and resolution of plasma diagnostics have driven more advanced force balance models as well as new approaches to equilibrium reconstruction, such as Bayesian inference techniques [1]. In this work we focus on the development of two MHD force balance models, discuss their constraint to laboratory data, and scope their potential to describe astrophysical phenomena.

EFIT TENSOR is an implementation of the axis-symmetric guiding centre plasma which supports pressure anisotropy and toroidal flow.[2] The free functions are one dimensional poloidal flux functions and all non-linear contributions to the toroidal current density are treated iteratively. The parallel heat flow approximation chosen for the model is that parallel temperature is a flux function and that both parallel and perpendicular pressures may be described using parallel and perpendicular temperatures. In addition to benchmark tests, the code has been applied to MAST data in cases with significant anisotropy.

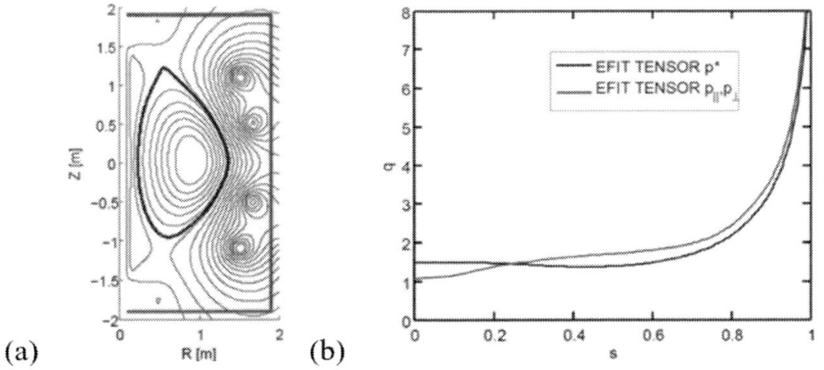


Figure 1: (a) EFIT TENSOR reconstrution for #29221 at 190ms, (b) EFIT TENSOR $q$ profile reconstructions for plasmas with isotropic (p*) and anisotropic ($p_{||}$, $p_{\perp}$) pressure profiles.

Figure 1(a) shows a MAST reconstruction for #29221 at 190ms. This discharge featured 1.6MW of NBI heating with a plasma current of 900kA, and featured a range of MHD activity. Calculations from TRANSP reveal the beam plus thermal population anisotropy is

$p_\perp / p_{||} \sim 1.7$. As shown in Fig. 1(b) reconstructions for this discharge that use the isotropic assumption $p^* = (p_{||} + p_\perp)/2$ have a significantly different $q$ profile to those with $p_\perp / p_{||} \sim 1.7$.

We have also developed a single adiabatic compressional stability model in which $p_{||}$ and $p_\perp$ do joint work. This model, together with the double adiabatic (CGL) model for stability has been implemented in the continuous spectrum code CSMIS and the global stability code MISHKA-A. Figure 2(a) and 2(b) show the difference in the n=1, $\gamma$ =0 continuum for the isotropic and anisotropic reconstruction, while Figs. 2(c) and (d) show the difference in global mode structure for the two modes. The TAE gap for the anisotropic continuum is much more open than the isotropic case, and the mode peak more inboard than the isotropic case. The implications of this differnt radial structure on wave-particle transport are being currently explored by coupling the EFIT TENSOR and MISHKA-A codes to the wave-particle interaction code HAGIS.

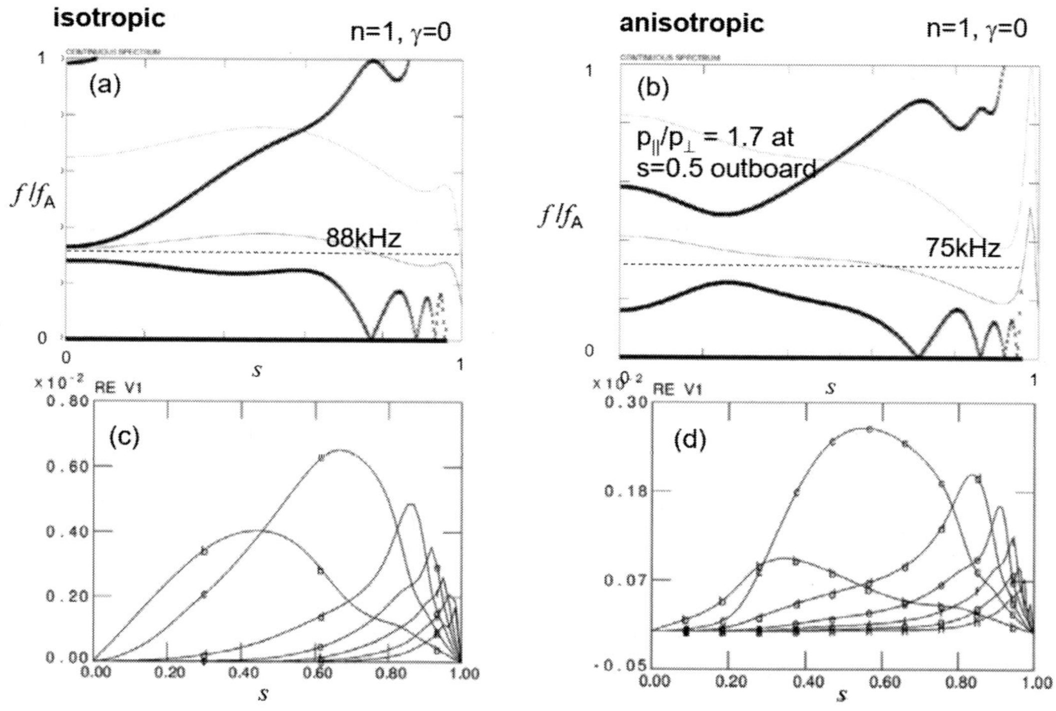


Figure 2: (a) and (b) continuous spectrum for isotropic and ansitropic plasmas. Panels (c) and (d) show global TAE modes for (a) and (b) respectively.

We also report on the development and numerical implementation of multi-region relaxed MHD variational principle MRXMHD, which comprises different Taylor relaxed regions separated by singular currents at ideal MHD interfaces. This model which supports stepped pressure profiles, enables the resolution of chaotic fields and magnetic islands which occur in toroidal asymmetric plasmas.

In MRxMHD invariant tori act as ideal MHD barriers to relaxation, so that Taylor constraints are localised to subregions. For such a system the energy functional can be written

$$W = \sum_{l=1}^{N} (U_l - \mu_l H_l / 2)$$

for a Lagrange multiplier $\mu_l$ and with the internal energy $U_l$ and helicity $H_l$ of each region $R_i$ given by

$$U_l = \int_{R_l} \left( \frac{B_l^2}{2\mu_0} + \frac{P_l}{\gamma - 1} \right) d\tau^3 \qquad H_l = \int_{R_l} (\mathbf{A}_l \cdot \mathbf{B}_l) d\tau^3$$

Solutions of this system of equations provide a Beltrami field, $\nabla\times\boldsymbol{B}_l = \mu_l \boldsymbol{B}_l$ in each plasma region, with a pressure jump condition $[[P_l + B_l^2/2\mu_0]]=0$ across each interface. We have verified that MRxMHD reproduces MHD in the limit of continuously nested flux surfaces.[3] Extended MRxMHD to include non-zero plasma flow [4]and plasma anisotropy[5], Generalized straight field line coordinates concept to fully 3D plasmas[6], Related helical bifurcation of a Taylor relaxed state to a tearing mode[7,8] and developed techniques to establish pressure jumps a surface can support.[9]

In this work we use the MRxMHD code SPEC[10] to describe the emergence of the quasi-single helicity state in RFX-mod.[11] We have used a two barrier MRxMHD model as a minimal model to describe the formation of double and single helical axis states in the reverse field pinch (RFP). We have compared the plasma energy of the MRxMHD (fully 3D and axisymmetric solutions), as well as ideal MHD (fully 3D and axisymmetric) solutions as a function of barrier position. As the position of the barrier is moved inboard from the plasma edge the fully 3D MRxMHD state diverges from the axisymmetric solution, and has lower energy, showing that the RFP bifurcated state has lower energy (preferred) than the comparable axisymmetric state. This offers an energy principle reason for the formation of single helical axis and double helical axis states. Figure 3 shows experimental[12] and MRxMHD computed Poincare plots for the double helical axis (DAX) and single helical axis (SHAX) states. This MRxMHD computed Poincare plot shows the formation of separate magnetic axes as well as magnetic islands and stochastic regions.

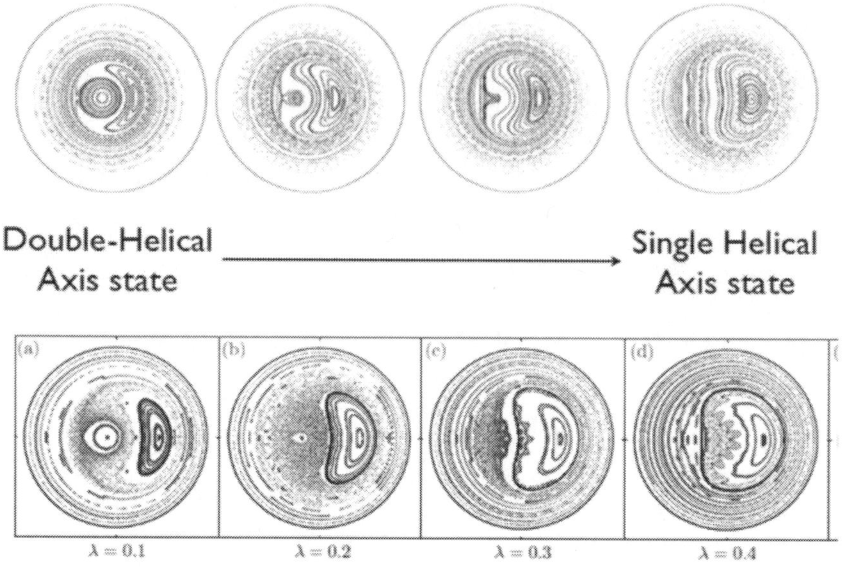


Figure 3: The upper row shows the experimental Poincaré plot with increasing plasma current, while the lower row shows MRxMHD computed Poincaré plots.

In ongoing work we plan to develop SPEC to handle free boundary equilibria, including the vacuum region and external conductors, as well as include toroidal flow.

**Acknowledgements:**

This work was supported by the Australian Research Council through grant FT0991899, DP110102881, and by the RCUK Energy Programme under Grant No EP/I501045 and the European Communities under the contract of Association between EURATOM and CCFE. The views and opinions expressed herein do not necessarily reflect those of the European Commission.

---

[1] G T von Nessi, M J Hole, J Svensson, L Appel, Phys. Plas. 19, 013106 (2012).
[2] M. Fitzgerald, L. C. Appel, M. Hole, Nucl. Fusion **53** (2013) 113040
[3] G. Dennis *et al*, Phys. Plas. ,032509, 2013
[4] G.R. Dennis, S.R. Hudson, R.L. Dewar, M.J. Hole, Phys. Plas. **21**, 042501 (2014)
[5] G.R. Dennis, S.R. Hudson, R.L. Dewar, M.J. Hole, Phys. Plas. **21**, 072512 (2014)
[6] R. L. Dewar, S. R. Hudson, A. Gibson, *Plasma Phys. Control. Fusion,* **55,** 014004, 2013
[7] Z. Yoshida and R. L. Dewar , *J. Phys. A: Math. Theor.* **45**, 365502, 2012
[8] M. J. Hole, R. Mills, S. R. Hudson and R. L. Dewar Nucl. Fusion 49 (2009) 065019
[9] M. McGann, ANU PhD thesis, 2013
[10] S. R. Hudson, R. L. Dewar, G. Dennis, M. J. Hole, M. McGann, G. von Nessi and S. Lazerson, Phys. Plasmas 19, 112502, 2012
[11] . R. Dennis, S. R. Hudson, D. Terranova, P. Franz, R. L. Dewar, and M. J. Hole, Physical Review Letters, 111, 055003 (2013)
[12] Fig. 6 of P. Martin et al., Nuclear Fusion 49, 104019 (2009)

# Cherenkov diagnostic observations of fast electron losses in FTU and interpretation with gyrokinetic simulations

F. Causa[1], S. Briguglio[1], P. Buratti[1], B. Esposito[1], G. Fogaccia[1], V. Fusco[1],
E. Giovannozzi[1], G. Pucella[1], G. Vlad[1] and the FTU Team*
L. Jakubowski[2], K. Malinowski[2], M. Rabinski[2], M. J. Sadowski[2], J. Zebrowski[2]
[1] *Unità Tecnica Fusione - ENEA C. R. Frascati - via E. Fermi 45, 00044 Frascati (Roma), Italy*
* *See the appendix of G. Pucella et al., Proc. 25th IAEA, St Petersburg, Russia, 2014*
[2] *National Centre for Nuclear Research, 7 Andrzeja Soltana Str., 05-400 Otwock, Poland*

## Introduction

Runaway electrons (REs) in tokamaks can reach energies in the range of tens of mega-electron-volts, [1], especially during plasma disruptions, representing a major concern for the safety of large tokamaks, [2]. Uncontrolled rapid loss of runaway current is detrimental, severely damaging plasma-facing components. Understanding how spontaneous or applied magnetic perturbations affect RE transport is crucial to evaluate the possibility of controlling and mitigating the RE current, [3]. Recently, an optical diagnostic system based on the Cherenkov effect, [4], was sucessfully installed and tested on the FTU tokamak, [5] to study the dynamics of non-thermal electrons. Emphasys is given here to results of the diagnostics in scenarios involving magnetic islands because they are more readily controllable than disruptions and, therefore, can be useful to identify key mechanisms of REs interacting with magnetic perturbations. The detailed analysis of the experimental scenarios is completemented by simulations obtained using the HMGC code, [6], in which a given RE distribution is evolved in an electormagnetic field representing the island perturbation to a FTU-like equilibrium.

## Relevant diagnostics

The Cherenkov probe, [4], [5], consists of a diamond detector mounted on a TZM (99% Mo, 0.5% Ti, 0.1% Zr) head inserted in the FTU vessel at a variable radial position in the limiter shadow, on the equatorial plane of FTU, at $\phi = 150$ deg, Figure 1. Data acquisition rate was 2.5 MHz over the duration of the discharge (up to 2 s), thus resulting in a high temporal resolution comparable to or better than that of other relevant existing FTU diagnostics. The probe is not sensitive to visible radiation, as verified experimentally by the lack of signal from the Cherenkov probe in the presence of a plasma with strong visible emission but no REs.

In FTU, a liquid organic proton recoil scintillator (NE213) and a proportional counter (BF3) are used to monitor, respectively, gamma and neutron emissions, [1]. The NE213 scintillator is located outside the cryostat on the equatorial plane, at $\phi = 210$ deg, and is sensitive to both neutron and gamma rays. Six BF3 proportional counters are used, positioned above the tokamak, insensitive to gamma rays. NE213 and BF3 have been cross-calibrated such that their outputs are identical in the absence of gamma rays. The gamma ray contribution due to the presence of REs is obtained by subtracting the neutron baseline (BF3) from the neutron and gamma rays integrated signal (NE213). The plasma magnetic activity is monitored by means of a set of poloidal field pick-up (Mirnov) coils. In particular, the poloidal ($m$) and toroidal ($n$) periodicity numbers of helical magnetic islands chains generated by tearing instabilities are determined by cross-phase measurements. Further information on the internal island structure is given by temperature profile oscillations, as measured by a twelve-channel Electron Cyclotron Emission (ECE) diagnostics. The ECE sightline is on the equatorial plane, at 90 deg toroidal distance from the poloidal limiter and 60 deg from the Cherenkov probe, Figure 1.

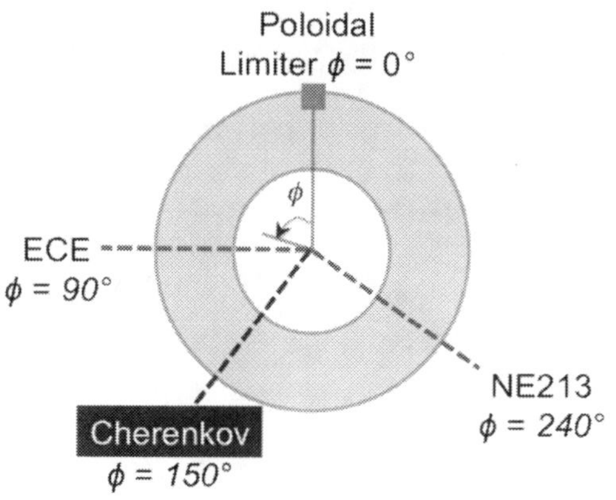


Figure 1: *Top view, simplified schematic of the FTU equatorial section with positions of relevant diagnostics.*

**Results**

The interest here is in scenarios with tearing modes developing magnetic islands where enhanced RE losses can be observed, in correlation with the island O-point transit in front of the probe, to study the dynamics of REs in the presence of magnetic islands. In the presence of MHD activity consisting in the growth and locking of a magnetic island with poloidal and toroidal periodicity numbers $(m, n) = (2, 1)$, a clear phase-relation was identified between Cherenkov, ECE and NE213 signals, showing that the modulation of the Cherenkov signal is due to the rotation of the magnetic island, Figure 2. The phase lags correspond to the toroidal separations between diagnostics. The scenario was simulated by evolving a model distribution of REs in a given electromagnetic field representing the island perturbation to a FTU-like equilibrium, using a particle-in-cell module of the HMGC code, [6]. The particle distribution satisfies the non-linear Vlasov equation in the low-$\beta$ drift-kinetic limit: perpendicular components of the vector potential and finite Larmor radius effect are neglected, while magnetic

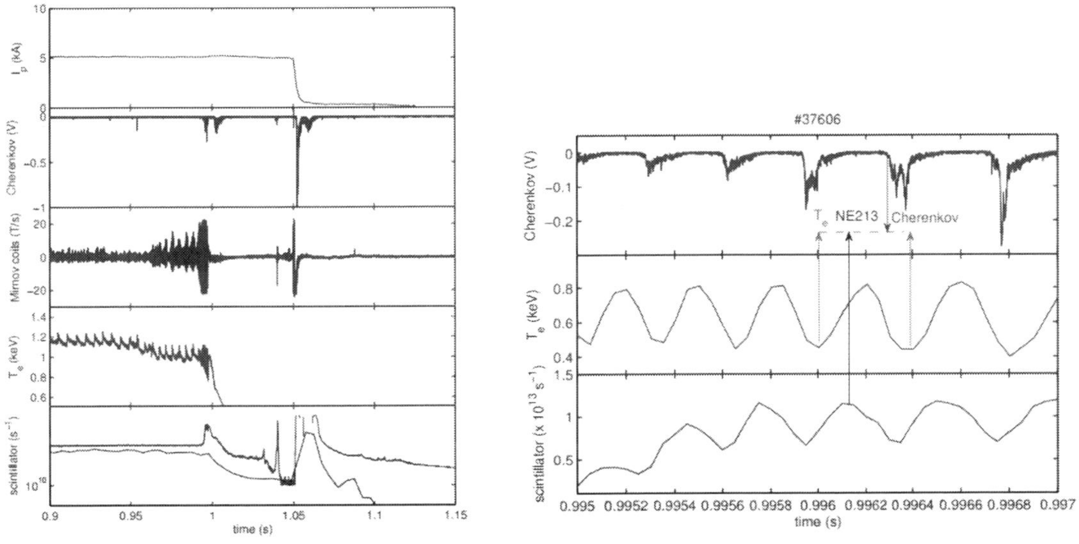


Figure 2: *Time traces of (Left) relevant quantities, note mode locking at t = 1 s and disruption at 1.05 s and (Right) zoom-in of Cherenkov, ECE (channel 10) and gamma-ray signals during island rotation in FTU pulse 37606 (B = 5.3 T, I = 520 kA).*

drift orbit widths are fully retained. The initial RE distribution function was assumed to be anisotropic Maxwellian with different parallel and perpendicular widths corresponding to temperatures $T_{\|} = 1Mev$ and $T_{\perp} = 50keV$. The RE density profile was assumed to be vanishing outside magnetic surface $\psi_{out}/\psi_{edge} = 0.76$, here assumed to be the limiter position, near the plasma edge. The effect of a static (zero frequency and fixed amplitude) perturbation to the poloidal flux, with poloidal and toroidal periodicity numbers $(m, n) = (2, 1)$, on the RE distribution was calculated to investigate whether a detectable particle flux is produced in the outer region, where the initial RE density vanishes. The simulations show that in the outer region, the particle density perturbation exhibits the same poloidal and toroidal periodicity of the magnetic island, Figure 3 (Left), consistently with the experimentally observed correlation of enhanced RE losses and the passage of the island O-point in front of the Cherenkov probe. Further, the outward flux is almost exclusively due to trapped particles. In addition, the simulation results yield a quadratic scaling of the number of particles displaced in the outer region with respect to the amplitude of the perturbed poloidal magnetic field $\delta B_{\theta}/B_{\theta}$ in the outer region ($\psi/\psi_{edge} = 0.8$), Figure 3 (Right). The non-linear scaling obtained with the HMGC simulations is consistent with the experimental results, although the latter show that the scaling is nearly quadratic for small perturbations, becoming higher-order in the presence of larger perturbations.

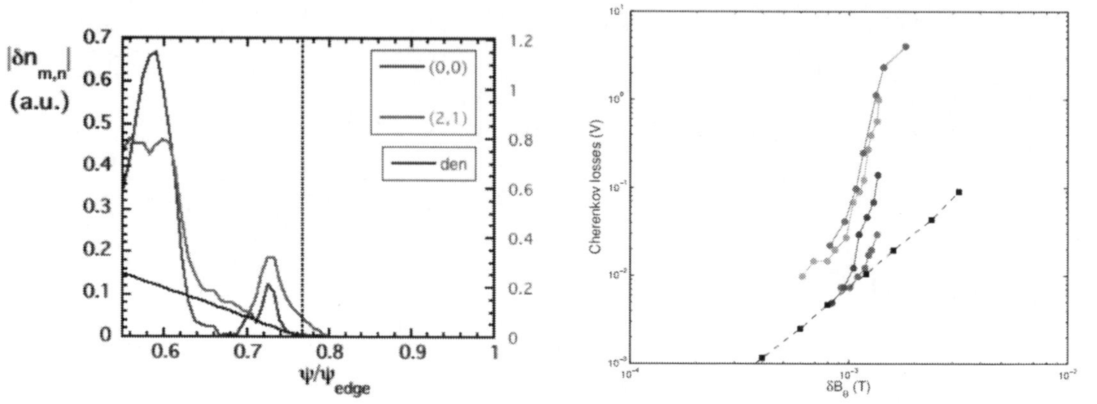


Figure 3: *(Left) HMGC simulated results showing the modulus of Fourier harmonics* $\delta n_{(2,1)}$ *(red) and* $\delta n_{(0,0)}$ *(blue) vs* $\psi/\psi_{edge}$*. The density perturbation outside the* $\psi_{out}$ *surface (indicated by the dashed line) is dominated by the (2,1) component. [Initial density profile for reference (black)]. (Right) Simulated (black, square markers) and experimental (coloured curves) losses vs magnetic perturbation amplitude. (Experimental curves: series of four discharges at B = 5.3 T and I = 520 kA).*

## Conclusions

Scenarios involving RE losses driven by magnetic islands were analysed in detail to gain a deeper insight into the mechanisms governing RE transport from plasma core to plasma edge. The experimental results show that the Cherenkov signal modulation is due to the island rotation and growth, and that the phase lags are due to the toroidal separations between diagnostics. The HMGC-PIC simulations provide a useful insight into RE dynamics in the presence of magnetic islands, showing that the outward flux is due to displaced trapped particles. Importantly, although simplified, the HMGC-PIC captures the non-linear scaling of the outward flux of particles with respect to the amplitude of the perturbed magnetic field representing the island.

## References

[1] B. Esposito et al., Dynamics of high energy runaway electrons in the Frascati Tokamak Upgrade, Phys. Plasmas 10 (2003).

[2] V.V. Plyusnin, Study of runaway electron generation during major disruptions in JET, Nucl. Fusion 46 (2006).

[3] B. Esposito et al., Runaway electron experiments in FTU, This conference.

[4] L. Jakubowski et al., Cherenkov-type diamond detectors for measurements of fast electrons in the TORE-SUPRA tokamak, Rev. Sci. Inst. 81 (2010).

[5] F. Causa et al., Cherenkov emission provides detailed picture of non-thermal electron dynamics in the presence of magnetic islands, Poster EX/P2-49, 25th IAEA Fusion Energy Conference, St Petersburg (Russia), 13-18th October 2014.

[6] S. Briguglio et al., Phys. Plasmas, 5, 3287 (1998).

# Modeling of disruption mitigation by massive gas injection in JET with JOREK and IMAGINE

A. Fil[1], E. Nardon[1], M. Hoelzl[2], G.T.A Huijsmans[3], F. Orain[2], M. Becoulet[1], P. Beyer[4], G. Dif-Pradalier[1], R. Guirlet[1], H.R. Koslowski[5], M. Lehnen[3], J. Morales[7], S. Pamela[6], C. Reux[1], F. Saint-Laurent[1], C. Sommariva[1] and JET contributors[6,*]

[1] *CEA, IRFM, F-13108 Saint-Paul-Lez-Durance, France*

[2] *Max Planck Institute for Plasma Physics, Boltzmannstr. 2, 85748 Garching b. M., Germany*

[3] *ITER Organization, Route de Vinon sur Verdon, 13115 Saint Paul Lez Durance, France*

[4] *Aix-Marseille University, CNRS, PIIM UMR 7345, 13397 Marseille Cedex 20, France*

[5] *Forschungszentrum Juelich GmbH, Institute of Energy and Climate Research Plasma Physics (IEK-4), Trilateral Euregio Cluster, 52425 Juelich, Germany*

[6] *EUROfusion Consortium, JET, Culham Science Centre, Abingdon, OX14 3DB, UK*

[7] *Ecole Polytechnique Federale de Lausanne, Centre de Recherches en Physique des Plasmas, 1015 Lausanne, Switzerland*

[*] *See the Appendix of F. Romanelli et al., Proceedings of the 25th IAEA Fusion Energy Conference 2014, Saint Petersburg, Russia*

## Introduction

A Disruption Mitigation System (DMS) is mandatory in ITER in order to reduce electromagnetic forces, mitigate heat loads and avoid Runaway Electrons (RE) [1]. These combined objectives make the design of the DMS a complex and challenging task, for which substantial input from both experiments and modeling is needed. We present here modeling results on disruption mitigation by Massive Gas Injection (MGI), which is one of the main methods considered for the DMS of ITER.

This article is divided as follows: the first part is devoted to the study of the gas penetration into the plasma with the first-principle based 1D code IMAGINE. The second part presents simulations of MGI-triggered disruptions in JET with the 3D non-linear MHD code JOREK.

## First principle modeling of neutral gas penetration during massive gas injection

IMAGINE is a 1D code in the radial direction which includes a complete model of atomic physics with ADAS coefficients. Neutral transport is convective, in agreement with first principles. The equations for a deuterium MGI derive from the mass, momentum and energy conservation, with sources taking atomic processes into account:

$$\partial_t n_e = n_e n_0 I - n_e^2 R + \partial_x (D \partial_x n_e) \quad (1)$$

$$\frac{3}{2}n_e\partial_t T_e = -n_e n_0 I(E_{ion} + \frac{3}{2}T_e) - n_e n_0 L + \partial_x(\chi n_e \partial_x T_e) - \frac{3}{2}T_e\partial_x(D\partial_x n_e) \quad (2)$$

$$\frac{3}{2}n_e\partial_t T_i = -\frac{3}{2}n_e(I + <\sigma v>_{cx})(n_0 T_i - P_0/e) + \partial_x(\chi n_e \partial_x T_i) - \frac{3}{2}T_i\partial_x(D\partial_x n_e) \quad (3)$$

$$\partial_t n_0 = -\partial_x(n_0 V_0) - n_e n_0 I + n_e^2 R \quad (4)$$

$$m_0 n_0 \partial_t V_0 = -m_0 n_0 V_0 \partial_x V_0 - \partial_x P_0 - m_0(n_0 n_e <\sigma v>_{cx} + n_e^2 R)V_0 \quad (5)$$

$$\partial_t P_0 = -V_0\partial_x P_0 - \gamma P_0 \partial_x V_0 + n_e <\sigma v>_{cx}(e n_0 T_i - P_0) - n_e I P_0 + n_e^2 R e T_i \quad (6)$$

$n_e$ and $T_e$ are the electron density and temperature (in eV) of the plasma, $T_i$ is the ion temperature (in eV) of the plasma, $n_0$ is the neutral density, $P_0$ and $V_0$ are the pressure and radial velocity of the neutrals. We simulate the JET shot 86887, starting from experimental $(n_e, T_e)$ plasma profiles and a realistic initial number of particles in the gas reservoir. In the simulation domain, the vacuum injection tube which links the gas reservoir and the plasma edge is also included. First, a rarefaction wave is propagating in the injection tube and we obtain the same gas flow as in laboratory experiments [2]. After 0.9 ms, the neutral gas reaches the plasma edge and starts penetrating into the plasma. Figure 1 presents the evolution of the electron temperature profile without (left) and with (right) the inclusion of charge-exchange. The slower gas penetration in the latter case is attributed to the very fast charge-exchange heating of the neutrals, which creates a shock wave slowing down the gas (see Figure 2 which presents the evolution of the neutral density profile). It is thus found that the energy transfer by charge-exchange plays a major role in the gas penetration into the plasma and is a key ingredient to recover a realistic pre-TQ time as well as a realistic increase of the plasma density.

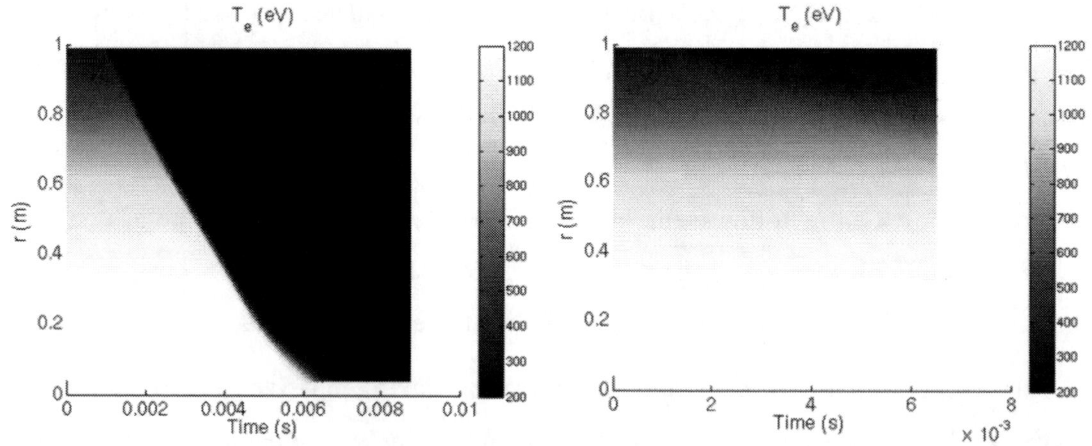


Figure 1: *Electron temperature evolution, left/right: without/with energy transfer by charge-exchange*

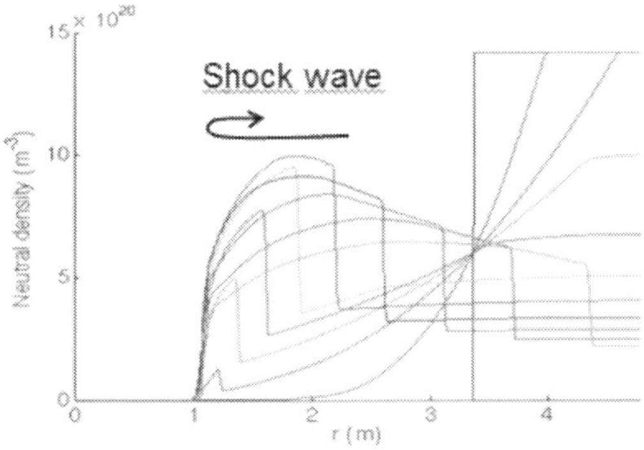


Figure 2: *Neutral density at different time steps for the simulation including energy transfer by charge-exchange*

**3D non-linear MHD modeling of MGI-triggered disruptions**

We now present simulations with the 3D non-linear MHD code JOREK. The code calculates the evolution of the neutral and plasma densities, plasma temperature, plasma velocity and electromagnetic field, giving access to key quantities which are not directly measurable, e.g. the toroidal electric field responsible for the creation of REs. The aim at present is to validate the code, for which purpose we simulate a relatively simple case: MGI of pure $D_2$ during the same recent JET shot (86887) as simulated with IMAGINE. The appropriate atomic physics for a $D_2$ MGI (see equations below) is included but the modeling of the gas injection and the transport of neutrals is simplified compared to IMAGINE. The MGI is treated as a volumetric source term $S_n$ which is calibrated so as to match the experimental density measurements using synthetic interferometers.

$$\frac{\partial\rho}{\partial t} = -\boldsymbol{\nabla}\cdot(\rho\boldsymbol{v}) + \boldsymbol{\nabla}\cdot(D_\perp\boldsymbol{\nabla}_\perp\rho + D_\parallel\boldsymbol{\nabla}_\parallel\rho) + \rho\rho_n S_{ion}(T) - \rho^2\alpha_{rec}(T) \quad (7)$$

$$\begin{aligned}\frac{\partial(\rho T)}{\partial t} = &-\boldsymbol{v}\cdot\boldsymbol{\nabla}(\rho T) - \gamma\rho T\boldsymbol{\nabla}\cdot\boldsymbol{v} + \boldsymbol{\nabla}\cdot(\kappa_\perp\boldsymbol{\nabla}_\perp T + \kappa_\parallel\boldsymbol{\nabla}_\parallel T) + \frac{2}{3R^2}\eta_{\text{Spitzer}}(T)j^2 \\ &-\xi_{ion}\rho\rho_n S_{ion}(T) - \rho\rho_n L_{lines}(T) - \rho^2 L_{brem}(T)\end{aligned} \quad (8)$$

$$\frac{\partial\rho_n}{\partial t} = \boldsymbol{\nabla}\cdot(\boldsymbol{D}_n : \boldsymbol{\nabla}\rho_n) - \rho\rho_n S_{ion}(T) + \rho^2\alpha_{rec}(T) + S_n \quad (9)$$

Other synthetic diagnostics have been implemented in JOREK and the results of the simulations are compared to the experiment in [4]. The MHD activity is dominated by internal kink modes and tearing modes ($m/n = 2/1$ and $3/2$) and is qualitatively similar to the experiment. Compared to the results presented in [4], a much higher MHD activity is now observed due to the reduction of the hyperresistivity used in the simulations. In Figure 3, magnetic fluctuations are plotted and an important burst of MHD is observed. At the peak of MHD activity, small scale structures of toroidal current density are observed (middle) and the Poincare plot (right) shows that the magnetic field is stochastic across the whole plasma (weakly stochastic in the core). The evolution of the electron temperature at the center is also plotted and shows that the temperature is still high (few hundreds of eV) after the MHD burst, even including a background of impurities, indicating that the simulations do not reproduce a complete thermal quench. This may come from a too low parallel heat conductivity in these simulations. This hypothesis is currently being tested.

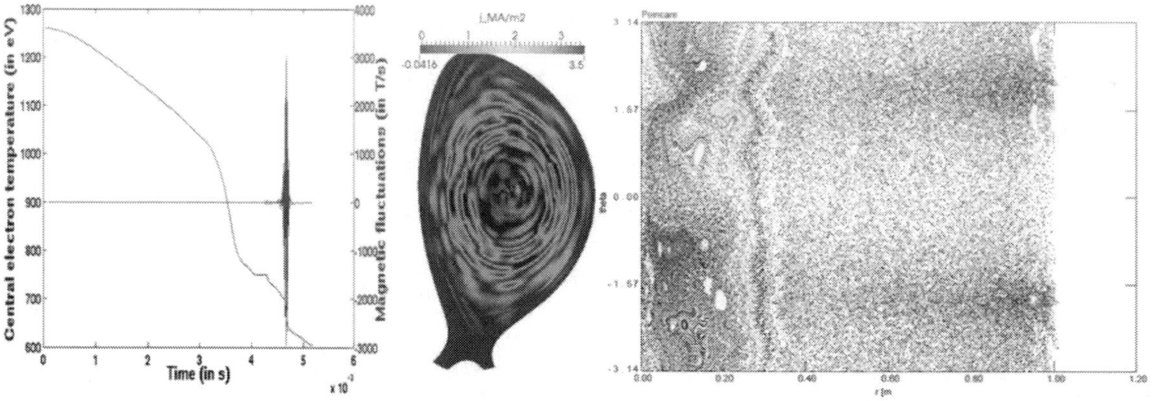


Figure 3: *Central electron temperature and magnetic fluctuations evolution (left). Poloidal cross-section of the toroidal current density (middle) and Poincare plot (right) at the peak of MHD activity.*

**Acknowledgement**

We thank G. Pautasso and E. Fable for useful suggestions. This work has been carried out within the framework of the EUROfusion Consortium and has received funding from the Euratom research and training programme 2014-2018 under grant agreement No 633053. A part of this work was carried out using the HELIOS supercomputer system (IFERC-CSC), Aomori, Japan, under the Broader Approach collaboration, implemented by Fusion for Energy and JAEA, and using the CURIE supercomputer, operated into the TGCC by CEA, France, in the framework of GENCI and PRACE projects. The views and opinions expressed herein do not necessarily reflect those of the European Commission or the ITER Organization.

**References**

[1] M. Lehnen et al., J. Nucl. Mater., (2014), `http://dx.doi.org/10.1016/j.jnucmat.2014.10.075`

[2] S.A. Bozhenkov, M. Lehnen, K.H. Finken, G. Bertschinger, H.R. Koslowski, D. Reiter, R.C. Wolf and TEXTOR Team, Nucl. Fusion **51**, 083033 (2011).

[3] G.T.A. Huysmans and O. Czarny, Nucl. Fusion **47**, 659 (2007).

[4] A. Fil, E. Nardon et al., Physics of Plasmas, Accepted and published in July, (2015).

# Runaway electron mitigation by resonant and non-resonant magnetic perturbations in RFX-mod tokamak discharges

M.Gobbin[1], M.Valisa[1], R.B.White[2], L.Marrelli[1], P.Piovesan[1], M.E.Puiatti[1]

[1]*Consorzio RFX, Corso Stati Uniti 4, 35127 Padova, Italy*
[2]*Princeton Plasma Physics Laboratory, Princeton, New Jersey, USA*

***Introduction***. Understanding the mechanisms behind Runaway Electron (RE) generation and the ways RE formation can be prevented is of paramount importance for future fusion reactors. In fact, RE with energies of several MeV generated during the sudden cooling involved in disruptions may cause severe damage to the plasma facing components and vacuum vessel wall. The injection of massive gas into the plasma is the main solution presently envisaged to avoid the RE avalanche. Also other strategies have been developed to enhance the RE de-correlation as the application of external magnetic perturbations (MP) [1]. The RFX-mod device [2] (major radius $R_0$=2m, minor radius a =0.459m, maximum toroidal field on axis $B_0$=0.6T) which can be run as a Tokamak, is equipped with an advanced system for the control of Magnetohydrodynamic modes based on 192 saddle coils independently fed and on a state of the art real-time hardware and software architecture [3]; for this reason, it is well suited to study the possible RE de-confinement in response of applied magnetic perturbations with different modal numbers and amplitude.

In this paper the conditions when a significant number of high energy electrons is generated during the flattop phase of tokamak RFX-mod plasmas are analyzed, both in terms of density and of loop voltage ($V_{loop}$), thus contributing to the scaling studies reported in [4]. Magnetic perturbations have been applied in order to enhance fast electron losses preventing their acceleration to higher energies; results are discussed and interpreted by numerical simulations with the Hamiltonian guiding center code ORBIT [5], recently updated to a relativistic version.

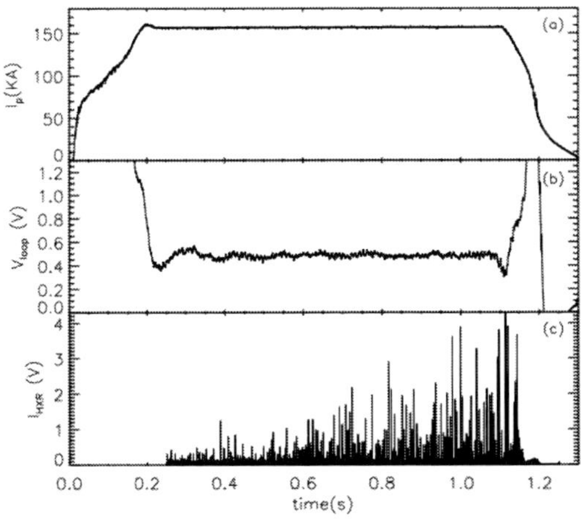


**Figure 1.** *Example of Tokamak discharge (#37936) in RFX-mod: (a) plasma current, (b) loop voltage and (c) signal from the scintillators detecting RE events.*

***RE generation/suppression in RFX-mod***. The high energy electrons dynamics in RFX-mod is investigated thanks to the soft-x-ray tomography and to 2 scintillators, which allow the detection of hard-x-rays produced when the electrons are lost from the plasma and impact on the first wall. The scintillators are placed at two different toroidal positions ($\phi_1$=187.5°, $\phi_2$=262.5°) on the equatorial plane. Fig.1-(a) shows the plasma current evolution in a typical RFX-mod tokamak discharge at low density ($n_e<2\cdot10^{18}m^{-3}$, corresponding to $n_e/n_G<0.1$) with safety factor at the edge q(a) < 2; in (b) the corresponding $V_{loop}$ is reported. The scintillator signal $I_{HXR}$ is shown in panel (c) as a function of time: each event (spike) corresponds to hard-x-ray (HXR) emission from a lost electron hitting the first wall and the amplitude of the signal is proportional to the energy carried by the electron itself. As it is clear from panel (c), on average, the energy of the lost electrons increases with time as they are continuously accelerated by the toroidal electric field before drifting out of the plasma. A statistical study has been carried out over many discharges (~150, similar to the one reported in Fig. 1 but with different electron density), dividing the time scale into intervals of 5ms; for each of them the total number of RE events with an amplitude greater than the background noise (≈ 0.2 V) has been determined together with the averaged values of other quantities like density and $V_{loop}$. The results of this analysis are summarized in the contour plot shown in Fig.2 reporting the percentage fraction of RE events at a given $n_e$ (x-axis) and accelerating toroidal electric field (i.e $E=V_{loop}/(2\pi R_0)$, y-axis). Most of the events occur at density lower than $n_e<4\cdot10^{18}m^{-3}$ with electric field in the range 0.03-0.1V/m. Nevertheless, few events can be observed at higher density (~$10^{19}$ $m^{-3}$) too at greater amplitudes of the electric field.

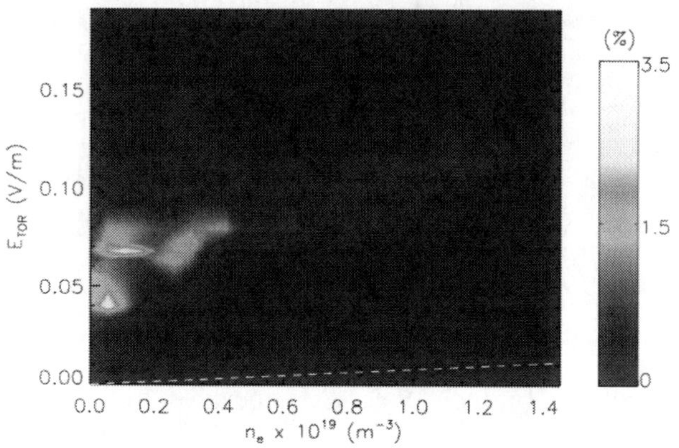


**Figure 2.** *Percentage fraction of RE events as a function of the electron density and toroidal electric field in RFX-mod Tokamak discharges.*

As reported in [4] for several devices, also in RFX-mod the critical toroidal electric field, required to observe RE events at a given density, exceeds the theoretical one predicted taking into account only of the friction effect due to collisionality (dotted white line in the plot). This confirms, also for RFX-mod, that other yet unknown loss mechanisms might play a role in the RE generation/suppression mechanisms.

***Effect of magnetic perturbations on RE losses.*** The presence of small radial magnetic fields can prevent the electrons to reach high energy before being lost. In fact, when magnetic perturbations are applied, depending on their amplitude and on the equilibrium configuration, partially chaotic regions may develop. Due to fast radial diffusion within these domains, electrons are lost in a shorter time and with lower energy with respect to an unperturbed configuration. Hence, the energy of the lost electrons is expected to be a decreasing function of the applied perturbation amplitude.

To investigate this issue, MPs have been applied during the flattop phase of low density Tokamak discharges similar to those described in the previous section. An example of MP evolution is shown in Fig. 3-(a) where a 10Hz rotating (2,1) mode is applied to a plasma with q(a) < 2. The amplitude of the MP increases linearly to a maximum value and then decreases to zero. The corresponding signal from the scintillator is reported in panel (b). It is worth to note that the maximum of RE energy is observed to occur about 0.15-0.2s after the beginning of the MP. Similarly there is a delay of 0.1-0.15s between the maximum of the MP and the minimum of the scintillator signal. Such a behavior can be explained by considering that during the MP application, the toroidal electric field is still accelerating the electrons and a finite time is required for electrons to diffuse till the edge of the plasma. This phenomenology can be roughly described by assuming that the variation of energy $dU$ for electrons in a partially chaotic domain during a time interval $dt$ is given by:

$$dU = Evdt - \alpha[b_r(t)]^2 dt \qquad (1)$$

where the first term takes into account of the electric field ($E$) acceleration and the second one represents an "effective" friction force due to stochastic diffusion, here assumed to be proportional (by a costant α) to the square of the amplitude $b_r$ of the mode. Collisional effects are neglected. By solving this differential equation (writing the velocity $v$ as a function of $U$: $v=(U/2m)^{1/2}$ ) the energy time evolution $U(t)$ is obtained as reported in panel (b) with a continuous red line and shows the same delay (hereafter called $\Delta$) observed experimentally between the scintillator signal and the MP. This simple model has been used to interpret also

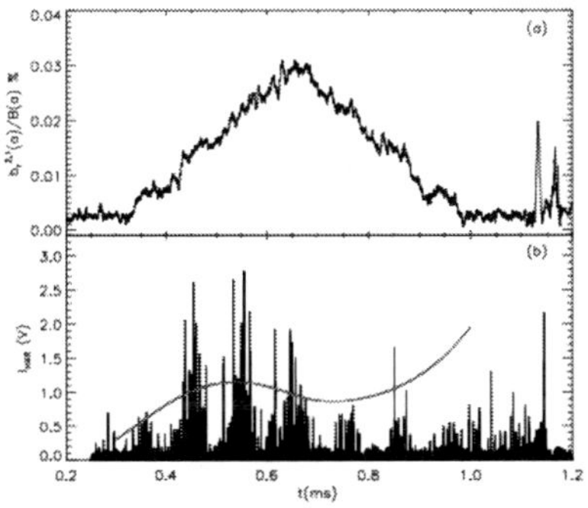


**Figure 3.** *(a) (2,1) rotating magnetic perturbation applied with varying amplitude (#37938);(b) scintillator signal in black ; in red the theoretical prediction.*

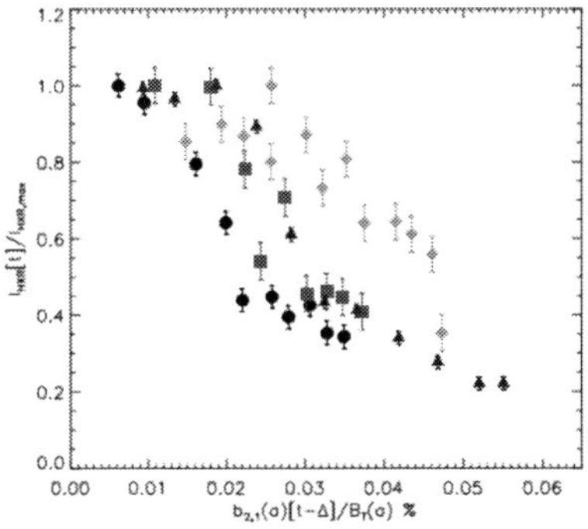

**Figure 4.** *Scintillator signal as a function of MP amplitude at the time t-Δ. Each color is a different shot.*

similar discharges: it is found that Δ mainly depends on the slope of the MP ramp and on the applied electric field magnitude. Fig.4 shows that the scintillator signal at the time *t* has a decreasing trend when plotted vs the amplitude of the MP evaluated at *t*-*Δ*.

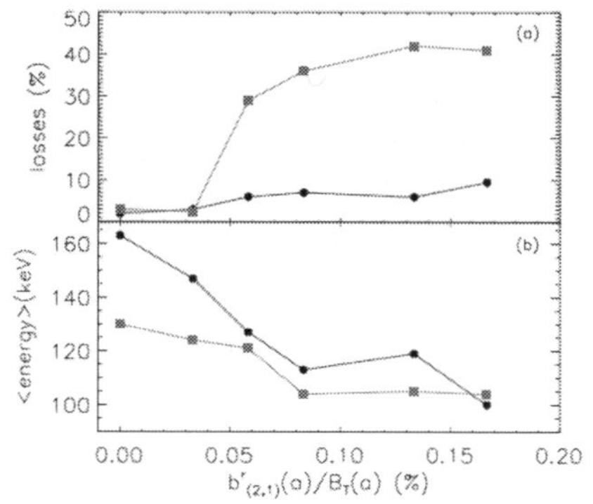


**Figure 5.** *Percentage fraction of RE lost from the plasma (a) and of their average final energy (b) from ORBIT numerical simulation. In black: q(a)=1.8; in red: q(a)=2.3*

***Numerical simulations.*** The decreasing of the energy of the lost fast electrons at higher values of MP has been also observed in numerical studies performed with the code ORBIT. In these simulations electrons have an initial energy of 100keV; at t=0 a constant perturbation (2,1) is applied; the effect of the toroidal electric field (*E=0.05V/m*) is implemented as well as slowing down collisions (assuming $n_e=2\cdot 10^{18}m^{-3}$). Fig.5 (a) shows the number of lost electrons vs the MP amplitude after 8ms while their average energy is reported in Fig.5(b). Two RFX-mod equilibria have been considered, one with q(a)=1.8 (black line) and another one with q(a)=2.3 (red line). In the latter case a small amplitude of the perturbation is enough to increase the fraction of lost electrons to about 30-40% of the initial population; on the contrary, in the simulation with a q(a)<2 equilibrium, the fraction of lost electrons still depends on $b^r_{21}(a)/B_T(a)$ but remains under 10%. These results can be explained by the higher level of stochasticity developing in the scenario where the mode is resonant. In both cases the final energy of lost electrons decreases at higher MP with respect to an unperturbed equilibrium.

***Conclusions.*** Dedicated experiments to study the RE dynamic in low density tokamak discharges have been recently performed for the first time in the RFX-mod device. Both experiments and simulations show that the application of MPs partially increases the losses of RE before they reach high energies. Further studies are currently in progress in order to understand which kind of MPs should be applied in order to de-correlate also the remaining confined electrons which still continue to be accelerated, especially in the core of the plasma.

**Acknowledgment.** *This project has received funding from the European Union's Horizon 2020 research and innovation programme under grant agreement number 633053. The views and opinions expressed herein do not necessarily reflect those of the European Commission.*

**References**

[1] Lehnen M. et al. 2008 PRL **100** 255003
[2] Sonato P. et al. 2003 Fusion Eng. Des. **66** 161
[3] Marrelli L. et al. 2007 Plasma Phys. Control. Fusion **49** B359
[4] Granetz R.S. et al. 2014 Phys. Plasmas **21** 072506
[5] White R.B. et al. 1984 Phys. Fluids **27** 2455

# Longitudinal laser ion acceleration in gas jets: experimental optimization on the Titan laser facility and numerical investigation of the ultra-high intensity limit

E. d'Humières[1], S. Chen[2], P. Antici[3], M. Bailly-Grandvaux[1], T. Gangolf[2], M. Lobet[4], G. Revet[2], A.M. Schroer[5], O. Willi[5], J. Santos[1], V. Tikhonchuk[1], H. Pépin[3], J. Fuchs[2]

[1]*Univ. Bordeaux, CNRS, CEA, CELIA, UMR 5107,33400 Talence, France*

[2]*LULI, École Polytechnique, CNRS, CEA, UPMC, route de Saclay, 91128 Palaiseau, France*

[3]*INRS-ENT, Varennes, Québec, Canada*

[4]*CEA, DAM, DIF, 91297 Arpajon, France*

[5]*Institut für Laser-und Plasmaphysik Heinrich Heine Universität Düsseldorf*

Intense research is being conducted on sources of laser-accelerated ions and their applications [1]. Several experiments on the low density regime of laser ion acceleration [2] were performed in recent years. In this regime, volume effects are expected to dominate, while for solid foils, ion acceleration is directly related to the electron surface density and the number of accelerated ions is limited. Simulations therefore show that it is possible to reach high ion energies with a high number of accelerated ions and a high conversion efficiency [3]. This scheme also leads to less debris than solid foils and is more adapted to high repetition lasers.

Very promising results were recently obtained at LULI [4] and on the LLNL Titan laser in this regime [5]. In both experiments, a first ns pulse was focused on a thin target to explode it and a second laser with a high intensity was focused on the exploded foil. The delay between two lasers allowed to control the density gradient seen by the second laser pulse. The transition between various laser ion acceleration regimes depending on the density gradient length was studied. With a laser energy of a few Joules, protons with energies close to the energies of TNSA accelerated protons were obtained for various exploded foils configurations. In the high energy regime (~180 J), protons with energies significantly higher than the ones of TNSA accelerated protons were obtained when exploding the foil while keeping a good beam quality. These results demonstrate that low-density targets are promising candidates for an efficient proton source that can be optimized by choosing appropriate plasma conditions.

We report on new experiments performed on the Titan laser in 2014 using compact and dense gas jets made available recently by important progresses on nozzle optimization [6]. The experimental optimization of longitudinal laser ion acceleration is discussed using this setup, which is very promising for several applications of laser ion acceleration even if important progresses are still required on the maximum energy of the beam.

The Titan short pulse with a 0.7 ps duration, a wavelength of 1054 nm, a 26 microns FWHM was focused on a hydrogen gas jet with a characteristic full width at half maximum of 300 microns on the propagation axis. The laser energy was varied from 50 to 180 J (corresponding to laser intensities up to $7x10^{19}$ W/cm$^2$). By varying the pressure from 180 bars to 900 bars, the maximum density of the gas jet ranged from 0.5 $n_c$ to 2.5 $n_c$. Three Thomson Parabolas at 0°, 16° and 32° from the short pulse propagation axis were used as well as a proton spectrometer at 90°. Depending on the maximum density of the gas jet, pure highly collimated proton beams with quasi-monoenergetic structures were measured for the first time at this wavelength.

Preliminary 1D Particle-In-Cell simulations were performed with the code PICLS [7]. The laser parameters are the same as above and the density profile corresponds to a lineout of the density measured in the experimental configuration. Proton energy spectra measured in the increasing part of the density gradient after 4 ps for two laser intensities are plotted in Figure 1. A peaked structure, similar to the one observed in the Titan experiments, is visible between 2 and 4 MeV in both cases. An analysis of the phase spaces shows that this structure corresponds to the electrostatic shock launched in the increasing gradient by the laser pulse.

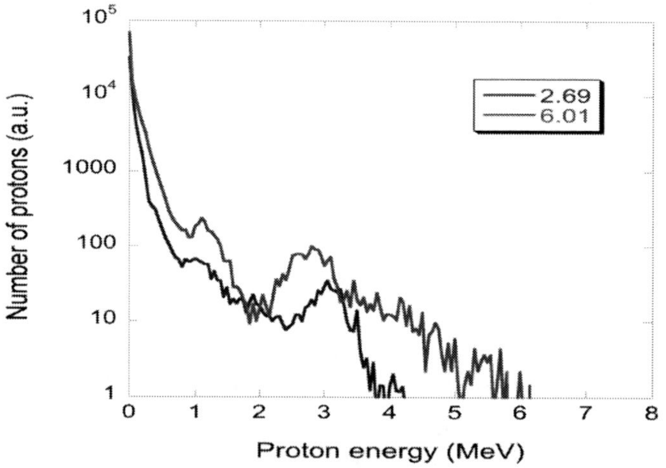


Figure 1: Proton energy spectra measured in the increasing part of the density gradient after 4 ps for $a_0$=2.69 (blue line) and for $a_0$=6.01 (red line).

Scaling shock acceleration in the low density regime to ultra high intensities (>$10^{22}$ W/cm$^2$) is a challenge as radiation losses and electron positron pair production change the optimization of the shock process. Using large-scale Particle-In-Cell simulations including these effects, we have investigated and modeled the transition to this regime in which intense beams of relativistic ions can be produced.

Following Refs. [8,9], the PIC code CALDER [10] has been enriched with numerical models of synchrotron radiation and Breit-Wheeler pair production [11]. In the classical case, the low-energy photon emission is treated using Sokolov's radiation friction model with a quantum-corrected radiated power [12]. In the quantum regime, Monte Carlo schemes are used to describe the discrete emission of gamma photons and their annihilation into e–e+ pairs. The physical parameters are chosen so as to reproduce laser and target conditions that could be tested with near-future laser installations like Apollon in France and the Extreme Light Infrastructure in Europe. The linearly-polarized laser pulse has a 1 μm wavelength, a 32 fs (60 $\omega_0^{-1}$, where $\omega_0$ is the laser frequency) FWHM duration and a peak intensity ranging from $I_0 = 10^{22}$ to $8.9\times10^{23}$ Wcm$^{-2}$ (i.e., a normalized field amplitude $a_0 = eE_0/m_ec\omega_0 = 85$ to 800). The focal spot FWHM is 10 μm. The target is taken to be a fully-ionized H plasma with a cosine squared density profile, a maximum electron density $n_e = 2\ n_c$ and a total thickness ranging from 100 to 400 μm.

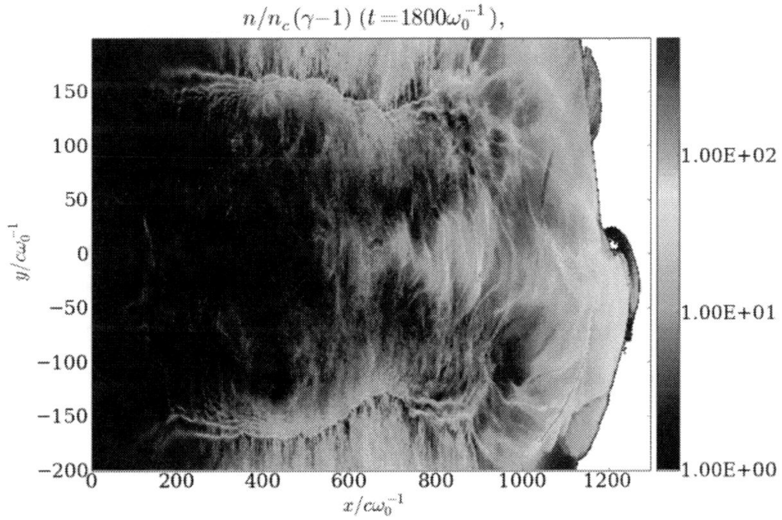


Figure 2: 2D map of the proton energy density after 1800 $\omega_0^{-1}$ for I=$10^{22}$ W/cm$^2$ for a 2 $n_c$, 150 microns long cos$^2$ target. The laser comes from the left side of the simulation box.

Figure 2 shows the 2D map of the proton energy density after 1800 $\omega_0^{-1}$ for I=$10^{22}$ W/cm$^2$ for a 2

$n_c$, 150 microns long $\cos^2$ target. As already described in [5], the high concentration of ion energy density and the complexity of the high energy ion front which is followed by high energy filamentary structures is clearly visible. Proton energies of more than 900 MeV are measured in these simulations.

In summary very promising results were recently obtained in the low density regime of laser ion acceleration using new nozzles able to deliver denser and more compact gas jets. The experimental optimization of longitudinal laser ion acceleration can be achieved using this setup, which is very promising for several applications of laser ion acceleration even if important progresses are still required on the maximum energy of the beam.

Using large-scale Particle-In-Cell simulations including these effects, we have investigated and modeled the transition to the ultra high intensity regime of laser ion acceleration in low density targets in which intense beams of relativistic ions can be produced. These relativistic ion beams are of great interest for high-energy laboratory astrophysics.

[1] A. Macchi et al., Rev. Mod. Phys. 85, 751 (2013); E. d'Humières (2012). Ion Acceleration by High Intensity Short Pulse Lasers, Laser Pulses - Theory, Technology, and Applications, Prof. Igor Peshko (Ed.), ISBN: 978-953-51-0796-5, InTech, DOI: 10.5772/46137.

[2] L. Willingale et al. Phys. Rev. Lett. 96, 245002 (2006); P. Antici et al. New J. Phys. 11, 023038 (2009); C. Palmer et al., Phys. Rev. Lett. 106, 014801 (2011); D. Harberger et al., Nature Physics 8, 95 (2012).

[3] E. d'Humières et al., J. Phys.: Conf. Ser. 244, 042023 (2010); F. Fiuza et al. Phys. Rev. Lett. 109, 215001 (2012); E. d'Humières et al., Phys. Plasmas 20, 023103 (2013).

[4] M. Gauthier et al. Phys. Plasmas 21, 013102 (2014).

[5] E. d'Humières, et al., Plasma Phys. Controlled Fusion 55, 124025 (2013).

[6] F. Sylla et al., Rev. of Sci. Inst. 83, 033507 (2012); S.N. Chen et al., Nuclear Instruments and Methods in Physics Research Section A: Accelerators, Spectrometers, Detectors and Associated Equipment 740, 105-106 (2014).

[7] Y. Sentoku, E. d'Humières, L. Romagnani, P. Audebert, J. Fuchs, Phys. Rev. Lett. 107, 135005 (2011).

[8] R. Duclous R., J.G. Kirk, and A.R. Bell, Plasma Phys. Control. Fus., 53, 015009 (2011).

[9] N.V. Elkina, A.M. Fedotov, I.Y. Kostyukov, et al., Phys. Rev. ST Accel. Beams 14, 054401 (2011).

[10] E. Lefebvre et al., Nuclear Fusion 43, 629 (2003).

[11] M. Lobet, E. d'Humières, M. Grech, et al., arXiv preprint arXiv:1311.1107 (2013).

[12] I.V. Sokolov, N.M. Naumova, J.A. Nees, et al., Phys. Plasmas 16, 093115 (2009).

# N-graphene synthesis using $N_2$-Ar remote plasmas

A. Dias[1], E. Tatarova[1], J. Henriques[1], F.M. Dias[1], M. Abrashev[2], U. Cvelbar[3], G. Filipič[3], Zh. Dimitrov[2], Zh. Kissovski[2]

[1] *Instituto de Plasmas e Fusão Nuclear, Instituto Superior Técnico, Universidade de Lisboa, Portugal*

[2] *Faculty of Physics, University of Sofia, 1164, Sofia, Bulgaria*

[3] *Department of Surface Engineering and Optoelectronics - F4, Jozef Stefan Institute, Slovenia*

## 1. Introduction

Plasma environments constitute powerful tools in materials science by allowing the creation of innovative materials and the enhancement of long existing materials that would not otherwise be achievable[1]. In particular, $N_2$-Ar large-scale plasma sources driven by microwaves demonstrate advantageous properties for plasma processing techniques. In the present work graphene sheets were N-doped using a $N_2$-Ar large-scale remote plasma treatment. The samples were treated at different compositions of $N_2$-Ar gas mixture. The N-doped graphene sheets have been analysed applying scanning electron microscopy (SEM), Raman and Fourier transform infrared (FT-IR) spectroscopy. Optical emission spectroscopy was also performed to detect the emission of excited nitrogen atoms.

## 2. Experimental

A slot antenna excited, surface wave sustained $N_2$-Ar plasma source described in [2,3] has been used (see Figure 1). The discharge vessel consists in a tube with ~70 cm of length and closed at the top by a quartz window. Gases are injected in the chamber, considering a total flow of tens of sccm. The slot-antennas are cut in the broad wall of the feeding rectangular waveguide shown in the figure. Microwave power of 600 W at 2.45 GHz is fed to the chamber. The plasma is created by the energy carried by surface waves propagating radially and azimuthally along the interface between the plasma and a quartz dielectric plate. A pure $TM_{140}$ surface mode, i.e. an evanescent wave along z, is excited. The graphene sheets with few atomic layers (1-5), synthesized previously applying microwave plasma-based method as described in [4,5], are placed in a remote plasma zone to be doped with nitrogen.

Two set of samples were produced to study the influence of $N_2$ percentage in $N_2$-Ar gas mixture (5-20 %) on the process of graphene doping.

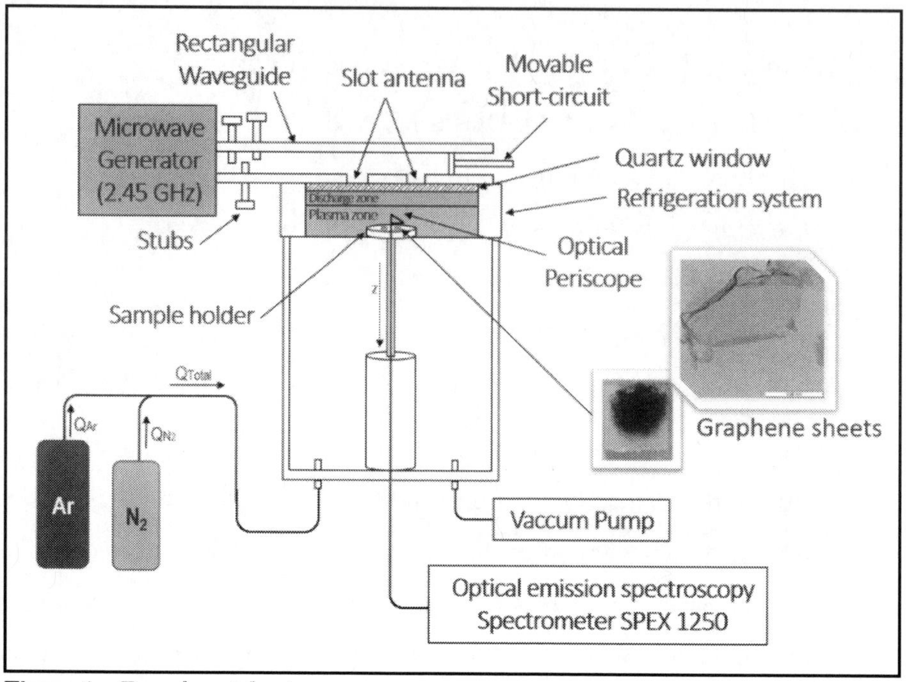


**Figure 1** – Experimental setup.

## 3. Results and discussion

The plasma emission spectra in the 300-900 nm range detected at the axial distance where the substrate with the free-standing graphene sheets were placed are shown in Fig. 2. As observed the spectrum is dominated by Ar atomic lines. Intense nitrogen atomic lines at 821.6 nm and 822.3 nm belonging to the transition N (3p→3s) have been detected. The lines intensity decreases as the $N_2$ percentage in the $N_2$-Ar gas mixture increases from 5-20 %.

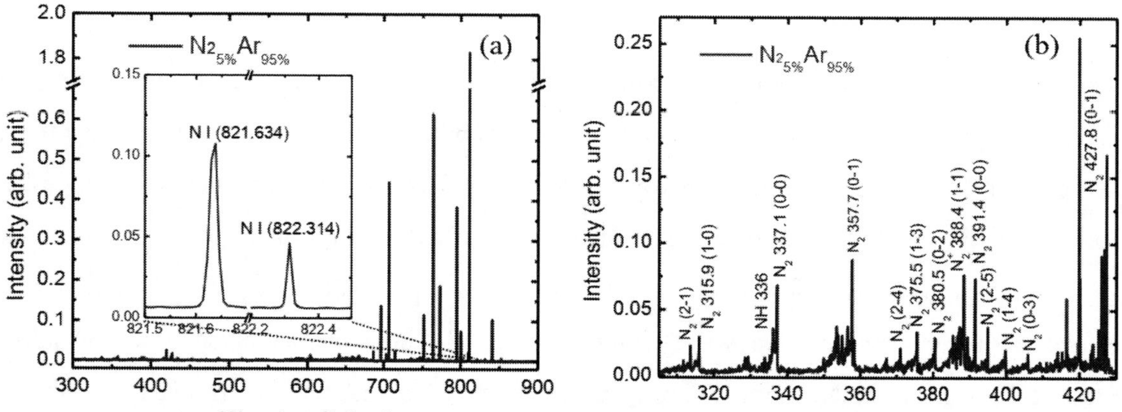


**Figure 2** – (a) Optical emission spectra emitted by $N_2$-Ar plasma with 5% of $N_2$ in the gas mixture. Experimental conditions: P= 600 W, p= 1 mbar; (inset) Nitrogen atomic lines. (b) Emission spectra of the $N_2$ first negative system.

The rotational temperatures corresponding to the $N_2$ first negative system are commonly used as a measure of the gas temperature. To this end, the rovibronic bands in the 389-391 nm range, have been used. The rotational temperature has been determined using LIFBASE code [6]. The gas temperature varies in the limits 900 K-1400 K when the percentage of $N_2$ in the $N_2$-Ar gas mixture increases from 5 to 20%. Therefore, energetic heavy particles could be present at the location of samples processing.

Fig. 3 shows the Raman spectra of 'pure' graphene sheets and the evolution of the Raman spectrum with the nitrogen percentage in $N_2$-Ar remote plasma. As seen there are three dominant peaks in the spectra. The three peaks are attributed to the G band at ~ 1585 $cm^{-1}$, 2D band at ~ 2662 $cm^{-1}$, and D band at ~1333 $cm^{-1}$. Taking into account the ratio between the 2D and G peak integral intensities (~2) and the full width at half maximum of the 2D-band (~45 $cm^{-1}$) of graphene sheets before plasma treatment the result is an indication of few-layer graphene sheets. As seen from Fig. 3(a) the D/G peaks intensity ratio increases with the $N_2$ percentage (from ~0.76 to ~1.19). Fig. 3(b) shows a SEM image of $N_2$-Ar plasma treated graphene sheets at 5% of $N_2$ in the gas mixture.

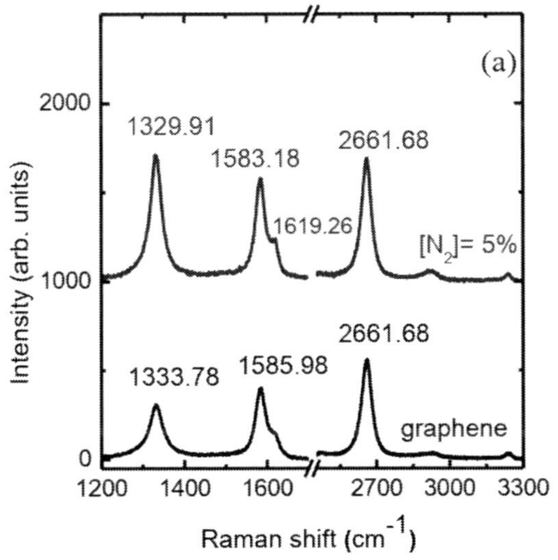

**Figure 3** – (a) Raman spectra of graphene and $N_2$-Ar plasma treated graphene sheets with 5% of $N_2$ in the gas mixture. (b) SEM image of graphene sheets treated with a $N_2$-Ar plasma with 5% of $N_2$ in the gas mixture. ( P = 600 W, p = 1mbar.

FT-IR spectra were used to investigate the bonding difference between graphene and N-graphene (see Fig. 4). The FT-IR spectrum of graphene shows some characteristic bands of surface oxidized groups at ~1733 $cm^{-1}$ which most likely corresponds to C=O bond. The peak at ~2360 $cm^{-1}$ corresponds to the C–O bonds and the features between 2841 and 3000 $cm^{-1}$ are consistent with C–$H_x$ stretching vibrations of chemisorbed hydrogen of various types present in all carbon films. In the case of the N-graphene sample a new peak at ~1244 $cm^{-1}$, corresponding to the C–N stretching vibrations appears.

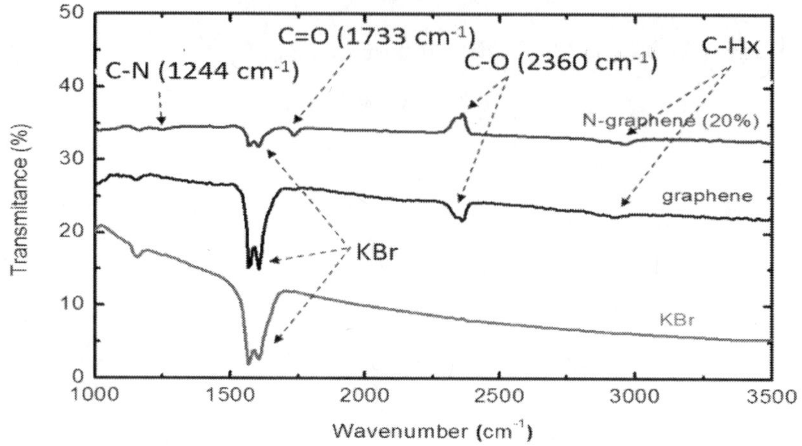


**Figure 4** – FT-IR spectra of KBr, graphene and $N_2$-Ar plasma treated graphene sheets with 20 % of $N_2$ in the mixture of $N_2$-Ar (P = 600 W, p = 1mbar).

## 4. Conclusions

It was demonstrated the production of N-graphene by means of microwave $N_2$-Ar plasma based nitrogen doping of free standing graphene sheets. The plasma treatment of the graphene sheets was performed at different $N_2$ percentages in the $N_2$-Ar gas mixture while keeping constant the microwave power (P = 600 W), pressure (p = 1 mbar), exposure time and spatial position in the remote plasma zone. The results obtained demonstrate that the level of doping and type of functional groups attached to graphene scaffold can be controlled by percentage of $N_2$ in the $N_2$-Ar gas mixture. The work is in progress to make further optimization of the process by tailoring the plasma environment and fluxes of nitrogen atoms towards targeted nanostructures.

**Acknowledgements.** Work partially funded by Portuguese FCT - Fundação para a Ciência e a Tecnologia, under Project UID/FIS/50010/2013 and grant SFRH/BD/52413/2013 (PD-F APPLAuSE).

## References

[1] E Tatarova, N Bundaleska, J Ph Sarrette and C M Ferreira; Plasma Sources Sci. Technol. **23** (2014) 063002.

[2] E Tatarova, J P Henriques, F M Dias and C M Ferreira, J. Phys. D: Appl. Phys. **39**, 2747-2753 (2006).

[3] J P Henriques, E Tatarova, F M Dias and C M Ferreira, J. Appl. Phys. **103**, 103304 (2008).

[4] E Tatarova, J Henriques, C C Luhrs, A Dias et al; Appl. Phys. Lett. **103** (2013) 134101-5.

[5] E Tatarova, A Dias, J Henriques, A M Botelho do Rego, A M Ferraria, M V Abrashev, C C Luhrs, J Phillips, F M Dias and C M Ferreira; J. Phys. D: Appl. Phys. **47** (2014) 385501.

[6] J Luque and D R Crosley, Menlo Park, CA: SRI International, LIFBASE Version 2.0. Available at www.sri.com/psd/lifbase (1999).

# Modelling and Optimization of an Electrohydrodynamic (EHD) Thruster

Victor H. Granados[1], Mario J. Pinheiro[2] and Paulo A. Sá[1]

[1] *Departamento de Engenharia Física, Faculdade de Engenharia da Universidade do Porto*
*Rua Doutor Roberto Frias s/n 4200-465, Porto, Portugal*
[2] *Departamento de Física, Instituto Superior Técnico - IST*
*Universidade de Lisboa, Av. Rovisco Pais, 1049-001 Lisboa, Portugal*

**Abstract**

*A theoretical self-consistent two dimensional model of an axisymmetric electrostatic plasma actuator is presented. The computational model solves the governing equations for a plasma kinetic model using the software* COMSOL MULTIPHYSICS *for discharges using three different gases independently: argon, nitrogen and oxygen at a pressure of 0.5 Torr. In the argon scenario, we include the following heavy (neutral and charged) species: Ar(X), Ar* and $Ar^+$ while the nitrogen scenario includes: N, $N^+$, $N_2$, $N_2^+$ and $N_4^+$ and the oxygen scenario considers: O, $O^+$, $O^-$, $O_2$, $O_2^+$ and $O_2^-$.*

## Introduction

Since the discovery of the ionic wind effect in 1709 by Francis Hauksbee [1] there is a rising interest in EHD thrusters as a useful mean to convert electric energy into kinetic energy. The mechanism of a EHD thruster consists in an applied electric field that accelerates the electrically charged particles while collisional coupling ensures the momentum transfer between ions and neutrals [2]. The electrically neutral particles are ionized in the region surrounding the electrode with smaller curvature radius (anode on the present case) due to the higher electric field on its surface, which is the well-known Corona Effect [3]. The charged particles flow towards the electrodes, depending on their charge, where they are collected. For the discharge to be sustained, a secondary electron emission must occur at the cathode, that is due to the impact of ions on its surface and its effect is essential for the stability of the mechanism.

## Thruster model

The model consists of a set of differential equations modelling the different phenomena taking place in a DC-discharge. The electrostatic field is computed using the Poisson equation ($\varepsilon\nabla^2 V = -\rho$, where $\varepsilon$ is the plasma permitivity ($\varepsilon = \varepsilon_0\,\varepsilon_r$), $V$ is the electric potential and $\rho$ is the space charge density), which is computed through the plasma chemistry by the equation:

$$\rho = e\left(\sum_{j=1}^{N} Z_j n_j - n_e\right) \tag{1}$$

where $Z_j$ is the charge number of ions, $n_j$ and $n_e$ are the ions and electron densities, respectively. For the electron-impact reactions considered on the model, we assumed a Maxwellian distribution function and computed the rate coefficients from the equation:

$$k_j = \left(\frac{2e}{m_e}\right)^{1/2} \int_0^{\infty} u\sigma_j(u) f(u) du \tag{2}$$

where $u$ is the electron energy (eV) and $\sigma_j(u)$ represents the elastic and inelastic (excitation and ionization) cross sections considered for each reaction.

The fluid is asumed to be quasi-steady, with a small Mach number ($M < 0.3$), and althought it is often account as incompresible, for simulation purposes we considered the compressible fluid Navier-Stokes equations described by the following equation, which includes variations in the space charge density as depicted by the continuity equation:

$$\begin{gathered} \rho\frac{\partial \mathbf{u}}{\partial t} + \rho(\mathbf{u}\cdot\nabla)\mathbf{u} = -\nabla p + \nabla\cdot\left[\mu\left(\nabla\mathbf{u} + (\nabla\mathbf{u})^T\right) - \frac{2}{3}\mu(\nabla\cdot\mathbf{u})\mathbf{I}\right] + \mathbf{F} \\ \frac{\partial \rho}{\partial t} + \nabla\cdot(\rho\mathbf{u}) = 0 \end{gathered} \tag{3}$$

where $\mathbf{u}$ is the fluid velocity, $p$ is the pressure, $\mu$ is the dynamic viscosity, $\mathbf{I}$ is the indentity matrix and $\mathbf{F}$ is the volume force vector, which comes from the integration of the force density ($\mathbf{f} = \rho\mathbf{E}$) all over the volume. A RC coupling circuit is used in series with the DC source in order to avoid the arc regime between its electrodes.

Table 1: Rate coefficients

(a) Argon discharge

| Reaction | Rate coefficient $[m^3/s$ or $m^6/s]$ | Ref. |
|---|---|---|
| $e + Ar \rightarrow e + Ar$ | $\sigma(u)$ | [4] |
| $e + Ar \rightarrow e + Ar^*$ | $\sigma(u)$ | [4] |
| $e + Ar \rightarrow 2e + Ar^+$ | $\sigma(u)$ | [4] |
| $e + Ar^* \rightarrow 2e + Ar^+$ | $\sigma(u)$ | [4] |
| $Ar^* + Ar^* \rightarrow e + Ar + Ar^+$ | $6.2 \times 10^{-16}$ | [5], [6] |
| $Ar^* + Ar \rightarrow Ar + Ar$ | $3.0 \times 10^{-21}$ | [5], [6] |

(b) Nitrogen discharge

| Reaction | Rate coefficient $[m^3/s$ or $m^6/s]$ | Ref. |
|---|---|---|
| $e + N_2 \rightarrow e + N_2$ | $\sigma(u)$ | [7] |
| $e + N_2 \rightarrow 2e + N_2^+$ | $\sigma(u)$ | [7] |
| $e + N_2 \rightarrow e + 2N$ | $\sigma(u)$ | [8] |
| $e + N_2^+ \rightarrow 2N$ | $2.8 \times 10^{-13}(300/T_e)^{0.5}$ | [9] |
| $e + N_2^+ + N_2 \rightarrow 2N_2$ | $2.6 \times 10^{-39}(300/T_e)^{1.5}$ | [9] |
| $e + N_2^+ \rightarrow N_2$ | $1 \times 10^{-25}(T_g/T_e)^{4.5}$ | [9] |
| $e + N_4^+ \rightarrow N_2 + N_2$ | $2 \times 10^{-12}(T_g/T_e)^{0.5}$ | [9] |
| $N_2^+ + N \rightarrow N^+ + N_2$ | $7.2 \times 10^{-19} exp(T_g/300)$ | [9] |
| $N_4^+ + N_2 \rightarrow N_2^+ + 2N_2$ | $2.1 \times 10^{-22} exp(T_g/121)$ | [9] |
| $N_2^+ + 2N_2 \rightarrow N_4^+ + N_2$ | $5.0 \times 10^{-41}$ | [9] |

(c) Oxygen discharge

| Reaction | Rate coefficient $[m^3/s$ or $m^6/s]$ | Ref. |
|---|---|---|
| $e + O \rightarrow e + O$ | $\sigma(u)$ | [4] |
| $e + O_2 \rightarrow e + O_2$ | $\sigma(u)$ | [4] |
| $e + O_2 \rightarrow e + 2O$ | $\sigma(u)$ | [4] |
| $e + O_2 \rightarrow 2e + O_2^+$ | $\sigma(u)$ | [4] |
| $e + O \rightarrow 2e + O^+$ | $\sigma(u)$ | [4] |
| $e + O_2 \rightarrow O_2^-$ | $\sigma(u)$ | [4] |
| $2e + O_2^+ \rightarrow e + O_2$ | $1 \times 10^{-31}(300/T_e)^{4.5}$ | [9] |
| $e + O_2^+ + O_2 \rightarrow 2O_2$ | $6 \times 10^{-39}(300/T_e)^{1.5}$ | [9] |
| $e + O_2^+ \rightarrow O + O$ | $2 \times 10^{-13}(300/T_e)$ | [9] |
| $e + O + O_2 \rightarrow O + O_2^-$ | $1 \times 10^{-43}$ | [9] |
| $e + O + O_2 \rightarrow O^- + O_2$ | $1 \times 10^{-43}$ | [9] |
| $O^+ + O + O_2 \rightarrow O_2^+ + O_2$ | $1 \times 10^{-41}$ | [9] |
| $O^+ + O_2 \rightarrow O_2^+ + O$ | $3.3 \times 10^{-17} exp(-0.00169 T_g)$ | [9] |
| $O_2^- + O \rightarrow O^- + O_2$ | $3.3 \times 10^{-16}$ | [10] |
| $O^- + O \rightarrow O_2 + e$ | $1.4 \times 10^{-16}$ | [10] |
| $O_2^- + O_2 \rightarrow 2O_2 + e$ | $2.7 \times 10^{-16}(T_g/300)^{0.5} exp(-5590/T_g)$ | [10] |

We considered self-consistent kinetic models for each gas, which includes electron-impact reactions (elastic, excitation and ionization), and chemical reactions in order to model the behaviour of the gas during the discharge. The reaction rates of the kinetic models are detailed in Table 1. The simulation conditions were: $P = 0.5\ Torr$, $T_g = 300\ K$, $V = 250\ V$, $R = [20\ -\ 1500]\ k\Omega$ and $C = 1\ pF$.

**Results and discussions**

Figure 1 shows the electron density distributions for the three gasses considered. Nitrogen showed at least two orders of magnitude greater than oxygen and than argon.

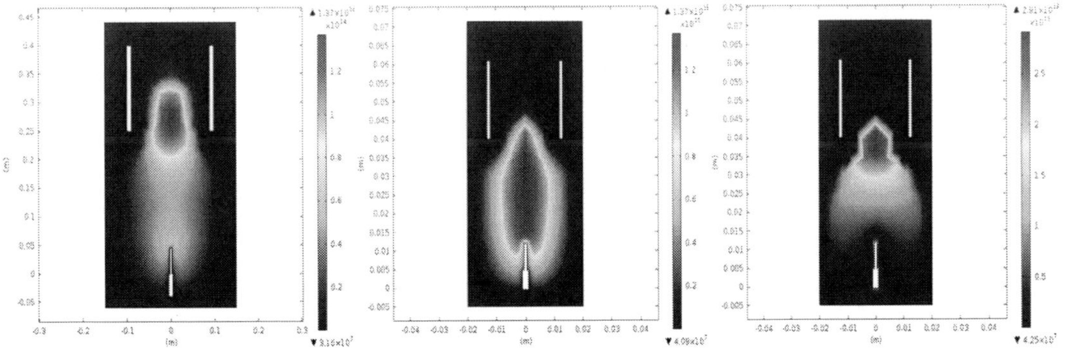

(a) Argon (b) Nitrogen (c) Oxygen

Figure 1: Electron density distributions ($m^{-3}$)

Figure 2 shows the fluid speed distributions for the three gases. In argon, the speed is two orders of magnitude less than in nitrogen and the configuration shows a very different profile which is due to a less inefficient momentum transfer. Nitrogen gas presents the highest peak speed (12.5 $cm/s$) which indicates a more efficient ionic wind phenomenon.

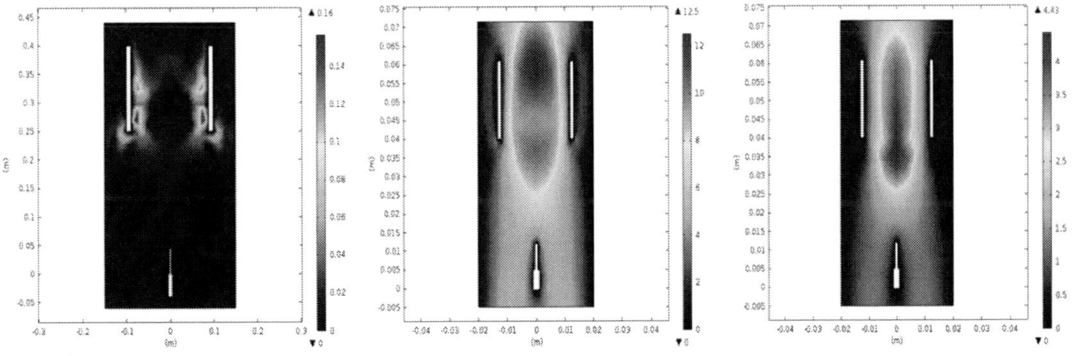

(a) Argon (b) Nitrogen (c) Oxygen

Figure 2: Fluid velocity distributions ($cm/s$)

In order to calculate the total thrust produced by the flux of gas going through the EHD thruster, we integrate over the output of the thruster. Table 2 shows the net thrust produced using the different gasses in the thruster's chamber, along with the power consumption and the

Thrust-to-Power (F/P) ratio, a known parameter to determine the ammount of thrust obtained for a given power input. According to Masuyama et al. [3]: "EHD thrusters have documented thrust-to-power ratios as high as 26 $mN/W$, which is orders of magnitude greater than electrostatic propulsion systems used in space"

Table 2: Net thrust produced by EHD thruster using different gases

| Gas | Thrust ($nN$) | Power ($mW$) | Thrust-to-Power ratio (nN/W) |
|---|---|---|---|
| $Ar$ | $9.77 \times 10^{-4}$ | 38.1 | $2.56 \times 10^{-2}$ |
| $N_2$ | 3.86 | 183.1 | 21.1 |
| $O_2$ | 0.331 | 10.4 | 31.8 |

Nitrogen gas presents the highest thrust production among the gases considered (ten times higher than oxygen), however, oxygen presented a higher Thrust-to-Power ratio which indicates a higher efficiency in transforming electric energy into thrust. Argon gas produced the fewest amount of thrust, as well as the lowest Thrust-to-Power ratio.

## Conclusions and future work

We conclude that argon contributes with the smallest thrust due to its mono-atomicity nature while nitrogen produces the highest thrust due to the $N_2^+$ ions, the most effective conveyors of momentum to the neutrals. The oxygen was more efficient in transforming power to thrust but generated a total net thrust four times smaller than nitrogen's. We corroborated that plasma actuators produce thrust in presence of different gases ($Ar$, $N_2$ and $O_2$) with high energy-conversion performance and thus consists in a attractive technology for near future aerospace applications. Althought our thurster is far from a commercial application, we proved the feasibility of simulating the ionic wind phenomena which will lead us to an optimization process until we reach a better EHD thruster.

## References

[1] Hauksbee, F., Physico-Mechanical Experiments on Various Subjects, London, (1709), 1st Edition, pp. 46-47.

[2] Alexandre A. Martins and Mario J. Pinheiro, Phys. Plasmas **18**, 033512 (2011) http://dx.doi.org/10.1063/1.3562874.

[3] Masuyama, K. and Barrett, S. R. H., On the performance of electrohydrodynamic propulsion. Proc. R Soc. A, (2013), **469**: 20120623.

[4] Morgan database, www.lxcat.net, retrieved on January 25, (2015).

[5] Bogaerts, A. and Gijbels, R., Modeling of metastable argon atoms in a direct-current glow discharge, Physical Review A. **52** (1995).

[6] Lam, S. K., Zheng, C-E, Lo, D., Dem'yanov, A. and Napartovich, A. P., Kinetics of $Ar_2^*$ in high-pressure pure argon, Journal of Physics D: Applied Physics **33**, (2000) 242.

[7] IST-Lisbon database, www.lxcat.net, retrieved on January 25, (2015).

[8] Itikawa database, www.lxcat.net, retrieved on January 25, (2015).

[9] Kossyi, I.A., Kostinsky, A.Y., Matveyev, A.A. and Silakov, V.P. Plasma Sources Sci. Technol. **1** (1992) 207.

[10] Tatarova, E., Dias, F. M., Ferreira, C. M., Guerra, V., Loureiro, J., Stoykova, E., Ghanashev, I. and Zhelyazkov, I., Self-consistent kinetic model of a surface-wave-sustained discharge in nitrogen. J. Appl. Phys. (1997), **30**, 2663-2676.

# Electron and positron acceleration in Pulsar Wind Nebulae

S. Giacchè[1], J.G. Kirk[1], T. Amano[2]

[1]*Max-Planck-Institut für Kernphysik, Heidelberg, Germany*

[2]*Department of Earth & Planetary Science, University of Tokyo, Japan*

## Introduction

Pulsars release their spin-down power in the form of a relativistic electron-positron wind characterised by a large magnetisation parameter $\sigma \gg 1$ [1], defined as the ratio of the Poynting flux to the particle kinetic energy flux. When the wind impacts with the slowly moving surrounding environment, a reverse or termination shock (TS) forms, where the Poynting flux is dissipated. At this location, the energy is converted into kinetic energy of charged particles, mostly electrons and positrons, which emit inverse Compton (IC) and synchrotron radiation that is observed as a Pulsar Wind Nebula (PWN) downstream of the shock. These particles are likely to be accelerated to high energies via a Fermi-like process in which they gain energy by elastically scattering off electromagnetic waves both upstream and downstream and performing many shock crossings.

The process able to perform the energy conversion at TS remains uncertain, even though recently it has been suggested [2] that the propagation of superluminal waves upstream of the shock might trigger the dissipation of the Poynting flux through the formation of instabilities and small-scale shocks. Superluminal waves can be generated at the TS when the pulsar frequency exceeds the proper plasma frequency $\omega_{p0} = \sqrt{\frac{8\pi ne^2}{m}}$, where $n$ is the density of the fluid and $m$ is the electron mass. In this case the incoming fluid is slowed down and heated up until a precursor with turbulent electromagnetic fields is established ahead of the shock. Particles encountering the turbulence can be reflected into the upstream, forming a population of pre-accelerated particles which can be considered to have been injected into a subsequent Fermi-type process.

are injected in the true Fermi-type mechanism.

We address this problem in the context of $\gamma$-ray binaries, in which a pulsar orbits a massive main sequence star. In this case, the wind launched by the companion star compresses the pulsar wind and dictates the location of the TS [3], which in turn affects the density in the pulsar wind. During the resulting orbital modulation of the local plasma frequency, the condition $\omega > \omega_{p0}$ for the formation of the shock precursor can be fulfilled, triggering the particle acceleration.

In this contribution we investigate the role played by the onset of this electromagnetically

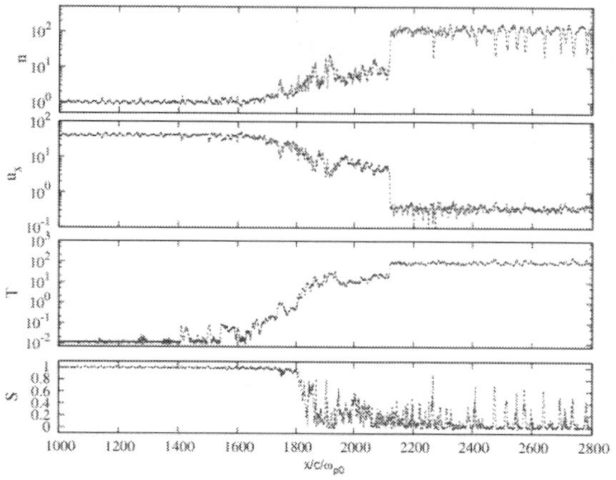

Figure 1: *From top to bottom: density, x component of the four velocity, temperature and normalised Poynting flux profiles of the plasma across the shock ($x_{sh} \sim 2120c/\omega_{p0}$) at $t = 2500/\omega_{p0}$. The interaction with the superluminal waves forms a precursor in the upstream where physical quantities smoothly vary before the sharp transition at the shock front.*

modified structure in the injection mechanism. The turbulent fields in the shock front are simulated with a 1-dimensional code and used to perform numerical integration of a large number of particle trajectories from which we determine the fraction of particles reflected at the shock, i.e. the fraction of particles which enters a Fermi acceleration process.

**Method**

We simulate the pulsar wind TS by mean of a 1-dimensional, relativistic, 2-fluid code supplemented by Maxwell's equations [2]. The pulsar striped-wind driven by the obliquely magnetised, rotating pulsar is treated in the upstream as a fully transverse, circularly polarised, sinusoidal magnetic shear (a "sheet-pinch") whose phase-averaged magnetic field is null. In the striped-wind model such a scenario is realised in the pulsar equatorial plane [4]. The initial conditions in the downstream are determined by the solution of the Rankine-Hugoniot relations, once the following parameters are provided: the Lorentz factor of the incoming fluid $\Gamma_s$, the magnetisation parameter $\sigma$ and the frequency of the pulsar $\omega$ in units of the upstream proper plasma frequency $\omega_{p0}$.

We use the background fields obtained with this set-up to numerically integrate trajectories of test particles (electrons and positrons) subject to the Lorentz force. We employ a form of the equations of motion which is suitable to avoid issues of momentum accumulation when only the magnetic field is present. The solution of our set of equations is advanced adopting a 4th order Runge-Kutta method with adaptive time step-size. To investigate whether this configuration of turbulent fields is able to inject particles in a Fermi-like acceleration process, we initiate the trajectories far upstream, with an isotropic distribution and almost at rest in the local fluid frame. We set two boundaries in the simulation box, one upstream and one downstream at distance $L_{up}$ and $L_{down}$ from the shock front respectively. Each trajectory is followed until it reaches the region between the boundaries and it then crosses either of the two. Particles registered at the upstream boundary, are labeled as reflected and contribute to the number of particles injected in

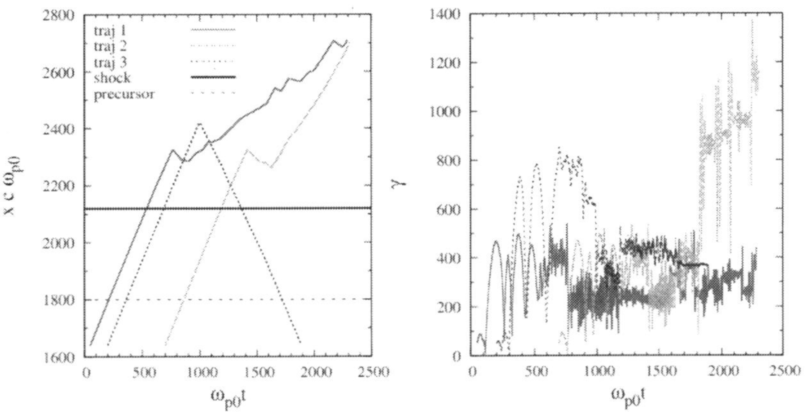


Figure 2: *Typical trajectories. Left panel: time evolution of the x coordinate of the motion. The solid black line represents the shock position and the dashed black line represents the approximate position of the border of the precursor in the steady state. Right panel: energy evolution of the corresponding particle.*

the acceleration process.

**Results**

We aimed to obtain a precursor which, after a first phase of expansion into the upstream, stabilises in a steady state for $\Delta t \gg 1/\omega_{p0} > 1/\omega$, namely comparable to, but still smaller than, the characteristic timescale of the orbit of the binary system. We obtain this configuration for $\Gamma_s = 40$, $\sigma = 10$ and $\omega = 1.2\omega_{p0}$. In Fig. 1 we show the profile of density, $x$ component of the 4-velocity, temperature and Poynting flux of the plasma across the shock front at $t = 2500/\omega_{p0}$, when the stationary precursor has already established. For the chosen configuration, physical quantities like density and temperature form a region of smooth transition, which is the precursor created through the interaction with electromagnetic waves, before undergoing a sharp transition at the location of the shock ($x_{sh} \sim 2120c/\omega_{p0}$) as it happens for hydrodynamical shock fronts. In addition, we notice in the fourth panel of Fig. 1 that the Poynting flux starts decreasing at the border of the precursor $1800c/\omega_{p0}$ and it is almost completely dissipated downstream the shock, a part from some residual field fluctuations. With our initial conditions, after $\Delta t \sim 1200/\omega_{p0}$ the precursor stops expanding into the upstream, reaching a stationary state during which its spatial extension remains constant. This phase lasts $\Delta t \sim 1200/\omega_{p0}$ before the simulations is stopped. Periodic boundary conditions of this steady state are created to increase the time interval available for numerical integration. We tested whether this implementation affected the trajectories and we found that the choice of the number of folding and of the time extension of the periodic boundary does not affect spectra and angular distributions of particles registered on both upstream and downstream boundaries.

Because of the interaction of the "sheet-pinch" with the superluminal waves, the precursor

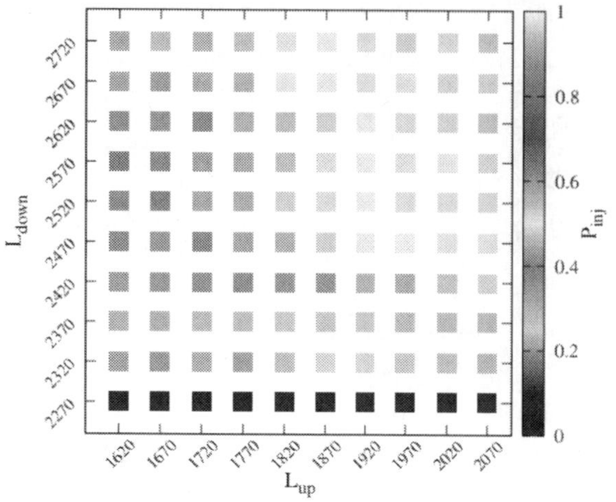


Figure 3: *Injection probability grid obtained counting the particles crossing the upstream boundary for different positions of the downstream and upstream boundaries.*

and the region immediately downstream the shock are characterised by turbulent fields which deflect and accelerate test particles. This is visible in Fig. 2, where the evolution of the $x$ coordinate and of the corresponding energy are plotted as a function of time for a sample of characteristic trajectories. The fraction of particles injected in a true acceleration process is computed by counting how many trajectories cross the upstream boundary travelling against the flow. We compute this number for many positions of $L_{up}$ at fixed position of $L_{down}$ and repeat the procedure for many values of $L_{down}$, thus building a grid of values for the injection probability. This is plotted in Fig. 3, where we show that as the two boundaries recede from the shock, $P_{inj}$ asymptotes to $\sim 40\%$. This value is large compared to that found for other relativistic shocks (e.g. $P_{inj} \sim 12\%$ for relativistic, oblique shocks [5]), suggesting that electromagnetically modified shocks are very effective in injecting particles in a Fermi mechanism.

## Acknowledgements

SG acknowledges the International Max Planck Research School for Astronomy and Cosmic Physics (IMPRS-HD) for financial support.

## References


[1] B. M. Gaensler and P. O. Slane, Annual Review of Astronomy & Astrophysics **44**, 1, 17 (2006)

[2] T. Amano and J.G. Kirk, The Astrophysical Journal **770**, 18 (2013)

[3] I. Mochol and J.G. Kirk, The Astrophysical Journal **776**, 40 (2013)

[4] F.V. Coroniti, The Astrophysical Journal **349**, 538 (1990)

[5] A. Achterberg, Y. Gallant, J.G. Kirk and A.W. Guthmann, Mon. Not. R. Astron. Soc. **328**, 393 (2001)

# KSTAR experimental results toward solving burning plasma issues

Jong-Gu Kwak and KSTAR teams

National Fusion Research Institute, Yusong, Daejeon, Korea

Aim of the nuclear fusion research is to realize the fusion reactor for solving energy problem in the world and ITER would be the big milestone for proving the scientific and engineering feasibility of nuclear fusion at burning plasma state. Moreover, the study on the reactor requires operation of the machine to be simple as well as producing high efficient fusion reaction which is to be in cost effective. In this regards, KSTAR has been looking forward to demonstrating the high performance long pulse discharge using highly advanced physics concept and steady state compatible superconducting magnets.[1] By highly elaborating engineering effort to operation of the superconducting machine based on the design concept of AT physics, KSTAR has shown the very outstanding experimental results which would impact on the design of the future reactor. While demonstrating the pulse-length of more than 40 sec H-mode flat-top in 2014 characterizing uniqueness of superconducting machine and achieving high beta($b_N$~4) above no-wall limit transiently by optimal $I_p/B_T$ and with $P_{ext}$ ~ 3 MW, KSTAR has also been exploring various means to achieve and sustain steady-state, ELM-suppressed/mitigated H-modes using versatile in-vessel control coils (IVCC), ECCD/ECH, and/or SMBI. The 40s sustaining time corresponds to more than 20 times of current diffusion time that is sufficient time to provide the current profile control by changing hardware of ECH actuator mirror mechanically. KSTAR also showed that the level of the intrinsic error field of the machine is very low and it is suggested that KSTAR may not need any error field correction coil to realize ITER baseline scenario.

In particular, taking advantage of the versatile 3-rows of IVCC which has similar structure as ITER, KSTAR accomplished both n=1 and n=2 RMP-driven, ELM-crash suppressed regimes that lasted up to 5 sec. The demonstration of n=1 ELM suppression was one of the unique results in world-wide tokamak researches. The suppression window in the safety factor q95 has also been extended from 6.5 to 3.9 depending on the coil configuration and the existence of multiple integer operational window of q95 for n=1 RMP ELM suppression indicating the strong impact of resonant component on the ELM suppression. Significant progress has been also made on the investigation of the underlying mechanism on RMP suppression using measurements of pedestal fluctuations and modeling of plasma response. In particular, recent success for longer ELM suppression using mid-coil only application opens the very promising possibility of RMP ELM suppression applicable to the burning plasma where the high neutron and thermal heat flux inside the vacuum vessel requires very careful design of in-vessel components such as coil in terms of material selection and neutron shielding because it should make the design to be simple as much as possible and avoid many engineering issue of manufacturing and installing coils inside the vessel.

For steady state operation, tokamak requires the plasma current drive non-inductively. In advanced scenarios with $q_{95} \geq 8.5$, the plasma current was lowered to 300-400 kA to increase the fraction of non-inductive current drive by increasing $\beta_p$ . considering auxiliary heating power limitation of present KSTAR resources. The plasma performance was increased after the transition but suddenly collapsed. However, the plasma quickly recovered from this event and the plasma performance was increased again even further up to $\beta_N$ of 3.0, $\beta_p$ of 3.5, $H_{89}$ of 2.0, and $H_{98}$ of 1.7. An interpretive ASTRA simulations reveal that the plasma current is driven non-inductively more than 100%. The toroidal magnetic field strength was 2 T. By adjusting plasma control scheme and adding fast radial plasma controller in 2015 campaign, it is expected that longer pulse would be possible. To support external off-axis current drive knob, the new innovative high efficiency current drive, so called helicon CD is developed. The helicon CD is known to have a high efficiency CD on the high density plasma so that if it is successfully validated on KSTAR, it would supplement LHCD in KSTAR. Figure 2(a) showed the installed mockup TWA(Travelling Wave Antenna) at KSTAR and 2(b) showed s-parameter measurement for the antenna. The working frequency is 500 MHz. For sustaining burning plasma, it is necessary to keep fusion output constantly and it is recognized to sustain the burning plasma condition by adjusting input gas. However, if the plasma wall is saturated and recycling is very high, the density control is very difficult in neutral beam dominated heating machine with graphite wall. So it is very important to see how long is the fuel retention at graphite wall in superconducting tokamak in addition to nuclear safety issue from fuel(tritium) accumulation.

Burning plasma at tokamak condition also requires continuous operation without abnormal event such as disruption because the ejected heat from stored energy in a short time is not acceptable to heat limit of PFC at disruption and subsequent halo-current and runaway electrons induce severe damages on the machine if it is not properly controlled. So forecasting the disruption event or mitigation of heat and runaway electron just after that is one of key research issues for ITER as well as future tokamak fusion reactor. KSTAR has equipped various 2D MHD image diagnostics such as ECEi and IR/bolometer TV to monitor pre-disruption state and thermal quench process. ECEi clearly showed that multiple disruption events happen after the mode locking due to the m/n=2/1 mode growth. In addition, one of KSTAR results showed the possibility of delay of starting point of disruption from the tearing mode by pre-application of NRMP to the plasma so that it enables to save time for the operator to counteract the disruption if not avoidable.[2] n=2 non-resonant field could hinder the growth of n=1 locked mode after locking. Burning plasma requires the plasma current as high as possible to enhance the confinement and this also increase halo-current at disruption event. It is well known that toroidal peaking in halo-currents induces non axisymmetric force on the tokamak vessel and induced the engineering issue for reactor design. Recent KSTAR results suggest that there might be some correlation between intrinsic error field and toroidal peaking of halo currents so that it might be possible to reinforce the vessel structure against the disruption selectively before full power operation. Figure 3(a) and 3(b) showed halo-current distribution on the upper and lower diverter and figure 3(c) showed direction of intrinsic error field on KSTAR.

In summary, KSTAR has successfully extended the operation boundary toward high-performance, long-pulse H-mode (~ 43 Sec at 0.6MA), transiently accessing to high beta ($\beta_N$ ~ 4) beyond no-wall limit. ELM suppression/mitigation physics study have been prioritized, expanding the operation range of q95 (4 – 6) in n=1 and n=2 and longer suppression (~5s) by using middle coil only in conjunction with long pulse sustainment at superconducting machine and IR TV measurement on the divertor. For in-depth study of disruption, overall process of disruption is visualized and past failures of passive plate is interpreted with observed toroidal asymmetry of Halo current and possibly intrinsic error field

References

[1] Jong-Gu Kwak el al., "An overview of KSTAR results", Nuclear Fusion 53(2013)104005

[2] J. Kim el al., PRL submitted (2015)

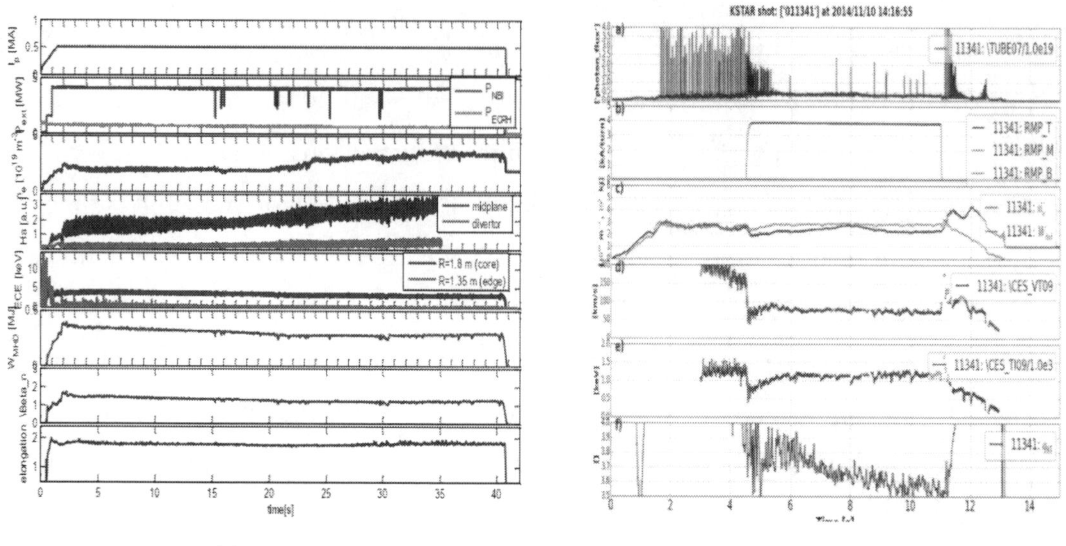

(a) (b)

Fig. 1. the time evolution of long pulse H-mode discharge(a) and typical shot of ELM suppression where the duration time is more than 5s and the middle coil is only used(b)

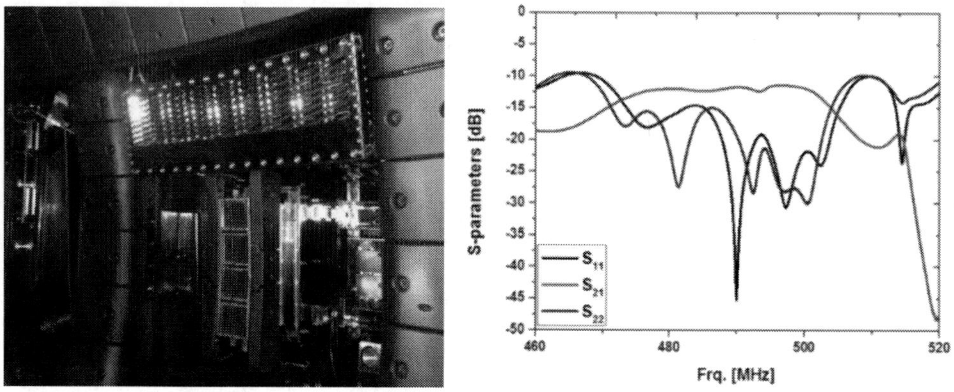

(a) (b)

Fig. 2. Helicon mockup antenna installed at KSTAR(a) where helicon antenna is positioned above LHCD antenna and measured s-parameter(b)

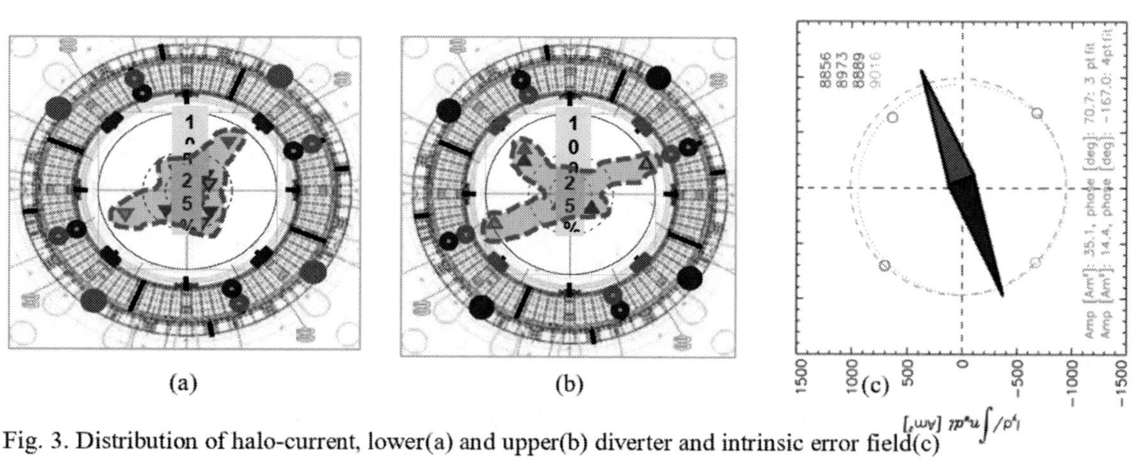

(a) (b) (c)

Fig. 3. Distribution of halo-current, lower(a) and upper(b) diverter and intrinsic error field(c)

# High-β steady-state scenarios in DIII-D with on-axis current drive

C.C. Petty[1], J.R. Ferron[1], F. Turco[2], J.M. Park[3], A.M. Garofalo[1], A.W. Hyatt[1], G.L. Jackson[1], T.C. Luce[1], C.T. Holcomb[4], E.J. Doyle[5] and W.M. Solomon[6]

[1]*General Atomics, P.O. Box 85608, San Diego, CA 92186-5608, USA*
[2]*Columbia University, 116th St and Broadway, New York, NY 10027, USA*
[3]*Oak Ridge National Laboratory, P.O. Box 2008, Oak Ridge, TN 37831, USA*
[4]*Lawrence Livermore National Laboratory, 700 East Ave, Livermore, CA 94550, USA*
[5]*University of California Los Angeles, Los Angeles, CA 90095-7099, USA*
[6]*Princeton Plasma Physics Laboratory, Princeton, NJ, USA*

**Abstract.** Experiments in the DIII-D tokamak have demonstrated potential new paths to fusion steady state based on peaked current profiles with $q_{min}\approx1$ that have high ideal stability limits, excellent confinement and benefit from efficient on-axis current drive. In the hybrid scenario, steady-state conditions ($V_{surf}$=0) using central ECCD and NBCD are achieved in 1.0 MA discharges with a beta value ($\beta_N$=3.6) that is 80%-90% of the ideal n=1 with-wall limit. Interestingly, the hybrid mechanism that anomalously broadens the current profile to maintain $q_{min}>1$ and prevents sawteeth continues to function despite the intense central current drive. In the "high $\ell_i$" scenario, the combination of broad pressure profile and a peaked current profile tailored to maximize $\ell_i$ allows $\beta_N\approx4.8$ and $H_{98y2}\approx1.8$ to be achieved transiently in a discharge that is overdriven ($V_{surf}$<0). The achieved $\beta_N$ is near the no-wall limit with the ideal-wall limit higher at $\beta_N\approx5$-6.

## I. Introduction

This paper discusses two high-β scenarios in DIII-D with on-axis current drive that are consistent with the Q=5 steady-state mission in ITER. While both cases have $q_{min}\approx1$, the hybrid scenario relies on the self-organized current profile produced by that regime, whereas the high $\ell_i$ scenario takes a more active approach to create an optimized current profile. Hybrid plasmas have the advantage of robustness and insensitivity to the current drive profile; the high $\ell_i$ scenario is more complex but has higher stability limits and higher confinement, as well as a less critical need for a high pedestal.

## II. Steady-State Hybrid Scenario

In DIII-D, experiments show that the beneficial characteristics of the hybrid scenario are maintained when central co-current drive is applied to increase the non-inductive fraction to ≈100% [1]. The advantages of the hybrid regime over the $q_{min}>2$ Advanced Tokamak (AT) regime are (1) good alignment between the current drive and plasma current profile is not necessary as poloidal magnetic flux pumping self-organizes the current density profile in hybrids with an m/n=3/2 tearing mode [2], and (2) the current drive in hybrids can be located near the plasma center where the current drive efficiency is highest. The high current drive efficiency can fully compensate for a lower bootstrap current fraction in the $q_{min}\approx1$ regime compared to the $q_{min}>2$ AT regime.

The natural attributes of the hybrid scenario make it a robust regime for high-β, steady-state plasmas, as shown in Fig. 1 where the surface loop voltage $V_{surf}$=0, thermal confinement factor $H_{98(y,2)}$=1.56 and normalized beta $\beta_N$=3.64 (toroidal beta $\beta_T$=3.1%) are sustained for the maximum duration of the beam pulse without exciting the deleterious m/n=2/1 tearing mode. Half of the plasma current is driven non-inductively near the plasma center by electron cyclotron current drive (ECCD) and neutral beam current drive (NBCD), with the other half

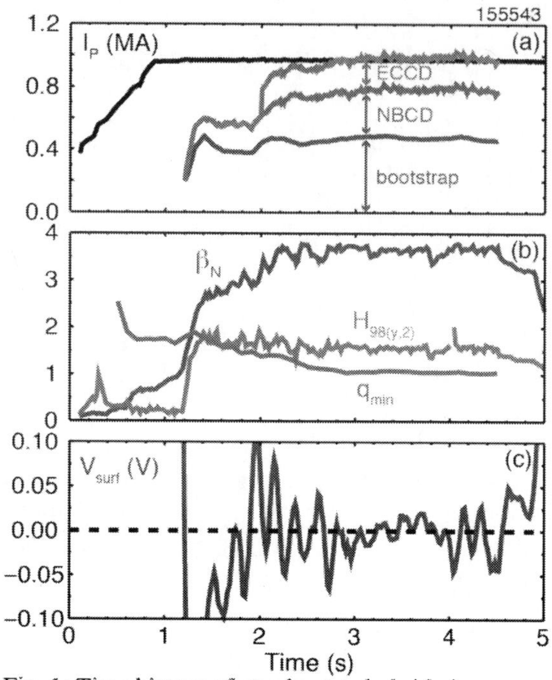

Fig. 1: Time history of steady-state hybrid plasma.

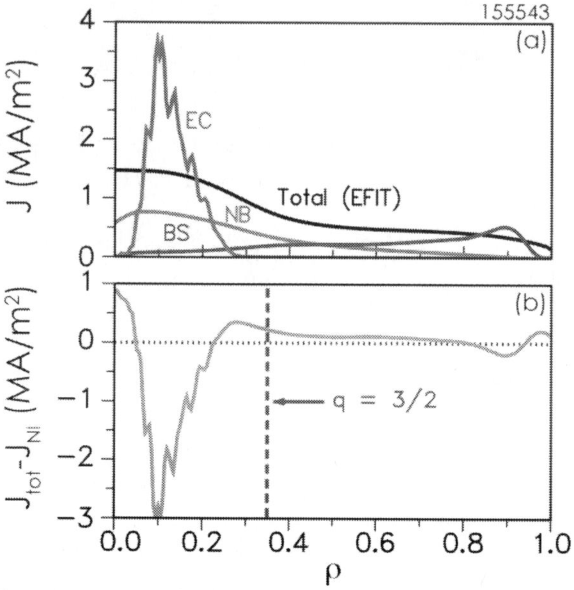

Fig. 2: (a) Total current density from EFIT, and modeled current densities from ECCD, NBCD and bootstrap. (b) Difference between total current density from EFIT and modeled non-inductive current densities.

generated by the bootstrap current. The non-inductive currents calculated by TRANSP are overlaid with the measured plasma current in Fig. 1a. While this demonstrates that $V_{surf}$=0 is consistent with the calculated non-inductive currents, Fig. 2a shows that the reconstructed current profile is not consistent with the sum of the driven current profiles. The current profile anomaly is displayed in Fig. 2b, where the calculated non-inductive current density is subtracted from the total current density determined by equilibrium reconstruction. Inside of the q=3/2 surface, the current profile is strongly overdriven and the current relaxation time is sufficiently short that $q_{min}$ should drop below 1 by the end of the discharge. The fact that $q_{min}$ remains above unity and sawteeth are absent shows that the hybrid scenario maintains an anomalously broad current profile even in the presence of strong central current drive.

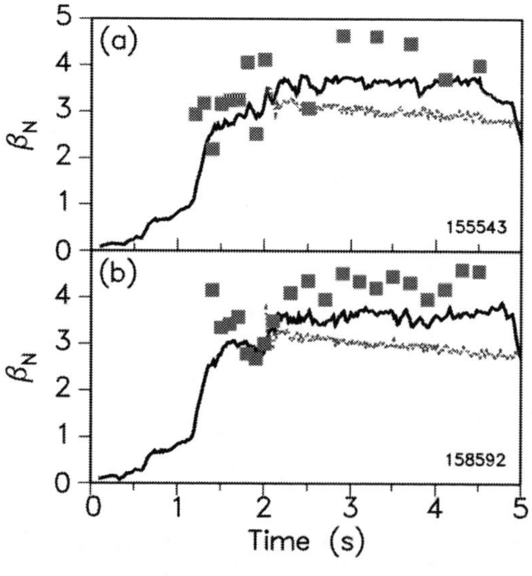

Fig. 3: Comparison of measured $\beta_N$ (solid black line), ideal-wall (red squares) and no-wall (dotted blue line) n=1 limits for (a) mixed on/off-axis beams and (b) on-axis beams.

The beta value obtained in steady-state hybrids is 80%-90% of the ideal n=1 with-wall limit. The theoretical stability limits are calculated by the DCON code using EFIT reconstructions constrained by the experimental pressure profile, MSE polarimetry and a neoclassical calculation of the pedestal bootstrap current density. These experiments utilized both on-axis and off-axis beam deposition (two of the six co-beams can inject off-axis) to affect the stability limit by varying the pressure profile peakedness ($f_P=P_0/\langle P \rangle$). Figure 3 shows that in hybrids with complete current drive, $\beta_N$=3.64 (i.e., $\beta_N=4.9\ell_i$) is reached and maintained for the duration of high power NBI. This value exceeds the no-wall n=1 stability limit (average DCON value $\beta_N=4\ell_i$) and is close to the ideal-wall n=1 limit. Using off-axis beam power reduces $f_P$ from ~3.4 to ~3.1, mainly by changing the fast ion pressure profile, with the largest systematic differences between on/off-axis injection occurring for $\beta_N$>3.3. On average the DCON calculated ideal-wall limit is ≈10% higher for hybrid plasmas with off-axis beam injection, but experimentally little difference is seen in the maximum beta ($\beta_N$=4.0 for 0.8 s) with or without off-axis beam power. Future plans have DIII-D increasing the number of off-axis beams from two to four, and the number of co-beams from six to eight, which may allow the theoretical advantages of a broader pressure profile to be more easily realized.

**III. High $\ell_i$ Scenario**

By taking a more active approach in tailoring the peaked current profile to maximize $\ell_i$, the ideal stability limit and confinement can be increased beyond the more passive approach described in the previous section. These performance improvements arise largely as a result of higher poloidal field in the discharge core and larger magnetic shear in the outer half of the plasma. The benefits of this "high $\ell_i$" scenario are seen in Fig. 4, where $\beta_N$≈4.8 and $H_{98(y,2)}$≈1.8 are achieved transiently at $\ell_i$≈1.3 [3]. To form the high $\ell_i$ target, the discharge begins with a long ohmic phase so that the electron temperature is low and the current density profile becomes peaked in the core with $q_{min}$≈1. Next, ECCD at ρ≈0.4 is added to increase the electron temperature and "freeze in" the peaked current profile, after which high power beam heating is applied to transition the plasma into ELMy H-mode and ramp up $\beta_N$. As described later, the ohmic current profile slowly becomes less peaked in time during the H-mode phase, resulting in the decreasing confinement time (and thus decreasing $\beta_N$ at fixed heating power) seen in Fig. 4 as $\ell_i$ evolves to a lower value.

Plasmas with high $\ell_i$ and $\beta_N$=4-5 are predicted to be stable to low-n ideal MHD instabilities even without the effect of a conducting wall. As seen in Fig. 5, the peak $\beta_N$ for the discharge shown in Fig. 4 is near the no-wall n=1 limit calculated by DCON, with the ideal-wall n=1 limit higher at $\beta_N$≈5-6. The ideal infinite-n ballooning mode stability limit calculated using the BALOO code is slightly below that of the ideal-wall n=1 mode. Consistent with these stability calculations, a global, pressure-limiting instability has not yet been clearly observed in the experiment. Instead, in cases where stability determines the limit to performance (such as when $\beta_N$ is raised above 5), the observed mode is most commonly a m/n=2/1 resistive tearing mode.

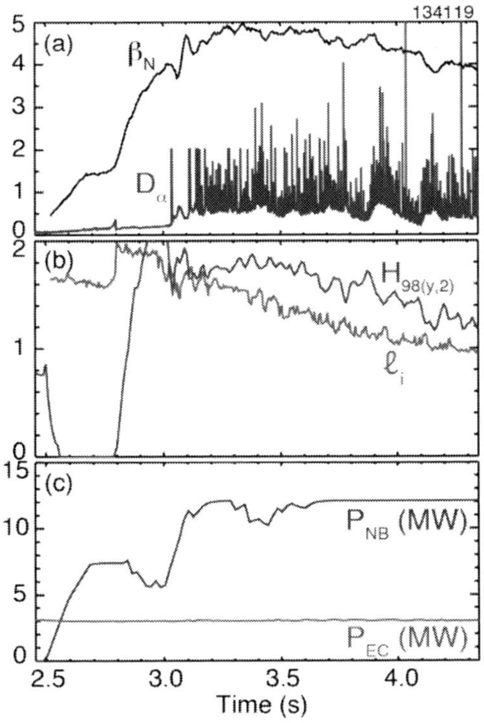

Fig. 4: Time history of high $\ell_i$ discharge.

Broadening of the pressure profile with increasing $\beta_N$ plays an important role in enabling stable access to high plasma pressure along with the elevated values of $\ell_i$. The broad pressure profiles obtained in these high $\ell_i$ plasmas, $f_P$~2.5, strongly increases the ideal-wall stability limit [4]. Interestingly, a low H-mode pedestal height, which can result from pedestal physics and/or the application of 3D magnetic fields, is not necessarily detrimental to this steady-state scenario as it works to raise $\ell_i$ and thus the ideal-wall limit. As a result, the high $\ell_i$ scenario is a promising option for the ITER Q=5 steady-state mission that can perform well in plasmas with a low pedestal height.

The elevated $\ell_i$ in present experiments is largely a result of the inductively-driven current profile. Figure 6 displays the individual current density components (bootstrap, NBCD, ECCD and ohmic),

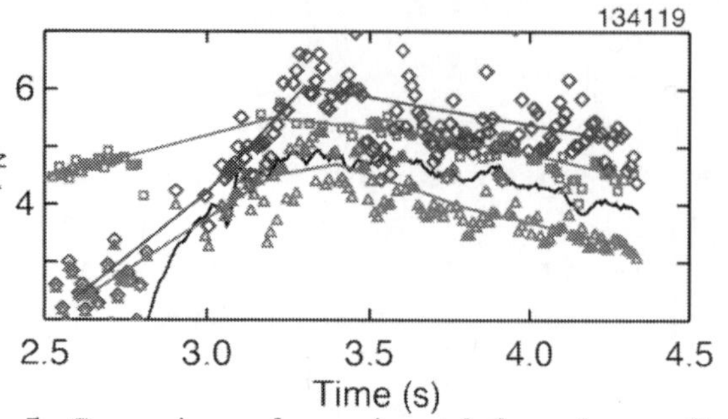


Fig. 5: Comparison of experimental $\beta_N$ and no-wall (triangles) and ideal-wall (diamonds) n=1 stability limits. The squares are the ideal infinite-n ballooning mode stability limit. The lines are sketches to indicate the location of various sets of data points.

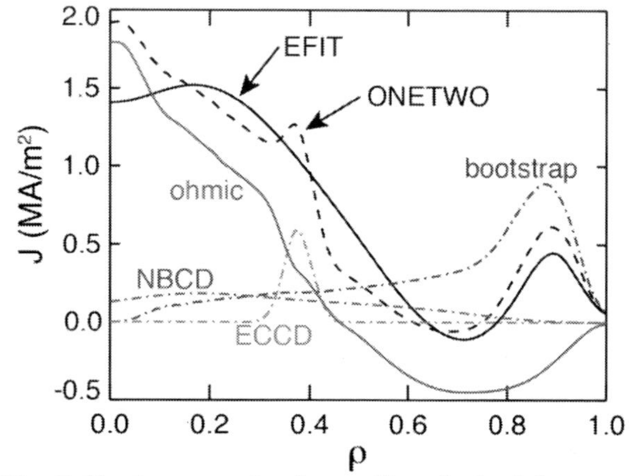


Fig. 6: Total current density profiles obtained from an equilibrium reconstruction and as calculated by ONETWO. The individual current density components from ONETWO are also plotted.

as well as the total current density, calculated by the ONETWO transport code for the discharge in Fig. 4 at 3.35 s; for reference, the total current density profile from a well-constrained equilibrium reconstruction is also plotted. The plasma current is actually overdriven with a bootstrap current fraction of ≈80% and $V_{surf}$<0. The negative surface loop voltage penetrates relatively quickly through the outer half of the plasma, with the negative inductively-driven current density outside ρ=0.5 offsetting some of the bootstrap current, helping to maintain an elevated value of $\ell_i$. The negative loop voltage penetrates slowly towards the axis during the high $\beta_N$ phase, decreasing the ohmic current in the plasma center and causing $\ell_i$ to drop over time. With a long enough pulse duration, this discharge would eventually evolve to a relatively low $\ell_i$ case because of the high pedestal bootstrap current.

To make the high $\ell_i$ scenario not only fully non-inductive but also stationary, the portion of the evolving ohmic current driven by the positive loop voltage in the core needs to be replaced with on-axis ECCD and NBCD (and higher plasma current is needed to eliminate the negative loop voltage). The maximum value of $\ell_i$ that can be obtained by tailoring the profile of the externally-driven current density will decrease with increasing bootstrap current in the outer half of the plasma. Therefore, the "optimized high $\ell_i$" equilibrium has about 50% bootstrap current and 50% external current drive near the axis. Transport code modeling of the steady-state current profile using TGLF [5] to self-consistently predict the temperature profiles examined the accessible current density and pressure profiles to maximize the stationary value of $\beta_N$ consistent with the calculated stability limits. The result of this modeling study is a stationary 1.1 MA plasma with $\ell_i$=1.07, $\beta_N$=4 and $H_{98(y,2)}$=1.1 using a combination of central ECCD (9 MW), mixed on/off-axis NBCD (20 MW total) and 50% bootstrap current. The ideal n=1 stability limit for this case is $\beta_N$=4.1 without a conducting wall and $\beta_N$=4.8 with a wall. This simulated high $\ell_i$ case is consistent with the planned upgrades to the DIII-D heating systems.

This work was supported in part by the US Department of Energy under DE-FC02-04ER54609, DE-FG02-04ER54761, DE-AC52-07NA27344, and DE-FG02-08ER54984.

[1] Turco F. *et al* 2015 *Phys. Plasmas* **22** 056113.
[2] Petty C.C. *et al* 2009 *Phys. Rev. Lett.* **102** 045005.
[3] Ferron J.R. *et al*, accepted for publication in *Nucl. Fusion.*
[4] Ferron J.R. *et al* 2005 *Phys. Plasmas* **12** 056126.
[5] Staebler G.M., Kinsey J.E., Waltz R.E. 2007 *Phys. Plasmas* **14** 055909.

# Preparing WEST tungsten divertor operation: Nitrogen seeded H-mode plasma simulation with SOLEDGE2D-EIRENE

H. Bufferand[1], G. Ciraolo[2], Y. Marandet[1], C. Bourdelle[2], J. Bucalossi[2], J. Denis[2], N. Fedorczak[2], D. Galassi[3], Ph. Ghendrih[2], R. Leybros[3], B. Pégourié[2], E. Serre[3], P. Tamain[2], E. Tsitrone[2]

*1. Aix-Marseille Université, CNRS, PIIM UMR 7345, F-13013 Marseille, France*

*2. CEA, IRFM, F-13108 Saint-Paul-Lez-Durance, France*

*3. Aix-Marseille Université, CNRS, Centrale Marseille, M2P2 UMR 7340, F-13451 Marseille, France*

## Introduction

In the perspective of operating WEST (W Environment Steady State) tokamak at Cadarache [1, 2], a dedicated effort has been made to develop the transport code SOLEDGE2D-EIRENE [3]. This code is used in a prospective way to investigate the operational domain of WEST, focusing on target heat loads and plasma conditions in the SOL. During the operation, the code will be used to analyze experiments, help interpreting edge plasma behavior (using synthetic diagnostics for instance [4]), and link WEST results to ITER. Using edge transport codes in a predictive way is a challenge since turbulent transport is not modeled self-consistently. Prescribed radial diffusion coefficients are usually adjusted to match experimental density and temperature profile in the outer mid-plane. Therefore, to investigate WEST plasma properties, we first simulate an H-mode plasma on ASDEX Upgrade which is a comparable medium size tokamak. The simulation relies on the transport coefficients determined in Ref. [5]. The same transport coefficients are then used to simulate edge plasma conditions for a WEST H-mode discharge.

## AUG H-mode simulation

In [5], a thorough study is carried out to match AUG H-mode discharge #17151 with SOLPS. The influence of the different transport code parameters on the solution is carefully studied. In this contribution, we propose to reproduce this work with SOLEDGE2D-EIRENE. The magnetic equilibrium for AUG shot #17151 at t=4.13s has been used to generate a SOLEDGE2D-EIRENE grid (resolution: 110 points in the radial direction, 180 points in the

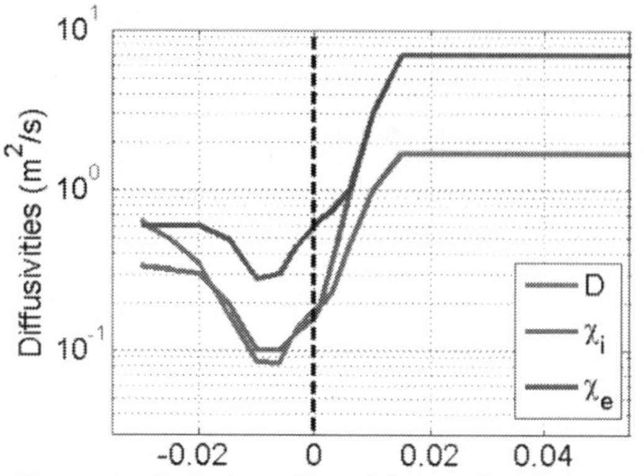


*Figure 1 : Radial profiles of flux surface averaged transport coeffcients used in SOLEDGE2D-EIRENE simulations inspired for [5]*

poloidal direction). The simulation domain is extended up to the first wall which is considered to be made of Carbon. The plasma simulated by the code is made of Deuterium with a small amount of Carbon. The Carbon is generated by wall sputtering. The sputtering yield used in the simulation is constant and set to $Y_C = 2.3\%$ including both physical and chemical sputtering and has been adjusted to match $P_{rad} = 1.15MW$ as measured in the experiment. At the core edge interface, the input power is set to $P_{SOL} = 5.4MW$ with an equal repartition between ions and electrons and the density is set to $n_e = 4.2 \cdot 10^{19} m^{-3}$. Transport coefficients are identical to those given by Chankin in [5]. They follow a $1/B$ dependence. The flux surface average values are reported on Figure 1, the H-mode transport barrier being created by a drop in diffusivities at the vicinity of the separatrix. Drifts are not taken into account.

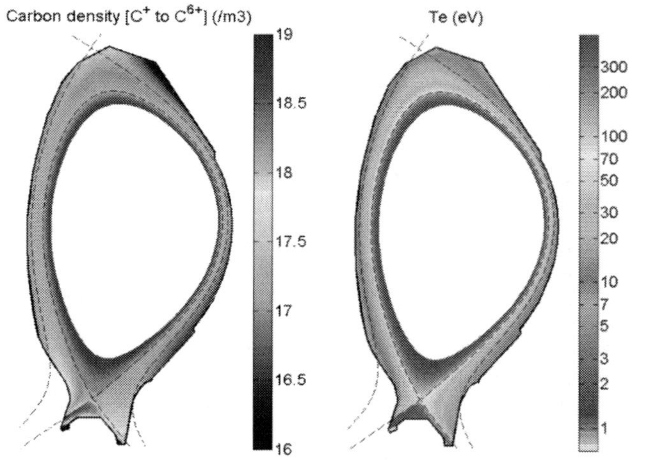


*Figure 2 : SOLEDGE2D-EIRENE simulation results for AUG #17151 H-mode plasma. Left: Carbon density (log scale), right: electron temperature.*

Typical simulation results for electron density, temperature and Carbon density are reported on Figure 2. Radial profiles in the outboard mid-plane are reported on Figure 3. A rather good agreement is found between

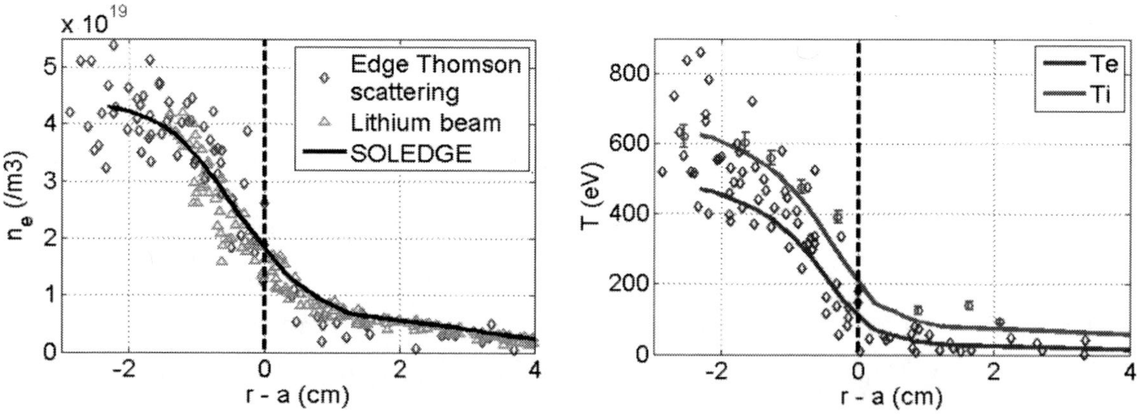


*Figure 3 : Radial profiles for plasma density (left) and ion and electron temperature (right). Comparison between measurements (symbols from [5]) and SOLEDGE2D-EIRENE simulation (lines).*

simulated and measured results for density and temperature, as expected. Indeed, the transport coefficients, which are the main free parameters of the code, have been adjusted in Ref. [5] so as to match experimental mid-plane profiles.

Figure 4 shows target profiles for density, temperature and heat flux on the outer target. The simulation results are in rather good agreement with Langmuir probes measurements and IR camera data.

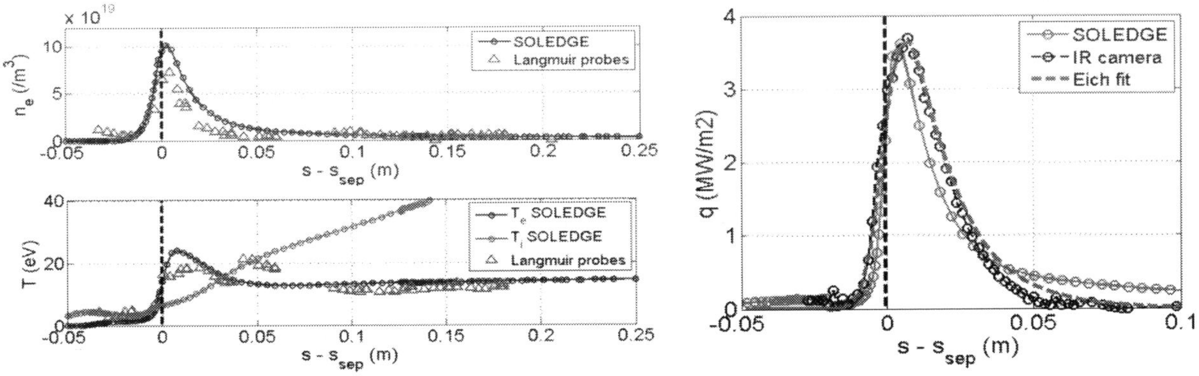


*Figure 4 : Comparison between measurements (symbols from [5]) and SOLEDGE2D-EIRENE simulation results for outer target profiles on AUG..*

On Figure 4 is also displayed a fit with Eich's function [6]. The fitting parameter found for this AUG case ($f_x = 4.65$) are: heat flux radial decay $\lambda_q = 3.5mm$, divertor spreading $S = 1.7mm$. The value found for $\lambda_q$ is compatible with the scaling law given by [7]. We consider that this case provides a reference set of transport coefficients for an H-mode plasma in a medium-size tokamak, and thus rely on this set for WEST simulations.

**WEST H-mode simulation**

The wall material for WEST is Tungsten and Tungsten sputtering is not taken into account in the simulation. The Carbon radiator sputtered from the wall in the AUG case is replaced by Nitrogen seeding in the lower divertor private flux region. Nitrogen is supposed to entirely recycle on the wall ($R_N^\star = 1$). The input power entering the SOL is set to $P_{SOL} = 8MW$ (equal repartition between ions and electrons), the electron density at the separatrix is set to $n_{sep} = 3.4 \cdot 10^{19} m^{-3}$ by adjusting core fueling and the Nitrogen puff rate is set to $\phi_N = 5.6 \cdot 10^{21} el/s$. Simulation results are summarized on Figure 5. The following values are found for heat flux width using Eich fit: $\lambda_q = 3mm$ and $S = 1.2$mm.

This simulation is part of an ongoing effort to map the WEST tokamak operation domain, which is aimed at providing guidance to prepare for WEST operation.

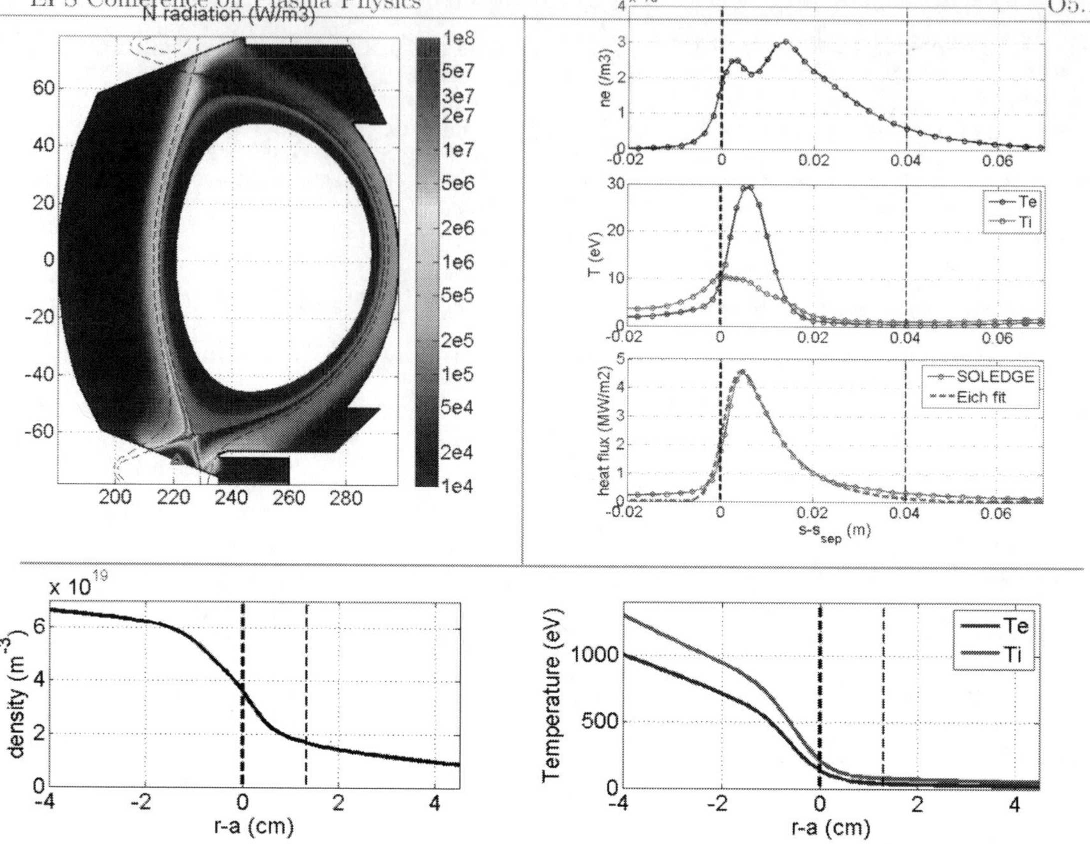


*Figure 5 : SOLEDGE2D-EIRENE simulation results for WEST H-mode discharge. Top Left-hand quarter: Nitrogen radiation (purple triangle show location of Nitrogen puff). Top Right-hand corner: Outer target profiles. Bottom: outer mid plane profiles.*


## Acknowledgments

This work was granted access to HPC resources of Aix-Marseille Université financed by the Equip@Meso (ANR-10-EQPX-29-01) of the program 'Investissements d'Avenir' supervised by the Agence Nationale pour la Recherche. We also acknowledge ASDEX Upgrade team for providing access to the database. Finally, part of this work has also been carried out within the framework of the French Research Federation for Fusion Studies, and of the EUROfusion Consortium (WP PFC) and has received funding from the Euratom research and training programme 2014-2018 under grant agreement No 633053. The views and opinions expressed herein do not necessarily reflect those of the European Commission.


## References


[1] C. Bourdelle et al., *Nucl Fusion* **55** (2015) 063017
[2] J. Bucalossi et al., *Fusion Eng. and Design* **10** (2014) 907-912
[3] H. Bufferand et al., *Nucl. Fusion* **55** (2015) 053025
[4] J. Rosato et al., *J. Quantitative Spectroscopy and Radiative Transfer* (2015)
[5] A. V. Chankin et al., *Plasma Phys. Control. Fusion* **48** (2006) 839-868
[6] T. Eich et al., *Phys. Rev. Lett.* **107** (2011) 215001
[7] T. Eich et al., *Nucl. Fusion* **53** (2013) 093031

# ECR plasma heating experiment in the GDT magnetic mirror

P. A. Bagryansky[1, 3], A. V. Anikeev[1, 3], A. A. Balakin[1, 2], G. G. Denisov[2], E. D. Gospodchikov[1, 2], A. A. Ivanov[1, 3], A. A. Lizunov[1], T. A. Khusainov[1, 2], Yu. V. Kovalenko[1, 3], V. I. Malygin[2], V. V. Maximov[1, 3], O. A. Korobeinikova[3], S. V. Murakhtin[1, 3], E. I. Pinzhenin[1], V. V. Prikhodko[1, 3], V. Ya. Savkin[1, 3], A. G. Shalashov[1, 2], O. B. Smolyakova[1, 3], E. I. Soldatkina[1, 3], A. L. Solomakhin[1, 3], D. V. Yakovlev[1, 3], K. V. Zaytsev[1]

[1]*Budker Institute of Nuclear Physics, Siberian Branch of RAS, Novosibirsk, Russia*

[2]*Institute of Applied Physics of Russian Academy of Sciences, Nizhny Novgorod, Russia*

[3]*Novosibirsk State University, Novosibirsk, Russia*

Open magnetic systems for plasma confinement are suitable for a number of nuclear fusion applications. The near term one is a high-power D-T fusion neutron source (see for example [1,2]) capable of producing neutron flux of several MW/m$^2$. Such neutron flux is required for full-scale neutron-material interaction research, aimed at design of first wall and other structural elements of future fusion reactors. Furthermore, high-power neutron source can be used as a driver for subcritical fission reactors including devices for burning of long lived radioactive wastes [3,4,5]. Finally, a series of studies has shown that open magnetic traps with reasonably improved axial confinement are completely consistent with reactor-level fusion projects with power gain factor $Q \gg 1$. Improvement of axial confinement can be achieved, for example, through the use of ambipolar plugs [6,7] or multimirror end-sections [8]. The most attractive from engineering and physical standpoint are axisymmetric magnetic mirrors. The simplicity of design, intrinsic capability of sustaining high beta plasmas ($\beta \approx 1$), natural channel of impurities and thermonuclear ashes removal, possibility of direct power conversion with near-unity efficiency are crucial benefits of open magnetic systems for plasma confinement.

However, lack of experimental data for the electron temperatures suitable for nuclear fusion applications significantly diminishes the advantages of open systems mentioned above. Low electron temperature, most probably, is caused by high level of axial energy losses which was observed in many experiments. This circumstance accounts for today`s relatively low level of research activity in the field of open systems for magnetic plasma confinement.

This paper presents the first results of auxiliary electron cyclotron plasma heating (ECRH) experiment, which is presently under way on the gas-dynamic trap (GDT, Budker Institute, Novosibirsk) machine – an axisymmetric magnetic mirror with high mirror ratio [9,10]. The plasma confined in GDT consists of two components with different mean energies. First one is so-called warm ions with isotropic Maxwell velocity distribution. These ions are confined in a gas-like (gas-dynamic) regime and thus have isotropic speed distribution due to high collision frequency. The second component is a population of hot ions, which is produced as a result of oblique injection of hydrogen or deuterium neutral beams into the plasma. The hot ions are confined in adiabatic regime which means that their movement is governed by conservation of adiabatic invariants. As a result they are bouncing within the region between two turning points near the magnetic mirrors.

One of the main issues is the magneto-hydro-dynamic (MHD) stability of the bulk plasma, for which the current configuration of the GDT is inherently unstable. To suppress the anomalous transverse transport caused mainly by flute MHD modes, we use a novel technique, which we call "vortex confinement" [11]. Strictly speaking, this method is not meant to suppress MHD modes, but rather to saturate them at a relatively low level by the differential rotation of outer plasma layers induced by an externally applied radial electric field. This produces a vortex-like structure with essentially closed flux lines. In the GDT, the vortex confinement technique is realized by applying a biasing potential between the ring-shaped radial plasma limiters and the central sections of plasma-facing end plates. The vortex confinement results in a stable confinement of hot plasma in the central core region, which appears to be unaffected by peripheral convection. The main conclusion based on the theoretical analysis [11] and experimental results [12] is that the transverse power losses can be limited to the level of 10–15% of the longitudinal (gas-dynamic) losses. Implementation of the vortex confinement in GDT allowed achieving a record value, for axisymmetric traps, of $\beta \approx 0.6$ [12, 13]. To provide an optimal confinement the biasing potential should be of the order of $T_e$. This condition is easily violated during the fast rise of the electron temperature in ECRH experiments. Thus, to stabilize plasma we need to apply additional voltage to the plasma periphery during the ECRH phase.

Energy confinement times of hot ions as well as their velocity spread are determined mostly by the collisional slowing-down on the bulk electrons. Since the collisional time $\tau_{ei} \propto T_e^{3/2}$, the electron drag force is rapidly decreasing with increasing electron temperature. For this reason, electron temperature is the main factor limiting the

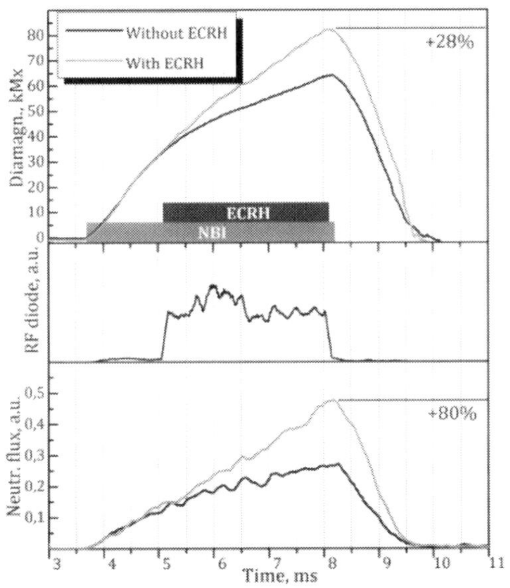


**Fig. 1.** Diamagnetic signal of fast ions with and without ECRH in magnetic configuration with broad power deposition (top). Stray radiation signal (middle) illustrating steady absorption of microwave power. D-D neutron flux signal (bottom) [18].

confinement time of fast ions and thus the power efficiency of a beam-driven fusion reactor based on a magnetic mirror. To significantly raise the electron temperature and thus the overall energy confinement time of the hot ion component, an electron cyclotron resonance (ECR) heating system based on two 450 kW/54.5 GHz gyrotrons was designed and installed. The crucial part of the heating scheme [14] is the strong refraction of the microwave beam in the inhomogeneous magnetized GDT plasma, which allows the beam to reach the ECR region. Technical details of ECRH system at the GDT are described in Ref. [15].

Two distinct modes of ECR heating were established which we call the "core heating" and the "broad heating". In the core heating mode the electron temperature increased only within a few centimetres from the device axis and the ECR heating started after 2 ms of NB injection. After approximately 0.3 ms the electron temperature, measured by laser scattering on the axis of the machine, increased from 200 eV (without ECR) to 400 eV, while the scattered spectrum indicated that the electron velocity distribution remained Maxwellian [16]. Shortly after this we observed an MHD activity, which dramatically increased the radial losses. In the broad heating mode the ECR surface had been shifted by 3 cm towards the mirror and the heating started 1 ms earlier. This resulted in a much wider power deposition and much more stable plasma (was stable during the whole time of NB operation). There was also a strong rise both in diamagnetic signal and in the neutron flux from D-D reactions (see Fig.1.), which might be of impotance in thermonuclear applications of the GDT device. In the next experiment the core heating mode was essentially stabilized by appling an additional (two-step) biasing voltage to the limiter to provide more optimal regime of vortex confinement. The sceme allowed adding limiter voltage during the shot at a specified time moment. The time of stable ECR heating was increased up to 0.6 ms what resulted in the record electron temperature more than 900 eV registered at the GDT to date. Plasma density in this mode was $0.65 \cdot 10^{19}$ m$^{-3}$. Even though

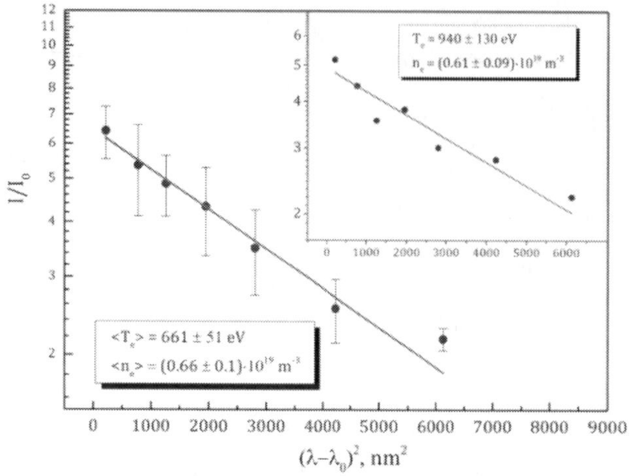


**Fig. 2.** Electron energy spectrum measured by Thomson scattering on the axis and averaged over 7 consecutive shots. Fit of these data suggests a Maxwellian electron distribution function with an electron temperature of $660 \pm 50$ eV and a density $(0.66 \pm 0.10) \times 10^{19}$ m$^{-3}$. The same data for one of the shots with an electron temperature above 900 eV is shown in the insert [17,18].

the MHD activity is still present, it no longer leads to the dramatical loss of the entire plasma and tends to become less destructive with the increase of the limiter potential jump. The energy spectrum of heated electrons is close to Maxwellian (see Fig. 2.). We find that our theoretical understanding of resonant plasma heating has been proven experimentally, thus the proposed novel ECRH scheme works quite robustly. The measured increase of electron temperature to nearly 1 keV along with the previous experiments, which demonstrated plasma confinement with $\beta \approx 60\%$, provide a firm basis for extrapolating the gas dynamic trap concept to fusion-relevant applications [17,18].

The work is supported by the Russian Scientific Foundation (Project No. 14-12-01007).

[1] Coengsen F.H., et al., J. Fusion Energy **8** (1989) 237.

[2] Bagryansky P.A., et al., Fusion Eng. Des. **70** (2004) 13.

[3] Simonen T.C., et al., Nucl. Fusion **53** (2013) 063002.

[4] Noack K., et al., Ann. Nucl. Energy **35** (2008) 1216.

[5] Yurov D.V., et al., Fusion Eng. Des. **87** (2012) 1684.

[6] Dimov G.I., Phys. Usp. **48**(11) (2005) 1129 (in Russian).

[7] Fowler T.K. and Logan B.G., Comments on Plasma Physics. Controlled Fusion **2** (1977) 167.

[8] Beklemishev A., et al., Fusion Sci. Technol. **63** 1T (2013) 46.

[9] Mirnov V.V. and Ryutov D.D., Sov. Tech. Phys. Lett. 5 (1979) 279 (in Russian).

[10] Ivanov A.A. and Prikhodko V.V., Plasma Phys. Control. Fusion **55** (2013) 063001.

[11] Beklemishev A.D., et al., Fusion Sci. Technol. **57** (2010) 351.

[12] P. A. Bagryansky et al., Fusion Sci. Technol. **59**(1t), 31 (2011).

[13] Simonen T.C., et al., J. Fusion Energy **29** (2010) 558.

[14] Shalashov A.G., et al., Phys. Plasmas **19** (2012) 052503.

[15] Bagryansky P.A., et al., Fusion Sci. Technol. **63** (2013) 340.

[16] Bagryansky P.A., et al., Nucl. Fusion, **54** (2014) 082001.

[17] Bagryansky P.A., et al., Phys. Rev. Letters, **114**, 205001 (2015).

[18] Bagryansky P.A., et al., Nucl. Fusion **55,** (2015) 053009.

# Time-resolved measurements of ion energy distribution function in cold low-density plasma induced by nanosecond pulses of extreme ultraviolet radiation

A. Abrikosov[1,2], O.Yakushev[1,3], D.Lopaev[1,4]

[1] *EUV-Labs, Ltd., Moscow, Troitsk, Russia*

[2] *Moscow Institute of Physics and Technology, Moscow, Russia*

[3] *Lebedev Physics Institute, Moscow, Russia*

[4] *Skobeltsyn Institute for Nuclear Physics, Moscow State University, Moscow, Russia*

Extreme ultraviolet lithography (EUVL) is a stage of semiconductor chip manufacturing which is expected to be introduced at technical nodes 22 nm and below [1]. An EUVL tool consists of pulsed light source and mirror optical system operating at a wavelength 13.5 nm. Since such radiation is strongly absorbed by any gases, the EUVL tool is maintained under ultra-high vacuum conditions. However, due to technical characteristics, various background gases, mainly hydrogen and argon, are present in the vacuum vessel of an EUVL tool at low pressures of ones to tens pascal. During operation these gases are ionized by both the high-energy EUV photons and the collisions with the secondary photoelectrons emitted by the mirrors. Thus, the plasma is formed within the EUVL tool.

Such plasma is known to cause various physical and chemical processes at the optical surfaces. These include, for example, etching the carbon contaminations that can accumulate on the mirrors due to hydrocarbon decomposition [2]. Other well-known effect is the delamination of the uppermost layers of the mirror, the so-called blistering [3]. The latter leads to the decrease in mirror reflectivity, which is undesirable for the commercial viability of EUVL tools. In order to reduce the degradation caused by the plasma, methods of controlling such plasma are required. This, in turn, calls for the development of diagnostic methods.

In this paper we present the unique experimental setup which allows to reproduce the experimental conditions close to those of an actual EUVL tool [4]. This consists of the pulsed EUV light source, the collector mirror and the ultra-high vacuum clean chamber. The clean chamber can be pumped for high vacuum and pressurized with various background gases. In this chamber the sample holder is mounted thus imitating the EUV-irradiated surface. We implement a retarding field energy analyzer (RFEA) as a diagnostic technique. This technique allows measuring the ion energy distribution function, a very important quantity since it is precisely what determines the rates of the surface processes invoked by the plasma. The results of the measurements are presented at the end of the paper.

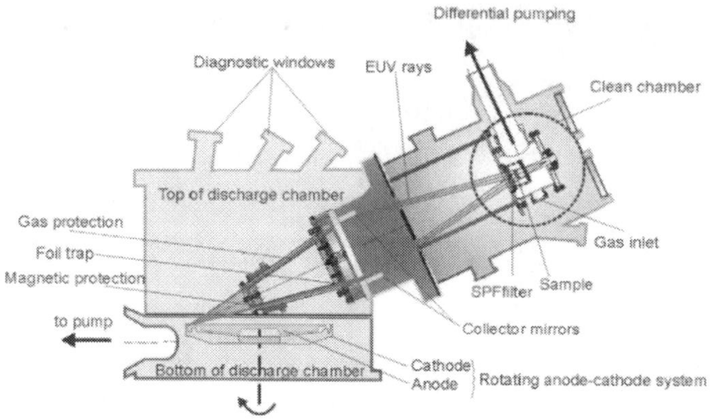


Fig. 1. Schematic view of the experimental setup.

The experimental consists of an EUV source based on the laser-ignited z-pinch discharge in tin vapor. It emits radiation pulses with the repetition rate of 2 kHz. The radiation spectrum in the EUV range has a strong peak at 13.5 nm. The duration of EUV pulse is about 40 ns at half-maximum. The light from the pinch is collected by the grazing incidence collector and then focused into THE clean chamber. To filter out the visible and deep UV radiation the light passes through the spectral purity filter which is a 150 nm thick zirconium foil covering the entrance window of the chamber. The light entering the chamber consists mostly of wavelengths in the range 10-20 nm. In the focal plane of the collector a sample holder for irradiated surface is installed. It is a copper electrode with a mount for 1 inch wafers, which in our experiments we replaced with the RFEA.

The clean chamber can be pressurized with the background gas with the pressures up to 100 Pa. In our measurements we use hydrogen as it is the primary gas present in the EUVL tool. In order to enhance ionization, a negative bias voltage can be applied to the sample. This creates an electric field which accelerates secondary electrons emitted from the surface by photoelectric effect. The bias voltage also varies the energy of ions impinging onto the surface. Opposite to the sample, a grounded cylindrical electrode is mounted. This electrode focuses the field lines and confines the plasma within its volume. The conditions within the clean chamber are close to those of actual EUV lithography tool.

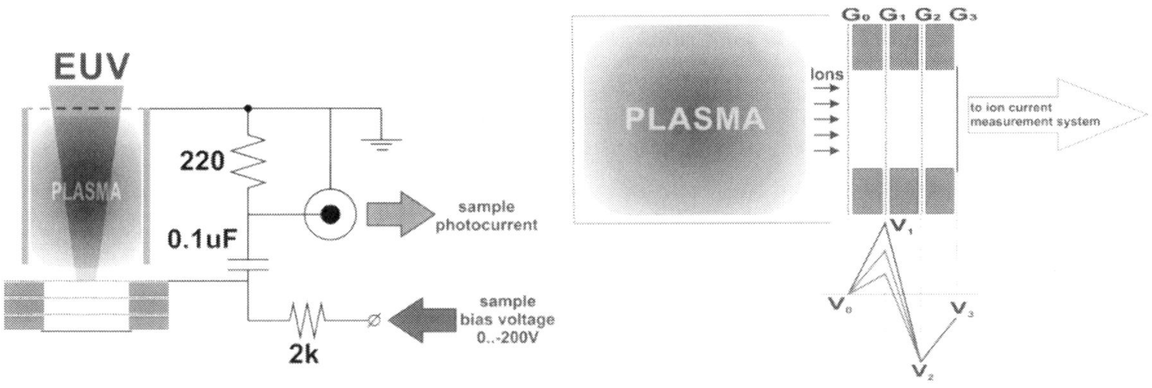


Fig. 2. Schematic view of the clean chamber rig(left), internal structure of the retarding field analyzer (right).

The structure of the retarding field energy analyzer is shown in fig. 2. It consists of three nickel grids $G_0$, $G_1$ and $G_2$ each with the transparency of about 40%, separated by ceramic insulator plates of 0.5 mm thickness. It is terminated with a copper collector plate $G_3$ which is connected with the current measurement system. The grid $G_0$ is facing the plasma and EUV radiation and is biased to the desired sample potential of 0 to -200 volt. All the other grids are biased with respect to $G_0$. The grid $G_2$ is biased to the potential -60 volt to repel secondary electrons, the collector plate $G_3$ has a potential of -23 volt to attract ions entering the analyzer. The grid $G_1$ acts as a potential barrier for ions. During the measurements, the potentials $V_0$, $V_2$ and $V_3$ are fixed constant while $V_1$ is swept above $V_0$. Such voltage sweep leads to the decrease in ion current so that only the ions with energies $E>eV_1$ pass the barrier and contribute to the collector current. The energy distribution function is therefore obtained by taking the derivative $dI/dV_1$ of the ion current.

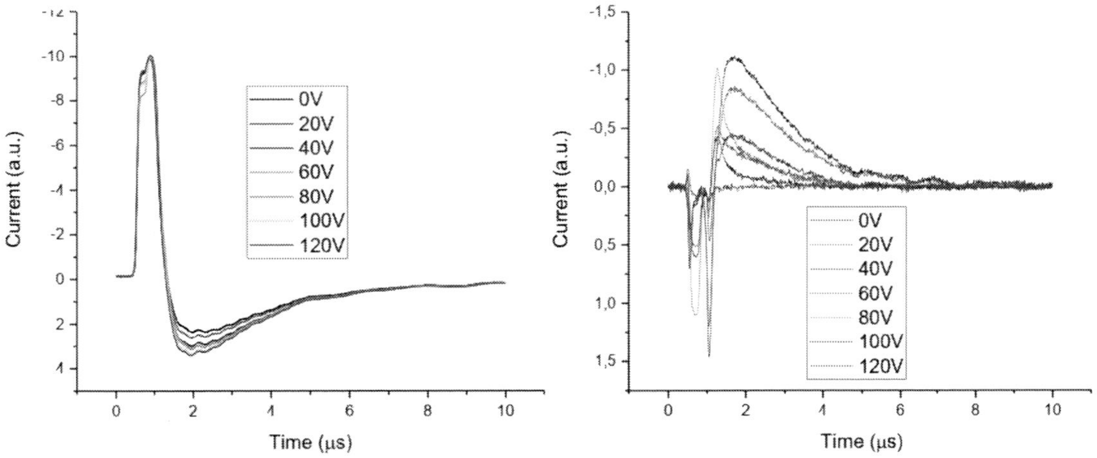


Fig. 3. Ion current at different $V_1$ values as recorded (left) and with the vacuum baseline subtracted (right)

Fig. 3 depicts the shape of the analyzer current at different values of $V_1$. All the measurements were performed in 30 Pa of $H_2$ with the upper grid potential $V_0$=-100V. The signal has a distinct baseline which is due to the displacement current within the analyzer. This can be eliminated by subtracting the pulse in vacuum. During the first 1.5 microseconds the useful signal can't be extracted because of the current amplifier oscillations. After 6 microseconds the signals drop below the noise level. Fig. 4 depicts the current-voltage curves extracted from the signals at various times. These are smoothed and differentiated in order to obtain the ion energy distribution function.

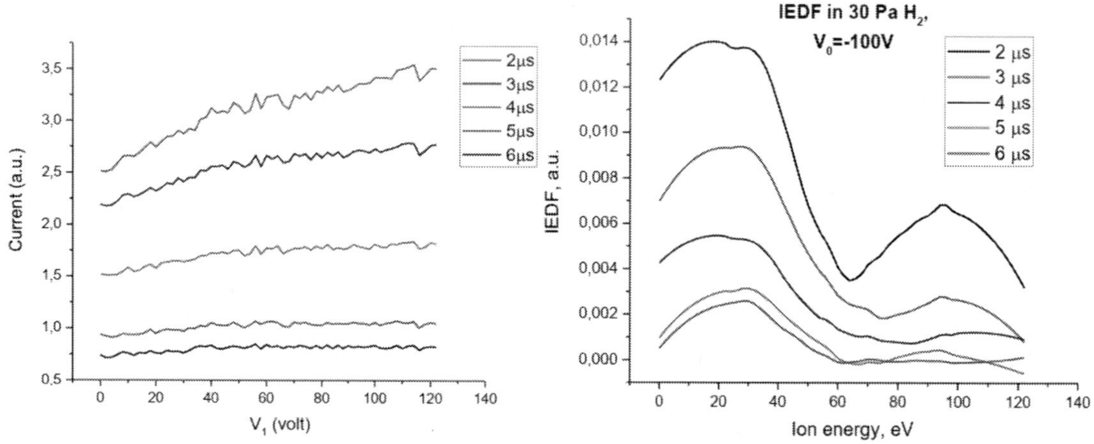


Fig. 4. IEDF integral as obtained from the current pulses (left), IEDF at different time moments (right)

We see that the ion spectrum is bimodal with two distinct peaks. However, the peak around 20 volt is not genuine but due to inaccurate performance of the analyzer at low $V_1$. Such is caused by ionization within the analyzer and, therefore, the distortion of the retarding potential. The second peak is due to acceleration of ions by the upper grid potential $V_0$. Therefore, the IEDF is peaked around 100 eV.

**References**

[1] Bakshi V 2006 *EUV Sources for Lithography* (Bellingham,WA: SPIE press)

[2] Chen J Q *et al* 2009 *Opt. Express* **17** 16969–79

[3] Kuznetsov A S, Gleeson M A and Bijkerk F 2012 *J. Phys.Condens. Matter* **24** 052203

[4] A Dolgov *et al* 2015 *Plasma Sources Sci. Technol.* **24** 035003

# Progress in understanding W control using ICRH in the JET-ILW tokamak

P.Mantica[1], F.J.Casson[2], M.Valisa[3], C.Angioni[4], R.Bilato[4], E.Belonohy[4], C.Giroud[2], M.Goniche[5], N.Hawkes[2], E. Lerche[6] and JET contributors*

EUROfusion Consortium, JET, Culham Science Centre, Abingdon, OX14 3DB, UK

[1] *Istituto di Fisica del Plasma "P.Caldirola", Consiglio Nazionale delle Ricerche, Milano, Italy*
[2] *CCFE, Culham Science Centre, Abingdon, OX14 3DB, UK*
[3] *Consorzio RFX, Consiglio Nazionale delle Ricerche, Padova, Italy*
[4] *Max Planck Institut fur Plasmaphysik, Garching, Germany*
[5] *CEA-Cadarache, St. Paul Lez Durance, France*
[6] *LPP-ERM-KMS, Belgian State, TEC partner, Brussels, Belgium*
[*] *See the Appendix of F. Romanelli et al., Proceedings of the 25th IAEA Fusion Energy Conference 2014, Saint Petersburg, Russia*

W transport has been studied intensively in JET with ITER-like wall (ILW), since high performance Hybrid scenarios are presently limited in duration by the onset of W accumulation, whilst Baseline H-modes feature reduced pedestal performance possibly linked with the high gas puff rates needed to keep the W influx low.

In high NBI power JET plasmas, neoclassical W transport is enhanced by poloidal asymmetries due to centrifugal effects and was found to be the dominant player in the central region, giving rise to W accumulation as soon as the $n_e$ profile gets peaked [1]. In addition, at the high collisionalities of baseline scenarios, the poloidal asymmetries cause a reduction of the temperature screening, with partial loss of its beneficial effects against $n_e$ peaking [2]. Baseline scenarios can nevertheless reach stationary conditions due to the W expelling effect of sawteeth and less pronounced $n_e$ peaking. Hybrid scenarios, instead, although reaching better performance in the initial phase when $n_e$ is flat, get rapidly (2-3 s) polluted by W accumulation.

Extrapolation of both scenarios [3, 4] to the high current and 33 MW NBI power expected in next campaigns and in D-T experiments indicates that 1) at the high $n_e$ achievable in baseline scenarios the NBI deposition is shifted towards the outer region, thereby decreasing the central $n_e$ peaking and mitigating the W accumulation; 2) in Hybrids at $n_{e0}<1\ 10^{20}\ m^{-3}$ (required for core transport improvement) the NBI deposition is still central and contributes to peaked $n_e$ profiles and high rotation, both enhancing W accumulation. Therefore the use of ICRH in Hybrid scenarios as a control tool for core W transport is mandatory. First analysis of the effects of central ICRH power on W transport [5] indicates various mechanisms with beneficial effects: reduction of $n_e$ peaking, of rotation, and possible peaking of $T_i$ (all counteracting neoclassical inward convection), increased turbulent transport, additional screening from the fast ion temperature gradient, and poloidal asymmetries due to fast ion temperature anisotropy [6], which may lead to outward neoclassical convection. All these mechanisms require a localized central deposition, maximizing the gradients of the fast ions.

In this paper, experimental studies and related modelling of W transport in H-mode Baseline and Hybrid scenarios with/without ICRH at different radial locations is reported. The main diagnostic used for W is the SXR tomography, as described in detail in [1]. Mo Laser Blow-Off and the associated SXR time behaviour was also used to identify separately the diffusion and convection terms and so provide an even more stringent comparison with theory than by steady-state profiles only. The ICRH deposition has been evaluated with TORIC-SSFPQL [7] and simulations with NEO [8] and linear GKW [9] have been made to study the theory predicted effect of ICRH on W and compare with experimental results, as described in [1,2,5].

Figs.1-6 summarize results for two Baseline discharges at 2.8 T/ 2.5 MA, 85308 with 19 MW NBI and 85307 with 14.7 MW NBI and 4.9 MW ICRH (42.5 MHz, $\rho_{dep}$~+0.03, $n_H/n_D$~ 5.5%). One can observe in Fig.1 in presence of ICRH the flattening of $n_e$ and reduction of $v_{tor}$, decreasing the neoclassical inward convection, the peaking of $T_e$ increasing turbulent transport, and the peaked fast ion profile with $T_\perp/T_\parallel$~ 8. We note that with respect to the ICRH modeling of the same discharge in [5], a lower value of $n_H/n_D$ is used, after an experimental reassessment of the available measurements. Experimentally the discharge with ICRH does not exhibit the strong accumulation seen with NBI only (Figs. 2b and 3c) and has a hollow $n_W$ profile. The NEO+GKW calculated W diffusion (D) and convection (V) profiles in Figs.5-6 indeed show a strong inward convection (and consequently peaked $n_W$ - Fig.4) in the NBI only case, whilst ICRH is able to reverse the sign of the central V, leading to hollow $n_W$ profiles. This effect is already present when considering only the effective fast ion temperature $T_{eff}=(T_{\|}+2T_{\perp R0})/3$, but it is much larger when anisotropy is included. The 2D SXR W emission (Fig.3a) shows the hollow pattern with LFS poloidal asymmetry due to centrifugal effects, as in experiment. We note from Fig.4 that in this case the hollow W profile is due to the direct ICRH fast ion effects and not to the ICRH induced plasma profile modifications (case $T_H=T_i$), which attenuate the peaking but are not enough to reach the reversal.

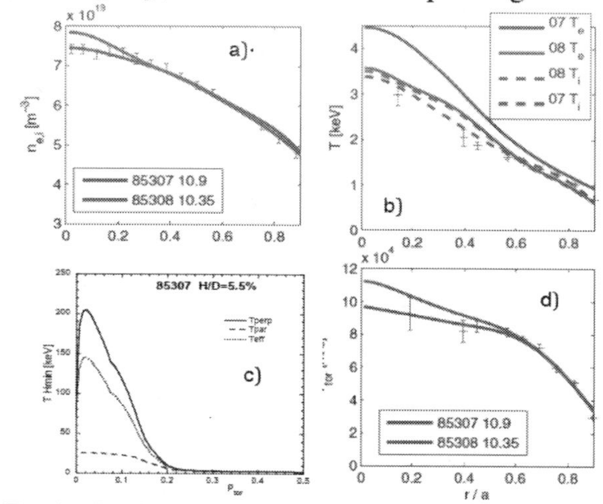


*Fig.1: Profile comparison of $n_e$, $T_i$, $T_e$, $v_{tor}$ and ICRH fast ion $T_H$ for baseline shots 85308 (NBI only) and 85307 (NBI+ICRH, H/D=5.5%)*

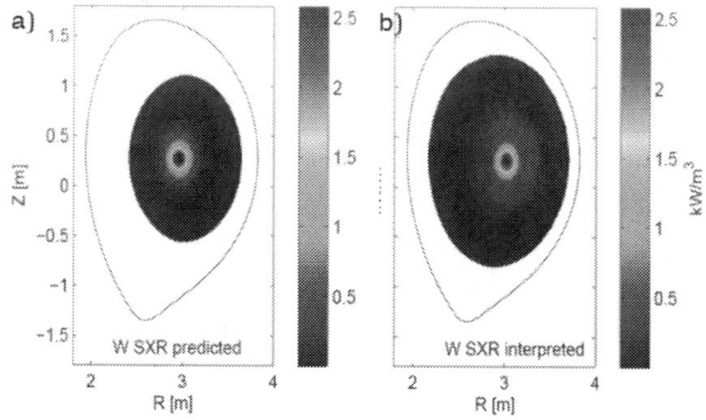


*Fig.2: 2D SXR emission due to W in shot 85308 (NBI only) at t=10.35 s: a) NEO+GKW simulation; b) experimental.*

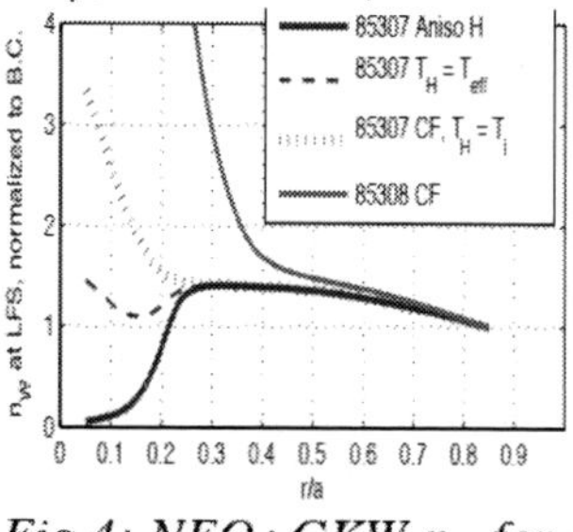


*Fig.4: NEO+GKW $n_W$ for 85307–85308. CF = only centrifugal effects*

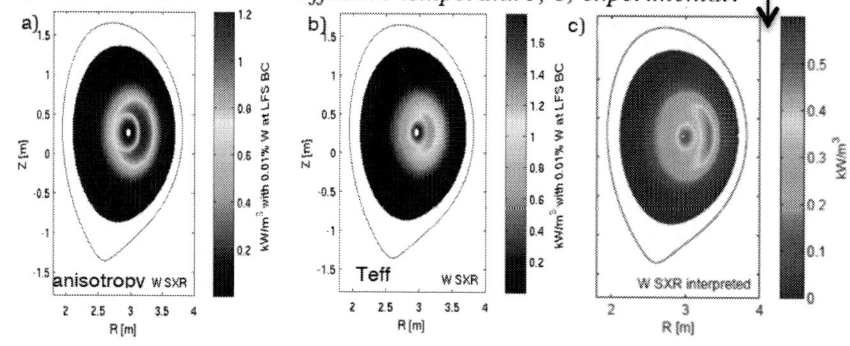


*Fig.3: 2D SXR emission due to W in shot 85307 (NBI +ICRH, H/D=5.5%), t=10.9s: a) NEO+ GKW simulation including fast ion anisotropy; b) NEO+GKW simulation with only fast ion effective temperature; c) experimental.*

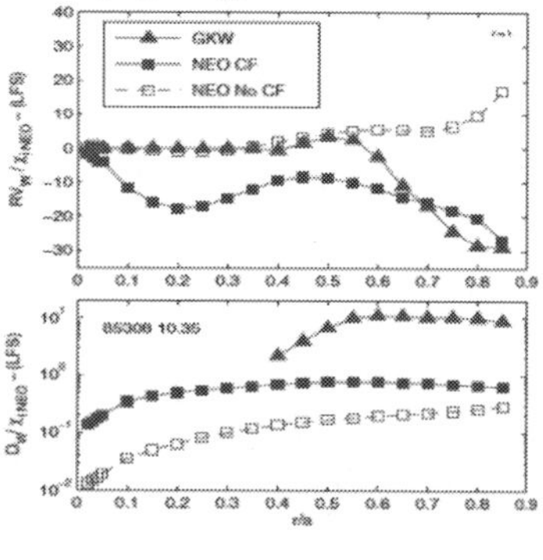


*Fig.5: NEO and GKW W D and V profiles for shot 85308*

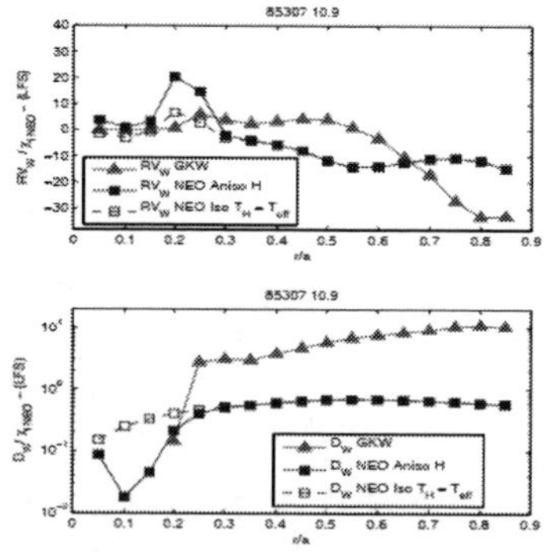


*Fig.6: NEO and GKW W D and V profiles for shot 85307*

In addition to good agreement between simulated and experimental steady 2D W density profiles, the NEO+ GKW flux surface averaged (FSA) Mo transport coefficients calculated for both shots have been used in time dependent 1D simulations with JETTO-SANCO [10] and UTC SXR post-processor to reproduce the time variation of the SXR line integrals following Mo LBO. As shown in Fig.7, very good

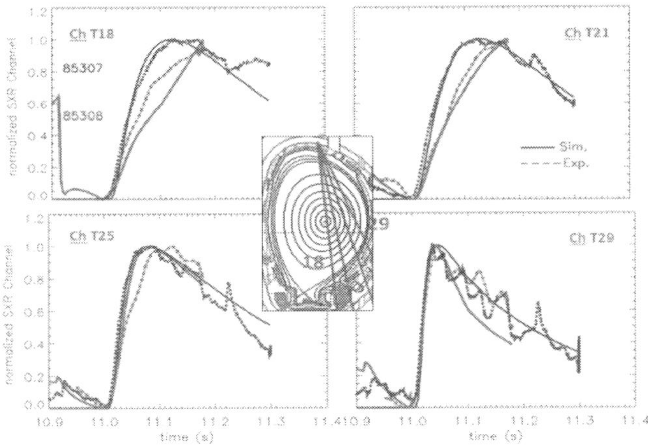


agreement is obtained when anisotropy is included. This was not the case for the simulations in [5] at $n_H/n_D$=9.4%, which shows the sensitivity of the modelling to the ICRH details.

*Fig.7: Experimental and NEO+GKW (with anisotropy) time evolution of a few selected channels of the SXR vertical camera during a Mo LBO for discharges 87307 (blue, NBI+ICRH) and 87308 (red, NBI).*

For Hybrid scenarios, we consider two discharges at 2.9 T/2.5 MA, 87253 with 22 MW NBI and 4 MW ICRH (46.2 MHz, $\rho_{dep}$~-0.14, $n_H/n_D$~ 4%) and 87252 with 22 MW NBI and 4 MW ICRH (51.4 MHz, $\rho_{dep}$~-0.4, $n_H/n_D$~ 4%). It is seen from the experimental SXR tomography (Fig.9) that only the discharge with central ICRH presents hollow W profiles, indicating that W accumulation can be controlled by ICRH in order to extend the scenario duration. Profiles for both shots are shown in Fig.8. In Hybrids due to non-linear e.m. stabilization [11] and low collisionality, $T_i$ can be significantly more peaked than $T_e$. Unfortunately due to the very low level of C in the central region, the $T_i$ profiles must be obtained from Be CX and are available with large uncertainties and only outside $\rho_{tor}$~0.3, whilst the central $T_i$ gradient is a key ingredient for this analysis. The choice was to use $T_i = T_e*2.5/(1.5 + \rho_{tor})$ which is shown in Fig.8b to lie within the uncertainties of the Be CX profiles where available. It is clear that the large freedom in the $T_i$ profile whose gradient plays a key role in determining neoclassical transport leaves a lot of room in the final W profiles obtained in the simulations. So our purpose here is mainly to investigate which of the various stabilization mechanisms associated to ICRH are at work in these Hybrid shots, and whether within the experimental uncertainties we can reach agreement with the experimental observations.

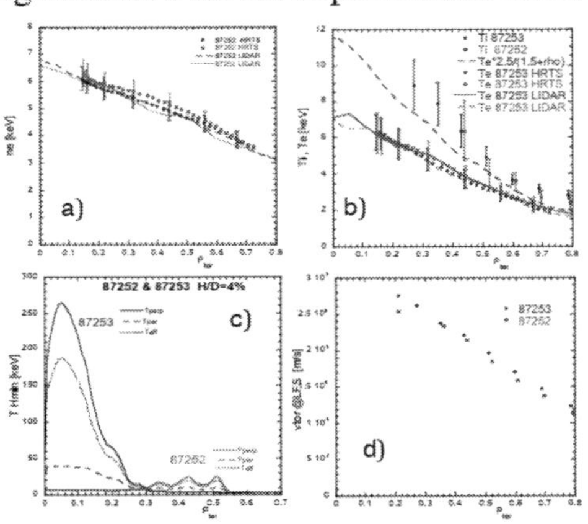


*Fig.8: Profiles of $n_e$, $T_i$, $T_e$, $v_{tor}$, ICRH fast ion temperatures for Hybrid 87253 (blue, ICRH on-axis) and 87252 (red, ICRH off-axis)*

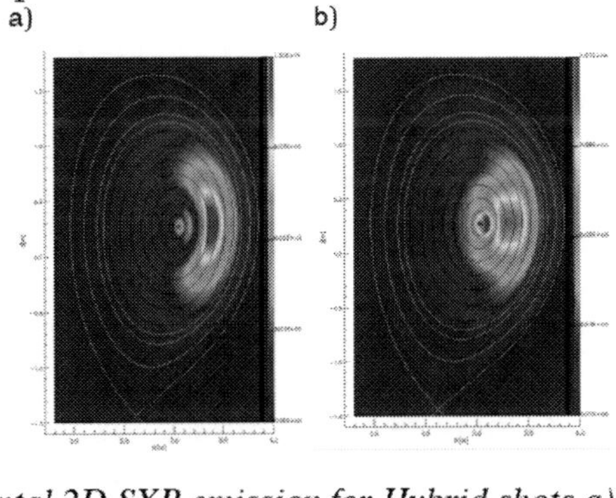

*Fig.9: Experimental 2D SXR emission for Hybrid shots a) 87253 (ICRH on-axis) and b) 87252 (ICRH off-axis) at t=6.4s*

Fig.8 shows that ICRH has no effect on $n_e$ and $T_e$, contrary to what seen in the Baseline case, which had a larger ICRH/NBI power ratio. Also the effect on rotation is negligible. The $T_i$ peaking is not expected to change much since ICRH in (H)-D is mainly electron heating. Instead the fast ion $T_{eff}$ and the anisotropy have sharp profiles for the on-axis ICRH case, whilst they are not expected to have any significant effects in the off-axis case (Fig.8c). Therefore we expect that the only mechanism at work to reduce W accumulation in Hybrids is the one associated to direct ICRH effects through fast ion screening and anisotropy. How general this conclusion is will be investigated in future experiments. Fig.10 shows the NEO+GKW calculated W D and V profiles. We note that since the two shots have identical plasma profiles and off-axis ICRH plays an insignificant role, the curves for 87252 practically coincide with those for 87253 with no ICRH effects included. Fig.11 shows the simulated 2D W SXR emission, which matches well the experimental one, yielding hollow $n_W$ only for the on-axis ICRH shot and only when anisotropy effects are included (Fig.12).

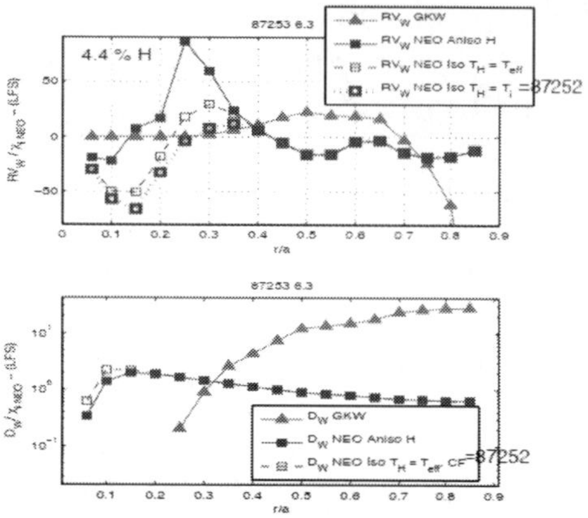


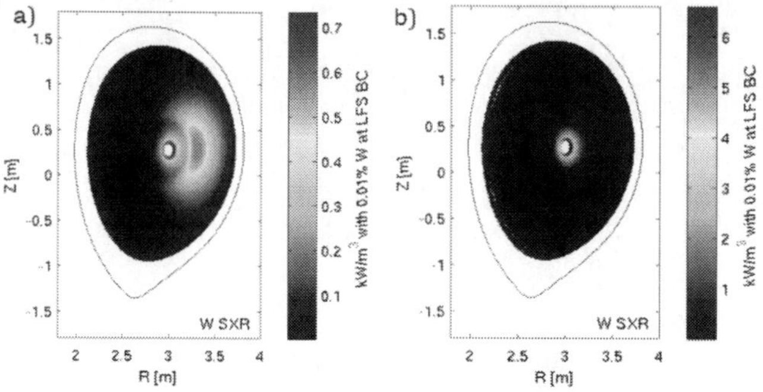


*Fig.11: NEO+GKW (with anisotropy) 2D W SXR emission for Hybrid shots a) 87253 and b) 87252*

*Fig.10: NEO+GKW W D and V profiles for shot 87253 (87252 coincides with 87253 $T_H$=$T_i$ or CF)*

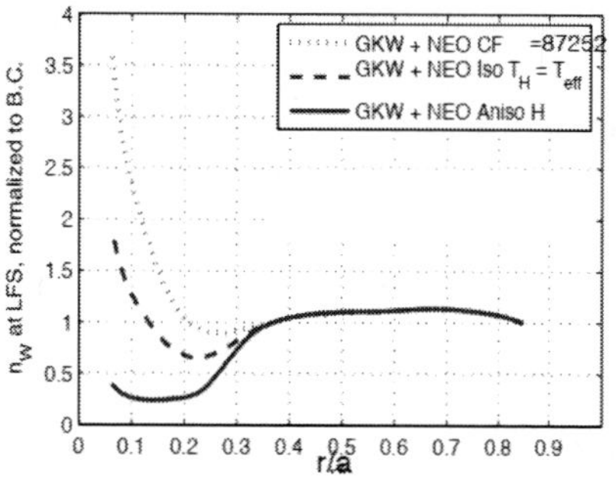


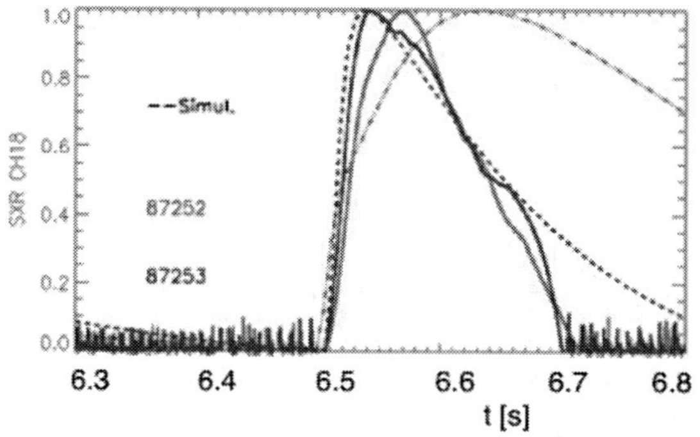


*Fig.12: NEO+GKW $n_W$ for 87253 (ICRH on-axis). 87252 (ICRH off-axis) coincides with 87253 CF.*

*Fig.13: Experimental (full) and simulated (dashed) time evolution of one selected central channel of the SXR vertical camera during a Mo LBO for discharge 87253 (blue, ICRH on-axis) and 87252 (red, ICRH off-axis).*

The NEO+GKW FSA Mo transport coefficients have been used in JETTO-SANCO simulations of Mo LBO. In Fig.13 the experimental and simulated SXR time traces indicate good agreement in 87253, whilst accumulation seems overestimated in 87252 (which is visible also in Fig.9b vs Fig.11b), requiring some sensitivity studies within the experimental uncertainties. A scan in minority concentration in simulations (not shown) indicates that when decreasing $n_H/n_D$ the fast ion screening is reduced (as it depends on minority density) and anisotropy effects get comparatively more weight, so overall intermediate $n_H/n_D$ ~4-5% are to be preferred. Concluding, it is found both in experiments and in the modelling that both in Baseline and Hybrid scenarios ICRH with narrow central deposition plays a major role in counteracting W accumulation, with a key role played by fast ion screening and anisotropy, and appears a promising tool to extend the duration of the high performance phase in next JET campaigns.

"This work has been carried out within the framework of the EUROfusion Consortium and has received funding from the Euratom research and training programme 2014-2018 under grant agreement No 633053. The views and opinions expressed herein do not necessarily reflect those of the European Commission."

[1] C.Angioni et al, Nucl. Fusion **54**, 083028 (2014).
[2] C.Angioni et al, Phys. Plasmas **22**, 055902 (2015)
[3] L.Garzotti, private communication
[4] J.Garcia, private communication
[5] F.Casson et al, Plasma Phys. Control. Fusion **57** (2015) 014031
[6] R.Bilato et al., Nucl. Fusion **54**, 072003 (2014)
[7] M. Brambilla and R. Bilato, Nuclear Fusion, **49**, 085004 (2009)
[8] E.Belli and J.Candy, Plasma Phys. Control. Fusion, **50**, 095010 (2008)
[9] A.G. Peeters et al, Comput. Phys. Commun. **180**, 2650 (2009)
[10] Lauro Taroni L. et al, Proc. 21st EPS Conf. on Controlled Fusion and Plasma Physics (Montpellier, France, 1994) vol 1, p 102
[11] J.Citrin et al., PRL **111**, 155001 (2013)

# The role of carbon on the H-mode confinement in ASDEX Upgrade with a metal wall

M.N.A. Beurskens[1,Previously 2], M.G. Dunne[3], L. Frassinetti[4], M Bernert[3], M. Cavedon[3], R. Fischer[3], A. Järvinen[5], A Kallenbach[3], F. M. Laggner[6], R. M. McDermott[3], G. Tardini[3], E. Viezzer[3], E. Wolfrum[3], the ASDEX Upgrade Team and the EUROfusion MST1 team*.

[1] *Max-Planck-Institut für Plasmaphysik, D-17491 Greifswald, Germany*
[2] *CCFE Fusion Association, Culham Science Centre, Abingdon, OX14 3DB, UK*
[3] *Max-Planck-Institut für Plasmaphysik, D-85748 Garching, Germany*
[4] *Division of Fusion Plasma Physics, Association VR, KTH, Sweden*
[5] *Aalto University, Tekes, P.O.Box 4100, 02015 Espoo, Finland*
[6] *Institute of Applied Physics, TU Wien, Fusion-ÖAW, 1040 Vienna, Austria*

**Section 1: Introduction**

Future fusion devices such as ITER and DEMO will have metal plasma facing wall components. This is because the conventional use of carbon as a first wall material is prohibited due to its high tritium retention properties. Seeding light impurities (like $N_2$) is needed to protect plasma facing components from steady state heat loads. An unexpected beneficial effect of this recipe is an improvement of confinement stemming from the pedestal region and resulting in confinement quality previously observed often only in carbon devices [1, 2, 3]. This observation raises the question as to whether nitrogen seeding in metal devices substitutes the presence of carbon as an intrinsic impurity, and which physics mechanisms are at play in the confinement improvement.

**Section 2: Experiment.**

In AUG the most extensive $N_2$ seeding studies aiming at confinement optimisation have been conducted in plasmas with a plasma current of $I_p$ = 1MA, $B_T$ =2.5T at low triangularity ($\delta$=0.2) and an edge safety factor of $q_{95}$ = 4.6 [1, 4]. In this paper, this reference scenario has been selected for comparing the effect of $N_2$ and $CD_4$ seeding on plasma confinement as part of an ongoing wider study [5,6,7] in which the potential of various seeding gases (He, $N_2$, $CD_4$, Ne, Ar and Kr) to optimise the confinement in combination with reduced divertor heat load mitigation, is reviewed.

A control experiment comparing $N_2$ and $CD_4$ seeding has been conducted with:

1) a reference plasma without seeding – (#30513) – $\Gamma_D \sim 2.5*10^{22}$ elec/s,

2) a plasma with $N_2$ seeding – (#30507) – $\Gamma_D \sim 2.5*10^{22}$ and $\Gamma_N \sim 1*10^{22}$ elec/s,

3) a plasma with $CD_4$ seeding – (#30518) – $\Gamma_D \sim 1.3*10^{22}$ and $\Gamma_{CD4} \sim 1.5*10^{22}$ elec/s.

Halfway through the pulse the upper triangularity is varied from $\delta_{up}\sim 0.05$ to $\delta_{up}\sim 0.3$ (lower triangularity fixed at $\delta_{low}\sim 0.45$). Plasma $I_p/B_T$ = 1MA/2.5T, input power $P_{tot}\sim$13MW ($P_{NBI}$=10MW and $P_{ECRH}$=1.5MW, $P_{Ohm}$=0.5MW)

Figure 1 shows that the plasma stored energy $W_{MHD}$, and normalised confinement $H_{98,y2}$ increase by 20-30%, for both $N_2$ and $CD_4$ seeded plasmas (numbering 1-3 as above). The plasma density remains unaffected by the seeding gases and is only affected by the change in plasma shaping where the density normalised to the Greenwald density increases from $f_{GW} \sim$ 0.6 to 0.75. The neutral particle density in the divertor $n_{0\text{-divertor}}$ and main chamber $n_{n\text{-main}}$ remain largely unaffected as does the divertor compression ratio $n_{0\text{-divertor}}/n_{0\text{-main}}$ ~200-230, a typical range for AUG. The confinement benefit of seeding is due to an increase in pedestal temperature by 20-30%, Figure 2&3, while the pedestal density remains unchanged.

*See http://www.euro-fusionscipub.org/mst1.

It should be noted that none of the shots discussed here suffers from confinement degrading 2/1 or 3/2 MHD activity in the plasma centre

The seeding experiment has led to the same impurity concentration, Figure 4, as measured with core and edge charge exchange system: $c_{N7+}$ and $c_{C6+}$ ~1.5-2% going from the plasma centre, and due to the similar charge of $C^{6+}$ and $N^{7+}$ , also similar derived $Z_{eff}$ profiles. The plasma radiation is affected by the introduction of the light-impurity gases. Figure 5 shows the tomographic reconstruction of the AUG bolometric data; as both N and C have their peak in radiation below T=100eV, mainly the divertor radiation is affected. The local radiation pattern of N and C are indeed very similar, with the increase in divertor radiation being the most apparent. The outer divertor remains in an attached regime and moves into a state of high recycling. This is seen from inter-ELM (50-90% of ELM cycle) Langmuir probe measurements (Figure 6) which show a strong increase in the ion saturation current and a slight decrease of the electron temperature for both seeding gases, typical for a transition to a high recycling regime. Note that this experiment does not aim at maximising radiation and detachment like in [6,7].

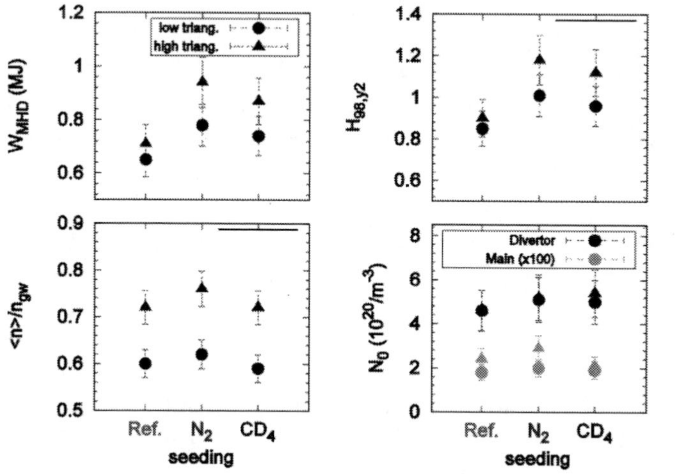


*Figure 1: Time averaged parameters for low triangularity phase (t=3-3.2s) and high triangularity phase (t=5-5.2s) for a) plasma stored energy; b) normalised confinement $H_{98,y2}$; c) Greenwald density fraction; d) divertor and outer-mid-plane main chamber neutral pressure*

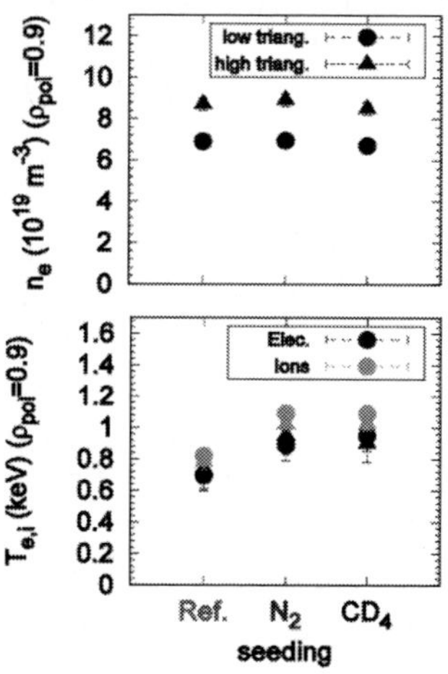


*Figure 2: Pedestal density and electron/ion temperature at $\rho_{pol}$=0.9 for low and high triangularity plasmas*

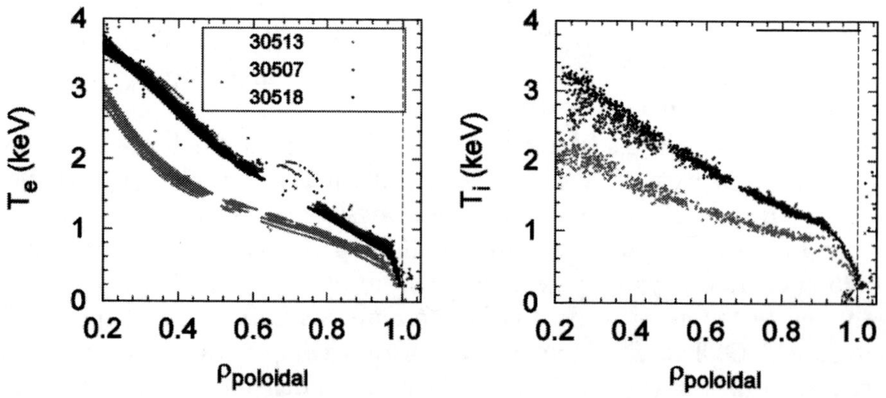


*Figure 3: a)$T_e$ from ECE b) $T_i$ from Charge exchange system for low triangularity phase*

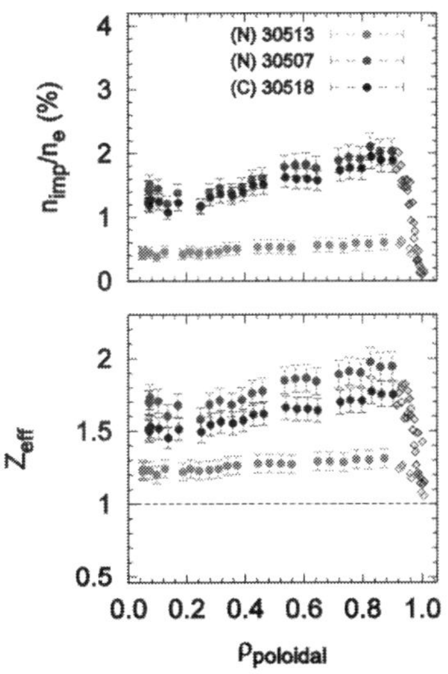


*Figure 4: impurity concentration of $N^{7+}/C^{6}$ and derived $Z_{eff}$ for the low $\delta$ phase (t=3-3.2s)*

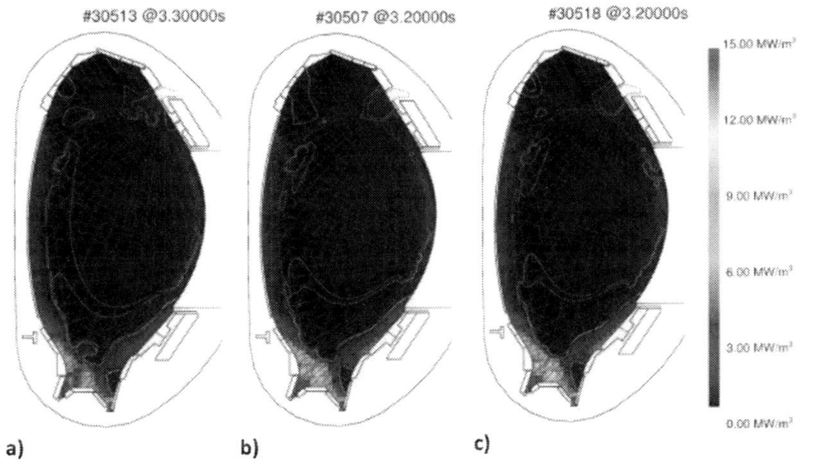


*Figure 5a) Reference, b) $N_2$ seeding c) $CD_4$ seeding: Radiated power density from tomographic reconstructions of bolometer lines of sight for the low triangularity phases. (high triangularity phases are not shown here, but show similar results)*

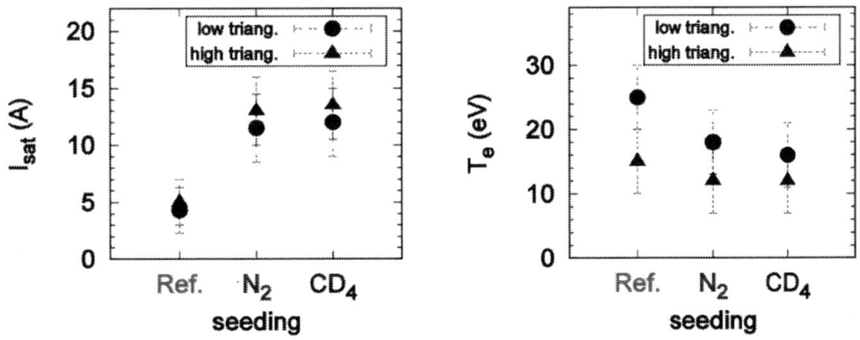


*Figure 6a) ion saturation current and b) target temperature as measured by Langmuir probes in between ELMs (50-90% of ELM cycle)*

**Section 3: Discussion and conclusions**

It is observed that seeding either nitrogen or carbon leads to a similar increase of the pedestal temperature by ~20-30%. In other aspects the plasmas look very similar too;

a) the radiation intensity and location are similar
b) the effect on divertor detachment state is similar
c) and the ELMs collapse time is reduced by equal amounts (not shown)

The experiment therefore shows that in a metal wall device, where the presence of carbon is disallowed due to tritium retention issues, nitrogen can be used to replace it as a radiator. Moreover it shows that the lower confinement often found in metal devices compared to carbon devices can (at least partly) be recovered by extrinsic nitrogen seeding. The outstanding question remains: What causes this confinement benefit driven by the pedestal temperature increase?

*Ion dilution:* The confinement benefit cannot stem from a dilution effect in combination with a hollow $Z_{eff}$ profile. Although figure 4 indeed shows that the $Z_{eff}$ profile is hollow, this is insufficient to cause an 20-30% confinement increase, as quantified in [3].
*Core transport:* The core profiles are stiff, like in [1, 3] so the confinement benefit cannot be caused by reduced core transport.
*Reduced particle exhaust:* both carbon and nitrogen are 'sticking' impurities [8], and could therefore influence the outgassing of deuterium from the first wall. However no evidence of a reduced neutral particle pressure is observed.
*Pedestal stability:* The seeded and unseeded pedestals are consistent with peeling-ballooning pedestal stability [5]. This as such does not explain why a higher pedestal temperature is obtained. The seeded plasmas have a higher $\beta_N$ than the unseeded one, which allows for reaching the ballooning limit at elevated pressure gradient. However, this does not explain the dynamics how the plasma reaches the increased $\beta_N$ in the first place
Another related possibility is that the impurity seeding affects the ratio of pressure gradient versus bootstrap current [5] in a way that allows a route to larger pressure gradients. This indeed can happen, but pedestal modelling shows that this can only cover a ~5% increase in $p_{ped}$ in a static situation.
*Separatrix temperature:* Finally, an option is presented where $N_2$ (or $CD_4$) seeding cools the separatrix temperature which effectively then shifts the steep pressure gradient region inwards to a region in the magnetic equilibrium where increased pressure gradients can be sustained. Simulations for JET have shown that a reduction of $T_{sep}$ from 100eV to 80eV, combined with a rigid shift of a modelled tangent hyperbolic pedestal profile, can indeed lead to an increase of the pedestal top pressure by 20-30% [9]. This would also require a reduction of the separatrix density, and therefore it is not certain that this is what happens in experiment presented here. The difficulty here is that to proof this mechanism experimentally, the profiles position with respect to the magnetic equilibrium needs to be known well within the current error margin of ~1cm.
In conclusion, the experiment presented here shows that the intrinsic presence of carbon in devices with a carbon plasma facing wall, can (at least in some cases) be replaced by nitrogen in metal devices. The two gases have very similar radiation potential and radiate mostly at temperatures below $T_e$=100eV, [10,11], which at similar impurity concentrations, leads to similar radiation and detachment behaviour. Some promising mechanisms have thusfar been presented to explain the confinement benefit, but a full understanding remains outstanding.

***Ackowledgements***
*This work has been carried out within the framework of the EUROfusion Consortium and has received funding from the Euratom research and training programme 2014-2018 under grant agreement No 633053. The views and opinions expressed herein do not necessarily reflect those of the European Commission.*

**References**
[1] J. Schweinzer, et al. Nuclear Fusion, 51(11), 2011
[2] C. Giroud, et al. Nuclear Fusion, 53(11), 2013
[3] M. Beurskens, et al Plasma Phys. Control. Fusion 55 (2013)
[4] A. Kallenbach, et al. Plasma Physics and Controlled Fusion, 55(12), 2013.
[5] M. Dunne, et al. In 42nd EPS Conference on Plasma Physics, Lisbon, 2015.
[6] A. Kallenbach et al 2015 Nucl. Fusion 55 053026
[7] F. Reimold et al 2015 Nucl. Fusion 55 033004
[8] S Brezinszek, et al, Journal of Nuclear Mat. 2014
[9] S Saarelma, accepted for pub. Physics of plasmas
[10] A. Järvinen, to be submitted to Nuclear fusion
[11] Giroud, C. et al. Plasma Phys. Control. Fusion 57 (2015) 035004

# MHD simulations of impurity transport during sawteeth for ASDEX-Upgrade plasmas

J.-H. Ahn[1], X. Garbet[1], R. Guirlet[1], O. Février[1], H. Lütjens[2], M. Sertoli[3],
the ASDEX-Upgrade Team and the EUROfusion MST1 Team*

[1] *CEA, IRFM, F-13108 Saint Paul-lez-Durance, France*

[2] *Centre de Physique Théorique, Ecole Polytechnique, CNRS, F-91128 Palaiseau, France*

[3] *Max Planck Institut für Plasmaphysik, Boltzmannstraße 2, D-85748 Garching, Germany*

## Introduction

Several tokamaks are now equipped with Tungsten plasma facing components (PFCs) to prepare the operation of ITER. It has been observed that in some conditions, the accumulation of Tungsten in core plasmas may lead to a radiative collapse due to excessive cooling. Hence, the control of heavy impurity transport has become a subject of renewed interest. RF heating appears to be a useful tool for preventing accumulation. However, in core plasmas, the precise mechanism has not been fully understood yet as it involves neoclassical and turbulent transport combined with sawtooth dynamics. In a number of experiments, it has already been shown that a sawtooth collapse induces a sudden radial transport of impurities, more specifically an impurity flux between edge and core plasma [1].

This paper presents intermediate steps towards the modelling of impurity behaviour during sawtooth cycles observed in an ASDEX-Upgrade discharge with a central ECRH deposition [2], performed during the EUROfusion MST1 2014 campaign. Fluid equations for impurities in the high collisional regime have been derived to describe the effects of neoclassical transport. This model has been implemented in the XTOR-2F code [3] in order to be coupled with nonlinear MHD equations for plasma electrons and ions. Also, preliminary results on the effects of a RF-like heat source on impurity transport are presented.

## Impurity transport model coupled with MHD equations

Heavy impurities are in the Pfirsch-Schlüter collisional regime due to their high charge numbers, while the main ion species remains in the banana-plateau collisional regime in core plasmas. In this framework, the neoclassical theory predicts that the ion temperature gradient drives an outward convection, and is thus responsible for a favorable screening effect [4].

The fluid equations (Eq. (1) and (2)) for impurities corresponding to this case have then been implemented in the XTOR-2F code

$$\partial_t n_z + \nabla.(n_z\mathbf{V}_z) + \frac{\nabla p_z}{Ze}.\nabla\times\frac{\mathbf{B}}{B^2} = \nabla.\left(D_z^{turb}\nabla n_z - \mathbf{V}_{p,z}^{turb} n_z\right) + S_z \tag{1}$$

$$\partial_t V_{\parallel,z} = -\frac{\nabla_\parallel p_z}{m_z n_z} - \frac{Ze\nabla_\parallel\phi}{m_z} - \nu_{zi}\left(V_{\parallel,z} - V_{\parallel,i}\right) + \frac{3}{5}\frac{Z^2}{m_z}\nabla_\parallel T_i \tag{2}$$

in addition to the fluid equations that describe the plasma electrons and ions [3]. The diffusion

*See *http://www.euro-fusionscipub.org/mst1*

coefficient $D_z^{turb}$ and the pinch velocity $\mathbf{V}_{p,z}^{turb}$ on the R.H.S. of Equation (1) correspond to the contribution of turbulent transport, and are defined as radial functions constant in time. It can be shown that this model reproduces well the neoclassical impurity transport, including the thermal screening. In fact, the term corresponding to the thermal screening can be derived from the expression of the thermal force between the main ion species and impurity. This model allows one to study the combined effect of collisional transport and sawtooth-induced transport of impurities in the Pfirsch-Schlüter regime.

Since the neoclassical time scales are larger than the time scales of MHD instabilities, neoclassical quantities are averaged on magnetic flux surfaces, denoted with angled brackets. With the previous fluid equations (Eq. (1) and (2)), the radial projection of neoclassical impurity flux is derived and can be written as

$$\langle \Gamma_z . \nabla r \rangle = -D_z^{neo} n_z (\partial_r \ln n_z - Z \partial_r \ln n_i - ZH \partial_r \ln T_i) \qquad (3)$$

where the constant $H \simeq -0.5$ for impurities in the Pfirsch-Schlüter regime and the main ion species in the banana-plateau regime [5].

**XTOR-2F simulation results**

In a first stage, simulations without kink, in a circular cross section, have been performed in order to verify the neoclassical transport model which has been implemented in fluid equations. To facilitate the monitoring of impurity transport, a flat profile was chosen for the initial impurity density. Also, analytical initial profiles of ion density and temperature are chosen and represented in Figure 1. It is expected that only the screening effect due to the temperature gradient appears in the region $r/a < 0.4$, where the ion density profile is flat. Figure 2-a shows that the model without the thermal force induces an inward flux of impurity due to the ion density gradient. However, the thermal force in the fluid equations clearly produces an outward impurity flux (Fig. 2-b). This test confirms that the fluid equations analytically derived and implemented in the XTOR-2F code reproduces the well known temperature screening effect as predicted by the neoclassical theory.

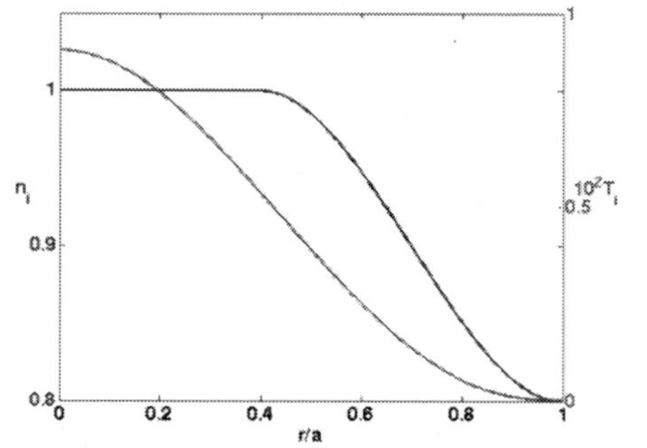


Figure 1: The initial ion density profile (blue) and temperature profile (green) in XTOR-2F units

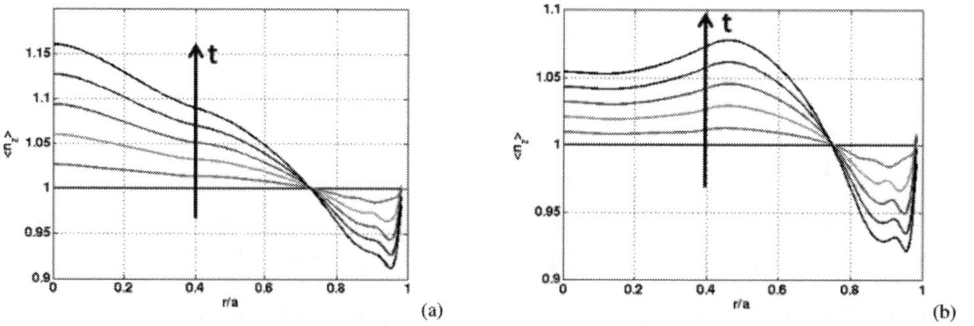


Figure 2: Time evolution of the surface averaged impurity density (a) without and (b) with thermal force in the model (Each colour corresponds to a time point).

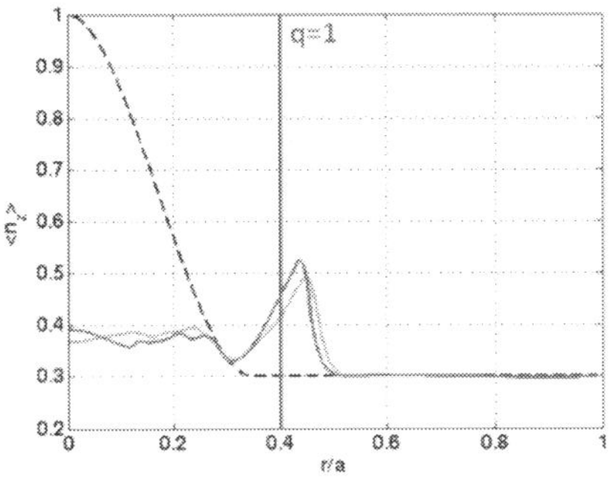


Figure 3: Surface averaged impurity density before (dashed lines) and after (solid lines) sawtooth crash for model without (green) and with (red) temperature screening.

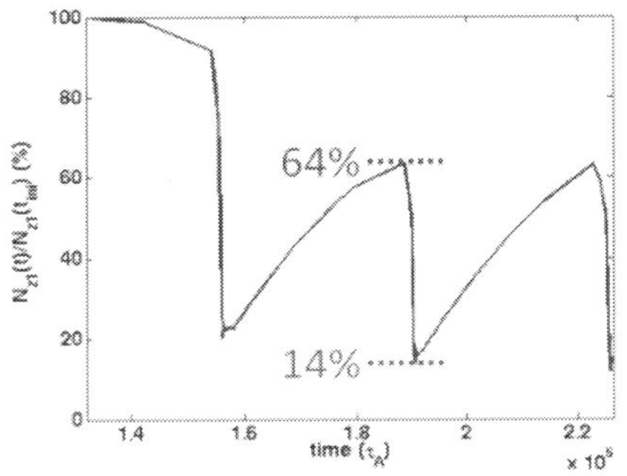


Figure 4: Time Evolution of the relative impurity quantity inside the volume $V_1$ delimited by the $q = 1$ surface (*cf.* Eq. (4)).

Figure 3 shows the impurity density evolution during a sawtooth crash from XTOR-2F simulations with the implemented model. An arbitrary peaked impurity profile is chosen at the beginning to highlight the temperature screening effect. The comparison of two cases, without and with temperature screening, reflects no remarkable difference in the impurity behaviour during the sawtooth crash, since the MHD time scale is much smaller than the neoclassical transport time scale. Figure 4 shows the time evolution of the ratio $N_{z1}(t)/N_{z1}(t_{ini})$, where the quantity $N_{z1}(t)$ is defined as

$$N_{z1}(t) = \int_{V_1} \left(n_z(\mathbf{r},t) - n_z(r_{q=1},t_{ini})\right) d^3\mathbf{r} \tag{4}$$

to assess the impurity content inside the plasma volume $V_1$ delimited by the $q = 1$ surface [6]. This quantity allows to trace and quantify the effects of the neoclassical transport which will appear after several sawtooth cycles in ongoing simulations.

In the XTOR-2F code, a heat source is added in the energy equation to model RF heating in the plasma. The propagation model of the heat source is identical to the one implemented for current drive [7]. However, this additional heat source does not precisely reflect ECRH nor ICRH deposition in experiments, as the actual XTOR-2F energy equation is written for total pressure. Thus, the heat source is considered to be equally distributed to plasma electrons and ions. A simulation without kink with a Gaussian heat source on the magnetic axis was performed, in order to test its effects on the impurity transport. Again, a flat profile was chosen for the initial impurity density to monitor the impurity transport. Figure 5 shows that, when an on-axis heat source is applied, impurities are depleted from the near-axis region of the plasma compared to the case without heat source (Fig. 2-b). Further analysis is necessary in order to understand how the applied heat source affects the MHD quantities and the impurity transport properties.

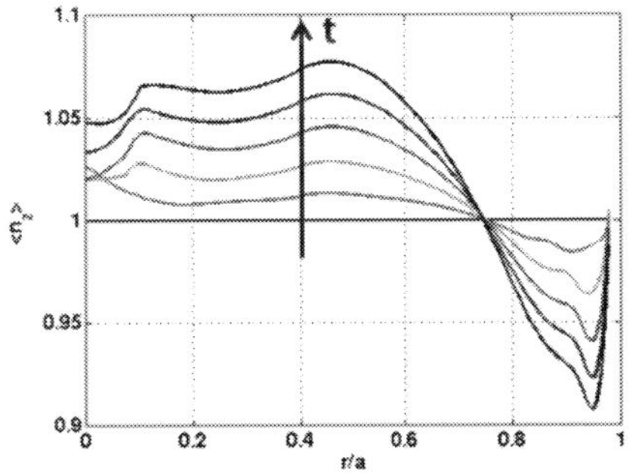


Figure 5: Time evolution of surface averaged impurity density when a heat source is applied on the magnetic axis (Each colour corresponds to a time point).

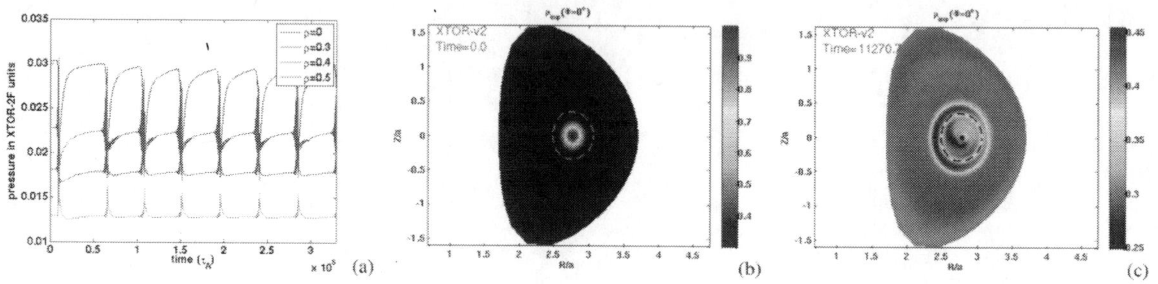

Figure 6: XTOR-2F simulation of sawteeth in an elongated geometry ($\kappa = 1.6$, $\delta = 0.4$). (a) Time evolution of pressure in different radial positions. Impurity density contours (b) before and (c) after sawtooth crash.

The XTOR-2F code has been shown to simulate sawtooth cycles in fair agreement with some experiments in circular cross section [8]. However, simulations of sawtooth cycles in a non-circular cross section have not been done yet. It is well known that the plasma elongation strongly destabilizes the kink mode [9]. A full scan of XTOR-2F input parameters has then been done to simulate stationary sawtooth cycles in an elongated cross section (Fig. 6-a). It is shown that the implemented model (Eq. (1) and (2)) gives the same impurity behaviour in an elongated cross section (Fig. 6-b and 6-c) as simulations in a circular cross section. Simulations in an elongated cross section will allow to compare the model with ASDEX-Upgrade discharges.

**Conclusion and prospect**

The XTOR-2F code has been modified to model impurity transport in the Pfirsch-Schlüter regime. Simulations show that the model reproduces the temperature screening effect as predicted by the neoclassical theory. Impurity transport is dominated by MHD convection during the sawtooth crash phase, as an effect of collisional transport is expected on a longer time scale. Thus, simulations of a large number of successive sawtooth cycles are ongoing to assess the interplay between the neoclassical transport and sawteeth. A full scan of input parameters allowed the XTOR-2F code to simulate sawteeth in an elongated cross section. It is also observed in simulations that the presence of a heat source tends to deplete core impurities.

With all these ingredients, plasma scenarios similar to the referred ASDEX-Upgrade discharge [2] will be simulated. Theses studies will allow to provide more physical understanding of the impurity dynamics and control for an efficient operation in future fusion devices.

➤*Numerical resources were provided by CINES, IDRIS and TGCC of GENCI (x2015056348), Mésocentre of Aix-Marseille University (15b025), Helios of IFERC-CSC (MaCoToP).*
➤*This work has been carried out within the framework of the EUROfusion Consortium and has received funding from the Euratom research and training programme 2014-2018 under grant agreement No 633053. The views and opinions expressed herein do not necessarily reflect those of the European Commission.*

**References**

[1] R. Dux et al., *Nucl. Fus.* **39** 1509 (1999)
[2] M. Sertoli et al., *this conference*, O4.129
[3] H. Lütjens and J.-F. Luciani, *J. Comp. Phys.* **229** 8130 (2010)
[4] P. Helander and D.J. Sigmar, *Cambridge Univ. Press* (2002)
[5] S.P. Hirshman and D.J. Sigmar, *Nucl. Fus.* **21** 1079 (1981)
[6] T. Nicolas et al., *Phys. Plasmas* **21** 012507 (2014)
[7] O. Février et al., *this conference*, P1.104
[8] T. Nicolas et al., *Phys. Plasmas* **19** 112305 (2012)
[9] H. Lütjens et al., *Nucl. Fus.* **32** 1625 (1992)

# Modeling of the impact of ECCD sweeping on NTM stability in ASDEX-Upgrade

O. Février[1]*, P. Maget[1], H. Lütjens[2], G. Giruzzi[1], M. Reich[3], O. Sauter[4], P. Beyer[5],
the ASDEX Upgrade team and the EUROfusion MST1 Team.†
[1]*CEA, IRFM, F-13108 Saint Paul-lez-Durance, France.*
[2]*Centre de Physique Théorique, Ecole Polytechnique, CNRS, France.*
[3]*Max-Planck-Institut für Plasmaphysik, D-85748 Garching, Germany.*
[4] *Ecole Polytechnique Fédérale de Lausanne (EPFC), Centre de Recherches en Physique des Plasmas, Station 13, CH-1015 Lausanne,Switzerland.*
[5]*Aix-Marseille Université, CNRS, PIIM UMR 7345, 13397 Marseille Cedex 20, France.*

**Introduction**

By reducing the maximum achievable $\beta$, Neoclassical tearing modes (NTMs) limit the performances of fusion devices. They can be controlled using Electron Cyclotron Current Drive (ECCD) as demonstrated in many tokamaks [1]. A new method, relying on sweeping ECCD deposition around the estimated position of the mode has been successfully tested on TCV[3] and applied to ASDEX-Upgrade during the 2014 MST-1 campaign [2]. It relies on the idea that the mean effect of the sweeping will still be favorable for the mode reduction, which relaxes the need for precise rational surface location. This allows the development of robust control systems. This paper focuses on the effects of ECCD sweeping across the rational surface on the stability index $\Delta'$, in particular to quantify the potential destabilizing effect. The stabilizing role of localized ECCD inside the island is also discussed and preliminary results are presented. The analysis is performed using the 3D full MHD code XTOR-2F [4], which includes a current source term modeling the ECCD in the Ohm's law [10].

**Brief description of the experiment**

An example of successful application of the sweeping strategy can be found in AUG pulse #30594 (figure 1 of [2]). A (3,2)-NTM is triggered by rising NBI power up to 17 MW. Once the mode is detected, the NBI power is reduced to 10 MW. The plasma current is about 1 MA, and $\beta_N \approx 1.8$. A first gyrotron is switched on with a power of about 750 kW. Mirrors orientations are then modified such that the current deposition radial location oscillates around the rational surface. This ensures that current is periodically deposited inside the O-Point. A second gyrotron, and then a third gyrotron are switched on, leading to the mode disappearance. The sweeping strategy hence proves its capability to successfully suppress an NTM.

*olivier.fevrier@cea.fr
†See http://www.euro-fusionscipub.org/mst1.

**Physical model and equilibrium with neoclassical physics**

The nonlinear MHD code XTOR-2F [4] solves the two fluid 3D MHD equations in a torus. The neoclassical effects are implemented via the neoclassical viscous stress tensors [5]. The current induced by the RF source implementation is detailed in [10]. We use an equilibrium issued from AUG pulse #30594 at $t = 4.0s$, that is after a (3,2)-NTM has appeared on the $q = 3/2$-surface, at $\rho \approx 0.55$, where $\rho$ is the normalized poloidal flux. For simplicity, the shape of the separatrix is modified to be up-down symmetric. Pressure and density profiles are fitted from AUG pulse #30594 using IDA (*Integrated Data Analysis*) at $t = 4.0s$. The central density is $n_i(0) = 9.4 \times 10^{19}$ m$^{-3}$, the magnetic field on the axis is $B_0 = 2.5$ T and the central temperature $T_e^0 \approx 3.6$ keV. At equilibrium, the bootstrap current computed by XTOR is in good agreement with predictions of the Sauter's formula [6] (figure 1).

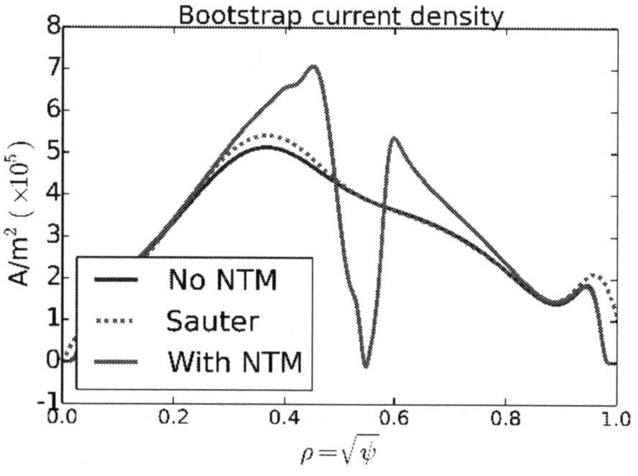


Figure 1: Bootstrap current density averaged over flux surfaces, as obtained with XTOR-2F and using Sauter's formula.

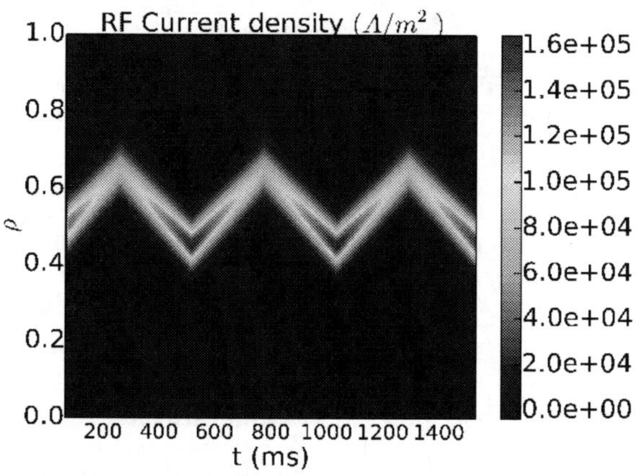


Figure 2: Evolution of the RF Current density

**Effect of ECCD sweeping on the linear stability**

One of the concerns arising from the sweeping procedure is that it could have deleterious effects on stability through equilibrium modifications, by increasing the mode stability index $\Delta'$ when the current deposition is off the rational surface [7]. In the following, we consider the evolution of the stability index $\Delta'_{(3,2)}$ of the (3,2) mode for a current injection (12 kA, corresponding to approximatively 3 gyrotrons) at different locations around the rational surface, using the one-fluid MHD model. The radial width of the current source has been determined using data from TORBEAM[11] at $t = 4.0$ s. As expected, we retrieve a stabilizing effect for $\rho_{ECCD} \geq \rho_{q=3/2}$, and a destabilizing one otherwise. This is plotted on figure 3 where the evolution of $\Delta'_{(3,2)}$ during a triangular sweep around the surface $q = 3/2$ is also plotted, with the experimental frequency of 2 Hz and also with a frequency of 4 Hz. One can see that the sweeping process induces a hysteresis-like modification of the equilibrium. On figure 2, the temporal evolution of the current density is plotted for a sweeping frequency of 2 Hz. The amplitude of the $\Delta'_{(3,2)}$ modification, is not enough to destabilize the mode, as it is compensated by curvature stabilization, whose value is approximatively 23, following [9]. It can also be noted that increasing the sweeping frequency reduces the deleterious effects on equilibrium, as it gives less time for the current to locally modify the equilibrium.

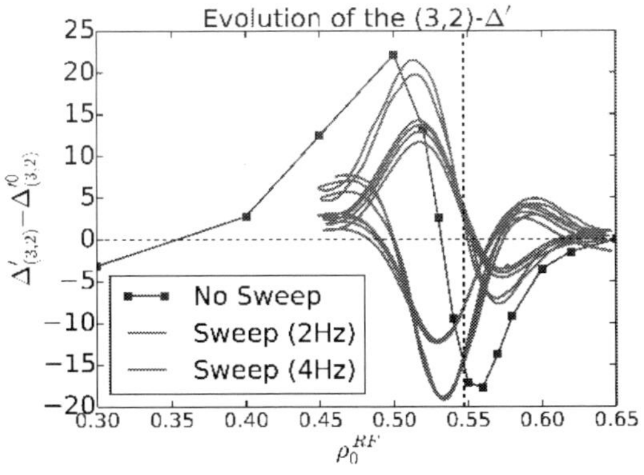


Figure 3: Effect of ECCD on $\Delta'_{(3,2)}$. The vertical dashed line indicates the $q = 3/2$-surface position.

We introduce a simple analytical model to reproduce this result, described in (equation 1), where $T$ is a triangle wave function, centered on $\rho = 0.55$, with a total amplitude of 0.2, a frequency $f$ and $\Delta'_s(\rho_{RF})$ the final value of $\Delta'_{(3,2)}$ for a source at fixed position $\rho_{RF}$.

$$\begin{cases} \frac{\partial \Delta'}{\partial t} = \frac{1}{\tau}\left(\Delta'_s(\rho_{RF}(t)) - \Delta'\right) \\ \rho_{RF}(t) = T(0.55, f, 0.2) \end{cases} \tag{1}$$

This model is able to capture the hysteresis phenomenon, as shown on figure 4. We retrieve the decrease of the impact on equilibrium as the frequency is increased (figure 5). (It should be noted that in experiments, this frequency is limited by the time necessary for the mirrors to move to their prescribed positions). To conclude, the effect of sweeping on linear stability is not negligible, although not critical in the present experiment.

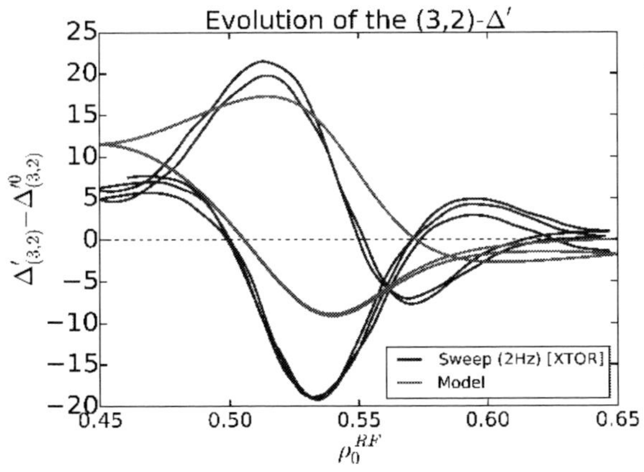


Figure 4: Evolution of $\Delta'_{(3,2)}$ procuded by the analytical model (equation 1) and XTOR.

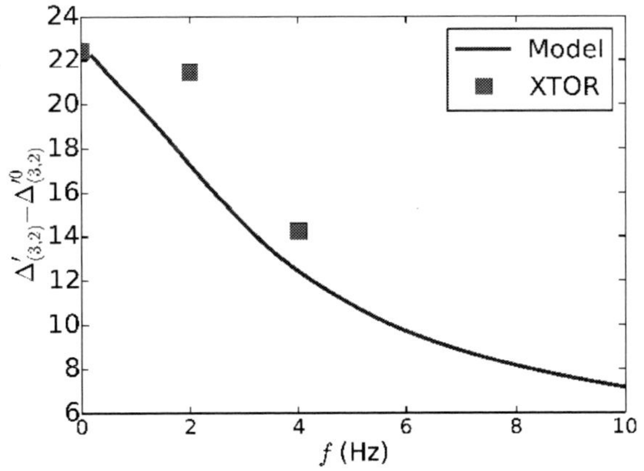


Figure 5: Variation of the maximum reached $\Delta'_{(3,2)}$ for different frequencies of the sweeping.

## Toward NTM stabilization by ECCD Current

The previous section shows that the most important effects for the island stabilization are the 3D effects [8]. We consider the situation where an NTM is present during the ECCD application. Simulations are carried out using a two-fluids neoclassical model. A (3,2)-NTM is driven by inducing a current perturbation on the rational surface $q = 3/2$, using the current source term. Once the island is properly setup and the bootstrap current nullified inside (figure 1), we proceed to drive current on the rational surface $q = 3/2$, using a 1D source term. In these preliminary simulations, we have a large ergodic region in the vicinity of the magnetic island. This leads to a a large broadening of the RF-driven current density (as shown on figure 6), reducing the efficiency $\eta_{RF}$ [8] of the control.

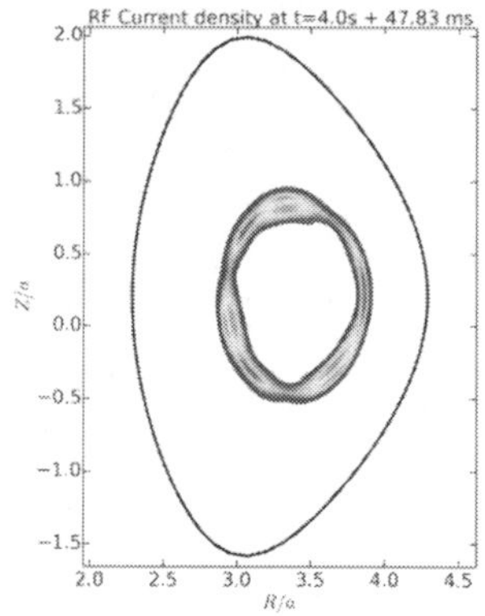


Figure 6: Distribution of the RF Current density.

## Conclusion

In this work, the impact on the sweeping strategy on the linear stability of the (3,2)-mode has been investigated. It is shown that the sweeping method can, when all the gyrotrons are switched on, have a significantly impact on the linear stability. It is also shown that the sweeping process induces an hysteresis-like modification of the equilibrium, due to the finite-timescale needed to perturb the equilibrium. Increasing the sweep frequency helps mitigating these effects. The work regarding the (3,2)-NTM stabilization has been initiated using the full bi-fluid and neoclassical MHD mode.

## Acknowledgement

*This work is part of the AMICI project (ANR-14-CE32-0004-01) funded by the Agence Nationale pour la Recherche. Numerical resources were provided by CINES, IDRIS and TGCC of GENCI (x2015056348), Mésocentre of Aix-Marseille University (15b025), Helios of IFERC-CSC (MaCoToP).This work has been carried out within the framework of the EUROfusion Consortium and has received funding from the Euratom research and training programme 2014-2018 under grant agreement No 633053. The views and opinions expressed herein do not necessarily reflect those of the European Commission.*

## References

[1] M. Maraschek *Nucl. Fusion* **52**, 074007 (2012) [2] M. Maraschek *42st EPS Conference on Plasma Physics, Lisbon* (2015) [3] F. Felici *Nucl. Fusion* **52**, 074001 (2012) [4] H. Lütjens and J.-F. Luciani *J. Comp. Phys.* **229**, 8130 (2010) [5] N. Mellet *Nucl. Fusion* **53**, 043022 (2013) [6] O. Sauter, *Phys. Plasmas* , 2834 (1999) [7] A. Pletzer and F. W. Perkins *Phys. Plasmas* **6**, 1589 (1999) [8] C. C. Hegna *Phys. Plasmas* **4**, 2940 (1997) [9] H. Lütjens *Phys. Plasmas* **8**, 4267 (2001) [10] O. Fevrier *41st EPS Conference on Plasma Physics, Berlin* (2014) [11] E. Poli et al. , Comput. Phys. Commun. **136**, 90 (2001)

# First principle numerical simulations of the SOL in ASDEX Upgrade

A.H. Nielsen[1], L. Aho-Mantila[2], J. Madsen[1], H.W. Müller[3], N. Vianello[4], H. Radhakrishnan[5], J. Signoret[6], M. Wischmeier[7], B. Sieglin[7], D. Carralero[7], G. Birkenmeier[7], R. Coelho[8], J. Juul Rasmussen[1], V. Naulin[1], M. Løiten[1], L. Tophøj[1], R. Schrittwieser[9], S. Costea[9], the EUROfusion MST1 Team[10], the EU-IM Team[11] and ASDEX Upgrade Team

[1]*Tech. Uni. of Denmark, Kgs. Lyngby, Denmark,* [2]*VTT Tech. Research Centre of Finland,* [3]*Uni. of Vienna, Austria,* [4]*EPFL, CRPP, Ch-1015 Lausanne, Switzerland,* [5]*University of Cyprus, Cypres* [6]*CEA, IRFM, France,* [7]*Max-Planck-Institut für Plasmaphysik, Garching, Germany,* [8]*Instituto de Plasmas e FusĂo Nuclear, Lisboa, Portugal,* [9]*Institute for Ion Physics and Applied Physics, Innsbruck, Austria,* [10]*www.euro-fusionscipub.org/mst1,* [11]*www.euro-fusionscipub.org/eu-im*

## Introduction

The turbulent transport in the edge and scrape-off-layer (SOL) regions at the outboard mid-plane of ASDEX Upgrade is investigated numerically and experimentally. Experimental data obtained during the MST1 campaign from a number of L-mode discharges were analyzed, including L-mode detached discharges induced by Nitrogen seeding.

We use the numerical code HESEL, which simulates interchange turbulence on outboard midplane in 2D perpendicular to the magnetic field and where parallel dynamics has been parametrized. In the numerical simulations the divertor conditions are mimicked by use of a sheath boundary condition term in the generalized vorticity equation. This condition affects the dynamics of the cross field motion of blobs. As energy, particles and momentum are mostly transported into the SOL by these blobs, changing the sheath condition will change the overall dynamics. To facilitate comparison of experimental and numerical results we have implemented the Stuttgart probe[1] as a synthetic diagnostics into the HESEL code. We observe reasonable agreement for the radial momentum transport using floating potential as a proxy for the plasma potential. Using plasma potential provides significantly different results in HESEL. We also numerically investigate the energy flux, not accessible experimentally. The energy flux is split into three components: the convective, the conductive, and the triple product. In the near SOL region all three components are equally important whereas the triple product dominates the energy flux in the far SOL region.

A standardised Kepler workflow, with HESEL embedded, is under development within the Integrated Modelling framework by the Code Development EUROfusion project. The workflow allow direct access to diagnostic data and shot parameters from the database at ASDEX

Upgrade.

**Model**

HESEL is an energy conserving four-field model based on the Braginskii equations[2] governing the dynamics of a quasi-neutral, simple plasma. It describes interchange-driven, low-frequency turbulence in a plane perpendicular to the magnetic field at the outboard midplane. In the limit of constant ion pressure the model reduces to the ESEL model, which has successfully modeled fluctuations and profiles in JET, MAST, EAST and TCV, see [3]. The HESEL model includes the transition from the confined region to the region of open field lines (SOL) and the full development of the profiles across the last closed flux surface (LCFS). The model is solved in a local slab geometry with the unit vector $\hat{\boldsymbol{z}}$ along the inhomogeneous toroidal magnetic field. The generalized vorticity is the manifestation of the polarization current in the model and describes charge separation due to the inertia in the ion response to changes in the $\boldsymbol{E} \times \boldsymbol{B}$ and ion diamagnetic drifts.

In a cooperation between the task Turbulence and synthetic diagnostics workflows under WPCD and MST1 a standardized workflow has been developed which is now being tested. In the work flow experimental data including profiles of density $n_e$ and electron temperature $T_e$, $q_{95}$, magnetic field, $B_0$, are read directly from the AUG database and passed to the HESEL actor. The HESEL actor then submits a job to the Gateway Linux cluster. Data from the synthetic diagnostics, such as Langmuir probes and Lithium beam implemented in the HESEL code, together with key parameters are automatically stored for later analysis.

**Results**

In Fig. 1a we display the profiles from a HESEL simulation using parameters obtained from AUG shot #30301. The first notable thing is that the density and temperature profiles all have sharp gradients close to the Last Closed Flux Surface (LCFS) with density and electron temperature approaching a background value already within 1 cm and ion temperature $2-3$ cm from the LCFS. However we can observe significant ion and electron energy fluxes, $\langle n_e T_\alpha \tilde{v}_r \rangle$ , $\alpha = \{i, e\}$, far into the SOL. These fluxes are not transported by convective, $\langle T_\alpha \rangle \langle \tilde{n}_e \tilde{v}_r \rangle$, nor conductive, $\langle n_e \rangle \langle \tilde{T}_\alpha \tilde{v}_r \rangle$, components as these to a large degree follow the mean of the density, $\langle n_e \rangle$ and temperatures, $\langle T_e \rangle$ and $\langle T_i \rangle$, but of the so-called triple product, $\langle \tilde{n}_e \tilde{T}_\alpha \tilde{v}_r \rangle$. This is a consequence of the very intermittent blob dynamics responsible for the energy transport in the SOL.

Figure 1b displays the time evolution of the radial power distribution. Each magnetic field line has been mapped from the outboard midplane in HESEL domain onto the divertor. In mapping the electron and ion energies we assume that the plasma is convected with the local

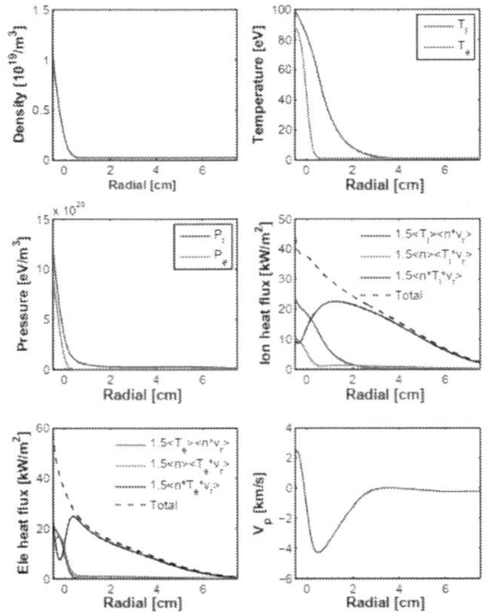


(a) Profiles.

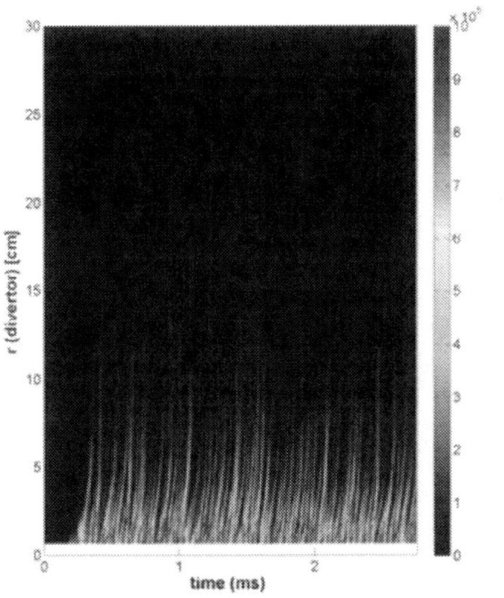


(b) Power deposition on the divertor.

Figure 1: HESEL simulation simulation using parameters obtained from AUG #30301.

ion sound velocity; $c_s(r,\theta,t) = \sqrt{(T_e(r,\theta,r)+T_i(r,\theta,t))/m_i}$ and a poloidal average has been performed. The individual traces from blob evolutions are clearly visible.

Figure 2 displays the Stuttgart probe head with 14 individual pins each measure either ion saturations current, $I_s = eA_{ip}n\sqrt{(T_e+T_i)/m_i}$ or floating potential, $V_f = V_p - \alpha T_e$ where $V_p$ is the plasma potential, $\alpha \sim 3.2$ for a deuterium plasma. The configuration of the individual pins varies between shots. In HESEL we have implemented a synthetic Stuttgart probe head at different stationary radial locations.

As an example of a comparison between the Stuttgart probe and the synthetic probe Fig. 3 displays the momentum transport at a radial position 1 cm from LCFS. The radial and poloidal velocities are approximated as; $v_r \sim -(V_f^{12} - V_f^{9})/(\delta\theta B_0)$ and $v_\theta = (V_f^{12} - 0.75V_f^{9} - 0.25V_f^{7})/(\delta r B_0)$. We notice that the floating potential and therefore also the velocities will be strongly influenced by electron temperature fluctuations. Also, pin separations and misalignment with respect to a local $r_\theta$ coordinates will affect the velocity calculation.

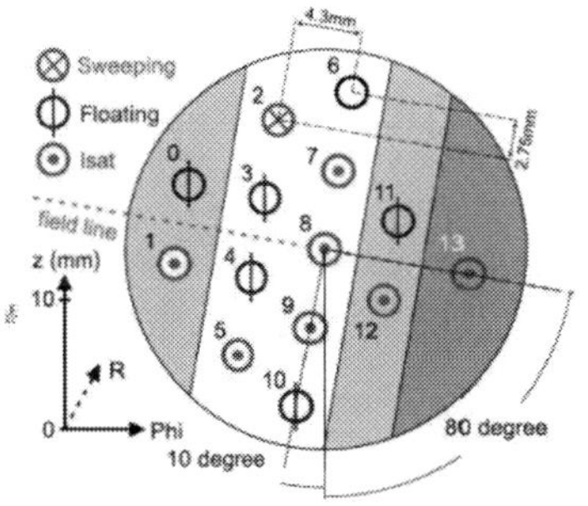


Figure 2: The Stuttgart probe used on the MEM at AUG.

The momentum transport, $M_r$, is divided into contributions from the Reynolds stress, $Re =< n >< \tilde{v}_p\tilde{v}_r >$, the convective term, $< v_p >< \tilde{n}\tilde{v}_r >$ and the so called triple product, $< \tilde{n}\tilde{v}_p\tilde{v}_r >$, where $< f >= \int_0^t f(t)dt$. Based on results from the HESEL we conclude that if the velocities are based on plasma po-

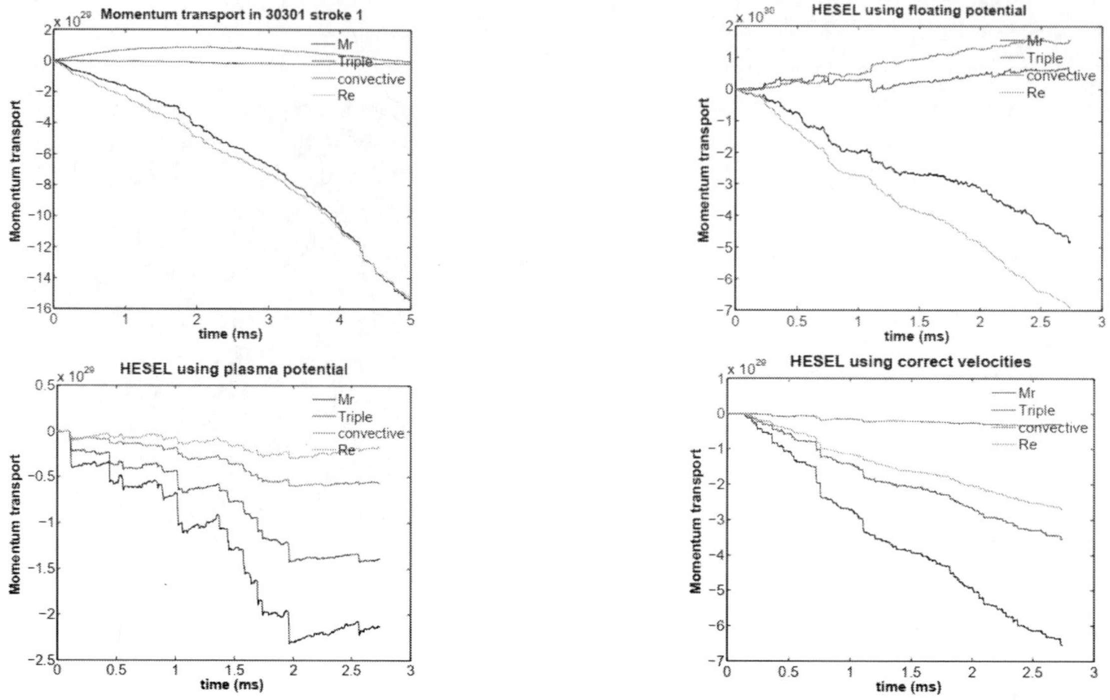


Figure 3: Momentum transport calculated from the Stuttgart probe and synthetic data.

tential the derived momentum are overall in agreement with the results based on the correct velocities. If, on the other hand, velocities are based on floating potential using cold probes we obtain strong deviations. It should be noted that the synthetic probe data applying the floating potential indeed are in close agreement with the measured ones. Thus, one may apply the simulated data to provide a more correct estimates for the momentum fluxes.

## Acknowledgement

This work has been carried out within the framework of the EUROfusion Consortium and has received funding from the Euratom research and training programme 2014-2018 under grant agreement No 633053. The views and opinions expressed herein do not necessarily reflect those of the European Commission.

## References

[1] B. Nold *et al.* New Journal of Phys. 14 (2012) 063022.

[2] S.I. Braginskii. *In Review of Plasma Physics, (ed. Leontovich, M. A.). New York, NY: Consultants Bureau*, Vol. 1, p. 205 (1965).

[3] W. Fundamenski *et al.* Nucl. Fusion, **47**, 417 (2007); F. Militello *et al.* Plasma Phys. Controlled Fusion, **55**, 025005 (2013); N. Yan *et al.* Plasma Phys. Controlled Fusion, **55**, 115007 (2013); O.E. Garcia *et al.* Phys. Plasmas, **12**, 062309 (2005).

# Determining fast-ion velocity-space distribution functions using velocity-space tomography

A.S. Jacobsen[1], M. Salewski[1], B. Geiger[2], L. Stagner[3], J. Eriksson[4], S.K. Nielsen[1], W.W. Heidbrink[3], S.B. Korsholm[1], F. Leipold[1], J. Rasmussen[1], M. Stejner[1], M. Weiland[2]
and the ASDEX Upgrade Team

[1] *Technical University of Denmark, Department of Physics, Kgs. Lyngby, Denmark*
[2] *Max Planck Institute for Plasma Physics, Garching, Germany*
[3] *University of California Irvine, Irvine, CA, USA*
[4] *Department of Physics and Astronomy, Uppsala University, Sweden*

## Introduction

In a burning plasma the heating comes from the very fast fusion products themselves. To ensure effective heating, the fusion products must be sufficiently confined until they have transferred their energy to the bulk plasma. Therefore, studies of the confinement of fast ions in current fusion devices are important. This includes studies of the interplay between fast ions and plasma instabilities.

Using velocity-space tomography it is possible to infer the fast-ion velocity-space distribution function directly from measurements [1, 2, 3, 4]. This is done by combining measurements from several fast-ion detectors measuring in the same location in position space. Velocity-space tomography requires knowledge about the velocity-space sensitivity of the fast-ion diagnostics. This sensitivity is described by so-called velocity-space weight functions. These have been formulated for fast-ion $D_\alpha$ (FIDA) spectroscopy [5, 6], collective Thomson scattering (CTS) [7], neutron emission spectrometry (NES) [8, 9], $\gamma$-ray spectroscopy (GRS) [10] and neutral particle scattering (NPA) [11]. A weight function, $w$, relates a fast-ion measurement, $s$, to the full fast-ion velocity distribution function, $f$.

$$s = \iint w f \, \mathrm{d}E \mathrm{d}p \,, \tag{1}$$

where the integration is with respect to energy and pitch. The pitch is here defined as $\frac{v_\parallel}{v}$ with positive sign in the direction of positive toroidal current. $v_\parallel$ is the velocity parallel to the magnetic field and $v$ is the speed of the ion. Equation (1) can be discretized giving a linear set of equations:

$$S = WF \,. \tag{2}$$

The goal is now to calculate $F$ from equation (2). However, this is mathematically an ill-posed inverse problem which can only give sensible results if the problem is regularized. There exist many different ways to regularize inverse problems. In the example shown in this paper we use truncated singular value decomposition. This is effectively a filtering method which removes the finer details in the solution suppressing the effect of the noise in the measurements.

**Tomography of a sawtooth crash**

FIDA spectroscopy measures Doppler-shifted light emitted by excited fast neutrals right after they have undergone charge exchange with injected NBI ions. Here, we present results of velocity-space tomographies using coherently averaged measurements taken just before and just after sawtooth crashes in discharge #30815 in the ASDEX Upgrade tokamak [2]. In this discharge we measured FIDA light in four different views. Each view consists of several lines of sight measuring in different locations in position space. Each view has a line of sight that intersects the NBI beam at the same location in the plasma center, each with a different angle to the magnetic field. The angle between the local magnetic field and the line of sight determines the projection direction of the velocity distribution and therefore determines the velocity-space sensitivity and thus the shape of the weight functions [6]. The measurement volume is located around $\rho_t = 0.1$, well within the sawtooth inversion radius located near $\rho_t = 0.3$. Figure 1(a) shows a tomography of the fast-ion velocity distribution function calculated using FIDA spectra measured right before a sawtooth crash. Figure 1(c) shows a theoretical distribution calculated using TRANSP/NUBEAM [12]. The shaded region at low energies correspond to low Doppler shifts in the FIDA spectra. Here the FIDA light cannot be observed as the halo, the beam emis-

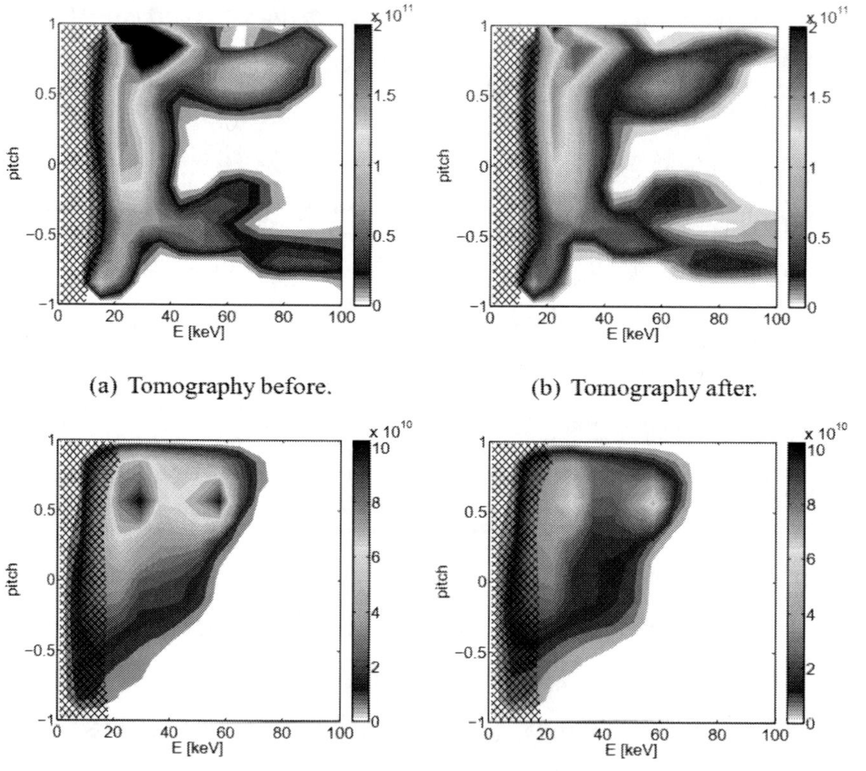


(a) Tomography before.
(b) Tomography after.
(c) TRANSP before.
(d) TRANSP after.

Figure 1: Central fast-ion velocity distribution functions before and after a sawtooth crash in units of [ions/keV/cm$^3$].

sion and the cold D-alpha light are much stronger [13]. Several similarities between figures 1(a) and 1(c) are evident. Most ions have a positive pitch and the NBI full-energy peak at 60 keV can be seen. However, several discrepancies are evident as well, especially the $E > 40$ keV feature at a pitch around -0.5. This is most likely an artefact introduced by the inversion. The FIDA data in this discharge contain impurity lines which obscure parts of the spectra with large Doppler-shifts which might have been able to suppress this artefact through a measured lack of FIDA light.

Figure 1(b) shows a tomography calculated using data measured right after the sawtooth crash. Figure 1(d) shows the corresponding TRANSP/NUBEAM calculation. A clear drop in the distributions can be seen. Furthermore, the drop in the fast-ion distribution is larger for passing ions compared to trapped ions. To investigate this effect further, the relative change is calculated as $\frac{F_{before}-F_{after}}{F_{before}}$. Figure 2 shows the relative change as a function of pitch for two different energies. Trapped ions with pitch values around 0 are redistributed less than passing ions with pitch close to 1. This behaviour has also been observed on other machines [14]. For comparison, the relative change predicted by the Kadomtsev model implemented in TRANSP is also shown.

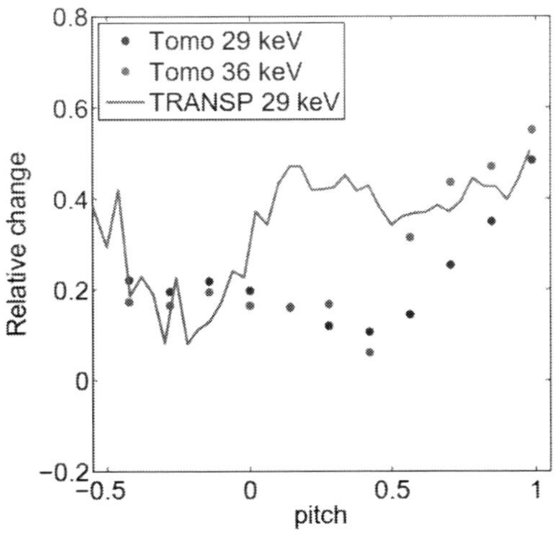


Figure 2: Relative change in fast ion density as a function of pitch for two different energies.

**Discussion**

As we mentioned in the introduction, the regularization of an inverse problem is essential in order to achieve sensible results. Many different regularization methods exists. The result depends on the choice of regularization and it should therefore preferably be made based on knowledge or assumptions about the true solution. A comparison of inversion methods for velocity-space tomography will be presented elsewhere.

The inversions can be improved by combining several different types of fast-ion diagnostics. For example, the FIDA views at ASDEX Upgrade could be combined with the CTS, NPA and NES diagnostics. In addition to the benefits gained by including more measurements, the different diagnostics yield complimentary information about fast-ion velocity space even if their projection angle is the same. Examples of weight functions for FIDA, CTS and NES are shown in figure 3, all calculated with a projection angle of 14°. The weight functions are plotted as a function of $v_{\|}$ and $v_{\perp}$ as the differences are more clear in these coordinates. The FIDA and CTS weight functions have triangular shapes while the NES weight function is circular.

## Conclusion

We have presented results of velocity-space tomography on a sawtooth crash at ASDEX Upgrade as an example of an application of the method. Using velocity-space tomography it is possible to recreate the full fast-ion velocity distribution function based on a combination of measurements from different detectors, each only measuring a projection of the distribution function. Our tomographic inversion suggests that passing ions are more affected by the sawtooth crash than trapped ions.

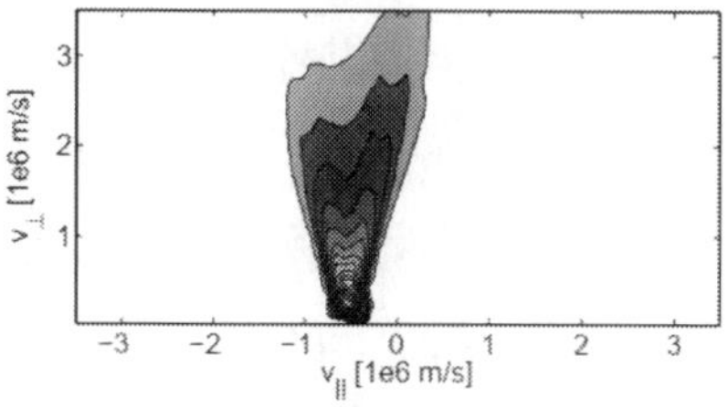


(a) FIDA.

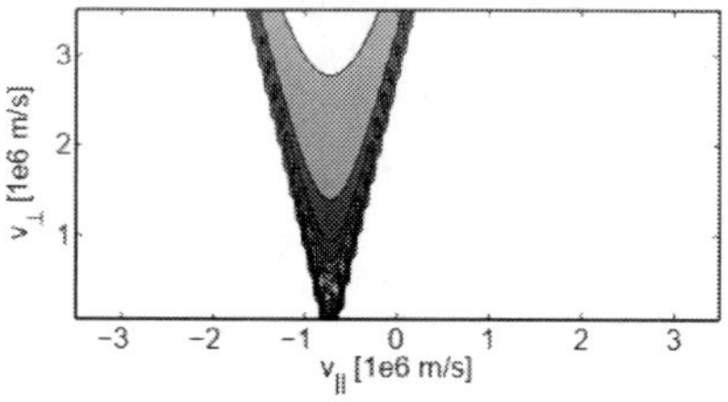


(b) CTS.

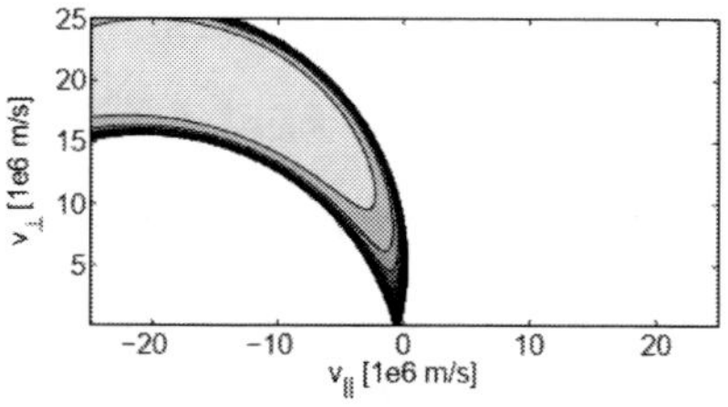


(c) NES.

Figure 3: Examples of FIDA, CTS and NES weight functions.

## Acknowledgement

This work has been carried out within the framework of the EUROfusion Consortium and has received funding from the Euratom research and training programme 2014-2018 under grant agreement No 633053. The views and opinions expressed herein do not necessarily reflect those of the European Commission.

## References

[1] M. Salewski et al., Nuclear Fusion **54**, 023005 (2014)

[2] B. Geiger et al., accepted in Nuclear Fusion, title: *Fast-ion transport and neutral beam current drive in ASDEX Upgrade*, (2015)

[3] M. Salewski et al., Nuclear Fusion **52**, 103008 (2012)

[4] M. Salewski et al., Nuclear Fusion **53**, 063019 (2013)

[5] W.W. Heidbrink et al., Plasma Physics and Controlled Fusion **49**, 1457-1475 (2007)

[6] M. Salewski et al., Plasma Physics and Controlled Fusion **56**, 105005 (2014)

[7] M. Salewski et al., Nuclear Fusion **51**, 083014 (2011)

[8] A.S. Jacobsen et al., Nuclear Fusion **55**, 053013 (2015)

[9] A.S. Jacobsen et al., Review of Scientific Instruments **85**, 11E103 (2014)

[10] M. Salewski et al., submitted, title: *Velocity-space sensitivity of high-resolution two-step reaction γ-ray spectroscopy*

[11] W.W. Heidbrink et al., Plasma Physics and Controlled Fusion **56**, 095030 (2014)

[12] A. Pankin et al., Computer Physics Communications **159**, 157-184 (2004)

[13] B. Geiger et al., Review of Scientific Instruments **84**, 113502 (2013)

[14] S.K. Nielsen et al., Nuclear Fusion **51**, 063014 (2011)

# Effects of 3D magnetic perturbation on divertor heat load redistribution on ASDEX Upgrade

Y. Gao[1], M. Rack[1], Y. Liang[1], B. Sieglin[2], P. Denner[1], W. Suttrop[2], A. Kirk[3], E. Wolfrum[2], the EUROfusion MST1 Team* and the ASDEX Upgrade Team

[1] *Forschungszentrum Jülich GmbH, Institut für Energie- und Klimaforschung – Plasmaphysik, 52425 Jülich, Germany*

[2] *Max-Planck-Institut für Plasmaphysik, D-85748 Garching, Germany*

[3] *CCFE, Culham Science Centre, Abingdon, Oxon, OX14 3DB, UK*

For the ITER design, one of the biggest challenges is the power handling with respect to the safety of in-vessel components. The application of magnetic perturbations (MPs) is a promising technique for controlling or optimizing plasma-wall interactions [1]. ASDEX Upgrade is equipped with toroidally distributed in-vessel MP coils (8 above and 8 below the mid-plane), which can perform a rigid rotation or differential phase scan with toroidal mode numbers $n = 1,2$.

On ASDEX Upgrade, ELM mitigation has been achieved using MPs. In low collisionality experiments ($\nu^*_{\mathrm{ped}} < 0.5$), the MPs have been shown to reduce the heat flux factor [2, Eq. 1] by 75 %, at the cost of an MHD confined energy drop of $\sim 30\,\%$ [3, 4]. In high collisionality with strong gas puffing, ELM mitigation can be obtained by the change from Type-I to small ELMs [5].

The heat load distribution on the outer divertor plate has been analysed in detail, by means of infra-red (IR) camera measurement with a high spatial resolution of $\sim 0.6$ mm and a frame rate of $\sim 3.3$ kHz. In both collisionality regimes, clear splitting of the outer strike line has been observed due to the application of MPs.

Fig. 1 shows the footprints at the target plate of Divertor-III.BG1 during a rigid rotation of an $n = 2$ MP at low collisionality for AUG discharge #30839. Here, the phase between the field patterns in the upper and lower MP coils has been kept constant ($\Delta\Phi = 0°$, "even" parity). The rotation has been performed with a frequency of 0.5 Hz. Except for the first toroidally symmetric strike line, two additional strike lines on the outer divertor have been observed to move with the rotating MPs. The appearance of these additional strike lines is due to the new channels for particle and energy transport in the magnetic flux tubes generated by the MPs.

In both the two cycles of the rotation with differing heating power (4.8 MW and 7.1 MW), the entire 3D features of the magnetic topology at the plasma edge induced by MPs manifest

*See http://www.euro-fusionscipub.org/mst1

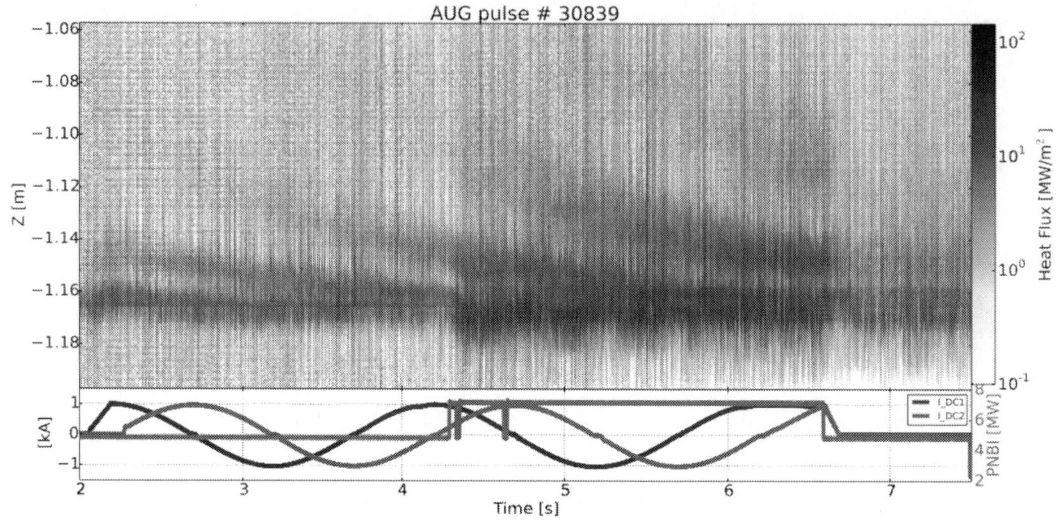


**Figure 1:** The heat flux map showing the movement of secondary strike lines with rigid rotating MPs. The lower part of the figure are the time evolution of the MP coils currents (blue, red) and NBI power (violet).

themselves as the helically striated divertor footprint patterns. After each cycle of the rotation with constant plasma parameters and IR monitoring at one toroidal angle, the whole toroidal distribution of the heat load can be obtained. The amplitude of the heat flux has been observed to differ at different lobes on the target. It may be due to the different connection length of the field lines in the 3D magnetic lobes intersected by the target plate. This could be modelled and compared with experimental results in the future using field line tracing codes.

In a high collisionality regime, gas puffing scans are performed with the application of MPs. In Fig. 2, the heat flux distribution over time for discharge # 30589 is presented (toroidal magnetic field $B_T = -2.5\,\mathrm{T}$ and plasma current $I_P = 800\,\mathrm{kA}$). The lower panel shows a ramping up phase of the gas puffing rate within the flattop of MP coil currents. The signal of divertor current $I_{\mathrm{div}}$ presented here is used as ELM indicator, which shows the transition of Type-I ELMs to more frequent small ELMs starting at $\sim$ 6.97 s due to the increase of the gas puffing rate. The gas inlet system consists of several valves at different toroidal angles but all fuelling from the lower private flux region. It can be seen obviously that the strike line splitting appears for both low and high gas puffing time periods within the flattop of MPs as shown in Fig. 2. Before the flat-top of MP coil currents, the splitting cannot be seen. After the turning off of MP coil currents, the splitting fades away gradually. This indicates that the MPs modify the edge magnetic topology and generate additional channels for particle and energy transport.

In order to study the gas puffing effect on heat flux redistribution between the first strike line and splitting patterns, two time periods are selected and compared in detail. From 5.705 s to

6.089 s the gas puffing rate is at the constant lowest value of $5 \times 10^{21}$ D/s, while from 7.238 s to 7.499 s the gas puffing rate is at the constant highest value of $30 \times 10^{21}$ D/s. In Fig. 3, the two time periods have been zoomed in and analysed in detail. At the low gas puffing rate, 40 Type-I ELMs are identified with crash duration of $\sim 3$ ms and the peak heat flux up to 85 MW m$^{-2}$. At

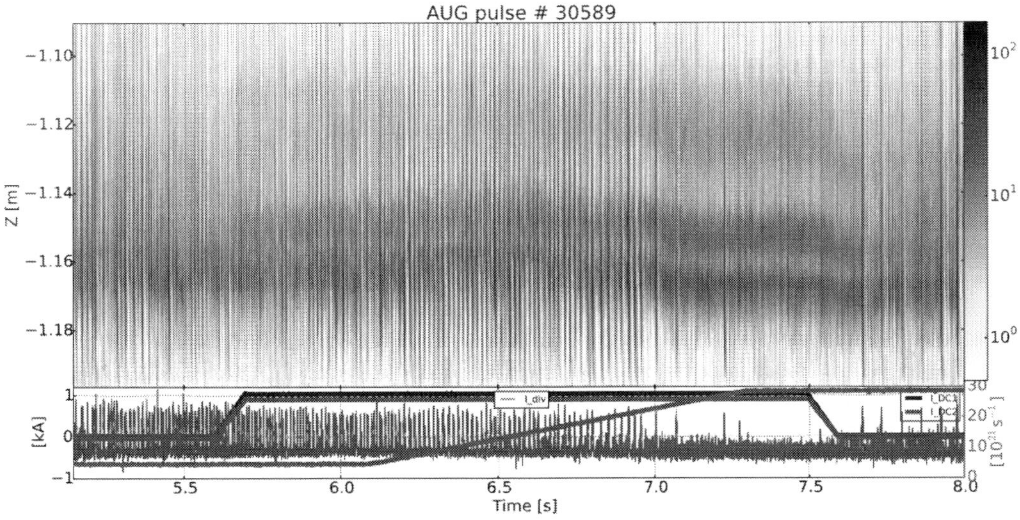


**Figure 2:** The time evolution of heat flux distribution on the outer target. The lower panel shows the gas puffing rate (violet), the MP coils currents (blue, red) and the dievertor current signal (green).

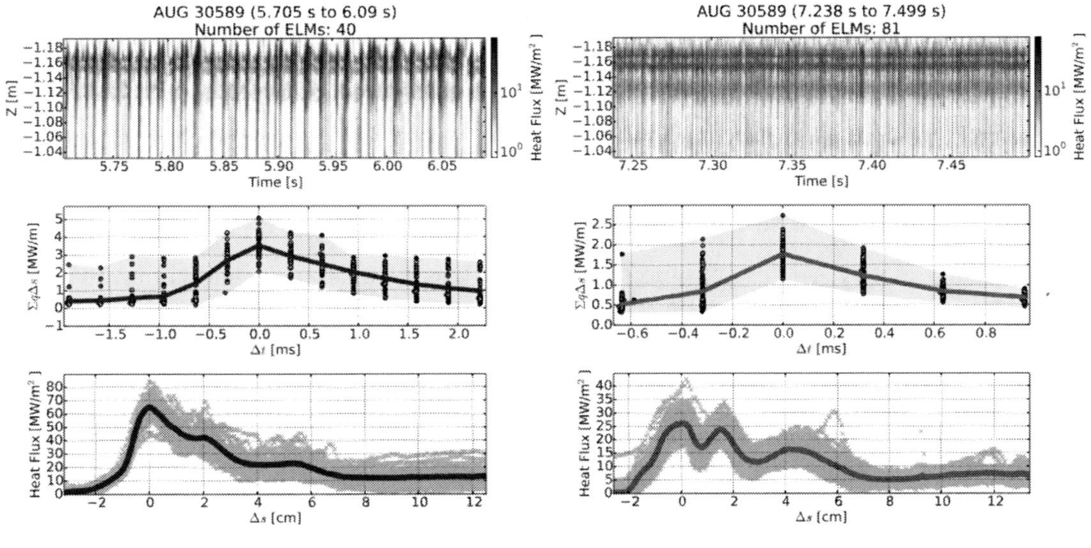


**Figure 3:** The comparison of the heat flux distribution on the outer target between low (left) and high gas puffing (right). The first row shows the time evolution of the heat flux distribution on the target. The second row shows the spatially integrated heat flux profile with $\Delta t = 0$ referring to the peak ELM crash moment. The bottom row shows the heat flux profiles on the target for every ELM at $\Delta t = 0$ as well as the average of these profiles (bold curve). The $\Delta s$ is correlated to the first peak.

the high gas puffing rate, 81 small ELMs are detected with power decay time reduced to $\sim$ 1 ms and the peak heat flux lower than 45 MW m$^{-2}$. In Fig. 4, the comparison of the averaged heat flux profile is presented. All the profiles are normalized to the peak of the ELM crash heat flux profile at the low gas puffing rate. The strike line splitting has be seen in all the profiles. But the distribution of the heat load in different lobes could be modified by the gas puffing rate. With increasing gas puffing the peak heat flux is reduced by 60 % during the ELM crash phase, while the ratio of the secondary to the first peak in the heat flux profile can be enhanced by a factor of 1.8. The averaged pre-ELM profiles are generated using the same phases in the discharge as the ELM crash profiles. During the pre-ELM phase, the heat flux is higher in all the lobes at high gas puffing rate.

At ELM crash phase, the effect of ELM mitigation by gas puffing is obvious because of the reduction of the heat load at the first toroidally symmetric strike line. However the ratio of the secondary to the first heat flux peak is enhanced. The formation of the secondary peaks is considered to be the parallel transport of particles along the perturbed field lines in the flux tubes induced by the MPs. Depend on the penetration depth of the MPs, these perturbed field lines may lead hot particles from confined region to the divertor with short connection length. Another explanation could be the increased perpendicular transport in high collisionality plasma, which could populate more particles into the outer lobes. This result is qualitatively consistent with the observations on EAST, which may provide an attractive approach to actively redistribute the heat load [6].

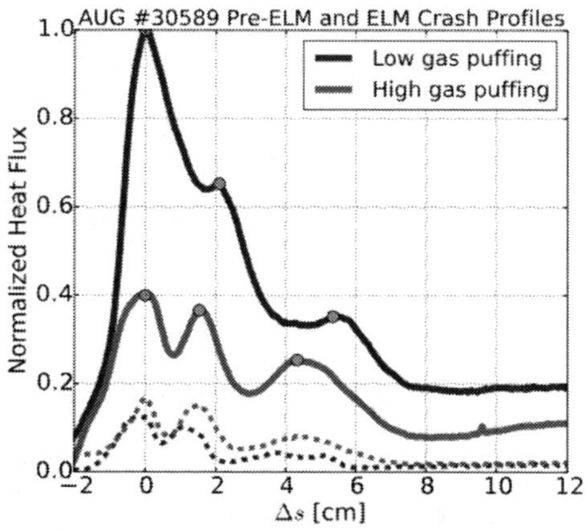


**Figure 4:** The comparison of averaged heat flux profile between low (blue) and high gas puffing rate (red). The dashed lines are pre-ELM footprints.

## Acknowledgement

This work has been carried out within the framework of the EUROfusion Consortium and has received funding from the Euratom research and training programme 2014-2018 under grant agreement No 633053. The views and opinions expressed herein do not necessarily reflect those of the European Commission.

## References

[1] A. Loarte *et al. Nuclear Fusion*, **54** (3), 033007 (2014).
[2] A. Herrmann. *Plasma Physics and Controlled Fusion*, **44** (6), 883 (2002).
[3] W. Suttrop *et al.* In *25th IAEA Fusion Energy Conference (FEC 2014)* (2014).
[4] A. Kirk *et al. Nuclear Fusion*, **55** (4), 043011 (2015).
[5] E. Wolfrum *et al.* In *41st EPS Conference on Plasma Physics* (2014).
[6] Y. Gao *et al. Plasma Science and Technology*, **16** (2), 93 (2014).

# Stable Operation at Disruptive Limits by Means of EC at ASDEX Upgrade

G.Granucci[1], B.Esposito[2], M.Maraschek[3], L.Giannone[3], A.Gude[3], E.Poli[3], C.J.Rapson[3], M.Reich[3], J.Stober[3], W.Treutterer[3], M.Willensdorfer[3], C. Cianfarani[2], G. Pucella[2], C. Sozzi[1], C. Galperti[4], S. Nowak[1], ECRH team[1], ASDEX Upgrade team[1]and MST1 team[*]

[1]*IFP-CNR, Via R. Cozzi 53, 20125 Milan, Italy*

[2]*ENEA for EUROFusion, Via E. Fermi 45, 00044-Frascati (Roma), Italy*

[3]*Max-Planck Institut für Plasmaphysik, Boltzmannstr. 2, 85748 Garching, Germany*

[4]*Centre de Recherches en Physique des Plasmas, EPFL, 1015 Lausanne, Switzerland*

**Introduction.** The technique of disruption avoidance based on the use of electron cyclotron (EC) waves has been established in recent years with experiments in FTU [1], ASDEX Upgrade [2] and DIII-D [3] in various types of disruptive scenarios. The proposed scheme uses a combination of precursors to trigger the Electron Cyclotron (EC) power directed at a q rational surface to reduce amplitude of the main MHD mode, delaying or even avoiding the current quench. In view of the application to ITER, the disruption avoidance technique should be complemented by a strategy of safe plasma shut-down using ECCD instead of ECRH as required for NTM control. As plasma disruption is definitely accompanied by MHD activity, whatever is the initial reason for instability start up, the EC wave can be effective, exploiting its high localization and local power density, as well done in the MHD control experiments [4]. This paper reports on experiments aiming to demonstrate the capability of EC power to avoid disruptions in AUG triggered in high performance plasma close to $\beta_N$ or density limits.

**Experimental Setup.** Two different experimental setups were used to obtain disruptions. The first scenario was an H-mode with 7MW of NBI at 1MA and 2.2 T to destabilize NTM mode at high $\beta_N$ limit; the second was an L-mode plasma at 0.6 MA, 2.5 T with pre-set gas puffing to push density above the limit. In both cases EC power was triggered by precursors, related to MHD presence and initiation of current quench. The EC launcher steering was slaved to a reference aiming to inject the power onto q=2 rational surface, using the real time (RT) reconstruction of equilibrium [5] and an accelerated TORBEAM ray tracing [6], based on density profile reconstruction. Approaching the disruption reconstructed profiles and the diagnostics exhibit fast variations, easily leading to errors in RT reconstructions inducing

[*] See http://www.euro-fusionscipub.org/mst1

steering mirrors to wrong positions, this has been addressed by real-time TORBEAM calculation. In the experiments the new AUG ECRH system was used, at the frequency of 140 GHz with power up to 1.3 MW (3 gyrotrons).

**High Beta limit Experiments and results**. The most challenging target was the high $\beta_N$ disruption scenario, in which the Lock Mode signal (LM) and Vloop were used to trigger the EC power at q=2 surface, where (2.1) mode develops and grows. Both ECRH and ECCD were tested on a repetitive target, obtaining complete disruption avoidance when the power was injected in the island. The EC power, (see Fig.1 left), is effective to maintain plasma stable also if the mode is still locked. During the EC pulse the same RT algorithm used in [4] maintained power at q=2 rational surface. At EC switch off the mode rotates again, leading to disruption in the Ip ramp down. A radial scan was performed (Fig.1 right) by adding a fixed offset to the deposition reference, aiming to demonstrate that it is the action on the main instability that avoids the disruption, and not an increase of the total power in the plasma.

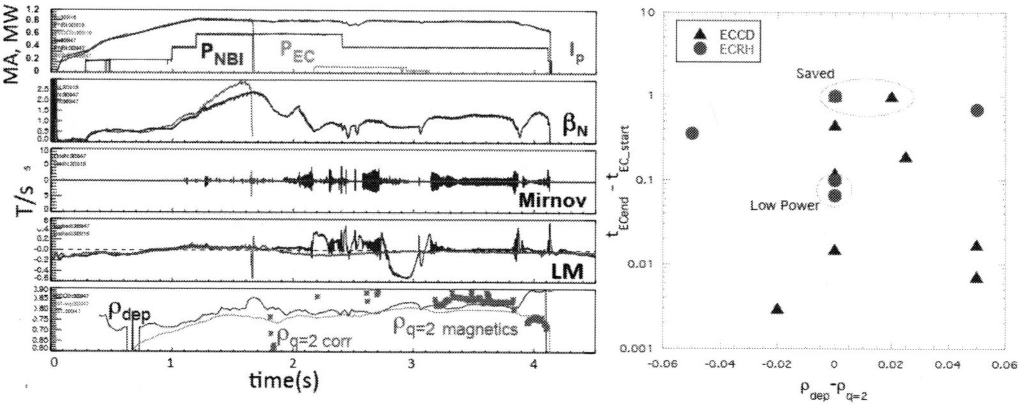


*Figure1: Left: Plasma evolution during high $\beta_N$ experiment. In red signals of disruptive target (#30918) in blue a plasma saved (#30947) by ECCD (toroidal injection of 9°). Right: high $\beta_N$ disruption radial scan with ECRH and ECCD results in term of surviving time of the plasma, expressed with duration of EC pulse (max length pre-set at 1s). Close to q=2 surface spread of results is due to errors in RT reconstruction. EC power: 0.6 - 1.3MW.*

This has obtained in both cases (CD or pure heating) although a greater precision on the deposition location was needed in the ECCD case (the most effective with NTM), for which the proper deposition is crucial to avoid disruption. In fact CD deposition inside the resonant surface even accelerate the disruption, in agreement with Rutherford equation [7]. This represents the challenge of such a technique, as the real time reconstruction of the target (mode position and beam localization) during a disruptive phase might be difficult and not reliable, at least with the actual status of diagnostics. A further RT tool to detect MHD onset and mode analysis is under study and was successfully applied off-line to data. It is based on SVD technique (Singular Value Deconvolution) described in [8] and capable to detect the most relevant mode using combination of Mirnov coils signals. Such an analysis has shown that in many disruption avoidance experiments after a successful mitigation of the (2,1) dominant mode, the plasma is killed by one external mode (typically a (3,1) or (4,1) and not by an internal (3,2) mode. This observation suggests a strategy for future experiments, using several gyrotrons aiming at different rational surfaces.

**Density limit Experiments and results.** The EC has been also used to avoid disruptions occurring at the density limit. In this case a L-mode at 0.6 MA, 2.5 T was chosen as scenario on which increase electron density by pre-programmed gas puffing up to density limit (0.6 Greenwald) and automatically switch on EC power just before the current quench: the most reliable precursor to trigger EC power was found to be the Vloop (threshold value 1.5 V).

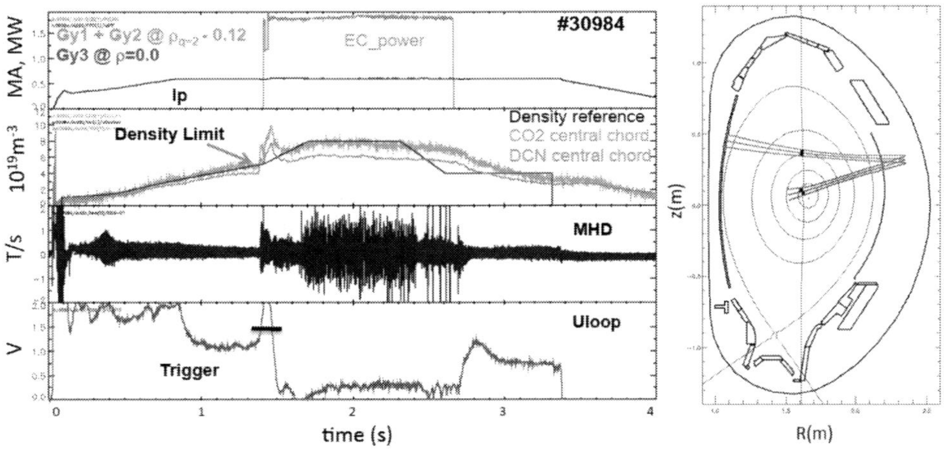


*Figure2: Left: Plasma evolution of main parameters during density limit disruption avoidance experiment (#30984). ECRH power is triggered by Vloop threshold at 1.5 V, the line averaged density ($CO_2$ and DCN central chords are shown) is pre-programmed at 8 $10^{19}m^{-3}$, two times the onset of disruption (start of MARFE). Right: the beam tracing of the used gyrotrons in the experiment. All the beams are in pure heating scheme.*

The system reaction time is around 7 ms. In these experiments the RT tracking of q=2 resonant surface was used to compensate the RF beam refraction. This effect was the limit of the previous experiments with no RT reported in ref [2] in which the EC deposition radius was moving away as density increases. This technique makes it possible, as shown in Fig.2 left, to maintain stable a discharge with a density 1.5 times the disruptive limit (MARFE, locked 2/1-mode). It is necessary to mention that in this scenario seems to be crucial to have a good core profile with sawteeth, obtained with EC central heating. This has to be further investigated in the future. In Fig.2 right, the used launching scheme is represented with two gyrotrons at q=2 surface (green path) and a third one aimed at the plasma centre (red path). All the gyrotrons were triggered at the same time. The analysis of the density evolution, during the phase above the limit, has demonstrated that the profiles maintain their peaking factor, suggesting the confinement of particle has been ameliorated. Further detailed analysis of confinement and transport are on going.

**Conclusion and future work** The technique of disruption avoidance based on the use of EC waves with RT control of steering mirrors has been successfully applied to ASDEX Upgrade in two types of disruptive scenarios: H-mode with high $\beta_N$ and L-mode at density limit. The power, automatically triggered by precursor signals and directed on the resonant q=2 surface, is able sustain the plasma current for the duration of EC pulse, and in some case also afterwards. The ECCD has been demonstrated to be effective in disruptions caused by $\beta_N$ limit, driven by NTM: compared to ECRH (which is also effective) a higher precision in power localization is necessary. For operation above the density limit ECRH has been used. In both the cases a proper RT reconstruction of equilibrium and ray tracing is at the basis of the success of the technique as well as the proper choice of the disruption precursors. The presented experiments were performed exploiting the LM detector and Vloop, while in future experiments a combination Mirnov coils signal could be used to detect mode onset and its kind. In view of the application to ITER, the disruption avoidance technique, more promising the case of high $\beta_N$ for the less power required, should be complemented by a more general strategy of safe plasma shut-down, developing a routine tool to be used in plasma operations.

## References

[1] B.Esposito, G.Granucci_et al. Phys.Rev.Let., **100** (4), 045006 (2008)

[2] B.Esposito, G.Granucci, M.Maraschek et al, Plasma Physics and Contr. Fusion, **53**, 124035 (2011)

[3] F.Volpe et al., Physics of Plasmas, **16**, 102502 (2009)

[4] M.Maraschek – This EPS conference, P1.112

[5] L. Giannone et al., Fusion Engineering and Design **88**, 3299 (2013).

[6] M. Reich et al., Fusion Eng. Design, online: http://dx.doi.org/10.1016/j.fusengdes.2014.04.024

[7] S. Nowak et *al.*, Physics of Plasmas, **19**, 092508 (2012)

[8] C. Galperti et al, Plasma Phys. Contr. Fusion **56**, 114012 (2014)

**Acknowledgement** This work has been carried out within the framework of the EUROfusion Consortium and has received funding from the Euratom research and training programme 2014-2018 under grant agreement No 633053. The views and opinions expressed herein do not necessarily reflect those of the European Commission.

# Dynamics of filaments in the vicinity of the separatrix during ELM cycles

J. Vicente[1], G.D. Conway[2], L. Meneses[1], C. Silva[1], M.E. Manso[1], the ASDEX Upgrade Team[2], and JET Contributors*

EUROfusion Consortium, JET, Culham Science Centre, Abingdon, OX14 3DB, UK

[1]*Instituto de Plasmas e Fusão Nuclear, Instituto Superior Técnico, Universidade de Lisboa, 1049-001 Lisboa, Portugal*

[2]*Max-Planck-Institut für Plasmaphysik, 85748 Garching, Germany*

**Introduction**

Theory suggests that ELMs grow explosively by radial expansion and ejection of plasma filaments into the scrape-off layer (SOL). Filaments have been observed and studied in the past years, mainly in the SOL during the ELM crash phases. Despite the large amount of data collected in those late stages of the filament life-cycle (i.e. at the SOL) the early stages of filaments are scarcely documented due to the lack of diagnostics able to measure with high time and spatial resolutions in their birth and development regions. As a consequence, the role of filaments in the transport of particles and energy during ELMs is not yet fully understood. MHD simulations indicate that either conductive or convective losses (the later *via* ejection of plasma filaments) can become dominant loss mechanisms depending on the ELM size, which highlights the importance of filaments and opens possibilities for alternative ITER scenarios such as the Type-III ELMy H-mode.

**Reflectometry Techniques**

Reflectometry is a diagnostic technique able to probe broad regions of the plasma edge with good time/spatial resolutions. Electromagnetic waves launched into the plasma are reflected at a plasma cut-off layer determined by the wave frequency $F$ for ordinary (O-mode) polarization, or by both $F$ and the local magnetic field for extraordinary (X) mode. According to the 1D geometric optics approximation the phase shift $\Delta\varphi$ of the O-mode reflected wave is proportional to the radial displacement $\Delta r$ of the reflecting density layer, such that $\Delta r = \Delta\varphi c / 4\pi F$, where c is the velocity of light. For small radial displacements, such that the magnetic field changes are negligible, the above relation should also hold for X-mode polarization. Thus, for fixed probing frequency, the time derivative of the phase provides the radial velocity

*See the Appendix of F. Romanelli et al., Proceedings of the 25th IAEA Fusion Energy Conference 2014, Saint Petersburg, Russia

$V_r = \Delta r/\Delta t$ of the probed density layer, either in O- or X-mode. Using reflectometry, a technique for filament detection was developed based on the fast movements of density layers induced by filaments [2]. A $V_r$ threshold criterion was employed with a typical value of $|V_r| >$ 1.2km/s, which was also suitable for this work.

**Reflectometry Systems (AUG and JET)**

At AUG two O-mode reflectometry systems operate in the Q (33-49.2 GHz) and V (49.4-72 GHz) bands respectively, using monostatic antennae configurations located on the low field side (LFS). The Q-band antenna is very close to the mid-plane and the V-band antenna is displaced downwards towards the X-point in lower single null magnetic configurations. At JET, a radial correlation reflectometer can be used in fixed frequency where two X-mode channels operate in the W band (73.4-110 GHz) sharing a bi-static antennae configuration. The antennae are located at the LFS close to the magnetic axis. The data acquisition rate for both the JET and AUG systems is 2 MHz.

**The Type-I ELM Cycle at JET**

The techniques developed for filament studies at AUG [2] were applied for the first time at JET in Type-I ELMy H-modes. In discharge #84704 the magnetic field at the flat-top phase reached $B_t$ = 2.4T. The cut-off (density layer) for the lowest probing frequency (73.4 GHz) was located at the plasma edge, below the pedestal top, just inside of the separatrix. For higher magnetic field values the location of the cut-off would be further out. Filament detection was applied over a period of fairly stable plasma parameters [46.5-47.0s] and the $V_r$ peak detections have been ELM synchronized ($t_0$=0 was taken at the ELM peaks observed in the Be emission line signal). In Fig. 1 it is displayed the probability distribution function (PDF) of the detection-times along the ELM cycle.

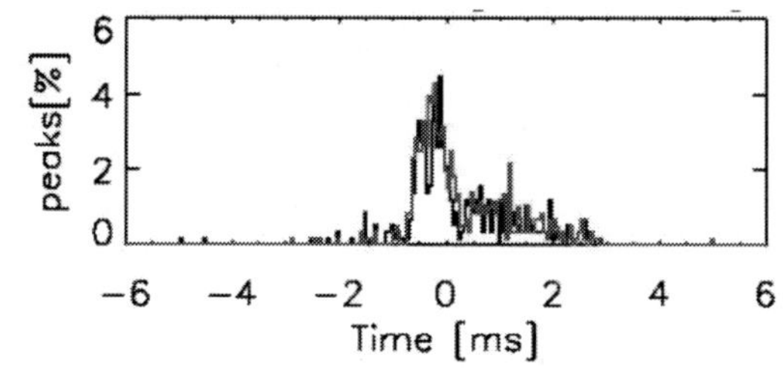


Figure 1: *Filament activity along a Type-I ELM cycle in JET. PDFs of detection times of positive (black) and negative (red) $V_r$ peaks, F=73.44GHz.*

It is observed that the occurrence of filaments is largely enhanced at the ELM onset period, some hundred μs before the ELM peak ($t_0$=0). In Fig. 2 are displayed the signatures of filaments on the reflectometry signals (two closely spaced density layers were probed) during inter-

ELM periods, obtained with conditional averaging (CA).

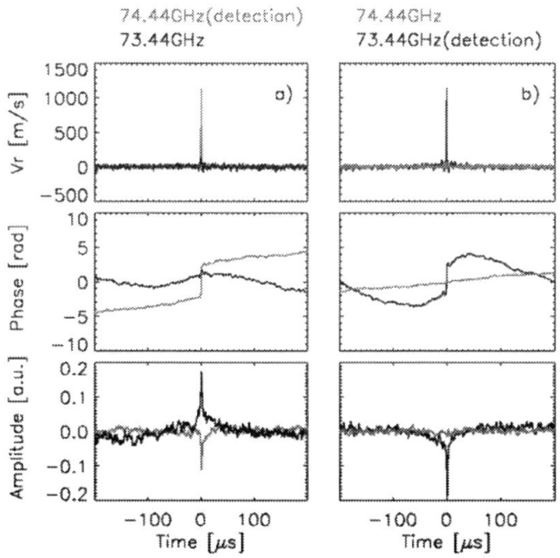


Figure 2: *Inter-ELM filament signatures on reflectometry signals at JET using CA and detection signals from F=74.44GHz a) and F=73.44GHz b).*

Both channels were used to provide primary and secondary signals. It is observed that when performing detections using the signal reflected at the highest density, correlated amplitude peaks are observed in the signal reflected from the lower density layer. Since there is only a small smooth phase oscillation associated (of much lower *frequency* <5kHz than any possible ELM precursor mode), it is likely that these signatures correspond to large (poloidally) filaments propagating with a dominant poloidal velocity. In general, the signatures in the detection channel, are similar to those obtained previously at AUG: $V_r$ peaks associated with amplitude dips and jumps in the phase signals. This strengthens the suggestion that filaments display similar dynamics in both machines.

**The Type-III ELM Cycle at AUG**

In AUG discharge #27016, a period with 20 type-III ELMs was considered and filament detection was again ELM synchronized ($t_0$=0 taken at the ELM peaks in divertor currents). Li beam and Thomson scattering diagnostics allowed to locate the reflecting density layers of each channel operating at their lowest frequencies: $\rho_{pol} \approx 0.99$ (V-band) and $\rho_{pol} \approx 1.00$ (Q-band). Results in Fig.3 show the enhancement in filament activity at the ELM onset period as seen before in type-I ELMs at both JET and AUG.

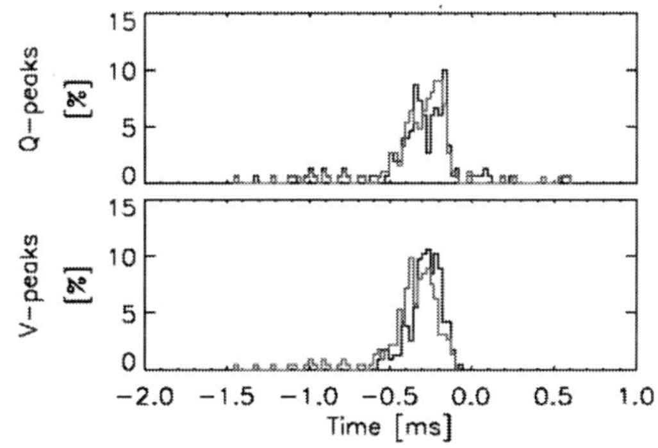


Figure 3: *Filament activity along the type-III ELM cycle at AUG. PDFs of times of detection of positive (black) and negative (red) $V_r$ peaks with both the Q and V bands.*

After the ELM peak, filamentary activity is strongly reduced, contrary to what was observed in type-I ELMs where some activity remained for some ms (even if filaments displayed smaller size and/or amplitude). Differences in the PDF shapes from the two channels prevent a time delay analysis (between poloidally displaced measurements) as was the case for type-I ELMs, where a dominant poloidal velocity component of filaments at the ELM onset period was found [2].

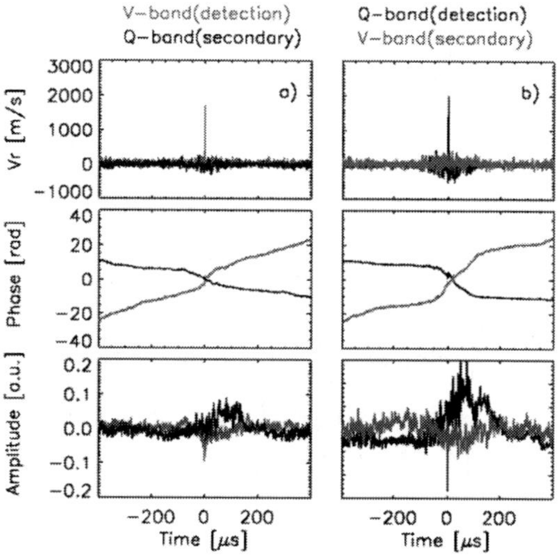


Figure 4: *Signatures obtained using the V-band a) and the Q-band b) signals for detection at AUG.*

The signatures obtained (see Fig.4) display large phase *jumps* (10-20 radians) anti-correlated in the two channels and an amplitude *hill* in the signal reflected from the lowest density probed layer. These features suggest that the density perturbations have large poloidal width (displacement between measuring spots is ≈20cm) and are likely propagating along a significant radial length.

**Conclusions**

Filaments were studied for the first time at JET in type-I ELMy H-modes using X-mode reflectometry. Results suggest that filament activity is enhanced at the ELM onset with dominant propagation in the poloidal direction, as seen before at AUG. Studies at AUG in Type-III ELMy H-mode reveal that at the ELM onset, filament activity is enhanced but does not display a preferred poloidal propagation direction and is suppressed more rapidly than in type-I ELMs (in JET or AUG).

**Acknowledgments**

This work has been carried out within the framework of the EUROfusion Consortium and has received funding from the Euratom research and training programme 2014-2018 under grant agreement No 633053. IPFN activities also received financial support from "Fundação para a Ciência e Tecnologia" through project UID/FIS/50010/2013. The views and opinions expressed herein do not necessarily reflect those of the European Commission.

**References**

[1] G.T.A. Huijsmans and A. Loarte, Nucl. Fusion **53** (2013) 123023

[2] J. Vicente *et al.*, Plasma Phys. Control. Fusion **56** (2014) 125019

# Dependence of poloidal asymmetries of the edge turbulence on the magnetic configurations in ASDEX Upgrade

V. Nikolaeva[1,2,3], P. Manz[2,3], L. Guimarais[1], G. D. Conway[2], R. Fischer[2], M. E. Manso[1], T. T. Ribeiro[2], F. Ryter[2], C. Silva[1], U.Stroth[2,3], E. Wolfrum[2] and the ASDEX Upgrade team

[1] *Instituto de Plasmas e Fusão Nuclear, Instituto Superior Técnico, Universidade Lisboa, PT*

[2] *Max Planck Institute for Plasma Physics, Boltzmannstr. 2, D-85748 Garching, Germany*

[3] *Physik-Department E28, Technische Universität München, 85748 Garching, Germany*

**Introduction**

Understanding plasma turbulence and transport properties in edge and SOL plasma regions is relevant to improve reliability of a fusion reactor. Most edge measurements are made on the low-field side (LFS) because of easier diagnostics access but since SOL/edge properties are not poloidally symmetric data from the high-field side (HFS) are essential.

**Experimental set-up**

At ASDEX Upgrade (AUG), the frequency modulated continuous wave (FMCW) reflectometry system provides simultaneous measurements on the HFS and LFS, that is dedicated to density fluctuation measurements in a fixed frequency regime. Operation in O mode provides independent of the magnetic field measurements that is essential for study with different magnetic configurations. Four channels (K, Ka, Q and V bands) on each inner and outer sides of tokamak featuring individual single antennae (located approximately at the tokamak midplane), cover the probing density range from 0.3 to 7 x $10^{19}$ $m^{-3}$. The diagnostic is equipped with homodyne detection and phase is extracted by applying the Hilbert transform method. Validation of the method is done by comparison with overlapping data (for the same probing

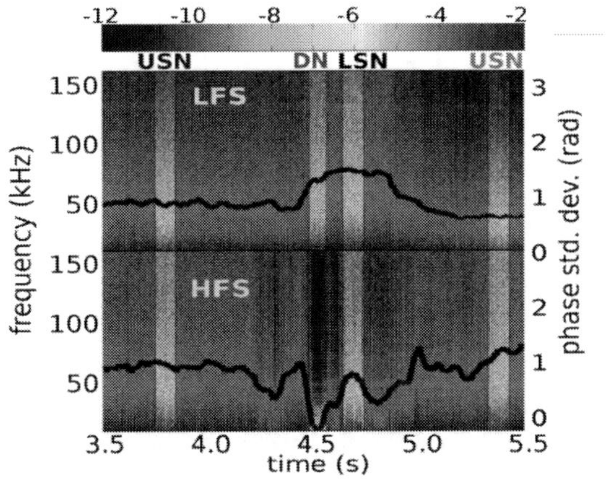


Fig. 1. Spectrograms of phase of reflectometers signals (K band, 0.4 x $10^{19}$ $m^{-3}$) at SOL on the LFS (top) and the HFS (bottom) through the L-mode discharge. The four bright zones correspond to different magnetic configurations, from left to right – USN, DN, LSN and back to USN. Results in Figs 2 and 3 refer to those four zones. Black curves are standard deviation of signals phases at the LFS and the HFS.

densities) obtained by hopping reflectometer with detection enabling to extract phase directly.

**Experimental results**

The results from reflectometry in L-mode discharge #31633 with the magnetic configuration transitions from upper single null (USN) through double null (DN) to lower single null (LSN) and back to USN are presented in Fig.1. Standard deviation of phase is proportional to the density fluctuation level δn/n [1]. Spectrograms of phase and level of standard deviation of phase display the δn/n changes observed in the SOL for different magnetic configurations

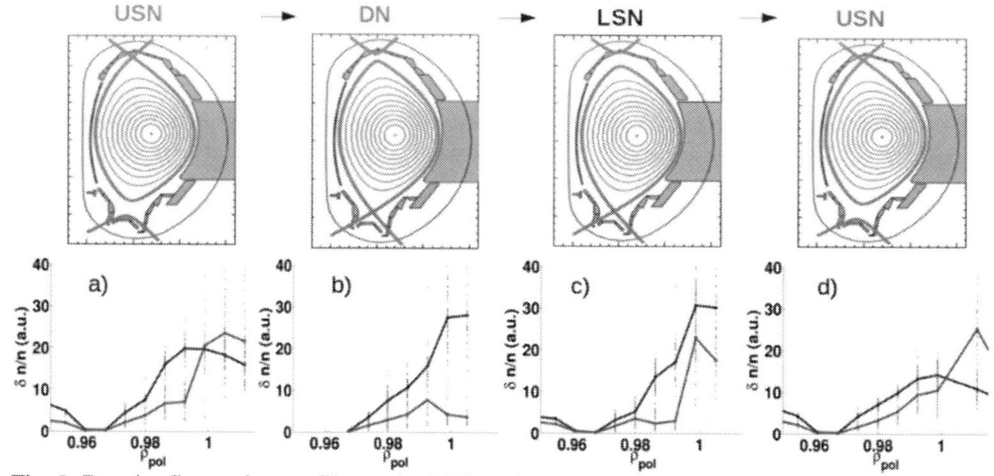


Fig. 2. Density fluctuation profiles at the HFS (red color) and the LFS (blue color) for different magnetic configurations: a) USN at the beginning of the discharge, b) DN, c) LSN and d) USN at the end of the discharge.

(Fig. 1), which are more pronounced on HFS. Radial profiles of δn/n at the HFS and the LFS shown in Fig.2 for the four highlighted zones in Fig. 1, were obtained using 1D C. Fanack model [2]. The effect of magnetic configuration on the poloidal asymmetries of density fluctuations is more pronounced outside the separatrix (Fig. 2 and Fig. 3). A δn/n drop inside the separatrix at around ρ = 0.96 is observed in the region of strong radial electric field Er shear for all configurations (Fig. 3), that is in agreement with the previous experimental results at AUG [3] and turbulence stabilization theories [4]. In LSN configuration the turbulence level is found to be much lower on the edge and SOL HFS (Fig. 2c) in comparison with the LFS. This is expected from common interchange instability mechanism and widely observed on many tokamaks. The highest levels of density fluctuations are observed in LSN and DN configurations on LFS. The lowest fluctuation levels are observed in the DN configuration (Fig. 2b) in SOL HFS, that corresponds to this region becoming isolated from the LFS SOL. On the LFS DN δn/n profiles behaviour is similar with the LSN, see Fig. 3. The strongest HFS/LFS asymmetry of fluctuations occurs outside the separatrix in DN plasma shape that was also seen with three dimensional electromagnetic gyrofluid computations from the GEMR code [5]. In USN configurations it has been observed that the inboard SOL turbulence level exceeds those of the outboard (Fig. 2a, Fig. 2d). That is surprisingly as the common interchange instability is driven by the bad curvature which is located at the outboard side. One possible explanation may come from GEMR simulations.

**Comparison of experimental results and GEMR simulations**

Recent GEMR simulations have shown the importance of another instability in the near SOL, the so-called conducting wall instability (CWI) [6]. Here temperature fluctuations are driven by the temperature gradient and translated to potential fluctuations. A simulation of the

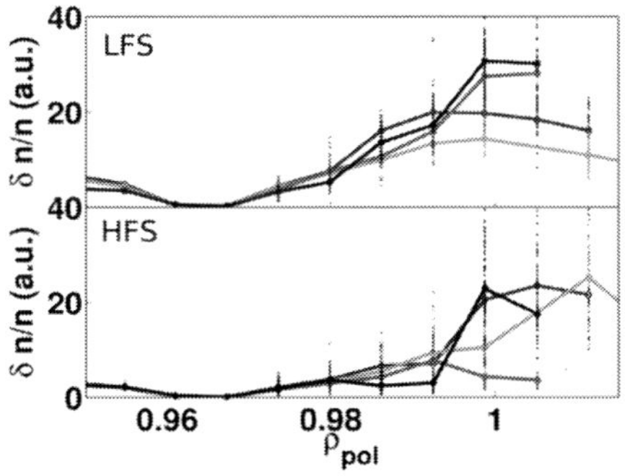


Fig. 3. Density fluctuation $\delta n/n$ profiles on the LFS (top) and the HFS (bottom) for different magnetic configurations – USN at the beginning of the discharge (blue), DN (red), LSN (black) and USN at the end of the discharge (green). USN $\delta n/n$ profiles have wider ρ range due to corresponding shifts of density profiles.

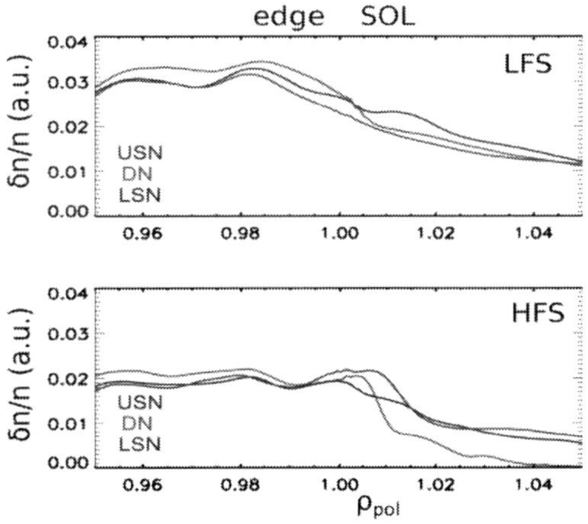


Fig. 4. Density fluctuation profiles $\delta n/n$ obtained by GEMR simulations on the LFS (top) and the HFS (bottom) for different magnetic configurations – USN (blue), DN (red) and LSN (black).

experiment was performed with the gyrofluid GEMR code with the same parameters for the three configurations. Even though the simulations were done in a circular plasma geometry with varying limiter positions only and therefore without taking the X-point geometry into account, the three configuration are called LSN (black), USN (blue) and DN (red) for easier comparison with the experiment. As the CWI is driven by the temperature gradient a setup with a strong electron temperature gradient $L_n=3\,L_T$ was chosen with the drawback that density fluctuations are smaller in comparison to cases with steeper density gradient.

Although the simulations are done in a circular geometry and effects of the triangularity are not taken into account, the turbulence levels show differences in the near SOL (Fig.4). In the region close to the separatrix $1 \leq \rho \leq 1.005$ the $\delta n/n$ is higher for LSN in comparison with

USN on the LFS and vise versa on the HFS that is in agreement with the experimental results. The $\delta n/n$ for DN configuration is closer to LSN on the LFS and to USN on the HFS. At $\rho >$ 1.005 the $\delta n/n$ in the DN configuration strongly drops at HFS. Between $1.005 \le \rho \le 1.015$ the largest difference between configurations is observed. The turbulence level on the HFS is similar to the turbulence level on the LFS in the USN configuration as observed in the experiment. The cross-phase between potential and density fluctuations $\alpha(\widetilde{\varphi},\widetilde{n}_e)$ is close to zero at around $\rho > 1.005$ in USN, showing its similarity to a drift-wave instability. At that location, the estimated cross-phase between potential and temperature fluctuations $\alpha(\widetilde{\varphi},\widetilde{T}_e)$ is close to zero, which shows that the underlying instability mechanism is the CWI.

**Conclusions**

Effect of magnetic configuration on the poloidal asymmetries of density fluctuations is mostly pronounced outside the separatrix. The strongest HFS/LFS asymmetry of fluctuations in DN SOL, corresponds to HFS SOL being isolated from LFS. In USN configurations, HFS SOL turbulence increases above the LFS level. This behavior is currently under investigation and might be induced by the CWI driven by different temperature gradients at these locations that are configuration dependent. Drop of density fluctuations inside the separatrix both at LFS and HFS observed in region of strong radial electric field shear for all configurations.

**Acknowledgements:** This work was carried out within the framework of the Erasmus Mundus International Doctoral College in Fusion Science and Engineering (FUSION-DC) and the EUROfusion Consortium. It has received funding from the Euratom research and training programme 2014-2018 under grant agreement No 633053. IST activities also received financial support from "Fundação para a Ciência e Tecnologia" through project UID/FIS/50010/2013. The views and opinions expressed herein do not necessarily reflect those of the European Commission.

**References**

[1] R. Nazikian and E. Mazzucato, Rev. Sci. Instrum. **66 (1)**, 392 (1995)

[2] C. Fanack et al., Plasma Phys. Control. Fusion **38**, 1915 (1996)

[3] J. Schirmer, PhD thesis (2005)

[4] H. Biglari et al., Phys. Fluids **B2**, 1 (1990)

[5] T. T. Ribeiro and B. D. Scott, Plasma Phys. Control. Fusion **50**, 055007 (2008)

[6] P.Manz et al., Phys. Plasmas **22**, 022308 (2015)

# LFS/HFS edge density profile dynamics on ASDEX-Upgrade

L. Guimarais[1], V. Nikolaeva[1,2], S. Potzel[3], F. Reimold[3], M. Bernert[3], D. Carralero[3], G. D. Conway[3], M. E. Manso[1], J. Santos[1], A. Silva[1], C. Silva[1], E. Wolfrum[3] and the ASDEX-Upgrade Team

[1] *Instituto de Plasmas e Fusão Nuclear, Instituto Superior Técnico, Universidade de Lisboa, Portugal,* [2] *Technische Universität München, James-Franck-Str. 1, 85748, Garching, Germany,* [3] *Max-Planck-Institut für Plasmaphysik, Boltzmannstr. 2, 85748, Garching, Germany*

## Introduction

Next generation fusion devices are required to operate in regimes with high energy confinement and high density in order to achieve large fusion power. They will explore new regimes of boundary plasma physics with competing needs of plasma performance and plasma facing components. A reduction of the power loads on the plasma facing components will be required due to the narrow width of the scrape-off layer (SOL) power flux profile characteristic of the H-mode. This necessitates high density and low temperature SOL plasmas detached from the divertor. The development of a High Field Side High High Density Front (HFSHD) associated with the inner divertor detachment has been reported [1, 2], which is located in the SOL at the height of the X-point. In these regimes the edge plasma and the plasma-wall interaction exhibits a complex dynamic that needs to be understood.

In this work the dynamics of the midplane edge electron density profiles and its relation to divertor detachment is studied on ASDEX-Upgrade (AUG) both in L- and H-mode, taking advantage of the high resolution High-Field-Side (HFS)/Low-Field-Side (LFS) O-mode reflectometry measurements.

## L-mode density ramp discharges

It was previously shown[1] that as plasma density increases, divertor detachment undergoes three distinct states: onset state (OS), fluctuating state (FS) and complete detachment state (CDS) [1]. Figure 1a,b shows the temporal evolution of the HFS/LFS iso-density contours from reflectometry together with other relevant plasma parameters. Gas fuelling was ramped up leading to an increase in the plasma density until the density limit occurs. Transition to the FS is observed at 2.5 s (determined by the onset of X-point fluctuations), while the CDS (characterized by the end of the X-point fluctuations and density and temperature reduction at the outer strike-point) occurs at 3.2 s. As illustrated in Figure 1a,b, at low densities (below $3.5\times10^{19}\,m^{-3}$) edge density profiles measured by reflectometry at the midplane are roughly HFS/LFS symmetric. However, around $t = 2.2$ s, well before the transition to the FS, a density increase is observed in front of the inner wall leading to strong poloidal asymmetries in the SOL density. Divertor plasma is often observed to oscillate back and forth between the OS and the FS [3]. Around 2.5 s, these oscillations reach the HFS midplane leading to a strong modulation of the HFS density profiles but are not detected in the LFS. At the transition to the FS the HFS/LFS density asymmetry decreases, increasing again for $t > 2.8$ s when the CDS is approached. Even in the presence of resonant magnetic perturbations, seen in Figure 1c as $I_{RMP1}$ and $I_{RMP2}$, density profiles change little until core density starts to increase, half a second later.

The high density at the midplane may be related to the so-called HFSHD observed on the HFS far-SOL divertor region. The divertor HFSHD, characterized by high density in the divertor volume measured by Stark broadening, is seen to move upwards from the strike-point to the X-point as detachment develops (Figure 1d). The HFS high density region observed at the midplane continues to develop as the gas fuelling is ramped up and the density limit is approached, suggesting that the HFSHD reaches the midplane. However, the AXUV measurements do not indicate the existence of a significant radiation outside the divertor region. The evolution of the LFS profiles is significantly different from that at HFS. As the gas fuelling is ramped, the LFS edge density increases across the entire region scanned by the reflectometry diagnostic and particularly outside the separatrix leading to a broad SOL. This broad density profile at the LFS is most probably a result of an increased radial transport at the LFS [4]. Note that the midplane HFSHD is formed before any significant SOL broadening is observed at the LFS.

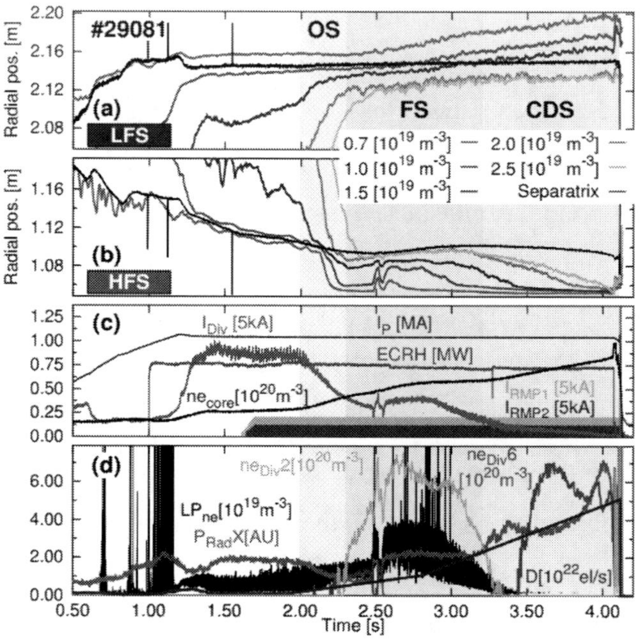


Figure 1: *Temporal evolution of the radial position of different density layers for an L-mode (29081) at the LFS (a) and HFS (b) obtained from reflectometry density profiles, together with plasma current, applied ECRH power, divertor shunt current, core line-averaged density (c), divertor target density $LP_{ne}$, volume density at lower ($ne_{Div2}$) and upper ($ne_{Div6}$) inner divertor, radiated power in the X-point region and D fuelling (d).*

## H-mode discharges

In high power H-mode discharges at AUG, a high density front also develops in the far SOL of the HFS divertor. Figure 2 shows the temporal evolution of the HFS/LFS iso-density contours of an H-mode discharge with nitrogen seeding together with the time traces of relevant plasma parameters. In the first phase (up to 2.8 s) moderate NBI power and gas fuelling are applied resulting in large amplitude low frequency ELMs. In a second phase (3−3.5 s) both NBI power and gas fueling rate are increased, leading to faster ELMs. Finally, at 3.5 s, nitrogen is puffed at constant rate to cool the divertor. The behaviour of the HFSHD changes strongly despite the roughly constant line-averaged core density. The increase of both heating power and fuelling leads to the development of the HFSHD that is then reduced when nitrogen is injected.

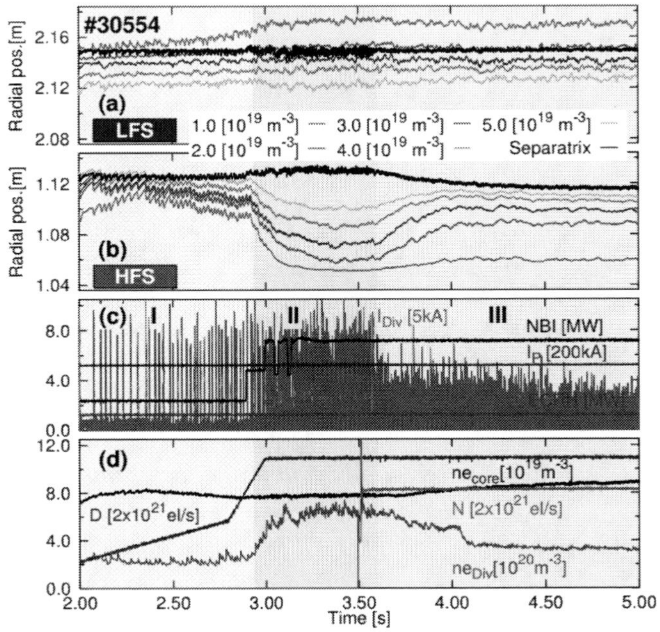


Figure 2: *Temporal evolution of the radial position of different density layers along a H-mode discharge (30554) for LFS (a) and HFS (b). Plasma current, applied ECRH and NBI power, divertor shunt current (c), core line-averaged density, divertor volume density and D and N fuelling (d).*

Reflectometry data shows the HFSHD extending all the way up to the HFS midplane, but not to the LFS, leading to strong poloidal asymmetries in the edge density profiles. At the midplane, the HFSHD is located in the SOL just in front of the inner wall with densities greater than $6\times10^{19}\,\mathrm{m}^{-3}$ (diagnostic limit at the HFS). A delay ($\approx 25\,\mathrm{ms}$) is observed between the density increase in the divertor volume (HFSHD formation) and that seen at the midplane. The ELM dynamics are found to be strongly influenced by the HFSHD. The conditional averaging of edge density profiles to study their evolution during ELMs can be seen in Figure 3 for the first and second discharge phases. In the case of low frequency Type-I ELMs (Figure 3a-c) the pre-ELM density profile is recovered after 2 ms at the LFS, while at the HFS a HFSHD is observed during the ELM that then persists for about 5 ms. During this period the particle flux to the inner divertor is reduced, as confirmed by divertor probes.

The power radiated in the inner divertor is also reduced as indicated by divertor AXUV measurements. These results suggest that in the period just after the ELM no HFSHD exists in the divertor region. The divertor target density and temperature are observed to increase when the midplane HFSHD disappears 5 ms after the ELM. Both inner and outer divertor shunt currents show an ELM signature lasting 2 ms, compatible with LFS profile data. In the second discharge phase (Figure 3d-f), the ELM frequency increases and the HFSHD is continuously observed at the midplane although the density in front of the inner wall increases further during ELMs. The formation of the midplane HFSHD during ELMs may be related with an enhancement of the plasma-wall interaction at the inner wall caused by the broadening of the plasma profiles.

## Summary

The formation of a HFSHD at midplane has been observed with reflectometry in both L- and H-mode regimes associated with divertor detachment. In L-mode, the midplane HFSHD develops continuously as the discharge density is increased while in H-mode, the midplane HFSHD is essentially observed during ELMs for low ELM frequency and it is continuously observed when the ELM frequency is increased. Results indicate that for low frequency ELMs the HFSHD builds up at the midplane while it vanishes in the divertor region.

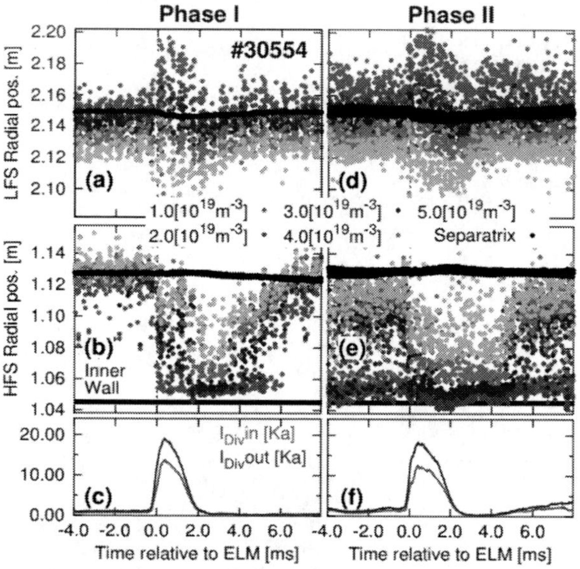


Figure 3: *Temporal evolution of the radial position of different density layers during low* ($1.7-2.8$ s) *(a,b) and high* ($3.0-3.5$ s) *(d,e) frequency Type-I ELMs obtained using Conditional Averaging, together with the divertor shunt current (c,f).*

## Acknowledgements

This work has been carried out within the framework of the EUROfusion Consortium and has received funding from the Euratom research and training programme 2014-2018 under grant agreement No 633053. The views and opinions expressed herein do not necessarily reflect those of the European Commission. IPFN activities also received financial support from "Fundação para a Ciência e Tecnologia" through project UID/FIS/50010/2013 and grant SFRH/BD/87738/2012.

## References

[1] S. Potzel et al, Nuclear Fusion **54** (2014) 013001

[2] S. Potzel et al, Journal of Nuclear Materials (2015), in press

[3] S. Potzel et al, Journal of Nuclear Materials 438 (2013) S285-S290

[4] D. Carralero et al, Nuclear Fusion **54** (2014) 123005

# Real-time control of MHD instabilities using ECCD in ASDEX Upgrade

M.Maraschek[1], M.Reich[1], K.Behler[1], L.Giannone[1], A.Gude[1], E.Poli[1],
C.J.Rapson[1], O.Sauter[2], J.Stober[1], W.Treutterer[1], H.Zohm[1],
ASDEX Upgrade team[1] and EUROfusion MST1 team*

[1] *Max-Planck Institut für Plasmaphysik, Boltzmannstr. 2, 85748 Garching, Germany*
[2] *Centre de Recherches en Physique des Plasmas, EPFL, 1015 Lausanne, Switzerland*

**Introduction** Future tokamaks, like ITER and DEMO, must be capable of controlling neoclassical tearing modes (NTMs) to ensure high $\beta_N$ plasma operation. These MHD instabilities are performance limiting and in the worst case they are destructive when they trigger a disruption. A proven actuator with which these instabilities can be controlled is electron cyclotron heating and current drive (ECCD) at the rational surface. This requires a precise control of the radial deposition location [1].

In order to achieve this control on ASDEX Upgrade, a large number of real-time (rt) diagnostics and intelligent controllers work in unison, all coordinated by the discharge control system (DCS). The system automatically detects the excitation of magnetic islands and identifies their mode number from magnetic probe signals. It uses electron cyclotron emission (ECE) and magnetic probe signals in combination to infer their radial position. The DCS aims the ECCD at the appropriate rational surfaces in order to stabilize the modes.

A newly introduced method originating at TCV that uses radial sweeps of the ECCD deposition alleviates the precision requirements for ECCD deposition such that inaccurately determined rational surfaces are not prohibitive for NTM stabilization and prevention [2].

**Technical upgrades** In addition to the essential rt equilibrium reconstruction (typically 4ms, 2ms possible data rate) [3] the controllers require density profile (1ms) [4], detection of MHD marker positions, such as rational surfaces from MHD modes from the electron cyclotron emission (6ms) [5], or the detection of sawteeth [6], ECCD beam-tracing in real-time (15-25ms) [7] and global plasma parameters (such as $I_p$ or $\beta_p$ from rt-equilibrium) to make their decisions.

**Mode detection** For the detection of a specific $(m/n)$-mode with poloidal mode number $m$ and toroidal mode number $n$ a subset of magnetic signals, i.e. $dB_\theta/dt = \dot{B}_\theta$-coils, on the low field (LFS, $\theta = 0^\circ$) and on the high field side (HFS, $\theta = 180^\circ$), are used. From these a set of hardware generated weighted sums, i.e. spatially Fourier filtered signals of the form

$$\dot{B}_{m/n} = \left( \beta_m^{LFS} \frac{\sum_i \alpha_{n,i}^{LFS} \cdot \dot{B}_i^{LFS}}{\sum_i |\alpha_{n,i}^{LFS}|} + \beta_m^{HFS} \frac{\sum_j \alpha_{n,j}^{HFS} \cdot \dot{B}_j^{HFS}}{\sum_j |\alpha_{n,j}^{HFS}|} \right) \Big/ \left( |\beta_m^{LFS}| + |\beta_m^{HFS}| \right)$$

are created. The $\alpha_{n,i}^{xFS}$ define the toroidal mode number $n$ to which the $\dot{B}_{m/n}$ is sensitive, the $\beta_m^{xFS}$ the dominating odd or even poloidal mode number $m$. In this way eight frequency independent signals $[n=1]$, $[n=2]$, $[n=\text{odd}]$ (sum of two coils on LFS 180° apart), $[n=\text{even}]$ (difference of two coils on LFS 180° apart), $[n=\text{odd}/m=\text{even}]$, $[n=\text{even}/m=\text{odd}]$, $[n=1/m=\text{even}]$ and $[n=2/m=\text{odd}]$ are derived in real-time.

These hardware generated $\dot{B}_{m/n}$ signals are used on a rt computer to calculate their amplitude using the envelope and communicating the results to the DCS on a 2ms sample rate. For constant mode frequency, $f_{m/n}(t) = f_{m/n}$, this amplitude is proportional to the island perturbation field.

*See http://www.euro-fusionscipub.org/mst1

The DCS can detect the existence of a (3/2) or a (2/1)-mode by any of these signals and initiate the stabilization attempt at the corresponding rational surface, $\rho_{m/n}$. Locked $n = 1$ modes are identified independently by a set of saddle coils [8]. After the mode has been removed, the gyrotrons can be turned off and are available for other modes. In the future the DCS will decide automatically, which gyrotrons can be most efficiently steered to the resonant surface [9].

Additionally, these mode signals are also connected to the ECE data acquisition system, where they are sampled together with the ECE signals. A correlation analysis is performed to determine the radial location of the phase jump of $\approx \pi$ occurring at the position of the X-point of the island [5]. Together with the rt-equilibrium this provides the radial location, $\rho_{m/n-NTM}(t)$, of the island. For each of the magnetic signals an independent fast Fourier transformation is performed and the frequency, $f_{m/n}(t)$, and amplitude, $A(f_{m/n}(t))$, is determined for $f > 2$kHz. Hence the time evolution of $B_{m/n}(t) \sim A(f_{m/n}(t))/f_{m/n}(t)$ becomes available.

**ECCD deposition control** The real-time implementation [10, 7] of the beam-tracing code TORBEAM calculates the radial deposition, $\rho_{dep}(t)$, of the ECCD for each of the four gyrotrons individually. The code requires the rt-equilibrium and the rt density profile.

Based on $\rho_{dep}$, the DCS steers the launching mirrors towards $\rho^{target}_{m/n}$. As the most efficient radial localization, $\rho_{dep}(t)$, of the ECCD with respect to the real resonant surface, $\rho_{m/n}(t)$, and the required accuracy are still open issues, the system provides the possibility to pre-program for each (m/n)-mode a time dependent offset to the radial target,

$$\rho^{target}_{m/n}(t) = \rho_{m/n}(t) + \Delta\rho_{m/n}(t).$$

It is possible to select the radial target trajectory, $\rho_{m/n}(t)$, calculated in real-time, for each gyrotron independently before the discharge. From the rt-equilibrium the radius of the resonant surface, $\rho_{m/n-eq}(t)$, the detection radius from the rt ECE analysis, $\rho_{m/n-NTM}(t)$ and a generic, mode number independent detection radius $\rho_{NTM}(t)$ from the correlation analysis can be selected.

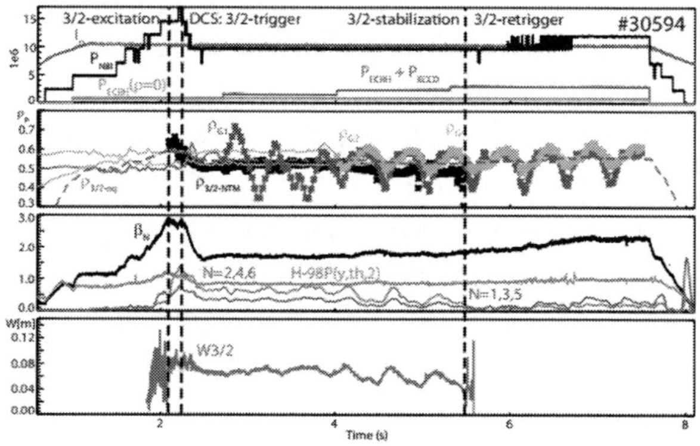


Figure 1: *NTM stabilization using sweeping with 3 $P_{ECCD}$[MW] steps and attempt to re-trigger mode. first box: $I_p$[MA] (red), $P_{NBI}$[MW] (black), central $P_{ECRH}$[MW] (green) and total $P_{ECRH} + P_{ECCD}$[MW] (blue). second box: $\rho_{3/2-NTM}(t)$ (black symbols), $\rho^{target}_{3/2}(t)$ (blue/orange/cyan lines), $\rho_{ECCD}(t)$ (blue/orange/cyan symbols) from rt-TORBEAM, $\rho_{3/2-eq}(t)$ from rt-equilibrium. third box: $\dot{B}_{n=odd}$ (blue), $\dot{B}_{n=even}$ (red) $\beta_N$ (black) and H-98P(y,th,2)-factor (green). forth box: island width $W_{3/2}$[m] (red)*

**Island width** The island width, $W_{m/n-ECE,ref}$ in $m$, is calculated from the Fourier filtered ECE amplitude and phase profile at the mode's frequency, $f_{m/n,ref}$[Hz], according to [11]. The profile is considered versus the minor plasma radius, $r$[m], on the low field side via the equilibrium. With this information from a reference discharge at a reference time (index $ref$) and the corresponding magnetic measurements, i.e. $\dot{B}_{m/n,ref}$ [T/s], the island width becomes available for discharges with self-similar $q$-profiles. With the tracked frequency, $f_{m/n}(t)$ [Hz], of a (m/n)-NTM on a single Mirnov coil and its amplitude, $\dot{B}(f_{m/n}(t))$ [T/s], from any discharges the island width

$$W_{m/n}(t) = \left(W_{m/n-ECE,ref} \Big/ \sqrt{\dot{B}_{m/n,ref}/f_{m/n,ref}}\right) \cdot \sqrt{\dot{B}(f_{m/n}(t))/f_{m/n}(t)}$$

is calculated offline for further analysis. This approach is possible, as in all present experiments a plasma current of $I_p = 1.0$MA and a toroidal field of $B_t = -2.5$T have been used.

**Experimental setup and scenarios** The stabilization experiments have been performed in the improved H-mode or hybrid scenario with $I_p = 1$MA and $B_t = -2.6$T with up to 17.5MW of NBI heating power. One gyrotron is aimed centrally ($\rho_{dep} = 0$), in order to avoid unwanted impurity peaking. This leaves three gyrotrons available for NTM control experiments.

**Stabilization scheme** For the stabilization scheme, the NBI power is initially ramped up to maximal 17.5MW in order to trigger an NTM. Typically, a (3/2), and occasionally also a (2/1)-NTM is triggered. The natural onset is characterized by $\beta_N^{onset,int}$ and $P_{NBI}^{onset,int}$. Any excited mode triggers the DCS, which switches to different program branch or activates a controller. It reduces the NBI power to typically $P_{NBI}^{stab}$ = 10MW for both the (3/2) and the (2/1)-NTM. Only with $P_{NBI} \leq 10$MW the presently available ECCD power (3 gyrotrons with up to $\approx 2.1$MW) for mode stabilization is seen to be able to remove the NTM.

For preemptive NTM avoidance in the same discharge scenario the NBI power is ramped up to an initial $P_{NBI}$ and subsequently the ECCD is applied at the resonant surface, where the first mode is expected, i.e. at the $q = 3/2$-surface. With the preemptive $P_{ECCD}^{pre}$ applied, the NBI power is increased until an NTM is triggered at $P_{NBI}^{onset,pre}$ and $\beta_N^{onset,pre}$.

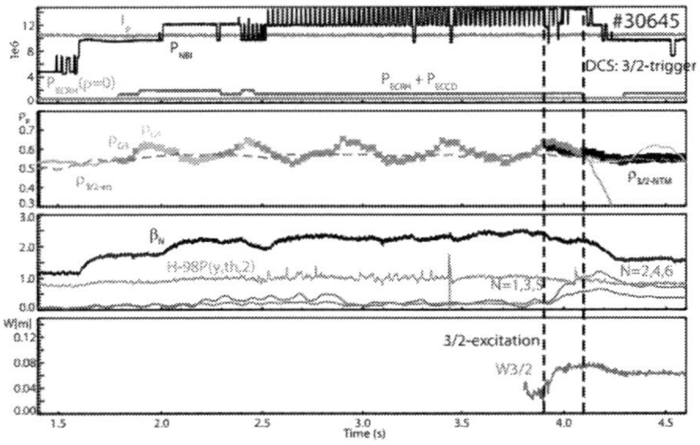


Figure 2: *Preemptive NTM-avoidance using sweeping with one gyrotron ($P_{ECCD}^{pre} \approx$ 0.65MW). Traces and colour coding similar as in figure 1.*

**Deposition scheme** In previous experiments the main aim was to retrieve the optimal value for $\Delta\rho_{m/n}(t) = \Delta\rho_{m/n}$ for the stabilization of NTMs ([12] and references therein). Typically, a slow sweep has been performed to get an estimate for $\Delta\rho_{m/n}$ for subsequent discharges. This approach suffers from the need of a very accurate and reliable determination of $\rho_{m/n}(t)$. The new approach originating at TCV is the use of a triangular sweeping of $\Delta\rho_{m/n}(t)$ centred around the resonant surface estimated from the rt-equilibrium, $\rho_{m/n-eq}(t)$, during the stabilization process. In this case there is always an instant during the sweep when the gyrotron is located optimally for stabilization. This can be identified by the largest island size reduction rate, $\tau_{decay} \sim dW/dt(t) < 0$, during the sweep for small $P_{ECCD}$. For $P_{ECCD} \geq P_{ECCD}^{stab}$ the mode disappears during the sweep. For an efficient stabilization the sweeprate $\tau_{sweep}$ has to be sufficiently large compared to the typical decay time $\tau_{decay}$ for optimal radial localization.

For preemptive avoidance of NTMs, sweeping the deposition might be the only possibility. Without the determination of the radial localization, $\rho_{m/n-NTM}(t)$, of an existing (m/n)-mode, only the equilibrium prediction, $\rho_{m/n-eq}(t)$, is available. However, $\rho_{m/n-eq}(t)$ suffers at present from the absence of internal constraints providing information on the $q$-profile.

Possible caveats of the sweeping technique are a longer required duration of the stabilization compared to an immediately optimal deposition. Also a larger amount of ECCD power, $P_{ECCD}^{stab}$, might be required, as the time integrated driven current within the island is lower.

**Stabilization of a (3/2)-NTM with sweeping of $\rho_{m/n}^{target}(t)$** In discharge #30594 (figure 1) three gyrotrons are subsequently activated with a delay of 1.3s in the intermediate $P_{NBI} = 10$MW phase. A (3/2)-NTM is triggered in the high $P_{NBI}^{onset,int} = 17.5$MW phase at $\beta_N^{onset,int} = 2.8$. During the 1.3s period with constant $P_{ECCD}$ the sweeping of $\rho_{ECCD}$ is applied to all three gyrotrons and therefore a three step power scan has been achieved. Only with the maximal $P_{ECCD}^{stab} = 2.1$MW the (3/2)-NTM disappears during the first sweep of $\rho_{ECCD}$ over $\rho_{3/2-NTM}$. After the removal of the (3/2)-NTM a maximal $\beta_N = 1.8$ has been achieved with $P_{NBI} = 10$MW. A subsequent rise of $P_{NBI}$ up to 12.5MW and $\beta_N = 2.3$ did not retrigger an NTM, as the ECCD in its sweeping mode was preemptively avoiding the reexcitation.

**Preemption of (3/2)-NTM with sweeping** Compared to previous preemptive experiments [3], figure 2 shows discharge #30645, where with only one radially sweeping gyrotron, i.e. $P_{ECCD}^{pre} = 0.65$MW the excitation of an NTM with $P_{NBI} = 12.5$MW could be avoided. Only during a $P_{NBI}$-ramp at $P_{NBI}^{onset,pre} = 15$MW at $\beta_N^{onset,pre} = 2.45$ a (3/2)-NTM gets excited. The otherwise identical comparison discharge without ECCD showing an earlier NTM excitation at lower $P_{NBI}$ and $\beta_N$ is still missing. However, typically at these $\beta_N$ values NTMs are excited.

**Summary and outlook** Several updates, which have been included in the setup for MHD control at ASDEX Upgrade have been reported. A major new development has been a continuous sweep of the ECCD deposition around the resonant surface, $\rho_{m/m-eq}$, both for NTM stabilization and preemptive avoidance. This technique relaxes the requirements on the accuracy of the localization of the ECCD. It has enabled us to perform NTM avoidance in new scenarios, without predetermination of the exact $q = m/n$ position based on preparation discharges. In the future, a definition of a generic figure of merit is required, to make the efficiency of stabilization and avoidance and also modes with different mode numbers directly comparable.

Stabilization of multiple modes, amplitude based stabilization, sawtooth tailoring and disruption avoidance are further examples of the generic capabilities of specialized controllers of the ASDEX Upgrade DCS, which will be further optimized for its joint application in ITER.

**References**

[1] C. C. Hegna and J. D. Callen, Phys. Plasmas **4**, 2940 (1997).

[2] D. Kim et al., in *55th An. APS Meeting of the Div. of Plasma Physics*, page PP8.00073, 2013.

[3] L. Giannone et al., Fusion Engineering and Design **88**, 3299 (2013).

[4] A. Mlynek et al., Nuclear Fusion **51**, 043002 (2011).

[5] M. Reich et al., Fusion Eng. Design **61**, 309 (2012).

[6] A. Gude et al., in *Proc. of the 42nd EPS Conf. on Plasma Physics*, page P1.122, 2015.

[7] M. Reich et al., in *Proc. of the 25th IAEA Conference Fusion Energy, 2014*, pages IAEA–CN–221/PPC/P1–267, 2014.

[8] M. Maraschek et al., in *Proc. of 40th EPS Conf. on Plasma Phys.*, page P4.127, 2013.

[9] C. J. Rapson et al., Fusion Eng. Design , online: http://dx.doi.org/10.1016/j.fusengdes.2015.01.007 (2015).

[10] M. Reich et al., Fusion Eng. Design , online: http://dx.doi.org/10.1016/j.fusengdes.2015.04.024 (2015).

[11] J. P. Meskat et al., Plasma Physics and Controlled Fusion **43**, 1325 (2001).

[12] M. Reich et al., in *Proc. of the 40th EPS Conf. on Plasma Physics*, page P2.151, 2013.

**Acknowledgement** *This work has been carried out within the framework of the EUROfusion Consortium and has received funding from the Euratom research and training program 2014-2018 under grant agreement No 633053. The views and opinions expressed herein do not necessarily reflect those of the European Commission.*

# On the role of the edge density profile for the L-H transition power threshold in ASDEX Upgrade

L.M. Shao[1,2], E. Wolfrum[1], F. Ryter[1], G. Birkenmeier [1,3], F.M. Laggner[4],
E. Viezzer[1], R. Fischer[1], M. Willensdorfer[1], B. Kurzan[1] and the ASDEX Upgrade Team
[1] *Max-Planck-Institut für Plasmaphysik, 85748 Garching, Germany*
[2] *Institute of Plasma Physics, Chinese Academy of Sciences, Hefei 230031, China*
[3] *Physik-Department E28, Technische Universität München, 85748 Garching, Germany*
[4] *Institute of Applied Physics, TU Wien, Fusion@ÖAW, 1040 Vienna, Austria*

On ASDEX Upgrade, the carbon tiles were replaced stepwise by tungsten (W) plasma facing components (PFCs) starting from 2003 and ending in 2007 [1]. Since 1999, the so-called 'H-mode standard shot' has been run as the first plasma discharge of each operation day at ASDEX Upgrade. The L-H transition is induced by a power ramp of neutral beam injection (NBI) achieved by beam modulation with increasing duty-cycle. A decrease of the power threshold ($P_{L-H}$) of about 25% for the H-mode standard shots was found when the full tungsten coverage of PFCs was reached [2, 3]. Neoclassical theory predicts that the edge radial electric field ($E_r$) is mainly driven by the main ion species [4, 5]. At ASDEX Upgrade it has been shown that, at fixed $B_T$ and magnetic configuration, the minimum in the neoclassical radial electric field $E_{r,neo}$ at the L-H transition is almost independent of density in the full tungsten wall [6].

The 'H-mode standard shot' was originally run at $|B_T| = 2.0$ T, but this had to be increased to $|B_T| = 2.5$ T when the tungsten coverage of the vacuum vessel increased [3]. The complete dataset at $|B_T| = 2.5$ T consists of 11 carbon wall (69% W coating PFCs) discharges with core line-averaged densities of about $4.5 \times 10^{19}$ m$^{-3}$ (2 shots) and $5 \times 10^{19}$ m$^{-3}$ (9 shots), and 14 discharges with full tungsten PFCs with a core line-averaged density around $5 \times 10^{19}$ m$^{-3}$. The dataset varies in plasma shape and location of gas fuelling (from the main chamber or lower divertor region). By comparing various shapes in both PFCs as well as dedicated experiments in the full tungsten wall discharges, in which we used both main chamber and divertor fuelling, we verified that these variations have no impact on $P_{L-H}$.

In figure 1(a)-(c) we compare the values of edge plasma quantities at $\rho_{pol} = 0.98$ in carbon and full tungsten walls just before the L-H transition by plotting these quantities versus $P_{L-H}$. These quantities are electron densities (taken in a time interval of 40 ms), temperatures (taken in a time interval of $\sim 60$ ms) and pressures. The power threshold values are obtained from

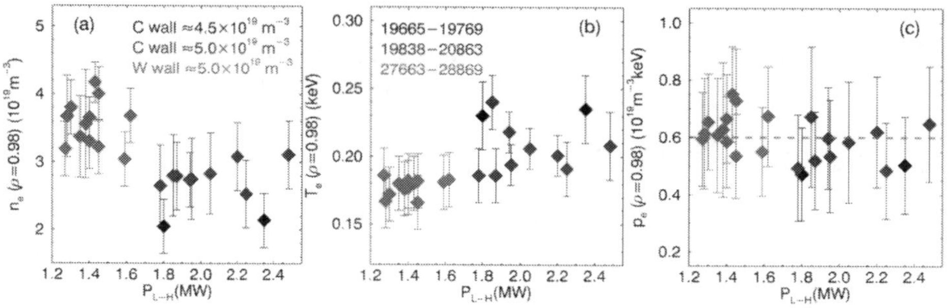


Figure 1: *Edge (a) electron densities, (b) electron temperatures and (c) electron pressures versus power thresholds at* $\rho_{pol} = 0.98$.

the loss power, $P_{loss} = P_{heat} - dW/dt$ at the L-H transition, where $P_{heat}$ is the sum of ohmic and auxiliary heating powers, and $W$ is the plasma stored energy. Figure 1 indicates that $P_{L-H}$ is clearly lower with the full tungsten wall than with the carbon wall. Higher electron densities and lower temperatures at $\rho_{pol} = 0.98$ are found in the discharges with full tungsten PFCs. The pedestal top electron pressures seem comparable in both PFCs.

Figure 2 shows a comparison of typical edge electron density profiles with the same core line-averaged density for two H-mode standard discharges within different plasma facing materials. The time slices for the edge density profile data are taken in a time interval of 20 ms just before the L-H transition. The discharge #28869, run with full metal wall, exhibits a higher $n_e$ at $\rho_{pol} = 0.98$ and a steeper edge density gradient $\nabla n_e$ compared to #20393 with carbon PFCs. The black curves are edge density profiles fitted with a modified hyperbolic tangent (MTanh) function.

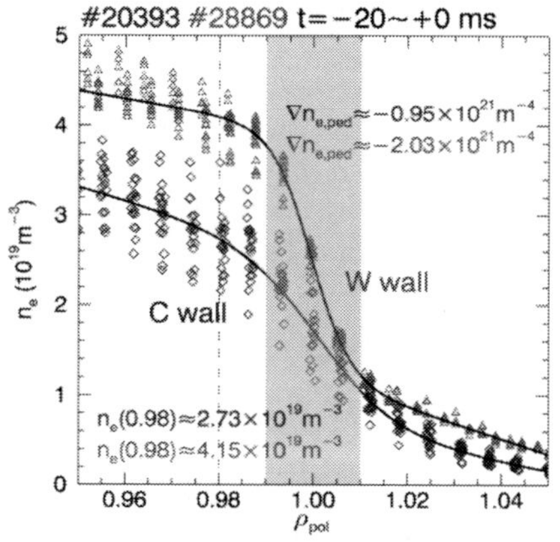


Figure 2: *Edge density profiles comparison of plasmas with the same core line-averaged density but different PFCs.*

Figure 3 shows the edge density gradients taken in a time interval of about 40 ms versus $P_{L-H}$ just before the L-H transition for our dataset. The edge density gradients are evaluated by fitting the edge density profiles applying the MTanh function and then averaging the gradients in $\rho_{pol} = 0.99 - 1.01$ (see the cyan shadowed region in figure 2) (filled diamonds) or applying the least absolute deviation (LAD) function (open circles). The values calculated by these two methods are consistent. With values of about $-1.5 \times 10^{21}$ m$^{-4}$ , steeper edge density gradients are found in the full tungsten wall discharges corresponding to the lower

$P_{L-H}$. In the carbon wall discharges, the edge density gradients are flatter, with values around $-0.95 \times 10^{21}$ m$^{-4}$.

Following the results from reference [6], we investigate whether the minimum in $E_{r,neo}$ just before the L-H transition is the same with the carbon and full tungsten walls in our dataset. The expression of $E_{r,neo}$ is

$$E_{r,neo} = E_{r,dia} - \alpha \frac{\nabla T_i}{e} + \frac{B}{B_t} B_\theta < V_{i\parallel} > . \quad (1)$$

$$E_{r,dia} = \frac{\nabla P_i}{n_i e} = \frac{\nabla n_e}{n_e}\frac{T_i}{e} + \frac{\nabla T_i}{e}. \quad (2)$$

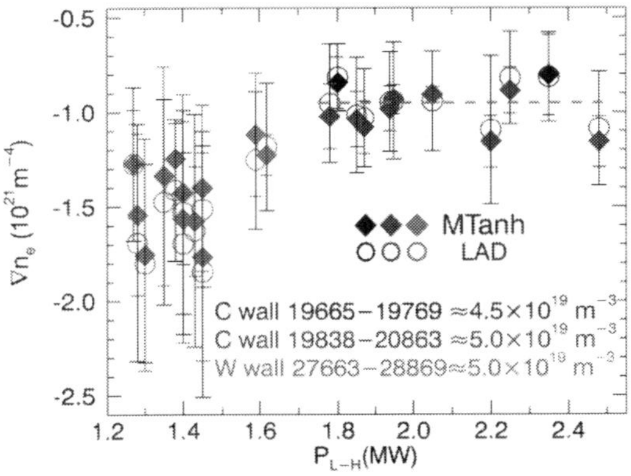


Figure 3: *Edge density gradients versus power thresholds. The edge density gradients are fitted by applying the MTanh function (filled diamonds) and the LAD function (open circles).*

Here, the toroidal field $B_t \approx 1.93$ T and poloidal magnetic field $B_\theta \approx 0.34$ T at the low-field side plasma edge of our dataset is determined by the equilibrium code CLISTE [7]. $\frac{B}{B_t} B_\theta < V_{i\parallel} > \approx 0.35 < V_{i\parallel} >$ and we assume $n_i = n_e$ at the plasma edge in ASDEX Upgrade. Combining equation (1) and equation (2), the neoclassical $E_r$ equation can be written as

$$E_{r,neo} = \frac{\nabla n_e}{n_e}\frac{T_i}{e} + (1-\alpha)\frac{\nabla T_i}{e} + 0.35 < V_{i\parallel} > . \quad (3)$$

In standard H-mode discharges, the mean impurities parallel velocity is typically $< 10^4$ m/s at the plasma edge before the L-H transition. Thus, the third term $0.35 < V_{i\parallel} >$ in equation 3 is less than $3.5kV/m$. This value is much less than the former two terms and neglected in the following analysis. As the edge CXRS diagnostic was installed in 2009, the direct measurement of the $E_r$ profiles is unavailable for our carbon PFCs dataset. However, $E_{r,neo}$ can be estimated from equation (3) using the routinely measured electron edge density and temperature profiles. According to [8, 9], $E_{r,min}$ is always located just inside the separatrix, therefore $\rho_{pol} = 0.995$ is artificially chosen as the radial position where the minimum of the $E_{r,neo}$ is expected.

The inverse density gradient length $\nabla n_e/n_e$ can simply be calculated by dividing the density gradient values (applying LAD method) in figure 3 by the corresponding edge density at $\rho_{pol} = 0.995$. $T_e$ and $T_i$ at the pedestal top are close to each other for $n_{e,ped} \geq 2 \times 10^{19}$ m$^{-3}$ and heating powers in the range of $P_{L-H}$ [6], but at the separatrix $T_i$ is somewhat larger than $T_e$. Therefore, we assume $T_i = T_e$ at $\rho_{pol} = 0.98$ and $T_i = 0.14$ keV (measured $T_e \sim 0.10$

keV) at the separatrix. The ion temperature profile is almost linear at the plasma edge, thus the edge ion temperature gradients can simply be estimated as $\nabla T_i = \frac{T_i(0.98)-T_i(1.00)}{r(0.98)-r(1.00)}$ and $T_i(0.995) = \frac{T_i(0.98)+3T_i(1.00)}{4}$. Since the collisionality $\nu_{*i}$ [10] at $\rho_{pol} = 0.995$ varies between 1.40 and 4.52 before the L-H transitions in the dataset, the factor $\alpha$ in equation (3) is in the plateau-collisional regime [4]. According to equations (2) and (3) and the terms $\frac{\nabla n_e}{n_e}(0.995)$, $T_i(0.995)$, $\nabla T_i$ and $\alpha$ described above, the minimum radial electric fields $E_{r,dia}(0.995)$ and $E_{r,neo} - \frac{B}{B_t} B_\theta < V_{i\|} > (0.995)$ are estimated.

The minimum of the $E_r$ well is plotted in figure 4 versus $P_{L-H}$. The values of the minimum of $E_{r,neo}$ (neglecting the third term $0.35 < V_{i\|} >$ in equation 3) are constant around -13 kV/m within the errorbars in a wide range of power thresholds and consistent with $E_{r,neo} \approx -15$ kV/m found for a wide range of pedestal top densities in reference [6]. We also compare the evaluated $E_{r,dia}(0.995)$ taking $T_i = 0.14$ keV at the separatrix ($T_i = T_e$ at $\rho_{pol} = 0.98$) and $T_i = T_e$ at the plasma edge. It shows that when choosing $T_i = 0.14$ keV at the separatrix, the calculated $E_{r,dia}(0.995)$ are much closer in the discharges with both PFCs.

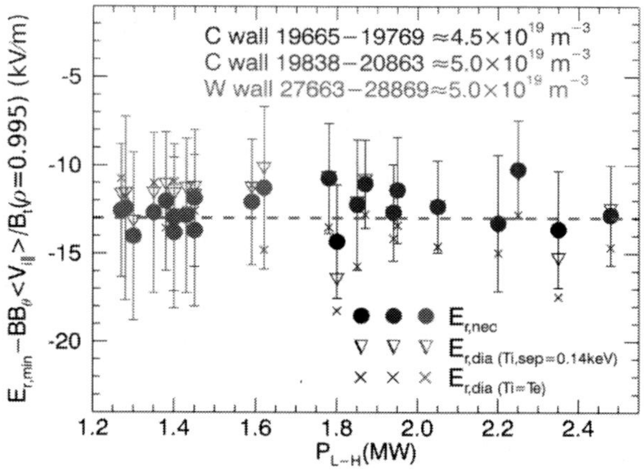

Figure 4: *Radial electric fields at $\rho_{pol} = 0.995$ versus power threshold evaluated according to equations (2) and (3).*

These results suggest that the minimum edge $E_{r,neo}$, just before the L-H transition, seems to be a constant value in different PFCs despite different threshold power. The lower $P_{L-H}$ in full tungsten PFCs can be attributed to steeper edge density profiles.

## References


[1] Neu R, *et al.* 2007 *J. Nucl. Mat.* **367** 1497-502.

[2] Neu R, *et al.* 2013 *J. Nucl. Mat.* **438** S34-S41.

[3] Ryter F, *et al.* 2013 *Nucl. Fusion* **53** 113003.

[4] Hinton F and Hazeltine R 1976 *Rev. Mod. Phys.* **48** 239-308.

[5] Stroth U, Manz P and Ramisch M 2011 *Plasma Phys. Control. Fusion* **53** 024006.

[6] Sauter P, *et al.* 2012 *Nucl. Fusion* **52** 012001.

[7] Schneider W, *et al.* 2000 *Fusion Eng. Des.* **48** 127-34.

[8] Conway G D, Angioni C, Ryter F, Sauter P and Vicente J 2011 *Phys. Rev. Lett.* **106** 065001.

[9] Viezzer E, *et al.* 2013 *Nucl. Fusion* **53** 053005.

[10] Sauter O, Angioni C and Liu-Liu Y R 1999 *Phys. Plasmas* **6** 2834.

# Beta limit and plasma response to n=1 perturbations in ASDEX Upgrade

V. Igochine[1], P. Lauber[1], P. Piovesan[2], M. Schneller[1], E. Strumberger[1], W. Suttrop[1], D.Yadykin[3], P.Bettini[2], A. Bogomolov[4], T. Bolzonella[2], I.G.J. Classen[4], M. Dunne[1], A. Gude[1], M. Maraschek[1], L. Marrelli[2], R. McDermott[1], M. Reich[1], G. Tardini[1], the EUROfusion MST1 Team* and the ASDEX Upgrade Team

[1] *Max Planck Institute for Plasma Physics, Boltzmannstr. 2, 85748 Garching, Germany*
[2] *Consorzio RFX, Padua, Italy*
[3] *Chalmers Univ., Gothenburg, Sweden*
[4] *DIFFER, Eindhoven, The Netherlands*

Tokamak plasmas with high normalized pressure are subject to various resistive and ideal MHD instabilities. The actual limit depends on several factors, including the stabilizing influence of the conducting components facing the plasma surface, kinetic interaction between the plasma and the marginally stable/unstable MHD modes, existence of low order rational surfaces and external actuators (heating, external perturbations, current drive). It was shown in previous work that the plasma response to n=1 perturbations from B-coils [1] in ASDEX Upgrade tokamak increases with increase of normalized beta $\beta_N$ ($\beta_N = \beta a B_t / I_p, \beta = 2\mu_0\langle p\rangle/\langle B\rangle$). This is a typical indication of the proximity to the so-called "no wall" beta limit [2]. In the present work we continue to study different effects influencing plasma stability to global n=1 modes which allows us to extend the achievable $\beta_N$.

**Kinetic interaction between n=1 mode and NBI particles**

High $\beta_N$ discharges were performed with dominant NBI heating in ASDEX Upgrade. The resulting plasma has high rotation and resonant interaction between the Doppler shifted mode frequency ($\omega_{E\times B} - \omega_{n=1}$) and plasma particles becomes possible. The analytical expression for changes of the mode energy $\delta W$ gives clear ideas about possible resonant frequencies [3]:

$$\delta W \sim \sum_{l=-\infty}^{\infty} \frac{(\omega_{n=1} + i\gamma_{n=1} - n\omega_{E\times B})\frac{\partial f_j}{\partial \varepsilon} - \frac{1}{eZ_j}\frac{\partial f_j}{\partial \Psi}}{\langle \omega_d^j \rangle + l\omega_b^j - i\nu_{eff}^j + n\omega_{E\times B} - \omega_{n=1} - i\gamma_{n=1}}$$

where $f_j$ is the distribution function of the particles $j$, ε is the particle energy, $\omega_{n=1}$ is the mode frequency in the plasma frame, $\gamma_{n=1}$ is the mode growth rate, Ψ is the magnetic flux,

* See http://www.euro-fusionscipub.org/mst1

$\nu_{eff}^{j}$ is the collision frequency, $Z_j$ is the effective charge. The first four frequencies in the denominator are: the precession drift frequency: $\omega_d = \frac{\rho_L}{r}\frac{v_{th}}{2qR_0}$ (for pitch angle $\Lambda = 1$); the bounce frequency: $\omega_b = \left(\frac{r}{2R_0}\right)^{1/2}\frac{v_\perp}{qR_0}$ (for $\Lambda = 1$); the collision frequency $\nu_{eff}$, and the $E\times B$ frequency, where $v_{th}$ is the thermal velocity, $\rho_L$ is the Larmor radius, $q$ is the safety factor value. The $E\times B$ frequency is $\omega_{E\times B} = \omega_\phi - \omega_{*i}$, where $\omega_\phi$ is the toroidal plasma rotation frequency and $\omega_{*i}$ is the ion diamagnetic frequency. All these frequencies as a function of $\rho$ are shown in figure 1a assuming experimental profiles. The presented frequencies show the upper estimation for possible resonances with passing particles for discharge #29100 with an unstable n=1 kink mode (pitch angle $\Lambda = 1$, no geometrical effects, etc.). Figure 1b shows results of HAGIS code [4] simulations for energy exchange between an $n = 1$ structure with multiple poloidal mode numbers (linear MHD code MARS) and a realistic distribution function (transport code TRANSP). HAGIS calculations of $\delta W$ include resonant and non-resonant interactions in realistic plasma geometry for the same discharge as in figure 1a. The results are shown in figure 1b. The real resonance frequencies are downshifted with respect to our simplified estimations, which is an expectable result for a realistic situation.

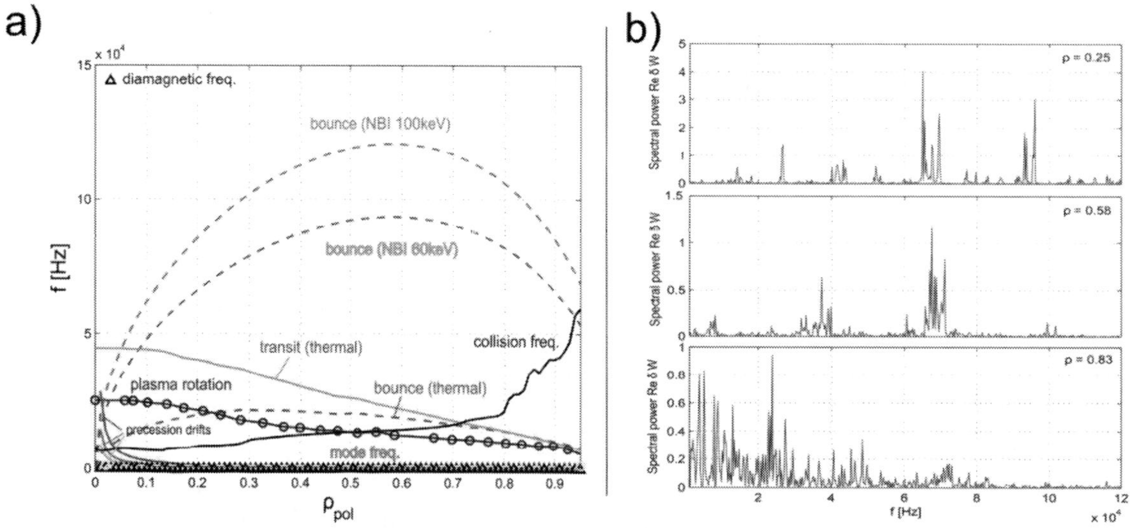


**Figure 1. Analysis of discharge #29100 a) Main resonance frequencies and plasma rotation frequency are shown. b) Spectral analysis of $Re(\delta W)$ for different radial positions in the plasma. $Re(\delta W)$ was calculated with HAGIS code taking into account realistic particle distribution and mode structure.**

The main interactions in the plasma core ($\rho = 0.25, \rho = 0.58$) are at high frequencies where the bounce resonances with NBI particles are important. The low frequency resonances become important close to the plasma boundary ($\rho = 0.83$). Spectral analysis of different particle species demonstrates the main players in the interaction of NBI particles with an $n = 1$ mode (figure 2). The spectral power density $Re(\delta W)$ at different radii for the full distribution function from TRANSP code is shown in figure 2a. (This is the same calculations as in figure 1b.) The other figures consider only part of the full distribution function from TRANSP for $\delta W$: only co-passing particles are considered in figure 2b; only counter-passing particles are taken in figure2c; only trapped particles are considered in figure 2d.

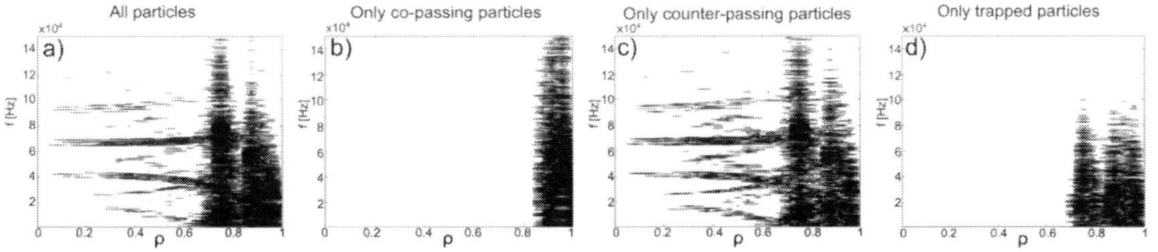


**Figure 2. Spectral analysis of $Re(\delta W)$ from HAGIS code for #29100. a) All particles are considered (the same case as in figure 1b); b) Only co-passing particles are taken into account; c) Only counter-passing particles are taken into account; d) Only trapped particles are taken into account.**

These figures show that co-passing particles are not important. Counter-passing particles are important at all radii and at different resonant frequencies. (Thus, radial NBI beams 1, 2 and 5, which produce the largest amount of counter-passing particles, play the dominant role.) Trapped particles are important for $\rho \geq 0.7$ (NBI 6 and 7). The next step will be an optimization of NBI start-up sequence to ensure maximal stabilization influence from NBI particles on the n=1 mode which should extend the achievable beta limit.

**Influence of external n=1 rotated perturbations from B-coils**

B-coils installed in ASDEX Upgrade were used for application of oppositely rotated perturbations in upper and lower row of coils. This produces constant changes of the pitch angle keeping the n=1 helicity of the perturbations unchanged. Reaction of the plasma to these perturbations becomes visible above the critical value $\beta_{N,crit} \approx 2.3$ (dashed line in figure 3a). Plasma displacement, estimated from temperature profile measurements ($\xi = -\widetilde{T_e}/\nabla\langle T_e\rangle$), shows visible changes above this $\beta_{N,crit}$ for differentially rotated n=1 perturbations (figure 3d). The same threshold is observed in experiments with rigid rotation of n=1 perturbations in similar plasma discharges. Independent measurements of the displacement close to the plasma boundary with 2D ECE Imaging system show amplitudes of the perturbations around 1 cm, which is in good agreement with our results [5]. The plasma rotation changes strongly through the radius as shown in figure 3c. Corresponding changes of the phase of the perturbation is shown in figure 3b. The dashed lines indicate maximal values of the rotation which corresponds to the same phase of n=1 perturbation. Explanation of such changes in the plasma rotation is not straightforward and probably related to neoclassical toroidal viscosity (NTV) torque difference for different pitch angles on n=1 perturbations. Detailed modelling with realistic profiles is foreseen for quantitative understanding of the effect. At the same time, the global nature of this effect and it dependence on normalized plasma pressure is clearly seen experimentaly.

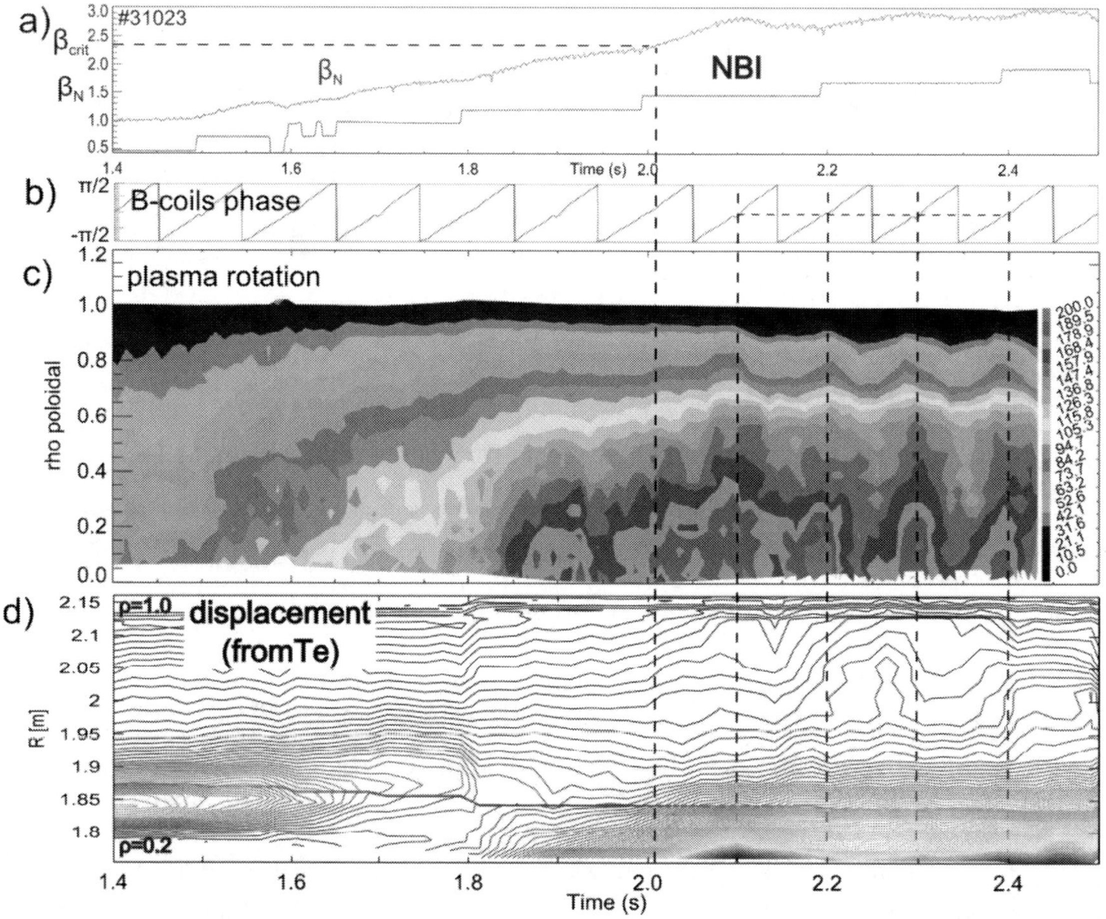


**Figure 3. #31023 a) Normalized beta and NBI power; b) phase of n=1 perturbation from B-coils shows changes in pitch angle; c) plasma rotation measurements; d) displacement from ECE.**

**Acknowlegments**

*This work has been carried out within the framework of the EUROfusion Consortium and has received funding from the Euratom research and training programme 2014-2018 under grant agreement No 633053. The views and opinions expressed herein do not necessarily reflect those of the European Commission.*

**References**

[1]W. Suttrop et. al., 2009 Fusion Eng. Des. 84, 290
[2] V. Igochine et. al., 40th EPS Conference, 2013, P5.147 (http://eps2013.aalto.fi/?page=proceedings)
[3] B. Hu, R. Betti, J. Manickam, Physics of Plasmas 12, 057301 (2005)
[4] S. Pinches et.al. Computer Physics Communications, 111, (1998), p.133
[5] P. Piovesan et. al., this conference, P1.144

# Effect of magnetic perturbations on pedestal, SOL and divertor in high collisionality plasmas at ASDEX Upgrade

E. Wolfrum[1], M. Faitsch[1], G. Fuchert[1], L. Barrera-Orte[2], M. Bernert[1], G. Birkenmeier[1,3], D. Carralero[1], T. Eich[1], Y.Gao[4], B. Kurzan[1], F.M. Laggner[5], H.W. Mueller[1,6], S. Potzel[1], M. Rack[4], B. Sieglin[1], H.J. Sun[1], W. Suttrop[1], M. Willensdorfer[1], the EUROfusion MST1 Team[7], and the ASDEX Upgrade Team

*1 Max-Planck-Institute for Plasma Physics, 85748 Garching, Germany*
*2 Eurofusion PMU, 85748 Garching, Germany*
*3 Physik-Department E28, Technische Universität München, 85748 Garching, Germany*
*4 Institut für Energie und Klimaforschung – Plasmaphysik, Forschungszentrum Jülich, 52425 Jülich, Germany*
*5 Institute for Applied Physics, TU Wien, Fusion@ÖAW, 1040 Vienna, Austria*
*6 Institute of Materials Chemistry and Research, University of Vienna, 1090 Vienna, Austria*
*7 See http://www.euro-fusionscipub.org/mst1*

Edge localized modes (ELMs) appear in plasmas with steep pressure gradients in the edge transport barrier. These ELMs lead to high heat loads onto the divertor. Magnetic perturbations (MPs) are applied in order to achieve a reduction in the ELM energy fluence to the target. At the same time MPs affect radial transport in the scrape-off layer due to the additional magnetic field component in radial direction. In high collisionality discharges at ASDEX Upgrade application of MPs lead to the disappearance of large type-I ELMs as soon as a critical density is achieved [1]. The critical density is higher in plasmas with high heating power, indicating a dependence on collisionality [2]. New experiments were carried out with n=2 MPs in discharges with three different plasma currents, 600 kA, 800 kA and 1 MA, in which the density was raised continuously to span Greenwald fractions (f_GW) from 0.5 to 0.75. While at the lower densities both ELM types, large type-I and small ones, are found simultaneously, the large ELMs disappear at high f_GW. The influence of the density ramp on ELM behaviour, the structure of the midplane kinetic profiles from the pedestal top into the scrape-off layer (SOL), the thermographic footprint in the divertor as well as on the fluctuations in the SOL are presented exemplarily for the 600 kA case.

In figure 1 time traces are presented of ASDEX Upgrade (AUG) discharge #30590, which was run at -2.5 T and 600 kA. The top panel shows the total heating power, the stored energy (W_mhd), the radiated power (P_rad) and the current through the MP coils (I_coil). W_mhd drops from 0.38 to 0.32 MJ during the first MP interval and from 0.41 to 0.40 MJ during the second one. In the second panel the gas fuelling ramps (green) as well as the reaction of the core and edge interferometry channels are plotted. The bottom panel shows the shunt current measurements at the divertor plates which is used as ELM indicator.

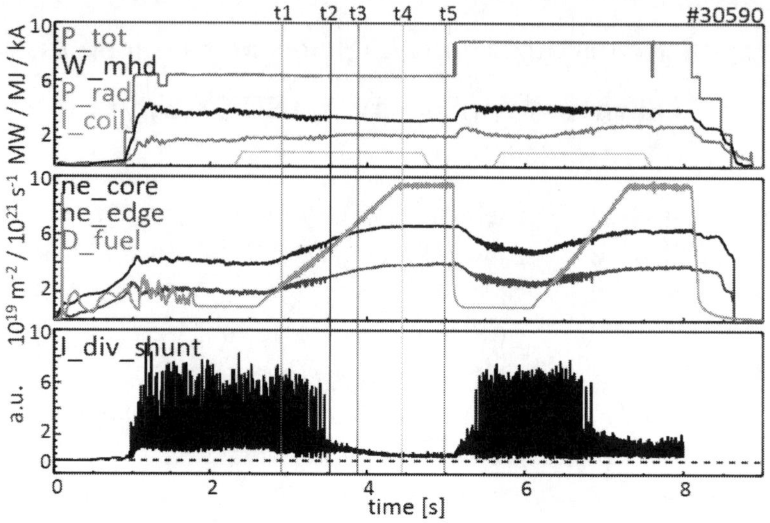


Figure 1: Time traces of #30590. Gas fuelling ramps are applied at two different heating powers. At the time points t1-t5 kinetic profiles are shown in figure 2 and outer divertor heat flux profiles in figure 3.

Several typical features for this kind of discharges can be observed from the time traces: (i) In the two phases with differing heating power the large type-I ELMs disappear during the density rise. A higher transition density can be observed with the higher heating power, indicating that collisionality might play a role. (ii) The small ELMs persist even after the MP coils are switched off (compare t4 and t5). No significant difference in the ELM appearance in the divertor current measurements can be seen, when MP coils are switched off. This observation is not true in the discharges with higher current: both in the 800 kA as well as in the 1 MA discharge the density drops slightly when the MP coils are switched off and a few type-I ELMs appear amongst the small ELMs. As has been shown in a previous study of high density discharges with and without MP [3] this might be due to the additional fuelling effect connected to the high density front at the HFS [4]. Infrared data of this 600 kA discharge is used to compare the footprint of small ELMs with and without MPs as well as the change of the footprint with rising density. Figure 2 shows the lobe structures of the heat flux on the outer target plate, where the measurements are time averaged during phases with small ELMs only or inter-ELM (t1). With increasing density the outer lobes are more and more pronounced, while the lobe closest to the separatrix shows a lower heat flux. The outer lobes also become wider. The change of the measured lobe pattern might be attributed to rotation of the lobe pattern caused by an altered plasma current near the separatrix. Another possible explanation for the widening of the lobes could be increased perpendicular transport. Also shown is a comparison of the heat flux structure of the small ELMs without the application of MPs for the highest density value (t5). Without MPs the lobe structure disappears and is replaced by a single wide feature. While single small ELMs also sometimes show a lobe-like structure, the position of these lobes is not locked, so that integration leads to the broad heat flux deposition profile.

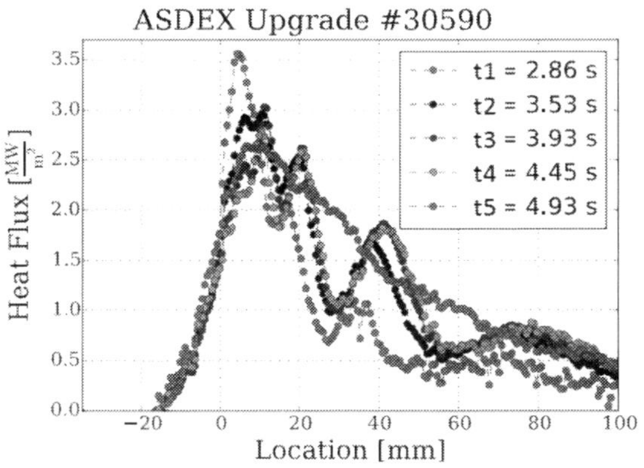


Figure 2: Heat flux profiles at time points indicated in figure 1. t1-t4 with MP, t5 without MP.

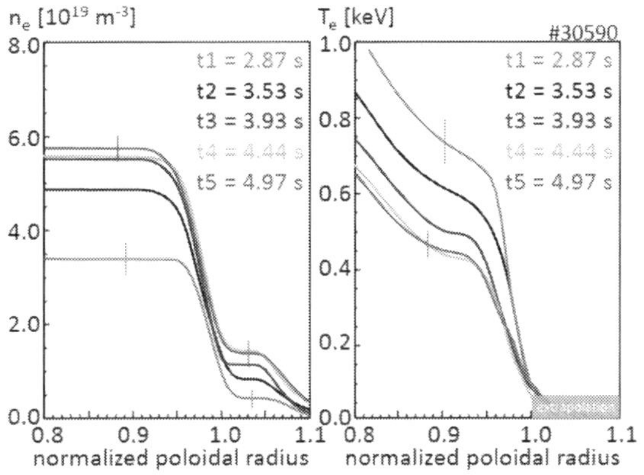


Figure 3: Edge electron density and temperature profiles of time points t1-t5. SOL $T_e$ values are only extrapolated (grey shaded). The $n_e$ SOL 'shoulder' increases with gas fuelling rate.

In figure 3 the corresponding edge electron density ($n_e$) and temperature ($T_e$) profiles are shown for the selected time points with increasing density. The data are derived from Lithium beam and DCN diagnostics for $n_e$ and from ECE for $T_e$ via integrated data analysis [5]. The $n_e$ pedestal gradient does not change within the error bars. With the rising density the pedestal top temperature is reduced from 700 eV to 400 eV. From time point t2 on ($T_{e,ped}$ = 600 eV) only mitigated ELMs appear, which is at remarkably high pedestal temperatures. If collisionality is the determining factor, then this could be explained by the high $q_{95}$ value for this 600 kA discharge. In the SOL $n_e$ increases with rising fuelling levels. The typical long $n_e$ fall-off lengths associated with increased filamentary transport, connected to high collisionalities in the divertor [6], are observed. When the MP coils are switched off in the high density period, no significant changes in the pedestal profiles can be observed (compare t4 and t5). The $T_e$ values in the SOL are greyed out, as ECE data carry no information about the local $T_e$ in this area. An analysis of the near SOL decay lengths from Thomson scattering data reveals a widening of both, $T_e$ and $n_e$, with increasing collisionality (not shown).

The filamentary transport was measured with Langmuir probes on the mid-plane manipulator. Figure 4 shows filaments measured at the position R = 2.167 m, z = 0.313 m, corresponding to $\rho_{pol}$ = 1.09, i.e. well in the limiter shadow of the inner heat shield. The bottom trace shows the ion saturation current, while the top two panels show $T_e$ and $n_e$. All values are normalized to their mean. At time point t1 a period without type-I ELMs is chosen, so that the small filaments occurring between type-I ELMs can be compared with the small ELMs in the mitigated phase, which appear as large filaments at time t4. While the average $T_e$ of both phases is very similar with 14 and 16 eV, the average $n_e$ increases by a factor of 2. The values are consistent with the densities of the SOL shoulder shown in figure 3 and reflect the fact that the formation of the SOL $n_e$ shoulder coincides with increased filament activity across the whole SOL.

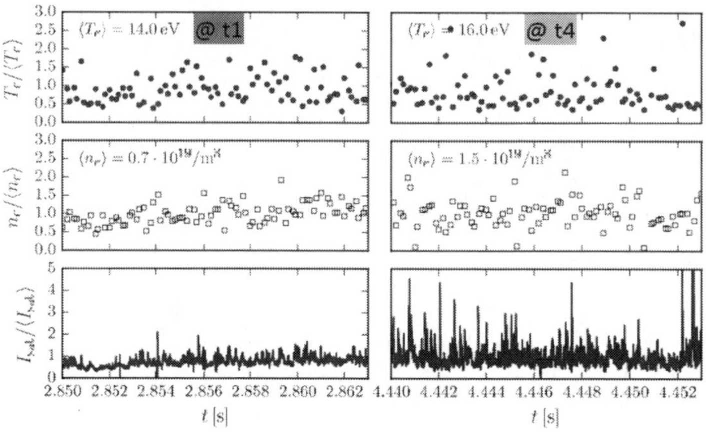

Figure 4: Langmuir probe data, $T_e$ (top), $n_e$ (middle) and $I_{sat}$ (bottom), from two different time periods: left, time point t1 but without type-I ELMs, right: t4 is a mitigated phase at high fuelling level. All values are normalized to their mean.

Taken all observations together the following picture of the effect of increasing density on SOL and divertor can be drawn: Filaments are born near the separatrix and expelled into the SOL. With increasing collisionality these filaments become larger and contribute to the $n_e$ shoulder in the SOL [6]. From a certain density or collisionality threshold on no type-I ELMs are triggered anymore. Magnetic perturbation coils create a pattern in the SOL magnetic field, giving rise to a lobe structure on the divertor surface. The larger filaments occur together with a wider near SOL $T_e$ decay length at the midplane. This cross field transport could populate the lobes which are positioned further away from the nominal undisturbed strike line, so that these lobes carry more heat flux and become wider. At the same time the lobe closest to the nominal strike line carries less heat flux. This reduction of the temperature causes an increase in collisionality in front of the LFS divertor, which has been linked to the onset of the SOL $n_e$ shoulder.

References:

[1] W. Suttrop et al, Phys.Rev.Lett. 106 (2011) 225004
[2] E. Wolfrum et al, EPS Conf. on Plasma Physics (2014), Berlin, Germany, P2.006
[3] E. Wolfrum et al. EPS Conf. on Plasma Physics (2013), Espoo, Finland, P4.105
[4] S. Potzel et al, J. Nucl. Mat. (2015) in press
[5] R. Fischer et al, Fus. Sci. Techn. 58 (2010) 675
[6] D. Carralero et al, Nucl. Fusion 54, 12 (2014) 123005

This work has been carried out within the framework of the EUROfusion Consortium and has received funding from the Euratom research and training programme 2014-2018 under grant agreement No 633053. The views and opinions expressed herein do not necessarily reflect those of the European Commission.

# Measurement and interpretation of active Balmer alpha spectra at ASDEX Upgrade

B. Geiger[1], R. Dux[1], M. Dunne [1], A. Lebschy[1], R. McDermott[1], R. Reimer[2], M. Salewski[3], G. Tardini[1], M. Weiland[1] and the ASDEX Upgrade team[1]

[1]*Max-Planck-Institute for Plasma Physics, Garching, Germany,* [2]*Max-Planck-Institute for Plasma Physics, Greifswald, Germany,* [3]*Technical University of Denmark, Lyngby, Denmark*

## Introduction

Active Balmer alpha spectroscopy has become an important diagnostic technique in fusion devices because it allows investigation of the main-ion temperature and rotation [1], the fast-ion density [2] and the magnetic field structure [3]. The technique is based on the observation of Balmer alpha radiation (656.1 nm, n=3->2), emitted by deuterium neutrals that are present along neutral beam injection (NBI) lines. Besides the injected neutrals, also main ions and fast ions can be analyzed because these get neutralized by charge exchange reactions and subsequently emit light. An example spectrum is plotted in figure 1, which shows the active components of Balmer alpha spectra (passive radiation, measured when neutral beam injection was turned off, has been subtracted). In addition to the measurement, simulated contributions are illustrated (from FIDASIM ).

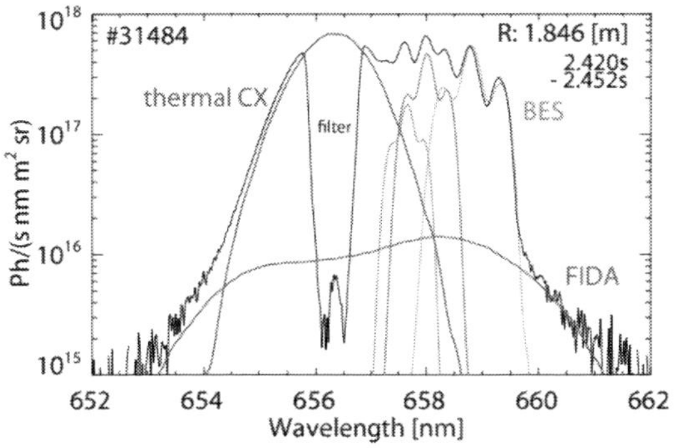


Figure 1: *Measured and simulated active Balmer alpha spectrum.*

The predicted beam emission (BES) is plotted in shades of orange. It is emitted by the injected neutrals after excitation through collisions with electrons and ions and can be subdivided into three energy components with different Doppler shifts: positive neutral beams, as applied here, inject full, half and one third energy neutrals. The spectral shape of the individual energy components is dominated by the Stark effect since the injected neutrals experience a strong $\mathbf{v}_{NBI} \times \mathbf{B}$ Lorentz electric field. This splits the radiation into three major peaks. The distance between these peaks depends on the orientation and strength of the magnetic field and, hence, allows to obtain information on the magnetic equilibrium. In blue is illustrated the simulated main ion charge exchange radiation, which can be described very well by a Gaussian curve. By analyzing the spectral width and Doppler shift of this component, main ion temperatures and rotations can be determined. The predicted charge exchange radiation emitted by fast beam ions (also called fast-ion D-alpha (FIDA) radiation) is illustrated in red in figure 1. A TRANSP [5] predicted fast-ion distribution function has been input to FIDASIM to consistently model the radiation. As can be seen, the FIDA component has broad wings and an asymmetric shape which contains information on fast-ion densities and the fast-ion velocity space distribution.

## Diagnostic setup

In the ASDEX Upgrade tokamak, a new Balmer alpha diagnostic has been installed that can address all topics mentioned above. The diagnostic has been built in the framework of a major charge exchange recombination spectroscopy (CXRS) diagnostic installation that makes use of an array of 70 lines of sight (LOS). Figure 2 shows a sketch of these LOS, which intersect the 93 keV heating beam, labeled NBI Q8, at various positions from the low-field-side to the high-field-side.

21 of the LOS are foreseen for beam emission spectroscopy (BES) measurements while the others are used for the CXRS analysis of impurity ions. The BES spectrometer setup is shown in the lower right of figure 2. It is based on the Czerny turner approach with a 2000 lines/mm reflection grating and uses two 200 mm, F/2.0 lenses instead of mirrors. 22 fibers with 400$\mu m$ diameter can be connected at its entrance and the application of two more lenses (56 mm, F/1.4 and 42 mm, F/0.95) allows us to block passive radiation at 656.1 nm, which would lead to saturation of the attached CCD camera. The additional lenses focus an intermediate image, of which parts are blocked by a wire, slightly de-magnified on the CCD camera. The effect of the wire can be seen in figure 1 since the measured spectrum is close to zero between 655 nm and 657 nm. The spectral range covered by this spectrometer approach is 15 nm and the spectral resolution is 0.05 nm. The latter is determined mainly by the entrance slit, which is set to a width of 50 $\mu m$.

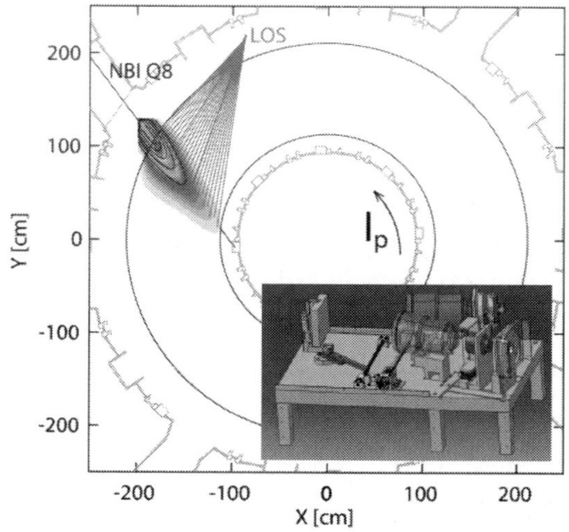


Figure 2: *Top-down view showing the LOS setup in red and a contour plot of the neutral density during operation of NBI Q8.*

## Study of the magnetic field structure

The good spectral resolution of the diagnostic allows us to investigate the beam emission very accurately. Figure 3 shows a contour plot of the beam emission measured in discharge #31484 ($B_t$=-2.5 T, $I_p$=800 kA). Clearly, the beam emission is attenuated towards smaller radii R where it exhibits larger Doppler shifts and a stronger Stark splitting. The larger Doppler shifts are explained by the viewing geometry of the diagnostic. The angles between LOS and NBI Q8 become less perpendicular at lower radii, i.e. change from 83 degree to 77 degree. The strong splitting at low radii is well explained by the 1/R dependence of the toroidal field and allows us to clearly identify the independent Stark lines. In contrast

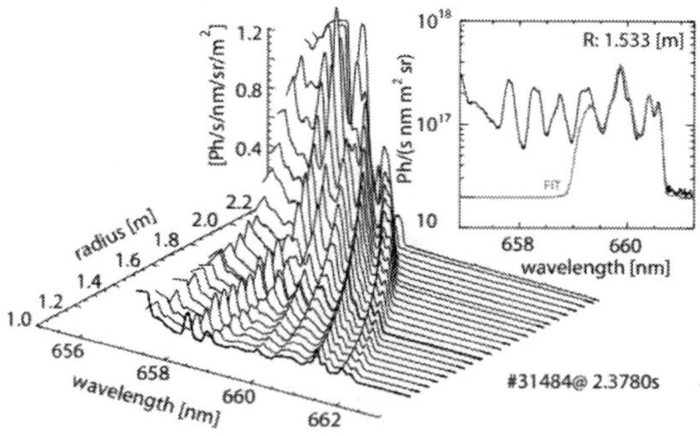


Figure 3: *Measured beam emission as a function of radius and wavelength.*

to the observed 3 major peaks per energy-component on the low-field side (see figure 1), the beam emission observed at the high-field-side allows us to study the Stark spectrum with its 9 transitions [6] (see upper right plot of figure 3). By fitting the spectra with an accurate model including the detailed shape of each Stark line [6], the distance of the individual lines and, therefore, the strength of the $\mathbf{v}_{NBI} \times \mathbf{B}$ electric field can be determined.

Figure 4 compares a measured radial $\mathbf{v}_{NBI} \times \mathbf{B}$ profile with profiles resulting from different equilibrium reconstructions (the corresponding q-profiles are displayed in the sub-plot). The error-bars of the measurement represent the statistical noise withing 5 subsequent time points, measured with 5 ms exposure time. Clearly, the equilibrium with the very low q-profile plotted in blue does not represent the measurement. This already shows that the BES measurement may allow us to improve future equilibrium reconstructions within a Bayesian framework.

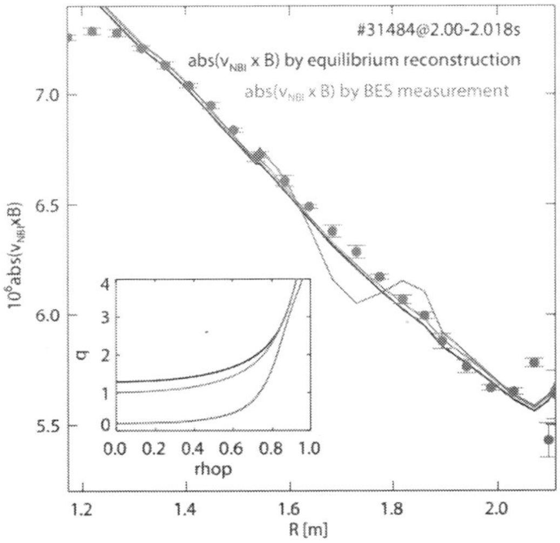


Figure 4: *Measured $\mathbf{v}_{NBI} \times \mathbf{B}$ profile compared with equilibrium reconstructions.*

## Main ion $T_i$ and $v_{tor}$

The main ion temperature, $T_i$, and rotation, $v_{tor}$, can be determined by fitting the thermal charge exchange component with a Gaussian curve. This is well within the abilities of our diagnostic setup because the beam emission is, in particular on the high-field-side, shifted far away.

In figure 5, we show radial profiles of the measured main-ion temperature and main-ion rotation with error-bars that represent the statistical noise. No data is shown for R<1.35 m and R>1.95 m because there, too low active signals and too strong overlap with the beam-emission prevent the analysis, respectively. The measured profiles have been corrected by cross-section effects, velocity space effects and a reduced radial resolution using the FIDASIM code: Based on the measured apparent [1] temperature and rotation, we calculated synthetic main-ion spectra and fitted them with a Gaussian curve. The ratios between the profiles input to FIDASIM and the fitted apparent temperature and rotation profiles have been used to correct the measurement. In addition to the direct measurement, figure 5 illustrates in black the temperature and rotation of nitrogen from which main-ion profiles have been determined neoclassically using TRANSP (blue). In particular in the plasma center, the main-ion measurement exceeds the temperatures from the nitrogen measurement and better agreement is found with the measured electron temperature profile, plotted

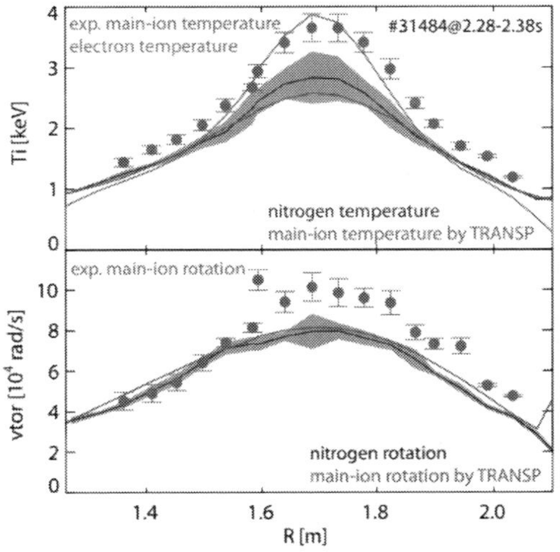


Figure 5: *Temperature and rotation profiles.*

in green. The directly measured main-ion rotations roughly agree within the error-bars with the ones determined using nitrogen but are systematically larger. The origin of this discrepancy is not yet clear and will be the subject to more detailed investigations.

**Fast-ion D-alpha**

The presence of the beam emission and the main-ion radiation in the spectra restricts the observable fast-ion radiation to large Doppler shifts. The velocity space observed at such wavelengths is displayed in color in figure 6 for a central LOS and for two wavelength ranges. In addition, the TRANSP predicted fast-ion velocity space distribution for discharge #30841 (similar to #31484) is displayed, which mainly exhibits positive pitch values ($v_{||}/v_{tot}$), well explained by the co-current NBI geometry. The overlap between the fast-ion distribution function and the weight function for wavelengths between 651 nm and 652 nm is very small. This explains the absence of FIDA radiation in figure 1 in this wavelength range. Better overlap with the weight function is observed for wavelengths between 661 nm and 662 nm, which is in agreement with the red-shifted FIDA contribution in figure 1. Radial profiles of the FIDA radiation, integrated between 661 nm and 662 nm and normalized by the measured level of the beam emission, are shown in figure 7. The comparison with the FIDASIM+TRANSP prediction shows good agreement, indicating that fast-ions are consistently modeled.

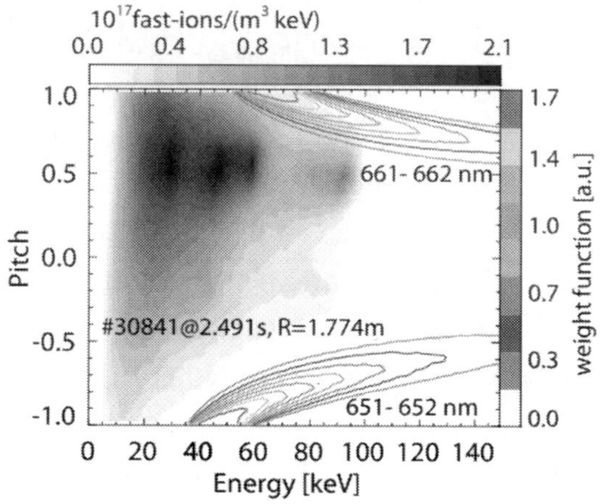


Figure 6: *Weight functions and fast-ion distribution function.*

**Acknowledgement**

This work has been carried out within the framework of the EUROfusion Consortium and has received funding from the Euratom research and training programme 2014-2018 under grant agreement No 633053. The views and opinions expressed herein do not necessarily reflect those of the European Commission.

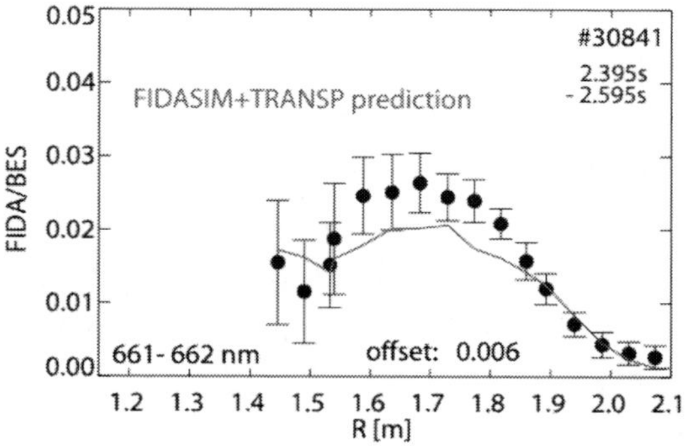


Figure 7: *Radial FIDA profile, normalized by the beam emission intensity and compared with the prediction from FIDASIM+TRANSP*

**References**

[1] GRIERSON, B. A. et al., RSI **81** (2010)

[2] HEIDBRINK, W. W. et al., PPCF **46** (2004) 1855

[3] WOLF, R. et al., Nuclear Fusion **33** (1993) 1835

[4] HEIDBRINK, W. et al., Commun. Comput. Phys. **10** (2011) 716

[5] PANKIN, A. et al., Computer Physics Communications **159** (2004) 157

[6] DUX, R. et al., this proceeding **P1.121** (2015)

# Effect of measured toroidal flows on tokamak equilibria

R. Fischer, L. Giannone, K. Lackner, R. M. McDermott, E. Viezzer, H.P. Zehrfeld,
and the ASDEX Upgrade Team
[1] *Max-Planck-Institut für Plasmaphysik, EURATOM Association, Boltzmannstr. 2, D-85748 Garching, Germany*

The magnetic confinement of a tokamak plasma is routinely calculated by solving the Grad-Shafranov equation describing the ideal magneto-hydrodynamic (MHD) equilibrium for the poloidal flux function for axisymmetric geometry. A basic assumption usually is that small effects of the centrifugal force due to toroidal and poloidal plasma rotation can be neglected. But in present-days unbalanced high-power beam-heated plasmas the toroidal plasma rotation is typically rather large. Toroidal plasma rotation velocities up to 300 km/s have been measured at the ASDEX Upgrade tokamak which corresponds to Mach numbers up to 0.5 in the plasma center. For other tokamaks (MAST, DIII-D, NSTX, JET) sonic Mach numbers approaching unity have also been reported [1] (and references therein). For such large Mach numbers the centrifugal force results in a redistribution of the particles on the flux surface. The surfaces of constant pressure do no longer coincide with the flux surfaces and significantly modified flux surface topology can arise. Reasonably accurate equilibrium reconstructions for moderate and high Mach numbers are necessary to study plasma stability since it is known that strong flows may considerably damp plasma instability [2]. Additionally, for an improvement of the flux surface geometry by multiple temperature measurements on the same flux surface (iso-flux constraint) [3] flows might have to be considered. If the underlying equilibrium equation is incomplete, a constraint from the measurement of the geometry of flux surfaces can lead to erroneous current distributions trying to compensate for the missing terms in the equilibrium equation.

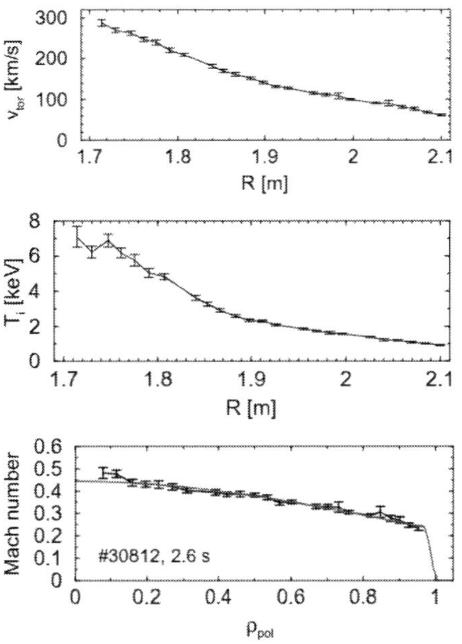


Figure 1: *Profiles of the toroidal rotation velocity and ion temperature as a function of the major radius for #30812, 2.6 s, and the corresponding Mach number profile as a function of* $\rho_{\mathrm{pol}}$ *and a parabolic fit.*

Various theoretical studies of toroidal and poloidal mass flow modifications to axisymmetric ideal MHD equilibria exist [4, 1] (and references therein). The purpose of the present paper is to study the effects of the toroidal flow on various equilibrium parameters quantitatively based on flow velocity profiles determined from measurements of the ion temperature and toroidal rotation velocity profiles at ASDEX Upgrade. The poloidal flow is not considered in the present work since it is typically much smaller and, therefore, only of minor importance [5].

The Grad-Shafranov equation for the poloidal flux function $\psi$ describing ideal magneto-hydrodynamic equilibrium in two-dimensional tokamak geometry with purely toroidal flow is:

$$\left(R\frac{\partial}{\partial R}\frac{1}{R}\frac{\partial}{\partial R}+\frac{\partial^2}{\partial z^2}\right)\psi(R,z)=-4\pi^2\mu_0 R^2\left(\frac{\partial p_s}{\partial\psi}+\frac{p_s R^2}{2R_a^2}\frac{\partial M_t^2}{\partial\psi}\right)\exp\left(\frac{M_t^2R^2}{2R_a^2}\right)-\mu_0 F\frac{\partial F}{\partial\psi} \quad (1)$$

$F(\psi)$ is the poloidal diamagnetic current. Assuming rigid rotation on a flux surface, the angular velocity $\Omega$ can be evaluated from the measured toroidal rotation velocity $v_t$: $\Omega(\psi)=v_t/R$. The centrifugal force re-distributes particles on a flux surface according to a Boltzmann distribution resulting in a modified plasma pressure $p(R,\psi)=p_s\exp\left(\frac{m\Omega^2R^2}{2k_BT}\right)=p_s\exp\left(\frac{M_t^2R^2}{2R_a^2}\right)$ where $M_t^2:=\frac{m\Omega^2R_a^2}{k_BT}$ and $R_a$ is the major radius of the magnetic axis. With temperature $T$, ion mass $m$, and $c_s^2=\frac{k_BT}{m}$ defining the velocity of sound, the toroidal Mach number is $M_t=\frac{R_a}{R}\frac{v_t}{c_s}$. Please note that $p_s$, $T$, $\Omega$ and, therefore, $M_t$ are assumed to be constant on a poloidal flux surface whereas the pressure $p$ is not a flux quantity due to the particle re-distribution on a flux surface.

The extension of the Grad-Shafranov equation with purely toroidal flow is included in the Grad-Shafranov solver IDE (Integrated Data analysis Equilibrium) [3] which is a free-boundary equilibrium code. The code can be run in an interpretative mode of reconstructing an equilibrium from measured data, or, to avoid effects of the plasma control, it can be run in a predictive mode with prescribed source terms of pressure and diamagnetic current. For the predictive mode to stabilize the inherent vertical instability of (1) an additional magnetic field from currents in external coils (V2o and V2u at ASDEX Upgrade) is applied which program-internally adjusts itself so that the magnetic axis is at a prescribed position ($R_a$, $z_a$). To study the effect of the toroidal rotation on the equilibrium, the predictive run can be performed in two different ways: 1. Keep the axis position constant for various Mach number profiles. The centrifugal force has to be balanced with the additional magnetic field, resulting in a shift of the last-closed flux surface (LCFS) towards the inner vessel wall [8]. 2. Since an experimental control of the magnetic axis position is not realistic, the second approach is to get rid of the appearing inward shift: In an additional iteration the stabilizing axis position is adjusted such that the currents in the external

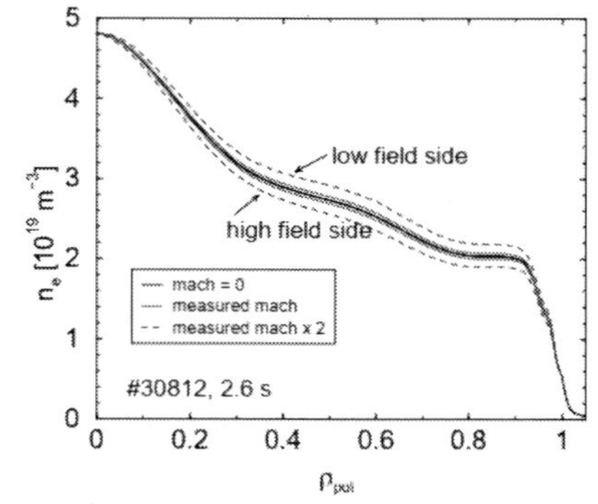


Figure 2: *Variation of the particle density for the static case (black line), the measured Mach number profile (red line), and multiplied by a factor of 2 (blue dashed line).*

coils vanish. This gives the *natural* equilibrium not corrupted by the stabilizing condition.

$T$ and $v_t$ are measured at ASDEX Upgrade with a high-resolution charge exchange system [6]. Fig. 1 shows the measured $T$ and $v_t$ profiles for discharge #30812, 2.6 s, chosen for its large toroidal rotation velocity (1 MA, 2.6 T, $q_{95} = 4.5$, $n_e = 5.8 \times 10^{19}$ m$^{-3}$, 2.5 MW NBI and 0.6 MW ECRH heating), and the corresponding Mach number profile including a parabolic fit. To study the effect of the various Mach number profiles, we start from the source profiles of a reconstructed equilibrium in interpretative mode employing the measured Mach number profile. The Mach number profile is then scaled from 0 (static, no flow) to 2. A factor of 2 results in Mach numbers close to 1 at the magnetic axis and close to 0.8 at mid radius. Fig. 2 shows the variation of the particle density as a function of $\rho_{pol}$ for the static case ($M_t = 0$), and due to the redistribution on the flux surface for the measured Mach number profile (red line), and its scaling by a factor of 2. In contrast to the density profile without flow being constant on a flux surface, the density decreases on the high-field-side (hfs) and increases on the low-field-side (lfs) of the flux surface with increasing Mach number. As the deviation is rather small for the measured Mach number of about 0.4 at mid-radius, the deviation increases approximately quadratically with the Mach number.

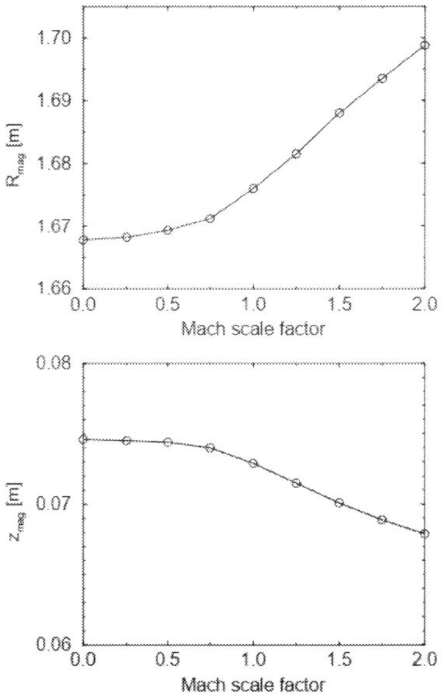


Figure 3: *Magnetic axis position as a function of the Mach number scaling.*

Fig. 3 shows the variation of the magnetic axis position ($R_{mag}$, $z_{mag}$) as a function of the scaling factor. The magnetic axis shows outward shifts of 0.8 cm for the measured Mach number profile and of 3.1 cm for a two-times larger Mach number profile. The radial shift starts with an approximately quadratic dependence on the Mach number with a transition into linearity for

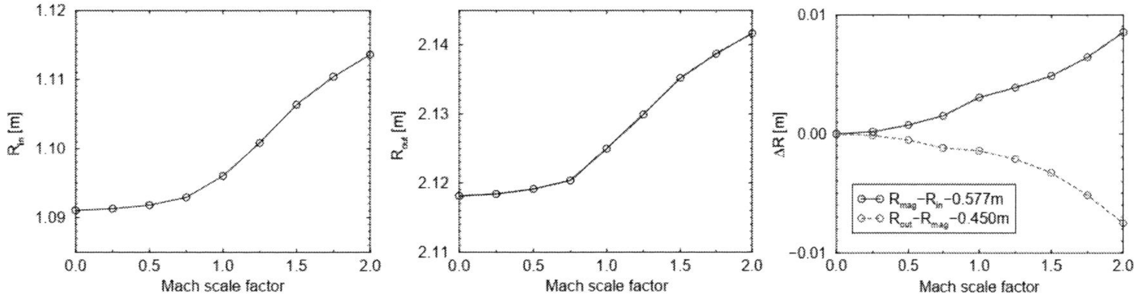


Figure 4: Variation of the innermost $R_{in}$ and outermost $R_{aus}$ position of the LCFS as a function of the Mach number scaling factor, and change of difference $R_{mag} - R_{in}$ and $R_{out} - R_{mag}$.

larger values resembling the observation of a theoretical study [7]. The shift for the vertical position of the magnetic axis is much smaller (∼mm) because the centrifugal force acts in radial direction. Fig. 4 shows the variation of the innermost $R_{\text{in}}$ and outermost $R_{\text{out}}$ position of the LCFS as a function of the Mach number scaling factor. Both quantities increase with Mach number resulting in an overall outward shift of the LCFS.

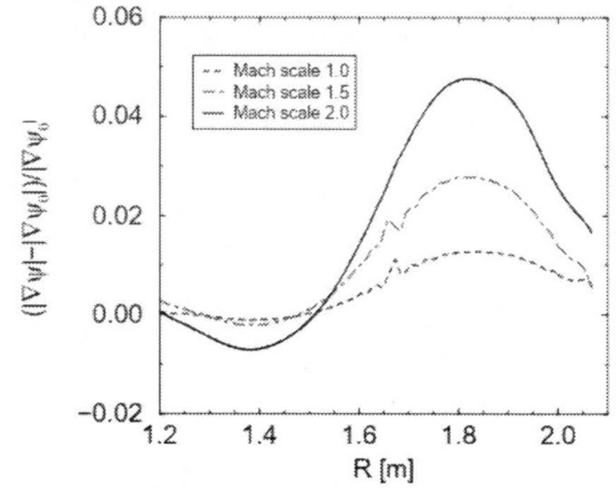


Figure 5: *Relative change of the mid-plane flux gradient with flow with respect to the static case.*

This effect can neither be seen in the interpretative mode of the equilibrium reconstruction because the magnetic measurements determine the LCFS position robustly, nor in the predictive mode with magnetic axis position held constant. An artificial scaling of the Mach number profile in the interpretative mode results in modifications of the source profiles in the plasma center only. In the predictive mode with magnetic axis position held constant the LCFS shifts to the inner vessel wall [8]. This is consistent with the increase of the distance $R_{\text{mag}} - R_{\text{in}}$ shown in the right panel of Fig. 4. In contrast $R_{\text{out}} - R_{\text{mag}}$ decreases with Mach number showing an increased flux gradient on the lfs. Fig. 5 shows the relative change of the gradient of the flux in the midplane with respect to the gradient in the static case. The largest change of the flux gradient is at the lfs (outer) midplane at mid radius. The change of the gradient increases approximately quadratically up to about 5% for a Mach number of 0.8 at mid radius (scale 2.0). Although the plasma shows a shift of the LCFS to the lfs, a compression on the lfs and a stretching on the hfs, there is nearly no effect on the plasma volume, plasma surface, horizontal or vertical extension, or the ellipticity.

This work has been carried out within the framework of the EUROfusion Consortium and has received funding from the Euratom research and training programme 2014-2018 under grant agreement No 633053. The views and opinions expressed herein do not necessarily reflect those of the European Commission.

## References


[1] A. Thyagaraja and K.G. McClements. *Physics of Plasmas*, 13:062502, 2006.

[2] E. Strumberger, H.P. Zehrfeld, and S. Günter. In *28th EPS Conference on Controlled Fusion*, volume 25A, page P5.036. 2001.

[3] R. Fischer, et al. In V. Naulin, C. Angioni, and M. Borghesi et al., editors, *40th EPS Conference on Plasma Physics*, volume 37D, page P2.139. European Physical Society, Geneva, 2013.

[4] B.J. Green and H.P. Zehrfeld. *Nucl. Fusion*, 13:750, 1973.

[5] T. Takeda and S. Tokuda. *J. Comput. Phys.*, 93:1–107, 1991.

[6] E. Viezzer, T. Pütterich, R. Dux, R. M. McDermott, and the ASDEX Upgrade Team. *Rev. Sci. Instrum.*, 83:103501, 2012.

[7] W. Kerner and O. Jandl. *Computer Physics Communications*, 31:269–285, 1984.

[8] J. Geiger. PhD thesis, Technische Univ. München, 1992.

# Diffusive and non-diffusive contributions to momentum transport in AUG experiments and the European intrinsic rotation database

R. M. McDermott[1], T. Odstrcil[1], C. Angioni[1], B. P. Duval[2], A. N. Karpushov[2], R. Fischer[1], A. Salmi[3], T. Tala[3], G. Tardini[1], E. Viezzer[1] and the ASDEX Upgrade Team

[1]*Max-Planck-Institut für Plasmaphysik, Garching, Germany* [2] *Ecole Polytechnique Federale de Lausanne (EPFC), Centre de Recherches en Physique des Plasmas, Station 13* [3]*VTT, P.O. Box 1000, FIN-02044 VTT, FInland*

## Introduction

Momentum transport has become an active area of research within the tokamak community due to the realization that significant rotation and rotation shear is needed to tame turbulent transport and ensure plasma stability. Moreover, knowledge of the rotation profile is needed to predict impurity particle transport. As such, the goal of the rotation community is to gain a fundamental understanding of the sources, sinks and transport of momentum in fusion plasmas in order to accurately predict the rotation profiles. At ASDEX Upgrade (AUG) significant effort has been dedicated to this via a wide variety of experimental techniques.

## Database analysis of type-I ELMy H-mode plasmas

One such technique, a profile database approach pioneered by Weisen *et al* [1, 2], is able to disentangle the diffusive momentum flux from all of the non-diffusive fluxes. This approach has an advantage over traditional torque modulation analyses in that it does not have to assume a negligible residual stress. It does, however, require the non-diffusive transport within the dataset to be similar, as it returns only a single averaged non-diffusive flux profile. The approach employs a momentum conservation equation re-written in a linear form ($y = mx + b$) such that the unknowns are encapsulated entirely in the slope and the offset.

$$\frac{R}{L_\omega} = \frac{R}{\omega_\phi}\frac{\partial \omega_\phi}{\partial r} = \frac{\chi_i}{\chi_\phi}\left(\frac{R\Pi_\phi}{n_i m_i R \chi_i \omega_\phi} + \frac{R\Gamma_n}{\chi_i n_i}\right) + \left(\frac{Rv_\phi}{\chi_\phi} + \frac{R\Pi_{RS}}{n_i m_i R \chi_\phi}\right) \quad (1)$$

In Eqn. 1 the left hand side is the normalised gradient of the toroidal rotation frequency ($R/L_\omega$), which can be obtained directly from measurements. The first set of brackets on the right hand side contains the normalised NBI torque and the particle flux ($t^* + \Gamma_n^*$), both of which can be calculated from transport codes such as TRANSP. There remain only two unknowns: the Prandtl number, $P_r = \chi_\phi/\chi_i$, which is the inverse of the slope in equation 1, and the non-diffusive contributions (the pinch number $V_N = RV_\phi/\chi_\phi$ and the normalised residual stress momentum flux $\Pi_{RS}^*$), which together correspond to the y-intercept.

Using this equation and a sufficiently large and varied database of observations, an averaged Prandtl and pinch number profile can be obtained by applying linear fits to the dataset at multiple radial locations. This is illustrated for one radial position on the left-hand side of Fig. 1 for a database of steady, NBI-heated, AUG H-mode plasmas [3]. Here, the measured $R/L_\omega$ is plotted against the normalised NBI torque. The dashed line corresponds to the case of $P_r = 1$ and zero non-diffusive flux. The vertical offset of the data points from this line clearly demonstrates the presence of a non-diffusive flux. The blue shaded region indicates the uncertainty in the analysis and translates directly to the error bars shown in Fig. 1b. Here, the profiles of the resultant non-diffusive momentum flux and Prandtl number are shown. The Prandtl number is approximately 1 everywhere (within the error bars), while the non-diffusive flux is small in the core and increases toward the edge, qualitatively consistent with a Coriolis momentum pinch [4]. However, the pinch numbers obtained from GKW, quasi-linear, gy-

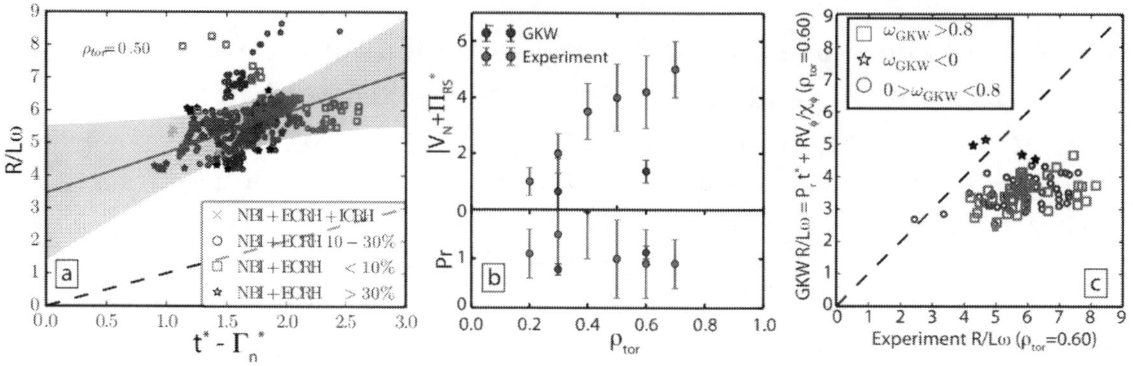


**Figure 1:** (left) Normalised toroidal rotation gradient vs. the normalised torque plus particle flux for a database of NBI heated AUG H-mode plasmas at $\rho_\phi$ =0.5. (right) Profile of the non-diffusive momentum flux derived from the linear analysis plus the pinch numbers calculated by GKW at two radial positions.

rokinetic simulations at $k_y\rho_i = 0.3$ of a representative subset of the database are significantly smaller than the experimentally determined flux. The resulting GKW predicted rotation gradients are a factor of two smaller than the experimentally measured values. This can be seen in Fig. 1c where the $R/L_\omega$ values predicted by GKW (the sum of the diffusive and convective fluxes) at $\rho_{tor} = 0.6$ are plotted against the experimentally measured values. This discrepancy suggests the presence of a co-current directed residual stress momentum flux to resolve the difference. Unfortunately, within the uncertainties of the measurements, nothing quantitative can be concluded.

It is potentially possible to gain information on the physics at play by looking at the parameter dependences of the experimental and theoretical data. To this end, the experimental $R/L_\omega$ data between $\rho_\phi = 0.2$ and $0.7$ were regressed against the parameters found to be relevant in an analogous work performed for a database of JET discharges [2].

$$R/L_\omega = 0.6(t^* - \Gamma_n^*) - 0.9T_e/T_i - 0.3q + 19\epsilon^{\frac{1}{2}} - 0.5 \qquad (2)$$

The standard deviations associated with the regression variables are $[0.1, 0.3, 0.1, 2, 0.7]$. Unsurprisingly, the radial coordinate $\epsilon = r/R$ is the most relevant parameter in the regression followed by the diffusive contribution to the transport (applied torque), $T_e/T_i$ and the safety factor $q$. In contrast to the JET results, $R/L_{ne}$ is not found to be important in this regression. However, if $\epsilon$ is removed from the regression and the data at individual radii examined, then $R/L_{ne}$ becomes the second most statistically relevant parameter after the applied torque. A dependence of $R/L_\omega$ on $R/L_{ne}$ is expected due to the contribution of the Coriolis momentum pinch [4]. Indeed, in the regressions of the GKW predicted pinch numbers, this dependence shows up clearly. The lack of an $R/L_{ne}$ dependence in the experimental $R/L_\omega$ when $\epsilon$ is included is, perhaps, not too surprising given that the GKW results under-predict the measured rotation gradients by a factor of two; i.e. the missing half of the non-diffusive flux may exhibit a different $\epsilon$ (and other) dependence that masks the contribution. Taken together, this data indicates that there is a significant, non-diffusive, co-current directed, momentum flux that is not taken into account in the modeling. This is qualitatively consistent with the evidence for a co-current intrinsic torque in NBI heated plasmas at DIII-D [5].

## Comparison of ASDEX Upgrade and TCV intrinsic rotation profiles

Given that intrinsically driven momentum fluxes can play a significant role in determining the rotation in fusion plasmas, it is important to gain a better understanding of the mechanisms

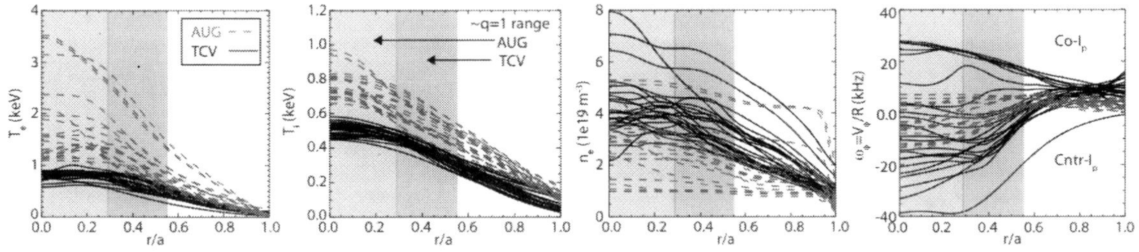


**Figure 2:** Electron and ion temperature, electron density and intrinsic toroidal rotation profiles from TCV (black solid) and AUG (red dashed) LSN, favorable drift, positive triangularity Ohmic L-mode discharges.

driving these fluxes and their dependences on plasma parameters. Work in this direction has been performed at AUG which showed that the intrinsic rotation is strongly linked to the electron density gradient and to the plasma density and temperature through the collisionality [6, 7]. More recently, through the framework of the European intrinsic rotation database, AUG and TCV intrinsic rotation data from Ohmic L-mode plasmas have been compared for the first time using the same analysis tools and methodologies. The AUG dataset is limited to LSN, favourable drift discharges with positive upper and lower triangularity, which is also the planned ITER configuration. Only TCV profiles from plasmas matching these criteria were used for the comparison, limiting the TCV dataset to 23 samples. For the AUG data, a subset of 20 samples that spanned the available parameter space was selected. The plasma profiles corresponding to the selected samples are shown in Fig. 2. The electron and ion temperatures tend to be higher in the AUG dataset. This is partly explained by the higher plasma currents and partly by the AUG data extending to a lower density range.

Interestingly, the toroidal velocity profiles (not shown) from the two machines are, at first glance, remarkably similar. They both display co-current edge rotations of roughly 10km/s and demonstrate a wide variety of profile shapes in the plasma core resulting in central toroidal rotations between -20 and +20km/s. However, the toroidal rotation frequency profiles (right hand side of Fig. 2), which account for machine size, show differences. The TCV profiles cover a wider range in $\omega$ and, more importantly, a wider range in the normalised toroidal rotation gradient, $u' = -(R^2/v_{th,i})\, d\omega/dr$, where $v_{th,i} = \sqrt{2T_i/m_i}$ is the main ion thermal velocity. The profiles display other differences. The "flattening" point of the profile, i.e. the point inside of which the gradient is effectively zero, differs between the two devices as a function of $r/a$, $\rho_{tor}$ and $\rho_\psi$. The region of steep rotation gradient for the AUG profiles extends further inside ($0.25 < r/a < 0.65$) compared to TCV ($0.45 < r/a < 0.65$). The hypothesis that the flattening of the profiles, hence the limiting of the gradient region, is caused by sawteeth is supported by the relative positions of the q=1 surfaces, see Fig. 2. The rather large difference in q=1 surface position persists even for plasmas with similar $q_{95}$ values. The TCV dataset includes $q_{95}$ values between 2.7 and 4.9, while the AUG data ranges between 3.7 and 8.5, providing only a small region of overlap around $q_{95} = 4$. However, if the aspect ratio difference between the two devices ($\bar{\epsilon} = 0.31$ for AUG and $0.26$ for TCV) is taken into account, the positional agreement of the q=1 surfaces in terms of $\epsilon$ improves and one finds that the regions of strongest rotation gradient overlap between $0.1 < \bar{\epsilon} < 0.175$.

A comparison of the parameter dependences of the observed rotation gradients in this $\epsilon$ region yields encouragingly similar results. In Fig. 3 the normalised rotation gradients from TCV and AUG are shown as a function of the dominant parameter dependences found in [7] as well as $\rho^*$. With the exception of a small group of points with positive rotation gradients (co-$I_p$ profiles), the data from the two machines overlay quite well. The trend with $R/L_{ne}$

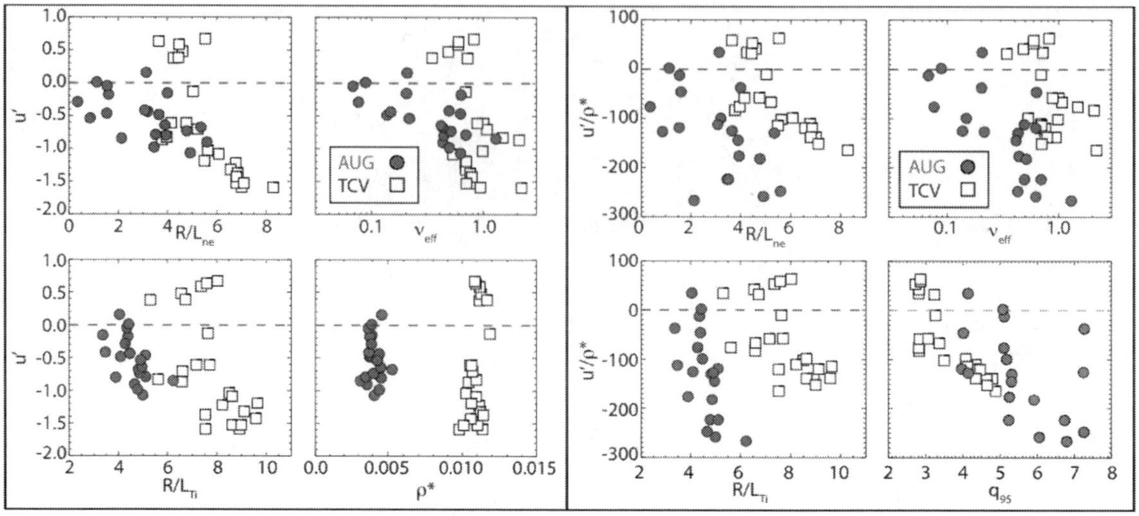


**Figure 3:** Parameter dependences of $u'$ (left) and $u'/\rho^*$ (right) averaged over $0.1 < \epsilon < 0.175$.(right)

is particularly striking. These datasets also show strong trends with other parameters such as $q_{95}$, that is itself strongly correlated with $R/L_{ne}$ (not shown). Most core localised residual stress momentum fluxes are expected to scale with $\rho^*$. In Fig. 3 it is clear that the available range of $\rho^*$ from either machine is too small to see any kind of scaling. Together though, the data spans over a factor of two in $\rho^*$. Multi-variable regressions of the combined datasets that include $\rho^*$ always yield $\rho^*$ as the dominant parameter dependence with a positive coefficient, and explain the co-current TCV rotation points as having the highest $\rho^*$ combined with the lowest normalised gradients, all of which have negative coefficients. These co-current points cause the regressions with $\rho^*$ to have a lower RMSE error than those without.

However, one would expect the magnitude of the residual stress, $\Pi_{RS}$, to scale as $\rho^*$ multiplied by some function with other parameter dependences, i.e. $\Pi_{RS} \sim \rho^* F(R/L_{ne}, R/L_{Ti}, ...)$. In this analysis, as $u'$ is being used as a proxy for $\Pi_{RS}$, we would expect $u'/\rho^*$ to demonstrate the residual stress parameter dependences more clearly than $u'$ alone. Accordingly, the $u'/\rho^*$ values of the database were also examined and regressed. The right hand side of Fig. 3 shows the results for the same subset of parameters. This normalisation significantly reduces the previously observed correlations, although it improves the correlation of the data with $q_{95}$, shown only for the $u'/\rho^*$ dataset. Additionally, the ability of a regression to reproduce the experimental $u'/\rho^*$ is not improved over $u'$ alone. Based on the analyses performed thus far on the combined AUG and TCV datasets, there is no clear evidence of a dominant $\rho^*$ dependence in the intrinsic rotation. Ideally, the intrinsic rotation profiles between the two machines should be compared at matched dimensionless parameters (excluding $\rho^*$). Unfortunately, this data does not yet exist within the database, but will be the focus of future experimental effort.

## References

[1] H. Weisen et al, Nucl. Fusion 52, 042001 (2012)
[2] H. Weisen et al, Nucl. Fusion 52, 114024 (2012)
[3] T. Odstrcil, Masters Thesis University of Stuttgart (2013)
[4] A. G. Peeters et al, Phys. Rev. Lett. 98, 265003 (2007)
[5] W. Solomon et al, Nucl. Fusion 49, 085005 (2009)
[6] C. Angioni et al, Phys. Rev. Lett. 107, 215003 (2011)
[7] R.M. McDermott et al, Nucl. Fusion 54, 043009 (2014)

**Acknowledgements:** This work has been carried out within the framework of the EUROfusion Consortium and has received funding from the Euratom research and training programme 2014-2018 under grant agreement No 633053. The views and opinions expressed herein do not necessarily reflect those of the European Commission.

# Erosion of metal plasma facing components by arcing at the inner divertor baffle of ASDEX Upgrade

V.Rohde, M.Balden, and the ASDEX Upgrade Team

*Max Planck Institut für Plasmaphysik, Boltzmannstr. 2,85478 Garching, Germany*

## INTRODUCTION:

In the standard picture of plasma wall interaction (PWI) arcs are not taken into account as erosion source: in the $80^{th}$ it was shown that for carbon physical and chemical sputtering dominate and arcs are only triggered during unstable plasma phases [1]. On the other hand arc traces are observed in all present fusion devices. Recently, investigations in ASDEX Upgrade (AUG) show that ELMs may trigger arcs at some locations [2] and the morphology of dust collected can be explained by droplet production in arcs [3].

Whereas previous investigations use the typical AUG PFCs, i.e. tungsten (W) coating of 5 micron thickness on a carbon substrate, bulk material was investigated this time. Polished probes were used as the surface roughness of technical materials is in the micron range, which is the typical width of arc traces. Moreover, scratches produced during the fabrication process influence the direction of arcs and even hide them. Two different materials were selected: W as material of the ITER divertor and P92, martensitic-ferritic high temperature steel similar to EUROFER, which is under discussion for DEMO.

## ARCING:

In contrast to sputter processes, arcs are localized and the amount of arc traces varies strongly locally even on a single tile. Markers, made out of polished W and P92 steel, were installed at the inner divertor baffle region, which is prone for arcing [2, 4] to investigate the erosion. In contrast to physical sputtering, arcs produce not only ions and neutral atoms, but also droplets. In a simplified picture arcs are a localised heat source, driven by the ohmic heat of the current in the arc. As the typical spot size is less than a micron, localised melting produces the characteristic traces. Due to the current, arcs have to move in a magnetic field perpendicular to the field direction. The arc jumps from one to another single spot, resulting in the typical arc traces, which can be seen on a macroscopic scale.

Whereas arcing is observed at many tiles in AUG, significant erosion was found only in deposition dominated regions [2]. At the inner baffle region, layers consisting of C, B, W,O and hydrogen isotopes reach a thickness of typical 2-3 μm [5]. Due to the shallow angle of the magnetic field direction and shadowing effects, the thickness of the deposits varies even on one tile. The role of deposits for the arcing process is not completely understood, but it is assumed that the bad electrical conductivity of the deposits reduces the velocity of the arc movement. As the arc burns for a longer time at the same location, it causes stronger erosion.

Local melting and splashing of the molten material, i.e. droplet production, is an important erosion process by arcing. Indeed huge amounts of W droplets are observed in the dust collected at AUG. Taking the average shot time a flux of $10^6$ droplets/s with a typical size of 2 μm can be calculated, which results in about 50 % of the dust collected [2].

The erosion by arcing depends on the melting temperature and the heat effusivity of the material used. As the melting temperature of W (3695 K) is much higher than for P92 (1800 K) and the effusivity is also higher (21 vs 9 $kW/(m^2K)$) a higher erosion is expected for the steel. Measurements of the erosion rate are mostly from the $80^{th}$ and use a normalisation to the charge of the arcs. Taking the average of different papers [6] a slightly higher erosion of W compared to Fe is expected (78 vs 61 μg/C), which is due to the higher specific mass.

**EXPERIMENTAL:**

The probes were exposed for one experimental campaign (Dec 13 till Oct 14). They were installed at inner divertor baffle (Bgr.6) of sector 9. After the opening of the vessel the tiles were removed and a part of the probes was cleaned by wiping using deionised water to remove the deposited layers. A confocal laser scanning microscope was used to image the surface by normal optical light microscopy (Fig. 1a and 1c) and to obtain 3D topography data (Fig. 1b and 1d). These 3D data allow determining the material loss. Underneath the deposited layer, a large fraction of the polished surface is unaffected, which is observed at cleaned areas. Any depression beneath the polished surface is assigned as eroded. The volume of these depressions regarding a reference plane parallel to the surface and their areal fraction are shown in Fig. 2.

**TUNGSTEN PROBES:**

A view on a typical region of the cleaned W surface is shown in Fig 1a. The polished surface below the deposition layer is partially affected by arcing, the dark structures with the direction from top to bottom. The arcs removed locally the complete deposition layer of typical 3 μm thickness (not shown in this figure). Two different kinds of traces are observed: single craters and complete traces indicating different burning conditions for the arcing. Derived from the optical picture, about 30 % of the surface is affected by arcing, but the depth profile (Fig 1b) is needed to determine the erosion. Most of the traces seen in the optical picture are also found at the depth profile, but the typical erosion is less than 0.2 μm. Additionally, a significant amount of molten W is deposited above the polished level (yellow) close to the erosion region. Only some craters of up to 5 μm depths (blue) are found.

**STEEL PROBES:**

An equivalent region as for W is analysed and shown in Fig. 1(c,d) for P92 steel. Again the arcs erode the substrate below the deposition layers. From optical observations about 20 % of the surface is affected by arcing. The number of arc traces is slightly lower than for W. Beside the long arc traces also craters are present (Fig. 1(c)). But in contrast to the W surface the respective craters are much bigger with about 100 μm diameter and about 30 μm depth. These craters dominate the erosion. An interesting detail of the measurements are the direction of the arc traces. As the steel used can be magnetized, the local magnetic field is influenced by the probes themselves. For this reason the direction of the arc traces, which are orientated perpendicular to the magnetic field, bends at the edges of the probes.

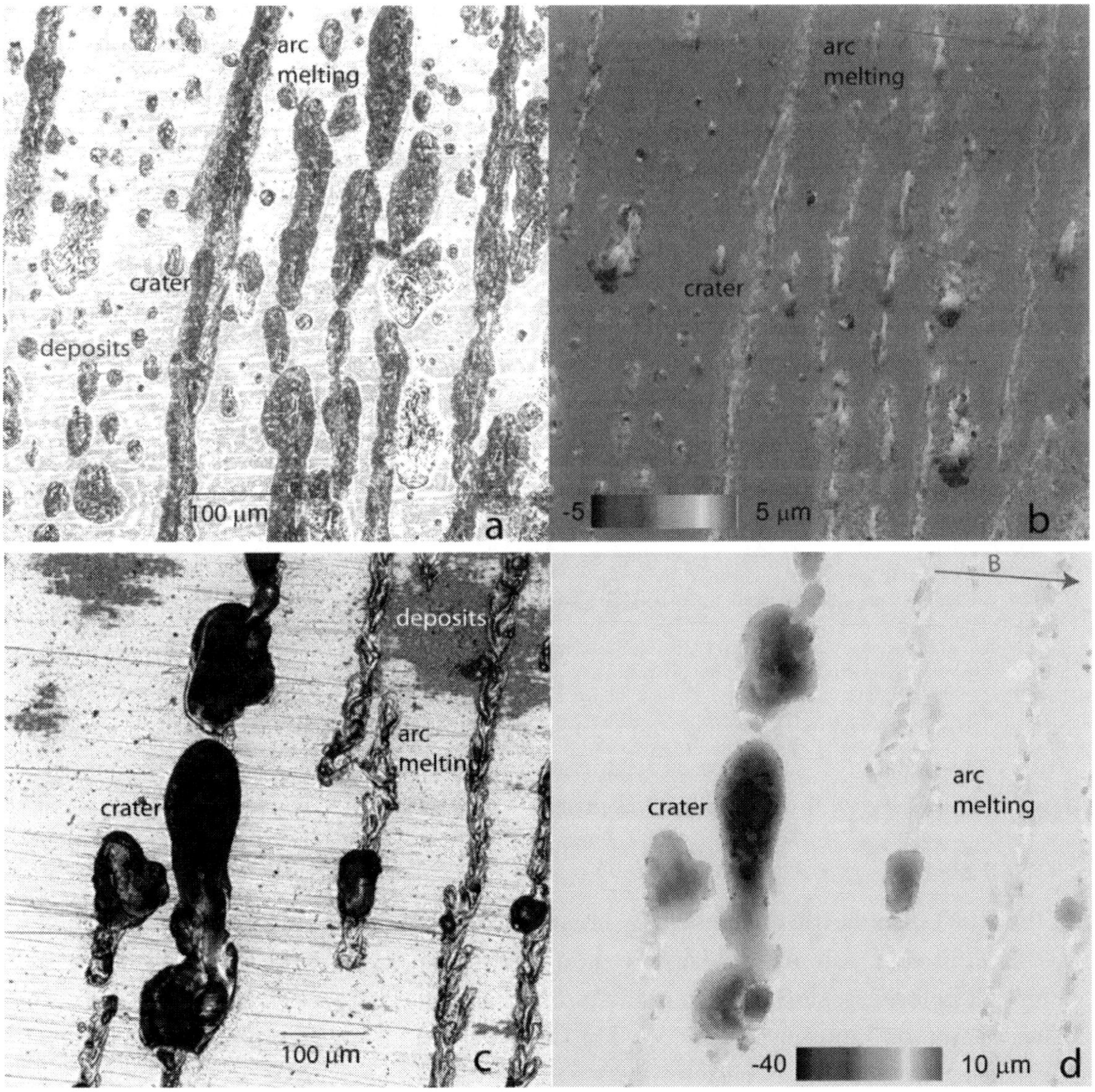


Figure 1: Detail of the investigated region showing typical arc traces on W (a, b) and P92 steel (c, d): light microscope images (a, c) and the respective 3D topography data (b, d). The brownish areas in (a, c) are remains from cleaning off the deposited layers. Note the different height scale of the two 3D topography data (b, d).

## DISCUSSION:

To compare the erosion for both materials, depth profiles of a 1.8*1.8 mm area obtained at the same toroidal and equivalent poloidal positions are used. The volumes of the depressions below a reference plane parallel to the surface were determined for various distances to the surface level. These volumes normalised to the total area analysed, presenting the average erosion, are shown together with the respective area faction in Fig 2. To measure the total erosion a reference level, presenting the unaffected surface has to be defined. The accuracy of this level is limited by deposits, melting from arcing and the accuracy of the polished surface. To overcome this problem only erosion 0.2 µm below the reference level was taken into account. Using this level only 7.3% of the W surface is eroded by arcing, but 15.6 % of the steel one. Normalizing the total eroded volume to the surface area yields an average erosion of ~30 nm for W and ~920 nm for P92, i.e. the erosion of steel is 30 times higher compared to

W. The mean erosion rate for the campaign (7125 s of plasma) is $2.7*10^{13}$ at $cm^{-2}s^{-1}$ for W and $1.1*10^{15}$ at $cm^{-2}s^{-1}$ for P92. Even as the position of the probes is deposition-dominated, significant erosion is observed, which can be compared with the erosion at the outer divertor strike line of $7*10^{14}$ at $cm^{-2}s^{-2}$ for W [5].

The big difference of the erosion for W and P92 is not expected from the erosion yield for arcing given in literature [6]. All literature data are for clean surfaces but in AUG the strong erosion is correlated with deposits. The influence of these layers is still not complete understood, and laboratory investigations would be beneficial. A hypothesis to explain the high erosion for P92 is the production of droplets, which increases strongly, as the material temperature reaches the melting point [6]. The lower effusivity and melting temperature for P92 may explain this behaviour. To understand the droplet production, a set of different polished probes was installed for the next experimental campaign of AUG. To scan different melting points and effusivity, this set comprises probes from Al, Cu, Cr, SS, P92, Mo and W.

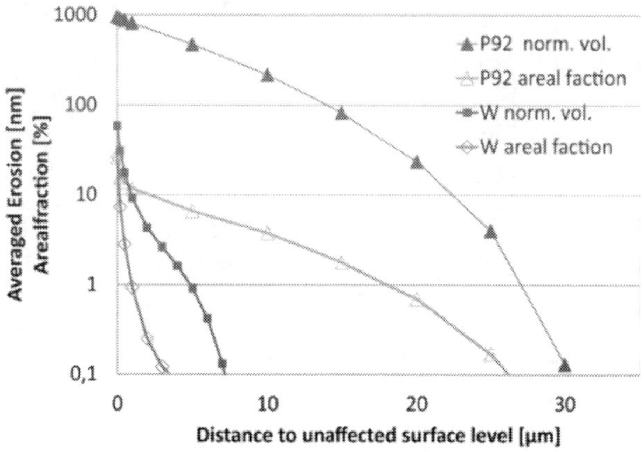


Figure 2: Normalized volume and areal fraction of depression versus distance to the unaffected, originally polished surface.

**CONCLUSION:**

At the AUG inner baffle the erosion by arcing on P92 steel is about 30 times higher than for W, i.e. the steel erosion at the baffle region is comparable to W erosion at the outer strike point. Therefore, erosion by arcing has to be taken into account to determine the optimal material mix for future fusion devices. For Be, designed for the inner baffle region of ITER, the melting temperature is close to the steel, but the effusivity is even higher than for W, which complicates predictions. Further investigations, using different materials are started at AUG to disentangle the different effects and to allow estimating the droplet production by arcing.

[1] McCracken, J.Nucl.Mat., 93, (1980), 3
[2] Rohde et al., J.Nucl.Mat., 438, (2013), 800
[3] Balden et al., Nucl.Fusion, 54, (2014), 073010
[4] Herrmann et al., J.Nucl.Mat., 390, (2009), 747
[5] Mayer et al., Phys.Scr., T138, (2009), 014039
[6] Post, Behrisch, Physics of Plasma-Wall interactions in Controlled Fusion, ISBN 0-306-42097-XJ

This work has been carried out within the framework of the EUROfusion Consortium and has received funding from the Euratom research and training programme 2014-2018 under grant agreement No 633053. The views and opinions expressed herein do not necessarily reflect those of the European Commission.

# Electron density and temperature structures during edge localized modes in the edge plasma of ASDEX Upgrade

B. Kurzan, L. Giannone, and ASDEX Upgrade team

*Max-Planck-Institut für Plasmaphysik, 85748 Garching, Germany*

## Introduction

Past investigations of electron density and temperature ($n_e$, $T_e$) structures of type I edge localized modes (ELMs) [1] were hampered by the low frequency of hitting an ELM with the lasers of the Thomson scattering diagnostic, because the duration of an ELM was only about 1 ms. The ELMs in the discharges presented here, have a duration of up to 7 ms, and are hit by the lasers much more frequently. A zoo of 2D electron density and temperature structures is observed now during the ELM. Additionally indications for high temperature, low density filaments are found when a limiter is introduced in the scrape-off layer (SOL).

## Setup of edge TS system and 2D snapshots of electron density and temperature

The edge TS system [2] is equipped with 6 lasers with a repetition rate of 20 Hz each. The scattering volumes of length 25 mm in the vertical (z) direction and diameter 1.5 mm of the 6 lasers and 11 spatial channels form a $6\times11$ (R, z) matrix in the plasma. When the 6 lasers are fired within 2.5 $\mu$s in burst mode operation, 2D snapshots of the electron density and temperature are obtained. The position of the separatrix is determined by fitting a 1.5D heat transport model [3] to an ensemble of 6 electron temperature profiles mapped by a single magnetic equilibrium to the outer mid-plane. The obtained separatrix position is now more accurate ($\pm1$ mm), than the value obtained from magnetic equilibrium reconstruction ($\pm5$ mm). For obtaining relative electron density and temperature structures $\Delta n_e$, $\Delta T_e$ the fitted functions (1.5D heat transport model for $T_e$ and a modified hyperbolic tangent function for $n_e$) are subtracted from the $T_e$, $n_e$ values. The significance of the correlations in $\Delta n_e$, $\Delta T_e$ structures, is given by the covariance $\Delta n_e \Delta T_e/(\sigma n_e \sigma T_e)$, with $\sigma n_e$, $\sigma T_e$ as the errors of measurement of $n_e$, $T_e$. In the 2D snapshots flux surfaces of the magnetic equilibrium are overlaid: the position of the separatrix, is indicated in blue, and the positions of the first wall contacts of the flux surfaces on the high and low field side are marked in green and red respectively.

## Discharge parameters

The investigated discharges are hydrogen plasmas with hydrogen neutral beam injection and toroidal magnetic field $B_t$=-2.5 T: For the discharge #31721 the plasma parameters are $\bar{n}_e=5\times10^{19}\mathrm{m}^{-3}$, $I_p$=0.8 MA, $P_{NBI}$=3.5 MW, $P_{ECRH}$=2-3 MW, and for the discharges #31726/7,

$\bar{n}_e$=7×10$^{19}$m$^{-3}$, I$_p$=1.0 MA, P$_{NBI}$=5.4 MW, P$_{ECRH}$=2.1 MW.

**Electron density and temperature structures during and in-between ELMs**

During the ELM a non-zero halo current (signal $I_{polsola}$) flows between the inner and outer divertor. It is used to define the beginning and end of an ELM. The according time points are indicated by red and green vertical line in the signal traces of $I_{polsola}$ in fig. 1. The D$_\alpha$-radiation in the divertor usually shows a peak before the rise of the poloidal halo current and is low during the ELM (see fig. 1). The electron density and temperature structures observed during the ELM, are: A hole in both the electron density, and temperature exists at the same position inside the separatrix (fig. 1a). Additionally to the first case also an electron temperature blob outside the separatrix is correlated with a hole in the electron density (fig. 1b). A density blob at the separatrix and a density hole in the SOL can be both anti-correlated with the electron temperature (fig. 1c). A hole in the electron temperature can exist inside the separatrix, which is not correlated with a corresponding structure in the electron density (fig. 1d). In inter-ELM phases, with peaking $D_\alpha$ radiation in the divertor, the observed structures are: At the separatrix small scale structures with higher quasi-mode number are found, where the blobs and holes in the electron density and temperature are in sync (fig. 1e). Some cases exist, where anti-correlated blobs and holes in electron density, and temperature exist over a radial distance of about 12 cm (fig. 1f).

**Hot filaments in the far SOL at and behind the limiter**

A limiter is introduced at a position in the SOL, where the open flux surfaces do not connect to the divertor plates, but to wall elements outside the divertor. Here the flux surfaces connect to the middle of the inner heat shield and the lower passive stabilizing loop. Local halo current measurements indicate that during the ELM plasma is transported into the limiter shadow. Usually a background plasma with electron density $n_e \approx 3\times10^{18}$m$^{-3}$ and temperature $T_e \approx$ (30±18) eV is observed here (fig. 2a). Sometimes hot filamentary structures with electron density $n_e \approx 3\times10^{18}$m$^{-3}$ and temperature $T_e \approx$ (100±60) eV are observed in the limiter shadow (fig. 2b). Note that the relative error of measurement of the electron temperature is $\sigma T_e/T_e$= 60% for both the background plasma and the hot filament. The relative error scales with the electron density, which is about the same for both cases.

When the whole SOL connects to the divertor (discharge #31721), hot filaments can also be observed. They are further away from the separatrix near the flux surfaces, which connect to the lower PSL and the lower part of the inner heat shield.

**Summary and Discussion**

The 2D structures of electron density and temperature observed during an ELM suggest,

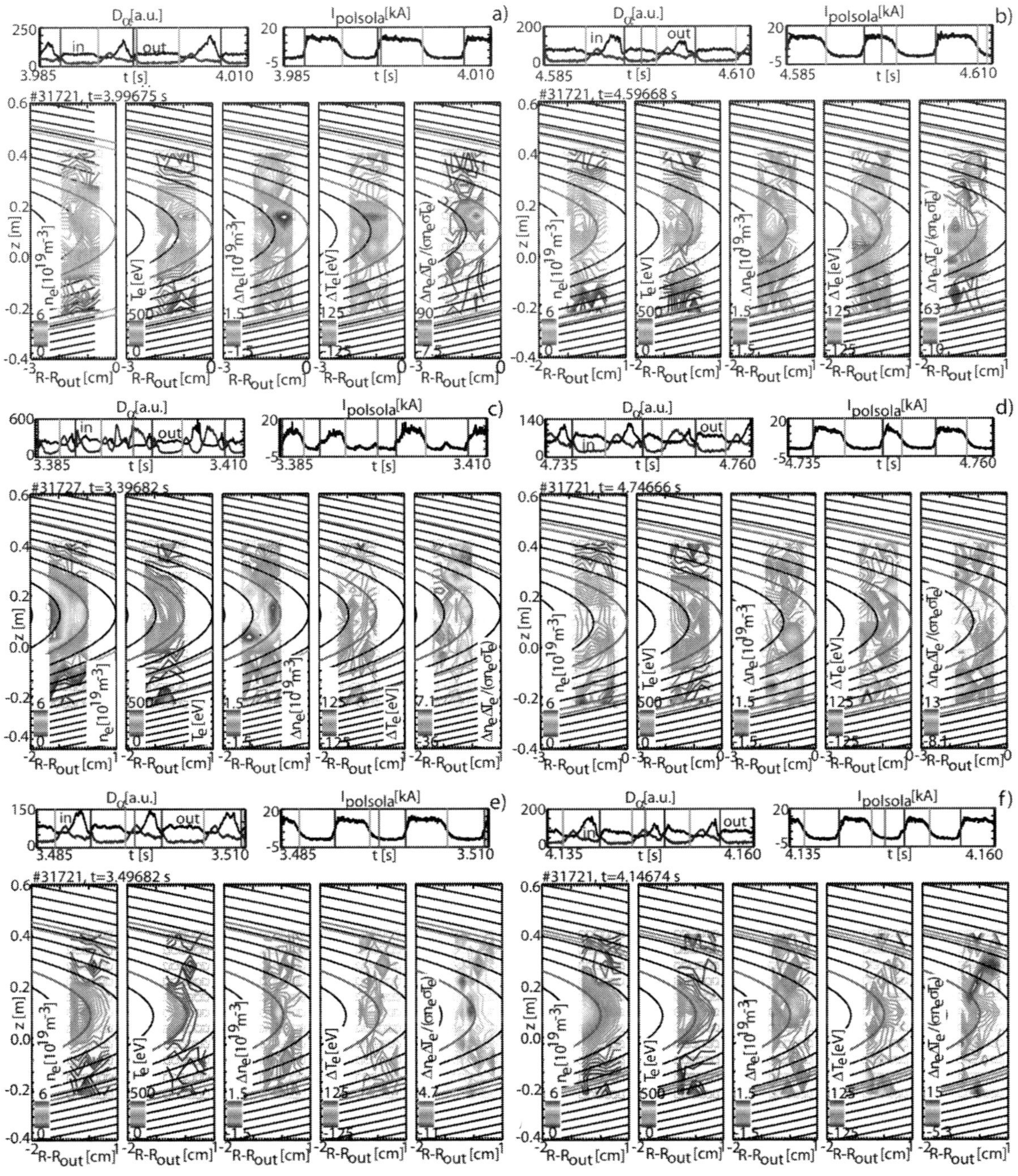

Figure 1: *Electron density and temperature structures during ELMs, with non-zero* $I_{polsola}$*, a) - d), and in-between ELMs, with peaking* $D_{\alpha}$*, e) - f). For details see the main text.*

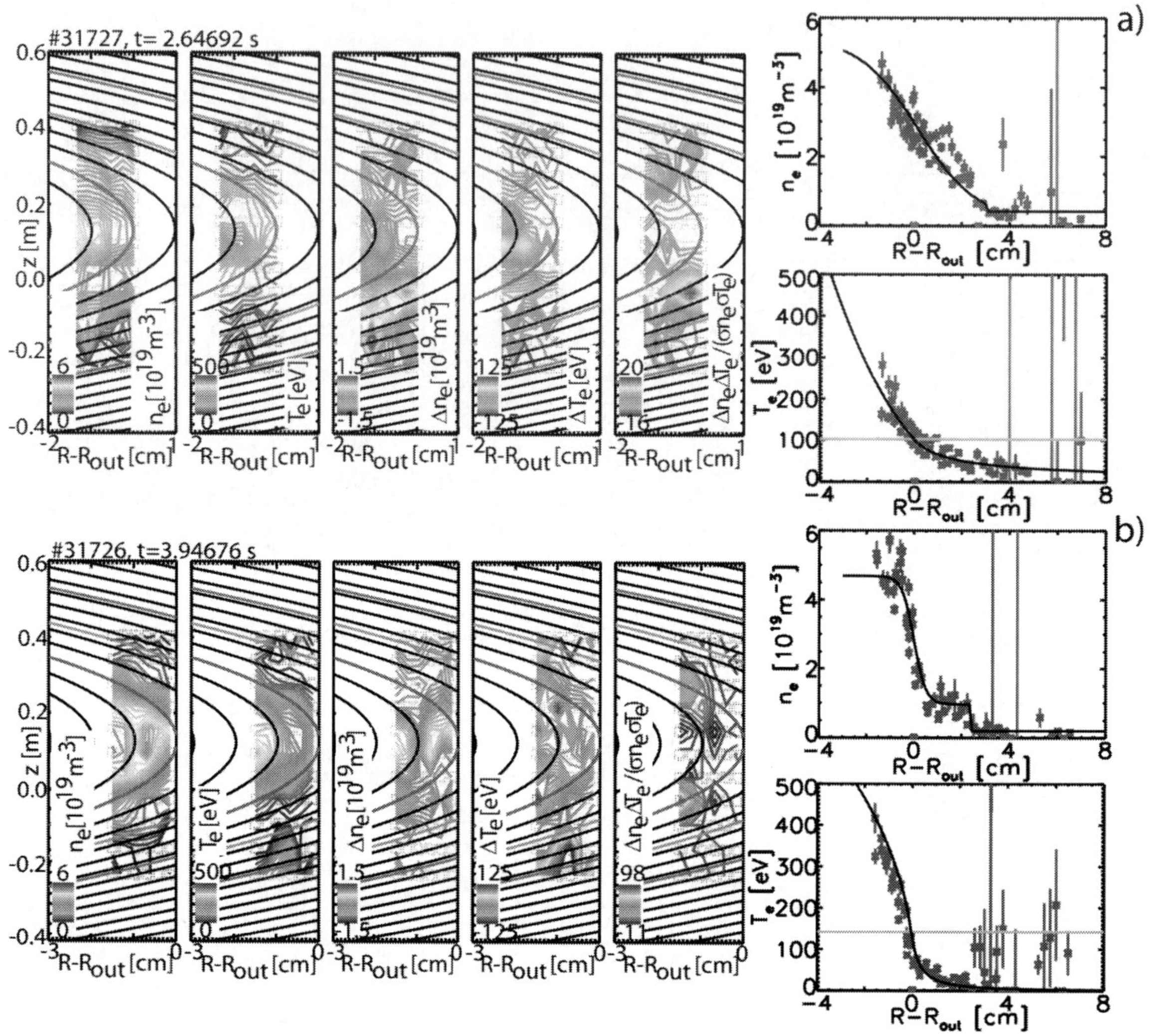


Figure 2: *Observed profiles and structures in the far SOL: cold background profile (a), and hot filaments at and behind the limiter (b).*

that the ELM is not only made up of electron density and temperature blobs in sync, which transport particles and heat. There exist also anti-correlated and uncorrelated electron density and temperature structures. The physics behind the high temperature and low density filaments close and behind the limiter in the SOL is not yet understood.

## References


[1] B. Kurzan *et al.*, Phys. Rev. Lett. **95**, 145001 (2005)

[2] B. Kurzan *et al.*, Rev. Sci. Instrum. **82**, 103501 (2011)

[3] V. Dose *et al.*, Nucl. Fus. **41**, 1671 (2001)

# A forward model for beam emission spectroscopy at ASDEX Upgrade

R. Dux, B. Geiger, R. M. McDermott, E. Viezzer and the ASDEX Upgrade team
*Max-Planck-Institut für Plasmaphysik, Garching, Germany*

The ASDEX Upgrade tokamak is equipped with a new charge exchange recombination spectroscopy (CXRS) system, which observes the neutral beam #8 on a dense grid of radially distributed lines-of-sight. The lines-of-sight have a toroidal orientation and measurements are made across the plasma centre up to the high field side. A dedicated spectrometer has been installed to monitor the $D_\alpha$ emission of the beam atoms, which can deliver an almost direct measure of the neutral beam density [1]. Since the uncertainty of the neutral beam density from beam attenuation calculations increases with the path length of the beam in the plasma, a direct measure is especially important for the measurements on the high field side.

For the above application, a model of the beam emission spectrum has been developed in order to simplify the fit of the complicated spectral feature, where each beam species (full, half and third energy component) contributes with a $D_\alpha$ multiplet consisting of 9 spectral lines. In addition, the spectrum is overlapped by the $D_\alpha$ radiation of the beam halo and a CII multiplet. The model was compared with first measurements.

**Beam Emission Spectrum**

The standard value for the acceleration voltage of beam 8 is 93 keV for deuterium operation. The atoms with 93 keV energy carry about 63% of the beam power, while the atoms at half and third energy contribute 28% and 9% respectively. At these energies, the line splitting of the $D_\alpha$ spectrum in the magnetic field of the tokamak is determined by the Stark effect due to the electric field $\vec{F} = \vec{v} \times \vec{B}$ in the rest frame of the fast atoms (motional Stark effect, MSE). It produces a symmetric multiplet of 15 lines, where the very weak 6 outermost lines can be neglected. The line splitting is proportional to the electric field $F$, i.e. the velocity v of the atom, the magnetic field strength $B$ and the angle $\varphi$ between $\vec{v}$ and $\vec{B}$: $\Delta\lambda_{\mathrm{MSE}}$[nm]= $2.76\times10^{-8}\Delta k F$[V/m] with $\Delta k$=-4,-3,...,4 and $F = \mathrm{v}B\sin\varphi$. The line strength varies with the angle $\psi$ between $\vec{F}$ and the direction $\vec{k}$ of the emitted radiation: for the 3 innermost $\sigma$-lines $L_\sigma = L_{\sigma,90}(1+\cos^2\psi)$ and for the remaining $\pi$-lines $L_\pi = L_{\pi,90}(1-\cos^2\psi)$. For $\psi$=90$^o$ and equal population of all upper states, the line ratios $L_{\sigma,0}$:$L_{\sigma,\pm1}$:$L_{\pi,\pm2}$:$L_{\pi,\pm3}$:$L_{\pi,\pm4}$ are 5490:1936:729:2304:1681 [2]. However in tokamak plasmas, deviations from these line ratios have been found [3] and were modelled to be caused by a non-statistical population of the sub-levels of the upper $n=3$ states [4]. When denoting the upper states with $|nkm\rangle$, i.e. with the electric quantum number $k$ and the magnetic quantum number $m$ for the angular momentum about the direction of $\vec{F}$, the modelling found equal population of states with same values of $|k|$ and $|m|$. Thus, there are three independent occupation numbers $b_{|k|,|m|}$, which describe the non-statistical population, e.g. $b_{0,2}$, $b_{1,1}$ and $b_{2,0}$, while the forth number $b_{0,0}$ follows from the normalisation condition: $n^2 = b_{0,0} + 4b_{1,1} + 2(b_{2,0} + b_{0,2})$. The above line ratios are then (882$b_{0,0}$+4608$b_{0,2}$):1936 $b_{1,1}$:729 $b_{0,0}$:2304 $b_{1,1}$:1681 $b_{2,0}$. For our measurements on the high-field side, the $D_\alpha$ multiplet is very well resolved and first fits (see below) give typical values of $b_{0,0}$=1.05, $b_{0,2}$=0.84, $b_{1,1}$=1.00, and $b_{2,0}$=1.13, which leads to a suppression of the central $\sigma$0-line and an increase of the outer $\pi$4-lines. In the model function, the spectra of all beam species are calculated with the same $b$-values, which presently are fit parameters but might be fixed later. For the atoms at 31 keV, the inclusion of the Paschen-Back effect starts to produce deviations from the above simple $\psi$-dependence. For the lines-of-sight on the high-field side, the $\sigma$-lines are reduced by up to about 5% and the $\pi$-lines are increased by up to about 7%. The line shift is still well represented by the pure MSE-formula. Finally, the line shift due to the Doppler effect has to be taken into account which depends on the angle $\alpha$ between $\vec{v}$ and $\vec{k}$: $\Delta\lambda_{\mathrm{Dop}} = -\lambda_0 v \cos\alpha / c$.

For the calculation of the spectrum, the velocity distribution $F(\vec{v},\vec{x})$ of the beam atoms for all points along the line-of-sight $\vec{x} = \vec{x}_0 + l\vec{k}$ is needed. It is usually calculated with Monte Carlo models. Here, we use the same basic assumptions as in the Monte Carlo models but replace the individual holes (774 for the beams at ASDEX Upgrade) of the extraction grid by a continuously emitting surface, which leads to rather simple analytic expressions. Each point on a grid emits the same number of atoms per second. The mean starting velocity is directed to a horizontal and vertical focus and the velocity direction has an angular spread, which is described by a Gaussian distribution. Thus, the velocity distribution function just depends on the orientation $\vec{u} = \vec{v}/v$ of the velocity while the absolute value is fixed. Furthermore, it is the product of two distributions functions in the two directions perpendicular to the beam direction, which we denote here by $x$ and $y$, i.e. $F(\vec{v},\vec{x}) = F(u_x,u_y,\vec{x}) = F_x(u_x,x,z)F_y(u_y,y,z)$. For a certain transition $n \longrightarrow m$ of the MSE multiplet, the line shape function $S_{n\rightarrow m}(\lambda)$ at wavelength $\lambda$ is given by an integration over the LOS and the velocity orientations.

$$S_{n\rightarrow m}(\lambda) = \int_{-\infty}^{\infty} dl \int_{-\infty}^{\infty} du_x \int_{-\infty}^{\infty} du_y F(u_x,u_y,l) w_{n\rightarrow m}(u_x,u_y,l)\,\delta[\lambda - \lambda_{n\rightarrow m}(v,u_x,u_y,l)]$$

Here, $\lambda_{n\rightarrow m}(v,u_x,u_y,l)$ is the wavelength of the $D_\alpha$ line being shifted due to the Doppler effect and the motional Stark effect, $w_{n\rightarrow m}(u_x,u_y,l)$ is a weighting function, which takes variations of the beam attenuation, beam excitation, and the emission into direction $\vec{k}$ into account and $\delta[\lambda - \lambda_{n\rightarrow m}]$ represents the $\delta$-function.

The above integral for the line shape has to be numerically calculated. Fig.1 illustrates the main elements, which contribute to the integral for 2 lines-of-sight. The sightline 'COR-67' (left column) intersects the beam on the high-field side at a major radius $R$=1.22 m and 'COR-25' (right column) at the low-field side at $R$=1.82 m. The upper three boxes of each column show profiles along the sightline coordinate $l$, where we use the distance $\Delta l$ to the point with maximum $\mathrm{D}_\alpha$ emission. The profile of the $\mathrm{D}_\alpha$ emission $\varepsilon$ into all directions is displayed in the upper box. It has been calculated with a collisional radiative model yielding the beam attenuation and the beam excitation along the beam. It is wider for the LOS on the high-field side since the beam is intersected at a smaller angle. The product of $\varepsilon$ with the $\psi$-term from above yields the weight $w$ of the integral. However, the variation of this $\psi$-factor is not so large and the shape of $w$ is dominated by $\varepsilon$. The next lower boxes show the profiles, which determine the line shifts. $\Delta\lambda_{\mathrm{MSE}}$ is proportional to the magnetic field strength $B$, which is normalised to its value at $\Delta l$=0. It does not vary a lot along the intersection region and is about 50% larger on the LOS at the high-field side. All following quantities depend on $\vec{v}$ and only the corresponding mean values of the velocity distribution at this point are plotted. The $\sin\varphi$-term contributing to $\Delta\lambda_{\mathrm{MSE}}$ shows a considerable variation, especially on the high-field side, where it changes by about a factor of 2. The variation of the $\cos\alpha$-term of the Doppler shift is 27% larger on the low-field side LOS since it is more perpendicular to the beam. The boxes on the next line display the average wavelength difference of three lines with respect to the wavelength at the emission maximum ($\Delta l$=0). The central $\sigma$0-line is only affected by the Doppler shift and the part of the LOS, which is closer to the observer ($\Delta l$ <0), is blue shifted compared to $\Delta l$=0, while it is red shifted for $\Delta l$ >0. For the outermost $\pi$-4 line on the blue side of the multiplet, the variation of $\Delta\lambda_{MSE}$ and $\Delta\lambda_{Dop}$ across the intersection region add up and the blue/red-shift with respect to the emission at $\Delta l$=0 increases, especially, for the LOS on the high-field side. For the outermost $\pi$4 line on the red side of the multiplet, the MSE- and Doppler-shift variation have opposite sign. For 'COR-25', the total line shift variation is just somewhat reduced. However, on the high-field side, the MSE-shift is so strong, that the far side of the LOS now contributes to the blue line wing, while at the near side, the average line shift is almost constant with a maximum of $\langle\Delta\lambda\rangle\approx 0$ nm. In the lowest boxes, the line shape function $S_\lambda$ (normalised to area 1) is plotted versus the wavelength difference $\Delta\lambda$ to the average wavelength at the emission maximum. For the low-field side, the shapes are still Gaussian like, while the outermost $\pi$-lines on the high-field side develop asymmetrical non-Gaussian shapes.

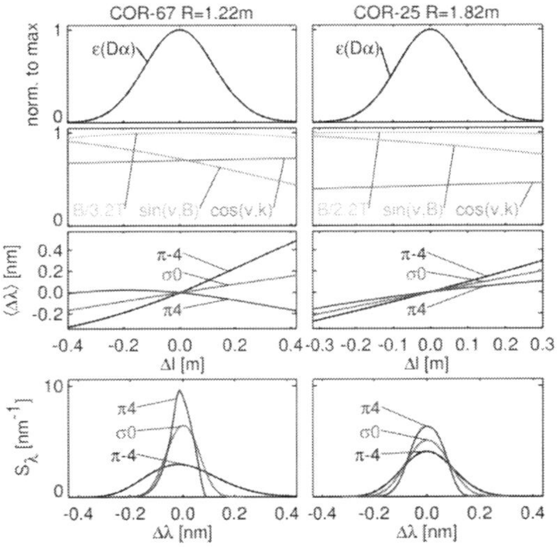


**Fig.1:** *The spectral shape of some lines from the $D_\alpha$ multiplet (lower boxes) and profiles along the sightline, which lead to these shapes (upper three boxes) for 2 lines-of-sight: left on the high-field side and right on the low-field side.*

For Fig.1, the atom energy was set to 93 keV. The shifts scale linearly with the velocity v and the line shapes of the other beam species are just a factor v more narrow since the weighting function $w$ is very weakly depending on v. Thus, the line shapes are only calculated for one beam energy and rescaled with the corresponding beam velocity during the fit. Furthermore, it turns out, that besides the product $B\sin\varphi$ at $\Delta l$=0, all other discharge dependent parameters have only a minor influence on $w$ or the change of $B\sin\varphi$ across the intersection region, such that the line shapes can be calculated before the fit for all LOS and MSE lines on a grid of values of $B\sin\varphi$ at $\Delta l$=0 and saved in a lookup table, which is then used to compute the actual line shape by linear interpolation.

The shape of the MSE spectrum is thus fixed and only the line strength due to the three energy components of the beam are free parameters in the fit function. An integral part of the beam emission spectrum is the broad line due to the beam halo, whose radiance is a few times stronger than the MSE features. The halo emission is unpolarised and well described by a shifted Gaussian. At present, we either use $T_i$ and $v_{rot}$ from the CXRS diagnostic to define the line position and shape or leave it together with the radiance of the halo as free parameters in the fit. The CII doublet $2s^2 3p\,^2P_{3/2,1/2} \rightarrow 2s^2 3s\,^2S_{1/2}$ at 657.8 nm and 658.3 nm is another prominent feature overlapping with the beam emission spectra. We fix the normalised flux surface coordinate of the emission region to $\rho_{pol}$=1.02, i.e. in the scrape-off layer, and calculate for the two intersections of the LOS and the corresponding flux surface the $B$ field and the angle between $\vec{B}$ and $\vec{k}$. This defines the Zeeman splitting and the relative weight of $\sigma$- and $\pi$-lines of the Zeeman multiplet. Each multiplet line is given a Gaussian shape corresponding to a low ion temperature around 10 eV. The spectra of the two intersection points are added up, normalised to an area of 1, and multiplied with the line radiance of the total CII doublet, which is a further fit parameter. Finally, the spectrum is folded with the instrumental line shape and a constant background is the last fit parameter. At present, we neglect the FIDA emission in the model function.

Fig.2 shows two fits of the model to measured spectra on the same LOS as previously used in Fig.1. The spectra are a temporal average from 2-2.1s from #31484, an H-mode plasma with $B_T$=2.5 T, $I_p$=0.8 MA, and line averaged density $n_e$=5.3×10$^{19}$ m$^{-3}$. The data are taken with a preliminary spectrometer setup, where the spectral region does not extend up to the blue side of the unshifted $D_\alpha$ line, such that the halo is not well seen in the data. The upper spectrum is from the high-field side, where the maximum of the MSE emission is at $\rho_{pol,max}$=0.93 while the lower graph is from the low-field side with $\rho_{pol,max}$=0.32. In this low density discharge, the MSE features can nicely be measured on the high-field side (note, the change of the scale of the spectral radiance!). The data are shown by the grey crosses, while the fit is shown as a black line. The model spectrum of the beam species with full, half and third energy is depicted by a blue, green and red line. The CII-multiplet is given by a cyan line and the halo is shown in magenta. The basic behaviour of the line shapes of the MSE feature is well described by the fit. A close inspection of the line widths yields that the model fits the measured widths only within ±10%, i.e. too wide for the $\sigma$0-line, but too narrow for the $\pi$-lines on the red side of the spectrum for the high-field side LOS and again too wide for the same lines at the low-field side LOS. Given the restricted knowledge about the velocity distribution function of the beam and the simplifying beam model, this fit is still considered to describe the spectra reasonably well. When applying the fit to all LOS, the line radiance of $D_\alpha$ of all beam species can be well fitted up to the innermost LOS on the HFS and will be used to study LFS/HFS impurity density asymmetries based on an experimentally verified beam attenuation.

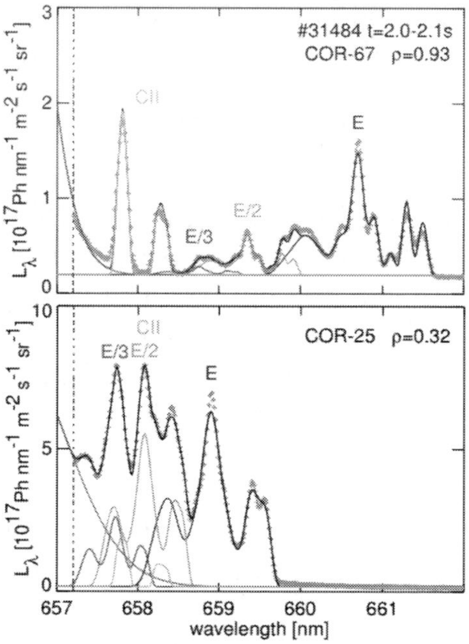


**Fig.2:** *Measured an fitted spectral radiances for the same LOS as in Fig.1.*

Acknowledgement

This work has been carried out within the framework of the EUROfusion Consortium and has received funding from the Euratom research and training programme 2014-2018 under grant agreement No 633053. The views and opinions expressed herein do not necessarily reflect those of the European Commission.

**References**

[1] B. Geiger et al, this proceeding (2015) P1.116.
[2] E. Schrödinger et al, Ann. d. Phys. **80**(1926)468.
[3] E. Delabie et al, Plasm. Phys. Control. Fusion **52**(2010)125008.
[4] O. Marchuk et al, J. Phys. B: At. Mol. Opt. Phys. **43** (2010) 011002.

# Automated sawtooth detection with multi-signal analysis

A. Gude, M. Maraschek, P. Eulenberg, V. Igochine, ASDEX Upgrade Team
*Max Planck Institute for Plasma Physics, Boltzmannstr. 2, D-85748 Garching, Germany*

**Introduction** Sawtooth crashes (SCs) are periodic relaxations of the core plasma which occur when the central safety factor, $q_0$, is below 1. They consist of a ramp-phase during which temperatures and particle densities evolve slowly, and a fast crash that basically flattens these quantities inside the mixing radius, $\rho_{mix}$. For peaked pre-crash profiles, the SC decreases the values close to the magnetic axis and increases them just inside $\rho_{mix}$. At the inversion radius, $\rho_{inv}$, the values remain unchanged. SCs decrease the core confinement, but they also help to avoid impurity accumulation in the hot plasma centre, and to get rid of helium ash in a future fusion reactor. Large sawtooth crashes can trigger neoclassical tearing modes (NTMs), which gets more likely for long sawtooth periods. Sawtooth control therefore does not necessarily aim at avoiding sawteeth but at keeping them small and frequent.

**Motivation** Sawtooth control requires reliable SC detection. Suitable diagnostics are ECE for electron temperature, $T_e$, and Soft X-ray (SXR) detectors for radiation. The advantages of ECE are its local measurement and the dependence on just one plasma quantity, but the plasma coverage is often incomplete and ECE signals are prone to failure when $n_e$ is close to/above the cut-off density. The line-integration of SXR signals affects the observed $\rho_{inv}$. Combined with the dependence on various quantities, especially the non-linear dependence on $T_e$, SXR signals are more difficult to interpret than those of ECE. In the presence of tungsten (W) impurities and central heating, SXR profiles are often hollow (local minimum in the centre) because of hollow tungsten density, $n_W$, profiles. A SC then leads to increase of $n_W$ on axis and decrease further out in such cases [1]. With peaked $T_e$ profiles, there are competing crash effects: central SXR emissivity is increased with $n_W$ but decreased with $T_e$, leading to a large variety of SC characteristics. In ITER, with a W divertor and central $\alpha$ particle heat-

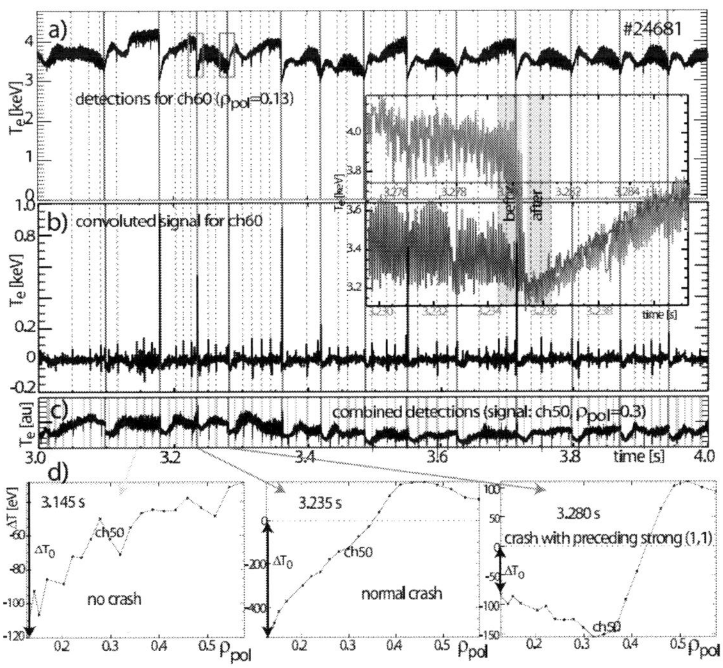


Figure 1: *a) Time trace and edge detections for most central ECE channel. Detected SCs are marked in green, missed ones is red. b) convolution of (a), c) combined detections using all channels shown in (d), d) crash profiles for an ELM, a normal and a mode-preceded crash. Small inset: zoom of boxes in a).*

ing, such conditions can be expected.

Moreover, in ASDEX Upgrade in plasmas with central wave heating, it has been observed, that strong (m=1,n=1) MHD instabilities can affect $T_e$ before the SC [2], leading to strongly reduced drops (figure 1 a)). In order to detect such SCs, the detection threshold has to be low, which will lead to detection of more events that are not SCs (false positives: FPs). Figure 2 b) shows that losses in the most central ECE channel for FPs are of the same order as for mode-preceded crashes, for which the highest losses occur at larger $\rho_{pol}$ (figure 1 d). These circumstances require a suitable discrimination between SCs and other detected events. We focus on SXR signals because of their availability in most plasma conditions and the excellent plasma coverage. The goal is reliable SC detection for statistical analysis or real-time application.

**SC detection** The necessary quantity that a sawtooth detection must deliver with sufficient precision is the crash time point, $t_{SC}$. Once this is known, other SC characteristics can be extracted, e.g., inversion positions or amplitude. Imprecise $t_{SC}$ can lead to wrong extracted values. The acceptable deviation ($\approx 0.3\,\text{ms}$), is slightly larger than the crash duration (mostly $\leq 0.1\,\text{ms}$) to account for precursor effects. The first task is to detect fast changes (edges) in detector signals observing the SC. These edges can be jumps both to smaller or to larger values, depending on the measurement position (inside or outside $\rho_{inv}$) and on the SC type (normal or inverted crash). Edge detection is a task often used in image processing [3]. Our task is simpler in the sense that we only consider edges in one dimension (time). Since the crash appears (almost) simultaneously in all channels, we can use the second dimension (space) to improve the detection reliability. Figure 1 a) shows that detection in the central channel alone misses some SCs, which are found when a larger range of signals is used (fig 1 c). Furthermore, there are other events that cause abrupt changes in the signals, which must be discriminated. Therefore, pure edge detection, sophisticated as it may be, is not sufficient for reliable detection of the whole range of SCs. We have to use crash characteristics for the decision if an event is likely to be a SC or not. Our SC detection is performed in three steps: 1) Edge detection in individual signals, 2) Correlation of detected events in a range of signals, 3) Analysis of crash profile (difference of radial distributions after minus before the crash) for a wide signal range to discriminate FPs.

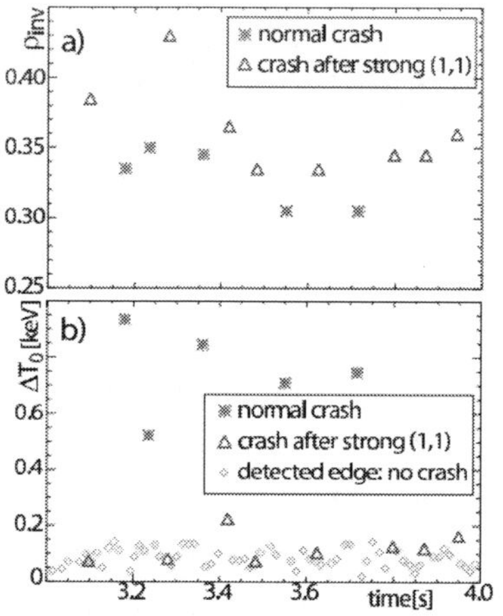


Figure 2: *a) $\rho_{inv}$ of crashes of figure 1, b) $T_0$ for all detected edges of figure 1 c).*

**Edge detection** The edge detection consists of two parts. First the signal is convoluted with a suitable kernel function. Following [3], we chose the first derivative of a Gaussian with width $\sigma_k$, (cut off at $\pm 3.5\sigma_k$). This transfers edges in the signal to peaks in the convoluted signal, (see figure 1 a),b)). The peak heights are equal to the edge heights in the original signal for a normalized kernel. Then the decision has to be made, which of the local extrema in the convoluted signals are due to edges and which are due to remaining noise. A threshold for the peak height has to be defined. This must be self-adapting to the discharge conditions for automated detection. One can use the noise level of the original signal or of the convoluted one for this purpose. The estimation of the noise level from the signal level itself is error-prone, because MHD modes and other events, including the crash precursor, a (1,1) mode, itself, which have to be regarded as noise for SC detection, are not caught.

We present a self-adaptive threshold for peak discrimination using the kurtosis, a measure for the peakedness of a statistical distribution $X = x_1, ..., x_i, ..., x_n$, defined as: $\kappa(x) = 1/n \sum((x_i - \mu)/\sigma_x)^4$, where $\mu$ is the mean value of $X$ and $\sigma_x$ the square root of the variance of $X$. It is calculated for slices of the convoluted signal, called kurtosis windows. For kurtosis windows with only Gaussian noise, $\kappa = 3$. For $X$ with many data close to $\mu$ and also far away from $\mu$ (e.g., a signal with a large peak and some noise in the remaining kurtosis window), $\kappa \gg 3$. Since $\kappa$ is invariant to changed order of data points, the position of the peak has to be determined beforehand. We do this by searching for the absolute extremum in each kurtosis window and shift the window in time such that the extremum is in its middle.

**Edge detection parameters** The kurtosis window has to be larger than the peak width, which is determined by the kernel width. Otherwise there is no noise contribution and $\kappa < 3$. Analysis of the ratio of kurtosis over edge height in an otherwise constant (but noisy) signal reveals that reliable edge detection is possible only for kurtosis windows that are at least 8 to 10 times longer than the kernel. The kernel width must be large enough for effective noise filtering. On the other hand, it must be small in order to detect the crash but not the recovery phase. The latter can be much shorter than the sawtooth period, but in ASDEX Upgrade it is not shorter than 2 ms. We chose a full kernel width of 1.5 ms, corresponding to $\sigma_k \approx 0.214$ ms. For NBI heated discharges with plasma rotation frequencies around 10 kHz this is long enough to smooth precursor oscillations. This choice requires a kurtosis window of at least 12 ms ($\pm 6$ ms around the peak centre), allowing to detect SCs with minimum distance $\approx 6.5$ ms. The kurtosis threshold, $\kappa_{min}$, has to reject most noise peaks. Because of statistical scatter and other perturbations, it must be significantly larger than 3. We chose $\kappa_{min} = 4.5$, which is rather low and results in a few more false detections in this first step, but we will discriminate those in the last step.

**Correlation of detected events** The central time points of all kurtosis windows for which the threshold is exceeded, are analysed for all detection channels together. Events that are less than 0.3 ms apart are combined to one event. Events detected in only one signal are ignored. The average over all contributing time points, weighted with the peak height, gives the combined $t_{SC}$. For a larger maximum time distance, detections around the crash can lead to imprecise $t_{SC}$.

**Event discrimination** Last, we have to decide whether a detected edge results from a SC. Typical assumptions are that a SC reduces $T_e$ and radiation in the centre and that $\rho_{inv}$ is close to the $q = 1$ surface and therefore does not change rapidly. Both criteria are no longer applicable in the presence of tungsten impurities and central heating: A SC can also lead to positive edges in central channels, and figure 2 a) shows that $\rho_{inv}$ can vary strongly from crash to crash. For SXR, the variation of $\rho_{inv}$ is even stronger and more common.

In any case, a SC causes redistribution. Therefore, we analyse the crash profile, which is represented by the convoluted signal values $(c)$ at $t_{SC}$ $(p := c(t_{SC}))$. We chose a profile range inside $\rho_m := \rho_{pol,max} = 0.7$, in order to include the full mixing region. The crash profiles, originally only defined at the measurement positions, are interpolated to an equidistant grid in $\rho$.

The first parameter represents the significance of a crash event: $A_{norm} = A_{crash}/A_{rad}$ with

$A_{crash} = \int_{-\rho_m}^{\rho_m} |p(\rho)| d\rho$ and $A_{rad} = \int_{-\rho_m}^{\rho_m} s(t_{SC}, \rho) d\rho$ ($s$: original signal). $A_{norm}$ might also serve as SC amplitude, however, it depends on the precrash profile shape, and a proper quantification is still to be studied. We found values as low as 2 % for inverted crashes in SXR. The second parameter is determined by the redistribution: $A_{net} = |\int_{-\rho_m}^{\rho_m} p(\rho) d\rho| / A_{crash}$. Redistribution means $A_{net} < 1$. For the case in figure 1, this criterion is sufficient to reject all FPs but accept all mode-preceded crashes. Another type of FPs arises from slow (1,1) mode oscillations with large growth rate. Since the (1,1) mode has opposite phase on different sides of the magnetic axis, crash profiles are strongly asymmetric for mode induced detections. The third parameter, therefore quantifies the profile asymmetry: $p_{asym} = \sqrt{\int_0^{\rho_m} (p(\rho) - p(-\rho))^2 d\rho} / A_{crash}$. Mode detections are rejected by the requirement $p_{asym} \leq 1$. A drawback is that also SCs that are preceded by a slow (1,1) mode can be rejected, depending on the mode phase at $t_{SC}$.

**Results and Outlook** Figure 3 shows an example of the SC detection with inverted crashes and strong variations. In the 37 discharge seconds studied in detail so far, using only one of the SXR cameras, we achieved very good rejection of FPs (1091) and acceptance of SCs, including inverted (1195), with only few accepted FPs (34, most of which have $A_{norm} < 3\%$ and are much smaller than the surrounding SCs) and rejected SCs (15). Crashes missed in the first step are not yet counted but scarce, and events that are difficult to decide (e.g., a fishbone that causes SC-like losses) are ignored. Note that despite a very broad range of heating and plasma parameters, we did not adjust any of our chosen parameters. These are: kernel length: 1.5 ms, kurtosis window: 12 ms, kurtosis threshold: 4.5, profile width: $\rho_{pol} \leq 0.7$, and for sawtooth crash acceptance: $A_{norm} \geq 2\%$, $A_{net} \leq 0.9$ and $p_{asym} \leq 1$. Rejected SCs so far were all preceded by a slow (1,1) mode, which in ASDEX Upgrade only occurs with low NBI heating power. Since the rejection depends on the mode phase, such crashes are accepted from SXR cameras with different viewing angles. Combining 2 or more cameras will therefore be a next step. Possible further steps are profile parameter refinement, replacing fixed acceptance thresholds by probability variation, preprocessing of signals, and an extended statistical investigation of detection reliability, including missed detections. Real-time implementation and application on tomographic reconstructions are foreseen for the future.

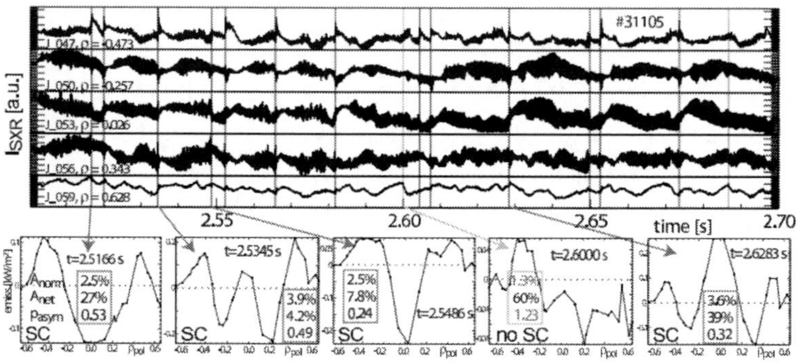


Figure 3: *Some SXR time traces with detected SCs (green), rejected FPs (yellow) and 1 missed SC (red, too close to next SC); Below: some crash profiles with profile parameters.*

[1] A. GUDE et al., *Proc. 37th EPS Conf. Plasma Physics, Europhys. Conf. Abstr.*, **34A** P4.124,2010
[2] M. SERTOLI et al., Plasma Physics and Controlled Fusion **57** 07500 (2015)
[3] J. CANNY, *IEEE Trans. on Pattern Analysis and Machine Intelligence*, **PAMI-8**, NO. 6, Nov. 1986

This work has been carried out within the framework of the EUROfusion Consortium and has received funding from the Euratom research and training programme 2014-2018 under grant agreement No 633053. The views and opinions expressed herein do not necessarily reflect those of the European Commission.

# Study of suppressed tearing modes seeded with non-axsisymmetric magnetic perturbation fields at the ASDEX Upgrade tokamak

S. Fietz[1], R. Coelho[2], I. Classen[3], M. Maraschek[1], W. Suttrop[1], H. Zohm[1],
the EUROfusion MST1 Team[4] and the ASDEX Upgrade Team

[1] *Max-Planck-Institut für Plasmaphysik, Garching, Deutschland*
[2] *Instituto de Plasmas e Fusão Nuclear, IST, Lisboa, Portugal*
[3] *FOM Institute DIFFER, Nieuwegein, The Netherlands*
[4] *See http://www.euro-fusionscipub.org/mst1*

## 1. Introduction

Non-axisymmetric magnetic perturbation (MP) fields arise in every tokamak e.g. due to misalignments in the positioning of the external field coils. Many tokamaks are equipped with an additional set of coils, which produce a 3D MP field, which is mainly used to either compensate for the intrinsic error field or to mitigate ELMs.

These perturbation fields give rise to a global plasma response. Depending on plasma parameters such as rotation and resistivity, they can either deform the plasma, mainly at the boundary, or lead to the formation of magnetic islands (mode penetration) by reconnection, and ultimately to a slowing down, locking to the vessel wall and the occurrence of a disruption. This can be avoided due to the fact that magnetic islands normally rotate together with the plasma. This hampers magnetic islands to lock and screens the effect of the external MP field. However beside driving magnetic reconnection, MP fields additionally exert torques to the plasma, which slow down the plasma rotation. This rotation damping again increases the sensitivity to MP field induced effects and favours the locking of islands.

Several studies exist where the formation of islands and rotation damping due to MP fields are investigated and modelled. These modelling results can explain experimental phenomena, but no direct quantitative comparison with experimental data has been done. In this paper dedicated experiments are presented which have been performed in order to investigate the evolution of the rotation effects during the mode penetration process in detail. In addition a direct modelling of these experimental data using a non-linear error-field penetration model [1] of Fitzpatrick is presented. This model is developed for low beta, circular plasmas, and describes the influence of MP fields on the plasma response, including magnetic island formation and induced torques. The study has been performed at the ASDEX Upgrade (AUG) tokamak in low collisionality ($\nu^*_{\mathrm{core}} \approx 0.01$) L-mode plasmas. AUG is equipped with a set of 16 in-vessel saddle coils [2]. These B-coils enable the generation of resonant and non-resonant, static and rotating MP fields.

## 2. Basic theory

The mode penetration process is mainly dominated by resonant MP field effects. Since the non-resonant effects, like the NTV, are assumed to be small at AUG only the resonant contributions are considered in the following calculation.

In a conducting plasma the resonant MP field components induce helical shielding currents at the corresponding resonant surfaces ($r_s$). Depending on the phase of the shielding current, it can influence the plasma or a magnetic island in two ways. The sine component of this current leads to an electromagnetic torque in the presence of a current at the resonant surface, which influences the rotation frequency $\omega$. Secondly the cosine component of the current modifies the island stability and hence the evolution of the island width $W$. This can be expressed in two equations based on a model derived by Fitzptarick [1].

$$\frac{dW}{dt} \approx 2m\left(\frac{W_{\mathrm{vac}}}{W}\right)^2 \cos(\Delta\phi) - \Delta' + \Delta_{\mathrm{pol}} \tag{1}$$

$$\frac{d\omega}{dt} \approx (W\,W_{\mathrm{vac}})^2 \sin(\Delta\phi)\,\delta(r-r_s) + T_{\mathrm{intrinsic}}(r) + T_{\mathrm{viscous}}(r) \tag{2}$$

The evolution of the full island width $W$ is described by the modified Rutherford equation (MRE) (eq. 1). The equation of motion (EOM) (eq. 2) describes the evolution of the island or plasma frequency $\omega$ at the resonant surface. In this model the amplitude and phase of the resonant MP field is parameterised using the concept of a vacuum island $W_{vac}$, which is the island resulting from the superposition of the equilibrium and the perturbation field. The parameter $\Delta\phi$ is the phase difference of the O-points of the vacuum island and the actual magnetic island. In addition to the resonant MP field effects also the Delta prime term ($r_s\Delta' = -2m$) and the polarisation current effect ($\Delta_{\mathrm{pol}} \propto -1/W^3$) according to [3] are included in the MRE. The EOM is extended (see also [4]) compared to [3] in order to included a radially dependent viscous and intrinsic torque. Both are adjusted to match the initial velocity profile without MPs. Equations 1 and 2 are a coupled system of non-linear differential equations which describes the slowing down, locking and growth of a pre-existing island [5] but also the mode penetration by the MP field.

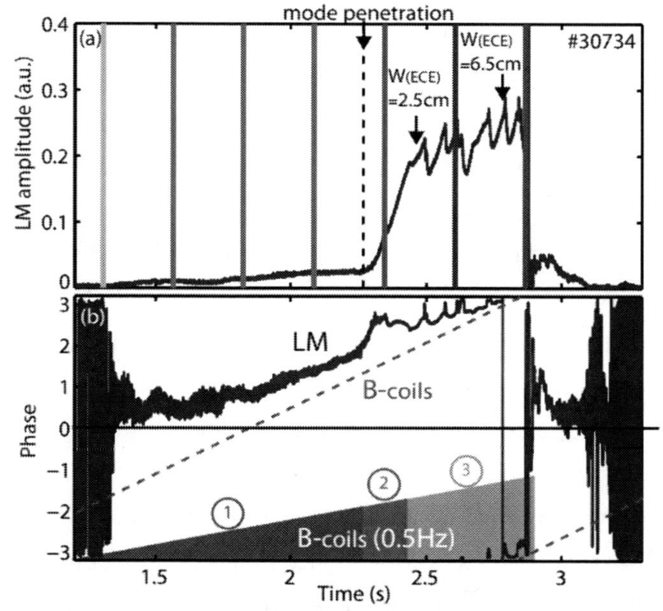


Figure 1: (a) Amplitude and (b) phase of the n=1 plasma response measured with saddle coils at the HFS [6]. In (b) also the programed phase of the rotating B-coils is indicated in blue.

## 3. Experimental results

In AUG mode penetration caused by resonant MP fields is only observed in low density ($< 2.5 \times 10^{19}\ 1/m^3$), low rotation discharges. Therefore the toroidal plasma rotation was only measured within NBI beam blips. In addition the MP field frequency was restricted to 0.5 Hz due a strong shielding of the MP field amplitude at rotation frequencies larger than 1.5 Hz. This is caused by shielding currents induced in the copper structure on which the MP coils are mounted.

By ramping up a slowly rotating MP field three phases can be distinguished (figure 1):
The phase of "linear" plasma response [6] (blue region), where the MP field induces a deformation of the plasma but the amplitude is not high enough to drive full magnetic reconnection. In this phase the plasma response follows the MP field amplitude. In the second phase of mode penetration (red region) the MP field exceeds a threshold which initiates full forced reconnection at the q=2 surface. The related non-linear formation of a (2/1) magnetic island is observed in the magnetic data and in the electron temperature. In the third phase (orange region) a new 3D equilibrium is established which is interrupted by some minor disruption events which flush out part of the plasma energy. In this phase the island is further growing to a size larger than $W_{vac}$ ($\approx$4.6cm). It is also visible that the island phase decouples from the MP field (different slope of phase in (b)). This means that the island rotates slightly slower than the MP field, due to additional torques acting on the island.

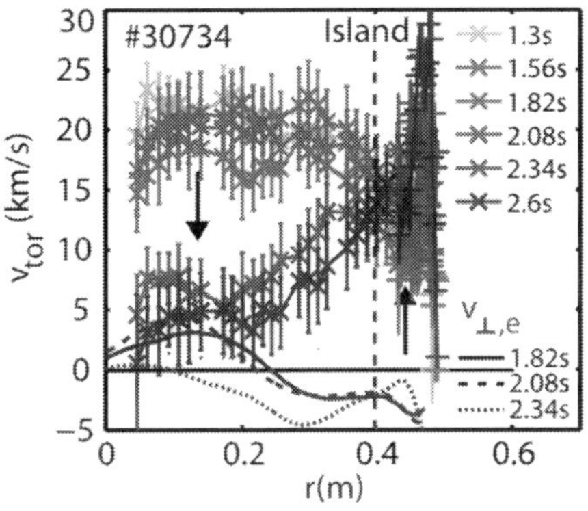


Figure 2: Evolution of the measured toroidal rotation profile and the calculated electron perpendicular velocity.

During the mode penetration process (fig. 2) the core toroidal rotation decreases up to the point of mode penetration. After the mode penetration it drops suddenly and stays almost constant during the locked island phase. In contrast the toroidal rotation at the edge seems to increase like observed at TEXTOR [7]. As theoretically expected the perpendicular electron velocity is around zero. However when a magnetic island of a sufficient size is present also the toroidal plasma rotation at the island surface is expected to be zero. One explanation for this observation is that against expectations the kinetic profiles are not completely flattened inside the island.

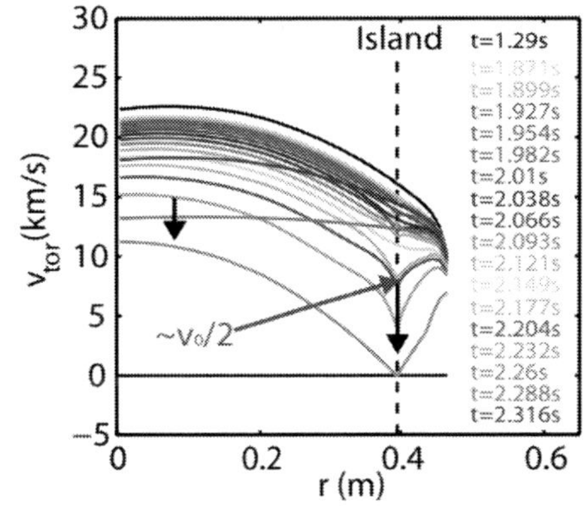


Figure 3: Evolution of the modelled toroidal rotation profile.

In any case these experiments confirm the predicted slow decrease of the plasma rotation towards the time of mode penetration and the small electron perpendicular velocity when an island is formed.

## 4. Modelling

To model the mode penetration equations (1) and (2) have to be solved simultaneously. Since this is not possible for $W \to 0$ the following substitution has to be made: $W^3 \to Y$ and $(1/W^2)\, dW/dt \to dY/dt$ [3]. In the modelling all quantities are taken from the experiment to be able to make a qualitative but also quantitative comparison. The effects at different resonant surfaces are treated by modelling the radial rotation profile and are coupled via the radially dependent viscous torque in the EOM, whereas the MRE is solved for each surface (q=1,2,3) individually.

Taking the experimental vacuum values for $W_{vac}$ (7, 5 and 3cm at q=1,2,3) results in a slowing down of the plasma rotation (fig. 4 (blue)). Small islands are induced at the three surfaces but the overall influence of the MP field on the island stability and plasma rotation is too small to achieve mode penetration. To achieve mode penetration and a concomitant stronger slowing down of the toroidal plasma rotation, $W_{vac}$ at q=2 had to be increase by 40 % (fig. 4 (black)) to 7cm. With this the theoretically expected time variation of the $\Delta\Phi$ and $W$ before the mode penetration ((c) and (d)) and the later abrupt growth of a (2/1) island are well reproduced. In the modelling the (2/1) island grows to maximum $W_{vac}$ due to the choice of $\Delta'$ and $\Delta_{pol}$ and the missing bootstrap drive in the model. Only small islands with $W \ll W_{vac}$ are induced at q=1 and 3 which decay as soon as the MP field is switched off. Also the large island at q=2 decays without MP field due to the missing bootstrap drive. The global toroidal plasma rotation (fig. 3) decreases strongly and stops at the q=2 surface. This is in contradiction to the experiment where only $v_{\perp,e}$ is around zero but $v_{tor}$ remains around 10 km/s.

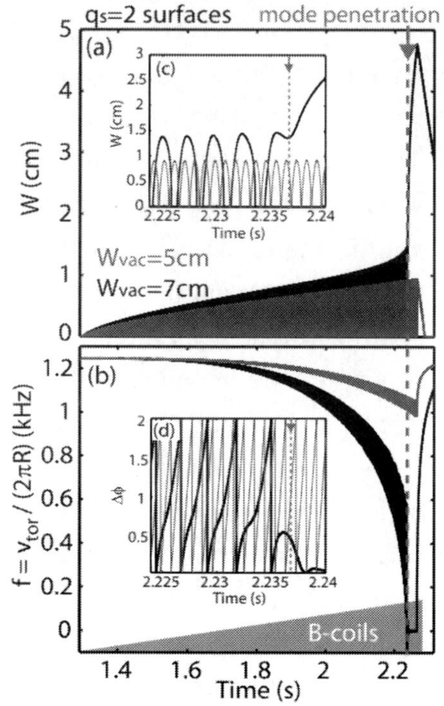


Figure 4: Modelled (a) size and (b) frequency at $q = 2$ for two different $W_{vac}(q = 2)$. (c) Detailed evolution of the island width and (d) differential phase.

In both modelled cases the contribution of $\Delta'$ and $\Delta_{pol}$ are negligibly small compared to the influence of the resonant MP field, which means that the plasma is in the Rutherford regime.

## 5. Conclusion and Outlook

Mode penetration is achieved in low density, low rotation L-mode discharges at AUG. Plasma rotation measurements confirm the predicted decrease of plasma rotation before the mode penetration. The mode penetration process with the concomitant slowing down of the plasma rotation can be modelled but higher MP fields are required compared to the experiment. This indicates that effects are missing in the model e.g. the non-resonant braking by NTV, which enhance mainly the effect on the plasma rotation. Therefore additional effects are planned to be implemented in the basic model and also the difference in the evolution of the toroidal rotation in the modelling and the experiment will be investigated in more detail..

**Acknowlegement** This work has been carried out within the framework of the EUROfusion Consortium and has received funding from the Euratom research and training programme 2014-2018 under grant agreement No 633053. The views and opinions expressed herein do not necessarily reflect those of the European Commission. The support from the Helmholtz-Postdoc-Fellowship is gratefully acknowledged.

## References


[1] R. Fitzpatrick, NF **33**, 7 (1993)
[2] W. Suttrop et al., Fus. Eng. Des. **84** (2009)
[3] R. Fitzpatrick, PPCF **54** (2012)
[4] I. Classen, PhD thesis, Eindhoven (2007)
[5] S. Fietz et al., NF **55** (2015)
[6] M. Maraschek, EPS, Helsinki (2013) P4.127
[7] M. De Bock, PhD thesis, Eindhoven (2007)

# Ion heat and momentum transport in the edge pedestal of ASDEX Upgrade

E. Viezzer[1], E. Fable[1], M. Cavedon[1,2], F. M. Laggner[3], C. Angioni[1], M. G. Dunne[1], R. Dux[1], M. Bernert[1], A. Burckhart[1], R. M. McDermott[1], T. Odstrcil[1,2], T. Pütterich[1], F. Ryter[1], E. Wolfrum[1], the ASDEX Upgrade Team and the EUROfusion MST1 Team*

[1] *Max Planck Institute for Plasma Physics, Garching, Germany*

[2] *Physik-Department E28, Technische Universität München, Garching, Germany*

[3] *Institute of Applied Physics, TU Wien, Fusion@ÖAW, Vienna, Austria*

## Introduction

Edge localized modes (ELMs) are magnetohydrodynamic instabilities that occur at the plasma edge of high confinement (H-mode) plasmas. These bursty edge perturbations expel particles and energy from the pedestal region, leading to a transient degradation of the H-mode transport barrier. ELMs are thought to be driven by a complex interplay between the edge pressure gradient $\nabla p$ and current density $j$. The energy exhaust during an ELM (which lasts typically for 1–2 ms) can account for up to 30 % of the total stored energy in the plasma.

In this paper we present the temporal evolution of the ion and electron profiles during the ELM cycle. A comparison of the toroidal momentum and thermal energy losses is shown for a variety of H-mode plasmas with different conditions. Analysis of the ion heat transport at the plasma edge shows that the ion heat conductivity is close to the neoclassical level in high collisionality plasmas, while at lower collisionality the ion heat transport is observed to be a factor of $\sim$2 higher in the pedestal region.

## Profile evolution during an ELM cycle

The unique edge diagnostic suite available at ASDEX Upgrade (AUG) allows us to measure the edge kinetic profiles on a sub-ms to ms time-scale and with a spatial resolution of less than 5 mm, making it ideal to study the recovery of the profiles after an ELM crash. The temporal evolutions of the edge temperature, density and toroidal rotation profiles during the ELM cycle were analyzed in various H-mode discharges with low ELM frequency ($<$130 Hz).The parameter range for the analyzed dataset are toroidal magnetic field on-axis $B_t$ = 2.4–2.5 T, plasma current $I_p$ = 0.8–1.0 MA, triangularity $\delta$ = 0.2–0.34, heating power 4.0–9.5 MW, fuelling level $\Gamma_D$ = 1.6–11.5 $\times$ $10^21$ $e^-$/s and pedestal ion collisionalities (at $\rho_{pol}$ = 0.97) $\nu_i^*$ = 0.2–2.5.

In figure 1 example profiles of the electron temperature ($T_e$), density ($n_e$), ion temperature ($T_i$) and impurity toroidal rotation ($v_{tor}$, usually measured on boron at AUG) for two different pedestal ion collisionalities, $\nu_i^*$=0.32 (black) and $\nu_i^*$=1.45 (red), are shown. The low collisionality discharges were obtained with low D fuelling (1.5–3.0 $\times$ $10^21$ $e^-$/s), high $\delta$ ($\sim$ 0.34) and

*See http://www.euro-fusionscipub.org/eu-im

medium heating power level (9 MW). Only data measured -4.5 ms up to -1.0 ms before the ELM crash are plotted. While the electron density is only slightly lower (~20 %) at low $v_i^*$, the core electron and ion temperatures are roughly 50 % higher. Interestingly, $v_{tor}$ changes sign at the plasma edge ($\rho_{pol} > 0.95$) and is in the counter-current direction at low collisionality (see figure 1(d)). As shown, the H-mode 'dip' feature in the edge toroidal rotation profile [1] is maintained, but is shifted to negative values. The feature also appears to be wider, concurrent with a wider $T_i$ pedestal (see figure 1(c)). Using the ELM synchronization technique, which sorts the measured data with respect to the onset of an ELM, the temporal behaviour of the profiles can be tracked during the entire ELM cycle. In figure 2 the values of $n_e$, $T_e$, $T_i$ and $v_{tor}$ at the radial position $\rho_{pol}$ = 0.97 are plotted against the ELM onset time. Note that for $n_e$ and $T_e$ the measurements evaluated with the integrated data analysis approach [2] with a temporal resolution below 1 ms are used, while $T_i$ and $v_{tor}$ are measured with the standard edge CXRS system with a time resolution of 2.3 ms.

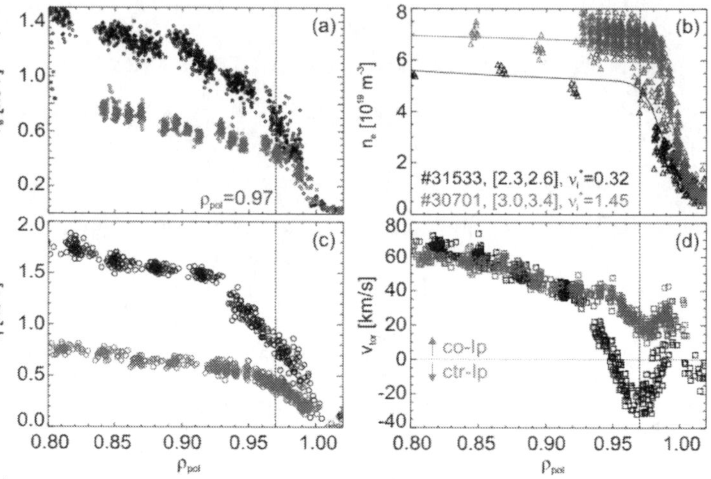


Figure 1: Plasma profiles at low (black) and high (red) $v_i^*$: (a) $T_e$, (b) $n_e$, (c) $T_i$, (d) $v_{tor}$.

Previous ELM cycle studies revealed that the electron density and temperature exhibit different recovery timescales [3]. During the ELM both the $n_e$ and $T_e$ gradient are small, followed by five distinct phases in the $T_e$ recovery, while the $n_e$ recovery shows three phases [3]. After the initial $\nabla T_e$ recovery, the density recovers rapidly and typically reaches its pre-ELM value 4–5 ms after the ELM crash. The full $T_e$ recovery usually evolves on a longer timescale. Figure 2 shows a similar behaviour of $n_e$ and $T_e$ as observed in [3] in both discharges. While $T_e$ and $T_i$ evolve on similar timescales in both discharges, $n_e$ shows an initial fast recovery phase (3–4 ms after the ELM onset). The $T_i$ crash seems to evolve on a slower timescale compared to $T_e$, however, note that the drop could be obscurred by the temporal resolution of the edge CXRS system. At low $v_i^*$, $n_e$ exhibits a second slow recovery phase to reach its pre-ELM value [4]. In this

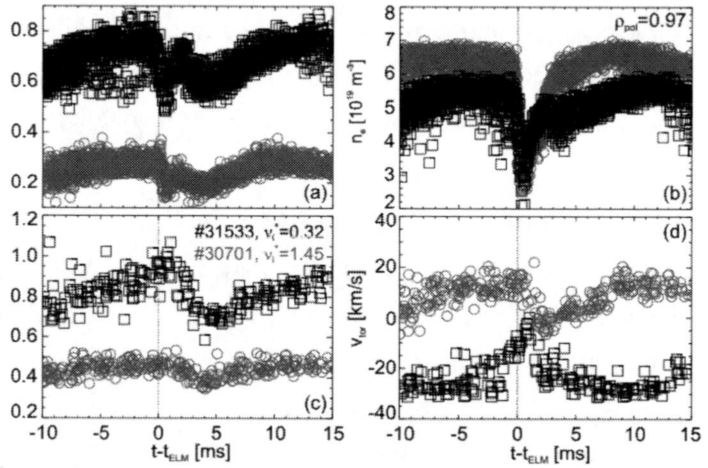


Figure 2: Temporal evolution of (a) $T_e$, (b) $n_e$, (c) $T_i$ and (d) $v_{tor}$ at $\rho_{pol}$=0.97 during the ELM cycle. t-$t_{ELM}$=0 denotes the ELM onset.

case, also the recovery of the edge impurity toroidal rotation to its pre-ELM value appears to be faster than the ion temperature recovery. At both collisionalities, the ELM crash induces a breaking of the edge plasma. Nonlinear modelling is required to study the dominant mechanism that is responsible for a faster restoration of the edge toroidal momentum compared to the heat transport. At higher $\nu_i^*$, the recovery times of $n_e$, $T_e$, $T_i$ and $v_{tor}$ to reach their pre-ELM values are similar and are on the order of ~6 ms [3, 5]. Note that at low $\nu_i^*$ the average ELM energy loss per ELM is larger (by a factor of ~2–3) compared to high $\nu_i^*$ plasmas.

Evaluating the relative losses in $n_e$, $T_e$, $T_i$ and $v_{tor}$ shows that the toroidal rotation exhibits a larger relative drop after the ELM compared to the drop in temperature and density profiles. The toroidal momentum and thermal energy losses are calculated by volume integrating the pre- and post-ELM profiles, i.e. only data where the full exposure time of the frame lies in the time windows [-4.5, -1.0] ms and [+0.5, +3.5] ms are used, respectively. Note that here the volume integral extends from $\rho_{pol}$ = 0.9 to 1.0, i.e. the region which is mostly affected by the ELM. In all analyzed discharges, at both low and high $\nu_i^*$, the ejected relative thermal energy loss is smaller than the relative toroidal momentum loss (see figure 3), in agreement with previous analysis at JET with the carbon wall [6]. A possible explanation for the different behaviour in the ELM induced energy and momentum losses was given by the presence of neutrals in the pedestal and SOL region [7]. Dedicated experiments will be performed in the upcoming AUG campaign to test this hypothesis in an all-metal device.

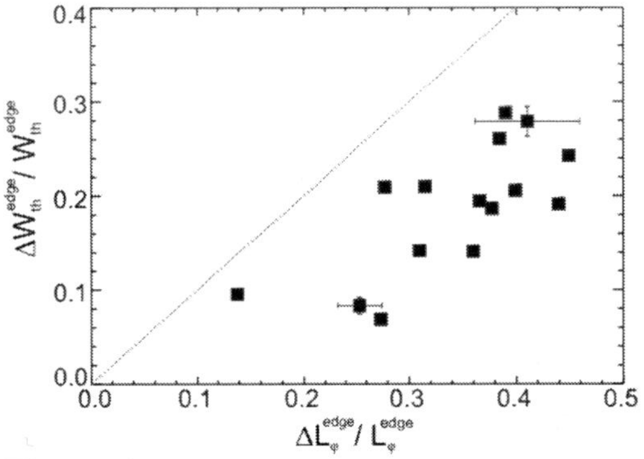


Figure 3: Relative loss in thermal energy plotted versus the relative loss in edge toroidal momentum.

**Ion heat transport analysis**

Extensive studies of the heat transport in the plasma core have been carried out in the past years, however, little focus was directed to the pedestal region due to lack of high-resolution diagnostics. Previous analysis of the temperature gradients measured at AUG were compared to simulations using the 2D fluid code B2.5 and showed that the pedestal ion heat transport coefficient in-between ELMs are consistent with neoclassical theory [8, 9].

In this work, the ion heat transport is studied using power balance analysis with ASTRA [10]. For the analysis dedicated discharges at low collisionality, where the ion and electron heat channels are not too strongly coupled, were selected. For the power balance analysis, the effects due to an ELM are excluded and only data from -5 ms up to -1 ms before the ELM onset are used. The analysis yields an edge ion heat conductivity on the order of $\chi_i^{PB} \sim 0.7$ m$^2$/s (see figure 4(a)) and exhibits similar values as the electron heat conductivity ($\chi_e^{PB}$). Note that the temporal evolution of $T_i$ is taken into account for the evaluation of the heat flux and the $\chi^{PB}$ profiles shown in figure 4 correspond to the temporal average in the analyzed time window. Comparison to the neoclassical profile calculated with NEO [11] (red) shows that at low $\nu_i^*$ the ion heat transport is approximately a factor of ~2 higher than the neoclassical level at the plasma edge. Note that $\chi_i$ could be influenced by a possible residual effect of the ELM crash since at low $\nu_i^*$ the $T_i$ profile is still increasing until the next ELM occurs (see figure 2(c)).

At higher $\nu_i^*$ (see figure 4(b)), the ion heat transport approaches the neoclassical level in the pedestal region, while further inwards it exceeds the neoclassical value by a factor of 4–5. Note that here the uncertainties are larger due to a stronger coupling between ion and electron heat channels. To quantify the differences between low and high $\nu_i^*$ plasmas, this analysis will be extended to the upgraded edge CXRS diagnostic [13] which enables measurements with a time resolution down to 30 $\mu$s, thus allowing for detailed studies of the ion heat transport just after the onset of an ELM.

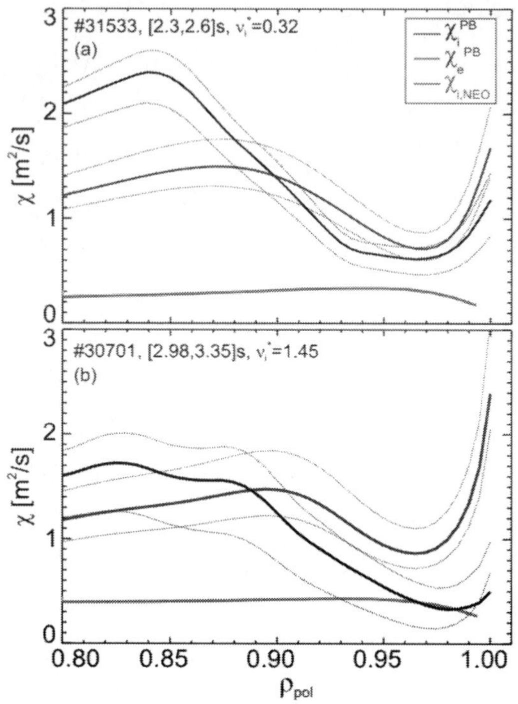


Figure 4: Ion (black) and electron (blue) heat conductivities determined via power balance. The neoclassical prediction is shown in red.

## Conclusions

Analysis of the ELM synchronized profiles shows that the ejected relative toroidal momentum loss due to an ELM is larger than the relative loss in thermal energy. At low $\nu_i^*$ the edge toroidal rotation recovers faster than the ion temperature after an ELM crash, indicating that first the momentum transport level is restored, followed by the heat transport level. Power balance analysis show that the ion heat transport in the pedestal region is near the neoclassical level at high $\nu_i^*$, while at low $\nu_i^*$, $\chi_i$ is a factor of 2 higher compared to the neoclassical value. A possible residual effect of the ELM may have an impact on $\chi_i$ at low $\nu_i^*$. Comparison to other neoclassical codes such as NEOART [12] are the subject of future work.

**Acknowledgments** This work has been carried out within the framework of the EUROfusion Consortium and has received funding from the Euratom research and training programme 2014-2018 under grant agreement No 633053. The views and opinions expressed herein do not necessarily reflect those of the European Commission. The support from the EUROfusion Researcher Fellowship programme under grant number WP14-FRF-IPP/Viezzer is gratefully acknowledged.

## References

[1] T. Pütterich *et al.*, PRL 102 025001 (2009)
[2] R. Fischer *et al.*, FST 58 675 (2010)
[3] A. Burckhart *et al.*, PPCF 52 105010 (2010)
[4] F. M. Laggner *et al.*, P1.147 this conference
[5] E. Viezzer *et al.*, PPCF 53 053005 (2013)
[6] T. W. Versloot *et al.*, PPCF 52 045014 (2010)
[7] T. W. Versloot *et al.*, PPCF 53 065017 (2011)
[8] A. V. Chankin *et al.*, PPCF 48 839 (2006)
[9] E. Wolfrum *et al.*, EPS conference (2007)
[10] G. V. Pereverzev *et al.*, IPP Report (2002)
[11] E. A. Belli *et al.*, PPCF 54 015015 (2012)
[12] A. G. Peeters *et al.*, PoP 7 268 (2000)
[13] M. Cavedon *et al.*, P1.130 this conference

# Characterisation of Pulsations Close to The L-H Transition in AUG

G. Birkenmeier[1,2], M. Cavedon[1,2], G.D. Conway[2], P. Manz[1,2], G. Fuchert[2],
F.M. Laggner[3], T. Happel[2], A. Medvedeva[1,2], V. Nikolaeva[1,4], D. Prisiazhniuk[1,2],
L.M. Shao[2], M. Maraschek[2], T. Pütterich[2], F. Ryter[2],
M. Willensdorfer[2], E. Wolfrum[2], U. Stroth[1,2], H. Zohm[2], and the ASDEX Upgrade Team.

[1] *Physik-Department E28, Technische Universität München, 85748 Garching, Germany*

[2] *Max Planck Institute for Plasma Physics, Boltzmannstr. 2, 85748 Garching, Germany*

[3] *Institute of Applied Physics, TU Wien, Fusion@ÖAW, Wiedner Hauptstr. 8-10, 1040 Vienna, Austria*

[4] *Instituto de Plasmas e Fusão Nuclear, Instituto Superior Técnico, Universidade Técnica de Lisboa, 1049-001 Lisboa, Portugal*

In fusion plasmas, a regular pulsation of the perpendicular flow velocity $\mathbf{u}_\perp$ and the density fluctuations $\tilde{n}$ can occur at the transition from low to high confinement regimes (L-H transition). These pulsations, sometimes referred to as limit-cycle oscillations [1], I-phase [2] or dithering cycles [3], have frequencies in the low kilohertz range and are associated with an improved confinement with respect to the L-mode.

It is discussed whether an interaction between edge turbulence and zonal flows [4, 5] regulate the I-phase pulsation. However, there are also indications that the Reynolds stress drive of zonal flows is too weak in order to explain the measured flows in some cases [6]. Therefore, it is still an open question whether the turbulence-flow interaction is the dominant mechanism to explain the experimental findings or whether there are other mechanisms at work as proposed in theoretical models [7, 8]. The experimental results presented in the following indicate that the I-phase pulsations are accompanied by electromagnetic phenomena which point to physics beyond the flow-turbulence interaction picture.

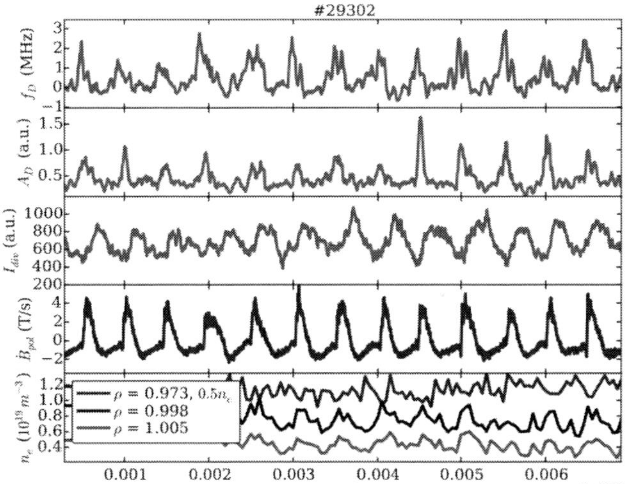


Figure 1: *The appearance of the I-phase pulsation in the Doppler shift $f_D \sim \mathbf{u}_\perp$ and amplitude $A_D \sim \tilde{n}$ of the Doppler reflectometer signal, the divertor shunt current $I_{div}$, the poloidal magnetic field coil $\dot{B}_\theta$, and the electron density $n_e$ at three different radial positions.*

The I-phase is typically studied with Doppler reflectometry [2] since the pulsation of the turbulence amplitude $\tilde{n}$ and the flow velocity $\mathbf{u}_\perp$ during this intermediate phase between L-mode and H-mode can easily be detected by the Doppler shift $f_D \sim u_\perp$ and the amplitude $A_D \sim \tilde{n}$ of the backscattered microwave. An example of an I-phase 6 ms after the L-I transition at medium density is shown in Fig. 1 (first two rows). During the phase of high turbulence level (peaks in $A_D$) the particle transport is increased resulting in a modulation of the divertor shunt current $I_{div}$ (third row in Fig. 1) with the same frequency of about 2 kHz. As already described in Ref. [3], the pulsations are accompanied by a magnetic response visible in signals of magnetic pick up coils measuring $\dot{B}_\theta$ (Fig. 1, second last row). In lower single null configuration, this magnetic response seems to originate from the X-point region and propagates along the high-field side from bottom to top. The outer part of the electron density profiles reconstructed from lithium beam emission spectrocscopy (Li-BES) data every 50 $\mu$s is modulated during an I-phase cycle. The separatrix and scrape-off layer density is increased while the inner parts of the profile ($\rho_{pol} < 0.98$) stay unchanged or are slightly reduced during the turbulence burst of the I-phase (last row in Fig. 1). This indicates that the density gradient is reduced during the high turbulence phase.

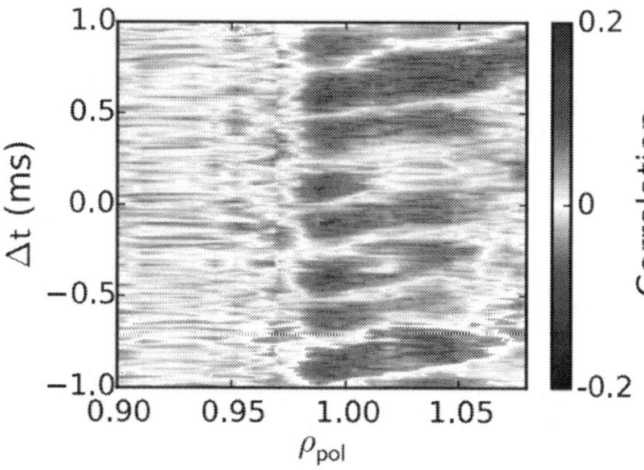


Figure 2: *Correlation between Doppler reflectometry velocity signal and radial channels of Li-BES rawdata ($\sim n_e$).*

The modulation of the electron density profile seems to propagate from inside to outside and is strongest slightly inside the sparatrix. This can be deduced from the slightly oblique stripes in the spatio-temporal cross-correlation of the Doppler velocity signal $f_D$ with different radial channels of the Li-BES system (Fig.2). The highest correlation is found at $\rho_{pol} = 0.99$ (green plus sign) and the time shift of the maximum is negative $\Delta t \approx -100$ $\mu$s indicating that the flow velocity at the probing position rises before the density profile change takes place.

The frequency of the I-phase pulsation in individual discharges decreases linearly with the density and temperature in the edge during the evolution from L-mode to type-I ELMy H-modes indicating a weaker drive or a stronger damping of the pulsation at higher density and temperature. In a series of discharges with varying magnetic field $B$, plasma current $I_p$ and density $n_e$ the I-phase frequency was studied in order to find a determining quantity for the pulsation frequency across various discharges. The model presented in Ref. [7] predicts an I-phase frequency scaling proportional to the neoclassical poloidal damping rate $\nu_{pol}$.

The data set from ASDEX Upgrade, however, does not follow this trend as shown in Fig. 3 (upper left box). Different colors indicate the covered I-phase frequency range in different discharges. The frequency likewise does not clearly scale with the poloidal Alfvén speed $v_A \sim I_p/\sqrt{n}$ as proposed by Solano *et al.* [9] based on JET data (lower right box). Furthermore, it does not scale with single parameters (e.g. $B$, $T_e$, $n_e$) and shows a weak dependence on the normalized Larmor radius $\rho^*$ (lower left box) and the normalized pedestal pressure $\beta_{ped} = 2\mu_0 p_{ped}/B^2$ (upper right box). It is therefore unclear by which quantity the I-phase frequency is determined.

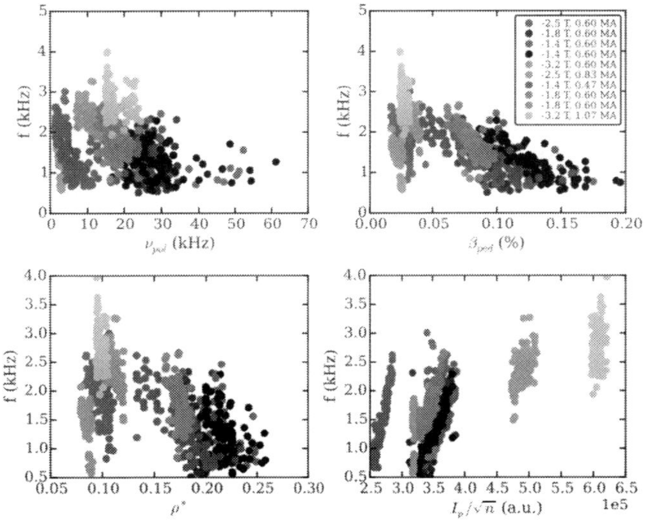


Figure 3: *I-phase frequency $f$ versus poloidal damping rate $\nu_{pol}$, normalized pedestal pressure $\beta_{ped}$, normalized Larmor radius $\rho^*$, and poloidal Alfvén speed $v_A \sim I_p/\sqrt{n}$.*

In the very late phase of the I-phase, the frequency is typically very low due to high density and temperature, and the pulsation becomes very intermittent and spiky. In this phase, each spike is accompanied by a magnetic precursor clearly visible in magnetic pick up coils measuring $\dot{B}_r$ sharing the characteristics of type-III ELM precursors. Its frequency ranges from 50 to 100 Hz as shown in the spectrogram of $\dot{B}_r$ (Fig. 4). Even before the precursor appears in the $\dot{B}_r$-signal, it is already detectable in the phase signal of the reflectometer which probes the (movement of the) cutoff-layer close to the separatrix (Fig. 4, third row).

In the early I-phase, magnetic precursors are hardly distinguishable. In the phase signal of the reflectometer, however, very small but significant quasi-coherent oscillations can be found (Fig. 5, third row). We interpret these findings as follows: each I-phase pulse has a precursor similar to a type-III ELM which is, however, too weak to be detected with the magnetic pick up coils in the early I-phase but always existing and in principle detectable as a modulation of the separatrix density. The existence of precursors in all phases of the pulsation points to a dynamic as it is known from magnetohydrodynamic instabilities which grow in a linear phase and result in a crash after a certain amplitude is exceeded [10].

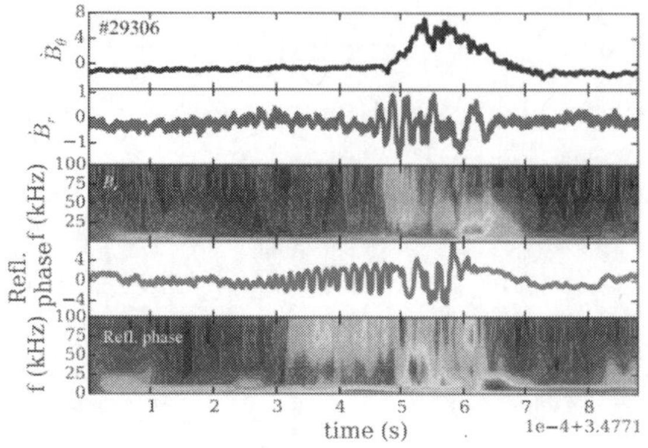


Figure 4: *Late I-phase: precursor modes are clearly visible in the $\dot{B}_r$-signal (red and spectrogram below) and even earlier in the reflectometer phase signal (purple and spectrogram below) prior to each I-phase pulse (black).*

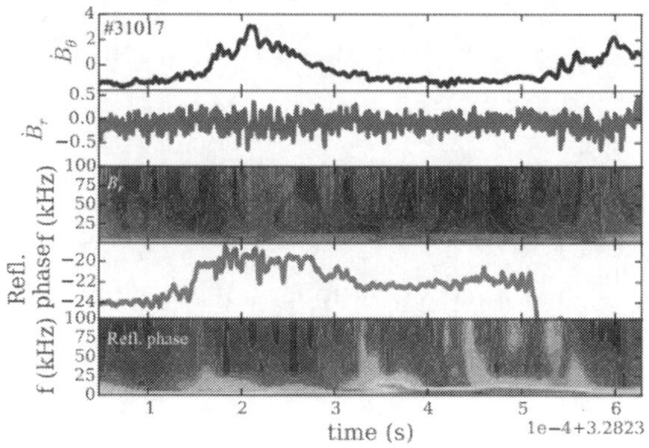


Figure 5: *Early I-phase: There are no precursors in the $\dot{B}_r$-signal but small oscillations in the reflectometer phase indicating a small precursor activity already at the beginning of the I-phase.*

**Acknowledgement**

This work has been carried out within the framework of the EUROfusion Consortium and has received funding from the Euratom research and training programme 2014-2018 under grant agreement No 633053. The views and opinions expressed herein do not necessarily reflect those of the European Commission.

**References**

[1] K. Itoh, S.-I. Itoh, A. Fujisawa, Plasma and Fusion Research, **8**, 1102168 (2013)

[2] G. D. Conway, C. Angioni, F. Ryter, P. Sauter, J. Vicente, and ASDEX Upgrade Team, Phys. Rev. Lett. **106**, 065001 (2011)

[3] H. Zohm *et al.*, Plasma Phys. Control. Fusion **37**, 437-446 (1995)

[4] P. H. Diamond, S.-I. Itoh, K. Itoh, and T. S. Hahm, Plasma Phys. Controll. Fusion **47**, R35 (2005).

[5] L. Schmitz *et al.*, Nucl. Fusion **54**, 073012 (2014)

[6] T. Kobayashi *et al.*, Nucl. Fusion **54**, 073017 (2014)

[7] G.M. Staebler, R.J. Groebner, Plasma Phys. Control. Fusion **57**, 014025 (2015)

[8] H. Zohm *et al.*, Phys. Rev. Lett. **72**, No. 2 (1994)

[9] E. Solano *et al.*, Proceeding of the 40th EPS Conference on Plasma Physics, Espoo (2013)

[10] H. Zohm, Plasma Phys. Control. Fusion **38**, 105-128 (1996)

# Characterisation of a Compact Neutron Spectrometer and unfolded neutron emission spectra in ASDEX Upgrade

G. Tardini[1], R. Fischer[1], F. Gagnon-Moisan[2,3], M. Weiland[1], A. Zimbal[2]
and the ASDEX Upgrade Team

[1] Max-Planck-Institut für Plasmaphysik, Boltzmannstr. 2, 85748 Garching, Germany
[2] Physikalisch-Technische Bundesanstalt, Bundesallee 100, 38116 Braunschweig, Germany
[3] Paul Scherrer Institut, 5232 Villigen PSI, Switzerland

## 1 Introduction

The physics of fast ions is a crucial issue in future burning plasmas like ITER due to their potential damage of first wall components as well as possible confinement deterioration. Fast ion generation and transport have been therefore subject of extensive investigations and diagnostic development in the last decade.
The energy spectrum of fusion neutrons is a footprint of the fast ion distribution, because the kinetic energy of the deuterons affects the energy of the fusion neutron, on top of the energy of 2.452 MeV carried by a neutron from the d-d reaction at rest.
A Compact Neutron Spectrometer (CNS) is installed at the ASDEX Upgrade tokamak [1][2]. It measures Pulse Height Spectra (PHS), resulting from the convolution of the neutron energy spectrum with the instrumental response function. While the PHS can provide evidence for higher energetic tails of the fast ion distribution [3], a quantitative evaluation requires the unfolded Neutron Emission Spectra (NES).
In this paper we discuss the characterisation of the detector and its response function. Moreover, the first unfolded NES are presented. Two phases of the same Neutral Beam Injection (NBI) heated discharge are compared: one with Ion Cyclotron Range of Frequency (ICRF) on top, and one without. A direct measurement of the NES tails and a comparison with Line-Of-Sight (LOS) simulated spectra provide a quantitative evaluation of the fast ion energy distribution.

## 2 Characterisation of the detector

For a given incoming neutron energy, the CNS detects a full energy spectrum of recoil protons, due to the random scattering angle. The scintillator properties and geometry, as well as the detector efficiency, make this response individual from detector to detector. Therefore, the response of the CNS to neutrons for different known energies has to be determined in order to unfold the experimental PHS.
Such responses were measured at the facility of the Physikalisch-Technische Bundesanstalt (PTB), with a white neutron field in the interval 1-17 MeV. As the distance of the detector to the neutron source is fixed, we can select narrow intervals of the neutron energy via their time-of-flight, and measure the associated PHS. A set of PHS is simulated with the NRESP code [4] for the same energies. Such synthetic PHS are then fitted to the experimental ones, assuming an energy dependent Gaussian broadening and using the maximum entropy of the solution as an optimisation criterium [5]. The results are shown in Fig. 1 .

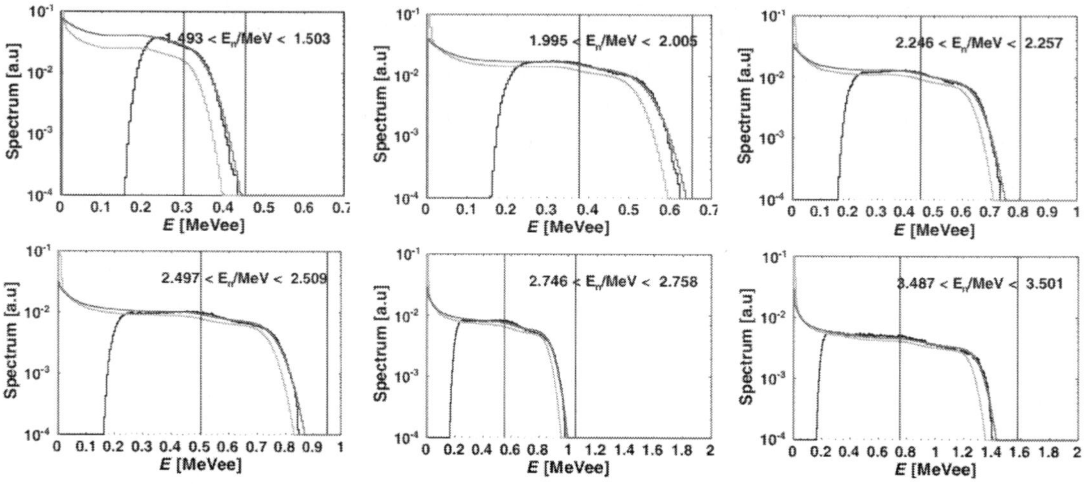


*Figure 1 . Detector response at several energies. Measured (black), simulated (green), optimised (red). The blue vertical lines delimit the region used for the fit.*

The fit in Fig. 1 matches the experimental results accurately over several orders of magnitude, while the statistical noise of the measurements is avoided. This is needed for accurate unfolding, so the fitted response matrix is the one used for deconvolution purposes.

The response for selected energy intervals is compared to that of monoenergetic neutrons, generated with the cyclotron of the PTB facility. A deuterium beam is accelerated towards a deuterium gas target, while the detector measures the neutrons from the D-d fusion. The comparison is shown in Fig. 2 .

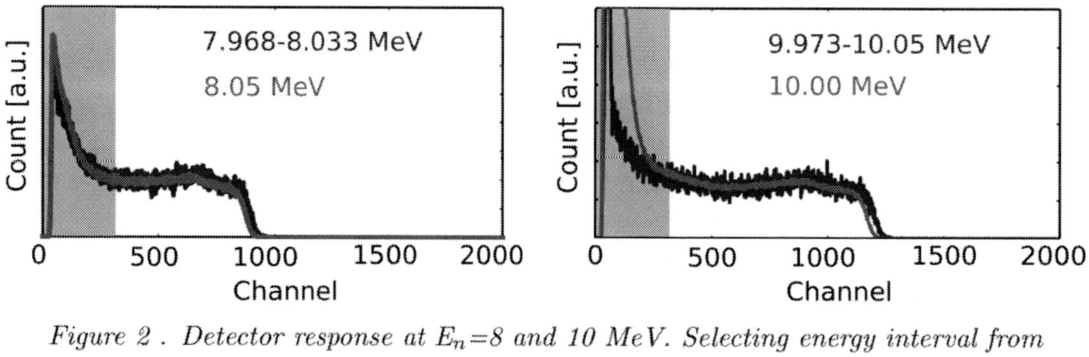


*Figure 2 . Detector response at $E_n$=8 and 10 MeV. Selecting energy interval from time-of-flight measurement (blue), monoenergetic (red).*

The shapes of the responses for 8 and 10 MeV in Fig. 2 agree well with the time-of-flight PHS concerning the slope, while the position of the edge is slightly lower for the monoenergetic neutrons. For low channels, the PHS of monoenergetic neutrons is dominated by other reactions and it should not be considered for the comparison.

## 3 Detection of ICRF 2nd harmonic D-acceleration

Two different PHS have been measured for the H-mode discharge # 29795 in a phase with NBI heating only and a later phase where ICRF was added (see [3]). Now the detector's response function is used to unfold both measured PHS, using a method based on maximum entropy [6]. The outcome is displayed in Fig. 3 (a).

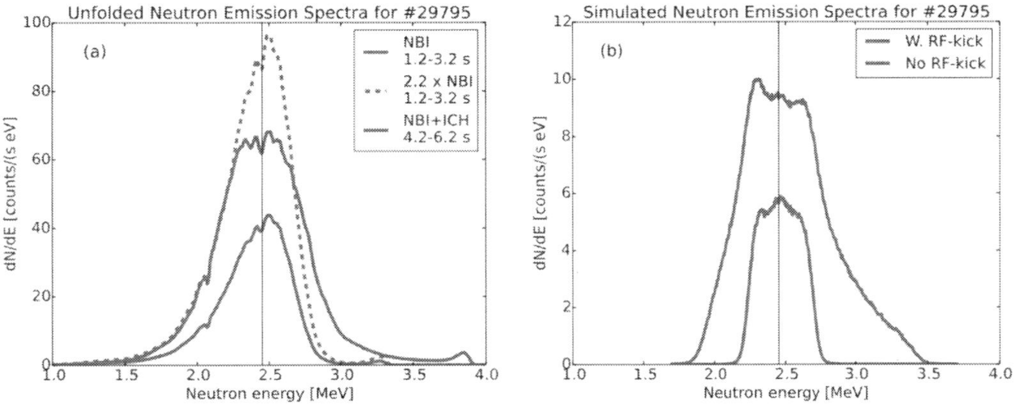


*Figure 3 . NES of discharge # 29795 (a) Unfolding the measured PHS. NBI-only (continous red), NBI-only multiplied by 2.2 for shape comparison (red dashed), NBI+ICRF (green). (b) LoS simulation: without (red) and with the RF-kick operator (green).*

The strong energetic tail in case of NBI+ICRF is clearly visible in Fig. 3 (a), as the green curve extends way beyond the red dashed (NES of the NBI-only phase multiplied by an ad-hoc factor). The maximum energy of the fast ions can be estimated by looking at the maximum $\Delta E_n$ with respect to the reference value of 2.452 MeV (black vertical line). In fact, from simple kinematic considerations one has $\Delta E_n \approx \sqrt{E_{fus} E_{f.i.max}}$, where $E_{fus} =$ 2.452 MeV is the reference d-d neutron energy and $E_{f.i.max}$ the maximum fast ion energy in the considered plasma. Substituting $E_{f.i.max} = 60$ keV (the NBI energy applied in this discharge) one obtains $\Delta E_n \approx 0.4$ MeV, which is observed for the red curve of Fig. 3 (a). A $\Delta E_n = 1$ MeV, exhibited by the green curve in Fig. 3 (a) corresponds to $E_{f.i.max} \approx$ 400 keV.

The effect of the synergy between ICRF and NBI can be simulated [11][3], in particular in the TRANSP code [8] since a RF-kick operator was implemented [9]. In Fig. 4 the NBI+ICRF is simulated with and without RF-kick operator, while keeping the same background plasma as simulation input.

When the RF-kick operator is included, fast deuterons are accelerated up to $\sim 500$ keV (see Fig. 4 (b)). The resulting distribution function is used as input in the GENESIS code [10], which calculates the NES in the proper LOS of the detector. The code prediction for both cases is shown in Fig. 3 (b), with the same colour code. There is a qualitative agreement with the experimental unfolded NES in Fig. 3 (a). However, it appears that the high energy tail of the neutron distribution is overestimated, while the low energy end the tail is less pronounced.

**4 Conclusions**

The CNS installed at ASDEX Upgrade is now fully characterised. The response matrix is obtained by fitting simulated PHS to the measured response at given neutron energies, selected by the neutron time-of-flight. This is accurate and smooth enough to unfold the experimental PHS [5].

The first examples of unfolded NES for ASDEX Upgrade are presented and discussed

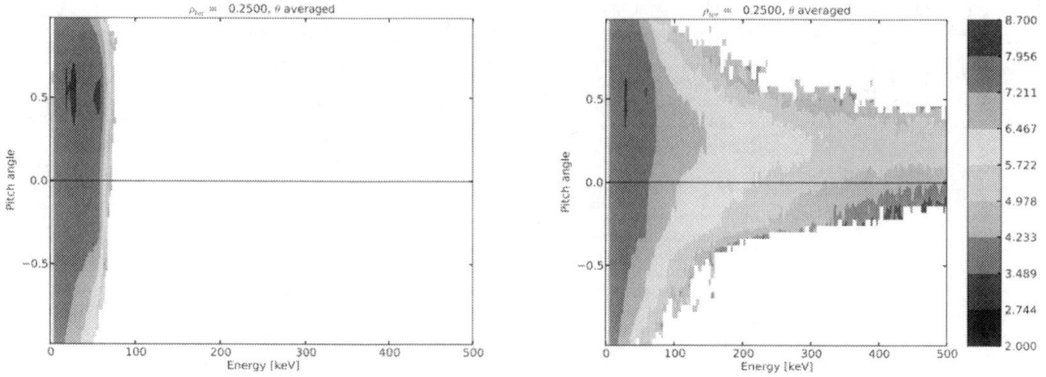


*Figure 4 . Logarithmic TRANSP fast ions distribution functions as a function of energy and pitch angle at $\rho_{tor} = 0.25$. D with higher energy is predicted in presence of ICRF (right plot).*

here, providing clear evidence of fast deuterium acceleration by direct ICRF 2nd harmonic heating. An upper limit of the fast ions present in the tokamak can be derived directly from the unfolded spectrum, amounting roughly to 1 MeV. The experimental NES can be compared to the LOS prediction obtained coupling the fast-ions orbit following code TRANSP and the GENESIS code, calculating fusion reactions along a LOS. The prediction is qualitatively good and the enhancement due to the RF-kick is in quantitative agreement with the unfolded experimental NES. The simulated neutron spectrum has a a slightly different shape, with a stronger tail at the high energy end but less pronounced at the low energy end. A comparison with the prediction of the TORIC6 code, coupled to a Fokker-Planck solver [11], is underway.

Weight functions of the diagnostic have been calculated, in order to combine several fast ions diagnostics in overlapping regions of the veocity space.

**References**

[1] G. Tardini *et al.*, Journal of Instrumentation **7** (2012) C03004

[2] L. Giacomelli *et al.*, Review of Scientific Instrument **82** (2011) 123504

[3] G. Tardini *et al.*, 41th EPS Conference on Plasma Physics, O2.103 (2014)

[4] G. Dietzke and H. Klein, Physikalisch-Technische Bundesanstalt Report PTB-ND-22, Braunschweig, Germany (1982), ISSN 0572-7170

[5] F. Gagnon-Moisan *et al.*, Journal of Instrumentation **7** (2012) C02023

[6] M. Reginatto, A. Zimbal, Rev. Sci. Instruments **79** (2008) 023505 [11] R. Bilato *et al.*, Nuclear Fusion **51** (2011) 103034

[8] A. Pankin, D. McCune, R. Andre *et al*, Comp. Phys. Comm. **159**, No. 3 (2004) 157

[9] B. H. Park *et al.*, Paper JP8.00122, Bull. Am. Phys. Soc. 54, (2012)

[10] M. Nocente *et al.*, Nuclear Fusion **51** (2011) 063011

[11] R. Bilato *et al.*, Nuclear Fusion **51** (2011) 103034

**Acknowlegment:** This work has been carried out within the framework of the EUROfusion Consortium and has received funding from the Euratom research and training programme 2014-2018 under grant agreement No 633053. The views and opinions expressed herein do not necessarily reflect those of the European Commission.

# Controlled fuelling of high density scenarios at ASDEX Upgrade in support of ITER and DEMO

P. T. Lang, L. Casali, Ch. Day[1], M. Dunne, E. Fable, R. Fischer, O. Kardaun, G. Kocsis[2], B. Kurzan, R. McDermott, V. Mertens, A. Mlynek, B. Ploeckl, Ch. J. Rapson, I. Sharov[3], T. Szepesi[2], W. Treutterer, E. Viezzer, H. Zohm, ASDEX Upgrade Team

MPI für Plasmaphysik, EURATOM Association., Boltzmannstr. 2, 85748 Garching, Germany
*1) Karlsruhe Institute of Technology, 76021 Karlsruhe, Germany*
*2) WIGNER RCP RMKI, POB 49, 1525 Budapest, Hungary*
*3) State Polytechnical University, 29 Politechnicheskaya st., 195251 Saint Petersburg, Russia*

## INTRODUCTION

Standard operation in a future fusion reactor will aim to establish a high core density in order to harvest a maximum of output power. In addition, a high separatrix density is essential for achieving the necessary detached divertor scenario [1]. Therefore, in the ITER project and in European DEMO studies [2], core fuelling to planned target densities is envisaged to be performed entirely by injecting pellets formed from frozen fuel and launched from the torus inboard side. On the other hand, the separatrix density will most likely be adjustable by gas puff in these devices, leading to a certain degree of freedom in the control scheme of core and separatrix (edge) density. The development of relevant high-density plasma scenarios and related control strategies capable to cope with pellet actuation became hence an important task within the ASDEX Upgrade (AUG) program [3]. AUG, equipped with a pellet system adapted to efficient inboard fuelling, as well as with a sophisticated discharge control system for flexible real time plasma control and also a versatile set of diagnostics, is very suited for this kind of investigations. Accordingly, a summary of the efforts undertaken will be reported.

## SET UP: ASDEX UPGRADE, PELLET LAUNCHER AND CONTROL SYSTEM

ASDEX Upgrade [3] is a divertor tokamak with all plasma facing components completely covered with tungsten (W). In-vessel saddle coils can produce non-axisymmetric magnetic perturbations, presently 16 B-coils (each 8 upper and lower ones, referred to as Bu- and Bl-coils, respectively) are installed which can create a mainly radial field with toroidal mode numbers up to n = 4 [4]. Investigations reported here where performed operating with the newly installed massive tungsten divertor III configuration.

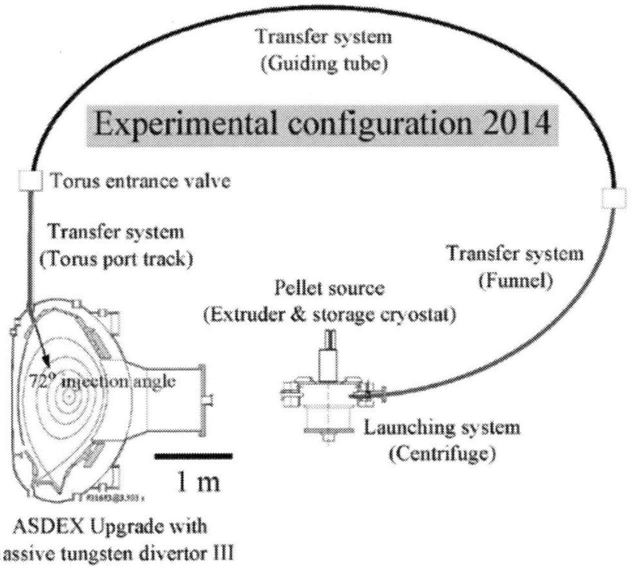


*Figure 1: Set up as operated in the campaign 2014. AUG with an all W cladded wall and now also equipped with the massive tungsten divertor III and the full set of B coils. The revitalized pellet launching system was optimized and commissioned for efficient particle fuelling from the torus inboard at high speed and high particle throughput. As well, admixing different gases to the main fuelling species (mostly deuterium) is now facilitated.*

The pellet system based on a centrifuge accelerator and a looping transfer line has been optimized and commissioned for efficient core particle fuelling. Presently, the system is capable of delivering pellets with a nominal particle content ranging from 1.5 – 3.7 x $10^{20}$ D in the velocity range 240 – 1040 m/s from the magnetic high field side of the torus with repetition rates of up to 70 Hz. Within a given pellet train launched into a discharge both pellet speed and size are fixed. The pellet imposed particle flux applied for fuelling can be adjusted by changing the pellet repetition rate. However, due to the centrifuge principle only rates equal to an integer fraction of the centrifuge revolution frequency can be chosen. Recently, the pellet source was modified now allowing admixing small amounts (up to 5%) of other gas species to the main pellet carrier gas (usually deuterium). Hence, investigations became possible with pellets composed essentially from fuel but with some admixtures [5]. The pellet observation system was also upgraded to include two ultra-fast CMOS cameras and is now capable of fast individual pellet tracking up to 1 Mframe/s.
The discharge control system (DCS) [6] is our tool for fully-automated and controlled execution of the plasma discharges. It supplies sophisticated real-time control methods comprising real-time diagnostic integration, dynamically adaptable multivariable feedback schemes, actuator management and a powerful monitoring and pulse supervision concept based on segment scheduling and exception handling. Therefore, the DCS requires reliable and valid information about the plasma density. Usually measured by a DCN laser interferometer, data can become invalid due to the strong pellet imposed local perturbation. Therefore, a sophisticated performance evaluator electron density module was designed. It computes the density from different DCN and $CO_2$ laser interferometer channels and a specially adapted Bremsstrahlung channel. Results are auto-corrected if possible (e.g. by detecting and eliminating DCN fringe jumps) or set invalid. A replacement strategy for invalidated interferometer densities tries to compute a substitution value from the average of the remaining valid interferometer channels. If no valid interferometers signals remain the value BRT calculated from the Bremsstrahlung (density assumed proportional to square root of signal) are adopted. In order to avoid offset errors, the BRT signal is calibrated against the last trusted density measurement from the interferometer.

**CONDITIONING OF LAUNCHING AND CONTROL SYSTEM**

The newly set up density module first was carefully tested and commissioned to ensure reliable and safe machine operation with the pellet actuator. It turned out helpful to provide the DCS information of an envisaged pellet arrival as "a priory" information for the fringe correction and substitution algorithm and to take eventually according control actuation by e.g. pausing ECR heating during the pellet ablation ("notching technique"). The pellet launching system was optimized with respect to the fuelling performance taking into account the pellet production process [7] and all pellet parameters. The best setting found, applied for essentially all fuelling experiments is at a speed of about 560 m/s for the largest available pellets. Thus, at a maximum rate of 70 Hz a pellet particle flux of about 2.6 x $10^{22}$ D/s could be established.

**FUELLING OF THE NITROGEN SEEDED IMPROVED H-MODE SCENARIO**

In the all-metal wall configuration, considerable energy confinement improvements can be realized in some plasma scenarios by applying N seeding [8]. This is in particular the case for the scenario applied for our recent investigations of the high density regime. Here, pellet fuelling easily achieved core densities far beyond the Greenwald limit $n_{Gw}$ with high density operation resulting in benign ELM behaviour without affecting the energy confinement [9]. Steady state phases could be achieved but only at a moderate value of H98 $\approx$ 0.76. Incorporating N seeding at an appropriate level enhanced this value to about 1 for cases applying sole gas (D and N) puffing reaching densities of about 0.7 x $n_{Gw}$. In the approach

reported here it was aimed to combine pellet fuelling and N seeding in order to establish high density high confinement operation if possible still emerging benign ELM behaviour.
Ultimately, simultaneous real-time control of the edge and the core density will be requested. Here, the purpose was to establish a high central electron density $n_e^0$ beyond $n_{Gw}$ while keeping the separatrix density $n_e^{sep}$ constant in order to prevent edge-induced confinement degradation. For this, pellet injection and conventional gas puffing are employed as actuators; the pellet sequence pre-programmed and the gas puff in feedback control. Standard $n_e^0$ and $n_e^{sep}$ measurements are not compatible with pellet actuation. However, unambiguously correlated substitutes have been identified utilisable for the investigated high density scenario. The divertor neutral gas pressure $n_0^{div}$, measured by a pressure gauge, was found to be directly correlated with $n_e^{sep}$. For $n_e^0$ control, the validated line-averaged plasma density $\overline{n_e}$ as provided by the DCS density module can be employed. By employing these surrogate control parameters, core densities up to 1.9 x $n_{Gw}$ have been established by pellet injection while $n_e^{sep}$ was kept stable by controlled reduction of the initial gas puff rate. An example is displayed in figure 2. There, $n_e^{sep}$ is kept by controlling $n_0^{div}$ via actuation on the gas puff while a steady train of pellets evokes a strong core density rise.

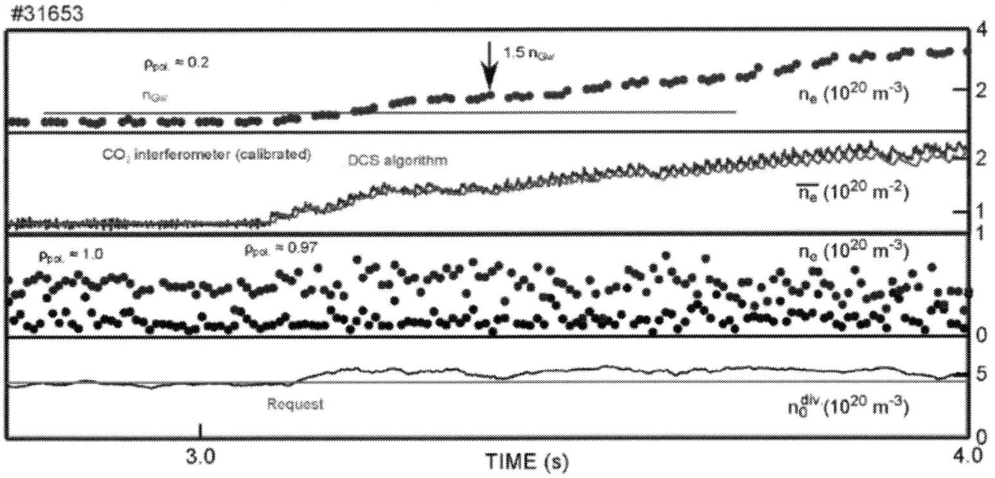


*Figure 2: Controlling the divertor neutral gas pressure keeps the local edge density (close to separatrix at $\rho_{pol} \approx 1.0$ and pedestal top at $\rho_{pol} \approx 0.97$) constant. The increase of the validated line average density correlates well with the evolution of the core density at $\rho_{pol} \approx 0.2$.*

Obviously, high density operation is achieved straightforward with pellet fuelling in the N seeding scenario as well. However, a strong N pump out was observed for peaked density profiles generated by pellet injection. Depletion of N resulted in an ensuing loss of the N induced confinement surplus, ultimately causing a setback of confinement to values typical for unseeded discharges. As a countermeasure, enhanced N puffing was applied during the pellet phase. Such the depletion effect can be mitigated, however too strong puffing resulted in impurity accumulation causing finally a radiative collapse. The best case achieved so far applying simple feed forward N puffing is shown in figure 3. It also shows the discharge from figure 2 with a steady single pellet train and feedback controlled edge density.
During the pellet induced high density phase, ELM amelioration is still observed. Nitrogen puffing during the initial phase results in a significant confinement surplus with respect to the unseeded reference. The N depletion in the high density phase is here partially compensated by an added extra to the reference puffing rate. However, this causes obviously an accumulation of heavy impurities (particularly W, concentration $c_W$ displayed) towards the end of the high density phase finally resulting in a strong confinement reduction due to the increasing radiative loss power. Nevertheless it becomes obvious high density operation is feasible with at least some of the N induced extra confinement still preserved. A first step

consequently envisaged for the campaign 2015/16 is to extend feedback control also on pellet actuator. Aiming to control $n_e^0$ at a reasonable level by accordingly reducing the pellet rate/flux is expect to result in less N depletion and confinement surplus loss. Next, N seeding in feedback control will be implemented to avoid unbearable radiative losses.

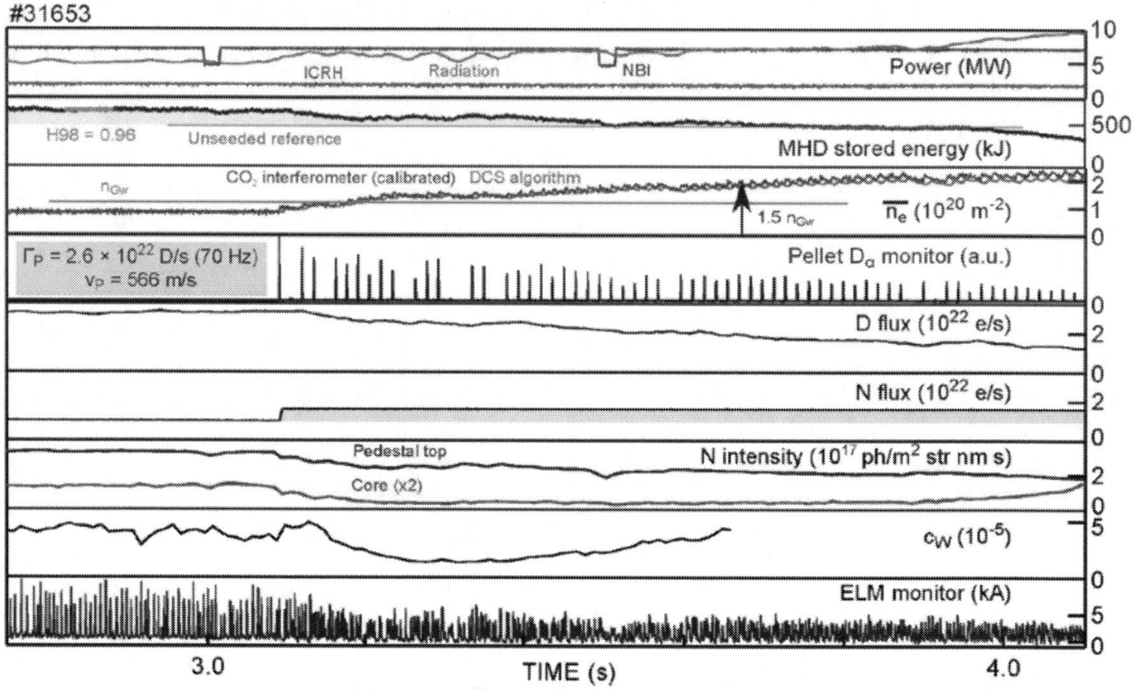


*Figure 3: The N pump out can be mitigated by an increased N puffing rate during the high density phase. Thus, high density operation with benign ELM behavior and still higher confinement compared to the unseeded reference can be achieved.*

## PELLETS WITH NITROGEN ADMIXED

Motivated by the higher fuelling efficiency of the pellets with respect to gas puffing and the prediction N admixed to D ice can enhance the mechanical stability [10], we performed an attempt to supply the N for seeding at least partially by the pellets. Indeed, by forming pellets from a gas mixture containing 1% N, an N contends of 0.8% in the pellets could be detected. As well, an enhancement of the N concentration in the plasma during the pellet phase was confirmed by spectroscopy. However, although delivering of N was found more efficient than by gas puffing, the N flux carried by the pellets was still insufficient to gain significant impact on the confinement. In addition, no positive influence on the mechanical stability of the pellets could be found. Contrary to these expectations, less density increase was observed when injecting N admixed pellets compared a train of pure D pellets. This seems to indicate for higher pellet mass losses during the transfer and hence a reduction of mechanical stability.

**REFERENCES:**

[1] H. Zohm et al., Nucl. Fusion 53, 073019 (2013)
[2] P.T. Lang et al., Fusion Eng. Des., in press (2015)
http://www.sciencedirect.com/science/article/pii/S0920379615002434
[3] U. Stroth et al., Nucl. Fusion **53,** 104003 (2013)
[4] W. A. Suttrop et al., Proc. 40th EPS conference (Espoo), P4.117 (2013)
[5] B. Ploeckl et al., Fusion Eng. Des., in press (2015)
http://www.sciencedirect.com/science/article/pii/S0920379615000241
[6] G. Raupp et al Fusion Eng. Des. **84,** 1575 (2009)
[7] B. Plöckl and P. T. Lang, Rev. Sci. Inst. **84,** 103509 (2013)
[8] J. Schweinzer et al., Nucl. Fusion **51,** 113003 (2011)
[9] P.T. Lang et al., Nucl. Fusion **54,** 083009 (2014)
[10] L.A. Alekseeva et al, Physics of the Solid State **48,** 1513 (2006)

# Non-linear simulations of ELMs in ASDEX Upgrade including diamagnetic drift effects

A. Lessig[1], M. Hölzl[1], F. Orain[1], I. Krebs[1,5], S. Günter[1], E. Franck[4], J. Morales[2], M. Becoulet[2], G. Huysmans[3], M. Dunne[1], ASDEX Upgrade Team[1]

[1] *Max-Planck-Institut für Plasmaphysik, Boltzmannstrasse 2, 85748 Garching, Germany*

[2] *CEA-IRFM, Cadarache, 13108 Saint-Paul-Lez-Durance, France*

[3] *ITER Organization, 13067 Saint-Paul-Lez-Durance, France*

[4] *INRIA Nancy Grand Est, Tonus Team & IRMA, Strasbourg University, France*

[5] *Max-Planck/Princeton Center for Plasma Physics*

## Introduction

Edge-localized modes (ELMs) are relaxation-like oscillatory instabilities at the boundary of H-mode plasmas. They eject particles and energy into the scrape-off layer (SOL) on very short time-scales and are driven both by the steep pressure gradient as well as the high bootstrap current in the pedestal region of an H-mode plasma. While they are a mean to control impurity and particle density, extrapolations of experimental data indicate that in devices like ITER the high transient heat loads on plasma-facing components and in particular the divertor targets are beyond the tolerance level of present-day materials. It is therefore important to study ELMs both theoretically and experimentally in order to reveal their fundamental properties which is necessary for the prediction of ELMs and the design of ELM mitigation systems in ITER and DEMO. We present first ELM simulations for ASDEX Upgrade (AUG) including diamagnetic drift effects and investigate the influence of the diamagnetic terms onto the evolution of the toroidal mode spectrum.

## Model & Simulations

The model used in our simulations are the reduced resistive MHD equations in toroidal geometry [1] supplemented with ideal wall boundary conditions except for the divertor targets where Bohm boundary conditions are applied. They are solved using the non-linear MHD code JOREK [2, 3]. It uses a flux surface aligned isoparametric Bezier finite element grid in the poloidal plane and a pseudo-spectral discretization in toroidal direction. The flux surface aligned grid is constructed by solving the Grad-Shafranov equation using input-profiles for $T, \rho, FF'$ and $\Psi|_{\text{bnd}}$ created from experimental data using the CLISTE equilibrium reconstruction code [4]. The time stepping is fully implicit leading to large sparse matrices which have to be solved in each time step causing a high memory consumption. This is particularly true for many toroidal harmonics and low resistivity which requires a high grid resolution to resolve the thin current structures formed during an ELM.

## Simulation Results

*Full ELM crash in AUG*

In this section, we present a simulation of a full ELM crash in AUG without diamagnetic drift effects ($\tau_{IC} = 0$). The simulation is based on a typical ELMy H-mode discharge (shot #29342@4.25s) and involves a large number of toroidal harmonics (n=1-22). This allows to study the non-linear evolution of the ELM but (as mentioned above) requires a very large amount of CPU memory which limits the resolution of the poloidal grid and the accessible values for the resistivity. In the simulation at hand the core resistivity is $2.5 \times 10^{-7}$ $\Omega$m which is approximately one order of magnitude larger than the realistic Spitzer value. Figure 1 shows the time evolution of the normalized magnetic energy for the different toroidal harmonics. At the onset of the ELM high toroidal mode numbers are clearly dominant whereas for later times intermediate and low mode numbers become more and more important. Similar transitions of high-n to low-n components have also been observed experimentally [5, 6] and are driven by quadratic mode coupling [7].

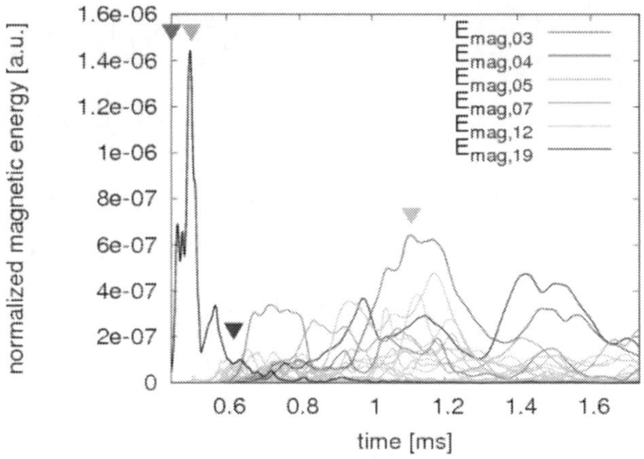


Figure 1: *Time-trace of normalized magnetic energy for different toroidal harmonics during the ELM crash*

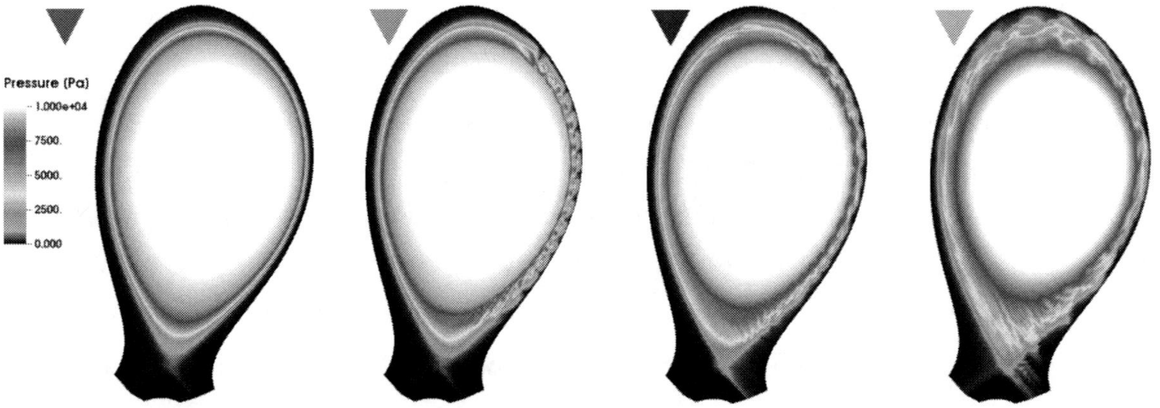


Figure 2: *Pressure distribution in the poloidal plane during the ELM crash*

Figure 2 shows the pressure distribution in the poloidal plane at four different points in time which are indicated by colored triangles in figure 1. The evolution of the pressure profile in the pedestal region (outward midplane) is shown in figure 3 using a coloring corresponding to the triangles in figures 1 and 2, respectively. One

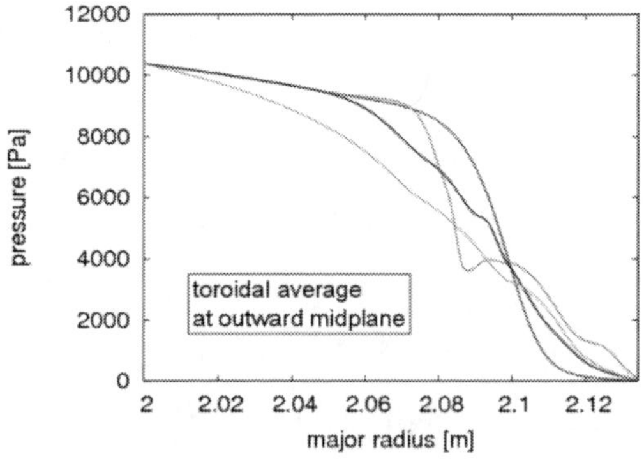


Figure 3: *Pedestal collapse during the ELM crash*

can clearly see how the pedestal collapses at the onset of the ELM and eventually flattens as the crash continues. Again this is in qualitative agreement with experimental findings [8]. Unfortunately, the pedestal does not build back up again in the simulation. Instead we observe continuous strong magnetic activity ('ballooning turbulence') after the crash (see figures 1 and 2) which prevents the pedestal from recovering.

*Diamagnetic Stabilization*

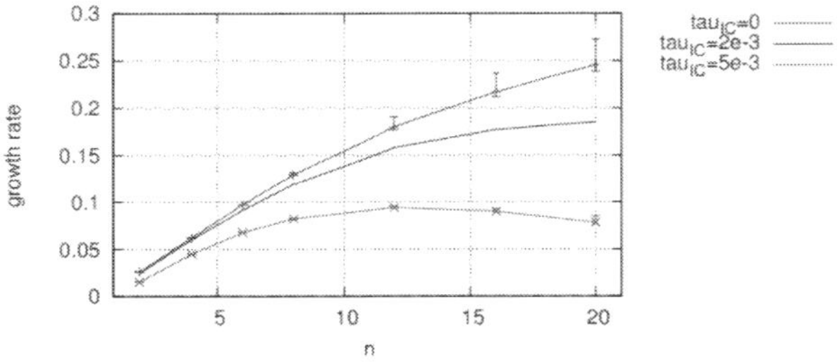

Figure 4: *Diamagnetic stabilization of high-n components*

In this section we show first results regarding simulations of ELMs in AUG including diamagnetic drift effects. The simulations are based on AUG discharge #23221@4.25s with a core resistivity of $7.5 \times 10^{-7}$ $\Omega$m. In order to study the influence of the diamagnetic terms on the toroidal mode spectrum we have performed several quasi-linear runs including only a single toroidal harmonic and mode numbers ranging from $n = 2$ to $n = 20$. Figure 4 shows the linear growth rate of the instability versus mode number both with ($\tau_{IC} = 5 \times 10^{-3}$) and without ($\tau_{IC} = 0$) diamagnetic drift effects as well as for an intermediate value of the diamagnetic parameter ($\tau_{IC} = 2 \times 10^{-3}$). One can clearly see a stabilization of high-n components through the diamagnetic drift terms leading to a maximum of the growth rate at intermediate mode numbers. This is in qualitative agreement with experimental observations [11]. The simulations above have been conducted with an intermediate resolution of the poloidal grid. In order to estimate the error bars in the plot, we have performed resolution scans of the growth rate of the individual toroidal harmonics for $\tau_{IC} = 0$ and $\tau_{IC} = 5 \times 10^{-3}$ as shown in figure 5. The growth

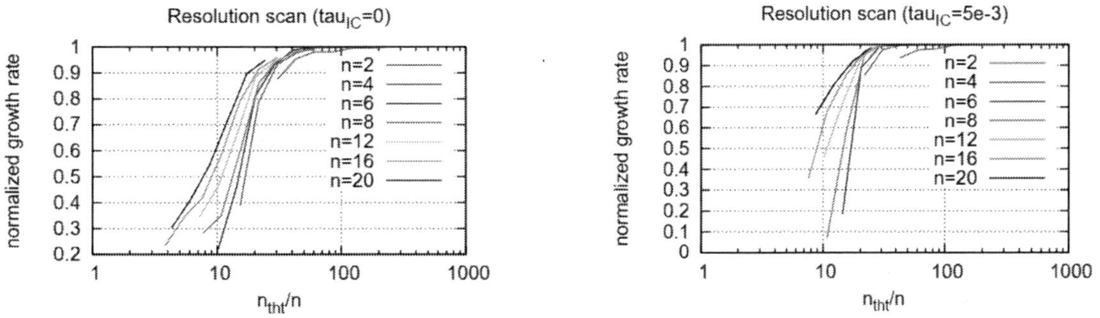


Figure 5: *Resolution scan with and without diamagnetic drift effects for different toroidal harmonics*

rates have been normalized to their saturation value and plotted against the number of poloidal grid points divided by the mode number. This choice seems reasonable since, although radial and poloidal resolution of the grid have been varied simultaneously, the poloidal resolution is the limiting one due to the fine filamentary structures whose typical length scale is inversely proportional to the mode number. Consequently high mode numbers are difficult to resolve as is clearly visible from the plot.

Figure 6 shows two simulations of a full ELM crash (for simplicity including only $n = 8$) with and without diamagnetic drift effects. In the former case, the ballooning turbulence is clearly suppressed which is a key ingredient for the simulation of ELM cycles [9].

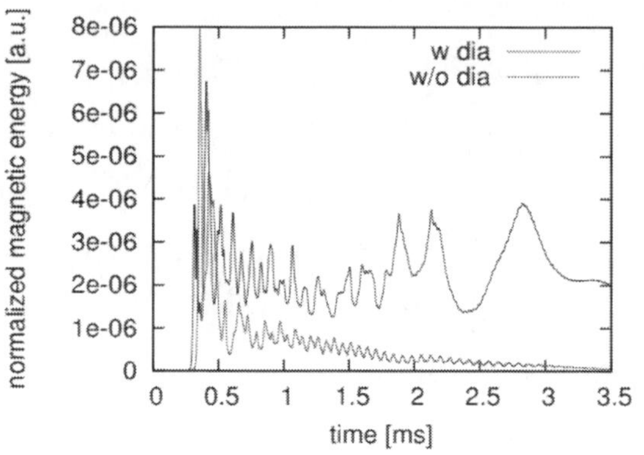


Figure 6: *Suppression of ballooning turbulence*

## Summary and Outlook

We have performed simulations of full ELM crashes in AUG with and without diamagnetic drift effects. The simulations without diamagnetic drift are in qualitative agreement with experimental findings regarding the evolution of toroidal Fourier spectrum and pedestal. However, multi-ELM simulations are not possible due to a persistent ballooning turbulence after the first crash. Simulations including diamagnetic drift effects have confirmed the stabilization of high toroidal mode numbers as well as the suppression of ballooning turbulence after the first crash. They are thus a key ingredient for multi-ELM simulations.

Next steps will be to perform multi-ELM simulations for AUG as well as to include and verify a model for a consistent bootstrap current which has recently been implemented in JOREK [10]. On the long term we aim to identify different ELM types as observed in experiments [11] and to compare the results to experimental observations, e.g., regarding pedestal evolution and heat deposition patterns.

## Acknowledgments

This work has been carried out within the framework of the EUROfusion Consortium and has received funding from the Euratom research and training programme 2014-2018 under grant agreement number 633053. The views and opinions expressed herein do not necessarily reflect those of the European Commission.

## References

[1] F. Orain et al., Phys Plasmas **20**, 102510 (2013)
[2] G. Huysmans et al., Nucl Fusion **47**, 659 (2007)
[3] O. Czarny et al., J Comput Phys **227**, 7423 (2008)
[4] M. Dunne, private communication (2014)
[5] J. E. Boom et al., Nucl Fusion **51**, 103039 (2011)
[6] R. P. Wenninger et al., Nucl Fusion **53**, 113004 (2013)
[7] I. Krebs et al., Phys Plasmas **20**, 082506 (2013)
[8] P. Schneider et al., Plasma Phys Control Fusion **56**, 025011 (2014)
[9] F. Orain et al., Plasma Phys Control Fusion **57**, 014020 (2015)
[10] S. Pamela, Invited talk at EPS 2015, Lisbon (2015)
[11] H. Zohm, Plasma Phys Control Fusion **38**, 105 (1996)

# Efficiency of surface shaping for protection of leading edges in the divertor of ASDEX Upgrade from direct plasma exposure

K. Krieger[1,*], J.W. Coenen[2], D. Brida[1], Y. Gao[2], J.P. Gunn[3], A. Hakola[4], A. Herrmann[1], T. Lunt[1], S. Potzel[1], B. Sieglin[1], ASDEX Upgrade Team, EUROfusion MST1 Team[*]

[1]*Max-Planck-Institut für Plasmaphysik, Garching, Germany*
[2]*Forschungszentrum Jülich, IEK - Plasmaphysik, Jülich, Germany*
[3]*CEA, IRFM, Saint Paul Lez Durance, France*
[4]*VTT Technical Research Centre of Finland Ltd, P.O.Box 1000, VTT, Finland*

Melting of tungsten plasma-facing components (PFCs) in tokamaks may occur either by transient heat flux excursions due to plasma instabilities or by excessive power loads to leading edges at gap side faces of PFCs exposed to the parallel power flux. Minimising potential melt damage requires, apart from mitigation of excessive transient power loads, a PFC design with optimised surface shaping to shadow PFC and castellation side faces from the parallel power flux. Code predictions of the shadowing efficiency based on 3D modelling of single particle trajectories, including gyro orbit motion but neglecting the sheath electric field [1], indicate that for typical castellation shadowing depths < 0.5 mm, hot ions, which become expelled during an ELM and flow down to the divertor target plates, may penetrate a geometrically shadowed area because their gyro orbit radius is comparable to the shadow depth (e.g. in ITER, with ELM energies $E_{i,ELM}$ = 5 keV and a toroidal field of $B_t$ = 6 T, the gyro radius is $r_L$ = 2.4 mm while typical shaping elevations are ≈ 0.5 mm). For typical ITER ELM parameters the resulting transient power flux to otherwise shadowed edges could, according to these simulations, exceed the tungsten damage thresholds.

To assess the validity of the simulations' underlying physics model, the predicted shadow penetration effect was studied in ASDEX Upgrade using ITER baseline type-I ELMy H-mode discharges, which provide ELM ion energies of 400-700 eV according to their typical plasma pedestal temperatures [2]. Despite their lower energy, their gyro radius is similar to that of corresponding ITER ELM ions because the toroidal magnetic field in the ASDEX Upgrade discharges is $B_t$ = 2 T so that for $E_{i,ELM}$ = 0.4 keV, $r_L$ = 2 mm. Due to the toroidally tilted installation of the outer divertor target tiles with an upstream edge elevation of 0.8 mm ASDEX Upgrade provides shadowed zones at the outer divertor with a toroidal width of ≈ 8 mm (Figure 1) and a depth similar to planned shaping steps at the ITER target castellations. Finally, the magnetic field angle of incidence is also comparable in both ITER and ASDEX Upgrade (Figure 2). Code predictions for the ASDEX Upgrade geometry and plasma conditions using the 3D ion orbit model suggested observable penetration effects during type I ELMs. To validate the predicted effect, the power flux to a customised target

[*]*see http://www.euro-fusionscipub.org/mst1*

tile was measured by infra-red thermography with sub-mm lateral resolution (Figure 3) [3]. In addition, two adjacent sample tiles equipped with a 20 nm W-marker layer were exposed using the divertor manipulator DIM-II [4]. This allowed quantifying the lateral distribution of W erosion both inside and outside of the shadowed zone by Rutherford backscattering analysis of the layer thickness pre- and post-exposure. The poloidal variation of W gross erosion was measured spectroscopically by quantitative observation of the WI 400.9 nm spectral line emission at the sample tile surface.

The thermal response of the tile surface was measured by the IR camera with a frame rate of 1 kHz and an exposure time of 13 μs per frame. Because the ELMs in these discharges have a duration of about 1 ms, analysis of a single ELM event could not provide conclusive evidence. Therefore, in a first analysis step, the IR response at three representative toroidal image positions was quantified by coherent averaging over all ELMs in the discharge #31697. Figure 4 shows the resulting time traces of the IR response ≈ 25 mm above the strike line for a point in the plasma exposed part of the surface, a point just inside the shadowed area of the surface and, for comparison, for a point inside the gap between adjacent tiles. The signal originating from the tile gap should not show any effect due to ELM impact safe for IR volume emission from the plasma fan above the target plate. Indeed, one observes significant IR emission following the actual ELM impact with a duration well beyond that of the actual ELM with a duration of ≈4 ms. This contribution is also seen in the signals from the actual tile surface and is approximately toroidally uniform. It is attributed to bremsstrahlung emitted by the high density cool divertor plasma built up transiently by the recycling particles of the ELM pulse. The effect is further amplified by the IR system's viewing geometry with a line of sight approximately tangentially along the divertor plasma fan. In contrast, the IR transient due to the actual ELM is in the shadowed surface zone by a factor of 6-10 lower than in the plasma exposed surface fraction. Still, Figure 4 clearly shows evidence for penetration of hot ELM ions into the magnetically shadowed zone, however, at a significantly lower fraction than predicted by the numerical simulations based on the simple 3D single particle model. Figure 4 also demonstrates that the IR time resolution allows to analyse the IR response to ELM impact by comparing single thermography image frames before an ELM and just after ELM impact. An example is given in Figure 5, which shows the toroidal profile of the surface temperature increase right after ELM onset at a time of t = 1.994 s. Also shown is the expected temperature increase according to the ion orbit simulations (red curve) as well as the extent of the geometric shadow (blue curve). While the simulation predicts a transient power flux due to hot ELM ions at approximately 40% of the power flux delivered to the plasma exposed surface fraction, the IR measurement yields an upper limit of about 10%. This is in line with the measured thickness change of the W-marker tile, which shows clear erosion near the strike point zone in the plasma exposed tile area, while in the shadowed area no erosion is seen within the experimental error.

For the discrepancy there are a number of potential causes. One reason could be the lack of electrical field drifts in the 3D orbit model. The influence of that simplification was studied for the ITER geometry and parameter space by PIC simulations [5], which also took into account possible collective mechanisms and sheath effects not included in the single particle model. However, these simulations revealed no significant discrepancies to the simple 3D orbit model. Another effect leading to the observed weaker penetration fraction might be the energy transfer from ELM electrons to ions during their passage from upstream to the divertor, which, in the free streaming ELM model [6], is required to maintain quasi-neutrality [7]. If the ions arriving at the target plate have significantly larger parallel energy than their perpendicular thermal energy, a correspondingly larger fraction will have passed the shadowed zone before their gyromotion leads to impact at the shadowed surface.

In conclusion the ASDEX Upgrade experiments demonstrate that hot ELM ions indeed can penetrate magnetically shadowed zones at shaped plasma facing surfaces. However, the resulting transient power loads to shadowed areas are smaller than predicted in simulations. Further experimental analysis is on the way to assess the consequences for predictive modelling of the power loads at ITER target plate module concepts with shaped castellations.

**References**

[1] J. Gunn et al., *Surface Heat Loads on Tungsten Monoblocks in the ITER Divertor*, Proc. of the 25th IAEA Fusion Energy Conf. 2014, FIP/1-2, 499, Saint Petersburg, Russia.

[2] J. Schweinzer et al., *Development of the Q=10 Scenario for ITER on ASDEX Upgrade*, Proc. of the 25th IAEA Fusion Energy Conf. 2014, EX/9-4, 169, Saint Petersburg, Russia.

[3] B. Sieglin et al., *Real Time Capable IR Thermography for ASDEX Upgrade*, subm. to Rev. Sci. Instr.

[4] A. Herrmann et al., *A large divertor manipulator for ASDEX Upgrade*, Fus. Eng. and Design, in press (2015), doi:10.1016/j.fusengdes.2015.02.007.

[5] M. Komm et al., *Particle-in-cell simulations of heat loads on ITER divertor monoblocks during steady-state and transient plasma conditions*, 15th Int. Conf. on Plasma Facing Mat. & Comp., Aix-en-Provence, 2015.

[6] T. Eich et al., J. of Nucl. Mat. **390-391** (2009) 760-763.

[7] Ch. Guillemot et al., *Experimental estimation of tungsten impurity sputtering due to Type I ELMs in JET-ITER-like Wall*, 15th Int. Conf. on Plasma Facing Mat. & Comp., Aix-en-Provence, 2015.

**Acknowledgement**

*This work has been carried out within the framework of the EUROfusion Consortium and has received funding from the Euratom research and training programme 2014-2018 under grant agreement No 633053. The views and opinions expressed herein do not necessarily reflect those of the European Commission.*

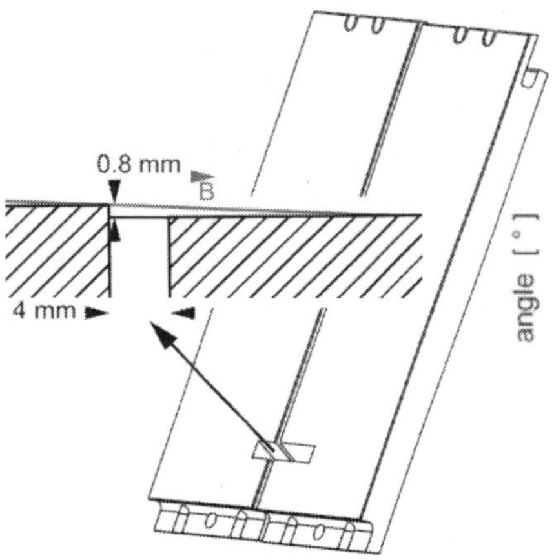


Figure 1: *Outer divertor tiles with cross-section showing magnetic shadowing.*

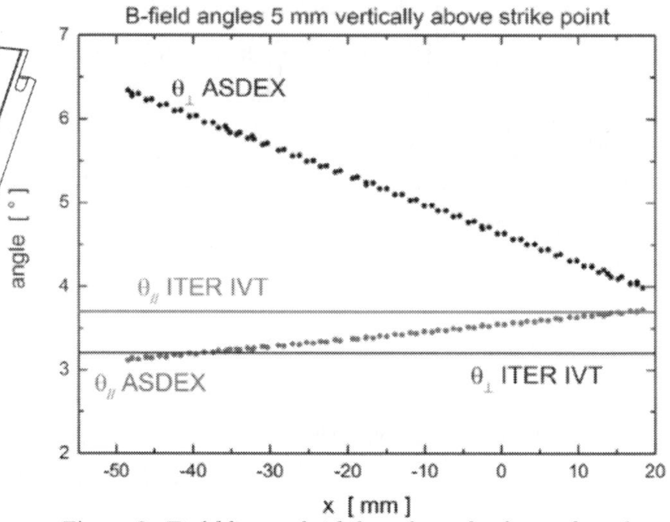


Figure 2: *Field line poloidal pitch angle, $\theta_{||}$, and angle to tile surface, $\theta_{\perp}$, for ITER and ASDEX Upgrade #31697.*

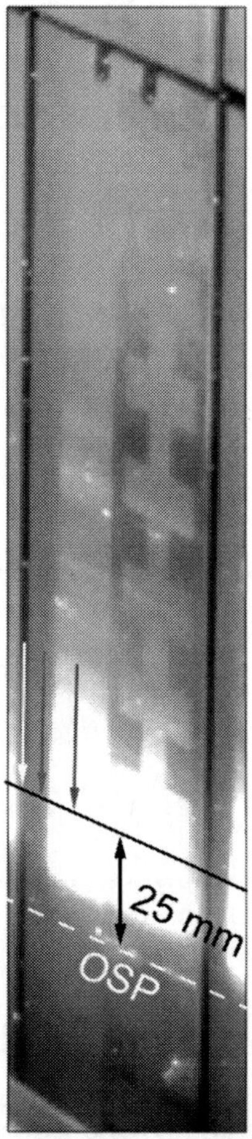


Figure 3: *Outer divertor tile observed by thermography camera with locations for coherent average of ELMs (Figure 4).*

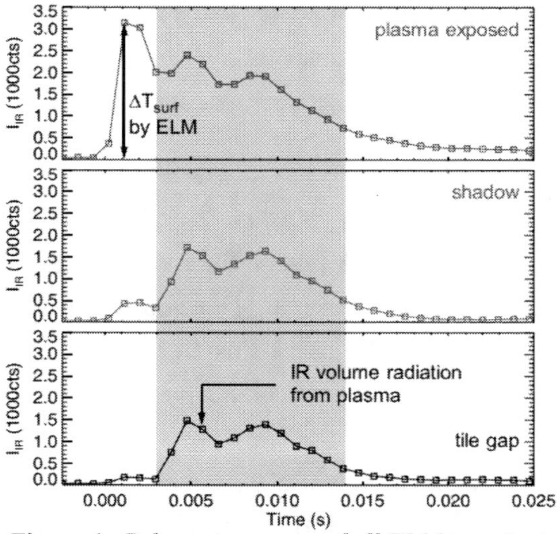


Figure 4: *Coherent average of all ELM transients in #31697 at 3 points 25 mm above strike line.*

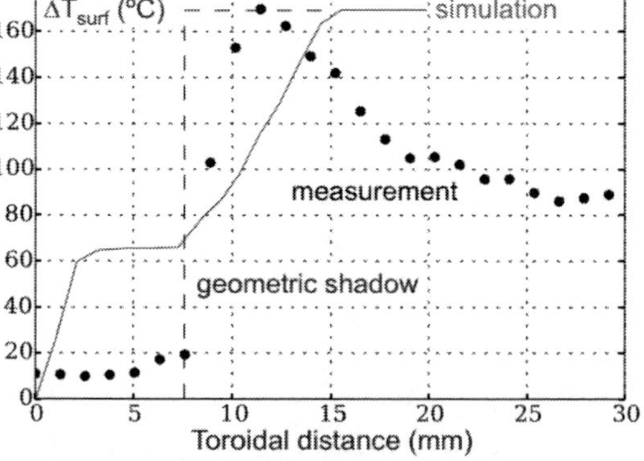


Figure 5: *Toroidal profile of ΔTsurf by ELM at t=1.994 s, 10 mm above strike line in #31697.*

# Study of the edge radial electric field during the L-H transition at ASDEX Upgrade

M. Cavedon[1,2], T. Pütterich[1], E. Viezzer[1], G. Birkenmeier[1,2], T. Happel[1],
F. Laggner[3], F. Ryter[1], U. Stroth[1,2] and the ASDEX Upgrade Team
[1] *Max Planck Institute for Plasma Physics, 85748 Garching, Germany*
[2] *Physik-Department E28, Technische Universität München, 85748 Garching, Germany*
[3] *Institute of Applied Physics, TU Wien, Fusion@ÖAW, 1040 Vienna, Austria*

In 1982, a regime with improved confinement (H-mode) was discovered at ASDEX [1]. The H-mode exhibits an edge transport barrier (ETB), a region of reduced particle and energy transport at the plasma edge. It is widely accepted that **E**×**B** velocity shear, caused by the local radial electric field $E_r$, is responsible for the suppression of turbulence and therefore for the establishment of the ETB. The formation of the ETB coincides with the transition from the low confinement mode (L-mode) to the H-mode and it is triggered when heating power exceeds a certain threshold $P_{\mathrm{thr}}$, scaling in [2].

Neoclassical theory predicts $E_r$ to be dominated by the diamagnetic velocity of the main plasma ions, as also suggested by experimental observations in H-mode [3] but not assessed yet in the transition from L-mode to H-mode. Recent observations at ASDEX Upgrade (AUG) have shown a correlation between the edge ion heat channel and the H-mode onset [4]. In line with that, a threshold for the $E_r$ minimum, a proxy of $\nabla E_r$, has been observed and is independent of density for a magnetic field of $B_t = 2.5\,\mathrm{T}$ highlighting the importance of $\nabla p_i/(Z_i n_i e)$ in the L-H transition mechanism [5]. However, the triggering mechanism in the transition process is unclear. A pulsating phase of the edge $E_r$ and of the turbulence amplitude, originally called dithering H-mode [6] and more recently limit cycle oscillation (LCO) or "I-phase" [7, 8] is often observed close the L-H transition where the role of turbulence induced flow shear is under investigation [8, 9]. Extensive effort has been directed to understanding the L-H transition, however, a first principle model which reproduces the experimental observations and predicts quantitatively the power threshold has not been developed yet. The aim of the present study is to provide a better understanding of the background **E**×**B** velocity shear necessary for the H-mode onset together with an analysis of the interplay between turbulence, flows and kinetic profiles in the dynamic evolution towards the L-H transition.

For this purpose, the edge charge exchange system at AUG has been upgraded. The new system has 30 toroidally (blue) and 20 poloidally (red) aligned lines of sight (LOS) focused at the low field side pedestal region (fig. 1) and permits accurate measurements of impurity density, velocity and temperature profiles.

The toroidal and poloidal projections of the LOS are shown respectively in figures 1 a and b. Through the ion radial force balance ($E_r = \nabla p_\alpha/(Z_\alpha n_\alpha) - v_{\theta,\alpha}B_\phi + v_{\phi,\alpha}B_\theta$, $\alpha$ : impurity species) it is possible to reconstruct the $E_r$ profile from the impurity CXRS measurements, thus allowing the interaction between flows and kinetic profiles to be studied. Moreover a new spectrometer has been designed which, by means of an interference filter, allows the measurements of up to nine LOS at a sampling rate of 20 kHz. Such a time resolution is a fundamental requirement to investigate the fast dynamics close to the L-H transition which typically shows frequencies of a few kHz [6–9].

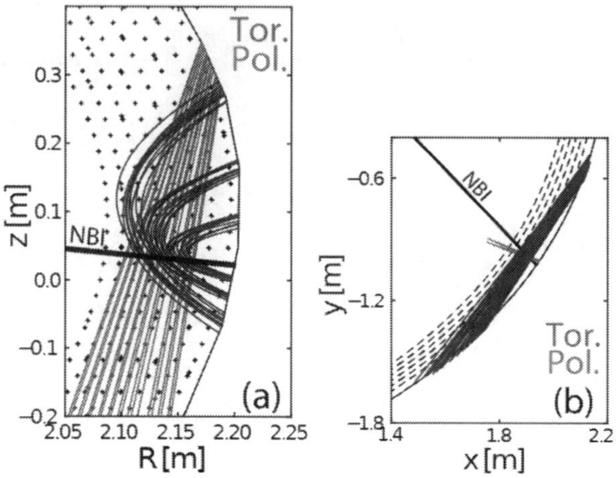


Figure 1: *Poloidal* (a) *and toroidal* (b) *projection of the upgraded edge charge exchange system at AUG*

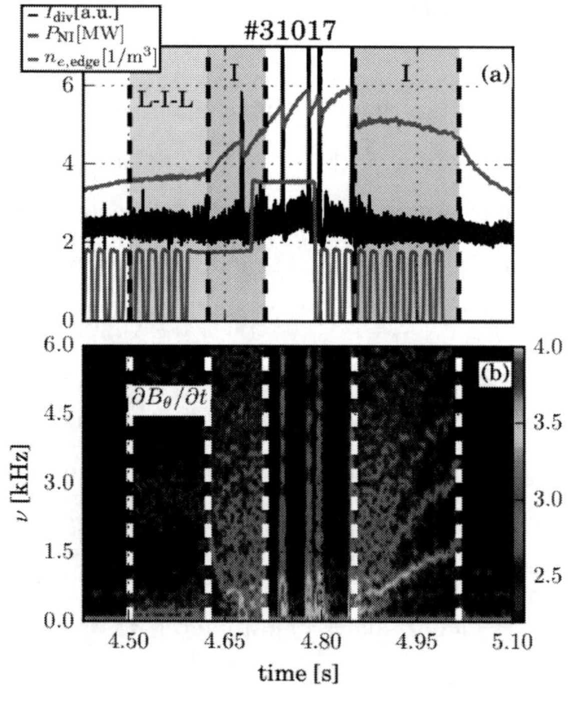


Figure 2: *Example of typical L-H transition:* (*a*) *time traces of the $n_{e,edge}$ (green), $P_{NI}$ (red) and $I_{div}$ (black),* (*b*) *spectrogram of the bottom inward Mirnov coil.*

In order to address the background **E**×**B** velocity shear condition for the H-mode onset, it is necessary to understand the dynamics of the L-H transition process and to define a time point where the plasma changes its confinement to enable a comparison of different discharges. In particular, correlations between the $E_r$ profile just before the confinement change and the power threshold have to be investigated in order to experimentally relate the macroscopic effects on $P_{thr}$ to the microscopic **E**×**B** shear of turbulence. A series of discharges with different plasma parameters has been performed where several L-H and H-L transitions have been induced by slowly ramping power up and down to allow active charge exchange measurements. In figure 2a, the time traces of the input power $P_{\mathrm{NI}}$ (red), the line integrated edge electron density $n_{\mathrm{e,edge}}$ (green) and the divertor current $I_{\mathrm{div}}$ (black) are shown.

Close to the L-H and H-L transition the I-phase has been identified (fig. 2a, blue shaded area). This phase is clearly visible in the spectrogram of a poloidal Mirnov at the low inboard side coil (fig. 2b) and in the divertor shunt current $I_{\mathrm{div}}$ (black, fig. 2a) where the typical fluctuations in the range of a few kHz are detected. The appearance of the dithering phase at the back transition and their effect on $\partial B_\theta/\partial t$ has already been reported in other machines [10–12]. A zoom-in of this phase is plotted in figure 3a where the turbulence amplitude measured by Doppler reflectometry (black) shows the typical bursts which are correlated with the spikes in $\partial B_\theta/\partial t$. These spikes result in harmonics in the Fourier space (fig. 2b) and they show the same characteristics when the spectrogram is more blurry. In several of the analyzed transitions, the I-phase also extends in between type-I ELMs and it is connected with a density rise (fig. 2, blue shaded area) denoting a confinement closer to H-mode rather than L-mode. This is further confirmed by looking at the H-L transition where the plasma goes back to L-mode only when these fluctuations are vanishing.

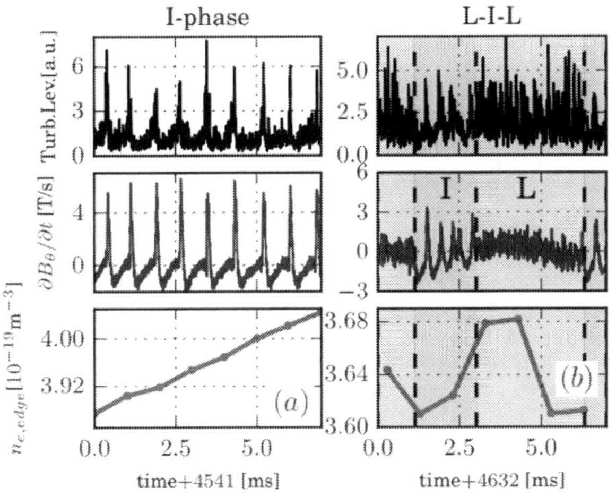


Figure 3: *Zoom-in of the I-phase (a) and of the L-I-L oscillations (b) comparing the turbulence level measured by Doppler reflectometry (black), $\partial B_\theta/\partial t$ from a Mirnov coil (blue) and the integrated edge electron density (green).*

Just before the onset of the I-phase another type of fluctuations is visible with a frequency of around 150 Hz (fig. 2, red shaded area). A zoom-in in one period (fig. 3b) shows that the plasma is shortly transiting from L-mode into the I-phase, indicated by a few bursts, and then it goes back to L-mode where the turbulence level (black) is higher. This rapid L-I-L transitions modulate also the edge electron density (green) which is increasing during the dithering and drops shortly after the plasma goes back into L-mode. These oscillations are only observed for electron densities above the minimum of the dependence of the power threshold on $n_e$. This suggests that the plasma goes back to L-mode due to the increased $n_e$ and consequently increased $P_{\mathrm{thr}}$. The final transition to the I-phase happens only when the input power is further increased, in this case only when the NI is switched on for a longer time (fig. 2).

In order to understand the background conditions for the H-mode onset, the radial electric field has been compared after the L-I-L oscillations but before the I-phase (in this case $t = 4.624$ s). It is however crucial to also understand the trigger of the H-mode by comparing the fast interplay of flows, kinetic gradients and turbulence during these fluctuations. Self induced turbulence flows, also called zonal flows (ZFs), are debated to be the actuator of these intermediate phases [8, 9]. Due to the improved temporal resolution of the edge CXRS diagnostic (a few hundreds of $\mu$s) and by applying conditional averaging, the measured $\mathbf{E}\times\mathbf{B}$ velocity has been compared to the diamagnetic velocity of the main ions ($v_{\text{dia,i}} = \nabla p_i/(n_i B)$) just before the L-I and I-L transition and few bursts after the L-I transition. No strong deviation inside the error bar was found between the two values in these phases limiting a possible contribution from ZFs exactly at L-I transition and/or during the initial bursts. This is indeed where self induced turbulence flows are reported to be most active.

The $E_r$ minimum just before the L-I transition is presented as a function of the magnetic field $B$ in figure 4 for discharges with different $B_t$ performed in deuterium (blue) and in hydrogen (red). $E_r$ scales roughly linearly with $B$ suggesting that a critical $\mathbf{E}\times\mathbf{B}$ velocity shear is necessary to get into H-mode. This also seems to be consistent with the almost linear dependence of the $P_{\text{thr}}$ on $B_t$ [2]. The minimum in the $E_r$ measured in hydrogen also aligns with those measured in deuterium even if the power threshold changes by a factor of two between the two isotopes [13]. However, considering that the turbulence characteristics at the edge are similar in hydrogen and deuterium [14], this further supports the idea of a critical $\mathbf{E}\times\mathbf{B}$ velocity shear for L-H transition. These observations combined with those reported in [5] and [4] indicate that the background conditions for the H-mode onset are dominated by the diamagnetic velocity of the main ions.

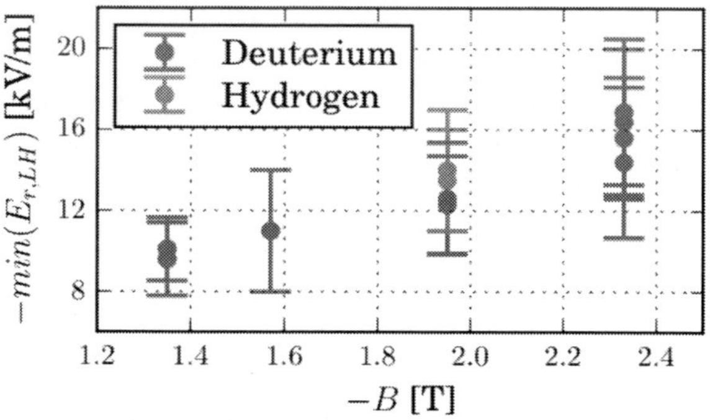


Figure 4: *Minimum of $E_r$ just before the final L-I transition in function of B for deuterium (blue) and hydrogen (red) plasmas*

## References

[1] Wagner, F. et al. *Phys. Rev. Lett.* (1982).

[2] Martin, Y. R., , et al. *Journ. of Phys.* (2008).

[3] Viezzer, E. et al. *Nucl. Fusion* (2014).

[4] Ryter, F. et al. *Nucl. Fusion* (2014).

[5] Sauter, P. et al. *Nucl. Fusion* (2012).

[6] Zohm, H. *Phys. Rev. Lett.* (1994).

[7] Conway, G. D. et al. *Phys. Rev. Lett.* (2011).

[8] Schmitz, L. et al. *Phys. Rev. Lett.* (2012).

[9] Kobayashi, T. et al. *Phys. Rev. Lett.* (2013).

[10] Zohm, H. et al. *PPCF* (1995).

[11] Schmitz, L. et al. *Nucl. Fusion* (2014).

[12] Cheng, J. et al. *Phys. Rev. Lett.* (2013).

[13] Righi, E. et al. *Nuclear Fusion* (1999).

[14] Hennequin, P. et al. In *EPS*, (2015).

This work has been carried out within the framework of the EUROfusion Consortium and has received funding from the Euratom research and training programme 2014-2018 under grant agreement No 633053. The views and opinions expressed herein do not necessarily reflect those of the European Commission.

# Evolution of ELMy H-mode performance in presence of core radiation on ASDEX Upgrade

R. Ochoukov[1], V. Bobkov[1], C. Angioni[1], M. Bernert[1], M. Dunne[1], R. Dux[1], R. Neu[1,2], J.-M. Noterdaeme[1,3], T. Odstrčil[1,2], T. Pütterich[1], F. Reimold[1], the AUG Team, and the EUROfusion MST1 Team

[1]*Max-Planck-Institut für Plasmaphysik, Boltzmannstr. 2, 85748 Garching, Germany*

[2]*Technische Universität München, Boltzmannstr. 15, 85748 Garching, Germany*

[3]*UGent, Applied Physics Department, 9000 Gent, Belgium*

## 1. Introduction

High performance, high temperature tokamak plasmas require that a large fraction of the thermal power that crosses the separatrix from the core ($P_{sep}$) is radiated away in the form of photons to mitigate heat loads on and erosion of plasma facing components (PFCs). It is common to use nitrogen and/or noble gases, such as neon, argon, or krypton, as a radiating species to transfer the plasma thermal energy into electromagnetic radiation. However, in case of a low-Z number of the radiating species, relatively high concentrations of the radiator must be used, which can impact the fusion reaction rate of the plasma core through dilution. High-Z radiators, on the other hand, are capable of radiating a large fraction of the plasma thermal power, while being present in low ($\sim 10^{-5}$-$10^{-4}$) concentrations in the plasma core. But, unlike low-Z radiators, it is crucial to maintain control over the core radiation profile to prevent central radiation peaking and subsequent degradation of the plasma performance and the discharge stability. The goal of the study is to determine the feasibility of using controlled core radiation to reduce $P_{sep}$, and hence, heat loads on PFCs, while maintaining high plasma performance in a stable discharge.

The early results on ASDEX Upgrade show that the core plasma performance parameters – the stored plasma energy ($W_{MHD}$) and the confinement factor (H98) – remain robust (Fig. 1 (a) and (b)) as the core radiating power fraction is increased (with tungsten (W) being the dominant radiating impurity). We remind that the radiated power is not included in the definition of H98. While the radial electron temperature ($T_e$) profiles show a slight drop inside $\rho_{tor}$ <0.5 in the high radiation case, the plasma density ($n_e$) profile is little changed in the same region resulting in the plasma stored energy ($W_{MHD} \sim n_eT_e$) approximately

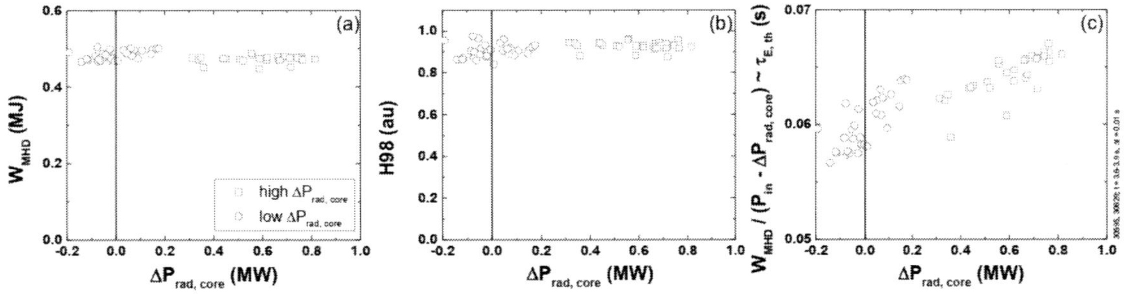


*Figure 1: Changes in (a) the plasma stored energy $W_{MHD}$, (b) the confinement factor H98, and (c) the thermal energy confinement time $\tau_{E,\,th}$ as a function of the radiated power from the plasma core $\Delta P_{rad,\,core}$. The "base" $P_{rad}$ value (when $\Delta P_{rad,\,core}$ = 0 MW) was 3.5 MW.*

constant before and during the high core radiation discharges. A combination of a constant $W_{MHD}$ and an increase of the core radiating power fraction suggests that the thermal energy confinement time ($\tau_{E,\,th} \sim W_{MHD}/(P_{in} - \Delta P_{rad,\,core})$), where $P_{in}$ is the total incident heating power, constant for these discharges) increases in the high core radiation discharge (Fig. 1 (c)). Data analysis using a time-dependent tokamak transport code TRANSP reveals that the electron heat diffusivity ($\chi_e$) decreases across the plasma core during the high radiation regime, consistent with the observed improvement in $\tau_{E,\,th}$. Note, that the high radiation regime remains stable for ~1 s (equal to the duration of the Ion Cyclotron Range-of Frequencies (ICRF) power pulse in non-standard phasing [1]). The radial tungsten concentration ($c_W$) profile across the core remains uniform without central peaking. Overall, the early results show a promising (robust) plasma performance in ELMy H-modes on AUG with a high core radiating fraction warrantying further study of this scenario either on its own or in combination with other high-Z radiators such as krypton [2].

**2. Experimental Method**

The discharge parameters for this study were: the toroidal magnetic field $B_T = -2.5$ T and the plasma current $I_p = 0.8$ MA. The edge safety factor $q_{95}$ was 5.4. The main ion species was deuterium and the core radiating impurity was tungsten, "seeded" from the active ICRF antenna limiters by operating antennas in non-standard phasing [1], while maintaining a constant ICRF power of 1.5 MW. The plasma shape was a lower single null. The H-mode was maintained by 4.9 MW of neutral beam power. Additionally, 1.4 MW of core electron cyclotron resonance heating (ECRH) was used to prevent tungsten concentration ($c_W$) from centrally peaking. $P_{in}$ (including ohmic heating) was ~8 MW. Core radiation was monitored with poloidal arrays of soft x-ray (SXR) detectors, sensitive to photons with energies >1.8 keV [3]. The $c_W$ radial profiles were reconstructed from the SXR emissivity ($\varepsilon_{SXR}$) profiles, assuming that the dominant radiating species in the core was W [4]. The data from these discharges were subsequently analysed using TRANSP to determine the effect of core radiation on electron and ion thermal diffusivities ($\chi_e$ and $\chi_i$) across the plasma core.

**3. Experimental Results and Discussion**

As a result of central ECRH heating, the discharges showed large sawtooth (ST) oscillations with $\Delta T_e \sim 1$ keV. In order to account for the observed ST oscillations in these discharges, the plasma profiles were examined just before the ST crash and just after the ST crash. The radial $T_e$ profiles inside the $q = 1$ region showed the largest modulation during an ST cycle (Fig. 2 (a)), while the sawtooth induced variations in $T_i$ and $n_e$ profiles were small (Fig. 2 (b) and (c)). In the discharges with high core radiation, the $T_e$ profile was slightly reduced for $\rho_{tor}$ <0.5, with the ST amplitude also being reduced (Fig. 2 (a)). The reduction of $T_i$ was observed further in, inside the ST inversion radius at $\rho_{tor}$ <0.31 (Fig. 2 (b)). The $n_e$ profile showed a minimal change across the plasma core in the high radiation regime (Fig. 2 (c)). The overall plasma stored energy ($W_{MHD} \sim n_eT_e$), which is a volume-integrated quantity,

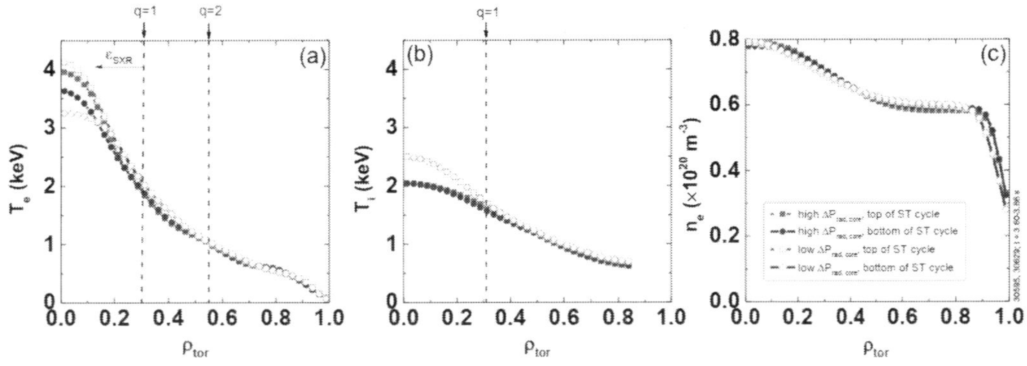


*Figure 2: The radial profiles of (a) the electron temperature ($T_e$), (b) the ion temperature ($T_i$), and (c) the plasma density ($n_e$) before and during the discharges with high core radiation. The data is shown for the top and the bottom of a sawtooth (ST) cycle. The radial locations of q = 1 and q = 2 rational surfaces are shown. The radial location of the soft X-ray emission region ($\varepsilon_{SXR}$) is also shown.*

remained approximately unchanged before and during the discharges with high core radiation (Fig. 1 (a)), as the observed $T_e$ and $T_i$ changes occurred over a small plasma volume.

The SXR diagnostic on ASDEX Upgrade is sensitive to photons typically in the energy range of >1.8 keV [3], which happened to coincide with the $T_e$ and $T_i$ values inside the ST inversion radius (Figs. 2 (a) and (b)). The $\varepsilon_{SXR}$ profiles showed that the radiation increase in these discharges originated from the plasma core and was influenced by the phase of the ST cycle (Fig. 3 (a)). Note, that the data is shown for four consecutive ST cycles and the $c_W$ profiles, averaged over a ST cycle, remained uniform across the plasma core without evidence for peaking (Fig. 3 (b)). The modulation of $c_W$ during an ST cycle was "in-phase" with the $T_e$ modulation (Fig. 3 (b)). In all cases the levels of $c_W$ remained $\leq 10^{-3}$ in the core.

The lack of a drop in the plasma performance with the increase in the core radiated power (Fig. 1 (a) and (b)) indicates that the transport of thermal energy across the plasma drops (and, hence, improves) with core radiation. Such trend is observed in the thermal energy transport coefficients for the electrons (Fig. 4 (a)) and, to some degree, for the main ions (Fig. 4 (b)), as estimated by TRANSP. The radial $\chi_e$ values are reduced across the entire confined plasma profile (Fig. 4 (a)), while the $\chi_i$ values are enhanced inside the sawtooth inversion

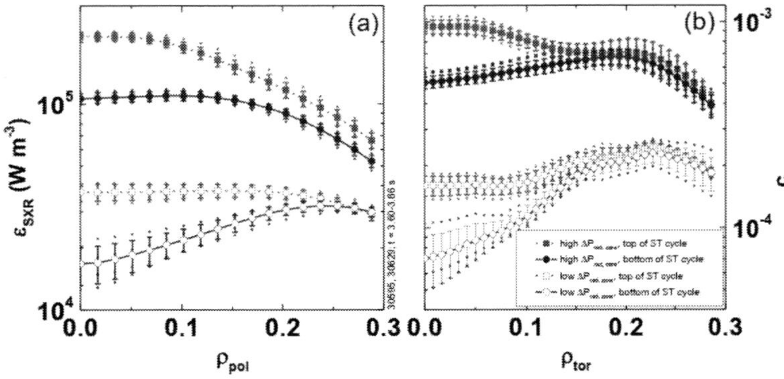


*Figure 3: The radial profiles of (a) the measured soft X-ray emissivity ($\varepsilon_{SXR}$) and (b) reconstructed tungsten concentration ($c_W$) inside the sawtooth (ST) inversion radius $\rho_{tor}$ = 0.31. The data is shown for four consecutive sawtooth cycles.*

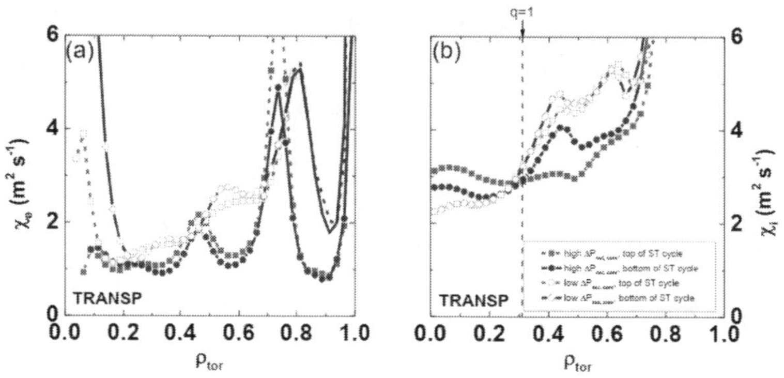


*Figure 4: The radial profiles of (a) the electron heat diffusion coefficient ($\chi_e$) and (b) the ion heat diffusion coefficient ($\chi_i$) before and during the discharges with high core radiation. The values are deduced by TRANSP.*

radius but are reduced outside the q = 1 region (Fig. 4 (b)) in the presence of core radiation. The observed variation in $\chi_e$ is expected as the core radiation directly impacts the $T_e$ profile and, most importantly, the gradient of the $T_e$ profile ($\nabla T_e$). $\nabla T_e$, in turn, sets the electron heat diffusivity across the core. The stiffness of the $T_e$ and $n_e$ pedestal profiles has been analysed previously [5] and shown to be equally stiff before and during the discharges with high core radiation. This further constrains the radial plasma pressure profile to remain constant, and, in turn, the heat transport coefficients to become reduced. The situation with $\chi_i$ is more complex, as the $T_i$ profiles depend on the strength of the electron-ion coupling. Note that the opposite effect – no change in $W_{MHD}$ with an increase in core ECRH heating – was previously observed on AUG [6] and is linked to $\nabla T_e$, $\nabla T_i$, and $n_e$ via collisionality [6].

## 4. Conclusion

Discharges with high core radiation fraction were examined to determine the applicability of core radiation as a means of mitigating thermal heat loads on the plasma facing components, while maintaining stable, high-performance core plasma. W "seeded" by an active ICRF antenna operated in non-standard phasing was used to change the core radiated power. The early results reveal that $W_{MHD}$ and H98 factor remain robust in the discharges with high core radiation. Since the increase in $P_{rad}$ originates from the core due to the presence of W, it implies that $\tau_{E,\,th}$ is improved in the discharges with high core radiation.

## 5. Acknowledgements

This work has been carried out within the framework of the EUROfusion Consortium and has received funding from the Euratom research and training programme 2014-2018 under grant agreement No 633053. The views and opinions expressed herein do not necessarily reflect those of the European Commission.

## 6. References

[1] V. Bobkov et al., 21st Topical Conference on RF Power in Plasmas, May 2015.
[2] M. Bernert et al., these proceedings, 2015.
[3] V. Igochine et al., Internal IPP Report "IPP 1/338", May 2010.
[4] M. Sertoli et al., Plasma Phys. Control. Fusion **57** 075004, 2015.
[5] R. Ochoukov et al., 41st EPS Conference on Plasma Physics, June 2014.
[6] A. Manini et al., Plasma Phys. Control. Fusion **46** 1723, 2004.

# Kinetic profiles and plasma response during application of 3D magnetic perturbations fields

M. Willensdorfer[1], W. Suttrop[1], E. Strumberger[1], M. Cavedon[1,2], I. G. J. Classen[3], M. Dunne [1], S. Fietz[1], R. Fischer[1], C. Fuchs[1], A. Kirk[4], F. M. Laggner[5], E. Wolfrum[1], N. C. Luhmann Jr.[6], B. Vanovac [3], the EUROfusion MST1 Team and the ASDEX Upgrade Team

[1] *Max Planck Institute for Plasma Physics, 85748 Garching, Germany,*

[2] *Physik-Department E28, Technische Universität München, 85748 Garching, Germany,*

[3]*FOM-Institute DIFFER, Dutch Institute for Fundamental Energy Research,*

[4] *CCFE, Culham Science Centre, Abingdon, Oxon, OX14 3DB, UK,*

[5] *Institute of Applied Physics, TU Wien, Fusion@ÖAW,*

[6] *University of California at Davis, Davis, CA 95616, USA*

## Introduction

The H-mode is accompanied by edge localized modes (ELMs), which may imply intolerable power loads on first wall materials in large fusion devices. One method to suppress or to mitigate such ELMs is the application of external magnetic perturbation (MP)-fields. This technique has been demonstrated to suppress large ELMs in ASDEX Upgrade (AUG), DIII-D, KSTAR and to mitigate them in JET, MAST, NSTX [2].

Recent experiments demonstrated ELM mitigation at low collisionality $\nu^{\star}$ in AUG [1, 3]. Accompanied with ELM mitigation is a decrease of density, the so-called density pump-out, whereas the temperature is not strongly affected. Towards ELM mitigation, the ELM frequency can reach several 100 Hz making an analysis of inter-ELM kinetic profiles and their evolution very challenging. Additionally, the MP-fields introduce a toroidal asymmetry, which makes the analysis and its interpretation of toroidally localized measurements even more complicated. In this paper, we investigate changes in kinetic profiles towards ELM mitigation and the effect of toroidal asymmetry on the plasma using toroidally localized diagnostics and rotating MP-fields.

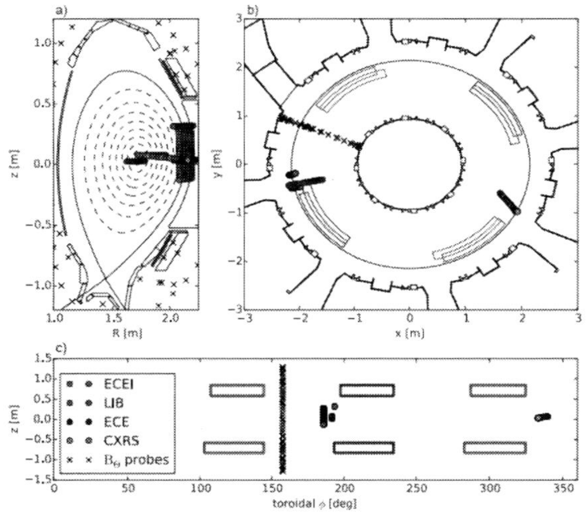


Figure 1: (a) poloidal plane, (b) toroidal plane, (c) toroidal angle versus z

## Experimental setup

Recent diagnostic upgrades of electron cyclotron emission (ECE) ($T_e$), charge exchange recombination spectroscopy (CXRS) ($T_i$) and the lithium beam (LIB) ($n_e$) diagnostics allow us to resolve inter-ELM kinetic profiles during ELM mitigation, although the ELM frequency amounts to up to $\sim$ 800 Hz. Figure 1 shows poloidal and toroidal positions of several used diagnostics at AUG. Additionally to the standard set of edge diagnostics for kinetic profiles, we also used ECE imaging (ECEI), which can resolve poloidal structures. ECEI and ECE was only used at medium q ($q_{95} \sim 5.2$) discharges due to the higher magnetic field ($|B_T| \sim 2.5$ T) and their limited frequency range.

The MP-field in AUG is generated by 16 perturbation coils and its versatile power supply allows rigid rotations of the MP-field using a coil configuration with a toroidal mode number n of one (n=1) or n=2 . Fig. 1 shows the position of the coils and the colors indicate an n=2 perturbation.

**Kinetic profiles during ELM mitigation**

Figure 2 shows ELM mitigation experiments at medium q. Time traces of Fig. 2(a) show two phases (labelled as MP I an II), when an n=1 MP-field was applied. These two phases differ only by the applied MP-field alignment caused by a toroidal phase shift between the upper and lower coil set. Although the strength of the currents were the same, only the second phase achieved ELM mitigation indicating the importance of the MP-field alignment on conditions for ELM mitigation (more details in [4]). Additionally, the ELM mitigation was induced by a radial sweep, which led to a larger perturbation field at the plasma edge due to the smaller distance between the plasma and the MP-coils.

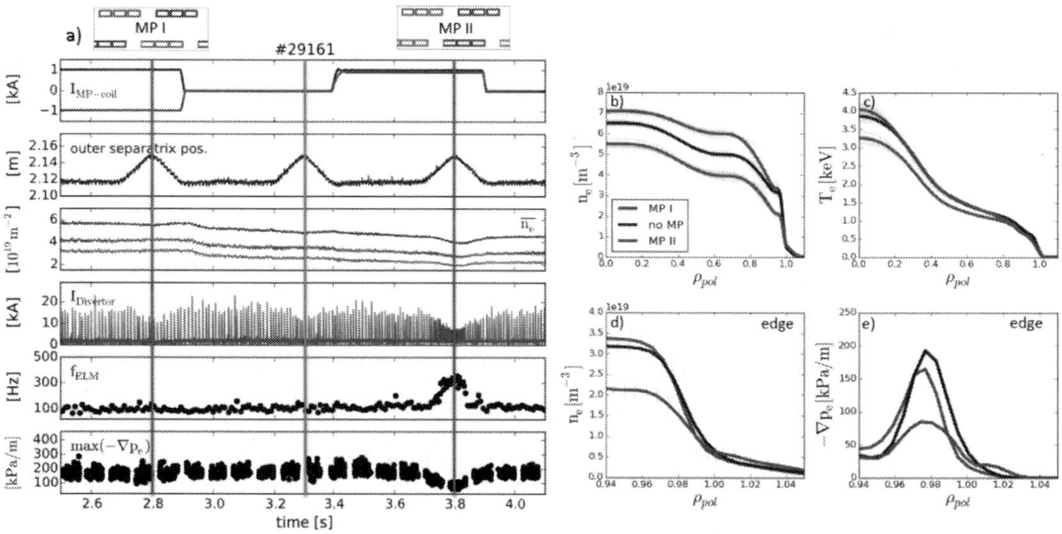


Figure 2: a) from top to bottom: MP coil current, outer separatrix position, line integrated density, divertor current, ELM frequency, maximum edge pressure gradient, (b) electron density profile, (c) electron temperature profile, d) edge electron profile, (e) electron pressure gradient. The color scaling of the vertical lines in (a) corresponds to the time windows used in (b-e).

Accompanied with ELM mitigation is a decrease of density, the so-called density pump-out (Fig. 2(b)), whereas the electron temperature is not strongly affected (Fig. 2(c)). Measurements indicate that the density pump-out comes along with smaller edge gradients within the pedestal region in the electron density profile (Fig. 2(d)) as well as electron pressure profile (Fig. 2(e)). The increase in ELM frequency and the lower critical electron pressure gradient indicate a change in ELM stability boundary, since a shallower pressure gradient is usually more stable to peeling-ballooning modes. Here, three dimensional effects caused by the MP could play a role, which are not considered in this analysis.

**Toroidal asymmetry of kinetic profiles during MP field induced ELM mitigation**

A three dimensional distortion of the plasma boundary is well documented [5, 6]. To analyze toroidal asymmetries of kinetic profiles, we rotated the MP field toroidally [5] and used toroidally localized diagnostics.

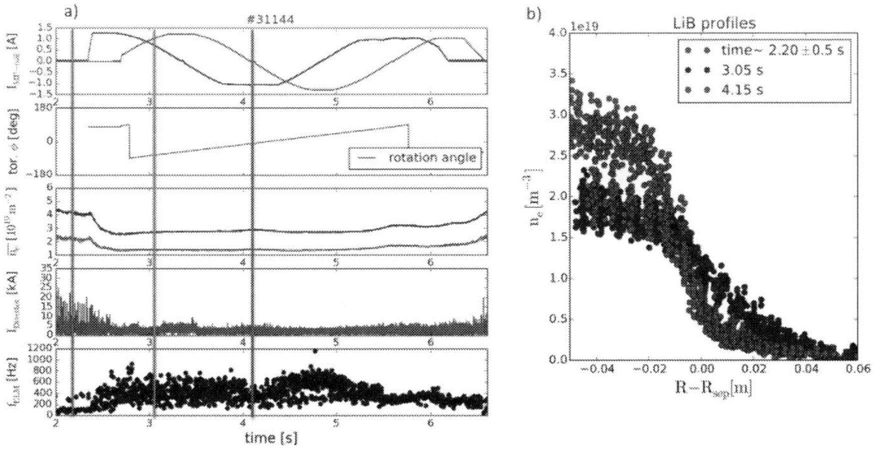


Figure 3: ELM mitigation experiment, (a) from top to bottom: MP coil currents, toroidal angle of rotation, line integrated density, divertor current, ELM frequency (b) edge electron density profile from LIB. LIB profiles are aligned with each other. The color scaling of the vertical lines in (a) corresponds to time windows used in (b).

Figure 3(a) shows time traces of an experiment at low q ($q_{95} \sim 3.8$) using a rotating n=2 MP field. The application of this perturbation resulted in ELM mitigation, which is sustained throughout the rotation until the MP coils were switched off. A rigid rotation of the MP pattern was applied meaning a constant phase between the upper and lower coil set. The MP field was rotated by about one full turn, thus, by 180 degrees due to the n=2 configuration. As before, ELM mitigation, density pump-out and an increase of ELM frequency up to 600 Hz are observed. To resolve inter ELM profiles, we reconstructed LIB profiles with a temporal resolution of 100 $\mu$s. In Fig. 3(b), LIB electron density profiles prior to the MP onset (red) and at two different time-points i.e. toroidal rotation angles (blue and green) during the rotation of the MP-field are shown. The pedestal top values of the density profiles decrease, when the MP-field is turned on. Furthermore, the density gradients (in real space) vary with toroidal rotation angle within the steep gradient region, whereas the pedestal top values remain constant. The edge gradients change by 50%. The ion temperature profiles from CXRS also show alteration in the edge gradients, but less pronounced. This indicates a toroidal variation of the magnetic structure due to the MP-field e.g. flux surface compression/expansion.

**Plasma response measurements**

To measure the plasma response due to a MP-field, the electron temperature is the best quantity. Due to the high electron heat transport along the field lines, the electron temperature is assumed to be constant and therefore, changes in the electron temperature profile reflect changes in magnetic topology. We used ECEI measurements during one discharge applying rigid rotating MP-field with even parity and n=2 configuration at medium q to detect variations in the magnetic topology (shown in Fig. 4). An influence of the MP-field on the ELM behavior could not be observed. Figure 4(a) shows the ECEI channel array (crosses) and the color scaling indicates the mean temperature during the rotation. The time traces using inter ELM data of the analyzed channels, which are close to the rational $q \sim 5$ surface, are shown in Fig. 4(b). During one full turn, ECEI measurements reveal a poloidal propagating n=2 structure. From the poloidal velocity and the given toroidal velocity and mode number (n=2), one can determine the poloidal mode number m. Preliminary evaluation of the poloidal number m indicates a pitch aligned response. But due to the high uncertainties in the position of the ECEI radiation, forward modeling of ECE radiation transport is required and will be employed. Furthermore, the measurements show a poloidal asymmetry of the amplitude, which is not expected from ideal MHD codes. But amplitude measurements of radially and toroidally localized diagnostics could be compromised by the feedback of the plasma shape and position control due to the rotating MP-field. In principle, this could also cause a poloidal asymmetry and will be investigated in more detail.

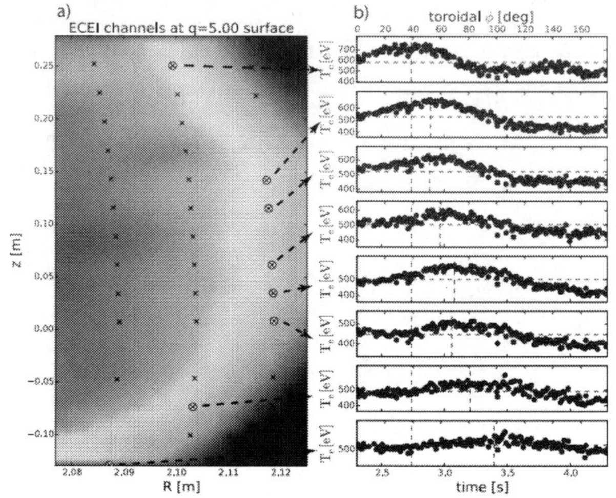


Figure 4: (a) The ECEI channel array is shown (cold resonance). The white circle indicate the ECEI channels around the $q \sim 5$ surface. (b) Time traces of the corresponding ECEI channels.

[1] W. Suttrop *et al*, 25th IAEA Int. Conf. on Fusion Energy, EX/P1-23, 2014
[2] M.E. Fenstermacher *et al*, 23rd IAEA Int. Conf. on Fusion Energy, ITR/P1-30, 2010
[3] M. Garcia-Munoz *et al*, Plasma Phys. Control. Fusion **53**, (2013) 124014
[4] A. Kirk *et al*, Nucl. Fusion 55, 043011, 2015.
[5] C. F. Fuchs *et al*, 41st EPS Conference on Plasma Physics, P-2004, 2014
[6] I. T. Chapman *et al*, Nucl. Fusion **54**, (2014) 123003

This work has been carried out within the framework of the EUROfusion Consortium and has received funding from the Euratom research and training programme 2014-2018 under grant agreement No 633053. The views and opinions expressed herein do not necessarily reflect those of the European Commission.

# SXR tomography of tungsten radiation patterns with and without MHD

T. Odstrčil[1,2], T.Pütterich[1], M. Weiland[1], M. Odstrčil[3], A. Gude[1] and the ASDEX Upgrade Team

[1] *Max-Planck-Instiüt für Plasmaphysik, D-85748 Garching ,Germany*

[2] *Physik-Department E28, Technische Universität Munchen, Garching, Germany*

[3] *Optoelectronic Research Center, University of Southampton, Southampton, UK*

## Introduction

The soft X-ray (SXR) radiation in combination with tomographic reconstruction allows to monitor the local plasma emissivity and the impurity content in the plasma core with good temporal and spatial resolution. This is particularly important in the ASDEX Upgrade tokamak (AUG), where the core SXR radiation is dominated by tungsten (W) ions released from the first wall. A specific feature of the W SXR radiation are the large gradients caused by neoclassical impurity transport and steep dependence of the W radiation on the electron temperature.

AUG is equipped with seven cameras, providing in total of 224 lines of sight (LoS) and allow for a spatial resolution of 2 cm horizontally and 4 cm vertically (see Fig. 1). The cameras H and J at the low field side have three heads with separated diode arrays and the top camera K has two heads. Plasma radiation with energy below $\sim$1.8 keV is absorbed in a bended 75 $\mu$m thick Be filter. Signal to noise ratio in H-mode discharges is up to 100 for cameras H-K and lower, about 20, for G, L and M cameras observing plasma through the gaps in limiter/divertor.

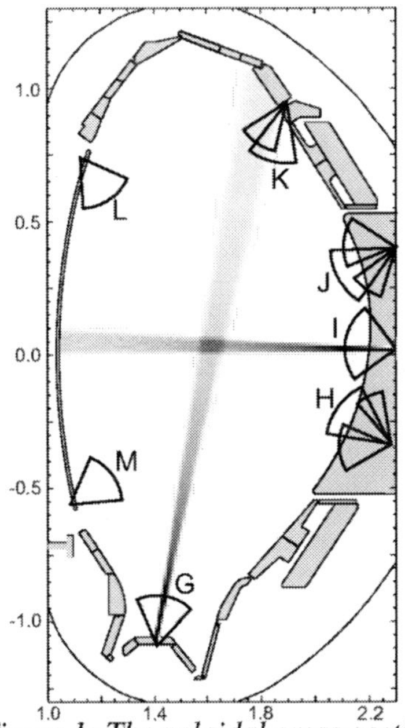

*Figure 1: The poloidal cross-section of the SXR diagnostic setup in AUG. An example for real profiles of lines of sight (LoS) is shown for one horizontal and one vertical LoS.*

Algorithms accurate and flexible tomography are required for a reliable reconstruction of the radiation contours. The standard methods like Cormack-Bessel or Tikhonov-Phillips method regularized by second order derivatives are not able to provide the required accuracy due to significant ringing artifacts, negative emissivity regions and underestimation of peaks. Therefore, we propose a new nonlinear regularization method and we will demonstrate its superiority to the Minimum Fisher Regularization (MFR) [1].

## Tomographic inversion of SXR measurements

The investigated algorithm belongs to the class of pixel based methods where the unknown emissivity profile is discretized into $N$ rectangular pixels $g_i$ and consequently the response of $j$-th detector $f_j$ on $i$-th pixel $g_i$ is described by the contribution matrix $\mathbb{T}_{ij}$. This task leads to a system of linear equations which is singular or ill-posed. A common method to find a unique solution is the Philips-Tikhonov regularization that minimizes the misfit together with a regularization functional $O(g)$, i.e $\min_g \left(\|\mathbb{T}g - f\|^2 + \lambda O(g)\right)$. The equation is solved by SVD of integrated projection matrix [2]. The solution is obtained in the form $g = \mathbb{V}\mathbb{D}^{-1}\mathbb{W}(\lambda)\mathbb{U}^T f$, where the matrix $\mathbb{U}$ is mapping the measurements to a sparse representation. The inverse of $\mathbb{D}$ is a diagonal matrix amplifying small features suppressed by the line integration and columns of $\mathbb{V}$ are the basis vectors in the reconstruction space. Finally, $\mathbb{W}$ is a filtering matrix suppressing the noise dominated dimension defined as $W_{ii} = (1 + \lambda/D_{ii}^2)^{-1}$. Optimal value of $\lambda$ parameters was selected by cross-validation PRESS method [6]

## Nonlinear regularization functional

A remarkable improvement of the SXR emissivity reconstruction can be achieved if the regularization is based on MFR. The key of its effectiveness is a nonlinear adaptation of the reconstruction basis $\mathbb{V}$ to the observed solution. The emissivity profile $g$ is consequently represented by a lower number of basis vectors. The compression of an energy spectrum $(U^T f)^2$ between the first and the last iteration of MFR is shown in Fig. 2. The number of basis vectors was reduced from ~140 to ~30 leading to a factor of 5 noise sensitivity reduction.

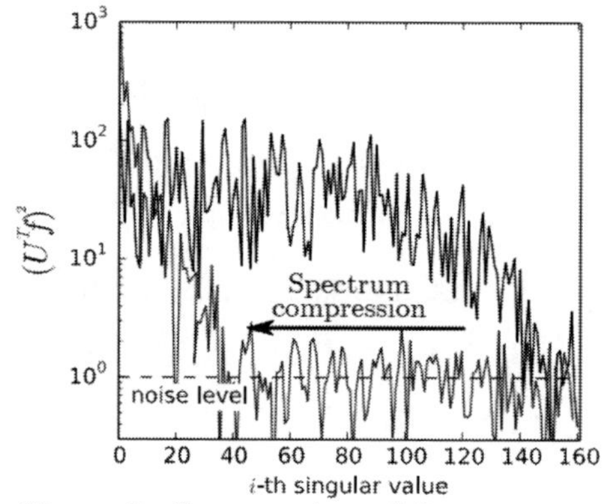


*Figure 2: Compression of the energy spectrum during nonlinear iterations. First iteration is in black, last in blue.*

Motivated by a frequent observation of exponentially peaking SXR radiation profiles, we propose a new regularization functional that results in minimal variation of the normalized gradient of emissivity: $O(g) = \sum_{i,j=1}^{2} \int \left(\frac{\partial^2 \ln(g)}{\partial x_i \partial x_j}\right)^2 dS$. We will call this method Minimum Normalized Gradient Regularization (MNGR). The discretization scheme is implemented analogously to MFR i.e. $O(g) = \sum_{i,j=1}^{2} g^T D_i^T W D_j D_i^T W D_j g$, where $\mathbb{W}_{ii} = 1/g_i$ and $D_i$ is a finite difference operator.

The new isotropic MNGR method was compared with anisotropic (i.e. smoothing along magnetic surfaces [3]) and isotropic MFR on four different artificial profiles called *phantoms*. The first profile called "Gaussian" is a moderately peaked profile defined by $\exp(-\rho^2/2)$. The second "Peaked" profile is defined as sum of this Gaussian (simulating the bremsstrahlung) and a peaked function $(\rho^2 + 0.02)^{-2}$. The third phantom "snake", representing impurity accumulation in the O-point of the mode, is obtained by displacing the peak in the previous profile radially and adding rotation in poloidal direction. And finally the "Hollow" phantom simulating hollow profile observed in discharges with ECRH heating and saturated 1/1 mode. The comparison is shown in the following table:

| | Gaussian | Peaked | Snake | Hollow |
|---|---|---|---|---|
| Anisotropic MFR | **1.4%** | 8.5% | 17% | **2.8%** |
| Isotropic MFR | 3.0% | 19% | 15% | 7.5% |
| Isotropic MNGR | 4.3% | **6.7%** | **9%** | 5.3% |

*Table 1: Relative differences ($\|g-g_0\|_2/\|g_0\|_2$) of the reconstructed profile with respect to phantoms*

These comparisons clearly show the advantage of the anisotropic MFR for profiles with a high level of poloidal symmetry. However if this assumption is not correct, like in the case of a snake mode, or snake-like sawtooth crash, the MNGR method is clearly better. Also for strongly peaked profiles the MNGR method performs slightly better than anisotropic MFR. The reconstruction quality is illustrated by Fig. 3 for a peaked profile. Isotropic MFR shows significant artifacts along the LoS, but contours of anisotropic MFR and new MNGR are almost indistinguishable despite no a priori knowledge about the magnetic equilibrium in MNGR is used.

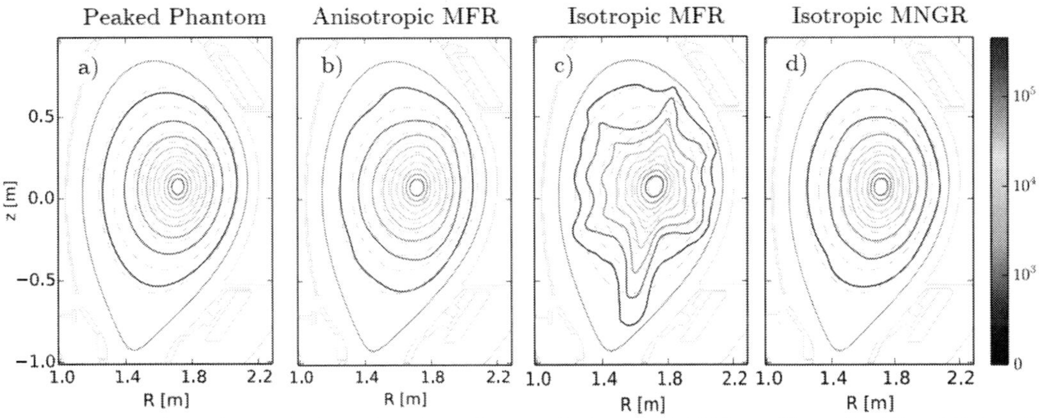


*Figure 3: a) a significantly peaked artificial profile simulating real AUG profiles with 1% additional noise was reconstructed with the b) anisotropic MFR, c) isotropic MFR and d) new isotropic MNGR method. Contours are shown in a nonlinear scale.*

## Multichannel harmonics filtering of SXR data

An issue closely related to the tomography of magneto-hydrodynamic (MHD) modes and the subsequent computation of 2D tungsten density profiles [4] is the treatment of noise in the signal. A common technique is based on bi-orthogonal filtering of the input data [5] or reconstructed profiles [1]. The first approach enhances the most significant spatial-temporal coherence but due to latter amplification in tomography small but important features will be lost. The post-processing approach is also not optimal because of an irreversible loss of reconstruction details caused by spatial filtering in the tomography algorithm applied on noisy data. We propose an alternative method for the filtering of quasi-stationary signals. A scheme of the filter is shown in the Fig. 4.

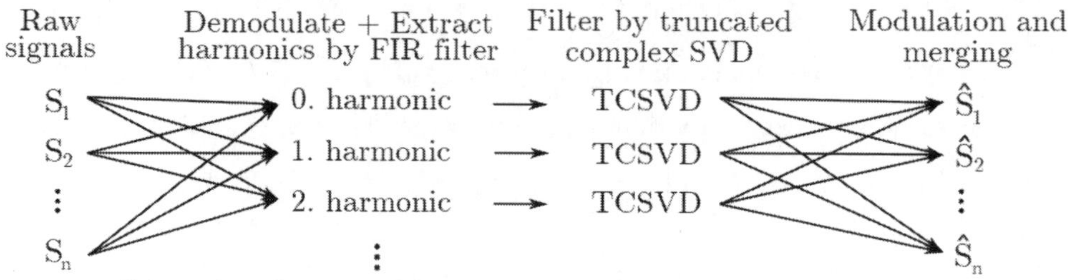


*Figure 4: A diagram of the multichannel filter algorithm architecture*

At first, the signals are demodulated by a complex exponential modulated on $M$-order of the fundamental frequency of the selected MHD mode. Further, the harmonic modes are extracted by the finite impulse response (FIR) filter designed using the Hamming window. The order of the filter enables the control of the bandwidth and the temporal resolution. This procedure results in slowly evolving complex signals describing phase and amplitude of the harmonics. This set of signals can be very effectively filtered by the truncated complex SVD (TCSVD) to suppress uncorrelated variation of the signals. Each channel is weighted according to its uncertainty in order to reduce influence of damaged or very noisy channels. Finally, the harmonics are modulated back and merged into the filtered signal. The obtained signals are virtually noiseless and therefore only a low spatial filtering will by introduced later by the tomography. Moreover, if multiple MHD modes with non-overlapping harmonics occur this method allows to isolate and reconstruct them separately.

## Conclusions

We have introduced a novel non-linear regularization method for tomography allowing to reconstruct a peaked radiation profiles. A comparison with a standard MFR method is showing significant improvement with respect to isotropic regularized version mainly due to reduction of the artifacts in the profiles with high poloidal asymmetries.

Moreover a very efficient method for the filtering of quasi-stationary signals from SXR was presented. Thanks to these methods, it is possible to recover more details of the MHD modes that are necessary for instance in the case of 2D W density profiles reconstructions [4] or reconstructions of a very weak modes which would be lost in noise.

## Acknowledgment

This work has been carried out within the framework of the EUROfusion Consortium and has received funding from the Euratom research and training programme 2014-2018 under grant agreement No 633053. The views and opinions expressed herein do not necessarily reflect those of the European Commission.

## References

[1] M. Anton et al., PPCF **38** (1996) 1849

[2] N. Terasaki, et al. Fus. Eng. Des. 34 (1997)686

[3] M. Odstrcil, et al., Nuc.Mth.in Phys. 2012 686

[4] M. Sertoli et al., this conference O4.129

[5] C. Nordone at al., PPCF **34**, (1992) 1447

[6] G.Golub, at al., Technometrics 21.2 (1979)

# Generation and suppression of runaway electrons in ASDEX Upgrade disruptions

G. Pautasso, P.J. McCarthy (*), C. Fuchs, R. Dux, S. Potzel, G. Papp, M. Sertoli, G. Tardini, K. Lackner, A. Scarabosio, A. Mlynek, L. Giannone, the ASDEX Upgrade team the EUROfusion MST team (+),
Max-Planck-Institut für Plasmaphysik, 85748 Garching, Germany,
(*) Department of Physics, University College Cork, Cork, Ireland, (+) http://www.euro-fusionscipub.org/mst1

## Introduction

Runaway electrons (REs) are expected to be generated in ITER disruptions at any significant plasma current. While the generation mechanisms of REs are believed to be understood, there is not a general consensus on which loss mechanisms are dominant in present experiments, and which one will prevail in ITER. This makes the calculation of the expected RE current and energy - and the consequent damage to the plasma facing components - uncertain and the design of the ITER disruption mitigation system challenging.

ASDEX Upgrade (AUG) disruptions, typically of diverted and elongated plasmas, do not exhibit formation of RE beams for the following reasons: Usually, a disrupting plasma has a density large enough to suppress the Dreicer generation; the avalanche generation is not expected to be significant in AUG; during the current quench, an elongated plasma becomes vertically unstable, moves up- or downwards, becomes limited on the lower or upper divertor and develops a halo, i.e. a region of open flux surfaces from which the REs escape, within milliseconds. Therefore, in order to allow studies of RE generation and dissipation during disruptions, a scenario for RE generation during the thermal quench had to be - and was - established during the 2014 campaign.

This paper presents the analysis and results of a series of RE-dedicated discharges. The magnetic equilibrium and the free-electron density profile reconstruction, the evaluation of the argon concentration during the whole RE beam, which allow the comparison between measured and computed friction force on the REs, are the main results of the work presented in the following.

## Experimental scenario and measurements

RE generation. REs have been generated by injecting between 0.07 and 0.2 bar×litre of argon in a low target density (volume averaged $n_e \leq 2\times10^{19}$ m$^{-3}$) circular plasma, with a current of 0.8 MA, a toroidal magnetic field of 2 T and 2 MW of ECRH heating. The injected argon induces a fast quench of the current carried by thermal electrons, followed by a long-lived runaway beam, forming a toroidal current of up to 300 kA, and lasting up to 380 ms. Although the target plasmas were similar (repeats), different initial RE current were observed, independently of the amount of argon injected – for reasons not yet identified.

RE suppression. In some of the RE beams generated, argon was injected a second time (0.17 and 0.7 bar×litre with a second valve) with a delay of 70 ms from the first argon puff. The RE current decay rates, after both the first and the second argon puff, show a clear dependence on the argon amount injected – i.e. the more gas the faster the decay, as illustrated by fig. 1 - suggesting that RE energy loss by collision (friction and radiation induced) with the electrons and the high Z impurity atoms is a significant RE dissipation mechanism. In the following section, the expected – from density measurements – and observed collisional damping are discussed.

Gas inventory. The amount of deuterium gas puffed into the vessel to fuel the plasmas under analysis, integrated over the whole discharge, ranges between $3\times10^{20}$ and $8\times10^{20}$ atoms. The amount of argon injected into the vessel and plasma, within a few ms after the opening of the fast valve, amounts to $2.4\times10^{21}$ argon atoms per bar originally in the valve reservoir. Since the valve reservoir pressure was varied between 0.7 and 7 bar (and the 7 bar injection followed a 1 bar), a total amount between

$1.7\times10^{21}$ and $1.9\times10^{22}$ argon atoms was injected into the vessel in a single discharge. A uniform distribution of these argon amounts within an empty vessel volume of 39.7 $m^3$ would result in neutral densities ranging between $4.3\times10^{19}$ and $4.8\times10^{20}$ $m^{-3}$. The ratio of the argon versus deuterium amount injected varies from 2 to 65.
The maximum vessel pressure, measured by two Baratron capacitance manometers (absolutely calibrated but too slow to measure the time evolution of the pressure) are in agreement with the total amount of injected gas. The fast pressure gauges were unfortunately saturated during the RE beam.

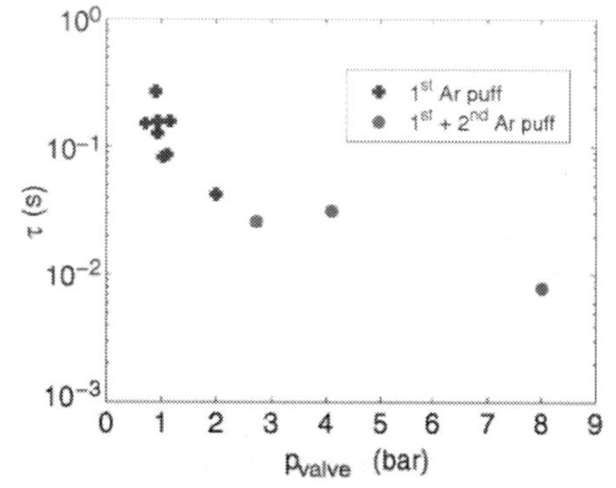


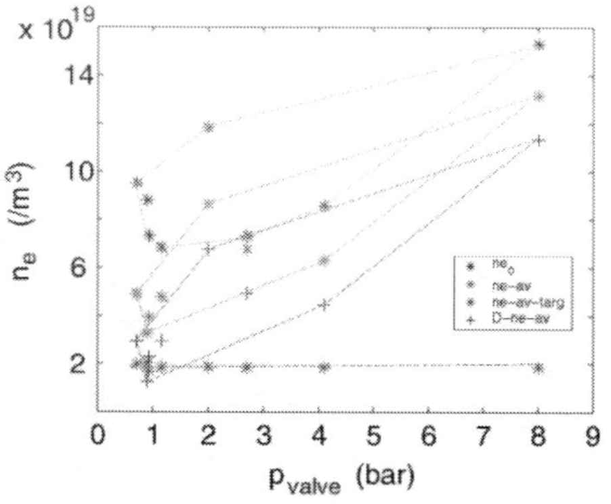


Left: fig. 1. *Decay time,* $\tau = I_p/(dI_p/dt) \sim \langle I_p\rangle/(\Delta I_p/\Delta t)$ *of the RE current versus argon pressure in the valve.*
Right: fig. 2. *Central ($ne_0$) and volume averaged electron (ne-av) density after argon injection; volume averaged electron density in target plasma and increase (D-ne-av) of the density after argon injection.*

Electron density. Line integrated measurements of the electron density, $n_e$, are available during the whole RE beam lifetime from 5 DCN chords and from 2 $CO_2$ chords. The $n_e$ profiles could be reconstructed for all the discharges during most of the RE beam because the magnetic equilibria could be calculated. The first and second argon injection increase significantly the volume averaged $n_e$: Fig. 2 reports an increase of a factor between 2 and 6 calculated from density measurements recorded 25 ms after the first argon puff ($p_{valve} < 2.5$ bar) and 10 ms after the second argon puff ($p_{valve} > 2.5$ bar).
Plasma position and its control. The circular plasma is vertically stable; the position and the shape of the plasma are in agreement with Soft X-Ray profiles. The plasma control system does not contribute to the vertical stability because the current in the control coils is part or all of the time saturated. A slow vertical displacement sets-in only in a few cases and does not affect the overall current decay rate of the RE beam.
Current control. The plasma current is ramped down by the control system; after injection of "small" amounts of argon, the controlled current follows the pre-programmed time trace; larger amounts of argon increase the RE losses, accelerate the current decay and the OH system cannot keep the actual current at the level of the reference value.
Current decay. The RE current does not exhibit a regular monotonic decay but rather a "bumpy" time behaviour correlated to sporadic MHD events, which are under investigation. The final phase of the current is often characterized by a faster loss of 30–100 kA, followed again by the slower decay of some tens of kA. Alternatively the current goes to zero with a constant slope.
The value of the edge safety factor ($q_{edge}$) at the time of the rapid current drop is not known in general, because the equilibrium reconstruction fails and radiation emission from the beam is not strong enough to allow its localization. Linear time extrapolation of $q_{edge}$, toward the end of the RE beam, indicates that its value will not drop below 5 at the time of the final current decay.
Radiated power. The foil bolometers measure a total radiated energy of the order of magnitude of the initial kinetic and magnetic energy of the RE beam. This suggests that radiation plays a significant role in the RE energy dissipation. Measurements of Bremsstrahlung and synchrotron radiation are not

available for this dedicated experiment.

HXR spectrum. Time-resolved measurements of gamma spectra from a gamma/neutron spectrometer are available and can give indication of the RE energy spectra, if interpreted with a model of RE-matter interaction. The spectrometer line of view crosses the beam on the midplane and intersects the inner wall, where the RE beam is limited and slowly scraped off. Therefore most of the photons should be generated by the interaction of the REs with the wall material. The gamma spectrum is peaked at energies of ~ 100 keV and does not change its form significantly as the RE current decays.

**RE collisional damping**

REs lose energy through different known mechanisms: By inelastic collision mainly with bound and free electrons, through small angle scattering and transfer of energy to the matter in which they travel; by collision with ions and Bremsstrahlung radiation; by cyclotron and synchrotron radiation. The first mechanisms, i.e. the friction force of the electrons on the REs, is discussed in this contribution, since the other mechanisms are evaluated to be smaller.

The RE current time evolution (with time constant $\tau_r$) is expected to be described by

$$1/\tau_r = dn_r/dt \ 1/n_r \simeq e(E_\phi - E_c)/p_r \qquad \text{Eq.(1)}$$

where $E_\phi$ is the toroidal electric field, $E_c$ is the so-called critical electric field, $n_r$ is the RE density and "$p_r$" is the average RE momentum.

Electric field. The evaluation of $E_\phi$ in the plasma requires a self consistent modelling of the plasma current evolution, which is mainly carried by the REs in the analysed discharges. Numerical modelling has not been carried out yet and therefore $E_\phi$ is not precisely known. Nevertheless, measurements of the loop voltage ($U_{loop}$) at different poloidal positions around the plasma are available. The closest – to the plasma – $U_{loop}$ coil is inside the vessel, behind the inboard shield, some 10 cm far from the the plasma surface; the measurement from this coil is equipped with a 30 Hz low-pass filter, which limits the validity of the measurements to slow transients (RE plateau). Faster measurements (30 kHz bandwidth) are available under the lower divertor and on the LFS, further away from plasma. The values of the poloidal flux from equilibrium reconstruction are also available and, in principle, allow the calculation of the electric field at the plasma surface. In reality, since this calculation requires the time derivative of the separatrix-coil flux difference, the resultant $E_\phi$ is very noisy.

An average $E_\phi$ can also be calculated knowing the plasma self-inductance and the multual-inductance between the plasma and other coils/structures:

$$2\pi R_c E_\phi \simeq L_p dI_p/dt + M_{OH-p} dI_{OH}/dt + M_{vv-p} dI_{vv}/dt \qquad \text{Eq.(2)}$$

where $M_{OH\text{-}p}$ and $M_{vv\text{-}p}$ are the mutual inductances between the OH coils and the plasma, and between the vessel structures and the plasma, respectively.

The fast and slow $U_{loop}$ measurements and the estimations with Eq. (2) give the following consistent picture:

- during the CQ, transient loop voltages of 40-50 V are reached;
- during the RE plateau, the loop voltage decreases from 10 to a few V in all cases;
- during the fast RE termination caused by the second argon injection, transient loop voltages up to 10 V are generated.

$E_\phi = U_{loop} / (2 \pi R_c) \sim U_{loop} / 10$, since the current major radius, $R_c \sim 1.6$ m.

The critical electric field. $E_c$ is a linear function of the free and bound electron density; therefore its estimation requires knowledge of the plasma atomic species and their ionization state:

$$E_c = e^3 n_e \ln(\Lambda_{e,free}) \ / \ (4\pi\epsilon^2 m_e c^2), \qquad n_e = n_{e,free} + \ln(\Lambda_{e,bound})/\ln(\Lambda_{e,free})\, n_{e,bound} \qquad \text{Eq.(3)}$$

Impurities other than argon are neglected here. Fig. 3 shows $E_c$ as function of single-ionized argon density.

Argon concentration. A total of 25 + 17 line integrated measurements of light emitted by $Ar^+$ at 401.39 nm are available during the RE beam lifetime with a high time resolution. They allow a spatially resolved calculation of the $Ar^+$ density, when the electron temperature and density are known, since the line radiance is given by the following integral along the line of sight:

$$L = 1/4\pi \int n_e \, n_{Ar} \, 0.343 \, f_1 \, X_{eff}(T_e, n_e) \, dl \qquad \text{Eq.(4)}$$

The photon emissivity coefficient $X_{eff}$ for the $Ar^+$ transition $3p^4(3P)4p\ ^4D_{7/2}$ -> $3p^4(3P)3d\ ^4D_{7/2}$ at 401.39 nm has been calculated from a collisional radiative model. The model does only consider the population of the upper states by electron excitation collisions. For the solution of the collisional radiative model, the ADAS208-code has been used. It gives the photon emission rate density per $Ar^+$ ion density and electron density for the whole multiplet. Within the multiplet, the specific line has a relative line strength of 0.343 when assuming LS-Coupling. The $Ar^+$ fractional abundance $f_1$ has been calculated from the ADAS data sets for rate coefficients of ionisation and recombination assuming a balance of ionisation and recombination rates; ($f_1 X_{eff}$) has a maximum at $T_e$ = 2.4 eV.
The argon concentration in the target plasma, calculated with Eq. (4), assuming a constant $T_e$ = 2.2 eV (argon singly ionized) is e.g. less than 35 % at 50 ms after the first argon injection.
Fig. 4 summarises the comparison between the measured $E_\phi$ and $E_c$ induced from argon and electron density measurements (t = 1.05 s) under different assumptions: argon density derived from spectroscopic measurements (spect); density increase after 1st argon puff all due to $Ar^+$ (dne); argon density given by a homogeneous distribution in the torus of the injected argon (inj). Even assuming that neutral argon or ArIII (not yet measured) play a role in the RE slowing down, the friction force of the free and bound electrons is smaller than the electric force on the REs and can not account for the observed RE current decay.

## Conclusions

The main finding of the described RE experiment is that the quantities of argon injected, to create and then suppress the REs, has a clear influence on the time decay of the RE beam, i.e. the more impurities, the faster the RE current decay. Nevertheless the critical electric field, calculated from the electron density and the concentration of singly ionized argon is 3-10 times lower than the toroidal electric field sustaining the RE beam, and cannot account for the RE current decay observed.

Acknowledgment. Michael Beck and Wolfgang Weisbart get the credit for maintaining and updating the AUG DMS. This work has been carried out within the framework of the EUROfusion Consortium and has received funding from the Euratom research and training programme 2014-2018 under grant agreement No 633053. The views and opinions expressed herein do not necessarily reflect those of the European Commission."

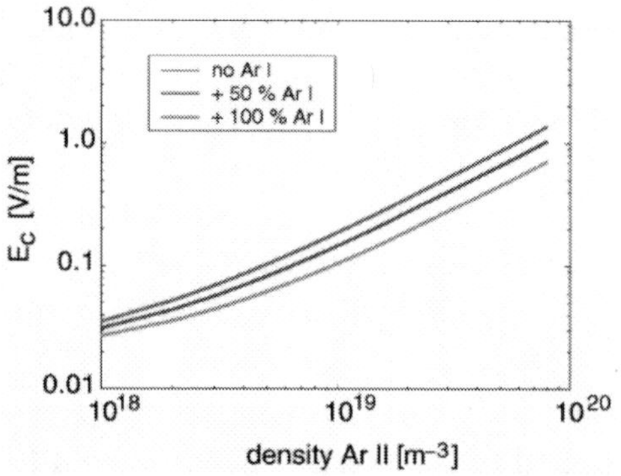


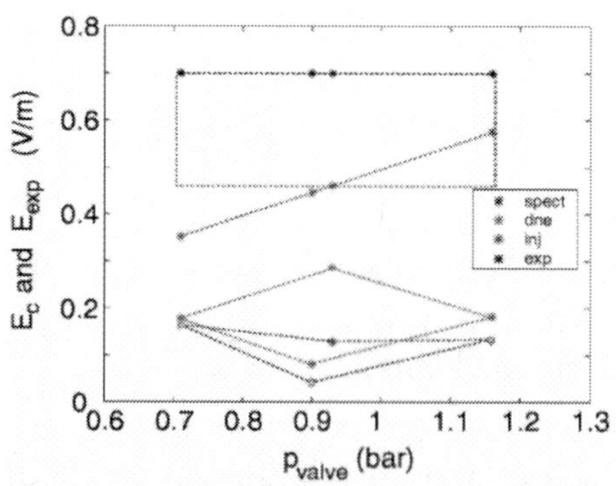


*Left: fig. 3. Critical electric field, $E_c$ (Eq. 3), as function of the Ar II density. A fully-ionized deuterium density of $2\times10^{19}$ $m^{-3}$ has been assumed in the background. The influence of neutral argon (Ar I) on $E_c$ is also shown. Right: fig. 4. Critical electric field, $E_c$, calculated under different assumptions (see text), and $E_\phi$ (Eexp) deduced from $U_{loop}$ measurements.*

# High radiation scenarios with radiation inside the confined region at ASDEX Upgrade

M. Bernert[1], F. Reimold[1,2], R. Dux[1], T. Eich[1], S. Glöggler[1], A. Kallenbach[1], B. Lipschultz[3], M. Wischmeier[1], the EUROfusion MST1 team* and the ASDEX Upgrade team

[1]Max-Planck-Institut für Plasmaphysik, Boltzmannstraße 2, 85748 Garching, Germany
[2]Forschungszentrum Juelich GmbH, Wilhelm-Johnen-Straße, 52425 Juelich, Germany
[3]University of York, York Plasma Institute, Heslington, York, YO1 9DD, United Kingdom

## Introduction

For future reactors, plasma scenarios with high radiated power fractions are required to reduce the power flux to the divertor target plates. While for ITER about 85-90% of the exhaust power has to be dissipated, for DEMO more than 95% are required to meet the material limits. The power dissipation consists of perpendicular transport, charge exchange losses and radiation losses, the latter being the biggest contribution [1, 2]. The radiated power can be increased by introducing seed impurities. These can be chosen to increase primarily the core (e.g. Kr, W) or scrape-off-layer radiation (e.g. N, Ne). To retain the enhanced energy confinement of the H-mode, the power flux across the separatrix has to stay above the H-L threshold. Therefore, at ITER additional radiation in the core is not feasible, while for DEMO the expected excess of alpha heating above the H-mode threshold allows (and requires) significant core radiation. [3]

## High radiation scenarios at ASDEX Upgrade

High radiation scenarios are investigated at the fully tungsten covered ASDEX Upgrade tokamak (AUG). These include divertor and core radiation, e.g. the double radiative feedback as demonstrated in [4]. In 2014, a pronounced detachment [5] of the outer target was achieved at highest heating powers ($P_{heat}/R \approx 13\,\mathrm{MW/m}$). In these scenarios up to 17 MW of neutral beam heating, 4 MW ion cyclotron resonance heating and 1 MW of electron cyclotron resonance heating (in O-2 mode) was applied. Significant central microwave heating is required to avoid the accumulation of tungsten. To facilitate detachment and mitigate magnetic modes, a neutral gas pressure of more than 2 Pa is required, which is achieved by reducing the cryo pump capacity. The divertor detachment is induced by intense impurity seeding of either krypton or nitrogen. With Kr about 75% of the heating power is radiated. With N seeding it is possible to achieve a radiated fraction of about 90%. These two scenarios are discussed here.

## Predominant edge radiation (N seeding)

The pronounced divertor detachment can be achieved at heating powers of about 20 MW with intense seeding of nitrogen ($\Gamma_N \geq 1.5 \cdot 10^{22} e^-/s$). Figure 2 shows the tomographic reconstructions of the radiation before and during the detached phase of discharge AUG #30506 (see also Fig. 1a). A poloidally strongly localized radiator around the X-point evolves with the detachment. Despite this radiator, the plasma is in a quasi-stable regime (withstanding ELMs and trips of heating power), unlike in the operation with carbon as first wall material [6].

*See http://www.euro-fusionscipub.org/mst1.

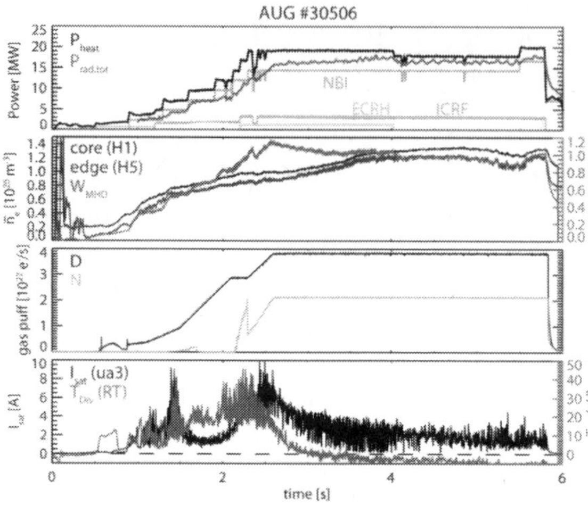


(a) Only N seeding

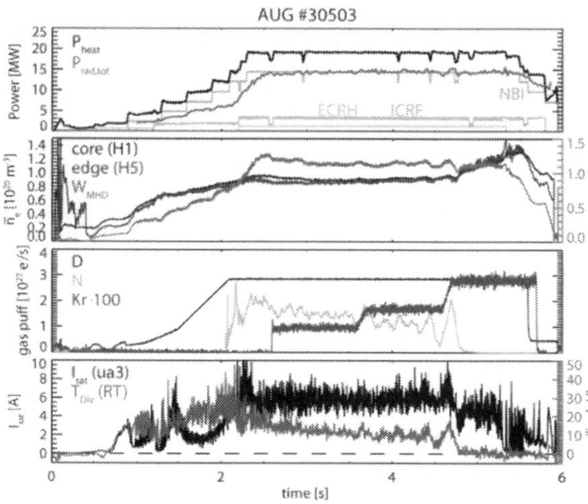


(b) Kr seeding (stepwise replacing N seeding)

Figure 1: Time traces of two high radiation discharges.

In the presented discharge, a constant N puff leads to a slow increase of the central N concentration ($c_N = 2\% \rightarrow 2.5\%$). With this evolution, the X-point radiator moves upwards inside the confined region. The time scale of this evolution is around 2 s.

When the X-point radiator moved significantly inside the confined region, intense deuterium line radiation can be observed in the region between the X-point and the radiator. This indicates electron temperatures ($T_e$) in the range of a few eV and high neutral densities inside the confined region. The pedestal top $T_e$ is reduced to about $400-500$ eV [7]. The $T_e$ measured in the midplane and the observation of deuterium line radiation at the X-point suggest the presence of parallel $T_e$ gradients. Such gradients combined with a X-point radiator were also reproduced by SOLPS modeling [7].

The reduction of pedestal top temperature does only marginally affect the overall confinement, as the pressure stays constant within $\rho_{pol} < 0.8$ [8]. Thus, the (volume integrated) energy confinement time is reduced by about 10% in comparison to the attached case, whereas the core pressure is not affected.

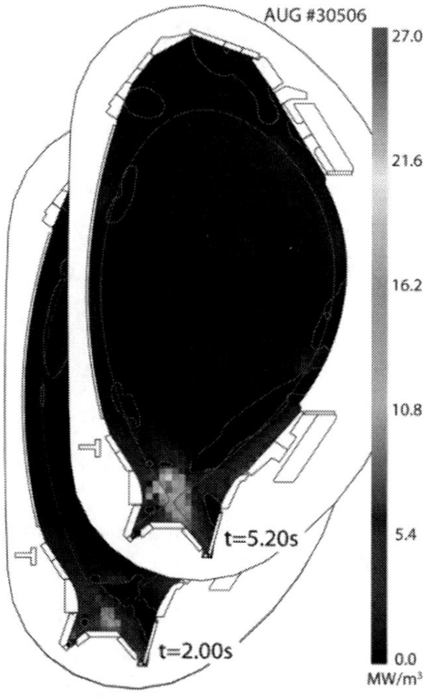


Figure 2: Evolution of the radiation for N induced detachment

**Predominant core radiation (Kr seeding)**

Krypton radiates predominantly at electron temperatures of about $300-500$ eV. Thus, it is mainly radiating in the pedestal region for standard scenarios at AUG.

Figure 1b shows a discharge, where Kr is step-wise substituting the (feedback-controlled) N. At the first two Kr levels, the plasma parameters change only marginally, Kr is compensating the effects of N (i.e. total radiated power, impact on confinement). At the highest seeding level ($\Gamma_{Kr} = 3 \cdot 10^{20} e^- / s$) a radiating ring in the pedestal region comes up (see Fig. 3).

The power flux into the divertor is reduced, the outer divertor detaches. This nonlinear response of the plasma on the Kr seeding indicates a radiation condensation: If sufficient Kr is present in the pedestal, it cools down the plasma to temperatures where Kr radiates more efficiently. By a density increase, which is regularly observed at the detachment, the pedestal radiation is even further increased.

Such a radiation condensation is observed to either keep the confinement constant for several energy confinement times (AUG #30503, $P_{heat} \approx 19$ MW) or reduce it (e.g. AUG #31645 & #31648, $P_{heat} \approx 10.5$ MW). The different responses of the plasma depend most likely on the resulting temperature profile and the dominant radiation location of Kr. The change of the radiation profiles before and after the radiation condensation (see Fig. 4) show that at a higher heating power, the dominant Kr radiation stays in the pedestal region, while in the lower heated discharges, the radiation moves further inside the confined region. In the first case, the temperature reduction is compensated by a density increase. At lower heating power, the density only changes marginally, leading to the reduced confinement.

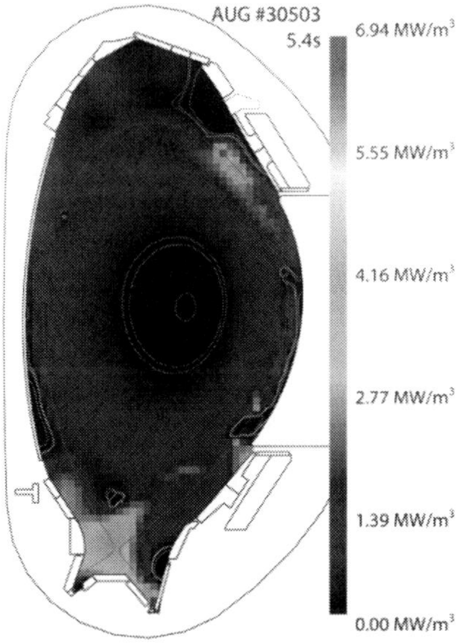


Figure 3: Reconstruction of total radiation in case of Kr induced detachment

**ELM transport:** The pedestal radiation with Kr is strongly influenced by ELMs. Measurements of AXUV diodes mainly observing the pedestal or the scrape-off layer (SOL) radiation at the beginning of Kr seeding are shown in Figure 5. The pedestal radiation increases with the Kr seeding but is strongly modulated by ELMs. Kr is transported into the pedestal in between ELMs and flushed out by ELMs [10]. The absence of some ELMs and the resulting strong increase of the pedestal radiation indicate a potential runaway situation. Therefore, high frequent ELMs are required to remain in a quasi-stable operation.

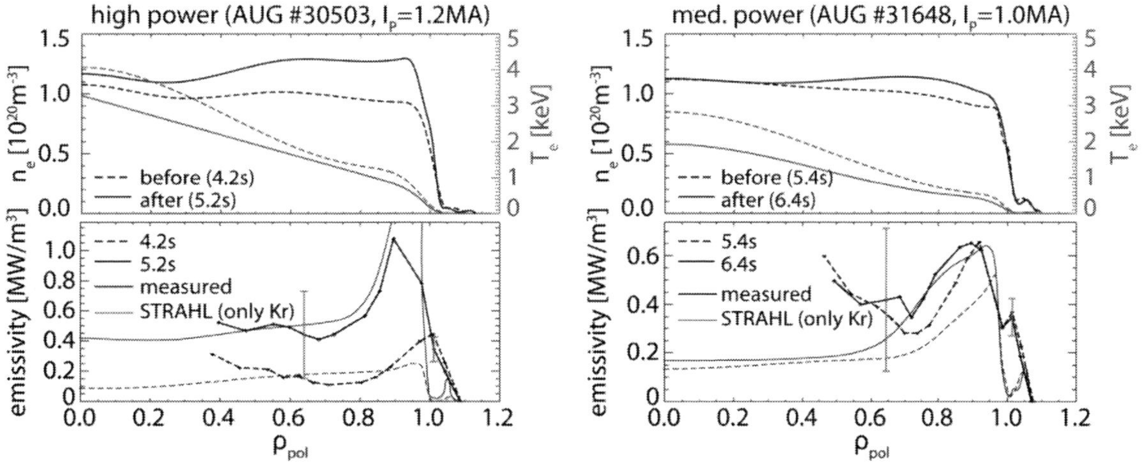


Figure 4: Density and temperature profiles and emissivity (measured by bolometry or simulated with STRAHL [9]) of two Kr seeded discharges before and after the radiation condensation.

The ELM frequency is reducing with increasing Kr radiation. This is in line with the type-I ELM frequency dependency on the heating power or the power flux over the separatrix [11], respectively. At the time of the radiation condensation, the ELM frequency drops drastically, which might lead to an unstable regime. However, at the same time the ELM type changes (most likely type-III), the ELM frequency increases and the plasma is back in a quasi-stable operation.

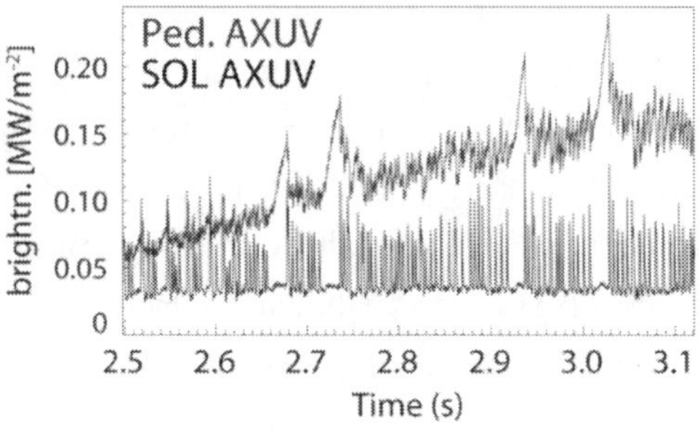


Figure 5: Influence of ELMs on pedestal radiation in presence of Kr.

## Discussion

For both scenarios (N or Kr seeding), the dominant radiation originates in the detached state from the confined region. While this is expected from krypton, the radiative loss parameter $L_z$ of nitrogen would suggest dominant divertor radiation. Hence, the radiation has for both scenarios a strong impact on the plasma parameters, resulting in the radiation condensation. The latter can occur due to a negative temperature dependence of $L_z$. The radiation condensation is for N poloidally localized above the X-point, where radiation losses compete with parallel heat conduction [12]. For Kr this condensation occurs more symmetrically distributed around the pedestal top and perpendicular heat transport must play an important role. Both scenarios are comparable to poloidally detached plasmas [12], reducing significantly the divertor heat load.

The presented scenarios were not yet operated in a steady-state condition. The long term stability is not shown, but both scenarios are stable versus transients, such as ELMs or beam trips. Further development and advanced control might be necessary. In case of the N seeded discharge, real-time algorithms, which control the position of the X-point radiator, should be feasible. The time scale of the evolution of the discharge (~2 s) is much slower than the typical transport times of the nitrogen into the pedestal (~ 50 - 100 ms). For Kr seeding the ELM behaviour plays an important role and further experiments are required.

## Outlook

The presented experiments demonstrate possible exhaust scenarios at reactor relevant values of $P_{heat}/R$. The optimal impurity mix has to be found in order to increase the fusion gain. While for current experiments, N and Kr are be adequate impurities to create the required radiation pattern, future experiments might have to use other impurities, e.g. xenon. However, the observations presented here highlight, that the intricate response of the plasma on the radiation has to be taken into account.

## Acknowledgement

This work has been carried out within the framework of the EUROfusion Consortium and has received funding from the Euratom research and training programme 2014-2018 under grant agreement No 633053. The views and opinions expressed herein do not necessarily reflect those of the European Commission.

## References

[1] M. Wischmeier et al, J. Nucl. Mat. **463**, 22 (2015) and references therein
[2] A. S. Kukushkin et al, J. Nucl. Mat. **438**, S203 (2013)
[3] H. Zohm et al, Nucl. Fusion **53**, 073019 (2013)
[4] A. Kallenbach et al, Nucl. Fusion **52**, 122003 (2012)
[5] A. Kallenbach et al, Nucl. Fusion **55**, 053026 (2015)
[6] J.A. Wesson et al., Nucl. Fusion **29**, 641 (1989)
[7] F. Reimold et al, J. Nucl. Mat. **463**, 128 (2015)
[8] F. Reimold et al, Nucl. Fusion **55**, 033004 (2015)
[9] R. Dux, STRAHL, annual IPP report **10/30** (2006)
[10] R. Dux et al, Nucl. Fusion **51**, 053002 (2011)
[11] H. Zohm et al, Plasma Phys. Control. Fusion **38**, 105 (1996)
[12] B. Lipschultz et al, Phys. Rev. Letters **81**, 1007 (1998) and references therein

# On the high density in the HFS far SOL at ASDEX Upgrade and its impact on plasma confinement

S. Potzel[1], M. Dunne[1], R. Dux[1], L. Guimarais[2], F. Reimold[3], A. Scarabosio[1], M. Wischmeier[1]
the ASDEX Upgrade team and the EUROfusion MST1 Team*

[1]*Max-Planck-Institut für Plasmaphysik, Garching, Germany*
[2]*Instituto de Plasmas e Fusao Nuclear, Universidade de Lisboa, Lisboa, Portugal*
[3] *Forschungszentrum Juelich, Juelich, Germany*

## Introduction

Detachment studies at AUG in L- and H-mode revealed that, when the inner divertor is strongly detached while the outer one remains attached, the ionization front close to the inboard separatrix has moved to the X-point and electron densities of the order of $10^{20}\,\mathrm{m}^{-3}$ are measured spectroscopically at the X-point. Simultaneously, radiative fluctuations appear in the high field side (HFS) divertor close to the X-point and a second high density region ($\sim 10^{20}\,\mathrm{m}^{-3}$) forms in the scrape-off layer (SOL) at the HFS [1]. In H-mode, this so called HFS high density (HFSHD) spans out to $\rho_{pol} \approx 1.15$ (or 10 times the power decay length $\lambda_q$) into the far SOL and extends well above the X-point, measured independently by spectroscopy and interferometry [2]. Furthermore, in the all-tungsten AUG, high performance discharges are usually conducted with medium to high gas fuelling rates where the HFSHD is always present. Seeding of impurities has been shown to improve the plasma confinement [3, 4], and also reduces the HFSHD and the corresponding neutral flux density $\Gamma_{n,\mathrm{HFS}}$ (typically by $50-70\,\%$).

Here, we will discuss the influence of the heating power, edge density, fuelling rate and plasma current on $n_e$ and the spatial extent of the HFSHD and show that there is a correlation between the $n_{e,HFSHD}$ reduction and confinement improvement due to impurity seeding.

## Discharge and diagnosic overview

The main diagnostics used to investigate the HFSHD are shown in figure 1. The local $n_e$ in the inner divertor and HFS SOL is determined by Stark broadening analysis (SBD) of the spectroscopically measured Balmer $\mathrm{D}_\varepsilon$ line [6]. While the spectroscopic system covers the entire divertor region, we focus here only on two lines of sight (LOS), named *DIV* and *HFS* view. The two LOS intersect also the LFS SOL, but the $\mathrm{D}_\varepsilon$ emission in this region is more than one order of magnitude lower than the emission in the inner divertor and thus negligible. In addition, two reflectometers measure the $n_e$ profiles at the LFS and HFS mid-plane [5]. Neutral fluxes are measured with AUG pressure gauges. The gauges used here are situated at the entrance to the inner divertor, at the outer mid-plane and below the divertor private flux region (PFR).

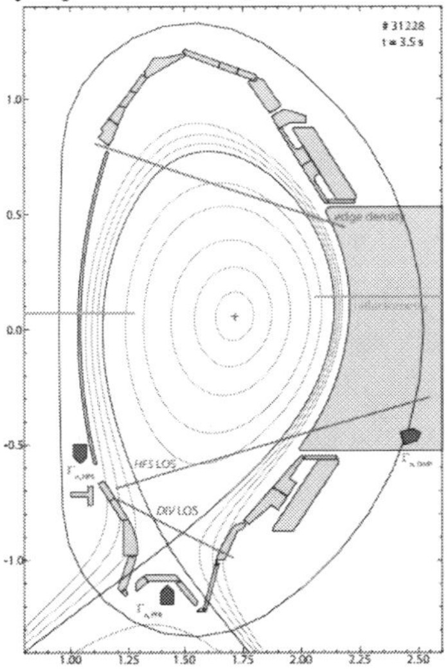

Figure 1: *DIV* & *HFS* LOS, neutral pressure gauges (red), reflectometry (cyan) and edge density interferometer (purple)

For this work, a set of 27 lower single null H-mode discharges with a magnetic field of $B_T = -2.5\,\mathrm{T}$ and a plasma current of 0.8 or 1.0 MA has been analyzed. The inner divertor plasma is always detached at the vertical target component (flat ion saturation current profile) whereas the outer divertor plasma remains attached. The total additional heating power $P_{TOT}$ in this database

*See http://www.euro-fusionscipub.org/mst1

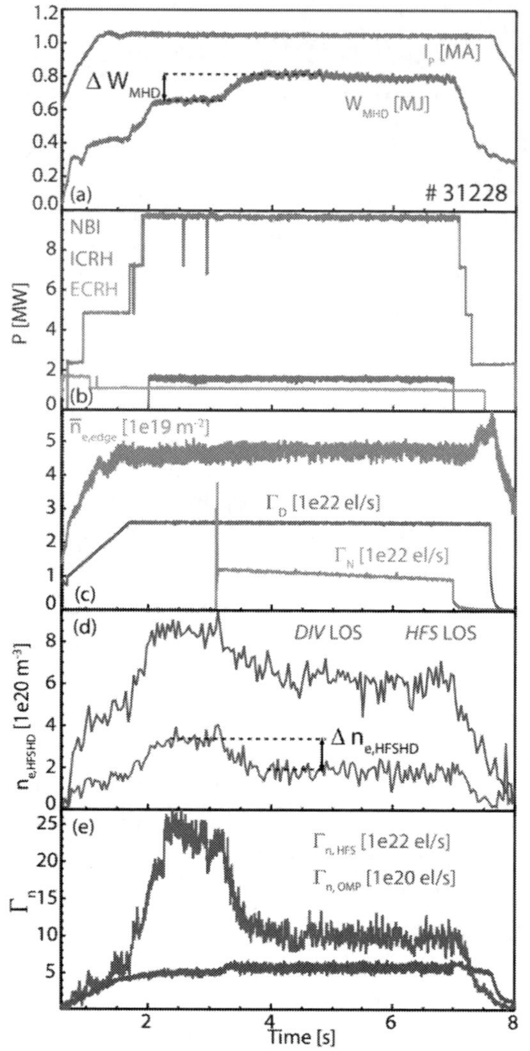


Figure 2: Time traces of a typical discharge of the database. See text for explanation

spans from 3 – 16 MW, consisting mainly of neutral beam heating (NBI), electron cyclotron heating $P_{ECH} \leq 2$ MW and, in some case, ion cyclotron heating $P_{ICH} \leq 4$ MW. The feed forward gas fuelling rate varies from $6 \times 10^{20} - 3.4 \times 10^{22}$ el/s. Figure 2 shows time traces of a typical discharge from the database. During the current ramp up phase of the discharge, the plasma fuelling and the heating power are simultaneously increased. In the flat-top phase of the discharge, the fuelling is kept constant, resulting in an approximately constant main plasma density and divertor neutral pressure, and the heating power is increased further. Correlated with the increase of the heating power the HFSHD develops. During the highest heating phase, densities of $8.5 \times 10^{20}\,\mathrm{m}^{-3}$ and $3.4 \times 10^{20}\,\mathrm{m}^{-3}$ are measured with the *DIV* and *HFS* LOS (see Fig. 1), respectively, which is more than one order of magnitude larger than the separatrix density. The neutral fluxes in this region, $\Gamma_{n,HFS}$ increase up to $2.5 \times 10^{23}$ el/s, which is about 500 times the neutral fluxes at the outer mid-plane $\Gamma_{n,mp}$, see Figure 2e. Moreover, the $n_e$ profiles at the HFS and LFS mid-plane are similar before the HFSHD formation with a separatrix density of $n_{e,sep} \approx 3 \times 10^{19}\,\mathrm{m}^{-3}$. When the HFSHD is present, the LFS $n_e$ profile is almost unchanged, but at the HFS a density of $n_e = 6 \times 10^{19}\,\mathrm{m}^{-3}$ (which is the maximal measurable density of the reflectometry system) is measured at a normalized poloidal radius of $\rho_{pol} \approx 1.09$. For details see [5]. Later in the discharge, nitrogen (sometimes also krypton in addition to nitrogen) is injected into the divertor private flux region while the heating power is kept constant. The impurity seeding leads to an increase of the plasma stored energy $W_{MHD}$ from 600 to 800 kJ and also reduces $n_e$ and $\Gamma_{n,HFS}$ in the HFSHD region.

**High field side high density scaling**

This section evaluates the dependence of the HFSHD on $P_{TOT}$, $I_p$ (or $q_{95}$ as $B_T$ is constant), the plasma fuelling and the main plasma density. Figures 3a-d shows $n_e$ measured with the *DIV* and *HFS* LOS versus $P_{TOT}$ for $I_p = 0.8$ MA and $I_p = 1.0$ MA, respectively. The color scale represents the applied deuterium fuelling. The increase of the HFSHD with $P_{TOT}$ is clearly visible in both LOS and both currents. For the *DIV* LOS, i.e. at the entrance of the inner divertor, it is at a given $P_{TOT}$ approximately independent of $I_p$ ($q_{95}$), see Fig. 3a,b. In contrast, in the HFS SOL (*HFS* LOS), $n_e$ is higher in $I_p = 0.8$ MA discharges at a given $P_{TOT}$. Thus, the spatial extent of the HFSHD from the entrance to the inner divertor towards the HFS mid-plane is larger with lower $I_p$ (higher $q_{95}$). This is revealed in Fig. 3e, where $n_{e,HFSHD}(DIV) - n_{e,HFSHD}(HFS)$ is shown. Furthermore, the poloidal gradient (as well as the absolute value) of $n_{e,HFSHD}$ towards the HFS mid-plane increases with heating power, independent of $I_p$ ($q_{95}$). In order to disentangle the dependence of the fuelling and plasma density on the HFSHD, we first need to discuss the effect of $\Gamma_D$ on $n_{e,edge}$. In 1.0MA discharges, the plasma density does not change with fuelling levels $\Gamma_D \geq 0.5 \times 10^{22}$ el/s (Fig. 3f). Hence, the increase of the HFSHD at a given $P_{TOT}$ (Fig. 3a, c) is due to the fuelling rate $\Gamma_D$ and independent of $n_{e,edge}$. However, with $I_p = 0.8$ MA, there is a clear increase of $n_{e,edge}$ with increasing $\Gamma_D$ until $\Gamma_D \approx 1.8 \times 10^{22}$ el/s, then $n_{e,edge}$ almost saturates. The increase of the HFSHD at a given $P_{TOT} \leq 7$ MW (Fig. 3b,d) is thus caused by the increase of $n_{e,edge}$. Above $P_{TOT} = 7$ MW, the increase of the HFSHD at a given $P_{TOT}$ is smaller and mainly caused by the increased fuelling rate. To sum this up, as long as $n_{e,edge}$ can be increased with the fuelling rate $\Gamma_D$, there is a strong, indirect influence of $\Gamma_D$ on the HFSHD via the $n_{e,edge}$ increase. Otherwise, the HFSHD still increases with $\Gamma_D$, but this is a smaller effect. Finally, there is a clear correlation between the HFSHD and the neutral fluxes in HFS SOL $\Gamma_{n,HFS}$, as shown in Figure 3g,h. The increase of the HFSHD with $\Gamma_{n,HFS}$ is independent of $P_{TOT}$, $n_{e,edge}$ and the fuelling rate $\Gamma_D$, but there is a different scaling for different $I_p$ ($q_{95}$). Moreover, at a given $\Gamma_{n,HFS}$, the HFSHD extent is larger at lower $I_p$ (higher $q_{95}$). It is important to note that there is no correlation between the HFSHD and the neutral fluxes in the PFR, i.e. the divertor neutral pressure (not shown).

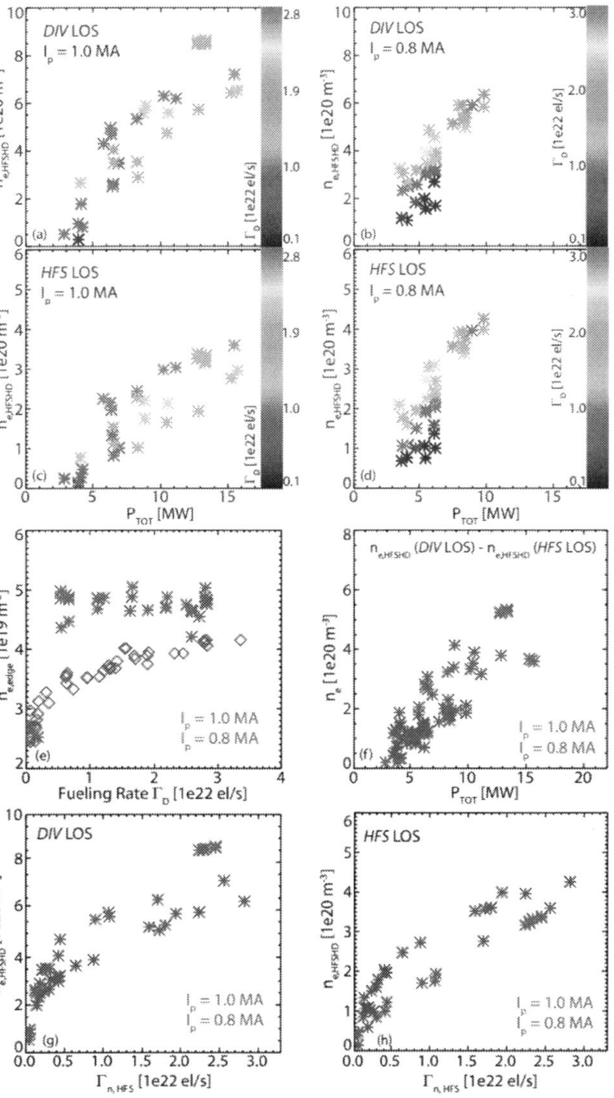


Figure 3: (a-d) HFSHD versus total heating power, (e) edge density versus fuelling rate, (f) difference of *DIV* & *HFS* LOS versus total heating power, (g-h) HFSHD versus neutral fluxes in the HFS SOL.

**Correlation between the reduction of the HFSHD and confinement improvement due to impurity seeding**

As already mentioned above, seeding of impurities leads to a reduction of the HFSHD and also to an increase of the stored energy $W_{MHD}$. Figures 4a,b reveal their correlation, where the HFSHD reduction $\Delta n_{e,HFSHD}$ is plotted against the confinement improvement $\Delta W_{MHD}$, which are the differences of the two quantities before and during the impurity seeding (see Fig. 2). Independent of the impurity species, $\Delta W_{MHD}$ increases clearly with the absolute value of $\Delta n_{e,HFSHD}$ for both, the *DIV* and *HFS* LOS. Moreover, the more $N_2$ is injected, the stronger is $\Delta n_{e,HFSHD}$, which is in Fig. 4c, only shown for the *HFS* LOS, but the *DIV* LOS is similar. Note that the abscissa shows the injected impurity atoms per second. If on top of $N_2$ Kr is injected ($\Gamma_{Kr} \leq 3.5 \times 10^{19}$ atoms/s), the HFSHD is reduced even further, which reveals that Kr has the similar effect as $N_2$ on the HFSHD. Moreover, as about 2 orders of magnitude less Kr atoms are

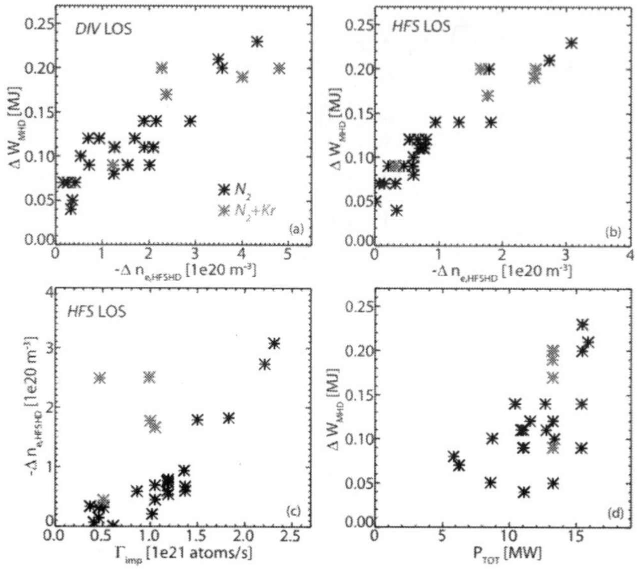


Figure 4: (a,b) Confinement improvement versus HFSHD reduction for the *DIV* & *HFS* LOS, (c) HFSHD reduction versus the impurity seeding rate in atoms per second for the *DIV* LOS, (d) Confinement improvement versus total heating power.

injected compared to N indicates that the amount of atoms (e.g. $Z_{eff}$) is not what matters for the $\Delta n_{e,HFSHD}$ and $\Delta W_{MHD}$ increase, but the amount of power which can be radiated per impurity atom ($L_Z$). Finally, at higher heating power a stronger confinement improvement can be achieved (Fig. 4f), depending on the impurity injection rate. This is consistent with the observations that the HFSHD is larger at higher $P_{TOT}$. Therefore, a higher $\Delta n_{e,HFSHD}$ and hence $\Delta W_{MHD}$ can be reached, given that enough impurities are injected. It must be noted that the outer divertor always remains attached in this database and, thus, at higher $P_{TOT}$ it is also possible to inject more impurities.

## Conclusions

All together, this indicates that the HFSHD and the correlated neutral fluxes $\Gamma_{n,HFS}$ are governed by the local recycling in the inner divertor, which is mainly determined by $n_{e,edge}$ and the power reaching the HFS SOL. The recycling fluxes under these conditions are about $1-2$ orders of magnitude larger than the gas fuelling rate, which has only a weaker effect on the HFSHD. Moreover, the spatial extent of the HFSHD is smaller at higher $I_p$ ($q_{95}$).

Finally, there is a clear correlation between the reduction of the HFSHD $\Delta n_{e,HFSHD}$ and the confinement improvement $\Delta W_{MHD}$ due to impurity seeding. The higher $\Delta n_{e,HFSHD}$, the stronger $\Delta W_{MHD}$, irrespective of the impurity species. Consequently, at higher $P_{TOT}$ and hence higher $n_{e,HFSHD}$, a higher $\Delta W_{MHD}$ can be achieved, provided that there is enough impurity injection, which radiates power in the SOL, to reach a high $\Delta n_{e,HFSHD}$. This is all consistent with the fact that in the old AUG and JET C-machines the confinement was generally better due to different operational boundary conditions such as the fuelling [7]. Similar to N, C radiates mainly in the SOL and is always present in the SOL as it is produced even at low temperatures via chemical sputtering. Hence, the HFSHD, which was also observed in AUG-C [8], could be intrinsically reduced in C-machines, leading to a better confinement. The correlation between $\Delta n_{e,HFSHD}$ and $\Delta W_{MHD}$ due to impurity seeding can only be quantitatively investigated, however, with extensive 2D simulation codes such as SOLPS, where the HFSHD has not been successfully reproduced yet.

## Acknowledgment

This work has been carried out within the framework of the EUROfusion Consortium and has received funding from the Euratom research and training programme 2014-2018 under grant agreement No 633053. The views and opinions expressed herein do not necessarily reflect those of the European Commission.

## References

[1] S. Potzel, et al., NF **54** 0130011 (2014)
[2] S. Potzel, et al., JNM, (2015)
[3] J. Schweinzer, et al., NF **51** 113003 (2011)
[4] M. Dunne, et al., *This conference*
[5] L. Guimarais, et al., *This conference*
[6] S. Potzel, et al., PPCF, **56**, 025010 (2014)
[7] M. Beurskens, et al., PPCF, **55**, 124043 (2013)
[8] K. McCormick, et al., JNM, **390**, S465 (2009)

# Indirect measurement of the poloidal rotation in the core of ASDEX Upgrade plasmas with charge exchange recombination spectroscopy

A. Lebschy[1,2], R. M. McDermott[1], B. Geiger[1], M. Cavedon[1,2], M. G. Dunne[1], R. Dux[1], R. Fischer[1], E. Viezzer[1] and the ASDEX Upgrade Team

[1] *Max-Planck-Institut für Plasmaphysik, Boltzmannstr. 2, D-85748 Garching, Germany*

[2] *Physik-Department E28, Technische Universität München, D-85748 Garching, Germany*

The radial electric field ($E_r$) is a critical parameter to understand the physics of microinstabilities and turbulent transport in fusion plasmas [1]. While difficult to measure, it can be calculated by evaluating the radial force balance equation for a given ion species $\alpha$:

$$E_r = \frac{\nabla p_\alpha}{e Z_\alpha n_\alpha} - u_{\mathrm{pol},\alpha} B_{\mathrm{tor}} + u_{\mathrm{tor},\alpha} B_{\mathrm{pol}}, \quad (1)$$

where $\nabla p_\alpha$ denotes the radial pressure gradient, $eZ_\alpha$ the charge, $n_\alpha$ the density and $u_{\mathrm{pol},\alpha}$ and $u_{\mathrm{tor},\alpha}$ the poloidal and toroidal rotation velocities of the species $\alpha$. $B_{\mathrm{pol}}$ and $B_{\mathrm{tor}}$ are the poloidal and toroidal magnetic field components determined from the magnetic equilibrium reconstruction. One of the most commonly used diagnostic to evaluate the ion quantities is active charge exchange recombination spectroscopy (CXRS), which measures the line-emission resulting from the charge exchange process between fully stripped impurity ions with neutrals from the highly energetic neutral beam injection (NBI) sources.

At the plasma edge, where the temperatures are low, the direct observation of $u_{\mathrm{pol}}$ is a standard technique (see e.g. [2]). In the plasma core, however, the direct measurement of $u_{\mathrm{pol}}$ is more challenging due to practical geometry and signal to noise issues associated with the poloidal lines-of-sight (LOS) as well as atomic physics issues. The energy dependence of the charge exchange cross-sections [3] in combination with the finite lifetime of excited atomic states lead to an apparent rotation in the poloidal direction [4] exceeding the desired poloidal rotation at high temperatures and magnetic fields.

**Indirect measurement principle**

A promising method to obtain more accurate measurements of the poloidal rotation, which is less sensitive to the atomic physics effects, is the evaluation of the inboard-outboard asymmetry of $u_{\mathrm{tor}}$. This method is based on the neoclassic description of the plasma flow, i.e. the plasma flow can be expressed as a sum of a component parallel to the magnetic field and a purely solid body rotation:

$$\vec{u} = \hat{u}(\psi)\vec{B} + \hat{\omega}(\psi) R \vec{e}_{\mathrm{tor}}, \quad (2)$$

where $\hat{u}$ and $\hat{\omega}$ are flux functions [5]. This means that by evaluating the plasma rotation, preferably $u_{\mathrm{tor}}$, at two distinct locations on the same flux surface, $\hat{u}$ and $\hat{\omega}$ can be evaluated and, therefore, the poloidal rotation can be measured indirectly. In order to have a precise estimation of the rigid body component, the two points with the largest separation are ideally taken, i.e. on the low-field side (LFS) and the high-field side (HFS). In reality, the measurement points are given by the injection geometry of the NBI sources. A big advantage of this measurement technique is additionally that the toroidal rotation asymmetry caused by a poloidal flow is amplified

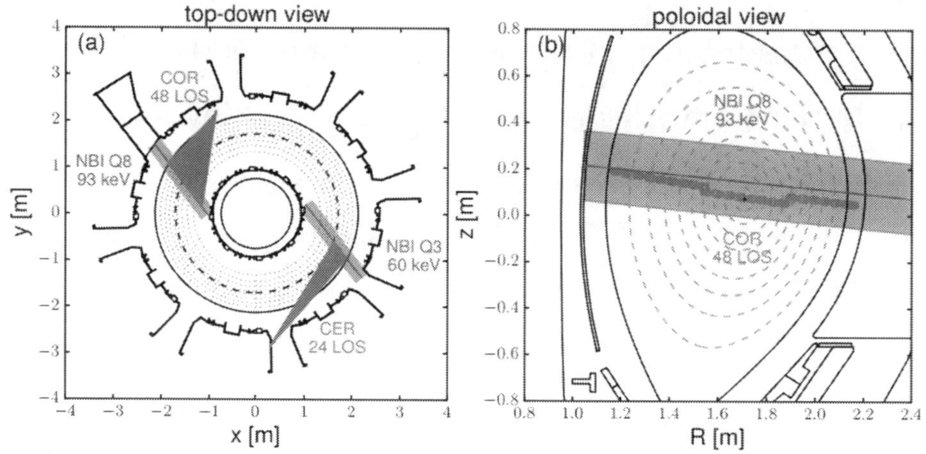

Figure 1: Upgrade of the core CXRS systems at ASDEX Upgrade.

by the safety factor $q$ [8].

Equation (2) has been derived under the assumption that $|u_{\text{tor}}| \ll u_{\text{therm}}$, i.e. without HFS-LFS asymmetries in the impurity density ($n_{\text{imp}}$) profile so that $n_{\text{imp}}$ is a flux function. If the toroidal rotation approaches the thermal velocity then a more general form of the plasma flow has to be considered [6]. However, in this work only plasmas with an impurity mach number below 0.5 were considered so that $n_{\text{imp}}$ can be assumed to be a flux function.

The described indirect measurement technique has been applied as well at DIII-D [7] and TCV [8]. Compared to the direct measurement technique they were able to reduce the uncertainty in $u_{\text{pol}}$ by a factor of 2 (from $2\,\text{km}\,\text{s}^{-1}$ to $1\,\text{km}\,\text{s}^{-1}$). At higher ion collisionalities ($\nu_i^* > 0.2$), good agreement between the experimentally observed poloidal rotation and the expected one from neoclassics are reported. However, in discharges with an internal transport barrier (ITB) or at very low ion collisionalities, deviations of $u_{\text{pol}}$ from neoclassic are reported [9]. These findings are also consistent with previous results reported from direct measurement of $u_{\text{pol}}$ [3, 10–13].

**Upgrade of the core CXRS systems at AUG and initial profile comparison**

A new core CXRS system (COR, see figure 1) has been installed at ASDEX Upgrade (AUG) in order to measure the HFS-LFS profiles of the CXRS quantities and, thus, enable the indirect measurement of $u_{\text{pol}}$. This new system has in total 48 toroidal LOS aligned along the NBI source 8, which has an injection energy of 93 keV. The system measures the ion temperature ($T_i$), $u_{\text{tor}}$ and $n_{\text{imp}}$ from the pedestal top on the LFS to the pedestal top on the HFS. Figure 1a shows a top-down view of AUG and illustrates the geometry of the two NBI sources being used for core CXRS measurements as well as the toroidal LOS geometry of the new core CXRS diagnostic and the pre-existing one (CER). Figure 1b illustrates the alignment of the new CXRS system along NBI source 8 in the poloidal plane. All 48 LOS are imaged simultaneously by two F/2.8 flexible wavelength spectrometers. The systems use Princeton Instruments ProEM $512 \times 512$ 16 μm pixel cameras that, with the 25 regions of interest defined on each camera, are capable of a maximum readout rate of 2.4 ms. The standard time resolution, however, is between 5 and 10 ms due to signal to noise considerations.

Figure 2 compares the measured profiles of $T_i$ and $u_{\text{tor}}$ for the two systems (CER in black; COR in blue and red) in a typical H-mode plasma with a core line-integrated density of

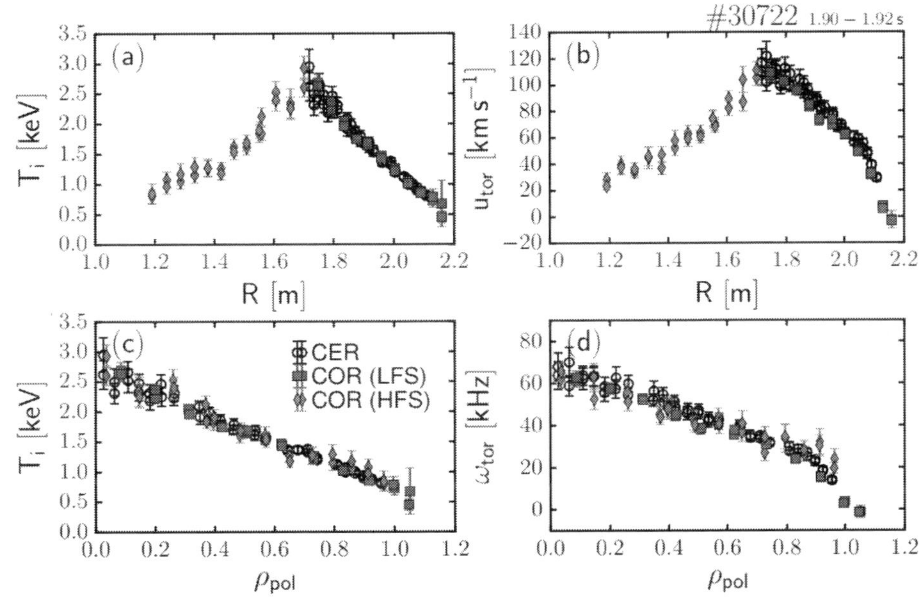


Figure 2: Comparison of HFS-LFS CXRS profiles.

$n_e = 7.4 \times 10^{19}\,\mathrm{m}^{-2}$ and 5 MW of NBI heating. A maximum of 3.0 keV for $T_i$ and $120\,\mathrm{km\,s}^{-1}$ for $u_{tor}$ were measured by both diagnostics in the plasma core. These profiles are shown in figure 2a and b as function of the major radius $R$. There is overlap between the CER and COR on the LFS such that the measurements of these two systems can be compared directly in this region. There is a very good agreement of the ion temperature on the LFS (see figure 2a) indicating that the dispersion of the spectrometers are well understood. The toroidal rotation of the two systems also agrees within error bars on the LFS.

When mapping the measurement locations to the normalized poloidal flux function $\rho_{pol}$, good agreement between the HFS and LFS ion temperatures is generally found. In the 2014 campaign, however, there were some cases where the LFS and HFS $T_i$ did not agree completely. This could be due to uncertainties in the magnetic equilibrium which effects the mapping of these measurements in flux coordinates. This possibility is being investigated.

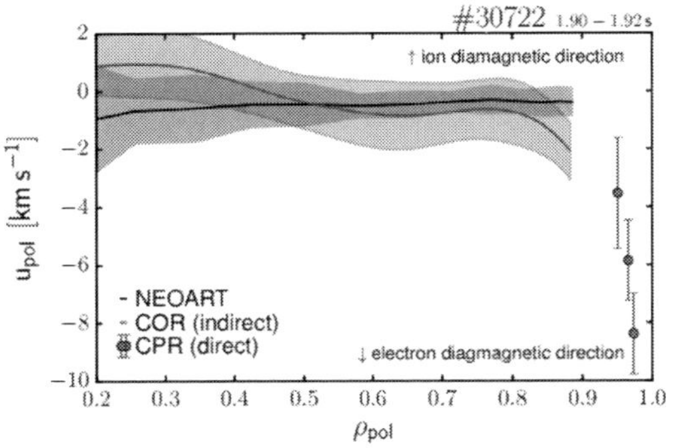


Figure 3: Initial $u_{pol}$ profile.

Comparing $u_{tor}$ (see figure 2b) on HFS and LFS, one can see that $u_{tor}$ is higher on the LFS. This is because the rotation is, to first order, a solid body rotation, meaning that the toroidal rotation frequency $\omega_{tor} = u_{tor}/R$, is the flux function and not the rotation velocity. By plotting $\omega_{tor}$ on HFS and LFS (see figure 2d) one can identify deviations away from a solid body rotation and, hence, identify the presence of a poloidal flow. In the plasma shown, the good agreement of $\omega_{tor}$ on HFS and LFS indicates that $u_{pol}$ is small in the plasma core.

Figure 3 shows the calculated poloidal rotation (blue curve) from the asymmetry in the rotation frequency shown in figure 2d. The method reveals a poloidal rotation in the electron diamagnetic direction in the region $0.4 < \rho_{pol} < 0.9$. The magnitude and direction of $u_{pol}$ for that area is in very good agreement with the neoclassical predictions from NEOART [14]. Towards the plasma center, the indirect method reveals a change in the sign of the poloidal rotation. In this region, the reconstructed rotation agrees within the error bars with the neoclassical prediction. In figure 3 also direct measurements of the impurity poloidal velocity from the edge CXRS system (CPR) are displayed. While the directions and magnitudes measured by the core and edge systems agree, they can not be compared directly as there is no radial overlap between the diagnostics.

The errors on the poloidal rotation are calculated by a Monte-Carlo approach. The measured toroidal rotation and measurement location of the LOS are varied within their uncertainties and their resolutions, respectively. Each "new" profile is fit with a spline to create a distribution of calculated poloidal velocities which provides the error estimate of this quantity. In total this measurement technique can resolve poloidal flows with an accuracy of $\pm 1\,\mathrm{km\,s^{-1}}$.

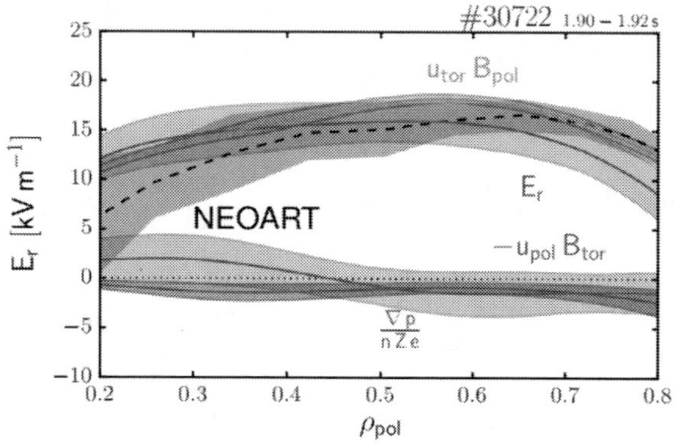

Figure 4: Initial $E_r$ profile and comparison to the neoclassical prediction from NEOART.

Being able to measure $T_i$, $u_{tor}$, $u_{pol}$ and $n_{imp}$, one can evaluate the radial force balance equation and calculate $E_r$. Figure 4 shows the calculated $E_r$ in purple and the single contributions to the radial electric field shown in equation (1) in blue, green and red. It reaches a maximum value of $18\,\mathrm{kV\,m^{-1}}$ at $\rho_{pol} \sim 0.6$ and is dominated by the contribution of the toroidal rotation times the poloidal magnetic field component.

## Summary

The new core CXRS system measures $T_i$, $u_{tor}$ and $n_{imp}$ from the pedestal top on the LFS to the pedestal top on the HFS. This enables first HFS-LFS comparisons of the CXRS quantities in the core of ASDEX Upgrade plasmas. In general, good agreement between the LFS and HFS profiles is found. The poloidal rotation can be measured indirectly by evaluating the asymmetry of $u_{tor}$ on LFS and HFS, which means that this new diagnostic can provide measurements of $E_r$ in the core of AUG plasmas.

## Acknowledgment

This work has been carried out within the framework of the EUROfusion Consortium and has received funding from the Euratom research and training programme 2014-2018 under grant agreement No 633053. The views and opinions expressed herein do not necessarily reflect those of the European Commission.

## References

[1] G. Dif-Pradalier et al, PRL **103**, 065002 (2009)
[2] E. Viezzer et al, Nucl. Fusion **53**, 053005 (2013)
[3] R. E. Bell et al, Physics of Plasmas **17**, 082507 (2010)
[4] R. E. Bell and E. J. Synakowski, AIP Conf. Proc. **547**, 39 (2000)
[5] P. Helander, *Collisional Transport in Magnetized Plasmas*, Camebridge: Cambridge University Press, 2002
[6] K. H. Burrell and C. Chrystal, Phys. Plasmas **21**, 072316 (2014)
[7] C. Chrystal et al, Rev. of Sc. Instruments **83**, 10D501 (2012)
[8] A. Bortolon et al, Nucl. Fusion **53**, 023002 (2013)
[9] B. A. Grierson et al, Nucl. Fusion **53**, 063010 (2013)
[10] K. Crombé et al, PRL **95** 155003 (2005)
[11] A. R. Field et al, PPCF **51** 105002 (2009)
[12] K. D. Marr et al, PPCF **52** 055010 (2010)
[13] W. M. Solomon et al, Phys. of Plasmas **13**, 056116 (2006)
[14] A. G. Peeters, Phys. of Plasmas **7**, 268 (2000)

# Experimental characterisation of sawtooth crash precursors on ASDEX Upgrade via Soft X-Ray tomography

D. Vezinet[1], V. Igochine[1], A. Gude[1], D. Meshcheriakov[1]
and the ASDEX Upgrade Team and the EUROfusion MST1 Team
[1] *Max Planck Institute for Plasma Physics, Boltzmannstr. 2, 85748 Garching, Germany*

**Introduction** Magnetic reconnection in the core of tokamak plasmas is responsible for a quasi-periodic relaxation process that significantly affects energy and particle confinement [1, 2]. Named *sawtooth crashes* - after the characteristic shape they entail on the time traces of the core electron temperature and Soft X-Ray (SXR) measurements - this type of event has become both a classic of tokamak operation and a subject of continued theoretical and experimental interest. Indeed, the exact mechanisms governing the reconnection time and the triggering of the crash remain partially unexplained [3, 4], and the recent addition of heavy impurities in some Tokamaks complicates further both the physics and the data interpretation [5].

In an effort to provide both return on experience for tokamak control and experimental input for MHD model validation, this paper describes the systematic quantitative characterisation, using SXR tomography, of approximately 50 sawtooth crash precursors drawn from 9 shots of the ASDEX Upgrade database. In particular, the time evolution of the growth rate of the internal m=1 kink mode is robustly estimated up to the crash. The growth rate is found to remain relatively constant during a preliminary phase of a typical duration of 5000-30000 times the poloidal Alfven time scale $\tau_A = \frac{r_1\sqrt{\rho\mu_0}}{B_P(q=1)} = 0.2 - 0.4\ \mu s$, (where $r_1$ is the $q = 1$ surface minor radius) at a value around or below 0.1 % of $\frac{1}{\tau_A}$. It then surges in less than 2000 $\tau_A$ to values which are typically of the 0.2-0.5 % of $\frac{1}{\tau_A}$ (but the error bars in this short phase are larger). Also observed (but not quantitatively characterised) is the systematic absence of phase shift (i.e. poloidal position) of the SXR hot core at the time of the crash as would be expected from complete reconnection and redistribution of all impurities with plasma flux, and the existence of a post-cursor after the collapse, that keeps on rotating as though it were a diminished continuation of the precursor.

In the following, the criteria governing the selection of the presented cases are briefly explained, then a brief description of the methodology of analysis is provided and exemplified in a particular case, and finally the results of the systematic analysis of all the crashes are presented and discussed.

**Data selection and methodology** The zoology of the effects of sawtooth crashes on SXR measurements is large, and in order to stick to cases that can be interpreted beyond reasonable doubt, we dismissed all cases for which trajectory of the SXR maximum was not thought to give a reasonable estimation of the trajectory of the initial hot core. Hence, cases like hollow or flat profiles, profiles flattened by visible locked modes, very asymmetric profiles, compound crashes or very slow crashes were all dismissed. Even for peaked profiles, the SXR maximum may be significantly shifted from the hot core in high rotation plasmas due to tungsten [5]. This may cause a phase shift on the line-integrated measurements, but it should not affect significantly the estimation of the displacement growth rate since the trajectory is shifted but not significantly distorted.

The methodoly of analysis, inspired from [6], is as follows: SXR tomographic inversions are performed for all time steps of a 200 kHz-sampled interval encompassing the sawtooth precursor and collapse. At each time step the position of the profile maximum is robustly determined by calculating the center of mass of the top 5 % of the profile. Several different regularisation methods have been tested, ranging from 1st order linear regularisation to linearised minimum Fisher regularisation, and two different mesh sizes were tried. While the shape of the profile may be significantly affected by the regularisation method (the minimum Fisher being more prone to reconstructing peaked profiles), the position of the profile maximum was found to be a quite robust feature, with typical differences below the mesh size.

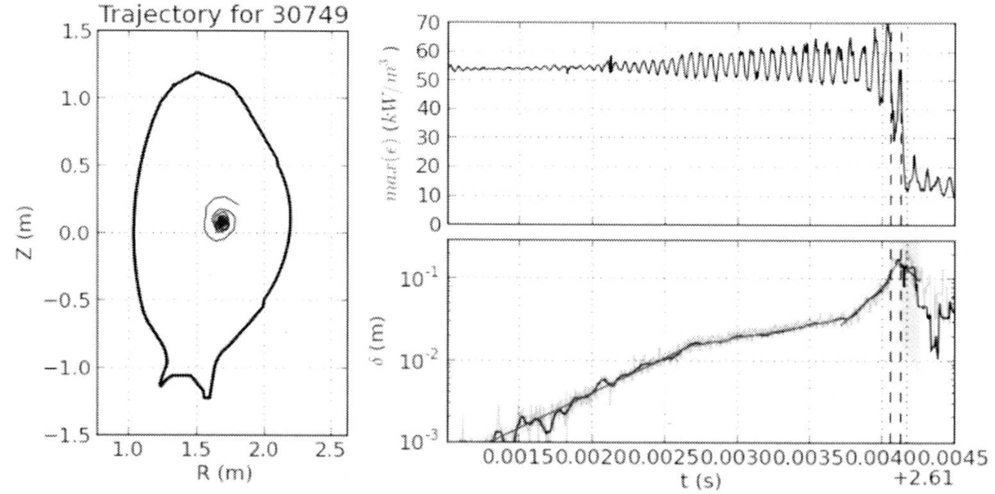


Figure 1: *(left) Trajectory in the poloidal cross section of the hot core during the precursor oscillations (upper right) time trace of the SXR emissivity inside the hot core (lower right) time trace of the estimated displacement, in semilog scale, from which piece-wise constant growth rate can be derived. The temporary dampening of the growth rate in the intermediate phase may be due to fishbones*

From the maximum trajectory tracking, which reveals an outward spiraling hot core, the mode frequency can be estimated and then used to compute a robust estimate of the time evolution of the core displacement using a geometrical method based on a one-period averaging of the trajectory. The time trace of the displacement can then be fitted by exponentials - since it can be seen in 1 that the growth rate is piece-wise constant. Both the displacement and the growth rate can then be normalised by the poloidal Alfven time scale to facilitate comparisons with literature [7].

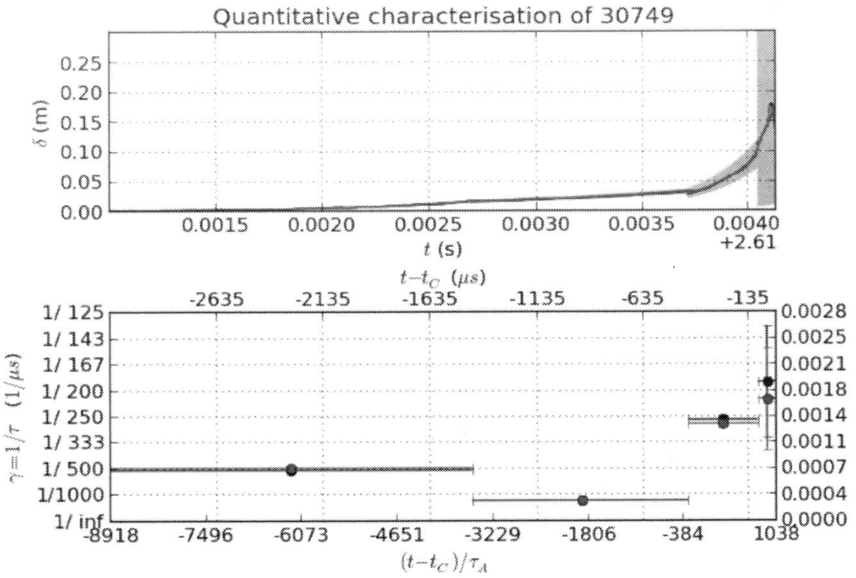


Figure 2: *(top) Time trace of the core displacement, computed with two different regularisation functionals (black and red). Shaded areas indicate the fit quality (bottom) Associated estimated piece-wise constant growth rates, with horizontal error bars showing the fitting interval, and vertical error bars indicate the fitting quality. Axis are doubled to show absolute and normalised values*

**Systematic analysis** Applying this methodology to the other crashes reveals good reproducibility of the features observed in the previous example. Indeed, one can clearly identify a long preliminary phase of slow and relatively constant growth and a second much shorter phase of sudden increase of the growth rate. The quantitative agreement of the normalised displacements and growth rates in fig.3 is also remarkable. Indeed, these shots have significantly varying plasma parameters, since the central elecron temperature is $1.5-3.5\ keV$, the electron density $0.4-1.10^{20}\ m^{-3}$, the plasma rotation $100-250\ km/s$, the plasma current $0.8-1\ MA$ and the toroidal magnetic field $1.8-2.8\ T$. One can see that once normalised they all offer a quite similar picture. Some points show a large intial growth rate. They correspond to rare cases for which the beginning of the growth could be clearly seen and these cases are not numerous enough, in the presented results, to check the reproducibility of this behaviour.

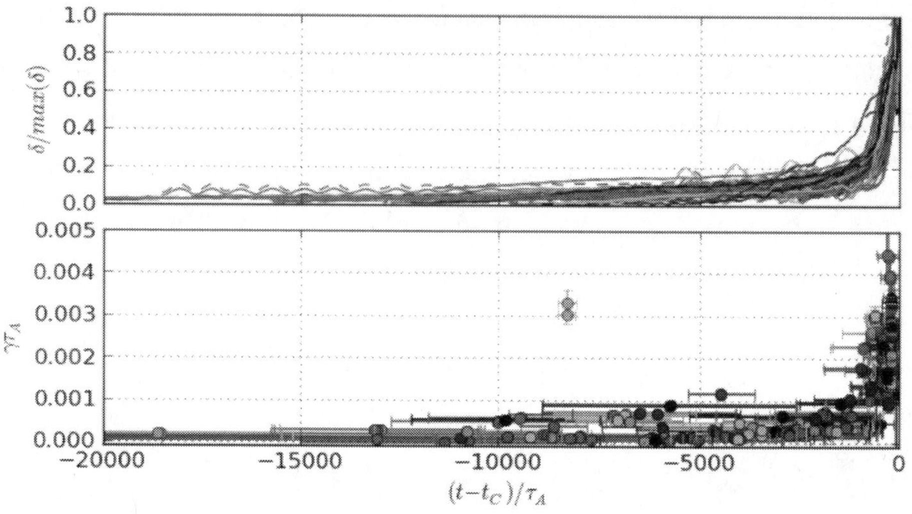


Figure 3: *(top) Time trace of the normalised core displacement for all the crashes (bottom) Associated estimated piece-wise constant growth rates*

Though the quantitative characterisation of the postcursor will not be shown here, it must be said that all the crashes presented here do have a postcursor structure which, on the SXR at least, resembles a diminished continuation of the precursor (no phase shift is observed at the collapse time). Also, the methods and hypotheses (regarding W redistribution during the mode growth [5]) used for this analysis do not permit to guarantee that a growth rate superior to the mode frequency can be correctly quantified, hence, all such cases were removed from the presented results.

This work has been carried out within the framework of the EUROfusion Consortium and has received funding from the Euratom research and training programme 2014-2018 under grant agreement No 633053. The views and opinions expressed herein do not necessarily reflect those of the European Commission. This project was supported by an EFDA Fellowship for Fusion Research 2014-2016.

## References

[1] T. Nicolas et al., Physics of Plasma **21**, 012507 (2014)

[2] J-H. Ahn et al., Poster to this conference

[3] R. J. Hastie, Astrophysics Space Science **256**, 117-204 (1997)

[4] I. T. Chapman, Plasma Phys. Control. Fusion **53**, 013001 (2011)

[5] M. Weiland, Plasma Phys. Control. Fusion, Sp. is.: MHD Stab. Contr., 2015

[6] J. A. Wesson, Nucl. Fus. **30**, 2545 (1990)

[7] A. Y. Aydemir, Phys. Fluids B **4**, 3469 (1992)

# Quantitative prediction of the blob generation rate

G. Fuchert[1], D. Carralero[1], P. Manz[2], U. Stroth[1,2], E. Wolfrum[1], ASDEX Upgrade team
[1] Max-Planck-Institut für Plasmaphysik, 85748 Garching, Germany
[2] Physik-Department E28, Technische Universität München, 85748 Garching, Germany

## Introduction

Filaments of increased pressure ("blobs") that transport particles and energy towards the wall are commonly observed in the scrape-off layer (SOL) of magnetically confined fusion experiments. They can lead to an enhanced perpendicular energy and particle transport, wall deterioration, and increased impurity densities. However, there is still no model to predict the generation rate of blobs, which would be of major interest for understanding and modeling the SOL physics and to ensure the safe and reliable operation of future fusion power plants.

In this novel approach, the generation rate of blobs is related to the dispersion of the underlying turbulence and the stability of the blobs themselves. Predictions from this model are compared to experimental values from ASDEX Upgrade.

## Prediction

In many fusion experiments it is observed that blobs are generated around a specific radial position ("generation region"). Inside of this region, usually a micro instability is present. According to many experimental observations (e. g. from Alcator C-Mod [1]), blobs detach from quasi-coherent structures of this instability (*seed fluctuations*). In ASDEX Upgrade this region is observed close to the separatrix. Probe measurements in L-mode plasmas indicate that the plasma turbulence in this generation region is dominated by a drift-wave-like instability. Hence, it is assumed that the generation rate is given by the frequency of this dominant edge instability. Since the frequency is linked to the spatial structure size by the dispersion relation it is necessary to know on which size scale the blob generation takes place. In ASDEX Upgrade the characteristic size of coherent structures remains constant when crossing the separatrix [2]. Hence, the blob size can be used as a proxy for the scale of the generating instability.

For a fixed observer the characteristic rate of seed fluctuations is then given by the background $\mathbf{E}\times\mathbf{B}$-drift velocity $v_{\mathrm{E}_r\times\mathrm{B}}$ and the phase velocity $v_{\mathrm{ph}}$ of the instability:

$$f_{\mathrm{g}} = (v_{\mathrm{ph}} + v_{\mathrm{E}_r\times\mathrm{B}})/\lambda_{\perp}\,. \tag{1}$$

For ASDEX Upgrade L-mode discharges it is assumed that in the SOL $E_r = -3\nabla T_{\mathrm{e}}/e$ [3] and, as discussed above, that the seed instability is drift-wave-like. Hence, $v_{\mathrm{ph}}$ is approximated by the diamagnetic velocity $v_{\mathrm{dia}}$ and it follows:

$$f_{\mathrm{g}} = \left| \frac{1}{3\delta_\theta} \cdot \left( \frac{\nabla p_{\mathrm{e}}}{e n_{\mathrm{e}} B} - 3\frac{\nabla T_{\mathrm{e}}}{eB} \right) \right| , \tag{2}$$

with the electron density $n_{\mathrm{e}}$, pressure $p_{\mathrm{e}}$ and temperature $T_{\mathrm{e}}$, the elementary charge $e$ and magnetic field strength $B$. Here, the conversion between the poloidal blob size $\delta_\theta$, defined as the full (poloidal) width at half maximum (FWHM) of the density perturbation, and the poloidal wave length $\lambda_\perp$ of the dominant edge instability $\lambda_\perp = 3\delta_\theta$ has been used, since the FWHM of the positive part of a sine wave is approximately $\lambda/3$.

Due to the small fraction of large amplitude events and the experimental focus on these, most of the generated structures are not detected. The detection rate $f_{\mathrm{d}}$ can be predicted as follows: The seed instability generates blobs at a rate $f_{\mathrm{g}}$. They propagate radially outward, where a fraction $\gamma_{\mathrm{d}}$ of them is detected, depending on the blob amplitude distribution function and threshold amplitude $a_{\mathrm{thresh}}$:

$$f_{\mathrm{d}} = f_{\mathrm{g}} \gamma_{\mathrm{d}} . \tag{3}$$

In principle $\gamma_{\mathrm{d}}$ could be calculated from a given amplitude distribution. Since this distribution is not yet known, $\gamma_{\mathrm{d}}$ will be determined from experimental data below.

This line of argument is only valid as long as the blob transit time at a given position $\tau_{\mathrm{d}}$ is small compared to the characteristic waiting time $\tau_{\mathrm{w}} = 1/f_{\mathrm{g}}$. In a simple model, for $\tau_{\mathrm{d}} \geq 1.7\tau_{\mathrm{w}}$, a typical blob is a compilation of $(\tau_{\mathrm{d}}/\tau_{\mathrm{w}}) + 0.3$ single events (these relations can be derived from Fig. 1, the calculations are not shown here). This reduces the number of clearly distinguishable blobs. In this case $f_{\mathrm{d}}$ is predicted as

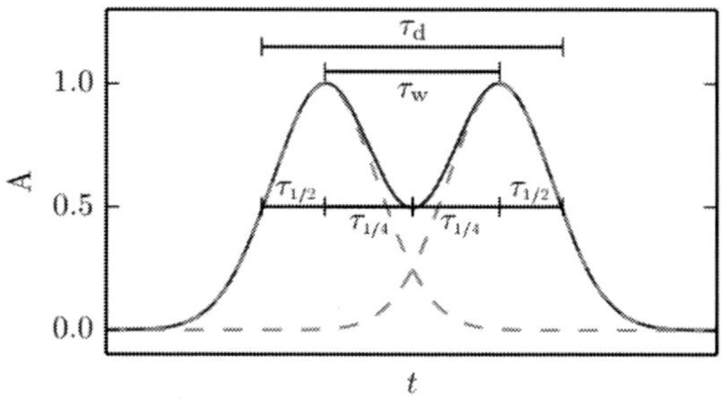


Figure 1: *In a pressure sensitive quantity A, single blobs are not properly resolved for $\tau_{\mathrm{d}} \geq 1.7\tau_{\mathrm{w}}$. A characteristic structure (closed line) is a compilation of $(\tau_{\mathrm{d}}/\tau_{\mathrm{w}}) + 0.3$ single events (dashed lines).*

$$f_{\mathrm{d}} = \frac{f_{\mathrm{g}}}{\tau_{\mathrm{d}}/\tau_{\mathrm{w}} + 0.3} \gamma_{\mathrm{d}} = \frac{1}{\tau_{\mathrm{d}} + 0.3\tau_{\mathrm{w}}} \gamma_{\mathrm{d}} . \tag{4}$$

For the comparison with experimental data, a circular blob shape is assumed, i. e. the radial size is equal to the poloidal one ($\delta_r = \delta_\theta$).

**Comparison with experimental data**

Blob detection rates were measured in L-mode discharges at ASDEX Upgrade (#29321-#29326 and #29887, toroidal magnetic field $B_{\mathrm{t}} = -2.5\,\mathrm{T}$, safety factor $q_{95} = 5.32$, plasma current $I_p = 800\,\text{kA}$, and ECRH power of 600 kW) for different fueling levels. Using probes, the detection rate $f_d$ was measured at the radial position $\rho_{pol} = 1.025$. The kinetic profiles required for evaluating the model equations are deduced from the standard edge diagnostics [4].

From these data and Eqs. (3) and (4), the unknown $\gamma_d$ can be estimated. For the evaluation, $\tau_d$ is identified with the radial auto-correlation time of the probe signal. A good agreement is obtained for $\gamma_d = 0.035$, as is shown in Fig. 2, where the circles correspond to the predictions of Eq. (3) for $\tau_d/\tau_w < 1.7$ and the squares to $\tau_d/\tau_w \geq 1.7$, Eq. (4) (evaluated at the separatrix). The observed detection rate scales linearly with the prediction. However, the slope is smaller than one. Hence, the aforementioned assumptions should be revised in future experiments. Nevertheless, the majority of points is predicted correctly within a factor of two.

The successful prediction of $f_d$ included experimental data for $\delta_\theta$ and $\tau_d$ from probe measurements. A predictive model, however, should only depend on either measured or simulated density and temperature profiles. Now, the model equations can be evaluated by taking the most stable blob size $\delta_b$ as the poloidal blob size and estimating $\tau_d \approx \delta_b/v_{b,r}$ (assumption of a circular blob shape).

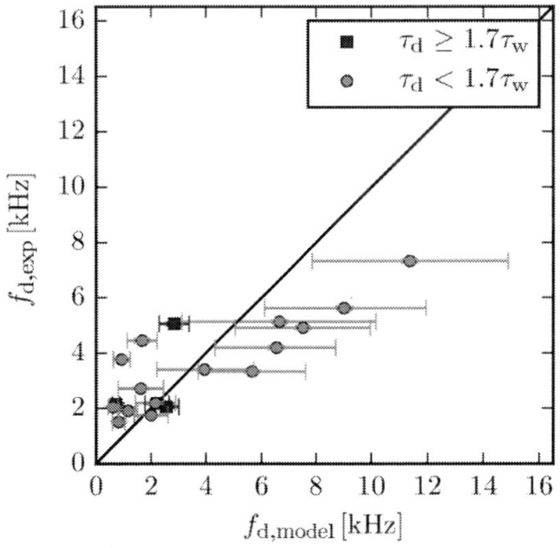


Figure 2: *Experimental blob detection rate $f_{d,exp}$ around $\rho_{pol} = 1.025$ compared to a prediction from Eq. (3) for $\tau_d < 1.7\tau_w$ and Eq. (4) for $\tau_d \geq 1.7\tau_w$, respectively with $\gamma_d = 0.035$.*

At ASDEX Upgrade, the identification of the relevant blob regimes is subject of present research [5]. There are at least two different regimes relevant for typical L-mode discharges depending on the connection of the plasma to the target plates. For low edge densities the size and velocity of blobs seem to scale as predicted by the well-studied sheath limited regime, probably influenced by a finite ion temperature $T_i$.

Following Ref. [6] and assuming that in the SOL $T_i \approx 3T_e$, $l_\parallel = 70\,\text{m}$ (parallel connection length) and $R = 2.155\,\text{m}$ (major radius of the separatrix), the following scaling laws are obtained:

$$\delta_b = 4\rho_s \left(\frac{l_\parallel^2}{\rho_s R}\right)^{1/5}, \quad v_{b,r} = 4c_s \left(\frac{\rho_s}{\delta_b/2}\right)^2 \frac{l_\parallel}{R}. \tag{5}$$

Above a threshold line-integrated edge density of about $2.5 \cdot 10^{19}\,\text{m}^{-3}$ the blob size is observed to scale linearly with the density. The identification of this regime with one of the cases described in the literature is ongoing. Therefore, in this regime a constant blob velocity of $v_{b,r} \approx 600\,\mathrm{m/s}$ and an empirical scaling for the blob size

$$\delta_b[\mathrm{m}] = 0.047 \cdot (\langle n_{\mathrm{edge}} \rangle [10^{19}\,\mathrm{m}^{-3}] - 2.5) + 0.030 \tag{6}$$

are assumed in accordance with the experimental values reported in Ref. [5]. Using these equations and $\gamma_{\mathrm{d}} = 0.035$ the prediction of $f_{\mathrm{d}}$ is repeated (shown in Fig. 3).

The absolute values are again predicted correctly within a factor of two for most data points. Improving the scaling laws for $\delta_{\mathrm{b}}$ and $v_{\mathrm{b},r}$ in the high-density regime may improve this agreement further.

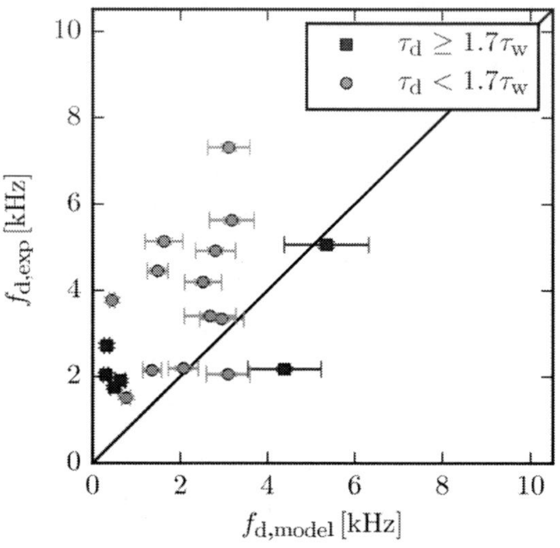


Figure 3: *Same representation as Fig. 2, but using analytical scaling laws for $\delta_{\mathrm{b}}$ and $\tau_{\mathrm{d}}$.*

## Summary and conclusion

A phenomenological model has been derived from the experimental findings that blobs are generated in a distinct region close to the separatrix by a dominant edge instability and that there seems to be a relation between this instability and the blobs in terms of detection frequency and size.

A comparison with experimental detection rates has been done for ASDEX Upgrade L-mode discharges. A good overall agreement was achieved, using the kinetic profiles as only input.

## References

[1] J. L. Terry *et al*, Phys. Plasmas **10**, 1739 (2003)

[2] B. Nold *et al*, Plasma Phys. Control. Fusion **52**, 065005 (2010)

[3] P. C. Stangeby, The Plasma Boundary of Magnetic Fusion Devices, Taylor and Francis Group, NY (2000)

[4] R. Fischer *et al*, Fus. Sci. Techn. **58**, 675 (2010)

[5] D. Carralero *et al*, Nucl. Fusion **54**, (2014)

[6] P. Manz *et al*, Phys. Plasmas **20**, 102307 (2013)

This work has been carried out within the framework of the EUROfusion Consortium and has received funding from the Euratom research and training programme 2014-2018 under grant agreement No 633053. The views and opinions expressed herein do not necessarily reflect those of the European Commission. Furthermore, this work is supported by Agence Nationale pour la Recherche, contract ANR-11-BS09-023-03 (SEDIBA).

# Numerical Investigation of Impurity Seeded Radiation Enhancement in the Divertor region with Magnetic Perturbations in ASDEX Upgrade

D. Brida[1,2], T. Lunt[1], M. Wischmeier[1], D. Carralero[1], M. Faitsch[1], Y. Feng[3],
B. Kurzan[1], A. Scarabosio[1], B. Sieglin[1], E. Wolfrum[1], W. Zhang[1],
the ASDEX Upgrade Team and the EUROfusion MST1 Team *

[1] *Max-Planck-Institut für Plasmaphysik, Boltzmannstr. 2., D-85748 Garching bei München*
[2] *Physik-Department E28, Technische Universtität München, D-85747 Garching bei München*
[3] *Max-Planck-Institut für Plasmaphysik, Wendelsteinstr. 1., D-17491 Greifswald*

## Introduction

Currently one of the most promising concepts for future fusion devices, such as ITER and DEMO, is the divertor tokamak, in which power and particles leaving the confined region are transported to the divertor targets. The power flux to the targets can locally exceed the tolerable target heat flux limit and it is therefore required that a large fraction of the power is dissipated before reaching the target. It has been shown in various experiments that this goal can be achieved by seeding low to medium Z impurities [2]. In [3] an empirical scaling formula for the power radiated by the nitrogen impurities has been found that depends linearly on the power decay length $\lambda_q$ [1]. One way to increase $\lambda_q$ could be the application of Magnetic Perturbations (MPs). MP coils were introduced at ASDEX Upgrade (AUG) and a number of other tokamaks to study their mitigating effect on Edge Localized Modes (ELMs) [4], intermittent high power flux bursts which can pose a serious risk for the targets. MPs can change the magnetic topology strongly and lead to the formation of stochastic regions inside the separatrix and finger-like deformations of the separatrix, so called lobes. To study numerically how the application of MPs affects the impurity radiation in the SOL and divertor, the edge code EMC3-Eirene [5] was employed to simulate an AUG L-mode deuterium bulk plasma with and without MPs. Subsequently nitrogen seeding simulations were performed on these bulk plasmas and the impurity radiation distribution between the MP-on and off cases considered in detail.

## EMC3-Eirene simulations and deuterium bulk plasma

EMC3-Eirene is a code package consisting of the plasma fluid code EMC3 and the kinetic neutral code Eirene. Both, the fluid as well as the neutral part, employ a Monte-Carlo scheme to solve the underlying equations. This approach allows that the grid used for the simulations does

*See http://www.euro-fusionscipub.org/mst1.

not have to be flux aligned, making it possible to simulate the transport in complex 3D field geometries, such as those occurring in magnetically perturbed plasmas. For the perturbation field a vacuum approach was assumed, i.e. the MP field was added linearly to the equilibrium field and no response of the plasma was taken into account, which could shield the MPs partially [7]. The discharge used to model the bulk plasma was the low-density AUG L-mode 30639. In this discharge rigidly rotated MPs with a toroidal mode number of $n = 2$ were switched on between $t = 2.0s$ and $t = 4.4s$. For the MP-off and MP-on case the time points $t = 1.95s$ and $t = 4.3s$ were chosen, respectively. To create the simulation grid, which stores the information about the magnetic configuration, the equilibrium from the 2D equilibrium code CLISTE [6] for each time point was used. Due to the n=2 symmetry the grid for the MP-on case covered 180°. In the radial direction the grids ranged from $\rho_p \equiv \sqrt{\psi_N} \approx 0.89$ to $\rho_p \approx 1.06$, with $\psi_N$ being the normalized poloidal flux. The input power in the simulations was given by $P_{net} = P_{tot} - P_{rad}$, where $P_{tot}$ is the total heating power of the discharge at the respective time point and $P_{rad}$ is the radiated power, and for the density the averaged separatrix density $n_{sep}$ served as a boundary condition. The corresponding values used were $P_{net} = 0.84MW$, $n_{sep} = 0.7 \cdot 10^{19} m^{-3}$ for the MP-off case and $P_{net} = 0.72MW$, $n_{sep} = 0.6 \cdot 10^{19} m^{-3}$ for the MP-on case. For both cases the same constant particle and heat cross-field diffusion coefficients $D_\perp = 0.15 m^2 s^{-1}$ and $\chi_{e,\perp} = \chi_{i,\perp} = 0.9 m^2 s^{-1}$ were assumed. The nitrogen impurity simulations were carried out in a semi self-consistent way: An impurity tracing simulation was performed, in which a simplified set of fluid equations consisting of particle and momentum conservation equations, was solved. The impurity radiation is then taken into account in the plasma energy equation. Since it is not yet possible to include a fully realistic seeding impurity source in the simulations, it was assumed that nitrogen is sputtered by the main ions with a spatially constant sputtering yield $Y = -\Gamma_z / \Gamma_D$, where $\Gamma_z$ is the impurity flux and $\Gamma_D$ is the target deuterium flux. The power radiated by the impurities was constrained to $P_Z^{rad} = 0.2 P_{net}$, which is achieved by the code by adapting $Y$ iteratively. From the impurity radiation distribution one can draw conclusions about the effectiveness of the radiation in the divertor region.

**Results**

In Fig.1a the outer midplane temperature (top) and density profiles (bottom) for the bulk plasmas obtained from the simulation without (left) and with MPs (right) in dependence of $\rho_p$ are shown. The simulation results are both compared to the experimental data from core and edge Thomson scattering. Here the edge Thomson scattering profiles were shifted by +1cm and those of the core by -4cm, a procedure always carried out in the present AUG campaign in order to align the data with the equilibrium. The simulation profiles for both cases agree

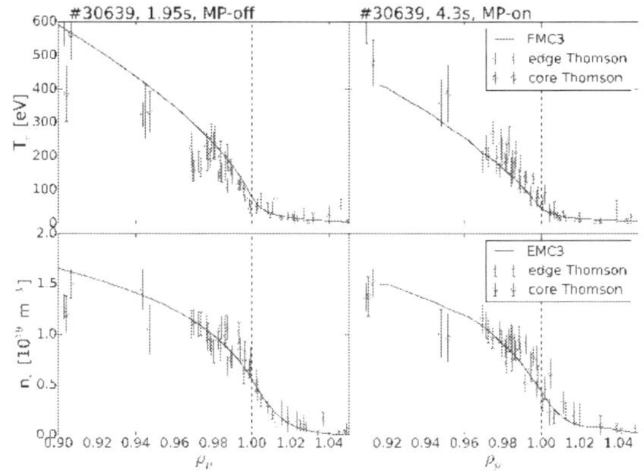


(a)

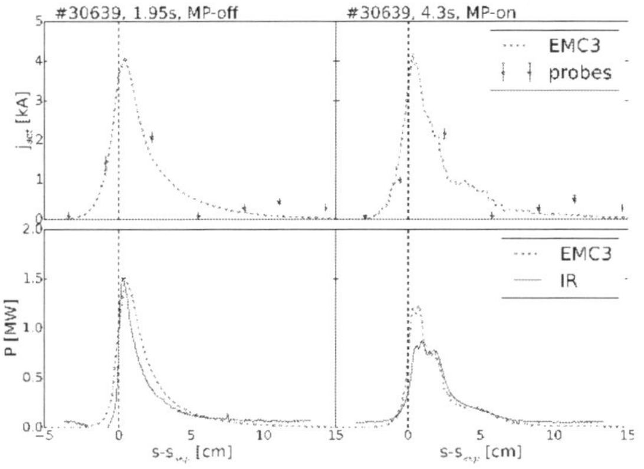


(b)

Figure 1: *a) Temperature and density outer midplane profiles at* $\phi = 11.25°$ *for the case without MPs (left) and with MPs (right). b) Target saturation current* $j_{sat}$ *at* $\phi = 168.8°$ *and power deposition profile at* $\phi = 11.25°$ *for the case without MPs (left) and with MPs (right).*

roughly within the errorbars with the Thomson scattering data, however, the temperature for the MP-on case is reduced by about 30% compared to the MP-off case in the simulations. In Fig.1b top the EMC3 outer target particle flux is compared to the saturation ion current density measured by the Langmuir probes at $\phi \approx 169°$. The EMC3 outer target power flux is shown in the bottom row of Fig.1b together with the experimental power flux obtained from the 2D IR diagnostics at $\phi \approx 11.25°$. For the MP-off case there is good agreement between the simulation results for the target heatflux and the experimental measurements, while for the MP-on case the EMC3 heat flux is higher near the separatrix than the experimentally observed one. It should be noted, however, that due to inaccuracies in the equilibrium reconstruction the IR profile was shifted along the poloidal direction by -2.3cm, such that a good agreement with the simulation results was achieved. For the case with MPs a modulation of the heat flux can be seen in the power deposition profiles of the simulation and the IR measurements. This footprint pattern on the target is a consequence of the magnetic lobes, and it has already been shown previously that EMC3-Eirene is able to reproduce these patterns [8].

In Fig.2 poloidal cross-sections of the impurity radiation without (left) and with MPs (right) are shown. For the MP-off case 48% of the total impurity radiation are radiated in the core, 47% in the SOL and 5% in the Private Flux Region (PFR). For the MP-on case 35% are radiated in the core, 58% in the SOL and 7% in the PFR. With other words, for the considered scenario and parameters, the effectiveness of the radiation in the SOL increases with MPs in the simulations which leads to a reduction of the core impurity radiation of about 25%.

## Summary, Outlook

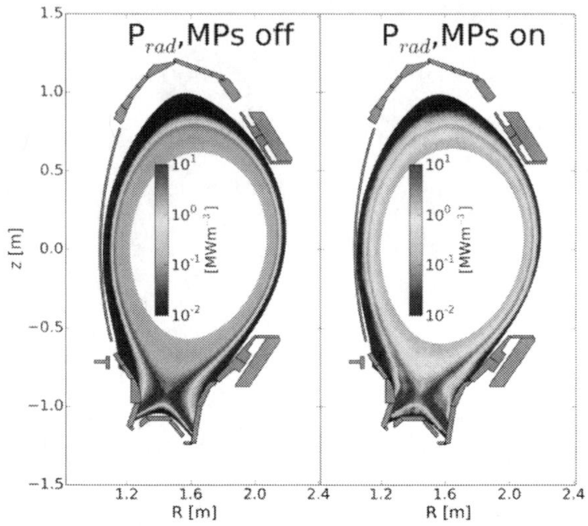


Figure 2: *Nitrogen impurity radiation distribution for the case without MPs (left) and with MPs (right).*

Deuterium bulk plasmas for a low density L-mode with and without MPs were simulated, assuming the vacuum approach. With the bulk plasmas, impurity simulations were carried out in a semi self-consistent way and the radiation distribution was investigated. For the case with MPs the radiation in the core was about 25% lower than for the case without MPs, showing that the effective change in transport associated with MPs could indeed enhance the impurity radiation in the divertor region. In the future L- and H-mode experiments with impurity seeding and MPs will be conducted at AUG and analyzed in detail with the help of EMC3-Eirene. Furthermore it is foreseen to develop a more realistic impurity model, containing point sources and impurity recycling. Moreover, screening currents will be included in the simulation to mimic the plasma response, which may be the cause for the reduced temperature and density found in the outer midplane profile for the MP case.

## Acknowledgment

This work has been carried out within the framework of the EUROfusion Consortium and has received funding from the Euratom research and training programme 2014-2018 under grant agreement No 633053. The views and opinions expressed herein do not necessarily reflect those of the European Commission.

## References

[1] T. Eich, et. al., Phys. Rev. Lett., **107**, (2011)
[2] A. Kallenbach et al., Plasma Phys. Control. Fusion, **52**, (2010)
[3] A. Kallenbach et al., IAEA, (2012)
[4] W. Suttrop et al., Phys. Rev. Lett., **106**, (2011)
[5] Y. Feng, et. al.. Contrib. Plasma Phys. **44**, 57-69 (2004)
[6] P. J. Carthy, Phys. Plasmas **6**, 3554 (1999)
[7] M. Becoulet, et. al., Nucl. Fusion **52**, (2012)
[8] T. Lunt, et. al., Nucl. Fusion **52**, (2011)

# Influence of ion cyclotron heating and MHD instabilities on the fast-ion distribution in ASDEX Upgrade

M. Weiland[1], B. Geiger[1], R. Bilato[1], A. Jacobsen[2], M. Salewski[2], P. A. Schneider[1], G. Tardini[1] and the ASDEX Upgrade team[1]

[1]*Max-Planck-Institut für Plasmaphsyik, Garching, Germany*

[2]*Technical University of Denmark, Dk-2800 Kgs. Lyngby, Denmark*

## Introduction

Fast ions are generated in present-day tokamaks by neutral beam injection and ion cyclotron resonance heating, and are used for plasma heating and current drive. In future fusion devices, energetic $\alpha$-particles from fusion reactions will be the main heating mechanism. Hence, it is crucial to understand the transport behavior and creation mechanisms of fast ions.

A way to measure the fast-ion distribution is the FIDA (fast-ion D-alpha) diagnostic, which analyzes the Doppler-shifted D-alpha radiation of neutralized fast-ions spectroscopically. With lines of sight at different radial positions, radial fast-ion density profiles can be measured. At the same time, the shape of the Doppler spectrum contains information about the 2D velocity space distribution $f(E, v_{\|}/v)$. Observation from different viewing angles allows consequently a tomographic reconstruction of $f(E, v_{\|}/v)$. For this purpose, the FIDA diagnostic at AUG has been upgraded from two to five views for the 2014 campaign, and the spectrometer has been upgraded to measure blue and red Doppler shifts simultaneously.

These recently developed diagnostic capabilities are used to study the effect of sawtooth crashes on the fast-ion velocity distribution and the further acceleration of fast D ions by 2nd harmonic ion cyclotron heating.

## Diagnostic setup

The FIDA diagnostic at ASDEX Upgrade consists now of five viewing arrays (see fig. 1). Each view consists of several radially distributed lines of sight, and has a different angle towards the magnetic field $\Phi$. The latter determines the observed region in velocity space, and hence an equal distribution of angles (in the plasma center) has been chosen: $\Phi \approx 10°, -20°, -50°, 70°$ and $85°$. The FIDA spectrometers are capable of measuring the blue- and red-shifted part of the D-alpha spectrum simultaneously, and have a capacity of up to 37 viewing channels. Each wavelength $\lambda$ in the FIDA

Figure 1: *3D visualization of the five FIDA line of sight arrays.*

spectrum can be interpreted as a weighted integral of the fast-ion velocity distribution $f(E,\xi)$:

$$FIDA(\lambda) = \int_0^\infty \int_{-1}^{+1} W(\lambda, E, \xi) f(E, \xi)\, dE d\xi \quad (1)$$

Hereby, the pitch is denoted with $\xi = v_{\parallel}/v$. The so called weight functions $W$ can be calculated with the FIDASIM code [1, 2]. In fig. 2, weight functions are shown for five FIDA views in the plasma center and two wavelengths corresponding to Doppler-shifts $\Delta\lambda = \pm 3.9$ nm. Weight functions at different Doppler-shifts have typically similar shapes, but are shifted with respect to the energy. It can be seen that the velocity space is well covered. This allows a tomographic reconstruction of $f(E,\xi)$. Therefore, eq. 1 is discretized to a matrix equation $\vec{s} = W\vec{f}$. The tomography can then be calculated from the FIDA signals $\vec{s}$ by $\vec{f} = W^+\vec{s}$, where we calculate $W^+$ by first order Tikhonov regularization as described in [3]. This is equivalent to calculating a least-squares fit, with the additional condition, that the solution $f$ should have small gradients.

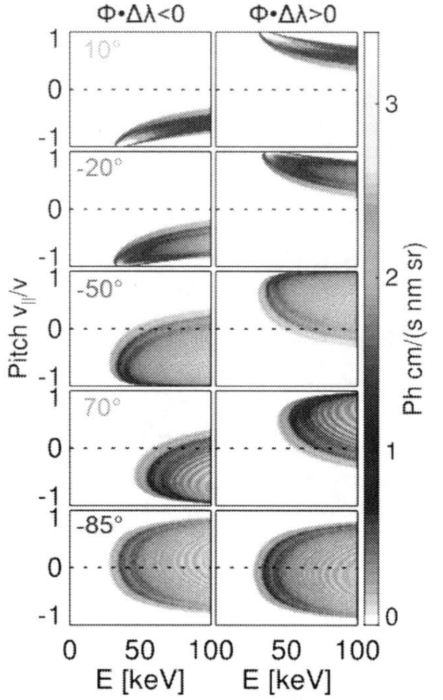


Figure 2: *Weight functions of the five FIDA views and two Doppler-shifts* $\Delta\lambda = \pm 3.9$ nm.

**Influence of RF heating on the beam ion distribution**

The coupling of radio frequency (RF) waves to ions is an important physics aspect for future fusion devices. In ITER, 2nd harmonic ion cyclotron resonance heating of tritium is one of the foreseen ICRF schemes, along with He-3 minority heating. The 2nd harmonic heating has the benefit, that it can accelerate the main ion species directly, however it is only efficient for ions with large Larmor radii (with respect to the RF wave length). At ASDEX Upgrade, the Larmor radii of D beam ions (i.e. from 60 keV NBI) are large enough for effective 2nd harmonic absorption.

We have analyzed discharge #30809, which has $B_t = -2.4$ T and $I_p = 1.0$ MA. The ICRF frequency is 36.5 MHz and the resonance layer is hence located at $R \approx 1.69$ m, i.e. very close to the magnetic axis. We have compared $t = 4.60$ s with 2.4 MW NBI + 1.8 MW ICRF and $t = 4.48$ s with 2.4 MW NBI only. FIDA data from four views are available for this discharge. We have calculated tomographies from central lines of sight ($R \approx 1.75 - 1.78$ m), well behind the ICRF resonance layer.

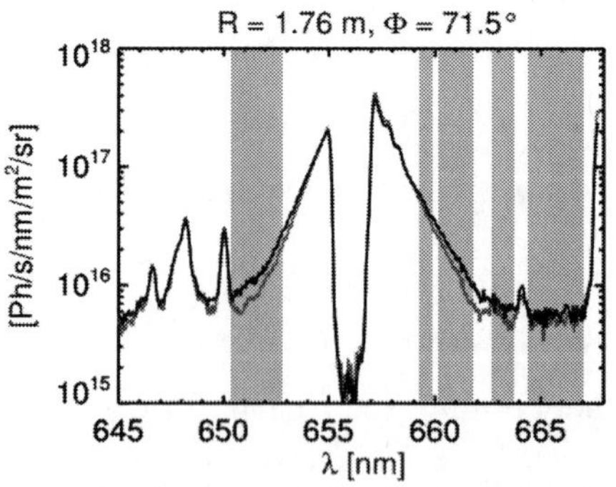


Figure 3: *FIDA spectrum for* 4.48 s *(red) and* 4.60 s *(black). The blue shaded area is used for the tomography.*

The spectra of the $\Phi = 72°$ view are shown in fig. 3. It can be seen directly, that the FIDA spectrum is broader for $t = 4.60$ s, which indicates the presence of more energetic fast ions. The tomography is shown in fig. 4. It has to be noted, that in the presence of ICRF also fast H (minority) ions are expected to contribute to the Balmer-Alpha spectrum. The weight functions for hydrogen are almost identical with respect to the velocity (or $E/m$). Therefore, we have plotted the tomography as function of $E/m$ and interpret it as the sum of fast D and fast H.

The NBI injection energy (30 keV/u) is indicated with a dotted line. At $t = 4.48$ s, the tomography yields mainly fast ions with energies below the injection energy, as expected. In the presence of ICRF ($t = 4.60$ s), two high energy tails are clearly seen. The stronger tail appears at pitches $\approx 0.7$, and can be identified as beam ions, which are further accelerated. A second, weaker high energy tail is seen at pitches $-0.3$ to $0.0$. It could originate from trapped ICRF-accelerated deuterium ions, which have the inside of their banana orbit at our measurement position with negative pitches, and from fast H ions, which are accelerated to high perpendicular velocities by ICRH. The tomography results can be directly compared to theoretical predictions or analytical models of the H fast-ion distribution function [4], and this will remain for future work.

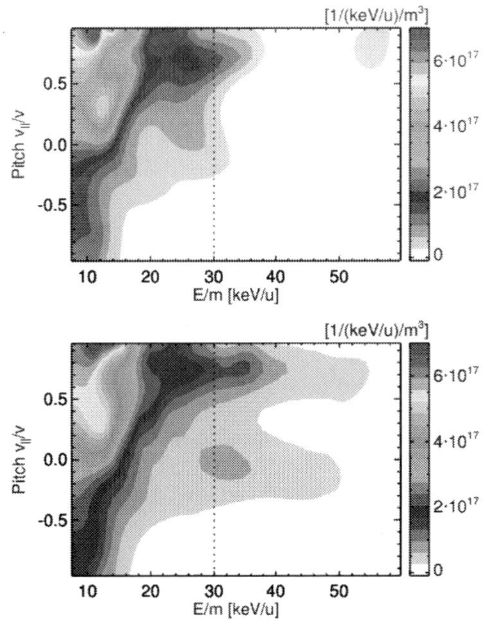


Figure 4: *FIDA tomography in the presence of 30 keV/u NBI and NBI+ICRF (down).*

**Fast-ion redistribution by sawtooth crashes**

Sawtooth crashes can strongly redistribute fast ions, which has been demonstrated e.g. from FIDA [5] or neutron measurements [6]. The improved FIDA setup and the additional views can be used now to study the velocity dependence of this redistribution in detail.

We have analyzed discharge #31557. The discharge was run with a magnetic field of -2.6 T, a plasma current of 1 MA and 2.5 MW NBI. Sawtooth crashes are directly visible e.g. in time traces of electron temperate and the FIDA raw signals.

We have calculated reconstructions before and after the sawtooth at 2.24s from five FIDA views, which measure all approximately at the same radial position at $\rho_{pol} \approx 0.10$, and inside the sawtooth inversion radius ($\rho_{pol} \approx 0.45$). Regions with impurity lines and beam emission are excluded from the input spectra.

The result of the tomography is shown in fig. 5. From the reconstruction, it is possible to determine the total fast-ion density (of fast-ions above 20 keV), and thus quantify the effect of the sawtooth: It causes a 25% drop of fast-ion density in the plasma center. With a cut of the fast-ion distribution function at constant energy, we can estimate, how this total density drop is distributed along different pitches (fig. 6). It can be seen that fast-ions with high pitches are much stronger distributed (-50%), while more strongly gyrating fast ions with pitches close to 0 are much less affected by the sawtooth. This is in accordance with [7], where a FIDA tomography is calculated with singular value decomposition from four FIDA views.

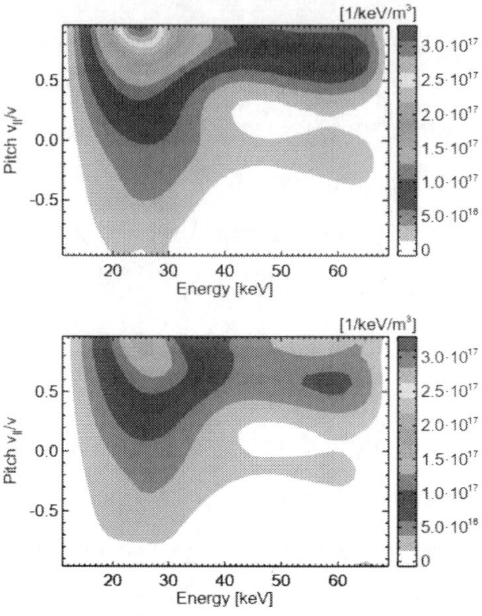


Figure 5: *FIDA tomography before (top) and after a sawtooth crash.*

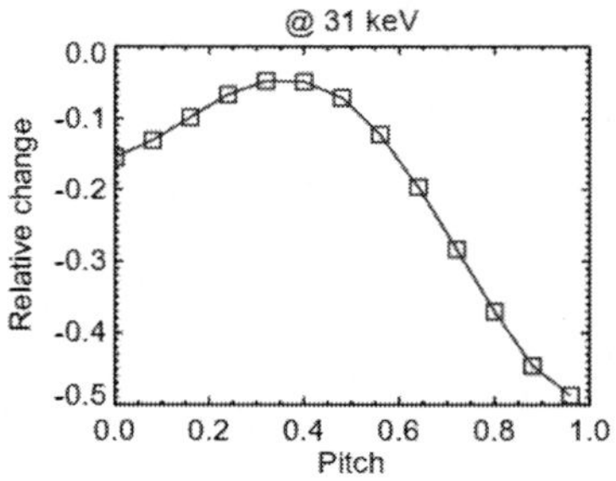


Figure 6: *Pitch distribution of the relative change of* $f(E = 31\ \text{keV}, v_{\parallel}/v)$.

## Summary and conclusion

The recent upgrades of the FIDA diagnostic at ASDEX Upgrade allow a tomographic reconstruction of the 2D fast-ion velocity distribution and hence velocity-space resolved fast-ion studies. In the presence of sawtooth crashes, a 25% overall decrease of central fast-ion density is observed, whereby the decrease is found to be much stronger (up to 50%) for fast ions with large pitches $v_{\parallel}/v$. The effect of ICRF heating on a beam ion distribution is clearly seen by high energetic tails in the tomography above the NBI injection energy. The pitch distribution of the energetic tails can be resolved, and two contributions at $\frac{v\|}{v} \approx 0.7$ (approx. same as NBI) and at $\frac{v\|}{v} \approx -0.15$ are found.

**Acknowledgment**

This work has been carried out within the framework of the EUROfusion Consortium and has received funding from the Euratom research and training programme 2014-2018 under grant agreement No 633053. The views and opinions expressed herein do not necessarily reflect those of the European Commission.

## References

[1] W.W. Heidbrink et al., Comm. Comp. Physics, 8, 2010.
[2] M. Salewski et al., Plasma Physics and Controlled Fusion, 56(10):105005, 2014.
[3] M. Anton et al., Plasma Physics and Controlled Fusion, 38(11):1849, 1996.
[4] R. Dendy et al., Physics of Plasmas, 2, 1623-1636 (1995)
[5] B. Geiger et al., Plasma Physics and Controlled Fusion, 57(1):014018, 2015.
[6] J.A. Lovberg et al., Physics of Fluids B: Plasma Physics, 1(4):874–892, 1989.
[7] B. Geiger et. al., Nuclear Fusion, 55(10):accepted, 2015.

# Design of the new Imaging Motional Stark Effect Diagnostic at ASDEX Upgrade

A. Burckhart[1], O. Ford[1], M. Reich[1], R. Wolf[1], ASDEX Upgrade Team[1]

[1] *Max Planck Institut für Plasmaphysik, Garching and Greifswald, Germany*

## Introduction

Motional Stark Effect (MSE) diagnostics are some of the few diagnostics capable of providing information on the current distribution in magnetically confined fusion plasmas. However, they are typically very difficult to calibrate and to maintain due to the complexity of the systems. Imaging MSE (IMSE), on the other hand, offers several advantages over the classical approach [1, 2], for example a simpler, more economical setup and at least one order of magnitude more spacial measurements points. Like classical MSE diagnostics, IMSE systems also utilise the Stark-split D-$\alpha$ light from neutral beams injected into fusion plasmas. However, once collected the light is lead through a series of birefringent plates and a polariser before being imaged onto a CCD chip, forming an interference pattern. A prototype of the new diagnostic was successfully tested at ASDEX Upgrade (AUG) in 2013 [3], motivating a permanent IMSE system, which was installed for the 2015 experimental campaign. The new system has a specially designed set of optics that maximises the collected light throughput under unusual mechanical restrictions while enabling three neutral beam sources to be viewed simultaneously. In this contribution, after a brief description of the IMSE diagnostic principle, the design of the new setup at AUG will be presented.

## MSE and IMSE principle

Neutral beams injection (NBI) of fast deuterium particles is used for heating and diagnostic purposes in many fusion plasmas. The particles are excited and emit D-$\alpha$ radiation. The line

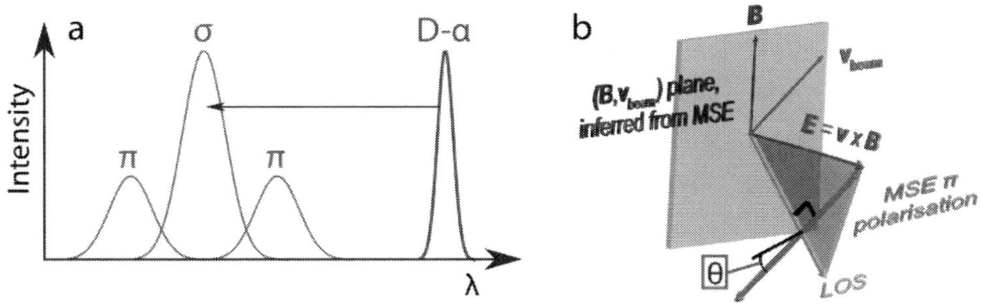


Figure 1: a: $\pi$ and $\sigma$ components of the MSE spectrum. b: 3D visualisation of the MSE principle: the $\pi$ polarisation is the projection of **E** onto the plane perpendicular to the LOS. Knowing **E** lets us calculate the (**B**, $\mathbf{v}_{beam}$) plane

emission is Doppler shifted by the particle's velocity $\mathbf{v}_{beam}$) and Stark split by the electric field in the rest frame of the atom ($\mathbf{E} = \mathbf{v}_{beam} \times \mathbf{B}$) into $\pi$ and $\sigma$ components (fig 1a). These are polarised parallel and perpendicular to the projection of **E** in the plane perpendicular to the line of sight (LOS) [4]. The measurement of the polarisation angle provides the direction of **E** and, therefore, the plane common to the beam and **B** (fig 1b). Depending on the direction of the LOS, a change in the direction of B (magnetic pitch angle) will have a larger or smaller impact on $\theta$. We call this the pitch angle sensitivity of the diagnostic.

Conventional MSE polarimeters measure $\theta$ by spectrally selecting one component with a narrow filter and use a pair of photo-elastic modulators to analyse the polarisation. Because of the varying Doppler shift, each spatial point requires a very finely tuned filter, optics, sensor and digitiser. Typically, the beam is observed on 10 to 20 LOS. Unlike conventional MSE systems, the IMSE diagnostic utilises all spectral components, leading to a brighter signal. Two birefringent plates modulate the image from the beam with orthogonal interference patterns (fig. 2a). The difference in wavelength of the $\pi$ and $\sigma$ components is exploited by including a delay plate in the setup, which enables the components to be added constructively in the interference pattern, maximising the signal intensity, despite their orthogonal polarisations.

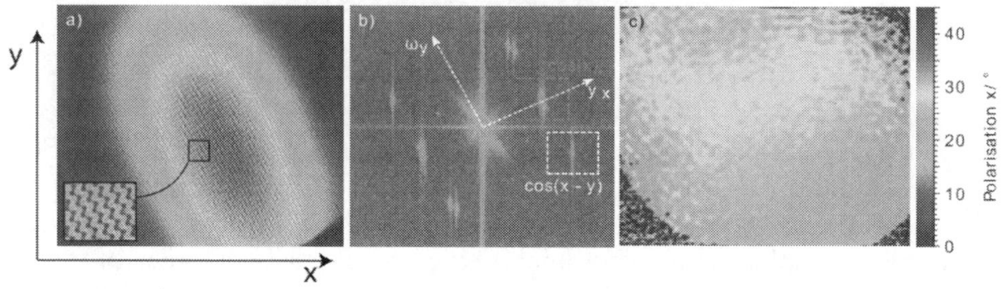


Figure 2: (a) interference pattern on the camera chip, (b) Fourier transform of the image and (c) MSE polarisation angle in 2D

The intensity on the chip is proportional to

$$\zeta cos(2\theta) + \zeta sin(2\theta) cos(x+y) + \zeta sin(2\theta) cos(x-y), \quad (1)$$

where $\theta$ is the polarisation angle, x and y are the axes on the camera chip and the fringe contrast $\zeta$ is, in a first approximation, a constant. The three components are isolated from the 2D Fourier transform (fig. 2b) and a 2D image of the MSE polarisation angle can be calculated (fig. 2c).

Compared to the classical MSE, the IMSE setup is simpler and more economical, only one calibration is necessary rather than calibrating all channels independently. The 2D nature of the system simplifies the identification of reflections and provides inherent position calibration as the vessel components on the image can easily be compared to the CAD model of the machine. It also provides more information on the current distribution [5], as the system performs roughly 200 simultaneous angle measurements.

**Optical design**

Several toroidal and poloidal locations were considered for the permanent IMSE system installed at AUG and evaluated according to multiple criteria, such as the lack of view obstruction. The LOS must not be obstructed by limiters, tiles, or other diagnostics. A large Doppler separa-

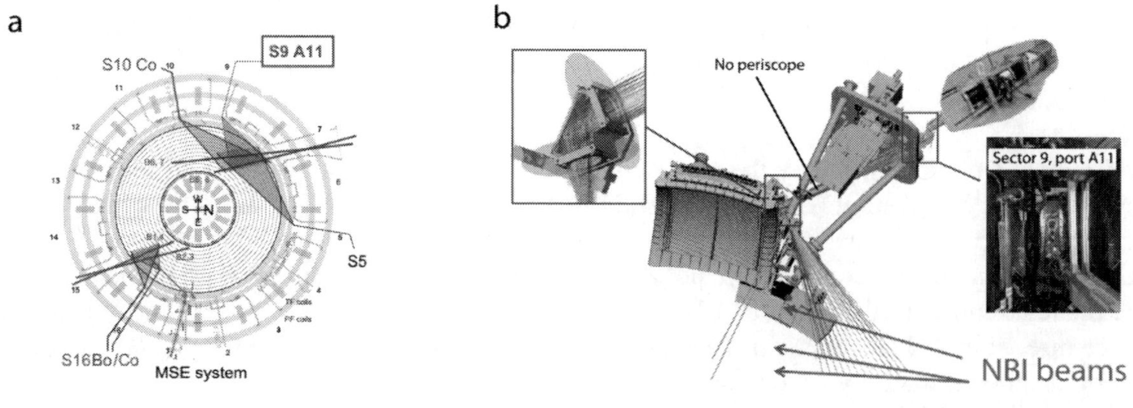


Figure 3: (a) Options for the diagnostic location, (b) 3D model of the optical relay system

tion was also a design criterion, as the MSE signal should be far from the D-$\alpha$ background. Blue shifted light is better than red, as this avoids contamination of the spectrum by carbon impurity lines. The location also impacts the spatial resolution. If the LOS is tangential to a flux surface, most of the light integrated over the beam width originates from that flux surface, which leads to a good radial resolution of the diagnostic. However, the pitch angle sensitivity of the diagnostic is higher if the view is not toroidal.

The best location (sector 9 port A11, see figure 3a) comes with strong space restrictions. The presence of other diagnostics in the same port (ECE, ECE imaging, heat shield temperature monitor) limits the size and positioning freedom of the mirror box, and prevents the installation of a periscope, limiting the maximum light throughput (fig. 3b). The size of the existing vacuum port was increased, albeit to only 100mm diameter. Outside the vacuum vessel, the poloidal field coil support structure and the backend of other diagnostics restrict the light path, creating the need for external mirrors.

Two boundary conditions were the starting point for the optimisation of the optical relay system, finding a compromise between field of view and total collected solid angle, and the requirement to set best resolution to the plasma centre.

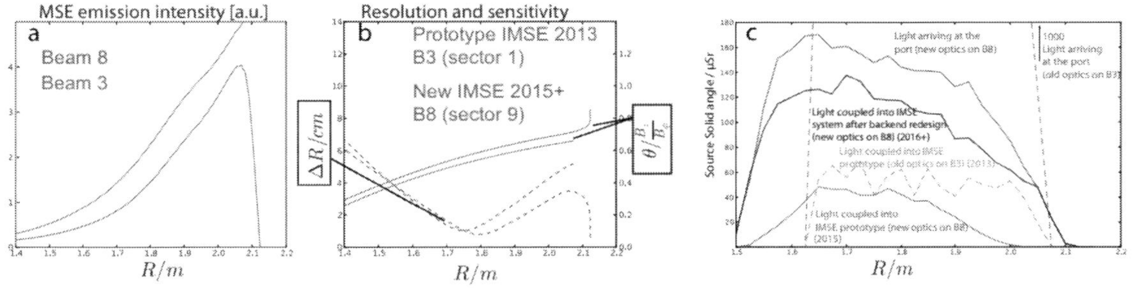


Figure 4: (a) Intensity of the beam signal at different radial locations, (b) resolution and pitch angle sensitivity versus plasma radius, (c) light cone reaching the IMSE backend

The field of view was chosen to see 3 beams, after which the system was optimised for maximum light throughput, given the space restrictions. Without considering the finite imaging capabilities of the lenses and the IMSE backend, the resolution at the plasma centre is limited to 1cm by the beam width and the diagnostic geometry (fig. 4b). The pitch angle sensitivity, the ratio of polarisation angle and pitch angle, is roughly 0.5, while the speed of the optics is f/2.7. The beam emission intensity of source 8 is higher than source 3 (fig. 4a), but the light cone coupled by the new system is slightly smaller than in the prototype tests of 2013 on the old MSE optics (fig. 4c). Therefore, the initial performance will be similar to the prototype tests, which yielded less noisy and faster measurements than the conventional MSE [3]. The amount of coupled light is expected to double after the planned redesign of the IMSE backend.

The imaging properties of the optical relay system were optimised with a custom 3D ray tracer. After several thousand optimisation steps in different considered scenarios, the best compromise between cost and imaging quality consisted of six custom lenses, of which two were aspherical to correct for spherical aberration. With this configuration, the optics achieve a resolution of 0.5cm at the beam. Since this is better than the 1cm resolution imposed by the beam width (cf. fig 4b), further optimisation would not improve the diagnostic resolution.

Two lamps were installed in the field of view of the diagnostic for absolute angle calibration [6]. The beam from a wavelength tunable laser is led into the vacuum vessel via optical fibres, then through a collimating lens and finally a polarising filter to provide a source of polarised light.

**Materials used in the permanent system**

Since most glasses are subject to Faraday rotation, an effect that rotates the polarisation angle of polarised light in the presence of a magnetic field parallel to light path, the lenses must be made of a special material. Most existing MSE systems use Schott SFL6 because of its low Verdet constant. However, SFL6 is not available any more. We tested different samples of Schott and Ohara glass at magnetic fields up to 80mT and calculated their Verdet constants at 656nm:

| | | |
|---|---|---|
| SCHOTT NSF6 | 28°/(T m) | ± 1°/(T m) |
| OHARA BSL7Y | 869°/(T m) | ± 150°/(T m) |
| OHARA S-FSL5 | -675°/(T m) | ± 9°/(T m) |
| OHARA TIH6 | 8°/(T m) | ± 17°/(T m) |

Since the samples provided were not ideal for optical measurements, some measurement errors are relatively high. However, the Faraday rotation in TIH6 is clearly the lowest. It was chosen to substitute Schott SFL6 for the lenses and a polished and optically coated cylindrical sample will be tested at magnetic fields up to 2T in the near future to characterise it more accurately. However, TIH6 is very susceptible to thermal stress and tends to crack with fast temperature variations. Therefore, a 5mm thick fused silica protection cover was mounted in front of the first lens to absorb most of the radiation from the plasma. The vacuum window is also composed of fused silica, and a coil was mounted at the port such that the magnetic field can be directly measured and the Faraday rotation included in the analysis code, as fused silica has a relatively high Verdet constant of 200°/(T m) so that the varying vertical and radial control fields must also be considered.

The ratio of p/s polarisation intensity reflected by the three mirrors must be close to 100% for the IMSE system to measure the correct polarisation angle. Furthermore, no phase shift must exist between the p and s components. The quality of the mirror is more crucial than for the existing MSE system, as three mirrors had to be installed. The theoretical p/s ratio of the installed mirrors is 99.97% and within our measurement capabilities a perfect ratio was measured in tests.

**Conclusions and outlook**

A new 2D imaging MSE system was designed, built and installed at the AUG tokamak. The optical design was optimised such that three beams are visible in its field of view and light throughput was maximised within the strong space restrictions. A sharp image is obtained by correcting aberrations with custom aspherical lenses. An alternative glass with low Faraday rotation, TIH6, was found to replace the obsolete Schott SFL6. An accuracy of roughly 0.3° in pitch angle is expected for the new system. First measurements are expected in the first weeks of the 2015 campaign of AUG.

Next steps include the redesign of the IMSE backend. Larger filters and crystals, as well as a custom collimator lens will be used in order to couple more light. If the light throughput is sufficient, a faster camera will be installed to increase the time resolution of the diagnostic, currently 10ms. Furthermore, the calibration lamps and the mirror box will be equipped with shutters during the next opening of the AUG vessel to prevent the coating of optical components during boronisations and the resultant drift of the calibration.

**Acknowledgements**

This work has been carried out within the framework of the EUROfusion Consortium and has received funding from the Euratom research and training programme 2014-2018 under grant agreement No 633053. The views and opinions expressed herein do not necessarily reflect those of the European Commission.

**References**

[1] J. Howard, PPCF, **50**, 125003 (2008)
[2] O. Ford *et al*, 40th EPS conference, P6.006 (2013)
[3] O. Ford *et al*, 40th EPS conference, O2.110 (2013)
[4] F.M. Levinton, PRL, **63**, 2060-2063 (1989)
[5] J. Svensson, PPCF, **50**, 085002 (2008)
[6] F.T. Löbhard, Diploma Thesis, TU München, (2011)

# High-resolution internal measurements of 3D plasma response for model validation in high-β plasmas

P. Piovesan[1], V. Igochine[2], Y.Q. Liu[3], M. Maraschek[2], L. Marrelli[1], D.A. Ryan[3,4], W. Suttrop[2], M. Weiland[2], D. Yadykin[5], M. Gobbin[1], D. Terranova[1], N.M. Ferraro[6], M.J. Lanctot[6], A. Bogomolov[7], I. Classen[7], M. Dunne[2], A. Gude[2], R. McDermott[2], M. Reich[2], E. Wolfrum[2], the ASDEX Upgrade Team and EUROfusion MST1 Team*

*[1]Consorzio RFX, Padova, Italy; [2]MPI für Plasmaphysik, Garching, Germany; [3]CCFE, Abingdon, UK; [4]University of York, York, UK; [5]Chalmers University, Gothenburg, Sweden; [6]General Atomics, San Diego, CA, USA; [7]DIFFER, Nieuwegein, The Netherlands.*
* See *http://www.euro-fusionscipub.org/mst1.*

The plasma response to external 3D fields, due to error fields or intentionally applied by non-axisymmetric coils, can impact tokamak performance. Helical distortions of flux surfaces can affect heat, particle, momentum, and fast ion transport, as well as tearing, ELM, and Alfvén eigenmode stability. Such effects are usually more prominent at high normalized pressure $\beta_N = \beta_T a B_T / I_P$ approaching the ideal MHD no-wall limit. Here the ideal kink mode is only marginally stable and can largely amplify 3D fields resonant with it. The paper presents an experimental and theoretical investigation of the plasma response to applied 3D fields in ASDEX Upgrade high-β hybrid plasmas [1]. These experiments were made possible by the new availability in 2014 of AC power supplies for the non-axisymmetric B-coils, two rows of 8 internal coils placed above and below the midplane.

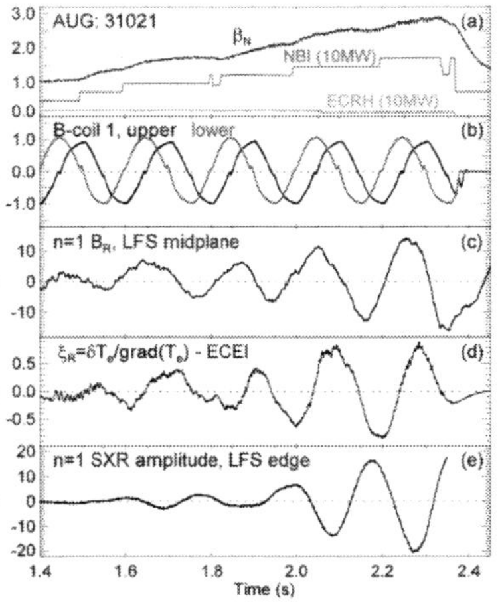


Fig. 1. Perturbation of different edge quantities in response to a rotating n=1 field during a $\beta_N$ ramp in AUG: (c) total n=1 $B_r$ (vacuum n=1 field ≈2.5G not subtracted), (d) radial displacement from ECE-I at pedestal top, (e) SXR edge channel at LFS.

Previous similar studies in DIII-D [2] focused on advanced tokamak plasmas, which feature lower no-wall limits than hybrids due to elevated safety factor with $q_{min}$>1.5. The n=1 response in this case is dominated by the external kink and is localized in the outer region of the plasma. On the other hand, the hybrid plasmas investigated here have $q_{min}$≈1. We find that this has a significant effect on the n=1 response profile, largely increasing the internal kink component with respect to the external one. Helical core displacements of the order of 1cm are observed, which may have consequences on hybrid operation, as discussed below. These measurements also provide an interesting, new test case to validate MHD simulations.

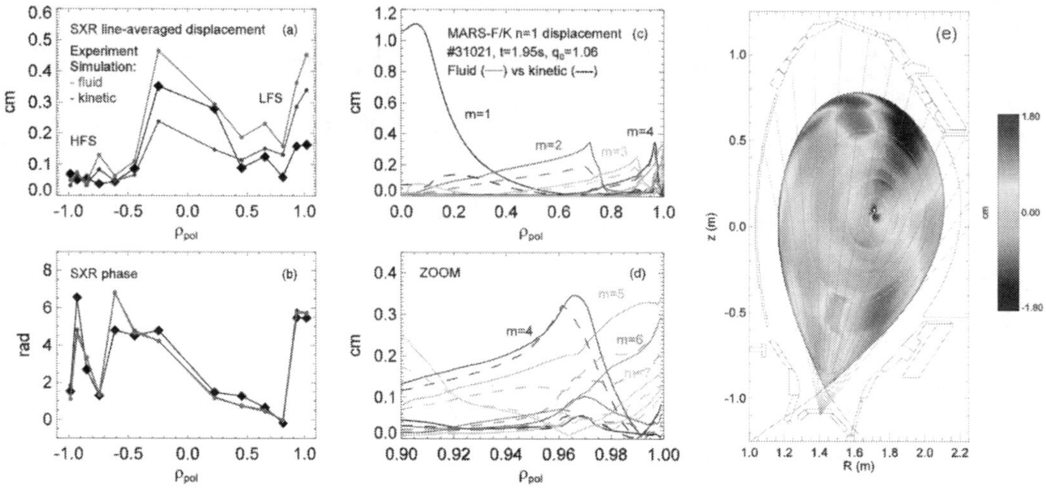


Fig. 2. (a) Amplitude and (b) phase of n=1 SXR line-averaged displacement in experiment (black diamonds) and simulated with MARS-F fluid (red) and MARS-K kinetic eigenfunctions (blue), shown in (c,d) for discharge #31021, t=1.95s. (e) SXR chords used in the analysis and contour of MARS-F n=1 displacement.

The plasma response to slowly rotating (5-27Hz) n=1 magnetic perturbations was measured with a wide set of edge and internal diagnostics during $\beta_N$ ramps obtained by increasing the NBI power up to 17MW, Fig. 1(a). A benign 3/2 NTM always appears during the $\beta_N$-ramp. The poloidal spectrum of the applied n=1 field is determined by the differential phase among upper and lower B-coil rows, $\Delta\phi^{U/L}$. The discharge reported in Figs. 1 to 3 has $\Delta\phi^{U/L}$=315°, corresponding to an almost pitch non-resonant vacuum field. Fig. 1 shows the resulting n=1 perturbation measured by $B_r$ pick-up probes at LFS midplane, Fig. 1(c), n=1 radial displacement from ECE-Imaging channels across the LFS pedestal, Fig. 1(d), and a SXR chord tangent to the separatrix on LFS, Fig. 1(e). In all diagnostics the n=1 perturbation increases with $\beta_N$, indicating amplification from a marginally stable, pressure-driven mode.

The entire n=1 response profile was measured from HFS to LFS by two SXR cameras with identical line of sight geometry, toroidally separated by 135°, Fig. 2. The difference between the two cameras allows extracting the n=1 response out of variations due to ELMs that otherwise dominate the signals. This difference may pick-up n≠1 contributions, expected in any case to be small. The amplitude and phase of the n=1 SXR line-averaged displacement are shown in Fig. 2(a,b) for discharge #31021 at t=1.95s, when $\beta_N$≈2 is below the no-wall limit, varying in these plasmas among 2.5 and 2.9. The n=1 eigenfunctions calculated by MARS-F fluid and MARS-K [3] kinetic simulations are shown in Fig. 2(c,d). These simulations used experimental profiles of resistivity and toroidal rotation and high-resolution CLISTE [4] equilibria with various constraints (kinetic profiles, fast ion pressure from TRANSP, 3/2 NTM position, SOL currents, no MSE). The SXR measurements perturbed by

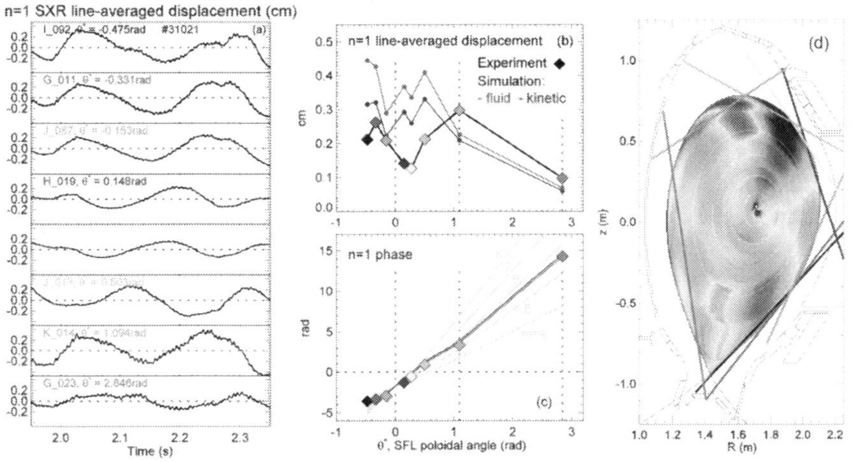


Fig. 3 (a) The n=1 radial displacement from SXR channels almost tangent to the separatrix is in phase with the applied n=1 field. (b) n=1 displacement and (c) phase in experiment (diamonds) and MARS-F/K simulations (red/blue) *vs* SFL poloidal angle. (d) Edge SXR channels used in the analysis and n=1 MARS-F displacement.

MARS-F/K eigenfunctions were simulated taking into account the exact 3D geometry of the SXR volumes of sight [5] and are shown in red/blue in Fig. 2(a,b). MHD predictions are in good agreement with experiment: in both cases a large core displacement is present, mainly due to the m=1 internal kink response, while the edge is dominated by external kink/peeling harmonics with m>nq. MARS-F predicts large sensitivity of the internal kink response with respect to changes in $q_0$ in the range 1.0-1.1 within the equilibrium reconstruction uncertainty ($q_0$ scan not shown). The simulation results reported here correspond to $q_0$=1.06, at which the internal kink response is maximum. On the other hand, drift-kinetic damping can also sensibly change the internal kink response. It must be thus concluded that, to quantitatively validate drift-kinetic effects in the core, internal q profile measurements are necessary.

The edge poloidal structure of the n=1 response predicted by MARS-F/K agrees with edge measurements at different toroidal/poloidal angles, including eight SXR channels almost tangent to the separatrix, Fig. 3(d), ECE-Imaging and Lithium beam. A poloidally propagating structure at the n=1 field frequency is evident when comparing the SXR displacement at different poloidal angles, Fig. 3(a). The phase of these oscillations has a linear dependence on the straight field line poloidal angle, $\theta^*$, a linear fit giving m=5.5±0.2. The n=1 displacement is larger at LFS than HFS, confirming the ballooning nature of the n=1 response. In this case, adding kinetic effects really improves agreement with experiment, since the edge q profile has much lower uncertainty than in the core. All these evidences confirm that the edge n=1 response is mainly due to the ideal external kink/peeling mode. No evidence of magnetic islands was found, though this may be due to diagnostic limits, e.g. line integration in SXR, low optical thickness for ECE-Imaging channels near the separatrix.

To probe the kink-resonant nature of the n=1 response, the m-spectrum of the applied n=1 field was scanned by varying the differential phase $\Delta\phi^{U/L}$ at constant $\beta_N$. Fig. 4 shows the

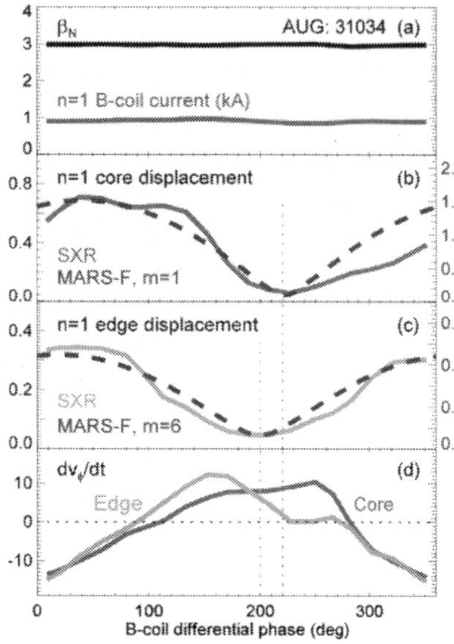


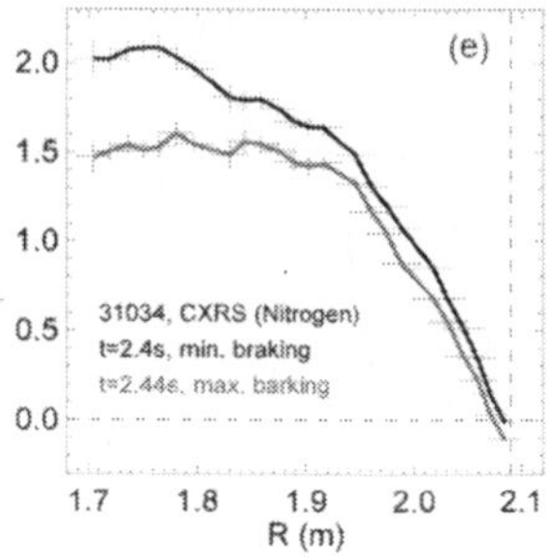


Fig. 4. (a) $\beta_N$ and n=1 B-coil current, (b) n=1 SXR core and (c) edge displacement *vs* MARS-K displacement, (d) time derivative of core/edge rotation averaged over three $\Delta\phi^{U/L}$ scans. (e) Toroidal rotation profiles at minimum and maximum n=1 response.

core/edge LFS SXR displacement *vs* $\Delta\phi^{U/L}$, compared to the time derivative of core/edge rotation, indicative of rotation braking. All quantities are averaged over three consecutive $\Delta\phi^{U/L}$ scans. The n=1 response dependence on $\Delta\phi^{U/L}$ agrees with MARS-K (dashed blue lines). The core/edge responses are minimum at different $\Delta\phi^{U/L}$ in agreement with MARS-K. The n=1 response causes global rotation braking largest in the core, where the response also peaks, as shown by the rotation profiles in Fig. 4(e) at minimum (black) and maximum (red) n=1 response. Edge rotation reverses to counter-$I_P$, a small but robust effect confirmed by accurate error estimates. Evaluation of the torques due to the n=1 displacement, i.e. NTV, e.m., and changes in NBI torque due to fast-ion transport, is ongoing.

The internal response measurements reported here provide an excellent basis to validate MHD codes. More experiments are needed to fully validate kinetic effects, in particular knowledge of core q is necessary. Work is ongoing to extend code validation to M3D-C$^1$ with 2-fluid effects [6] and the free-boundary 3D equilibrium code V3FIT [7].

This work has shown that hybrid plasmas are highly sensitive to 3D fields as the no-wall limit is approached. Their distinctive feature is an n=1 response peaking in the core due to $q_{min}\approx 1$. This causes significant rotation braking and may also increase core thermal and fast ion transport, with possible consequences on hybrid discharge performance. Such effects may become an issue in presence of error fields or 3D fields for ELM mitigation and should be carefully evaluated. In future experiments it will be also interesting to extend these studies to lower rotation, where 3D field penetration is easier. Work is ongoing to compare these results to other tokamaks equipped with non axi-symmetric coils such as DIII-D.


[1] O. Gruber *et al.*, Nucl. Fusion **49**, 115014 (2009).
[2] M.J. Lanctot *et al.*, Phys. Plasmas **18**, 056121 (2011).
[3] Y. Liu *et al.*, Phys. Plasmas **15**, 112503 (2008).
[4] P.J. Mc Carthy, Phys. Plasmas **6**, 3554 (1999).
[5] M. Weiland *et al.*, accepted for publication in PPCF.
[6] N.M. Ferraro *et al.*, Phys. Plasmas **17**, 102508 (2010).
[7] J.D. Hanson *et al.*, Nucl. Fusion **49**, 075031 (2009).


This work has been carried out within the framework of the EUROfusion Consortium and has received funding from the Euratom research and training programme 2014-2018 under grant agreement No 633053. The views and opinions expressed herein do not necessarily reflect those of the European Commission.

# Real-time simulation of internal profiles in the presence of sawteeth using the RAPTOR code and applications to ASDEX Upgrade and RFX-mod

C. Piron[1], F. Felici[2], D. Kim[3], C.J. Rapson[4], M. Reich[4], O. Sauter[3], T. Goodman[3], W. Treutterer[4], H. van den Brand[5], I. Chapman[6], the ASDEX Upgrade Team, C. Finotti[1], G. Manduchi[1], G.Marchiori[1], L. Marrelli[1], O. Kudlacek[1], P. Piovesan[1], C. Taliercio[1], the RFX-mod Team and the EUROfusion MST1 Team[7]

[1] *Consorzio RFX, Padua, Italy*

[2] *Technische Universiteit Eindhoven, Eindhoven, The Netherlands*

[3] *CRPP-EPFL, Lausanne, Switzerland*

[4] *Max-Planck-Institut für Plasmaphysik, Garching, Germany*

[5] *FOM-DIFFER, Eindhoven, The Netherlands*

[6] *EURATOM/CCFE Fusion Association,Culham Science Centre, Abingdon, UK*

[7] *See http://www.euro-fusionscipub.org/mst1*

The high performance of future thermonuclear fusion reactors will require an integrated model-based approach to estimate and control the plasma state and its evolution in real-time. As an example, if an accurate control of the plasma profiles is well coordinated with the active control of MHD instabilities, disruptions can be detected in time and avoided with preventive actions using multiple actuators.

The supervision and the active control of the plasma state can be achieved in real-time by combining the RApid Transport simulatOR – RAPTOR code [1] with a state observer. In so doing, a rapid yet accurate plasma profile evolution estimate can be obtained every few ms by solving two coupled 1D diffusion equations involving the electron temperature ($T_e$) and the poloidal magnetic flux.

Recently, RAPTOR has been upgraded to include the Porcelli's sawtooth model [2] to reproduce the effects of the sawtooth instability on the safety factor (q) and the $T_e$ profiles in real-time. The implemented heuristic model predicts a sawtooth crash whenever the magnetic shear on the q=1 surface ($s_1$) overcomes a critical threshold, being $s_1(crit.)\approx0.2$ in most plasmas of interest. The localization of the q=1 surface is performed with a linear interpolation on the iota ($\iota=1/q$) profile, that is mapped over a dense spatial grid (1001 points in the present work) for this purpose. If the $\iota$ profile is not monotonic, the code takes into account the outer radius where the equality $\iota=1$ is satisfied. Once a sawtooth crash is triggered, the plasma profiles evolve according to either the Kadomtsev's complete magnetic reconnection model [3] or to the incomplete one [2]. Eventually, the new profiles are down sampled back to the spatial grid of the real-time model that is usually 100 times less dense than the sawtooth model's one such that the code execution time is compatible with the time scale of the plasma dynamics.

The RAPTOR-based state observer, originally developed on TCV [1], has also been recently embedded in the ASDEX-Upgrade [4, 5] and RFX-mod real-time [6] control systems.

In ASDEX-Upgrade RAPTOR has also been used in offline simulations to model operational scenarios in sawtoothing plasma, providing a better understanding of sawtooth control and locking experiments, like the one that is reported in Fig.1. This figure shows the time evolution of the NBI power (a), the EC one (b) and its corresponding deposition locations (c), as the color code suggests. In this experiment an EC launcher (red) is pointed well inside the q = 1 surface, while the pulsed power of the remaining ones (blue and green) is deposited just outside this surface with the aim to stabilize and lock sawteeth. In this experiment the sawtooth locking period is increased in consecutive time windows in the range $T_{LOCK}$ = [70, 100, 140]ms, while the duty cycle is fixed at 80%. The last two panels compare the core ECE $T_e$ (e) and the measured sawtooth period (f) (blue) with the RAPTOR predictions (red). The agreement is remarkably good, sawtooth locking almost succeeds only with $T_{LOCK}$=100ms both in the experiment and in RAPTOR simulation as the near overlap with the green dashed line suggests. Offline RAPTOR simulations with different pre-programmed EC square

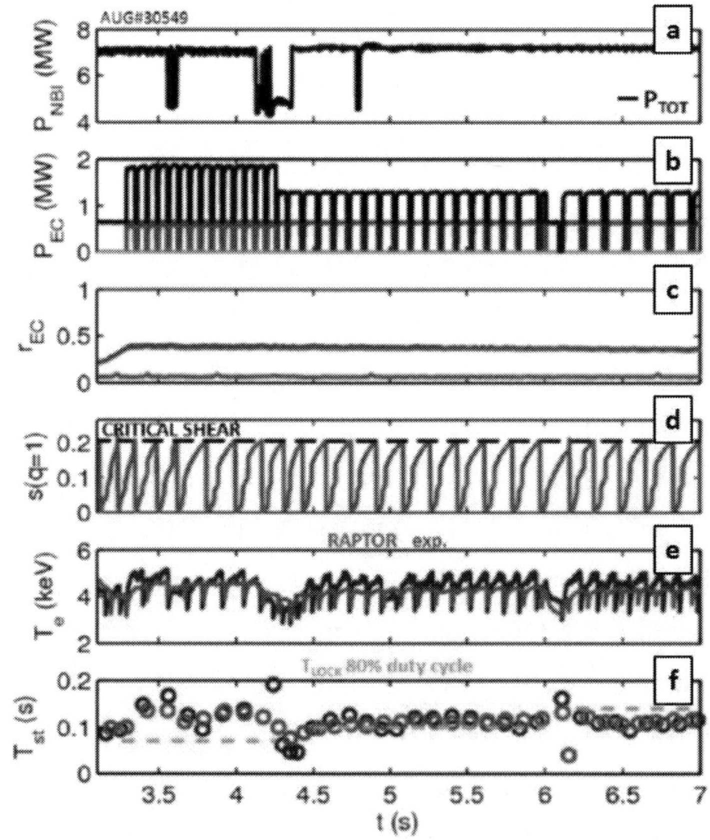


*Fig.1 RAPTOR validation in an ASDEX-Upgrade sawtooth locking experiment. a) Neutral beam injected power, b) Electron Cyclotron - EC injected power, c) radial deposition of the EC power (colour code corresponds to panel b)), d) magnetic shear on the q=1 surface predicted by RAPTOR (red) and critical threshold (dashed black line), e) measured (blue) and RAPTOR (red) core electron temperature, f) measured (blue) and RAPTOR (red) sawtooth period, expected period from locking power injection (dashed green line).*

waveforms and power deposition profiles suggest that sawtooth locking is very sensitive to the deposition radius for a given locking period and duty cycle, thus this code is a valuable to optimize the design of such experiments before their execution.

In RFX-mod a RAPTOR-based state observer has been embedded for the first time in the real-time MARTe framework and it has been used to model low-q edge sawtoothing tokamak plasmas. The final aim is a proof of concept experiment where a plasma disruption will be predicted on the basis of the sawtooth period rather than from direct magnetic measurements, that have not been included in this code yet. Sawteeth disappear very reproducibly in RFX-mod q(a)<2 plasmas as a current driven m=2, n=1 Resistive Wall Mode grows in amplitude prior a plasma disruption, as it can be seen in Fig.2 (on the left), but the same evidence is documented also in q(a)>2 plasmas that are terminated by a locked m=2, n=1 Tearing Mode (TM) (on the right). RAPTOR, that is unaware of this MHD dynamics, would predict stationary sawteeth thus the real-time comparison of the measured sawtooth period with the RAPTOR one will possibly predict the occurrence of a disruption a few tens ms early in both these operational regimes. A preparatory experiment for this task is reported in Fig.3. In this experiment, that validates RAPTOR in real-time in a circular tokamak plasma at q(a)=1.7, the pre-programmed gas puffing injection is switched off at t=0.6s to slowly ramp down the electron density (b) and as a result the sawtooth period shortens (d). The core electron temperature and sawtooth period predicted in real-time by RAPTOR (c), d) in red) follow the experimental trend (blue), though they bear the discretization due to the finite cycling time of the code. The latter was fixed to 2ms in these experiments to resolve changes in the sawtooth

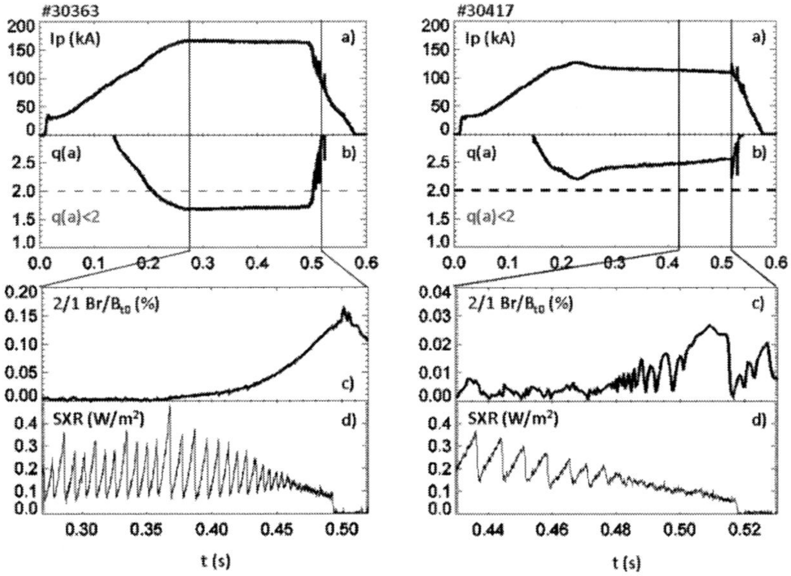


*Fig.2 RWM (on the left) and TM (on the right) mitigating effect on sawtreeth in RFX-mod. Time history of a) the plasma current, b) the edge safety factor and c) the m = 2, n = 1 amplitude of the radial magnetic field normalized to the equilibrium field.*

period in the order of tens ms. The spatial grid was reduced accordingly to 6 points, in order the execution time to be safely shorter than the cycling time. With this configuration the sawtooth model takes around 1.5ms to execute at every crash (e). Since a 25% margin in the execution time is still left, an optimization of the code will likely allow increasing the time resolution in future experiments, improving the accuracy of the real-time predictions.

The disruption avoidance technique that will be tested in RFX-mod will be a demonstrative example of a possible integrated control application with RAPTOR. The real-time comparison of the experimental dynamics with the modelled one can be exploited to contribute to the development of robust disruption avoidance schemes in larger tokamaks, like ASDEX Upgrade or JET, where other instabilities, like locked TMs or radiative ones, can trigger plasma disruptions.

**References**:

[1] F. Felici *et al* 2011 *Nucl. Fusion* **51** 083052

[2] F. Porcelli *et al* 1996 *Plasma Phys. Control. Fusion* **38** 2163

[3] B.B. Kadomtsev 1987 *Rep. Prog. Phys.* **50** 115

[4] F. Felici *et al* 2014 *Proceedings of the 41st EPS Conference on Plasma Physics, Berlin, Germany*

[5] F. Felici *et al* 2015 O4.127 *this conference*

[6] G. Manduchi *et al* 2014 *Fusion Engineering and Design* **89** 224

**Acknowledgment.** *This work has been carried out within the framework of the EUROfusion Consortium and has received funding from the Euratom research and training programme 2014-2018 under grant agreement No 633053. The views and opinions expressed herein do not necessarily reflect those of the European Commission.*

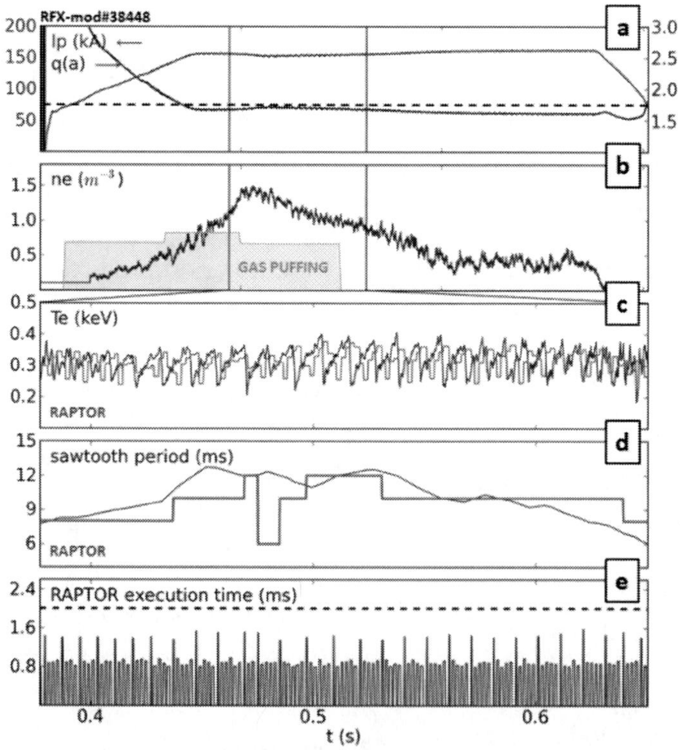


*Fig.3 RAPTOR validation in a RFX-mod low-q sawtoothing plasma. a) Plasma current (blue), edge safety factor (black) and q(a)=2 stability limit (black dashed line), b) electron density (blue) and the pre-programmed gas puffing injection (orange), c) measured (blue) and real-time RAPTOR predicted (red) core electron temperature, d) measured (blue) and real-time RAPTOR predicted (red) sawtooth period, e) RAPTOR execution time in MARTe (blue) and RAPTOR cycling time (black dashed line).*

# Evaluation of the neoclassical toroidal viscous torque in ASDEX Upgrade

A. F. Martitsch[1], S. V. Kasilov[1,2], W. Kernbichler[1], M. F. Heyn[1], E. Strumberger[3], S. Fietz[3], W. Suttrop[3], A. Kirk[4], the ASDEX Upgrade Team[3] and the EUROfusion MST1 Team *

[1] *Fusion@ÖAW, Institut für Theoretische Physik - Computational Physics, TU Graz, Petersgasse 16, A–8010 Graz, Austria*

[2] *Institute of Plasma Physics, National Science Center "Kharkov Institute of Physics and Technology", Akademicheskaya Str. 1, 61108 Kharkov, Ukraine*

[3] *Max-Planck Institut für Plasmaphysik, D-85748 Garching, Germany*

[4] *CCFE, Culham Science Centre, Abingdon, OX14 3DB, UK*

## Introduction

Non-axisymmetric non-resonant magnetic perturbations in a tokamak (e.g., toroidal field (TF) ripple, error fields, coils for ELM mitigation purposes) give rise to the so-called neoclassical toroidal viscous (NTV) torque, which can have a significant impact on the plasma rotation. The effect of the NTV torque on the plasma rotation has been observed in experiments [1]. For the evaluation of the NTV several analytical and semi-analytical approaches [2] are presently used, which make simplifying assumptions concerning geometry and collision operators. Numerical calculations of the NTV torque without such simplifications are presented here for the TF ripple and ELM mitigation coils in ASDEX Upgrade, and are compared to analytical models. The results are obtained from a quasilinear version of the code NEO-2, which is based on a numerical approach given in [3]. The only assumption is that the perturbations are small enough such that the particle motion within the perturbed flux surface is only weakly affected by the perturbation field (quasilinear approach). Since non-resonant magnetic perturbations are well described by the ideal MHD theory, the 3D equilibria of the ASDEX Upgrade discharges can be reconstructed using the code NEMEC [4]. These equilibria are used as an input for the upgraded code NEO-2.

## Definitions

In Boozer coordinate system $(r, \vartheta, \varphi)$ with re-defined flux surface label [3], $\langle|\nabla r|\rangle = 1$ and metric determinant $\sqrt{g}$, the flux surface averaged NTV torque density $T_\varphi^{\text{NA}}$ can be computed directly from the flux-force relation [2, 3],

$$T_\varphi^{\text{NA}} = -\frac{\sqrt{g}B^\vartheta}{c}\sum_\alpha e_\alpha \Gamma_\alpha^{\text{NA}}, \tag{1}$$

where $\sqrt{g}B^\vartheta = \partial\psi_{\text{pol}}/\partial r$ is the radial derivative of the poloidal flux and $c$, $e_\alpha$ and $\Gamma_\alpha^{\text{NA}}$ denote speed of light, charge and non-ambipolar particle flux density of species $\alpha$, respectively. Here a co-variant notation is used for magnetic field, velocity and torque density components. The non-ambipolar particle fluxes are expressed through transport coefficients $D_{ij}^{\text{NA}}$ and thermodynamic forces $A_j$,

$$\Gamma_\alpha^{\text{NA}} = -n_\alpha\left(D_{11}^{\text{NA}}A_1 + D_{12}^{\text{NA}}A_2\right), \tag{2}$$

where $n_\alpha$ is $\alpha$ species density, and transport coefficients are evaluated by NEO-2. The thermodynamic forces are specified by

$$A_1 = \frac{1}{n_\alpha}\frac{\partial n_\alpha}{\partial r} - \frac{e_\alpha E_r}{T_\alpha} - \frac{3}{2T_\alpha}\frac{\partial T_\alpha}{\partial r}, \qquad A_2 = \frac{1}{T_\alpha}\frac{\partial T_\alpha}{\partial r}, \tag{3}$$

*See http://www.euro-fusionscipub.org/mst1

where $T_\alpha$ and $E_r$ are $\alpha$ species temperature and radial electric field, respectively. The radial electric field can be calculated from the toroidal rotation frequency of ions via the relation

$$V^\varphi = \frac{c}{\sqrt{g}B^\vartheta}\left(E_r - \frac{1}{e_i n_i}\frac{\partial (n_i T_i)}{\partial r}\right) + qV^\vartheta, \qquad V^\vartheta = \frac{ckB_\varphi}{e_i\sqrt{g}\langle B^2\rangle}\frac{\partial T_i}{\partial r}, \tag{4}$$

where $q$ and $\langle\ldots\rangle$ are safety factor and the neoclassical flux surface average, respectively. Here the coefficient $k = 5/2 - D_{32}/D_{31}$ is determined by the parallel ion flow obtained from the NEO-2 solution for the unperturbed, axisymmetric problem,

$$\langle V_{\|i}B\rangle = B_\varphi\left(V^\varphi - qV^\vartheta\right) + \frac{ckB_\varphi}{e_i\sqrt{g}B^\vartheta}\frac{\partial T_i}{\partial r} = -\left(D_{31}A_1 + D_{32}A_2\right). \tag{5}$$

**Results**

Evaluation of the NTV torque by NEO-2 and analytical models is performed here for the ASDEX Upgrade equilibrium #30835@3200ms provided by the code NEMEC. In Fig. 1 the experimentally measured profiles of density $n_e$, temperatures $T_i$ and $T_e$, toroidal ion rotation frequency $V^\varphi$, and computed toroidal Mach number of the $\mathbf{E}\times\mathbf{B}$ rotation $M_t = cRE_r(v_T\sqrt{g}B^\vartheta)^{-1}$ as well as the safety factor for the corresponding shot are shown as functions of the normalized poloidal radius $\rho_{\mathrm{pol}} = (\psi_{\mathrm{pol}}/\psi^a_{\mathrm{pol}})^{1/2}$ where $\psi_{\mathrm{pol}} = 0$ on the magnetic axis. In the pre-

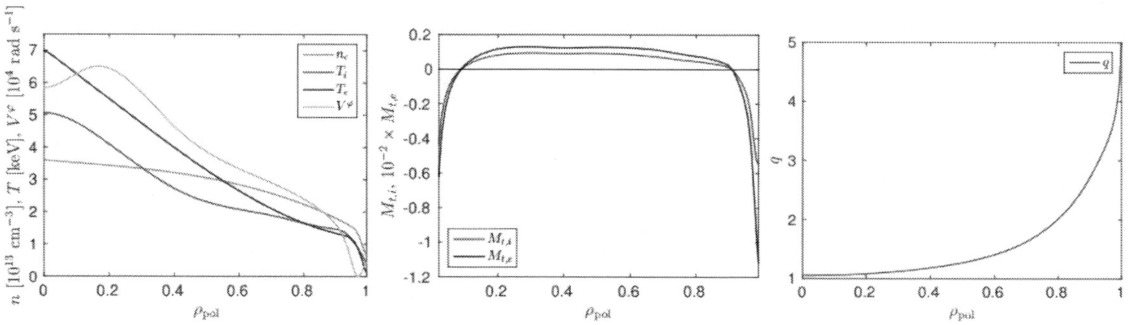


**Figure 1:** *Radial profiles of density, temperatures, toroidal rotation frequency (left), toroidal Mach number (center) and safety factor (right) for ASDEX Upgrade shot #30835@3200ms.*

sented modeling of the NTV torque non-resonant magnetic perturbations due to both, TF ripples (short-scale perturbation, n=16) and ELM mitigation coils (medium scale perturbation, n=2,6), are considered. For the TF ripple a comparison of NEO-2 results with analytical estimates of the torque density and integral torque,

$$T_\varphi^{\mathrm{NA,int}} = \int_{V(\rho_{\mathrm{pol}})} \mathrm{d}^3r\, T_\varphi^{\mathrm{NA}}, \tag{6}$$

where $V(\rho_{\mathrm{pol}})$ is the volume limited by the flux surface with given $\rho_{\mathrm{pol}}$, is shown in Fig. 2. It can be seen that the NTV torque acts in the direction opposite to the experimentally measured plasma rotation velocity and the integral torque is about -0.8 Nm. The NTV torque produced by TF ripples is mainly applied to ions and corresponds to the ripple-plateau regime [5] in most of the plasma volume. The analytical estimate, used here for comparison, can be obtained from a general expression for the particle flux in the ripple-plateau regime (Eq. (45) of Ref. [5]) by replacing in this derivation the simplified magnetic field with the more general form,

$$B(r,\vartheta,\varphi) = B_0(r,\vartheta) + B_n(r,\vartheta)\cos(n\varphi), \tag{7}$$

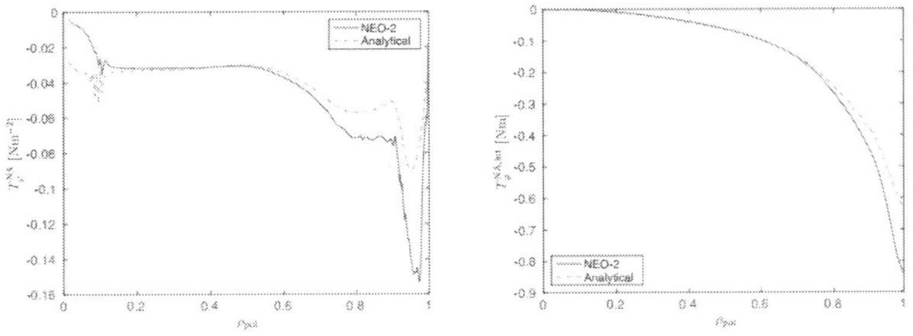


**Figure 2:** *Radial profiles of the NTV torque density (left) and the integral torque (right) produced by the TF ripple.*

where $B_0$ is the unperturbed magnetic field and $B_n$ is the perturbation field amplitude. The non-ambipolar diffusion coefficients valid for a general tokamak geometry are then

$$D_{11}^{\mathrm{NA}} = \frac{\sqrt{\pi}}{4} \frac{n m_i^2 c^2 v_T^3 B_0^2}{e_i^2 g (B_0^\vartheta)^2 B_0^\varphi} \left( \int_0^{2\pi} \frac{\mathrm{d}\vartheta}{B_0^2} \right)^{-1} \int_0^{2\pi} \frac{\mathrm{d}\vartheta}{B_0^3} \left( \frac{B_n}{B_0} \right)^2, \quad D_{12}^{\mathrm{NA}} = 3 D_{11}^{\mathrm{NA}}, \tag{8}$$

where the notation is the same as in Ref. [3]. The difference in the integral torque between the NEO-2 result and the analytical estimate is less than 25%. In Fig. 3 NEO-2 results for the NTV torque density and integral torque produced by ELM mitigation coils are compared to the bounce-averaged model of Shaing [2, 3]. The integral torque produced by ELM mitigation coils is -0.5 Nm, and adds up to the torque produced by the TF ripple giving a total torque of -1.3 Nm. For the NTV torque, not only ions but also electrons make a significant contribution, which is in the direction of (positive) plasma rotation and which partly balances the negative ion torque. Note that the electron torque agrees by order of magnitude with the result of the asymptotical model [2]. The observed discrepancies can be attributed to the rather small aspect ratio where the analytical model of Shaing [2] can significantly deviate from accurate computations [3]. However, this is not the case for ions where NEO-2 results exceed the results of the asymptotical model [2] significantly. It should also be noted that the torque density profile exhibits distinctive substructures in the vicinity of resonant surfaces, which are indicated by the vertical lines. A rather peculiar point in this profile is the resonant surface $(m,n) = (6,2)$ (light blue vertical line) which almost coincides with the zero of the electric field. The increased electron torque density around this point is due to the fact that for small values of the electric field $1/\nu$ transport is dominant in absence of resonant regimes. In order to determine relevant

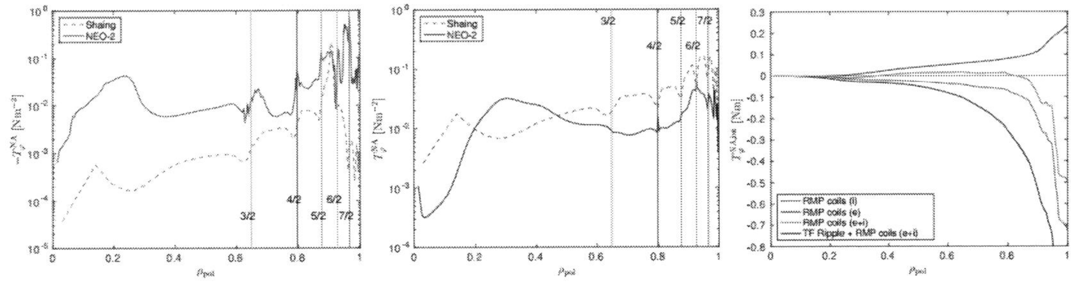


**Figure 3:** *Ion (left) and electron (center) contribution of the ELM mitigation coils to the NTV torque density and the integral torque (right) computed by NEO-2 as functions of the normalized poloidal radius. Vertical lines indicate the positions of resonant surfaces with $q(\rho_{\mathrm{pol}}) = m/n$, where $m$ and $n$ are the poloidal and toroidal mode numbers, respectively.*

transport regimes, a scan of the diffusion coefficient $D_{11}^{\mathrm{NA}}$ over collisionality and otherwise the same parameters as in the experimental profile has been performed at $\rho_{\rm pol} = 0.5$, see Fig. 4. It is seen that electrons there are at the onset of $\nu - \sqrt{\nu}$ regime while the plateau like behavior of ion coefficient indicates the resonant diffusion regime. A drop of ion torque near the zero of the electric field corresponds to the reduced resonant contribution which at relatively low Mach numbers increases with $M_t$ [6].

**Discussion**

The total value of the integral NTV torque for AUG shot #30835 as computed by NEO-2 is -1.3 Nm, which is less than the NBI torque value of +3.9 Nm. As it is known (see, e.g., Ref. [3]) the integral rotational moment is a conserved quantity and therefore missing balance between the NTV and NBI torques indicates the presence of other momentum sources. One obvious momentum source unaccounted here is the resonant torque [7] , which is applied mainly to the electrons (and is positive) in resonant layers around rational flux surfaces where the ideal MHD theory is not valid. Other possible reasons for discrepancies could be the NTV torque produced by error fields and fields from eddy currents in the wall induced by intrinsic MHD modes, which are not taken into account in this computation and which can have both, resonant and non-resonant contributions. A further possible source is the momentum exchange through the plasma boundary. As it is known, particle and heat fluxes and also the toroidal momentum fluxes to the inner and outer divertor plates are generally not the same. In particular, this effect is responsible for the toroidal plasma spin up by gas puff [8]. Finally, losses of fast particles produced by NBI [9] can produce a significant negative torque. Thus, a study is of interest of the discharges where RMPs are absent and the rest volume sources of the momentum (NTV from TF ripple and error fields, and fast particle losses) are taken into account since the balance of these sources should be met with sources at the edge.

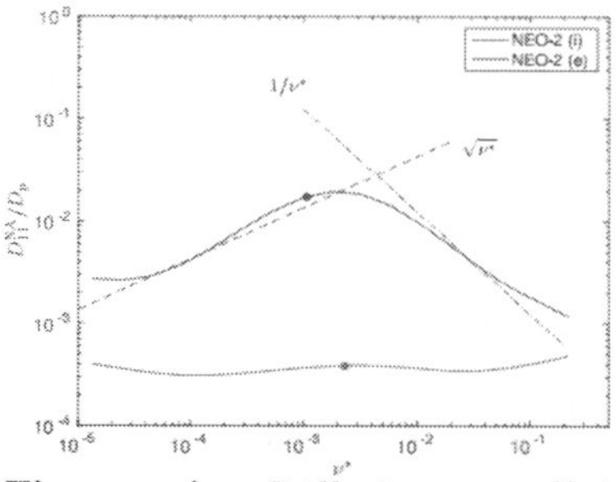


**Figure 4:** *Diffusion coefficient normalized by the plateau diffusion coefficient* $D_p = \pi q v_T \rho_L^2 (16R)^{-1}$ *for* $\rho_{\rm pol} = 0.5$ *as a function of the collisionality parameter* $\nu^\star = 2\nu q R v_T^{-1}$. *Actual collisionalities for the poloidal radius are shown by filled circles. Note that the normalizing plateau diffusion coefficient is by a a square-root of mass ratio smaller for electrons than for ions.*

**Acknowledgement**

This work has been carried out within the framework of the EUROfusion Consortium and has received funding from the Euratom research and training programme 2014-2018 under grant agreement No 633053. The views and opinions expressed herein do not necessarily reflect those of the European Commission. The authors gratefully acknowledge support from NAWI Graz.

**References**


[1] W. Zhu *et al*, Phys. Rev. Lett. **96**, 225002 (2006); A. M. Garofalo *et al*, Phys. Plasmas **16**, 056119 (2009); Y. Sun *et al*, Nucl. Fusion **52**, 083007 (2012).

[2] K. C. Shaing *et al*, Nucl. Fusion, **50**, 025022 (2010); J. Park *et al*, Phys. Rev. Lett. **102**, 065002 (2009); Y. Sun *et al*, Phys. Rev. Lett. **105**, 145002 (2010).

[3] S. V. Kasilov *et al*, Phys. Plasmas **21**, 092506 (2014).

[4] S. P. Hirshman *et al*, Comput. Phys. Commun. **43**(1), 143-155 (1986).

[5] A. H. Boozer, Phys. Fluids, **23**, 2283, 1980.

[6] C. G. Albert *et al*, this conference, P1.183.

[7] M. F. Heyn *et al*, Nucl. Fusion **54**, 064005 (2014).

[8] V. Rozhansky *et al*, J. Nucl. Mater. **291**, 337-339 (2005).

[9] T. Kurki-Suonio *et al*, Plasma Phys. Contr. Fusion **42**, 277-282 (2000).

# The effects of triangularity and main ion species on the inter-ELM profile evolution in ASDEX Upgrade

F. M. Laggner[1], E. Wolfrum[2], M. G. Dunne[2], M. Cavedon[2,3], P. Manz[2,3], G. Birkenmeier[2,3], R. Fischer[2], E. Viezzer[2], M. Willensdorfer[2], F. Aumayr[1], the EUROfusion MST1 Team* and the ASDEX Upgrade Team

[1] *Institute of Applied Physics, TU Wien, Fusion@ÖAW, 1040 Vienna, Austria*

[2] *Max Planck Institute for Plasma Physics, 85748 Garching, Germany*

[3] *Physik-Department E28, Technische Universität München, 85748 Garching, Germany*

In high confinement mode (H-mode) plasmas the steep gradients at the plasma edge (called pedestal) are quasi-periodically destabilised by edge localised modes (ELMs). Since the pedestal strongly affects the global plasma performance, further understanding of the temporal approach towards the pedestal stability limit (also known as pedestal recovery) is needed to optimise plasma scenarios. Previous studies have shown different recovery timescales of the electron density $n_e$ and temperature $T_e$ pedestal in between ELM crashes [1]. In these experiments the $n_e$ pedestal recovered first, while the $T_e$ pedestal started to recover after the $n_e$ pedestal was established. Gyrokinetic modelling attributed the recovery timescales to the domination of different turbulent modes [2]. In the phase of the $n_e$ recovery trapped electron modes were found in the pedestal, whereas in the pre-ELM pedestal after the $T_e$ recovery, especially on large scales, microtearing modes and kinetic-ballooning modes were dominant.

Experiments recently conducted at ASDEX Upgrade extended the investigated parameter space in plasma triangularity $\delta$, since this parameter strongly influences the edge stability. Additionally, to quantitatively compare the temporal development of the pedestal in plasmas with different main ion species and to test edge stability codes, a pedestal top $n_e$, $T_e$-match in deuterium (D) and hydrogen (H) was performed. All presented discharges were conducted at a plasma current of 1 MA, $-2.5$ T toroidal magnetic field in lower single null configuration. A medium $\delta \sim 0.23$ and a low $\delta \sim 0.19$ shape was established. Characteristic pre-ELM edge profiles, evaluated by integrated data analysis [3], for the two shapes (heating power $P_{\mathrm{Heat}} \sim 5.25$ MW

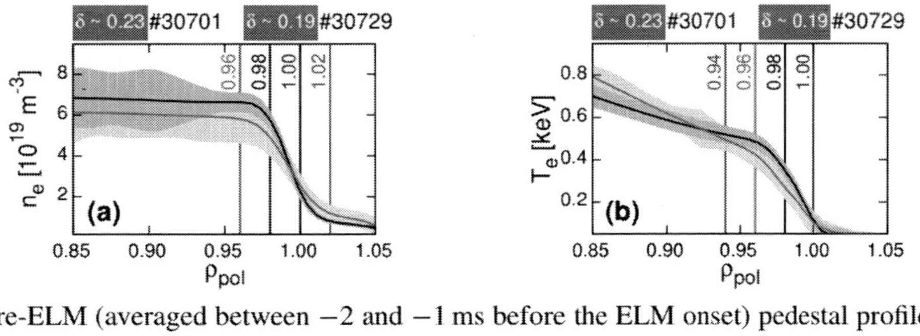


Figure 1: Pre-ELM (averaged between $-2$ and $-1$ ms before the ELM onset) pedestal profiles of $n_e$ (a) and $T_e$ (b) for medium ($\delta \sim 0.23$, black) and low ($\delta \sim 0.19$, magenta) triangularity. Owing to the different plasma shape the pedestal gradients of $n_e$ and $T_e$ are shallower at low $\delta$ compared to medium $\delta$.

* See http://www.euro-fusionscipub.org/mst1.

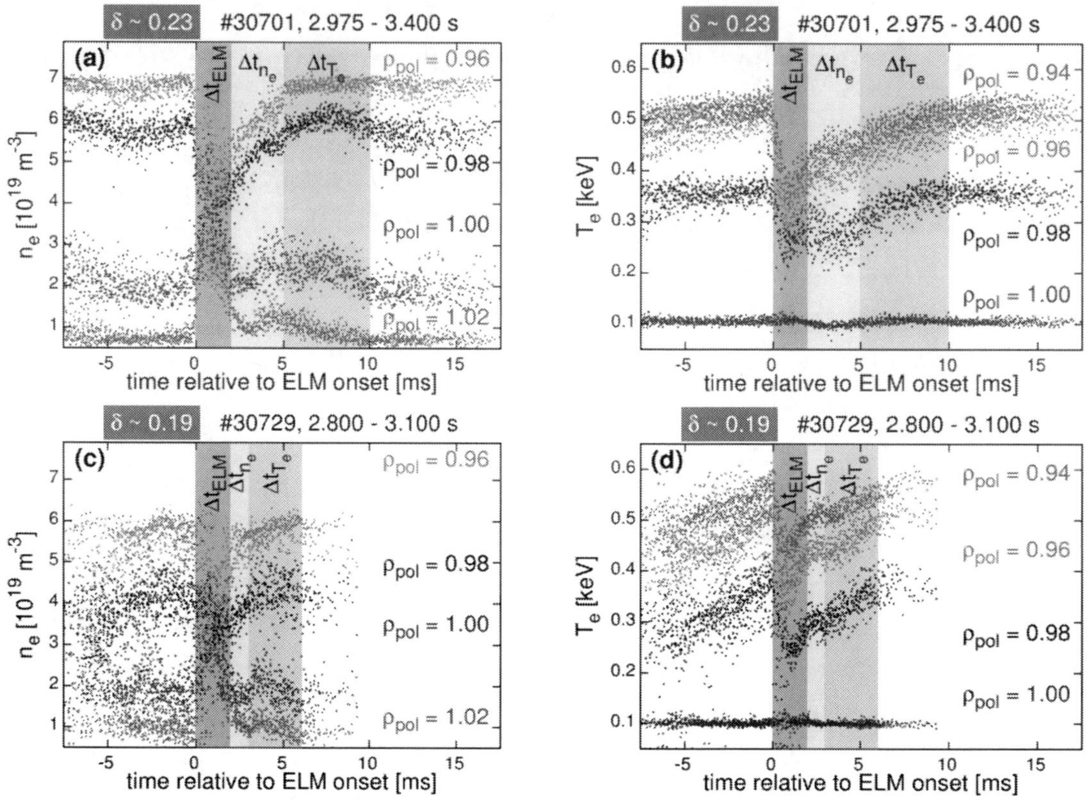


Figure 2: Temporal evolution of the ELM synchronised $n_e$ and $T_e$ at certain radial locations ($\rho_{pol}$) for medium $\delta \sim 0.23$ (a,b) and low $\delta \sim 0.19$ (c,d). In the low $\delta$ case $f_{\mathrm{ELM}}$ is roughly a factor 2.5 larger, the pedestal $T_e$ gradient recovers faster and does not saturate in comparison to medium $\delta$.

at medium $\delta$, $P_{\mathrm{Heat}} \sim 8.20\,\mathrm{MW}$ at low $\delta$) are presented in figure 1. The low $\delta$ discharge (magenta) has shallower pedestal gradients at similar pedestal width compared to the medium $\delta$ discharge (black), which is consistent with previous observations [4]. Several steps of heating power were applied to both low and medium triangularity discharges. In all cases the recovery of the pedestal after the ELM crash exhibited the same phases for both plasma shapes. As examples, figure 2 compares the inter-ELM evolution of ELM synchronised $n_e$ and $T_e$ profiles for medium (a,b) and low (c,d) $\delta$. The ELM frequency $f_{\mathrm{ELM}}$ in the low $\delta$ discharge is $\sim 180\,\mathrm{Hz}$, which is $\sim 2.5$ times larger than in the medium $\delta$ discharge ($\sim 70\,\mathrm{Hz}$), a change which can be partially attributed to the higher $P_{\mathrm{Heat}}$. The ELM duration ($\Delta t_{\mathrm{ELM}}$), here defined as the period with enhanced transport, visible in the divertor shunt current measurements (c.f. figure 3, black curve), is $\sim 2\,\mathrm{ms}$ for both discharges. In the low $\delta$ scenario two types of ELM cycles (data of both included in figure 2c,d) were observed (with similar $\Delta t_{\mathrm{ELM}}$), in which the 'fast' ELM crashes occurred immediately after the $n_e$ pedestal recovery and before the full $T_e$ pedestal recovery (compared to the 'slow' ELM cycle). For both $\delta$ cases the $n_e$ pedestal recovers faster than the $T_e$ pedestal. The $n_e$ pedestal reaches its pre-ELM state faster (see/compare $\Delta t_{n_e}$) in the low $\delta$ than in the medium $\delta$ discharge. Additionally, a second $n_e$ peak in the scrape off layer ($1.00 < \rho_{pol} < 1.02$) at 3 to 4 ms after the ELM onset is observed in both shapes. Its occurrence is correlated with the divertor neutral pressure. During the phase of the $n_e$ recovery the pedestal

$\nabla T_e$ (between $\rho_{pol} = 0.98$ and $\rho_{pol} = 1.00$) does not evolve (see figure 2b,d). The $T_e$ pedestal recovery time ($\Delta t_{T_e}$) at low $\delta$ is faster ($\sim 3\,\text{ms}$) than at medium $\delta$ ($\sim 5\,\text{ms}$), but this might be caused by the larger $P_{\text{Heat}}$ in the low $\delta$ discharge. Nevertheless, the pedestal $\nabla T_e$ (between $\rho_{pol} = 0.98$ and $\rho_{pol} = 1.00$) recovery saturates approximately 10 ms after the ELM onset in the medium $\delta$ discharge. This saturation is accompanied by the onset of radial magnetic fluctuations ($\partial B_r/\partial t$) with frequencies $\sim 240\,\text{kHz}$ (figure 3) measured at the low field side midplane. This behaviour may be connected to the onset of a pedestal limiting instability as described in [5]. Because these fluctuations disappear with each ELM and set on as soon as the $\nabla T_e$ is recovered, it is very likely that the modes are located in the steep gradient region.

The isotope pedestal top $n_e$, $T_e$-match required roughly a factor of 2 higher heating power as well as a factor of almost 10 higher fuelling rates in the H plasma to achieve similar pedestal top parameters. The pre-ELM $n_e$ and $T_e$ profiles of both plasmas are presented in figure 4. The pedestal top values agree well, however, the pedestal top ion temperature $T_i$ is noticeably higher in the H plasma. The comparison of the $n_e$ profiles (figure 4a) reveals shallower pedestal gradients in H plasmas. These are probably caused by the change of particle confinement for different main ion species, since a different ionisation source, further inside the plasma in H (at similar particle confinement), would lead to higher pedestal top $n_e$. Figure 5 compares the inter-ELM recovery of $n_e$ (a,c) and $T_e$ (b,d) for both main ion species. Both plasmas show qualitatively similar recovery phases, i.e. the pedestal $n_e$ recovers on faster timescales than the pedestal $T_e$. $\Delta t_{\text{ELM}}$ is in the range of $\sim 1.5\,\text{ms}$ for both main ion species. The $n_e$ pedestal takes longer to recover in the D plasma ($\Delta t_{n_e} \sim 6.5\,\text{ms}$) in comparison to the H plasma ($\Delta t_{n_e} \sim 3.5\,\text{ms}$), which possibly has a connection to the 10 times higher fuelling rate in H. In the H plasma a significantly higher $f_{\text{ELM}}$ ($\sim 100\,\text{Hz}$) is observed than in

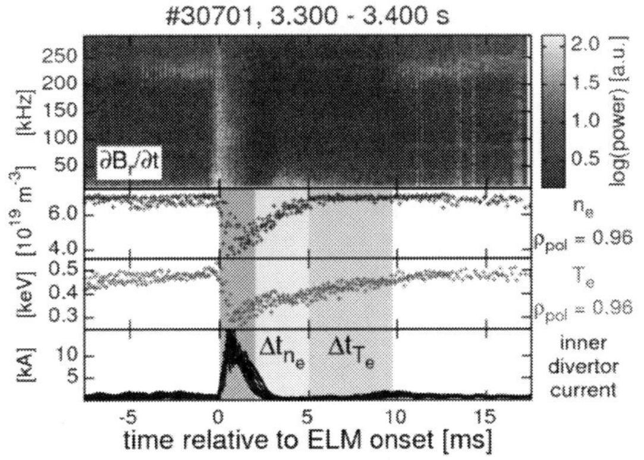


Figure 3: ELM synchronised spectrogram of radial magnetic fluctuations ($\partial B_r/\partial t$). When a certain pedestal top $T_e$ (red) is reached frequencies of $\sim 240\,\text{kHz}$ get dominant, correlating with a saturation of the pedestal $\nabla T_e$ (c.f. figure 2b).

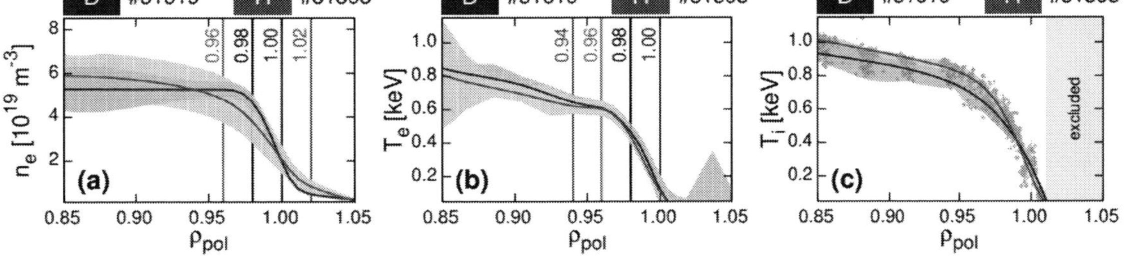


Figure 4: Pre-ELM (averaged between $-2$ and $-1$ ms before the ELM onset) profiles of $n_e$ (a), $T_e$ (b) and $T_i$ (c) for D (blue) and H (red). The gradient of the $n_e$ profile in H is shallower than in D, leading to a wider pedestal and a reduction of the achievable pressure gradient.

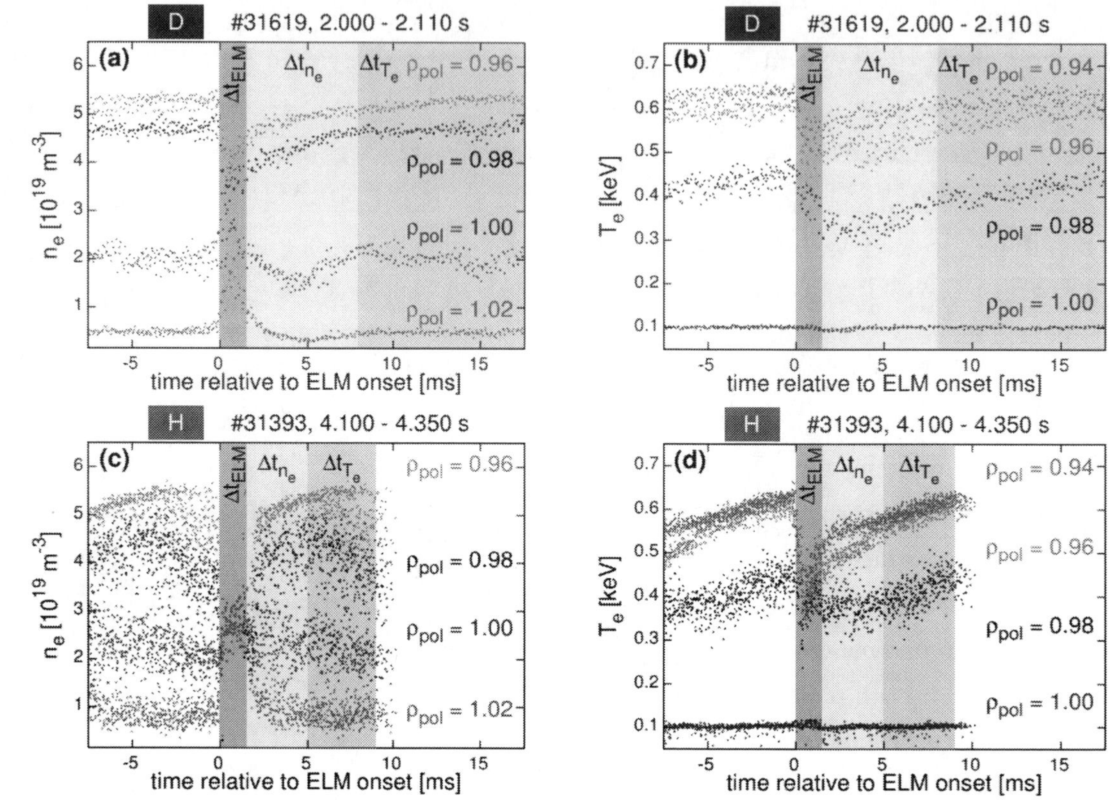


Figure 5: Evolution of $n_e$ (a,c) and $T_e$ (b,d) at certain positions ($\rho_{pol}$) for the D (a,b) and H (c,d) plasma. The $T_e$ pedestal in H recovers faster than in D owing to the higher $P_{Heat}$.

D ($\sim$ 50 Hz). The ELM energy and particle losses are higher in the H plasma, but the fraction of power losses caused by ELMs $P_{ELM}/(P_{Heat} - P_{rad,core})$ is approximately similar in both plasmas. Linear MHD stability analyses indicate that the pre-ELM pedestals are peeling-ballooning limited for both main ion species. Nevertheless, the operational point of the H plasma is at shallower pressure gradients caused by the shallower pedestal $\nabla n_e$.

From the presented observations following conclusions can be drawn: The different inter-ELM recovery timescales for the $n_e$ and $T_e$ pedestal ($n_e$ pedestal recovers faster than the $T_e$ pedestal) are robust and observed at different $\delta$ and in plasmas with different main ion species. In cases when the pedestal $\nabla T_e$ saturates the onset of radial magnetic fluctuations with frequencies larger then 200 kHz is found, which are probably the signature of a pedestal gradient limiting instability. The D and H plasma comparison found much shallower gradients in the H $n_e$ pedestal, which might be related to different particle confinement.

## References

[1] A.Burckhart, et al. Plasma Physics and Controlled Fusion **52**, 10 (2010)
[2] D. R. Hatch, et al. Nuclear Fusion **55**, 6 (2015)
[3] R. Fischer, et al. Fusion Science and Technology **58** 2 (2010)
[4] J. Stober, et al. Plasma Physics and Controlled Fusion **42**, 5 (2000)
[5] A. Diallo, et al. Physical Review Letters **112**, 11 (2014)

## Acknowledgement

This work has been carried out within the framework of the EUROfusion Consortium and has received funding from the Euratom research and training programme 2014-2018 under grant agreement No 633053. The views and opinions expressed herein do not necessarily reflect those of the European Commission.

F. M. Laggner is a fellow of the Friedrich Schiedel Foundation for Energy Technology.

# Fast changes in the mode structure of chirping energetic particle driven modes

L. Horváth[1], G. Papp[2,3], G. I. Pokol[1], Ph. Lauber[3], G. Por[1],
A. Gude[3], V. Igochine[3] and the ASDEX Upgrade Team[3]
[1] *Institute of Nuclear Techniques, Budapest University of Technology and Economics, Hungary*
[2] *Max-Planck/Princeton Center for Plasma Physics*
[3] *Max Planck Institute for Plasma Physics, Garching, Germany*

Supra-thermal energetic particles (EPs) in tokamak plasmas can excite various instabilities which can lead to an enhanced transport of fast particles. These instabilities constitute a nonlinear system where kinetic and MHD non-linearities can both be important making it difficult to describe the phenomenon [1]. Several types of non-linear behaviour of the mode amplitude and frequency are observed on present-day tokamaks. This behaviour ranges from steady-state – where the amplitude saturates at nearly constant frequency – to an explosive growth. The type of behaviour is expected to significantly influence the impact of the instabilities on the fast particle transport. Thus, their thorough understanding from both experimental and theoretical side is essential.

Bursting beta induced Alfvén eigenmodes (BAEs) and EP-driven geodesic acoustic modes (EGAMs) are routinely driven unstable in the ramp-up phase of off-axis NBI heated plasmas in ASDEX Upgrade (AUG) [2]. The investigation of these bursting modes can provide important information about the underlying physics of the non-linear behaviour of EP-driven modes.

Our goal is to experimentally investigate the rapid changes in the radial structure of bursting EP-driven modes during the non-linear chirping phase. Due to diagnostic and data analysis complexities this task has never been accomplished before. Some modes are expected to retain their radial structure, while others would be expected to change. Even qualitative results can provide important information about the underlying physics and strengthen (or challenge) our present theoretical understanding. Furthermore, an analysis based on the methods developed here can serve as the basis of comparison with numerical codes which simulate the time evolution of EP-driven modes.

**Analysis principles** The only fluctuation measurement which has spatial resolution and these modes were visible in its signal was the SXR diagnostic. The ECE and the ECE imaging (ECEI) diagnostics would be good candidates due to their high time- and spatial resolution, however, BAEs and EGAMs were not visible by them in the investigated shots. Since the SXR diagnostic is a line-integrated measurement, it is not straightforward to reconstruct the radial structure of the mode, but many line-of-sights (LOSs) are available which makes it possible to qualitatively investigate the time evolution of the radial structure.

In general, the signal-to-noise ratio of the SXR measurements was poor. In many cases, modes which were clearly visible on the magnetic spectrogram were only visible on one or two LOSs of SXR. The strategy was to find cases where the mode is observable on at least 3 adjacent LOSs of a particular SXR camera. Finally, 3 cases for BAEs and 5 cases for EGAMs were found where the oscillation amplitude on the SXR signals was sufficiently high to identify them visually on the SXR spectrograms.

To examine the changes in the radial eigenfunctions, a radial mapping of the oscillation amplitudes was constructed. This means that each LOS was labelled with the normalized poloidal flux of the magnetic flux surface to which the LOS was tangential. The oscillation amplitude was calculated for each LOS and it is plotted as the function of the radial coordinate. This way the radial mapping of the oscillation amplitudes can be evaluated at any time instant. The oscillation amplitude were not normalized with the gradient of the background SXR radiation, because it did not change significantly during a chirp. The rapidly changing mode frequency and the low signal-to-noise ratio were handled with an advanced time-frequency transform based method [3].

**Beta-induced Alfvén eigenmodes** In the gaps of the shear Alfvén continuum weakly damped normal modes can exists. These gap modes can be easily excited by the EP population. The BAE gap is introduced by the coupling between compressible acoustic waves and the shear Alfvén continuum [4]. BAEs are driven by the radial gradient in the EP distribution function [5].

A magnetic spectrogram with downchirping BAEs is shown in figure 1a. A thorough mode number analysis [6] was carried out by using the magnetic pick-up coil system of AUG which showed that the observed BAEs have $n = 1$ toroidal and $m = -3$ poloidal mode number.

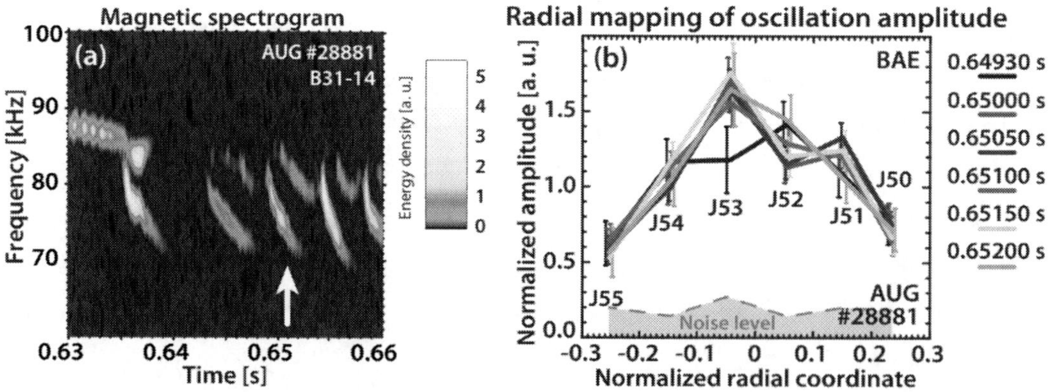


Figure 1: **(a)** *Downchirping BAEs in the range of* 70 − 85 *kHz are visible on the magnetic spectrogram.* **(b)** *The radial mapping of the oscillation amplitude of BAEs on the different SXR LOSs.*

The radial mapping for the chirp indicated with white arrow in figure 1a is presented in figure 1b. Since this work focuses on the relative changes in the radial structure of the modes, each curve of the radial mapping was normalized with its integral. The background noise level is estimated from time-frequency intervals of the SXR spectrograms where no coherent mode appeared. The average of the background noise level of the different channels is also plotted in the radial mappings. The error bars were estimated from the background noise level by taking into account the error propagation in the amplitude reconstruction formula [7].

The time evolution of the radial structure is similar in all 3 cases to the one presented in figure 1b, i.e the results do not show significant changes in the radial distribution of the oscillation amplitudes. This suggests that if there is any change in the radial eigenfunction it is smaller than the uncertainty of our measurement. This observation is consistent with the physical picture that BAE is a normal mode of the plasma and its radial structure strongly depends on the background plasma parameters rather than on the EP distribution.

**Energetic particle driven GAMs** EGAM is an energetic particle mode which can be excited if the EP pressure is sufficiently high compared to the thermal pressure and the EP drive can overcome the continuum damping. Their mode structure is independent from the shear Alfvén continuum, and sensitively depends on the EP distribution. Their frequency usually correspond to a characteristic frequency of the EP orbital motion.

In figure 2a a magnetic spectrogram with upward chirping EGAMs is shown. In principle EGAMs should only be detectable with density fluctuation measurements because it is an electrostatic mode, however, due to sideband coupling it is clearly visible on magnetic fluctuation measurements as well [8]. According to the mode number analysis carried out by using the magnetic probe measurements the mode coupled to EGAMs (which allows to detect the mode in the magnetic fluctuation signals) have $n = 0$ toroidal and $m = -2$ poloidal mode number.

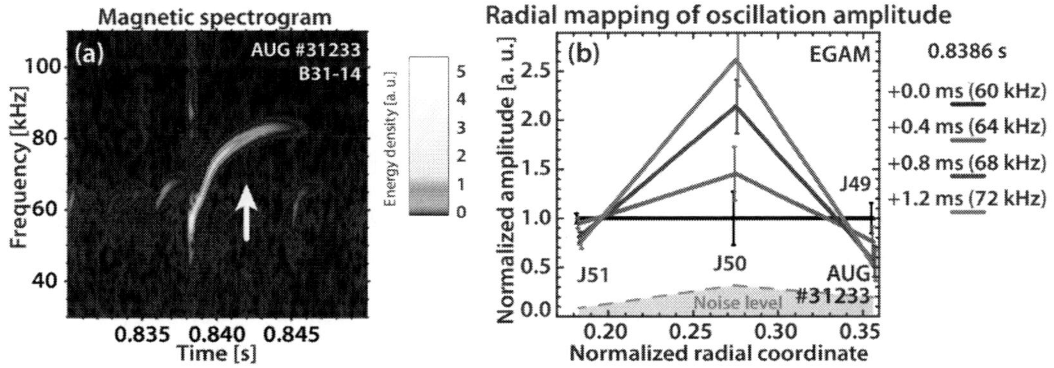


Figure 2: **(a)** *Upchirping EGAMs in the range of 40 - 90 kHz are visible on the magnetic spectrogram.* **(b)** *The radial mappings of the oscillation amplitude of EGAMs in 3 SXR LOSs which are located above the magnetic axis.*

The radial mapping for the chirp indicated with white arrow in figure 2a is presented in figure 2b. In this case the reconstructed amplitudes were normalized to the amplitude at the beginning of the chirp. Figure 2b shows the results of LOSs located above the magnetic axis. On LOSs located below the magnetic axis, radial mappings do not show any significant change in the mode structure. However, on LOSs located above the magnetic axis (figure 2b), a shrinkage of the mode is visible, since as time evolves the relative amplitude in the middle channel is rising compared to the outer channels. This shrinkage of the EGAM radial structure was significant in all 5 investigated cases.

The observed shrinkage of EGAMs during the non-linear chirping phase is consistent with our present theoretical understanding. According to the EGAM resonance condition ($\omega - \omega_t = 0$), the mode frequency is equal to the transit frequency of the interacting ions. The EGAM drive is proportional to the velocity phase space gradient in the EP distribution function. The coordinates of the steepest gradient determine the velocity phase space coordinates of the interacting particles. The mode frequency is rapidly increasing as it is visible in figure 2a. This means that as time evolves, the resonance condition changes and shifts to particles with higher transit frequency. The simulated EP distribution (TRANSP [9]) showed that the resonance condition moves toward more passing particles during the non-linear phase. Since the radial structure of EGAMs is determined by the orbit of the interacting ions and more passing particles have a narrower orbit width, the experimentally observed shrinkage of the radial mode structure can be explained.

**Conclusions** Our analysis showed that in case of the observed downward chirping BAEs the changes in the radial structure were smaller than the uncertainty of the measurement. This behaviour is consistent with that the radial structure of BAEs – as normal modes – strongly depends on the background plasma parameters rather than on the EP distribution.

In case of rapidly upward chirping EGAMs the analysis consistently shows shrinkage of the mode structure. This can be explained by the changing resonance condition in the velocity phase space of EPs. The mode structure of EGAMs is sensitive to the EP distribution. The rising frequency of the mode indicates that, as time evolves, the EGAM is driven by more passing particles which have narrower orbit width. This leads to the experimentally observed shrinkage of the mode structure.

*Authors at BME NTI acknowledge the support of the Foundation for Nuclear Engineers (JNEA) and Hungarian State grant NTP-TDK-14-0022. This work has been carried out with support of FuseNet, the European Fusion Education Network. This work has been carried out within the framework of the EUROfusion Consortium and has received funding from the Euratom research and training programme 2014-2018 under grant agreement No 633053. The views and opinions expressed herein do not necessarily reflect those of the European Commission.*

**References**

[1] W. W. Heidbrink, Physics of Plasmas **15**, 055501 (2008)
[2] Ph. Lauber et al., Presented at the 13th IAEA TCM, Beijing, China (2013)
[3] L. Horváth et al. 41st EPS Conference on Plasma Physics **38F**, P2.008 (2014)
[4] A. D. Turnbull et al., Physics of Fluids B, **5**, 2546-2553 (1993)
[5] C. Nguyen et al., Plasma Physics and Controlled Fusion **51**, 095002 (2009)
[6] L. Horváth, *Reducing systematic errors in time-frequency resolved mode number analysis*, submitted to Plasma Physics and Controlled Fusion, arXiv:1506.04864 (2015)
[7] L. Horváth, *Analysis of fast ion induced instabilities in tokamak plasmas*, MSc thesis, Budapest University of Technology and Economics, arXiv:1506.05134 (2015)
[8] Ph. Lauber, Physics Reports, **533**, 33-68 (2013)
[9] A. Pankin, Computer Physics Communications, **159**, 157-184 (2004)

# MHD modes analysis in high-beta ASDEX Upgrade configurations including a 3D model of conductors

S. Mastrostefano[1], P. Bettini[2,3], T. Bolzonella[3], V. Igochine[4], Y. Liu[5], F. Villone[1] and D. Yadykin[6]

[1] *CREATE, DIEI, Università di Cassino e del Lazio Meridionale, Cassino (FR), Italy*

[2] *Consorzio RFX, Padova, Italy*

[3] *DII, Università di Padova, Padova, Italy*

[4] *Max Planck Institute for Plasma Physics, EURATOM Association, Garching, Germany*

[5] *CCFE Fusion Association, Culham Science Centre, Abingdon, OX14 3DB, UK*

[6] *Uppsala University, Uppsala, Sweden*

## 1. Introduction

The main goal of this work is to analyze MHD modes in high-beta configurations of the ASDEX-Upgrade device, evaluating and quantifying the effect of 3D conductors, e.g. in order to find possible extension of operation space in case of modification of the various conducting structures. Assuming that resistive instabilities (e.g. tearing modes) are suitably prevented, the maximum achievable value of beta is limited by the onset of the ideal external kink instability. The actual limit depends mainly on the stabilizing effect of the conducting components around the plasma: any plasma perturbation induces eddy currents in the conducting structures, which tend to counteract the instability itself. If the conductors are close enough, the instability is slowed down from fast Alfvén times to the electromagnetic times, giving rise to the so-called Resistive Wall Modes (RWM). Several different codes and models can be used to analyse this problem, ranging from linearized axisymmetric codes to nonlinear three-dimensional models. In this paper, we use the CarMa code [1], which is able to analyse RWM in presence of 3D volumetric conducting structures.

## 2. Reference plasma configuration and geometry

The AUG reference configuration [2] used in this paper is described by a safety factor on axis $q_0 = 1.17$, major radius 1.64 m, toroidal magnetic field at axis $B_\phi = 2.38$ T and toroidal plasma current $I_p = 1$ MA. In the following, the plasma beta has been scanned keeping plasma current and safety factor constant. Fig. 1 reports the plasma boundary and the q profile. For this AUG configuration, the vacuum vessel is relatively far from the plasma - close to or even beyond the limit position, as it will be discussed later. In such situation, a sophisticated and detailed description of the structures is necessary to quantify the actual stabilizing effect [3]. The CarMa code [1] is used to analyse the effect of the main 3D structures (Fig. 2), specifically including the vacuum vessel with port extensions and bellows and the PSL, made of two toroidally continuous copper rings connected by poloidal bridges.

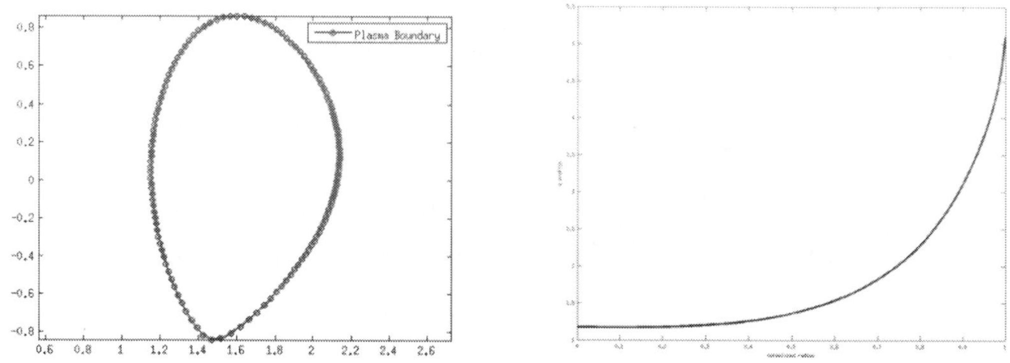


Fig. 1. Plasma boundary and q profile

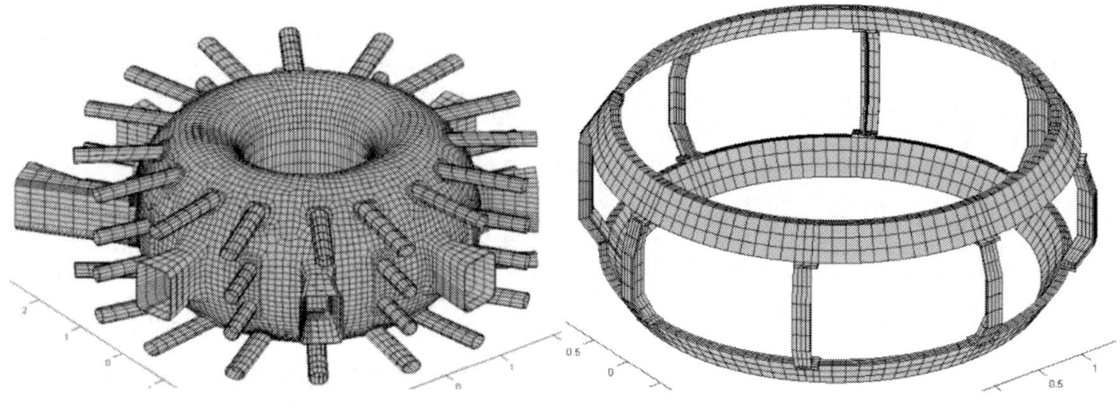

(a) (b)

Fig. 2. 3D conducting structures considered: (a) vacuum vessel; (b) PSL

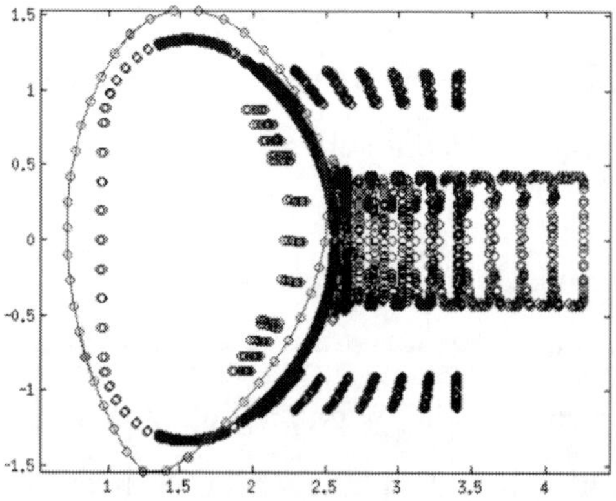


Fig. 3. Ideal wall limit position (red) compared to actual conducting structures (blue)

### 3. Results

First of all, the no-wall beta limit has been computed as $\beta_{nowall} = 3.06$. A beta value above this value ($\beta_N = 3.21$) has been selected, so as to destabilize the ideal pressure-driven external kink n=1 mode. For this configuration, the limiting ideal wall position has been computed for an axisymmetric wall conformal to the plasma boundary: any conducting structure located outside this position is not able to slow down the instability to electromagnetic times. Figure 3 shows the relative position of the ideal wall limit position with respect to actual structures, showing that the vessel is expected to be only partially effective, while the PSL is close enough to plasma. Indeed, considering an axisymmetric conformal wall located at the radial position of the PSL (Fig. 4a: gray+blue+red), a RWM appears with a normalized growth rate $\gamma\ \tau_w = 0.47$. This value was positively compared against MARS-F [4], which provides the same value within approximately 6%. Fig. 4b shows the corresponding current density pattern. In order to evaluate the effect of the partial poloidal coverage of the PSL, other two cases have been considered. The first one corresponds to rings connected by toroidally continuous bridges (red+blue), providing the same result within a few percents, while the second corresponds to the rings only (red), providing an instability on Alfvén time scale. A computation carried out on the true 3D geometry (Fig. 2) provides again an ideal instability: in other words, considering its realistic geometry, the PSL is not able to stabilize the plasma. A similar consideration applies also to the vessel alone.

Conversely, if both the vessel and the PSL are considered together, the instability is slowed down to electromagnetic time (growth rate $\gamma \approx 163\ s^{-1}$). Figure 5a shows the corresponding current density pattern. We conclude that there is a synergistic effect of PSL and vessel: in a sense, the PSL is close enough to plasma, while the vessel guarantees poloidal continuity.

The current density pattern of Fig. 5a clearly shows some interruptions between consecutive sectors, which are interpreted as being due to the presence of high-resistance bellows. To prove this conjecture, we consider a fictitious case in which the bellows have been removed: the current density pattern (Fig. 5b) shows no interruptions (except for port extensions, obviously) and the growth rate drops to $\gamma \approx 104\ s^{-1}$.

The same analysis has been carried out for a higher beta value ($\beta_N = 3.66$ [2]). The ideal wall limit position in this case is completely inside the vessel, so that the vessel does not contribute to stabilization. A 2D wall located at the PSL position again stabilizes the plasma, but the realistic 3D structure (PSL with bridges + vessel with bellows) cannot slow down the instability to electromagnetic times.

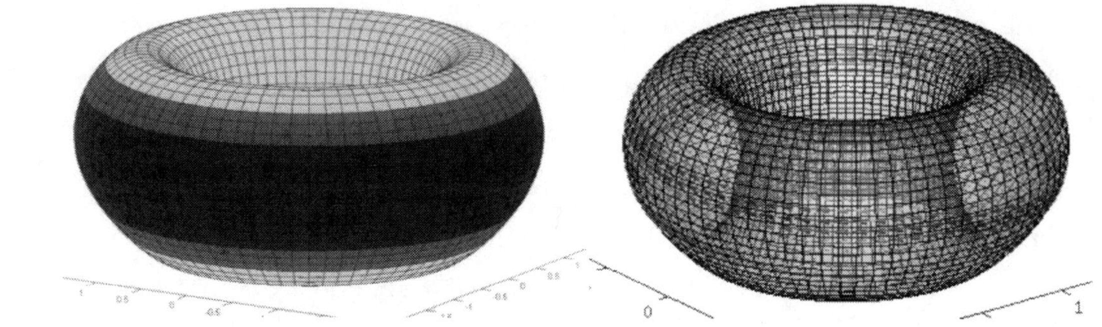

(a) (b)

Fig. 4. 2D wall located at the same position of the PSL: (a) geometry, (b) current density pattern

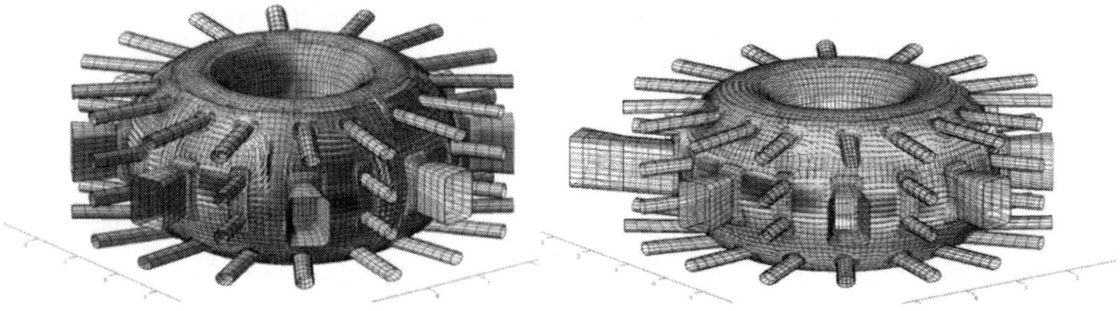

(a) (b)

Fig. 5. Current density patterns in the vessel: (a) reference; (b) no bellows

**4. Conclusions**

A detailed 3D analysis of the external kink in AUG has been carried out, investigating the 3D effects of conducting structures surrounding the plasma, investigating in particular the effects of partial poloidal coverage of PSL and the connection of rings through bridges, and the influence of the high-resistance bellows between sectors. The main conclusion is that 3D effects may be significant: a 2D wall located at the PSL position may be too optimistic, while the realistic 3D wall is able to stabilize configurations with a normalized beta not much above the no-wall limit. These conclusions may not necessarily apply to other equilibrium configurations. This work was supported in part by Italian MIUR under PRIN grant#2008E7J7A3 and in part by EUROFusion MST1 Work Package.

[1] A.Portone, F.Villone, Y. Liu, R. Albanese, G. Rubinacci, *Plasma Phys. Control. Fusion* **50** (2008) 085004

[2] V. Igochine et al., *40$^{th}$ EPS conference on Plasma Physics* (2013) paper P5.147

[3] E. Strumberger et al., *Phys. Plasmas* **15** 056110 (2008)

[4] Y. Liu et al., *Phys. Plasmas* **7** 3681–90 (2000)

# NBI modelling by upgraded TASK3D-a code in preparation of LHD deuterium campaigns

P. Vincenzi[1,2], S. Murakami[3], M. Osakabe[4], R. Seki[4], M. Yokoyama[4], T. Bolzonella[1]
[1]*Consorzio RFX, Corso Stati Uniti 4 - 35127 Padova, Italy*
[2]*Università degli Studi di Padova, 35122 Padova, Italy*
[3]*Department of Nuclear Engineering, Kyoto University, Kyoto 606-8501, Japan*
[4]*National Institute for Fusion Science, Toki, 509-5292, Japan*

LHD (Large Helical Device) [1] is a superconducting helical device (heliotron type) capable of studying current-less plasmas. The experiment is equipped with 5 hydrogen neutral beam injectors: 2 perpendicular positive-NBIs (40-50 keV, up to 12MW) and 3 tangential (co- and counter-current) negative NBIs (180-190 keV, up to 16MW). Currently hydrogen plasma experiments are run at LHD, but the use of deuterium (D) is planned in the future campaigns, together with the upgrade of at least 3 NBIs to D injection. In view of a helical fusion reactor it is therefore important to investigate the isotope effect and understand the energetic ion confinement in helical devices. In tokamak devices it is already known that the increase of plasma ion mass leads to better confinement properties [2,3].

Interpretative analyses of the LHD experiments are currently carried out with TASK3D-a integrated transport code [4], capable of treating hydrogen and multi-ion species plasma [5,6]. Effort has been recently put to start the upgrade of TASK3D-a code to extend the analysis capabilities to D plasmas with D NBIs. One of TASK3D-a routines is named FIT3D and it has been developed to evaluate radial profiles of NBI absorbed power, beam pressure, beam source and induced momentum. This paper presents the first upgrades proposed for FIT3D routine in order to allow the complete modelling of D NBI injection and the interaction with D plasma. Two main modifications have been proposed and here illustrated: the first regards the evaluation of the beam D particle ionization in the D plasma, while the second regards the implementation of a module which estimates the neutron rate production from D-D reactions, both from thermal and beam-plasma reactions.

**Beam stopping cross section for D NBI ionization process in D plasmas**

A central part of FIT3D routine is the calculation of the so-called "beam stopping cross section". The beam attenuation process is described by the stopping cross section $\sigma_s = 1/(n_e\lambda)$, where $n_e$ is the plasma density and $\lambda$ is the attenuation length. Currently FIT3D code uses an analytical formula for the beam stopping cross section proposed by Janev [7]. The formula was originally calculated for neutral hydrogen injection in H plasma; however changing the relative atomic mass of the beam particles *u* in the input $E_{NBI}$ /u we can obtain the attenuation of a deuterium beam. Janev's formula is calculated for 100$\leqslant$ $E_{NBI}$ (keV/u)$\leqslant$ $10^4$. Considering LHD hydrogen NBIs, the two positive NBIs are out of this range. In case of D beams, the beam energy divided by atomic weight of deuterium (keV/u parameter) will be less than 100 kev/u for all the NBI lines, being therefore out of the range of Janev's calculation. For this reason we proposed the use of the beam stopping cross section formula provided by Suzuki [8], which represents an updated and extended alternative to Janev's formulation. This formula is calculated for a wider $E_{NBI}$/u range and it uses different fitting coefficients in case of H, D and T background plasmas, while Janev did not distinguish different background plasma cases. Given the possibility of choosing D plasma background and given the wider $E_{NBI}$/u range of fit's validity, the use of Suzuki formulation is preferable in the analysis of D plasma and NBI with TASK3D-a tool. Figure 1 compares the beam stopping cross section for NBI D 80 keV particles (i.e. 40keV/u - typical LHD NBI energy) calculated with Janev and Suzuki formulas, with an arbitrary $Z_{eff}$ of 1.5 and different plasma densities and temperatures. We can note some discrepancies, especially at high density: as remarked 40 keV/u is not in the range of Janev's formula.

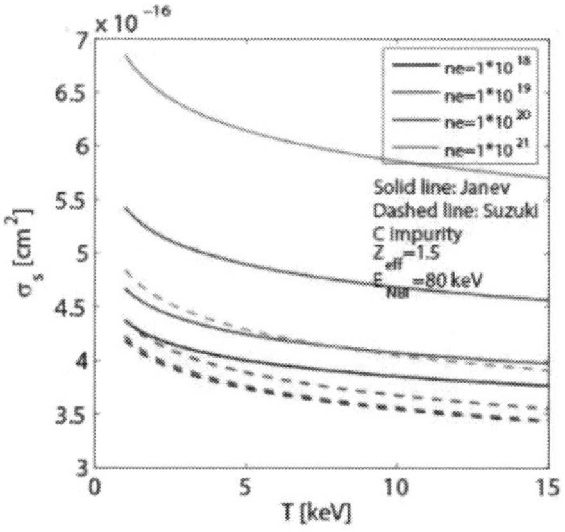


**Figure 1**: Comparison of beam stopping cross sections for D LHD NBI with 80 keV, calculated by Janev (solid line) and Suzuki (dashed line) formulas

**Neutron production from thermal plasma and NBI-plasma fusion reactions**

In D plasmas a significant neutron production is expected due to D-D fusion reactions, especially if energetic D particles are injected by the beams. D-D fusion reactions are:

1. D + D → p (3.02MeV) + T (1.01 MeV) → proton production
2. D + D → n (2.45 MeV) + $^3$He (0.82 MeV) → neutron production

Reactions originate in background (thermal) plasma, from beam-plasma interactions and beam-beam interactions. In order to have an evaluation of the neutron rate and fusion power production, we implemented a new routine in FIT3D code. In the present work beam-beam fusion reactions are neglected, being a minor fusion source and quite complicated to be treated. For thermal fusion reactions we decided to implement the Bosch formula for the fusion reactivity <σv> for a Maxwellian plasma [9]. Regarding the beam-plasma fusion source, the integral for <σv> has to be calculated taking into account both plasma and beam particle distribution function. D-D fusion cross section formula has been taken from Bosch [9] and the beam particle distribution function f(v) from Rome [10]. In order to simplify the Rome's formula, we decided to neglect the anisotropy in the velocity distribution function, therefore not considering any dependence on pitch angle $\xi$, and to consider only stationary solutions, therefore not considering any dependence on t. The resulting approximation depends only on particle velocity, takes into account slowing down of fast ions and f(v) becomes (in case of negligible density of background neutrals) $f(v) = \frac{\tau_S}{4\pi(v^3+v_c^3)}U(v_0 - v)$, where $v_0$ is the velocity of beam particles with energy $E_{NBI}$ and $v_c = \sqrt{2E_c/m_f}$ is the critical velocity of fast ions with mass $m_f$ (see Stix's paper [11] for definition of critical energy $E_c$). A Maxwellian distribution function is considered in the integral for the (isotropic) background plasma, and the integral calculation follows the procedure adopted for the code FIT3D_DD developed by S. Murakami and described in the work of M. Homma and S. Murakami [12], except that the fusion cross sections are calculated by the Bosch fit [9] and not by the outdated Duane's formulation [13]. The code FIT3D_DD described in [12] is not integrated in TASK3D-a, and therefore we decided to implement a fusion source routine directly in TASK3D-a. Figure 2 shows a comparison of neutron rate calculation between the code FIT3D_DD and the upgraded FIT3D routine of TASK3D-a transport analysis suite.

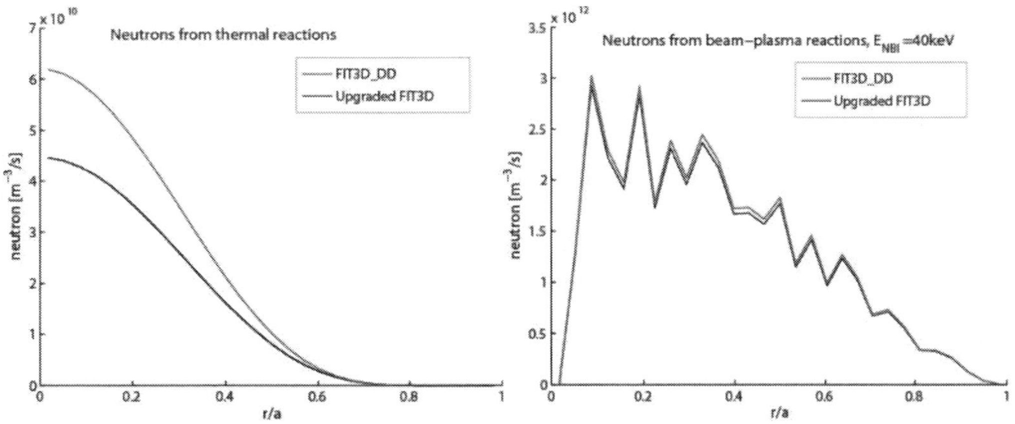


**Figure 2**: comparison between upgraded FIT3D code and FIT3D_DD code by S. Murakami [12] of neutron rate from thermal D plasma (left) and D NBI (40 keV) – D plasma interactions (right), obtained with arbitrary density ($n_{e,0}$=3e19 $m^{-3}$) and temperature profiles ($T_{e,0}$=1keV).

**Conclusions**

In future plans of LHD experiment, D plasma with D NBI injection discharges are scheduled in order to investigate the isotope effect and confinement properties in the path towards fusion helical reactors. Recently the upgrade of TAKS3D-a transport analysis suite has started, in order to expand the analysis capabilities for LHD experiments from H discharges to D discharges. Interaction between NBI and plasma is considerably affected by the gas change, and the present work introduces the first modifications implemented in the code in order to allow the complete modelling of D neutral beam injection in D plasma. In particular the neutral particle ionization routine has been modified, with the implementation of beam stopping cross section formula more suitable for LHD D cases. Moreover the code has been extended with the implementation of a routine which estimates the neutron rate and fusion power production both from thermal background plasma and beam-plasma interactions.

**Acknowledgements**

This project has received funding from the European Union's Horizon 2020 research and innovation programme under grant agreement number 633053. The views and opinions expressed herein do not necessarily reflect those of the European Commission. One of the authors (PV) gratefully acknowledges the hospitality of NIFS institute, Toki-city (Japan).

**References**

[1] O. Motojima et al, Nucl. Fusion 40 (2000) 599
[2] J. G. Cordey et al., Plasma Phys. Control. Fusion 42 (2000) A127-A132
[3] H. Urano et al., Nucl. Fusion 52 (2012) 114021
[4] M. Yokoyama et al., Plasma Fusion Res. 8 (2013) 2403016
[5] A. Sakai et al., Plasma Fusion Res. 9, (2014) 3403124
[6] H. Yamaguchi and S. Murakami, Plasma Fusion Res. 9 (2014) 3403127
[7] R.K. Janev et al, Nucl. Fusion 29 (1989) 2125
[8] S Suzuki et al, Plasma Phys. Control. Fusion 40 (1998) 2097
[9] H.-S. Bosch and G.M. Hale, Nuclear Fusion 32 (1992) 611-631
[10] Rome et al, Nucl. Fusion 16 (1976) 1
[11] T H Stix, Plasma Phys. 14 (1972) 367
[12] M. Homma et al, Plasma Fusion Res. 10 (2015) 3403050
[13] Duane B.H., "Fusion cross section theory", in Annual Report on CTR Techn. 1972 (Wolkenhauer W.C., Ed)

# Investigation on radial transport of perpendicular momentum in the SOL of AUG during L-I-H transition

S. Costea[1], J.J. Rasmussen[2] , N. Vianello[3] , H.W. Müller[4,5] , V. Naulin[2], R. Schrittwieser[1] , A.H. Nielsen[2], J. Madsen[2] , C. Ionita[1] , M. Spolaore[6] , D. Carralero[4] , F. Mehlmann[1] and the ASDEX Upgrade Team[4]

*1 Institute for Ion Physics and Applied Physics, Innsbruck, Austria*

*2 Technical University of Denmark, Kongens Lyngby, Denmark*

*3 EPFL-CRPP, Lausanne, Switzerland*

*4 Max Planck Institute for Plasma Physics, Garching, Germany*

*5 Institute of Materials Chemistry and Research, Vienna, Austria*

*6 Consorzio RFX, Padova, Italy*

Transport in the edge and scrape-off layer (SOL) region of magnetically confined plasmas is dominantly turbulent, caused by strong events in the form of filamentary structures that originate from the outboard midplane and are elongated along the magnetic field [1]. In addition to particles and energy also momentum is transported radially by turbulence.

We have investigated the radial transport of momentum perpendicular to the total magnetic field ($\mathbf{B}_{total}$) in the SOL of ASDEX Upgrade during L-mode (*Fig.1*), using the Innsbruck-Padova probe head (*Fig.2*). The radial transport of perpendicular momentum ($M_r$) has been estimated using electric fields derived from the potential difference of floating pins and the plasma density estimated from the ion saturation current of negatively biased pin.

$$M_r = n\, v_r\, v_\perp$$

$$\approx n\,(-E_\perp B_{total}^{-1})\,(E_r B_{total}^{-1})$$

$$\approx n\,(V_{fl,2}-V_{fl,5})\,(d_{\perp,2\text{-}5}B_{total})^{-1}\,(V_{fl,1}-V_{fl,2})\,(d_{r,2\text{-}1}B_{total})^{-1}$$

$$n \approx I_{is,4}\,(e\,A_{p,i})^{-1}\,(m_i\,/\,T_e)^{1/2}$$

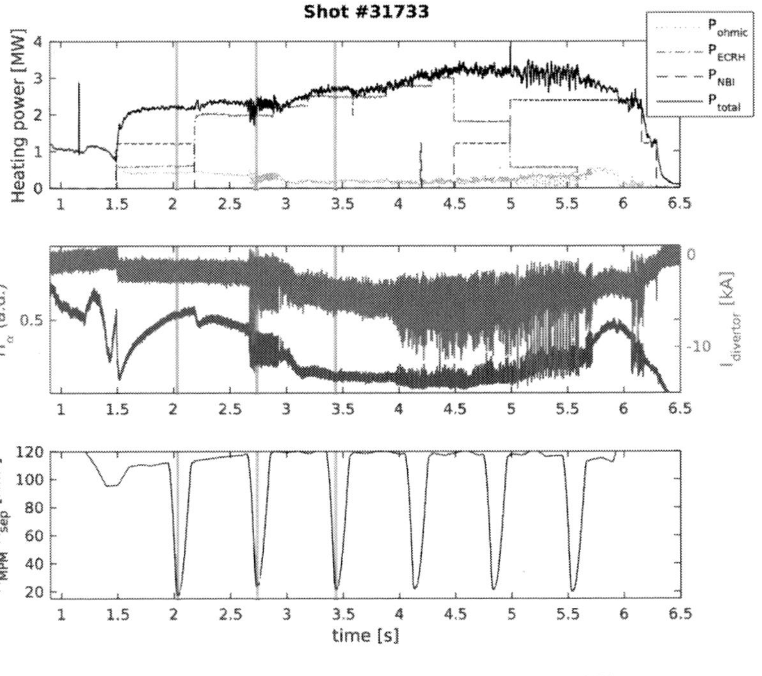


***Fig.1**: Hydrogen plasma shot in AUG. Top: heating scheme. Middle: light emission and divertor current indicating the different confinement regimes: L-mode <2,6s <I-phase <2,9s < H-mode Bottom: Position of the midplane manipulator (MPM), on which the Innsbruck-Padova probe head was mounted, with respect to the separatrix (located by magnetic reconstruction). Vertical grey bars: 20 ms time intervals used in the probe data evaluation*

We split all quantities into stationary and fluctuating components:

$n = n_0 + n_f$; $v_r = v_{r,0} + v_{r,f}$; $v_\perp = v_{\perp,0} + v_{\perp f}$

$M_r \approx n_0\, v_{r,f}\, v_{\perp f} + n_f\, v_{r,f}\, v_{\perp,0} + n_f\, v_{r,f}\, v_{\perp f}$

$M_r \approx Re + v_{\perp,0}\, \Gamma_{r,f} + n_f\, v_{r,f}\, v_{\perp f}$

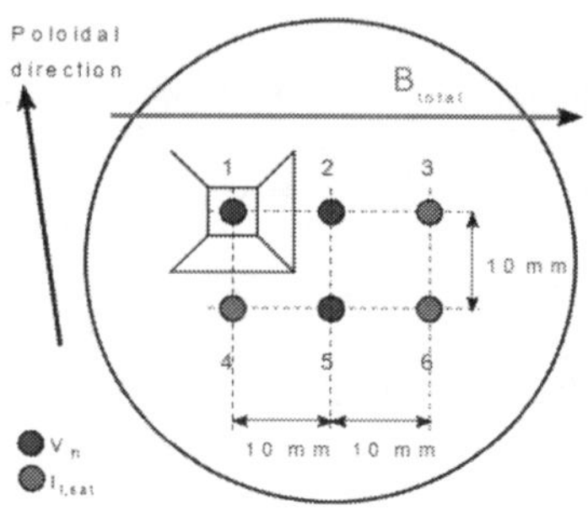


*Fig.2: Innsbruck-Padova probe head: six graphite pins, 1 mm diameter, 2 mm length. Pin #1 radially protruding by 3 mm, i.e. deeper into the plasma. Probe head aligned to the direction of the total magnetic field. Blue: floating potential. Red: ion saturation current*

The plasma density estimated from ion saturation current and radial transport of perpendicular momentum ($M_r$) and its subcomponents are presented in *Fig. 3*, for L-mode, L-I-H transitions and H-mode.

Since the charged particles also carry heat from the edge of the closed-field region of plasma, it is expected for density fluctuations ($n_f$) to be in phase with electron temperature fluctuations ($T_{e,f}$) and plasma potential fluctuations ($V_{p,f}$) [2]. It is known that the floating potential is affected by electron temperature: $V_{fl}=V_p-\alpha T_e$, $\alpha\approx 3$. In this perspective, the phase relation between $n$ estimated by ion saturation current and floating potential is investigated.

By conditionally averaging samples of all recorded signals during L-mode, triggered by events higher than 2 standard deviations in the estimated density from pins 3 and 4 (*Fig.4*), we observe that $n_f$ is in phase with $V_{fl,f}$ for pins located on the same magnetic field line (e.g. pins 1-2-3). Although this result appears to be in contradiction with [2], it might be caused by

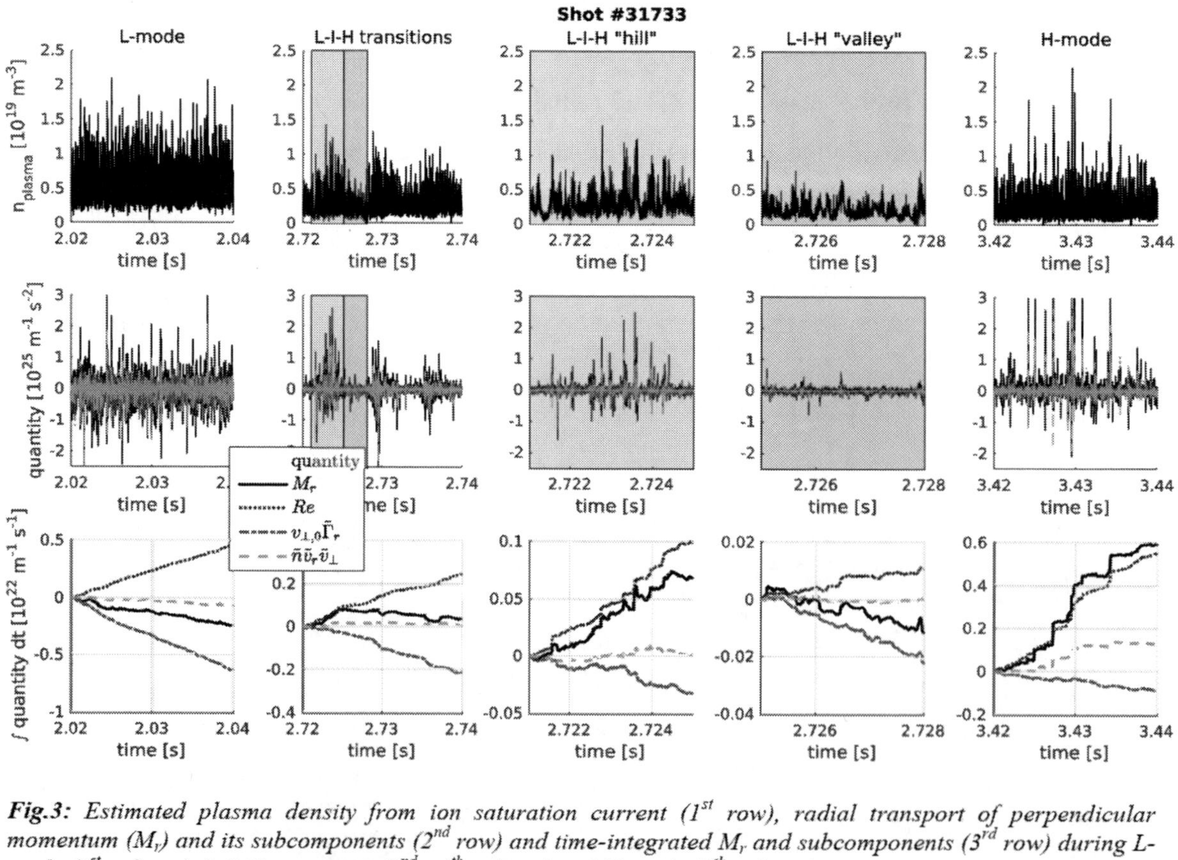


*Fig.3: Estimated plasma density from ion saturation current (1st row), radial transport of perpendicular momentum ($M_r$) and its subcomponents (2nd row) and time-integrated $M_r$ and subcomponents (3rd row) during L-mode (1st column), L-I-H transitions (2nd - 4th column) and H-mode (5th column).*

smaller values $T_{e,f}$ compared to $V_{p,f}$, and so the phase between $V_{fl,f}$ and $V_{p,f}$ is preserved but the amplitude of $V_{fl,f}$ is reduced. For the pins sitting on a different magnetic field line than the trigger, we observe a delay of ~10 μs which corresponds to a typical $v_\perp$=1 km/s flow. Since the perpendicular pin separation is comparable to blob size [1], in the case of floating pins not aligned with the trigger along $\mathbf{B}_{total}$ (e.g. pin 1 and 2, 2nd column) the average of samples is not a defined peak (as pin 1 and 2 in the 1st column) but a combination of positive and negative peaks, indicating the ambipolar structure of blobs, as also seen here [3].

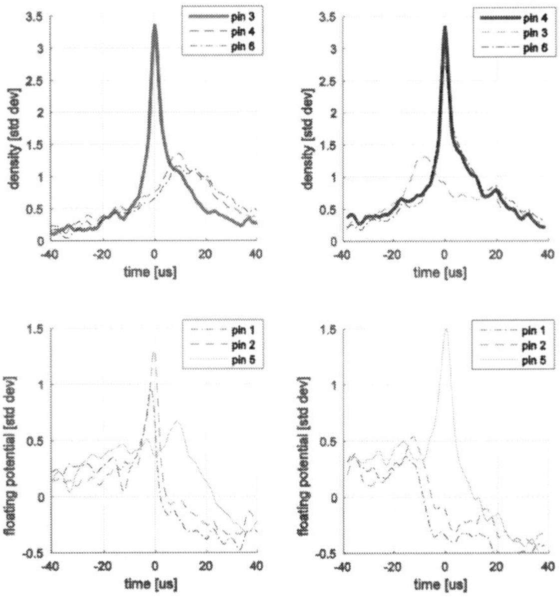


*Fig.4: Conditionally averaged samples of all recorded signals triggered by events in the estimated density from pin 3 (1st column) and pin 4 (2nd column).*

To test the validity of using the floating potential instead of the plasma potential, in the absence of localized electron temperature fluctuation measurements, a comparison with simulation has been performed. ESEL is a 2-dimensional inter-change turbulence code which simulates the edge-SOL region on the outboard midplane of a tokamak in the plane perpendicular to the total magnetic field (slab geometry). It makes use of an electrostatic fluid model which solves the equations for continuity, momentum and electron temperature [4]. A virtual probe head consisting of separated synthetic point-pins was placed 1 cm away from the separatrix in order to simulate the experimental probe head and the recorded signals during a L-mode discharge. The synthetic signals provided by the simulation on each synthetic pin are: plasma potential, plasma density, electron temperature and perpendicular and radial flow velocities. From these, the synthetic ion saturation current and the synthetic floating potential signals are constructed.

Results of synthetic diagnostic data (cold Langmuir pins, ion saturation pins) are in very good agreement with the experimental data (*Fig.5*, column 1).

If we compare the synthetic diagnostic data with simulation results (density, radial and perpendicular velocities, *Fig.5* column 3), we see that in the cold pin case the convective momentum has opposite direction with respect to the simulated one. If the cold pins are replaced by highly emissive pins ($V_{floating}=V_{plasma}$) then the convective momentum ($v_{\perp,0}\,\Gamma_{r,f}$) of the synthetic emissive probe diagnostic agrees with the simulated data but the Reynolds stress

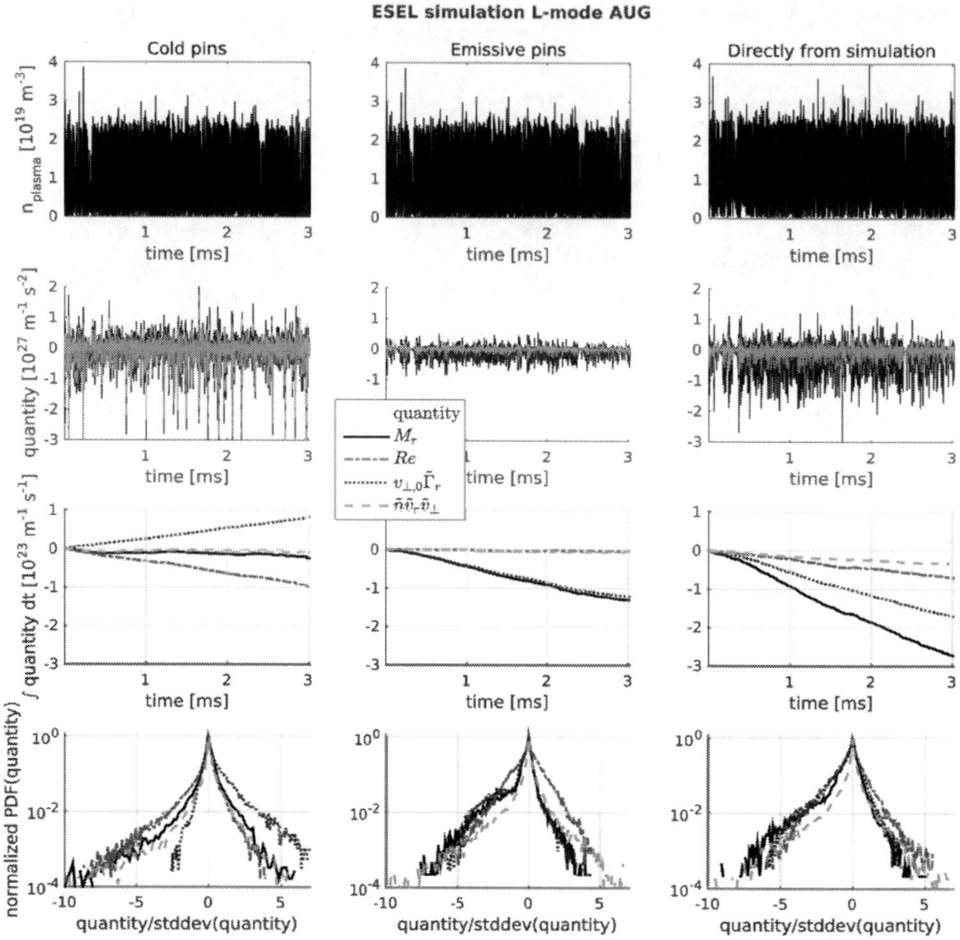


***Fig.5:*** *Analysis on $M_r$ from synthetic probe measurements simulating the experimental probe head in L-mode plasma.* *Column 1: simulated Langmuir cold pins and ion saturation pins.* *Column 2: simulated emissive pins ($V_{floating}$= $V_{plasma}$) and ion saturation pins.* *Column 3: density and velocities directly from simulation.*

is underestimated. Furthermore, the emissive pins preserve the probability density function (*Fig.5*, 4$^{th}$ row), of the simulated momentum transport and its subcomponents, even though the overall amplitude of the signals is reduced, possibly due to the pin separation.

In conclusion, we see a clear difference in the radial transport of perpendicular momentum and its subcomponents between the L-mode and H-mode. During L-I-H transitions the transport seems to be an alternation between L- and H-mode transport. The simulation could recreate the experimental time-integrated momentum transport but it indicates that one should take care when using cold floating potential as estimation of plasma potential in the presence of electron temperature fluctuations.

**Acknowledgement**

This work has been carried out within the framework of the EUROfusion Consortium and has received funding from the Euratom research and training programme 2014-2018 under grant agreement No 633053. The views and opinions expressed herein do not necessarily reflect those of the European Commission. This work was also supported by the Friedrich Schiedel Foundation for Energy Technology.

**References:**

[1] H.W. Muller *et al*, *Nucl. Fusion* **51** (2011) 073023

[2] B. Nold *et al*, *New Journ. Phys*. **14** (2012) 063022

[3] D. Carralero *et al.*, *Nucl. Fusion* **54** (2014) 123005

[4] W. Fundamenski *et al.*, Nucl. Fusion **47** (2007) 417

# Turbulence stabilization due to high beta and fast ions in high-performance plasmas at ASDEX Upgrade and JET

H. Doerk[1], C. Challis[2], J. Citrin[3,4], M. Dunne[1], J. Garcia[3], T. Goerler[1], F. Jenko[5], F. Ryter[1], P. A. Schneider[1], E. Wolfrum [1], the ASDEX Upgrade Team[1] and JET Contributors*

*EUROfusion Consortium, JET, Culham Science Centre, Abingdon, OX14 3DB, UK*

[1] *Max-Planck-Institut für Plasmaphysik, D-85748 Garching, Germany*

[2] *CCFE, Culham Science Centre, Abingdon, OX14 3DB, UK*

[3] *CEA, IRFM, F-13108 Saint-Paul-lez-Durance, France*

[4] *FOM Institute DIFFER P.O. Box 1207 NL-3430 BE Nieuwegein, The Netherlands*

[5] *Dept of Physics and Astronomy, University of California, Los Angeles, CA 90095, USA*

**See the Appendix of F. Romanelli et al., Proceedings of the 25th IAEA Fusion Energy Conference 2014, Saint Petersburg, Russia*

High-performance fusion plasmas are desired to reach high plasma beta. Yet, the dependence of the thermal confinement time on this important parameter is unclear: Dedicated experiments yielded inconclusive results [1] and most theoretical results are obtained in simplified setups (overview e.g. in Ref. [2]). Using high accuracy plasma parameter measurements and realistic geometry for recent H-mode discharges at ASDEX Upgrade and JET (with ITER-like-wall), turbulent transport is studied by means of GENE gyrokinetic simulations in the plasma core.

## Electromagnetic effects in plasma microturbulence

The gyrokinetic simulation code GENE is developed for studying microturbulence in strongly magnetized plasmas, such as fusion plasmas. In this work, we use the local flux-tube version of GENE [3]. Finite values of $\beta = 8\pi p_0/B_0^2$ allow for electromagnetic effects, which are characterized into two types, dynamic and geometric. Self-consistent perpendicular ($\tilde{A}_{\|}$) and parallel ($\tilde{B}_{\|}$) magnetic fluctuations account for dynamical $\beta$ effects. Geometric $\beta$ effects are related to pressure-induced changes in the magnetic equilibrium and thus the magnetic drifts. When the fast ion pressure $p_{\mathrm{fast}}$ constitutes a significant fraction of the total pressure, contributions to both electromagnetic effects are expected.

## A pair of ASDEX Upgrade discharges varying $\beta$

In the ASDEX Upgrade discharges #29197 (case A, $\beta_N = 1.67$) and #29224 (case B, $\beta_N = 2.6$), $\beta_e$ varies by a factor of two at mid-radius, while changes in other dimensionless parameters, such as $\rho^\star$ and $\nu^\star$, as well as magnetic geometry are by far less pronounced [2].

At the reference position of $\rho_{\text{tor}} = 0.5$, both plasmas appear unstable to ion temperature gradient (ITG) modes and microtearing modes (MTMs). As seen in Fig. 1(a) for case B, ITG is stabilized and MTMs become more unstable at higher $\beta_e$. Kinetic ballooning modes (KBM) are not expected to play a role for transport, since the experimental value of $\beta$ is at 40% of the KBM threshold for case B and 20% in case A. In Fig. 1(b), the gyrokinetic results on thermal transport are compared to the power balance computed with the TRANSP code [5]. In these simulations, magnetic flutter transport (that would be expected from MTM turbulence) is negligible, even though ITG and MTM can exhibit comparable growth rates. Thus, ITG turbulence reduction is the dominant $\beta$ effect. The resulting nonlinear up-shift of $a/L_{T,\text{crit}}$ increases with beta, which is in agreement with previous results obtained for simplified setups [4]. Due to the high stiffness in the simulation results, these findings are within the experimental uncertainty. However, the trend of higher gradient at higher beta is consistent between experiment and gyrokinetic modelling in the plasma core. The experimental finding of degrading global confinement $\tau_E B_0 \sim \beta^{-0.2}$ (although weaker than the IPB(y,2) result $\tau_{E98y2} B_0 \sim \beta^{-0.9}$) is thus likely attributed to pedestal physics.

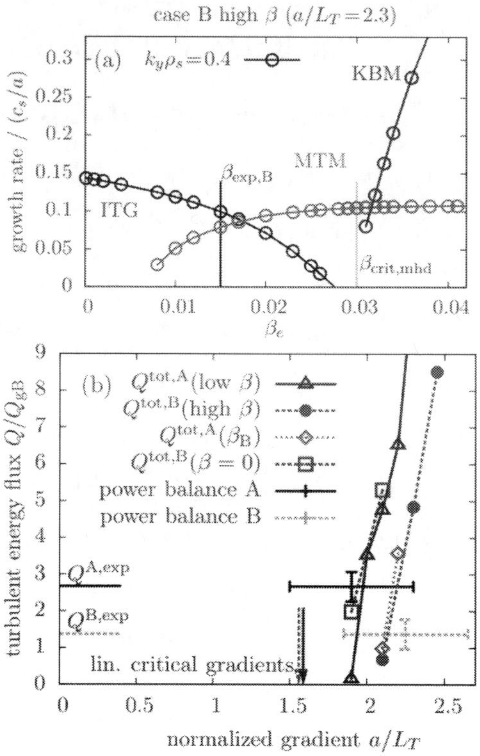


Figure 1: *(a) Growth rates as a function of $\beta_e$ for a selected $k_y$ (case B) at $\rho_{\text{tor}} = 0.5$. (b) Turbulent heat transport from GENE simulations compared to the power balance results.*

**A power scan at JET-ILW in hybrid mode**

In the following, we analyze two discharges from a power scan in a low triangularity hybrid configuration at JET: #84798 (low power LP) and #84792 (high power HP). To explain the measured weak power degradation of confinement and increasingly peaked ion temperature in the core, a fast ion induced positive feedback between edge and core has been proposed [6, 7]: Increased core pressure due to $\beta$-stabilized turbulence improves pedestal stability by an increased Shafvanov-shift, which in turn elevates the core profiles and further increases $\beta$. The virtuous cycle is stopped once $\beta_{\text{crit}}$ for the instability of KBM (or fast particle driven modes, like beta induced Alfvén eigenmodes BAE) is reached and strong fast particle transport sets in.

Figure 2 shows the growth rate of the most unstable mode for both JET plasmas at $\rho_{\text{tor}} = 0.33$.

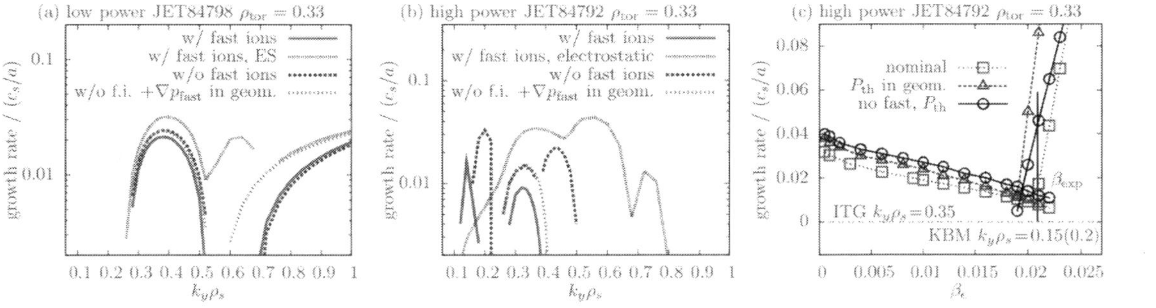


Figure 2: *Wavenumber spectra for JET discharges 84798 (a) and 84792 (b) at* $\rho_{tor} = 0.33$*. (c)* $\beta_e$ *scan in the HP case, we select* $k_y\rho_s = 0.35$ *for ITG and, to capture* $\beta_{crit}$*,* $k_y\rho_s = 0.15$ *for nominal case KBM and* $k_y\rho_s = 0.2$ *otherwise.*

The nominal case is compared to the following reduced setups: (1) electrostatic (with fast ions), (2) neglected fast ions in dynamics and geometry, (3) neglecting fast ions in dynamics, but retaining their pressure contribution to the equilibrium. Indeed, while electromagnetic effects are somewhat stabilizing for the low power case and fast ions play a minor role, these have greater impact for the high power case. Here, the nominal $\beta$ is close to $\beta_{crit}$, which is rather low due to low magnetic shear, and ITG growth rates are strongly reduced due to mostly dynamic $\beta$ effects. The ratio $\beta/\beta_{crit}$ has been demonstrated to be a good measure for the degree of ITG turbulence reduction [8]. A transition to higher harmonic tearing parity ITG [9] is not observed, but may be triggered by even larger $\nabla p_{fast}$. In the present case, geometric stabilization due to $p_{fast}$ has little effect on ITG, but shifts $\beta_{crit}$ upwards by about 10%, as depicted in Fig. 2(c).

Figure 3 shows an $a/L_{Ti}$ scan of GENE turbulent ion heat transport in comparison to power balance results. For the high power discharge, the experimental operation point lies between the strongly stabilized ITG regime and KBM/BAE turbulence. In the KBM/BAE turbulence regime, thermal transport for fast particles and electrons generally is large and more stiff than ion thermal transport, which is expected to set a limit on the electron and fast particle temperature gradients.

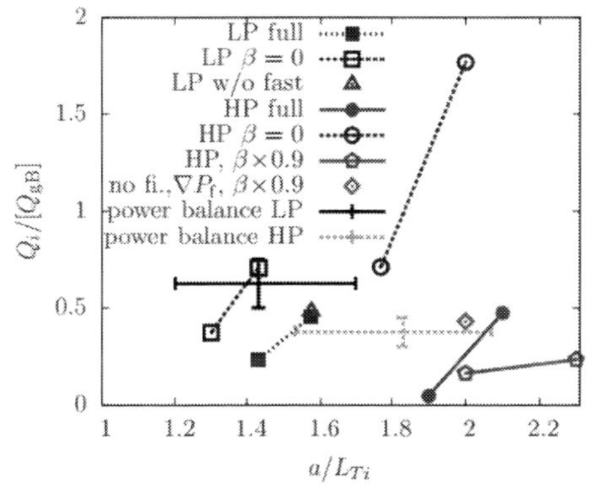


Figure 3: *GENE turbulent transport as a function of* $a/L_{Ti}$*, compared to power balance results. for JET #84798 (low power) and #84792 (high power).*

In view of future experimental validation beyond the match of macroscopic quantities (like $Q_i$), GENE turbulence data is characterized in greater detail. Distinctions between turbulence types are made by real frequency (or phase velocity) and cross phase relations. As an example, we observe the $n \times \phi$ phase angle close to zero for electrostatic ITG and about $\pi$ for KBM turbulence. Idealized KBM (interchange) turbulence exhibits a cross phase of $\pi/2$ only [10, 11]. Furthermore, it will be helpful to compare turbulence correlation lengths and times with reflectometry measurements.

**Summary and Conclusions**

Electromagnetic effects in fusion plasmas play an increasingly important role, as $\beta_{\rm crit}$ is approached. In the present ASDEX Upgrade beta scaling experiment in standard H-mode, $\beta/\beta_{\rm crit}$ does not exceed 40%. Also the threshold for strong MTM transport is not overcome, despite substantial growth rates in the linearized system. Thus, increased core confinement is expected due to ITG turbulence reduction at higher beta. The global confinement time is degrading weakly $\tau_E B \sim \beta^{-0.2}$ in this set of experiments, however.

In the JET hybrid H-mode power scan, one finds $\beta \approx \beta_{\rm crit}$ at high power, especially in the very core of the plasma, where magnetic shear is low. Including a fast ion dynamic species contributes to dynamic electromagnetic ITG turbulence reduction and this effect is very strong in the high power case. These results support the hypothesis of a fast particle induced positive feedback between core and edge stability towards high beta, which has been invoked to explain weak power degradation also in comparable C-wall JET hybrid plasmas.

Overall, our results indicate that turbulence stabilization due to increased beta and/or fast particles contributes to improved plasma confinement, providing a tool for the optimising future experiments.

*This work has been carried out within the framework of the EUROfusion Consortium and has received funding from the Euratom research and training programme 2014-2018 under grant agreement No 633053. The views and opinions expressed herein do not necessarily reflect those of the European Commission.*

**References**

[1] C. C. Petty Phys. Plasmas **15**, 080501 (2008)
[2] H. Doerk et al., Phys. Plasmas **22**, 042503 (2015)
[3] F. Jenko et al., Phys. Plasmas **7**, 1904 (2000)
[4] M. J. Pueschel and F. Jenko, Phys. Plasmas **17**, 062307 (2010)
[5] A. Pankin et al., Computer Phys. Comm. **159**, 157 (2004)
[6] C.D. Challis et al. Nucl. Fusion **55**, 053031 (2015)
[7] J. Garcia et al. Nucl. Fusion **55**, 053007 (2015)
[8] J. Citrin et al. Plasma Phys. Controll. Fusion **57**, 014032 (2015)
[9] M. Romanelli et al., Plasma Phys. Controll. Fusion **52**, 045007 (2010)
[10] B. D. Scott Phys. Plasmas **12**, 062314 (2005)
[11] P. Manz et al. Plasma Phys. Control. Fusion **56**, 035010 (2014)

# Development and first operations of the IShTAR test facility

R. D'Inca[1], R. Ochoukov[1], K. Crombe[2], I. Morgal[4], J. Jacquot[1], F. Louche[2], D. Van Eester[2], S. Heuraux[3], S. Devaux[3], J. Moritz[3], E. Faudot[3], F. Fischer[1], G. Siegl[1], H. Fünfgelder[1], H. Faugel[1], J-M. Noterdaeme[1,4]

[1] *Max-Planck-Institut für Plasmaphysik, Garching, Germany*

[2] *LPP-ERM-KMS, TEC partner, Brussels, Belgium*

[3] *Institut Jean Lamour UMR 7198 CNRS-Université de Lorraine, Nancy, France*

[4] *Department of Applied Physics, Ghent University, Belgium*

**Abstract**

Tthe IShTAR testbed facility, used for the study of RF waves/plasma interactions, is now in a fully operational status, equipped with an external plasma helicon source and an ICRF antenna. We present the first measurements of the density and of the helicon and ICRF wave structures in the plasma.

## Experimental setup and objectives

The IShTAR (Ion cyclotron Sheath Test ARrangement) is a linear magnetized testbed dedicated to the analysis of plasma/ICRF waves interactions in controlled conditions [2]. Detailed description of the testbed assembly and analysis of the first produced plasma have been previously presented [3]. The setup consists of (cf. Fig 1): the plasma source composed of 4 magnetic coils and a helicon antenna ($B = 0.1\ T$, $P = 3\ kW$, $f = 11\ MHz$), the main vessel with two coils operating up to 8 kA and the ICRF single-strap antenna; its operational instrumentation comprises a cylindrical 0.8 mm diameter Langmuir probe and two $\dot{B}$-probes (oriented parallel and perpendicular to the static magnetic field), all of them mounted on a manipulator arm moving between the wall and the vessel center.. The objective of this study is to characterize the plasma produced by the auxiliary plasma source and the behaviour of the ICRF antenna in presence of this plasma (plasma density profile, position inside the vessel, the regime of power coupling: capacitive C, inductive H or helicon W [4], the wave profiles from helicon and ICRF antennas, and the instabilities). These preliminary results are required firstly to provide inputs for theoretical models and for the antenna design; secondly, they will help to refine the requirements on the future plasma instrumentation (compensated probes, E-field measurement); and, thirdly, they support the optimization of the plasma source operation (by reaching the helicon mode [1]).

## Plasma characterization

The purpose is to reach the highest possible plasma densities and to get a uniformal profile to be representative of a tokamak edge (so that the dipsersion relations of the wave propagating

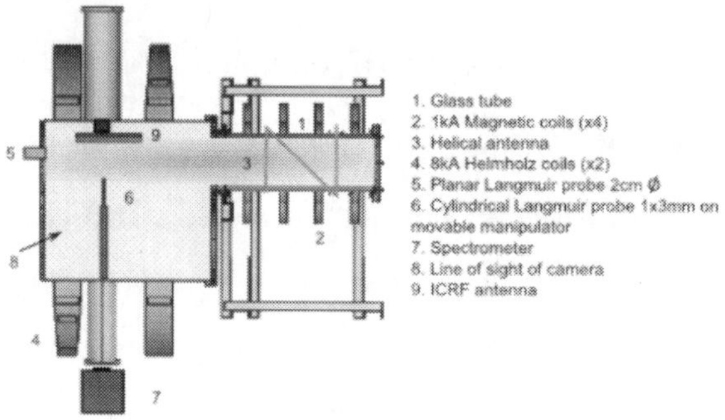


Figure 1: Cutaway (top view) of ISHTAR with its main components

inside the plasma are similar). Measurements of the plasma density made3 during a power scan by the planar Langmuir probe (which can detect smaller densities) reveals a jump between regime 1 and 2, characteristic of a transition from capacitive to inductive coupling or from inductive to helicon. A drop in the floating and plasma potential confirms this transition. There is no clear evidence yet of the type of mode reached in regime 2: it could be helicon (but there is no observation of a first jump from capacitive mode at lower power) or inductive. The calculation of the dispersion relation shows that for $B = 0.1\ T$, the helicon mode $m = 1$ should start to propagate at a density of $10^{16}\ m^{-3}$.

The cylindrical Langmuir probe shows a strongly oscillating signal in the vessel center for some discharges, confirming the presence of an instability. The density fluctuates from $1.8 \cdot 10^{18}\ m^{-3}$ to below $10^{17}\ m^{-3}$, lower than for stable discharges at the same power (cf. Fig. 3). As confirmed by video recordings, it is a sign of bursts of plasma moving over the probe. The floating potential from discharges with this instability reveals a different behavior than from stable discharges: at the beginning and during low power (<750 W) phases, the potential decreases and is negative instead of increasing and being positive. This would indicate that the instability could be triggered by changes in the initial conditions of the discharge.

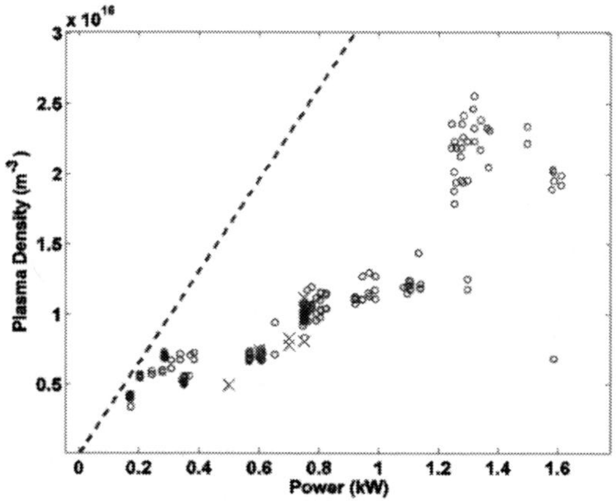


Figure 2: Density as a function of the injected power for a series of discharges . Black line is the maximum density due to losses.

The behavior of the plasma is conditioned by the mechanisms of ionization and coupling between the helicon wave and the plasma. A first step is to measure the wave spatial structure with the $\dot{B}$ probes when the plasma source is operating alone. The evolution of the fields (parallel and perpendicular) intensity during a power ramp up shows on Fig. 4 a) a mode transition between 300 and 400 W. On Fig. 4 b), the spatial profile perpendicular to the magnetic field is represented (perpendicular field) for an injected power of 750 W: a bump is visible and could correspond to the position of the plasma, which has to be confirmed by Langmuir probes measurements.

Yet, we notice that the change in amplitude along the manipulator axis is not important: the wave "fills in" the vessel, which can be explained by the low frequency (11 MHz) relatively to the characteristic dimension (1 $m$) of the testbed. The absence of a spatial localization of the wave on the coupling with the plasma still has to be evaluated.

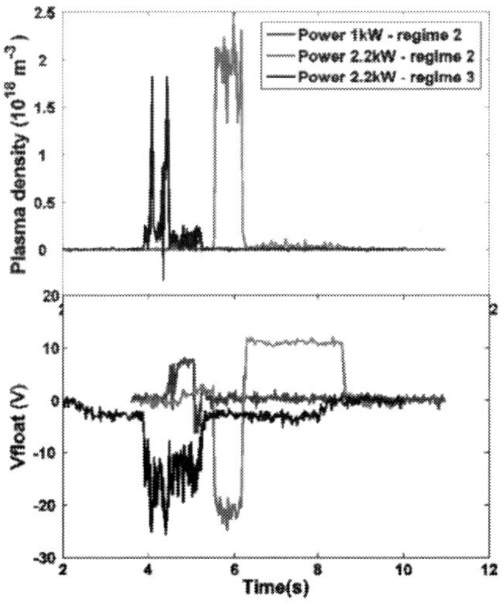


Figure 3: Comparison of density (top) and floating potential (bottom) for discharge in stable and unstable regimes.

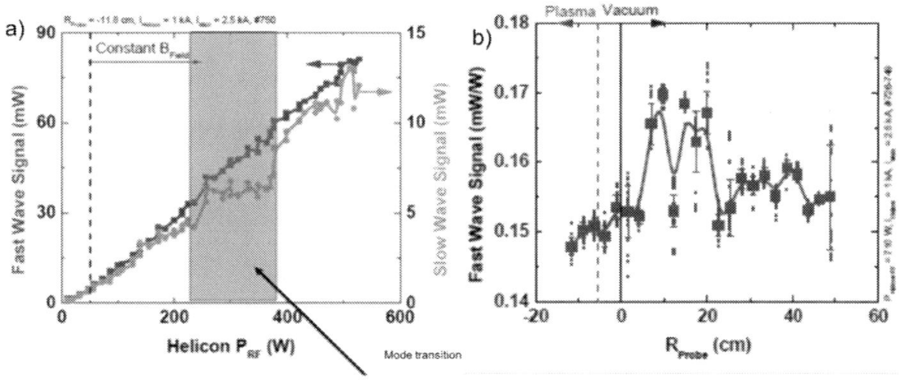


Figure 4: a) Evolution of helicon wave amplitude (perpendicular and parallel) with injected power. b) Helicon wave spatial structure perpendicular to the magnetic field at a constant power of 750 W.

## ICRF operations

A simple strap ICRF antenna has been operated for the first time on the testbed in presence of the plasma (either self-generated by the strap or coming from the auxiliary plasma source). No major RF breakdown has been observed and the system has delivered a nominal power of up to 400 W at 5 MHz. An evaluation of the RF field profile perpendilar to the magnetic field has been done in vacuum ($P = 10^{-5}\ mbar$) and with Argon without external plasma. The comparison of the cases on Fig. 5 shows that the field amplitude decreases with the self-generated plasma, probably indicating a coupling/absorption of the wave within the plasma. A bump near the wall is also visible in the plasma case, in the region where the plasma density is the lowest. But, as in the case of the helicon wave, the spatial variations of the field amplitude are small, indicating that the field "fills in" the vessel. A numerical analysis with COMSOL is ongoing to check the impact of this uniform wave structure on the representativity of the sheath with respect to the tokamak case. Further tests at higher power and different frequencies will provide further experimental data.

## Conclusion

A preliminary characterization of the plasma generated by the helicon source and of the RF fields from the ICRF antenna has been carried out. A plasma mode transition is observed but still has to be characterized. A Further steps involve the optimization of the operating parameters like the plasma density by adding more power to the helicon source, aligning it with the magnetic field and using other gas types (Helium,Hydrogen).

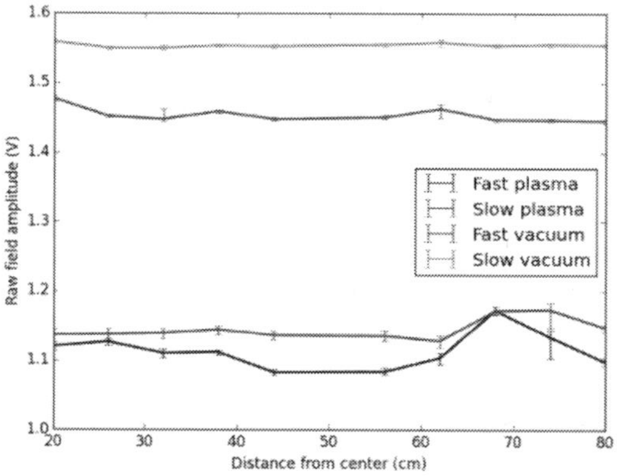


Figure 5: Measurement of field (perpendicular:fast and parallel:slow)profiles in vacuum and with self-generated plasma.

## Acknowledgments

This work has been carried out within the framework of the EUROfusion Consortium and has received funding from the Euratom research and training programme 2014-2018 under grant agreement No 633053. The views and opinions expressed herein do not necessarily reflect those of the European Commission.

## References

[1] CHEN, F. F. Physics of helicon discharges. *Physics of Plasmas (1994-present) 3*, 5 (1996), 1783–1793.

[2] CROMBE ET AL, K. Studies of rf sheaths and diagnostics on ishtar. *RF Conference 2015* (2015).

[3] D'INCA ET AL, R. Ishtar: a test facility to investigate sheath effects during ion cyclotron resonance heating. *To be published* (2015).

[4] ELLINGBOE, A., AND BOSWELL, R. Capacitive, inductive and helicon-wave modes of operation of a helicon plasma source. *Physics of Plasmas (1994-present) 3*, 7 (1996), 2797–2804.

# Influence of radiation on the energy spectrum of runaway electrons during tokamak disruptions

L. Zeng[1], H.R. Koslowski[2], Y. Liang[2], A. Lvovskiy[2], K. Wongrach[3], K.H. Finken[3], X. Gao[1]

[1] *Institute of Plasma Physics, Chinese Academy of Sciences, 230031 Hefei, China*

[2] *Forschungszentrum Jülich GmbH, Institute of Energy and Climate Research - Plasma Physics (IEK-4), 52425 Jülich, Germany*

[3] *Institut für Laser- und Plasmaphysik, Heinrich-Heine-Universität Düsseldorf, 40225 Düsseldorf, Germany*

*E-mail: zenglong@ipp.ac.cn*

## 1. Introduction

Runaway electron (RE) currents of several mega amperes are expected to be generated in ITER disruptions due to avalanche multiplication [1]. An uncontrolled loss of these high-energy electrons to the plasma facing components might cause serious damage [2]. Impurity injection on RE plateau during disruptions, as a candidate method for ITER, is proposed for mitigating REs. Experimental results on DIII-D and Tore-Supra show large quantities of high-Z gas are injected into the RE plateau, resulting in decay of the RE current. We present here observations on the TEXTOR tokamak, which indicates influence of impurity injection on the energy spectrum of REs.

## 2. Experimental observations

Disruptions are deliberately triggered by injection of large amounts of Argon using a fast disruption mitigation valve (DMV) at 2.0s on TEXTOR. A RE current plateau forms during the following current quench phase. Then the second DMV was triggered on the RE plateau.

Figure 1 compares two discharges, #117528 with neon injection by the second DMV at 2.01s, 10ms after the first DMV, and #117533 without impurity injection during the RE plateau. The time evolution of plasma current shows that argon injection by the second DMV causes an enhanced decay rate of the RE current. For #117528, just several milliseconds after the second DMV triggering, the signals of H$\alpha$ and SXR rise and then decrease slowly, which is clearly attributed to the increase of collisionnality. Meanwhile, no clear change is found in the ECE and synchrotron radiation.

Helium injection by the second DMV has also been tried on TEXTOR, shown in Fig. 2. Similar to #117993 which is without impurity injection during the RE plateau, no clear change of the RE current is found in #117999, in which helium is injected at 2.01s by the second DMV. The signals of H$\alpha$, ECE, and SXR raise after helium injection and then quickly decrease, even less than #117993.

Compared with these two cases, injection of amount of middle-Z (or high-Z) gas into RE plateau can cause enhanced current decay but no sufficient effect is observed with low-Z gas injection. The difference of radiation signals, such as Hα, ECE, and SXR, suggests that radiation caused by impurity injection may play an important role on the evolution of RE plateau.

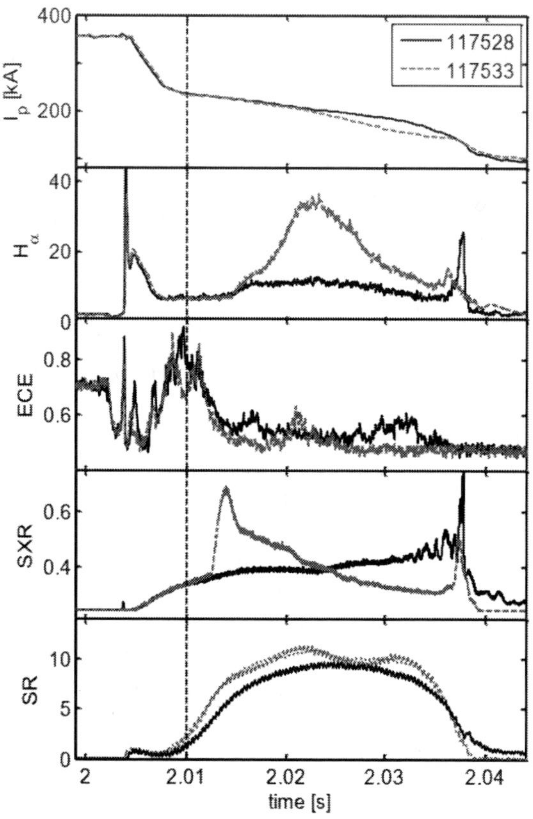


*Fig. 1 Time traces from two discharges showing (a) plasma current, (b) Hα, (c) ECE, (d) SXR, and (e) synchrotron radiation shot 117849. The dashed line indicates the triggering moment of the second DMV.*

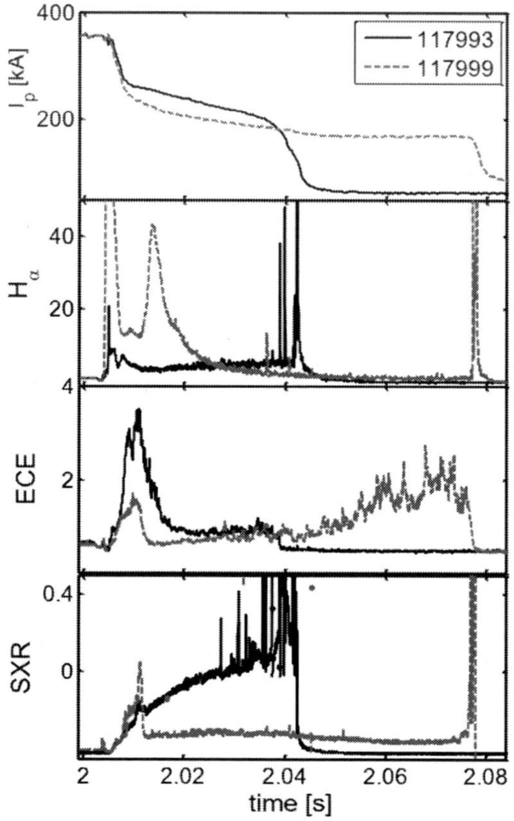


*Fig. 2 Time traces from two discharges showing (a) plasma current, (b) Hα, (c) ECE, and (d) synchrotron radiation shot 117849.*

### 3. Discussion

Experimental evidence from several tokamaks indicates that the electric field strength necessary for RE generation could in fact be several times larger than the critical electric field. Besides electric field and Coulomb collisions, many other potential mechanisms, such as bremsstrahlung [3] and synchrotron radiation [4, 5] have been considered as possible explanations for these observations.

The energy loss rate due to the collisions of a RE with the plasma particles is given by

$$P_{\mathrm{coll}} = (dW/dt)_{\mathrm{coll}} = n_e e^4 \ln\Lambda / (4\pi\varepsilon_0^2 m_e v), \tag{1}$$

The energy gain in the electric field is $P_E = eE_{\|}v_{\|}$. The relativistic collisional theory of runaway electrons predict that the threshold electric field for runaway generation $E_{\text{th}}$ is set by the condition that $P_E = eE_{\|}v_{\|}$ should be at least larger than such minimum value of $P_{\text{coll}}$ when $v \to c$. Actually, runaway electrons not only lose energy due to collisions but they also lose energy in the form of bremsstrahlung and synchrotron radiation at high energy. The energy loss due to bremsstrahlung when they are deflected by the nuclei of atoms is given by

$$P_B = Z^2\, n_i T_e^{0.5} / (7.69\times10^{18})^2 * (1+k_B T_e/m_e c^2), \quad (2)$$

The synchrotron radiation losses is

$$P_{SR} = 2/3\, r_e m_e c^3\, (v/c)^4 \gamma^4 <1/R^2>. \quad (3)$$

The total stopping power is obtained by adding the collision, bremsstrahlung and synchrotron radiation losses, $P_{\text{loss}} = P_{\text{coll}} + P_B + P_{SR}$. When $P_E$ exceeds $P_{\text{loss}}$, REs are generated and the corresponding electric field is the new critical electric field $E_{\text{th}}$.

The new energy limit for REs is plotted in Fig. 3. The total stopping power and contribution of Collision, bremsstrahlung, and synchrotron radiation are also shown. For the curve of the total stopping power, the low energy region is dominated by collision while the bremsstrahlung and synchrotron radiation losses dominate at high energies. Fig. 3 indicates that including the bremsstrahlung and synchrotron radiation losses, the new threshold electric field for RE generation must be larger than prediction of the relativistic collisional theory of REs. As illustrated in Fig. 3, for a given electric field large than the threshold electric field, energy balance ($P_E = P_{\text{loss}}$) is found at two critical points, $P_{th1}$ and $P_{th2}$. An electron with an energy large than $P_{th1}$ will be accelerated by the electric field to run away until a balance is achieved due to the bremsstrahlung and synchrotron radiation losses.

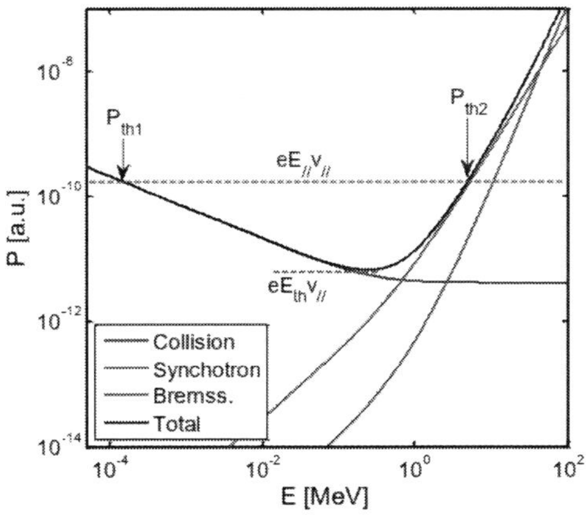


*Fig. 3 Collision, bremsstrahlung, synchrotron radiation, and total stopping powers for an electron. . The energy gain in the electric field crosses the curve of the total stopping power in two points $P_{th1}$ and $P_{th2}$.*

Basing on the method, the effect of impurity injection on RE plateau can be discussed here. The electron density will increase due to impurity injection will increase the electron density ($n_Z$) and $Z_{\text{eff}}$. The energy losses caused by the collision and synchrotron radiation are both not sensitive to the noble gas species too much, so the difference of radiation in the experiments could be caused by the bremsstrahlung.

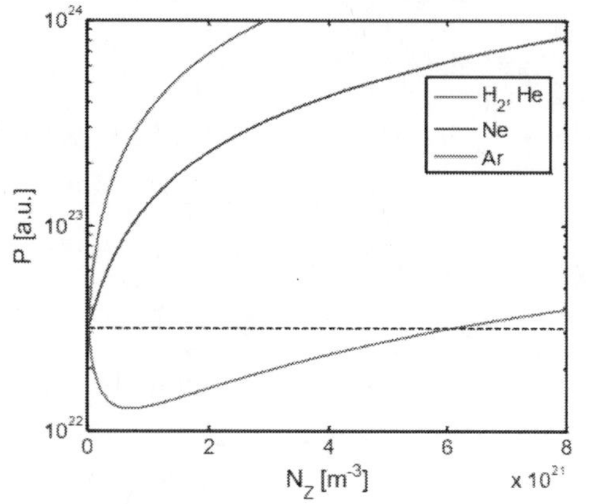


*Fig. 4 bremsstrahlung due to impurity injection by the second DMV. Both low-Z ($H_2$, He) and middle-Z (Ne, Ar) are considered here.*

Assume $n_Z = 10^{20}$ m$^{-3}$ and $Z_{eff} = 18$ for the background plasma after disruptions triggers by DMV. Simulation results including impurity injection by the second DMV are shown in Fig. 4. It is clear that for $H_2$ and He, the bremsstrahlung decreases when the injection amount is less than a value but for Ne and Ar, the bremsstrahlung increases continually. The effective charge number of $H_2$ and He is less than the background plasma and mixture of them together will decrease the final $Z_{eff}$ if the injection amount is limited. Impurity injection by the second DMV changes the bremsstrahlung and then the total energy balance. Simulation results (Fig. 5) show that when helium is injected to the RE beams and the bremsstrahlung will decrease, the new $P_{th2}$ ($P_{th2}$') is large than before, which means that REs with the previous energy $P_{th2}$ will be accelerated to $P_{th2}$'. When argon is injected and the bremsstrahlung will increase, the new $P_{th2}$ ($P_{th2}$') is less than before, which means that REs with the previous energy $P_{th2}$ will be reduced to $P_{th2}$'. These results are consistent with the results in TEXTOR and support the experimental observations that middle-Z (or high-Z) gas injection by the second valve is more effective at reducing RE current in DIII-D and Tore Supra disruptions.

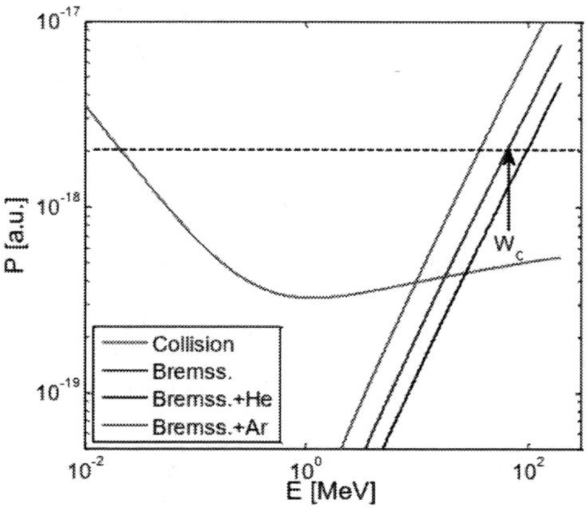


*Fig. 4 the energy balance changed due to impurity injection by the second DMV. Both low-Z (He) and middle-Z (Ar) are considered here.*

*Acknowledgement.* —Support from the National Magnetic Confinement Fusion Science Program of China under Contracts No. 2013GB106003 and 2014GB106004 and the National Natural Science Foundation of China under Contract No. 11475221 is gratefully acknowledged.

*References.*

[1] T. C. Hender *et al* Nucl. Fusion **47**, S128 (2007)
[2] M. Lehnen *et al* Nucl. Mater. **390–391**, 740 (2009)
[3] M. Bakhtiari *et al* Phys. Rev. Lett. **94**, 215003 (2005)
[4] J.R. Martín-Solís, R. Sánchez, and B. Esposito, Phys. Rev. Lett. **105**, 185002 (2010)
[5] K. Wongrach *et al* submitted to Nucl. Fusion (2015)

# An investigation of the plasma response to applied RMPs on TEXTOR

P Denner[1], Ph Drews[1], Y Liang[1], Y Yang[2] and the TEXTOR team

[1] *Forschungszentrum Jülich GmbH, Institut für Energie- und Klimaforschung – Plasmaphysik, 52425 Jülich, Germany*

[2] *Institute of Plasma Physics, Chinese Academy of Sciences, 230031 Hefei, China*

## Introduction

In order to avoid damage to plasma-facing components on ITER, type-I edge-localized modes (ELMs) must be either mitigated or suppressed [1]. The application of resonant magnetic perturbations (RMPs) to the plasma provides a promising method of ELM mitigation or suppression. However, in order to understand the mechanism by which RMPs mitigate or suppress ELMs, it is necessary to understand the plasma response to the application of RMPs.

RMPs have been applied to TEXTOR plasmas using the Dynamic Ergodic Divertor (DED) [2]. TEXTOR was also equipped with a fast movable magnetic probe (FMMP) capable of measuring the magnetic field in the edge of TEXTOR plasmas with applied RMPs. By subtracting the vacuum field, direct measurements of the plasma response to RMPs have been obtained.

## Experimental set-up

TEXTOR's DED consisted of sixteen helical coils on the high-field side of TEXTOR. It could be configured to produce fields with mode number $m/n = 3/1$, $6/2$ or $12/4$. The results presented here were obtained in $3/1$ and $6/2$ configurations. The DED frequencies available were $\pm 1\,\mathrm{kHz}$ and $\pm 5\,\mathrm{kHz}$ for the $3/1$ configuration and $\pm 1.4\,\mathrm{kHz}$ for the $6/2$ configuration, where positive frequencies represent a rotation of the field in the counter-current (electron diamagnetic drift) direction and negative frequencies correspond to the co-current direction.

The FMMP was located at the midplane on the low-field side of TEXTOR and could be plunged into the plasma edge in order to obtain radial profiles of the magnetic field. The probe contains three groups of three coils. Within every group, one coil is oriented in each of the radial, toroidal and poloidal directions so that every component of the magnetic field can be measured at three locations simultaneously.

When the probe was plunged into the plasma, radial profiles of the magnetic field were obtained. The measured field is correlated with the DED signal, and the Fourier component corresponding to the DED frequency is selected in order to distinguish the effect of the DED field from the background equilibrium plasma. This fluctuating part of the magnetic field is labelled $\delta B$. The same procedure is carried out for the magnetic field measured in a vacuum shot. This vacuum field is then subtracted from the data, and the remaining field is considered to be generated by the plasma as a response to the RMPs. This process is outlined in more detail in [3].

The duration of the probe plunge is much longer than the DED time period, so many DED cycles occur during a single plunge. If probe measurements taken at different radial locations but at the same point in the DED cycle are compared, then any difference in the amplitude or phase of these measurements is most likely due to radial variation in $\delta B$. Therefore, if $\delta B$ is plotted as a function of radius and time point in the DED cycle, radial variations in the amplitude or phase of $\delta B$ should appear.

Figure 1 shows three examples of such plots. An example for a vacuum shot is shown in figure 1 (*a*). In this case, there is no measureable radial variation in the DED field. Figure 1 (*b*) and (*c*) show similar plots but with the addition of a TEXTOR plasma. In figure 1 (*b*), there is a clear $\sim 180°$ jump in the phase of $\delta B_\theta$ at $r \approx 45$ cm. This is interpreted as being caused by the presence of a screening current at this radial location, which should correspond to a resonant surface. Figure 1 (*c*) shows no such phase jump, but a resonant surface is expected to exist within the range of $r$ covered by the probe. This is interpreted as penetration of the RMP field and destruction of the screening current on this resonant surface.

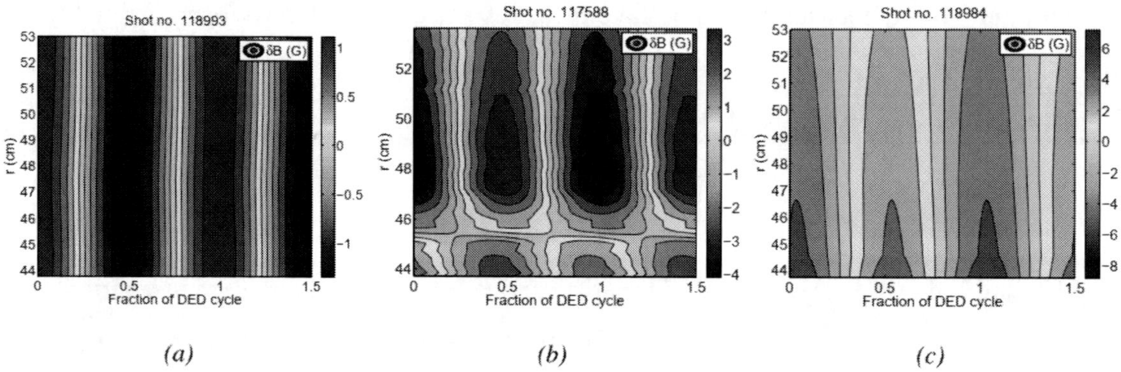


Figure 1: Example contour plots for (*a*) a vacuum shot, (*b*) screening and (*c*) field penetration.

## Screening and field penetration

For the $3/1$ configuration, a transition from screening of the applied perturbation to field penetration as the DED current increases has previously been reported and described in detail in [3]. However, for the $6/2$ configuration, only screening has been observed, an example of which can be seen in figure 2. The perturbation spectrum for the $6/2$ configuration is weaker than for the $3/1$ configuration and is very localized at the plasma edge with a narrow resonant window, as can be seen in figure 3. As a result, for $q_a$ values of 4.25 or above, $n = 2$ RMPs have not been observed to have any effect on the plasma that can be measured by the FMMP.

One possible explanation for why field penetration has not been observed with $n = 2$ RMPs – even at lower values of $q_a$ and high values of $I_{\mathrm{DED}}$ – is that the perturbation field was simply not strong enough for field penetration to occur. However, an analysis of the data from an array of in-vessel Mirnov coils revealed that field penetration in the 3/1 configuration was accompanied by the appearance of a 2/1 mode that locked to the DED frequency [3]. Therefore, another possible explanation is that this 2/1 mode, which is absent in the case of $n = 2$ RMPs, is required for field penetration. On the other hand, it could well be that the field penetration causes or at least enables the growth of the 2/1 mode.

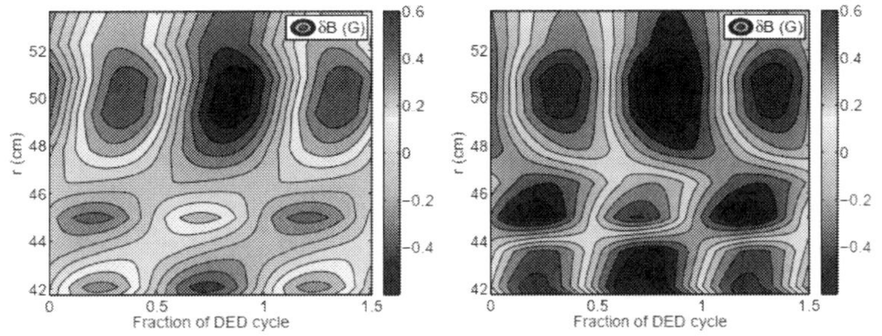


(a) (b)

Figure 2: Screening currents on the $q = 7/2$ and $q = 4$ surfaces with $n = 2$ RMPs for (a) $I_{\mathrm{DED}}$ =1.2 kA and (b) $I_{\mathrm{DED}}$ =2.3 kA.

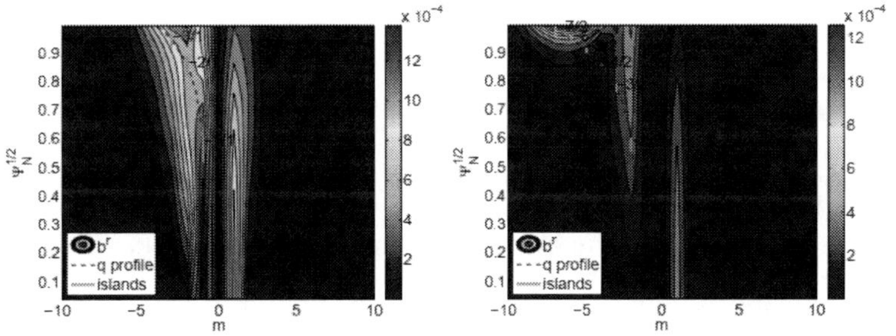


(a) (b)

Figure 3: Perturbation spectra for (a) the 3/1 configuration with $I_{\mathrm{DED}}$ =0.8 kA and (b) the 6/2 configuration with $I_{\mathrm{DED}}$ =1.8 kA. Despite having more than double the DED current, the $n = 2$ perturbation is weaker, and it is very localized at the plasma edge with a narrow resonant window.

**Hodographs**

A hodograph is the locus of one end of a variable vector, with the other end at the origin. Hodographs enable us to combine measurements of two or all three components of the magnetic field in one plot and provide a new way of looking at the data.

The example shown in figure 4 shows a phase change in the plasma response to $n = 1$ RMPs that occurs for DED currents above a certain value when there is a large difference in rotation between the plasma and the perturbation field but not when there is a smaller difference in rotation. This phase change indicates that screening currents are formed on rational surfaces in the plasma edge. The fact that the phase change is not observed when there is a smaller difference between the rotation of the plasma and that of the perturbation field suggests that the

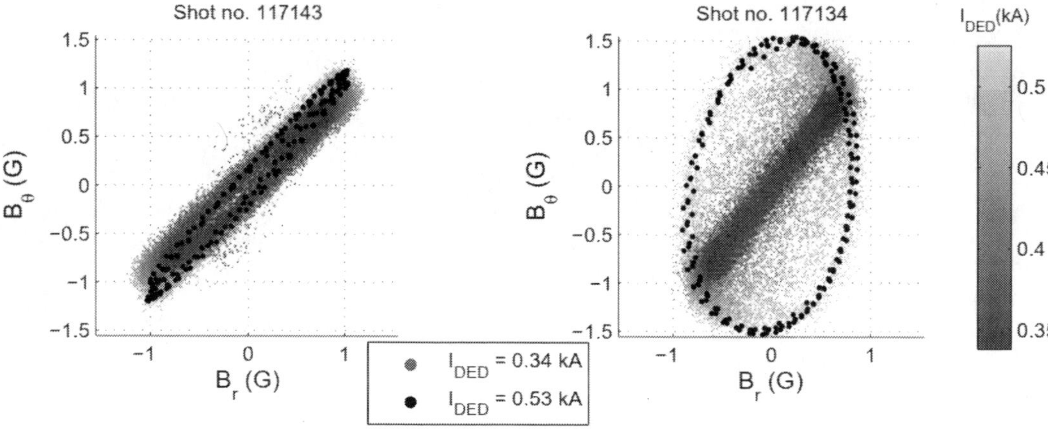


Figure 4: Hodographs of the plasma response with different values of plasma rotation, showing a constant phase (left) and a phase change (right) as $I_{\rm DED}$ exceeds a certain value.

RMPs are not screened when their rotation is close to that of the electron fluid, i.e. when the perturbation is almost static in the reference frame of the electron fluid.

**Summary and conclusions**

Screening currents on multiple surfaces have been observed for both $n = 1$ and $n = 2$ perturbation fields. Furthermore, a transition from screening to field penetration at higher values of $I_{\rm DED}$ has been observed for $n = 1$ perturbations. However, this field penetration has not been reproduced with $n = 2$ RMPs. This is possibly due to the fact that the $n = 2$ perturbations are weaker and have a narrower resonant window, or possibly because the field penetration requires the presence of a $2/1$ mode.

Hodographs represent a promising new method of analysing the data that can combine two or all three components of the magnetic field in one plot. A systematic analysis of the magnetic topology using hodographs is currently ongoing.

The FMMP has now been deployed on other tokamaks.

**Acknowledgments**

The financial support from the Helmholtz Association in the framework of the Helmholtz–University Young Investigators Group VH-NG-410 is gratefully acknowledged.

**References**

[1] Loarte A *et al.* 2003 *Plasma Phys. Control. Fusion* **45** 1549

[2] Finken K H 1997 *Fusion Eng. Des.* **37** 445

[3] Denner P *et al.* 2014 *Nucl. Fusion* **54** 064003

# Effect of magnetic field geometry on blob structure and dynamics in TJ-K

S. Garland, M. Ramisch, T. Hirth

*IGVP, Universität Stuttgart, 70569 Stuttgart, Germany*

## Introduction

A detailed understanding of scrape-off layer (SOL) turbulence is of high importance to the design of future fusion reactors. Density fluctuations in the SOL of most magnetic confinement devices exhibit positive skewness, which is generally associated with relatively dense and hot mesoscale structures elongated along the magnetic field lines, otherwise known as blobs [1]. These structures transport particles and heat through the SOL to the reactor first wall, potentially resulting in high heat loads on plasma facing components. Knowledge of the dynamics and structure of blobs is therefore crucial. For this contribution, investigations have been carried out into the effect of magnetic field curvature and shear on poloidal blob velocity and blob 3D structure respectively.

## Experimental setup

Experiments were conducted in the small-sized torsatron TJ-K [2], which has a major radius of 0.6 m and a minor plasma radius of 0.1 m. Electron densities are of the order of $10^{17}\,\mathrm{m}^{-3}$ whilst electron temperatures are typically around 10 eV and those of ions 1 eV, allowing Langmuir probe access to the entire plasma volume. Measurements can therefore be taken with the necessary spatio-temporal resolution to study plasma turbulence in detail.

A range of Langmuir probe diagnostics are available for use at TJ-K, with a data acquisition frequency of up to 1 MHz. For the present contribution two main probe diagnostics were used: a 2D scanning probe unit, capable of sampling fluctuations within a poloidal cross section; a probe matrix comprising an equally spaced square 8x8 probe grid covering $49\,\mathrm{cm}^2$. In typical TJ-K discharges, fluctuations in ion saturation current, $\tilde{I}_{i,sat}$, can be assumed proportional to those of density due to the lack of temperature fluctuations [3].

In order to create a controlled environment for blob studies, a pair of poloidal limiters were installed at the toroidal angles $\phi = 150^\circ$ and $270^\circ$, giving an enlarged SOL as well as an extended region of approximately uniform magnetic field line connection length.

## Geodesic curvature drive

Blob propagation is generally thought to be driven by charge separating drifts, e.g. curvature drifts, which polarise blob filaments perpendicular to the magnetic field. The subsequent transverse electric field results in the ExB advection of the blob through the SOL [5, 4]. Due to

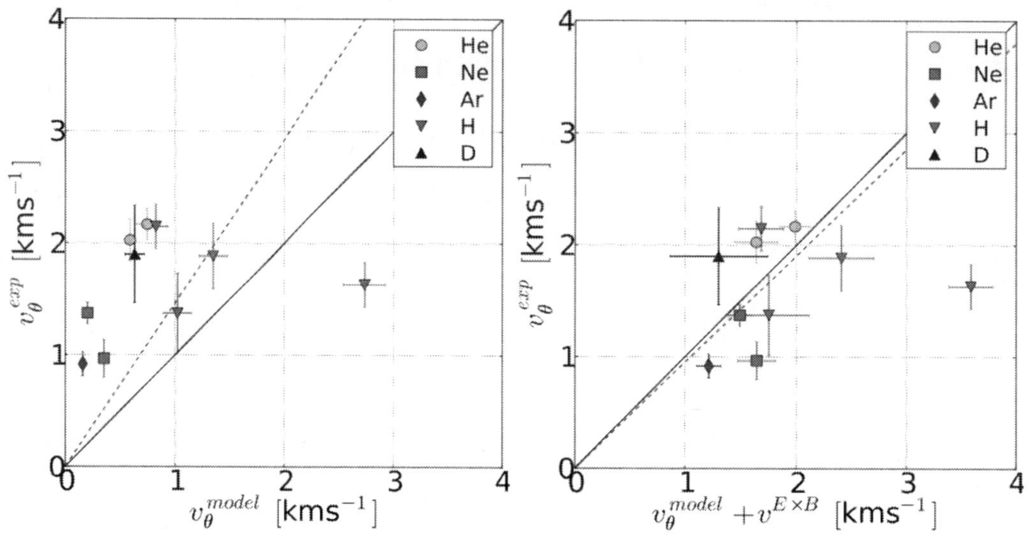


Figure 1: *Experimental poloidal blob velocities compared to predictions from the analytical model with (right) and without (left) the addition of background ExB flow speeds.*

diagnostic limitations in many magnetic confinement devices, the study of blob velocity in the perpendicular direction has mostly been limited to the radial component, thought to be driven by normal curvature. It is predicted, however, that non-zero geodesic curvature gives rise to a poloidal velocity component.

An analytical model [4, 5] has previously been found to compare well to experimental radial blob velocities in TJ-K in the so-called inertial regime [6]. In this regime the radial blob velocity is related to the blob size $\delta_b$, the field line curvature $1/R$, and the blob-background density ratio $\hat{n} = (n_{blob} - n_0)/n_0$, via $v_r = \sqrt{2\delta_b/R}c_s\hat{n}$ [6, 7], where $c_s$ is the speed of sound. Within the same approximation, the model can be used to express blob velocity, $\mathbf{v}_b$, in terms of the field line curvature vector for an arbitrary coordinate system,

$$\mathbf{v}_b = -2c_s^2\gamma^{-1}\hat{n}\boldsymbol{\kappa} \tag{1}$$

where $\gamma = \sqrt{2/(R\delta_b)}c_s$, and $\boldsymbol{\kappa}$ is the curvature vector. Thus, the radial and poloidal velocity components can be expressed approximately in terms of the normal and geodesic curvature,

$$\begin{aligned} v_{r,b} &= -2c_s^2\gamma^{-1}\hat{n}\kappa_n \\ v_{\theta,b} &= -2c_s^2\gamma^{-1}\hat{n}\kappa_g \end{aligned} \tag{2}$$

Blob velocity components were determined experimentally by calculating blob centre of mass (CoM) displacements from 2D conditionally averaged probe data, obtained on a 1 cm discretised grid with a fixed reference probe. The blob detection trigger condition applied to the reference

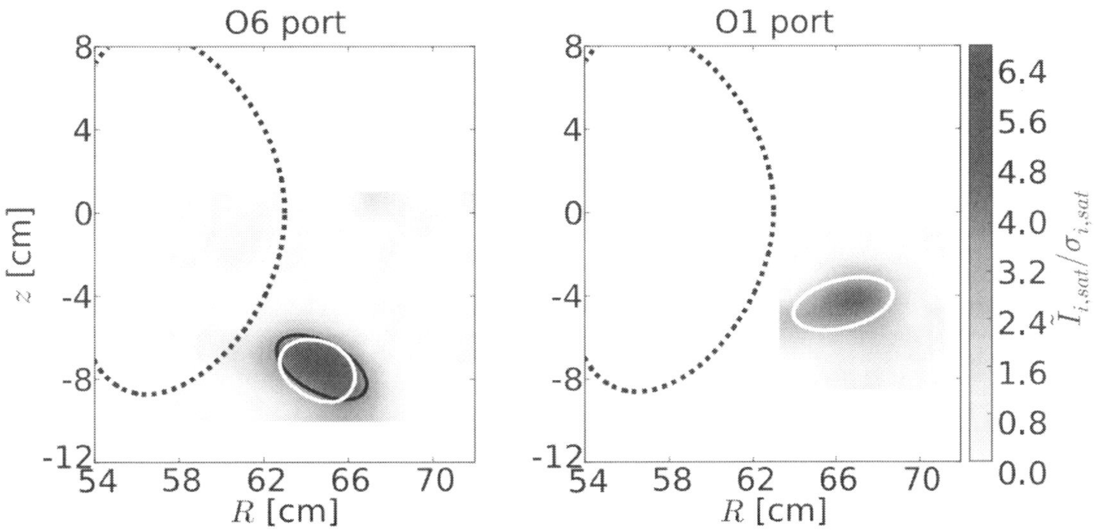


Figure 2: *The conditional average of $I_{i,sat}$ fluctuations normalised by $\sigma_{I_{i,isat}}$ at two toroidal positions, O6 (left) and O1 (right), separated by 60°. In the left hand plot, the white ellipse is fitted to the cross section of the blob structure, and in the right hand plot the black ellipse is fitted to the struture whilst the white shape corresponds to the field line traced ellipse.*

signal was $2\sigma_{I_{i,isat}}$. CoM coordinates were also used to extract local parameters for input into the analytical model. In order to vary the parameter space, the gases H, D, He, Ne and Ar were used at field strengths ranging between approximately 60 and 85 mT.

The left hand plot of figure 1 shows the experimental poloidal velocity components compared to the predictions of the analytical model in the inertial regime, as per equation 1. The linear fit of the data, given by the blue dashed line, indicates a substantial deviation of the experimental data from the prediction of the analytical model.

Radial plasma potential gradients in the SOL of TJ-K lead to background ExB flow which may also affect the poloidal blob propagation. In order to determine the background flow, emissive probes were used to directly measure plasma potential profiles for each shot, whilst the magnetic field components were determined using a field line tracing code. As can be seen from the right hand graph of figure 1, the poloidal velocity components compare well to the model when the background ExB flow speed is also accounted for. This result is consistent with the geodesic curvature drive of poloidal blob propagation.

**The 3D structure of blobs**

Due to fast parallel transport, blobs are generally assumed to be field-aligned structures. A consequence of this is that they are influenced by magnetic field shear, becoming deformed in regions of high shear (e.g. [8]).

The three dimensional structure of blobs has been investigated using $\tilde{I}_{i,sat}$ measurements in two different poloidal cross sections, separated by 60°. At one toroidal position (port O1, $\phi = 30°$) fluctuations were measured using the 8x8 probe matrix, whilst at the other (port O6, $\phi = 330°$), the 2D scanning probe was used to take simultaneous measurements. An additional fixed probe was installed in order to conditionally average all fluctuations with a common reference signal, and the same trigger condition of $2\sigma_{I_{i,isat}}$ was used.

Blob structures occuring in the conditionally averaged probe matrix data were fitted with ellipses, the coordinates of which were then followed using a field line tracing code along the vacuum magnetic field lines to the poloidal plane containing the 2D scanning probe (O1$\longmapsto$ O6). Figure 2 shows the conditionally averaged fluctuations at the two different toroidal positions, for time $t = 0\,\mu$s in the conditional average time series. The data was taken during a He discharge at $B \approx 70\,$mT.

It can be seen from the figure that the traced ellipse (white), deformed through field line tracing by magnetic shear, is roughly aligned with ellipse fitted to the blob structure at O6 (black). This indicates that blob structure is influenced strongly by magnetic shear. Future work will investigate this effect further by extending the region of study into higher shear regions.

## References


[1] D. A. D'Ippolito et al., Phys. Plasmas **18**, 060501 (2011)

[2] U. Stroth et al., Phys. Plasmas **11**, 2558 (2004)

[3] N. Mahdizadeh et al., Plasma. Phys. Contr. Fusion **47**, 569 (2005)

[4] S. I. Krasheninnikov et al., J. Plasma Phys. **74**, 679 (2008)

[5] D. A. D'Ippolito et al., Phys. Plasmas **9**, 222 (2002)

[6] G. Fuchert, PhD thesis, IGVP Stuttgart (2013)

[7] C. Theiler et al., Phys. Rev. Lett. **103**, 065001 (2009)

[8] J. L. Terry et al., J. Nucl. Mater. **390-391**, 339-342 (2009)

# Investigation of the energy transfer to zonal flows at the stellarator TJ-K

B. Schmid[1], P. Manz[2,3], M. Ramisch[1], U. Stroth[2,3]

[1] *IGVP, Universität Stuttgart, 70569 Stuttgart, Germany*

[2] *Max-Planck-Institut für Plasmaphysik, 85748 Garching, Germany*

[3] *Physik-Department E28, Technische Universität München, 85747 Garching, Germany*

**Introduction** In fusion devices, an increased zonal flow activity is found when the bifurcation to high confinement (H-mode) is reached. For zonal flows, it has been shown that they can reduce turbulent transport [1]. With their $m = n = 0$ topology and finite radial wave number $k_r$, the zonal flow is intrinsically connected to a poloidal shear flow. Drift-wave eddies are tilted and drive the shear flow, which leads to a self-amplification of the zonal flow. The zonal flow energy is then dissipated through collisional losses or transferred via the geodesic transfer effect to the geodesic acoustic mode. A key parameter in the drift wave - zonal flow system is the collisionality. It determines the coupling strength between the density and potential. By changing the collisionality, the scaling of the spectral energy transfer into the different channels can be studied.

**Stellarator TJ-K** The experiments were carried out in the low-temperature plasmas of the stellarator TJ-K. To study the energy transfer from the drift waves to the zonal flow and to the geodesic acoustic mode (GAM), a poloidal Langmuir-probe array was used for the measurements (Figure 1). The array covers four neighbouring flux surfaces in the edge of the confinement region, whereby it is possible to investigate the energy transfer directly in wave-number space. Simultaneous measurements of density and potential fluctuations allow to also address the nonlinear cross-coupling in drift-wave turbulence [2].

Figure 1: *Poloidal Reynolds-stress array with Langmuir probes on four flux surfaces (dotted white lines) just inside the separatrix (dashed white line). The detail illustrates how the electric field components are measured.*

With respect to parameter settings, the experiment has a high flexibility, which allows an operation in different collisionality regimes. Mainly by changing the ion mass and magnetic field strength, the collisionality can be varied by about two orders of magnitude. Thereby it is possible to study the transition from the hydrodynamic case ($C > 1$) to the adiabatic case ($C < 1$).

**Scaling of the energy transfer to the zonal flow** It was shown that the turbulence in the stellarator TJ-K is drift-wave dominated [3] and for the analysis it is assumed that it satisfies the nonlinear wave-coupling equation. Using the modified Ritz method, developed by Kim et al [4], which incorporates moments of fourth-order, it is then possible to estimate the spectral transfer of density fluctuation activity, which includes the nonlinearity of the $E \times B$-drift. A conditional averaging technique is used to calculate the temporal ensemble average. For the zonal flow as

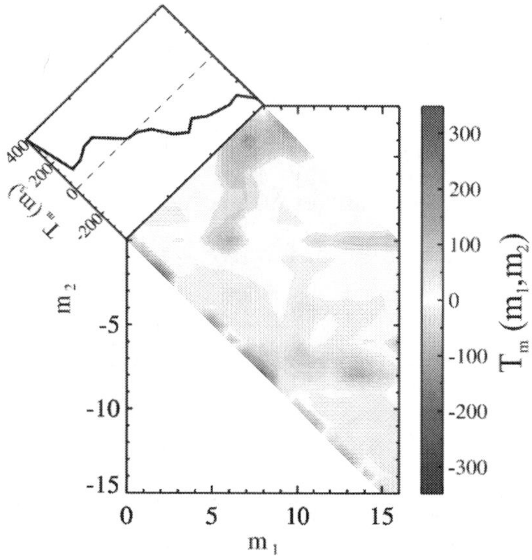


Figure 2: *Nonlinear power transfer function in wavenumber space (contour plot) and the integrated spectrum (line plot) showing the energy transfer to the $m_3 = 0$ mode.*

trigger event, an energy transfer from small-scale drift waves to the $m_3 = 0$ potential mode is found, corresponding to an nonlocal invers energy transfer (Figure 2). This energy transfer to the zonal flow is calculated for various measurements with different collisionality. The collisionality is thereby given as

$$C = \frac{\hat{\nu}}{\hat{k}_{\parallel}^2} \propto \frac{B\, n_e}{k_{\parallel}^2\, m_i\, T_e^{5/2}},$$

where $\hat{\nu}$ is the normalised collision frequency and $\hat{k}_{\parallel}$ is the normalised parallel wavelength. With the collision frequency as inverse of the collision time, one gets the shown dependencies on ion mass, magnetic field, electron density and temperature. The relative zonal flow power $P_{ZF}/P_{total}$ as a function of the collisionality is shown on the left hand side of figure 3. In the limit of the adiabatic case ($C \to 0$), the zonal flow contribution to the complete spectrum strongly increases. On the right hand side of figure 3, the scaling of the energy transfer to the zonal flow ($m_3 = 0$) is plotted. For low collisionality the energy transfer to the zonal flow also increases, showing the increased drive by the drift waves. The collisionality is a key parameter in drift-wave turbulence.

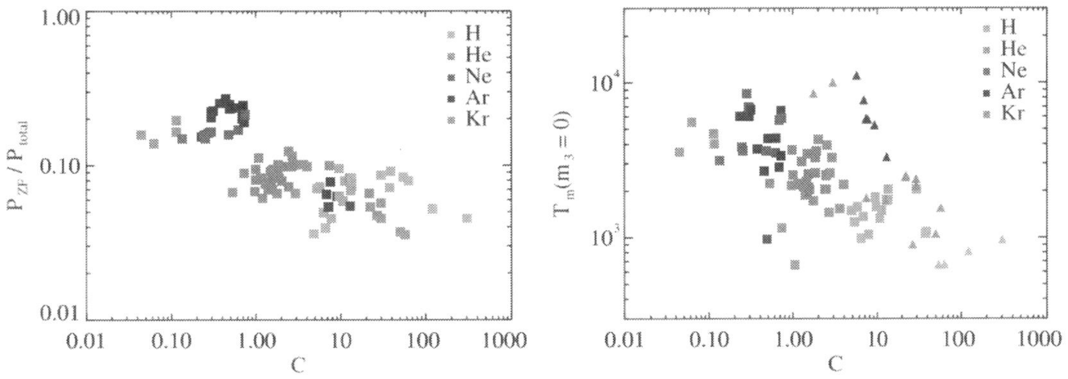


Figure 3: *Scaling with collisionality of relative spectral zonal flow power (*left*) and energy transfer to the zonal flow (*right*). Measurements with high magnetic field are drawn as triangles.*

In the two dimensional Hasegawa-Wakatani equations, as a model for drift-wave turbulence, it determines the coupling of the density and the potential field,

$$\partial_t n + \{\phi, n\} + \kappa_n \partial_y \phi = C^{-1}(\phi - n) \,,$$
$$\partial_t \Omega + \{\phi, \Omega\} = C^{-1}(\phi - n) \,.$$

Here $n$, $\phi$ and $\Omega$ denote the density, the potential and the vorticity, respectively. $\kappa_n$ is the normalised background gradient and $\{.,.\}$ represents Poisson brackets. For high collisionality, the two equation decouple and for low collisionality the density and potential act similar. Since the zonal flow is a pure potential structure, but the drift waves are sheared in the density, the drive is more efficient for higher coupling.To further investigate this coupling, also measurements with a different bias setting of the poloidal probe array were performed. For this setting, blocks of probes measuring ion-saturation current alternate with probes on floating potential when going around the circumference. So it is possible to calculate the Reynolds stress from the density fluctuations, the so called pseudo-Reynolds stress, together with the Reynolds stress from potential fluctuations [5]. Since Reynolds stress and pseudo-Reynolds stress can be measured on two fluxsurfaces, the corresponding zonal flow drive $-\partial_r(\langle \mathrm{R} \rangle_{\mathrm{fs}})$ of both quantities is obtained. The scaling of the maximum crosscorrelation between Reynolds-stress drive and pseudo Reynolds-stress drive with collisionality is shown in figure 4. For lower collisionality the corre-

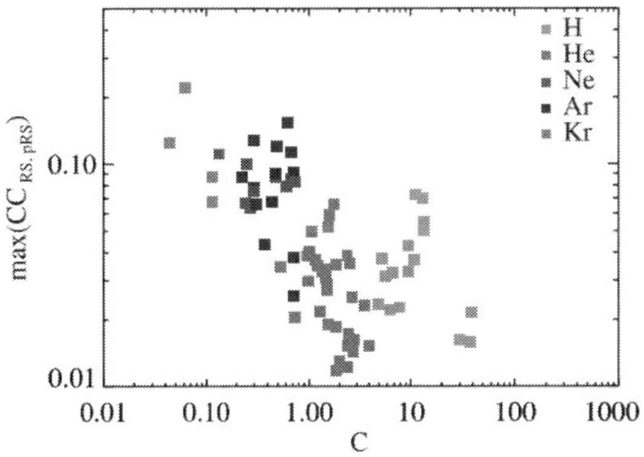


Figure 4: *Scaling of the maximum cross-correlation between Reynolds-stress drive and pseudo Reynolds-stress drive with the collisionality.*

lation between both increases, pointing to an increased coupling between density and potential.

**Energy transfer to the GAM** For measurements with high magnetic field (triangles in figure 3), and therefore low $\rho_s = \sqrt{m_i T_e}/eB$, an increased coupling between zonal flow and $m_1 = m_2 = 6$ density mode is found. The scaling of the energy transfer to this $m = 6$ mode is shown in figure 5. For decreasing collisionality more energy is transferred from the zonal flow ($m_3 = 0$ potential mode) to the $m = 6$ density mode, where the trend is similar as the scaling shown in figure 3. As stated above, energy can be transferred from the zonal flow to the geodesic acoustic mode. Due to the asymmetric poloidal flow structure, density will be advected into regions with high geodesic curvature. Therefore, the GAM in tokamak geometry has a $m = 1, n = 0$ density mode structure. Since the stellarator TJ-K has a sixfold magnetic field symmetry, this mode could indicate a GAM whose mode structure reflects the magnetic field geometry of the stellarator.

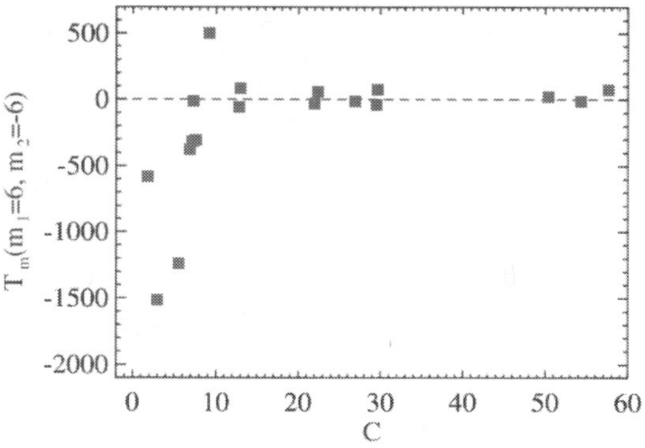


Figure 5: *Energy transfer to the $m = 6$ density mode for measurements with high magnetic field, i.e. low $\rho_s$.*

**Conclusion** Various measurements at different collisionalities have been carried out at the stellarator TJ-K to study the scaling of the energy transfer to zonal flows. For low collisionality an increased energy transfer to the zonal flow was found. Along with this, also the relative power of the zonal flow in the total spectrum strongly increases. With a different measurement setting, also Reynolds stress and density-based pseudo-Reynolds stress are obtained. The correlation between these two quantities increases with lower collisionality, pointing to a higher coupling of density and potential. For shots with high magnetic field, a prominent energy transfer to a $m = 6$ density mode is observed, showing the same scaling behaviour as the zonal flow. This mode could be the geodesic acoustic mode, reflecting the magnetic field structure of the stellarator.

**References**


[1] G. Birkenmeier et al., Phys. Rev. Lett. **110**, 145004 (2013).

[2] P. Manz et al., Plasma Phys. Control. Fusion **53**, 095001 (2011).

[3] U. Stroth et al., Phys. Plasmas **11**, 2558 (2004).

[4] J. S. Kim et al., Phys. Plasmas **3**, 3998 (1996).

[5] S. B. Korsholm et al., Plasma Phys. Control. Fusion **43**, 1377 (2001).

# A hybrid fluid–kinetic approach for 3D plasma edge transport in He-plasma at Wendelstein 7-X

M. Rack[1], D. Reiter[1], Y. Feng[2], P. Börner[1] and F. Hasenbeck[1]

[1]*Forschungszentrum Jülich GmbH, Institut für Energie- und Klimaforschung – Plasmaphysik, 52425 Jülich, Germany*

[2]*Max-Planck-Institute für Plasmaphysik, 17491 Greifswald / 85748 Garching, Germany*

## Introduction

The complex three-dimensional structure of the optimised stellarator Wendelstein 7-X can only partially be accessed by means of experimental measurements, as a consequence of the only quite limited number of measurements at distinct locations with different plasma cross-sections and properties. This highlights the importance of precise modelling (i.e. numerical bookkeeping) as an extended method for understanding the three-dimensional nature.

A plasma edge simulation code for treating complex magnetic configurations is the fluid plasma edge Monte-Carlo code in three dimensions (EMC3) [1] coupled to the kinetic (neutral) transport code EIRENE [2, 3]. EMC3 bases on a Monte-Carlo algorithm [4, Ch. 4.1] for a reduced set of Braginskii equations [5] formulated in a Fokker-Planck scheme while EIRENE solves extended Boltzmann equations. A detailed ab initio derivation of the EMC3 model equations can be found in [6, Ch. 2]. EMC3-EIRENE has been continuously improved and verified, but one remaining restriction was that the bulk ion species is limited to hydrogen isotopes, with higher-Z ions being treated as trace-impurities.

However, in initial operation phases Wendelstein 7-X (and probably also ITER) will be operated with helium plasma. For a more direct interpretation of first measurements, computational quantification of helium plasma flows in the edge is required. Therefore, an approach is presented and initiated for the simulation of helium plasma by slightly extending EMC3 to facilitate a treatment of the main plasma species with $Z \neq 1$ and expanding the use of EIRENE features within EMC3, e.g. its existing non-linear mode of operation, metastable resolved reactions and, most important, currently strongly reduced kinetic trace ion transport module.

## Generalization of the EMC3 model equations for $Z \neq 1$

The simulation of $He^{++}$ with EMC3 requires a generalization of the considered fluid equation to $Z \neq 1$. In the simply modified equations the particle balance reads

$$\nabla \cdot \left[ nV_{i\|}\boldsymbol{b}_{\|} - D\boldsymbol{b}_{\perp}\boldsymbol{b}_{\perp} \cdot \nabla n \right] = S_{\text{ion}} , \tag{1}$$

with the plasma density $n = n_i = n_e/Z$, the parallel ion flow velocity $V_{i\|}$, the normalized parallel and perpendicular magnetic field vectors $\boldsymbol{b}_\|$ and $\boldsymbol{b}_\perp$, and the ionization source $S_{ion}$ due to electron impact collisions with neutrals. The momentum balance reads

$$\nabla \cdot \left[m_i n V_{i\|} V_{i\|} \boldsymbol{b}_\| - \eta_\| \boldsymbol{b}_\| \boldsymbol{b}_\| \cdot \nabla V_{i\|} - D \boldsymbol{b}_\perp \boldsymbol{b}_\perp \cdot \nabla \left(m_i n V_{i\|}\right)\right] = -\boldsymbol{b}_\| \cdot \nabla p + S_m \,, \tag{2}$$

with the ion mass $m_i$, the parallel ion viscosity coefficient $\eta_\|$ (as defined in [5, Eq. 2.22]), the perpendicular diffusion coefficient $D$, the plasma pressure $p = n_e T_e + n_i T_i$, and the momentum source $S_m$ due to charge-exchange and elastic collisions with neutrals. The electron and ion energy balances read

$$\begin{aligned} \nabla \cdot \left[\frac{5}{2} Z n T_e V_{i\|} \boldsymbol{b}_\| - \kappa_e \boldsymbol{b}_\| \boldsymbol{b}_\| \cdot \nabla T_e - \frac{5}{2} Z T_e D \boldsymbol{b}_\perp \boldsymbol{b}_\perp \cdot \nabla n - Z \chi_e n \boldsymbol{b}_\perp \boldsymbol{b}_\perp \cdot \nabla T_e\right] \\ = k\left(T_i - T_e\right) + S_{e,e} + S_{imp} \,, \end{aligned} \tag{3}$$

$$\begin{aligned} \nabla \cdot \left[\frac{5}{2} n T_i V_{i\|} \boldsymbol{b}_\| - \kappa_i \boldsymbol{b}_\| \boldsymbol{b}_\| \cdot \nabla T_i - \frac{5}{2} T_i D \boldsymbol{b}_\perp \boldsymbol{b}_\perp \cdot \nabla n - \chi_i n \boldsymbol{b}_\perp \boldsymbol{b}_\perp \cdot \nabla T_i\right] \\ = k\left(T_e - T_i\right) + S_{e,i} \,, \end{aligned} \tag{4}$$

with the electron and ion temperature $T_e$ and $T_i$, the parallel electron and ion thermal conductivity $\kappa_e$ and $\kappa_i$ (based on [5, Eq. 2.12 and 2.15]), the parallel electron and ion heat diffusivity $\chi_e$ and $\chi_i$, $k = 3 m_e n_e / (m_i \tau_{ee})$ with $\tau_{ee}$ being the electron collision time (as defined in [5, Eq. 2.5e]), the energy source of electrons $S_{e,e}$ and ions $S_{e,i}$ due to plasma-neutral interaction (inclusive radiation), and the energy source $S_{imp}$ due to impurity radiation.

The boundary conditions are adapted by generalizing the sound speed $c_s = \sqrt{(ZT_e + T_i)/m_i}$.

**Estimates for a typical helium ionisation and population distribution**

The ion temperature of a typical plasma edge in a medium-sized fusion device ranges from a few eV to several 100 eV at densities in the order of $1 \times 10^{13}\,\mathrm{cm}^{-3}$. In a detached divertor, plasma densities in the order of $1 \times 10^{15}\,\mathrm{cm}^{-3}$ can be reached. A zero-dimensional estimate under non-local-thermodynamic-equilibrium conditions (using FLYCHK [8, 7]) provides first indications that a reasonable treatment of helium plasma as a single-fluid system with helium in the second ionisation state is justified. The graph in fig. 1 shows that inde-

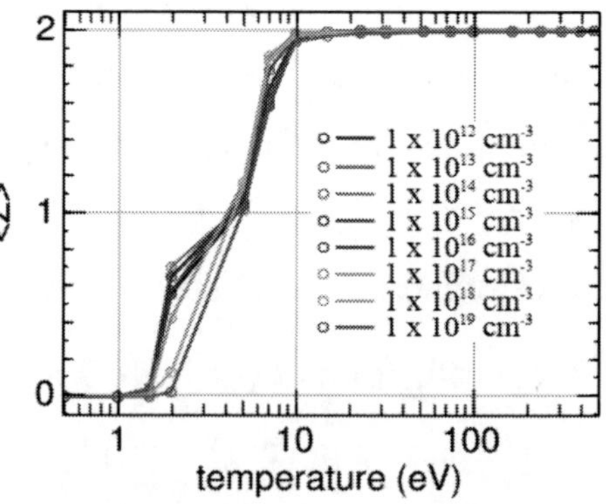


**Figure 1:** Helium averaged charge state for electron densities $1 \times 10^{12}\,\mathrm{cm}^{-3}$ to $1 \times 10^{19}\,\mathrm{cm}^{-3}$. Based on [7].

pendent of the density from a temperature of about 5 eV the dominant part of the helium ions is in the second charged state. At a temperature above 10 eV a fully ionized helium plasma is present.

A second estimate on the correct treatment of helium plasma can be made based on EIRENE test simulations by comparing the recycling flux from the targets with fluxes from volume-recombination. The simulations show that the dominant source for $He^+$ under the given conditions is ionisation from He rather than volume-recombination from $He^{++}$. Thus, it leads to the same conclusion as the zero-dimensional estimate under non-local-thermodynamic-equilibrium conditions: helium in the second ionisation state is dominant in the bulk of the plasma edge. Furthermore, this can be strengthened by comparing the simulated spatial distribution of $He^{++}$, neutral helium and $He^+$ near the location of the limiter in a typical limiter discharge at Wendelstein 7-X ($n_{i,up} = 1 \times 10^{13}\,cm^{-3}$, $P = 2\,MW$), see fig. 2. Only in the vicinity of the limiter the neutral helium density (dashed) is dominant in the scrape-off layer (SOL) but decreases rapidly in the plasma edge region, magnitudes below the $He^{++}$ density (solid). A similar behaviour is seen for $He^+$ (dashed).

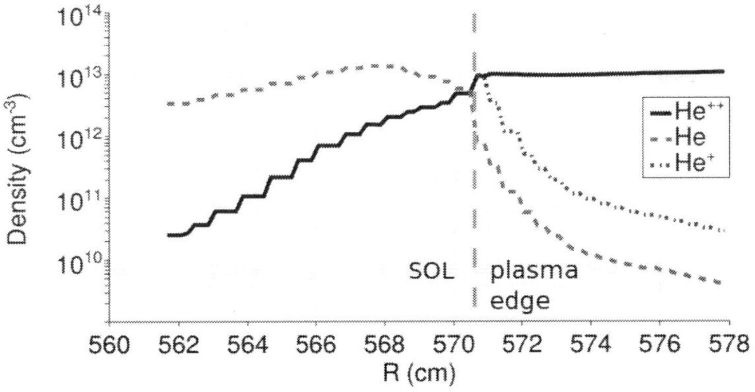


**Figure 2:** Comparison of density distributions of $He^{++}$, neutral helium, and $He^+$ calculated for typical Wendelstein 7-X limiter plasma with an upstream density of $n_{i,up} = 1 \times 10^{13}\,cm^{-3}$ and a heating power of $P = 2\,MW$. The radial profiles are taken along the mid-plane near the limiter.

## Numerical treatment of $He^+$ and $He^{++}$

A treatment of $He^+$ with a fluid approach is not very appropriate because the short life time of $He^+$ ions prevents thermalisation and therefore a relaxation towards a Maxwellian distribution of the $He^+$ ions. An adequate treatment of $He^+$ is either quasi-static, when the characteristic transport lengths are much shorter than the length scale of interest, or even kinetic, otherwise. In the quasi-static approach the ion is stopping at its place of birth until it gets further ionised or recombines. The simplest approach for a kinetic treatment is the movement along the magnetic field lines which is already implemented in EIRENE. Additional effects like ion drifts are neglected.

The atomic and molecular collision data as originally presented by Janev *et. al* [9] and Fujimoto [10] (for helium) are extended and stored in the HYDHEL [11] and AMJUEL [12] database to treat the different atomic processes within EIRENE simulations (for further details on the treatment of helium reactions see [13]). The following atomic processes are considered:

$$e + \mathrm{He}(1|1\mathrm{S}) \rightarrow e + \mathrm{He}^{+} + e \qquad [12, \mathrm{rx\ H.4\ 2.3.9a}],$$
$$\mathrm{He}^{++} + \mathrm{He} \rightarrow \mathrm{He} + \mathrm{He}^{++} \qquad [11, \mathrm{rx\ H.1/H.3\ 6.3.1}],$$
$$e + \mathrm{He}^{+}(1\mathrm{s}) \rightarrow \mathrm{He}(1|1\mathrm{S}) + \ldots \qquad [12, \mathrm{rx\ H.4\ 2.3.13a}],$$
$$e + \mathrm{He}^{+} \rightarrow \mathrm{He}^{++} + e + e \qquad [12, \mathrm{rx\ H.4\ 2.2C}].$$

The dominant ion species $He^{++}$ is treated with the fluid approach using the generalized model for EMC3 as presented above.

## Summary and outlook

Using simple estimates it is shown that the dominant helium species in the plasma edge region is $He^{++}$ which can be treated with a fluid approach, simulated by the extended model of the plasma fluid code EMC3. For a sufficient consideration of plasma wall interactions a treatment of neutral helium and $He^{+}$ is also required which can be performed using the kinetic transport code EIRENE.

Future studies need to quantify the applicability of a quasi-static over a kinetic approach for $He^{+}$ and the use of metastable resolved over unresolved transport. Processes currently neglected like $He^{++} \rightarrow H^{+}$ recombination [12, rx H.4 2.3.2C] and He–$He^{+}$ charge-exchange [11, rx H.1/H.3 5.3.1] will be considered via the non-linear collision model of EIRENE.

## Acknowledgement

This work has been carried out within the framework of the EUROfusion Consortium and has received funding from the Euratom research and training programme 2014-2018 under grant agreement No 633053. The views and opinions expressed herein do not necessarily reflect those of the European Commission. The authors gratefully acknowledge the computing time granted on the supercomputer JUROPA at Jülich Supercomputing Centre (JSC).

## References

[1] Y. Feng *et al.* *Journal of Nuclear Materials*, **266-269**, 812 (1999).
[2] EIRENE. http://www.eirene.de.
[3] D. Reiter *et al.* *Fusion Science and Technology*, **47** (2), 172 (2005).
[4] H. Risken. *The Fokker-Planck Equation: Methods Of Solution And Applications: Methods of Solutions and Applications*. Springer-Verlag, 2nd edition (1989).
[5] S. I. Braginskii. *Reviews of Plasma Physics*, **1**, 205 (1965).
[6] D. Harting *et al.* Technical Report Jül–4173, Forschungszentrum Jülich GmbH (2005).
[7] FLYCHK. https://www-amdis.iaea.org/FLYCHK/.
[8] H.-K. Chung *et al.* *High Energy Density Physics*, **1** (1), 3 (2005).
[9] R. Janev *et al.* *Elementary Processes in Hydrogen-Helium Plasmas*. Springer-Verlag, 1st edition (1987).
[10] T. Fujimoto. *Journal of Quantitative Spectroscopy and Radiative Transfer*, **21** (5), 439 (1979).
[11] HYDHEL datafile. http://www.eirene.de/html/hydhel.html. Based on [9].
[12] AMJUEL datafile. http://www.eirene.de/html/amjuel.html.
[13] D. Reiter *et al.* Technical Report Jül–2605, Forschungszentrum Jülich GmbH (1992). Based on [10].

# Estimation of Wendelstein 7-X magnetic field perturbation due to permeable materials

T. Andreeva[1], R. Brakel[1], J.Kißlinger[2], M. Köppen[1]
[1] *Max-Planck-Institut für Plasmaphysik, EURATOM Association, Teilinstitut Greifswald, Wendelsteinstraße 1, 17491 Greifswald, Germany*
[2] *Max-Planck Institut für Plasmaphysik, EURATOM Association, Boltzmannstraße 2, 85748 Garching, Germany*

Wendelstein 7-X (W7-X), currently under commissioning at the Max-Planck-Institut für Plasmaphysik in Greifswald, Germany, is a helical advanced stellarator, combining the modular coil concept with optimized plasma properties. To achieve the designed magnetic field structure and reliable operation of the machine not only a high accuracy of coil manufacturing and assembly is required, but also a careful assessment of the influence of all permeable materials in the vicinity of the plasma. This paper discusses the simplified modelling of the permeable material influence on the magnetic field perturbation which can be relevant also for other devices aiming to assess more accurate the magnetic field by consideration of the impact of permeable materials and presents a typical example of such calculations for Wendelstein 7-X.

**Keywords**: Wendelstein 7-X, stellarator, permeable materials, magnetic field perturbation

**Introduction**

During a construction of Wendelstein 7-X (Figure 1) a multitude of different materials is used, which should be compatible with the experimental conditions on one hand and ensure a proper operation on the other hand. To guarantee the unperturbed topology of the magnetic field the handling of permeable materials in the W7-X hall and boundary conditions for their application are strictly regulated by the corresponding technical guideline. It requires to use materials with magnetic permeability $< 1.01$ within the outer vessel area. Only for the welding seams this limit is relaxed up to 1.05. In all cases deviating from these specified requirements the impact on the magnetic field perturbation must be accurately estimated taking into account the exact location and the specific geometry of a permeable machine component or its pieces. This analysis is a basis for the further release procedure of a permeable material, registered afterwards in the corresponding data base, which allows one to follow the details of the application whenever necessary. The obligatory condition for the allowance to use a permeable material is that the magnetic field perturbation $B_s/B_0$ should be less than $1\cdot10^{-4}$, where $B_s$ – is a perturbed magnetic field, $B_0$ – an unperturbed magnetic field at the location of interest/ computation point (plasma edge).

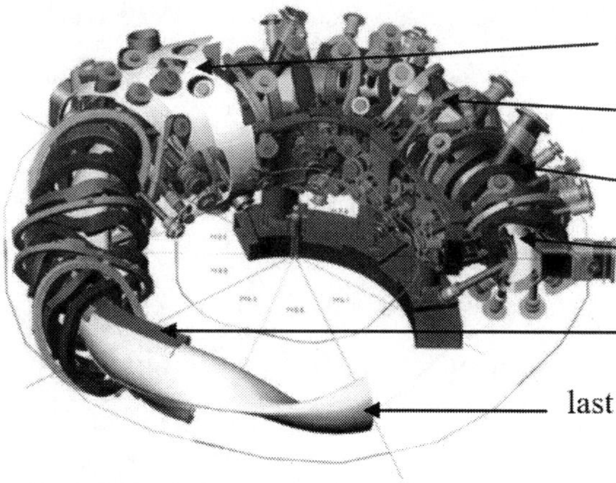


*Figure 1: Schematic view of Wendelstein 7-X.*

**Modeling methods**

There are several possibilities to estimate the magnetic field perturbation $B_s/B_0$. The simplest method, applicable for a small permeable component located far away from the plasma, is to use the analytical approximation of one permeable sphere [1]:

$B_s/B_0 = 2\ B_k/B_0\ (\mu_r - 1)/\ (\mu_r + 2)\ \cdot(r_k/d)^3$, (1)

where $B_k$ – magnetic field at the location of the permeable sphere, $\mu_r$ – relative permeability of the considered material, $r_k$ – radius of the permeable sphere with the volume equal to the volume of the considered permeable component, d – distance between the permeable sphere and the location of interest. In this approximation the angle between the local magnetic field and the vector along the distance d is neglected.

Another possibility is to use finite element (FE) calculations, for example with help of ANSYS code. This method should be used when considering the non-linear dependencies of the magnetic permeability and for complicated geometries of the permeable components especially when an accurate modelling of the magnetic field perturbation near the allowable limit is necessary. The disadvantage of this method is that FE calculations are very time consuming.

To find a compromise between both needs (to perform calculation in reasonable time and to avoid too rough approximation) the method of mutually interacted spheres was developed for W7-X tasks and realised in the program "W7-X permeabilities" (author:

M. Köppen) being used for standard W7-X applications. The mathematical basis is deduced from methods of classical electrodynamics described in [2].

The magnetization M of a permeable sphere in an external homogeneous magnetic field $B_0$ is $M = \frac{3}{\mu 0}(\frac{\mu-\mu 0}{\mu+2\mu 0})B_0$, where μ - permeability of the sphere, $\mu_0$ – permeability of vacuum. The magnetic scalar potential is given by $\phi_M = \frac{1}{3} M a^3 \frac{cos\emptyset}{r^2}$, where a - radius of the sphere, r – distance between the sphere and computation point, ∅ - angle between the magnetic field of the sphere and the distance vector between the sphere and the computation point. The vector of the magnetic field strength is given by $\mathbf{H} = -\nabla\phi_M$. The magnetic field components are defined now from the vector equation: $\mathbf{B} = \mu_0 \mathbf{H}$.

The initial magnetic field $B_0$ can be calculated by Biot–Savart law. Then the perturbed magnetic field is calculated for every out of N mutually interacting spheres, which gives 3N equations written above. Afterwards the magnetic field perturbation at each computation point is found as a superposition of the local initial field $B_0$ and a sum of local perturbations from all spheres. This algorithm is realised in MATLAB package which allows a "matrix-division" (a possibility to derive vector **x** like **x**= A**b** from the equation A**x**=**b**, where A is a matrix and **b** is a vector).

The complex geometries can be modelled in this method with help of many spheres, where each sphere is small enough to consider the field $B_0$ as homogenous. The method of mutually interacting spheres is much more accurate than the approximation (1) and is extremely fast in comparison to FE calculations (100s against 2 days for a task with 2500 spheres). This allows checking the magnetic field perturbation for all reference W7-X configurations [3] and estimating its maximum value in a reasonable time. Result inaccuracies depend on how conservative is the geometry model.

**One example for Wendelstein 7-X applications**

For the user convenience the MATLAB program realizing the method described above includes a graphical user interface (GUI). The input data consist of the geometry description of permeable components approximated with spheres, the radii of these spheres and their magnetic permeability.

One typical task during Wendelstein 7-X construction was the estimation of the magnetic field perturbation caused by magnetic parts of cooling circuits with an increased permeability. Each circuit part was represented by several spheres distributed between its ends. The radii of the spheres were calculated from the corresponding volumes of the circuits. The magnetic permeability value was chosen conservative, equal to 2000. The maximum magnetic field perturbation was chosen among calculations for 18 operational cases (9 reference operational cases x 2 directions of currents in trim coils). The resulting perturbation value of $5.45 \cdot 10^{-7}$ was significantly less than the specified allowable limit of $10^{-4}$.

Figure 2 represents the GUI interface during the calculation showing the modeled geometry and the magnet system of W7-X. Figure 3 shows the value of the maximum magnetic field perturbation and its distribution at the plasma edge.

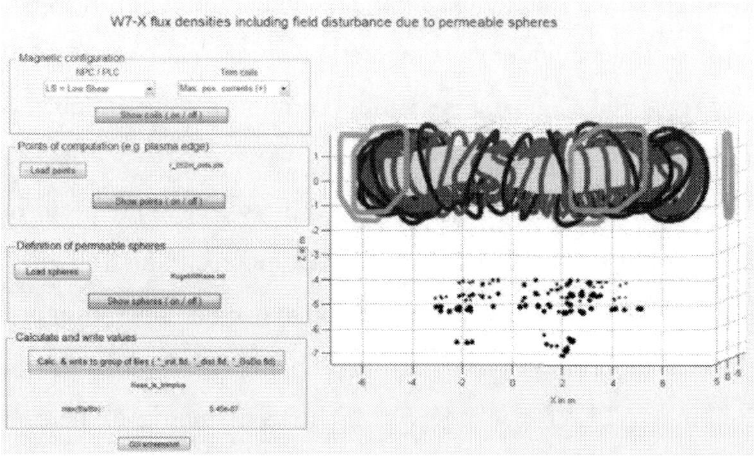


*Figure 2: Modeling of the influence of permeable cooling circuit parts for W7-X. NPC are shown is red, PC – in blue and trim coils –in green color. Plasma edge is shown in yellow. Black points correspond to spheres representing circuits.*

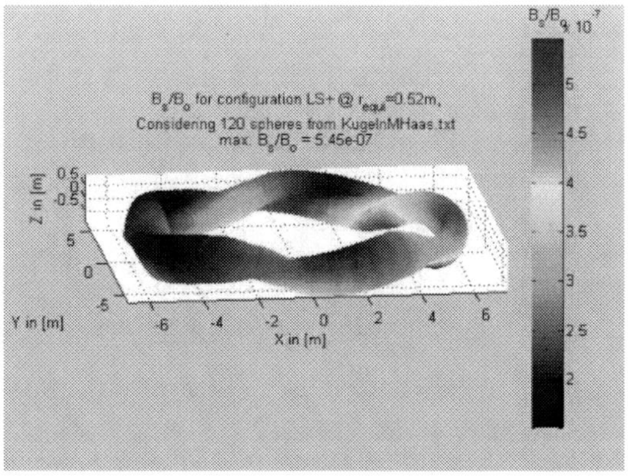


*Figure 3: Maximum magnetic field perturbation for the task with permeable cooling circuit parts.*

**Summary**

To achieve the designed magnetic field structure and reliable operation of the Wendelstein 7-X careful assessments of the influence of all permeable materials in the vicinity of plasma are necessary. For this purpose the official procedure for handling the permeable materials was established based on the corresponding technical guideline. The method of mutually interacting spheres is a reasonable compromise between analytical approximations and FE calculations, and allows performing quick and reliable assessments. All results of calculations are stored in the W7-X documentation system. Material data are collected in the corresponding data base, giving an overview about main properties of any specific component, where the magnetic permeability does not comply with guideline requirements.

**References**

1. W. Weizel, "Lehrbuch der theoretischen Physik", Springer-Verlag, 1963.

2. J. D. Jackson, "Classical electrodynamics", John Wiley & Sons Inc., 1999.

3. T. Andreeva, et al., "Characteristics of main configurations of Wendelstein 7-X", Problems of Atomic Science and Technology, Series: Plasma Physics, Vol. 4 (2002) 45-47.

# Reflectometry fluctuation diagnostics at Wendelstein 7-X

T. Windisch[1], T. Estrada[2], M. Hirsch[1], W. Kasparek[3], A. Krämer-Flecken[4], C. Lechte[3], S. Wolf[3], P. Xanthopoulos[1] and T. Klinger[1]

[1] *Max-Planck-Institute for Plasma Physics, Greifswald, Germany*

[2] *Centro de Investigationes, Medioambientales y Technologicas (CIEMAT), Madrid, Spain*

[3] *Inst. für Grenzflächenverfahrenstechnik und Plasmatechnologie, Univ. Stuttgart, Germany*

[4] *Institut für Energie- und Klimaforschung, Forschungszentrum Jülich, Germany*

Wendelstein 7-X, the first superconducting stellarator along the five-fold symmetric HELIAS line, will start operation in fall this year. In its first commissioning operation phase (OP1.1) plasma energy is limited to 2 MJ, which corresponds to a pulse length of 0.4 s at 5 MW ECR-heating. The plasma boundary with open flux surfaces is produced by five graphite inboard limiters in order to protect in-vessel components. In the second operation phase (OP1.2), an inertially cooled island divertor and a water cooled first wall allow 10 s operation at 10 MW heating power. Finally, in OP2, scheduled to start in 2019, 1800s pulses with 10 MW are envisaged using water cooled island divertors.

From the beginning, a set of microwave reflectometry systems dedicated to measure the spatio-temporal evolution of density fluctuations are operated, i.e, a V-band Doppler reflectometer (DR, provided by CIEMAT) and a Ka-band poloidal correlation reflectometer (PCR, provided by FZ Jülich). In the later operation phases, the DR will be extended to the W-band, and a novel steerable Doppler reflectometer (SDR, developed at IVGP Stuttgart) will be installed. The systems are briefly described in the following. All reflectometer systems are designed to measure fluctuations localized around the last closed flux surface (LCFS), where large variations

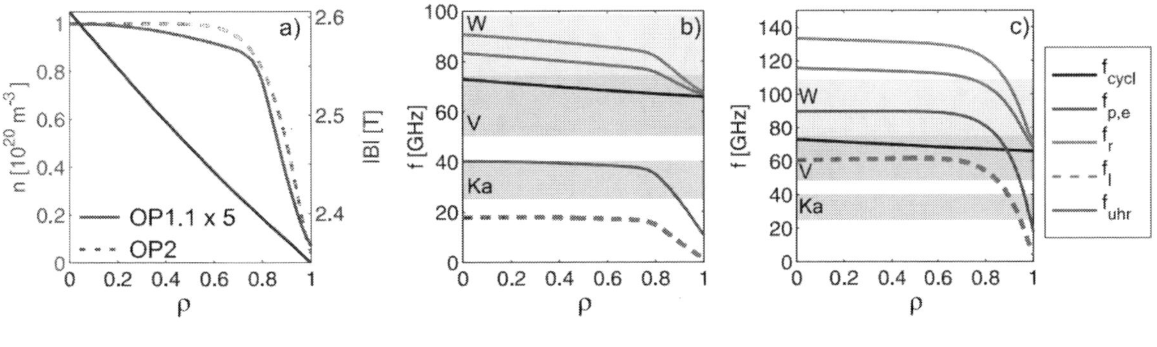


Figure 1: (a) Assumed radial profiles of $n$ and |B| (solid: scenario OP1.1, dashed: scenario OP2), and (b-c) corresponding resonance and cut-off frequencies for sc. OP1.1 and sc. OP2. The used microwave bands are shown as colored boxes.

in the perpendicular $E_r \times B_t$-flow are expected.

Assumed density profiles for two operation scenarios (scenario OP1.1, scenario OP2) are shown in Fig. 1a. In scenario OP2, the expected peak density is approx. five times larger than in scenario OP1.1. The reflectometer systems are located in one of the five highly elongated corners of W7-X, the so-called bean shaped plane ($\phi = 72°$). Typical flux-surfaces and the design beam paths of the DR and PCR are shown in Fig. 2. The design cut-off positions are around the outer midplane. In contrast to a tokamak, the outer midplane is not a bad curvature region for all toroidal locations per se. For the bean shaped and triangular shaped planes of W7-X, the normal magnetic field curvature $\kappa$ is shown in Fig. 3 along the magnetic field lines crossing the respective planes at poloidal angle $\theta = 0$ (blue line). Also shown are the mode structures associated with ITG fluctuations as obtained from linear GENE simulations with adiabatic electrons (red lines). Around the outer midplane ($\theta = 0$) the poloidal structure of the fluctuations differs significantly. In the case of the bean shaped plane the fluctuations peak at the design cut-off position ($z = 0$). This theoretical prediction could be proven directly utilizing the reflectometry systems.

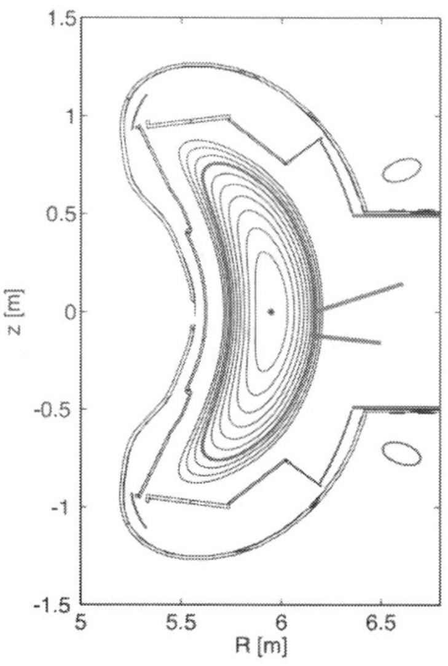


Figure 2: Flux surfaces and DR (blue) and PCR (green) beam paths.

The broadband DR system operates in V- and W-band ($50-110$GHZ) using o- or x-mode polarization. The corresponding cut-off positions for OP1.1 and OP2 are shown in Fig. 1b-c. The measurement principle relies on the Bragg backscattering mechanism. Via the Bragg condition $k_\perp = 2k_0 \sin(\theta_{\rm tilt})$ fluctuations with $k_\perp$ are selected out of the turbulent spectrum at

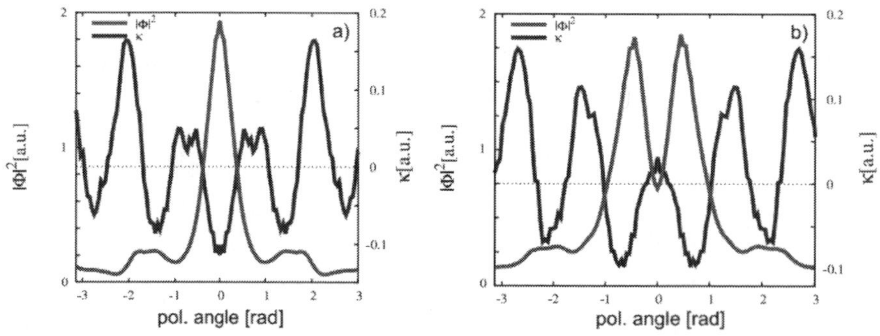


Figure 3: Normal magnetic field curvature $\kappa$ (blue) and ITG potential fluctuations (red) on a flux surface for the (a) bean and (b) triangular shaped plane as obtained from linear GENE runs (taken from Ref. [1]).

the cutoff-layer. The bistatic system has a fixed incidence angle to the flux surface normal $\theta_{tilt} = \pm 18°$. The design cutoff is located at $\rho = 0.85$, which is 4 cm inside the LCFS.

This flux surface is highlighted in Fig. 2 as orange line . The chosen tilt angle is a compromise between path length (beam waist, damping), suppression of the strong $m = 0$ scattering component and desired wavenumber. The chosen angle corresponds for $f = 80\,\text{GHz}$ to $\lambda_\perp = 0.6\,\text{cm}$. The accessible wavenumber range as computed with TRAVIS [2] is shown in Fig. 4. Since the measured signal is Doppler shifted, it allows to estimate the perpendicular velocity of the fluctuations (provided $k_\parallel \ll k_\perp$). In order to enhance the wavenumber resolution and to ensure a broadband operation of the DR system, special attention has been paid to the Gaussian antenna system, which consists of six mirrors. Details about the optimization strategy can be found in Ref. [3]. The O-mode PCR system consists of five antennae, i.e. one launching and four receiving antennae [4]. With a cut-off density $n_c = 2 \cdot 10^{19}\,\text{m}^{-3}$ it measures in sc. OP2 around the LCFS $\rho = 0.96 - 0.98$. Via correlation techniques the poloidal propagation and correlation lengths and times of the turbulent structures can be analyzed. In addition the occurrence of (quasi-)coherent modes may be measured directly and the pitch angle of the local magnetic field can be estimated.

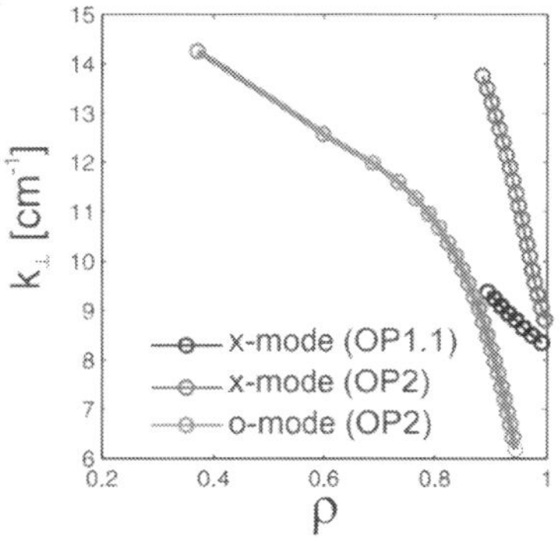


Figure 4: Accessible perpendicular wavenumber of the DR system.

The SDR system allows to change the incidence angle $\theta_{tilt}$ and thereby scanning a part of the wavenumber spectra without any movable parts in the vacuum. This is achieved by a phased-array antenna consisting of 32 stacked H-plane sectoral horns [5]. The phased feed array allows to tilt angle between $\pm 20°$ by a small frequency change $\Delta f \approx 700\,\text{MHz}$ at around 15 frequencies in the W-band. The offset tilt angle is chosen to allow a direct comparison with the DR system. The accessible radial and wavenumber range as estimated with TRAVIS is shown in Fig. 5.

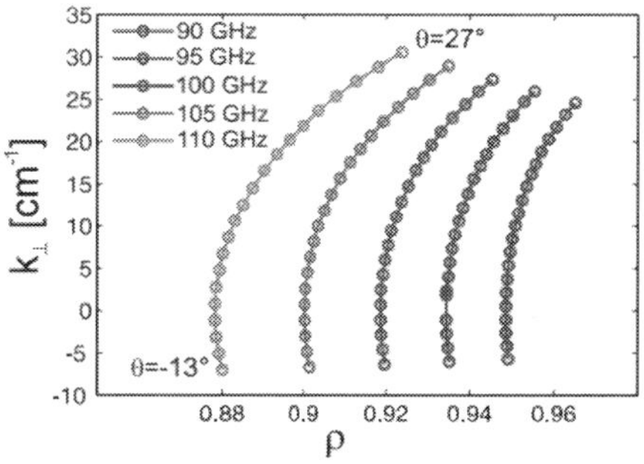


Figure 5: Accessible perpendicular wavenumber of the SDR system.

In general the interpretation of the measurement signals is often quite challenging due to the complex interplay between microwave reflection, scattering (multiple), refraction and mode coupling, especially for larger fluctuation amplitudes. Here, full wave calculations can be used

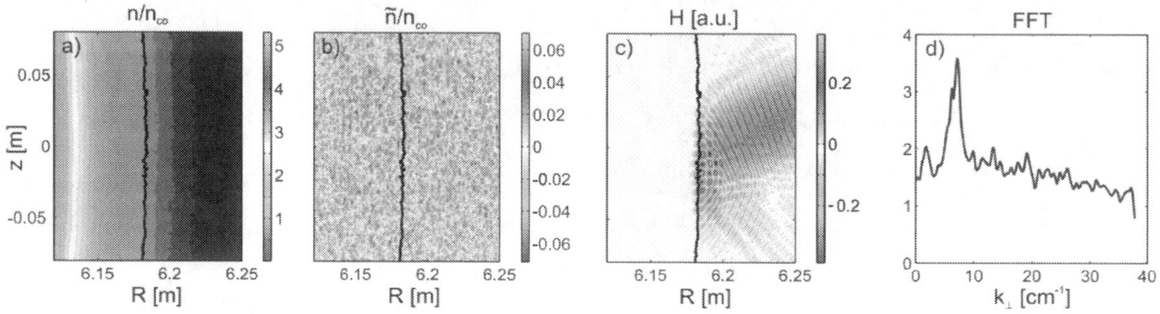


Figure 6: Normalized density (a) and density fluctuations (b) as obtained from GENE simulations, full wave simulation (c), and radially averaged spectra of density fluctuations (d).

to gain a better understanding. For this purpose we use the IPF-FD3D code developed at the University of Stuttgart [6]. The input density fluctuations for the full wave code is taken from flux-surface GENE simulations [7] at $\rho = 0.8$, which then defines the cut-off layer position and microwave frequency, respectively. Note, that in the domain of the flux surface simulations ($x = 264\,\rho_s \sim 25\,\mathrm{cm}$ for $T_e = 254\,\mathrm{eV}$) the radial density and temperature gradients are assumed to be constant. The density and fluctuations in the interesting full wave domain are shown in Fig. 6a,b. Both are normalized to the cut-off density $n_{co}$ for $f = 72.8\,\mathrm{GHz}$. The black line indicates the nominal corrugated cut-off layer for this frequency. The fluctuation amplitude is $\sim 4\%$. The result of the full wave calculation for the upper DR antenna ($\theta_{tilt} = -18°$) is shown in Fig. 6c. The spectrum of the simulated density fluctuations (Fig. 6d) has a distinct peak at $k_\perp \sim 7\,\mathrm{cm}^{-1}$, which corresponds to $k_\perp \rho_s \sim 0.7$ . This is close to the design value obtained from the Bragg condition for this frequency, $k_\perp^d = 9\,\mathrm{cm}^{-1}$. The simulation of the dynamic response of the reflected microwave signal is currently ongoing.

## References

[1] Proll J, Trapped particle instabilities in quasi-aerodynamics stellarators, PhD thesis, EMAU Greifswald (2013).

[2] Marushchenko NB, Turkin Y, Maassberg H, Comp. Phys. Comm. **185**, 165 (2014).

[3] Hirsch M *et al.*, Nucl. Fusion **46** S853 (2006).

[4] Krämer-Flecken A, Soldatov S, Vowinkel B et al., Rev. Sci. Instrum. **81**, 113502 (2010).

[5] Rohmann P. et al. IEEE Int. Symposium on Phased Array Systems & Technology, 559 (2013).

[6] Lechte C, IEEE Trans. Plasma Sci., **37**, 1099 (2009).

[7] P. Xanthopoulos, H. Mynick, P. Helander, Y. Turkin *et al.*, Phys. Rev. Lett. **113**, 155001 (2014).

# Modeling of electron cyclotron current drive for finite collisionality plasmas in Wendelstein 7-X

G. Kapper[1], W. Kernbichler[1], S.V. Kasilov[1,2], N.B. Marushchenko[3]

[1] *Fusion@ÖAW, Institute of Theoretical and Computational Physics, TU Graz, Austria*

[2] *Institute of Plasma Physics, National Science Center "Kharkov Institute of Physics and Technology", Ukraine*

[3] *Max Planck Institute for Plasma Physics, EURATOM Association, Germany*

## Introduction

The low-shear concept of Wendelstein 7-X avoids low-order rational flux surfaces in the plasma core since large magnetic islands can be formed there. In turn, such a resonant surface is placed at the plasma edge, thus producing an island divertor configuration. In order to preserve these features, which can be destroyed even by the arise of a relatively small bootstrap current, this current should be compensated by electron cyclotron current drive (ECCD) [1]. In modern fusion devices the evaluation of the generalized Spitzer function (GSF), which serves as an ECCD efficiency, is usually performed in the low collisionality limit typical for these devices, where one can effectively use the bounce averaged kinetic equation. At the same time, at high densities and mild electron temperatures, which can be expected at the initial stage of the device operation and at the start up of plasma discharge, finite collisionality effects should be taken into account. Such an approach requires the solution of the high dimensional (4D) drift kinetic equation, which is a more difficult task as compared to the 2D bounce averaged equation. In this work, the impact of finite collisionality effects on ECCD in a high-mirror configuration of Wendelstein 7-X is studied using a combination of the codes NEO-2 [2], a 4D drift kinetic equation solver for plasmas with finite collisionality using the full linearized Coulomb collision operator and realistic device geometry, and TRAVIS [3], a ray-tracing code developed for studies of electron cyclotron current drive, heating and emission. Recently, this code combination was used to study the influence of finite collisionality on ECCD in tokamaks, where the generalized Spitzer problem at a given magnetic flux surface is essentially 3D [2]. The 4D problem of stellarators led to the development of a new interface between the codes based on precomputed data of NEO-2 for a realistic magnetic model from an equilibrium code and plasma parameter profiles from a transport code. As in Ref. [2], the non-relativistic limit is used here for the GSF, which is compared to asymptotic limits of low and high plasma collisionality.

**Code combination within the adjoint approach**

A commonly used technique for current drive modeling is the adjoint approach [4], where the flux surface averaged parallel current density,

$$\langle j_{\|}B\rangle = e\,l_c\left\langle \int d^3p\,\frac{\partial \bar{g}}{\partial \mathbf{p}}\cdot\mathbf{\Gamma}_{\mathrm{RF}}\right\rangle, \tag{1}$$

is expressed by an adjoint GSF $\bar{g}$ computed by NEO-2 and a microwave induced quasilinear diffusion flux in phase space $\mathbf{\Gamma}_{\mathrm{RF}}$ computed by TRAVIS, where $\langle\dots\rangle$ denotes flux surface average, $e$ is the electron charge, $B$ is the magnetic field module and $l_c$ is the mean free path [2]. The adjoint function is related by $\bar{g}(v_{\|}) = -g(-v_{\|})$ to the usual GSF $g$, which is a solution to the conductivity problem written in terms of integrals of motion in velocity space as follows,

$$v_{\|}\mathbf{h}\cdot\nabla f_{\mathrm{M}}g - \hat{L}_{\mathrm{CL}}f_{\mathrm{M}}g = \frac{B}{l_c}v_{\|}f_{\mathrm{M}}. \tag{2}$$

Here, $v_{\|}$ is the parallel velocity, $\mathbf{h}$ is a unit vector along the magnetic field, $f_{\mathrm{M}}$ is a Maxwellian and $\hat{L}_{\mathrm{CL}}$ is a linearized collision integral. Cross-field rotation has been ignored in Eq. (2) because it has small effect on the GSF in finite collisionality regimes as well as at low collisionalities where the electrons are mostly in the $1/\nu$ regime. Therefore, a 4D problem (2) can be reduced to a 3D problem for a single field line (position on field line is given by a line parameter $\varphi_s$) which is long enough to cover the given flux surface (labeled by the normalized toroidal flux value $s$) densely. Representing the energy dependence of the GSF $g(\mathbf{r},\mathbf{p})$ by a series expansion over normalized Laguerre polynomials of the order $3/2$ (Sonine polynomials) $L_m^{(3/2)}(z)$, the solution to Eq. (2) is

$$g(\mathbf{r},\mathbf{p}) = \sum_{m=0}^{M} g_m(\varphi_s,\eta)L_m^{(3/2)}(z), \tag{3}$$

where $\eta = p_{\perp}^2/(p^2B)$ and $z = p^2/(2mT) = mv^2/(2T)$ are the normalized perpendicular adiabatic invariant and energy, respectively. The GSF expansion coefficients $g_m(\varphi_s,\eta)$ are computed by NEO-2 using a conservative finite difference scheme on a 2D grid over $\varphi_s$ and $\eta$ with an adaptively refined $\eta$-grid in order to resolve the trapped-passing boundary layer. The spatial dependence of $g$ on Boozer coordinates $(s,\vartheta,\varphi)$ required by TRAVIS is reconstructed by an interface with the help of linear interpolation on a non-uniform 3D grid formed by (discretized over $\varphi_s$) segments of (long enough) field lines representing a prescribed set of flux surfaces.

**Results**

In this work, two ECCD scenarii in W7-X, namely the X2- and the O2-scenario, are investigated for three different collision models, i.e., long mean free path (LMFP) limit by the internal

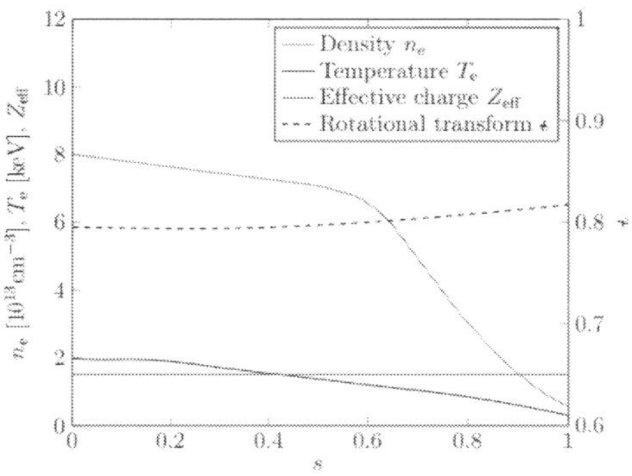


**Figure 1:** *Temperature, density, effective charge and rotational transform profiles over the normalized toroidal flux.*

TRAVIS model, finite collisionality by NEO-2 and the high-collisional limit using the classical Spitzer function [5]. Profiles of plasma parameters shown in Fig. 1 correspond to the onset of the LMFP regime (collisionality is roughly three times smaller than needed for the plateau regime). The microwave beam is launched off axis ($Z = -0.1$ m) in the horizontal plane with the poloidal launch angle $\alpha = 5.5$ deg (angle between this plane and the beam). Dependencies of the total current generated in the plasma on the toroidal launch angle $\beta$ (angle between the beam and the meridian plane containing the launcher) are shown in Fig. 2. It can be seen that in the standard X2-scenario the current is by 25% different from the LMFP limit, as it can be expected, while for the advanced O2-scenario the current significantly differs from both, LMFP and high collisionality limits. In the latter case, characterized by weak absorption and respectively smaller current, a significant part of this current is produced by particles from the trapped region. This is better seen from Fig. 3 where the distributions of the absorbed power and generated current along the microwave beam are shown for $\beta = 18.5$ deg. Due to the strong absorption in the X2 case, most power is absorbed by strongly passing particles, whose distribution function is not significantly different from the GSF in the LMFP limit, before the beam reaches the cold resonance. In the O2-scenario weak absorption results in power absorption and respective counter-current by passing particles on the other side of cold resonance, as well as in a significant amount of wave energy absorbed by trapped particles whose current is mainly determined by the position of the absorption region but not significantly by the parallel wave number [6, 2]. This can be used to increase the total toroidal current.

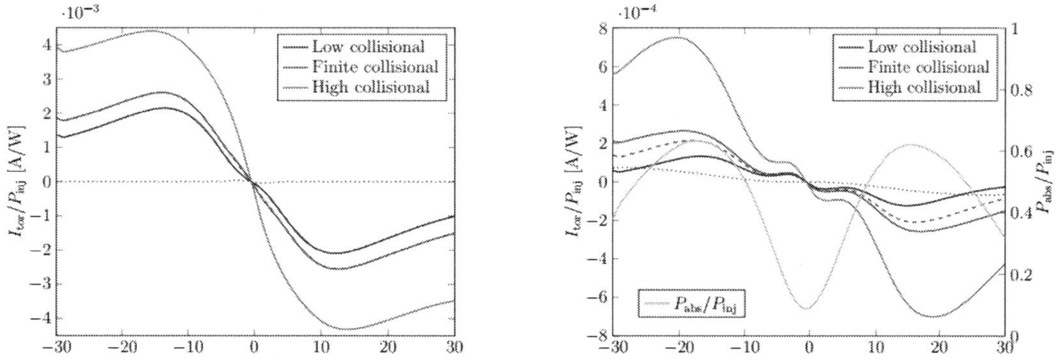


**Figure 2:** *Total toroidal current by passing (dashed), trapped (dotted) and all (solid) particles as function of the toroidal launch angle $\beta$ for X2-mode (left) and O2-mode (right). The fraction of the absorbed power $P_{abs}/P_{inj}$ is shown for the O2-mode only, since for the X2-mode there is total absorption.*

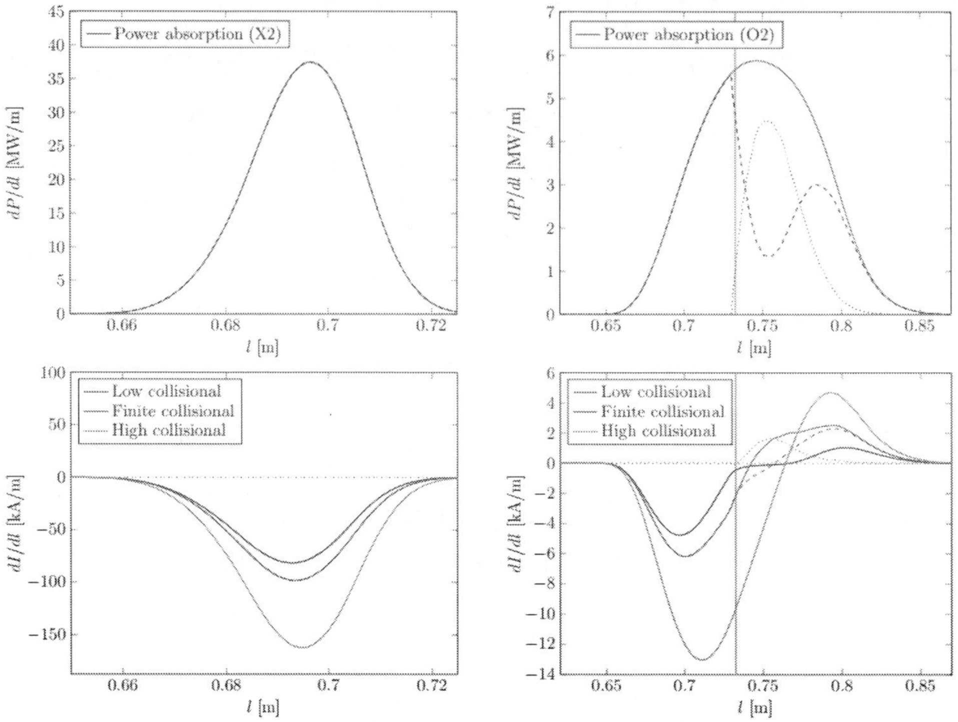


**Figure 3:** *Local power absorption (upper plots) and local current generation (lower plots) for X2-mode (left) and O2-mode (right) by passing (dashed), trapped (dotted) and all (solid) particles along the ray trajectory. Cold resonance position at $l = 0.73$ m is marked by a vertical (cyan) line.*

## Conclusion

This report extends the ECCD studies for tokamaks [2] to stellarator geometry. The studies of the X2- and O2-scenarii in a high-mirror configuration of W7-X at the onset of the LMFP regime show that in the O2-scenario finite collisionality effects play a significant role on ECCD current, which cannot be approximated by asymptotical collisionality limits. This effect is significant also in the X2-scenario, where the current is increased by 25%. Properties of ECCD on trapped particles can be used to improve the efficiency in advanced (O2, X3) scenarii, which are only available at high densities, where the cut-off prohibits the standard (O1, X2) scenarii.

## Acknowledgments

This work has been carried out within the framework of the EUROfusion Consortium and has received funding from the Euratom research and training programme 2014-2018 under grant agreement No 633053. The views and opinions expressed herein do not necessarily reflect those of the European Commission. The authors gratefully acknowledge support from NAWI Graz.

## References

[1] M. Romé, *et al.*, *Plasma Phys. Control. Fusion* **40**(4), 511 (1998)
[2] W. Kernbichler, *et al.*, *EPJ Web of Conferences* **87**, 01006 (2015)
[3] N. B. Marushchenko, *et al.*, *Comput. Phys. Commun.* **185**(1), 165 (2014)
[4] T. M. Antonsen and K. R. Chu, *Phys. Fluids* **25**(8), 1295 (1982)
[5] L. Spitzer, Jr. and R. Härm, *Phys. Rev.* **89**(5), 977 (1953)
[6] P. Helander and P. J. Catto, *Phys. Plasmas* **8**(5), 1988 (2001)

# The interaction between neutral particles and turbulent plasma in the tokamak SOL

C. Wersal, P. Ricci, F.D. Halpern, F. Riva

*Ecole Polytechnique Fdŕale de Lausanne (EPFL), Centre de Recherches en Physique des Plasmas (CRPP), Lausanne, Switzerland*

**Abstract**

The first-principle understanding of the processes in the Scrape-Off-Layer (SOL) of a tokamak is crucial for the development of a thermonuclear reactor. Since the plasma temperature in the SOL is rather low, the plasma is typically not fully ionized, and the neutral atoms play an important role in determining the SOL regimes. We have derived a kinetic model for neutral atoms in the SOL that contains the fundamental elements of neutral dynamics, while remaining relatively simple. The model has been coupled to the drift-reduced Braginskii equations and is implemented in GBS [1], a three-dimensional numerical code developed to simulate SOL turbulence. Details of the neutral model and the interactions with the plasma are given and we present first results indicating a transition into a regime that shows typical signatures of the conduction limited regime, e.g. significant parallel temperature gradients.

**The neutral and plasma model**

We describe the dynamics of the distribution function of a single mono-atomic neutral species, $f_{\rm n}$, by using the following kinetic equation

$$\frac{\partial f_{\rm n}}{\partial t} + \vec{v} \cdot \frac{\partial f_{\rm n}}{\partial \vec{x}} = -\nu_{\rm iz} f_{\rm n} - \nu_{\rm cx}\left(f_{\rm n} - \frac{n_{\rm n}}{n_{\rm i}} f_{\rm i}\right) \tag{1}$$

being $f_{\rm i}$, $n_{\rm n}$, and $n_{\rm i}$ the ion distribution function, the neutral density, and the ion density, respectively. The ionization and charge-exchange processes are described, respectively, through the use of Krook operators with collision frequencies defined as $\nu_{\rm iz} = n_{\rm e}\langle v_{\rm e}\sigma_{\rm iz}(v_{\rm e})\rangle$ and $\nu_{\rm cx} = n_{\rm i}\langle v_{\rm i}\sigma_{\rm cx}(v_{\rm i})\rangle$, where $\langle\rangle$ is the average over the electron and ion distribution function, respectively. In the present work, the effective reaction rates, $\langle v\sigma\rangle$, are taken from the OpenADAS database (http://open.adas.ac.uk).

We consider a single ion species plasma. We start our derivation of the drift-reduced Braginskii equations from the kinetic Boltzmann equation of ions and electrons, where we include collision terms in the form of Krook operators to describe the interaction with the neutrals. The kinetic equation for the ions is

$$\frac{\partial f_{\rm i}}{\partial t} + \vec{v} \cdot \frac{\partial f_{\rm i}}{\partial \vec{x}} + \vec{a} \cdot \frac{\partial f_{\rm i}}{\partial \vec{v}} = \nu_{\rm iz} f_{\rm n} - \nu_{\rm cx}\left(\frac{n_{\rm n}}{n_{\rm i}} f_{\rm i} - f_{\rm n}\right) + C(f_{\rm i}) \tag{2}$$

while the kinetic equation for the electrons is

$$\frac{\partial f_e}{\partial t} + \vec{v} \cdot \frac{\partial f_e}{\partial \vec{x}} + \vec{a} \cdot \frac{\partial f_e}{\partial \vec{v}} = \nu_{iz} n_n \left[ 2\Phi_e(\vec{v}_n, T_{e,iz}) - \frac{f_e}{n_e} \right] + C(f_e) \tag{3}$$

where $\vec{a}$ is the particle acceleration due to the Lorentz force, $\Phi_e(\vec{v}, T)$ is a Maxwellian distribution function for electrons, and $C(f_i)$ and $C(f_e)$ are the Coulomb collision operators. To derive the ionization operator in Eq. (3), we assume that a fast electron is lost, and two lower-energy electrons appear with a Maxwellian distribution function isotropically in the neutral frame of reference, and a temeprature, $T_{e,iz}$, so that the total electron kinetic energy is reduced by the ionization energy, $E_{iz}$, giving $T_{e,iz} = T_e/2 - E_{iz}/3 + m_e v_e^2/6 - m_e v_n^2/3$, where $T_e$ and $v_e$ are the local electron temperature and fluid velocity respectively.

Following the work of Braginskii [3] and, e.g., Zeiler [4], we take the first three moments of the electron and ion kinetic equations in the limit $\omega_c \tau \gg 1$, where $\omega_c = qB/m$ is the gyro-frequency and $\tau$ the typical Coulomb collision time. Then, we simplify the hereby derived Braginskii equations in the drift limit, observing that $d/dt \ll \omega_{ci}$ for typical SOL turbulence. The resulting drift-reduced Braginskii equations are

$$\frac{\partial n}{\partial t} = -\frac{1}{B}[\phi, n] + \frac{2}{eB}[C(p_e) - enC(\phi)] - \nabla_\parallel(n v_{\parallel e}) + \mathscr{D}_n(n) + S_n + n_n \nu_{iz} \tag{4a}$$

$$\frac{\partial \tilde{\omega}}{\partial t} = -\frac{1}{B}[\phi, \tilde{\omega}] - v_{\parallel i}\nabla_\parallel \tilde{\omega} + \frac{m_i \omega_{ci}^2}{e^2 n}\nabla_\parallel j_\parallel + \frac{2B}{c m_i n}C(p) + \mathscr{D}_{\tilde{\omega}}(\tilde{\omega}) - \frac{n_n}{n}\nu_{cx}\tilde{\omega} \tag{4b}$$

$$\begin{aligned}\frac{\partial v_{\parallel e}}{\partial t} + \frac{e}{m_e c n}\frac{\partial \Psi}{\partial t} &= -\frac{1}{B}[\phi, v_{\parallel e}] - v_{\parallel e}\nabla_\parallel v_{\parallel e} + \frac{e}{\sigma_\parallel m_e} j_\parallel + \frac{e}{m_e}\nabla_\parallel \phi - \frac{T_e}{m_e n}\nabla_\parallel n \\ &- \frac{1.71}{m_e}\nabla_\parallel T_e + \mathscr{D}_{v_{\parallel e}}(v_{\parallel e}) + 2\frac{n_n}{n}\nu_{iz}(v_{\parallel n} - v_{\parallel e})\end{aligned} \tag{4c}$$

$$\frac{\partial v_{\parallel i}}{\partial t} = -\frac{1}{B}[\phi, v_{\parallel i}] - v_{\parallel i}\nabla_\parallel v_{\parallel i} - \frac{1}{m_i n}\nabla_\parallel p + \mathscr{D}_{v_{\parallel i}}(v_{\parallel i}) + \frac{n_n}{n}(\nu_{iz} + \nu_{cx})(v_{\parallel n} - v_{\parallel i}) \tag{4d}$$

$$\begin{aligned}\frac{\partial T_e}{\partial t} &= -\frac{1}{B}[\phi, T_e] - v_{\parallel e}\nabla_\parallel T_e + \frac{4T_e}{3eB}\left[\frac{T_e}{n}C(n) + \frac{7}{2}C(T_e) - eC(\phi)\right] + \frac{2T_e}{3e}\left[\frac{0.71}{en}\nabla_\parallel j_\parallel - \nabla_\parallel v_{\parallel e}\right] \\ &+ \mathscr{D}_{T_e}(T_e) + \mathscr{D}_{T_e}^{\parallel}(T_e) + S_{T_e} + \frac{n_n}{n}\nu_{iz}\left[-\frac{2}{3}E_{iz} - T_e + m_e v_{\parallel e}\left(v_{\parallel e} - \frac{4}{3}v_{\parallel n}\right)\right]\end{aligned} \tag{4e}$$

$$\begin{aligned}\frac{\partial T_i}{\partial t} &= -\frac{1}{B}[\phi, T_i] - v_{\parallel i}\nabla_\parallel T_i + \frac{4T_i}{3eB}\left[\frac{T_e}{n}C(n) - \frac{10}{3}C(T_i) - eC(\phi)\right] + \frac{2T_i}{3e}\left[(v_{\parallel i} - v_{\parallel e})\frac{\nabla_\parallel n}{n} - \nabla_\parallel v_{\parallel e}\right] \\ &+ \mathscr{D}_{T_i}(T_i) + \mathscr{D}_{T_i}^{\parallel}(T_i) + S_{T_i} + \frac{n_n}{n}(\nu_{iz} + \nu_{cx})\left[T_n - T_i + \frac{1}{3}(v_{\parallel n} - v_{\parallel i})^2\right]\end{aligned} \tag{4f}$$

with $p = n(T_e + T_i)$ the total pressure. The generalized vorticity, $\tilde{\omega} = \omega + 1/e\nabla_\perp^2 T_i$, is related to the electrostatic potential by $\nabla_\perp^2 \phi = \omega$, while $\nabla_\perp^2 \Psi = j_\parallel$. The following operators have been introduced $\nabla_\parallel f = \mathbf{b}_0 \cdot \nabla f$, $[A, B] = \hat{b} \cdot (\nabla A \times \nabla B)$, and $C(A) = \frac{B}{2}\left(\nabla \times \frac{\hat{b}}{B}\right) \cdot \nabla A$. The details of the model are presented in Ref. [2], where the boundary conditions and aditional collision processes are discussed.

**Solution of the neutral kinetic equation in typical SOL relevant parameters**

We now solve the kinetic advection equation for the neutrals, Eq. (1), by using the method of characteristics, under the assumption that plasma-related quantities are known. We consider two approximations, valid in the typical SOL parameter regime, which considerably simplify the solution, that is $\tau_{\text{neutral flight time}} < \tau_{\text{turbulence}}$ and $\lambda_{\text{neutral mean free path}} \ll 1/k_{\parallel,\text{plasma}}$, leading to a steady state solution of the neutral kinetic equation independently on each poloidal plane. Within these two approximations, the formal solution of the neutral kinetic equation, Eq. (1), is

$$f_n(\vec{x}_\perp,\vec{v}) = \int_0^{r_{\perp b}} \frac{\nu_{cx}(\vec{x}'_\perp)n_n(\vec{x}'_\perp)\Phi_i(\vec{x}'_\perp,\vec{v})}{v_\perp} \exp\left[-\frac{1}{v_\perp}\int_0^{r'_\perp} \nu_{cx}(\vec{x}''_\perp)+\nu_{iz}(\vec{x}''_\perp)dr''_\perp\right]dr'_\perp \quad (5)$$

where $\vec{x}'_\perp = \vec{x} - r'_\perp\hat{\Omega}_\perp$ and $\hat{\Omega}_\perp = \vec{v}_\perp/v_\perp$. For better readability, we do not include the explicit notation of the $t$ and $x_\parallel$ parametric dependencies. A linear integral equation for $n_n(\vec{x}_\perp)$ is obtained by integrating Eq. (5) in velocity space, which is

$$n_n(\vec{x}_\perp) = \int_D n_n(\vec{x}'_\perp)\nu_{cx}(\vec{x}'_\perp)K_{p\to p}(\vec{x}_\perp,\vec{x}'_\perp)dA', \quad (6)$$

where we have rearranged the integrals, $dA'$ is the infinitesimal area of $D$, which is the optically to $\vec{x}_\perp$ connected domain, and where the following kernel function has been defined

$$K_{p\to p}(\vec{x}_\perp,\vec{x}'_\perp) = \int_0^\infty \frac{1}{r'_\perp}\Phi_{\perp i}(\vec{x}'_\perp,\vec{v}_\perp)e^{-\frac{1}{v_\perp}\int_0^{r'_\perp}\nu_{\text{eff}}(\vec{x}''_\perp)dr''_\perp}dv_\perp, \quad (7)$$

in which $\Phi_{\perp i}(\vec{x}_\perp,\vec{v}_\perp) = \int \Phi_i(\vec{x}_\perp,\vec{v})dv_\parallel$. We remark that the kernel does neither depend on $f_n(\vec{x}_\perp,\vec{v})$, nor on any of its moments. After numerically solving Eq. (6), the distribution function of the neutral atoms, $f_n(\vec{x}_\perp,\vec{v})$, can be readily evaluated by using Eq. (5).

**First plasma turbulence simulations with self-consistent neutral dynamics**

The neutral model derived in this paper has been used to perform the first simulations of SOL plasma turbulence that include self-consistently the neutral dynamics. A newly developed version of the GBS code has been used for this purpose. We compare here a low plasma density simulation, $n_0 = 5\cdot10^{18}\text{m}^{-3}$, with a high plasma density simulation, $n_0 = 5\cdot10^{19}\text{m}^{-3}$. Both simulations consider a limited SOL geometry, with a toroidal limiter on the high field equatorial midplane, $R/\rho_{s0} = 500$, $m_i/m_e = 400$, $2\pi a = 800\rho_{s0}$, $a$ being the minor radius, and $T_{e0} = 10\text{eV}$.

The poloidal dependence of the relevant plasma quantities for the low- and high-density simulations are shown in Fig. 1. The displayed profiles are averaged over a time window of 5 $R/c_{s0}$, over the full toroidal angle, and over a radial region extending for 20 $\rho_{s0}$, just outside the separatrix. We point out a few interesting differences between the high- and low-density simulations. The normalized density in the high-density simulation is flatter than in the low-density simulation. The plasma source due to ionization inside the SOL is much higher in the

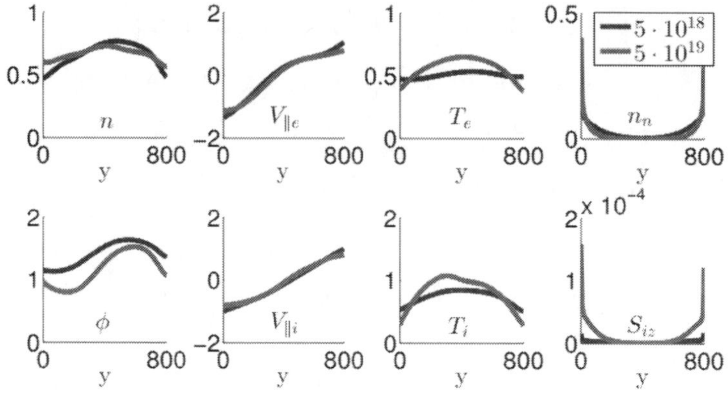


Figure 1: Time-averaged poloidal profiles of $n$, $\Phi$, $V_{\|\mathrm{e}}$, $V_{\|\mathrm{i}}$, $T_\mathrm{e}$, $T_\mathrm{i}$, $n_\mathrm{n}$, and $S_\mathrm{iz}$ for the low (blue) and high (red) plasma density scenario.

high-density simulation, so that there is a higher plasma source close to the limiter, preventing the plasma density to drop in this region. The parallel velocity profiles (which are expected to be approximately linear if the plasma source is poloidally constant) are somewhat flatter close to the limiters in the high-density scenario, however, the flattening is not particularly significant, because a relatively large fraction of the plasma density source is still due to the poloidally constant outflow of particles from the core. Both electron and ion parallel temperature gradients increase in the high-density scenario, which results in a higher amount of heat conduction as compared to the low-density simulation.

## Acknowledgments

Part of the simulations presented herein were carried out using the HELIOS supercomputer system at Computational Simulation Centre of International Fusion Energy Research Centre (IFERC-CSC), Aomori, Japan, under the Broader Approach collaboration between Euratom and Japan, implemented by Fusion for Energy and JAEA; and part were carried out at the Swiss National Supercomputing Centre (CSCS) under Project ID s549. This work has been carried out within the framework of the EUROfusion Consortium and has received funding from the Euratom research and training programme 2014-2018 under grant agreement No 633053. The views and opinions expressed herein do not necessarily reflect those of the European Commission.

## References

[1] P. Ricci, et al., Plasma Phys. Control. Fusion **54**, 124047 (2012)

[2] C. Wersal, et al., Submitted to Nuclear Fusion, May 2015

[3] S. I. Braginskii, Reviews of Plasma Physics **1** 205 (1965)

[4] A. Zeiler, et al., Phys. Plasmas, **4**(6) 2134 (1997)

# Real-time capable first principle based modelling of tokamak turbulent transport

J. Citrin[1,2], S. Breton[2], F. Felici[3], F. Imbeaux[2]
T. Aniel[2], J.F. Artaud[2], B. Baiocchi[4], C. Bourdelle[2], Y. Camenen[5], J. Garcia[2]
[1]*FOM Institute DIFFER – Dutch Institute for Fundamental Energy Research, P.O. box 6336, 5600 HH Eindhoven, The Netherlands*
[2]*CEA, IRFM, F-13108 Saint Paul Lez Durance, France*
[3]*Eindhoven University of Technology, Department of Mechanical Engineering, Control Systems Technology Group, P.O. Box 513, 5600MB, Eindhoven, The Netherlands*
[4]*Istituto di Fisica del Plasma "P. Caldirola", Associazione Euratom-ENEA-CNR, Milano, Italy*
[5]*Aix-Marseille Université, CNRS, PIIM UMR 7345, 13397 Marseille, France*

An accurate predictive model for turbulent transport fluxes driven by microinstabilities is vital for the interpretation and optimization of present-day experiments, and extrapolation to and control of future machines. However, the computational cost of direct numerical simulation with massively parallel nonlinear gyrokinetic codes, $10^4 - 10^5$ CPUh for fluxes at a single radius, precludes their use for integrated tokamak transport simulations.

Increased tractability is gained by applying transport models based on the quasilinear approximation, largely valid in the core of tokamak plasmas [1, 2]. These are validated by comparison to nonlinear simulations, and have proven successful in reproducing experimental profiles in many cases. Compared to nonlinear simulations, a $\sim$6 orders of magnitude speedup is gained. However, their computational speed is still insufficient for applications such as convenient large-scale scenario development, trajectory optimization, and simulations for developing real-time controllers.

We suggest an approach to overcome these challenges. The central point is to emulate the original transport model with a neural network (NN) nonlinear regression of quasilinear fluxes previously compiled in a database. The neural network emulator is then orders of magnitude faster than original flux calculation, and can be used for real-time applications. Neural networks have been previously applied for regression of DIII-D heat fluxes from experimental power balance databases [3]. In this work, we suggest to apply the same methodology for code output.

The QuaLiKiz quasilinear gyrokinetic transport model [4, 5] was employed in this work. The computational time for the QuaLiKiz eigenvalue solver at a single wavenumber is $\sim$1 s.

A database of QuaLiKiz solutions was constructed, in the ion temperature gradient (ITG) instability regime. The code was run with adiabatic electrons for simplicity. The database covers

Table 1: Summary of input parameters for the QuaLiKiz adiabatic electron ITG database employed in this work

| Parameter | Min value | Max value | No. of points |
|---|---|---|---|
| $R/L_{Ti}$ | 2 | 12 | 30 |
| $T_i/T_e$ | 0.3 | 3 | 20 |
| $q$ | 1 | 5 | 20 |
| $\hat{s}$ | 0.1 | 3 | 20 |
| $k_\theta \rho_s$ | 0.05 | 0.8 | 16 |
| Total no. of points | | | 3 840 000 |

five input parameters: the driving normalized logarithmic ion temperature gradient $R/L_{Ti}$, the ion to electron temperature ratio $T_i/T_e$, the safety-factor $q$, the magnetic shear $\hat{s} \equiv \frac{r}{q}\frac{dq}{dr}$, and the normalized wavenumber $k_\theta \rho_s$, where $\rho_s \equiv \sqrt{T_e m_i}/(Z_i q_e B)$. The database content is summarized in table 1. The training sets for the neural network were sifted from this database. Outputs include growth rates, frequencies, and ion heat flux. We concentrate on the analysis of the ion heat flux output, which involves a summation over the wavenumbers in the construction of the quasilinear saturation rule.

A multilayer perceptron neural network is used, which is a nonlinear function with tunable variables (weights and biases), with the property of universal approximation [6]. Linear combinations of the inputs and biases are propagated through a series of nonlinear transfer function vectors (named 'hidden layers'), until eventually linearly combined to an output layer. With two hidden layers and a single output value (as used in this work), this is represented as:

$$y = b_3 + \sum_i^N w_i^2 g\left(b_i^2 + \sum_j^M w_{ij}^1 g\left(b_j^1 + \sum_k^I w_{jk}^{in} x_k\right)\right) \tag{1}$$

Where $y$ is the output 'neuron' containing the output value (e.g. ion heat flux), $x_k$ the vector of input values, $b^n$ the bias vectors, $w^{in}$ the M$\times$I weight matrix connecting the input vector to the 1st hidden layer, $w^1$ the N$\times$M weight matrix connecting the two hidden layers, and $w^2$ the weight vector connecting the 2nd hidden layer to the output neuron. $g$ is the nonlinear transfer function, defined as a sigmoid in this work:

$$g(x) = \frac{2}{1+e^{-2x}} - 1 \tag{2}$$

Following a series of optimization tests, two hidden layers, as shown in equation 1, were employed here. The hidden layer sizes M and N were set to 40. The input layer size, I, is 4 for ion heat fluxes, and 5 for growth rates and frequencies.

The key stage is the determination of the optimized values of the weights and biases through fitting a 'training set' of predetermined mappings, taken from a subset of 35000 QuaLiKiz database outputs. Following training, the network output then emulates the original model within the database input parameter envelope. This is validated by comparison to validation sets sifted from the database, chosen different from the training set. To avoid overfitting the data, regularization techniques were used in the optimization.

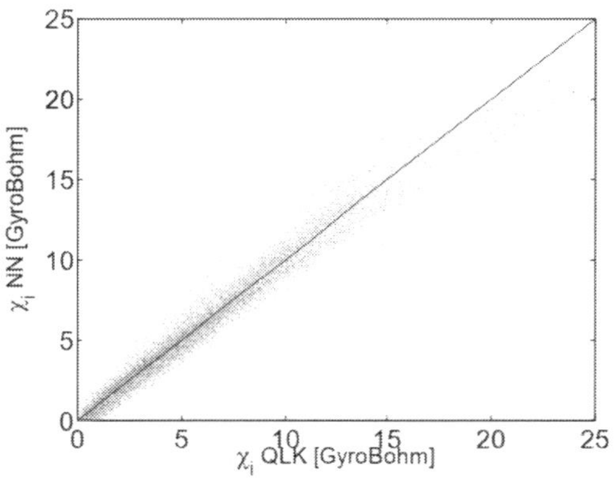


Figure 1: *Comparison between normalized ion heat fluxes obtained directly from QuaLiKiz (x-axis) and those from its NN regression (y-axis)*

A comparison between the regression NN and QuaLiKiz outputs for a validation set of 10000 unstable cases is shown in figure 1. The regression network has an RMS error of 0.77 in gyroBohm units ($\chi_{iGB} = \frac{T_i^{3/2} m_i^{1/2}}{(Z_i q_e B)^2 r}$) when compared to the validation set. This RMS error is primarily due to the regularization constraint. The impact of this error on the simulated profiles is minor, due to stiffness, and corresponds to a $\Delta(R/L_{Ti}) = 0.29$.

The typical quality of the fits can be seen in figure 2, displaying scans of the 4 separate input parameters while the others remained fixed. Each NN output is calculated on a sub 10 $\mu s$ timescale in MATLAB on a Intel(R) Xeon(R) E5450 CPU @ 3.00GHz. This is a 5 order of magnitude speedup compared to the original QuaLiKiz calculations.

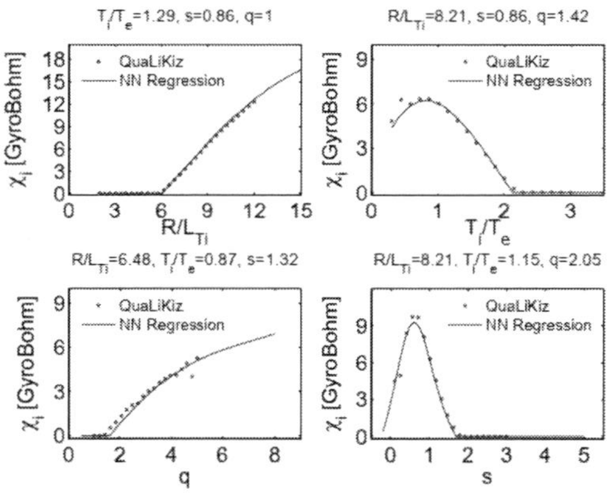


Figure 2: *Comparison of NN parameter scans (blue solid lines) vs the original QuaLiKiz ion heat flux calculations (red dots). The scans are in $R/L_{Ti}$ (top left panel), $T_i/T_e$ (top right panel), q (bottom left panel) and $\hat{s}$ (bottom right panel)*

A transport model based on the trained neural network was implemented both in the CRONOS [7] and RAPTOR [8] integrated modelling codes. We focus here on the real-time simulation capabilities offered in RAPTOR.

Presently, RAPTOR only models electron heat transport. The NN model output was thus modified to roughly approximate ITG regime electron heat transport by setting a constant $q_i/q_e = 3$. This is based on typical nonlinear and quasilinear observations in the ITG regime [9, 2].

In figure 3, we compare a RAPTOR simulation of an ITER hybrid scenario, using the QuaLiKiz NN model for electron heat transport, with a simulation of the same case originally carried out [10] using CRONOS and the GLF23 [11] transport model. A RAPTOR simulation of an entire 300 s ITER discharge took 10 s on a single CPU. This combination of simulation speed and first-principle modelling is unprecedented. With CRONOS/GLF23, the simulation took 24 hours.

This computational speed opens up many new possibilities for real-time controller design and validation, scenario preparation and optimization, and real-time discharge supervision. This initial QuaLiKiz neural network emulator is a proof-of-principle, and there remains much scope for expanding the number of input dimensions in the databases used for the fits, as well as employing slower yet more complete linear gyrokinetic codes for populating the database. This work is ongoing.

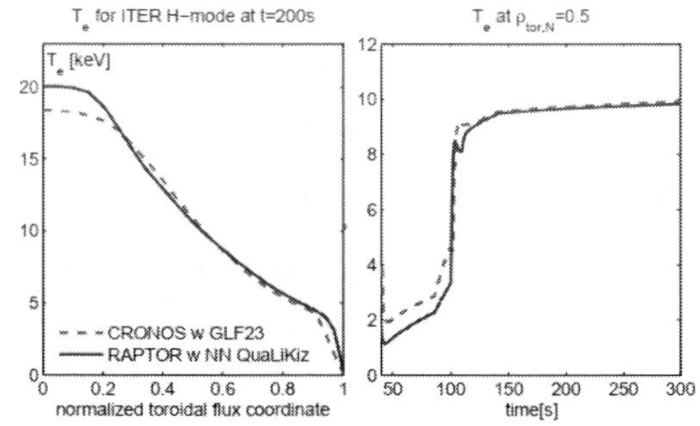


Figure 3: *Comparison between $T_e$ predictions for an ITER hybrid discharge carried out with CRONOS/GLF23 [10] (red curve) and a RAPTOR simulation using the QuaLiKiz NN transport model (blue curve). A typical H-mode profile (left panel) and time dependence at mid-radius (right panel) are shown. The LH transition was set at 100 s.*

*Acknowledgments.–* This work is part of the research programme 'Fellowships for Young Energy Scientists' (YES!) of the Foundation for Fundamental Research on Matter (FOM), which is financially supported by the Netherlands Organisation for Scientific Research (NWO). The authors greatly thank O. Meneghini, S. Smith, U. von Toussaint and P. Xanthopolous for inspiring discussions.

## References

[1] T. Dannert and Frank Jenko *Phys. Plasmas* **12** 072309 (2005)

[2] A. Casati *et al. Nucl. Fusion* **49** 085012 (2009)

[3] O. Meneghini, C. J. Luna, S. P. Smith, and L. L. Lao *Phys. Plasmas* **21** 060702 (2014)

[4] C. Bourdelle *et al. Phys. Plasmas* **14** 112501 (2007)

[5] C. Bourdelle. Turbulent Transport in Tokamak Plasmas: bridging theory and experiment. Physics. Aix Marseille Université, (2015). https://tel.archives-ouvertes.fr/tel-01113299

[6] Christopher M Bishop *Neural networks for pattern recognition.* Clarendon press Oxford (1995)

[7] J.F. Artaud *et al. Nucl. Fusion* **50** 043001 (2010)

[8] F. Felici and O. Sauter *Plasma Phys. Control. Fusion* **54** 025002 (2012)

[9] J.E Kinsey, R.E. Waltz and J. Candy *Phys. Plasmas* **12** 062302 (2005)

[10] J. Citrin *et al., Nucl. Fusion* **50** 115007 (2010)

[11] R.E. Waltz, G.M. Staebler, W. Dorland and G.W. Hammett *Phys. Plasmas* **4** 2482 (1997)

# The influence of hot neutrals in simulations of gas puff modulation

A.S. Christensen[1], J. Madsen[1], V. Naulin[1], J.J. Rasmussen[1],
A. Salmi[2], L.E.H. Tophøj[1] and JET contributors*[3]
[1] *PPFE, DTU Physics, Technical University of Denmark, Denmark,* [2] *VTT, Espoo, Finland,*
[3] *EUROfusion Consortium, JET, Culham Science Centre, Abingdon, OX14 3DB, UK*

## Introduction

Experiments utilizing the gas puff modulation technique have been carried out in JET for the purpose of studying plasma fueling as well as particle transport [1]. It is of great interest to simulate such experiments, as this allows for deeper insight in the mechanics that determines the fueling rate when applying gas puffing, as well as providing knowledge of the dynamics of neutral particles. This has to some extend been achieved in numerical simulations, which have proved successful in reproducing profiles for the modulation experiments in [1].

In this paper we make progress towards a simple to use model which allows to explore parameter space in a dynamical fashion and can treat modulation experiments in a meaningful way. We present numerical investigations on how the gas puff modulation characteristics are affected by the neutral particle model. Two 1D models are studied; the first model includes only one species of neutral atoms originating from the gas puffing, whereas the second model also accounts for the hot neutrals created in charge exchange collisions between hot plasma and the cold neutrals ($ion_{\mathrm{hot}} + neutral_{\mathrm{cold}} \underset{\mathrm{cx}}{\rightarrow} neutral_{\mathrm{hot}} + ion_{\mathrm{cold}}$).

It is found that the neutral model does not affect the transport of the modulated signal significantly, but the amplitude of the modulation is several times larger in the two-species neutral model. Also, the hot neutrals are seen to penetrate deeper into the edge, which along with the increased amplitude for the hot neutral model indicates a significant difference in fueling in the two models.

## Models

To simulate the transport of heat and particles, we apply a system of 1D transport equations [2]. The density equation for the quasi-neutral plasma species reads

$$\partial_t n_{\mathrm{p}} + \tfrac{1}{r}\partial_r \left(r\Gamma_{\mathrm{p}}\right) = S_{\mathrm{p}}, \qquad \Gamma_{\mathrm{p}} = -D_{\mathrm{p}}\partial_r n_{\mathrm{p}} + v n_{\mathrm{p}}, \tag{1}$$

*See the Appendix of F. Romanelli et al., Proceedings of the 25th IAEA Fusion Energy Conference 2014, Saint Petersburg, Russia

where $n_p$ is the plasma density, $v$ is vthe convective velocity, $D_\mathrm{p} = D_\mathrm{p}(\chi_\mathrm{e})$ is the diffusion coefficient and $S_p$ is the plasma particle source. The electron temperature equation reads

$$\tfrac{3}{2}\partial_t\left(n_\mathrm{p}T_\mathrm{e}\right) + \tfrac{1}{r}\partial_r\left(r\left[q_\mathrm{e} + \tfrac{5}{2}T_\mathrm{e}\Gamma_\mathrm{p}\right]\right) = Q_\mathrm{e} + \frac{\Gamma_\mathrm{p}}{n_\mathrm{p}}\partial_r\left(n_\mathrm{p}T_\mathrm{e}\right), \qquad \frac{q_\mathrm{e}}{n_\mathrm{p}} = -\chi_\mathrm{e}\partial_r T_\mathrm{e} + uT_\mathrm{e}, \quad (2)$$

where $u$ is the electron heat convection velocity, $\chi_\mathrm{e}$ the heat diffusivity and $Q_\mathrm{e}$ is the heat source. In order to account for turbulent effects, the electron heat diffusivity is described by a critical electron temperature gradient model [3], such that the heat diffusion grows when the electron temperature gradient exceeds a critical level $\kappa$

$$\chi_\mathrm{e} = \chi_{\mathrm{e}0} + \lambda T_\mathrm{e}^\alpha \left[\frac{|\nabla T_\mathrm{e}|}{T_\mathrm{e}} - \kappa\right]^\beta H\left(\frac{|\nabla T_\mathrm{e}|}{T_\mathrm{e}} - \kappa\right), \quad (3)$$

where $H$ is the Heaviside step function, $\chi_{\mathrm{e}0}$ is the collisional heat diffusion coefficient and $\lambda$ a coefficient.

Two models for the neutral particles are applied. In the two-species model the neutral population consists of a cold neutral species which enters the system by gas puffing, and a hot neutral species which is created in charge exchange collisions between hot plasma and cold neutrals. Both hot and cold neutral densities $n_{\mathrm{n_{c,h}}}$ are described by a diffusion model

$$\partial_t n_{\mathrm{n_{c,h}}} + \tfrac{1}{r}\partial_r\left(r\Gamma_{\mathrm{n_{c,h}}}\right) = S_{\mathrm{n_{c,h}}}, \qquad \Gamma_{\mathrm{n_{c,h}}} = -D_{\mathrm{n_{c,h}}}\partial_r n_{\mathrm{n_{c,h}}}, \quad (4)$$

where $D_{\mathrm{n_{c,h}}}$ are the diffusion coefficients.

The plasma and neutral equations are coupled through the sources

$$S_\mathrm{p} = \left(n_\mathrm{n}k_\mathrm{iz} - n_{\mathrm{n_c}}k_\mathrm{cx}\right)n_\mathrm{p}, \qquad Q_\mathrm{e} = -k_\mathrm{iz}n_\mathrm{n}T_\mathrm{iz}, \quad (5)$$

$$S_{\mathrm{n_c}} = -\left(k_\mathrm{cx} + k_\mathrm{iz}\right)n_{\mathrm{n_c}}n_\mathrm{p}, \qquad S_{\mathrm{n_h}} = \left(n_{\mathrm{n_c}}k_\mathrm{cx} - n_{\mathrm{n_h}}k_\mathrm{iz}\right)n_\mathrm{p}, \quad (6)$$

where $n_\mathrm{n} \equiv n_{\mathrm{n_c}} + n_{\mathrm{n_h}}$, $T_{iz} = 13.6\,\mathrm{eV}$ is the ionization energy and $k_\mathrm{iz}$ and $k_\mathrm{cx}$ are the ionization and charge exchange rates which both depend on the plasma temperature [4]. External plasma and cold neutral sources are imposed through boundary conditions, and in particular the gas puffing is simulated by modulating the cold neutral source on the wall.

The one-species model for neutrals is similar to the above, but the charge exchange reaction rate $k_\mathrm{cx}$ is quenched such that no hot neutrals are created.

**Results**

By solving the system of equations (1)-(6) numerically we obtain the neutral profiles shown in Figure 1, modulated plasma density profiles in Figure 2 and phase shift and amplitude for the modulation in Figure 3. The plots for experimental data shown in Figures 2 and 3 are constructed from a measured plasma density profile in phase 1 of JET shot #87425 with GIM7 modulation (for details, see [5, 6]). These, along with experimental electron temperature data, were also used when the free parameters of the model were estimated.

In Figure 1 it is seen, that in both models the cold neutrals have a small penetration depth and the densities vanish inside the last-closed-flux-surface (LCFS) which lies at approximately at $R = 3.80\,\mathrm{m}$. The hot neutrals have a much lower density at the wall (at $R = 3.864\,\mathrm{m}$), but penetrate far deeper into the plasma edge. This indicates that hot neutrals may play a significant role in plasma fueling.

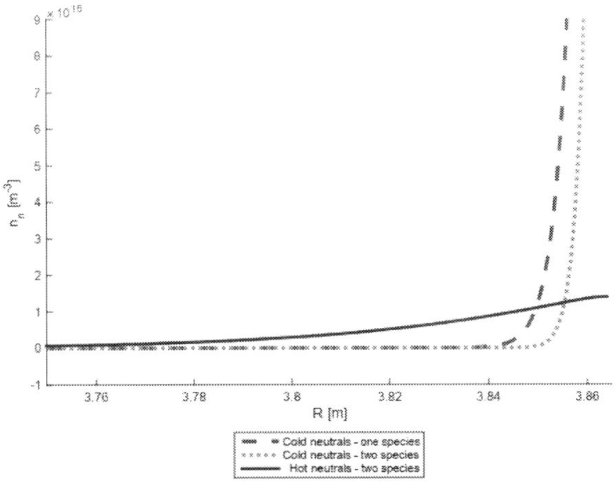


Figure 1: Neutral profiles in steady state, shown both for the one-species model (dashed), and for the two-species model with cold (dotted) and hot (full) neutrals for JET geometry.

Figure 2 shows the modulation of the density as a function of time. Both simulations mimic the plasma behavior to gas puffing well by reproducing a modulation phase shift which increases, and an amplitude which decreases when approaching the center. The phase shift and amplitude are depicted in Figure 3. This shows that the calculated phase shift only depends weakly on which of the two models is applied, and that the ionization source region is deeper for the model which includes hot

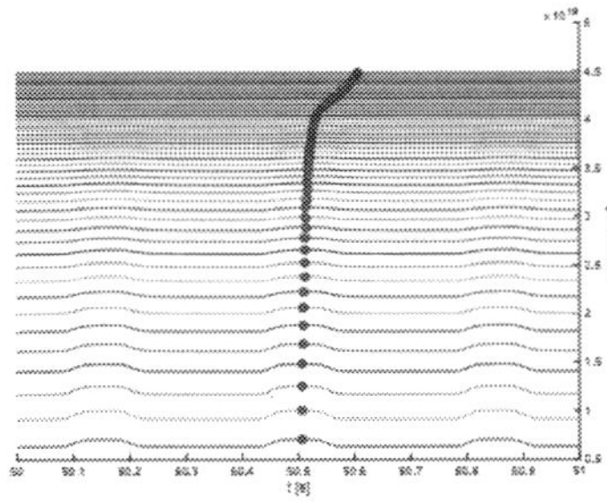

(a) One neutral species.

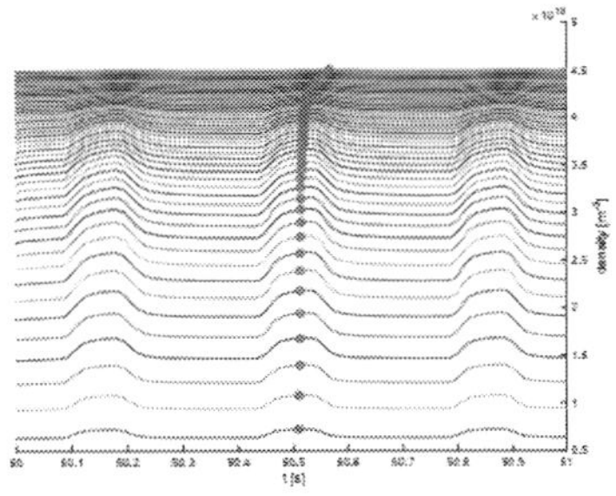

(b) Two neutral species.

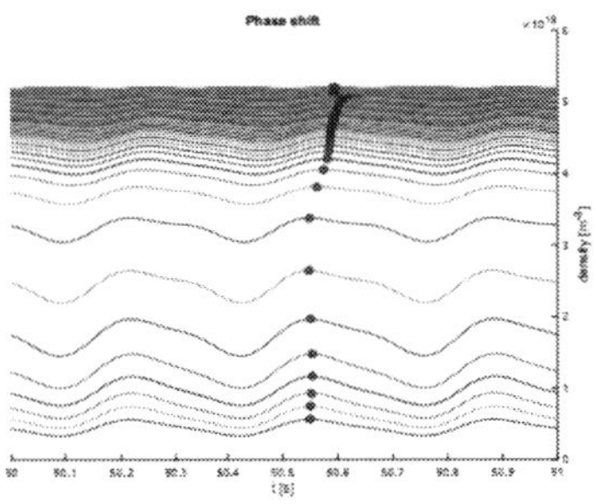


(c) Experimental data.

Figure 2: Simulations of and experimental data for gas puff modulation of plasma density. One of the perturbations is tracked and marked by colored circles.

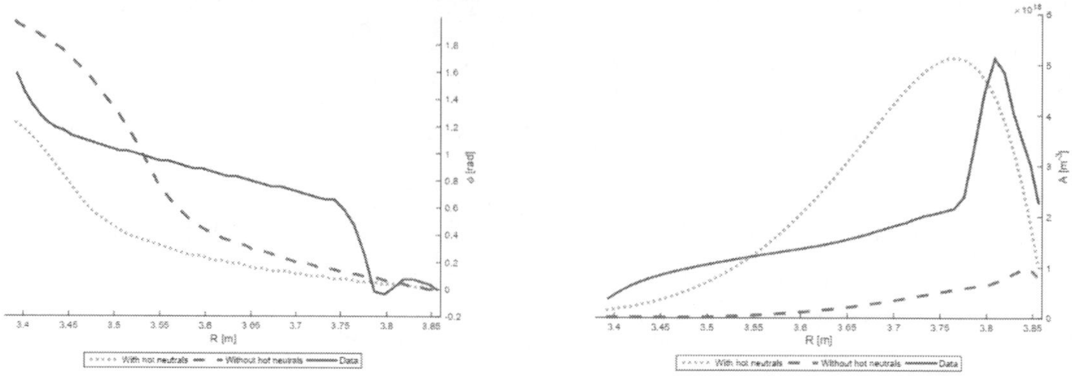


(a) Phase shift. (b) Amplitude.

Figure 3: Phase shift and amplitude of plasma density modulation for one-species model (dashed), two-species model (dotted) and experimental data (full).

neutrals. The latter is concluded from the flatter phase shift of the two-neutral-species model in the outer region. It is also observed that the amplitude of the modulation is several times larger for the model that includes hot neutrals. This effect is due to the deeper penetration of the hot neutrals inside LCFS.

**Discussion and conclusion**

The effect on the gas puff modulation characteristics are investigated for two different models for neutral particles. A comparison shows that the phase shift in the plasma density modulation is similar for the two neutral models, whereas the amplitude of the modulation is significantly larger for the model which includes hot neutrals produced in charge exchange collisions.

The weak influence on the phase shift suggests that the choice of neutral model is not significant for the plasma particle transport. In contrast the strong influence that the choice of neutral model has on the amplitude of the density modulation indicates that the fueling rate inside LCFS is very dependent on whether hot neutrals are included or not.

*This work has been carried out within the framework of the EUROfusion Consortium and has received funding from the Euratom research and training programme 2014-2018 under grant agreement No 633053. The views and opinions expressed herein do not necessarily reflect those of the European Commission.*

**References**

[1] A. Salmi et al. 41st EPS conference on Plasma Physics, Berlin, 2014, P1.008

[2] P. Mantica, F. Ryter, C. R. Physique 7 (2006)

[3] F. Imbeaux et al, Plasma Phys. Control. Fusion 43 (2001)

[4] P.C. Stangeby, IoP Publishing, (2000)

[5] A. Salmi et al. 42nd EPS conference on Plasma Physics, Lisbon, 2015, conference poster

[6] T.T.J. Tala, 20th joint US-EU TTF workshop, Salem, 2015, conference oral presentation

# Kinetic resonances effect on magnetic braking in tokamaks

I.G. Miron

*National Institute for Lasers, Plasma and Radiation Physics, Magurele-Bucharest, Romania*

The investigation of tokamak plasma kinetic effects such as the particles precession and bounce frequencies resonance effect on the dynamics of the plasma rotation under different collisionality regimes is the goal of this work.

It has been found that the kinetic processes mentioned above increase the plasma neoclassical toroidal viscosity (NTV) effect on the toroidal rotation of the plasma when peculiar conditions are fulfilled. NTV is due to the breaking of the magnetic field axisymmetry that enhances the radial nonambipolar particle transport and hence increases the NTV braking torque that finally damps the plasma rotation. When the particles electric, magnetic and bounce frequencies resonate, the nonambipolar radial particle flux increases and the plasma toroidal rotation is affected. Starting from the CGL expression of the pressure tensor, $\mathbf{p} = p_{\|}\mathbf{nn} + p_{\perp}(\mathbf{I}-\mathbf{nn}) = p\mathbf{I} + \pi$ (**n**=**B**/B, **B** is the equilibrium magnetic field), we have derived the 3D parallel (to **B**) perturbed pressure tensor term :

$$p_{\|} = \sum_{\mathrm{j=e,i}}\sum_{m,n} \exp[\mathrm{i}(m\theta - n\varphi) + \gamma t] \int_0^1 (1-\Lambda B/B_{z0})^{1/2}\Big\{\alpha_{1j}^{mn}(1-\Lambda B/B_{z0}) + \alpha_{2j}^{mn}(\Lambda B/B_{z0}) + \sum_{l}\left[\beta_{1j}^{mn}(1-\Lambda B/B_{z0}) + \beta_{2j}^{mn}\Lambda B/B_{z0}\right]\left[\gamma J_{5/2}^{lnj} + \gamma_{1j}^{mn}(1-\Lambda B/B_{z0})^{1/2} J_3^{lnj}\right]\Big\} d\kappa^2 \qquad (1)$$

A similar expression is obtained for the perpendicular pressure tensor, $p_{\perp}$. The $\alpha$ terms are pure fluid contributions, whereas the terms under the $l$ summation are due to the particles precession and bounce oscillations. $m$, $n$ and $l$ are the harmonic numbers in the poloidal, toroidal and trapped particles bounce orbit expansion, respectively. $\gamma$ is the perturbation growth rate. $B_{z0}$ is the equilibrium toroidal magnetic field in the major axis. $\Lambda = 1/[1+\varepsilon(2\kappa^2-1)]$ is a parameter depending on the trapped bouncing particles normalized pitch angle $\kappa$ [1] for a low plasma inverse aspect ratio $\varepsilon = a/R_0$. $\gamma_{1j}^{mn} = inB/(R_0B_{z0})\sqrt{2/M_j}$ ($M_j$ is the ion or the electron mass). $\alpha_{1,2j}^{mn}$ and $\beta_{1,2j}^{mn}$ are analytically derived expressions of plasma fluid and kinetic parameters, respectively. $J_{5/2,3}^{lnj}$ are defined by:

$$J_{5/2,3}^{lnj} = \int_0^\infty E^{\,5/2,3}\exp[-E/(K_BT_j)]\frac{dE}{in\Omega_z + \nu_k + \gamma_2^l\sqrt{\Lambda EB/B_{z0}} + \gamma_3^{mn}\Lambda EB/B_{z0}} \qquad (2)$$

$K_B$ is the Boltzmann constant, $T_j$ the $j$-particle temperature and $E$ the particle total energy. $\Omega_z$ is the plasma toroidal rotation frequency. $\omega_b = -(\mathrm{i}\gamma_2^l/l)\sqrt{\Lambda EB/B_{z0}}$ and $\omega_D = \mathrm{i}\gamma_3^{mn}\Lambda EB/(nB_{z0})$ are the 3D derived particle bounce frequency and bounce averaged toroidal precession frequency, respectively, where $\gamma_2^l$ and $\gamma_3^{mn}$ are analytically obtained parameters. $\nu_k \sim \nu E^{-3/2}$ is the Krook collisional parameter [2], with $\nu$ the particle collision frequency. Depending on the range of the plasma rotation frequency, the above integral is different for specific collisional regimes. In the non-resonant regimes ($1/\nu$ and $\nu - \sqrt{\nu}$) [3] the particles collision frequency influences the NTV magnitude. The $1/\nu$ regime also includes the trapped-circulating particle boundary layer physics [4]. Park et al. model [5] that involves the Krook collision operator, does not consider any $1/\nu$ regime. The above mentioned model has been recently validated [6] for the weak collisional regime close to the dangerous superbanana plateau regime ($SB-P$). We are interested to describe this regime under kinetic resonance conditions. In a lower collisionality case, the main contribution for the NTV magnitude is due to the resonant particles. For very low collisionality, the integral (2) over the particle energy becomes improper. For the case we are interested in, the $l \neq 0$ case, the toroidal precession frequency is significantly lower compared to the trapped particle bounce frequency and can be neglected. On the other hand, at usually high plasma rotation levels, no resonance occurs in the absence of the bounce oscillations, the precession rotation level being too low to resonate. The energy integral that measures the resonance effect is derived at different collisional regimes: a high collisional regime, a low collisional near the superbanana regime and an intermediate regime of collisionality. It should be noted that the use of a Krook collisional operator cannot include the collisional boundary layer introduced by Shaing et al. [4] to overcome the singularity of a pitch angle integral at the trapped-circulating particles boundary limit. However, at very low collisionality regimes we are interested in, the results using the Krook operator are valid. For the $1/\nu$ collisional, $SB-P$ and $1/\nu \to SB-P$ regimes we have analitically obtained specific Cauchy principal values for the integral quantities given by (2). With the aid of it we get a final analytic form of the NTV torque

$$\langle \mathbf{B}_{\text{z-total}} \cdot \nabla \pi \rangle = \sum_{\mathrm{m,n}} \Big\{ \left[ f_{\mathrm{mn}}(r)\Phi^{\mathrm{mn}} + g_{\mathrm{mn}}(r)\Phi'^{\mathrm{mn}} \right] S^0_{\mathrm{n}} \left[ (2-s)\Phi^{\mathrm{mn}*} + r\Phi'^{\mathrm{mn}*} \right]$$
$$+ \sum_{\substack{\mathrm{j}=-3 \\ j\neq 0}}^{3} h_{\mathrm{mnj}}(r, S^0_{\mathrm{n}}, S^2_{\mathrm{n}})\Phi'^{\mathrm{mn}} \left[ (2-s)\Phi^{\mathrm{m+j,n}*} + r\Phi'^{\mathrm{m+j,n}*} \right] + \bar{h}_{\mathrm{mnj}}(r, \Lambda_{\mathrm{j}}, \Lambda'_{\mathrm{j}}, \Lambda''_{\mathrm{j}}) S^0_{\mathrm{n}} \Phi'^{\mathrm{mn}} \Phi^{\mathrm{m+j,n}*} \Big\} \quad (3)$$

$f_{\mathrm{mn}}$, $g_{\mathrm{mn}}$ and $\bar{h}_{\mathrm{mnj}}$ are exactly derived coefficients. $\pi$ is the traceless stress tensor and $\mathbf{B}_{\text{total}} = \mathbf{B} + \mathbf{b}$, where the perturbed magnetic field is parameterized as $\mathbf{b} = \nabla \times [(1/B)\nabla\Phi \times \mathbf{B}] \times \mathbf{B}$. $\Phi$ is the perturbed normalized scalar electric potential. $s$ is the local magnetic shear. $'$ means radial (flux) derivative and $*$ indicates the hermitian conjugation. $\Lambda_{\mathrm{j}} = \Delta\delta_{|\mathrm{j}|,1} + E\delta_{|\mathrm{j}|,2} + T\delta_{|\mathrm{j}|,3}$ is the plasma shaping parameter (depending on local Shafranov shift $\Delta$, ellipticity $E$ and triangularity $T$) and $\delta_{|\mathrm{j}|,\mathrm{i}}$ is the Kronecker delta. $S^0_{\mathrm{n}}$, $S^2_{\mathrm{n}}$ and $c_{\mathrm{ln}}$ are exactly derived kinetic coefficients. To estimate the NTV torque quantity, the Fourier components of the perturbed magnetic field are found. In terms of $\Phi^{\mathrm{mn}}$ and $\Phi'^{\mathrm{mn}}$ we have analytically solved the perturbed magnetic field equations (perturbed plasma, vacuum and feedback circuit equations) obtaining

$$\Phi^{l_i} = \frac{1}{\bar{\Delta}} \sum_{\substack{l_1,..,l_{2L}=1 \\ \text{distinct}}}^{2L} sgn(l_1,..,l_{2L}) \prod_{p=1}^{2L} \Gamma^{p}_{l_p l_i}, \quad \Phi'^{l_i} = \frac{1}{\bar{\Delta}} \sum_{\substack{l_1,..,l_{2L}=1 \\ \text{distinct}}}^{2L} sgn(l_1,..,l_{2L}) \prod_{p=1}^{2L} \tilde{\Gamma}^{p}_{l_p l_i} \qquad (4)$$

They already depend on the kinetic effects. The new index ordering is explained in [7]. The $\Gamma^{p}_{l_p l_i}$, $\tilde{\Gamma}^{p}_{l_p l_i}$ and $\bar{\Delta}$ coefficients depend on plasma, wall and feedback circuit parameters.

In Fig. 1 the calculated normalized NTV for $1/\nu$ and the transition regime to superbanana plateau regime are plotted. The dependence of ion-ion collisional operator is showed. The profiles agree with those obtained with the Park-Boozer formula, for the same collisionality regimes [8]. No assumption about the trapped-circulating particles boundary layer regime is made.

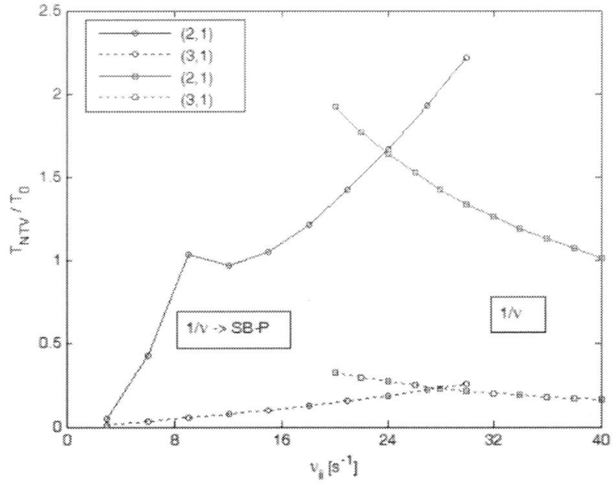


Figure 1: *NTV torque for non-resonant collisional regimes as a function of ion-ion collisional frequency.*

Fig. 2 shows the dependence of the NTV torque at resonance conditions on the error field components amplitude. A higher error field drives a higher NTV quantity. The NTV is calculated at the level of two magnetic surfaces: inside and close to plasma boundary. As expected, the NTV close to the plasma surface is lower. Being calculated at resonance, at very low collisionality, it is clear that a higher plasma rotation is a condition to avoid the resonance.

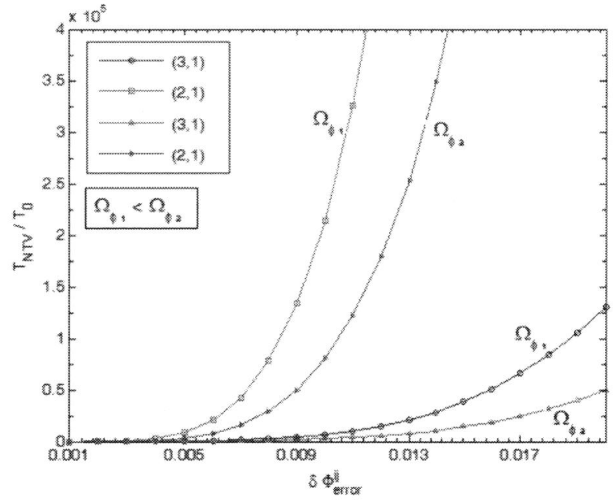


Figure 2: *Error field dependence of NTV at resonance for different plasma toroidal rotations.*

Fig. 3 shows the bounce harmonic number resonant contributions to the calculated NTV for the resonant regime, at very low collisional frequency. It is obvious the significant increase of the NTV at rotation levels that satisfy the resonance condition.

For each *l* harmonic number in the trapped particles bounce orbit expansion, there is a peak of the NTV torque, at a different plasma rotation. The higher *l* is, the lower NTV peak is obtained at a higher corresponding plasma toroidal rotation.

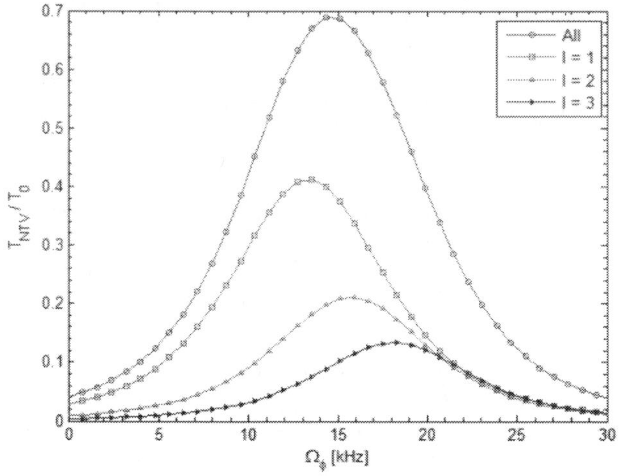


Figure 3: *Calculated NTV torque as a function of plasma toroidal rotation. The l harmonic number contributions.*

To resume, a 3D perturbed particle distribution function has been obtained as a function of the magnetic field perturbations. The latter are derived by solving the linearized perturbed complete system of the multimode magnetic field equations. The magnetic perturbations are not inserted parametrically into the calculated NTV but they are calculated (4), already enclosing the kinetic resonance influence.

In summary, a model to provide an analytic form of the NTV torque (3) is built. It describes and proves the destabilizing influence of the particles kinetic resonances effect on magnetic braking of plasma rotation in tokamak plasmas of specific collisionality regimes.

## Acknowledgments

This work was supported by Euratom and carried out within the framework of the European Fusion Development Agreement. This work has also been carried out within the framework of the EUROfusion Consortium and has received funding from the Euratom research and training programme 2014-2018 under grant agreement No 633053. The views and opinions expressed herein do not necessarily reflect those of the European Commission.

## References

[1] K.C. Shaing, Phys. Plasmas **10**, 1443 (2003)
[2] H. E. Mynick, Nucl. Fusion **26**, 491 (1986).
[3] Y. Sun, Y. Liang, K.C. Shaing et al., Nucl. Fusion **51**, 053015 (2011)
[4] K.C. Shaing et al., Phys. Plasmas **15**, 082506 (2008)
[5] J-k. Park, A.H. Boozer and J.E. Menard, Phys. Rev. Letters **102**, 065002 (2009)
[6] J-k. Park, Y.M. Jeon, J.E. Menard et al., Phys. Rev. Letters **111**, 095002 (2013)
[7] I.G. Miron, Plasma Phys. Control. Fusion **50**, 095003 (2008)
[8] S. Satake, J.-K. Park, H. Sugama, and R. Kanno, Phys. Rev. Letters **107**, 055001 (2011)

# Impact of rotating magnetic islands on profile flattening and turbulence

D. Zarzoso[1,2], W. A. Hornsby[2], F. J. Casson[3], E. Poli[2], A. G. Peeters[4], S. Nasr[1]

[1] *Aix-Marseille Université, CNRS, PIIM, UMR 7345 Marseille, France.*

[2] *Max Planck Institute for Plasma Physics, Boltzmannstr. 2, 85748 Garching, Germany*

[3] *CCFE, Culham Science Centre, Abingdon, Oxon, OX14 3DB, UK*

[4] *Theoretical Physics V, Dept. of Physics, Universitaet Bayreuth, Bayreuth, Germany, D-95447*

The presence of magnetic islands in nuclear fusion devices has strong consequences for the core confinement and can eventually lead to a disruption. It is therefore essential to analyse the stability of magnetic islands, which is mostly determined by the bootstrap current, in particular by the density gradient [1]. In the absence of turbulence, small magnetic islands rotating in the ion diamagnetic direction flatten the density due to the adiabatic response of trapped ions, reducing the stabilizing effect of the bootstrap current [2]. We elucidate in the present work whether the adiabatic response is still able to flatten the profile or not in the presence of turbulence. We analyse also the impact of rotating islands on turbulent transport and propose other mechanisms additional to the standard suppression of the linear drive. For this purpose we use the gyrokinetic code GKW [3], which solves the gyrokinetic equation for both ions and electrons in toroidal geometry, coupled to the quasi-neutrality condition. The equations are solved in the flux-tube approximation with a linearized collision operator consisting only of the pitch-angle scattering part, essential for the trapping-detrapping physics. We do not solve Ampère's law, i.e. we impose the island width and rotation frequency. The parameters of the simulations are: $R/L_n = 2.2$, $R/L_{T,i} = R/L_{T,e} = 6.9$, $T_e/T_i = 1$, $\varepsilon = 0.19$, $q = 1.5$ and $\hat{s} = 0.16$. We present results using sufficiently high numerical resolution to describe turbulent regimes as well as the physics around the separatrix of the magnetic island [4]. We perform a scan on the island width and the island rotation frequency. The island widths we consider are $W = [2, 6, 9, 12, 18]$, normalized to the ion Larmor radius. We perform also a simulation without island to have a turbulent reference case for comparison. This allows us to quantify the impact of the magnetic island on turbulence.

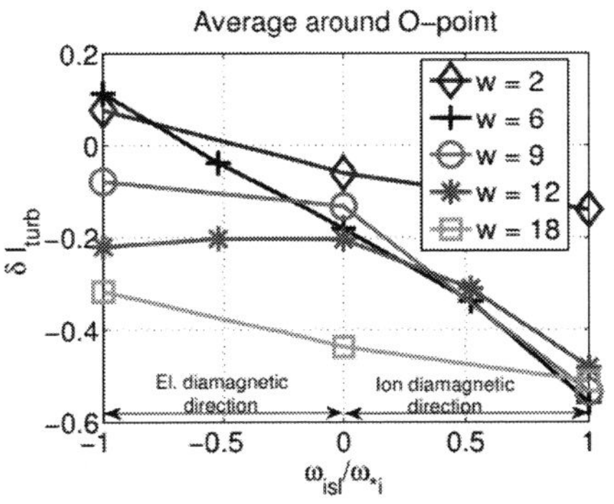


Figure 1: *Turbulence modification with an magnetic island as a function of the island rotation frequency and width.*

In figure 1 we give the relative modification of the turbulent intensity with respect to the reference simulation as a function of the island width and rotation frequency, $\delta I_{\text{turb}}$, averaged in time over the saturated phase and in space around the O-point. The presence of an island tends to stabilize turbulence around the O-point. This general result is in agreement with previous gyrokinetic results [5], where the island was large enough to lead to a rather significant flattening of radial profiles and in addition it was a static island, i.e. there was no rotation frequency. The reduction of turbulence was due to the flattening of radial profiles leading to a reduction of the turbulence linear drive. However, in our case, we consider smaller islands and the flattening is not necessarily as significant as in the case of larger islands. We also observe that islands rotating in the ion diamagnetic direction are more efficient to reduce turbulence than islands rotating in the electron diamagnetic direction. Finally, very small islands ($W \approx 2$) seem to impact turbulence in a less significant way than larger islands, characterized by $W \geq 18$. It is clear that the impact on turbulence due to intermediate islands ($W \approx 6-9$) depends more strongly on the island rotation frequency than the impact of larger islands. In particular, the width $W = 6-9$ seems to be the optimal range for the rotation dependence. Beyond those widths, the dependence on the island rotation frequency tends to disappear and the island reduces the turbulence independently of the direction of the rotation. To determine whether the reduction comes from the suppresion of the linear drive, we analyse the impact of the island on the radial profiles.

On the left-hand side of figure 2 we plot the normalized density gradient averaged around the O-point. The island $W = 6$ rotating in the ion diamagnetic direction is able to flatten the density profile and this flattening level is reduced when increasing the island width. This might be due to the adiabatic response of trapped ions [2], which is analysed later. On the right-hand side of figure 2 we plot the normalized ion temperature gradient as a function of the island rotation frequency and the island width. Also in this figure, for reference, we give the value of $R/L_{T,i}$ for a static island $W = 12$ when the turbulent modes are filtered out in the simulation. The suppression of the linear drive seems to be the dominating mechanism for turbulence reduction when the island is large enough. However, for intermediate magnetic islands, another mechanism should be considered, since the linear drive is increased.

This mechanism can be related to the increase of a radial electric shear around the separatrix when the island is rotating with respect to the plasma. The transfer of energy between modes is expected to be more efficient when island and turbulence structures are co-rotating because the time of interaction between large-scale (zonal flows and island) and small-scale (turbulence) structures is maximized. In addition to the time interaction, one must consider also the scale lengths. In particular when the shear of the radial electric field exhibits scale lengths close to the decorrelation length of turbulence, the transfer of energy can lead to a reduction of turbulence.

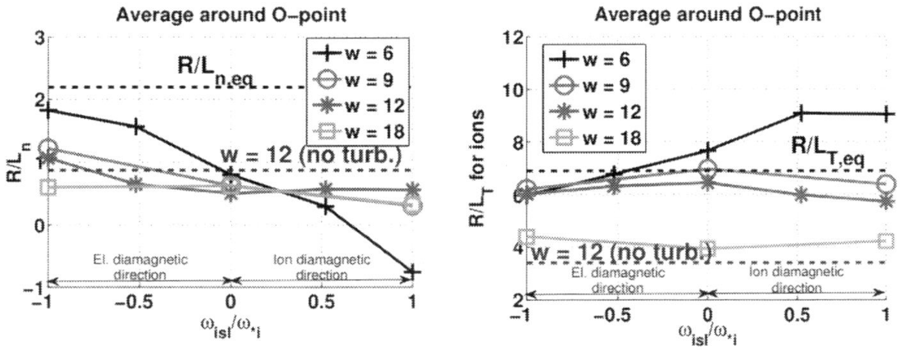


Figure 2: *Normalized density (left) and ion temperature (right) gradients as a function of the island rotation frequency and width.*

In figure 3 we plot for the island $W = 6$ the radial profile of the radial electric shear through the O-point for two rotation frequencies: ion (solid line) and electron (dashed line) diamagnetic frequencies. The island separatrix is highlighted by two vertical lines. It can be observed that the shear is clearly increased when the island is rotating in the ion direction. To elucidate whether the flattening when the island rotates in the ion direction is due to trapped ions or not in the presence of turbulence, we analyse the perturbed ion density and compare it to that obtained in the absence of turbulence. In the top panel of figure 4 we give the perturbed ion density in velocity space on the separatrix of an island of width $W = 6$ in the absence of turbulence.

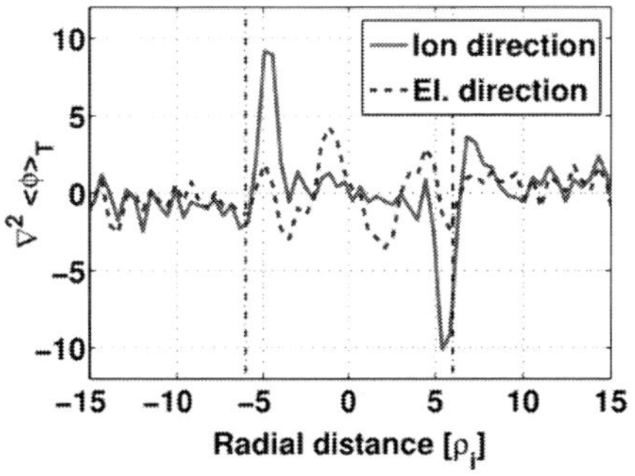


Figure 3: *For $W = 6$, radial profile through the O-point of the shear of the radial electric field for two island rotation directions: electron (dashed line) and ion (solid line).*

The left-hand side of the panel represents an island rotating in the ion diamagnetic direction and the right-hand side an island rotating in the electron diamagnetic direction. The solid black lines separate the population of trapped particles from that of passing particles. It can be observed that the resonant particles are mainly the trapped ones. The perturbed ion density is entirely positive (resp. negative), meaning that they contribute to the flattening (anti-flattening) of the profile when the island is rotating in the ion (resp. electron) diamagnetic direction. Since the response changes its sign when changing the sign of the rotation frequency, these results imply that trapped ions are responsible for the flattening and that their response exhibit an adiabatic behaviour.

We compare these results with the results obtained with an island of the same size in the presence of turbulence. This is illustrated in the bottom panel of figure 4, for an island rotating in the ion (left-hand side) and electron (right-hand side) directions. We observe that the perturbed density is also mainly localized in the trapped region. However, when the island rotates in the electron diamagnetic direction, the response of particles does not exhibit the opposite behaviour to the one observed when the island rotates in the ion diamagnetic direction. This clearly shows that, although the flattening is mainly due to trapped ions (the response is localized within the trapped region), their adiabatic response cannot be considered as the only reason for the flattening of density profile.

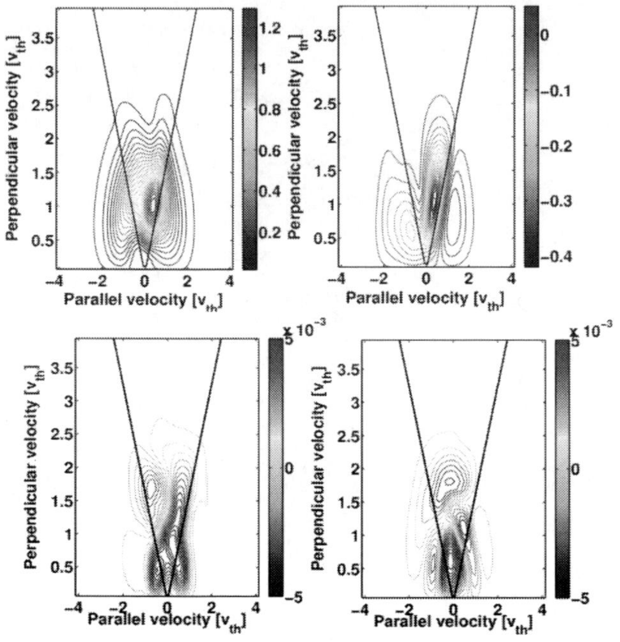


Figure 4: *Perturbed ion density on the separatrix as a function of the parallel and perpendicular velocities in the absence (top) and in the presence (bottom) of ITG turbulence. Left:* $\omega_{\rm isl} = \omega_{*i}$. *Right:* $\omega_{\rm isl} = \omega_{*e}$

This work has been carried out within the framework of the EUROfusion Consortium and has received funding from the European Union's Horizon 2020 research and innovation programme under grant agreement number 633053, from the RCUK Energy Programme [grant number EP/I501045] and from the A*MIDEX project (no. ANR-11-IDEX-0001-02) funded by the "Investissements d'Avenir" French Govern- ment program, managed by the French National Research Agency (ANR). The views and opinions expressed herein do not necessarily reflect those of the European Commission. Simulations were performed on the IFERC-CSC Helios super-computer within the framework of the TURBISLE project.

## References

[1] O. Sauter et al, Phys. Plasmas (1994-present), **6**, 7 (1999)

[2] A. Bergmann et al, Phys. Plasmas (1994- present), **16**, 9 (2009)

[3] A. G. Peeters et al, Computer Physics Communications,**180**, 12 (2009)

[4] D. Zarzoso et al, Phys. Plasmas (1994-present) (2015)

[5] W. A. Hornsby et al, Phys. Plasmas (1994- present), **17**, 9 (2010)

# Towards self-consistent runaway electron modeling

G. Papp[1,2], A. Stahl[3], M.Drevlak[1], T. Fülöp[3], Ph. W. Lauber[2] and G. I. Pokol[4]

[1]*Max Planck Institute for Plasma Physics, Garching/Greifswald, Germany.*
[2]*Max-Planck/Princeton Center for Plasma Physics.* [3]*Dept. of Applied Physics, Chalmers University of Technology, Gothenburg, Sweden.* [4]*Institute of Nuclear Techniques, Budapest University of Technology and Economics, Budapest, Hungary.*

**Introduction** Reliable runaway electron (RE) mitigation after disruptions is one of the most important challenges for safe ITER operation [1]. A proper understanding of the generation and losses of REs is therefore essential. Recent experimental [2, 3] and theoretical [4] results suggest that knowledge of the electron distribution function is essential to properly describe the evolution of post-disruption runaway beams. One way to model the electron distribution function is by solving the Fokker-Planck equation [5, 6]. However, in order to properly account for the time variation of plasma parameters, a self-consistent evolution of the background plasma (temperature and density profiles, impurity composition etc) is necessary. This paper reports on the first results of our self-consistent combined fluid-type and Fokker-Planck simulations of post-disruption runaway evolution.

**Numerics** Here we consider two Fokker-Planck solvers. **CODE** [5] (COllisional Distribution of Electrons) is a fast 2-V continuum (Eulerian) fully implicit Fokker-Planck solver with arbitrary electric field and collisions, using a relativistic momentum-conserving collision operator valid for arbitrary energies. It also takes the effect of synchrotron radiation into account [4], which can significantly change the runaway growth rate under certain plasma conditions. In the future we plan on using an even more sophisticated tool, LUKE [6], which, among other things, includes toroidicity effects, electron heating & current drive, etc. In this paper, only results using CODE are discussed.

To follow the evolution of the plasma background we use the **GO** code [7]. GO calculates runaway electron generation in disruptions using a self-consistent fluid model of the electric field and other important plasma parameters. It includes an ADAS-based, temperature and density dependent collisional-radiative atomic physics model for an arbitrary ion composition. In the future, GO could be replaced by e.g. a self-consistent 3D MHD (such as JOREK or NIMROD) tool calculating the whole disruption phase including impurities. However, the relative simplicity and orders of magnitude smaller computational requirements of GO will always be an advantage.

Schematically the coupling is relatively simple: a separate distribution function is maintained for each radial point in GO, and in each GO time step, CODE is invoked to evolve the distribution. CODE determines the necessary number of sub-iterations to evolve the distribution, based on the plasma parameters. In order to save CPU time, each CODE call has an independent setting of numerical parameters (energy grid, number of Legendre polynomials, number of iterations, etc). CODE utilizes dynamic grids, where the grid is continuously adapting based on the distribution function and the plasma parameters. This gives roughly a factor of ~100x (or more) speedup compared to a fixed grid approach. Load balancing is of particular importance: each (radial) CODE run has to finish before the global plasma parameters can be evolved and the amount of time it takes to evolve the distribution can differ by orders of magnitudes for the different radial grid points. Paralellisation is utilized in an "MPI+openMP"-style approach. Each CODE call is a separate worker while the optimum number of threads is determined by the size of the sparse CODE matrix and the number of necessary sub-iterations. We observe good sparse scaling as a function of grid sizes and number of iterations. We believe a factor of 1.5-2x speedup could be achieved by further optimizing the parallellisation scheme, however it is uncertain if the potential improvement justifies the necessary time and effort. In general, the bottleneck is memory bandwidth, which might be improved by utilizing high memory bandwidth Intel Xeon Phi accelerator cards.

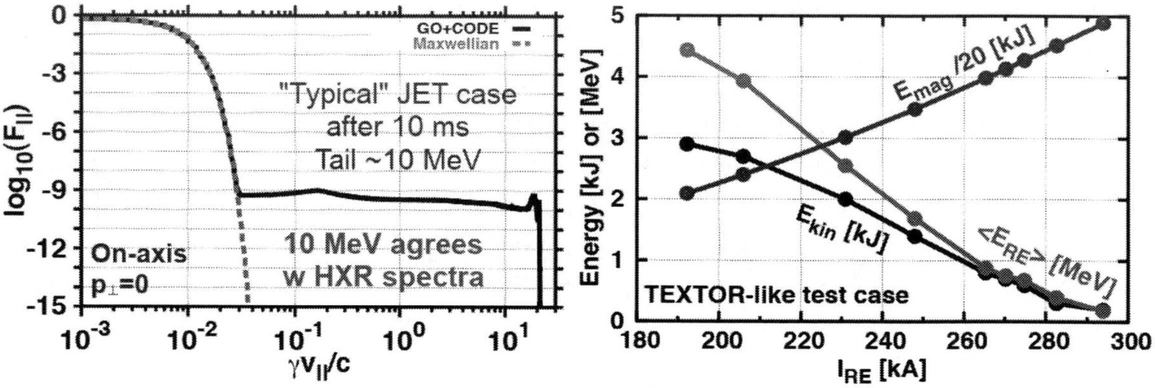


**Figure 1:** *(a) Typical spectrum evolution from GO+CODE for JET parameter range. The maximum runaway energy agrees well with the energy usually inferred from HXR spectral reconstruction. (b) Magnetic energy, total kinetic energy and mean runaway energy as a function of post-CQ runaway current. In higher RE current scenarios the electric field decays faster and consequently the kinetic energy carried by the runaways is lower.*

**First results** The initial test simulations were conducted with an "as simple as possible" approach. We use a TEXTOR-like plasma with $T_e = 1.3$ keV predisruption central electron temperature, $n_e = 1.7 \cdot 10^{19}$ m$^{-3}$ central electron density, $B_0 = 2.1$ T, $R = 1.8$ m, $I_0 = 305$ kA. The atomic physics module in GO is inactive, the thermal quench is instead initiated by a forced exponential collapse of the electron temperature. In the figures presented in this paper the post-thermal-quench central electron temperature was 5 eV.

Figure 1b shows the magnetic energy, total kinetic energy and mean runaway energy as a function of post-CQ runaway current (scanned by changing the thermal quench time). We note that in scenarios that produce a larger post-thermal-quench runaway current, the electric field decays faster (as it is being "consumed" by the REs) and consequently the kinetic energy carried by the runaways is lower. This also means that mitigation of already formed runaway beams by high-Z material injection might be more effective in higher runaway current scenarios.

If we compare the saturation runaway currents between GO+CODE and GO we observe a 35% (~60 kA for these parameters) extra production of runaways, as is illustrated in figure 2. Deciding which model is closer to reality is a nontrivial task. In general, CODE shows a higher initial burst of runaways in the early stages of evolution compared to the analytical primary generation formulas [8, 9].

However, it should be noted that the Dreicer generation formula was derived for quasi-steady-state distributions. We will clarify the picture by comparing to other Fokker-Planck solvers [6]. In conclusion we can say that the coupling between GO and CODE ***works***, and the first results are encouraging. However, there is still further optimization, testing and validation to experiments [3] to be done.

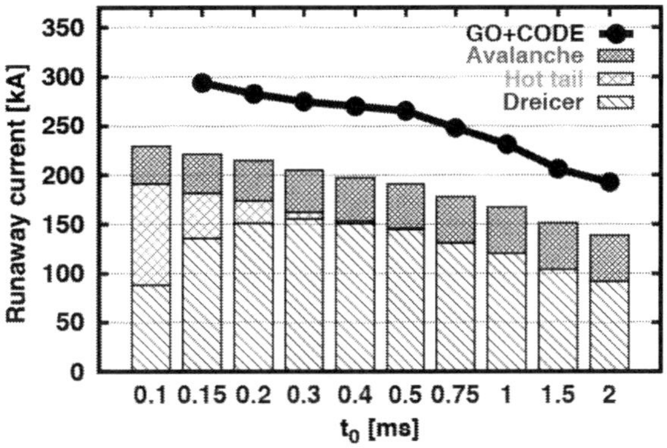


**Figure 2:** *Comparison of GO+CODE and GO: ∼35% difference in runaway current is observed.*

## The effect of Resonant Magnetic Perturbations

In the past few years we have analysed in detail the effect of Resonant Magnetic Perturbations (RMP) on runaway electron losses in ITER (see [10] and references therein). The most important conclusion was that runaways closer to the edge (r/a = 0.7) can be quickly lost under the influence of RMPs, however, core particles are expected to be well confined. However, earlier computations were only using a test particle approach [10] and therefore we could not take into account the effect of RMPs on runaway generation with a self-consistently calculated electric field evolution. Figure 3a illustrates the runaway losses under the effect of RMPs. A flat runaway profile of 1 MeV particles is initialized and then followed up to 100 ms. Figure 3a shows the evolution of the runaway density profile as a function of time and normalized flux. Close to the edge particles can be lost as soon as after 1$\mu$s, while particles in the central region are well confined.

The usual 1 dimensional description of transport caused by magnetic perturbations is the Rechester-Rosenbluth diffusion [11], which scales with $(\delta B/B)^2$. If we now introduce RR diffusion into GO [7] with the $\delta B(r)$ profile calculated for the ITER ELM perturbation coils [10], we arrive to the profile evolution presented in figure 3b. If we compare figures 3a-b (mind the logarithmic time scale on the y-axis) we immediately see that the RR diffusion does not reproduce the test particle results obtained with ANTS. We also found that not only RR diffusion is not applicable, but the transport itself is not diffusive. Diffusive transport would imply a flux proportional to the local gradient, but we found that the flux is proportional to the local density, which is an indication of a convective type loss [12]. Even though the individual particle orbits are chaotic, the ensemble behavior is smooth (figure 3a), therefore we approximated the particle losses with an $N_{\text{lost}}(t) \propto 1-\exp\{-t/\tau(\psi)\}$ trend, where $\tau(\psi)$ is the characteristic loss time associated with a certain initial $\psi$ position for the particles. We have performed fits of $\tau(\psi)$ to the evolution of fast electron density obtained by the 3D test particle simulations for various cases. We note that the dependence of $\tau(\psi)$ on the radial coordinate is relatively smooth and is well approximated by an exponential dependence $\tau(\psi) \propto \exp\{-\psi/\psi_0\}$, where $\psi_0 \propto \delta B/B$. Figure 3c shows the density profile evolution based on the exponential losses using the fitted $\tau(\psi)$ values. It clearly resembles the particle evolution calculated by ANTS.

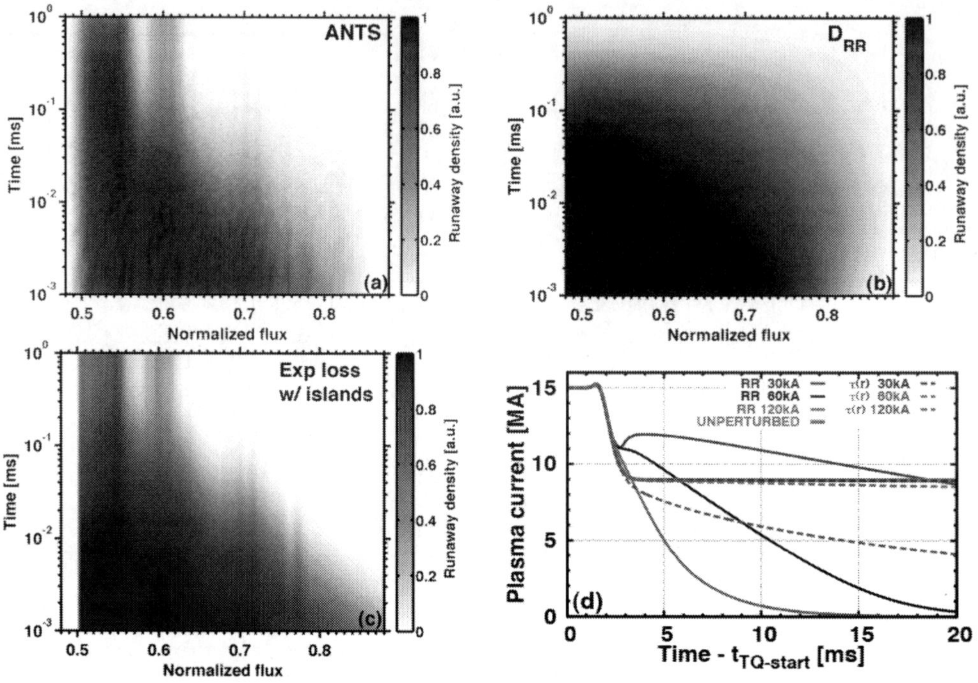


**Figure 3:** *(a) Evolution of a flat runaway density profile in 3D ANTS simulations. (b) Evolution of the density profile under RR diffusion and (c) with the exponential loss model. (d) The RR diffusion suggests that RMPs can suppress runaway formation with ITER, it is not the case with the loss model based on the ANTS results [10, 12].*

To illustrate the differences between the implications of the diffusive and exponential-loss models, we have also implemented the latter in GO. Figure 3d shows a disruption simulation for the same ITER scenario that was used for the 3D simulations. Without losses a runaway beam of ~9 MA would form. Using the RR diffusion model we get suppression of the runaways after 20 ms at $I_{\mathrm{RMP}} = 60$ kA and after 10 ms if $I_{\mathrm{RMP}} = 120$ kA. As the dashed lines clearly illustrate, **virtually no runaway suppression is achieved** with the exponential loss model. Although the perturbation penetrates the core, the test particle simulations show that the transport is essentially untouched inside the radius $r/a = 0.7$ for $I_{\mathrm{RMP}} = 60$ kA. Therefore, even if the transport is largely increased towards the edge region in the exponential loss description, this does not contribute much to runaway suppression. The reason is that most of the runaways are formed in the plasma core, where there is good runaway confinement according to the test particle simulations.

**Acknowledgment** This work was funded by the Max-Planck/Princeton Center for Plasma Physics. This work has been carried out within the framework of the EUROfusion Consortium and has received funding from the Euratom research and training programme 2014-2018 under grant agreement No 633053. The views and opinions expressed herein do not necessarily reflect those of the European Commission.

## References

[1] E. M. Hollmann *et al.*, PoP **22** 021802 (2015)
[2] G. Pautasso *et al.*, EPS 2015 P1.134
[3] E. M. Hollmann *et al.*, PoP, **22** 056108 (2015)
[4] A. Stahl *et al.*, PRL **114** 115002 (2015)
[5] M. Landreman *et al.*, CPC **185** 847 (2014)
[6] J. Decker *et al.*, PSFC/RR-05-3 (2005)
[7] G. Papp *et al.*, NF **53** 123017 (2013)
[8] J. Connor and J. Hastie, NF **15** 415 (1975)
[9] H. M. Smith *et al.*, PoP **15** 072502 (2008)
[10] G. Papp *et al.*, PPCF **54** 125008 (2012)
[11] A. B. Rechester, *et al.*, PRL **40** 38 (1978)
[12] G. Papp *et al.*, JPP (a.), arXiv:1504.05074

# Scattering of electromagnetic wave beams from density fluctuations in tokamak plasmas

E. Poli[1], O. Maj[1], H. Weber[1], E. V. Sysoeva[2], E. Z. Gusakov[2]

[1] *Max-Planck-Institut für Plasmaphysik, Garching, Germany*

[2] *Ioffe Institute, St. Petersburg, Russia*

Well collimated electron cyclotron (EC) wave beams are routinely used in present-day tokamaks for heating, current drive, and for diagnostic purposes. In several applications, turbulent fluctuations of the density can potentially affect the desired result, broadening the power deposition profile or altering the spectrum of the probing beam. The theoretical treatment of the problem is made difficult by the fact that, in general, turbulent fluctuations do not satisfy the WKB ordering on which most of the tools usually employed for the description of short-wavelength waves in fusion plasmas rely. On the other hand, an increasing number of theoretical estimates indicates that beam scattering can be important in particular on reactor-size tokamaks, where small deviations of the beam trajectory can propagate over large distances and have thus sizeable effects. In this paper, an approach to this problem based on the solution of the wave kinetic equation, for which a rigorous treatment of the scattering term has been developed, is presented, and the corresponding numerical implementation in the new code WKBeam is described. First results for a model distribution of the turbulence are discussed.

## Theoretical background

We look for a solution of the wave equation for the wave field $E(\omega,x)$,

$$\nabla \times \nabla \times E(\omega,x) - \kappa^2 \hat{\varepsilon} E(\omega,x) = 0, \tag{1}$$

in the short-wavelength (WKB) limit $\kappa \gg 1$, where $\kappa = Lk_0$, with $L$ the inhomogeneity scale of the medium and $k_0 = \omega/c$ the vacuum wavevector of a wave of fixed frequency $\omega$. In the previous equation, the coordinate $x$ is normalized to $L$. The wave equation can be written in the symbolic form $DE = 0$, where $D$ is the dispersion tensor. The wave kinetic equation is constructed starting from the equation for the field correlation $DEE^\dagger = 0$ by applying the Weyl symbol map. In the case of the field correlation function, the result of the Weyl symbol map is the Wigner function

$$W(x,N) = \int e^{-i\kappa N\cdot s} E\left(\omega, x+s/2\right) E^\dagger\left(\omega, x-s/2\right) \mathrm{d}s$$

(properly speaking, one should talk about a Wigner matrix, due to the vector nature of the electric field, but this distinction will be ignored here, as in the following we will treat different modes of propagation separately; see [1] for a complete discussion).

The Wigner function satisfies an equation equivalent to Eq. (1), which in the limit $\kappa \to \infty$ leads to:

$$\begin{aligned} &\mathscr{O}(1): && D(x,N)W(x,N) = 0, \\ &\mathscr{O}(\kappa^{-1}): && -\frac{i}{2\kappa}D(x,N)\left(\overleftarrow{\partial_x}\cdot\overrightarrow{\partial_N} - \overleftarrow{\partial_N}\cdot\overrightarrow{\partial_x}\right)W(x,N) \equiv \frac{i}{2\kappa}\{D,W\} = 0, \end{aligned}$$

where in the last step the Poisson brackets have been introduced. Imposing the condition of non-trivial solutions for $W$ on the first equation leads to the dispersion relation. The second equation has the form of a steady state kinetic equation and is used to determined the evolution of the Wigner function, starting from the antenna plate. If fluctuating and dissipative media are considered, adopting the ordering $D \to D_0 + \kappa^{-1/2}D_{\text{fluct}} + \kappa^{-1}D_{\text{abs}}$ (where $D_0$ is the unperturbed, cold-plasma dispersion operator) leads to the form of the wave kinetic equation considered in this work:

$$\{D_0, W\} = -2\gamma W + S(\Gamma, W), \tag{2}$$

where $\gamma = e^*\varepsilon^a e$ is the absorption coefficient and $S$ the scattering term, which is described in some more detail below and depends on the density correlation upon its Wigner transform $\Gamma$. The choice of the ordering of the fluctuating term simply reflects the fact that the fluctuation term appears quadratically in the final result.

The derivation of the scattering operator follows McDonald [2] and relies on the approach of Ref. [3] which is briefly sketched here, since it represents the most important approximation of the present approach. Write the wave equation in the compact form $(D_0+D_1)E = 0$, where the ensemble average of the fluctuations part vanishes, $\langle D_1\rangle = 0$. If $E_0$ is a solution of $D_0E_0 = 0$, the solution of the wave equation can be written as $E = E_0 - (D_0^{-1}D_1)E$. We look for an iterative solution, $E = E_0 - (D_0^{-1}D_1)E_0 + (D_0^{-1}D_1D_0^{-1}D_1)E_0 + \cdots$. Truncating at the lowest significant order (*Born approximation*) and applying ensemble average yields

$$D_0\langle E\rangle = \langle D_1D_0^{-1}D_1\rangle E_0$$

where to lowest order one can replace $E_0$ with $\langle E\rangle$. Applying the same procedure to $(D_0 + D_1)EE^\dagger(D_0^\dagger + D_1) = 0$ yields

$$D_0\langle W\rangle = \langle D_1W_0^{-1}D_1\rangle\,(D_0)^{-1} + \langle D_1D_0^{-1}D_1\rangle W_0,$$

where the right-hand side leads (applying the Weyl symbol map and the WKB limit) to a scattering term in the form

$$S = \int\left[\sigma(x,N,N')W(x,N') - \sigma(x,N',N)W(x,N)\right]\mathrm{d}N'$$

where the scattering cross section $\sigma$ is

$$\sigma(x,N,N') = \frac{\kappa^4}{(2\pi)^3}\left|e^\dagger(x,N)(\mathbf{I}-\varepsilon)e(x,N')\right|^2 \delta(H(x,N'))\Gamma(x,N-N').$$

Note that $\Gamma$ varies on the scale length of the equilibrium scale (typically the scale of the gradients driving the turbulence), so that the application of the WKB ordering is justified.

**Benchmarks**

The previous equations are implemented in the code WKBeam [4], which evolves the Wigner function along Hamiltonian orbits, including scattering and absorption. Diffraction effects are accounted for by initializing the rays at the antenna plane according to the Wigner function. Scattering events are generated along each ray according to the scattering amplitude $\sigma$. Relevant quantities (e.g. power deposition profiles) are reconstructed through the corresponding Monte Carlo estimators (binning in phase space). A detailed code verification in the absence of turbulence has been performed, i.e. against analytic solutions (free space, lens-like medium, linear-layer medium) and numerical solutions in tokamak geometry (beam tracing code TORBEAM [5]). See Ref. [1] for some examples. Here we present the first results of a verification exercise that includes turbulence and relies on the two-dimensional model of Sysoeva *et al.* [6], which is based on the same assumptions (Born approximation) as employed in the derivation of the scattering term implemented in WKBeam.

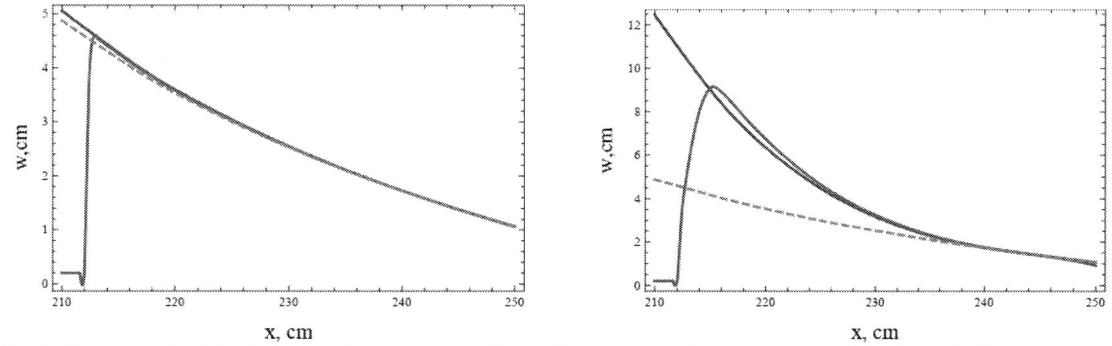


Fig. 1. *Benchmark of the beam width as obtained from WKBeam (red) against Eq.(11) of Ref. [6] (blue). The green dashed line shows the result in the absence of turbulence. The fluctuation level $\delta n_e$ is 0.5% of the cutoff density in the left panel, 5% in the right panel.*

It is indeed possible to show that the equation (Eq.(11) of Ref. [6]) for the Fourier-transformed amplitude $|a_{k_y}|^2$ of the electric field, which is connected to the Wigner function through the relation

$$|a_{k_y}|^2 = \frac{L^2\sqrt{k^2-k_y^2}}{2\pi}\int W(x,N)\mathrm{d}y\mathrm{d}k_x,$$

can be derived exactly from the wave kinetic equation (2). Fig. 1 shows a very good agreement between the beam width resulting from a numerical solution of the equation for $|a_{k_y}|^2$ and that obtained from WKBeam in slab geometry.

**Results**

Preliminary results obtained with the WKBeam code have been presented in [7]. For ITER, it has been found that a turbulence layer with a fluctuation level $\delta n_e/n_e \approx 10\%$ located around the plasma edge can lead to a broadening of the absorption profile by about a factor of two for parameters typical of the upper EC antenna, confirming previous estimates.

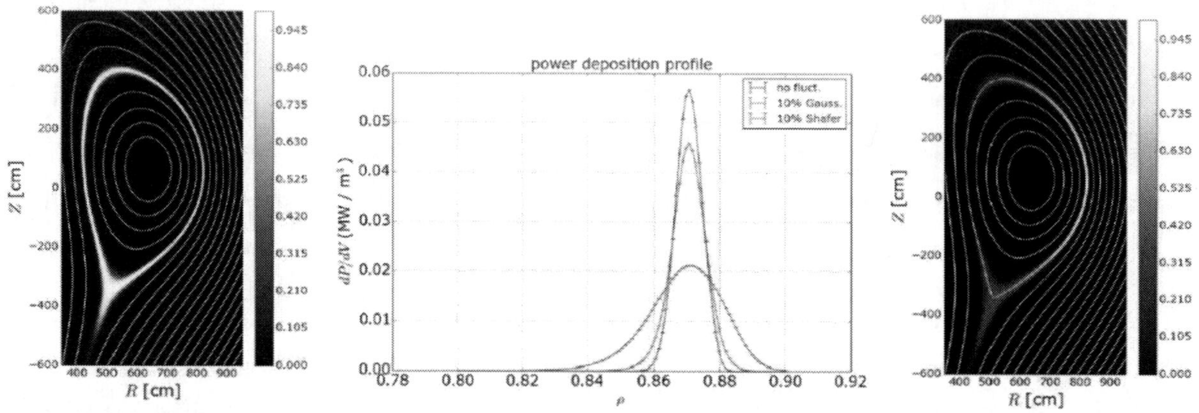


Fig. 2. *Power profile broadening in ITER, homogeneous poloidal distribution of the turbulence (left) vs. the model of Ref. [8] (right).*

Here, we present as an example a comparison of the power deposition profile for ITER mentioned above (homogeneous poloidal distribution of the turbulence) with a model which takes into account the poloidally inhomogenous distribution of the turbulence [8]). For the ITER upper launcher, which is located well above the midplane, the effect of turbulence "ballooning" reduces significantly the impact of the fluctuation on the deposition profile. This result stresses the importance of a realistic desription of the fluctuations for this analysis.

**Note**: This work received funding from Fusion for Energy under Grants 161 and 615. The views expressed in this publication are the sole responsibility of the author and do not necessarily reflect the views of Fusion for Energy.

**References**

[1] H. Weber, IPP Report IPP 5/134 (2013), http://edoc.mpg.de
[2] S. W. McDonald, Phys. Rev. A **43**,4484 (1991).
[3] F. C. Karal, J. B. Keller, J. Math. Phys. **5**, 537 (1964).
[4] H. Weber, O. Maj, E. Poli, Comp. Phys. Comm. (to be submitted).
[5] E. Poli, A. G. Peeters, G. V. Pereverzev, Comp. Phys. Comm. **136**, 90 (2001).
[6] E. V. Sysoeva *et al.*, Nucl. Fusion **55**, 033016 (2015).
[7] H. Weber, O. Maj, E. Poli, EPJ Web of Conf. **87**, 01002 (2015).
[8] M. W. Shafer *et al.*, Phys. Plasmas **19**, 032504 (2012).

# Representation of finite amplitude ideal magnetohydrodynamic modes

D. Meshcheriakov[1], V. Igochine[1], H. Zohm[1], A. Gude[1], M. Maraschek[1], D. Moseev[1], D. Vezinet[1], M. Dunne[1] and the ASDEX Upgrade Team

[1] *Max-Planck-Institut fur Plasmaphysik Boltzmannstrasse 2, 85748, Garching bei Muenchen, Germany*

## Introduction

The ideal modes are described by the equations of ideal magnetohydrodynamics (MHD). One of the most important properties of ideal MHD is that there can be no change in topology of magnetic flux surfaces. The perturbation of the equilibrium field through $\delta\vec{B} = \nabla \times (\vec{\xi} \times \vec{B})$ is the conventional way to represent ideal MHD modes. It was shown [1] that the original topology is only conserved if the perturbed field is calculated to all orders in displacement $\vec{\xi}$. The simple representation of the MHD code results as a superposition of equilibrium and linearized perturbed field will result in an island structure which is unphysical for the ideal instabilities [2]. Our method is shown to conserve the magnetic field topology even at experiment relevant perturbation amplitude. In order to illustrate possible practical applications, the developed approach is applied to a real ASDEX Upgrade equilibrium.

## Analytical expression for the covariant component of the Lundquist identity

The analytical expression for the change of the magnetic field connected with a finite displacement of the ideally conducting incompressible liquid was derived by Lundquist [3] in 1951. It's derivation is based on the consideration of the magnetic flux tube deformation by an arbitrary perturbation and the conservation of the magnetic flux through any element of a surface wholly composed of lines of force. Finally this expression casts into the so-called Lundquist identity:

$$\vec{B}(\vec{r}_0,t) = \vec{B}(\vec{r}_0 - \vec{\xi},0) + [\vec{B}(\vec{r}_0 - \vec{\xi},0) \cdot \nabla]\vec{\xi}(\vec{r}_0,t) \quad (1)$$

Advantages of use of this expression over the standard method of perturbing an equilibrium magnetic field through $\delta\vec{B} = \nabla \times (\vec{\xi} \times \vec{B})$ were clearly demonstrated in [1]. In the present work we generalize this approach to arbitrarily shaped plasmas and perturbations.

First we derived a general expression for the covariant representation of the perturbed magnetic field:

$$B(\vec{r}_0,t)_j = B(\vec{r}_0-\vec{\xi},0)_j + \frac{1}{\sqrt{g_{ii}}} B(\vec{r}_0-\vec{\xi},0)_i \frac{\partial \xi_j}{\partial x_i} - B(\vec{r}_0-\vec{\xi},0)_i \xi_k \Gamma^k_{j,i} \tag{2}$$

where $g_{ij}$ is a metric tensor of the coordinate system for which the magnetic field needs to be calculated and $\Gamma^k_{j,i}$ are Christoffel symbols, which are defined in the following way:

$$\Gamma^k_{j,i} = \frac{1}{2}\sum_s g^{ks}\left(\frac{\partial g_{is}}{\partial x^j} + \frac{\partial g_{js}}{\partial x^i} + \frac{\partial g_{ij}}{\partial x^s}\right) \tag{3}$$

**Calculation of the perturbed ASDEX Upgrade field**

The choice of the most convenient coordinate system for description of the perturbation arises from the variation of the poloidal and toroidal components of magnetic field on the equilibrium flux surface due to toroidicity and shaping of the plasma. As discussed in [4], there is a variation of the pitch angle along the field line in a general toroidal equilibrium and thus MHD modes can no longer be described by a single Fourier harmonic in poloidal angle. The proposed solution of this problem is a mapping of a general equilibrium to a screw pinch like geometry with a so-called straight field line angle $\theta^*$ instead of a poloidal angle. Such coordinate system, where coordinates are $\rho$ - normalized poloidal flux, $\theta^*$ -straight field line angle and $\phi$ - toroidal angle, is chosen for the present studies.

In order to demonstrate consistency of the presented method at realistic plasma shapes, the equilibrium reconstruction of ASDEX Upgrade discharge ($AUG28746, t = 1.7s$) ) is chosen as a test bed. The equilibrium reconstruction was done with CLISTE interpretative code [5].

A radial component of the plasma perturbation corresponding to a single ideal mode can be written as:

$$\xi(\rho) = \xi_0(\rho)\cdot\cos(m\theta^* - n\phi + \Phi_0) \tag{4}$$

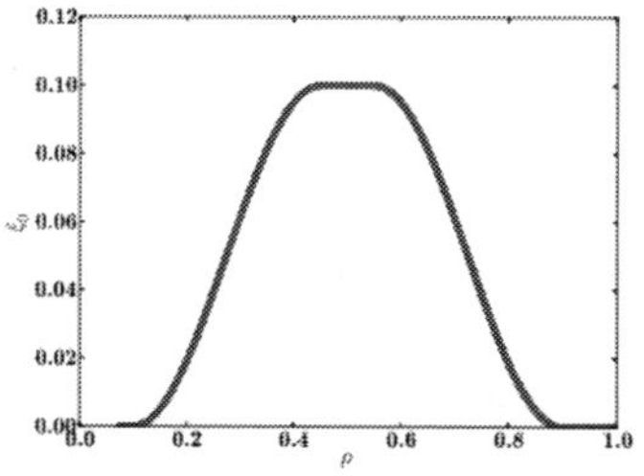


Figure 1: *Normalized test perturbation profile*

where $\Phi_0$ is the initial phase. The poloidal and toroidal components of $\vec{\xi}$ can be determined from the additional constraints of incompressibility of the plasma ($\nabla\cdot\xi = 0$) and the slip motion condition ($\nabla\times(\vec{\xi}\times\vec{B}_\phi) = 0$):

$$\vec{\xi} = \begin{pmatrix} \xi_\rho \\ \xi_{\theta^*} \\ \xi_\phi \end{pmatrix} = \begin{pmatrix} \xi_0(\rho)\cdot\cos(m\theta^* - n\phi + \Phi_0) \\ -\frac{1}{m}(\xi_0(\rho) + \rho\xi_0'(\rho))\cdot\sin(m\theta^* - n\phi + \Phi_0) \\ 0 \end{pmatrix} \tag{5}$$

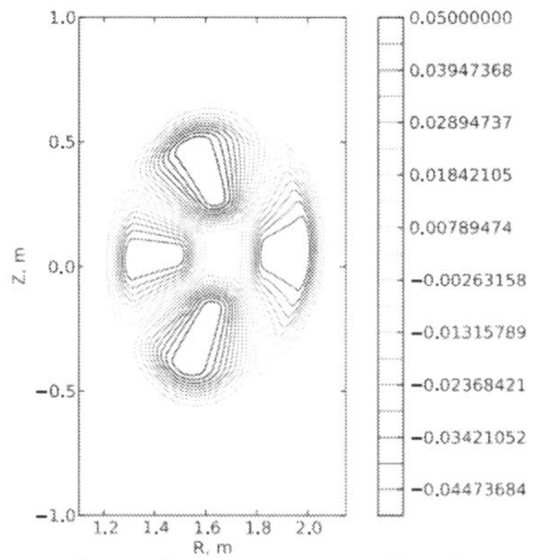


Figure 2: *Contours of the radial component($\xi_\rho$) of the test perturbation*

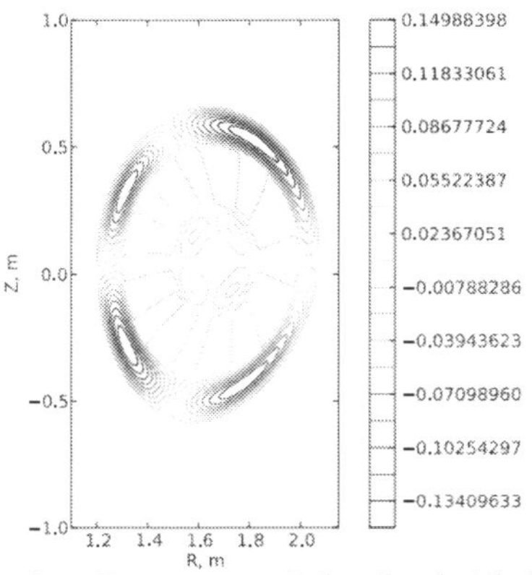


Figure 3: *Contours of the "poloidal"($\xi_{\theta^*}$) component of the test perturbation*

In order to simulate an impact of a localized mode, the radial mode structure shown in Figure 1 was assumed. It consist of a region with a constant displacement and two gradient regions which represent vanishing of the mode far from the resonant surface. Since the expression for the $\theta^*$ component of the perturbation contains the first derivative of $\xi_0(\rho)$, our function should be continuously differentiable. Therefore, the shape of the gradient regions was chosen to be a squared cosine half-wave. The 2D structures of the radial and straight field line angle components of the $m = 2$ perturbation are shown on Figure 2 and Figure 3, respectively.

$$\xi_0(\rho) = \begin{cases} 0 & \rho \in [0, \rho_1[ \\ \xi_0 \cos^2\left(\frac{\pi}{2}\frac{\rho-\rho_2}{\rho_2-\rho_1}\right) & \rho \in [\rho_1, \rho_2] \\ \xi_0 & \rho \in ]\rho_2, \rho_3] \\ \xi_0 \cos^2\left(\frac{\pi}{2}\frac{\rho-\rho_3}{\rho_4-\rho_3}\right) & \rho \in ]\rho_3, \rho_4] \\ 0 & \rho \in ]\rho_4, 1] \end{cases} \tag{6}$$

In this expression, $\xi_0$ defines the perturbation amplitude and the parameters $\rho_1, \rho_2, \rho_3, \rho_4$ the location of the mode, it's width and the width of the gradient regions. It is important to mention here that the perturbation amplitude has to be smaller than the width of the gradient regions. Otherwise, according to Equation 1, flux surfaces can be moved directly to the region with $\xi_0(\rho) = 0$, where the convective part of the Lundquist identity is zero and thus the displacement is not compensated in the gradient regions.

The implementation of our method is performed in three major steps: first, components of the displacement vector are calculated in the straight field line coordinates, then a coordinate transformation is made in order to get it's components in the cylindrical coordinate system $(R, Z, \phi)$ and finally, the cylindrical version of the Lundquist identity is applied. The special

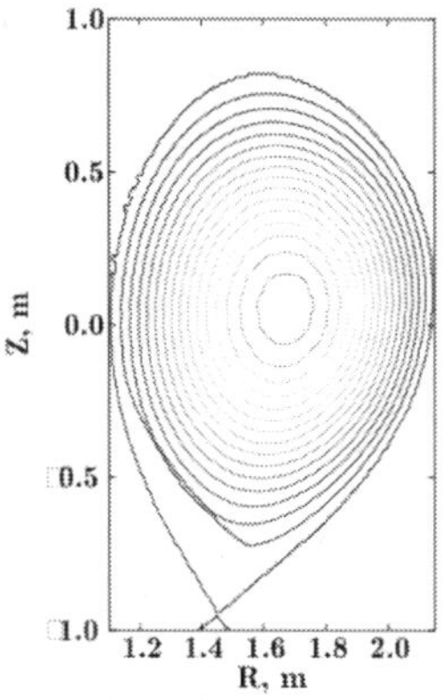


Figure 4: *Flux surfaces of the equilibrium magnetic field*

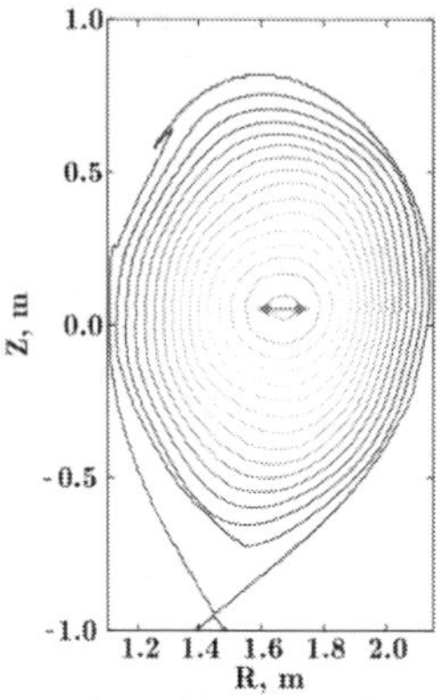


Figure 5: *Flux surfaces of the magnetic field with $m = 2$ perturbation. $\xi_0 = 0.05$*

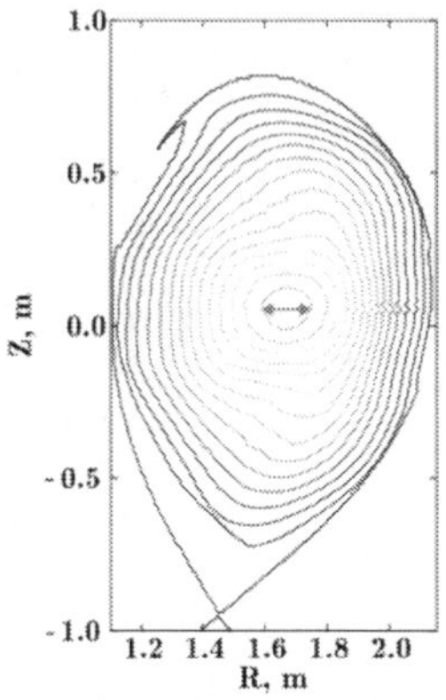


Figure 6: *Flux surfaces of the magnetic field with $m = 2$ perturbation. $\xi_0 = 0.1$*

case of an $m = 1$ mode requires additional treatment since it involves a special point $\rho = 0$, where the Jacobian is 0 and thus the coordinate transform is not defined.

Isocontours of the equilibrium and perturbed poloidal flux are shown on Figures 4 5 and 6. When the perturbation with $m = 2$ is applied, one sees the expected compression of the flux surfaces in the horizontal direction and stretching in the vertical one. In the regions with zero perturbation amplitude, flux tubes remain unperturbed in the case with $\xi_0 = 0.05$. However, when the amplitude is increased to $\xi_0 = 0.1$, the width of the gradient regions seems to be insufficient to screen the displacement completely. It is important to point out that no islands are observed and thus the topology is conserved.

## Conclusions

An alternative method of calculating the perturbed magnetic field has been presented. It includes both an analytical expression for the covariant component of the perturbed magnetic field and a numerical tool for modification of the ASDEX Upgrade equilibrium field in the presence of a perturbation. The ability to apply finite (up to experiment relevant amplitudes) perturbations without breaking the topology is demonstrated.

**Acknowlegments**

*This work has been carried out within the framework of the EUROfusion Consortium and has received funding from the Euratom research and training programme 2014-2018 under grant agreement No 633053. The views and opinions expressed herein do not necessarily reflect those of the European Commission.*

## References

[1] *R. B. White, Physics of Plasmas* **20**, *(2013).*
[2] *A. D. Turnbull et al., Physics of Plasmas* **20**, *(2013).*
[3] *S. Lundquist, Physical Review* **83**, *(1951).*
[4] *H. Zohm,* Magnetohydrodynamic Stability of Tokamaks *(Wiley-VCH, Weinheim, Germany, 2014), pp. 34–37.*
[5] *P. McCarthy and al., Phys. Plasmas* **6**, *3554 (1999).*

# Relativistic runaway electrons in a near-threshold electric field

Pavel Aleynikov[1] and Boris N. Breizman[2]

[1]*Max-Planck-Institut für Plasmaphysik, Greifswald, Germany*

[2]*Institute for Fusion Studies, The University of Texas, Austin, TX 78712, USA*

Runaway electrons (RE) generated via the avalanche mechanism [1] during disruption events in tokamaks have to be mitigated to minimize the detrimental impact from their losses to the wall, especially in future tokamaks such as ITER. Even if the inductive electric field is initially very large, it is expected to drop to the avalanche threshold level after a transient phase of fast RE generation. The subsequent longer-lasting behaviour of RE is governed by the near-threshold electric field. This regime is of particular interest for the mitigation of the RE current and energy.

In this work we first discuss a kinetic theory for relativistic runaway electrons in the near-threshold regime [2]. In a uniform magnetic field $B$ and constant electric field $E$ along the magnetic field lines the RE distribution function $F$ satisfies the relativistic Fokker-Planck equation

$$\frac{\partial F}{\partial s}+\frac{\partial}{\partial p}\left[E\cos\theta-1-\frac{1}{p^2}-\frac{p\sqrt{1+p^2}}{\bar{\tau}_{rad}}\sin^2\theta\right]F=$$
$$\frac{1}{\sin\theta}\frac{\partial}{\partial\theta}\sin\theta\left[E\frac{\sin\theta}{p}F+\frac{(Z+1)}{2}\frac{\sqrt{p^2+1}}{p^3}\frac{\partial F}{\partial\theta}+\frac{1}{\bar{\tau}_{rad}}\frac{\cos\theta\sin\theta}{\sqrt{1+p^2}}F\right], \quad (1)$$

where $p$ is the particle momentum (normalized to $mc$), $\theta$ is the pitch-angle, $s$ is the time variable (normalized to $\tau\equiv\frac{4\pi\varepsilon_0^2m_0^2c^3}{e^4n_e\ln\Lambda}$), $E$ is the electric field (normalized to $\frac{mc}{e\tau}\equiv E_c$), and $\bar{\tau}_{rad}\equiv\frac{\tau_{rad}}{\tau}=\frac{6\pi\varepsilon_0m_0^3c^3}{e^4B^2}\frac{1}{\tau}$ is the normalized time of synchrotron losses. The normalization of the distribution function is given by $\int F dp\sin\theta d\theta=1$. In fully ionized plasmas $Z$ is the ion charge, whereas in cold post-disruption plasmas with impurities, $Z$ should be adjusted to capture the effects of fast-electron scattering on impurity ions and atomic nuclei. Also, the expression for $\tau$ needs to be generalised to take into account collisions with bound electrons (as pointed out in [1]).

In the near-threshold regime the pitch-angular relaxation timescale is shorter than the momentum evolution timescale in (1). In this case, a small parameter $\varepsilon=\frac{E-E_a}{E_a}$ (where $E_a$ is the avalanche threshold as introduced in [2]) enters the equation allowing us to find the angular distribution to the lowest order. The average pitch angle of this distribution is

$$\cos\theta_{av}=\frac{1}{\tanh\left[A(p)\right]}-\frac{1}{A(p)}, \quad (2)$$

where

$$A(p)\equiv\frac{2E}{(Z+1)}\frac{p^2}{\sqrt{p^2+1}}. \quad (3)$$

We use this angular distribution to integrate Eq. (1) over all pitch-angles, which gives a one-dimensional continuity equation,

$$\frac{\partial G}{\partial s}+\frac{\partial}{\partial p}U(p)G=0, \tag{4}$$

for the momentum distribution function $G(s;p)$ with the "flow velocity"

$$U(p)\equiv E\cos\theta_{av}-1-\frac{1}{p^2}-\frac{Z+1}{E\bar{\tau}_{rad}}\frac{p^2+1}{p}\cos\theta_{av}. \tag{5}$$

The profile of the flow velocity $U(p)$ (shown in Fig. 1) predicts peaking of the runaway distribution at $p=p_{max}$ in phase space. This feature is consistent with the prediction obtained from a truncated dynamical model in Ref. [3]. Note that it is straightforward to generalize this solution to include bremsstrahlung effects. This will preserve the trend for peaking of the runaway distribution [4].

To verify this analytical result we solve Eq. (1) numerically (using FiPy Finite Volume PDE Solver [5]). Fig. 2 shows an example of calculation for the parameters of Fig. 1. The initial distribution in Fig. 2 consists of two separated RE populations (left plot), one with the momentum between $p_{min}$ and $p_{max}$ and the other with the momentum higher than $p_{max}$. The right plot in Fig. 2 shows the distribution function $F$ at the time $s=10$. We note that both populations have gathered at the anticipated $p_{max}\approx 8$ location. The angular part of the equilibrated distribution function $F$ also agrees reasonably with Eq. (2).

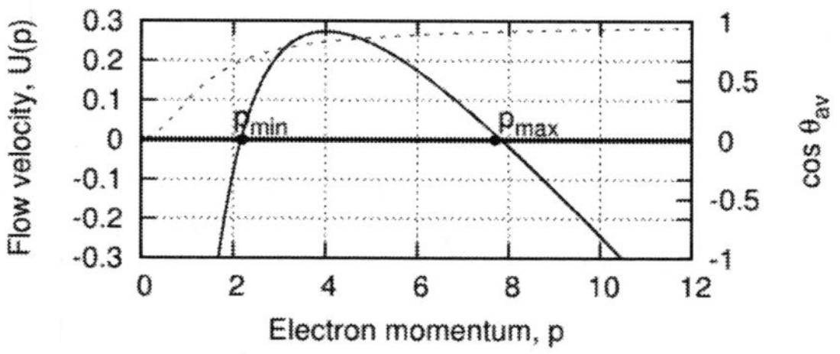


*Figure 1. The flow velocity $U(p)$ defined by Eq. (5) for $Z=2$, $E=2.2$ and $\bar{\tau}_{rad}=10$ (solid line) and $\cos\theta_{av}$ defined by Eq. (2) (dashed line).*

Note that Eq. (4) predicts contraction of the momentum distribution into a delta-function. This is not observed in the numerical calculations, where such contraction eventually stops and the established distribution has a finite width. This width reflects the limitation of accuracy of the lowest order solution obtained in [2].

With the lowering of the electric field the root $p_{min}$ increases and $p_{max}$ decreases. They merge when the electric field has a certain value $E_0$, which is referred to as the "sustainment threshold" in [2]. If the electric field is less than $E_0$, then all electrons decelerate, which precludes sustainment of the RE population. The equality $p_{min}=p_{max}$ serves as a formal condition for determining $E_0$. A convenient analytic fit for $E_0$ is given by Eq. (8) in [2]:

$$E_0\approx 1+\frac{(Z+1)}{\sqrt{\bar{\tau}_{\rm rad}}}\left(\sqrt[6]{\frac{1}{8}+\frac{(Z+1)^2}{\bar{\tau}_{\rm rad}}}\right)^{-1}. \tag{6}$$

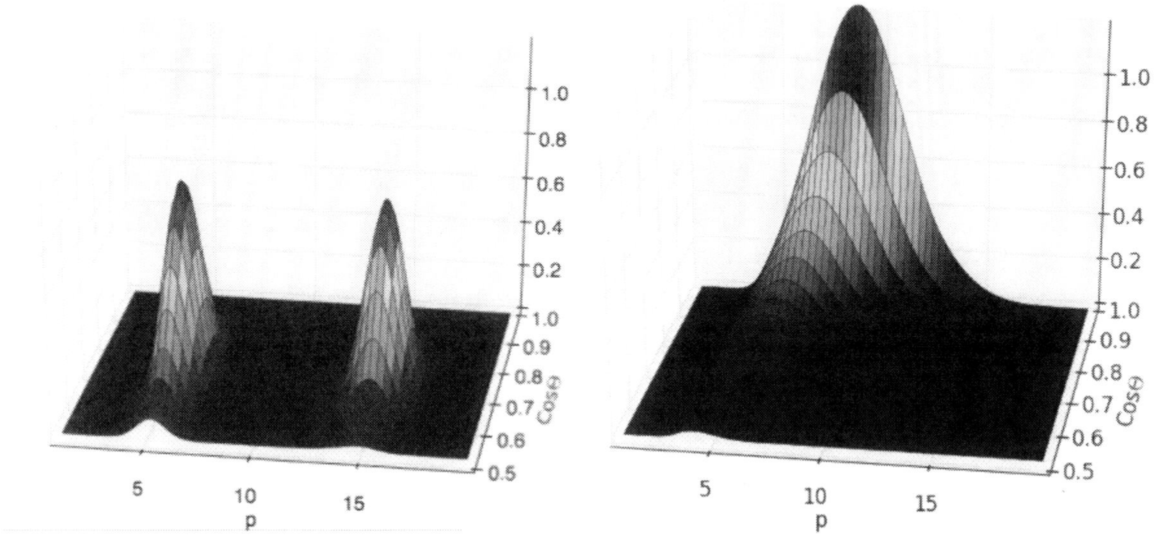


*Figure 2. Numerical calculations of the distribution function F. Left figure - initial condition. Right figure - equilibrated distribution function F at the time $s = 10$.*

This value of the sustainment threshold $E_0$ agrees with the experimental observations of the fast loss of the RE population [6]. For the cases shown in Fig. 10 of [6], Eq. (6) predicts $E_0 \approx 2 - 2.5$, which is in a reasonable agreement with the position of the fast drop of the HXR signal in Fig. 10(c) of [6] (i.e. $E_\phi/E_c \approx 2.7$). Such effect of fast loss of the RE population is miscalculated in a recent work [7] due to a simplified description of the secondary electron source, which assumes extremely high energies of the primary electrons.

Note that $E_0$ is close to unity when $\bar{\tau}_{rad}$ is high, and that the critical momentum $p_{min}(E_0) \equiv p_{max}(E_0)$ has relatively weak dependence on $\bar{\tau}_{rad}$, as shown in Fig. 3.

The near-threshold regime for runaway electrons determines the decay time of the toroidal current during runaway mitigation in tokamaks. To verify that the electric field does drop to the $E_0$ value during the current decay we compare our estimates with the self-consistent Monte Carlo modelling of the RE and the plasma current evolution [8]. In these calculations, a high-Z impurity (Ar) is introduced at the "plateau" stage of RE evolution, to mimic the Massive Gas Injection. Thus instantly changes $E_c$, $Z$ and $\bar{\tau}_{rad}$ values.

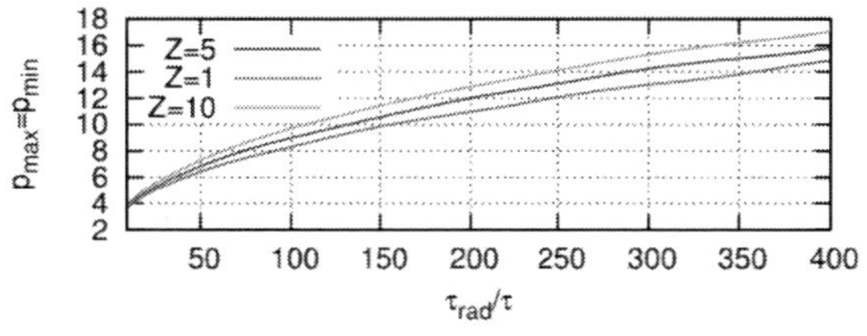


*Figure 3. Dependance of the stable point $p_{min} = p_{max}$ at $E = E_0$ on the plasma parameters.*

Figure 4 shows an example of such comparison. In this example, a prescribed seed RE current initiates the avalanche, which reaches saturation by $\approx 15ms$. The injection of Argon with $n_{Ar} = 3 \cdot 10^{20} m^{-3}$ takes place at $\approx 30ms$. The subsequent evolution of the RE current is apparently linear. This is explained by the hysteresis phenomenon introduced in [2], which involves the avalanche threshold field $E_a > E_0$, thus bounding the inductive electric field to $E_a > E > E_0$ ($E_a$ is only somewhat higher than $E_0$).

The green curves in Fig. 4 show the values expected from the analytical solution of the Eq. 1. The inductive electric field, self-consistently calculated from the Monte Carlo modelling of the RE kinetic equation with the current channel evolution, indeed, saturates at a value of $E_0 = 5V/m$, as predicted by Eq. (6). The average RE energy is also in a reasonable agreement with the $p_{max}$ value, given that a certain time is required for equilibration of the distribution function. This timescale (the time required for the bump formation) can be estimated from the value of the flow velocity between $p_{min}$ and $p_{max}$ and the width of this interval.

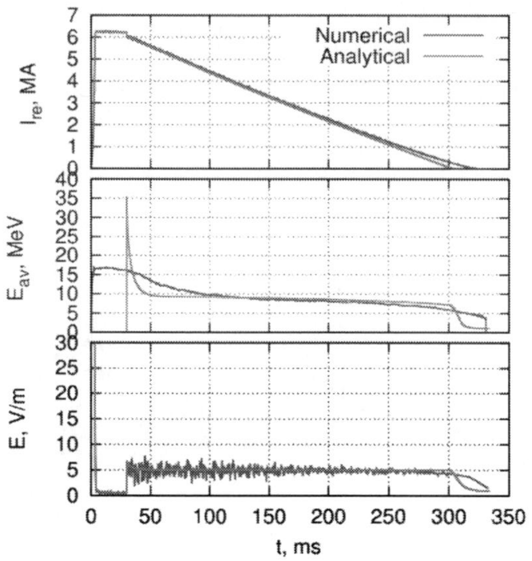


*Figure 4. The comparison of RE current evolution (top), RE average energy ($p_{max}$) (middle) and inductive electric field ($E_0$) (bottom) between self-consistent calculations [8] and reduced model [2].*

In summary, we have compared our approximate analytical solution of the kinetic equation [2] with the numerical results. This comparison confirms the peaking of the RE distribution function at the phase-space attractor near the expected $p_{max}$ value. Self-consistent Monte Carlo modelling of the RE and the plasma current gives the electric field that provides sustainment of the RE. This electric field agrees with the theoretical predictions of the runaway sustainment threshold $E_0$. The sustainment threshold governs the passive RE current decay (RE mitigation timescale), which appears to be consistent with the numerical calculations as well as with what is seen in the mitigation experiments [9].

The authors are grateful to Dr. Per Helander for constructive comments. This work was supported by the U.S. Department of Energy Contract No. DEFG02-04ER54742.

## References

[1] M. N. Rosenbluth, S. V. Putvinski, Nucl. Fusion **37**, 1355 (1997)
[2] P. Aleynikov, B. Breizman, Phys. Rev. Lett., **114**, 155001 (2015)
[3] J. R. Martin-Solis et. al., Phys. Plasmas **5**, 2370 (1998)
[4] M. Bakhtiari, et al. Phys. Rev. Lett., **94**, 215003 (2005)
[5] J. E. Guyer, D. Wheeler, J. A. Warren, Computing in Science & Engineering **11**(3) (2009)
[6] C. Paz-Soldan et al., Phys. Plasmas **21**, 022514 (2014)
[7] A. Stahl, E. Hirvijoki, J. Decker, O. Embreus, and T. Fülöp PRL **114**, 115002 (2015)
[8] P. Aleynikov, et al. 25th IAEA Fusion Energy Conference,TH/P3-38 (2014)
[9] E. Hollmann et al., Nucl. Fusion **53**, 083004 (2013).

# Helical Mirrors for Active Plasma Flow Suppression in Linear Magnetic Traps

A. V. Sudnikov[1,2], A. D. Beklemishev[1,2], I. A. Ivanov[1,2], V. V. Postupaev[1,2]

[1] *Budker Institute of Nuclear Physics, Novosibirsk, Russia*

[2] *Novosibirsk State University, Novosibirsk, Russia*

## Introduction

Recent advances in physics and technology of linear magnetic traps led to the achievement of relatively high plasma parameters. In a quasistationary collisional regime, the relative plasma pressure up to $\beta \approx 60\%$ was demonstrated in GDT device [1] at mean energy of hot ions of 12 keV. The electron temperature up to 0.9 keV was obtained recently with an additional ECR heating [2]. A simple interpolation of GDT parameters to higher NBI injection energy and D-T fuel resulted in a project of GDT-based neutron source with Q = 0.02 that is sufficient for a material test facility with neutron flux of 2 MW/m$^2$ [3].

Suppression of the longitudinal losses is the key issue in increasing plasma parameters. Multiple-mirror [4] or ambipolar [5] passive barriers are used to suppress axial plasma flow. Recently, a new idea of active plasma counterflow pumping with the combination of helicoidal magnetic and radial electric fields was proposed [6]. Plasma rotates due to $\boldsymbol{E}\times\boldsymbol{B}$ drift. Periodical variations of magnetic field move along the flow in its reference frame, inducing longitudinal force acting on the trapped particles. Velocity could be estimated as

$$V_z \approx \frac{hcE_r}{2\pi r B_z},$$

where $h$ is the helicity period, $r$ is the plasma radius, $E_r$ is the radial electric field and $B_z$ is the longitudinal magnetic field. Up- or downstream direction of this force depends on the directions of the electric and the magnetic fields and on the helical structure. Theory predicts exponential dependence of the flow suppression on the magnetic structure length, that is more favorable than the linear dependence in passive confinement systems.

This paper describes the layout of an experiment for the verification for this theory, namely the required plasma parameters and the optimization of the helicoidal magnetic field.

## Plasma parameters

The most effective suppression of the longitudinal plasma flow corresponds to the conditions when mean free path of the ion $\lambda$ is of the order of the period of the helicity:

$$h \sim \lambda \sim \frac{T_i^2}{4\pi\Lambda e^4 Z^4 n_i} \qquad\qquad n_i \sim \frac{T_i^2}{4\pi\Lambda e^4 Z^4 h}$$

Plasma should be magnetized to suppress the transversal conductivity:

$$\rho_B \ll r,\ \rho_B \ll \frac{\lambda}{2\pi} \sim \frac{h}{2\pi},\ r \sim \frac{h}{2\pi} \qquad\qquad B_z[T] \gg \frac{2\sqrt{2}\pi c}{e\sqrt{m_i}} \frac{\sqrt{T_i}}{h}$$

Velocity of the magnetic perturbation $V_z$ should be greater than the thermal velocity $V_i$, therefore the radial electric field:

$$E_r > B_z \frac{2\pi r}{h} \frac{V_i}{c} \gg \frac{2}{em_i} \frac{T_i}{r}$$

To achieve an appropriate concentration of the trapped particles the mean corrugation of the magnetic field along the field line in the plasma cross-section should be $R \sim 1.5$–$2$.

Duration of the experiment $\tau$ should be greater than the ion transit time to achieve the steady state. Length of the device is $N\times h$, where $N \sim 10$ is the number of periods, therefore:

$$\tau \gg \frac{N\times h}{V_i} \sim \frac{\sqrt{m_i}}{4\sqrt{2}\pi\Lambda e^4 Z^4} \frac{T_i^{\frac{3}{2}}}{n_i} N$$

Summarizing the estimations above the required plasma parameters are:

| | | |
|---|---|---|
| $n_i \sim 10^{19}\ m^{-3}$ | $E_r \sim 100\ {}^{V}\!/_{cm}$ | $h \sim 20\ cm$ |
| $T_i \sim 10-100\ eV$ | $\tau \sim 0.1\ s$ | $N = 12$ |
| $B_{max} = 0.1-0.3\ T$ | $r \sim 5\ cm$ | $R_{mean} \sim 1.5-2$ |

**Magnetic field structure**

Magnetic system of the device includes the plasma gun coils, entrance and exit expander coils, solenoid with helicoidal field and interconnections between these parts — see Fig. 1. The main goal of the optimization is to maximize the variation of the magnetic field along the field line in the cross-section of the plasma. We note that the mirror ratio R is equal to unity on the magnetic axis in a helically symmetric magnetic system. Plasma

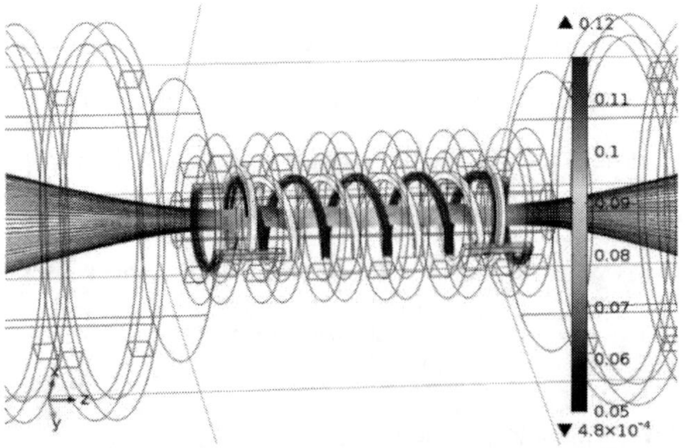

Fig. 1. Configuration of the magnetic system with a helicoidal section for control of the axial plasma flow.

gun and expander coils are axisymmetric and therefore need no complicated optimization. Combined solenoid is composed of two parts: strait axial component of the helicoidal field is created by the set of the flat coils; helicity is induced by the even number of spiral bars with counter-flowing currents. Two bars provide the most favorable distribution of the magnetic field variation along the radius [7]. The free variables are the period of the helicity to the diameter of the spiral bars ratio $h/d$ and helical to strait magnetic field ratio $B_{hel}/B_{str}$. The range of variables where detailed calculations were performed was $h/d = 0.6$–$1.2$, $B_{hel}/B_{str} = 0.4$–$0.8$. Field lines were assumed to be lost if they cross the cylinder with the diameter $d_c = 0.8\ d$.

Optimal values of the parameters were $B_{hel}/B_{str} \sim 0.75$–$0.8$, $h/d \sim 0.8$–$0.9$ (Fig. 2).

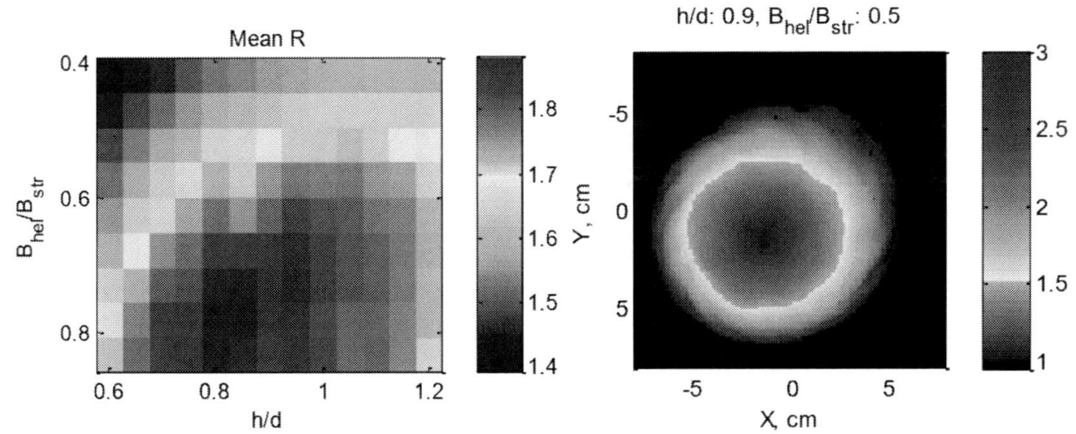


Fig. 2. Left: mean variation of the magnetic field along the field line $R$ in parameter space. Right: distribution of the $R$ value along the cross-section near the optimal parameters.

Magnetic axis in helical magnetic field is sufficiently 3-dimensional. Saddle-like correction coils were used in interconnections between the combined solenoid and expanders to match it with the geometrical axis and, therefore, plasma gun and radial electric field source. Axes match when the current in the correction coil is proportional to the current in spiral bars, $I_{corr} = 0.23\ I_{spiral}$. (Fig. 3). The ends of the helicoidal winding create parasite mirrors at the ends of the combined solenoid. The mirrors are eliminated by the additional planar error correction coil with the current $I_{planar} = 0.57\ I_{spiral}$.

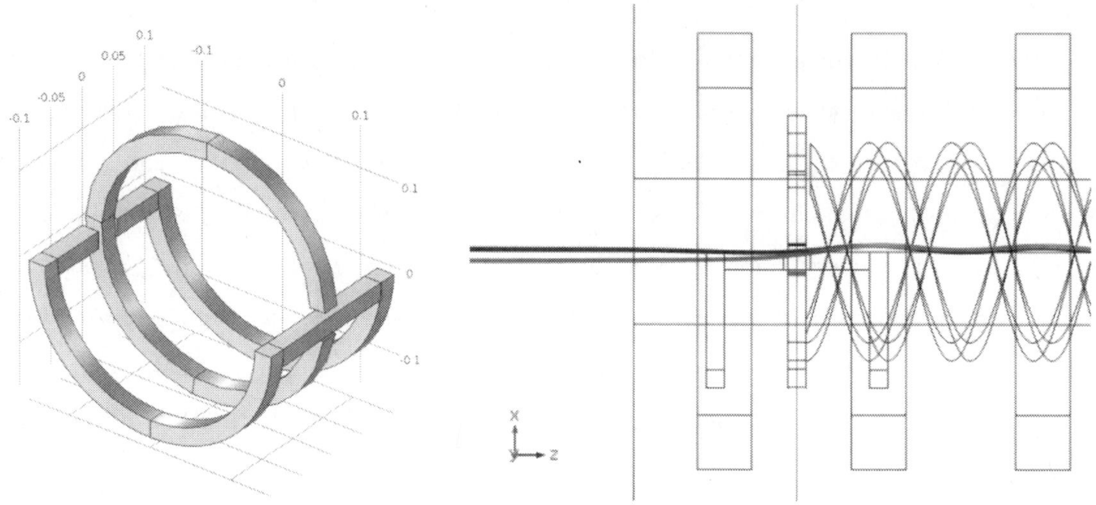


Fig. 3. Left: correction coil layout. Right: magnetic axis before (red) and after (blue) correction.

## Summary

Plasma parameters and magnetic field configuration that are required to verify the helicoidal confinement theory are appropriate for a moderate-scale experimental device of the concept exploration category. Such a device is now planned in BINP.

## Acknowledgements

This work was supported by Russian Science Foundation (project 14-50-00080).

[1] T.C. Simonen, et al., *J. Fusion Energy* **29**, 558 (2010)

[2] P.A. Bagryansky, et al., *Nuclear Fusion* **55**, 053009 (2015)

[3] A.A. Ivanov and V.V. Prikhodko, *Plasma Phys. Control. Fusion* **55**, 063001 (2013)

[4] A. Burdakov, et al., *Fusion Sci. Technol.* **51** (No. 2T), 106 (2007)

[5] T. Imai et al, *Fusion Sci. Technol.* **63** (No. 1T), 8 (2013)

[6] A. D. Beklemishev, *Fusion Sci. Technol.* **63** (No. 1T), 355 (2013)

[7] Masahiro Wakatani, *Stellarator and heliotron devices*, Oxford: Clarendon Press, 27 (1998)

# Mirror based fusion neutron source: R&D status and applications

A.V. Anikeev[1,2], P.A. Bagryansky[1,2], A.D. Beklemishev[1,2], A.A. Ivanov[1,2],
E.Yu. Kolesnikov[1], M.S. Korzhavina[1], O.A. Korobeinikova[1,2], A.A. Lizunov[1,2],
V.V. Maximov[1,2], S.V. Murakhtin[1,2], E.I. Pinzhenin[1], V.V. Prikhodko[1,2], E.I. Soldatkina[1,2],
A.L. Solomakhin[1,2], Yu.A. Tsidulko[1], D.V.Yakovlev[1,2], D.V. Yurov[1]
[1] *Budker Institute of Nuclear Physics SB RAS, Novosibirsk, Russia*
[2] *Novosibirsk State University, Novosibirsk, Russia*

## 1. Introduction

The Budker Institute of Nuclear Physics in worldwide collaboration develops a project of a 14 MeV neutron source for fusion material studies and other applications [1,2]. The projected neutron source of plasma type is based on the gas dynamic trap (GDT), which is a special magnetic mirror system for plasma confinement [3]. Essential progress in plasma parameters was performed in recent experiments at the GDT facility in the Budker Institute, which is a hydrogen (deuterium) prototype of the source. Stable confinement of hot-ion plasmas with the relative pressure exceeding 0.5 was demonstrated. The electron temperature was increased up to 0.9 keV in the regime with additional ECRH of a moderate power [4]. These parameters are the record for axisymmetric open mirror traps. These achievements shift the projects of a GDT-based neutron source on a higher level of competitive ability and make possible today to construct a source with reasonable parameters, suitable for materials testing. The first part of this paper presents a brief review of basic experimental results obtained on the GDT device in recent years. The second part of the paper focuses on numerical simulations of the GDT neutron source and its possible applications including a fusion material test facility and a fusion-fission hybrid system.

## 2. Resent results of the GDT experiment

Fig.1 shows the layout of the GDT experimental device with quarter-section. This is a 7 m long axisymmetric mirror trap with high mirror ratio ($B_0 = 0.3$ T, $B_m$ up to 15 T) for plasma confinement. Warm maxwellian plasma is confined in a gas dynamic regime, which is characterized by collisional particle losses into the end chambers of the device. An inclined

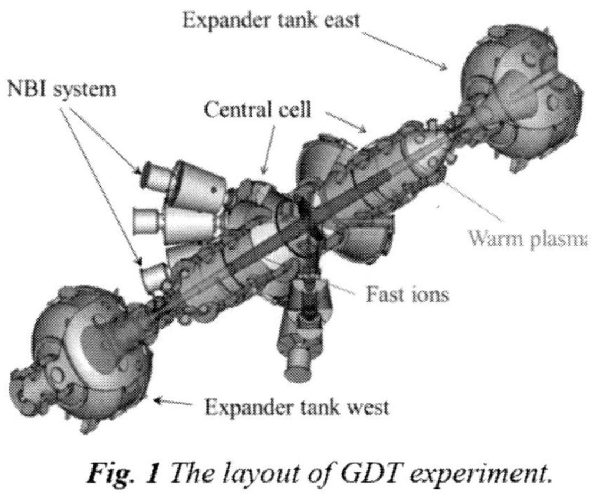


***Fig. 1** The layout of GDT experiment.*

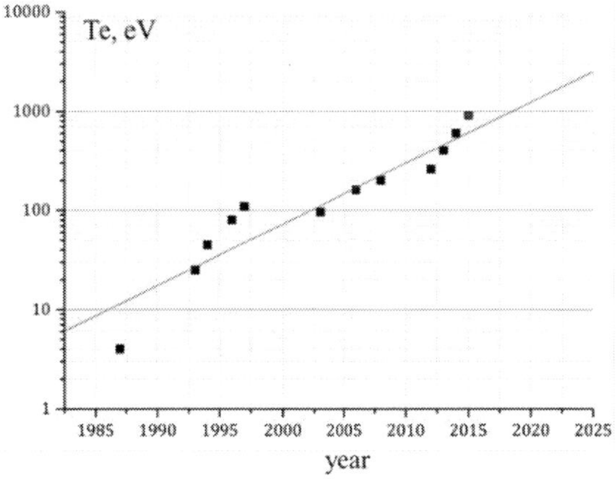


***Fig. 2** Electron temperature in GDT experiments vs. year of achievement.*

injection of eight deuterium atom beams (with energy 20-25 keV and total power 5 MW) produces fast ions oscillating back and forth between the hills of the magnetic field. The peaks of the fast ion density appearing near to their turning points represent the volumes of high plasma pressure and intense fusion neutron production.

Fig. 2 presents the rise of the electron temperature in GDT experiments over recent 40 years. The maximal electron temperature obtained in this year (red point) exceeds 0.9 keV that is correspond to electron temperature in first tokamaks. This result was achievable due to the additional ECR heating by the new ECRH system (two 54.5 GHz gyrotrons with 0.4 MW power each + beam lines) [5].

Obtained in the GDT experiment electron temperature substantially exceeds previously predicted [2] limit for $T_e$ in a magnetic mirror trap with neutral beam injection: $T_e \sim 0.01\ E_{inj}$, were $E_{inj}$ is the energy of injected neutral atoms. Thus it is possible to abandon this prediction and use in GDT-NS modeling a self-consistent value of the electron temperature.

In GDT experiments the relative plasma pressure β exceed 0.5 that corresponds to fast ion density up to $5\text{x}10^{19}\ \text{m}^{-3}$ with $\langle E_i \rangle$=10 keV. All these results were obtained by using a new efficient method of transverse plasma confinement, so called "vortex confinement" [6]. Shear flows, driven via biased end plates and limiters, in combination with finite-Larmor-radius effects are shown to be efficient in confining high-beta plasmas even with a magnetic hill on axis.

The GDT experimental achievements make possible today to construct a source with reasonable parameters, suitable for materials testing. The next sections of this paper are devoted to the numerical simulation of several neutron sources based on the achieved experimental data.

## 3. Results of the GDT neutron source simulations

During past few years several transport codes have been developed and applied for computational studies of GDT-NS in parallel to the experimental research. The plasma physics calculations of the neutron source's parameters have been performed by the Integrated Transport Code System (ITCS) [7]. ITCS includes different modules for plasma, particles transport and neutron production modelling. The main 3D Monte-Carlo module for fast ion transport MCFIT+ has been substantially upgraded for adequate simulation of the various versions of the mirror based neutron source. The experimental and theoretical foundations of these phenomena were obtained in the GDT-U experimental facility in the Budker Institute. Brief simulations of GDT plasma parameters and optimization researches were made by the one-dimensional plasma code DOL that is developing in Budker Institute for fast calculation of main plasma parameters evaluation in the mirror trap [8].

Table 1 presents a summary of simulation results for several mirror based neutron source projects. The first column shows achieved GDT experimental results for comparison with parameters of the projected devices. The GDT experiment regime was also simulated by the numerical codes described above and results of this simulation are in a good agreement with experimental data that is the best proof of our numerical tools.

**TABLE 1.** Main results of neutron source simulations:

| Parameters | GDT exp. | GDT-U (next) | GDT-NS | GDMT NS | Mirror NS |
|---|---|---|---|---|---|
| Magnetic field, $B_0$ / $B_m$ [T] | 0.34/12 | 0.5/15 | 1/15 | 1/9 | 2/15 |
| Effective mirror ratio, $k$ | 35 | 30 | 15 | 10…100 | 7.5 |
| NB injected/heat power, [MW] | 5/3 | 9.6/7.2 | 40/30 | 40/30 | 100/90 |
| NBI energy, $E_{inj}$ [keV] | 25 | 20 | 65 | 65 | 80 |
| Pulse duration, [s] | 0.005 | 0.03 | continuous | continuous | continuous |
| Warm ion density, $n_w$ [$10^{20}$ m$^{-3}$] | 0.3 | 0.5 | 0.2 | 0.3 | 0.08 |
| Fast ion density, $n_f$ [$10^{20}$ m$^{-3}$] | 0.5 | 0.7 | 2.5 | 3.5 | 1 |
| Mean ion energy, $T_i$ [keV] | 10 | 10 | 35 | 30 | 60 |
| Electron temperature, $T_e$ [keV] | 0.25 / 0.9* | 0.4 / 1* | 0.7 | 1.5 | 6 |
| Relative plasma pressure, β | 0.6 | 0.5 | 0.5 | 0.5 | 0.5 |
| Fusion energy gain factor, $Q_{fus}$ | - | - | 0.05 | 0.1 | 0.5 |
| Fusion neutron power, $P_n$ [MW] | - | - | 1.5 | 3 | 45 |

* with additional ECRH

The second column presents parameters of the GDT-U project – next step of the GDT experiments. A strong modernization of the GDT magnetic system for 0.5 T in midplane and up to 15 T in mirrors is planned. Also a new NBI system includes four modules with 2.4 MW 20 keV D-beam each with 30 ms pulse duration and corresponded power supply will be realized in near future. The main goals of the update are the demonstration of increasing mirror plasma parameters closed to moderate GDT neutron source project and achieving steady-state confinement. Fig.3 shows a time dependence of the ion ($T_i$) and electron ($T_e$) temperatures in GDT-U during NBI heating only. As it's seem the 30 ms power beam injection is enough for a physical steady-state regime. An additional ECRH can increases the electron temperature over 1 keV for parameters in Table 1.

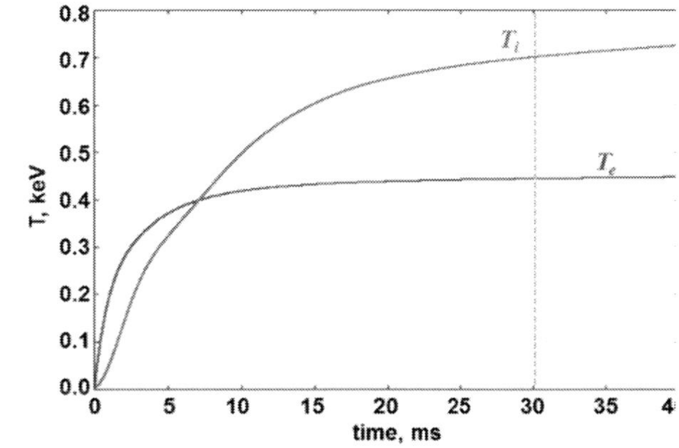


*Fig.3. Simulated temperature in GDT-U vs. time.*

The number of DT fusion neutron source projects were simulated on the base of the achieved GDT experimental results ($T_e \sim 0.6$ keV, β ~0.5). First of them is a model of GDT-NS proposed for fusion material research [2]. It is an axially symmetric mirror machine of the GDT type, 10 m long and with magnetic field $B_0 = 1$ T and mirror ratio of 15. The source parameters are presented in the column 3 of Table 1. The Gas-Dynamic Multiple-mirror Trap (GDMT) based on GDT and multiple-mirror trap GOL-3 results may be a near-term substitute [9]. GDMT based neutron source has improved axial confinement with the effective mirror ratio $k$ up to 100 and high $T_e$. The simulated GDT-NS parameters (col.4 in the Table) allow to propose this neutron source as a basic for different applications including fusion material test facilities and moderate fusion driven (hybrid) systems (FDS).

The first analysis of possibility to use the GDT based neutron source as a driver in the sub-critical system (FDS) was made in [10]. It has shown necessity to optimize the GDT-NS parameters. Optimized mirror based NS with $Q_{fus} = 0.5$ uses 100 MW of 80 keV NBIs and has 6 m long n-zone with up to $1.4 \times 10^{19}$ n/s production is presented in last column of Table 1. It assumes a kinetic regime of axial confinement, vortex transverse confinement and warm maxwellian plasma minority for the DCLC stabilization. This powerful NS is proposed for basic of future FDS (hybrids) for MA burning or for Th-reactor realizing (see Fig 4 and 5).

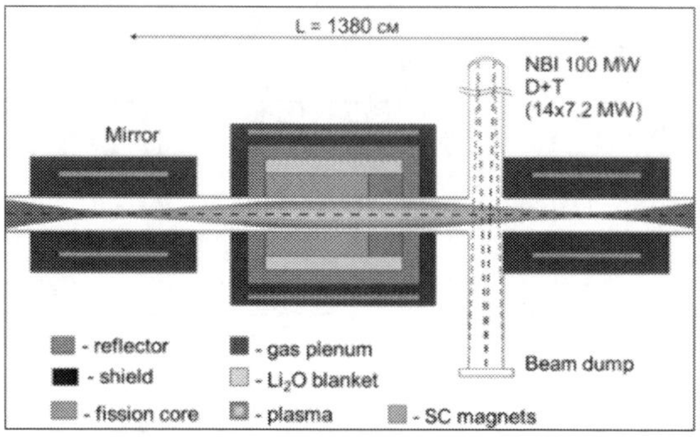


*Fig.4. General layout of the mirror based FDS*

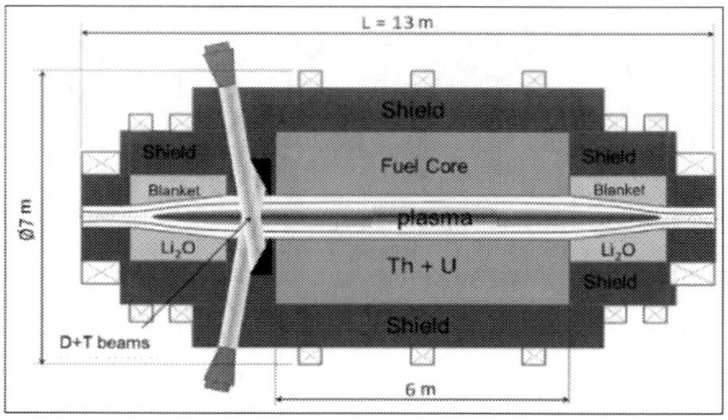


*Fig.5 Mirror based hybrid reactor with Th+U fuel*

## 4. Conclusions

Resent GDT experiments results show the possibility of realizing competitive neutron sources based on axisymmetric mirror cell.

The next step of GDT experiment was simulated. The results show the possibility of achieving steady state confinement with high plasma parameters.

Neutron sources based on GDT and multi-mirror GDMT were simulated. Proposed neutron sources with $Q_{fus}$ up to 0.1 can be used as a basic for fusion material test facilities and moderate fusion driven systems (FDS).

Numerical optimization of the mirror based neutron source for the driver in FDS hybrid system was made. Proposed source with $Q_{fus} = 0.5$ can be used for effective MA burning or for Th-reactor realizing.

This research was supported by the Russian Scientific Foundation Grant №14-50-00080.

**References:**

[1] A.Ivanov, E. Kryglyakov, Yu. Tsidulko, *Journal of Nuclear Materials* **307-311,** 1701(2002).
[2] P.A. Bagryansky, et al., *Fusion Engineering and Design* **70,** 13-33(2004).
[3] A.A.Ivanov and V.V. Prikhodko, *Plasma Phys. Control. Fusion* **55,** 063001 (2013).
[4] P. A. Bagryansky, et al., *Phys. Rev. Lett.* **114**, 205001 (2015).
[5] P A Bagryansky, et al., *Nuclear Fusion* **54,** 082001(2014).
[6] A. D. Beclemishcev, et al., *Fusion Science and Technology* **57**, Issue 4, 351-360 (2010).
[7] A.V. Anikeev, et al., *Trans. Fusion Technology* **39,** 183(2001).
[8] D.V. Yurov, et al. *Fusion engineering and Design* **87** 1684-1692 (2012).
[9] A. D. Beklemishev, et al., *Fusion Science and Technology* **63**, 1T, 46-51 (2013).
[10] K. Noack et al., *Annals of Nuclear Energy* **35** (2008) 1216-1222.

# Plasma surface currents in the presence of a resistive wall and their connection to Halo currents during disruptions in tokamaks

V.V. Yanovskiy and R. Paccagnella

*Consorzio RFX, Associazione EURATOM-ENEA sulla Fusione, Padova, 35127, Italy*

**1. Introduction.** Major disruptions and VDEs result in halo and eddy currents flowing in the conducting structures surrounding plasma [1–4]. These currents can produce unacceptable forces on vacuum vessel and in-vessel components, and their careful evaluation is necessary. Recent analytical theory [5, 6] predicts that the ideal plasma surface current can contribute to the halo, especially when the plasma boundary almost coincides with the rational surface. Here we speculate that such a particular case must not be considered in the frame of the ideal MHD, because it leads to a singularity for the plasma displacement. Treating the cold post-disruption plasma edge as a resistive layer we derive a dispersion relation for the mode growth rate and frequency, expressions for the resistive plasma "surface" current and eddy currents in the wall. Our approach has some similarities with that in [7, 8].

**2. Formulation of the problem.** We consider a cylindrical plasma of radius $a$ surrounded by a coaxial magnetically thin resistive wall with radius $b$, thickness $d$ and uniform resistivity $\eta_w$. Assuming that the rational surface is sufficiently close, but still outside of the plasma boundary we treat the plasma edge as a thin resistive layer with given thickness $W$ and resistivity $\eta_{pl}$, while the rest of the plasma is supposed to be ideal and inertialess. The plasma-wall gap and space behind the wall are treated as a vacuum. In this case, the equations governing the evolution of the system are:

$$\frac{\partial}{r\partial r}\left(r\frac{\partial\psi}{\partial r}\right)-\frac{m^2}{r^2}\psi-\frac{1}{(\mu-n/m)}\frac{\partial}{r\partial r}\left[\frac{\partial}{r\partial r}(r^2\mu)\right]\psi=0 \quad (1)$$

for the ideal plasma and vacuum regions;

$$\frac{\partial\psi}{\partial t}-i\Omega\psi=\frac{\eta_{pl}}{\mu_0}\Delta\psi \quad \text{with} \quad \Omega\equiv\frac{nV_z(a)}{R}-\frac{mV_\theta(a)}{r} \quad (2)$$

for the resistive plasma layer (at $\mu-n/m<<1$);

$$\frac{\mu_0}{\eta_w}\frac{\partial\psi}{\partial t}=\Delta\psi \quad (3)$$

for the resistive wall. These are derived in the cylindrical geometry $(r,\theta,z)$ for the $(m,n)$ mode of the magnetic perturbation $\mathbf{b}=\nabla\psi\times\mathbf{e}_z$, assuming also $nb/(mR)<<1$, where $R$ is the major radius, $\mu \equiv 1/q = RB_\theta / rB_z$, $B_\theta$ and $B_z$ are equilibrium poloidal and toroidal magnetic fields, $V_\theta(a)$ and $V_z(a)$ are equilibrium poloidal and toroidal edge velocities, respectively. For the time dependence of the form $\exp(\gamma t + i\omega t)$ with the growth rate $\gamma$, and the mode rotational frequency $\omega$, integration of Eqs. (2) and (3) across the resistive layer and the wall under the constant - $\psi$ assumption yields the following boundary conditions

$$\tau_\eta \left[\gamma + i(\omega - \Omega)\right] = \left.\frac{r\psi'}{\psi}\right|_a - \frac{a}{r_-}\left.\frac{r\psi'}{\psi}\right|_{r-} \quad \text{and} \quad \tau_w(\gamma + i\omega) = \left.\frac{r\psi'}{\psi}\right|_{b-}^{b+}, \tag{4}$$

where $\tau_\eta \equiv \mu_0 aW / \eta_{pl}$, $\tau_w \equiv \mu_0 bd / \eta_w$, $r_- = a - W$ and prime means the radial derivative. For the parabolic distribution of the equilibrium current

$$j_z = j_0\left(1 - \alpha_j \frac{r^2}{a^2}\right), \tag{5}$$

the solution of Eq. (1) within the ideal plasma region can be found in terms of hypergeometric functions $\psi_{pl} \propto z^{m/2}(1-z)^{-1}F(A,B;A+B+1;z)$ [9], so that

$$\left.\frac{r\psi'}{\psi}\right|_{r-} = m + 2z(\ln F(A,B;A+B+1;z))'_z\Big|_{r-} \quad \text{with} \quad z = \frac{\alpha_j r^2}{2a^2(1 - nq_0/m)}, \tag{6}$$

where $A = (m - \sqrt{m^2+8})/2$, $B = (m + \sqrt{m^2+8})/2$ and $q_0$ is the safety factor at the magnetic axis. The solutions in the vacuum gap and external vacuum are $\psi_{gap} \propto Cr^m + Dr^{-m}$ and $\psi_{out} \propto r^{-m}$, where $C$ and $D$ are arbitrary constants. Applying the second of Eqs. (4) we have

$$\left.\frac{r\psi'}{\psi}\right|_a = -m\frac{2m + \tau_w(\gamma + i\omega)(1 + (a/b)^{2m})}{2m + \tau_w(\gamma + i\omega)(1 - (a/b)^{2m})}. \tag{7}$$

Substitution of (6) and (7) into the first of Eqs. (4) provides us with a dispersion relation.

Once one knows the mode growth rate and frequency the currents in the resistive layer and wall can be calculated by

$$\mu_0 i_\eta = -\psi'\big|_{r-}^{a} = -\tau_\eta\left[\gamma + i(\omega - \Omega)\right]\psi(a) \quad \text{and} \quad \mu_0 i_w = -\psi'\big|_{b-}^{b+} = -\tau_w(\gamma + i\omega)\psi(b). \tag{8}$$

**3. Computation results and discussion.** To provide a basis for the comparison Fig. 1 shows the results of [5, 6] where the whole plasma is treated as ideal. The region within which the rational surface is close to the plasma boundary is designated by the vertical lines. Predictions of our model are presented in Figs. 2-6 showing the dependences of the normalized growth rate on the edge safety factor for the $m/n = 2/1$ mode. In Fig. 2 the growth rate of the locked mode

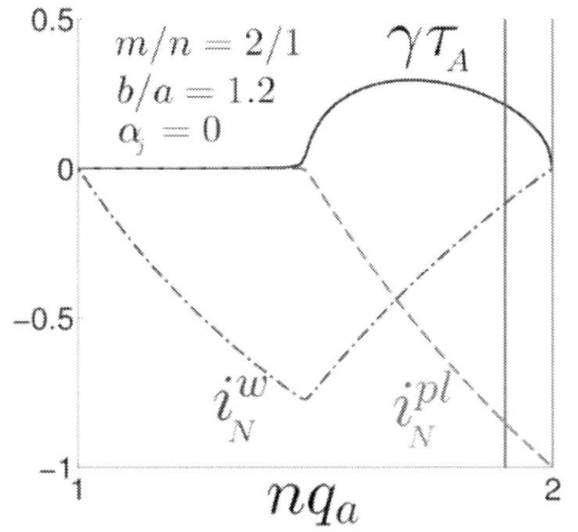


Fig. 1. Normalized growth rate, ideal plasma surface and resistive wall eddy currents vs. edge safety factor for flat current profile.

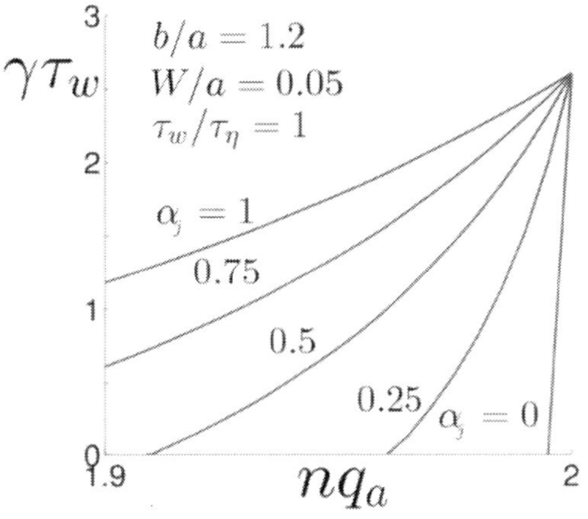


Fig. 2. Normalized growth rate of the locked mode vs. edge safety factor for different current profiles.

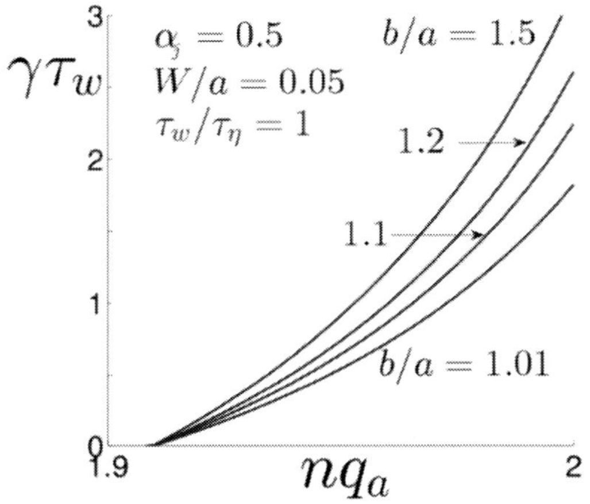


Fig. 3. Normalized growth rate of the locked mode vs. edge safety factor for different wall positions.

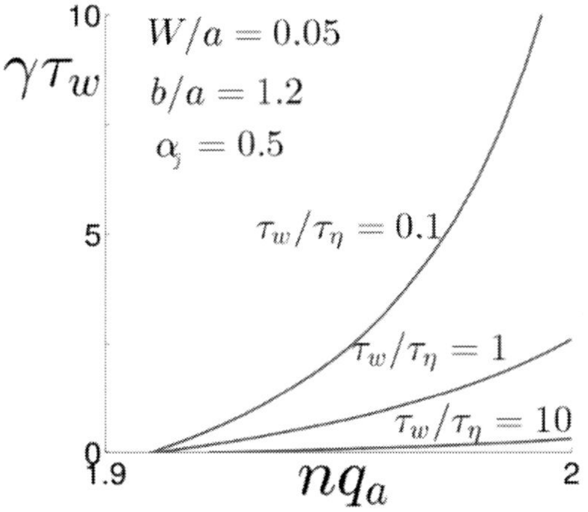


Fig. 4. Normalized growth rate of the locked mode vs. edge safety factor for different ratios of the wall and resistive layer times.

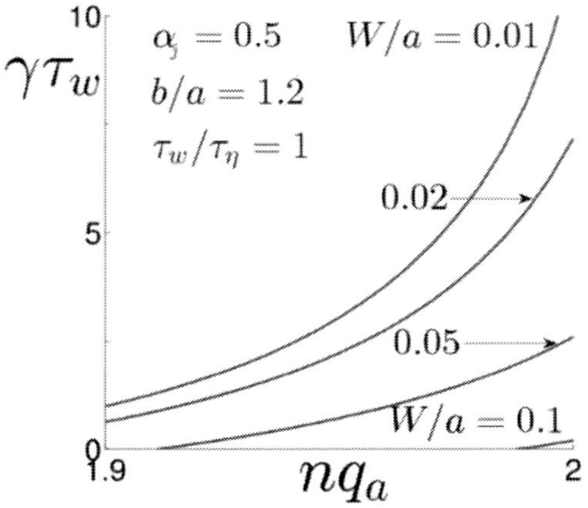


Fig. 5. Normalized growth rate of the locked mode vs. edge safety factor for the different thickness of the resistive layer.

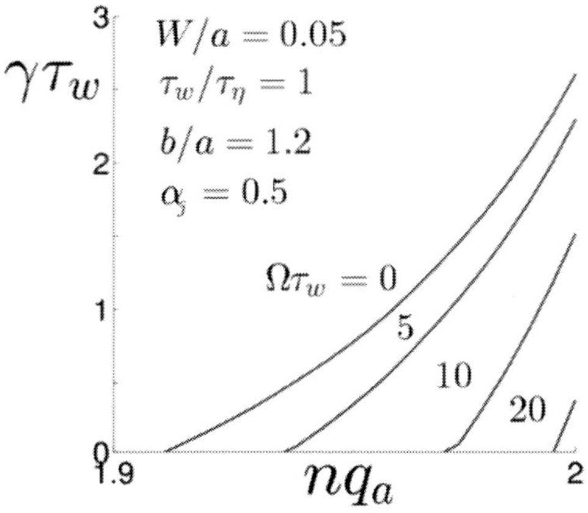


Fig. 6. Normalized growth rate of the rotating mode vs. edge safety factor for different plasma velocities.

is plotted for different current profiles from flat $\alpha_j = 0$ to the purely parabolic $\alpha_j = 1$. At the larger current gradient the mode becomes more unstable. Fig. 3 illustrates the stabilizing influence of the wall on the perturbation, which is slower when the wall is closer to the plasma. After the thermal quench the plasma is relatively cold, so the resistive effects becomes stronger as can be seen in Fig. 4 highlighting the dependence on the ratio $\tau_w / \tau_\eta$. The thickness of the plasma resistive layer $W$ determines the plasma resistive time $\tau_\eta$ and also enters relation (6), Fig. 5 shows that for larger $W$ the mode growth slows down. Adding of the plasma rotation to the model makes the dispersion relation complex that results in non-zero mode frequency, which in turn provides stabilizing effect on the perturbation. According to Fig.6 even moderate plasma rotation can completely suppress the mode.

Eqs. (8) allow to estimate edge plasma and wall eddy currents. For locked modes they are simply proportional to the mode growth rate and the amplitude of the perturbed poloidal magnetic flux at the plasma edge and the wall, respectively. In this case they are related as $i_\eta / i_w = \tau_\eta \psi(a) / \tau_w \psi(b)$.

**4. Conclusion.** In contrast with results in [5, 6] the model predicts increase of the perturbed plasma edge current for steeper profiles of the equilibrium current. For locked modes it becomes larger with a more distant wall, colder plasma and thinner resistive layer. Plasma rotation provides mode stabilization, to address its influence on the perturbed edge current dynamics further analysis is needed. Our consideration is free from the assumption that the mass density has a jump at the plasma boundary, which is the main reason leading to the existence of the surface currents within the ideal MHD.

**Acknowledgments.** One of the authors (VVY) is grateful to V.D. Pustovitov for continuous support. This work was supported by the RFX-mod team.

[1] ITER Physics Expert Group on Disruptions, Plasma Control, and MHD and ITER Physics Basis Editors, Nucl. Fusion **39**, 2251 (1999).

[2] T. C. Hender *et al.,* Nucl. Fusion **47**, S128 (2007).

[3] R. Paccagnella *et.al.,* Nucl. Fusion **49**, 035003 (2009).

[4] H. Strauss *et.al.,* Phys. of Plasmas **17**, 082505 (2010).

[5] L. E. Zakharov, Phys. Plasmas **15**, 062507 (2008).

[6] L. E. Zakharov, Phys. Plasmas **18**, 062503 (2011).

[7] J. M. Finn, Phys. Plasmas **2**, 198 (1995).

[8] J. M. Finn, Phys. Plasmas **2**, 3782 (1995).

[9] A. B. Mikhailovskii, *Instabilities in a Confined Plasma* (Institute of Physics, Bristol, 1998).

# Multiple domain scheme for heat transport analysis in plasmas with magnetic islands: a first study.

Y. Zhang[1], F. Auriemma[1], A. Fassina[1], D. Lopez-Bruna[2], R. Lorenzini[1], E. Martines[1], B. Momo[1], F. Sattin[1] and D. Terranova[1]

[1] *Consorzio RFX (CNR, ENEA, INFN, Università di Padova, Acciaierie Venete Spa), Corso Stati Uniti, 4, 35127, Padova, Italy*

[2] *Laboratorio Nacional de Fusión, Ciemat, 28040, Madrid, Spain*

## 1. Introduction

In magnetically confined plasma experiments aiming to nuclear fusion, the existence of rational surfaces leads to the formation of so-called magnetic islands observed in magnetic configurations like tokamak, stellarator and reversed field pinch (RFP). The formation of islands modifies magnetic topology by breaking the nested magnetic surfaces, forming their own, individual axis as well as a separatrix contained inside certain flux surfaces around which the stochastic layer develops [1,2]. As a consequence, the presence of magnetic islands strongly modifies particle and heat transport, which affect the plasma confinement. Therefore, the transport under magnetic configuration with multi-axes needs to be studied further.

Transport research benefits of the availability of several well-developed 1.5D transport codes, which have been widely used in single axis magnetic configuration for decades while it is beyond their capability to treat situations with the presence of magnetic islands. Indeed in plasma configurations with multi-axes a monotonic radial coordinate is undefinable in the whole plasma volume, which, together with the correct metrics (the spatial derivative of the volume and the first element of the metric tensor, denoted as $V'$ and $G_1$ in this paper, respectively), are essential to solve the transport equations. Facing this issue, a new approach capable of studying transport under multi-axis configuration is presented.

In this paper we show a preliminary study of the energy transport in RFX-mod [3], solving the equation $\frac{3}{2}n_e\frac{\partial T_e}{\partial t}+\frac{1}{V'}\frac{\partial \Gamma_e}{\partial \rho}=S_e$ under both single and multiple magnetic axes configurations, where $\Gamma_e=-V'G_1 n_e \chi\frac{\partial T_e}{\partial \rho}$ is the energy flux, $n_e, T_e$ are the local plasma density and temperature, $S_e$ is the energy source and $\chi$ is the thermal diffusivity. To this end, a new scheme with multi-axes (dubbed Multi Domain Scheme - MDS), together with a new tool, named DAx_transp3, is presented. DAx_transp3 is a Fortran routine capable of studying transport in both single axis scheme and MDS. It works as a proxy for ASTRA [4], giving the opportunity of testing the feasibility of the multi-axes domain in a controlled environment. DAx_transp3 routine has been successfully benchmarked with ASTRA in single axis regime based on the experimental data obtained on RFX-mod, and the results are shown in Section 2. Detailed information of DAx_transp3 routine and its preliminary results on multi-axes regime are also presented in Section 3.

## 2. Single domain: benchmark with ASTRA

Two electron temperature profiles, measured by Thomson scattering (TS) in shot #30068 at t=0.09s and t=0.11s, are shown in Figure 1, as a function of the minor radius r. The hot region is shifted around 10cm far from the geometrical center. These thermal structures are small in

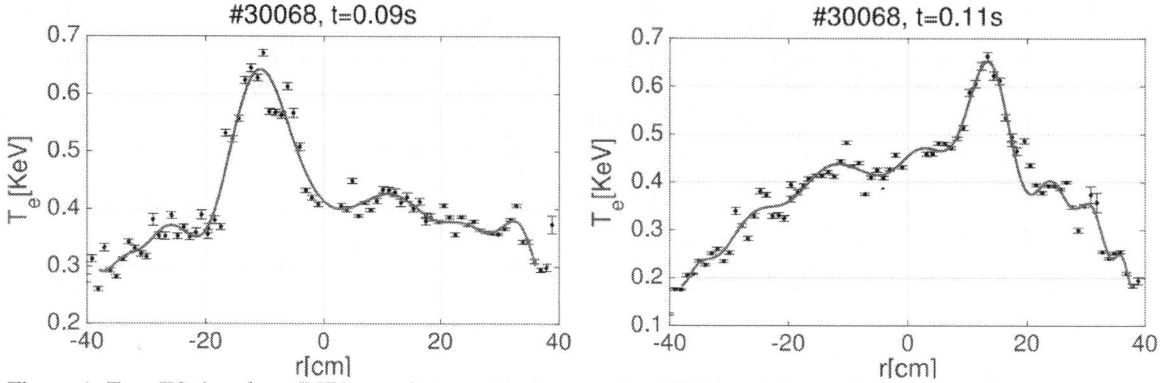


Figure 1. Two TS data from RFX experiment with shot number 30086 at different time. r is the minor radius.

the sense that they exist only on one side of the geometric center, and we dub these states as $SHAx_i$ state, in order to distinguish them from Single Helical Axis (SHAx) state [5,6], in which the dimension of the thermal structure is big enough to cross the geometric center of the vacuum chamber. It is important to notice that in both $SHAx_i$ and SHAx states only one magnetic helical axis is present, which means a global monotonic flux coordinate can be found. Figure 2 top shows the same $T_e$ profiles, plotted as a function of the flux coordinate $\rho = \sqrt{V_\rho/(2\pi^2 R)}$, where $V_\rho$ is the helical volume enclosed by each flux surface labeled by $\rho$ and $R = 2m$ is the major radius of RFX. From the center towards the edge, a temperature gradient, which is related to electron internal transport barrier (eITB) [6], could be identified between the hot core and outer region, corresponding to the marked region in Figure 2.

The benchmark work was performed between DAx_transp3 routine and ASTRA, based on a database of about 20 shots with no separatrix. SHEq code [7] is capable of calculating the magnetic equilibrium as well as the flux surface average of the kinetic quantities. In particular, here it is used to provide $\rho$, $T_e(\rho), V'$ and $G_1$. Afterwards, these quantities, together with a selection of $\chi$ profiles, are provided as input information to both codes to find the best temperature profile matching between numerical results and experimental data. For DAx_transp3, the selection of best $\chi$ is done by hand while for ASTRA, it is done

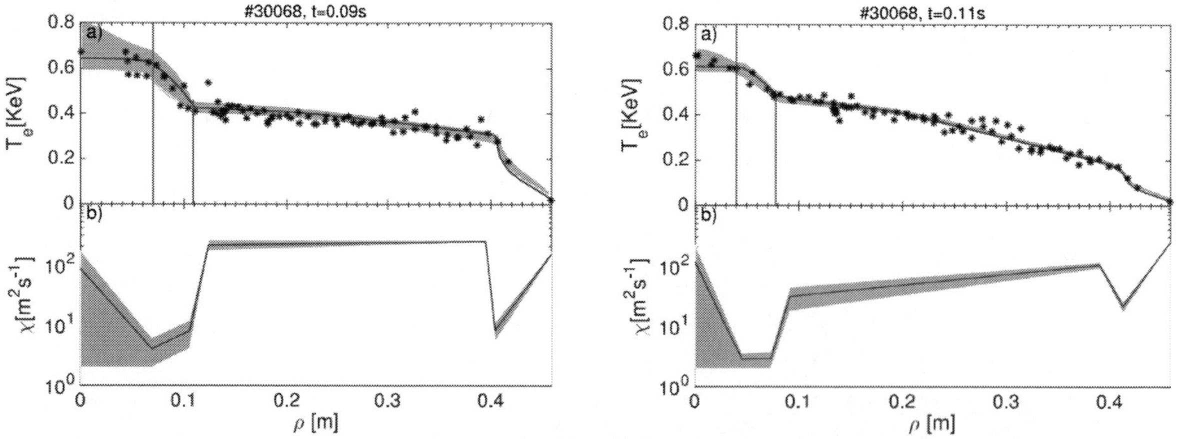


Figure 2. Benchmark result corresponding to the two cases shown in Figure 1. Upper panels (a) report the experimental $T_e$ (asterisks), the numerical profile computed with the DAx_transp3 routine (red line) and the CI's computed by ASTRA (blue shadow area). The lower panels (b) show the corresponding energy diffusivities (red line for the values used in the DAx_transp3 routine, blue area for ASTRA ones).

automatically using Genetic Algorithm [8], which generates a serials of "best choice", also named confidence interval (CI's). The benchmarking results are shown in Figure 2. In the upper two pictures, the black asterisks are the experimental electron temperature data while the red lines are the numerically computed $T_e$ profile from DAx_transp3. The blue shadow area in Figure 2 shows the CI's of $\chi$ (bottom plots) and of the corresponding $T_e$ profile computed by ASTRA. In these two cases, numerical results of both $T_e$ and $\chi$ profiles from DAx_transp3 routine are in good agreement with ones from ASTRA, which indicates the benchmarking between DAx_transp3 routine and ASTRA is successful. Moreover the CI's of the resulting numerical $T_e$ profiles are within the error of TS data, showing that the $\chi$ profiles correctly describes the energy transport level in the two cases.

## 3. Multiple Domain Scheme

Figure 3 shows a cross section of a so-called Double Axes (DAx)[7] magnetic configuration, in which there is a magnetic island with its own axis. The contour of the magnetic surfaces is plotted with the island separatrix represented by the red line. Region I is formed by nearly circular surfaces in the plasma core enclosed by the separatrix and nested around the main magnetic axis. The bean-shaped magnetic surfaces nested around the island O-point form region II. The surfaces outside the separatrix form region III. These regions are the three domains of MDS, in each of which a monotonic coordinate can be found and the kinetic quantities can be described as flux functions (i.e. they are constant over the flux surfaces).

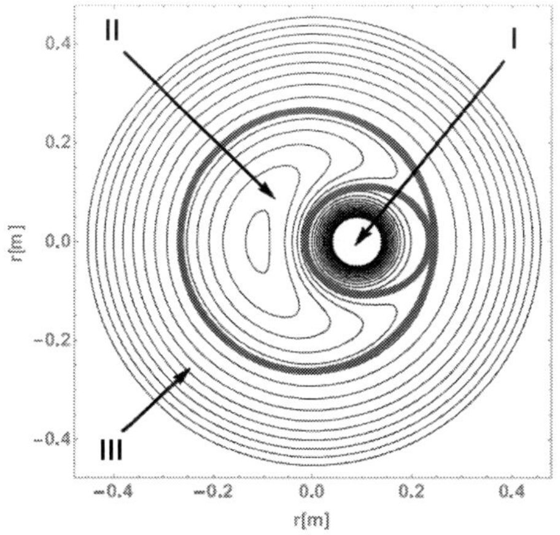

Figure 3. Poloidal cross-section of a DAx with the three domains of the MDS.

The three domains are coupled via common boundary conditions on fluxes and plasma quantities on the mutual interfaces. Under this regards a particular effort has been devoted to designing the coupling through the separatrix, which models the fluxes in the separatrix region located at the island boundary.

To this end, at each time-step, the DAx_transp3 routine solves the energy transport equation for a given diffusivity profile $\chi$. The boundary conditions for zone I and II are: $dT_e/d\rho=0$ at $\rho=0$ (i.e. on the core and island axes), $T_e=T_{sep}$ at the edge (interface with the separatrix). The temperature profile in zone III is computed solving the same equation with different boundary condition: at the separatrix interface the temperature is fixed at $T_{sep}$ whereas the edge temperature (interface with the plasma wall) is given by the experimental value $T_a$. After each time step the fluxes leaving (or entering, depending on the $\nabla T_e$ sign) the three zones and entering the stochastic layer are computed, being respectively $\Gamma_I$, $\Gamma_{II}$ and $\Gamma_{III}$. Then the evolution of the separatrix temperature is estimated with an explicit scheme: $\frac{\partial T_{sep}}{\partial t} = \frac{2}{3n_e}\frac{(\Gamma_I+\Gamma_{II}+\Gamma_{III})}{V_{sep}}$, being $V_{sep}$ the volume of the stochastic layer surrounding the separatrix. $V_{sep}$ has been chosen as the volume of the last cell of the island numerical grid (zone II). The simulation ends when convergence has been obtained in the three zones and the fluxes in the stochastic layer are mutually compensated, keeping constant the $T_{sep}$.

Figure 4 reports an example of DAx case studied with the DAx_transp3 routine: the mapping of the experimental $T_e$ in the three zones are reported with asterisks (upper panels) and with the continuous line for the numerical $T_e$ profile. The lower ones show the corresponding

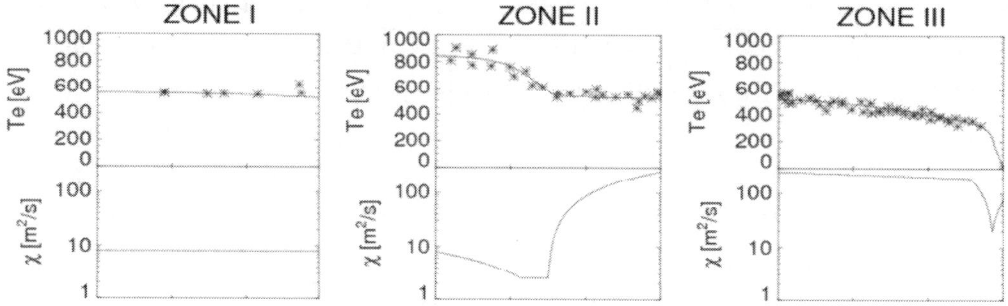


Figure 4. Experimental $T_e$ in the three regions for shot 37386 at 60ms (asterisks) mapped over the local $\rho$. The continuous line represents the numerical temperature profile. The lower panels show the corresponding diffusivity profiles.

diffusivity. The zone I and III have higher transport than the barrier region that exhibits a transport level comparable with the $SHAx_i$ eITB, in this example.

## 4. Conclusion and future work

The development of a new scheme to study transport in a magnetic configuration with islands is ongoing. A new tool, DAx_transp3 routine, has been developed and successfully benchmarked with ASTRA in QSH state without separatrix. The same tool has been tested on a DAx case, as shown. The next step is to build a larger database of QSH with separatrix, analyzing the influence of the presence of islands on plasma transport. Moreover, it is important to mention that besides the scheme presented in this paper, at least another scheme is under study: in the TJ-II stellarator [9] we are approaching the problem merging together zone I and III, finding a convenient flux coordinate and solving the transport equation by means of ASTRA. At the separatrix position an external routine has to be called in order to solve the transport in the zone II (island) and provide to the transport code the additional local energy source (or sink) to be included in the transport calculation.

## Acknowledgement

This project has received funding from the European Union's Horizon 2020 research and innovation programme under grant agreement number 633053. The views and opinions expressed herein do not necessarily reflect those of the European Commission.

**References**

[1]Fitzpatrick R. 1995 *Physics of Plasmas* **2** 825–38
[2]Hölzl M. *et al.* 2007 *Phys. Plasmas* **14** 052501
[3]Sonato P. et al. 2003 Fusion Eng. Des. 66 161
[4]Pereverzev G.V. *et al.* 2002 IPP Technical Report 02 Max-Planck-Institut für Plasmaphysik
[5]Lorenzini R. *et al.* 2012 *Nucl. Fusion* **52** 062004
[6]Lorenzini R. *et al.* 2009 *Nature Physics* **5** 570–4
[7]Martines E. *et al.* 2011 *Plasma Phys. Control Fusion* **53**(035015)
[8]Auriemma F. *et al.* P4.157 this conference
[9]Alejaldre C. *et al.* 1990 Fusion Technol. 13, 521

# Toroidal rotation in resonant regimes of tokamak plasmas due to non-axisymmetric perturbations in the action-angle formalism

C. G. Albert[1], M. F. Heyn[1], S. V. Kasilov[1,2], W. Kernbichler[1], A. F. Martitsch[1]

[1] *Fusion@ÖAW, Institut für Theoretische Physik - Computational Physics, TU Graz, Petersgasse 16, 8010 Graz, Austria*
[2] *Institute of Plasma Physics, National Science Center "Kharkov Institute of Physics and Technology", Akademicheskaya Str. 1, 61108 Kharkov, Ukraine*

## Introduction

Toroidal torque generated via neoclassical toroidal viscosity (NTV) [1, 2, 3, 4] caused by external non-resonant, non-axisymmetric magnetic perturbations (TF-ripples, error fields, perturbations caused by ELM mitigation coils away from resonant surfaces) has a significant effect on toroidal plasma rotation in tokamaks. Besides collisional transport regimes, an important role (in particular, in ASDEX-Upgrade [4]) is played by resonant transport regimes such as superbanana-plateau [1] or bounce and bounce-transit resonance [2] regimes where transport coefficients are independent of the (small) collision frequency. Here, a universal approach by the canonical Hamiltonian quasilinear formalism in action-angle variables [5] is presented. This approach is well developed for the general case of small electromagnetic perturbations in tokamaks (see e.g. [6]). The described treatment covers both trapped and passing orbits in a unified way and does not require any simplifications of device geometry. Numerical results of NTV evaluation in resonant transport regimes are presented and compared to references [1, 3].

## Toroidal torque and radial transport within canonical Hamiltonian quasilinear theory

Within the action-angle formalism, the quasilinear kinetic equation describing the evolution of the slowly varying averaged part of the distribution function $f_0 = f_0(\mathbf{J})$ over the canonical angles can be cast to the form

$$\frac{\partial f_0}{\partial t} = \sum_{\mathbf{m}} m_i \frac{\partial}{\partial J_i} Q_{\mathbf{m}}, \qquad Q_{\mathbf{m}} = Q_{\mathbf{m}}(\mathbf{J}) = \frac{\pi}{2} |H_{\mathbf{m}}|^2 \delta\left(m_j \Omega^j - \omega\right) m_k \frac{\partial f_0}{\partial J_k}, \tag{1}$$

where $J_k$ are the actions, $\omega$ is the perturbation frequency, $\delta(\dots)$ is the Dirac delta-function. Summation is assumed over repeated indices $i, j$ and $k$, and bold typesetting denotes the full set of three quantities, $\mathbf{J} = (J_1, J_2, J_3)$. Splitting $H = H_0 + \tilde{H}$ of the Hamiltonian $H$ into an unperturbed (averaged over angles) part $H_0 = H_0(\mathbf{J})$ and a perturbation part $\tilde{H}$ leads to the following definitions of the canonical frequencies $\Omega^j$ and of the amplitudes $H_{\mathbf{m}} = H_{\mathbf{m}}(\mathbf{J})$ of the Fourier expansion of $\tilde{H}$ over the canonical angles $\theta^k$:

$$\Omega^i = \frac{\partial H_0}{\partial J_i}, \qquad \tilde{H} = \mathrm{Re} \sum_{\mathbf{m}} H_{\mathbf{m}} e^{i\left(m_k \theta^k - \omega t\right)}, \tag{2}$$

where summation is performed over all three Fourier indices $\mathbf{m} = (m_1, m_2, m_3)$. In a tokamak, two of the actions are the perpendicular adiabatic invariant (up to a constant factor the magnetic moment) $J_1 = J_\perp$ and the canonical toroidal momentum $J_3 = p_\varphi$ respectively given by

$$J_\perp \approx \frac{m_\alpha v_\perp^2}{2\omega_{c\alpha}}, \qquad p_\varphi = m_\alpha v_\varphi + \frac{e_\alpha}{c} A_\varphi \approx m_\alpha v_\parallel h_\varphi + \frac{e_\alpha}{c} A_\varphi, \tag{3}$$

where $m_\alpha$, $e_\alpha$ and $\omega_{c\alpha}$ are $\alpha$-species mass, charge and cyclotron frequency, respectively, $v_\varphi$, $h_\varphi$ and $A_\varphi$ stand for the toroidal co-variant components of particle velocity, unit vector along the magnetic field and vector potential, respectively, and $v_\perp$ and $v_\parallel$ are perpendicular and parallel velocities (all quantities are defined for the unperturbed magnetic field). In flux (Boozer) coordinates $(s,\vartheta,\varphi)$ where $s=\psi_{\rm tor}/\psi_{\rm tor}^a$ is the toroidal flux normalized by its edge value, $A_\varphi=-\psi_{\rm pol}(s)$ is a flux function (poloidal flux). Introducing $s_\varphi=s_\varphi(p_\varphi)$, which is the solution to the second equation in (3) with $v_\parallel$ set to zero (banana tips in case of trapped orbits), the poloidal action $J_2=J_\vartheta$ is defined for both, trapped and passing particles by

$$J_\vartheta=\frac{e_\alpha}{c}A_\vartheta(s_\varphi)\delta_{t-p}+J_\parallel,\qquad \text{trapped: } \delta_{t-p}=0,\quad \text{passing: } \delta_{t-p}=1, \tag{4}$$

where $A_\vartheta(s)=s\psi_{\rm tor}^a$ and the parallel adiabatic invariant is defined via the bounce average as

$$J_\parallel=\frac{m_\alpha\tau_b}{2\pi}\langle v_\parallel^2\rangle_b,\qquad \langle a(\vartheta)\rangle_b\equiv\frac{1}{\tau_b}\int\limits_0^{\tau_b}{\rm d}\tau\, a\left(\vartheta_{\rm orb}(\vartheta_0,\tau)\right). \tag{5}$$

Here $a(\vartheta)$ is any function of poloidal angle and integrals of motion $(J_\perp,H_0,s_\varphi)$, $\vartheta_{\rm orb}(\vartheta_0,\tau)$ is the (periodic) solution of the unperturbed guiding center (orbit) equations starting at $\vartheta_0$ $(B(\vartheta_0)=B_{\rm min})$, and $\tau_b$ is the bounce time, $\vartheta_{\rm orb}(\vartheta_0,\tau_b)=\vartheta_{\rm orb}(\vartheta_0,0)=\vartheta_0$. Since $\Omega^1=\langle\omega_{c\alpha}\rangle_b$ is much larger then the other frequencies, for quasi-static $(\omega\to 0)$ magnetic perturbations with perpendicular scale much larger than Larmor radius and banana width only $m_1=0$ contributes significantly in (1). Such Fourier amplitudes correspond to a gyroaverage because the gyrophase is a linear function of the first canonical angle, $\phi=\theta^1+\Delta\phi(\theta^2,\mathbf{J})$, while the remaining guiding center variables are independent of $\theta^1$. For a similar reason the Fourier amplitudes of the Hamiltonian perturbation in the form of a single toroidal harmonic $\propto\exp(in\varphi)$ differ from zero only for $m_3=n$, since the toroidal angle $\varphi$ is the only coordinate depending (linearly) on the canonical angle $\theta^3$, as given by a variable transformation in linear order over the Larmor radius,

$$\theta^2=\Omega^2\tau,\qquad \varphi=\theta^3-q\theta^2\delta_{t-p}+q\vartheta_{\rm orb}(\vartheta_0,\tau). \tag{6}$$

$\Omega^2=\Omega^2(\mathbf{J})=\omega_b=2\pi\tau_b^{-1}$ is the bounce frequency and $q=q(s_\varphi)$ the safety factor. Namely, within ideal MHD quasistatic electromagnetic perturbations are fully described by perturbations of the $B$ module on perturbed flux surfaces, $B=B_0(\vartheta)+{\rm Re}\left(B_n(\vartheta)\exp(in\varphi)\right)$, which results in

$$H_{\mathbf{m}}=\left\langle\left(m_\alpha v_\parallel^2(\vartheta)+J_\perp\omega_{c\alpha}(\vartheta)\right)\frac{B_n(\vartheta)}{B_0(\vartheta)}{\rm e}^{inq\vartheta-i\left(m_2+nq\delta_{t-p}\right)\omega_b\tau}\right\rangle_b, \tag{7}$$

where $\vartheta=\vartheta_{\rm orb}(\vartheta_0,\tau)$ in all functions of $\vartheta$ and only $\mathbf{m}=(0,m_2,n)$ with various $m_2$ contributing to the result. Thus, the resonance condition given by the $\delta$-function in Eq. (1) is reduced to

$$m_j\Omega^j-\omega=0\quad\to\quad m_2\omega_b+n\Omega^3=0, \tag{8}$$

$$\Omega^3=q\omega_b\delta_{t-p}+\Omega_{\rm tor},\qquad \Omega_{\rm tor}\equiv-\omega_b\frac{\partial J_\parallel}{\partial p_\varphi}=\Omega_{tE}+\langle\Omega_{tB}\rangle_b,\qquad \Omega_{tE}=-\frac{cq}{\psi_{\rm tor}^a}\frac{{\rm d}\Phi}{{\rm d}s}, \tag{9}$$

with $\Phi$ being the equilibrium electrostatic potential. In Eq. (9) only the contribution of the $\mathbf{E}\times\mathbf{B}$ drift to the cross-field toroidal rotation frequency $\Omega_{\rm tor}$ is given explicitly because $\Omega_{tB}$ is rather complicated. Condition (8) describes all regimes of interest here: The resonance $m_2=0$ for trapped particles corresponds to the superbanana-plateau resonance, $m_2=m$ for passing with poloidal mode $m$ gives a transit resonance, which is the only one surviving in the infinite aspect ratio limit where it reduces to a Cherenkov (TTMP) resonance. Finite mode numbers $m_2$ correspond to bounce and bounce-transit resonances for trapped and passing particles, respectively.

In the case of small enough perturbations considered here, quasilinear effects are weak so that $f_0$ is close to a local Maxwellian. Then Eq. (1) can be replaced by a set of radial transport equations for the moments of $f_0$. A transport equation of interest is the conservation law of generalized toroidal momentum, which is obtained by multiplying Eq. (1) by $p_\varphi\delta\left(s-s_{\rm can}(\theta,\mathbf{J})\right)$, integrating the product over phase space and dividing the result by $\mathrm{d}V/\mathrm{d}s$,

$$\frac{\partial}{\partial t}\left\langle \int \mathrm{d}^3 p\, p_\varphi f_0 \right\rangle + \frac{\mathrm{d}s}{\mathrm{d}V}\frac{\partial}{\partial s}\left(\frac{\mathrm{d}V}{\mathrm{d}s}F_{p_\varphi}\right) = T_\varphi, \tag{10}$$

where $s_{\mathrm{can}}$ is a variable change law, $V$ is the volume within the flux surface and $\langle \dots \rangle$ is the neoclassical "flux surface" average. The flux surface averaged toroidal torque density $T_\varphi$, which dominates the transport term containing the momentum flux density $F_{p_\varphi}$, is

$$T_\varphi = -n\frac{\mathrm{d}s}{\mathrm{d}V}\int \mathrm{d}^3\theta \int \mathrm{d}^3 J\, \delta\left(s - s_{\mathrm{can}}\right)\sum_{m_2} Q_{\mathbf{m}} \approx -2\pi^2 n m_\alpha^3 \frac{\mathrm{d}\psi_{\mathrm{pol}}}{\mathrm{d}V}\int_0^\infty \mathrm{d}v\, v^3 \int_0^{1/B_{\mathrm{min}}} \mathrm{d}\eta\, \tau_b \sum_{m_2} Q_{\mathbf{m}}, \tag{11}$$

where setting $s_{\mathrm{can}} \approx s_\varphi$ (ignoring the FLR effects) and using the $\delta$-function made the integration over angles and $J_3$ trivial. The remaining integration variables were changed from $J_\perp$ and $J_\vartheta$ to $\eta = v_\perp^2/(v^2 B_0)$ and velocity module $v$. Substituting $f_0$ in (1) by a drifting Maxwellian, $f_0 = (2\pi m_\alpha T_\alpha)^{-3/2} n_\alpha \exp((e_\alpha \Phi - H_0)/T_\alpha)$, where $n_\alpha = n_\alpha(s_\varphi)$ as well as $T_\alpha$ and $\Phi$, for $\mathbf{J}$ satisfying the resonance condition one obtains

$$m_k \frac{\partial f_0}{\partial J_k} = -\frac{n c f_0 \left(A_1 + A_2 u^2\right)}{e_\alpha \langle |\nabla \psi_{\mathrm{pol}}| \rangle}, \qquad \frac{A_1}{\langle |\nabla s| \rangle} = \frac{1}{n_\alpha}\frac{\partial n_\alpha}{\partial s} + \frac{e_\alpha}{T_\alpha}\frac{\partial \Phi}{\partial s} - \frac{3}{2T_\alpha}\frac{\partial T_\alpha}{\partial s}, \qquad \frac{A_2}{\langle |\nabla s| \rangle} = \frac{1}{T_\alpha}\frac{\partial T_\alpha}{\partial s},$$

where $u = v/v_T$ and $v_T = \sqrt{2T_\alpha/m_\alpha}$. Relating $T_\varphi$ to the flux-surface averaged particle flux density $\Gamma = -n_\alpha(D_{11}A_1 + D_{12}A_2)$ via the flux-force relation [1] $T_\varphi = -e_\alpha c^{-1}\langle |\nabla \psi_{\mathrm{pol}}| \rangle \Gamma$, resonant transport coefficients follow as

$$D_{11} = \frac{\pi^{3/2} n^2 c^2 q v_T}{e_\alpha^2 \langle |\nabla s| \rangle^2 \psi_{\mathrm{tor}}^a}\frac{\mathrm{d}s}{\mathrm{d}V}\int_0^\infty \mathrm{d}u\, u^3 \mathrm{e}^{-u^2}\sum_{m_2}\sum_{\mathrm{res}}\left(\tau_b |H_{\mathbf{m}}|^2 \left| m_2 \frac{\partial \omega_b}{\partial \eta} + n\frac{\partial \Omega^3}{\partial \eta}\right|^{-1}\right)_{\eta = \eta_{\mathrm{res}}}, \tag{12}$$

and $D_{12}$ containing an extra factor $u^2$ in the sub-integrand. Here $\eta_{\mathrm{res}} = \eta_{\mathrm{res}}(u)$ are (generally multiple) roots of Eq. (8) resolved with respect to $\eta$.

**Numerical implementation, benchmarking results and discussion**

Coefficients (12) are computed numerically allowing for the general case of a perturbed tokamak magnetic field in Boozer coordinates. Bounce averages are performed via numerical time integration of zero order guiding center orbits and an effective numerical procedure for root finding for Eq. (8) is realized using the scalings $\omega_b(u,\eta) = u\bar{\omega}_b(\eta)$ and $\Omega_{tB}(u,\eta) = u^2\bar{\Omega}_{tB}(\eta)$ and pre-computation of $\bar{\omega}_b$ and $\bar{\Omega}_{tB}$ on an adaptive $\eta$-grid. For testing, a circular concentric flux surface tokamak configuration is used with safety factor shown in Fig. 1. The perturbation field amplitude in Eq. (7) is taken as $B_n(\vartheta) = \varepsilon_M B_0(\vartheta)\exp(im\vartheta)$ with (formally infinitesimal) $\varepsilon_M$ set to 1 in all plots. Numerical results for $D_{11}$ are normalized below by the mono-energetic plateau value $D_p = \pi q v_T^3/(16R\bar{\omega}_{c\alpha}^2)$ where $R$ is the major radius, the (0,0) Boozer harmonic is used for the reference frequency $\bar{\omega}_{c\alpha}$ and perturbation wavenumbers $(m,n)$ are indicated in the titles. In Fig. 1 results of Eq. (12) for relatively weak radial electric fields where the superbanana-plateau regime is possible are compared to the analytical large aspect ratio formula of Shaing [1] for a range of aspect ra-

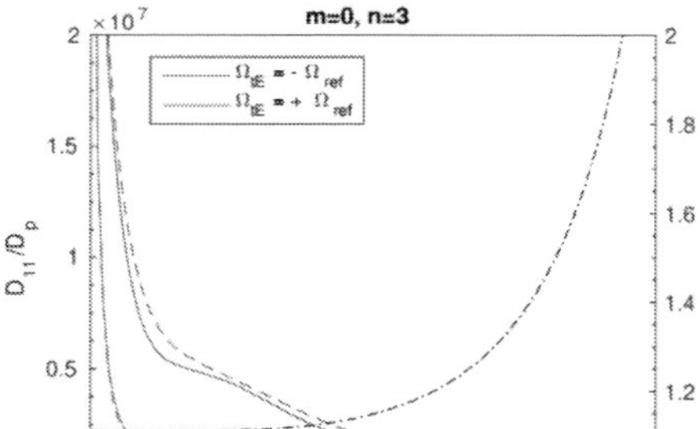


*Figure 1: Superbanana plateau - results of Eq. (12) (solid) and analytical formula [1] (dashed). Here $\Omega_{\mathrm{ref}} = cT_\alpha/(e_\alpha \psi_{\mathrm{tor}}^a)$ and $q$ is the safety factor (dash-dotted).*

tio values $A$. Earlier this formula was shown to be in agreement with computations by the quasilinear version of the drift kinetic equation solver NEO-2 [3] for $A = 10$ and the same

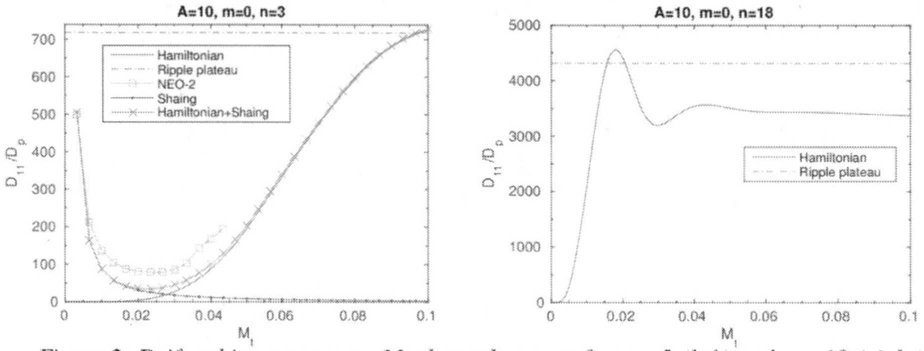

*Figure 2: Drift-orbit resonances - Mach number scan for $n = 3$ (left) and $n = 18$ (right).*

parameters as in Fig. 1. In Fig. (2) Mach number scans of $D_{11}$ for $A = 10$ are shown for $\Omega_{tB} = 0$ what is valid for relatively "large" Mach numbers $M_t = \Omega_{tE}R/v_T$. Since $m = 0$, passing particles contribute little, which corresponds to the regime of bounce resonances.

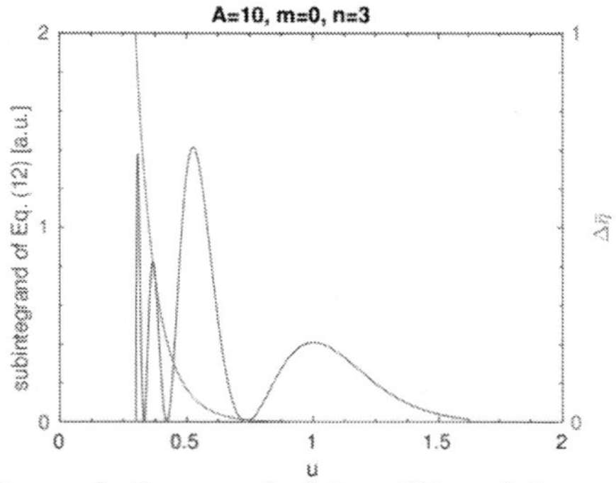

*Figure 3: Integrand of Eq. (12) and the resonance line for $M_t = 0.02$ and $m_2 = 1$.*

Results are compared to the ripple-plateau regime [7] where $D_{11} = 4\pi^{-1/2}nqA^2\varepsilon_M^2 D_p$, to the universal formula of Shaing for collisional transport in bounce-averaged regimes [1] and to results of NEO-2. The last two computations were done for very low $\nu^* = 5\cdot 10^{-5}$ and $n = 3$. The sum of resonant and bounce averaged coefficient is seen to be a factor 2 smaller than the NEO-2 result at intermediate Mach numbers, $M_t \sim 0.02$. The reason can be seen from Fig. 3 where the integrand of Eq. (12) and the normalized resonance condition $\Delta\bar{\eta} = (\eta_{\text{res}}(u)B_{\max} - 1)/(B_{\max}/B_{\min} - 1)$ are plotted for the mainly contributing first bounce harmonic $m_2 = 1$. Significant contributions lie around $u = 1$ where the distance between resonance and trapped-passing boundary is very small, $\Delta\bar{\eta} \sim \exp(-2\pi m_2 u/(qM_t\sqrt{2A}))$.

## Conclusion

The proposed unified approach treats all resonant NTV regimes in tokamaks in the same manner. A numerical code has been developed for NTV calculations in these regimes without using model geometry simplifications. A validation for the superbanana plateau regime shows good agreement to Ref. [1]. For drift-orbit resonances, the toroidal torque from NEO-2 [3] is significantly higher than the sum of collisional bounce averaged [1] and resonant regimes. This is due to strong contributions near the trapped-passing boundary. Therefore, collisional boundary layer analysis will be additionally required even at very low $\nu^*$.

## Acknowledgments

This work has been carried out within the framework of the EUROfusion Consortium and has received funding from the Euratom research and training programme 2014-2018 under grant agreement No 633053. The views and opinions expressed herein do not necessarily reflect those of the European Commission. The authors gratefully acknowledge support from NAWI Graz.

## References

[1] K. C. Shaing et al., Nucl. Fusion **50**, 025022 (2010)
[2] K. C. Shaing et al., Plasma Phys. Control. Fusion **51**, 075015 (2009)
[3] S. V. Kasilov et al., Phys. Plasmas **21**, 092506 (2014)
[4] A. F. Martitsch et al., this conference, P1.146
[5] A. N. Kaufman, Phys. Fluids **15**, 1063 (1972)
[6] A. Bécoulet et al., Phys. Fluids B **3**, 137 (1991)
[7] A. H. Boozer, Phys. Fluids **23**, 2283 (1980)

# Simultaneous control of plasma profiles and neoclassical tearing modes with actuator management in tokamaks

E. Maljaars[1], H. van den Brand[1,2], F. Felici[1], C.J. Rapson[3], W. Treutterer[3], O. Sauter[4], M.R. de Baar[1,2]

[1] *Eindhoven University of Technology, Mechanical Engineering, Control Systems Technology, P.O. Box 513, 5600MB Eindhoven, The Netherlands*

[2] *FOM Institute DIFFER, Eindhoven, The Netherlands*

[3] *Max Planck Institute for Plasma Physics, Garching, Germany*

[3] *Centre de Recherches en Physique des Plasmas, EPFL, Lausanne, Switzerland*

## Introduction

Tokamaks require a plasma control system (PCS) to help ensure that physics goals are met while remaining within operational and machine limits. Especially in future long pulse devices (e.g. ITER), several control tasks need to be executed simultaneously that share a limited set of actuators, where the priority of control tasks and the availability of actuators may change suddenly due to plasma or hardware events [1].

PCS architectural design has recently gained much attention in the literature in view of future long pulse tokamaks (e.g. ITER [2] and Tore Supra WEST [3]) and currently operational tokamaks (e.g. AUG [4]). In the PCS design, actuator management is needed to resolve the conflicting requests of multiple control tasks that share a limited set of actuators [1].

Actuator management for the Electron Cyclotron (EC) beams at AUG is considered in [5], where gyrotrons are optimally allocated to low level controllers for gyrotron power and launcher steering, based on higher level power requests with corresponding importance and effectiveness.

In this work we show simultaneous control of plasma profiles and suppression of neoclassical tearing modes (NTMs) by using a shared amount of available EC-power that is often used for each of these individual control tasks. The proposed PCS architecture contains a high level actuator management layer that allocates resources to the profile and NTM-controller before execution of these control tasks. As such, an intelligent profile controller can be used to ensure satisfaction of (time-varying) operational limits, which would be impossible if calculated actuator commands would be modified afterwards.

## Control architecture

Based on the existing (quite similar) PCS architecture designs in [2, 3, 4], we propose a control architecture to facilitate the simultaneous control of profiles and NTMs, that can be extended to more tasks. The control architecture is presented in Figure 1.

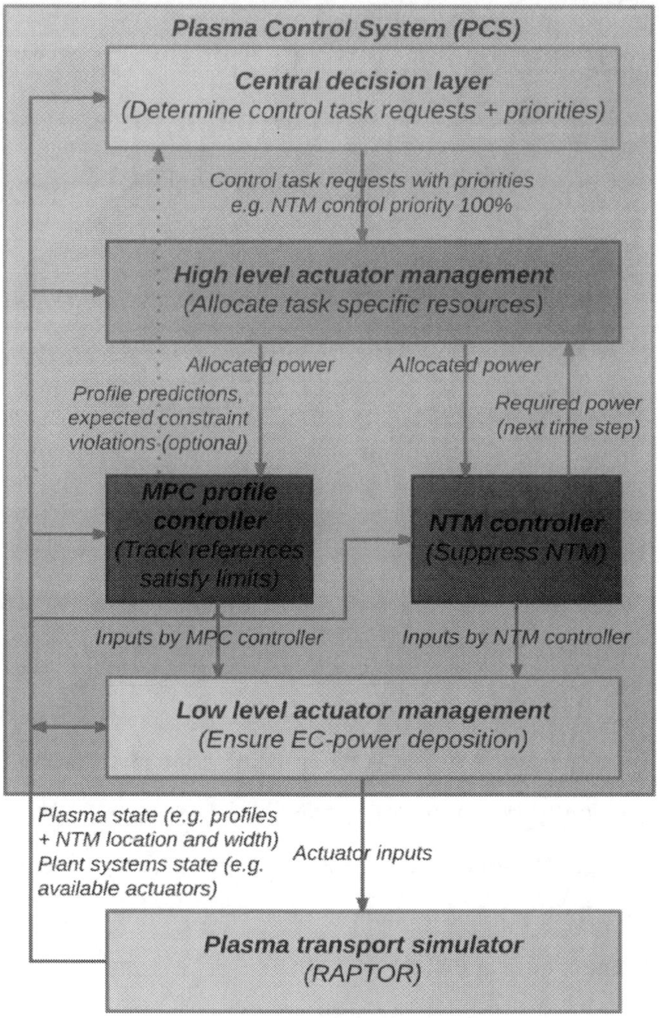

Fig. 1: PCS architecture for control of NTMs and profiles.

A central decision layer sets control task priorities based on the state of plasma and status of hardware (top). These priorities are used by the high level actuator management layer to allocate a limited set of resources to each control task, such that these controllers are aware of their resources. The NTM controller provides the actuator input for the next step and the desired allocated power for the next step is given back to the high level actuator management. The MPC controller computes at every time instant the future optimal control inputs for 120 time steps ahead, based on profile evolution models such that a $q$-profile reference is tracked and actuator and operational limits are satisfied [6]. The MPC controller could also provide profile predictions and warnings for expected constraint violations back to the central decision layer (not in this work). The inputs as given by the NTM and MPC controllers are in this work simply joined with feedforward inputs in the low level actuator management block, whereas functionality in this block could be extended based on the work in [5]. Finally the actuator inputs are given to the plasma transport simulator RAPTOR [7].

**Simulation results**

A simulation environment is built in Simulink [8] containing the designed PCS, connected to RAPTOR. RAPTOR has been extended to include the effect of NTMs on the plasma profiles, where the local heat transport is enhanced proportional to the island width, while the island width evolution is modeled by the Modified Rutherford Equation [9]. An ITER Hybrid scenario simulation in RAPTOR as described in [10] has been modified and has now a nominal flat top current of $I_p = 12$ MA, 33 MW NBI, and 25.7 MW EC. Actuator trajectories are optimized to achieve stationarity of profiles at the beginning of the flat-top (100 s) using the method

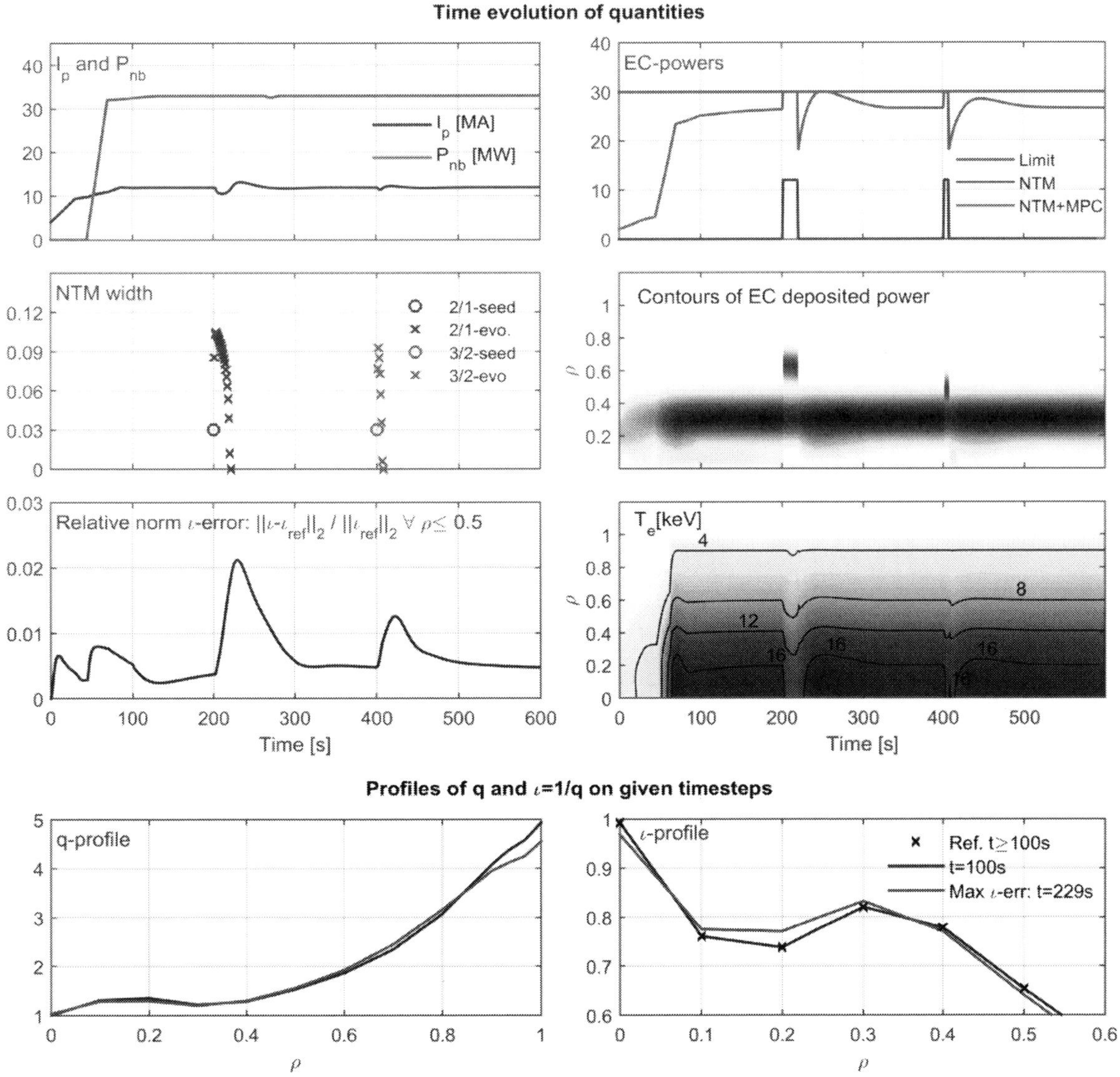


Fig. 2: Simulation results of effective simultaneous control of NTMs and profiles. Seed islands created for 2/1 NTM (200s) and 3/2 NTM (400s), resulting first in a significant drop in $T_e$. NTMs are fully suppressed while MPC profile controller uses plasma current $I_p$ and remaining EC-power to reduce $q$-profile tracking error and keep $q \geq 1$.

in [7, 10]. The MPC-controller can request EC-power deposition at three fixed $\rho$-locations: $\rho_{\text{tor}} = [0.05, 0.2, 0.3]$, where $\rho$ is the normalized toroidal magnetic flux coordinate. The NTM controller can request EC-power deposition at the location of the NTMs if present, assuming perfect alignment and no power modulation. In practice, a low level control system will ensure that gyrotrons and steering mirrors deliver the requested EC-power deposition. The total available EC-power is 30MW, which exceeds the power presently planned for ITER, but note that no IC-power is used in these simulations. Priorities are set such that in case of an NTM, the NTM controller will request 12MW per NTM (2/1 or 3/2). With the remaining available power, the MPC-control objective is to track the nominal $q$-profile in the region $\rho \leq 0.5$ while ensuring

$q \geq 1$, where it can use the power to the three EC-beams, NBI power, and plasma current $I_p$.

The simulation results are presented in Figure 2. Seed islands of 3 cm are created after 200 s for the 2/1 NTM and after 400 s for the 3/2 NTM, respectively. The 2/1 NTM is suppressed in 16 s and the 3/2 NTM in 6 s. At the times of maximum island widths, a peak can be noticed clearly in the deposited EC-power $V'P_{ec}$, whereas a drop in $T_e$ is visible. By modifying the inputs, the MPC controller can limit the growth of the relative error in the $q$-profile during the presence of the NTMs, while keeping $q \geq 1$. An important side effect of the limited changes in the $q$-profile is that the $\rho$-positions of the NTMs do not change significantly.

## Conclusion and outlook

We have shown simulations of the simultaneous operation of a profile controller and an NTM controller within a PCS architecture where these controllers are aware of their resources, allocated by a high level actuator management layer. Closed-loop simulations for the ITER tokamak show that the proposed design can effectively respond to the occurrence of an NTM by suppressing it, while at the same time the MPC profile controller maintains the safety factor profile close to its reference and within the operational limits.

This preliminary work can be continued in multiple ways. The simultaneous control of profiles and NTMs can be tested on the newly developed Plasma Control System Simulation Platform (PCSSP) [11]. Other control tasks sharing the same actuators could be added. The provided expected profile predictions and constraint violations can be used by the central decision layer to improve decision making. Extending the MPC controller to track also $\beta_{pol}$ may enable temporarily lowering $\beta_{pol}$ for faster suppression of NTMs. Also more advanced NTM-controllers could be implented, and the low level actuator management can be extended, e.g. based on [5].

## References


[1] D. Humphreys *et al*, Phys. Plasmas **22** 2 (2015)
[2] W. Treutterer *et al*, Fus. Eng. and Design **89** 5 512 – 517 (2014)
[3] N. Ravenel *et al*, Fus. Eng. and Design **89** 5 (2014)
[4] W. Treutterer *et al*, Fus. Eng. and Design **89** 5 146 – 154 (2014)
[5] C. J. Rapson *et al*, Fus. Eng. and Design *In press, online February 7th 2015* (2015)
[6] E. Maljaars *et al*, Nuclear Fusion **55** 2 2015
[7] F. Felici *et al*, Plasma Phys. Control. Fusion **54** 2 (2012)
[8] Simulink Release 2015a, The Mathworks, `www.mathworks.com`
[9] O. Sauter *et al*, Plasma Phys. Control. Fusion **44** 9 (2002)
[10] J. van Dongen *et al*, Plasma Phys. Control. Fusion **56** 12 (2014)
[11] M. L. Walker *et al*, Fus. Eng. and Design *In press, online February 16th 2015* (2015)


*This work has been carried out within the framework of the EUROfusion Consortium and has received funding from the Euratom research and training programme 2014-2018 under grant agreement No 633053. The views and opinions expressed herein do not necessarily reflect those of the European Commission.*

# Mimetic spectral element method for the Grad-Shafranov equation

A. Palha[1], F. Felici[1], B. Koren[1]

[1] *Eindhoven University of Technology, P.O. Box 513, 5600 MB Eindhoven, The Netherlands*

## Motivation

The computation of magnetohydrodynamic (MHD) equilibrium of toroidal plasmas plays a central role in tokamak fusion studies, not only as a fundamental tool for the computation of transport and stability calculations but also for the development of advanced tokamak control systems. Under perfect axisymmetric conditions, the vector calculus formulation of the Grad-Shafranov equation is the one most commonly used, [1, 2]:

$$-\nabla\cdot\left(\frac{1}{\mu r}\nabla\psi\right) = r\frac{\mathrm{d}p}{\mathrm{d}\psi} + \frac{1}{\mu_0 r}f\frac{\mathrm{d}f}{\mathrm{d}\psi} := j_t\,. \tag{1}$$

This equation can be written as a system of three equations:

$$\begin{cases}\nabla\psi = \vec{u}\\ \nabla\cdot\vec{h} = \tilde{j}_t\end{cases} \quad\text{and}\quad \mathbb{K}\vec{u} = \vec{h} \quad\text{with}\quad \mathbb{K} := \begin{bmatrix}\frac{1}{\mu_0 r} & 0\\ 0 & \frac{1}{\mu_0 r}\end{bmatrix}. \tag{2}$$

The two equations on the left define differential (topological) relations and the third one establishes a constitutive relation between $\vec{u}$ (the gradient of $\psi$) and the flux vector $\vec{h}$. The direct discretization of this formulation using standard techniques presents two challenges:

- Conservation of current at the discrete level ($\int_\Omega \nabla\cdot\vec{h}\,\mathrm{d}\Omega = \int_\Omega j_t \mathrm{d}\Omega$).
- Extension to curved meshes.

The first one is an important requirement for coupling with transport equations, the second enables optimal high order approximation in curved domains.

## Differential Geometry Formulation

In this work, we propose to use the differential geometric formulation of this equation:

$$\mathrm{d}\star_{\mathbb{K}}\mathrm{d}\star\tilde{\psi}^{(2)} = \tilde{j}_t^{(2)}\,. \tag{3}$$

This corresponds to two topological relations (left) and two constitutive relations (right):

$$\begin{cases}\mathrm{d}\psi^{(0)} = u^{(1)}\\ \mathrm{d}\tilde{h}^{(1)} = \tilde{j}_t^{(2)}\end{cases} \quad\text{and}\quad \begin{cases}\star\tilde{\psi}^{(2)} = \psi^{(0)}\\ \mathbb{K}\star u^{(1)} = \tilde{h}^{(1)}\end{cases}, \tag{4}$$

where each physical quantity is either an inner, $\Lambda^k$, or outer, $\tilde{\Lambda}^k$, oriented a $k$-form:

$$\psi^{(0)} \in \Lambda^0, \quad u^{(1)} \in \Lambda^1, \quad \tilde{h}^{(1)} \in \tilde{\Lambda}^1, \quad \tilde{j}_t^{(2)}, \tilde{\psi}^{(2)} \in \tilde{\Lambda}^2.$$

Here d is the exterior derivative and $\star$ is the Hodge-$\star$ operator that transforms $k$-forms into $(k+1)$-forms and changes their orientation from inner to outer and vice-versa:

$$\star : \Lambda^k \to \tilde{\Lambda}^{k+1} \quad \text{or} \quad \star : \tilde{\Lambda}^k \to \Lambda^{k+1}.$$

With these three operators, $(\mathrm{d}, \star, \mathbb{K})$, it is possible to construct the following double de Rham complex

$$\begin{array}{ccccc} \Lambda^0(\Omega) & \xrightarrow{\mathrm{d}} & \Lambda^1(\Omega) & \xrightarrow{\mathrm{d}} & \Lambda^2(\Omega) \\ \star\updownarrow & & \mathbb{K}\star \downarrow\uparrow \star\mathbb{K}^{-1} & & \star\updownarrow \\ \tilde{\Lambda}^2(\Omega) & \xleftarrow{\mathrm{d}} & \tilde{\Lambda}^1(\Omega) & \xleftarrow{\mathrm{d}} & \tilde{\Lambda}^0(\Omega) \end{array} \tag{5}$$

These spaces of inner and outer oriented $k$-forms are intimately related to the fact that $k$-forms naturally integrate over $k$-dimensional domains, $\Omega_k$. For example, $h^{(1)}$ is a flux quantity and therefore it naturally integrates across lines (since these equations are defined in a two dimensional space).

This formal classification of physical quantities into inner and outer oriented $k$-forms, although seeming a mere mathematical artefact, has deep impacts in the formulation of the discretization of the equations. This association is important because a physical system generally preserves integral quantities and these quantities are topological in the sense that they do not depend on a particular coordinate system or metric employed. Integration of differential forms is independent of any metric notions in space; no Riemann metric, dot products or $k$-dimensional volume elements are required. Integrals are of paramount importance in mimetic discretizations, because we cannot exactly represent differential forms and the operations on and between differential forms in a finite dimensional setting. On the other hand we can represent integrals of differential forms and the integral relations between them exactly.

Using the de Rham complex, (5), the inner product between inner oriented $k$-forms ($\alpha^{(k)}, \beta^{(k)} \in \Lambda^k$) becomes

$$\left(\alpha^{(i)}, \beta^{(i)}\right) := \int_\Omega \alpha^{(i)} \wedge \star\beta^{(i)}, \quad i = \{0,2\} \quad \text{and} \quad \left(\alpha^{(1)}, \beta^{(1)}\right) := \int_\Omega \alpha^{(1)} \wedge \mathbb{K}\star\beta^{(1)},$$

and for outer oriented $k$-forms ($\tilde{\alpha}^{(k)}, \tilde{\beta}^{(k)} \in \tilde{\Lambda}^k$) yields

$$\left(\tilde{\alpha}^{(i)}, \tilde{\beta}^{(i)}\right) := \int_\Omega \tilde{\alpha}^{(i)} \wedge \star\tilde{\beta}^{(i)}, \quad i = \{0,2\} \quad \text{and} \quad \left(\tilde{\alpha}^{(1)}, \tilde{\beta}^{(1)}\right) := \int_\Omega \tilde{\alpha}^{(1)} \wedge \star\mathbb{K}^{-1}\tilde{\beta}^{(1)}.$$

This formulation particularly contrasts with the one obtained with traditional vector calculus for the inner product between 1-forms (vectors in the vector calculus formulation):

$$\left(\vec{a},\vec{b}\right) := \int_{\Omega} \vec{a}\cdot\vec{b}\,\mathrm{d}\Omega .$$

It is possible to note that neither $\mathbb{K}$ nor $\mathbb{K}^{-1}$ appear in the vector calculus inner product. This results in a different finite element formulation, as will be seen below.

**Mimetic Discretization**

The mimetic spectral element method used in this work follows directly from the framework introduced in [4]. The key ideas of this method are: (i) special basis functions capable of representing the integral character of $k$-forms and (ii) a weak formulation that takes into account the relations established in the de Rham complex, (5). These basis functions, $\tilde{\varepsilon}_i^{(k)}$, satisfy the following properties:

$$\tilde{\varepsilon}_i^{(0)}(\tilde{\tau}_{(0),j}) = \delta_{ij}, \quad \int_{\tilde{\tau}_{(1),j}} \tilde{\varepsilon}_i^{(1)} = \delta_{ij} \quad \text{and} \quad \int_{\tilde{\tau}_{(2),j}} \tilde{\varepsilon}_i^{(2)} = \delta_{ij},$$

where $\tilde{\tau}_{(0),j}$, $\tilde{\tau}_{(1),j}$ and $\tilde{\tau}_{(2),j}$ are the points, edges and surfaces of the spectral element mesh.

The weak problem of (4), then reduces to finding the approximate solution, $\tilde{\psi}_h \in \tilde{\Lambda}^2$, such that:

$$\begin{cases} \left(\mathbb{K}\star\mathrm{d}\star\tilde{\psi}_h^{(2)},\tilde{\varepsilon}_j^{(1)}\right) = \left(\tilde{h}_h^{(1)},\tilde{\varepsilon}_j^{(1)}\right) & \forall\tilde{\varepsilon}_j^{(1)} \\ \left(\mathrm{d}\tilde{h}_h^{(1)},\tilde{\varepsilon}_j^{(2)}\right) = \left(\tilde{j}_t^{(2)},\tilde{\varepsilon}_j^{(2)}\right) & \forall\tilde{\varepsilon}_j^{(2)}. \end{cases}$$

Using integration by parts we can rewrite this as:

$$\begin{cases} \int_{\partial\Omega}\tilde{\psi}_h^{(2)}\wedge\tilde{\varepsilon}_j^{(1)} - \left(\tilde{\psi}_h^{(2)},\mathrm{d}\tilde{\varepsilon}_j^{(1)}\right) = \left(\tilde{h}_h^{(1)},\tilde{\varepsilon}_j^{(1)}\right) & \forall\tilde{\varepsilon}_j^{(1)} \\ \left(\mathrm{d}\tilde{h}_h^{(1)},\tilde{\varepsilon}_j^{(2)}\right) = \left(\tilde{j}_t^{(2)},\tilde{\varepsilon}_j^{(2)}\right) & \forall\tilde{\varepsilon}_j^{(2)}. \end{cases}$$

In doing this, the discrete analog of the flux operator ($\mathbb{K}\star\mathrm{d}\star$) is the negative adjoint of the discrete divergence (d).

**Numerical Results**

This formulation was applied to a Soloviev solution of the Grad-Shafranov equation, [3], corresponding to:

$$\psi(r,z) = \frac{f_0 R_0^2 a^2}{2}\left[1-\left(\frac{z}{a}\right)^2-\left(\frac{r-R_0}{a}+\frac{(r-R_0)^2}{2aR_0}\right)^2\right], \quad p(\psi) = \frac{f_0}{\mu_0}\psi + p_0, \quad f\frac{\mathrm{d}f}{\mathrm{d}\psi} = f_0 R_0^2,$$

with $a = 0.5$, $f_0 = 1.0$ and $R_0 = 1.1$. To assess this method on meshes of different curvature, the following space transformation that maps the space $(\xi,\eta)\in[-1,1]\times[-1,1]$ into a curved domain $(x,y) \in [x_a, x_b] \times [y_a, y_b]$ is used:

$$\begin{cases} x(\xi,\eta) = (\xi + c\sin(\pi\xi)\sin(\pi\eta) + 1)\,\frac{x_b - x_a}{2} \\ y(\xi,\eta) = (\eta + c\sin(\pi\xi)\sin(\pi\eta) + 1)\,\frac{y_b - y_a}{2} \end{cases}$$

An example of these results, together with the error, are presented in Figure 1. With this formulation, optimal $h$- and $p$-convergence are obtained regardless of the curvature of the mesh, as can be observed in Figure 2. This shows the potential of this method for the accurate solution of the Grad-Shafranov equation on curved meshes.

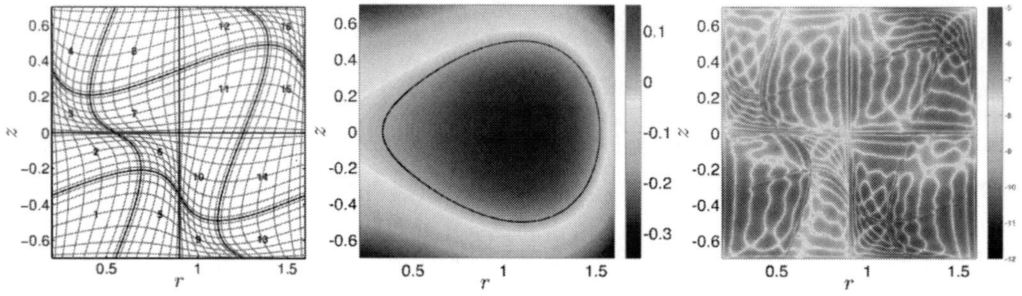

Figure 1: Example of a four by four element mesh with $c = 0.2$ (left), approximate Soloviev solution, $\psi_h$, using basis functions with polynomial degree $p = 8$ (center) and associated numerical error, $\log|\psi_h - \psi|$, with respect to the analytical solution $\psi$ (right).

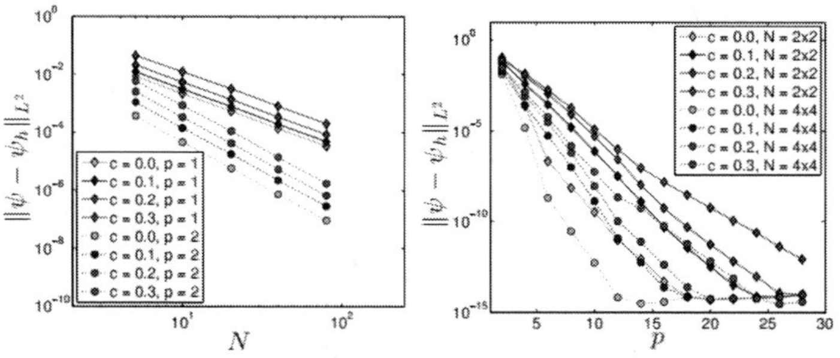


Figure 2: Error convergence plots for $h$- and $p$-refinement, left and right respectively. Optimal algebraic convergence of order $(p+1)$ (left) and exponential convergence (right) on curved meshes.

## References

[1] H. Grad and H. Rubin, Proceedings of the Second United Nations Conference on the Peaceful Uses of Atomic Energy **31** (1958)

[2] V.D. Shafranov, Soviet Journal of Experimental and Theoretical Physics **6**, 545 (1958)

[3] Howell, E. C., Sovinec, C. R., Computer Physics Communications, **185**, 1415-1421 (2014)

[4] A. Palha, P.P. Rebelo, R. Hiemstra, J. Kreeft, M. Gerritsma, Journal of Computational Physics **257** Part B (2014)

# Numerical Action-Angle transform for Guiding-Center particle motion in Axisymmetric Tokamak equilibria.

P.A. Zestanakis[1], Y. Kominis[2], G. Anastassiou[1], K. Hizanidis[1]

[1]*School of Electrical and Computer Engineering,*

[2] *School of Applied and Physical Science,*

*National Technical University of Athens, Greece*

## Introduction

The Lagrangian of the Guiding Centre (GC) motion of a charged particle is $\mathscr{L} = (\mathbf{A} + \rho_{\|} \cdot \mathbf{B}) \cdot \mathbf{v} + \mu\dot{\xi} - H$, where $\mathbf{A}$ and $\mathbf{B}$ are the vector potential and the magnetic field respectively, $\mathbf{v}$ is the guiding centre velocity, $\mu$ the magnetic moment, $\xi$, the gyrophase, $\rho_{\|}$ the parallel velocity to the magnetic field, normalized with $B$ and $H = \rho_{\|}^2 B^2/2 + \mu B + \Phi$ the Hamiltonian, with $\Phi$ the electric potential [1]. All quantities are evaluated at the guiding centre and normalized with respect to the nominal magnetic axis gyrofrequency and the major radius $R$. It has been shown that, when the magnetic coordinate system is appropriately chosen, the dynamical system is Hamiltonian and one can define $P_\tau$ and $P_\chi$ to be the canonical poloidal and toroidal momenta respectively [2]. In axisymmetric equilibria, the canonical position $\chi$ is ignorable and $P_\chi$ is conserved, so that the dynamical system, being reduced to one Degree Of Freedom (DOF), is integrable. However, the motion in phase space is non–trivial and there is no straightforward way to predict the behaviour of the system when perturbations are introduced and the integrability is lost.

In this paper we present a method for transforming any dynamical system describing the GC motion of a given axisymmetric equilibrium to an action angle (AA) phase space, where the study of the dynamics is significantly simplified. We demonstrate how this procedure makes it remarkably easy to pinpoint the location and the resonances in the presence of perturbations, as well as determine the conditions for nonlinear interaction due to stochastization, synergy between different perturbations, and confinement loss.

## The Action Angle Transform

The conserved canonical momenta $P_\chi$ and $\mu$ are already the actions of the toroidal motion and the gyromotion respectively. The AA pair $(J, \theta)$ of the poloidal motion is found by integrating along a closed orbit in the poloidal plane. By virtue of the Liouville–Mineur–Arnlold theorem, such a transform always exists locally. In particular we can cover all phase space with a measurable set of AA transforms, one for each phase space region that is bounded by a separatrix. From now on we shall call such a region a *continent*, while the set of transforms for all regions is called an *atlas*. In each continent such a transform is implicitly generated by a generating function of the form $F_2 = F(\tau, J, P_\chi, \mu)$. Dependence of $F_2$ on $P_\chi$ and $\mu$ implies that $\chi$ is also transformed to an angle variable

$$\bar{\chi} = \chi - f_\chi(J, \theta, P_\chi, \mu) \tag{1}$$

and so does $\xi$. Therefore, this procedure generates the transform to the AA pairs $(P_\chi, \bar{\chi})$ and $(\mu, \bar{\xi})$. In the AA phase space, the actions $J$, $P_\chi$ and $\mu$ remain constant, while the angles $\theta$, $\bar{\chi}$ and $\xi$ evolve linearly in time, with frequencies $\omega_\theta$, $\omega_{\bar{\chi}}$ and $\Omega_c$ respectively.

**Orbital spectrum analysis**

A perturbation of the form $\delta\mathbf{B} = \nabla \times \sigma\mathbf{B}$ can be straightforwardly included in the guiding centre Hamiltonian as $H = (\rho_c - \sigma)^2 B^2/2 + \mu B + \Phi$, $\rho_c = \rho_\| + \sigma$. This introduces a first and second order perturbation Hamiltonian term. Any perturbation with physical meaning should be given in terms of the magnetic coordinates, or even the lab coordinates. The first order Hamiltonian $H_1$ is proportional to $\sigma$. Due to the nonlinear depencence of $\bar{\chi}$ on $\theta$ (eq. 1), a monochromatic mode $\sigma = A_{m,n}(\psi)\exp(i(m\chi + n\tau - \omega t))$ in the magnetic coordinates gives an infinite series of modes in the AA phase space, so that

$$H_1 = \sum_s \mathscr{H}^1_{s,m}(J, P_\chi)\ e^{i(m\bar{\chi} + s\theta - \omega t)}, \tag{2}$$

where

$$\mathscr{H}^1_{s,m}(J, P_\chi) = \frac{1}{2\pi} \oint_{J, P_\chi = \text{const.}} H^1_{m,n}(\psi)\ e^{i(n\tau + n f_\chi(J, P_\chi, \theta) - s\theta)} d\theta. \tag{3}$$

As equation eq. 2 indicates, the resonances of the perturbation are located in action space at the points where the resonance condition

$$m\ \omega_{\bar{\chi}}(J, P_\chi, \mu) + s\ \omega_\theta(J, P_\chi, \mu) - \omega\ = 0 \tag{4}$$

is met. The location of the resonances in action space depends only on the spectral parameters $m$ and $\omega$. The actual profile of the perturbation, i.e the depencence on $n$ or $\psi$, is relevant in defining the amplitude of the resonant terms, but not in pinpointing their location in the orbital spectrum. Since $s$ can take on any integer value, each bounded continent contains an infinite number of such frequencies, most of which are located in the narrow chaotic sea near the separatrix, where $\omega_\theta$ approaches zero. In the bulk of each continent there are only a few, if any, sites where eq. 4

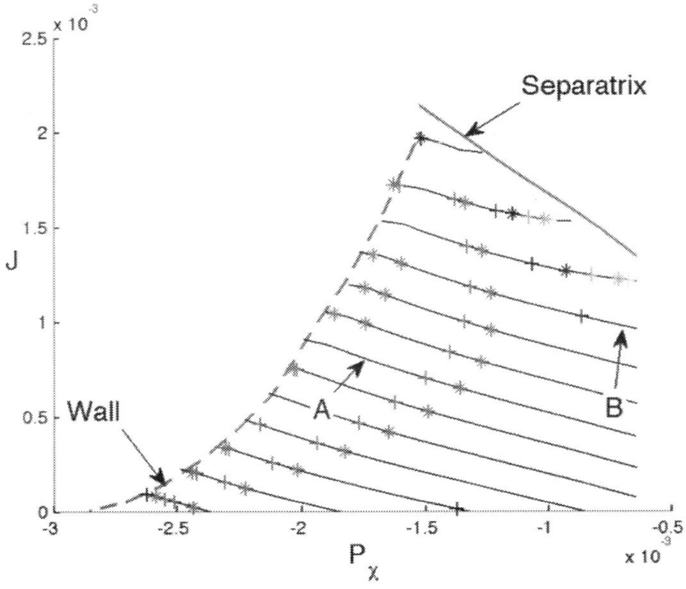


(a) Resonance chart. The solid black lines depict the energy surfaces, crosses and stars correspond to resonances with $m = 10$ and $m = 8$ respectively.

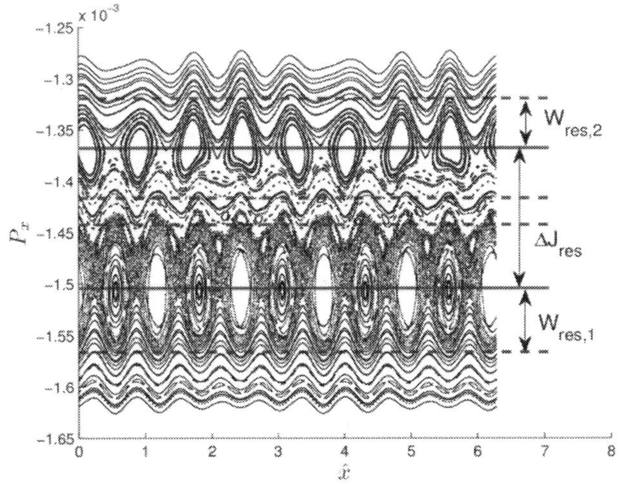


(b) Poincare plot on the surface A of Fig 1a for two modes with subcritical amplitude. The semi-analitically calculated positions of the resonances as well as their widths are denoted with solid and dashed lines respectively.

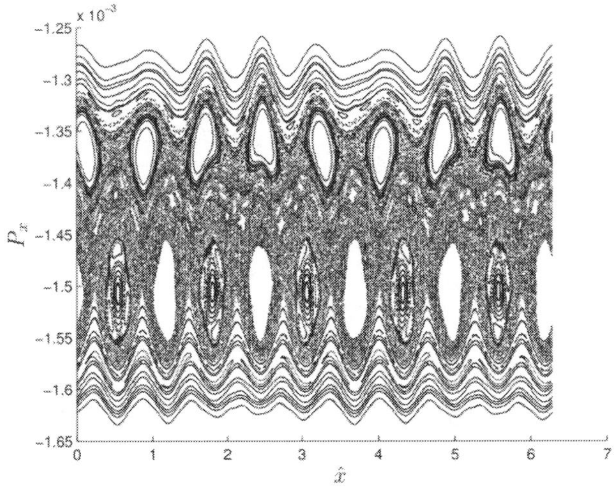


(c) The same Poincare plot for perturbations with critical amplitude. KAM lines between the two resonances have been destroyed and significant redistribution can take place.

Figure 1: Inspection of the resonance chart can reveal the phase space regions when mode synergy can be significant. The analytically calculated resonance positions, width and overlap conditions are in excellent agreement with the simulations.

is satisfied.

The case of time independent perturbations is particularly simple, but rather indicative of the power of the AA transform and the orbital spectrum transform. Near a particular resonance $m\ \omega_{\bar{\chi}} + s\ \omega_\theta = 0$, the dynamics follow a pendulum– like Hamiltonian and a trapped area of width proportional to the square root of the perturbation amplitude is formed. The width depends on $\mathscr{H}_{s,m}(J, P_\chi)$ and can be easily calculated once the AA transform has been performed. The Hamiltonian is conserved and the quantity $P_\chi - m/sJ$ is an adiabatic invariant. The cases where the ratio $m/s$ or $s/m$ becomes very large are of little interest, since the adiabatic invariant coincides with one of the actions, so that no significant redistribution takes place.

Coexistence of more than one resonances on the same energy surface can lead to chaotic redistribution, due to destruction of the adiabatic invariant. Chiricov criterion, which relies on resonance overlap, provides a very good estimation of the required conditions for chaotic motion. Figure 1a depicts the chart of the $m = 10$ and $m = 8$ resonances in a banana continent of a peaked large aspect ratio equilibrium. As demonstrated by Figs. 1b, 1c, inspection of the resonance chart and employment of the orbital spectrum transform (Eq. 3) provides useful insight on the dynamics of the perturbed phase space.

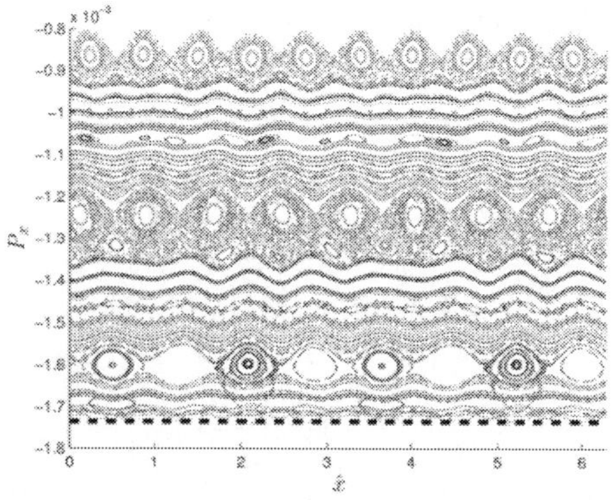

(a) Poincare cut for the energy surface B of Fig 1a and subcritical amplitude 0.08 $a_{\text{Chirikov}}$. Only two of the resonances have partially overlaped.

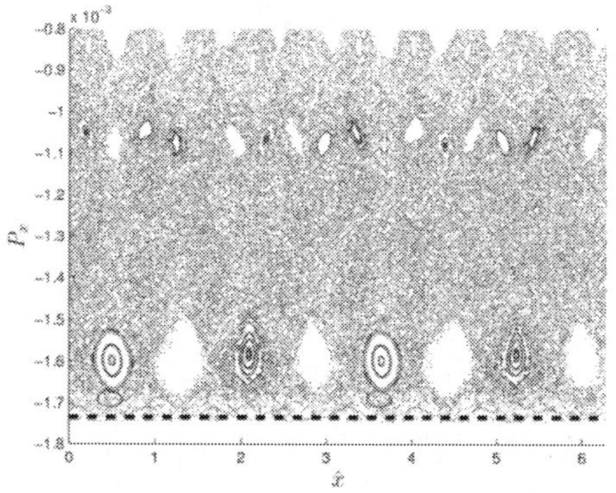

(b) The same, with amplitude 0.3 $a_{\text{Chirikov}}$. Although, this is still subctitical, the KAM surfaces have been destroyed. Chirikov criterion overestimates the critical amplitude, by ignoring higher order resonances.

Figure 2: The AA transform as a tool for estimating conditions for confinement loss. The outer closed flux surface is marked with a thick dashed line.

The case of the energy surface B in Fig 1a is of particular importance, because there is a line of resonances linking a deeply trapped part of phase space to the plasma wall (dashed blue line), provided that all consecutive resonances overlap. Chirikov criterion determines the critical amplitude at $a_{crit} = 1.2 \cdot 10^{-3}$,which is an overestimation, due to the strong presence of higher order resonances, as demonstrated in Fig. 2.

## References

[1] R. G. Littlejohn. *J. Plasma Physics* 29, 2 1983.

[2] R. B. White and M. S. Chance. *Phys. Fluid* 27(10), 1984.

# Multiple scattering of radio frequency waves by blobs: Homogenization of a mixture of blobs and the Waterman-Truell approach(*)

K. Hizanidis[1], S. I. Valvis[1], F. Bairaktaris[1] , Gr. Zouros[1], G. Kokkorakis[1], J. Roumeliotis[1]

[1]*National Technical University of Athens, Athens, Greece*

A. K. Ram[2]

[2]*Plasma Science and Fusion Center, MIT, Cambridge MA, USA*

Radio frequency waves are of tantamount importance for heating and current drive in magnetized fusion plasmas. The scattering process of these waves by a multitude of density fluctuations, such as blobs in the edge region, is studied by homogenizing the part of the edge region which is populated by an ensemble of ellipsoidal blobs immersed in the ambient plasma. For that region, the effective permittivity tensor is formulated on the basis of the depolarization dyadic. The ambient density is considered as step-wise constant. In general, the interfaces between the slab characterized by the effective permittivity tensor and the ambient plasma are not necessarily aligned with the ambient magnetic field which is considered as homogeneous.

From now on, the scattering problem of an incident wave is amenable to two different approaches: (a) by considering the scattering and transmission through a composite slab surrounded by the ambient plasma and (b) by a suitably adapted version of the Waterman-Truell approximation formula for the effective wave number in conjunction with the fulfillment of a self consistency condition for the mean field.

*(a) Scattering and transmission through a composite slab*

From this point of view, the waves propagate in three regions. The first one is homogeneous and anisotropic and consists of ions with number density $n_1$. Then, the incident wave propagates in an intermediate dielectric mixture plasma region which consists of a homogeneous anisotropic environment in which the ions have number density $n_3$ and anisotropic spherical inclusions (blobs) with ion number density $n_B$. After getting through this intermediate region, the wave propagates in a homogeneous anisotropic region consisting of ions with number density $n_2$. Furthermore, the axis of the intermediate region is not necessarily parallel to the external imposed magnetic field $\vec{\mathbf{B}}_0$ (figure 1).

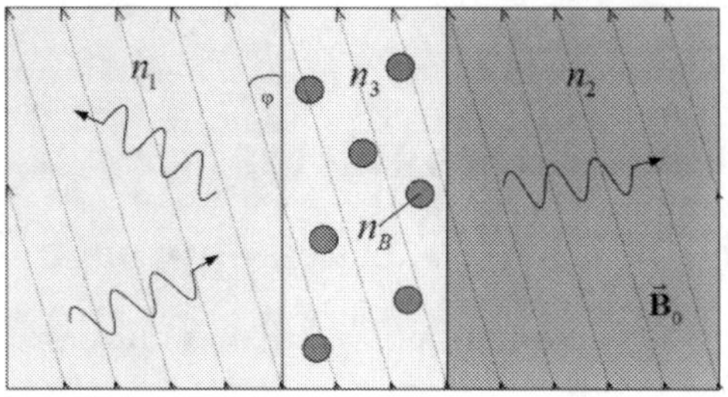


Figure 1: A simplified depiction of the problem – the three propagation regions

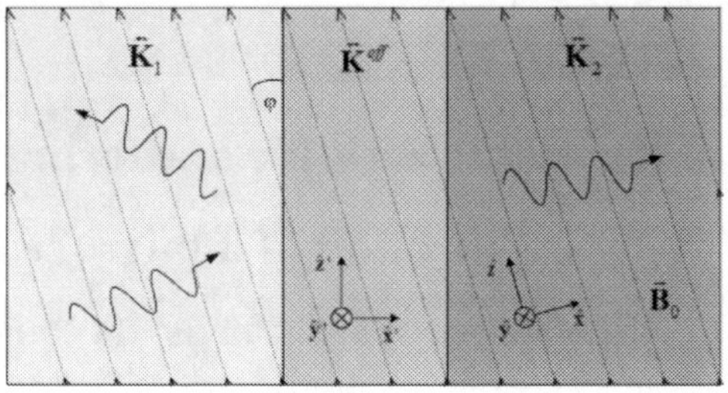


Figure 2: Intermediate region is considered as a dielectric region with calculated effective tensor $\ddot{\mathbf{K}}^{eff}$

So, in the intermediate propagation region, there are anisotropic spherical inclusions with an arbitrary permittivity tensor $\ddot{\mathbf{K}}^{B}$ located in an homogeneous anisotropic environment with a permittivity tensor $\ddot{\mathbf{K}}^{P}$. This problem, can be simplified by replacing the region of plasma with spherical blobs with a dielectric region with effective permittivity tensor $\ddot{\mathbf{K}}^{eff}$ (figure 2) [1], [2].

$$\ddot{\mathbf{K}}^{eff} = \tilde{\ddot{\mathbf{I}}} \cdot \ddot{\mathbf{K}}^{P}, \quad \tilde{\ddot{\mathbf{I}}} \equiv \ddot{\mathbf{I}} + f\left(\ddot{\mathbf{K}}^{B} - \ddot{\mathbf{K}}^{P}\right) \cdot \left[\ddot{\mathbf{K}}^{P} + (1-f)\ddot{\mathbf{N}} \cdot \left(\ddot{\mathbf{K}}^{B} - \ddot{\mathbf{K}}^{P}\right)\right]^{-1}$$

where $\ddot{\mathbf{K}}^{eff,B,P} = \begin{pmatrix} K_{\perp}^{eff,B,P} & -jK_{\times}^{eff,B,P} & 0 \\ jK_{\times}^{eff,B,P} & K_{\perp}^{eff,B,P} & 0 \\ 0 & 0 & K_{\parallel}^{eff,B,P} \end{pmatrix}$ and $\tilde{\ddot{\mathbf{I}}} = \begin{pmatrix} \tilde{I}_{\perp} & -j\tilde{I}_{\times} & 0 \\ j\tilde{I}_{\times} & \tilde{I}_{\perp} & 0 \\ 0 & 0 & \tilde{I}_{\parallel} \end{pmatrix}$.

The depolarization matrix is $\ddot{\mathbf{N}} = \begin{pmatrix} N_{\perp} & 0 & 0 \\ 0 & N_{\perp} & 0 \\ 0 & 0 & N_{\parallel} \end{pmatrix}$ with $N_{\perp} = \left(1 - N_{\parallel}\right)/2$ and $f$ is the volume fraction that the inclusions occupy, while:

I.) $K_{\parallel}^{P} > K_{\perp}^{P}$ : $N_{\parallel} = \dfrac{K_{\parallel}^{P}/K_{\perp}^{P}}{\sqrt{K_{\parallel}^{P}/K_{\perp}^{P} - 1}^{3}}\left(\sqrt{K_{\parallel}^{P}/K_{\perp}^{P} - 1} - \arctan\sqrt{K_{\parallel}^{P}/K_{\perp}^{P} - 1}\right)$

II.) $K_{\parallel}^{P} < K_{\perp}^{P}$ : $N_{\parallel} = \dfrac{K_{\parallel}^{P}/K_{\perp}^{P}}{2\sqrt{1 - K_{\parallel}^{P}/K_{\perp}^{P}}^{3}}\left(\ln\dfrac{1+\sqrt{1-K_{\parallel}^{P}/K_{\perp}^{P}}}{1-\sqrt{1-K_{\parallel}^{P}/K_{\perp}^{P}}} - 2\sqrt{1-K_{\parallel}^{P}/K_{\perp}^{P}}\right)$

and

$$\tilde{I}_\perp = 1 + f\frac{K_\perp^P\left(K_\perp^B-K_\perp^P\right) - K_\times^P\left(K_\times^B-K_\times^P\right) + (1-f)N_\perp\left[\left(K_\perp^B-K_\perp^P\right)^2-\left(K_\times^B-K_\times^P\right)^2\right]}{\left[K_\perp^P+(1-f)N_\perp\left(K_\perp^B-K_\perp^P\right)\right]^2-\left[K_\times^P+(1-f)N_\perp\left(K_\times^B-K_\times^P\right)\right]^2}$$

$$\tilde{I}_\parallel = 1 + f\frac{K_\parallel^B-K_\parallel^P}{K_\parallel^P+(1-f)N_\parallel\left(K_\parallel^B-K_\parallel^P\right)},\quad \tilde{I}_\times = f\frac{K_\perp^P K_\times^B-K_\times^P K_\perp^B}{\left[K_\perp^P+(1-f)N_\perp\left(K_\perp^B-K_\perp^P\right)\right]^2-\left[K_\times^P+(1-f)N_\perp\left(K_\times^B-K_\times^P\right)\right]^2}$$

Also, in a more realistic approach, the initial spheroidal shape of the scattering regions is distorted into ellipsoidal during an affine transformation necessary to keep the potential outside
distorted into ellipsoidal during an affine transformation necessary to keep the potential outside
the sphere obeying the Laplace equation [2]. In this case, the depolarization matrix components are given by the integral $N_x = \frac{a_x a_y a_z}{2}\int_0^\infty \left(s+a_x^2\right)^{-1}\left[\left(s+a_x^2\right)\left(s+a_y^2\right)\left(s+a_z^2\right)\right]^{-\frac{1}{2}} ds$, where $a_x, a_y, a_z$ are the semiaxes of the ellipsoidal inclusion. This hyperelliptic integral can be expressed as function of known elliptic integrals [3].

In a coordinate system (*x, y, z*) with the *z*-axis aligned with the ambient magnetic field, one needs to rotate around *y* to the one aligned with the surfaces of the slab (*x', y', z'*) using the rotation matrix $\vec{\vec{\mathbf{R}}}$, via $\left(\vec{\vec{\mathbf{K}}}^{B,P,eff}\right)' = \vec{\vec{\mathbf{R}}}\cdot\vec{\vec{\mathbf{K}}}^{B,P,eff}\cdot\vec{\vec{\mathbf{R}}}^T$. Furthermore, the propagation vector of the incident wave must be expressed in the new aligned coordinate system. Choosing the plane of incidence to be the (x-z) plane [the containing the blobs slab is infinite in the (y-z) directions] and following Berreman [4], one may write the two vectorial Maxwell's equations for the phasors of the electromagnetic field as $\left(\vec{\vec{\mathbf{M}}} + j\frac{\omega}{c}\vec{\vec{\mathbf{C}}}\right)\cdot\vec{\mathbf{F}} = 0$, where $\vec{\vec{\mathbf{M}}}$ consists of spatial derivatives with respect to *x, z,* $\vec{\vec{\mathbf{C}}}$ contains the permittivity tensor components and the rotation angle $\varphi$ and $\vec{\mathbf{F}}$ includes the six components of the electric and the magnetic field. Then, since the properties of the various regions involved change along the *x*-direction, one may Fourier transform the differential equation in the *z*-direction accordingly [exp($jk_\parallel z$)] and therefore $\left(\tilde{\vec{\vec{\mathbf{M}}}} + j\vec{\vec{\mathbf{C}}}\right)\cdot\tilde{\vec{\mathbf{F}}} = 0$ where the arc on top signifies Fourier transform in the *z*-direction so that the differentiation with respect to $z$ changes to a simple algebraic operation related to the $z$ component of the wavevector. The latter linear problem is solved in the three domains of Fig. 2 where the density can be taken as piece-wise constant.

*(b) Waterman-Truell approach*

This approach is limited to spherical blobs. However, it is capable of incorporating multiple scattering effects. In its algorithmic implementation involves an iterative procedure where the previous method for the calculation of the effective dielectric tensor of the effective medium is required. It also involves the solution of two scattering problems, namely the propagation of the incident wave in the effective medium and scattered by (1) a spherical blob of density which is equal to the equivalent to the composite slab (solved previously) and (2) by a spherical blob of blob density. After solving these two scattering problems using the method of Ram and Hizanidis [5], a test takes place to find out if a consistency condition referring to the mean fields and the wavenumbers is satisfied. If it is, the procedure is considered completed and it ends. If it's not, then the wave vector is being replaced with a more exact using the Waterman and Truell method [6], the two scattering problems are being solved again and the procedure repeats itself until the consistency condition is satisfied.

(*)Supported in part by the Hellenic National Programme on Controlled Thermonuclear Fusion associated with the EUROfusion Consortium

[1] A. K. Ram et al., Phys. Plasmas 20, 056110 (2013).

[2] Ari Sihvola, Homogenization of a dielectric mixture with anisotropic spheres in anisotropic background (1996), Department of Electroscience, Electromagnetic Theory, Lund Institute of Technology, Sweden

[3] Gradshteyn / Ryzhik, Table of Integrals, Series, and Products (1980), Academic Press Inc., New York, USA

[4] D. W. Berreman (1972), Optics in Stratified and Anisotropic Media: 4x4-Matrix Formulation, Journal of the Optical Society of America, 62:4, 502-510, DOI:10.1364/JOSA.62.000502

[5] Abhay K. Ram & Kyriakos Hizanidis (2013) Scattering of electromagnetic waves by a plasma sphere embedded in a magnetized plasma, Radiation Effects and Defects in Solids: Incorporating Plasma Science and Plasma Technology, 168:10, 759-775, DOI: 10.1080/10420150.2013.835633

[6] A. Lakhtakia (1993) Application of the Waterman-Truell approach for chiral composites, International Journal of Electronics, 75:6, 1243-1249, DOI: 10.1080/00207219308907198

# Advanced divertor equilibrium calculations with finite current density on the plasma boundary.

P.J. Mc Carthy,[1] K. Lackner,[2] H. Zohm,[2] T. Lunt,[2] J.C. Fuchs,[2]
L. Giannone,[2] W. Suttrop,[2] and The ASDEX Upgrade Team[2]

[1] *Department of Physics, University College Cork, Cork, Ireland.*

[2] *Max Planck Institut für Plasmaphysik, Boltzmannstrasse 2, D-85748 Garching, Germany*

## Introduction

The high exhaust power fluxes that are expected to reach critical values in the narrow wetted area on the ITER and DEMO divertor plates (implying the need for a large core radiation fraction for the latter) have motivated the development of the X-divertor (XD) [1] and Snowflake (SF) [2] advanced divertor configurations. The aim of the XD is to expand the poloidal flux in the proximity of the target plates. Expansion of the flux surfaces, primarily around the X-point (XP), can also be achieved by the SF, where a second order null in the poloidal magnetic field (both the field and its first derivatives vanish) results in a configuration exhibiting local six-fold symmetry. The benefits of strong flaring in the SF configuration are conditional on the divertor plates being in the vicinity of the SF origin, however the increased scrape-off layer (SOL) volume associated with the SF will independently have the beneficial effect of increasing radiative losses. Here we focus on the ideal snowflake and its variants already introduced in [2], namely the 'snowflake plus' ($SF^+$) and 'snowflake minus' ($SF^-$) where the second order null is replaced by two closely positioned conventional X-points. In the case of the $SF^+$ the secondary XP lies in the private flux region (PFR) of the primary XP, while for the $SF^-$ the XP lies in the primary SOL. In what follows we consider (i) the effect of finite current density on the ideal SF configuration, (ii) the fitting of decay lengths for currents in the SOL and PFR and (iii) $SF^-$ configurations that are experimentally feasible on ASDEX Upgrade.

## Snowflake calculations with finite current density on the plasma boundary

Since its introduction in [2], work on the SF configuration reported so far (e.g. [3]) has assumed a vanishing toroidal current density at the separatrix, although Ryutov et al. [4] have considered finite $j_\phi$ in the SOL for $SF^+$ and $SF^-$ cases. More recently, the volume of the scrape-off layer has been proposed as a new figure of merit for advanced divertor configurations [5]. The predictive/interpretive CLISTE equilibrium code [6] has added functionality to calculate snowflake-like equilibria under the realistic assumption that the current density profile extends across the separatrix into the SOL and PFR. CLISTE can compute an ideal snowflake at any feasible location by assigning values to four of the poloidal field coil currents so as to satisfy four conditions at the specified location of the SF origin:

$$\psi_R = \psi_Z = \psi_{RR} = \psi_{RZ} = 0 \tag{1}$$

where $\psi_R = \partial\psi/\partial R$, etc. and the condition $\psi_{ZZ} = 0$ is redundant provided the toroidal current density vanishes on the plasma boundary. For the realistic case of $j_{\phi,\,boundary} \neq 0$ where at a first order null the Grad Shafranov equation simplifies to

$$\psi_{RR} + \psi_{ZZ} = -\mu_0 R j_\phi \tag{2}$$

a second order null and hence an ideal snowflake is not possible. To study the transition in the SF configuration from $j_\phi = 0$ to $j_\phi \gg 0$ the condition $\psi_{RR} = 0$ in (1) is replaced by $\psi_{RR} - \psi_{ZZ} = 0$ which imposes a second order null for $j_\phi = 0$ where the poloidal flux takes the form $\psi(R, Z) - \psi_{XP} = a((R - R_x)^2 - (Z - Z_x)^2)$ (in suitably rotated coordinates) at the location $R_x, Z_x$ of a single null, while at the same time allowing a circular current density distribution ($\psi_{RR} = \psi_{ZZ} \neq 0$) for $j_\phi > 0$. Contour plots for a $20 \times 20\,\mathrm{cm}^2$ region centred on the (nominal) SF origin for three cases using the ASDEX Upgrade coil configuration are shown in figure 1.

FIG. 1: Contour plot detail (20 cm × 20 cm centred on nominal SF origin) of poloidal flux for three $I_p$ = 800 kA SF calculations; (a) ideal SF with $j_\phi = 0$ at the origin, (b) $j_\phi = 25\,\mathrm{kA\,m^{-2}}$, (c) $j_\phi = 100\,\mathrm{kA\,m^{-2}}$.

As $j_\phi$ increases, the second order null of the ideal SF is replaced by a triangular region bounded by three X-points and an O-point corresponding to a shallow extremum in flux. In the right-hand plot in figure 1 this region is (approximately) an equilateral triangle of side 10 cm where the O-point is at $\rho_{pol} = 1 - 0.00015$ and $j_\phi$ is approximately constant (∼5% variation) taking a value ($\simeq 100\,\mathrm{kA\ m^{-2}}$) which is ∼ 5% of the maximum $j_\phi$ value in the plasma core.

**Experimental determination of current density decay lengths in the SOL and PFR**

A prerequisite for the potential experimental relevance of the finite $j_\phi$ SF-like configurations is that current density in the PFR should have a sufficiently long decay length. ASDEX Upgrade is well equipped with robust divertor tile current measurements of SOL poloidal currents whose dependence on plasma parameters has previously been reported [7]. The resolution and bandwidth for individual tile signals was improved in 2012, and subsequent discharges which include a slow horizontal sweep of the separatrix very occasionally generate data where a tile extremity transits from well outside the SOL (in the deep PFR) to well inside the SOL under quasi stationary plasma conditions ($I_p$, $\bar{n}_e$, $W_{\mathrm{mhd}}$). Two such discharges were analyzed, and an exponential decay in Ipol (and, by assumption, $j_\phi$) versus $\delta\rho_{pol}$, the separation of the separatrix from the tile extremity, was found to fit the data extremely well for the first discharge (#29779, $I_p = 1\,\mathrm{MA}$, $\bar{n}_e = 2.7 \times 10^{19}\,\mathrm{m^{-3}}$, $W_{\mathrm{mhd}} = 100\,\mathrm{kJ}$, L-mode/I-phase) and reasonably so for a second discharge (#29729, $I_p = 1\,\mathrm{MA}$, $\bar{n}_e = 5.5 \times 10^{19}\,\mathrm{m^{-3}}$, $W_{\mathrm{mhd}} = 750\,\mathrm{kJ}$, H-mode) where averaging over high frequency ELMs was necessary. Separate regressions of $\mathrm{Ipol_{tot}} - \mathrm{Ipol_{roof}}$ vs. $\delta\rho_{pol}$ were made for time points with positive $\delta\rho_{pol}$, and of $\mathrm{Ipol_{PFR}}$ vs. $\delta\rho_{pol}$ for negative $\delta\rho_{pol}$, where $\mathrm{Ipol_{tot}}$ is the total Ipol flowing in the inner divertor. The results shown in figures 2 and 3 indicate decay lengths (expressed in terms of $\rho_{pol}$) for Ipol in the PFR of $\lambda_{\mathrm{PFR}} = 0.0007$, 0.0011 for #29779 and #29729, in both cases comfortably exceeding the range of $\delta\rho_{pol}$ in the SF-like triangular region in figure 1(c). The corresponding SOL values are $\lambda_{\mathrm{SOL}} = 0.0046$, 0.0071 which equate to outer midplane spatial decay lengths of 3.2 and 4.4 mm, respectively.

**Search for feasible SF$^-$ configurations for ASDEX Upgrade**

Transport simulations and experimental results on TCV [8] indicate that the SF$^-$ configuration is more promising than the SF$^+$ for reducing the power flux at the divertor plates. Most ASDEX Upgrade coils are remote from the plasma, requiring large currents to generate SF configurations which generally require a substantial reduction in the plasma current to avoid exceeding coil current and/or vertical force limits. CLISTE in predictive mode has the facility to search parameter space using Monte Carlo methods and hence identify optimal configurations satisfying pre-specified criteria. Two such configurations are shown in figure 4. The $I_p = 800\,\mathrm{kA}$ case, with a spatially remote secondary XP, does not differ strongly from the standard XP configuration even though it satisfies the SF$^-$ criterion of magnetic proximity: the outer midplane equivalent separation is ∼ 1 mm. The second configuration exhibits large flux expansion at the divertor plates, but at the price of a much reduced plasma current of 300 kA.

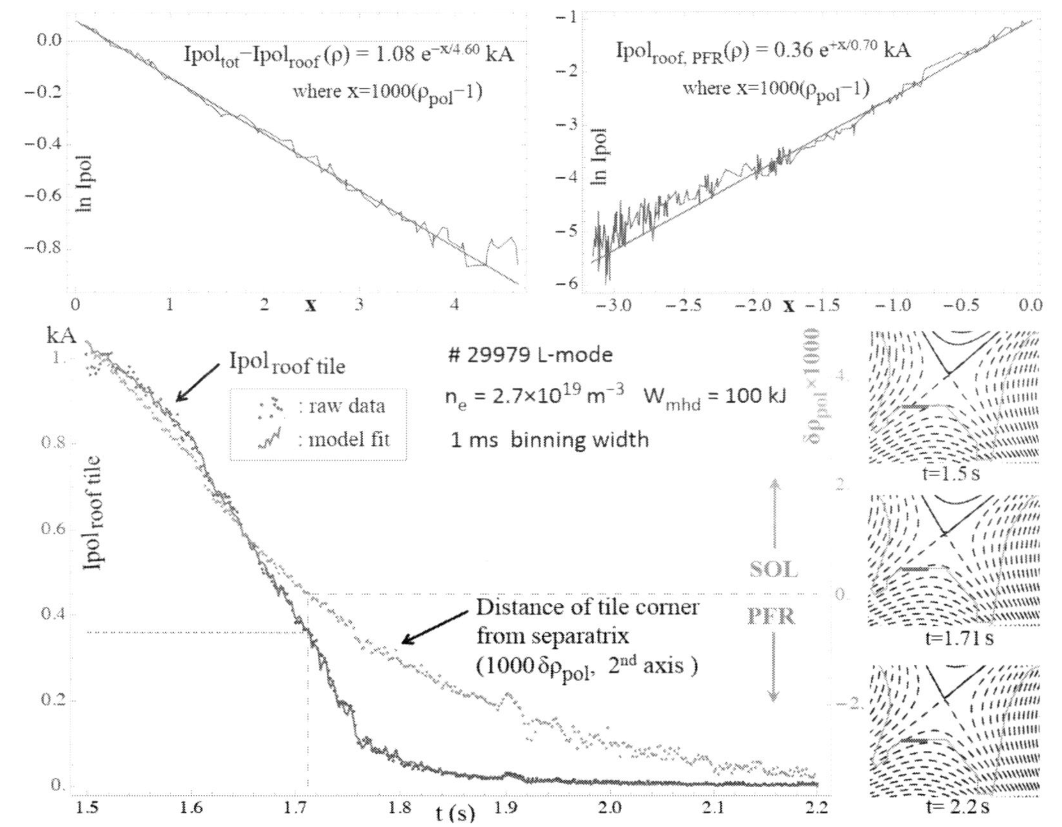


FIG. 2: Fits to poloidal current through roof baffle tile 14DUMoi in ASDEX Upgrade L-mode/I-Phase shot #29779 versus $\delta\rho_{pol}$ distance from separatrix. The upper plot pair show $\ln(\mathrm{Ipol_{tot}} - \mathrm{Ipol_{roof}})$ vs. $\delta\rho_{pol}$ data and regression line for $\delta\rho_{pol} > 0$ (left) and $\ln \mathrm{Ipol_{roof,PFR}}$ vs. $\delta\rho_{pol}$ for $\delta\rho_{pol} < 0$ (right). The lower plot shows the raw (blue points) and fitted (wine-red curve) data vs. time. The light brown dots are the values (2nd axis) of $1000\,\delta\rho_{pol}$ obtained from CLISTE equilibrium fits to magnetic data *not* including the tile currents. The separatrix leg departs from the tile at $t = 1.71$ s (dotted lines). The tile geometry (red) and flux surfaces at the sweep extremes and the crossing point are shown on the right.

## Summary

Ideal snowflake configurations are not possible when a finite current density flows on and outside the separatrix, as is observed experimentally. CLISTE computations of the evolution of the topology of the SF flux surfaces with increasing $j_\phi$ on the separatrix show that a triangular region with quasi-uniform $j_\phi$ and bounded by three X-points is established. Initial analysis of several divertor swept ASDEX Upgrade discharges with favorable geometry and information-rich tile current data yielded private flux region decay lengths for the poloidal current and, by assumption, $j_\phi$ of order $\Delta\rho_{pol} = 0.001$. Feasible $SF^-$ configurations using the existing ASDEX Upgrade coils have been found by a Monte Carlo scan of parameter space.

[1] PM Valanju et al, Phys.Plasmas **16** 056110 (2009).
[2] DD Ryutov et al, Phys. Plasmas **15** 092501 (2008).
[3] K Lackner, H Zohm Fus. Sci. Tech. **63** (2012) 43-48.
[4] DD Ryutov, MA Makowski and MV Umansky, Plasma Phys. Control. Fusion **52** 015001 (2010).
[5] K Lackner, H Zohm, *private communication*, (2014).
[6] PJ Mc Carthy and ASDEX Upgrade Team, Plasma Phys. Control. Fusion **54** 015010 (2012).
[7] A Kallenbach et al., J. Nucl. Materials **290-293** (2001) 639.
[8] T Lunt et al., Plasma Phys. Control. Fusion **56** 035009 (2014).

*This work has been carried out within the framework of the EUROfusion Consortium and has received funding from the Euratom research and training programme 2014-2018 under grant agreement No 633053. The views and opinions expressed herein do not necessarily reflect those of the European Commission.*

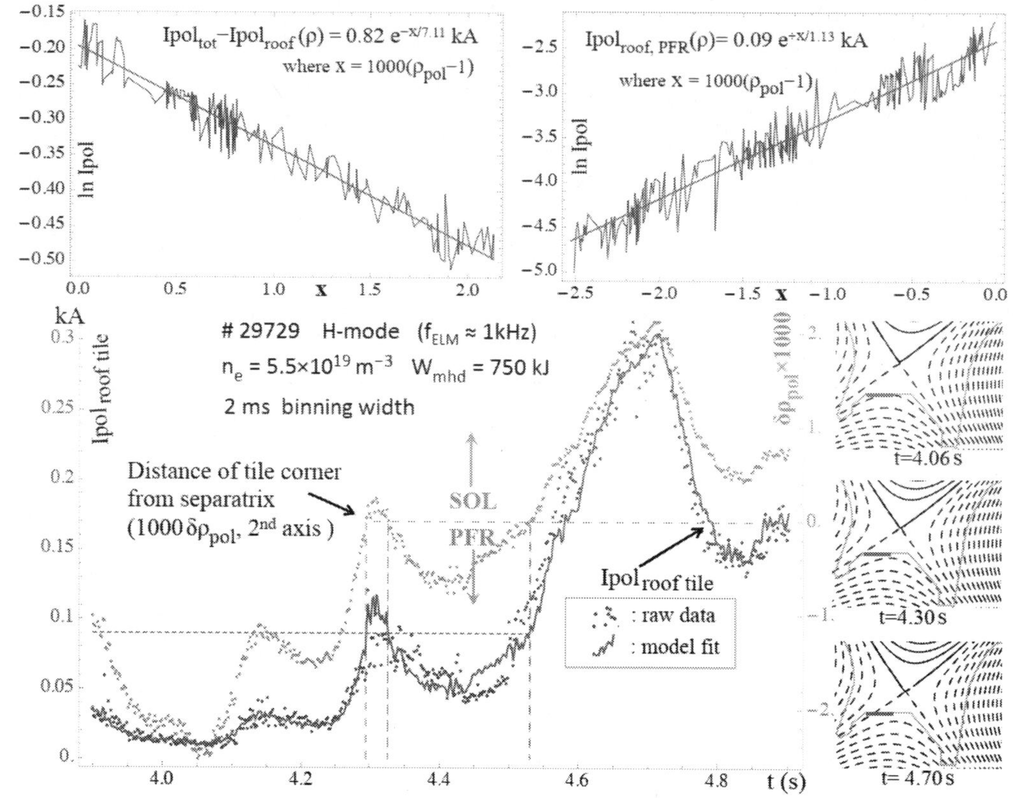


FIG. 3: Fits to poloidal current through roof baffle tile 14DUMoi in ASDEX Upgrade H-mode shot #29729 versus $\delta\rho_{pol}$ distance from separatrix. See previous caption for details.

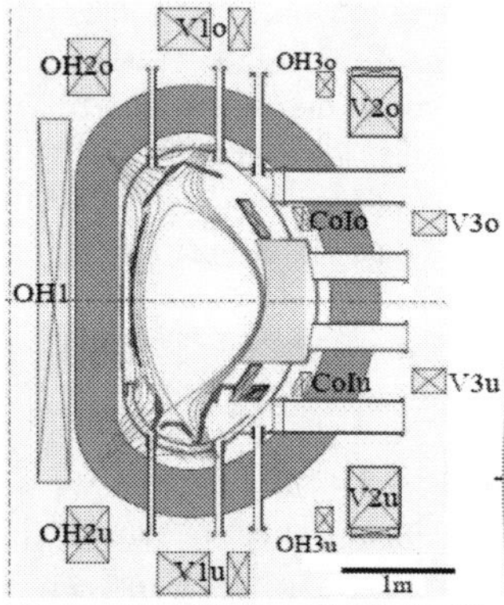


| Coil | Current | Limit | $F_V/F_{V,max}$ |
|---|---|---|---|
| V1o | −14.4 | 59 kA | 0.08 |
| V1u | −9.0 | 59 kA | 0.05 |
| V2o | 18.1 | 49 kA | 0.43 |
| V2u | 16.6 | 49 kA | 0.43 |
| V3o | −46.4 | 55 kA | 0.52 |
| V3u | −46.4 | 55 kA | 0.49 |
| CoIo | 1.9 | 30 kA | 0.04 |
| CoIu | −8.2 | 30 kA | 0.20 |
| OH1,3 | 0.0 | 47 kA | 0.00 |
| OH2o | 29.6 | 47 kA | 0.74 |
| OH2u | 41.9 | 47 kA | 0.32 |

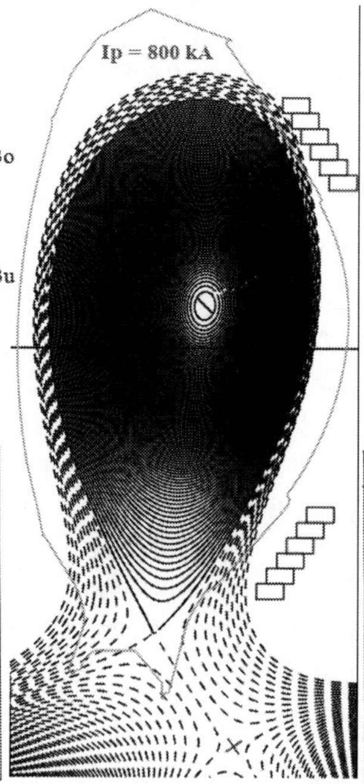


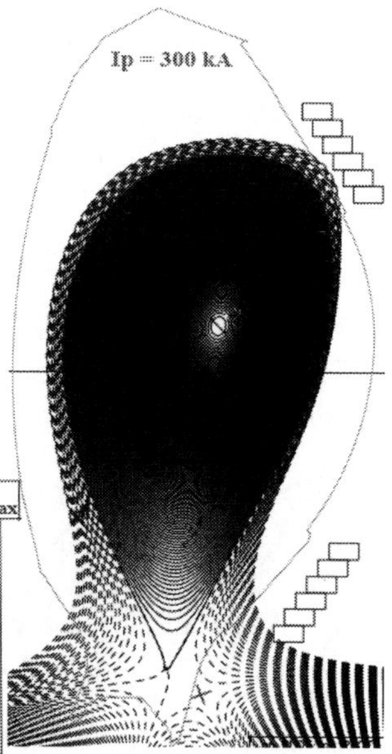


| Coil | Current | $F_V/F_{V,max}$ |
|---|---|---|
| V1o | −15.4 kA | 0.06 |
| V1u | −13.4 kA | 0.07 |
| V2o | 11.0 kA | 0.13 |
| V2u | 16.0 kA | 0.25 |
| V3o | −28.0 kA | 0.16 |
| V3u | −28.0 kA | 0.25 |
| CoIo | 24.3 kA | 0.29 |
| CoIu | −23.7 kA | 0.55 |
| OH1,3 | 0.0 kA | 0.00 |
| OH2o | 26.3 kA | 0.66 |
| OH2u | 30.9 kA | 0.32 |

FIG. 4: Feasible $SF^-$ configurations for ASDEX Upgrade including a table of coil currents per turn, maximum values and vertical forces expressed as a fraction of the maximum permissible force on each coil for (a) remote and (b) close secondary X-points, in each case with magnetic separation equivalent to a gap of ~1 mm in the outer midplane. Case (b) is possible only at the cost of a reduced plasma current due to the greater demands on the coil currents to achieve closer proximity of the X-points and also relies on significant currents in the CoIo, CoIu position control coils.

# Perturbations of microwaves due to plasma density fluctuations

A. Köhn[1,2], M. Thomas[3], R.G.L. Vann[3], E. Holzhauer[1], J. Leddy[3]

[1] *IGVP, University of Stuttgart, Germany*

[2] *Max Planck Institute for Plasma Physics, Garching, Germany*

[3] *York Plasma Institute, Department of Physics, York, U.K.*

Electromagnetic waves in the microwave regime play an important role in present plasma experiments as they are used for heating [1] and diagnostics purposes [2]. In both cases, the microwaves have to traverse the plasma boundary. In fusion related plasma, substantial density fluctuation levels are known to occur there [3] which can significantly distort the traversing microwave beam and thus lead to reduced coupling efficiencies to the plasma or to ambiguous diagnostics results. This work aims to investigate the average perturbation of a fluctuating plasma on a microwave beam by means of full-wave simulations. The advantage of a full-wave treatment is that no a-priori assumptions about the strength or size of the plasma density perturbations need to be made. The 2D full-wave code IPF-FDMC [4] is used to study the effect of a fluctuating plasma density on a propagating microwave beam. It is a finite-difference time-domain code based on a cold plasma description.

Ensemble averaging over a number of density profiles is required in order to be statistically relevant. The density fluctuations appear to be frozen in the frame of the microwave since they vary on the kHz scale [3], whereas the microwave oscillates on the GHz scale. The fluctuations itself are created by a Hasegawa-Wakatani drift-wave turbulence model within the BOUT++ framework [5]. A slice of fluctuating plasma density is then superimposed onto a homogeneous background density $n_0$ with a value of $n_0 = 0.5\,n_{\text{e,cut-off}}$, where $n_{\text{e,cut-off}}$ is the corresponding cut-off density of the injected microwave.

The numerical grid has a size of $10\,\lambda_0$ in the $x$-direction, where $\lambda_0$ is the vacuum wavelength of the injected microwave, and $5\,\lambda_0$ into the $y$-direction. It is surrounded by absorbing (non-radiating) boundaries. The fluctuating density slice extents along the whole $x$-direction but only from $y = 1\ldots 4\,\lambda_0$. The microwave beam is injected along $y = 0$ and along $y = 5\,\lambda_0$, a receiving antenna is placed. The injected beam has a Gaussian shape with the beam waist located in the antenna plane having a radius of $w_0 = 2\,\lambda_0$. The polarization is such that an O-mode is injected and the background magnetic field points into the direction perpendicular to the 2D numerical grid with a constant strength of $Y = \omega_{ce}/\omega_0 = 0.5$, where $\omega_{ce}$ is the electron cyclotron frequency.

Figure 1 shows two snapshots of the absolute value of the wave electric field taken after the

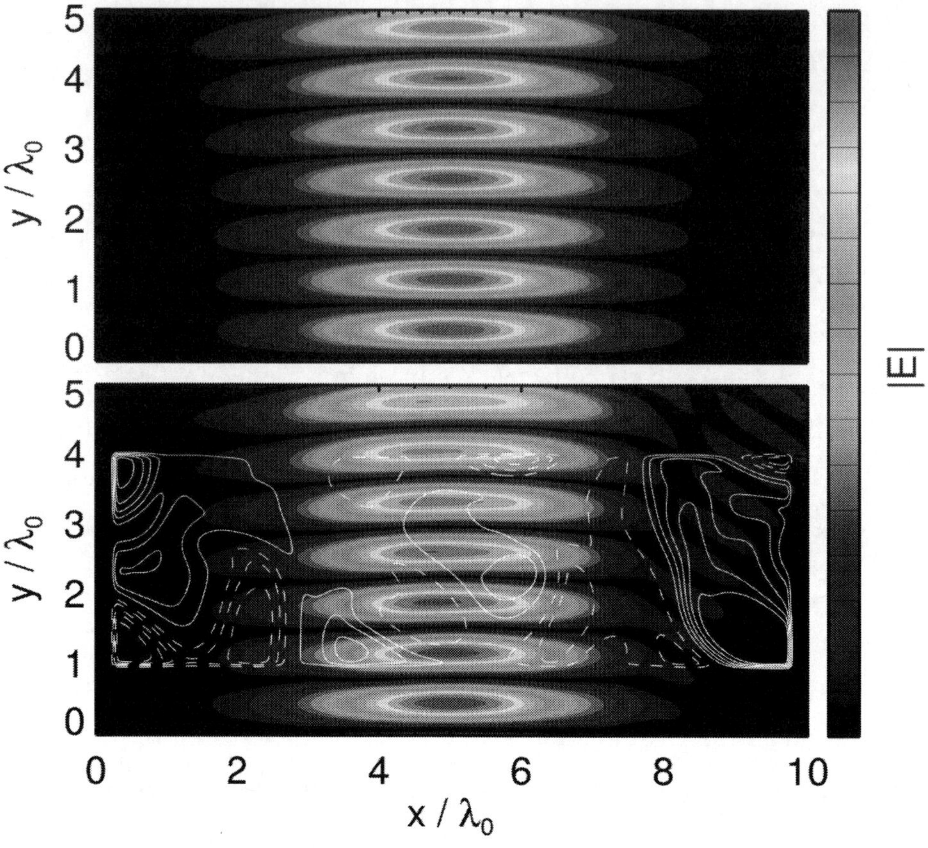


Figure 1: *Snapshots of the absolute value of the wave electric field after the steady state solution is achieved. Top: for the case of a homogeneous plasma with* $n_0 = 0.5\, n_{e,cut\text{-}off}$ *everywhere; bottom: for the case of a turbulence slice with an average structure size of* $c_L \approx 0.5\,\lambda_0$ *(white solid and dashed lines correspond to positive and negative density perturbations, respectively).*

steady state solution is achieved. The top snapshot shows the case for a homogeneous plasma with a density of $n_0 = 0.5\, n_{\text{e,cut-off}}$ and the bottom snapshot for a fluctuating plasma with an average size of the density structures of $c_L \approx 0.5\,\lambda_0$ and an average fluctuation level of $\sigma \approx 4\,\%$. The average size $c_L$ is taken as the distance where the spatial auto-correlation drops to a value of 0.5 and the fluctuation level is taken as the standard deviation of the mean-free density fluctuations. Comparing the two snapshots, it can be seen how the microwave beam is perturbed by the plasma density fluctuations. In this case, the position of the maximum signal is slightly shifted to the left and a sidelobe at the very right hand side occurs.

To quantitatively investigate the effect of the fluctuations on the microwave beam, the rms-value of the wave electric field $\tilde{\mathbf{E}}$ is continuously recorded at the detector antenna as a function of the coordinate $x$: $\tilde{E}_{\text{rms}} = \sum_t \left(|\tilde{\mathbf{E}}|/\sqrt{T}\right)$, where $t$ is the time coordinate and $T$ the wave period.

This signal is then compared with the signal of the homogeneous case, $E_{\mathrm{rms}}$. Figure 2 shows this comparison for the case of the average structure size which yielded the largest perturbation. The effect of the density fluctuations is obvious. Note, that for this plot only a few of the realizations from the ensemble of density profiles are shown. A scattering parameter $\alpha$ is defined as the sum of the squared deviations of the rms-signals, allowing to quantify the result of one full-wave run with a single value:

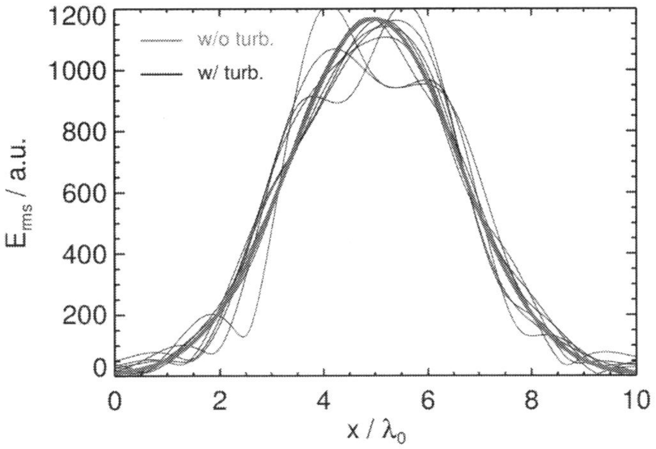


Figure 2: *Detector antenna signal for the homogeneous case (red) and for a few turbulence slices with an average structure size of $c_L \approx 0.5\,\lambda_0$.*

$$\alpha = \frac{\sum_x \left(\tilde{E}_{\mathrm{rms}} - E_{\mathrm{rms}}\right)^2}{\sum_x E_{\mathrm{rms}}^2}. \tag{1}$$

Looking at Fig. 2, the necessity of sufficient ensemble averaging is apparent. To check the quality of the averaging process, the average position of the maximum signal in the detector antenna plane, $\tilde{x}_{\mathrm{max}}$, is compared with the position for the homogeneous case, $x_{\mathrm{max}}$. The difference of these two values should vanish, $\Delta = \tilde{x}_{\mathrm{max}} - x_{\mathrm{max}} \stackrel{!}{=} 0$. Only then, the results will be statistically relevant. The average value of $\Delta$ is plotted in Fig. 3 as a function of the number of ensembles $N$. It can bee seen that $N$ is sufficiently large, since an asymptotic value of $\Delta = 0$ is reached (within the errorbars).

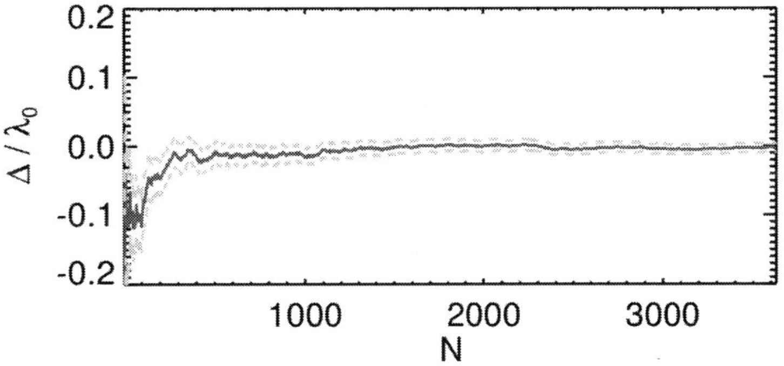


Figure 3: *Average value of the spatial difference of the maximum signal at the detector antenna (compared to the position of the maximum signal for the homogeneous case) as a function of the number of ensembles for an average structure size of $c_L \approx 0.5\,\lambda_0$ (the red curve corresponds to the average and the dashed orange curves to the error).*

The average size of the density structure is varied from $c_L \approx 0.2\,\lambda_0 \ldots 1\,\lambda_0$. With increasing structure size $c_L$, the number of ensembles $N$ is also increased as a smaller number of density structures fit into one turbulence slice. Figure 4 shows the scattering as a function of the average structure size. A maximum scattering at a size of $c_L \approx 0.5\,\lambda_0$ can be clearly seen. For decreasing and increasing sizes, the average scattering is decreasing.

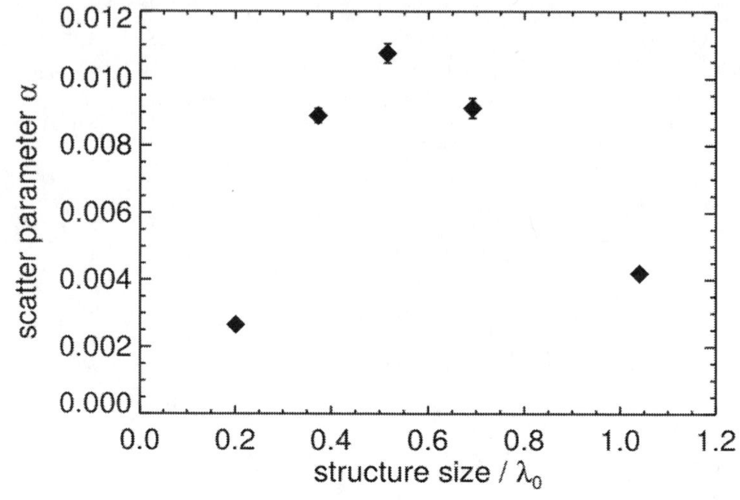


Figure 4: *Scattering parameter* $\alpha$ *(defined by Eq. (1)) as a function of the average size of the density structures for an average fluctuation level of* $\sigma \approx 4\,\%$.

Note that in a previous study, the influence of a single blob-like structure on a traversing microwave beam was studied [6], and the deterioration was found to peak at a larger structure size. This is, however, not in contradiction to the result presented here, as we are interested in the *average* perturbation whereas in the previous study we were interested in the *maximum* perturbation.

Further investigations are performed at the moment in order to elaborate the role of the strength of the fluctuations and of the size of the turbulence region.

## References

[1] V. Erckmann and U. Gasparino *Plasma Phys. Control. Fusion* **36** 1869 (1994)

[2] H.-J. Hartfuss and T. Geist *Fusion Plasma Diagnostics with mm-Waves* (Wiley-VCH, Weinheim, 2014)

[3] S. J. Zweben *et al.*, *Plasma Phys. Control. Fusion* **49** S1 (2007)

[4] A. Köhn *et al.*, *Plasma Phys. Control. Fusion* **50** 085018 (2008)

[5] B Dudson *et al.*, *Comp. Phys. Comm.* **180** 1467 (2009)

[6] T. Williams *et al.*, *Plasma Phys. Control. Fusion* **56** 075010 (2014)

**Acknowledgements** Part of the simulations were performed on the HELIOS supercomputer system at Computational Simulation Centre of International Fusion Energy Research Centre (IFERC-CSC), Aomori, Japan, under the Broader Approach collaboration between Euratom and Japan, implemented by Fusion for Energy and JAEA.

# Statistical analysis of the oscillations preceding ELM instabilities

S. Zoletnik[1], L. Zsuga[1,2], TEXTOR Team[3]

[1] *Wigner Research Centre for Physics, Budapest, Hungary*

[2] *Budapest University of Technology and Economics, Budapest, Hungary*

[3] *Institute of Energy and Climate Research, Forschungszentrum Jülich, Germany*

## Introduction

Former Lithium Beam Emission Spectroscopy (Li-BES) measurements in limiter H-mode plasmas on TEXTOR[1] showed the appearance of growing oscillations before ELM instabilities[2]. These precursor structures were noticeable only in the middle of the pedestal region a few 10 μs-s before ELMs with about 40kHz frequency. With the purpose of getting more information on the connection of ELM triggering and precursor structures a statistical analysis of Li-BES measurements on TEXTOR are presented in this paper. The ELM frequency in these TEXTOR discharges is around 1 kHz and they are likely type III.

## Measurement technique

The measurements were made with the 35 keV Lithium beam diagnostic[3]. At the edge of the plasma the beam light emission is nearly proportional to the local electron density, thus these measurements can give information about density changes in the pedestal region. The light signals were measured by 16 Avalanche Photodiode (APD) detectors with 1 MHz analog bandwidth and 2.5 MHz sampling rate. The APDs measure the light signal with 1 cm radial resolution. The signal to noise ratio was in the 10-20 range. Although the APD system uses a 2 nm narrow band optical filter the background light can be substantial, especially during ELMs. To separate the background from the beam emission the fast chopping technique[3] was used which modulates the beam with 250 kHz. Averaging the signal in the beam-on and beam-off times one gets both the background, and beam+background signals with 4 μs time resolution. After interpolation the background can be subtracted thus a clean Li-beam signal is obtained with 4 μs time resolution. Channels from 9 to 14 examined the pedestal area and the Scrape off layer as shown in *Fig.1a*. The effect of ELMs on the measurement signals deeper in the plasma was negligible.

## Detecting ELMs in the signals

The background light emission increases during ELMs because of the increased plasma-wall interaction. This change allows us to locate the starting time of ELMs. An ELM reference signal was created by summing up the background signals from all channels. An approximate ELM onset time was chosen by hand in this reference signal. Well separated ELMs were selected, ELMs too close to each other were neglected. A trigger level was set at about 10% of an average ELM pulse height and a program selected the time around each approximate ELM onset time where the reference signal increases above the trigger level. These times were used as the accurate onset time of the ELMs as shown in *Fig.1b*.

Time interval of precursors

The analysis was done on three shots. Several hundred ELM onset times were chosen in each with the algorithm described above. The background corrected Li-BES signals were examined in the [-100, 0] μs interval relative to the ELM onset time. Overplotting them showed tendencies as seen in *Fig.1c*. In the last 50μs before the ELM onset the amplitude of the oscillations in Li-beam signals 9-11 increases remarkably, thus the [-50,0] μs interval is called "precursor time window". In the [-100, -50] μs interval the oscillations seem to have constant amplitude, therefore this time interval was used as a reference time window in order to estimate the noise level. The source of this noise is the signal amplifier, photon statistics and plasma activity unrelated to ELMs.

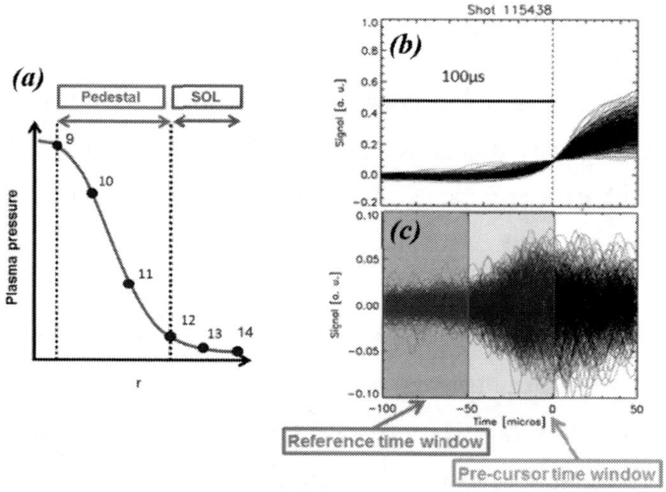


*Fig. 1. (a) Cartoon of the location of the analysed measurement channels relative to the pedestal. (b) ELM reference signals overplotted around the determined exact ELM onset times. The reference signal crosses the trigger level at the red dashed line. (c) Background corrected and 10-70 kHz bandpass filtered Li-beam signal in channel 11 overplotted around the exact ELM onset times.*

Statistical analysis

As the first step the frequency distribution of the precursor oscillations was studied. As their amplitude changes on a 100 μs timescale the frequency resolution cannot be better than 10 kHz. The background corrected Li-beam signals were digitally filtered with a series of 10 kHz wide bandpass filters covering the 0-100 kHz frequency range. The mean energy of the oscillations in the $[t_1,t_2]$ time interval of the $\tilde{S}_c^f$ filtered signals was calculated as

$$E(c,f,t_1,t_2) = \frac{\sum_{n_{ELM}} \int_{t_1}^{t_2} \left(\tilde{S}_c^f - \langle \tilde{S}_c^f \rangle\right)^2}{n_{ELM}},$$

where $t_1$, $t_2$ are the start and end times of the time window in μs, $n_{ELM}$ is the number of processed ELMs, *c* stands for the channel (*c* = 9, 10, …, 14) and *f* stands for the frequency band of the filter (*f* = 0-10 kHz, 10-20 kHz, …, 90-100 kHz). These mean energies were calculated in both the reference and precursor time windows. $E(c, f, -100, -50)$ was used as the noise level of the precursor energy calculation. As the noise energy is expected to be statistically independent of the precursor oscillations, the precursor mean energy can be calculated as

$$E_p(c, f) = E(c, f, -50, 0) - E(c, f, -100, -50).$$

$E_p(c, f)$ were depicted for all three shots on *Fig. 2a*. The graphs suggest that the precursor energy has a maximum in the pedestal region between 10-40 kHz frequency. In order to study the spatial distribution of precursors in more detail the signals were filtered for the

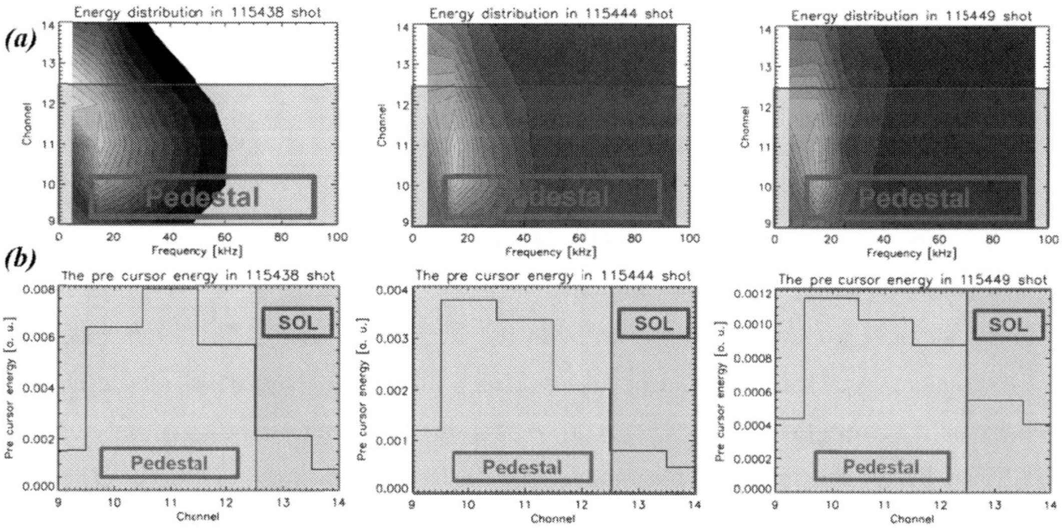


*Fig 2. (a) Distributions of the precursor mean energy $E_p(c, f)$ in the three studied shots. (b) Spatial distribution of the precursor mean energy in the 10-40 kHz frequency band.*

10-40 kHz band and the spatial distribution of $E_p(c, 10 - 40\ kHz)$ is plotted in *Fig. 2b*. This plot clearly places the precursor structures in the pedestal area of the plasma.

The above investigations only gave information about the precursor activity averaged for all ELMs. The next step was to investigate to what extent the precursor activity changes from ELM to ELM. Therefore the histogram of the precursor energy of each ELM was calculated for each channel and time window for all three shots separately. An example is shown in *Fig. 3*. The energy of the oscillations in the reference time window is limited to small values while in the precursor time window much higher energies are observed. The noise level of the precursor detection was determined from the histogram in the reference time window in a

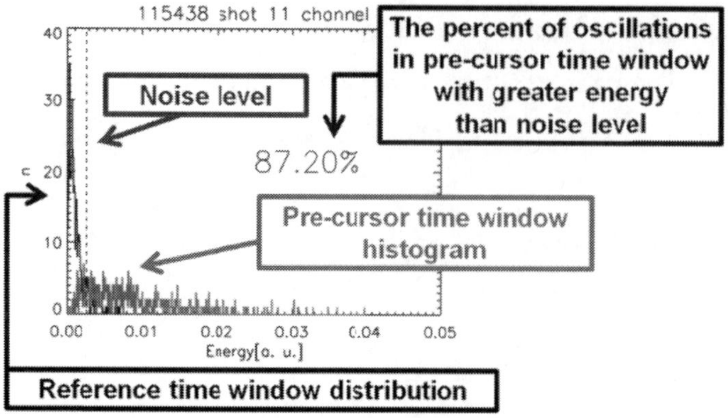


*Fig. 3. Histogram of precursor energy in shot 115438, channel 11. The black curve shows the histogram in the reference time window, while the red curve is the histogram in the precursor time window. The percentage of ELMs with precursor energy above the noise level is 87% in this case.*

way that 90% of ELMs have oscillations in the reference time window with energy below this value marked by a dashed vertical line. As the final step of the analysis the percentage of ELMs was calculated which have higher precursor energy than the noise level. These percentages are summarized in the table.

## Conclusions

| | Shot | | |
|---|---|---|---|
| Channel | 115438 | 115444 | 115449 |
| 9 | 50 % | 34 % | 18 % |
| 10 | 77 % | 45 % | 26 % |
| 11 | 87 % | 38 % | 23 % |
| 12 | 79 % | 33 % | 25 % |
| 13 | 53 % | 37 % | 27 % |
| 14 | 37 % | 30 % | 24 % |

The statistics showed that the ELM precursor activity is concentrated to the middle of the pedestal area with 10-40 kHz frequency. In one shot (115438) nearly 90% of ELMs have clear precursor energy above the noise level. In the two other analysed shots the tendency is less clear. Although in the pedestal area a significant increase is seen in the average energy of oscillations 50 μs before the ELM only for about 30% of ELMs can this precursor activity be detected above the noise level. The global parameters ($I_p$, $B_t$, $q_a$, $n_e$) are nearly the same for all shots, the difference lies in the ELM activity. In the shot where precursor oscillations are clearly distinguishable ELMs are quasi-periodic ($f \sim 1$ kHz) and they are preceded by a quiet period. In the other shots ELMs happen irregularly. Because of the absence of the quiet period it is more difficult to determine the ELM onset time and the noise level calculated from the reference time interval is also higher.

## References

[1] S. Soldatov, *et al*, *Plasma Phys. Control. Fusion* **52** 085001 (2010)
[2] S. Zoletnik, *et al*, *Int. Congress on Plasma Physics. ECA* **36F** O3.107 (2012)
[3] D. Dunai, *et al*, *36th EPS Conference on Plasma Phys. ECA* **33E** P-1.182 (2009)

# About dynamics of shocks induced by complex spot laser irradiation of layered targets

A. Aliverdiev [1,2*], D. Batani [3], A. Amirova [4], R. Benocci [5], R. Dezulian [5], E. Krousky [6], M. Pfeifer [6], J. Ullschmied [6], J. Skala [6], and K. Jakubowska [7,3]

[1] *IGR DSC RAS - Pr. Shamilya 30A, 367030, Makhachkala, Russia*

[2] *Dagestan State University -Ul. M. Gadzhieva 43A, 367025, Makhachkala, Russia*

[3] *Université Bordeaux, CEA, CNRS, CELIA (Centre Laser Intense at Applications), UMR 5107,F-33405 Talence, France*

[4] *IP DSC RAS - ul.Yaragskogo 94, 367003, Makhachkala, Russia*

[5] *Dipartimento di Fisica G.Occhialini, Universita di Milano Bicocca, Milan, Italy*

[6] *PALS Research Centre, Za Slovankou 3, 18221 Prague 8, Czech Republic*

[7] *Institute of Plasma Physics and Laser Microfusion, 23 Hery Str., 01-497 Warsaw, Poland*

* *aliverdi@mail.ru*

The aim of the present work is the analysis of laser inducted shock behavior in experiments with double (or complex) focal spot, in particular, the cumulative effect in double layer foam-metal targets.

Experiments were realized using the PALS (Prague Asterix Laser System) iodine laser. By splitting the laser beam in two equal parts with a prism we could obtain two focal spots, which gives very large non-uniformity. A photographic objective has been employed to image the target rear face onto a streak. [1] One of the main experimental results is the difference in time-resolved rear-side self-emission images obtained with the streak camera for aluminium and aluminium-foam targets. Namely, we obtained the bright area between two spots for the double-layer targets (look at fig. 1a), which is not have place in the experiments with monolayer Al target (look at fig. 1b).

In order to provide qualitative simulations with a 2D code [2,3], we assumed an axial-symmetric approximation, i.e. we have used a coaxial striped spatial profile for the laser spot. Some experimental data and preliminary analysis (including simulations with "mono-ring" and "central-spot+ring" 2D simulations) was recently published in [4]. In a case of double-layer targets the shock generated at the interface spreads back into the foam, continues to compress the foam-base, and finally produce a central region with a much higher pressure, as compared to the case of the simple Al foils.

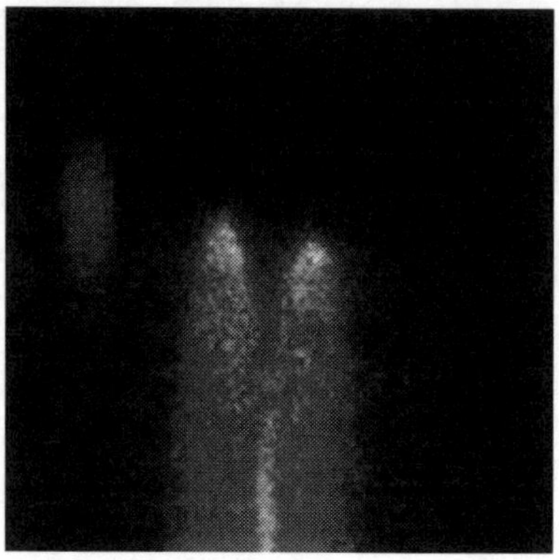

a

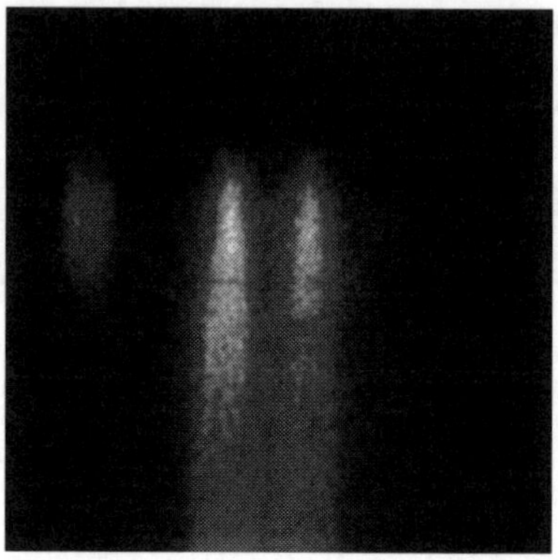

b

**Fig. 1(colour on-line).** Rear side optical streak camera experimental images (in arbitrary units): a) target with 10 μm Al+50 μm foam ($\rho_{foam}$ = 50 mg/cm$^3$); b) 10 μm simple Al target. The total time window is 1600 ps (vertical) and the imaged region is 1330 μm wide (horizontal). The laser energy is the same for both shots: E = 52 J. Time flows from top to bottom. The signal on the upper right part of the image is the time fiducial indicating the arrival of the laser pulse on the front side of the target.

The 2D simulations with an axial symmetry are not completely adapted to reproduce our experiment. Indeed, the convergence of the shocks towards the axis can cause an artificial increase in pressure. Also the region around the centre of the cylinder is critical for possible artefacts in discretization. Hence, in order to acquire more confidence in our results, we realized simulations with another spatial profile, i.e. we used a double-ring spatial profile to see a shock collision dynamics and additional central spot with the same size for additional comparison with simulations of one-spot dynamics. The diameter of the central spot and the thickness of rings as well as the distance between rings were chosen close to experimental spot diameters. The diameters of bands and the distance between central spot to the first ring were chosen much larger than the ring thickness.

We have analyzed the dependence of shock behaviour from laser intensity and foam density. Some results will be presented in the poster and will be published soon. Fig. 2 demonstrates the temperature of the rear side (the last cell of the target) as a function of time for the Al-foam ($\rho_{foam}$ = 50 mg/cm$^3$, left) and the Al (right) targets for a set of laser intensities.

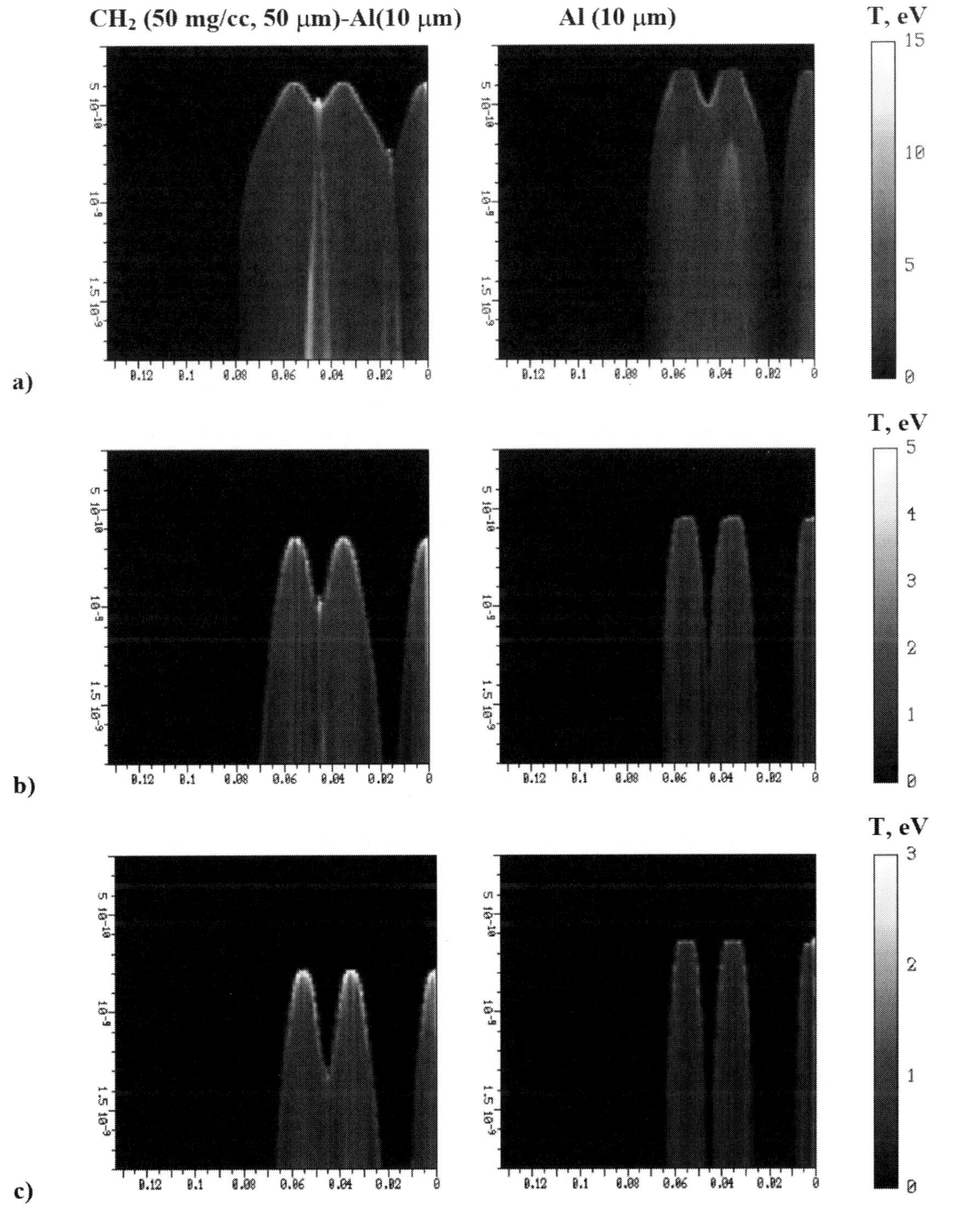


**Fig. 2 (colour on-line).** The temperature of the rear side (the last cell of the target) as a function of time for the Al-foam ($\rho_{foam} = 50$ mg/cm$^3$, left) and the simple Al (right) targets for different laser intensities: (a) $10^{15}$ W/cm$^2$; (b) $10^{14}$ W/cm$^2$; (c ) 5 $10^{13}$ W/cm$^2$. The total time window is 1600 ps (vertical) and the imaged region is 1330 μm wide (horizontal) as in fig. 1.

Finally, we have experimentally observed and numerically analyzed the behaviour of shocks generated by a "double laser spot" on foam-Al layered targets. We have observed the shock breakout on the target rear side and we saw that the region between the two spots becomes brighter that the spots themselves. This is explained as a result of the collision between the two individual shocks, generating larger pressure in the collision region. We performed hydrodynamic simulations in 2D (in "double-ring" format), which despite the assumption of axial symmetry seem to reproduce the qualitative feature of our results.

**Acknowledgments.** We warmly acknowledge technical staff of PALS. A.A. is also grateful to ESF (program SILMI), COST (STSM visit grant in the framework of COST Action MP1208 "Developing the physics and the scientific community for inertial fusion"), RFBR (12-01-96500, 12-01-96501), European Physical Society and "Dynasty" Foundation of D. Zimin for support.

## References

[1] R. Benocci, et al. 2009 *Phys. Plasmas* **16** 012703

[2] R. Ramis, et al., 2009 *Computer Physics Communications* **180**, 977

[3] A. Aliverdiev, et al. 2014 *Phys. Rev E* **89**, 053101

[4] A. Aliverdiev, et al. 2015 *Nukleinika* **60** (2) 213

# Neutron spectra and temperature diagnostics of imploded ICF plasma: models and 1D and 2D simulations

S. Atzeni and G. Rossi

*Dipartimento SBAI, Università di Roma "La Sapienza", Roma, Italy*

## 1. Introduction

The spectrum of neutrons released by an inertially confined fusion (ICF) plasma provides valuable information on the emitting plasma [1]. This includes, but is not limited to, ion temperature and confinement parameter $\rho R$. For instance, a neutron-averaged ion temperature is obtained from the width of unscattered portion of neutron spectra, $T \propto \Delta E_{\mathrm{FWHM}}$ [2]; see below. The confinement parameter of ignition-size targets is obtained from the down-scattered-ratio ($DSR$), i.e the ratio of the 14.1 MeV DT neutrons scattered in the 10–12 MeV range to the unscattered neutrons (neutrons in the 13–15 MeV range); approximately, $\rho R[\mathrm{g/cm^2}] = 21DSR$.

General expressions for the production spectrum of neutrons (and fast charged particles) can be found in the literature [3, 4]. It is confirmed that energy distribution $f(E)$ of the neutrons emitted by a uniform source, at rest, with Maxwellian ion velocity distribution, is very well approximated by a Gaussian, $f(E) \propto \exp[-(E- <E>)^2/2\sigma^2]$, with average energy and rms deviation, respectively [2],

$$<E>= \frac{3}{2}T + \frac{m_o}{m+m_o}(Q+ <K>); \qquad \sigma = \left(\frac{2m}{m+m_o}T <E>\right)^{1/2},$$

where $m$ and $m_o$ are the masses of the neutron and of the other fusion product, respectively, $Q$ is the reaction yield, $<K>$ the mean relative energy of the reacting nuclei, and $T$ is the ion temperature in units of energy. (Slightly more accurate expressions, as well expressions for higher moments of the distribution can be found in [5]). In practice $\sigma \simeq [TQ(2mm_o/(m+m_o)^2)]^{1/2}$, since typically both $T$ and $<K>$ are much smaller than $Q$. The relations between FWHM of the energy distribution ($\Delta E_{\mathrm{FWHM}}$) and the reaction-weighted ion temperatures for DT and neutron branch of DD reactions follow immediately:

$$T_{\mathrm{DT}} = (\Delta E_{\mathrm{FWHM}}/176.8)^2; \qquad T_{\mathrm{DDn}} = (\Delta E_{\mathrm{FWHM}}/82.4)^2, \tag{1}$$

where both neutron energies and ion temperatures are measured in keV.

However, ICF imploded plasmas are space- and time-dependent, plasma motion effects can also be important (see the next section), and emitted neutrons can be scattered; see, e.g. Fig. 1, showing the spectrum emitted by a NIF-size ignited target. As a first step in the development of

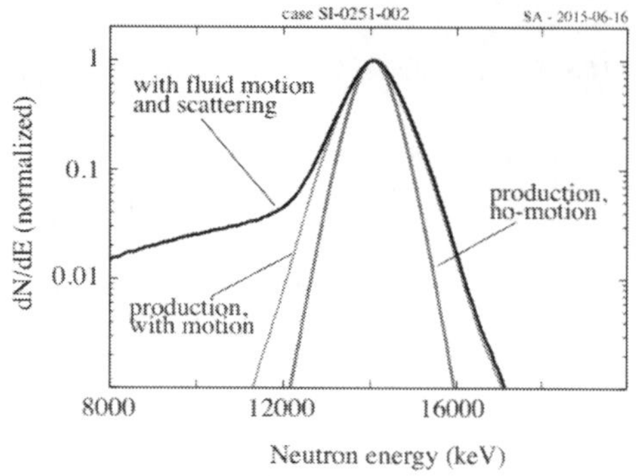


Figure 1: *DT neutron spectrum released by a reactor-size, shock-ignited target (see Sec. 3.3).*

a synthetic neutron diagnostics, we have included a neutron Monte Carlo package in the 2D ICF code DUED [6]. The source term accounts for both ion motion and bulk fluid motion. Scattering is also included, while, at the moment, secondary reactions and reactions by knocked-on nuclei are neglected.

## 2. Fluid motion effects

In most cases, scattering does not affect the width of the distribution, which provides information on ion temperature. Instead, fluid motion effects can be significant. Insight is provided by simple analysis, applying to both the angularly integrated spectrum from a point source in motion with velocity $u$, and to the spectrum released by a homogeneous and spherically symmetric plasma imploding/exploding with velocity $u$. We have tested that the production spectrum obtained from Eq. (35) of ref. [3] is accurately reproduced by the convolution of a Gaussian (with rms deviation $\sigma$) accounting for thermal motion only, and a distribution accounting for fluid motion only. The latter is a constant for energy in the interval $[E_0(1-w)^2]; E_0(1+w)^2])$ and is zero elsewhere, with $E_0 = (1/2)mv_0^2 = Qm_o/(m+m_o)$ the neutron energy in the fluid system, and $w = u/v_0$ and $v_0$ neutron velocity in the fluid system.

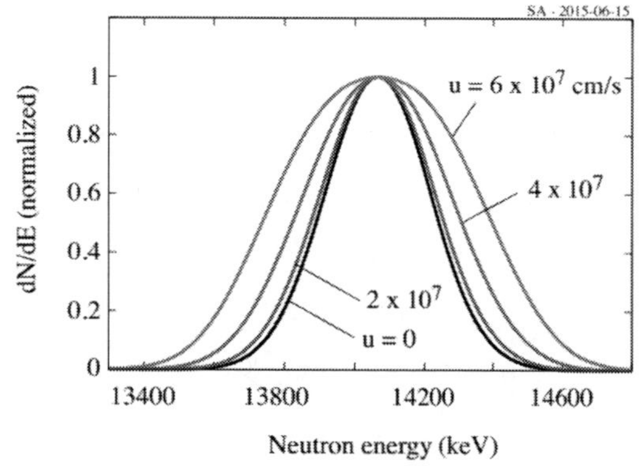


Figure 2: *DT neutron production spectrum, for a 4 keV plasma, and different values of the fluid velocity u.*

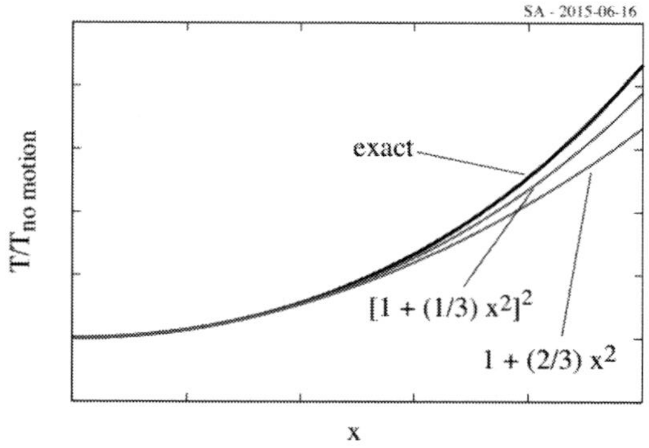


Figure 3: *Ratio* $T/T_{\text{no motion}}$ *vs the dimensionless parameter* $x = 2(u/v_0)(<E>/\sigma)$.

Distributions of DT neutrons from a 4 keV plasma are presented in Fig. 2 for different values of the fluid velocity $u$. In this case, motional effects become appreciable for velocities in excess of $2\times 10^7$ cm/s. Simple analysis (based on power expansion of the convolution to third order in $x$; details will be given elsewhere) provides an approximate expression for the FWHM of the spectrum, as a function of $T$ and the parameter $x = 2 <E> w/\sigma = 0.229(u_7/\sqrt{T})\sqrt{A_n + A_o}$, where $u_7$ is the fluid velocity in units of $10^7$ cm/s, and $A_n$ and $A_o$ are the mass numbers of the reaction products. Using Eq. (1) to obtain a neutron-weighted ion temperature we can write

$$T \simeq T_{\text{no motion}}(1 + \frac{1}{3}x^2)^2 \simeq T_{\text{no motion}}(1 + \frac{2}{3}x^2). \quad (2)$$

Such expressions are a few per cent accurate for $x < 0.6$ (i. e. $u_7 < 1.2\sqrt{T}$, for DT), as can be seen from Fig. 3. Practical expressions for DT- and DD-weighted temperatures are, respectively $T_{DT} \simeq T_{DT\text{no motion}}(1 + 0.176u_7^2/T)$ and $T_{DD} \simeq T_{DD\text{no motion}}(1 + 0.14u_7^2/T)$. From these expressions one can expect significant fluid motion contribution to neutron-weighted temperatures in exploding pushers (where reactions occur when a strong shock hits the center) and in ignited targets (which explode with velocity well in excess of $10^8$ cm/s), and only minor effects in conventional close-to-ignition targets (in which reactions occurs at implosion stagnation).

**3. Coupled DUED hydrodynamics and neutron Monte Carlo simulations**

We now discuss neutron spectra generated by DUED radiation hydrodynamics simulations with the Monte Carlo neutron package.

3.1 - High-density $D^3He$-filled exploding pusher implosions at OMEGA [7]. At low gas-fill density, 1D simulations largely overestimated both yields and temperatures, evidencing the role of kinetic effects [7]. At the highest densities (few mg/cm$^3$), fluid simulations were closer to experiments. Yields were overestimated by a factor 2–3, and temperatures were not far from the experimental values. In particular, experimental DD-weighted temperature were about 12 keV, while the temperature inferred from DUED-MC is about 10.3 keV; 8.8 keV when fluid motion effects are neglected. This is consistent with an average velocity of about $3.5 \times 10^7$ cm/s at the time of neutron emission.

3.2 -NIF-like, high-foot implosion implosions. We have performed 1D simulations of direct-drive targets with approximately same fuel mass (0.2 mg), implosion velocity ($3.1 \times 10^7$ cm/s) and adiabat ($\alpha \simeq 2.5$) as the NIF high-foot experiments [8]. A typical spectrum is shown in Fig. 4. From such a spectrum we obtain $DSR = 0.045$ and DT weighted temperature $T_{DT} = 4$ keV (with a contribution smaller than 150 eV from fluid motion; the curves with motion included and neglected indeed nearly

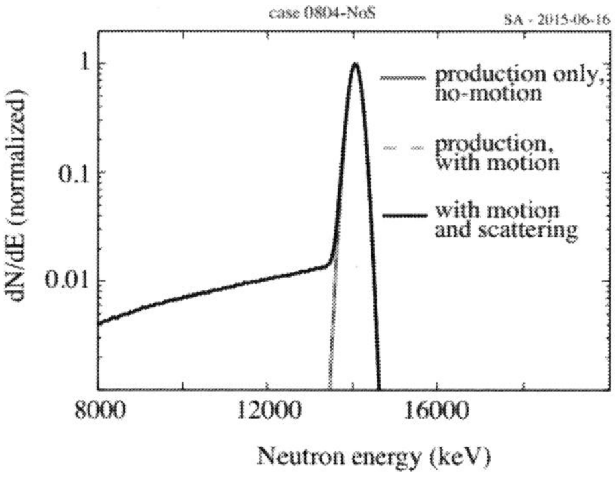


Figure 4: *Spectrum from a NIF-like high foot implosion.*

overlap in the figure). Such values, as well as other hot spot parameters are in reasonable agreement with experimental values. A significant discrepancy (also reported in Ref. [8]), instead concerns the difference $\Delta T = T_{DT} - T_{DD}$ of DT- and DDn-weighted temperatures. The simulations give $\Delta T = 200$-300 eV, only about one half of the experimental values. According to the results of Sec. 2, fluid motion (even turbulent motion) cannot account for such a larger $\Delta T$ difference. Such effects as fluid layering or species separation [9, 10, 11], not modelled by DUED, could instead contribute to increasing $\Delta T$.

3.3 - High-gain shock-ignition targets. In particular, we refer to the *robust R* cases presented in Sec. 5 of Ref. [12]. Figure 1 shows the spectrum released by a target driven by an 830 kJ laser pulse, and achieving gain about 100 (85 MJ yield). In this case one notices the very large broadening due to fluid motion during thermonuclear burn (increasing the DT-weighted temperature from 25 keV to 45 keV). 2D simulations of targets failing to ignite due to excessive hot spot asymmetries (as the case shown in the lhs panels of Fig. 10 of Ref. [12]) show signficant effects of fluid motion, too, yielding $T_{DT} = 6.5$–7 keV and $T_{DT\,\mathrm{no\ motion}} \simeq 5$ keV. A possible explanation is that asymmetric implosions result in incomplete conversion of kinetic energy into internal energy.

Work supported by the Italian project MIUR PRIN2012AY5LEL. I thank M. Gatu-Johnson, J. Frenje, G. Gorini, R. Petrasso, M. Rosenberg and A. Schiavi for useful discussions.

**References**


[1] J. Frenje *et al.*, Nucl. Fusion **53**, 043014 (2013)

[2] H. Brysk, Plasma Phys. **15**, 611 (1973)

[3] B. Appelbe and J. Chittenden, Plasma Phys. Controll. Fusion **53**, 045002 (2011)

[4] B. Appelbe and J. Chittenden, High Energy Density Physics **11**, 30 (2014)

[5] L. Ballabio, J Kallne and G. Gorini, Nucl. Fusion **38**, 1723 (1998)

[6] S. Atzeni *et al.*, Comput. Phys. Commun. **169** (2005) 153

[7] M. Rosenberg *et al.,* Phys. Rev. Lett. **112**, 185001 (2014)

[8] O. Hurricane *et al.*, Nature **506**, 343 (2014)

[9] D. T. Casey *et al.*, Phys. Rev. Lett. 108 075002 (2012)

[10] C. Bellei *et al.*, Phys. Plasmas **20** 012701 (2013)

[11] A. Inglebert *et al.*, EPL **107** 65003 (2014)

[12] S. Atzeni, A. Marocchino and A. Schiavi, Plasma Phys. Controll. Fusion **57**, 014022 (2015)

# Initialization of charged particle beam in OSIRIS

A. Balzarini[1,2], R.A. Fonseca[2,3], J. Vieira[2], E. Boella[2], L.O. Silva[2]

[1] *Dipartimento Energia, Politecnico di Torino, Turin, Italy*

[2] *GoLP/Instituto de Plasmas e Fusão Nuclear - Laboratório Associado, Instituto Superior Técnico, Lisboa, Portugal*

[3] *ISCTE - Instituto Universitário de Lisboa, Lisboa, Portugal*

## Introduction

There are several phenomena that can be investigated looking at the interaction between a particle beam and a plasma. One important example is that of beam plasma accelerators, where a particle beam acts as a driver for wakefield acceleration in the plasma. The tools of choice for modeling these scenarios are particle-in-cell (PIC) codes such as OSIRIS[1], that is a fully relativistic, massively parallel, PIC code. The current algorithm for initializing the EM fields associated with the beam implemented in OSIRIS works by gradually accelerating the beam, and has some limitations, in particular regarding the computational time required to converge to a solution. In this work we present a new algorithm that overcomes this problem, leading to an improvement of the code performance and of the accuracy of the initial fields.

## Initialization of EM-fields in PIC code

Electromagnetic PIC codes generally work by iteratively advancing particles and fields in time starting from a self-consistent description of the system. The default is to initialize it assuming that it is charge and current neutral, with no EM fields present. This is not suitable for simulating a charged particle beam, where beam current and charge are expected to generate both electric and magnetic fields. To initialize the values of fields associated with the beam, the current algorithm in OSIRIS would accelerate the beam, from zero to the velocity correspondent to the required Lorentz factor $\gamma$, through a finite number of time-steps[1]. During this initialization stage, the beam would not react to simulation fields, that would gradually converge to a value close to the real one, especially for large number of initialization time-steps. Although this scheme was easy to implement, it had some problems: the algorithm converges very slowly, so the initialization stage requires significant time that is not negligible when compared with the total simulation time. The number iterations required also grows on the size of the simulation box, further slowing larger simulations.

[1]This mimics the gradual acceleration of a beam in a constant field.

## New algorithm

The goal of the new scheme is to evaluate, at time $t = 0$, the EM field associated with a particle beam that is moving in $x_1$ direction with an high $\gamma$. Following the existing OSIRIS code, the new algorithm is parallelized[2] and it is written for 2D-cartesian, 3D-cartesian and 2D-cylindrical coordinates (axial-symmetrical). The steps that the new algorithm does are:

**1**. The charge density, the velocity of the beam and the geometrical parameters of the system are moved from the laboratory reference frame, where the beam has its $\gamma$, to the beam reference frame, where its velocity is 0, using the Lorentz transformation. With the hypothesis that the particles are moving in $x_1$ direction, the result of the transformation is only a stretch of the space along $x_1$ by a factor $\gamma$ and a consequent reduction of the charge density, $\rho$.

$$x_1' = \gamma x_1; \quad x_2' = x_2; \quad x_3' = x_3; \quad \rho' = \frac{\rho}{\gamma}; \tag{1}$$

Since we are dealing with beams that are ultra relativistic, with $\gamma \gg 1$, the cell shape in the beam reference frame will be significantly stretched along the $x_1$ direction leading to $dx_1' \gg dx_2' \simeq dx_3'$. As result, we can consider a cell in this frame like an infinite rod along the $x_1$ direction.

**2**. On the beam reference frame the velocity is zero, making this an electrostatic problem. Therefore it is possible to evaluate the electric field, **E**, using Gauss's law, that, expressed in OSIRIS normalized units, is:

$$\int \mathbf{E} \cdot \mathbf{dS} = \int \rho dV \tag{2}$$

Considering that each cell is an infinite rod we get for a 2D-cartesian case:

$$E_2 = \frac{\rho' \cdot dx_2'}{2} \tag{3}$$

And for 2D-cylindrical/3D-cartesian one:

$$E(r) = \frac{\rho' \cdot dx_2' \cdot dx_3'}{2 \cdot \pi \cdot r} \tag{4}$$

Due to the stretching of the grid, two grid-points that lie in $x_1$ direction are too far to influence each other. As consequence the E-field is only evaluated in a plane perpendicular to $x_1$, the direction of the motion of the charge, while is set to zero elsewhere. Then, since each charged cell is approximated to an infinite rod, there is no $E_1$ component of the field, and it is set to zero everywhere. All this cumulative effects lead to a speed-up of the code.

**3**. Once all the components of the electric field are evaluated, it is possible to come back to the laboratory frame of reference using again the Lorentz transformation applied to the fields:

$$\mathbf{E}_\perp = \gamma \cdot \mathbf{E}_\perp'; \quad \mathbf{B}_\perp = \frac{\mathbf{v}}{c} \times \mathbf{E}; \tag{5}$$

[2]For distributed memory systems, using MPI.

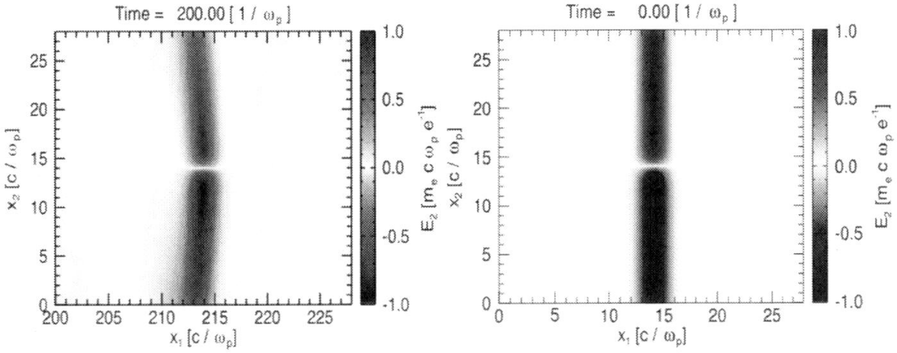


Figure 1: $E_2$ of a 2D gaussian positron beam propagating in $x_1$ direction obtained with the traditional algorithm (left) and the new one (right). $512 \times 512$ cells, $2 \times 2$ particles per cell, 4 nodes. The time required to obtain the result on the left is 165 s, 10000 time-steps, while for the other one is less than 0,1 s.

Giving the staggering of field values in the Yee mesh used by OSIRIS, we need to center some of them, which we do by linear interpolation. A comparison between the new algorithm presented here and the traditional one for a 2D case, can be seen in fig. 1.

**Parallelization**

OSIRIS was designed from scratch to be massively parallel, so it could tackle large simulation problems. The traditional algorithm required no additional effort to parallelize, as it only involved the accelerating local particles. In the new algorithm however, careful attention needs to be given to this issue, as the charge of the beam in one particular parallel node may influence the fields in other ones. One possible approach would be broadcast the charge from a given nodes to all the nodes and then calculate the corresponding fields locally, repeating for all source nodes. However, given that the field lines lie in planes perpendicular to the direction $x_1$, so they have the same $x_1$ coordinate of the charges that contribute to their generation, this would be inefficient, as most of the nodes would not be doing any work.

Considering our specific problem it is possible to improve the parallelization efficiency by dividing the domain using the MPI communicators, so that these will have inside only the nodes characterized by the same $x_1$ coordinate. In this way the information about the charge inside a communicator is only broadcast to nodes that are all also inside, but not to all the others. As result we have a significant speedup of the code, leading to an almost linear performance increase with the number of nodes in $x_1$ direction.

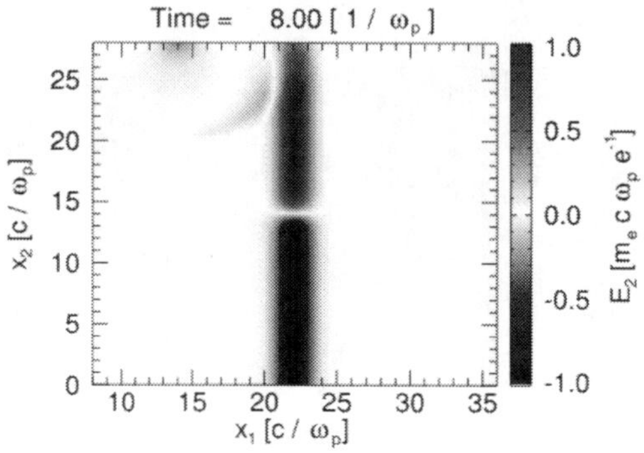


Figure 2: $E_2$ component of a 2D gaussian positron beam propagating in $x_1$ direction, it is possible to see the difference between the upper boundary (without PML initialization) and the lower one (with PML initialization).

**Absorbing boundary conditions**

Beam plasma simulations are generally performed using some form of absorbing boundary conditions for the fields, to simulate open space boundaries. In OSIRIS perfectly matched layer (PML) absorbing boundary conditions[2] are used, implementing in particular the version described here [3]. Inside the PML, we use a modified version of Maxwell's equation that define a medium with vacuum impedance but with gradual damping, so that EM waves leave the simulation box and are damped without being reflected. To avoid spurious reflection in these boundaries, we also need to initialize the beam field values inside the PML, matching not only the particle beam conditions, but also the specific medium properties in this region. Fig. 2 illustrates the effect of this initialization.

**Conclusion**

We developed a new algorithm in OSIRIS allowing for the initialization of the EM-fields associated to a highly relativistic charged particle beam. The new scheme is parallelized and can work with PML boundary conditions. The main advantage of the new scheme is an improvement in the speed of the code that can be, depending on the problem, up to three order of magnitude faster. Moreover, while the previous algorithm gives only an approximation of the EM-field values, the new algorithm evaluates their exact values at time zero.

**References**

[1] R. A. Fonseca *et al.*, Lecture Notes in Computer Science, **2331**, 2002, 342-351

[2] A. Taflove, S. C. Hagness Computational electrodynamics:the finite-difference time-domain method, Artech house, Norwood, 2000

[3] J. Vay, Journal of computational physics, **165**, 2000, 511-521.

# Pellet ignition using ions shock accelerated in the corona

R.A. Cairns[1], E. Boella[2], M. Vranic[2], L.O. Silva[2], R. Trines[3], P. Norreys[3,4], R. Bingham[3,5]

[1] *University of St Andrews, St Andrews, UK*
[2] *Instituto Superior Técnico, Lisbon, Portugal*
[3] *STFC Rutherford Appleton Laboratory, Didcot, UK*
[4] *University of Oxford, Oxford, UK*
[5] *University of Strathclyde, Glasgow, UK*

Fast ignition, using a beam of high energy electrons generated by a short laser pulse and directed into a previously compressed fusion pellet, was suggested some time ago as a means of reducing the total energy needed to achieve ignition [1]. More recently attention has been given to the use of ions rather than electrons. One suggested method is to use a separate target to generate the ions by the familiar target normal surface acceleration process, as described in a recent review by Fernandez et al [2]. Another is to use ions driven by the ponderomotive force of a very intense laser beam [3]. The requirement for this would be for a very tightly focused beam at an intensity of around $10^{22}$ W/cm$^2$. Recently we suggested that another possible scheme is to use a collisionless shock generated in the subcritical corona region of a compressed pellet to generate fast ions that would penetrate to the dense core and be absorbed there [4]. In this paper we expand on this idea and show that, while further detailed work is needed, the scheme appears feasible and worth further investigation. We shall first give a brief account of some of the relevant properties of collisionless shocks, then present some calculations of ion stopping power in dense plasmas to ascertain the approximate energy range needed. With the aid of some shock simulations we conclude that shock acceleration of ions from the corona to the required energy does indeed appear feasible.

Turning first to the nature of collisionless shocks in an unmagnetized plasma, a key paper with early simulations was published more than forty years ago by Forslund and Freidberg [5]. They showed that at low Mach numbers there is a smooth structure with a potential ramp followed by downstream potential oscillations. In this regime the potential jump is not sufficient to reflect a large number of the upstream ions. We have recently developed an analytic theory to describe this regime [6] and shown that the structures so formed may be relevant to a number of laser plasma experiments. At higher Mach numbers (above around 1.5) there is a change to a more complicated structure in which a large fraction of the upstream ions are reflected and there is a population of trapped ions in the downstream region. The low Mach number shocks can give rise to a very narrow accelerated ion energy spectrum, since only a small part of the tail of the ion distribution is reflected. We have shown that this effect can well explain some experiments on ion acceleration [7]. In the higher Mach number regime there is a broader energy spectrum, but many more accelerated ions [6,7]. For the purpose of pellet ignition a narrow energy spectrum does not seem necessary, but rather the aim should be to obtain as large an ion flux as possible so that the higher Mach number regime is appropriate. A rough estimate of the ion energy can be obtained by noting that in the laboratory frame the speed of the reflected ions is twice the shock speed. Thus, if we approximate the ion sound speed by $(T_e/m_i)^{1/2}$, the ion energy is around $2M^2T_i$ in a single species plasma.

Turning now to the ion energy needed for fast ignition, we look at the range of ions in a plasma, using a formula for stopping power given by Li and Petrasso [9]. This work takes account of small and large angle scattering from ions and electrons, quantum effects and excitation of collective plasma oscillations. Estimates of the relevant Coulomb logarithms are given. Here we take the value 3 for ion-electron collisions and 10 for ion-ion collisions, typical of the range of values given by Li and Petrasso. To obtain an idea of the ion range we have looked at a range of parameters, looking at the trajectory of an ion in a plasma with a density ramp increasing by one order of magnitude.

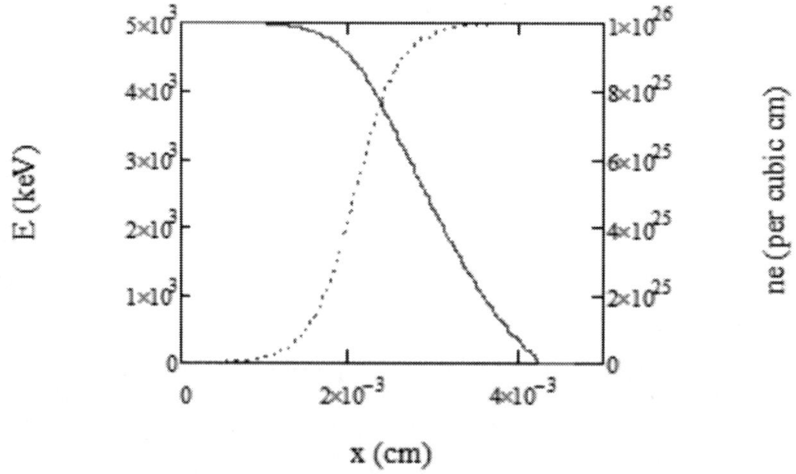


Figure 1. Ion energy (red) in the density ramp shown in blue.

Figure 1 shows the energy loss of a 5MeV hydrogen ion going into a plasma consisting of equal concentrations of deuterium and tritium where the density goes up to $10^{26}$ $cm^{-3}$ and the temperature is 5keV. At the highest density it can be seen that the ion range becomes of the order of a few tens of μm. A similar plot for a density two orders of magnitude lower is shown in Figure 2.

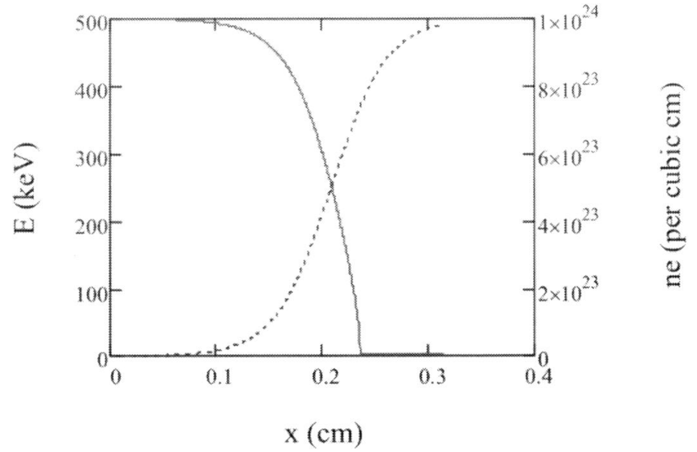


Figure 2. As for Figure 1 but with lower density.

Now the particle range has gone up to be of the order of mm. From calculations like this we estimate that ions of around 5-10 MeV generated in the corona will be able to penetrate to the dense plasma core and be sufficiently rapidly slowed down so as to deposit most of their energy there.

Finally we look at some simulations using OSIRIS [10] of a shock generated in a D-T plasma density ramp by a $CO_2$ laser pulse of intensity $10^{18}$ $W/cm^2$ and duration 14ps.

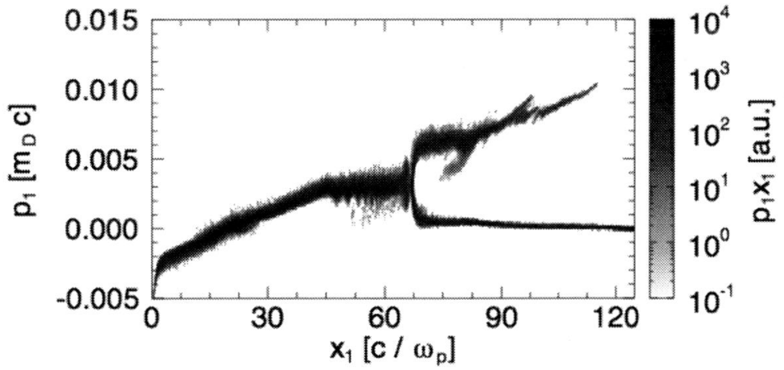


Figure 3. Deuterium phase space.

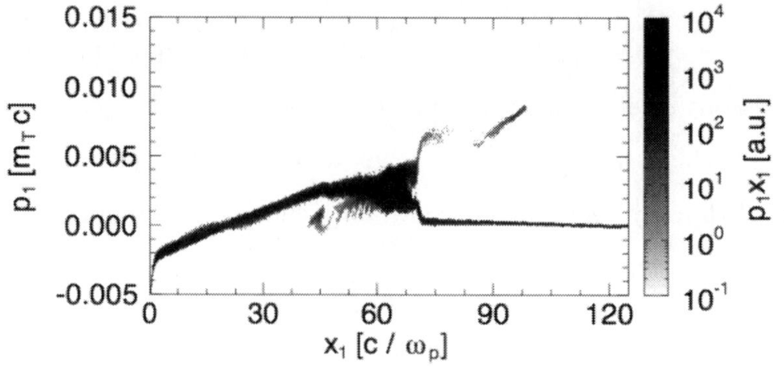


Figure 4. Correponding tritium phase space.

Clearly the lighter ions are preferentially accelerated and it has been shown that energies up to a few 10s of MeV can be obtained. In the corona of a plastic coated target we would have a carbon/hydrogen or possibly carbon/deuterium plasma, but these preliminary simulations indicate that the lighter component can be picked out and very effectively accelerated by a collisionless shock.

In conclusion, our results indicate that collisionless shock launched in the corona of a compressed target by a short intense laser pulse can generate ions of the energy needed to penetrate into the dense core and deposit their energy there.

This work was supported by the UK Engineering and Science Research Council.

## References

1. M. Tabak et al, Phys. Plasmas **1**, 1626 (1994)
2. J.C. Fernandez et al, Nucl. Fusion **54**, 054006 (2014)
3. N. Naumova et al, Phys. Rev. Lett. **102**, 025002 (2009)
4. R.A. Cairns et al, Plasma Phys. Control. Fusion **57**, 044008 (2015)
5. D.W. Forslund and J.P. Freiberg, Phys. Rev. Lett. **27**, 1189 (1971)
6. R.A. Cairns et al, Phys. Plasmas **21**, 022112 (2014)
7. D. Haberberger et al, Nat. Phys. **8**, 95 (2012)
8. F. Fiuza et al, Phys. Rev. Lett. **109**, 215001 (2012)
9. C-K. Li and R. Petrasso, Phys. Rev. Lett. **70**, 3059 (1993)
10. R.A. Fonseca et al, Lecture Notes in Computer Science, **2331**, 342 (2002)

# Investigation of spontaneous magnetic fields generated under different regimes of laser-produced plasma creation

T. Pisarczyk[1], S. Gus'kov[2,3], R. Dudzak[5,6], T. Chodukowski[1], J. Dostal[5,6], N.N. Demchenko[2,3], P. Korneev[3], Z. Kalinowska[1], M. Kalal[4], O. Renner[6], M. Shmid[6], S. Borodziuk[1], E. Krousky[5,6], J.Ullschmied[5,6], J. Hrebicek[5,6], T. Medrik[5,6], J. Golasovski[5,6], M. Pfeifer[5,6], J. Skala[5,6], P. Pisarczyk[7]

[1] *Institute of Plasma Physics and Laser Microfusion, Warsaw, Poland*
[2] *P.N. Lebedev Physical Institute of RAS, Moscow, Russian Federation*
[3]*National Research Nuclear University (Moscow Eng. Phys. Inst.), Russian Federation*
[4] *Czech Technical University in Prague, FNSPE, Prague, Czech Republic*
[5] *Institute of Plasma Physics ASCR, Prague, Czech Republic*
[6] *Institute of Physics ASCR, Prague, Czech Republic*
[7] *Warsaw University of Technology, ICS, Warsaw, Poland*

Shock ignition (SI) approach [1] currently becomes one of the most promising scenarios of target ignition in the inertial confinement fusion (ICF). One of the main research aims related to the SI is the understanding of ablation pressure generation mechanisms at different irradiation conditions. The corresponding investigations were carried out on PALS facility with different diagnostics [2-6]. From the SI point of view, particularly interesting is the initial phase of the expansion, related to the interaction of the laser pulse with the generated plasma. In this phase different processes of anomalous absorption, responsible for the generation of fast electrons, appear and the electric and magnetic fields are created. These fields can significantly modify plasma parameters and transport of fast electrons to the ablation surface affecting the formation of the shock wave ablation pressure.

For the experimental studies of these processes, the interferometry and polarimetry are highly demanded at the early phase of the laser plasma expansion, provided a sufficiently short diagnostic pulse would become available. This requirement was fulfilled by a two-channel polaro-interferometer, sketched in Fig. 1, irradiated by 40 fs Ti:Sa laser working at the wavelength of 808 nm.

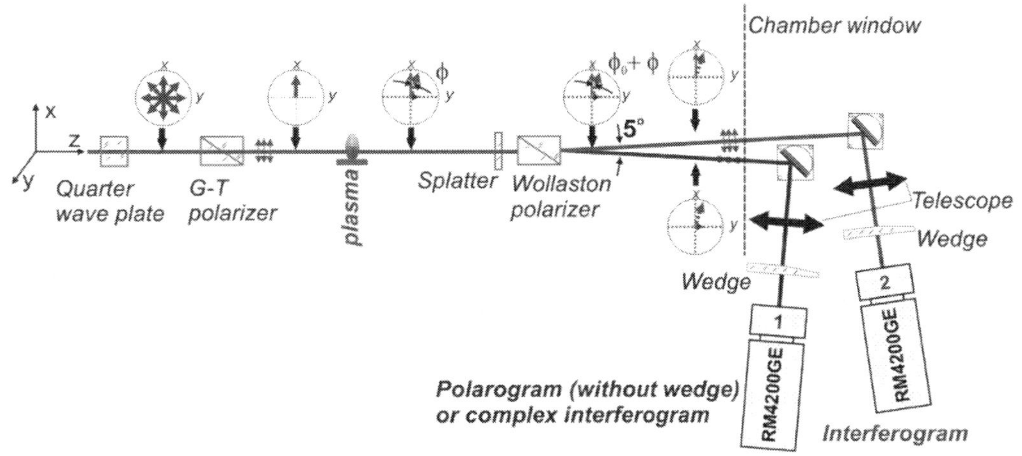


Fig. 1 Optical scheme of the two-channel polaro-interferometer.

Two-channel polaro-interferometer consists of two channels which provide a possibility to measure the magnetic fields in two ways depending on the optical configuration of the polarimetric channel. In the first case, a polarogram and an interferogram are recorded in each channel and magnetic field distribution in plasma are calculated on the basis of the Faraday rotation angle and the electron density distributions. In an alternate case, instead of the polarogram, a complex interferogram can be registered via an initial rotation of the polarization plane with a wedge added in the polarimetric channel (see Fig. 1). Information about magnetic field distribution can be obtained directly on the basis of the amplitude-phase analysis of the interferogram [7].

In the presented preliminary studies planar targets made of materials with different atomic numbers (plastic and Cu) were illuminated by the 1ω laser beam with energy range of 250-500 J, focused to the minimum possible focal spot radius - $R_L$=50 μm. A methodology presented in [8-10] was used for polaro-interferometric measurements. To ensure the optimal registration conditions in the polarimentric channel, the measurements at the initial angle of rotation of the polarizer of **$\varphi_0$ = 2°** were carried. The polarograms with corresponding interferograms illustrate the interaction of the 1ω laser beam at the energy about 250 J with the plastic targets are shown in Fig. 2a, while in the case of the Cu targets and the same experimental conditions are presented Fig. 2b.

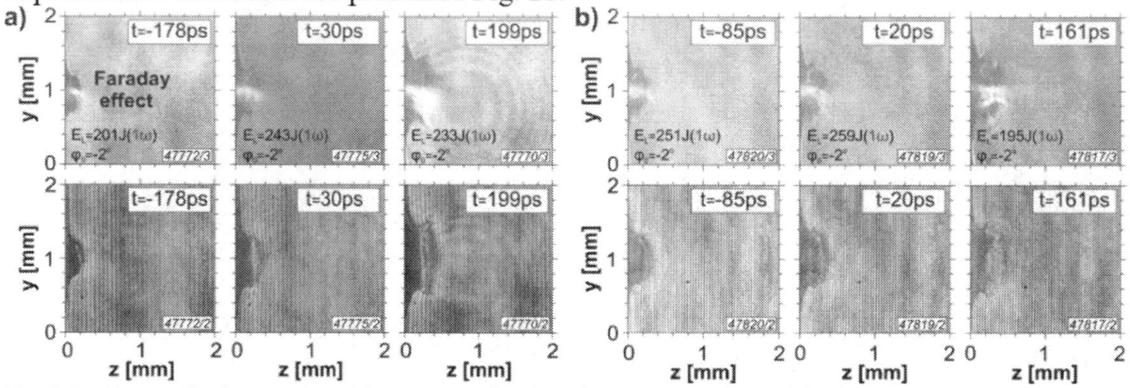


Fig. 2 Sequence of polarograms with corresponding interferograms registered for the 1ω laser beam interaction with the planar massive targets made of: a) plastic and b) Cu.

The Faraday effect, clearly visible in both polarograms, demonstrates a proper functioning of the polaro-interferometer. The effect is visible only in the bottom half polarograms, which proves that the spontaneous magnetic fields (SPM) have azimuthal symmetry.

One of the goals of such research was to prove the influence of the character of the ablative plasma expansion on the SPM structure, namely: (i) the spherical expansion of the fast component in the case of the light target material (plastic) and (ii) planar (axial) expansion which is enforced by a heavy Cu plasma [11]. To obtain the distribution of the magnetic field, the methods of analysis described in papers [6] have been applied. In the case of the axial plasma symmetry, the equations describing the relations between the Faraday rotation angle (φ), the distribution of phases (**δ**) and the plasma parameters are applied, namely:

$$\varphi(y) = 5.24 \cdot 10^{-17} \cdot \lambda^2 \int_y^R \frac{B_\varphi(r) n_e(r) y dr}{\sqrt{r^2-y^2}} \tag{1}$$

$$\delta(y) = 8.92 \cdot 10^{-14} \cdot \lambda \int_y^R \frac{n_e(r) dr}{\sqrt{r^2-y^2}} \tag{2}$$

where: $B_\varphi(r)$ - the azimuthal magnetic field distribution, $n_e(r)$-electron density distribution, λ – the wavelength of a probe beam.

Expressions (1) and (2) reduce simply to an integral Abel equation [10]:

$$S(y) = 2 \int_y^1 \frac{f(r) dr}{\sqrt{r^2-y^2}} \tag{3}$$

For the case (1) the transformation is given by:

$$f_B(r) = 2.62 \cdot 10^{-17} \cdot \lambda^2 R \left[\frac{B_\varphi(r) n_e(r)}{r}\right] \quad \text{and} \quad S_B(y) = \frac{\varphi(y)}{y} \tag{4}$$

and for (2) it is:

$$f_n(r) = 4.46 \cdot 10^{-14} \cdot \lambda R n_e(r) \qquad \text{and} \qquad S_n(y) = \delta(y) \qquad (5)$$

Substitution of $n_e(r)$ from (4) to (5) gives for $B_\varphi(r)$:

$$B_\varphi(r) = \frac{1.7\cdot 10^3}{\lambda} \cdot \left[\frac{r f_B(r)}{f_n(r)}\right] \qquad (6)$$

To solve this equation and find the magnetic field distribution $B_\varphi(r)$, the FFT Abel inversion method was used. The SPM distributions for the different expansion time moments, as a result of the 1ω laser beam interaction with the planar massive plastic target are presented in Fig. 3a, while Fig. 3b shows distributions which correspond to Cu target for the same laser pulse parameters.

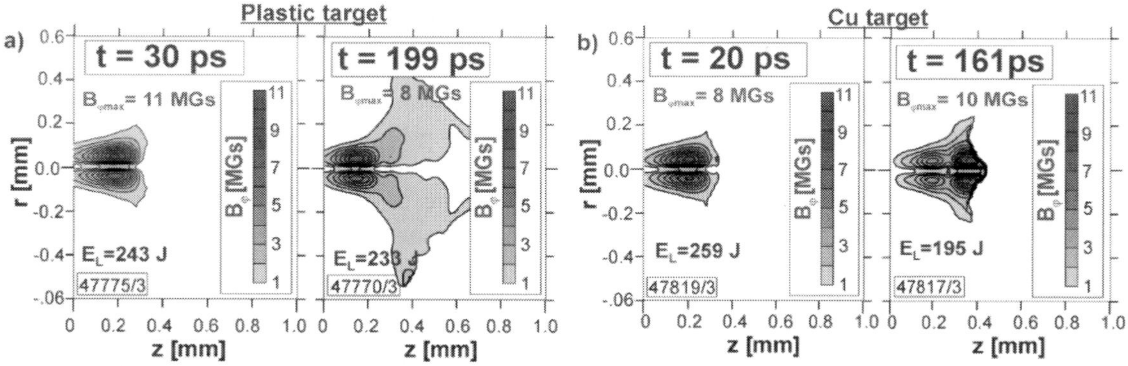


Fig. 3 2D distributions of the SPM in the ablative plasma during the 1ω laser beam interaction with: a) the plastic and b) Cu target for two different expansion times related to the maximum intensity of the laser pulse.

From the Fig. 3a maximum value of SPM about 10 MGs is seen in the vicinity of the target surface does not significantly change during the action of the laser pulse. However, the distribution of the SPM changes distinctly with the time, while the ablative plasma expands. The spherical expansion of the more fast light plasma from a plastic target causes that the distribution of SPM is becoming clearly divergent at the expansion time moment t=199 ps *(it resembles the effect of fountain stream)*. The SPM distributions generated at different expansion times, as a result of the 1ω laser beam interaction with the planar Cu massive target, are shown in Fig. 3b. The influence of the planar plasma expansion is clearly seen in this case. In contrast with the plastic plasma, total current cross-section flows in a plasma column within diameter not exceeding 0.5 mm. Moreover, the distribution of the SPM in the plasma column clearly differs from the distribution for the case of light plastic plasma. These distributions prove that the axial expansion of the heavy plasma significantly reduces the divergence angle of the *fountain flow* of electrons from the front part of the plasma column. Due to this fact, the maximum amplitude of the SPM at the front of the plasma column is larger than the one in the vicinity of the target and reaches a value of about 10 MGs.

For a more consistent theoretical analysis of the expanding plasma parameters and the SMF, the general plasma transport relations may be useful. Following [12], the equation for the magnetic field balance with plasma parameters may be written as follows (the small items are omitted for simplicity):

$$\frac{\partial \vec{B}}{\partial t} + \frac{c^2}{4\pi}\nabla \times \frac{\nabla\times\vec{B}}{\sigma} - \nabla \times \left(\vec{v}\times\vec{B}\right) - \frac{c}{e}\nabla \times \frac{\nabla p + \vec{R_T}}{\partial t} = 0, \qquad (7)$$

where $t$ is time, $c$ is the light velocity, $\sigma$ is the plasma conductivity, $v$ is the plasma flow velocity, $e$ is the electron charge absolute value, $p \approx neTe$ is the plasma pressure, $\vec{R}_T \sim n_e \nabla T_e$ is the so-called thermo-force, $T_e$ is the electron temperature. From Fig.2 it follows that the characteristic distances are ~30 μm and times ~100 ps, measured in the experiment;

measured densities of plasma flow $n_e \approx 3\times10^{19}$ cm$^{-3}$ (not shown in figures), and the plasma temperature for the considered laser parameters has the order of $T_e \sim U_p \sim 3$ keV, where $U_p$ is the ponderomotive energy. Considering estimations for $\vec{R}_T$ and σ from [12], for the third and fourth items in eq. (7) we find the scaling $\sim 10^{16..17}$ Gauss/sec. Then, the first item $\frac{\partial \vec{B}}{\partial t} \sim \vec{B}/10^{10}\ sec^{-1}$ leads to the estmation value of magnetic field as several Megagauss. While the third item in (7) regulated the relation between the flow velocity and SPM, the main effect for the SPM generation comes from the last item in (7), which contains the gradient of the electron temperature $T_e$.

**Conclusions:** It should be emphasized that the realized investigations of the SPM on PALS experiment by means of the femtosecond polaro-interferometry are greatly innovative if we take into account that: (i) as far as we know the space-time SPM distributions during the laser pulse interaction with the target have been measured with the high resolution for the first time, and (ii) such detailed information about SPM distributions offer major cognitive possibilities, especially in the SI concept.

The hitherto studies on SI performed by the authors [3, 4, 6, 11] show that investigations of the expansion regimes play a central role in the knowledge of absorption processes and ablation mechanisms responsible for the ablation pressure formation, with the participation of fast electron. In addition, they showed that the character of the plasma expansion depends not only on the target material but also on the irradiation conditions of the target, namely: the focal spot radius and the wavelength of the laser beam. It is expected that wider range of a diagnostic possibilities for the measuring the SPM distributions can be particularly useful for further studies of the SI scenario.

**Acknowledgments:** This paper was supported in part by the Access to Research Infrastructure activity in the 7th Framework Program of the EU Contract No.284464, Laserlab Europe III, by the Czech Republic's Ministry of Education, Youth and Sports under PALS RI-LM 2010014 and by the ToIFE Project of the EUROfusion Consortium and by the French-Polish bilateral collaboration program POLONIUM and of the COST Action MP1208 "Developing the physics and the Scientific Community for Inertial Fusion by Russian Foundation for Basic Research (Projects No. 14-02-00010 and No. 13-02-00295).

**References:**

1. R. Betti et al, *Phys. Rev. Lett.* **98**, 155001 (2007).
2. P. Koester et al, *Plasma Physics and Controlled Fusion* **55**, 124045(2013).
3. S.Yu. Gus'kov et al, *Laser and Particle Beams* **32**, 177-915 (2014).
4. T. Pisarczyk et al, *Phys. Plasmas* **21**, 012708.(2014).
5. D. Batani et al, *Phys. Plasmas* **21**, 032710 (2014).
6. T. Pisarczyk et al, *Laser and Particle Beams* **33**, *221-236 (2015).*
7. M. Kalal, *Czechoslovak J. Phys*. **41**, 743 (1991).
8. N.G. Basov, J. Wołowski, E.G. Gamali, S. Denus, T. Pisarczyk, A.A. Rupasov, G.S. Sarkisov, G.B. Sklizkov, V.T. Tikhonchuk, A.S. Shikanov. *Detection of spontaneous magnetic field in a laser plasma in the "Delfin-1" device.* JETP Lett. **45** (4), (1987).
9. T. Pisarczyk et al, *Journal of Soviet Laser Research* **11** (1) (1990).
10. T. Pisarczyk et al, *Laser and Particle Beams* **12** (3), 549-561 (1994).
11. A.Kasperczuk et al, Laser and Particle Beams **27**, 415-427 (2009).
12. Braginsky, S.I. (1965). Transport processes in a plasma, in Reviews of Plasma Physics, Vol. 1, ed. M.A. Leontovich (Consultants Bureau, New York)

# *GEMpix detector: a new diagnostic for Laser Produced Plasmas monitoring*

G. Claps[1], D. Pacella[1], A. Romano[1], L. Gabellieri[1], F. Causa[1], R. De Angelis[1], P.L. Andreoli[1], G. Cristofari[1], F. Consoli[1], F. Murtas[2]

[1]Unità Tecnica Fusione, ENEA, C.R. Frascati, Via E. Fermi, 45 - 00044 Frascati, Rome, Italy

[2]Istituto Nazionale di Fisica Nucleare, Via E. Fermi, 40, 00044 Frascati, Rome, Italy

## *Introduction*

A common characteristic of Laser produced plasmas (LPPs) is their intense emission in the X-ray spectral region, in short time scale. These plasmas can reach electron temperatures of several keV, electron density up to $10^{23}$ electrons/cm$^3$ and therefore they can be considered as pulsed sources of X-rays. They can be considered for several applications. Inertial Confinement Fusion (ICF) is one of the most relevant. In this case, X-rays play a role not only as a diagnostic mean of the physical properties of the plasma, but also to drive the implosion, in case of indirect scheme. In addition laser plasma could be used also as X-rays sources in many technological applications like X-ray microscopy, nano-lithography, dynamic studies of mechanical properties of materials, X-ray time resolved radiography. For these reasons the development of new X-ray imaging devices is relevant even beyond the field of laser produced plasmas. X-ray imaging diagnostics from LPPs are realized so far mainly by X-ray CCDs , Micro Channel Plates and Imaging Plates. This paper presents a new detector in the soft X-ray range, the "GEMpix", which can have several advantages compared to the conventional imaging devices. It is a gas detector based on the GEM technology [1,2] with a quad-medipix chip as read-out electronics. Its potential as LPP imaging diagnostic has been preliminary demonstrated through some measurements performed on the ABC facility (ENEA, Frascati) [3,4].

## *The GEMpix detector*

The GEMpix detector [5] is a micro-pattern gas detector based on Gas Electron Multiplier technology (GEM). It has been realized like a chamber with a cathode, an anode and three GEM foils. The cathode is made by an thin aluminate mylar foil (12 μm) and defines also the detector window. The active volume is represented by the gas layer between the cathode and the first GEM foil. Triple-GEM structure allows a gain from 1 to $10^4$ without discharge problems. The anode is a quad-medipix chip that has 512 x 512 pixels, each one with an area of 55 x 55 μm$^2$. It is based on a C-MOS technology and consists of four medipix chips hold together without the silicon layer which, in this case, is substituted by the triple-GEM gas chamber. It can work both in counting mode and in Time over Threshold (ToT) mode. In the second case, each pixel is used in a Wilkinson type ADC, so that the registered digital counts are proportional to time in which the signal is above the threshold, which in turn is proportional to the charge collected by the pixel.

This charge can be originated by a single photon interaction when the particle flux is not very high. An estimate of this flux limit (few tens of MHz) results mainly by the front-end-electronics and the electron drift velocity, giving an intrinsic time resolution of about 100 ns. Charge collected at higher fluxes implies the contemporary detection of many photons, whose signals are overlapped, corresponding to a charge integration mode. In both cases the gain provided by GEMs can be properly set in order to evaluate the released charge without saturation of the read-out electronics. Then GEMpix is able to work on a wide dynamic range: the estimated value is four orders of magnitude higher respect to the medipix detector which uses a silicon layer as active material. Dynamic range has been studied in the New Imaging X-ray Techniques (NIXT) Laboratory (ENEA) using mono-energetic photons of three different energies (fig. 2): 3.7 keV (Ca), 6.4 keV (Fe) and 8.0 keV (Cu), in the regime of single photon detection.

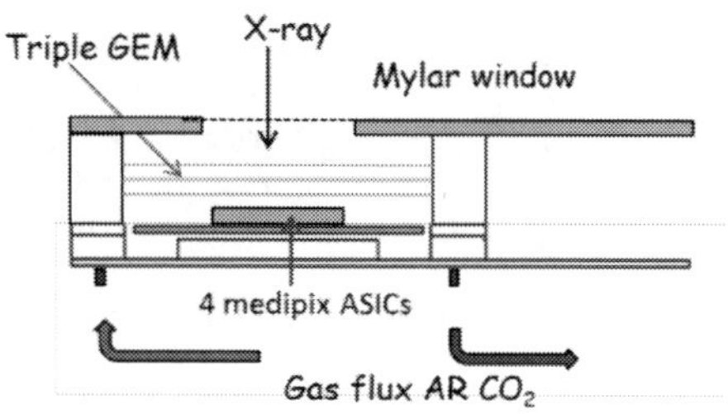


***Figure 1:*** *schematic layout of a GEMpix detector; it is evidenced the detector window, the triple-GEM and the quad-medipix chip. In and out gas connectors are also evidenced.*

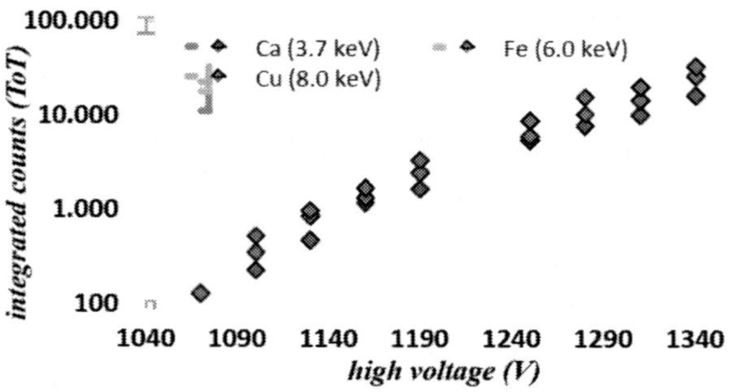


***Figure 2:*** *Integrated counts per incident photon from three different X-ray sources (Ca, Fe and Cu) when the gain voltage is varied from 1070 to 1340 V for $ArCO_2CF_4$ (45/15/40) gas mixture.*

It is clear from the plot that when the gain voltage is increased, there is a corresponding increase of the total charge collected on the pixels and then digitized in counts. The transverse size of the electron cloud collected on the pixels depends on the GEM gain. The detected photon appears as a blob whose diameter is greater and greater when the applied voltage is increased (fig. 3). Then working at low gain voltages, smaller blobs are observed and photon interacting on different points of the active area can be distinguished, if the photon fluence is low, according to the observations made before. Since the time resolution of GEMpix is about 100 ns, as previously mentioned, counting mode can be used up to a few MHz, while charge integration mode (ToT configuration) will be used for detecting X-rays burst, like in the case of LPPs, whose time width is of the order of few ns, or even shorter, if femtosecond lasers are used. As a result, charge is produced by more than one photon. Characterization in laboratory in regime of single photon detection has been done, in particular to study the intensity response and the spatial resolution (blob size). In order to check the imaging capabilities of this detector, experiments have been done with real laser produced plasmas. Some preliminary tests have been then performed on the ABC test facility (ENEA, Frascati).

***ABC tests***

ABC laser facility is dedicated to the study of Inertial Confined Fusion (ICF) and is equipped with two beam lasers which can deliver up to 100 J in pulses few nanoseconds long with a wavelength of 1.054 μm. Interaction of laser with target happens in a high vacuum spherical tank. A test measurement with GEMpix was conducted during two experimental campaigns. In both cases, laser pulse had time duration of 3 ns and the experiments consisted in the interaction of a single laser beam with a thin plane target made of different materials. In the first one, a beryllium window 5 mm in diameter and 50 μm in thickness was mounted on a tank port, without any pinhole. This was a very preliminary measurement aimed to test the GEMpix on the X-ray coming from this type of plasmas. Fig. 4 shows the result obtained on a target of lithium 6 Fluoride ($^6$LiF) having a thickness of 334 nm on a substrate of C28 of 127 μm. The acquired image is uniform, as expected, and the transverse size is equal to the Be window diameter.

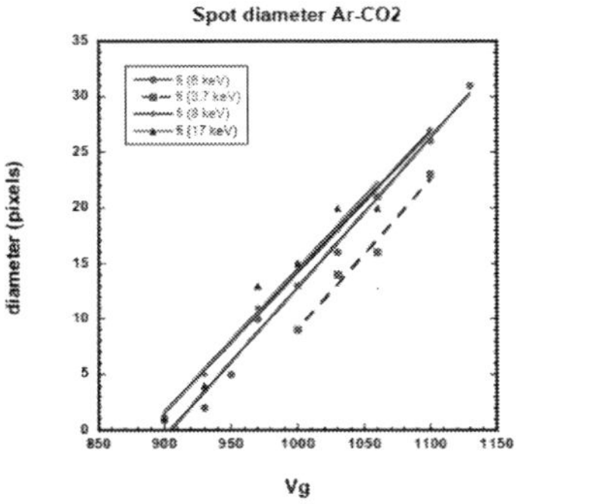


*Figure 3: Blob mean diameter per incident photon from four different X-ray sources (Ca, Fe, Cu and Mo) when the gain voltage is varied from 900 to 1140 V for $ArCO_2$ (70/30) gas mixture.*

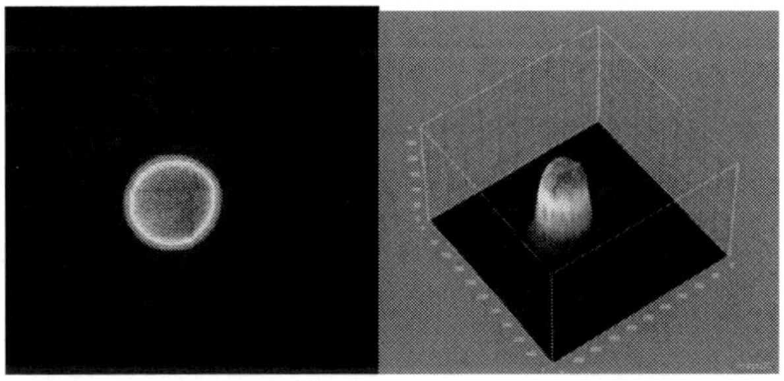

***Figure 4: X-ray spot coming from a beryllium window 5 mm in diameter and 50 μm in thickness when the ABC laser interacted with a $^6$LiF target (334 μm) on a C28 substrate (127 μm). An $ArCO_2$ (70/30) gas mixture was used with a gain of about 440.***

The result of this measurement has demonstrated that GEMpix can work well in ToT mode with high photon fluence and good uniformity is observed, taking into account that the spatial resolution is about 50 μm In this case the GEM voltage was 900 V with a gain about 400. The following step was the addition of a pin-hole, inside the vacuum chamber, between the plasma and the window, giving a magnification of 1.5 in order to make an X-ray imaging of the plasma plume coming from the target after laser interaction. The pin-hole was located almost at a right angle respect to the incident laser direction. Results are shown in fig. 5 in the case of an aluminum target (7 μm thick).

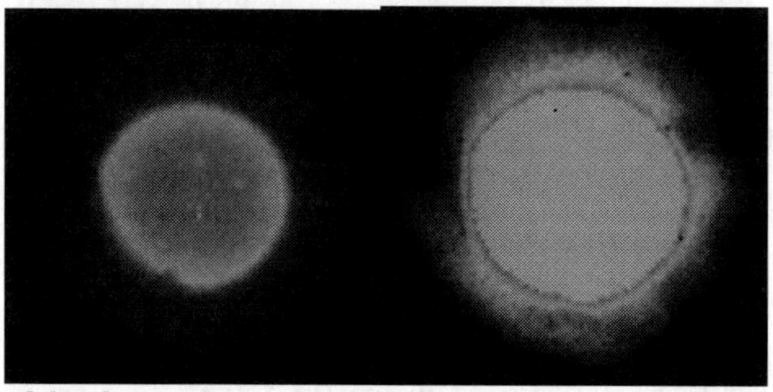

*Figure 1: X-ray image of the plasma plume seen through a pinhole with 1.5 of magnitude when the ABC laser interacted with a Al target (7 μm). An $ArCO_2$ (70/30) gas mixture was used with a gain of about 440.*

Image on the left gives information on the integrated charge due to all the incident photons. On the right, instead, intensity scale (colors) has been selected to saturate the core, but showing the presence of a corona around the plasma core. This corona exhibits clearly poloidal modulations, imaged with a spatial resolution of about 50 μm. On one side it is possible to observe a cut, which identifies the target.

***Conclusions***

GEMpix detector has been proposed as soft X-ray imaging detector for laser produced plasmas and the preliminary results obtained from ABC Frascati laser facility has been discussed and they appear promising. In fact the adjustable sensitivity of the detector allowed to obtain physical response without signal saturation, while electronic noise is practically absent. Imaging capabilities have been demonstrated, in a pinhole configuration, showing the shaping of the core emissions and the poloidal modulations of the corona, with good spatial resolution. Based on these results, suited configurations could be defined in order to assess the spectral distribution of the detected X-rays, thanks to the proportionality with the photon energy. Further calibrations and characterization will be done at the ultra-short laser facility Eclipse (Bordeaux) in July 2015.

[1] F. Sauli, *"Nucl. Instr. and Meth."* A 386 (1987) 531.

[2] Pacella, D., Bellazzini, R., Brez, A., Pizzicaroli, G., *Fast 2-D soft X-ray imaging device based on micro pattern gas detector* (2003) Nuclear Physics B - Proceedings Supplements, 125, pp. 332-336.

[3] Inertial Fusion Progress RT/ERG/FUS/93/67, ISSN/1120-5598, ENEA, 1994.

[4] C. Strangio, A. Caruso, LaserPart.Beams16(1998), 45.

[5] F. Murtas, 3rd INTERNATIONAL CONFERENCE ON MICROPATTERN GASEOUS DETECTORS, 1–6 JULY, 2013, Zaragoza, Spain.

# Hydrodynamic analysis of self-similar radiative ablation flows

J.-M. Clarisse[1], J.-L. Pfister[2], S. Gauthier[1], C. Boudesocque-Dubois[1]

[1] *CEA, DAM, DIF, F-91297 Arpajon, France*

[2] *ENS Cachan, F-94235 Cachan, France*

Hydrodynamic instabilities, and in particular ablation flow instabilities, are a key and withstanding issue for laser-driven inertial confinement fusion (ICF). In an ongoing effort to obtain a better description of the early-irradiation flow instabilities of an ICF pellet implosion (IPI), a dedicated approach using, as mean flows, self-similar ablative heat-wave solutions of the gas dynamics equations with nonlinear heat conduction [1] in slab symmetry, has been developed and applied to direct laser irradiation configurations [2, 3]. This approach consists in computing, with a spectrally accurate numerical method [4], semi-infinite slab mean flows and their time-dependent linear perturbations. Key features of real ablation flows such as unsteadiness, confinement, compressibility that may be critical to the flow stability, are thus fully and accurately taken into account. Under the assumption of radiative conduction and for positive pressures and radiation fluxes at the slab external boundary, such self-similar solutions can be held as being representative of the ablation of an ICF-pellet shell driven by a gas-filled hohlraum radiation. A logical and preliminary step prior to any hydrodynamic stability study is then to perform a hydrodynamic analysis of such mean flows. In doing so, a framework for analyzing more general (*i.e.* not necessarily self-similar) unsteady ablative mean flows is provided.

## Dimensionless equations of motion

The proposed model flow considers the motion of an inviscid, heat-conducting fluid with a polytropic equation of state, $p = R\rho T$, $\mathscr{E} = RT/(\gamma-1)$, and a nonlinear heat conductivity of the form $\kappa = \kappa_0(\rho_0/\rho)^{\mu}(T/T_0)^{\nu}$, $\kappa_0 > 0$, $\mu \geq 0$, $\nu > 0$, where $\rho_0$ and $T_0$ are homogenization constants. The fluid which is initially at rest, of uniform finite density $\rho_I$, and occupying the half-space $x \geq 0$, is set into motion by time-increasing pressure and heat flux applied at its external boundary surface $\Gamma$. Relying on the physical governing parameters $\gamma$, $R$, $\kappa_0$, $\rho_I$, characteristic boundary pressure $p_\Gamma$ and heat flux $\varphi_\Gamma$, supplemented by a reference time $t_c$, two dimensionless formulations of the equations of motion in Lagrangian coordinates are proposed:

**$(\mathscr{B}_p, \mathscr{B}_\varphi)$-formulation [2]**

$$\begin{cases} \partial_{\bar t}[1/\bar\rho] - \partial_{\bar m}\bar v_x = 0, \quad \partial_{\bar t}\bar v_x + \partial_{\bar m}\bar p = 0, \\ \partial_{\bar t}\left[\frac{1}{2}\bar v_x^2 + \frac{1}{\gamma-1}\bar T\right] + \partial_{\bar m}[\bar p\bar v_x + \bar\varphi_x] = 0, \\ \bar\varphi_x = -\bar\rho\,\bar\kappa\,\partial_{\bar m}\bar T, \end{cases}$$

**$(\mathscr{M}, \mathscr{P}e)$-formulation**

$$\begin{cases} \partial_{\bar t}[1/\bar\rho] - \partial_{\bar m}\bar v_x = 0, \quad \partial_{\bar t}\bar v_x + \frac{1}{\gamma\mathscr{M}^2}\partial_{\bar m}\bar p = 0, \\ \partial_{\bar t}\left[\frac{1}{2}\bar v_x^2 + \frac{1}{\gamma\mathscr{M}^2}\frac{1}{\gamma-1}\bar T\right] + \frac{1}{\gamma\mathscr{M}^2}\partial_{\bar m}[\bar p\bar v_x + \bar\varphi_x] = 0, \\ \bar\varphi_x = -\frac{\gamma}{\gamma-1}\frac{1}{\mathscr{P}e}\bar\rho\,\bar\kappa\,\partial_{\bar m}\bar T, \end{cases}$$

with $\bar{p} = \bar{\rho}\bar{T}$ and $\bar{\kappa} = \bar{\rho}^{-\mu}\bar{T}^{\nu}$, and where $\bar{m}$ is such that $\mathrm{d}\bar{m} = \bar{\rho}\mathrm{d}\bar{x}$. Driving dimensionless parameters, apart from $\gamma$, are ascribed with the former formulation to the sole boundary condition parameters $\mathscr{B}_p$ and $\mathscr{B}_\varphi$ (*cf.* Tab. 1), while they are, with the latter, enchased in the governing equations under the form of Mach and Péclet number scales $\mathscr{M}$ and $\mathscr{Pe}$. For initial and boundary conditions compatible with a reduction of variables (Tab. 1), these formulations lead to two complete classifications of self-similar solutions (*cf.* Tab. 2).

| | $\bar{\rho}(\bar{m},0)$ | $\bar{v}_x(\bar{m},0)$ | $\bar{T}(\bar{m},0)$ | $\bar{p}(0,\bar{t})$ | $\bar{\varphi}_x(0,\bar{t})$ |
|---|---|---|---|---|---|
| $(\mathscr{B}_p,\mathscr{B}_\varphi)$ | 1 | 0 | 0 | $\mathscr{B}_p\bar{t}^{2(\alpha-1)}$ | $\mathscr{B}_\varphi\bar{t}^{3(\alpha-1)}$ |
| $(\mathscr{M},\mathscr{Pe})$ | 1 | 0 | 0 | $\bar{t}^{2(\alpha-1)}$ | $\bar{t}^{3(\alpha-1)}$ |

Table 1: Invariant-density self-similar solutions. Initial and boundary conditions.

| $\mathscr{B}_p$ | $\mathscr{B}_\varphi$ | $\mathscr{M}^2$ | $\mathscr{Pe}$ |
|---|---|---|---|
| $(\gamma-1)\gamma^{-2}\,\mathscr{Pe}\mathscr{M}^{-2}$ | $(\gamma-1)^{3/2}\gamma^{-5/2}\,\mathscr{Pe}^{3/2}\mathscr{M}^{-2}$ | $\gamma^{-1}\,\mathscr{B}_\varphi^2\mathscr{B}_p^{-3}$ | $\gamma(\gamma-1)^{-1}\,\mathscr{B}_\varphi^2\mathscr{B}_p^{-2}$ |

Table 2: Relations between the driving parameters $(\mathscr{B}_p,\mathscr{B}_\varphi)$ and $(\mathscr{M},\mathscr{Pe})$.

**Invariant-density self-similar radiative ablation flows**

Self-similar solutions leaving invariant the flow density correspond to the similarity exponent definition $\alpha = (2\nu-1)/2(\nu-1)$ and the similarity laws of Tab. 3, the systems of PDEs being replaced by systems of ODEs for the flow reduced variables, with initial and boundary conditions of Tab. 1 being transformed, respectivley, into boundary conditions at $\xi = +\infty$ and at $\xi = 0$ [2]. For sufficiently low values of the external boundary heat flux, the heated fluid region may be considered to be bounded by a non-isothermal shock-wave discontinuity [1, § C] and the boundary conditions at $\xi = +\infty$ replaced by the relevant Rankine–Hugoniot jump relations at the shock-front location (*e.g.* [2]). Self-similar radiative ablation flows within this bounding shock-wave approximation have been investigated — for a monatomic gas ($\gamma = 5/3$) and the fully ionized gas model of Kramers ($\mu = 2$, $\nu = 13/2$) — by means of an extensive exploration of the space of parameters $(\mathscr{B}_p,\mathscr{B}_\varphi)$ for solutions computable with the numerical method detailed in [4]. A set of 631 ablative solutions, presenting a large variety of spatial pro-

| $\bar{m}$ | $\bar{\rho}(\bar{m},\bar{t})$ | $\bar{v}_x(\bar{m},\bar{t})$ | $\bar{T}(\bar{m},\bar{t})$ | $\bar{p}(\bar{m},\bar{t})$ | $\bar{x}(\bar{m},\bar{t})$ |
|---|---|---|---|---|---|
| $\bar{t}^{\alpha}\xi$ | $G(\xi)$ | $\bar{t}^{\alpha-1}V(\xi)$ | $\bar{t}^{2(\alpha-1)}\Theta(\xi)$ | $\bar{t}^{2(\alpha-1)}P(\xi)$ | $\bar{t}^{\alpha}X(\xi) = \bar{t}^{\alpha}[V(\xi)/\alpha + \xi/G(\xi)]$ |

Table 3: Invariant-density self-similar solutions. Dependent/independent-variable reduction.

files (Figs. 1$a$, $c$, $e$, $g$) and including a solution extrapolated from an IPI simulation with an ICF hydrodynamics code (IPI-like flow), has thus been obtained.

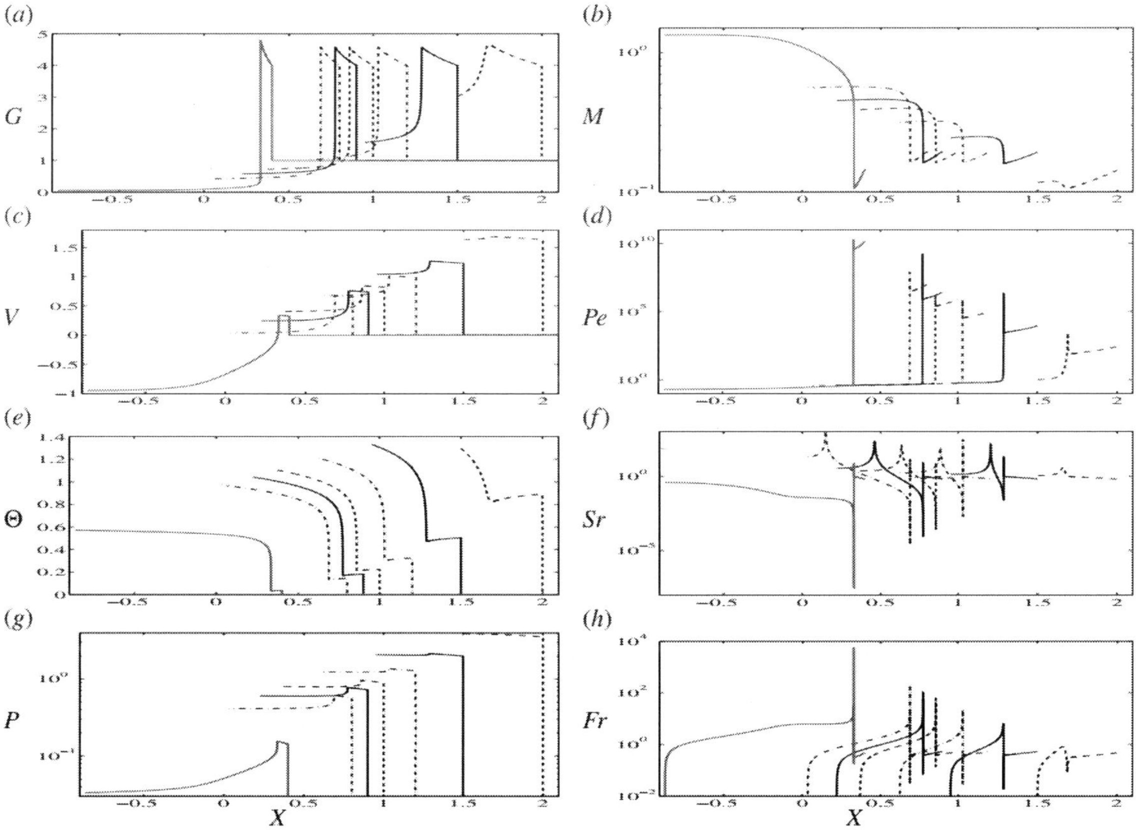

Figure 1: Reduced flow variables ($a$, $c$, $e$, $g$) and hydrodynamic characteristic numbers ($b$, $d$, $f$, $h$) *vs* the reduced coordinate $X$ (Tab. 3), for a particular flow sample (IPI-like flow in red).

## Hydrodynamic analysis

The analysis of these ablative flows has been performed in the reference frame of the ablation front (af) — here ascribed to be the locus of the minimum temperature-gradient length, $\ell_T = \min_x \ell_{\nabla T}$ — leading to the definition of local Mach, Péclet, stratification and Froude numbers in terms of the fluid velocity relative to the ablation front $v'_x$, the ablation-front acceleration $a^{\rm e}_{\rm af}$, and proper local gradient lengths (Tab. 4). Completed by the flow-region relative lengths $\ell_{\rm cond}/\ell_{\rm tot}$ (conduction region), $\ell_T/\ell_{\rm tot}$ (ablation-front stiffness), and $(1-\ell_{\rm cond}/\ell_{\rm tot})$

| | Mach | Péclet | stratification | Froude |
|---|---|---|---|---|
| flow analysis | $M = \frac{\lvert v'_x \rvert}{\sqrt{\gamma p/\rho}}$ | $Pe = \frac{\lvert v'_x \rvert \ell_{\nabla T}}{\kappa/\rho R}$ | $Sr = \frac{\ell_{\nabla p} \lvert a^{\rm e}_{\rm af} \rvert}{p/\rho}$ | $Fr = \frac{\lvert v'_x \rvert}{\sqrt{\lvert a^{\rm e}_{\rm af} \rvert \ell_{\nabla v'}}}$ |
| LM criteria | $\gamma M^2 \ll 1$ | $\frac{\gamma^2}{\gamma-1} \frac{M^2}{Pe} \ll 1$ | $Sr' = \frac{\ell_{\nabla v'} \lvert a^{\rm e}_{\rm af} \rvert}{p/\rho}$ | $\frac{\gamma M^2}{Fr^2}\ (= Sr') \ll 1$ |

Table 4: Hydrodynamic characteristic number definitions and LM approximation criteria.

(compressed-fluid region), this analysis yields a global and detailed description of the flow hydrodynamic features. For example, strong compressibility effects — fast expansions of the heated fluid ($\ell_{\rm cond}/\ell_{\rm tot} \approx 1$, Figs. 2*a*, *b*), isentropic Chapman–Jouguet (CJ) points ($M = 1$, Figs. 1*b* and 2*c*, *d*), ablation fronts with high stiffnesses and strong dominance of pressure gradients over inertial forces ($Sr_{\rm af} \ll 1$, Fig. 1*f*) — are evidenced for sufficiently large values of the Péclet scale $\mathscr{P}e$ ($\propto \mathscr{B}_\varphi^2/\mathscr{B}_p^2$), and in particular for the IPI-like flow. Regarding the

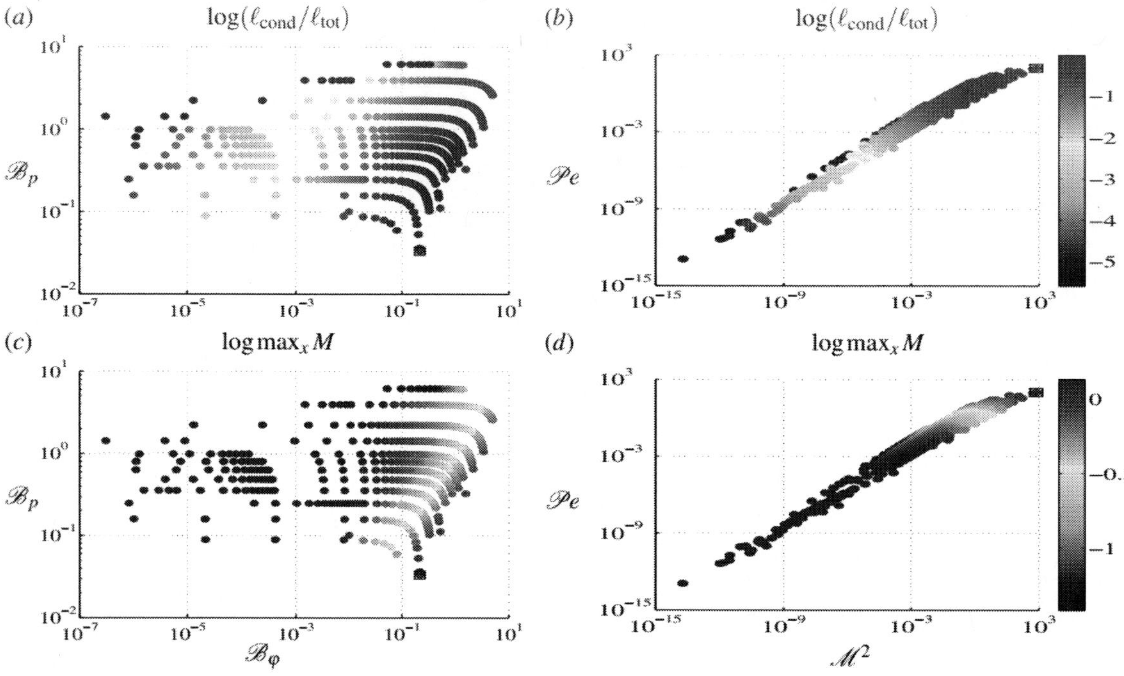


Figure 2: Radiative self-similar flow. Hydrodynamic characteristic numbers (IPI-like flow □).

low-Mach-number (LM) approximation commonly used in ablative Rayleigh–Taylor instability modeling, none of the flows satisfies the Froude criterion $\gamma M^2/Fr^2$ $(= Sr') \ll 1$ (see Tab. 4) in the vicinity of the external surface and of the ablation front (Fig. 1*h*), while flows with CJ points obviously do not fulfill the low-Mach criterion over part of their conduction region (Fig. 1*b*). For arbitrary time-dependent flows, the same set of characteristic numbers is equally relevant, the corresponding analysis having to be performed repeatedly over time.

**References**

[1] R. Marshak, Phys. Fluids **1**, 24 (1958)

[2] C. Boudesocque-Dubois, S. Gauthier and J.-M. Clarisse, J. Fluid Mech. **603**, 151 (2008)

[3] J.-M. Clarisse, C. Boudesocque-Dubois and S. Gauthier, J. Fluid Mech. **609**, 1 (2008)

[4] C. Boudesocque-Dubois, V. Lombard, S. Gauthier and J.-M. Clarisse, J. Comp. Phys. **235**, 723 (2013)

# Experimental and numerical investigation of the early time dynamics of single wire plasma explosions

I. Fitilis[1,2], A. Skoulakis[1,2], E. Kaselouris[1,2], I.K. Nikolos[3], M. Bakarezos[1,2], N.A. Papadogiannis[1,2], V. Dimitriou[1,2] and M. Tatarakis[1,2*]

[1]*Centre for Plasma Physics & Lasers ( CPPL), Technological Educational Institute of Crete, Chania & Rethymnon, Greece*

[2]*School of Applied Sciences, Technological Educational Institute of Crete, Chania, Greece*

[3]*School of Production Engineering & Management, Technical University of Crete, Chania, Greece*

**Corresponding Author: m.tatarakis@chania.teicrete.gr*

**ABSTRACT**

The aim of this work is to investigate the initial stages of the Z-pinch explosion as the phase changes from solid to plasma. Experiments of single wire explosion were carried out using a Z-pinch device implementing a current pulse of 35 kA peak with rise time of 60 ns (10%-90%). Experimental results concerning the expansion dynamics of the exploded material were obtained from a modified Fraunhofer diffraction pattern method and also shadowgraphic and interferometric techniques utilizing a 150 ps pulsed laser. Furthermore, a coupled transient electromagnetic-thermal-structural multiphysics computational FEA (Finite Element Analysis) using a single material ALE (Arbitrary Lagrangian-Eulerian) element is carried out. In order to simulate the matter's response, a strength material model along with an equation of state for pressure and an equation of state for electrical conductivity are used. Temperature-dependent material properties are also taken into account. The comparison of experimental and numerical results aims to contribute towards the comprehension of transition from solid to plasma phase of the wire.

The Z-pinch plasma device is a topic with continues interest due to the important emerging applications and the fascinating fundamental physics involved [1]. The physical processes at the initial stages of the plasma formation along the wire, flowing by high current pulse, are very important for the plasma evolution on Z-pinch devices. [2] The early time dynamics have been proven to be important for the development of the MHD instabilities in plasma. [2,3] So as to study such dynamics including the crucial study of the phase change conditions from the thermoelastic to the melting and plasma regimes, thick metallic wires have been investigated through experiments and simulations in this work.

For the experimental investigation of the initial times of wire explosion a Z-pinch pulsed power device is used, powered from a Marx-bank of 600 Joule energy capacity after passing through a water-filled pulse forming line (PFL) and a self-breaking SF6 switch. A copper wire of 300 um diameter and 15.2 mm length is fixed by soldering at the conical shaped electrodes inside a vacuum chamber of $10^{-4}$ mbar. A V-dot probe measure the derivative of the voltage at the PFL and a Rogowski groove measure the derivative of the current passing through the wire. For the optical probing of the fast phenomena during the wire explosion a SBS-compressed Nd:YAG Q-switch laser (EKSPLA, SL312) with 150 ps pulse duration and 532 nm wavelength is utilized. Shadowgraphy, schlieren and interferometric imaging techniques are implemented to record the early times of the plasma formation. The schlieren stop is a knife-edge oriented parallel to the wire which enable the distinguish of the light deviation caused by plasma from that caused by neutral Cu vapors.[4] The imaging of the Fraunhofer diffraction pattern at the focus of a lens is employed as a method to measure the expansion of the wire at times before the plasma formation.[5]

A coupled 3D transient electromagnetic-thermal-structural multiphysics computational FEA analysis using a single material ALE (Arbitrary Lagrangian Eulerian) element is carried out. Electromagnetic fields are computed by solving the Maxwell equations in the eddy-current approximation. When the electromagnetic fields have been computed, the Lorentz force term $\mathbf{F}=\mathbf{j}\mathrm{x}\mathbf{B}$, where **j** is the current density and **B** the magnetic field, is evaluated and added to the mechanical solver, which computes the deformation of the wire. Moreover, the Joule heating term ($j^2/\sigma$, where $\sigma$ the electrical conductivity) is added to the thermal solver in order to update temperature. Furthermore, the skin depth effect of the conducting wire is also taken into account.[6]

The ALE algorithm consists of a classical Lagrangian step in which the mesh moves along with the modeled material, a rezone step in which the mesh is modified to preserve good quality through the computation of the Eulerian time step and a remapping step in which the solution is conservatively transferred from the old mesh to the new rezoned one.

The hydrodynamic and deviatoric behavior of the metal is taken into account by using an equation of state combined with a strength material model. Analytical Gruneisen equation of state [7] coupled with Johnson-Cook [8] strength material model are used for this purpose. Furthermore, electrical conductivity versus temperature and density is computed using

Burgess equation of state [9]. Regarding the boundary conditions, the ends of the wire are fixed at environmental temperature, at 27 °C. An important feature of the developed simulation is that the ALE mesh is adaptively refined in order to accurately simulate the dynamic phase changes of matter. The loading source term is the alternating current as measured and recorded during the real experiments.

A series of interferometric and schlieren images, taken at different times after the current start, showing that the plasma formation time is about 150 ns ± 10 ns, it doesn't occur simultaneously along the wire and the transition from neutral vapors to plasma is very fast. At later time the corona plasma occupies all the wire length. Figure 1 shows two instances from the initiation of the coronal plasma. The upwards/downwards fringe shifts at the interferograms, as well as the left/right lighten at the schlieren images, is attributed respectively to the plasma/ vapors domination on the refractive index near the wire.

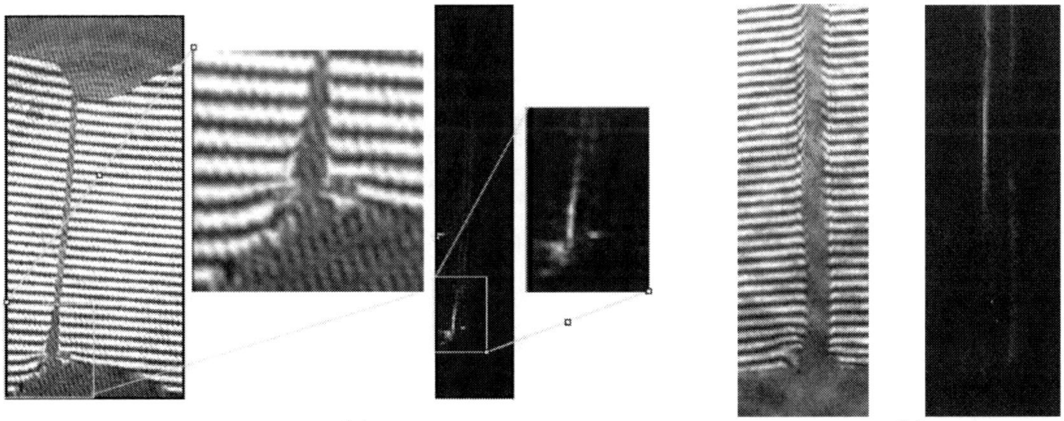

(a) (b)

Figure 1. Interferograms and schlieren images from the initiation of coronal plasma forming at a) 150 ns and b) 162 ns (different shot) after the current start.

For the measurements of the wire expansion (solid or melted) before the plasma formation the Fraunhofer diffraction pattern of the laser pulse at the focus of a lens is recorded and the wire diameter is calculated from the fitting of $(\text{sinc}(x))^2$ function at the line out plot along the fringes axis (Figure 2). An average radial expansion rate of 65 m/s results from the experimental measurements, while a radial expansion rate of 60 m/s is computed from the simulations.

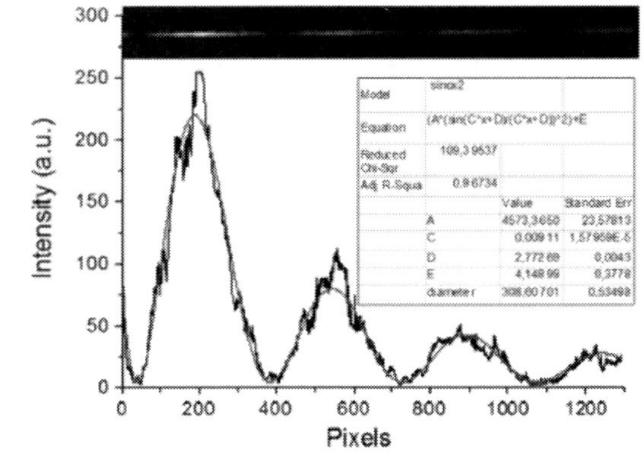


Figure 2. Line out of the diffraction pattern (top image) and calculation of the wire's diameter.

In Figure 3 the temperature and density simulation results in relation to the time after the current start are presented. The vaporization of the copper at a time of 160 ns is in consistence with the indication of vapors in the experimental images at these times.

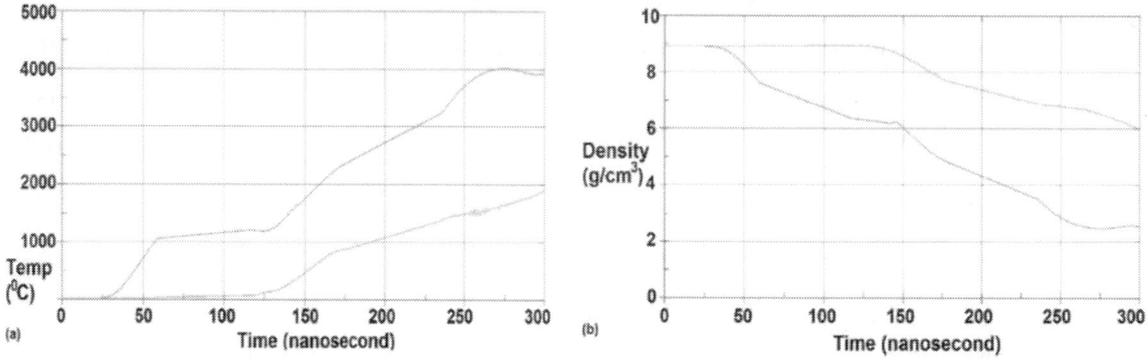


Figure 3. Finite element numerical results of a) temperature and b) density in relation to time, for elements 150 μm away from the wire's axis (outer region, red curve) and 5 μm far from the axis (core region, green curve).

The combination of the multiphysics FEM model and the experimental method is capable to describe the wire expansion dynamics. Further numerical simulations that use a multiphase equation of state instead of the analytical Gruneisen along with new experiments are still under development in order to monitor the conditions for coronal plasma formation.

**Acknowledgments**

The authors acknowledge financial support through the Action "National Research Infrastructure for HiPER" MIS 376841 (co-funded by the European Union and Hellenic National funds within the Operational Programme "Competitiveness and Entrepreneurship").

## References

[1] M.G. Haines, Plasma Phys. Control. Fusion **53**, 093001 (2011)

[2] X. Zhu et al., IEEE T Plasma Sci. **40**, 3329-3333 (2012)

[3] M. Tatarakis et al., Phys. Plasmas **5** (3), 682-691 (1998)

[4] S. I. Tkachenko et al., Plasma Phys Rep. **35**, 734-753 (2009)

[5] S.A. Khodier, Optics & Laser Technology **36,** 63-67 (2004)

[6] G. Le Blanc et al., "Ramp wave compression in a copper strip line: comparison between MHD numerical simulations (LS-DYNA) and experimental results (GEPI device)"10th International LS-DYNA Conference, Dearborn, Michigan, 8-10 June 2008.

[7] K. Nagayama, "Introduction to the Gruneisen Equation of State and Shock Thermodynamics", Kindle Edition, (2011)

[8] G.R. Johnson and W.H. Cook, Eng. Fract. Mech **21**, pp. 31-48 (1985)

[9] T. Burgess, "Electrical resistivity model of metals", 4th International Conference on Megagauss Magnetic-field generation and related topics, Santa Fe, NM, 14-17 July 1986

# Combined high frequency and arc plasma sources for contacts application

A.V. Leonovych[1, 2], V.S. Taran [2]

1 V.N. Karazin Kharkiv National University, Kharkiv, Ukraine,

2 Institute of Plasma Physics, NSC KhIPT, Kharkiv, Ukraine

The use of plasma sources for metal coatings is very common. Feeding large constant displacement to the metal substrate can obtain high quality coatings with excellent adhesion properties. But in this case the temperature of the sample during the coating will also be very high. This option is great in case of massive samples of solid refractory alloys such as for various kinds of cutting tools. But what if you have to work with the finer details where there is a greater probability of overheating areas? Many materials can not be used as a substrate at such temperatures because they will be destroyed. In the case of dielectric, constant displacement can not be fed in the usual manner.

So in our case there was a problem of creating the metal contacts for CVD diamond detectors of ionizing radiation. The material of the substrate is dielectric material and very sensitive to overheating. Since operating conditions can lead to heating above T = 120°C at which standard adhesive contact starts degradation it was decided to use a connection by soldering. Due to inert properties of chromium, dual-layer contacts (chrome-stainless steel) were obtained and investigated.

EXPERIMENTAL SETUP

The application of metal contacts on to CVD diamond samples was carried out on Bulat-6 type device, additionally equipped with a high frequency generator. Fig. 1 schematically shows the device for HF-cleaning and ion-plasma assisted deposition.

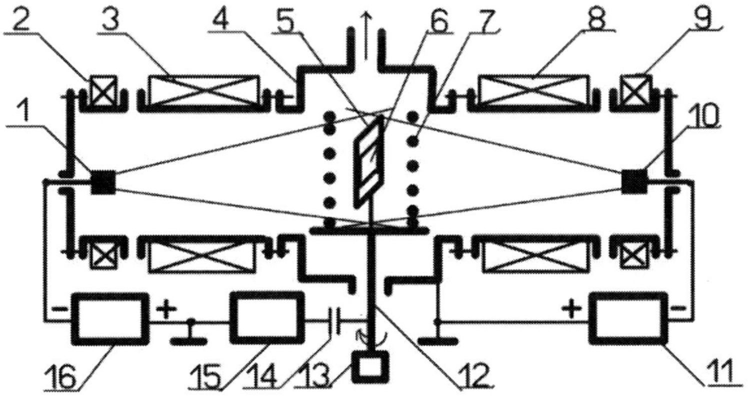


Fig. 1. Scheme of the experimental setup:

1, 10- cathode, 2, 9 - stabilizing reels, 3, 8 - focusing reels, 4 - vacuum chamber, 5 - container (lining), 6 - sample, 7 - HF antenna, 11, 16 – arcs power source, 12 - rotating device, 13 - reducer, 14 - capacitor,15 - HF generator.

Samples used for studies are polycrystalline CVD diamond on a 8×8×2 mm$^3$ substrate of single-crystal Si. The important point is not to allow ingress of evaporator material to the samples side surfaces. Otherwise leakage currents can occur, which is unacceptable. Therefore, the samples were placed in a fluoroplastic holder, jointly with the masks of PTFE or aluminum foils in titanium polyacetal special container. Designed container helps to fulfill the technical requirements and eventually get metal contacts of the desired geometry [1].

The sample container and the electrode with HF coils are mounted on a rotating device in the center of the vacuum chamber against the Cr and SS arc evaporators. Rotating table was connected to the HF generator via a capacitance.

Plasma method of ion cleaning is technically simple to implement but it helps us to carry out uniformity processing of any difficult form details. Also using assisted HF potential during dielectrics application helps us to eliminate the accumulation of over-discharge-surface and hence inhibition of ions that occurs at long constant-displacement potentials. An important fact is a decrease of droplet fraction formations during application due to additional ionization.

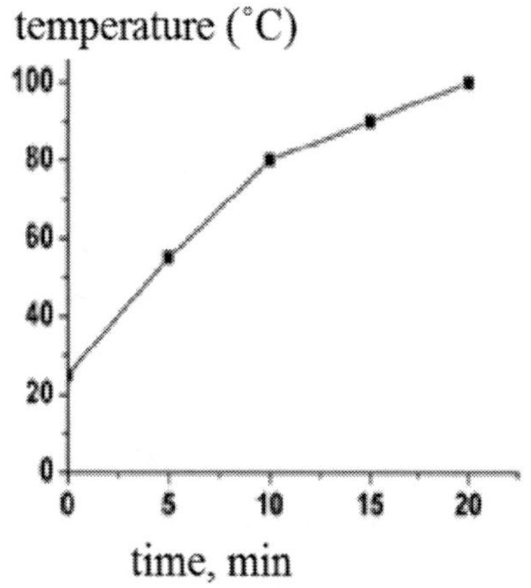


Fig. 2. Samples temperature during coatings application

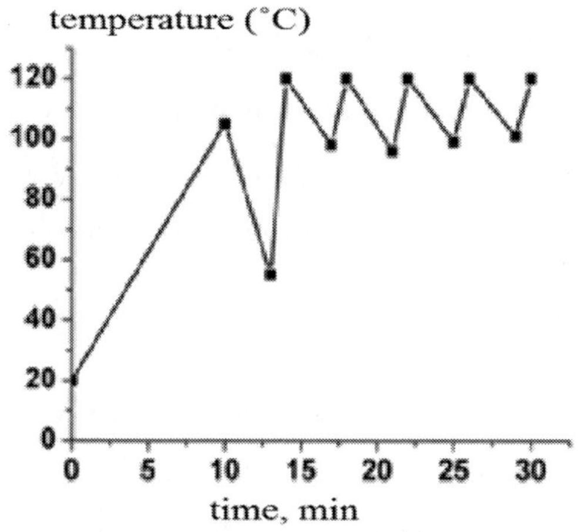


Fig. 3. The temperature of the samples during applying coatings

The important point on cleaning and applying stage is temperature monitoring. Electro and spectrometric characteristics of semiconductor detectors of ionizing radiation depends on many factors, such as, compliance with the stoichiometry of the composition, perfection of macro and microstructure, the presence of defects in a thin surface layer, the degree of perfection and electrical resistance of metal-semiconductor contacts. Exceeding the critical temperature can negatively affect all of the above.

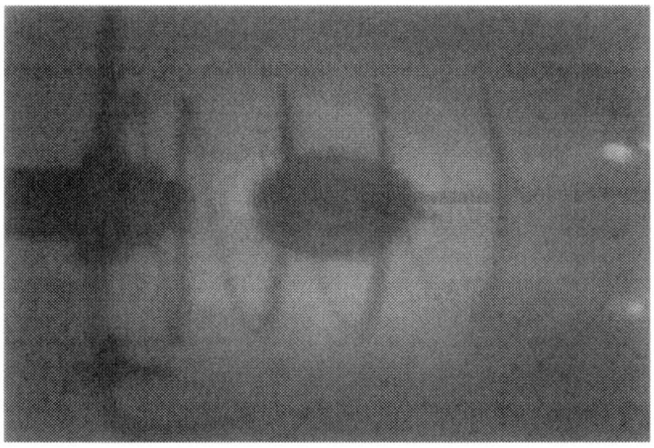

Fig. 4. Finishing ion cleaning with HF-generator

The cleaning process starts simultaneously with rotary apparatus (Fig. 4). Neutral Ar gas inlet and supplies HF voltage to the target (sample container, HF-coil). Approximate cleaning mode: the shifting voltage $U_{bias}$ = -(700….900) V; pressure $P_{Ar}$=2 $10^{-1}$Pa. The cleaning took 3…5 minutes, and temperature was not higher than 100 °C (Fig. 2).

The application of coatings starts immediately after cleaning. Cr and then SS contacts applies in a single technological cycle. A characteristic feature is pulsed operation of the evaporator on a particular program. The plasma ion method lead to heating the substrate (container with the sample) up to a temperature above permissible, therefore, there was designed a pulse mode of operation of the evaporator with switched HF voltage (assisted sputtering). In such a mode the temperature do not exceed 120 °C (Fig. 3). The metal layers thickness was 0.2….0.3 μm. Parameters for metal layers applying: arc current – 100 A; pulsed evaporator mode - 2 min, pause duration – 3 min, the number of cycles - 10; the shift voltage – 100 V; focusing coil current - 1.4 A; frequency of HF generator – 5 MHz. The total time of the coating is determined on the basis of the required thickness of the contact area (200…450 nm) [1,2].

RESULTS AND DISCUSSION

Table 1 shows the electrical characteristics to 7 detectors with double-layer contacts.

Tab. 1. The electrical characteristics of detectors

| № | The thickness h, microns | R at 400 V after application double-layer contacts, Ohm | counting rate, counts / s |
|---|---|---|---|
| 1 | 224 | 2,0E+12 | 13,2 |
| 2 | 191 | 2,0E+12 | 9,6 |
| 3 | 220 | 2,4E+12 | 8,6 |

| | | | |
|---|---|---|---|
| 4 | 216 | 4,7E+11 | 4,8 |
| 5 | 182 | 1,5E+12 | 9,8 |
| 6 | 150 | 3,2E+12 | 13 |
| 7 | 229 | 2,2E+12 | 15,4 |

In Fig. 5 it is shown the CVC of two samples with the smallest and largest value of electrical resistance.

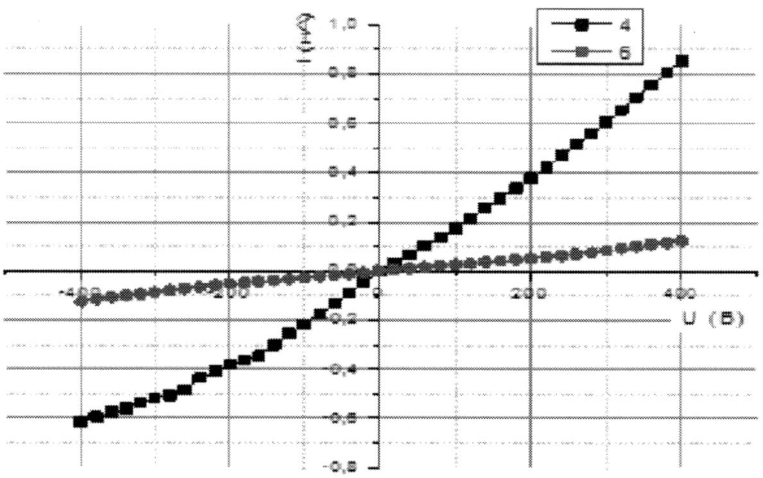


Fig. 5. CVC of two samples with the smallest and largest value of electrical resistance.

Samples that have a linear CVC with minimal leakage current can be used as detectors of ionizing radiation. Registration of the pulse-height distributions of the detector signal under α irradi-ation from $^{239}$Pu source performed with the following parameters of spectrometric tract: $U_{bias}$ = 400V, the shaping amplifier gain - 1000, shaping time τ = 1 microsecond, acquisition live time - 60 minutes.

REFERENCES

1. R.M. Muratov, A.A. Vierovkin, V.E. Kutny, Yr.N. Nezovibat'ko, A.V Rybka, V.S. Taran.Development and application of metal contacts on polycrystalline diamond films using combination of HF and ARC plasma sources // InternationalConference-School on Plasma Physics and Controlled Fusion,Kharkov, 2014, p.158.
2. R.M. Muratov, A.A. Vierovkin, V.E. Kutny, Yr.N. Nezovibat'ko, A.V Rybka, V.S. Taran.Development and application of metal contacts on polycrystalline diamond films using combination of HF and ARC sources // VANT "Series Plasma Physics(), №7, pp.188-190, 2014.

# Numerical calculation of ion runaway distributions

O. Embréus[1], S. Newton[1,2], A. Stahl[1], E. Hirvijoki[1], T. Fülöp[1]

[1] *Applied Physics, Chalmers University of Technology, SE-412 96 Göteborg, Sweden*

[2] *CCFE, Culham Science Centre, Abingdon, Oxon, OX14 3DB, United Kingdom*

**Background** The phenomenon of particle runaway in a plasma is well known, arising because the friction force experienced by a fast charged particle decreases with particle energy, so a sufficiently strong electric field can allow the particle to be accelerated - or *run away* - to high energy. Ion runaway has long been of interest in the astrophysical community, where it is thought to contribute to the observed abundance of high energy ions in solar flares [1]. It has also been used to study the behavior of lightning channels [2] and was observed in magnetic confinement devices [3, 4]. The detailed mechanism of ion runaway differs from that of electron runaway. Friction with the electrons drifting in the electric field acts to cancel a portion of the accelerating field. This effect is captured by replacing the electric field with the effective electric field $E^* = (1 - Z_i/Z_{\text{eff}})E$, with the consequence that highly charged impurities may be accelerated anti-parallel to the electric field. The remaining friction against the bulk has a non-monotonic velocity dependence, depending on composition, with friction against electrons limiting the maximum energy attainable in moderate electric fields.

Earlier analytic treatments of runaway ions were focused on the initial stages of the process [3, 5], or limited to the high-electric field limit and unable to provide details of the runaway tail formation and its features [6]. Here we describe the formulation and implementation of an efficient, open-source [7], finite-difference–spectral-method tool, CODION (COllisional Distribution of IONs), that solves the two-dimensional momentum space ion kinetic equation in a homogeneous plasma. Using CODION we obtain illustrative two-dimensional ion velocity space distributions, which demonstrate the typical behaviour of runaway ions in a variety of physical scenarios.

**Formulation** We consider the problem of ion acceleration by induced electric fields parallel to the magnetic field in a plasma. We assume straight field line geometry and a homogeneous background plasma. The time evolution of the ion distribution is then given by the Fokker-Planck equation, $\partial_t f_i + \frac{Z_i e}{m_i} E^* \left(\xi \partial_v + (1-\xi^2)/v\, \partial_\xi\right) f_i = \sum_s C_{is}\{f_i\}$, where $\xi = v_\parallel / v$ and $C_{is}$ is the collision operator. We restrict the study to fully ionized plasmas. We consider the case where an electric field appears in what was previously an equilibrium state, and may linearize the distribution around a Maxwellian $f_M$ if we only consider cases of small total runaway density, or the initial phase of the process. The linearized collision operator for self-collisions is given by $C^l_{ii}\{f_i\} = C_{ii}\{f_i, f_{Mi}\} + C_{ii}\{f_{Mi}, f_i\}$, where the first, the test-particle operator, describes collisions of the perturbed distribution with the bulk, while the second, the field-particle operator, describes the response of the bulk to the perturbation, for which we use the model described in Ref. [8]. Assuming the other ion species to remain near equilibrium, the collision operator is

$$\sum_s C_{is}\{f_i\} = \frac{1}{\tau_{ie}} \sum_s \frac{n_s Z_s^2}{n_e} \left\{ \frac{\phi(x_s) - G(x_s)}{2x_i^3} \frac{\partial}{\partial \xi} \left[ (1-\xi^2) \frac{\partial f_i}{\partial \xi} \right] + \right.$$

$$\left. + \frac{1}{x_i^2} \frac{\partial}{\partial x_i} \left[ 2 \frac{T_i}{T_s} x_i^2 G(x_s) f_i + x_i G(x_s) \frac{\partial f_i}{\partial x_i} \right] \right\} + \frac{1}{\tau_{ii}} \left[ 2\nu_s(v) x_i \xi u_\parallel + \nu_E(v) x_i^2 Q \right] f_{Mi},$$

where $\tau_{is}^{-1} = n_s e^4 Z_i^2 Z_s^2 \ln\Lambda / 4\pi\varepsilon_0^2 m_i^2 v_{Ti}^3$, $x_s = v/v_{Ts} = \sqrt{m_s v^2/2T_s}$ and the usual error function $\phi(x) = (2/\sqrt{\pi}) \int_0^x \mathrm{d}y\, e^{-y^2}$ and Chandrasekhar function $G(x) = [\phi(x) - x\phi'(x)]/2x^2$ appear. The moments $u_\parallel$ and $Q$ of the distribution function appearing in the momentum and energy restoring terms of the self-collision operator are $u_\parallel\{f_i\} = \frac{3}{2} v_{Ti} \left( \int \mathrm{d}^3v\, \nu_s(v) v_\parallel f_i \right) / \left( \int \mathrm{d}^3v\, \nu_s(v) v^2 f_{Mi} \right)$ and $Q\{f_i\} = v_{Ti}^2 \left( \int \mathrm{d}^3v\, \nu_E(v) v^2 f_i \right) / \left( \int \mathrm{d}^3v\, \nu_E(v) v^4 f_{Mi} \right)$, with the scattering frequencies given by $\nu_s(v) = 4G(x_i)/x_i$ , and $\nu_E(v) = 8G(x_i)/x_i - 2\phi(x_i)/x_i^3$.

Important features of the runaway behaviour can be demonstrated by reducing to the simpler case of an ion test particle, neglecting the effect of diffusion. The ion equation of motion takes the form $m_i \mathrm{d}v_\parallel/\mathrm{d}t = Z_i e E^* + F_i^{\text{test}}$ [2], where the collisional friction on a test particle is given by $F_i^{\text{test}}(v) = -Z_i^2 e E_D \sum_s \frac{n_s Z_s^2}{n_e} \frac{T_e}{T_s} (1 + m_s/m_i) G(x_s)$, and $E_D = n_e e^3 \ln\Lambda / 4\pi\varepsilon_0^2 T_e$ is the Dreicer field. Thus ion acceleration can occur when $E/E_D > F_i^{\text{test}}/Z_i e E_D |1 - Z_i/Z_{\text{eff}}|$, where $Z_{\text{eff}} = \sum_i Z_i^2 n_i/n_e$ is the effective charge. The electric field required to accelerate a test ion with given velocity is illustrated in Fig. 1 for low and high $Z_{\text{eff}}$. The non-monotonic friction shows two maxima, near the ion and electron thermal velocities, and also shows a strong dependence on charge and composition. This qualitative behaviour is what allows a runaway ion population to form in the presence of a sufficiently strong electric field.

**CODION** The discretization scheme used in CODION is based on an expansion of the distribution function in Legendre polynomials, $f_i(t, p, \xi) = \sum_l f_l(t, p) P_l(\xi)$, as in CODE [9]. A description of the scheme, together with convergence tests and more, can be found in Refs. [10, 11].

Figure 2(a) shows a typical example of the evolution of the ion distribution for a case where the electric field is above the minimum required for runaway acceleration. The plasma parameters used are characteristic for tokamaks with a hot bulk deuterium plasma at 1 keV and fully ionized native carbon impurities. A contour plot of the distribution in velocity space when steady state is reached is shown in Fig. 2(b). The required fields of order 1 V/m are, however,

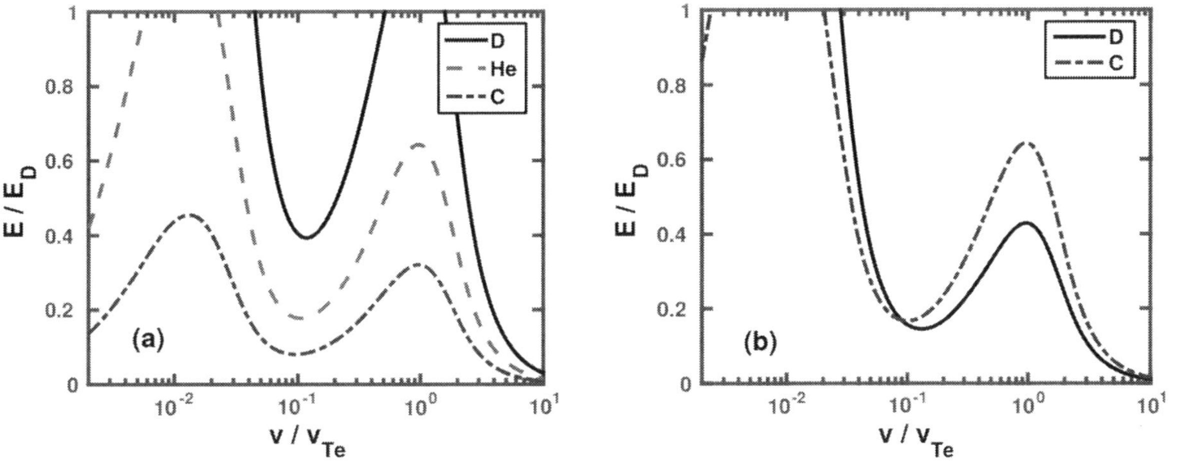


Figure 1: Electric field needed to accelerate fully ionized test particles of various ion species in impure deuterium plasmas. (a) $n_{\mathrm{C}}/n_{\mathrm{D}} = 0.4\%$, $n_{\mathrm{He}}/n_{\mathrm{D}} = 5\%$, $Z_{\mathrm{eff}} = 1.2$ and (b) $n_{\mathrm{C}}/n_{\mathrm{D}} = 4\%$, $n_{\mathrm{He}}/n_{\mathrm{D}} = 5\%$, $Z_{\mathrm{eff}} = 2$. All particle species are taken to be at the same temperature.

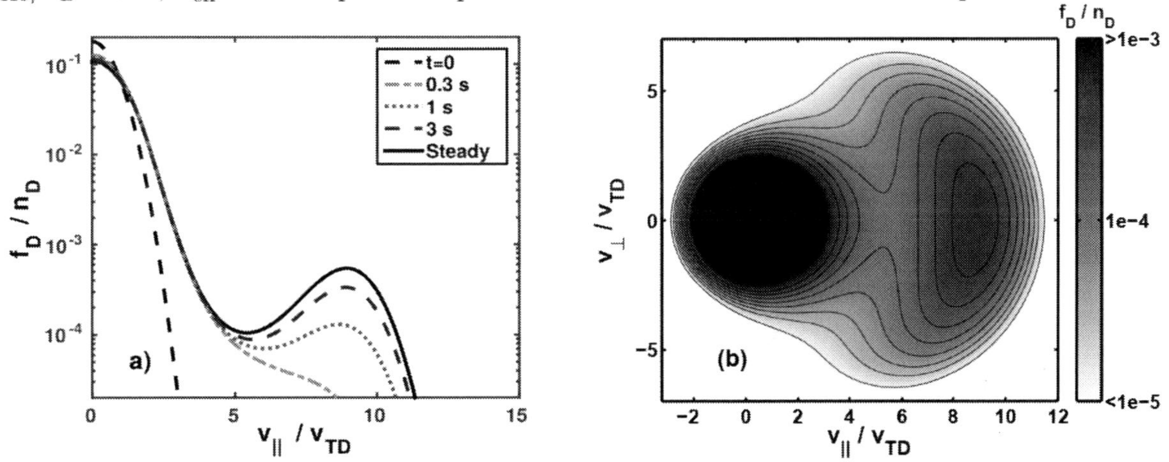


Figure 2: Deuterium distribution function in a plasma characterized by $T = 1\,\mathrm{keV}$ for all particle species, $n_e = 3 \cdot 10^{19}\,\mathrm{m}^{-3}$, $Z_{\mathrm{eff}} = 2$ due to fully ionized carbon impurities with $n_C/n_D = 4\%$, and $E = 1.64\,\mathrm{V/m}$. (a) Time evolution of the $\xi = 1$ cut of the distribution and (b) contour plot of the steady state distribution, established after 20 s.

unlikely to occur in operational tokamak plasmas for long enough that an appreciable runaway ion population has time to form.

Large electric fields are induced during tokamak disruptions and are well-known to generate runaway electrons [12]. However, low-frequency magnetic fluctuations in the frequency range $f \approx 60 - 260$ kHz have been observed in disruptions induced by massive gas injection (MGI) [13]. The observed frequency range is consistent with low-mode number Toroidal Alfvén Eigenmodes (TAEs) [14], which can be resonantly driven by fast ions satisfying $v_{\|} = \{v_A/3, v_A\}$, where $v_A = B/\sqrt{\mu_0 \rho_m}$ is the Alfvén speed and $\rho_m$ the mass density.

A limited analytic study of the potential for runaway ions to drive such modes was given in Ref. [5]. CODION allows us to investigate the scenario in detail. With the same plasma composition as used in the previous case, but with a lower temperature of $T = 10\,\mathrm{eV}$, we may solve for the ion runaway distribution in a post-disruption plasma after a characteristic time scale of 2 ms, assuming a Maxwellian initial ion distribution. The resulting distribution functions are illustrated in Fig. 3 for various electric fields. The resonant velocity is $v_A/3 \sim 30\text{-}40\, v_{TD}$, indicating that runaway ions typically have too low energy to drive TAEs, and also require higher electric fields than expected of typical post-disruption fields according to models [15]. Note, however, that fast ion populations present due to heating schemes in use before the disruption may not be completely expelled, and it is therefore not certain that the initial distribution is accurately described by a Maxwellian as assumed here.

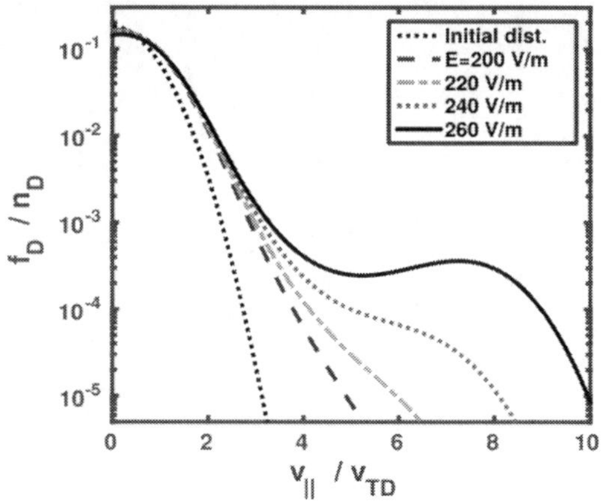

Figure 3: Deuterium distribution function resulting from various electric fields after 2 ms of acceleration from an initial Maxwellian in a post-disruption plasma.

**Summary** A numerical solver of the 2D ion Fokker-Planck equation has been presented, and its usefulness demonstrated in investigating the conditions required for ion runaway in cold and hot tokamak plasmas. It is shown that Alfvénic activity observed in disruption experiments are unlikely to be explained by the runaway mechanism alone.

## References

[1] G. D. Holman, Astrophys. J. **452**, 451 (1995).

[2] T. Fülöp, M. Landreman, Phys. Rev. Lett. **111**, 015006 (2013).

[3] P. Helander *et al.*, Phys. Rev. Lett. **89**, 235002 (2002).

[4] S. Eilerman *et al.*, Phys. Plasmas **22**, 020702 (2015).

[5] T. Fülöp, S. Newton, Phys. Plasmas **21**, 080702 (2014).

[6] H. P. Furth, P. H. Rutherford, Phys. Rev. Lett. **28**, 545 (1972).

[7] CODION is available at `https://github.com/Embreus/CODION`.

[8] I. G. Abel *et al.*, Phys. Plasmas **15**, 122509 (2008).

[9] M. Landreman, A. Stahl, T. Fülöp, Comp. Phys. Comm. **185**, 847 (2014).

[10] O. Embréus *et al.*, Phys. Plasmas **22**, 052122 (2015).

[11] O. Embréus, MSc thesis, Chalmers University of Technology (2014).

[12] E. M. Hollmann *et al.*, Nucl. Fusion **53**, 083004 (2013).

[13] L. Zeng *et al.*, Phys. Rev. Lett. **110**, 235003 (2013).

[14] C. Z. Cheng *et al.*, Phys. Fluids **29**, 3695 (1986).

[15] H. Smith *et al.*, Phys. Plasmas **13**, 102502, (2006).

# Energy partition in kinetic turbulence in strongly magnetized plasmas

L.M. Fazendeiro[1], N.F. Loureiro[1]

[1] *Instituto de Plasmas e Fusão Nuclear, IST, Universidade de Lisboa, 1049-001 Lisboa, Portugal*

**Introduction:** Turbulence is one of the fundamental unsolved problems in classical physics, ubiquitous in both laboratory and astrophysical plasmas [1]. In many cases of interest (e.g., the solar corona, the interstellar medium or the core of modern-day tokamaks) the collisional frequency is so small compared to the dynamic frequencies of interest that a fluid approach is not justified and a kinetic description is thus required. Due to the huge computational costs involved in performing fully kinetic ($6D$) simulations of turbulent plasmas, there is great advantage in utilizing reduced models that can still accurately capture the main relevant aspects of the problem. In addition, our physical understanding may also be greatly enhanced by them. One such fluid-kinetic model is KREHM (Kinetic Reduced Electron Heating Model), a rigorous asymptotic reduction of gyrokinetics in the limit of electron plasma beta $\beta_e \sim m_e/m_i$ [2]. The model has been numerically implemented in the `Viriato` code [3, 4], used in this work, in which we simulate decaying kinetic Alfvénic turbulence in $3D$, starting from an Orszag-Tang (OT) initial condition [5].

**Model and code benchmarks:** In KREHM the perturbed electron distribution function is defined as $\delta f_e = g_e + (\delta n_e/n_{0e} + 2v_{\|}u_{\|e}/v_{the}^2)F_{0e}$, where $F_{0e}$ is the equilibrium Maxwellian, $v_{\rm the} = \sqrt{2T_{0e}/m_e}$ the electron thermal speed, $v_{\|}$ the velocity coordinate (parallel to the magnetic guide field, $B_0$), $\delta n_e$ is the electron density perturbation (the zeroth moment of $\delta f_e$), $u_{\|e} = (e/cm_e)d_e^2\nabla_\perp^2 A_\|$ is the parallel electron flow (the first moment of $\delta f_e$), $A_\|$ is the parallel component of the vector potential, and $d_e = c/\omega_{pe}$ is the electron skin depth, with $\omega_{pe} = \sqrt{4\pi n e^2/m_e}$ the electron plasma frequency. The KREHM equations are [2]:

$$\frac{1}{n_{0e}}\frac{d\delta n_e}{dt} = -\hat{\boldsymbol{b}}\cdot\nabla\frac{e}{cm_e}d_e^2\nabla_\perp^2 A_\|, \tag{1}$$

$$\frac{d}{dt}\left(A_\| - d_e^2\nabla_\perp^2 A_\|\right) = \eta\nabla_\perp^2 A_\| - c\frac{\partial\varphi}{\partial z} + \frac{cT_{0e}}{e}\hat{\boldsymbol{b}}\cdot\nabla\left(\frac{\delta n_e}{n_{0e}} + \frac{\delta T_{\|e}}{T_{0e}}\right), \tag{2}$$

$$\frac{dg_e}{dt} + v_\|\hat{\boldsymbol{b}}\cdot\nabla\left(g_e - \frac{\delta T_{\|e}}{T_{0e}}F_{0e}\right) = C[g_e] + \left(1 - \frac{2v_\|^2}{v_{\rm the}^2}\right)F_{0e}\hat{\boldsymbol{b}}\cdot\nabla\frac{e}{cm_e}d_e^2\nabla_\perp^2 A_\|, \tag{3}$$

where $\frac{d}{dt} = \frac{\partial}{\partial t} + \frac{c}{B_0}[\varphi,\ldots]$, $\hat{\boldsymbol{b}}\cdot\nabla = \frac{\partial}{\partial z} - \frac{1}{B_0}[A_\|,\ldots]$ and $[\ldots,\ldots]$ denotes the Poisson bracket. $C[g_e]$ is the collision operator and $\eta$ the Ohmic resistivity. The electrostatic potential, $\varphi$, is given by the gyrokinetic Poisson law $\frac{\delta n_e}{n_{0e}} = \frac{Z}{\tau}\left(\hat{\Gamma}_0 - 1\right)\frac{e\varphi}{T_{0e}}$, with $\tau = T_{0i}/T_{0e}$ and $\hat{\Gamma}_0$ denoting the inverse Fourier transform of $\Gamma_0(\alpha) = I_0(\alpha)e^{-\alpha}$; $I_0$ is the modified Bessel function and $\alpha = k_\perp^2\rho_i^2/2$, with $\rho_i$ the ion Larmor radius. Equation (3) is a kinetic equation for the reduced electron distribution function $g_e(x,y,z,v_\|,v_\perp,t)$, with no explicit dependence on $v_\perp$. If such a dependence is not introduced by the collision operator $C[g_e]$ [2], then $v_\perp$ can be integrated out and $g_e$ effectively becomes only $4D$, $g_e(x,y,z,v_\|)$. In `Viriato` $g_e$ is expanded in Hermite polynomials, which transforms the electron drift-kinetic eq. (3), into a coupled set of $M$

fluid-like equations for each of the coefficients ($g_m$) of the Hermite polynomials of order $m$ [2].

In Fig. 1 we show two of the many benchmarks performed with `Viriato` [4]. The left plot is a direct comparison with the gyrokinetic code `AstroGK` [6], where the linear growth rate of the tearing mode is computed for several values of the Lundquist number, $S = av_A/\eta$, where $a$ is a typical equilibium length scale and $v_A = B_0/\sqrt{4\pi n_0 m_i}$ the Alfvén speed based on the guide-field. The parallel Alfvén time is defined as $\tau_A = L_\parallel/v_A$, where $L_\parallel$ is the reference length-scale, parallel to the guide field. We note that for $\beta_e \sim 30m_e/m_i$ the agreement is already quite good, suggesting that KREHM may well be a valid description of plasma dynamics outside its strict asymptotic limit of validity. The right plot shows a calculation of the frequency and damping rate of the kinetic Alfvén wave (KAW), whose dispersion relation can be obtained by linearizing eqs. (1–3), showing an excellent agreement between the analytical prediction and the numerical results obtained with `Viriato`.

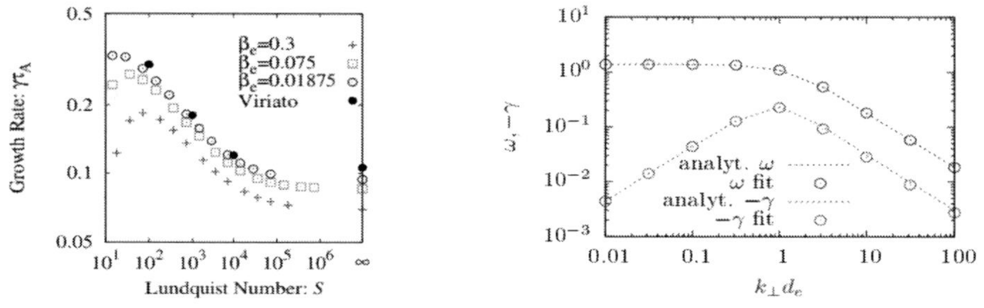


Figure 1: Left: Tearing mode growth rate as a function of the Lundquist number; results obtained both with the gyrokinetic code `AstroGK` [6], for varying values of $\beta_e$, and `Viriato`. Right: Frequency and damping rate of the linear kinetic Alfvén wave (KAW), in units of $\tau_A$, at fixed $k_\perp \rho_i = 1$, $M = 20$, as a function of $d_e$. Lines are the exact solution of the analytical dispersion relation, data points were obtained with `Viriato`.

**Main results and discussion:** In this work, the KREHM equations are solved in a $3D$ periodic cubic box of size $2\pi$ in each direction and $256^3$ points. In these simulations there are two available channels for energy dissipation: one via hyperdiffusion in coordinate space ($k-$space in `Viriato`) and the other (if non-isothermal electrons, and thus $g_e \neq 0$ are considered) via hypercollisions in the $m-$space of the Hermite polynomials. Thus, and similar to the energy cascade occurring from small $k$ (where energy is injected) to large $k$ (where dissipation takes over), in the familiar Richardson picture, there is now also transfer of energy occurring from small to large $m$, i.e., phase-mixing in velocity space.

In Fig. 2 we show two energy spectra for the case of decaying $3D$ OT kinetic turbulence, with parameters $\rho_i = 2$, $d_e = 0.3$ (in code units) and non-isothermal electrons, with $M = 40$ Hermite polynomials, at two different times. The slopes indicated refer to power laws that have been widely discussed in the literature. In particular, we see that there is a sharp separation between electric and magnetic energy scalings, occurring at around $(k_\perp/2\pi)\rho_i \sim 1$, in agreement with solar wind observations [7]. Also shown is a $-2.8$ slope often reported in observations of the solar wind (e.g., [8]).

In Fig. 3 we show the result of the competition between the two energy dissipation channels, for

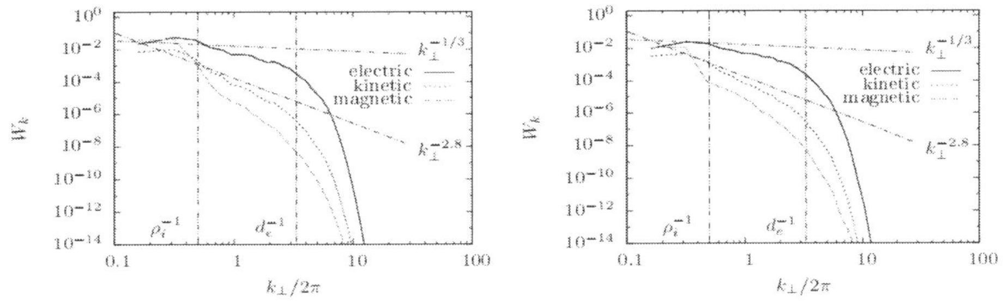


Figure 2: Spectra for decaying turbulence with $\rho_i = 2$, $d_e = 0.3$ and $M = 40$ at $t/\tau_A = 1.5$ (left) and 3.0 (right). Blue line represents perpendicular electric energy; red line kinetic energy, green the perpendicular magnetic field energy. See text for discussion of the power laws.

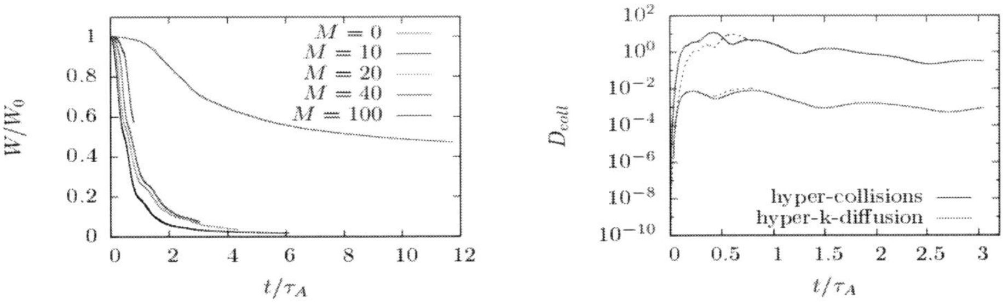


Figure 3: Left: Total energy, normalized to its initial value, $W_0$, for a $256^3$ system, with $\rho_i = 2$, $d_e = 0.3$. Note the dramatic decrease in energy decay rate when the $m's$ are switched off ($M = 0$). Right: Energy dissipation channels for the $M = 40$ (full lines) and $M = 100$ (dotted lines) runs. The predominance of the hyper-collisions (in $m$ space) in both cases is clearly displayed.

the same parameters of Fig. 2. The left plot shows the values of total energy, normalized to its initial value, for a case with isothermal electrons ($M = 0$, i.e., $g_e = 0$) and for a successively higher number of Hermite polynomials. The different durations of the runs are due to the fact that they become much more computationally expensive as $M$ increases. The right plot shows the total values for each of the two dissipation channels: hyperdiffusion in $k-$space and hypercollisions, corresponding to the energy cascade in velocity space (linear phase-mixing). This is shown for two runs only ($M = 40, 100$) but the $M = 10, 20$ runs also display the same behaviour. Note also that the run with $M = 100$ (dotted lines) takes longer to reach the same value of energy dissipation via hyper-collisions, in accordance with its slower energy decay rate, shown in the left plot.

Finally, for completeness, we show in Fig. 4 (left) energy spectra in $m-$space for this set of runs, at $t = 0.73\tau_A$. This plot clearly illustrates how the inertial range develops as the number of Hermite polynomials increases and the system subsequentely takes longer to reach the dissipation range. Also shown is a $-1/2$ power law, deduced in [2] for the inertial range slope of linear phase-mixing KAWs. The right plot shows energy spectra for $M = 40$ at the same times as Fig. 2. The power law seems to be obeyed, well beyond its regime of validity, even after most of the initial energy has been dissipated.

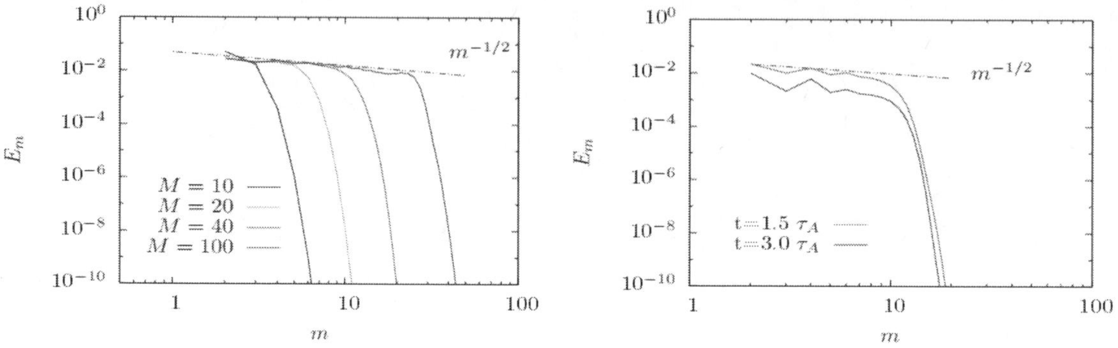


Figure 4: $E_m = |g_m^2|/2$ energy spectra in $m-$space, with a $-1/2$ power law [2]. Left: same values of $M \neq 0$ shown in Fig. 3 (left), at the same time, $t = 0.73\tau_A$. Right: Spectra for the $M = 40$ run, at two different times, the same as shown in Fig. 2.

**Conclusions and future work:** In this work we show preliminary results suggesting that linear phase-mixing may play a dominant role in decaying kinetic turbulence energy dissipation. We note that these results are still not conclusive and further parameter scans are in progress. In particular, the non-linearity of the system only becomes effective at about $t \sim \tau_A$ (see Fig. 3, left, $M = 0$ run). This means that for non-isothermal electrons ($g_e \neq 0$, i.e., $M \neq 0$) linear phase-mixing begins dissipating energy well before the other channel (in $k$-space) can do so. However, there is a strong indication (from the results presented here as well as from other, smaller runs) that this may indeed be the dominant energy dissipation mechanism in these systems, a topic which will be the main subject of a forthcoming publication.

Other questions of interest include ascertaining the exact role played by current sheets in energy dissipation [9, 10]. In order to quantify this we intend to perform numerical simulations where turbulent mixing (due to nonlinearity) and linear phase-mixing in velocity space are taking place simultaneously, which in turn will require increasing the number of Hermite polynomials even further.

**Acknowledgements:** This work was partly supported by Fundação para a Ciência e Tecnologia via Grants UID/FIS/50010/2013 and IF/00530/2013. Simulations were carried out at Helios (IFERC) and Stampede (TACC).

## References


[1] A.A. Schekochihin, S.C. Cowley, W. Dorland *et al*, Astrophys. J. Suppl. S., **182**, 310-377 (2009).

[2] A. Zocco and A.A. Schekochihin, Phys. Plasmas, **18**, 2309 (2011).

[3] N.F. Loureiro, A.A. Schekochihin and A. Zocco, Phys. Rev. Lett. **111**, 025002 (2013).

[4] N.F. Loureiro, W. Dorland, L.M. Fazendeiro *et al.*, arXiv:1505.02649, submitted to J. Comput. Phys., (2015).

[5] S.A. Orszag and C.-M. Tang, J. Fluid Mech., **90**, 01, (1979).

[6] R. Numata, G.G. Howes, T. Tatsuno *et al.*, J. Comp. Phys., **229**, 24, (2010).

[7] S. Bale, P. Kellogg, F. Mozer *et al.*, Phys. Rev. Lett., **94**, 21, (2005).

[8] O. Alexandrova, J. Saut, C. Lacombe *et al.*, Phys. Rev. Lett., **103**, 16, (2009).

[9] V. Zhdankin, D.A. Uzdensky and S. Boldyrev, Phys. Rev. Lett., **114**, 065002, (2015).

[10] K.D. Makwana, V. Zhdankin, H. Li *et al.*, Phys. Plasmas, **22**, 042902, (2015).

# Preliminary study of differential effects in fast magnetic compression experiments

R. Gueroult[1], N.J. Fisch[1,2]

[1] *Princeton Plasma Physics Laboratory, Princeton University, Princeton, NJ, USA*

[2] *Department of Astrophysical Sciences, Princeton University, Princeton, NJ, USA*

Fast magnetic plasma compression has been suggested for time resolved plasma wave properties control [1]. In such concepts, an axial magnetic field $\mathbf{B} = B(t)\hat{\mathbf{z}}$ changes with time, in order to compress a pre-magnetized plasma, and hence control the plasma density under the assumption that $n \propto |B|$.

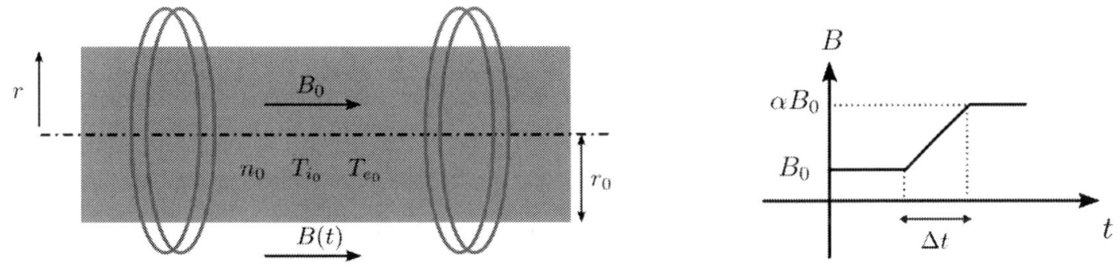


(a) Geometry considered

(b) Time evolution of the control magnetic field

Figure 1: Fast magnetic compression of a pre-magnetized plasma. A background magnetic field $\mathbf{B} = B_0\hat{\mathbf{z}}$ is embedded in the plasma. The axial magnetic field outside the plasma $B(t)$ is ramped up over a time $\Delta t$. For fast compression, $\Delta t < 2\pi/\omega_{ci}$.

## Simple picture

When the compression time $\Delta t$ is large compared to the ion gyro-period $\tau_{ci} = 2\pi/\omega_{ci}$, with $\omega_{ci} = q_i B_0/m_i$, both ions and electrons are magnetized. The charged particles motion then consists essentially in the radial cross field drift $E_\phi \times B_z$, with $E_\phi(r) = -r\dot{B}/2$ the induced azimuthal electric field.

On the other hand, when the compression time is shorter than the ion gyro-period, but larger than the electron gyro-period, electrons and ions motion differ. Magnetized electrons drift radially inward, similarly to what is observed for slower compression. Contrarily, ions are essentially unmagnetized on this time-scale, and ions motion is primarily dictated by the induced electric field $E_\phi(r) = -r\dot{B}/2$. Since the ions motion is largely azimuthal, radial charge separation occurs, which is the source of a radial electric field $E_r$. When this radial electric field becomes comparable or larger than the azimuthal field, ions move radially inward. Another consequence of this radial electric field is the azimuthal electron drift $-E_r \times B_z$, which can momentarily lead to large azimuthal currents.

Interestingly, experimental evidence of ion separation has been reported in the case of magnetic field penetration in a non-magnetized plasma on similar time-scales [2]. In light of the very simplified fast compression picture depicted above, one can ask whether the ion separation could be the result of differential rotation effects [4, 5]. To better address these questions, a numerical modeling effort has been initiated.

**Preliminary results from numerical modeling**

In order to avoid additional complications associated with cylindrical geometry, the model considered here is 1D planar. As illustrated in Fig. 4, we consider the interaction of an electromagnetic pulse propagating in vacuum and a magnetized plasma slab of finite width using a fully electromagnetic relativistic PIC code. A background field $\mathbf{B} = B_0\hat{\mathbf{z}}$ is present everywhere in the domain. The pulse amplitude is $\delta B$, with $\delta B/B_0$ typically equal to 1%, and the pulse width is $\pi c/\omega$, with $\omega = 2.5\omega_{ci}$. The plasma compositions considered are either an electron-proton mixture, or an electron-proton-He$^+$ mixture, in which case $n_{0_{\mathrm{He}^+}} = n_{0_{\mathrm{H}^+}} = n_0/2$. Real mass ratios are used, with $m_{\mathrm{H}^+}/m_e = 1836$ and $m_{\mathrm{He}^+}/m_{\mathrm{H}^+} = 4$. In all cases, the ion cyclotron frequency of reference $\omega_{ci}$ is the proton cyclotron frequency.

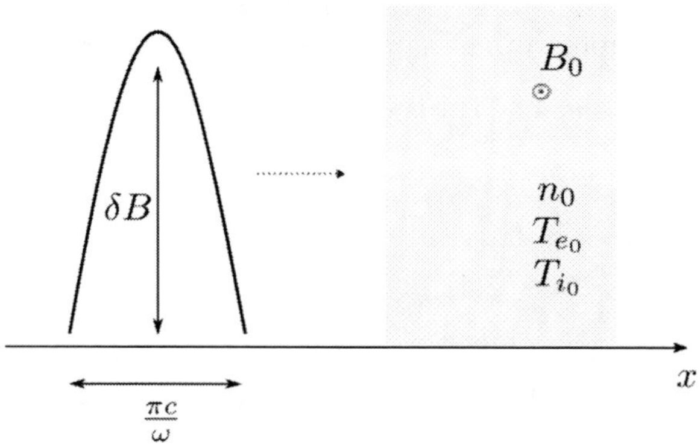


Figure 2: Geometry modeled with a fully electromagnetic relativistic PIC code. $\delta B/B_0$ is typically 1%, and $\omega = 2.5\omega_{ci}$. Background field $B_0 = 1$ T, $n_0 \sim 10^{12}$ cm$^{-3}$, $T_{e_0} = 3$ eV, $T_{i_0} = 0.1$ eV.

Upon interaction with the plasma slab, the incident pulse energy,

$$\varepsilon_p = \frac{\pi c}{5\mu_0\omega_{ci}}\delta B^2,$$

is divided into three components. First, a large fraction of the incident energy is reflected in the form of a reflected pulse. Second, part of the energy is transferred to plasma kinetic energy.

Third, some of the incident energy is transferred to plasma waves. The existence of incident, reflected and transmitted pulses is illustrated through the magnetic field spatial and temporal evolution given in Fig. 3, and obtained for an electron-proton plasma. For the plasma parameters and pulse shape considered here, the energy transmitted to the plasma wave is about 12%. From Fig. 3(b), the phase velocity can be estimated to be about $2.1 \pm 0.2\ 10^7$ m.s$^{-1}$. This is consistent with the phase velocity of a compressional Alfven wave, which for the small temperatures considered here propagates at the Alfven speed $v_A = B/\sqrt{\mu_0 n_0 m_i} \sim 2.2\ 10^7$ m.s$^{-1}$.

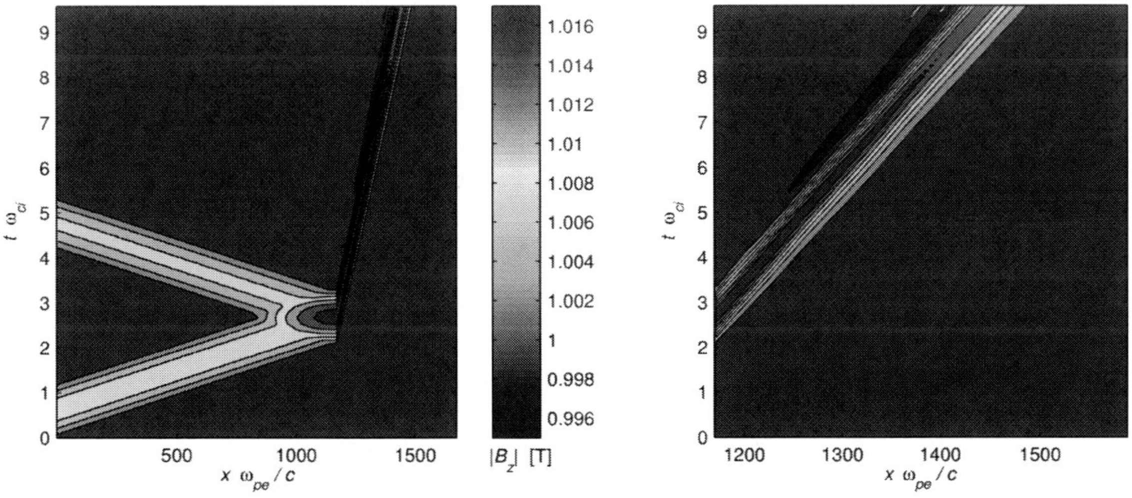


(a) Whole simulation domain

(b) Zoom on the initial plasma region

Figure 3: $Z$ component of the magnetic field in the simulation domain for an electron-proton plasma. The plasma initially extends from 1170 $c/\omega_{pe}$ to 1588 $c/\omega_{pe}$. The incident pulse amplitude is $\delta B/B_0 = 1\%$, and the initial density is $n_0 = 10^{12}$ cm$^{-3}$.

Comparison of the time evolution of the energy content obtained for an electron-proton plasma and a multi-ions species plasma indicates evidence of differential heating effects. For the electron-proton case, Fig. 4(a) shows that after an initial particle acceleration phase, the wave propagates with no dissipation. The wave energy is conserved, and particles do not acquire energy from the wave. On the other hand, for a multi-ions species plasma, wave damping is observed as illustrated in Fig. 4(b). This result is consistent with previous results obtained for magnetosonic pulses [3], *i. e.* larger compression times. The initial phase is similar to the electron-proton plasma case, with most of the energy transferred to the ions axial motion. For the $H^+ - He^+$ mixture considered, the energy initially transferred to the light and heavy species is of the order of $m_{H^+}/(m_{H^+} + m_{He^+})$ and $m_{He^+}/(m_{H^+} + m_{He^+})$, respectively. After this initial phase, the propagating plasma wave transfers energy to the particles, leading to wave damping. Interestingly, the initial energy transfer is preferentially towards the light ions (protons), while the wave damping seems to preferentially heat heavy ions ($He^+$).

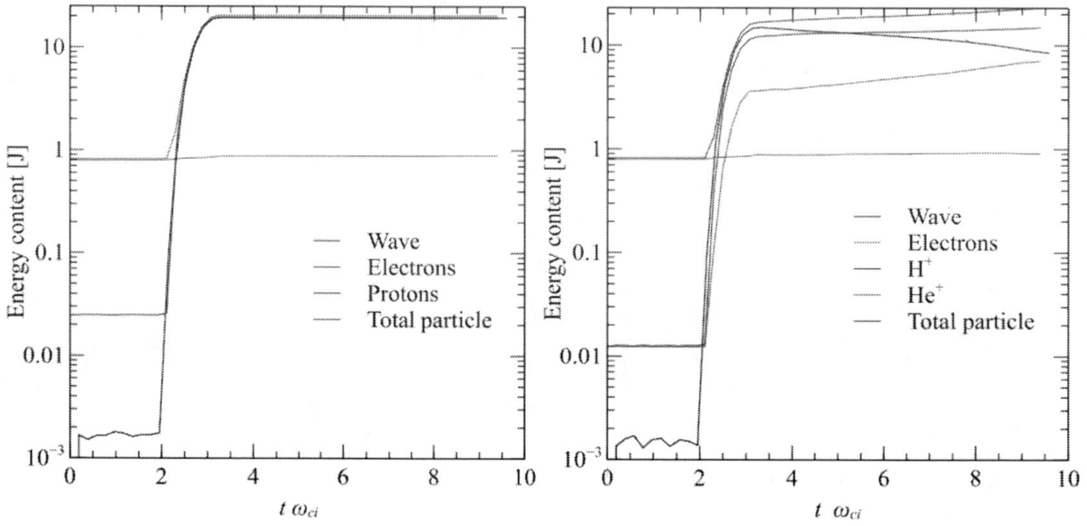


(a) $e^- - H^+$ (b) $e^- - H^+ - He^+$

Figure 4: Time evolution of plasma wave and particle kinetic energy content for two different plasma compositions. Wave damping is observed for multi-ions species plasma, but not for the electron-proton plasma. In both cases, the incident pulse energy $\varepsilon_p \sim 156$ J.

## Discussion and Caveat

The results presented here offer only a preliminary understanding of the physical mechanisms through which differential effects might be observed in fast magnetic compression experiments. To fully understand this effect, it remains to extend the simulations here to different relative abundances of different ions, as well as to consider a variety of different plasma parameters. To compare to experiments, it would be necessary as well to consider cylindrical geometry.

**Acknowledgments** This work is supported under the NNSA SSAA DE-FG5208NA28553, DOE 67350-9960 (Prime # DOE DENA0001836), DOE DE-FG02-06ER54851, and DOE DE-AC02-09CH11466.

## References

[1] P. F. Schmidt and N. J. Fisch, Phys. Rev. Lett. **109**, 255003 (2012)
[2] A. Weingarten, R. Arad, Y. Maron and A. Fruchtman, Phys. Rev. Lett. **87**, 115004 (2001)
[3] Daiju Dogen, Mieko Toida and Yukiharu Ohsawa, Phys Plasmas **5**, 1298 (1998)
[4] R. C. Davidson, Physics of Nonneutral Plasmas (2001)
[5] J.M. Rax, A. Fruchtman, R. Gueroult and N. J. Fisch, *in preparation*

# Gaussian Radial Basis Functions for Plasma Physics: Numerical Aspects

E. Hirvijoki[1], J. Candy[2], E. Belli[2], and O. Embréus[1]

[1] *Applied Physics, Chalmers University of Technology, Göteborg, Sweden*

[2] *General Atomics, San Diego, USA*

A magnetized plasma is described by the Vlasov equation and the non-linear Fokker-Planck collision operator [1]. The Vlasov part describes phase-space advection and the collision operator adds dissipation due to collisional energy and momentum exchange. Numerical discretization of the collision operator, however, is far from trivial. Recently, we have developed a new approach [2] to address this issue. The new approach is based on an expansion in Gaussian Radial Basis Functions (RBFs), a method widely used in neural network calculations [3]. In this paper, we discusss useful details regarding the numerical implementation of the RBF method.

**Introduction** In Ref. [2], we adressed the kinetic equation and the collision operator by expressing the distribution function as a sum of shifted Maxwellians, i.e., as Gaussian RBFs. In the continuous summation limit, an equivalent expression for the distribution becomes

$$f_a(\mathbf{v},t) = \int \left(\frac{\gamma}{\pi}\right)^{3/2} \exp\left[-\gamma\left(\mathbf{v}-\mathbf{u}\right)^2\right] W_a(\mathbf{u},\gamma,t)\, d\mathbf{u}\, d\gamma,$$

where the weight function $W_a(\mathbf{u},\gamma,t)$ is normalized to the density of the species $a$ according to $\int W_a(\mathbf{u},\gamma,t)\, d\mathbf{u}\, d\gamma = n_a(t)$. The Rosenbluth potentials, which are needed to describe the non-linear collision operator, can then be calculated analytically according to

$$\varphi_b(\mathbf{v},t) = -\frac{1}{4\pi}\int \gamma^{1/2}\Phi(\gamma^{1/2}|\mathbf{v}-\mathbf{u}|)\, W_b(\mathbf{u},\gamma,t)\, d\mathbf{u}\, d\gamma,$$
$$\psi_b(\mathbf{v},t) = -\frac{1}{8\pi}\int \gamma^{-1/2}\Psi(\gamma^{1/2}|\mathbf{v}-\mathbf{u}|)\, W_b(\mathbf{u},\gamma,t)\, d\mathbf{u}\, d\gamma,$$

where $\Phi(s) = \mathrm{erf}(s)/s$, $\Psi(s) = [s+1/(2s)]\mathrm{erf}(s) + \exp(-s^2)/\sqrt{\pi}$, and $\mathrm{erf}(s)$ is the error function.

For numerical considerations, the integration is changed into a discrete sum by defining the weight-function to be a sum of delta-functions: in 3D velocity space one chooses $W_a(\mathbf{u},\gamma,t) = \sum_i w_a^i(t)\,\delta(\mathbf{u}-\mathbf{v}_i)\,\delta(\gamma-\gamma_a^i)$ and in the axisymmetric case one can choose cylindrical velocity space coordinates and $W_a(\mathbf{u}) = \sum_i w_a^i(t)\frac{1}{2\pi v_{\perp i}}\delta(u_\parallel - v_{\parallel i})\delta(u_\perp - v_{\perp i})\delta(\gamma-\gamma_a^i)$. As was shown in Ref. [2] the collisional part of the Fokker-Planck equation then reduces to

$$\sum_i \frac{\partial w_a^i}{\partial t} F_a^i = \sum_{b,k,\ell} w_b^k(t)\, w_a^\ell(t) L_{ab}\left[\frac{m_a}{m_b}F_b^k F_a^\ell + \mu_{ab}\frac{\partial \varphi_b^k}{\partial \mathbf{v}}\cdot\frac{\partial F_a^\ell}{\partial \mathbf{v}} - \frac{\partial^2 \psi_b^k}{\partial \mathbf{v}\partial \mathbf{v}} : \frac{\partial^2 F_a^\ell}{\partial \mathbf{v}\partial \mathbf{v}}\right], \quad (1)$$

where $\mu_{ab} = m_a/m_b - 1$, $L_{ab} = (e_a e_b/m_a\varepsilon_0)^2 \ln\Lambda_{ab}$, $\ln\Lambda_{ab}$ is the Coulomb logarithm, and $w_a^i$ are the expansion weights. The RBF basis functions become $F_a^i = \left(\gamma_a^i/\pi\right)^{3/2} \exp\left[-\gamma_a^i(\mathbf{v}-\mathbf{v}_a^i)^2\right]$, for the 3D weight-function, and $F_a^i = (\gamma_a^i/\pi)^{3/2} I_0(2\gamma_a^i v_{i,\perp} v_\perp) e^{-\gamma_a^i(v_\parallel - v_{i,\parallel})^2 - \gamma_a^i(v_\perp^2 + v_{i,\perp}^2)}$, corresponding to the 2D axisymmetric weight function. $I_0(x)$ is the zeroth modified Bessel function of the second kind. The only notable difference between the 3D and 2D axisymmetric case is that in 3D, the weight function completely eliminates the **u**-integral in the potential functions, while in 2D the potentials $\varphi_b^i$ and $\psi_b^i$ have to be computed numerically as *ring-like* averages of the functions $\Phi$ and $\Psi$.

In this paper, we provide a detailed discussion of some of the key aspects needed for an efficient implementation of Eq. (1). We will focus on the evaluation of the right-hand-side and also on the specific case of axisymmtry.

**Equation for the weights** Even after the expansion, the Fokker-Planck equation, Eq. (1), has a velocity space dependence. In order to obtain an equation for the weights only, this dependence has to be eliminated.

Multiplying Eq. (1) with a test function $\Theta^i$ and integrating over velocity space gives

$$\sum_j M_a^{ij} \frac{\partial w_a^j}{\partial t} = \sum_{b,k,\ell} w_b^k(t)\, w_a^\ell(t) C_{ab}^{ik\ell}, \tag{2}$$

where the so-called *mass* matrix $M_a^{ij} = \int d\mathbf{v}\, \Theta^i F_a^j$ and the tensor like collision term $C_{ab}^{ik\ell}$ is

$$C_{ab}^{ik\ell} = \int d\mathbf{v}\, \Theta^i L_{ab} \left[ \frac{m_a}{m_b} F_b^k F_a^\ell + \mu_{ab} \frac{\partial \varphi_b^k}{\partial \mathbf{v}} \cdot \frac{\partial F_a^\ell}{\partial \mathbf{v}} - \frac{\partial^2 \psi_b^k}{\partial \mathbf{v}\partial \mathbf{v}} : \frac{\partial^2 F_a^\ell}{\partial \mathbf{v}\partial \mathbf{v}} \right] \equiv \int d\mathbf{v}\, \Theta^i C_{ab}^{k\ell}.$$

The traditional Galerkin projection is obtained by choosing $\Theta^i = F_a^i$, and the center collocation method by choosing $\Theta^i = \delta(\mathbf{v} - \mathbf{v}_i)$. In the center collocation, the matrix is thus simply $M_a^{ij} = F_a^j(\mathbf{v}_i)$ and the collision tensor becomes $C_{ab}^{ik\ell} = C_{ab}^{k\ell}(\mathbf{v}_i)$.

The true challenge in solving Eq. (2) is that the right hand side contains a rank-3 tensor, which is not sparse: introducing the Gaussian RBF-basis is advantageous in the sense that the expression for the collision operator in 3D velocity space becomes analytic but, simultaneously, each basis function extends to infinity, making both the matrix and the tensor full. The specific structure of the collision tensor, however, can be used to reduce the memory demands significantly, especially if the center collocation method is used.

Consider the evaluation of the collision operator at a set of points $\{\mathbf{v}_i\}_{i=1}^N$. Since $C_{ab}^{k\ell}(\mathbf{v})$ can be expressed as a sum of products of two different functions we can compute the value of the collision operator at the points $\mathbf{v}_i$ according to

$$\boldsymbol{C} = L_{ab} \left[ \frac{m_a}{m_b} ([\boldsymbol{F}_b]\boldsymbol{w}_b)^T([\boldsymbol{F}_a]\boldsymbol{w}_a) + \mu_{ab}([\boldsymbol{\partial}\boldsymbol{\varphi}_b^\sigma]\boldsymbol{w}_b)^T([\boldsymbol{\partial}\boldsymbol{F}_a^\sigma]\boldsymbol{w}_a) - ([\boldsymbol{\partial\partial}\boldsymbol{\psi}_b^{\sigma\nu}]\boldsymbol{w}_b)^T([\boldsymbol{\partial\partial}\boldsymbol{F}_a^{\sigma\nu}]\boldsymbol{w}_a) \right],$$

where summation over repeated indices $\sigma$ and $\nu$ is assumed, $\boldsymbol{w}_a$ is a vector containing the weights, $\boldsymbol{F}_a$ is a matrix with components $(\boldsymbol{F}_a)_{ij} = F_a^j(\mathbf{v}_i)$, the derivative matrices in Cartesian 3D velocity space have the components

$$[\boldsymbol{\partial\varphi}_b^\sigma]_{ij} = \frac{\partial \varphi_b^j(\mathbf{v}_i)}{\partial v^\sigma}, \qquad [\boldsymbol{\partial\partial\psi}_b^{\sigma\nu}]_{ij} = \frac{\partial^2 \psi_b^j(\mathbf{v}_i)}{\partial v^\sigma \partial v^\nu},$$

and similarly for the distribution function. Later we give the expressions in cylindrical coordinates. The upper index $^T$ denotes a matrix transpose. Thus, instead of storing a full rank-3 tensor to evaluate the right hand side in Eq. (2), it is enough to compute and store the matrices $[\boldsymbol{F}_a]$, $[\boldsymbol{\partial\varphi}_a^\sigma]$, $[\boldsymbol{\partial F}_a^\sigma]$, $[\boldsymbol{\partial\partial\psi}_a^{\sigma\nu}]$, and $[\boldsymbol{\partial\partial F}_a^{\sigma\nu}]$ for each of the plasma species of interest.

In the center collocation method, the matrix sizes are $N \times N$ for $N$ basis functions, and the discretized equation for the weights can then be expressed in a matrix form as

$$[\boldsymbol{F}_a]\frac{\partial \boldsymbol{w}_a}{\partial t} = \boldsymbol{C}(\boldsymbol{w}_a, \boldsymbol{w}_b).$$

In the Galerkin projection, one has to compute the integrals over the velocity space but, as it has to be done numerically, it can be formulated into a similar form

$$[\boldsymbol{M}_a]\frac{\partial \boldsymbol{w}_a}{\partial t} = \boldsymbol{K}(\boldsymbol{w}_a, \boldsymbol{w}_b),$$

where the mass matrix and the integral of the collision operator are defined by

$$[\boldsymbol{M}_a]_{ij} = \sum_\ell b_\ell F_a^i(\mathbf{v}_\ell) F_a^j(\mathbf{v}_\ell), \qquad \boldsymbol{K}_i(\boldsymbol{w}_a, \boldsymbol{w}_b) = \sum_\ell b_\ell F_a^i(\mathbf{v}_\ell) C_\ell(\boldsymbol{w}_a, \boldsymbol{w}_b),$$

and $b_\ell$ are the integration weights. Using the described method one avoids storing a full rank-3 tensor.

**Velocity space axisymmetry** Using cylindrical velocity space coordinates $(v_\parallel, v_\perp)$ the relative magnitude of two vectors $\mathbf{v}$ and $\mathbf{u}$ is $|\mathbf{v}-\mathbf{u}| = \sqrt{(v_\parallel - u_\parallel)^2 + v_\perp^2 + u_\perp^2 - 2v_\perp u_\perp \cos(\theta_v - \theta_u)}$ from which it is easy to observe that $(\partial_{\theta_v} + \partial_{\theta_u})|\mathbf{v}-\mathbf{u}| = 0$, and then further show that the Rosenbluth potentials become independent of the cylindrical angle $\theta_v$ when the 3D potentials are integrated over the angle $\theta_u$. In order to compute the axisymmetric collision operator in cylindrical velocity-space coordinates, we thus need the expressions

$$\begin{aligned}
\frac{\partial \varphi_b^k}{\partial \mathbf{v}} \cdot \frac{\partial F_a^\ell}{\partial \mathbf{v}} &= \frac{\partial \varphi_b^k}{\partial v_\parallel}\frac{\partial F_a^\ell}{\partial v_\parallel} + \frac{\partial \varphi_b^k}{\partial v_\perp}\frac{\partial F_a^\ell}{\partial v_\perp}, \\
\frac{\partial^2 \psi_b^k}{\partial \mathbf{v}\partial \mathbf{v}} : \frac{\partial^2 F_a^\ell}{\partial \mathbf{v}\partial \mathbf{v}} &= \frac{\partial^2 \psi_b^k}{\partial v_\perp^2}\frac{\partial^2 F_a^\ell}{\partial v_\perp^2} + 2\frac{\partial^2 \psi_b^k}{\partial v_\perp \partial v_\parallel}\frac{\partial^2 F_a^\ell}{\partial v_\perp \partial v_\parallel} + \frac{1}{v_\perp^2}\frac{\psi_b^k}{\partial v_\perp}\frac{\partial F_a^\ell}{\partial v_\perp} + \frac{\partial^2 \psi_b^k}{\partial v_\parallel^2}\frac{\partial^2 F_a^\ell}{\partial v_\parallel^2}.
\end{aligned}$$

As these expressions are also summations of products of two functions the equation for the weights is obtained in the same manner as was described above.

Derivatives of the distribution function are simple to calculate analytically and the derivatives of the potential functions $\psi_i$ become

$$\frac{\partial^2 \psi_i}{\partial v_\parallel^2} = -\frac{\gamma_i^{1/2}}{8\pi}\left\langle \frac{\Psi'(\gamma_i^{1/2}|\mathbf{v}-\mathbf{v}_i|)}{\gamma_i^{1/2}|\mathbf{v}-\mathbf{v}_i|} + \gamma_i(v_\parallel - v_{i,\parallel})^2\left(\frac{\Psi''(\gamma_i^{1/2}|\mathbf{v}-\mathbf{v}_i|)}{\gamma_i|\mathbf{v}-\mathbf{v}_i|^2} - \frac{\Psi'(\gamma_i^{1/2}|\mathbf{v}-\mathbf{v}_i|)}{\gamma_i^{3/2}|\mathbf{v}-\mathbf{v}_i|^3}\right)\right\rangle_{\theta_i},$$

$$\frac{\partial \psi_i}{\partial v_\perp} = -\frac{1}{8\pi}\left\langle \gamma_i^{1/2}\left[v_\perp - v_{i,\perp}\cos(\theta-\theta_i)\right]\frac{\Psi'(\gamma_i^{1/2}|\mathbf{v}-\mathbf{v}_i|)}{\gamma_i^{1/2}|\mathbf{v}-\mathbf{v}_i|}\right\rangle_{\theta_i},$$

$$\frac{\partial^2 \psi_i}{\partial v_\perp^2} = -\frac{\gamma_i^{1/2}}{8\pi}\left\langle \frac{\Psi'(\gamma_i^{1/2}|\mathbf{v}-\mathbf{v}_i|)}{\gamma_i^{1/2}|\mathbf{v}-\mathbf{v}_i|} + \gamma_i(v_\perp - v_{i,\perp}\cos(\theta-\theta_i))^2\left[\frac{\Psi''(\gamma_i^{1/2}|\mathbf{v}-\mathbf{v}_i|)}{\gamma_i|\mathbf{v}-\mathbf{v}_i|^2} - \frac{\Psi'(\gamma_i^{1/2}|\mathbf{v}-\mathbf{v}_i|)}{\gamma_i^{3/2}|\mathbf{v}-\mathbf{v}_i|^3}\right]\right\rangle_{\theta_i},$$

$$\frac{\partial^2 \psi_i}{\partial v_\perp \partial v_\parallel} = -\frac{\gamma_i^{1/2}}{8\pi}\left\langle \gamma_i(v_\parallel - v_{i,\parallel})\left[v_\perp - v_{i,\perp}\cos(\theta-\theta_i)\right]\left[\frac{\Psi''(\gamma_i^{1/2}|\mathbf{v}-\mathbf{v}_i|)}{\gamma_i|\mathbf{v}-\mathbf{v}_i|^2} - \frac{\Psi'(\gamma_i^{1/2}|\mathbf{v}-\mathbf{v}_i|)}{\gamma_i^{3/2}|\mathbf{v}-\mathbf{v}_i|^3}\right]\right\rangle_{\theta_i},$$

where $\langle\cdot\rangle_\theta$ refers to integration over the angle $\theta$. Similarly one would calculate the derivatives for the potential $\varphi_i$. For the results presented in Ref. [2] these integrals were evaluated numerically using the Simpson's rule which is a particularly good method for periodic integrands. High level of accuracy can be obtained already with 100 integration points, as the integrands are periodic with respect to the angle $\theta$.

**Summary** We have described in detail the numerical implementation of the Gaussian radial basis function method to discretize the Fokker-Planck collision operator in both full 3D and in axisymmetric 2D velocity space. Especially, we focused explicitly on decomposing the rank-3 tensor that appears in the collision operator into a form that only requires storing rank-2 matrices. This approach saves both memory and computation time. In addition, we gave explicit formulas for evaluating the axisymmetric velocity space derivatives of the Rosenbluth potentials that are needed for the axisymmetric implementation.

**Acknowledgments** We are grateful for the encouraging comments and insightful discussions with Prof. Tünde Fülöp, Dr. Istvan Pusztai, and Mr. Adam Stahl. Especially, without the help from Prof. Fülöp, the collaboration between Chalmers University of Technology and General Atomics would not have been possible. Funding for the collaboration was received also from the Chalmers Foundation.

## References

[1] M. N. Rosenbluth, W. M. MacDonald, and D. L. Judd. Fokker-Planck Equation for an Inverse-Square Force. *Phys. Rev.*, 107:1–6, Jul 1957.

[2] E. Hirvijoki, J. Candy, E. Belli, and O. Embréus. The Gaussian Radial Basis Function Method for Plasma Kinetic Theory. *arXiv:1503.06312*, 2015.

[3] J. Park and I. W. Sandberg. Universal approximation using radial-basis-function networks. *Neural Comput.*, 3(2):246–257, June 1991.

# The influence of electron-electron scattering on the asymptotic behaviour of the strongly degenerate plasma electrical conductivity

V.S. Karakhtanov

*Institute of Problems of Chemical Physics, Russian Academy of Sciences, Chernogolovka, Russia*

The dc conductivity of fully ionized plasma in the strongly degenerate limit is investigated using the linear response theory. An analytical expression for the temperature dependence of the difference between conductivities of plasma with and without electron-electron scattering is obtained. It is found that this difference is proportional to the fourth power of the degeneracy parameter, if we neglect the ion-ion correlations. Accounting for ion-ion correlations in the approximation of the Debye-Hückel type leads to the third power of dependency above. The result is applicable in constructing interpolation formulas for calculating the plasma conductivity in the wide temperature and density regions on the basis of the Lorentz model.

## Introduction

The temperature behaviour of the electron-electron scattering influence on plasma transport properties is interesting both for the description of experiments and the construction of interpolation formulas for the conductivity [1, 2, 3]. The electron-electron scattering makes a significant contribution to the conductivity of low-density non-degenerate plasma [4], where the systematic treatment of it with the dynamical character of dielectric screening is possible on the base of linear response theory [5]. For technical details and asymptotic behaviour of conductivity of low-density non-degenerate plasma see [6]. In strongly degenerate electron systems the electron-ion scattering is dominated and the conductivity is well defined without the electron-electron collisions account [7, 8]. Some works were aimed at the constructions of interpolation formulas, linking the real plasma conductivity with the Lorentz one by introducing the correction factor between them. In particular, two different correction factors were introduced in [9] and [10]. In order to create more suitable correction factor and interpolation formula for conductivity, it would be useful to investigate the possibilities to determine an asymptotic behaviour of it in both limits on degeneracy. In the present work we consider the limit of high degeneracy.

## Basic approximations

Following [11], consider a neutral two-component plasma consisting of free singly charged particles with charges $e_i$ (ions) and $e_e$ (electrons, $e_i = -e_e = e$) at temperature $T$ and density $n = n_{\rm e} = n_{\rm i}$ in the adiabatic limit (the electronic mass is much less than the ionic mass), interacting via Coulomb forces. Introduce degeneracy parameter $\Theta = (2m_e k_B T/\hbar^2)(3\pi^2 n_e)^{-2/3}$ and the electron-ion coupling constant $\Gamma = (e^2/4\pi\varepsilon_0 k_B T)(4\pi n_e/3)^{1/3}$. For the strongly degenerate plasma $\Theta \ll 1$. Within the linear response theory in the formulation of Zubarev [5], the transport properties are expressed via force-force correlation functions [12, 13, 14]. The resulting expression for conductivity is

$$\sigma = -\frac{e^2}{\Omega \det(d)} \begin{vmatrix} 0 & N_0 \\ \overline{N_0} & d \end{vmatrix}, \tag{1}$$

$$N_n = \begin{pmatrix} N_{n0} & N_{n1} & \dots & N_{nl} \end{pmatrix}, \tag{2}$$

$$\overline{N_n} = \begin{pmatrix} N_{n0} \\ N_{n1} \\ \vdots \\ N_{nl} \end{pmatrix}, d = \begin{pmatrix} d_{00} & d_{01} & \dots & d_{0l} \\ d_{10} & d_{11} & \dots & d_{1l} \\ \vdots & \vdots & \ddots & \vdots \\ d_{l0} & d_{l1} & \dots & d_{ll} \end{pmatrix}. \tag{3}$$

In (1)-(3) $\Omega$ - the system volume, $N_{mn}, d_{mn}$ are correlation functions for the thermodynamic equilibrium, $N_e$ - the number of electrons and $\beta = (k_B T)^{-1}$ . The dimension of the matrix $d$ coincides with the number of moments in the corresponding relevant statistical operator (see [5]). In the adiabatic limit we can omit the ion flux [15] and obtain for Eq.(3)

$$d_{mn} = d_{mn}^{ei} + d_{mn}^{ee}, \tag{4}$$

$$N_{mn} = N_e \frac{\Gamma(m+n+5/2)}{\Gamma(5/2)} \frac{I_{m+n+1/2}(\beta\mu_e^{id})}{I_{1/2}(\beta\mu_e^{id})}, \tag{5}$$

with $I_\nu(y)$ - the Fermi integrals, $\mu_e^{id}$- the ideal part of the electronic chemical potential.

The correlation functions $d_{mn}$ are evaluated using thermodynamic Green's functions. As in [11], we restrict ourselves to $l = 1$ in (2), (3).

In the adiabatic limit

$$d_{mn}^{ei} = \frac{4m_e^2}{3\pi^2\beta^2\hbar^3} \int_0^\infty dx x^{n+m+2} f_k^e (1 - f_k^e) Q_{ei}(x), \tag{6}$$

$$Q_{ei}(x) = \frac{\beta^2\Omega^2}{16\pi x^2} \int_0^{2k} \left| \frac{V(q)}{\varepsilon_e(q,0)} \right|^2 S_{ii}(q) q^3 dq, \tag{7}$$

$$\varepsilon_e(q,0) = 1 + \Omega V(q)(1 - G_e(q))\chi_e^{(0)}(q,0), \tag{8}$$

with $x = \frac{\beta\hbar^2 k^2}{2m_e}$, $Q_{ei}(x)$ - the transport cross-section for electron-ion scattering, $\varepsilon_e(q,0)$ - the effective static electronic dielectric function, $S_{ii}(q)$ - the ion-ion structure factor, $\chi_e^{(0)}(q,\omega)$ the free-electron polarizability, $G_e(q)$ - the static electronic local field correction.

The correlation function $d^{ei}_{mn}$ for the electron-electron scattering in the strongly degenerate limit was evaluated in [11]:

$$d^{ee}_{11} = 17.3\Theta^{7/2}k_F d \int_0^{2k_F} \frac{dq}{q^2\varepsilon_e^2(q,0)}, \tag{9}$$

where $d = \frac{8}{3}\frac{m_e^{1/2}e^4N_e^2\beta^{3/2}}{\Omega(4\pi\varepsilon_0)^2}$.

Representing $d^{ei}_{m,n-m} = \int_0^\infty f(E)d((\beta E)^n U_{ei}(E))$, we obtain

$$d^{ei}_{00} = U_{ei}(E_F) + \frac{\pi^2}{6}T^2U''_{ei}(E_F) + \frac{7\pi^4}{360}T^4U^{IV}_{ei}(E_F) + \ldots, \tag{10}$$

$$d^{ei}_{01} = \Theta^{-1}(d^{ei}_{00} + \delta_1\Theta^2 + \delta_2\Theta^4 + \ldots), \tag{11}$$

$$d^{ei}_{11} = \Theta^{-2}(d^{ei}_{00} + \varepsilon_1\Theta^2 + \varepsilon_2\Theta^4 + \ldots), \tag{12}$$

$$N_{00} = N_e, N_{01} = N_e\Theta^{-1}(1 + \alpha_1\Theta^2 + \alpha_2\Theta^4 + \ldots), \tag{13}$$

where $E_F$ is the Fermi energy.

For $d^{ee}_{11}$ we suggest $\varepsilon_{ei} = d^{ee}_{11}/d^{ei}_{11} = o(\Theta^2)$. As the analysis shows, $\varepsilon_{ei} \sim \Theta^4$ for $S_{ii} = 1$, $\varepsilon_{ei} \sim \Theta^3$ for the Debye-Hückel model, and has a tendency to $\Theta^3 ln\Theta$ for HNC procedure.

**Results and discussion**

Substituting all the expansions into (1), we obtain:

$$\sigma = \sigma_{Lorentz}\left(1 - \varepsilon_{ei}\left(\frac{E_F U'_{ei}(E_F)}{U_{ei}(E_F)} - \frac{3}{2}\right)^2\right). \tag{14}$$

Compare the result obtained with the approximations:

$$\sigma_S = \sigma_{Lorentz}(1 - 0.0125\Theta), \tag{15}$$

$$\sigma_F = \sigma_{Lorentz}\left(1 - 0.209\Theta^2\right) \tag{16}$$

of [9] and [10], correspondingly.

For $\varepsilon_e(q,0) = 1 + \frac{\kappa^2}{q^2}$ (Thomas-Fermi approximation for the electronic dielectric function, $\kappa^2 = \frac{4k_F}{\pi a_0}$, $k_F$ - Fermi wave number, $a_0$ - Bohr radius)

$$d^{ee}_{11} = 8.65\Theta^{7/2}d\left(\sqrt{k_s}\arctan\sqrt{k_s} - \frac{k_s}{1+k_s}\right), \tag{17}$$

where $k_s = \frac{4k_F^2}{\kappa^2} = \pi k_F a_0 = \frac{6.029}{r_s}$, $(a_0 r_s)^3 = \frac{3}{4\pi n_e}$.

If $S_{ii}(q) = 1$,

$$d^{ei}_{00} = \frac{3\sqrt{\pi}}{8}\Theta^{3/2}d\left(ln(1+k_s) - \frac{k_s}{1+k_s}\right), \tag{18}$$

and $\varepsilon_{ei} \sim \Theta^4$.

More realistic structure factor is obtained from linearization of HNC procedure, or equivalent contour integral treatment of electron-ion collisions dynamical screening in adiabatic approximation [16]. For the Thomas-Fermi dielectric function (Debye-Hückel model)

$$S_{ii}(q) = \frac{q^2+\kappa^2}{q^2+\kappa^2\left(1+\frac{2}{3\Theta}\right)}. \tag{19}$$

This structure factor approximates well the results of HNC procedure for $q$ up to $2k_F$ and $\Gamma \sim 1$, and it can be used for estimates up to $r_s \sim 2\Theta$.

In this approximation we obtain

$$\varepsilon_{ei} = 17.35\Theta^3 \frac{\sqrt{k_s}\arctan\sqrt{k_s} - \frac{k_s}{1+k_s}}{k_s - ln(1+k_s)}, \frac{E_F U'_{ei}(E_F)}{U_{ei}(E_F)} = \frac{k_s^2}{(1+k_s)(k_s - ln(1+k_s))}. \tag{20}$$

Further analysis can be carried out using the numerical procedures for determining the structure factor. HNC procedure shows the tendency to $\Theta^3 ln\Theta$ for $\varepsilon_{ei}$ behaviour. The value $U'_{ei}(E_F)$ is very sensitive to the details of $S_{ii}(q)$ and needs further research.

Unlike (15) and (16), correction factors to Lorentz plasma conductivity show other behaviour on $\Theta$ and are dependent on electronic density. These results should be considered in further constructions of interpolation formulas for the conductivity

## References

[1] A. Esser, and G. Röpke, Phys. Rev. E **58**, 2446 (1998)

[2] A. Esser, R. Redmer, G. Röpke, Contrib. Plasma Phys. **43**, 33 (2003)

[3] S. Ichimaru, S. Tanaka, Phys. Rev. A **32**, 1790 (1985)

[4] L. Spitzer and R. Härm, Phys. Rev. **89**, 977 (1953)

[5] D. N. Zubarev, *Nonequilibrium Statistical Mechanics* (Plenum, New York, 1974)

[6] V. S. Karakhtanov, R. Redmer, H. Reinholz, and G. Röpke, Contrib. Plasma Phys. **53**, 639 (2013)

[7] J. M. Ziman, Philos. Mag. **6**, 1013 (1961)

[8] T. E. Faber, *Introductions to the Theory of Liquid Metals* (Cambridge University Press, Cambridge, England, 1974)

[9] W. Stygar, G. Gerdin, and D. Fehl, Phys. Rev. E **66**, 046417 (2002)

[10] V. E. Fortov, V. Ya. Ternovoi, M. V. Zhernokletov, M. A. Mochalov, A. L. Mikhailov, A. S. Filimonov, A. A Pyalling, V. B. Mintsev, V. K. Gryaznov, and I. I. Iosilevski , JETP **97**, 259 (2003)

[11] V. S. Karakhtanov, J. Phys: Conf. Series **529**, 012017 (2014)

[12] C.-V. Meister and G. Röpke, Ann. Phys. A **494** 133 (1982)

[13] G. Röpke, Physica A **121**, 92 (1983)

[14] G. Röpke, Phys. Rev. A **38**, 3001 (1988)

[15] H. Reinholz, R. Redmer, and D. Tamme, Contrib. Plasma Phys. **29**, 395 (1989)

[16] V. S. Karakhtanov, R. Redmer, H. Reinholz, and G. Röpke, Contrib. Plasma Phys. **51**, 355 (2011)

# Bohm-criterion approximation versus optimal matched solution for a spherical probe in radial-motion theory

Siegbert Kuhn[1], Alif Din[2]

[1] *Institute for Theoretical Physics, University of Innsbruck, Technikerstrasse 21A, A-6020 Innsbruck, Austria*

[2] *Theoretical Physics Division, PINSTECH, P. O. Nilore, 44000, Islamabad, Pakistan*

**Abstract.** It is demonstrated that in the large-negative-bias regime the probe characteristics obtained for the correct potential distribution ("M solution", cf. A. Din and S. Kuhn, Phys. Plasmas **21**, 103509 (2014)) differs substantially from those obtained by application of the Bohm criterion. This raises questions about the correctness of probe theories relying on merely applying the Bohm criterion.

**1. Introduction and overview.** The theory of positive-ion collection by a spherical probe immersed in a low-pressure plasma was reviewed and extended by Allen et al. [1]. Numerical extensions were given later by Allen and Turrin [2] and Chen [3]. In all of these treatments, a nonneutral solution (governed by the Poisson equation, potential profile $V_{nn}^{(m)}(r;r_m)$) was matched to the quasineutral solution (governed by the quasineutrality condition, potential profile $V_{qn}(r)$) at some "matching" radius $r_m$, resulting in the "matched" global solution ("M solution", potential profile $V^{(m)}(r;r_m)$). An efficient numerical procedure for systematically determining the optimal matching radius $r_{mo}$ was presented recently by Din and Kuhn [4]. There it was also demonstrated explicitly that the optimal M solution (i.e., the M solution with $r_m = r_{mo}$) differs substantially from the "B solution" obtained by merely applying the Bohm criterion, which holds at $r = r_{qn,s}$, the singularity point of the quasineutral solution.

In the present work we extend this comparison by comparing the current-voltage probe characteristics ensuing from the B solution and the optimal $(r_m = r_{mo})$ M solution.

**2. Comparison of the M and B solutions; applicability of the Bohm-criterion approach**

A frequently used method for approximately solving plasma-wall transition problems is the "Bohm-criterion approach" [5], [6], in which the quasineutral or presheath solution is used all the way down to the singularity point, where the Bohm criterion is satisfied and the sheath (which on the presheath scale is infinitesimally thin) is located, cf. Fig. 1(b) of [6]. Let us now discuss how, for our spherical probe problem, the approximate solution following from this Bohm-criterion approach ("B solution") compares with our correct solution based on the matching approach ("M solution").

Let us start out from the fairly general quasineutrality condition presented in eq. (2.17) of [6], which in our case (spherical symmetry, collisionless cold ions, Boltzmann-distributed electrons) reads

$$\left[m^i\left(u^i\right)^2 - kT^e\right]\frac{dn^e}{dr} = \frac{-2m^i\left(u^i\right)^2 n^e}{r}. \tag{1}$$

Since the right-hand side vanishes nowhere, the singularity $dn^e/dr \to -\infty$ clearly occurs where the "cold-ion Bohm criterion"

$$\left|u^i\right| = \sqrt{\frac{kT^e}{m^i}} \tag{2}$$

is satisfied. Applying the normalization scheme $\tilde{Q} = Q/Q_{ref}$ with the reference quantities

$$V_{ref} = kT^e/e,\ \ n_{ref} = n_0,\ \ r_{ref} = \lambda^e_{D0} = \sqrt{\varepsilon_0 kT^e/\left(n_0 e^2\right)},\ \ E_{ref} = \sqrt{n_0 kT^e/\varepsilon_0},$$

$$v_{ref} = c_s := \sqrt{kT^e/m^i},\ \ I_{ref} = c_s := 4\pi\left(\lambda^e_{D0}\right)^2 e n_0\sqrt{2kT^e/m^i}, \tag{3}$$

the normalized forms of eqs. (1) and (2) read

$$\left[\left(\tilde{u}^i\right)^2 - 1\right]\frac{d\tilde{n}^e}{d\tilde{r}} = \frac{-2\left(\tilde{u}^i\right)^2 \tilde{n}^e}{\tilde{r}} \tag{4}$$

and

$$\left|\tilde{u}^i\right| - 1, \tag{5}$$

respectively.

In Fig. 1 (taken from [4]) we compare the B and M solutions for $\tilde{I}^i$ = -200. The B solution looks qualitatively like the solution in Fig. 1(b) of [6] and is totally different from our (numerically approximate but conceptually correct) optimal $\left(\tilde{r}_m = \tilde{r}_{mo}\right)$ M solution. In particular, on the presheath scale shown here the sheath of the B solution is infinitesimally thin and for $\tilde{r}_p < \tilde{r} < \tilde{r}_{qn,s}$ cannot reach the probe surface, whereas the nonneutral ("sheath") region of our M solution obviously is of finite width, extending all the way down to the probe surface.

In summary, the B solution turns out to be visibly wrong in our case: The Bohm-criterion approach is absolutely inapplicable for $\tilde{r}_p < \tilde{r} < \tilde{r}_{qn,s}$ and yields an approximation to the correct solution which is extremely poor for $\tilde{r}$ slightly above $\tilde{r}_p$ and tends towards the correct solution only for. $\tilde{r} \to \infty$. The fundamental reason for this failure was discussed in Ref. [4].

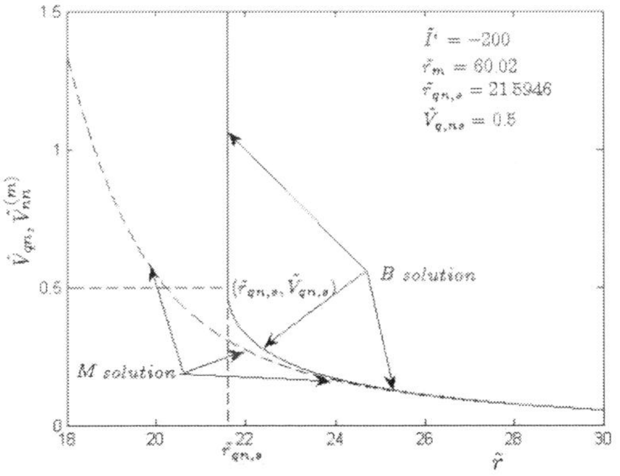


Figure 1: Comparison of potential profiles resulting from our optimal matching approach (M solution) and the Bohm-criterion approach (B solution) [4].

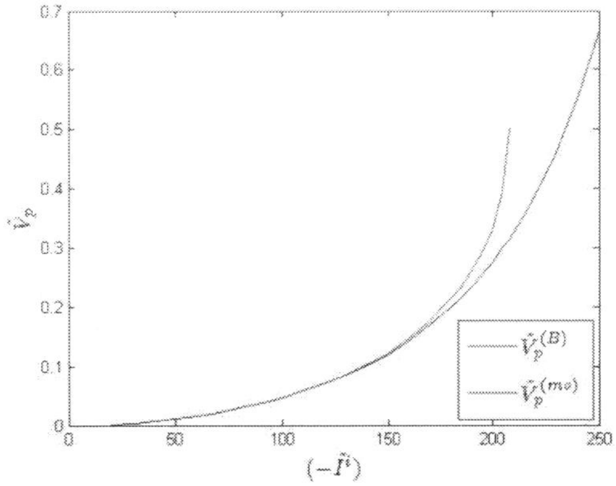

Figure 2: Comparison of the $\tilde{V}_p vs.\left(-\tilde{I}^i\right)$ probe characteristics following from the B solution (upper curve) and the optimal M solution for probe radius $\tilde{r}_p = 22$.

## 3. Comparison of the $\tilde{V}_p vs.\left(-\tilde{I}^i\right)$ probe characteristics for the M and B solutions

From potential distributions $\tilde{V}(\tilde{r})$ obtained for various values of $\tilde{I}^i$ (as exemplified by Fig. 1) we can extract the $\tilde{V}_p vs.\left(-\tilde{I}^i\right)$ probe characteristics in the large-negative-bias regime for a given value of the probe radius $\tilde{r}_p$ (as exemplified by Fig. 2 for $\tilde{r}_p = 22$) as follows. As a representative example consider Fig. 1, which shows the B and optimal M solutions for $\tilde{I}^i =$ = -200, in which case the singularity of the quasineutral solution occurs at $\tilde{r}_{qn,s}$ = 21,594 and the optimal matching radius has been found to be $\tilde{r}_{mo}$ = 60:02. For $\tilde{r}_p < \tilde{r} < \tilde{r}_{qn,s}$ the B solution is inexistent and hence yields no value of the probe potential at all, whereas the M solution yields some value of $\tilde{V}_p$ p for any chosen value of $\tilde{r}_p$: In the region of existence of the B solution, $\tilde{r}_{qn,s} \le \tilde{r}_p < \tilde{r}$, the potential distribution corresponding to the B solution always lies above the one corresponding to the Mo solution and tends towards the latter as $\tilde{r} \to \infty$. From this kind of reasoning we obtain characteristics as exemplified in Fig. 2 for . $\tilde{r}_p$ = 22; which (together with the potential distributions exemplified by Fig. 1) show that the Bohm-criterion approximation is absolutely inadequate for $\tilde{r}_p < \tilde{r} < \tilde{r}_{qn,s}$ and very poor for $\tilde{r}$ slightly above but near the singularity point, but becomes more and more acceptable with increasing $\tilde{r}_p$.

**4. Conclusion.** The fact that the (very approximate) Bohm-criterion approach leads to probe characteristics that differ strongly from the characteristics following from the (essentially correct) matching approach raises questions about the correctness of probe theories relying on merely applying the Bohm criterion. Further investigations into this problem will be conducted by the present authors.

**Acknowledgment.** This work was partially supported by the Pakistan Atomic Energy Commission (PAEC). The views and opinions expressed herein do not necessarily reflect those of the PAEC.

**References**

[1] JE Allen, R. L. F. Boyd, P. Reynolds, Proc. Phys. Soc. 70, 297 (1957).

[2] J. E. Allen and A. Turrin. Proc. Phys. Soc. 83, 177 (1964).

[3] F. F. Chen, J. Nucl. Energy 7, 47 (1965).

[4] A. Din and S. Kuhn, Phys. Plasmas 21, 103509 (2014).

[5] K.-U. Riemann, J. Phys. D: Appl. Phys. 24, 493 (1991).

[6] K.-U. Riemann, J. Tech. Phys. 41, 89 (2000).

# Energy partition in kinetic turbulent reconnection

R. Numata[1], N.F. Loureiro[2]

[1] *Graduate School of Simulation Studies, University of Hyogo, Kobe, Hyogo 650-0047, Japan*

[2] *Instituto de Plasmas e Fusão Nuclear, Instituto Superior Técnico, Universidade de Lisboa, 1049-001 Lisboa, Portugal*

## Introduction

Magnetic reconnection is a commonly observed fundamental process in both astrophysical and fusion plasmas. It allows topological change of magnetic field lines, and converts the free energy in the magnetic field into various forms of energy, such as bulk plasma flows, plasma heating, or non-thermal particle acceleration.

In weakly collisional plasmas, phase mixing processes caused by kinetic effects, such as Landau damping and finite Larmor radius (FLR) effects, create oscillatory structures in velocity space, which must eventually be regularized by collisions. Therefore, even if collisions are infrequent, energy dissipation and resulting plasma heating can be significant. This heating mechanism due to phase mixing has been demonstrated using both a reduced [1] and a fully gyrokinetic model [2]. It has also been found that, for high beta gyrokinetic plasmas, the current sheet becomes unstable, and the resultant plasmoid (secondary island) can significantly alter the efficiency of electron and ion heating. This finding suggests that more efficient energy conversion may take place in turbulent reconnection, which is likely to be present in natural environment.

In this paper, we present a preliminary study of plasma heating in turbulent magnetic reconnection in strongly magnetized plasmas.

## Ion and electron heating via phase mixing

In this section, we summarize the previous result of ion and electron heating due to phase mixing during magnetic reconnection by means of gyrokinetic simulations. (Full details are described in [2].)

We consider magnetic reconnection of strongly magnetized plasmas in a two-dimensional doubly periodic slab domain. We initialize the system by a tearing unstable magnetic field configuration. The equilibrium magnetic field profile is $\boldsymbol{B} = B_{z0}\hat{z} + B_y^{\mathrm{eq}}(x)\hat{y}$ ($B_{z0} \gg B_y^{\mathrm{eq}}$), where $B_{z0}$ is the background guide magnetic field and $B_y^{\mathrm{eq}}$ is the in-plane, reconnecting component, related to the parallel vector potential by $B_y^{\mathrm{eq}}(x) = \partial A_\parallel^{\mathrm{eq}}/\partial x$, and $A_\parallel^{\mathrm{eq}}(x) \propto A_{\parallel 0}^{\mathrm{eq}} \cosh^{-2}(x/a)$ ($A_{\parallel 0}^{\mathrm{eq}}$ is a constant and $a$ is the magnetic shear length). The maximum of $B_y^{\mathrm{eq}}$ is used to define the Alfvén time $\tau_A$, which characterizes the time scale of the system. We impose a sinusoidal perturbation to initiate magnetic reconnection. Simulations are performed in the collisionless tearing-mode regime where the frozen-flux condition is not broken by collisions.

During magnetic reconnection process, the initial magnetic energy is redistributed. To estimate plasma heating, we measure the collisional energy dissipation rate of species $s$, $D_s$. Without collisions, the gyrokinetic equation conserves the generalized energy $W$, and it is dissipated by collisions as $dW/dt = -\sum_s D_s$. The collisional dissipation increases the entropy, and is turned into heat [3].

Figure 1, left column shows the temporal evolution of the reconnection rate (top), energy components (middle), and dissipation rate (bottom) for a high-beta case ($\beta_e = 1$). Looking at the reconnection rate measured by the electric field at the center of the domain ($E_X$), we observe that magnetic reconnection occurs around $t/\tau_A = 30 \sim 40$, followed by a plasmoid formation ($t/\tau_A = 40$) and ejection ($t/\tau_A = 50$). (Note that, after the plasmoid formation, $E_X$ is not a proper measure of the reconnection rate since the center of the domain becomes an $O$ point.)

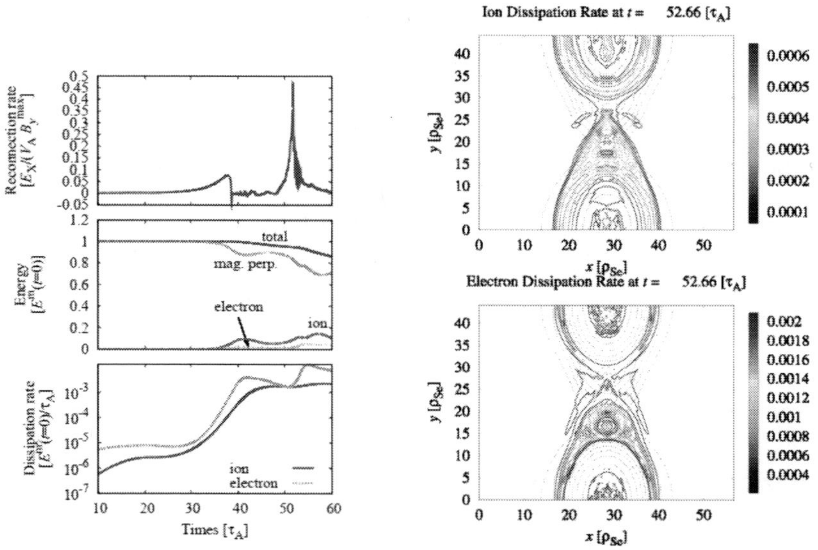


Figure 1: Left: Temporal evolution of reconnection rate, energy components, and dissipation rate. Right: spatial distribution of dissipation rate of ions and electrons at the plasmoid ejection. Length is normalized by the ion sound Larmor radius $\rho_{Se} = 0.177a$.

In the initial reconnection phase ($t/\tau_A = 30 \sim 40$), the energy dissipation rate gradually increases, and an appreciable amount of energy is converted into ion and electron heating. To detail the heating process, we plot the spatial distribution of the dissipation rate in Fig. 1, right column. The plots show the dissipation rate of ions and electrons at the plasmoid ejection phase. If we look at the distribution function at the place where strong dissipation is occurring, we see clear phase mixing structure in velocity space [2]. We also found that the energy dissipation of electrons are significantly enhanced at the newly formed $X$ point when a plasmoid is formed [2].

**Driven simulation**

To study efficient energy conversion by turbulence during magnetic reconnection, we drive turbulence in the system. We have implemented and tested two types of driving forces:

- Oscillating Langevin Antenna [4]

  The oscillating Langevin antenna is already implemented in `AstroGK`, which drives a current in the $z$ direction (perpendicular to the reconnection plane). It consists of the driven/damped oscillator and white noise:

$$\frac{\mathrm{d}A_{\parallel,\boldsymbol{k}_\mathrm{a}}}{\mathrm{d}t} = -\mathrm{i}\omega_\mathrm{a} A_{\parallel,\boldsymbol{k}_\mathrm{a}} + F_\mathrm{a}, \tag{1}$$

  where $\omega_\mathrm{a}$ is a complex frequency, $F_\mathrm{a}$ is a random number, and $\boldsymbol{k}_\mathrm{a}$ is the wave number of the mode that we wish to drive.

- Direct forcing

  We can also add direct driving force in the gyrokinetic equation evolving the non-Boltzmann part of the distribution function $h_s$,

$$\frac{\partial h_s}{\partial t} = \cdots + F f_{0s}, \tag{2}$$

  where $F$ is a random noise, and is independent of velocity, $f_{0s}$ is the Maxwellian distribution function. This forcing induces perturbations of density and electrostatic potential, thus results in driving perpendicular flows due to the $\boldsymbol{E}\times\boldsymbol{B}$ drift.

Figure 2, top panel shows the reconnection for the cases with i) the antenna drive with $\omega_\mathrm{a}\tau_A = 800$, $\boldsymbol{k}_\mathrm{a}\rho_\mathrm{Se} = (2.5/\sqrt{2}, 1.6/\sqrt{2})$ for $\beta_\mathrm{e} = 1, 0.01$, and ii) direct forcing for $\beta_\mathrm{e} = 0.01$. By the antenna with $\omega_\mathrm{a}\tau_A = 800$, a thin current layer is supported for long time, and higher reconnection rate is achieved. However, the current sheet structure shown in Fig. 2, bottom-left panel, is rather laminar, and turbulence seems not to be excited. The direct forcing, on the other hand, can efficiently drive turbulence as shown in Fig. 2, bottom-right panel.

**Conclusion**

Magnetic reconnection can efficiently convert energy to heat via phase mixing in strongly magnetized plasmas, as has been proved by gyrokinetic simulations. In this study, we have considered two different methods to drive turbulence in gyrokinetic magnetic reconnection simulation to investigate effects of turbulence on plasma heating. It seems the direct drive is advantageous because i) it can effectively drive turbulence in two-dimensional case, ii) it is directly

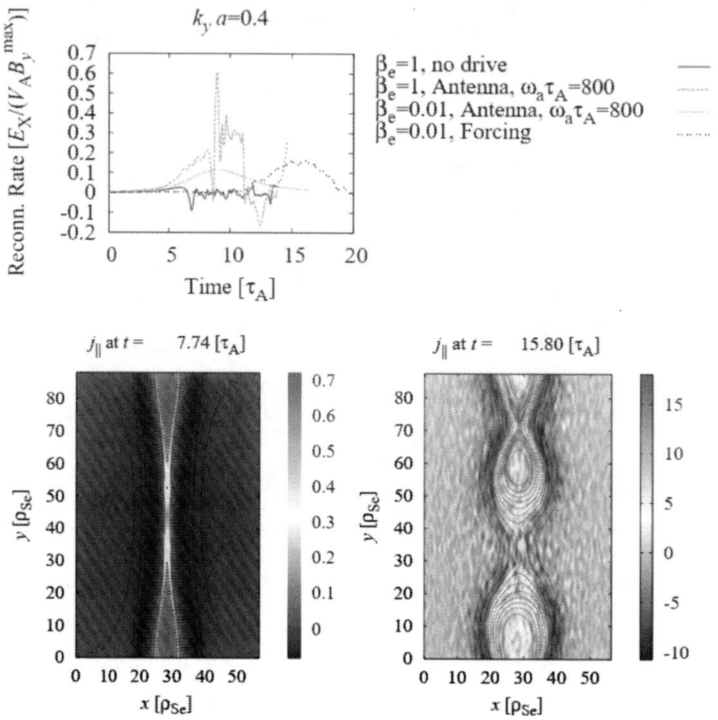


Figure 2: Top: Reconnection rate for the cases with various driving forces. Bottom: Current distribution for $\beta_e = 1$, $\omega_a \tau_A = 800$ case (Left) and $\beta_e = 0.01$, direct forcing case (Right).

comparable with the similar study in the framework of MHD [5]. The antenna drive may not be adequate for the current study since it directly changes reconnecting magnetic field. More thorough and systematic survey on the effects of external forcing will be the subject of a future publication.

## Acknowledgments

This work was supported by JSPS KAKENHI Grant Number 24740373. NFL was supported by grant No. IF/00530/2013 from Fundação para a Ciência e a Tecnologia. All simulations are performed using the HELIOS supercomputer system at IFERC-CSC.

## References


[1] N. F. Loureiro, A. A. Schekochihin, and A. Zocco, Phys. Rev. Lett. **111**, 025002 (2013).

[2] R. Numata and N. F. Loureiro, J. Plasma Phys. **81**, 305810201 (2014).

[3] G. G. Howes *et al.*, Astrophys. J. **651**, 590 (2006).

[4] J. M. TenBarge, *et al.*, Comput. Phys. Commun. **185**, 578 (2014).

[5] N. F. Loureiro, *et al.*, Mon. Not. R. Astron. Soc. **399**, L146 (2009).

# Interpretation of partially detached divertor operation with and without impurity seeding in JET with EDGE2D-EIRENE

A. E. Jaervinen[1], S. Brezinsek[2], C. Giroud[3], M. Groth[1], C. Guillemaut[3], P. Belo[3], G. Corrigan[3], P. Drewelow[4], D. Harting[2], A. Huber[2], K. Lawson[2], B. Lipschultz[5], C. Maggi[3], G.F. Matthews[3], A.G. Meigs[3], D. Moulton[1], M.F. Stamp[3], S. Wiesen[2], and JET Contributors[1]

*EUROfusion Consortium, JET, Culham Science Centre, Abingdon, OX14 3DB, UK.*

[1] *Aalto University, Espoo, Finland.*

[2] *Forschungszentrum Jülich GmbH, Institut für Energie- und Klimaforschung – Plasmaphysik, 52425 Jülich, Germany.*

[3] *CCFE, Culham Science Centre, Abingdon, OX13 3DB, UK.*

[4] *Max-Planck Institute for Plasma Physics, Garching, Germany.*

[5] *York Plasma Institute, University of York, Heslington, York YO10 5DD, UK.*

Radiative power exhaust with extrinsic impurities is necessary in ITER to maintain the divertor surface heat fluxes below 10 MW/m$^2$ while producing 500 MW of fusion power [1]. Reliable prediction of these scenarios for the next step devices relies on model validation with existing experimental facilities. To address these needs, partially detached divertor operation with $N_2$ and Ne injection in high-triangularity, high confinement mode JET plasmas with both strike points on vertical targets have been experimentally investigated at toroidal magnetic field of 2.7 T and plasma current of 2.5 MA with about 16 – 25 MW of NBI and 3 – 5 MW of ICRH heating [2, 3]. In this study, the divertor characteristics in these plasmas are further dissected and simulated with the multi-fluid code package EDGE2D-EIRENE [4, 5, 6]. The EDGE2D-EIRENE setup, including cross-field transport coefficients, is similar to that used in [7]. The simulation grid in this study corresponds to a high triangularity, vertical target configuration, whereas the grid in [7] corresponded to a high triangularity equilibrium with the low-field side (LFS) strike point on the horizontal target. Further details about the configurations can be found in [2]. Deuterium is injected into the computational domain at the high-field side (HFS) divertor into the common SOL with the puff rate adjusted to provide a LFS mid-plane separatrix electron density of 4.1e19 m$^{-3}$, at which the model captures the measured LFS divertor saturation currents and strike point electron temperatures in the unseeded plasmas at medium deuterium fuelling rates (about 2.2 – 2.6e22 electrons/s). As in the experiment, the deuterium puff level is not changed when increasing the impurity injection rate in the model. With this setup, the predicted LFS mid-plane electron density and

[1] *See the appendix of F.Romanelli et al., 25th IAEA FEC, St. Petersburg, Russia, 2014.*

temperature profiles are consistent with the JET high resolutions Thomson scattering measurements in the unseeded plasmas [8]. However, the scrape-off layer (SOL) densities further than 0.5 cm away from the separatrix are underestimated by a factor of 2 compared to the JET Lithium beam (Li-beam) measurements [9]. This is presumably caused by unaccounted SOL transport processes as well as underestimated deuterium ionization sources in the far SOL. These mechanisms are subject to further studies in forthcoming publications. In this contribution, overlapping scans with N and Ne injection, based on this unseeded reference plasma simulation, are conducted and compared to experimental observations. Both impurity species are assumed fully recycling.

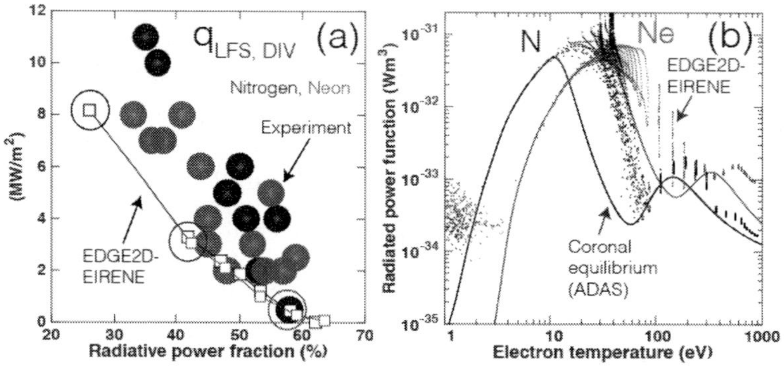


Figure 1. (a) Measured (solid circles) and predicted (connected hollow squares) LFS divertor peak heat flux as a function of radiated power fraction. Since the EDGE2D-EIRENE simulation grid includes only the pedestal and SOL regions, the radiated power fraction used as the x-axis for the simulations is calculated according to the formula $f_{rad}^{EDGE2D\text{-}EIRENE} = (P_{rad}^{EDGE2D\text{-}EIRENE} + 4\ MW)/18MW$, where 4 MW stands for the unaccounted core radiation contribution. The simulation cases representing the unseeded (black), medium-$N_2$/Ne (blue) and high-$N_2$/Ne (red) injection cases are circulated with color-coded circles b) The radiated power function for N (black), and Ne (red), as a function of electron temperature in coronal equilibrium (solid lines) and in EDGE2D-EIRENE simulations (dots) in the medium-$N_2$/Ne injection cases.

The simulations show that, changing only the radiating species while maintaining all the other input parameters unchanged, the LFS divertor peak heat flux, $q_{LFS,DIV}$, is reduced in similar fashion with either $N_2$ or Ne injection (fig. 1a). This result is qualitatively consistent with experimental observations and obtained in the simulations when adjusting the impurity injection rate to reproduce the experimentally measured levels of radiation. However, whereas the radiative potential of nitrogen peaks around electron temperatures of 10 – 40 eV, reducing to low values at temperatures above this region, the peaking region for neon extends from 20 eV up to 100 eV (fig. 1b). Furthermore, for electron temperatures inside the pedestal ($T_e > 500$ eV), the radiative potential of neon is about a factor of 5 – 10 higher than that of nitrogen. As a result, at impurity injection rates producing partially detached divertor conditions at the LFS divertor plate (the cases with the red circle in the fig. 1a), more than 90% of nitrogen radiation in the computational domain, which extends about 12 cm inside the separatrix at the LFS mid-plane, is predicted to occur in the divertor chamber below the vertical height of Z = -1.2 m, and more than 92% outside the separatrix. Therefore, nitrogen seeding is predicted to enable seeded, partially detached divertor operation in JET with a very minor reduction of the power crossing the separatrix, $P_{SOL}$. Increasing the nitrogen injection further in the simulations, leads to a transition to a radiation-condensation, X-point MARFE, formation inside the X-point. Once this formation occurs, the nitrogen radiation zone shifts inside the X-point and more than 50% of nitrogen radiation occurs inside the separatrix. However, due to numerical difficulties, these simulations do not reach a converged state. On the other hand, in partially detached divertor conditions with neon injection, less than 50% of radiation is predicted to occur below the vertical height of Z = -1.2 m, and less than 60% outside the separatrix. As a result, at neon injection rates required for substantial reduction of $q_{LFS,DIV}$, substantial reduction of $P_{SOL}$ is also predicted. These findings are qualitatively consistent with experimental observations, where in partially detached nitrogen seeded plasmas (JPN 85274) about 43% of the total radiated power is estimated occur below the vertical height of Z = -1.2m. In contrast, in partially detached neon seeded plasmas (JPN 87181) about 16% of the total radiated power is estimated to occur in this region. Estimating the radiation inside and outside the separatrix, the difference between two seeding species becomes smaller: 34% outside the separatrix with N, 30% with Ne. However, this type of analysis is susceptible to large error-bars due to spatial uncertainties of the bolometric reconstruction in the vicinity of the X-point. The exact divertor radiation fractions differ between the model and the experiment due to the missing core radiation contribution inside the pedestal in the EDGE2D-EIRENE simulations, due to spatial uncertainties in the experimentally estimated radiation distribution, as well as due to possible disagreement in the detailed radiation distribution between the model and the experiment. Nevertheless, a qualitative agreement between the experiment and the model is obtained, and overall these findings are also consistent with recent ITER prediction, conducted with the SOLPS code package, where the operational space with neon seeding is predicted to be more conservative than the operational space with nitrogen seeding, due to the proximity of $P_{SOL}$ in ITER to the H-to-L back transition threshold [10].

When adjusting the impurity injection rate to reproduce the measured radiated power, the simulations capture the experimentally observed particle and heat flux reduction at the LFS divertor plate (fig. 2 a,b). With both nitrogen and neon injection, the dominant particle flux reduction in the detachment assisted by impurities is caused by reduced deuterium ionization and recycling rate, which is possible due to reduced power reaching the deuterium ionization front. The recombination sink is less than 10% of the LFS ionization source in all of the simulated cases. However, the divertor $D_\alpha$-emission is underestimated by about a factor of 2 in the unseeded plasmas, and about a factor of 3 – 5 in the partially detached conditions, even though the LFS divertor particle fluxes are consistent with measured values. This indicates a shortfall in the deuterium radiation, which is presumably caused by underestimated recombination emission, unaccounted emission related to deuterium molecules, as well as due to other unaccounted $D_\alpha$-emission channels in the post processing routine in EDGE2D-EIRENE. While in the unseeded plasmas, the overall divertor radiation is underestimated by a factor of 2, with increasing seeding rate the overall measured divertor radiation levels are obtained with both nitrogen and neon. However, as this happens, it is very likely that the deuterium radiation contribution is underestimated throughout the impurity injection scan, as is also indicated by the underestimated deuterium alpha emission levels.

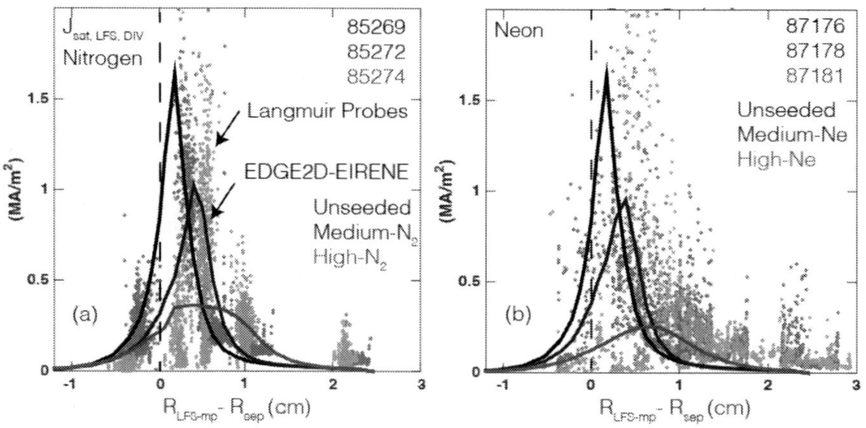


Figure 2. Measured (dots) and predicted (solid lines) LFS divertor saturation currents in the unseeded (black), medium seeding rate (blue), and high seeding rate (red) plasmas with nitrogen (a) and neon (b) injection. The illustrated simulation cases are circled in the figure 1a.

*This work has been carried out within the framework of the EUROfusion Consortium and has received funding from the Euratom research and training programme 2014-2018 under grant agreement No 633053. The views and opinions expressed herein do not necessarily reflect those of the European Commision.*

[1] PIPB editors, *Nucl. Fusion* **47** (2007)
[2] C. Giroud, *et al. Plasma Phys. Cont. Fus.* **57** (2015) 035004
[3] C. Giroud, *et al. Proc. of the 25th IAEA FEC St. Petersburg Russia, October* 2014, EX/P5-25
[4] R. Simonini, *et al. Contrib. Plasma Phys.* **34** (1994) 368 – 373
[5] D. Reiter, *et al. Jour. Nucl. Mater.* **196 – 198** (1992) 80 – 89
[6] S. Wiesen, *EDGE2D/EIRENE code interface report, JET ITC-Report,* http://www.eirene.de/e2deir_report_30jun06.pdf, *2006*
[7] A. E. Jaervinen, *et al. Jour. Nucl. Mater. in Press* (2014) doi:10.1016/j.jnucmat.2014.10.047
[8] M. N. A. Beurskens, *et al. Nucl. Fusion* **48** (2008) 095004.
[9] M. Brix, *et al. Rev. Sci. Instrum.* **83** (2012) 10D533.
[10] H. D. Pacher, *et al. Jour. Nucl. Mater. in Press* (2014) doi:10.1016/j.jnucmat.2014.11.104

# Improvement of neutron yield predictions in JET with ASCOT

T. Koskela[1], A. Snicker[1,7], S. Sipilä[1], A. Salmi[2], T. Johnson[3], C. Marchetto[4], M. Schneider[5] M. Romanelli[6], H. Weisen[8], and JET contributors*

*EUROfusion Consortium, JET, Culham Science Centre, Abingdon, OX14 3DB, UK*
*[1] Aalto University School of Science, Espoo, Finland; [2] VTT Technical Research Centre, Finland; [3] Department of Fusion Plasma Physics, School of Electrical Engineering, KTH Royal Institute of Technology, SE-10044 Stockholm, Sweden; [4] Instituto di Fisica del Plasma, CNR, Milano, Italy; [5] CEA, IRFM, F-13108 Saint Paul-lez-Durance, France; [6] CCFE, Culham Science Centre, Abingdon,OX14 3DB, UK; [7] Max-Planck-Institut für Plasma Physik, Boltzmannstr.2, 85748 Garching, Germany; [8] CRPP, Ecole Polytechnique Federale de Lausanne, CH-1015 Lausanne, Switzerland*

There is a large scatter between measured and predicted neutron yields in JET plasmas. The measured neutron yield in neutral beam (NBI)-heated plasmas often falls below predictions by commonly used simulation tools that assume neoclassical transport of fast ions. This is often referred to as a “neutron deficit” [1]. In radiofrequency (ICRH)-heated plasmas the opposite is often observed. Due to the acceleration of NBI ions by the ICRH, measured neutron yields tend to exceed predicted values [2]. In this contribution we present improvements to the Monte Carlo fast ion code ASCOT [3], aimed at tackling these deficiencies in modelling.

**Can fast ion redistribution due to NTMs explain the neutron deficit?** The neutron deficit can be mitigated by prescribing an ad-hoc anomalous diffusion fast ion coefficient in simulations [4]. This suggests additional fast ion transport is the cause of the deficit, but it does not explain the source of the transport, and has little predictive capability. A possible source of anomalous transport are long-lived MHD modes, such as neoclassical tearing modes (NTM)s. Earlier simulations with ASCOT [5], using realistic island parameters ($w = 10$ cm, $\Omega = 15$ kHz), derived from electron cyclotron emission (ECE) and Mirnov coil measurements, have shown that the redistribution of fast ions by a single stationary NTM reduces the predicted neutron yield but is not sufficient to explain the neutron deficit. This is illustrated in Figure 1a that shows the DD neutron rate from an interpretive JETTO/ASCOT simulation as a function of the width of a 3/2 island in JET discharge 77269 previously analyzed in [5]. In this work, we fix the temperature and density profiles to the ones measured by Thomson Scattering and focus only on the transport of fast ions. The profiles do not respond to changes in the NBI heating profile, for example. It is seen that even an unrealistically large island width is not enough to bring down the prediction to the measured level. We also include the rotation of the island, neglected in [5]. The result of a scan over the rotation frequency is shown in 1b. We find that the rotation of the island has a negligible effect on the predicted neutron yield. However, its effect in redistributing the fast ions can be clearly seen in the neutron profile, shown in Figure 1b. We can conclude that the local effect of a single island is not sufficient to significantly alter the total neutron yield. The temperature and density gradients of the background plasma are small due

*See the Appendix of F. Romanelli et al., Proceedings of the 25th IAEA FEC 2014, St. Petersburg, Russia

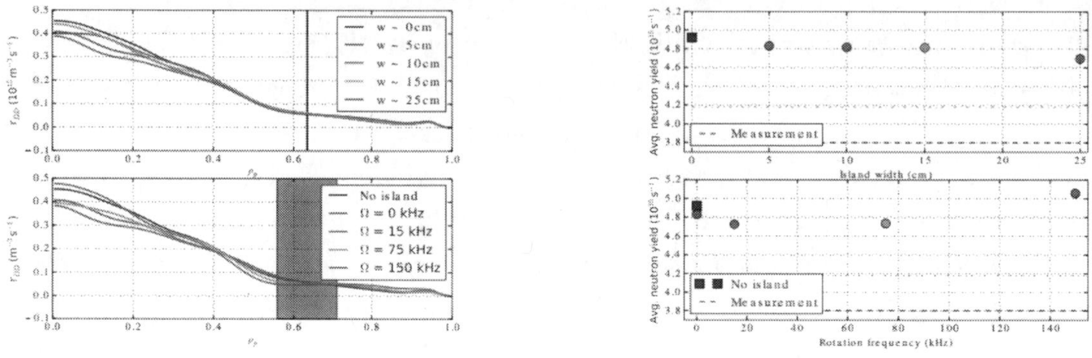


Figure 1: Scan of predicted (a) radial neutron profile and (b) volume integrated neutron rate in the width and rotation frequency of a 3/2 island. The experimentally measured width was approximately 10 cm and the frequency was approximately 15 kHz. The measured neutron rate is indicated by the dashed red line in (b).

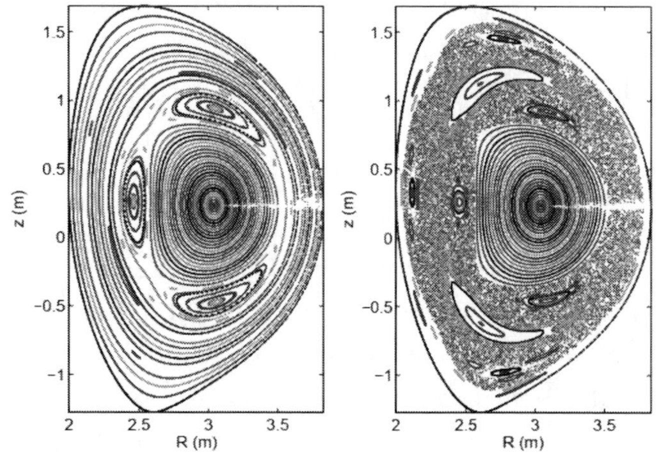


Figure 2: Poincare plots of magnetic field lines using (a) a signle 3/2 island and (b) a partially overlapping chain of three islands (3/2,2/1,3/1).

to the island and transporting fast ions across the island will not, therefore, significantly change their probablity of undergoing fusion reactions with the background. We therefore repeat the excercise with a string of 3/2, 2/1 and 3/1 islands connecting each other to create a channel for the fast ions to be transported from inside mid-radius to the top of the pedestal. A poincare plot of the magnetic geometry, obtained by field line tracing, is shown in Figure 2. It shows that the islands overlap in large parts of the plasma cross-section, creating stochastic field regions that would, in reality, lead to the termination of the discharge before developing to this stage. Therefore, these results most likely overestimate any realistic decrease in neutron yield due to magnetic islands. Even so, we find the total simulated neutron yield still overestimates the measurement. Figure 3 shows a comparison of simulated density, neutron and fast ion JxB torque profiles between simulations with no island, single island and three islands. The chain of three islands reduces the neutron deficit from 25 % to 18 %, and including rotation further reduces it

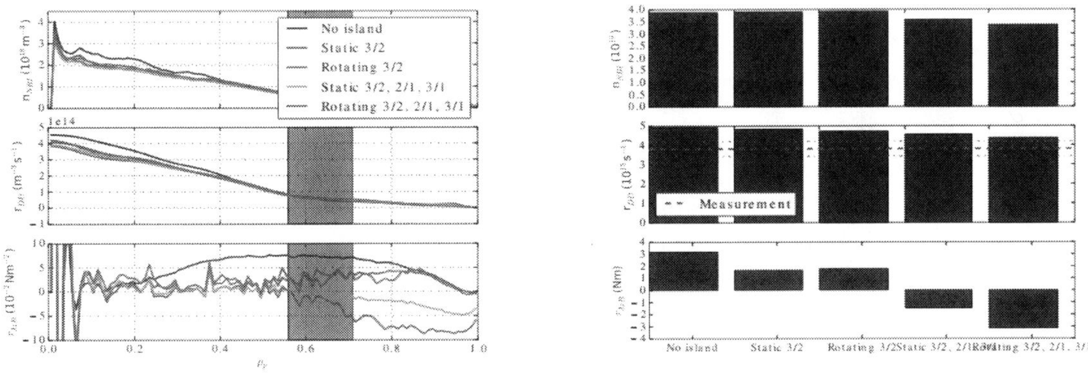


Figure 3: Scan with one or mutliple static and rotating islands. (a) profiles of beam ion density, DD neutron rate and beam JxB torque in simulations with static and rotating islands. (b) Volume-integrated values of the profiles. The measured neutron rate is indicated by the dashed red line in the middle panel.

to 13 %. However, even this extreme case still leaves half of the deficit unaccounted for. There is, however, a redistibution of fast ions outside mid-radius due to the islands that reverses the JxB torque of the NBI ions and induces a loss fraction of 10 %. In the extreme case of three islands, this reversal is strong enough to cancel the collisional component of the NBI torque.

**First results of modelling NBI/ICRH coupling with ASCOT/RFOF** The ASCOT code has been coupled to the RFOF library [6] to calculate RF heating by a Monte Carlo "kick" operator that resolves the acceleration of both thermal and NBI ions. The advantage of a Monte Carlo method is that it can include any number of particle species with different source profiles and resolve higher harmonic frequency heating schemes. In this work, we present ASCOT/RFOF modelling results of a 3rd harmonic D heated JET discharge 86459, with 3MW ICRH power and 4.5MW NBI power. The NBI deposition is calculated with the BBNBI module [7] and the RF deposition is simultaneously calculated by RFOF. The ASCOT simulation has been performed without coupling to a full wave-field solver for simplicity, the PION code has been used to determine the power absorbed by the ions and the amplitude of the accelerating electric field is assumed constant over the plasma cross-section. The simulated fast ion distribution, compared to the SPOT and PION codes, is shown in Figure 4a. We note that ASCOT predicts a slightly lower gradient for the distribution in the 1 MeV to 2 MeV range, but there is reasonably good agreement on the cut-off of the distribution at 2.2 MeV. The analyzed discharge was designed to study fusion products and, therefore, exhibits a high neutron rate. Due to the heating mix, the measured neutron yield significantly exceed what would be predicted by NBI or ICRH modelling alone, as is illustrated in Figure 4b. It shows that a neutron rate of $1 \cdot 10^{15} s^{-1}$ is predicted by modelling the NBI heating with the PENCIL code, and a neutron rate of $2 \cdot 10^{15} s^{-1}$ is predicted by modelling the ICRH heating only with ASCOT/RFOF. The measured neutron rate, $6 \cdot 10^{15} s^{-1}$, is well matched by an ASCOT/RFOF simulation of both NBI and ICRH simultaneously. These calculations do not include the beam-beam component of the neutron rate,

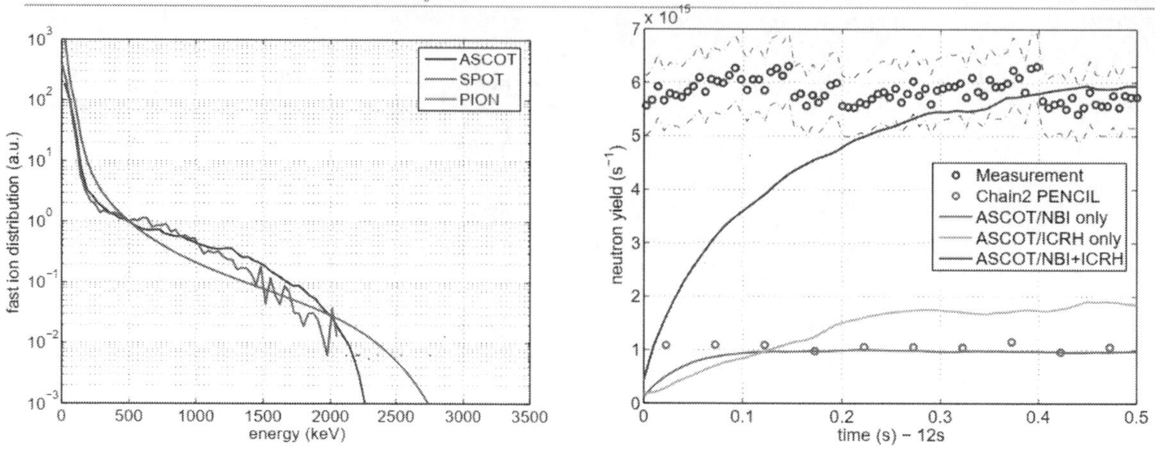


Figure 4: (a) Comparison of normalized fast ion energy distribution between ASCOT/RFOF, SPOT/RFOF and PION. (b) Measured neutron yield in 86459 and ASCOT simulations with NBI, ICRH and NBI+ICRH.

including it would most likely lead to a slight overestimation of the neutron yield, consistently with the NBI results.

**Conclusions** ASCOT can be used to simulate complex fast-ion dynamics, including 3D effects and wave-particle interactions. In this work we have applied it to neutron rate calculations in JET, which will be important in the upcoming DT campaign. We have found that magnetic islands due to NTMs can cause local redistribution of fast ions, large enough to reverse the sign of the JxB torque due to the radial ion flux. However, the redistribution does not significantly change the integrated neutron yield, and is unable to explain the neutron deficit observed in JET experiments. In ICRH and NBI heated plasmas, ASCOT can reproduce the measured neutron rate by self-consistently modelling both ICRH and NBI heating. Further benchmarking is ongoing.

**Acknowledgements** This work has been carried out within the framework of the EUROfusion Consortium and has received funding from the Euratom research and training programme 2014-2018 under grant agreement No 633053. The views and opinions expressed herein do not necessarily reflect those of the European Commission. Financial support from the Emil Aaltonen foundation is gratefully acknowledged. The supercomputing resources of CSC - IT center for science were utilized in the studies.

**References**

[1] H. Weisen, et. al., This Conference
[2] G. Tardini, et. al., 41th EPS Conference on Plasma Physics (2014), O2.103
[3] E. Hirvijoki, et. al., Computer Physics Communications 185 (2014), pp. 1310-1321
[4] Y. Baranov et al, Plasma Phys. Control. Fusion 51 (2009) 044004
[5] C. Marchetto, et. al., 41th EPS Conference on Plasma Physics (2014), P1.018
[6] T. Johnson et. al., AIP Conference Proceedings 1406 (2011), p.373-376
[7] O. Asunta et. al., Computer Physics Communications 188 (2015), pp. 33-46

# Density fluctuations measured with fast frequency swept reflectometry

F. Clairet[1], A. Sirinelli[2], L. Meneses[3] and JET contributors*

EUROfusion Consortium, JET, Culham Science Centre, Abingdon, OX14 3DB, UK

[1] *CEA, IRFM, F-13108 St. Paul-lez-Durance cedex, France*

[2] *ITER Organisation, 13067 Saint-Paul-lez-Durance, France*

[3] *IPFN, IST, Universidade de Lisboa, 1049-001 Lisboa, Portugal*

Plasma turbulence is known to induce an anomalous heat and particle transport responsible of a strong degradation of the plasma energy confinement, and we present recent achievements to provide precise measurements of density fluctuations using the microwave reflectometry diagnostic. In this work, the turbulence measurements are retrieved from the fluctuations of the reflected signals of frequency swept systems. This method, first tested on Tore Supra [1], offers the great advantage to provide a localized and continuous radial determination of the density fluctuation level from their radial wavenumber power spectra $S(k_r)$. The fast sweep capability offers the possibility to study the plasma turbulent events with a high time resolution. Using the complete set of the JET X-mode fast sweeping heterodyne reflectometers (KG10) [2] we can provide fluctuation profiles from the edge to the center over a wide variety of plasma discharges in terms of density and magnetic field.

**Determination of the density fluctuation profile**

While the phase of the reflected signal is commonly used to calculate a radial density profile, the plasma density fluctuations are extracted from its phase fluctuations. Thanks to the profile reconstruction we can retrieve the radial dependency of these phase fluctuations $\delta\phi(r)$ and calculate their wavenumber spectra by performing a FFT through a radial window $\Delta r$ (providing a minimum accessible wavenumber $k_{min} = 2\pi/\Delta r$). The relationship, or transfer function, between the phase fluctuations and the density fluctuations spectra is given by a 1D Helmholtz code which simulates the propagation of the probing electromagnetic wave into fluctuating plasma given the experimental density and magnetic field profiles of the discharge. The reflected signal records the plasma fluctuations through Bragg backscattering physical processes, where $k_{fluc} = 2k_{wave}$, of the probing electromagnetic wave with $k_{wave}$ varying from $k_0$ ($2\pi F/c$) at the edge to 0 at the cut-off (Fig. 1).

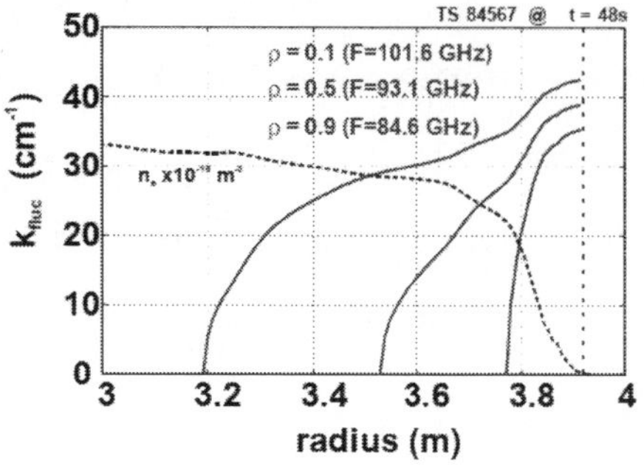


Fig. 1 : Probing wavenumbers of the plasma fluctuations according the probing frequency.

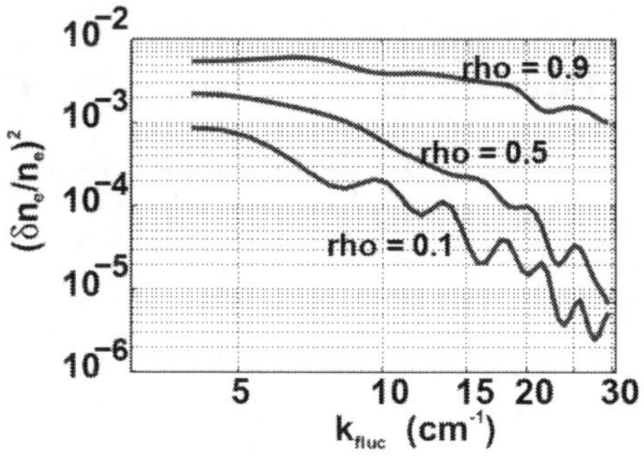


Fig. 2 : Wavenumber density fluctuation spectra calculated at the cut-off positions of Fig. 1 with an integrated radial window of 3.14 cm ($k_{min}$=2cm$^{-1}$).

The fluctuation levels are calculated from the integrated spectra (Fig. 2) for $k_{min} < k_{fluc} <$ 10cm$^{-1}$ ($k_{min}$ depends on the integrated radial window) in order to preserve the radial localisation of the measurement.

## H-mode discharge

This technique has been applied to the JET 84567 discharge (Fig. 3) with a time resolution of 300 μs (phase spectra averaged from 5 sweeps) and a radial integration of Δr=1.57 cm. It exhibits the time evolution of the density fluctuation radial profile around the LH transition in the entire plasma, from the edge to the center (Fig. 4). The turbulence reduction arises at

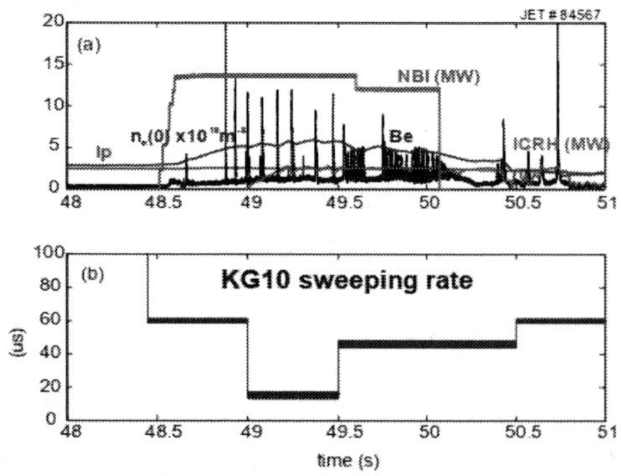


Fig. 3 : (a) Time trace of the H-mode discharge scenario and (b) delay between consecutive frequency sweeps of the reflectometers.

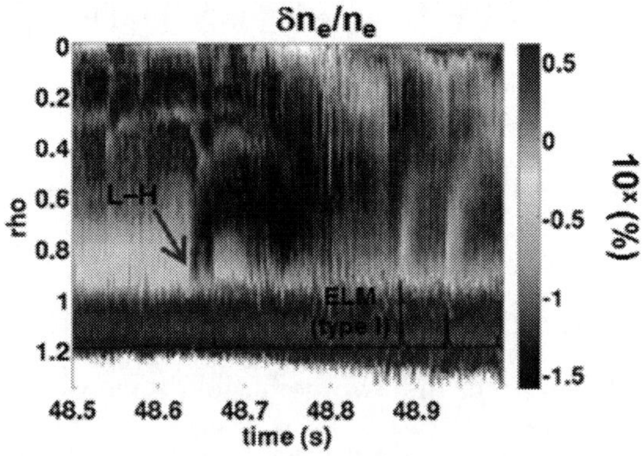


Fig. 4 : Time evolution of the density fluctuation radial profile. The black line accounts for edge Be emission.

about 48.65s and has been evaluated to occur first near the edge (ρ ~ 0.95) which then propagates toward the center. Much of the turbulence reduction is noticeable up to mid-radius and has no noticeable influence further in the plasma center.

## Wavenumber spectra

Along with a decrease of the turbulence in H-mode, a substantial steepening of the spectra is clearly evidenced (Fig. 5) and can be seen from the edge to the center (Fig. 6). The modification of the spectral slope can be related to non-linear effects of the density fluctuations onto the phase fluctuations of the reflected signal. This is particularly true at the edge where the plasma turbulence is high, however, the linear relationship between phase and density fluctuations should be more solid toward the center where the fluctuation level is low ($\delta n/n < 1\%$).

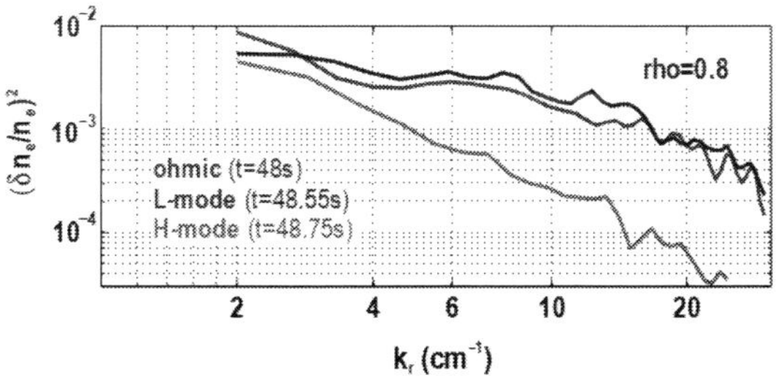


Fig. 5 : Wavenumber density fluctuation spectra comparison between ohmic, L-mode and H-mode.

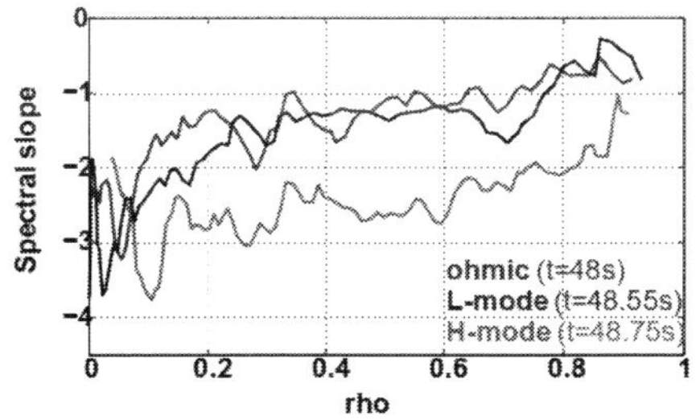


Fig. 6 : Spectral slope comparison between ohmic, L-mode and H-mode. The slopes are calculated between 2 and 10 $cm^{-1}$

## ELM crash

An ELM of type I (Fig. 7 and 8) provides a strong plasma perturbation with considerable fluxes hitting the inner wall. It offers the advantage of having a long inter ELM

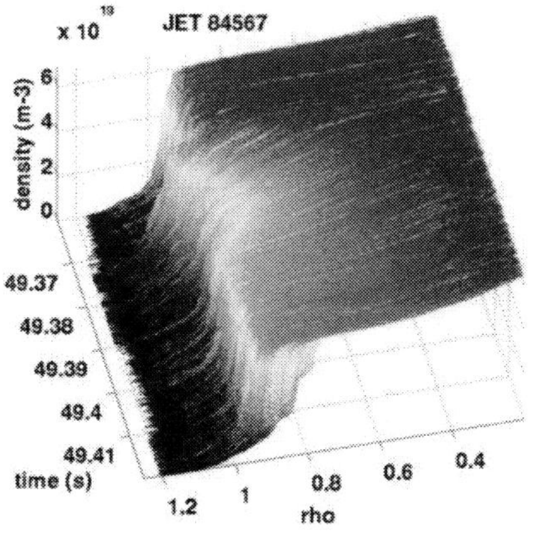


Fig. 7 : Density profile evolution during a type I ELM event.

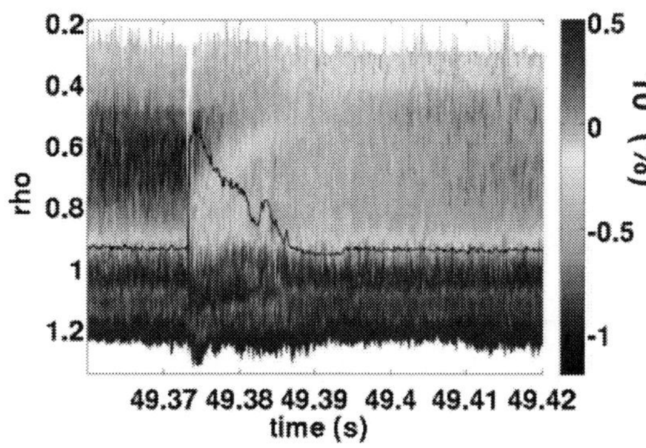


Fig. 8 : Time evolution of the density fluctuation radial profile. The black line account for edge Be emission.

period and can be thus more easily studied individually. Before the ELM crash, no precursor is observed. At the crash onset, a substantial increase of the density fluctuations is observed starting from the edge accompanying the strong plasma expulsion toward the edge. During the crash a turbulent wave front propagates toward the plasma center at about 30m/s accompanying a density collapse. Once the turbulent wave front disappears, the plasma density relaxes to its former state.

**Conclusion**

In this work, we demonstrate that the fast swept reflectometry is a powerful tool to analyze the plasma density fluctuations. While fixed frequency systems provide a substantially better signal to noise ratio, the sweeping systems provide a continuous radial measurement of the plasma turbulence from the edge to the center. Thanks to the fast sweeping capabilities the turbulent changes during the L-H transition as well as the ELM crash which can be extremely well temporally observed. The density fluctuation level can be determined with very few sweeps (~ 5 to 10), the wavenumber spectra require much more (100 to 200) and is still subject to substantial nonlinear interaction between the probing electromagnetic wave and the plasma fluctuations which may alter its spectral shape.

**Acknowledgments**

This work has been carried out within the framework of the EUROfusion Consortium and has received funding from the Euratom research and training programme 2014-2018 under grant agreement No 633053. The views and opinions expressed herein do not necessarily reflect those of the European Commission.

**See the Appendix of F. Romanelli et al., Proceedings of the 25th IAEA Fusion Energy Conference 2014, Saint Petersburg, Russia*

[1] L. Vermare et al. Nucl. Fusion 46, S743–S759 (2006)

[2] A. Sirinelli et al. Rev. Sci. Instrum. 81, 10D939 (2010)

# Role of ELMs in the production and concentration of Tungsten in JET discharges

P. Devynck[1], N. Fedorczak[1], O. Meyer[1] and JET Contributors*

EUROfusion Consortium, JET, Culham Science Centre, Abingdon, OX14 3DB, UK

[1]CEA-IRFM F-13108 Saint-Paul-lez-Durance, France

Abstract

We report the analysis of a series of 230 shots taken randomly in the JET ITER-Like Wall (ILW) database. The goal of the analysis is to study the effects of the Edge-Localized-Mode (ELM) crash on the Tungsten source eroded in the outer divertor. The main findings are that the mean Tungsten source per ELM is proportional to the mean density drop of the pedestal during the crash. During Nitrogen seeding, the tungsten source per ELM is increased further. The frequency of the ELMs is found to be driven by the Power crossing the separatrix but is also fond to depend inversely on the density drop of the pedestal. The scaling for the frequency applies for seeded discharges as well. The role of the temperature pedestal in the process has not been tested and is still unclear. The ratio of Tungsten in the bulk over tungsten source during ELMs is a decreasing function of the ELMs frequency.

Database and signal processing

The only criteria applied for the shots selection is that the chosen time window should be long enough to include several ELMs, with no obvious change in ELM regime. The main plasma parameters should be stationary during the time window. The database includes all kinds of

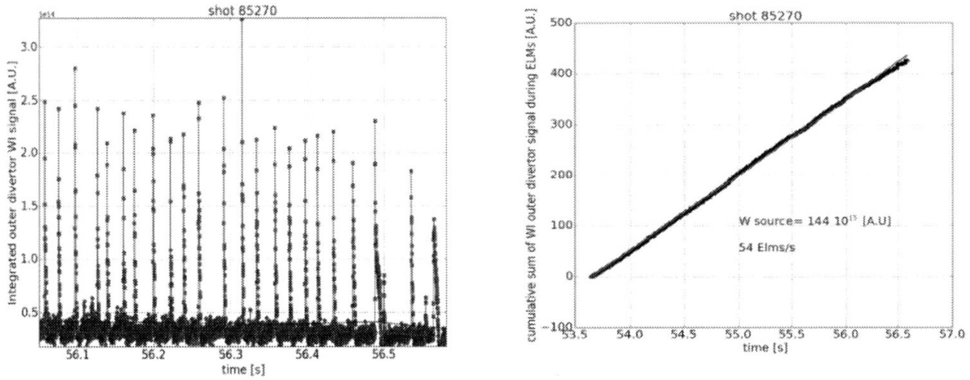


**Figure 1**
Left : example of ELM filtering on WI outer divertor Signal.
Right : Cumulative sum of filtered ELMs as a function of time.

plasmas during ILW operation, including Nitrogen and Neon seeding shots. The total power applied to the plasma ranges from 6 to 26 MW, Toroidal field from 1.3 to 2.87 T, Plasma current from 1.3 to 3MA, NL and $N_{pedestal}$ are measured using lines of sight of the Interferometry diagnostic, NL ranges from 0.9 to 2.4 $10^{20}$ $m^{-2}$. The gas fuelling was also varied and the data include shots with kicks. The ELM filtering program is applied on the

*See the Appendix of F. Romanelli et al., Proceedings of the 25th IAEA Fusion Energy Conference 2014, Saint Petersburg, Russia

integrated outer divertor signal measuring the neutral source of Tungsten WI. This signal is composed by summing up 10 spectroscopic lines of sight monitoring the outer divertor. No calibration is applied to this signal. An example of filtering is shown in Figure 1 left. The mean number of ELMs per second is measured (ELMs of all sizes are included in the count). The cumulative sum of the WI signal during ELMs is computed as a function of time and displayed in Figure 1 Right. The slope of this curve gives a value proportional to the number of Tungsten atoms emitted per second. The mean number of Tungsten atoms per ELM (uncalibrated) is also calculated.

Scaling of the tungsten source produced by the ELMs

Figure 2 left: The mean tungsten source per ELM (outer divertor) is found to be proportional to the mean density drop of the pedestal during ELM crash. Dependence with pedestal

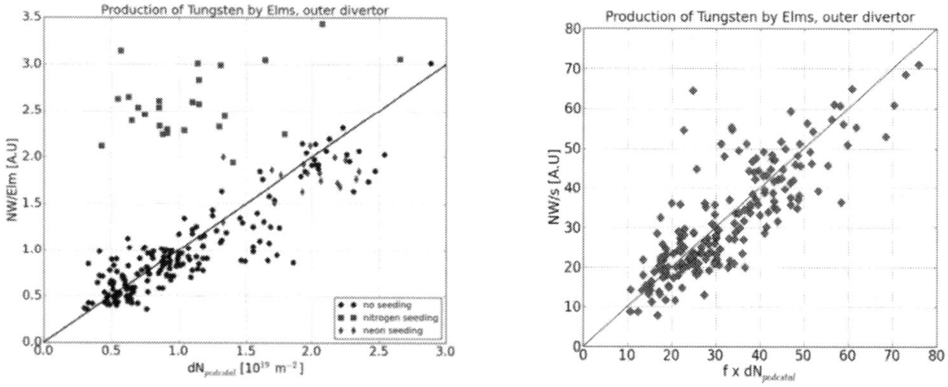


**Figure 2**

Left : Mean number of tungsten atoms per ELM as a function of mean pedestal density drop during ELM crash. Right : Dependence of mean number of tungsten atoms per second with mean pedestal density drop during ELM crash and ELM frequency ( No seeding).

temperature has not been checked so far and will be investigated in the near future. However the data is already rather well aligned (standard deviation 0.25). We have not observed any effect of the position of the outer strike point though its position was scanned from the center horizontal target over the corner to the vertical target. The result presented here is not incompatible with the idea that physical erosion by ELMs occurs in a range of particles energies where the cross section for erosion is constant [1-4]. As far as the nature of particles responsible for the tungsten erosion during ELM crash is concerned, it is necessary to estimate the concentration of the different impurities to give an answer. This concentration is not calculated for these shots though concentrations of about 1% of Beryllium have been estimated in ILW plasmas [5]. With such a concentration Deuterium should dominate the physical erosion. We notice however that the case with nitrogen seeding stands out and shows an increase of the tungsten source by about a factor of 2 relative to the scaling. Surprisingly no increase of the tungsten source with neon seeding is found. Figure 2 right is simply the consequence of the previous finding. The source rate during ELMs $N_{tungsten}/s$ in the case where no nitrogen seeding is applied is simply proportional to the mean drop of the pedestal density multiplied by the ELMs frequency.

Scaling of the ELMs frequency

The frequency for type 1 ELMs has been found to be an increasing function of the power crossing the_separatrix [6]. Here we reexamine this dependency, keeping in mind that the ELMs encountered so far in our database with the ILW regardless of their size have the characteristics of type 1 ELMs. The result is shown in Figure 3. The frequency of the ELMs is found to be increasing with the power crossing the separatrix and decreasing with the density drop of the pedestal during ELM crash. The standard deviation of the data around the fitting curve is 8.5 Hz. This scaling is in agreement with the usual observation that the ELM size

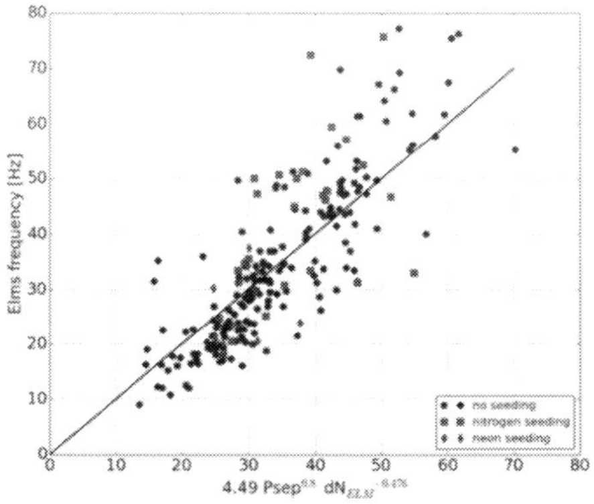


**Figure 3**
Scaling of ELMs frequency with Power through separatrix and mean density drop during ELMs

($dN_{ELM}$) decreases with the frequency but by inverting the scaling it can be seen that $dN_{ELM}$ is also an increasing function of the power through the separatrix. The result here suggests that the time to refill the pedestal density may be a factor driving the frequency of the ELMs. We notice also that the data for nitrogen and neon seeding follows the same scaling. The same process probably drives the ELM frequency in seeded discharges. We plan to extend the analysis to type 1 ELMS in Carbon discharges and check type 2 ELMS in JET Carbon and ILW discharges against the scaling presented here.

Dependences of the amount of tungsten in the discharge

The amount of tungsten in the bulk of the discharge is roughly estimated with the help of the bulk radiated power and mean density of the plasma. This supposes that the radiation from the bulk comes only from tungsten. In this case the relative invariance of the radiative function of tungsten with electron temperature [7] allows writing the amount of tungsten in the discharge as:

$$\iiint Nw \, dV = Prad_{bulk}/(<Ne> L_w^{eff})$$

The ratio of Number of tungsten atoms estimated in this way over tungsten source during ELMs is plotted in Figure 4 as a function of ELMs frequency. This ratio is found to be a decreasing function of the ELMs frequency indicating that the flushing effect of the ELMs is real. At frequencies below 40 Hz many shots have a high ratio of bulk tungsten over source during ELMs. Individual observation of the two terms forming the ratio indicates that the amount of tungsten in the discharge is high while the source of tungsten by ELMs is low for these shots. In fact the amount of tungsten eroded by the ELMs continues obeying the scaling displayed in figure 2 left but the level of radiated power is unusually high. We will investigate if the increase in radiated power correlates with the tungsten bundle line measured in the bulk or if some other type of impurity must be invoked to explain the high level of radiated power measured for these shots. In the latter the ratio displayed in Figure 4 for these shots would be wrong. Another possibility is a change of the tungsten influx produced by changes in gas flow and SOL properties.

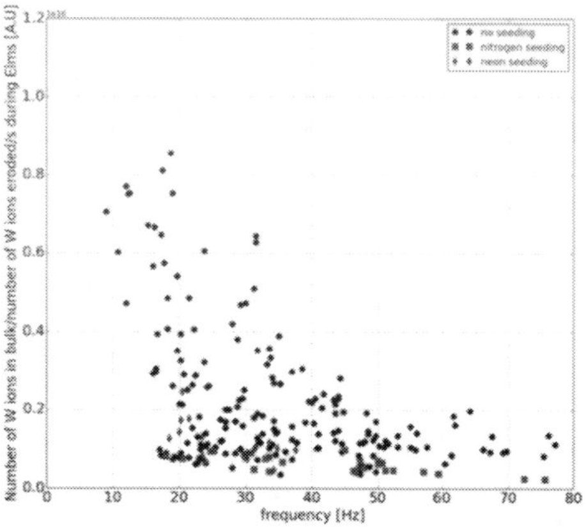


**Figure 4**

Ratio of number of tungsten ions in the bulk of discharge over number of tungsten atoms eroded by the ELMs /s. The ratio is shown as a function of ELMs frequency.

## Conclusions

A database of 230 shots in ILW has allowed drawing some conclusions on the effect of tungsten erosion by ELMs. One of the strongest results is that the mean tungsten source per ELM is directly proportional to the mean pedestal density drop during the ELMs. Some results are also presented showing that the ELMs frequency is a function of both the power through separatrix but also of mean pedestal density drop during ELM crash. The number of shots seems to be sufficient to draw firm conclusions about these two items. However, the situation is less clear about the flushing effect of the ELMs on tungsten where the observed dependencies are rather fragile. In particular the estimate of the amount of tungsten in the discharge using the bulk radiated power has not been correlated with observations of spectroscopic lines of tungsten. To really confirm the flushing effect of the ELMs on tungsten, it is necessary to increase considerably the number of shots in the database. It is also necessary to investigate how the temperature of the pedestal modifies (or not) these results.

Acknowledgment:
This work has been carried out within the framework of the EUROfusion Consortium and has received funding from the Euratom research and training programme 2014-2018 under grant agreement No 633053. The views and opinions expressed herein do not necessarily reflect those of the European Commission.

*References:*

[1] C. Guillemaut et al. Submitted to Plasma Phys. and Control. Fusion.
[2] W. Fundamenski et al., Plasma Phys. Control. Fusion **48** (2006) 109–156
[3] T. Eich et al., J. Nucl. Mater. **390–391** (2009) 760–763
[4] D. Moulton et al., Plasma Phys. Control. Fusion **55** (2013) 085003
[5] S. Brezinsek et al., PSI 2014
[6] H.Zohm, Plasma Phys. Control. Fusion, **38**,105
[7] Neu R *et al* 2010 *Proc. 24th IAEA Fusion Energy Conf. (Daejon, Republic of Korea)*

# Global vs local gyrokinetic studies of core microinstabilities in JET hybrid discharges with ITER like wall

S. Moradi[1], I. Pusztai[2], J. Citrin[3,4], Hyun-Tae Kim[5], I. Lupelli[5], WJ. Kettlewell[1], and the JET contributors *

*EUROfusion Consortium, JET, Culham Science Centre, Abingdon, OX14 3DB, UK*
[1] *Fluid and Plasma Dynamics, Université Libre de Bruxelles, Brussels, Belgium*
[2] *Applied Physics, Chalmers University of Technology, Göteborg 41296, Sweden*
[3] *FOM Institute DIFFER, P.O. box 6336, 5600 HH Eindhoven, The Netherlands*
[4] *CEA, IRFM, F-13108 Saint Paul Lez Durance, France*
[5] *CCFE, Culham Science Centre, OX14 3DB Abingdon (UK)*

## Introduction

In recent years there has been an increasing worldwide effort to further the development of the so-called hybrid or improved H-mode scenarios. In these plasmas, representing a hybrid between "baseline" and "advanced tokamak" scenarios, enhanced normalized confinement is achieved (compared to the ITER baseline ELMy H-mode) through current profile optimization, avoiding deleterious MHD, and access to high $\beta_N$. We present a study on the characteristics of core microinstabilities in two hybrid plasmas from recent JET experiments with ITER like wall (selected from Ref. [1]) based on local and global gyrokinetic simulations with the *GYRO* code [2]. In particular, we study the role of various plasma parameters such as normalized plasma pressure and Shafranov shift on the onset and stability of the microinstabilities. The time evolution of the selected plasma discharges from the ILW power scan experiments: #84545 (high-$\delta$) and #84792 (low-$\delta$) are shown in Fig. 1. In these discharges, a "current overshoot" before the Neutral Beam Injection (NBI) pulse allows the formation of a wide region of low magnetic shear in the plasma core which allows access to the hybrid conditions with high $\beta_N$ and $H_{98} > 1$ [3].

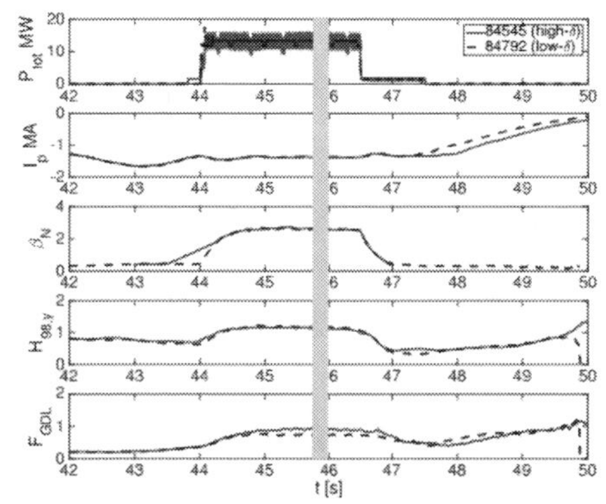


Figure 1: *Time traces of main plasma parameters for the selected plasmas.*

## Results of gyrokinetic simulations

For our microinstability analysis three radial positions are selected: $\rho = 0.3,\ 0.45$ and $0.6$, where $\rho$ is the normalised toroidal flux (as defined in [2]). Magnetic geometry parameters are taken from EFIT constrained by MSE measurements, and the profiles calculated by TRANSP are averaged over a time window of 0.5 s over the period of the highest $H_{98}(y,2)$ factor with nearly stationary temperatures and density (shown by the grey field in Fig. 1).

* See the Appendix of F. Romanelli et al., *Proceedings of the 25th IAEA Fusion Energy Conference 2014, Saint Petersburg, Russia*

Figures 2(a-f) show the result of a $\beta_e$ scans for the three selected radial positions of #84545 (high-$\delta$) discharge computed by Local (solid red lines) and Global (dashed-dotted blue lines) linear simulations. A significant difference between the two types of simulation, as seen in this figure, is the lower KBM threshold in the local runs in all three radii, while the global scans show no onset of KBMs throughout the $\beta_e$ scan range. In #84792 (low-$\delta$) discharge we did not find any significant difference in the overall results and similar trends were observed. At the experimental levels of $\beta_e$ ($\rho = 0.3 \rightarrow \beta_e \sim 0.015$, $\rho = 0.45 \rightarrow \beta_e \sim 0.01$ and $\rho = 0.6 \rightarrow \beta_e \sim 0.006$ ), ITGs are found to be the dominant instability in Global simulations at all the considered radii. Here, as $\beta_e$ is increased the ITGs are completely stabilised without any transition to a KBM branch of instabilities. Note, that the modes with almost zero real frequency are potentially numerical rather than physical modes as we find ITGs to be completely stabilised here.

Comparing Local and Global results, the largest difference is observed for the inner core radius at $\rho = 0.3$ as can be seen in Fig. 2 (a and b), while for the outer core radii i.e. $\rho = 0.45$ and $0.6$, the Local and Global simulation results are closer and specially for the experimental values of $\beta_e$ they overlap (see Fig. 2 (c-f)). Such a trend is expected because for the inner core due to very low magnetic shear the Local assumptions are less valid while as the magnetic shear is increased by moving away from the core we expect the Local assumptions to hold and therefore the Global and Local results to be close. However here, the impact of profile variation still plays an important role and therefore there are further differences found as $\beta_e$ is increased further beyond the experimental values.

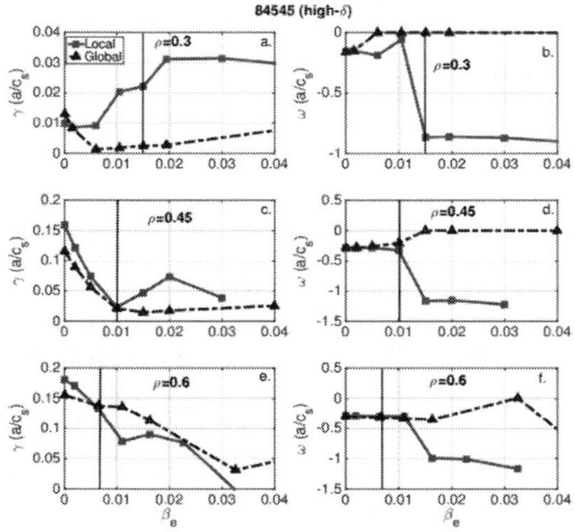


Figure 2: *Imaginary and real frequency of the most unstable modes as functions of $\beta_e$ with self-consistent variation of the $\alpha_{MHD}$ at $\rho = 0.3$ (a and b), $\rho = 0.45$ (c and d) and $\rho = 0.6$ (e and f). Note that negative frequencies are in the ion diamagnetic direction. Black vertical lines represent the experimental values of $\beta_e$.*

We have examined the stabilisation role of Shafranov shift on the unstable modes with Local and Global simulations and the results are presented in Figs. 3. Here the variation of Shafranov shift is applied by varying the parameter $c_p$ defined as $\alpha_{MHD} = -q^2 R_0 8\pi/(B_{unit}^2)\frac{dp}{dr}c_p$, where $R_0$ is the major radius of the centroid of the flux surface, and $p = \sum_a n_a T_a$ is the total plasma pressure. $c_p$ is the geometric pressure gradient scaling parameter which allows an artificial adjustment of $\alpha_{MHD}$ without modifying the background gradients as presented in Ref. [5]. As

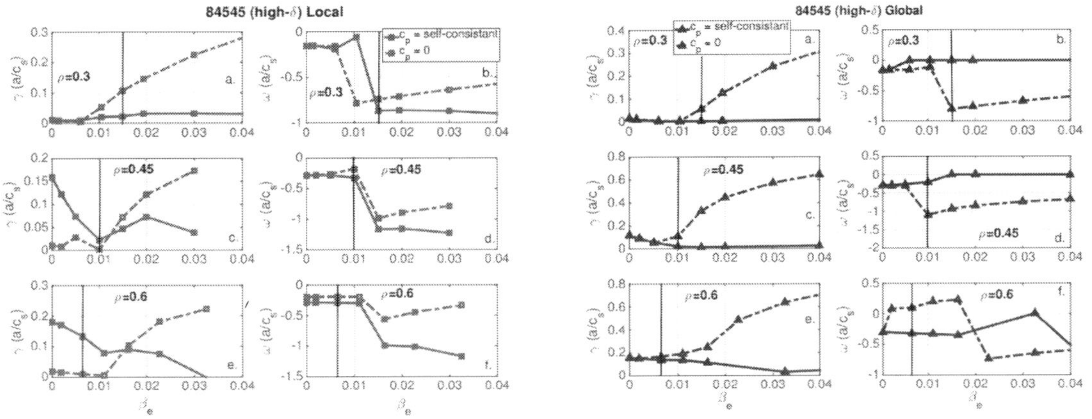


Figure 3: *Imaginary and real frequency of the most unstable modes as functions of $\beta_e$ with self-consistent variation of the $\alpha_{MHD}$ and without i.e. $c_p = 0$ at $\rho = 0.3$ (a and b), $\rho = 0.45$ (c and d) and $\rho = 0.6$ (e and f). Left figures are Local and right figures are Global results, with black vertical lines representing the experimental values of $\beta_e$.*

seen in this figure in both Local and Global simulations we observe a strong stabilisation effect of Shafranov shift on the growth rates at large $\beta_e$, while the effect is minimal or destabilising e. g. in the case of Local runs at $\rho = 0.45$ and $\rho = 0.6$ (see Figs. 3 left (c-f)). A significant impact of the Shafranov shift is on the ITG/KBM transition which is found to be strongly shifted towards lower $\beta_e$ values in Global simulations and without it we find KBMs to be dominant unstable mode at the experimental values of $\beta_e$ for $\rho = 0.3$ and $\rho = 0.45$. Furthermore we investigate the role of magnetic shear on the stability characteristics in the two simulation types. Figures 4 illustrate the results of $\beta_e$ scans with the experimental values of magnetic shear (solid lines) and with a 30% increase in the magnetic shear (dashed-dotted lines). As can be seen in this figure in the case of Local simulations (left figures) an increase in the magnetic shear results in a strong stabilisation of the unstable modes for the whole range of considered $\beta_e$ and at all radii. Here we find that such an increase will result in the complete stabilisation of KBMs at high $\beta_e$ and the appearance of the subdominant TEM due to complete stabilisation of the dominant ITG by increasing $\beta_e$. In the Global simulations however, a 30% increase in magnetic shear does not impact the overall growth rate values for all radii, but here also we observe the appearance of the TEM as the ITG is stabilised at high $\beta_e$.

**Summary and conclusions**

In a series of $\beta_e$ scans for the two selected hybrid discharge we have shown that indeed there is a significant difference in the characteristics of the dominant instabilities between the Local and Global assumptions. The most significant difference is the strong reduction of the ITG growth rate and the absence of the ITG/KBM transition in the Global simulations. The reason for this strong ITG stabilisation is found to be a combination of Shafranov shift and electromagnetic effects. A 30% increase in the magnetic shear is found to impact the Local results more strongly than the Global results however such an increase is expected to further

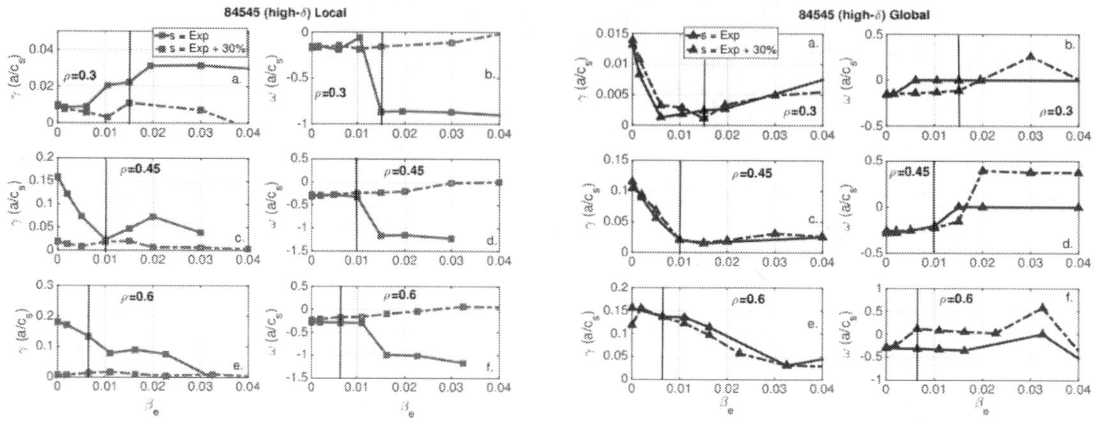


Figure 4: *Imaginary and real frequency of the most unstable modes as functions of $\beta_e$ with self-consistent variation of the $\alpha_{MHD}$ with experimental (solid lines) and +30% (dashed-dotted lines) values of magnetic shear, at $\rho = 0.3$ (a and b), $\rho = 0.45$ (c and d) and $\rho = 0.6$ (e and f). Left figures are Local and right figures are Global results, with black vertical lines representing the experimental values of $\beta_e$.*

stabilise the ITG instability in the core. Our analysis suggest that electron directed modes such as TEM/MTM are subdominantly present in these plasmas and as the combination of high $\beta$ and Shafranov shift effects result in complete stabilisation of the ITG modes, these modes become more visible since they are less affected by these effects. Strong reduction in KBMs growth rate, to an almost marginally unstable state, in global in global (in contrast to local) simulations was previously observed in Ref. [4] for the inner core of the C-wall hybrid plasmas. The importance of non-linear $\beta$ stabilisation of the ITG modes in the core of the JET hybrids in both C-wall and ILW were shown in Refs. [6, 7] through local non-linear gyrokinetic simulations. Here it was presented that the $\beta/\beta_{crit}$ ratio where $\beta_{crit}$ is the critical $\beta$ value for the onset of electromagnetic modes such as KBMs, is an important factor for the effectiveness of the $\beta$ stabilisation of the ITGs. Our results shows the importance of global simulations on obtaining an accurate $\beta_{crit}$ for the onset of KBMs in high-$\beta$ regimes which are found to be significantly higher than what was found in local simulations. Hence, our findings suggest a possible rise of the upper power limit for regimes showing a lack of plasma confinement degradation [1], which may be restricted by destabilisation of KBMs as $\beta$ is increased.

**Acknowledgments**

The authors would like to thank J Candy for providing the *GYRO* code. This work has been carried out within the framework of the EUROfusion Consortium and has received funding from the Euratom research and training programme 2014-2018 under grant agreement No 633053. The views and opinions expressed herein do not necessarily reflect those of the European Commission. Sara Moradi has benefited from a mobility grant funded by the Belgian Federal Science Policy Office and the MSCA of the European Commission (FP7-PEOPLE-COFUND-2008 n 246540).

**References**

[1] C. Challis et al, Proceedings 25th IAEA Fusion Energy Conference, 2014, St. Petersburg, Russia
[2] J. Candy and E. Belli, General Atomics Report GA-A26818 (2011)
[3] J. Hobirk et al *Plasma Phys. Control. Fusion* **54** 095001 (2012).
[4] S. Moradi et al, Nucl. Fusion **54** 123016 (2014)
[5] E. A. Belli and J. Candy, *Phys. Plasmas*, **17** 112314 (2010).
[6] J. Citrin et al, Plasma Phys. Control. Fusion **57** 014032 (2015)
[7] J. Garcia et al, Proceedings 25th IAEA Fusion Energy Conference, 2014, St. Petersburg, Russia

# Gyrokinetic modelling of baseline H-mode JET plasmas with C Wall and ITER-like wall

D. Tegnered[1], P. Strand[1], H. Nordman[1], C. Giroud[2], Hyun-Tae Kim[2], G.P. Maddison[2], M. Romanelli[2], G. Szepesi[2] and JET Contributors*

*EUROfusion Consortium, JET, Culham Science Centre, Abingdon, OX14 3DB, UK*

[1]*Department of Earth and Space Sciences, Chalmers University of Technology, SE-412 96 Göteborg, Sweden.* [2]*CCFE, Culham Science Centre, Abingdon, OX14 3DB, UK*

## Introduction

A deterioration in global confinement is observed at JET in baseline H-mode experiments following the change from a carbon wall (CW) to an ITER-like wall (ILW) with beryllium and tungsten [1]. One cause is the high deuterium gas puffing rate necessary in ILW discharges in order to mitigate W accumulation. For low triangularity plasmas, this degradation of confinement with fuelling level was also observed for CW discharges [2]. The deterioration is correlated by a degradation of pedestal confinement with lower electron temperatures at the top of the edge barrier region. This leads to a lower electron temperature in the core, thereby changing the NBI heat deposition profiles in the core. As a result, the core energy confinement time is influenced with lower electron energy confinement time and similar ion confinement time in the ILW case [1]. To study the effect of the ILW on confinement, a database has been created comprising a set of JET discharges with ILW and matched CW discharges using the same criteria as in [3]. In the present work, transport due to Ion Temperature Gradient (ITG)/Trapped Electron Mode (TEM) turbulence is calculated for similar CW and ILW discharges using the gyrokinetic code GENE [4], in order to assess the differences seen in core energy confinement.

## GENE simulations setup and discharge parameters

GENE solves the nonlinear gyrokinetic Vlasov equation coupled with Maxwell's equations. Collisions are modelled using a linearised Landau-Boltzmann collision operator [5]. Magnetic fluctuations are included in all simulations. The Miller geometry model [6] is used in a flux tube domain. Miller parameters are extracted from numerical geometries reconstructed by the EFIT code [7]. For the linear simulations both an initial value solver and an eigenvalue solver that can find subdominant modes are used. Two ILW discharges and two CW discharges with global parameters matched as closely as possible are analysed. The matched global parameters

*See the Appendix of F. Romanelli et al., Proceedings of the 25th IAEA Fusion Energy Conference 2014, Saint Petersburg, Russia

are the plasma current, the toroidal magnetic field, applied NBI power, average electron density, safety factor, and triangularity. The discharges are baseline H-mode with ion temperature and rotation measurements available through charge exchange spectroscopy. Discharge parameters are taken from TRANSP runs [8] performed with electron density and temperature profiles from high resolution Thomson scattering measurements. One impurity species is included in the simulation, 1.9% carbon for the CW discharges and 0.4% beryllium for the ILW discharges. The impurity density is calculated from $Z_{eff}$, assumed constant over the whole radius [1]. The four discharges are analysed at $\rho = 0.5$ where $\rho$ is the normalized toroidal flux coordinate. The discharges are pair wise 74313 (CW), 85407 (ILW), 74324 (CW) and 85406 (ILW). The data is averaged over a one second time window and smoothed in the radial direction.

**Linear results**

Due to the large uncertainty in the parameter $R/L_{T_i}$, the linear and nonlinear simulations of the ILW and CW discharges are performed as scans over $R/L_{T_i}$. The turbulence is ITG dominated for $R/L_{T_i} > 4$ for the ILW discharges and TEM dominated for lower $R/L_{T_i}$ while for the CW discharges the TE mode is not excited. Thus, the discharges are all ITG dominated at experimental $R/L_{T_i}$. The ITG threshold is slightly lower for the ILW discharges and the normalized growth rates are larger at the same $R/L_{T_i}$. The ILW versus CW pairs considered are not perfectly matched with respect to dimensionless parameters. This leads to differences in linear stability of the main instabilities in the discharges. The reason for the mismatch in many parameters is related to the difference in pedestal height. This difference in the edge region translates into differences in the core of key parameters like $\beta$, Shafranov shift, and collisionality. These differences are expected to disappear if the pedestal confinement is recovered, e.g. through N seeding [9]. The difference in impurity content between the pairs leads to a slightly more stable situation in the CW case which should remain even if the pedestals are similar. In Figure 1, the effect of the difference in dimensionless parameters on the linear stability is summarized. The figure shows the relative change in the ITG growth rate when the values of the parameters in one discharge is changed to that of the corresponding paired discharge. As seen, the mismatch in $\beta$, Shafranov shift, magnetic shear, and electron temperature gradient serve to destabilize the ILW discharges relative to the CW discharges while the mismatch in collisionality and ion to electron temperature ratio tend to stabilize the ILW discharges. The difference in the safety factor and triangularity does

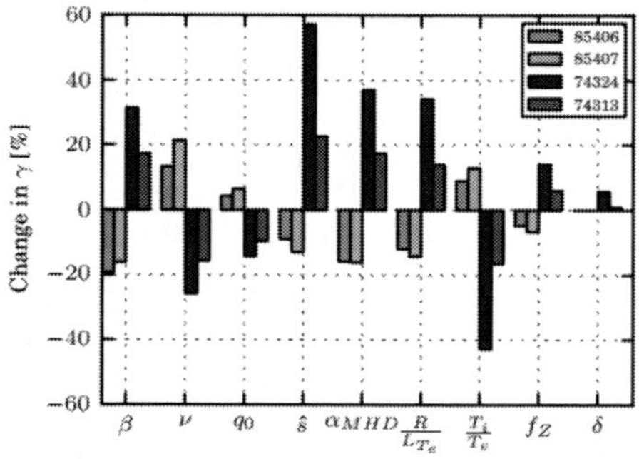


**Figure 1:** Growth rate change at $k_y\rho_s$=0.3

not substantially change the linear stability properties.

**Nonlinear results**

For the nonlinear GENE simulations, a simulation domain in the perpendicular plane of $[L_x, L_y] = [146, 126]$ is used, with a resolution of $\left[n_x, n_{k_y}, n_z, n_{v_\parallel}, n_\mu\right] = [96, 48, 32, 64, 16]$. The resolution and simulation domain are checked through convergence tests. In order to quantify the effects of rotation, it was included in simulations of discharge 74313 and 74324. As can be seen in Figure 2a, this resulted in a reduction in the ion heat flux of around 20%. For these simulations, both the effect from the toroidal shear and Coriolis and centrifugal forces are included. Figures 2a and 2b show the scaling of ion and electron heat flux with $R/L_{T_i}$ in normalized gyroBohm units. The electron temperature gradient is here fixed at the experimental value. An estimate of the stiffness is obtained from these normalized fluxes. As observed, the stiffness of the ILW discharges is larger than the matched CW-discharges. In non normalized units the heat flux for all the four discharges is comparable at the same $R/L_{T_i}$. The ion heat flux is larger than the electron heat flux as expected for ITG dominated discharges. In Figure 2a, the ion heat flux at $\rho = 0.5$ taken from the corresponding TRANSP runs is also shown. For the discharges at lower $R/L_{T_i}$ the experimental heat flux is comparable with the simulated flux while for the discharges at higher $R/L_{T_i}$, the simulated ion heat flux is up to a factor $\sim 3$ higher. The discrepancy between the experimental and simulated fluxes can be explained by the uncertainty in the input parameters, in particular the uncertainty in $R/L_{T_i}$ is large for the ILW discharges. The results follow the linear trends in that the linearly more unstable ILW discharges show significantly larger normalized fluxes. The electron energy confinement times are shorter for the ILW discharges while the ion energy confinement times are similar, as seen in Figures 2c and 2d, in line with the experimental analysis of [1]. The change is due to the difference in NBI heating power deposited to the electrons and ions in the ILW versus CW

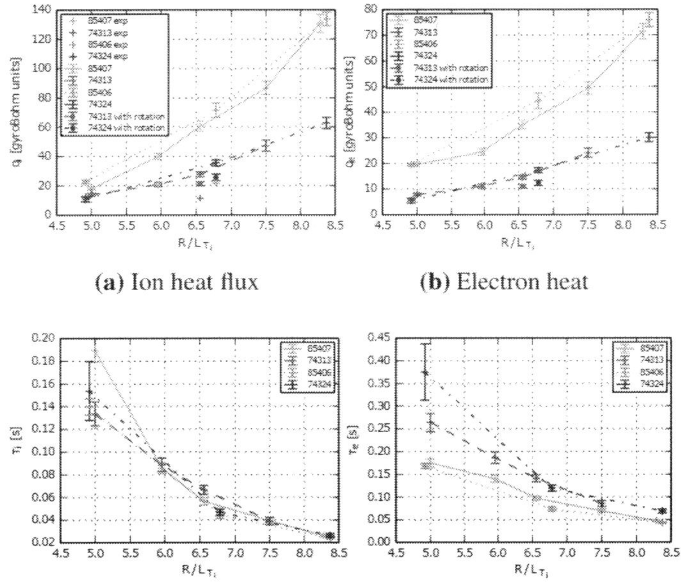


(a) Ion heat flux (b) Electron heat

(c) Ion energy confinement times (d) Electron energy confinement times

**Figure 2:** Nonlinear $R/L_{T_i}$ scans, electron and ion heat flux and energy confinement times

cases. The fraction of total NBI power deposited to the electrons is larger for ILW discharges as compared to the CW discharges. This is a result of the lower edge $T_e$ in the ILW discharges.

## Conclusion

The linear sensitivity scans showed that the relative change in key plasma parameters between the ILW and CW discharges had a significant effect on the ITG mode stability. The total effect of these parameter mismatches was that the ILW discharges were destabilized compared to the CW discharges at all $k_y\rho_s$. The nonlinear results followed the linear ones in that the ILW discharges showed higher normalized heat fluxes at both comparable and experimental $R/L_{T_i}$. The ion energy confinement times were similar, comparing the CW and ILW discharges while the electron energy confinement times were shorter for the ILW discharges which was in line with experimental analysis. These results indicate that the core confinement in the ILW discharges was affected by changes in key plasma parameters due to the degradation of the edge pedestal if compared to CW discharges. Hence, we expect the core confinement in the ILW discharges to be improved if the edge pedestals were recovered.

## Acknowledgements and references

The simulations were performed on resources provided by the Swedish National Infrastructure for Computing (SNIC) at PDC Centre for High Performance Computing (PDC-HPC), on the HELIOS supercomputer system at Computational Simulation Centre of International Fusion Energy Research Centre (IFERC-CSC), Aomori, Japan, under the Broader Approach collaboration between Euratom and Japan, implemented by Fusion for Energy and JAEA, and on the supercomputer JUROPA at Jülich Supercomputing Centre (JSC). This work was funded by a grant from The Swedish Research Council (C0338001). This work has been carried out within the framework of the EUROfusion Consortium and has received funding from the Euratom research and training programme 2014-2018 under grant agreement No 633053. The views and opinions expressed herein do not necessarily reflect those of the European Commission.

[1] Hyun-Tae Kim, M Romanelli, I Voitsekhovitch, T Koskela, J Conboy, C Giroud, G Maddison, E Joffrin, et al. *Plasma Physics and Controlled Fusion*, 57(6):065002, 2015.

[2] R Neu, G Arnoux, M Beurskens, V Bobkov, S Brezinsek, J Bucalossi, G Calabro, C Challis, JW Coenen, E De La Luna, et al. *Physics of Plasmas (1994-present)*, 20(5):056111, 2013.

[3] MNA Beurskens, J Schweinzer, C Angioni, A Burckhart, CD Challis, I Chapman, R Fischer, J Flanagan, L Frassinetti, C Giroud, et al. *Plasma Physics and Controlled Fusion*, 55(12):124043, 2013.

[4] F Jenko, W Dorland, M Kotschenreuther, and BN Rogers. *Physics of Plasmas (1994-present)*, 7(5):1904–1910, 2000.

[5] F Merz. *Gyrokinetic simulation of multimode plasma turbulence.* PhD thesis, University of Münster, 2009.

[6] RL Miller, MS Chu, JM Greene, YR Lin-Liu, and R. E. Waltz. *Physics of Plasmas (1994-present)*, 5(4):973–978, 1998.

[7] LL Lao, H St. John, RD Stambaugh, AG Kellman, and W Pfeiffer. *Nuclear Fusion*, 25(11):1611, 1985.

[8] RJ Hawryluk et al. *Physics of plasmas close to thermonuclear conditions*, 1:19–46, 1980.

[9] C Giroud, GP Maddison, S Jachmich, F Rimini, MNA Beurskens, I Balboa, S Brezinsek, R Coelho, JW Coenen, Lorenzo Frassinetti, et al. *Nuclear Fusion*, 53(11):113025, 2013.

# The effect of nitrogen seeding on ELM filaments

A.V. Bogomolov[1], I.G.J. Classen[1], A.J.H. Donné[1,2,3], H. Meyer[4], M. Dunne[3], P.A. Schneider[3], E. Wolfrum[3], B. Vanovac[1], R. Fischer[5], L.Frassinetti[6], N.C. Luhmann Jr.[7] and the ASDEX Upgrade Team, EUROfusion MST1 Team.

*[1]FOM-Institute DIFFER, Dutch Institute for Fundamental Energy Research, 5600 HH Eindhoven, The Netherlands*
*[2]Eindhoven University of Technology, 5600 MB Eindhoven, The Netherlands*
*[3]EUROfusion, Boltzmannstraße 2, 85748 Garching, Germany*
*[4] Culham Centre for Fusion Energy,Culham Science Centre, Oxfordshire, OX14 3DB, UK*
*[5]Max-Planck-Institut für Plasmaphysik, Boltzmannstraße 2, 85748 Garching, Germany*
*[6]Division of Fusion Plasma Physics, Association EURATOM-VR, KTH, Sweden*
*[7]University of California at Davis, Davis, CA 95616, USA*

E-mail: A.Bogomolov@differ.nl

## Introduction

Edge localised modes (ELMs) are strongly influenced by nitrogen seeding: in comparison to non-seeded discharges nitrogen seeded ELMs are often smaller and have a higher frequency [1]. Using the Electron Cyclotron Emission Imaging (ECEI) diagnostic the spatial - temporal evolution of the electron temperature for nitrogen seeded and normal type I ELMs has been studied in dedicated discharges on ASDEX Upgrade (AUG). The AUG ECEI system is a 2D diagnostic which consists of a linear array of 16 detectors, looking at different vertical positions in plasma. Each detector acts as a standard (1D) ECE radiometer with 8 radial channels. In total ECEI has 128 channels which cover an observational area of 10x40 cm [2]. The difference in ELM duration is connected with additional filaments being ejected in the longer ELMs during an extra phase in the crash identified before [1]. The duration of this phase is correlated with the number of filaments ejected each filament carrying away a certain of energy and particles leading to the higher pressure drop of the longer ELMs. In the following section the measurements and a possible explanation are discussed.

## Evolution of ELMs during nitrogen seeding

Introducing nitrogen into the plasma gradually changes the ELM behaviour. From Fig. 1c it is seen, that the ELMs can be separated in two groups according to their duration: a 'long' ELMs group, with a duration in the range of 2-3ms (red stars) and a 'short' ELMs group (green dots), with a duration of about 1 ms [3]. In the absence of nitrogen both groups simultaneously coexist with the long ELMs prevailing. At 2.3 s the first step of the $N_2$ puff begins. Due

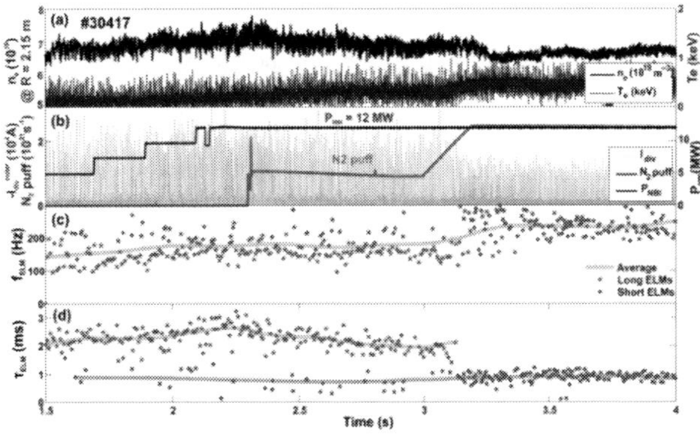


Figure 1. ELM frequency evolution related to the nitrogen seeding. a) Density and temperature time traces, b) Divertor current, NBI heating power and nitrogen puff rate, c) ELM frequency, d) ELM duration. The ELMs with a duration of more than 1.5 ms are considered to be 'long' and are marked with red stars, while the ones with a duration of less than 1.5ms are 'short' and are marked with green dots.

to the low puffing rate, it does not lead to any significant changes in the ELM behaviour. However, after the second step which starts at 3.0 s, the ELMs change: long ELMs disappear and are completely substituted by short ELMs.

**ELM duration and filaments**

The comparison between the short ELMs in the presence of nitrogen and the conventional type-I ELMs was performed on AUG and presented in ref. [3]. Three different phases of an ELM crash have been identified (see Fig. 2). (I) The ELM crash itself with two subphases: (Ia) the rise of the divertor thermal current signal, (Ib) the flat phase of the divertor current; (II) the decay of the divertor current, (III) the inter-ELM phase, during which the pedestal parameters begin to recover [3]. Short nitrogen seeded ELMs differ from type-I ELMs by the absence of the (Ib) phase, where the divertor current is flat. Skipping the (Ib) phase makes nitrogen seeded ELMs to be on average 1 ms shorter than type-I ELMs. The duration of phase (Ib) is correlated with the existence of additional filaments. Here by filaments we consider relatively hot and dense structures in comparison to background, which are elongated in poloidal direction. These structures, opposite to coherent and regular structure of ELM onset modes, are singular events and are only observed outside the separatrix (see Fig. 2b,c). Due to limitations of ECEI only the filaments passing the viewing window with high enough density and the size more than 2.5 $cm^2$ can be seen [2].

Fig. 3a shows the upper mentioned dependency: short ELMs never have more than one filament and in most cases they have no filaments at all. The ELMs which produce two or more filaments are always long. There is also a correlation between the number of filaments and both the temperature and density drop, caused by an ELM. Fig. 3b,c shows these dependencies, with the temperature and density drops taken at $\rho_{pol} = 0.95$. The ELMs with two or more filaments have a tendency to reduce the plasma kinetic profiles more than the ELMs with only one filament or with no filaments at all.

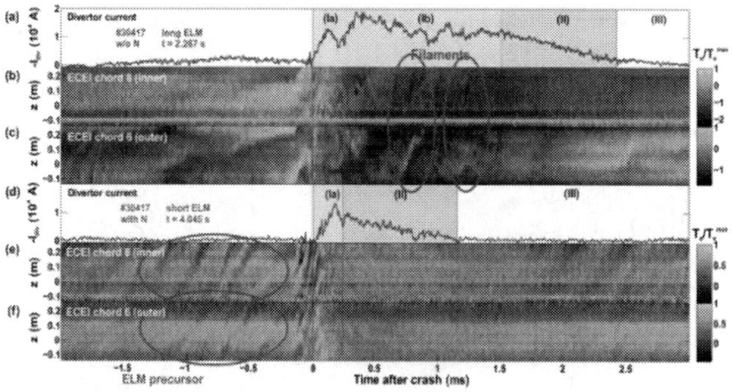


Figure 2: Long (in red) and short (N2 seeded, in green) ELMs as seen by the divertor current (a,d) and ECE Imaging (b,c;e,f). a) Divertor current for an ELM without nitrogen. b) ECEI inner vertical chord. c) ECEI outer vertical chord, two filaments are marked by the red circles d) Divertor current for a nitrogen seeded ELM. e) ECEI inner vertical chord for the ELM without nitrogen, ELM precursor is marked. f) ECEI outer vertical chord.

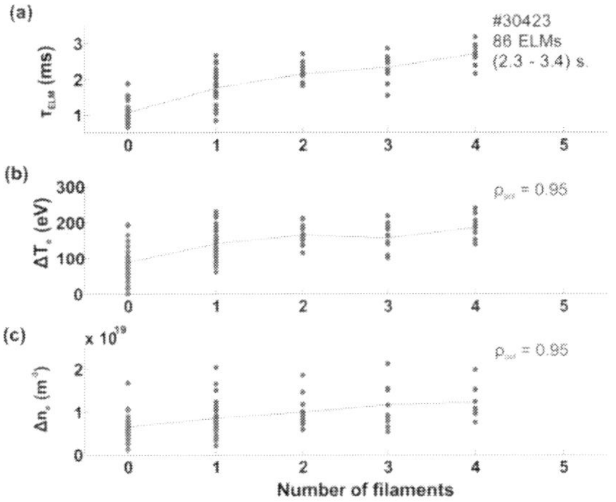


Figure 3: The effect of filaments seen by ECEI on: a) ELM duration b) Electron temperature drop c) Density drop. Long ELMs are marked with the red stars, short ELMs are marked with the green dots. The lines indicating the mean of the groups.

The less filaments are expelled by an ELM, the shorter the ELM is in terms of divertor current signal. As nitrogen seeded ELMs are usually short, a hypothesis is that nitrogen reduces the number of filaments expelled by ELMs. As it is shown in ref. [4], nitrogen seeding enhances the radiation around the X-point, which reduces the temperature in the SOL relative to non-seeded discharges. This likely increases the resistivity as well as the viscosity in the SOL. According to JOREK simulations for JET, lower SOL temperatures lead to a lower number of filament sets ejected [5]. Assuming that nitrogen seeding has a similar effect on the SOL as modeled in ref. [4] in our plasmas, the reduction of filaments due to higher SOL resistivity and viscosity may well explain the differences we observe. The study of the SOL changes and possible JOREK modeling of the AUG experiments is subject to future work.

**Filaments related temperature and density drop**

Fig. 4 shows the averaged density and temperature drops, caused by the ELMs with different number of filaments. The data is obtained from the profiles, delivered by Integrated Data Analysis (IDA) [6]. There is a trend seen, which shows, that the crashes with a higher number of filaments lead to higher drops in both the density and the temperature profiles. Each filament causes, on average, 5 kJ of energy loss, compared to 30 kJ for an averaged loss caused by a type-I ELM in the considered discharge.

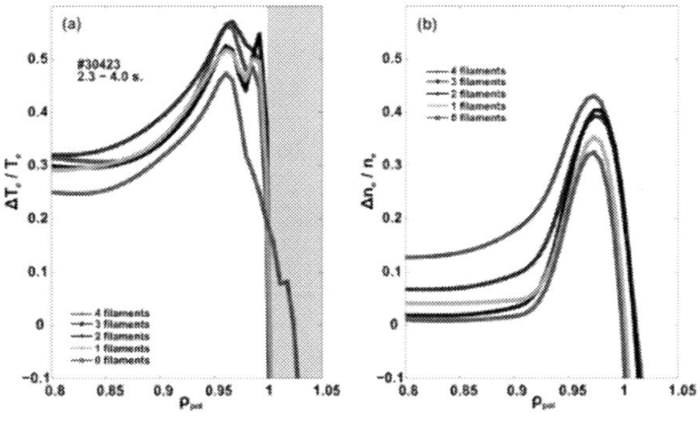


Figure 4: Relative temperature and density drops averaged by the different number of filaments, expelled by the ELMs. The data in the shaded area on the temperature plot is outside of the separatrix and is not reliable.

**Discussion**

In the experiments conducted on AUG both long and short ELMs take place simultaneously, with the domination of long type-I ELMs, before nitrogen seeding is applied. When the nitrogen puff is high enough (~80% of the $D_2$ gas puff rate), the long type-I ELMs completely disappear. The ELM size is usually determined by the ELM duration and/or the pressure drop caused by it. As is shown, both the duration and the temperature and density drop are correlated to the number of filaments expelled by an ELM. With some assumptions a possible mechanism, explaining this ELM behaviour is the following. Nitrogen changes SOL parameters, also reducing the temperature in the divertor and SOL regions [1][4][7] which can lead to reduction of the number of expelled filaments [5]. As soon as there is enough nitrogen in the plasma, the number of filaments expelled by each ELM reduces and subsequently ELMs become shorter. The mechanism of the additional losses in the absence of nitrogen had not been identified for a long time, however, filaments were one of the likely candidates [3]. With the help of ECEI on AUG it became possible to prove, that filaments are a determining factor for these losses.

**Acknowledgment**

This work has been carried out within the framework of the EUROfusion Consortium and has received funding from the Euratom research and training programme 2014-2018 under grant agreement No 633053. The views and opinions expressed herein do not necessarily reflect those of the European Commission.

**References**

[1] A. Kallenbach et al. Nucl. Fusion 49 (2009) 045007

[2] I.G.J. Classen et al. Rev. Sci. Instrum. 81 (2010) 10D929

[3] P. Schneider et al. Plasma Phys. Control. Fusion 56 (2014) 025011

[4] C. Fuchs et al. Journal of Nuclear Materials 415 (2011) S852-S855

[5] S. Pamela et al. Effect of SOL Temperature on Filament Dynamics in MAST Proc. 41st EPS Conf. on Plasma Physics Berlin, Germany, 23-27 June 2014 P5.061

[6] R. Fischer et al. Fusion Science and Technology 58/2 (2010) 675-684

[7] O. Gruber et al. Nucl. Fusion 49 (2009) 115014

# Gyrokinetic modelling of light impurity peaking in JET baseline and hybrid H-modes: a missing ingredient?

P. Manas[1], Y. Camenen[1], S. Benkadda[1], H. Weisen[2], C. Angioni[3],
F. J. Casson[4], C. Giroud[4] , M. Gelfusa[5] and JET contributors*
*EUROfusion Consortium, JET, Culham Science Centre, Abingdon, OX14 3DB, UK*
[1] *Aix-Marseille Université, CNRS, PIIM UMR7345, 13397 Marseille, France*
[2] *CRPP, École polytechnique fédérale de Lausanne (EPFL), CH-1015 Lausanne*
[3] *Max-Planck-Institut für Plasmaphysik, D-85748 Garching, Germany*
[4] *CCFE, Culham Science Centre, Abingdon, OX14 3DB, UK*
[5] *Department of Industrial Engineering, University of Rome "Tor Vergata", via del Politecnico 1, Roma, Italy*

In this paper, quasilinear gyrokinetic modelling of the Carbon peaking factor in JET baseline and hybrid H-modes is presented. The study is based on a database of $\sim$ 1800 shots (during the carbon wall era) covering a large range of plasma parameters: $I_p < 2.6MA, q(r) = 1.5-4, n_e = 2-6\cdot 10^{19}m^{-3}, P_{NBI} = 5-22MW, P_{ICRH} = 0-4MW$. High toroidal rotation cases with improved magnetic equilibrium reconstruction have been extracted ($\sim$ 156 shots) leading to a reduced database previously used for momentum transport studies [1]. The objective of the present work, based on the reduced set, is twofold: can the modelling approach be validated against experimental data and is toroidal rotation an important ingredient for light impurity transport as shown on ASDEX Upgrade [2].

Before addressing these questions, the robustness of the carbon density profiles, provided by the Charge Exchange Recombination Spectroscopy (CXRS) diagnostic [3] (using the C5+ N=8 to 7 line), and of their radial gradients is assessed by performing multiple fits using gaussian noise with experimental error bars as standard deviation. A small deviation for the radial coordinate has also been used. In Fig. 1, the corresponding fitted profiles of carbon density and logarithmic density gradient ($R/L_{n_c}$) are shown using or not the outer localised radial experimental value. The choice of including or not this last experimental point is motivated by the larger uncertainties on the neutral beam attenuation (linked to the pedestal width and position) and hence on the deduced carbon density. Furthermore, it has been observed that taking into account this last point leads to a strong correlation between $R/L_{n_c}$ and the shot number over physically relevant parameters for cases close to the edge ($\rho \geq 0.8$ ). In the following, a detailed comparison between modelled and experimental carbon peaking factor is presented at $\rho = 0.65$ where $R/L_{n_c}$ is negative (outward pinch velocity) and the fitted gradients are robust with respect to fitting hypotheses.

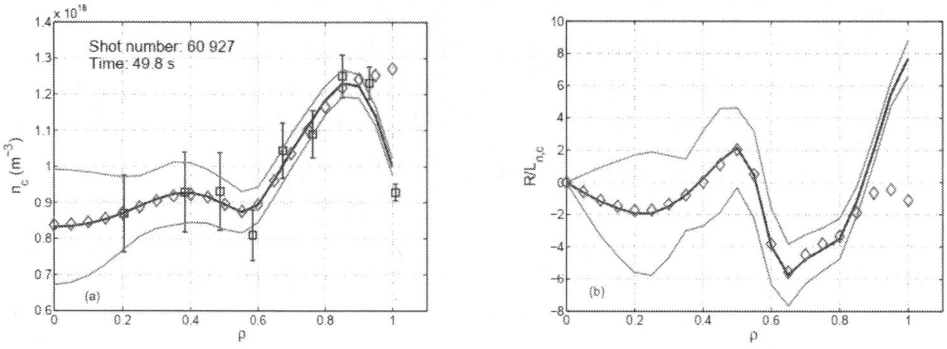


Figure 1: Carbon density (a) and logarithmic density gradient (b) profiles. Fitted profiles using the outer localised radial experimental point (blue line) or not (red diamonds) are shown together with error bars for the former (pink lines). Experimental data are represented by blue squares.

The modelling of the carbon peaking factor is performed using the flux-tube version of the gyrokinetic code GKW [4] which includes centrifugal effects and realistic magnetic equilibrium to compute the turbulent impurity flux. Neoclassical contributions to this peaking factor are evaluated using the neoclassical code NEO and found to be negligible compared to turbulent transport (a factor of $\sim 100$ between the neoclassical and turbulent diffusivities). A quasilinear approach is chosen to tackle the large number of experimental data. Using a mixing length rule [5], the quasilinear particle flux can be written as follows:

$$\Gamma_{QL} = \sum_{k_\theta} \Gamma_{k_\theta} \frac{\gamma_{k\theta}}{< k_\perp^2 >} \qquad \text{with} \qquad < k_\perp^2 > = \frac{\int |\phi|^2 k_\perp^2 \,\mathrm{d}s}{\int |\phi|^2 \mathrm{d}s} \tag{1}$$

with $\gamma_{k\theta}$ the growth rate of the linear instability at the poloidal wave number $k_\theta$, $\phi$ the electrostatic potential, $s$ the field aligned coordinate and $k_\perp$ the local perpendicular wave vector. The normalised impurity flux is decomposed in a diffusive and a convective part:

$$\frac{R\Gamma_{QL}}{n} = D_{QL}\frac{R}{L_n} + RV_{QL} = D_{QL}\left(\frac{R}{L_n} + C_{T,QL}\frac{R}{L_T} + C_{u,QL}u' + C_{p,QL}\right) \tag{2}$$

with $R$ the major radius, $u'$ the normalised gradient of toroidal rotation frequency ($R^2\nabla\Omega/v_{th,i}$), $D$ a diffusion coefficient and $V$ the pinch velocity consisting of three terms: thermo-diffusion ($C_T\frac{R}{L_T}$), roto-diffusion ($C_u u'$) and a constant ($C_p$). Since there is no source of carbon in the core, the steady state carbon particle flux is zero, hence giving a direct relation for the quasilinear logarithmic density gradient $R/L_{n_c} = -RV_{QL}/D_{QL}$.

In Fig. 2 experimental and modelled carbon peaking factor are represented against the normalised gradient of toroidal rotation frequency $u'$ and the normalised collisionality $\nu_*$ for cases without ICRH at $\rho = 0.65$ (ICRH can induce significant carbon poloidal asymmetries and these cases are left for another study). The predicted $R/L_{n_c}$ is also shown neglecting roto-diffusion, underlining the importance of this term in recovering negative experimental carbon peaking factor at high $u'$. In Fig. 2b, a marked dependency of experimental $R/L_{n,c}$ against the neoclassical collisionality $\nu_*$ is observed which does not appear for the modelled $R/L_{n,c}$. Furthermore, while modelled and experimental carbon peaking factor are in agreement within experimental error bars at low $\nu_*$ ($< 0.03$), a clear difference is seen at higher $\nu_*$. Variations of physical parameters used in gyrokinetic simulations such as the safety factor ($\pm 50-100\%$), the magnetic shear ($\pm 50\%$), the toroidal velocity ($\pm 50-100\%$) and its gradient ($\pm 50-100\%$), the logarithmic temperature ($\pm 15\%$) and density gradients ($\pm 20\%$) do not make it possible to recover such high negative values of the carbon peaking factor. Several mixing length rules have also been tested together with finite radial wave numbers $k_x$, changing the spectral shape to increase higher or lower $k_\theta \rho_i$ contributions.

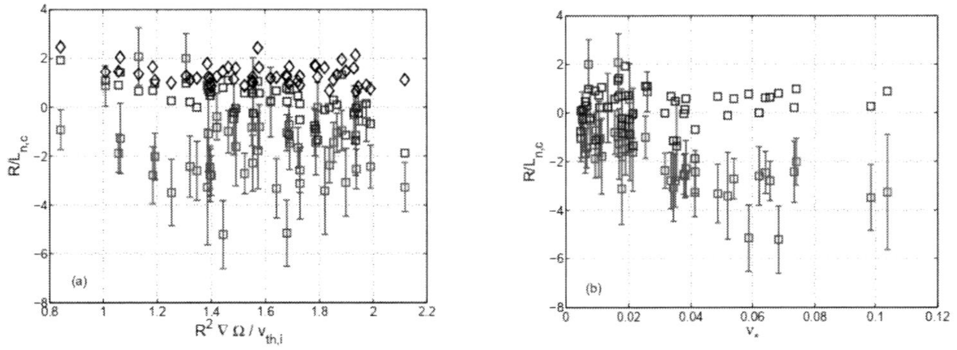


Figure 2: Modelled carbon peaking factor (blue squares) and experimental values (red squares) versus the normalised gradient of toroidal rotation frequency (a) and $\nu_*$ (b). Cases without roto-diffusion are also represented (black diamonds).

To better characterise this $\nu_*$ dependency, multilinear regressions are performed on experimental and modelled $R/L_{n_c}$ (Fig. 3) with the following physically relevant set of parameters: $R/L_n$, $R/L_{T_i}$, $R/L_{T_e}$, $T_i/T_e$, $u'$, $R\Omega/v_{th,i}$, $q$, $\hat{s}$, $Z_{eff}$, $\nu_*$, $\beta$. The regressed $R/L_{n_c}$ is computed using 3 variables with their coefficients $b$, uncertainty $\delta b$ and the statistical significance (STS) $b/\delta b$ (see [1] for more details). The expression for the regressed $R/L_{n_c}$ is then of the form:

$$\frac{R}{L_{n_c}} regressed = b_1 var_1 + b_2 var_2 + b_3 var_3 + constant \qquad (3)$$

Using this decomposition, a set of fits are obtained with increasing standard deviation $\sigma$. In Fig. 3 regressions for only one set of 3 variables is represented. For the 5 best fits (minimum standard deviation) recurring parameters for both modelled and experimental $R/L_{n_c}$, $u'$ and $R/L_n$ are always obtained (Fig. 3a, 3c) while the third variable can be either $T_i/T_e$, $R/L_{T_i}$, $R\Omega/v_{th,i}$ or the magnetic shear. This dependence in $u'$ and $R/L_n$ is very similar in magnitude (linear coefficients b) with a lower value for modelled $R/L_{n_c}$. In Fig. 3b and 3d, $R/L_n$ has been suppressed from the set of parameters unraveling the strong dependency of the experimental carbon peaking factor and $\nu_*$ ($\nu_*$ and $R/L_n$ are strongly correlated). This dependence in $\nu_*$ is less pronounced for modelled $R/L_{n_c}$ which is consistent with results of Fig. 2. It is also interesting to note that the coefficients in front of $R/L_{T_i}$ (linked to thermo-diffusion) are similar for experimental and modelled carbon peaking factor.

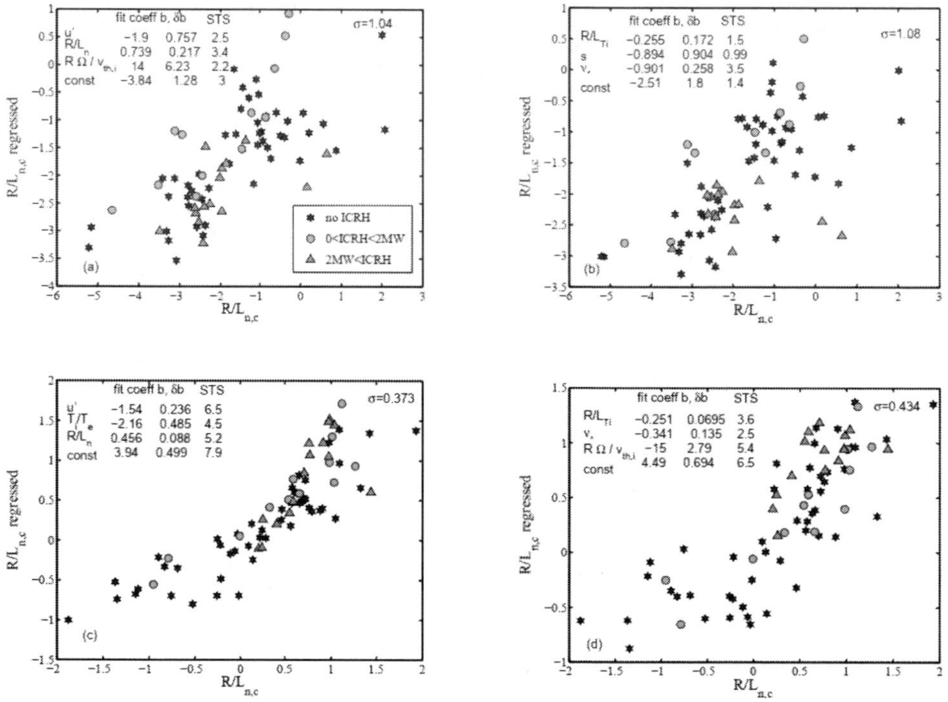


Figure 3: Multilinear regression of the experimental (a,b) and modelled (c,d) carbon peaking factor at $\rho = 0.65$.

To conclude, provided roto-diffusion is included, a rather good prediction of the carbon peaking factor at high gradient of toroidal rotation frequency and low $\nu_*$ is observed for this JET H-modes database, using quasilinear gyrokinetic simulations (neoclassical contributions are negligible). However the calculations fail to predict high negative experimental $R/L_{n_c}$ at high $\nu_*$. Multilinear regressions of modelled and experimental $R/L_{n_c}$ help identifying the relevant parameters and confirm a stronger dependence on $\nu_*$ for experimental $R/L_{n_c}$ than in the modelling. This work has been carried out within the framework of the EUROfusion Consortium and has received funding from the Euratom research and training programme 2014-2018 under grant agreement No 633053. The views and opinions expressed herein do not necessarily reflect those of the European Commission.

**References**

[1] H. Weisen *et al*, Nuclear Fusion **52**, 042001 (2012)
[2] F. J. Casson *et al*, Nuclear Fusion **53**, 063026 (2013)
[3] C. Giroud *et al*, Review of Scientific Instruments **79**, 10F525 (2008)
[4] A. G. Peeters *et al*, Computer Physics Communications **180**, 2650 (2009)
[5] F. Jenko, T. Dannert and C. Angioni, Plasma Phys. Control. Fusion **47**, B195 (2005)

*See the Appendix of F. Romanelli et al., Proceedings of the 25th IAEA Fusion Energy Conference 2014, Saint Petersburg, Russia

# The "neutron deficit" in JET baseline H-modes and hybrid regimes

H. Weisen[1], Hyun-Tae Kim[2], J. Strachan[3], S. Scott[3], Y. Baranov[2], J. Buchanan[2], M. Fitzgerald[2], D. Keeling[2], D. King[2], L. Giacomelli[4], T. Koskela[5], M.J. Weisen[6], C. Giroud[2], W.G. Core[2], K.-D. Zastrow[2] and JET contributors*

*Eurofusion Consortium JET, Culham Science Centre, Abingdon, OX14 3DB, UK*
*[1]CRPP, Ecole Polytechnique Fédérale de Lausanne, Switzerland, [2]Culham Centre for Fusion Energy, Abingdon, [3]PPPL, Princeton University, USA, [4]IFP-CNR, Milano, Italy, [5]Aalto University School of Science, Espoo, Finland, [6]University of Strathclyde, Glasgow, UK*
**see the Appendix of F. Romanelli et al, Proc 25th IAEA Fusion Energy Conference, 2014, St. Petersburg, Russia*

## 1. INTRODUCTION.

We refer as «neutron deficit» to a situation where the measured neutron rate falls short of expectations based on codes such as TRANSP, assuming only collisional fast ion orbit diffusion or no radial fast ion transport at all. The neutron deficit affects JET-C and JET-ILW discharges alike and can range from 0 to 50% typically. Frustratingly, despite being well characterised by the study here presented, the neutron deficit has so far eluded all attempts at identifying its causes. In JET, unlike ITER, neutrons are primarily from beam-thermal reactions and the causes of the neutron deficit are believed to due to processes affecting the fast ion-thermal reactivity, such as fuel dilution, NBI deposition and fast ion transport. Previous studies of the neutron deficit in JET have focussed on the trace Tritium campaign in 2003 [1,2]. The study presented here is based on Deuterium discharges, mostly from the JET carbon phase under EFDA (JET-C) and cover a wide range of plasma conditions in baseline H-mode, hybrid scenarios, together with a few L-modes. A subset of 317 samples, presented here, covers the following ranges:

$0.8\text{MA} \le I_p \le 4\text{MA}$, $1\text{T} \le B_T \le 3.4\text{T}$, $2\text{ MW} < P_{NBI} < 23\text{MW}$, $2.1 \le q_{95} \le 4.7$,

$1.5\times10^{19} \le \langle n_e \rangle \le 9.4\times10^{19}\text{m}^{-3}$, $0.002 \le \langle n_C \rangle / \langle n_e \rangle \le 0.06$, $1.4 \le Z_{eff}(VB) \le 4$, $0.06 \le \tau_E \le 0.5\text{s}$,

$0.47 \le H98 \le 1.4$

Here $n_C$ refers to the carbon density from CXRS, the brackets refer to volume averages, $Z_{eff}(VB)$ is the effective charge as measured by visible bremstrahlung, $\tau_E$ is the energy confinement time based on the kinetic stored energy calculated from the plasma profiles and H98 is $\tau_E$ normalised to the IPB98(y,2) scaling. Discharges with $P_{ICRH}/P_{NBI}>0.1$, increased toroidal ripple, 3D fields from internal coils and the entire trace T campaign were excluded. Neutron rates were measured using the JET fission chambers, as recently recalibrated [3].

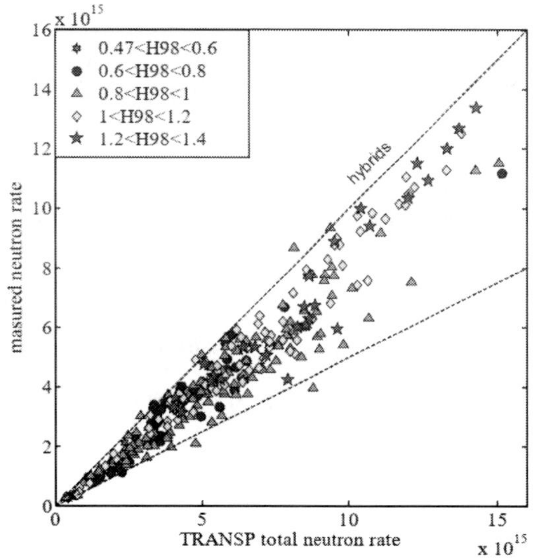


Fig.1 Measured v expected neutron rates in JET-C

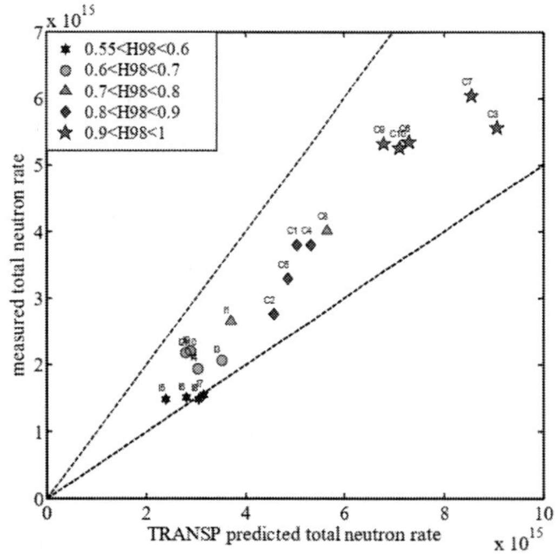


Fig.2 Measured versus expected neutron rates for 10 pairs of discharges with members from JET-C (labelled C1-C10) and JET-ILW (labelled I1-I10). I1,I2 and I3 were nitrogen seeded with $Z_{eff}$ not exceeding that of their JET-C counterparts.

TRANSP simulations for the 317 samples from JET-C were produced using $T_i$, the toroidal rotation $\omega_\phi$ and the dilution inferred from $n_C$ from CXRS and electron density and temperature from LIDAR Thomson scattering. Importantly, no ad-hoc adjustments to the measurements were made with the purpose of improving agreement between calculations and measurements. The measured neutron rates are in the range 50-100% of the total neutron rate expected from TRANSP, as seen in fig.1 The symbols refer to classes of H98.

Fig.2 shows that the neutron deficit also affects JET-ILW (i.e. after JET was equipped with Tungsten and Be PFC's) plasmas by comparing pairs of baseline H-modes in JET-C and JET-ILW with matching $I_p$=2.5MA, $B_T$=2.7T, $\langle n_e \rangle$ and $P_{NBI}$ (in the range 14-17MW) and triangularity, as detailed in [4]. CXRS ion temperature measurements were not available for the JET-ILW discharges, however they were chosen to have high enough density ($7\times10^{19}m^{-3}<\langle n_e \rangle<10^{20}m^{-3}$) to safely assume $T_i$=$T_e$. The JET-ILW discharges had a clearly lower confinement (H98≈0.63 on average) than the JET-C cases (H98≈0.9). The deficit was also somewhat larger, with the ratio of measured to expected neutrons rates, $R_N/R_{N\ TRANSP}$≈0.7 for the 10 JET-C samples and $R_N/R_{N\ TRANSP}$≈0.64 on average for the 10 JET-ILW samples.

## 2. PARAMETER DEPENDENCIES AND MODELLING

The deficit correlates with plasma parameters, being smallest or absent in discharges with high $T_e$, $T_i$, and $\beta_N$, such as hybrid scenarios. Contrary to widespread belief, $Z_{eff}$ and dilution appear to play at best a minor role. The peakedness of the NBI power deposition profile, expressed as $\langle q_{NB}(\rho<1/3)\rangle/\langle q_{NB}(\rho<1)\rangle$, ranging from 0.4 (hollow) to 5 (very peaked), is uncorrelated with the deficit, showing that potential errors in the beam stopping calculations do not explain the neutron deficit. Simple linear regressions of the ratio $R_N/R_{N\ TRANSP}$ (measured/expected), as shown in fig.3, using a small number (3-4) of plasma parameters achieve standard deviations near 10%, well within the best expectation, given that the key parameters governing the neutron rate (e.g. $T_e$, $T_i$, $P_{NBI}$, $n_e$, $n_C$) are quoted with errors of the order of 10%. They demonstrate that the neutron deficit is at least to a large part a matter of plasma physics, but do not exclude the possibility that systematic experimental errors, such as calibration errors, may also play a role. The parameters in the example regression are $\beta_N$, $T_i(\rho=1/3)$, $T_{slow}(\rho=1/3)/\tau_E$ and $\beta_{fast}= W_{fast}/(B^2/2\mu_0)$. The latter has a negative coefficient, suggesting that fast ion pressure driven modes may play a role. $T_{slow}(\rho=1/3)$ is the fast ion slowing down time from the birth energy to thermal evaluated at 1/3 of the minor radius. Note that these parameters should not be interpreted as being causally related to the deficit.

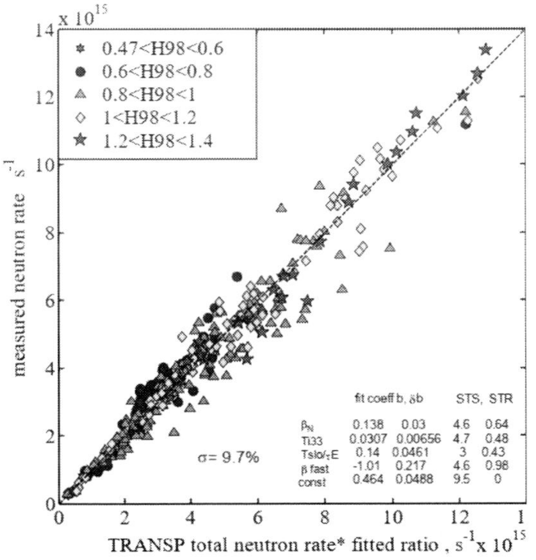


Fig.3 Measured neutron rate versus TRANSP prediction corrected with a regression. STS is the statistical significance for a 90% confidence interval and STR is the normalised statistical relevance.

A simple model was used for assessing a hypothetical relation between fast ion transport and thermal energy transport. It calculates the radial diffusion of fast ions, assuming $D_f \propto \chi$, together with the fast ion slowing down and associated neutron production. Neutron rates are reduced by transport because of the increased (unproductive) slowing down on the electrons at the lower $T_e$ experienced by fast ions having moved outwards, as well as due to the reduced $\langle \sigma v \rangle_{DD}$ at lower $T_i$. Fig.4 shows the observed ratio $R_N/R_{N\ TRANSP}$ versus the ratio obtained from assuming $D_f=\chi_i$ from the power balance and $D_f=0$ using the model. It is plain that the observed dependencies are not reproduced by the model. In particular, ions with

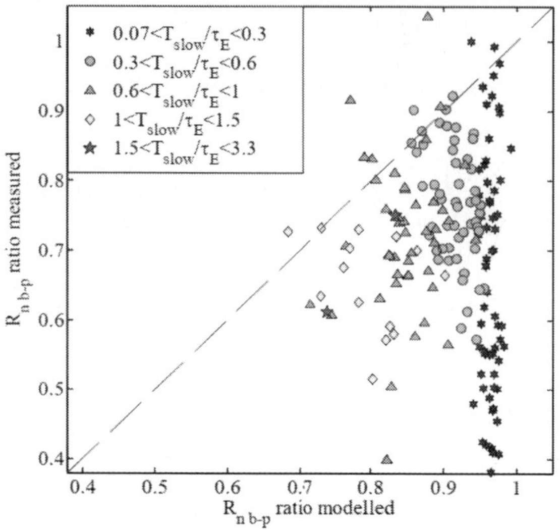


Fig.4 Measured versus modelled beam-thermal neutron deficit, assuming $D_f=\chi_i$ and $D_f=0$

$T_{slow}(\rho=1/3)/\tau_E<0.3$ (blue stars) should be to a large degree immune to transport at the time scale of $\tau_E$, but the results show, to the contrary, that this class of data spans the whole range, from no deficit to the worst in the database.

3. MHD

Fast ion redistribution by a variety of MHD modes, such as sawteeth, NTMs and fishbones, provides a potential explanation. Neutron rates in most baseline H-modes are too low for measuring sawteeth using the JET neutron camera. This drawback has been overcome by the boxcar-ing of 2600 quasi-identical sawteeth with 0.5ms time resolution from 132 repeated pulses, using a central ECE signal for crash timing. The result shows that neutron sawteeth (reflecting those of the underlying fast ions) are very similar to those of other plasma parameters such as $T_e$, with a ~16% crash for the central channel and inverted neutron sawteeth outside the inversion radius. Ad-hoc modelling of the mixing, as well as the (selectable) sawtooth model in TRANSP, show that mixing by sawteeth can account at best for a few % of the neutron deficit on a sawtooth cycle averaged basis, when sawteeth are present at all. Boxcar-ing of 18000 ELMs has revealed no discernible effect of the ELMs on the neutron rates. A still ongoing analysis of MHD activity based on the JET high resolution toroidal array and spanning the mode number range $-10<N<10$ has so far revealed no significant correlation between mode activity and the neutron deficit. It is tempting to see this as supportive of ASCOT simulations of fast ions in the 3D fields of NTM's, which show that NTM's can only account for a modest reduction in neutron rates below expectations and are far from being able to account for the largest deficits, even for island chains spanning the minor radius [5]. However, as the observed modes have not yet been categorised into types such as NTM's, fishbones, kinks etc, the lack of a correlation between MHD activity and the neutron deficit may simply indicate that the modes producing the largest signals at the probes may not be the most potent ones at causing fast ion transport.

## References

[1] Y. Baranov et al, PPCF 51 (2009) 044004
[2] M. Nocente et al, Nucl. Fusion 54 (2014) 104010
[3] D. B. Syme et al, ISFNT-11, Barcelona, 16-20.9.2013
[4] Hyun-Tae Kim et al, PPCF 57, 2015 065002
[5] T. Koskela et al, this conference, P2.106

# MHD in JET hybrid plasmas with the ITER Like Wall

P Buratti[1], M Baruzzo[2], C D Challis[3], D Frigione[1], J Garcia[4], E Giovannozzi[1], T C Hender[3], J Hobirk[5], E Joffrin[4], J Mailloux[3], G Pucella[1] and JET Contributors*

*EUROfusion Consortium, JET, Culham Science Centre, Abingdon, OX14 3DB, UK*

*[1]ENEA for EUROfusion Via E. Fermi 45, 00044 Frascati (Roma), [2]RFX, Corso Stati Uniti 4, Padova, Italy, [3]CCFE, Culham Science Centre, Abingdon, OX14 3DB, UK, [4]CEA, IRFM, F-13108 Saint-Paul-lez-Durance, France, [5]Max-Planck-Institut für Plasmaphysik, D-85748 Garching, Germany, *See the Appendix of F. Romanelli et al, Proc. of the 25th IAEA FEC, Saint Petersburg, Russia, 2014*

The achievable performance in terms of beta and duration of hybrid scenarios is often limited by the onset of MHD modes that degrade confinement or even induce disruptions. The most dangerous modes have tearing structure with poloidal and toroidal mode numbers $(m, n)$ = (2, 1), (3, 2), (4, 3). In addition, studies in JET plasmas with the ITER-like wall (JET-ILW), which includes tungsten divertor surfaces, revealed that also the internal (1, 1) mode can assume tearing character [1] and that tearing modes can foster central accumulation of tungsten [2].

This paper is aimed at a statistical survey of MHD instabilities that can affect the performance of JET hybrid discharges, with particular attention to onset conditions and saturated amplitude in terms of the normalised beta ($\beta_N$); other issues such as the influence on impurity accumulation and on energy confinement will be reported elsewhere [3].

The database includes about 260 pulses for JET-ILW and 180 selected pulses for the previous carbon wall (JET-C). Modes appearing after radiated power exceeds 70% of input power are discarded; this eliminates instabilities associated with heavy impurity influxes and hollow temperature profiles. Modes with $n = 4$ or higher are not considered. Toroidal numbers and mode amplitudes (zero-to-peak poloidal field oscillation) are evaluated by cross-spectral analysis of a toroidal array of Mirnov coils. Poloidal numbers are identified from the mode location as inferred by matching the mode frequency against the plasma rotation profile [4], e.g. the $n = 2$ mode location is close to the $q = 1.5$ and then its poloidal number is $m = 3$.

**Statistical overview**

The maximum amplitude reached by the $n = 2$ mode during each pulse is shown in Fig. 1 for JET-C and JET-ILW datasets separately. A similar trend to increase mode amplitude with normalised beta can be observed in both cases, with about 30% larger values for JET-C. Data clustered near zero amplitude correspond to weak (1-2 cm island width) and generally transient modes that are not relevant to the present analysis. These modes are discarded in the following by imposing a threshold of $3\times10^{-6}$ T s on the time-integrated amplitude. Maximum amplitudes of $m = 1$, $n = 1$ and $n = 3$ modes with the ILW are shown in Fig. 2. Threshold values of $10^{-5}$ T s for $n = 1$ and $10^{-6}$ T s for $n = 3$ are used in the following. The $m = 2$, $n = 1$ occurs rarely; it is easily recognized for having location at $q = 2$ and large amplitude (> $1.5\times10^{-3}$ T typical).

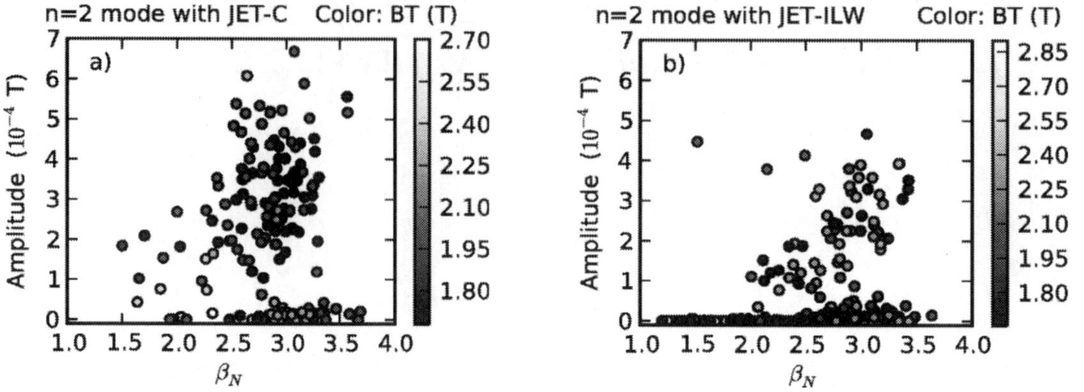


Figure 1. Maximum amplitude of the $m = 3$, $n = 2$ mode during each pulse as a function of normalised beta a) with the C-wall and b) with the ILW. The toroidal field is shown by color code. Outliers above $3\times10^{-4}$ T at $\beta_N < 2.5$ in b) are from pulses with early NBI timing at $q_{95} = 3$.

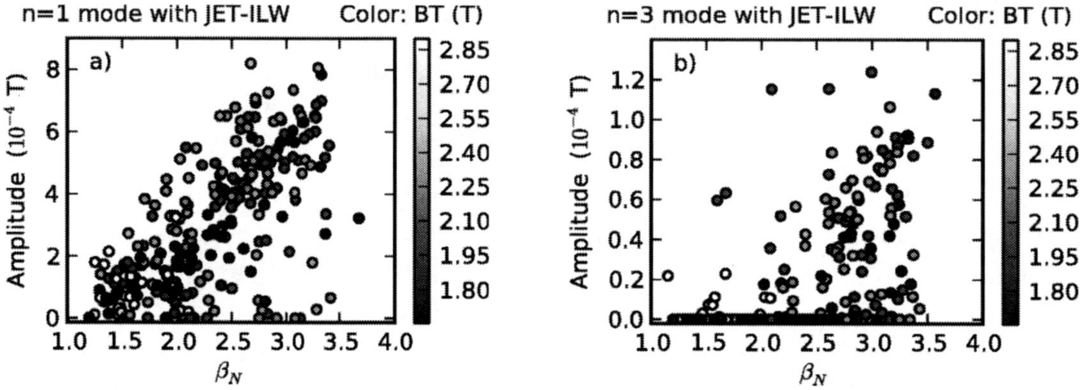


Figure 2. Maximum mode amplitude detected during each pulse with the ILW as a function of normalised beta a) for $m = 1$, $n = 1$ and b) for $m = 4$ (or 5), $n = 3$. Color code refers to toroidal field.

The incidence of modes above threshold with $n = 1$, 2 or 3 is compared in Fig. 3 for the JET-C and JET-ILW. The $n = 1$ activity is in the core ($q \approx 1$) region; cases with $m = 2$, $n = 1$ activity are rare and have been ignored to stay with three-sets Venn diagrams; four-sets ones will be shown elsewhere [3]. Generally the $n = 3$ activity is at the $q = 4/3$ surface, but in some pulses it is at the $q = 5/3$ surface. The incidence of pulses without MHD activity or with core $n = 1$ activity alone is much larger in the ILW database.

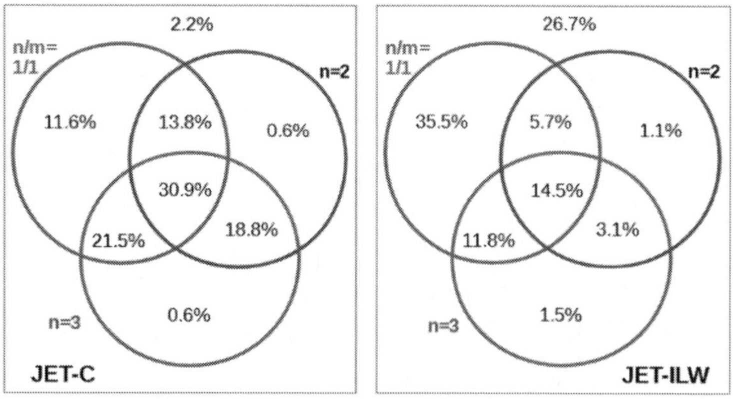


Figure 3. Venn diagrams of mode incidence observed in pulses with JET-C (left) and with JET-ILW (right). Data outside circles indicate pulses without modes above threshold.

From Fig. 4 it can be observed that a possible reason for the differences is that the proportion of pulses with lower $\beta_N$ is much higher with the ILW. The systematic difference in density (due to operation constraints to avoid impurity influx with the ILW) could also play a role by increasing collisionality, which probably affects the density peaking and then the bootstrap drive with it. Another difference is the earlier NBI timing (< 4 s) in several JET-C pulses at low toroidal field (1.7 T): the 1/1 mode is very rare and other modes are very frequent in these pulses.

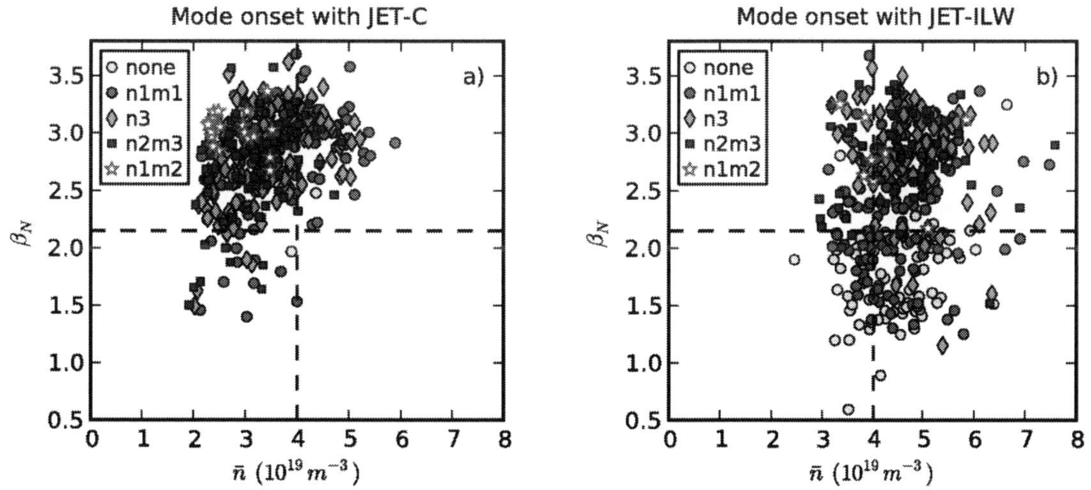


Figure 4. Normalised beta as a function of line-average density at mode onset, i.e. at threshold crossing time, or at time of maximum beta for mode-free cases (with "none" label). Cross-hairs are to help the comparison.

**Mode onset**

The first question addressed in this section is how long the high performance phase of hybrid pulses lasts before the onset of MHD activities. From Fig. 5 it can be seen that presently achieved durations from NBI start to mode onset time with JET-ILW can only exceed 3 s at $\beta_N$< 2.5, while longer durations were obtained with the JET-C at any $\beta_N$. However, the main reasons for the systematic lack of long high power pulses with the ILW were heat load and impurity problems. The second question, how $q_{95}$ affects the maximum $\beta_N$ that can be reached before the onset of MHD activities with $n/m > 1$, is addressed in Fig. 6, from which it can be seen that the maximum $\beta_N$ decreases from 3.5 to 3 as $q_{95}$ is lowered from 4 to 3. Also the mode-free region at low $\beta_N$ tends to disappear at $q_{95} \approx 3$.

The last addressed question concerns the role of sawtooth crashes (or of strong fishbones) as triggers of the observed modes. Different shapes of temperature profile variation inside the sawtooth inversion radius were observed, ranging from the usual flattening to partial flattening to unperturbed central temperature (annular crash). Two onset types, with trigger by ordinary (i.e. with central temperature drop) and by annular sawtooth respectively, were classified in order to recognize the occurrence of the annular shape. The onset of modes in the presence of frequent fishbones, for which no clear-cut time coincidence can be appreciated, was classified as uncertain. Last, there are mode onsets that occur far from sawtooth crashes or even in sawtooth-free pulses, and were then classified as not sawtooth triggered. From Fig. 7 it can be seen that, for the $n = 2$ mode, ordinary sawtooth triggers are dominant at lower $\beta_N$ and higher $q_{95}$, while triggers by annular crashes appear at higher $\beta_N$. Modes without sawtooth trigger are more frequent at high $\beta_N$ or at low $q_{95}$; cases at low $\beta_N$ and low $q_{95}$ are from pulses in which the minimum $q$ was raised and the sawtooth was suppressed by advancing NBI timing. Qualitatively similar results are found for $n = 3$ modes.

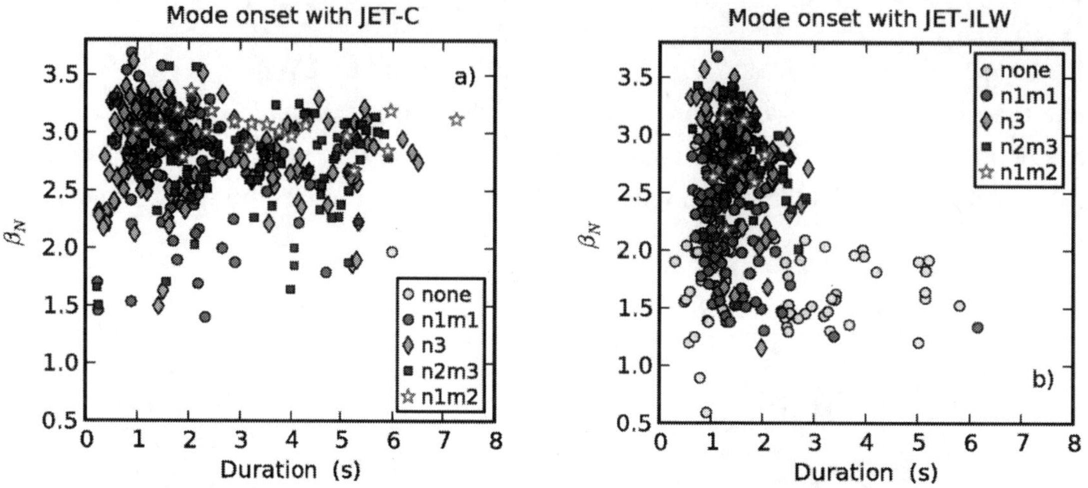


Figure 5. Normalised beta at mode onset as a function of duration from NBI start to mode onset time. Duration before radiation exceeds 70% of input power is used for pulses without MHD (grey dots).

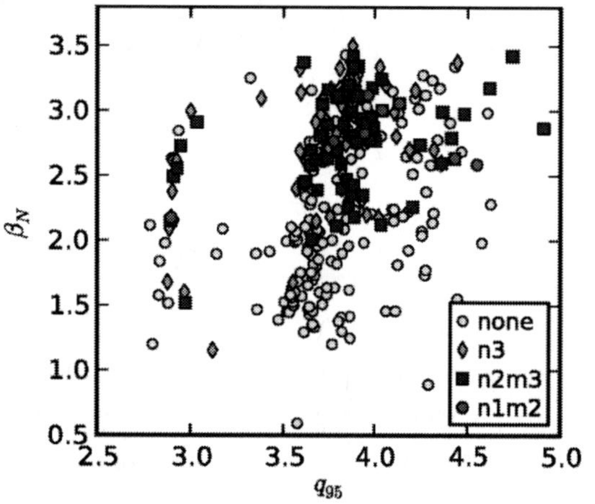


Figure 6. Normalised beta at the onset of instabilities with m/n = 2/1 (red), 3/2 (blue) and 4/3 or 5/3 (green) in ILW hybrid plasmas. Grey dots represent the maximum beta in discharges without the mentioned modes.

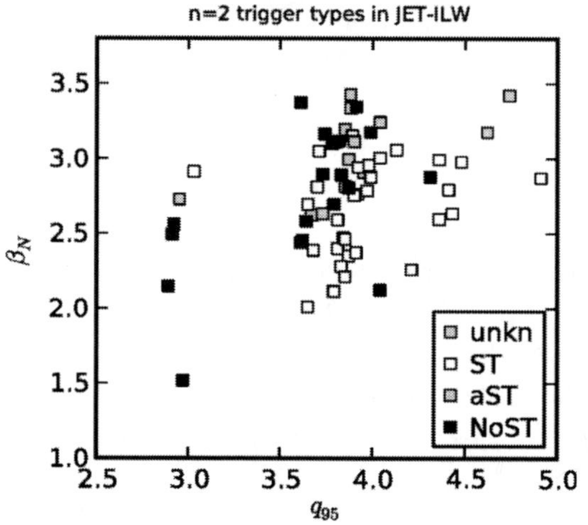


Figure 7. Onset types for the $m = 3$, $n = 2$ mode in ILW, ordinary sawtooth trigger in white, annular sawtooth trigger in yellow, uncertain in grey, not triggered by sawtooth in black.

## Summary

High $\beta_N$ pulses with the ILW develop MHD instabilities after 2-3 s of NBI heating; presently, pulses with MHD-free duration in excess of 5 s only exist at $\beta_N < 2$. Short durations at $\beta_N > 3$, with the ILW, appear to be caused by intrinsically less stable profiles, in fact sawtooth triggers seldom occur in this range. It is important to remark that, rather than to MHD onset, the systematic lack of long high power pulses with the ILW was due to heat load and impurity problems. Improved impurity control should also lead back towards the previous JET-C stability conditions.

[1] M. Baruzzo *et al 40th EPS Conf.* (2013) P5.161
[2] T.C. Hender *et al 41th EPS Conf.* (2014) P1.011
[3] T.C. Hender *et al* in preparation.
[4] P Buratti *et al 41th EPS Conf.* (2014) P1.014

Acknowledgments. *This work has been carried out within the framework of the EUROfusion Consortium and has received funding from the Euratom research and training programme 2014-2018 under grant agreement No 633053. The views and opinions expressed herein do not necessarily reflect those of the European Commission.*

# Overview of hybrid development in JET with ITER-Like Wall

D Frigione[1], C Challis[2], J Garcia,[3] J Hobirk[4], B Alper[2], G Artaserse[1], C Angioni[4], M Baruzzo[5], E Belonohy[4], R Bilato[4], A Brett[2], P Buratti[1], I Carvallho[6] , F Casson[2], L Frassinetti[7], L Garzotti[2], E Giovannozzi[1], D Grist[2], NC Hawkes[2], E Lerche[8], J Mailloux[2], P Mantica[9], M Maslov[2], J Morris[2], G Pucella[1], E Rachlew[10], M Reich[4], F Rimini[2], ACC Sips[10], MF Stamp[2], P Tamain[3], M Tsalas[11], M Valisa[5], H Weisen[12] and JET contributors*

EUROfusion Consortium, JET, Culham Science Centre, Abingdon, OX14 3DB, UK
**1**)ENEA for EUROfusion, via E. Fermi 45, 00044 Frascati (Roma), Italy. **2**)CCFE, Culham Science Centre, Abingdon, OX14 3DB, UK. **3**)CEA, IRFM, F-13108 Saint-Paul-lez-Durance, France. **4**)Max-Planck-Institut für Plasmaphysik, D-85748 Garching, Germany. **5**)Consorzio RFX, corso Stati Uniti 4, 35127 Padova, Italy. **6**) Instituto de Plasmas e Fusão Nuclear, Instituto Superior Técnico, Universidade de Lisboa, 1049-001 Lisboa, Portugal. **7**)Fusion Plasma Physics, EES, KTH, SE-10044 Stockholm, Sweden. **8**)Laboratory for Plasma Physics Koninklijke Militaire School - Ecole Royale Militaire Renaissancelaan 30 Avenue de la Renaissance B-1000, Brussels, Belgium, 9) Istituto di Fisica del Plasma, CNR/ENEA, Milano, Italy. **10**)European Commission, B-1049 Brussels, Belgium. **11**)FOM Institute DIFFER P.O. Box 1207 NL-3430 BE Nieuwegein, The Netherlands. **12**)Ecole Polytechnique F.d.rale de Lausanne (EPFL), CRPP, CH-1015 Lausanne, Switzerland
**See the Appendix of F. Romanelli et al., Proc. 25th IAEA Fusion Energy Conference 2014*

Since 2011 JET has operated with a Tungsten (W) divertor and a Beryllium (Be) main chamber to test first wall materials adopted for ITER [1]. The hybrid scenario [2], originally characterized by a low central shear q-profile and aiming at a moderate density (Greenwald fraction ~0.6), $q_{95}$~4, $\beta_N$>2.5 and $H_{98y2}$ > 1.2, has been studied and developed with this new ITER like wall (ILW). The main difference with the former Carbon wall (C-wall) is the lack of an intrinsic edge radiator and the presence of metals radiating from the core. Tungsten accumulation is a major issue in Hybrid: it has recently been shown that the phenomenon is mainly associated with the peaking of the electron density profiles, typical of this regime, plus additional contributions coming from MHD and poloidal asymmetries [3]. Gas puffing and central ICRH heating are used to mitigate this effect which can degrade the performance or even lead to a disruption: the first increases the ELM frequency which helps in flushing out impurities and the latter has beneficial effects on neoclassical particle pinch via density pump out [4]. An optimization of the resonance position and minority concentration is needed for ICRH to be effective in impurity control: while this work has been successfully done in baseline H-mode, it needs further investigation for the hybrid case.

A reduced pedestal confinement is generally observed with the present metal wall. In the hybrid regime, this is partially compensated by a more peaked pressure profile which allows to recover typical C-wall performance ( $H_{98y2}$ ~1.2-1.4 with $\beta_N \sim 3$ for 2-3 s) [5]. Dedicated power scans, performed at constant current, field, shape and density, confirm that the power degradation is weaker than expected from $H_{98y2}$ scaling [6]. A closer inspection of the data shows that this result is obtained only when the gas rate is reduced at the minimum level required for impurity control. A wider database, including plasmas with different engineering parameters, seems to indicate that the normalized pressure ($\beta_N$) is more relevant for high confinement than the input power per se, though disentangling the role of these two parameters is not trivial. In a first analysis, a dependence of the kind $H_{98y2} \propto \beta_N^{0.63}$ is found to best fit data for Ip > 1.8 MA and $1.1 < \beta_N < 2.4$ (fig. 1). The indication that best performance is obtained at high beta (and/or power) is also theoretically justified and reproduced in modeling activity as a virtuous core-edge feedback that reduces micro-turbulence and MHD effects [7]. A comparison between hybrid (tailored q-profile) and baseline (relaxed q-profile) run at same 'engineering' parameters such as density, current, field and power does not show any major difference in confinement, beta, neutrons and MHD stability [8]. This observation, together with the task of maximizing the equivalent fusion performance for future DT experiments, motivated the removal of the current overshoot introduced in the past to shape the q-profile [9, 10]. In fact, the no-overshoot start-up avoids low q transient phases and fast current ramps when pushing to high performance that require maximum current and field in the flattop. JET experience shows that a H factor in excess of 1 can be achieved also without the overshoot though this issue needs further optimization. A main heating timing scan, performed to optimize the q-profile, points as well to the same conclusion: figure 2 shows the case of two discharges where the main heating start differs by 1.5 s showing only small differences in performance. The early heating features less radiation losses and a higher ELM frequency.

When moving to lower $q_{95}$, the threshold for NTM onset decreases as well [11]. A comparison with baseline plasmas shows that the hybrid behavior is not more stable than baseline's when operating in the same $\beta_N$ and $q_{95}$ range (fig. 3). In the last campaigns, a substantial effort was devoted to push the hybrid to high absolute performance i.e. to high current and field: the strategy was to increase the current first at $q_{95}$~4 (I=2.5 MA; B=2.9 T) and then lower $q_{95}$ at constant field. This study was performed at low trinagularity (δ~0.25) to take advantage of the lower operating density that maximizes the fusion performance at constant pressure. With these parameters, a new ILW neutron yield record has been established of $2.3x10^{16}$ n/s at $\beta_N$= 2.1 and $H_{98y2}$= 1.1. In spite of a limited amount of available power, it was possible to achieve a good peak performance, while obtaining stationarity was far more difficult. As seen in figure 3, these plasmas were heavily affected by impurity accumulation. A gas scan performed at 2.5 MA, 2.9 T and 23-26 MW of additional power, revealed a strong sensitivity of ELM frequency to small changes of gas rate not seen at lower current and power: this made it difficult to find a good compromise between impurity control and performance degradation. Figure 4 shows the two extremes of a five shot scan in which the gas rate was gradually increased: a 40% gas reduction is seen to cause a dramatic effect on performance and impurity behavior. An approach to $q_{95}$=3 operation performed at constant field (2.9 T) and increasing current not only confirmed that, at lower $q_{95}$, MHD instabilities onset have a lower $\beta_N$ threshold but also that more gas is needed to keep the same ELM frequency for impurity control: this counteracts the expected confinement improvement due to the higher current. Further investigation will be needed to optimize $q_{95}$ for best fusion performance within the power limits envisaged for the JET DT campaign. In view of this, a modeling activity to predict the DT performance is being carried out including extrapolations to higher current and power. In the coming campaigns a substantial experimental time will be devoted to hybrid studies to focus on various aspects of scenario integration and optimize the target for DT experiments.

*This work has been carried out within the framework of the EUROfusion Consortium and has received funding from the Euratom research and training programme 2014-2018 under grant agreement No 633053. The views and opinions expressed herein do not necessarily reflect those of the European Commission.*

References

**(1)** G Matthews et al., Phys. Scr. T145 (2011) **(2)** TC Luce. et al 2003 Nucl. Fusion 43 321 **(3)** Angioni et al. Nucl. Fusion 54 (2014) 083028 **(4)** P.Jacquet et al., Phys. Plasmas 21 , 061510 (2014) **(5)** M Beurskens et al., Nucl. Fusion 54 (2014) 043001 **(6)** C Challis et al., Nucl. Fusion 55 (2015) 053031 **(7)** J Garcia et al. , Nucl. Fusion 55 (2015) 053007 **(8)** J Mailloux et al., 41st EPS conference on Plasma Physics, Berlin 2014, O4.127 **(9)** J Hobirk Plasma Phys. Control. Fusion 54 (2012) 095001 **(10)** E Joffrin Nucl. Fusion 54 (2014) 013011 **(11)** P Buratti et al. this conference P2.115

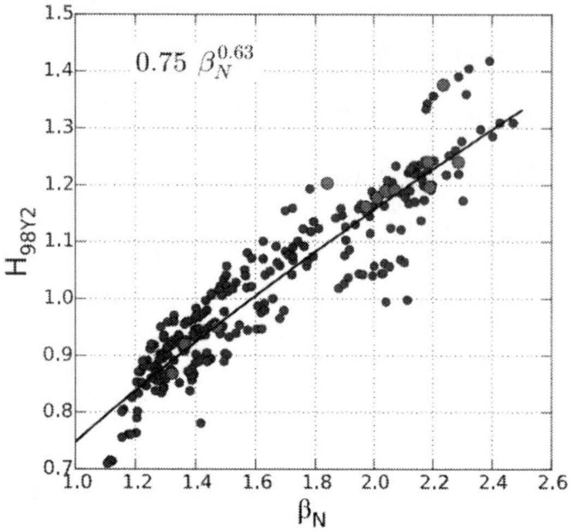


Figure 1: H98y2 versus $\beta_N$ for Hybrid with $I_p$> 1.8 MA. Red symbols refer to plasmas with constant ϱ* (5.0-5.1 x$10^{-3}$).

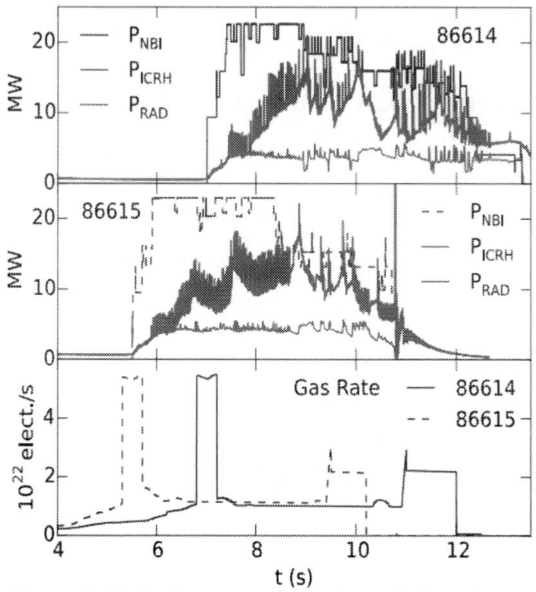


Figure 2. Tailoring q-profiles via main heating timing. Earlier start of NBI power (86615), giving a flatter q profile, produces more frequent ELMs (see spikes on $P_{RAD}$ signal) thus mitigating the impurity accumulation ($P_{RAD}$) at same gas rate. The performance ($\beta_N$, $H_{98y2}$) remains virtually unchanged.

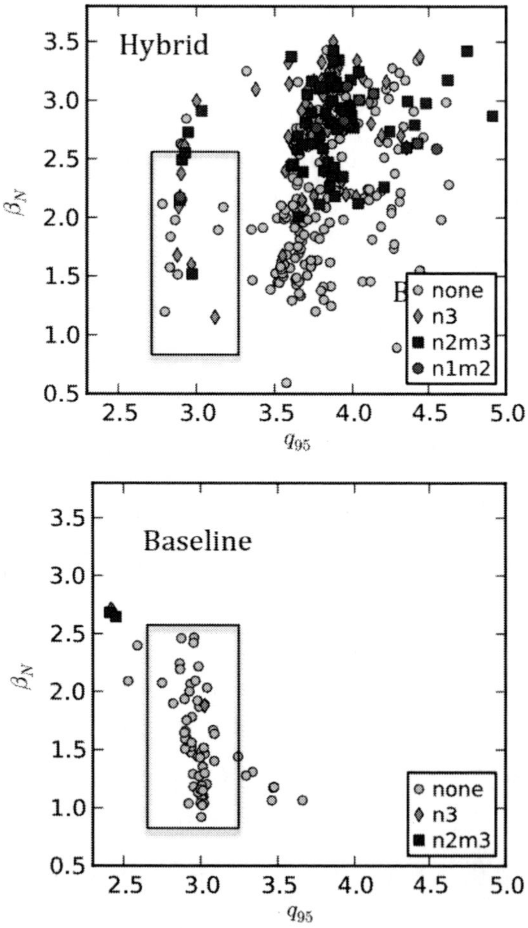


Figure 3: $\beta_N$ at mode onset versus $q_{95}$. Rectangles: comparison range.

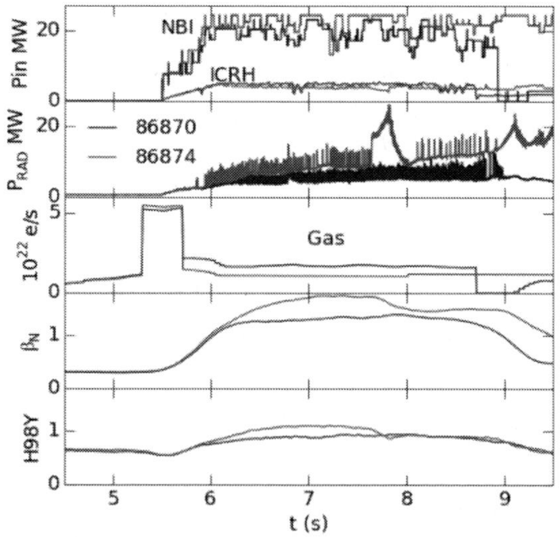


Figure 4. Gas scan at 2.5 MA, 2.9 T, low triangularity, tile 6. Shot 86874 (red) with 40% less gas exhibits a better performance at the cost of more radiation, lower ELM frequency and poorer stability.

# Fundamental cyclotron $^3$He minority ICRF heating experiments in H plasmas in JET in presence of the ILW

D. Van Eester[1], E. Lerche[1], Y. Kazakov[1], P. Jacquet[2], M. Santala[2], M. Goniche[3], L. Colas[3], A. Shumack[1], R. Dumont[3], K. Crombé[1] and JET contributors*

*EUROfusion Consortium, JET, Culham Science Centre, Abingdon, OX14 3DB, UK*

[1]*LPP-ERM/KMS, EUROfusion Consortium Member - Trilateral Euregio Cluster, Brussels, Belgium*

[2]*CCFE, Culham Science Centre, Abingdon, UK*

[3]*CEA, IRFM, F-13108 Saint-Paul-Lez-Durance, France*

**See the Appendix of F. Romanelli et al., Proceedings of the 25th IAEA Fusion Energy Conference 2014, Saint Petersburg, Russia*

## Introduction

Efficient plasma heating schemes are a prerequisite for reaching fusion relevant temperatures in fusion machines. On the road to reaching ignition, non-activated scenarios - such as $(^3He)-H$ plasmas - will first be used to characterise plasma dynamics. Moreover, $^3He$ minority heating is intended to be adopted during the ramp-up phase of $D-T$ pulses in ITER to increase the fraction of bulk ion heating and to boost the fusion reactivity. Core fundamental $^3He$ minority cyclotron heating experiments were performed in hydrogen majority plasmas during the 2014 JET hydrogen campaign to help determining the road ITER should take.

## Experimental results

Though the accent of the here discussed experiments was on obtaining insight in RF (radio frequency) heating dynamics, the parameters were chosen in a more general context, namely to help developing baseline heating scenarios in all-metal-wall fusion machines and allowing to ensure high confinement (H-) modes while preventing heavy impurity accumulation in the core: (i) The operating field and current were fixed at $B_o = 3.25T$ and $I_P = 2.5MA$. (ii) Up to 5MW of ICRF power were coupled using the A2 antennas in dipole phasing. To place the $^3He$ cyclotron layer in the core, the experiments were performed using $f = 33MHz$. (iii) Densities of the order of $n_{e,o} \approx 4 \times 10^{19}/m^3$ were used.

Compared to earlier $(^3He)-H$ experiments (see e.g. Mayoral et al. [1] and Van Eester et al. [2]), $n_{e,o}$ was higher by a factor of $1.3-1.4$. Operation at higher density reduces the temperature that can be reached. Whereas core electron temperatures up to $T_{e,o} \approx 4.5keV$ were observed when applying $5MW$ of RF power under similar conditions in the past, the maximum $T_e$ reached now was merely $2.7keV$. No ion temperature profile is available but a 1-point guess at minor radius $\rho \approx 0.3m$ locally yields $T_i \approx 2.2-2.3keV$ while $T_e(0.3) \approx 1.8-2keV$. The left subplot of Fig.1 depicts the electron temperature. The present profiles are clipped while they were strongly peaked in the past. In absence of sawtoothing and any other anomalous transport enhancements, $T_{e,o} \approx 3.4keV$ could have been reached (see the dashed curve in Fig.1-left). For *identical* energy, the relative fraction of the peak temperatures would be consistent with the imposed density change ($4.5/3.4 \approx 1.3$) but the fact that the plasma energy was even *higher* by roughly a factor 1.5 in the recent experiments points to the differing role of energy source/loss channels. Obvious candidates that come to mind are transport phenomena of diffusive/convective type in between sawtooth crashes and shorter timescale mechanisms associated with the crashes themselves. On the other hand, one may wonder if the energy sources were less efficient or radiation more dominating.

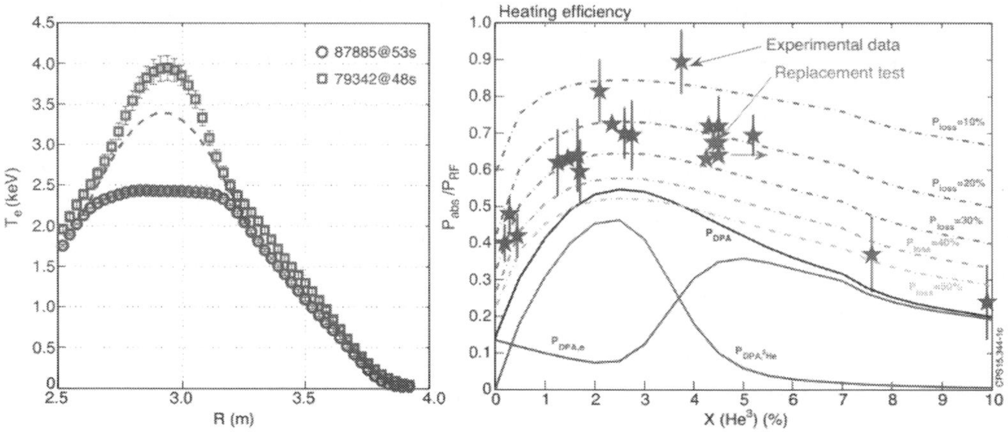

Figure 1: *Electron temperature profiles for recent ITER-like wall shot #87885 and comparison with Carbon wall shot #79342 both with $P_{RF} = 4MW$ (left), and experimental (stars) and theoretically predicted (lines) heating efficiency as a function of the $^3He$ minority concentration (right).*

The global RF induced energy increase was $0.25MJ/MW$, comparable to past results. It was shown through break-in-slope (BIS) analysis of the energy that up to 80% of the RF power launched was retrieved in the main plasma. In line with preparatory modelling and analysis, optimal heating was achieved at a $^3He$ minority concentration of around 3% (see Fig.1-right). Recall that the BIS method identifies instantaneous modification and does not capture transport effects. Bolometry shows that 20 to 30% of the launched power is radiated as bulk radiation. Compared to the earlier experiments, the peaking of the radiation was similar; no signs of impurity accumulation were seen in the main heating phase. The total bulk radiation was 60% higher but the energy in the recent experiments was already higher by as much as $0.4MJ$ before the RF heating was switched on (operating at higher $I_p$ yields higher Ohmic power). At higher powers, the radiated power increases but the fraction stays roughly unchanged. The fraction of the radiated power is high at very low concentrations and re-increases beyond the $X[^3He]$ at which heating is optimal. Correcting the theoretically obtained prediction for the heating efficiency (labeled as DPA i.e. double pass absorption in the right subfigure of Fig.1 and obtained by the 1D wave propagation and damping code TOMCAT [3]) for the observed - radiative - losses yields efficiencies in fair agreement with the heating efficiencies found via BIS analysis. The total double pass absorption is the sum of the power absorbed by the $^3He$ minority and the electrons. To the exception of the region of negligibly small minority concentrations ($X[^3He] < 0.5\%$) $^3He$ minority fundamental cyclotron heating dominates the absorption until the mode conversion regime is reached at $X[^3He] = 3-4\%$, after which electron absorption is stronger than ion absorption. The overall double pass absorption dies away slowly beyond $X[^3He] \approx 5\%$. RF induced fast minority ion tails were observed. The tails are most prominent at modest concentrations but require a minimum of $^3He$ to be present, consistent with what is expected from theory. The left subfigure of Fig.2 shows the fast ion flux data obtained from the neutral particle analyser for the energy range between 200 and $400keV$.

Whereas the global heating efficiency is explained satisfactorily invoking global radiation losses, the inability to rebuild a peaked core electron temperature profile after a sawtooth crash is striking. Intended for preparing H-mode studies to be performed when topping up the RF power by neutral beam injection, the adopted value of $I_P$ for the chosen $B_o$ could not prevent $q_o$ from dipping under 1 at the adopted auxiliary power level. One aspect that helps understanding why the temperature profile cannot recover is closely related to fast particle dynamics. Fig.2-right shows the redistributed power deposition profile as obtained from the 2D wave solver CYRANO [4] and the 1D Fokker-Planck solver StixRedist [5]; transport phenomena such as equipartition between bulk ions and electrons - more important at higher densities - were *not* accounted for in this simulation. Most of the power absorbed centrally by the $^3He$ minority is Coulomb collisionally redistributed to the bulk *ions*. Simulations made for earlier experiments (plot not included) yield another, more usual result: the fast tail transferring most of its energy to the *electrons*. This difference - mainly due to the higher collisionality - makes that a smaller fraction of the RF power is available to help $T_e$ recover from sawtooth crashes if they occur. Note also that the deposition profile has a finite extent. In view of the error bars on various quantities, a fraction of the RF power might not even be available inside $q = 1$ to help rebuilding $T_e$.

The clipping limits the potential of ICRH to build up a strong electron temperature gradient. Since the sawtooth frequency $f_{ST}$ scales as $T^{-3/2}$ [6], low $T_e$ yields high sawtooth frequencies and hence tends to keep the core electron temperature low through frequent flushing. It requires a sufficient amount of electron heating to overcome this effect. Transport studies need to be undertaken to determine if the observed $T_e$ dynamics can be understood.

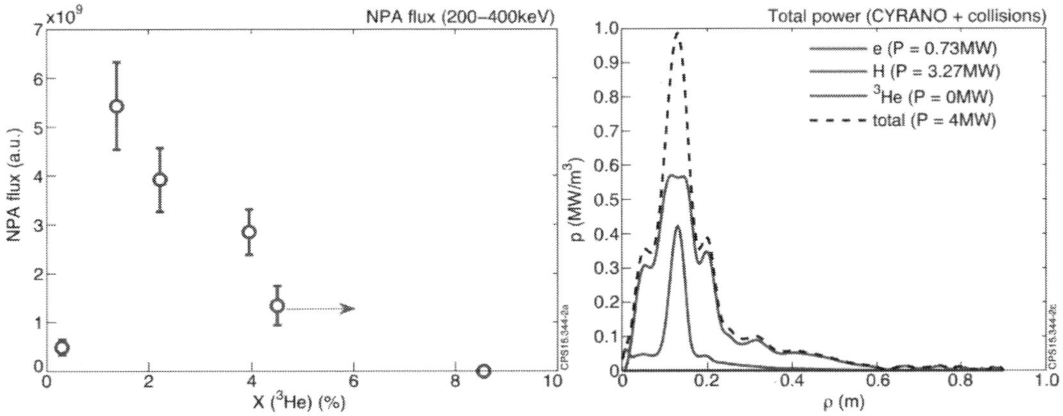


Figure 2: *Neutral particle analyser fast ion flux as a function of the* $^3He$ *minority concentration (left), and power deposition profile after Coulomb collisional redistribution of the RF induced minority tail energy to bulk ions and electrons as a function of the magnetic surface labeling parameter* $\rho$ *(right).*

$^3He$ being an extremely expensive gas [9], reducing its consumption is desirable. From the RF point of view, the key role of a minority population is to affect the polarization of the wave, hereby increasing or reducing the potential of a wave to transfer its energy to the plasma. Exploiting this dependence, Kazakov proposed replacement schemes whereby not readily available gases can partly be substituted by more common ones while not affecting the wave characteristics significantly [10]. Some first tests were performed to assess the potential of a $^3He$ replacement scheme relying on puffing $N_2$, one of the $D$-like impurities, into the $H$ plasma. Although further testing is needed, these first results were promising. Two indications are (i) the purple stars added to the heating efficiency left subplot of Fig.2, and (ii) the red circle added to the right subfigure of Fig.2. Both data confirm the expected trend, although the $N$ injection was modest.

**Conclusions**

$^3He$ minority fundamental cyclotron heating experiments were performed in JET in a H majority plasma in the framework of optimising scenarii for the nonactivated phase of future fusion machines. As is characteristic for so-called inverted schemes, optimal heating occurs at $^3He$ minority concentrations of a few percent, and the heating efficiency dwindles when $X[^3He]$ goes beyond $6-7\%$. The heating itself was fairly efficient provided the proper $X[^3He]$ was used (the optimal efficiency reached was $\approx 80\%$ at $X[^3He] \approx 3\%$); the obtained net heating performance crudely matched earlier results: $0.15MJ/MW$. Operating at - for L-mode - high density successfully mitigated the typically high $P_{rad}$ and $W$ content observed in lower $n_{e,o}$ L-modes despite the low central temperatures reached, hinting at reduced edge high Z impurity sources. However, inside the $q = 1$ surface, the electron temperature is flat. The combined effect of high collisionality, radiated power and sawtoothing activity at - too - modest auxiliary heating power is thought to be responsible for the observed phenomenology but more detailed modelling is needed for better understanding. This illustrates the challenge of reaching high core (electron) temperatures in high collisionality plasmas (such as high $I_p$ baseline H-modes) and confirms the need for high power ICRF or NBI preheating as well as the right choice of the q-profile. One directly RF-related contributing factor is that - opposite to what usually is the case and due to the chosen operation point - the minority tail Coulomb collisionally transfers the bulk of the wave energy it absorbs to the thermal majority ions and hence does not significantly contribute to the rebuilding of the core electron temperature. Optimising a heating scenario not just requires finding suitable parameters guaranteeing efficient absorption of the auxiliary energy, but also necessitates studying the losses and inefficiencies associated with these parameters.

**Acknowledgements**

This work has been carried out within the framework of the EUROfusion Consortium and has received funding from the Euratom research and training programme 2014-2018 under grant agreement No 633053. The views and opinions expressed herein do not necessarily reflect those of the European Commission.

**References**

[1] M.-L. Mayoral et al., *Nucl. Fusion* **46** (2006) S550-S563
[2] D. Van Eester et. al, *Plasma Phys. Control. Fusion* **54** (2012) 074009
[3] D. Van Eester and R. Koch, *Plasma Phys. Control. Fusion* **40** (1998) 1949-1975
[4] P.U. Lamalle, *Nonlocal theoretical generalization and tridimensional study of the coupling of an ICRH antenna to a tokamak plasma*, PhD Thesis, LPP-ERM/KMS Report **101** (1994) Université de Mons
[5] D. Van Eester and E.A. Lerche, *Plasma Phys. Control. Fusion* **53** (2011) 092001
[6] W. Park and D.A. Monticello, *Nucl. Fusion* **30** (1990) 2413
[7] D. Van Eester et al., Proc. *41st EPS Conference on Plasma Physics*, Berlin, 23rd June 2014
[8] M. Valisa et al., *Proc. 25th IAEA Fusion Energy Conference*, St Petersburg, Russia (2014) EFDA-JET-CP(14)06/29
[9] D. Shea and D. Morgan, *The Helium-3 Shortage: Supply, Demand, and Options for Congress*, Congressional Research Service, CRS Report for Congress (2010); on-line: http://www.fas.org/sgp/crs/misc/R41419.pdf
[10] Ye.O. Kazakov, T. Fülöp and D. Van Eester, *Nucl. Fusion* **53** (2013) 053014

# Isotope exchange experiments on ITER-like wall in JET

T. Wauters[1], D. Douai[2], T. Loarer[2], S. Brezinsek[3], JET Contributors*

*EUROfusion Consortium, JET, Culham Science Centre, Abingdon, OX14 3DB, UK*

[1] *Laboratory for Plasma Physics, ERM/KMS, 1000 Brussels, Belgium, TEC partner*

[2] *CEA, IRFM, F-13108 St-Paul-Lez-Durance, France*

[3] *Forschungszentrum Jülich GmbH, Institut für Energie und Klimaforschung - Plasmaphysik, 52425 Jülich, Germany*

* *See the Appendix of F. Romanelli et al., Proceedings of the 25th IAEA Fusion Energy Conference 2014, Saint Petersburg, Russia*

## Introduction

While the permanent retention rate of fuel on the JET-ITER like wall (ILW, beryllium main wall and tungsten divertor) is reduced by a factor of $\sim 18$ with respect to operation with the carbon wall [1, 2, 3], the rates may still lead to an unacceptable Tritium inventory build-up when extrapolated to ITER. Uncertainties remain on the absolute rates. Post mortem analysis finds permanent retention rates of $5.7 \cdot 10^{18}\,\mathrm{D.s^{-1}}$ [1] while equally well established gas balance analysis techniques result in rates of $0.2 - 1.5 \cdot 10^{20}\,\mathrm{D.s^{-1}}$ [2, 3]. Long term outgassing may give reason for these difference [2]. For this paper a worst case scenario for the permanent retention rate is considered, and used as a figure of merit in the presented investigation. Extrapolated from JET to ITER, considering a 4 times larger surface area on ITER and a 50% Tritium content in the plasma, one quickly arrives at a permanent retention of 0.5 gT within one 400 s D:T pulse on ITER. The safety limit of 640 gT in ITER [4] may as such be reached within the first few years of ITER D:T operation.

Fuel recovery experiments relying on isotopic exchange by (i) tokamak plasmas, hereafter named $I_p$-plasma, (ii) ion cyclotron wall conditioning plasmas (ICWC) and (iii) DC glow discharges (GDC), have been performed on JET. The experiments, exchanging the stored fuel content in the plasma facing components (PFC), provide insight on the sizes of the accessible fuel reservoir for each of the techniques. In this contribution the results, based on gas balance analysis consisting of combined volumetric analysis and gas chromatography, are revisited and complemented by in-vessel Be deposition patterns obtained by ex-situ surface analysis of retrieved wall and divertor tiles. The study aims at assessing the performance of the isotopic exchange techniques in mitigating the fuel inventory build-up.

## Transient storage to permanent storage during isotopic exchange by $I_p$-plasma operation

The change-over experiment by $I_p$-plasma, described in detail in [3], was carried out through 13 consecutive D discharges (with only $D_2$ gas injection) on a wall pre-conditionned with H. For all these pulses, a total plasma duration of $\sim 150$ s was performed in limiter configuration and about the same duration $\sim 150$ s in X point configuration with $I_p = 2.0$ MA, $B_T = 2.4$ T, $\langle n_e \rangle \approx 4.5 \cdot 10^{19}\,\mathrm{m^{-3}}$, constant gas injection of $3.0 \cdot 10^{21}\,\mathrm{D.s^{-1}}$ and 0.5 MW of auxiliary heating by ICRH in L-mode. The plasma isotopic ratio was exchanged from 95% in a reference H pulse before the experiment to 5% H in the final D pulse. The gas balance of the experiment shows that throughout $3 \cdot 10^{22}$ D atoms were pumped out of the vessel (see Table 1).

The key processes governing fuel recycling by plasma discharges are identified in [3]. The species in the plasma scrape of layer (SOL) are fuelled both by gas injection and wall release. Transport of ionised particles in the SOL determines the dominant areas for interaction between

Table 1: Gas balance and derived accessible fuel reservoir for different isotopic exchange techniques on JET-ILW. The unit 'atoms' designates the initial dominant isotope specie in the PFC's.

| | Removed (gas balance) | Permanent retenion rate | Accessible, maximum |
|---|---|---|---|
| $I_p$-plasma | $3 \cdot 10^{22}$ atoms | $1.5 \cdot 10^{20}$ atoms.s$^{-1}$ | $3 \cdot 10^{22} + 0.5 \cdot 10^{22}$ atoms |
| ICWC | $6.2 \cdot 10^{22}$ atoms | $1.5 \cdot 10^{20}$ atoms.s$^{-1}$ | $6.2 \cdot 10^{22} + 0.4 \cdot 10^{22}$ atoms |
| GDC | $10 \cdot 10^{22}$ atoms | *none* | $10 \cdot 10^{22}$ atoms |

plasma and wall (PWI). As main processes in this interaction one distinguishes (i) implantation of fuel for which the implanted species remain mostly accessible to the recycling process, (iia) codeposition of fuel with eroded wall materials, mostly with Be but also with W, in areas accessible for re-erosion, and (iib) long term retention in form of codeposits located in areas inaccessible to the used plasma type or plasma configuration. The continuous operation of the vacuum pumps removes steadily a fraction of the present neutral gas.

We relate process (iib) to the earlier quoted permanent retention rate. Considering the worst case scenario for the permanent retention rate in $I_p$-plasma discharges (Table 1) and knowing the averaged isotopic ratio for each pulse (from [3]) we can estimate the amount of D that was removed from the first wall in the isotopic exchange process, but then stored permanently into deposits (by process iib) instead of being pumped out of the vessel by the equation:

$$N_{pr,D} = \sum_{pulse} Q_{pr} \Delta t \rho_D \tag{1}$$

with $Q_{pr}$ the retention rate, $\Delta t$ the pulse length, $\rho_D$ the isotopic ratio for D and $N_{pr}$ the total stored D in permanent deposits. For the above experiment this amounts to $0.5 \cdot 10^{22}$ D atoms. This number adds to the pumped amount of D atoms to deliver the amount of fuel stored in the vessel that can be accessed in principle by isotopic exchange with $I_p$-plasma (Table 1).

From this analysis it is concluded that $I_p$-plasma removes fuel from the transient reservoir and stores a fraction of it (13% estimated) into deposits where it remains inaccessible for further isotopic exchange by $I_p$-plasma. This unwanted effect may seem less severe than expected, which is due to the fact that for the JET-ILW the plasma isotopic ratio is quickly (less than one minute [3]) dominated by the injected gas. The permanent deposits that are continuously formed contain as such only small fractions of the isotopes of interest (T in the case of ITER).

**Transient storage to permanent storage during isotopic exchange by ICWC**

Several experiments on isotopic exchange by ion cyclotron wall conditioning (ICWC) have been performed on JET [5]. For this contribution we use as reference the JET-ILW $H_2$-ICWC experiment on walls "naturally" loaded with D throughout the JET-ILW D campaigns. As described in detail in [6], the JET ICRF antennas were operated at 25 MHz with toroidal field values of 1.65 T simulating on JET the ITER half field case (2.65 T/40 MHz) with on axis location of the fundamental $H^+$ resonance layer. Powers of 50 – 240 kW could be coupled to the low density ICRF plasma of $0.2 - 2.4 \cdot 10^{17}\,m^{-3}$. 20 pulses with variable duration of 2 – 20 s resulted in $\sim 218$ s of plasma exposure. The gas balance of the experiment shows that throughout $6.2 \cdot 10^{22}$ D atoms were pumped out of the vessel (see Table 1) while $8.9 \cdot 10^{22}$ of the injected H atoms remained retained in the vessel.

Similar to the previous section the amount of fuel that can be accessed by ICWC is estimated. Via similar processes as (i), (iia) and (iib) a part of the D will have migrated into permanent storage, remaining herafter inaccessible to ICWC. In a crude approach we take the surplus of retention as permanent retention. This corresponds then to a 1 to 1 exchange in ICWC accessible areas and storage of $N_{pr,H} = 8.9 \cdot 10^{22}$-$6.2 \cdot 10^{22}$ H atoms in permanent deposits. Using again equation (1), relying on averaged isotopic ratios per pulse and the pulse lengths while assuming that the ICWC permanent retention rate is a constant, we arrive equally at a permanent retention rate of $Q_{pr,ICWC} = 1.5 \cdot 10^{20}\,s^{-1}$. Finally the amount of D that was removed from the first wall in the isotopic exchange process, but stored permanently into deposits (iib) instead of being pumped out of the vessel, is estimated by equation (1) as $0.4 \cdot 10^{22}$ D. As such, the reservoir accessible in principle by ICWC amounts to $6.6 \cdot 10^{22}$ atoms which are $3.1 \cdot 10^{22}$ atoms more than accessible by $I_p$-plasma (see Table 1).

From this analysis it is concluded that ICWC-plasma removes fuel (supposedly $3.1 \cdot 10^{22}$ atoms) from areas that are not accessible by $I_p$-plasma, as judged by the larger accessible reservoir. ICWC migrates also less released wall isotopes to permanent storage areas relative to the pumped amount of wall isotopes. From previous studies it is believed that the migration of D from transient to long term retention in ICWC can be even further reduced by optimising the RF duty cycle [7].

Table 2: Summary of the retained D total amounts from JET-ILW 2010-2012 adapted from [1]

| JET vessel area | %D stored | Comment |
|---|---|---|
| Upper Dump plates | 5% | Accessible by ICWC |
| Outer Poloidal and Inner Wall Guard Limiters | 22% | Accessible by ICWC |
| Divertor Apron (Tile 0 and Tile 1) | 33% | Accessible by ICWC |
| Inner Divertor (without Tile 0) | 18% | Possible to rely on divertor baking |
| Outer Divertor | 13% | Possible to rely on divertor baking |
| Remote areas | 11% | / |

**Fuel removal by $I_p$-plasma vs. ICWC (in view of T-recovery on ITER)**

To assess the effectiveness of ICWC and $I_p$-plasma for T-recovery, based on the JET results, a 400 s H-plasma is assumed *on JET*. The pulse is expected to retain (worst case) $6.0 \cdot 10^{22}$ H atoms in permanent deposits, and as well $3.5 \cdot 10^{22}$ H atoms that are accessible to $I_p$-plasma.

Using the above described $I_p$-plasma procedure to recover the wall isotopes, the $3.5 \cdot 10^{22}$ H atoms will be mobilised from the accessible areas. Of these atoms $3.0 \cdot 10^{22}$ are pumped out of the vessel while $0.5 \cdot 10^{22}$ H atoms will be stored into the permanent deposits, increasing the permanent reservoir to $6.5 \cdot 10^{22}$ H atoms. This results from the fact that from fuel removal point of view, the PWI interaction areas are not sufficiently different from those in interaction during normal operation. For this reason it seems not a good strategy to apply $I_p$-plasma pulses for the mere purpose of T removal by isotopic exchange between ITER D:T shots (see also [3]).

Using the above described ICWC procedure to recover the wall isotopes may also mobilise the $3.5 \cdot 10^{22}$ H atoms from the $I_p$-plasma accessible areas. This is supported by the demonstrated complete isotopic change-over by ICWC in a follow up experiment on JET-ILW [8], confirmed by tokamak plasma measurements. The additional removed $3.1 \cdot 10^{22}$ H atoms are stemming then from the $I_p$-plasma permanent deposit areas. For ITER, this means that 52% of the permanently stored H in one 400 s pulse can be recovered from the vessel, taking that the small fraction of extra retention is mitigated by applying optimal RF duty cycles.

**ICWC plasma wetted areas and JET codeposition patterns from surface analysis**

ICWC plasmas on JET feature clear recycling radiation areas at the Outer Poloidal and Inner Wall Guard Limiters (low field and high field side resp.) and visually reach the Upper Dump plates and Divertor Apron (meaning tile 0 and 1 following [1]) by applying a small vertical magnetic field ($B_V/B_T \approx 1\%$) [9]. It is therefore preliminary considered that ICWC optimally conditions the low and high field side areas, while removal at the top of the vessel and nearing the divertor can be maximized by applying poloidal magnetic fields. The cleaning efficiency inside the divertor may be modest. Langmuir probes sensitive to ICWC plasma fluxes that could confirm these statements are not available on JET.

Codeposition patterns from post mortem (PM) surface analysis [1] learn that the deposits located on areas accessible by ICWC, supposedly the Outer Poloidal Limiters, Inner Wall Guard Limiters, Upper Dump plates and Divertor Apron comprise 60% of the total retention (Table 2). This means that the present ICWC procedure may prove already capable of removing a significant fraction of the permanent retention in the main vessel that is accumulated in one 400 s pulse. In addition from PM analysis it is found that the contribution of impurities to retention on JET is most important in the divertor. With less impurities on ITER, e.g. less residual C, less deposits will be inaccessible by ICWC. The impurities play a minor role in the overall gas balance [10].

**Removal by glow discharge conditioning**

The maximum frequency of de-energising and re-energising of the toroidal field coils on ITER is set at once a week. As GDC can't be operated in presence of magnetic fields the maximum GDC frequency on ITER is likewise set at once a week. Isotopic exchange experiments by $H_2$-GDC on JET have shown that the technique is able to remove $10 \cdot 10^{22}$ D atoms from the first wall [5]. This result is interpreted as GDC having the largest interaction area of the 3 compared isotopic exchange techniques. Moreover the exchange by GDC occurs such that the permanent retention caused is negligible [5].

To appreciate the applicability of GDC for T-recovery on ITER we consider 5 operation days with a modest number of 10 H pulses of 400s each operated *on JET*. In such an operation week $3 \cdot 10^{24}$ H atoms would be retained in permanent deposits (worst case) while $3.5 \cdot 10^{22}$ H atoms are located in the accessible reservoir for $I_p$-plasma. Operating $D_2$-GDC would remove in total $10 \cdot 10^{22}$ H from the vessel, representing supposedly $3.5 \cdot 10^{22}$ H atoms stemming from the transient storage area and $6.5 \cdot 10^{22}$ H from the $I_p$-plasma permanent deposits. The latter amount is a factor 50 smaller than the worst case accumulated H over the considered operation days. Employing GDC for the purpose of T removal on ITER seems therefore not a secure strategy, which is merely a consequence of the need for $B_T = 0$ T.

## Conclusion

The analysis of fuel recovery experiments by isotopic exchange on JET-ILW learns that $I_p$-*plasmas* do not remove fuel from permanent codeposition areas. This is a mere consequence of the PWI interaction areas being not sufficiently different from those in interaction during normal operation, be-it from fuel removal point of view. Moreover, amounts of the $I_p$-plasma accessible fuel migrates through PWI and SOL transport to permanent storage areas (13% in worst case). Therefore, for T inventory control on ITER, applying solely $I_p$-plasmas exchanging the wall isotopic ratio is not a viable scenario (It may still apply for removal from the divertor).

In the case of *ion cyclotron wall conditioning*, it is found that the ICWC PWI area for fuel removal must include a part of the $I_p$-plasma permanent codeposition areas, judged from $\sim 2\times$ larger interaction reservoir than for $I_p$-plasma. It is estimated that 52% of the permanent retention by a 400s plasma can be recovered by ICWC operation. This amount corresponds to almost the complete expected retention in the main vessel including the divertor apron, based on post-mortem surface analysis data. This suggests that operation of ICWC between D:T pulses is a promising strategy for T inventory control on ITER. Although hard to operate, "*divertor baking*" is relied upon to remove the remaining deposits inside the divertor. Divertor baking at 350°C or 240°C (slower) features an expected removal efficiency of more than 90% of T from the divertor [5].

*Glow discharge conditioning* features the largest interaction area of the 3 compared techniques. Nevertheless GDC may prove unable to play an important role in T-recovery on ITER as it can be applied only at maximum frequency of once a week. The latter is far too less in case of worst case retention rates. It is however recommended, adopting a "good house keeping approach" for ITER, to foresee regular GDC operation.

Finally, the presented JET experiment are performed with JET-ILW at 200°C, whereas ITER will operate ICWC and $I_p$-plasma at 70°C. As shown in [11, 12], one key factor determining isotopic exchange is surface temperature. Complementary ICWC experiments are therefore suggested in the JET-ILW with walls at 70°C.

## Acknowledgments

This work has been carried out within the framework of the EUROfusion Consortium and has received funding from the Euratom research and training programme 2014-2018 under grant agreement No 633053. The views and opinions expressed herein do not necessarily reflect those of the European Commission.

## References

[1] Heinola et al., J. Nucl. Mater. **463** (2015) 961-965
[2] S. Brezinsek et al., Nucl. Fusion **53** (2013) 083023
[3] T. Loarer et al., Nucl. Fusion **55** (2015) 043021 (8pp)
[4] M. Shimada, R.A. Pitts, J. Nucl. Mat. **415** (2011) S1013-S1016
[5] D. Douai et al., J. Nucl. Mat. **463** (2015) 150-156
[6] T. Wauters et al., J. Nucl. Mat. **463** (2015) 1104-1108
[7] T. Wauters et al., J. Nucl. Mat. **415** (2011) S1033-S1036
[8] T. Wauters et al., Proc. of the 25th FEC, Oktober 2014, Saint Petersburg, Russia, EX/P5-21
[9] D. Douai et al., J. Nucl. Mat. **415** (2011) S1021-S1028
[10] S. Brezinsek et al., J. Nucl. Mat. **463** (2015) 11-21
[11] D. Kogut et al., Proc. of the 15th PFMC, May 2015, Aix-En Provence, France, submitted to Phys. Scrypta.
[12] D. Douai et al., 2015, this conf.

# ELM mitigation using *n=2* magnetic perturbations on JET with an ITER-like wall and comparison with a C-wall

Y. Liang[1], S. Jachmich[2], P. Lomas[3], P. Denner[1], M. Rack[1], Y. Gao[1], N. Wang[1], L. Li[1], I. T. Chapman[3], and JET contributors*.

*EUROfusion Consortium, JET, Culham Science Centre, Abingdon, OX14 3DB, UK*

*1Forschungszentrum Jülich GmbH, Institut für Energie- und Klimaforschung – Plasmaphysik, 52425 Jülich, Germany;*

*[2] Laboratory for Plasma Physics Koninklijke Militaire School - Ecole Royale Militaire Renaissancelaan 30 Avenue de la Renaissance B-1000, Brussels, Belgium;*

*[3] CCFE, Culham Science Centre, Abingdon, Oxon, OX14 3DB, UK;*

Active control of edge-localized modes (ELMs) by externally applied fields offers an attractive method for next-generation tokamaks, e.g. ITER. To date, most ELM mitigation/suppression experiments have been performed on devices with a carbon-wall (C-wall)[1, 2]. However, a carbon-free tungsten divertor is foreseen for the phase active on ITER, while the main chamber blanket modules in ITER are protected by shaped (limiter-like) beryllium panels. Therefore, it is urgent and important to prove the applicability of ELM control/suppression with magnetic perturbations in plasmas with ITER-like PFCs and to perform a systematic comparison with previous results with a C-wall.

On JET, the previous experiments have shown that type-I ELMs can be mitigated in a relatively wide operational domain ($q_{95}$, pedestal collisionality, triangularity, configurations …) with a C-wall by applying static low *n* external magnetic perturbation fields [2,3]. Since 2012, the ITER-like wall (ILW) has been installed on JET to replace the previous C-wall, and the EFCC system is upgraded to allow a maximum coil current twice as large as the previous one. To date, JET is the only device capable of studying ELM mitigation with magnetic perturbations in plasmas with an ILW and comparison with a C-wall. Recently, mitigation of type-I ELMs was observed with an $n = 2$ field on JET with the ILW [4] in a relative wide range of the pedestal collisionality ($0.3 < \nu_e^* < 2.5$) and plasma density ($0.55 < n_e/n_e^{GW} < 0.9$) as shown in figure 1. Here, $n_e^{GW}$ is the Greenwald density limit. Several new findings with the ILW were identified and contrasted with the previous C-wall results for comparable conditions.

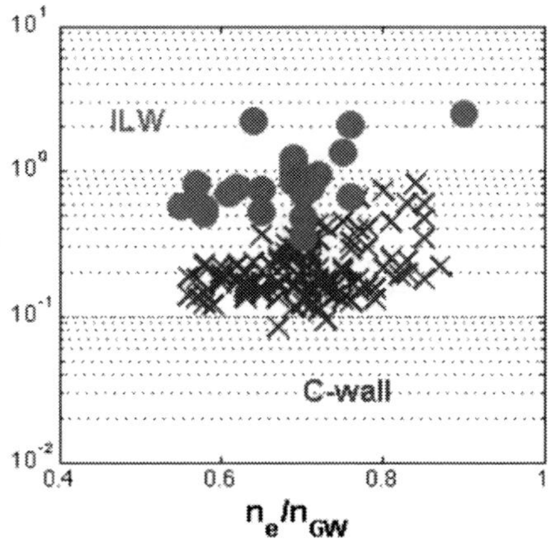


Figure 1 Experimentally determined access condition in terms of pedestal collisionality ($\nu^*_e$) versus pedestal density as a fraction of the Greenwald density ($n_e/n_{GW}$) for mitigation of type-I ELMs using EFCC on JET with the C-wall ($n$=1 and 2 EFCC) and the ILW ($n$=2 EFCC only)

In the high collisionality ($\nu^*_{e,ped} \geq 2.0$) regime, no clear ELM mitigation or suppression was observed during an application of the $n = 1$ or 2 fields with a EFCC current up to 48kAt on JET

*see the Appendix of F. Romanelli et al, Proc 25th IAEA Fusion Energy Conference 2014, St. Petersburg, Russia

with the C-wall, where the target plasma has a relatively high ELM frequency (few 100 Hz). However, with the ILW, a strong mitigation of type-I ELMs was observed in both high and low triangularity H-mode plasmas when the $n = 2$ field was applied as shown in figure 2. In this experiment, the EFCC coil current was kept at 80kAt for 2 seconds, which is about ten times the plasma energy conferment time. The large regular type-I ELM becomes small in size and irregular in frequency in the low triangularity case (# 87511), while it almost disappeared when the plasma triangularity increases from ~0.3 up to ~0.45 during the flattop of the EFCC current (# 87516). The plasma shapes used in this experiment are named as V5OH and HT3R configurations on JET for the low and high triangularity plasmas, respectively. In the high collisionality regime with the ILW, no density drop (so called pump-out effect) was observed, and splitting of the outer strike point was observed in the H-mode plasma during the application of the $n = 2$ field on JET. With an increase in NBI heating power, the effect of the $n = 2$ fields on ELMs becomes much less pronounced in H-mode plasmas with the HT3R configuration as shown in figure 2 (bottom plot). The large type-I ELM re-appears when the NBI power increases from 3MW to 6MW in the H-mode plasma. The ELM frequency slightly increases with a further increase in the NBI power from 6MW to 9MW. This agrees with the typical power dependence of the Type-I ELM.

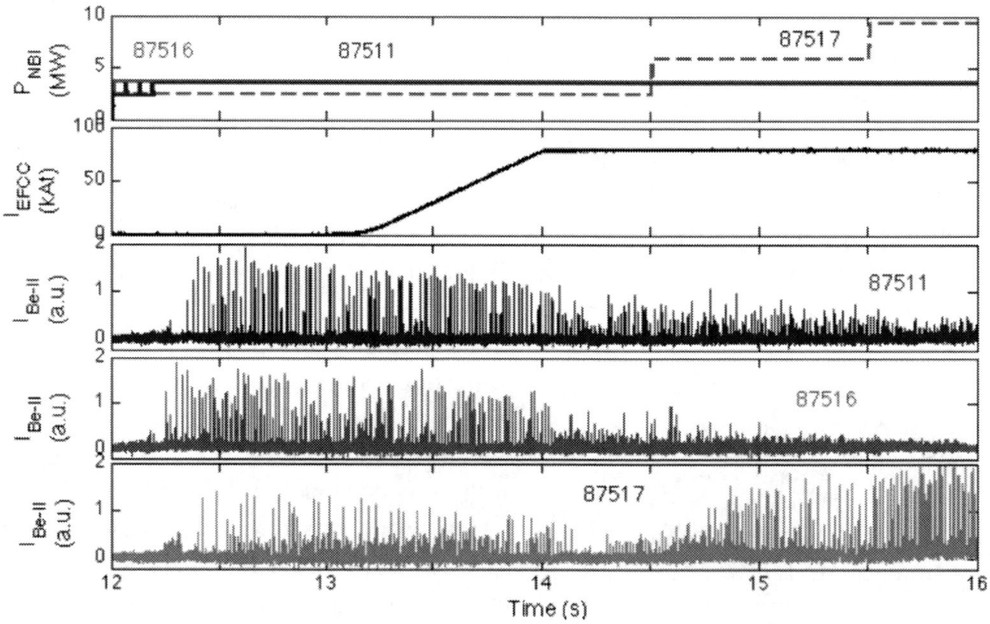


**Figure 2 Time evlutions of NBI power, EFCC current and intensity of Be-II emissions for the H-mode plasma with a low (87511) and high (87516, 87517) triangularities, respective.**

In the moderate collisionality ($0.3 < \nu^*_{e,ped} < 2.0$) regime, ELM mitigation with an increasing ELM frequency up to 6 times of the unmitigated case with the $n = 1$ fields and up to 5 times with the $n = 2$ fields has been achieved in different plasma configurations and a wide range of $q_{95}$ on JET with the C-wall. Since 2012, ELM mitigation has been observed in the moderate collisionality regime on JET with the ILW using the $n = 2$ fields. A saturation effect of ELM mitigation and a reduction in the maximal ELM peak heat load, due to the splitting of the outer

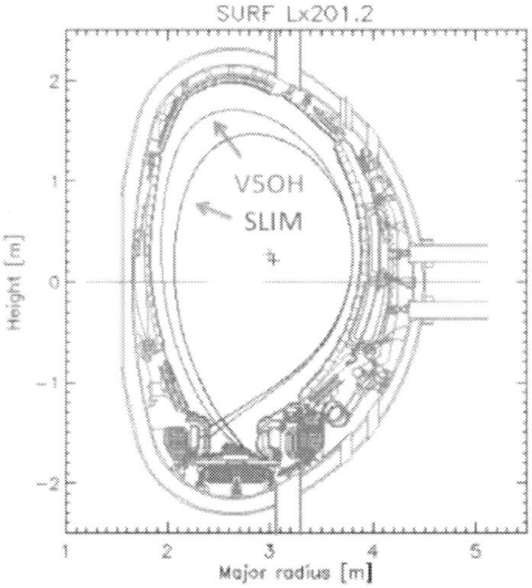


Figure 3 Comparison the plasma shape between SLIM (blue) and V5OH (red) configurations.

strike point, were observed during the application of the $n$ = 2 fields. With the $n$ = 2 fields, clear pre-ELM structures were observed on the outer divertor plate, and its dependence on $q_{95}$ [4].

Recently, the plasma shape dependence of the ELM mitigation using the $n$ = 2 field has been observed on JET with the ILW. Two different plasma configurations, V5OH and SLIM as shown in figure 3, were chosen as the target plasmas in this experiment. Both of these two configurations have a low plasma triangularity, but different significantly in the inner and top gap between plasma and first wall. Therefore, the SLIM plasma has a larger edge clearance and smaller plasma volume comparing with the V5OH plasma. Although both the SLIM and the V5OH configurations have a similar divertor configuration, but the outer strike point of the SLIM configuration is closer to the divertor pumping port, and the pumping efficiency of the SLIM configuration is better.

With the V5OH configuration mixed small and large ELMs appear during an application of the $n$ = 2 field in the H-mode plasmas with the pedestal electron collisionlity of 1.1 as shown in figure 4. The plasma core density drops from $4.0\times10^{19}m^{-3}$ to $3.5\times10^{19}m^{-3}$, while the core electron temperature reduces as well. With an increasing in the NBI heating power from 6MW to 9MW and 12MW, the pedestal collosionality reduces down to 0.7 and 0.4, respectively, and the small ELMs disappeared and the more regular Type-I ELMs are remaining in the H-mode plasma. The remaining Type-I ELMs during the flattop of the $n$ = 2 EFCC phase have an ELM frequency of few 10 Hz, even lower than that before applying the $n$ = 2 field, however, the ELM peak heat flux was reduced.

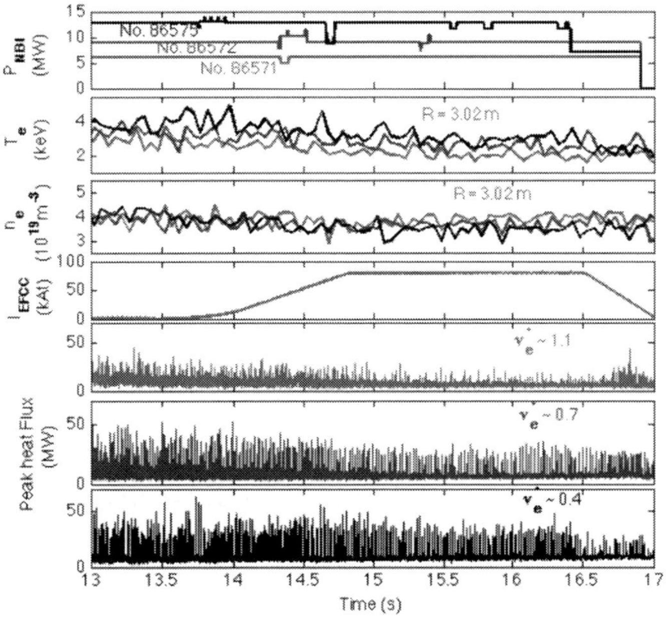


Figure 4 Overview of the ELM mitigation using n=2 fields in the H-mode plasmas with different NBI heating power of 6MW (86571), 9MW (86571) and 12 MW (86572). The signal from top to bottom are NBI power, plasma electron temperature and density measured by TS at the plasma core, EFCC coil current and Peak heat flux measured by the IR camera viewing outer divertor plate.

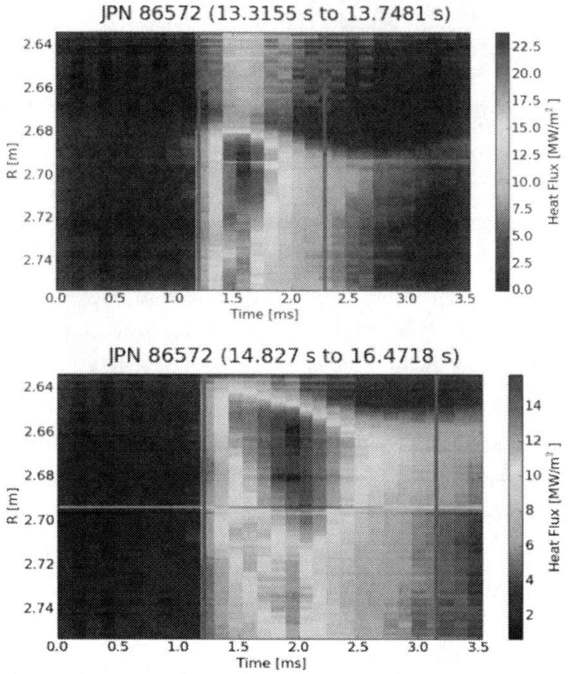


Figure 5 Averaged ELM heat flux distribution on the outer divertor measured without (upper) and with (bottom) an application of an n = 2 field.

Figure 4 shows a comparison of the averaged ELM heat flux distributions on the outer divertor plate before and after the application of the $n$ = 2 fields. With the $n$ = 2 field, the ELM peak heat flux reduces from 22 MW/m$^2$ to 15 MW/m$^2$, and splitting of the outer strike point has been seen during the ELM crash. The heat flux at the secondary strike point is about 12 MW/m$^2$, which is ~75% of the ELM peak heat flux. Both the secondary and the original strike points are moving slowly out along the divertor plate after the ELM crash, and the heat load decay time of the mitigated ELM is about 2ms, which is by a factor of two longer than that of the non-mitigated ELM.

To summarize, in the high collisionality regime, significant ELM mitigation using the $n$ = 2 fields has been observed on JET with the ILW, while no/weak ELMs mitigation was observed in previous experiments with a C-wall. Very small and frequent ELMs were observed during the application of the $n$ = 2 fields in both, low (V5OH) and high triangularity (HT3R) plasmas with a low NBI power, and there is no density pump-out was observed.

In moderate collisionality plasmas, the influence of plasma configuration on the collisionality dependence of ELM mitigation with the $n$ = 2 fields has been observed on JET with the ILW. With the $n$ = 2 fields, an increasing in ELM frequency by a factor of 4 are observed in the plasma with the SLIM configuration, while the ELM frequency even slightly drops in the plasma with the V5OH configuration. Reduction in the maximal ELM peak heat load, due to the splitting of the outer strike point, were observed during the application of the $n$ = 2 fields. The JET experimental observations indicate that the edge boundary condition (or the wall effect) is important for the ELM mitigation/suppression with magnetic perturbations on ITER.

"This work has been carried out within the framework of the EUROfusion Consortium and has received funding from the Euratom research and training programme 2014-2018 under grant agreement No 633053. The views and opinions expressed herein do not necessarily reflect those of the European Commission."

[1] T. Evans et al., Phys. Rev. Lett. 92, 235003 (2004). [2] Liang Y, *et al* , Phys. Rev. Lett. 98 265004 (2007), [3] Liang Y, *et al.*, Plasma Phys. Control. Fusion 49 B581 (2007). [4] Liang Y, *et al.*, Nucl. Fusion **53** 073036 (2013).

# Screening of external magnetic perturbation fields due to sheared flow

L. Li[1,2*], Y. Liang[1,2], Y. Q. Liu[3], N. Wang[2,4] and F. C. Zhong[1]
[1]*Donghua University, Shanghai, China*
[2]*Forschungszentrum Jülich GmbH, Institut für Energie- und Klimaforschung – Plasmaphysik, Jülich, Germany;*
[3]*CCFE, Culham Science Centre, Abingdon, OX14 3DB, UK*
[4]*Huazhong University of Science and Technology, Wuhan, China*

## 1. Introduction

Plasma response of the external magnetic field plays an important role in suppression or mitigation of the edge localized modes (ELMs) in tokamaks [1-4]. It is known that the single fluid theory predicts various screening regimes [5, 6] when plasma responds to the externally applied 3D magnetic field perturbations. One particular regime of screening, which occurs only at very slow flow, is associated with toroidal averaged curvature effect and referred as the GGJ-regime [7]. It has been shown in Ref. 6 that this effect can induce additional screening of the field pitch aligned components of the RMP field. From the experimental viewpoint, this can be an interesting regime, when the plasma flow locally vanishes near one rational surface in the pedestal region of the H-mode plasma.

While investigating the effect of RMP field penetration under the condition of vanishing flow near the rational surface, in a toroidal plasma with finite equilibrium beta, it is unavoidable to enter into the GGJ screening regime. Therefore, in this work, much of our effects will be devoted to study how the radial profile of the plasma flow affect the GGJ screening. We shall also compare the screening in the so-called resistive-inertial (RI) regime [5], which occurs at faster flow than the GGJ regime.

## 2. The MARS-F plasma response model

We used the MARS-F code [8] to compute the plasma response to the external magnetic field under the dc condition, while prescribing a radial profile for the plasma toroidal flow speed $\mathbf{V}_0 = R\Omega\hat{\phi}$. We refer to [8] for detailed description of the computational models for the linear plasma response.

In order to better understand the numerical results from the code, we shall consider a simple equilibrium, as that described in Ref. 6. All the results in Ref. 6 were obtained

* E-mail: l.li@fz-juelich.de

assuming a uniform equilibrium rotation profile. In this work, we consider sheared flow, with special choices of the rotation profiles. More specifically, we define two types of shear flow profiles.

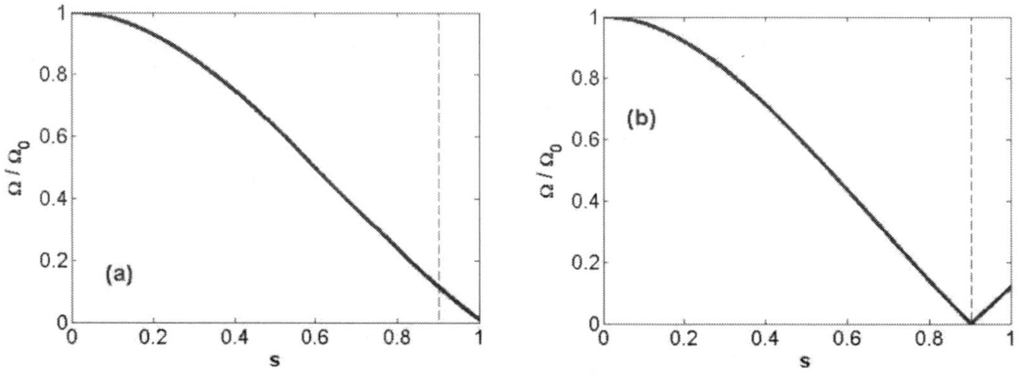


Figure 1. Radial profiles of the equilibrium toroidal rotation frequency, with the dash lines indicating the location of the $q = 2$ rational surface, which is the only rational surface for the chosen equilibrium and the chosen n=1 magnetic perturbation. (a) without, and (b) with, a strong local variation of flow speed near the rational surface.

*2.1. A sheared rotation profile without strong local variation near the rational surface*

Such a profile is prescribed by the following model

$$\Omega = (\Omega_0 - \Omega_1)(1 - 2s^2 + s^3) + \Omega_1 \quad (1)$$

where $\Omega_0$ is the amplitude of the plasma rotation frequency at the magnetic axis, and $\Omega_1$ is defined as the rotation frequency at the plasma surface [8]. An example of such a profile, normalized to unity at the magnetic axis, is shown in figure 1(a).

*2.2.Flow profiles with strong local variation near rational surface*

In order to investigate the relative effect of the flow shear and the flow amplitude, at the rational surface, on the plasma response to the external magnetic field, we also consider another family of profiles, with nearly vanishing flow speed at the rational surface. The profile is defined as

$$\Omega = \left| \frac{(1 - 2s^2 + s^3) - (1 - 2s_q^2 + s_q^3)}{2s_q^2 - s_q^3} \Omega_0 \right| + \Omega_q \quad (2)$$

where $s_q$ is the radial position of the rational surface $q = 2$, and $\Omega_q$ is the plasma rotation frequency at this surface. Figure 1(b) shows one example of the radial equilibrium plasma rotation frequency profile, where again $\Omega$ is normalized to unity at the magnetic axis. In further numerical simulations, the on-axis rotation frequency $\Omega_0$ is one plasma parameter that scans.

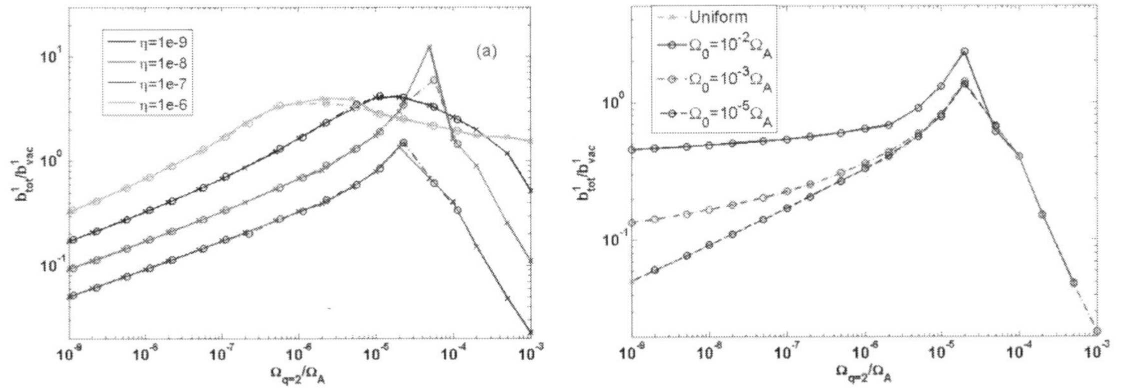


Figure 2. The effect of different rotation profiles on the plasma response to the n=1 external field. (a) comparison of the plasma response between the uniform plasma flow profile (solid) and that defined in Section 2.1 (dash-dotted), while assuming various values for the normalized plasma resistive (the inverse of the Lundquist number). (b) plasma response with the flow profile as defined in Section 2.2., with various choices of the flow amplitude, compared also with the uniform flow case. Note a good overlap between the uniform flow case and the case with very slow flow in (b). The plasma response is measured by the m/n=2/1 resonant harmonic of the radial field, computed at the q=2 surface and normalized by the corresponding quantity for the vacuum field.

## 3. Numerical results

Extensive modelling and analysis efforts have been carried out, in order to identify the screening physics associated with the local variation of the plasma flow speed near the rational surface. The results are summarized in Fig.2. Figure 2(a) compare the modelled plasma response, assuming two types of flow profiles – one is the uniform flow, the other is a flow profile as defined in Section 2.1 and shown in Fig. 1(a). The key point here is that, we plot the response, defined as the amplitude of the m/n=2/1 resonant harmonic of the perturbed radial magnetic field component at the q=2 surface, against the *local* rotation frequency at the same q=2 surface, for different types of the flow profile.

The fact that the curves, obtained with two types of flow profiles, nearly overlap for various choices of the plasma resistivity, indicates that the plasma screening (in both the GGJ-regime and the RI-regime) of the external magnetic field is quantitatively determined by the magnitude of the local plasma speed at the rational surface. No strong variation of the local flow shear has been allowed in this type of scans shown in Fig. 2(a).

A subtle difference, however, appears, if we do allow a strong local variation of the flow near the q=2 surface, as shown in Fig. 1(b). The computed plasma response from this flow scan is summarized in Fig. 2(b). The separation between the GGJ screening regime and the RI-regime is again separated by the *local* flow speed at the rational surface, as expected. On the other hand, when the ratio between the on-axis flow and the local flow speed becomes very large (enhanced by orders of magnitude), the global radial shape of the plasma flow also starts to affect the screening in the GGJ-regime, essentially reducing the GGJ-screening.

## 4. Conclusion and discussion

Based on a simple toroidal equilibrium, we try to computationally understand the screening of the external 3D magnetic fields by the toroidal flow. In particular, we investigate how the screening changes when the local flow speed approaches zero at the rational surface. According to the single fluid theory, the relevant screening regime at slow flow is the GGJ-regime. Without a strong local variation of the flow profile, the *local* magnitude of the flow at the rational surface is the dominant factor affecting the GGJ screening. With a strong local variation of the flow (and thus a large variation of the global flow profile), the GGJ screening is reduced.

## Acknowledge

The work was partially supported by the National Natural Science Foundation of China (Grant No. 11405029). This work is also funded by the China Scholarship Council.

## References

[1] Evans T E et al 2006 Nat. Phys. **2** 419
[2] Liang Y et al 2007 Phys. Rev. Lett. **98** 265004
[3] Kirk A et al 2011 Plasma Phys. Controlled Fusion **53** 065011
[4] Suttrop W et al 2011 Phys. Rev. Lett. **106** 225004
[5] Fitzpatrick R 1998 Phys. Plasma **5** 3325
[6] Liu Y Q et al 2012 Phys. Plasma **19** 072509
[7] Glasser A H et al 1975 Phys. Fluids **7** 875
[8] Liu Y Q et at 2010 Phys. Plasma **17** 122502

# Measurement of plasma response to n=2 fields using saddle loops on JET

N. Wang[1,2,*] Y. Liang[1], S. Gerasimov[3], P. Denner[1], P. Drews[1], S. Jachmich[4], L. Li[1], P. Lomas[3], and the JET Contributors[†]

*EUROfusion Consortium, JET, Culham Science Centre, Abingdon, OX14 3DB, UK*

[1]*Forschungszentrum Jülich GmbH, Institut für Energie- und Klimaforschung - Plasmaphysik, 52425 Jülich, Germany*

[2]*Huazhong University of Science and Technology, 430074 Wuhan, China*

[3]*CCFE, Culham Science Centre, Abingdon, OX14 3DB, UK*

[4]*Koninklijke Militaire School-Ecole Royale Militaire, B-1000 Brussels, Belgium*

## Introduction

Active control of Type-I edge-localized mode (ELM) using 3D fields has been demonstrated in several tokamaks [1-5], and it provides an attractive method for ELM control in ITER. To date, more and more experimental results show the plasma response to the 3D perturbation fields is important for understanding the mechanism of ELM control. On JET, it has been found that the plasma response to $n$ = 2 fields measured by saddle loops was saturated during the ramp-up of the current flowing in the error field correction coils (EFCCs) [6]. Recently, the experiment in DIII-D shows rapid increase in the inner-wall magnetic response to an $n$ = 2 field when the plasma transitions into ELM suppression [7].

To further investigate the plasma response to $n$ = 2 fields on JET, the eddy currents in the vacuum vessel induced by the varying of EFCC currents should be taken into account. Since the former work [6] hasn't considered the eddy effect, this paper will focus on the analysis of eddy effect on measured $n$ = 2 field based on a vacuum pulse.

## Experimental setup and eddy current model

JET is equipped with four error field correction coils (EFCCs) outside the vacuum vessel. A set of saddle loops, with 14 toroidal rows and 8 loops in each row, are fitted against the ex-vessel wall. A sketch of EFCCs and saddle loops is shown in Figure 1. Only one low field side row of saddle loops, which will be used for the analysis of eddy effect in this paper, are shown. They are named from S101 to S801, based on their toroidal locations. Using the radial magnetic fields measured by these 8 saddle loops, two orthogonal components of $n$ = 2 field can be calculated, i.e. $b_r^{\cos} = (b_{S101} - b_{S301} + b_{S501} - b_{S701})/4$ and $b_r^{\sin} = (b_{S201} - b_{S401} + b_{S601} - b_{S801})/4$. Then the amplitude of $n$ = 2 field is calculated via $b_r^{n=2} = \sqrt{(b_r^{\sin})^2 + (b_r^{\cos})^2}$ , and the phase is $\phi^{n=2} = \mathrm{atan}(b_r^{\sin}/b_r^{\cos})$.

* E-mail: *wangnc@hust.edu.cn*

† See the Appendix of F. Romanelli et al., *Proceedings of the 25th IAEA Fusion Energy Conference 2014*, Saint Petersburg, Russia

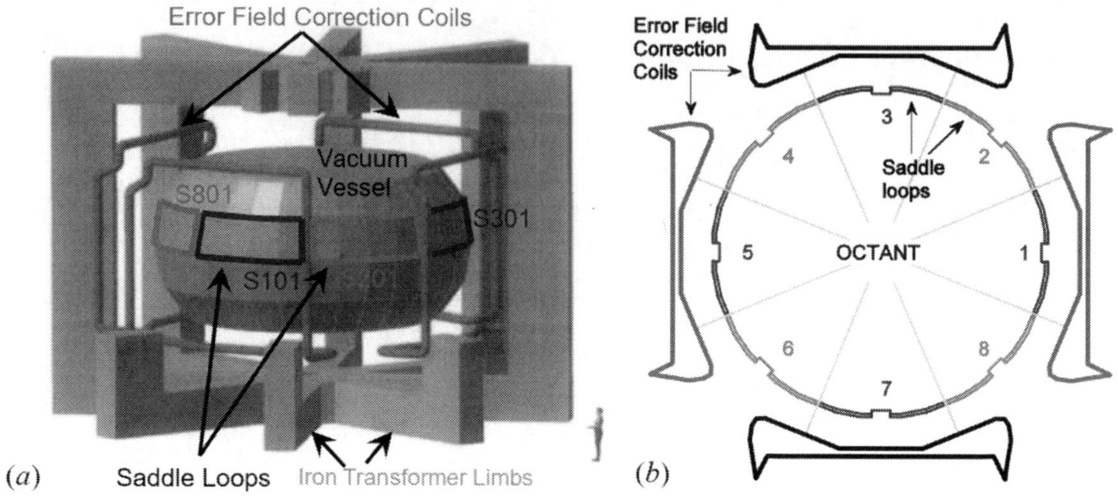


Figure 1 *Sketch of the EFCCs and the low field side saddle loops in (a) 3D view and (b) plan view.*

In order to obtain the plasma response to applied $n = 2$ EFCC field, the vacuum field must be removed from the signals of plasma pulses. Since the saddle loops in the odd octants are exactly under the EFCCs as shown in Figure 1(*b*), the $b_r^{cos}$ signal can measure the vacuum fields induced by $n = 2$ EFCCs, which are significantly larger than the $n = 2$ field from plasma response. Except for the direct coupling between saddle loops and EFCCs, the eddy currents excited in the vessel by $dI_{EFCC}/dt$ can also contribute to the measured $n = 2$ field. An analytical model has been used to describe the eddy effect on saddle loop measurement on JET. This model was developed to analysis the eddy effect on a single saddle loop on the J-TEXT tokamak [8]. As shown in Figure 3 of Ref. 8, the eddy currents flowing in the vacuum vessel are equivalent to several *L-R* circuits which couple with the EFCCs and saddle loops. Neglecting the mutual inductance between eddy circuits and assuming zero state of the system, the analytical solution of this model can be written as

$$b_r = M_{sl} I_s(t) - \sum_{j=1}^{k} H_e^j \left[ I_s(t) - \left(\exp\left(-t/\tau_e^j\right)/\tau_e^j\right) \int_0^t I_s(t) \exp\left(t/\tau_e^j\right) dt \right], \qquad (1)$$

where $M$ is the mutual inductance, $H$ is a parameter relevant with inductances as defined in Ref. 8, $I$ is the source current, and $\tau$ is the time constant of eddy circuit. The subscripts $s$, $l$, $e$ denote the EFCCs, saddle loop, and eddy circuits, and $k$ is the number of eddy circuits used. The vacuum pick-up induced by the EFCCs and the eddy currents can be identified by fitting Eq. (1) to the signals measured by the saddle loops in a vacuum reference pulse (no plasma).

## Results

JET pulse 82563 is shown in Figure 2 as an example for the application of Eq. (1). After the plasma in pulse 82563 was landed, EFCC current was ramped up at a rate of 1.9 kA/s to the 1.5 kA flattop, and then oscillations with frequencies (amplitudes) of 10 Hz (0.45 kA), 30

Hz (0.15 kA) and 50 Hz (0.05 kA) were superimposed to the flattop. Both the measured $b_r^{\cos}(n=2)$ and $b_r^{\sin}(n=2)$ are fitted to the function of $I_{EFCC}$ as described by Eq. (1). The parameters $M$, $H$ and $\tau$ can be calculated using the nonlinear least square fitting.

The *cosine* component of vacuum $n$ = 2 field from the EFCC coils, $M\times I_{EFCC}$, is shown by the blue line in Figure 2(*a*), while the vacuum field from all the eddy currents is shown by the magenta line. The sum of these two parts makes up the fitted signal (red line in Figure 2(*a*)). As a comparison, the original measured signal is shown by the black line. The fitting error between measured and fitted signals is shown in figure 2(*c*). The eddy current part, shown in magenta line of Figure 2(*a*), is contributed from 3 eddy circuits with different time constants, i.e. 9.2 ms, 54 ms, and 1.09 s, respectively. The three parts are plotted in Figure 2(*b*). All the signals are analyzed and plotted in the original acquired frequency of 500 Hz, and haven't been smoothed or filtered. This indicates that the fitting error is so small that it is at the same order as the noise. Hence it is shown that the 3 assumed eddy currents are enough to describe the eddy effect.

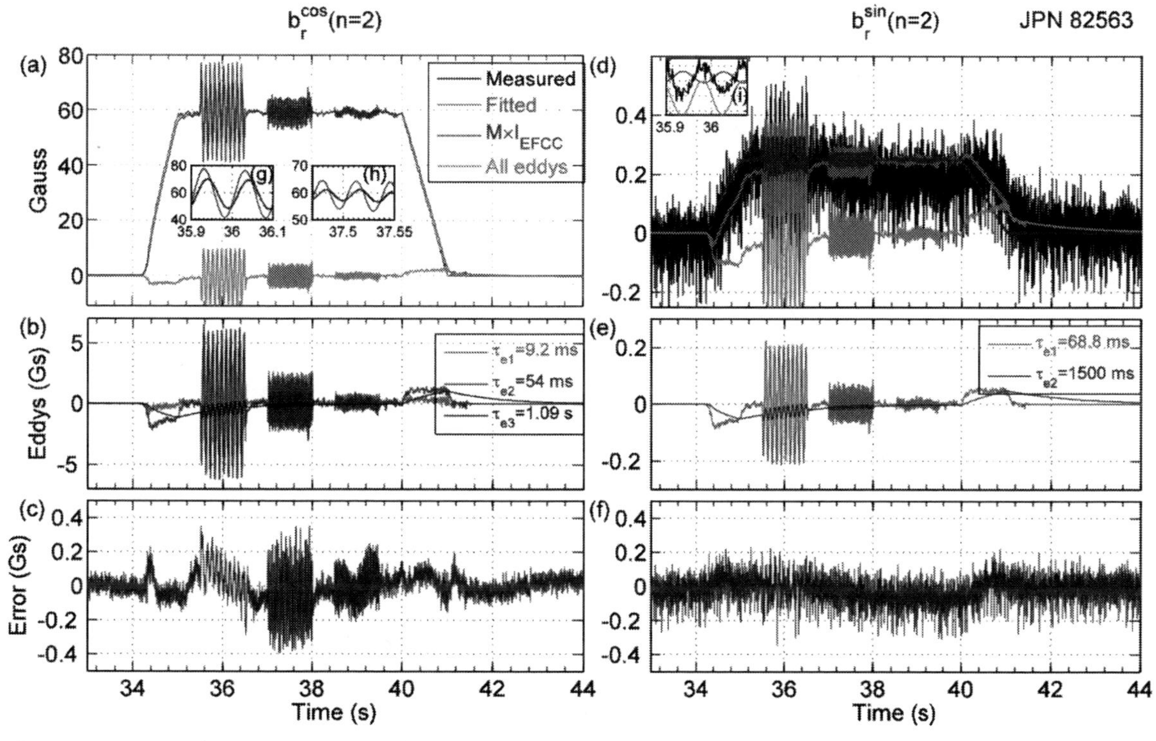


Figure 2 *Measured vacuum n = 2 field and the analysis results in a pulse 82563 with n = 2 EFCC.*

Figure 2(*g*) and (*h*) show the expanded views of measured signals, fitted signals and $M\times I_{EFCC}$ with 10 Hz and 30 Hz oscillations. The significant differences between measured signals and $M\times I_{EFCC}$ show the large contributions from eddy currents again. The total $n$ = 2 field from all the eddy currents can be as large as 2.5 Gauss during the ramp-up/down of $I_{EFCC}$ at around 1.9 kA/s, and around10 Gauss during the 10 Hz/0.45 kA oscillations.

Figure 2 (*d*) through (*f*) show the results for $b_r^{\sin}(n=2)$ with the same color coding as (*a*) through (*c*), and Figure 2 (*i*) shows the expanded view of Figure 2(*d*) during the 10 Hz oscillations. Due to the location of even saddle loops and the configuration of EFCC current directions, $b_r^{\sin}(n=2)$ shows a very small vacuum field during the application of the EFCCs, and two eddy circuits are enough to describe the eddy effect, as shown in Figure 2(*e*).

Considering the smaller oscillation current at 30 Hz, the fitting error during the 30 Hz oscillation is significantly larger than that of 10 Hz oscillation. Since the higher frequency oscillations can induce eddy current with faster time constant, fast eddy current will be needed in the model. But if the time constant is faster than 10 ms, the integration, expressed by the second part in right hand side of Eq.(1), will result in an unacceptable numerical error for data acquired at 500 Hz. So the time constants are restricted to be larger than 10 ms in the fitting. This may be the reason why the errors for higher frequency oscillations are larger. A higher data acquisition rate is needed to provide better fitting for high frequency oscillations.

## Conclusion and Acknowledgments

An analytical model [8] has been successfully used in analyzing the eddy effect during the application of the EFCC field. The vacuum $n$ = 2 field generated by the eddy current with $dI_{EFCC}/dt \sim$ 1.5 kA/s is around 2.5 Gauss, and ~10 Gauss during the 10 Hz/0.45 kA oscillation. The eddy effect can be removed with an error less than ±0.3 Gauss. In low-beta H-mode plasmas, the eddy effect due to the conducting vacuum vessel has a similar amplitude as the measured plasma response.

This work has been carried out within the framework of the EUROfusion Consortium and has received funding from the Euratom research and training programme 2014-2018 under grant agreement No 633053. The views and opinions expressed herein do not necessarily reflect those of the European Commission. N.W. is thankful for the support from the China Scholarship Council.

## References

[1] T. E. Evans *et al.*, *Phys. Rev. Lett.* **92** (2004) 235003.
[2] Y. Liang *et al.*, *Plasma Phys. Controlled Fusion* **49** (2007) B581.
[3] A. Kirk *et al.*, *Plasma Phys. Controlled Fusion* **53** (2011) 065011.
[4] W. Suttrop *et al.*, *Phys. Rev. Lett.* **106** (2011) 225004.
[5] Y. Jeon *et al.*, *Phys. Rev. Lett.* **109** (2012) 035004.
[6] Y. Liang *et al.*, *Nucl. Fusion* **53** (2013) 073036.
[7] R. Nazikian *et al.*, *Phys. Rev. Lett.* **114** (2015) 105002.
[8] Y. Ding, N. Wang, *et al.*, *Rev. Sci. Instrum.* **85** (2014) 043502.

# Impact of electron scale modes on electron heat transport in the JET tokamak

N. Bonanomi[1,2], J. Citrin[3,4], P.Mantica[1] and JET contributors*

EUROfusion Consortium, JET, Culham Science Centre, Abingdon, OX14 3DB, UK

[1] *Istituto di Fisica del Plasma "P.Caldirola", Consiglio Nazionale delle Ricerche, Milano, Italy*
[2] *Università di Milano-Bicocca, Milano, Italy*
[3] *CEA-Cadarache, St. Paul Lez Durance, France*
[4] *FOM Institute DIFFER, P.O. box 6336, 5600 HH Eindhoven, The Netherlands*
**See the Appendix of F. Romanelli et al., Proceedings of the 25th IAEA Fusion Energy Conference 2014, Saint Petersburg, Russia*

In dedicated electron heat transport experiments in JET L-mode plasmas [1], lower values of $R/L_{Te}$ are observed, at the same level of gyro-Bohm normalized electron heat flux, in the presence of significant NBI (Neutral Beam Injection) power with respect to discharges with pure ICRH (Ion Cyclotron Resonance Heating) applied in mode conversion (MC) scheme yielding dominant electron heating. The discharges studied in this paper were made with C-wall and with $B_0$~3.45 $T_{e,0}$~5 keV, $T_{i,0}$~2.5 – 5 keV, $n_{e,0}$~(2 – 3)$10^{19}$ $m^{-3}$ and $I_p$~(1.8 – 3) MA with $I_p$ overshoot, ramp-up and ramp-down.
As seen in Fig.1, the $R/L_{Te}$ decrease is due to both a decrease in inverse critical gradient length and an increase in stiffness. This is in contrast with the strong reduction of ion stiffness observed in presence of NBI (Fig.1a, [2]), which was interpreted as due to non-linear electromagnetic stabilization of ITG modes by fast ion pressure gradient [3].

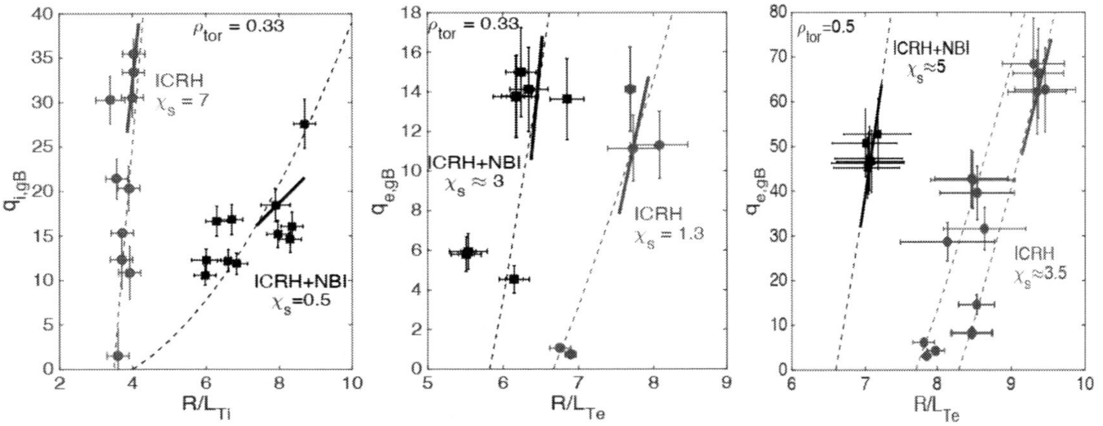


**Figure 1: Effect of the presence of NBI heating on ions in L-mode JET discharges at $\rho_{tor}$=0.33 (left panel, reproduced from [2]) and on electrons at $\rho_{tor}$=0.33 (centre panel) and $\rho_{tor}$=0.5 (right panel) [1].**

The main differences in NBI heated plasmas with respect to pure ICRH-MC plasmas are lower values of $T_e/T_i$, higher values of $R/L_{Ti}$, the presence of additional fast ions and higher toroidal rotation. Due to the stabilization effects of fast ions on ITG, the effects of higher $R/L_{Ti}$ are not expected to be significant. Possible effects of $T_e/T_i$ on TEM modes thresholds have been analysed with linear gyrokinetic simulations and the results suggest that it can't explain the experimental observation [1]. However, one possible effect of lower values of $T_e/T_i$ is an increase of the electron heat flux carried by ETG modes, for which a stabilizing effect of $\tau=Z_{eff}T_e/T_i$ is expected [4].

In this work we investigate the presence of ETG modes in these JET discharges and their effects on the electron heat flux using linear and nonlinear gyro-kinetic simulations with the gyro-kinetic code GENE in the local limit [5].

To study properly the impact of ETG modes on transport, multiscale gyro-kinetic simulations including both electron and ions scales are necessary: ion scale zonal flows can provide a mechanism for ETG streamer saturation; ETG modes can affect the ion scales through non-linear coupling mechanisms, increasing the level of heat transport carried by TEM/ITG modes [6,7,8,9]. However, such simulations demand exceeding computational resources ($10^6 - 10^7$ CPUh per run) and could not be afforded for this work. Instead, we carried out separate scale simulations. In all simulations, Miller geometry, collisions, kinetic electrons and experimental input parameters varying within their error range were used. In all ion-scale simulations, a carbon impurity was included, at a level consistent with the experimental values of $Z_{eff.}$ Fast ions and electromagnetic effects were retained in the NBI case. In the ETG simulations, we used adiabatic ions and included the measured external flow shear. This leads to ETG streamer saturation. We assume here that the external flow shear leads to a similar ETG saturation level as the ion scale zonal flows would have done in a multi-scale simulation. However, validating this assumption is out of the scope of this work. Extensive convergence tests were made for both linear and non-linear cases.

Linear gyro-kinetic simulations were carried out to establish the effect of $\tau$ on ETG linear threshold within the experimental parameters range. The results obtained at $\rho_{tor}$=0.5 are shown in Fig. 2.

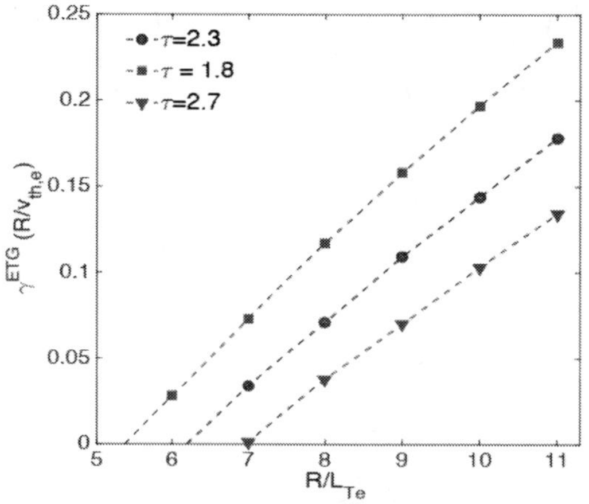


**Figure 2: ETG linear growth rates vs $R/LT_e$ for different values of $\tau$.**

The linear threshold of ETG modes decreases for lower values of $\tau$, i. e. for lower values of $T_e/T_i$. The values found are below the experimental values of $R/L_{Te}$, indicating that ETG modes are unstable for the experimental set of parameters in both ICRH and ICRH+NBI cases.

Regarding non-linear simulations, we used the parameters of JET discharge n. 78834 for the pure ICRH heating case and of JET discharge n. 78842 for the ICRH+NBI heating case. In the TEM/ITG nonlinear runs, we used a box size of $[L_x,L_y]=[100,125]\rho_i$ , with a numerical resolution of [128,24,48,48,12] points in $[x,y,z,v_{/\!/}, \mu]$ and $0.05 \le k_y\rho_i \le 1.2$. In the ETG case, we used a box size of $[L_x,L_y]=[195,125]\rho_e$, with a numerical resolution of [256,24,48,48,12] points in $[x,y,z,v_{/\!/}, \mu]$ and $0.05 \le k_y\rho_e \le 1.2$. x, y, z, $v_{/\!/}$, $\mu$ indicate respectively the radial direction, the binormal direction, the parallel direction, parallel velocity and magnetic moments. $k_y$ is the binormal mode number and $\rho_{i/e}$ is the ion/electron Larmor radius.

We made a scan in $R/L_{Te}$ of the electron heat flux in order to compare the levels of the heat flux and of the electron stiffness with the experimental values. The results obtained at $\rho_{tor}$=0.5

for the electron heat flux are shown in Fig. 3. In Fig. 4 the results obtained at $\varrho_{tor}$=0.5 for ions are compared with the experimental values. All the fluxes are normalized to gyro-Bohm units using $q_{e(i),gB} = q_{e(i)}/(T_e n_e \rho_s^{*2} c_s)$, where $c_s = \sqrt{T_e/m_i}$ and $\rho_s^* = c_s m_i/ReB$.

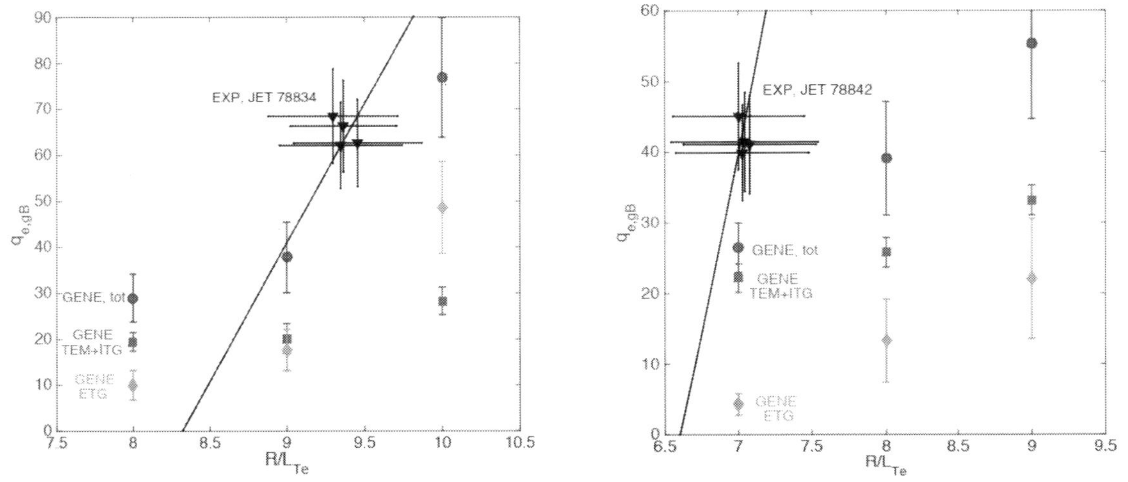


**Figure 3: Normalized electron heat flux vs R/L$_{Te}$. Experimental flux (black triangles), GENE TEM/ITG flux (red squares), GENE ETG flux (green diamonds) and GENE TEM/ITG+ETG flux ( blue circles) for discharges with pure ICRH heating (left) and for discharges with ICRH+NBI heating (right). The black lines indicate the experimental slope (stiffness) of the electron heat flux.**

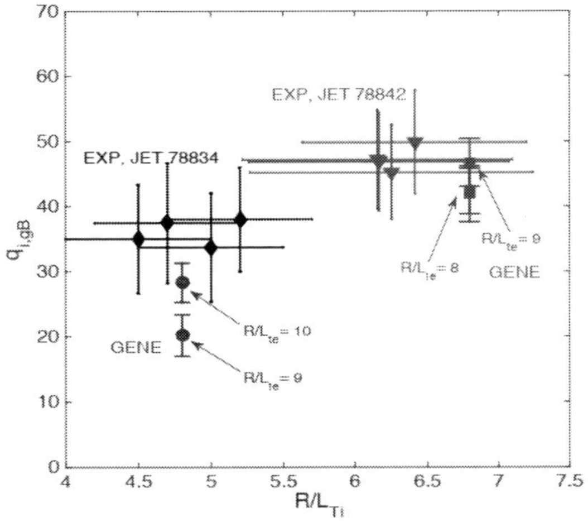


**Figure 4: Normalized ion heat flux vs R/L$_{Ti}$. Experimental flux of the ICRH case (black diamonds) and of the ICRH+NBI case (purple triangles); GENE TEM/ITG flux for the ICRH case (blue circles) and GENE TEM/ITG flux for the ICRH+NBI case (red squares).**

The experimental normalized ion heat flux remains unchanged in the two cases despite the differences in R/L$_{Ti}$ and this is reproduced quite well in the simulations using fast ions and electromagnetic effects, confirming what found in previous works [3]. The fact that the ion heat flux is reproduced is an indication of the consistency of our simulations.

Regarding the electron heat flux, the simulations indicate that a considerable amount of flux is independent of R/L$_{Te}$ and is carried by non-diagonal terms such as R/L$_n$ TEM modes and especially ITG modes (~25% of the experimental flux in the ICRH case and ~40% of the experimental flux in the NBI case). The flux carried by ion scale modes is in both cases too low to reproduce the experimental flux: in both cases we can reproduce the ~50% of the experimental flux with TEM-ITG modes. The scan in R/L$_{Te}$ allows also a comparison with the experimental slope of the flux and also in this case the TEM/ITG contribution to the flux cannot alone reproduce the experimental slope. In the NBI case, the experimental values of $R/L_{Te}$ are also very close to the nonlinear threshold of $\nabla T_e$ TEM modes: this could indicate that an other kind of instability, such as ETG, is carrying the remaining part of the flux. In both cases ETG modes are unstable: as mentioned, the amount of ETG flux calculated is just indicative as multiscale scale simulations would be needed, but in both cases we can't reproduce the experimental values and especially the experimental slope of the electron heat flux without retaining the ETG flux. This suggests that ETG modes could play an important role for electron heat flux for our experimental range of parameters and can help to explain the higher electron stiffness and the lower threshold values found experimentally in the NBI case. Also, a small reduction of the TEM modes threshold due to lower collisionality in the NBI case can contribute to explain the experimental observations.

**Conclusions**

This work provides a comparison between experiments and gyrokinetic simulations for JET L-mode discharges with and without substantial ion heating provided by NBI. It indicates that a significant amount of electron heat flux (~25%) can be carried by non-diagonal terms such as ITG modes. Using ion scale modes alone it's difficult to reproduce the experimental slope and the experimental electron heat flux, reaching only the ~50% of the experimental values. Electron scale modes can help to reproduce the experimental fluxes in both ICRH and ICRH+NBI cases. Furthermore, being more unstable in presence of substantial ion heating due to lower values of $T_e/T_i$, ETG modes can help to explain why, in presence of NBI heating, we observe lower values of $R/L_{Te}$: the electron heat flux carried by ions scale modes doesn't increase significantly in the NBI case, but in this case there is almost the same amount of electron heat flux carried by ETG modes at lower values of $R/L_{Te}$. It is important to underline again that these results are only indicative. Complete scale simulations should be done in order to properly consider the nonlinear interactions between different scales modes.

**Acknowledgements**

The authors would like to thank Tobias Görler, Daniel Told and Frank Jenko for precious advice on GENE simulations.

The authors are also grateful to D. R. Mikkelsen for assistance.

This research used resources of the National Energy Research Scientific Computing Center, a DOE Office of Science User Facility supported by the Office of Science of the U.S. Department of Energy under Contract No. DE-AC02-05CH11231. A part of this work was carried out using the HELIOS supercomputer system at Computational Simulation Centre of International Fusion Energy Research Centre (IFERC-CSC), Aomori, Japan, under the Broader Approach collaboration between Euratom and Japan, implemented by Fusion for Energy and JAEA.

This work has been carried out within the framework of the EUROfusion Consortium and has received funding from the Euratom research and training programme 2014-2018 under grant agreement No 633053. The views and opinions expressed herein do not necessarily reflect those of the European Commission.

**References**

[1] N. Bonanomi et al., submitted to Nuclear Fusion
[2] P. Mantica et al., Phys. Rev.Lett. **107**, 135004 (2011)
[3] J. Citrin et al., Phys. Rev.Lett. **111**, 155001 (2013)
[4] F. Jenko et al., Phys. Plasmas **8**, 4096 (2001)
[5] F. Jenko et al., Phys. Plasmas **7**, 1904 (2000)
[6] N. T. Howard et al., Phys. Plasmas **21**, 112510 (2014)
[6] Candy et al., JPCS **78**, 012008 (2007)
[7] T. Gorler, F. Jenko, Phys. Rev.Lett. **100**, 185002 (2008)
[8] R. E. Waltz et al., Phys. Plasmas **14**, 056116 (2007)

# MHD analysis of Beta Collapses in AT JET discharges

E.Alessi[1], L.Piron[2], P.Buratti[3], J. Mailloux[2], N.Hawkes[2], E.Joffrin[4], Y.Baranov[2], E.Giovannozzi[3], M.P. Gryaznevich[5,6], T.C.Hender[2], T.C.Luce[7], G.Pucella[3], C.Sozzi[1] and JET contributors*

EUROfusion Consortium, JET, Culham Science Centre, Abingdon, OX14 3DB, UK

[1] *Istituto di Fisica del Plasma, C.N.R., EURATOM-ENEA Association, Milan, Italy*

[2] *CCFE, Culham Science Centre, Abingdon, UK*

[3] *C.R. Frascati, EURATOM-ENEA Association, Frascati (Rome), Italy*

[4]*CEA, IRFM, F-13018 Saint-Paul-lez-Durance, France*

[5]*Tokamak Energy Ltd, Culham Science Centre, Abingdon, OX14 3DB UK*

[6]*DTU, Fysikvej, DK - 2800 Kgs. Lyngby, Denmark*

[7]*General Atomics, PO Box 85608, San Diego, CA 92186–5608, USA*

**See the Appendix of F. Romanelli et al, Proc. of the 25th IAEA FEC, Saint Petersburg, Russia, 2014*

**Introduction.** Advanced Tokamak (AT) experiments aim at identifying favorable conditions for stable plasmas at high normalised pressure ($\beta_N$>3) in view of extrapolating to next step fusion devices. Higher values of $\beta_N$ are limited by the onset of MHD activity [1, 2]. In experiments performed at JET with Carbon Wall (CW), large collapse events [3] have been observed to suddenly reduce the $\beta_N$ value, and were referred to as beta collapses (BC). BCs were observed to be ELM-like events but involving a larger plasma volume and triggered by a global kink-like MHD mode [3].

Recent experiments were performed in JET with ITER-Like Wall (ILW) using up to 26MW NBI power in plasmas with $B_T$ from 1.7T to 2.4T, with $I_P$ from 1.9MA to 1.5MA ($q_{95}$ from 3.5 to 5.5). The initial q-profile was varied by changing the start time of the high heating phase, with $q_{min}$ from 1.2 to 2.5. All q-profiles have positive magnetic shear. Contrary to the observations in CW, BCs found in ILW experiment did not show any clear oscillatory precursor in the usual fft based analysis of magnetic signals. An analysis of the MHD activity before the BC is here addressed by application of Singular Value Decomposition (SVD) [4,5] to a toroidal array of Mirnov coils.

Resonant Field Amplification (RFA) measurements were performed in JET experiments by application of non-axisymmmetric magnetic field from four Error Field Correction Coils (EFCCs) arranged symmetrically around the outside of the vacuum vessel, each spanning

70° toroidally. This measure allows to probe the plasma stability, using the so-called MHD spectroscopy technique [6].

**Singular Value Decomposition.** Magnetic crashes associated with BC are very fast events, with durations of ~0.5ms similar to ELM ones. MHD analysis is here carried out by application of SVD to a toroidal array of six Mirnov coils in time windows of 0.5ms. SVD [4] decomposes a matrix in a collection of singular values $\{SV_i\}$ (the strength of the mode), of Principal Components $\{PC_i\}$ (the mode timeline), and of Principal Axes $\{PA_i\}$ (the spatial periodicity of the mode). In order to evaluate the toroidal order number n of the dominant SVD mode, the n's likelihoods $L_n$ [5] are calculated as: $L_n = |PA_1 \cdot V_n|^2$; where $V_n$ is the expected $PA_1$ calculated from numerical signals with n periodicity [5]. Amplitudes $A_n$ of perturbations with n periodicity is calculated from the amplitude $A_x$ of the fluctuations sensed by a given Mirnov coil, by

$$A_n = \frac{A_x \sum_i SV_i^2 |PA_i \cdot V_n|^2}{\sum_i SV_i^2}.$$

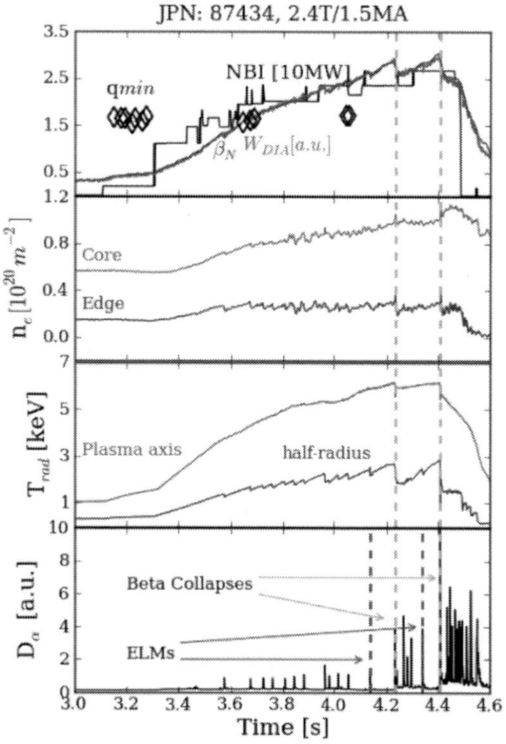


**Figure 1**. From top to bottom: $q_{min}$ values from MSE constrained EFIT, NBI power normalized to 10MW, $W_{DIA}$ normalized to 1.76MJ; core and edge line integrated electron density; electron temperature near plasma axis and at half radius; Dalpha. Beta Collapse events are indicated by vertical dashed orange lines while ELMs analyzed in Fig.2 by red dashed lines.

**Detailed analysis of JET ILW #87434.** The time traces for this pulse are reported in Fig.1 where two BCs occurring at t=4.23s and 4.41s are indicated by vertical dashed orange lines. Differently from ELMs (dashed red lines in Fig.1, bottom panel) such events affect the electron temperature profile from edge to near the plasma axis, (3rd panel in Fig.1) and produce sudden relative losses of Wdia (stored diamagnetic energy) of 10 and 17% respectively, reducing $\beta_N$ from 3 to less than 2.5.

In Fig.2, the SVD results for 40ms around each event highlighted in Fig.1 by vertical dashed lines are shown. For each case the peak frequency of the $PC_1$, the $L_n$ values, and the $A_n$ are shown from top to bottom. $L_n$>0.5 indicates the presence of a mode with a given n order number.

A weak n=4 mode is present both before the ELM at 4.14s (Fig.2a) and before the BC at 4.23s (Fig.2b), while in the larger BC at 4.41s (Fig.2d) a n=5 mode is found to suddenly grow in about 20ms and to undergo to three frequency-amplitude cycles before the BC. Such a mode is located at R~3.44m (CXS), where q~1.8±0.2 from MSE-EFIT.

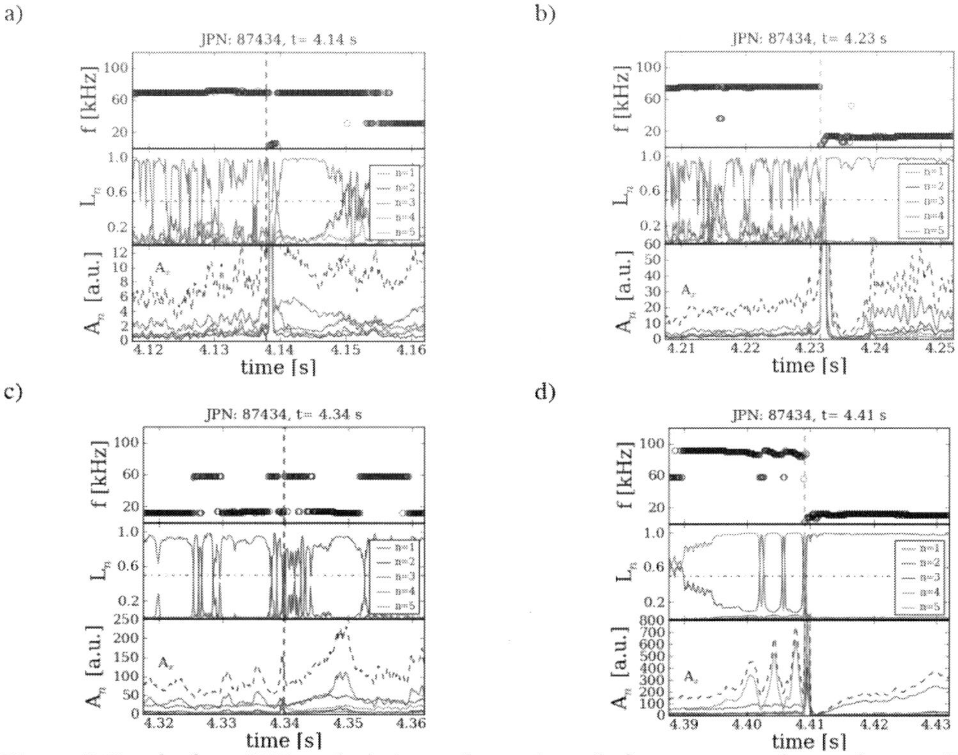


**Figure 2.** Results from SVD analysis (top to bottom): mode frequency; n order number; amplitudes of the modes. Results are shown in four cases: a) ELM before the first BC; b) first BC; c) ELM before the second BC; d) second BC (see also figure 1).

In both cases shown in figure 2, SVD is not capable of identifying a clear toroidal periodicity when the crash happens, but in all cases it indicates the presence of a low frequency n=1 mode immediately after the crashes. The main difference between ELMs and BCs is the presence of an n=1 mode that develops as a 2/1 mode.

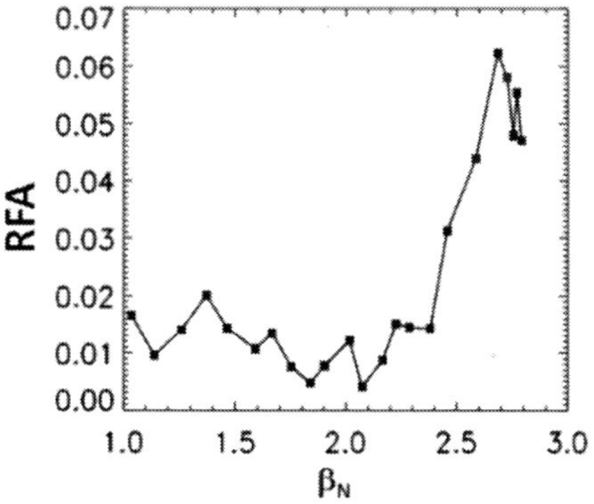


**Figure 3.** RFA measurements in pulse #87434.

Results of RFA measurements are shown in Fig.3 as a function of $\beta_N$. Toroidally opposite EFCCs are connected in series with oppositely directed currents to produce a n=1 magnetic field. RFA is defined as

$$RFA = \frac{B_r|_{rsensor} - B_{r,ext}}{B_{r,ext}}.$$

In Fig.3 for $\beta_N>2.5$ the strengthening of the RFA is a sign that the plasma pressure is approaching the no-wall beta limit.

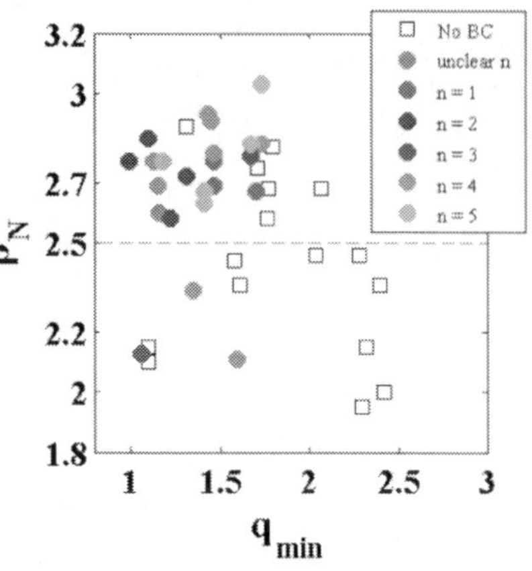


**Figure 4**. Maximum $\beta_N$ in AT pulses with (circles) and without (squares) Beta Collapses. Colors refer to the dominant mode n resulting from SVD analysis. $q_{min}$ from MSE-EFIT.

**SVD results**. SVD analysis has been performed in 24 events found in 19 JET AT ILW pulses. Such events were selected only taking into account the relative losses of $\Delta W_{dia}$>8%. Results of such analysis are shown in the $q_{min}$-$\beta_N$ diagram of Fig.4. Circles in Fig.4 are the $q_{min}$-$\beta_N$ values just before the BC, colors are related with the n of the dominant SVD mode before the BC. Values of n from 1 o 5 are found before BCs, but only in some cases with n=3,5 such activities behave as a precursor growing exponentially in few tens of ms before the BC. n=1,2,4 modes are usually already present and with constant amplitude before the BC. A weak dependence is found between $q_{min}$ and n (n=2 for $q_{min}$=1.45, and n=3,5 for 1.45<$q_{min}$<1.75).

It is noticeable that BCs are mainly concentrated in ($q_{min}$<1.75, $\beta_N$>2.5) region of operational parameters. Furthermore, in 20 cases out of 21 for $\beta_N$>2.5 a strong 2/1 is triggered after the BC, while only 1 out of 3 for $\beta_N$<2.5. The black squares in Fig.4 refer to ($q_{min}(\beta_{Nmax})$, $\beta_{Nmax}$) of 16 pulses clearly unaffected by BC with $\beta_N$>1.9 and NBI>20MW.

**Summary**. MHD analysis before BCs has been carried out by SVD. Such analysis showed that modes with high n up to 5 are present before the BC and in certain cases they act as precursor (Fig.2 and 4) showing a fast growth in ~10ms. This is in contrast with previous CW observations of a BC triggering by a n=1 kink like mode [1,3]. RFA enhancement (Fig.3) for $\beta_N$>2.5 shows that plasma pressure is approaching the no-wall beta limit. This is found in good agreement with the concentration of BC in the $q_{min}$-$\beta_N$ diagram (Fig.4).

**Acknowledgments**
This work has been carried out within the framework of the EUROfusion Consortium and has received funding from the Euratom research and training programme 2014-2018 under grant agreement No 633053. The views and opinions expressed herein do not necessarily reflect those of the European Commission.

**References**

[1] Buratti P., et al. 2012 *Nucl. Fusion* **52** 023006
[2] Hender T.C., et al. 2007 *Nucl. Fusion* **47** S128
[3] Baranov Yu.F., et al. 2012 *Nucl. Fusion* **52** 023018
[4] Nardone C., 1992 *Plasma Phys. Control. Fusion* **34 9** 1447
[5] Galperti C., 2014 *Plasma Phys. Control. Fusion* **56** 114012
[6] Reimerdes H., 2005 *Nucl. Fusion* **45** 368

# Measuring Intrinsic Rotation in the JET tokamak

M.F. F. Nave[1], J. Bernardo[1], R. Coelho[1], A. Czarnecka[2], J. Ferreira[1], A. Figueiredo[1], C. Giroud[3], J. Hillesheim[3], V. Parail[3], A. Salmi[4], K-D. Zastrow[3] and JET Contributors*

*EUROfusion Consortium, JET, Culham Science Centre, Abingdon, OX14 3DB, UK*

[1]*Instituto de Plasmas e Fusão Nuclear, Inst. Superior Técnico, Univ. de Lisboa, Portugal*

[2] *Institute of Plasma Physics and Laser Microfusion, Warsaw, Poland*

[3] *Culham Centre for Fusion Energy, Abingdon, Oxfordshire, UK*

[4]*VTT Technical Research Centre of Finland, Espoo, Finland*

**See the Appendix of F. Romanelli et al., Proc. 25th IAEA Fusion Energy Conference 2014, St Petersburg, Russia*

**Introduction** - Intrinsic rotation in JET plasmas has been determined from X-ray Crystal Spectroscopy (XCS) [1], Charge eXchange Recombination Spectroscopy (CXRS) [2, 3] and from the observed frequency of Magneto Hydro-Dynamics (MHD) modes [3]. Intrinsic rotation, observed in plasmas with Ohmic, lower-hybrid frequencies and Ion-Cyclotron Resonance Frequencies (ICRF) heating, has quite low amplitude, typically one order of magnitude lower than rotation in plasmas with Neutral Beam Injection (NBI). At JET this turned out to be difficult to measure since often it is at the limit of detection. Data analysis methods for all three diagnostic techniques had to be improved. Careful and dedicated data analysis for JET C-wall' experiments have shown that these three methods of detecting the toroidal rotation give comparable rotation velocities and consistent determination of the direction of rotation.

**CXRS** - The most common method for obtaining radial profiles of toroidal angular frequency, as well as ion temperature is through analysis of CXRS of $C^{+6}$ with neutral D measured during NBI blips (short pulses of NBI with $P_{NBI}$=1-2 MW). The spectral analysis requires the fitting of a 'cold' component, the passive, due to charge exchange between $C^{+6}$ ions and thermal deuterium at the edge, and a 'hot', active, component that arises from the interaction of $C^{+6}$ with the fast neutral beam. The velocity of the toroidal rotation is determined from the Doppler shift of the resulting active spectrum of $C^{+5}$ with respect to a $Be^{+1}$ emission line at 527.063 nm that originates in the cold scrape-off layer (SOL), that is used as reference assuming rotation in this layer to be negligible. (A schematic view of JET with indication of the relevant layers for CXRS analysis is shown in fig. 1. The NBI setup for CXRS measurements is shown in fig. 2.) For Ohmic and other low temperature plasmas, the spectral lines of active and passive components are of the same order thus can be difficult to separate (figure 3). In this case the spectrum of the passive taken just before the application of beam power is subtracted from the spectrum measured at the beginning of the NBI blip. The subtracted spectrum is analyzed by fitting it to only one Gaussian component. A recent assessment on how to improve the spectral fitting of the passive component has been performed. Among the questions considered were: improving the passive fitting by adding data from several passive frames; the effect that edge and core MHD events, such as ELMs and sawteeth, occurring during the time interval taken for the passive analysis, may have on the measurements of both rotation and ion temperature at the beginning of the blip; and whether the $Be^{+1}$ line rotates. For the later it has been found, that the $Be^{+1}$ does rotate during H-modes with ICRH, with significant effect on the rotation near the edge. Based on this assessment, the intrinsic rotation database is being re-analyzed taking 10 passive frames before the blip and making sure that the reference $Be^{+1}$ is always chosen during an Ohmic phase.

Since NBI provides a toroidal momentum source, the question of whether the time resolution of the CXRS measurements, at present 10 ms, was suitable to measure intrinsic rotation has been addressed in two ways: (i) extrapolation of measurements during the NBI to a time before the blip and, (ii) modelling the rotation produced by the NBI blip. An example of linear extrapolation to time t=0 ms is shown in figure 4 for the angular frequency and the ion temperature profiles, that in both cases are very close to the measured profiles at t=10 ms. Modelling was performed with the code JETTO for different models of momentum diffusion, using as input the NBI time dependent torque calculated with the code ASCOT. The NBI rotation was found to be negligible in the first 10-50 ms, confirming that rotation measurements taken at the first 10 ms are indicative of the rotation before the NBI blip.

**XCS** - The toroidal angular frequency in the plasma core can be obtained with a high-resolution x-ray crystal spectrometer (XCS) that observes the spectrum around the resonance line of the helium-like nickel ion $Ni^{26+}$. This is a technique ideal for measuring rotation in plasmas with no NBI momentum; however, it gives information at only one radial position. The location of the emission shell depends on the plasma temperature, in the range of 3-8keV. As a reference line for Doppler shift measurement is not available, the wavelength scale is calibrated by comparison with the angular frequency derived during NBI from CXRS of C [1]. Figure 5 shows a typical discharge used for XCS calibration. The figure shows the intensity of $Ni^{26+}$ measured with an integration time of 30 ms, while in the bottom box are the $Ni^{26+}$ toroidal angular frequencies, determined during NBI and in the Ohmic phase before the NBI. For the Ohmic plasma the spectra has been added over extended phases lasting up to 2 s. The reference wavelength has been determined by comparing CXS and CXRS measurements of 20 discharges with NBI power varying from 2.5 to up to 11 MW (figure 6). The CXS and CXRS measurements were matched at the radii of maximum CXS emission. Figure 5 shows that the Ohmic plasma central rotation is in the direction counter to the plasma current. The NBI phases are co-rotating. The direction and amplitude of rotation measured with XCS and with CXRS in Ohmic and in plasmas with ICRH have been found to be consistent.

**Conclusion** – Careful analysis of data from charge exchange spectroscopy and X-ray crystal spectroscopy have provided information of intrinsic toroidal velocities of JET plasmas with C-wall. CXRS intrinsic rotation measurements have been routinely compared with the observed frequency and direction of propagation of MHD modes with low n-mode numbers, such as sawtooth precursors and NTMs, which have confirmed the direction of rotation and measured rotation frequencies in the core of the plasma. A benchmark of XCS, CXRS and

MHD observations for ICRF L-mode plasmas has shown that all three methods give comparable angular rotation velocities. CXRS measurements of C-velocities with the JET ILW are more demanding to analyse as the available C has been reduced. Measuring Neon rotation is being considered and experiments with Ne puffs for CXRS analysis are planned for the re-start phase of the next JET experimental campaign later in 2015. A new tool for measuring rotation in JET is Doppler backscattering (DBS), which has recently started being used [5]. DBS measures the lab frame perpendicular propagation velocity of turbulent structures, which in the core of tokamaks is often dominated by the D toroidal velocity even when there is no external momentum input. Commissioning and cross-checking against CXRS measurements is also planned for the re-start of the next JET campaign.

**References**

[1] Eriksson, L-G, Righi E and K-D Zastrow 1997 Plasma Phys. Control. Fusion 39 27
[2] Noterdaeme, J-M et al 2003 Nucl. Fusion 43 274
[3] Nave, M. F. F. et al. 2012 Plasma Phys. Control. Fusion 54 074006
[4] Giroud C et al 2008 Rev. Sci. Instrum. 79 10F525
[5] Hillesheim, J.C., International Reflectometry Workshop , Juelich, Germany, May, 2015

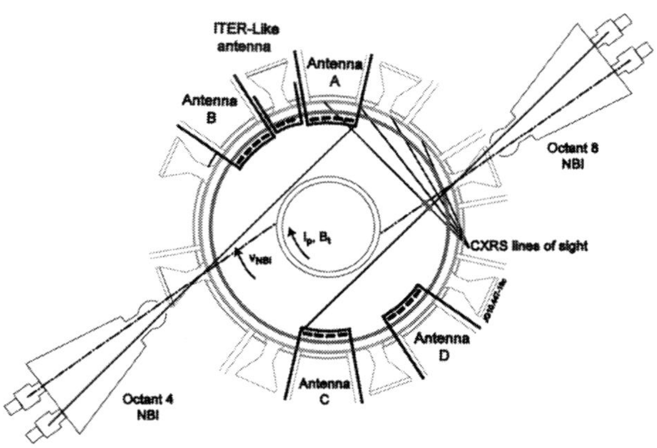


Fig 1. Schematic view from the top of JET showing the two NBI injectors, NBI lines and, the CXRS lines of sight. (also shows the ICRH antenas). CXRS emission layers which contribute to the measured spectra are indicated as the cold non-rotating SOL in green being the zone of the reference $B^{1+}$ line, the passive warm and slowly rotating layer in blue and the active hot and strongly rotating region which is shown by red dots in the intersection of beam lines and CXRS lines of sight. Also indicated in the figure are the usual directions of the toroidal field $B_T$, the plasma current $I_p$ and the direction of the rotation induced by NBI, $v_{NBI}$.

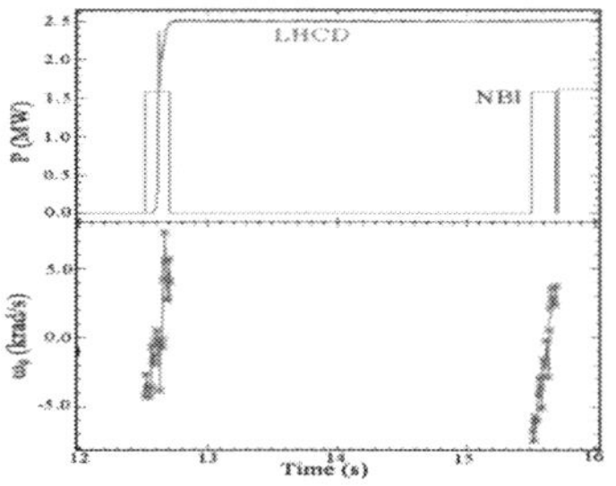


Fig. 2. NBI setup for CXRS measurements (pulse # 74866) with NBI blips lasting 200 ms in an Ohmic phase (beginning of 1st blip) and a phase with 2.5 MW of Lower Hybrid (2nd blip). On the bottom, the time evolution of C angular frequency from a central channel (R=3.1 m).

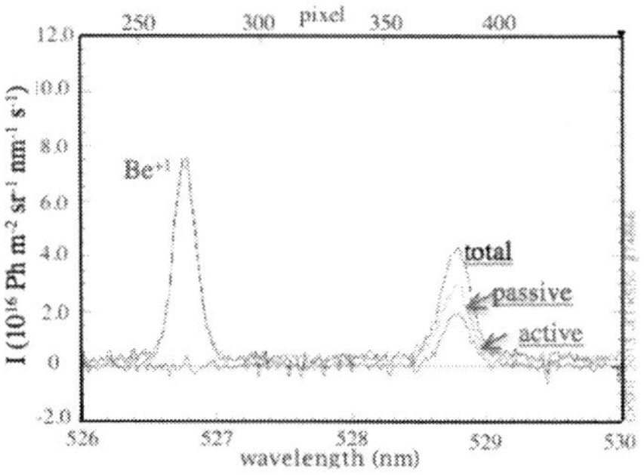


Figure 3. Measured $C^{+5}$ spectral line at beginning of 2nd blip, compared with passive line measured immediately before the blip. Active line, i.e. total with passive subtracted.

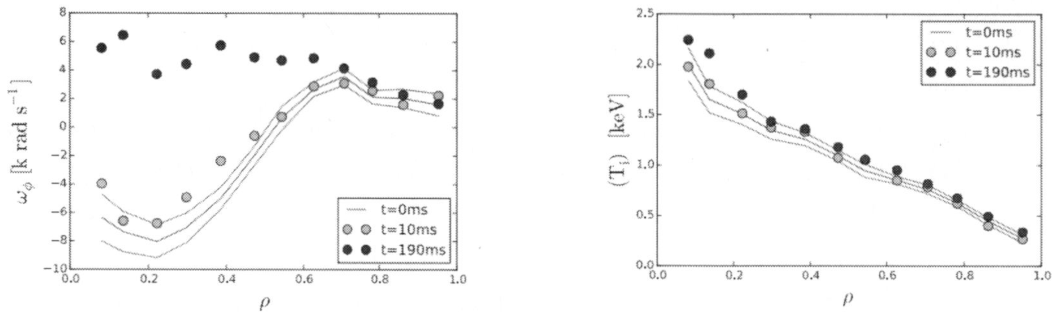


Figure 4- Profile measurements of (a) toroidal angular frequency and (b) ion temperature for the 2nd blip in fig. 2. Orange dots for measurements at the beginning of NBI at 10 ms, blue dots at the end of the blip at t=190 ms, red lines are the linear extrapolation to time t=0 ms. Magenta lines are the linear extrapolation of statistical errors in the spectral fitting, determined from the error in the least-squares fit [4].

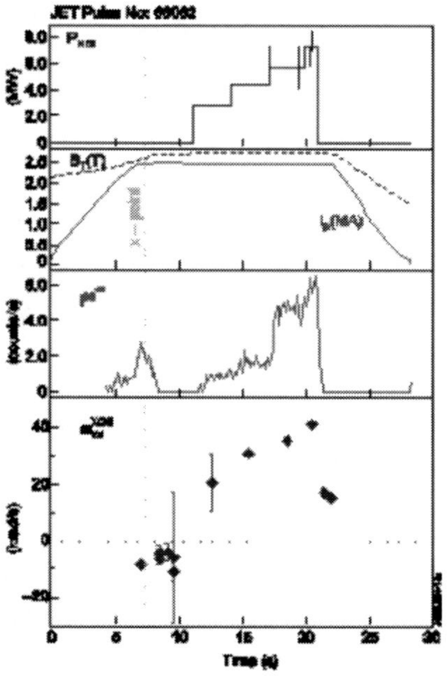

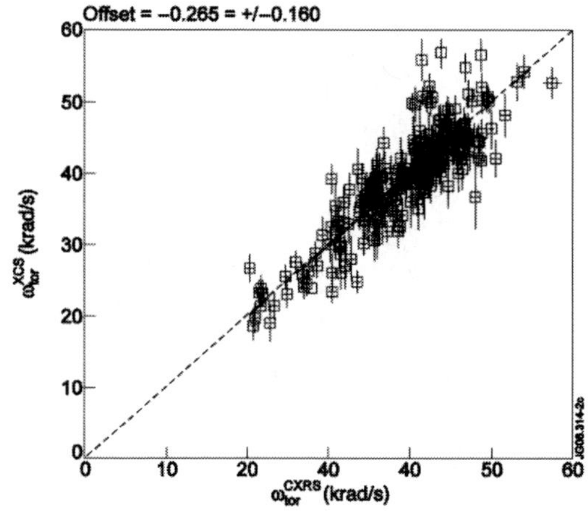


Fig. 5 -Typical discharge used for XCS calibration. (a) $P_{NBI}$, (b) $I_p$ and $B_T$, (c) Intensity of $Ni^{26+}$ with integration time 30 ms; (d) Ni core toroidal angular frequency, $w_f$, versus time determined from XCS (spectra added over extended phases lasting up to 2 s)

Fig. 6 The XCS wavelength scale was calibrated by comparison with the angular frequency obtained during NBI with CXRS using 20 discharges ($B_T$ =2.7T, $I_p$=2.5 MA) with NBI heating ($P_{NBI}$=2.5-11 MW)

**Acknowledgements** - This work has been carried out within the framework of the EUROfusion Consortium and has received funding from the Euratom research and training programme 2014-2018 under grant agreement No 633053. IST activities also received financial support from "Fundação para a Ciência e Tecnologia" through project UID/FIS/50010/2013. The view and opinions expresses herein do not necessarily reflect those of the European Commission.

# A stability study of α-particle driven Alfvén eigenmodes in JET D-T plasmas

J. Ferreira[1], D. Borba[1], R. Coelho[1], L. Fazendeiro[1], A.C.A. Figueiredo[1],
N. F. Loureiro[1], F. Nabais[1], P. Rodrigues[1], J. Conboy[2], M. Fitzgerald[2],
S.E. Sharapov[2], I. Voitsekhovitch[3], and JET Contributors[§]

*EUROfusion Consortium, JET, Culham Science Centre, Abingdon, OX14 3DB, UK*
*[1]Instituto de Plasmas e Fusão Nuclear, Instituto Superior Técnico,*
*Universidade de Lisboa, 1049-001 Lisboa, Portugal*
*[2] CCFE, Culham Science Centre, Abingdon OX14 3DB, United Kingdom*
*[3] EuroFusion Consortium Programme Unit, Boltzmannstr. 2, D-85748, Garching, Germany*

**Abstract**

As a contribution to the effort towards the assessment of future scenarios with tritium and deuterium-tritium (D-T) mixture plasmas in the Joint European Torus (JET), a scientific workflow that has recently been developed for the systematic linear-stability assessment of Alfvén eigenmodes in the presence of fusion born α-particles, is here applied to a representative JET D-T scenario.

## Introduction

Fusion-born α-particles are ubiquitous in D-T burning plasmas as they comprise a considerable part of the total heating energy. Although being in minority they are born with high energies ~3.5 MeV for a deuterium-tritium reaction, and being thus super-Alfvénic they can potentially destabilize global MHD modes such as the Alfvén eigemodes (AEs). In some conditions these instabilities can strongly enhanced the radial transport of this energetic population with serious consequences on burning performance and wall damage [Sharapov]. These are crucial problems, especially for future machines like ITER or DEMO, and a good understanding of the energetic particle physics supported by validated modelling tools is therefore of particular importance in order to better predict and plan burning plasma scenarios.

Many experiments and theoretical studies done during the last 20 years have improved our knowledge, pushed the development of better diagnostic techniques, and allowed the opportunity to enhance our numerical tools. Notwithstanding these great efforts further studies can be done, specially taking advantage of the forthcoming JET D-T campaign, to extend our understanding of energetic particle-wave interaction physics in plasmas surrounded by metallic walls. The

§See the Appendix of F. Romanelli et al., Proceedings of the 25th IAEA Fusion Energy Conference 2014, Saint Petersburg, Russia preparation of the next DT campaign is therefore of crucial importance and motivates the work here presented. As a contribution to the assessment of future DT scenarios, a scientific workflow that has recently been developed for the systematic linear-stability assessment of Alfvén eigenmodes (AE) in the presence of fusion born $\alpha$-particles is here used to analyze a representative JET DT scenario.

**Linear stability analysis of Alfvén eigenmodes**

In this work we follow a similar procedure that has been used to assess the stability of AEs in ITER burning-plasma scenarios [Rodrigues 2014, 2015; Figueiredo 2015]. As outlined in Figure 1, from an equilibrium picked from a TRANSP [Goldston 1981] run of a typical JET D-T discharge, which is then refined by CHEASE [Luetjens 1996] and/or HELENA [Huysmans 1991] equilibrium codes, all possible AEs are systematically searched using the ideal-MHD code MISHKA [Mikhailovskii 1999] by intensively scanning over a frequency and wave-number range. The linear stability of these modes is then evaluated by computing the energy exchange between these AEs and the fusion $\alpha$s, thermal D, T and electron populations, using the drift-kinetic code CASTOR-K [Borba 1999; Nabais 2015]. The $\alpha$-particle drive and background damping of these modes are then analysed. Also of significant importance are the radiative damping, continuum damping and collisional AE damping, which are not taking into account here but will be part of a future publication. Also, the mechanisms responsible for wave damping on the impurities and He ash populations are neglected in the present study owing to the fact that is reasonable to assume that they are low when compared with the ion and electron damping

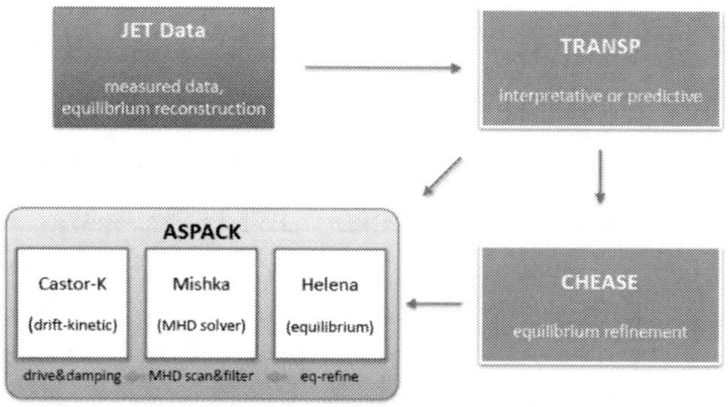


Figure 1 - Diagram of the workflow followed in the present study for the AE linear-stability assessment by the $\alpha$-drive.

**Results and Discussion**

A high-performance fusion JET D-D experimental pulse was selected for this study and is used as a reference for a predictive TRANSP analysis to extrapolate that discharge to a mixed D-T plasma scenario, and therefore estimate the $\alpha$ content that could be reached in future JET D-T experiments. Relevant operational parameters for the selected discharge are: toroidal magnetic field of $B_T$ = 3.46 T at the magnetic axis, plasma current of $I_{plasma}$ = 3MA, central ion and electron temperatures of $T_{i0} \sim 20$ keV and $T_{e0} \sim 7$ keV, and a D-T mixture of the order 60%/40% at the plasma core. From this run a time slice is chosen during the burning phase for the systematic linear assessment of the AEs. This is illustrated in figure 2 where ion and electron temperature and density profiles, together with the $\alpha$ content expressed by the $\beta_\alpha$ normalized pressure and radial derivative profiles, as well as the safety factor and ion mass profiles are shown for t = 46.38s, around the time $\alpha$ particles start to destabilize AEs. A dynamical study to evaluate the impact of current diffusion (evolution of safety-factor profile) and the increase of $\alpha$ content on the stability of AEs along the discharge is left as future work.

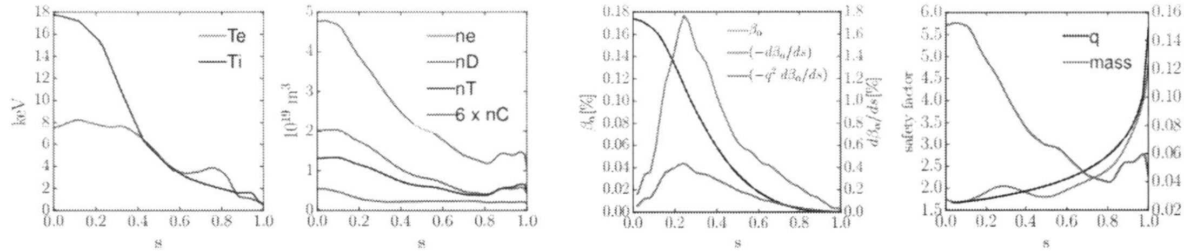


Figure 2 – Plasma profiles taken at t=46.38s from the TRANSP extrapolation to DT of a JET DD pulse #40214. From left to right, figure shows ion and electron temperature profiles, density profiles for different species, normalized $\alpha$-pressure ($\beta\alpha$) and radial derivatives, ion mass and safety-factor profiles (from TRANSP in red, EFIT reconstruction from DD in black), as function of "s", the square root of the normalized poloidal flux.

As illustrated in the rightmost plot of figure 2, the safety-factor (q) profile for the reconstructed equilibrium by EFIT (in black) for the reference DD pulse is monotonic, contrary to the self-consistently equilibrium calculated by TRANSP where current diffusion is solved. Without loss of generality we have chosen to use the simpler monotonic-q equilibrium for this preliminary analysis. The non-monotonic self-consistent TRANSP equilibrium will be assessed in future work.

As described earlier, for the selected equilibrium a set of allowed AEs is systematically found and each mode is then processed by CASTOR-K to calculate the drive due to interaction with $\alpha$-particles and wave damping on the thermal D ions, T ions and electrons. The total effective normalized growth rate is then given by $\gamma/\omega = (\gamma_\alpha + \gamma_D + \gamma_T + \gamma_e)/\omega$, where $\gamma_s$ is the AE growth rate for the interaction with the population "s", and $\omega$ the angular AE frequency. Positive growth rates quantify an energy transfer to the wave (drive), whereas negative values are related with wave damping. In figure 3 we show the ideal MHD continuum spectra for the aforementioned equilibrium, and

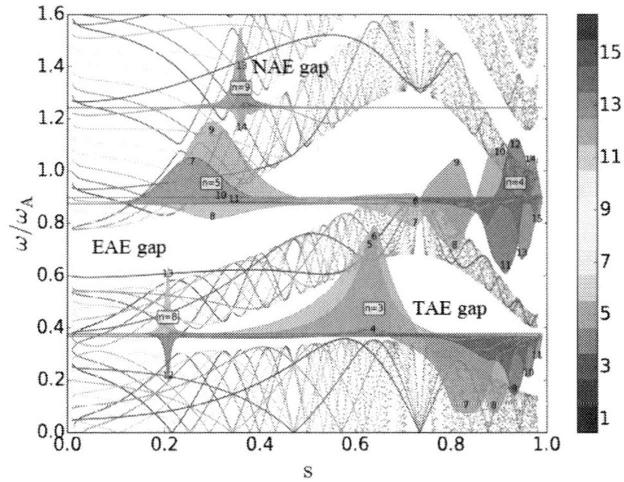


Figure 3 - ideal MHD continuum spectra for n up to 16, as function of square root of the normalized poloidal flux (s). AE modes are shown within the TAE, EAE and NAE gaps.

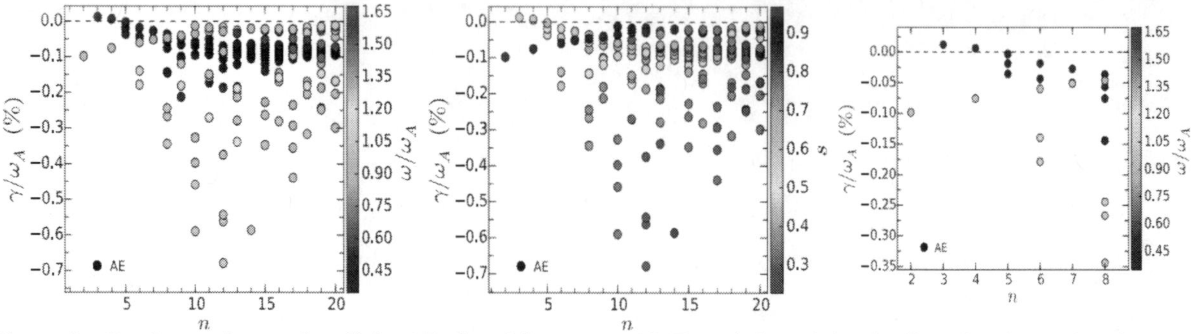

Figure 4 – Total growth rates for all the AEs found for n up to 20. From left to right, the first plot shows that the most unstable modes are low-n TAEs, but many of the high-n EAEs ($\omega/\omega_\alpha \sim 1$) that are found close to the marginal stability threshold can start to have a deleterious role with the increase of $\beta_\alpha$. In the middle plot each mode is colored by the radial location of its peak amplitude. A careful analysis shows that modes closer to the unstable threshold are localized in the outer half of the plasma, where the ion damping is significantly lower than in the very hot inner core. Lastly, in the rightmost plot, we plot the low-n AEs in the region of interest, where the only unstable modes were found.

the structure and localization of some of the AEs that can be found within the gaps. The stability threshold for each mode (more than 400 AE where found) is determined by balancing the energetic ion drive against the dominant damping mechanisms. In figure 4 we show the total growth rates for all AEs found. A comprehensive analysis of these results will be left for a future publication. As a summary of this preliminary assessment we found that the outer radial low-n TAEs, localized in a lower ion-damping region, are more easily destabilized in spite of being driven by fewer $\alpha$-particles. The high ion Landau damping in the very hot inner half of the plasma may overcome the $\alpha$-drive. Future work will try to assess if core localized AEs can be destabilized in JET D-T hot-ion plasmas.

**Acknowledgements**

This work has been carried out within the framework of the EUROfusion Consortium and has received funding from the Euratom research and training programme 2014-2018 under grant agreement No 633053. The views and opinions expressed herein do not necessarily reflect those of the IST, FCT, CCFE, or the European Commission. IST activities received financial support from "Fundação para a Ciência e Tecnologia" (FCT) through project UID/FIS/50010/2013. All computations were carried out using the HELIOS supercomputer system at the Computational Simulation Centre of the International Fusion Energy Research Centre (IFERC-CSC) in Aomori, Japan, under the Broader Approach collaboration between Euratom and Japan implemented by Fusion for Energy and JAEA. PR was supported by EUROfusion Consortium grant no. WP14-FRF-IST/Rodrigues and NFL was supported by FCT grant no. IF/00530/2013.

**References**

**[Borba 1999]** D. Borba and W. Kerner, J. Comput. Phys. 153 (1999) 101. **[Figueiredo 2015]** A. C. A. Figueiredo *et al.* "Comprehensive evaluation of the linear stability of Alfvén eigenmodes driven by alpha particles in an ITER baseline scenario", to be submitted to Nucl. Fusion (2015), also see this conference P4.123. **[Goldston 1981]** R. J. Goldston, D. C. McCune, *et al.*, J. Comput. Phys. 43, (1981) p. 61. **[Horton 2015]** L. Horton and JET Contributors, 12th International Symposium on Fusion Nuclear Technology, ICC JEJU, Jejun Island Korea, oral presentation (2015). **[Huysmans 1991]** G. Huysmans *et al.*, Int. J. Mod. Phys. C 2 (1991) 371. **[Luetjens 1996]** H. Luetjens, A. Bondeson, O. Sauter, Comput. Phys. Commun. 97 (1996) 219 **[Mikhailovskii 1999]** A.B. Mikhailovskii *et al.*, Plasma Phys. Rep. 23, (1997), p. 844. **[Nabais 2015]** F. Nabais *et al.*, Plasma Sci. Technol. 17 (2015), p. 89. **[Rodrigues 2014]** P. Rodrigues *et al.*, Proceedings of the 25th IAEA Fusion Energy Conference, Saint Petersburg, Russia, TH/P3-25 (2014), arXiv:1410.2744. **[Rodrigues 2015]** P. Rodrigues *et al.*, "Systematic linear-stability assessment of Alfvén eigenmodes in the presence of fusion $\alpha$-particles for ITER-like equilibria", accepted by Nucl. Fusion (2015), also see this conference P4.124. **[Sharapov 2011]** S.E. Sharapov, I. Voitsekhovitch, *et al.*, 38th EPS on Plasma Physics, Strasbourg, France, P5.117, (2011). **[Sharapov 2012]** S.E. Sharapov, *et al.*, IAEA, San Diego, United States of America, OV/4-3, (2012). **[Sharapov 2013]** S. E. Sharapov *et al.*, Nucl. Fusion 53 (2013) 104022

# Observation of fast chirping modes at frequencies above the Alfvén frequency in JET

F. Nabais[1], D. Borba[1], R. Coelho[1], L. Fazendeiro[1] J. Ferreira[1], A. Figueiredo[1], M. Fitzgerald[2], P. Rodrigues[1], S. Sharapov[2] and JET Contributors[§]

*EUROfusion Consortium, JET, Culham Science Centre, Abingdon, OX14 3DB, UK*
*[1]Instituto de Plasmas e Fusão Nuclear, Instituto Superior Técnico, Universidade de Lisboa, 1049-001 Lisboa, Portugal.*
*[2]CCFE Fusion Association, Culham Science Centre, Abingdon OX14 3DB, UK.*

[§]See the Appendix of F. Romanelli et al., Proceedings of the 25th IAEA Fusion Energy Conference 2014, Saint Petersburg, Russia

## Introduction

Modes propagating with frequencies above the Alfvén frequency ($\omega>\omega_A$) are not normally observed in currently operating tokamaks as the plasma on these machines contains only a modest number of energetic ions with velocities above the Alfvén velocity ($v>v_A$). Notable exceptions are the spherical tokamaks NSTX and MAST, where the injection of super-Alfvénic beams produced significant populations of ions verifying the condition ($v>v_A$) [1], [2]. In these spherical tokamaks, modes with frequencies in the range $\omega_A<\omega<\omega_C$ are routinely observed. Here, $\omega_C$ is the cyclotron frequency of beam ions. These modes have been identified as CAE (Compressional Alfvén Eigenmodes) and in few cases as GAE (Global Alfvén Eigenmodes) or a mix of CAE and GAE. The distinction between CAE and GAE is usually quite difficult. Experiments carried out in the conventional tokamak DIIID using low magnetic fields so the injected beam was super-Alfvénic allowed to observe similar MHD activity [3]. CAE/GAE are likely to be destabilized by alpha particles in ITER and they may be important for stochastic heating of thermal ions [4] as well as be used for alpha particles diagnostics [5].
This paper reports on the observation of modes with frequencies in the range $\omega>\omega_A$ which occurred in JET experiments. Contrary to the previously reported cases, in JET, the modes in the super-Alfvénic range of frequencies were destabilized by ICRH accelerated ions and not by beam-injected ions. JET experiments used low plasma densities and high ICRH power, which allowed a significant population of energetic ions in the MeV range of energies to build up in the plasma [6].

## MHD activity in the frequency range $\omega>\omega_A$

Figure 1 shows two distinct spectrograms obtained with Mirnov Coils using an acquisition frequency of 1 MHz during JET low plasma density high ICRH power experiments. The nearly constant frequency modes observed between 200 kHz and 250 kHz are TAE, the modes with slowly decreasing frequencies observed in the same range of frequencies are tornado modes, i.e. TAE localized inside the q=1 surface. The fast chirping modes (FCM) are observed to cover nearly all the frequencies ranges shown in figure 1.

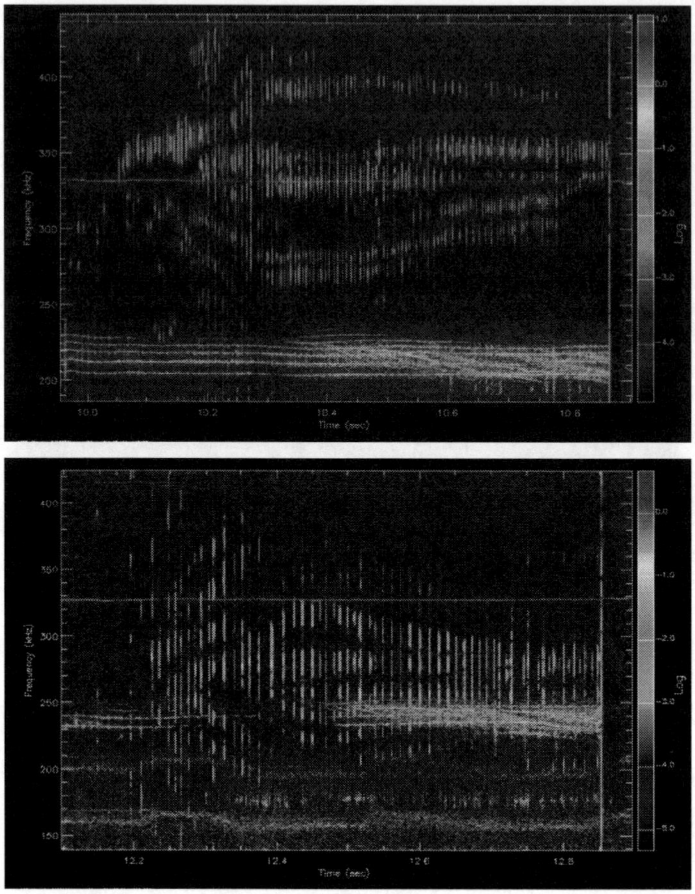


*Figure 1: Spectrogram of MHD activity obtained with Mirnov Coils using an acquisition frequency of 1 MHz.*

All these modes are suppressed by the monster sawtooth crashes observed near the end of the represented time periods (vertical lines). The FCM present a bursting behavior and are composed by several different frequency bands that may change slowly in frequency as time evolves. In particular, the FCM seem to be affected by the presence of tornado modes. In this set of experiments, a few Mirnov coils were acquiring at a frequency of 2 MHz (see figure 2). Spectrograms made using signals acquired at 2 MHz show the FCM at frequencies between 750 and 800 kHz. This means the FCM observed in figure 1 are aliases (signals that are undistinguishable from the real signals and that are observed at lower frequencies due to an insufficient sampling rate). The Alfvén frequency in the plasma core in these experiments was typically around 450 kHz, which is well below the FCM frequency. The sub cyclotronic modes in the range of frequencies $\omega_A<\omega<\omega_C$ observed in DIIID, NSTX and MAST may span over a large frequency range, in some cases the modes are observed up to around the cyclotron frequency.

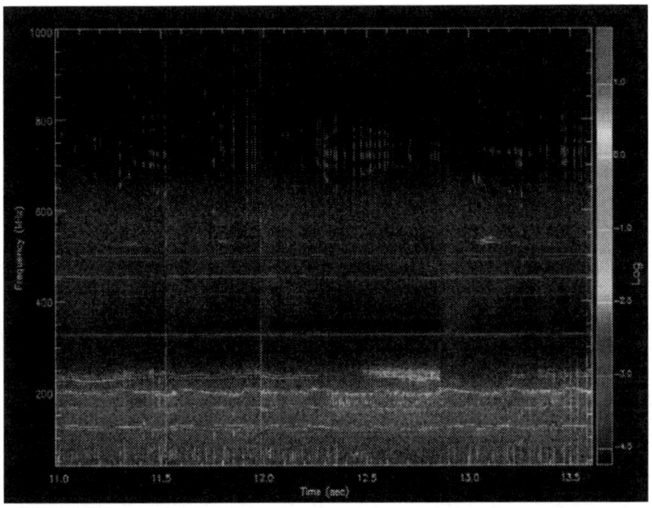

*Figure 2: Spectrogram of MHD activity obtained with Mirnov Coils using an acquisition frequency of 2 MHz.*

In JET, the anti-aliasing system prevents higher frequency modes to be observed in the magnetic spectrograms (which would be observed as aliases), so there is no information regarding the higher frequency limit of the FCM. Despite the presence of JET anti-aliasing system, in the analysis it is considered the possibility of the mode observed around 800 kHz still be an alias.
FCM are observed both with positive and negative toroidal mode numbers n, this is, they propagate both in the co and counter current directions. This suggests the drive of FCM to be associated with the free energy in the velocity space, implying the distribution in energy must have a local "bump-on-tail" like distribution ($\partial F/\partial E>0$). The loss of fast ions from the plasma in this set of experiments was measured by a scintillator plate. These measurements indicated that FCM have little or no influence on the loss of fast ions.

## Comparison of FCM with CAE/GAE in NSTX, MAST and DIIID

The sub cyclotronic modes in the range of frequencies $\omega_A<\omega<\omega_C$ observed in DIIID [3], NSTX [1][7] and MAST [2][8] were identified as CAE. In NSTX and DIIID, a mixture of CAE and GAE were sometimes observed. A bursting behavior of the modes, similar to JET FCM, has been observed in all three tokamaks, though sometimes, modes were observed as continuous (not bursting) modes. In DIIID, NSTX and MAST, groups of modes at different frequency ranges were observed. In JET only one group, near the lower frequency limit, was observed but in case there were other groups of modes present in the plasma they wouldn't be observed due to the inexistence of appropriate diagnostics acquiring at a sufficiently high sampling rate. Each of these groups of modes presented a finer frequency splitting and a presence of "bands". FCM in JET also showed the presence of bands. In all three mentioned tokamaks, modes propagating both in the co and counter current direction were observed (in DIIID, toroidal mode numbers were inferred, not measured), consistent with the drive of both CAE and GAE which is associated with a local positive gradient of the fast ion distribution in energy.

## Summary and discussion

The observation of fast chirping modes with frequencies above the Alfvén in JET is reported. Contrary to the modes observed in this ranges of frequencies in other tokamaks, namely DIIID, MAST and NSTX, in JET the modes were not destabilized by the injection of super-Alfvénic beams into the plasma but by high power ICRH applied on a low density plasma. Alfvénic modes known to exist in the range of frequencies $\omega_A<\omega<\omega_C$ are the Compressional Alfvén Eigenmodes and the Global Alfvén Eigenmodes. While all the aspects mentioned in this and previous section suggest the fast chirping modes observed in JET to be CAE/GAE modes, there is still an open issue related with the mode frequency. The lower frequency limit of CAE/GAE in DIIID, MAST and NSTX deviate significantly from the cyclotron frequency of the energetic ions. Similarly, in JET the FCM frequency deviates significantly from the cyclotron frequency. However, when CAE/GAE are destabilized by super-Alfvénic beams, the ions can resonate with the lower frequency CAE/GAE through the Doppler shifted wave particle resonance because the parallel velocity of the energetic ions is large. This is not the case of ICRH accelerated ions. So, if the fast chirping modes observed in JET are indeed CAE/GAE, the ICRH accelerated ions must possess a significant drift velocity.

### References

[1] – E.D. Fredrickson *et al.* Phys. Rev. Lett. **87**, 145001 (2001)
[2] – L.C. Appel *et al.* Plasma Phys. Control. Fusion **50**, 115011 (2008)
[3] – W.W. Heidbrink *et al.* Nucl. Fusion **46**, 324 (2006)
[4] – N.N. Gorelenkov *et al.* Nucl. Fusion **43**, 228 (2003)
[5] – G. A. Cottrell *et al.* Nucl. Fusion **33**, 1365 (1993)
[6] – F. Nabais *et al.* Nucl. Fusion **52**, 083021 (2012)
[7] – N.N. Gorelenkov *et al.* Nucl. Fusion **42**, 977 (2002)
[8] – S.E. Sharapov *et al.* Plasma Phys. Control. Fusion **21**, 082501 (2014)

### Acknowledgements

The views and opinions expressed herein do not necessarily reflect those of IST, FCT, or the CCFE Association. IST activities received financial support from "Fundação para a Ciência e Tecnologia" through project UID/FIS/50010/2013. This work has been carried out within the framework of the EUROfusion Consortium and has received funding from the Euratom research and training programme 2014-2018 under grant agreement No 633053. The views and opinions expressed herein do not necessarily reflect those of the European Commission. Paulo Rodrigues was supported by EUROfusion Consortium grant no. WP14-FRF-IST/Rodrigues.

# Parameters and Stability of Runaway Electron Dominating Discharge in JET with ITER-Like Wall

V.V. Plyusnin(1), V.G. Kiptily(2), A.E. Shevelev(3), E.M. Khilkevitch(3), M. Brix(2), S. Gerasimov(2), G. F. Matthews(2) and JET contributors*

*EUROfusion Consortium, JET, Culham Science Centre, Abingdon, OX14 3DB, UK*

*(1) Instituto de Plasmas e Fusão Nuclear, Instituto Superior Tecnico, Universidade de Lisboa, Lisboa, Portugal;*
*(2) CCFE, Culham Science Centre, Abingdon, OX14 3DB, UK*
*(3) Ioffe Institute, St. Petersburg, 194021, Russia;*
**See the Appendix of F. Romanelli et al., Proceedings of the 25th IAEA Fusion Energy Conference 2014, Saint Petersburg, Russia*

**Introduction** Potentially severe generation of high-energy runaway electrons (RE) in tokamaks could occur on any stage of discharges: during breakdown and plasma current ramp-up, during stationary stages and during disruptions. Uncontrolled generation of RE currents up to several MA during disruptions is considered as most dangerous for the device first wall [1]. Usually the RE generation is detected with the aid of HXR diagnostics and photo-neutrons. Relativistic RE interacting with background plasma, neutrals or surrounding plasma-facing components (PFCs) produce a bremsstrahlung hard X-ray emission (HXR) in the MeV energy range and photo-neutrons. Measurements of these radiations provide the information on energy characteristics of RE. For the device protection and RE studies 5 scintillation time-resolved HXR monitors and $^{235}U$ and $^{238}U$ fission chamber neutron rate monitors at 3 different locations operate in a current mode with 0.1 ms time resolution. The HXR emission has been measured with the set of horizontally and vertically viewing NaI(Tl), $Bi_4GeO_{12}$ (usually referred as BGO) and $LaBr_3$ spectrometers. The data on spatial distribution of HXR emission sources in the plasma has been obtained with the JET neutron/gamma profile monitor routinely used for neutron and gamma rays measurements. The HXR raw data measured by BGO spectrometer has been processed with de-convolution procedure using the DeGaSum code [2]. This procedure allowed reconstruction of the energy spectra of generated RE populations. The RE distribution function (REDF) is important characteristic describing the RE generation process. Evolution of REDF can characterize different stages of this process such as acceleration of electrons and their scattering on background plasma and neutrals, interaction with first wall, etc. Sometimes the RE were detected in-flight during the transient stages of discharge breakdown and plasma current ramp-up in JET with ITER-Like Wall (JET-ILW) [3]. Events of RE generation during flat top stages of discharges in JET constitute a subject of highly rated interest, first of all due to the rare possibility to study the onset, growth and reasons for loss of relativistic electrons (REs) [4]. In this report we present a study of RE parameters generated and detected during quasi-stationary stage of the JET-ILW discharge following the large plasma density decrease. In order to avoid an ambiguous interpretation of the processes occurring on this stage the evolution of measured plasma parameters (density, electron temperature, loop voltage, plasma current) have been numerically processed in frames of conventional theory for RE generation.

**Experimental data and discussion** During the flattop stage of ordinary JET-ILW pulse (JPN #86078) a sudden loss of control on inlet of working gas resulted in decrease of plasma density to extremely low value: $\langle n_e \rangle \leq 2.E18\ m^{-3}$ (from approx. t = 18.4 sec in Figure 1). This regime with low density has been sustaining approximately 8 seconds. The neutron diagnostic registered almost complete disappearance of fusion neutron yield from the deuterium plasma following this density decrease. Plasma current during this low-density stage was kept almost constant ($I_{pl} \cong 1.8$ MA) till to discharge termination (Figure 1). The measured loop voltage ($V_{loop}$) at the beginning of density decrease was $V_{loop} \cong 0.65$ V, that corresponds to $\langle T_e \rangle \cong 1.5$

keV calculated from classical resistivity with neo-classical correction on fraction of trapped particles. Obtained $<T_e>$ value is very close to $T_e$ data provided by other JET diagnostics. In low-density phase the mean (drift) current velocity increases up to $u_0 = (I_{pl}/\pi a_{pl}^2)/(en_e) \leq$ 3.0E6 m/sec, where $a_{pl}$ is plasma radius (from EFIT) and $<n_e>$ is measured plasma density. Large drift velocity ($u_0 \cong 0.1 v_{Te}$) inevitably will lead to asymmetry of the electron distribution function ($f_e(v_e) \sim n_e * exp(-(v_e - u_0)^2/v_{te}^2)$). Simple calculation shows that appeared asymmetry causes a significant increase of the number of electrons (up to ~10%) moving in the direction of acceleration. It results in increase of number of electrons with velocities higher than the critical one: $v_e > v_{Dr} = \sqrt{\frac{e^3 n_e \ln\Lambda\,(2+Z_{eff})}{4\pi\varepsilon_0^2 m_e E}}$, where $v_{Dr}$ is a critical velocity for generation of primary (Dreicer) runaway electrons (RE). According to the theory of RE generation two mechanisms are responsible for generation of RE. The primary mechanism, when acceleration of electrons by external E-field is higher than frictional drag of Coulomb collisions with plasma particles. And the secondary avalanching, when existing RE (collisionless on their origin) transfer a significant energy (higher than threshold for runaway acceleration) to surrounding thermal electrons via close distance collisions thus enabling multiplication of the number of RE. It is clear that secondary avalanching could occur only if the primary generation will provide the substantial RE flux with sufficient energy. Note, that the close distance collisions are less frequent by the factor $8*ln\Lambda$ than usual Coulomb (distant) collisions. Dynamics of both processes is determined by the runaway growth rate, which is mainly depending on the ratio of externally applied electric field ($E_0$) to the threshold field, above of which a free acceleration is possible: $\alpha = E_0/E_{DR}$, where $E_{DR} = \frac{e^3 n_e \ln\Lambda\,(2+Z_{eff})}{4\pi\varepsilon_0^2 m_e v_{Te}^2}$ – is a threshold field for Dreicer process, and $\beta = E_0/E_{CR}$, where $E_{CR} = \frac{e^3 n_e \ln\Lambda\,(2+Z_{eff})}{4\pi\varepsilon_0^2 m_e c^2}$ - is the critical field for avalanching.

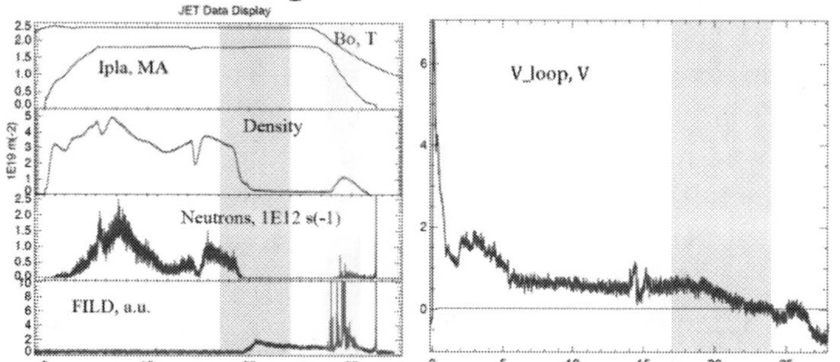

*Figure 1. JPN#86078: General view of discharge parameters evolution (left chart). Right chart - loop voltage time evolution till to beginning of pulse termination. Studied stage (RE) is highlighted in pink (both charts), in yellow – RE instability phase*

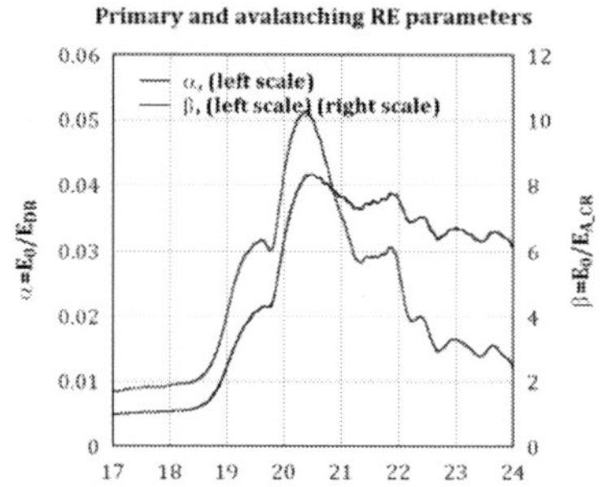

*Figure 3. Evolution of RE generation parameters ($E/E_{DR/CR}$) for primary and avalanching processes in JPN#86078*

Figure 3 presents evolutions of α and β during low-density stage. RE generation has been modelled with conventional equation for evolution of RE density $\frac{dn_{RE}}{dt} = \lambda_R - \frac{n_{RE}}{\tau_R} + \frac{n_{RE}}{t_0}$, where

$\lambda_R = C(Z_{eff}) \cdot n_e v_e \varepsilon^{-\frac{3(Z_{eff}+1)}{16}} \cdot \exp\left(-\frac{0.25}{\varepsilon} - \sqrt{\frac{Z_{eff}+1}{\varepsilon}}\right)$ is the primary generation rate, and

$t_0 = \frac{4\pi\varepsilon_0^2 m_e^2 c^3}{e^4 n_e}\sqrt{\frac{3}{\pi}}(Z_{eff}+5)\cdot(\beta-1)^{-1}$. So, that $1/t_0$ is the secondary avalanching growth rate and confinement time $\tau_R$ is infinity. Modelling revealed large increase of runaway growth rate and simultaneous decrease of runaway critical energy ($W_{cr}$) when density decreased below 1E19 m$^{-2}$, i.e. at t ≥ 19.5 sec (Figure 4). It is important to note, that contribution from runaway avalanching source to total RE population was less than expected from theory since the Dreicer acceleration and EDF asymmetry provided nearly total contribution into RE population ($n_{RE}$~1.E15 $m^{-3}$ at t=24 s), that corresponds to RE current fraction $I_{RE} \leq 200$ $kA/m^2$. Evolution of REDF indicates that further sustainment of RE stage is provided by avalanching. Gradual decrease of loop voltage indicates not only increase of $<T_e>$, but also increase of RE fraction in total plasma current $I_{pl}$~1.8 MA. Loop voltage becomes negative at large RE current fraction. After t = 24 s the numerical analysis of experimental data fails due to inversion of loop voltage signal. Simultaneously to increase of runaway growth rate the vertical and horizontal HXR detectors registered the increase of HXR signals from the plasma centre (Figure 5). The scintillator of the FILD detector (Fast Ions Loss Diagnostics, Figure 1) also registered increase of the signal, which is associated with the presence of HXR emission caused by RE in plasma. RE produce electromagnetic radiation in collision with plasma particles and when they hit the surrounding PFC. In both cases the predominant radiation production mechanism is bremsstrahlung. Figure 6 presents the measurements of HXR radiation performed using the JET $\gamma$-camera with vertically and horizontally viewing sets of detectors in direction perpendicular to the direction of RE motion (along the toroidal field lines).

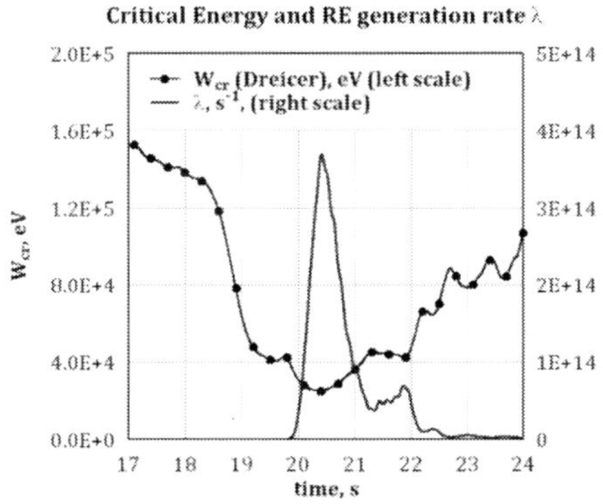


*Figure 4. Evolution of critical energy $W_{cr}$ for runaway process and RE generation rates.*

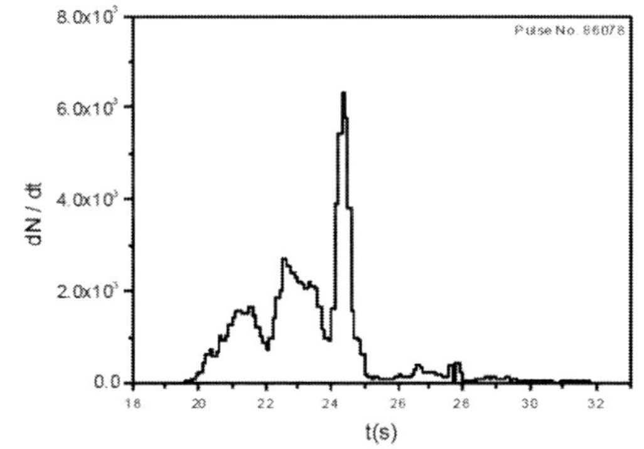


*Figure 5. Signal from HXR detector during low-density phase of JPN#86078*

HXR spectrometry has been used to study the dynamics of RE generation process. Energy spectra of RE have been measured. A study of RE energy spectra revealed the generation of electrons with maximal energies up to 7 MeV and characteristic energy 2.5 MeV (Figure 7).

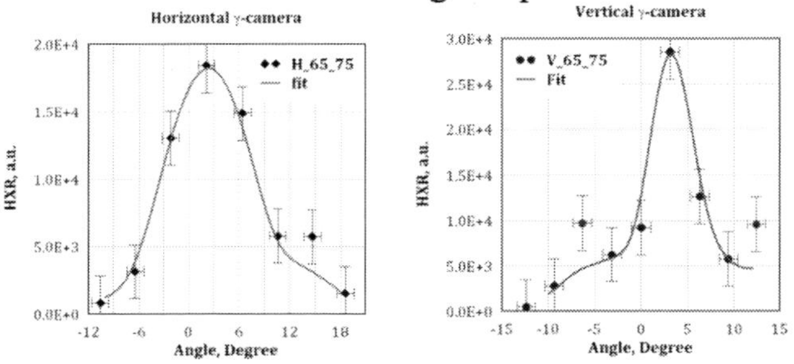


*Figure 6. HXR bremsstrahlung measured by different channels of gamma camera. Each channel is characterized by unique angle of inclination to vertical of horizontal plane.*

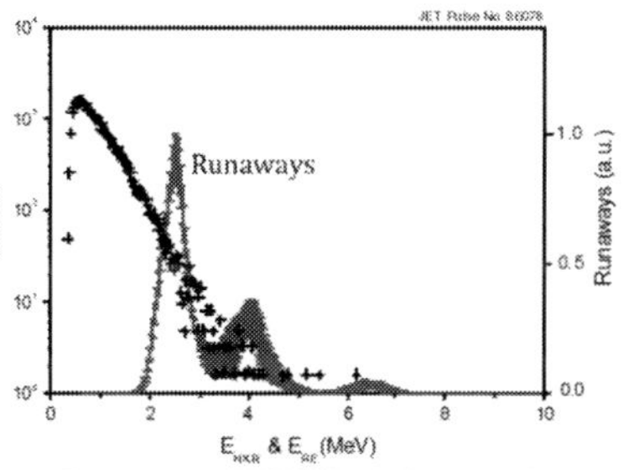


*Figure 7 JPN#86078: HXR spectrum and reconstructed REDF at beginning RE stage (t=20.2-22 s)*

2D map of HXR emissivity source in Figure 8 is created by a combination of data from two gamma cameras. Taking into account that HXR bremsstrahlung is radiated in direction of RE motion during scattering and it has very narrow forward radiation pattern with angular half-width $\theta \sim {}^{1}/_{\gamma}$, $\gamma$–relativistic parameter [5], one can conclude that RE are scattered by ions during one act of scattering and this scattering is by large angles $\geq \pi/2$ (close distance collisions). Analysis of HXR radiation patterns for vertical and horizontal cameras plotted in polar coordinates (Fig. 9) reveal the correspondence of RE energy determined with HXR spectroscopy. Two narrow patterns with $\theta \sim 0.1$ have been detected in the vicinity of q=1.

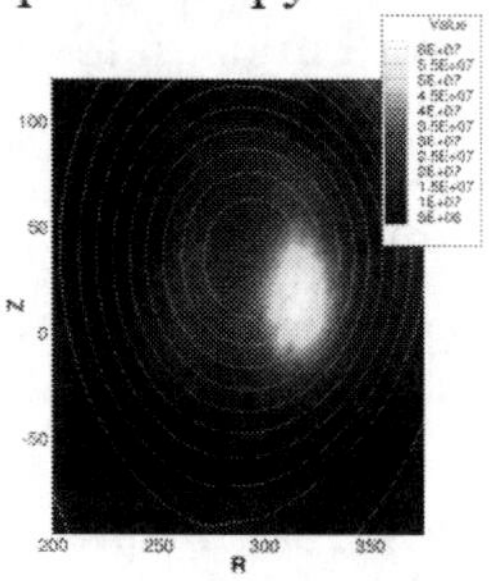

*Figure 8. JPN#86078: 2D map of HXR emission sources during RE stage*

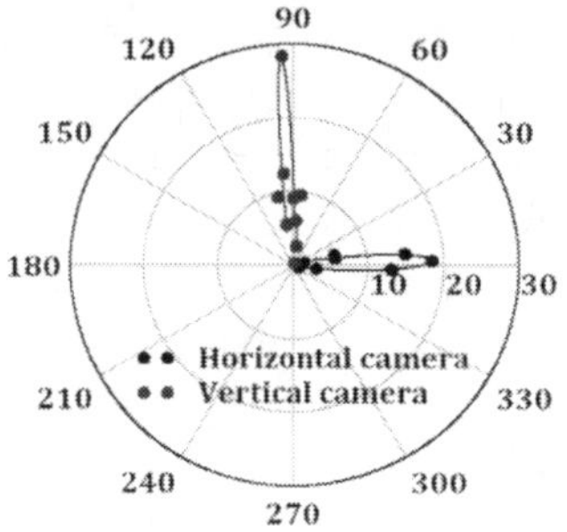


*Figure 9. HXR radiation intensity patterns in polar coordinates for vertical and horizontal γ-cameras: pole – is a point of maximum of emission in Fig. 8.*

The kinetic instability driven by RE (fan instability) arises due to the anomalous Doppler effect [6]. The magnetized Langmuir oscillations in plasma could be generated if the certain conditions on runaway beam velocity and density are satisfied: $V_{beam} > 3V_{cr}*(\omega_{ce}/\omega_{pe})^{3/2}$, where $V_{cr}=V_{Te}\alpha^{-1/2}$ ($\alpha=E_0/E_{DR}$); and $\nu_{eff} > \nu_e$, where $\nu_e=2.91*10^{-6}\ln\Lambda*n_e Z_{eff}T_e^{-3/2}$ and $\nu_{eff}$ is the effective collision frequency, which characterizes the enhancement of collisions due to electrostatic oscillations: $\nu_{eff}\approx\pi^{1/2}\omega_{pe}(\omega_{pe}/\omega_{ce})K(Zeff)*\varepsilon^{-3(Z+1)/16-1.5}\exp\{-1/4\varepsilon-((Z+1)/\varepsilon)^{1/2}\}$.

First analysis shows that this instability is the first candidate to explain intense burst of photo-neutrons and signal from FILD (Fig. 1, time range is highlighted in yellow). This data will be studied more in detail to provide the information on possible methods for destabilization of RE beams.

**Summary**

- RE generation process was detected on flattop stage of JET discharge using HXR diagnostics and identified. Main parameters of RE generation (critical fields, energies etc.) have been evaluated. Temporal, spatial and energy characteristics of RE generation process were studied.
- HXR spectrometry measured the HXR spectra, and RE with maximal energies up to 7 MeV and mean energy of 2.5 MeV were detected.
- It is found that RE current fraction could achieve half of the total plasma current, while the lower energy electron population (with several tens keV) could carry the other current part.
- Scattering of RE on plasma ions due close distance collisions was identified using the analysis of HXR bremsstrahlung.
- The evidence for kinetic instability (characteristic HXR bursts) has been observed only during plasma discharge ramp-down and significant decrease of magnetic field. Parameter space for instability has been analyzed.

**Acknowledgement** *"This work has been carried out within the framework of the EUROfusion Consortium and has received funding from the Euratom research and training programme 2014-2018 under grant agreement No 633053. IST activities also received financial support from "Fundação para a Ciência e Tecnologia" through project UID/FIS/50010/2013. A.E. Shevelev and E.M. Khilkevitch are grateful for financial support from the Ministry of Education and Science of the Russian Federation (Agreement No.14.619.21.0001, 15.08.2014, id RFMEFI61914X0001). The views and opinions expressed herein do not necessarily reflect those of the European Commission."*

**References:**

[1] M. Lehnen et al, Nuclear Fusion 53 (2013) 093007;
[2] A. Shevelev et al, Nuclear Fusion 53 (2013) 123004
[3] V. Kiptily, "Runaways measurements on JET" ITPA MHD & EP TGs Meeting, April 2013, Abingdon, UK
[4] R. S. Granetz et al. Phys. of Plasmas 21, 072506 (2014)
[5] J.D. Jackson. Classical Electrodynamics, John Wiley &Sons, Inc., Hoboken, New Jersey, 1962.
[6] Parail V V and Pogutse O P, 1978 Nuclear Fusion 18 303

# Measurement of radiation asymmetries during disruption mitigation at JET

P. Drewelow[1], U. Kruezi[2], H.R.Koslowski[3], M.Lehnen[4], G.F. Matthews[2] and JET Contributors*

*EUROfusion Consortium, JET, Culham Science Centre, Abingdon, OX14 3DB, UK*

[1] *Max-Planck Institute für Plasma Physik, Euratom Association, 17491 Greifswald, Germany*

[2] *CCFE, Culham Sience Centre, Abingdon OX14 3DB, UK*

[3] *Forschungszentrum Jülich GmbH, 52425 Jülich, Germany*

[4] *ITER Organization, Route de Vinon-sur-Verdon - CS 90 046,13067 St Paul Lez Durance Cedex, France*

Disruptions, the accidental loss of current and confined energy in Tokamak plasmas can cause damage to the in-vessel components and vital plant and support structures. Thus, the ability to predict, to avoid and if inevitable to mitigate this harm is crucial for existing fusion experiments and for future fusion reactors like ITER. Massive Gas Injection (MGI) is a possible approach to be used as a disruption mitigation system (DMS) enabling radiative dissipation of the stored energy. However, toroidal asymmetries in the radiation can diminish the energy dissipation and it is expected that in the case of ITER such asymmetries in the radiation could cause melting.

## Setup at JET

At JET, the Tokamak experiment closest to ITER in terms of operating parameters and size, a DMS based on two fast disruption mitigation valves (DMV) has been established and is routinely used in close loop operation [1]. The toroidal location of these valves and their approximated injection trajectories are shown in figure 1 (DMV1: orange, DMV2: green). In order to study the mitigation efficiency at JET, bolometric measurements are used to assess the radiation level and distribution. The overview bolometer KB5 consisting of two cameras (red and blue in figure 2) allows a reconstruction of the poloidal radiation pattern by tomography. However, this analysis of the vertical and horizontal bolometer data assumes toroidal symmetry of the radiation and is known to produce artefacts in case this assumption does not hold. Thus, a set of

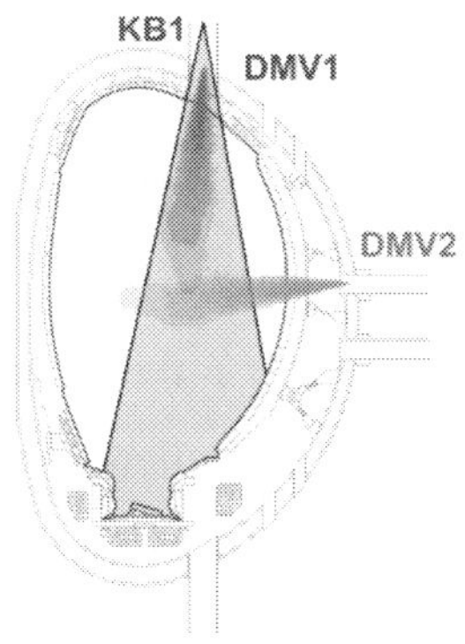


Figure 1: *Poloidal cross section JET torus*

*See the Appendix of F. Romanelli et al., Proceedings of the 25th IAEA Fusion Energy Conference 2014, Saint Petersburg, Russia

four single channel bolometers (KB1) at different toroidal positions with an equivalent field of view (FOV) was recently re-installed (purple in figure 1 and 2).

**Experiments and data analysis strategy**

Due to limitations in the data acquisition of the single channel bolometer KB1, a mitigated disruption was repeated four times and the toroidal radiation distribution measured with each of the four bolometers consecutively. The main plasma parameters prior to the disruption (central density $n_e = 6 \cdot 10^{19} m^{-3}$ and temperature $T_e = 1.8\text{keV}$, plasma current $I_p = 1.5\text{MA}$, total radiation $P_{rad} = 3\text{MW}$ and input heating power $P_{NBI} = 14\text{MW}$) are all comparable during the pulse. Tomographic reconstructions show a similar poloidal radiation pattern in the pre-disruption plasma. All time vectors are corrected by the time $t_{\text{DMV}}$ when the DMV puff reaches the plasma. The DMV delivered 10bar l of a gas mixture of 10% Argon and 90% Deuterium.

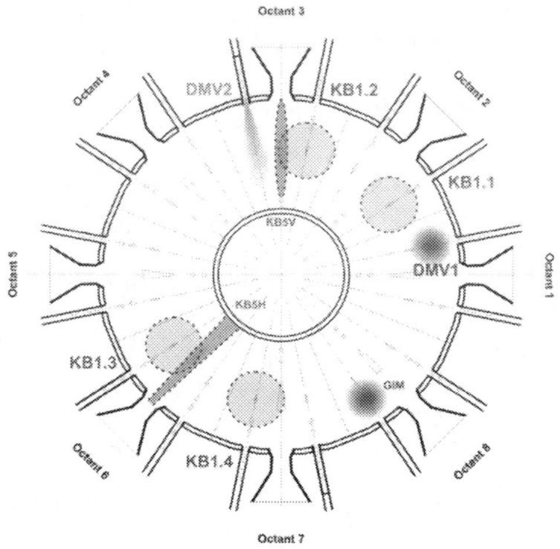


Figure 2: *Horizontal overview of measurement setup at JET*

The radiated power $P_{\text{rad, i}}$ measured by each KB1 bolometer $i$ is derived using the bolometer equation

$$P_{\text{rad,i}} = \frac{\tau_{\text{i}} G_{\text{i}}}{S_{\text{i}} g} \times \left( \frac{dU_{\text{i}}}{dt} + \frac{U_{\text{i}}}{\tau_{\text{i}}} \right) \tag{1}$$

with the voltage signal of KB1 $U_{\text{i}}$, the cooling time of the detector foil $\tau_{\text{i}}$, the amplification factor of the data acquisition $g$, the detector sensitivity $S_{\text{i}}$ and the etendue of the field of view $G_{\text{i}}$. Most of these parameters are known, except for $S_{\text{i}}$ and $G_{\text{i}}$. They are derived using a cross calibration to the total power $P_{\text{rad,tot}}$ measured by the vertical camera of the KB5 bolometer in octant 3 during a phase of complete toroidal symmetry of the plasma (limiter start-up or a divertor fuelled symmetrically heated phase).

The cross calibration $P_{\text{rad,o}} = k_{\text{geo}} \times P_{\text{rad,i}} = \frac{1}{8} P_{\text{rad,tot}}$ can be used to derive an estimate for the radiation $P_{\text{rad,o}}$ in each octant $o$, assuming a constant geometry factor $k_{\text{geo}}$ that represents the ratio between the radiation within the FOV of KB1 and that within the total volume of one octant. This assumption may not hold for a plasma with asymmetry even within one octant or when a large fraction of the radiation is not within the poloidal FOV of KB1 (e.g. torus inboard side, see figure 1). Dividing $\frac{\max[P_{\text{rad,o}}]}{\text{mean}[P_{\text{rad,o}}]} = f_{\text{TRPF}}$ results in the toroidal radiation peaking factor which is unaffected by the geometric factors. In the following, an average $\langle f_{\text{TRPF}} \rangle$ is shown that

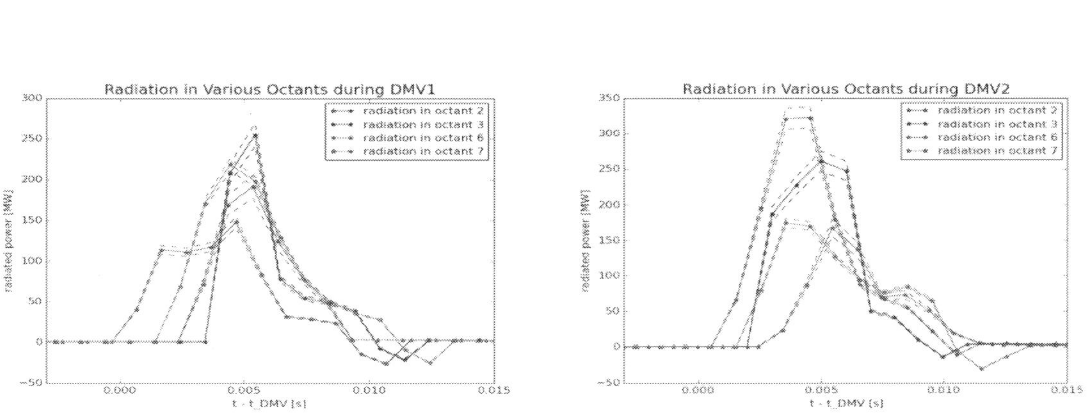


(a) (b)

Figure 3: *Radiated power during the application of (a): DMV 1 (vertical in octant 3) and (b): DMV 2 (horizontal in octant 6)*

is derived from the integral radiated energy.

The toroidal asymmetry measurement was validated with the integrated radiated energy during a pulse fuelled by a Gas Inlet Module (GIM) in octant 8 (see figure 4). The peaking at octant 7 agrees with asymmetry induced artefacts observed in the tomographic reconstruction.

**Radiation characteristics during disruption mitigation**

Using the vertical injection valve in octant 1 (DMV1) during a disruption resulted in a radiation that was detectable in the octant order: 2, 7, 6 and 3 (see figure 3 (a)). Since KB1.3 ($\Delta\varphi = -157.5°$) is reached before KB1.2 ($\Delta\varphi = +67.5°$), the clockwise expansion of the radiation cloud seems about three times as fast as in anti-clockwise direction. When integrating the radiation during the disruption (neglecting the negative drop in figure 3 (a)), the main radiation was emitted in octant 7 rather than in 2 (see figure 4, green line). This indicates a preferred clockwise gas or impurity transport that leads to an average toroidal peaking of $\langle f_{\mathrm{TRPF}} \rangle \approx 1.3$ in octant 7. Mitigating the disruption with a horizontal gas injection at octant 3 (DMV2), the radiation occurs in the order octant 2, 7, 3 and 6 (see figure 3 (b)). Due to its proximity, the radiation signal would be expected first in octant 3 (see figure 2). It is possible that it only seems delayed after the octant 2 radiation because

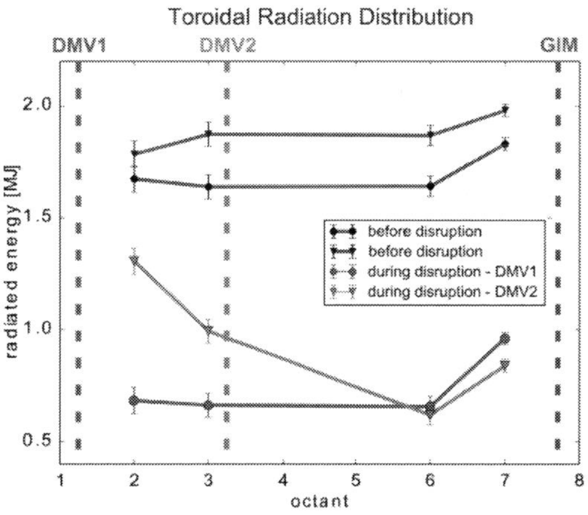


Figure 4: *Toroidal distribution of radiated energy before and during the disruption mitigated by different DMV1 and DMV2*

poloidally the radiation is limited to the blind angle of KB1.2 (see figure 1) in the first phase of the disruption. The integral radiation during the disruption peaks at octant 2 with $\langle f_{\mathrm{TRPF}} \rangle \approx 1.4$. Apart from the slightly higher $\langle f_{\mathrm{TRPF}} \rangle$ value it should also be noted that the radiation peak extends further around the torus with horizontal injection (ca. over octant 7-8-1-2-3), while it only covers up to three octants (7-8-1) during the vertical gas injection (see figure 4).

## Conclusion

Measuring the toroidal radiation distribution with a set of four toroidally distributed bolometers revealed differences in the radiation peaking during a disruption mitigated by horizontal or vertical massive gas injection. Both injections resulted in a similar peaking factor (vertically 1.3, horizontally 1.4). The peak location seems shifted in both cases by $\Delta\varphi_{\mathrm{peak}} \approx 20^{\circ}$ to $\Delta\varphi_{\mathrm{peak}} \approx 50^{\circ}$ toroidally. This stands in contrast to simulations at DIII-D [3], predicting the radiation peak on the opposite torus side due to a gas puff induced 1/1 MHD mode. The average radiated energy peaking is also of a lower level than the previously measured transient peaking factor in the pre-thermal-quench phase of 8 [2] indicating that the thermal-quench radiation compensates by far the initial radiation located at the gas injection point. The injection methods differ in the toroidal width of the total radiated energy peak. Vertical injection confines the peaking to maximal three octants, while horizontal injection causes a broad peak extending over at least 5 octants. This difference might be related to a wider gas diffusion in the larger volume outside the plasma that is available at the horizontal injection point.

Future work will focus on a more systematic investigation into the interaction of the gas puff with pre-existing MHD modes or plasma rotations and on the further analysis of the time resolved peaking in different phases of the disruption.

## Acknowledgement

This work has been carried out within the framework of the Contract for the Operation of the JET Facilities and has received funding from the European Union's Horizon 2020 research and innovation programme. The views and opinions expressed in this paper do not necessarily reflect those of the European Commission or the ITER Organization.

## References

[1] U. Kruezi et al, Proceedings of 28th SOFT, EFD-C(14)05/22 Dia ref. CPS14.691 (2014)

[2] M. Lehnen et al, Nuclear Fusion **51**, 123010 (2011)

[3] Izzo V.A., Physics of Plasmas **20**, 056107 (2013)

# Confinement and pedestal in dimensionless collisionality scans of low triangularity H-mode plasmas in JET-ILW

L. Frassinetti[1], M. N. A. Beurskens[2], S. Saarelma[2], J. E. Boom[3], E. Delabie[4], J. Flanagan[2], C. M. Kempenaars[2], Giroud[2], P. Lomas[2], L. Meneses[5], S. Menmuir[2], I. Nunes[5], F. Rimini[2], and JET contributors*

EUROfusion Consortium, JET, Culham Science Centre, Abingdon, OX14 3DB, UK

[1]Division of Fusion Plasma Physics, KTH Royal Institute of Technology, Stockholm SE

[2]CCFE, Culham Science Centre, Abingdon,OX14 3DB, UK

[3]Max-Planck-Institut für Plasma Physik, Boltzmannstr.2, 85748 Garching, Germany

[4]Oak Ridge National Laboratory, Oak Ridge, TN 37831-6169 USA

[5]Instituto de Plasmas e Fusão Nuclear, Universidade de Lisboa, 1049-001 Lisboa, Portugal

** See the Appendix of F. Romanelli et al., Proceedings of the 25th IAEA Fusion Energy Conference 2014, Saint Petersburg, Russia*

## INTRODUCTION

The baseline type I ELMy H-mode scenario with H=1 has been re-established in JET with the new tungsten divertor and beryllium main wall (JET-ILW) in 2011. Comparing carbon wall (JET-C) discharges in similar conditions, a degradation of the confinement has been observed in JET-ILW, with the reduction mainly driven by a lower pedestal pressure [1]. Results presented in [2,3,4,5] show that JET-ILW at low collisionality ($\nu^*$) reaches a confinement comparable to the JET-C. The present work describes the results on the confinement and on the pedestal of a dimensionless collisionality scan.

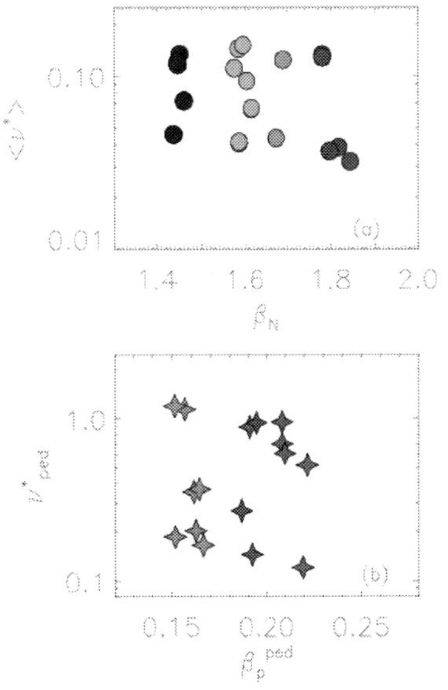


*Figure 1. Range of dimensionless parameters used. The colours highlight data with similar $\beta_N$ (a) and similar pedestal $\beta_\theta$ (b).*

## THE DATA SET

The collisionality scan is obtained in baseline JET-ILW plasmas by allowing to vary power, current and gas in the range $P_{NBI}\approx11\text{-}22$MW, $I_p\approx1.5\text{-}2.5$MA and $\Gamma_{D2}\approx1\text{-}6\cdot10^{22}$ (e/s). The safety factor and the triangularity are kept constant at $q_{95}\approx3$ and $\delta\approx0.22$. The volume averaged collisionality $\langle\nu^*\rangle$, defined as the ion-electron collision frequency normalized to the bounce frequency, $\nu^*=6.9\cdot 10^{-18} n_e R q_{95} Z_{eff} \ln\Lambda/(\varepsilon^{3/2} T_e^2)$, is varied by a factor 5. Four collisionality scans obtained at four different $\beta_N$ values have been performed. The parameters range is shown in figure 1(a). The volume averaged normalized ion poloidal Larmor radius is kept constant at $\langle\rho^*\rangle\approx 2.4$-$2.7\%$. This dataset is used to investigate the dependence of the global confinement on the collisionality. However, the change in collisionality affects the density peaking, therefore, a dataset with constant $\beta_N$ has different pedestal beta. For this reason, figure 1(b) shows a dataset [whose shots are mostly the same as those in figure 1(a)] in which the plasmas with constant pedestal $\beta_\theta$ have been highlighted. The corresponding $\nu^*$ varies by a factor 10 at the pedestal, while $\rho^*$ at the pedestal is constant at $\rho^*_{ped}\approx 1.6$-$1.8\%$. This dataset is used when discussing the pedestal structure.

**CONFINEMENT**

The confinement factor versus the normalized beta is shown in figure 2(a). The dashed line shows the typical trend between $H_{98}$ and $\beta_N$ in JET-C. The $\nu^*$ scans of the JET-ILW data show a more complex behaviour. The high $\nu^*$ JET-ILW plasmas have low confinement ($H_{98}\approx 0.8$), while at low $\nu^*$ JET-ILW reaches confinement comparable to JET-C. The improvement from high to low $\nu^*$ is more evident at high $\beta_N$ than at low $\beta_N$ (compare the blue and the red dots in figure 1a). The trend of $H_{98}$ and total stored energy versus $\langle\nu^*\rangle$ are shown in figures 2(b) and 2(c) respectively. The increase of $H_{98}$ and $W_{th}$ with the reduction of collisionality is clear.

The increase in $W_{th}$ is due to both the increase of the pedestal pressure and of the core pressure, as described in figure 3. Figure 3(a) shows the pedestal electron temperature $T_e^{ped}$ versus the pedestal electron density $n_e^{ped}$. Reducing $\nu^*$ in the JET-ILW dimensionless collisionality scan produces a decrease of $n_e^{ped}$ and an increase of $T_e^{ped}$. The $T_e^{ped}$ increase is stronger than the $n_e^{ped}$

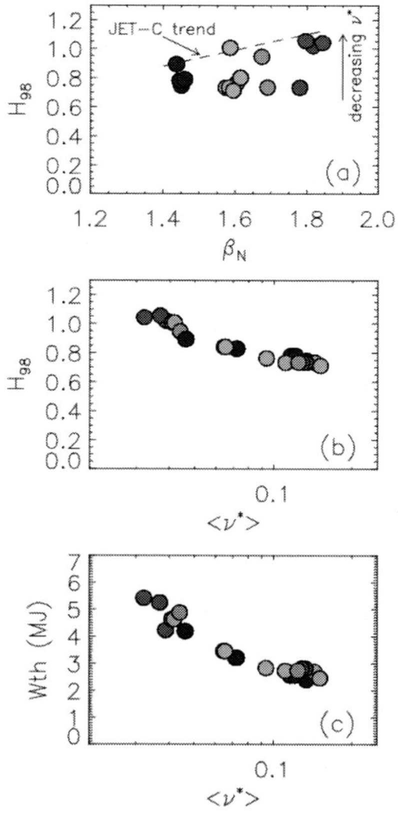


*Figure 2. Confinement factor versus $\beta_N$ (a) and versus collisionality (b). Total thermal energy versus collisionality. Colour code as in figure 1(a).*

reduction, so a higher pedestal pressure is obtained ($P_e^{ped}\approx$4kPa at high $\nu$* and $P_e^{ped}\approx$6kPa at low $\nu$*). In the core, figure 3(b), the density does not change significantly, while the temperature increases from ≈2.5keV to ≈5keV and consequently the pressure from ≈25kPa to ≈45-50kPa.

The change in collisionality strongly affects the peaking [2,6], as shown in figure 4(a). The $T_e$ peaking remains constant, as shown in figure 4(b), so the pressure peaking increases at low collisionality.

Therefore, the increase of the $W_{th}$ with the $\nu$* reduction is due to (1) the increase in the pedestal $W_{th}$ due to the increase in $P_e^{ped}$ (≈50%) and to (2) the increase in the core $W_{th}$ (≈75%) due to the increase in $P_e$ peaking and to the $T_e$ profile stiffness.

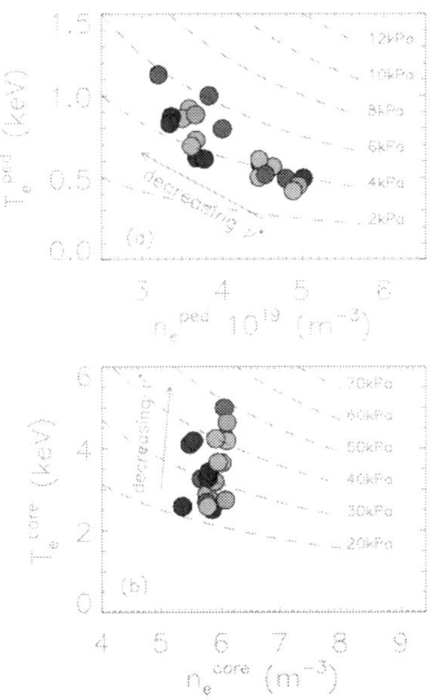


*Figure 3. Electron temperature versus electron density at the pedestal (a) and in the core (b). Dashed lines highlight the isobar curves. Colour code as in figure 1(a).*

**PEDESTAL STRUCTURE and EPED RESULTS**

Recent experimental results [4] show that the pedestal pressure width $w_{pe}$ in JET-ILW scales with the pedestal $\beta_\theta$., as observed in other devices and predicted by the KBM model. In this work, $w_{pe}$ is estimated as the average between $T_e$ and $n_e$ pedestal widths, in agreement with the definition in EPED. Figure 5(a) shows the correlation of $w_{pe}$ with $\nu$*. A positive trend is observed. This is in reasonable agreement with what recently described in [7], where the width is observed to increase at high gas level. The present data at low $\nu$* are in agreement with the EPED results based on the KBM predictions [$w_{pe}=0.076(\beta_\theta^{ped})^{0.5}\approx0.03$] as shown in figure 5(b). The deviation from the prediction at high $\nu$* suggests that the present models are not sufficient to describe in detail the pedestal width behaviour.

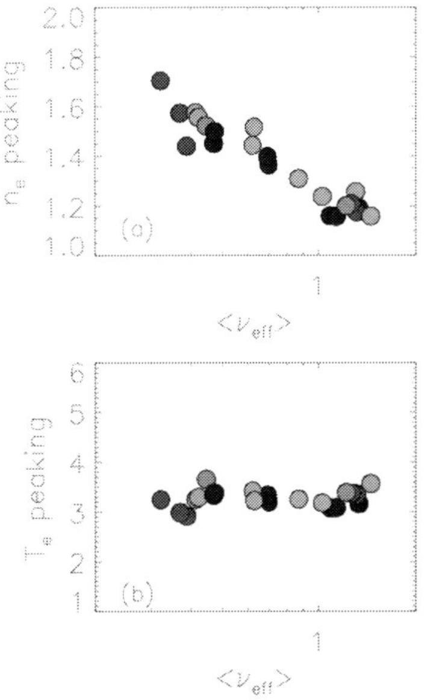


*Figure 4. Density peaking (a) and temperature peaking (b) versus effective collisionality. Colour code as in fig 1(a).*

The EPED results for the pedestal temperature is shown in figure 6. The EPED modelling has been performed in order to match the $\rho^*_{ped}$ (≈1.7%) and $\beta_\theta^{ped}$ (≈0.15) of the JET-ILW ν* scan. EPED correctly reproduces the qualitative behaviour of the pedestal height with the ν*, but overestimates the experimental $T_e^{ped}$ by ≈10% at high ν* and ≈15-20% at low ν*. The EPED results are consistent with the fact that the present JET-ILW plasmas are far from the stability boundary, figure 6(b).

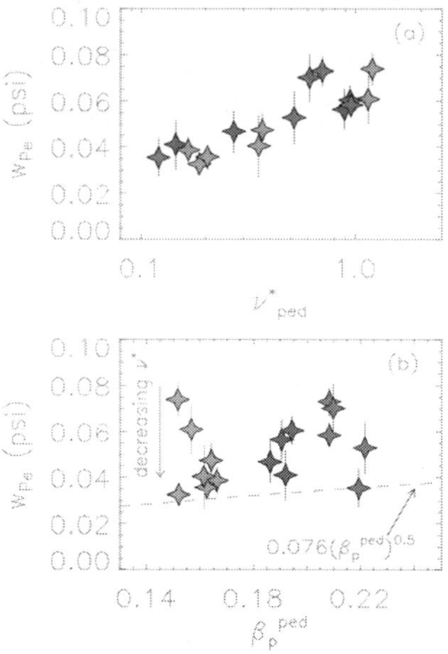


*Figure 5. Pressure pedestal width versus pedestal collisionality (a) and beta poloidal (a). Colour code as in figure 1(b).*

## CONCLUSIONS

The work shows that JET-ILW at low collisionality can reach confinement comparable to JET-C. The high stored energy is achieved by both the increased pedestal pressure (≈50%) and by the increased core pressure (≈75%). It was not possible to find a perfect match between the present data set and JET-C (in terms of ν*,ρ*,$\beta_\theta$ and $q_{95}$), so a quantitative comparison cannot be done, but, qualitatively, JET-C and JET-ILW have similar trends with collisionality.

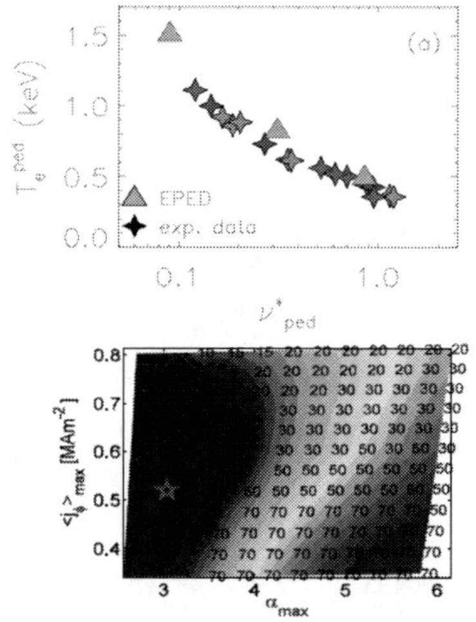


*Figure 6. Pedestal temperature versus $\nu^*_{ped}$ for experimental data and EPED simulations (a). Stability diagram for a low ν*plasma (b).*


## ACKNOWLEDGEMENT

This work has been carried out within the framework of the EUROfusion Consortium and has received funding from the Euratom research and training programme 2014-2018 under grant agreement No 633053. The views and opinions expressed herein do not necessarily reflect those of the European Commission."


## REFERENCES


[1] C. Giroud et al., Nucl. Fusion **53** 113025 (2013)
[2] M. Beurskens et al., Nucl. Fusion **54** 043001 (2014)
[3] C. Challis et al., Nucl. Fusion **55** 053031 (2015)
[4] C. Maggi et al., submitted to Nucl. Fusion
[5] I. Nunes et al., 25th IAEA Fusion Energy Conf. (S. Petersburg, 2014) EX/9-2
[6] M. Maslov et al., Nucl. Fusion **49** 075037 (2009)
[7] M. Leyland et al., Nucl Fusion **55** 013019 (2015)

# Confinement and pedestal structure in high performance scenarios in JET-ILW and comparison with JET-C

E. Stefanikova[1], L. Frassinetti[1], P. Lomas[2], I. Nunes[3], M. Baruzzo[4], F. Rimini[2], S. Saarelma[2], S. Wiesen[5], M. Peterka[6,7], and JET contributors*

*EUROfusion Consortium, JET, Culham Science Centre, Abingdon, OX14 3DB, UK*

*[1]Division of Fusion Plasma Physics, KTH Royal Institute of Technology, Stockholm SE*

*[2]CCFE, Culham Science Centre, Abingdon, OX14 3DB, UK; [3]Centro de Fusao Nuclear, IST, Lisboa, Portugal;*

*[4]RFX, Corso Stati Uniti 4, Padova, Italy; [5]Forschungszentrum Jülich GmbH, D-52425, Jülich, Germany;*

*[6]Institute of Plasma Physics AS CR, Prague, Czech Republic,*

*[7]Charles University in Prague, Faculty of Mathematics and Physics, Prague, Czech Republic*

** See the Appendix of F. Romanelli et al., Proceedings of the 25th IAEA Fusion Energy Conference 2014, Saint Petersburg, Russia*

## INTRODUCTION

High plasma current and field is one way to develop high performance scenarios with high thermal energy ($W_{th}$ > 10MJ) and temperature ($T_e^{core}$ ≈ 7-15keV). Recently, JET-ILW has reached 4.0MA in the baseline scenarios with low triangularity in quasi stationary-conditions, avoiding W accumulation and controlling the divertor heat loads. The stored energy in JET-ILW tends to be lower than in JET-C [1]. This work studies the role of the pedestal in the confinement of JET-ILW baseline plasmas of a current scan and discusses the differences with the JET-C. The dataset used has $I_p$ in the range 2-4.5MA, $P_{NBI}$ ≈ 4-26MW, gas level $\Gamma_{D2}$ ≈ 0-10·$10^{22}$ e/s.

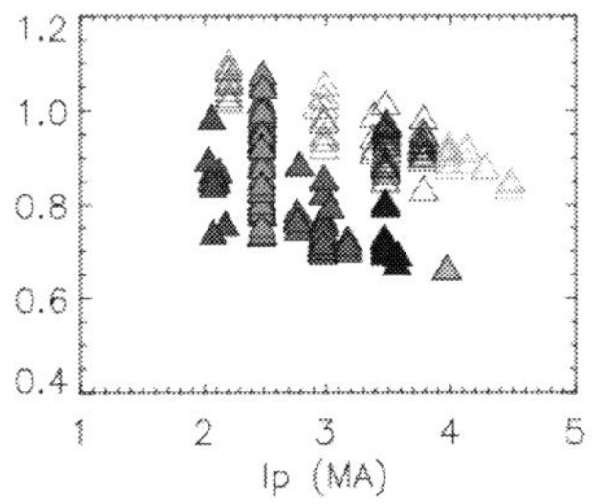

**Figure 1.** Range of $H_{98}$ covered by JET-C (open symbols) and JET-ILW (full symbols) shots for different currents.

## GLOBAL CONFINEMENT

The confinement enhancement factor covered by the analysed shots is $H_{98}$ ≈ 0.6-1.1, figure 1. For plasma current higher than 2.5MA, JET-ILW tends to have lower $H_{98}$ than JET-C. For lower currents, JET-ILW can reach $H_{98}$ comparable to JET-C.

Figure 2(a) shows the total stored thermal energy $W_{th}$ versus the plasma current. An increase of $W_{th}$ in both JET-C and JET-ILW with $I_p$ is observed [1]. However, the increase is stronger in JET-C at high $I_p$. The stored energy in JET-ILW tends to be lower than in JET-C up to 20% for high $I_p$. Comparable $W_{th}$ is obtained only for $I_p \leq$ 2.5MA. The present JET-ILW dataset at high

$I_p$ tends to have higher radiation from the bulk plasma but the lower $W_{th}$ cannot be ascribed to a lower power through the separatrix. Figure 2(b) shows $W_{th}$ versus $P_{sep}= P_{in}$ - $dW_{th}/dt$ - $P_{rad,bulk}$. At high current [light and dark blue color in the figure 2(b)] the stored energy in JET-ILW is lower than in JET-C, with $P_{sep}$ in the range 15-25MW. The lower $W_{th}$ in JET-ILW is related to a reduction in both the core $W_{th}$ and in the pedestal $W_{th}$, as shown in figures 2(c) and 2(d) respectively.

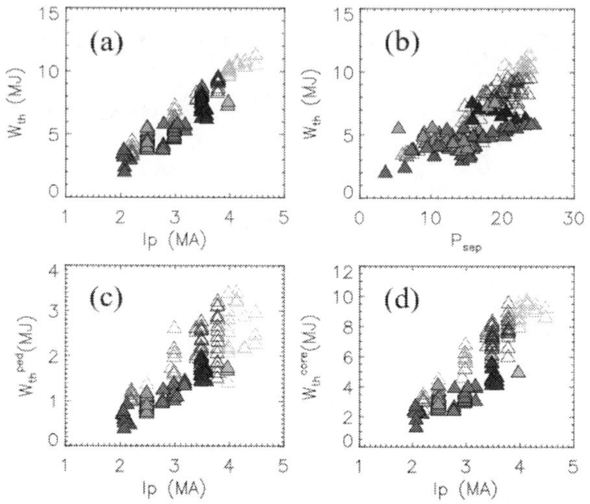


**Figure 2.** (a) Scaling of the total stored thermal energy with $I_p$; (b) total stored thermal energy versus power through the separatrix; (c) core energy vs $I_p$; (d) pedestal energy vs $I_p$. Open symbols – JET-C, full symbols – JET-ILW.

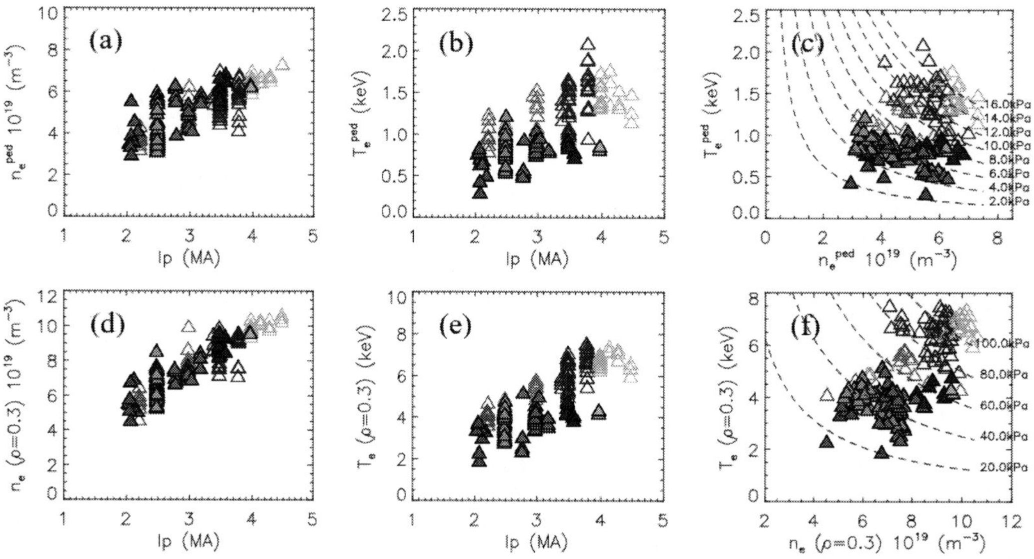


**Figure 3.** (a) Scaling of the pedestal density with current; (b) scaling of the pedestal temperature with current; (c) pedestal temperature vs pedestal density; (d) scaling of the core density with current; (e) scaling of the core temperature with current; (f) core temperature vs core density. Dashed lines indicate constant pressure. Open symbols – JET-C, full symbols – JET-ILW.

## CORE AND PEDESTAL ELECTRON DENSITY AND TEMPERATURE

Figures 3(a) and 3(b) illustrate the scaling of pedestal density ($n_e^{ped}$) and temperature ($T_e^{ped}$) with $I_p$. In JET-C, both $n_e^{ped}$ and $T_e^{ped}$ increase with $I_p$. In JET-ILW, $n_e^{ped}$ increases with $I_p$, and a good overlap with JET-C is observed. Instead, $T_e^{ped}$ increases only until $I_p \approx 2.5$MA. For $I_p >$ 2.5MA, $T_e^{ped}$ is relatively constant, as already observed in [1]. Therefore, the low $W_{th}^{ped}$ in JET-ILW is mainly due to a low $T_e^{ped}$. The difference in behaviour with $I_p$ might be related to

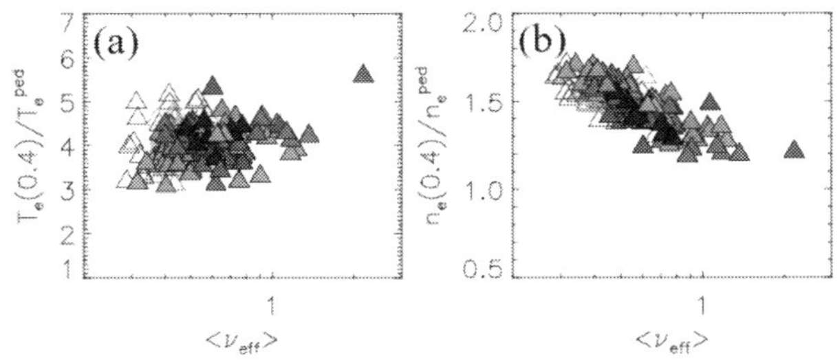


**Figure 4.** (a) $T_e$ peaking vs $\nu_{eff}$; (b) $n_e$ peaking as a function of $\nu_{eff}$. Open symbols – JET-C, full symbols – JET-ILW. Colors correspond to levels of $I_p$.

the larger gas fuelling used in JET-ILW in order to control W accumulation [1]. The pedestal pressure ($P_e^{ped}$) in JET-ILW at high $I_p$ does not reach values comparable to JET-C, figure 3(c). In the core, figures 3(d), 3(e) and 3(f), a similar behaviour to the pedestal is observed.

Figure 4 shows the dependence of profile peaking with the effective collisionality ($\nu_{eff}$). Figure 4(a) shows that $T_e$ peaking has a weak or no trend with $\nu_{eff}$. Therefore a low $T_e^{ped}$ is related to a low $T_e^{core}$ due to the $T_e$ profile stiffness. Figure 4(b) follows a decreasing trend of $n_e$ peaking with $\nu_{eff}$ in agreement with [2]. JET-C and JET-ILW show the same trend of the peaking versus $\nu_{eff}$. But, due to lower $T_e$ at high current, JET-ILW tends to have higher $\nu_{eff}$ and therefore lower $n_e$ peaking than JET-C. This further reduces the core contribution to the JET-ILW $W_{th}$.

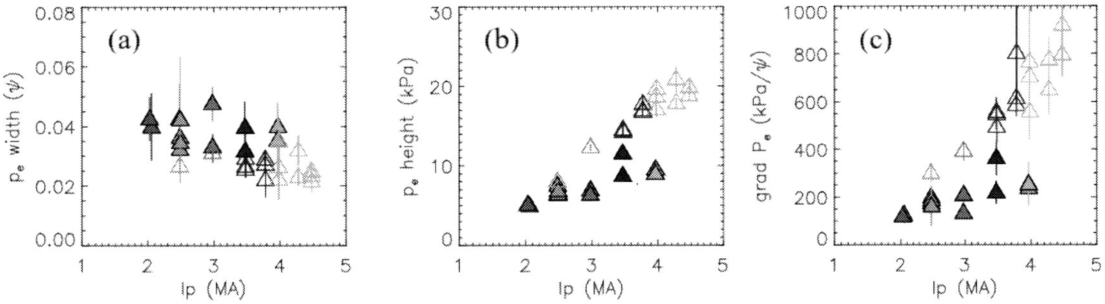


**Figure 5.** (a) $p_e$ pedestal widths vs $I_p$; (b) $p_e$ pedestal heights vs $I_p$; (c) grad $p_e$ vs $I_p$. Open symbols – JET-C, full symbols – JET-ILW.

## PEDESTAL WIDTH AND PEDESTAL GRADIENT

The pedestal width analysis has been performed at each $I_p$ level for the shots with the highest $H_{98}$, both in JET-C and JET-ILW. Figure 5(a) shows pedestal $p_e$ widths versus $I_p$. No clear trend with $I_p$ is observed, but JET-C tends to have a lower pedestal width than JET-ILW. This might be related to the higher gas level used in JET-ILW [3]. Figure 5(b) shows the pedestal $p_e$ versus $I_p$ and figure 5(c) shows the correlation of the pedestal pressure gradient with $I_p$. A significant reduction, by a factor 2 or higher, for JET-ILW is observed.

## STABILITY ANALYSIS

The stability analysis has been performed using the MISHKA and ELITE codes for the shot with the best confinement at each current level, in both JET-ILW and JET-C. The JET-C shots are near or slightly beyond the stability boundary at each current level. Figure 6(a) and 6(b) show the stability diagram at $I_p$ = 2.5MA and $I_p$ = 4MA for JET-C. The JET-ILW shots are not far from the stability boundary only at low current ($I_p$ < 2.5MA), while at high current JET-ILW is deep in the stable region, far from the boundary. Figure 6(c) and 6(d) show the stability diagram at $I_p$ = 2.5MA and $I_p$ = 4MA for JET-ILW. This is consistent with the fact that JET-ILW at high current has not reached performances comparable to the JET-C.

## CONSLUSIONS

From the results above it can be concluded that low performance in JET-ILW (compared to JET-C) at high currents is due to a reduction in both core and pedestal stored energy and this is caused by a reduction of the $T_e$. Low $T_e^{ped}$ produces low $T_e^{core}$ due to $T_e$ profile stiffness. Moreover, the low $n_e$ peaking also reduces the contribution to $W_{th}^{core}$. Pedestal $p_e$ widths from the selected subset of shots do not follow any obvious trend with $I_p$. But JET-ILW reaches comparable or larger pedestal widths than JET-C. Also the pressure gradient at the pedestal is strongly reduced in JET-ILW with high $I_p$. This is in accordance with stability analysis from the MISHKA code which showed that at high $I_p$ JET-ILW is far from the stability boundary.

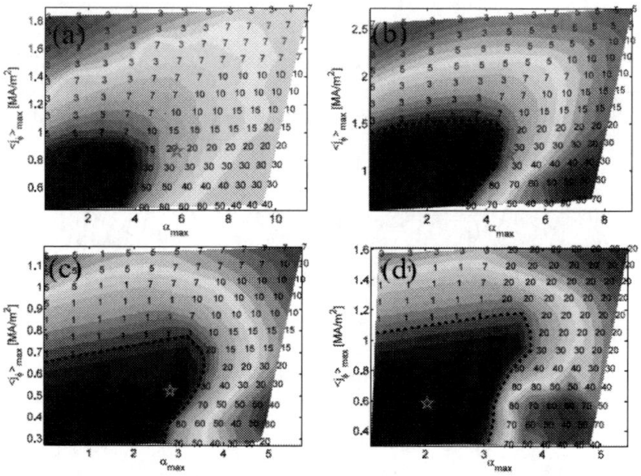


**Figure 6.** Stability diagram at (a) $I_p$ = 2.5MA, JET-C, (b) $I_p$ = 4MA, JET-ILW, (c) $I_p$ = 2.5MA, JET-ILW, (d) $I_p$ = 4MA, JET-ILW.

## ACKNOWLEDGEMENTS

This work has been carried out within the framework of the EUROfusion Consortium and has received funding from the Euratom research and training programme 2014-2018 under grant agreement No 633053. The views and opinions expressed herein do not necessarily reflect those of the European Commission.

## REFERENCES

[1] I Nunes et al., EX/9-2 Proceedings of the 25th IAEA FEC 2014, Saint Petersburg, Russia

[2] M. Beurskens et al.,Nucl. Fusion **54**, 043001 (2014)

[3] M. Leyland, et al.,Nucl. Fusion **53**, 083028 (2013)

# Alpha heating in JET plasmas with sawteeth

R.V. Budny[a] and JET contributors[b]

*EUROfusion Consortium, JET, Culham Science Centre, Abingdon, OX14 3DB, UK*

Alpha heating is essential for practical energy production from DT fusion reactions. Experiments to detect alpha heating were performed in TFTR (1994) [1] and in JET (1997) [2]. Measurements of alpha heating were difficult since the alpha heating power $p_\alpha$ was small relative to the external heating $p_{\rm ext}$, and since the experiments were not accurately reproducible. It was difficult to realize comparable DD or TT discharges for comparison. One cause of irreproducibility in the JET discharges was the presence of sawteeth.

This paper studies effects of sawteeth in the discharges featured in [2]. They had $I_p$=3.8MA and $B_{\rm tor}$=3.6T, and density 35% of the Greenwald empirical limit. The mix of D and T was varied from discharge to discharge by changing the D and T in the neutral beam injection while maintaining their total powers approximately constant ($P_{\rm NB} \simeq$ 10.3MW). Several of them had global fusion power to heating power ratio $Q_{\rm DT}$=0.7, and core values (defined in [3]) of 1.2. These are near the highest values seen in JET. The fraction of T to total beam power $f_{\rm T}=P_{\rm T-NB}/(P_{\rm NB})$ is an important metric for the scans. Here, one with D-only, three with D and T, and two with T-only NB [2] are used. Summary values are in Table I.

Sawtooth crashes cause rapid drops in the core electron temperature $T_e$. Minor disruptions also cause rapid crashes, but sawtooth crashes have simultaneous rapid increases in $T_e$ outside the mixing radius (past the radius where the safety factor is unity). Crash times were determined from core electron cyclotron measurements of $T_e$. Examples are shown in Fig. 1-a). The pattern of frequent small sawteeth crashes petering off after the start of NB is typically seen in JET discharges. Even large sawtooth crashes within the first second of NB did not appear to have major effects on the later performance. However, larger crashes occurring later in the phase of stored energy ramp-up did have significant impacts, discussed below. Sawtooth crashes were generally seen in TFTR supershots continuing briefly into the NB phase, but they were usually suppressed later during the NB phase. Effects of the sawtooth crashes in the JET scan can be seen on the total neutron emission rate $S_n$ and on the central neutron emission rate $s_n(0)$ observed in fits to neutron camera measurements in Fig. 1-b).

The electron density profiles $n_e$ were sometimes hollow, as often seen during the JET DT campaign The general trend was for the $n_e$ shape to very from peaked for H to flatish for D to increasing peakedness in DT and T H-mode plasmas [4]. For this reason, and also since the heating directly effects the stored energy, it is insightful to focus on the electron and ion energy densities $w_e$ and $w_i$.

At least in the core, $w_e$ and $n_e$ increased approximately linearly in time from the start of NB for at least 1.6s. Plots of $w_e$ volume-averaged from the magnetic axis to x($\equiv$ square-root of the normalized toroidal flux) = 0.2 in Fig. 1-c) show $w_e$ increasing with approximately the same rate except for abrupt decreases at sawtooth crashes. They do not show a clear trend in $s_n(0)$ or $f_T$ suggesting that $p_\alpha$ is not playing a significant role, perhaps due to systematic changes in confinement or energy loss rates compensating for increased $p_\alpha$. The values for $w_i$ shown in Fig. 1-d) do show a trend of increasing $f_T$ suggesting that an isotopic effect might be playing a role if 43011 is excluded . This discharge had the most hollow $n_e$ and recycling rates 50% higher than the others. The $n_e$ pedestal was the lowest and the $T_i$ and $T_e$ pedestals the highest in the scan. This discharge was pivotal for the argument against isotopic mass effects in [2].

The $T_e$ profiles also are similar across the scan at the same time, and do not exhibit consistent trends in $f_T$ or $s_n(0)$ up to the times of the first significant sawtooth crash as shown in Figs. 2-a). At later times sawtooth crashes break the tight clustering of rates and the discharges that have not crashed continue to increase at the same growth rates. By 14s a clear separation of $T_e$ and $T_i$ profiles appear with the high $s_n(0)$ shots, still before sawtooth crashes, having the highest $T_e$ and $T_i$. Profiles of $T_i$ shown in Figs. 2-b) and Figs. 2-d) show greater separation. Central values of $T_e$ shown in [2] just before the last sawtooth crashes did have clear separation with $S_n$. This separation of $T_e$ can be attributed to the sawtooth crashes and ion-electron coupling $p_{ie}$. Increased sawtooth period with increasing $f_T$ has been reported [5].

The discharges were modeled using the TRANSP code to study heating and transport. The modeling achieves approximate agreement with $S_n$ and $s_n(0)$, as well as with the EFIT diamagnetic energy indicating that the modeling is approximately accurate. The largest discrepancy between the simulated and measured $s_n(0)$ differed by at most 12% at the peak, indicating that $p_\alpha$ is credible. The sawtooth crashes are modeled assuming Kadomtsev helical mixing of current and fast ions at the observed sawtooth crash times. Sawteeth crashes have been observed to mix fast beam and alpha ions [6]. The core volume-averaged $p_\alpha$, fast alpha density $n_\alpha$ and pressure $\beta_\alpha$ also increase approximately linearly in time, except for sharp breaks at sawtooth crashes. At 13.6s the predicted peak core averaged $p_\alpha$ is half the maximum calculated for this scan. In 42856 at 14.1s the peak predicted ratio $p_\alpha/p_{ext}$ is near 0.2.

The electron channel time-integrated the beam fueling and electron heating source rates during the energy ramp-up phase. The rates of increases of the volume-integrated $n_e$(x) and the volume- and time-integrated NB electron source rates within x are nearly equal. The agreement of rates holds both for the core (x$\simeq$0.2) and larger x, suggesting that the TRANSP beam deposition calculation is accurate and the electron flux is negligible, or that an inward pinch and outward diffusivity transport were canceling each other.

Similarly the volume integrated electron energy $\int_0^1 dV w_e$ increased at a constant rate $\simeq$2MW throughout the ramp-up. TRANSP uses measurements to calculate the heating and loss profiles. The dominant terms for electrons are $p_e \equiv p_{be} + p_\alpha + p_{oh} + p_{ie} - p_{cond_e} - p_{conv_e}$. The radiation emission profile $p_{rad}$ is not available, so TRANSP is used to predict the line, bremsstrahlung, and synchrotron emission. Their total volume-integrated rates scale in time approximately as the measured total emission, but are lower by about a factor of 0.8. The convected and conducted power losses $p_{\mathrm{cond_e}}$ and $p_{\mathrm{conv_e}}$ are TRANSP outputs. To study $p_\alpha$ consider core $\int_0^x dV w_e \simeq \int_0^x \int_{12}^t dt dV p_e$ integrated to x=0.2 or 0.3. This is dominated by $p_{\mathrm{be}}$. During the end of the ramp-up and into the flat-top phases the terms $p_{\mathrm{oh}}$, $p_{\mathrm{rad}}$, $p_{\mathrm{ie}}$, $p_{\mathrm{cond_e}}$, $p_{\mathrm{conv_e}}$, and $p_\alpha$ contribute roughly equally, except for the $f_{\mathrm{T}}$=0 discharge where $p_\alpha$ is negligible. This makes it is difficult to accurately quantify $p_\alpha$. The total thermal ion stored energy $\int dV w_{\mathrm{i}}$ increased at the same rate, also $\simeq$2MW. The dominate terms are: $p_{\mathrm{i}} \equiv p_{\mathrm{bi}}$ - $p_{\mathrm{ie}}$ - $p_{\mathrm{cx}}$ -$p_{\mathrm{cond_i}}$ - $p_{\mathrm{conv_i}}$. One large uncertainty is the charge-exchange loss term $p_{\mathrm{cx}}$ which is estimated (from the estimated recycling rate) to be large near the edge. The $p_{\mathrm{cond_i}}$ and $p_{\mathrm{conv_i}}$ terms are large. In the mid-radius region near 13.6s $\chi_i$ is around 0.5 $\mathrm{m}^2/s$ and $\chi_e$ is less.

An important parameter from the modeling is the hydrogenic isotopic mass $< A >$ defined as ($n_{\mathrm{H}}$+2$n_{\mathrm{D}}$+3$n_{\mathrm{T}}$)/($n_{\mathrm{H}}$+$n_{\mathrm{D}}$+$n_{\mathrm{T}}$). This is calculated from species conservation using the beam neutral ionization rates and the wall recycling and gas fueling rates [7]. The core <A> is dominated by the beam neutral ionization rates. Across the scan $< A > \simeq f_{\mathrm{T}} + 2$. The time delay $\delta_t$ is another important parameter. Values in Table I tend to increase with $f_{\mathrm{T}}$ and $< A >$ as shown in Figs. 3-a). Neither $S_{\mathrm{n}}$ nor $p_\alpha$ are linearly correlated with $\delta_t$ as shown in Figs. 3-b). This suggests that increasing core isotopic mass is the cause of longer $\delta_t$, and thus contributing to the higher core $T_e$ and $T_i$.

In conclusion $< A >$ is strongly correlated with the increase of $\delta_t$, and the delayed crashes allowed $T_e$ and $T_i$ to obtain higher values. At constant time $T_i$ was higher for discharges with higher $< A >$. Thus longer $\delta_t$ and $p_{\mathrm{ie}}$ could explain the higher $T_e$. The computed $p_\alpha / p_{\mathrm{ext}}$ is up to 0.2, but electron power balance in the core shows that multiple terms with uncertainties are comparable to $p_\alpha$.

Work supported in part by the US DoE contract No. DE-ACO2-76-CHO3073, and carried out within the framework of the EUROfusion Consortium, and has received funding from the Euratom research and training programme 2014-2018 under grant agreement No 633053. The views and opinions expressed herein do not necessarily reflect those of the European Commission.

[a] *Princeton University, Princeton, NJ 08543, USA*

[b] See the Appendix of F. Romanelli, *et al.*, Proceedings of the 25th IAEA Fusion Energy Conference 2014, Saint Petersburg, Russia.

[1] G.Taylor, J.D.Strachan, R.V.Budny, and D.R.Ernst, Phys. Review Lett. **76** 2722 (1996).
[2] P.R. Thomas, P Andrew, B.Balet, D. Bartlett, J. Bull, B.de Esch, *et al.*, Phys. Review Lett. **80** 5548 (1998)
[3] R.V.Budny, J.G.Cordey, "Core fusion power gain and alpha heating in JET, TFTR, and ITER", to be submitted
[4] R.V.Budny, D.R. Ernst, T.S.Hahm, D.C.McCune, J.G.Cordey, *et al.*,, Phys of Plasmas, **7** 5038 (2000)
[5] M.F.F. Nave, N.N. Gorelenkov, K.G. McClements, S.J. Allfrey, *et al.*,, Nucl. Fusion **42** 281 (2001)
[6] G McKee, R.Fonck, B.Stratton, R.Bell, R.Budny, C. Bush, B. Grek, *et al.*, Phys. Review Lett. **75** 649 (1995)
[7] R.V.Budny, M.G.Bell, D.K.Mansfield, J.D.Strachan, S.Zweben, *et al.*,, $21^{st}$ EPS, Montpelier, France (1994)

| Discharge | $f_T$ | $\delta_t$ [s] | $< A >$ (14.0s) | $p_\alpha$ [$10^{-2}$MW/m$^3$] | $n_\alpha$ [$10^{17}/m^3$] |
|---|---|---|---|---|---|
| 40365 | 0.0 | 0.9026 | 1.99 | 0.0 | 0.0 |
| 42870 | 0.27 | 0.9515 | 2.24 | 2.1 | 0.38 |
| 42856 | 0.52 | 1.4565 | 2.55 | 4.7 | 1.02 |
| 42847 | 0.88 | 1.4015 | 2.49 | 3.6 | 0.88 |
| 42840 | 1.00 | 1.6100 | 2.77 | 2.1 | 0.40 |
| 43011 | 1.00 | 1.6217 | 2.91 | 0.7 | 0.17 |

Table 1: $f_T$, *time delay* $\delta_t$ *to 1st significant crash, core (volume-averaged to x=0.2) hydrogenic mass,* $p_\alpha$ *and number of fast alpha ions before the 1st significant crash.*

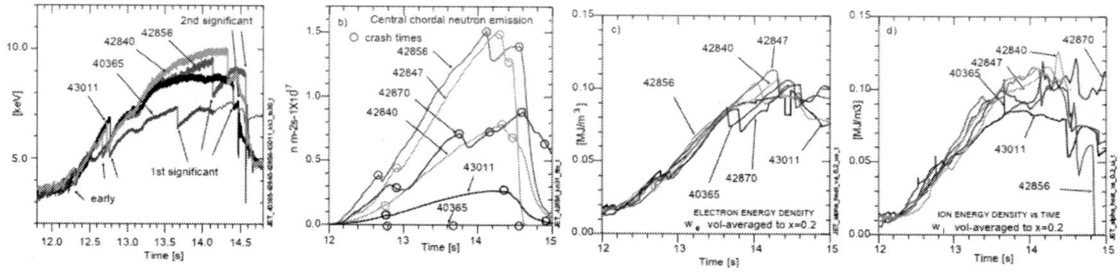


Figure 1: a) Examples of sawtooth / minor disruption crashes in ECE signals of the discharges with $f_T$ = 0.0, 0.52, and 1.0 from the alpha heating scan; significant crashes / minor disruptions seen after 13.6s; b) Sawtooth crash effects in the measured central chordal neutron emission rates; Time evolution of the core volume-averaged c) electron and d) thermal ion energies.

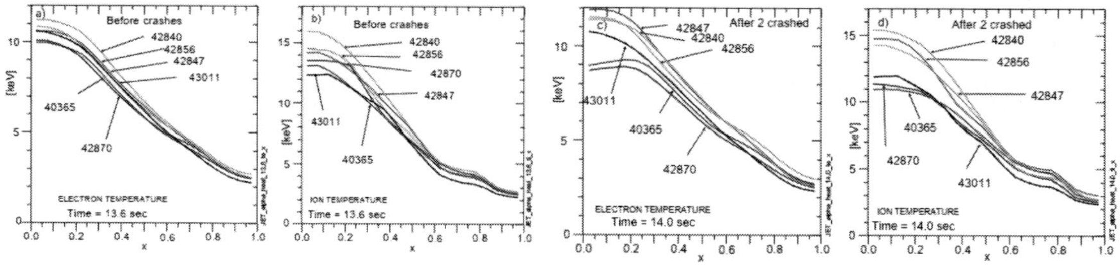


Figure 2: Profiles of $T_e$ and $T_i$ profiles at 13.6s (before the first significant sawtooth crash) and 14.0s (after two of the discharges experienced their first significant sawtooth crash).

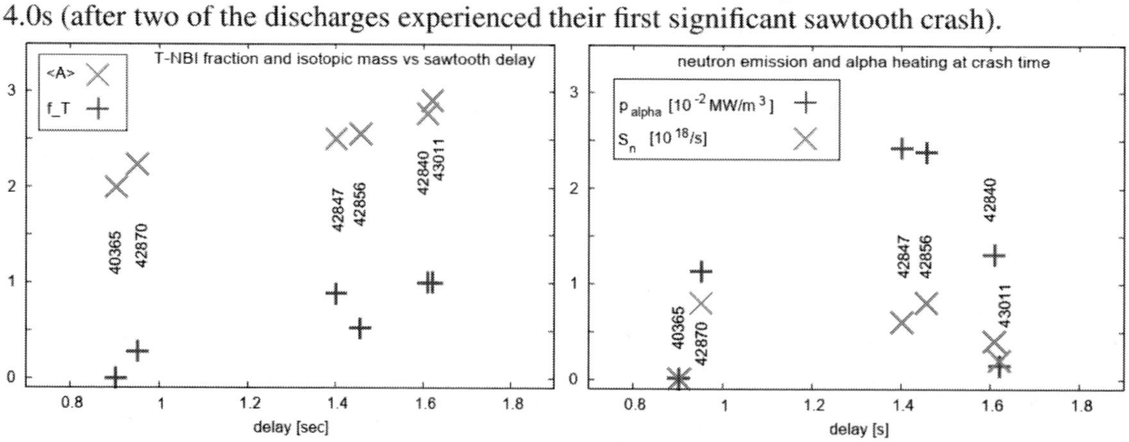


Figure 3: Correlations of delay times to the first significant sawtooth crash with a) core hydrogenic mass and tritium fraction of NB power; b) neutron emission and core alpha heating at the 1st significant crash times.

# Experimental characterization of *M-Mode* in JET tokamak

N. Vianello[1], E. R. Solano[2,3], E. Delabie[4], J. Hillesheim[5], D. Refy[6] , S. Zoletnik[6]
P. Buratti[7], J. E. Boom[8], R. Coelho[9], A. Figueiredo[9], E. Lerche[10], L. Meneses[9], F. Rimini[5]
ACC Sips[11], G. Artaserse[7], E. Belohony[3] and the JET Contributors *
*EUROfusion Consortium, JET, Culham Science Centre, Abingdon, OX14 3DB, UK, [1] EPFL CRPP CH-1015 Lausanne, Switzerland [2]EUROfusion PMU, Culham Science Centre, Abingdon, [3] Laboratorio Nacional de Fusion, CIEMAT, Madrid, Spain [4]ORNL, Oak Ridge, TN, USA [5] CCFE, Culham Science Centre, Abingdon UK, [6]Wigner Research Centre for Physics, Budapest, Hungary, [7]ENEA for EUROfusion, Frascati, Italy, [8] IPP Garching, Germany [9]Instituto de Plasmas e Fusao Nuclear IST, Portugal, [10]LPP-ERM/KMS, Brussels, Belgium [11] European Commission, Brussels, Belgium, * See the Appendix of F Romanelli et al., Proc. 25th IAEA Fusion Energy Conference 2014, St Petersburg, Russia*

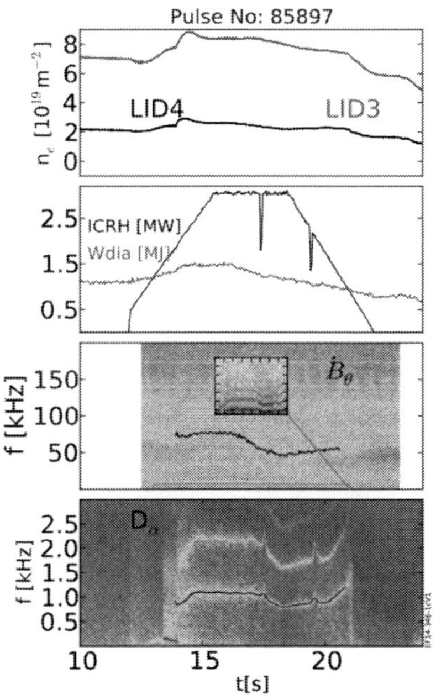

Figure 1: From top to bottom: Core(red) and edge(black) line integrated density, ICRH power (black) and $W_{dia}$ (red), spectrogram of $\dot{B}_\theta$ and spectrogram of $D_\alpha$

In recent years detailed investigations on the L-H transition in tokamaks have revealed that at power close to the L-H power threshold an intermediate oscillatory phase between the L and H phases exists [1]. This phase, generally appearing prior to a final transition to a proper H-Mode, is characterized by *dithering cycles* with periodic oscillations between high and low confinement regime and it has been confirmed in a variety of devices [2–5]. In all these devices the temporal dynamics has been often association to limit cycle oscillations (LCO) describing the interaction between turbulence (generally electrostatic turbulence is considered) and $\mathbf{E}\times\mathbf{B}$ flow, although some of the machines have been observing also a magnetic oscillation during these phases [2, 6, 7]. It is in any case accompanied by periodic oscillations of the pedestal profiles. A weak ELM-free H-Mode regimes, dubbed *M-Mode* has been observed at JET [8], with a clear increase of pedestal density and a weak temperature pedestal. A peculiarity of the M-Mode is the presence of a clear low-frequency coherent magnetic oscillation, with periodicity $(n,m) = (0,1)$ around 1-2 kHz. It has been observed in the pedestal region, as confirmed by ECE, Reflectometer and Li-Beam diagnostic, but it also modulates particle and heat flux at the target as confirmed by $D_\alpha$, Langmuir probes and fast infrared cameras.

The present contribution will continue the description of the experimental observation associated to this state, highlighting the existence of an associated high frequency magnetic oscillations and the observed modulation of edge and SOL density gradient. In Figure 1 a typical ICRH heated pulse with the appearance of the M-Mode is shown. The M-mode oscillation is clearly visible in the spectrogram of the $D_\alpha$ in the bottom panel where the dominant frequency around 1 kHz and the second harmonic are shown. The mode appears whenever the ICRH power is increased and it is accompanied by a modest increase in the energy content and in the edge and core density. In the same plot the third panel shows the spectrogram of a magnetic pick-up coil located in the HFS. The inset amplify the evolution of the spectrogram in the lower frequency branch where the M-Mode appears but also a second branch at higher frequency (50-80 kHz) is clearly evident. This high frequency branch (HFB) exhibits similar spectral properties with respect to the dominant low frequency oscillation as shown in Figure 2. The HFB has a clear $n = 0$ toroidal periodicity and it is worth noting that these broad-band oscillation is present even before the appearance of the M-mode, marked with a vertical white line. At the appearance of the M-Mode the HFB branch starts pulsating at the characteristic frequency of the M-mode as shown in the bottom panel, where the band-pass filtered $\dot{b}$ signal obtained from two poloidally separated probes are shown. We have marked the typical periodicity of the M-Mode with an arrow and in the smaller inset we provide a zoom on the two signal to highlight that also the HFB maintain the $m = 1$ periodicity. As stated at the beginning the M-Mode is generally observable in different diagnostic apart from the magnetic, both in the pedestal and in the Divertor region. In the top panel of figure 3 the Power Spectral Density (PSD) of the magnetic field (in the HFS of the tokamak, above the equator), is shown together with the same quantity calculated for a fast ECE channel located at the top of the pedestal the reflected phase of the correlation reflectometry $\Phi_{rfl}$. The dominant frequency of the M-Mode, located near 1.5 kHz for this pulse, is clearly visible in all signals together with higher harmonics. The proper phase between magnetic and kinetic fluctuations (i.e density and temperature) has been analyzed in the frequency

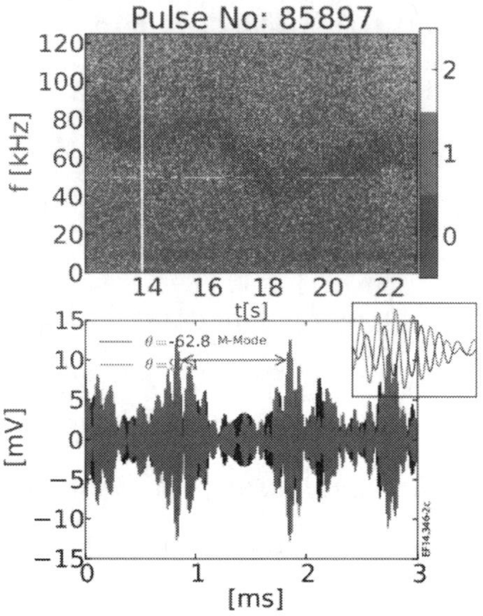


Figure 2: Top:Toroidal mode spectrum. Bottom: zoom on two $\dot{b}$ probe, located at different poloidal position, pass-band filtered between 60 and 80 kHz. The inset magnify the two signals to highlight the poloidal periodicity

domain and shown in the same figure 3.

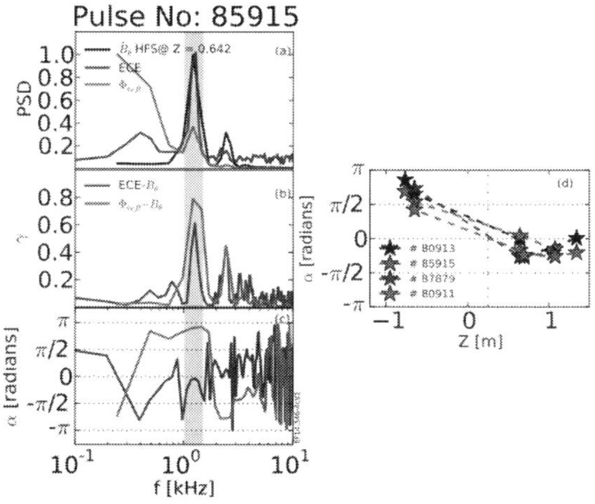


Figure 3: (a) PSD of $\dot{B}_\theta$ (black), ECE Channel (blue), $\Phi_{refl}$(red). Coherence (b) and Phase (c) between integrated magnetic and ECE (blue) and magnetic Reflectometer (red). (d) Phase between ECE and Magnetic field at different Z computed at the M-Mode frequency

The magnetic signal has been integrated with a proper high-pass filtering above 50 Hz and coherence and phase with the ECE and Reflectometer have been computed. The observed high coherence at the mode frequency between magnetic and kinetic signals, ensures that the phase estimate is reliable in this frequency range. Actually, consistently with the idea that the M-Mode is associated to a small up-down motion of pedestal plasma surface [8], the phase is found to depend on the $Z$ value of the chosen probe. In the same Fig. 3 panel (c) we show the phase, between ECE and magnetic field, computed at the M-Mode frequency as a function of the vertical position for a series of pulses. Considering the ECE LOS is located at $Z = 0.248$ we can conclude that, within the uncertainty due to the poloidal resolution, the phase between $T_e$ and $b_p$ is close to zero. We have checked that the same consideration holds for the density oscillations suggesting we have a pressure perturbation in phase with $B_\theta$.

Investigation with a Li-Beam reveals that the entire pedestal gradient is oscillating at the frequency of the M-Mode. The time evolution of the entire pedestal and SOL density profile has been computed using Li-Beam data and a combination of hyperbolic tangent and linear fit (corresponding to the pedestal and SOL region respectively). It has been observed that there is no modulation of the pedestal position neither of the pedestal height, whereas the gradient is found to oscillate at the M-Mode frequency, as confirmed by the figure 4 where phase and coherence between $\nabla n_e$ and the integrated magnetic field computed at different vertical position are shown. As done for the temperature and density we have

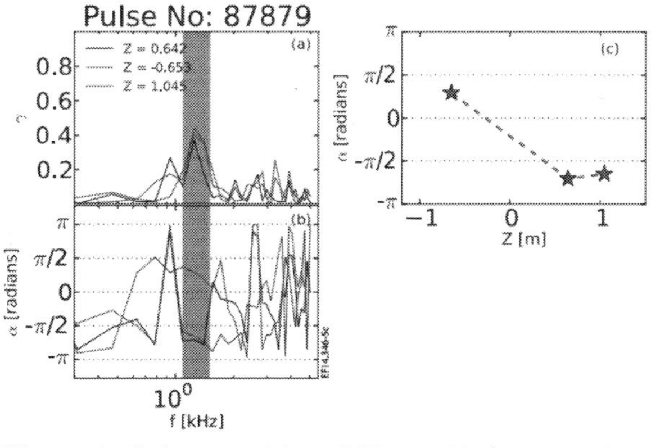


Figure 4: Coherence (a) and Phase (b) between density gradient and magnetic field computed at different poloidal position. (c) Phase computed at the M-Mode frequency as a function of vertical position

also computed the phase at the M-Mode frequency, and 4 (c) shows this phase as a function of the vertical position of the probes. We can conclude that the density gradient is almost in quadrature with the magnetic field.

Finally, hints of the scaling of the Mode frequency with the poloidal Alfvén speed have already been provided [8]. A further consistent observation has been obtained by comparing similar discharges obtained during L-H threshold investigation in Hydrogen and Deuterium plasmas [9]. The two sets have been obtained comparing similar discharges with $I_p = 1.7$MA and $B_T$=1.8T. The mode frequency has been tracked in time and the scaling with the time evolution of the poloidal alfven speed as computed at $\rho = 0.97$ is shown in the bottom panel of figure 5. Apart from a slight deviation at lower frequency, where we should take into account the feedback system reacting to vertical oscillation at JET, a linear trend is visible, consistent with the poloidal Alfvén speed scaling with an appropriate isotope dependence.

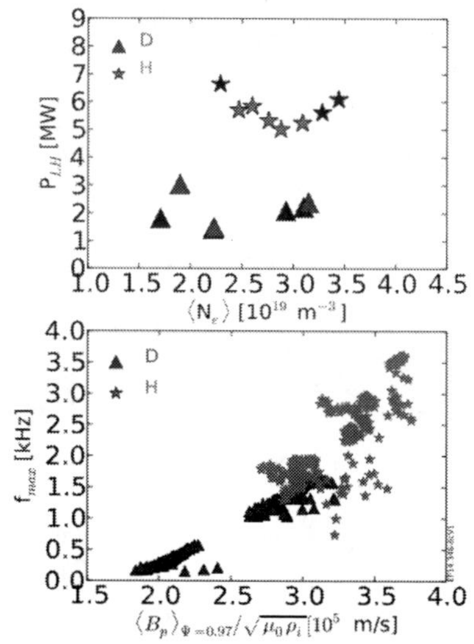


Figure 5: Top: L-H Power threshold vs density for Hydrogen and Deuterium discharges. Bottom: scaling of the M-Mode frequency vs Poloidal alfven speed. The color code is consistent with the top panel

Concluding further investigation of the M-Mode on JET tokamak is reported. The oscillation of the poloidal magnetic field is in phase with density and temperature. An higher frequency broadband fluctuation is observed with the same symmetry of the M-Mode. At the appearance of the M-Mode this HFB starts pulsating. The pedestal gradient is found to oscillate at the mode frequency, resembling similar observation in the dithering or I-phase. The dynamical interplay with the density gradient is presently under investigation. Finally by considering an isotope scaling the dependence of the M-Mode on the poloidal alfven speed has been confirmed.

This work has been carried out within the framework of the EUROfusion Consortium and has received funding from the Euratom research and training programme 2014-2018 under grant agreement No 633053. The views and opinions expressed herein do not necessarily reflect those of the European Commission

[1] L Schmitz et al., Physical Review Letters **108**, 155002 (2012).

[2] G. S. Xu et al., Nuclear Fusion **54**, 103002 (2014).

[3] J Cheng et al., Nuclear Fusion **54**, 114004 (2014).

[4] S. H. Müller et al., Physics of Plasmas **21**, 042301 (2014).

[5] T Estrada et al., Nuclear Fusion **55**, 063005 (2015).

[6] K. H. Burrell et al., Physics of Fluids B: Plasma Physics **2**, 1405–7 (1990).

[7] Y Xu et al., Plasma Physics and Controlled Fusion **57**, 014028 (2015).

[8] E. R. Solano et al., in 40th EPS Conference on Plasma Physics (2013), P4.111.

[9] E Delabie et al., in Proceedings of the 24th iaea fec conference (2014), EX/P5–24.

# ELM occurrence times in relation to the phase evolution of global measurements in JET

S C Chapman[1], R O Dendy[2,1], N W Watkins[1,3], T N Todd[2], F A Calderon[1] and JET Contributors*

*EUROfusion Consortium, JET, Culham Science Centre, Abingdon OX14 3DB, UK*

[1]*Centre for Fusion, Space and Astrophysics, Department of Physics, Warwick University, Coventry CV4 7AL, UK*

[2]*CCFE, Culham Science Centre, Abingdon, Oxfordshire OX14 3DB, UK*

[3]*Centre for Analysis of Time Series, London School of Economics, London WC2A 2AE, UK*

## 1. Introduction

Quantitative characterisation of ELMing processes using measured sequences of ELM occurrence times and inter-ELM time intervals is novel[1-6] and shows unexpected structure. Recently[4-6] we found that the signals from system scale diagnostics, the toroidally integrating full flux loops VLD2 and VLD3 in the divertor region of JET, contain statistically significant information on the occurrence times of intrinsic ELMs seen in the BeII signal. The ELMs occur preferentially when the full flux loop signals are at a specific temporal phase. Here we show, see also [6], that these signal phases contain information on the global build-up to an intrinsic ELM.

## 2. Results

We determine[4,6] each ELM occurrence time from the peak of the Be II signal. After each ELM there is a large amplitude oscillatory response in the VLD2 and 3 signals. There is a class of prompt ELMs[4] that occur within this initial response to the previous ELM. Non-prompt ELMs occur when the system's response to the previous ELM, as measured by VLD2 and 3, has died away. Fig.1 plots standardised raw timeseries for a prompt (left) and non-prompt (right) ELM.

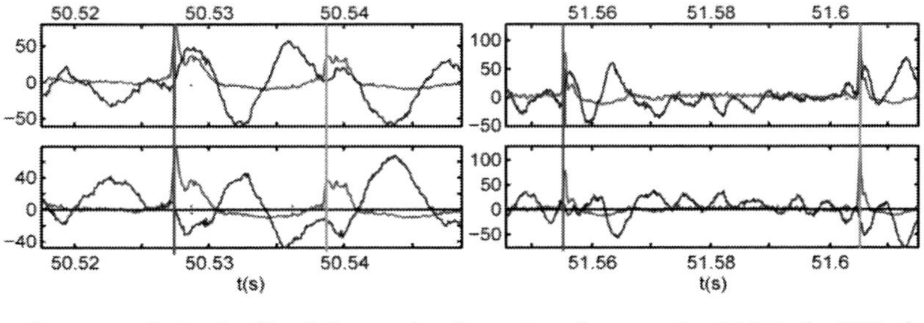


Fig.1: Examples of measured standardised timeseries for pairs of successive ELMs in JET plasma 83773: Be II intensity (red) and full flux loops (blue) VLD3 (upper panels) and VLD2 (lower panels). Left panels show a prompt ELM, and right panels a non-prompt ELM. To facilitate comparison we standardise the signal amplitudes, and the sign convention of the VLD2 and VLD3 signals assigns them opposite polarity. ELM occurrence times are marked by vertical red and green lines.

*See the Appendix of F Romanelli *et al.*, Proc. 25th IAEA Fusion Energy Conf. 2014, St Petersburg, Russia

Figure 2 shows the VLD3 signal traces for all prompt ELMs in JET plasma 83773 with inter-ELM time interval $\Delta t < 0.015$ s. For comparison, the VLD3 traces are shifted in time such that $t = 0$ is the occurrence time of the first ELM, and in amplitude such that the trace is at zero when $t = 0$. We mark on these traces the occurrence times of the first (red) and second (green) ELMs. Ordering the data in this manner demonstrates clustering of ELM occurrence times with respect to the phase of the VLD3 signal. These prompt ELMs all arrive after ~1½ to 1¾ oscillations of the VLD3 response to the previous ELM.

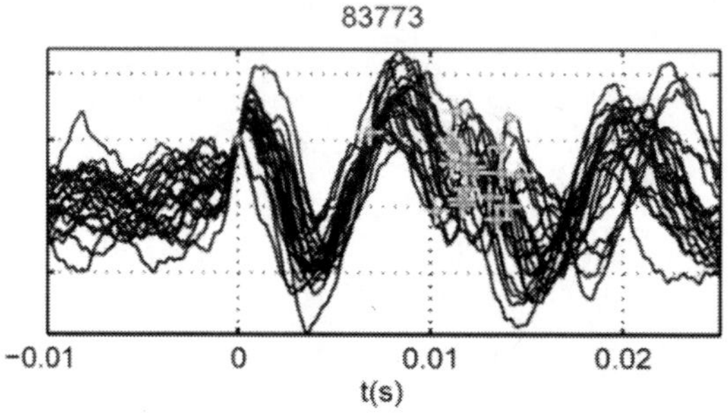


Fig.2: Prompt ELM occurrence times (green circles/yellow crosses) superimposed on corresponding VLD3 traces for ELMs in JET plasma 83773. All ELM pairs and traces are shifted such that $t = 0$ when the first ELM (yellow-filled red circles) of a pair occurs, and the VLD3 amplitude (normalised as in [6]) is zero at $t = 0$.

Figure 3 (left panel) plots the instantaneous amplitude and phase of the VLD3 full flux loop signal versus time for all the ELMs in JET plasma 83773. The instantaneous phase is determined from the measured signal as in [4]: the timeseries $S(t)$ is decomposed by Hilbert transform into an instantaneous amplitude $A(t)$ and phase $\phi(t)$ such that $S(t) = A(t)\exp[i\phi(t)]$. We choose the zero time for this plot as in Fig.2. The full flux loop phase difference is taken from the time of the first ELM in the entire sequence. There is clear phase bunching, which we find for *all* non-prompt ELMs during the flat-top H-mode in JET plasmas 83769-83775. ELMs are more likely to occur when the phase of the full flux loop signals is at a specific value. This phase relationship is stable throughout the plasma, that is for all ELMs that occur between 48.5s and 53.8s. We have established that this is not a trivial correlation by testing against randomly shuffled ELM arrival times and VLD2 and 3 timeseries in [4,6].

The upper panels of Fig.3 plot (green circles) the instantaneous amplitude and phase at the time of the second ELM as determined by the peak in the Be II signal. The rise time of the VLD3 response to the ELM is fast enough that it is possible that VLD3 is already responding to the ELM by the time the Be II reaches its peak. The VLD3 phase at the ELM time could then be dominated by its response to the ELM. To test this, we also plot on Fig.3 the amplitudes and phases of the VLD3 signal at times $dt$ before the second ELM; $dt$ is increased as we move down the figure. We then see that at $dt = 0$ (the Be II peak time) the instantaneous amplitude is for some ELMs comparable to the response to the previous ELM. At $dt = -1ms$ that is, 1ms before the ELM, the VLD3 amplitude is for most ELMs closer to the unperturbed level and for $dt = -2ms$ the VLD3 amplitude is generally at the unperturbed level. This confirms that the response to the ELM in the VLD3 occurs within 1-$2ms$ just before the ELM time as determined by the peak in the Be II. At this time, the non-prompt VLD3 phases are bunched, suggesting that there is an indication of the build-up to an ELM in these global, toroidally integrating signals.

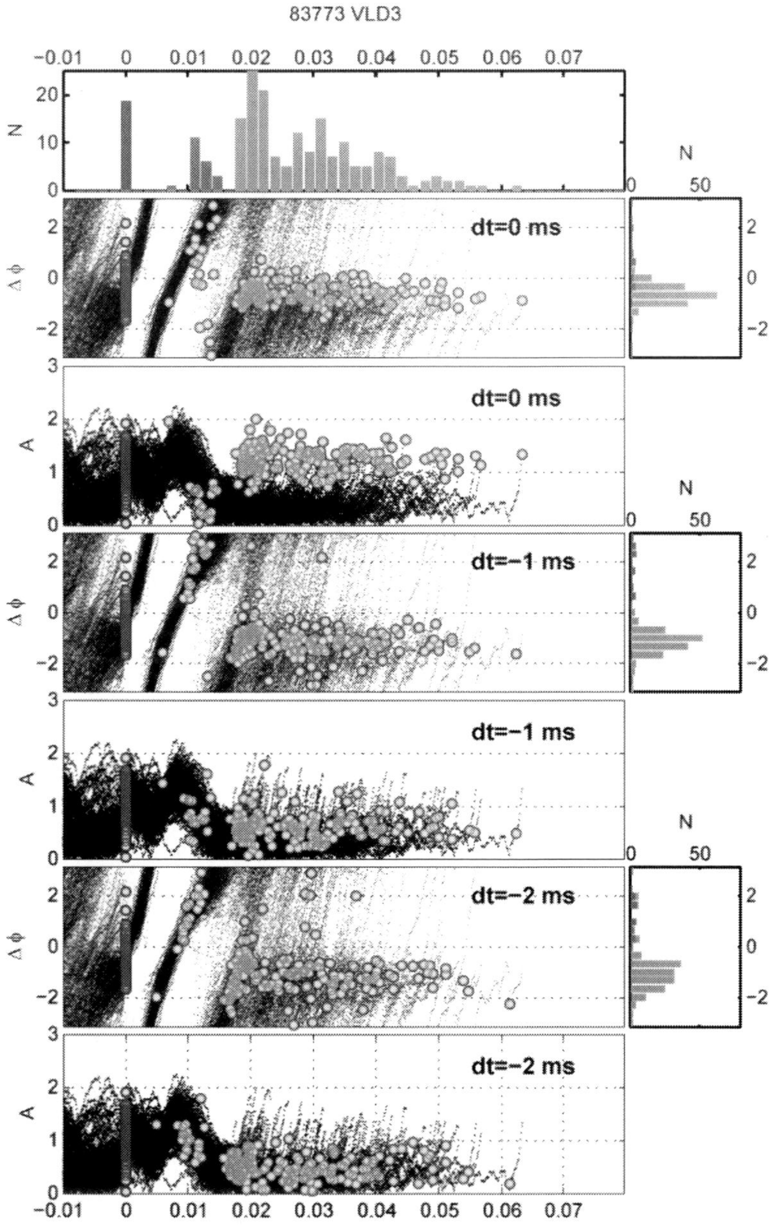


Fig 3: ELM occurrence times and VLD3 instantaneous amplitude and phase in the flat-top of JET plasma 83773. Each set of panels is as follows. Main Panel: VLD3 instantaneous amplitude $A$ (standardised units) and phase $\Delta\phi$, (modulo $2\pi$), plotted as functions of time, up to the occurrence time of the next ELM. The coordinates are time, and phase difference from the occurrence time of the first ELM in the sequence. ELM occurrence times are marked on each VLD3 trace with yellow-filled red circles (first ELM) and green circles (second ELM). Right Panel: histogram of VLD3 phase differences at the time of the second ELMs, for inter-ELM time intervals $\Delta t < 0.015$ (magenta) and for all ELMs (green). Top Panel: histogram of inter-ELM time intervals for prompt ELMs (magenta) and non-prompt ELMs (green). The instantaneous amplitude and phase are plotted at three times $dt$ before the second ELM.

The timeseries in Fig.2 also show that prompt ELMs occur at a specific phase in the VLD3 response to the previous ELM. This phase relationship is seen in the panel of Fig.3 with $dt$ = $2ms$. The prompt ELMs occur at a different phase to that of the non-prompt ELMs.

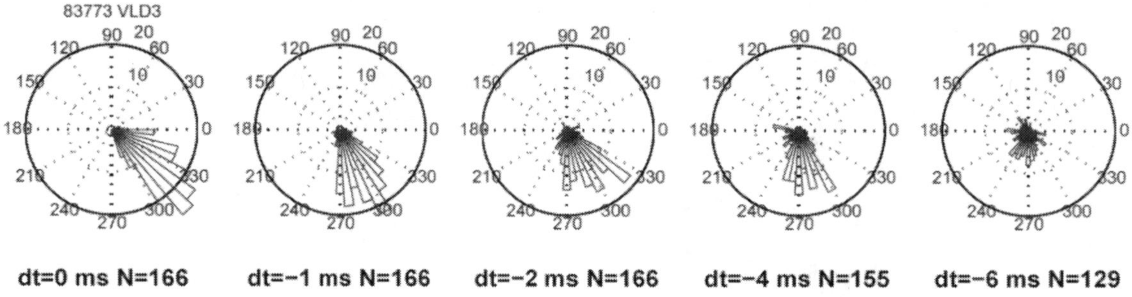


Fig 4: Polar histograms of VLD3 instantaneous phase in the flat-top of JET plasma 83773 at time $dt$ before the ELM time; from left to right, progressively earlier in time before the ELM. ELMs at $\Delta t - dt < 0.015s$ are excluded. $N$ is the number of ELMs remaining in the histogram.

Fig. 4 shows VLD3 phase at times $dt$ progressively before the ELM time (time of Be II peak). Phase bunching is still seen at $4ms$ before the ELM. Detailed analysis[6] shows that up to ~$5ms$ before the ELM, there is statistically significant phase bunching. This identifies a global build-up to an ELM as seen in the VLD3 full flux loop signal, which we also find in VLD2.

## 3. Conclusions

Here we show that the phase of the *global* full flux loop signal contains precursor information for non-prompt ELMs: during ~2-$5ms$ before each ELM, it aligns to the same value. Our results suggest a scenario for intrinsic ELMing where the plasma and its interacting environment, including the control system, together self-generate a global plasma perturbation such that the plasma is magnetically 'self-kicked'. This then leads[5] to an ELM if this global perturbation is sufficient to modify conditions at the plasma edge for instability, as first proposed in [5], and subsequently investigated in [7]. Prompt ELMs[4] occur at a specific phase within the initial full flux loop response to the previous ELM. This suggests an additional testable hypothesis: that compound ELMs may be a pattern of successive prompt ELMs. This is consistent with the observed[2] narrow spread in the time intervals between successive component ELMs in a compound ELM sequence. One would then expect a well-defined phase relationship between high time resolution full flux loop signals and the burst occurrence times within compound ELMs.

We thank the MPIPKS Dresden. This work has been carried out within the framework of the EUROfusion Consortium and has received funding from the Euratom research and training programme 2014-2018 under grant agreement No 633053 and from the RCUK Energy Programme [grant number EP/I501045]. The views and opinions expressed herein do not necessarily reflect those of the European Commission.

[1] J Greenhough, S C Chapman, R O Dendy and D J Ward, Plasma Phys. Control. Fusion **45** 747 (2003)
[2] F A Calderon, R O Dendy, S C Chapman, A J Webster, B Alper *et al.*, Phys. Plasmas **20** 042306 (2013)
[3] A J Webster, R O Dendy, F A Calderon, S C Chapman *et al.*, Plasma Phys. Cont. Fusion **56** 075017 (2014)
[4] S C Chapman, R O Dendy, T N Todd, N W Watkins *et al.*, Phys. Plasmas **21** 062302 (2014)
[5] S C Chapman, R O Dendy *et al.*, 41st EPS Conf. Plasma Phys. (2014) paper P1.101
[6] S C Chapman, R O Dendy *et al.*, Phys. Plasmas, submitted (2014)
[7] A J Webster *et al*, Symp. Fus. Technol. 2014 paper EFDA–JET–CP(14)05/06

# Particle source and edge transport studies in JET H-mode gas puff modulation experiments

A. Salmi[1], T. Tala[1], P. Mantica[2], A. Järvinen[3], L. Meneses[4], S. Mordijck[5], V. Naulin[6], J. Juul Rasmussen[6], J. Svensson[7], L. Giacomelli[2], R. Gomes[4], M. Groth[3], T. Koskela[3], C. Maggi[8], M. Maslov[8], G. Sips[9], H. Weisen[10] and JET contributors*

EUROfusion Consortium, JET, Culham Science Centre, Abingdon, OX14 3DB, UK
*[1]VTT, Espoo, Finland; [2]IFP, CNR-ENEA, Milan, Italy; [3]Aalto University, Helsinki, Finland; [4]IPFN, IST, Universidade de Lisboa, Portugal; [5]College of William & Mary, Virginia, USA; [6]DTU Physics, Lyngby, Denmark; [7]IPP, Greifswald, Germany; [8]CCFE, Abingdon, UK; [9]EFDA CSU, Culham, UK; [10]CRPP, Lausanne, Switzerland.*

Gas modulation experiments in H-mode plasmas featuring a scan in collisionality to study particle transport and sources at the plasma edge have been carried out on JET [1,2]. The local electron density response to the gas injection was measured with reflectometer and Thompson scattering diagnostics close to the midplane. Modulation amplitudes below 1% (in the core) are reliably measured thus allowing minimal plasma disturbance and the possibility to use data from multiple harmonics. The linearity of the electron density response was verified in identical plasmas by having different gas modulation amplitudes.

The set of experiments were run in a corner configuration with strike points near the pumping ducts and gas modulation from top of the machine approximately from a point source. Figure 1 illustrates the plasma shape and density modulation due to the gas injection for the shot #87420 ($B_T$=2.67T, $I_P$=2MA). The propagation of the perturbation is clearly seen in the reflectometer signal up to mid radius yielding data for up to three harmonics.

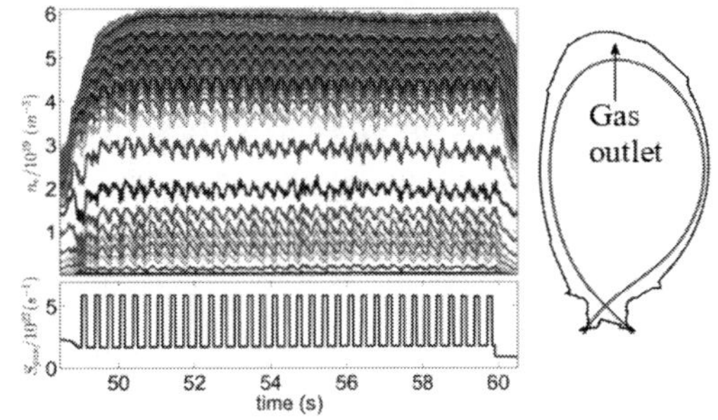


Figure 1 Experimental gas waveform and resulting density modulation. Top location (GIM7) was used for gas injection #87420.

Time resolved electron density measurements together with 1.5D transport modelling are used to clarify the mechanisms responsible for fuelling of the plasma. In particular we aim to understand whether plasma fuelling is due to inward convection or whether even in H-mode there could be significant source penetrating pedestal top. We assume that electron density obeys the following 1.5D equation with geometrical shape corrections included

$$\frac{\partial n_e}{\partial t} = \frac{1}{V'}\frac{\partial}{\partial \rho}\left[V'\left(\langle(\nabla\rho)^2\rangle D\frac{\partial n_e}{\partial \rho} - \langle\nabla\rho\rangle v n_e\right)\right] + S(\rho,t)$$

where $\rho$ is the radial coordinate (=square root of normalised poloidal flux), $\langle\ \ \rangle$ denotes flux surface average, S is the particle source (including both cold neutrals and neutral beam (NB) ions) and $V'$ is the radial derivate of the plasma volume. We assume time independent radial profiles for the diffusion D and convection $v$. Here, the electron density profile $n_e$ is measured as a function of time and the geometrical factors $\langle(\nabla\rho)^2\rangle$, $\langle\nabla\rho\rangle$ and $V'$ are obtainded from EFIT equilibrium solver. The unknown $\boldsymbol{D}, \boldsymbol{v}$ and $\boldsymbol{S}$ are solved using an iterative scheme where they are varied inside a non-linear optimisation routine until the simulated electron density best matches the experimental data.

The 3 Hz rectangular gas modulation with ~35% duty cycle yields density perturbation above the noise level for up to 3rd harmonic. Experimental data thus provides us one real valued equation from the steady state and three complex valued equations from the three harmonics totalling to seven equations (we effectively use only 5 equations). As we have only three unknown profiles we have more than sufficient information to solve the unknowns simultaneously. The steady state equation provides us a relation between $D, v$ and $S$ thus eliminating one unknown:

$$v = \frac{\langle(\nabla\rho)^2\rangle}{\langle\nabla\rho\rangle}\frac{\frac{\partial \bar{n}_e}{\partial\rho}}{\bar{n}_e}D + \frac{\int_0^\rho V'\bar{S}d\rho}{\bar{n}_e V'\langle\nabla\rho\rangle}$$

We use the following parametrisation for $D$ and $S$

$$D(\rho) = \frac{1}{2}(a_1 + a_2\rho^{a_3})\left(1 - \tanh\frac{\rho - a_4}{a_5}\right) + a_6$$

$$S(\rho, t) = S_{NBI} + a_7 \exp\left(\frac{1-\rho}{a_8}\right)(1 + a_9 S_t(t - a_{10}))$$

where the ten scalar coefficients $a_x$ are optimised. The experimental gas waveform $S_t(t)$ is normalised such that $\bar{S_t} = 1$ and $\Delta S_t = 1$.The neutral beam particle source $S_{NBI}$ is calculated with PENCIL. Figure 2 gives a graphical illustration of the free parameters.

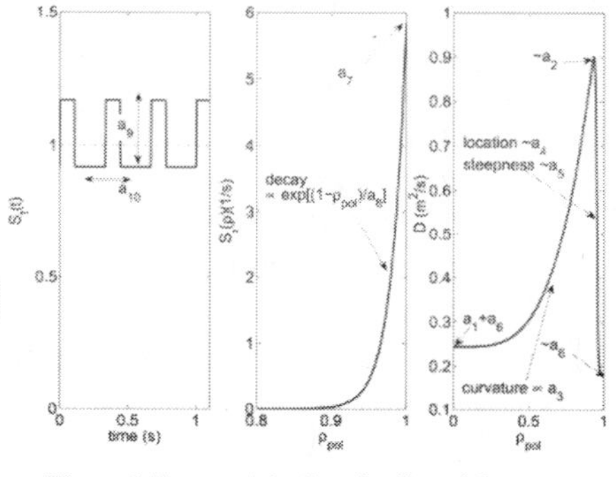

Figure 2 Parametrisation for D and S.

Figure 3 gives the result of the fitting when all parameters are allowed to be optimised (full black line) and when the source decay length (i.e. $a_8$) forced to be 8 $cm$ (dashed black line). We use reflectometer data for the three harmonic amplitudes and Thompson scattering diagnostic for the first harmonic phase profile. This is because at the moment we are still in the process of validating the phase profiles from processed reflectometer data. With this data set we are able to obtain a fit that reproduces the experimental data within the error bars ($\chi^2/N \sim 1$) thus improving on previous analysis [2]. Note that the experimental error bars include only statistical error of the data mapped on a fixed a coordinate system but not those that might arise due to mapping itself, other data processing or from other unknown sources. The small diffusion inside $\rho_{pol} < 0.6$ radius is essentially due to the rapidly decaying perturbation amplitude which is not reproduced with $D \sim 1\ m^2/s$. The resulting small diffusive flux $\Gamma_D$ is then mostly balanced by the NB particle flux thus eliminating the need for inward convection to explain the peaked density profile. For more details please see [1].

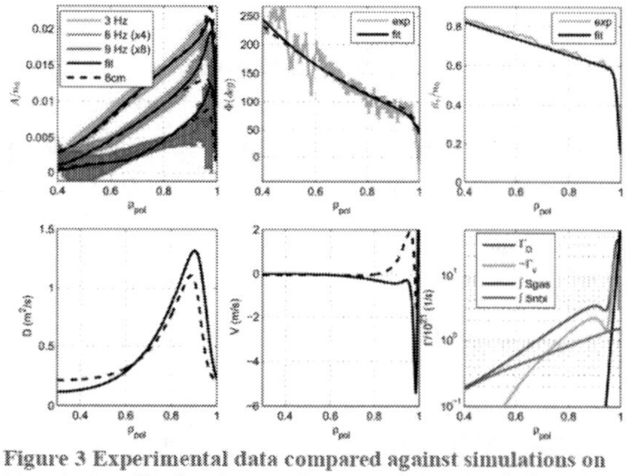

Figure 3 Experimental data compared against simulations on top row (amplitude, phase and average density) and the corresponding diffusion and convection profiles together with the assosciated fluxes on the bottom row. Full black line is for narrow source (1cm) and dashed line for deep source (8cm).

The best overall fit is found with a relatively narrow cold neutral ionisation profile ($a_8 \sim 1\ cm$) which in turn requires inward convection at the edge. However, ionisation source widths of up to 8 $cm$ are equally well fitted inside $\rho_{pol} < 0.9$ while outside they start to deviate. Interestingly, when source width is fixed to a value larger than $\sim 2\ cm$ the additional source going across the pedestal cannot be compensated by increasing diffusion but outward convection is required as shown in dashed line for the 8 $cm$ case in Figure 3. As outward convection appears less likely than edge localised inward convection this could indicate that the cold neutral ionisation source is not significantly penetrating the pedestal. Further gas modulation experiments in plasmas with more opaque SOL could potentially clarify this question. Regardless of the cold neutral source width it is the NB flux that contributes mostly to the density peaking in the core plasma. With the narrow source we found that the modulated incoming flux crossing the separatrix needed to fit the experiment is about 20% of the injected modulated gas and that the steady state flux is about twice the amount of the total gas fuelling. Typically these numbers have factor of 2 uncertainties.

In addition to core diagnostics the neutral sources were diagnosed with visible light divertor and wide angle cameras behind $D_\alpha$ and $D_\beta$ filters. They were operated at maximum available framerate ($60\,Hz$) for the full duration of the discharge. Figure 4 shows the wide angle view of the $D_\beta$ radiation. The bright spot at the top of the vessel is the gas injection location. One can observe that it is not only the actual injection point that features oscillations but that significant $3\,Hz$ component is present everywhere and especially at the divertor indicating that fuelling indeed modulates the recycling quite strongly. Furthermore, frame b) provides new insight on the time scales. The radiation peak and thus potentially also the maximum ionisation rate at the divertor takes place roughly $50\,ms$ after it appears at the top of the vessel. The time trace e), the spectrum f) and the error indicator d) illustrate periodicity and noise in the data. For long modulations, like in here, the time scales can possibly be estimated to less than $5\,ms$ accuracy.

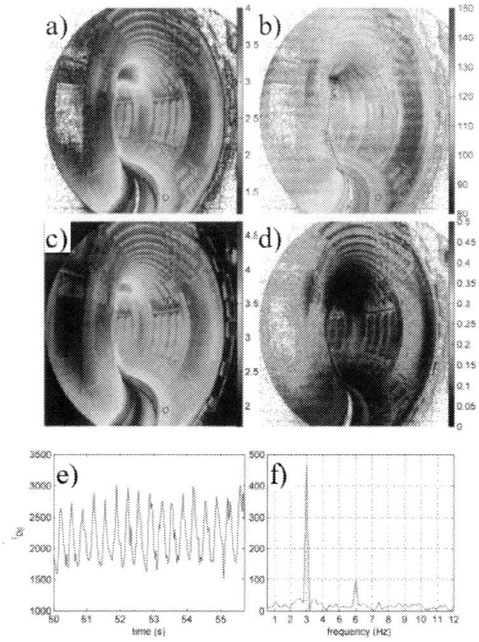

Figure 4 Wide angle camera $D_\beta$ radiation. a) 3Hz amplitude, b) 3Hz phase, c) steady state, d) error indication $\delta A/A$, e) intensity at the red circle and f) spectrum at red circle.

Figure 5 shows the tomographic inversion [3,4] of the divertor camera $D_\alpha$ image on a 2D plane. Also here amplitude and average intensity are plotted on log scale. While the recycling is highest at the targets one can also see a 'plume' on the inner divertor apron clearly synchronised with the gas modulation that cycles the upstream conditions. It also appears that the recycling at this location is generating some higher energy neutrals that are able to penetrate deeper into the confined plasma. Furthermore, the frame b) shows that on the inside of the X-point the emission is decreasing when modulated gas is on which indicates that the inner strike point leg periodically detaches creating very cold and dense plasma during the added gas. Moreover, it is seen that $D_\alpha$ radiation peaks on the outer divertor $10-20\,ms$ after it peaks on the inner divertor. With a connection length of the order of $50\,m$ this would correspond to a sound speed of a $0.1\,eV$ deuteron which shows that parallel flow only cannot explain such a long difference. The coherent phase and the

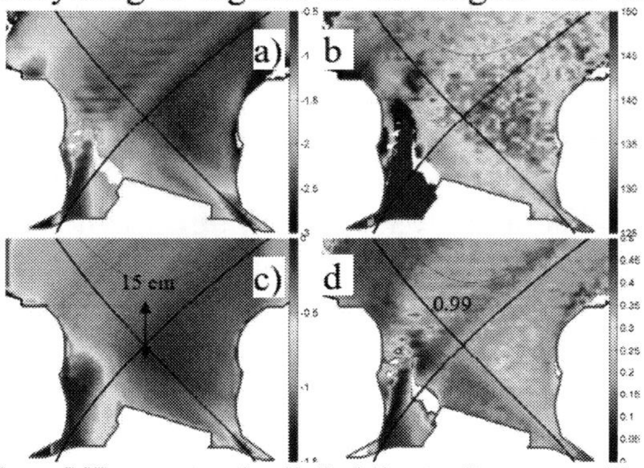

Figure 5 2D reconstruction [3,4] of divertor $D_\alpha$ camera with $\log_{10}$ colour scale. a) 3Hz amplitude, b) 3Hz phase, c) steady state, d) error $\delta A/A$. $\rho = 0.99$ surface is 1 cm from separatrix at midplane.

separatrix aligned amplitude together seem to suggest that cold neutrals are able to penetrate $10-15\ cm$ across the separatrix near the X-point, mainly on the LFS. However, as the plasma profiles steepen towards the midplane the penetration depth will also be reduced.

To better understand the recycling and ionisation profiles time dependent EDGE2D-EIRENE modelling was performed for roughly one modulation cycle. Figure 6 summarises the findings from this simulation. The underlying statistical noise and the lack of synthetic $D_\alpha$ and $D_\beta$ diagnostics are obviously presenting challenges for comparisons against the camera data. Nevertheless, one can notice that modelling suggests that ionisation inside the separatrix is strongly concentrated on the low field side of the plasma (Figure 6a). However, these simulations do not include cross-field drifts or currents, and, therefore, the in-out divertor asymmetries are likely to be underestimated. As a result, the fractional contribution of LFS in the pedestal fuelling profile is probably overestimated relative to the HFS fuelling. The width of the flux surface averaged ionisation profile inside the separatrix is estimated to be roughly 1 $cm$ on the outboard mid plane being consistent with both the divertor camera and the 1.5D modelling. The dynamics of the simulated ionisation EDGE2D-EIRENE are rather complicated as seen on the phase plot (Figure 6b) and can vary significantly between neighbouring cells. Generally, they also seem quite fast compared to wide angle camera images. One possible contributing factor here could be that the actual source is essentially a point source but when communicated from EIRENE to EDGE2D it is made axisymmetric. Thus, 3D effects might need to be included to be able to reproduce the measured emissions and their relevant time scales.

To summarise: we have presented experimental data and modelling from gas modulation

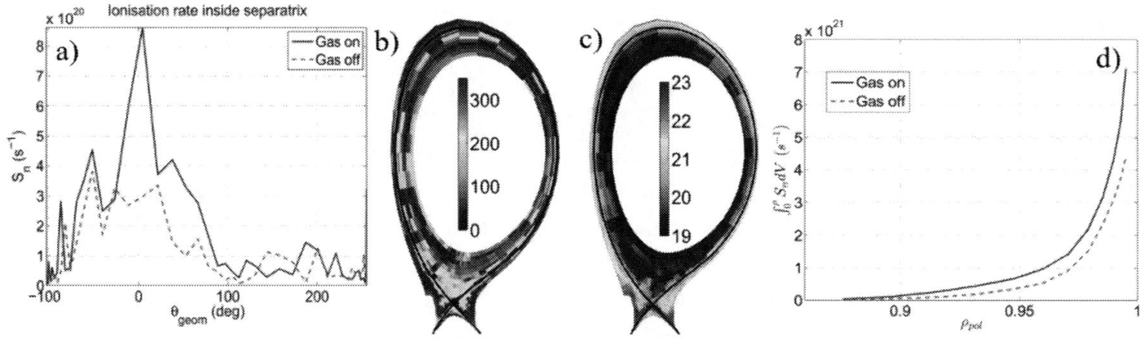


**Figure 6 Ionisation rates and their 5 Hz phase and amplitude 2D profiles from EDGE2D-EIRENE simulations : a) ionisation rate inside separatrix, b) phase delay w.r.t. gas on time, c) $\log_{10}$ amplitude and d) radial ionisation profile**

experiments that suggest that particle source inside separatrix is fairly narrow and it does not contribute much inside the pedestal top. Inward convection of the order of $5\ m/s$ at the plasma edge is needed to sustain the steep pedestal while the convection quickly drops to zero towards the core where small diffusion and NB source are responsible for the density peaking. But we also remind that as the error bars at the edge are large and potentially underestimated the wider source profiles cannot yet be ruled out with certainty.

[1] T. Tala et al. 2015 EPS, Portugal [2] A. Salmi et al. 2014 EPS, Germany
[3] J. Svensson 2011 EFDA–JET–PR(11)24 [4] J. Svensson et al 2007 Proc. IEEE WISP

* *See the Appendix of F. Romanelli et al., Proc. 25th IAEA Fusion Energy Conference 2014, St Petersburg, Russia*

# Dimensionless Collisionality Scans for Core Particle Transport in JET

T. Tala[1], P. Mantica[4], A. Salmi[1], C. Bourdelle[3], C. Giroud[4], J. Hillesheim[4], C. Maggi[4], L. Meneses[5], M. Maslov[4], S. Menmuir[4], S. Moradi[6], S. Mordijck[7], V. Naulin[8], H. Nordman[9], J. Juul Rasmussen[8], G. Sips[10], A. Sirinelli[3], M. Tsalas[11], H. Weisen[12] and JET contributors

EUROfusion Consortium, JET, Culham Science Centre, Abingdon, OX14 3DB, UK
[1]VTT, P.O. Box 1000, FI-02044 VTT, Espoo, Finland
[2]Istituto di Fisica del Plasma, via Cozzi 53, 20125 Milano, Italy
[3]CEA, IRFM, F-13108 Saint-Paul-lez-Durance, France
[4]CCFE, Culham Science Centre, Abingdon, OX14 3DB, UK
[5]Instituto de Plasmas e Fusão Nuclear, IST, Universidade de Lisboa, Portugal, IST, Lisbon, Portugal
[6]Ecole Polytechnique, CNRS UMR7648, LPP, F-91128, Palaiseau, France
[7]College of William & Mary, Virginia, USA
[8]Danish Technical University Physics, Lyngby, Denmark
[9]Chalmers University of Technology, Göteborg 41296, Sweden
[10]European Commission, Brussels, Belgium
[11]DIFFER, Nieuwegein, Netherlands
[12]CRPP, Lausanne, Switzerland
*See the Appendix of F. Romanelli et al., Proc. of the 25th IAEA Fusion Energy Conference 2014, Saint Petersburg, Russia

## 1. Introduction

Particle transport and fuelling are major open questions in understanding the ITER physics basics [1]. Extensive database studies on density peaking on several tokamaks showed that in the H-mode plasmas, collisionality is the most important parameter determining the density peaking [2,3]. The neutral beam fuelling source was found to be the second most important factor. On the other hand in L-mode, the plasma internal inductance $l_i$ was to found to be the key factor in determining the density peaking [4].

To verify the trends found in references [2-4], particle transport has been studied in JET by performing a dimensionless collisionality scan both in H-mode (ITER like wall) and L-mode (carbon wall) plasma. Gas puff modulation technique was exploited in all plasma discharges to obtain perturbative particle transport coefficients. Furthermore, density perturbation will shed light on the question from where this peaking is coming to understand details of the beam and neutral particle source. However, the analysis of the neutral particle source and the edge and SOL particle transport studies are reported in another paper in this conference [5].

## 2. Experimental Set-Up for Collisionality Scans and Gas Puff Modulation Technique

The H-mode discharges were run with the strike point in the corner (not located at vertical nor horizontal divertor tiles) so that the operational space to vary density was largest. The typical plasma parameters in this low triangularity baseline scenario plasmas were $B_t$=2.7T, $I_p$=2.0MA, $n_{e0}$=6x10$^{19}$m$^{-3}$, $T_{e0}$=3.5keV and $Z_{eff}$~1.3. The gas puff modulation was performed with a gas valve at the top of the machine at 3Hz frequency using rectangular waveform, the rate varying from 0 to 6x10$^{22}$s$^{-1}$ at 30% duty cycle. Another gas injection module at the divertor location was used to keep the volume averaged density constant.

In order to test the linearity of the gas puff modulation technique, the modulated gas puff rate was doubled for one of the discharges. The phase profile stayed the same while the amplitude doubled, thus giving confidence that fuelling from the gas puff modulation at this level is linear and does not perturb the plasma background transport. Therefore, in determining the transport coefficients, the assumption that the background is time-independent seems valid.

The plasma response (electron density modulation) is measured with a multi-band profile reflectometer capable of good spatial and temporal resolution [6] and high resolution Thomson Scattering. This data was found to agree well with respect to steady-state profiles and the amplitude, but the there is a significant difference in the modulation phase profile between the reflectometer and Thomson scattering diagnostics. This is issue is being reviewed for the time being.

## 3. Results from the Collisionality Scans

3-point collisionality υ* scans were performed both in JET H-mode and L-mode plasmas. In each scan, roughly a factor of 5 in υ* was achieved by scanning $I_p$ from 1.7MA to 2.5 MA and $B_t$ from 2.0T to 3.0T, respectively. The dimensionless parameters, $q$, $\rho^*$, $\beta_n$ and $T_i/T_e$ were matched very well within the H-mode scan, the difference between the shots being only a few % (<10% in worst case). This is illustrated in figure 1. Even the temperature gradient length $R/L_T$ was constant within the scan. The volume averaged density is very similar, within 5% between the three discharges, operationally keeping density constant when scanning the plasma current to such an extent is in general far from trivial in all metal device. While the volume averaged $n_e$ is very similar, the density profile shape is significantly more peaked at low υ*, as shown in figure 2 (left frame). The density peaking factor, defined here as $R/Ln_e$, increases in the inner core ($r/a$=0.3) from 0.3 to 2.7 and in the outer core ($r/a$=0.8) from 1.6 to 3.5 when the volume averaged υ* decreases from 0.47 to 0.09. Similar dimensionless 3-point υ* scan was executed with the carbon wall in L-mode by scanning υ* by a factor of 4. Unlike in H-mode, no change in $R/Ln_e$ was observed, indicating same particle transport coefficients in each shot, shown in figure 2 (right frame).

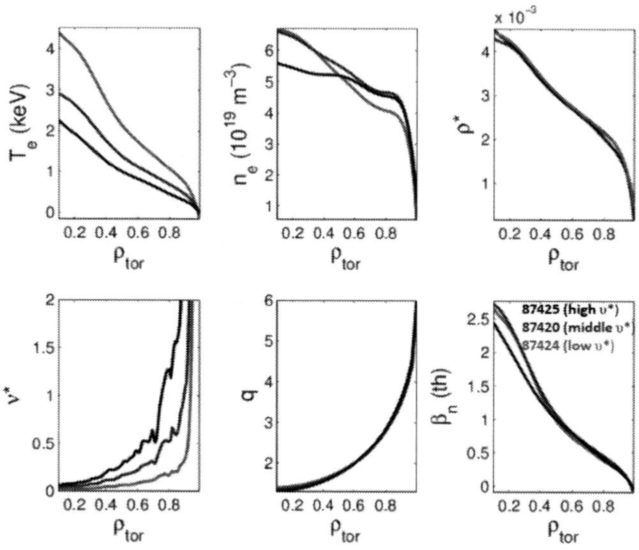


Figure 1. The main parameters in the JET dimensionless 3-point H-mode collisioality scan.

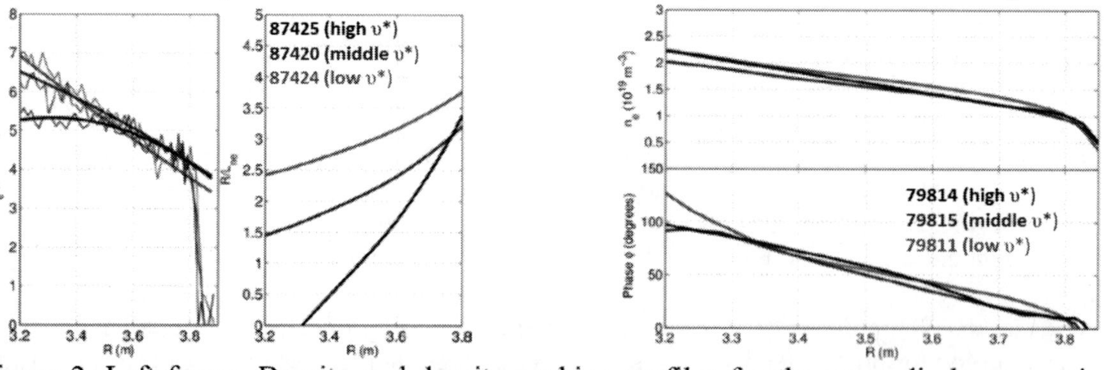


Figure 2. Left frame: Density and density peaking profiles for the same discharges as in figure 1. Right frame: Density profiles and the phase profiles of the density modulation for the dimensionless L-mode collisionality scan.

The results from both of the dedicated H-mode and L-mode scans are consistent with the earlier density database results in L-mode [4] and H-mode [2,3] with respect to collisionality dependence of density peaking. Opposed to JET, on DIII-D no $\upsilon^*$ dependence on collisionality was reported either H (or L-mode) within the same range of the volume averaged $\upsilon^*$ [7]. It is to be noted that the scan covered only a factor of 2 and besides NBI heating, additional ECH was used in accessing the low $\upsilon^*$ point.

4. Particle Transport Coefficients and a Comparison to Theory

The key question is what the fraction of inward convection versus NBI particle source and neutral particle source is in contributing to the observed relatively strong density peaking. One can point out that the NBI fuelling rate increases from $0.8 \times 10^{21}$ 1/s to $2.3 \times 10^{21}$ 1/s within the collisionality scan. Therefore, one has to obtain the particle transport coefficients from the gas puff modulation data to distinguish the relative roles of fuelling versus transport. The experimental data from the gas puff modulation analysis indicates that a narrow neutral fuelling profile and peaked at the edge is most consistent with all transport analyses and thus does not contribute to density peaking even at $\rho$=0.8 [5]. SOLPS modelling is on-going for more precise neutral source calculation and edge particle transport analysis.

The transport coefficients have been determined using ASTRA transport code by choosing the diffusion D and convection v in such way that the least square error between the simulated and experimental steady-state $n_e$, and the modulation amplitudes $A(n_e)$ and phases $\Phi(n_e)$ is minimized. This technique is analogous to the analysis method of the JET NBI modulation experiment [8]. The beam particle source is calculated with PENCIL. The results are shown in figure 3 for JET pulse 87420 which is the intermediate shot in the collisionality scan. Steady-state and modulation amplitude $A(n_e)$ of $n_e$ are well reproduced with the inferred D and v on the bottom right panel. Also the phase of $n_e$ modulation $\Phi(n_e)$ is well reproduced, but only against HRTS data. No combination of D and v can reproduce the phase from the reflectometry. There is an issue identified with the analysis of the profile reflectometry data and this is being assessed right now. The same set of D and v reproduces also the second harmonic data and is thus an extra confidence on the accuracy of the analysis. To quantify the relative level of density peaking originating from the fuelling sources versus inward pinch, the density peaking factor Rv/D by using the determined D and v are compared to the measured peaking $R/L_{ne}$. These D and v from Astra optimization include the NBI fuelling source term and therefore, they give the peaking factor due to only transport effects. This is shown in figure 4.

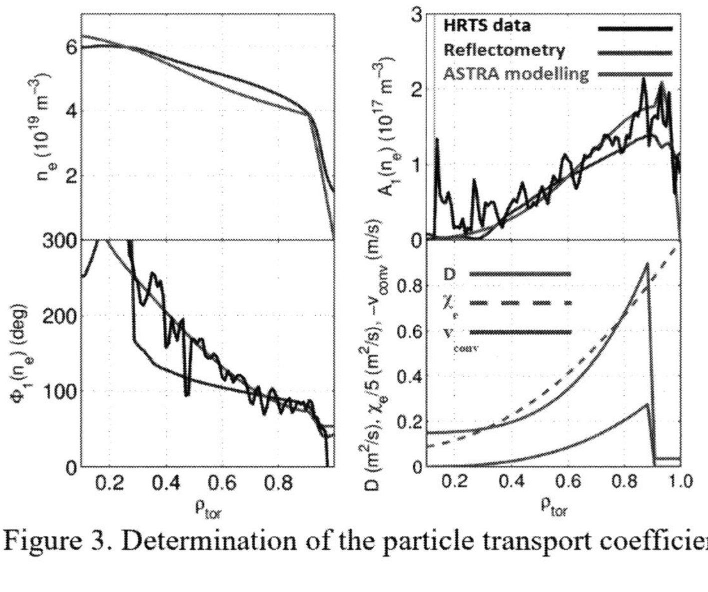


Figure 3. Determination of the particle transport coefficients for pulse no 87420.

Furthermore, the density peaking factor from linear gyro-kinetic with GYRO and fluid runs with the Weiland model were added for comparison. GYRO runs were linear, electro-magnetic, and with drift-kinetic electrons at $k_y$=0.3. The results turned out to be sensitive to the choice of $k_y$ [9]. Weiland fluid runs were also at $k_y$=0.3. $R/L_n$ was searched for the root of the interpolated normalized electron flux as a function of $a/L_{ne}$ by assuming zero particle flux.

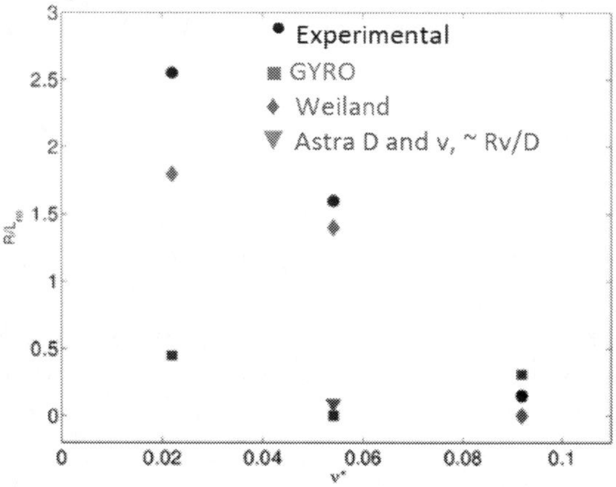


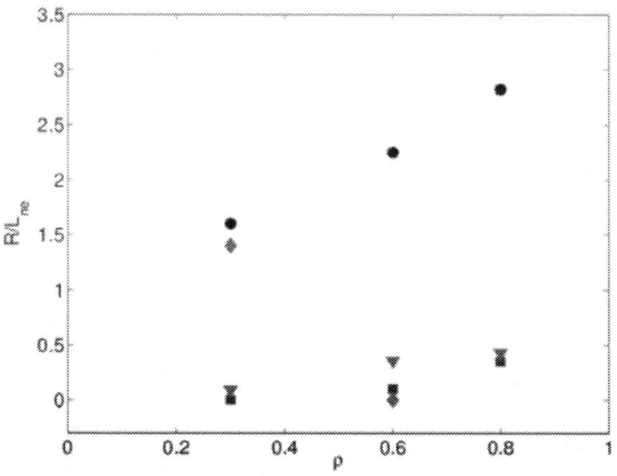


Figure 4. Left frame: Density peaking as a function of $\upsilon^*$ from the 3-point $\upsilon^*$ scan at $\rho$=0.3 and right frame: density peaking for JET pulse no 87420. Both frames include the experimental, Astra inferred and the modelled GYRO and Weiland density peaking profiles.

There are two very important messages in figure 4. Firstly, the difference between the measured experimental $R/L_{ne}$ and the ones from Astra that include the NBI fuelling contribution show that the NBI fuelling seems to play a major role in contributing to the density peaking. In fact, for this JET, fuelling is the dominant factor as the part below the green triangles are due to inward pinch and the part above with respect to the experimental ones (black points) are due to NBI fuelling. Secondly, GYRO with the assumptions discussed above predicts very low density peaking, which is actually consistent with the density peaking factors from the transport modelling (green triangles).

## 5. Summary

Density peaking increases in the inner core ($r/a$=0.3) from 0.3 to 2.7 and in the outer core ($r/a$=0.8) from 1.6 to 3.5 when $\upsilon^*$ decreases from 0.47 to 0.09 in JET H-mode plasmas while density peaking does not depend on $\upsilon^*$ in JET L-mode plasma. For this particular scan, experimental evidence indicates that a dominant part of this peaking originates from NBI fuelling and inward pinch is a subdominant fraction. This is supported by the simple linear runs with GYRO although much more work is needed here to make a conclusion. More discharges are needed to quantify more precisely the fractions of these two contributions and also much more work is required on the modelling front to make exact comparisons.

Acknowledgement. This work has been carried out within the framework of the EUROfusion Consortium and has received funding from the Euratom research and training programme 2014-2018 under grant agreement No 633053. The views and opinions expressed herein do not necessarily reflect those of the European Commission.

[1] A. Loarte *et al.*, Nucl. Fusion **53**, 083031 (2013).
[2] H. Weisen *et al.*, Plasma Phys. Control. Fusion **48**, A457 (2006).
[3] C. Angioni *et al.*, Nucl. Fusion **47** 1326–1335 (2007).
[4] H. Weisen *et al.*, Plasma Phys. Control. Fusion **46**, 751 (2004).
[5] A. Salmi *et al.*, 42nd EPS Conference, Lisbon, Portugal, 22-26 June 2015.
[6] A. Sirinelli *et al.*, Review of Scientific Instruments **81**, 10D939 (2010).
[7] E. Doyle *et al.*, 24th IAEA Fusion Energy Conference, San Diego, USA, October 8-13, 2012.
[8] T. Tala *et al.* Phys. Rev. Lett. **102**, 075001 (2009).
[9] C. Bourdelle *et al.*, Phys. Plasmas **14**, 112501 (2007).

# A comparative multivariate analysis of disruption classes between JET and AUG

R. Aledda[1], B. Cannas[1], A. Fanni[1], A Murari[2], A. Pau[1], G. Sias[1], the ASDEX Upgrade Team[3], the JET Contributors* and the EUROfusion MST1 Team**

*EUROfusion Consortium, JET, Culham Science Centre, Abingdon, OX14 3DB, UK*

*[1] Electrical and Electronic Engineering Dept- University of Cagliari, Cagliari, Italy*

*[2] Consorzio RFX-Associazione - EURATOM ENEA per la Fusione, Padova, Italy*

*[3] Max-Planck-Institut für Plasmaphysik, D-85748 Garching, Germany*

## I. Introduction

Disruptive events still represent one of the main concerns for safe operation of large size tokamak devices, in particular in relation to the preservation of the structural integrity of the machine. This aspect plays a key role in the design and running of the next step experimental devices as ITER and motivates the need of developing methods and techniques aimed to minimize both the number and the severity of disruptions. Furthermore, when a disruption occurs it would be particularly important to be able to distinguish among its different types to improve avoidance and mitigation schemes. In order to extrapolate results from existing devices to the next step ones, it is crucial to interpret the multi-machine plasma confined experiments data with a firm physical basis. The definition of common basis and criteria for cross-machine analysis would facilitate the development of portable systems for disruption prediction and classification, which is becoming of increasingly importance for the real-time plasma control and operation.

Recently, criteria for manual disruption classification have been proposed both for JET [1] and AUG [2]. It has been noted that several physics instabilities in AUG [2] are also usual disruption precursors in JET [1]. In this paper, two data bases have been used containing 116 flat-top disruptions occurred at JET with ITER Like Wall (ILW) from 2011 to 2012 and 102 flat-top disruptions occurred at ASDEX Upgrade (AUG) from 2011 to 2014. A manual classification of disruptions has been done for the AUG database. In particular, the analyzed AUG discharges have been clustered in different classes following the chain of events as proposed in [1] for JET. The manual classification for JET was already available [3]. The distribution of the disrupted pulses in the classes is very similar in the two machines, even if the relative percentage of each class of disruption is strongly influenced by the scientific program and/or by the number of sessions devoted to a given program. A smaller percentage of disruptions during impurity accumulation (RPK) has been found in AUG with respect to

* See the Appendix of F. Romanelli et al., Proceedings of the 25th IAEA Fusion Energy Conference 2014, Saint Petersburg, Russia.
** See http://www.euro-fusionscipub.org/mst1.

JET; this reflects the capability of this device to centrally heat the plasma with ECRH in most of the experimental scenarios [2].

The availability of common disruption classes for the two machines allows us to perform an analysis aiming to identify a robust set of dimensionless parameters, with particular reference to basic quantities available in real-time in different machines. The raw signals have been processed in the time and frequency domains in order to synthesize non-dimensional indicators fitting both the considered devices. In this paper, the procedure followed to build such indicators has been shown with reference to the locked mode (LM) signal. The validity of the proposed indicator has been assessed by using it as disruption predictor. Its performance has been compared with that of the LM raw signal, which is commonly used both at AUG and JET to trigger the mitigation system.

## II. Physics-based indicators

An indicator can synthesize more signals in order to describe a complex phenomenology, or can be based on a single signal when the signal itself is intrinsically representative of a disruptive behaviour, as for example in the case of the locked mode. In most of the cases, in order to maximize the information content in the signal, it could be required to remove noise or unwanted spikes, or simply to extract the trend of the signal filtering transient phenomena. After the signal-processing step, in order to allow an eventual comparison among different machines it is required to find suitable scaling factors and to normalize the signal.

Even if the amplitude of the LM signal scales with several quantities, in this work the normalization has been done with respect to the amplitude of the plasma current. To remove the trend and the offset from this signal (sampling frequency of 10 kHz), its maximum deviation from the mean over a sliding window of 3.2 ms is evaluated (Fig. 1c). Moreover, a FFT of the normalized LM signal is performed over a sliding window of 51.2 ms, and the DC component is removed. Then, the standard deviation of the FFT (Fig. 1d) and the sum of the FFT components, weighted with respect to the frequencies themselves (Fig. 1e), are evaluated. The final indicator obtained multiplying these three signals has been plotted in logarithmic scale to better visualize its behaviour (Fig. 1f). In order to statistically assess its effectiveness, such indicator has been tested as a disruption predictor both on the considered JET and AUG databases and its performance have been compared to those ones obtained with the simple raw signal. In both the cases, the probability density function of the indicator for samples belonging to the safe and the disruptive phases of a subset of discharge (training set) have been compared in order to identify a range of values that potentially discriminate the two phases. In that range, a threshold has been optimized maximizing the successful prediction rate on the training set, and then a test over a separated set of discharges (test set) has been performed. Testing the indicator on both JET and AUG, a significant improvement of about 10% of the successful predictions on the test sets is achieved, passing from 69% to 77% at JET and from 65% to 74% at AUG. A disruption is considered successfully predicted when triggered between 1.5 s and 10 ms before the disruption time for JET, and between 500 ms and 2 ms before the disruption time for AUG.

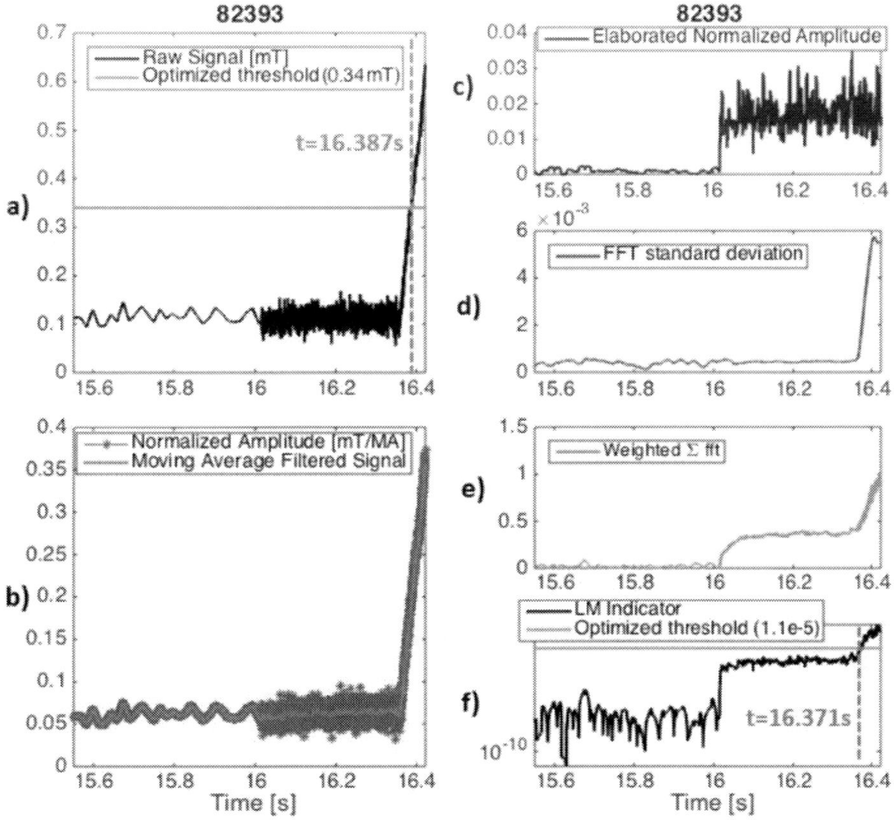


**Figure 1: Construction of the Locked Mode indicator for JET**

As it has been introduced before, an indicator can be built not only from a single signal, but also properly combining information and sequence of events described by more signals. In Fig. 2, a set of dimensionless parameters is reported which represents some of the most relevant quantities describing the accumulation of high-Z impurities. Depending on the plasma underlying conditions, the accumulation of high-Z impurities can lead to disruption both in JET and in AUG. Even if the time scales are different, a robust calculation of some peaking factors of basic quantities, as radiation and temperature profile, can be well correlated with the slow impurity profile peaking, as well as the final reduction or even suppression of the sawtooth activity, followed by the onset of strong tearing modes up to the final mode locking.

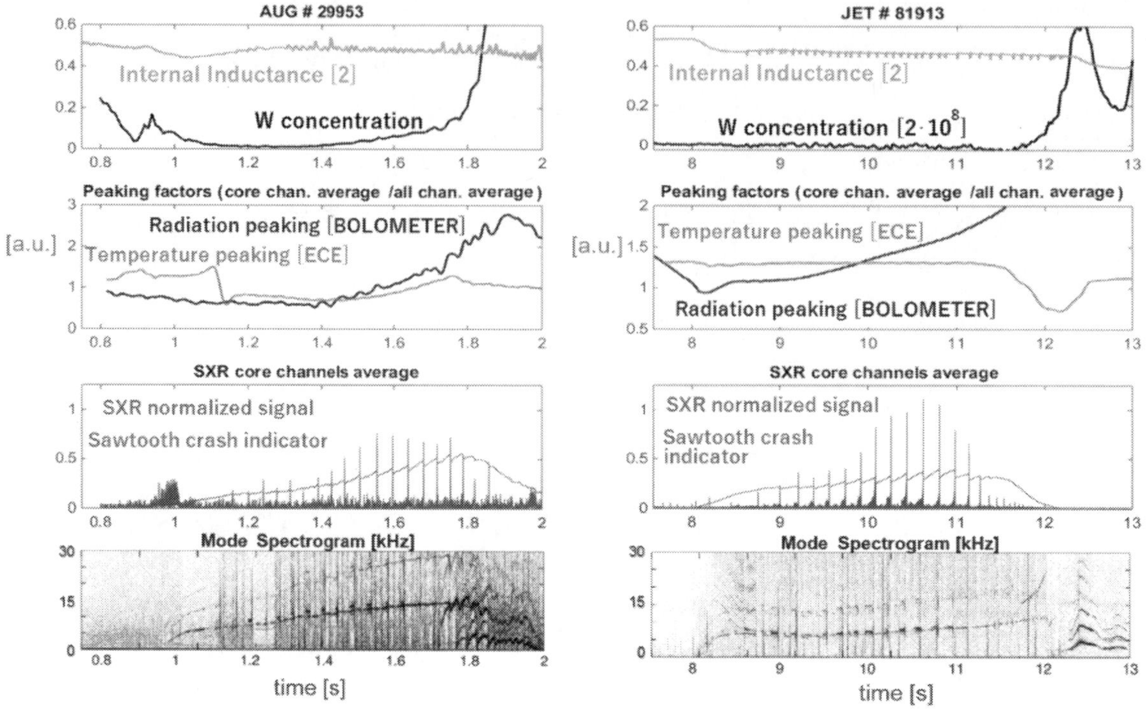


**Figure 2: Dimensionless parameters describing impurity accumulation on AUG and JET.**

## III. Conclusions

In order to get reliable and comparable dimensionless parameters, the steps of signal processing and feature extraction are of primary importance as well as the investigation of suitable scaling factors for normalization. This requires an extensive modelling of past experiments, concerning in particular the definition of thresholds for the obtained indicators, and a connection to the operational space where the plasma is evolving. This link can be provided in real-time through mapping by means of manifold learning techniques, as described in [4].

## Acknowledgement

This work has been carried out within the framework of the EUROfusion Consortium and has received funding from the Euratom research and training programme 2014-2018 under grant agreement No 633053. The views and opinions expressed herein do not necessarily reflect those of the European Commission.

## References

[1] de Vries P.C. *et al.* (2011) Nucl. Fusion 51 53018.
[2] Pautasso G. *et al.* (2014) 41th EPS Conference on plasma physics, Berlin 2014.
[3] de Vries P.C. *et al.* (2012) Plasma Phys. Control. Fusion 54 124032.
[4] Cannas B. *et al.* (2013) Nucl. Fusion 53 (2013) 093023.

# Edge plasma – lower hybrid wave interaction and current drive efficiency

M.Goniche[1], K.Kirov[2], J.Mailloux[2], G.Antar[3], B.J.Ding[4], A.Ekedahl[1], M.H.Li[4], Y.Peysson[1]
and JET Contributors*

*EUROfusion Consortium, JET, Culham Science Centre, Abingdon, OX14 3DB, UK*

[1] *CEA, IRFM, F-13108 Saint-Paul-lez-Durance, France.*

[2]*CCFE, Culham Science Centre, Abingdon, OX14 3DB, UK*

[3]*American University of Beirut, Riad el-Solh, Beirut 1107-2020, Lebanon*

[4] *Institute of Plasma Physics, Chinese Academy of Sciences, Hefei 230031, P. R. China*

**See the Appendix of F Romanelli et al., Proc. 25th IAEA Fusion Energy Conference 2014, St Petersburg, Russia*

## Introduction

Lower Hybrid Current Drive (LHCD) has the unique capability to drive current with high efficiency in the outer part of the plasma (r/a>0.5) and to shape the current profile required for the fully non-inductive steady state scenario. So far, fully or quasi-fully non inductive discharges have been carried out at low or medium plasma density ($\bar{n}_e$<5x10$^{19}$m$^{-3}$) whereas the ITER steady state scenario assumes $\bar{n}_e \sim 7\times10^{19}$m$^{-3}$.

Experiments have been performed on JET [1], Alcator C-Mod [2], FTU [3] and Tore Supra [4] at ITER-relevant plasma densities. The fast electron tail and the LHCD efficiency were estimated from the bremsstrahlung emission by the Hard X-ray (HXR) diagnostic (40-200keV) for the last 3 facilities and from the Electron Cyclotron Emission of the fast electron population ($ECE_{non\ th.}$) for JET and Tore Supra. When the density is ramped up at constant LHCD power, the HXR or ECE signal decays very fast typically like $\bar{n}_e^{-k}$ with k~3 on Tore Supra, k~4.5 on C-Mod. The HXR signal is satisfactory reconstructed when ad-hoc broadening of the wave index spectrum is included in the C3PO/LUKE ray-tracing/Fokker Planck code [5, 6]. The fast electron tail can be increased by more than on order of magnitude by additional heating provided by ICRH (C-Mod [7]) or by changing the particle recycling wall obtained by lithium layer deposition (FTU [3], EAST [8]). Whereas the first technique can both increase the core and edge electron temperature, the second one can be assumed to mostly affect the edge temperature. Low edge temperature may lead to parasitic wave absorption (collisional absorption) or wave spectrum modification (wave scattering, parametric decay). We here investigate the link between the frequency spectrum broadening, related to the coherent quasi-mode at the ion sound frequency (~MHz), and the fast electron tail.

**Experimental results**

The effect of LHCD power on spectrum broadening was investigated on JET. The frequency spectrum was measured outside the vacuum vessel by a magnetic loop or from the reflected RF power. When the LH power is increased from 0.1MW to 3.2MW, the pump broadening, measured 30dB below the maximum, increases linearly and doubles without observed threshold.

$ECE_{non\ th.}$ and HXR measurements were performed, on JET and EAST respectively, during L-mode discharges whose line-averaged density $\overline{n_e}$ was ramped-up from 1.5 (EAST)-1.9 (JET)$\times 10^{19}m^{-3}$ to 2.8 (JET)-3.5 (EAST) $\times 10^{19}m^{-3}$ by gas injection. The toroidal field was 1.95-2.43T on JET and 2.24T on EAST. Following the Stix-Golant accessibility condition (no propagation effect), the wave can penetrate to the very edge only where the density gradient is large: typically for a density of $\overline{n_e}$=2.5x$10^{19}m^{-3}$ the wave penetrates up to r/a~0.99 at 2.1T and r/a~0.90 at 2.4T on JETwith a parallel wave index was used $N_{//}$=1.8. On EAST ($N_{//}$=2.1), the wave can penetrate much further in, up to r/a~0.6.

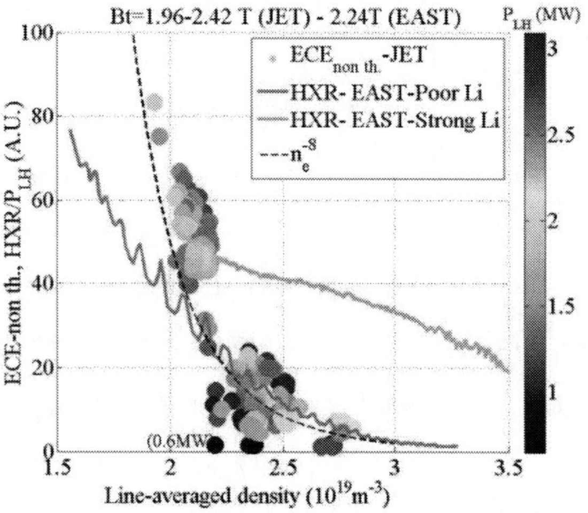


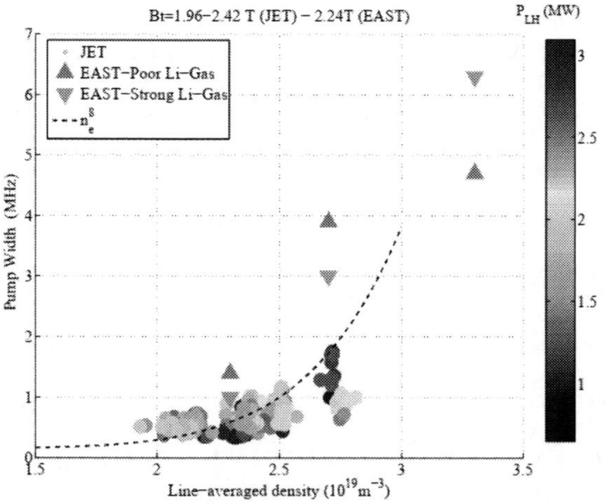


Figure 1. a) $ECE_{non\ th.}$(JET) and HXR (EAST) b) Pump width at -30dB (JET) , -20dB (EAST) as a function of $\overline{n_e}$ . Bt=2.22-2.43T, $N_{//}$=1.8 (JET), Bt=2.24T, $N_{//}$=2.1 (EAST).

For this range of magnetic fields, a very strong decrease of the signal (k=-8) is observed on both machines when $\overline{n_e}$ exceeds ~2.2x$10^{19}m^{-3}$ (figure 1-a). Although the $ECE_{non\ th.}$ and HXR signals are normalized to LHCD power, the fast electron emission still increases with power, consistent with a beneficial effect on the generation of the fast electron tail of higher electron temperature from the higher injected power. After lithium deposition on the vessel walls in EAST, $1/\bar{n}_e$ scaling is found up to $\bar{n}_e$=2.6x10$^{19}$m$^{-3}$. For the same discharges, a weak increase of the pump width $\Delta f_{-30dB}$ (from ~0.5MHz to ~1MHz) is observed on JET with increasing density whereas $\Delta f_{30dB}$ increases from 3.8MHz to 8.7MHz on EAST (1.4MHz to 4.7MHz when measured at -20dB, figure 1-b). After lithium deposition, a strong increase of $\Delta F$ is observed with even a larger value for the highest density case when compared to the high recycling case.

From the same database, the $ECE_{non\ th.}$ signal is now plotted as a function of the density normalized to the accessible density at r/a=0.9 (figure 2.a). The scattering of the fit is significantly reduced (RMS goes from 7 to 4) and the k exponent is reduced from 8 to 5. It should be noted that for an edge density exceeding the accessible density by ~70%, there is still a fast electron tail. The EAST data, performed with a higher wave index ($N_{//}$=2.1), have a much lower range of normalized density (0.4-0.75) and do not scale the same way as the JET data, keeping the normalizing factor of the ECE data with the HXR data of figure 1. Parametric decay is expected to affect strongly the LHCD efficiency when the ratio of the lower hybrid frequency $\omega_{LH}$ to the operating frequency $\omega$ approaches ~0.5 [9]. Here $\omega_{LH}/\omega$ lies between 0.13 and 0.16 for the JET data and between 0.18 and 0.21 for the EAST data (Figure 2.b). The EAST $\Delta f$ data are now more in the line of the JET ones although there is still a difference when considering that the $\Delta f_{30dB}$ is about twice $\Delta f_{20dB}$.

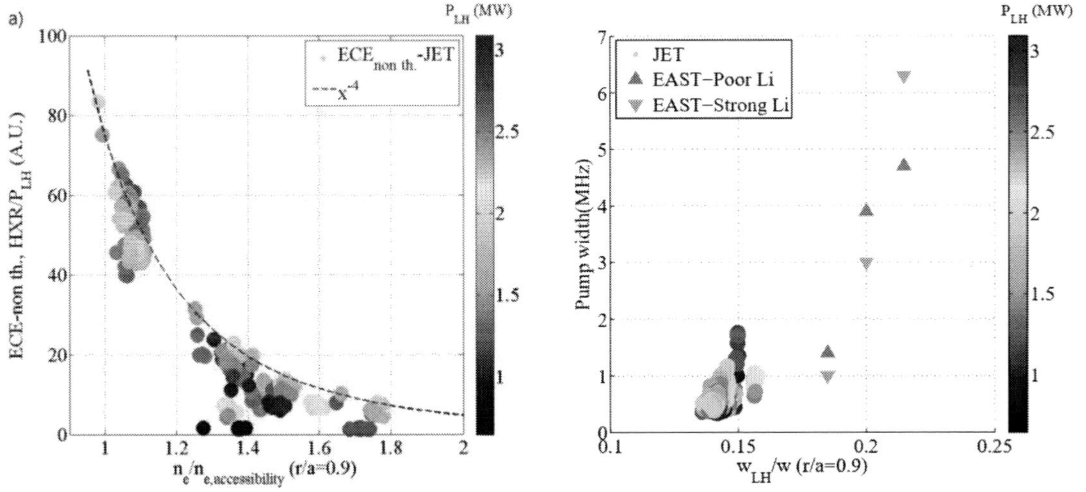


Figure 2. a) $ECE_{non\ th.}$(JET) as a function of $n_e/n_{e,\ acess}$. b) Pump width at -30dB (JET) , -20dB (EAST) as a function of $\omega_{LH}/\omega$. Bt=1.96-2.43T, $N_{//}$=1.8 (JET), Bt=2.24T, $N_{//}$=2.1 (EAST).

**Discussion and conclusions**

The link between parametric instabilities triggered at the plasma edge and LHCD efficiency have been studied on a data base of 56 JET discharges and 2 EAST discharges with a density ramp. On JET, for these weakly accessible conditions, the fast electron tail decays, as it does on C-Mod or Tore Supra (k=3-5) when the density is normalized to $n_{e,accessibility}$. However the amplitude of the ion-sound quasi-mode, derived from $\Delta f_{30dB}$, increases quite modestly with density and is poorly related to the fast electron tail decay. The HXR data of EAST ('poor lithium' case) indicates a similar decay of the fast electron tail for $\bar{n}_e > 2\times10^{19}m^{-3}$ although the wave can penetrate deeper in the plasma. The pump width is much larger than for the JET discharges for similar density and the $\omega_{LH}/\omega$ scaling is more relevant when comparing the results of the two machines. However when comparing the results before (high $n_e$, low $T_e$ at the edge) and after (low $n_e$, high$T_e$) Li deposition on the walls , there is clearly a discrepancy between the frequency broadening at high density ($\Delta f$ is wider after Li deposition) and fast electron tail (larger by more than one order of magnitude after Li deposition).

For this range of densities, JET and EAST data do not show evidence that parametric decay in ion-sound wave is the dominant mechanism to explain the decay of the fast electron tail with density. On JET the very poor accessibility of the wave could be the main cause of the strong decay of the fast electron tail with density.

**Acknowledgements**

This work has been carried out within the framework of the EUROfusion Consortium and has received funding from the Euratom research and training programme 2014-2018 under grant agreement No 633053. The views and opinions expressed herein do not necessarily reflect those of the European Commission. The ITPA/IOS is acknowledged for initiating this work.

## References

[1] K. Kirov et al 2012, Plasma Phys. Control. Fusion 54 074003
[2] G.Wallace et al, Phys.Plasmas, **19**, 062505 (2012), doi: 10.1063/1.4729734]
[3] R.Cesario et al., 2010 Nature Commun. **1** 55
[4] M.Goniche et al., Nucl. Fusion 53 (2013) 033010 (9pp) doi:10.1088/0029-5515/53/3/033010
[5] J.Decker, Y.Peysson et al., Physics of Plasmas, **21**, 092504 (2014); doi: 10.1063/1.4894749
[6] Y.Peysson, J.Decker, et al., I5-O12, this conference.
[7] S. Shiraiwa et al, Nuclear Fusion **53**, 113028 (2013).
[8] B.J.Ding et al., 2013 Nucl. Fusion **53** 113027
[9] Y. Takase et al, Phys. Fluids **28**, 983 1985

# The effects of resonant magnetic perturbations on fast ion confinement in the Mega Amp Spherical Tokamak

K.G. McClements[1], R.J. Akers[1], W.U. Boeglin[2], M. Cecconello[3], D. Keeling[1], O.M. Jones[1,4], A. Kirk[1], I. Klimek[3], R.V. Perez[2], K. Shinohara[5], K. Tani[6]

[1] *CCFE, Culham Science Centre, Abingdon, Oxfordshire, OX14 3DB, UK*

[2] *Department of Physics, Florida International University, Miami, Florida 33199, USA*

[3] *Department of Physics and Astronomy, Uppsala University, SE-751 05, Uppsala, Sweden*

[4] *Department of Physics, Durham University, South Road, Durham DH1 3LE, UK*

[5] *Japan Atomic Energy Agency, Naka, Ibaraki 311-0193, Japan*

[6] *Tokyo Institute of Technology, 2-12-1 Ookayama, Meguro-ku, Tokyo 152-8550, Japan*

## 1. Introduction

Resonant magnetic perturbations (RMPs) are frequently superimposed on the axisymmetric fields in tokamak plasmas to suppress or ameliorate edge localised modes (ELMs). In several devices RMPs have been found to degrade energetic ion confinement [1-3]. In this paper we report evidence for RMP-induced beam ion transport in the Mega Amp Spherical Tokamak (MAST); further details can be found in [4].

## 2. Experimental evidence for RMP-induced fast ion transport in MAST

Fig. 1 shows time traces from two single-null diverted (SND), MAST pulses with plasma current $I_p$ = 400kA: 30086 (RMPs with toroidal mode number $n$ = 3) and 30090 (no RMPs). The application of RMPs coincided with a substantial drop in total neutron rate $S_n$ that was not matched in pulse 30090, which had no RMPs. As was usual in MAST, the application of RMPs caused the ELMs to have somewhat lower amplitudes and to occur more frequently [Fig. 1(c)]. The NUBEAM module of the TRANSP plasma simulation code is used to model the collisional evolution of beam ions, with the option of adding *ad hoc* non-collisional transport [5]. Fig. 2(a) again shows $S_n$ in pulse 30086 (black curve), here with synthetic rates computed from NUBEAM simulations with purely collisional transport (green dashed-dotted curve) and with additional diffusivity $D_a$ = 1m$^2$s$^{-1}$ (blue dotted curve). The synthetic rate from the simulation with anomalous fast ion transport tracks the measured rate fairly well until RMPs are applied, then remains high while the measured rate drops by nearly 50%. Shortly after the application of RMPs in 30086, the amplitude of MHD activity dropped considerably [Fig 2(b)], which suggests that it would be appropriate to set $D_a = 0$ during this phase of the pulse, were it not for the presence of RMP-induced transport.

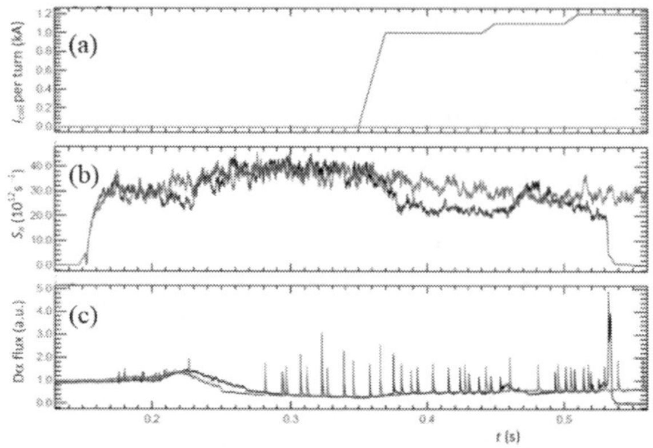


**Fig. 1** Time traces from MAST pulses 30086 (black) and 30090 (red) showing (a) RMP coil current per turn, (b) neutron rate $S_n$, and (c) Dα emission.

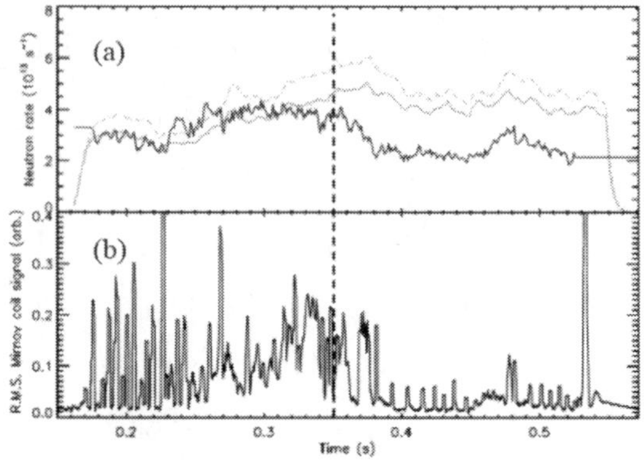


**Fig. 2** (a) Measured neutron rate (black) and rates simulated using TRANSP with $D_a = 0$ (green curve) and $D_a = 1\text{m}^2\text{s}^{-1}$ (blue curve) in MAST pulse 30086. (b) Root mean square signal measured using outboard Mirnov coil. The dashed vertical line indicates the time at which RMPs were first applied.

Fig. 3 shows rates of fusion protons recorded in four channels, corresponding to unconfined particle trajectories crossing the midplane at four different major radii (indicated on the figure) for pulse 30086. A similar trend can be seen in all channels and in $S_n$, with drops in the measured rates occurring shortly after the end of the RMP ramp-up.

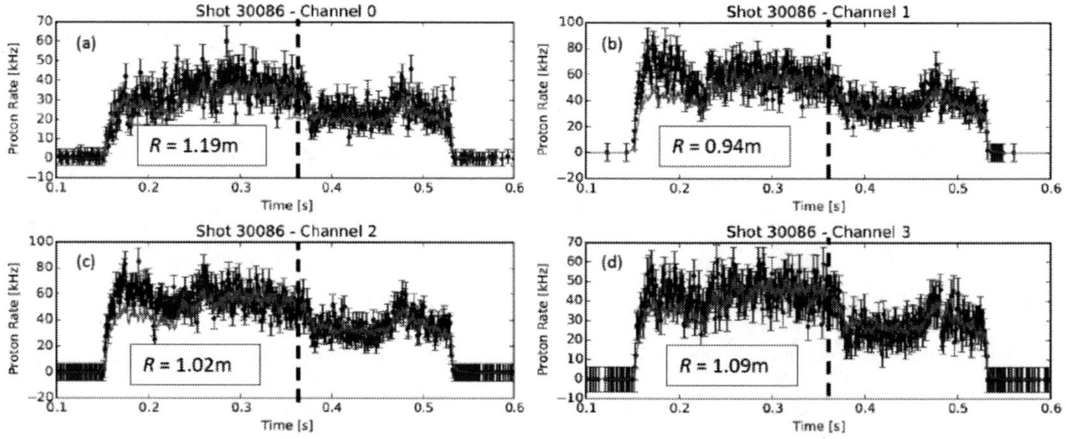


**Fig. 3** Blue circles: rates of fusion protons recorded in four detector channels during MAST pulse 30086. Red curves: $S_n$ normalised to same scale as proton rates. The dashed lines indicate the time at which the RMP coil current reached its maximum value. The values of $R$ are the average major radii at which protons detected in the four channels crossed the midplane.

Fig. 4 shows fast ion Dα (FIDA) emission in the two pulses discussed above. The left-hand and right-hand plots show data for two different values of $R$, here representing the average major radius at which the FIDA line of sight intersected one of the beamlines. Higher Doppler shifts from the Dα rest wavelength in air (656.1 nm) correspond to higher line-of-sight velocities, hence higher minimum beam ion energies, which are indicated on the figure. When the RMPs are applied the emission drops sharply at both values of $R$ and in every wavelength range, suggesting that fast ions with a wide range of energies are affected.

Fig. 5 shows total neutron rates and neutron camera (NC) data from two similar pulses, with and without RMPs. In the former the RMP coil current was ramped up gradually then held constant [Fig 5(a)]. In Figs. 5(c-e) the parameter $p$ is the tangency major radius of the NC line of sight while $Z$ is the vertical position of the line of sight at the tangency point. The total neutron rates and the rates in all NC channels clearly diverge as the coil currents are ramped up, with many fewer neutrons being produced in the pulse with RMPs.

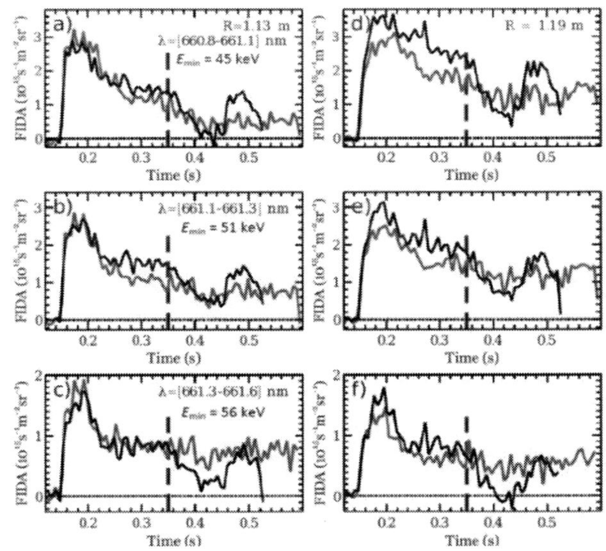


**Fig. 4** FIDA emission versus time in MAST pulses 30086 (black) and 30090 (red) for three wavelength ranges, corresponding to different minimum fast ion energies, and two major radii, $R$. Dashed lines indicate the time at which RMPs were applied in pulse 30086.

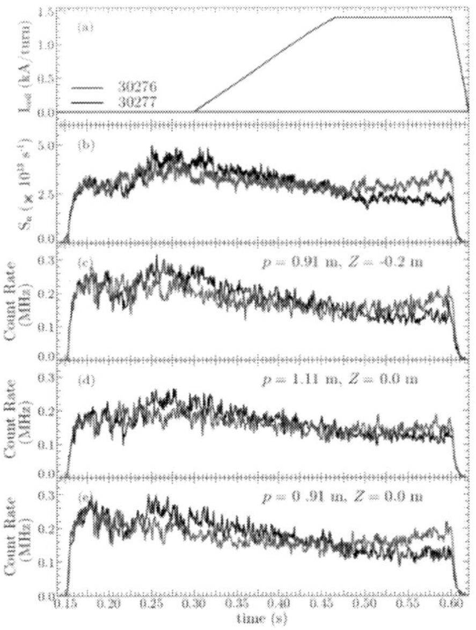

**Fig. 5** Time traces from MAST pulses 30276 (red) and 30277 (black) showing (a) RMP coil current per turn, (b) $S_n$, and (c-e) neutron rates in three different NC channels.

## 3. First principles-based modelling

Simulations of beam ions in MAST plasmas with and without RMPs have been carried out using the F3D-OFMC code [6]. Beam ion orbits were tracked from birth to thermalization, taking into account collisions with bulk ions and electrons. The simulated RMPs had $n = 6$, and the equilibrium and plasma profiles were based on SND MAST pulse 27205, with $I_p$ = 600 kA. Two alternative representations of the first wall (at which particles were considered to be lost) were used, designated as W1 and W2: see Fig. 6 in [4]. Both MAST neutral beams were modelled in the simulations, with primary birth energies set equal to 65 keV and 51 keV. Table 1 lists the figures obtained in the simulations for the total loss of neutral beam heating power due to fast ion losses. It is apparent that the RMP-induced loss of beam heating power (~11%) does not depend on either the choice of first wall or finite Larmor radius effects.

| magnetic field | orbit-following scheme | wall | Total loss of beam power (%) | loss of power due to RMPs (%) |
|---|---|---|---|---|
| axisymmetric | full orbit | W1 | 5.7 | - |
| $n = 6$ RMPs | full orbit | W1 | 16.5 | 10.8 |
| $n = 6$ RMPs | guiding centre | W1 | 16.2 | 10.5 |
| axisymmetric | full orbit | W2 | 11.5 | - |
| $n = 6$ RMPs | full orbit | W2 | 22.6 | 11.1 |
| $n = 6$ RMPs | guiding centre | W2 | 22.1 | 10.6 |

**Table 1.** Beam power losses found in F3D-OFMC simulations of MAST plasmas. RMP-induced losses were obtained by subtracting results for full orbit losses in axisymmetric plasmas from total figures.

### 4. Summary

The effects of RMPs on beam ion confinement in MAST have been studied using measurements of neutrons, fusion protons and FIDA emission. In SND MAST pulses with $I_p$ = 400 kA, the total neutron emission dropped by around a factor of two when $n$ = 3 RMPs were applied. The measured neutron rate during RMPs was much lower than that calculated using TRANSP, even when non-classical (but axisymmetric) *ad hoc* fast ion transport was taken into account. Sharp drops in spatially-resolved neutron rates, fusion proton rates and FIDA emission were also observed. First principles-based simulations of RMP-induced fast ion transport in MAST, using the F3D-OFMC code, show similar losses for two alternative representations of the MAST first wall, with and without full orbit effects taken into account; for $n$ = 6 RMPs in a 600 kA plasma, the additional loss of beam power due to the RMPs was found in the simulations to be approximately 11%.

This work has received funding from the RCUK Energy Programme [grant number EP/I501045] and from Euratom. The views and opinions expressed herein do not necessarily reflect those of the European Commission.

[1] M. Garcia-Munoz et al., Plasma Phys. Control. Fusion **55**, 124014 (2013)
[2] M. Garcia-Munoz et al., Nucl. Fusion **53**, 123008 (2013)
[3] M.A. Van Zeeland et al, Plasma Phys. Control. Fusion **56**, 015009 (2014)
[4] K.G. McClements et al., Plasma Phys. Control. Fusion **57**, 075003 (2015)
[5] A. Pankin et al., Comput. Phys. Commun. **159**, 157 (2004)
[6] K. Tani et al., J. Phys. Soc. Japan **50**, 1726 (1981)

# Searching for Enhanced RF Field Amplitudes in the SOL Using a Simplified Cold-Plasma Model

R.J. Perkins[1], J. C. Hosea[1], N. Bertelli[1], E.F. Jaeger[2], C. K. Phillips[1], G. Taylor[1], J. R. Wilson[1]

[1] *Princeton Plasma Physics Laboratory, Princeton, United States*

[2] *XCEL Engineering Inc., Oak Ridge, United States*

NSTX uses a high-harmonic fast-wave (HHFW) antenna for plasma heating and current drive, but a significant amount of the HHFW can be lost along open magnetic fields lines in the SOL to the divertor regions under certain conditions [1,2]. The underlying cause is related to fast-wave propagation in the scrape-off layer (SOL), as the heating efficiency is strongly correlated with the location of the righthand cutoff. Recent results with the full-wave simulation AORSA suggest that the RF electric field amplitude grows to large values in the SOL when the SOL density is large enough that there is no cutoff region in front of the antenna [3]. However, interpretation of the AORSA results have not been straightforward due to the complicated vessel and magnetic geometry [4]. In this paper, we employ a cylindrical cold-plasma model to determine which features of the AORSA results can be reproduced in this simplified geometry. We do indeed find modes that conduct a significant fraction of power in a low density periphery plasma surrounding the dense core plasma, but more work is needed to quantitatively link these results to those from experiments and from AORSA.

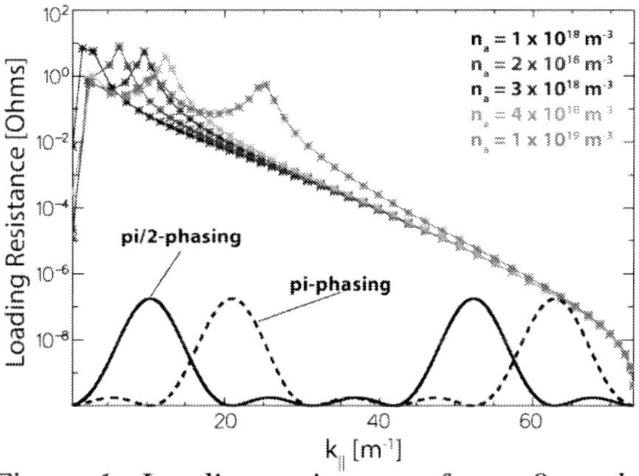


Figure 1: *Loading resistance of $m = 0$ modes peaks at a $k_{||}$ that depends on $n_a$. Plotted for reference are antenna spectra for $\pi/2$ and $\pi$ phasing.*

The cylindrical cold-plasma model, based on previous work [5], consists of a cylindrical core plasma of radius $r_c$ and density $n_c$ surrounded by a lower-density annulus of radius $r_a$ and density $n_a$, which is in turn surrounded by a vacuum region with a perfectly conducting vessel of radius $r_w$. A uniform axial magnetic field is used, which would correspond to a "zero-pitch" tokamak. For each azimuthal wavenumber $m$, the system dispersion (the system of equations obtained by matching solutions in each region to boundary conditions) is solved for a set of discrete radial modes, denoted by the index $n$. Each mode has a single axial wavenumber $k_{||}$ and three corresponding radial wavenumbers $k_{\perp}$, one for each region. In the core and annulus, $k_{\perp}$ is given by the local cold-plasma dispersion, and while both slow and fast waves are included in the model, in this paper $k_\perp$ will denote the fast-wave branch, the slow wave always being cutoff and confined to a very narrow radial region at the core-annulus and annulus-vacuum interface. In the vacuum region, the modes are always cutoff, with $k_\perp^{\text{vac}} = (k_\parallel^2 - \omega^2/c^2)^{1/2} \approx k_\parallel$. We typically model the antenna using four current straps situated at $r = r_s$ with arbitrary phasing, an inter-strap distance 0.15 m, and uniform azimuthal current distribution over an angle of $\pi/2$. A Faraday screen at $r = r_F$ shorts out the $E_z$-mode from the antenna. We chose parameters to mimic the conditions in NSTX: $n_c = 5\times10^{19}$ m$^{-3}$, $f = 30$ MHz, $B = 0.323$ T which gives $\omega/\omega_{ci} = 12$, $r_c = 0.515$ m, $r_a = 0.575$ m, $r_F = 0.600$ m, $r_s = 0.650$ m, and $r_w = 0.700$ m.

We begin by analyzing azimuthally symmetric modes ($m = 0$). The loading resistance for each $m = 0$ mode is plotted against the mode axial wavenumber in Fig. 1 for different annulus densities. Loading resistance $R_{m,n}$ is related to the wave power $P_{m,n}$ conducted by each mode via $R_{m,n}I_{\text{ant}}^2/2 = P_{m,n}$, with $I_{\text{ant}}$ the current in the antenna strap. $P_{m,n}$ is calculated either via the power output of the antenna

$$P_{m,n} = \int \langle \mathbf{E}_{m,n}\cdot\mathbf{J}_{\text{ant}}\rangle\, dV, \qquad (1)$$

or via the axial Poynting flux integrated over the cross-section of the cylinder

$$P_{m,n} = \int_{r=0}^{r=r_w}\int_{\theta=0}^{\theta=2\pi} \langle \mathbf{E}_{m,n}\times\mathbf{H}_{m,n}\rangle\, dr d\theta, \qquad (2)$$

where $< \ldots >$ denotes time-averaging of complex quantities. Both methods for computing the mode power give the same result. The mode powers are computed for a single-strap antenna, which uniformly excites all $k_\parallel$, to get a sense of the mode structure without complication from the particular choice of an antenna spectrum. For reference, the spectral shapes for $\pi/2$ and $\pi$ phasing of a four-strap antenna are plotted in Fig. (1). In general, loading resistance decreases with $k_\parallel$, as the wave fields become more evanscent in the vacuum region with a cutoff wavenumber $k_\perp^{\text{vac}} \approx k_\parallel$. However, a peak in loading resistance is clearly seen, and the $k_\parallel$ of the peak increases with the annulus density. The peak always occurs at a $k_\parallel$ lower than the righthand cutoff $c^2k_\parallel^2/\omega^2 = R$.

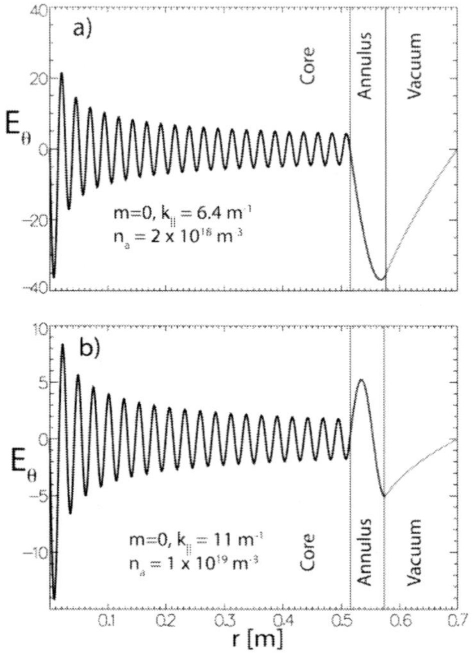


*Figure 2: $E_\theta(r)$ for peak modes is large in the annulus and has an approximate 1/4 or 3/4 wavelength structure there.*

Figure 2a shows the radial profile of $E_\theta$ for the peak $m = 0$ mode for $n_a = 2.0\times10^{18}$ m$^{-3}$ (e.g. peak of red curve in Fig. (1)). At the core-annulus interface, $E_\theta \approx 0$, and there is approximately one quarter wavelength in the annulus region. In the annulus, $E_\theta$ is a linear combination of $J_1(k_\perp r)$ and $Y_1(k_\perp r)$; since $k_\perp = 29.2\ \mathrm{m}^{-1}$ in the annulus for this mode, the Bessel functions are well-approximated by their asymptotic forms $\sin(k_\perp r - 3\pi/4)/(k_\perp r)^{1/2}$ and $\cos(k_\perp r - 3\pi/4)/(k_\perp r)^{1/2}$, so we can compute the phase difference across the annulus as $k_\perp(r_a - r_c) = 1.754$, which is indeed just above the quarter-wavelength value of $\pi/2 = 1.571$. Table 1 shows the values of $k_\parallel$, $k_\perp$, and $k_\perp(r_a - r_c)$ for each peak in Fig. 1. While $k_\parallel$ increases markedly with density, $k_\perp$ remains relatively constant. Finally, note that fitting three quarter wavelengths into the annulus would require $k_\perp \approx 79\ \mathrm{m}^{-1}$, which is above the maximum value that $k_\perp$ can attain for typical SOL densities on NSTX. However, if we raise the annulus density to very large values, we do indeed observe multiple peaks; at $n_a = 1\times10^{19}\ \mathrm{m}^{-3}$ (Fig. 1), the peak mode at $k_\parallel = 11\ \mathrm{m}^{-1}$ has $k_\perp = 78.5\ \mathrm{m}^{-1}$ and a radial $E_\theta$ profile given by Fig. 2b.

Table 1: Wavenumbers for peak mode as annulus density increases.

| $n_a[\mathrm{m}^{-3}]$ | $k_\parallel\ [\mathrm{m}^{-1}]$ | $k_\perp\ [\mathrm{m}^{-1}]$ | $k_\perp(r_a - r_c)$ |
|---|---|---|---|
| $1\times10^{18}$ | 1.80 | 33.1 | 1.99 |
| $2\times10^{18}$ | 6.43 | 29.2 | 1.75 |
| $3\times10^{18}$ | 9.68 | 29.8 | 1.99 |
| $4\times10^{18}$ | 12.4 | 30.7 | 1.84 |
| $1\times10^{19}$ | 25.3 | 33.5 | 2.01 |
| $1\times10^{19}$ | 11.0 | 78.5 | 4.71 |

For a mode to be relevant to the SOL-loss studies on NSTX, it should conduct a significant fraction of the wave power through the annulus. Figure 3 partitions the wave power $P_{m,n}$ (evaluated for unit antenna current) carried by each $m = 0$ mode with $n_a = 2\times10^{18}\ \mathrm{m}^{-3}$ (red curve of Fig. (1)) into power conducted in the core, annulus, and vacuum region by integrating the axial Poynting flux over the appropriate cross-section. Most modes conduct the majority of their power in the core, but the peak mode carries 54% of its power in the annulus, with the average Poynting flux there being four times larger than in the core. Further, this mode accounts for 58% of the total wave power, making such modes relevant to the NSTX studies in that they conduct a significant fraction of the total power through the annulus. Since the antenna spectrum favorably weights modes with $k_\parallel$ that closely match the antenna, one can conceivably explain the SOL loss on NSTX by selecting SOL parameters such that these peak modes which conduct significant power in the annulus lie under the antenna spectrum peak.

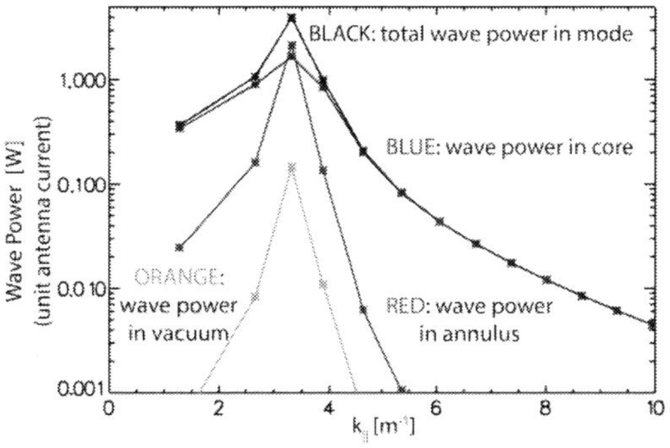


Figure 3: *Wave power conducted in core/annulus/vacuum. $m = 0$ and $n_a = 2\times10^{18}\ \mathrm{m}^{-3}$.*

We extend the analysis to non-axisymmetric modes in Figure 4a, which plots loading resistance on a linear scale from modes with $m = -3$ to $m = +3$. Modes are weighted by $|c_m|^2$, with $c_m$ the azimuthal Fourier component for an antenna of $\pi/2$ span: $|c_0|^2 = 1/16$ and $|c_m|^2 = \sin^2(m\pi/2)/(\pi m)^2$ for $m \neq 0$. We include the azimuthal spectral weighting but not the axial weighting because the azimuthal weighting is essentially fixed and does not change with the choice of antenna phase. Similar to the $m = 0$ case of Fig. 1, for each $m$ loading resistance peaks for a handful of modes. However, the $k_{\parallel}$ of these peak increase with $m$. Thus, the condition for maximal loading, one-quarter wavelength fitting in the annulus, holds only very appoximately for $m \neq 0$ (Fig. 4b), and this dependence on $m$ will be the subject of future investigation.

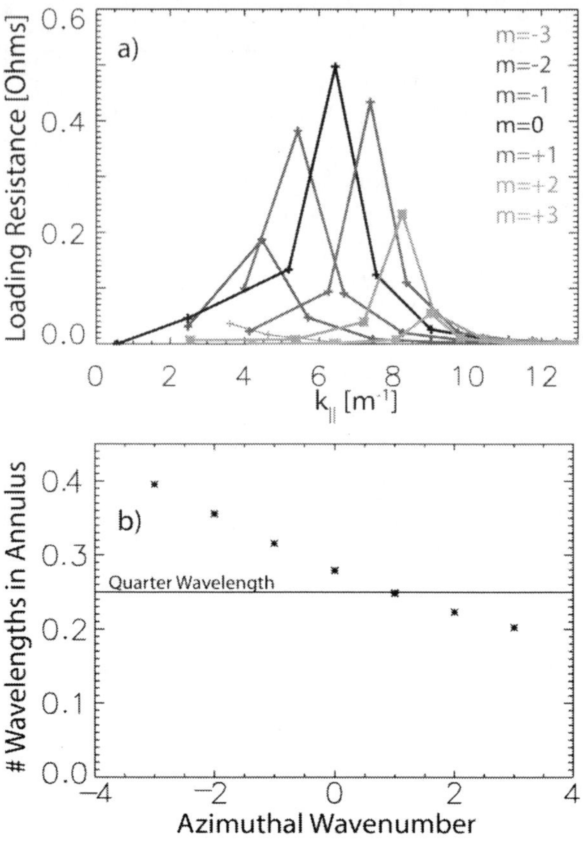


Figure 4: *a) Loading resistance for various m. b) Numer of wavelengths in the annulus for various m.* $n_a = 2 \times 10^{18}$ m$^{-3}$.

The cylindrical cold-plasma model contains modes that conduct a signficant fraction of power in the lower-density annulus region and that carry a significant fraction of the total power. Such modes are good candidates for explaining the SOL losses in NSTX. A clear next step would be to test if the Poynting flux is field-aligned, as observed in experiments. This requires a rather detailed study, as the Poynting flux will vary both in $z$ and $\phi$ due to interference of modes. Ultimately, this work is part of an effort to include the proper edge damping into full-wave codes so that they can reproduce the losses observed in NSTX and predict their importance for ITER.

This work was supported by DOE Contract No. DE-AC02-09CH11466.

## References

[1] J.C. Hosea *et al.*, Phys. Plasma, **15**, 056104 (2008)
[2] R.J. Perkins *et al.*, Phys. Rev. Lett **109**, 045001 (2012)
[3] N. Bertelli *et al*, Nucl. Fusion **54** 083004 (2014)
[4] N. Bertelli, to appear in AIP Conf. Proc.
[5] J.C. Hosea and R.M. Sinclair, Phys. Rev. Lett. **23** 3 (1969)

# Stabilization of Electron-Scale Turbulence by Electron Density Gradient in NSTX

J. Ruiz Ruiz[1], Y. Ren[2], W. Guttenfelder[2], Anne E. White[1], S. M. Kaye[2], B. P. Leblanc[2], E. Mazzucato[2], K. C. Lee[3], C. W. Domier[4], D. R. Smith[5], and H.Yuh[6]

[1]*MIT-Plasma Science and Fusion Center, Cambridge, Massachusetts 02139, USA*
[2]*Princeton Plasma Physics Laboratory, Princeton, New Jersey 08543, USA*
[3]*National Fusion Research Institute, Daejeon, South Korea*
[4]*University of California at Davis, Davis, California 95616, USA*
[5]*University of Wisconsin-Madison, Madison, Wisconsin 53706, USA*
[6]*Nova Photonics, Inc., Princeton, New Jersey 08540, USA*

Theory and experiments have shown that electron temperature gradient (ETG) turbulence on the electron gyro-scale, $k_\perp \rho_e \lesssim 1$, can be responsible for anomalous electron thermal transport in NSTX [1, 2]. Electron scale (high-k) turbulence is diagnosed with a high-k microwave scattering system [3] in NSTX. Here we report on the stabilization effects of the electron density gradient on electron-scale density fluctuations in a set of NSTX NBI-heated H-mode plasmas, following previous work by Y. Ren *et al.* [4].

Electron density fluctuations on the electron gyro-scale ($k_\perp \rho_e \lesssim 1$) were measured with a 280 GHz microwave coherent scattering system implemented on NSTX [3]. Five collection channels simultaneously measure five different wave numbers in the range $5 \lesssim k_\perp \lesssim 30$ cm$^{-1}$. Measured wave vectors are primarily radial $k_r$, with a small binormal component $k_b$ satisfying $k_b/k_r \approx 0.2-0.3$ ($k_r$ is perpendicular to the flux surface and $k_b$ is perpendicular to the local magnetic field and inside the flux surface). The near mid-plane trajectory of the probe beam and the k-response are computed using a ray tracing code. In the scattering experiment presented here the high-k diagnostic is sensitive to fluctuations taking place at $R \approx 135-136$ cm ($R_{maj} = 0.85$ m, r/a $\sim$ 0.7-0.8). Here we focus on a particular experiment of an NSTX NBI-heated H-mode discharge (shot 141767) on a time range after t = 0.3 s.

In figure 1 is plotted the frequency fluctuation spectrogram from channel 1 in $a$), the total scattered power (integrated in frequency) in $b$), the difference between the experimental and critical electron temperature gradient at the scattering location in $c$), and the critical electron temperature gradient formula ($cf.$ [5]) in $d$). The critical gradient formula derived in [5] is given by $(R_0/L_{T_e})_{crit} = max\{0.8R_0/L_{n_e}, (1+\tau)(1.33+1.99\hat{s}/q)(1-1.5\varepsilon)(1+0.3\varepsilon d\kappa/d\varepsilon)\}$, where $\tau = Z_{eff}T_e/T_i$. The two-terms in the $max$ function are plotted in $d$). $R_0/L_X$ are normalized gradients of the quantity $X$, such that $R_0/L_X = -R_0/(dlnX/dr)$, and $R_0$ is the corresponding flux surface center. The critical gradient was also explicitly computed using GS2 linear simulations (triangles in 1.$d$)), and a good agreement is found with the Jenko critical gradient formula despite the high-$\beta$, tight aspect ratio condition of NSTX. High values of the electron density gradient will make the $0.8R_0/L_{n_e}$ term dominant, and it will determine the value of the critical gradient which should have a stabilizing effect on the ETG instability.

For early times ($t \lesssim 0.33$ s), the experimental temperature gradient is at marginal stability levels with respect to the critical gradient (1.$c$)), and no high-k fluctuations are observed in Fig. 1.$a$) $-$ $b$). The intense spike at $t \approx 292$ ms corresponds to an ELM event and is independent of ETG fluctuations. At $t \sim 330$ ms, $R/L_{T_e}^{exp} - R/L_{T_{ec}}^{Jenko}$ increases in Fig. 1.$c$), and ETG becomes unstable. At that time, electron density fluctuations start to develop in Fig. 1.$a$) $-$ $b$). During the time span $t \sim 0.36 - 0.53$ s, a very similar value of $R/L_{T_e}^{exp} - R/L_{T_{ec}}^{Jenko}$ gives rise to very different density fluctuation amplitudes in 1.$a$) $-$ $b$). This suggests a nonlinear dependence of the fluctuation amplitude with different plasma parameters. We aim to show the nonlinear dependence of the density fluctuation amplitude on the equilibrium electron density gradient.

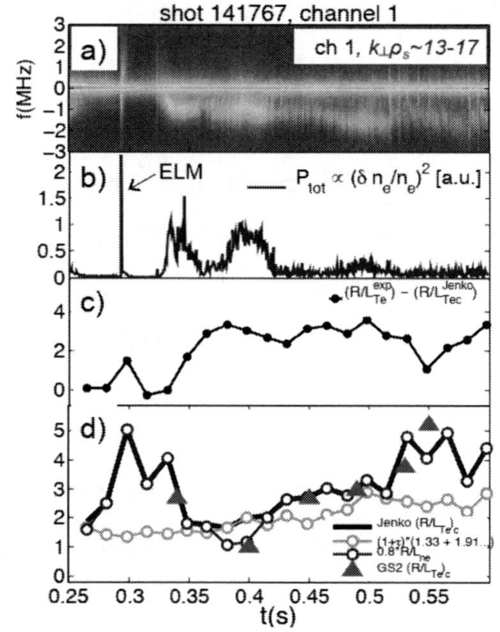


Figure 1: $a$) *Fluctuation spectrogram.* $b$)*Total scattered power from channel 1.* $c$) $R/L_{T_e}^{exp} - R/L_{T_{ec}}^{Jenko}$. $d$) *Jenko critical gradient (black), both terms in the max-function, and GS2 computed critical gradient.*

The electron density gradient term (blue curve in Fig. 1.$d$)) is dominant for most part of the time range of interest (t $>$ 0.3 s). High values of $R/L_{n_e}$ during $t \lesssim 330$ ms raise the critical gradient for ETG to the experimental temperature gradient value; ETG is marginally stable and no high-k fluctuations are detected (Fig. 1.$a) - b)$). As the electron density gradient drops at $t \approx 330$ ms, $R/L_{T_e}^{exp} - R/L_{T_eC}^{Jenko}$ increases and ETG is unstable; high-k fluctuations (Fig. 1.$a$)-$b$)) start to develop and the scattered power increases. ETG remains unstable and high-k fluctuations persist until the end of the discharge. Between $360 \lesssim t \lesssim 410$ ms, the electron density gradient term becomes subdominant and an enhancement of high-k fluctuations is observed in figure 1.$a$)-$b$). After a sudden increase in the electron density gradient at $t \approx 520$ ms, $R/L_{T_e}^{exp} - R/L_{T_{ec}}^{Jenko}$ is reduced and high-k fluctuations mitigate.

Linear gyrokinetic simulations carried out with the gyrokinetic code GS2 [6] show that the electron density gradient has a strong impact on the wavenumber corresponding to the maximum linear growth rate and the real frequency. The radial component $k_r$ was set to 0 to find the most unstable mode. Electromagnetic effects and collisions were also included in these linear runs, and used local Miller equilibrium.

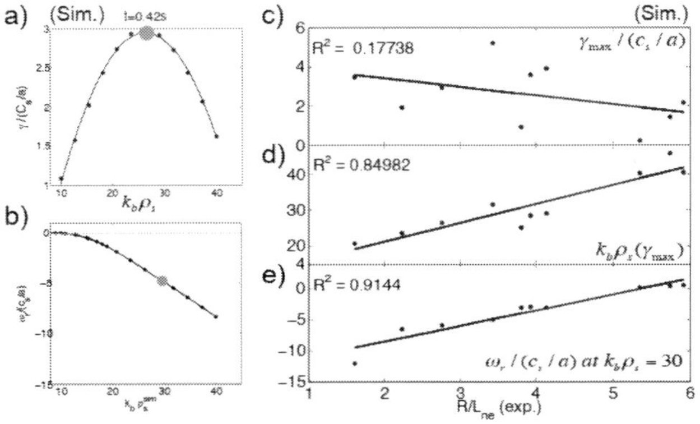


Figure 2: *Orange dots indicate values picked $\gamma_{max}/(c_s/a)$, $k_b\rho_s(\gamma_{max})$ and $\omega_r(k_b\rho_s = 30)$ to plot against $R_0/L_{n_e}$.*

At each time, the maximum linear growth rate ($\gamma_{max}/(c_s/a)$) and the wavenumber corresponding to the maximum linear growth rate $k_b\rho_s(\gamma_{max})$ are calculated. These two quantities are then computed for several times, and are plotted in Fig. 2.$c) - d)$ with respect to the local value of the density gradient $R_0/L_{n_e}$. No correlation is observed between $\gamma_{max}/(c_s/a)$ and $R_0/L_{n_e}$, however a clear correlation can be established between the density gradient and the wavenumber at the maximum linear growth rate $k_b\rho_s(\gamma_{max})$. To study the real frequency, we pick a fixed wavenumber ($k_b\rho_s = 30$), and calculate the frequency corresponding to that fixed wavenumber (Fig. 2.$b$)). A very clear correlation is observed between the real frequency $\omega_r(k_b\rho_s = 30)$ and the local value of $R_0/L_{n_e}$. A local scan on the electron density gradient (not shown here) was carried out using GS2 and completely confirms all correlations established here.

In figure 3 we make a very similar analysis, but this time using experimental measurements. A clear parallel between Fig. 2 and 3 can be made if linear growth rates in 2.$a$) are substituted with the wavenumber spectrum of fluctuations from experiment in 3.$a$) and the real frequency from GS2 in 2.$b$) is replaced with the Doppler subtracted, plasma frame frequency of fluctuations in 3.$b$). The maximum fluctuation level $(\delta n_e/n_e)^2_{max}$ in Fig. 3.$c$) is found to have a weak correlation with $R_0/L_{n_e}$, but the wavenumber corresponding to the maximum fluctuation level in Fig. 3.$d$) is found to have a very strong correlation with density gradient, similar to that shown in Fig. 2.$d$). The Doppler subtracted, plasma frame frequency corresponding to a fixed wavenumber ($\omega_p(k_\perp \rho_s = 13.2)$) is shown to have a notable dependence on density gradient. The same trends found from the linear GS2 simulation are reproduced from experimental data.

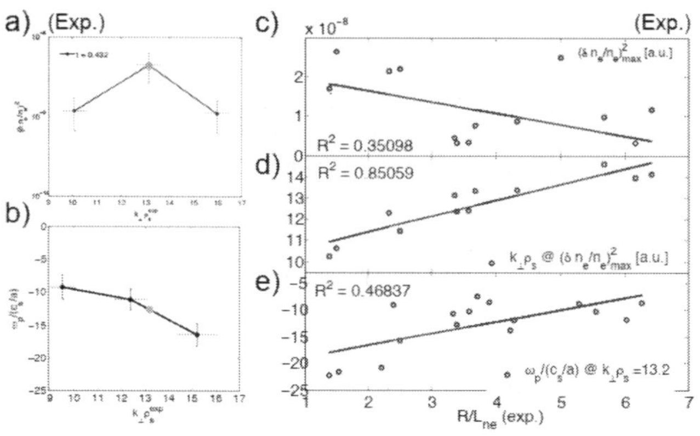


Figure 3: $(\delta n_e/n_e)^2_{max}$, $k_\perp \rho_s((\delta n_e/n_e)^2_{max})$ *and* $\omega_p(k_\perp \rho_s = 13.2)$ *from a), b) plotted against* $R_0/L_{n_e}$ *in c), d), e).*

In summary, the electron density gradient is shown to affect high-k ETG turbulence (Fig. 1) through the linear threshold determined by the Jenko critical gradient expression, as well as a nonlinear stabilizing effect as shown in 1. Simulations (Fig. 2) and experiment (Fig. 3) suggest that the electron density gradient is shifting ETG turbulence to even higher wavenumbers and at the same time reducing the frequency of the instability. This work has been supported by US. D.O.E. contract DE-AC02-09CH11466.

**References**


[1] W. Horton. *et al*, Nucl. Fusion **45**, (2005) 976-985.

[2] E. Mazzucato *et al.*, Nucl. Fusion **49**, 055001 (2009).

[3] D. R. Smith *et al.*, Rev. Sci. Instrum., **79**, 123501 (2008).

[4] Y. Ren *et al*, Phys. Rev. Lett. **106** 165005 (2011).

[5] F. Jenko *et al*, Phys. Plasmas **8**, 4096 (2001).

[6] M. Kotschenreuther *et al*, Comput. Phys. Commun. **88** 128 (1995).

# Measurements of the natural plasma flow during the precursor of TCABR density limit disruptions

F. Salzedas[1,2], G. Ronchi[3], J.H.F. Severo[3], R.M.O. Galvão[3], E.K. Sanada[3]

[1] *Universidade do Porto, Faculdade de Engenharia, 4200-465 Porto, Portugal*

[2] *Instituto de Plasmas e Fusão Nuclear, Instituto Superior Técnico, Universidade de Lisboa, 1049-001 Lisboa, Portugal*

[3] *Instituto de Física, Universidade de São Paulo, São Paulo CEP 05508-090; Brazil*

## Introduction

Magnetic islands are known to degrade confinement of energy and particles in tokamak plasmas. They are also present in the precursor of density limit disruptions being observed until the onset of the energy quench of the disruption. The subject of the effect of plasma flow on tearing mode stability has been studied both theoretically [1] and experimentally [2]. However the experimental work has focused on plasmas with Neutral Beam Injection (NBI), which adds high momentum to both the plasma and the island, deviating them from their natural velocities. It is important to know the plasma/island rotation behavior without NBI, not just to know their natural values but because future large tokamaks, like ITER, are expected to have low plasma rotation due to both large plasma inertia and low applied torque. In this work we will show experimental values of the ion velocity and the magnetic island velocity measured in TCABR ohmic plasmas, without NBI, and during the precursor of a density limit disruption. In this way it was possible to compare the natural plasma velocity with the island velocity. The ion velocity was inferred from Doppler spectroscopy of the impurity lines of CVI. Details of this diagnostic system and an error analysis of the measured signal can be found in [3]. With the present set up, the toroidal velocity can be measured with time resolution equivalent to 600 Hz and precision somewhat better than 5 km/s.

## Plasma parameters and experimental conditions

The events described here were observed in several TCABR hydrogen plasmas with the following typical range of parameters, 60 kA $\leq I_p \leq$ 80 kA, $B_\phi = 1.06$ T, $a = 0.18$ m, $R_0 = 0.61$ m, and $q_a \approx 3.5$. The electron density was ramped up, to provoke a density limit disruption. During the plasma current flat top a magnetic island with $m/n = 2/1$, where $m$ and $n$ are the poloidal and toroidal mode numbers, exhibited fast growth to saturation with its amplitude remaining constant for approximately 20 milliseconds before the plasma disrupted (see plasma #30026 in Fig. 1 (d)). This type of behavior, commonly observed in tokamaks, has the peculiar feature

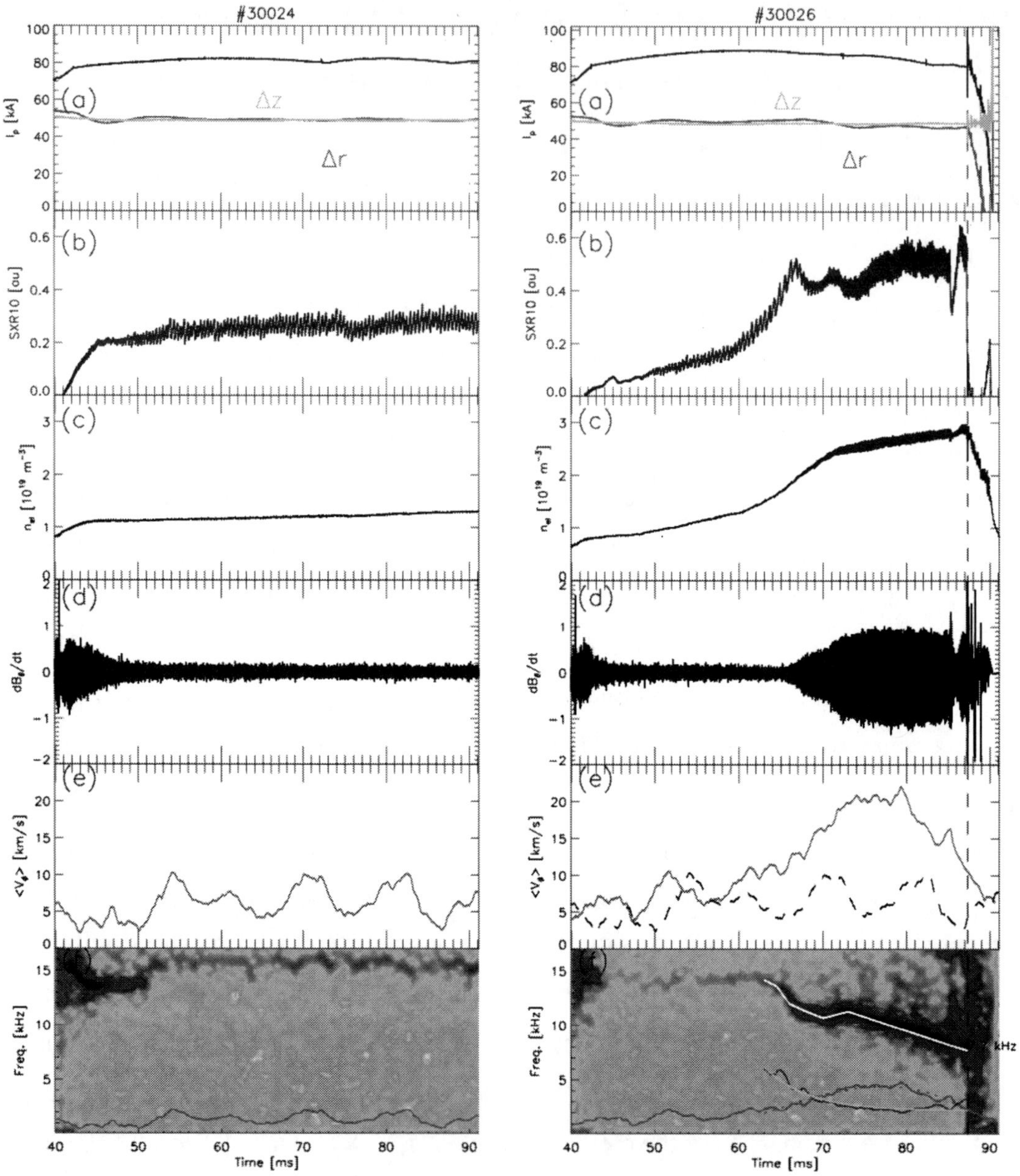


Figure 1: (a) plasma current, $I_p$, radial, $\Delta r$ and vertical, $\Delta z$ displacements of the magnetic axis, (b) SXR central chord, (c) central electron density, $n_e$ at $r = 0.5$ cm, (d) $dB_\theta/dt$, (e) intrinsic average toroidal rotation velocity $< v_\phi >$ measured at $r = 12$ cm, (f) spectrogram of $dB_\theta/dt$ with the toroidal rotation shown at (e) expressed in Hz (red line). For plasma #30026 is also shown the ratio $f_{island}/f_\phi$ (blue line).

in which the mode stays with very large and constant amplitude during several milliseconds until the disruption occurs. This clearly points out that the mode amplitude *per si* is not a sufficient condition to trigger the disruption. As is also well known, in parallel with the amplitude evolution is observed a reduction in the mode rotation frequency (Fig. 1 (f)) mainly due to the interaction of the mode with the metallic wall. For the plasmas in this study the disruptions occurred when the island rotation was $7.5 \leq f_{MHD} \leq 7.9$ kHz. Each tokamak has its characteristic range of frequencies for which disruptions occur. In large ones disruptions occur when the mode stops rotating and locks to the wall. It is not known why the disruption occurs at these frequencies, however the fact that it does reveals that somehow rotation can be part of a disruption trigger. We will show new input to this question, where it was found that at the disruption the plasma and the island toroidal angular acceleration, $\alpha$ tend to the same value.

TCABR operates with the plasma current opposite to the toroidal magnetic field $B_\phi$ and the intrinsic toroidal rotation is in the direction of $B_\phi$ or in counter current direction. The plasma toroidal velocity, $v_\phi$ was measured at the radial position of the mode's rational surface, $q = 2$ at $r_{island} = 12$ cm. The measurements were done in a plasma with a 2/1 island (see #30026 in Fig.1(e)) and in a plasma without an island (see #30024 in Fig.1(e)). The plasma toroidal rotation is also displayed in Fig.1(d) in units of Hz for better comparison with the island rotation frequency.

**Island and plasma rotation**

As the mode amplitude suddenly increases its rotation frequency drops from approximately 14 kHz to 11 kHz. The plasma toroidal rotation frequency, that is in the same direction as that of the island, increases slightly from 2.3 kHz to 2.6 kHz. So while the plasma experiences a small acceleration the island is decelerating. The island rotation keep's on decelerating until reaching the lowest rotation speed of $\approx 7.7$ kHz at the disruption.

However the plasma will only accelerate for about more 10 ms until it's rotation frequency reaches the maximum value of 4.5 kHz (or 20.9 km/s), at around 76 ms. From this time on the plasma will also decelerate. In Fig.2 (a) are plotted both the plasma and island toroidal angular acceleration. These curves were obtained from the derivative of a 3rd order polynomial fit of the velocity measurements also shown Fig.2 (b). The vertical dotted line indicates the time of the disruption. At this time the plasma is rotating at 2.3 kHz which is about three times smaller than the rotation of the magnetic island.

For comparison is also shown in Fig.1 discharge #30024 without magnetic island. For this discharge all parameters were kept the same as for #30026, with the exception of the density that was maintained constant at about $1.1 \times 10^{19}$ m$^{-3}$. The plasma toroidal rotation also measured at $r = 12$ cm is observed to evolve at around 5 km/s throughout the discharge. This is in sharp contrast with the behavior in plasma #30026 where $< v_\phi >$ reaches a maximum value of

20 km/s, when the island amplitude was saturated, which is four times higher than the speed without island. We are still analyzing the behavior of the plasma poloidal rotation that is more difficult to interpret because considering the predictions of neoclassical theory, the main ions poloidal velocity can be quite smaller than that of the impurities, in particular close to the plasma edge [4]. Even so it is clear that the island rotates at a quite different velocity from the plasma with the island apparently being able to drag the plasma. This indicates the island slips-trough the plasma, contrary to what is commonly assumed in most theoretical models.

The non linear evolution of a magnetic island that rotates with angular velocity $\Omega = \omega/\omega_\phi$ relatively to the plasma angular velocity $\omega_\phi$, is described in Eq.(88-89) of[1]. When the island reaches a constant amplitude there is a solution for these equations in which $\Omega$ decays exponentially in time. The yellow line in #30026 Fig.1(f) is a non-linear least squares fit to the data with the function

$$\Omega(t) = \frac{f_{island}}{f_\phi} = 2.22 + 8.60 \times 10^5 e^{-\frac{t-63}{5.13}},$$

indicative of an exponentially decay in time.

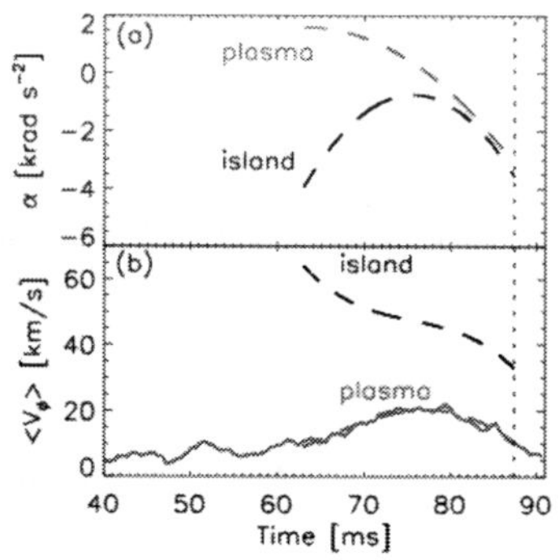


Figure 2: (a) Angular toroidal acceleration and (b) toroidal velocity for the period since the island saturates to disruption.

## Conclusions

We have shown for the first time experimental evidence from TCABR high density plasmas, of synchronization between plasma and $2/1$ magnetic island natural toroidal angular acceleration, $\alpha$, just before the disruption occurs. The data also indicates the island slips-trough the plasma, contrary to what is commonly assumed in most theoretical models.

This work has been supported by different grants from FAPESP (Fundação de Amparo à Pesquisa do Estado de São Paulo) and CNPq (Conselho Nacional de Desenvolvimento Científico e Tecnológico) and by the University of São Paulo and University of Porto Collaboration Agreement. IPFN activities received financial support from Fundação para a Ciência e Tecnologia through project UID/FIS/50010/2013.

## References

[1] A. Smolyakov, et al, *Phys. Plasmas* **2**, 1581 (1995).

[2] R. J. La Haye, et al, *Phys. Plasmas* **17**, 056110 (2010).

[3] J.H.F. Severo, et al, *Nuc. Fusion* **49**, 115026 (2009).

[4] Y.B. Kim, et al, *Phys. of Fluids B* **3**, 2050 (1991).

# Characterization of toroidal intrinsic rotation with MHD activity in the TCABR tokamak

G. Ronchi[1], J.H.F. Severo[1], F. Salzedas[2,3], R.M.O. Galvão[1]

[1] *Instituto de Física, Universidade de São Paulo, São Paulo CEP 05508-090; Brazil*

[2] *Universidade do Porto, Faculdade de Engenharia, 4200-465 Porto, Portugal*

[3] *Instituto de Plasmas e Fusão Nuclear, Instituto Superior Técnico, Universidade de Lisboa, 1049-001 Lisboa, Portugal.*

## Introduction

Plasma rotation has an important play in stabilization of MHD modes and reducing turbulent transport of particles and energy. Because in fusion reactors it is expected the torque provided by external sources will be small, the intrinsic (or spontaneous) rotation is of great interest[1, 2, 3]. Furthermore, the origin and physics of plasma rotation is also an important issue by itself.

The behavior of the intrinsic toroidal rotation during the growth and saturation of m/n = 2/1 magnetic islands, triggered by programmed density ramp up, has been investigated in L-mode ohmic discharges in the TCABR tokamak. In those discharges $R = 0.61$ m, $a = 0.18$ m, $I_p \approx 80$ kA, $B_t = 1.07$ T, $q(a) \approx 3.5$ and the toroidal spontaneous rotation of the plasma core is in the counter-current direction. The results show that the plasma is accelerated as the island starts to grow, while the island frequency slows down. And, as the island saturates, the toroidal rotation decreases quite rapidly (faster than the island), and the discharge is followed by a major disruption. In some discharges, where the density decreases after the island saturation (and thus, avoiding the plasma disruption), the MHD instability becomes smaller until it vanishes, and the toroidal rotation slows down to its original value before the gas injection.

## Diagnostic system

The plasma rotation in the TCABR is measured with 2 photomultiplier (PMT), per optical chord, at the axial and lateral output of a single monochromator, as shown in figure 1. They are adjusted to measure slightly different wavelengths by means of a beam splitter inside the monochromator output. Thus, in this scheme, the plasma rotation $v$ can be estimated by the Doppler shift of an impurity emission line with central wavelength $\lambda_0$: $v = c\frac{\Delta\lambda}{\lambda_0} = R(\Delta\lambda)$ where $\Delta\lambda$ is the Doppler shift of the spectral line due to the rotation, $c$ is the speed of light and $R(\Delta\lambda)$ is the ratio of the signal from the two slits as a function of the Doppler shift. The $R(\Delta\lambda)$ is obtained from calibration, where all optical ports are set to view the central chord of the plasma poloidal cross section (where the rotation is negligible), and the ratio of the impurity emission signal is recorded as a function of wavelength.

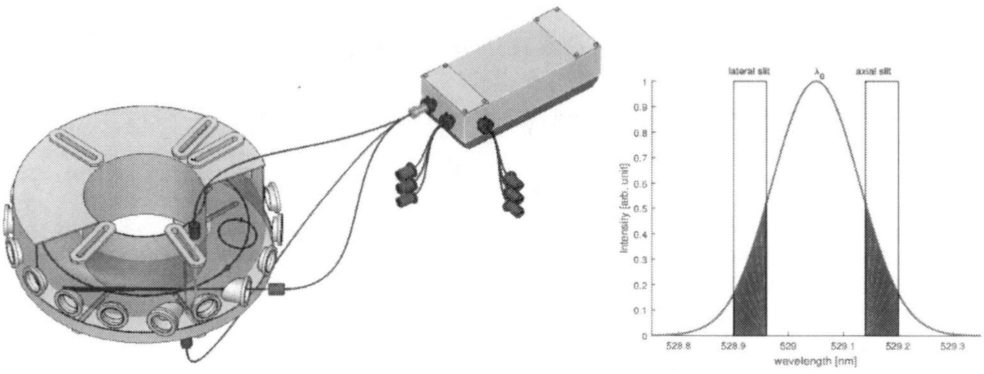


Figure 1: *Experimental setup: the light from the poloidal chords and toroidal chord arrive in the entrance slit of the monochromator, and each signal is measured in the PMT at the axial and lateral output slit. Currently, there are 3 optical ports, one for toroidal rotation, and 2 for poloidal rotation.*

**Results**

In those discharges, the carbon line CVI (529.05 nm) was employed to measure plasma rotation velocity. In the figure 2 are shown the main signals from discharges with and without disruption where, in both cases, the density was ramped up to induce the MHD activity. The mode m/n = 2/1 is the dominant one, as estimated from spatial FFT from the Mirnov coils (figure 3).

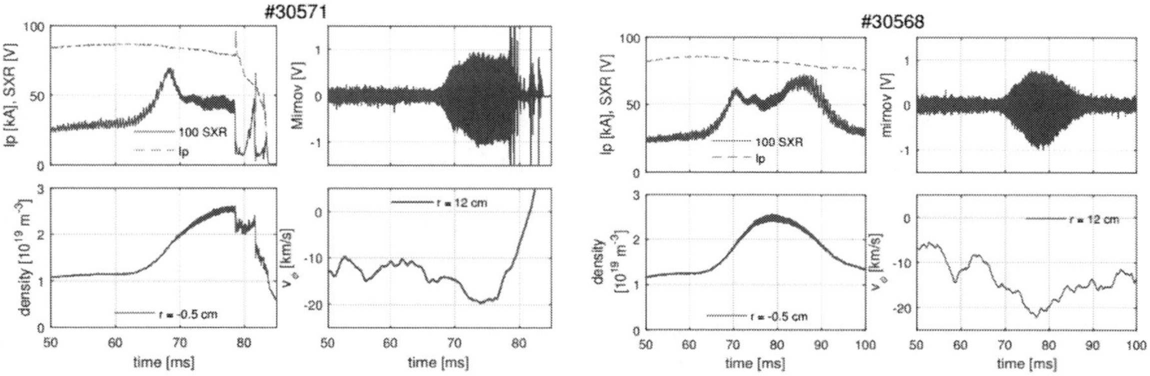


Figure 2: *Plasma current, soft X-rays (SXR), line-average electron density, Mirnov signal and toroidal velocity for discharges with (#30571) and without disruption (#30568)*

In these discharges the line-averaged electron density is slowly increased from $1 \times 10^{19}$ m$^{-3}$ to $2.8 \times 10^{19}$ m$^{-3}$. In this process, the sawtooth instability, as seen in the SXR diagnostic, fades out just before a 2/1 magnetic island is destabilized, indicating an internal change of the safety factor profile.

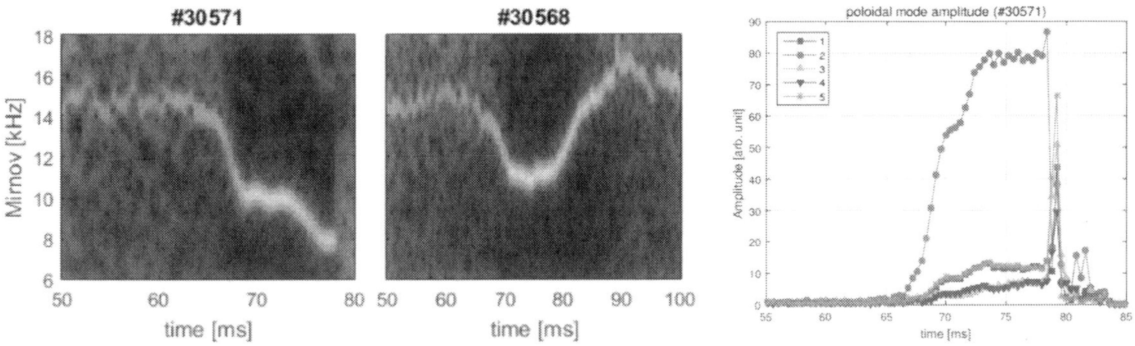


Figure 3: *Spectrogram of the Mirnov oscillations for discharges with (#30571) and without disruption (#30568), and the poloidal amplitude for the discharge #30571.*

At this point, the plasma core velocity increases and, a few milliseconds later, the mode amplitude saturates, while the toroidal velocity continues to increase. If density continues to increase, the strong interaction between the magnetic island and the mirror currents in the plasma wall reduces the MHD frequency, and in a few milliseconds later, the toroidal velocity starts to decrease rapidly and before it changes its sign, a major disruption occurs.

The radial profile of the line-integrated toroidal rotation velocity for plasmas both with and without MHD is shown in figure 4: the average velocities during approximately 50 ms to 65 ms (without MHD), and the maximum velocity in the presence of MHD (usually at $\sim$ 77 ms, just after the mode m/n = 2/1 saturation). The rotation in the plasma core is in the counter current direction, while at the edge, usually it is in co-current direction.

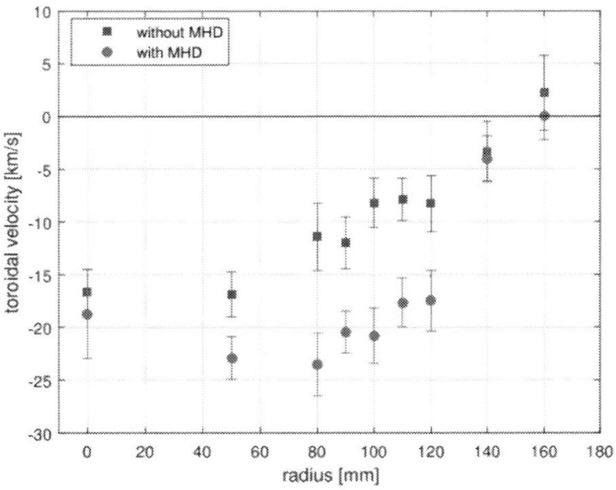


Figure 4: *The radial rotation profile with and without m/n=2/1 activity.*

## Discussion

Even in the absence of MHD activity, there is a natural breaking of axisymetry due to toroidal field ripple and magnetic error fields, leading to a spontaneous rotation as it has been extensively investigated [4, 5]. The increase in toroidal velocity may be explained due to enhanced transport (from breaking axisymmetry), leading to an internal rearrangement of toroidal momentum [6].

Further investigation must be done with electrostatic probes at the scrape-of-layer to determine the radial flux change in the presence of 2/1 tearing mode. Also, a proper inversion of this signal is still necessary to estimate the actual rotation value, particularly, in the plasma core.

A similar behaviour, although not presented here, is also observed in during mode coupling (n=2, 3) in discharges with small plasma vertical displacement ($\sim$ 1 cm), but in this case the interpretation is not straightforward.

## Acknowledgements

This work has been supported by different grants from The State of São Paulo Research Foundation and CNPq, and by the University of São Paulo and University of Porto Collaboration Agreement.

## References

[1] F.I. Parra *et al. Phys. Rev. Lett.* 108, 095001 (2012)

[2] Severo J. H. F. *et al. Nucl. Fusion* 43 1047 (2003)

[3] Severo J. H. F. *et al. Nucl. Fusion* 49 115026 (2009)

[4] M. F. F. Nave *et al. Phys. Rev. Lett.* **105** 105005 (2010)

[5] A. M. Garofalo *et al. Phys. Rev. Lett.* **101**, 195005 (2008).

[6] J S deGrassie. *Plasma Phys. Control. Fusion* **51** 124047 (2009).

# Tokamak equilibrium reconstruction code LIUQE: implementation, applications, developments

J.-M. Moret[1], P. Bettini[3], S. Coda[1], B.P. Duval[1], F. Felici[2], H.B. Le[1], H. Reimerdes[1], O. Sauter[1], the TCV Team[1][1] and the EUROfusion MST1 Team[4]

[1]*CRPP - EPFL, Centre de Recherches en Physique des Plasmas, École Polytechnique Fédérale de Lausanne, Lausanne, Switzerland;* [2]*Eindhoven University of Technology, Eindhoven, The Netherlands;* [3]*Consorzio RFX, Padova, Italy;* [4]*http://www.euro-fusionscipub.org/mst1*

Equilibrium reconstruction consists of identifying, from experimental measurements, a plasma current distribution that satisfies the pressure balance. Elongated equilibria in TCV impose specific requirements on its reconstruction code LIUQE [2], mainly the stabilisation of the vertical position, the influence of vessel eddy currents and for real time (RT) control a cycle time compatible with the actuator response time. Coupling with RAPTOR [3] is under development, in which 1D transport equations are solved to obtain physics based profiles for the pressure and the current density.

**Vertical stabilisation.** LIUQE adopts a method based on an iterative solution of the Poisson equation for the poloidal flux $\psi$, $\Delta^*\psi = -2\pi\mu_0 r j_\varphi$, coupled with a linear parametrisation of the plasma current density $j_\varphi$. This algorithm is unstable against vertical motion for elongated shapes. The growth rate is given by $1 - \partial_z^2\psi_{eA}/\partial_z^2\psi_A$ [2] where $\psi_e$ is the flux produced by the external coils, and the subscript $A$ tags quantities evaluated on the magnetic axis; this growth rate is thus directly related to the curvature index of the shaping field, and reaches 2 for high elongations. To stabilise the algorithm, it is assumed that the available measurements contain the information necessary to determine the equilibrium vertical position; thus a free parameter $\delta z$ is added when fitting the plasma current distribution, seeking for solutions of the form $j_\varphi = 2\pi\left(rp'(\psi(r, z+\delta z)) + \frac{TT'(\psi(r, z+\delta z))}{\mu_0 r}\right)$. The specific response to $\delta z$ of a magnetic measurement located at $(r_m, z_m)$ can be conveniently evaluated using the relation $\int G j_\varphi(r, z+\delta z)drdz \cong \ldots + \delta z\int G\partial_z j_\varphi(r, z)drdz = \ldots + \delta z\int \partial_{z_m} G j_\varphi(r, z)drdz$ where $G(r_m, z_m, r, z)$ is the appropriate Green's function. The cumbersome calculation of $\partial_z j_\varphi$, discontinuous on the LCFS, is thus avoided. The Poisson equation can be solved with the unshifted plasma current distribution but with shifted external currents, $\Delta^*\psi_+(r, z) = -2\pi\mu_0 r(j_\varphi(r, z) + j_e(r, z-\delta z))$, which implies a correction on the boundary conditions given by $-\delta z\int \partial_z M(r, z, r', z')j_e(r', z')dr'dz'$ where $M$ is the mutual inductance. The final solution writes $\psi = \psi_+ + \delta z\partial_z\psi_+$. This particular treatment does not impair the algorithm efficiency and accuracy, since it relies only on the smooth derivative $\partial_z \psi_+$ and on precalculated $\partial_{z_m} G$ and $\partial_z M$.

**Vessel currents.** A vacuum vessel with low resistance designed to enhance passive stabilisation of the vertical position or a conducting shell that stabilises MHD modes give rise to eddy currents with a sizable influence on the equilibrium reconstruction. These are taken into account with dedicated measurements or with an appropriate electromagnetic modelling of the conducting structures. In the case of TCV [2], the vacuum vessel is poloidally divided in 38 segments, each of them being equipped with a flux loop. The current in the segments is estimated with $I_s = -R_s^{-1} U_s$ where $U_s$ is the voltage measured by the loops. The segment resistance $R_s$ is experimentally obtained by applying long current ramps $\frac{dI_a}{dt}$ in the poloidal field coils during which the relation $I_s = -R_s^{-1} M_{sa} \cdot \frac{dI_a}{dt}$ holds. The magnetic measurements $X_m$ can then be used to evaluate $R_s$, solving $X_m - G_{ma} \cdot I_a = -G_{ms} \cdot R_s^{-1} M_{sa} \cdot \frac{dI_a}{dt}$ in a least square sense.

**Error analysis.** During the measurement fitting step, the current distribution is parametrised with base functions $g_g(\psi)$ as follows: $j_\varphi = \sum_g a_g r^{\nu_g} g_g(\psi)$, with $\nu_g = \pm 1$ for the $p'$ and $TT'$ terms respectively. A covariance analysis of this procedure shows that, using the available measurements, i.e. magnetic measurements including the diamagnetic flux, the covariance between the coefficients $a_g$ typically reaches 0.99 in absolute value when 3 base functions are used, while it is maintained to 0.80 for 2 base functions. In the latter case, error propagation yields an accuracy of 3% for the central pressure current density. This limit in the number of relevant base functions is compatible with the measurement set, the total plasma current giving a constraint on the base function sum, and the diamagnetic flux on the $TT'$ terms.

**Coupling with RAPTOR.** The goal of coupling LIUQE and RAPTOR is to obtain equilibrium reconstruction using physics based profiles for the pressure and the current density, obtained by solving with RAPTOR 1D transport equations constraint by measurements. This alleviates the arbitrary choice in the base functions and allows, for example, steep or non-monotonic profiles. The flux surface averaged (FSA) equations involve flux contour integrals in the form $Q_{\mu\nu} = (2\pi)^{-\mu} \oint |\nabla\psi|^{\mu} r^{\nu} dl_\theta$, a particular case being the safety factor $q = \frac{TQ_{-1,-1}}{2\pi}$. Based on the contour description as the distance from the magnetic axis $r^*(\rho, \theta)$ for a given normalised flux $\rho^2$ and poloidal angle $\theta$, one can rewrite $Q_{\mu\nu}$ as $Q_{-1,\nu} \sim \oint_0^{2\pi} r^\nu \rho^{-1} \partial_\rho r^{*2} d\theta$ and $Q_{1,\nu} \sim \oint_0^{2\pi} r^\nu \left(r^{*2} + \frac{1}{4} r^{*-2} (\partial_\theta r^{*2})^2\right) \rho (\partial_\rho r^{*2})^{-1} d\theta$. Near the magnetics axis, a second order Taylor expansion of the flux yields the analytical formula given in [2]. For a divertor equilibrium, the limit values on the separatrix are $Q_{-1,\nu}/Q_{-1,1} = r_X^{\nu-1}$ and $Q_{1,\nu} = 0$, $r_X$ being the radial position of the active X point.

In between an efficient determination of the flux contours is necessary. It is based on a bilinear interpolation of the flux $\hat{\psi}$ in the grid cell. A Newton method is used to search for the intersection of the flux contour at level $\psi_\rho$ and the radius at angle $\theta$, iteratively applying the update: $r^* \leftarrow r^* - (\hat{\psi} - \psi_\rho)/(\nabla\hat{\psi} \cdot \nabla r^*)$ where $\hat{\psi}$ is evaluated at $(r_A - r^* cos\theta, z_A + r^* sin\theta)$. Each Newton iteration is interleaved with a reconstruction iteration; the latest values for $\psi$ and $(r_A, z_A)$ are used while the cell containing the contour point at the previous iteration is used for the interpolation. Note finally that for an accurate evaluation of the derivative $\partial_\rho r^{*2}$ in $Q_{\mu\nu}$, a non equidistant $\rho$ grid is chosen, especially near an X point.

In its simplest form the coupling between LIUQE and RAPTOR is achieved by interleaving a LIUQE iteration for each RAPTOR iteration: RAPTOR is fed with the 2D equilibrium geometry obtained from LIUQE, while LIUQE uses as base functions for $p'$ and $TT'$ the profiles calculated by RAPTOR using a FSA Grad-Shafranov equation.

**Implementation.** To facilitate its collaborative development and its application to other devices, LIUQE is coded in MATLAB, with a generic diagnostic description and an interface with the ITM data structure. It has been successfully used to analyse RFX tokamak equilibria.

In its RT implementation, the number of iterations is reduced to one, while the value for the current distribution in the Poisson equation is taken from the previous time sample, assuming that a sufficiently short sampling time is used so that the equilibrium does not change significantly between two samples. The RT version has been implemented in TCV's control system [4], coded in its SIMULINK standard environment [5]. Running on one single core of an INTEL i7 processor overclocked at 5 GHz, a cycle time of 200 μs was achieved for an $(r, z)$ grid of 28 by 65, using all 133 measurements and including the FSA quantities necessary for the RT RAPTOR on a $(\rho, \theta)$ grid of 16 by 32. Amongst the choice of numerical methods and code optimisation techniques that lead to this performance, we can cite the followings:

- In fitting the currents to the measurements, the latter are grouped in the measurements $Y_e$ sensitive only to the external currents $I_e$, those $Y_i$ sensitive only to the plasma current distribution, i.e. the parameters $a_j = \{a_g, \delta z\}$, and those $Y_r$ sensitive to both. The fitting is then obtained by solving in a least square sense the equations $\begin{bmatrix} Y_r - W_{re} \cdot J_e \\ Y_i \end{bmatrix} = \begin{bmatrix} W_{rj} \\ W_{ij} \end{bmatrix} \cdot a_j$ and $\begin{bmatrix} Y_r - W_{rj} \cdot a_j \\ Y_e \end{bmatrix} = \begin{bmatrix} W_{re} \\ W_{ee} \end{bmatrix} \cdot J_e$ where $W$ are the corresponding response matrices. This system can be solved iteratively first for $a_j$ and then for $J_e$; in doing so the only matrix which depends on the equilibrium and cannot be precalculated and inverted has the size of $a_j$, at most 4 by 4, and is inverted in RT using an unrolled LDL decomposition.

- Quantities depending on $(r, z)$ are stored with $z$ as the fastest varying index, leading to contiguous indexing for the numerous operations along the $z$ direction.
- For loop intensive operations are coded in C. They pervade the Poisson solver, based on a direct tridiagonal resolution in the $r$ direction and a cyclic reduction in the $z$ direction, the search for the X points and the magnetic axis, and the identification of the plasma domain.
- Matrix and vector multiplications use the *Intel Math Kernel Library (MKL)*.
- Code vectorisation using the specific instruction set of the target processor is automatically achieved with the compiler option *-ax*. This vectorisation is further facilitated by using integer and floating point numbers with the same binary length.
- All floating point operations are carried out in single precision, since none of the algorithm steps depends unduly on rounding errors. The required memory for variables and parameters then fits in the 8 Mbytes of the L3 cache, minimising cache faults.

**Conclusion.** The equilibrium reconstruction code LIUQE features specific requirements to handle highly shaped equilibria in TCV. It has been translated in MATLAB and SIMULINK to facilitate future collaborative developments and applications to other devices. Its RT implementation runs on the TCV control system with a 200 μs cycle time and serves as a basis for iso-flux shape control [6] and for sawtooth and NTM control using ECH [1]. Coupling with RAPTOR offers a promising way to introduce physics based source profiles, including steep or non monotonic profiles often encountered in advanced scenarios.

**Acknowledgments.** This work was supported in part by the Swiss National Science Foundation and the Netherlands Organisation for Scientific Research, and has been carried out within the framework of the EUROfusion Consortium and has received funding from the Euratom research and training programme 2014-2018 under grant agreement No 633053. The views and opinions expressed herein do not necessarily reflect those of the European Commission.

[1] S. Coda for the TCV Team, Nucl. Fusion 55 82015) 104004

[2] J.-M. Moret, B.P. Duval, H.B. Le, S. Coda, F. Felici, H. Reimerdes, Fusion Eng. Des. 91 (2015) 1

[3] F. Felici, O. Sauter, S. Coda. B.P. Duval, T.P. Goodman, J.-M. Moret, et al., Nucl. Fusion 51 (2011) 08052

[4] H.B. Le, F. Felici, J.I. Paley, B.P. Duval, J.-M. Moret, S. Coda, et al., Fusion Eng. Des. 89 (2014) 155

[5] F. Felici, H.B. Le, J.I. Paley, B.P. Duval, S. Coda, J.-M. Moret, et al., Fusion Eng. Des. 89 (2014) 165

[6] H. Amand, J.-M. Moret, F. Felici, S. Coda, H.B. Le, 41st EPS Conf. on Plasma physics (2014) P5.069

# Investigating the double scale length of limited plasmas with nonlinear simulations of the TCV Scrape-Off Layer

F. Nespoli[1], B. Labit[1], I. Furno[1], F.D. Halpern[1], P. Ricci[1]

[1] *Ecole Polytechnique Fédérale de Lausanne, Centre de Recherches en Physique des Plasmas, 1015 Lausanne, Switzerland*

## Introduction

Heat loads onto the limiter for inboard limited plasmas have been measured recently in TCV using infrared thermography [1]. The parallel heat flux profile in the Scrape-Off Layer (SOL) exhibits a double scale length, confirming previous results from other tokamaks [2]. A short scale length of the order of several millimeters is always present in the vicinity of the Last Closed Flux Surface (LCFS), while in the main SOL the heat flux decay length is typically ten times longer. Even though the ITER first wall design was recently changed to handle the extra heat flux associated with such a narrow feature [3], the underlying physics is not well understood yet.

## Nonlinear simulations of the TCV SOL

To improve our understanding, numerical simulations of the TCV SOL are performed using the GBS code [4]. This code solves the drift-reduced Braginskii equations in a 3D geometry. The resulting plasma turbulence determines self-consistently both the equilibrium profiles and their fluctuations. The simulations include the effects of magnetic shear $\hat{s} = 1.5$, finite aspect ratio $\epsilon = 0.24$ and ion temperature $\tau = 1$. The reference case is based on experimental parameters from the TCV discharge #49170, the electron density $n_{e,0} = 5\ 10^{18} m^{-3}$ and temperature $T_{e,0} = 25$ eV at the LCFS are estimated from flush mounted Langmuir probe (LP) data, located on the central column of TCV, acting as the limiter. These values determine $\rho^* = \rho_s/R_{ax}$, and therefore the size of the simulation, and the normalized Spitzer resistivity $\nu = e^2 n_e R_{ax}/(m_i \sigma_{\|} c_s)$, where $\rho_s$ is the ion Larmor radius computed with the sound speed $c_s$, and $R_{ax} = 0.84$ m is the position of the magnetic axis. The toroidal field on axis is $B_t = 1.45$ T. The value of $q_{edge} = 3.2$ is computed from the equilibrium reconstruction. The geometry of the toroidal limiter in the simulations is shown in Fig.1, together with the main simulation parameters and a snapshot of the resulting electron density. Also, in this first simulation the

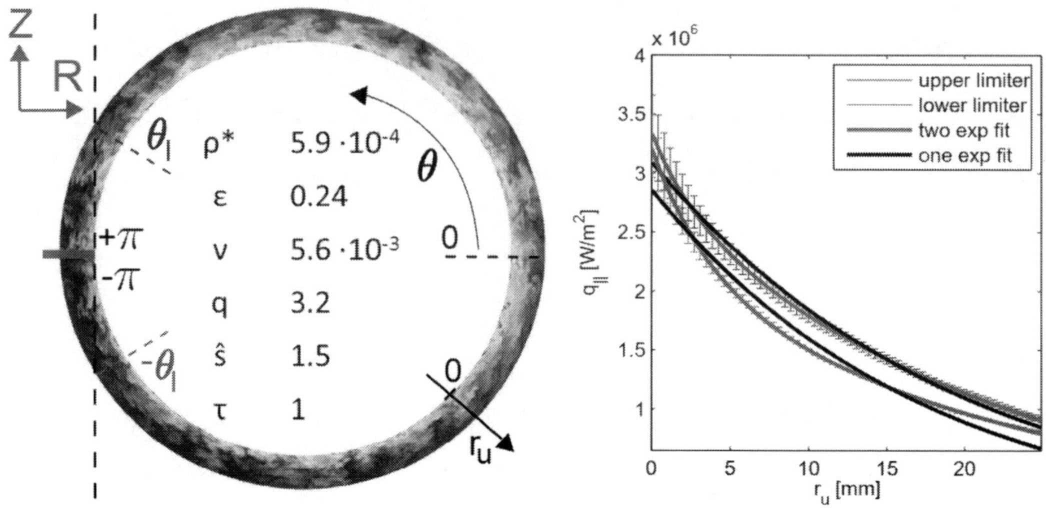


Figure 1: Left: snapshot of the electron density resulting from the GBS simulation, together with simulation (red) and TCV (black, dashed) limiter geometry. Right: fit of radial $q_{||}$ profiles taken along the limiters (blue, green) with a single exponential (black) and with a sum of two exponentials (red).

magnetic field and plasma current are antiparallel, contrary to the experiment.

**Comparison with experimental data**

The parallel heat flux is computed from the plasma density $n_e$, the electron and ion temperature $T_e, T_i$ as $q_{||} = \gamma n_e \sqrt{\frac{T_e + T_i}{m_i}} T_e$, where all the quantities are averaged in time and in the toroidal direction, and $\gamma = 7$ is the sheath power transmission factor. In order to compare with experimental data, $q_{||}(r_u, \theta)$ has been averaged for $\theta_l < \theta < \pi$ and $-\pi < \theta < -\theta_l$, where $\theta_l$ is the angle where the whole GBS profile crosses the TCV-like limiter, resulting in two radial profiles, for the upper and lower parts of the limiter respectively. The narrow feature observed in the experiments is also seen in the simulations. Indeed, the limiter profiles are well fitted by the sum of two exponentials

$$q_{||}(r_u) = q_s \exp(-r_u/\lambda_s) + q_l \exp(-r_u/\lambda_l) \tag{1}$$

while the fit with a single exponential is unsatisfactory as shown in Fig.1. The fitted values of $\lambda_s = 2.7, 3.7$ mm and $\lambda_l = 22.3, 25.3$ mm for the upper and lower limiters respectively are similar with the experimental ones $\lambda_{s,exp} = 2.9$ mm , $\lambda_{l,exp} = 36.7$ mm. Furthermore, $q_{||}(r_u)$ has been analyzed at every fixed $\theta$, revealing a strong poloidal variation of the fitted parameters (Fig.2). Above the midplane ($\theta \geq 0$), the narrow feature is relatively less

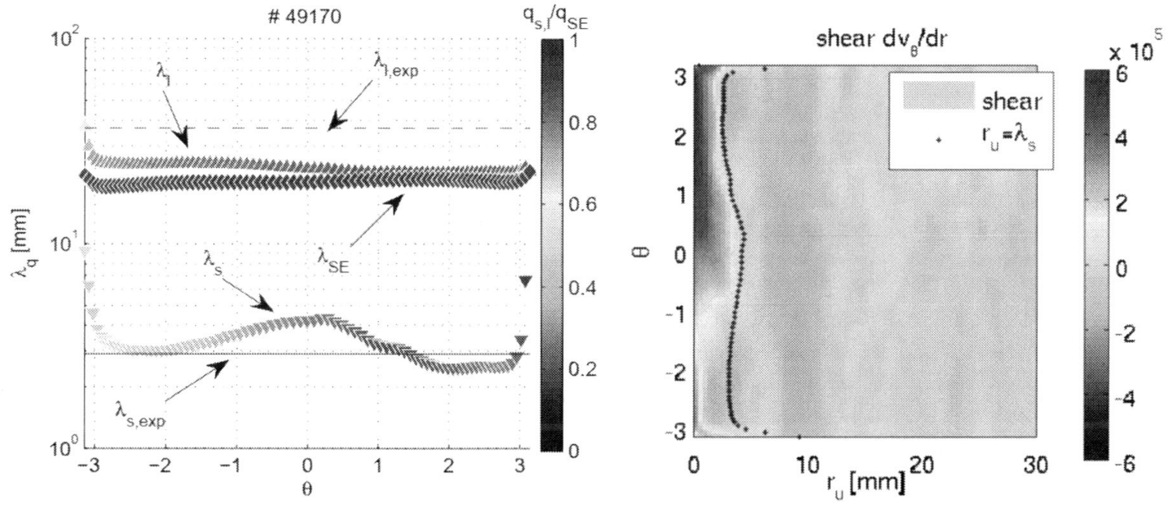


Figure 2: Left: poloidal variation of the fitted decay lengths of the heat flux, color-coded with their relative magnitude. The value resulting from a single exponential fit is shown, together with the experimental values. Right: velocity shear in the polidal plane, displayed with the short scale length of the heat flux profile.

strong since $q_s/q_{SE}$ decreases, where $q_{SE}$ is the heat flux magnitude resulting from a single exponential fit. Such an asymmetry might be due to the $\mathbf{E}\times\mathbf{B}$ drift velocity, whose main component is in the positive $\theta$ direction. Also, a strong correlation is found between the velocity shear $dv_\theta/dr$ of the $\mathbf{E}\times\mathbf{B}$ velocity and the steeper part of the profile, as shown in Fig. 2. The current in the SOL flowing to the limiter is computed from the simulation as $j_{||}=q_e n_e(v_{i,||}-v_{e,||})$, where $v_{i,||}, v_{e,||}$ are the ion and electron parallel velocities. Two profiles are obtained for the limiter with the same procedure used for $q_{||}$. The result is compared in Fig. 3 with the current measured by the LPs biased at the limiter potential. The presence of electronic currents in the region $r_u \leq \lambda_s$ is found in both experiments and simulations, suggesting a correlation between the narrow feature and such non-ambipolar currents, as already observed in COMPASS [5].

**The role of resistivity**

Experimentally, it has been shown that the excess heat load in the SOL scales with $\Delta P_{SOL} \equiv 4\pi R_{LCFS}B_\theta/B_\phi \int_0^\infty [q_{||}(r_u)-q_{||,l}(r_u)]dr_u \propto T_e^{1.43}n_e^{-1.01}$ [1], i.e. approximately with $1/\nu$. Three more simulations have been performed, in which the resistivity is increased by a factor 10, 20, 40 respectively. Due to time constraints, these simulations are not fully converged yet.

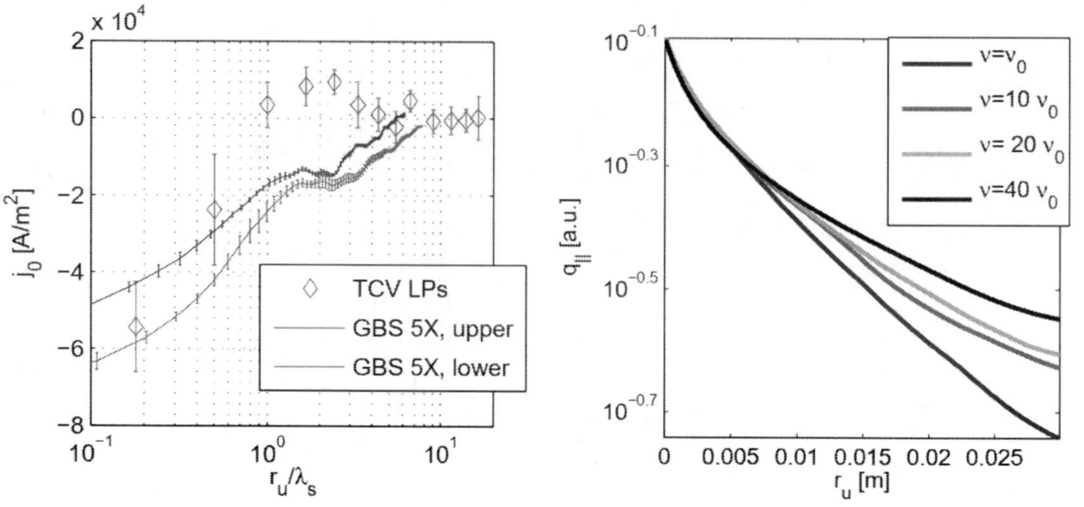


Figure 3: Left: density current flowing to the grounded limiter measured by LP in TCV discharge #49170 (red diamonds), and average profiles from the GBS simulation (5X magnification) for the upper (blue) and lower (green) limiter. Rigth: change in the (poloidally averaged) $q_{||}$ profiles increasing the resistivity.

Preliminary results are shown in Fig.3, where the poloidally averaged profiles of $q_{||}$ are displayed. Qualitatively, increasing $\nu$ flattens the profiles in the main SOL, while the narrow feature is still present. A quantitative analysis is foreseen, once an increased statistics will be available, together with a more detailed analysis of the statistical properties of plasma density and potential fluctuations. Moreover, another simulation with a different safety factor ($q_{edge} = 5.2$) is being performed and experiments in TCV are planned for the MST-1 campaign at the end of 2015. In particular, experiments in helium plasmas are foreseen.

This work has been carried out within the framework of the EUROfusion Consortium and has received funding from the Euratom research and training programme 2014-2018 under grant agreement No 633053. The views and opinions expressed herein do not necessarily reflect those of the European Commission.

[1] F. Nespoli et al., J. Nucl. Mater. 463 (2015)

[2] G. Arnoux et al., Nucl. Fusion 53 (2013) 073016

[3] M. Kocan, R.A. Pitts, et al., Nucl. Fusion 55 (2015) 033019

[4] P. Ricci et al., Plasma Phys. Controlled Fusion 54 (2012) 124047

[5] R. Dejarnac et al., J. Nucl. Mater. 463 (2015)

# Advanced divertor research on the TCV tokamak

W.A.J. Vijvers[1,2,3], G.P. Canal[2], B.P. Duval[2], B. Labit[2], B. Lipschultz[3], T. Lunt[4], F. Nespoli[2], H. Reimerdes[2], U.A. Sheikh[2], C. Theiler[2], C. Tsui[5], K.H.A. Verhaegh[3], and the TCV team

[1]*FOM Institute DIFFER - Dutch Institute for Fundamental Energy Research, Eindhoven, the Netherlands;* [2]*Ecole Polytechnique Fédérale de Lausanne (EPFL), Centre de Recherches en Physique des Plasmas (CRPP), Lausanne, Switzerland;* [3]*York Plasma Institute, University of York, United Kingdom;* [4]*Max Planck Institute for Plasma Physics, Garching, Germany;* [5]*UCSD, San Diego, USA*

## Introduction

Continuous operation of a power plant-scale fusion reactor (e.g. DEMO) will require the safe exhaust of $10^8$ W of power via a $10^{-3}$ m thin scrape-off layer (SOL) surrounding the confined plasma. Although expanded in the divertor, the SOL thickness (set by plasma transport physics) usually limits the volume available for radiative and collisional energy losses, as well as the surface area of the reactor-wall usable for exhaust. Several concepts have been proposed to improve upon the conventional magnetic geometry of the single-null divertor (SND). These aim to enlarge the SOL volume ($V_{SOL}$) and plasma-wetted wall area ($A_{wet}$), and thereby lower the peak heat fluxes to the walls. Since the flux expansion at the target in any divertor configuration is limited by the minimum acceptable total fieldline angle, the alternative concepts aim to increase $A_{wet}$ by enhancing the cross-field transport. TCV contributes to a DEMO exhaust solution by investigating the exhaust physics of a wide range of alternative divertor concepts. In previous work, cross-field transport was shown to be significantly enhanced in the Snowflake divertor (SFD) [1], possibly in part due to instabilities at high $\beta_{pol}$ (normalized plasma pressure) [2]. This paper highlights recent TCV results on how diffusion and drifts affect the target profiles in the SFD, and compares the spatial distribution of radiation in the X-divertor (XD), and X-point Target Divertor (XTD) with that of the better known SND.

## Scrape-off layer plasma transport in the Snowflake divertor

SFD configurations generally feature a second x-point, $X_2$, (and associated separatrix) in the vessel and in the proximity of the primary one. The lower poloidal field ($B_p$) in the null region locally increases $V_{SOL}$ and the connection length ($L_{//}$) compared to the SND, which might facilitate access to a detached divertor state. In TCV, $V_{SOL}$ and $L_{//}$ only increase in the near-SOL, whereas the full SOL benefits in larger tokamaks [3]. The manifestation of any potential benefits and the spatial target profiles depend strongly on the location of $X_2$ relative to the primary x-point, see e.g. [4, 1] and the experimental results below.

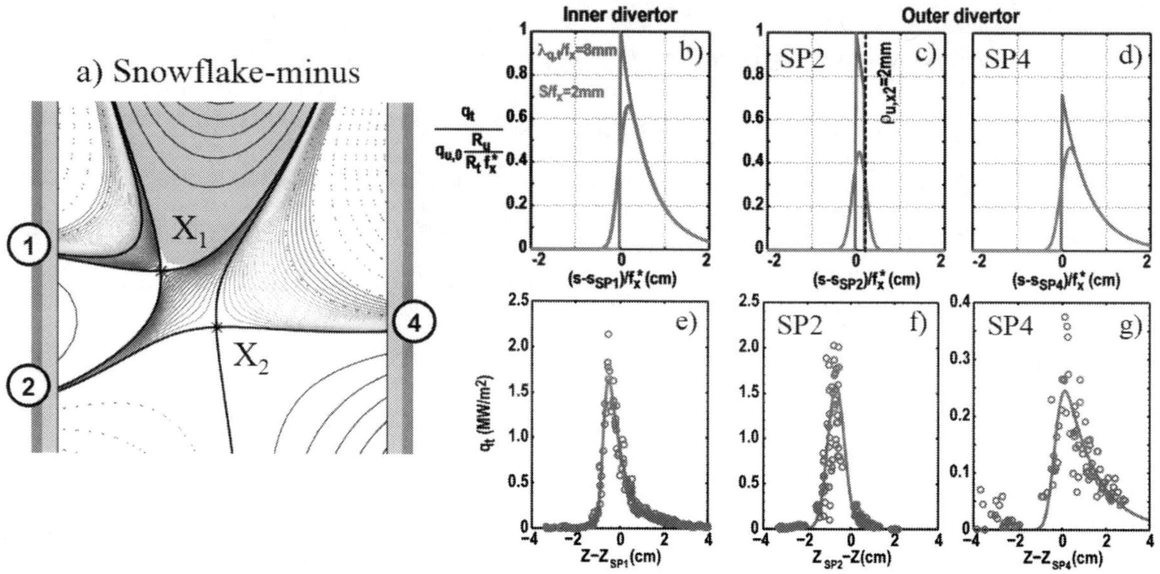


**Figure 1:** *a) SOL geometry of the Snowflake-minus configuration. The secondary separatrix forks the SOL and diverts parts of the plasma to two strike points. Colors indicate the assumed heat flux density. b-d) Calculated target profiles at SP1,2,4 (mapped upstream and normalized to the upstream heat flux) without (blue) and with (red) cross-field diffusion modeled by convoluting an exponential SOL power density profile with a Gaussian distribution, following [5]. e-g) Target profiles measured by Langmuir probes near the same SPs of the SF- configuration, fitted with an Eich-profile.*

If $X_2$ is placed in the common flux region, i.e. in the SF- configuration, the upstream heat flux profile on one side of the SOL splits into two strike points (SPs), SP2 and SP4 for the configuration in Fig. 2. The target profiles depend on the fraction of the SOL width inside the secondary separatrix. A simple analytical model for the target profiles in the SND was presented in [5], approximating the cross-field transport in the divertor by a convolution of an exponential upstream profile of the heat flux in the poloidal direction with a Gaussian. In a similar way, we truncated the exponential profiles of the SF- and convoluted them with a Gausssian. Fig. 1c shows that, with realistic values for the SOL width and the S-parameter (Gaussian spreading), the creation of two additional steep perpendicular gradients in the SOL enables a significant reduction of the peak heat flux by cross-field diffusion. Figs. 1e-g show the experimental target heat flux profiles (measured with Langmuir probes) accurately fitted with the model function. A more detailed numerical study of cross-field diffusion in the SF- using the transport code EMC3-Eirene indicates an even stronger reduction of the *parallel* heat flux by up to a factor 2 [6]. Note that neither model assumes increased effective cross-field diffusivities in the SFD.

When $X_2$ resides in the private flux region (a SF+ configuration), particle and heat fluxes to the extra strike points only occur through cross-field transport from the SOL. These fluxes are significantly greater than that explained by modelled cross-field diffusion alone, both in L-mode and (particularly) during ELMs [7, 1]. One of the foci of our research is to identify any non-diffusive transport mechanisms. If we assume that parallel density and temperature profiles remain unchanged, the poloidal gradients of these quantities, and the resulting ExB drifts, are enhanced in the SFD (compared to a SND). The parallel and perpendicular components of this convective flow are illustrated in Fig. 2a. Fig. 2b shows that the target profile measured at the inner primary strike point is significantly altered in the SF+, i.e. without changing the SOL topology, even leading to a second peak. This double peak disappears when any ExB drifts are reversed by reversing the toroidal field (Fig. 2c), demonstrating qualitative consistency with the drift model [8]. This and other cross-field transport mechanisms will be further explored with experiments and modelling.

**Radiation peaking in the XD and XTD**

DEMO operation will likely require detachment of direct plasma fluxes from the material surfaces. This requires significant volumetric energy losses upstream of the targets, notably in the form of impurity radiation. Ideally, a detachment front should be stabilized away from the core to reduce the risk of impurities affecting the performance. To this end, it is investigated how poloidal flux expansion ($f_x$), flux surface flaring, and total flux expansion (due to the target radius) affect detachment front stability. In [9], detachment was studied at up to $f_x = 9$, with $D_\alpha$ emissions suggesting that recombination occurs closer to the target and further from the separatrix at larger $f_x$. Recent experiments extended the range of TCV configurations with X-divertor [10] (x-point outside wall) and X-point Target Divertor [11] (x-point inside vessel) configurations, which give yet larger $f_x$ ($\sim$10–40) and flux surface flaring (c.f. Fig. 3). At a density below the detachment threshold ($n_e^{avg} \approx 3e19\ m^{-3}$), we observe in the tomographically inverted bolometer data that the total emissivity peaks near the target at increased $f_x$, and further towards the high-field side than in [9]. The difference between the XD and XTD is insignificant in this particular experi-

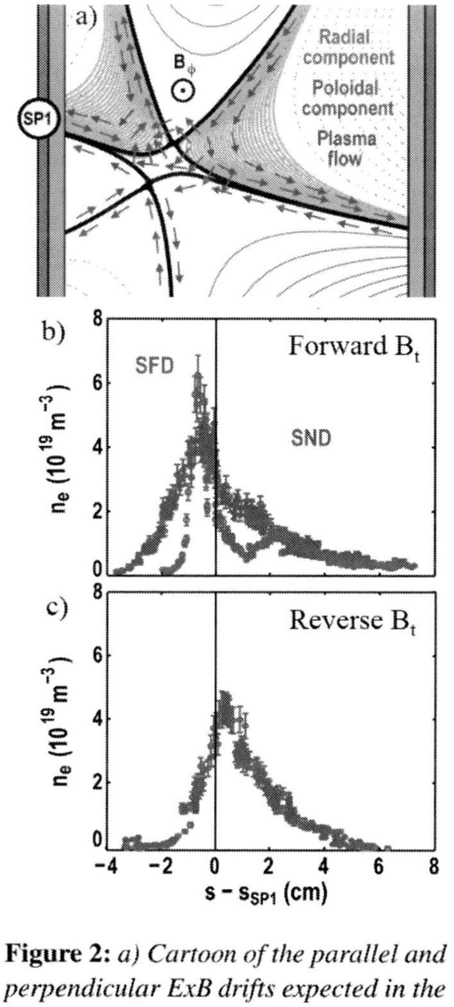


**Figure 2:** *a) Cartoon of the parallel and perpendicular ExB drifts expected in the SF+ divertor configuration. b) Target density profiles at the inner strike point in forward toroidal field in the SND (blue) and SFD (red) configurations. c) Target density profile at SP1 at the SFD in reversed $B_t$.*

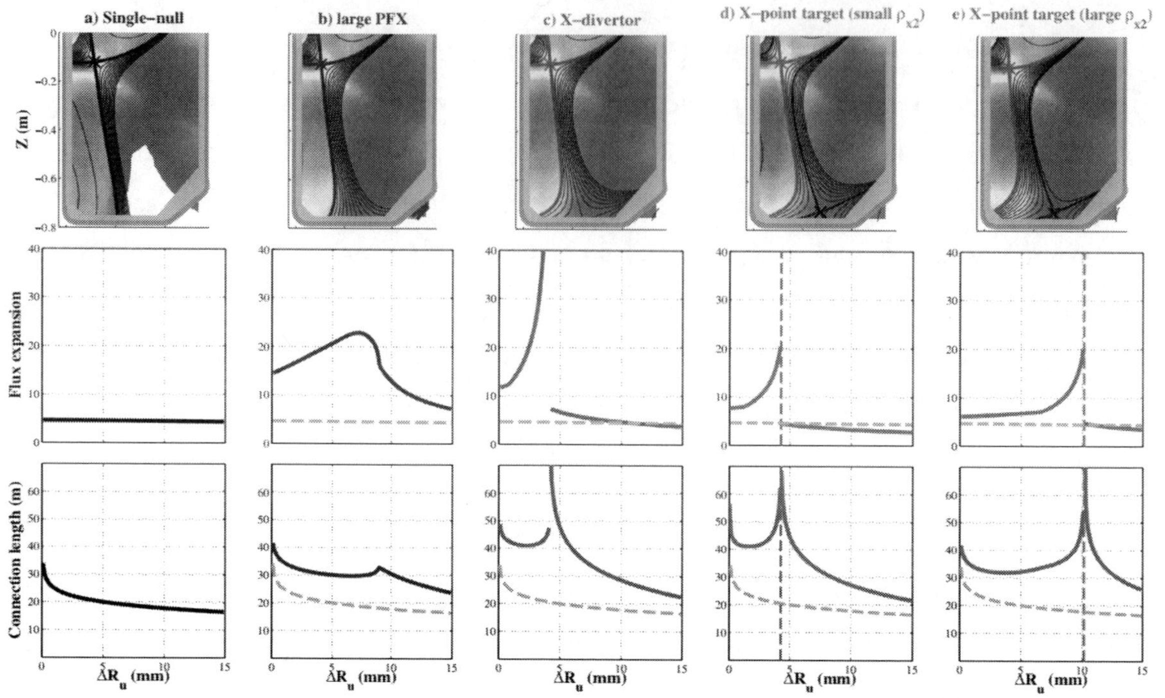


**Figure 3:** *Emissivity distributions on a common color scale (top), target flux expansion (middle) and connection length (bottom) for five configurations with large flux surface flaring in the divertor.*

ment. We speculate that the locally enhanced connection length (factor 1.5–2, see Fig. 3), which increases the residence time of charged particles near the target, and favorable poloidal angle of the magnetic flux surfaces with respect to the neutral recycling flux are responsible for the enhanced volumetric losses near the target. Establishing a neutral recycling flux opposite to the main poloidal plasma flow, while operating at the minimum magnetic field angle, could provide an important advantage over SND configurations with an inclined target (e.g. vertical target divertor).

*This work has been carried out within the framework of the EUROfusion Consortium and has received funding from the European Union's Horizon 2020 research and innovation programme under grant agreement number 633053. The views and opinions expressed herein do not necessarily reflect those of the European Commission. This work was supported in part by the Swiss National Science Foundation and NWO.*

## References

[1] W.A.J. Vijvers, et al, Nucl. Fus. **54**, 023009 (2014)
[2] D. Ryutov, et al, Physica Scripta **89**, 088002 (2014)
[3] H. Reimerdes, et al, Plasma Phys. Control. Fusion **55**, 124027 (2013)
[4] D. Ryutov, et al, Plasma Phys. Control. Fusion **52**, 105001 (2010)
[5] T. Eich, et al, Phys. Rev. Lett. **107**, 215001 (2011)
[6] T. Lunt, this conference and submitted to Plasma Phys. Control. Fusion
[7] T. Lunt et al, Plasma Phys. Control. Fusion **56**, 035009 (2014)
[8] G.P. Canal, et al, submitted to Nucl. Fus.
[9] R.A. Pitts, et al, Journal of Nuclear Materials **290-293**, 940-946 (2001)
[10] M. Kotschenreuther, et al, Phys. Plasmas **14**, 072502 (2007)
[11] B. LaBombard, B.C. Second and D. Third, APS 2013, Denver

# Model-based reconstruction and feedback control of the plasma particle density in tokamaks

T.C. Blanken[1], F. Felici[2], M.R. de Baar[1,2], W.P.M.H. Heemels[2] and the TCV team[3]

[1] *FOM-institute DIFFER, Dutch Institute for Fundamental Energy Research, Association EURATOM-FOM, PO Box 1207, 3430 BE Nieuwegein, The Netherlands*

[2] *Eindhoven University of Technology, The Netherlands*

[3] *École Polytechnique Fédérale de Lausanne, CRPP-EPFL, Lausanne, Switzerland*

We present a new model-based approach for real-time reconstruction and feedback control of the plasma particle density. Accurate knowledge and control of the plasma density is necessary to achieve the required plasma pressure, staying below stability limits (e.g. Greenwald limit) and e.g. ECRH/ECE cutoff limits. A control-oriented, physics-based model is presented, which is then employed to design a dynamic profile reconstruction algorithm and a feedback controller. This approach to real-time profile reconstruction is similar to recent work on estimation of temperature and current density profiles [1] and extends this work for the particle density profile.

## Physics-based model for control

We present a physics-based model of the tokamak density transport for observer and controller design that is flexible to adapt for multiple devices with multiple diagnostics and actuators and takes physical parameters of the plasma into account. The model is based on a 1D PDE for radial particle transport in the plasma [2] (describing flux-surface averaged electron density $n_e(\rho,t)$, where $\rho = \sqrt{\Phi/\Phi_{LCFS}}$ is the spatial variable and $\Phi$ is the toroidal magnetic flux) and two ODEs of the wall and vacuum particle inventories, $N_w(t)$ and $N_v(t)$. See Figure 1 for the modeled transport. The radial plasma particle transport is governed by

$$\frac{\partial}{\partial t}(n_e V') = \frac{\partial}{\partial \rho}\left(V' G_1 \chi \frac{\partial n_e}{\partial \rho} + V' G_2 v n_e\right) + V' S \qquad (1)$$

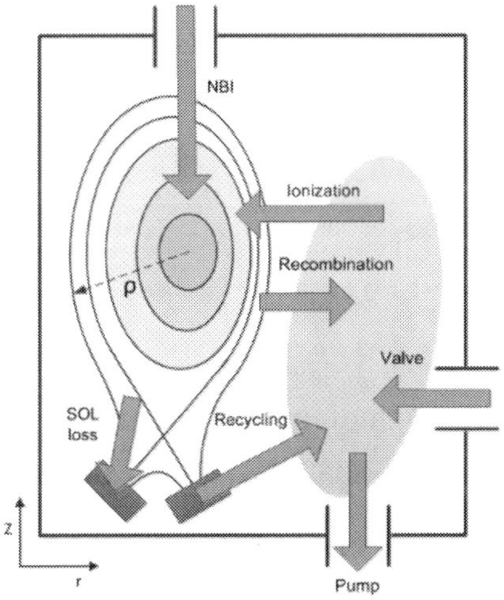


**Figure 1:** *Graphical representation of the plasma, the wall components, the neutral vacuum and the modeled particle flows in the tokamak.*

where $V' = \frac{\partial V}{\partial \rho}$ and $G_1, G_2$ are geometric terms which depend on the plasma equilibrium. We choose to model transport in an empirical fashion and therefore set the diffusion coefficient $\chi$ and drift velocity $v$ as simple functions of $\rho$. The net plasma particle source $S$ is modeled as

$S = S_{ion\&rec} + S_{inj} - S_{SOL\rightarrow wall}$, and the wall and vacuum inventory balances are modeled as

$$\frac{dN_w}{dt} = \Gamma_{SOL\rightarrow wall}(t) - \Gamma_{recycling}(t) \quad (2)$$

$$\frac{dN_v}{dt} = \Gamma_{valve}(t) + \Gamma_{recycling}(t) - \Gamma_{ion\&rec} - \Gamma_{pump}(t) \quad (3)$$

where $\Gamma_{SOL\rightarrow wall} = \int_{V_p} S_{SOL\rightarrow wall} dV$, $\Gamma_{ion\&rec} = \int_{V_p} S_{ion\&rec} dV$ and $V_p = \int V' d\rho$ is the plasma volume. We model the sources and flows as

$$S_{ion\&rec} = \langle\sigma v\rangle_{ion}(T_e)\, n_n n_e - \langle\sigma v\rangle_{rec}(T_e)\, n_e^2$$

$$S_{inj} = \Lambda_{NBI}(\rho)\Gamma_{NBI}(t) + \Lambda_{pellet}(\rho)\Gamma_{pellet}(t) \qquad S_{SOL\rightarrow wall} = \frac{n_e}{\tau_{SOL}}\delta_{\rho\in SOL}$$

$$\Gamma_{recycling} = \frac{N_w(t) - c_w V_{v,0} n_n}{\tau_{release}} + \frac{N_w(t)}{N_{sat}}\Gamma_{SOL\rightarrow wall} \qquad \Gamma_{pump} = \frac{V_{v,0} n_n}{\tau_{pump}}$$

where $n_n$ is the neutral vacuum density, $\tau_{release}$ is the outward wall diffusion time constant, $c_w$ is a dimensionless balance constant and $\Gamma_{NBI}$ and $\Gamma_{pellet}$ are the NBI and pellet fuelling rates.

We have included the influence of the LCFS electron temperature $T_{e,b}(t) = T_e|_{\rho=1}$, plasma current, 2D equilibrium, and operational modes (limited or diverted plasma $c_D \in \{\mathrm{lim}, \mathrm{div}\}$, L- or H-mode $c_H \in \{\mathrm{L}, \mathrm{H}\}$) on various transport coefficients as follows. We choose $v = v_0(\rho)\frac{I_p}{I_{p,0}}$ to represent the increase of pinch at higher current. An H-mode density pedestal implies a reduction of transport in the plasma edge and is reproduced by lower edge diffusion $\chi(\rho \in \mathrm{SOL})|_{c_H=\mathrm{H}} \ll \chi(\rho)|_{c_H=\mathrm{L}}$ and a lower drift velocity $v_0(\rho)|_{c_H=\mathrm{H}} < v_0(\rho)|_{c_H=\mathrm{L}}$. Furthermore, $\tau_{SOL}(c_D)$ is the time constant of particle loss in the scrape-of layer (chosen to increase for a diverted plasma), $N_{sat}(c_D, c_H)$ is the wall saturation level (chosen to increase on a limiter-to-divertor transition but decrease on a L-to-H transition), and $\tau_{pump}(c_D)$ is the pumping time constant (may depend on divertor strike point locations). We define the time-varying external input parameter $p(t) = \left[\begin{array}{cccccc} T_{e,b} & I_p & V' & \psi & c_D & c_H \end{array}\right]$ and we assume that these values are known in real time through diagnostics or real-time equilibrium reconstruction.

The neutral vacuum density outside the plasma is approximated as $n_n \approx \frac{N_v}{V_r - V_p}$ where $V_r$ is the vessel volume. We choose ad-hoc approximations of the spatial dependency of the neutral vacuum density $n_n$ and the electron temperature $T_e(\rho, t)$ inside the plasma. These approximations are parametrized using the vacuum density $\frac{N_v}{V_r - V_p}$ and the LCFS electron temperature $T_{e,b}(t)$, respectively, based on known spatial distributions of ionization and recombination near the edge. We choose the deposition functions $\Lambda_{NBI}(\rho)$ and $\Lambda_{pellet}(\rho)$ empirically based on known deposition locations from detailed physics analysis. We have nominal constants: the nominal vacuum volume $V_{v,0}$ and the nominal plasma current $I_{p,0}$, which are device-specific and define the operational point of the model.

Using a spatial discretization of $n_e$ and a time discretization, the system (1)-(3) is written as

$$x_k = f(p_{k-1}, x_{k-1}) + Bu_{k-1} \quad (4)$$

where $x_k = x(t_k)$ consists of $N_w(t_k)$, $N_v(t_k)$ and a parameterization of $n_e(\rho, t_k)$, the inputs are $u_k = \begin{bmatrix} \Gamma_{\text{valve}}(t_k) & \Gamma_{\text{NBI}}(t_k) & \Gamma_{\text{pellet}}(t_k) \end{bmatrix}^T$ and $p_k = p(t_k)$. The outputs $y_k$ of the interferometry system are the line-integrated electron density along multiple chords and depend on the 2D equilibrium

$$y_k = \left[ \int_{L_1} n_e(\rho, t_k) dL \cdots \int_{L_N} n_e(\rho, t_k) dL \right]^T \quad (5)$$

A forward model of the line-integrals is given by

$$y_k = C_k(p_k) x_k \quad (6)$$

**Dynamic state observer for density profile reconstruction**

We design a dynamic state observer, or Extended Kalman filter to estimate the density profile in real-time. It recursively fuses data of multiple diagnostic channels with model information. More precisely, the estimate $\hat{x}_k$ of the state $x_k$ at time $t_k$ is a linear combination of (a) a one-step ahead prediction $\hat{x}_k^p$ given the previous estimate $\hat{x}_{k-1}$ (using (4)) and (b) the innovation $z_k = y_k - C_k(p_k)\hat{x}_k^p$, which is the difference between measurements (5) and predicted line-integrated density (using (6)). Furthermore, the observer is employed to estimate systematic modeling errors and disturbances as slow-moving deviations from the one-step predictions.

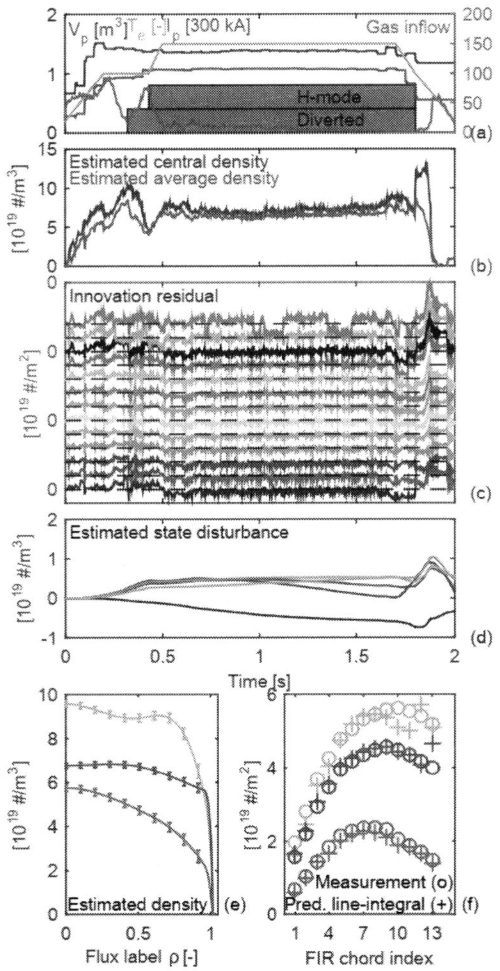


**Figure 2:** *Offline observer results for TCV shot #41953. The gas valve and parameter data are shown in (a). We use artificial temperature data. The estimated central and average density are shown in (b). The elements of the innovation $z_k$ are shown in (c), individually offset at intervals of $10^{19}\,\#/m^2$. The estimated disturbance is shown in (d). The estimated density profiles at $t = 0.12$ (magenta), $t = 0.3$ (light blue) and $t = 1$ (red) are shown in (e) with confidence bounds. The measurements and predicted line-integrated density at these time instants are shown in (f).*

In Figure 2, offline reconstruction results of the observer on interferometry and equilibrium reconstruction data of TCV are shown. The conformity of measurements and line-integrated density predictions (see Figure 2(f)) indicates accurate estimation of the profile.

## Feedback control design using robust control theory

We use the model (4) for designing feedback controllers for the density that are robust against model uncertainties and disturbances [4]. With the description of dynamics in different modes (4), we design controllers for each mode combination ($c_D \times c_H$) and switch between the controllers according mode transitions. Using the MATLAB Robust Control Toolbox [5], we design a linear switching feedback controller $K_{c_D,c_H}$ to track a predefined reference signal $r(t)$ for the volume-averaged density $\bar{n}_e(t) = \frac{1}{V_p}\int_{V_p} n_e dV$ using the gas valve. Additionally, we apply an anti-windup strategy to prevent the controller from integrating when the gas valve saturates.

$$\Gamma_{\text{valve}} = K_{c_D,c_H}(r - \bar{n}_e) \qquad (7)$$

In Figure 3, results of a simulation of the closed-loop interconnection of the plant model ((4) and (6)) with the observer and switching controller (7) are shown. In order to verify the robustness of the observer and controller, various coefficients ($\chi, v_0, \tau_{SOL}, N_{sat}$) have been doubled or halved, whereas the simulation model uses unperturbed coefficients. The controller is able to track representative reference signals, with the performance limited by the inability to rapidly decrease the density.

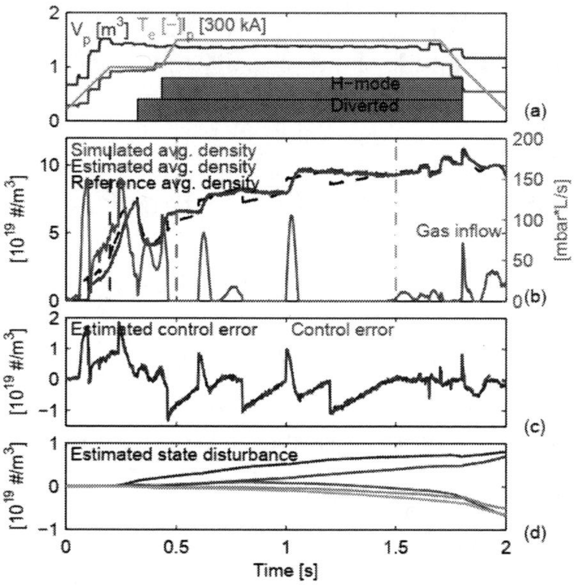


**Figure 3:** *Closed-loop simulation results for a high density H-mode plasma. The parameter data (a) is taken from TCV shot #41953. The controller is able to let the average density (green in (b), obtained by the observer) evolve along the reference trajectory (black in (b)). Negative steps are not tracked: the valve closes and the density decays autonomously.*

## Outlook on future work

Diagnostic faults such as fringe jumps or misbehaviour of the plasma (preceding a disruption) may be detected by the observer as inconsistencies between measurements and model-based predictions. Also, the method can be used for control of the profile shape, provided that the actuators allow such control.

## References

[1] F. Felici *et al*, American Control Conference, Portland, USA (2014)

[2] F. Hinton and R. Hazeltine, 'Theory of plasma transport in toroidal confinement systems', Rev. Mod. Phys. (1976)

[3] W. Vijvers *et al*, 39th EPS Conference on Plasma Physics, (2012)

[4] M. Morari and E. Zafiriou, 'Robust Process Control', Prentice Hall (1989)

[5] *www.mathworks.com/products/robust/*

# Runaway electrons in non-disruptive Tore Supra plasmas

E. Nilsson, Y. Peysson, J. Decker, J.F. Artaud, T. Aniel, M. Irishkin,
D. Mazon, F. Saint-Laurent and the Tore Supra team
*CEA, IRFM, F-13108, Saint-Paul-lez-Durance, France*

Runaway electrons can form in tokamak plasmas when the toroidal electric field exceeds some critical value ($E_c$) proportional to the electron density and may cause serious damage to plasma facing components in ITER [1]. These relativistic runaway electrons may appear during plasma disruptions [2], current ramp-up or ramp-down and even in the current flattop in quiescent low density plasmas as observed in several tokamaks [3]. In this work we perform interpretative kinetic modelling of the runaway formation in the current flattop of scenarios in the Tore Supra (TS) tokamak.

Runaway electrons are detected in the TS tokamak in low line averaged density discharges ($< n_e >< 10^{19}\ m^{-3}$) in the current ramp-down by hard X-ray (HXR), photo-neutron and electron cyclotron emission (ECE) measurements. Such signatures are found in discharge #40719 (Fig. 1), where the electron density in the current flattop is $< n_e >= 0.64 \cdot 10^{19} m^{-3}$, corresponding to $E/E_c \approx 8$, or $E/E_D \approx 0.06$, where $E_D = E_c m_e c^2/T_e$. However, in a similar discharge (#40721), with two times higher density, no runaways are observed even though the electric field exceeds the critical electric field ($E/E_c \approx 4$, or $E/E_D \approx 0.02$). This does not guarantee that no runaways or suprathermals are formed, it could simply be a consequence of that the population is too small to be detected by the diagnostics. Another signature of the runaway population is the remaining current ($\sim 50$kA) at the termination of the plasma as seen in Fig. 1(c) (t=$15.7 - 16$s). This current plateau is believed to originate from a beam of well confined runaway electrons. Such a plateau is not seen in the higher density discharge (#40721).

The runaway electron formation is simulated using the LUKE code [4], a solver of the 3-D linearized bounce-averaged relativistic electron Fokker-Planck equation in a toroidal geometry. Global plasma parameters such as parallel electric field and plasma equilibrium are prescribed by the fast integrated modelling code METIS [6]. With this work we aim to contribute to understanding of runaway electron formation processes, and provide information beyond the experimental measurements.

**Modelling of non-disruptive scenarios with LUKE/METIS**

In order to understand the different outcome of the two non-disruptive TS scenarios, the formation of runaway electrons from the combined effect of Dreicer and avalanche is studied with the LUKE code. Temperature and density profiles are prescribed in METIS by fitting the

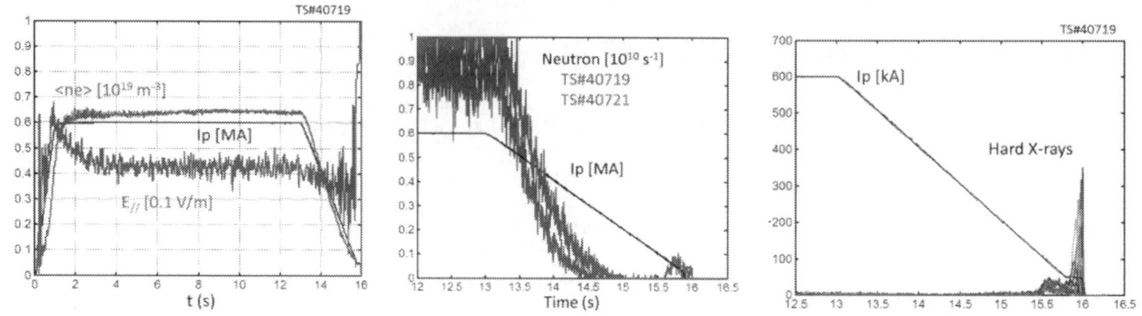


(a) (b) (c)

Figure 1: (a) Line averaged electron density, plasma current and parallel electric field strength in discharge #40719. (b) A peak in the neutron signal is observed at the plasma termination for the low density shot (40719) but none for the higher density shot (40721). (c) HXR data from the vertical camera (channels 1-21) in the energy range $E_{HXR} = 20 - 200\,\mathrm{keV}$ observed in the current ramp down in 40719.

combined results of several diagnostics including Thomson scattering, ECE, reflectometry and interferometry, weighted with Bayesian analysis. Given the METIS equilibria, the LUKE code calculates the effect on the electron distribution function of runaway processes including the Dreicer and avalanche effect through a recently implemented knock-on source term [5].

For #40719 the METIS/LUKE simulations show that runaways are progressively formed during the current flattop (Fig. 2), concentrated near the magnetic axis (Fig. 3(a)). Even though the density is lower off-axis and the E-field profile rather flat, $E/E_D$ decreases with the radius due to the temperature profile. This would explain the slower Dreicer generation off the magnetic axis. Also, the increase of magnetic trapping effects off-axis contributes to a reduced runaway rate [5].

Figure 3(a) shows the calculated current density profile as carried by external runaways (with kinetic energy $E_k > 1\,\mathrm{MeV}$), when assuming that they move at the speed of light, compared to the current density profile of the bulk. The calculated Ohmic plasma current in the flattop at 13 s is 0.54 MA, compared to an experimental plasma current of 0.55 MA. To match the current, the effective charge had to be reduced from the predictions by METIS ($Z_{eff} = 3.9$) to $Z_{eff} = 2.6$. At the end of the current flattop the predicted current carried by external runaway electrons ($E_k > 1\,\mathrm{MeV}$) is 33 kA, which is the approximate level of remaining current ($I_p = 50\,\mathrm{kA}$) likely carried by a well confined beam of runaways, as seen in the termination of the plasma (Fig. 1(c)).

The simulations are compared to experimental measurements through reconstructed HXR

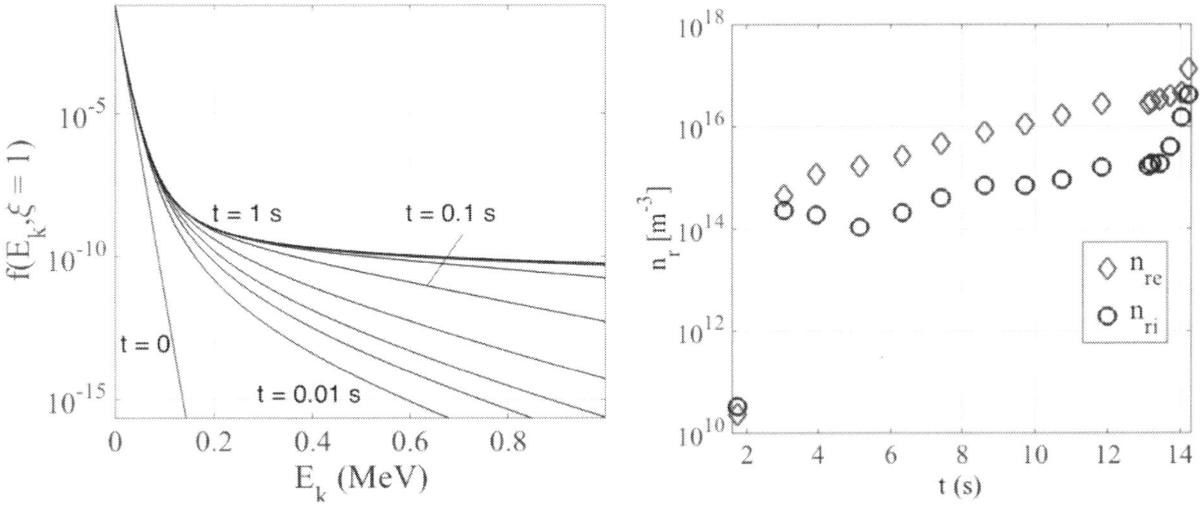


(a) (b)

Figure 2: (a) Electron distribution function at $\xi = 1$ in TS discharge #40719 in current flat-top during 1 s. (b) Evolution of internal ($p_c < p < p(E_k = 1\,\mathrm{MeV})$) and external ($p > p(E_k = 1\,\mathrm{MeV})$) runaway electrons.

tomography emission based on the electron distribution function with the R5X2 code [7], that calculates the fast electron Bremsstrahlung (FEB) cross-section and integrates the emission along the lines of sight, accounting for the response function of the detectors. The HXR signals from a vertical (chords $1-21$) and horizontal (chords $22-59$) cameras provide information about the suprathermal electron population. The emission as measured by each detector is described by the count rate in the energy range $50-110\,\mathrm{keV}$ (see Fig. 3(b)). LUKE simulations predict a runaway population concentrated near the magnetic axis. The shape and amplitude of reconstructed FEB emission agree fairly well with measurements.

With METIS/LUKE simulations we can investigate whether any suprathermals are formed also in the discharge without runaway signature (#40721), where $E > E_c$. The simulation of #40721 confirms the experimental observations; the runaway production is negligible as the simulations show the population of suprathermal electrons is formed during this pulse is negligible. The external runaway population, i.e. electrons with kinetic energy $E_k > 1\,\mathrm{MeV}$, is so small that it would only carry a current of around $5\,\mathrm{mA}$ at the end of the 10 seconds long current flattop. These results are in line with the parametric study of runaway formation performed in Ref. [5], where it is found that 10 seconds of $E/E_c \approx 2.5$, which is the local central electric field strength, is not sufficient for a significant runaway population to form in a $3\,\mathrm{keV}$ plasma. Furthermore, these results support the experimental observations of Ref. [3] where at least $E/E_c \sim 3-12$ is required to generate a detectable population of runaway electrons in

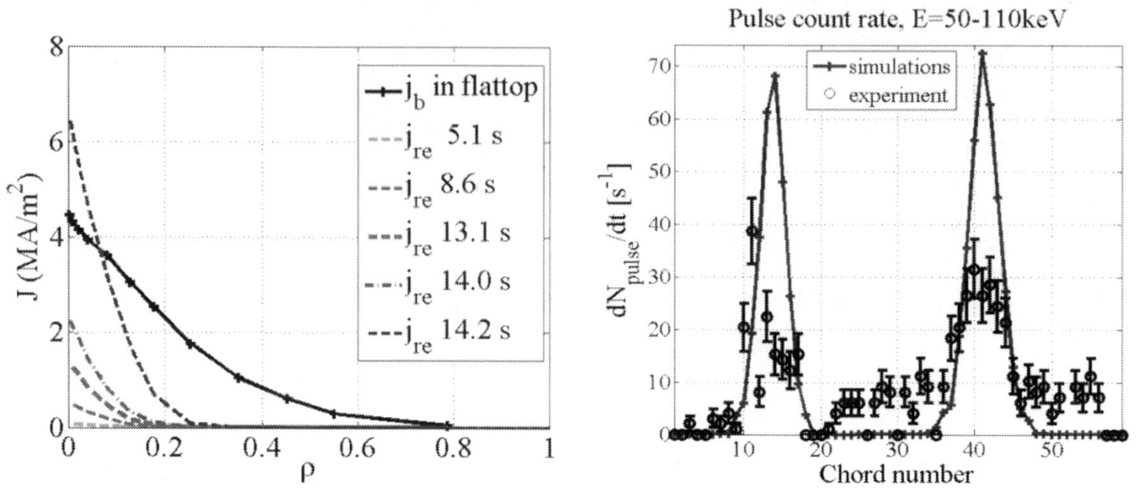


(a) (b)

Figure 3: (a) Current density profile of bulk (black) and current density carried by runaways ($E_k > 1\,\mathrm{MeV}$). (b) Reconstructed FEB profile, compared to measured FEB emission from HXR cameras.

various tokamaks.

## Conclusion

The Fokker-Planck solver LUKE, is used to model runaway electron formation through Dreicer and avalanche mechanisms in non-disruptive Tore Supra scenarios in near-critical E-field. Simulations reveal progressive build-up of a suprathermal population during the flattop in the discharge where runaways are detected (#40719), but not in the higher density discharge (#40721) where $E/E_c \approx 4$. These results agree with experimental observations from various other tokamaks [3] where at least $E/E_c \approx 3-12$ is required for a detectable runaway population to form. The order of magnitude of the current carried by runaway correspond well to experimental indications. The magnitude of the reconstructed FEB emission from suprathermal electrons is well reproduced but the profile is more centrally peaked than experimental measurements.

## References


[1] T. C. Hender et al., Nucl. Fusion **47** (2007)

[2] F. Saint-Laurent et al., Fusion Science and Technology **64** (2013)

[3] R. S. Granetz et al., Phys. of Plasmas **21** (2014)

[4] J. Decker, Y. Peysson, EUR-CEA-FC-1736, Euratom-CEA (2004)

[5] E. Nilsson et al., Submitted to Plasma Physics and Controlled Fusion (2015)

[6] J.F. Artaud et al., Nucl. Fusion **50** (2010)

[7] Y. Peysson and J. Decker, Phys. of Plasmas **15** (2008)

# Long pulse RF heating and current drive scenarios for WEST

A. Ekedahl, C. Bourdelle, J.-F. Artaud, L. Colas, J. Decker, L. Delpech, R. Dumont, M. Goniche, W. Helou, J. Hillairet, G. Lombard, P. Mollard, E. Nardon, Y. Peysson, E. Tsitrone and the Tore Supra / WEST Team

*CEA, IRFM, F-13108 Saint Paul-lez-Durance, France.*

**Introduction.** The longstanding expertise of the Tore Supra team in long pulse radiofrequency (RF) heating and current drive systems will now be exploited in the WEST device (tungsten-W Environment in Steady-state Tokamak) [1]. WEST will allow an integrated long pulse tokamak programme for testing W-divertor components at ITER-relevant heat flux (10-20 MW/m$^2$). At the same time, crucial aspects relevant for ITER-operation will be treated, such as avoidance of W-accumulation in long discharges, monitoring and control of heat fluxes on metallic plasma facing components (PFCs) and coupling of RF waves in H-mode plasmas. To fulfil its programmatic goals, the WEST research plan has been structured in two main headlines, namely: "ITER grade PFC tests" and "Towards long pulse H mode and steady-state operation" [2]. The WEST device and research plan is open to all ITER partners.

The transformation of Tore Supra to WEST involves replacing the carbon PFCs by a combination of bulk and coated tungsten (W) elements and inserting upper and lower divertor coils in order to produce a divertor plasma configuration. The plasma volume in WEST is therefore reduced to 15 m$^3$, compared to 25 m$^3$ in Tore Supra. With major radius R $\approx$ 2.5 m and minor radius a $\approx$ 0.5 m, the aspect ratio of WEST is fairly large (A $\approx$ 5), as compared to other tokamaks. The expected L-H power threshold in WEST has been assessed with different scaling laws [2], yielding an expected power threshold below 5 MW at $n_e = 8\times10^{19}$ m$^{-3}$ and $B_T$ = 3.6 T. Considering the installed additional heating power capability in WEST, access to H-mode appears feasible with good margin.

**Heating and current drive in WEST.** The additional heating and current drive power in WEST is provided by RF heating systems, namely Ion Cyclotron Resonance Heating (ICRH) and Lower Hybrid Current Drive (LHCD), delivering up to 9 MW of ICRH power and 7 MW of LHCD power. Successful long pulse operation with LHCD + ICRH, leading to 1 GJ injected energy, has been achieved in Tore Supra L-mode plasmas [3, 4], but the WEST programme now brings new challenges. The main plasma heating is provided by ICRH and scenarios compatible with high ICRH power, without causing W-accumulation, must be developed. The LHCD system needs to provide sufficient non-inductive current in high density H-mode plasmas in order to sustain long pulse scenarios required for testing the W-divertor under ITER-relevant particle fluence (~$10^{27}$ D/m$^2$) and time scale (>100 s).

To allow coupling to H-mode plasmas, three new ELM-resilient ICRH antennas have been designed for WEST [5, 6]. They are provided by ASIPP (Hefei, China) as in kind contribution. The antennas are based on a load resilient antenna concept already tested on Tore Supra. Starting from this design, a number modifications have been made, e.g. improving the coupling by optimising the antenna front face design and adding cooling throughout the antenna for long pulse operation (3 MW / 1000 s). The ICRH generator has been upgraded to allow high power operation (9 MW / 30 s) at high reflected voltage. The WEST ICRH system is thus the first ever ICRH system combining continuous wave (CW) operation and load tolerance capability for coupling on H-mode. The nominal operating frequencies are 53 ± 2 MHz and 55.5 ± 2 MHz, in order to allow flexibility in the resonance layer location around the magnetic axis.

A high power (9 MW at the generator), CW LHCD system is already available for WEST, consisting of 16 high power CW klystrons (f = 3.7 GHz), feeding two actively cooled multijunction launchers [7]. The main modification of the system involves a reshaping of the toroidal profile of the launcher mouth on the Full-Active-Multijunction launcher, in order to guarantee the LH wave coupling on WEST plasmas [7]. The Passive-Active-Multijunction (PAM) launcher will not be modified in the first instance, since this ITER-relevant concept already offers good coupling properties at low edge density [8].

Local gas injection will be added to all five antennas [9]. The gas injections are poloidally distributed along the antennas, the radial location being ~15 cm behind the protection limiters and the toroidal location being ~15 cm outside the protection limiters. According to recent 3D SOL-modelling [10], these choices for the gas injection locations should allow poloidally homogeneous SOL density at the antennas. In addition, 2D modelling of the WEST SOL has been carried out with SolEdge2D-EIRENE [11], showing that gas injection from the outer mid-plane indeed increases the SOL density at the radial location of the antennas.

| H-mode scenario ($B_T$ = 3.6 T) | High power | Standard | High fluence |
|---|---|---|---|
| Plasma current | 0.8 MA | 0.6 MA | 0.6 MA |
| Line averaged density | 8 $10^{19}$m$^{-3}$ | 7 $10^{19}$m$^{-3}$ | 7 $10^{19}$m$^{-3}$ |
| Greenwald fraction | 64% | 75% | 75% |
| **Total RF power** | **15 MW** | **12 MW** | **9.3 MW** |
| Lower Hybrid Current Drive | 6 MW | 6 MW | 6.3 MW |
| Ion Cyclotron Resonance Heating | 9 MW | 6 MW | 3 MW |
| **Pulse length** | **30 s** | **60 s** | **1000 s** |

**Table 1:** Reference scenarios for WEST operation (from METIS).

**Modelling of WEST scenarios**. Based on the RF heating and current drive capability, three reference plasma scenarios are targeted (Table 1). These scenarios have been constructed from integrated modelling with the 0D version of the CRONOS code, called METIS [12], by imposing several constraints, such as maximum tolerable electron temperature on the divertor target and energy confinement factor (H = 1). Well established scaling laws are used for the density profile peaking and radiated power, etc. The "Standard scenario" (12 MW / 60 s) is targeted for testing ITER-like divertor components at relevant heat flux (10-20 MW/m$^2$) and demonstrating integrated long pulse H-mode operation (~60 s). The "High fluence" scenario is envisaged for plasma-wall interaction studies (up to 1000 s at 10 MW). Finally, a "High power" scenario (15 MW / 30 s) is developed for high performance discharges.

The ICRH power deposition profile is obtained from the full wave solver EVE, with the Fokker-Planck module AQL [13]. Figure 1a illustrates an EVE/AQL simulation for 6 MW ICRH at 55.5 MHz and hydrogen minority concentration $n_H/n_e = 6\%$. The power is deposited within a normalised radius of 0.4, essentially on minority hydrogen ions (~4.8 MW). It is then redistributed through collisional relaxation of the fast minority ions between electrons (~2.6 MW) and bulk ions (~1.1 MW). The missing power (~ 1.1 MW) is partly due to the losses of fast ions due to the magnetic ripple (2% at the outer mid-plane in WEST). Some flexibility in terms of electron/ion heating can be obtained by varying the ICRH power level, as seen in Figure 1b, or the hydrogen minority concentration.

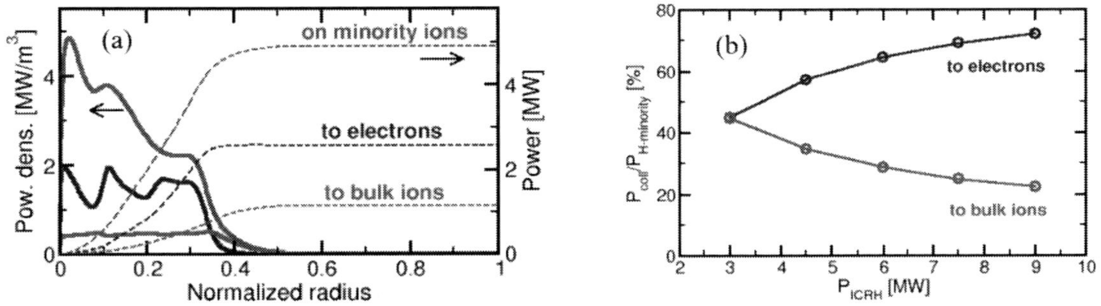


**Figure 1:** EVE/AQL modelling of a WEST scenario for 6 MW ICRH at 55.5 MHz and $n_H/n_e = 6\%$. Power density absorbed by species (a), power split between electrons and bulk ions as a function of ICRH power (b).

The LHCD modelling is carried out with ray-tracing + Fokker-Planck codes C3PO/LUKE [14]. The three reference scenarios listed in Table 1 have been modelled using $N_{//}$-spectrum peaked at $N_{//0} = 2.0$, to improve the accessibility of the LH waves to the plasma. Figure 2a shows the electron density and temperature profiles from METIS for the "High fluence" scenario ($I_P = 0.6$MA, $n_e = 7\times10^{19}$m$^{-3}$, $P_{Add} = 9$ MW), while Figure 2b shows the ray trajectory for a ray with $N_{//0} = 2.0$. The LH power deposition and current profiles have been modelled with the recent "Tail LH" model in LUKE [15], which has proven to reproduce well the experimental results on Tore Supra and EAST [14]. The LH power deposition and current profiles are rather central (Figure 2c) and yield a current drive efficiency of $\eta = 0.12\times10^{20}$ $Am^{-2}W^{-1}$. It should be notes that the current drive efficiency predicted in WEST is higher than that obtained in the Tore Supra configuration. This is attributed to the X-point configuration that introduces an $N_{//}$-upshift that facilitates the propagation of the waves to the plasma core.

Based on the ICRH and LHCD simulations presented above, together with the METIS simulations, it is expected that the LHCD power available to the plasma (7 MW) should be sufficient to sustain long pulse operation in WEST. Indeed, the simulations show that the loop voltage ($V_{Loop}$) is only ~6 mV in the "High fluence" scenario, when using 6.3 MW of LHCD power + 3 MW ICRH, which should allow reaching 1000 s pulse length.

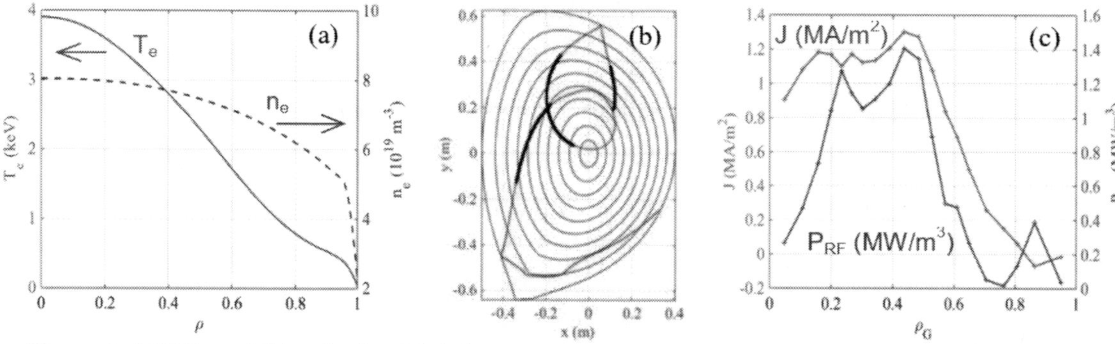


**Figure 2:** LHCD modelling for the "High fluence" scenario: electron density and temperature profiles (a), poloidal ray trajectory for a ray with $N_{//0} = 2.0$ (b), current density and absorbed power profile (c).

## References


[1] J. Bucalossi et al., Fusion Eng. Des. **89**, 907 (2014).

[2] C. Bourdelle et al., Nucl. Fusion **55**, 063017 (2015).

[3] R. Dumont et al., Plasma Phys. Control. Fusion **56**, 075020 (2014).

[4] M. Goniche et al., Phys. Plasmas **21**, 061515 (2014).

[5] Z. Chen et al., Fusion Eng. Des. **94**, 82 (2015).

[6] J. Hillairet et al., Proc. 21st Topical Conf. on RF Power in Plasmas, Lake Arrowhead, CA (2015).

[7] L. Delpech et al., Fusion Eng. Des. (2015), http://dx.doi.org/10.1016/j.fusengdes.2015.01.020

[8] A. Ekedahl et al., Nucl. Fusion **50**, 112002 (2010).

[9] A. Ekedahl et al., Proc. 21st Topical Conf. on RF Power in Plasmas, Lake Arrowhead, CA (2015).

[10] W. Zhang et al., Proc. 21st Topical Conf. on RF Power in Plasmas, Lake Arrowhead, CA (2015).

[11] H. Bufferand et al., This conference, paper O5.139.

[12] J.-F. Artaud el al., Nucl. Fusion **50**, 043001 (2010).

[13] R. J. Dumont and D. Zarzoso, Nucl. Fusion **53**, 013002 (2013).

[14] Y. Peysson et al., This conference, paper I5.012, to be submitted to Plasma Phys. Control. Fusion (2015).

[15] J. Decker et al., Phys. Plasmas **21**, 092504 (2014).

# Progress with the ENEA-ORNL high-speed four barrels pellet injector

A. Frattolillo[1], S. Migliori[1], S. Podda[1], F. Bombarda[1], G. D'Elia[1], F. Poggi[1],
S. K. Combs[2], C. R. Foust[2], S. J. Meitner[2], D. T. Fehling[2], L. R. Baylor[2] and G. Roveta[3]
[1] *ENEA, C. R. Frascati, Frascati, Italy*
[2] *Oak Ridge National Laboratory, Oak Ridge, YN. USA*
[3] *Criotec Impianti, Chivasso, Italy*

**Introduction.**

Pellet injection is a well proven technology representing to date the most efficient method for core fuelling of fusion plasmas. It is believed that, due to favourable drift effects across the plasma column, injection from the inboard torus side via curved guide tubes can provide high fuelling efficiency, in spite of restrictions imposed to the projectiles speed by such a scheme. Extensive tests carried out at ORNL suggest that the speed will need to be limited to a few hundreds of m/s, for reliable delivery of intact pellets [1]. Simulation codes, however, indicate that the fuelling efficiency increases with the injection speed [2], while recent tests at Asdex Upgrade demonstrated that $D_2$ ice pellets can survive at velocities of ~ 1 km/s inside a guiding system, provided this is very carefully designed [3]. Moreover, when injection from the inboard torus side is impractical due to tight space (as will be for instance the case of Ignitor and JT60-SA), pellets have to be necessarily injected from the outboard side at speeds ≥ 3 km/s, in order to achieve sufficient fuel penetration in spite of the unfavourable drift.

ENEA and ORNL are therefore collaborating on the development of a high-speed four-barrel "pipe gun" injector for the Ignitor experiment (fig. 1), designed to launch $D_2$ ice pellets (1.9, 2.6, 3.2 and 4.4 mm in size) at speeds up to 4 km/s, using two-stage light gas guns[4].

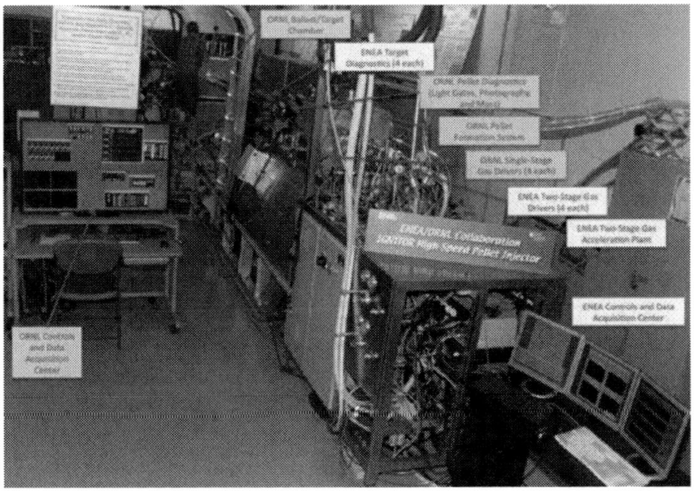

Figure 1. The Ignitor high-speed four barrel pellet injector at ORNL

**Innovative features of the ENEA-ORNL injector.**

ENEA and ORNL have been collaborating on high-speed pellet injectors since early 90's. A previous joint experimental effort was dedicated to the development of a high-speed repeating injector, by coupling an existing ORNL piston extruder and an ENEA two-stage gun capable of operating at a repetition rate of 1 Hz. During preliminary tests with this device, the very low break-away pressure of extruded pellets (as compared to that of "in-situ" frozen projectiles) caused them to start moving rather unpredictably, resulting in a wide spread and in a significant reduction of speed performance (◘ 2 km/s). A spring actuated check valve, opening at a well-defined upstream pressure threshold and designed to allow the dynamic transmission of the pressure pulse, was therefore developed by the ENEA team and fitted at the downstream end of the pump tube. This device, capable of suitably tailoring the shape of the rising edge of the pressure pulse, prevented premature movement of the projectiles and improved their acceleration, allowing to launch intact pellets at speeds up to 2.5 km/s, while strongly reducing the spread of experimental data [5], [6].

An upgraded (electromagnetically driven) version of this valve has been included in the design of the present "pipe gun" injector, with the aim of investigating whether it may be beneficial from the point of view of speed performance, as well as to improve both reliability and versatility of the whole system. As a matter of fact, this valve prevents the propellant gas, initially admitted at relatively low pressure in front of the piston, from escaping through the gun breech when the cut off valve (separating the gun and the barrel) is opened for firing, thus protecting the piston, which is prevented from hitting the end of the pump tube. Besides strongly reducing the risk of failures of the two-stage gun, this valve allows performing test shots even in the absence of a projectile inside the barrel. Moreover, the new design of the valve, which uses the magnetic field generated by a solenoid to produce the closing force, is much more reliable and versatile as compared to the previous spring loaded prototype, actually getting rid of issues related to the progressive yielding of the spring, while allowing to adjust the force acting on the shutter by simply varying the electric current flowing in the solenoid. The injector accommodates both an ORNL propellant valve and an ENEA two-stage gun on each barrel (fig. 2). A check valve is placed downstream of each propellant valve, preventing flow of hydrogen gas in the backward direction; this ensures that the propellant valves will not be exposed to the hot, high peak pressure pulses generated when

Figure 2. The new arrangement accommodating both a propellant valve and a two-stage gun on each barrel.

firing the two-stage guns. Several hundred shots have been performed, alternating single and two-stage gun operation, without any failure. In the future, this unique feature will provide great flexibility on a fusion experiment since the operator can switch seamlessly and at any time between standard and high-speed pellets, on any or all gun barrels.

**Latest improvements.**

The injector's cryostat can operate as a cryogen free device. In joint tests carried out up to date at ORNL, it was cooled down using a Cryomech PT810 two-stage pulse tube cryo-generator, already in use at ORNL. With this device, featuring a cooling power of 80 W @ 80K on the first stage and 14 W @ 20K on the second stage, the minimum temperatures reached inside the cryostat were ~100K at the thermal radiation shield, and ~10 K at the pellet freezing zone (nominal 8 K should be available on the second stage with no thermal load). In these conditions launching speeds up to 2.6 km/s have been easily obtained, but pellets were broken above 2 km/s. A more powerful pulse tube cryo-cooler (Cryomech PT415) has been procured by the ENEA team, and replaced the existing unit with the aim of enhancing pellet formation and producing better ice quality. This device, featuring a refrigerating power of 40 W @ 45 K on the first stage and 1.5 W @ 4.2 K on the second (nominal minimum temperature without thermal load is 2.8 K), was chosen also for its design, almost similar to that of the PT810, allowing replacement of the existing unit with minimal mechanical adjustments. Preliminary cooling tests with the PT415 showed a significant reduction of the temperatures achieved at both the shield (~50 K) and the freezing region (~7.5 K).

**Remote operation of the injector.**

Remote operation capability via Ethernet is a further distinctive feature of this injector. With computer security recently given higher priority than ever at ORNL, getting remote access to local computer has been really challenging. A remote control station has been set up at ENEA Frascati (figure 3), consisting of

**Figure 3.** The remote control station at ENEA Frascati

several computers that exactly replicate the configuration of the local control system at ORNL. A point to point secure connection enables each ENEA terminal talking with its corresponding local unit. Exceptions to the ORNL firewall had to be raised to allow the static IP addresses assigned to ENEA computers to get through, log into their own paired local unit and share their desktops, using RealVNC® software. Remote operators can thus get full access to the C&DAS of the injector, using the comprehensive LabView software running on local computers. To ensure an adequate level of IT security, access from Frascati to ORNL computers via VNC is protected by a password, but also subject to prior authorization by an operator at ORNL. The full Ethernet bandwidth (1 Gbit) is used to speed up the communication. Additional desktops are also provided for video conferencing, using Blue Jeans platform, and to control and monitor two Ethernet based video surveillance cameras, one installed at ORNL and the other at ENEA, so that visual contact can be maintained throughout each experimental session. Recently, preliminary tests using ORNL propellant valves have been successfully carried out. Solid deuterium pellets have been seamlessly and repeatedly formed and launched at moderate speeds (~1 km/s), using up to all four barrels, without coming across any operational issue or failure. Communication via 1 Gbit Ethernet proved fast enough, so that remote operators do not detect any substantial delay as compared to local operation. Actually, remote operation of the injector turns out to be as easy as local operation at ORNL. Experimental data, stored in the hard disk of ORNL computers, are also available to the ENEA team throughout the experimental session, however cloud based solution are being considered to make data accessible to both parties at any time, as well as live streaming of cameras that monitor and record the impact of pellets on the final targets.

**Acknowledgments**

The authors wish to thank Prof. B. Coppi of MIT for including the fast pellet injector in the Ignitor program. This work was supported in part by ENEA of Italy and by the US DoE.

**References.**

[1] S. K. Combs et al., Fusion Engineering and Design 75-79 (2005) 691

[2] P. T. Lang et al., Proc. Of the 28th Symposium on Fusion Technology (SOFT 2014), presentation O2A.2

[3] A. Lorenz et al., Fusion Eng. Design 58-59 (2001) 325

[4] A. Frattolillo et al., Proc. of the 37th EPS Conf. on Plasma Phys., ECA Vol. 34A, P-5.153 (2010)

[5] A. Frattolillo et al. Rev. Sci. Instrum. Vol. 67(5), May 1996, 1834-1841

[6] S. Migliori, A. Frattolillo, patent n. ITRM950100 (Aug. 20th, 1996)

# Predictions of neutral beam current drive in DEMO using BBNBI and ASCOT within the European Transport Simulator

O. Asunta[1], R. Coelho[2], D. Kalupin[3], T. Johnson[4], T. Franke[3,5], and EU-IM Team*

[1] *Department of Applied Physics, Aalto University, P.O. Box 14100, FI-00076 AALTO, Finland*

[2] *Instituto de Plasmas e Fusão Nuclear, Instituto Superior Técnico, Univ. de Lisboa, Portugal*

[3] *EUROfusion Consortium Programme Unit, Boltzmannstr. 2, D-85748, Garching, Germany*

[4] *KTH Royal Institute of Technology, 10044 Stockholm, Sweden*

[5] *Max-Planck-Institut für Plasmaphysik, Boltzmannstr. 2, D-85748 Garching, Germany*

## Introduction

Extensive predictive modelling is absolutely essential to avoid costly mistakes when designing future fusion devices. Modelling the entire burning plasma requires self-consistent treatment of a number of interconnected physics processes from magnetic equilibrium and auxiliary heating to radiation and transport of heat and particles. European Integrated Modelling (EU-IM) framework provides a platform for combining numerical tools that model the different physics processes in the plasma. It allows building complex integrated simulations (*workflows*) by connecting pre-compiled physics modules (*actors*) using standardized data structures known as *Consistent Physical Objects* (CPOs) [1].

Neutral beam injection (NBI) is used extensively in current fusion devices for, e.g., driving current and heating the plasma. It is also foreseen to be used in ITER and future fusion reactor, DEMO. In this work BBNBI [2] was used to model the injection and ionization of neutrals. The ionized particles were then followed using Monte Carlo code ASCOT [3] to model how they, e.g., heat the plasma and drive current. BBNBI and ASCOT were ran within one of the most advanced workflows developed on EU-IM framework, the European Transport Simulator (ETS) [4], to study neutral beam current drive (NBCD) in DEMO. In particular, the effect of horizontal and vertical tilting of the beam on NBCD and plasma heating was investigated. In addition to gaining valuable insight into the characteristics of NBI in DEMO, the objective of this work was to demonstrate that existing tools are capable of flexible and sophisticated modelling of NBI, for arbitrary injection geometries, as a part of an integrated transport simulation.

## Neutral beam injection in DEMO

Simulations were carried out for two discharge scenarios of EU DEMO1 (2012) design; one with a *peaked* and the other with *flat* density profile. The temperature and density profiles of

*See http://www.euro-fusionscipub.org/eu-im.

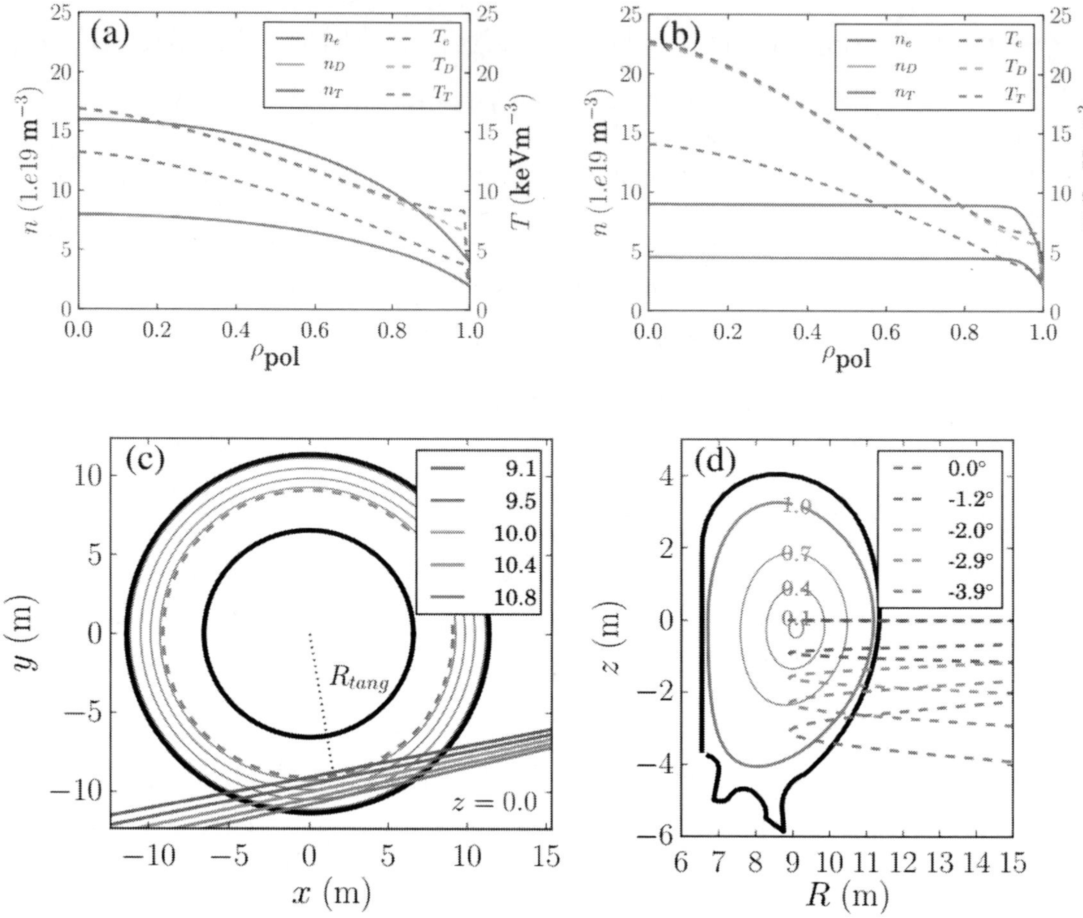


Figure 1: $T$ and $n$ profiles of the *peaked* (a) and *flat* (b) density DEMO discharge scenarios, and the beamlines of the scans performed by increasing the $R_{\text{tang}}$ (c) and tilting the beam downwards (d). A set of magnetic flux surfaces (grey), including the magnetic axis (dashed) and the last closed flux surface (bold), are shown. In (c) and (d), the red beamline with $R_{\text{tang}} = 9.1$ m and $\theta_{\text{tilt}} = 0.0°$, is the reference case.

these discharges are presented in Fig. 1(a) and (b), respectively. In both scenarios, $n_{\text{D}} = n_{\text{T}}$, $I_{\text{p}} =$ 16 MA, $B_\phi = 6.791$ T, and $P_{\text{NBI}} = 50$ MW. Magnetic equilibrium was calculated within the ETS using CHEASE [5] based on plasma boundary obtained from 2013 CREATE [6] simulations.

For the NB injector, a model of ITER injector was scaled up to DEMO dimensions. In the reference case, horizontal injection was assumed. The study consisted of a scan of three beam energies; 0.75 MeV, 1.0 MeV, and 1.5 MeV. When tilting the beam, the injector grid was held fixed at $R = 46.0$ m, $z = 0.0$ m, and the beamlines were aimed so that corresponding horizontally and vertically tilted beams had their tangency point at about the same value of $\rho_{\text{pol}}$. Many of the used beamlines would not, in reality, pass through the beam port. Therefore, the results present an integrated modelling exercise testing and verifying the physics actors and are not meant to address any decision making on the design and parameters of the DEMO NBI system.

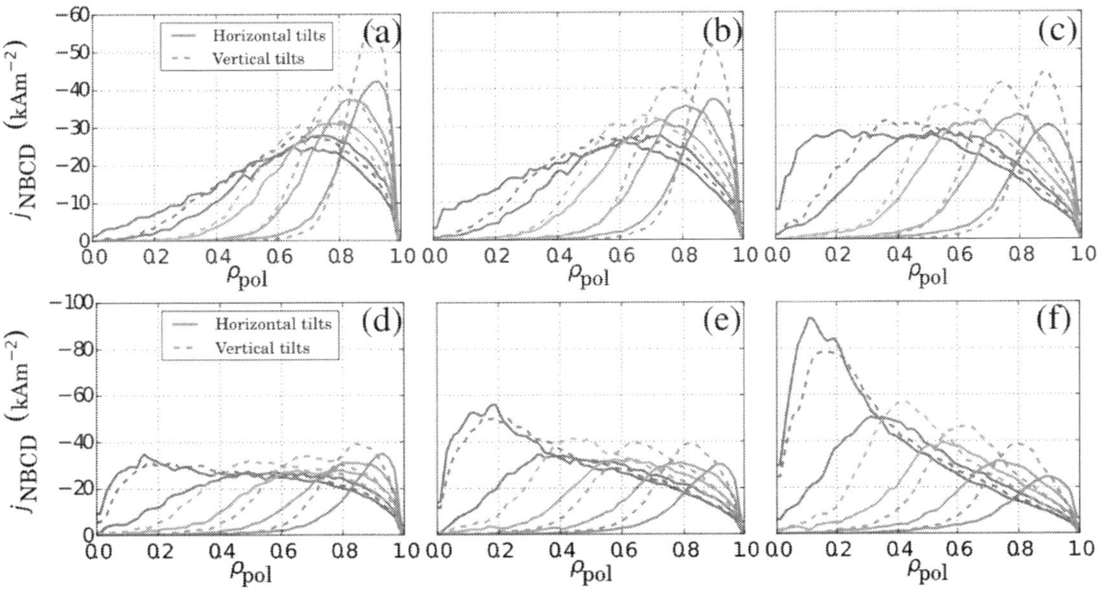


Figure 2: NBCD for horizontal (solid) and vertical (dashed) scans for the peaked (top) and flat density discharge (bottom). The beam energy was varied from 0.75 MeV (left) to 1.0 MeV (centre) and to 1.5 MeV (right). The red solid curve is the reference case, whereas the other colors are as indicated in Fig. 1.

The NBCD density profiles resulting from the beam scans are illustrated in Figs. 2(a)–(c) and (d)–(f) for the peaked and flat density discharges, respectively. The colors and line styles correspond to those of the beamlines in Fig. 1. The shape of $j_{\rm NBCD}$ depends strongly on beam aiming and energy. High energy beams reach the plasma core better than low energy beams, whereas tilting the beam horizontally or vertically moves the profile outwards. The power deposition from fast ions to the plasma (electrons and ions) follows the same trends as the driven current, with the exception of the very core where the fast ion current is strongly shielded.

Since the peaks of the resulting current density profiles are not at exactly the same locations in $\rho_{\rm pol}$, the radially increasing volume differentials make quantitative visual comparisons between them difficult. Therefore, the integrated values of power deposition to the plasma ($P_{\rm dep,tot}$) and the driven current ($j_{\rm NBCD,tot}$) for all the beamlines are plotted in Fig. 3. It is noteworthy how tilting the beam vertically, while driving current deeper in the plasma where the electron shielding is larger, still produces higher total beam driven currents compared to horizontal tilting. That is, from the current drive perspective, tilting the beam vertically is the better of the two options.

The largest tilts have most shine-through and orbit losses, resulting in significantly lower power deposition to the plasma. For the peaked density plasma, $j_{\rm NBCD,tot}$ does not decrease with increased tilting as strongly as the power deposition because the decrease in electron shielding towards the plasma edge counteracts the effect of increased shine-through and losses.

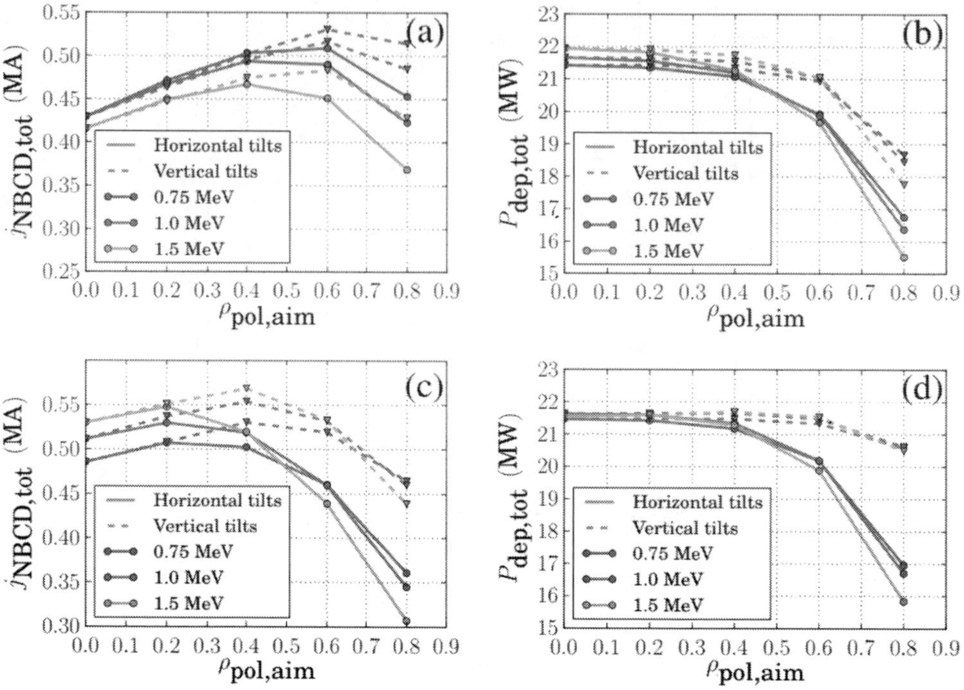


Figure 3: Total NBCD (left) and power deposition to plasma (right) in the peaked (top) and flat density (bottom) discharges as a function of flux surface ($\rho_{\mathrm{pol,aim}}$) at which the beam was aimed.

The results indicate that, in both scenarios, beams tangential at $\rho_{\mathrm{pol}} = 0.4$ ($R_{\mathrm{tang}} = 10.0$ m, $\theta = 0.0°$ and $R_{\mathrm{tang}} = 9.1$ m, $\theta = -2.0°$, orange in Fig. 1) are a good compromise to maximise $j_{\mathrm{NBCD,tot}}$ with a minimal penalty in $P_{\mathrm{dep,tot}}$. The desired current density profile shape and engineering constraints will eventually dictate how far out the beams should be tilted.

## Summary

It was demonstrated that BBNBI and ASCOT are capable of flexible and sophisticated modelling of NBI as a part of an ETS simulation. The simulations showed that tilting the DEMO beam vertically results in a few per cent higher current drive than comparable horizontal tilts.

**Acknowledgements**

This work has been carried out within the framework of the EUROfusion Consortium and has received funding from the Euratom research and training programme 2014-2018 under grant agreement number 633053 and from Tekes – the Finnish Funding Agency for Innovation under the FinnFusion Consortium. The views and opinions expressed herein do not necessarily reflect those of the European Commission.

## References

[1] F. Imbeaux et al, Comput. Phys. Commun. **181** (2010) 987–998
[2] O. Asunta et al., Comput. Phys. Commun. **188** (2015) 33–46
[3] E. Hirvijoki et al., Comput. Phys. Commun. **185** (2014) 1310–1321
[4] D. Kalupin et al., Nucl. Fusion **53** (2013) 123007
[5] H. Lütjens et al., Comput. Phys. Commun. **97** (1996) 219–260
[6] R. Albanese et al., Fusion Eng. Des. **66–68** (2003) 715–718

# Gyrokinetic investigation of isotope effect on flow oscillations in ohmic tokamak discharges

P. Niskala [1], T.P. Kiviniemi[1], S. Leerink[1], T. Korpilo[1]

[1]Department of Applied Physics, Aalto University, Espoo, Finland

## 1. Introduction

Improved confinement has been measured in deuterium plasmas compared to hydrogen discharges with similar parameters [1]. This "isotope effect" runs counter to the general understanding of heat and particle transport in plasmas. Turbulence is held as the main culrpit for anomalous transport observed in tokamaks, and one would expect the increased ion mass to boost transport and thus lead to deterioration of confinement.

On the other hand, many theories on L-H transition rely on radial flow shear to suppress the turbulent transport [2]. A variety of mechanisms affect the flows in plasmas, creating mean and zonal flows. The latter ones include the geodesic acoustic mode (GAM), a time-varying radial oscillation of potential. Given the connection between flows and the plasma confinement, possible effects of ion species on the GAM dynamics have gathered interest. We study the isotope effect on GAMs via gyrokinetic full-f simulations of hydrogen and deuterium discharges from the FT-2 tokamak. Here we present the analysis of two particular discharges from the computational point of view, while combining simulations with experimental observations is discussed by Gurchenko et al [3].

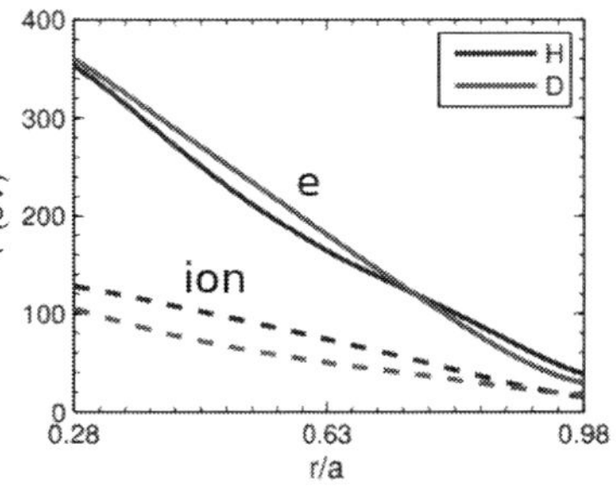


Figure 1: *Deuterium case had a higher $T_e/T_i$ ratio. The $O^{6+}$ impurities have same temperature as the main ions.*

## 2. Simulation setup

The simulations were conducted using ELMFIRE, a gyrokinetic code simulating coupled neoclassical and turbulent dynamics [5, 6, 7]. ELMFIRE uses the particle-in-cell method to model the evolution of full plasma distribution function plasma. The distribution is initialised as particles on a toroidal grid according to initial input profiles for density and temperature. A total of three particle species can be included in the calculations: electrons and two ions. Collisions between the particles have been implemented using a binary collision model.

At every time step, charge densities are distributed to the grid. Particle movement utilizes as a hybrid explicit-implicit pusher. The explicit step produces charge separation, which is then used

Table 1: *Simulation parameters.*

| Isotope | $B_\phi$ (T) | $I_0$ (kA) | $T_e$ (eV) | $T_i$ (eV) | $n_e$ ($10^{19}$ m$^{-3}$) | $Z_{\text{eff}}$ |
|---|---|---|---|---|---|---|
| H | 2.27 | 20.5 | 416 | 147 | 2.68 | 2.3 |
| D | 2.22 | 19.4 | 435 | 151 | 2.48 | 2.8 |

to solve the electric potential to ensure quasineutrality. The potential is interpolated to particles to calculate the polarisation drift on ions and parallel electric field acceleration on both ions and electrons in the implicit step.

The magnetic geometry of ELMFIRE is fixed as co-centric flux surfaces with no shifts. Simulation region is limited radially, resulting in a toroidal annulus. Electric field is set to zero at the inner boundary and potential to zero at the wall. Toroidal magnetic field on the magnetic axis and the current profile are given as an input to determine the total magnetic field and safety factor profile.

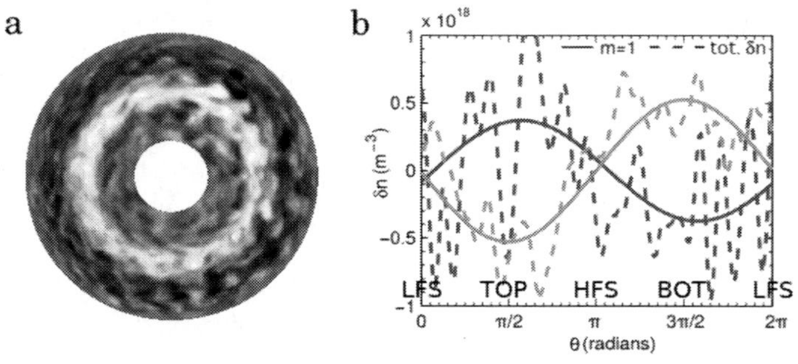


Figure 2: *The poloidal snapshot of (a) shows the characteristic temporally varying GAM oscillation, while the $m = 1$ density fluctuation is visible for opposite temporal phases in (b).*

The simulations are based on two plasma discharges from the FT-2 tokamak ($a = 0.08$ m, $R_0 = 0.55$ m), denoted here on as just hydrogen (H) and deuterium (D). The basic parameters are presented in table 1. Besides the isotope, main difference between the cases comes from the effective charge number $Z_{\text{eff}}$ and ion temperature, as seen in figure 1. Partially ionised oxygen ($O^{6+}$) was taken included as the impurity ions in the simulations, and the density profile was initialised according to the effective charge. Simulation grid spanned radially from $r = 0.02$ m to $r = 0.08$ m with 120 radial, 150 poloidal and 8 toroidal grid points. A total of approximately $10^9$ particles was initialised with a simulation time step of 0.03 microseconds.

### 3. Simulation results

The main features of geodesic acoustic mode are clearly visible in the simulations. Firstly, the expected radial oscillation is evident in the results for the electric potential, as seen in figure 2a. The temporal frequency for these dominating fluctuations is smaller for deuterium (below 50 kHz), as would be expected based on simplest analytical expressions for frequencies [4]. The frequency is largely constant in radial direction. Secondly, fluctuations are also present in the radial electric field, which couples the potential oscillation to a sinusoidal $m = 1$ mode in density. Figure 2b shows that this mode is clearly present in the simulations as density fluctuations at the characteristic GAM frequency.

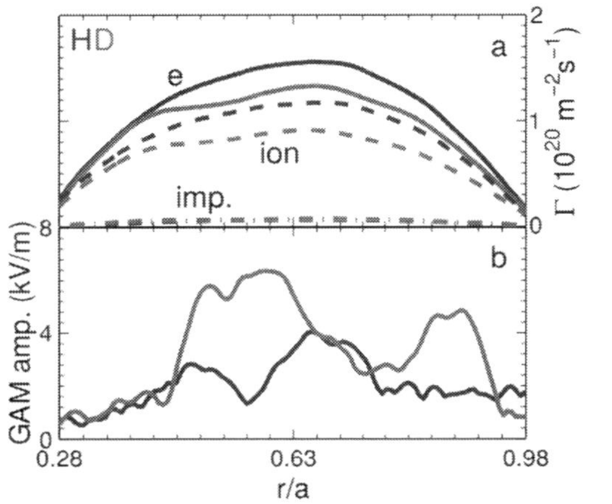


Figure 3: *Particle fluxes are larger for hydrogen (a), while $E_r$ oscillations at the GAM frequency have a higher amplitude for the deuterium case (b).*

The amplitude of the temporal $E_r$ oscillations is distinctly higher for deuterium, as figure 3b demonstrates. Similar trend has been observed for zonal flows at TEXTOR [8]. The amplitudes have large spatial variation, with the strongest oscillations located close to the middle plasma radius. The amplitude of the $m = 1$ density oscillations scales similarly, indicating that the top-bottom asymmetry is captured in the simulations. The radial wavelength of the oscillations also increases from deuterium to hydrogen, consistently with both earlier simulations and theoretical work [9, 10]. Based on the theoretical work by Hahm et al [11], higher amplitude combined with lower frequency could lead to higher effective shearing rate and stronger reduction of turbulence for deuterium.

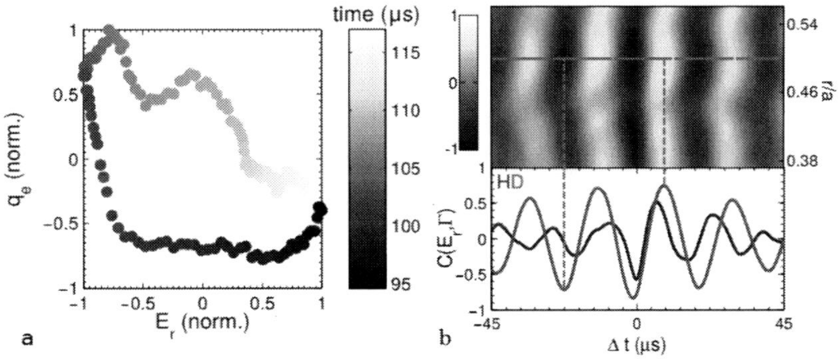


Figure 4: *Flux and $E_r$ oscillations have limit-cycle-like characteristics, especially for the deuterium case (a). Correlation between $E_r$ and particle flux is stronger and less localised for deuterium compared to hydrogen (b).*

Hydrogen simulation produces larger energy fluxes, but conductive heat fluxes are similar for both isotopes. The difference is explained by particle fluxes, which are seen as the temporal mean levels in figure 3a. Generally it is difficult to attribute decreased particle fluxes to the stronger $E_r$ oscillations of the deuterium case, but there is a slight dip in the region where $E_r$ has the highest amplitude. Furthermore, $E_r$ and transport oscillations have a predator-prey-like phase shift in the simulations, as seen in the single oscillation period of figure 4a . Figure 4b demonstrates that this correlation between the two is also more distinct for deuterium. Moreover, the correlation is less localised for deuterium. These observations are compatible with increasing GAM amplitude and radial wavelength, but temporal standard deviation of flux oscillations grows slightly as well, indicative of the relationship between transport and $E_r$.

**4. Discussion**

While the simulations show the deuterium plasma to have stronger geodesic acoustic mode activity and increased correlation with transport, there is room for more detailed investigations. The computational study can be expanded by including more cases to further establish the trends in transport and amplitudes. Additionally, GAM wavelength and propagation at FT-2 deviates from the theoretical predictions of Itoh et al. Investigating the characteristics and scaling of the GAMs observed in FT-2 via ELMFIRE simulations offers an appealing avenue for deepening the understanding of flow dynamics in tokamak plasmas, bridging the theoretical and experimental observations through advanced computations.

**Acknowledgements:** We thank A.D. Gurchenko for E.Z. Gusakov for providing experimental input data. This work was supported by grant 278487 of the Academy of Finland has received funding from Tekes – the Finnish Funding Agency for Innovation. Computational resources for the work were provided by CSC - IT Center for Science and IFERC.

**References**

[1] M. Bessenrodt-Weberpals et al, Nuclear Fusion **33**, 1205 (1993)
[2] P.W. Terry et al, Reviews of Modern Physics **72**, 109 (2000)
[3] A.D. Gurchenko et al, I5.J203 at this conference
[4] N. Winsor et al, Physics of Fluids **11**, 2448 (1968)
[5] J.A. Heikkinen et al, Journal of Computational Physics **227** (2008)
[6] S. Leerink et al, Physical Review Letters **109**, 165001 (2012)
[7] T.P. Kiviniemi et al, Plasma Physics and Controlled Fusion **56**, 075009 (2014)
[8] Y. Xu, Physical Review Letters **110**, 265005 (2013)
[9] P. Niskala et al, Nuclear Fusion **55**, 073012 (2015)
[10] K. Itoh et al, Plasma and Fusion Research **1**, 037 (2006)
[11] T.S. Hahm et al, Physics of Plasmas **6**, 922 (1999)

# Microinstabilities in the Confinement Zone near the Tokamak Edge

C M Roach[1], P Abdoul[2], M Anastopoulos[2] J W Connor[1], D Dickinson[2], R J Hastie[1],
S Saarelma[1], and H R Wilson[2]

[1] *CCFE, Culham Science Centre, Abingdon, OX14 3DB, UK*

[2] *York Plasma Inst., Dep't of Physics, University of York, Heslington, York, YO10 5DD, UK*

## Introduction

Microturbulence influences transport in the edge region of tokamak H-mode plasmas, and the pedestal characteristics that are crucial for overall plasma performance. Local gyrokinetic calculations for MAST [1, 2] find that, at perpendicular lengthscales close to $\rho_i$, electromagnetic microinstabilities dominate with a striking mode transition at the pedestal top: microtearing modes (MTMs) dominate in the plateau inboard of the pedestal top, while twisting parity kinetic ballooning modes (KBMs) dominate in the pedestal itself. This paper reviews recent findings on: mode transitions at the MAST pedestal top, physics impacting on the (marginal) stability of KBMs in the pedestal, and microtearing modes in the plateau.

## Mode Transitions at the Pedestal Top: ETG also unstable in Plateau

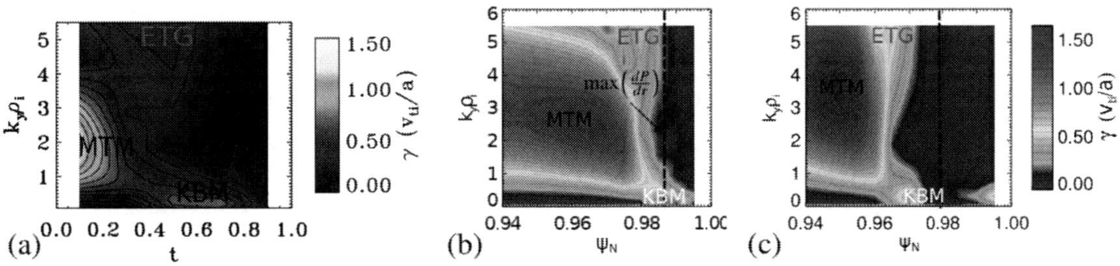


Figure 1: *(a) $\gamma(k_y\rho_i, t)$ of the dominant mode at the $\psi_n = 0.975$ surface (where t is normalised time during the ELM recovery). (b), (c) $\gamma(k_y\rho_i, \psi_n)$ at the start and end of the ELM recovery, respectively.*

As the MAST pedestal expands inwards during the ELM recovery [3], $dn_e/dr$ and $dP/dr$ increase on a plateau plasma surface as it joins the pedestal, which acts to stabilise MTMs until the onset of KBMs at higher $dP/dr$ [1]. Subsequent studies reveal that ETG is *also unstable* at higher $k_y\rho_i$ in the MAST plateau, and *also stabilised* as $dn_e/dr$ increases on a plateau surface joining the pedestal. Figure 1 shows that microstability spectra from [1, 2] *contain the low $k_y\rho_i$ tail of the ETG peak.* DBS and CPS measurements of $\delta n$ and $\delta B$ fluctuations at the MAST pedestal top during ELM recovery, appear consistent with linear ETG properties from GS2 [4].

## Impact of $\nu_*$ and Global Effects on KBMs in the Pedestal

KBMs are locally unstable, but close to marginal [1], in most of the collisional MAST pedestal studied in [1], but in lower $\nu_*$ MAST pedestals KBMs are locally stable across most of the

[1]ASDEX Upgrade pedestals are also close to the KBM limit [5].

steep gradient region [2], as has also been found for JET pedestals [6]. This increased stability at lower $\nu_*$ is due to higher bootstrap current density, $J_{bs}$, reducing magnetic shear, $s$, and improving access to 2nd stability. The sensitivity of the absolute local stability of KBMs to $J_{bs}$ will motivate the future use of more accurate bootstrap current models, e.g. [7].

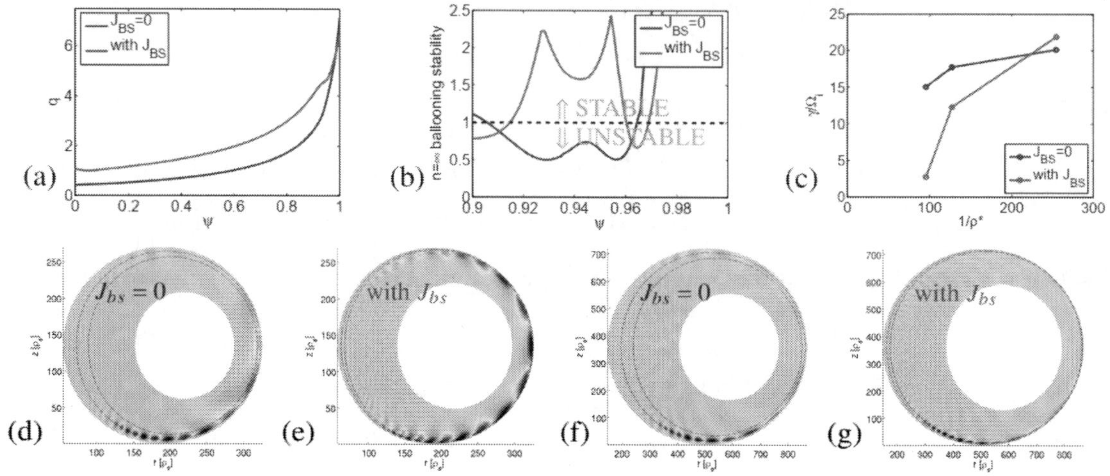


Figure 2: *(a) q profiles and (b) ideal $n=\infty$ ballooning stability index from circular MAST-like equilibria with/without $J_{bs}$. (c) Corresponding growth rates of $n = 10$ KBM, from ORB5, as functions of $1/\rho_*$. Eigenfunctions of electrostatic potential with,without $J_{bs}$ (d),(e) for $1/\rho_* = 95$ , and (f),(g) for $1/\rho_* = 254$.*

Global effects are also significant for ion scale instabilities in the pedestal as $\rho_i/L \not\ll 1$. Global mode properties can be determined by post-processing results from local gyrokinetic calculations (see [8, 9]). This reveals that global profile effects impact on the linear eigenmodes and reduce growth rates below those obtained from purely local analyses, as demonstrated for ITG in [9]. In future we will compare the local post-processing approach of [9] with linear global GK calculations of KBMs in the pedestal (using ORB5). Two circular MAST-like equilibria were generated with and without $J_{bs}$, with the $q$ profiles given in Figure 2(a). Figure 2(b) shows $n = \infty$ ballooning stability index profiles, indicating that the ballooning unstable region of the pedestal narrows with $J_{bs}$. Figure 2(c) shows linear growth rates for the $n = 10$ KBM from global GK calculations using ORB5 over a scan in $1/\rho_*$, and figures 2(d-g) show electrostatic potential eigenfunctions. At low $1/\rho_*$, where global effects are stronger, the KBM is more extended radially and less unstable. The global effects are more strongly stabilising in the case with $J_{bs}$ included, because the mode must extend into a locally stable region.

## Studies of Microtearing Modes

MTMs are unstable in the MAST plateau without collisions, and growth rates are sensitive to drifts [10]. [2] Linear collisionless local gyrokinetic calculations using GS2, GKW and GENE for

[2]MTMs with similar properties are also unstable in the JET plateau [6], and were recently found in global

a circular equilibrium surface based on MAST plateau equilibrium parameters, find dominant MTMs with similar growth rate and frequency spectra (see Figure 3).

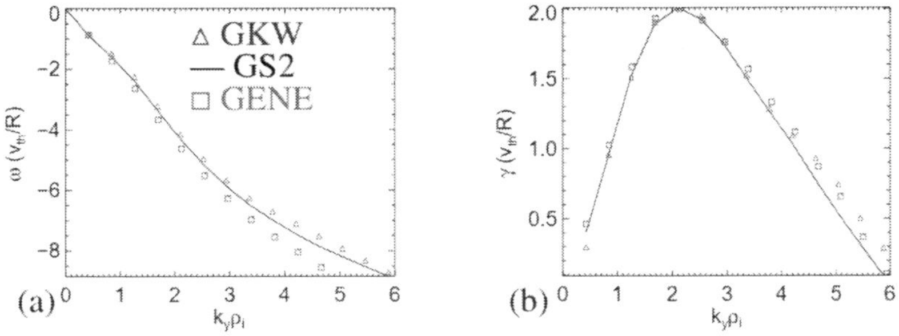


Figure 3: *(a) Real frequency, $\omega$, and (b) growth rate, $\gamma$, of dominant MTMs versus $k_y\rho_i$ from local linear collisionless GK calculations using GS2, GKW and GENE, for a circular MAST-like plateau surface.*

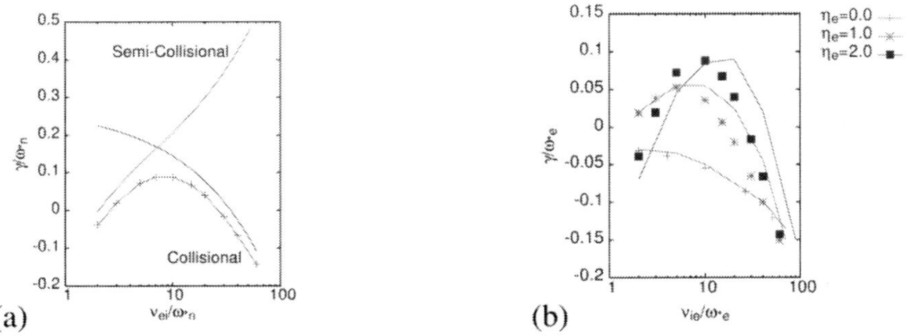


Figure 4: *MTM growth rates from GS2 versus electron collision frequency compared (a) with scalings in the collisional (blue) and semi-collisional (purple) limits obtained by Drake and Lee [12] at $\eta_e = 2$, and (b) at $\eta_e = 0, 1, 2$ with the model by Gladd* et al *[13].*

GS2 calculations of MTMs in slab geometry [14], with equilibrium parameters $\beta_e = 0.005$, $L_n/L_s = 0.05$, $k_y\rho_i = 0.05$ and various values of $\eta_e = L_{n_e}/L_{T_e}$, capture the trends predicted by theory for another microtearing instability mechanism that relies on collisions [12, 13]. Figure 4 shows MTM growth rates from GS2 versus electron collision frequency, comparing in Figure 4(a) with predictions by Drake and Lee [12], and in Figure 4(b) with calculations by Gladd *et al* [13] that more fully account for the electrostatic impact on the current layer.

Nonlinear MTM simulations at the MAST pedestal top predict significant electron heat flux, but convergence was only achieved with $\beta$ a factor 5 smaller than in experiment [15]. Edge fluctuation measurements in MAST [16] and ASDEX Upgrade [17], find type-I ELM precursors that propagate in the electron diamagnetic drift direction, and may be consistent with MTMs.

## Conclusions

Local gyrokinetic simulations suggest that electromagnetic microinstabilities play an important role at the edge of the confinement zone in MAST H-mode plasmas: MTMs and ETG dominate in the plateau region, and KBMs, which are sensitive to collisionality and global effects,

simulations of large aspect ratio tokamaks [11].

are close to marginal stability in the pedestal. Both ETG and MTM are stabilised as $dn_e/dr$ increases on a plateau surface as it joins the pedestal during the ELM recovery. The plateau MTM is unstable without collisions, and its linear mode properties have been verified, for circular shaping, by three local gyrokinetic codes. GS2 captures predictions for the scaling of MTM growth rates with $\nu_e$ in slab geometry where the MTM drive requires collisions. Nonlinear simulations are needed for more realistic plasma conditions in the edge plasma, to improve contact with edge fluctuation measurements.

**Acknowledgements**

This work has been carried out within the framework of the EUROfusion Consortium and has received funding from the Euratom research and training programme 2014-2018 under grant agreement No 633053 and from the RCUK Energy Programme [grant number EP/I501045]. The views and opinions expressed herein do not necessarily reflect those of the European Commission. GK calculations were carried out on supercomputers: HELIOS at IFERC, Aomori, Japan under the Broader Approach; ARCHER via the Plasma HEC Consortium [EPSRC Grant No.EP/L000237/1]; and HECToR [EPSRC Grant No. EP/H002081/1]. The authors are grateful for useful discussions with FJ Casson, H Doerk, J Hillesheim, A Kirk, R Scannell and D Told.

**References**

**[1]** D. Dickinson et al., Physical Review Letters **108**, 135002 (2012). **[2]** C. M. Roach et al., in Proceedings of 24th IAEA Fusion Energy Conference, San Diego, pp. (TH/5–1), 2012. **[3]** D. Dickinson et al., Plasma Phys. Controlled Fusion **53**, 115010 (2011). **[4]** J. C. Hillesheim et al., submitted to Plasma Phys. Controlled Fusion (2015). **[5]** D. Hatch et al., Nuclear Fusion **55**, 063028 (2015). **[6]** S. Saarelma et al., Nuclear Fusion **53**, 123012 (2013). **[7]** E. A. Belli et al., Plasma Phys. Controlled Fusion **56**, 045006 (2014). **[8]** D. Dickinson et al., Physics of Plasmas **21**, 010702 (2014). **[9]** P. A. Abdoul et al., Plasma Phys. Controlled Fusion **57**, 065004 (2015). **[10]** D. Dickinson et al., Plasma Phys. Controlled Fusion **55**, 074006 (2013). **[11]** A. K. Swamy et al., Physics of Plasmas **21**, 082513 (2014). **[12]** J. F. Drake and Y. C. Lee, Phys Fluids **20**, 1341 (1977). **[13]** N. T. Gladd et al., Phys Fluids **23**, 1182 (1980). **[14]** M. Anastopoulos, Mphys project dissertation, University of York, 2015. **[15]** D. Dickinson et al., in Proceedings of 40th EPS Conference on Plasma Physics, Strasbourg, France, Vol. ECA 37D, O3.105, 2013. **[16]** A. Kirk et al., Nuc. Fusion **54**, 114012 (2014). **[17]** P. Manz et al., Plasma Phys. Controlled Fusion **56**, 035010 (2014).

# Benchmarking Neutral Beam Injection codes within the European Integrated Modelling framework

M. Schneider[1], O. Asunta[2], T. Johnson[3], D. Kalupin[4], R. Coelho[5] and EU-IM team*

[1] *CEA, IRFM, F-13108 Saint-Paul-lez-Durance, France*

[2] *Department of Applied Physics, Aalto University, P.O. Box 14100, FI-00076 AALTO, Finland*

[3] *KTH Royal Institute of Technology, 10044 Stockholm, Sweden*

[4] *EuroFusion Cons. Prog. Unit, Boltzmannstr. 2, D-85748, Garching, Germany*

[5] *Instituto de Plasmas e Fusao Nuclear, IST, UTL 1049-001 Lisboa, Portugal*

## Introduction

Auxiliary heating systems are vital to reach fusion relevant temperatures in tokamaks. One of the most prevailing heating sources is neutral beam injection (NBI). Hence, reliable simulations of NBI sources in fusion plasmas are essential for preparing scenarios and predicting performances of future fusion devices.

The European Integrated Modelling (EU-IM) framework is developed for providing a standardized platform for the development of integrated modelling suits [1]. This framework offers a high level of modularity with standardized input and output enabling the selection between multiple combinations of simulation models.

NBI modelling requires two modelling steps: first, the injection and ionization of fast neutrals are estimated by beam deposition codes; secondly, the dynamics of NBI ions is simulated by Fokker-Planck solvers. At present two deposition codes are integrated in the EU-IM framework, the Monte Carlo code BBNBI [2] and the narrow-beam model NEMO [3]. They can be associated with four Fokker-Planck solvers, ASCOT [4], NBISIM [5], RISK [6] and SPOT [7], which are also integrated inside the EU-IM framework and used for the present benchmark. ASCOT and SPOT are Monte Carlo codes with a high level of accuracy including orbit width effects, while NBISIM is a simple 1D analytic model and RISK is a 2D Fokker-Planck code that combines finite elements and an eigenfunction expansion. Hence, the two latter are much faster but less accurate than ASCOT or SPOT since they operate in the zero-banana-width limit.

The Heating and Current Drive workflow combines codes for NBI, ICRH, ECRH and alpha-heating with a design that makes it easy to include synergies between the heating schemes. This workflow is used both as a standalone tool and to handle heating and current drive pro-

*See http://www.euro-fusionscipub.org/eu-im

cesses inside the European Transport Solver. The present study reports the results of the NBI benchmark activity using this Heating and Current Drive workflow and the codes mentioned above, covering all possible combinations between the deposition and Fokker-Planck codes, for ASDEX-Upgrade, JET and ITER and plasmas.

**Beam deposition and ionization**

Typical equilibrium and kinetic profiles for ASDEX-Upgrade, JET and ITER discharges have been used for this benchmark. The two deposition codes BBNBI and NEMO have been extensively benchmarked in order to verify their deposition in the multi-dimensional phase space. They are found to agree well, given their different approaches (Monte Carlo versus narrow beam model), as illustrated in Figures (1) and (2) showing the 1D and 2D profiles of the neutral beam deposition. Their total source rates are in very good agreement (below 1% difference, attributed to shine-through predictions).

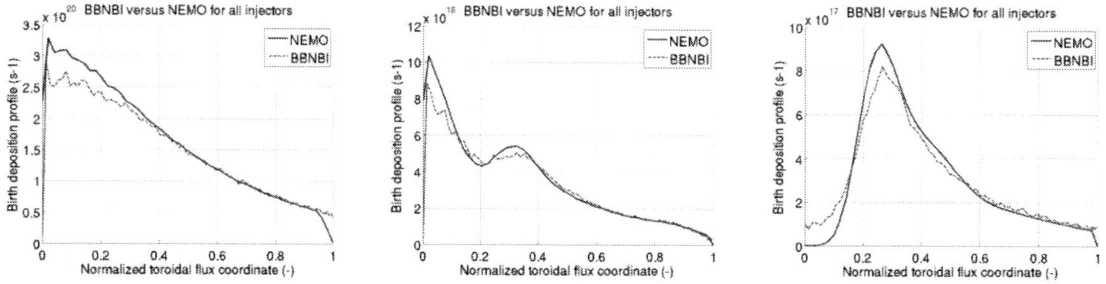


Figure 1: *BBNBI and NEMO 1D-ionization profiles, for ASDEX-Upgrade, JET and ITER plasmas, from left to right.*

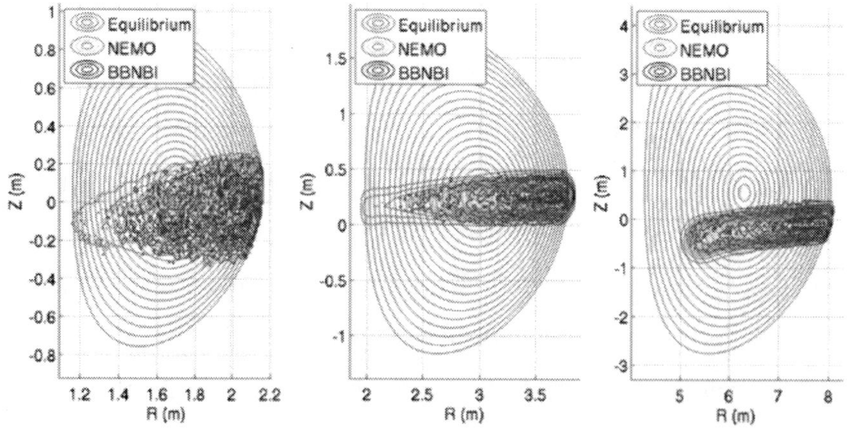


Figure 2: *BBNBI and NEMO 2D deposition profiles, for ASDEX-Upgrade, JET and ITER plasmas, from left to right.*

The second stage of NBI modelling after neutral beam deposition is the simulation of the dynamics of confined ions using Fokker-Planck codes, as presented in the next section.

## Power to the bulk and Neutral Beam Current Drive (NBCD)

All the combinations between the deposition and Fokker-Planck codes have been tested and compared in the fast ion slowing down steady-state. The resulting 1D profiles for the power to the bulk and the neutral beam current drive are presented in Figures (3) and (4) for ASDEX-Upgrade, JET and ITER. The total power to the bulk differs from less than 7% between all combinations while the total NBCD differs from less than 20%. As can be seen, RISK and NBISIM predict narrower profiles compared to SPOT and ASCOT along with small differences in the NBCD profiles. The reason is that orbit width effects are neglected both in RISK and NBISIM. Including an ad-hoc orbit width correction in RISK leads to a better agreement for power 1D-profiles compared to SPOT, as shown in Ref. [6]. However, NBCD profiles cannot be improved using this correction since a part of the NBCD is due to a diamagnetic current driven by fast ion finite orbit width and pressure gradients. It means caution is needed when using simple models like RISK for current drive modelling.

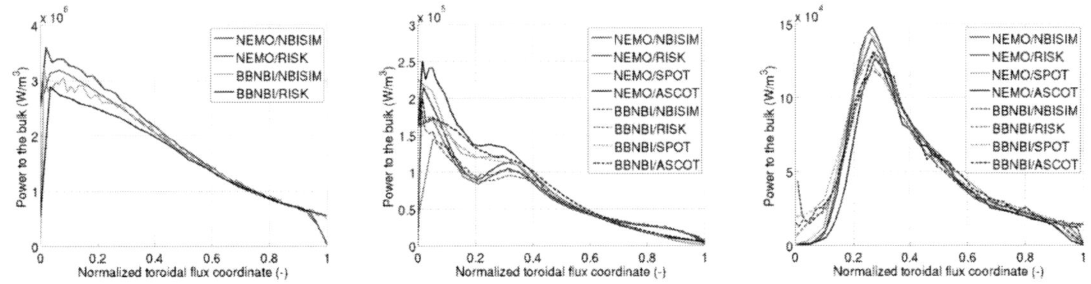


Figure 3: *Power to the bulk, for ASDEX-Upgrade, JET and ITER plasma, from left to right.*

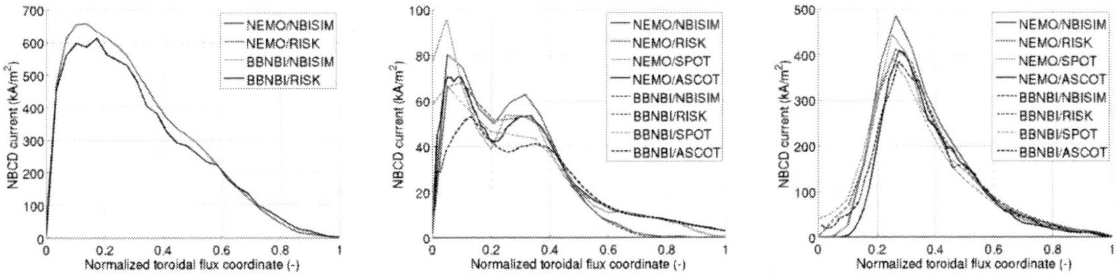


Figure 4: *NBCD profiles for ASDEX-Upgrade, JET and ITER plasmas, from left to right.*

## Ramp-up towards the fast ion steady-state

Accurately modelling transitory phases of fast ion heating and current drive is essential for predicting the full dynamics of a discharge inside a transport solver. The Fig. (5) shows the time evolution of the power to bulk electrons and ions for JET, as predicted by SPOT and RISK Fokker-Planck codes, showing an overall good agreement between the two codes.

## Conclusion

The comparison of the involved simulation codes for modelling heating and current drive induced by NBI fast ions evidences an overall good agreement between the physics models. The main discrepancies between the codes arise from orbit width effects that are not included in simple Fokker-Planck calculations.

This benchmark allows to trustfully choose the adapted NBI model according to the required accuracy and computation time for simulating a full plasma discharge in the context of integrated modelling frameworks, here within the European platform, and will be continued on the ITER Integrated Modelling and Analysis Suite [8].

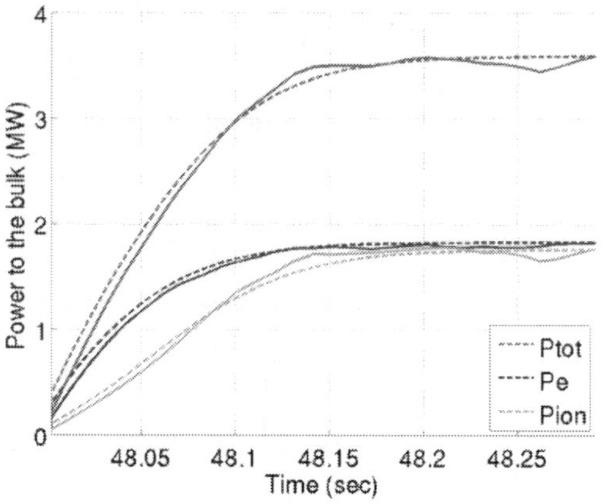


Figure 5: *Ramp-up of power to bulk electrons and ions for JET using SPOT (solid lines) and RISK (dashed lines) Fokker-Planck codes.*

## Aknowledgments

This work has been carried out within the framework of the EUROfusion Consortium and has received funding from the Euratom research and training programme 2014-2018 under grant agreement No 633053. The views and opinions expressed herein do not necessarily reflect those of the European Commission.

## References

[1] G.L. Falchetto et al, Nucl. Fusion 54 (2014) 043018

[2] O. Asunta et al, Computer Physics Comm. 188 (2015) 33-46

[3] M. Schneider et al, Nucl. Fusion 51 (2011) 063019

[4] E. Hirvijoki et al. Computer Physics Communications 185 (2014) 1310-1321

[5] From equation (5.4.12) of Wesson J. 2004 Tokamaks 3rd ed. (Oxford: Clarendon)

[6] M. Schneider et al, Nucl. Fusion 55 (2015) 013003

[7] M. Schneider et al, Plasma Phys. Control. Fusion 47 (2005) 2087-2106

[8] F. Imbeaux et al, Proc. 25th IAEA Fusion Energy Conference, TH/P3-41 - St-Petersburg (2014)

# AUTHOR INDEX

# AUTHOR INDEX

# AUTHOR INDEX

# AUTHOR INDEX

# AUTHOR INDEX

# AUTHOR INDEX

# AUTHOR INDEX

# AUTHOR INDEX

# AUTHOR INDEX

# AUTHOR INDEX

# AUTHOR INDEX

# AUTHOR INDEX

# AUTHOR INDEX

# AUTHOR INDEX

# AUTHOR INDEX

# AUTHOR INDEX

## AUTHOR INDEX

# AUTHOR INDEX

# 42nd EPS Conference on Plasma Physics (EPS 2015)

Europhysics Conference Abstracts Volume 39E

Lisbon, Portugal
22-26 June 2015

Part 2 of 2

**Editors:**

**R. Bingham**
**S. Atzeni**
**K. McClements**
**C. Silva**
**W. Suttrop**
**R. Foest**
**B. Goncalves**
**R. Coelho**

ISBN: 978-1-5108-2946-6

**Printed from e-media with permission by:**

Curran Associates, Inc.
57 Morehouse Lane
Red Hook, NY 12571

**Some format issues inherent in the e-media version may also appear in this print version.**



Printed by Curran Associates, Inc. (2016)



European Physical Society (EPS)
6 Rue des Freres Lumiere
68200 Mulhouse
France

Phone: 33 389 32 94 40
Fax: 33 389 32 94 49

www.eps.org

**Additional copies of this publication are available from:**

Curran Associates, Inc.
57 Morehouse Lane
Red Hook, NY 12571 USA
Phone: 845-758-0400
Fax: 845-758-2633
Email: curran@proceedings.com
Web: www.proceedings.com

# TABLE OF CONTENTS

## PART 1

## PART 2

# On the generation and saturation of the Zonal Flows

C. Norscini[1], Ph.Ghendrih[1],
T.Cartier-Michaud[1], G.Dif-Pradalier[1], X.Garbet[1], N.Nace[1], Y.Sarazin[1], P.Tamain[1]
[1] *CEA Cadarache, Saint Paul lez Durance, France*

## Introduction

In fusion plasmas heat transport is deleterious, therefore transport barrier (TB) generation is a key performance indicator. The spontaneous generation of a transport barrier is due to the interplay between turbulence and zonal flows (ZF). Turbulent eddies can transfer energy via non linear coupling to ZF that tend to store it for long time scale. On the other hand the efficiency of the energy transfer between turbulence and zonal flows can vary and different transport regimes have been observed. Two opposite regimes have been identified: (1) fully turbulent, where one can retrieve an homogeneous turbulence distribution and ZF are weak, (2) zonation, where the ZF dominate and the turbulence is damped. In many dynamical systems one can observe these various regimes, from oceans to planetary atmosphere [1].

We present here an analytical work on the interaction between turbulence and ZF reducing our model to a limited number of modes. The results are then compared with those obtained via a 2D fluid code. The role of non linear coupling in the ZF generation and saturation is defined and appears crucial in the transition from turbulent to zonation regime.

## Zonal flow source and sink: study of three modes coupling

Departing from the charge conservation equation, in an isothermal plasma, in the cold ions limit and in the case of absence of interchange forcing, the vorticity equation is derived as following

$$\partial_t W \quad + \quad [\phi, W] = \nu_\perp \Delta_\perp W + J \tag{1}$$

where $W$ is the vorticity $W = \Delta_\perp \Phi$, $x$ and $y$, the space coordinates corresponds to the poloidal angle and radial direction, $J$ is the parallel current loss. The convective turbulent transport $[\Phi, W] = \partial_x(W(-\partial_y \Phi)) + \partial_y(W(\partial_x \Phi))$ competes with the small scale diffusive transport with coefficient $\nu$. The loss term is defined as $< J >_y = 0$. Averaging along the flux surfaces, there is no current loss in parallel direction and the zonal flows $V_z = \partial_x < \Phi >_y$ evolution can be rewritten as

$$\partial_t V_z + \partial_x RS - \nu \Delta_x V_z = 0 \tag{2}$$

where $RS(x) =< \partial_x\tilde{\Phi}\partial_y\tilde{\Phi} >_y$ is the Reynold stress term and $\tilde{\Phi} = \Phi - <\Phi>_y$ is the potential fluctuations. The collisions term, in this case simplified through the viscosity term, acts as sink for ZF [3]. Conversely the RS term can act as source or sink of the zonal flows, according to the predator-prey scheme[2]. Namely, two possible scenarios have been identified: (a) Turbulence gives energy to ZF, i.e. the ZF are excited and turbulence is damped. ZF govern the large scale transport (in this case RS acting as source).(b) Once ZF source is drained, ZF start decaying and turbulence can grow until a turbulent burst can go through the barrier (in this case RS is a ZF sink). Once the turbulent transport becomes dominant, the cycle restarts. The RS term in the two different regimes has been studied in order to understand the mechanism that let the system cycle between both regimes (a) and (b). The interplay between zonal and turbulent modes can be addressed in the framework of non linear three mode coupling: the zonal flow, the streamer and the more homogeneous turbulent mode, respectively $\Phi_z(\kappa,0), \Phi_s(0,k_s), \Phi_t(\kappa,k_s)$. Two cases of study are proposed in this letter, comparable to the regimes previously mentioned:

(a) The condensation of the ZF, in the case the streamer $\Phi_s$ represents the equilibrium profile and $\Phi_z, \Phi_t$ are the small amplitude perturbations, if the growthrate is positive the turbulence is giving energy to ZF.

(b) The Kelvin Helmoltz (KH) instability, where the ZF modes is the equilibrium profile and $\Phi_z, \Phi_t$ are the perturbations, positive growthrate means that the ZF is giving energy to turbulence.

The Kelvin Helmoltz instability is the primary focus. We can study such instability in a more general way defining the perturbations as $\phi_l(k_{xl},k_y),\phi_{l+1}(k_{xl+1},k_y)$, where $k_{xl+1} = l\kappa$ with $l = [0,\inf]$, the dispersion relation can be rewritten as

$$[\gamma + \frac{\gamma_l+\gamma_{l+1}}{2}]^2 = -V_l V_l' k_y^2 \kappa^2 |\Phi_z|^2 - \gamma_l\gamma_{l+1} + \frac{\gamma_l+\gamma_{l+1}}{2}^2 \tag{3}$$

where $\gamma_i = \nu k_i^2 + \sigma/k_i^2, k_i^2 = k_{xi}^2 + k_y^2$ with $i = l, l+1$ and coupling terms are $V_l = \left(k_l^2 - \kappa^2\right)/k_{l+1}^2$, $V_l' = \left(k_{l+1}^2 - \kappa^2\right)/k_l^2$. The instability condition is then $-V_l V_l' k_y^2 \kappa^2 |\Phi_z|^2 - \gamma_l\gamma_{l+1} > 0$. If we define $R_l = -\frac{V_l V_l' k_y^2 \kappa^2 |\Phi_z|^2}{\gamma_l\gamma_{l+1}}$, then $R_l > 1$ becomes a necessary condition in order to have the coupled modes unstable, which lead to define three new variables $X_l, Y_l, Z_l$ such that $X^2 = \frac{k_{xl}^2+k_y^2}{K^2}$, $Y_l = \frac{k_{xl}}{K}$ and $Z_l^2 = X_l^2 - Y_l^2 = \frac{k_{yl}^2}{K^2}$. Let be $\bar{k}$ such that $\nu\bar{k}^2 = \frac{\sigma}{\bar{k}^2}$. In this way we rewrite $R_l = g^2 F(x,y,l)/G(X,y,l)$ such that

$$F = -V_l V_l' \left(\frac{|\Phi_z|\kappa^4}{k_l k_{l+1}}\right)^2 = (X_l^2 - Y_l^2)(X_l^2 - 1)(X_l^2 + 2Y_l) \tag{4a}$$

$$G = \gamma_1\gamma_2 \left(\frac{\sigma}{k_l k_{l+1}}\right)^2 = \left(\frac{k_l^4}{\bar{k}^4} + 1\right)\left(\frac{k_{l+1}^4}{\bar{k}^4} + 1\right) \; ; \; g^2 = |\Phi_z|^2 \frac{\kappa^8}{\sigma^2} \tag{4b}$$

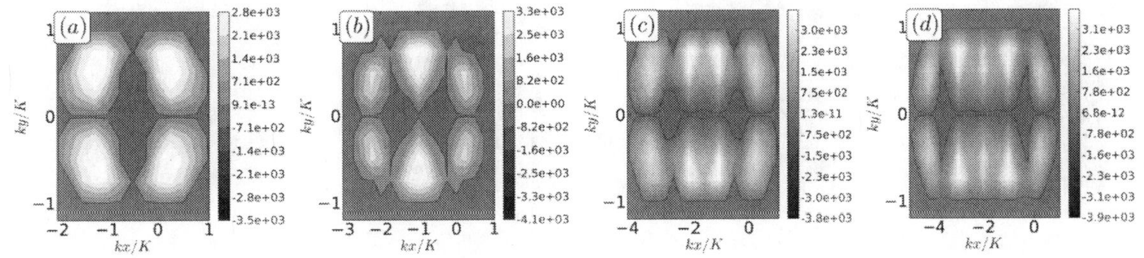

Figure 1: (a)the linear growthrate $\gamma\tau_{\parallel}$ in function of $k_y$ and $k_x$ in case of 2 coupled modes, (b) 3 modes, (c) 4 modes and (d) 5 modes

By definition $G$ and $g^2$ are positive and $X_l^2 \geq Y_l^2$, therefore $R_l$ is positive if $(X_l^2-1)(X_l^2+2Y_l) < 0$, which corresponds to two possible cases

$$Y_l^2+Z_l^2<1 \cap (Y_l+1)^2+Z_l^2>1 \quad or \quad Y_l^2+Z_l^2>1 \cap (Y_l+1)^2+Z_l^2<1 \tag{5}$$

In fig.1(a) the growthrate amplitude in function of the perturbation mode is plotted . Additionally, if we take in account more than one couple of modes, the following linear system is derived:

$$\begin{pmatrix} \Omega_1 & V_1 & 0 & \cdots & 0 \\ -V_1' & \Omega_2 & V_2 & \cdots & 0 \\ 0 & \ddots & \ddots & \ddots & 0 \\ 0 & 0 & \cdots & V_n' & \Omega_n \end{pmatrix} \begin{pmatrix} \Phi_l \\ \Phi_2 \\ \vdots \\ \Phi_n \end{pmatrix} = 0$$

Solving the linear system, we observe that the unstable spectrum region is expanded in $k_x$ if the number of perturbations coupled with $\phi_z$ increases, while the condition $\gamma > 0 => |k_y|/|\kappa| < 1$ is always true (see fig.1(b)-(c)-(d), where $\gamma\tau_{\parallel}$ is plotted in function of $k_x, k_y$ respectively in case of three, four and five modes contribution). Conversely, if the same procedure is repeated but considering the streamers as equilibrium condition and $\Phi_z, \Phi_t$ as perturbations (case (a)), the opposite condition is recovered: the region of spectrum of positive growthrate is constrained by $|k_y|/|\kappa| > 1$. To verify the analytical observation, the two regimes are now modeled. For (b) case, we assume a equilibrium profile $\Phi_z(\kappa = 0.024, 0)$ and two streamers are excited with different poloidal size, $\Phi_{s1}(0, \kappa/2), \Phi_{s2}(0, 7/4\kappa)$. From fig.2(a)-(b) we can observe that the perturbation $\Phi_{s1}$ is growing accordingly with the prevision of the linear analysis, i.e. the growthrate is approximately $\gamma\tau_{\parallel} \approx 2600$. On the other hand the streamers with poloidal size smaller than the ZF radial width is damped. If we consider the streamers like mode $\Phi_s(0, k_s = 0.024)$ as equilibrium condition and two zonal flows as perturbations $\Phi_{z1}(k_s/2, 0), \Phi_{z2}(7/4k_s, 0)$, we can observe that the zonal flow is excited only if the radial width of the zonal flow is larger than the poloidal width of the streamer, fig.2(c)-(d).

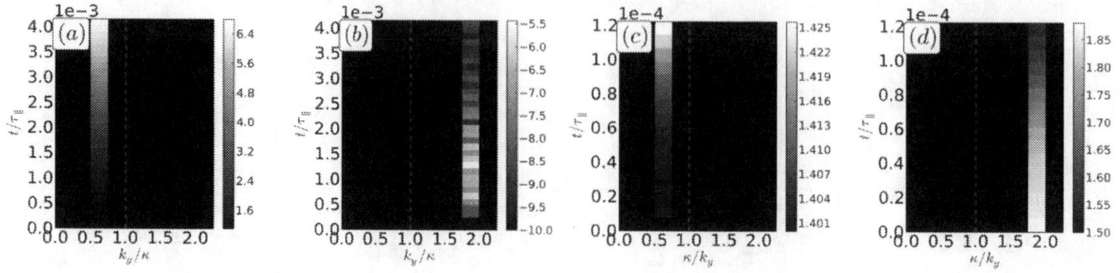

Figure 2: (a) evolution in time of the poloidal spectrum $k_y$ if a streamer $\Phi_{s1}$ or (b) $\Phi_{s2}$ is perturbed,(c) evolution in time of the radial spectrum $k_x$ if a zonal flow $\Phi_{z1}$ or (d) $\Phi_{z2}$ is perturbed.

## Conclusions

We can redefine the turbulent spectrum in two different regions called as (1)'B-modes region' and (2) 'S-modes region'. (1)'B' stays for big scales structure. This region of the turbulent spectrum, where the poloidal size of the turbulent structures is bigger than the radial size of ZF, cannot give but only receive energy from ZF through KH instability. Fundamentally this region is not source for ZF. (2) 'S' stays for small scale structures and define the region of the spectrum where the turbulence size is smaller than the radial size of ZF, namely S-modes transfer energy to the ZF and cannot act as a sink for ZF. Inserting the interchange forcing and the density equation we can simulate the selfconsistent zonal flows and turbulence interplay. According to the amplitude of the two regions $S-B$, we can recover different transport regimes[4]. If the S-modes dominates we are in the zonation regime, while if the B-modes dominates the system is fully turbulent.

## References


[1] GALPERIN, B., SUKORIANSKY, S., AND DIKOVSKAYA, N. Zonostrophic turbulence. *Physica Scripta 2008*, T132 (2008), 014034.

[2] KIM, E.-J., AND DIAMOND, P. H. Zonal flows and transient dynamics of the *l-h* transition. *Phys. Rev. Lett. 90* (May 2003), 185006.

[3] LIN, Z., HAHM, T. S., LEE, W. W., TANG, W. M., AND DIAMOND, P. H. Effects of collisional zonal flow damping on turbulent transport. *Phys. Rev. Lett. 83* (Nov 1999), 3645–3648.

[4] SHATS, M. G., XIA, H., AND PUNZMANN, H. Spectral condensation of turbulence in plasmas and fluids and its role in low-to-high phase transitions in toroidal plasma. *Phys. Rev. E 71* (Apr 2005), 046409.

# Neoclassical impurity transport in full-f global GYSELA simulations

D. Estève, X. Garbet, Y. Sarazin, S. Breton, Ph. Ghendrih, V. Grandgirard, T. Cartier-Michaud, G. Dif-Pradalier, C. Norscini
CEA, IRFM, F-13108 St.Paul-lez-Durance cedex, France

## Introduction

The computation of impurity transport in magnetized plasmas has gained a renewed interest for several reasons. Firstly an increasing number of tokamaks operate with tungsten plasma facing components in order to prepare the operation of Iter [1, 2]. Recent results from ASDEX-Upgrade [3] and JET [4] indicate that tungsten transport plays an essential role in the discharge history. Accumulation in the core may lead to reduced performances, and in some cases to radiative collapses. Secondly impurity seeding of edge plasmas with medium charge numbers $Z$ such as argon or neon is envisaged in Iter [5]. Inward penetration of these impurities should be prevented to avoid core dilution and a subsequent decrease of the fuel concentration. Finally fusion reactions will produce helium, which must be extracted efficiently, again to avoid core dilution. Neoclassical impurity transport has been investigated in details in the 80' [6, 7], and updated recently to account for poloidal asymetries due to centrifugal forces and/or RF heating [8, 9, 10]. Impurity fluxes are traditionally written as sums of turbulent and neoclassical contributions. However synergistic effects cannot be excluded so that it is desirable to compute these two contributions on an equal footing, at least to verify this hypothesis. The right framework for computing neoclassical and turbulent transport at the same time appears to be a set of gyrokinetic equations that account for collisions. This paper presents the first results of the GYSELA code with a new collision operator that has been implemented recently for this purpose.

## Model collision operator implemented in the GYSELA code

The collision operator that is implemented in the GYSELA code is of the form $C_a = \sum_b C_{ab}$ with

$$\begin{aligned} C_{ab}(\bar{F}_a) &= \frac{\partial}{\partial v_\parallel}\left[D_{d,ab}F_{M0a}\frac{\partial}{\partial v_\parallel}\left(\bar{f}_a - \frac{m_a v_\parallel U_{\parallel d,a}}{T_a}\right)\right] \\ &- \nu_{s,ab}\frac{m_a v_\parallel}{T_a}\left(U_{\parallel d,a} - U_{\parallel ba}\right)F_{M0a} \\ &+ \frac{2}{3}\frac{1}{N_a T_a}\left(Q_{ab} + V_{\parallel a}R_{\parallel ab} + q_{ab}\right)\left(\frac{m_a v^2}{2T_a} - \frac{3}{2}\right)F_{M0a} \end{aligned} \quad (1)$$

where $\bar{f}_a$ is the gyrocenter distribution $\bar{F}_a$ normalized to the Maxwellian $F_{M0a}$, $D_{d,ab}$ is the deflection diffusion coefficient and $\nu_{s,ab}$ is the slowing-down collision rate, $Q_{ab}$ is the energy collisional transfer rate, $R_{\parallel ab}$ is the friction force, $N_a$ the density, $m_a$ the mass, and $T_a$ the temperature. The function $U_{\parallel d,a}(v)$ and parameters $U_{\parallel ab}$ and $q_{ab}$ are adjusted to conserve momentum and energy, i.e.

$$\frac{m_a v}{T_a}U_{\parallel d,a}(v) = \frac{3}{4\pi}\int 2\pi d\xi\,\xi\,\bar{f}_a \quad (2)$$

$$U_{\parallel ab} = \frac{\left\langle \nu_{s,ab}v^2 U_{\parallel d,a}\right\rangle_a}{\left\langle \nu_{s,ab}v^2\right\rangle_a} \quad (3)$$

$$q_{ab} = N_a T_a \left\langle \sigma_{ab} \bar{f}_a \right\rangle_a \tag{4}$$

and $\xi = v_\parallel / v$ is the pitch-angle parameter and $v$ the velocity modulus. The bracket indicate an average over $F_{M0a}$, and the function $\sigma_{ab}(v)$ is given by

$$\sigma_{ab}(v) = -\nu_{d,ab} \frac{m_a v^2}{2T_a} v_\parallel \frac{\partial}{\partial v_\parallel} \ln \left( D_{d,ab} F_{M0a} v_\parallel \right) \tag{5}$$

This multi-species collision operator satisfies several constraints: particle, momentum and energy conservation, and it relaxes towards a Maxwellian in the direction parallel to the magnetic field.

**Numerical tests of the collision operator**

The operator Eq.(1) has been implemented in the GYSELA code. Its main properties have been verified, as detailed and illustrated in the following. The gyrokinetic equation is reduced to its collisional part only, i.e. the Lagrangian derivative along the particle trajectories is ignored. Keeping only one species allows a test of the conservation properties of the like-particle collision operator. In a second stage, a trace impurity is added and the transfer rates of momentum and energy from one species to another are computed. The main ion species is deuterium while impurities are helium, carbon and tungsten, in the trace limit $N_Z Z^2 \ll N_D$. A typical grid is ($N_r$, $N_\theta$, $N_\varphi$, $N_{v_\parallel}$,$N_\mu$) = (64,64,8,256,32).

Particle, momentum and energy conservation laws have been checked by running the code in the single particle case, with an initial distribution function that is a centered Maxwellian. The density, parallel velocity and temperature radial profiles are constant, and their normalized values are initialized to one. For each fluid quantity $X_D$, the absolute error $\delta X_D = |X_D(t) - X_D(t=0)|$ is calculated at mid-radius after a deuterium collision time $t = \tau_{DD}$. The normalized collisionality is $\nu_D^\star = (a/R_0)^{-3/2}\, \nu_D (qR_0/v_{TD}) = 2$. The errors are found to be $\delta N_D \simeq 10^{-9}$, $\delta V_{\parallel D} \simeq 10^{-12}$ and $\delta T_D \simeq 10^{-5}$. These errors are negligible when compared to the typical increments of the fluid quantities due to transport and sources during the same time lags.

Relaxation towards a Maxwellian is tested by initializing the code with a distribution function that is far from a Maxwellian in the parallel direction. Figure 1 shows that the initial state (fig. 1a) relaxes towards a Maxwellian distribution function (fig. 1b). The exchange of momen-

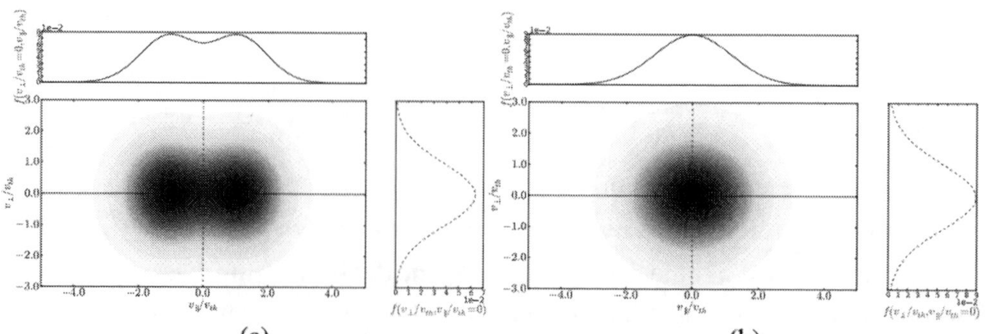

(a) (b)

Figure 1: 2D plot in the velocity space of the distribution function at initial (a) and final (b) time.

tum between the impurity and the main ions is investigated first. The code is initialized with a deuterium distribution function that is a shifted Maxwellian in the low Mach number limit, with $V_{\parallel i}/v_{Ti} = 0.05$. The initial distribution function of the impurity is a centered Maxwellian. Density and temperature profiles are flat with $T_D = T_Z$. The evolution of the mean parallel velocity is given by

$$\partial_t V_{\parallel a} = -\nu_{ab}(V_{\parallel a} - V_{\parallel b}) \tag{6}$$

where $\nu_{ab}$ is the momentum collisional transfer rate. In the case of thermal relaxation, the initial distribution function is a centered Maxwellian with constant density. The initial deuterium temperature is chosen larger than the impurity temperature, $T_D > T_Z$. The time evolution of each temperature is dictated by an equation similar to Eq.(6 ), with velocities replaced by temperatures. The time evolution of the mean velocity and temperature differences are shown on figure 2. The collisional damping rates are found to agree with the theoretical values.

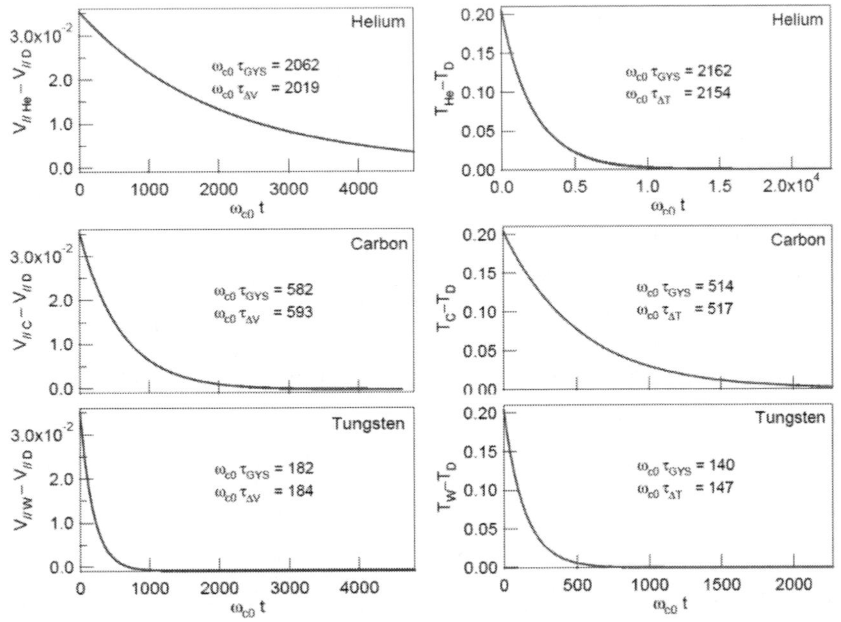


Figure 2: Time evolution of the difference between the parallel flow and temperature of the main ion species and those of helium, carbon and tungsten impurities.

## Recovering the theory of impurity neoclassical transport

A more challenging test consists in recovering the neoclassical theory of impurity transport. In this case, the Lagrangian derivatives along the particle trajectories are retained. We consider here the case where the main ion species is deuterium in banana regime, while the second species is a trace impurity in various collision regimes, depending on mass and charge. Gradients are below the instability threshold (no turbulence), so that collisional transport only is left. The expression of the impurity flux is known to be of the form

$$\Gamma_Z = -D_Z N_Z \left( \frac{\partial \ln N_Z}{\partial r} - Z \frac{\partial \ln N_i}{\partial r} - HZ \frac{\partial \ln T}{\partial r} \right) = -D_Z \frac{\partial N_Z}{\partial r} + V_Z N_Z$$

where $D_Z$ is diffusion coefficient, $V_Z$ the pinch velocity and $r$ the minor radius. The thermal screening factor $H$ is known to be equal to $-1/2$ in the Pfirsch-Schlüter regime. In a first step, all gradients are set to zero except the impurity density gradient. In that case diffusion only is expected. The diffusion coefficient is determined by plotting the impurity flux $\frac{\Gamma_Z}{N_Z}$ versus the density gradient $-\frac{\partial N_Z}{N_Z \partial r}$ for several values of the gradient. Points lay on a line and the slope is the diffusion coefficient $D_Z$. The diffusion coefficient agrees well with the analytical calculation and values given by the neoclassical code NEO [11], as shown in Fig.3a. A second test consists in setting a finite main ion density gradient. In that case, an impurity accumulation is expected. The ratio $\frac{V_Z}{D_Z}$ is found by plotting again the flux versus the gradient. It is equal to the value of $\frac{\partial N_Z}{N_Z \partial r}$ such that the flux vanishes. It is found to agree with the theoretical expectation $Z\frac{\partial N_i}{N_i \partial r}$, as shown in Fig.3b.

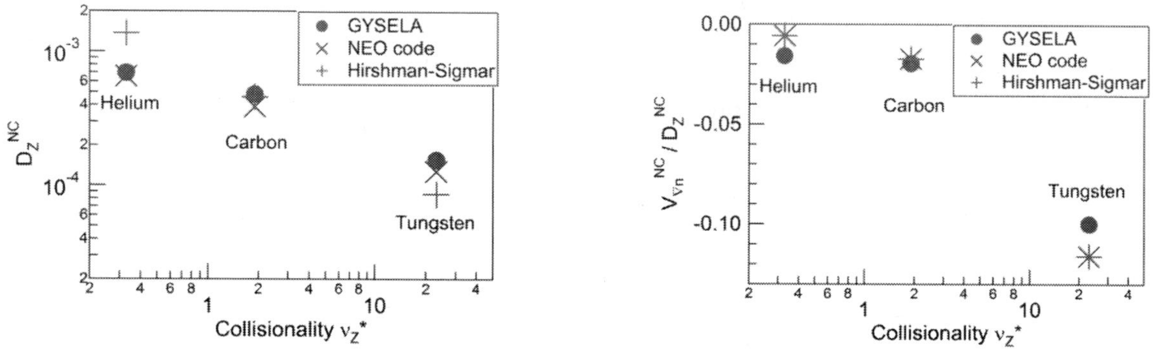


Figure 3: (a) Diffusion coefficient of helium, carbon and tungsten compared with the Hirshman-Sigmar analytical prediction and values calculated with the NEO code. (b) Same legend for the ratio $V_Z/D_Z$.

## Conclusion

A linearized multi-species collision operator has been implemented in the GYSELA gyrokinetic code. This operator has been numerically tested and is found to satisfy particle, momentum and energy conservation with an excellent accuracy. Also it relaxes towards a Maxwellian. Moreover the interspecies relaxation rates for momentum and energy well agree with the theoretical values. The code is also found to recover the main results of theory for impurity neoclassical transport. The diffusion coefficients are found to agree with analytical expressions and also with the NEO code. In the isothermal limit, accumulation is found due to the main ion density gradient. Thermal screening effect is being currently tested and first results are encouraging.

## References

[1] R. Neu, M. Balden, V. Bobkov, *et al.*, Plasma Phys. Control. Fusion **49**, B59 (2007).
[2] R. Neu, G. Arnoux, M. Beurskens, et al., Phys. Plasmas **20**, 056111 (2013).
[3] M. Sertoli, C. Angioni, R. Dux et al.,Plasma Phys. Control. Fusion **53**, 035024 (2011).
[4] C. Angioni, P. Mantica, T. Pütterich, *et al.*, Nucl. Fusion **54**, 083028 (2014).
[5] A. Loarte, B. Lipschultz, A.S. Kukushkin et al., Nucl. Fusion **47** S203 (2007).
[6] S.P. Hirshman and D.J. Sigmar, Physics of Fluids **20**, 418 (1977).
[7] S.P. Hirshman and D.J. Sigmar, Nucl. Fusion **21**, 1079 (1981).
[8] M. Romanelli and M. Ottaviani, Plasma Phys. Control. Fusion **40**, 1767 (1998).
[9] C. Angioni and P. Helander, Plasma Phys. Control. Fusion **56**, 124001 (2014).
[10] E. A. Belli, J. Candy and C. Angioni, Plasma Phys. Control. Fusion **56**, 124002 (2014).
[11] E. A. Belli and J. Candy Plasma Phys. Control. Fusion **50**, 095010 (2008).

# Kinetic Theory of Phase Space Plateaux

F. Eriksson[1], R. M. Nyqvist[1] and M. K. Lilley[2]

[1] *Dept. of Earth and Space Sciences, Chalmers Uni. Tech., 412 96 Göteborg, Sweden*

[2] *Phys. Dept., Imperial College, London, SW7 2AZ, UK*

Frequency sweeping signals are attributed to the formation and evolution of phase space holes and clumps in the non-thermal fast particle distribution [1, 2]. Recently their origin was shown to follow from the presence of a nearly unmodulated phase space plateau [3], centered at the wave-particle resonance, which supports a pair of shifted *edge modes* that destabilize due to dissipation in the background plasma and nonlinearly evolve into holes and clumps. The role of the plateau as a hole/clump breeding ground is further substantiated in this article via inclusion of fast particle collisions and sources. Also, it is demonstrated that relaxation of the plateau edge gradients has only a minor quantitative effect and does not change the plateau stability qualitatively.

We consider an electrostatic travelling wave with spatial period $\lambda$ and wave number $k = 2\pi/\lambda$ in a one-dimensional, uniform plasma equilibrium. The wave carrier frequency is assumed to be high enough that the plasma can be separated into a cold bulk, comprised of electrons and immobile ions, and a low density beam of energetic electrons that may interact resonantly with the wave and therefore need to be treated separately. The cold electrons respond linearly to the electric field $E(x,t)$, so their perturbed velocity $v_e$ satisfies the linear fluid equation

$$\frac{\partial v_e}{\partial t} = \frac{e}{m_e} E - 2\gamma_d v_e \,, \tag{1}$$

where the last term is a linear, dissipative friction force that damps the velocity perturbations exponentially in time. The fast electrons are described kinetically in terms of their distribution function $F(x,v,t)$, which evolves according to the kinetic equation

$$\frac{\partial F}{\partial t} + v\frac{\partial F}{\partial x} - \frac{e}{m_e} E \frac{\partial F}{\partial v} = \mathfrak{C}[F] + S(v) \ . \tag{2}$$

The right hand side represents fast particle collisions and sources, whose action is to relax $F$ towards an equilibrium distribution $F_0(v)$ that is taken as a constant, positive slope throughout the wave-particle resonance. It is modeled by the following combination of operators, commonly known as Krook-type collisions, collisional drag and velocity space diffusion,

$$\mathfrak{C}[F] + S(v) = -\beta(F - F_0) + \frac{\alpha^2}{k}\frac{\partial}{\partial v}(F - F_0) + \frac{v^3}{k^2}\frac{\partial^2}{\partial v^2}(F - F_0) \ . \tag{3}$$

The system is closed by Ampére's law

$$\frac{\partial E}{\partial t} = \frac{e}{\varepsilon_0}\left[ n_e v_e + \int v(F - F_0)\, dv \right] \,, \tag{4}$$

where $n_e$ is the unperturbed density of cold electrons. The plasma wave, with frequency $\omega_{pe}$, is driven at linear rate $\gamma_L$ by the fast electrons and weakly damped at rate $\gamma_d \leq \gamma_L \ll \omega_{pe}$ due to the friction force. All simulations are performed by means of a numerical algorithm, previously described in [4], that solves Eqs. (1) – (4) for the nonlinear evolution of $E(x,t)$ and $F(x,v,t)$.

Holes and clumps are nonlinear structures that extend in both real and velocity space and carry a particle deficit/surplus as compared to the surrounding distribution. Disregarding the effects of fast particle collisions and sources, they arise symmetrically shifted off the wave-particle resonance of a kinetically unstable bulk plasma eigenmode. Once firmly established, they tend to balance dissipative wave damping in the background plasma with the energy tapped by traversing fast particle phase space as coherent entities. Fig. 1 displays snapshots of the spatially averaged fast particle distribution that illustrate the presence of an intermediate plateau before the hole/clump pairs are created. At that stage, the initial mode is damped out and kinetically stable, but there are shifted resonances situated just inside the edges of the flat plateau region, responsible for small modulations of the plateau edge that begin to grow and eventually evolve into a hole/clump pair that detaches from the plateau.

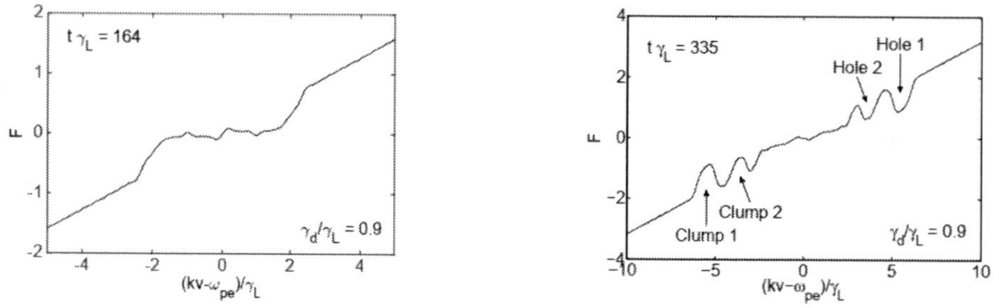


Figure 1: *Intermediate plateau and hole/clump pairs in the fast particle distribution.*

Inclusion of fast particle collisions and sources as given by Eq. (3) has significant impact on the formation (and evolution) of holes and clumps. Collisional relaxation, as mediated by velocity space diffusion or Krook-type collisions, inhibits hole/clump formation from the plasma wave resonance [5, 6]. Slowing down of the fast particles (drag), on the other hand, promotes it, in particular the formation and growth of holes [4].

The role of the intermediate plateau is now further substantiated via linear stability analysis of an unmodulated shelf with continuous edges (see Fig. 2a). Under the assumption that $k\Delta v \ll \omega_{pe}$ the resulting dispersion relation is

$$wz + i\pi\gamma + \frac{1-r}{r}\left[\log(1+z) - \log(1-z)\right] - \frac{1}{r}\left[\log(1-r+z) - \log(1-r-z)\right] = 0\,, \quad (5)$$

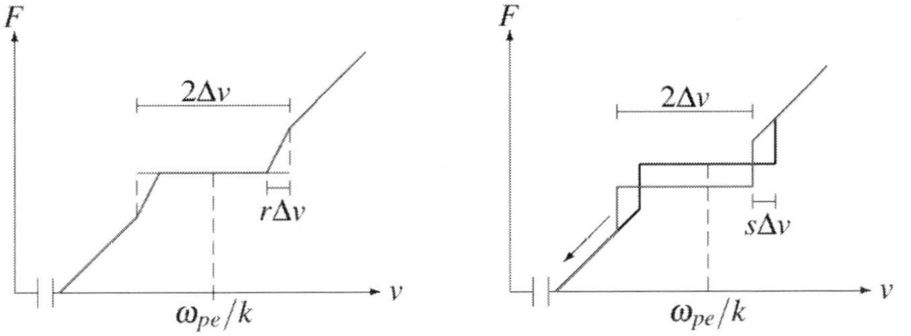


Figure 2: *Fast particle distribution for linear stability analysis of plateau state.*

where $r$ is defined in Fig. 2a, we have introduced the dimensionless variables

$$z \equiv \frac{\omega - \omega_{pe}}{k\Delta v}, \quad w \equiv \pi \frac{k\Delta v}{\gamma_L} \quad \text{and} \quad \gamma \equiv \frac{\gamma_d}{\gamma_L} \tag{6}$$

and it has been assumed that $|\mathrm{Re}[z]| < 1 - r$. A numerical analysis reveals that Eq. (5) has precisely three complex roots, of which one, dubbed $z_c$, sits at $\mathrm{Re}[z] = 0$ with a negative imaginary part that vanishes at $\gamma = 0$ and whose magnitude increases with $\gamma$. It is supplemented by a symmetric pair that bifurcates as $w$ varies, the bifurcation point and precise pattern depend on $r$ as well as $\gamma$ (see Figs. 3a-c). These modes tend asymptotically to the edges of the flat region, $\pm(1-r)$, as $w \to \infty$ and were hence dubbed *edge modes* (denoted $z_\pm$) in [3]. For a fixed $w$ the modes shift with the edges towards the plateau center as $r$ increases, as shown in Fig. 3d. Their imaginary parts remain almost completely constant, however, meaning that relaxation of the shelf edge within a narrow transition layer has a negligible effect on the plateau stability.

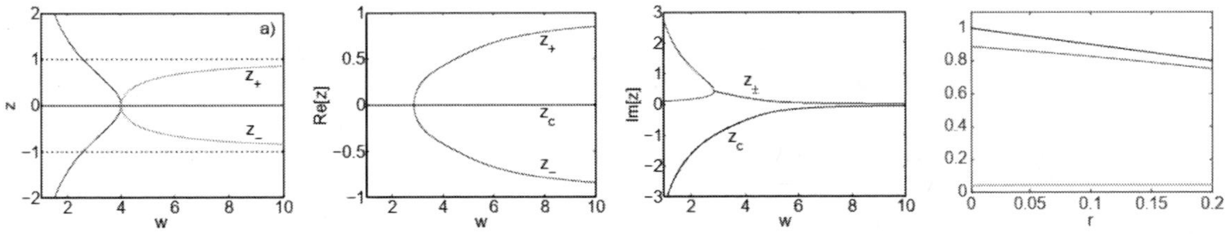


Figure 3: *a) Real (green) and imaginary (red) parts of roots to the dispersion relation* (5) *when* $r = 0$ *and* $\gamma = 0$. *b) Real and c) imaginary parts of roots to the dispersion relation when* $r = 0$ *and* $\gamma = 0.1$. *d) Edge modes* ($\pm\mathrm{Re}[z_\pm]$ *in red and* $10 \times \mathrm{Im}[z_\pm]$ *in green) when* $\gamma = 0.1$ *and* $w = 4\pi$. *The black line represents the edge of the flat region,* $|1 - r|$.

We proceed to further certify the plateau hypothesis by inclusion of fast particle collisions and sources. The Krook-type and velocity space diffusion operators both inhibit hole/clump production, as expected. For Krook-type collisions this is illustrated in Fig. 4 where hole/clump production ceases above roughly $\beta/\gamma_L = 0.01$ when $\gamma = 0.1$. For diffusive collisions the corresponding rate is $\nu/\gamma_L = 0.1$. Collisional drag, on the other hand, promotes hole/clump formation

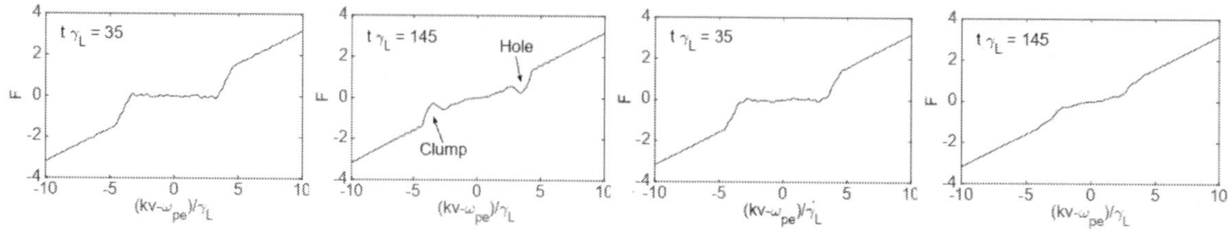


Figure 4: *Fast particle distribution in the presence of Krook-type collisions with rate* $\beta/\gamma_L = 0.005$ *(left two) and* $\beta/\gamma_L = 0.02$ *(right two) and* $\gamma = 0.1$.

and detachment. The effect on the plateau is a convection of the entire shelf down along the ambient linear distribution. Further, linear stability analysis of the shifted plateau in Fig. 2b sheds light on the previous result that collisional drag enhances holes and their sweeping rates but suppresses clumps. For large plateau widths, a power series in $1/w$ reveils the asymmetry

$$\mathrm{Im}\left[z_{\pm}\right] = \frac{\pi\gamma}{w^2}\left[1 + \frac{1}{w}(2\ln 2w + 1 \pm sw)\right] . \tag{7}$$

These results are further confirmed by nonlinear simulations of an initial plateau as a continuous shelf of tanh-type, cf. Fig. 5, where the plateau experiences a downward shift and the ensuing hole/clump pair evolves asymmetrically with a hole that grows faster than the clump.

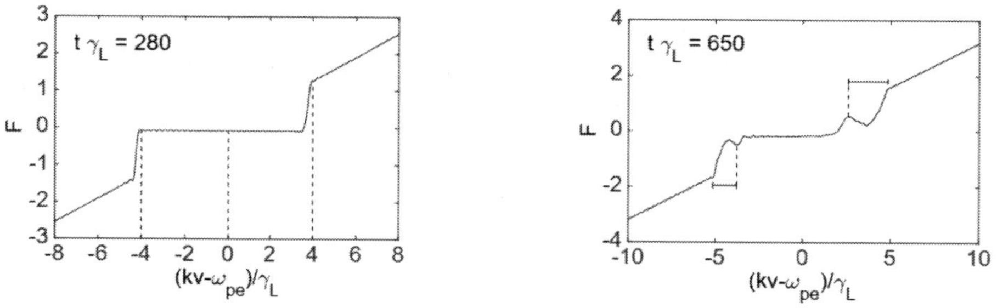


Figure 5: *Evolution of initial plateau when* $\alpha/\gamma_L = 0.03$ *and* $\gamma = 0.1$.

## References

[1] H. L. Berk *et al*, Phys. Lett. A **234**, 213 (1997)

[2] S. D. Pinches *et al*, Plasma Phys. Controlled Fusion **46**, S47 (2004).

[3] M. K. Lilley and R. M. Nyqvist, Phys. Rev. Lett. **112**, 155002 (2014).

[4] M. K. Lilley *et al*, Phys. Plasmas **17**, 092305 (2010).

[5] H. L. Berk *et al*, Phys. Rev. Lett. **76**, 8 (1996)

[6] B. N. Breizman *et al*, Phys. Plasmas **4**, (5) (1997).

# Tokamak GOLEM for fusion education - chapter 6

M. Dimitrova[2], O. Ficker[1,2], O. Grover[1], J. Kocman[1], J. Krbec[1,2], V. Loffelmann[1,2], L. Matena[1], J. Stockel[1,2], V. Svoboda[1], G. Vondrasek[1]

[1] *Faculty of Nuclear Sciences and Physical Engineering CTU in Prague, Prague, Czech. Rep.*

[2] *Institute of Plasma Physics CAS, v.v.i., Prague, Czech. Rep.*

GOLEM is the oldest operational tokamak in the world. It serves as an educational device at the Facuty of Nuclear Sciences and Physical Engineering at CTU in Prague. Several improvements of its diagnostic technology made during the last year as well as new experimental results are presented in this article.

## Microwave interferometry

No reliable density measurement method was available on the GOLEM tokamak until 2015. This absence seriously limited many experiments performed on the device. A microwave interferometer that had been used on CASTOR tokamak was recently made operational. The refurbishment also included upgrade of its electronic circuitry. The interferometer is of Wharton's type, meaning that the diagnostic wave frequency (around 70 GHz) is periodically swept by a sawtooth signal during operation (about 25 MHz), which produces lower frequency beats after the wave interferes with a reference. From the phase shift between the beats and the sweeping signal the plasma density can be calculated. The dependence on amplitude attenuation is eliminated by this method.

An analog circuit was developed as an alternative to a traditional computer algorithm. This circuit has higher temporal resolution than the digital calculation at the cost of higher demands on data acquisition systems (DAS). However, it seems that the analog evaluation should produce satisfactory results for most of the conducted experiments. Figure 1 shows comparison of both evaluation methods during one of the shots.

The interferometer only gives information on line-integrated density along the path of the wave (vertical direction). Average density is then calculated by estimating the distance passed through the plasma (limiter diameter). No better estimate of the plasma density is calculated at present.

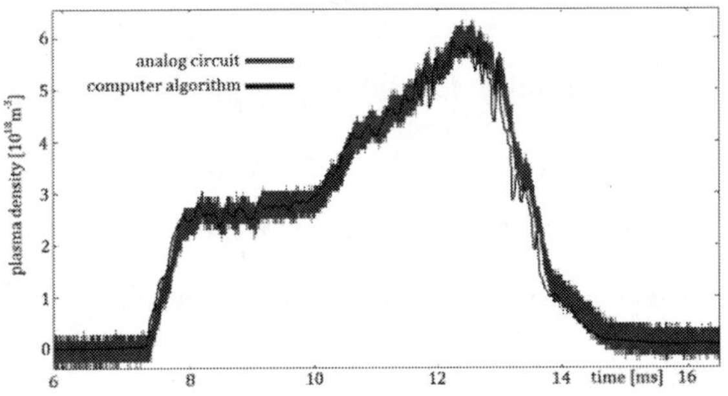


Figure 1: *Comparison of plasma density evolution as calculated by analog and digital means*

## Plasma position control

The GOLEM tokamak is equipped with four Mirnov coils used for plasma position (both vertical and radial) measurements and a set of positional stabilization coils. An electronic system has been developed which connects the sensors and actuators to a closed loop. First, the sensor signal is preprocessed using an analog integrator and subsequent digital filtering (removal of parasitic signals from the toroidal field and the field of feedback coils). The plasma position is then calculated using either a simplified Grad-Shafranov model or an even more simplified straight conductor approximation; The results are fed into a nonlinear proportional controller. Finally, the actuator signal is estimated as a suitable combination of the controller output and a predefined waveform (can be configured through the virtual control room) [1].

Both the closed-loop feedback and the predefined waveforms have been shown to be able to stabilize the plasma position evolution. This also leads to longer plasma duration.

## Runaway electrons

A new NaI(Tl) scintillation detector with a photomultiplier tube was installed for regular measurements. Although this detector is characterised by a relatively slow peak decay time, it can be useful in monitoring the runaway electrons (RE) losses via detection of the hard X-ray radiation released during their interaction with the wall. It seems that during standard operation (density $n_e = 3 - 8 \cdot 10^{18}$ m$^{-3}$), RE are created just during the plasma breakdown phase and/or during the plasma termination via some position instability (leads to creation of low density plasma in yet high electric field). The RE created during the breakdown drift to the low field side and are lost in few milliseconds, the loss of RE generated during partial plasma termination is usually faster and probably may occur on the high field side. Examples of both of these phenomena are shown in Fig. 2.

According to the analysis using the relation for primary RE generation rate (Kruskal-Bernstein), it is obvious that runaway electrons are created just during the breakdown (occurring at 12 ms) in the first shot presented in Fig. 2. In the second shot, the largest number of runaway electrons is created in the period of low plasma current and high loop voltage at the end of the discharge. The subsequent momentary recreation of plasma is probably caused by emission of secondary electrons from the wall after RE impact. The calculated RE densities seems to be reasonable regarding both plasma current and number of detected photons estimated from the HXR detector signal.

The tests of TimePix detector (hybrid silicon pixel detector) for direct fast electron measurements have continued. First measurements are planned for this year.

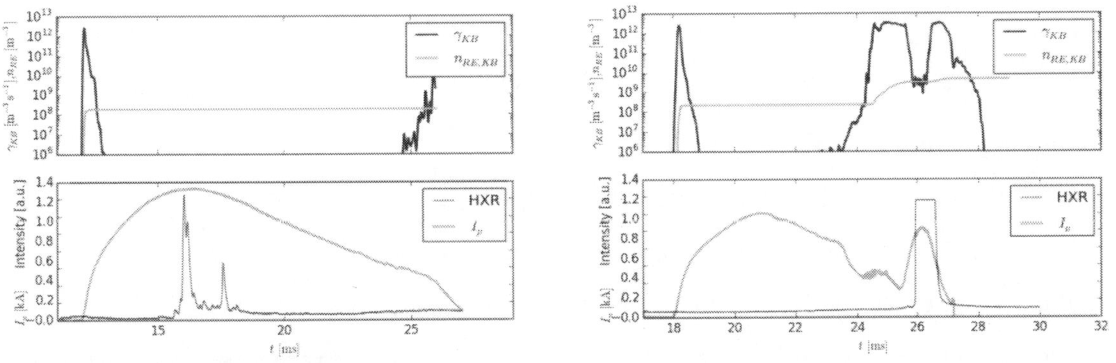


Figure 2: *Comparison of shots #18970 (left) and #18850, the latter one with a position instability; The figure shows the Kruskal-Bernstein rate (blue), an estimate of RE density (violet), plasma current (green) and the hard X-ray signal (red)*

## Probe measurements in microwave plasma

Pre-ionization of GOLEM discharges is performed by the microwave generator at f = 2.45 GHz. The resulting MW plasma is studied by means of the planar Langmuir probe located at the radius r/a = 0.8 (effective collecting area is 50 $mm^2$), and oriented perpendicularly to the magnetic field lines. Thanks to good reproducibility, the IV characteristics are measured on a shot to shot basis in 18 discharges (#18480 - #18497). The typical IV characteristic constructed at t = 12 ms is shown in the right panel in Fig. 3. It is seen that the saturation currents ($I_{+sat}$ and $I_{-sat}$) are not constant. Increase of $I_{+sat}$ is a consequence of the probe sheath expansion, however the decrease of $I_{-sat}$ with the probe voltage is not yet understood. The fitting function was used to calculate plasma parameters by using the technique proposed in [4].

The resulting evolutions of the electron temperature and the ion saturation current are plotted in the central panel of Fig. 3. It is evident that plasma is confined in the vessel long time after switching-of the MW power at t = 10.8 ms. Since that time, the ion saturation current,

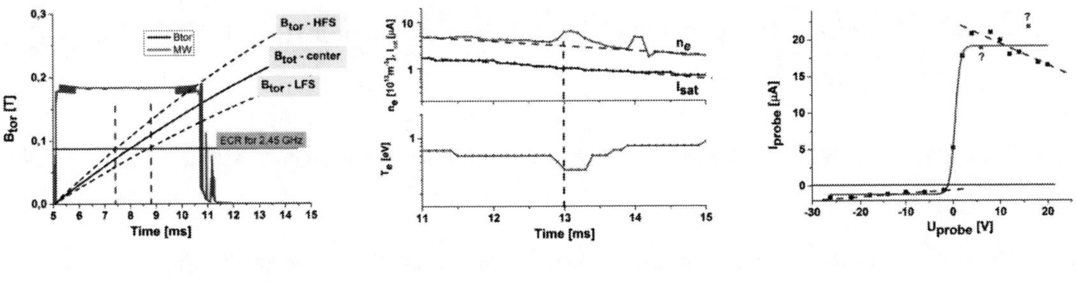


Figure 3: Left - *Temporal evolution of the toroidal magnetic field and the microwave power (the ECR resonance indicated);* Center - *Temporal evolution of selected plasma parameters after the MW power was switched off;* Right - *A selected IV characteristics ($t = 12$ ms, $B_{tor} = 0.188$ T)*

and electron density decay exponentially, with a time constant ~1.4 ms. Such a relatively long confinement of plasma is possible due to the low electron temperature, which remains constant at $T_e$ = 0.6-0.8 eV. Consequently, the centrifugal and E×B drift velocities are small.

**Visible light tomography**

The tokamak is equipped with two visible light cameras observing the plasma from two perpendicular directions. With use of rolling shutter effect, a frame rate of 40 000 fps can be reached while reducing the resolution to single line of 336 pixels.

An algebraic algorithm has been utilized to reconstruct the 2D profile of visible light emissivity of the plasma. The algorithm is based on linear parametrization of the space of possible solutions followed by solving a system of linear equations for the parameters. Because of ill-posedness of this problem, it is solved approximately by Tikhonov regularization (with an assumption of smoothness). The calculation is being done for all three color channels recorded by the cameras (red, green, blue) separately. A correction of the sensor nonlinearity is applied before the calculation.

Figure 4 shows two examples of the reconstructions. The latter one shows a hollow emissivity profile typical for most discharges; This probably corresponds to radiation coming from colder regions of the plasma. Many smaller fast-evolving structures can be seen in the reconstruction; These are probably artifacts caused by reflections from the first wall.

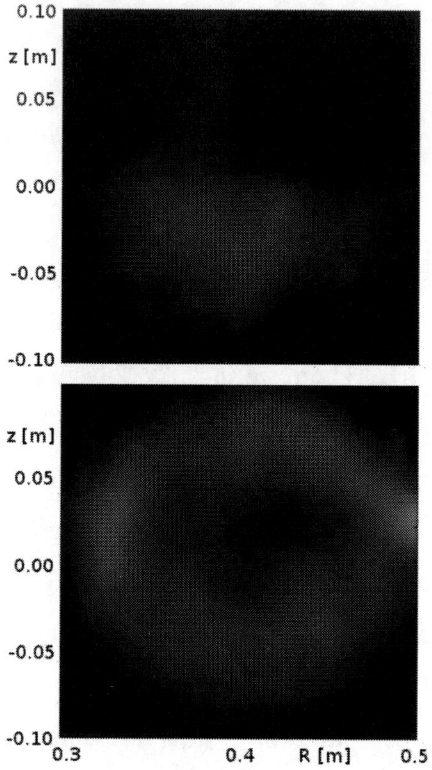


Figure 4: *Two tomographic reconstructions obtained during different phases of the same discharge*

**References**

[1] J Kocman, Řízení polohy plazmatického prstence na tokamaku GOLEM, Master Thesis, FNSPE CTU (2015)

[2] L Matena, Microwave interferometry on the GOLEM tokamak, Master Thesis, FNSPE CTU (2015)

[3] O Ficker, Generation, losses and detection of runaway electrons in tokamaks, Master Thesis, FNSPE CTU (2015)

[4] Ts Popov et al, Plasma Phys. Control. Fusion, 51 (2009) 065014

# Nonlinear simulations of the fishbone instability

M. Idouakass[1], M. Faganello[1], H. L. Berk[2], X. Garbet[3], S. Benkadda[1]

[1] *Aix Marseille Université, CNRS, PIIM UMR 7345, 13397 Marseille, France*

[2] *Institute for Fusion Studies, University of Texas, USA*

[3] *CEA, IRFM, 13115, St Paul lès Durance, France*

**Abstract**

We propose an extension of the precessional fishbone model developped by Odblöm et al. [1], which allows taking into account the MHD nonlinearities around the $q = 1$ surface, and a linear response for the energetic particles, which are well contained within the $q = 1$ surface. The fishbone mode, resulting from the interplay between the trapped energetic particles and the thermal plasma, is an electromagnetic mode characterized by bursts of activity and frequency down-chirping. It can cause enhanced transport and losses of energetic particles, such as ICRH- or NBI-produced particles in present tokamaks, or alpha particles in future fusion devices, and affect their energy deposition into the plasma.

## Introduction

The fishbone oscillation was first observed in the PDX tokamak with nearly perpendicular neutral beam injection active [2]. It was observed to have a dominant $m = 1$ poloidal mode number and $n = 1$ toroidal mode number, with a radial plasma displacement profile similar to the internal kink "top-hat" structure. Its frequency was close to the NBI-produced energetic particles precessional frequency, and was observed to decrease by a factor of about 2 during each burst.

This instability was later interpreted analytically in two different ways, corresponding to two different branches. One branch is a fishbone with a frequency close to the precessional drift frequency of the trapped energetic particles [3]. The other branch is a fishbone with a frequency close to the ion diamagnetic frequency [4]. In the case of the diamagnetic branch, there is a gap forming in the Alfvén continuum, within which the mode frequency is contained. In that case, the MHD resonances are essentially eliminated. In the case of the precessional branch, there is no gap forming in the Alfvén continuum, so that there are MHD resonances, leading to nonlinear MHD behavior, in addition to a kinetic nonlinear behavior.

Therefore, a nonlinear description of the precessional fishbone instability is a challenging problem, because of the interplay between the MHD nonlinearities around the $q = 1$ surface, and more precisely at the two Alfvén resonances [1], and the kinetic nonlinearities [5].

## Model

In order to study the competition between the nonlinear MHD and kinetic effects, while still being able to understand the origin of the different nonlinear dynamics, the model used is based on a domain separation.

The first domain is the core region centered on the magnetic axis, and extending up to a radius lower but close to that of the $q = 1$ surface. It is consedered that all energetic particles are contained within this region, so that the nonlinear kinetic evolution needs to be studied only in this region. At the same time only a linearized MHD response is kept for the plasma bulk, which behaves in an internal kink-like manner. The second domain is an annular region centered around the $q = 1$ surface, in which there are no energetic particles. It is considered that all MHD nonlinear effects take place in this region.

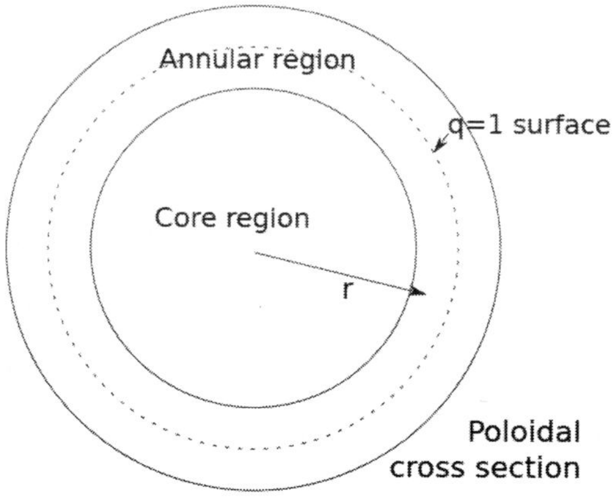


Figure 1: *Scheme of the domain separation*

## Annular region

Assuming that this region is a narrow layer centered around $r_*$, the radial location of the $q = 1$ surface, and considering a single helicity response, it is possible to simplify the geometry to a two dimensional slab. In the annular region, the nonlinear MHD dynamics are described by the two fluid Reduced MHD model, meaning that compressibility and anisotropic effects of the thermal pressure are neglected. The evolution is then given by the frozen-in equation and the vorticity equation :

$$\frac{\partial \psi}{\partial t} = [\psi, \phi] + \eta \Delta_\perp \psi \tag{1}$$

$$\frac{\partial}{\partial t}\Delta_\perp \phi = [\Delta_\perp \phi, \phi] + [\psi, \Delta_\perp \psi] + \nu \Delta_\perp \Delta_\perp \phi \tag{2}$$

where $\psi$ is the magnetic flux, proportional to the parallel vector potential, $\phi$ is the stream function, proportional to the electric potential, $\eta$ and $\nu$ represent small but finite resistivity and viscosity. The Poisson brackets are defined as $[f,g] = \partial_x f \partial_y g - \partial_y f \partial_x g$, where $x = r - r_*$ and $y$ represents the poloidal variable.

## Core region

In this region, where a kinetic description is used for the energetic particles, the fact that there is a time scale separation between the gyromotion, the bounce motion and the precessional drift of particles, is used to simplify the study. Indeed, the particles gyro-frequency is much higher than the bounce frequency, itself much higher than the precession frequency. Considering that experimentally, the fishbone mode has a frequency close to that of the precessional motion, a gyro and bounce averaging is done. Furthermore, only deeply trapped particles, with a velocity parallel to the magnetic field that is negligeable, are taken into account. This fact allows considering a simplified two dimensional phase-space, that is described in action-angle variables represented by the toroidal momentum $P_\alpha$ and the toroidal angle $\alpha$ respectively. A last simplification is to consider that the energetic particles are characterized by a single value of the magnetic moment : $\mu = \mu_*$. In this case, the Hamiltonian, separated in an equilibrium part and a perturbation, reads :

$$H(\mu, v_\parallel, P_\alpha, \alpha) = H_0(\mu_*, v_\parallel = 0, P_\alpha) + H_1(\alpha, P_\alpha) \tag{3}$$

In the Vlasov equation, the equilibrium contribution gives the precession frequency:

$$\frac{\partial H_{eq}}{\partial P_\alpha} = \omega_D \tag{4}$$

The perturbed Hamiltonian is given by the electric potential of the $m = n = 1$ mode, which has a kink-like structure : a rigid displacement of the whole core, characterized by $\phi/r$ nearly constant. The perturbed Hamiltonian reads

$$H_1(P_\alpha, \alpha) = er\frac{\phi(r_{bound}, \alpha)|_{n=1}}{r_{bound}} \tag{5}$$

where the radius $r = r(P_\alpha)$ is a function of the toroidal momentum and $r_{bound}$ defines the boundary between the core and the annular regions.

**Coupling**

The coupling between the particle dynamics and the bulk plasma evolution is done including the particle pressure in the linearized MHD response of the core. Integrating in space this response, from the magnetic axis to the boundary between the two domains, we obtain a time dependent boundary condition:

$$\left(\frac{\partial^2}{\partial t^2} + \omega_A^2\right) \frac{\partial \phi}{\partial r}\bigg|_{bound} = \frac{B_T}{r_{bound}^2 \rho_0} \frac{\partial}{\partial t} \int_0^{r_{bound}} dr r^2 (b \times \kappa) \cdot \nabla \left(\frac{P_{\perp,h}}{B_T}\right) \tag{6}$$

where $P_{\perp,h} = \int \mu B_T f d^3v$ is the energetic particle pressure.

**Linear Results**

Linearizing the Vlasov equation and the Reduced MHD equations, while taking into account the boundary condition coupling the two different dynamics, it is possible to obtain a dispersion relation. This is done in a way similar to that of Ödblom et al. [1], but taking different assumptions for the energetic particles distribution function. The differences come from the need to be able to describe, in the future, the nonlinear kinetic dynamics.

Taking a distribution function of the form $F_{eq} = AF(r)\delta(v_{\parallel})\delta(\frac{\mu}{\mu_*} - 1)$, the dispersion relation reads :

$$i = K\int_0^1 \frac{y^2 q(y)^2\left(-\frac{dF}{dy}\right)}{q - \omega T y}dy \quad ; \quad y = \frac{r}{r_*} \tag{7}$$

where $K$ is a normalized energetic content, and $T$ is a characteristic precession period

$$K = \frac{2\pi^2 \mu_* B_T \omega_c \Gamma}{2(1-q_0)\rho_0 k_\theta(r_*) v_{A,\theta}(r_*)} \quad ; \quad T = \frac{1}{\omega_D(r_*)} \tag{8}$$

Using this model, it is possible to get a threshold condition for the fishbone instability assuming that the distribution is:

- monotonically decreasing with radius, consistently with what would be expected from energetic particles in a fusion reactor
- going to a constant at the magnetic axis, and towards the $q = 1$ surface
- varying rapidly over a radial length small compared to $r_*$

Furthermore, the $q$-profile is considered to be nearly equal to 1 over all the core region.

With all these assumptions taken into account, the dispersion relation is simplified to :

$$i \simeq \frac{K y_0^2}{\omega T}\int_{-\infty}^{\infty} \frac{-\frac{dF}{dy}}{\frac{1}{\omega T} - y}dy \tag{9}$$

where $y_0$ is the normalized radius around which $\frac{dF}{dy}$ is peaked. This simplified dispersion relation gives, for the threshold condition :

$$K_{threshold} \simeq -\frac{1}{\pi y_0^3 F'|_{y_0}} \quad ; \quad \omega_{threshold} \simeq \frac{1}{y_0 T} = \omega_D(y_0) \tag{10}$$

These values are confirmed by numerical simulations based on the model presented in this paper.

This work was granted access to the HPC resources of Aix-Marseille Université financed by the project Equip@Meso (ANR-10-EQPX-29-01) of the program " Investissements d'Avenir " supervised by the Agence Nationale pour la Recherche.

## References

[1] A. Ödblom et al., Phys. of Plasmas **9**, 155 (2002)
[2] K. McGuire et al., Phys. Rev. Lett. **50**, 891 (1983)
[3] L. Chen et al., Phys. Rev. Lett. **52**, 1122 (1984)
[4] B. Coppi et al., Phys. Rev. Lett. **57**, 2272 (1986)
[5] B. Breizman et al., Phys. of Plasmas **4**, 1559 (1997)

# Influence of collisions on parametric instabilities produced by lower hybrid wave power

C. Castaldo[1], A. Di Siena[2], R. Fedele[2], F. Napoli[3], L. Amicucci[1], R. Cesario[1], G. Schettini[4]
[1]*ENEA, Unità Tecnica Fusione, C.R. ENEA Frascati, CP65, 00044 Frascati, Italy*
[2]*Dipartimento di Scienze Fisiche, Università Federico II and INFN, Napoli, Italy*
[3]*Lancaster Univerity, Engeenering Department, Lancaster, United Kingdom*
[4]*Università degli Studi Roma Tre, Dipartimento di Ingegneria, Roma, Italy*

The active control of the plasma current density profile is mandatory to achieve sufficient stability and energy confinement of the plasma in fusion reactors based on the tokamak concept. The lower hybrid current drive (LHCD) effect [1] has been proved as a flexible and efficient tool to achieve such goal and extend the duration of the inductive plasma current, which is intinsically transient. A proper design of the LHCD systems in fusion reactors requires to control the parametric instabilities (PI), which can occur in the plasma region located near the LHCD antenna mouth. The typical signature of PI, as detected by radiofrequency (rf) probes and frequency spectra analyzers, consists in a large broadening (up to about 40 MHz) at the operating line $f_o$ of the rf power source. PI were responsible for the failure of the early experiments of injection of lower hybrid waves in tokamaks to achieve the efficient ion heating predicted by the theory [2],[3],[4]. In the early LHCD experiments, operating at lower plasma density (up to line-averaged value $n_{e,av} \sim 0.4\ 10^{20}\ m^{-3}$), PI were comparably weak [5] and the rf power penetrated in the plasma core, successfully driving a plasma current [6]. However, PI play an important role also in LHCD scenarios. The observed weak broadening of the frequency spectrum due to PI should be indeed accompanied by a significant broadening of the spectrum in the parallel refractive index $N_{\|}$, as suggested in Ref. [7]. A full wave analysis of the propagation of a LH wave power pulse, performed in the frame of nonlinear fluid theory, predicted a spectral broadening $\Delta N_{\|}/N_{\|}$ of the order of unity for typical LH and plasma parameters [8]. Such spectral broadening has a profound effect on the the LHCD physics, concerning the power absorption and current drive in the plasma core. The latter are indeed based on the LHW interaction at the Landau resonance $v_{\|} = c/N_{\|}$ with a significant fraction of the plasma electrons, as given by the distribution function $f(v_{\|})$ in the velocity along the static magnetic field. Based on the numerical code $LH^{star}$ it was shown that the spectral broadening is an important contribution to bridge the $N_{\|}$ gap in LHCD [9,10]. The spectral gap problem in LHCD physics consists in explaining the mechanisms that broaden and up-shift the launched power spectrum in $N_{\parallel}$. A redistribution in the $N_{\parallel}$ spectrum coupled by LH antennas [11], is necessary for justifying the power absorption and the amount of current driven in the experiments, which should be otherwise negligible. A more remarkable success of the LHCD modeling taking into account the effects of PI is the prediction, confirmed by the experiments, that LHCD approaching reactor-graded high plasma densities can have success provided that a relatively high electron temperature in the outer plasma should be produced [12]. This allows to overcome the density limit observed in LHCD experiments, namely a threshold of the line-averaged density above which the current driven by the rf power injection becomes much lower than expected by the LHCD theory and sharply disappears. Important evidences of LH power penetration in the plasma core were obtained in the FTU tokamak at reactor relevant central plasma density $n_{e,o} \approx 5\ 10^{20}\ m^{-3}$ in low recycling regimes achieved with suitable plasma operations, e.g. performing wall lithization or injecting pellets to increase the core plasma density without perturbing the edge. This allowed to keep the edge electron temperature sufficiently high that the PI induced spectral broadening in the frequency domain was reduced by a factor two with respect to FTU plasma discharges with standard operations, in which, at even lower central plasma density $n_{e,o} \approx 3\ 10^{20}\ m^{-3}$, no LH wave-particle interactions in the plasma core are observed. This result suggests that in fusion reactor scenarios, such as ITER or DEMO, the edge electron temperature is sufficiently high to prevent the onset of PI. In this case, the LHCD power deposition and current drive can be controlled by two parameters of the coupled spectrum vs. the refractive index parallel to the static magnetic field, namely the peak value $N_{//o}$ and the width $\Delta N_{//o}$ of the main lobe [13]. These are determined, following the linear theory, only by the phasing of the waveguide array launcher, which can be varied electronically during the rf pulse. However, a robust and safe design of the LHCD system in fusion reactors requires that the LHCD antennas should be located in the shadow of a port of the vacuum vessel, where the rf power coupling is expected to occur in a cold, private plasma. In such conditions, strong PI are expected to occur [14]. As a main result of the present theoretical analysis, we have found that collisional suppression of PI allows such robust and safe design. Collisions are neglected in the standard modeling of PI in LHCD experiments [14], though they were included in the Miklos Porkolab's pioneering work on PI induced by LHW near the lower hybrid resonance [15] and their effects were confirmed by experiments [16]. A role of collisions on PI induced by LH waves, especially near the antenna mouth, can be expected in LHCD scenarios, observing that in such cold ($T_e \leq 10$ eV) and dense plasma ($n_e \sim 10^{18}\ m^{-3}$) the collision rates of the electrons are of the order of the frequency of the ion-acoustic quasi-modes.

The latter are involved in the main channel of the PI induced by LH power waves in current drive experiments. The present analysis is based on a specific collisional parametric dispersion equation derived to evaluate the growth rates of PI induced by LH wave power in the plasma region in front of the LH antenna mouth for LHCD scenarios. Details of the derivation will be proposed elsewhere, we limit the present discussion to the basic assumptions. Uniform and stationary plasma model is used, with Cartesian axis $Z$ aligned along the static magnetic field $\mathbf{B}_o$. The basic equations are given by the self-consistent system of the Maxwell equations coupled to the kinetic equation for the single particle distribution functions $f_\alpha$. Following [15], we adopt a particle-conserving BGK collision operator as an approximate model of the collision integral, $(\partial f_\alpha/\partial t)_{coll} \cong -\nu_\alpha[f_\alpha - f_\alpha^o(n_\alpha/n_\alpha^o)]$, where $f_\alpha^o$ is the unperturbed, local equilibrium, $\nu_\alpha$ are characteristic relaxation rates involving the plasma particle species $\alpha$ and $n_\alpha, n_\alpha^o$ are their number density corresponding respectively to $f_\alpha, f_\alpha^o$. We observe that the ion relaxation rates are two order of magnitude lower than the electron collision rates. As a result, we have found that the introduction of the collision operator into the ion kinetic equation has a negligible effect on the solutions of the parametric dispersion relation. The relaxation rates do not include the electron-neutral collisions as we are mainly interested to low recycling regimes and assume that the residual neutrals are ionized by the injected rf power. The plasma electrons are considered as completely magnetized, i.e. their motion can be represented by the 1D motion of their guide center along the direction of the magnetic field. The plasma ions do not contribute to the nonlinear dynamics. They are assumed fixed concerning the high frequency dynamics and completely magnetized concerning the low frequency dynamics. We observe that in the plasma region near LH antennas, with typical kinetic parameters $T_e \leq 10$ eV and $n_e \leq 5\ 10^{18}$ m$^{-3}$, the motion parallel to the magnetic field has been indeed recognized as the dominant dynamics concerning the standard collisionless parametric dispersion equation [17,18]. We use the spectral method and the perturbation theory to solve the self-consistent system obtained coupling the kinetic equation to the Maxwell equations. The expansion parameter is $\varepsilon = e|E_{z,o}|/m_e\omega_o v_{th,e} \ll 1$, where $e$ is the elementary charge, $|E_{z,o}|$ is the maximum amplitude of the z-component of the rf electric field, and $\omega_o = 2\pi f_o$ is the peak angular frequency of the injected rf wave power. We consider four waves interaction, involving the injected LH pump wave (subscript 0) two LH sidebands at lower (subscript 1) and higher (subscript 2) frequency, and a resonant mode or quasi-mode at low frequency (no subscript). The relevant selection rules for such interaction are given in standard notations by $\omega_1 = \omega - \omega_o, \boldsymbol{k}_1 = \boldsymbol{k} - \boldsymbol{k}_o, \omega_2 = \omega + \omega_o, \boldsymbol{k}_2 = \boldsymbol{k} + \boldsymbol{k}_o$. The parametric dispersion relation (PDR) thus obtained can be cast in the form $f_{PD}(\omega, k_{\parallel}, k_{\perp}, \delta, \omega_o, k_{\parallel,o}, \varepsilon, \sigma) = 0$ where $f_{PD}$ is the parametric dispersion function, which is defined by $f_{PD} = 1 - (\mu_1/\varepsilon_1^{hf} + \mu_2/\varepsilon_2^{hf})/\varepsilon^{lf}$ where $\varepsilon_{1,2}^{hf}$ and $\varepsilon^{lf}$ are, respectively, the linear dispersion functions for the low and high frequency fields and $\mu_{1,2} \propto \varepsilon^2$ are the nonlinear coupling parameters. Here the subsripts $\parallel$ ($\perp$) indicates components parallel (perpendicular) to $\mathbf{B}_o$, $\delta = \angle(\boldsymbol{k}_{\perp}, \boldsymbol{k}_{\perp,o})$ and $\sigma$ indicates the plasma parameters (density, temperatures of the electron and ion species and plasma composition). For fixed values of the subset of parameters $\{k_{\parallel}, k_{\perp}, \delta, \omega_o, k_{\parallel,o}, \varepsilon, \sigma\}$, the solution of the PDR thus consists of finding the zeros of the complex function $f_{PD}$ of the complex variable $\omega$. The imaginary part $\gamma$ of such zeros is identical to the imaginary part of $\omega_{1,2}$ and discriminates stable ($\gamma < 0$) and unstable ($\gamma > 0$) LH modes. The real and imaginary part of the zeros of the collisionless (a) and collisional (b) PDR, corresponding to the maximum values of $\gamma$ (with respect to $k_{\perp}$ and $\delta$) are plotted in figure 1 as a function of $N_{\parallel} = ck_{\parallel}/\omega_o$ for different LH pump frequencies $f_o$ and power density assumed proportional to $f_o$. The stabilizing effect of the collisions here shown is confirmed by an extensive analysis of PI in LHCD scenarios.

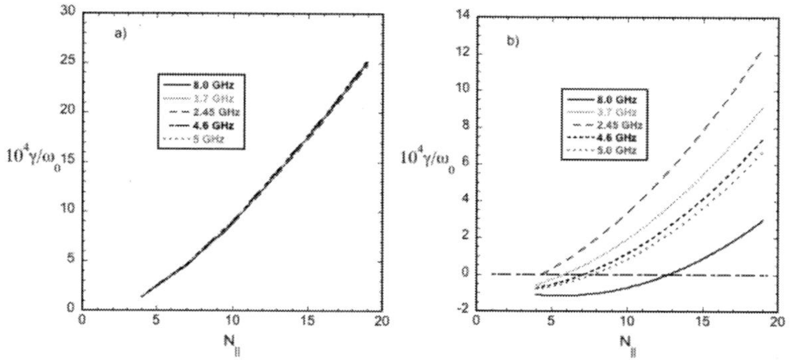


Fig. 1. Growth rates of LH sidebands for different frequencies $f_o$ of the LH pump waves in collisionless (a) and collisional (b) D plasmas, $n_e = 3\, n_{e,cut-off}$, $T_e = T_i = 5$ eV, $N_{\parallel,o} = 1.85$, P(kW/cm$^2$) = 0.6 $f_o$ (GHz).

[1] N. J. Fisch, *Rev. Mod. Phys.* **59** (1987) 175
[2] T. H. Stix, *Phys. Rev. Lett.* **15** (1965) 878
[3] C. F. Karney, *Phys. Fluids* **22** (1979) 2188
[4] M. Porkolab, et al. *Phys. Rev. Lett.* **31** (1977) 230
[5] Y. Takase, et al., *Phys. Fluids* **28** (1985) 983
[6] S. Bernabei, et al., *Phys. Rev. Lett.* **49** (1982) 1255
[7] Y. Takase, et al., *Phys. Rev. Lett.* **53** (1984) 274
[8] C. Castaldo, et al., *Phys. Letters A* **230** (1997) 336
[9] R. Cesario, et al., *Phys. Rev. Lett.* **92** (2004) 175002
[10] R. Cesario, et al., *Nucl. Fusion* **46** (2006) 462
[11] M. Brambilla, *Nucl. Fusion* **16** (1976) 11
[12] R. Cesario, et al., *Nature Comms*, **1** (2010) 55
[13] A. Cardinali, et al., *Journal of Physics: Conference Series* **561** (2014) 012002
[14] R. Cesario, et al., *Nucl. Fusion* **54** (2014) 043002
[15]M. Porkolab, *Phys. Fluids* **17** (1974) 1432
[16] H. Fujiyama, Y. Watanabe and M. Akazaki, *Journ. Phys. Soc. Japan*, **49** (1980) 1981
[17] F. Napoli, et al., *Plasma Phys. Contr. Fusion* **55** (2013) 095004
[18] A. Zhao, Z. Gao, *Nucl. Fusion* **53** (2013) 083015

# Predictive simulations of reactor-scale plasmas fuelled with multiple pellets with the European Transport Simulator

D. Kalupin [1], O.Asunta[2], R. Coelho[3], D. Coster [4], J. Ferreira[3], A. Figueiredo[3], K.Gal[5], I.Ivanova-Stanik[6],T. Jonnson[7], M. Owsiak[8], B. Palak[8], M.Schneider[9], R.Stankiewicz[6], P.Strand[10] *and the EU-IM team**

[1] EUROfusion Consortium Programme Management Unit, Boltzmannstr. 2, D-85748, Garching, Germany
[2] Department of Applied Physics, Aalto University, P.O. Box 14100, FI-00076 AALTO, Finland
[3] Instituto de Plasmas e Fusão Nuclear, Instituto Superior Técnico, Universidade de Lisboa 1049-001 Lisboa, Portugal
[4] Max-Planck-Institut für Plasmaphysik, Garching, Germany
[5] EuroFusion Consortium Programme Management Unit, Culham Science Centre, OX143DB, Abingdon, UK
[6] Institute of Plasma Physics and Laser Microfusion, Hery23, 01-497, Warsaw, Poland
[7] Department of Fusion Plasma Physics, School of Electrical Engineering, KTH Royal Institute of Technology, SE-10044 Stockholm, Sweden
[8] Poznan Supercomputing and Networking Center, IChB PAS, Noskowskiego 12/14, Poznan, Poland
[9] CEA, IRFM, F-13108 Saint-Paul-lez-Durance, France.
[10] Department of Earth and Space Sciences, Chalmers University of Technology, SE-41296 Göteborg, Sweden

## Introduction

Development of the operating scenario for fusion reactor requires integrated modelling addressing the critical reactor issues: plasma heating and fuelling, radiation from impurities, MHD stability, etc. Since all these issues are inter-linked, to demonstrate a successful operational scenario, the relevant physics need to be included in a single simulation. Thus, the environment used for scenario modelling should allow for the integration of multiple codes and physics modules into a single scientific workflow. The European Transport Simulator (ETS) [1] is an outstanding example of such an integrated workflow. The ETS workflow couples individual physics modules e.g. calculating the plasma magnetic equilibrium, deposition (by auxiliary heating systems) and transport of energy and particles, impurity radiation, and MHD. It offers several options of different fidelity for each physics component. Previously, ETS was verified against state-of-the-art transport codes and used to analyse data from existing tokamaks [2]. In this work, the capability of the ETS to simulate complex scenarios was used to study the possibility to control the plasma density in a reactor size machine by pellet injection.

## DEMO scenario simulations with ETS

Present simulations were configured to reproduce overall plasma scenario developed by the system code PROCESS [3]. ETS was configured to solve the transport equations

* See http://www.euro-fusionscipub.org/eu-im

for the current, the density (*for e, D, T, He, Ar and W*) and the temperature (*for e, D, T, He*). Temperatures for *Ar* and *W* were assumed to be equal to the temperature of deuterium ions. Transport coefficients were provided by the combination of Bohm-gyroBohm and neoclassical models, assuming the edge transport barrier at $\rho_{tor\_norm}=0.97$. Heat and current drive sources were calculated by the beam simulation package and α-heating module. Heat balance also included the radiation from impurities (*Ar* and *W*) and Synchrotron radiation.

The pellet module [4] used in simulations calculates the ablation rate of the pellet and provides flux surface averaged changes to density and temperature profiles, treated by the transport code as instantaneous event. In addition to the original model the current implementation includes *ExB* drifts following the scaling from ref.[5].

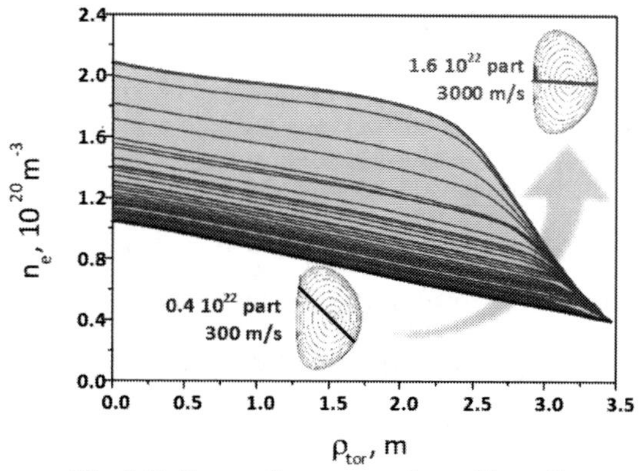


*Fig.1 Pellet cycle averaged profiles of electron density for different injection locations, mass and velocity of the pellet*

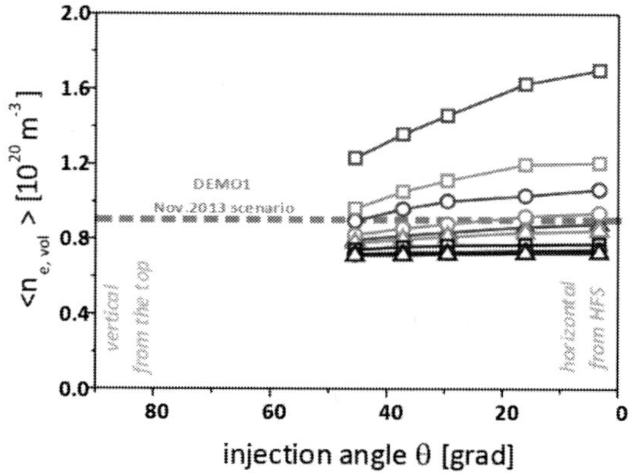


*Fig.2 Steady-state volume averaged electron density compared to the requirements by DEMO scenario*

Initial simulations were aimed in determining the domain of pellet parameters and injection locations, which might provide particle fuelling sufficient for maintaining the density in the reference DEMO scenario. Therefore, the simulations for five different poloidal angles of injection position, each with several pellet sizes and velocities were performed. To keep the particle throughput constant over the scan, the injection frequency was adjusted inversely proportional to the pellet size. As expected the fuelling efficiency increases over the scan with increasing velocity and size of injected pellets (see fig.1). This provides freedom on the pellet system optimization, where the reduction of the pellet size can be compensated by deeper penetration inside of the confined plasma. Nonetheless, a critical pellet size is found in simulations, below which the pellet injection system cannot provide the fuelling necessary to maintain the density required by the scenario considered for a fusion reactor. Figure 2 shows steady state volume averaged density versus the injection angle for the matrix of three different pellet sizes (*$4*10^{21}$, $6*10^{21}$ and $1.6*10^{22}$ particles*) by three different injection velocities (*triangles-300 m/s, cycles -1000 m/s, squares-3000 m/s*). The required density can not be achieved with the smallest pellets under any conditions. For those pellets the ablation occurs at a very peripheral region, the majority of the ablated materials is expelled from the plasma before it can contribute to the fuelling. At the same time, simulations with larger pellets offer several possible solutions where the required density is achieved.

Therefore for next sequence of simulations the intermediate pellet of $6*10^{21}$ particles and velocity of 1000 m/s was chosen. In this group of simulations the flexibility of pellet injector design in terms of injection angle was addressed. The controller for the pellet injection was configured to maintain the volume averaged density required by DEMO scenario. For the same entry point at the vessel, the injection angle (*with vertical axis*) was varied between 10° and 70°, see fig. 3. Results of simulations are presented in fig. 4.

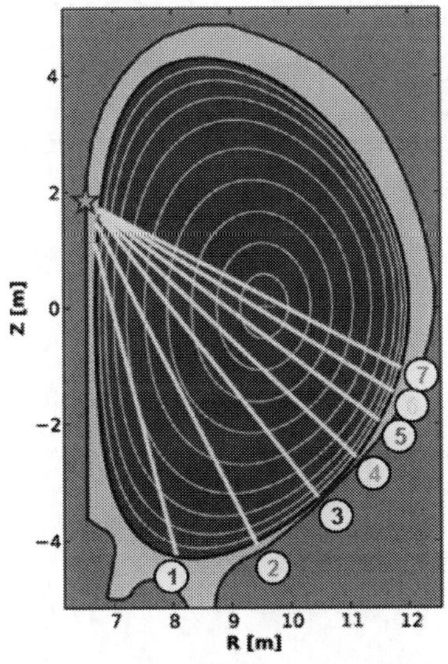


*Fig.3 Pellet injection geometry*

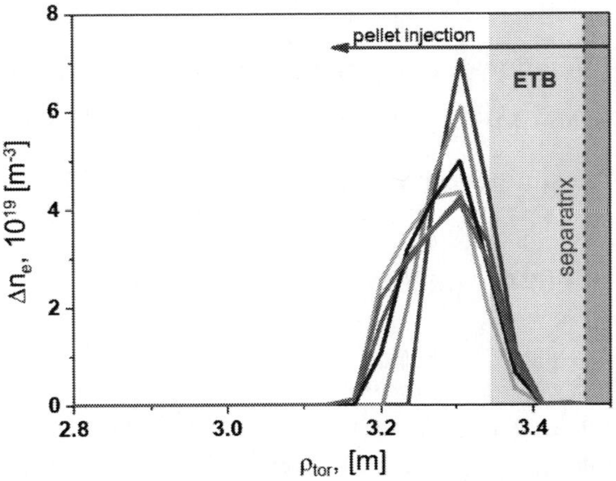


*Fig. 4 Pellet deposition profiles for different injection geometry*

The change of electron density profile caused by single pellet injection for different injection angles stays nearly the same for injection angles between 35° and 70°. For smaller angles the deposition shifts in the outward direction, when the fraction of the material deposited outside of the pedestal top increases. It makes fuelling less efficient and forces the controller to increase the frequency of the pellet injection in order to maintain the required density. Figure 5 shows the frequency of injection set up by controller at the steady state phase of simulated discharge (*when required plasma parameters are achieved*) as a function of injection angle. These results show the flexibility for the design of the pellet injection system for DEMO

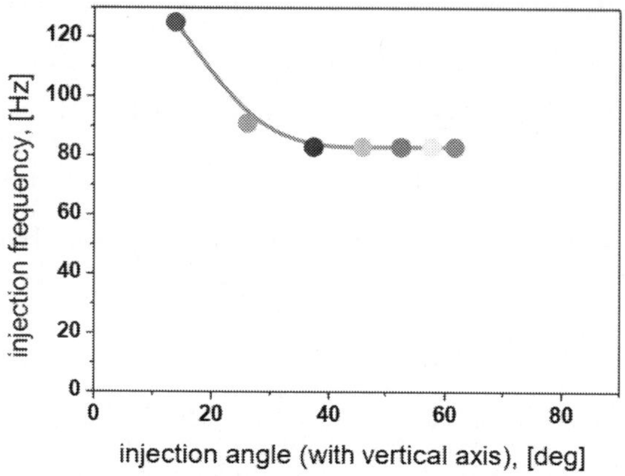


*Fig.5 Output from pellet injection controller: adjustment of the injection frequency depending on the injection angle*

**Conclusions & Remarks**

The European Transport Simulator has reached a mature state and can be applied for complex scenario simulations, eg for predictive simulation of reactor scale plasmas. The density for the foreseen DEMO operational scenario can be maintained by medium size pellets injected from the high field side. The injector design allows for some flexibility in choosing the pellet speed and injection geometry keeping the same fuelling efficiency. The computations presented do not include stability analysis of the MHD, whereas it might cause additional physics limitations in optimizing the pellet parameters, especially for largest pellets. Integration of the MHD stability chain into the ETS is foreseen as a next important development of the simulation package.


**Acknowledgment**

*This work has been carried out within the framework of the EUROfusion Consortium and has received funding from the Euratom research and training programme 2014-2018 under grant agreement No 633053. The views and opinions expressed herein do not necessarily reflect those of the European Commission.*


**References**


[1] D. Kalupin et al. Nucl. Fusion 53 (2013) 123007

[2] D. Kalupin et al. 38th EPS Conference on Plasma Phys. Strassbourg, 27 June-1 July 2011, Vol35G, P-4.111 (2011)

[3] N. P. Taylor (holder) and P. J. Knight, "PROCESS Reactor Systems Code", AEA Fusion Project Work File, F/RS/CIRE5523/PWF (1992)

[4] K. Gal et al, Plasma Phys. Control. Fusion **50** (2008) 055006

[5] F. Köchl 35th EPS Conference on Plasma Phys. Hersonissos, 9 - 13 June 2008 ECA Vol.32D, P-4.099 (2008)


The results presented here, obtained using the EU-IM framework, present an integrated modelling exercise testing and verifying the consistency of dedicated physics actors. They are not (yet) meant as a basis for decision making on the design and parameters of the DEMO PELLET system.

# Neoclassical transport in displaced tokamak helical cores

J. M. García-Regaña[1,2], H. Smith[3], Y. Turkin[3], M. Sertoli[2], C. Angioni[2], A. Bustos[3],
W. A. Cooper[5], J. Geiger[3], R. Kleiber[3], E. Strumberger[2] and ASDEX Upgrade Team.

[1] *EUROfusion PMU, Garching b. München, Boltzmannstr. 2 85748, Germany*
[2] *Max-Planck-Institut für Plasmaphysik, Boltzmannstr. 2 85748, Garching b. München, Germany*
[3] *Max-Planck-Institut für Plasmaphysik, Wendelsteinstr. 1 17491, Greifswald, Germany*
[4] *Departamento de Física, Universidad Carlos III, Leganés, 28911, Spain*
[5] *Centre de Recherches en Physique des Plasmas, EPFL, CH-1015 Lausanne, Switzerland*

## Introduction

Recent experiments in ASDEX Upgrade (AUG) have shown expulsion of Tungsten out of the helical core produced by internal (1,1) kink modes [1]. Given that impurities dilute the plasma fuel and cool it down due to their high radiative power, the identification of mechanisms that suppress or mitigate impurity accumulation is of central importance for the success of thermonuclear fusion.

Although tokamaks are nominally axi-symmetric, they are often subject to 3D perturbations, e.g. due to TF coil ripple, RMP coils used for ELM mitigation or, as in the aforementioned AUG experiments, developement of bifurcated helical MHD states. The loss of the toroidal symmetry of the magnetic field is recognized as potential overlap with the physics features and tools familiar to stellarators and a good basis for addressing the question of to what extent such physics features can show up in tokamaks. Following a similar reasoning, the closeness of stellarators to certain symmetries which make them exhibit tokamak properties like for instance large rotation, is currently an active research field [2]. In the present work we initiate the evaluation of some of these features in the frame of neoclassical theory, including the search for transport scaling at low collisionalities, estimation of the ambipolar electric field and potential variations. This preliminary assessment relies on the central motivation and mid-term objective of studying the dynamics of impurities in the aforementioned 3D tokamak equilibria.

## The 3D AUG equilibrium considered

3D equilibrium tokamak states can be generated with the Variational Moments Equilibrium Code VMEC [3], prescribing the plasma boundary, pressure and toroidal current profiles as well as an initial non-planar axis guess. If parameters are such that a weakly reversed shear $q$ profile with a minimum near unity is obtained and the axis guess is sufficiently distorted the output results in a 3D solution with helical core similar to a saturated $m/n = 1/1$ ideal internal

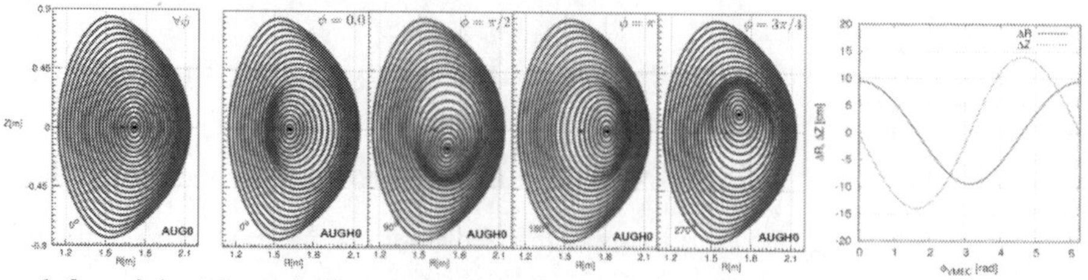

Figure 1: From left to right: Magnetic flux surface contours for the initial axi-symmetric AUG equilibrium (left-most figure); 3D final solution at 4 toroidal angles; and displacement of the axis along the cylindrical coordinates $R$ and $Z$ as a function of the toroidal VMEC coordinate.

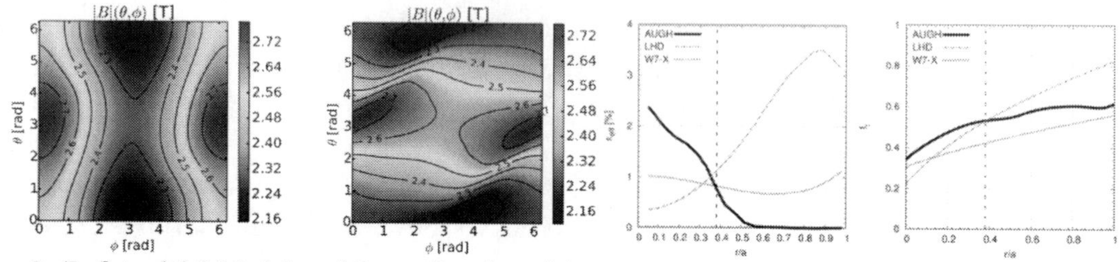

Figure 2: (Left to right) Modulus of $B$ as a function of the PEST angular coordinates $\theta$ and $\phi$ at $r/a = 0.1$ and $0.35$; $\varepsilon_{\rm eff}$ and $f_t$ radial profiles for the 3D AUG equilibrium, $Wendelstein7-X$ and LHD stellarator.

kink [4]. In the present study the plasma boundary corresponding to an perfectly axi-symmetric equilibrium of AUG has been considered (see left-most plot in fig. 1). The final developed 3D solution (see the 4 poloidal cross sections in fig. 1) shows a helical displaced core inside the $q = 1$ surface situated at the normalized effective radius $r/a \approx 0.4$. Finally the helical axis displacement amplitude (right-most plot of fig. 1) is approximately 10 and 12 cm along the usual cylindrical coordinates $Z$ and $R$ respectively.

Prior to quantitative neoclassical considerations, it is important to note that the equilibrium suffers a strong deviation from toroidal symmetry in the helical core, which is seen in the contour maps of the magnetic field strength in fig. 2. In addition, the fraction of trapped particles $f_t$ and the effective ripple $\varepsilon_{\rm eff}$ take values of the order of or larger than stellarators like LHD and W7-X in their standard configurations.

## Neoclassical simulations

Regarding the neoclassical simulations, these have been performed with the neoclassical version of the Monte-Carlo $\delta f$ PIC code EUTERPE [5, 6]. In the most general case EUTERPE solves the drift kinetic equation including the potential varying on the flux surface $\Phi_1$:

$$\frac{\partial f_1}{\partial t} + \dot{\mathbf{R}} \cdot \nabla f_1 + \dot{v}_\parallel \frac{\partial f_1}{\partial v_\parallel} = -f_{\rm M} \left(\mathbf{v}_{\rm d} + \mathbf{v}_{E1}\right) \cdot \nabla r \left[ \frac{n'}{n} + \frac{q}{T}\Phi_0' + \left( \frac{mv^2}{2T} - \frac{3}{2} + \frac{q}{T}\Phi_1 \right) \frac{T'}{T} \right],$$

with $' \equiv {\rm d}/{\rm d}r$, $f_1$ the departure from $f_0 = f_{\rm M} \exp(-Z|e|\Phi_1/T)$, the lowest order distribution function; $\Phi_0$ the ambipolar part of the electrostatic potential, $v_{E1}$ the $E \times B$ drift related to $\Phi_1$

and $v_d$ the magnetic drift. The characteristics at lowest order are for the guiding center $\mathbf{R}$, parallel velocity $v_\parallel$ and magnetic moment $\mu$: $\dot{\mathbf{R}} = v_\parallel \mathbf{b} - \frac{\nabla\Phi_0 \times \mathbf{B}}{B^2}$; $\dot{v}_\parallel = -\frac{Ze}{m}\mathbf{b}\cdot\nabla\Phi_1 - \mu\mathbf{b}\cdot\nabla B - \frac{v_\parallel}{B^2}(\mathbf{b}\times\nabla B)\cdot\nabla\Phi_0$; and $\dot{\mu} = 0$. $\Phi_1$ is obtained after requiring quasi-neutrality up to first order among the involved species. The assumption of adiabatic electrons considered here leads to the equation: $\Phi_1 = \frac{T_e}{e}\left(n_{0e} + n_{0i}\frac{T_e}{T_i}\right)^{-1} n_{1i}$, with $n_0$ and $n_1$ the equilibrium and perturbed densities, $T$ the temperature and $\{i,e\}$ the bulk ion and electron indices. The collisions (pitch angle scattering in the present work) are performed in each time-step applying a random kick in the markers' pitch angle after pushing them along the collisionless characteristics.

The collision frequency is defined as the sum over all the target species of the deflection collision frequency: $\nu = \sum_b \nu_D^{ab}$. To get a simplified picture of the physics, however, it can be set constant for all markers, which enables depicting the different collisional regimes participating in the transport processes. In particular for the above-mentioned AUG equilibrium and zero value of the radial electric field ($E_r = -\mathrm{d}\Phi_0/\mathrm{d}r$) the low collisional asymptotic $1/\nu$ regime shows up at the helical core for the transport matrix coefficient $L_{11}$ [7] (see fig. 3 left), while the usual *banana* scaling is recovered at a more external radius (see fig. 3 right).

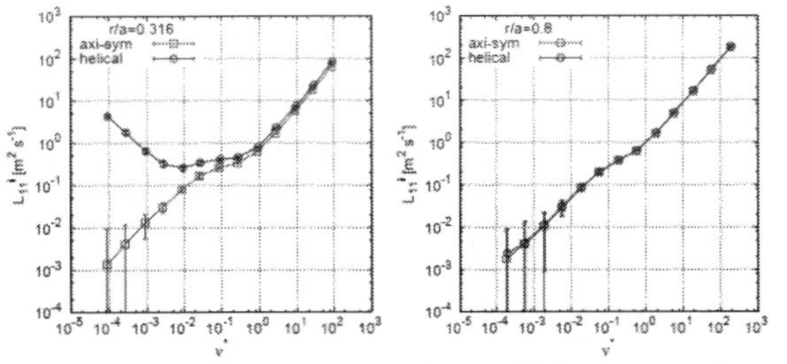

Figure 3: $L_{11}$ transport matrix coefficient as a function of the normalized thermal collision frequency $\nu^* = \nu R_0 q / v_{\mathrm{th}}$ inside the helical core (left) and far outside (right).

Along the same line-of-thought, considering a set of analytical temperature and density profiles similar to those reported for the experimental AUG discharge #31114 in [1], see fig. 4(left), the neoclassical radial transport of bulk ions ($H^+$) for $E_r = 0$ is significantly larger – between two and three orders of magnitude – in the helical core region compared with the outer axi-symmetric radii, regardless of the weaker gradients at the inner positions. This is observed in

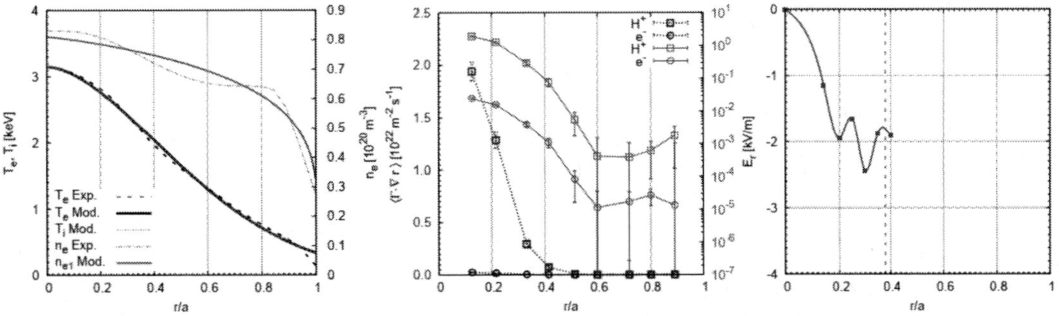

Figure 4: (Left) Plasma parameters fitted to AUG discharge #31114. (Center) Particle flux density of bulk ions ($H^+$) and electrons for $E_r = 0$ in logarithmic (red) and linear (blue) scales. (Right) Radial electric field estimation inside $q = 1$ surface.

fig. 4(center). Such large ion radial particle flux in comparison to that observed out of the helical core, which falls nearly to the noise level, can lead to the onset of an electric field in order to restore the subsequent strong lack of ambipolarity. The estimate of $E_r$ is shown at different radii for a pure Hydrogen plasma in the displaced core in fig. 4(right).

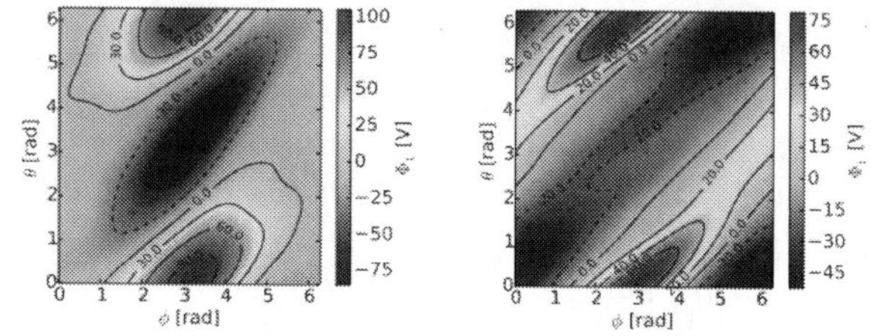

Figure 5: $\Phi_1(\theta,\phi)$ at $r/a = \{0.2, 0.35\}$.

Finally, in order to complete the picture for the electric field the potential variation within some flux surfaces at the displaced core has been obtained. This potential variation cannot be neglected, as is usually done, because the magnetic and electrostatic drifts become comparable as the charge state $Z$ increases, which gives rise to an electrostatic transport source that can compete with the grad-B and curvature drives. A similar thing happens for the ratio between the magnetic and electrostatic mirror terms. As $Z$ increases, electrostatic trapping becomes important, which affects the topology of the trapping boundaries set by the magnetic mirror term. In figure 5 the potential variation at $r/a = 0.2$ and 0.35 are represented. Considering the usual values obtained and measured [9, 10] in stellarators, these inside helical region are comparable and larger at similar collisionalities.

## Conclusions

Considering this preliminary assessment of the magnitudes and properties relevant for 3D devices like $\varepsilon_{\rm eff}$, presence of $1/\nu$ scaling, ambipolarity fulfillment and $\Phi_1$ amplitude, the conclusion can be drawn that it would be questionable to apply an axi-symmetric approach to the helical state. Further studies involving the impact on impurity transport are ongoing.

## Acknowledgements

This work has been carried out within the framework of the EUROfusion Consortium and has received funding from the Euratom research and training programme 2014-2018 under grant agreement No 633053. The views and opinions expressed herein do not necessarily reflect those of the European Commission.

## References

[1] M. Sertoli *et al.*, this conference O4.129 (2015).
[2] I. Calvo *et al.* Plasma Phys. Control. Fusion **55** 125014 (2013).
[3] S. P. Hirshman *et al.* J. Comput. Phys. **96**, 99 (1991).
[4] W.A. Cooper *et al.* Nucl. Fusion **53**, 073021 (2013).
[5] V. Kornilov *et al.* Nucl. Fusion **45**, 238 (2005).
[6] J. M. García-Regaña *et al.* Plasma Phys. Control. Fusion **55** 074008 (2013).
[7] C. D. Beidler *et al.* Nucl. Fusion **51**, 076001 (2011).
[8] M. Sertoli *et al.* Nucl. Fusion **53**, 053015 (2013).
[9] J. M. García-Regaña *et al.* to be submitted to Plasma Phys. Control. Fusion, preprint: http://arxiv.org/abs/1501.03967.
[10] M.A. Pedrosa *et al.* Nucl. Fusion **55** 052001 (2015).

# ICRH analysis of high-performance JET hybrid discharges using PION modelling and neutron spectrometry measurements

M.J. Mantsinen[1], J. Eriksson[2], C. Challis[3], D. Frigione[4], J. Garcia[5], C. Giroud[3], A. Hjalmarsson[2], D.B. King[3] and JET Contributors*

*EUROfusion Consortium, JET, Culham Science Centre, Abingdon, OX14 3DB, UK*

*[1]ICREA-Barcelona Supercomputing Center, Barcelona, Spain*

*[2]Uppsala University, Department of Physics and Astronomy, Sweden*

*[3]CCFE, Culham Science Centre, Abingdon, OX14 3DB, UK*

*[4]ENEA, Frascati, Italy*

*[5]CEA, IRFM, 13108 Saint-Paul-lez-Durance, France*

**See Appendix of F. Romanelli et al., Proc. 25th IAEA FEC 2014, Saint Petersburg, Russia*

**Introduction** The hybrid scenario is an operational tokamak plasma regime designed to achieve long pulse operation with a combination of inductive and non-inductive current drive. It has been suggested for ITER to allow operation at a high fusion power over 1000 s at a lower plasma current than for the inductive reference scenario. In the recent 2014 JET experimental campaign with the new ITER-like-wall (ILW) and deuterium as the main gas, high-performance hybrid discharges have been achieved with combined deuterium neutral beam injection (NBI) and ion cyclotron resonance heating (ICRH) [1]. ICRH was tuned to the fundamental cyclotron frequency of minority hydrogen ions which coincides with the second harmonic cyclotron frequency of deuterium ions in the plasma centre ($\omega \approx \omega_{cH} = 2\omega_{cD}$). In the present paper a number of JET high-performance hybrid discharges are analysed using the ICRH modelling code PION which takes into account NBI+ICRH synergy due to beam ions that are resonant with ICRH [2, 3]. Comparisons between the experimental results with modelling are presented, with special emphasis on the ICRH enhancement of the neutron rate measured with the neutron time-of-flight spectrometer TOFOR [4]. Furthermore, we analyse the ICRH power transfer to bulk ions and electrons and we extrapolate the best performing discharges to D-T.

**Experimental overview** The main parameters for JET hybrid discharges 86614, 87331 and 86871 analysed in the present paper are shown in Fig. 1. Discharges 86614 and 86871 were carried out at a toroidal magnetic field $B_T$ of 2.9 T and a plasma current $I_p$ of 2.5 MA, while a lower $B_T$ of 1.95 T and $I_p$ of 2.1 MA was used for discharge 87331. Up to 22 MW of D NBI was applied with injection energies in the range of 90-115 keV. In discharges 86614 and 86871 up to 5 MW of ICRH power was applied at a frequency of 42.5 MHz while in discharge 87331 up to 2.5 MW of ICRH power was used at 32.4 MHz. The $\omega \approx \omega_{cH} = 2\omega_{cD}$ resonance was located at $R_{res}$–$R_0$ of 2.5 cm, -33 cm and 9 cm at the peak performance for discharges 86614, 87331 and 86871, respectively. Here $R_{res}$ and $R_0$ are the major radii of the resonance and magnetic axis, respectively. The best performing discharge 86614 reached a normalized beta $\beta_N$ of 2.1 and a confinement factor H98 of 1.1. The duration of the high-power phase was limited by impurity accumulation [1].

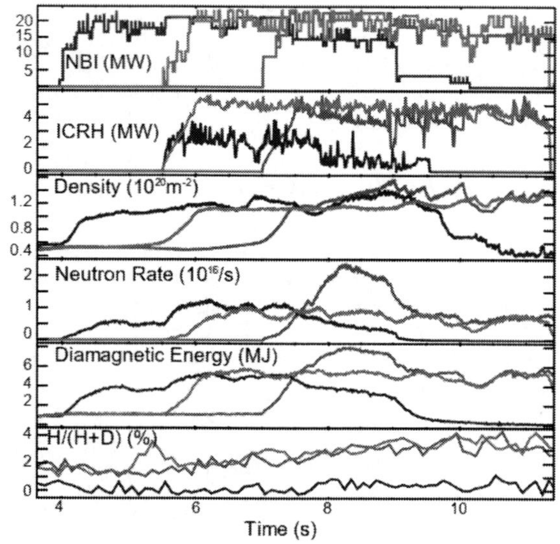


**Figure 1** NBI and ICRH power, line-integrated electron density, neutron yield, plasma diamagnetic energy and hydrogen concentration for JET hybrid discharges 86614 (red), 87331 (blue) and 86871 (magenta).

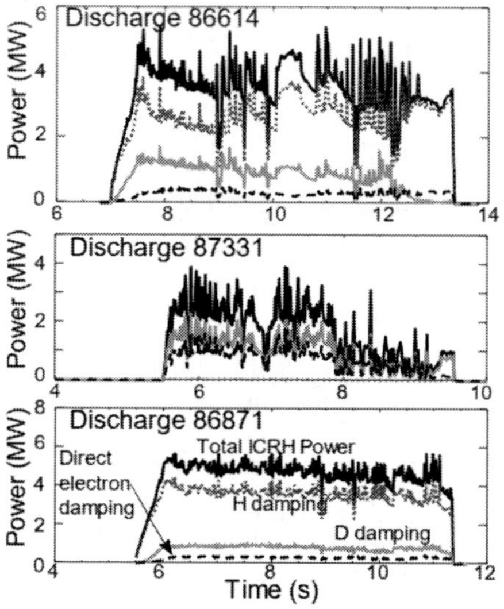


**Figure 2** Total ICRF power (solid black line), hydrogen damping (red), deuterium damping (green) and direct electron damping (dashed black line) as given by PION for discharges 86614, 87331 and 86871.

For the $\omega \approx \omega_{cH} = 2\omega_{cD}$ scenario, the ICRF power partitioning between the resonant hydrogen minority and deuterium ions depends on the hydrogen concentration $n_H/(n_H+n_D)$ [3]. The hydrogen concentration, deduced from the ratio of the $D_\alpha$ and $H_\alpha$ light collected along lines of sight through the plasma, was 2-3% in discharges 86614 and 86871 and below 0.5% in discharge 87331. Penning gauge spectroscopy in the divertor gave somewhat higher $n_H/(n_H+n_D)$ values of 2-4%, 3-5% and 1-2% for 86614, 86871 and 87331, respectively.

**Results from ICRF modelling and comparisons with experimental results** Figure 2 shows the ICRF power partition as given by PION for the three discharges. In discharges 86614 and 86871, up to ≈20%, 70-75% and 5-10% of the ICRH power is absorbed, respectively, by D ions, hydrogen minority ions and by direct electron damping by electron Landau damping and transit time magnetic pumping. In discharge 87331 damping by hydrogen minority ions is reduced due to a lower $n_H/(n_H+n_D)$. The best agreement with the measured data is obtained assuming a very low $n_H/(n_H+n_D)$ of < 0.1%. As result, the H damping is negligible, while direct electron and D damping take 45 and 55% of the ICRH power, respectively. The ICRH power transferred to the bulk ions was about 2, 2.5 and 1 MW for 86614, 86871 and 87331, respectively, the rest being transferred to the plasma electrons. The ICRH power was deposited centrally inside r/a < 0.5, where r/a is the normalised plasma minor radius.

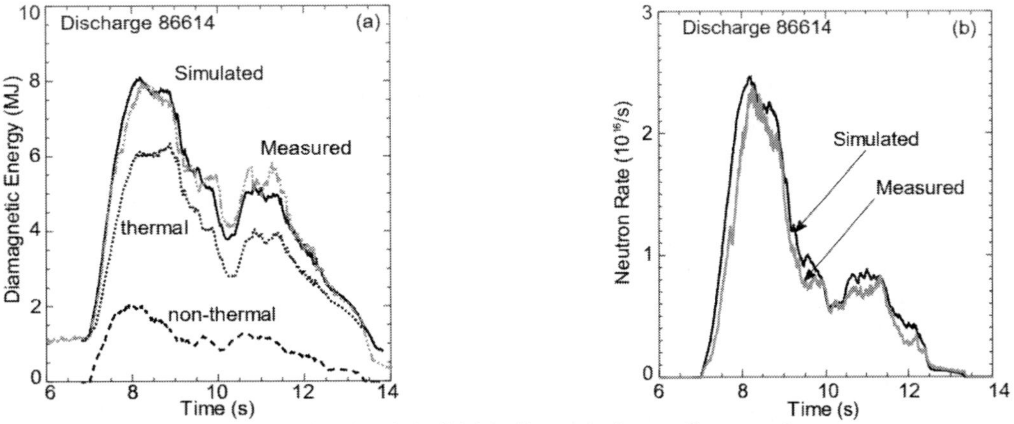


**Figure 3** (a) Measured (green) and simulated (solid black) total plasma diamagnetic energy content as given by PION and (b) measured (green) and simulated (black) total neutron rate as given by PION for discharge 86614. In (a) the measured thermal (dotted blue) and simulated nonthermal (black dashed) contributions are also shown.

The comparisons of the simulated and measured plasma diamagnetic energy content $W_{DIA}$ and total neutron rate $R_{NT}$ are shown in Fig. 3 for discharge 86614. The agreement between modelling and experimental results is remarkably good. Similar agreement is found for discharges 86871 and 87331. The thermal contribution to $W_{DIA}$ is about 75% while the non-thermal component is about 25%. Furthermore, according to PION, 5-20% of $R_{NT}$ is due to the second harmonic D acceleration (Fig. 4). The predicted ICRH enhancement of $R_{NT}$ has been compared with the results of the time-of-flight neutron spectrometer TOFOR [4]. TOFOR consists of two sets of plastic scintillator detectors that view the plasma vertically through the plasma centre. The energy of the impinging neutrons is deduced from the time of flight between these two sets of detectors. A measured TOFOR spectrum is separated into components due to thermonuclear (TH), beam-thermal (NBI), ICRH induced fusion reactions and scattered neutrons (Fig. 5). The spectral shape of the ICRH component is calculated from the D distributions given by PION while the NBI component is calculated from the slowing down distributions obtained with the NUBEAM code [5], which is part of the TRANSP package, and the TH component is a Gaussian with a width determined by the plasma temperature. An estimate of the ICRH enhancement is obtained from the fitted component intensities as $I_{ICRH}/(I_{TH} + I_{NBI})$. The ICRH enhancement estimated in this way is in good agreement with the PION simulations, as shown in Fig. 4.

**Conclusions and extrapolation to DT plasmas** Recent JET high-performance hybrid discharges with combined NBI and ICRH have been successfully modelled with the PION code. Good agreement between modeling and measured data was obtained, lending confidence in the simulations. The modeling shows that both central electron and ion heating is increased significantly with the application of ICRH. The extrapolation of discharges 87331 and 86614 to a 50%-50% D-T mixture yields a fusion power of about 4 and 7 MW, with an ICRH enhancement of the fusion yield of ≈5-10% and 15%, respectively. The goal for the JET hybrid scenario development in 2015-2016 is to more than double these values, i.e. to reach predicted fusion power of 15 MW with a pulse duration of at least 5s, in preparation of the JET deuterium-tritium campaign DTE2.

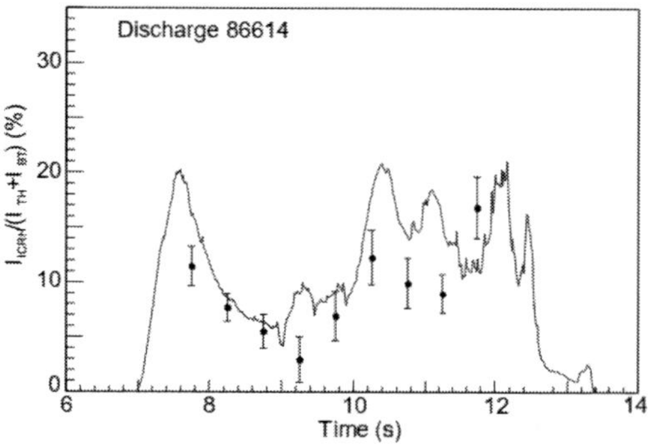


**Figure 4** Enhancement of the neutron rate due to second harmonic D damping according to PION (blue line) and TOFOR (points with error bards) for discharge 86614.

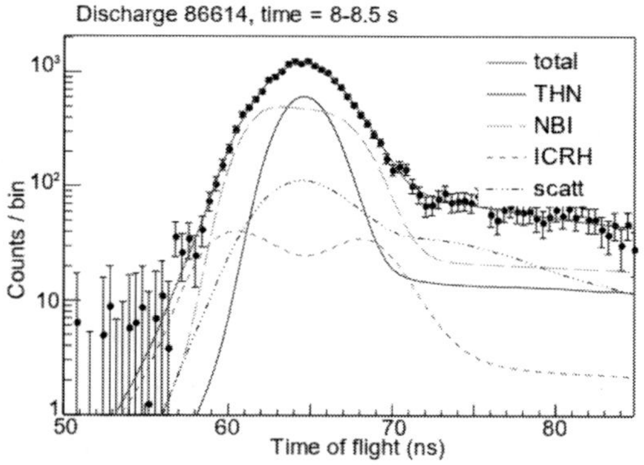


**Figure 5** TOFOR data points with error bars for discharge 86614 integrated over a time interval of t = 8-8.5 s. Also fitted components describing thermonuclear (TH), NBI, ICRH and scattered neutrons are shown.

**Acknowledgements** This work has been carried out within the framework of the EUROfusion Consortium and has received funding from the Euratom research and training programme 2014-2018 under grant agreement No 633053. The views and opinions expressed herein do not necessarily reflect those of the European Commission.

## References

[1] D. Frigione et al., this conference. [2] L.-G. Eriksson, T. Hellsten and U. Willén, Nuclear Fusion 33 (1993) 1037. [3] M.J. Mantsinen et al., Plasma Physics and Controlled Fusion 41 (1999) 843. [4] M. Gatu Johnson et al., Nucl. Instrum. Methods A, 591 (2008) 417. [5] A. Pankin et al., Comput. Phys. Commun. 159 (2004) 157.

# A novel gridded retarding field energy analyzer for IEDF measurement

Soumen Ghosh[1], P. K. Chattopadhyay[1], J. Ghosh[1], D. Bora[1]

[1]*Institute for Plasma research, Bhat, Gandhnagar-382428, India*

**Abstract**

A retarding field energy analyzer (RFEA) has been designed and developed to measure ion energy distribution function (IEDF) in helicon plasma. To design RFEA, determination of shape, size of the analyzer orifice, choice of grid and their separations are discussed. The four grid energy analyzer assembly provides easy access of all internal mechanical components and electrical connections. Fundamental problems regarding acquisition of correct RFEA data and their probable solutions are discussed.

## 1. Introduction

Conventional Langmuir probe measures electron energy distribution function (EEDF) in plasmas. RFEA is useful diagnostic for both ion energy distribution function (IEDF) and EEDF measurements. RFEA is widely used in expanding plasma systems, particularly where ion beam is generated [1]. RFEA has been used in capacitive and inductive RF plasma to measure IEDF and local plasma potential [2]. In this paper we describe the design of parallel plane four grid retarding field energy analyzer, which has been developed primarily to measure ion beam energy and plasma potential in helicon antenna produce radio frequency (RF) plasma. It is also shown that the design is flexible and provides easy mechanical and electrical access. Entering energetic electrons into the RFEA cavity, bias voltage scheme, offset adjustment, capacitive pick up problems are addressed.

## 2. RFEA principle and design

Charge particles are transmitted through an aperture, are analysed by the retardation of electric field established through bias potentials applied to the number of grids. The entrance slit must be wide enough to permit adequate flux transmission, yet sufficiently small such that the electrostatic sheath established around the slit edges be large enough to bridge the aperture width and hence shield the aperture from the bulk plasma. After the first entrance grid, second grid repels electrons (in ion mode operation) called repeller. Third grid samples different energetic ions, called discriminator. Forth grid suppresses the secondary electrons coming due to ion bombardment on the collector surface is the suppressor followed by the collector plate. A fraction of incident ion flux is transmitted through the slit. The slit entrance plate can be kept floating or biased negative to repel thermal electrons. The schematic of RFEA bias scheme is shown in figure (1).

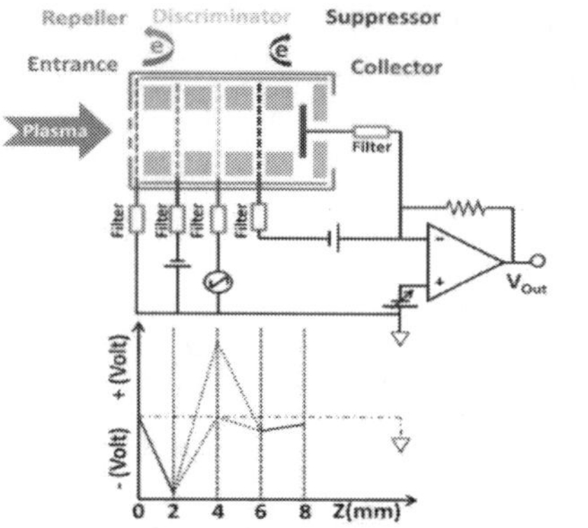


Fig 1. RFEA bias scheme for ion collection mode.

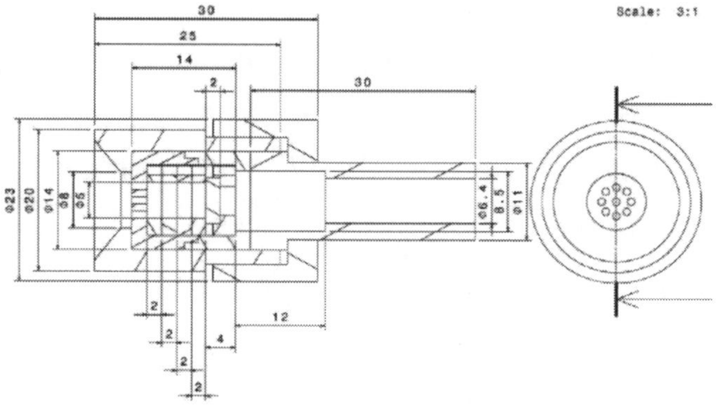


Fig 2. RFEA mechanical design, where all dimensions are in mm.

The ions of charge $e$, are assumed to enter from the main plasma with kinetic energy $E = eV_0$. They are retarded by the axially directed electrostatic field between the grid electrodes and collected at the collector. If $V_0 > V_d$ they will reach the plate and appear as collector current $I_c$; and if $V_0 < V_d$ they will be resisted. For discriminator voltage $V_d = 0$ to $V_S$, $I_c$ remain constant, means all ions gain a parallel energy of $eZ_iV_S$, where $eZ_i$ is the ion charge and $V_S$ is the sheath voltage, which is the plasma potential ($V_P$) with respect to the grounded chamber for a collision-less sheath. For $V_d > V_S$, $I_c$ starts to decrease. $I_c$ is expressed in equation (1)

$$I_C = eZ_iA_{eff}T_ET_1^4 \int_{v_{min}}^{+\infty} f(v)dv \; ...(1); \quad \frac{\partial I_c}{\partial V_d} = -\frac{e^2A_{eff}T_ET_1^4}{M_i} f\left(\sqrt{\frac{2eV_d}{M_i}}\right) \propto f(v) \; ...(2)$$

Where, $A_{eff}$ is the effective area of entrance apertures, $T_E$ is the transparency of the entrance slit plate, $T_1^4$ is the transparency of the total four mesh grid, $v_{min}=\sqrt{(2eV_d/M_i)}$, $v$ is the parallel ion velocity and $f(v)$ is the parallel ion velocity distribution. Ion energy distribution function (IEDF) as indicated in equation (2), which is proportional to the derivative of collector current and discriminator voltage. That will be derived from the RFEA ($I_c$ - $V_d$) characteristics.

Mechanical drawing of RFEA assembly is shown in figure (2). Distance between entrance grid to collector of RFEA is chosen to be less than ion neutral collision mean free path ($\lambda_{i-n}$) which is 8mm at $3.75\times10^{-3}$mbar. The inner diameter (circular plasma facing front side) of the RFEA is optimized to be 5mm, which accommodate array of holes, of diameter 0.9 mm each to make it close to satisfy the condition of electron entering into RFEA ($d\sim2\lambda_D$) [3]. RFEA energy resolution ($\Delta E/E$) = $D^2/16S^2$ depends on the inter grid separation (S) and grid wire separation (D) [4]. Large (~mm) grid wire separation increases energy resolution. However, this leads to very poor transmission (T) = $(D-2r)^2/D^2$, if the grid wire radius (r) is not very thin (1-10μm). We are using grids having D=110μm and 2r=33μm with grid transparency 0.49.

### 3. Results and discussion

Experiment is performed in helicon plasma system [5] [6], where plasma is produced by application of 200W RF power at $2x10^{-4}$mbar Argon fill pressure. Initial troubleshooting experiments are carried out by applying bias to all the grids with respect to the entrance grid, which is electrically connected with the Aluminium enclosure. While doing that we have also applied suppressor grid bias with respect to this common reference instead of suppressor grid bias with respect to collector as shown in figure (1). Keeping that bias configuration, the role of repeller bias or retarding potential on the RFEA characteristics is studied.

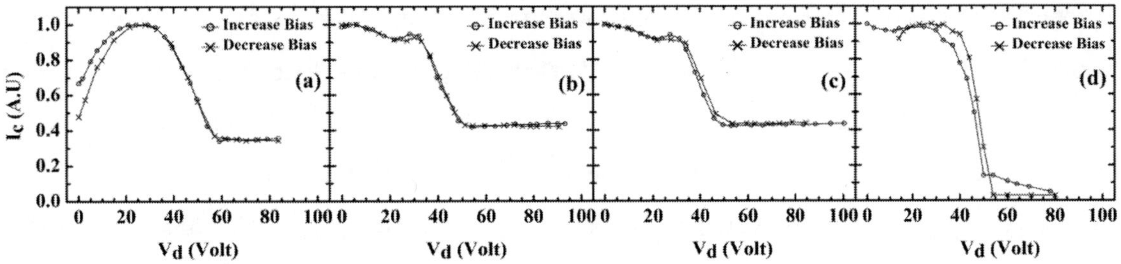


Fig 3. Collector current ($I_c$) versus discriminator bias ($V_d$) voltage with (a) Repeller bias ($V_R$) -60V, Suppressor bias ($V_S$) 0V, (b) $V_R$= -120V, and $V_S$= 0V, (c) $V_R$=-120V, $V_S$= -25V with respect to entrance grid and (d) $V_R$= -120V and $V_S$= -1.5V with respect to the collector.

Figure 3a shows that for $V_R$ = -60V, $I_c$ increases with $V_d$ from 0 to 15 V. Whereas, for $V_R$ = -120V (figure 3b), $I_c$ remains constant up to the plasma potential. In both the cases, the suppressor grid is kept at same potential. In figure (3c), $V_R$= -120V and $V_S$ =-25V, no significant change is observed in the characteristics figure (c) compared to figure (b). These results indicate that there is no significant role of $V_S$ for this biasing scheme. However, increased repeller bias significantly improves the characteristics by objecting high energy electrons entering into the RFEA. High energy electrons are restricted by the -120V repeller bias but not for -60V bias. The data presented in figure (3a, b, c), shows a dc off-set in all cases. To adjust this off-set the efficient way of using suppresser grid response for secondary emitted electrons, suppressor bias is applied with respect to collector as shown in figure (1). In that configuration, even a very small suppressor bias (-1.5V) with respect to the collector is sufficient to make the necessary off-set adjustment (figure 3d). In figure (3a, b, c and d), measurement of collector current for both increase and decrease bias voltage are found to be almost identical nature ($I_c$ - $V_d$) characteristics. This confirms that there is no such space charge accumulation between any grids.

After preliminary dc characterization and testing different measurement circuit schemes, a unipolar transistor amplifier ramp generated bias (0-120V) is applied to the discriminator grid to sample different energetic ions entering into the analyzer. This low frequency (~50mHz) ramp bias application to the discriminator essentially helps to reduce the capacitive pickup from the signal. At 50mHz ramp frequency, signal to capacitive pickup ratio is ~$10^3$. Therefore, the collector current signal is well resolved. Figure 4 shows the collector current to discriminator bias voltage along with its differentiation, which is then fitted with double Gaussian fit. First peak corresponds to the local plasma potential ($V_P$) and the second peak representation of ion beam ($V_b$). The ion beam energy detected by this RFEA is e($V_P$-$V_b$)~13eV, which is generated by the double layer like potential structure formation in this kind of expanding helicon plasma system.

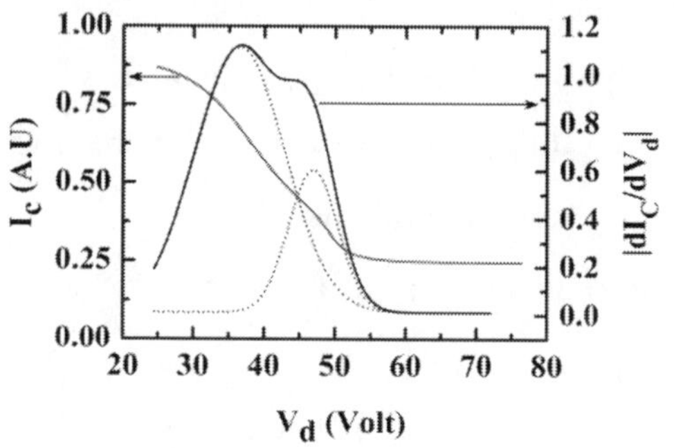


Fig. 4 Collector current ($I_c$) versus discriminator bias ($V_d$) with $V_R$= -100V with respect to the entrance grid and $V_S$= -5V with respect to the collector. Discriminator ramp bias frequency 50mHz.

## 4. Summary and conclusion

RFEA diagnostic has been designed and developed for measuring IEDF in helicon antenna produce plasma. The designed gives flexibility to access all the electrical components, grids and electrical connections. Initial dc characterization is carried out to understand suitable repeller bias voltage and bias scheme for ion collection mode operation. A low frequency (~50mHz) transistor amplifier circuit is used for application of bias to the discriminator and collector signal is acquired using an analog (*I to V*) convertor circuit, which detects the ion beam generated from the double layer like potential structure.

## References

[1] D. L. Rudakov et. al., *Rev. Sci. Instrum.* **70**, 476 (1999).
[2] K. Denieffe et. al., *J. Phys. D: Appl. Phys.* **44**, 075205 (2011).
[3] R. A. Pitts et. al., *Rev. Sci. Instrum.* **74**, 4644 (2003).
[4] J. Arol Simpson, *Rev. Sci. Instrum.* **32**, 1283 (1961).
[5] S. Ghosh et. al., *Plasma Sources Sci. Technol.* **24** 015017 (2015).
[6] K. K. Barada et. al., *Rev. Sci. Instrum.* **83** 063501(2012).

# High resolution equilibrium and stability calculations of pedestal and SOL plasma in tokamaks

S.Yu.Medvedev, A.A.Martynov, A.A.Ivanov, Yu.Yu.Poshekhonov

*Keldysh Institute, Russian Academy of Sciences, Moscow, Russia*

Tokamak edge plasma modelling relies upon computational grids aligned to magnetic surfaces because of highly anisotropic transport properties of the plasma with respect to magnetic field [1, 2]. The mapping of magnetic surfaces in the plasma and in the scrape-off layer (SOL) is usually performed based upon free boundary equilibrium data providing rather coarse resolution compared to very narrow pedestal and SOL widths predicted in ITER [3, 4]. The accuracy of the mapping procedure is certainly insufficient for MHD stability calculations and up to now only dedicated runs of nonlinear MHD code were capable to self-consistently treat the pedestal and SOL equilibrium and stability [5]. That is why fast and accurate equilibrium and stability calculations would significantly contribute to integrated modelling of edge plasmas.

**1 High resolution equilibrium with SOL** A starting point for high resolution fixed boundary equilibrium with plasma outside the separatrix is a free boundary equilibrium with poloidal flux function $\psi$ available outside the separatrix. The values of $\psi$ provide Dirichlet conditions at the boundary of SOL, private flux region and divertor plates for the solution of Grad-Shafranov equation while the position of the X-point and the separatrix are free to adjust. In the modified version of the adaptive grid CAXE code [6] plasma profiles can be independently prescribed in each sub-domain with nested flux surfaces formed by the separatrix branches, including a possibility of discontinuous toroidal current density across the separatrix. The same boundary conditions can be used for computation of a series of high resolution equilibria with varying profiles in the plasma and SOL. For the ITER Scenario 2 15MA plasma adding the high pressure gradient in the narrow SOL region corresponding to the separatrix pressure value $p_{sep} = 4.8$ kPa and the SOL drop-off length of $\sim 1.5$ mm [5] does not change the equilibrium much (Fig.1a). Note that despite high edge pressure gradient from the SOL side the separatrix angle is close to 90 degree.

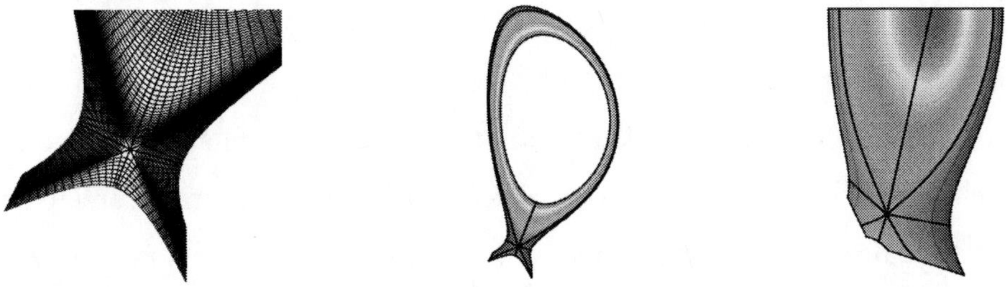

*Figure 1. (a) Fragment of the grid adapted to magnetic surfaces near the X-point for ITER 15MA equilibrium; (b) six subdomains for SN CAXE6-SOL version; (c) eight subdomains for DN CAXE8-SOL version.*

For a single null (SN) divertor plasma six sub-domains are used in the CAXE code (Fig. 1b). Another version of the CAXE code with eight sub-domains solves for up-down symmetric double null (DN) equilibrium (Fig. 1c).

**2 Ideal MHD stability with SOL plasma** The sheath compatible MHD boundary conditions on the open field lines [7] are applied to the variational formulation of the ideal MHD stability problem by setting $\xi_{||} \cdot n = 0, \quad \xi_{\perp} \cdot n = 0$, where $\xi$ is the vector of plasma displacement, $n$ is the normal to divertor plates. This formulation is almost identical to the standard variational principle with the exception of only one nonsymmetric term $\delta W_d = -1/2 \int_{S_d} A^* \cdot \delta j_s dS$, where $\delta j_s = n \times < \delta B >$ is a perturbed surface current at the divertor plates, $A$ is the vector potential of the perturbed magnetic field $\delta B_v = \nabla \times A$ in vacuum and the integral is taken over divertor plate surface. Assuming the absence of the surface current on divertor plates (e.g. due to toroidal gaps) we ended up with a traditional variational formulation with the only difference due to a special treatment of the boundary conditions at the divertor plates, where $B \cdot n \neq 0$. In the updated version of the KINX code [8] only the $\xi \cdot \nabla \psi$ component of displacement is free to vary at the divertor plates. It does not mean that perturbed magnetic field normal to divertor plates vanishes and it gives rise to the corresponding perturbation of the magnetic field in vacuum. However the self-consistent vacuum perturbation is not implemented in the KINX yet and $n \cdot \delta B_v = 0$ is assumed at the divertor plates instead, thus making the perturbed vacuum energy smaller and leading to just sufficient stability condition. The sensitivity to the way how the sheath compatible condition is implemented can be checked by setting even more stringent line tying condition $\xi = 0$ at the divertor plates making also $\xi \cdot \nabla \psi = 0$ and $n \cdot \delta B_v = 0$ with the vacuum energy treated consistently. It turns out that the difference between ideal MHD growth rates corresponding to the sheath compatible boundary condition or the line tying boundary condition (the latter leading also to $\delta W_d = 0$) for medium-n kink-ballooning modes localized at the outboard side of the plasma is rather weak.

In the unstructured grid stability code [9] the divertor plate boundary condition treatment is much easier. The ideal MHD stability condition $E \cdot B = 0$ is just replaced by the $E \cdot B \times n = 0$ there. The vacuum treatment is automatically consistent because of the tangential electric field continuity $< E \times n >= 0$ condition at the plasma-vacuum boundary which is natural for the vector Whitney basis functions used in the approximation.

The new versions of the ideal MHD stability codes were applied to preliminary studies of axisymmetric $n = 0$ mode driven by finite current density at the separatrix and to pedestal height limit gradient calculations taking into account high pressure gradients in narrow SOL for the ITER Scenario 2 15MA case.

The calculations of the $n = 0$ mode passive stabilization in tokamak plasma with finite edge current density reveal an existence of the mode localized near the X-points for DN divertor configurations once the global $m = 1$ mode has been stabilized by conducting wall (e.g. for negative triangularity plasmas [10]). Besides the value of current density at the edge, the proximity of the plasma boundary cutoff $\psi_b/\psi_{sep} \to 1$ to the separatrix is destabilizing. On the contrary, the deviation from plasma up-down symmetry in DN, for which perturbed surface current is induced only in case of finite edge current density, to SN configuration, is strongly stabilizing for the localized mode. All these features are consistent with the recent analytic results [11]. The calculation with the MHD_NX code taking into account plasma in SOL demonstrated that the "$n = 0$ peeling" mode is susceptible to the stabilization by even very narrow SOL plasma. The plasma displacement streamlines near the X-point, featuring reverse flow in SOL, are shown in Fig.2.

The issue of the pedestal stability of ITER plasmas taken into account high pressure gradient in SOL was recently revisited in [5] as the scaling [4] predicted very narrow SOL width $\sim 1$ mm with the corresponding pressure gradient exceeding several times the pedestal values. The dedicated calculations from [5] demonstrated quite expected fact that high but strongly localized pressure gradient cannot drive $n = 40$ mode unless the pedestal region is included in the analysis. This is because the unrestricted mode extends over the whole pedestal region despite high toroidal wave number. The updated KINX3-SOL code allows to estimate quantitative changes in the pedestal height limit due to the presence of the narrow SOL.

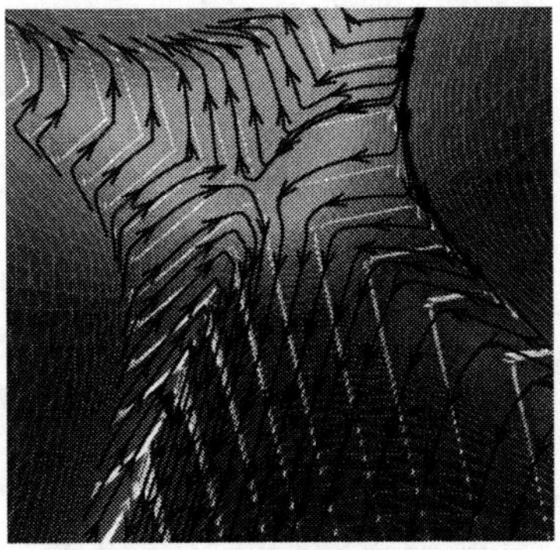

*Figure 2. Streamlines and vector plot of plasma displacement for "$n = 0$ peeling" mode in DN configuration. Color map corresponds to $E_\varphi$.*

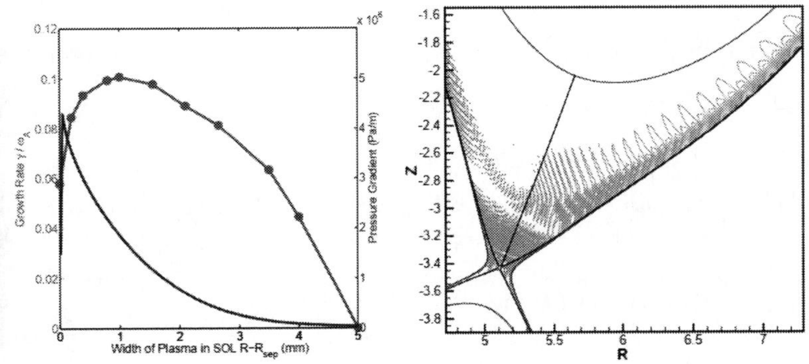


*Figure 3. (a) $n = 40$ mode growth rate and pressure gradient at the outboard vs the width of conducting plasma in SOL; (b) level lines of plasma displacement normal to magnetic surfaces, $\gamma/\omega_A = 0.098$, $R - R_{sep} = 1.56mm$.*

The same ITER Scenario 2 15MA plasmas with the pedestal height close to the marginal stability of $n = 40$ kink-ballooning mode was investigated. Fig. 3a shows the growth rate of the $n = 40$ mode and pressure gradient distribution (drop-off length 1.5 mm, separatrix pressure

4.8 kPa) versus the width of SOL at the outboard treated as ideally conducting plasma. Only moderate growth rate increase and eventual stabilization with increasing width of conducting plasma in SOL is demonstrated. For comparison the pedestal height limit was computed also with ideally conducting $p' = 0$ plasma in 1.5mm SOL by re-scaling the pedestal on the plasma side, which increases the limit by 2%. Almost the same increase in the pedestal height limit against the $n = 40$ mode was obtained due to re-scaling of $p'$ in SOL of the same width thus providing an evidence that the pressure at the top of the pedestal and not local pressure gradient value is the critical parameter for the pedestal stability.

**3 Discussion and future work** The developed high resolution equilibrium and stability codes with plasma outside the separatrix and open magnetic field lines at the divertor plates can be applied to a variety of edge and SOL plasma modelling applications. Preliminary results include the stabilization of the $n = 0$ mode driven by finite edge current density taking into account plasma outside the separatrix and estimates of the pedestal height limit set by kink-ballooning modes with high pressure gradient in narrow SOL.

Very localized "$n = 0$ peeling" mode which is unstable in the up-down symmetric DN divertor plasma seems to be susceptible to stabilization by a narrow layer of conducting plasma outside the separatrix.

The pressure at the top of the pedestal in ITER limited by kink-ballooning modes is determined by pressure value at the top of the pedestal also in case of high pressure gradient in SOL in accordance with existing scalings for the pedestal height (see e.g. [12]).

Another possible application of the developed suite of codes is coupling to the SOLPS [1] with a purpose to provide a new flexible interface to the equilibrium increasing the accuracy in magnetic surface aligned grid and to support self-consistent edge transport/equilibrium modelling. Gyrokinetic simulations of tokamak edge micro-instabilities could be also contingent on the uncertainties of the magnetic equilibrium [13].

[1] A.S. Kukushkin *et al.* Fusion Engineering and Design 2011 **86** 2865-2873
[2] R. Marchand and M. Dumberry. Computer Physics Communications 1996 **96** 232-246
[3] A.S. Kukushkin, H.D. Pacher, G.W. Pacher, V. Kotov, R.A. Pitts and D. Reiter. Nucl. Fusion 2013 **53** 123024
[4] R. Goldston. Nucl. Fusion 2012 **52** 013009
[5] A. Loarte *et al.* Journal of Nuclear Materials 2014 doi:10.1016/j.jnucmat.2014.11.122
[6] S.Yu.Medvedev, L.Villard, L.M.Degtyarev, A.A.Martynov, R.Gruber and F.Troyon. 20th EPS Conf. on Controlled Fusion and Plasma Phys., Lisbon, Proc. Contrib. Papers, Vol.17C, Part IV, 1279 (1993)
[7] H. R. Strauss. Physics of Plasmas 2014 **21** 032506
[8] L. Degtyarev *et al.* Comput. Phys. Commun. 1997 **103** 10
[9] S.Yu.Medvedev, A.A.Martynov, L.Villard, 40th EPS Conference on Plasma Phys., Espoo, Finland, 1 - 5 July, 2013 ECA Vol.37D, P5.145 (2013)
[10] S.Yu.Medvedev *et al.* Nuclear Fusion 2015 **55** 063013
[11] Z. T. Wang *et al.* Physics of Plasmas 2015 **22** 052107
[12] S. Yu. Medvedev *et al.* Contrib. Plasma Phys. 2010 **50** 1
[13] W. Wan *et al.* Physics of Plasmas 2015 **22** 062502

# Experimental campaign to test the capability of STARDUST-Upgrade diagnostics to investigate LOVA and LOCA conditions

L.A. Poggi[1], A. Malizia[1], J.F. Ciparisse[1], M. Gelfusa[1], A. Murari[2], S. Pierdiluca[1], E. Lo Re[1], and P. Gaudio[1]

[1] *Associazione EUROFUSION-ENEA, Department of Industrial Engineering, University of Rome "Tor Vergata", via del Politecnico 1, 00133, Rome, Italy*

[2] *Consorzio RFX-Associazione EUROFUSION-ENEA per la Fusione, I-35127 Padova, Italy*

## Introduction

An important issue related to future nuclear fusion reactors is the large amounts of dust produced by PMIs [1-7]. Re-suspension phenomena in case of LOVAs/LOCAs can cause serious hazard to the health of the operators (since particles are radioactive and of breathable size [9, 10-12]) and can cause explosions [13]. Experience achieved on reproducing thermo fluid-dynamic consequences of a LOVA [14-31] allowed the QEPM Research Group to develop an improved [32-34] facility ("STARDUST-Upgrade", Small Tank for Aerosol Removal and Dust – Upgrade) to reproduce dust re-suspension phenomena and to test diagnostics capability to investigate not only LOVAs but also LOCAs and their consequences as reported in the GSSR Report for ITER [35].

## 1. STARDUST-Upgrade facility overview and experimental methods

"STARDUST-Upgrade" facility (Small Tank for Aerosol Removal and Dust – Upgrade) is composed of a stainless steel vacuum chamber connected with diagnostics and a data acquisition system that ensure thermo fluid-dynamic conditions (pressure, temperature, local air velocity, pressurization rate) comparable to that expected in an ITER-like vacuum vessel during LOVA and LOCA events as reported in the GSSR Report [35]).

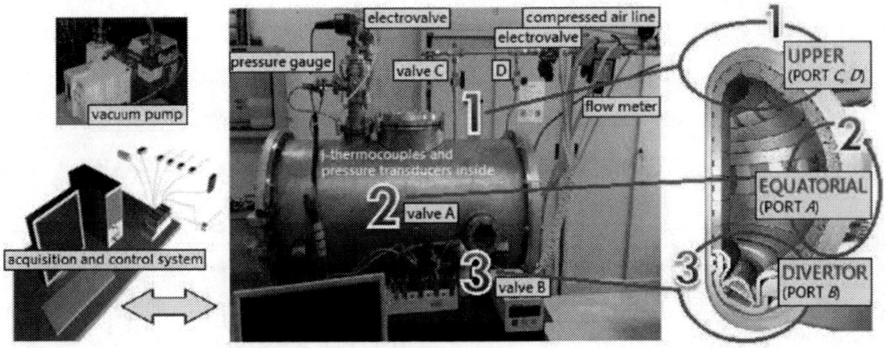


**Fig.1** "STARDUST-Upgrade" facility overview

During each vacuum failure accident experiment, the following signals have been acquired at a frequency of 50 Hz,: J-thermocouples temperature [°C], actual internal pressure [Pa], air flow rate [l/min], and differential pressure transducers output [Pa] used to calculate local air velocity [m/s] [8]. The "velocity transient time" (Table 1) was calculated in order to have information on the time range in which mobilization is expected. It was defined as the time at which the pressure transducer voltage signal returned to 0.50 mV, corresponding to its maximum acceptable zero point. This velocity, namely "v(0.50mV)", is equal to the minimum air velocity detectable by the system above the signal-to-noise ratio (SNR). Finally, the upper ports of "STARDUST-Upgrade" were used as inlet ports, reproducing LOVAs/LOCAs consequences from the upper ports of an ITER-like vacuum vessel. Pressurization experiments were performed at different air flow rates to achieve several pressurization rates from about 100 Pa/s to 400 Pa/s, including rates expected in GSSR Report in case of LOVA event [35].

## 2. Results

Internal pressure trends and corresponding pressurization rates are presented in Fig.2.

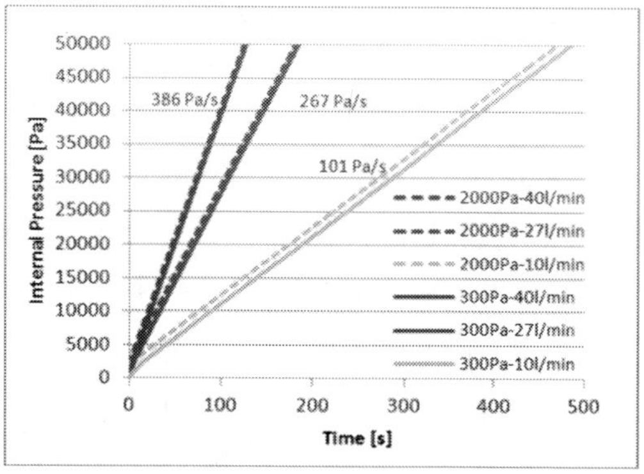


**Fig.2** Internal pressure trends measured by Pirani gauge inside "STARDUST-Upgrade" facility

Air velocity trends for first 25 seconds of air intake are reported in Fig.3 that shows a velocity peak in the first four seconds. The corresponding pressurization rates are that shown in Fig.2.

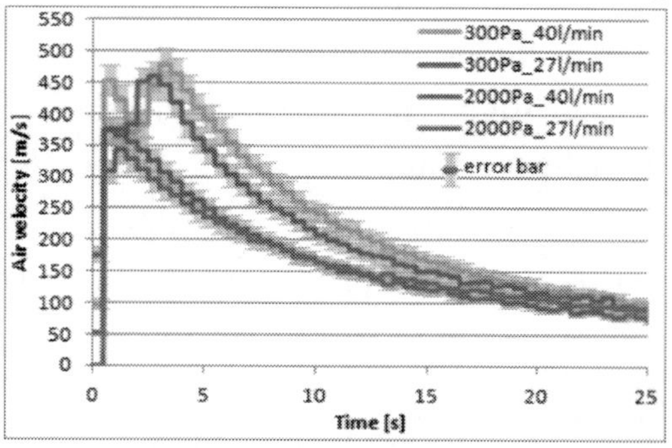


**Fig.3** Air velocity trends calculated for pressurization experiments in "STARDUST-Upgrade" facility

Table 1 summarizes the results on air velocity calculations. The maximum velocity value [m/s] is presented for each run along with corresponding time [s] at which the maximum was observed.

| | | | | | | | |
|---|---|---|---|---|---|---|---|
| *Initial Pressure* ***300*** *Pa* | Flow rate [l/min] | **40** | | | **27** | | |
| | Replication | *#1* | *#2* | *#3* | *#1* | *#2* | *#3* |
| | Max velocity [m/s] | 481.88 | 479.92 | 480.87 | 457.69 | 459.71 | 458.55 |
| | Time [s] | 3.36 | 3.36 | 3.36 | 2.80 | 2.80 | 2.80 |
| | Transient velocity time [s] | 58.69 | 57.78 | 55.56 | 45.97 | 49.17 | 48.20 |
| | v(0.50mV) | 48.32 | 37.68 | 40.23 | 56.35 | 49.12 | 45.93 |
| *Initial Pressure* ***2000*** *Pa* | Flow rate [l/min] | **40** | | | **27** | | |
| | Replication | *#1* | *#2* | *#3* | *#1* | *#2* | *#3* |
| | Max velocity [m/s] | 377.91 | 407.19 | 375.34 | 339.93 | 334.84 | 342.45 |
| | Time [s] | 1.40 | 1.40 | 1.40 | 1.40 | 1.40 | 1.40 |
| | Transient velocity time [s] | 44.54 | 66.53 | 44.06 | 52.50 | 51.36 | 54.55 |

| | | | | | | | |
|---|---|---|---|---|---|---|---|
| | v(0.50mV) | 40.03 | 30.52 | 40.18 | 42.92 | 49.67 | 42.84 |

**Table 1** Air velocity calculation data sheet for 27 l/min and 40 l/min air flow rate at two different initial internal pressures (300 Pa and 2000 Pa) of the chamber of "STARDUST-Upgrade" facility.

Fig.4 and Fig.5 compare pressurization curve and air velocity trends measured experimentally with CFD model predictions. Experimental and numerical results show substantial agreement for the first 20 seconds of pressurization. However, due to numerical viscosity effects, the numerical velocity predicted peak in the first seconds is lower than the experimental one.

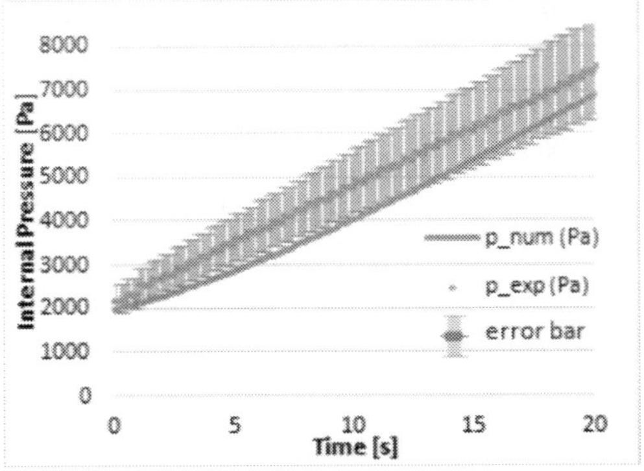


**Fig.4** Measured internal pressure of the chamber with error bars (p_exp) compared to model prediction (p_num) for air intake at 27 l/min at 2000 Pa initial internal pressure.

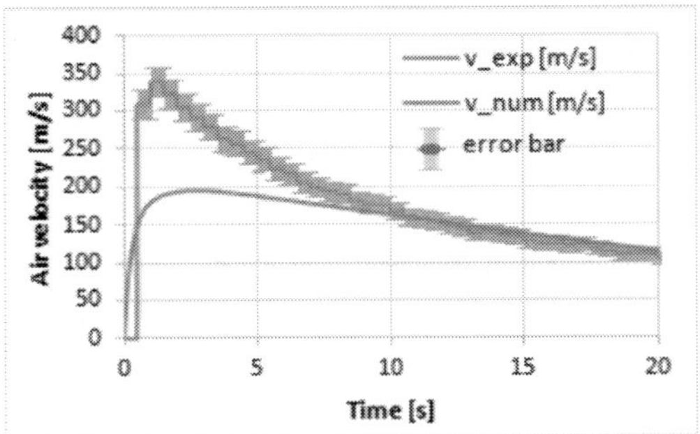


**Fig.5** Measured air velocity at the outlet of port C (v_exp) compared to model prediction (v_num) for air intake at 27 l/min at 2000 Pa initial internal pressure.

## 3. Conclusions

Pressurization of the chamber achieved through air intake from upper ports evidenced that "STARDUST-Upgrade" facility is able to reproduce a wide range of pressurization rates including what expected in GSSR Report [35], not only from the lower and equatorial level [1-7, 17] but also from the upper section of the vessel. Using upper ports also allows to replicate pressurization consequences of a LOCA from coolant system in the upper section of an experimental fusion plant, using air as model fluid. In the present work, a LOCA causing vacuum failure from the upper section is supposed to produce almost the same pressurization rates as LOVAs, but no experimental evidences are available at this time. The Transient velocity time shown in Table 1 resulted smaller than 60 seconds for all replications demonstrating that investigation of dust mobilization is crucial in the first seconds. Furthermore, Fig.4 and Fig.5 show a substantial agreement between numerical and experimental results. It is observed that the simulated air velocity trend does not reach the same measured peak value, and that the predicted pressurization rate is lower than the experimental one. This could be due to an insufficiently fine grid, which produces numerical viscosity. Future model development should improve the numerical accuracy. In conclusion, "STARDUST-Upgrade" facility is shown to be capable of reproducing the thermo fluid-dynamic consequences of a LOVA from lower, equatorial and upper part of the vessel, and the pressurization consequences of a LOCA from the upper part of the vessel. Since the fluid expected to enter the vessel during a LOCA could be different from air, next experimental campaigns will involve different fluids (e.g. steam) to properly reproduce a LOCA event.

**References**


1 J.P. Sharpe et al., Fusion Eng. Des. **63-64**, 153-163 (2002)
2 J. Winter, Phys. Plasmas **7**, 3862-3866 (2000)
3 A. Malizia et al., Adv. Mat. Sci. Eng. **2014**, Article ID 201831 pages 29 (2014)
4 A. Malizia, in *35th EPS Conference on Plasma Physics* 32, 696-699 (2008)
5 C. Bellecci et al., Fusion Eng. Des. **86**, (9-11) 2774-2778 (2011)
6 D. Di Giovanni et al., WSEAS Trans. Environ. Dev. **10**, 106-122 (2014)
7 A. Malizia et al., in *41st EPS Conference on Plasma Physics*, P. 5006 (2014)
8 A. Malizia, LAP LAMBERT Academic Publishing, Saarbrucken, Germany, 2014
9 K. Takase et al., Fusion Eng. Des. **42**, 83-88 (1998)
10 A. Malizia et al., Def. S and T Tech. Bull. **4**, (1) 64-76 (2011)
11 A. Malizia et al., Def. S and T Tech. Bull. **5**, (1) 36-45 (2012)
12 O. Cenciarelli et al., Def. S and T Tech. Bull. **6**, (1) 33-41 (2013)
13 J.P. Sharpe, P.W. Humrickhouse, Fusion Eng. Des. **81**, 1409-1415 (2006)
14 W.G. Brown, Int. J. Heat Mass Tran. **5**, 859-871 (1962)
15 K. Takase et al., Fusion Tech. **30**, 1459-1464 (1996)
16 K. Takase et al., Nucl. Sci. Eng. **125**, 223-231 (1997)
17 I. Lupelli et al., J. Fusion Energy ISSN 0164-0313, DOI: 10.1007/s10894-015-9905-8 (2015)
18 K. Takase, Fusion Eng. Des. **54**, 605-615 (2001)
19 P. Gaudio et al., In *Int. Conf. on Mathematical Models for Engineering Science*, 134-147 (2010)
20 P. Gaudio et al., Int. J. Syst. Appl. Eng. & Dev. **5**, 287-305 (2011)
21 M. Benedetti et al., Int. J. Syst. Appl. Eng. & Dev. **5**, 718-727 (2011)
22 M. Benedetti et al., Fusion Eng. Des. **88**, (9-10) 2665-2668 (2013)
23 C. Bellecci et al., Fusion Eng. Des. **88**, (9-11) 2774-2778 (2011)
24 C. Bellecci et al., Nucl. Fusion **51**, (5) 053017 (2011)
25 C. Bellecci et al., Fusion Eng. Des. **86**, (4-5) 330-340 (2011)
26 M. Benedetti et al., In: Mechanics RRi, 2nd Int. Conf. on FLUIDSHEAT'11, TAM'11; 142-147 (2011)
27 T. Pinna et al., Fusion Eng. Des. **85**, (7-9) 1410-1415 (2010)
28 C. Bellecci et al., In: *37th EPS Conference on Plasma Physics*; p. 703-706 (2010)
29 P. Gaudio et al., In: *Int. Conf. on Mathematical Models for Eng. Sci.* (MMES' 10) p. 134-147 (2010)
30 C. Bellecci et al., In: 33E1 ECA, editor. *36th EPS Conference on Plasma Physics* p. 266-269 (2009)
31 C. Bellecci et al., In: 32 EP, editor. *35th EPS Conference on Plasma Physics* 32; p. 9-13 (2008)
32 I. Lupelli et al., Fusion Eng. Des. **89**, (9-10) 2048-2052 (2014)
33 A. Malizia et al., Fusion Eng. Des. **89**, (9-10) 2098-2102 (2014)
34 P. Gaudio et al., In: *ICFDT 2013 Conference*, INFN, Frascati, Italy (2013)
35 IAEA. ITER Joint Central Team. *Generic Site Safety Report (GSSR)*
36 A. Malizia et al., Fusion Eng. Des. In Press. DOI:10.1016/j.fusengdes.2014.11.009

# Transport Barriers Formation and Properties Study Based on Bifurcation Concept

T. Onjun[1], B. Chatthong[1], R. Picha[2], N. Poolyarat[3] and J. Promping[2]

[1]*School of Manufacturing Systems and Mechanical Engineering, Sirindhorn International Institute of Technology, Thammasat University, Pathum Thani, Thailand*

[2]*Thailand Institute of Nuclear Technology, Bangkok, Thailand*

[3]*Department of Physics, Thammasat University, Pathum Thani, Thailand*

A coupled 2-fields bifurcation model is used to study the formations of both edge transport barrier (ETB) and internal transport barrier (ITB) in tokamak plasmas. In this work, the heat and particle transport equations can be written, respectively, in slab geometry as: $\frac{3}{2}\frac{\partial p}{\partial t}-\frac{\partial}{\partial x}[\chi_{neo}+\chi_{ano}f_s(v'_E,s)]\frac{\partial p}{\partial x}=H(x)$ and $\frac{\partial n}{\partial t}-\frac{\partial}{\partial x}[D_{neo}+D_{ano}f_s(v'_E,s)]\frac{\partial n}{\partial x}=S(x)$, where $p$ is the plasma pressure, $n$ is the plasma density, $\chi_{neo}$ and $D_{neo}$ are the heat and particle neoclassical transport coefficients, respectively, $\chi_{ano}$ and $D_{ano}$ are the heat and particle anomalous transport coefficients, respectively, $f_s$ is a suppression function which is assumed to depend on the flow shear $v'_E$ and the magnetic shear $s$, and $H$(x) and $S$(x) are the thermal and particle sources, localized at plasma center and edge, respectively. Furthermore, it is assumed that only the anomalous channel is suppressed by the two mechanisms with the suppression function of the form: $f_s(v'_E,s)=\frac{\beta|s|}{1+\alpha v'^2_E}\frac{1}{1+\gamma s^2}$, where $\alpha$, $\beta$, and $\gamma$ are arbitrarily chosen constants representing strengths of the suppression. This function is based on assumptions that the transports can be quenched or reduced by the flow shear and magnetic shear. Furthermore, the flow shear strength is also affected by the magnetic shear that is the magnetic topology also influences the flow in the plasma. The flow shear couples the two transport equations according to the force balance equation: $v'_E=c\frac{E'_r}{B}\,\square\,-\frac{c}{eBn^2}p'n'$, where $E_r$ is the radial electric field and $B$ is plasma magnetic field. Note that for simplicity, the effects of curvature, toroidal rotation, and poloidal rotation are neglected. The magnetic shear $s$ is calculated from the $q$-profile as follows: $s=\frac{x}{q}\frac{\partial q}{\partial x}$, $q=\frac{x}{R}\frac{B_\phi}{B_\theta}$, where $B_\phi$ and $B_\theta$ are toroidal and poloidal magnetic fields, respectively. A large aspect ratio tokamak of circular cross section approximation is assumed throughout the work. The toroidal magnetic field is assumed to be constant; while the poloidal magnetic field is calculated from the current using Ampere's law with current density of the form: $j(x) = j_0\left(1-(x-x_0)^2/a^2\right)^{\upsilon} + j_b$, where $x_0$ is the location of current density peak, $a$ is the plasma minor radius and $j_b$ is the bootstrap current which is locally proportional to pressure gradient( $j_b \propto -\nabla p$ ) [1]. Both transport equations can be decoupled using the method similar to that in Ref. [2]. Then, using the given current density and sources profiles the plasma pressure/density and their respective gradients can be numerically solved. All numerical results obtained are carried out using the same set of chosen constants; otherwise it is specified promptly. The transport coefficients are defined such that both thermal and particle anomalous diffusivities are around one order of magnitude over their neoclassical counterparts. The assumption is valid for ions, while for electrons the order of magnitude can be as high as two. In addition, the particle diffusivities are set to be roughly a quarter of thermal diffusivities [3]. Specifically, $\chi_{neo} = 1$, $\chi_{ano} = 10$, $D_{neo} = 0.25$, $D_{ano} = 2.5$, $\alpha = 0.1$, $\beta = 1$, and $\gamma = 0.1$. These values of transport coefficients are in a typical range of transport predicted by simulations using integrated predictive modeling codes using predictive core transport models [4-5].

**Plasma Profiles and Bifurcation Diagram**

Four scenarios are considered in this work. Scenarios 1 and 3 are the scenarios with the same heating which is less than the critical value required for *L-H* transition ($Q_L = 0.92Q_{L\rightarrow H}$), but with different location of current density peaking; at $x_0 = 0$ (plasma center), $x_0 = 0.2$ and $x_0 = 0.1$, respectively. On the other hand, scenarios 2 and 4 are the scenarios with the same heating which is more than the critical value required for *L-H* transition ($Q_H = 1.06Q_{L\rightarrow H}$), but with different location of current density peaking at; $x_0$=0 (plasma center), $x_0 = 0.2$ and $x_0 = 0.1$, respectively. It can be seen that each scenario yields the plasma with different performance. Firstly, there exists a heating threshold $Q_{L\rightarrow H}$ in which the plasma makes a transition from *L*-mode to *H*-mode with formation of an edge transport barrier (ETB) near plasma edge. In scenario 1, the plasma is set up with low heat source and no reverse shear profile (current profile peak at plasma center). This plasma remains in *L*-mode with central pressure and density equal to 0.69 and 0.77, respectively. Scenario 2 is setup with heat source greater than the critical value required for *L-H* transition ($Q_H > Q_{L\rightarrow H}$) and no reverse shear profile. In this scenario, the plasma makes a transition to *H*-mode with a pedestal width of 0.038 and central pressure and density of 1.24 and 1.42, respectively. Numerically, the heating in scenario 2 increases from that of scenario 1 by 15% but the central plasma pressure and density are almost doubled. This is a very significant enhancement which is why the *H*-

mode is preferred for tokamak operation. The bottom panels illustrate bifurcation diagram which mapped numerical results of each scenario into heat flux $Q$ versus pressure gradient $p'$ and particle flux $\Gamma$ versus density gradient $n'$ spaces. Theoretically, an example of a tradition $s$-curve bifurcation diagram can be found in Ref. [2] which shows a non-monotonic behavior. In figure 1, there appear discontinuities in pressure and density gradient profiles for scenario 2. They identify the locations of top of ETB.

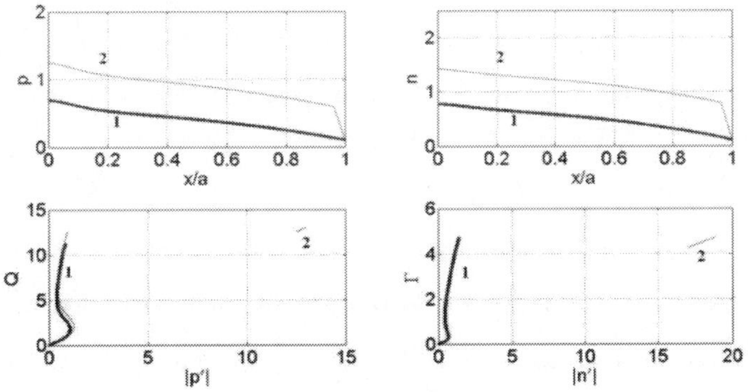


Figure 1: Pedestal Plasma pressure (top left) and density (top right) as a function of normalized minor radius $x/a$ for setup scenarios 1 (thick line) and 2 (thin line), and the results mapped onto heat flux versus pressure gradient (bottom left) and particle flux versus density gradient (bottom right).

Scenario 3 is the scenario with low heat source ($Q_L < Q_{L\to H}$) and current profile peaking at $x/a = 0.2$ ($x_0 = 0.2$) representing reversed shear profile. In this case, the plasma is still in $L$-mode but there is a formation of ITB near plasma core (see figure 2). The central pressure and density are around 1.23 and 1.36, which are increased from $L$-mode plasma without ITB by 78% and 77%, respectively. This implies that appropriate control of current profile in the plasma can lead to enhancement of plasma performance as well. The simulations in this work assume a form of current density distribution [3] with possibility to change the current peak location and with addition of bootstrap current effect. Scenario 4 is setup with high heat source ($Q_H > Q_{L\to H}$) and current profile peaking at $x/a = 0.2$. As expected based on the previous results, ETB and ITB formations can occur simultaneously. Note that the top of ITB appears to be closed to the location of current drive peak. The central pressure and density are increased to 1.94 and 2.06 which are 181% and 168% enhancement over $L$-mode performance, respectively. In this particular scenario, the jump in the gradients at lowest value of $Q$ corresponds to the top of ITB, next jump corresponds to the foot of ITB and the jump at highest value of $Q$ corresponds to the top of ETB.

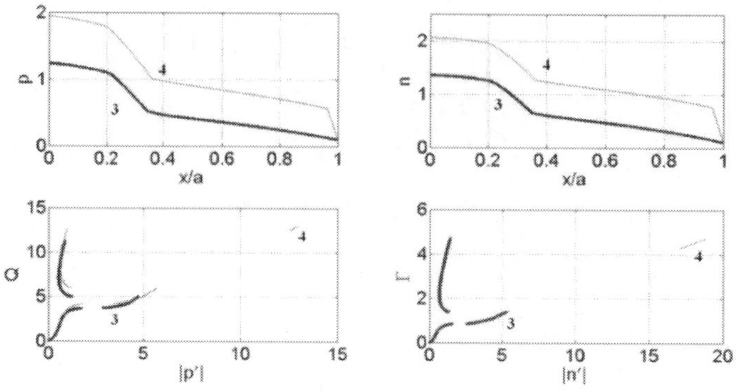


Figure 2: Plasma pressure (top left) and density (top right) as a function of normalized minor radius $x/a$ for setup scenarios 3 (thick line) and 4 (thin line), and the results mapped onto heat flux versus pressure gradient (bottom left) and particle flux versus density gradient (bottom right).

**Conclusions**

A coupled 2-fields bifurcation model is considered to analyze the formation and properties of ETB and ITB in tokamak plasmas. The transport equations for temperature and particle are self-consistently solved for the relation between local plasma gradients and corresponding fluxes in order to examine the ETB and ITB formations. It is found that the bifurcation nature can be observed when mapped onto fluxes versus gradients space in which abrupt changes in the gradients can be observed at plasma edge for ETB and plasma core for ITB. ETB formation depends sensitively on the heat flux. On the other hand, ITB formation is possible only with a presence of reverse magnetic shear profile with its width depends weakly on the heat flux.

**Acknowledgments**

This work was supported by the Commission on Higher Education (CHE) and the Thailand Research Fund (TRF) under Contract No.RSA5580041. B Chatthong thanks the Royal Thai Scholarship, Bangchak Petroleum PCL, and Thailand Institute of Nuclear Technology for financial support.

**References**

[1] R J Bickerton, J W Connor, and J.B. Taylor *Nature physical science*, **229** 110-112 (1971)

[2] M. A. Malkov, and P. H. Diamond *Phys. Plasmas*, **15** 12, 122301 (2008)

[3] John Wesson, Tokamaks, New York, USA: Clarendon Press 2004

[4] David Hannum, Glenn Bateman, Jon Kinsey *et al. Phys. Plasmas*, **8** 3, 964-974 (2001)

[5] Thawatchai Onjun, Glenn Bateman, Arnold H. Kritz *et al. Phys. Plasmas*, **8** 3, 975-985 (2001)

# Plasma Response in Tokamaks by Helical Current Sheet Application

A. C. Fraile Jr.[1,2], M. Roberto[1], I. L. Caldas[3]

[1] *Department of Physics,* Aeronautic Institute of Technology, *São José dos Campos, Brazil*

[2] *Aerothermodynamics and Hypersonic Division,* Advanced Studies Institute, *São José dos Campos, Brazil*

[3] *Institute of Physics, São Paulo Univesity, São Paulo, Brazil*

**ABSTRACT**

Resonant magnetic perturbations (RMPs) have been applied in tokamaks to mitigate or even suppress plasma edge localized modes (ELMs) [1]. For some RMPs, it is necessary to considere the plasma response to accurately calculate the perturbed magnetic field inside the plasma. Several models have been used to estimate the plasma response according to references [2-4]. In this work we consider an ergodic magnetic limiter to create a RMP [5], superimposed to large aspect ratio tokamak equilibrium, and a resonant current sheet inside the tokamak to simulate the plasma response. For a given perturbation the plasma response is estimated, as proposed in [3], by choosing the current sheet on the internal surface to produce a null radial magnetic field on that surface. Equilibrium and non-uniform escape of the chaotic field lines are analyzed under the influence of plasma response.

**1 – INTRODUCTION**

Several works have been used RMPs in tokamaks due to beneficial effects to mitigate or even suppress plasma edge localized modes in high confinement (H-mode) plasmas [1-3]. The literature shows the plasma reacts to RMPs modifying the magnetic field line transport [1].

**2 – RESONANT MAGNETIC PERTURBATIONS**

Resonant magnetic perturbations (RMPs) are widely employed in order to modify edge instabilities [5, 6]. In order to provide an analytic treatment to the inclusion of RMPs, the cylindrical coordinate system is used in this work, with contravariant components $(r,\theta,z)$ and covariant base $(\vec{e}_r,\vec{e}_\theta,\vec{e}_z)$ expressed as a function of Cartesian coordinates $(x,y,z)$ as described in ref. [5]. All equations in this work are derived considering an infinite cylinder with periodic length equal to $2\pi R_0$ along its symmetry axis.

Since the RMPs are originated from helical wires and helical current sheets, it is possible to define the following vector, which expresses the direction of a helix:

$$\vec{e}_h = \frac{\alpha}{\sqrt{b^2\alpha^2+1}}\vec{e}_\theta + \frac{1}{\sqrt{b^2\alpha^2+1}}\vec{e}_z \tag{1}$$

where $b$ is the cylinder radius and $\alpha = d\theta/dz$.

In order to analyze RMPs generated along helices, it is defined the parameter:

$$u_h = m_0\theta - n_0\left(\frac{z}{R_0}\right) \tag{2}$$

which is constant along a helix and defines its winding law, with mode numbers $m_0$ and $n_0$.

Considering a RMP that is resonant to the surface $r = r_0$ and its electrical current is located at the surface $r = r_j$, the perturbed magnetic field ($\vec{B}$) is calculated from Maxwell equations and it is associated to a scalar field:

$$\phi(r,\theta,z) = \begin{cases} \sum\limits_{k_z=-\infty}^{\infty}\sum\limits_{k_\theta=-\infty}^{\infty} C^i_{k_z,k_\theta} I_{k_\theta}(k_z\alpha r)e^{i(k_\theta\theta-k_z\alpha z)}, & \text{if } r \le r_j \\ \sum\limits_{k_z=-\infty}^{\infty}\sum\limits_{k_\theta=-\infty}^{\infty} C^e_{k_z,k_\theta} K_{k_\theta}(k_z\alpha r)e^{i(k_\theta\theta-k_z\alpha z)}, & \text{if } r > r_j \end{cases} \tag{3}$$

where $I_{k_\theta}(k_z\alpha r)$ and $K_{k_\theta}(k_z\alpha r)$ are, respectively, the modified Bessel function of the first and second kind with order $k_\theta$ and argument $k_z\alpha r$. The constants $C^i_{k_z,k_\theta}$ and $C^e_{k_z,k_\theta}$ must be calculated from the boundary condition expressing the magnetic field discontinuity along the surface $r = r_j$:

$$\Delta\vec{B}\Big|_i^e = \frac{\mu_0 j}{\sqrt{b^2\alpha^2+1}}\left(-b\alpha\vec{e}^z + b\vec{e}^\theta\right) \tag{4}$$

where $j$ is the surface current at $r = r_j$.

## 3 – MAGNETOHYDRODYNAMIC EQUILIBRIUM

The magnetic field associated to the magnetohydrodynamic (MHD) equilibrium can be obtained from the Grad-Shafranov equation [5], which provides the following covariant components:

$$\begin{cases} B_0^r = 0 \\ B_0^\theta = -\dfrac{\mu_0 I_p}{2r^2\pi}\left[1-\left(1-\dfrac{r^2}{a^2}\right)^{\gamma+1}\right] \\ B_0^z \cong -\dfrac{\mu_0 I_e}{2\pi R_0} \end{cases} \tag{5}$$

where $I_p$ is the plasma current, $I_e$ is the poloidal current and $a$ is the plasma column radius.

Table 1 shows numerical values for variables used in this work, corresponding to the geometry and the operation of the TCABR tokamak [7].

Tab. 1. Parameters values used to perform calculations.

| Parameter | Value |
|---|---|
| Plasma current: $I_p$ (A) | $7.0\times10^4$ |
| Poloidal current: $I_e$ (A) | $4.0\times10^6$ |
| Plasma column radius: $a$ (m) | 0.18 |
| Cylinder radius: $b$ (m) | 0.22 |
| Constant $\gamma$ | 3.0 |
| Characteristic length: $R_0$ (m) | 0.61 |

The safety factor ($q$) along the plasma cross section is shown in Fig 1.

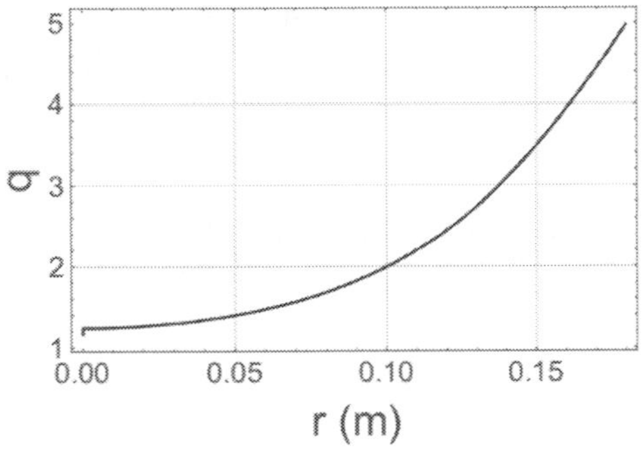


Fig. 1. Safety factor along the plasma cross section.

## 4 – HELICAL WIRES PERTURBATION

Two helical wires conducting electrical current $I_h$ in opposed senses are placed at the surface $r_j = b$ with the winding laws $0 = m_0\theta - n_0(z/R_0)$ and $\pi = m_0\theta - n_0(z/R_0)$. The resonant surface is located at $r_0 = 0.161$ m, corresponding to $m_0 = 4$ and $n_0 = 1$. Figure 2 shows Poincaré sections for $I_h = 0.1\% I_p$ and $I_h = 1.0\% I_p$ .

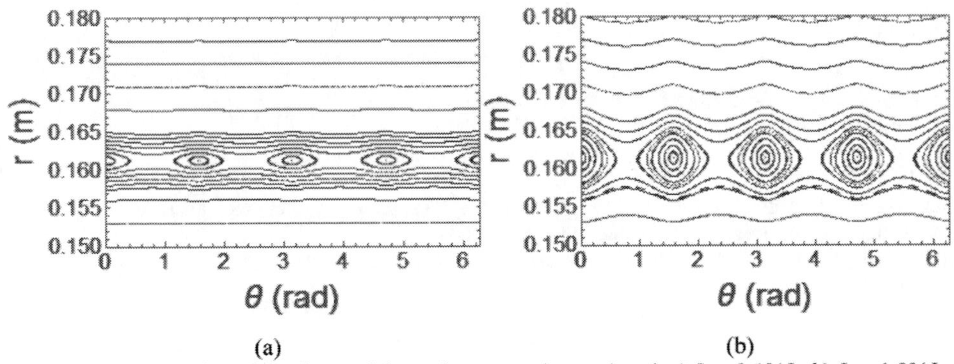


Fig. 2. Poincaré sections for modal numbers $m_0 = 4$, $n_0 = 1$ and: a) $I_h = 0.1\% I_p$; b) $I_h = 1.0\% I_p$.

## 5 – PLASMA RESPONSE

The effect of plasma response can be seen as magnifying or attenuating magnetic islands in Poincaré plots. In order to mimic the situation where the size of magnetic islands is reduced as a consequence of plasma response, it is imposed that the component of perturbed magnetic field orthogonal to the surface where a helical current sheet is located must be zero [3]:

$$\vec{B}\cdot\nabla r = 0 \tag{6}$$

where $\vec{B}$ includes the RMP generated by both the helical windings and the current sheet.

Figure 3 shows the effect of including the plasma response model to the Poincaré sections.

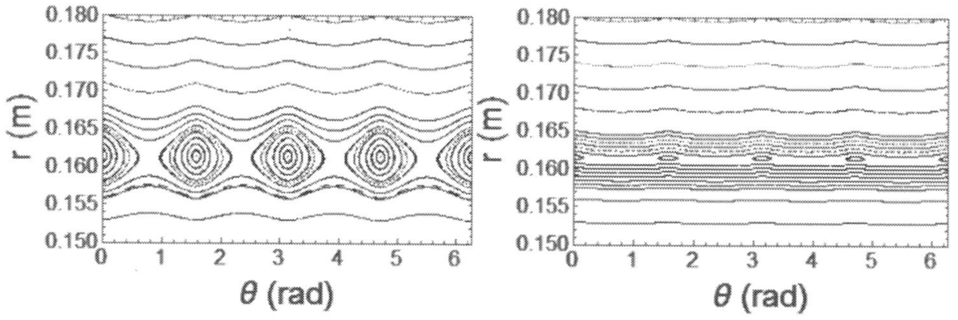


Fig. 3. Poincaré sections for modal numbers $m_0 = 4$ and $n_0 = 1$ and $I_h = 1.0\% I_p$: a) without plasma response model; b) with plasma response model.

## 6 – RESULTS AND CONCLUSIONS

Although the plasma response model presented in this work has been proposed to cylindrical plasma, it may be applied to any geometry. The next step to improve the model corresponds to the inclusion of a toroidal geometry through a polar toroidal coordinate system, which is capable of simulation Shafranov shift [5, 6].

## REFERENCES


[1] H. Frerichs, et al., *Physics of Plasmas* **19**, 052507 (2012)
[2] R. Fitzpatrick, *Physics of Plasmas* **17**, 112501 (2010)
[3] P. Cahyna, E. Nardon, *Journal of Nuclear Materials* **415** (2011)
[4] M. Becoulet, et. al., *Nuclear Fusion* **52**, 054003 (2012)
[5] E.C. Silva, I.L. Caldas, R.L. Viana, *IEEE Trans. on Plasma Science*, **29**, 4 (2001)
[6] M. Roberto, et al., *Physics of Plasmas* **11**, 214 (2004)
[7] E.C. Silva, I.L. Caldas, R.L. Viana, *Brazilian Journal of Physics* **32**, 39 (2002)

# Development of tokamak plasma vertical control system with neutral point taking into account

R.R. Khayrutdinov, M.L. Dubrov, V.E. Lukash, M.M. Sokolov, V.A. Khayrutdinova, V.N. Dokuka

*NRC Kurchatov Institute, Moscow, Russia*

## 1. Introduction

Since typical plasma in most of the existing tokamaks and future tokamak reactors, such as ITER, has a vertically elongated cross section, it is subject to vertical instability. Due to the size and therefore cost of such tokamaks, there will naturally be smaller margins in the poloidal field (PF) coil power supplies to provide the plasma stabilization by vertical feedback control system [1]. On the other hand, it is known, that the vertically elongated plasma has a neutral point (NP) in vertical direction [2, 3], above which the open loop plasma moves upward and correspondently the plasma moves downward if the plasma is located below NP.

Presented paper describes the design of linear models of T-15 tokamak vertically elongated plasma [4] (Fig. 1) together with development of vertical plasma control system. Here the connected in series fast poloidal field coils $10_{up}$ and $10_{dw}$ are used for plasma vertical control in present study. This system is optimized on the plasma position $Z_p$ relatively NP location. Evolution of plasma is simulated with non linear tokamak plasma model implemented in DINA code [5].

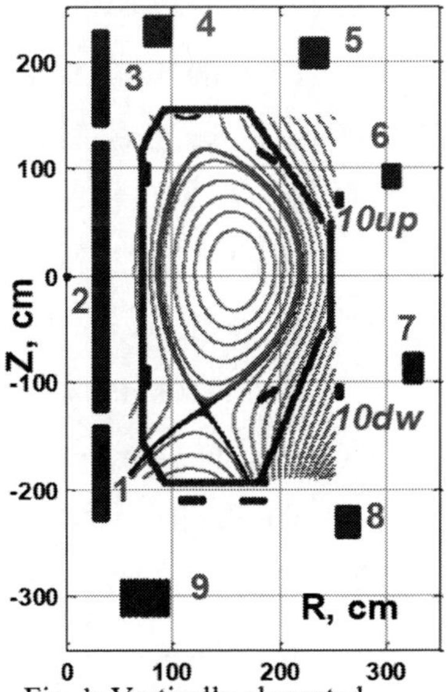

Fig. 1: Vertically elongated plasma in T-15

## 2. Linear plasma model

To develop control system to stabilize the plasma vertical position the linear model in following state space form

$$\begin{cases} dx(t)/dt = Ax(t) + Bu(t) \\ y(t) = Cx(t) + Du(t) \end{cases} \quad (1)$$

was created. Linear models have been obtained with use of DINA code by two methods: 1) direct linearization of non-linear plasma model at specified time moment [6] and 2) identification of numerical experiments on plasma movement in vertical direction [7]. Here $A$, $B$, $C$, $D$ are the matrixes of the model, $y$ is the output vertical position displacement $\delta Z$, the variable $u$ is the output voltage in the tenth PF coil. The vector $x$ is the state variable, which for T-15 plasma in case of direct linearization of non-linear model includes the variations of currents $\delta I$ in both active (10 pieces) and passive (54 pieces) coils. In canonical form the equation system (1) can be presented as

$$\begin{cases} d\delta I(t)/dt = A\delta I(t) + Bu(t) \\ y(t) = C\delta I(t) + Du(t) \end{cases} , \quad (2)$$

where $A=-L^{-1}R$, $B=-L^{-1}$, $D$=0. Inductivity matrix $L$, resistivity vector $R$ and matrix $C$ are calculated during linearization.

Identification case of linear model was calculated by applying of nonzero signal in one input together with zero signals in other inputs. Taking corresponding output signals the transfer functions between applied input and the all outputs are created. Transfer matrix, which is connected with $A$, $B$, $C$ and $D$ matrixes in (1) as $H(s)=C(sI-A)^{-1}B+D$ (here $s$ is the complex variable, $I$ is the identity matrix) is calculated by repeating of such routine for the every input. Then the inverse problem is solved to reconstruct the state space form (1) for identification case of linear model.

To choose the linear model for design of vertical position controller the comparison simulations of plasma evolution during several vertical displacement events (VDE) were carried out with use of three cases of linear simulation models and the non-linear plasma simulation model. Such VDEs were triggered by disturbance of voltage in one PF coil shown in Fig. 1 while the voltages in other coils were equal zero. Time averaged relative difference between $Z$ coordinate of plasma axis simulated with linear and identification plasma models for each case of PF coil disturbance are shown in Fig. 2 where $N_{coil}$ is a number of active coil. Here the three linear plasma model simulation cases are considered: 1) direct linearization of non-linear model with assumption of zero plasma resistance $R_p$, 2) direct linearization of non-linear model with assumption of T-15 plasma resistance equal $R_p$= 2.3280 mΩ and 3) identification case of non-linear model. One can see firstly that the including of plasma resistance in linear model decreases an error in response of non-linear plasma simulation and secondly the modeling results of plasma movement during VDE with use

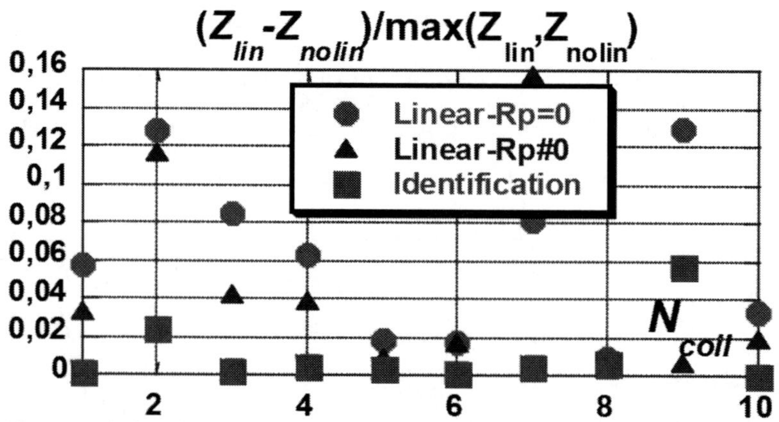


Fig. 2: Relative difference between Z coordinate of plasma axis simulated with linear and non-linear plasma models for each case of PF coil disturbance

of identification case is much closer to the non-linear modeling ones. Thus the plasma vertical controller for T-15 plasma movement was designed on the base of identification case of non-linear model.

**3. Design of vertical plasma position control system**

Design of T-15 plasma position control system is based on LQ-controller [8], in which the control signals are calculated as $u = -Kx$, where matrix $K$ is defined from minimizing of functional $I = \int_0^{+\infty} \left(x^T Qx + u^T Ru + 2x^T Nu\right)dt$. Here $Q$, $R$ are the diagonal weighting matrixes, matrix $N$ is selected to provide the acceptable control time and magnitude of the contrable impact. The ratio between the elements of R matrix determines the distribution of control in the coils. The relationship between the elements of Q matrix determines the difference between the deviations of the corresponding elements of the vector $x$ from zero.

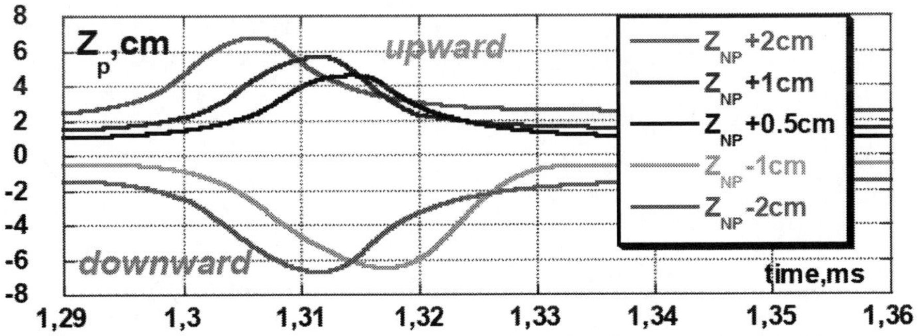


Fig. 3: Several upward and downward scenarios with plasma vertical position stabilization in T-15

**4. Simulation results**

Described in previous section plasma vertical position controller was used to study the T-15 plasma stabilization during both upward and downward VDEs with different set points of $Z$ plasma position relative to $Z_{NP}$, which is ≈ 0.5 cm for T-15 plasma. The five plasma stabilized scenarios in vertical direction were considered with following set points of $Z$: $Z_{sp1} = Z_{NP}+2$ cm, $Z_{sp2} = Z_{NP}+1$ cm, $Z_{sp3} = Z_{NP}+0.5$ cm, $Z_{sp4} = Z_{NP}-1$ cm, $Z_{sp5} = Z_{NP}$ -2 cm. The initial $Z$ positions of plasma in the beginning of these scenarios are chosen to be equal correspondently $Z_{p01}$ = 2.5 cm, $Z_{p02}$ = 1.5 cm, $Z_{p03}$ = 1.1 cm, $Z_{p04}$ = -0.5 cm, $Z_{p05}$ = -1.5 cm (see Fig. 3). The time moment $t_c$ when the plasma vertical position stabilization is

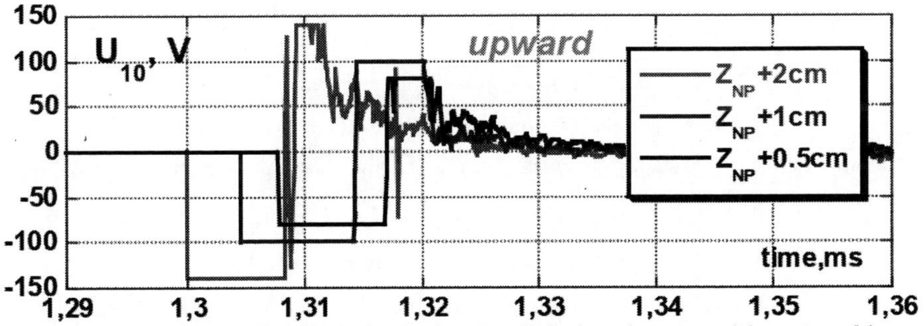


Fig. 4: Waveforms of voltages in 10-th P F coil during plasma position control in considered upward scenarios

switched on is defined when during VDE the value of $abs(Z_p(t_c)-Z_{p0}) \geq 2.5$ cm. At that time moment the voltage on 10-th PF coil is applying. The level of voltage $U_{max}$ is chosen to have the minimum its value to change the direction of uncontrolled vertical movement. In Fig. 4 the waveforms of $U_{10}$ are presented for all considered upward plasma movement VDE scenarios. One can see that the closer plasma is located to NP the smaller $|U_{max}|$ value is necessary to suppress the plasma vertical velocity. Such effect seems to be more effective in the case of plasma vertical movement towards X-point location (see Fig. 5). That means that the power supply of plasma vertical control system is lower if the controllable plasma is located closer to NP.

## 5. Conclusions

Simulation results of sensitivity study of plasma position control in T-15 have shown that the maximum value of voltage in control poloidal fields coils is decreasing if the plasma at the start of VDE locates closer to the neutral point. In that case there is possibility to decrease the power supply of plasma vertical control system. Such effect is more effective in the case of plasma vertical movement towards X-point location. This study is carried out with vertical control system based on T-15 plasma linear model designed with identification method of linearization obtained with use of DINA code.

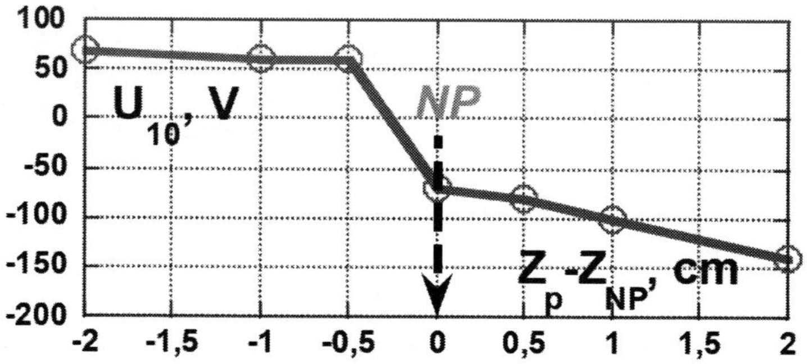

Fig. 5: Maximum value of voltage in 10-th PF coil depending on initial plasma position corresponding NP location

**Acknowledgments.** This work was supported by the Funds № 14-07-00483-а и № 14-07-00912-a of Russian Foundation for Basic Research.

### References

[1] J-Y Favez et al. Plasma Phys. Control. Fusion **47** (2005) 1709

[2] Y.Nakamura et al. // Nuclear Fusion 36 (1996) 643

[3] V.E. Lukash et al. // Plasma Phys. Control. Fusion 47 (2005) 2077-2086

[4] Azizov E.A. et al. Status of project of engineering-physical tokamak. — In: 23rd IAEA Fusion Energy Conf. Daejeon, Republic of Korea, 11—16 October 2010, FTP/P6-01

[5] R.R. Khayrutdinov and V.E. Lukash. J. of Comp. Physics, **109**, No. 2 (1993) 193-201

[6] Belyakov V., Kavin A. – In: Proc. International Conference «Physics and Control», St.-Petersburg, August 2003, p. 1019-1024

[7] Ljung L. System Identification – Theory for the User. New Jersey; Prentice Hall, Upper Saddle River. 2nd edition, 1999

[8] Belyakov, V. et al. Fusion Engineering and Design, Vol. 45, pp. 55-64, 1999

# Computation of Dust Particle Evolution in Divertor Plasma

N.Kh. Bastykova[1)], S.K. Kodanova[1)], T.S. Ramazanov[1)], S.A. Maiorov[2,3)]

[1] *Institute of Experimental and Theoretical Physics, Al-Farabi Kazakh National University, Almaty, Kazakhstan*

[2]*Prokhorov General Physics Institute, Russian Academy of Sciences, Moscow, Russia*

[3]*Joint Institute for High Temperatures, Russian Academy of Sciences, Moscow, Russia*

**Introduction.** Investigation of formation and evolution of dust particles in controlled fusion devices has become an important part of large-scale fusion plasma experiments [1-4]. The flux of hot particles from the central region to the walls of the reactor can cause destruction and evaporation of the wall surface, as a result of which solid particles of various shapes - from irregular forms to almost perfect spheres – can be injected into the plasma. These particles consist of materials used for divertor plates, the first inner wall and other structural elements, namely, graphite, titanium, tungsten, beryllium and steel.

As the speed of electron is much greater than that of ions the dust particles usually acquire a negative charge in the plasma. The mean value of the charge and its fluctuations are mainly determined by the ratio of masses of ions and electrons, their temperature and size of dust particles. A more detailed description of the charging process without a magnetic field may be found in [2, 5-7].

The magnetic field can also have a significant impact on the process of dust particle charging in the plasma. In order to study the impact of dust on the operation of fusion reactors it is necessary to take into account the influence of a strong magnetic field on the process of dust particle charging. In this paper, ion collisions with atoms were calculated using the Monte Carlo method and the dependence of the dust particle charge and fluxes of plasma particles on the dust particle on the magnetic field were obtained.

Migration of dust particles from the surface of the wall into the depth of the reactor and their evaporation largely determine the composition and characteristics of the edge plasma, and can therefore have a significant impact on the operation of the reactor. To take into account migration of dust particles in the reactor it is necessary to solve the equations of dust particle motion in the given electric and magnetic fields and equations of mass and energy balance.

**Dust particle charging.** At the first stage of this research the charge of a fixed, initially neutral dust particle with an infinite mass was calculated. The dust particle charge was calculated using the particle-in-cell method, and the number of collisions between ions and atoms was determined by the Monte Carlo method [5-6, 10-11].

The charge of dust particles was calculated for the following parameters of the divertor plasma [12]: the density of electrons and ions equal to $10^{14}$ $cm^{-3}$, the temperature of ions – $0.7$ $eV$ and electrons – $3$ $eV$. The charges of dust particles with radii of $0.5;1$ and $2\mu m$ for various values of the magnetic field $B \div (10\text{-}10^5)G$ were obtained. Fig. 1 shows the time dependence of the charge of the dust particle with a radius of $0,5$ $\mu m$, initially neutral, obtained using the particles-in-cell method and calculated according to the OML theory. Figure 2 shows the distribution of the dust particle charge as a function of time for different values of the magnetic field. In a plasma without a magnetic field most electrons reflect from the Coulomb barrier of the dust particle, and only a small part of fast electrons can reach the dust particle. On the contrary, ions are attracted by the dust particle and the collision cross-section becomes much larger than the geometrical cross section of the dust particle $\pi a^2$.

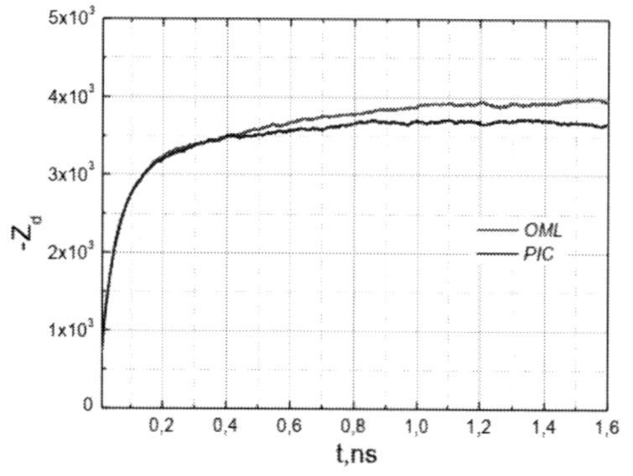


Fig. 1 – The distribution of the dust particle charge as a function of time

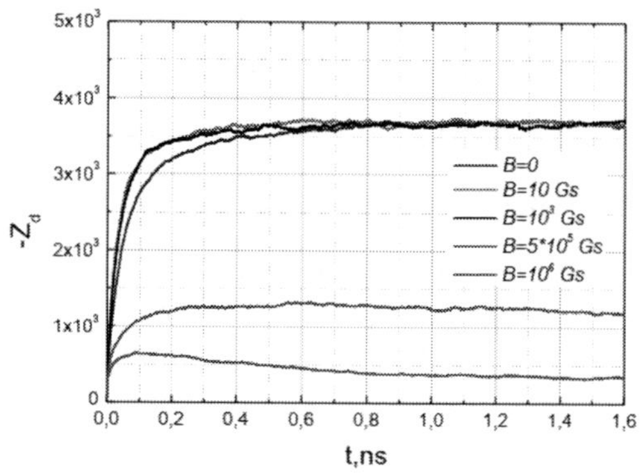


Рис. 2 – The distribution of the dust particle charge as a function of time (ns) for different values of the magnetic field

It was found out in [8] that the impact on the charge of the dust particle begins at a certain critical value of the magnetic field $B^e_{cr}(kG)a(\mu m) > 41.37\sqrt{T_e(eV)/3(eV)}$. In this paper it was found out that for the parameters of the divertor plasma and the dust particle with a radius of $a = 0.5\mu m$ the strength of the critical magnetic field is equal to $B^e_{cr} = 8.5 \cdot 10^4 G$. The magnetic field affects charging of the dust particle mainly due to the magnetization of electrons. In a weak magnetic field $B < B^e_{cr}$, when the gyro-radius of the electron is greater than the size of the dust particle, the influence of the magnetic field on the charge of the dust particle is very small. As the strength of the magnetic field increases to the values where the gyro-radius of electrons is equal to the radius of electron capture by the dust particle the picture changes. The electrons move along the magnetic field lines and can reach the surface of the dust particle only if the magnetic field line intersects it. However, the electrons with a low energy, as in the case without a magnetic field, are reflected from the Coulomb barrier in the reverse direction.

**Energy balance for a dust particle in the plasma.** In the plasma, the dust particle is subjected to the action of the flows of ions, electrons, atoms and radiation. The kinetic energies of incident electrons and ions, and the energy of their recombination on the surface of the dust particle contribute to the heating of the dust particle. In addition, on the surface of the dust particle there are such processes as heat emission during deposition, exothermic reactions, recombination of dissociated molecules, etc. (see a more detailed description of the model in [3, 4]).

The model of this research takes into account heating of the dust particle by the heat fluxes of electrons and ions, their recombination on its surface and cooling by radiation and evaporation. Every event of ion absorption on the dust particle leads to formation of an atom and the energy released in this case, is equal to the ionization potential of the gas atom. The loss of energy in the form of radiation is taken into account in accordance with the law of black body radiation, with the corrections of the Mie theory. As the temperature of the dust particle rises, its matter may undergo phase transitions. In our calculations, it first passes from the solid to the liquid state, and then, when its temperature reaches the boiling point, it rapidly loses its mass through evaporation. During the phase transition the temperature is assumed to remain constant.

The heating process of initially cold dust particles in a homogeneous deuterium plasma near the wall of the divertor was calculated. The following parameters were chosen for calculations [12], $T_e = 3.0eV, T_i = 0.7eV$, $T_a = 0.2eV$, $n_e = n_i = n_a = 2\times10^{14} cm^{-3}$, the initial temperature of dust particles $T_{d0} = 1000K$, radius $R_{d0} = 0.5\mu m$.

Fig. 3 shows the time dependence of the temperature and radius of the dust particle. In During the first 20 μs the dust particle temperature rises to $3695K$, then in the time interval from 20 μs and 100 μs the temperature does not change, as the incoming energy is spent on the process of melting of the dust particle. After the transition to the liquid state the temperature starts to rise quickly up to the boiling point of the dust particle at $T_{vaporation} = 4850K$. In the process of boiling the dust particle quickly loses its mass, and in 3000 μs its radius decreases to zero.

Fig. 4 shows the dependence of the lifetime of dust particles in a deuterium plasma on its density for different values of the electron temperature. It is seen that as the temperature of electrons increases, the lifetime of dust particles decreases, and the analysis shows that the main reason for this is an increase in the charge of dust particles.

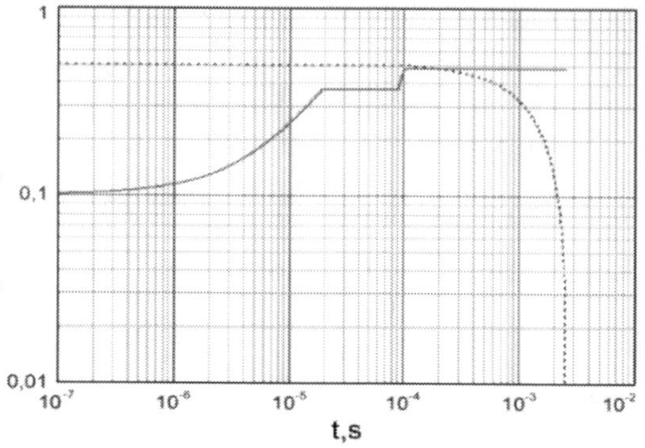


Fig. 3 –Time dependence of the dust particle temperature (solid line) and radius (dot line) during heating

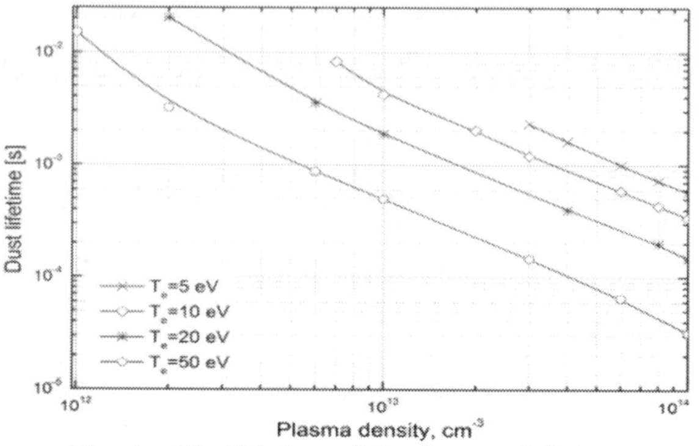


Fig. 4 – The lifetime of the dust particle in a homogeneous deuterium plasma

## Conclusions

The influence of the magnetic field on the charge, dynamics and lifetime of dust particles in the edge divertor plasma is investigated. A computational model based on the particle-in-cell and Monte Carlo methods was constructed and the time dependence of the dust particle charge and plasma fluxes on its surface was calculated. It has been shown that a strong magnetic field has a significant impact on the process of charging of dust particles in the divertor plasma. The numerical model describing heating and evaporation of dust particles in the plasma has been developed, and calculations that allow us to estimate the lifetime of the dust particle and its path in the edge divertor plasma have been made.

## References

[1]. V.N. Tsytovich and J. Winter, 1998 Phys.-Usp. **41,** 815.

[2]. V.E. Fortov, A.G. Khrapak, S.A. Khrapak, V.I. Molotkov, and O.F. Petrov, 2004 Phys. Usp. **47**, 447.

[3]. A. Yu. Pigarov, S. I. Krasheninnikov and et al., 2005 Physics of Plasmas **12**, 122508.

[4]. L. Vignitchouk, P. Tolias and S. Ratynskaia, 2014 Plasma Phys. Control. Fusion **56,** 095005.

[5]. S.A. Maiorov, S.V. Vladimirov, and N.F. Cramer, 2002 Plasma Physics Reports **28,** 946.

[6]. S.A. Maiorov, 2005 Plasma Physics Reports **31,** 749.

[7]. T.S. Ramazanov, S.K. Kodanova et al., 2011 Eur. Phys. Letters, **96,** 45004.

[8]. V. N. Tsytovich, N. Sato, G.E. Morfill, 2003 New Journal of Phys. **5**, 43.1-43.9.

[9]. Y. Tomita, G. Kawamura, T. Yamada, O. Ishihara, 2009 J. Plasma Fusion Ser. **8**, 273-276.

[10]. S.A. Maiorov, 2006 Plasma Physics Reports **32,** 802.

[11]. S.A. Maiorov, B.A. Klumov, 2013 Bullletin of the Lebedev Phys. Inst. **40,** 285.

[12]. F.Taccogna, R. Schneider, K. Matyash, S.Longo, M. Capitelli, D. Tskhakaya 2008 Contrib. Plasma Phys. **48**, 147-152.

# Effect of 3D Ferromagnetic Materials on Plasma Nonlinear Evolution in Fusion Devices

L. Barbato[1], S. Mastrostefano[1], G. Rubinacci[2], S. Ventre[1], F. Villone[1]

[1]*CREATE, DIEI, Università di Cassino e del Lazio Meridionale, 03043, Cassino, Italy*
[2]*CREATE, DIETI, Università Federico II di Napoli, Via Claudio 21, 80125, Napoli, Italy*

## 1. Introduction

Future magnetic confinement fusion devices, like ITER [1], have such high performances to require a special care in the dimensioning of various components, posing several challenges both from the physical and from the engineering point of view. The electromagnetic interaction of the plasma with the surrounding structures is fundamental for the dimensioning of many critical components (magnets, vacuum vessel, plasma facing components, etc.). This happens both during routine operation (e.g. magnetic control [2]) and in off-normal events (e.g. disruptions [3]).

One of the inherent difficulties in the modelling of such devices is related to the geometrical complexity of the metallic parts surrounding the plasma; another is due to the fact that both non-magnetic conductors and ferromagnetic materials may be present. Indeed, in next-generation devices, where a very high neutron fluence is expected, special reduced activation ferritic martensitic steels will be required for the shielding of the neutrons produced by fusion reactions. Also other applications of ferromagnetic materials in fusion devices can be envisaged, e.g. for the transformer core (as in JET), but also for the so-called ferromagnetic inserts to improve the toroidal field (TF) ripple.

Consequently, reliable computational tools are needed, able to describe such electromagnetic interaction in presence of complex ferromagnetic structures. In particular, recently [4] a novel computational tool has been presented, called CarMa0NL, with the unique capability of simultaneously considering three-dimensional effects of conductors surrounding the plasma and the inherent non-linearity of the plasma behaviour itself. This work reports an extension of this computational tool, namely regarding the inclusion of ferromagnetic materials (previously not considered) into the formulation.

## 2. Mathematical and numerical formulation

The CarMa0NL formulation, described in details in [4] if no magnetic materials are present, introduces a coupling surface $S$ in order to couple the nonlinear evolution of an axisimmetric plasma through equilibrium points, with 3D magneto-quasi-static equations valid in the conductors surrounding the plasma itself. Inside $S$, Grad-Shafranov nonlinear equations are solved in terms of the magnetic flux per radian $\underline{\psi}$ at the nodes of a triangular finite elements mesh, assuming as boundary condition the external flux on $S$, $\underline{\hat{\psi}}_e$ :

$$\underline{F}\left(\underline{\psi},\underline{\hat{\psi}}_e\right)=0 \; inside \; S \tag{1}$$

Resorting to the numerical formulation described in [4, 5], in the conductors $V_c$, discretized with a volumetric finite elements hexahedral mesh, the following equations hold:

$$\underline{\underline{L}}\frac{d\underline{I}}{dt}+\underline{\underline{R}}\,\underline{I}+\frac{d\underline{U}}{dt}+\underline{J}_m=\underline{V} \tag{2}$$

where the following terms are also present in the non-magnetic formulation [4]:

$$L_{i,j}=\frac{\mu_0}{4\pi}\int_{V_c}\int_{V_c}\frac{\nabla\times\mathbf{N}_i(\mathbf{r})\cdot\nabla\times\mathbf{N}_j(\mathbf{r}')}{|\mathbf{r}-\mathbf{r}'|}dV\,dV' \tag{3}$$

$$R_{i,j}=\int_{V_c}\nabla\times\mathbf{N}_i\cdot\eta\cdot\nabla\times\mathbf{N}_j dV \tag{4}$$

$$U_i=\int_{V_c}\nabla\times\mathbf{N}_i\cdot\mathbf{A}_{plasma}\,dV\,, \qquad \underline{U}=\underline{\underline{Y}}\,\underline{I}_{eq} \tag{5}$$

In these equations, $\underline{I}$ is the vector of degrees of freedom of 3D current density, $N_k$ are the edge elements basis functions, $\underline{V}$ is the vector of voltages applied to electrodes and the quantity $\mathbf{A}_{plasma}$ is the magnetic vector potential due to plasma currents, with the Coulomb gauge, so that $\underline{U}$ is the magnetic flux induced in 3D conductors by the plasma, which is represented by a set of suitable equivalent filamentary currents $\underline{I}_{eq}$ located on $S$. The term $\underline{J}_m$ is instead due to magnetization **M** in magnetic materials $V_f$, such that $\mathbf{B} = \mu_0\,(\mathbf{H}+\mathbf{M})$, which is discretized in terms of piecewise constant basis functions [5]:

$$\mathbf{M}(\mathbf{x},t)=\sum_k M_k(t)\,\mathbf{P}_k(\mathbf{x}) \tag{6}$$

$$\underline{J}_m=\underline{\underline{F}}\frac{d\underline{M}}{dt}, \quad F_{i,j}=\frac{\mu_0}{4\pi}\int_{V_c}\int_{V_f}\frac{\nabla\times\mathbf{N}_i(x)\cdot\mathbf{P}_j(x',t)\times(\mathbf{x}-\mathbf{x}')}{|\mathbf{x}-\mathbf{x}'|^3}dV'\,dV \tag{7}$$

The nonlinear characteristic of magnetic materials, $\mathbf{M} = \mathrm{G}(\mathbf{B})$, can be discretized as [5]:

$$\underline{\underline{D}}\,\underline{M}=\underline{G}(\underline{B}),\quad G_i(\underline{B})=\int_{V_f}\mathbf{P}_i(x',t)\cdot\left(\sum_k B_k(t)\,\mathbf{P}_k(\mathbf{x})\right)dV,\quad \underline{B}=\underline{\underline{D}}^{-1}\left(\underline{\underline{E}}\,\underline{M}+\underline{\underline{W}}\,\underline{I}_{eq}+\underline{\underline{F}}^T\underline{I}\right)+\underline{B}_c \tag{8}$$

where $\underline{B}_c$ is the magnetic field due to sources constant in time (e.g. TF coils) and

$$D_{ij} = \int_{V_f} \mathbf{P}_i \cdot \mathbf{P}_j dV, \quad W_{i,j} = \frac{\mu_0}{4\pi} \int_S \int_{V_f} \frac{\mathbf{P}_i(\mathbf{x}) \cdot \mathbf{J}_{eq}(\mathbf{x}',t) \times (\mathbf{x} - \mathbf{x}')}{|\mathbf{x} - \mathbf{x}'|^3} dS' dV$$
$$E_{ij} = \mu_0 \int_{V_f} \mathbf{P}_i \cdot \mathbf{P}_j dV - \frac{\mu_0}{4\pi} \int_{\partial V_{fi}} \int_{\partial V_{fj}} \frac{(\hat{\mathbf{n}} \cdot \mathbf{P}_i(\mathbf{x}'))(\hat{\mathbf{n}}' \cdot \mathbf{P}_j(\mathbf{x}))}{|x - x'|} dS' dS \tag{9}$$

We pose $\underline{M} = \underline{M}_c + \delta\underline{M}$, where $\underline{M}_c$ is the (time constant) magnetization corresponding to $\underline{B}_c$. We suppose that the perturbation around $\underline{M}_c$ is small; a linearization of the magnetic characteristic is performed, defining a local magnetic susceptivity $\chi$ and permeability $\mu$:

$$\delta\underline{M} = \underline{\underline{C}}\left(\underline{\underline{W}}\,\underline{I}_{eq} + \underline{\underline{F}}^T \underline{I}\right), \quad \underline{\underline{C}} = \left(\frac{\mu}{\chi}\underline{\underline{D}} - \underline{\underline{E}}\right)^{-1} \tag{10}$$

Substituting (6)-(10) in (2), we have:

$$\underline{\underline{L}}^* \frac{d\underline{I}}{dt} + \underline{\underline{R}}\,\underline{I} + \underline{\underline{Y}}^* \frac{d\underline{U}}{dt} = \underline{V}, \quad \underline{\underline{L}}^* = \underline{\underline{L}} + \underline{\underline{F}}\,\underline{\underline{C}}\,\underline{\underline{F}}^T, \quad \underline{\underline{Y}}^* = \underline{\underline{Y}} + \underline{\underline{F}}\,\underline{\underline{C}}\,\underline{\underline{W}} \tag{11}$$

which is formally analogous to the model without ferromagnetic materials [4], except for a redefinition of matrices, to take into account the presence of the magnetic materials. Similarly, the external flux on $S$ is modified as:

$$\underline{\psi}_e = \underline{\underline{Q}}\,\underline{I} + \underline{\underline{Q}}_m \underline{M} = \underline{\underline{Q}}\,\underline{I} + \underline{\underline{Q}}_m \underline{M}_c + \underline{\underline{Q}}_m \delta\underline{M} = \underline{\underline{Q}}_m \underline{M}_c + \left(\underline{\underline{Q}} + \underline{\underline{Q}}_m \underline{\underline{C}}\,\underline{\underline{F}}^T\right)\underline{I} + \underline{\underline{Q}}_m \underline{\underline{C}}\,\underline{\underline{W}}\,\underline{I}_{eq} \tag{12}$$

## 3. Results

A simple test case is presented, in which one equilibrium configuration of ITER is perturbed with a fictitious linear magnetic material placed in correspondence of the inner shell of the vessel (Fig. 1). It is also assumed that the ferromagnetic material is axisymmetric, so that, although a fully 3D mesh is given, a 2D code (CREATE_L [6]) can be used as benchmark. Figure 2 reports the $\underline{\underline{L}}^*$ and $\underline{\underline{Y}}^*$ matrices (11) as computed by CarMa0NL and CREATE_L, for 50 axisymmetric conductors representing the outer vessel shell and for 618 filamentary currents on the coupling surface, showing a good agreement. The perturbation of the matrices is concentrated close to the magnetic material, but the equilibrium configuration is affected globally (Fig 1c). The equilibrium solutions computed by CarMa0NL and CREATE_L differ only of a few percents, confirming the ability of CarMa0NL to correctly describe the effects of ferromagnetic materials on plasma equilibrium configurations.

## 4. Conclusions

In this paper we have presented an extension of the CarMa0NL computational tool, in order to account for 3D ferromagnetic conductors in the nonlinear evolutionary equilibrium of axisymmetric plasmas. Simple test cases have been solved, in which a one-to-one comparison with axisymmetric models is possible, showing good agreement. Future activity will be devoted to the analysis of the effect of realistic 3D ferromagnetic structures, like the ferromagnetic inserts. This work was supported in part by Italian MIUR under PRIN grant 2010SPS9B3.

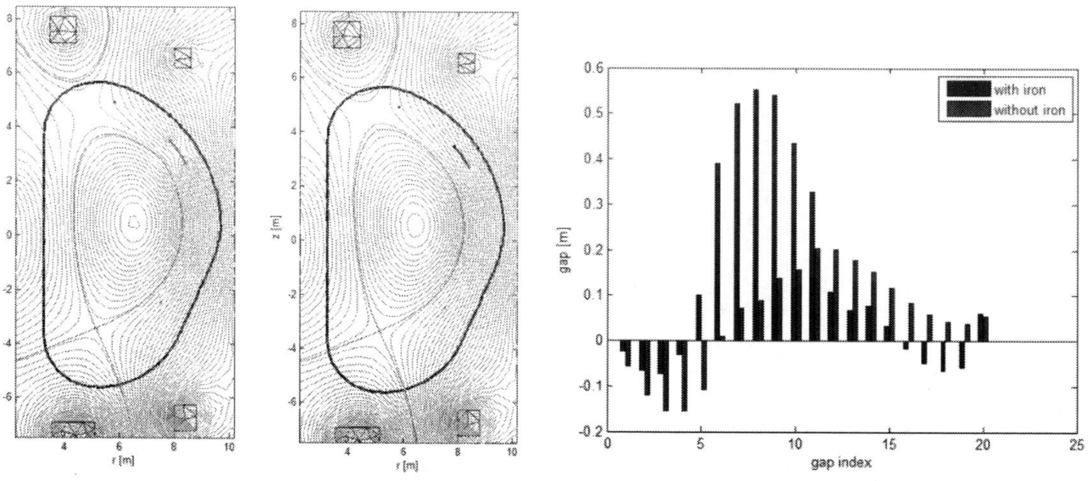


(a) (b) (c)

Fig. 1. Reference equilibrium with (a) and without (b) axisymmetric magnetic material and effect on gaps (c)

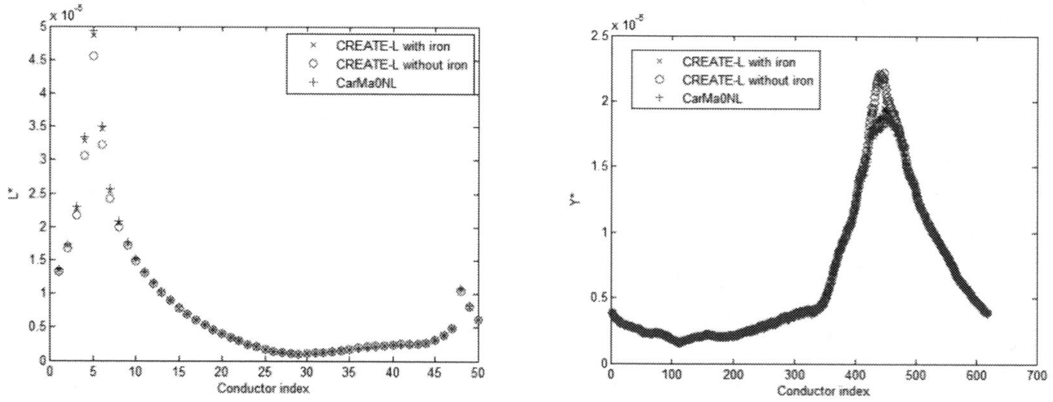


Fig. 2. Comparison of $\underline{\underline{L}}^*$ and $\underline{\underline{Y}}^*$ for test axisymmetric case.

[1] K. Ikeda, *Nucl. Fusion* **50** (2010) 014002

[2] Y. Gribov et al, *Nucl. Fusion* **47** (2007) S385

[3] T. Hender et al, *Nucl. Fusion* **47** (2007) S128

[4] F. Villone et al., *Plasma Phys. Control. Fusion* **55** (2013) 09500

[5] R. Albanese, G. Rubinacci, *Adv. Imaging El. Phys.* **102** (1998) 1-86

[6] R. Albanese and F. Villone, *Nucl. Fusion* **38** (1998) 723

# Design Options for Stellarator Divertors

**Alkesh Punjabi**[1], **Allen Boozer**[2]

[1]**Hampton University, Hampton, VA 23668, USA**

[2]**Columbia University, New York, NY 10227, USA**

ABSTRACT: Divertors are thought to be required to concentrate the plasma particle exhaust region on the walls for effective pumping while avoiding plasma degradation by impurities from the walls. Concentration eases particle handling but makes power handling impossible unless the power exhaust can be spread by electromagnetic radiation or by the geometry of the interception region with the walls. What is possible with divertors is determined by (1) the relative shape of the wall and the plasma and (2) the topological nature of the magnetic field lines near the plasma edge—magnetic surfaces, islands or stochastic regions. The range of possible topologies can be efficiently represented by parameters in magnetic field line maps. The range of shapes of the walls and plasma can be given as parametric functions. Using parameterizations of the topology and shape, the design options for stellarator divertors are being studied along with their sensitivity to the rotational transform and its shear. This work is supported by the US DOE grants DE-FG02-01ER54624 and DE-FG02-04ER54793 to Hampton University and DE-FG02-95ER54333 to Columbia University.

Generally, there are two design options for stellarator divertor designs. The first, dominated by magnetic islands as on W7-X, and second dominated by stochasticity as on LHD [1]. The two designs are based on different principles: islands on the plasma surface for W7-X, called a resonant divertor, and the use of sharp edges on the quasiaxisymmetric stellarator, called the nonresonant divertor. Primary purpose of a divertor is to concentrate the region on the wall in which the plasma particle exhaust occurs so pumps can be installed at this location. Covering the whole wall with pumps is not practical. Concentrating the particle exhaust is advantageous; concentrating the plasma energy exhaust is not. A fifth of the power released by DT fusion is released in alpha particles that heat the plasma and the rest in neutrons, which directly heat the walls. If none of the alpha power were lost from the plasma by electromagnetic radiation roughly 20% of the wall area would need to be devoted to the divertor region in order to handle the power exhaust. This area could be in a volume through which the plasma has access by a slot, which gives what is called a closed divertor. Generally it is assumed that most of the power can be radiated. Radiated power may become sufficiently large that the plasma recombines before it has contact with the surrounding structures. This is called plasma detachment. This gives a detached divertor. Some desirable features are: (1) a significant spatial separation between the region of good plasma containment and the intersection location of divertor field lines on the surrounding structures. This separation makes it more difficult of sputtered particles to enter the plasma volume. (2) A stable location for a slot for field lines to pass into a divertor chamber independent of the plasma state. (3) A controlled spreading of the heat load. The Hamiltonian equations for the field line are $d\psi_t/d\varphi = -\partial\psi_p/\partial\theta$ and $d\theta/d\varphi = \partial\psi_p/\partial\psi_t$. Position vector for field lines is given by $\vec{x}(\psi_t,\theta,\varphi) = R(\psi_t,\theta,\varphi)\hat{R}(\varphi) + Z(\psi_t,\theta,\varphi)\hat{Z}$. The standard position vector is given by $R = R_a(\varphi) + a\sqrt{\psi_t/\psi_a}\cos\theta, Z = Z_a(\varphi) + a\sqrt{\psi_t/\psi_a}\sin\theta$. Here we investigate the resonant divertor design option for stellarators. We use the NCSX parameters [2]. The total Hamiltonian is $\psi_p(\psi_t,\theta,\varphi) = \overline{\psi}_p(\psi_t) + \widetilde{\psi}_p(\theta,\varphi)$. The unperturbed Hamiltonian is $\overline{\psi}_P(\psi_t) = a\psi_t + b\psi_t^2$. The perturbation is $\widetilde{\psi}_P(\theta,\varphi) = \delta[\cos(m_1\theta - n_1\varphi) + \cos(m_2\theta - n_2\varphi)]$. The resonant map equations are $\psi_{j+1} = \psi_j + k\delta[\sin(m_1\theta_j - n_1\varphi_j) + \sin(m_2\theta_j - n_2\varphi_j)]$, and $\theta_{j+1} = \theta_j + k\bar{\iota}(\psi_{j+1})$. Both forward and the backward map are used. Resonant perturbation have modes (2,1)+(5,3) with amplitude = $10^{-3}$. Both modes are locked with no radial dependence. Footprints are calculated using continuous analogs of forward and backward maps. Lengths of lines, Liapunov lengths, and footprints are calculated. Two topologies for collecting surfaces; conformal tori and cylinders are studied. Here the tori are considered. The starting surface is shown in Fig. 4.

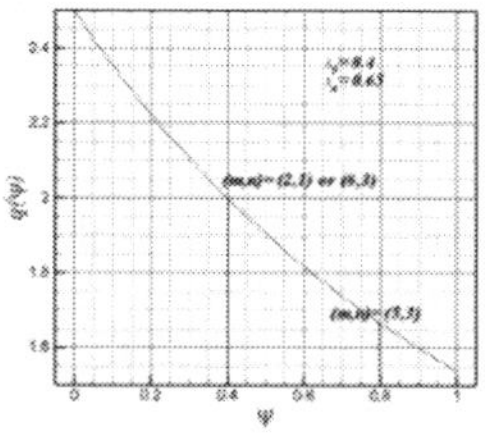


Fig. 1. Equilibrium safety factor for the NCSX.

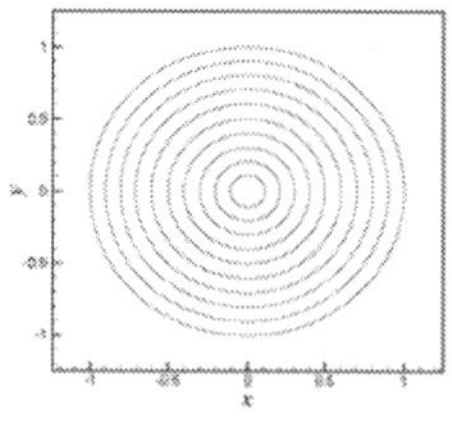


Fig. 2. The equilibrium magnetic surfaces in the NCSX generated by the map.

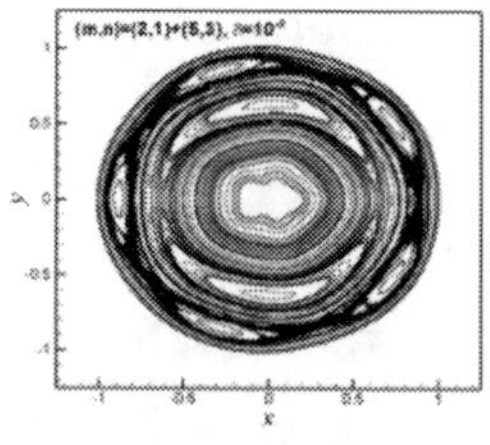


Fig. 3. Phase portrait for the resonant perturbation $(m,n)$=(2,1)+(5,3) with the amplitude $\delta=10^{-3}$ in the $\varphi=0$ plane.

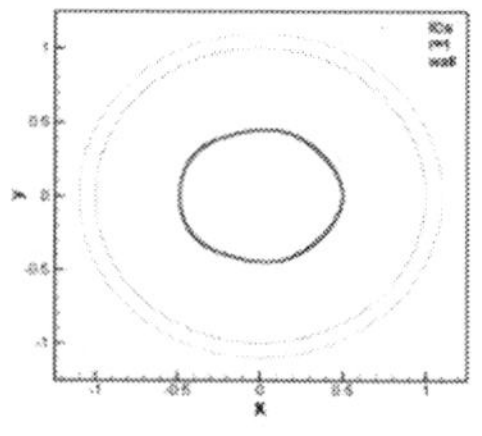


Fig. 4. The $r$=1/2 perturbed surface, the plasma boundary at minor radius $r$=1, and the wall at $r$=1.1 in the $\varphi$=0 plane.

The radially outward diffusion of field lines is implemented by the spiraling operator $r_{j+1} \rightarrow (0.9{+}0.2R_N D/N_p)r_{j+1}$. Phase portrait of starting surface from spiraling are shown in Fig. 5. The half-life, 1/e loss time, and the full confinement time all scale roughly as $1/D$.

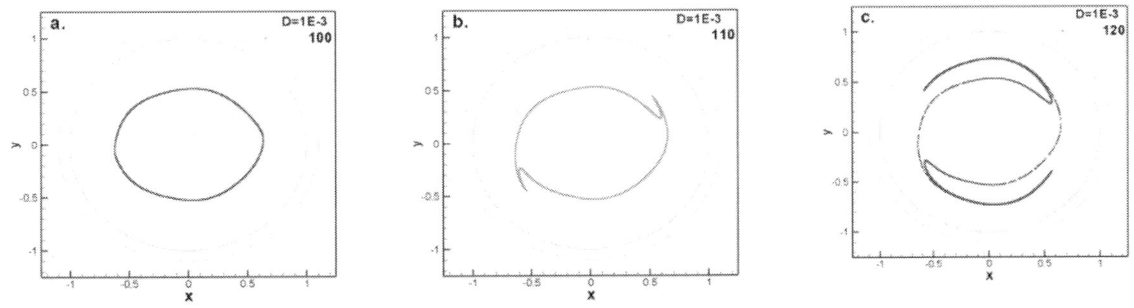


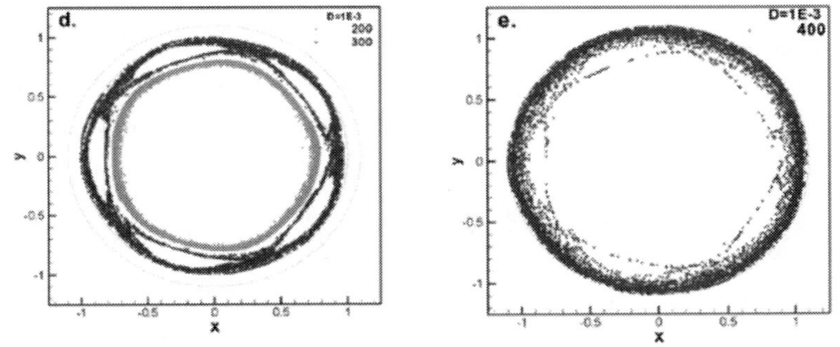


Fig. 5. Puncture plots of field lines in the $\varphi$=0 plane for spiraling coefficient $D$=1E-3 after (a) 100, (b) 110, (c) 120, (d) 200, 300, and (e) 400 circuits.

The important footprints are shown in Fig. 6.

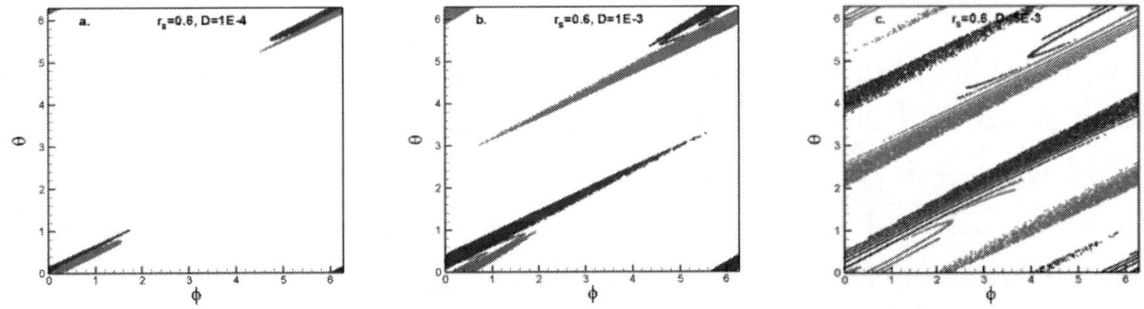

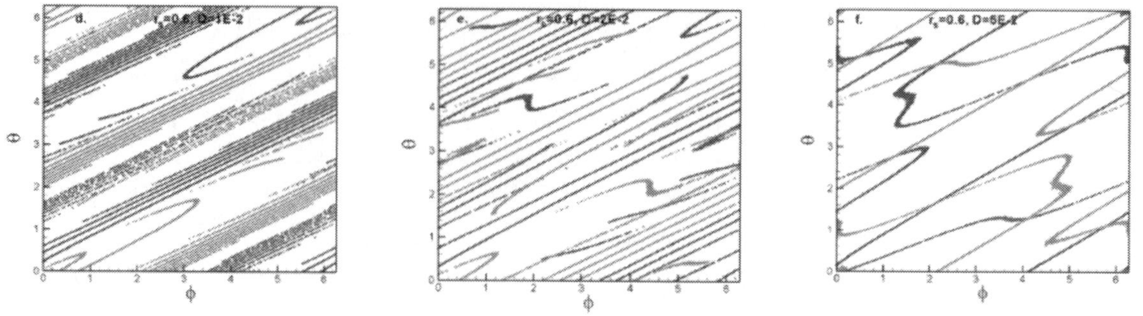


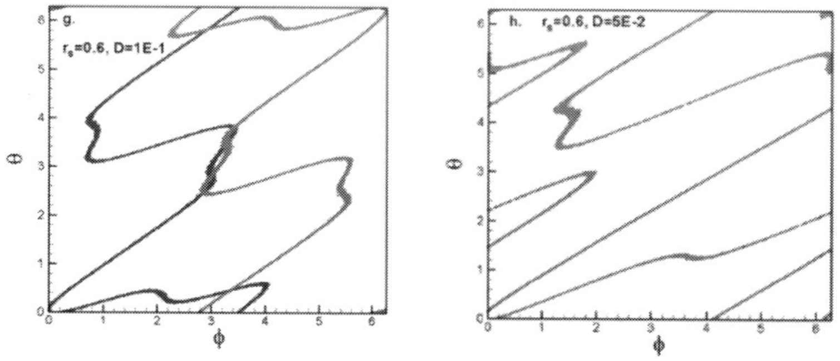


Fig. 6. $r_S$=0.6. (a) $D$=1E-4, (b) $D$ =1E-3, (c) $D$ =5E-3,(e) $D$ =1E-2, (f) $D$ =2E-2, (g) $D$ =1E-2, and (h) $D$ =5E-2 with the backward footprint superimposed on the forward footprint after the transformations $\varphi \to -\varphi + 2\pi$ and $\theta \to -\theta + 2\pi$. Forward footprint are shown in blue, and backward in green.

Generally the normalized area is larger for intercepting torus of larger radius. As $D$ increase from 1E-5 to about 1E-2, the area increases. The rise in the area can be roughly as much as an order of magnitude. The area reaches its maximum roughly when $D$=1E-2. On further increase in $D$, the area decreases. The average length of field lines $<l>$ scales roughly as $1/D$. The average Liapunov length $<l_L>$ of the lines varies from about 50 to 5000. The slowest variation in $<l_L>$ with $D$ is for $r_S$=1, and the fastest variation with $D$ is for $r_S$=0.8. $<l_L>$ scales very roughly as $1/D^{1/2}$. For innermost collecting surfaces $r_S$=0.6, 0.7, and 0.8 the maximum e-folds do not exceed roughly 10. But for outer intercepting tori, the jump in maximum e-folds is quite large; about 24 for $r_S$=0.9 and about 38 for $r_S$=1 and 1.1. For small enough $D$, for these outer surfaces, e-folds can exceed 20. Large e-folds spread the heat flux on intersecting surfaces [4]. Simple models can give important results [3].

[1] Y. Feng *et al, Plasma Phys. Control. Fusion* **53**, 024009 (2011).

[2] N. Pomphrey *et al, Fusion Science and Technology* **51**, 181 (2007).

[3] A. H. Boozer, *Nuclear Fusion* **55**, 025001 (2015).

[4] A. H. Boozer, *Plasma Physics and Controlled Fusion* **52**, 104001 (2010).

# Magnetic Shear and Stability of Golden Magnetic Barrier in Divertor Tokamak

Halima Ali[1] and Alkesh Punjabi[1]

[1]*Hampton University, Hampton, VA, 23668, USA*

ABSTRACT: In Hamiltonian chaos the last torus to be destroyed is the golden torus [1]. It has the most irrational winding number called the golden mean. An external perturbation can be designed to create an invariant torus inside chaos in Hamiltonian systems [2]. The required perturbation is an order of magnitude smaller than the perturbation that creates the chaos. A good magnetic surface can be created inside chaos in magnetically confined plasmas [2-7]. This is called a magnetic barrier. A magnetic barrier with golden value of winding number will be the most stable magnetic surface. Effect of magnetic shear on the stability of golden barrier is studied. Magnetic shear is an important parameter in magnetic confinement. A map for magnetic field lines with variable shear is used [5-6]. The question addressed is - how does the largest magnetic perturbation against which the golden barrier is stable vary with shear. How does the shear affect the confinement of plasma having the golden barrier is investigated. This work is supported by the US DOE grants DE-FG02-01ER54624 and DE-FG02-04ER54793. This research used resources of the NERSC, supported by the Office of Science, US DOE, under contract DE-AC02-05CH11231.

Recently, we used a control method and our method of maps to create invariant magnetic surface inside chaotic regions in magnetically confined plasmas [3-7] by adding to the poloidal flux an external magnetic perturbation that is an order of magnitude smaller than the perturbation that creates the chaotic regions. This invariant surface is called magnetic barrier. It effectively acts as "barrier" against field-line dispersion and, indirectly, particle diffusion, and can be beneficial for plasma confinement. The feasibility, in the DIII-D tokamak, of these magnetic barriers is confirmed in [7] using experimentally constrained EFIT [8] equilibria and vacuum field perturbations exerted by the internal I-coils, taking into account their actual geometry. We also introduced a new area-preserving logarithmic map in magnetic coordinates that calculates magnetic field line trajectories in single-null divertor tokamaks with variable shear. The map parameter denoted by $s$ controls the magnetic shear and is referred to as shear parameter. This map preserves the symplectic topological invariance of the Hamiltonian system. We used this map to study the resilience of golden magnetic surface in tokamaks against resonant magnetic perturbations [5-6]. In this paper, we use this map to study the effects of magnetic shear on the stability of magnetic barrier with golden value of safety factor with and without adding second order control term to the poloidal flux. We call this barrier the golden barrier. For magnetic perturbation, we choose two tearing modes with mode numbers $(m,n)=\{(4,3),(2,1)\}$, and $\chi_1(\theta,\phi)=\delta[\cos(4\theta-3\phi)+\cos(2\theta-\phi)]$. $\delta$ is the amplitude of the perturbation. The safety factor profile $q$ is given by $q(\psi)=1-s\ln(1-\psi)$. Here $\psi$ is normalized by $\psi_{SEP}$, the toroidal flux inside the separatrix surface. The value of safety factor at the magnetic axis, $q_0$, is equal to unity and $q\rightarrow\infty$ at the separatrix. We characterize the magnetic configuration by the value of edge safety factor, $q_{95}$, at the magnetic surface that encloses 95% of the poloidal flux compared to the poloidal magnetic flux inside the separatrix. Fig. 1 shows $q_{95}$ as a function of the shear parameter s. $q_{95}$ scales as $e^{2s/3}$. From Fig. 1, we also see that for $s$=1.0, $q_{95}$=3.1.

As the magnetic shear parameter $s$ is varied, the location of the golden magnetic surface changes. For each value of $s$, the magnetic perturbation given above is applied to these surfaces with increasing amplitude; and the critical levels of the perturbations, $\delta_{crit}$, when the surfaces break down are calculated. We found that as the shear parameter $s$ increases, $\delta_{crit}$ scales as $e^{-2s/3}$. Thus, $\delta_{crit}$ scales as $1/q_{95}$. See Fig. 2. Following [2-4], we add a second order magnetic perturbation denoted by $\chi_2$ to the poloidal flux $\chi=\chi_0+\chi_1$. $\chi_0$ is the equilibrium poloidal flux and $\chi_1$ is given above. We refer $\chi_2$ as the second order control term and we denote its location as $\psi_b$. Figure 3 shows the effects of $\chi_2$ on the phase portrait. Fig. 3a depicts the phase portrait when $\delta=2\times10^{-3}$ and $s=1$. This Fig. shows the $m$=4 islands and $m$=2 islands. The perturbation influences mainly the region in the vicinity of the resonant surface $\psi_{43}$. A large chaotic zone appears around this surface. The golden surface is broken into eight islands and there are no internal transport barriers within the chaotic region. The magnetic field lines close to the magnetic axis $\psi=0$ and the separatrix are regular. Fig. 3b shows the phase portrait with the same values of $\delta$ and $s$ as in Fig. 3a but with the second order control term $\chi_2$ added to the poloidal flux $\chi$. In this case, the phase portrait is more regular, the chaotic region around the resonant surface $\psi_{43}$ has disappeared and the golden magnetic barrier is created. Fig. 4 shows $\delta_{crit}$ as a function of the shear parameter $s$ with and without the second order control term $\chi_2$.

We define the magnetic shear of the magnetic field as $S=2\psi q'(\psi)/q(\psi)$, and the average magnetic shear as $\langle S\rangle=\frac{1}{\psi_{95}}\int_0^{\psi_{95}}S(\psi)d\psi$. The influences of magnetic shear and the edge safety factor in the stability and improvement of the quality of confinement in magnetically confined plasmas have been firmly established by the results of numerous theoretical and experimental studies using various configurations in the last decades [9-10]. Fig. 5 shows that the average magnetic shear <S> scales as $e^{2s/7}$. As the average magnetic shear increases, we found that the critical levels of the magnetic perturbations $\delta_{crit}$ when the golden surfaces break down generally decrease with or without the second order control term $\chi_2$. With the control term $\chi_2$, the values of $\delta_{crit}$ fluctuate as <S> increases from 1.1 to 1.3; then decrease with increasing <S>. We also found that, for any fixed value of <S>, the value of $\delta_{crit}$ required for the golden magnetic barrier to break down is about 45% to 50% higher with the second order control term than without. Fig. 6 shows these results.

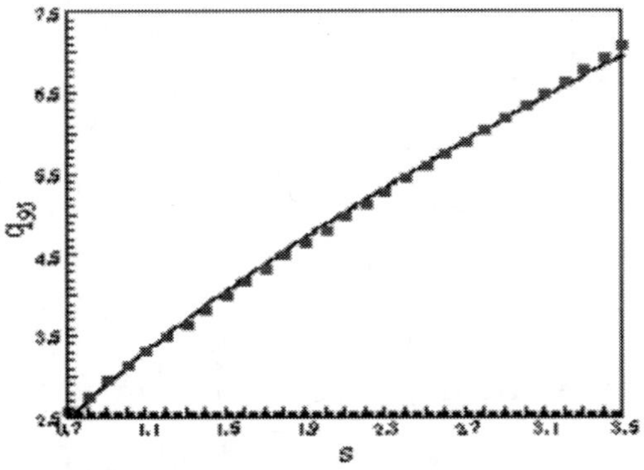


Fig. 1: The edge safety factor $q_{95}$ as a function of the shear parameter $s$. Curve through data points is a power fit: $q_{95}$ scales as $e^{2s/3}$.

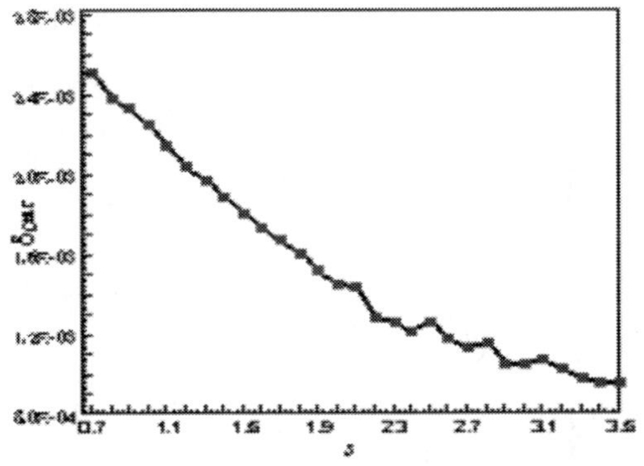


Fig. 2: $\delta_{crit}$ as a function of the shear parameter $s$. Curve through data points is a power fit: $\delta_{crit}$ scales as $e^{-2s/3}$.

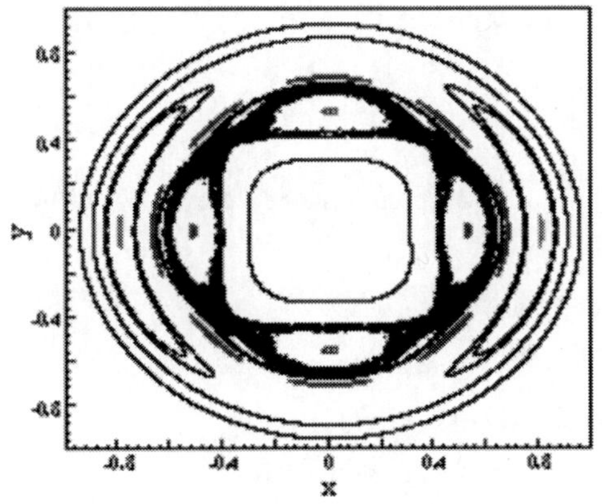


Fig. 3a shows the phase portrait when $\delta=2\times10^{-3}$ and $s=1$ without the control term $\chi_2$ at $\psi=\psi_{golden}$.

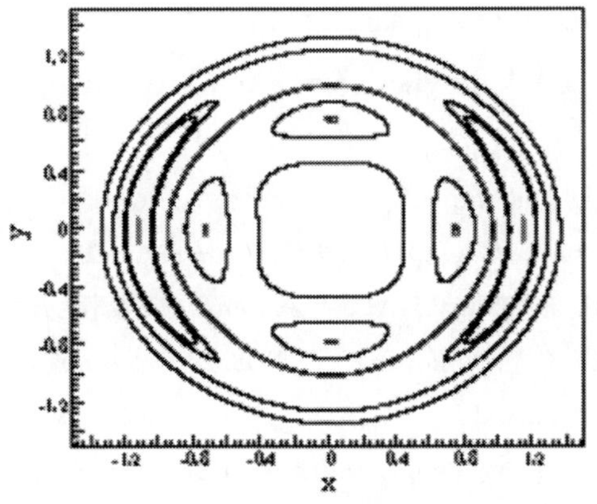


Fig. 3b shows the phase portrait with the same values of $\delta$ and $s$ with the control term $\chi_2$ at $\psi=\psi_{golden}$.

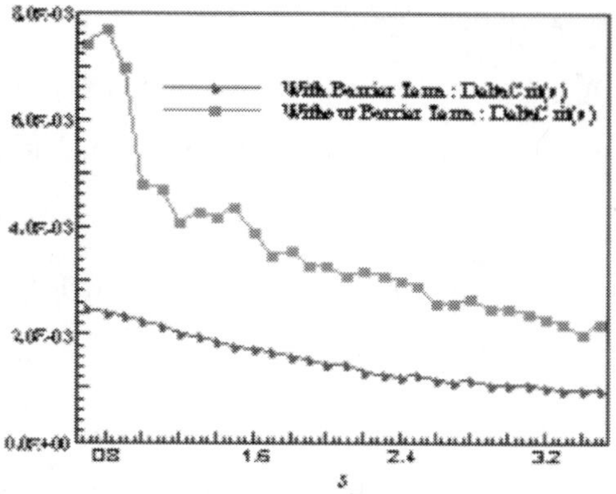


Fig. 4a: $\delta_{crit,}$ as a function of $s$ with and without the control term $\chi_2$ at $\psi = \psi_{golden}$.

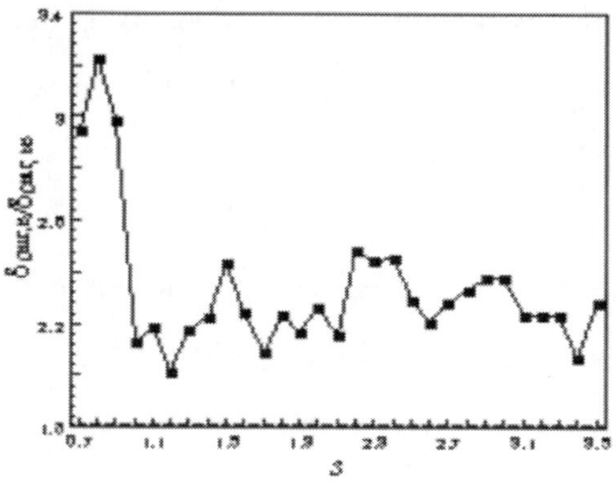


Fig. 4b: $\delta_{crit,\,B}/\ \delta_{crit,\,NB}$ as a function of $s$

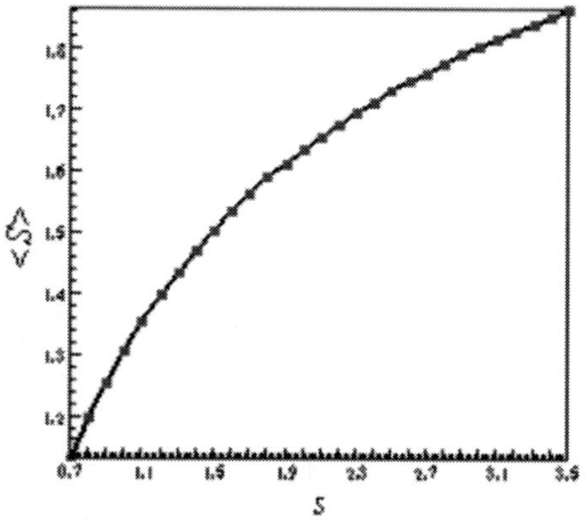


Fig. 5: $<S>$ as a function of the shear parameter $s$. Curve through data point is a power fit: $<S> \propto e^{2s/7}$.

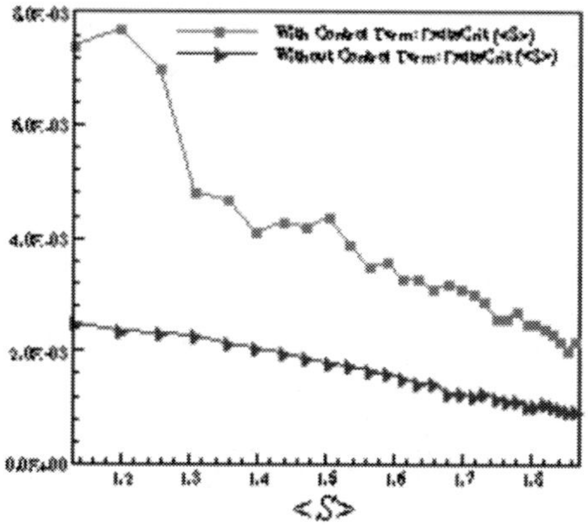


Fig. 6: $\delta_{crit,}$ as a function of $<S>$ with and without the control term $\chi_2$ at $\psi_{golden}$ added to the poloidal flux $\chi$.

Higher average shear makes the golden surface and the golden magnetic barrier weaker as they break down at lower levels of magnetic perturbation.

[1] Lichtenberg, A.J.; Liberman, M.A., *Regular and Chaotic Dynamics*, 2nd edn. Springer: New York (1991).

[2] Ciraolo, G. *et al*, *Physical Review E* **69**, 056213 (2004).

[3] Ali, H.; Punjabi, A., *Plasma Phys. Controlled Fusion* **49**, 1565 (2007).

[4] Ali, H. and Punjabi, A., *REDS: Inc. Plasma Sc. Plasma Tech.* **168**, 323 (2013).

[5] Ali, H. and Punjabi, *Europhy. Conf. Abs.* P5.137, Vol. **36F** ISBN 2914771797 (2013).

[6] Ali, H. and Punjabi, *Europhy. Conf. Abs.* P1.059,Vol. **38F** ISBN 2-914771-90-8 (2014).

[7] F.A. Volpe *et al*, *Nucl. Fus.* **52**, 054017 (2012).

[8] L. Lao, *et al.*, *Nucl. Fusion* **25**, 1611 (1985)

[9] [L.-G. Eriksson, *et. al.*, *Phys. Rev. Lett.*, **88**, 145001 (2002).

[10] O. Sauter, S.*et al*, *Phys. Rev. Lett.*, **94**, 105002 (2005).

# Statistical properties of poloidal transport in zonal flows with finite Larmor radius effects

Hector A. Velasco and Julio J. Martinell

*Instituto de Ciencias Nucleares, UNAM, A. Postal 70-543, México D.F., Mexico*

**I. Introduction.** A simple description of drift wave transport in presence of a zonal flow uses the dynamics of a test particle described by a Hamiltonian system which is liable to present chaotic behavior when the fluctuation amplitude is large. The zonal flow is manifested by the presence of a barrier to radial transport. To zeroth order in the particle Larmor radius the effect of the transport barrier in isolating the central plasma is erased for large enough fluctuation level. However, the inclusion of finite Larmor radius effects has been found to restore the barrier [1]. This is important for alpha particle confinement since it indicates that they are affected by a transport barrier even for large fluctuation levels. Another interesting problem is whether the type of transport is diffusive or not. This can sometimes be inferred from the statistical properties of transport since the variance of the particle distribution may exhibit a nonlinear dependence with time. In this work we have analyzed the statistical behavior of a large group of test particles in order to determine the resulting probability distribution function (PDF) for displacements in the poloidal direction, i.e. parallel to the transport barrier in non-monotonic zonal flows (non-twist Hamiltonian) taking into account finite Larmor radius (FLR) effects.

**II. Test particle model with FLR effects.**

Consider a test particle subject to the $E \times B$ drift velocity, with $B$ along the $z$ direction, which assumes that the particle follows the guiding center velocity. For finite gyroradii the particle actually samples spatial regions around the guiding center and this effect can be accounted for by taking the average of the drift equations over the gyroradius. Thus, the equations are

$$\frac{dx}{dt} = -\left\langle \frac{\partial \phi}{\partial y} \right\rangle_\theta, \qquad \frac{dy}{dt} = \left\langle \frac{\partial \phi}{\partial x} \right\rangle_\theta \tag{1}$$

where the gyroaverage, $\langle\rangle_\theta$, is defined as $\langle\Psi\rangle_\theta \equiv \frac{1}{2\pi}\int_0^{2\pi} \Psi(x+\rho\cos\theta, y+\rho\sin\theta)\,d\theta$ . $x$ and $y$ are radial and poloidal coordinates. This represents a Hamiltonian system with $\langle\phi\rangle_\theta$ as the Hamiltonian. In our model $\phi = \phi_0 + \phi_1$ where $\phi_0$ is the zonal flow, given by $\hat{u}_0(x) = \frac{d\phi_0}{dx} = \mathrm{sech}^2 x$, and the fluctuations $\phi_1$ are given by two modes of the Hasegawa-Mima equation, in the co-moving frame of one mode. Its gyroaverage is,

$$\langle\phi\rangle_\theta = J_{0\rho}(x) - \eta x + \varepsilon_1 J'_{k_1\rho}(x)\cos(k_1 y) + \varepsilon_2 J'_{k_2\rho}(x)\cos(k_2 y - \omega t), \tag{2}$$

where the prime denotes the derivative of the function $J_{k\rho}(x)$, which is defined by,

$$J_{k,\rho}(x) = \frac{1}{\pi}\int_0^{\pi} \tanh(x+\rho\cos\theta)\cos(k\rho\sin\theta)\,d\theta\,. \tag{3}$$

The equations 1 with 2 are integrated to determine the particle evolution. When $\varepsilon_2 = 0$ the system is time independent and the phase space (in our case the $x-y$ space) has integral surfaces. In general, it is characterized by two resonances on each side of the shearless point, where the mode velocity equals the flow velocity. For $\rho = 0$ the resonances can form two sepparate chains of islands (heteroclinic, when wave amplitude is low, see Fig. 1) or they can join in coupled chains (homoclinic, for large wave amplitude). For finite $\rho$ the island chains separation grows with $\rho$ and eventualy a bifurcation occurs that forms 4 island chains. When $\varepsilon_2 \neq 0$ the system is non integrable and chaotic regions appear near rational surfaces, which widen as $\varepsilon_2$ grows. But the shearless curve (between the islands) acts as a barrier that particles cannot cross until $\varepsilon_2$ exceeds a certain threshold that depends nonlinearly on $\rho$ [1]. Chaoric transport is thus limited by this transport barrier in radial ($x$) direction. On the other hand, poloidal ($y$) transport is not blocked by this effect and is thus interesting to study its properties in the non-twist Hamiltonian associated with the radial barrier. The statistical properties in the twist case (no barrier) were studied in [2] and in [3] including FLR effects.

**III. Transport properties.**

In order to study the nature of transport in $y$ direction we place a large number of particles centered around the hyperbolic point of Fig. 1 and let them evolve for a sufficiently long time. They comprise closed and open orbits in comparable proportions. Typically we use about $10^4$ particles but sometimes larger numbers are required to separate the different populations. The integrations are lengthy but we can take advantage of the independence of particle trajectories to perform parallel computations. Therefore we use GPU parallel programming based on CUDA in order to speed up the process. With this technique, execution times are found to improve by a factor of up to 100. In this way, the PDFs can be efficiently found. The behavior depends on the values of the parmeters $\varepsilon_1, \varepsilon_2$ and $\rho$. There are three types of particle trajectories: (1) trapped in the islands (2) free-streaming that flow unhindered along $y$ and (3) trapped-flight which alternate between the two (the most common, also called sticky-flight). For

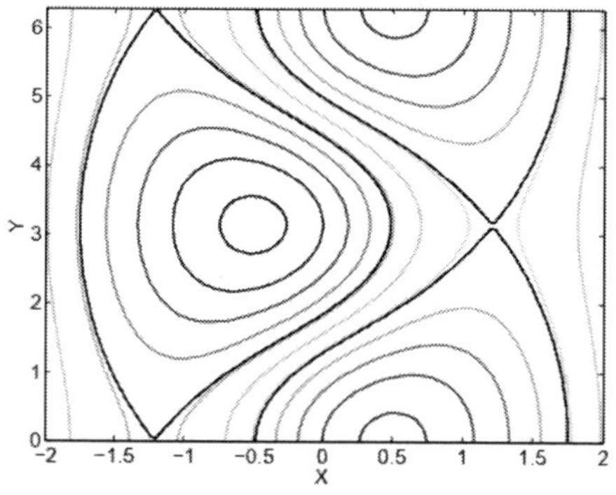


Figure 1: Topology of phase space.

non chaotic cases ($\varepsilon_2 = 0$) only the first two populations are present and are clearly separated; the orbits are deterministic. The distribution functions along $y$ are shown in Fig. 2. For $\varepsilon_2 \neq 0$ chaotic transport produce sticky-flight orbits and the distribution fuctions are wider as seen in Fig. 3.

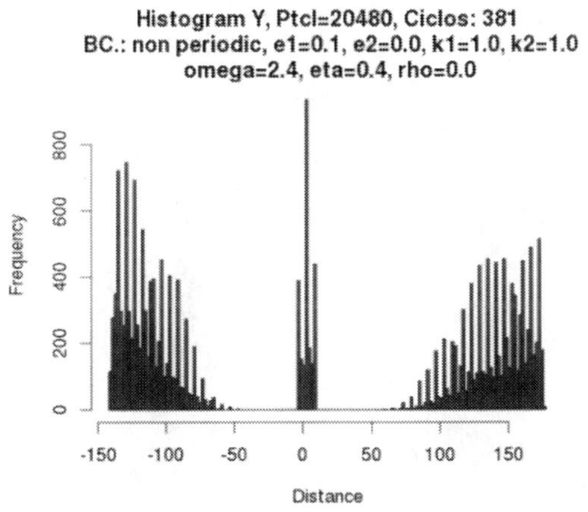


Figure 2: Distribution function of particles with non chaotic orbits.

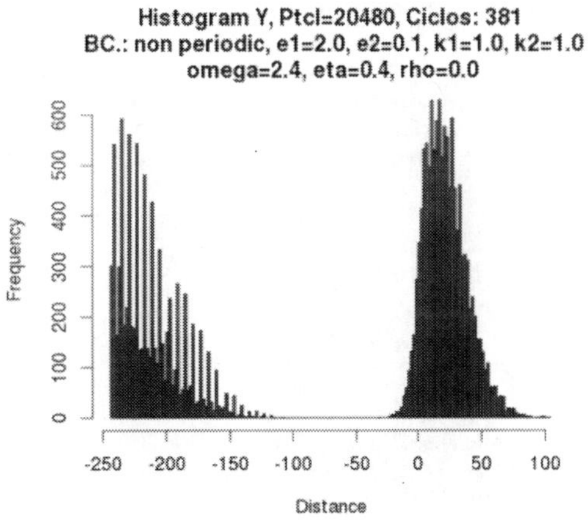


Figure 3: Patricle distribution function with chaotic orbits.

The population that stays near zero usually keeps the local coherence, while those that travel to the left or right display quite different behaviors. The figures presented give a sample of this variety. Usually one of the two extremes can have a more coherent shape and that is the one most commonly analyzed here. When the dispersion in particles looks wild the analysis cannot be performed.

The interesting property to follow is the width of the distribution as this is a meassure of the statistical spread that determines the transport type. Thus the variance $\sigma$ of the distribution of a single population is analyzed and its time dependence is determined. For diffusive transport the linear relation $\sigma \propto t$ is satisfied. However, nondiffusive behaviors have a nonlinear dependence of the type $\sigma \propto t^{\gamma}$. When this is done in the cases with chaos it is generally found that $\gamma \approx 2$ which is indicative of superdiffusive transport.

Another important charcteristic of the transport is the form of the probability distribution function (PDF). For diffusive transport the PDF is Gaussian. However, for some cases here when $\varepsilon_2 \neq 0$ they are not Gaussian, but present tails that decay more slowly. This is evidence for non-diffusive transport. This can be seen in Figs. 4 and 5 where the two types of distributions are obtained. The first case, with large $\varepsilon_2$ chaos level is high and the PDF is Gaussian since the information about trapped or free orbits is lost. In the second case the PDF is non Gaussian having a long tail on the right side. The tail is realted to the presence of long flights in some orbits. Trapping and circulating events in the particles trajectories are equivalent to Levy flights. It has been found that the long tails are described by Levy distributon functions [4] which asymptotically have the dependence $F \propto x^{-\mu}$. For a Levy-type PDF it is required that the exponent $\mu < 3$, so that the decay is slow enough. In the cases analyzed we have several situations with $\mu < 3$ although it is difficult to assign a value due to the complicated PDFs found.

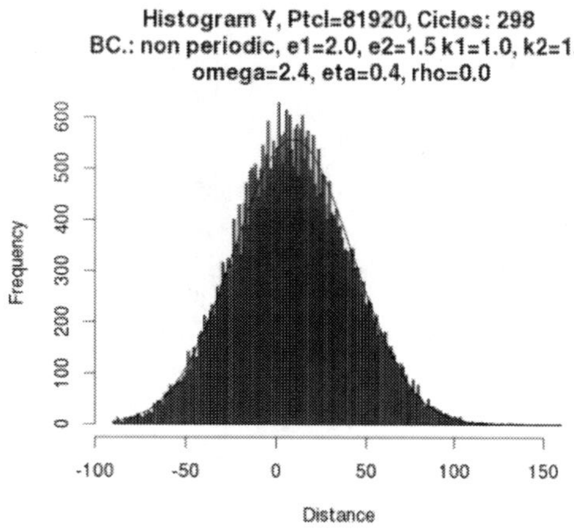


Figure 4: PDF with high level of chaos with a Gaussian-like behavior.

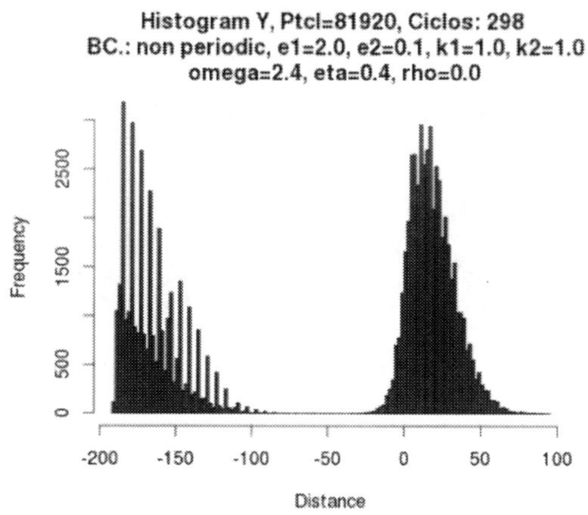


Figure 5: PDF with non-Gaussian distribution in the far population.

The effect of FLR is to increase the non-local effects, i.e. large values of $\rho$ usually have lower values of $\mu$. The results of having a bifurcation in the phase-space topology (for $\rho >$ 2.33) is to increase the degree of left right anisotropy. From the analysis we can conclude that the nondiffusive transport found for the cases of Hamiltonian systems with twist is preserved for non-twist systems. In particular the superdiffusive nature and the non-Gaussian PDF. FLR effects are not very noticeable although they modify quantitatively the transport.

**Acknowledgements.** This work was partially supported by projects DGAPA-UNAM IN106911, IN109115 and CONACyT 152905.

**References**

[1] J.J. Martinell and D. del-Castillo-Negrete, Phys. Plasmas **20**, 022303 (2013)

[2] D. del-Castillo-Negrete, Phys. Plasmas **7**, 1702 (2000).

[3] K. Gustafson, D. del-Castillo-Negrete and W. Dorland, Phys. Plasmas **15**, 102309 (2008).

[4] D. del-Castillo-Negrete, Phys. Fluids **10**, 576 (1998).

# Peculiar Transport Barriers Onset in Large Aspect Ratio Tokamaks

C. V. Abud[1], B. Bartoloni[2], I. L. Caldas[2]

*1- Federal University of Goiás, Catalão, Brazil*

*2- Institute of Physics, University of São Paulo, São Paulo, Brazil*

Internal transport barriers can reduce magnetic field line escape to the tokamak vessel and contribute to the plasma confinement improvement. The shape of the safety factor profile is among the factors responsible to the onset of such barriers. In this work, we investigate the onset of magnetic field line transport barriers for two kinds of peculiar safety factor profiles. For these profiles, we perform numerical simulations in the symplectic Ullmann map [1, 2] that describes the MHD equilibrium in large aspect-ratio tokamaks perturbed by a chaotic limiter.

Initially we take into account evidences that the formation of barriers may be correlated with some local flattening of the otherwise monotonic safety factor [3]. In this case, a local safety factor profile modification creates a low-shear zone where transport barriers can emerge [4]. Here, we show, for the considered symplectic Ullmann map, that a local flattened profile modifies the field lines topology and reduces the magnetic field lines escape.

For a large aspect ratio tokamak with ergodic magnetic limiter, we obtain the magnetic field line mapping from the Ullmann map. For the equilibrium the map is given by [1]:

$$r_{n+1} = \frac{r_n}{1 - a_1 sen\Theta_n}$$

$$\Theta_{n+1} = \Theta_n + \frac{2\pi}{q_0(r_{n+1})} + a_1 cos\Theta_n$$

where $a_1$ is introduced to take into account a correction for the toroidal geometry. The perturbation due to the effect of the ergodic limiter on the equilibrium configuration is given by [1]:

$$r_{n+1} = r_n^* - bC\left(\frac{r_n^*}{b}\right)^{m-1} sen(m\Theta_n^*)$$

$$\Theta_n^* = \Theta_{n+1} - C\left(\frac{r_n^*}{b}\right)^{m-2} cos(m\Theta_n^*)$$

where the control parameters are $q_0(0)$, $q_0(a)$, the perturbation amplitude C and the number of the wires m.

The equilibrium safety factor is given by

$$q_0 = \frac{rB_z}{R_0 B_\theta} \qquad (1)$$

where $B_\theta$ and $B_z$ are the poloidal and toroidal components, $R_0$ is the major radius (Fig. 1).

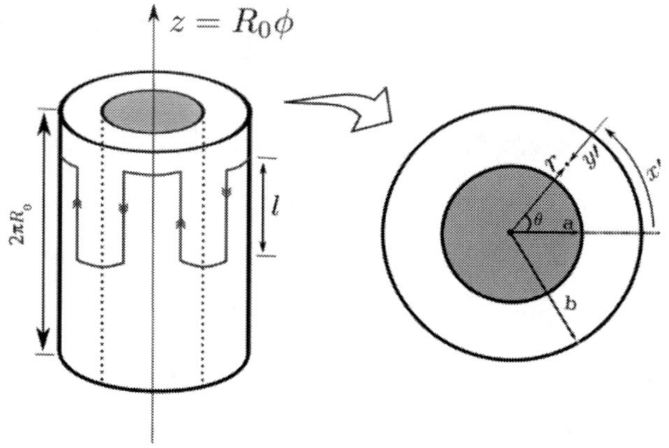


Figure 1: Coordinates $x'$, $y'$, major $R_0$ and minor $a$, $b$ radii.

For each current density profile, we will have a poloidal magnetic field and a correspondent safety factor profile. The perturbed safety factor is numerically calculated as:

$$q \equiv \lim_{k\to\infty} \frac{2\pi k}{\sum_{j=0}^{k}(\theta_{j+1}-\theta_j)} \qquad (2)$$

For a monotonic equilibrium safety factor profile with a local flatness point, shown in Fig. 2a, the perturbation map emerges from the effect of the ergodic limiter on the equilibrium configuration. Fig. 2b shows the map, for the considered $q_0$ profile, with an external invariant curve acting as a barrier in the low part of the map (small y) [5].

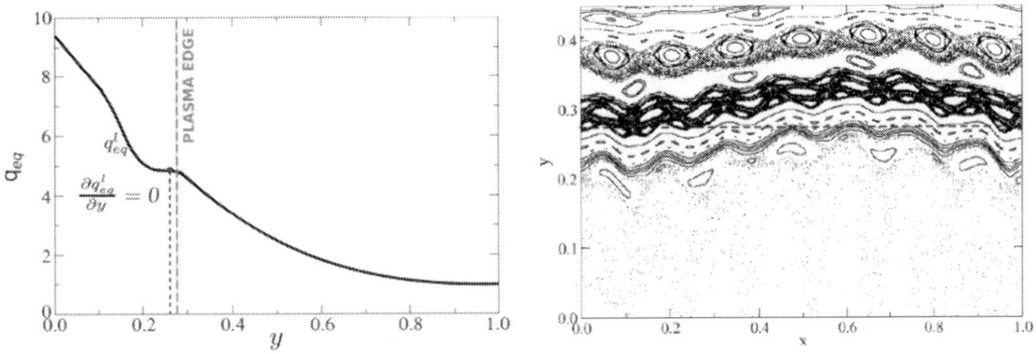


Figure 2: a) Safety factor profile. Red point indicates the local flatness. b) Phase space for the symplectic map monotonic profile with one flatness point ($q_0(0) = 0.83$, $q_0(a) = 5.0$, C=0.33).

The second considered safety profile is related to unperturbed plasma equilibrium with a radial hollow current profile (Fig. 3a).

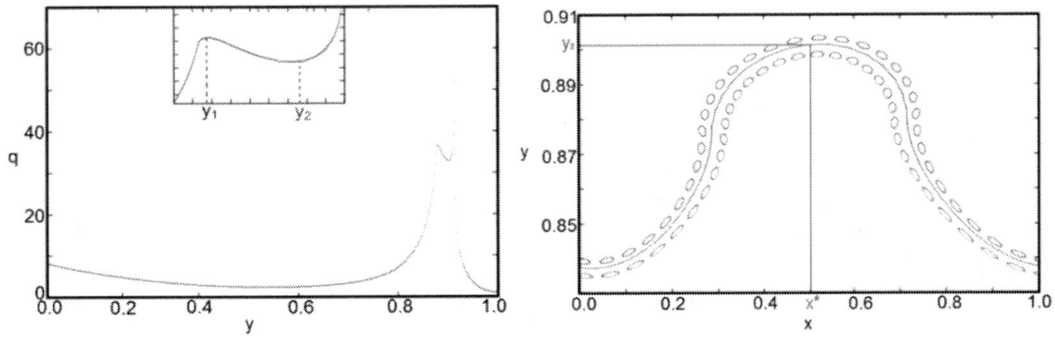


Figure 3: a) Numerical safety profile. In the inset, highlights of the maximum and minimum points. b) Shearless line and island chains ($q_0(0) = 0.83$, $q_0(a) = 5.0$, C=0.0).

Plasmas are confined in tokamaks by magnetic field lines on nested axisymmetric toroidal magnetic surfaces. However, some advanced confinement regimes with non inductive currents develop a radial hollow current profile, with a central hole where the current density is almost zero or even negative [6]. Solutions of the Grad-Shafranov equation describing magnetohydrodynamic equilibria with non nested magnetic surfaces have been proposed [7-10]. Here, we show that applying the Ullmann map for an equilibrium safety factor obtained from a central hollow electric current with negative density, gives rise to non-nested magnetic surfaces and two shearless invariant curves due to the current inversion (Fig. 3b). Furthermore, the considered equilibrium modified by the toroidal correction has a series of resonances with the corresponding islands and isolated chaotic regions (Fig. 4).

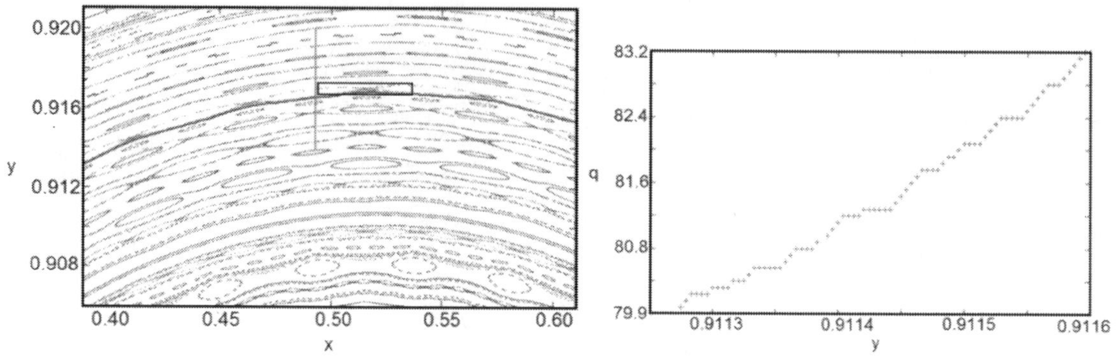


Figure 4: a) Poincaré section close to the divergence with many small island. b) Zoom of the numerical safety factor profile. Devil's stair indicates the presence of the islands. Same parameters of Fig. 3.

The described barriers are created by peculiar safety factor profiles. The first considered equilibrium has a local flat current profile at the plasma edge modifying the field lines topology and reducing the global field lines escape pattern [3]. The second equilibrium with non-nested magnetic surfaces presents two shearless invariant curves created by the current inversion reducing the field lines transport.

Acknowledgments

Work partially supported by the State of São Paulo Research Foundation (FAPESP, Brazil), under Grant N° 2011/19296-1, and the federal agencies CNPq, and CAPES.

References

1. K. Ullmann and I. L. Caldas, Chaos, Solit. and Fract. *11*, 2129 (2000).
2. J. S. E. Portela, I. L. Caldas, R. L, Viana. Europ. J. of Phys., *165*, 195 (2008).
3. K. A. Razumova et al., Plasma Phys. Rep. 27, 273 (2001).
4. D. Constantinescu, M.C. Firpo, Nucl. Fusion 52, 054006 (2012).
5. C. V. Abud, I. L. Caldas, to be published in Phys. Plasmas.
6. T. Fujita, Nucl. Fusion *50*, 113001 (2010).
7. A. A. Martynov, S. Yu Medvedev, and L. Villard, Phys. Rev. Lett. *91*, 085004 (2003).
8. P. Rodrigues, J. Bizarro, Phys. Rev. Letters, *95*, 015001 (2005).
9 . C. Martins, M. Roberto, I. L. Caldas, Phys. Plasmas, *18*, 082508 (2011).
10. D. Ciro Taborda, I. L. Caldas, Phys. Plasmas *20*, 102512 (2013).

# A quantitative model for heat pulse propagation in Large Helical Device plasmas using a travelling wave transformation

H Zhu[1], R O Dendy[2,1,3,4], S C Chapman[1,5,6] and S Inagaki[3,4]

[1]*Centre for Fusion, Space and Astrophysics, Department of Physics, University of Warwick, Coventry CV4 7AL, UK*

[2]*CCFE, Culham Science Centre, Abingdon, Oxfordshire OX14 3DB, UK*

[3]*Itoh Research Center for Plasma Turbulence, Kyushu University, Kasuga 816-8580, Japan*

[4]*Research Institute for Applied Mechanics, Kyushu University, Kasuga 816-8580, Japan*

[5]*Max Planck Institute for the Physics of Complex Systems, Dresden, Germany*

[6]*Department of Mathematics and Statistics, University of Tromsø, Tromsø, Norway*

**1. Introduction** The transport of energy across magnetically confined fusion plasmas, and the storage of energy within them, reflect a wide range of turbulent and nonlinear phenomenology. There is extensive experimental evidence for the transport phenomena that are non-diffusive and may be non-local. Here we focus on heat pulse experiment[1-6]. These heat pulses are typically initiated by injection into the edge plasma of ice pellets or supersonic molecule beams, by gas puffing, and by laser ablation. The zero-dimensional model is know to work at the best-diagnosed spatial location, capturing the time evolution of the pulse there, where $t = 0$ is defined to be the local arrival time of the initial impulsive perturbation. Therefore we conjecture that the zero-dimensional model should work at each location across the radial domain of the plasma within which the same physical processes determine the behaviour of the pulse. From this we infer that the model ought to apply in a frame that is co-moving radially with the heat pulse across this region. We then apply a travelling wave transformation by replacing $t$ by $\xi = x + v_0 t$; here $v_0$ is to be considered as proxy of the pulse propagation velocity in the radial direction $x$, and the sign convention adopted for $t$ assists consideration of inward propagation.

**2. Model description** The normalised zero-dimensional model[4] examined below is constructed in terms of the three key physical quantities that were measured[4] so as to characterise pulse propagation in LHD. These quantities are the deviation from steady state of the electron temperature gradient $\delta \nabla T_e$, the excess turbulent heat flux $\delta q_e$, and the deviation $\delta T_e$ of electron temperature from its steady-state value. The dimensionless counterpart of these variables are denoted by $x_1$, $x_2$ and $x_3$ respectively, as defined in Ref.[4]. The model of Ref.[4] shows quantative agreement between its outputs and experimental measurements of the time evolution. However, for given parameters and initial conditions, the model only simulates the time evolution of a passing heat pulse at a specific radius. The model equation (5) to (7) of Ref.[4] are as follows

$$\frac{dx_1}{dt} = \kappa_{T0}x_2 + x_1x_2\frac{\partial\kappa_T}{\partial x_1} + x_2x_3\frac{\partial\kappa_T}{\partial x_3} - \gamma_{L1}x_1$$

$$\frac{dx_2}{dt} = -\kappa_{Q0}x_1 + x_1^2\frac{\partial\kappa_Q}{\partial x_1} - x_1x_3\frac{\partial\kappa_Q}{\partial x_3} - \gamma_{L1}x_2$$

$$\frac{dx_3}{dt} = -\frac{1}{\tau_c}\frac{\eta}{\chi_0}x_2 - \gamma_{L2}x_3$$

Central to the present paper is the adoption of travelling wave transformations as a method for generating $(x,t)$-dependence from the $t$-dependent model above. We assume that

$$x_i(x,t) = y_i(\xi), \xi = x + v_0 t, i = 1,2,3$$

Here $x$ and $t$ are considered to be independent variables, and we refer to $v_0$ as the pseudo-velocity of the pulse. Then we get the following equations by substitution, leading order approximation and transposition:

$$v_0\frac{d^2y_1}{d\xi^2} - \gamma_{L1}\frac{dy_1}{d\xi} + \left(\frac{\kappa_{T0}}{v_0}\frac{\partial\kappa_Q}{\partial y_1} + \frac{\kappa_{Q0}}{v_0}\frac{\partial\kappa_T}{\partial y_1}\right)y_1^2 + \frac{\kappa_{T0}\kappa_{Q0}}{v_0}y_1 = -\frac{\kappa_{T0}\gamma_{L1}}{v_0}y_2 + \frac{\kappa_{T0}}{v_0}\frac{\partial\kappa_T}{\partial y_1}y_2^2$$

$$v_0\frac{d^2y_2}{d\xi^2} - \gamma_{L1}\frac{dy_2}{d\xi} + \frac{\kappa_{T0}\kappa_{Q0}}{v_0}y_2 = \frac{\kappa_{Q0}\gamma_{L1}}{v_0}y_1 - y_1y_2\left(\frac{\kappa_{Q0}}{v_0}\frac{\partial\kappa_T}{\partial y_1} + 2\frac{\kappa_{T0}}{v_0}\frac{\partial\kappa_Q}{\partial y_1}\right)$$

**3. Comparison of model outputs with LHD experimental data** We recall that $y_1$, $y_2$ and $y_3$ are the dimensionless counterpart of $\delta\nabla T_e$, $\delta q$ and $\delta T_e$. respectively. Solution of our model, embodied in the two coupled nonlinear ordinary differential equations above. The numerical values of parameters carry over identically from Ref.[4], to which we refer for the experimental motivation for these values.

Figure 1 compares time traces of the evolving electron temperature at multiple radial locations, obtained from the model and from experimental data (Te49708) for the R case. Several representative radii are marked by arrows on the right hand side. The model results are able to match experimental data from $\rho$ = 0.450 inward to the core, if we uniformly apply horizontal (+0.01) and vertical (+0.20) shifts, suggesting that the model applies over this broad radial range. It is also clear that electron temperature profiles from $\rho$ = 0.546 outward to $\rho$ = 0.703 are not simulated by the model. This suggests that different physics dominates heat pulse propagation in the outer region of this plasma.

Figure 2 demonstrates the comparison of model results and experimental data (Te49719) for the D case. In common with the R case shown Figure 1, the model results are good from $\rho$ = 0.450 inward to the core, with uniformly applied horizontal (+0.04) and vertical (+0.07) shifts, but not in the outer region of this plasma.

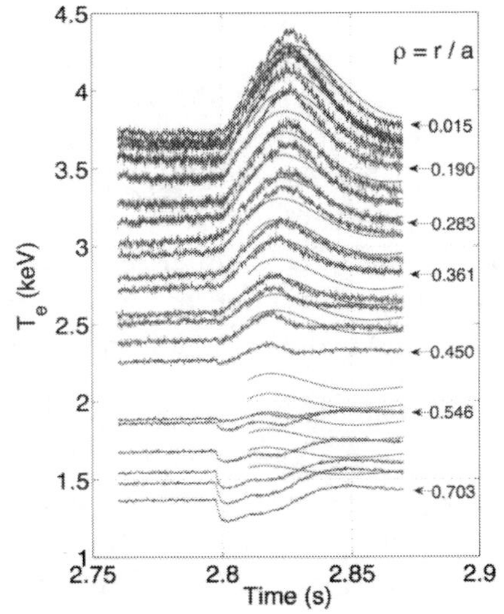


Figure 1. Time evolution of electron temperature at multiple radial locations, derived from LHD data(blue) and the model(red) for the core temperature rise(R) heat pulse propagation experiment in plasma 49708. Radial location rage from edge ($\rho$ = 0.703) to core ($\rho$ = 0.015), where $\rho = r/a$, $r$ is the radial co-ordinate and $a \sim$ 0.6m is minor radius of LHD. Model results match experimental data well from $\rho$ = 0.450 inwards to the plasma core, especially amplitudes and the time structure of pulse decay. The amplitude of model time traces increase from edge to core, as in the measured electron temperature profiles. Model results do not fit experimental data outwards from $\rho$ = 0.546 to $\rho$ = 0.703, implying that different physics applies in the outer LHD plasma. Reproduced from [6].

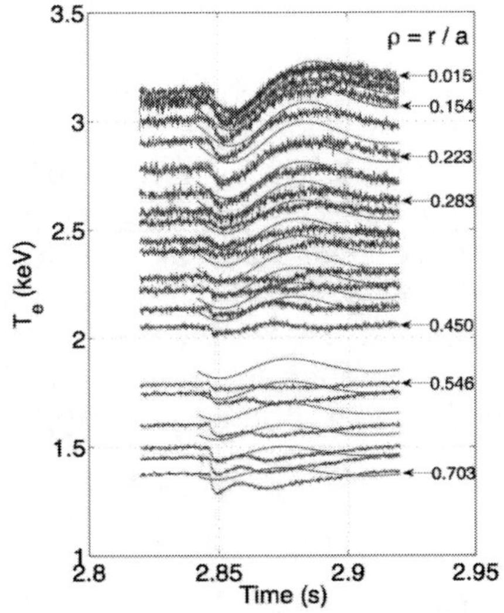


Figure 2. Time evolution of electron temperature at multiple radial locations, derived from LHD data(blue) and the model(red) for the core temperature drop(D) heat pulse propagation experiment in plasma 48719. Radial location rage from edge ($\rho$ = 0.703) to core ($\rho$ = 0.015), where $\rho = r/a$, $r$ is the radial co-ordinate and $a \sim$ 0.6m is minor radius of LHD. As in Figure 1, model results match experimental data well from $\rho$ = 0.450 inwards to the plasma core. Again, model results do not fie experimental data outwards from $\rho$ = 0.546 to $\rho$ = 0.703, reinforcing that different physics dominates in the outer LHD plasma. Reproduced from [6].

**4. Conclusions** We have derived from first principles a time-dependent model in one spatial dimension, which is able to describe quantitatively the radial inward propagation of heat pulses in the core of two plasmas in the Large Helical Device(LHD). In one plasma the central electron temperature rises, in the other it falls. This new model is derived from a travelling wave transformation of the zero-dimensional model of Ref.[4], which is known to capture the time-evolution of the heat pulse as it passes through a fixed radial location in these two plasmas. Comparison between model outputs and raw experimental data suggests that our model is able to describe heat pulse propagation well, within a broad radial range of the LHD core plasma from $r/a \approx 0.5$ to the centre.

The results of the present paper provide additional support to the physical proposals, which motivate the simple model equations above. Central to these proposals is the conjecture that heat pulses are structures, which involve plasma physics processes which are so strongly nonlinear that heat pulse evolution is primarily determined by the reactions of the perturbed heat flux and the perturbed temperature gradient on each other. In the present model, this is the dominant element of nonlinear physics, whereas turbulent transport plays a relatively minor dissipative role. It is this mutually coupled interaction between perturbed heat flux and temperature gradient that governs the local plasma dynamics of the heat pulse in space, equivalent to the local up-and-down dynamics of a water wave under gravity. We have shown in this paper that this coupling model lends itself readily to a travelling wave transformation, yielding spatio-temporal pulse propagation, and that the pulse velocity that emerges mathematically provides an adequate match to empirical results and expectations. This aspect of the analysis also provides guidance on a previously unanswered question, namely the generic character of the heat pulse: we have shown that it may be closely related to a Korteweg-de Vries-Burgers soliton.

This work was part-funded by the RCUK Energy Program under grant EP/I501045. This project has received funding from the European Union's Horizon 2020 research and innovation program under grant agreement number 633053. The views and opinions expressed herein do not necessarily reflect those of the European Commission. This work was also supported in part by the grant-in-aid for scientific research of JSPF, Japan (23244113) and KAKENHI (21224014) from JSPS, Japan. This work was also supported in part by the Grant-in-Aid for Scientific Research of JSPF, Japan (23360414) and by the collaboration programs of the National Institute for Fusion Science (NIFS) (NIFS13KOCT001) and the Research Institute for Applied Mechanics of Kyushu University.

[1] S Inagaki *et al*, Nucl. Fusion **46**, 133 (2006).
[2] P Mantica *et al*, Phys. Rev. Lett. **82**, 5048 (1999).
[3] S Inagaki *et al*, Plasma Phys. Control. Fusion **52** 075002 (2010).
[4] R Dendy *et al*, Plasma Phys. Control. Fusion **55**, 115009 (2013).
[5] S Inagaki *et al*, Plasma Phys. Control. Fusion **46** A71 (2004).
[6] H Zhu *et al*, submitted to Phys. Plasmas (2015).

# Scaling of the measured intensity of ion cyclotron emission with the concentration of energetic ions in large tokamak plasmas

R O Dendy[1,2], L Carbajal[2], S C Chapman[2] and J W S Cook[2]

[1]*CCFE, Culham Science Centre, Abingdon, Oxfordshire OX14 3DB, UK*

[2]*Centre for Fusion, Space and Astrophysics, Department of Physics, Warwick University, Coventry CV4 7AL, U.K.*

## 1. Introduction

Ion cyclotron emission (ICE) is highly suprathermal and its frequency spectrum displays narrow peaks at sequential cyclotron harmonics of energetic ions, at values which imply an emitting region localised to the outer mid-plane edge of the plasma. ICE was the first collective radiative instability driven by confined fusion-born ions that was observed from deuterium-tritium (D-T) plasmas in JET[1-3] and TFTR[4]. It is used as an energetic ion diagnostic on the largest contemporary tokamaks including DIII-D[5], ASDEX-Upgrade[6,7] and JT-60U[8,9], and in the large stellarator-heliotron LHD[10]. The most likely emission mechanism is the magnetoacoustic cyclotron instability (MCI)[11-18], whose role in relation to ICE observations was recently reviewed in Ref.[19]. The use of ICE as a diagnostic for confined energetic ion populations in ITER is under active consideration[20], and also for future D-T plasmas in JET. Diagnostic exploitation of ICE would be greatly assisted by theoretical understanding of the experimentally observed linear scaling, at low cyclotron harmonics, of its measured intensity with the number density concentration, $n_\alpha/n_i$, of fusion-born energetic ions $n_\alpha$ relative to the background thermal ions $n_i$. Here we provide a preliminary report of studies[21] which address this key observational aspect of ICE using a 1D3V hybrid code which simulates the self-consistent full orbit kinetics of energetic and thermal ions in the presence of a massless neutralising fluid of electrons. Importantly, these simulations are carried out in the fully nonlinear regime of the MCI, which has recently been shown[4] to generate, uniquely, the full observed ICE spectrum from the fundamental cyclotron frequency of the energetic ions upwards. We find that the experimentally observed linear scaling of ICE intensity with $n_\alpha/n_i$ at low harmonics is recovered by these simulations, and we relate this welcome outcome to the analytical theory of the MCI. These new results provide further support for, and understanding of, the MCI as the emission mechanism for ICE. This enhances the prospects for the successful exploitation of ICE as a diagnostic[20] for confined fusion products in future D-T plasmas in JET and ITER.

## 2. Hybrid simulation model results

The hybrid model resolves ion kinetics including gyromotion, and hence resolves the gyrophase-dependent physics of ion cyclotron harmonic resonance which is central to the MCI. The position of each ion evolves continuously in physical space and in velocity space under the self-consistent local Lorentz force. The magnetic field is updated using

Faraday's law, and the electric field using the momentum equation for the electron fluid in the limit of zero electron inertia[18,22]. This enables the study of fast instabilities such as the MCI that evolve on ion gyro timescales, and are rapid compared to gyro-averaged or fluid phenomena. In this preliminary account of the simulations to be reported more fully in Ref.[21], in addition to the majority thermal plasma there is an energetic minority alpha-particle population, whose concentration $\xi_\alpha$ relative to the thermal ions is in the range $n_\alpha/n_i = \xi_\alpha = 10^{-4}$ to $10^{-3}$, which forms a ring-beam distribution in velocity space with characteristic energy 3.5MeV. Plasma parameters are similar to those at the outer mid-plane edge of D-T plasmas in JET: background magnetic field strength $B_0$ = 2.1T, total number density $10^{19}m^{-3}$, electron temperature 1keV. The periodic 1D3V simulation domain comprises 8192 grid cells, each of order the deuteron Larmor radius so that ion gyromotion is fully resolved computationally, with 200 particles of each ion species per cell. Total energy is conserved within 3% of its initial value, and this change is smaller than any change in the energy of the ions and electromagnetic fields during the instability. Spectra obtained from our simulations in the fully nonlinear regime, for two different values of $\xi_\alpha$, are shown in Fig.1.

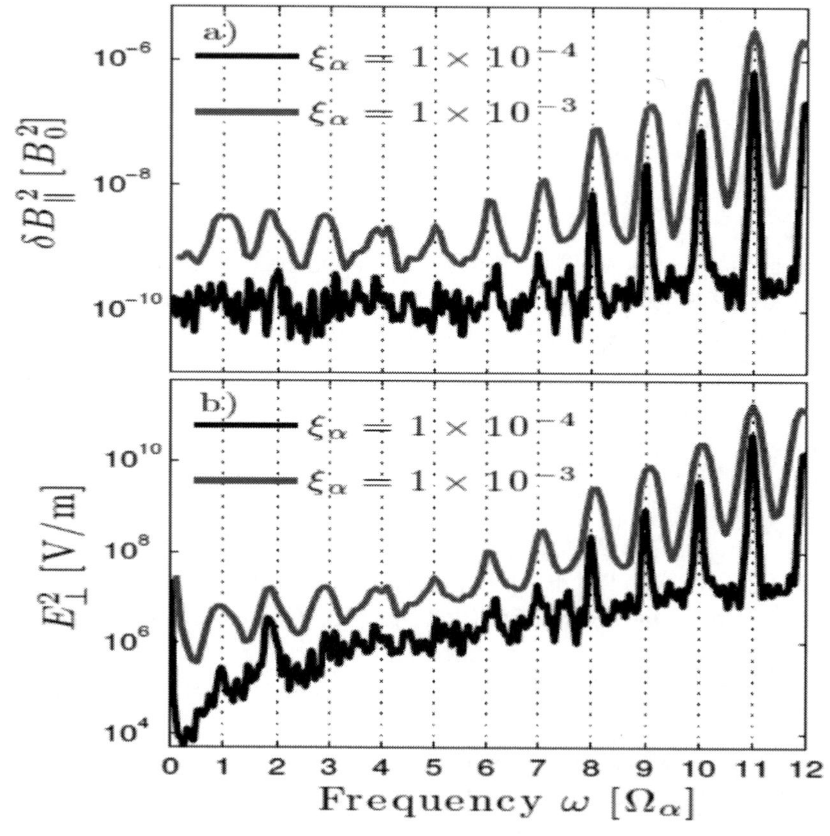


Fig.1 Log-normal frequency domain plot of the intensity of fields excited in hybrid simulations of deuterium plasmas containing an energetic minority alpha-particle population, undergoing the magnetoacoustic cyclotron instability: a) power density of the perturbed magnetic field for alpha-particle concentrations $\xi_\alpha = 10^{-3}$ (top red trace) and $\xi_\alpha = 10^{-4}$ (bottom black trace); b) power density of the transverse component of the electric field from the same simulations as panel a).

The simulation results are less intensive in their use of computational resources than those of the low-noise simulations to be reported in Ref.[21], but nevertheless enable us to investigate satisfactorily the scaling of the intensity of MCI-excited field energy with energetic ion concentration, in the fully nonlinear regime of the MCI. The objective is to obtain, from our simulation results at different alpha-particle concentrations, values for the conjectured scaling exponent η in the expression

$$(\delta B_{\parallel}/B_0)^2 \sim (\xi_\alpha)^\eta,$$

for comparison with the experimentally measured value of η for ICE in JET and TFTR,

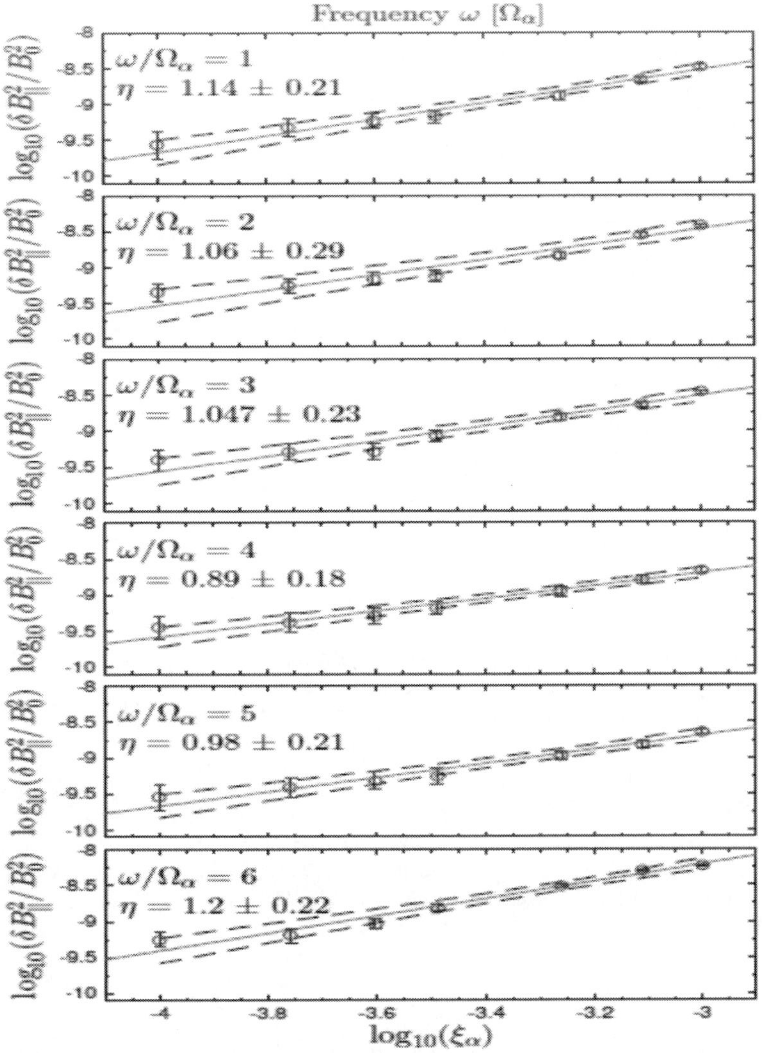


Fig.2 Scaling of the normalized magnetic field energy with alpha-particle concentration $\xi_\alpha$ in the hybrid simulations. The six panels display weighted least-squares linear fits for the base ten logarithmic plots of the normalized magnetic field energy $(\delta B_{\parallel}/B_0)^2$ versus $\xi_\alpha$ for the first six alpha-particle cyclotron harmonics $\omega = n\Omega_\alpha$. The black dashed lines in each of these panels represent the confidence interval of the weighted linear fit value of η, which is unity to within error bars at all six harmonics.

which is $\eta \approx 1$ as noted above[1,4]. Figure 2 shows that, to within error bars, our simulations indeed yield $\eta = 1$ for the six lowest cyclotron harmonics, and there are strong indications that this is the case also for higher harmonics.

**3. Conclusions**

The scaling of the MCI-excited magnetic field energy density with alpha-particle concentration in the fully nonlinear regime of these preliminary hybrid simulations is found to be linear. This conclusion includes the modes at low cyclotron harmonics, which are not predicted to be linearly unstable according to the analytical theory of the MCI. These simulation results are in good agreement with observations of ICE intensity in JET and TFTR. They further strengthen the capability to exploit ICE, through understanding of the emission mechanism, as a diagnostic for fusion-born alpha-particles in future D-T plasmas in JET and ITER. The relation between the simulation-derived scaling of $\eta$, examined here, and the scaling properties of the linear analytical growth rate of the MCI[14,15,4], will be discussed further in Ref.[21].

This work was supported in part by the RCUK Energy Programme, EURATOM, and the Mexican Council of Science and Technology (CONACyT). The views and opinions expressed herein do not necessarily reflect those of the European Commission.

[1] G A Cottrell, V P Bhatnagar *et al.*, Nucl. Fusion **33** 1365 (1993)
[2] R O Dendy, K G McClements *et al.*, Nucl. Fusion **35** 1733 (1995)
[3] K G McClements, C Hunt *et al.*, Phys. Rev. Lett. **82** 2099 (1999)
[4] S Cauffman, R Majeski *et al.*, Nucl. Fusion **35** 1597 (1995)
[5] W W Heidbrink *et al.*, Plasma Phys. Control. Fusion **53** 085028 (2011)
[6] R D'Inca, M Garcia-Munoz *et al.*, Proc. 38th EPS Conf. Plasma Phys. 2012 P1.053
[7] R D'Inca, PhD thesis, Max Planck Institute for Plasma Physics 2013
[8] M Ichimura, H Higaki *et al.*, Nucl. Fusion 48 035012 (2008)
[9] S Sato *et al.*, Plasma and Fusion Research **5** S2067 (2010)
[10] K Saito, R Kumazawa *et al.*, Plasma Sci. Technol. **15** 209 (2013)
[11] V S Belikov and Ya I Kolesnichenko, Sov. Phys. Tech. Phys. **20** 1146 (1976)
[12] R O Dendy, C N Lashmore-Davies and K F Kam, Phys. Fluids **B4** 3996 (1992)
[13] R O Dendy, C N Lashmore-Davies and K F Kam, Phys. Fluids **B5** 1937 (1993)
[14] R O Dendy, K G McClements *et al.*, Phys. Plasmas **1** 1918 (1994)
[15] K G McClements, R O Dendy *et al.*, Phys. Plasmas **3** 543 (1996)
[16] T Fülöp and M Lisak, Nucl. Fusion **38** 761 (1998)
[17] J W S Cook, R O Dendy and S C Chapman, Plasma Phys. Control. Fusion **55** 065003 (2013)
[18] L Carbajal, R O Dendy, S C Chapman and J W S Cook, Phys. Plasmas **21** 012106 (2014)
[19] R O Dendy and K G McClements, Plasma Phys. Control. Fusion **57** 044002 (2015)
[20] K G McClements, R D'Inca R O Dendy, L Carbajal, S C Chapman, J W S Cook, R W Harvey, W W Heidbrink and S D Pinches, Nucl. Fusion **55** 043013 (2015)
[21] L Carbajal *et al.*, to be submitted
[22] P Gingell, S C Chapman, R O Dendy and C S Brady, Plasma Phys. Control. Fusion **54** 065005 (2012)

# Interaction of explosive multiple filaments in magnetised plasmas

S.A. Henneberg[1], S.C. Cowley[2,3], H.R. Wilson[1]

[1] *York Plasma Institute, Department of Physics, University of York, Heslington, York, YO10 5DD, UK*

[2] *Culham Centre for Fusion Energy, Abingdon, Oxfordshire. OX14 3DB, UK*

[3] *Department of Physics, Imperial College, Prince Consort Road, London SW7 2BZ, UK*

E-mail: Sophia.Henneberg@york.ac.uk

## 1. Introduction

Two main operational regimes exist in tokamak fusion plasmas: the low-confinement mode (L-Mode) and the high-confinement mode (H-Mode). The H-Mode has a higher confinement than the L-Mode through a steep pressure gradient region at the edge, which creates a pedestal (Fig. 1). Fusion devices will likely operate in H-Mode because of its enhanced confinement. However the H-mode is vulnerable to violent plasma eruptions, called Edge Localised Modes (ELMs) which can be triggered by the steep edge pressure gradient. ELMs are quasi-periodic, filamentary instabilities which grow very rapidly [1]. They release a large amount of energy and particles which must be controlled to avoid excessive erosion of components on future fusion devices, such as ITER. Therefore it is very important to understand ELMs. Here, in particular, we are interested in how erupting plasma filaments with different heights influence each other's behaviour as this assists in understanding a more realistic early evolution of ELMs.

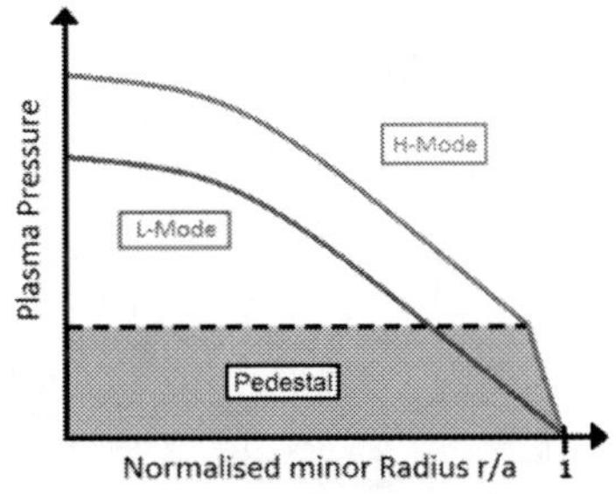


Figure 1: *Schematic pressure profile for L- and H-Mode*

## 2. Theoretical Model

### Background

We employ a slab geometry which is susceptible to the Rayleigh-Taylor instability, [2, 3], see Fig. 2. The magnetic field lines $\mathbf{B_0}$ are in the $z$-direction and $|\mathbf{B_0}|$ varies in the $x_0$-direction. The density $\rho_0$ and the pressure gradient $\nabla p_0$ are also only dependent on $x_0$. The gravity $\mathbf{g}$ points in the negative $x_0$-direction. The equilibrium is such that we have an unperturbed density and pressure at the $z$ boundaries: $z = 0$ and $z = L$. We assume periodic boundary conditions in the $y_0$ direction, and we adopt a constant temperature along the field lines as the thermal conduction is fast along them. Here $y_0$

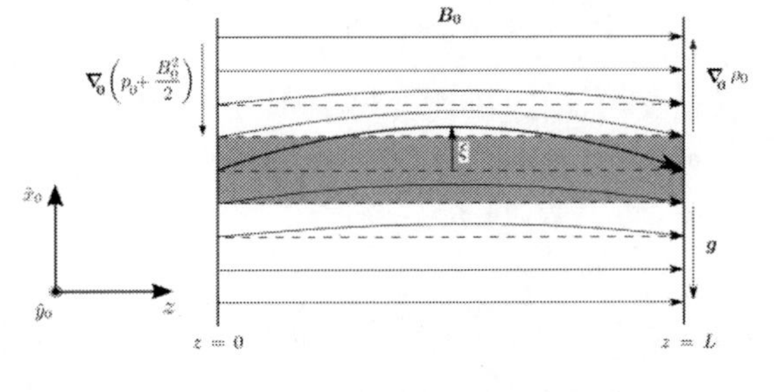


Figure 2: *Slab geometry: Straight field line represent the equilibrium.*

is equivalent to a poloidal angle in a tokamak plasma and $x_0$ is equivalent to the radial coordinate, $\psi$. The displacement perpendicular to the equilibrium magnetic field lines, $\xi$, of a field-aligned flux tube only depends on $x_0$, $y_0$ and $t$ and is described by non-linear ballooning theory, [2]-[7]:

$$\overbrace{C_0 \frac{\partial^2 \xi}{\partial t^2}}^{\text{Inertia Term}} = \overbrace{C_1\left(1-\frac{(x_0-x_{max})^2}{\Delta^2}\right)\xi}^{\text{Linear Instability Drive}} - \overbrace{C_2 \frac{\partial^2 u}{\partial x_0^2}}^{\text{Field Line Stability Term}} \tag{1}$$
$$+ \underbrace{C_3 \xi \frac{\partial^2 \overline{\xi^2}}{\partial x_0^2}}_{\text{Quasi-Linear-Nonlinearity Term}} + \underbrace{C_4\left(\xi^2-\overline{\xi^2}\right)}_{\text{Non-linear Growth Drive}} + \underbrace{\nu \frac{\partial^2}{\partial y_0^2}\frac{\partial \xi}{\partial t}}_{\text{Viscosity Term}}$$

where $\xi = \frac{\partial^2 u}{\partial y_0^2}$ and an overbar indicates an average over the periodic spatial direction, $y_0$. The 0-subscript indicates that it is a Lagrangian coordinate, where $\mathbf{r} = \mathbf{r_0} + \xi$ and therefore the $z$-direction corresponds to an Eulerian coordinate [2, 3]. The impact of different equilibrium properties is captured in the coefficients $C_0$-$C_4$, the viscosity $\nu$ and the parameter $\Delta$ - see Ref. [5], [6] and [7] for details. For some types of tokamaks the order of the time derivative can also change, but for strongly shaped poloidal cross section this second order time derivative is appropriate [7]. Therefore this equation describing early non-linear evolution of ballooning modes has basically the same form for different geometries and therefore our results are generic.

**Three Filament System**

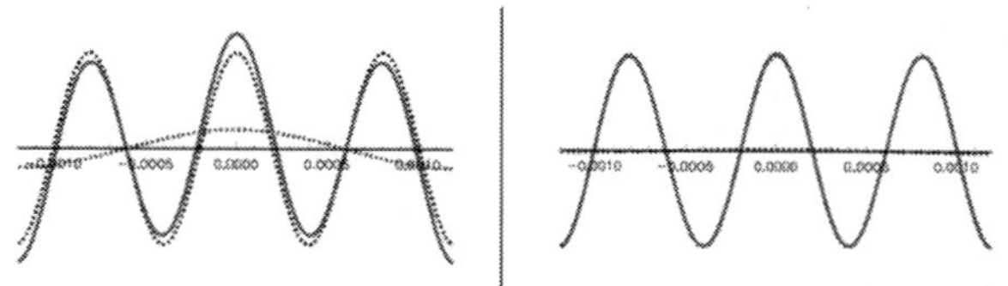

(a) *Superpostition (red) of two modes with $n_2 = 3n_1$ and $h_2 = 5h_1$. Blue: $n_1$ and black $n_2$-mode. In the middle is the main central filament with two side filaments.*

(b) *Superpostition (red) of two modes. This time with the actual heights $h_2 = 50h_1$ which produces a less than 2% larger main filament.*

Figure 3: *Initiation of the filaments for two different choices of relative amplitude.*

Solving the differential equation given by the linear terms of Eq. (1) which are highlighted in blue, we obtain the eigenmode solution which is used to initialise the simulation at sufficiently small amplitude $h_i$, so that the linear terms are initially dominant:

$$\xi_{init} = h_i \cos(n_i y_0) \exp\left(-\frac{(x_0-x_{max})^2}{2\sigma_i^2} + \gamma_i t\right) \tag{2}$$

with the toroidal mode number $n_i$, the linear growth rate $\gamma_i = -\frac{n_i^2 \nu}{2C_0} + \sqrt{\frac{C_1}{C_0} - \frac{\sqrt{C_1|C_2|}}{C_0 n_i \Delta} + \frac{n_i^4 \nu^2}{4C_0^2}}$ and the Gaussian width $\sigma_i = \sqrt{\frac{4\Delta}{n_i}\sqrt{\frac{|C_2|}{C_1}}}$. One can superimpose solutions with different mode numbers. Here we chose a two mode system with the following features: $n_2 = 3n_1$ and $h_2 = 50h_1$ which provides initial conditons with a main central filament which is only very slightly larger than the two side filaments, see Fig. 3a. In this case the difference in the filament amplitude is less than 2%, see Fig. 3b. A second simulation, which will be referred to as the single mode simulation rather than the two mode simulation, is performed with three equal sized filaments with mode number $n_2$.

### 3. Results and Discussion

Comparison between the two simulations described in the previous section shows that they evolve very similarly in the linear regime of the evolution, where the linear drive term of the ballooning mode envelope equation is dominant, see top Figs. 4a and 4b. However, after entering the non-linear regime,

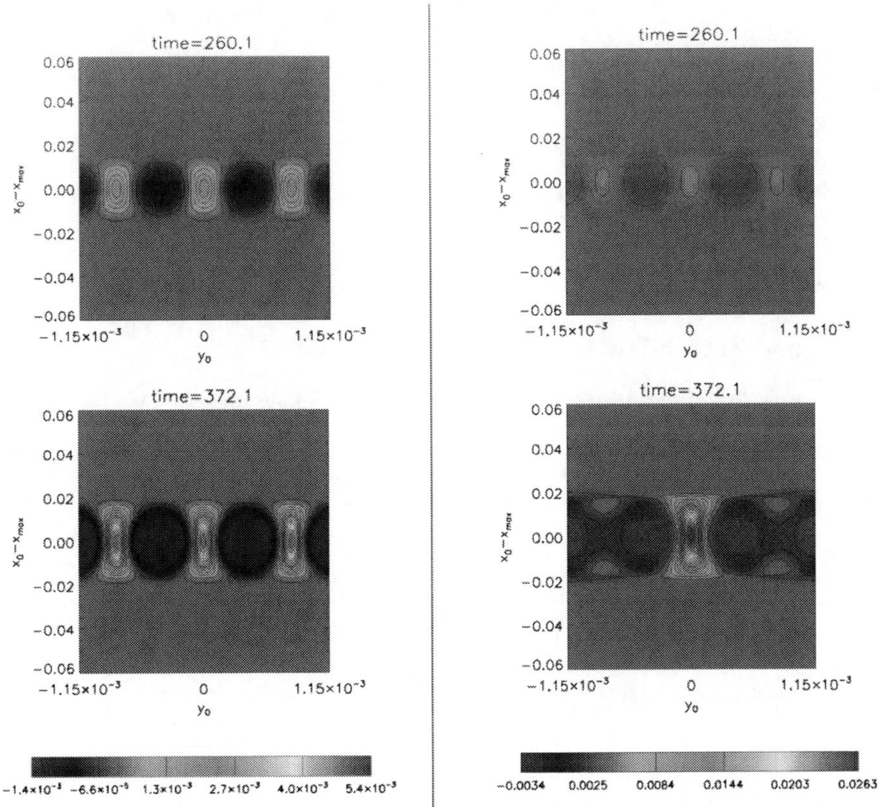


(a) *Initialised with three equal sized filaments. The top is at the beginning of the non-linear regime. The bottom shows the end of the simulation)*

(b) *Central, main filament is initialised slightly larger (less than 2%) than the two side filaments. Same times as in 4a.*

Figure 4: *Evolution of the filaments.*

the side filaments are slowed down and then suppressed, see Fig. 5. Comparing the height of the main filament of the two-mode simulation with the heights of the filaments of the single mode simulation, one finds the main filament is larger by a factor of approximately 5 at the end of the simulation, see bottom Figs. 4a and 4b. Therefore one can conclude that the main filament is gaining height from the interaction with the side filaments, which are themselves suppressed. The non-linear terms of the ballooning

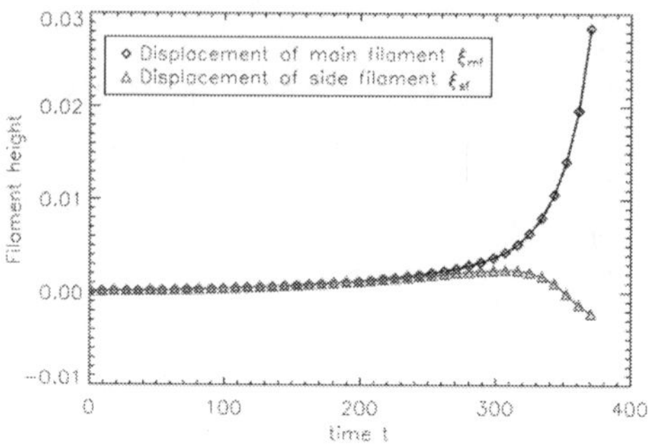


Figure 5: *Central filament (black) grows explosively. Side filaments (red) are suppressed in the non-linear regime.*

mode envelope equation are responsible for the enhanced growth of the main central filament and the suppression of the smaller side filaments. The quadratic nonlinearity has an averaged $\xi^2$ which arises from a low order incompressibility condition of the plasma. For the largest filament $\left(\xi^2 - \overline{\xi^2}\right) > 0$ and acts as a drive, while for the side filaments $\left(\xi^2 - \overline{\xi^2}\right) < 0$ acts to damp the eruption. This term acts as a down draft created by the main filament eruption, which pushes back down on the side filaments. The quasi-linear (cubic) nonlinearity stabilises the filaments at the most unstable height, $x_0 = x_{max}$, but is responsible for the widening of the main filament radially and driving wings of the side filaments, see Fig. 4b.

## 4. Conclusion and Future Work

We have shown how the interaction between filaments influences their evolution: Larger filaments gain amplitude from the interaction and grow faster than in the single mode case. Smaller filaments are suppressed in the non-linear regime. This interaction is caused by a down draft due to the incompressible plasma.

Future work is to calculate the coefficients $C_0$-$C_4$ and solve the non-linear ballooning equation, [7], for realistic tokamak geometry, including MAST and JET. This will allow a direct, quantitative comparison between theory and experimental observations.

## Acknowledgments

The support of the EPSRC for the FCDT (Grant EP/K504178/1) is gratefully acknowledged.

This work has been carried out within the framework of the EUROfusion Consortium and has received funding from the Euratom research and training programme 2014-2018 under grant agreement No 633053 and from the RCUK Energy Programme [grant number EP/I501045]. The views and opinions expressed herein do not necessarily reflect those of the European Commission.

## References

[1] A. Kirk et al, Phys. Rev. Lett., **96**, 185001 (2006)

[2] S.C. Cowley, B. Cowley, S.A. Henneberg and H.R. Wilson, arXiv:1411.7797, submitted to Proc. Roy. Soc.A

[3] S.A. Henneberg, S.C. Cowley and H.R. Wilson, submitted to Plasma Phys. Control. Fusion

[4] S.C. Cowley, M. Artun, and B. Albright, Phys. Plasmas **3**, 1848 (1996)

[5] O.A. Hurricane, B.H. Fong, and S.C. Cowley, Phys. Plasmas, **4**, 3565 (1997)

[6] S.C. Cowley, H. Wilson, O. Hurricane and B. Fong, Plasma Phys. Control. Fusion, 45, A31 (2003)

[7] H.R. Wilson and S.C. Cowley, Phys. Rev. Lett., **92**, 175006 (2004)

# Blob dynamics in TORPEX null point configurations

B. W. Shanahan[1], B. D. Dudson[1], F. Avino[2], I. Furno[2], P. Hill[1], and A. Fasoli[2]

[1]*York Plasma Institute, Department of Physics, University of York, York YO10 5DD, UK*

[2]*École Polytechnique Fédérale de Lausanne (EPFL), Centre de Recherches en Physique des Plasmas Association Euratom-Confédération Suisse, CH-1015 Lausanne, Switzerland*

## Introduction

Filaments, or blobs, are field aligned plasma structures which have been observed in the scrape of layer of magnetically confined plasmas. These filaments carry particles into the SOL and therefore play a role in determining the profiles during L-mode and inter-ELM H-mode scenarios. While there have been many investigations into the dynamics of such filaments [1, 2], few have concentrated on the explicit behavior near X-points. Magnetic tori such as the TORPEX device [3] replicate tokamak scrape off layer (SOL) scenarios while allowing straightforward diagnostic access. While filaments have been studied extensively experimentally within Torpex [4, 5, 6], no theoretical studies have yet explored the dynamics in X-point configurations [7]. The results presented here are a natural progression from previous studies in linear configurations [8], towards an ultimate aim of full tokamak edge simulations.

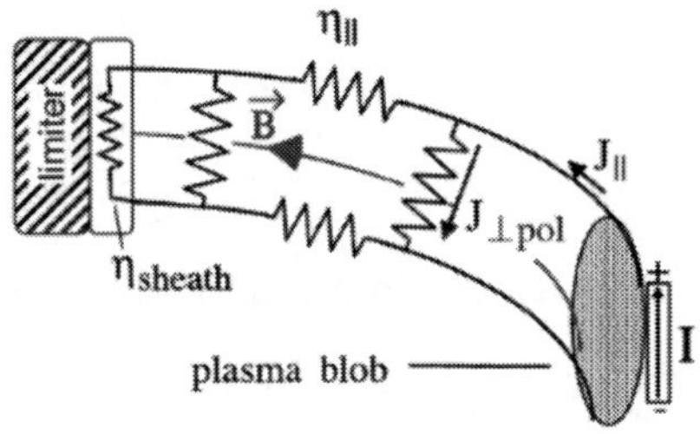


Figure 1: Circuit diagram indicating the mehanisms to resolve charge separation within a blob [2]

The dynamics of propogating filaments depends on the mechanism for charge dissipation within the blob and can be explained by considering the system as a circuit, Figure 1 [2]. If the charge separation caused by diamagnetic drifts is resolved primarily via the parallel current through the sheath, the filament is considered to be sheath-connected. If the connection length to the sheath is too high, or likewise the resistivity too large, charge is dissipated via the polarization current and the blob is considered to enter the inertially limited regime. The dependence of regime characterization can be detected in the perpendicular motion of the filament.

## Torpex null point scenarios

The aim of this work is to explore blob dynamics in the TORPEX device in X-point geometries [7]. The experimental magnetic geometry to be simulated is shown in Figure 2.

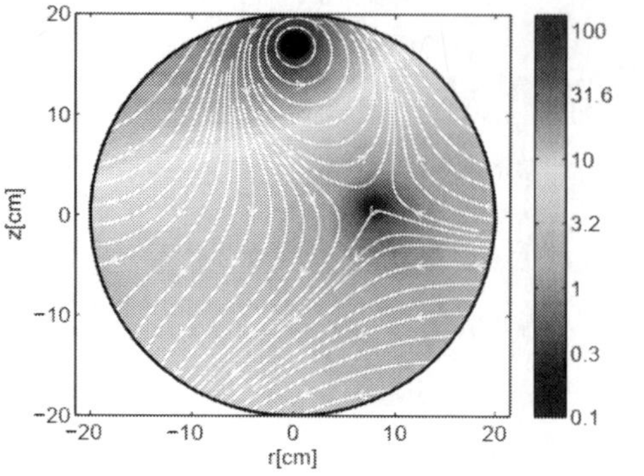


Figure 2: Magnetic geometry of Torpex X-point scenarios [7]

The TORPEX device has a major radius of 1m, minor radius of 20cm, and a toroidal magnetic field at about 75mT [7].

**Isothermal Model**

An isothermal cold-ion fluid model initially constructed for plasma blob studies [1, 9] has been extended for use in X-point scenarios [8]. The model is electrostatic, and inviscid; while the isothermal electron temperature $T_{e0}$ is 5eV. While this model includes several simplifications, it still captures relevant physics such as Kelvin-Helmholtz, interchange and driftwave turbulence, an important class of instabilities in tokamak edge plasmas [10], as it is a ubiquitous instability. The equations which are solved are given as follows in SI units:

$$\frac{dn}{dt} = 2c_s\rho_s\xi\cdot(\nabla n - n_0\nabla\phi) + \nabla_{\|}\frac{J_{\|}}{e} - n_0\nabla_{\|}u_{\|} \tag{1}$$

$$\rho_s^2 n_0\frac{d\Omega}{dt} = 2c_s\rho_s\xi\cdot\nabla n + \nabla_{\|}\frac{J_{\|}}{e} \tag{2}$$

$$\frac{du_{\|}}{dt} = -\frac{c_s^2}{n_0}\nabla_{\|}n \tag{3}$$

$$J_{\|} = \frac{\sigma_{\|}T_e}{en_0}(\nabla_{\|}n - n_0\nabla_{\|}\phi) \tag{4}$$

Where $\Omega \equiv \nabla_\perp^2\phi$ is vorticity, total derivatives are split via $\frac{d}{dt} = \frac{\partial}{\partial t} + \mathbf{u_E}\cdot\nabla + \mathbf{u}_{\|}\cdot\nabla$, and parallel derivatives are evaluated using $\nabla_{\|} = \mathbf{b}\cdot\nabla$. Curvature effects are included via the polarization vector $\xi \equiv \nabla\times\frac{\mathbf{b}}{B}$. In the above equations, $\rho_s = \frac{c_s}{\Omega_i}$ is the Bohm gyroradius. These equations are normalized such that density ($n$) is normalized to typical TORPEX values, $n_0 = 5\times10^{16}m^{-3}$, speeds are normalized to the sound speed, and $\phi = \frac{e\Phi}{T_{e0}}$ is the normalized electrostatic plasma potential. Boundary conditions for the vessel walls were set as the presheath boundary conditions for oblique magnetic fields [11]. Finally, the magnetic geometry shown in Figure 2 is implemented by specifying a function for the vector potential, $A_{ext}$, and modifying the parallel gradient operator such that: $\mathbf{b}\cdot\nabla f = \nabla_{\|}f - \left[\frac{A_{ext}}{B}, f\right]$.

This model differs from that used in reference [1] in that it incorporates parallel ion free streaming, $u_{\|}$, as parallel flows are vital when determining the effects of X-points. Additionally, energy conservation required the restriction that $n$ is considered constant ($n_0$) in terms where it is not differentiated, which is simply a limit of the imposed Boussinesque approximation which assumes that density fluctuations are small: $\nabla\times\left(n\frac{d\nabla_\perp\phi}{dt}\right) \approx n\frac{d}{dt}\nabla_\perp^2\phi$

**Simulation results**

Simulations were performed to validate the extension to toroidal geometries of the model described previously and to determine the nature of blob propagation within the TORPEX device.

As mentioned previously, we can determine the blob regime by examining the relative contributions of the given currents in a system. For the baseline TORPEX scenario($T_{e0} = 5eV, n_0 = 5 \times 10^{16} m^{-3}$), we found a peak parallel current density of $J_{\parallel} \sim 15A$ and a peak polarization current density of $J_{pol} \sim 800A$. This indicates that the filaments are inertially limited, as described in Figure 1. Previous studies on TORPEX filaments have also determined that blobs in Hydrogen plasmas to be inertially-limited [6].

To further analyze the filament dynamics, simulations were performed in various electron temperatures, as shown on the left in Figure 3. The center of mass velocity increases with increasing isothermal electron temperature. This result is expected as $E \times B$ velocity is proportional to electron temperature.

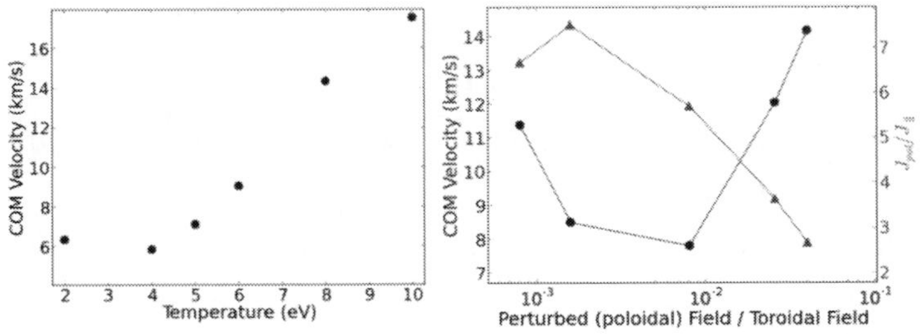


Figure 3: Parameter scans for blob propogation. Left: Center of mass velocity as a function of electron temperature. Right: COM velocity (blue) and Polarization current over parallel current (green) vs poloidal vector potential strength.

In addition, a scan of the poloidal magnetic field strength was conducted, as shown on the right in Figure 3.

The plot on the right in Figure 3 indicates somewhat unexpected results, in that there is a turnover in blob behavior at higher poloidal magnetic field strengths. One possible cause for this behavior is that as the poloidal field is increased, the parallel component of the poloidal blob propagation becomes more prominent, and therefore the blob is able to propagate faster.

**Experimental Comparison**

To further characterize the model and simulation methods, we have begun preliminary comparison to experimental measurements. Primarily, we have looked at time-averaged potential, as this is a straightforward and reliable

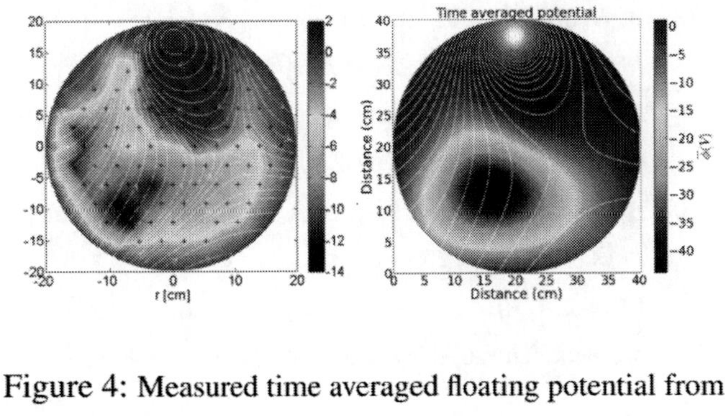


Figure 4: Measured time averaged floating potential from experiment (left), and simulations (right)

measurement within the TORPEX de-
vice. The results of this comparison are shown in Figure 4.

The discrepancy in the scales in Figure 4 are potentially due to the dependence of potential on temperature and density, and that our isothermal temperature assumption of 5eV should be reconsidered. While these are preliminary results, the experimental profiles share qualitative similarities to the results from simulation and merit further investigation.

## Conclusions and future work

We have successfully been able to model blob propagation in the TORPEX device. We have characterized the filament propagation regimes and established a promising correspondence with experimental measurements. The characteristics of the system can be explored further by varying the size of the initial seeding and implementing a realistic source within the system. We are also working toward extending our methods to full tokamak edge simulations in order to explore the mechanisms of blob propagation across the separatrix and null regions.

## References

[1] N R Walkden, B D Dudson, and G Fishpool. Characterization of 3d filament dynamics in a mast sol flux tube geometry. *Plasma Physics and Controlled Fusion*, 55(10):105005, 2013.

[2] D A D´Ippolito, J R Myra, and S J Zweben. Convective transport by intermittent blob-filaments: Comparison of theory and experiment. *Physics of Plasmas*, 18:060501, 2011.

[3] A Fasoli et al. Electrostatic instabilities, turbulence and fast ion interactions in the torpex device. *Plasma Physics and Controlled Fusion*, 52(12):124020, 2010.

[4] I. Furno, B. Labit, M. Podestà, A. Fasoli, S. H. Müller, F. M. Poli, P. Ricci, C. Theiler, S. Brunner, A. Diallo, and J. Graves. Experimental observation of the blob-generation mechanism from interchange waves in a plasma. *Phys. Rev. Lett.*, 100:055004, Feb 2008.

[5] M. Podestà, A. Fasoli, B. Labit, I. Furno, P. Ricci, F. M. Poli, A. Diallo, S. H. Müller, and C. Theiler. Cross-field transport by instabilities and blobs in a magnetized toroidal plasma. *Phys. Rev. Lett.*, 101:045001, Jul 2008.

[6] C. Theiler, I. Furno, P. Ricci, A. Fasoli, B. Labit, S. H. Müller, and G. Plyushchev. Cross-field motion of plasma blobs in an open magnetic field line configuration. *Phys. Rev. Lett.*, 103:065001, Aug 2009.

[7] F Avino, A Fasoli, and I Furno. The new torpex in-vessel toroidal conductor for the generation of a poloidal magnetic field. *Review of Scientific Instruments*, 85(3):–, 2014.

[8] B W Shanahan and B D Dudson. X-point modelling in linear configurations using bout++. *Journal of Physics: Conference Series*, 561(1):012015, 2014.

[9] J R Angus, M V Umansky, and S I Krasheninnikov. Effect of drift waves on plasma blob dynamics. *Physical Review Letters*, 108(21):215002, 2012.

[10] B D Scott. The nonlinear drift wave instability and its role in tokamak edge turbulence. *New Journal of Physics*, 4(1):52, 2002.

[11] J Loizu, P Ricci, F D Halpern, and S Jolliet. Boundary conditions for plasma fluid models at the magnetic presheath entrance. *Physics of Plasmas (1994-present)*, 19(12):–, 2012.

# Response of toroidal drift modes to profile evolution: a small-ELM model?

A. Bokshi[1], D. Dickinson[1], C. M. Roach[2], H. R. Wilson[1]

[1] *York Plasma Institute, Dept. of Physics, University of York, Heslington, York, YO10 5DD, UK*

[2] *CCFE Fusion Association, Culham Science Centre, Abingdon, Oxon, OX14 3DB, UK*

## Introduction

Toroidal drift instabilities are characterised by short wavelengths perpendicular to the magnetic field line and extended structure along it. And as a result of their radial localisation, these instabilities experience a weak radial equilibrium variation. The ballooning theory exploits this scale separation. To the lowest order in toroidal mode number $n$, the formalism provides the mode structure along the field line and local complex eigenvalue $\Omega_0(x) = \omega_0(x) + \mathrm{i}\gamma_0(x)$ ($x$ is a radial variable). To construct the full global mode structure and global (true) complex mode frequency $\Omega = \omega + \mathrm{i}\gamma$ from the leading order local results, we need to proceed to the next order in $n$. This higher-order theory then predicts two types of global mode structures depending on the equilibrium profiles for all toroidal microinstabilities [1, 2]: the *Isolated Mode* (IM) and the *General Mode* (GM). The IM exists for the special situation when the maxima in $\omega_0(x)$ and $\gamma_0(x)$ are co-located. This mode will typically balloon at the outboard midplane and have a strong global growth. The GM, however, does not have any constraint on $\Omega_0(x)$ and is therefore always accessible. It will peak at the top/bottom of the poloidal plane[1] and is much more stable.

## Model system

We demonstrate the essential physics with a global electrostatic toroidal fluid-ITG model for the perturbed potential $\phi_1(x,\theta)$ in a circular cross-section geometry with adiabatic electrons [3]:

$$\left[\rho_s^2\frac{\partial^2}{\partial x^2} - k_\theta^2\rho_s^2 - \frac{\sigma^2}{\Omega^2}\left(\frac{\partial}{\partial\theta} + \mathrm{i}nq'x\right)^2 - \frac{2\varepsilon_n}{\Omega}\left(\cos\theta + \mathrm{i}\frac{\sin\theta}{k_\theta}\frac{\partial}{\partial x}\right) - \frac{\Omega-1}{\Omega+\eta_s}\right]\phi_1(x,\theta) = 0. \tag{1}$$

The equilibrium parameters are defined in reference [3] (though note that $\Omega$ is normalised to the electron diamagnetic frequency $\omega_{*e}$). The parameters used in our simulations have been defined in table 1 and are broadly comparable to those found in the pedestal.

Table 1: *Equilibrium parameters used in simulations.*

| $a$ | $k_\theta\rho_i$ | $\hat{s}$ | $\varepsilon_n$ | $\tau$ | $q$ | $n$ | $m_0$ | $\gamma_E = \mathrm{d}\Omega_\phi/\mathrm{d}q$ |
|---|---|---|---|---|---|---|---|---|
| 0.5 | 0.2 | 25.0 | 0.08 | 1.0 | 1.4 | 50 | 70 | [-0.7, 0.7] |

[1]In general, the poloidal location where the IM/GM peaks would depend on profiles, shaping etc.

Eqn. 1 is solved by decomposing $\phi_1(x,\theta)$ into poloidal harmonics, i.e. $\phi_1(x,\theta) = \sum_m \phi_m(x)\exp(im\theta)$, and mapping the complex frequency $\Omega_m \to \mathrm{i}\partial/\partial t$ for each $\phi_m$. Toroidal flow-shear $\Omega'_\phi$ is included through a Doppler-shift, i.e. $\Omega \to \Omega + n\Omega'_\phi x$. An instantaneous complex mode frequency $\Omega_m(t) = \mathrm{i}\partial \ln \phi_m/\partial t$ is evaluated at the rational surfaces where the poloidal modes $m$ are expected to peak. For an eigenmode, we expect $\Omega_m(t) = \Omega$ for all $m$ and independent of time.

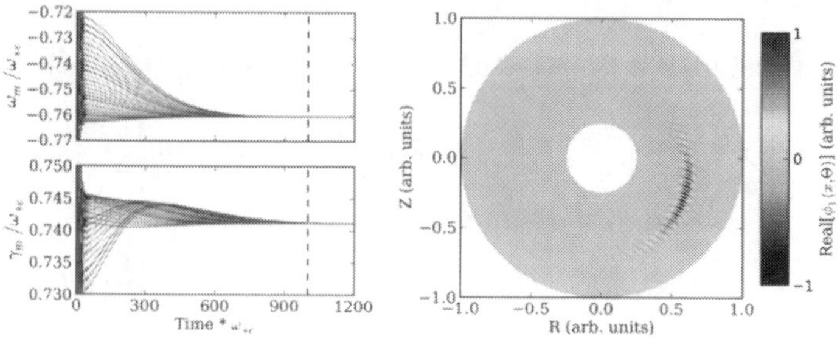


Figure 1: *(Left) Evolution of $\Omega_m(t)$. The instability is an eigenmode when $\Omega_m(t) \to \Omega$ (vertical line). (Right) The corresponding eigenfunction.*

**Stationary plasma profiles**

Simulations in this section were performed using the new initial-value code, holding all plasma profiles fixed in time. The simulations were initialised with noise, and after sufficient time, the solution converges to an eigenmode (Fig. 1).

**Effect of flow-shear**

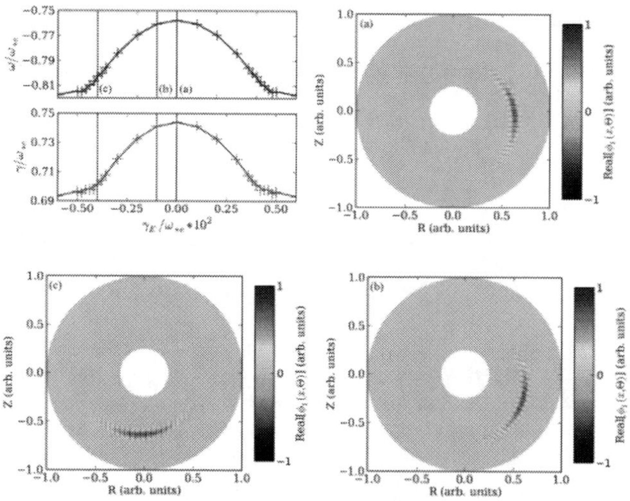


Figure 2: *Frame 1 shows the global $\Omega$ as a function of flow-shear $\gamma_E$. Subsequent frames show the evolution of IM into GM for indicated values of flow-shears.*

The IM is obtained for a quadratic ITG drive profile $\eta_s = \eta_g(1.0 - \eta_c x^2)$, with $\eta_g = 2.0$, $\eta_c = 62.5$ and flow-shear $\gamma_E = 0$. As we gradually increase $|\gamma_E|$, the peak in the real part of the local complex mode frequency $\omega_0(x)$ is shifted relative to $\gamma_0(x)$, and $\mathrm{d}\Omega_0(x)/\mathrm{d}x \neq 0$. The IM is then seen to smoothly evolve into the GM (Fig. 2), as also observed in [4, 5]. These results have been benchmarked against the eigenmode solution to eqn. 1 developed in [4] to within 0.1%.

**Time for eigenmode formation**

Starting with arbitrary initial conditions, the IM forms over $\mathcal{O}(10^2)$ and the GM structure forms over $\mathcal{O}(10^3)$ growth-times $\gamma T_{eig}$. These numbers suggest that non-linear regimes are likely to be entered before the linear mode structures are established. However, eqn. 1 is only valid for strongly unstable modes ($\eta_s \gg 1$).

If we increase $\eta_s$ by 100%, we find that the global growth rate has increased by $\sim 80\%$, whereas $T_{eig}$ only changes by 0.1%. Therefore close to marginal stability, it is possible that $\gamma T_{eig} \sim \mathcal{O}(1)$, and the structure and growth-rate of these linear modes may be important in determining non-linear transport. This should be investigated rigorously in the future using a more accurate plasma model.

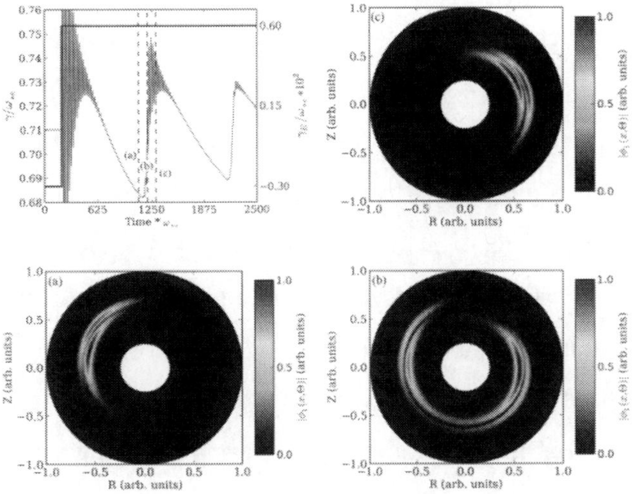


Figure 3: *Dashed vertical lines in frame 1 correspond to potential plots in subsequent frames (anti-clockwise).*

**Floquet Modes**

For plasma profiles held at high values of flow-shear ($\gamma_E \geq |0.4|$), we find that the linear mode rotates poloidally. It precesses rapidly through the bottom-half of the poloidal plane, slowing down in the top-half, performing many cycles, before eventually settling down at the top as a General Mode, with $\gamma_{FM}(t) \rightarrow \gamma_{GM}$ (Fig. 3). This is consistent with [6] which concludes that the Floquet form is generally a transient associated with starting conditions.

**Dynamic plasma profiles**

For results in this section, the flow-shear evolves over three time-scales, with all other plasma profiles held fixed.

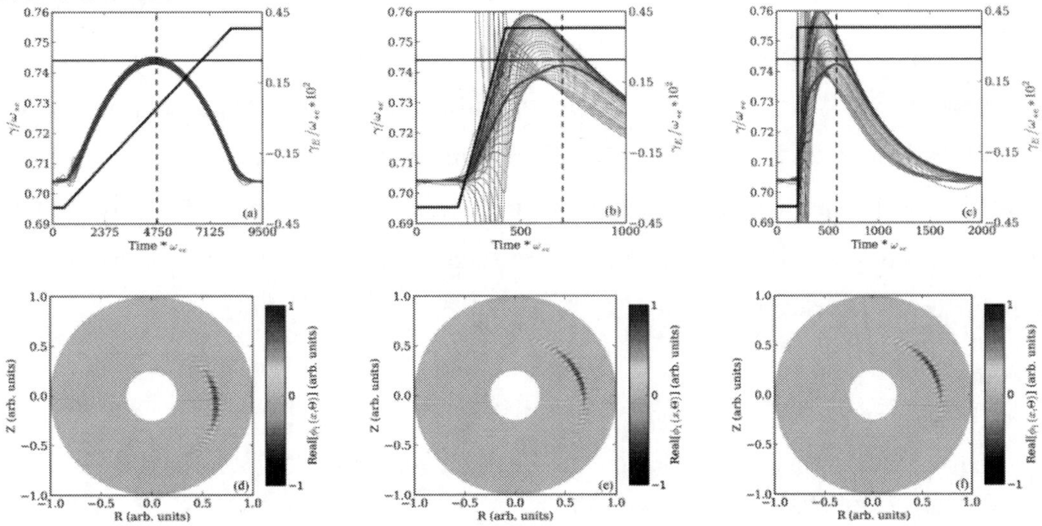


Figure 4: *Fig. 4a-4c show the evolution of the instability's global growth $\gamma$ (red) as a function of flow-shear $\gamma_E$ (blue). Fig. 4d-4f show the mode structure for the times indicated by the dashed-vertical lines in the frames above. The green-horizontal line indicates the IM growth-rate.*

**Eigenmode time-scale** For slowly varying profiles, the instability retains its eigenmode structure as it responds to the evolving flow (note that every $\gamma_m$ has the same value at each time - Fig. 4a). Assuming typical $\omega_{*e} = 10^5\ s^{-1}$, the flow-shear evolves over $\sim$ 100 ms time-scale.

**Small-ELM relevant time-scale** When profiles evolve over an $\mathcal{O}(1)$ ms time-scale, the eigenmode identity is lost. This is apparent since different Fourier harmonics grow at different rates (Fig. 4b). Nonetheless, $\gamma_{max} \sim \gamma_{\rm IM}$. However this $\gamma_{max}$ is realised $\sim$ 3 ms after the critical $\gamma_E = 0$ is passed (which would give the IM for flows held in time). Note this period is approximately the time it takes for the IM to form out of arbitrary initial conditions. Finally, we observe that the mode sits away from the outboard midplane at the time when $\gamma = \gamma_{max}$ (Fig. 4e).

**Instantaneous profile reversal** Starting with a mode sitting at the bottom of the poloidal plane, if the flow profile is immediately reversed such that the mode must balloon at the top, the instability again responds with a characteristic time, comparable to the time taken for the GM to form from initial noise. Further, all features identified for the small-ELM relevant ramp are recovered, and we additionally confirm that, independent of how rapidly the profiles change, the mode structure retains a *coherent* form as it gets convected poloidally (Fig. 4f).

## Discussion

Could then a GM-IM transition provide a burst of instability corresponding to a small-ELM? A number of issues need to be addressed before a conclusion can be drawn. First, for our fluid model, a high growth implies that non-linear regimes are likely to be entered much before the linear mode structure could respond during a GM-IM transition. It is important that we compare the GM-IM transition time to growth-times close to a more realistic, marginally-stable, GM stability boundary. Secondly, if a GM-IM transition does indeed provide a burst in transport, the profiles are likely to reset and the GM must form again before driving another crash in the next small-ELM cycle. For our ITG model parameters and typical values of $\omega_{*e}$, the GM formation time is significantly greater than the inter-small-ELM period ($\sim$ by a factor 10). Again one needs to ascertain this with a more realistic model of the pedestal, including kinetic ballooning mode physics, for example.

## References

[1] J W Connor, J B Taylor and H R Wilson, *Phys. Rev. Lett.* **70** (1993) 1803
[2] J B Taylor, J W Connor and H R Wilson, *Plasma Phys. Control. Fusion* **38** (1996) 243
[3] J W Connor and J B Taylor, *Phys. Fluids* **30** (1987) 3180
[4] D Dickinson, C M Roach, J M Skipp and H R Wilson, *Physics of Plasmas* **21** (2014) 010702
[5] P A Abdoul, D Dickinson, C M Roach and H R Wilson, *Plasma Phys. Control. Fusion* **57** (2015) 065004
[6] J B Taylor and H R Wilson, *Plasma Phys. Control. Fusion* **38** (1996) 1999

# The effect of clusters on the generation of high harmonics

M. Aladi, R. Bolla, I.B. Földes

*Wigner Research Centre for Physics of the Hungarian Academy of Sciences, Institute for Particle and Nuclear Physics, Department of Plasma Physics H-1525 Budapest, Hungary, POB 49.*

## 1. Introduction

High-harmonics as sources of attosecond pulses can be efficiently generated from cluster targets [1]. The problem of the use of cluster targets is that even for clusters ionization sets an upper limit for the applicable laser intensity [1,2] and thus, for the total harmonic energy. On the other hand, the generated high harmonics may become appropriate diagnostic tools for the internal structure of clusters.

The mechanism of high harmonic generation from clusters is still a subject of intense debates. A recent model of Ruf et al [3] suggests tunnel ionization from a partly delocalized electron wave function and recombination to this wave function, i.e. to the cluster itself.

Experiments were carried out using different noble gases from commercial valves with 1 to 12 bar backing pressures. He and Ne provided atomic gas targets, Xe was a source of clusters and Ar was intermediate depending on the pressure. The observation that in the presence of clusters the intensity of harmonics increases steeper with increasing pressure is in agreement with previous experiments [1] and it is a demonstration of the presence of clusters which is confirmed by Rayleigh scattering. High harmonics spectra show several interesting properties. Increasing the laser intensity results in a blue shift of the harmonics due to propagation effects both for atomic gases and for clusters. This is large enough for generating a continuously tuneable coherent light source in the EUV [2,4,5].

At modest intensities however - when free electrons in the interaction domain are rare – a spectral shift of opposite signature could be observed which increased with increasing pressure. Whereas the above-mentioned blue shift is attributed to the free electrons in the interaction region, this red shift is attributed to the nanoplasmas inside the clusters [5]. From the observed red shift we can even estimate the size and the density of the nanoparticles, thus it may even serve as a diagnostics of the clusters.

## 2. Experimental

A Ti:sapphire laser system was used delivering 4 mJ pulse energy of 40 fs pulse duration centered at a wavelength of 805 nm at 1 kHz repetition rate. The initial beam diameter was

9 mm. The linearly polarized laser beam was focused into the vacuum chamber using a plano-convex spherical lens of 30 cm focal length. Gas jets using a commercial valve (Parker series 9) were used with an additional nozzle. The density was characterized earlier using x-ray shadowgraphy [6]. The valve had an orifice of 0.99 mm diameter and it was opened for 1 ms duration. The backing pressure could be varied between 1 and 12 bar and the laser beam propagated 1 mm in front of the conical nozzle. The nozzle was situated some millimeters in front of the focal plane, thus the generating gas medium was in the diverging beam, preferring the short trajectories of electrons for the generation of harmonics.

A toroidal holographic grating (Jobin Yvon, 550 lines $mm^{-1}$) imaged the target onto the cathode of a microchannel plate (MCP) with a phosphor screen and the spectra were taken by a CCD. It is to be noted that the reflectivity of the grating decreases when the wavelength is shorter than 30 nm.

### 3. Results and Discussion

High harmonics were observed up to the 45$^{th}$ order with laser intensity increasing up to 1-2*$10^{14}$ W/cm$^2$ when saturation occurs due to the ionization of the media. Note that due to the relative high total gas density at the interaction range it occurs somewhat earlier than for low density gases. This saturation occurs both for atomic gases as He and Ne and for cluster targets as Xe. Fig. 1 illustrates the pressure dependence of the efficiency of harmonics in a typical case of 25ω.

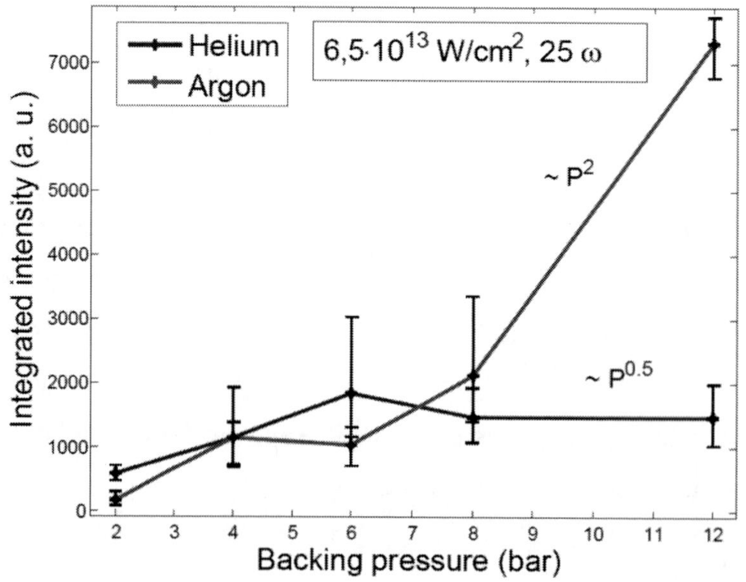


Fig. 1. Pressure dependence of the 25th harmonic in He and Ar for an intensity of 6.5×$10^{13}$ W/cm$^2$.

The conversion to harmonics shows that whereas in the atomic He the increase of harmonics intensity with pressure is modest and it saturates probably due to the increasing ionization, in the case of Ar a sharp increase sets in when – according to the Hagena-scaling [7] the clusters will consist of more than 1000 particles. This sharp increase is a well-known feature of harmonics generation from clusters [1].

An interesting phenomenon was observed in the case of relatively low, $6.5\times10^{13}$ W/cm$^2$ intensity in which case there are still not many free electrons in the interaction range. Harmonics spectra show a red shift of the harmonics, especially for Xe, in which case the clusters are relatively large. Increasing the pressure and consequently increasing the size of the clusters, this frequency shift reaches $\Delta\omega/\omega\approx10^{-2}$ for backing pressures above 8 bar. This corresponds – according to the Hagena-scaling [7] – to clusters consisting of more up to $10^5$ atoms. The more recent data of Dorchies et al [8] suggests – especially for the largest applied pressures - a slightly modified scaling which gives clusters consisting of $\sim6\times10^4$ atoms.

We can even use the observed wavelength shift to the independent estimation of the size and density of the clusters. According to Tisch [9] the phase mismatch due to the nanoplasmas in the clusters can be estimated as

$$\Delta k_{nanoplasma} = \frac{q\cdot 2\pi}{\lambda_\omega}\left(\sqrt{1-\frac{4\pi n_e r^3 n_{cl}}{3n_{qcrit}-n_e}} - \sqrt{1-\frac{4\pi n_e r^3 n_{cl}}{3n_{crit}-n_e}}\right).$$

Here $q$ is the order of the given harmonic, $\lambda_\omega$ is the laser wavelength, $n_e$ the electron density, $n_{cl}$ the cluster density, $r$ is the radius of the cluster (or nanoparticle), $n_{crit}$ and $n_{qcrit}$ are the critical densities for the laser and the given harmonic, respectively.

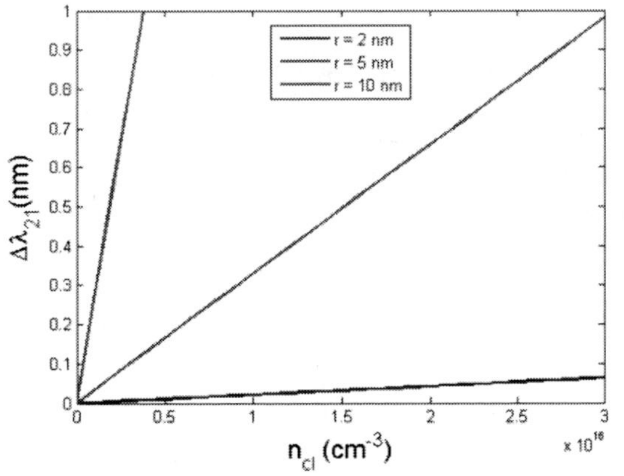


Fig. 2 Calculated wavelength shift of the 21st harmonic radiation as a function of cluster density with different assumed cluster radii..

Figure 2 compares the calculations with the experimental results as the wavelength shifts are illustrated depending on the density and radius of the cluster. We compare the calculations with the experimental results obtained for Xe with 12 bar backing pressure. The experiments gave a red shift of *0.5 nm* for the 21st harmonic in this case. Clearly, a very small radius of ~*2 nm* would require very high cluster density to reach the observed wavelength shift. In the case when the radius of the cluster is *10 nm*, then it consists of - taking account the van der Waals binding length for it to be 0.273 nm [10] – 60000 atoms in agreement with the scaling law [8]. In this case the cluster density is $1.9\times10^{15}\ cm^{-3}$ which corresponds to the earlier observed total atomic density of $10^{19}\ cm^{-3}$ [6]. If however a cluster radius of *5 nm* is assumed then the cluster consists of ~6000 atoms which is an order of magnitude less.

Our preliminary Rayleigh-scattering measurements using a 530 nm laser beam gave similarly, only some thousands of particles for the cluster sizes. Summarizing the results it can be concluded that measuring the detailed spectral properties of the generated harmonics gives not only a proof for the existence of nanoplasmas in the clusters but it also serves as an additional diagnostics for their structure.

**Acknowledgements**

This work has been carried out within the framework of the EUROfusion Consortium and has received funding from the Euratom research and training programme 2014-2018 under grant agreement No 633053. The views and opinions expressed herein do not necessarily reflect those of the European Commission. It was also supported by the COST MP1208 and MP1203 activities.

**References**

[1] T.D. Donnelly et al., *Phys. Rev. Lett.* **76**, 2471 (1996)
[2] M. Aladi and I.B. Földes, *Journal of Physics Conf. Ser.* **508**, 012016 (2014)
[3] H. Ruf et al., *Phys. Rev. Lett.* **110**, 083902 (2013)
[4] C. Altucci et al., *Phys. Rev. A* **61**, 021801(R) (1999)
[5] M. Aladi, I. Márton, P. Rácz, P. Dombi and I.B. Földes; High Power Laser Science and Engineering **2**, e32 (2014)
[6] R. Rakowski et al., *Nucl. Inst. Methods A*, **551**, 139 (2005)
[7] O.F. Hagena, *Rev. Sci Instrum.* **63**, 2374 (1992)
[8] F. Dorchies et al., *Phys. Rev.A* **68**, 023201 (2003)
[9] J. Tisch, *Phys. Rev. A* **62**, 041802 (R) (2000)
[10] G.M. Petrov et al., *Phys. Plasmas* **12**, 063103 (2005)

# Amplification of short laser pulses by Stimulated Brillouin Backscattering

T. Gangolf[1,2], L. Lancia[3], J.-R. Marquès[1], A. Giribono[3], K. Glize[1,4],
M. Blecher[2], L. Vassura[1,3], A. Frank[5], M. Quinn[6], M. Cerchez[2], C. Riconda[7],
S. Weber[6,8], M. Chiaramello[7], G. Mourou[7], O. Willi[2], J. Fuchs[1]

[1] *LULI, École polytechnique – CNRS – CEA – UPMC, 91128 Palaiseau, France*

[2] *ILPP, Heinrich-Heine Universität Düsseldorf, 40225 Dusseldorf, Germany*

[3] *Dept. SBAI, Università di Roma "La Sapienza", 00161 Rome, Italy*

[4] *CEA, Bruyères-le-Châtel, 91297 Arpajon, France*

[5] *GSI Helmholtzzentrum für Schwerionenforschung GmbH, 64291 Darmstadt, Germany*

[6] *IZEST, École polytechnique – CEA, 91128 Palaiseau, France*

[7] *LULI, Sorbonne Universités-UPMC–École polytechnique–CNRS–CEA, 75005 Paris, France*

[8] *Institute of Physics of the ASCR, ELI-Beamlines, 18221 Prague, Czech Republic*

## Introduction

In search for novel techniques for amplifying laser light to ever higher intensities, plasma amplification is being investigated by several groups, including the IZEST $C^3$ project [1]. A plasma-based approach benefits from the fact that a plasma can sustain much higher intensities than a solid state amplifier. In a plasma, energy can be transferred from one laser pulse (pump) to another (seed), either via a high-frequency plasma electron wave (stimulated Raman backscattering, SRS [2]) or by a low-frequency ion acoustic wave (stimulated Brillouin backscattering, SBS [3]). Especially, the strong coupling regime of SBS (sc-SBS) is of interest since seed pulses much shorter than the ion acoustic timescale $\frac{\lambda}{c_s} \approx 10$ ps can be amplified.

The strong coupling regime is reached when the the pump pulse $(\omega_0, k_0)$ is so intense that the plasma response is determined by the pump pulse. For this to happen, the pump must fulfill the threshold condition $(v_o/v_e)^2 = 4k_0 c_s \omega_0/\omega_{pe}^2$. Here, $v_o = eE/(\omega_0 m_e)$ is the electron quiver velocity in the laser electric field $E$, $v_e = \sqrt{kT_e/m_e}$ the electron thermal velocity, $c_s = \sqrt{ZkT_e/m_i}$ the ion sound velocity, and $\omega_{pe} = \sqrt{4\pi n_e e^2/m_e}$ is the electron plasma frequency. In more practical units,

$$I_{14}\lambda_{\mu m}^2 = 0.11\ T_{\text{keV}}^{3/2} \sqrt{\frac{Z}{A}\frac{n_c}{n_e}} \sqrt{1 - \frac{n_e}{n_c}}, \tag{1}$$

where $I_{14}$ is the intensity in $10^{14}\frac{W}{cm^2}$, $\lambda_{\mu m}$ is the pump wavelength in $\mu$m, $T_{\text{keV}}$ is the electron temperature in keV, $Z$ is the charge number, $A = m_i/m_e$ is the mass number, and $n_e$ is given in units of the critical density $n_c = \omega^2 \varepsilon_0 m_e/e^2$. Therefore, this regime is attained at high pump intensities, high plasma electron densities, and low plasma electron temperatures.

Then, the plasma wave is not a natural eigenmode of the plasma determined by the ion acoustic frequency but a quasi-mode with the dispersion relation $\omega_{sc} = (1+i\sqrt{3})/2\left(k_0^2 v_{osc}^2 \omega_{pi}^2/\omega_0\right)^{1/3}$ where $\omega_{pe} = \sqrt{4\pi Z^2 n_i e^2/m_i}$ is the ion plasma frequency so that the growth rate is $\gamma_{sc} = \Im(\omega_{sc})$ [4]. Advantages of amplification by sc-SBS compared to SRS are: The interaction length can be short, full pump depletion can be attained, pump and seed can have the same frequency, and the process is less sensitive to density inhomogeneities.

**Previous experiments**

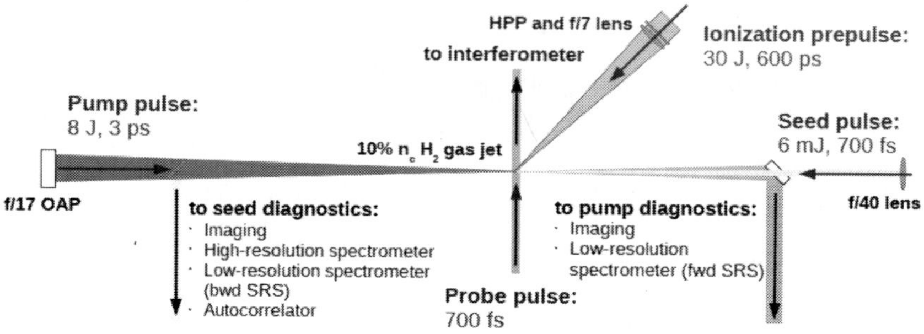


Figure 1: *Schematic setup of the 2013 experiment on SBS amplification at LULI.*

In the last experiments, the interaction between pump and seed was studied in a counterpropagating setup (Fig. 1) at the ELFIE laser facility at LULI ($\lambda$ = 1057 nm). The target was a preformed plasma, created using a long (ns) high energy laser pulse in order to create a plasma having an electron temperature of $T_e \approx 200$ eV and an electron density of $n_e \approx 0.1\, n_c$. Characterization of the plasma density by transverse interferometry did not allow to find the maximum density (a part of the interferogram was not accessible due to refraction of the probe), but the highest density observed can be given as a a lower limit for the plasma density: $10^{18}$ cm$^{-3}/p_{bar}$ where $p_{bar}$ is the nozzle backing pressure in bar ("meas." in Fig. 2).

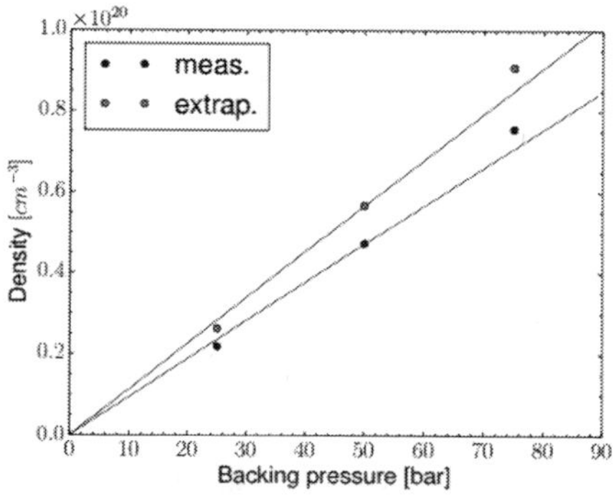


Figure 2: *Measured (blue circles) and extrapolated (green circles) on-shot plasma electron density from interferometry. Lines are linear fits.*

With these values and a pump intensity of $5 \times 10^{15}$ W/cm$^2$, the criterion in Eq. (1) is satisfied. The highest density observed can also be used to extrapolate the plasma density at the center of the gas jet using the values from ex-situ tomography, which yields $1.2 \times 10^{18}$ cm$^{-3}/p_{bar}$ ("extrap." in Fig. 2). This is 60 % of the expected value for full ionization, and we assume that we are close to the targeted values of up to $n_e \approx 0.1\, n_c$ .

In this beamtime, absolute amplification could be obtained [5]. These results are given in a companion contribution (Poster P2.224).

**Planned experiment: SBS amplification of high intensity pulses**

For future applications in the context of the $C^3$ scheme, it is planned to use the Brillouin backscattering process to amplify seed pulses up to intensities in the order of magnitude of $10^{18}\,\frac{\mathrm{W}}{\mathrm{cm}^2}$. As has been shown theoretically [6], amplification can be obtained even for seed pulses that are more intense than the pump pulse.

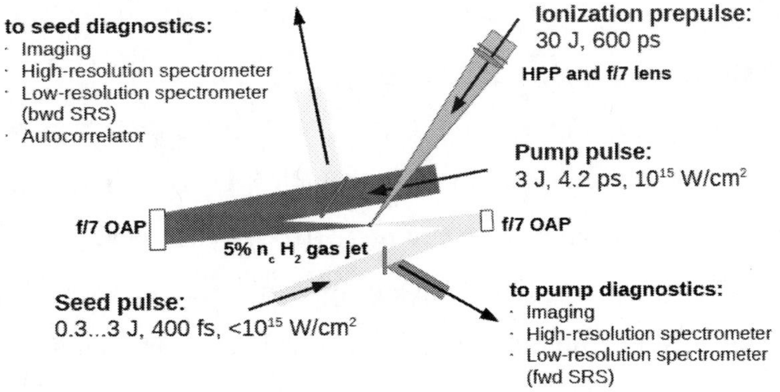


Figure 3: *Schematic setup of the planned experiment on SBS amplification at high seed intensities at the ELFIE laser facility at LULI.*

Therefore, we plan to study the process with an initial seed intensity as high as the pump intensity, mimicking a pulse that has already undergone some amplification (Fig. 3). A 3 J pump beam will counterpropagate against a 0.3 . . . 3 J seed. Being confocal with the same f number to maximize the overlap, the beams will be defocused to obtain different combinations of the intensities and spot sizes.

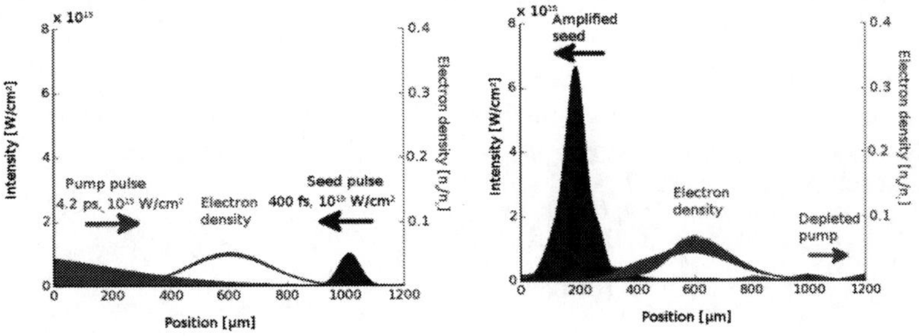


Figure 4: *1D 3V PIC simulation for a high intensity seed.*

The parameter space is limited by a maximum spot size of 90 $\mu$m (in order not to reduce the pump intensity below $10^{15}\ \frac{\mathrm{W}}{\mathrm{cm}^2}$) and a maximum seed intensity of $1 \times 10^{16}\ \frac{\mathrm{W}}{\mathrm{cm}^2}$ still with some defocus (60 $\mu$m spot size). This will allow us to study both the influence of the intensity and the spot size on the process.

The experiment is supported by simulations done by our group using the SMILEI PIC code. As shown in Fig. 4, the seed is amplified at high intensities and remains short.

**Planned experiment: SBS amplification of ultrashort pulses**

So far, SBS amplification has not been investigated experimentally for seed pulse durations $\tau_s < 400$ fs. This regime, however, is particularly interesting because the highest output intensities are easiest to achieve with shortest pulses as energy is limited. Furthermore, very short seed pulses are less subject to wavebreaking in the plasma [4]. On the other hand, the growth rate of sc-SBS limits amplification to a timescale $1/\gamma_{sc} = 215$ fs (for $1 \times 10^{16}\ \frac{\mathrm{W}}{\mathrm{cm}^2}$ pump). This will be investigated in an experiment (Fig. 5) at the ARCTURUS Ti:Sapphire laser facility (Düsseldorf, Germany). Initial seed durations down to 30 fs will allow us to verify up to which duration an initial seed remains short. A pump intensity up to $10^{16}\ \frac{\mathrm{W}}{\mathrm{cm}^2}$ (limited by the available energy) allows to remain in the strong coupling regime in spite of the shorter wavelength ($\lambda = 800$ nm). The diagnostics will be similar to those used in the previous experiments. In addition, they will include a SPIDER to obtain information on the spectral phase of the amplified seed.

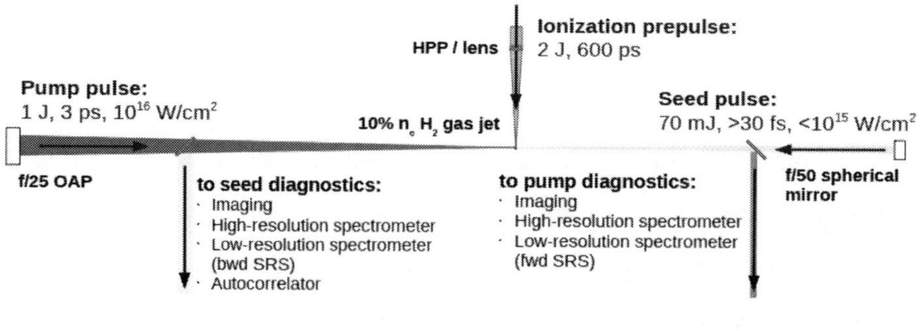


Figure 5: *Schematic setup of the planned experiment on SBS amplification of ultrashort pulses.*

## References


[1] G.A. Mourou et al., Optics Commun. **285**, 720 (2012)
[2] V. M. Malkin et al., Phys. Plasmas **7**, 2232 (2000)
[3] L. Lancia et al., Phys. Rev. Lett. **104**, 025001 (2010)
[4] C. Riconda et al., Phys. Plasmas **20**, 083115 (2013)
[5] L. Lancia et al., to be submitted
[6] S. Weber et al., Phys. Rev. Lett. **111**, 055004 (2013)

# Dynamical properties of dense plasma in inertial confinement fusion

T.S. Ramazanov[1], S.K. Kodanova[1], M.K. Issanova[1]

[1] *IETP, Al-Farabi Kazakh National University, Almaty 050040, Kazakhstan*

Recently, considerable attention of researchers is paid to the study of matter with high energy densities and, as a consequence, with high pressures and temperatures. The research in the field of inertial confinement fusion (ICF) on heavy ion beams takes a special place among the works on various aspects of this problem [1-2].

One of the most important parameters to describe the interaction of ions with matter is the energy of the projectiles. The stopping power is a parameter characterizing the rate of loss of the average energy of fast electrons or ions in plasma [3-4]:

$$\frac{dE}{dx} = 8\pi n\left(\frac{\mu_{\alpha\beta}}{m_\beta}\right)\cdot E_c \cdot b_\perp^2 \cdot \lambda_{\alpha\beta}\,, \quad (1)$$

here $E_c = \frac{1}{2} m_{\alpha\beta}\upsilon^2$ is the energy of the centre of mass, $\upsilon$ is the relative velocity of the scattered test particle, $b_\perp = \frac{Z_\alpha Z_\beta e^2}{2E_c}$, $\lambda_{\alpha\beta}$ is the Coulomb logarithm.

The Coulomb logarithm on the basis of the effective interaction potential of the particles is determined by the scattering angle of the pair Coulomb collisions. Introducing the centre of mass in the collision process the Coulomb logarithm reads [5-6]:

$$\lambda_{\alpha\beta} = \frac{1}{b_\perp^2}\int_0^{b_{\max}} \sin^2\left(\frac{\theta_c}{2}\right) b\,db\,, \quad (2)$$

The center-of-mass scattering angle can be obtained by the formula [5]:

$$\theta_c = \pi - 2b\int_{r_0}^{\infty} \frac{dr}{r^2}\left(1 - \frac{\Phi_{\alpha\beta}(r)}{E_c} - \frac{b^2}{r^2}\right)^{-1/2}, \quad (3)$$

here $m_{\alpha\beta} = m_\alpha m_\beta/(m_\alpha + m_\beta)$ is the reduced mass of the particles of kinds $\alpha$ and $\beta$; $b_\perp = Z_\alpha Z_\beta e^2/(m_{\alpha\beta}\upsilon^2)$, $b_{\min} = \max\{b_\perp, \lambda_{\alpha\beta}\}$ describes the minimum impact parameter, where $\lambda_{\alpha\beta} = \hbar/\sqrt{2\pi m_{\alpha\beta}k_B T}$ is the thermal de Broglie wave length.

In formula (3) $\Phi_{\alpha\beta}(r)$ is the interaction potential and $r_0$ is the distance of the closest approach for a given impact parameter b:

$$1-\frac{\Phi_{\alpha\beta}(r_0)}{E_c}-\frac{b^2}{r_0^2}=0. \tag{4}$$

It is known that in order to correctly describe static and dynamic properties of plasmas the collective screening effect is to be taken into account. In this work the dense plasma is considered for which quantum effects must be taken into account at short distances. Further, the effective interaction potential which including both charge screening at large distance and quantum effects at short distance will be used [7]:

$$\Phi_{\alpha\beta}(r)=\frac{Z_\alpha Z_\beta}{r\sqrt{1-\left(2\lambda_{\alpha\beta}/r_D\right)^2}}\left(\exp(-rB)-\exp(-rA)\right), \tag{5}$$

$$B^2=\frac{1}{2\lambda_{\alpha\beta}^2}\left(1-\sqrt{1-\left(\frac{2\lambda_{\alpha\beta}}{r_D}\right)^2}\right),\quad A^2=\frac{1}{2\lambda_{\alpha\beta}^2}\left(1+\sqrt{1-\left(\frac{2\lambda_{\alpha\beta}}{r_D}\right)^2}\right),$$

where $r_D$ is the Debye radius, $\lambda_{\alpha\beta}$ is the thermal wave length of the particles, and $m_{\alpha\beta}=m_\alpha m_\beta/(m_\alpha+m_\beta)$.

The basis of controlled nuclear fusion is to provide a fusion reaction of light nuclei. In this connection reaction involving the hydrogen isotopes deuterium and tritium (DT-cycle) are most important. The energy of the synthesis is arranged in the fuel. In case of DT reaction:

$$D+T\rightarrow\alpha(3{,}5M\text{э}B)+n\,(14{,}1M\text{э}B).$$

In this reaction, the total energy of 17.6 MeV is distributed between the $\alpha$-particle (3.5 MeV) and the neutron (14.1 MeV). To absorb the energy $\alpha$-particle, the size of the fuel has to exceed the range $\rho r$ [8]. The range of particles is determined as follows:

$$\rho r=\int_E^{E_0}\left(\frac{dE}{\rho dx}\right)^{-1}dE. \tag{6}$$

where the value of the stopping power $\frac{dE}{dx}$ is calculated in accordance with (1), $E_0$ is the initial energy of the particle. It should be noted the range as a product of mass density times distance, that means the unit of measurement is $g/cm^2$. For a more accurate description of the thermonuclear fusion also such parameters as mean deflection angle, stopping time, and depth of penetration of the ions have to be introduced. The stopping time of ion in DT plasma is determined by the following equation:

$$t_{dep}=\int_E^{E_0}\left(\frac{dE}{dt}\right)^{-1}dE. \tag{7}$$

The penetration depth of the ion with the initial energy $E_0$ can be calculated using equation:

$$\rho x = \int_{E}^{E_0} \langle \cos\theta \rangle \left( \frac{dE}{\rho dx} \right)^{-1} dE \,. \tag{8}$$

In the present paper the stopping time, the mean deflection angle, the depth of penetration, the effective range of the particles with different energies generated in the DT plasma. The stopping time of ion with initial energy $E_0 = 12\,MeV$ is shown in Fig. 1. The results presented in Fig. 1 show that the stopping time depends on the value of the initial energy as well as on density and temperature of the fuel. But the main problem is the stopping process of ions with energies up to 12 MeV during the initial time $t_{dep} \le 2\,ps$.

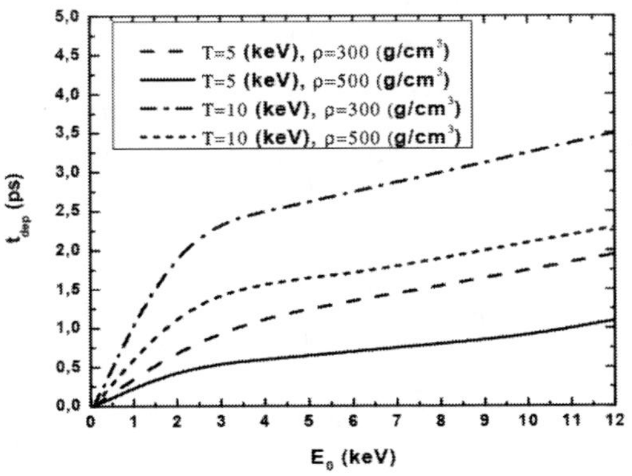


Fig. 1 – The stopping time of ions in DT plasma with initial energy $E_0 = 12\,MeV$.

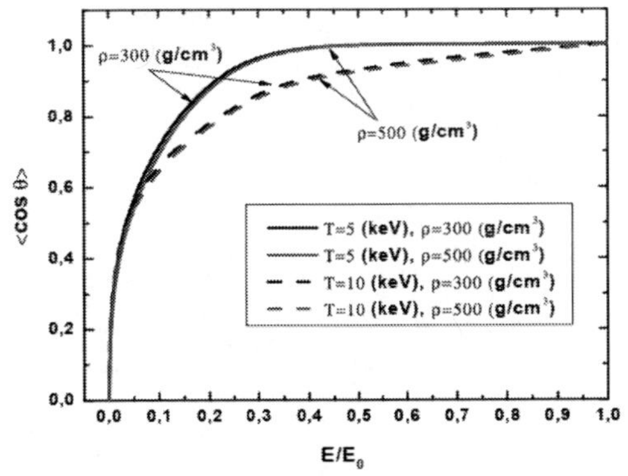


Fig. 2 – The mean deflection angle of ions in DT plasma with initial energy $E_0 = 12\,MeV$.

Figure 2 shows the mean deflection angle of the stopped ions with initial energy $E_0 = 12\,MeV$ in a DT plasma during the stopping process. The angle is presented as function of the residual energy after the collision process, normalized by the initial energy $E_0$. Fig. 2 shows that the incident ion is constantly changing the direction in the target while losing its energy. Thereat, the change of the mean deflection angle with the particle density is negligible. The influence of the temperature on the mean deflection angle is much more important. The mean angle increases if the temperature grows. The values of the stopping power and slowing down of the initial energy of the $\alpha$-particles under real conditions in the ICF target are presented in Fig. 3 and 4.

The range of $\alpha$-particles with an energy of 3.5 MeV in the plasma with a temperature of 30 keV is about $3\,g/cm^2$. Consequently, in order to realize an efficient self-heating due to the absorption of energy of the $\alpha$-particles by the fuel, the fuel must be brought to such conditions, where $\rho r > 3\,g/cm^2$.

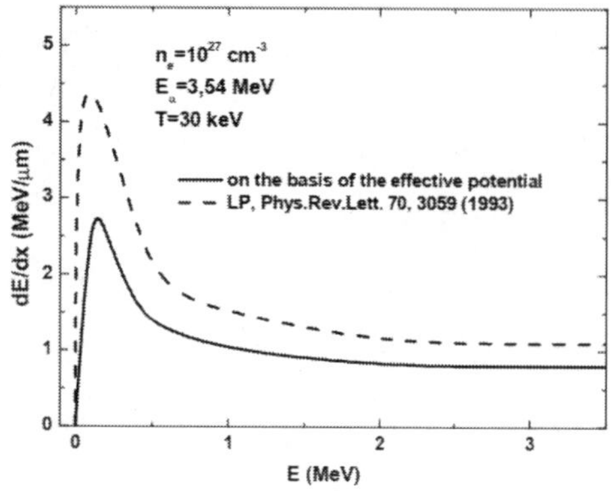


Fig. 3 – The stopping power of the $\alpha$ -particles with an initial energy $E_0 = 3.54\,MeV$ .

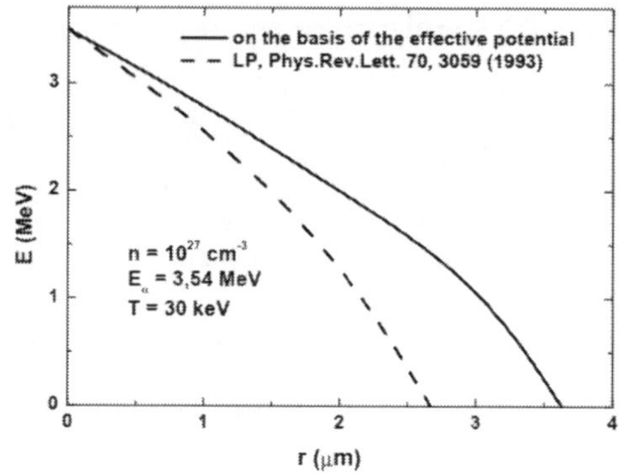


Fig. 4 – The slowing down energy of the $\alpha$ -particles with an initial energy $E_0 = 3.54\,MeV$ .

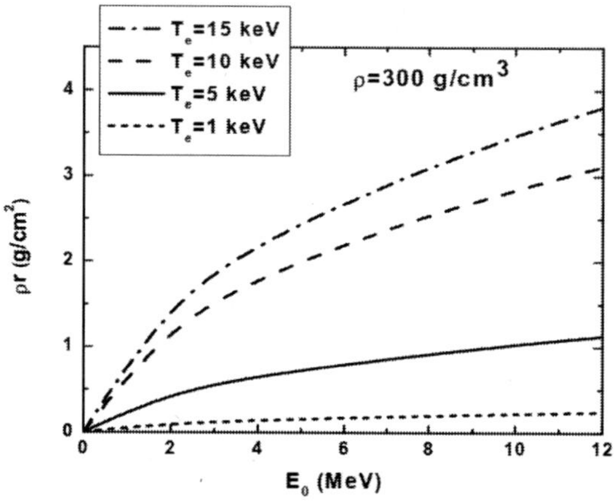


Fig. 5 – The range of protons in DT plasma with a density $\rho = 300\,g/cm^3$ .

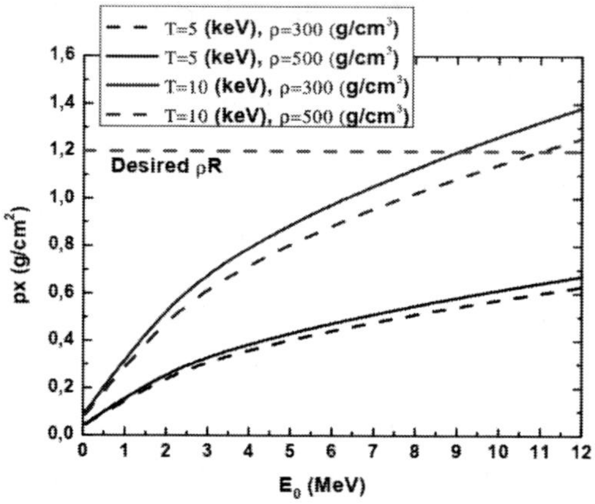


Fig. 6 – The penetration depth of the proton in DT plasma at different values density and temperature.

Figures 5-6 illustrate the dependence of the range and the penetration depth of the protons with different energies on the density and temperature of the target. The results show that at a lower target temperature of $T = 5\,\kappa eV$ , the target protons may conserve their energy inside the target if $\rho R < 1.2\,g/cm^2$. However, when the target is hotter $T = 10\,\kappa eV$ , the required initial energy of the proton is reduced to $E \leq 2\,MeV$ , in order to meet the required optimal deposition depth.

**References**


1. V.E. Fortov. Extreme states of matter. – Moscow, PHIZMATHLIT (2009).
2. V.E. Fortov. Extreme states of matter on Earth and in the Cosmos. – Berlin, Springer (2009).
3. C.A. Ordonez, M.I. Molina. Phys. Plasmas. 1, 2515 (1994).
4. T.S. Ramazanov, S.K. Kodanova. Phys. Plasmas 8, 5049 (2001).
5. G. Belyaev, M. Basko, et al. Phys. Rev. E 53, 2701 (1996).
6. A. Golubev, M. Basko. Phys. Rev E. 53, 2701-2707 (1996).
7. S.K. Kodanova, T.S. Ramazanov, et al. Contrib. Plasma Phys. 55, No. 2-3, 271 – 276 (2015).
8. M. Mahdavi and T. Koohrokhi. Physical Review E 85, 016405 (2012).

# Quasi monoenergetic electrons from shockwave injection

C. Koschitzki[1], M. Schnuerer[1], L. Ehrentraut[1]

[1] *Max Born Institut, Max Born Strasse 2a, 12489 Berlin, Germany*

## Quasi monoenergetic eletrons

It has been shown that injection in Laser Wakefiel Acceleration (LWFA) can be stimulated with density gradients of the plasma medium. As shown in [1] such a gradient can be introduced by placing a sharp edge into the flow of a super sonic nozzle (Fig. 4(b)). Energy spreads of the accelerated electrons of 3-5 MeV have been presented in the first experiment with maximum electron energies of 25 MeV. It was showen that the absolute energy spread does not increase with further acceleration [3]. At a 150 MeV the absolute energy spread remained at 3-5 MeV. As predicted in [2] we were able to reach electron beams with less than 1% energy spread (FWHM), when we accellerated shockwave injected beams to energies of 290 MeV.

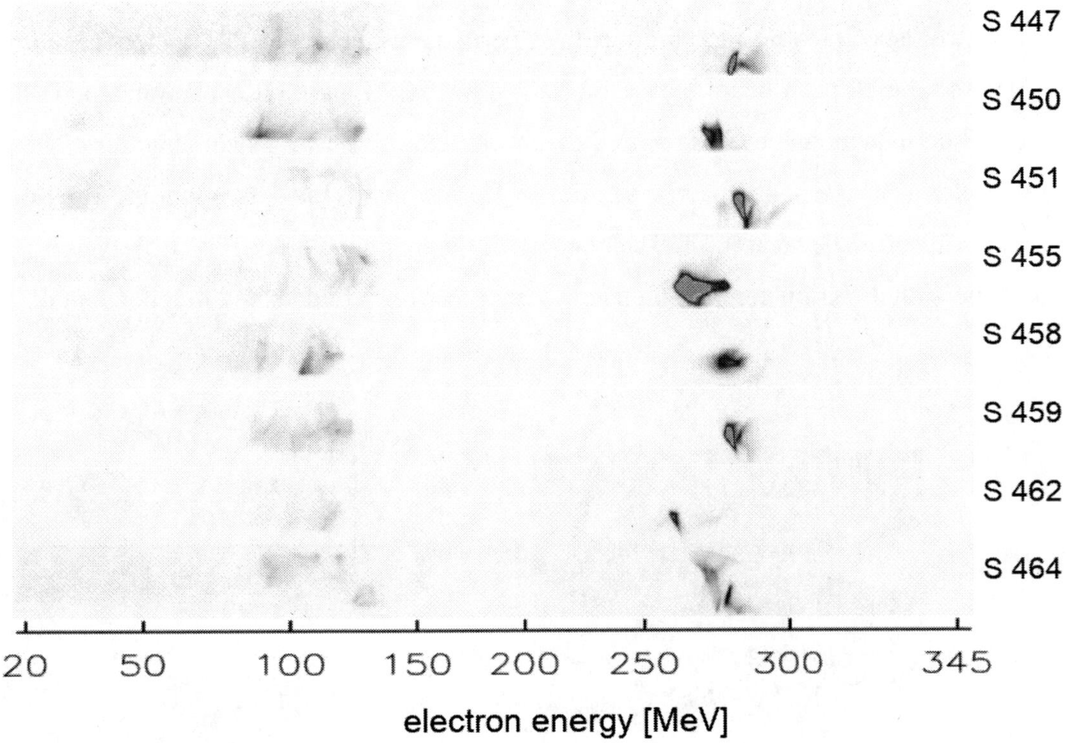


Figure 1: Camera images from magnet spectrometer. Corresponding shotnumbers are annotated to the right.

We used our Amplitude Pulsar 200 System beeing part of MBI's High Field Laser delivering 1.2 J on target with a pulse duration of 25 $fs$. The beam was focused with an f/15 parabola to a spotsize of $\sigma = 7.9\,\mu m$ , yielding a laser parameter $a_0 = 1.85$.

The quasi monoergetic beams have been accelerated above a C-D nozzle with an opening of 2 mm. The average density was $2 \times 10^{18}\,\mathrm{cm}^{-3}$. This yields a cold nonrelativistic wavebreaking limit of 135 GV/m which matches the achieved electron energies. High energy monoenergetic electrons can be achieved, when the razorblade reaches only 50 $\mu m$ into the opening of the nozzle. The razorblade is mounted at 100 $\mu m$ above the nozzle. 60 % of the measured spectra showed an energy spread of 2 % FWHM or less.

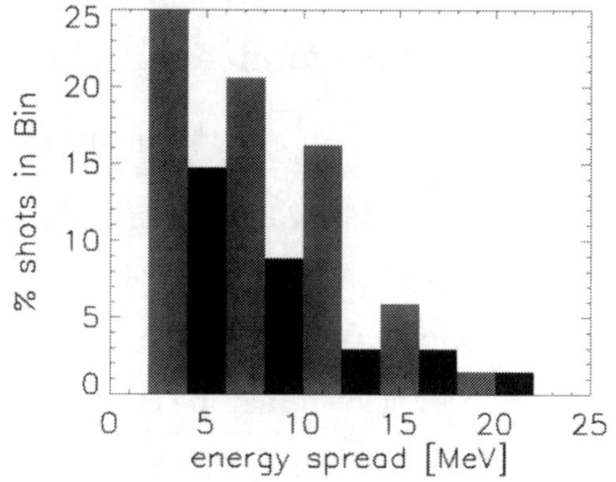


Figure 2: Energy spread distribution for 70 shots at 290 MeV. Bin size = 2MeV

**Magnet spectrometer**

Electron spectra were measured using a permanent magnet spectrometer (Fig. 3(a)), which dispersed the electron beam onto scintillating screens (Lanex Fast). The magnetic field of the NdFeB permanent magnets is focused to a peak field 0.7 T using steel chamfers. The peak field resembles a circular segment with a radius of 1.4$m$ which is the corresponding bending radius of a 1 GeV electron beam at 0.7 T. The gradient of the magnetic field (Fig. 3(b)) causes a focussing in dispersion direction creating a plane, where electrons are detected independent of their input

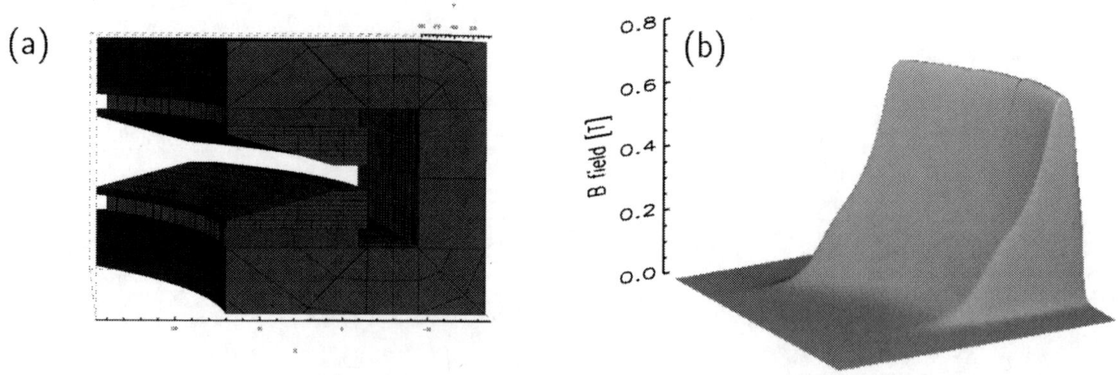


Figure 3: **(a)** - Model of the magnet spectrometer from 'Radia' field modelling software **(b)** - Measured magnetic field in the centered plane

angle, but only dependent on their respective energy. The scintillating screens are mounted directly onto the yoke on the side and the further end of the magnet spectrometer. Fig. 4(a) shows the deviation of the measured energy for electrons with a deviation angle of 1 mrad from the laser axis. An input aperture was placed in front of the magnet spectrometer accepting only electrons with deviations of $\pm 2.5$ mrad from the laser axis. To detect energy spreads of 4 MeV, the divergence angle of the electron bunches needs to be smaller then 2 mrad.

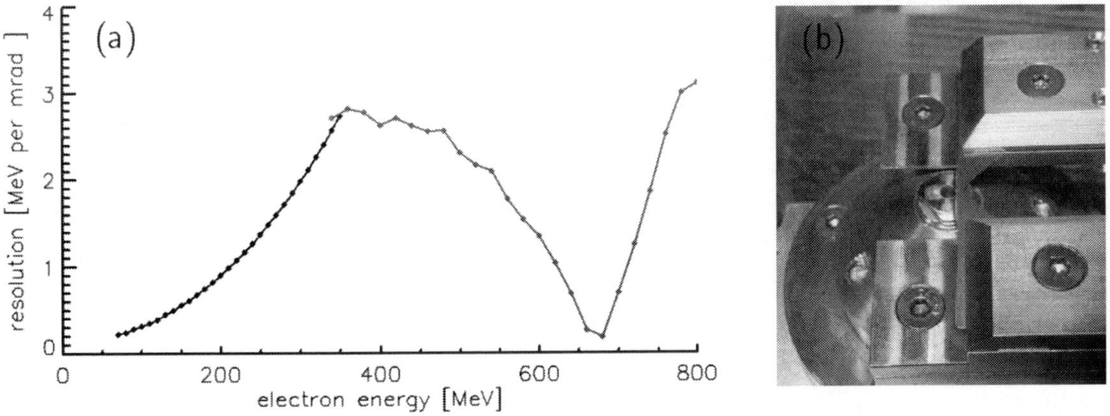


Figure 4: **(a)** - Plot of measured energy deviation per 1 mrad deviation of input angle **(b)** - Nozzle and razorblade

A LANEX Fast screen was placed into the electron beam path before the magnet spectrometer. The average beam divergence was measured at 1.5 mrad. This suggests that our measured spectra are actually resolution limited. The screen showed electrons for every shot and electron bunches with low divergence for 90% of all shots. The pointing has been measured to 3 mrad RMS. The pointing and an additional offset causes 60% of the accelerated electron bunches to miss the magnet spectrometer.

**Angular chirp**

While density gradients in the plasma target can cause a refractive steering of the laser and thus the electron beam. It has been showen that anuglar dispersion in the beam can also cause a steering and increased pointing of the electron beam [3]. It is suggested that angular chirp (AC) should be compensated to at least $0.2\,\mu\mathrm{rad/nm}$. In the horizontal plane this can usually be achieved by rotating the compressor gratings around the axis of the grooves. A detune of $\varepsilon$ then causes an angular chirp $d\theta/d\lambda = 2\varepsilon\tan\beta_0/(s\cos\alpha)$ with the groove spacing $s$, input angle $\alpha$ and dispersion angle $\beta_0$. This yields an AC of $0.2\,\mu\mathrm{rad/nm}$ per 0.1 mrad per detuning of our grating. Compensation of vertical AC can be achieved by rotating the grating around its surface normal. This rotation angle is not remotely controllable in our current compressor setup.

Hence we propose a different method to adjust the AC. As wedged windows cause a dispersive refraction as well, two identical wedges can be used to adjust the AC more precisely. The wedges are first placed to exactly compensate each other. As they are rotated in opposing directions their effects cancel out in one direction, but add up in the other. Thus with two fused silica wedges (wedge angle = 3°) we can apply an AC between negative and positive 2.1 $\mu$rad/nm by rotating the wedges over 180°. This option will be investigated in future experiments. The horizontal AC has been adjusted using the grating rotation to point electron beams through the aperture of the magnet spectrometer.

The research leading to these results has received funding from Deutsche Forschungsgemeinschaft within the program CRC/Transregio 18 and from LASERLABEUROPE (Grant Agreement No. 284464, EC Seventh Framework Program)

## References

[1] K. Schmid et al., Phys. Rev. ST Accel. Beams 13, **10**, 091301 (2010)
[2] A. Buck et al.,PRL 110 **11**, 185006 (2013)
[3] A. Popp et al.,PRL 105 **11**, 215001 (2010)

# Shock wave acceleration with short, high power laser pulses in two-ion-species plasma

Zs. Lecz[1], A. Andreev[1,2]

[1] *ELI-ALPS, ELI-HU Nkft, Szeged, Hungary*

[2] *Max-Born Institute, Berlin, Germany*

In the previous work [1] we have studied the proton acceleration in a proton plasma, which is basically produced by expanding a thin hydrogen foil initially at solid density. The dynamics of the plasma expansion together with the soliton propagation in this non-uniform medium results in the reflection of protons from the shock-front with relatively low energy spread. This mechanism has been proven for long pulses in near-critical plasma using long (picosecond) laser pulses at moderate intensity [2, 3]. In our study we have demonstrated that it works also with ultrashort high intensity pulses.

In the present work our goal is to investigate the same mechanism in the presence of multiple ion species, which is much easier to produce in real experiments by using CH targets, for instance. Thus the next step was to include a heavier ion beside the proton with density ratio $n_i/n_p$. The charge state $Z$ of heavy ions is a very important parameter, but we can include it in the parameter $n = Zn_i/n_p$, which gives the ratio of electrons originated from the heavy ion part and protons respectively. Another parameter, which needs to be included in a two-ion-species plasma is the mass ratio: $\alpha = m_i/(m_pZ)$. In the earlier work presented in Ref. [4] it was shown that the electric field acting on the light protons can be $\sqrt{\alpha}$ times greater than in the case of a plasma consisting only of protons (and electrons). Later we show that, in contrast with the results of Ref. [1], here the density scale length ($l_n$) at the rear side of the expanded plasma slab also becomes an important parameter.

Before we present the simulation results let us discuss theoretically the effect of these parameters: $n, \alpha$ and $l_n$. It is clear that the ion acoustic speed defines the dominant part in the velocity of reflected protons, which can be calculated from the dispersion relation of the plasma [5]:

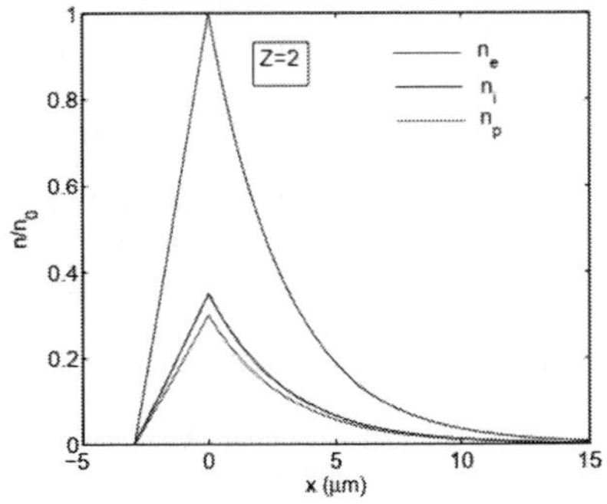


Figure 1: *Initial density distributions. The laser pulse comes from the left.*

$$C_{s,eff} = C_s \left( \frac{1+n/\alpha}{1+n} \right)^{1/2}, \quad (1)$$

where $C_s = \sqrt{T_h/m_p}$ corresponding to the proton plasma, where $n = 0$ and we obtain $C_s$ for the acoustic speed. The relation between the plasma length, density and pulse duration is the same as in [1]: the laser pulse has to be capable of heating all electrons in the target providing a nearly one-temperature hot plasma, which means: $l_f + l_n \approx ct_L$, where $l_f$ is the scale length at the plasma front side and $t_L$ is the laser pulse duration. In the presented simulations the electron and ion density profile is the same as in [1], only the absolute value has to fulfill the charge neutrality: $n_e = Zn_i + n_p$. The initial density distribution of different species is shown in Fig. 1.

The main difference, in comparison with the proton plasma case, is that the expansion of the background plasma is much slower, therefore $l_n$ can be considered constant. The velocity of protons is nearly uniform, because the electric field is also uniform [6]: $E_x = T_h/el_n$, and has the following form: $v_0(t) = F(\alpha)T_h t/(m_p l_n)$. The function $F(\alpha)$ is used to account for the effect of heavy ions [4] and we distinguish two regimes, where this function behaves as:

$$F(\alpha) = \begin{cases} \ln(n\sqrt{2\alpha}+\mathrm{e}), & \text{if } 0 \leq n \approx 1 \\ \sqrt{\alpha}, & \text{if } n \gg 1 \end{cases}$$

In the proton plasma case $n = 0$ has to be used, $l_n$ also depends on $v_0$ and in Ref. [1] we introduced $t_i$, which was the time when the reflection starts. In the one-ion-species plasma case we found that $v_0(t_i)$ does not depend significantly on $l_n$. Here the $l_n$ parameter comes also into play, which we assume to be constant during the acceleration, because the heavy ions can be considered immobile. This approximation is valid for large $n$ values, when the portion of protons in the plasma is small. Depending on $n$ and $\alpha$ parameters the reflection can occur before or after the proton solitary wave reaches the heavy ion front, which influences the energy.

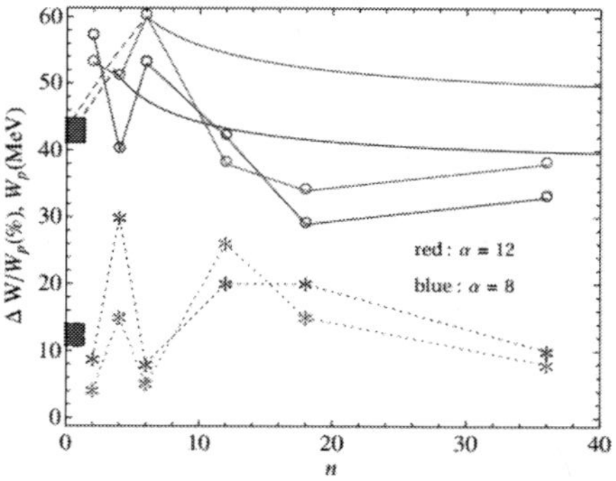


Figure 2: *Proton energy in the peak ($W_p$, joined circles) and relative energy spread ($\Delta W/W_p$, joined stars) as a function of n and $\alpha$ parameters. The full ( $n \gg 1$) and dashed (small n) lines show the results of analytical model described by Eq. (2). The green squares represent the result obtained in the case of proton plasma.*

Numerous simulations have been performed for different $l_n$, $\alpha$ and $n$ parameters. The laser intensity is always the same, $I_L = 9 \cdot 10^{20}$ W/cm$^2$, and uniform after 6 fs linear ramp. The pulse duration is 20 fs, the peak plasma density is $8n_{cr}$ and $Z = 2$ or 4 was used. The charge state, as a parameter, is not included explicitly in the modeling, therefore its value is not important: for any $Z$ value one can find mass and density ratios for a given $(n, \alpha)$ parameter-set. In Fig. 2 the results of a parameter-scan using PIC simulations are shown with $l_f = l_n = 3\mu$m. As we expected, the energy increases with $\alpha$, but for higher $n$ values, the energy is smaller. Another interesting result is that the energy spread is below 20 % in most of the cases.

Based on the knowledge gained from Ref. [1] we can derive an expression for the velocity of reflected protons. First we estimate the velocity of background plasma $v_0 = C_s\tau F(\alpha)$, where we introduce a dimensionless quantity: $\tau = C_s t/l_n$. For our simulation parameters it is easy to prove that $\tau$ is on the order of unity. Another parameter is introduced in order to simplify the final expression: $N_\alpha = ((1+n/\alpha)/(1+n))^{1/2}$, thus the velocity of reflected protons in the moving frame [7]: $v_r = 2M_{sh}C_sN_\alpha$, where $M_{sh}$ is the shock Mach number varying between 1 and 3 [8]. Now the final velocity in the laboratory frame reads:

$$\frac{v_f}{c} = \frac{C_s}{c}\frac{\tau F(\alpha) + 2M_{sh}N_\alpha}{1 + 2M_{sh}\tau N_\alpha F(\alpha)(C_s/c)^2}. \qquad (2)$$

This equation is the same as the one provided in Ref. [1] because in the limit of proton plasma ($n = 0$), when $F(\alpha) = 1, N_\alpha = 1$ and $M_{sh} = 1.6$, we obtain: $v_f = C_s(\tau + 3.2)/(1+3.2\tau(C_s/c)^2)$, where $\tau \approx 0.8$ was obtained [1]. By using the electron temperature formula from Ref. [1] we know that for our laser parameters $(C_s/c)^2 \approx 5\cdot 10^{-3}$, for the Mach number we can take $M_{sh} =$ 1.6 and for the acceleration time $\tau = 1$. In Fig. 2 these parameters are used, which shows qualitatively good agreement with the simulation results. At large $n$ values $\tau$ varies also, which makes the mechanism more complicated. This figure shows that the maximum proton energy can be increased by including also heavy ions, which provide a stronger acceleration of the background protons in the downstream plasma.

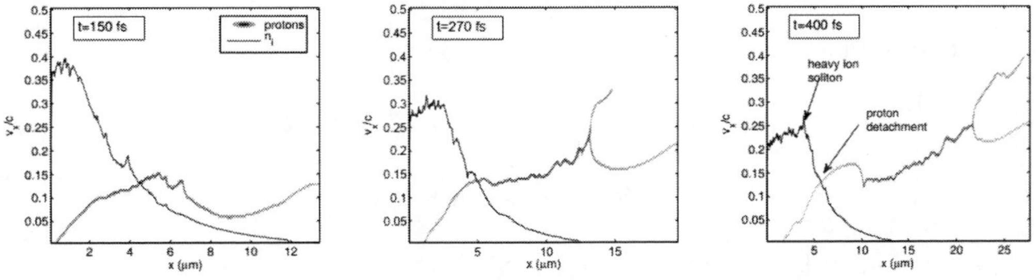


Figure 3: *Proton velocity phase-space (gray color) and heavy ion density profile (blue line) at three time instances for $l_n = 3\,\mu m$, $n = 6$, $\alpha = 12$. The ion density is normalized to $10^{28}$ $m^{-3}$.*

In order to get a deeper insight into the acceleration process we show the time evolution of the ion density profile and proton velocity phase-space in Fig. 3. First of all we see that the final velocity of reflected protons is mostly defined by the background velocity of protons, which is $v_0 \approx 0.17c$ at the beginning of reflection. During the whole acceleration process the density of heavy ions does not change significantly, only at the end a small peak appears, which corresponds to a heavy ion soliton. Another observable feature is the detachment of the protons from the plasma: the velocity and density profiles will not be continuous, but it is cut by the electric field associated with the heavy ion soliton (at $x \approx 5\,\mu$m). After the protons are pushed out from the plasma $v_0$ further increases due to self expansion and the proton reflection continues. Finally the reflected protons are cumulated in a small energy range around 60 MeV, which is about 40 % higher than the energy obtained in Ref. [1] with the same laser pulse.

In reality much higher charge states are present in heavy ion plasmas, which means higher $Z$ value. However, if we consider $Z \gg 1$, then the ions must be much heavier than protons and their density should be much smaller than the proton density in the material in order to obtain the results presented here. We have seen that the energy of proton bunch can be higher if heavier ions are also present, but it requires an optimal parameter-set, which is not trivial to achieve in experiments. The CH (or organic) materials do not seem to be adequate, because they do not meet these conditions, unless the low charge state is somehow maintained during the interaction. Preliminary 2D simulations show that high electron current is maintained by the protons moving relative to the heavy ion bulk, which in turn results in the development of Weibel instability [9, 10] for a time much longer than in the purely proton plasma [1]. This effect should be avoided to ensure the same proton beam quality as in 1D. In conclusion we can say that shock wave acceleration works also in multi-species plasmas with possible improvements.

**References**


[1] Zs. Lecz and A. Andreev, Phys. Plasmas **22**, 043103 (2015)

[2] Dan Haberberger et al., Nature Physics **8**, 95 (2012)

[3] Charlotte A. J. Palmer et al., PRL **106**, 014801 (2011)

[4] V. T. Thicknochuk et al., Plasma Phys Control. Fusion **47**, B869 (2005)

[5] Noah Hershkowitz and Young-C Ghim, Plasma Source Sci. Tech. **18**, 014018 (2009)

[6] T. Grismayer and P. Mora, POP **13**, 032103 (2006)

[7] A. Macchi et al., PRE **85**, 046402 (2012)

[8] F. Fiuza et al., Physics of Plasmas **20**, 056304 (2013)

[9] Peter H. Yoon and Ronald C. Davidson, Phys Rev A **35**, 2718 (1987)

[10] Y. Sentoku, K. Mima, S. Kojima and H Ruhl, Physics of Plasmas **7**, 689 (2000)

# Visible tomography of coherent modes in a linear magnetized plasma

P.A.S. David[1], A. Escarguel[1], Y. Camenen[1], J-P. Boeuf[2]

[1] *PIIM - UMR 7345 CNRS, Aix-Marseille Université, Marseille, France*

[2] *Université de Toulouse UPS, INPT / CNRS - LAPLACE, Toulouse, France*

## Introduction

In this work, we study low frequency instabilities regularly, called 'flute modes', rotating around thie axis of the linear magnetized plasma device MISTRAL [1], [2] (see Fig.1). Different diagnostics have been used to study them, such as a fast camera, Langmuir probes, and laser induced spectroscopy. They have shown that those instabilities are driven by the local electric field creating an $\vec{E} \times \vec{B}$ drift, but there might also be a ionization front due to the azimuthal propagation of a shock wave [3], [4]. A simple analytical model has also been developed to study the physical parameters that control the plasma rotating frequency[5]. However, the physics of the instability is not yet completely understood.

We will use a newly implemented tomographic diagnostic, giving a spatial resolution without intrusion, to complement previous measurements. Then results from PIC simulations with parameters very close to the experiment will be shown.

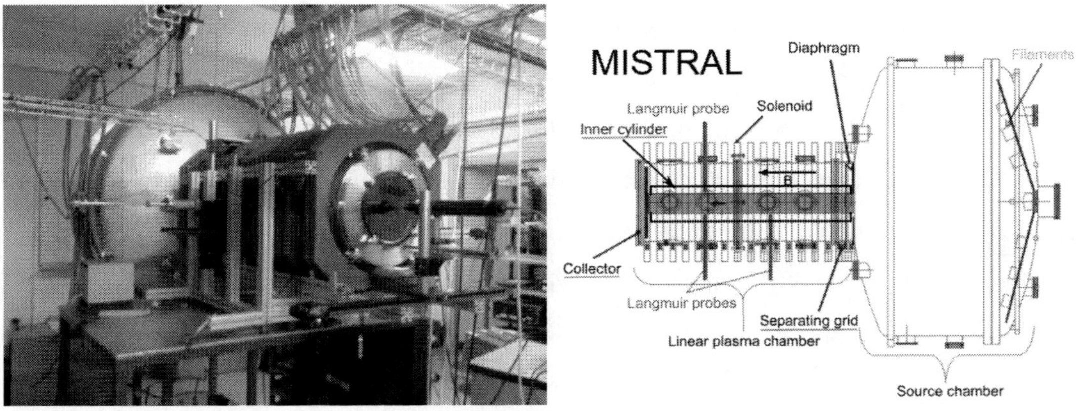


**Figure 1** – MISTRAL device - Items underlined in green can be individually polarized to extract or to further confine electrons or ions in both longitudinal and radial direction.

## Tomography diagnostic

In MISTRAL, the plasma is created in a large source chamber (1.4 m diameter, 1 m length) using a thermoionic discharge with 32 tungsten filaments and is then injected in a one meter long cylindrical magnetized interaction chamber. The plasma's behaviour is affected by several tunable parameters, such as the gas pressure ($1-6.10^{-4}$ mbar), the discharge energy ($30-40$ eV),

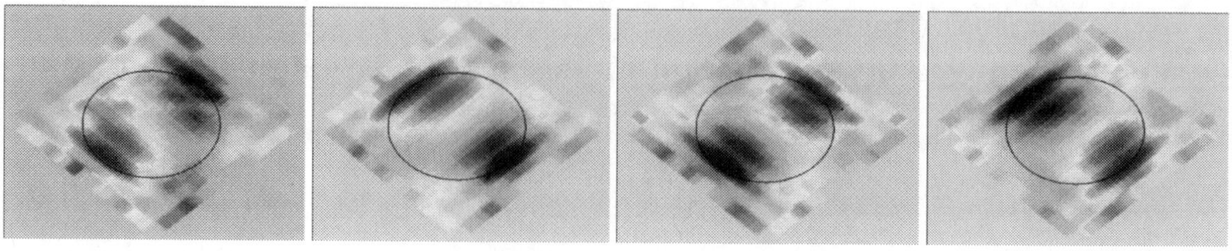

**Figure 2** – Normalized emissivity pertubation ($\sim \tilde{n}_e$) every 38.4 μs given by multiple-tomographic inversion on different pixels distributions (12x12 to 15x15 squares) for each experimental data. The black circle represent the position of the diaphragm.

the magnetic field intensity (150 Gauss), and the polarization of each end of the column (referred to as *separating grid* and *collector* in Fig.1) and of the inner cylinder (−25 to +25 V). The given windows of parameters are the one chosen to get stable modes rotating at 3 to 8 kHz.

We take advantage of the stability of those modes during several hours to simulates 32 simultaneous acquisitions using an optical single fibre moved to different positions and synchronizing the acquisitions with the plasma rotation. The fibre is collimated to collect light from a plasma cone between 8 and 13mm wide. A carbon based coating has been applied inside the vessel to limit excessive noise due to light reflections. Depending on the brightness of the plasma, two optical detectors can be used: a photomultiplier with a gain of about $10^{6-7}$, or a photodiode. Both detectors measure the visible light from the plasma without spectral filtering. The light mainly comes from the argon lines, which in the coronal model, is directly proportional to the electron density $n_e$. Different tomographic inversions are made to compare the quality of each numerical method, and especially the pixels shapes.

The tomographic inversion diagnostic consists in measuring the plasma from $n$ chords (collimated lines of sight), and using them to reconstruct the local plasma's emissivity. Here we use a finite-element technique [6] requiring to divide the cross-section of interest into $m$ pixels then to calculate the contribution of each pixel to each detector to write the linear system $S = L.E$, S and E being the vectors of the $n$ measurements, and of the $m$ unknown emissivities respectively, and the $n$ by $m$ matrix $L = (l_{ij})$, the volume of the $i^{th}$ chord inside the $j^{th}$ pixel. The main problems are that the matrix $L$ is very ill conditioned, and that having a large number of pixels a limited number of measurements makes the system under-defined. So instead of directly solving this system, we look for $E$ minimizing $\phi = \frac{1}{2}(L.E - S)^2 + \alpha\mathscr{R}$, with $\alpha$ a positive weighting parameter, and $\mathscr{R}$ a regularizing functional, here taken as the first order spatial derivative of the system [7]. No further hypothesis is made, for instance on the plasma shape or position.

Several parameters can cause numerical artefacts. The most important one is when measurement chords and pixels positioning are almost aligned, for instance with parallel chords (e.g. vertical and horizontal) and a grid of square pixels. This is due to the first order differential regularization that causes the numeric signal to be spread along the direction of the spatial derivative, which is amplified by the geometry of both the pixels and the chords. On the other hand, such a regular grid has very simple and well defined finite differences, thus, improving the numerical calculations. To overcome this problem the mesh grid is rotated so that the pixels' sides and distribution are not parallel to the measurements. Remaining artefacts being much less important and more random, they can be further mitigated by doing several inversions on the same experimental data, but with different numerical pixels configurations (pixels numbers and rotation angle). The fig. 2 shows the time evolution of the emissivity perturbation obtained from tomographic inversions on different rotated 12x12, 13x13, 14x14 and 15x15 square grids from which the time averaged emissivity over several rotations is subtracted and no spatial smoothing was used. The frequency of the rotation from the tomography (one full rotation in $\sim (4 \times 38.4\ \mu s)^{-1} \simeq 6.5$ kHz) corresponds to the frequency measured with a Langmuir probe. Additionally, we can see the density decreasing (blue region) on the opposite side of the expelled plasma arm (red region), which could not be measured with any other diagnostic and is consistent with numerical simulations showing the whole plasma center being dragged toward the instability.

Similar tests have been made with other configurations (up to seven time as many pixels as chords, different pixel shapes, other regularizing conditions) and can give similar results, but the previously described grid have given the most consistent ones. Finally, this study based on a single line of sight have been used to design a high resolution diagnostic to be installed during summer 2015 with a total of 128 lines of sight. The new set-up will first be tested and validated on global modes and then be used to study local turbulent modes.

## Numerical simulations

A circular cross section of a cylindrical magnetized plasma is simulated using a Particles-In-Cell Monte-Carlo-Collisions code developed at LAPLACE (Toulouse, France) [8]. The third dimension is solely used for the energetic electron injection and particles loss through the extremities of the column so only modes with axial symmetry can be described. Here, as in the experiments, the key parameter to generate rotation modes is to extract enough particles in the radial direction compared to the axial one, or similarly to have a better confinement in the axial direction. In those simulations, very close to the experiment, the time average over several rotations of the plasma shows that such a configuration creates a potential well at the edge of the diaphragm region (see fig. 3) which will force the excessive electrons to be radially expelled through an instability instead of freely drifting to the walls. The rotation movement of this instability then comes from a $\vec{E} \times \vec{B}$ configuration. There are ongoing studies to see whether the potential well can be experimentally measured on MISTRAL even though this variation only represents less than 25% of the maximum potential.

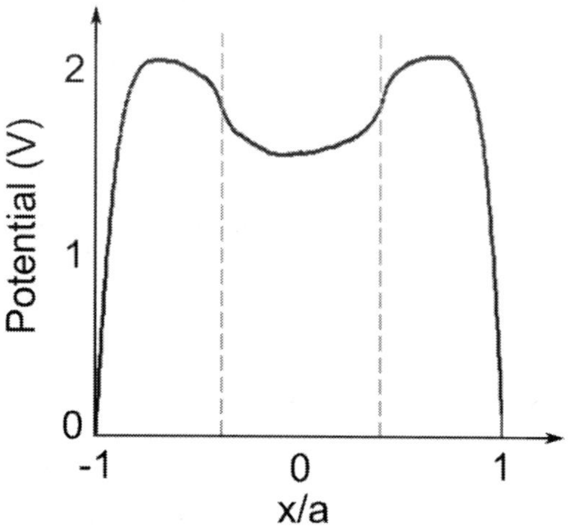


**Figure 3** – Radial distribution of the potential in PIC simulations. The dashed lines represent the diaphragm diameter.

However, there is an important difference between simulations and the experiment in the axial direction. In the former there are two distinct regions at the end of the column, the negatively polarized separating grid in the middle, and the grounded diaphragm around it whereas in the latter the grid takes the whole diameter in front of the diaphragm which cannot be use to extract drifting particles. The axial confinement is therefore different. This difference is being studied and new simulations are running with a more adapted geometry trying to reproduce this behaviour.

*This work has been carried out thanks to the support of the A∗MIDEX project (n○ ANR-11-IDEX-0001-02) funded by the "Investissements d'Avenir" French Government program, managed by the French National Research Agency (ANR)*

## References

[1] Th. Pierre, et al., Phys. Rev. Lett. **92**, 065004 (2004).
[2] A. Escarguel, Eur. Phys. J. D **56 (2)**, 209-214 (2010)
[3] S. Jaeger, et al., Phys. Plasmas **16**, 022304 (2009)
[4] C. Rebont, et al., Phys. Rev. Lett. **106(22)** (2011)
[5] B.M. Annaratone, Phys. Plasmas, **18 (3)** 032108 (2011)
[6] R.S. Granetz et P. Smeulders, Nuclear Fusion **28 (3)** (1988)
[7] M. Anton , et al., Plasma Physics and Controlled Fusion, **38 (11)**, 1849-1878 (1996)
[8] J.P. Boeuf et al., Phys. Rev. Lett. **111**, 155005 (2013)

# Spectroscopic diagnostics of weakly-ionized argon surface wave plasma based on CR model

H. Kawazoe[1], S. Takeuchi[1], G. Yamada[2], M. Setou[3], M. Matsuno[1]
[1] *Tottori University, Tottori, Japan*
[2] *Tokai University, Hiratsuka, Kanagawa, Japan*
[3] *Bando Chemical Industry, Kobe, Hyogo, Japan*

## 1. Introduction

Surface wave plasma of argon gas (Ar-SWP) has been utilized in such the material surface processing as etching, the plasma characteristics and the atomic elementary process although has not cleared up. Therefore, appropriate control of Ar-SWP is not achived. In the study an easy collisional-radiative model (CR model) was tried for analysing Ar-SWP characteristics. It has the virtue of non-intrusive diagnostics and has the potential of application to a low temperature and non-equilibrium plasma.

## 2. CR(Collisional-Radiative) model

It is necessary to know the elementary process of Ar-SWP for the establishment of a CR model. An elementary process will be modelled by estimating the sensitivity of electronic transition as indicated in Ref. [1]. However, this procedure cannot be adopted in the research, because there is no applicable model for the low-ionized and low temperature argon plasma of the research.

Therefore, we have tried to make a CR model by referring to such the forgoing research as Kano et al. for an argon atom[1], Fujimoto for a hydrogen atom [2], and so on[3][4]. As the results, we selected the level of $(4d+6s)^*$ to be a key. It means that the dominant effective quantum numbers become 5p* and >4f*. A CR model can decide both of electron temperature and electronic number density. The standard energy levels i=20 was selected for the electron

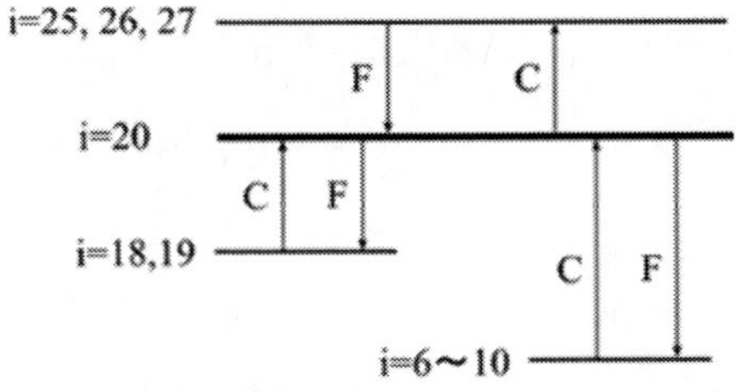


Level i=3, 5, 12- 17 were eliminated by the reason of no-account spectra in the experiment.

Fig.1 Atomic model of electronic transition for i=20

temperature of the Ar-SWP because this level has many electron transitions for sampling spectra. The schematic drawing of electronic transition is illustrated in Fig.1. Electron temperature was decided under the assumption of local equilibrium.

## 3. Experiments

A double probe and a plasma absorption probe (PAP) were made to measure electron temperature and electronic number density respectively. The spectroscope with an iced CCD camera was utilized for the analysis of spectra and for the development of a CR model The cylindrical chamber for the Ar-SWP generation is shown in Fig.2. The experimental conditions are the electric power supply of 0.4kW, the working gaseous pressure of 32Pa and the microwave frequency of 2.45GHz.

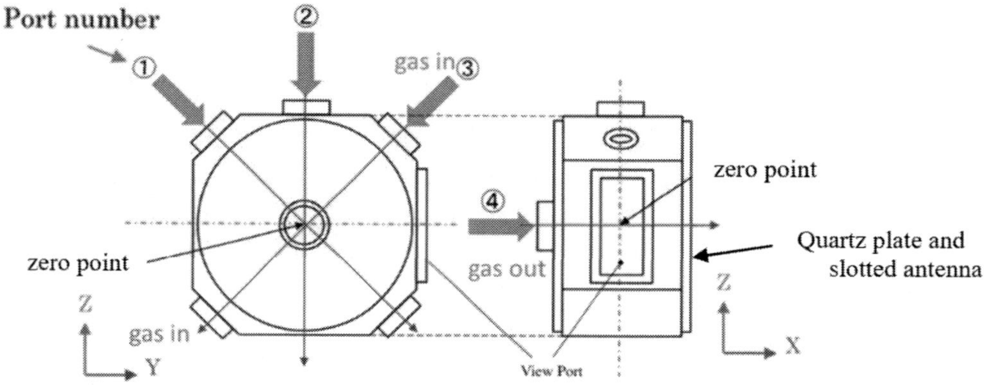


Fig.2 Details of the SWP chamber arrangement

## 4. Results and discussions

Figure 3 shows one of the spectra measured by the spectroscope for the CR model and it was applied to the equation (1), which represents the difference between the inflow and the outflow of the populations.

$$F(T_e) = [C_{6.20}N_6 + \ldots + C_{10.20}N_{10} + C_{18.20}N_{18} + C_{19.20}N_{19} + F_{25.20}N_{25} + F_{26.20}N_{26} + F_{27.20}N_{27}]$$
$$-[( C_{20.25} + C_{20.26} + C_{20.27} + F_{20.19} + F_{20.18} + F_{20.10} + \ldots + F_{20.6})N_{20} ] \quad \ldots(1)$$

The relation between $F(T_e)$ and $T_e$ in the result of Fig.3 is indicated in Fig.4. As assuming local equilibrium condition for the Ar-SWP, that is $F(T_e) = 0$, electron temperature becomes 2.3eV (about 27,000K) from Fig.4.

The electron number density measured by the double probe is shown in Fig.5, where the horizontal axis is the x-axis in Fig.2. The electron density is about $10^{10}$-$10^{12}$ counts/cm$^3$ and

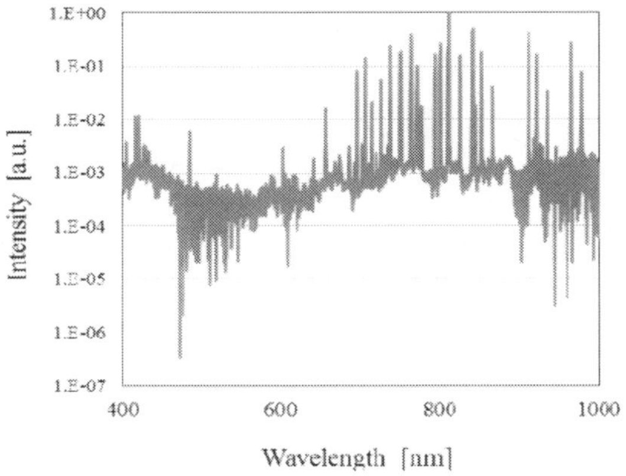


Fig.3 Spectra from the Ar-SWP

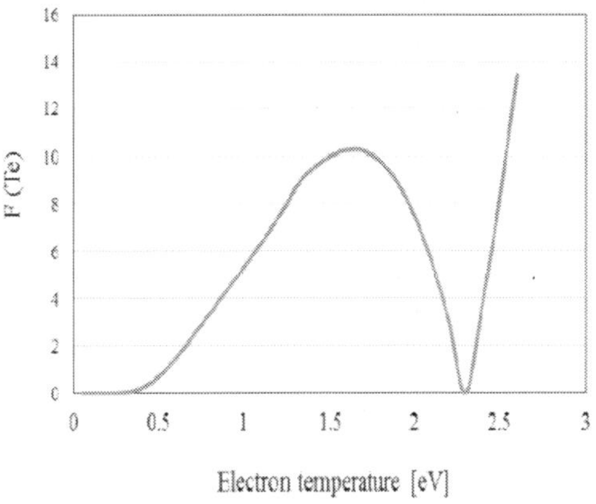


Fig.4 $F(T_e)$ with $T_e$ of the eq.1

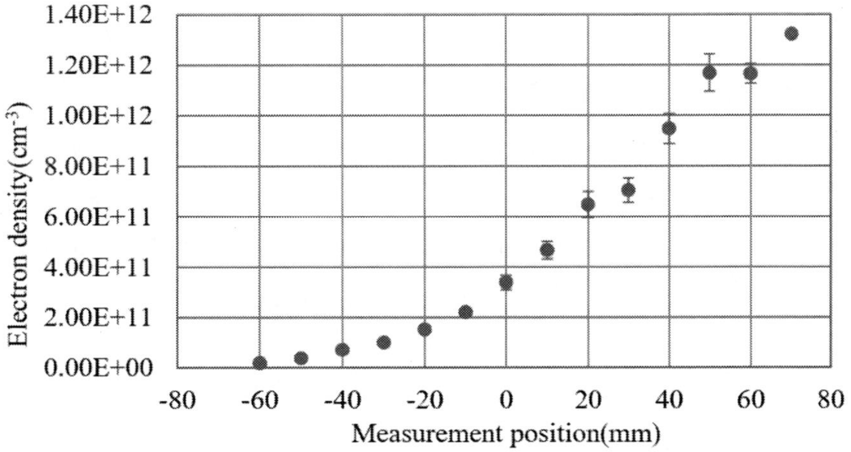


Fig.5 Electron number density along the cylindrical chamber axis measured through the port #4 in Fig.2

increases towards the quartz plate, from which a micro wave passes. The electron density on the yz-plane becomes maximum at the middle of the cylinder chamber, x=0.

Electron temperature measured by the double probe is the temperature of 2.44eV (about 28,000K), which is very close to that by the developed CR method. It was found that the CR model based on the energy level of i=20 is sure to involve the dominant elementary process of the Ar-SWP.

On the other hand, electrical excitation temperature of the plasma was about 4,500K as shown in Fig.6. These were derived on the basis of the Boltzmann plots. It is 23,000K lower than the results of the CR model and the PAP methods. The reason is considered that the electrical excitation temperature was estimated using the energy transition between 4s and 5p and it should not be dominant.

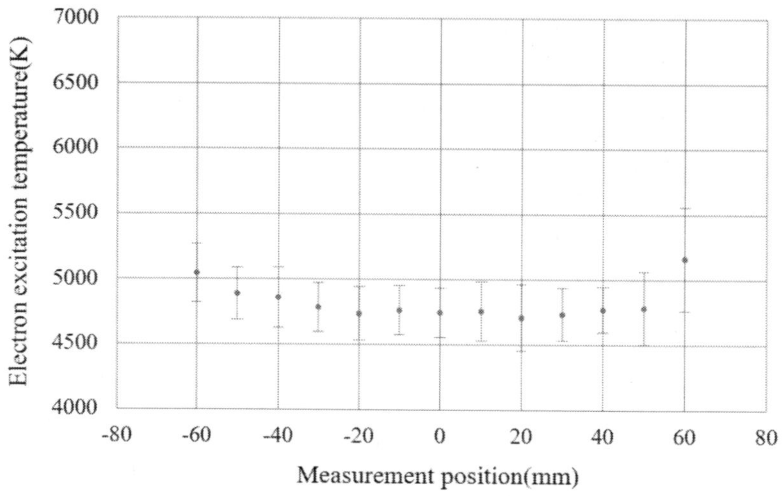


Fig.6 Electron excitation temperature, through the port #1

## 5. Conclusions

Collisional-radiative model has been developed to investigate the characteristics of argon surface wave plasma with low electron temperature and low electrical number density. It was applied to estimate the electron temperature and the obtained temperature is very close to the experimental one by the double probe method. It shows that the CR model involves the dominant elementary process of the argon surface wave plasma.


## Acknowledgement

This work was supported by Grant-in-Aid for Challenging Exploratory Research (No. 25630329), provided by the Japan Society for the Promotion of Science.


## References


[1] T. Fujimoto, J. Phys. Soc. Jpn. 47, 265 (1979).

[2] K. Kano, Spectroscopic Diagnostics of Argon Plasma and Application of the Collision-Radiative Model to Generation of Population inversion, Doctoral thesis of Tokyo Institute of Technology, 2001. (in Japanese).

[3] K. Tachibana and K. Fukuda, Jpn. J. Appl. Phys. 12, 895 (1973).

[4] A. Kimiura, H. Kobayashi, M. Nishida and P. Valentin, J. Quant. Spectrosc. Radiat. Transfer 34, 189 (1985).

# Multiresolution and fast multipole methods for streamer simulations

P. Bílek[1], Z. Bonaventura[1], M. Duarte[2] and A. Bourdon[3]

[1] *Department of Physical Electronics, Faculty of Science, Brno, Czech Republic*

[2] *CCSE, Lawrence Berkeley National Laboratory, Berkeley, USA*

[3] *LPP-CNRS, Ecole Polytechnique, Palaiseau, France*

## Introduction

This contribution presents a new numerical strategy for the computation of the electric field in multi-scale streamer discharge simulations. The electric field is determined by using an integral equation method to solve the Poisson's equation and accelerated by a fast multipole method (FMM) [1, 2, 3, 4]. An adaptive mesh refinement technique based on space adaptive multiresolution (MR) [5, 6] is implemented to obtain a hierarchy of embedded adapted grids (a binary tree in the case of one spatial dimension). To show the interest of the method, we present the simulation of a positive streamer discharge by means of a 1.5D model, where the spatio-temporal evolution of the charged particle densities is solved only along one spatial dimension and the electric field is determined in three dimensions using a disc method [7]. Even though a similar numerical strategy can be straightforwardly applied to higher spatial dimensions [8], the numerical efficiency of the electric field computation in a 1.5D model can be advantageous, as this model can be considered, for instance, to investigate thermal electron acceleration [9] or different kinetic schemes in streamer discharges [10]. All the details on the positive streamer model considered here as well as the numerical strategy to solve the spatio-temporal evolution of the particle densities can be found in [7], while for the sake of brevity here we focus only on the computation of the electric field.

## Computation of the electric field

The 1D computational domain $D$ is divided into a set of cells of different spatial resolution. The source distribution $\rho_i(y)$ is thus given on $N$ leaf nodes $D_i$. Hence, $D = \cup_{i=1}^{N} D_i$ and the electric field from a charged cylinder of radius $R$ at the target

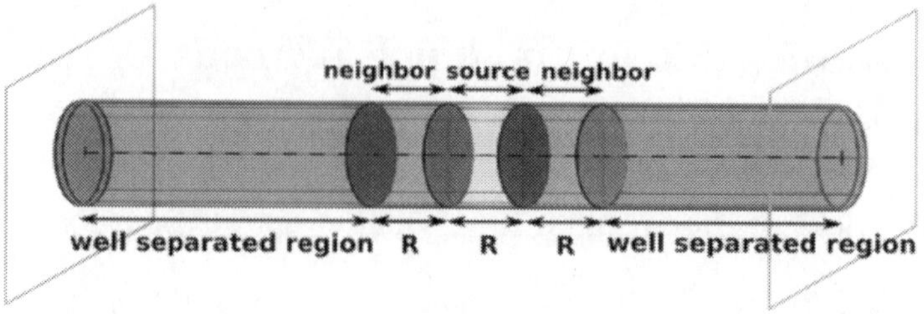


Figure 1: The multipole expansion is convergent in the well separated region from the source panel. $R$ is the radius of the cylinder.

point $x$ on the discharge axis is given by

$$E(x) = \frac{1}{2\varepsilon_0}\sum_{i=1}^{N}\rho_i \int_{D_i} K(x,y)dy, \tag{1}$$

where

$$K(x,y) = \mathrm{sign}(x-y) - \frac{x-y}{\sqrt{(x-y)^2+R^2}}. \tag{2}$$

A direct evaluation of (1) over a set of $N$ target points $x_j$ clearly requires $O(N^2)$ computations. The FMM relies on an approximation to the kernel $K(x,y)$ by a $p$-term expansion, with coefficients $\psi_k(R,y)$ depending on source points alone, and convergent in the well separated region from the source cylinder (see figure 1), i.e., for $|x| > \sqrt{y^2+R^2}$ [3, 4]. Then, we have

$$K(x,y) \approx \mathrm{sign}(x-y) - \mathrm{sign}(x)\left(1-\sum_{k=1}^{p}\psi_k(R,y)x^{-k}\right) = \mathrm{sign}(x-y) -$$
$$- \mathrm{sign}(x)\left(1-\frac{1}{2}R^2x^{-2} - yR^2x^{-3} - \frac{1}{8}(12y^2R^2-3R^4)x^{-4} - \dots\right). \tag{3}$$

The approximation to the electric field at $x$ can be then expressed as a superposition of local and far field contributions:

$$E(x) \approx E_{\mathrm{loc}}(x) + \frac{\mathrm{sign}(x)}{2\varepsilon_0}\sum_{k=1}^{p}\sum_{i=1}^{N_{\mathrm{ff}}}\rho_i\left(\int_{D_i}\psi_k(R,y)dy\right)x^{-k} = E_{\mathrm{loc}}(x) + \frac{\mathrm{sign}(x)}{2\varepsilon_0}\sum_{k=1}^{p}\alpha_k x^{-k}, \tag{4}$$

where the moments $\alpha_k$ depend only on the distribution of $N_{\mathrm{ff}}$ sources in well separated cells and can be precomputed, while $E_{\mathrm{loc}}(x)$ requires direct integration. Moreover, well

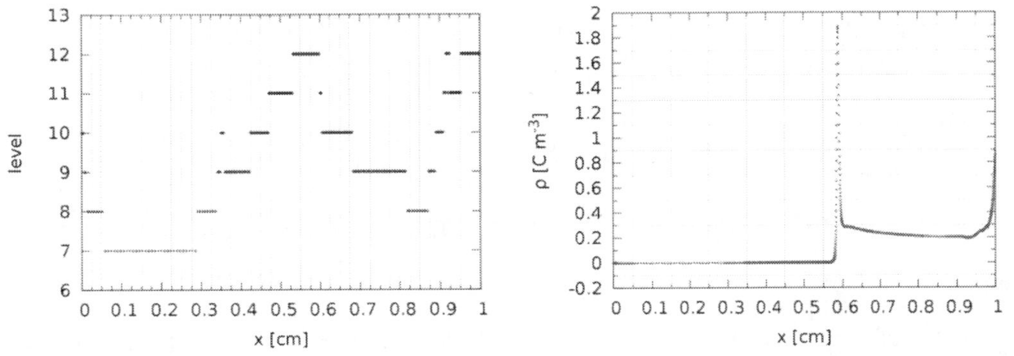


Figure 2: Adapted grid (left) corresponding to a given charge density (right) for an MR tolerance of $\eta = 10^{-6}$.

separated regions are taken into account on the coarsest level possible regarding the hierarchy of grid refinements. This allows in particular to take into account efficiently image charges required for Dirichlet boundary conditions. Consequently, the evaluation of $E$ at each leaf node involves $O(pN)$ operations.

**Results**

Positive streamers are simulated as in [7]. The electric field is computed by means of the FMM on an MR adapted grid. Grid levels corresponding to a given charge distribution at 6 ns are shown in figure 2, for an MR tolerance of $\eta = 10^{-6}$.

We compare the CPU time for the FMM, $t_{\mathrm{fmm}}$, with the one required in the direct computation, $t_{\mathrm{dir}}$ (evaluation of eq. (32) in [7]). Note that the FMM also involves direct computations, but only for the local contribution. Table 1 shows $t_{\mathrm{dir}}$ and $t_{\mathrm{fmm}}$ for several MR tolerances $\eta$ and $p = 25$, and thus adapted grids of different size #AG (#AG = 4096 corresponds to a uniform grid at level 12). The speedup obtained with the FMM depends on the number of cells in the adapted grid. Further analysis shows that the com-

Table 1: Computational results (CPU times $t_{\mathrm{dir}}$ and $t_{\mathrm{fmm}}$ in ms).

| $\eta$ | #AG | $t_{\mathrm{dir}}$ | $t_{\mathrm{fmm}}$ | speedup |
|---|---|---|---|---|
| $10^{-3}$ | 168 | 6 | 2.6 | 2.3 |
| $10^{-4}$ | 349 | 26 | 6.3 | 4.1 |
| $10^{-5}$ | 635 | 87 | 13 | 6.6 |
| $10^{-6}$ | 1064 | 244 | 26 | 9.4 |
| $10^{-7}$ | 2098 | 947 | 66 | 14.3 |
| $10^{-8}$ | 2809 | 1696 | 97 | 17.5 |
| $10^{-9}$ | 3232 | 2260 | 110 | 20.5 |
| $10^{-10}$ | 3900 | 3340 | 140 | 23.8 |
| — | 4096 | 3600 | 150 | 24.0 |

putational complexity of the direct method is indeed of $O(N^2)$, whereas the FMM behaves like $O(N^{1.3})$ with $p = 25$. This departure from linear complexity may be due to the relatively high number of local computations needed to ensure convergence depending on the streamer radius $R$. Nevertheless, computational speedup is significant even for cases with very low number of cells. This is achieved also thanks to the higher efficiency of the FMM when handling boundary conditions requiring image charges.

## Conclusion

A new approach was introduced to compute electric field in streamer simulations, where fast multipole acceleration was implemented on an adapted grid resulting from space adaptive multiresolution. Moreover, the FMM allows one to account for image charges needed for Dirichlet boundary conditions very efficiently. The same numerical strategy can be straightforwardly extended to higher spatial dimensions, where an even better computational efficiency is expected.

## Acknowledgements

This research has been supported by the Czech Science Foundation research project 15-04023S and by the project LO1411 (NPU I) funded by the Ministry of Education Youth and Sports of the Czech Republic. This work has been done partially within the LABEX Plas@par project, and received financial state aid managed by the Agence Nationale de la Recherche, as part of the programme "Investissements d'avenir" under the reference ANR-11-IDEX-0004-02.

## References

[1] L. Greengard, The rapid evaluation of potential fields in particle systems, MIT press (1988).
[2] L. Greengard, Science **265**, 909 (1994).
[3] R. Beatson, L. Greengard , Wavelets **1-37**, 37 (1997).
[4] F. Ethridge, Greengard L., SIAM J. Sci. Comput. **23**, 741 (2001).
[5] A. Harten, Commun. Pure Appl. Math. **48**, 1305 (1995).
[6] A. Cohen, S. Kaber, S. Müller, M. Postel, Math. Comput. **72**, 183 (2003).
[7] M. Duarte, Z. Bonaventura, M. Massot, A. Bourdon, S. Descombes, T. Dumont, J. Comput. Phys. **231**, 1002 (2012).
[8] M. Duarte, Z. Bonaventura, M. Massot, A. Bourdon, J. Comput. Phys. **289**, 129 (2015).
[9] O. Chanrion, Z. Bonaventura, D. Cinar, A. Bourdon, T. Neubert, Environ. Res. Lett. **9**, 055003 (2014).
[10] Z. Bonaventura, M. Simek, $32^{nd}$ ICPIG, July 26–31, 2015 Iași, Romania.

# A three dimensional simulation of inductively coupled plasma sources for TFT fabrications

Young-Jun Lee[1], HeeHwan Choe[1], Deuk-Chul Kwon[2]

[1] *Korea Aerospace University, Goyang, Korea*

[2] *Plasma Technology Research Center NRFI, Gunsan, Korea*

Recently, high-resolution display devices are widely used in TV, monitor and mobile devices. The number of pixels in ultra-high definition (UHD) devices are 4 or more times that of the usual high definition (HD) (1080p) ones [1, 2]. For higher resolution with a brighter screen, display devices need a thin film transistor (TFT) with fine design rule and a high mobility semiconductor material.

For the high mobility semiconductor materials, poly-Si or oxide semiconductors are adopted. For these materials, the $SiO_x$ is used for a dielectric layer instead of the usual $SiN_x$. Therefore, high-density plasma sources are required in the dry etching process for silicon-oxide film.

Because it is not difficult to obtain a high density ($\geq 10^{17}/m^3$), a good uniformity ($\leq 5$ %) at low pressure ($\leq$20 mTorr), ICP reactors are used for etching and deposition in these devices. Inductively coupled plasma reactors are widely used for etching or deposition processes in the microelectronics industries. In the fabrication processes of semiconductor devices, most discharge characteristics in plasma simulation can be obtained by assuming azimuthal symmetry, and plasma simulation in semiconductor processes have only been studied for small sizes ($\leq$ 500mm). However, for the fabrication processes of display devices such as TFT-LCD or AMOLED, a three dimensional studies are inevitable due to the absence of symmetry.

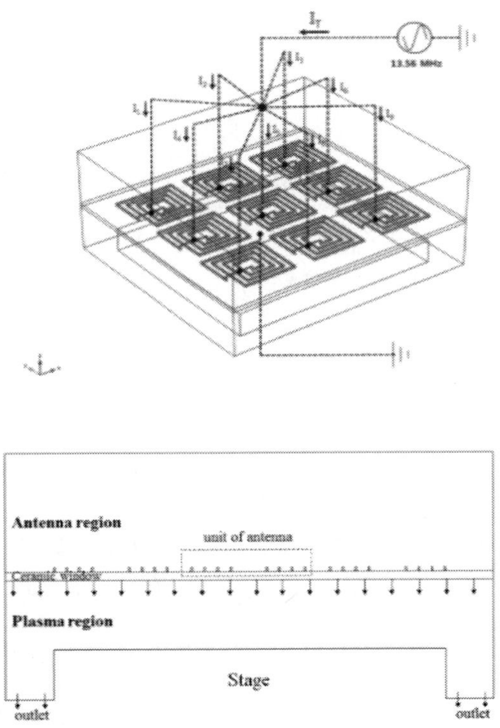


Figure 1: *(a) Schematic diagram of 8th G chamber and (b) cross sectional view in the yz plane*

The substrate size of the 8th generation is $2200 \times 2500$ mm and the process chamber is rectangular. This means that the process chamber has a large volume and no symmetry. Thus, a high cost for the experiment is inevitable to find process conditions. Plasma parameters are significant to find out optimized process conditions. Therefore, plasma discharge simulation may be an effective way for finding process conditions.

To apply these methods in display device processes, a large area without symmetry should be considered. In this paper, we developed a three dimensional 8th generation ICP simulation based on a fluid model without symmetry. $CF_4/O_2$ discharge was used to study the etching process of $SiO_x$ film. Three dimensional profiles of electron density, electron temperature, ion densities, and radical densities are investigated.

The simulation chamber is shown in fig.1. The antenna consists of 9 units. Each unit has a 4-turn coil. The input current (13.56MHz) flows through each antenna unit. A gas inlet is located beneath the ceramic window and an outlet is outside the stage.

The usual equation set for fluid description and ICP discharge are used for simulation [3, 4]. $CF_4/O_2$ chemical reactions are used from the data in ref [5, 6, 7]. The default process conditions are pressure in 20 mtorr, flow rate in 4000 sccm. The ratio of $(CF_4/(CF_4+O_2))$ is 0.1, 0.5, 0.7 and 0.9.

The profiles of electron density, $CF_3^+$, $CF_2^+$ and $F^+$ in the stage are shown in fig.2. $CF_3^+$ is dominantly produced by the collision between electron and feedstock $CF_4$. $CF_2^+$ is produced by the electron impact reaction of $CF_3$ and $CF_4$. As the difference of reaction rate constant, $CF_3^+$ has a better uniformity than $CF_2^+$ distribution. $F^+$ is produced by the collision between the electron and $CF_x$. As a result,

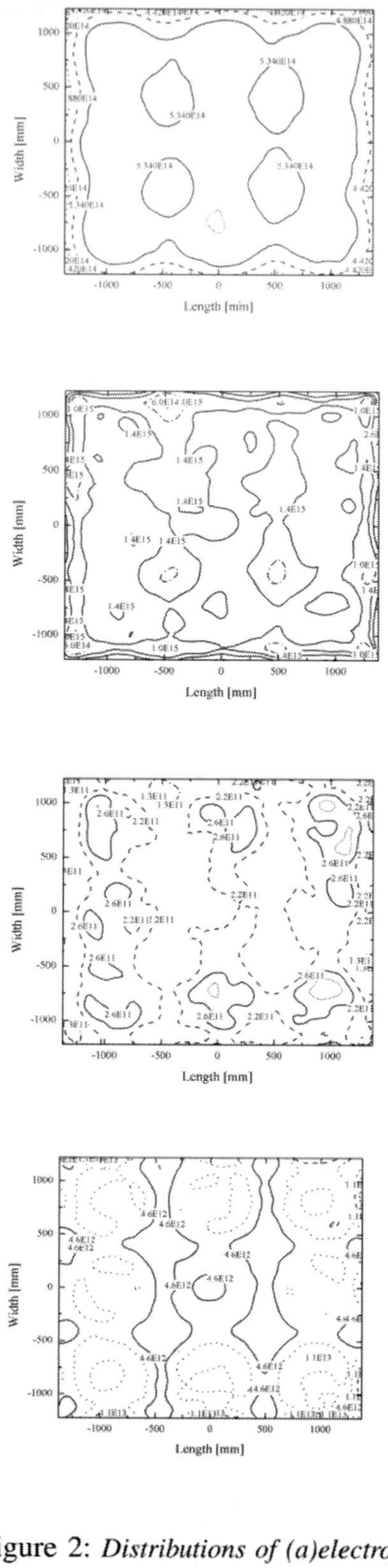


Figure 2: *Distributions of (a)electron density, (b)*$CF_3^+$*, (c)*$CF_2^+$ *and (d)*$F^+$

$F^+$ has a higher density than $CF_2^+$ and $CF^+$. The ions and neutrals coming to the substrate result in the etch rate of $SiO_2$ film. From the fluxes of ions and neutrals on the substrate, it is possible to obtain the uniformity of etch rate beside the average etch rate.

Fig.3 describes electron density, major radicals ($CF_4$, $CF_3$, $CF_2$, CF and F) and major ions ($CF_3^+$, $CF_2^+$, $CF^+$ and $F^+$) depending on $CF_4/O_2$ ratio. The addition of $O_2$ mole fraction decreased total neutral and positive ions. However, despite the increase of $O_2$ ratio, neutral fluorine has a region in which number density increases. F is produced to destroy $CF_x$ radicals by O atom. Therefore, the F number density is increased [8].

The etch rate of $SiO_2$ is known for relying on $\Gamma_{neutral}/\Gamma_{ion}$. This relation may or may not valid for $SiO_x$ film, but this relations is an useful guideline. The result like fig.4 was obtained by calculating $\Gamma_{neutral}/\Gamma_{ion}$ on substrate surface. The ratio of each species may increase or decrease, but the total ratio is important in etch rate. The ions move to the substrate by drift and diffusion. The neutrals are slower than the ions because they have diffusion only, but the ion-neutral collisions increase the velocity of neutrals. The ions and neutrals collide with the films on the substrate that there are two kinds of reactions which are physical sputtering and chemical reactions. As is well known, the simulataneous reactions increase the etch rate of film on the substrate. Since the physical sputtering and etching by chemical reactions are proceeded at the same time by ion and neutral flux, it is possible to obtain such a result [9].

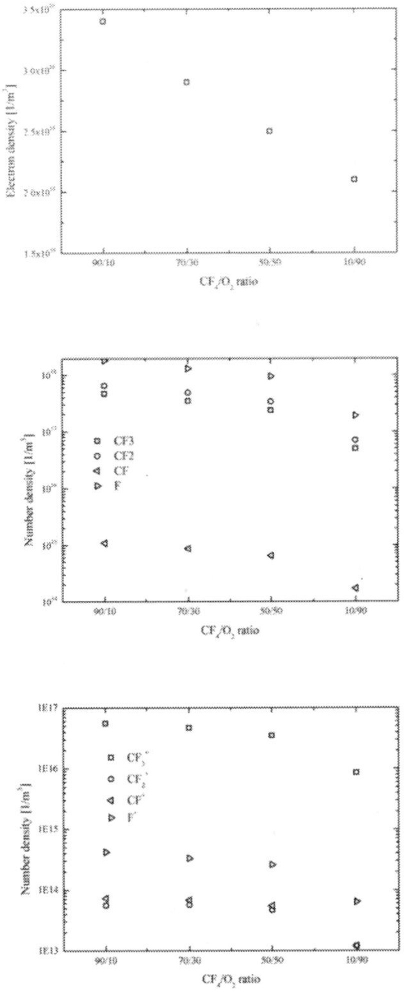


Figure 3: *Electron density, radicals and ions for* $CF_4/O_2$ *ratio*

In summary, a three dimensional ICP simulation with $CF_4/O_2$ discharge is conducted for an 8th generation plasma chamber in display device process. The profiles of radicals, ions and flux which is related to $SiO_x$ etch rate can be obtained. These reults may be used for the prediction of etch rate or uniformity. However, the detailed etching mechanism $SiO_x$ film (not $SiO_2$ film) are not well known. Therefore, to obtain reasonable predictions, the simulation results are expected to apply to process by matching the experiment result.

**Acknowledgements**

This work was supported in part by the International collaborative R&D program (N0000678) and by the Industrial Strategic Technology Development Program (10041681) funded by the Ministry of Trade, Industry and Energy of Korea(MOTIE, Korea).

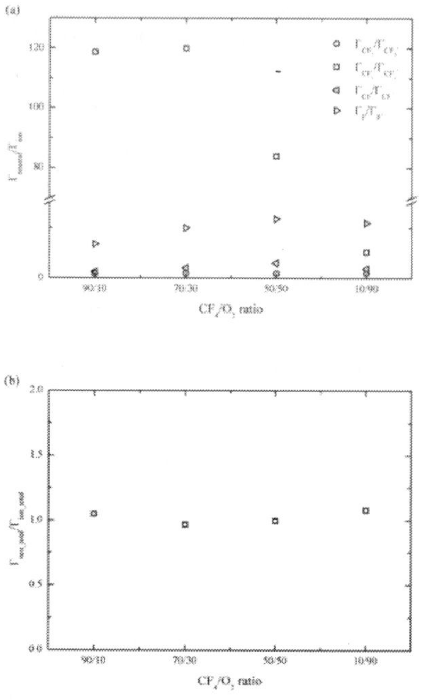


Figure 4: $\Gamma_{neutral}/\Gamma_{ion}$ *on substrate*

## References

[1] Jong-Sik Park, Journal of Semiconductor Technology and Science, **12**, 203 (2012)

[2] Harstead, E, Communications Magazine, IEEE, **50**, 218 (2012)

[3] J. D. Bukowski and D. B. Graves, Journal of Applied Physics, **80**, 2614(1996)

[4] Deuk Chul Kwon and N. S. Yoon, Journal of the Korean Physical Society, **47**, 163(2005)

[5] LIU Wenyao, Plasma Science and Technology, **15**, 885(2013)

[6] Hiroshi Fukumoto, Plasma Sources Sci. Technol. **18**, 1(2009)

[7] M.A. Liebermann, Principles of plasma discharge and materials processing, Wiley&Sons Inc., Hoboken, New Jersey (2005)

[8] Da Zhang and Mark J. Kushner, Journal of Vacuum Science Technology A, **19**, 524(2001)

[9] Jinyoung Son, Plasma Chemistry of Plasma Process, **Vol.34**, 239(2014)

# Fluid modelling of plasma at low pressures

J. Hromadka, T. Ibehej, R. Hrach

*Department of Surface and Plasma Science, Faculty of Mathematics and Physics, Charles University, Prague, Czech republic*

## Introduction

Thanks to the increasing performance of computers in recent years computer models brought new opportunities into plasma research. Nowadays, computer models of plasma are widely used to interpret experimental measurements and to better understand plasma behaviour in cases that are too complicated for theoretical description.

The most common computer modelling techniques used in plasma physics involve particle and fluid models. Whereas particle models are able to provide us with detailed microscopic information even about a system out of thermodynamic equilibrium, fluid models are much less demanding on the computational resources. Therefore, a lot of effort has been put into development of hybrid models that benefit from advantages of both, particle and fluid models. Hybrid models have already proven their usefulness [1]; on the other hand, they also suffer from several limitations. E.g. predictions of hybrid models based on combination of PIC method and drift-diffusion fluid model fail at low pressures where the assumptions of drift-diffusion approximation are violated. Since many technological applications of plasma require low-pressure regime (e. g. semiconductor processing by CCP discharges), it would be highly desirable to extend the range of applicability of hybrid models to the area of low pressures.

Our contribution deals with the question how the drift-diffusion model can be modified so that fluid modelling could be used also for low pressure plasma with sufficient accuracy. It was proven [2] that inclusion of full electron momentum equation can improve predictions of fluid models at low pressures for CCP discharges. We would like to discuss this topic more generally and propose a way how the area of applicability of fluid models could be extended.

## Effects of low pressure

We have used our 2D PIC computer model to investigate effects of low pressure on argon plasma sheath in the vicinity of a cylindrical probe with $2 \cdot 10^{-4}$ m radius. The motion of charged particles in electric field was resolved by Verlet algorithm, the electric field was obtained by finite difference method on square shaped computational mesh of $400 \times 400$ nodes. The computational domain of $4 \times 4$ cm dimensions was surrounded by sources of charged particles ($Ar^+$ ions and electrons) with Maxwellian velocity distribution. Scattering processes of charged particles with neutrals were treated by modified null collision method [3].

The computations were done for two biases of cylindrical probe with respect to the plasma potential ($\pm 5$ V) and for variable pressure of neutral gas in background ($1.33 \cdot 10^{-1} - 1.33 \cdot 10^{2}$ Pa). Number density of charged particles changed between $5.03 \cdot 10^{13}\,\mathrm{m}^{-3}$ and $1.59 \cdot 10^{15}\,\mathrm{m}^{-3}$ according to square root dependance on pressure. Temperature of $e^-$ in source of particles was set to 23 600 K, temperature of $Ar^+$ ions to 300 K. Cross sections of $e^-$, resp. $Ar^+$ scattering processes were considered according to [4], resp. [5].

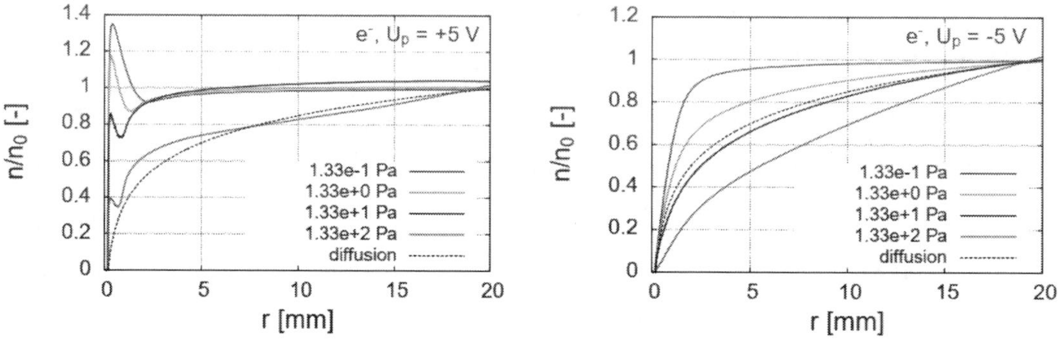


Figure 1: Normalized density of electrons in the vicinity of cylindrical probe for various values of neutral gas pressure and two probe biases.

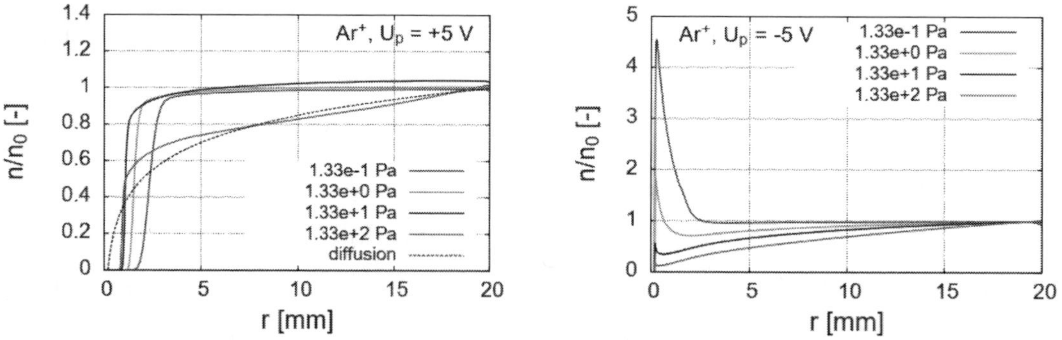


Figure 2: Normalized density of $Ar^+$ ions in the vicinity of cylindrical probe for various values of neutral gas pressure and two probe biases.

From Figures 1 and 2 it can be seen that with decreasing pressure of neutral gas in background charged particles that are attached by the probe bias cumulate near the probe since less collisions prevent them from getting into this area from bulk plasma. As a result, density profile of charged particles in the vicinity of the probe does not follow diffusion solution for low pressure.

Less charged particles are present in a volume unit at low pressure. For positively biased probe it has a consequence that potential is shielded out on larger distance at low pressure as

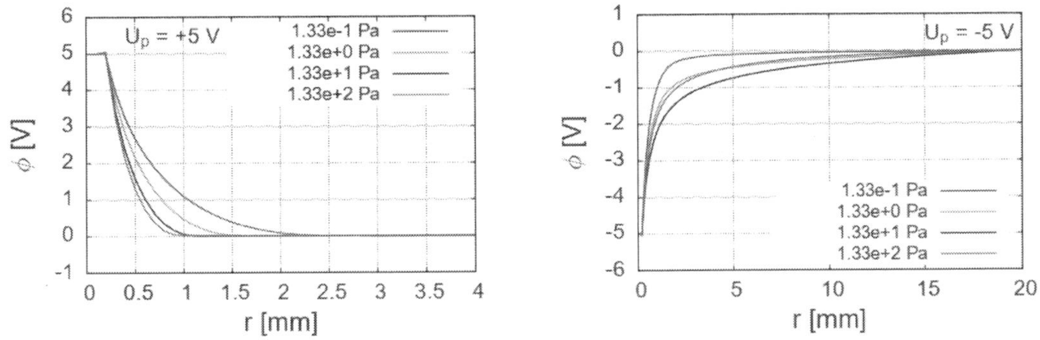


Figure 3: Electric potential in the vicinity of cylindrical probe for various values of neutral gas pressure and two probe biases.

can be seen from Figure 3. On the other hand, charged particles are cumulated near the probe at low pressure; thus, they are able to shield out the potential on shorter distance - this can be observed for negatively biased probe.

**Fluid modelling at low pressure**

The most frequently used fluid model of low temperature plasma is the drift-diffusion model

$$\boldsymbol{J} = \mu n \boldsymbol{E} - D\nabla n \tag{1}$$

which is derived from full-momentum equation

$$mn\left[\frac{\partial \boldsymbol{v}}{\partial t} + (\boldsymbol{v}\cdot\nabla)\boldsymbol{v}\right] = qn\boldsymbol{E} - \nabla p - mn\boldsymbol{v}\nu \tag{2}$$

by neglecting terms on the left hand side (LHS) of this equation. This approximation is only justifiable if we consider steady state $\left(\frac{\partial \boldsymbol{v}}{\partial t} = 0\right)$ and elastic collisions, whose effects are represented by the third term on right hand side (RHS) of Eq. (2), are so frequent that the second, inertial term on LHS can be neglected. Another assumptions of drift-difusion approximation: (i) Plasma can be treated as an ideal gas. (ii) The last term on RHS of Eq. (2) is only special case of velocity moment of the collisional term that emerges in Boltzmann equation. As a result, drift-diffusion model can be used to describe low temperature plasma at medium and high pressures. The usage of this approximation is discutable for positive ions because of their higher mass.

We have used the project FENICS [6] to implement 2D drift-diffusion model of cylindrical probe in argon plasma for different pressures. The results on Figure 4 show that drift-diffusion model is not able to capture effects of low pressure. Cumulation of charged particles near the probe with decreasing pressure can not be observed; diffusion character of density profile solution near the probe persists.

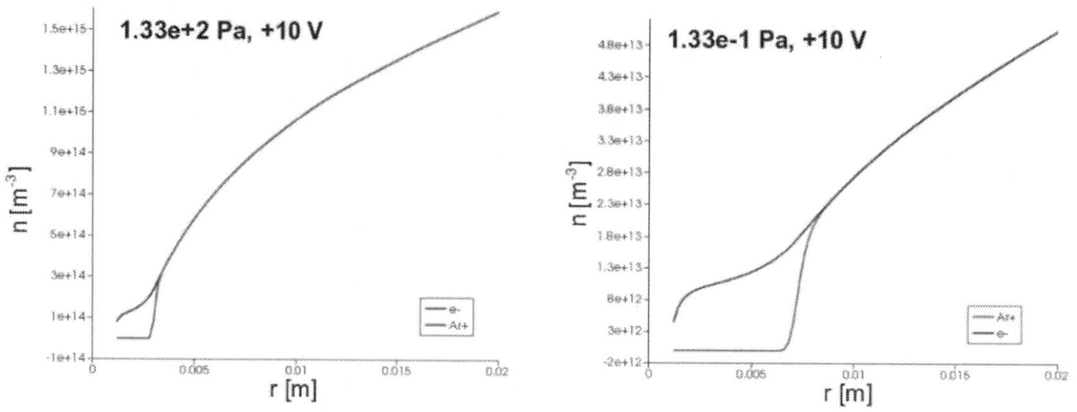


Figure 4: Number density of charged particles in the vicinity of the cylindrical probe for two different pressures. Results of 2D drift-diffusion fluid model. Probe bias: +10 V.

## Conclusion

It was proven that drift-diffusion model cannot capture the effects of low pressure. To propose fluid model that would be reliable also in low pressure area we have to move to the general momentum balance equation

$$mn\frac{\mathrm{d}\boldsymbol{v}}{\mathrm{d}t} = \pm en\boldsymbol{E} - \nabla p - \nabla\cdot\boldsymbol{\pi} + \boldsymbol{F_s} \tag{3}$$

where $\boldsymbol{P} = p\boldsymbol{I} + \boldsymbol{\pi}$ is the stress tensor, $\boldsymbol{\pi}$ is viscosity stress tensor, $\boldsymbol{F_s} = \sum \boldsymbol{F_{ss'}}$ is frictional force due to collisions and $\boldsymbol{P}$ and $\boldsymbol{F_{ss'}}$ are given by

$$\boldsymbol{P} = \int m\boldsymbol{w}\boldsymbol{w} f(\boldsymbol{r},\boldsymbol{v},t)\mathrm{d}^3\boldsymbol{v}, \qquad \boldsymbol{F_{ss'}} = \int m\boldsymbol{v}\left[\frac{\delta f}{\delta t}\right]_{ss'}\mathrm{d}^3\boldsymbol{v} \tag{4}$$

To find suitable closure relations for $\boldsymbol{P}$ and $\boldsymbol{F_{ss'}}$ for specific set of interactions between particles we can employ Chapman Enskog approach [7].

## References

[1] V. Hruby and R. Hrach, IEEE Transactions on Plasma Science **38**, 9 (2010)

[2] G. Chen and L. L. Raja, Journal of Applied Physics **96**, 11 (2004)

[3] S. Roucka and R. Hrach, IEEE Trans. Plasma Sci. **39**, 3244-3250 (2011).

[4] A. Phelps, http://jila.colorado.edu/avp/collision_data/, (2013).

[5] A. Bogaerts, R. Gijbels and W. Goedheer, Jpn. J. Appl. Phys., **38**, 4404-4415 (1999).

[6] A. Logg, K. A. Mardal, G. N. Wells et al., Automated Solution of Differential Equations by the Finite Element Method, Springer, (2012).

[7] S. Chapman and T. G. Cowling, The mathematical theory of non-uniform gases: an account of the kinetic theory of viscosity, thermal conduction, and diffusion in gases, Cambridge University Press, (1990).

# Simple model of the discharge for Hall thrusters

V.P. Shumilin[1], N.V. Shumilin[2], A.V. Shumilin[2]

[1] *Moscow Institute of Physics and Technology (State University), Dolgoprudnyi, Russia*

[2] *Moscow Institute of Electronics and Mathematics National Research University Higher School of Economics, Moscow, Russia*

For some specific types of Hall thrusters the comparison of dependencies of the current of accelerated ions on the discharge voltage, which have been calculated within the limits of simple theoretical model, with the experimental data is presented. It is shown, that the offered simple model gives results close to a reality both qualitatively and quantitatively.

In paper [1] a simple model for the determination of interrelation between integral characteristics of Hall thrusters with an anode layer is offered. We understand accelerating voltage, strength of a magnetic field, working gas flow rate, current of accelerated ions and geometrical sizes of the engine as integral characteristics. This model includes the equation of balance of the forces acting on the Hall current area, and the relation between the current of accelerated ions and working gas flow rate.

If the axial component of the magnetic field is either absent or weakly varying in the Hall current region, then the force exerted by the magnetic field on the "Hall current loop" is proportional to the difference between the magnetic pressures on both sides of the "loop" and the cross-sectional area of the ion beam. On the other hand, this force is proportional to the product of the current of accelerated ions and their momentum. This allows one to write an equation describing the balance of forces acting on the Hall current region. In so doing, it is assumed that the "Hall current loop" is imbedded in the equipotential anode plasma, while almost the entire potential difference is concentrated between the anode and cathode plasmas, i.e., the electric forces can be ignored.

Assuming that all ions pass across the total potential difference $\varphi_0$ applied to the discharge and have no angular scatter in velocities, the balance of forces can be expressed in the form

$$\frac{l}{c} J_H H_0 = \frac{M}{e} J_i \sqrt{\frac{2e\varphi_0}{M}},$$

where $M$ is the atomic mass of the working gas, $e$ is the elementary charge, and $J_i$ is the ion beam current, $c$ is the speed of light in vacuum, $J_H = \int j_H(r,z)dS \approx \Delta R \int_0^L j_H(z)dz$ is the total Hall current, $\Delta R = R_2 - R_1$, $R_2$ is the outer radius of the region occupied by the Hall current, $R_1$ is its inner radius, $l = \pi(R_1 + R_2)$ is the length of the midchord of the cross section of the region occupied by the Hall current, $L$ is the length of the region occupied by the Hall current along the $z$ axis, $H_0$ is the average magnetic field in the Hall current region.

In addition to this balance, another condition should be satisfied: the degree of ionization of working gas should be high enough to ensure the required value of the current of accelerated ions and, consequently, the above balance of forces.

It is obvious that the current of accelerated ions is proportional to the working gas flow rate

$$J_i = P_i Q,$$

where $Q$ is the flux of working gas in amperes. Coefficient of proportionality $P_i$ has the meaning of ionization probability of a working gas atom upon passing through the entire Hall current region. The high degree of ionization of the working gas in Hall thrusters is achieved because ionization is predominantly produced by the electrons forming the Hall current. We will assume that ionization is produced only by these electrons. The question of the ionization probability of neutral particles in a region with a known dependence of their depletion frequency on the coordinate has been considered in [2]. Using a simple model of gas discharge in crossed electric and magnetic fields, it is possible to find the relationship between the ionization frequency of the working gas and the integral characteristics of the thruster. Thus, the ionization probability turns out to depend on the ratio of the magnetic pressures on both sides of the "Hall current loop", the accelerating voltage, the sort of gas, and the magnetic field.

As a result, we have two relationships that interrelate the accelerating voltage, the square root of the ratio of the magnetic pressures on both sides of the "Hall current loop" (the demagnetization parameter), the magnetic field strength, the current of accelerated ions for a given sort of working gas and given geometrical sizes of the Hall thruster. Excluding from them, e.g., the demagnetization parameter, we obtain the required relationship between the integral characteristics of the Hall thruster. In particular, for a specific thruster, this allows one to plot the dependence of the current of accelerated ions on the accelerating voltage at fixed values of the magnetic field strength and the working gas flow rate, as well as the dependence of the current of accelerated ions on the magnetic field strength at fixed values of the accelerating voltage and the working gas flow rate, and to consider the question of the domain of existence of the Hall thruster discharge in the ($H_0, \varphi_0$) coordinates. In spite of the fact that the proposed simple model was initially intended for Hall thrusters with anode layer (TAL), its application to stationary plasma thrusters (SPT) yields acceptable results.

The dependences of the current of accelerated ions on discharge voltage at constant value of the magnetic field strength, fixed gas flow rate and temperature of working gas ($T = 300$ K) for TAL (Fig. 1A), SPT (Fig. 1B, 1C, 1D) and for hybrid Hall thruster (Fig. 1E) received using this method are presented in Fig. 1. The horizontal dotted straight lines correspond to the working gas flow rate, circles note their working points in the assumption of total ionization depletion of working gas. The curves 1 correspond to the hydrodynamic model of neutral particles motion; curves 2 - to the kinetic model with equal absolute values of initial velocities of neutral particles; curves 3 - to the kinetic model with Maxwellian distribution of initial velocities of neutral particles; curves 4 are the result of recalculation of experimental data from [3 - 8]. Thus, simple model offered in work [1] gives results close to a reality both qualitatively and quantitatively.

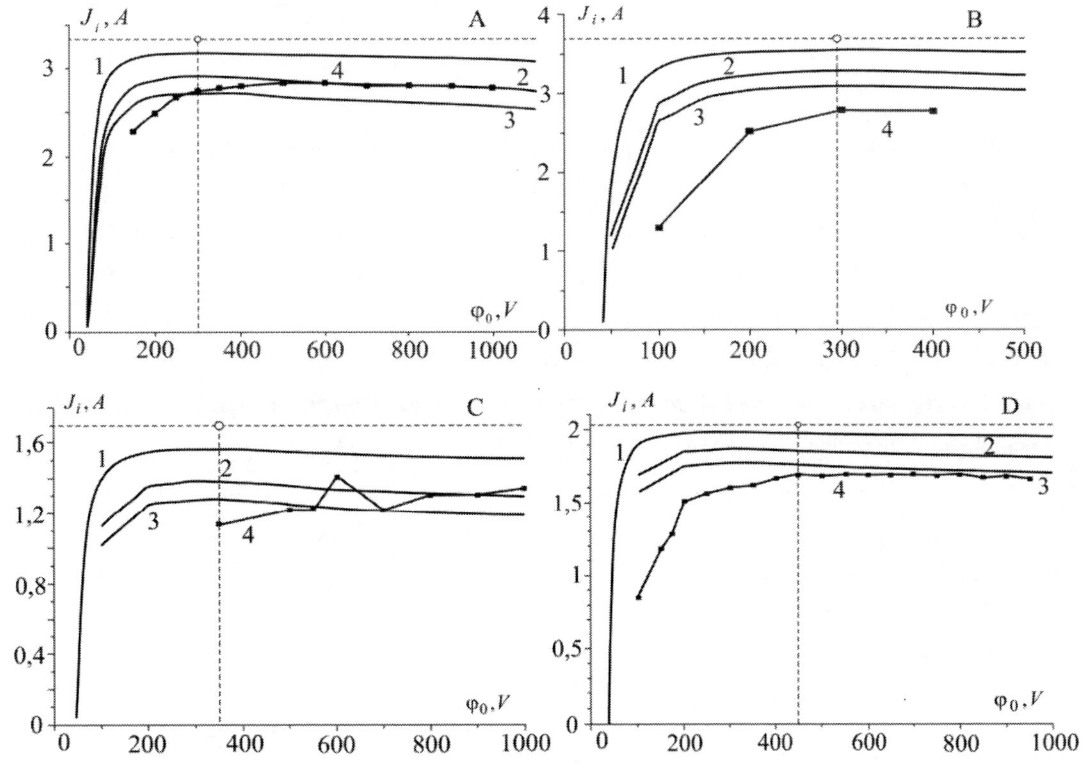

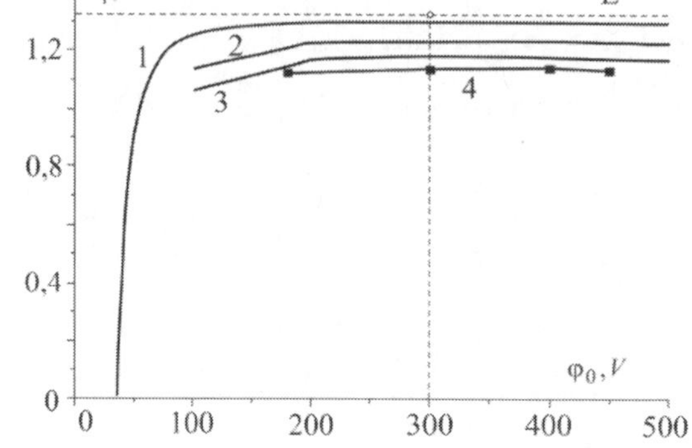


*Fig. 1. Dependence of the current of accelerated ions on discharge voltage at constant value of the magnetic field strength, fixed gas flow rate and temperature of working gas (xenon) ($T = 300$ K) for thrusters A) TAL-WSF ($H_0$ =223 G, $\dot{m} = 4.52$ mg/s) [3, 4]; B) SPT-100 ($H_0$ =233 G, $\dot{m} = 5.32$ mg/s) [5]; C) PPS-1350 ($H_0$ =214 G, $\dot{m} = 2.31$ mg/s) [6]; D) BHT-1000 ($H_0$ =200 G, $\dot{m} = 2.76$ mg/s) [7]; E) PlaS-40 ($H_0$ =250 G, $\dot{m} = 1.8$ mg/s) [8]. $\dot{m}$ - working gas mass flow rate.*

A simple model for the determination of interrelation between integral characteristics of Hall thrusters is proposed. Using this model, we obtained the dependence of the current of accelerated ions on the discharge voltage for some specific types of Hall thrusters. The proposed model can be used for preliminary estimates of the parameters of newly designed Hall thrusters.

**References:**


1. *V.P. Shumilin, A.V. Shumilin, and N.V. Shumilin.* Plasma Physics Reports. 2014. Vol. 40, No. 3, pp. 229-238.
2. *V.P. Shumilin, A.V. Shumilin, and N.V. Shumilin.* High Temperature. 2012. Vol. 50, No. 2, pp. 178-185.
3. *A. Semenkin.* External anode layer thruster parameters at a high specific impulse regimes. 25th International Electric Propulsion Conference, August 24-28, 1997, Cleveland, Ohio. IEPC-97-055. pp. 341-344. 1997.
4. *A. Semenkin, A. Kochergin, V. Garkusha, G. Chislov, A. Rusakov, S. Tverdokhlebov, C. Sota.* RHETT/EPDM flight anode layer thruster development. 25th International Electric Propulsion Conference, August 24-28, 1997, Cleveland, Ohio. IEPC-97-106. pp. 661-666. 1997.
5. *J. Bareilles, G. J. M. Hagelaar, L. Garrigues et al.* Physics of Plasmas. 2004. Vol. 11, No. 4, pp. 3035-3046.
6. *V. Vial, A. Lazurenko, A. Bouchoule et al.* PPS 1350-G in an extended operation domain: comparison between experimental and simulation results - AIAA 2004-3607, 2004.
7. *B. Pote and R. Tedrake.* Performance of a high specific impulse Hall thruster - IEPC-01-035, 2001.
8. *M.Yu. Potapenko, V.V. Gopanchuk.* Development and research of the plasma thruster with a hollow magnet anode PlaS-40 - IEPC-2013-52, 2013.

# The influence of anode temperature on the length of the anode plasma in a gas discharge under conditions of strong ionization depletion of neutral particles

V.P. Shumilin[1], A.V. Shumilin[2], N.V. Shumilin[2]

[1] *Moscow Institute of Physics and Technology (State University), Dolgoprudnyi, Russia*

[2] *Moscow Institute of Electronics and Mathematics National Research University Higher School of Economics, Moscow, Russia*

Although thrust of electric propulsion thrusters is much smaller than that of chemical thrusters, they can reach much higher propellant exhaust velocities. The most popular type of such thrusters are Hall thrusters that are widely used for correction of orbits of long-life spacecrafts and for controlling their orientation in space today. These thrusters are also perspective for distant space missions because of their ability for reaching high velocities and because of their large efficiency factor of working medium consumption. In electric ion propulsion thrusters of this type the gas discharge in the crossed electric and magnetic fields is used. Working gas is delivered into the discharge channel of the thruster, where it is exposed to ionization depletion under activity of Hall electrons. The ions, formed during ionization depletion, are accelerated by electric field and generate thrust of Hall thruster [1].

In [2,3] the problem of structure of gas discharge anode plasma with strong ionization depletion of neutral particles was considered. There was supposed that mean directed velocity of neutral particles was constant and equal to $\overline{v}/2=\sqrt{2T/\pi M}$, where $T$ is the temperature of neutral particles measured in energy units; $M$ is the mass of neutral particle.

As in the papers [2,3], we will consider only case of collisionless motion of ions in one-dimensional approximation and Boltzmann thermal electron distribution. Motion of neutral particles will be described using kinetic approximation. We will assume that neutral particles have an arbitrary semi-isotropic in velocity space velocity distribution function at anode surface [4].

The potential $\varphi(x)$ at some point of the anode plasma is at its maximum. Let us choose the origin of the coordinate and potential at this point, that is $\varphi(0)=0$, $d\varphi(x)/dx\big|_{x=0}=0$. A neutral gas is fed from the $x=x_a$ ($x_a<0$) anode plane, with the backward flow of neutral particles being absent. All of the ions generated in the region between this plane and the potential maximum turn back, recombine on the $x_a$ surface, and come back to the discharge region in the form of neutral particles. Thus, the flux of neutral particles from the $x = x_a$ plane exceeds the gas flow rate to which the neutral flux density at the potential maximum, $q_0$, corresponds. The ions are trapped in the region between $x_a$ and the potential maximum in a steady state. Indeed, in view of the conservation of the heavy component $eq(x) + j_i(x) = eq_0$, where $q(x)$ is the neutral flux density, $j_i(x)$ is the ion current density, $e$ is the elementary charge. Since $j_i(0) = 0$, then $q(0) = q_0$.

Let us introduce dimensionless potential $\eta = -e\varphi(x)/T_e$, where $T_e$ is the plasma electron temperature in energy units. Using continuity equation for neutral particles and quasi-neutrality equation we can obtain following expression for dependence of neutral flow density upon potential [2]

$$q(\eta)/q_0 = 1 \mp \varepsilon e^{-\eta}\mathrm{erfi}(\sqrt{\eta})/\sqrt{\pi}, \qquad (1)$$

where sign "-" corresponds to a region $x \le 0$, sign "+" corresponds to a region $x \ge 0$; $\varepsilon = (n_0/q_0)\sqrt{2T_e/M}$ is a depletion parameter; $n_0$ is slow-electron density at $x = 0$.

It should be noticed that anode surface is usually significantly heated by accelerated ions that move back from anode plasma and electrons. After neutralization at the anode surface these ions come back to the discharge region in the form of neutral particles. So we have two groups of neutral particles that obviously have different temperatures.

Let us consider the case when atoms of neutral gas that are supplied from anode surface and particles that recombine on the $x = x_a$ surface and come back to the anode plasma again as neutral particles have different semi-isotropic in velocity space velocity distribution functions. Let us assume that these distribution functions are characterized by functions of velocity modulus $F_1(v)$ and $F_2(v)$ correspondingly, that satisfy requirements $\int_0^\infty F_i(v)v^k dv < \infty$, $k = 0,1,2,3,\ldots$, $i = 1,2$.

In [4] expression for neutral gas flow density was obtained that accounts strong ionization depletion of neutral particles. It can be showed that in our case using results of paper [4] and expression (1) we can write for neutral gas flow density following expression

$$q(\eta)/q_0 = 2\int_0^\infty E_3\left(\frac{\varepsilon}{2}\frac{\bar{v}}{v}(s(\eta) - s_a)\right)\left[\frac{F_1(v)}{\int_0^\infty F_1(v)v^3 dv} + \frac{\varepsilon}{\varepsilon_*}\frac{F_2(v)}{\int_0^\infty F_2(v)v^3 dv}\right]v^3 dv, \qquad (2)$$

where $E_m(x)=\int_0^\infty \frac{\exp(-xw)}{w^m}dw = x^{m-1}\int_x^\infty \frac{\exp(-t)}{t^m}dt$, $m=1,2,3...$ is the integral exponent [5];

$\varepsilon_* = \pi\sqrt{\eta_0}$ is critical value of the depletion parameter at which the cathode plasma boundary goes to infinity and the entire flow of neutrals is depleted in the plasma region [2]; $\eta_0 \approx 0.854$ is potential at the plasma boundaries; $s = x/l$ is dimensionless coordinate; $l = (n_0\bar{v}/2\omega_0 q_0)\sqrt{2T_e/M}$ ; $\omega_0 = \text{const}$ is constant ionization depletion frequency of neutral particles.

It is easy to notice that if distribution functions of velocity modulus $F_1(v)$ and $F_2(v)$ coincide, then equation (2) takes the same form as equation that was obtained in paper [3].

Expressions (1) and (2) allow us to obtain dependence of potential upon coordinate for different neutral velocity distribution functions $F_1(v)$ and $F_2(v)$. Let us consider in particular cases when absolute values of the initial velocities of neutral particles are equal ($F_i(v) = \delta(v-\bar{v})$) and when neutral particles have Maxwellian velocity distribution ($F_i(v) = \exp(-4v^2/\pi\bar{v}^2)$) for different values of ratio of the anode temperature to the temperature of supplied gas ($T_a/T_0$).

Dependences of anode (1*a* and 1*c*) and cathode (1*b* and 1*d*) boundary coordinates of anode plasma upon depletion parameter are showed on figure 1 for different values of ratio of the anode temperature to the temperature of supplied gas ($T_a/T_0$). Figures 1*a* and 1*b* are plotted for kinetic model with equal absolute values of the initial velocities of neutral particles; figures 1*c* and 1*d* are plotted for kinetic model with Maxwellian velocity distribution of neutral particles. It can be seen from these figures that effect of anode temperature increases with an increase of depletion parameter.

It is obvious that decrease of average neutral particle lifetime in the anode plasma leads to increase of length of the anode plasma if depletion parameter is fixed. Therefore, taking into account both kinetic effects of predominant depletion of tangentially propagating and slow moving neutral particles while considering the flow of neutral particles [3] and heating of the anode surface leads to increase of length of the anode plasma.

It should be noticed that coordinate of cathode boundary of anode plasma rapidly increase with an increase of depletion parameter, and if depletion parameter is large enough then length of the anode plasma is approximately equal to distance from point of maximum value of anode plasma potential to its cathode boundary.

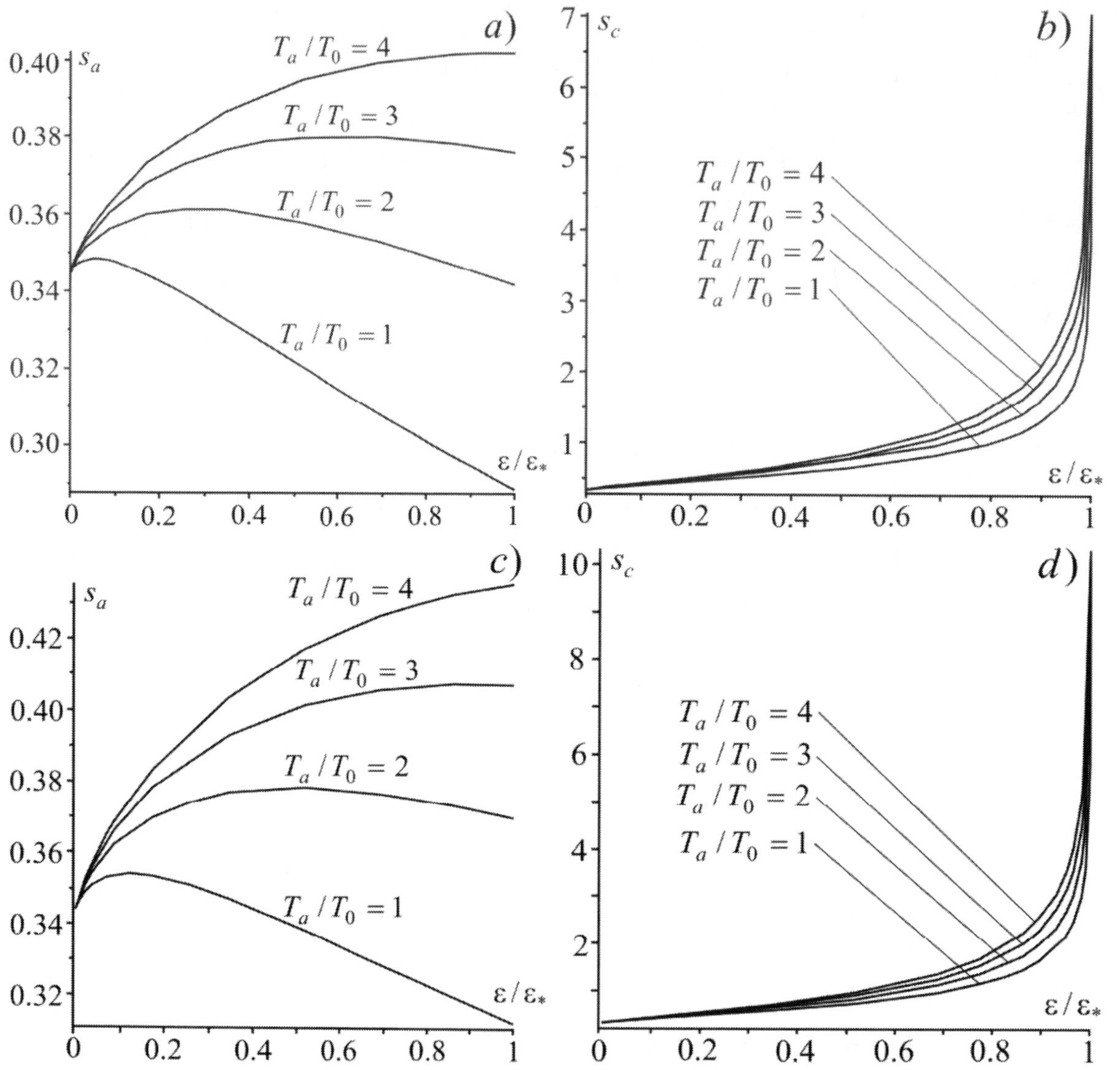


*Figure 1. Dependences of anode and cathode boundary coordinates of anode plasma upon depletion parameter for different values of ratio of the anode temperature to the temperature of supplied gas. (a), b) correspond to kinetic model with equal absolute values of the initial velocities of neutral particles; c), d) correspond to kinetic model with Maxwellian velocity distribution of neutral particles).*

**References:**

1. D.M. Goebel and I. Katz. Fundamentals of Electric Propulsion: Ion and Hall Thrusters. Wiley-Blackwell. (2008).

2. A.V. Zharinov and V.P. Shumilin, Journal of Experimental and Theoretical Physics, Vol. 103, №2, P. 324 (2006).

3. V.P. Shumilin, A.V. Shumilin and N.V. Shumilin, Kinetic effects in ionization depletion of neutral particles in the anode plasma of the gas discharge. ESCAMPIG XXII, Greifswald, Germany, July 15-19, 2014, P1-06-05.

4. V.P. Shumilin, A.V. Shumilin and N.V. Shumilin, High Temperature, Vol. 50, №2, P. 178 (2012).

5. A.F. Nikifirov and V.B. Uvarov. Special Functions of Mathematical Physics. Basel (Switzerland): Birkhauser. (1988).

# 2D Simulations and experiments on streaming instabilities in a Penning discharge

M. King[1], D.C. Speirs[1], T. Heelis[1], R. Bingham[1,2], C.G. Whyte[1], C.W. Robertson[1], A.D.R. Phelps[1], M.E. Koepke[1,3], S.L. McConville[1], R.A. Cairns[1,4], I. Vorgul[4], & K. Ronald[1]

[1] *Department of Physics, SUPA, University of Strathclyde, Glasgow, G4 0NG, Scotland*

[2] *STFC Rutherford Appleton Laboratory, Harwell, Oxford, Didcot, OX11 0QX, England*

[3] *Department of Physics, West Virginia University, Morgantown, WV 26506-6315, USA*

[4] *School of Mathematics and Statistics, University of St Andrews, KY16 9SS, Scotland*

Instabilities may arise when a plasma has a strongly non-thermal component in its distribution function. Laser injected 'spark-plug' electron beams are proposed as a method of heating the core of inertially confined fusion targets in schemes for 'fast ignition'. These schemes are attractive as they promise a very substantial reduction in the scale of the laser required for fuel compression. Streaming instabilities can contribute to the transfer of energy from these energetic electrons to heat the ions to the levels required to initiate fusion reactions [1]. Experiments have shown that more energy can be transferred to the plasma than can be accounted for from purely electron-ion collisions [2,3]. It has been proposed that the two-stream instability may be responsible as it can parametrical decay into ion-acoustic waves which are then damped by ion-ion collisions resulting in additional energy transfer to the ions [4].

Multidimensional kinetic computer simulations of these instabilities may be achieved using modern parallel Particle in Cell (PiC) codes [5]. In order to ensure the predictions of the simulations are accurate, it is desirable to test their output over as wide a range in parameter space as possible. We present 2D simulations of the interaction conducted using XOOPIC [6] between a moderately energetic electron beam (~10-100kV), with a narrow thermal distribution, propagating in a background plasma ~5cm in diameter and with a density of ~$5\text{-}9\times10^{16}\,m^{-3}$ with both the beam and plasma confined radially by a strong axial magnetic field. The two stream instability will originate from a point source disturbance within such a streaming system [7]. If a density fluctuation arises from this disturbance in one stream of particles, then the electric field will initiate a plasma oscillation at that location. However, these fields can modulate the charged particle densities of the second stream and the drift of these density modulations through each other can result in energy exchange. This leads to growth of the energy associated with the electric fields feeding from the energy of the initial particle streams.

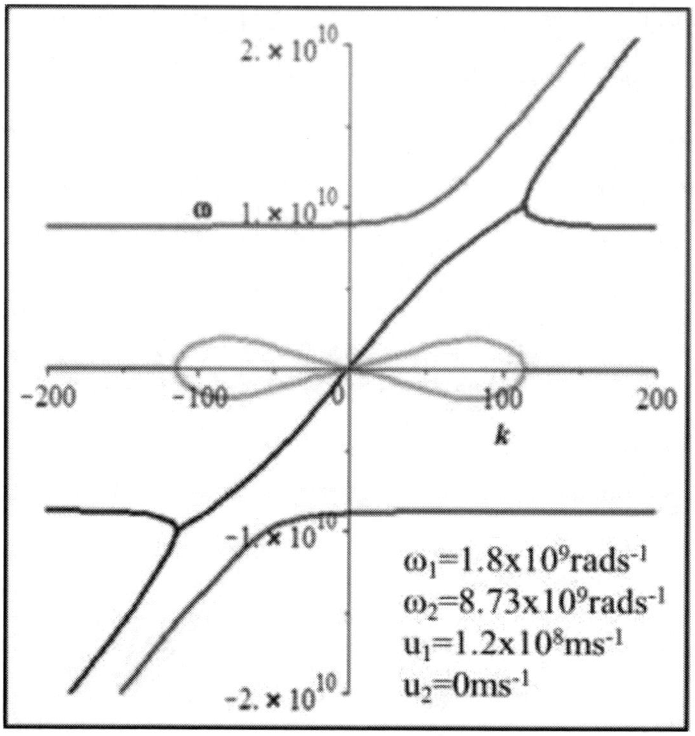


Fig 1 - Dispersion relation obtained from linear theory of the two-stream instability parameters used in the numerical simulations

Figure 1 shows the linear theory predictions of the temporal growth rate, the frequency and spatial scale of the oscillations induced by the streaming instability (assuming the insytability grows near the point where Im(ω) is maximum). Looking at figures 2 and 3, the simulations show clearly the growth and saturation of the two-stream instability, the evolution of phase space vortices and the trapping of beam electrons in axial velocity due to strong axial electric fields. The AC fields encourage the formation of axially aligned spatial structures in both the ion and electron densities. These structures act to inhibit the two stream process and drift at the ion acoustic velocity. The simulations were used to define the parameters for a laboratory experiment. A schematic of the apparatus is presented in figure 4, including the electron accelerator, the magnetic field system and the plasma column (based on a Penning discharge).

The electron beam is created using a Pierce-type electron accelerator geometry. A Penning trap type plasma discharge was used to create the necessary background plasma. Water-cooled solenoids were used to create the focusing magnetic field for the electron beam as well as the confining magnetic field for the plasma. These solenoids allowed the magnetic fields to be varied as required to maintain laminar beam transport and stable Penning discharge operation.

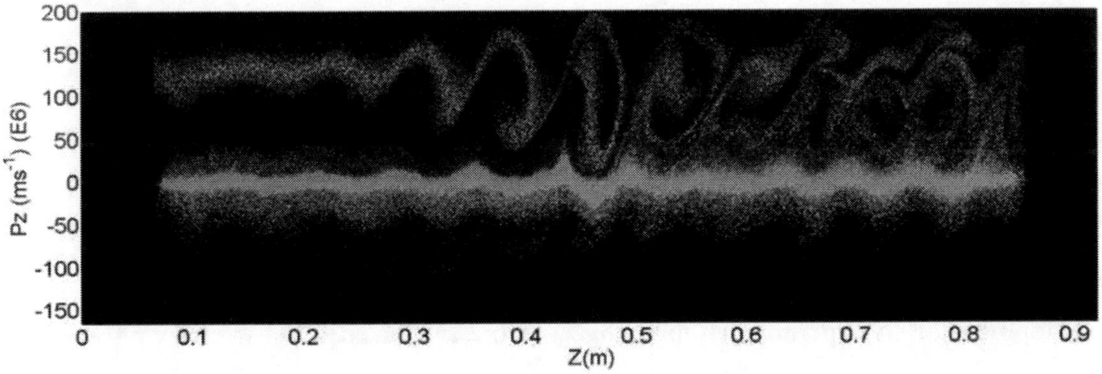


Fig 2 - axial-momentum against z of all particles in the simulation after 150ns (orange=beam electrons, green=plasma electrons, blue=plasma ions)

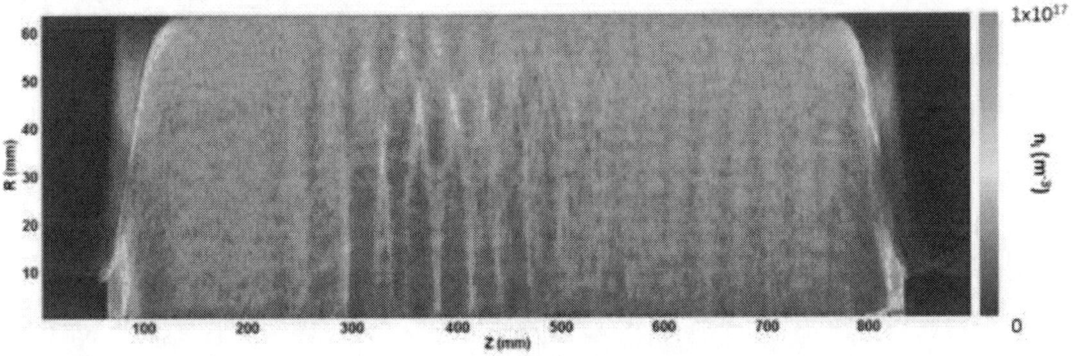


Fig 3 - Z-R plot of ion density after 150ns showing the formation and drift of ion density modulations- electrons follow similar behaviour, initial density $2.4\times10^{16}m^{-3}$, dia. 60mm

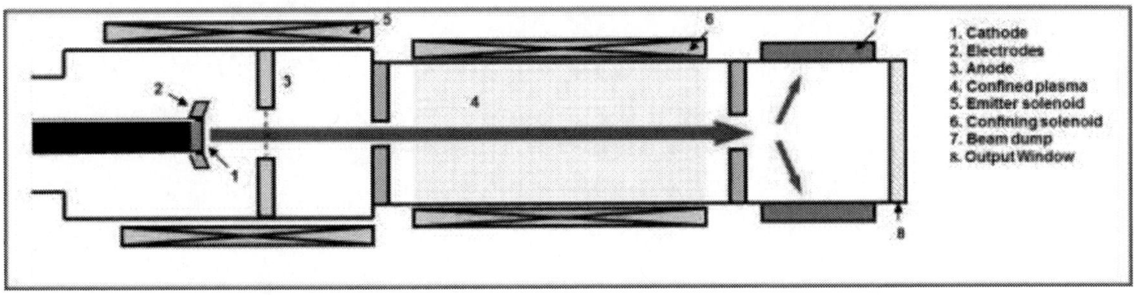


Fig 4 – Experimental configuration for two-stream experiments.

The high current Penning discharge trap features specially designed insulators at each end to suspend the anode in the centre of the system electrically separated from the main anode (which is the electrical cathode for the Penning discharge). Stable plasma conditions were obtained over the range 1mA-25mA in DC discharge operation current and pulsed operation was achieved to over 40mA. Helium gas was admitted to the discharge chamber at a pressure of $7x10^{-4}$mB in a vacuum of $5x10^{-6}$mB, whilst the magnetic field was typically 0.05-0.3T. The discharge was diagnosed by two distinct methods (Langmuir probe and interferometric microwave phase-evolution measurements). These diagnostics indicate plasma densities in the region of $3x10^{16}m^{-3}$ at a discharge current of 18mA (DC).

In summary long duration simulations have shown transfer of energy from the electrostatic oscillations excited by a streaming instability in a meter long magnetised plasma column enclosed in a metal tube to excite ion motion. The simulations have informed the development of an experimental apparatus which has created a plasma, having a density similar to that predicted by the simulations to facilitate the effect, through which an electron beam of ~50-100keV and a current of 1-50A may be propagated.

## References

[1] Zhuo, H.B. et al., Physical Review Letters, 112, 215003 (2014)

[2] R. Kodama et al., Nature (London) 412, 798 (2001)

[3] R. Kodama et al., Nature (London) 418, 933 (2002)

[4] J.T. Mendonça et al, Phys Rev Lett, 94, 245002 (2005)

[5] N.J. Sircombe et al, Plasma Phys. Control. Fusion, 50, 065005 (2008)

[6] J. P. Verboncoeur, A. B. Langdon and N. T. Gladd, Comp. Phys. Comm. 87, 199 (1995). Code available via http://ptsg.eecs.berkeley.edu

[7] Stix T.H., Waves in Plasmas, Springer (1992).

# Fast measurements of the electron temperature in divertor region of the COMPASS tokamak using ball-pen probe

J. Adamek[1], J. Seidl[1], R. Panek[1], M. Komm[1], P. Vondracek[1,2], J. Stöckel[1] and COMPASS team

[1] *Institute of Plasma Physics CAS, Prague, Czech Republic*

[2] *Faculty of Mathematics and Physics, Charles University, Prague, Czech Republic*

## 1 Introduction

Investigation of the electron temperature and consequently the parallel heat flux on the divertor target during Edge Localized Modes (ELMs) is of importance for the ITER tokamak in order to determine the physical sputtering and melting of the plasma facing components and, consequently, the life time of the divertor targets. The electron temperature in divertor regions of present-day devices is typically measured by swept Langmuir probes or triple probe technique [1]. However, the measurement of its evolution during ELM filaments requires high temporal resolution, which has not been achieved by these techniques yet. Alternative option how to determine the electron temperature with high time resolution is to use direct measurements of the plasma ($\Phi$) and floating ($V_{fl}$) potentials. The electron temperature $T_e$ is consequently determined from the difference of both potentials normalized by the proper coefficient for specific gas and magnetic field [2, 3, 4]. The floating potential is routinely measured by Langmuir probe (LP) and the plasma potential can be directly measured by ball-pen probe (BPP) [4]. The BPP was used to measure plasma potential in the edge plasma on COMPASS [3, 5], ASDEX Upgrade [2, 3, 5], MAST [6] and ISTTOK [7] using probe manipulator. Recently, a new array of BPPs and dome-shaped Langmuir probes was installed also in a divertor target of the COMPASS tokamak. The first results of fast measurements of the electron temperature on divertor during ELMy H-mode with sub-microsecond time resolution are reported in this paper.

## 2. Divertor BPP and LP construction and experimental set-up

The new array of BPPs (labelled A to L) has been implemented directly to the graphite divertor target on the COMPASS tokamak as shown in Fig. 1. Each ball-pen probe is made of stainless-steel collector with 3 mm diameter and boron-nitride shielding tube with 6 mm diameter and. The collector is retracted at $h \sim 0.7$ mm below the graphite surface. Four additional dome-shaped LPs (Q, N, R, P) are also mounted on the same divertor target as seen in Fig. 1. The LPs are made of graphite with 4 mm diameter and protruding 1.3 mm to the plasma. Toroidal distance between neighbouring LPs is roughly 6 cm and similar distance is applied also for BPP and LP. In current configuration, a set of probes BPP_H, LP_Q, LP_N is located at the

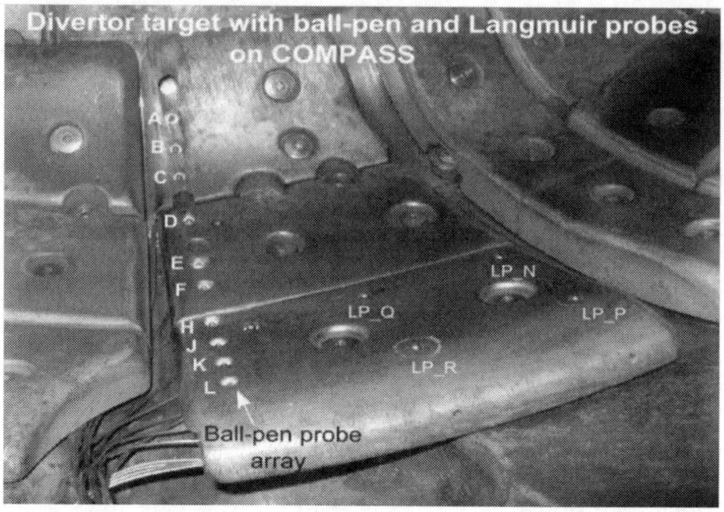


**Fig. 1:** Picture of the ball-pen probe (BPP) array and few dome-shaped Langmuir probes mounted on divertor target on COMPASS. The toroidal distance between BPP and LP is roughly 6 cm.

same poloidal position. The Langmuir probes can operate in floating or ion saturation current mode ($V_{probe}$= -270 V). The total exposed area of LP is approximately 30 mm$^2$. All signals are acquired with a sampling frequency rate $f$ = 5 MSPS. The BPPs are implemented to the divertor target as flush-mounted probes, which significantly reduce the direct impact of high heat flux on the probe material. On the other hand, the BPPs in the divertor are surrounded by large graphite (conducting) tiles, which is a new probe arrangement in contrast to the BPP mounted on probe manipulator. Therefore, in order to verify their *I-V* characteristics, we have applied swept voltage (+/-200 V, $f$= 1 kHz) on two BPPs (BPP_J, BPP_D) with significantly different poloidal locations. An example of resulting *I-V* characteristics of both probes obtained during the L-mode phase ($B_T$= 1.15 T, plasma current $I_P$ = 300 kA, density $n_e = 6{\cdot}10^{19}$ m$^{-3}$) of H-mode discharge #4259 are plotted in Fig. 2. It can be seen that both probes provide nearly symmetric *I-V* characteristics during the whole time interval. It is also seen that electron and ion branch of the *I-V* does not saturate and therefore the evaluation of the ratio of the electron and ion saturation current $R=I_{sat\_e}/I_{sat\_i}$ requires a linear extrapolation of both branches. The linear fit in Fig.2 is calculated for BPP voltage > +100V on the electron branch and for BPP voltage < -100V on the ion branch. If we assume the floating potential of BPP is close to the plasma potential $\Phi$ then the logarithm of the ratio $R$ is found close to ln($R$) ~ 0.8. The resulting value ln($R$) is close to the value obtained with BPP mounted on probe manipulator on AUG [2]. The determination of the ratio $R$ during ELMs is difficult because of very short time scale of ELM filaments. However, the estimation of the electron and ion saturation current during ELM is shown in Fig. 3. Two similar ELM events, ELM_1 and ELM_2, with the same D-alpha signal, were detected on swept BPP_J and BPP_D during ELMy H-mode phase of the discharge #4259

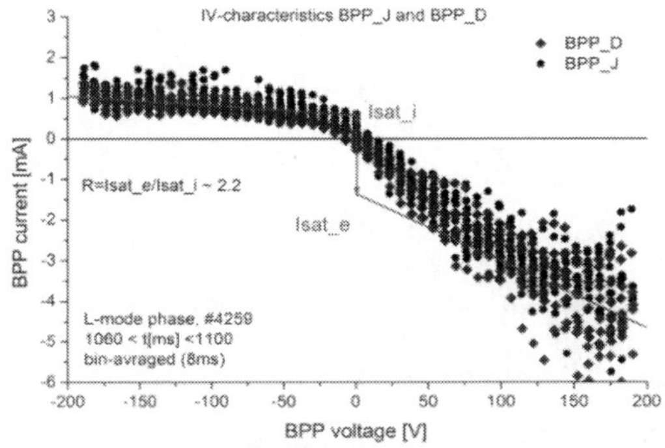


**Fig. 2:** The bin-averaged *I-V* characteristics of BPP_D (diamond) and BPP_J (circle) during whole L-mode phase of discharge #4259.

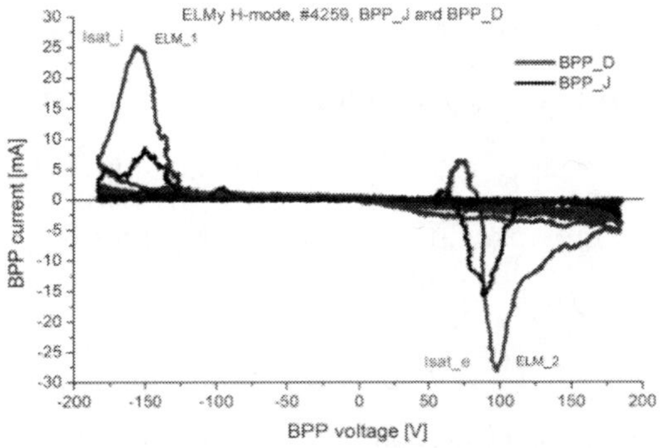


**Fig. 3:** The *I-V* characteristics of BPP_D (blue) and BPP_J (black) during ELMy H-mode #4259 with two similar ELM events incoming to both probes.

($n_e = 9{\cdot}10^{19}$ m$^{-3}$) at the moments with opposite value of applied voltage. This confirm that both values $I_{sat\_e}$ and $I_{sat\_i}$ are also roughly balanced within the ELM, with ratio $1< R <2$ independent on the probe position. Note, the investigation of ratio $R$ of BPP during ELMy H-mode can be also found in systematic measurements on ASDEX Upgrade [2].

**3 The electron temperature within ELM filaments**

The investigation of the electron temperature $T_e$ with a sub-microsecond temporal resolution is based on fast measurements of BPP and LP potentials, $\Phi^{BPP}$ and $V_{fl}$, and on the formula $T_e=(\Phi^{BPP} - V_{fl})/2.2$. The coefficient 2.2 is obtained from the ln($R$) = 2.8 of LP in deuterium plasmas (assuming $T_i = T_e$ and no secondary electron emission) minus the ln($R$) of BPP. The values of ln($R$) of BPP on COMPASS, shown in Fig. 2, and on AUG [2] are in agreement within the error bars. Therefore, we are using value of ln($R$) = 0.6 of BPP for both devices. Note, the same formula was already used for comparative measurements of the electron temperature using BPP/LP mounted on the reciprocating manipulator and Thomson scattering on COMPASS [3], AUG [3] and MAST [6] in L-mode plasmas with a very good agreement of both techniques. The first results of fast electron temperature measurements within the ELMs using BPP/LP mounted on horizontal manipulator were performed on COMPASS [8]. However, this is the first time measurement of the electron temperature within ELMs using BPP/LP on the divertor with sub-microsecond time resolution. At first, both Langmuir probes, LP_Q and LP_R, were connected as a floating probe in order to calculate the cross-correlation coefficient. It has been found that cross-correlation coefficient is within the range of 0.8-0.9 for L-mode, ELM free H-mode and ELMy H-mode phases of the discharge #7876. This confirms that both LPs and BPP_H are well magnetically connected. In the following H-mode discharge #7891, the LP_R was switched to the $I_{sat}$ regime ($V_{bias}$ = -270 V). It allowed us to provide simultaneous measurements of the electron temperature $T_e$ by combining BPP_H and LP_Q and ion saturation current $I_{sat}$. The time evolution of the electron temperature and ion saturation current during one ELM is plotted in Fig. 4. It is seen that the value of the electron temperature is strongly varying within the filament structure. However, it must be noted that a small part of resulting $T_e$, roughly 9% of total signal during H-mode phase, has values below 0 eV, because in some cases it is observed during ELM filaments that both signals $\Phi^{BPP}$ and $V_{fl}$ are slightly shifted in time. This can be caused by a small misalignment of BPP and LP with respect to the magnetic field lines or different dimensions of each probe. However, the major part of $T_e$ values can be used in order to estimate the value of parallel heat flux $q_{//}=7{*}T_e{*}I_{sat}/S_{LP}$, where 7 is the

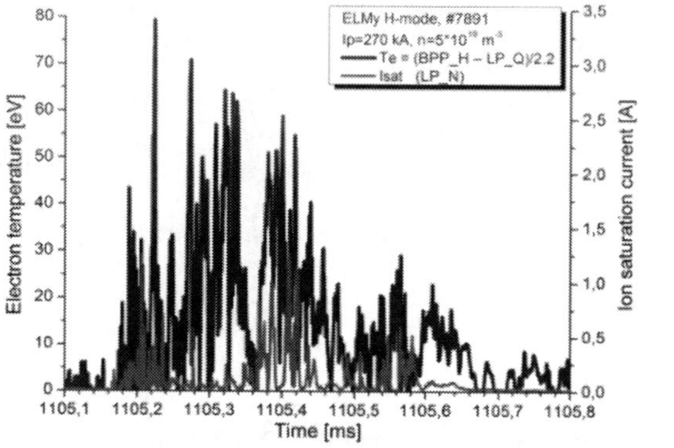


**Fig. 4:** Temporal evolution of the electron temperature and ion saturation current during one ELM event in H-mode #7891 on divertor on COMPASS.

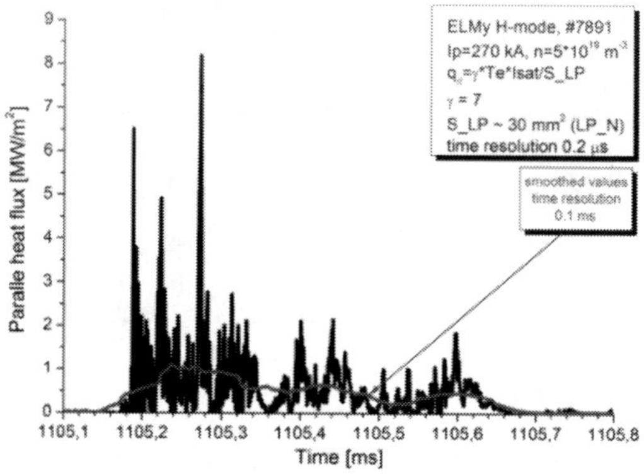


**Fig. 5:** Temporal evolution of parallel heat flux obtained from $T_e$ and $I_{sat}$ measurements and heat transmission coefficient $\gamma=7$ during one ELM event in H-mode #7891.

value of heat transmission coefficient [9] and $S_{LP}$ is the total exposed surface of LP_N. The time evolution of $q_{//}$ during one ELM is plotted in Fig. 5. It is seen that value of parallel heat flux during ELMs reaches a couple of MW/m$^2$. These values are obtained roughly 6 cm from outer strike point. The red line represents smoothed values of parallel heat flux to 0.1 ms (10 kHz), which is comparable to resolution of common IR camera. In such case, the resulting values of $q_{//}$ are below 1 MW/m$^2$, strongly underestimating the peak heat flux.

**Acknowledgements** This work was supported by the project of the Czech Science Foundation 15-10723S and MSMT Project #LM2011021. It has been carried out within the framework of the EUROfusion Consortium and has received funding from the Euratom research and training programme 2014-2018 under grant agreement No 633053. The views and opinions expressed herein do not necessarily reflect those of the European Commission.

**References**

[1] S. Jachmich et al., 36th EPS Conference on Plasma Phys., Sofia, P-2.159 (2009).
[2] J. Adamek et al., Contrib. Plasma Phys. 50, 854 (2010).
[3] J. Adámek, et al., 41st EPS Conference on Plasma Physics, Berlin, P2.011 (2014).
[4] J. Adamek et al., Czech. J. Phys. 54, C95-C99 (2004).
[5] J. Adamek et al., Contrib. Plasma Phys. 54, No. 3, 279 – 284 (2014).
[6] N. R. Walkden et al., Review of Scientific Instruments 86, 023510 (2015).
[7] C Silva et al., Plasma Phys. Control. Fusion 57 (2015) 025003 (9pp)
[8] J. Seidl et al., 41st EPS Conference on Plasma Physics, Berlin, P5.059 (2014).
[9] J.G. Watkins et al., Journal of Nuclear Materials 390–391 (2009) 839–842

# Effects of Plasma Control on Runaway Electrons in the COMPASS Tokamak

J. Mlynar[1], O. Ficker[1,2], M. Vlainic[1,3], V. Weinzettl[1], M. Imrisek[1,4], R. Paprok[1,4], M. Rabinski[5], M. Jakubowski[5] , M. Tomes[1,4], M. Peterka[1,4], R. Panek[1] and the COMPASS Team

[1]*Institute of Plasma Physics AS CR, Prague, Czech Republic*

[2]*Czech Technical University in Prague, Faculty of Nuclear Sciences and Physical Engineering, Czech Republic*

[3]*Department of Applied Physics, Ghent University, Ghent, Belgium*

[4]*Charles University in Prague, Faculty of Mathematics and Physics, Prague, Czech Republic*

[5] *National Centre for Nuclear Research (NCBJ), Otwock-Swierk, Poland*

## 1. Introduction

Runaway Electron (RE) mitigation after disruptions presents one of the key challenges for safe operation of future fusion reactors. In support of related research efforts (see e.g. [1]-[7]) systematic studies of RE generation and losses in the COMPASS tokamak started in early 2014 in the Framework of the EUROfusion WP14-MST2-9 research project. The COMPASS tokamak features ITER-relevant geometry with $R = 0.56$ m, $a = 0.23$ m, $I_p < 400$ kA, $B_T \sim 1.15$ T, up to 2x 350 kW of Neutral Beam Heating and typical pulse length around 300ms. It can operate with inner limiter or single null (divertor) plasma configuration. In the latter case, routine operation in the H-mode was achieved. Plasma edge studies and the related diagnostic development are in the main focus of the COMPASS research team [8]. The RE research relevant diagnostics is rather sparse at present (see Fig. 1), however, it has been extended in collaboration with the NCBJ Poland (development of the Cherenkov detector) and the Czech Technical University (NaI(Tl) detector, MediPix pinhole camera), while further investments are foreseen (e.g. the neutron detector). The hard X-ray (HXR) scintillation detectors present the key diagnostic tool in this work. These detectors measure non-collimated radiation that originates mostly from RE collisions with the tokamak inner wall

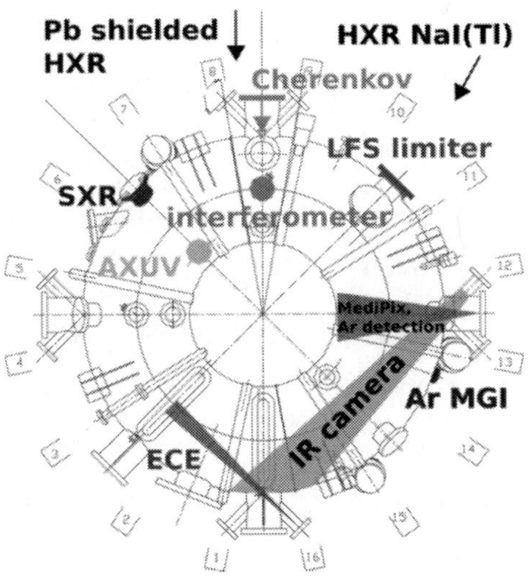


**Figure 1.** Setup of the diagnostic systems for the RE studies at COMPASS tokamak.

(bremsstrahlung and deexcitation in solid components). One HXR detector is located outside the tokamak so that its threshold for the HXR energy is above 50 keV. The second HXR detector is at a similar distance from the tokamak, however it is shielded by 10 cm Pb box; its lower energy threshold is therefore well above 500 keV. In the RE dominated COMPASS discharges, the RE generated synchrotron radiation was directly measured by infrared camera, see [9]. These observations indicate that the maximum energy of the Runaway Electrons (REs) is around 25 MeV. RE beams after argon triggered plasma disruptions were successfully observed in the current ramp-up phase of the COMPASS plasma discharge, which opened new prospects for ITER-relevant studies on RE control and mitigation at COMPASS [10]. In this contribution, it is demonstrated that the RE seed production during plasma breakdown plays a key role in formation of the RE population observed in the current plateau and current ramp-down of the discharge. Next, the observed dependences of the RE confinement on plasma shaping and plasma current are presented.

## 2. Plasma deuterium fuelling and generation of the Runaway Electrons

Experiments with plasma fuelling were aimed at improving our understanding of the impact of the initial (seed) RE generation on the overall properties of REs in the subsequent current plateau phase of the discharge. The results are also applicable to other COMPASS research programmes where the early RE mitigation may improve plasma parameters and/or diagnostic performance. In the experiments, the plasma border was defined by high-field side limiter and plasma density was feedback controlled in the current plateau at a level far from the RE dominated discharges in order to minimise RE production in the plateau phase. First, discharges with circular cross-sections and plasma density $n_e \sim 3.1 \cdot 10^{19}$ m$^{-3}$ were studied, see Fig. 2a. It can be seen that a single initial puff of fuel at plasma breakdown resulted in generation of an intense RE population, which is due to high $U_{loop}$ and low density in the current ramp-up phase. In the discharge plateau, however, REs got mitigated in a relatively high plasma density; this process is witnessed by the exponential decrease of the HXR radiation. Under the same conditions, a secondary fuel puff during the early current ramp-up completely mitigated the RE seed and, consequently, no HXR radiation was observed. In either case the Pb shielded HXR detector registered no useful signal. Second, discharges with elongated plasmas and density $n_e \sim 2.5 \cdot 10^{19}$ m$^{-3}$ were examined, see Fig. 2b. These experiments confirmed importance of the initial RE seed for the subsequent evolution of REs. Furthermore, acceleration (and possibly avalanching) of the RE population in the discharge plateau is testified by loss of highly energetic REs (as observed by the Pb shielded detector) in the current ramp-down phase with zero $U_{loop}$. Interestingly, the data also demonstrated that a low-amplitude secondary fuel puff in the early current ramp-up phase can hinder production of the high energy REs only.

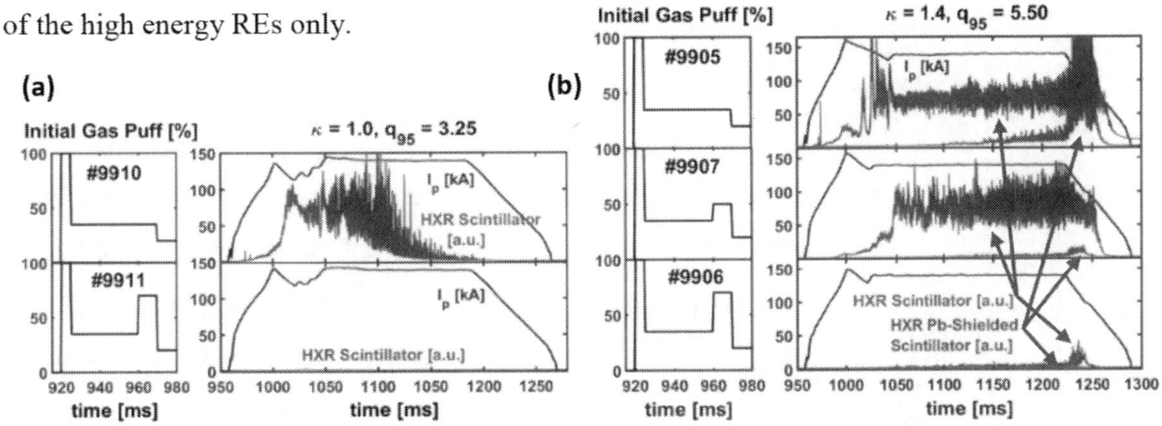


**Figure 2.** Fuelling scenarios at the plasma breakdown and its impact on RE population in the COMPASS tokamak: (a) circular plasma ($\kappa = 1$, $q_{95} = 3.25$) and (b) elongated plasma ($\kappa = 1.4$, $q_{95} = 5.50$) cross-sections.

## 3. Confinement of Runaway Electrons in shaped plasmas

In order to assess influence of plasma shaping on the RE confinement, HXR data from discharges with circular plasma and elongated plasma cross-sections were compared, see Fig. 3. Notice that in these studies, plasma current was kept constant in order to maintain similar plasma temperatures ($T_e \sim 550$ eV), and two levels of plasma density were examined. The HXR data give evidence of improved confinement in shaped plasmas. Indeed, in the low density case ($n_e \sim 1.9 \cdot 10^{19}$ m$^{-3}$, blue line) data from the shielded detector indicate continuous loss of the energetic REs in the case of circular plasmas, while in elongated plasmas the energetic REs are lost in a massive amount only at current ramp-down. At higher density ($n_e \sim 3.1 \cdot 10^{19}$ m$^{-3}$, red line) the evidence is even stronger, with early loss of all REs in the circular plasma, while some energetic REs survive in the elongated plasma.

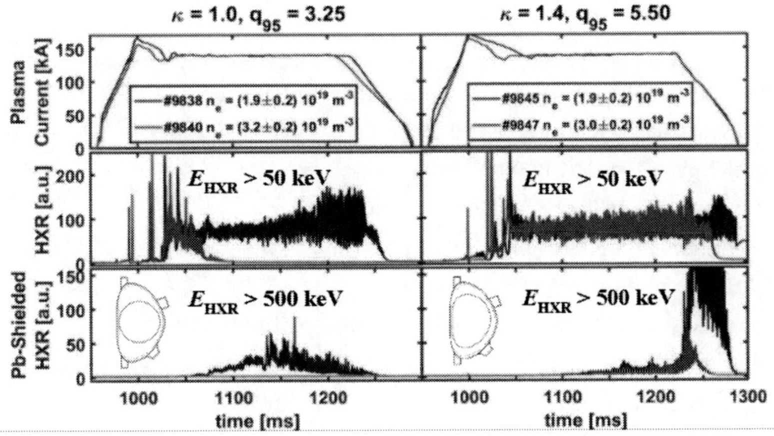


**Figure 3.** Plasma shaping and RE confinement in the COMPASS tokamak

However, interpretation of these observations is challenging and beyond available models. Among others, a possible interplay between the safety factor (magnetic field helicity) profile and the RE confinement have been discussed. Notice that at a constant plasma current the edge safety factor $q_{95}$ increases with plasma elongation, which may strongly influence MHD properties of the discharge. To this end, RE confinement in elongated COMPASS discharges was studied as a function of increasing plasma current at a constant plasma density ($n_e \sim 2.5 \cdot 10^{19}$ m$^{-3}$). The results confirmed that the RE losses tend to increase with decreasing edge safety factor $q_{95}$, see Fig. 4. However, the HXR radiation data also reflected rather sudden loss of the RE population; at the time of the loss, a clear onset of MHD activity was observed in several discharges.

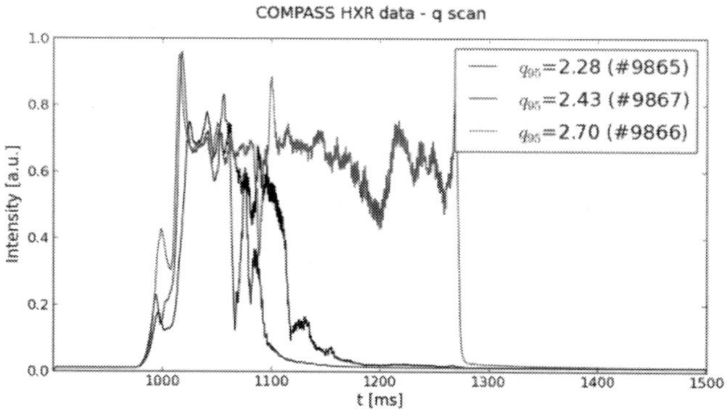


**Figure 4.** RE confinement in the COMPASS elongated plasmas with variable edge safety factor $q_{95}$

## 4. Conclusions

Dedicated experiments at COMPASS tokamak proved that initial fuelling of the discharge predetermines the RE population. Improved confinement of REs was observed in shaped plasmas; data indicate a significant role of the MHD effects. Future COMPASS RE campaigns shall be focused, among others, on benchmarking studies of the models, in particular on LUKE [11], and on field independent plasma heating via Neutral Beam Injector.

## Acknowledgments

This work has been carried out within the framework of the EUROfusion Consortium and has received funding from the Euratom research and training programme 2014-2018 under grant agreement No 633053. The COMPASS operation has been supported by the MSMT project LM2011021. The views and opinions expressed herein do not necessarily reflect those of the European Commission.

## References

[1] V. Plyusnin et al., Proc. 25th Fusion Energy Conference, St Petersburg 2015, EFDA-JET-CP(14)06/34
[2] A. Loarte et al., Nuclear Fusion 51 (2011) 073004
[3] B. Esposito et al, Phys Plasmas 10 (2003) 2350
[4] E. M. Hollmann et al, Nucl. Fusion 53 (2013) 083004
[5] F. Saint-Laurent, Proc. 36th EPS Conf. on Plasma Physics, Sofia 2009, ECA Vol. 33E, P4.205
[6] E.D. Frederickson, Nucl. Fusion 55 (2015) 013006
[7] R. Yoshino, S. Tokuda and Y. Kawano, Nuclear Fusion 39 (1999) 151
[8] R. Panek at al., invited talk at this conference, to be submitted to Plasma Phys. Control. Fusion
[9] M. Vlainic et al, poster presentation at this conference, P4.108
[10] M. Vlainic et al, accepted by J. Plasma Phys
[11] J. Decker, Y. Peysson, EUR-CEA-FC-1739 (2004)

# Observation of geodesic acoustic mode-like oscillations on COMPASS

J. Seidl[1], M. Hron[1], J. Adamek[1], P. Vondracek[1,2], J. Horacek[1], C. Hidalgo[3], A. Melnikov[4,5], L. Eliseev[4], T. Markovic[1,2], J. Stöckel[1], D. Basu [6], P. Hacek[1,2], J. Havlicek[1,2], M. Imríšek[1,2], K. Kovarik[1,2], V. Weinzettl[1], R. Panek[1] and COMPASS Team

[1] *Institute of Plasma Physics AS CR, Prague, Czech Republic*

[2] *Faculty of Mathematics and Physics, Charles University, Prague, Czech Republic*

[3] *Ciemat, Madrid, Spain*

[4] *National Research Centre "Kurchatov Institute", Moscow, Russia*

[5] *National Research Nuclear University MEPhI, Moscow, Russia*

[6] *Plasma Physics Laboratory, University of Saskatchewan, Saskatoon, Canada*

Geodesic acoustic mode (GAM) [1] is a turbulence-driven high-frequency branch of zonal flows, which can play a role in self-regulation of turbulent transport. It is a toroidaly symmetric, $n = 0$, oscillating flow with potential constant on a flux surface, $m_\phi = 0$. Due to geodesic curvature the flow couples to a pressure perturbation with $m_p = 1$ standing wave pattern. Even though GAM is mainly an electrostatic mode, its magnetic component has been recently described both in experiment and theory. While in the circular plasmas the magnetic component has $m_B = 2$ poloidal mode number, other Fourier components can be induced by plasma shaping [2]. The GAM oscillations are driven by non-linear three-wave coupling with ambient turbulent oscillations [3]. In practice, the coupling is typically detected using bicoherence analysis. Frequency of the mode scales with the ion sound speed $c_s$ and tokamak major radius $R_0$, $2\pi f_{GAM} = Gc_s/R_0$, where $G$ is a factor dependent on plasma shape. In ciruclar plasmas $G = (2+q^{-2})^{1/2}$ [1] but lower values are expected in elongated plasmas [2, 4].

In this paper, we present observation of GAM-like oscillations in D-shaped discharges on the COMPASS tokamak. The mode is detected in a form of long range correlations of plasma potential between pair of reciprocating probes and the oscillations are clearly correlated with oscillations of poloidal magnetic field.

## Experimental setup

COMPASS ($R$ = 0.56 m, $a$ = 0.2 m) is a tokamak with ITER-like divertor plasma configuration. It is equipped with two pneumatic reciprocating probe manipulators, located at the top of the vessel and at the low-field side (LFS) midplane, toroidally shifted by 22.5°. Typically the manipulators are not directly magnetically connected. Both probe heads consist of combination of Langmuir probes (LP) measuring floating potential $V_{fl}$ and/or ion saturation current $I_{sat}$, and ball-pen probes (BPP) providing fast measurement of a potential that can serve as a close proxy to the plasma potential $\phi$ [5]. Moreover, combination of LP and BPP allows fast sub-microsecond measurement of electron temperature $T_e = (V_{BPP} - V_{fl})/2.2$ [6]. All the probe measurements are sampled at 5 MHz. Both manipulators allow deep reciprocation inside the last closed flux surface, without overheating the probe head. However, position of the probes and EFIT reconstruction are not fully in agreement, with a shift at LFS of D-shaped plasmas of about $\sim 2$ cm. As both probes provide fast measurement of $T_e$, we therefore carry out surface labeling using an assumption that $T_e$ is constant over flux surface.

The probe diagnostics was supplemented by measurement of poloidal magnetic field $B_p$ fluctuations with 2 MHz sampling rate, using two poloidal rings of Mirnov coils and a set of saddle loops covering the whole toroidal angle.

**Mode properties**

Fig. 1 shows cross-coherence between $V_{fl}$ during simultaneous deep plunge of the probes. Correlated oscillations are detected in $V_{fl}$ as well as $\phi$ around $\sim 30$ kHz whenever both probes get inside the radius with zero radial electric field (i.e. local maximum of radial profile of plasma potential). When the probes are at the same flux surface, mutual phase shift is close to zero (Fig. 1c), consistent with $m_\phi = 0$ expected for GAMs. This is further supported by estimation of wavenumber-frequency spectra from two-point probe measurements [8] that rules out poloidal mode numbers with $m_\phi > 2$. $I_{sat}$ measurements at midplane show

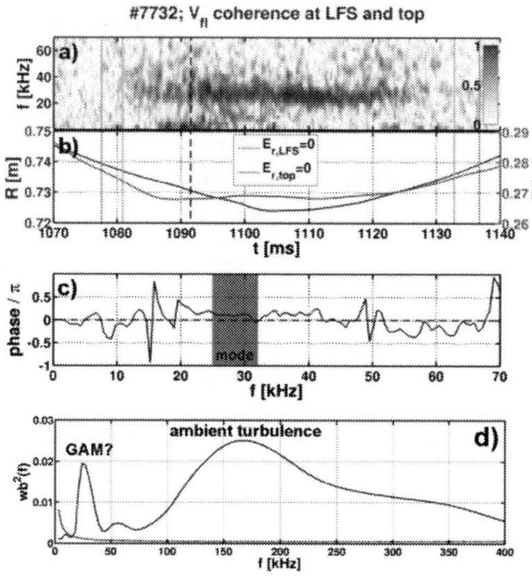


Figure 1: *Coherence of $V_{fl}$ at LFS and top during simultaneous probe reciprocation a), position of the probes in time b), phase shift c) between probes when crossing the same flux surface - marked by dashed black line in a). Summed squared wavelet bicoherence of $V_{fl}$ at LFS d), dashed black line shows level of statistical significance.*

no or very weak oscillations at this frequency, consistent with GAM oscillations, but the $m_p = 1$ structure needs to be further verified by $I_{sat}$ measurement at the top. Further, wavelet bicoherence [7] (Fig. 1d) confirms non-linear coupling of the mode to the ambient turbulent fluctuations in a broad 100-400 kHz range of frequencies. The turbulent drive of the mode may explain its observation in L-mode only, disappearing during L-H and reappearing shortly after the H-L transition. Note that in some cases we detected significant bicoherence also bellow 5 kHz, i.e. in the expected range of the low-frequency zonal flows. These fluctuations are, however, out of the scope of this paper.

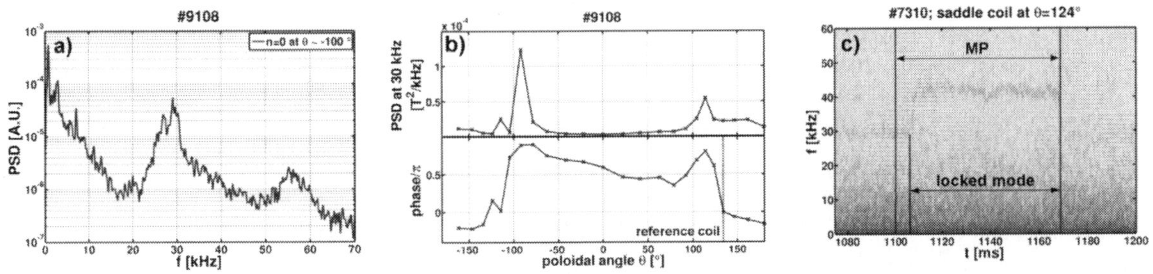


Figure 2: *Spectrum of toroidal $n_B = 0$ component a) and poloidal structure of magnetic fluctuations b). The frequency peak in a) is broadened by oscillations with sawtooth period. Change of mode frequency simultaneously with generation of locked mode phase by MP c).*

## Magnetic component

Magnetic sensors show oscillations of $B_p$ correlated with the potential oscillations measured by the probes. In toroidal direction, saddle coils spanning over the whole toroidal circumference confirm toroidal symmetry $n_B = 0$ (Fig. 2a) of the mode. Poloidally spaced coils confirm its non-rotating nature and show poloidal structure different from $m_B = 2$ (Fig. 2b). While $m_B = 2$ is expected for GAM in circular plasmas, [2] predicts that additional Fourier components can be excited due to plasma shaping.

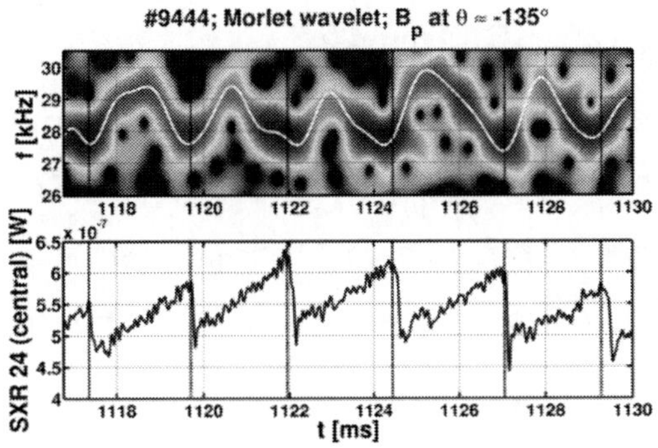


Figure 3: *Wavelet periodogram showing oscillations of frequency and amplitude of the mode compared with central soft-X ray (SXR) channel. White line marks instantaneous frequency estimated from Hilbert transform.*

## Mode frequency

Frequency of the mode in D-shaped discharges is observed in the range $\sim 25-35$ kHz. With typical edge temperature $T_e \gtrsim 30$ eV this is in the range expected for GAM frequency in circular plasmas provided $T_e = T_i$ and $\gamma_i = 1$, but empirical AUG scalings for D-shaped plasmas [4] would require rather high value of ratio $\gamma_i T_i / T_e \sim 7$, where $\gamma_i$ is ion specific heat ratio and $T_i$ is ion temperature. The frequency does not significantly change with radius within ~1-2 cm layer penetrated by the probes, consistent with non-local eigenmode structure reported on some devices, e.g. [9, 10]. However, in discharges with sawtooth activity (all shots shown here), the frequency peak of the mode is broadened by temporal oscillations of the frequency. This is best demonstrated on the magnetic component of the mode that shows clear periodic oscillations of the frequency by $\delta f \sim 2$ kHz with period similar to that of sawtooth instability (Fig. 3). This could be related to the temperature dependence of GAM frequency and to the heating of edge plasma by energy released from the core after each sawtooth crash, as observed e.g. on T-10 [11]. Note the relatively small dimensions of COMPASS plasma that allow non-negligible core-edge interactions.

## Magnetic perturbations

COMPASS is equipped with set of coils used for generation of $n = 2$ magnetic perturbation (MP) field [12]. While MP typically does not affect frequency of the GAM-like mode, Fig. 2c shows that when MP generated locked mode is present, the frequency sharply but still continuously rises by $\sim$15 kHz and returns back when MP coils are turned off and the locked mode disappears. Poloidal and toroidal structure of the magnetic fluctuations changes during presence of the locked mode towards dominant $n_B = 2$ and $m_B \geq 6$ components. While mechanism responsible for the transformation of the mode is still not fully clear, we note that very similar observation has been recently made during MP experiments on ASDEX Upgrade [13].

## Summary

An oscillatory mode has been observed at the edge of COMPASS plasma, inside of the radius with $E_r = 0$, that provides a reference point for the position of the edge sheared flows. Frequency of the mode is in the range of frequencies expected for GAMs and bicoherence analysis confirms non-linear coupling of the mode with broad spectrum of turbulent fluctuations. Also poloidal and toroidal structure of the magnetic component and poloidal mode number of the potential oscillations are in favor of conclusion that it could be classified as a GAM. Characteristic temperature dependence of GAM frequency needs, however, yet to be confirmed. We note that presence of the mode was detected also on several other diagnostics, such as soft-X ray detectors or divertor ball-pen probes, whose analysis will follow.

**Acknowledgements.** This work was supported by the projects GACR P205/12/2327, GACR 15-10723S and MSMT #LM2011021. The work was carried out within the framework of the EUROfusion Consortium and has received funding from the Euratom research and training programme 2014-2018 under grant agreement No 633053. The views and opinions expressed herein do not necessarily reflect those of the European Commission.

## References

[1] N. Winsor, et al., Phys. Fluids **11**, 2448 (1968)
[2] C. Wahlberg, et al., 41st EPS Conference Proceedings, P2.047 (2014)
[3] P. H. Diamond, et al., Plasma Phys. Control. Fusion **47**, R35 (2005)
[4] G. Conway, et al., Plasma Phys. Control. Fusion **50**, 055009 (2008)
[5] J. Adámek, et al., Contrib. Plasma Phys. **54**, 279 (2014)
[6] J. Adámek, et al., 41st EPS Conference Proceedings, P2.011 (2014)
[7] B. Ph. van Milligen, et al., Phys. Plasmas **2**, 3017 (1995)
[8] J. M. Beall, Jour. Apl. Phys. **53**, 3933 (1982)
[9] A. V. Melnikov, et al., Nuclear Fusion **55**, 063001 (2015)
[10] C. A. Meijere, et al., Plasma Phys. Control. Fusion **56**, 072001 (2014)
[11] A. V. Melnikov, et al., Plasma Phys. Control. Fusion **48**, S87 (2006)
[12] P. Cahyna, et al., Nucl. Fusion **49**, 055024 (2009)
[13] G. D. Conway, et al., Plasma Phys. Control. Fusion **57**, 014035 (2015)

# Alfvén-wave character oscillations in tokamak COMPASS plasma

T. Markovič[1,2], J. Seidl[1], A. Melnikov[3,4], P. Háček[1,2], J. Havlicek[1,2],
A. Havránek[1,5], M. Hron[1], O. Hronova[1]. M. Imríšek[1,2], F. Janky[1], K. Kovařík[1,2], O. Mikulín[1],
R. Pánek[1], R. Papřok[1,2], J. Pipek[1], P. Vondráček[1,2], V. Weinzettl[1]
[1] *Institute of Plasma Physics CAS, Prague, Czech Republic*
[2] *Faculty of Mathematics and Physics, Charles University in Prague, Prague, Czech Republic*
[3] *National Research Centre 'Kurchatov Institute', Moscow, Russian Federation*
[4] *National Research Nuclear University MEPhI, Moscow, Russian Federation*
[5] *Faculty of Electrical Engineering, Czech Technical University in Prague, Prague, Czech Republic*

In magnetic confinement fusion devices, high-energy particles originating e.g. from auxiliary-heating systems can drive instabilities by channeling the free energy (typically pressure gradients) to otherwise stable plasma eigenmode via resonance of particle velocity with plasma Alfvén velocity [1]. Consequently, the confinement of these particles is degraded by their interaction with the resulting shear Alfvén waves. Toroidal Alfvén Eigenmodes (TAE), caused by an interference of two counter-propagating Alfvén continuum waves [2], are of the most concern due to their low damping rate. In the recent work from the COMPASS tokamak [3], observations of possible TAE and BAE (Beta-induced Alfvén Eigenmodes) in H-mode plasmas in the spectral range of 50-250 kHz were reported. The present work introduces further observations of plasma fluctuations in the spectral range of 150-2500 kHz in both L-mode and H-mode plasmas showing their Alfvén-wave character.

**Experimental arrangement**

The COMPASS tokamak is a device with an ITER-like plasma geometry, having major radius $R_0 = 0.57$ m and minor radius about 0.2 m. In discharges described in this work, plasma current was $I_p < 0.3$ MA, toroidal magnetic field $B_0 = 1.15$ T and pulse duration $\Delta t < 0.4$ s. The tokamak is capable of H-mode plasma generation, either ohmic or NBI assisted ($2 \times 300$ kW deuterium beams, energy up to 40 keV). It is also equipped with a Resonant Magnetic Perturbation (RMP) system of variable spatial configuration [4]. Plasma eigenmode fluctuations are detected by two sets of Mirnov Coils (MC, 24 coils each) located at different toroidal positions measuring poloidal magnetic field fluctuations across the whole poloidal cross section (specifically, as coherence between two coils at the same poloidal position from different toroidal sets). Other detection coils are mounted on the U-probe manipulator [7] located in the bottom part of the chamber on Low-Field Side (LFS). The maximal detectable frequency is given by $f_{\mathrm{Nyquist}}$

of used data acquisitions, i.e. 1 MHz for MC coils and 2.5 MHz for U-probe coils, respectively.

**Frequency dependence on plasma parameters**

The first indication that observed oscillations have Alfvén wave character comes from a scaling of their frequency with global plasma parameters. An approximate spectral position of the center of the Alfvén-continuum gap is described in [2, 5]. In the scope of this work, a simplified scaling based on measured plasma parameters was used:

$$f_A(t) \sim \frac{B_0(t)}{q_{95}(t)\sqrt{n_e(t)}}, \quad (1)$$

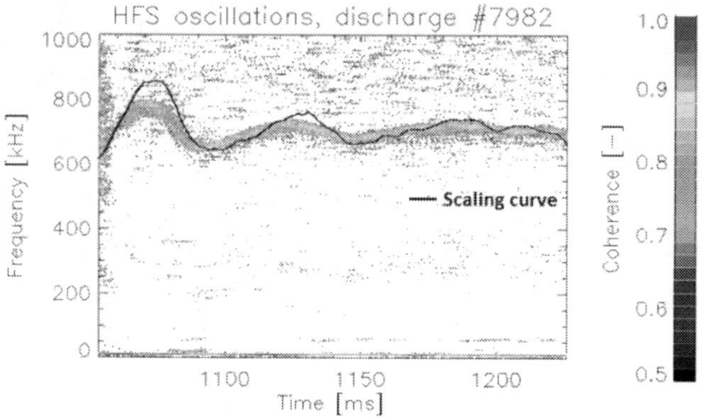


Figure 1: *High-frequency plasma oscillations on HFS during L-mode phase. Signal represents coherence between two selected MC coils. Black line represents eq. (1).*

with $n_e$ being line-averaged electron density and $q_{95}$ edge plasma safety factor. Experiments with ramps in each of these three parameters, as well as discharges featuring $n_e$ oscillations, have shown that high-frequency High-Field Side (HFS) oscillations typically follow eq. (1) in both limited and diverted plasmas (with some abberation at higher densities) – see Fig. 1. During H-mode phase, the trend of fluctuation frequencies is slightly different from the scaling, which is most likely attributed to changes in a plasma profile.

**Travelling wave character**

Alfvén waves are detected as high-frequency fluctuations due to high velocity at which they travel along helical field-lines of the confining field. In laboratory reference frame, plasma rotation attributes as a Doppler shift correction [6]:

$$f_{\text{tot}}(r) = f_A(r) + n \cdot f_{\text{plasma}}(r), \quad (2)$$

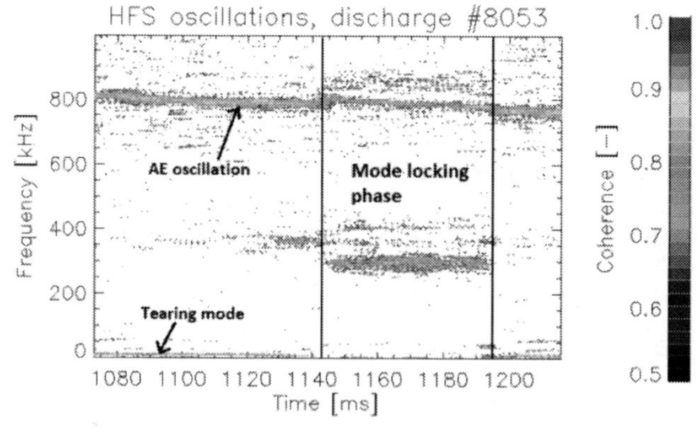


Figure 2: *High-frequency plasma oscillations on HFS. Signal represents coherence between two selected MC coils. Black lines show the start and the end of period in which mode locking occured.*

where $n$ is toroidal mode number (periodicity) of the oscillation and $f_{\text{plasma}}$ is frequency of plasma toroidal rotation at corresponding position. In the first approximation, rotation of low-frequency MHD instabilities of tokamak plasmas, e.g. tearing modes or kink modes, is fully attributed to the Doppler shift term. An interaction of magnetic islands with DC RMP field can cause radially localized braking of plasma rotation effectively mitigating the Doppler shift.

During mode locking phases of RMP discharges, it was seen that high-frequency oscillations localized on HFS persist, suggesting that these fluctuations are travelling-wave character shear Alfvén waves. This can be seen in Fig. 2, where the 800 kHz mode retains its frequency even after the 10 kHz tearing mode oscillation vanishes. Nevertheless, due to lack of rotation measurements, there is still some uncertainty in the radial extent of rotation braking that takes place, as well as in radial position of the studied high-frequency oscillation.

**Oscillations correlated with runaway electrons**

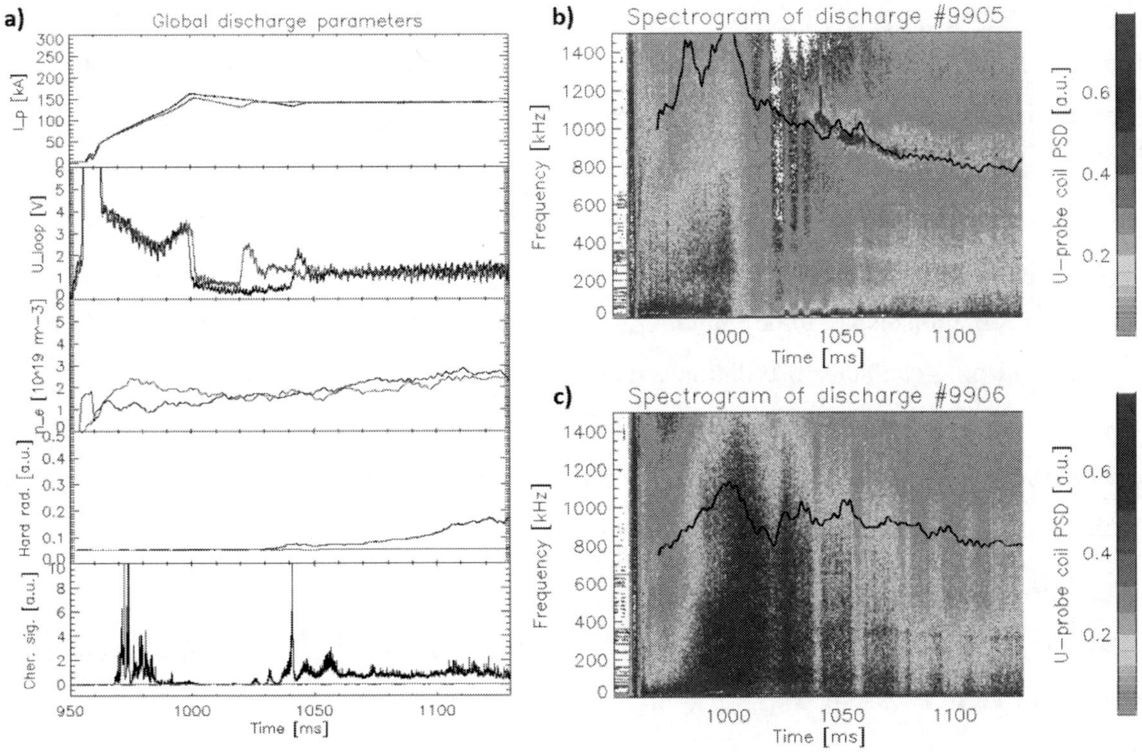


Figure 3: *a) Global parameters of shots no. 9005 (black) and no. 9906 (blue). $I_p$ – plasma current, $U_{\text{loop}}$ – loop voltage, $n_e$ – line-averaged electron density, Hard radiation – photo-neutron detector signal, Cher sig. – Cherenkov detector measuring lost electrons at 50-260 keV energies on LFS. b) Spectrogram of U-probe coil signal on LFS for shot no. 9905 with scaling curve from eq. (1). c) The same graph as b) for shot no. 9906.*

In the previous work done on COMPASS [3], a possible involvement of runaway electrons in driving observed plasma fluctuations was discussed. Although a subsequent investigation of discharges with high $U_{\text{loop}}$ and low $n_e$ (hence with increased runaway electron population) did not show any straightforward indications for modes described up to this point, a new type of very high-frequency oscillations (up to $f_{\mathrm{Nyquist}} = 2.5$ MHz) was observed. This tends to have a chirping character with frequency generally following eq. (1). Fig. 3 a) shows two identical discharges, the shot no. 9906 has a stronger gas puff during a plasma start-up phase than the one of no. 9905 – see the difference in $n_e$ there. The runaway seed was affected and hence production of runaway electrons mitigated, as it can be seen from the different behavior of both hard radiation and Cherenkov detector signals. The spectrogram in fig. 3 b) shows 0.8-1.2 MHz oscillation, whose frequency follows eq. (1). Absence of this fluctuation in the spectrogram in fig. 3 c) implies correlation between mode incidence and a significant runaway electron population.

## Conclusions

Series of the high-frequency plasma eigenmodes from 150 kHz up to $f_{\mathrm{Nyquist}} = 1$ MHz have been observed in the COMPASS tokamak plasmas by the magnetic diagnostics with an improved frequency coverage compared to the past work [3]. The eigenmode oscillations are present in both limiter and diverted plasmas, best visible on HFS and persist over L-H and H-L transitions. It was shown that their frequency follows Alfvén wave-like scaling described in [2, 5]. Moreover, discharges with RMP-induced plasma rotation braking suggest that the observed fluctuations rotate in the plasma-reference frame. Nonetheless, the driving mechanism of these oscillations is yet to be found. Discharges with increased runaway electron population contain chirping-character oscillations, which have frequency up to $f_{\mathrm{Nyquist}} = 2.5$ MHz that follows Alfvén-wave-like dependency. Future work will focus at specification of toroidal and poloidal mode numbers of the oscillations and their radial localization, further studies of RMP field effect and simulations of Alfvén continuum spectrum and its gaps by the KINX code [8].

## Acknowledgements

This study was supported by the projects GACR P205/12/2327, GACR 15-10723S and GA14-35260S and by grant RSF 14-22-00193.

## References

[1] N.N. Gorelenkov, S. D. Pinches, K. Toi, Nucl. Fuison **54**, 125001 (2014)
[2] W.W. Heidbrink, Physics of Plasmas **15**, 055501 (2008)
[3] A.V. Melnikov, T. Markovic, L.G. Eliseev, et al., Plasma Physics and Controlled Fusion **57**, 065006 (2015)
[4] P. Cahyna, et al, Nucl. Fusion **49**, 055024 (2009)
[5] D.A. D'Ippolito, J. P. Goedbloed, Plasma Phys. **22**, 1091 (1980)
[6] S.E. Sharapov, 11th CMSS on Plasma and Fusion Energy Physics proceedings (2013)
[7] K. Kovařík, et al., 41st EPS Conference Proceedings, P5.025 (2014)
[8] S. Medvedev, et al., Computer Phys. Communications **103**, S0010-4655 (1997)

# ELM and inter-ELM electromagnetic filaments in the COMPASS Scrape Off Layer

M. Spolaore[1], K. Kovařík[2,3], J. Stöckel[2], J. Adamek[2], I. Ďuran[2], M. Komm[2], E. Martines[1], J. Seidl[2], N. Vianello[1,4] and COMPASS Team

[1]*Consorzio RFX (CNR, ENEA, INFN, Università di Padova, Acciaierie Venete SpA), Corso Stati Uniti 4, 35127 Padova, Italy*
[2]*Institute of Plasma Physics AS CR, Za Slovankou 3, Prague, Czech Republic*
[3]*Charles University in Prague, Faculty of Mathematics and Physics, Prague, Czech Republic*
[4]*EPFL, CRPP, CH-1015 Lausanne, Switzerland*

Filamentary structures have been observed in all magnetic configurations with very similar features despite the difference in the magnetic geometry: theory and experiments suggest they exhibit a radial convective motion across the SOL, and the interest in blob dynamics is further motivated by their interaction with first wall and divertor.

Despite their possible different generation mechanisms, turbulent structures and Edge Localized Mode (ELM) filaments share some common physical features, as the localization in the cross-field plane and the associated parallel current, with a convective radial velocity component somehow related to their dimension.

The electromagnetic (EM) effects on filament structures deserve particular interest, among the others for the implication they could have for ELM, related for instance to their dynamics in the transition region between closed and open field lines or to the possibility, at high beta regimes, of causing line bending which could enhance the interaction of blobs with the first wall. In this contribution the presence of ELMs and inter-ELM EM filaments will be investigated in the COMPASS tokamak, where a new probe head was recently developed and successfully commissioned [1]. The diagnostic, based on the U-probe concept [2], allows the simultaneous measurements of electrostatic and magnetic fluctuations, with high time resolution suitable for the identification of EM features of filaments, providing in particular the direct measurement of the current density associated to filaments.

The probe head was inserted at different radial positions in the SOL of D-shaped diverted discharges. The COMPASS experiment was operated in these discharges in ohmic H-mode, with the clear presence of different type of ELMs. The COMPASS tokamak [3] is a compact experimental device (R = 0.56 m, a = 0.2 m) operated in divertor plasma configuration with ITER-like plasma cross-section. Presently, COMPASS operates with plasma current up to 400 kA and toroidal magnetic field in the range 0.9 – 1.8 T and elongation 1.8. Two neutral beam injectors provide power of 2 x 0.4 MW at the beam energy of 40 keV for additional plasma heating. Recently, an Ohmic as well as NBI assisted H-mode has been successfully achieved on the COMPASS tokamak after application of boronization of the vacuum vessel interior. The L-H transition is followed either by an ELM-free period or ELMs with frequency in the range of 80 – 1 000 Hz. During 2014 the system for application of magnetic perturbations at edge was successfully commissioned. In this paper the electromagnetic features of ELM and inter-ELM filaments observed in the Scrape Off Layer of Compass device are shown. In figure 1 the main plasma features of shot #9328 are shown. This shot was chosen as representative case, where an extended ohmic H-mode was obtained. In this case two H-mode phases are observed, from 1075 to 1120 ms and from 1170 to 1200 ms. As clearly seen in the $D_\alpha$ monitor, these phases are characterized by the sharp strong events with a frequency varying with time. In the right column of fig. 1 some quantities as measured by the Compass U-probe are shows. The probe is placed in a fixed position during the discharge in the bottom part of the device and at low field side. In this case the radial insertion of the probe head is 15 mm from the wall and corresponds to a distance of about 50 mm from the Last Closed Flux Surface (LCFS).

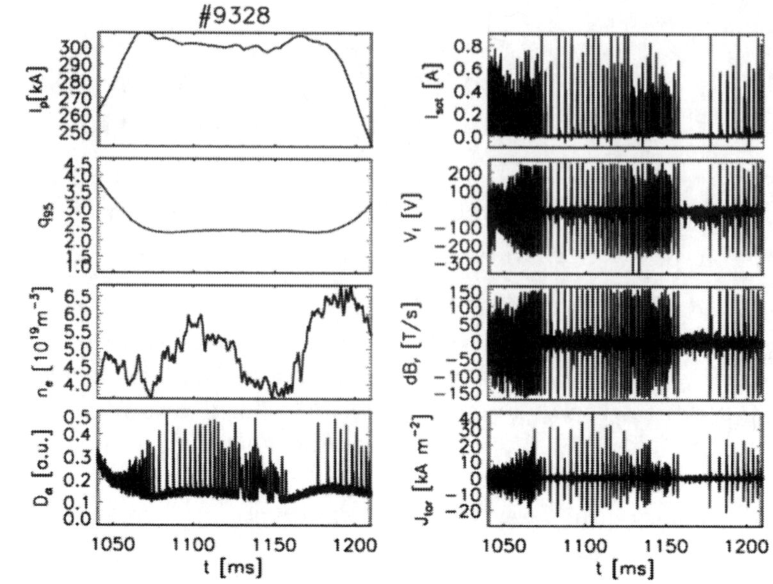


**Fig. 1** Time evolution of the main plasma parameters of shot #9328 (right column). Representative data from the Compass U-probe are shown in the left column.

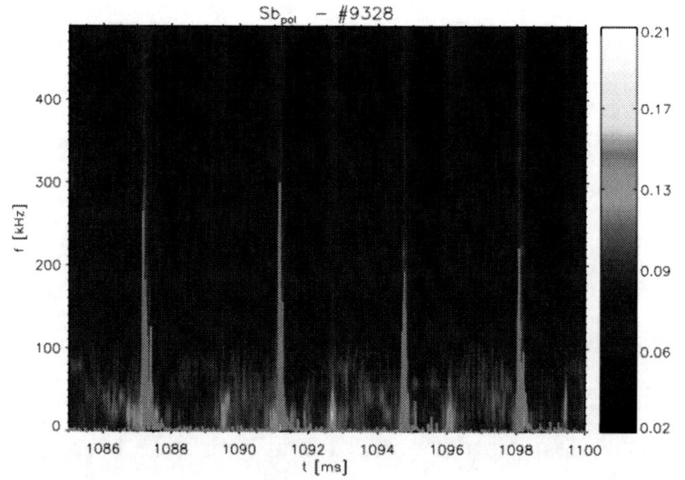


**Fig. 2** Spectrogram of 10 kHz high pass filtered $\delta b_{pol}$, during a ELMy phase. The local $I_{sat}$ [a.u.] measurement is overplotted for comparison (green line).

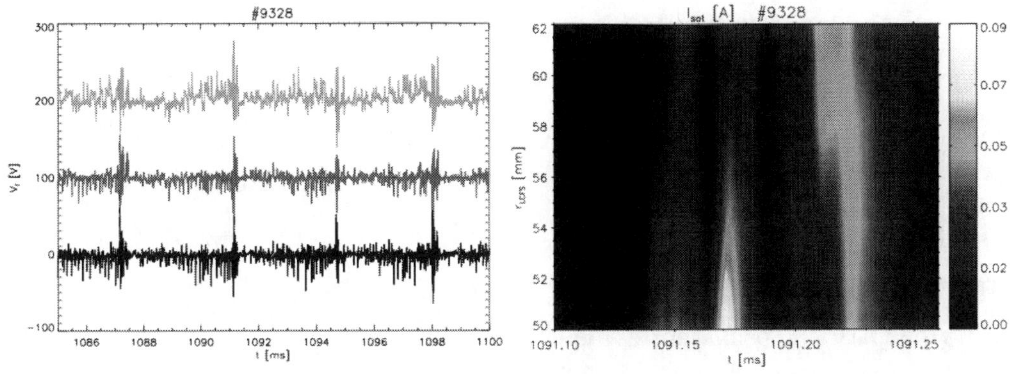


**Fig. 3** Time evolution during an ELMy H-mode phase of $V_f$ (left panel) measured at different radial positions (darker colors refer to outermost positions). $I_{sat}(r_{LCFS},t)$ during a single ELM event (right panel).

It can be observed that strong events on fluctuations of both $I_{sat}$ and $V_f$ are present and correspond to the $D_\alpha$ ELMs events. It is worth noting that also the magnetic fluctuations exhibit similar features, see the $dB_r(t)$ signals shown as representative, thus providing a clear indication of a EM feature of ELM filaments observed. The use of a 2D array of 3-axial coils allows also obtaining a direct monitor of the local parallel current density fluctuations, $J_{tor}$, measured along the whole discharge. It can be observed that strong events corresponding to ELMs are observed also in this case. In order to better explore this issue a focus on the ELMy phase is presented in the following. In particular, fig. 2 shows the spectrogram of the poloidal component of magnetic field fluctuation, $\delta b_{pol}$, during this phase. For comparison the local measurement of $I_{sat}$ time evolution is overlapped. Given that the strong $I_{sat}$ events, indicative of density bursts, can be considered as the signature of an ELM filament [4], it can be observed that also a clear magnetic activity is correlated to ELMs and is characterized by an abruptly spread spectrum. The dataset provided by the Compass U-probe includes also radially spaced information on $V_f$ and $I_{sat}$. Fig. 3 shows the time evolution, in the same phase seen in fig. 2, of $V_f$ measured in three different positions radially spaced by 8 mm. The ELM events involve all the $V_f$ signals, suggesting a radially extended potential structure. In the same figure the detail of the time evolution around a single event as seen by the radial

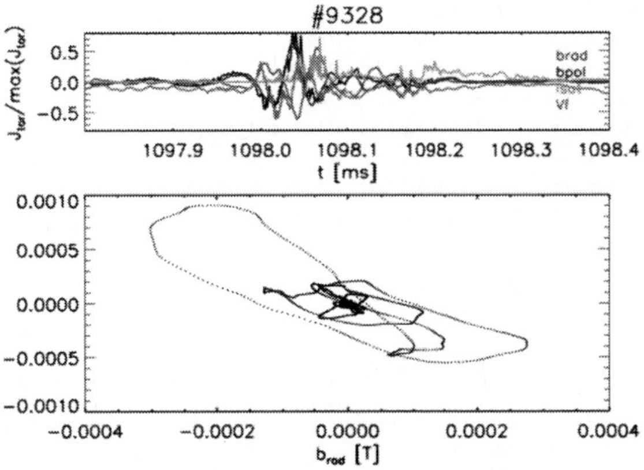


**Fig. 4** Time evolution during a single ELM event of $\delta J_{tor}$, $\delta b_{pol}$, $\delta b_{rad}$, $\delta I_{sat}$ and $\delta V_f$, normalized to their maximum value (top). Cross field pattern described by the $\delta b_{pol}$ and $\delta b_{rad}$ fluctuations during an ELM event (bottom).

array of $I_{sat}$ measurements is shown. In this case four measurements are used, radially spaced by 4 mm, the distance from the estimated position of the LCFS, $r_{LCFS}$, is shown. The revealed density structure associated to this event is complex: a radial extended structure appears at the beginning, then suddenly develops and radially propagates and exhibits multiple fragmentations. Fig. 4 shows details of a single ELM event according to locally measured magnetic and electrostatic quantities. The main $I_{sat}$ peak corresponds to a potential valley, furthermore, a main positive peak of parallel current density, $J_{tor}$, is observed. However, also negative secondary $J_{tor}$ peaks are evident, providing a nearly zero time integral of the $J_{tor}$ associated to the ELM filament. This behavior suggests a $J_{tor}$ current density pattern closing on itself. Its filamentary feature is confirmed by the closed patterns described by the $\delta b_{pol}$ and $\delta b_{rad}$ fluctuations in the cross-field plane [5,6], however a statistical analysis deserves in order to conclude on the detailed $J_{tor}$ topology. The ELM events result than characterized by a composite EM filamentary structure. A further detail can be provided by analyzing the inter-ELM phases. An example is shown in fig. 5, where the time behavior in between two ELMs is shown for $J_{tor}$, $I_{sat}$ and $V_f$ fluctuations. Several $I_{sat}$ events similar to the ELM ones are observed in these phases, but characterized by about one order of magnitude smaller amplitude and time scales. Also these inter-ELM events are characterized by associated $V_f$ valley and $J_{tor}$ filamentary structure. A zoom on one of them is shown as an example in fig. 5. Summarizing, ELM and inter-ELM electromagnetic filaments were measured during ohmic H-mode discharges in Compass device. Similar structures with different scales were observed in the two cases.

*This work has been carried out within the framework of the EUROfusion Consortium and has received funding from the Euratom research and training programme 2014-2018 under grant agreement No 633053 and received founds from MSMT CR #LM2011021.The views and opinions expressed herein do not necessarily reflect those of the European Commission.*

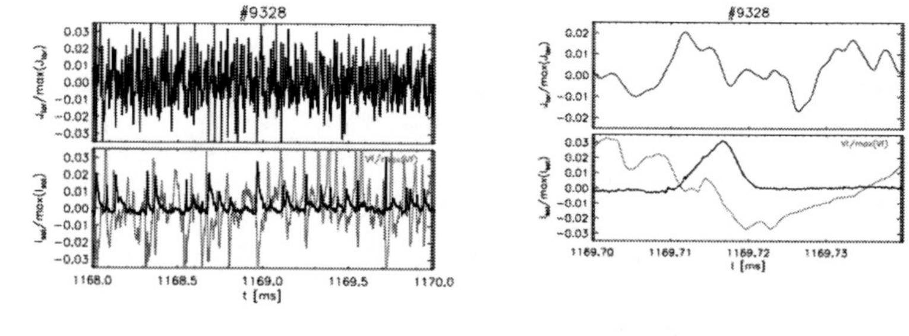


**Fig. 5** Time evolution during an inter-ELM phase of $\delta J_{tor}$, $\delta I_{sat}$ and $\delta V_f$, normalized to their maximum value (left), zoom of the same quantities on a single inter-ELM event detected on $I_{sat}$ (right).

[1] K. Kovarik, et al. 41st EPS Conference on Plasma Physics, Berlin, Germany 2014, P5.025
[2] M. Spolaore et al. Phys. Plasmas 22, 012310 (2015) http://dx.doi.org/10.1063/1.4906869
[3] R. Panek et al., 40th EPS Conference (2013) Vol. 37D, P4.103
[4] A. Leonard Phys. of Plasmas 21 (2014) 090501
[5] N. Vianello et al. Phys. Rev. Lett. 106, 125002 (2011)
[6] N. Yan et al., Plasma Phys. Control. Fusion 56 (2014) 095023

# Synchrotron Radiation from Runaway Electrons in COMPASS Tokamak

M. Vlainic[1,2], P. Vondracek[2,3], J. Mlynar[2], V. Weinzettl[2], O. Ficker[2,4], M. Varavin[2], R. Paprok[2,3], M. Imrisek[2,3], J. Havlicek[2,3], R. Panek[2], J.-M. Noterdaeme[1,5] and the COMPASS Team

[1] *Department of Applied Physics, Ghent University, Ghent, Belgium*

[2] *Institute of Plasma Physics AS CR, Prague, Czech Republic*

[3] *Charles University in Prague, Faculty of Mathematics and Physics, Prague, Czech Republic*

[4] *Czech Technical University in Prague, Faculty of Nuclear Sciences and Physical Engineering, Prague, Czech Republic*

[5] *Max Planck Institute for Plasma Physics, Garching, Germany*

**Introduction.** Runaway electron (RE) experiments are an important part of the ongoing ITER-relevant studies on COMPASS, as the REs could severely damage the plasma facing components in the future fusion reactors. The electron is said to "run away" (in the velocity space), when the collisional drag force $F_{coll}$ acting on it becomes smaller than the accelerating force $F_{acc}$ due to the toroidal electric field $E_{tor}$. This net accelerating force comes from the fact that the $F_{coll}$ decreases with approximately quadratic dependence on the velocity. The theory made by Dreicer [1, 2] defines a critical velocity $v_c$ above which one an electron diffusing in the velocity space becomes the RE (*primary/Dreicer* mechanism). Beside diffusing, a thermal electron can escape to the runaway region if it collides with the primary generated RE and both the electrons remain in the runaway region (*secondary/avalanche* mechanism).

A relativistic accelerated charged particle in the presence of the magnetic field emits Synchrotron Radiation (SR). As SR is emitted preferentially in the direction parallel to the RE motion (the so-called headlight effect) it should be observed from a tangential view. Consequently, the SR offers a valuable opportunity for measuring parameters of the confined high-energy REs directly from the plasma core.

In this paper it is demonstrated that the relative intensity of the infrared (IR) radiation is correlated with the critical energy $W_c$ for production of REs. Furthermore, analysis of the first direct observation of the RE beam in the COMPASS tokamak with the calibrated camera is presented.

**Experimental Setup.** The COMPASS tokamak [3] is a experimental fusion device with major radius $R_0 = 0.56\,\mathrm{m}$ and minor radius $a = 0.23\,\mathrm{m}$. Toroidal magnetic field $B_{tor}$ was $1.15\,\mathrm{T}$ for all discharges reported in this paper, the typical pulse length is 0.4 s, although the low current circular discharge dominated by REs can last approximately 1 s. Furthermore, the SR was successfully measured at the low line-averaged density discharges ($\bar{n}_e \leq 2.5 \times 10^{19}\,\mathrm{m}^{-3}$) for the wide range of the plasma current values 100 – 250 kA.

The SR falls in the mid-wavelength IR region for the REs generated in the COMPASS tokamak, therefore the bolometric IR camera with the wavelength range 7.5 – 13 $\mu$m was used. During these experiments, the IR camera was installed at a midplane tangential port. The diameter of the observed area of the plasma cross-section varied between the RE campaigns: 14.9, 16.9 and 15.5 cm for the first, the second and the third campaign, respectively. Fig. 1 shows the resulting differences in the observed picture.

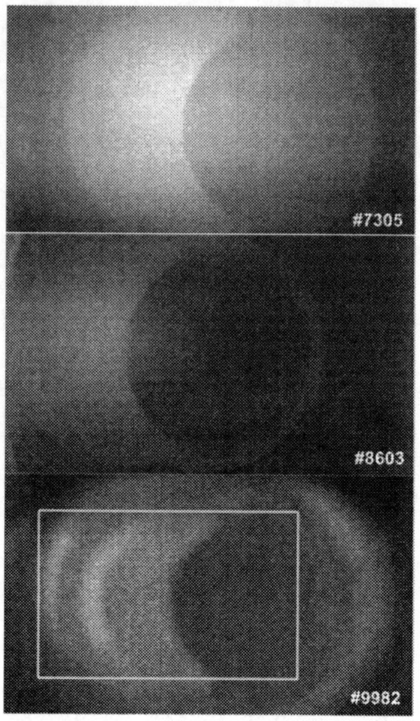


**Figure 1:** Different observations from the first (top) to the third (bottom) campaign. The moon-like shape is the carbon neutral beam injector dump. A yellow rectangle is the analysed area from Fig. 3.

**Results.** Even though the observed area was relatively large for monitoring of the plasma core in comparison to $a$, it seems that almost all of the recorded intensity is rather a reflection of SR from the vessel than a direct SR from the visible plasma volume. Nevertheless, it is possible to analyse the dependence of the relative synchrotron intensity as function of the critical energy $W_c$. Following the Dreicer runaway theory, the critical energy is given as [4]:

$$W_c = \frac{e^3 \ln \Lambda}{4\pi\varepsilon_0^2} \frac{n_e}{E_{tor}} \sqrt{2+Z_{eff}}, \quad (1)$$

where $Z_{eff}$ is the effective ion charge and $\ln \Lambda$ is the Coulomb logarithm. For the calculation given here $Z_{eff}$ is assumed to be constant and set to $Z_{eff} = 2$, therefore $W_c$ is proportional to the ratio between the electron density $n_e$ and the loop voltage $V_{loop}$, which are both measured quantities. In Fig. 2 the maximum of the relative intensity from the IR camera signal is plotted as a function of the averaged $W_c$ during the first 240 ms of the discharge. One can see that different set-ups have different maxima, but in general a margin around $\sim 40$ keV could be distinguished.

There are two main reasons why almost all discharges with the observation of the SR come from the reflection. First, the observed field of view is slightly farther from the region where the high-energy REs are presumably generated [5]. Moreover, even if the RE beam reaches proper radial position for the direct observation of the SR, the IR camera almost regularly disconnects. This is probably because the radiation is so strong that it saturates the camera's electronics. The cases when the camera disconnects are separately grouped and plotted on Fig. 2 and one can see that they occur only when $W_c < 37$ keV.

The evolution of the RE beam for the discharge #9814 is showed in Fig 3. The SYnchrotron spectra from RUnaway Particles (SYRUP) code [6] is used for the theoretical estimation of the synchrotron spectral power density $dP/d\lambda$ per RE. The SYRUP requires the maximum RE energy $W_{max}$ and the pitch angle $\theta = \arctan(v_\perp/v_\parallel)$ as inputs. The latter one can be measured from Fig. 3 [7], while $W_{max}$ has been calculated using the 0D-model from [8] that takes into account a power gain by the electric field $P_E$ and a power loss by the SR $P_{synch}$ for the highly relativistic electrons:

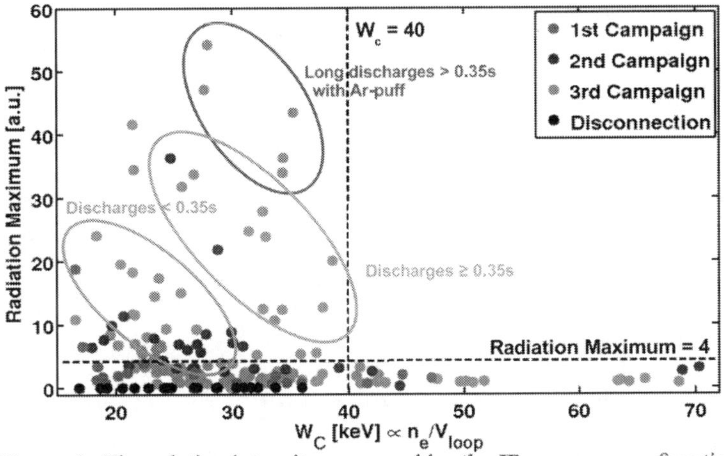


**Figure 2:** The relative intensity measured by the IR camera as a function of the averaged $W_c$ during the first 240 ms of the discharge. Discharges are grouped by the campaigns, while discharges when the IR camera disconnects during all campaigns are grouped separately.

$$\frac{dW_{max}}{dt} = P_E - P_{synch}, \quad P_E = ecE_{tor} = \frac{ecV_{loop}}{2\pi R}, \quad P_{synch} = \frac{2m_ec^3r_e\gamma^4}{3R_C^2} \tag{2}$$

where $r_e$ is the classical electron radius and $R_C$ is the curvature radius of the RE. The plasma parameters are necessary for the quasi-steady-state RE distribution function calculation implemented in SYRUP. For purpose of this paper, the necessary plasma parameters are taken to be constant: $\bar{n}_e = 1.6 \pm 0.2 \times 10^{19}\,\mathrm{m}^{-3}$, $T_e = 530 \pm 60\,\mathrm{eV}$, $Z_{eff} = 2$ and $E_{tor} = 0.33 \pm 0.04\,\mathrm{V/m}$. Furthermore, the calculation from Eq. 2 gives rising $W_{max}$ from 20 to 30 MeV at time when the SR was observed (Fig. 3).

The derivative $dP/d\lambda$ obtained from SYRUP should first be converted to an integral, measurable value such as brightness $B$ [8]:

$$B(\lambda, \theta, W_{max}) = \frac{dP}{d\lambda}\frac{2R}{\pi\theta}n_{re}^{synch}, \tag{3}$$

where $n_{re}^{synch}$ is the density of the observed REs. To get the comparable theoretical estimation

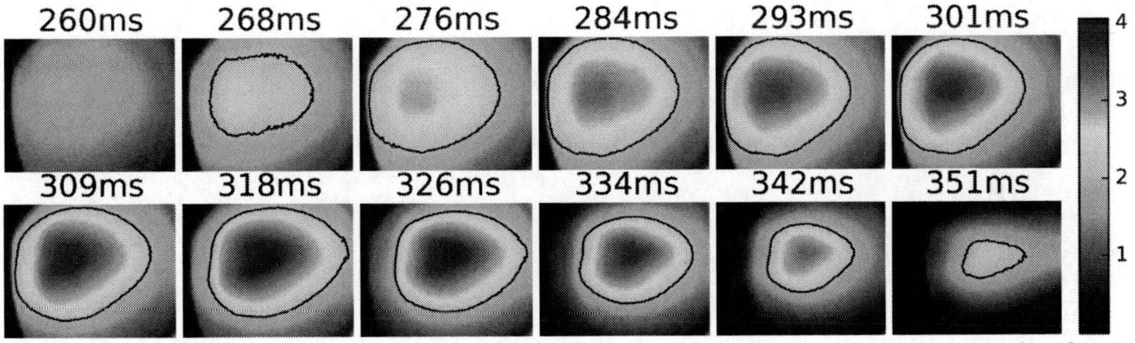


**Figure 3:** Time evolution of the RE beam observed with the IR camera (the color scale is in $\mathrm{W\,sr^{-1}\,m^{-2}}$). The black line represents 50% of the maximum intensity, which is taken as the margin of the RE beam. Note that in order to obtain this evolution, the radiation level as observed at 227 ms (the background thermal radiation) is subtracted from all the frames.

with the IR camera measurement, the brightness $B$ can be integrated over the camera wavelengths:

$$S(\theta, W_{max}) = \int B(\lambda, \theta, W_{max}) T(\lambda) \mathrm{d}\lambda, \quad (4)$$

where $T(\lambda)$ is a transparency of the optical path between the plasma and the IR camera. The pitch angle $\theta$ in the IR camera measurements varied from 0.15 to 0.30 rad, which leads to the corresponding RE density $n_{re}^{synch}$ values $0.65 - 1.7 \times 10^{15}\,\mathrm{m}^{-3}$. Note that this evaluated $n_{re}^{synch}$ gives only density of the high-energetic REs contributing to the SR.

**Discussion.** A method of extracting the relevant information from the reflected SR using an appropriate analysis is presented. From here the threshold of the critical energy $W_c = 40\,\mathrm{keV}$ during the discharge can be estimated. Note that the $W_c$ value is a dynamic quantity and that the value given is its average over the time range. Additionally, longer discharges have more intense synchrotron radiation, as expected due to the longer acceleration time.

One should note that SYRUP could be used to estimate the $W_{max}$ provided that the $n_{re}^{synch}$ is known, however calculation of $n_{re}$ is a demanding task by itself, which is out of the scope of this paper. Nevertheless, for the future work usage of the code name CODE [9] as an efficient numerical tool for approximation of the RE distribution function is planned for the $W_{max}$ estimation based on the observed SR.

**Acknowledgment.** This work has been carried out within the framework of the EUROfusion Consortium and has received funding from the Euratom research and training programme 2014-2018 under grant agreement No 633053. The views and opinions expressed herein do not necessarily reflect those of the European Commission. Next to thank is the project MSMT #LM2011021 from which the COMPASS operation is supported. Then, the authors would like to acknowledge work of the WP14-MST2-9 research project team. The "Joint Doctoral Programme in Nuclear Fusion Science and Engineering" is acknowledged by the first author for supporting the studies.

## References

[1] H. Dreicer, Phys. Rev. **115**, 238 (1959).

[2] H. Dreicer, Phys. Rev. **117**, 329 (1960).

[3] R. Panek *et al.*, Plasma Phys. Contr. Fusion - *To Be Submitted*; Invited talk at this conference (2015).

[4] I. Fernandez-Gomez, J. R. Martin-Solis, and R. Sanchez, Phys. Plasmas **19**, (2012).

[5] R. Gill *et al.*, Nucl. Fusion **42**, 1039 (2002).

[6] A. Stahl *et al.*, Phys. Plasmas **20**, 093302 (2013).

[7] R. Jaspers, Ph.D. thesis, Technische Universiteit van Eindhoven, Eindhoven, The Netherlands, 1995.

[8] J. H. Yu *et al.*, Phys. Plasmas **20**, 042113 (2013).

[9] M. Landreman, A. Stahl, and T. Fülöp, Comput. Phys. Commun. **185**, 847 (2014).

# Analysis of the potential of ICRF waves to heat bulk ions in DEMO

D. Gallart[1], M.J. Mantsinen [1,2], Y. Kazakov [3]

[1] *Barcelona Supercomputing Center, Barcelona, Spain*

[2] *Catalan Institution for Research and Advanced Studies, Barcelona, Spain*

[3] *Laboratory for Plasma Physics, LPP-ERM/KMS, Brussels, Belgium*

**Introduction** Among the auxiliary heating schemes presently envisaged for ITER, heating with waves in the ion cyclotron range of frequencies (ICRF) is the only scheme that can provide dominant bulk ion heating. Ion heating with ICRF waves results from the absorption of the wave power $P_{ICRF}$ by resonant ions $P_i$ which can subsequently transfer their energy to the fuel ions via collisions (Fig. 1). This effect is in competition with direct electron damping $P_e$ and other mechanisms which can also damp the wave power. Bulk ion heating dominates provided

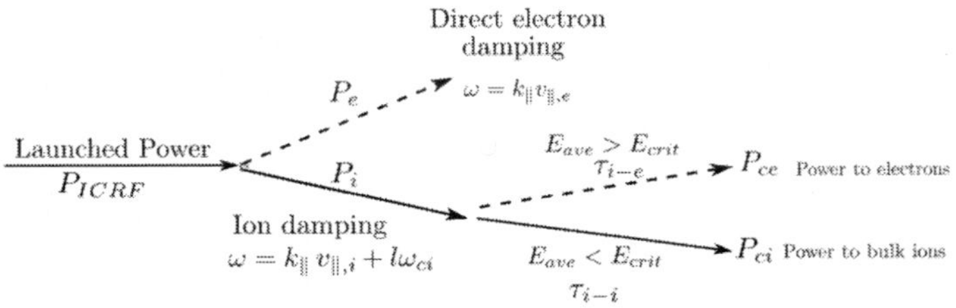


Figure 1: *Scheme of the power deposition in the ICRF heating process.*

that the energy of the resonant ions $E_{ave}$ is tailored to remain below the critical energy $E_{crit}$ at which they transfer energy equally to electrons $P_{ce}$ and ions $P_{ci}$ in collisions. Fast ions transfer their energy via collisions to bulk ions and electrons on the time scale of the slowing-down times on ions $\tau_{i-i}$ and electrons $\tau_{i-e}$. For fast $^3$He ions and tritons, which will be the main ICRF-resonant ion species in ITER, $\tau_{i-i}$ is 1 ms and 3 ms and $\tau_{i-e} =$ 1s and 3.4 s, respectively, at typical steady-state ITER parameters. The bulk ion heating capabilities of ICRF waves in ITER have been studied e.g. in Refs. [1, 2]. We analyze the potential of ICRF waves to heat the fuel D-T ions in DEMO. DEMO is a proposed nuclear fusion demonstration power plant that is expected to be built after ITER. While ITER's main purpose is to confirm the feasibility of nuclear fusion as an energy source, DEMO is planned as the first fusion reactor to produce net electrical energy. Our analysis is carried out for the DEMO 2015 (from now on DEMO2), using the ICRF modelling codes PION [3] and TORIC [4]. We also have analyzed the previous design DEMO 2013 (from now on DEMO1) see Ref. [5].

**Analysis of bulk ion heating in DEMO** We concentrate our studies on the second harmonic ICRF heating of tritium with and without $^3$He in a 50%:50% D-T plasma in DEMO. The ICRF scenario is for a standard low-field side (LFS) midplane launch with thermal plasma (there is

| Design | $R_0(m)$ | $a$ (m) | $B$ (T) | $I_p(MA)$ | $\kappa$ | $\delta$ | Plas. Vol. (m$^3$) | Fusion P. (MW) |
|---|---|---|---|---|---|---|---|---|
| DEMO1 | 9.25 | 2.64 | 6.8 | 18.6 | 1.52 | 0.33 | 2009 | 2119 |
| DEMO2 | 9.07 | 2.93 | 5.7 | 19.6 | 1.59 | 0.33 | 2502 | 2037 |

Table 1: DEMO1 and DEMO2 parameters.

no ICRF+NBI interaction) and a fixed toroidal mode number ($N_\phi \simeq Rk_\parallel$). $P_{ICRF} = 100\ MW$ is considered as a baseline value of the coupled ICRF power. The basic parameters of the ICRF system are summarized in Table 2 for both DEMO designs. The study has consisted of a scan of the electron density $n_e$ and the electron temperature $T_e$. In order to compare both DEMO designs from the bulk ion heating point of view, several simulations have been made for both designs at the same DEMO design point (DEMO1). Simulations show that direct electron damping grows as $n_e$ and $T_e$ increase but also and more strongly as the magnetic field $B$ decreases, see Ref. [6]. For the same $N_\phi$ DEMO1 showed higher bulk ion heating than DEMO2.

| Parameter | DEMO1 | DEMO2 |
|---|---|---|
| Toroidal magnetic field $B$ (T) | 6.8 | 5.7 |
| ICRF frequency (MHz) | 66, 70, 74 | 54, 57, 61 |
| Resonance location ($\frac{r_{res}}{a}$) | 0.2, -0.05, -0.2 | 0.2, -0.05, -0.2 |
| ICRF Power (MW) | 100 | 100 |
| Toroidal mode number $N_\phi$ | 50 | 50 |
| DEMO design point $n_e$ (m$^{-3}$), $T_e$ (keV) | $1.2 \cdot 10^{20}$, 30 | $1.0 \cdot 10^{20}$, 27.4 |

Table 2: ICRF parameters for DEMO.

**$^3$He minority heating** In this case, the frequency of the wave has been set equal to the fundamental harmonic of $^3$He, $\omega = \omega_{^3He} = 2\omega_T$, for central resonance $f = 57.6$ MHz. The

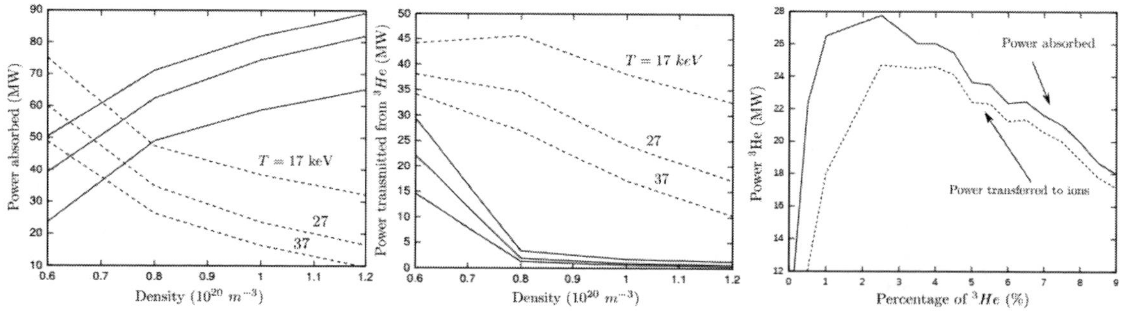


(a) (b) (c)

Figure 2: *For central resonance f = 57.6 MHz. (a) Scan in $T_e$ and $n_e$ of the power absorbed by $^3$He (dashed line) and electrons (solid line). (b) Power transferred by $^3$He to bulk ions (dashed) and to electrons (solid). (c) Power absorption by $^3$He and power transferred to ions by $^3$He for different concentrations at $n_e = 1.0 \cdot 10^{20}\ m^{-3}$.*

power absorbed by $^3$He tends to decrease as temperature $T_e$ and electron density $n_e$ increase. In fact, for densities $n_e \geq 0.8 \cdot 10^{20} m^{-3}$ direct electron damping clearly dominates (Fig. 2a). Only for low densities $n_e \sim 0.6 \cdot 10^{20}$ m$^{-3}$ the energy of fast ions is considerably high and relatively close to $E_{crit}$. The power transferred to electrons from $^3$He is only important for low densities (Fig. 2b), as $n_e$ increases the average fast ion energy is much smaller than $E_{crit}$ and therefore, for $n_e \geq 0.8 \cdot 10^{20}$ m$^{-3}$ the power transferred to electrons is negligible. At the DEMO design point the slowing-down times of fast $^3$He ions are $\tau_{i-i} = 8.5 \cdot 10^{-3}$ s and $\tau_{i-e} = 1.1$ s.

The dependence of bulk ion heating on the $^3$He concentration for the DEMO design point shows two different regions (Fig. 2c). Firstly, for low concentrations, the power transferred is considerably lower than the power absorbed as the energy of fast ions is high. In this region as the concentration of $^3$He increases the polarization of the wave is enhanced and therefore, the power absorbed and transferred becomes stronger. At $\sim 3\%$ of $^3$He concentration a maximum is reached. Beyond this maximum the wave polarization becomes increasingly unfavorable as the concentration increases, and consequently, the power absorbed and transferred by $^3$He decreases. However, the difference between the two curves is due to the average fast ion energy which is 500 keV and 131 keV for 0.5% and 5% concentrations, respectively.

**Second harmonic tritium scenario** This scenario has two main advantages in comparison with $^3$He minority heating. Firstly, that no $^3$He is required and secondly, there is no dilution by $^3$He. Therefore, the fusion rate D+T is higher than for $^3$He dilution (Fig. 3a). The power absorption (Fig. 3b) follows a similar trend as in the $^3$He minority scenario (Fig. 2a). The power transferred from tritium to bulk ions does not change considerably for different $n_e$ (Fig. 3c) because average fast ions energy changes substantially from low densities to high densities. For low densities the average fast ion energy is larger than $E_{crit}$ and the power is mainly transferred to electrons while for high densities it is lower than $E_{crit}$. This change in $E_{ave}$ with respect to $E_{crit}$, together with the decrease in absorbed power by tritons (Fig. 3b), results in a relatively constant power transfer to bulk ions as function of $n_e$. The power transferred by fast tritium ions to bulk ions for low densities is lower than the power transferred by $^3$He (Fig. 2b), 25 MW and 40 MW respectively. This difference is important for the start up where low densities are used and good bulk ion heating is needed. As the electron density grows the average fast ion energy decreases and bulk ion heating dominates for densities $n_e \geq 1.0 \cdot 10^{20}$ m$^{-3}$. At the DEMO design point the slowing-down times of fast tritium ions are $\tau_{i-i} = 3.6 \cdot 10^{-2}$ s and $\tau_{i-e} = 4.6$ s. The power transferred to bulk ions at the DEMO design point is 24.8 MW which is more than for minority heating scenario (22.4 MW).

**Off-axis heating** The results shown above are computed for a frequency of 57.6 MHz, which places the resonance in the center of the plasma. We have also carried out simulations with

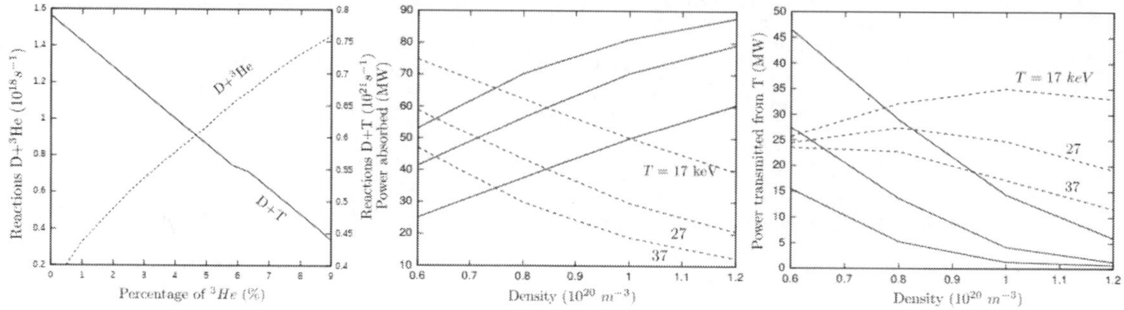


(a) (b) (c)

Figure 3: *For central resonance $f = 57.6$ MHz (a) Fusion reaction rates for different concentrations of $^3$He. (b) Scan in $T_e$ and $n_e$ of the power absorbed by tritium (dashed line) and electrons (solid line). (c) Power transferred by tritium to bulk ions (dashed) and to electrons (solid).*

the off-axis resonance locations, $r/a = 0.2$ and -0.2, corresponding to a wave frequency of 54.0 MHz and 61.7 MHz, respectively. The results are presented in Table 3. The bulk ion heating fraction by ICRF can be maximized in DEMO by placing the ICRF resonance slightly off-axis on the low-field side to minimize the competing direct electron damping. The $k_{||}$ used in this analysis is in line with earlier works on ICRF current drive for DEMO, cf. Ref. [7]. Further steps will consist of an analysis of lower $k_{||}$, where an increased power transfer to bulk ions is expected due to weaker direct electron damping, and studying the possibility to heat heavy impurity ions such as $^9$Be or $^7$Li via the three-ion ICRF mechanism suggested in Ref. [8].

| Composition | 54 MHz | 58 MHz | 62 MHz |
|---|---|---|---|
| $^3$He 5% | 42.9 MW | 22.4 MW | 14.3 MW |
| T | 40.8 MW | 24.8 MW | 8.6 MW |

Table 3: Power transferred to bulk ions ($P_{ICRF} = 100$ MW).

**Acknowledgments** This work has been carried out within the framework of the EUROfusion Consortium and has received funding from the Euratom research and training programme 2014-2018 under grant agreement No 633053. The views and opinions expressed herein do not necessarily reflect those of the European Commission. We are grateful to Dr. R. Wenninger and Dipl.-Ing T. Franke (PPPT, Garching) for DEMO parameters.

**References**

[1] V. Bergeaud, L.-G. Eriksson and D.F.H. Start, Nuclear Fusion **40**, 35 (2000) .

[2] R.J. Dumont and D. Zarzoso, Nuclear Fusion **53**, 013002 (2013).

[3] L.-G. Eriksson, T. Hellsten and U. Willén, Nuclear Fusion **33**, 1037 (1993) .

[4] M. Brambilla and R. Bilato, Nuclear Fusion **46**, S387 (2006).

[5] D. Gallart, M. J. Mantsinen, Y. Kazakov, Radio Frequency Power in Plasmas, AIP conf. proceed., (2015).

[6] A. Becoulet, Plasma Phys. Control. Fusion **38**, A1-A11 (1996).

[7] Ye.O. Kazakov et al., Plasma Phys. Control. Fusion **57**, 025014 (2015).

[8] Ye.O. Kazakov et al., Nucl. Fusion **55**, 032001 (2015).

# DEMO Exhaust Challenges Beyond ITER

R. Wenninger[1,2], R. Kemp[3], F. Maviglia[1], H. Zohm[2], R. Albanese[4], R. Ambrosino[5], F. Arbeiter[6], J. Aubert[7], C. Bachmann[1], W. Biel[8], E. Fable[2], G. Federici[1], J. Garcia[9], A. Loarte[10], Y. Martin[11], T. Pütterich[2], C. Reux[8], B. Sieglin[2], P. Vincenzi[12]

[1] *EUROfusion - Programme Management Unit, Garching, Germany*
[2] *Max-Planck-Institut für Plasmaphysik, Garching, Germany*
[3] *CCFE Culham, EURATOM Association, Culham, United Kingdom*
[4] *Università di Napoli Federico II, Naples, Italy*
[5] *Università di Napoli Parthenope, Naples, Italy*
[6] *Karlsruhe Institute of Technology, Karlsruhe, Germany*
[7] *CEA-Saclay, DEN/DM2S/SEMT, Gif-Sur-Yvette, France*
[8] *Institute of Energy- and Climate Research, Forschungszentrum Jülich GmbH*
[9] *CEA, IRFM, St Paul-Lez-Durance, France*
[10] *ITER Organization, Cadarache, Saint-Paul-Lez-Durance, France*
[11] *CRPP, EPFL, Lausanne, Switzerland*
[12] *Consorzio RFX, Padova, Italy*

The European fusion roadmap foresees a demonstration fusion reactor (DEMO) [1] to follow ITER with the aim of demonstrating the production of electricity in a fusion plant. Table 1 lists parameters of several recent EU designs: DEMO1 (pulsed, conservative physics and technology assumptions), DEMO2 (steady state, more optimistic physics and technology assumptions) designs. The designs are tokamaks in H-mode operation with a closed fuel cycle. In comparison to the ITER (Q = 10) design the European DEMO design options have significantly higher fusion power, higher $\beta_N$, higher temperature across the whole profile, higher fueling rate and higher core radiation fraction. Also there are more challenging engineering requirements (e.g. closed fuel-cycle) and consequently more restrictive engineering boundary conditions when compared to ITER. Due to this there is a number of areas, in which the physics challenges of DEMO - mostly related to the feasibility or the cost of the device - go considerably beyond the ones for ITER. This paper addresses the important subset of these additional challenges, which concentrates around the topic of power exhaust.

## DEMO Wall Loads

In the recent EU concept design analysis of the first wall of DEMO it is assumed to have W as armour material and water or He as coolant. In DEMO the requirements of electricity generation and tritium self-sufficiency lead to challenges for an efficient heat removal from the first wall. As the coolant temperature is considerably higher than in ITER and beyond the operational

Table 1: Overview of EU design parameters for DEMO1 and DEMO2 and an alternative DEMO2

| | EU DEMO1 2015 | EU DEMO2 2015 | EU DEMO2 2015 (Alt) |
|---|---|---|---|
| $R[m]$ | 9.1 | 7.5 | 7.5 |
| $A$ | 3.1 | 2.6 | 3.1 |
| $B_T$ [T] | 5.7 | 5.6 | 6.5 |
| $I_P$ [MA] | 20 | 22 | 18 |
| $H$ (rad. cor.) | 1.1 | 1.4 | 1.4 |
| $\beta_{N,tot}[\%]$ | 2.6 | 3.8 | 3.8 |
| $f_{bs}[\%]$ | 35 | 61 | 64 |
| $P_{sep}/R[MW/m]$ | 17 | 20 | 20 |
| $\tau_{burn}[h]$ | 2 | steady state | steady state |
| $P_{el,net}[MW]$ | 500 | 953 | 941 |

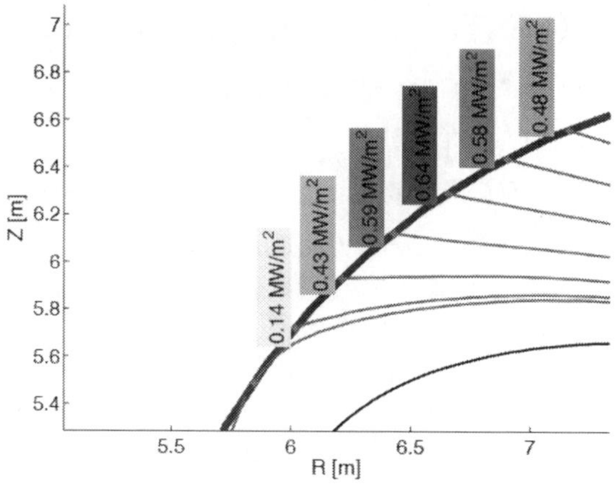


Figure 1: Power flux density to the inner top section of the DEMO first wall based on field line mapping: Vessel wall (blue) and intersecting field lines (red). It is assumed that $\lambda_q = 170mm$

window of Cu, EUROFER despite its relatively low thermal conductivity (~one tenth that of Cu) is currently foreseen as structural material in this area.

Considering these conditions leads to a power flux density limit of $1.5MW/m^2$ for water cooling and $1.0MW/m^2$ for He cooling respectively [2]. In comparison to this more than 50% of the first wall of ITER is specified for more than $3.5MW/m^2$ [3]. To identify at an early stage possible problems in this respect the development of the DEMO Wall Load Specification has been initiated. It has been reported that the peak power flux density during flat-top operation on the wall due to radiation can reach up to $0.45MW/m^2$ for DEMO relevant conditions [2], which is based on the assumption that in addition to the core radiation 100% of $P_{sep} = 150MW$ is radiated from the SOL.

Another important load component onto the first wall originates from thermal charged particles. To assess this, a heat flux density profile in the outer midplane has been mapped along the field lines to the first wall. It is assumed that the total $P_{sep} = 150MW$ is transfered by thermal charged particles from the outer midplane to the first wall. The main zone of wall interaction in the breeding area is in the inner top region of the device. Using the extrapolated midplane power fall off length $\lambda_q = 0.8mm$ [4] leads to negligible heat loads onto the first wall. Figure 1 displays the situation using $\lambda_q = 170mm$ [3] as most extreme value, which might apply in the mid to far SOL for DEMO plasmas with an intense blob activity [5]. This results in a peak power flux density of $0.64MW/m^2$. Similarly as in ITER [6], it has to be investigated by which factors the heat loads are enhanced when considering several types of deviations of the device and the plasma from the idealized situation assumed so far. The risk is recognized that this could lead to heat loads intolerable for EUROFER-based first wall technologies.

**ELMs in DEMO**

Similarly as in ITER, ELMs in DEMO can harm divertor PFCs by energy impact. The relevant limit is the melt limit of W, which is defined by the heat impact factor $\eta = \Delta W/(A\sqrt{t}) \approx 50MJ/m^{-2}s^{-0.5}$. Assuming that the heat impact duration is similar in ITER and DEMO, leads to an identical limit in the relative ELM energy density of $\Delta W/A \leq 0.5MJ/m^{-2}$. Figure 2 displays the dependence of $\Delta W/A$ on the relative ELM size $\Delta W/W$ for various assumptions. The most optimistic assumption (blue) is that the ELM broadening $b$ [7] has a constant value of 6. The most pessimistic assumption (red) is that $b \propto \Delta W/W$ for $\Delta W/W > 1\%$ and $b = 1$ below. These assumptions correspond to a limit in $\Delta W/W$ of 0.84% and 0.14% respectively for

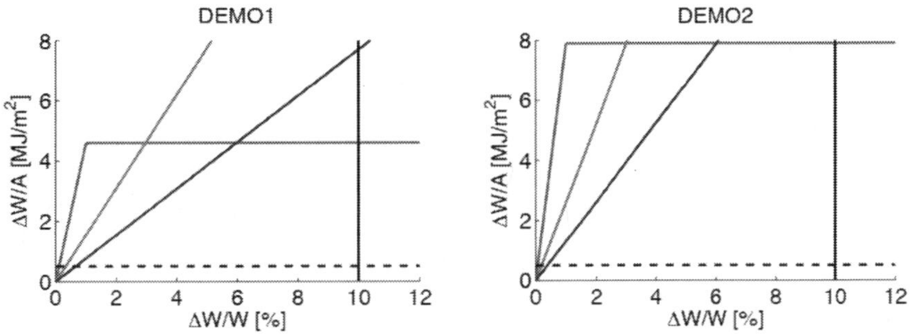


Figure 2: Dependence of the ELM energy density on the relative ELM size for various assumptions: Left: DEMO1, Right: DEMO2. Red: $b = 1$ and $b \propto \Delta W/W$, Magenta: $b = 3$, Blue: $b = 6$. The dashed line marks the limit of $0.5MJ/m^{-2}$. The vertical line represents the extrapolated natural relative ELM size for DEMO.

DEMO1. Considering the anticipated relative size of natural type I ELMs in DEMO of 10% [4], this means that a mitigation in terms of $\Delta W/W$ by a factor of 15 to 90 is required for DEMO1 (DEMO2: 25-150). Other limits, which should be addressed, are related to main chamber heat loads and ELM impurity flushing.

In view of this significant mitigation necessity a robust strategy to deal with ELMs needs to be a key component of any DEMO concept. No recent device is able to simultaneously operate in the DEMO range for normalized collisionality $\nu^\star$ and and Greenwald density fraction. Based on the assumption that pedestal physics dominates in this context, mitigation methods that are demonstrated for DEMO relevant collisionalities ($\nu^\star \approx 0.05$) need to be identified. For RMP ELM-mitigation [8] and the QH-mode [9] this is fulfilled, while the I-mode [10] has so far only been observed for $\nu^\star > 0.1$. Other alternatives to consider are ELM pacing by pellets or vertical kicks.

For all candidate methods for ELM mitigation in DEMO open R&D needs have to be addressed:

- RMP: Characterize penalty in confinement and investigate feasibility of coil integration
- QH-mode: Clarify access conditions at low torque
- I-mode: Determine the extent of the operation window between I-mode threshold and H-mode threshold
- ELM pacing by pellets / vertical kicks: Extrapolate $\Delta W(f_{ELM})$ and broadening; Expand data base and physics understanding; Investigate impact on gas balance (pellets) and control systems (vertical kicks)

**Key Size Drivers**

The size of DEMO is the main driver of its capital costs. It turns out that the major radius R is driven amongst others by parameters related to H-mode operation and divertor protection. Figure 3 shows the dependence of R on $f_{LH} = P_{sep}/P_{LH,scal}$ ($P_{LH,scal}$: [11]) and $P_{sep}/R$ from a calculation with PROCESS minimizing the major radius at fixed $P_{el,net} = 500MW$ and $\tau_{burn} = 2h$. For each $P_{sep}/R$, $f_{LH}$ has a minimum value corresponding to the highest achievable value of the magnetic field at the inner TF coil leg. To achive sufficient confinement quality and controllability of the plasma it might be necessary to control $f_{LH}$ towards a higher value, which would lead to a significant increase of R.

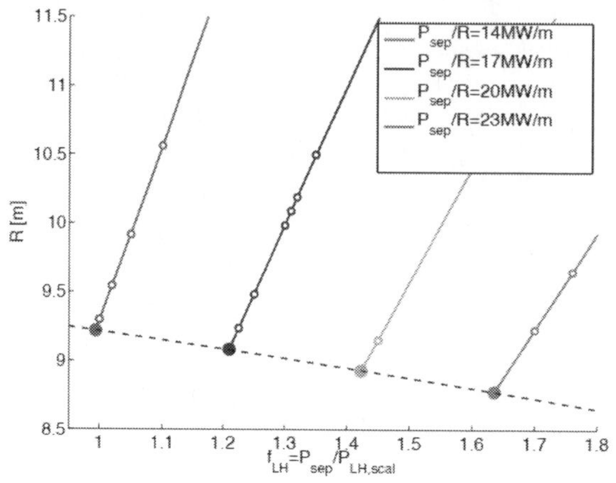


Figure 3: Dependence of the major radius on $f_{LH}$ and $P_{sep}/R$ from PROCESS calculations

The 95% confidence interval of the ITPA threshold scaling for ITER [11] ranges roughly from 50% to 200% of $P_{LH,scal}$. In DEMO $P_{LH} = 2P_{LH,scal}$ would roughly correspond to doubling the major radius. Consequently all possibilities to reduce this uncertainty need to be exhausted. This includes activities towards an improvement of the understanding of the LH-transition. Also it is necessary to develop a new scaling of the H-mode threshold power including more DEMO-relevant data (e.g. data from metal wall devices [12, 13] and high radiation plasmas). It might be helpful to include $Z_{eff}$ as a scaling parameter. Furthermore a better characterization of a possible dependence of $P_{LH}$ on the X-point height / shape [12] and identification of the key physics parameter governing this dependence is desirable especially for closed divertors, which are the option foreseen for DEMO.

## Acknowledgement

This work has been carried out within the framework of the EUROfusion Consortium and has received funding from the Euratom research and training programme 2014-2018 under grant agreement No 633053. The views and opinions expressed herein do not necessarily reflect those of the European Commission.

## References

[1] G. Federici *et al.*, Fusion Engineering and Design **89**, 882 (2014).

[2] R. Wenninger *et al.*, Nuclear Fusion **55**, 063003 (2015).

[3] R. Mitteau *et al.*, Journal of Nuclear Materials **415**, S969 (2011), proceedings of the 19th International Conference on Plasma-Surface Interactions in Controlled Fusion.

[4] R. Wenninger *et al.*, Nuclear Fusion **54**, 114003 (2014).

[5] D. Carralero *et al.*, Nuclear Fusion **54**, 123005 (2014).

[6] R. Mitteau *et al.*, Journal of Nuclear Materials - accepted (2015).

[7] H. Thomsen *et al.*, Nuclear Fusion **51**, 123001 (2011).

[8] A. Kirk *et al.*, Nuclear Fusion **55**, 043011 (2015).

[9] K. Burrell *et al.*, Nuclear Fusion **53**, 073038 (2013).

[10] D. Whyte *et al.*, Nuclear Fusion **50**, 105005 (2010).

[11] Y. R. Martin, T. Takizuka, and the ITPA CDBM H-mode Threshold Database Working Group, Journal of Physics: Conference Series **123**, 012033 (2008).

[12] C. Maggi *et al.*, Nuclear Fusion **54**, 023007 (2014).

[13] F. Ryter *et al.*, Nuclear Fusion **53**, 113003 (2013).

# Impurity Limits in a Reactor Grade Fusion Device

T. Pütterich[1], E. Fable[1], R. Dux[1], R. Neu[1,2], M.G. O'Mullane[3], R. Wenninger[4]

[1]*Max-Planck-Institut für Plasmaphysik, D-85748 Garching, Germany*

[2]*Technische Universität München, 85748 Garching, Germany*

[3]*CCFE, Culham Science Centre, Abingdon, Oxfordshire OX14 3DB, United Kingdom*

[4]*EUROfusion Programme Management Unit, 85748 Garching, Germany*

**Introduction and Production of Atomic Data** Impurity limits for fusion devices are a topic of recent interest, because of several reasons. First, the implementation of first walls with high-Z elements (e.g. ITER-like wall in JET, full W-wall in ASDEX Upgrade) demonstrated the importance of impurity control. Second, it is well known that a substantial fraction of the heating power of a fusion reactor must be radiated via seeded impurities before the power flux enters the regions of the scrape-off layer. Third, a low-Z radiator for the divertor region is required in order to further reduce the power fluxes to the divertor. The atomic data of the radiators are usually derived from different types of calculations such that an investigation on the Z-scaling may suffer from artifacts. In order to avoid such systematic artifacts, atomic data for all considered elements (cf. Fig. 1) were produced by one set of codes from ADAS [1].

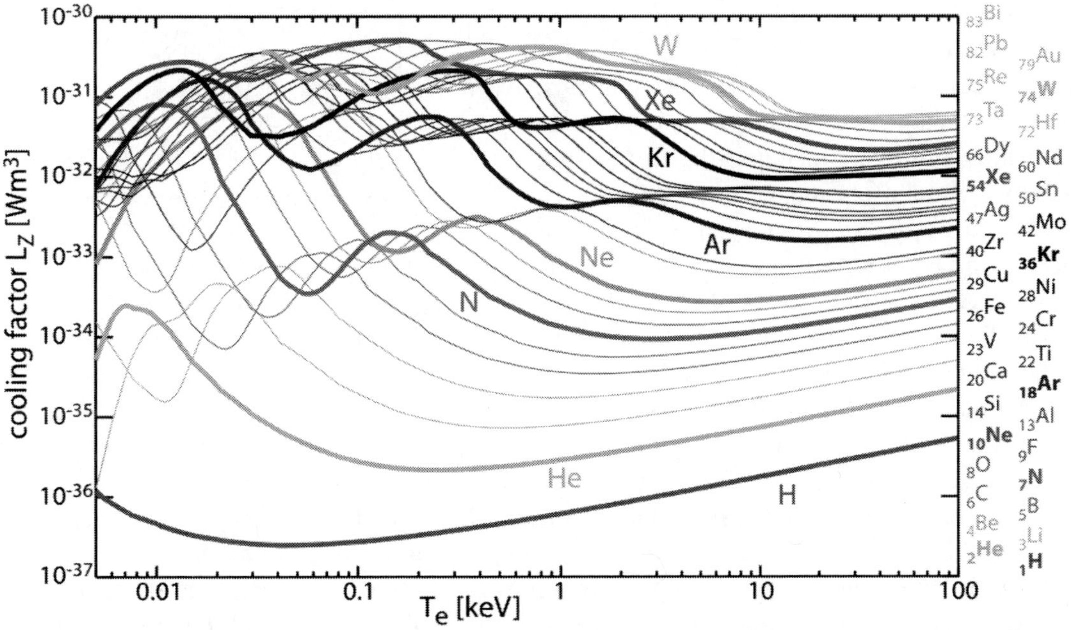


**Fig. 1:** *Calculated cooling factors for the considered elements from H to Bi.*

For all ion stages of the 35 elements under consideration, we used the Cowan code [2] via the infrastructure of ADAS to calculate the atomic/ionic structure and the electron impact collision cross sections (plane-wave Born). The spectra for each ion are then calculated using a collisional radiative model. For the cooling factor of the elements the configuration averaged scheme was used, as it is known to produce good radiation predictions [3]. For the determination of the recombination rates the 'type-A' implementation within the ADAS codes adas407/408 was used. For better results of these latter calculations all of the Cowan code calculations have been repeated using the intermediate coupling scheme. The latter also enables the prediction of detailed spectra for all elements. High-quality ionization rates were obtained via the configuration averaged distorted wave procedure of [4] using the ADAS implementation. As a result, we obtained a large set of radiation data for many elements including all elements that may exists in a fusion plasma.

**0D Reactor Model** The investigated 0D model corresponds to case 4 in section 3 of [5]. The model basically balances the heating by alphas with the loss terms due to radiation and transport. For the radiation the cooling factors are used, while the power lost via transport is described by $W/\tau_E$, where W is the stored thermal energy and $\tau_E$ is the energy confinement time. This balance provides the so-called burn condition and in the plane $nT\tau_E$ vs. $T$ it is matched on the so-called burn curve. In order to treat the helium (He) ash self-consistently, it is assumed that the He confinement time $\tau_{He}$ is proportional to $\tau_E$ (cf. [5]). In the 0D approach the ratio $\rho^* = \tau_{He}/\tau_E$ must stay below 15.7 in order for a burn condition to exists, while in experiment values down to $\approx 4$ have been observed [6]. In the present work the maximum impurity concentration is determined for a range of $\rho^*$ and each impurity. Beyond that maximum impurity concentration no burn condition exists.

**0.5D Reactor Model** In order to take profile effects into account, the above model was extended via the parameters $\frac{T_{core}}{\langle T\rangle}$ and $\frac{n_{core}}{\langle n\rangle}$. $\frac{T_{core}}{\langle T\rangle}$ is the ratio of the core temperature and the volume averaged temperature, $\frac{n_{core}}{\langle n\rangle}$ is the ratio of the core density and the volume averaged density. The profiles themselves are assumed to be linear along the radial coordinate while a circular cross section of the plasma is assumed. This is thought to roughly approximate real plasmas, while plasma shape and profile shape dependencies are only mildly influencing the final result due to the parametrization using $\langle T\rangle$ and $\langle n\rangle$. This is confirmed by studies on the plasma elongation $\epsilon$, which show that the influence of $\epsilon$ is negligible and thus, the effects of $\epsilon$ are not discussed further here. From the $T$- and $n$-profiles the $\alpha$-heating profile and the radiated power profile are derived. Note that for $\frac{T_{core}}{\langle T\rangle} = \frac{n_{core}}{\langle n\rangle} = \epsilon = 1$ the results of the 0D model are reproduced. In Fig. 2, the burn curves are shown for the 0D model (dashed, blue shaded lines) and 0.5D model (solid, red shaded lines). The 0.5D model was evaluated for $\frac{T_{core}}{\langle T\rangle} = 2.0$, $\frac{n_{core}}{\langle n\rangle} = 1.3$ and $\epsilon = 1.8$, which corresponds approximately to the EU DEMO1 2015 design. For each of the two models, lines are presented that correspond to impurity-free plasmas ($\rho^* = 0$, $c_{Xe} = 0$), plasmas with He (assuming $\rho^* = 5$) plus xenon at the maximum concentration 'x' and half that value. Clearly, the consideration of profiles reduces the amount of tolerable xenon - for the considered cases from $4.6 \cdot 10^{-4}$ to $1.0 \cdot 10^{-4}$. In order to investigate this effect as a function of $\frac{T_{core}}{\langle T\rangle}$, $\frac{n_{core}}{\langle n\rangle}$ and $\epsilon$ the whole parameter space has been investigated with both models. Throughout the parameter space the consideration of profiles makes the burn condition harder to achieve and less impurities of any kind can be tolerated. This also leads to a reduction of the maximum $\rho^*$ below of which a burn condition exists. The most unfavorable situation for a fixed $T_{core}$ is achieved, when $\frac{T_{core}}{\langle T\rangle}$ is maximal (investigated up to $\frac{T_{core}}{\langle T\rangle} = 2.5$) and $\frac{n_{core}}{\langle n\rangle}$ is minimal (investigated down to $\frac{n_{core}}{\langle n\rangle} = 1$). A strong temperature peaking leads to a reduction of the plasma volume in which the optimal temperature is achieved and consequently causes a reduction of the impurity limit. Conversely, density peaking emphasizes the hot plasma region where the fusion power exceeds the radiated power and de-emphasizes the colder plasma volume where radiation is more important (if $\frac{n_{core}}{\langle n\rangle} > 1$).

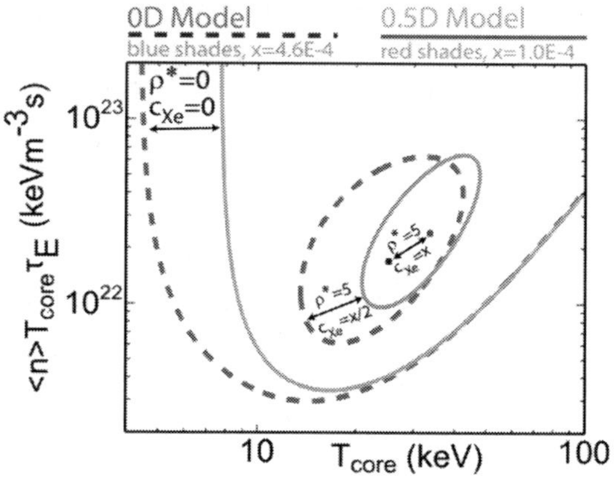


**Fig. 2:** *Burn curves for plasmas with no impurities and plasmas with He and Xe using the 0D-model (blue shades, dashed) and the 0.5D-model (red shades, solid).*

The above 0.5D model is not directly applicable to a specific reactor design, because within the 0.5D model the optimal $\langle n\rangle T_{core}\tau_E$ and $T_{core}$ is chosen for each impurity independently. In reality, certain design parameters of a reactor fix the plasma parameters and the question is what the impurity limits are for that specific reactor design (SRD), e.g. for a specific core temperature. Thus, the above 0.5D model is extended further by finding the maximal impurity density at a fixed temperature (0.5D-SRD).

**1D Reactor Model (ASTRA)** The 0.5D-SRD model (here with $T_{core} = 25\,keV$) can be compared to a proper 1D model of a reactor design similar to that presented in [7]. The 1D ASTRA simulation takes into account the radiation and heating effects on a radial grid resembling the EU DEMO1 2015 design [8, 9]. This 1D ASTRA model allows to set a boundary condition on the radiative cooling, which is important to protect the divertor. In ASTRA the boundary condition on the power flux exiting the plasma is set to 1.2 times the power threshold for H-mode, i.e. 160 MW, such that a continuous operation in H-mode can be guaranteed. Note that this ignores the additional radiative cooling in the divertor, which is still important and may require a low-Z radiator such as nitrogen. It still makes sense to compare the 0D and 0.5D models with $Q = \frac{P_{fusion}}{P_{aux-heating}} = \infty$ to the 1D ASTRA model: First,the synchrotron radiation not considered in the 0D and 0.5D models is about equal to the 50 MW of auxiliary heating taking place for the EU DEMO1 2015 design. Second, meeting the condition of 160 MW power crossing the separatrix in the EU DEMO1 2015 design requires a major reduction of the power flux, which is approximately resembled by the 0D and 0.5D models where also a major fraction of the produced power is radiated or lost via dilution. In Fig. 3(a), results from all models are compared for N and Xe versus $\rho^*$. The 0D model allows for larger impurity concentrations than the 0.5D, the 0.5D-SRD and the 1D ASTRA model. The 0.5D-SRD model reproduces the 1D ASTRA results within a factor of $\approx 3$ for $\rho^* < 6.5$. For $\rho^* \geq 6.5$ the 0.5D-SRD model results have not been evaluated, because the tolerable impurity content steeply drops, while in the 1D ASTRA simulation the impurities on top of He are more easily tolerated. However, in the 1D ASTRA model the fusion yield $Q$ (numbers nearby data points) drops at larger $\rho^*$, such that large $\rho^*$ must be avoided in a reactor for economical reasons. In fact, $Q$ is alarmingly low in the 1D ASTRA simulation even for $\rho^* < 6.5$. For a conservative estimate of $\rho^* = 5$, the impurity limits are presented in Fig. 3(b) and $Q$ ranges between 23 (using N) and 31 (using Xe). For $\rho^* = 5$, the 1D ASTRA model requires an impurity amount which is more than predicted by the 0.5D-SRD model and less than predicted by the 0D model. A major difference between the models is the resulting He-concentration, which is for the $Q = \infty$

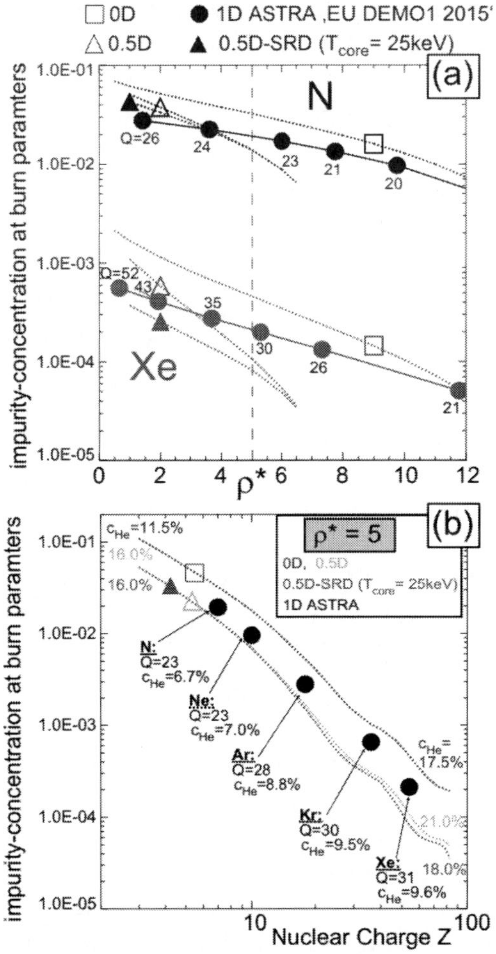


**Fig. 3:** *(a) Max. impurity concentration from all models for N, Ar, and Xe vs $\rho^*$. Numbers give Q for ASTRA runs. (b) Max. impurity concentration from all models vs. Z at $\rho^* =$ 5. Q and He concentrations are also given.*

models approximately 2 times larger than in the 1D ASTRA model. Probably this is caused by the finite $Q$ which allows for less He production at a certain $\tau_E$.

In Fig. 4 a possible remedy for a too low $Q$ is investigated. In Fig. 4(a), the effect of better confinement (red curves) and of an additional W-content of $c_W = 5 \cdot 10^{-5}$ (open symbols) on the tolerable/necessary Xe content is investigated with the 1D ASTRA model. Both the amount of the impurities and $Q$ are increasing for an improved confinement factor (H-factor from $\approx 1.1$ to $\approx 1.4$). For an additional W content of $c_W = 5 \cdot 10^{-5}$ the required Xe content is decreased, while a negligible change of $Q$ is observed. $Q$ is almost unchanged because both Xe and W cause mostly radiation but negligible dilution. However, if N is considered as a plasma impurity instead of Xe (cf. Fig. 4(b)) the partial replacement of N by W leads to an increased $Q$ due to the less severe dilution. Note that for all cases the density profile in the 1D ASTRA model is prescribed to match the EU DEMO1 2015 design, while a physics based model in ASTRA leads to a further reduction of Q. In conclusion, the low $Q$ values found in the 1D ASTRA model w.r.t. to the predicted $Q$ (i.e. 40 for the Xe case) by the DEMO1 design code requires further investigation. It hints to inconsistencies between the DEMO1 design code and the 1D ASTRA simulations. When comparing the 1D ASTRA model to the much simpler 0.5D models, the agreement for the tolerable impurity content are within a factor of 3 in the reactor relevant region of $\rho^* < 6.5$. The 0.5D model is largely independent of dimensions and geometry allowing for its application also in other devices such as stellarators or for a fast implementation in system codes. The 0D model gives always an upper limit to the impurity content.

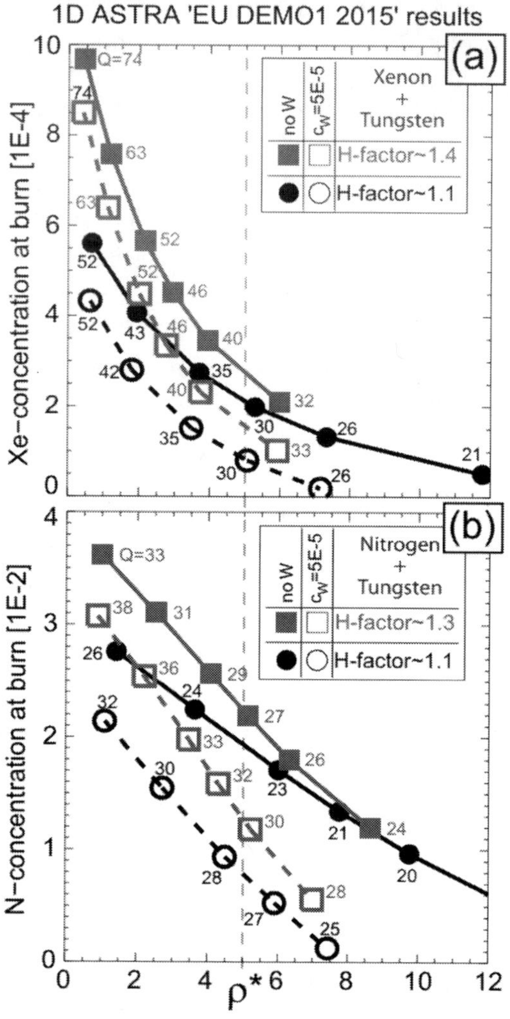


**Fig. 4:** *(a) Limit for Xe-concentrations vs. $\rho^*$ for standard (black, circles) and increased H-factor (red, squares), for the case without W (filled symbols) and with a W-concentration of $5 \cdot 10^{-5}$ (open symbols). (b) Limit for N-concentrations vs. $\rho^*$ for the same cases.*

## References

[1] H. P. Summers, The ADAS User Manual, version 2.6 http://adas.phys.strath.ac.uk (2004).
[2] R. D. Cowan, *The Theory of Atomic Structure and Spectra* (University of California Press, 1981).
[3] T. Pütterich *et al.*, NF **50**, 025012 (9pp) (2010).
[4] S. D. Loch *et al.*, PRA **72**, 052716 (2005).
[5] D. Reiter *et al.* NF **30**, 2141 (1990).
[6] H.-S. Bosch *et al.*, PPCF **39**, 1771 (1997).
[7] R. Wenninger *et al.*, NF **54**, 114003 (2014).
[8] R. Wenninger *et al.*, *this conference, P4-110.*
[9] R. Kemp, private communication, 2015.

**Acknowledgement** This work has been carried out within the framework of the EUROfusion Consortium and has received funding from the Euratom research and training programme 2014-2018 under grant agreement No 633053. The views and opinions expressed herein do not necessarily reflect those of the European Commission

# MHD limit cycles on FTU

G. Pucella, P. Buratti, C. Cianfarani, E. Giovannozzi, and FTU team*

*Unità Tecnica Fusione, C.R. ENEA Frascati, CP65, 00044 Frascati, Italy*

## Introduction

The development of large-amplitude tearing modes (TM) during the pre-programmed plasma density ramp-up in proximity of the density limit [1] or after injection of sufficient amounts of Ne gas [2] on FTU ($R_0 = 0.935$ m, $a = 0.30$ m, $B_T = 2$ - $8$ T, $I_p = 0.2$ -$1.6$ MA) shows a complex behavior that can be outlined in three stages. First stage: the magnetic island grows smoothly at constant rotation frequency. Second stage: amplitude and frequency feature large cycles of oscillations, with peak amplitude increasing progressively across cycles. Third stage: the island grows quickly to large amplitude and locks; this stage generally ends in a disruption. The amplitude and frequency oscillations show a well defined phase portrait, determining a so-called "limit cycle" on the Amplitude/Frequency plane. The existence of a stage with large amplitude and frequency modulations could be caused either by interaction between modes of different helicity or by island self-healing phenomena; resonant error field components could also play a role. Dedicated experiments were performed on FTU, in a wide range of plasma current and toroidal magnetic field values, to understand the origin of amplitude and frequency modulations of the (2, 1) TM and to obtain the scaling for transitions between different regimes (saturation, limit cycles, locking) in terms of plasma parameters and mode amplitude.

## MHD activity

The magnetic activity on FTU is analyzed by means of poloidal and toroidal arrays of magnetic pick-up coils installed inside the vacuum vessel. All the investigated pulses present a very similar MHD phenomenology [3], as illustrated in Figure 1, where the time traces of some relevant quantities are reported for a pulse with $B_T = 4.0$ T and $I_p = 700$ kA. The onset of a $m/n = 2/1$ TM ($m$ and $n$ are the poloidal and toroidal mode numbers, respectively) is identified at $t = 0.82$ s, when the frequency takes a well-defined value of 6.5 kHz. At first, the mode grows algebraically and its frequency remains

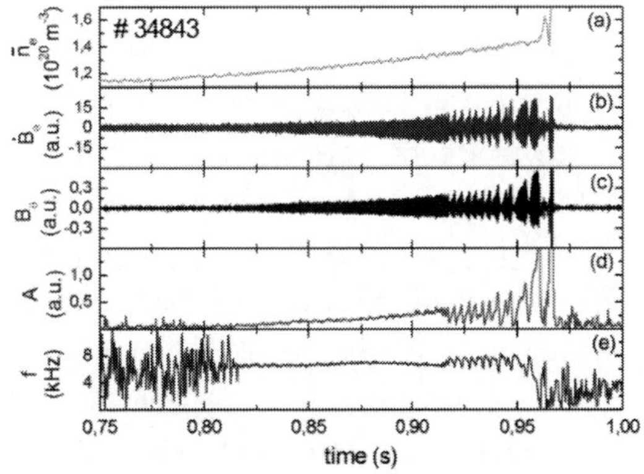


Figure 1. Time traces of some relevant quantities for the MHD activity for a specific pulse on FTU with $B_T = 4.0$ T and $I_p = 700$ kA: (a) central line-averaged density, (b) output from the pick-up coil, (c) poloidal magnetic perturbation, (d) mode amplitude, (e) mode frequency.

*See the appendix of G. Pucella et al., Proc. 25th IAEA Fusion Energy Conference, Saint Petersburg, Russia,2014

constant. Subsequently, the amplitude shows a quasi-periodic modulation, with the maximum value reached during each period still increasing algebraically. In the last phase, the mode growth speeds up and the frequency decreases to zero.

Regarding the signal processing, the output of the pick-up coils is integrated to obtain the poloidal magnetic perturbation associated to the MHD mode. In particular, a high-pass filter (1 kHz) is applied to eliminate slow drifting of the integral and a low-pass filter (20 kHz) is applied to eliminate noise. The Hilbert transform is used to calculate the amplitude and phase of signal starting from the poloidal magnetic perturbation and the derivative of the phase with respect to time is performed to obtain the mode frequency. Finally, a low pass filter is used on both amplitude and frequency to decrease the associated noise. We also considered that an air-core transformer was present on the MHD signal acquisition path. The resulting frequency dependent attenuation (particularly strong on lower frequencies) was analytically calculated and fitted against experimental data (as derived through comparison with a transformer-free acquisition chain), thus obtaining a correction that was applied to the final MHD amplitudes.

**Limit cycles**

As we can see from Figure 1, the temporal evolution of the TM during the density ramp-up is characterized by a “regular phase” where the mode amplitude increases at approximately constant frequency (from t = 0.82 s to t = 0.91 s in the Figure), and a “bursting phase” where the amplitude and the frequency of the mode oscillate (from t = 0.91 s to t = 0.96 s). Finally the mode amplitude grows up to high values and its frequency reduces rapidly up to the disruption. From a closer inspection of data it is possible to see that both the amplitude and frequency oscillations are higher for higher values of the line-averaged density (linearly proportional to the time for the discharge considered here). In particular, if we report the temporal evolution of the mode on the Amplitude/Frequency plane (see Figure 2), we can observe, after the regular phase (cyan lines), the formation of cycles with increasing area.

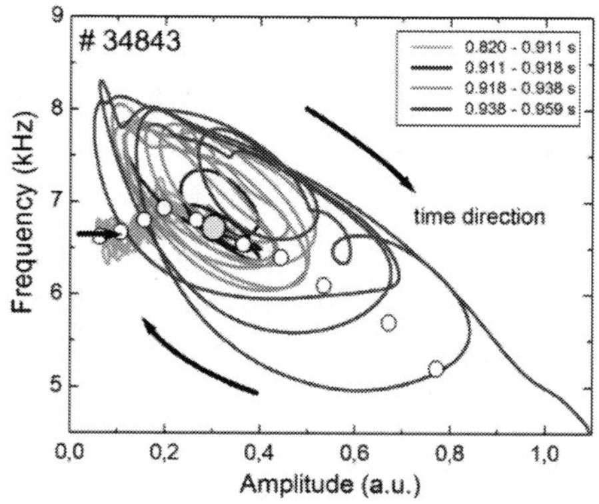

Figure 2. Time evolution of the tearing mode on the Ampltude/Frequency plane (see the legend for the meaning of colours). The evolution of the amplitude and frequency envelopes is also reported as open circles. The yellow circle corresponds to the critical mode amplitude for transition from smooth to cyclic behavior.

The critical mode amplitude for transition from smooth to cyclic behavior (reported as a yellow circle in Figure 2) was analyzed in terms of critical island width in a wide range of operational parameters, with toroidal field ranging from 4 to 8 T and plasma current from 0.5 to 0.9 MA. In particular, considering the well-known expression for the island width and employing the usual analytical estimate based on neglecting plasma current outside the mode resonant surface (good approximation going towards the density limit and for pulses with Ne gas injection), we have:

$$w = 4\sqrt{\left.\frac{rqB_{r1}}{mq'B_{\theta}}\right|_{r=r_s}} = 4\sqrt{\frac{Rr_s}{nB_z s_s}\cdot\left(\frac{r_c}{r_s}\right)^{m+1}\frac{1-(r_s/r_w)^{2m}}{1+(r_c/r_w)^{2m}}\cdot B_{\theta 1}(r_c)} \quad (1)$$

where $r_s$, $r_c$ and $r_w$ are the radii of the mode resonant surface, the pick-up coil and the wall, respectively. Critical values of the full island width for the onset of the limit cycles ranged from 1 cm to 2 cm, which are in the range of the critical island width for local temperature flattening by finite perpendicular heat conduction. The existence of a well defined threshold for the transition from smooth to cyclic behavior has been also confirmed in experiments of real time control of TM performed on FTU by using injection of electron cyclotron waves (ECW) inside the 2/1 magnetic island [4]. In pulses of this kind, a partial suppression of the 2/1 TM has been obtained during the ECW phase and a double transition from cyclic and smooth behavior was correspondingly observed (see Figure 3).

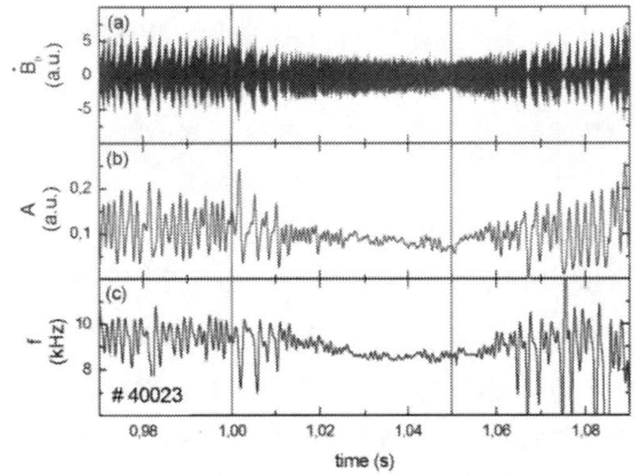

Figure 3. Example of stabilization of the 2/1 mode in FTU pulses at high density, by using injection of ECW. (a) Output from the pick-up coil, (b) mode amplitude, (c) mode frequency. Red bars indicate the ECW phase.

For a more detailed analysis of the MHD limit cycles, we report in Figure 4 the time traces of some relevant quantities for a given cycle of the MHD activity, while in Figure 5 we report the temporal evolution of the MHD mode on the Amplitude/Frequency plane. As we can see, we can distinguish four phases: a blue phase with increasing amplitude and approximately constant frequency, a navy phase with increasing amplitude and decreasing frequency, a green phase with decreasing amplitude and decreasing frequency, and an orange phase with

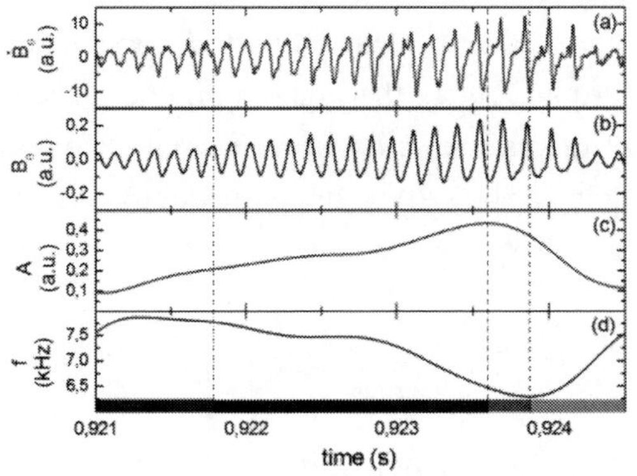

Figure 4. Time traces of some relevant quantities for a given cycle of the MHD activity.

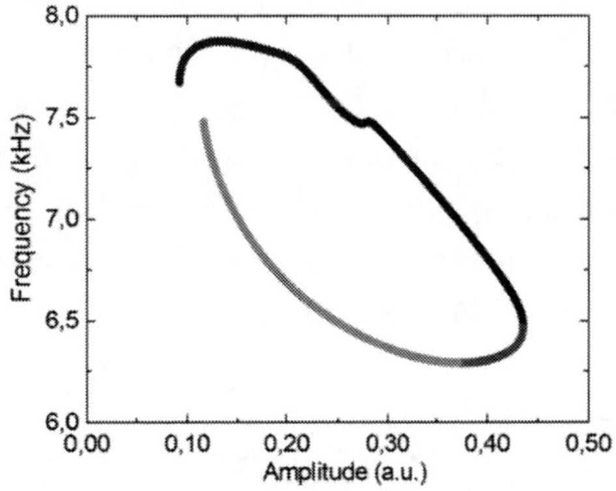

Figure 5. Temporal evolution (clockwise) of the MHD activity on the Amplitude/Frequency plane.

decreasing amplitude and increasing frequency. We noted that the trajectories of the mode on the (A, f) plane form closed curves (with area increasing with the line-averaged density), then the system exhibits, for a given density, an isolated closed trajectory, namely a so-called "limit cycle". For the purpose of this work we can try to understand the closed trajectories on the (A, f) plane by considering the diagrams produced on the phase-space from the oscillation in the four different phases of a limit cycle and comparing these diagrams with those produced by a harmonic oscillator (see Figure 6). As we can see, only on the first phase, characterized by an increasing amplitude and a constant frequency, the behavior of the tearing mode rotation is like a harmonic oscillator, whilst in the third phase, characterized by decreasing amplitude and decreasing frequency, we have a sort of "double system".

**Results and discussion**

The main results indicate that the presence of the q = 3 resonance in the plasma is necessary for the occurrence of deep and regular limit cycles for the (2, 1) mode. On the other hand, the presence of residual amplitude and frequency oscillations in pulses with $q_a < 3$ implies that other (weaker) modulation mechanisms are acting, such as interaction with the (1, 1) internal mode or with a non resonant (3, 1) surface kink. Stronger overtones are observed that can be justified by intra-wave modulation due to enhanced interaction with the resonant error field.

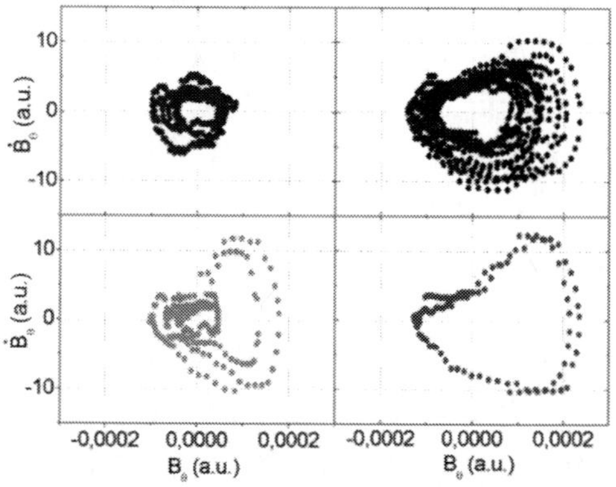

Figure 6. Diagrams in the phase-space for the oscillation in the four different phases of a limit cycle.

**Conclusions**

The MHD activity near the density limit or after injection of sufficient amounts of Ne gas shows an m/n = 2/1 tearing mode growing up to high amplitude, with a final phase characterized by amplitude and frequency oscillations, giving rise to limit cycles on the Amplitude/Frequency plane. The MHD cycles could be due to recursive island fragmentation (the island distortion increases before amplitude drops), but in the high-amplitude phases of the mode temporal evolution it is difficult to discriminate between a mode coupling mechanism and non-linear effects. A more accurate analysis will be done to better understand the possible physical effects responsible for the occurrence of deep and regular limit cycles for the (2, 1) tearing mode on FTU.

**References**

[1] G. Pucella, et al., Nucl. Fusion 53, 083002 (2013)
[2] A. Botrugno, et al., Proc. 12th Asia Pacific Phys. Conf. (Chiba), D1-PTu-02 (2012)
[3] G. Pucella, et al., Proc. 40th EPS Conf. on Plasma Phys. (Espoo), P5.139 (2013)
[4] C. Sozzi, et al., Proc. 25th IAEA FEC (S. Petersburg), P2.47 (2014)

# Runaway electron experiments in FTU

B. Esposito[1], D. Carnevale[2], J.R. Martin-Solis[3], L. Boncagni[1], Z. Popovic[3], F. Causa[1], M. Agostini[4], E. Alessi[5], G. Apruzzese[1], W. Bin[5], P. Buratti[1], C. Cianfarani[1], R. De Angelis[1], S. Galeani[2], C. Galperti[5,6], S. Garavaglia[5], M. Gospodarczyk[2], A. Grosso[1], G. Granucci[5], G. Maddaluno[1], D. Marocco[1], C. Mazzotta[1], L. Panaccione[1], A. Pensa[1], V. Piergotti[1], S. Podda[1], G. Pucella[1], G. Ramogida[1], M. Riva[1], G. Rocchi[1], M. Sassano[2], A. Sibio[1], C. Sozzi[5], B. Tilia[1], O. Tudisco[1], M. Valisa[4], ECRH team and FTU team

[1]*ENEA for EUROfusion, via E. Fermi 45, 00044 Frascati (Roma), Italy*
[2]*Dip. di Ing. Civile ed Informatica DICII, Università di Roma, Tor Vergata, Roma, Italy*
[3]*Universidad Carlos III de Madrid, Avda. Universidad 30, Leganes, 28911 Madrid, Spain*
[4]*Consorzio RFX, Padova, Italy*
[5]*IFP-CNR, Via R. Cozzi 53, 20125 Milano, Italy*
[6]*Centre de Recherches en Physique des Plasmas, EPFL, 1015 Lausanne, Switzerland*

**Introduction**

We present an overview of the FTU experimental results related to two important ITER physics issues concerning the mitigation of runaway electrons (RE): the measurement of the threshold electric field for RE generation and the position control and ramp-down of disruption-generated RE. A strong experimental program devoted to RE studies has been recently enabled in FTU by the integration of diagnostic, hardware and software tools in the real-time (RT) plasma control system. Several diagnostics are (being) upgraded to RT operation: the hard x-ray (HXR) radial profile monitor measuring in-plasma bremsstrahlung from RE, a set of Mirnov coils used for disruption prediction algorithms, the CO2 scanning interferometer for RE position control, a fission chamber monitoring the gamma-ray emission as a signature of RE interactions with the plasma facing components (PFC). In addition, a Cherenkov probe was installed to study the RE dynamics in presence of magnetic islands [1] and a RE imaging system (REIS) is being tested to detect visible/infrared synchrotron radiation emitted by RE.

**Measurement of threshold electric field for RE generation ($E_{thr}$)**

The aim of these experiments is to measure $E_{thr}$ at the RE *onset* and RE *suppression* occurring in the plasma current ($I_p$) flattop phase of deuterium ohmic discharges. The RE *onset* (see an example in Fig.1(right)) is obtained through a decreasing electron density ($n_e$) in the $I_p$ flattop. The RE *suppression* is achieved by starting a discharge with low gas prefill thus creating a RE population subsequently suppressed by a feedback-controlled phase of constant or increasing $n_e$. Key diagnostics are the $BF_3$ neutron detectors and the NE213 neutron/gamma scintillator, which give overlapping signals in the absence of RE [2]. The results (shown in Fig.1(left); more details can be found in [3]) essentially confirm the findings reported in [2] but refer to a wider database (56 discharges with $B_t$=2-7.2 T, $I_p$=0.35-0.9 MA, $Z_{eff}$=1.5-10). The measured $E_{thr}$ (obtained as $V_{loop}(t_C)/2\pi R_0$, where $t_C$ is the time when NE213 and BF3 traces diverge/converge in the *onset/suppression* experiments) is ~2-5 times larger than predicted by the classical collisional theory [4] and consistent with the new threshold calculated including synchrotron radiation losses [5]. The local central density is used to calculate $E_R^{rad}$ since RE are initially generated in the core of the plasma ([2] and below Fig.3(left)).

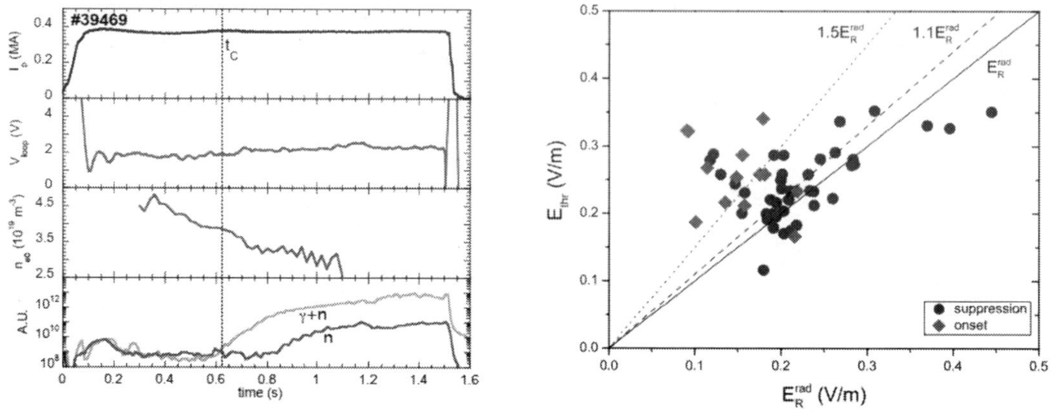


**Figure 1**. (left) *RE onset: time traces of $I_p$, loop voltage ($V_{loop}$), local central density, NE213 (γ+n) and $BF_3$ (n) signals.* (right) *Comparison of measured $E_{thr}$ with relativistic collisional+synchrotron radiation theory predictions ($E_R^{rad}$ [5]);*

## RE control

The aim of these experiments is the RT control of the horizontal position and current ramp-down of disruption-generated RE plateaus [6]. Two scenarios are adopted: a) a significant RE population is generated during the $I_p$ ramp-up/flattop by starting the discharge with low gas prefill and pre-programming a low $n_e$ reference (~$1.5 \times 10^{19}$ m$^{-3}$), followed by Ne injection to induce the disruption. b) a spontaneous disruption is induced during the $I_p$ ramp-up by setting extremely low gas prefill. A limited database is available as these scenarios do not produce RE plateaus routinely in all discharges, but only in a fraction of the attempts. The plasma parameters are $I_p$=0.36 MA, $B_t$=6 T. Key RT diagnostics for RE control are the fission chambers, the Mirnov coils and the CO2 scanning interferometer. The active coils used to control the position of the plasma and $I_p$ are T (transformer) to control $I_p$, V (vertical) and F (feedback) to control the plasma radial movements and elongation and H (horizontal) to control the plasma vertical position. The current of the V coil, which produces a vertical field similar to F but with a slower rate of change, is varied by the Current Allocator controller [7] in order to modify at run-time the F current and maintain the vertical field unchanged. An example of the results is shown in Fig.2: 1) the standard $I_p$ reference is substituted by a ramp-down that produces a lower $V_{loop}$ (see panel g), hence reducing the energy transferred by the central solenoid to the RE and, consequently, also the radial Shafranov shift. 2) the plasma outer radius ($R_{ext}$) is reduced dynamically at run-time (panel f) and this contributes to the reduction of the RE beam interaction with the PFC as indicated by the lower fission chamber signal (panel h). The RE control strategy is being upgraded by means of a control-oriented dynamical model for RE position and energy [8].

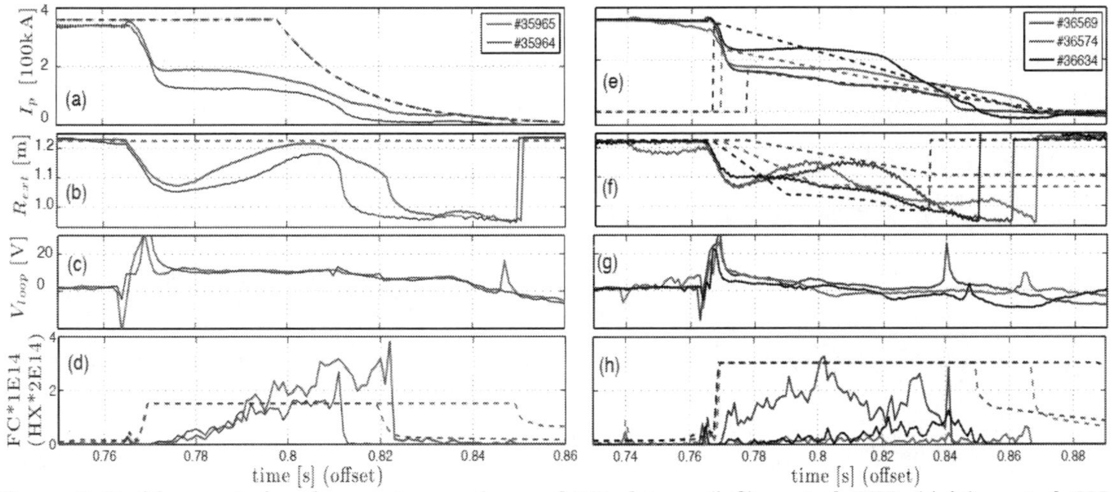


**Figure 2**. Position control and current ramp-down of RE plateau: (left) control OFF; (right) control ON. Dashed lines in panels (a), (b), (e) and (f) refer to reference traces, solid lines to actual measurements.

## RE diagnostics

Some results from two RE diagnostics recently implemented in FTU are presented below. Features of the HXR emitted perpendicularly to the magnetic field and produced by RE through bremsstrahlung in the plasma [9] are shown in Fig.3(left) (an increase of the RE energy is observed after the initial phase (0.6-1.0 s)) and Fig.3(right) (the RE beam originates in the plasma center and moves off-axis soon after). Fig.4(top) shows the (uncalibrated) spectra in the visible range obtained from the REIS: a wide-angle optical system collects the radiation from two plasma cross sections (corresponding respectively to the RE backward and forward view, see Fig.4(bottom left)) and transmits it to two visible spectrometers through an incoherent bundle of fibers. Optionally, an infrared spectrometer has been used in the forward view (Fig.4(bottom right)). These preliminary measurements (both discharges with $B_t$=4.1 T and $I_p$=0.5 MA) clearly indicate that the range of RE emission in FTU may extend from 500 nm to 2000 nm.

## Conclusions

The threshold electric field for RE generation was measured in the flattop of ohmic discharges and found to be larger by a factor ~2-5 than expected according to the purely collisional theory and very close to a new threshold that includes synchrotron radiation losses. This might imply a lower threshold density value to be achieved by means of massive gas injection (MGI) for RE suppression in ITER. The dissipation of the disruption–generated RE energy and population through RE horizontal position control was performed through feedback on the poloidal coils by exploiting a RT boundary reconstruction algorithm evaluated on magnetic moments. The RE beam was slowly ramped down within ≈100 ms with reduced interaction with the PFC. RE control can be therefore an alternative/complementary technique to MGI.

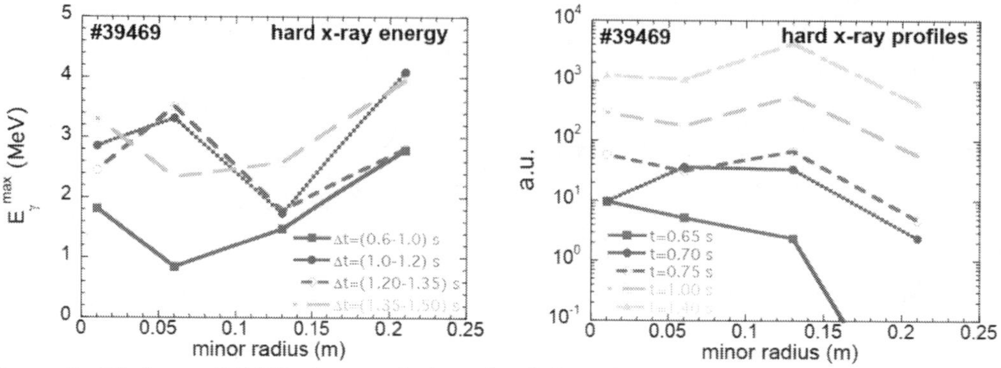


**Figure 3**. Discharge #39469: time evolution of radial profiles of HXR maximum energy (left) and line integrated HXR emission (right).

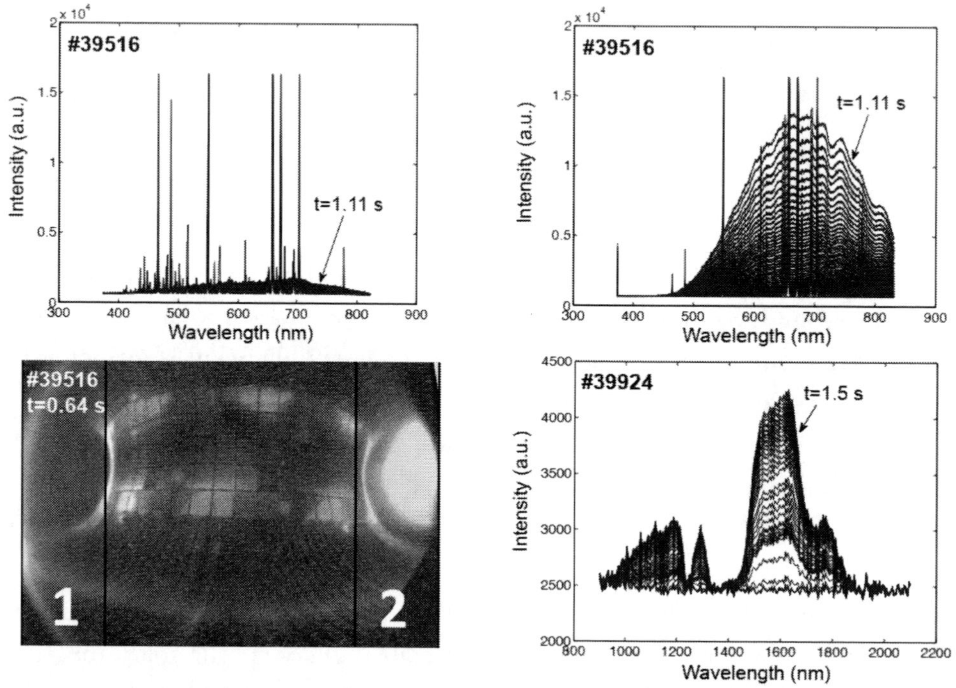


**Figure 4**. (top left) Backward and (top right) forward RE visible spectra (#39516); (bottom left) viewing layout of REIS: RE backward (1) and forward (2) view; (bottom right) forward RE infrared spectra (#39924): note the two OH-IR absorption bands (centered around 1.25 and 1.4 μm) due to the fiber bundle presence. Each spectral curve corresponds to a time slice: the last time (indicated by the arrow) is before the disruption.

## References

[1] F. Causa et al., paper O4.134, this conference.
[2] B. Esposito et al., Proc. 33rd IAEA Conference, St Petersburg (Russian Federation), EX/P2-50 (2014).
[3] Z. Popovic et al., paper P4.116, this conference.
[4] J. W. Connor and R. J. Hastie, Nucl. Fus., **15** (1975) 415.
[5] J. R. Martin-Solis et al., Phys. Rev. Lett., **105** (2010) 185002.
[6] D. Carnevale et al., Proc. 33rd IAEA Conference, St Petersburg (Russian Federation), EX/P2-48 (2014).
[7] L. Boncagni et al, Fus. Eng. Des., **88** (2013) 1109.
[8] D. Carnevale et al., paper P4.192, this conference.
[9] D. Marocco et al., Proc. 28th SOFT Conference, San Sebastian (Spain), (2014).

## Acknowledgements

This work was carried out within the framework of the EUROfusion Consortium and received funding from the Euratom research and training programme 2014-2018 under grant agreement No 633053 (WP14-MST2-9 and WP15-MST2-15) The views and opinions expressed herein do not necessarily reflect those of the European Commission. Funding was also received by Project No. ENE2012-317 (MINECO; Spain).

# Neon transport simulation for highly peaked density FTU plasma

C.Mazzotta[1], M.E.Puiatti[2], L.Gabellieri[1], G.Apruzzese[1], V.Dolci[1,3], L.Carraro[2], F.Causa[1], A.Romano[1] and the FTU team*

[1] *Unità Tecnica Fusione, ENEA C.R. Frascati, via E.Fermi 45, 00044 Frascati (Roma), Italy*

[2] *Consorzio RFX, Corso Stati Uniti 4, I-35127 Padova, Italy*

[3] *Dip. di Fisica, "Sapienza" Università di Roma, P.le Aldo Moro 5, 00185 Roma, Italy*

**See the appendix of G.Pucella et al., Proceedings of the 25th IAEA Fusion Energy Conference, Saint Petersburg, Russia, 2014*

## Introduction

It is well known that density profile peaking is mostly associated to better plasma performances. In particular, in the so called RI mode, which is an improved confinement regime obtained in Tokamaks by injecting controlled quantities of gaseous impurities (mainly noble gases such as Neon or Argon), it has been assessed that a stable radiative edge seeded with light impurities has beneficial effects on confinement and induces density peaking [1,2]. On the other hand, the risk associated with a too large an amount of impurities is to increase the plasma pollution, producing large values of the effective charge and of the radiated power from the plasma center, which can lead to plasma disruption. Indeed, general aim of impurity seeding experiments is a high radiation level from the edge avoiding impurity accumulation in the center. In this framework, impurities transport studies in reactor-grade plasmas are of outmost importance to deepen the understanding of their role in triggering such improved regimes. This paper addresses the issue of the behavior of Neon when used as a seeding impurity in FTU Tokamak experiments, where a significant increase of electron density peaking is observed [3].

## FTU Neon injection experiments

Neon injection experiments have been recently performed in FTU. Similar to TEXTOR, the first device where RI-modes have been well described [2], FTU has a nearly circular cross-section, a poloidal limiter and operates in L-mode. The Neon injections recently described in FTU (current I=360 kA, field B=5.2 T, density before injection $n_e=5\cdot10^{19}$ $m^{-3}$) can cause a spontaneous increase of the line average density by a factor 2 and the peaking factor ($n_{e0}/\langle n_e\rangle_{vol}$) easily reaches 3.5 and over [3]. From the point of view of transport, in particular of microturbolence analysis, first results lead to identify the mechanism of density peaking with the ITG growth rate, as calculated by Gyro-Kinetic codes [3,4], which decreases in consequence of Neon injection, resulting in a decrease of the outward particle flux and in the peaking of the density profile.

The applicability of these results to a machine with the size of e.g. ITER has to be further investigated, in particular deepening the understanding of the impurity transport mechanisms, in order to describe their role in the on-axis density increase. At this purpose a set of new experiments have been recently performed on FTU to extend the set of explored parameters (scan in current, starting densities, amount of impurity injected etc.). All these discharges need to be carefully modelled in terms of microstability analysis, and this is a work in progress for which the calculation of the impurity profile is decisive. In addition, the characterization of the Neon behavior is necessary to identify its contribution to the observed density peaking, and to investigate the influence of light impurities on particle transport [4,5,6,7]; in fact the result will be an input for the gyro-kinetic codes.

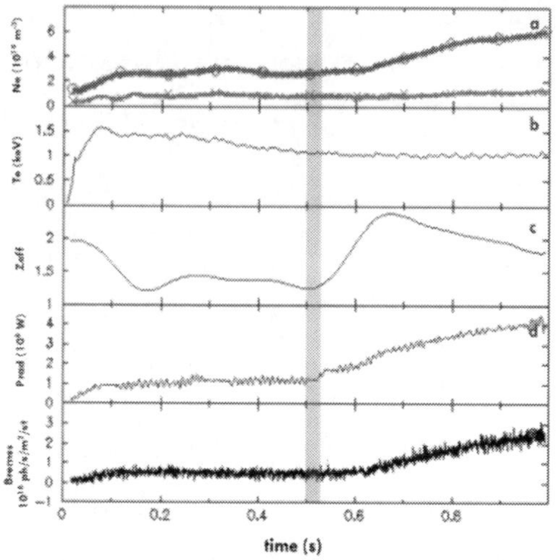


*Figure 1. time traces for #35907 discharge; the light red vertical box marks the Neon injection. a) line averaged electron density red for the central chord, pink for the peripheral one. b) Central Electron Temperature by ECE. c) Zeff time trace. d) Radiation power by bolometry. e) Central Bremsstrahlung signal.*

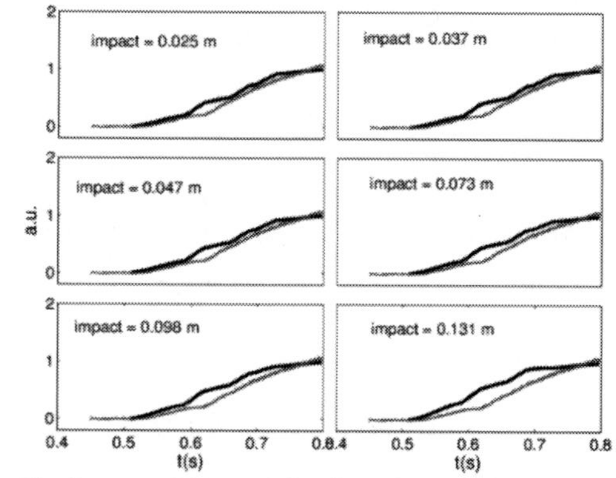


*Fig.2: experimental (red) and simulated (black) time evolution of SXR brightnesses corresponding to line of sights at different impact parameter.*

**Neon transport analysis**

In order to characterize Neon impurity ionization and transport subsequent to gas puff, a 1-dim collisional-radiative impurity transport code is used [8], which for each ion of charge Z solves the transport equation: $\frac{\partial n_Z}{\partial t} + \frac{1}{r}\frac{\partial}{\partial r}(\Gamma_Z) = S$ (1) where $S$ is a source term including the ionization and recombination balance and the neutral source. The flux of an ion of charge $Z$ is: $\Gamma_Z = -D(r)\frac{\partial n_Z(r)}{\partial r} + v(r)n_Z(r)$ with D(r) diffusion coefficient and v(r) radial pinch velocity. D(r) and v(r) are assumed to be the same for all ions. Profiles from high resolution interferometer for the electron density and from ECE for the temperature are used as input to the code, taking into account their temporal evolution.

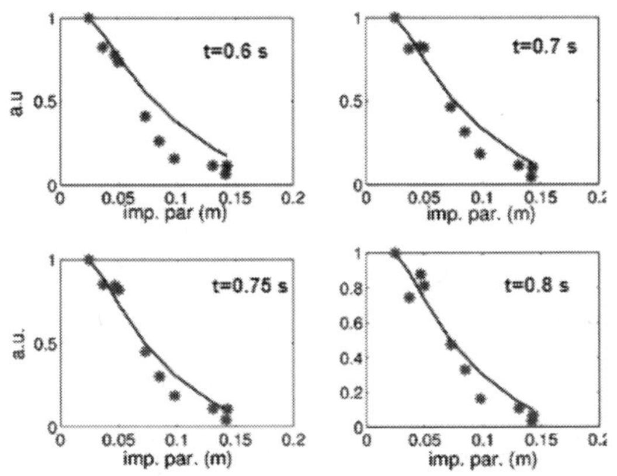


*Fig.3: comparison of simulated profiles SXR brightness profiles (blue) with experimental points (green) at different times.*

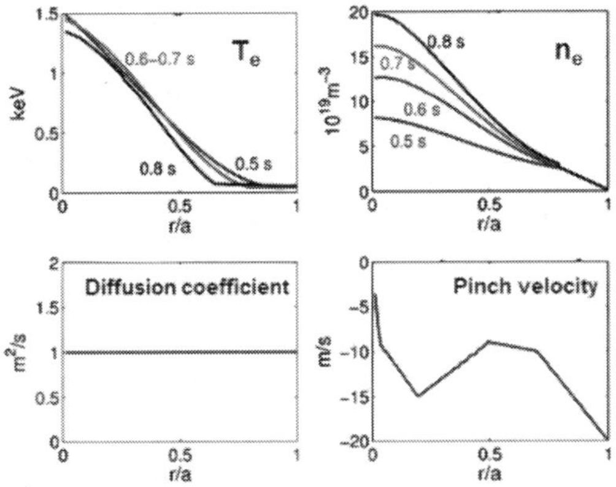


*Fig4: Temperature and density profiles used in the calculation (top); diffusion coefficient and pinch velocity resulting from the simulation.*

In eq. (1) D(r) and v(r) are assumed as free parameters within a procedure of minimization of the difference between the model output and the experimental data in terms of SXR emissivity (SXR being the radiation diagnostic characterized by the best time and space resolution). The result of the minimization procedure is then compared with other measurements, including effective charge, radiated power, spectral line time evolution. The analyzed discharge #35907 has a Neon injection starting at 0.5s (fig. 1, time traces of principal signals), which is viewed by SXR detectors and spectrometers 50ms later (fig. 2) due to the 1m distance of the gas valve from the plasma. The presence of heavy impurities is negligible in this discharge, so that most of the SXR brightness can be attributed to Neon. However, in the simulation the SXR brightness at t=0.5s (just before the Neon injection) are enhanced by a time-dependent factor proportional to the on-axis density increase and subtracted to the experimental signals, in order to isolate the Neon contribution. In figure 2 the comparison of SXR brightness time traces between simulation (black line) and measurement (red trace) at different impact parameters is shown. In figure 3 the experimental and simulated normalized profiles are reported at 4 times. At about 0.6s when the Neon enters the plasma, the density starts to grow up and peaks; at 0.8s the discharge reaches an electron density peaking of 3.2. Contrary to the electron density, the temperature does not change significantly in the central region (fig. 1 b), a modest cooling effect as density increases is observed at r/a > 0.5. Accordingly, the density and the temperature profile evolutions shown in fig.4

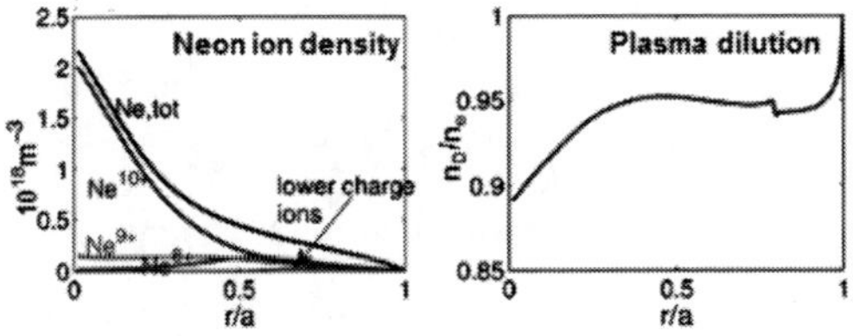


*Fig. 5: Neon ion densities (left) and plasma dilution (right).*

have been used in the simulation. The fig. 4 also shows the diffusion coefficient and pinch velocity resulting from the simulation, note that for D(r) an uniform profile was pre-assigned, and that the pinch velocity results being inward throughout the radius. As consequence the impurity presents a peaked density profile: the fig. 5 reports the profiles at different Neon ionization state. The SXR signals as shown in figs. 2-3 are compared in normalized units. To determine the absolute Neon density, the effective charge and total radiated power (fig. 1 c, d) have been compared between simulation and experiment, leading to the Neon ion densities and the total Neon density at t=0.8s as drawn in fig. 5. The Neon profile is peaked, but not much more than the same electron density. Indeed the plasma dilution due to Neon, defined as the ratio between deuterium ions and the total density, remains > 0.9 throughout the radius (fig. 5 at right), corresponding to a Neon contribution to the on-axis electron density of the order of $10^{19}m^{-3}$, much lower than the experimental increase > $10^{20}m^{-3}$. This is shown in fig.6: the black curve is the total density, the red is the total density with the Neon contribution subtracted, and the blue one is the difference between the electron density after and before the Neon injection. The increase of the effective charge integrated along the diameter is 0.6, consistent with the experimental findings.

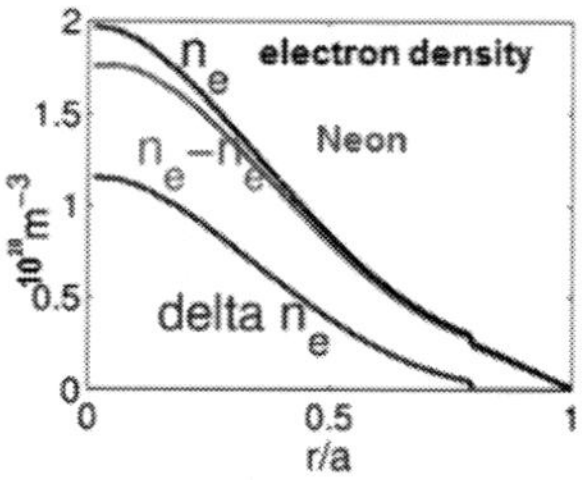

*Fig.6: black: total density; red: total density with subtracted Neon contribution; blue: total density increase after Neon injection.*

## Conclusions

The simulation of a Neon seeded FTU discharge, featuring strong peaking of the electron density, has shown that, though due to an inward pinch the Neon has a peaked profile, such peaking is not significantly greater than the density one. Indeed Neon is responsible only of a modest increase of the central density, since the dilution remains > 0.9, while in the time range considered, the on-axis electron density increases by a factor 2.5 in the simulation with an observed peaking factor that reaches 3.2.


[1] J. Ongena et al. 2001, Phys. Plasmas **8** 2188

[2] G.Telesca et al 2000 Nucl. Fusion **40** 1845

[3] C.Mazzotta 2014 proceeding at IAEA FEC 2014, 13-18 October, St Petersburg, Russia and accepted to NF

[4] G. Szepesi et al. 2013 Nucl. Fusion **53** 033007

[5] G.L.Jackson et al. 2002 PPCF **44**, 1893-1902

[6] M.E.Puiatti et al 2006 PoP **13**, 042501

[7] M. Romanelli et al. 2011 Nucl. Fusion **51** 103008

[8] M. Mattioli et al. 2001 J. Phys. B **34**, 127

# Experimental studies with the improved CTS diagnostics on FTU and evidences of scattered wave signals with short time-scale.

A. Bruschi[1], W. Bin[1], O. D'Arcangelo[2], E. Alessi[1], F. Belli[2], V. Cocilovo[2], L. Figini[1], C. Galperti[3], S. Garavaglia[1], G. Granucci[1], G. Grosso[1], S.B. Korsholm[4], M. Lontano[1], V. Mellera[1], A. Moro[1], S.K. Nielsen[4], G.Ramogida[2], U. Tartari[1]

[1] *Istituto di Fisica del Plasma - Consiglio Nazionale delle Ricerche, 20125 Milano, Italy*

[2] *ENEA for EUROfusion, via E. Fermi 45, 00044 Frascati (Roma), Italy*

[3] *Ecole Polytéchnique Fédérale de Lausanne, CRPP, CH-1015 Lausanne, Switzerland*

[4] *Technical University of Denmark, Department of Physics, DK-2800 Kgs Lyngby, Denmark*

The Collective Thomson Scattering (CTS) diagnostics is an important mean in ITER to determine the fast ions distribution function, inferred from the spectrum of the wave scattered around the frequency of a powerful mm-wave source. But the spectrum itself may be disturbed by some phenomena, as the recent observations [1, 2] possibly attributed to parametric decay instabilities (PDI) [3]. An EUROfusion Enabling Research project [4] aiming at investigating such phenomena in conditions close to those expected in ITER, i.e. at high density and with probe frequency lower than the electron cyclotron (EC) frequency in the plasma, is under implementation on the FTU tokamak. The investigation is done with the existing CTS diagnostics [5], now exploiting two lines of sight of the FTU EC launcher provided with mirrors movable in real-time [6, 7] and recently upgraded [8]. The experiments were performed at two toroidal magnetic fields, at 4.7 T (in which the fundamental EC resonance is on the high-field side) and in the ITER-like conditions at 7.2 T (with the fundamental EC layer located out of the plasma on the low-field side, actually in the equatorial port).

Using a symmetric arrangement of the probe and the receiving antennas, the scattering volume was usually located on the equatorial plane (Fig.1, left). Only in some special shots the volume was placed at the intersection of the q=2 surface with the EC resonance on the low-field side (Fig.1, center). The probe beam (500 kW/140 GHz, line 4 of the EC system) has been launched at a 5° toroidal angle in order to avoid direct reflection in the port. The receiving antenna was swept during the probe pulse with a toroidal forward movement from 1° to 9° and backward from 9° to 1°, in order to cross the probe beam and allow to discriminate the phenomena originated in the scattering volume. Different plasma conditions were tested, in particular in plasma shots with and without MHD activity.

**Common observations at $B_T$=7.2 T and $B_T$=4.7 T.**

a) Strong signals are seen at the end of the pulse, visible as lines at approximately 0.5 and 1 GHz from the gyrotron line (Fig.1, right): the possible explanation is in terms of gyrotron satellite lines at the end of the pulse (mode jumps).

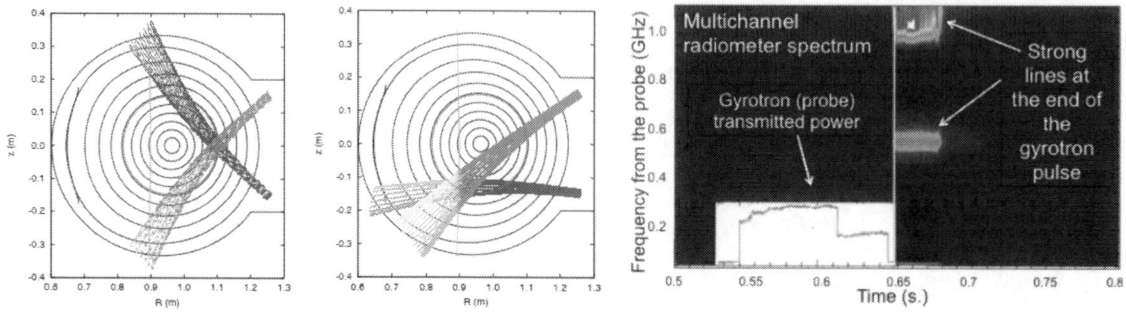


Fig.1: *Beam tracing at 4.7 T with trajectories crossing on the q=2 surface either on the equatorial plane, with symmetric launch (left) or on the resonance with asymmetric launch (center). Right: Multi-channel radiometer spectrum during probe pulse, showing strong emissions close to pulse end, possibly due to gyrotron mode jumps.*

b) Broad spectra are observed from 30 to 100 ms after the start of the gyrotron pulse (Fig.2). Possible explanation is the local breakdown that can be excited by the probe power at the location of the 1st (at 7.2 T) or 2nd (at 4.7 T) harmonic of the EC frequency, positioned close to the launching mirror. Breakdown plasma can propagate close to the receiving antenna.

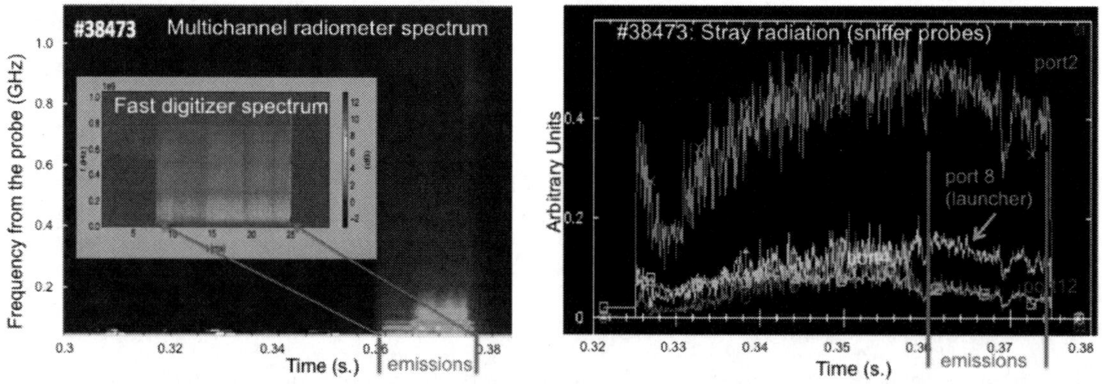


Fig.2: Left: Multi-channel radiometer and fast digitizer spectra during pulses. Right: Sniffer probes signals show a relatively higher stray radiation measured at the launcher port 8 in coincidence with intense spectral emissions.

**Observations at $B_T$=7.2 T:**

This is the case in which the fundamental EC resonance is out of plasma, as in the CTS of ITER. Signals appear as broad lines at frequencies rapidly changing in time, in the 0.6-1.1 GHz range away from the probe frequency, seen only with the time resolution allowed by the digitizer (Fig.3). The signals appear in shot with a very low MHD activity, as seen from the MHD probe (Mirnov coil) signals. The explanation for these signals is presently not clear.

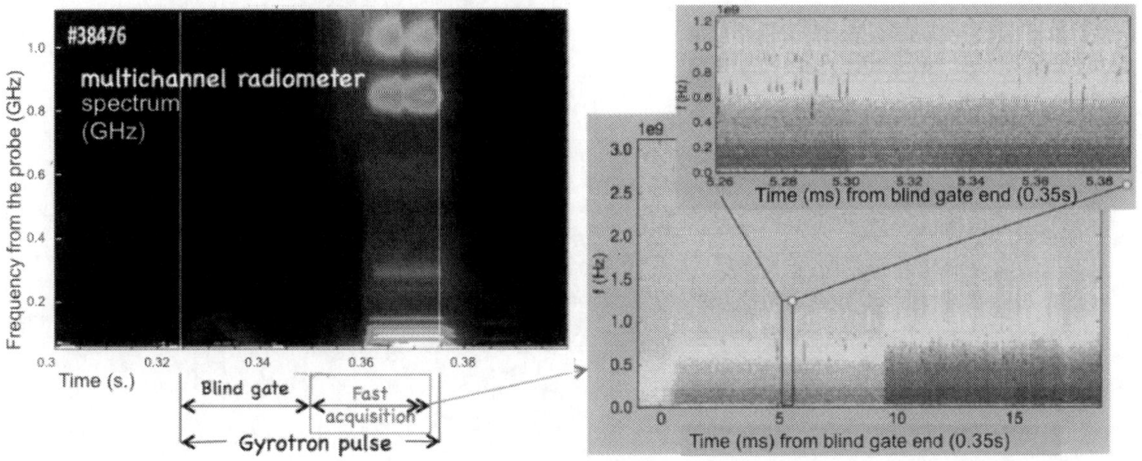


Fig.3: Received spectrum with the multi-channel radiometer (left) and with the fast digitizer (right): the increased resolution of the latter allows to see signals appearing as lines at frequencies rapidly changing in time.

**Observations at $B_T$=4.7 T (asymmetric trajectory)**:

In this case the 1st EC harmonic is in the plasma on the high field side, the scattering volume is at the intersection of between the resonance and an m:n=2:1 magnetic island. Burst of emissions are detected in particular when the MHD island is slowly rotating, and are apparently stronger when the beam trajectories overlap, at 5° toroidal angle (Fig.4).

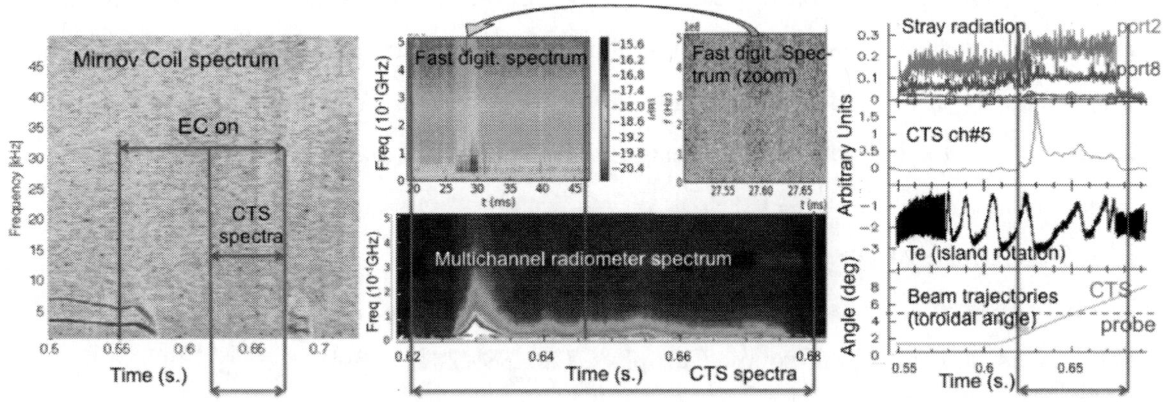


Fig.4: Left: Mirnov coil spectrum showing a MHD island with rotation speed slowing down to zero during pulse. Center: Spectra showing emissions that can be resolved as single bursts. Right: stray radiation correlates with ch#5 emission, ch#5 intensity correlates with beam overlapping, and bursts in ch#5 correlate with island rotation.

**Observations at $B_T$=4.7 T (symmetric trajectory)**:

In this case the EC resonance is in the plasma on the high field side, and the scattering volume is located where the q=2 surface crosses the equatorial plane on the low field side (see Fig.5). In this case magnetic islands are absent. The evidence is an emission in short bursts with spikes repeating at high frequencies, with no MHD activity seen by magnetic probes.

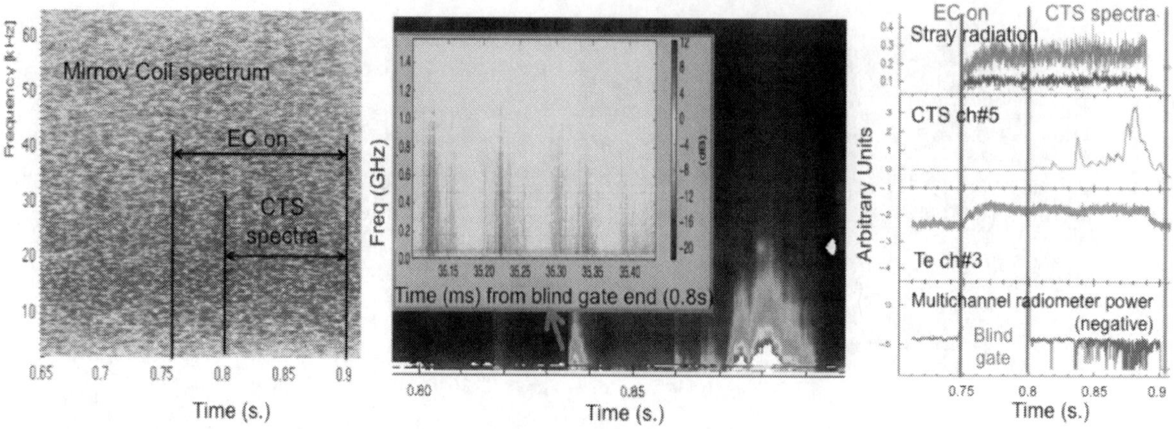


Fig.5: Left: No MHD activity is present during pulse. Center: radiometer (bottom) and digitizer (top left) spectra, showing emissions in bursts shorter than 10$\mu$s. Right: intensity is not correlated to stray radiation or temperature.

**Conclusions**

Anomalous signals were detected by the receiver in a frequency band of 1.2 GHz around the probe frequency, either as spectral lines or as continuum. The analysis of the data recorded with an ultra-fast digitizer allowed to detect rapid spectral changes with high frequency resolution. The MHD activity can only explain some instances of anomalous scattering while in other cases signals are not synchronous with magnetic island rotation or MHD is not relevant at all. Other spectral features may be attributed to plasma formation at the resonances close to the antennas during the gyrotron pulses, causing also gyrotron mode jumps possibly due to reflections. These results are particularly important and worth to be further investigated in relation to the design of the ITER CTS diagnostic system.


[1] E. Westerhof et al., Phys. Rev. Lett. 103 (2009) 125001

[2] S K Nielsen et al., Plasma Phys. Control. Fusion 55 (2013) 115003

[3] E.Z. Gusakov and A.Yu. Popov, Europhys. Lett. 99 (2012) 15001

[4] https://www2.euro-fusion.org/erwiki/index.php?title=ER15-ENEA-06

[5] U.Tartari et al, Nucl. Fusion 46 (2006) N.11, pp. 928–940

[6] A.Bruschi et al., Fusion Sci. Technol. Vol. 55 (2009) pp 94-107

[7] W.Bin et al., Fusion Eng. Des. Vol. 84 (2009) pp 451-456

[8] W. Bin, et al., Fusion Eng. Des. (2015), http://dx.doi.org/10.1016/j.fusengdes.2015.05.022



*This work has been carried out within the framework of the EUROfusion Consortium and has received funding from the Euratom research and training programme 2014-2018 under grant agreement No 633053. The views and opinions expressed herein do not necessarily reflect those of the European Commission.*

# Runaway Electron Dynamics and Critical Electric Field for Runaway Generation in the FTU Tokamak

Z. Popovic [1], B. Esposito [2], J. R. Martin-Solis [1], W. Bin [3], L. Boncagni [2], D. Carnevale [4], M. Gospodarczyk [4], D. Marocco [2], G. Ramogida [2] and M. Riva [2]

[1] *Universidad Carlos III de Madrid, Avda.Universidad 30, 28911-Madrid, Spain.*
[2] *ENEA for EUROfusion, via E. Fermi 45, 00044 Frascati (Roma), Italy.*
[3] *Istituto di Fisica del Plasma - CNR, Via R. Cozzi 53, 20125 Milano, Italy.*
[4] *Dip. di Ingegneria Civile ed Informatica DICII, Universita di Roma, Tor Vergata, Via del Politecnico 1, 00133 Roma, Italy.*

**1. Introduction** Large amounts of MeV runaway electrons (REs) generated during disruptions pose a significant danger for the operation of next-step devices like ITER. RE suppression by Massive Gas Injection (MGI) relies on increasing the electron density (and, hence, the electric field) above their critical values where no REs can be generated. Here, we report on experiments designed to evaluate the threshold electric field for runaway generation during the flat-top of ohmic (OH) discharges in the FTU tokamak. The measured threshold electric field is found to be significantly larger ($\sim 2-5$ times) than predicted by the classical collisional theory and can be explained to a great extent by the electron synchrotron radiation losses. Simulations of the RE dynamics (generation and energy evolution) during these experiments will be also presented.

**2. Experiments and data analysis** The relativistic collisional theory predicts that for runaway creation a threshold electric field $E_R = n_e\, e^3\, ln\Lambda/(4\pi\, \varepsilon_0^2\, m_e\, c^2)$ should be exceeded. In FTU, the results have been consistent with an increase of $E_R$ due to the electron synchrotron radiation losses [1]:

$$\frac{E_R^{rad}}{E_R} \cong 1 + C(Z_{eff})\, F_{gy}^{\alpha},$$

where $\alpha = 0.45 \pm 0.03$, $F_{gy} \equiv 2\varepsilon_0\, B_0^2/(3n_e\, ln\Lambda\, m_e)$, $C(Z_{eff}) \cong 1.64 + 0.53 Z_{eff} - 0.015 Z_{eff}^2$.

Here, the critical conditions have been tested during the flat-top phase of OH discharges in two experimental scenarios, as illustrated in Fig. 1: RE *onset* experiments (left, Fig. 1) in which the density is decreased until REs are generated and RE *suppression* (right, Fig. 1) of existing REs (created during a low density start-up) by gas injection. The presence of the REs is detected by the divergence of the $BF_3$ neutron detector and the NE213 neutron/gamma scintillator signals which allow to determine the critical time $t_c$ at which the REs appear/disappear in the onset/suppression experiments [2]. The threshold field is estimated as $E_{thr} \sim V_{loop}(t_c)/2\pi R_0$.

The database of analysed discharges covers a wide range of plasma parameters ($B_t = 2 - 7.2\,$T, $I_p = 0.35 - 0.9\,$MA and $Z_{eff} = 1.5 - 10$). The results (Fig. 2) are consistent with previous findings [1]: $E_{thr}$ is several times ($\sim 2-5$) larger than the relativistic

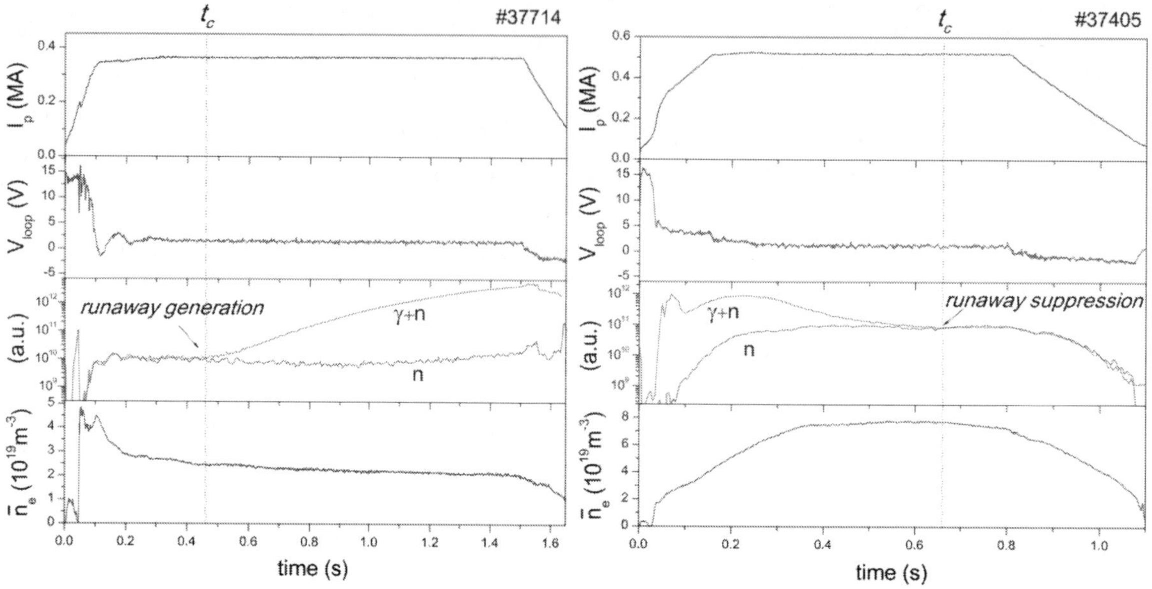


Figure 1: *Time traces of the plasma current ($I_p$), loop voltage ($V_{loop}$), NE213 ($\gamma+n$) and $BF_3$ ($n$) signals and line-averaged central density for representative discharges: RE onset (**left**) and RE suppression (**right**).*

collisional theory predictions, $E_R$ (Fig.2, left) and falls well within the range of critical fields predicted including the effect of the electron synchrotron radiation losses, $E_R^{rad}$ (Fig. 2, right). The local central density is used to calculate $E_R$ and $E_R^{rad}$ since REs are observed to be mostly generated in the centre of the plasma [1]. These findings are in agreement with a recent ITPA joint experiment in several tokamaks [3] and suggest that REs might be suppressed at lower densities than expected which could have important implications for RE mitigation in ITER.

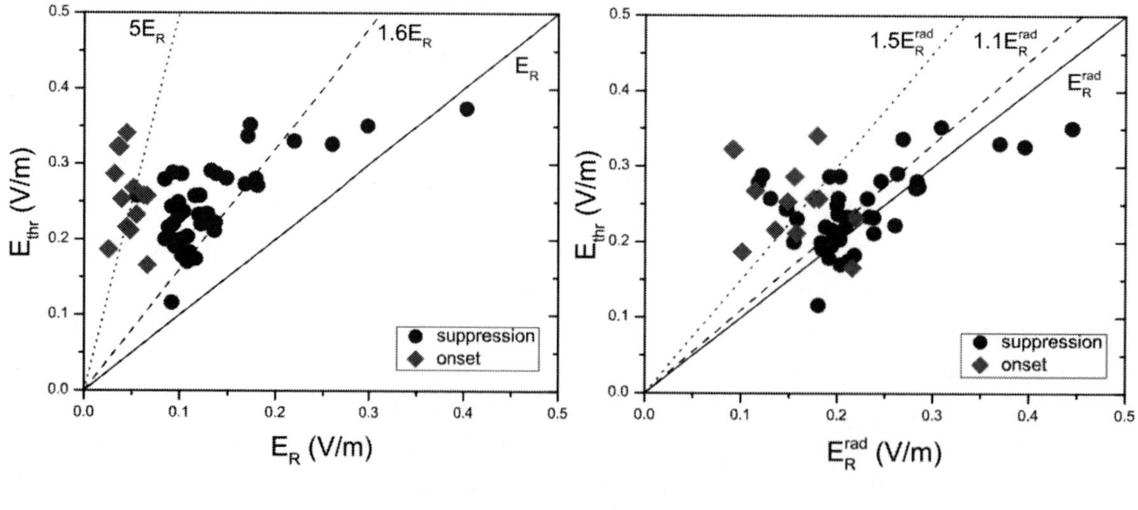


Figure 2: *Comparison of $E_{thr}$ with the relativistic collisional theory, $E_R$, (**left**) and synchrotron radiation theory $E_R^{rad}$ (**right**). Each point corresponds to a single discharge.*

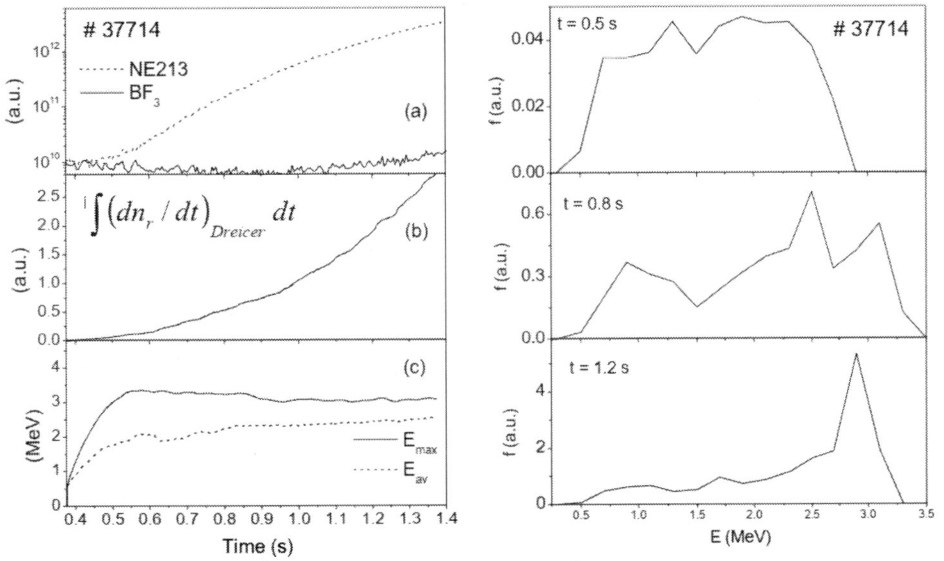


Figure 3: *For RE onset discharge # 37714:* ***Left:*** *Time evolution of $BF_3$ and NE213 scintillator signals (a), estimated runaway production (b), maximum and average runaway energies (c);* ***Right:*** *Calculated runaway distribution function at 0.5, 0.8 and 1.2 s.*

**3. Runaway Dynamics** The runaway dynamics for the RE onset and suppression experiments of Fig. 1 is illustrated in Figs. 3 and 4, respectively. For the case of the RE onset discharge (Fig. 3), the runaway production [trace (b)], calculated assuming to be dominated by the Dreicer mechanism, as typically observed in FTU OH discharges [3], is qualitatively consistent with the runaway measurements [traces picture (a)]. The runaway energy distribution function, $f(E)$, has been estimated calculating the energy evolution of the generated runaway electrons by means of a simple test particle model for the runaway dynamics [4] including the electric field acceleration, collisions with the plasma particles and deceleration due to synchrotron radiation losses. Then, the runaway distribution function is formally given by:

$$f(E,t) = \int_0^t \frac{\mathrm{d}n_r}{\mathrm{d}t'}(E_0,t')\,\mathrm{d}t', \tag{1}$$

where $t = 0$ denotes the start of the runaway generation and the integration is carried out over the times $t'$ for which an electron generated with energy $E_0$ would have gained, according to the test particle equations, an energy $E$ at time $t$. The resulting distribution function at three different times during the discharge is plotted in right Fig. 3, whereas the time evolution of the maximum and average electron energies ($E_{max}$ and $E_{av}$) is plotted in Fig. 3 (c). The distribution function is broad, extending up to the maximum runaway energy at each time step, and showing a trend along the discharge to accumulate close to the limiting energy $\sim 3\,\mathrm{MeV}$.

The RE suppression experiment is presented in Fig. 4. REs are generated in the beginning of the discharge (due to the large electric field and low density, Fig. 1) and the runaway production stops soon during the current ramp-up at $\sim 0.15\,\mathrm{s}$ [trace (b)].

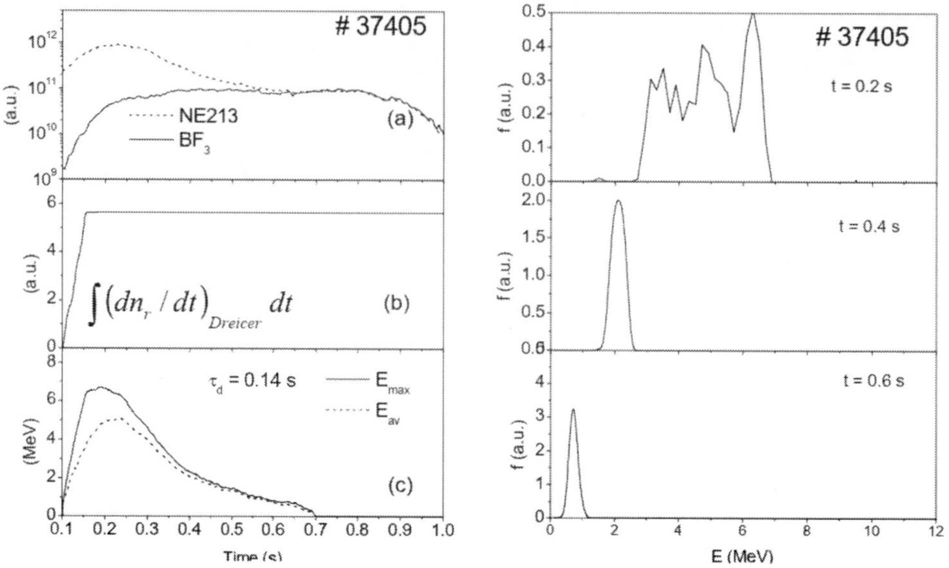


Figure 4: *For RE suppression discharge # 37405:* ***Left:*** *Time evolution of $BF_3$ and NE213 scintillator signals (a), estimated runaway production (b), maximum and average runaway energies (c);* ***Right:*** *Calculated distribution function at 0.2, 0.4 and 0.6 s.*

The runaway energy increases up to $\sim 7\,\text{MeV}$ [trace (c)] and decreases later on along the discharge until the REs are suppressed. As the runaway production stops early in the discharge ($\sim 0.15\,\text{s}$), the distribution function, initially broad (t = 0.2 s in right figure), evolves towards a mono-energetic runaway beam (t = 0.4 and 0.6 s in right Fig.4) with an average energy equal to the maximum runaway energy [Fig.4 (c)]. For most of the analysed RE suppression discharges, the electron radiation cannot account for all the runaway losses (typically $E_{thr} > E_R^{rad}$, as shown in Fig. 2) and additional energy dissipation mechanisms must be invoked to explain the observed RE suppression (in Fig. 4, a characteristic dissipation time $\tau_d \sim 0.14\,\text{s}$ has been assumed in the simulations to describe such unknown processes).

## Acknowledgements

This work was done under financial support from Dirección General de Investigación, Científica y Técnica, Project No.ENE2012-31753 and within the framework of the EUROfusion Consortium and has received funding from the Euratom research and training programme 2014-2018 under grant agreement No 633053. The views and opinions expressed herein do not necessarily reflect those of the European Commission.

## References


[1] B. Esposito et al., in Fusion Energy 2014 (Proc. 25th Int. Conf., St. Petersburg, 2014) (Vienna: IAEA) CD-ROM file EX/P2-50.

[2] B. Esposito et al., Phys. Plasmas **10** (2003) 2350.

[3] R. Granetz et al., Phys. Plasmas **21** (2014) 072506.

[4] J.R. Martin-Solis et al., Phys. Plasmas **5** (1998) 2370.

# Towards an Assessment of Alternative Divertor Solutions for DEMO

H. Reimerdes[1], L. Aho-Mantila[2], R. Albanese[3], R. Ambrosino[3], W. Arter[4], S. Brezinsek[5], H. Bufferand[6], G. Calabro[7], G. Ciraolo[6], D. Coster[8], H. Fernandes[9], J. Harrison[4], I. Kaldre[10], K. Lackner[8], O. Lielausis[10], J. Loureiro[9], T. Lunt[8], G. Mazzitelli[7], S. McIntosh[4], F. Militello[4], T. Morgan[11], N. Pelekasis[12], G. Pelka[13], V. Pericoli[7], V. Philipps[4], F. Subba[14], F. Tabares[15], B. Viola[7], R. Wenninger[8], R. Zagorski[13] and H. Zohm[8]

[1] *EPFL, Lausanne, Switzerland,* [2] *VTT, Finland,* [3] *Università di Napoli, Italy,* [4] *CCFE, Culham, UK,* [5] *Forschungszentrum Jülich, Germany,* [6] *CEA, St. Paul-lez-Durance, France,* [7] *ENEA-Frascati, Italy,* [8] *Max-Planck-Institut für Plasma Physik, Garching, Germany,* [9] *Universidade de Lisboa, IST, Portugal,* [10] *University of Latvia, Salaspils, Latvia,* [11] *FOM Institute DIFFER, Niewegein, The Netherlands,* [12] *University of Thessaly, Volos, Greece,* [13] *IPPLM, Warsaw, Poland,* [14] *Politecnico di Torino, Italy,* [15] *Ciemat, Madrid, Spain*

## Introduction

The European roadmap for fusion energy [1] has identified heat exhaust as a major challenge towards the realisation of magnetic confinement fusion. Since the scrape off layer (SOL) width is believed to be independent of the size of the device, the exhaust power increases faster than the area where the plasma interacts with the wall. A reactor must, therefore, harness an even greater heat flux than ITER. At the same time the higher particle and neutron fluence in a reactor imposes stronger constraints on the target materials and the admissible erosion. In the current baseline scenario, which foresees a single null magnetic configuration and divertor targets made out of tungsten (W) this must be achieved by maintaining a detached divertor and primarily increasing the radiation power. It is, however, uncertain whether a higher radiation fraction is compatible with the required energy confinement and whether transients can be sufficiently suppressed to maintain a detached divertor. In addition there are uncertainties in the predicted power decay length, which may modify the exhaust requirements. In order to mitigate the risk that the baseline scenario does not extrapolate from ITER to a DEMO reactor, the EUROfusion consortium is assessing alternative divertor solutions. The assessment focuses on solutions that have a high potential to be fully developed for a DEMO in the 2040s.

In the assessment it is first evaluated whether a proposed alternative could likely be implemented in a DEMO on the foreseen timescale. The assessment then seeks to quantify the potential benefits and costs of the alternatives compared to the baseline solution. The comparison with the baseline solution should decrease the effect of any systematic errors in the extrapolation from present day devices to a reactor. Considered alternatives can be divided into non-standard magnetic configurations, which may decrease the heat and particle flux to the divertor targets, and liquid metal target armor, which may increase their exhaust capabilities.

## Alternative magnetic configurations

The main geometric modification of the divertor are the increase of the flux expansion at the target, an increase of the major radius of the target and a reduction of the gradient of the poloidal field in the null point, associated with the *X divertor* (XD) [2], *Super-X divertor* (SXD) [3] and *Snowflake divertor* (SFD) [4] concepts, respectively. All configurations have at least partially

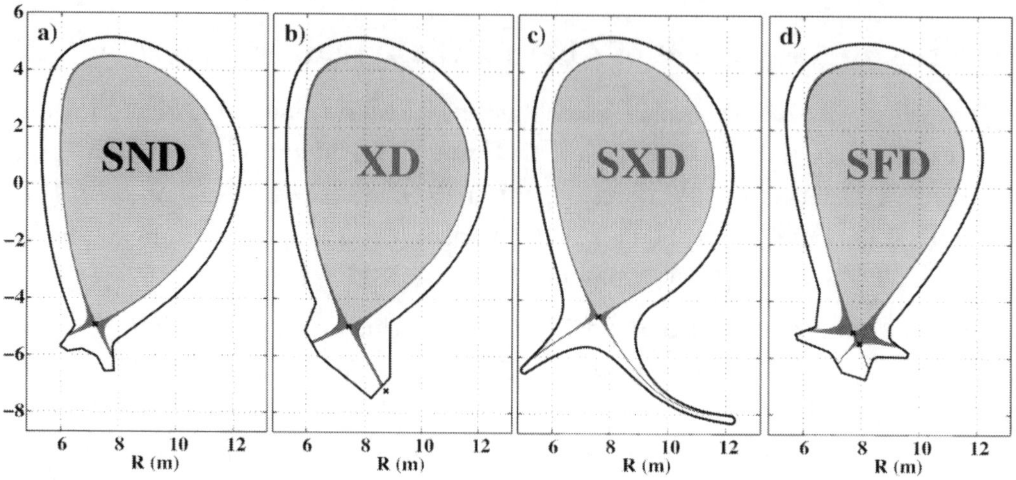


Figure 1: *(a) EU A=3.1 DEMO reference configuration (2014) and prototypes for the (b) X divertor, (c) Super-X divertor and (d) Snowflake divertor.*

been realised in experiments - in the case of the XD even before the concept was named and fully formulated. The SXD is presently only considered as a solution for the outer, usually heavier loaded, divertor leg.

The assessment is carried out in two steps. In the first step a DEMO size prototype of each alternative configuration is generated. The prototype configurations possess the key characteristics of the proposed alternatives, but are not yet optimised, Fig. 1. Core parameters and plasma shape are kept similar to the reference configuration, which has a single null divertor (SND) and an aspect ratio, $A = 3.1$, and is designed for an electric power output of $P_{\mathrm{elec}} = 500\,\mathrm{MW}$ (major radius, $R = 8.8\,\mathrm{m}$, toroidal field, $B_t = 5.8\,\mathrm{T}$, plasma current, $I_{\mathrm{P}} = 20\,\mathrm{MA}$). The axisymmetric targets are inclined for recycling neutrals to reflect towards the separatrix, also referred to as *closed* divertor configurations, but maintain a minimum grazing angle of magnetic field lines of $1.5°$. Coils are always placed to allow for at least 80 cm of neutron shielding. The XD and SXD prototypes only include an alternative solution for the outer divertor leg.

In the second step, which is presently carried out, the prototypes are optimised to reduce their costs and increase the potential benefit using a systematic approach [5]. Considered beneficial geometric quantities are the connection length, the SOL volume, a flaring of the flux surfaces towards the target quantified by the ratio of flux expansion at the target and the minimum flux expansion along the divertor leg and the ratio of the target radius and the null point radius. Considered costs are the total poloidal field coil currents (weighted with their major radius as a proxy for the required conductor volume), the forces on the poloidal field coils, the current of poloidal field coils located inside the toroidal field coils (weighted with their major radius) and the ratio of toroidal field coil volume and the plasma volume.

The extrapolation to DEMO must include a physics model as present day devices are not capable of testing alternative configurations under DEMO relevant divertor conditions. The performance of the alternative configurations is, therefore, evaluated using a set of boundary codes with various degrees of complexity. Technical challenges include the grid generation for the new SFD topologies and an increased run time of the codes due the size of the configurations.

Nevertheless, SFD grids were already generated for the EMC3-Eirene and SOLEDGE2D codes. Also the run time of the various codes has been manageable, but drifts were so far neglected and a kinetic descriptions of neutrals only used within EMC3-Eirene. In order to compare the ability of the proposed alternatives to meet the exhaust challenge several metrics have been proposed: 1. plasma density at the onset of detachment, 2. SOL impurity concentration for the required high divertor power loss and 3. the maximum divertor power loss before the loss of stability.

While various aspects of the selected models have been validated within the fusion community, all of the models include only a diffusive cross-field transport with empirical diffusion coefficient. It is well known that convective transport is important, in particular, in the high density detached operating regimes [6]. It is, furthermore, speculated that the low poloidal field in the SFD increases turbulent cross-field transport. However, models of turbulent transport are presently not capable of addressing novel magnetic configurations and the assessment can, therefore, not be comprehensive.

**Liquid metal plasma facing components**

A variety of liquid metal target concepts are proposed. They can be distinguished by their principle heat removal mechanism, which can be conduction, convection or evaporation. While concepts that are based on convection or evaporation promise a superior heat removal capability, they also face additional challenges such as the stability of the flowing layer and particle exhaust. The assessment of liquid metals, therefore, focuses on static liquids, whose feasibility is a prerequisite for more advanced concepts. A static liquid can be stabilised in capillary porous system (CPS), made out of a solid material [7]. Considered metals are lithium (Li), due to its compatibility with good plasma performance and tin (Sn), due to its high evaporation temperature. In addition Li/Sn alloys are also investigated as they promise the advantageous properties of both components. Before weighing benefits against costs, the physics and technological feasibility of a liquid metal based concept has to be established.

The heat removal capability is greatly determined by the temperature difference between the target surface and the coolant. The surface temperature is limited by material erosion, which is enhanced at higher temperature. In addition to temperature the effective erosion also depends on the surface texture and the plasma pressure. Lastly prompt redeposition due to the applied magnetic field can significantly reduce the effective erosion rate. The heat removal capability also depends on the thickness of the structure. Since liquid metals self heal eroded material, the thickness of the armor can be reduced over a solid target. However, structural components and coolant pipes must remain similar. The capability to self heal increases not only the lifetime of the divertor targets, but also opens the possibility to tolerate transients that cause some erosion provided that the substrate is not exposed. The tolerance to transient depends on the effectiveness of vapour shielding and on the film thickness, which in turn is limited by the surface stability.

A modelling effort has begun to predict the wetting and temperature evolution of a CPS under DEMO-relevant heat and particle loads. The tolerable increase of the impurity concentration in the SOL is modelled with transport codes with adapted boundary conditions.

With the predictive modelling being limited to various aspects of the CPS based liquid metal solution, the technical feasibility of stationary power and particle exhaust has to be demon-

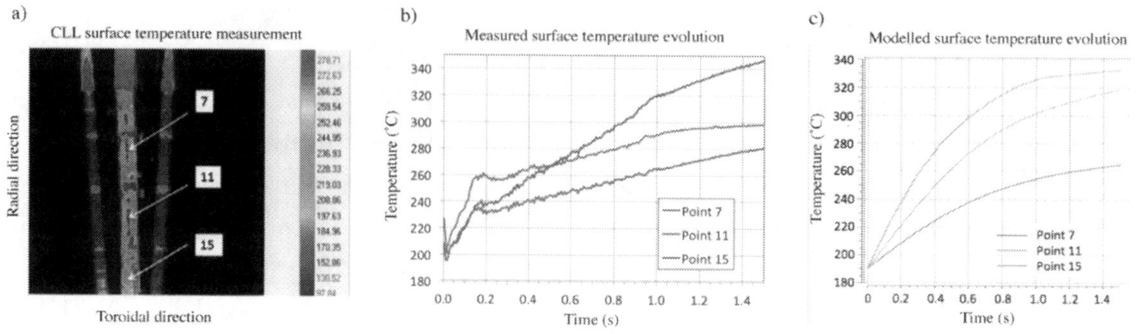


Figure 2: *(a) IR measurements of the CLL surface temperature and (b) its evolution are compared with (c) the surface temperature evolution modelled using ANSYS.*

strated in experiments. Effects are investigated separately in laboratory experiments and linear plasma devices as well as combined in the FTU tokamak, where experiments with a recently installed cooled liquid Li limiter (CLL) have commenced [8]. Upon conclusion of the CLL experiments it is planned to replace the CLL with a cooled liquid Sn limiter. FTU experiments have shown stability of the CPS system under heat flux of approximately 2 MW/m$^2$ in a 1.5 s discharge, but the target temperature and, hence, the heat removal is not stationary, Fig. 2. Present efforts, therefore, aim at an extension of the FTU pulse length to 5 s, which should suffice to demonstrate stationary power exhaust and evaluate the effect on the plasma boundary.

## Conclusion

The EUROfusion consortium is evaluating unconventional power and particle exhaust concepts as an alternative for their DEMO reactor. The assessment has identified the X divertor, Super-X divertor and snowflake divertor configurations as potential alternatives. In addition it is considering a liquid Li, Sn or Li/Sn alloy target based on a capillary porous system. Feasibility, costs and benefits of the alternatives are evaluated and compared to the conventional divertor solution. The assessment result is expected to inform the decision on a European divertor test tokamak facility.

## Acknowledgements

This work has been carried out within the framework of the EUROfusion Consortium and has received funding from the Euratom research and training programme 2014-2018 under grant agreement No 633053. The views and opinions expressed herein do not necessarily reflect those of the European Commission.

## References

[1] "Fusion Electricity - A roadmap to the realisation of fusion energy", F. Romanelli, et al., EFDA, 2012.
[2] M. Kotschenreuther, et al., *Phys. Plasmas* **14** (2007) 72502.
[3] P.M. Valanju, et al., *Phys. Plasmas* **16** (2009) 056110.
[4] D.D. Ryutov, *Phys. Plasmas* **14** (2007) 064502.
[5] R. Albanese, R. Ambrosino, M. Mattei, *Plasma Phys. Control. Fusion* **56** (2014) 035008.
[6] H.W. Müller, et al., *J. Nucl. Mater.* **463** (2015) 739.
[7] L.G. Golubchikov, et al., *J. Nucl. Mater.* **233** (1996) 667.
[8] G. Mazzitelli, et al., *J. Nucl. Mater.* **463** (2015) 1152.

# Impact of localized ECRH on Alfven Eigenmodes in the ASDEX Upgrade tokamak

M. Garcia-Munoz[1,2], M. A. Van Zeeland[3], S. Sharapov[4], I. G. J. Classen[5], B. Bobkov[2], J. Galdon-Quiroga[1], B. Geiger[2], V. Igochine[2], P. Lauber[2], N. Lazanyi[6], F. Nabais[7], V. Nikoleva[2], D. C. Pace[3], M. Rodriguez-Ramos[1], L. Sanchis-Sanchez[1], M. Schneller[2], A. Snicker[8], J. Stober[2] and the ASDEX Upgrade Team

[1] *FAMN Department, Faculty of Physics, University of Seville, Seville, Spain*

[2] *Max Planck Institut fur Plasmaphysik, Garching, Germany*

[3] *General Atomics, PO Box 85608, San Diego, CA 92186-5608, USA*

[4] *Culham Center for Fusion Energy, Culham Science Center, Abingdon, Oxfordshire, UK*

[5] *FOM-Institute DIFFER, Nieuwegein, The Netherlands*

[6] *BME NTI, Budapest, Hungary*

[7] *Instituto de Plasmas e Fusão Nuclear, Instituto Superior Técnico, Portugal*

[8] *Aalto University, Espoo, Finland*

## Introduction

Future burning plasma experiments such as ITER may be subject to the excitation of Alfven eigenmode (AE) instabilities by 3.5MeV fusion born alpha particles as well as fast-ions created by auxiliary heating systems. If allowed to grow unabated, these instabilities have the potential to cause fast-ion redistribution and loss leading to a degradation of the performance, heating and current drive efficiencies as well as to possible serious damage of first wall components. Of special importance to magnetically confined fusion plasmas are toroidal Alfven eigenmodes (TAEs) and reverse shear Alfven eigenmodes (RSAEs) because of their potential to eject fast-ions before their thermalization.

External actuators are commonly used to mitigate or even suppress magnetohydrodynamic (MHD) fluctuations of diverse nature. Among others, externally applied resonant magnetic perturbations (RMPs) and localised Electron Cyclotron Resonant Heating (ECRH) are the most extended methods to mitigate some of the most deleterious MHD perturbations in present tokamaks, i.e. Edge Localised Modes (ELMs) and Neoclassical Tearing Modes

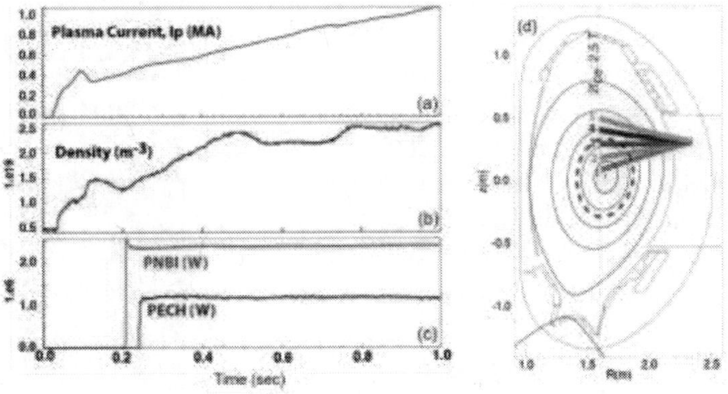


Fig.1. Overview of the main plasma and heating parameters in AUG discharges for NBI driven AE mitigation. Temporal evolution of (a) plasma current, (b) core line integrated density and (c) external heating. (d) ECRH deposition volumes.

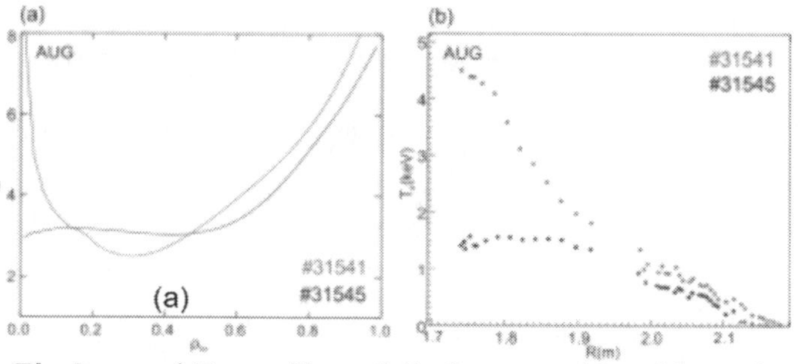


Fig.2. q and $T_e$ profiles of discharges with mitigated (black) and non-mitigated (red) AEs at t=0.4 sec.

(NTMs) respectively. The successful mitigation of ELMs and NTMs though RMPs and localised ECRH has, however, a limited operational window and sometimes important side effects on some plasma parameters that can affect the resulting plasma confinement, e.g. density pump-out and fast-ion losses caused by RMPs in ELM mitigated discharges [1]. A throughout understanding of the physics mechanism underlying the effect that external actuators have on plasma stability and eventually on additional transport is thus mandatory.

Experiments in DIII-D have shown the dramatic effect that localized ECRH can have on neutral beam driven Alfvén Eigenmode (AEs) activity in reversed magnetic shear plasmas [2]. The most commonly observed effect is a shift in the dominant observed modes from a mix of reversed shear Alfvén eigenmodes (RSAEs) and toroidicity induced Alfvén eigenmodes (TAEs) to a spectrum of weaker TAEs when ECRH is deposited near the shear reversal point ($q_{min}$). ECRH deposition near the magnetic axis typically increases the unstable mode amplitudes and resultant fast ion transport. It is found that for many of the intervals with minimal RSAE activity, or RSAEs with a much reduced frequency sweep range, that the geodesic acoustic mode (GAM) frequency at $q_{min}$ is very near or above the nominal TAE frequency - suggesting the so-called beta suppression mechanism [3] might explain some of the experimental observations. Although it has been observed that localised ECRH has a clear impact on AE activity not only in tokamaks but also in stellerators, a robust explanation of the experimental findings is still missing.

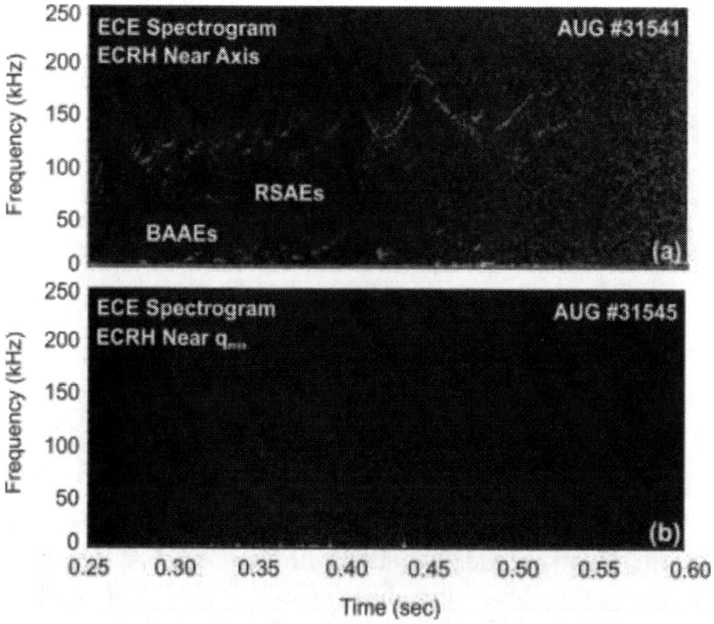


Fig.3. ECE spectrograms showing AE activity near $q_{min}$ in (a) #31541 and (b) 31545 AUG discharges.

The ASDEX Upgrade (AUG) tokamak is especially well suited to test the validity of this AE mitigation method due to its powerful and versatile heating mechanisms as well as internal fluctuation and fast-ion diagnostics. Recent experiments have been carried out to study the impact that localised ECRH has on AEs driven by fast-ions of Neutral Beam Injection (NBI) and Ion Cyclotron Resonant Heating (ICRH) origin. RSAEs driven by sub-alfvenic NBI ions have been successfully mitigated applying localised ECRH near $q_{min}$ (RSAE location) though almost no effect is observed on AEs driven unstable by supra-alfvenic ICRH ions.

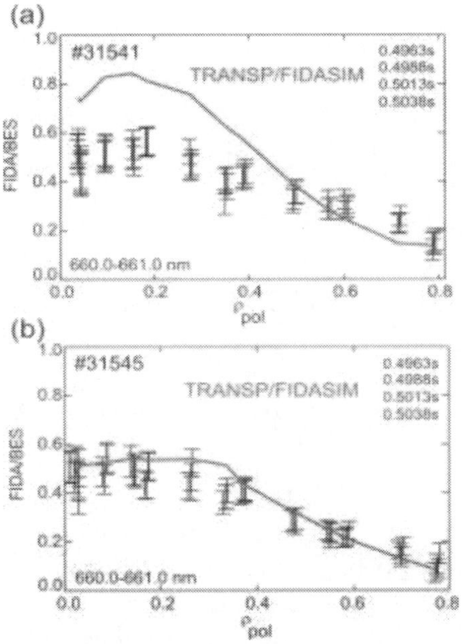


Fig.4. FIDA profiles measured in (a) a discharge with Alfvenic activity, #31541, and (b) without Alfvenic activity #31545.

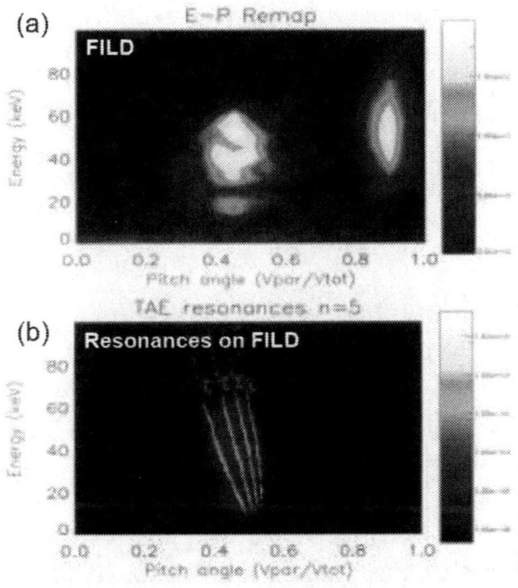


Fig.5. Velocity-space of fast-ion losses measured with FILD in a discharge with NBI driven AEs. (a) Remapped scintillator plate on orthogonal energy – pitch-angle coordinates. (b) Calculated resonances between escaping ions and n=5 TAE.

**ECRH impact on NBI driven AEs.**

NBI driven Alfvén Eigenmodes (AEs) are routinely obtained in the ASDEX Upgrade (AUG) tokamak during the current ramp-up phase with early NBI heating (starting at around 0.2 sec) and reversed q-profile. ECRH is typically applied tens of ms after the starting of the NB to maintain a reversed q-profile as long as possible and affect the observed AE activity. Fig.1-(a) shows the most relevant plasma and heating parameters in a typical NBI driven AE mitigation discharge. The ECRH deposition location is scanned in a shot-to-shot basis as shown in Fig.1-(b). Most commonly observed AEs comprise Reversed Shear Alfven Eigenmodes (RSAEs) and Toroidal Alfven Eigenmodes (TAEs). The q- and $T_e$-profiles of the studied AE discharges are presented in Fig.2. Fig.3-(a) shows an ECE spectrogram with the observed RSAE activity near $q_{min}$. At low frequencies, Beta Induced Alfven-Acoustic Eigenmodes (BAAEs) are also observed.For ECRH injection near $q_{min}$, (RSAEs location) the overall mode activity is significantly reduced. While in DIII-D, the unstable modes shift from strong RSAEs to weaker global TAEs and reminiscent RSAEs are visible only at their highest frequencies during their transition to TAEs [1], in AUG, almost all mode activity disappears with ECRH injection near $q_{min}$, see Fig.3-(b). With reduced or no mode activity, the fast-ion confinement is significantly improved, in fact, the fast-ion profile measured with Fast-Ion D-Alpha (FIDA) spectroscopy matches classical TRANSP predictions. Fig.4 shows the fast-ion profiles measured in a discharge with AE activity, Fig.4-(a) and in a discharge without AE activity Fig.4-(b). In agreement with this observation, and with classical fast-ion profiles, e.g. discharge #31545, no fast-ion losses induced by AEs are observed by the Fast-Ion Loss Detectors (FILD) systems. In contrast to this, clear convective losses are measured in discharges with NBI driven AE activity.

Fig.5-(a) shows the velocity-space of the escaping ions measured during the AE activity in a typical AUG discharge. Orit simulations show that only the lost ions with pitch angles between 0.3 and 0.6 can fulfil the AE resonance conditions, Fig.5-(b) show the possible resonances for a n=5 TAE.

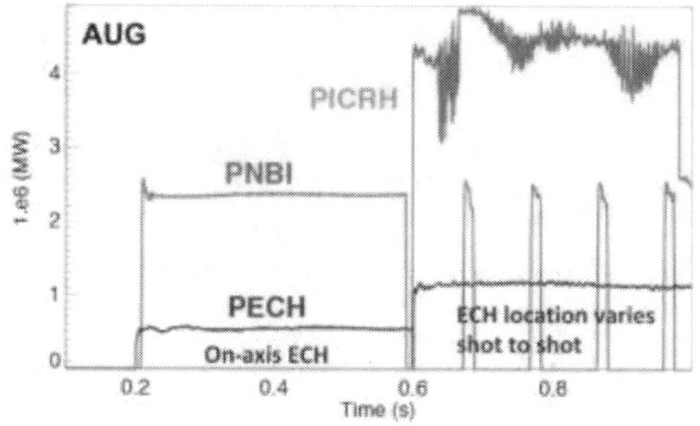


Fig.6. Heating timing / power in AUG discharges for ICRH driven AE mitigation.

**ECRH impact on ICRH driven AEs.** Similar experiments have been carried out to study the ECRH impact on AEs driven unstable by supra-alfvenic ICRH ions. Early on-axis ECRH and NBI is applied to obtain a high and reversed q-profile. ICRH heating is applied at ~0.6 sec with some NBI blips for diagnostics purposes and an ECRH deposition location scan in a shot-to-shot basis. Fig.6. shows the heating timing and power used in these discharges. ICRH driven AEs do not exhibit the same reduction in RSAE activity with ECRH near $q_{min}$. In contrast to this, no RSAEs are observed with ECRH on-axis while in discharges with ECRH applied near $q_{min}$ a rich pattern of RSAEs and TAEs driven by ICRH ions is clearly visible. Fig.7 shows the AE activity observed near $q_{min}$ in ECE fluctuations with ECRH near magnetic axis Fig.7-(a) and ECRH deposition near $q_{min}$, Fig.7-(b) and –(c). In all cases clear convective losses due to AEs are observed by the FILD systems. Future sensitivity experiments and modelling will focus on understanding this difference and the overall impact that the slightly ECRH modified q-profile, $T_e$ and fast-ion pressure may have on the observed AE activity as well as other possible mitigation mechanisms.

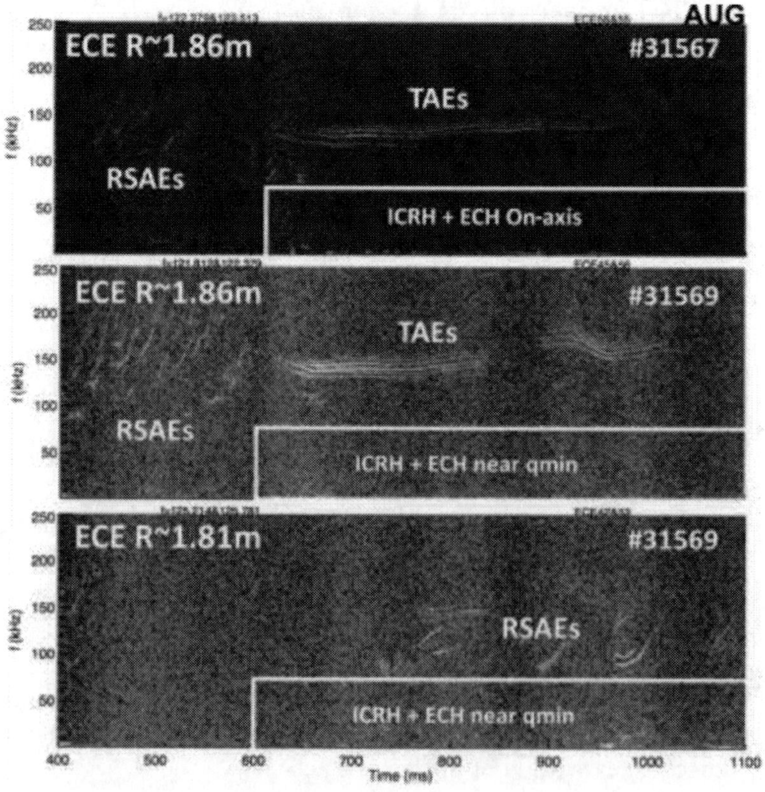


Fig.7. ECE spectrograms showing the ICRH driven AE activity with ECRH on-axis (a), and near $q_{min}$ (b) and (c).

**References**

[1] M. Garcia-Munoz et al., Plasma Phys. Control. Fusion **55,** 124014 (2013)

[2] M.A. Van Zeeland, et. al., Plasma Phys. Control. Fusion **50,** 035009 (2008)

[3] E. Fredrickson et al., Phys. Plasmas **14**, 102510 (2007)

*This work has been carried out within the framework of the EUROfusion Consortium and has received funding from the Euratom research and training programme 2014-2018 under grant agreement No 633053. The views and opinions expressed herein do not necessarily reflect those of the European Commission.*

# Coupled core/SOL modelling of fuelling requirements during the current ramp-up of ITER L-mode plasmas

P Belo[1], M Romanelli[1], V Parail[1], G. Corrigan[1], D Harting[1], L Garzotti[1], F Koechl[2], E. Militello-Asp[1], M Mattei[3], R Ambrosino[3], A Loarte[4], A S Kukushkin[4,*], R Sartori[5], M Cavinato[5]

[1]*CCFE, Culham Science Centre, Abingdon, OX14 3DB, UK*

[2]*Technical University Wien, 1020 Vienna, Austria*

[3]*CREATE, University of Naples "Federico II", Italy*

[4]*ITER Organization, Route de Vinon sur Verdon, 13067 St Paul Lez Durance, France*

[5]*Fusion For Energy Joint Undertaking, Josep Pla 2, 08019, Barcelona, Spain*

An initial assessment of fuelling of ITER L-mode plasmas indicated that it could be difficult to reach sufficient density for unrestricted use of neutral beam injection (NBI; 16.5 MW per injector; ion energy 1 MeV) by gas puff alone [1]. That assessment was carried out with the integrated Core/SOL suite of codes JINTRAC [2] by performing a gas rate scan, incrementally increasing the gas rate until the divertor plasma detached. The highest core line average density ($n_{eav}$) obtained for 15 MA / 5.3 T plasmas with 20 MW of Electron Cyclotron Heating (ECRH) marginally exceeded the density required for unrestricted use of NBI heating with acceptable shine-through loads on ITER's first wall $n_{eNBI} \sim$ 2.5-3.0 $10^{19}$ $m^{-3}$. For lower plasma currents the maximum density achieved by gas puffing was found to be lower than $n_{eNBI}$, which means that NBI modulation or operation at lower NBI energy would be required for acceptable shine-through loads. In this paper we therefore concentrate on the fuelling of lower current ITER H-modes as one option presently foreseen is to access the H-mode in the Q = 10 scenario in the ramp-up phase at a current of 10 MA which will require the use of NBI injection. Regarding SOL modelling assumptions, we have used the optimized EDGE2D/EIRENE parameter settings obtained from a dedicated benchmark between SOLPS and EDGE2D [2]. The perpendicular transport coefficients in the SOL are described by a tanh-fit decay from the core values to the asymptotic values of $\chi_{i,e}$=1.0$m^2$/s, $D_\perp$=0.3$m^2$/s in the far SOL. In the core, the plasma was modelling with JETTO with anomalous transport being described by the empirical Bohm/gryo-Bohm (BgB) model transport found to describe L-mode plasmas satisfactorily [3]. The ECRH power deposition was described by a Gaussian profile with its maximum at normalised $\rho_{tor}$ = 0.2, $\Delta\rho_{tor}$ = 0.26 and total of 10 MW.

We have carried out a series of L-mode simulations with different gas puffing rates from 1.0 $10^{21}$ $s^{-1}$ to 1.3 $10^{22}$ $s^{-1}$; at the highest fuelling rate the divertor plasma was found to detach with electron temperatures, $T_e$ < 1 eV for in a region 50 cm wide upwards of the separatrix strike point corresponding to 5 cm from the separatrix

[*] Present addresses : NRC "Kurchatov Institute", Kurchatov sq. 1, 123098 Moscow, Russia; NRNU MEPhI, Kashirskoye sh. 31, 115409 Moscow, Russia

mapped to the outer midplane. Applying higher gas puffing rates leads to density instabilities associated with oscillatory behaviour of plasma detachment in the simulations. In order to ensure that the confined plasma has reached steady state conditions the simulations were carried out for 10s of real plasma time until the core density (and temperature) profiles reached stationary conditions. The maximum plasma density achieved in these gas fuelled simulations for 10 MA plasmas in ITER is 1.8 $10^{19}$ $m^{-3}$, which is noticeably lower than $n_{eNBI}$.

As the plasma density achievable by gas fuelling is limited by divertor detachment, an additional heating scan was carried out by increasing the power to 20 and 40 MW with the highest fuelling rate explored at 10 MW (1.3 $10^{22}$ $s^{-1}$). This leads to an increase of the $n_{eav}$ to 2.0 $10^{19}$ $m^{-3}$ and 2.6 $10^{19}$ $m^{-3}$, respectively (Fig. 1). The increase of the main plasma density is caused by the increase of the separatrix density which is possible due to the re-attachment of the divertor plasma as the power entering the SOL increases. It is important to note that, as described in [1], neutral penetration into the confined plasma is very inefficient so that there is a direct relation between the separatrix density $n_{sep}$ and the main plasma density $n_{eav}$ which depends on core plasma physics and not edge neutral ionization. These results show that 40 MW of additional heating are required to achieve a density by gas fuelling which is marginally sufficient for unrestricted heating with NBI. Although this level of RF heating is in principle possible in ITER (20 MW ECRH and 20 MW of ICRH are available), additional heating at that power level is already in excess of that required for the plasma to enter the H-mode in these conditions ($P_{L-H}$ = 25 MW), from the scaling in [4]. Access of H-mode with RF heating only can pose complications due to the change of coupling efficiency of the ICRH heating due to changes in the edge plasma parameters due to the L-H transition itself.

A more robust alternative for ITER is to achieve higher density in L-mode by pellet fuelling so that the H-mode transition can be triggered by the application of NBI. The pellets in our modelling are assumed to have a 50-50 DT mixed and they enter the plasma by the high field side with an injection velocity of 300 m/s. The deposition of the pellet mass in the plasma is modelled with the neutral gas and plasma shield (NGPS) model [5]. To evaluate the impact of pellet size on the core density profile and on the divertor parameters a scan of the pellet size has been carried out at a constant fuelling rate of ~ 1.5 $10^{21}$ $s^{-1}$ by adjusting the pellet size to 2.5 $10^{20}$, 8.4 $10^{20}$, 1.4 $10^{21}$ and 3.0 $10^{21}$ DT particles per pellet. In addition, gas puffing with a rate of 5.0 $10^{21}$ $s^{-1}$ was applied together with a power of 10 MW by ECRH. Fig. 2 shows the result of this scan demonstrating final density achieved in the plasma is independent of the pellet size. It is important to note that because the modelled plasma is in L-mode with low edge temperatures the ablation of the pellet occurs relatively deep in the plasma (peak of pellet ablation at normalized toroidal $\rho_{toroidal} = 0.75$). On the contrary, the perturbation caused by the pellets to the divertor plasmas increases strongly with pellets size so that for the largest pellet in the scan the divertor plasma is fully detached after the injection of the pellet.

Due to a lack of a solid physics basis for SOL transport that can be extrapolated to ITER, we have performed a scan of SOL transport assumptions to determine how the L-mode fuelling results depend on them. Simulations with EDGE2D/EIRENE stand-alone of JET L-mode plasmas in the high recycling regime required transport coefficients, $\chi_{i,e}$=0.5m$^2$/s, $D_\perp$=0.15m$^2$/s, in order to match the density and temperature profiles at the outer mid plane separatrix and at the outer divertor target [6,7]. Based on this and on the observations from other tokamaks, the SOL transport coefficients were varied within a wide range of assumptions including : a) a tanh decay of the core values towards the far SOL asymptotic values $\chi_{i,e}$=1.0m$^2$/s, $D_\perp$=0.3m$^2$/s, b) constant transport coefficients in the SOL with the following coefficients: $\chi_{i,e}$=1.0m$^2$/s - $D_\perp$=0.3m$^2$/s; $\chi_{i,e}$=0.5m$^2$/s - $D_\perp$=0.15m$^2$/s; and $\chi_{i,e}$=0.25m$^2$/s - $D_\perp$=0.07m$^2$/s. In all the simulations pellets of a size 8.4 $10^{20}$ particles where injected with a frequency of 3.3 Hz (2.8 $10^{21}$ s$^{-1}$ pellet fuelling rate) in addition to gas puffing with a rate of 5.0 $10^{21}$ s$^{-1}$. As shown in Fig. 3, higher SOL transport leads to a decrease of $n_{eav}$ through the reduction in separatrix density for a given fuelling rate. This is in agreement with previous simulations performed with gas fuelling only and is reproduced in the density of the plasma achieved before pellet injection starts in Fig. 3 (a gas fuelling rate of $10^{22}$ s$^{-1}$ was applied before the start of pellet fuelling in these simulations). As shown in Fig. 4, the use of pellet fuelling breaks the link between separatrix density and line average density found in gas puffing by the direct deposition of particles in the plasma leading to higher densities being reached with pellets than with gas puffing. Here it should be noted that decreasing SOL transport increases the peak divertor heat flux and thus the separatrix density that can be achieved before the divertor detaches and thus allows a higher density to be reached by gas puffing or pellet fuelling. For gas puffing $n_{eav}$ can only reach or exceed $n_{eNBI}$ for the case with lowest SOL transport while for pellet fuelling this value is reached/exceeded for wider range of SOL transport assumptions. The density that can be reached by pellet fuelling can be further increased by reducing gas fuelling (from 5.0 $10^{21}$s$^{-1}$ to 1.0 $10^{21}$s$^{-1}$) and increasing pellet fuelling (8.4 $10^{20}$ to 2.0 $10^{21}$ particles per pellet injected with frequency of 3.3 Hz). This allows $n_{eav}$ to reaching $n_{eNBI}$ even for the most pessimistic SOL transport assumption (see Fig. 5).

*This work was funded jointly by the RCUK Energy Programme and F4E Grant 502 (under ITER Task Agreement C19TD51FE). The views and opinions expressed herein do not necessarily reflect those of the ITER Organization or those of Fusion for Energy.*

*[1] Kukushkin, A.S., et al, Nucl. Fusion* ***45*** *(2005) 608, [2] Romanelli, M., et al., Nucl. Fusion, to be published, [3] Erba, M., et al., Plasma Phys. Control. Fusion 37 (1995) 1249, [4] Martin, Y.R., et al ,J.*

*Phys.: Conf. Ser.* ***123*** *(2008) 012033, [4] Garzotti L., et al, Nucl. Fusion* ***37*** *(1997) 1167, [5] Groth, M., et al, 38$^{th}$ EPS, Strasbourg, France 2011, [6] Wiesen, S., et al, 19$^{th}$ PSI San Diego, USA, 2010.*

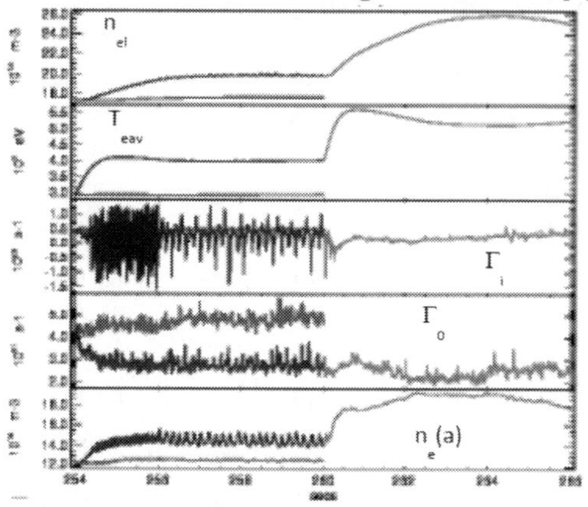


Figure 1: a) Line average core density; b) average electron temperature; c) main ion flux; d) neutral main ion flux and electron density at the separatrix time traces of the simulations with 1.3 $10^{22}$ $s^{-1}$ gas rate and different input ECRH power of ECRH: 10 MW; 20 MW; 40 MW

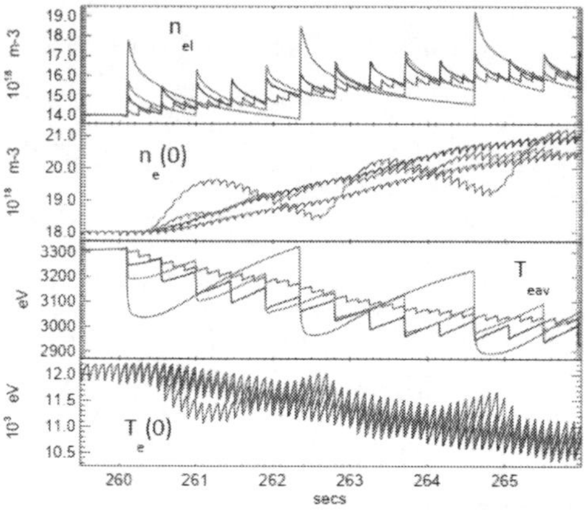


Figure 2: a) Line average core density; b) core axis electron density; c) average electron temperature; d) axis electron temperature time traces of the simulations with 5.0 $10^{21}$ $s^{-1}$ gas rate and different pellet size and frequency: 2.5 $10^{20}$ part. 6.7 Hz, 8.4 $10^{20}$ part. 2.2 Hz; 1.4 $10^{21}$ part. 1.1 Hz; 2.3.0 $10^{21}$ part. 0.44 Hz.

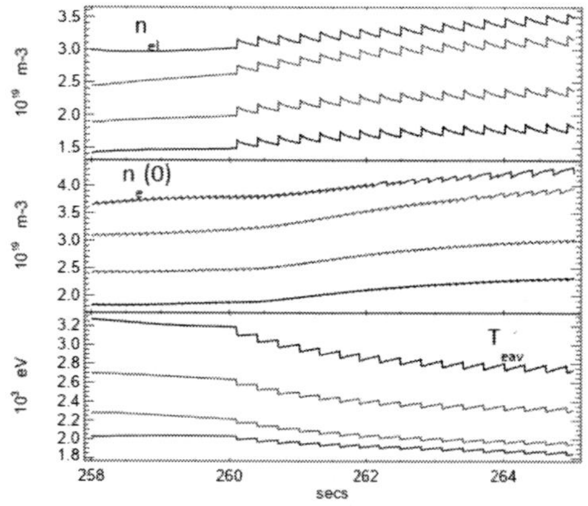


Figure 3: a) Line average core density; b) core axis electron density; c) average electron temperature; time traces of the simulations with 5.0e21 p/s gas rate, 1.5 mm pellet size 10/3 Hz pellet frequency and different SOL perpendicular transport coefficients: Tanh decay to $\chi_{i,e}$ = 1.0 $m^2/s$, $D_\perp$ = 0.3 $m^2/s$; constant $\chi_{i,e}$ = 1.0 $m^2/s$, $D_\perp$ = 0.3 $m^2/s$; $\chi_{i,e}$ = 0.5 $m^2/s$, $D_\perp$ = 0.15 $m^2/s$; $\chi_{i,e}$ = 0.25 $m^2/s$, $D_\perp$ = 0.07 $m^2/s$

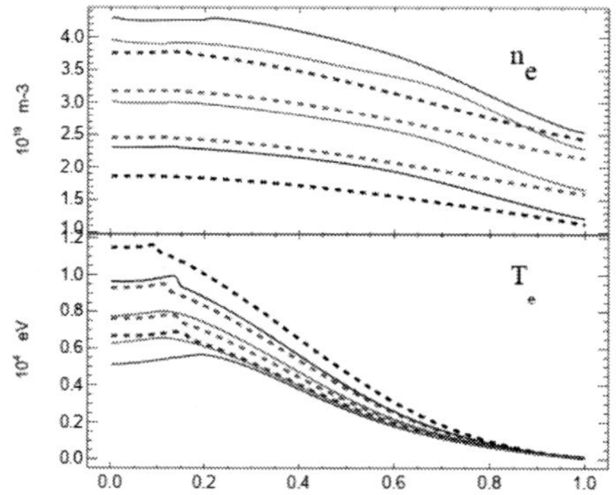


Figure 4: a) core density; b) core electron temperature profiles of the simulations with 5.0e21 p/s gas rate, 1.5 mm pellet size 10/3 Hz pellet frequency (continuous line); gas puff only of 1.0e22 p/s (dashed line) and different SOL perpendicular transport coefficients: Tanh decay to $\chi_{i,e}$ = 1.0 $m^2/s$, $D_\perp$ = 0.3 $m^2/s$; constant $\chi_{i,e}$ = 1.0 $m^2/s$, $D_\perp$ = 0.3 $m^2/s$; $\chi_{i,e}$ = 0.5 $m^2/s$, $D_\perp$ = 0.15 $m^2/s$; $\chi_{i,e}$ = 0.25 $m^2/s$, $D_\perp$ = 0.07 $m^2/s$

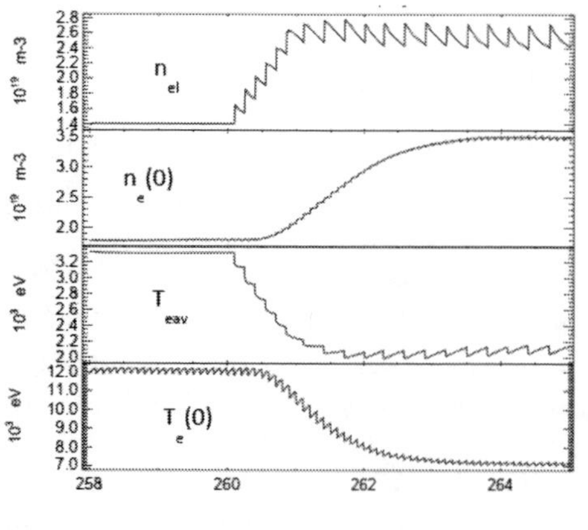


Figure 5: a) Line average core density; b) core axis electron density; c) average electron temperature; d) axis electron temperature time traces for the simulations with 1.0 $10^{21}$ $s^{-1}$ gas puffing rate and pellet injection with 2.0 $10^{21}$ particles per pellet injected with frequency of 3.3 Hz rate. SOL perpendicular transport coefficients described by tanh decay to $\chi_{i,e}$ = 1.0 $m^2/s$, $D_\perp$ = 0.3 $m^2/s$ in far SOL.

# Simulation of the influence of the operation conditions of GDC and ICWC on isotope exchange in Be layers as mean of T inventory control in ITER

D. Douai[1], D. Kogut[2], S. Brezinsek[3], T. Wauters[4] and JET Contributors[5]

[1]*CEA, IRFM, Association Euratom-CEA, 13108 St Paul lez Durance, France*
[2]*Aix-Marseille Université, CNRS, PIIM, UMR 7345, 13013 Marseille, France*
[3]*Forschungszentrum Jülich GmbH, Institut für Energie- und Klimaforschung, 52425 Jülich, Germany*
[4]*Laboratory for Plasma Physics, ERM/KMS, 1000 Brussels, Belgium, TEC partner*
[5]*EUROfusion Consortium, JET, Culham Science Centre, Abingdon, OX14 3DB*
*See the Appendix of F. Romanelli et al., Proceedings of the 25th IAEA Fusion Energy Conference 2014, Saint Petersburg, Russia*

## 1. Introduction

In the nuclear phase of ITER, wall conditioning will certainly contribute to the control of the tritium inventory - a major safety issue since the agreed safety limit inside the vessel must be kept under 640g during D:T operation [1] - by depleting tritium from the walls and in particular from that co-deposited with beryllium. Fuel retention experiments and post-mortem analysis in JET with ITER-like Wall (ILW) suggests that fuel implantation in Be and W dominates at first followed by fuel co-deposition with Be on top of the divertor tiles [2, 3]. In this aim, isotopic exchange is a good candidate to control tritium inventory, by replacing co-deposited or implanted tritium by deuterium in Glow Discharge Conditioning (GDC) or Ion Cyclotron Wall Conditioning (ICWC) plasmas. High efficiency has been reported, in particular on JET-ILW [1, 4, 5]. The isotopic exchange mechanisms at play in beryllium layers have been investigated by a 1D Diffusion Trapping Model for Isotopic eXchange – DITMIX [6, 7] validated against experimental profiles of hydrogen isotopes in beryllium layers deposited and irradiated in laboratory [6]. It reveals that the key factors influencing isotope exchange in such layers are flux, fluence and temperature. It is here applied to extrapolate trends on T removal efficiency vs temperature, expected flux and fluences during GDC or ICWC operation in ITER.

## 1. Isotope exchange experiments in JET-ILW

The efficiency of fuel removal by isotopic exchange with GDC and ICWC has been assessed in the JET-ILW [1, 4, 5]. Table 1 gives particle balances, integrated over the cumulated discharge durations in each experiment.

Table 1 : particle balance in GDC and ICWC experiments on JET

| | $H_2$-GDC | $D_2$-ICWC | $H_2$-ICWC |
|---|---|---|---|
| **RF cumulated duration** | 5000 sec., continuous | 85 sec. | 218 sec. |
| | | 5-10 sec. pulse, dwell~30 min. | |
| **Removal (atoms)** | $10.10^{22}$ H | $2,9.\ 10^{22}$ H | $6,2.10^{22}$ D |
| **Retention (atoms)** | $10.10^{22}$ D | $2.5.\ 10^{22}$ D | $8,9.10^{22}$ H |

Particle balances evidenced an accessible reservoir for isotopic exchange, i.e. the amount of wall atoms that can be replaced by the ICWC discharge specie, up to $10^{23}$ atoms for GDC operated nearly one and a half hour, and $6.10^{22}$ for ICWC operated in multi pulse mode [8], with discharges typically lasting 5-10 sec. and dwell time of 30 min. Roughly extrapolated to ITER, between a tenth of a mole and a mole of T could be removed by GDC or ICWC for similar durations [1].

Whereas GDC can only be operated in the absence of the toroidal magnetic field, ICWC discharges are produced in its presence, making them particularly interesting for T removal between D:T plasmas in ITER [1]. Isotope exchange with ICWC and GDC have been operated in JET-ILW with walls at 473K. in ITER, the inlet temperature of the pressurized cooling loop will be is 343K during plasmas and ICWC, the First Wall and the divertor being bakeable for conditioning up to 513K and to 623K, respectively.

## 2. Application of DITMIX to GDC and ICWC operated in JET and ITER

DITMIX is based on transport equations of hydrogen in metals such as used in TMAP and in other codes [9, 10]. It solves a system of partial differential equations and computes a 1D time-dependent solution for concentrations of dissolved and trapped hydrogen isotopes in beryllium [6]. The model considers two hydrogen isotopes in a-Be:H layers, either as soluted (mobile) or trapped (immobile) particles. The hydrogen transport in the film is governed by diffusion, and the model accounts for implantation, trapping, thermal and kinetic detrapping, as well as "swapping" between isotopes, introduced to simulate enhanced isotopic exchange all over the depth of the Be:H:D layer [6].

Hydrogen profiles from DITMIX were found in excellent agreement with profiles measured by $^{15}$N-NRA on pre-characterized 600 nm thick Be:H layers, with H/Be=0,04, which were irradiated by D ions with well-defined fluxes and energies, for different fluences and surface temperatures. The model provides a qualitative understanding of the isotope exchange mechanisms, although modelled and measured D profiles show less agreement in the bulk. DITMIX shows that the main factors determining isotopic exchange are the irradiation flux and fluence and the surface temperature [6].

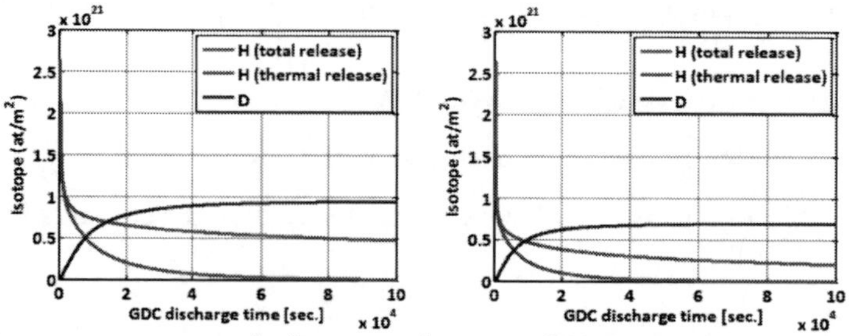


*Figure 1. isotope content in the layer as a function of GDC operation time at 473K and 513K.*

DITMIX was applied to GDC to simulate isotopic exchange in a 600 nm thick Be:H layer, with initial H/Be=0,04, for surface temperature of 473K and 513K. Typical $D^+$ ion energy of 500 eV and flux $\Gamma_D = 10^{17}$ D/m$^2$/s have been used for irradiation, switched on after a 500 s hold time to ensure a stable initial depth profile and switched off at $t = 10^5$ s. Hence, the total fluence of deuterons $D = 10^{22}$ D/m$^2$.

Figure 1 shows the calculated H content in the layer as a function of time at 473 and 513K. The green curve corresponds to a pure thermal H release, while the red one corresponds to the total release of H atoms from the layer exposed to the ion flux of GDC. Although baking alone is able to remove 85–90% of the retained hydrogen, the remaining amount is removed only if GDC is applied simultaneously. After $10^5$ s $D_2$-GDC, the Be:H layer is almost empty from hydrogen and only 0.04% of the initial H content is left in case of GDC at 513 K. Figure 2 shows the evolution of the isotope content in the layer during $D_2$-ICWC with $\Gamma_D = 10^{19}$ D/m$^2$/s, based on 0D simulations of ICWC plasma delivering an averaged flux over the entire first wall [11], and energy of 70 eV. This ion energy is probably overestimated, as ion impact energies of about 10–50 eV on the wall are expected. However the implantation depth varies very little in this energy range.

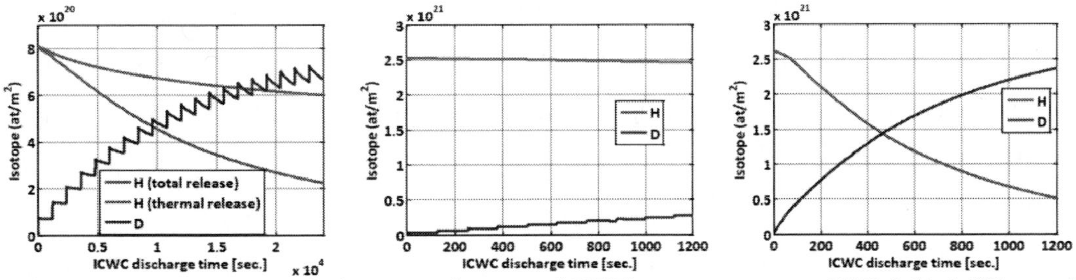

*Figure 1. isotope content in the layer as a function of ICWC operation time at 473K (left), 343K in multi pulse operation (middle) and 343K in continuous operation (right).*

Considering typical ICWC duty cycles in JET (Table 1), a sequence of 20 ICWC pulses was simulated at $T_{surf} = 473$K. In ITER, ~1200 sec. are available for conditioning between D:T pulses [1]. A sequence of 10 ICWC pulses with a duty cycle of 5/120 sec. ON/OFF has been therefore simulated and compared with continuous ICWC. Figure 2 (left) shows that at 473K, hydrogen (green and red curves) is continuously removed, while deuterium is implanted stepwise, part of it being released in the post-discharge. Decreasing $T_{surf}$ to 343K leads to a drastic reduction of thermally activated hydrogen transport and detrapping in beryllium, thus limiting isotopic exchange efficiency. At this temperature and flux, efficiency is recovered only if ICWC is applied continuously.

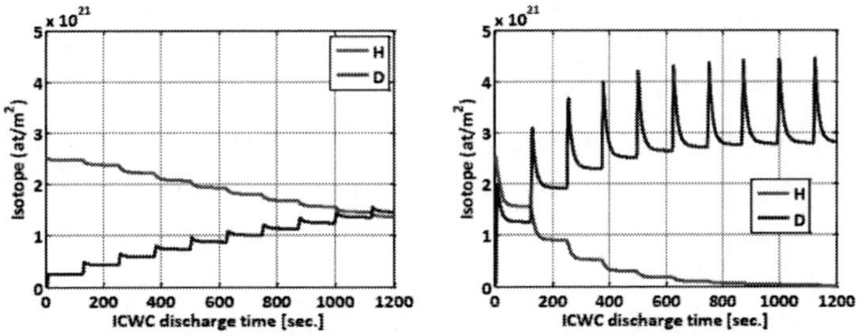

*Figure 3. isotope content in the layer as a function of ICWC operation time at 343K in multi pulse operation, for $\Gamma_D = 10^{20}$ D/m$^2$/s (left) and $10^{21}$ D/m$^2$/s (right).*

However, experimental evidence from ICWC discharges in Tore Supra [1] and in TEXTOR [12] suggest that the ion flux on limiters reaches $10^{21}$ D/m$^2$/s, this flux quickly decaying in the areas shadowed by the limiters. Fluxes of the order of $10^{20}$–$10^{21}$ D/m$^2$/s can be considered as an upper limit in ICWC discharges on the fully shaped ITER First Wall, without limiting surfaces. The temporal evolution of the isotope content for such fluxes is given in Figure 3. In the first case almost half of the initially retained H is removed after 10 pulses of 5 s, and a comparable amount of D is implanted in the co-deposit. For $\Gamma_D = 10^{21}$ D/m$^2$/s, the initial amount of H in the layers is decreased by nearly a factor 200 and the D content reaches saturation, i.e. the isotopic exchange is almost complete. It should be also noted that at this flux, erosion of Be layers by incident deuterons may not be longer negligible, leading to a possible re-deposition of the eroded material and a consequent H re-trapping.

## 3. Concluding remarks

Isotopic exchange with GDC or ICWC is potentially an efficient mean for the control of the tritium inventory in ITER. A 1D Diffusion Trapping Model for Isotopic eXchange –DITMIX - has been used to predict trends on the efficiency vs. expected GDC or ICWC operation parameters. The model is applied on a 600 nm thick a-Be:H layer, with H/Be=0.04, for expected wall temperatures, fluxes or fluences in ITER GDC or ICWC. Complete exchange by pulsed ICWC discharges is predicted for an ion flux to the wall $\Gamma_D = 10^{19}$ D/m$^2$/s impinging at E=70 eV, at $T_{surf}$=473K, the operation temperature in JET-ILW experiments. At $T_{surf}$=343K, the operation temperature of ICWC in ITER, and for the same ion flux and energy, isotope exchange is drastically reduced. Hence, one key factor determining isotopic exchange efficiency is surface temperature [6]. Complementary ICWC experiments are therefore suggested in the JET-ILW with walls at 343K, as it will be the case in ITER. For ion fluxes of $10^{20}$ and $10^{21}$ D/m$^2$/s, which may be considered as an upper limit in ICWC discharges on the ITER First Wall, the efficiency of pulsed ICWC is recovered at 343 K. Although complete exchange is predicted at $T_{surf}$=513K, $\Gamma_D = 10^{17}$ D/m$^2$/s and E=500 eV, representative conditions of GDC, the latter requires the absence of toroidal magnetic field, and it can therefore not be used in the silent time between ITER plasma pulses to control T inventory in ITER. In this respect GDC is not as attractive as ICWC for T-removal.

## References

[1] D. Douai *et al,* J. Nucl. Mater. (2015), doi:10.1016/j.jnucmat.2014.12.034
[2] S. Brezinsek *et al.*, Nucl. Fusion 53, 083023 (2013)
[3] A. Baron-Wiechec *et al.*, J. Nucl. Mater. (2015), doi:10.1016/j.jnucmat.2015.01.038
[4] T. Wauters *et al.*, J. Nucl. Mater. (2015), doi:10.1016/j.jnucmat.2014.12.097
[5] T. Wauters *et al.*, this conf.
[6] D. Kogut *et al.*, Proc. of the 15th PFMC, Aix-en-Provence, May 2015, to be published in Phys. Scr.
[7] D. Kogut, « Study of wall conditioning in tokamaks with application to ITER, PhD Thesis, 2014
[8] D. Douai *et al.*, Journal of Nuclear Materials 415 (2011) S1021– S1028
[9] M. Baldwin, T. Schwarz-Selinger, and R. Doerner, Nucl. Fusion 54, 073005 (2014)
[10] O. V. Ogorodnikova et al., J. Nucl. Mater. 273, 66 (1999)
[11] T. Wauters et al., Plasma Phys. Control. Fusion 53 (2011) 125003 (20pp)
[12] S. Möller *et al.*, . Nucl. Mater. (2015), doi:10.1016/j.jnucmat.2015.01.022

# Comprehensive evaluation of the linear stability of Alfvén eigenmodes driven by alpha particles in an ITER baseline scenario

A. C. A. Figueiredo[1], P. Rodrigues[1], D. Borba[1], R. Coelho[1], L. Fazendeiro[1], J. Ferreira[1], N. F. Loureiro[1], F. Nabais[1], S. D. Pinches[2], A. Polevoi[2] and S. E. Sharapov[3]

[1] *Instituto de Plasmas e Fusão Nuclear, Instituto Superior Técnico, Universidade de Lisboa, 1049-001 Lisboa, Portugal*

[2] *ITER Organization, Route de Vinon-sur-Verdon, CS 90 046, 13067 St Paul-lez-Durance Cedex, France*

[3] *CCFE, Culham Science Centre, Abingdon OX14 3DB, United Kingdom*

## Introduction

In ITER, the performance of burning plasmas will depend on the population of alpha particles being well confined within the plasma core, as the heating of the DT plasma will then rely mainly on the energy of these suprathermal particles that are produced by core fusion reactions. A phenomenon that can potentially hinder the successful operation of ITER is therefore the destabilization of Alfvén eigenmodes (AEs) by alpha particles [Fu 1989], whereby an increased radial transport of the latter could degrade the conditions necessary to sustain the fusion process and carry damaging heat loads to the tokamak wall [Sharapov 2013]. In this work the stability of AEs is systematically addressed for the 15 MA ELMy H-mode ITER baseline scenario [Polevoi 2002, Pinches 2015, Lauber 2015, Rodrigues 2015] making use of a recently introduced framework [Rodrigues 2015] that is based on the hybrid MHD drift-kinetic code Castor-K [Borba 1999, Nabais 2015].

## Results and Discussion

We consider two variants of the aforementioned ITER scenario with plasma current $I_p = 15$ MA, toroidal magnetic field $B_0 = 5.3$ T, and plasma major and minor radii $R_0 = 6.4$ m and $a = 2$ m, respectively[1]. The kinetic profiles in Figure 1 are plotted versus the radial coordinate $s$, the square root of the poloidal magnetic flux normalized to its boundary value. Both scenario variants have approximately the same electron density $n_e$ and impurity content $n_Z$, which is essentially Be, and the fuel density $n_{DT} = n_D + n_T$ is for an optimal mix of D and T, i.e., $n_D = n_T$. The main difference between the two variants is that in the one on the left of Figure 1, hereafter called *LoPed*, electron and ion temperatures $T_e$ and $T_i$, respectively, are much lower at the pedestal and much higher at the core than in the *HiPed* variant on the right. Naturally, the higher core temperatures go with a much higher density of alpha particles $n_\alpha$ and helium ashes (thermalized alpha particles) $n_{He}$ in *LoPed*. For convenience, differences in the safety factor $q$ are discussed apropos Figure 4.

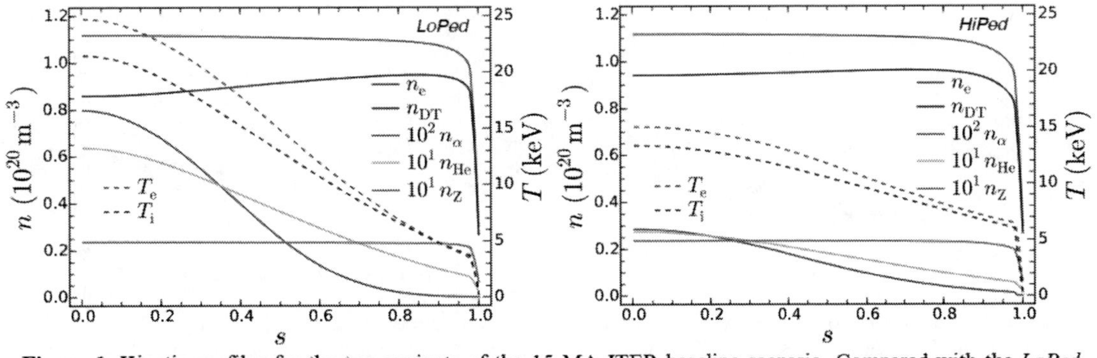


**Figure 1.** Kinetic profiles for the two variants of the 15 MA ITER baseline scenario. Compared with the *LoPed* scenario, *HiPed* is characterised by higher temperatures at the pedestal and lower core temperatures, together with a lower density of alpha particles and He ashes.

[1] Here, $R_0$ is the position of the magnetic axis.

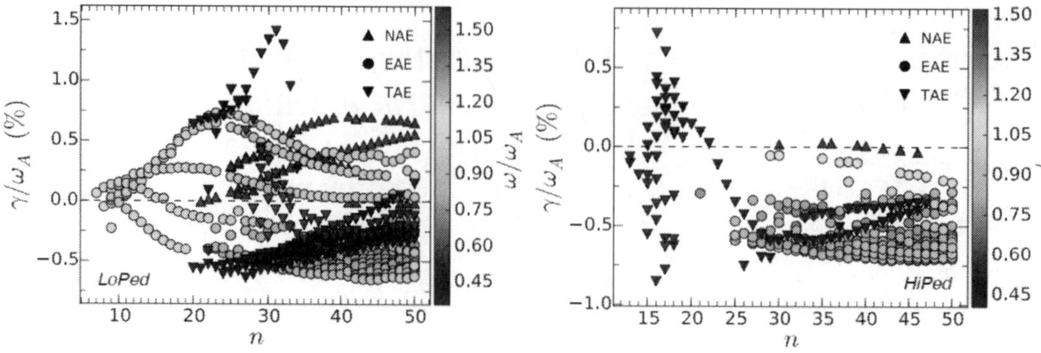


**Figure 2.** Linear growth rates $\gamma$ normalized to the Alfvén frequency $\omega_A = v_A/R_0$, where the Alfvén velocity at the magnetic axis is $v_A \approx 7.1\times10^6$ m/s in *LoPed* and $v_A \approx 7.0\times10^6$ m/s in *HiPed*, versus toroidal mode number $n$ and colored by AE frequency for the two variants of the 15 MA ITER baseline scenario. Different symbols are used for Toroidicity induced AEs (TAEs), Ellipticity induced AEs (EAEs), and Non-circular triangularity-induced AEs (NAEs). TAEs appear in dark-blue patterns corresponding to frequencies around $\omega_A/2$ (the middle of the TAE gap) and are the most unstable. No valid AEs have been found for $n < 7$ in *LoPed* and $n < 13$ in *Hiped*.

Besides Castor-K, which assesses the stability of AE by computing their linear growth-rates, our suite of numerical codes comprises Helena [Huysmans 1991] to obtain equilibria and Mishka [Mikhailovskii 1997] to calculate eigenmodes. AEs have been extensively considered in our methodical approach, a necessity that stems from our purpose of forecasting experimental results, rather than interpret them. Here, we focus on values of $n$ ranging from 1 to 50 in order to stay within the limits of the drift-kinetic ordering for the alpha particles [Rodrigues 2015]. All possible AEs have been determined by scanning the mode frequency $\omega/\omega_A$ from 0.01 to 2 in steps of $2\times10^{-5}$ and inputting $(\omega/\omega_A)^2$ to Mishka as a guess of the eigenvalue that, upon convergence, is returned together with the mode eigenfunction. A selection of valid AEs has then been made based on two criteria: the mode frequency cannot match the Alfvén continuum at any radial position where the mode amplitude is above 1% of its maximum, and the eigenfunction must be well-resolved radially with a numerical grid index $h_g < 0.3$ [Rodrigues 2015]. Around 2300 AEs were selected for *LoPed* and 400 for *HiPed*. These AEs were then processed by Castor-K, which calculates the energy $\delta W_P$ exchanged between a mode and a given population (P) of plasma particles as well as the associated growth rate $\gamma_P = \mathrm{Im}(\delta W_P)/(2\omega W_k)$, where $W_k$ is the kinetic energy of the mode perturbation [Borba 1999]. For each mode, four Castor-K runs were done to calculate the drive due to the alpha particles

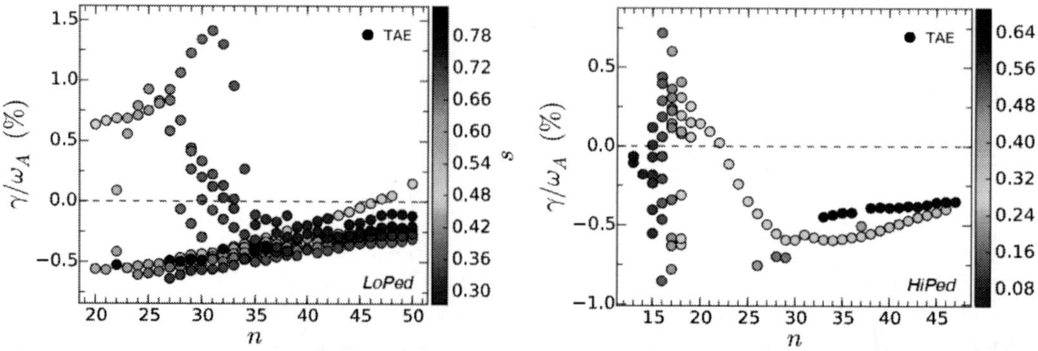


**Figure 3.** In both analysed scenarios the most unstable TAEs are core-localized. As seen from the $q$ profile in Figure 4, in *HiPed* these TAEs are inside the $q = 1$ surface, so they are 'tornado modes' [Sharapov 2013]. Due to the closing of the TAE gap few modes exist for $n > 20$ in *HiPed* — only those located at the bottom of the gap that avoid matching the continuous spectrum, as seen in Figure 5. Such is not the case in *LoPed* for which more modes exist for high $n$ due to the extended low magnetic-shear region that makes gaps close at an outer position, as seen in Figure 4, and because in *LoPed* higher-$n$ modes are located at inner positions than lower-$n$ modes. Moreover, the higher magnetic shear in *HiPed* does not allow the existence of Low Shear TAEs (LSTAEs) beyond $s \approx 0.3$.

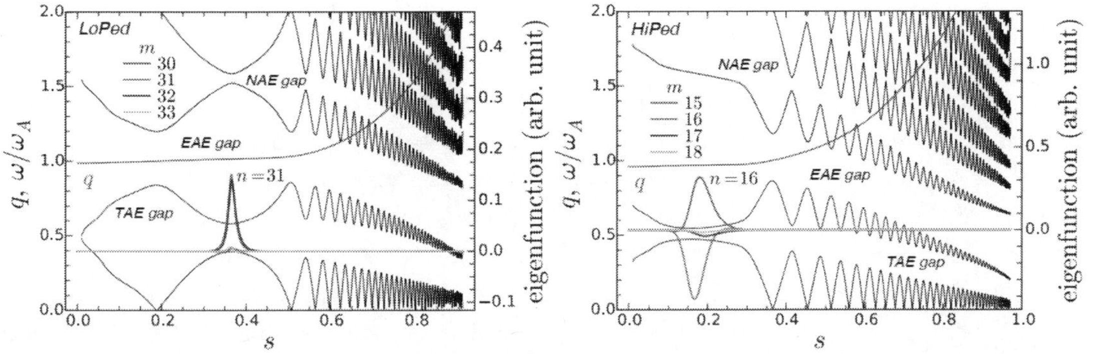


**Figure 4.** The eigenfunctions of the most unstable AEs represented by the colored lines have been calculated by Mishka using 17 harmonics with poloidal mode number starting at $m = n - 1$, but only the strongest 4 are shown. In both scenario variants the modes fall within the TAE gap of the Alfvén continuous spectrum. Notice that the baseline of the eigenfunction (zero value on the right axis) marks the mode frequency on the left axis. The on-axis value of the safety factor in *HiPed* is $q(0) = 0.96$, around 3% lower than $q(0) = 0.987$ in *LoPed*. While in *HiPed* the low magnetic-shear region ranges from the axis to $s \approx 0.3$, in *LoPed* it extends to $s \approx 0.5$. The $q = 1$ surface is located at $s \approx 0.2$ in *LoPed* and $s \approx 0.37$ in *HiPed*.

($\alpha$) and the damping due to the interaction with the bulk ions (DT), the electrons (e), and the helium ashes (He), thus obtaining the total growth rate $\gamma = \gamma_\alpha + \gamma_{DT} + \gamma_e + \gamma_{He}$ of the AE. The alpha-particle population has been described by a radius-independent slowing-down energy distribution with a crossover energy $E_c \approx 730$ keV in *LoPed* and $E_c \approx 585$ keV in *HiPed*, and a dispersion around the alpha-particle birth energy $\Delta_E \approx 50$ keV in both cases [Rodrigues 2015]. Radiative damping, which has a non-negligible contribution to $\gamma$ in *LoPed* [Rodrigues 2015] will be included in a forthcoming article [Figueiredo 2015].

The growth rates[2] shown in Figures 2 and 3 are larger in *LoPed* than in *HiPed*. Although EAEs and NAEs have positive growth rates in *LoPed*, all markedly unstable modes are TAEs in the vicinity of $n = 30$ in *LoPed* and $n = 15$ in *HiPed*. Figure 3 shows a difference in the evolution of TAE localization as $n$ increases: while in *HiPed* modes become progressively located at outer positions, the opposite occurs in *LoPed*. This results directly from $q$ being below 1 in *HiPed* and above 1 in *LoPed* within the region $0.2 < s < 0.37$. Notice that the unstable modes in Figure 3 are LSTAEs, well-localized and within the low magnetic-shear region of the plasma core. As shown in Figure 4, in *LoPed* the most unstable mode is a $n = 31$ TAE located at $s \approx 0.37$ near the bottom of the TAE gap with a growth rate $\gamma/\omega_A \approx 1.41\%$, whereas in *HiPed* the most unstable mode is a $n = 16$ TAE at $s \approx 0.17$ near the top of the TAE gap with $\gamma/\omega_A \approx 0.72\%$. While in *LoPed* the most unstable mode is close to $s = 0.38$ where the gradient of $n_\alpha$ is highest, that is not possible in *HiPed* because the maximum $n_\alpha$ gradient occurs at $s \approx 0.44$ where the higher magnetic shear only allows non-local modes that interact strongly with the Alfvén continuum. At this point it is interesting to verify the estimate of $n$ for the most driven AE that is based on matching the width of passing alpha-particle orbits and the TAE width [Pinches 2015,

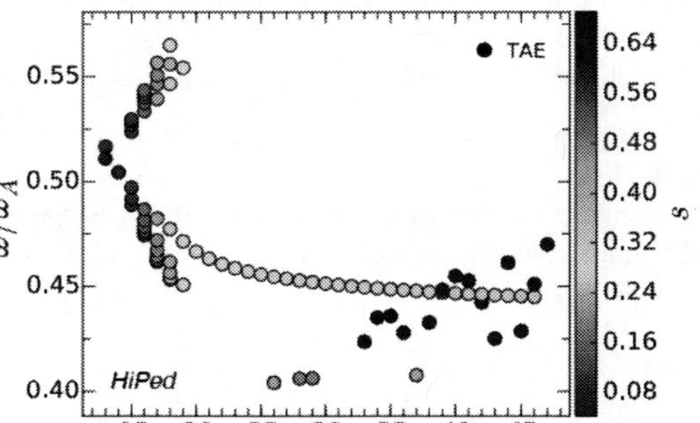


**Figure 5.** Frequency distribution of "tornado modes" inside the TAE gap in *HiPed*. The lower frequency branch is formed by symmetric modes at the bottom of the gap, while the modes in the upper branch are anti-symmetric TAEs located at the top of the gap.

[2] Notice the normalization to the Alfvén frequency instead of the more usual mode frequency.

Rodrigues 2015], which is given by $n \approx s/q^2 \times a/R_0 \times \Omega_\alpha/\omega_A$. The cyclotron frequency of the alpha particles is $\Omega_\alpha \approx 2.5\times10^8$ rad/s and $q(s)$ is taken at the mode location. Using $q(0.37) \approx 1.016$ we arrive at $n \approx 26$ in *LoPed*, while in *HiPed* $q(0.17) \approx 0.968$ leads to $n \approx 13$.

In Figure 5 the frequency of TAEs is plotted versus $n$ for *HiPed*. Two main branches are evident inside the TAE gap: an upper branch made of modes that rise in frequency as $n$ increases, and a lower branch with decreasing frequency modes. The lower-frequency branch is made of symmetric modes, while the modes in the upper branch are anti-symmetric [Pinches 2015, Lauber 2015]. Moreover, for a given $n$ the number of oscillations in the poloidal harmonics of these modes increases as the frequency increases in the upper branch, or as the frequency decreases in the lower branch. The long line that chirps down until $n \approx 45$ has the simplest symmetric TAEs with a single peak per $m$ (no oscillations). Its 'mirror' line in the upper branch ends abruptly at $n \approx 20$. This occurs because the TAE gap closes at an inner $s$ for modes at the top of the gap than at its bottom, causing the missing modes to be strongly affected by continuum damping.

**Summary**

The linear-stability of AEs for two variants of the 15 MA ELMy H-mode ITER baseline scenario has been analysed using the hybrid MHD drift-kinetic code Castor-K. The main difference between the two variants *LoPed* and *HiPed* is in the temperature profiles, particularly at the edge pedestal. Results show that the most unstable modes are TAEs with $n$ around 30 in *LoPed* and 15 in *HiPed*. In both variants of the ITER scenario these unstable TAEs are localized in a low magnetic-shear region of the plasma core. In the *HiPed* case a clear frequency distribution of symmetric and anti-symmetric TAEs, which are in fact 'tornado modes', has been found within the TAE gap that agrees with recent studies on the same ITER scenario [Pinches 2015, Lauber 2015]. Calculation of radiative damping is underway and will be part of a following publication. While radiative damping has been shown to somewhat reduce the highest *LoPed* growth rates [Rodrigues 2015], it should not alter our results for *HiPed* significantly since its most unstable mode is an anti-symmetric TAE, which is therefore practically unaffected by radiative damping [Nyqvist 2012].

**Acknowledgments**

This work has been carried out within the framework of the EUROfusion Consortium and has received funding from the Euratom research and training programme 2014-2018 under grant agreement No. 633053. IST activities received financial support from "Fundação para a Ciência e Tecnologia" (FCT) through project UID/FIS/50010/2013. The views and opinions expressed herein do not necessarily reflect those of the European Commission, IST, FCT, or the ITER Organization. All computations were carried out using the HELIOS supercomputer system at the Computational Simulation Centre of the International Fusion Energy Research Centre (IFERC-CSC) in Aomori, Japan, under the Broader Approach collaboration between Euratom and Japan implemented by Fusion for Energy and JAEA. PR was supported by EUROfusion Consortium grant no. WP14-FRF-IST/Rodrigues and NFL was supported by FCT grant no. IF/00530/2013.

**References**

[Borba 1999] D. Borba and W. Kerner, J. Comput. Phys. **153** (1999) 101.

[Figueiredo 2015] A. C. A. Figueiredo *et al.* "Comprehensive evaluation of the linear stability of Alfvén eigenmodes driven by alpha particles in an ITER baseline scenario", to be submitted to Nucl. Fusion (2015).

[Fu 1989] G. Y. Fu and J. W. Van Dam, Phys. Fluids B **1** (1989) 1949.

[Huysmans 1991] G. Huysmans *et al.*, Int. J. Mod. Phys. C **2** (1991) 371.

[Lauber 2015] P. Lauber, Plasma Phys. Control. Fusion **57** (2015) 054011.

[Mikhailovskii 1997] A. Mikhailovskii *et al.*, Plasma Phys. Rep. **23** (1997) 844.

[Nabais 2015] F. Nabais *et at.*, Plasma Sci. Technol. **17** (2015) 89.

[Nyqvist 2012] R. M. Nyqvist and S. E. Sharapov, Phys. Plasmas **19** (2012) 082517.

[Pinches 2015] S. D. Pinches *et al.*, Phys. Plasmas **22** (2015) 021807.

[Polevoi 2002] A. R. Polevoi *et al.*, J. Plasma Fusion Res. SERIES **5** (2002) 82.

[Rodrigues 2015] P. Rodrigues *et al.* "Systematic linear-stability assessment of Alfvén eigenmodes in the presence of fusion α-particles for ITER-like equilibria", accepted by Nucl. Fusion (2015).

[Sharapov 2013] S. E. Sharapov *et al.* Nucl. Fusion **53** (2013) 104022.

# Linear stability of $\alpha$-particle–driven Alfvén Eigenmodes in ITER burning-plasma scenarios

P. Rodrigues[1], A. C. A. Figueiredo[1], L. Fazendeiro[1] J. Ferreira[1], R. Coelho[1], F. Nabais[1], D. Borba[1], N. F. Loureiro[1], A. Polevoi[2], S. D. Pinches[2], and S. E. Sharapov[3]

[1]*Instituto de Plasmas e Fusão Nuclear,*

*Instituto Superior Técnico, Universidade de Lisboa, 1049-001 Lisboa, Portugal.*

[2]*ITER Organization,*

*Route de Vinon-sur-Verdon, CS 90 046, 13067 St Paul-lez-Durance Cedex, France.*

[3]*CCFE, Culham Science Centre, Abingdon OX14 3DB, United Kingdom.*

Plasma heating during the burning regime in tokamaks will rely upon the energy of fusion-born $\alpha$-particles which must be kept confined to keep the plasma hot and prevent wall damage. However, such particles can drive Alfvén Eigenmodes (AEs) unstable and can thus be transported away from the burning core. Before experiments on ITER are performed we rely upon extrapolation from present machines and simple predictive models to determine which AEs will interact most with fusion $\alpha$'s for a given scenario. On the other hand, *ab initio* simulations able to provide a self-consistent solution of the interaction between $\alpha$-particles and the bulk plasma are still computationally expensive for routine use in experiment design and planning. Therefore, more efficient approaches must be devised so that large swaths of parameter space can be explored, identifying the most unstable cases for later analysis with more specialized tools.

In this work an approach is presented that systematically evaluates the linear stability of all possible AEs for a given equilibrium [1]: these are found by intensively scanning over a frequency and wave-number range with the ideal-MHD code `MISHKA` [2], while the energy transfer between them and the plasma species (fusion $\alpha$'s, DT fuel, electrons and He ash) is evaluated with the drift-kinetic code `CASTOR-K` [3, 4]. The computational efficiency of the `MISHKA/CASTOR-K` pair is the key to handle the very large number of AEs involved in systematic stability assessments. The ability to produce a list of possible AEs ranked by their linear growth rate is illustrated with an ITER baseline scenario [5, 6]. Small variations of the safety factor $q$ are then considered, which result from slightly changing the plasma current. We shall find that such small changes can have a large influence on the linear-stability properties of AEs for this ITER baseline scenario.

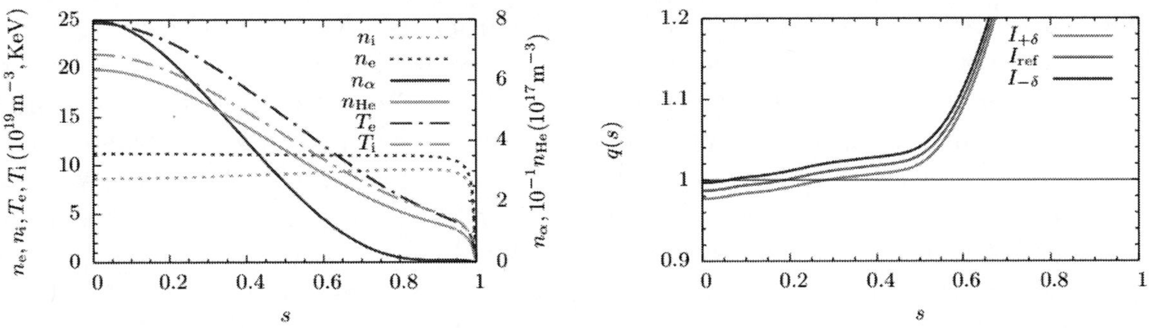


Figure 1: Kinetic (left) and safety-factor profiles (right) corresponding to three values of $I_p$.

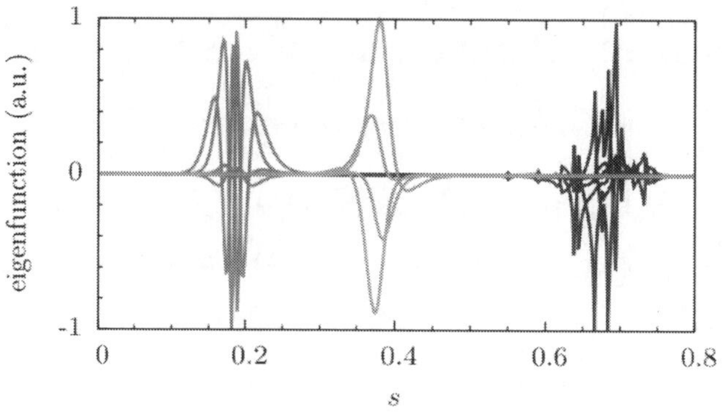


Figure 2: Examples of AEs with grid index $h_g = 0.005$ (well resolved, centre), 0.3 (marginally resolved, left), and 0.95 (poorly resolved, right).

Figure 1 displays the kinetic profiles of the aforementioned ITER scenario, with $I_p = 15\,\mathrm{MA}$, on-axis magnetic field $B_0 = 5.3\,\mathrm{T}$, minor radius $a = 2\,\mathrm{m}$, and magnetic axis at $R_0 = 6.4\,\mathrm{m}$. The DT fuel mix ratio is $n_D/n_T = 1$ and their combined density is $n_i = n_D + n_T$. The $q$ profile is also plotted in Figure 1, with $q(0) = 0.987$. In both plots, $s^2 = \psi/\psi_b$, $\psi$ is the poloidal flux, and $\psi_b$ is its value at the boundary. Two other profiles correspond to small variations of the reference value ($I_{ref} = 15\,\mathrm{MA}$) for $I_p$ and shall be discussed latter.

The linear-stability assessment starts with an equilibrium computed by HELENA [7], using the kinetic profiles and the $I_p$ value, which is then employed by MISHKA to find the frequency and radial structure of all AEs for a toroidal number $n$ in the range $1 \leqslant n \leqslant 50$ and poloidal harmonics $n-1 \leqslant m \leqslant n+15$. The upper limit for $n$ is set by $k_\perp \rho_\alpha \lesssim 1$, whence $n \lesssim (s/q)/(\rho_\alpha/a) \approx 50$, with $\rho_\alpha/a \approx 1/100$ the normalized $\alpha$-particle gyro-radius, $k_\perp \approx nq/(as)$, $q \approx 1$, and $s \approx 0.5$. For each $n$, the frequency range $0 \leqslant \omega/\omega_A \leqslant 1$ [where $\omega_A = V_A(0)/R_0$ and $V_A(0)$ is the on-axis Alfvén velocity] is sampled in small steps of size $2 \times 10^{-5}$ and each sample becomes MISHKA's initial guess. If the resulting AE solution does not converge in 5 iterations, it is discarded and the next sample is tried. Converged AEs are next filtered to remove spurious solutions with radial wavelengths shorter than the radial mesh size. AEs with sharp variations display a large grid index $h_g$ while a lesser one corresponds to smaller discretization errors [1], as illustrated in Figure 2. Only AEs with $h_g \leqslant 0.3$ are considered. AEs matching the Alfvén continuum at radial positions where their amplitude exceeds $\frac{1}{100}$ of its maximum value are also discarded. Next, CASTOR-K evaluates the energy exchange $\delta W_j$ between every selected AE and each species $j$: the $\alpha$-particles described by a radius-independent slowing-down energy distribution (with crossover energy $E_c = 730\,\mathrm{KeV}$ and dispersion $\Delta_E = 50\,\mathrm{KeV}$ around the birth energy) and then three thermal species (DT ions, electrons, and He ash), all described by local Maxwellians [1].

Although other sources of energetic particles are foreseen for the ITER scenario concerned (e.g., 40 MW of combined NBI and ECRH power) [6], only fusion-born $\alpha$-particles are considered in this work. Radiative-damping evaluation is computationally costly and performed only for the most unstable ones [1], but here it will be left to future work. So, the net linear growth rate is $\gamma = \gamma_\alpha + \gamma_{DT} + \gamma_e + \gamma_{He}$, with $\gamma_j = \mathrm{Im}(\delta W_j)/(2\omega W_k)$ and $W_k$ the AE's kinetic energy [3].

Three different magnetic equilibria are next considered, with $I_p$ assuming the reference value $I_{ref} = 15\,\mathrm{MA}$ and then two small perturbations of the latter, respectively $I_{+\delta}$ and $I_{-\delta}$, where $\delta = 0.16\,\mathrm{MA}$. The linear-stability assessment of these three cases is summarized in Figure 3,

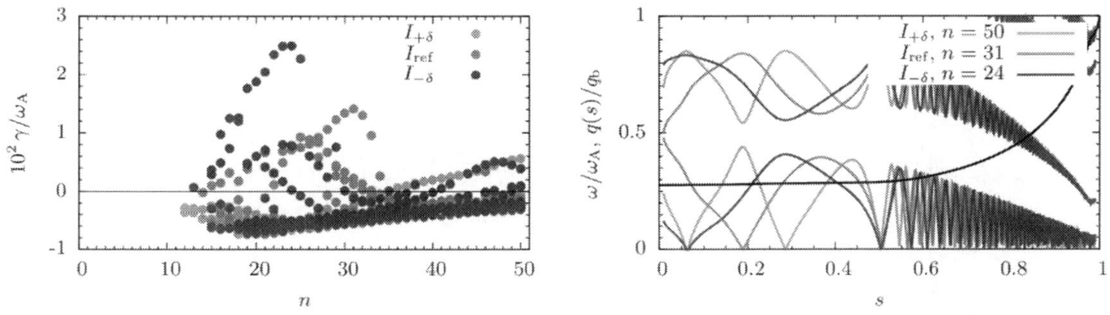


Figure 3: Linear growth rate versus $n$ (left), Alfvén continuum corresponding to the most unstable $n$ (right), and $q$ profile for $I_p = I_{\rm ref}$ normalized to its boundary value $q_b = 3.57$.

where only Toroidicity-induced AEs (TAEs) are shown for clarity. The reference case has its most unstable AEs around $n \approx 31$, corresponding to highly localized low-shear TAEs (LSTAEs) within the region $s \lesssim 0.5$, where the large radial separation between frequency gaps allows them to be excited. Actually, the most unstable AEs arises in the frequency gap located at $s \approx 0.37$ and depicted in Figure 3. Non-localized AEs coupling different frequency gaps only show in the plasma outer half, where the shear is higher and the frequency-gap separation is smaller. However, because the TAE gap closes towards the edge, most non-localized AEs forcibly interact with the Alfvén continuum and are assumed to be completely damped [1]. Changing $I_p$ by the small amount $\pm\delta$, whilst keeping the same kinetic profiles, results in new magnetic equilibria whose $q$ profiles are depicted in Figure 1 and for which $q(0)$ changes by $\pm 1\%$ of the reference value. The consequences are noticeable: The toroidal mode number $n$ of the most unstable AEs drops by 20% whilst the linear growth rate almost doubles for the lower current $I_{-\delta}$ [and thus higher $q(0)$]; Conversely, for the higher current value $I_{+\delta}$ [corresponding to lower $q(0)$], the majority of AEs become stabilized except some large-$n$ AEs ($n \gtrsim 35$). These proceed from the second frequency gap located near $s \approx 0.45$ (Figure 3) and have their linear growth rate reduced roughly by half with respect to the reference case. For all these three $I_p$ values, the most unstable modes are always even LSTAEs whose dominant poloidal harmonics are $m = n$ and $m = n+1$, so that their resonant surface has a radial location established by $q_{n,n} = 1 + 1/(2n)$.

The plots in Figure 4 are useful to bring some insight into these results. They clearly indicate that lowering $I_p$ also moves the radial position of the most unstable AEs inwards, and vice versa. The relation between the AEs location and $I_p$ is better understood if the latter is replaced by the on-axis value $q_0 \equiv q(0)$ and the safety factor is approximated by $q(s) = q_0 + \sigma s$ in the low-shear region, whence $\sigma s = 1 - q_0 + 1/(2n)$. Recalling that for efficient energy transfer the most unstable AEs follow $k_\perp \Delta_{\rm orb} \sim 1$, with $\Delta_{\rm orb} \sim aq/(\varepsilon\tilde{\Omega})$ the $\alpha$-particle orbit width, $\tilde{\Omega}$ its gyrofrequency normalized to $\omega_A$, and $\varepsilon = a/R_0$, one finds $n + (1-2\zeta)/(4n) + 1 = \zeta(1-q_0)$, where $\zeta \equiv \varepsilon\tilde{\Omega}/\sigma = (q/\sigma)(\Delta_{\rm orb}/a)^{-1}$. Subtracting the former relation and its evaluation with the values $n_{\rm ref}$ and $q_{\rm ref}$, which correspond to the reference case, gives

$$\left(1 + \frac{2\zeta - 1}{4n_{\rm ref}n}\right)(n - n_{\rm ref}) = -\zeta\,(q_0 - q_{\rm ref}). \tag{1}$$

ITER parameters are $\sigma \approx 0.07$, $\varepsilon \approx 0.3$, and $\tilde{\Omega} \approx 230$, whence $\zeta \approx 10^3$. Therefore, increasing $q_0$ above $q_{\rm ref}$ lowers $n$ below $n_{\rm ref}$ and conversely, as observed in Figures 3 and 4. On the other hand, the rise of $\gamma/\omega_A$ as $q_0$ increases (or $I_p$ drops) is due to the larger number of $\alpha$-particles found as AEs move inwards within $0.2 \lesssim s \lesssim 0.6$, where $dn_\alpha/ds$ is almost constant.

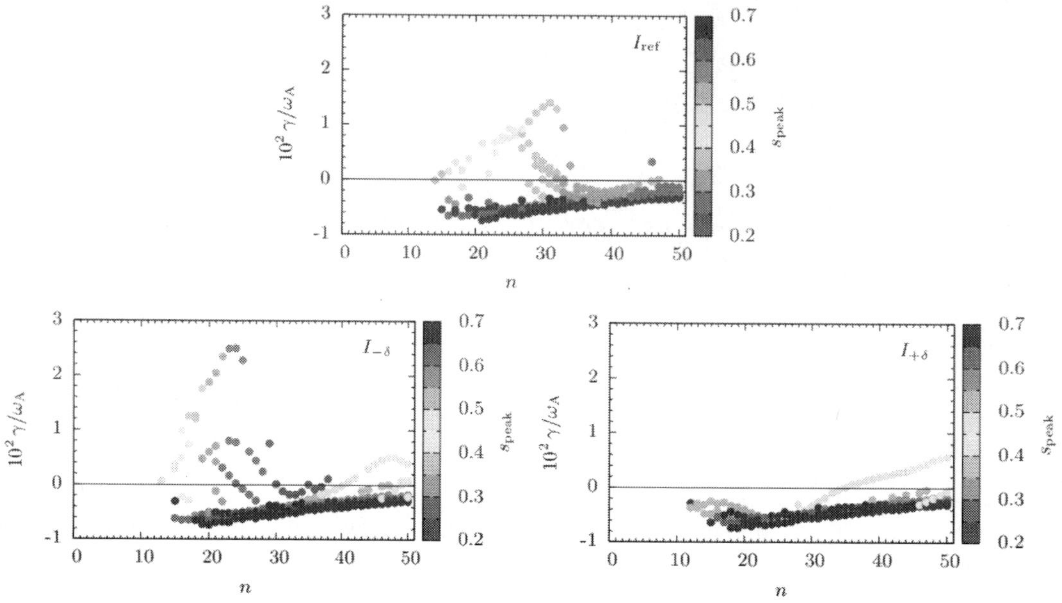


Figure 4: $\gamma/\omega_A$ versus $n$, with each AE colored by the radial location of its peak amplitude.

In summary, a computationally efficient approach for systematically assessing the linear stability of $\alpha$-driven AEs has been developed and used with an ITER baseline scenario. Predicted values for $\gamma/\omega_A$ and $n$ of the most linearly unstable AEs were found to be highly sensitive to small changes in $I_p$ (or $q_0$). Moreover, this sensitivity was shown to proceed from the large value attained by $\zeta$ in Eq. (1) due to the very low magnetic shear in the plasma core, denoted by having $\sigma \ll 1$, and very large gyrofrequency $\tilde{\Omega} \gg 1$.

**Acknowledgments**

This work was carried out within the framework of the EUROfusion Consortium and received funding from the Euratom research and training programme 2014-2018 under grant agreement no. 633053. IST activities received financial support from "Fundação para a Ciência e Tecnologia" (FCT) through project UID/FIS/50010/2013. The views and opinions expressed herein do not necessarily reflect those of the European Commission, IST, CCFE, or the ITER Organization. All computations were carried out using the HELIOS supercomputer system at the Computational Simulation Centre of the International Fusion Energy Research Centre (IFERC-CSC) in Aomori, Japan, under the Broader Approach collaboration between Euratom and Japan implemented by Fusion for Energy and JAEA. PR was supported by EUROfusion Consortium grant no. WP14-FRF-IST/Rodrigues and NFL was supported by FCT grant no. IF/00530/2013.

**References**

[1] P. Rodrigues *et al.* "Systematic linear-stability assessment of Alfvén eigenmodes in the presence of fusion $\alpha$-particles for ITER-like equilibria", accepted to Nucl. Fusion (2015).
[2] A. Mikhailovskii *et al.*, Plasma Phys. Rep. **23**, 844 (1997).
[3] D. Borba and W. Kerner, J. Comput. Phys. **153**, 101 (1999).
[4] F. Nabais *et al.*, Plasma Sci. Technol. **17**, 89 (2015).
[5] A. R. Polevoi *et al.*, J. Plasma Fusion Res. SERIES **5**, 82 (2002).
[6] S. D. Pinches *et al.*, Phys. Plasmas **22**, 021807 (2015).
[7] G. Huysmans et al., Int. J. Mod. Phys. C 2 **371**, (1991).
[8] Ph. Lauber, Plasma Phys. Control. Fusion 57 **054011**, (2015).

# Study of ELM triggering by axisymmetric in-vessel coils

Y. Gribov[1], A.A. Kavin[2], V.E. Lukash[3], R.R. Khayrutdinov[3], G.T.A. Huijsmans[1], A. Loarte[1]

[1] *ITER Organization, Route de Vinon sur Verdon, CS 90 046, 13067 St Paul Lez Durance Cedex, France*

[2] *Joint Stock Company "D.V.Efremov Institute of Electrophysical Apparatus", Saint Petersburg, Russia*

[3] *National Research Centre "Kurchatov Institute", Moscow, Russia*

## 1. Introduction

Periodic vertical movements of the plasma have been used to trigger ELMs in a number of tokamaks and can be realized in ITER by the use of the in-vessel coils dedicated for plasma vertical stabilization (VS coils shown in Fig. 1 by the black filled circles). The upper and lower VS coils have 4 turns each. The maximum value of current in each turn is 60 kA and the maximum value of voltage produced by the power supply is 2.3 kV. Earlier studies [1] showed that the plasma periodic vertical movements has potential for ELM triggering to provide tungsten exhaust from the main plasma during the initial phase of ITER operation in H-mode plasmas for which the plasma current will be typically about 7.5 MA.

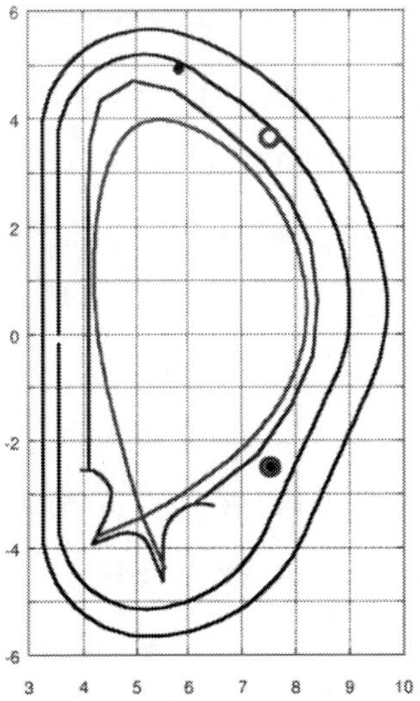


**Fig. 1**. Location of the VS coils: black filled circles –Baseline coils, red open circles – hypothetical Symmetric coils

On the basis that ELM triggering by vertical plasma position can be explained by the associated oscillation of the plasma current at the plasma edge [2], we use the following indexes to characterize the performance of the ITER VS coils for ELMs triggering:

1) $\Delta \boldsymbol{I}_{\mathbf{ped}} / \boldsymbol{I}_{\mathbf{ped}}$ -variation of the total current in the pedestal region $0.975 \leq \rho \leq 1$, where $\rho = \sqrt{\Phi/\Phi_b}$, $\Phi$ and $\Phi_b$ are the fluxes of toroidal magnetic field inside a given magnetic surface and inside the plasma boundary, respectively, and
2) $\Delta \boldsymbol{j}_{\mathbf{tot}}(\boldsymbol{\rho} = \mathbf{1}) / \boldsymbol{j}_{\mathbf{tot}}(\boldsymbol{\rho} = \mathbf{1})$ - variation of the plasma current density at the separatrix.

## 2. In-vessel coils connected in anti-series (standard configuration)

Simulations of an ITER 7.5 MA / 2.65 T H-mode scenario, performed with the DINA code [3] with feedback and feedforward control of the plasma current, position and shape, have demonstrated promising results of ELMs triggering at a frequency of up to 20 Hz by oscillations of the plasma vertical position using the VS coils connected in *anti-series* (i.e. producing a magnetic field predominantly in the *radial* direction) [4, 5]. It has been shown that with the power supply voltage $V_{VS} = \pm 2.3$ kV, the displacement of the plasma current center, $\Delta Z_p$, is about ±4 cm and this is sufficient to affect significantly the plasma current in the pedestal region leading to the triggering of ELMs. During the oscillations of the current in the VS coils, the plasma current in the region $0.975 \leq \rho \leq 1$ is changed by about 30% (index $\Delta I_{ped} / I_{ped}$) and the plasma current density at the separatrix $j_{tot}(\rho = 1)$ varies by about 60% (index $\Delta j_{tot}(\rho = 1) / j_{tot}(\rho = 1)$). The peak values of the coils current (40 kA) required to achieve this are within the engineering limits.

Two factors lead to the generation of plasma current in the pedestal region during the plasma vertical oscillations by the VS coils in ITER: 1) the up/down asymmetry of the ITER PF system, vacuum vessel and plasma, and 2) the up/down asymmetry of the VS coils in ITER, which are not located symmetrically with respect to the plasma midplane due to integration issues. Further DINA simulations have been carried out to determine to which degree these two factors contribute to the edge plasma induced current. The performance indexes calculated using the up/down asymmetric *Baseline* in-vessel coils adopted in the present ITER design (position shown in Fig. 1 by black filled circles) were compared with those calculated using a hypothetical up/down *Symmetric* in-vessel coils (position shown in Fig. 1 by red open circles). This comparison has shown that it is the up/down asymmetry of the PF system, vacuum vessel and plasma what dominates the generation of edge plasma current and not the asymmetry of the VS coils. For the *Symmetric* coils connected in *anti-series*, the index $\Delta I_{ped} / I_{ped}$ is about 20% and the index $\Delta j_{tot}(\rho = 1) / j_{tot}(\rho = 1)$ is about 50% to be compared with 30% and 50% respectively for the *Baseline* in-vessel coils.

### 3. In-vessel coils connected in series

Oscillations of the edge plasma current can also be achieved by fast oscillations of the axisymmetric vertical magnetic field, which can be produced in ITER by the in-vessel coils, if they are connected in *series* (for production of a predominantly *vertical* magnetic field). Such oscillations result in oscillations of the poloidal magnetic flux linked with the plasma leading to the oscillations of the edge plasma current (skin current). Unfortunately, the oscillations of the vertical magnetic field lead also to plasma radial oscillations, which significantly reduce the amplitude of the oscillations of the poloidal magnetic flux linked to the plasma, leading to the reduction of the amplitude of the edge plasma current oscillations. In addition the in-vessel coils connected in *series* cannot provide the radial magnetic field required for vertical stabilization, which in this case of the DINA simulations was performed only by the VS1 circuit varying the differential currents in the PF2+PF3 and PF4+ PF5 coils [4, 5]. Taking into account that the plasma radial inward and vertical displacements, caused by the variation of current in the in-vessel coils connected in *series*, complicate plasma vertical stabilization, in the DINA simulations it was assumed a 9 kV engineering limit on the VS1 power supply. It should be noted that in the present design the voltage limit is 6 kV, although, if necessary, it can be increased to 9 kV within the insulation limit for the PF2 – PF5 coils.

Three configurations of the in-vessel coils connected in *series* were used in the DINA simulations:

1) the hypothetical up/down *Symmetric* configuration with 4 turns in the upper coil and 4 turns in the lower coil (position shown in Fig. 1 by red open circles).
2) the present *Baseline* configuration with 4 turns in the upper coil and 4 turns in the lower coil (position shown in Fig. 1 by the black filled circles),
3) the *Modified Baseline* configuration with 4 turns in the upper coil and 2 turns in the lower coil (position shown in Fig. 1 by the black filled circles),

The *Modified Baseline* configuration was chosen for this sensitivity study with the goal of maximization of the performance indexes with the present location of the in-vessel coils.

For a given frequency of the oscillations (20 Hz), the amplitude of the current in the in-vessel coils depends on the chosen voltage produced by the coils power supply (simulated as a chopper). In the DINA simulations the voltage produced by the in-vessel coils power supply was increased (taking into account the 2.3 kV limit) until the plasma vertical stabilization performed by the VS1 feedback circuit is lost. This maximum allowable value the voltage, $V_{VS}$, and the corresponding amplitude of the current in the in-vessel coils, $I_{VS}$, are shown in Table 1. Table 1 also shows the values of the performance indexes obtained in the simulations.

**Table 1.**
Performance indexes characterizing generation of current in the pedestal region ($\Delta I_{ped} / I_{ped}$ and $\Delta j_{tot}(\rho = 1) / j_{tot}(\rho = 1)$), peak value of the coil currents ($I_{VS}$) and values of the applied voltages ($V_{VS}$)

| Coils configuration | $\frac{\Delta I_{ped}}{I_{ped}}$ | $\frac{\Delta j_{tot}(\rho = 1)}{j_{tot}(\rho = 1)}$ | $I_{VS}$ kA | $V_{VS}$ kV |
|---|---|---|---|---|
| **Baseline** coils connected in **anti-series** | 30% | 60% | 40 | ±2.3 |
| **Symmetric** coils connected in **anti-series** | 20% | 50% | 35 | ±2.3 |
| **Baseline** coils connected in **series** | 9% | 25% | 25 | ±1.6 |
| **Modified Baseline** coils connected in **series** | 16% | 40% | 40 | ±1.15 |
| **Symmetric** coils connected in **series** | 8% | 19% | 40 | ±2.3 |

It should be noted that the oscillations of the vertical magnetic field leads to two effects: 1) oscillations of the plasma current density at *all* values of $\rho$ due to the "frozen" profile of $q(\rho)$ caused by the associated radial plasma position oscillations, leading to periodic adiabatic compression of the plasma, and 2) oscillations of the plasma *skin* current – oscillations of the current in the pedestal region caused by the oscillations of the poloidal magnetic flux linked with the plasma.

The hypothetical up/down *Symmetric* in-vessel coils have the smallest value of the indexes $\Delta I_{ped} / I_{ped}$ and $\Delta j_{tot}(\rho = 1) / j_{tot}(\rho = 1)$ among the configurations with the coils connected in *series*. The present *Baseline* coils have higher values of these indexes than the *Symmetric* coils. The highest values of the indexes have the *Modified Baseline* coils, where the number of turns in the lower coil was reduced from 4 to 2. In this case the performance indexes "$\Delta I_{ped} / I_{ped}$" increases to 16% and the index "$\Delta j_{tot}(\rho = 1) / j_{tot}(\rho = 1)$" increases to 40%. However these values are lower than those obtained in the simulations when the *Baseline* in-vessel coils were connected in *anti-series* (the indexes are 30% and 60%, respectively).

## 4. Conclusions

The performance indexes characterizing generation of current in the plasma edge during oscillations of coil currents in the in-vessel VS coils have been calculated in DINA simulations of 7.5 MA / 2.65 T H-mode ITER plasmas with feedback control of the plasma current, position and shape, taking into account the design limits imposed on the engineering systems (coils, power supplies, etc.). The simulations have been performed for different configurations of the in-vessel coils for an oscillation frequency of 20 Hz. The highest values of the performance indexes are obtained for the *Baseline* configuration of the in-vessel coils connected in *anti-series* for plasma vertical stabilization: $\Delta I_{ped} / I_{ped} = 30\%$, $\Delta j_{tot}(\rho = 1) / j_{tot}(\rho = 1) = 60\%$. This indicates that the baseline configuration with its foreseen operational mode is optimum for ELM triggering by this technique for low current H-mode ITER operation.

Disclaimer: The views and opinions expressed herein do not necessarily reflect those of the ITER Organization.

**References**

[1] A.Loarte, et al., Nucl. Fusion, **54** (2014) 033007

[2] P.B. Snyder et al., Nucl. Fusion, **47** (2007) 961–968.

[3] R.R.Khayrutdinov, V.E.Lukash, J. Comput. Physics, **109** (1993) 193

[4] Y.Gribov, et al., $25^{th}$ Fusion Energy Conference, Saint Petersburg, Russia, 13-18 October 2014, PPC/P3-21

[5] Y.Gribov, et al., Plasma vertical stabilization in ITER, accepted by Nuclear Fusion in 2015

# Assessment of neutron emission from DD to DT operation of ITER

A.R. Polevoi[1], A. Loarte[1], R. Bilato[2], L. Bertalot[1], R.V. Budny[3], E. Fable[2], M. Gorelenkova[3], M. Loughlin[1], E. Polunovskiy[1]

[1]*ITER Organization, Route de Vinon-sur-Verdon, 13067 St Paul Lez Durance, France*

[2]*Max-Planck-Inst. für Plasmaphysik, Boltzmannstr.2, 85748 Garching, Germany,*

[3]*PPPL, Princeton University, Princeton, NJ 08543-0451, USA*

## 1. Introduction

An assessment of the 2.45 and 14 MeV neutron yield and neutron emission spatial distributions for the DD and DT phases of ITER operation is needed to define the details of a set of dedicated experiments to be carried out in order to establish the reliability of the neutronics calculations early in ITER nuclear operations [1] and to evaluate the capabilities and performance of the ITER neutron diagnostics in different phases/scenarios during ITER nuclear operation. The main challenge for these two issues is that the evaluations of the capability/reliability are foreseen to be carried out during the ITER DD operational phase (or early in the DT phase) when the neutron flux and fluence will be orders of magnitude lower than those expected at Q = 10.

## 2. Transport model

In our analysis we use self-consistent 1.5D simulations in the frame of ASTRA [2] with the scaling based transport model [3]. The boundary conditions are defined from SOLPS predictions for ITER operation with the tungsten divertor, and the pedestal parameters for H-mode plasmas are evaluated with the EPED1 model and the SOLPS-based boundary conditions [4]. For each set of operational conditions the electron and ion heat conductivities were fitted to the appropriate scaling: Ohmic [5], L-mode and H-mode [6]. For particle diffusivity we used the same assumption for all species, $D = D_e = 0.2\,\chi_{eff} = 0.1\,(\chi_e + \chi_i)$, based on JET experiments for the ITER-like conditions [7]. As noted in the ref. [7], this choice of D is lower than the tritium diffusivity obtained in the JET tritium trace experiments, ($D_T \sim 0.3 - 2\,\chi_{eff}$) or in the analysis of H-mode density profiles from multimachine databases [8] ($D_T \sim 0.4 - 0.7\,\chi_{eff}$). Thus, our estimate of tritium accumulation in DD plasma is conservative. In the simulations, together with the reactions between the thermal D and T ions, we take into account interaction of the fast D ions from the 1 MeV ITER NBI and the 1 MeV tritium produced from the DD reaction with the thermal D and T ions in approximation of finite electron and ion temperatures. The probability of the fusion reaction of a fast ion with energy $E_A$ with a thermal specie "B" during slowing down is given by $F_{AB}(E_A,T_i,T_e)= n_e \int<\sigma_{AB}V> dt = n_e \int<\sigma_{AB}V> E^{-1} \tau(E,T_i,T_e)\, dE$, where $\tau(E,T_i,T_e) \sim 1/n_e$ is the slowing down time, $dE/dt = -E/\tau(E,T_i,T_e)$, and $<\ldots>$ denotes averaging over the thermal species Maxwellian distributions. The source of the 14 MeV neutrons, $S_{n14}$, from DT reaction, $D+T \rightarrow n + {}^4He$, is given by $S_{n14} = n_{Tth}n_{Dth} S_{DT} + S_T F_{TD} n_{Dth}/n_e + S_{DNB} F_{DT} n_{Tth}/n_e$, where $S_{DNB}[10^{19}/s]=P_{NB}[MW]/1.6$ with $P_{NB}$ is the 1 MeV NBI absorbed power, $n_{Tth}$ , $n_{Dth}$, $n_e$ are the densities of the thermal T and D ions and electrons and $S_T$ is the source of 1.008 MeV T from the DD reaction both from reaction of thermal D-D and from D-fast D from NBI DD1: D+D (50%) $\rightarrow$ T + p ($S_T = 0.5\, n_{Dth}^2 S_{DD1} + S_{DNB} F_{DD1} n_{Dth}/n_e$). The source of 2.45 MeV neutrons, $S_{n2.45}$, from the DD (thermal DD + thermal-fast NBI D) reaction DD2: D+D (50%) $\rightarrow n_{2.45} + {}^3He$, is given by $S_{n2.45} = 0.5\, n_{Dth}^2 S_{DD2} + S_{DNB} F_{DD2} n_{Dth}/n_e$. The temperature dependence of the fusion rates and the probabilities of burn during fast particles slowing down are shown in Fig. 1 (a, b) with the expected increase of the ratio of thermal to slowing-down fusion rates $S_{DT}/F_{DT}$ with temperature (Fig.1 (a)). The fusion probabilities for fast particles take into account the finite value of the ion temperature both for DD and DT. To evaluate them we have extended the technique, proposed in [9] for DD, to DT plasmas keeping all terms of order $\varepsilon = T_i/E_A << 1$ omitted in [9]. This improved formulation leads to different fusion probabilities for fast particles at high $T_i$ in comparison with those evaluated for cold ions (see Fig. 1 (c)), as expected.

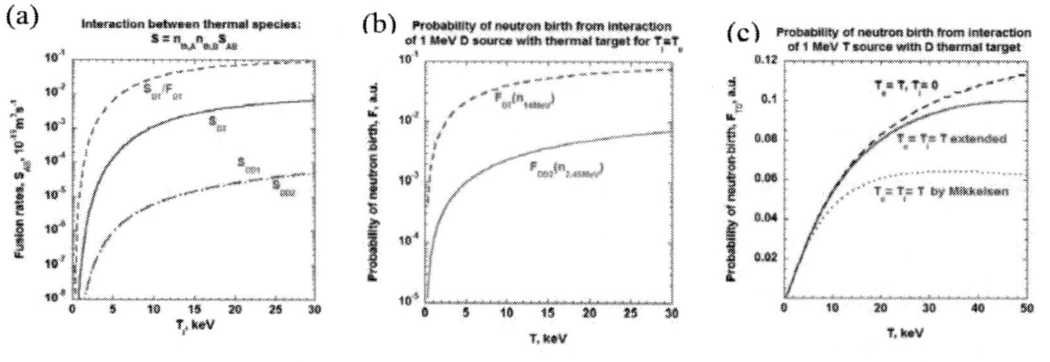


*Figure 1. Fusion rates for thermal species (a), probabilities of the fast ion reactions with thermal species during the slowing down (b), probability of fast ion reactions during slowing down of 1.008 MeV T on thermal D ions.*

### 3. Code benchmarking

The simulations with ASTRA [2] and a simplified NBI description [10] have been successfully benchmarked with simulations from TRANSP+NUBEAM [11], which uses a Monte-Carlo technique for description of fast ions, as shown in Fig. 2. (a, b) for a 12 MA ITER plasma with 33 MW of NBI power and $n_D = n_T$. In these conditions the 14 MeV neutron source from the fast D(NBI)-thermal T reaction is negligible compared to the thermal DT source while the 2.45 MeV neutron source from fast D(NBI)-thermal D reactions is comparable to that of the thermal DD reaction in these conditions (Fig. 2 (c)).

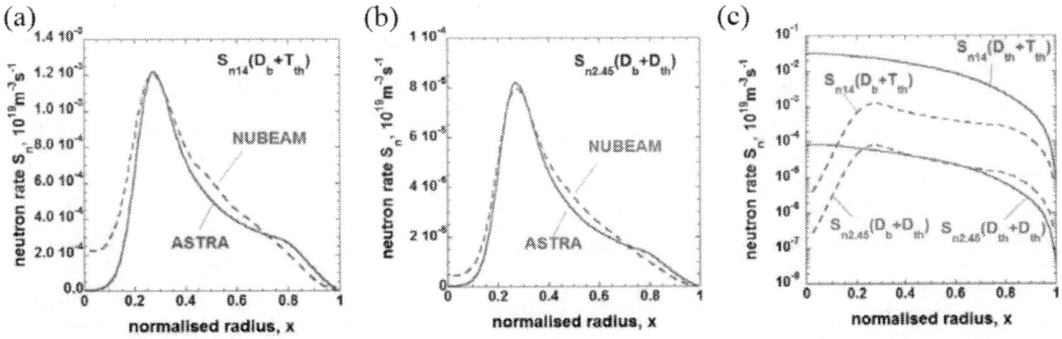


*Figure 2. Comparison of the neutron sources predicted by ASTRA and TRANSP+NUBEAM for ITER plasma with 33 MW of 1 MeV D-NBI and $n_D$=$n_T$. N neutron sources from the DT reaction and DD reactions: $S_{n14}$(Fig. 2 (a)) and $S_{n2.45}$(Fig. 2 (b)). Neutron sources predicted by TRANSP+NUBEAM for the thermal-thermal and fast-thermal fusion reactions (Fig.2 (c)).*

## 4. ITER simulations

Our assessment includes ITER DD and DT plasmas in Ohmic, L-mode and H-mode with full (5.3T) and half (2.65T) toroidal field and plasma currents 7.5 – 17 MA.

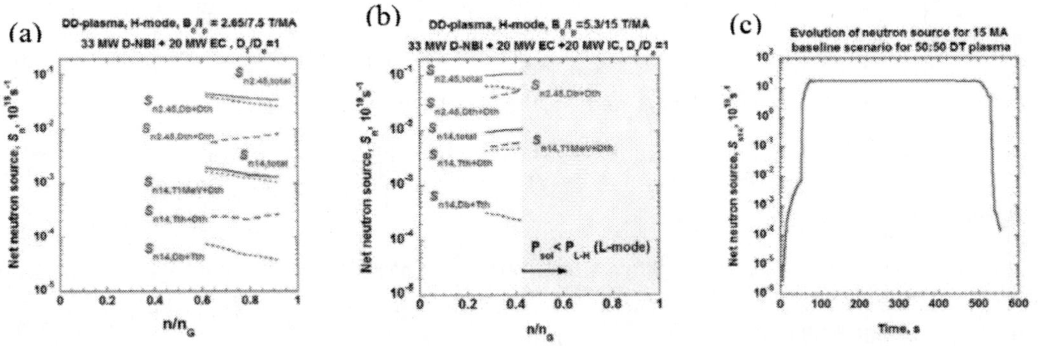


*Figure 3. Neutron production versus normalized plasma density for DD H-mode plasmas with $I_p$ = 7.5 MA (Fig. 3(a)), $I_p$= 15 MA (Fig.3.(b)) and 14 MeV neutron production rate versus time for the 15 MA Q=10 DT scenario (Fig. 3(c)).*

The DD plasma results for $S_n = \int S_n dV$ shown in Fig. 3 (a, b) assume zero T recycling because of the low amount of T produced. Increasing T recycling in these conditions up to 100% leads to a doubling of the edge T density and to an increase of the total T content by 30% in H-mode. This leads to an increase by 30% of the thermal T sources for 14 MW neutrons compared to the values in Fig.3. (a b). Increasing the T thermal diffusivity by a factor K = $D_t/D$ = 1.5 – 10, as observed in the JET T-trace experiments [7], decreases the thermal T content and thermal 14 MeV neutron sources as 1/K compared to the values in Fig.3. (a b). As shown in Fig. 3.a, for 7.5MA/2.65T DD plasmas the dominant 14 MeV neutron source is caused by the burn of 1 MeV T produced in the DD reaction (Fig. 1(a)) so that assumptions on T transport and recycling have little effect on the 14 MeV neutron rate in these conditions. The same applies to the 2.45 MeV neutron source which is dominated by the 1 MeV NBI-D reaction with the thermal D. For 15MA/5.3T DD plasmas, which have higher temperatures, the contribution from the thermal reactions the 14 MeV and 2.45 MeV neutrons increase and are comparable to those from fast particle-thermal reactions (Fig. 3 (b)) as expected from Fig. 1 (a) (see $S_{DT}/F_{DT}$). The typical neutron yields in DD and DT ITER Ohmic and H-mode plasmas are summarized in Table 1.

**Table 1. Neutron yield at plasma current flat-top**

| $B/I_p$,T/MA | $n_e,10^{19}m^{-3}$ | $S_n,10^{19}s^{-1}$ | $S_{n14}/S_n$,% | $n_T/n_e$,% | $P_{heat}/P_\alpha$,MW | mode |
|---|---|---|---|---|---|---|
| 2.65/7.5 | 3 | $1.06\ 10^{-5}$ | 0.8 | $2.74\ 10^{-6}$ | $7.5/4.64\ 10^{-7}$ | OH/DD |
| 2.65/7.5 | 3.6 | 0.0468 | 4 | 0.006 | 53/0.011 | H/DD |
| 5.3/15 | 3.3 | 0.1093 | 8.5 | 0.019 | 73/0.053 | H/DD |
| 5.3/15 | 11.2 | 17.45 | ~100 | 39.8 | 50/98 | H/DT |
| 5.3/17 | 12.5 | 25 | ~100 | 38.5 | 50/140 | H/DT |



**References**

[1] M. Loughlin, et al, Fusion Engineering and Design **89** (2014) 1865–1869
[2] G.V. Pereverzev, P.N. Yushmanov, "ASTRA Automated System for TRansport Analysis", IPP-Report IPP 5/98 (2002)
[3] A.R. Polevoi, et al, Plasma Phys. and Contr. Fusion **48** (2006) A449
[4] A.R. Polevoi, et al, Nucl. Fusion **55** (2015) 063019
[5] G. Bracco and K. Thomsen, Nucl. Fusion **37** (1997) 759
[6] ITER Physics Basis, 39 (1999) 2208
[7] M. Valovic, et al, Nucl. Fusion **47** (2007) 196-200
[8] C. Angioni, et al , Nucl. Fusion **47** (2007) 1326
[9] D.R. Mikkelsen, Nucl. Fusion **29** (1989) p. 1113
[10] A.Polevoi, H.Shirai, T.Takizuka,"Benchmark of the NBI Block in ASTRA Code versus the OFMC Calculations", JAERI-DATA-Code 97-014, JAERI, March 1997
[11] A. Pankin, et al., Computer Phys. Communications, Vol. 159, No. 3 (2004) 157-184

# Simulations of ITER and DEMO plasmas during pellet operation using BALDUR code with a theory-based toroidal velocity model

A. Wisitsorasak[1], P. Klaywittaphat[2], B. Chatthong[2], T. Onjun[2]

[1] *King Mongkut's University of Technology Thonburi, Bangkok, Thailand*

[2] *Sirindhorn International Institute of Technology, Thammasat University, Pathum Thani, Thailand*

## Introduction

The injection of small solid deuterium-tritium pellets has been demonstrated in various tokamak experiments that the pellet injection can give rise to the enhancement of the central plasma density and yield fusion performance to be in a more desirable range [1]. Moreover the transition from L-mode to H-mode can be accessed through pellet operation [2]. The pellet injection, as it entered into a hot plasma, also perturbs surrounding environment. In this work we investigate the modification of the toroidal rotation and the radial electric field after the pellet injection in ITER and EU-DEMO using 1.5D BALDUR integrated predictive modeling code [3].

## Neutral Gas Shielding Model

The neutral gas shielding model for ablation of frozen hydrogenic pellets was developed by Milora and colleagues and has been integrated into BALDUR code [4]. The model accounts for the effects of 1) an initial Maxwellian distribution of incident electron energies; 2) a cold plasma shield outside the neutral shield; 3) energetic neutral beam ions and alpha particles; and 4) self-limiting electron ablation in the collisionless plasma limit. The pellets in this study were launched into a tokamak from low field side (LFS). As it traveled into the targeted plasma, the pellet particles near the surface are subsequently ionized due to collision with background ion and electron plasma and neutral particles from neutral beam injection heating. This leads the formation of neutral cloud around the pellet material. The rate of the ablation process is approximately given by [4]

$$\dot{r}_p = -1.25\frac{\int_{r_p}^{r_0} n^0 dl}{2r_p n_s^m}\left(q^0 r_p \frac{\gamma-1}{2}\right)^{1/3} \tag{1}$$

where $q^0$ is the net energy flux, $\gamma = 7/5$ the ratio of specific heats for hydrogen gas, $m_p$ the mass of a proton, $A_p$ the atomic mass number of the pellet species, $n^0$ the atomic density in the neutral cloud, $r_0$ the radius of the neutral cloud. The detailed calculation was presented in [4].

## Toroidal Rotation Model

The toroidal rotation of the plasma is mainly contributed by two parts: neoclassical toroidal viscosity (NTV) and ion cyclotron resonance frequency (ICRF) heating, in which the auxiliary heating includes the effects of neutral beam injection (NBI), ion cyclotron resonance frequency (ICRF) heating, and lower hybrid (LH) heating. The toroidal rotation due to the auxiliary heating, in general, is much smaller than the other component from NTV; therefore, it was excluded in this study. The toroidal velocity component from NTV is caused by symmetry breaking of a non-axissymmetric field and has the form [5]:

$$v_{\phi,NTV} = \frac{1}{eZB_\theta}\left(3.54 - \left(\frac{r}{R}\right)K_l\right)\frac{dT_i}{dr} \tag{2}$$

## Results and Discussion

The BALDUR integrated predictive transport modeling code with is used to carry out the simulations for two designs: ITER with the designed parameter ($R = 6.2$ m, $a = 2.0$ m, $I_p = 15$ MA, $B_T = 5.3$ T, $\kappa_{95} = 1.70$, $\delta_{95} = 0.33$ and $n_l = 1.0 \times 10^{20}$ m$^{-3}$) [6], and EU-DEMO with the designed parameters ($R = 9.0$ m, $a = 2.25$ m, $I_p = 30.5$ MA, $B_T = 7.36$ T, $\kappa_{95} = 1.70$, $\delta_{95} = 0.33$ and $n_l = 1.03 \times 10^{20}$ m$^{-3}$) [7]. The pellet ablation is described by NGS model, while an anomalous transport is calculated using the MMM95 transport model. The boundary condition are computed through the pedestal model based on magnetic flow shear stabilization.

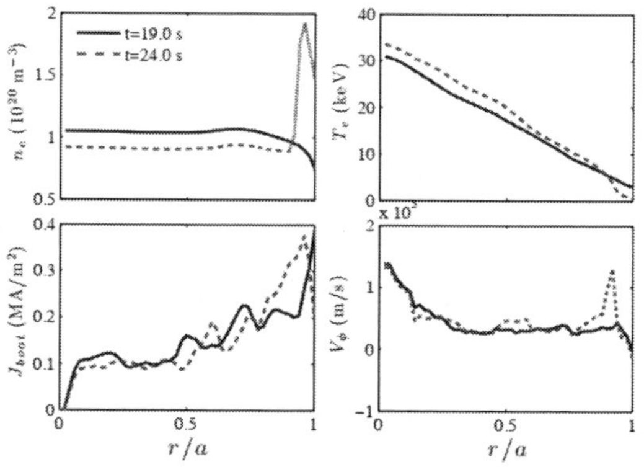


Figure 1: Comparison of the plasma profiles when pellets enter the target plasma in ITER.

Figure 1 illustrates the simulation results of the electron density $n_e$, the electron temperature $T_e$, the bootstrap current $J_{boot}$, and the toroidal velocity $V_\phi$ as predicted for ITER. The blue solid lines and the red dashed lines respectively represent the profiles before and after the pellet operation. The deuterium pellets of radius 3 mm and speed of 300 m/s were launched from LFS during time of $t = 20 - 70$ s. The launching frequency is 8 Hz. We will refer this injection scheme as Scenario I. After the pellets were released, one can see an overshoot in the density profile due to mass deposition of the pellets. The electron temperature also drops significantly near the edge, while it slightly increases in the central core area. The increased particle density also leads to an increase of the bootstrap current density near the edge of the plasma. The toroidal current is slightly suppressed near the edge of the plasma because of the increasing particle density in the local region. We also observed a peak of the toroidal current where the gradient of the electron temperature has the largest value.

While injection of small pellets at low speed is feasible for our current technology, launching the pellets of very large size and high speed into dense plasma is still scientifically interesting. In this part, we simulate the plasma profiles after the pellet injection as obtained by using EU-DEMO design. The pellets of radius 9 mm and speed of 8 km/s are injected from LFS plasma. The frequency of the pellet injection is 1 Hz. The time period of the inject is $t = 20 - 70$ s. This injection procedure will be called as Scenario II. The toroidal speed also shows larger value than the other case.

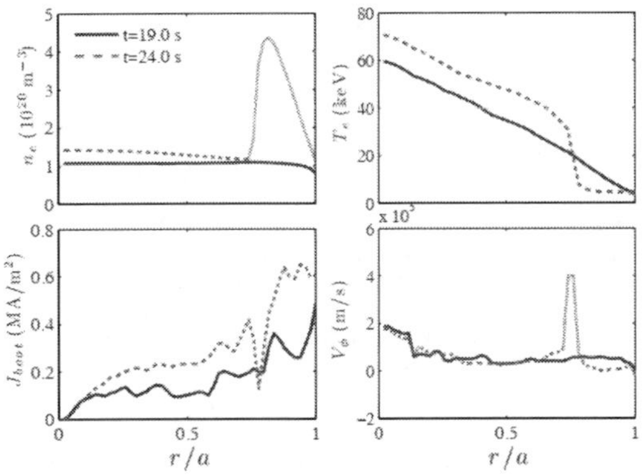


Figure 2: Plasma profiles in EU-DEMO before ($t = 19.0$ s) and 4 seconds after ($t = 24.0$ s) pellet injection. Note that the pellet size and speed are 9 mm and 8 km/s. The injection frequency is 1 Hz.

Figure 2 shows the plasma profiles before and after the pellet injection. Under this circumstance, the pellet operation results in an increase of the electron density and the electron temperature at the central core region. The size of the pellet used in Scenario II is substantially larger than that in Scenario I. Thus pellets are able to penetrate the plasma deeper and deposit larger amount of pellet material. This also leads to enhancement of the bootstrap current throughout the plasma region.

It is known that the toroidal rotation plays a crucial role in the mechanism of flow shear suppression for ITB formation. One quantity that plays crucial role such formation is the radial electric field which can be computed from the radial force balance equation:

$$E_r = v_{\phi i} B_\theta - v_{\theta i} B_\phi + \frac{Z_i}{n_i} \frac{dp_i}{dr}. \quad (3)$$

The contribution from each term fo the radial electric field as a function of a minor radius is presented in Figure 3 for EU-DEMO and Figure 4 for ITER plasma. As can be seen in Figure 4, the radial electric field $E_r$ is dominated by the toroidal velocity component. The poloidal velocity and the neoclassical

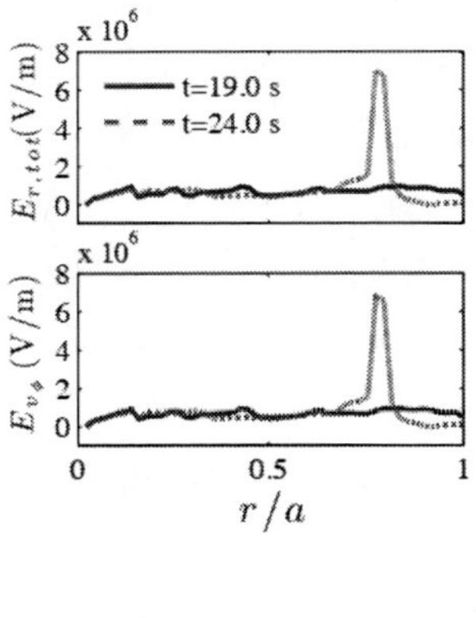


Figure 3: The radial electric field as simulated from EU-DEMO.

temrs are comparable. Consequently they tend to cancel each other. The radial electric field from the toroidal velocity exhibits similar behavior as the toroidal velocity. Both of them show significant reduction near the plasma edge, and a peak at the same location of the density profile.

## Conclusion

The impacts of pellets injected from the LFS on the plasma core and edge in ITER, and EU-DEMO are carried out using the 1.5D BALDUR integrated predictive modeling code combined with the NTV model and the NGS pellet ablation model. It is found that injection of pellets from LFS can enhance an increase in the plasma average density in the tokamak devices. As the particle density deposited in the plasma, the bootstrap current fraction increases. The toroidal rotation is suppressed near the plasma edge, but shows an increasing value at the same location of density peak due to the pellet injection. The change in the toroidal velocity leads to a local perturbation in the radial electric field through the radial force balance equation which is dominated by the toroidal velocity term.

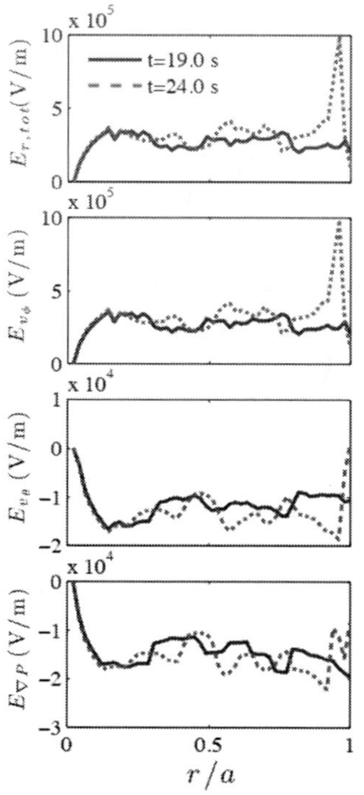


Figure 4: The radial electric field and its contribution from each term as determined from the radial force balanced equation. Data are simulated for ITER plasma after the pellet operation.

## References

[1] S. Milora, W. Houlberg, L. Lengyel, and V. Mertens, *Nuclear Fusion*, vol. 35, no. 6, pp. 657–754, 1995.

[2] M. Valovič, K. Axon, L. Garzotti, S. Saarelma, A. Thyagaraja, R. Akers, C. Gurl, A. Kirk, B. Lloyd, G. Maddison, *et al.*, *Nuclear Fusion*, vol. 48, no. 7, p. 075006, 2008.

[3] C. Singer, D. Post, D. Mikkelsen, M. Redi, A. McKenney, A. Silverman, F. Seidl, P. Rutherford, R. Hawryluk, W. Langer, *et al.*, *Computer physics communications*, vol. 49, no. 2, pp. 275–398, 1988.

[4] W. Houlberg, S. Milora, and S. Attenberger, *Nuclear fusion*, vol. 28, no. 4, p. 595, 1988.

[5] M. Kikuchi, *38th EPS Conference on Plasma Physics*, P4.115, 2011.

[6] R. Aymar, P. Barabaschi, and Y. Shimomura, *Plasma Physics and Controlled Fusion*, vol. 44, no. 5, p. 519, 2002.

[7] G. Federici, R. Kemp, D. Ward, C. Bachmann, T. Franke, S. Gonzalez, C. Lowry, M. Gadomska, J. Harman, B. Meszaros, *et al.*, '*Fusion Engineering and Design*, vol. 89, no. 7, pp. 882–889, 2014.

# Linear ideal and resistive MHD stability analysis of post-disruption plasmas with runaway electrons in ITER

K. Aleynikova[1,2,3], G.T.A. Huijsmans[2], P. Aleynikov[1,2]

[1]*Max-Planck-Institut für Plasmaphysik, EURATOM Association, Greifswald, Germany*

[2]*ITER Organization, Route de Vinon-sur-Verdon, St Paul-lez-Durance Cedex, France*

[3]*Moscow Institute of Physics and Technology, Dolgoprudny, Russian Federation*

Most of the plasma current can be replaced by a runaway electron (RE) current during plasma disruptions in ITER [1]. In this case, the post-disruption plasma current profile is likely to be more peaked than the pre-disruption profile [2]. Such plasmas may become MHD unstable. This would affect the runaway electron generation and confinement and the dynamics of the plasma position evolution (Vertical Displacement Event (VDE)), possibly limiting the time frame for RE and disruption mitigation.

For the evaluation of possible MHD instabilities in post-disruption plasmas in ITER, we calculate the self-consistent evolution of the plasma equilibrium during a VDE event with the free boundary equilibrium code DINA [4], starting with a prescribed seed RE current profile and a prescribed impurity profile. The linear ideal and resistive MHD stability of the resulting sequence of equilibria is then analyzed with the codes MISHKA [5] and CASTOR [6].

We start the DINA VDE calculations from an equilibrium of the 15MA Q=10 ITER scenario. We prescribe the Thermal Quench (TQ) event to happen, leading to a plasma temperature drop and a flattening of the q-profile inside q=2 flux surface. We also assume that a constant and flat Ar impurity density of $0.25 \cdot 10^{20} m^{-3}$ was introduced in the pre-TQ phase to mitigate thermal loads. This amount of Ar is enough to keep the post-disruption temperature low and provide a high inductive electric field to initiate a RE avalanche. In the present study, we side-step the questions of primary RE generation and prescribe the seed RE current profile. We use a Gaussian profile for the RE seed current, and our parametric scan involve a variation of the Gaussian profile width ($w$) and amplitude:

$$j_{seed} = j_0 e^{-\frac{\rho^2}{w^2}},$$

where $\rho$ is a minor radius.

The subsequent evolution of the plasma equilibrium is calculated self-consistently with the RE generation (via avalanche mechanism [1]), resulting in the evolution of the total plasma current profile as well as the whole equilibrium.

An example of such evolution (for the RE seed current of 22kA) is plotted in Figure 1. In this case the TQ occurs at 84ms, and then the avalanche quickly multiplies the RE current profile resulting in a total RE current of 8MA by 95ms. During this evolution, due to the contraction of the total current profile (Fig. 1-2)) the internal inductance $l_i$ grows from the initial value of 0.6 to 1.4, while the value of $q$ at the magnetic axis and at the boundary drops significantly.

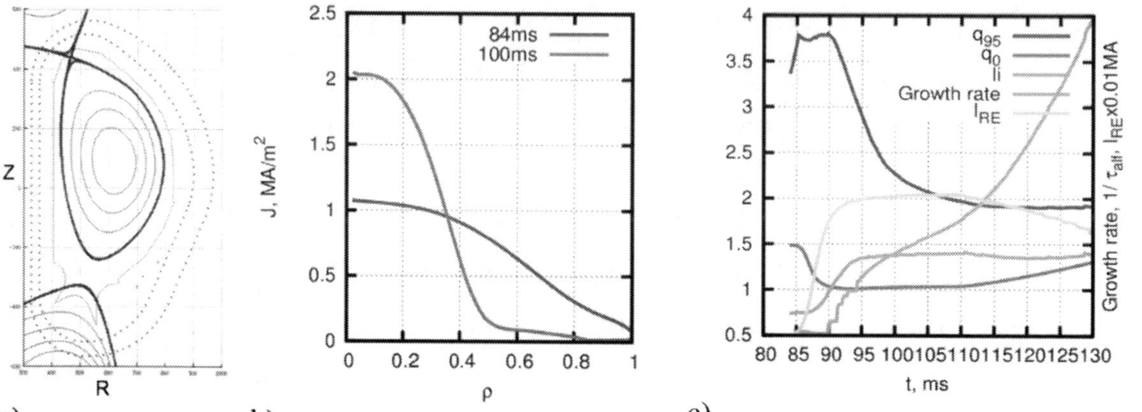


a) b) c)

*Fig. 1. The evolution of the plasma equilibrium for the RE seed current of 22kA: a) plasma shape b) "shrinking" of the total current profile c) time evolution of the the internal kink growth rate, $q$, $q_o$ , $l_i$ and $I_{RE}$*

It is reasonable to expect that $q \leq 1$ at the axis will give a rise to a sawtooth-like mode and integer $q$ at the boundary will result in an external kink instability. For brevity, we will provide only $q - l_i$ plots of other studied cases (see figure 2). In most of these cases $q_a$ approaches 1.

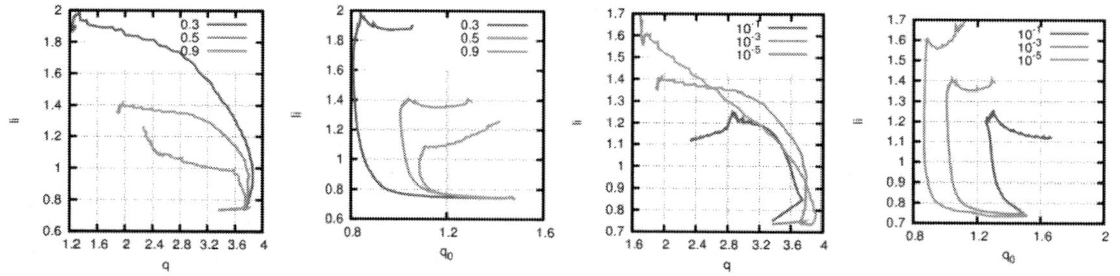


*a) b) c) d)*

*Fig. 2. a) $q_{95}$-li and b) $q_0$-li plots for different RE current profile width (w is indicated in the legends) and $I_0$ =22kA c) $q_{95}$-li and d) $q_0$-li plots for different on RE seed amount ($j_0$ is indicated in the legends) and w=0.5.*

Whilst ideal MHD modes are insensitive to the nature of plasma current, the resistive MHD instability in the runaway plasmas was investigated in [3]. It was shown that linear growth rate of resistive modes in a runaway-dominated plasma is approximately the same as in the plasma without runaways. We thus rely on these findings and proceed to a linear ideal and resistive MHD stability analysis of the resulting sequence of equilibria with MISHKA [5] and CASTOR [6] codes. MISHKA is an ideal, incompressible MHD code. In CASTOR, the linearized resistive MHD equations are solved in a general toroidal geometry, where the equilibria are assumed to be axisymmetric and static. We calculate the linear growth-rates of ideal MDH modes with both codes for crosscheck.

Figure 3a shows the evolution of the linear growth rate for the cases shown previously in Figs. 1 and Fig. 2. In all these cases, the dominant mode is an internal resistive kink mode with n=m=1. An example of the evolution of the structure of this mode is shown in Fig.3b.

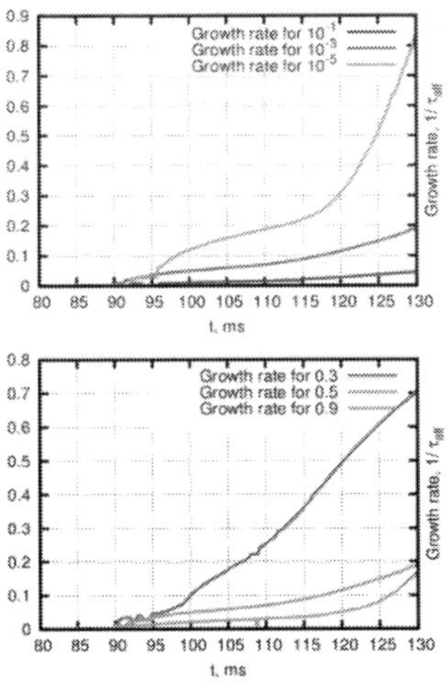


*Fig.3a Time evolution of the internal kink growth rate dependence on the RE seed amplitude, w=0.5 (up) and RE current profile width, $I_0$=22kA (down).*

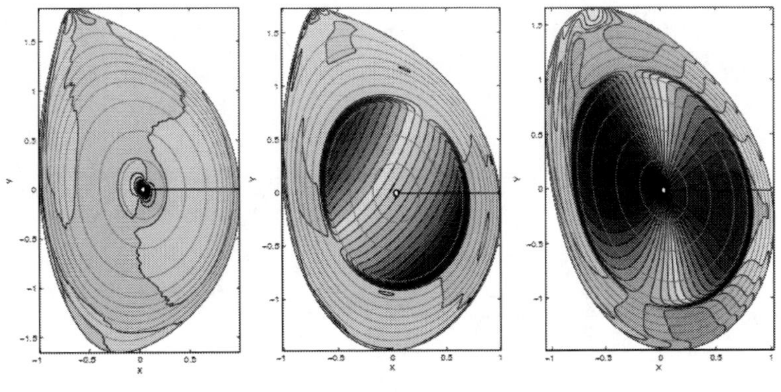

*Fig.3b The evolution of the internal kink mode during a VDE $I_0$=22kA, w=0.5. (shown is structure of the displacement of the perturbation in the poloidal plane.)*

This mode is present (and dominant) in most of the analyzed cases, and although it has relatively high growth rate it is not expected to significantly affect the course of VDE [7]. This mode will most likely result in a flattening of the current profile (this effect is not taken into account self-consistently in the current study). Omitting this mode from the consideration, we find that a variety of other modes develop in during the profile evolution. The growth rate and a selection of mode structures are shown in Fig.4a. Two extreme cases (narrow profile *w*=0.3 case and low seed current $I_0$=0.2kA case) result in a very peaked current profile (with no current on the periphery) giving rise to internal modes (Fig.4b -3,6). On the other hand, a smooth and high-seed RE profile (*w*=0.9 and $I_0$=2.2MA cases) provides more current at the boundary accompanied by external kink and q=2 tearing modes (Fig.4b - 1,4) which may result in fast bursts and RE losses [8]. The intermediate case ($I_0$=22kA, *w*=0.5) shows moderate external kink instability when $q_{95}$ approaches integers (Fig.4b -2,5).

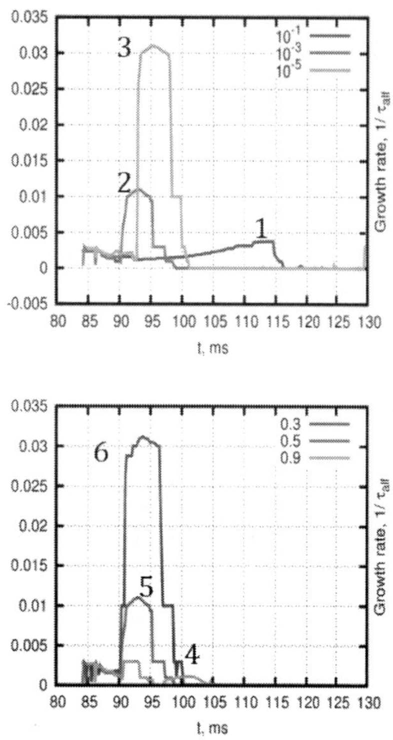


*Fig.4a Time evolution of growth rate dependence on the RE seed amplitude, w=0.5 (up) and RE current profile width, $j_0$=22kA (down) excluding m=n=1 mode.*

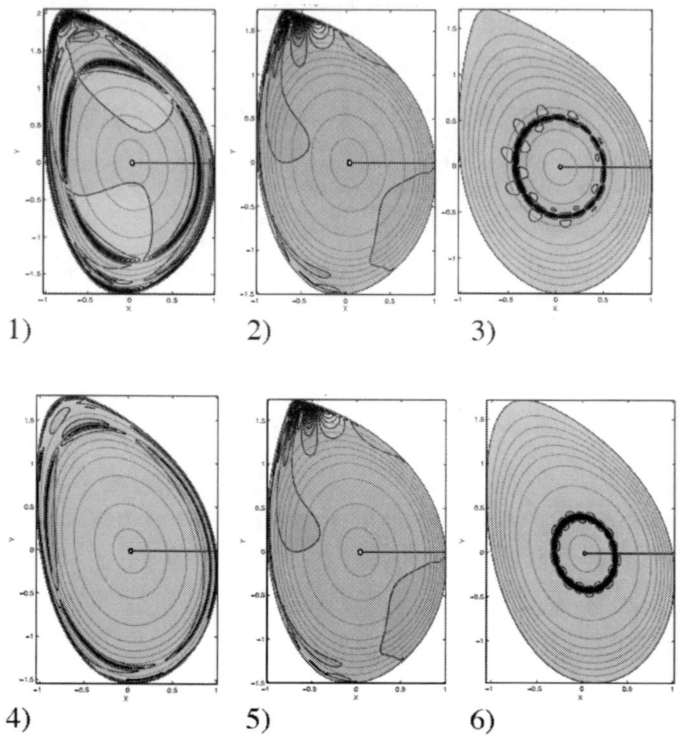


*Fig.4b Structure of the modes (excluding m=n=1 mode) which can develop during the evolution of the RE current profile in ITER disruptions. Dependence on RE seed amplitude, w=0.5 (up, 1-3) and RE current profile width $j_0$=22kA (down, 4-6)*

In summary, we conclude that both ideal and resistive modes (internal kink modes, tearing modes and external kink modes) can develop during the evolution of the RE current profile in ITER disruptions. These modes can significantly affect the RE confinement and a course of the VDE. External modes are more pronounced for higher and broader initial RE seeds, tearing modes are present in both extreme cases (while absent in the intermediate case), and almost any initially peaked RE seed profile causes an internal m=n=1 mode. Further investigations are required to derive a desirable disruption scenario for ITER, and non-linear MHD dynamics needs to be taken into account for the VDE evolution.



**References**


[1] M. N. Rosenbluth, S. V. Putvinski, Nucl. Fusion 37, 1355, (1997)

[2] L.G. Eriksson, P. Helander, F. Andersson, D. Andersson and M. Lisak,Phys. Rev. Lett. 92, 205004 (2004).

[3] Helander P., Grasso D., Hastie R.J. and Perona A. 2007 Phys. Plasmas 14 122102

[4] R. Khayrutdinov and V. Lukash, J. Comput. Physics 109 (1993) 193-201

[5] A. Mikhailovskii, G. T. A. Huysmans, W. O. K. Kerner, and S. Sharapov, Plasma Phys. Rep. 23, 844 1997

[6] W Kerner, JP Goedbloed, GTA Huysmans, S Poedts, E Schwarz. J. of comput. Phys. 142(2),271-303, (1998)

[7] Huishan Cai and Guoyong Fu, Nucl. Fusion 55, 022001, 2015

[8] Smith H.M., Feher T., Fu¨lo¨p T., Gal K. and Verwichte E. 2009 Plasma Phys. Control. Fusion 51 124008

# Optimization of plasma current ramp-up in DINA simulations of 15 MA ITER scenario with JINTRAC plasma transport modelling

V.E. Lukash[1], A.A. Kavin[2], Y. Gribov[3], R.R. Khayrutdinov[1], F. Koechl[4], A. Loarte[3]

[1]*National Research Centre "Kurchatov Institute", Moscow, Russia*

[2]*Joint Stock Company "D.V. Efremov Institute of Electrophysical Apparatus", Saint Petersburg, Russia*

[3]*ITER Organization, Route de Vinon sur Verdon, CS 90 046, 13067 St Paul Lez Durance Cedex, France*

[4]*Association EURATOM-ÖAW/ATI, Atominstitut, TU Wien, 1020 Vienna, Austria*

## 1. Introduction

Developed in 2014, a new version of the DINA code (DINA with prescribed kinetic profiles) [1] allows validation of the capability of ITER PF system to support scenarios obtained with other codes. These other codes may have more sophisticated plasma transport models, but do not have the capability of: 1) detailed simulation of plasma magnetic control, 2) correct calculation of the magnetic flux linked with the plasma from plasma initiation and 3) detailed integration of all engineering limitations. The simulation of ITER 15 MA $Q \approx 10$ scenario described in [1] was performed using the JINTRAC code suite to model plasma transport [2] for the fastest plasma current ramp-up (lasting 50 s) in ITER, which is limited by the voltage produced by the power supply of the Central Solenoid central modules (CS1U, CS1L) connected in series. The results of the JINTRAC modelling were then incorporated into a DINA simulation which showed that, in this scenario, due to the low value of the plasma internal inductance, $l_i$, at the start of current flattop (SOF), the maximum value of the magnetic field on the PF6 coil conductor, $max(B_{PF6})$, during the burn is significantly higher than the design limit (7.3 T vs. 6.4 T). It should be noted that for a given separatrix shape the maximum current in the PF6 coil (and the magnetic field on the coil conductor) in the flat top phase decreases with: 1) the increase of $l_i$ and 2) the increase of the current in the CS1U-L coils flowing opposite to the plasma current (i.e. with the increase of the poloidal magnetic flux consumption during plasma current ramp-up). Therefore the peak value of the magnetic field on the PF6 conductor can be reduced by increasing the plasma current ramp-up time, $t_{ramp}$, leading to the increase of the resistive and inductive magnetic flux consumptions in this phase. The corresponding increase of $l_i$ during the current flattop and the increase of negative current in the CS1U-L coils at the SOF lead to a reduction in the PF6 current and magnetic field on the PF6 conductor.

This paper presents the results of an optimization study for the duration of the plasma current ramp-up in the ITER 15 MA Q ≈ 10 scenario simulated with the DINA code using the JINTRAC plasma transport model. The purpose of the optimization was the reduction of the maximum magnetic field on the PF6 conductor to values lower than the design limit while maintaining an appropriate burn length as required for ITER (> 300 s). The simulations were performed for two values of the plasma current ramp-up time of 70 s and 80 s and include the modelling of plasma current, position and shape control taking into account: 1) engineering limits imposed on coils and on their power supplies (including the switching network units and converters), 2) eddy currents induced in the vacuum vessel and 3) allowable values of the plasma-wall gaps and positions of the separatrix strike points on the inner and outer divertor target plates. The simulations were carried out from the start of CS discharge (plasma initiation) to the end of burn (EOB), defined as the state when current in the CS1U-L coils approaches the engineering limit ($I_{CS1U\text{-}L}$(EOB) = -44.2 kA for an engineering limit of 45 kA).

## 2. Simulation model

The DINA simulations combined a 2-D free-boundary equilibrium evolution solver and 1-D calculation of the poloidal magnetic flux diffusion together with the JINTRAC plasma transport modelling for these plasma conditions.

### *JINTRAC plasma transport model*

JINTRAC is a system of codes for integrated simulations of tokamak scenarios [2]. It includes the 1.5D core transport code JETTO+SANCO and the multi-fluid SOL-divertor code EDGE2D-EIRENE. In JETTO+SANCO, the transport equations are solved for plasma current, pressures, main ion and impurity densities and toroidal momentum. In the ITER baseline scenario simulations for DINA, anomalous transport coefficients in the core are determined by GLF23 in H-mode and by the Bohm/gyroBohm model in L-mode. If the net heat flux, $P_{net}$, at the separatrix exceeds the L-H transition power threshold, $P_{L\text{-}H}$, anomalous transport within the ETB is gradually suppressed as a function of $exp(-(P_{net}- P_{L\text{-}H})/P_{L\text{-}H})$. The pedestal pressure, ETB width and boundary conditions at the separatrix are derived from EPED1-SOLPS calculations [3].

### *Steps in scenario simulations*

Plasma initiation and early current ramp-up to 1.6 MA (3.5 s after the start of CS discharge) were simulated using the standard version of the DINA code with a 0-D plasma transport model similar to that described in [4]. The plasma current of the first free-boundary plasma equilibrium is 0.1 MA ($t$ = 1.2 s) [1]. When plasma current exceeds 1.6 MA

($t > 3.5$ s), the simulation is performed with the newly developed DINA version using the kinetic profiles obtained in the corresponding JINTRAC simulations. The simulations of scenarios were performed in three steps:

- At the *first* step, the time sequence of the sets of ($R$, $Z$) coordinates of a number of points located on the plasma boundary and the waveform of the plasma current, used as input data in the JINTRAC code, were calculated by standard version of the DINA code utilizing the following approach: before the X-point formation (at 11.4 s and with $I_p = 3.4$ MA) the rate of plasma current ramp-up was kept the same as it was in the simulation with the 50 s plasma current ramp-up [1], after the X-point formation the rate of plasma current ramp-up was decreased to achieve the SOF at $t = 70$ s or at $t = 80$ s.
- At the *second* step, the sets of plasma boundaries and the waveform of plasma current obtained in the DINA simulation at step 1 were used in the JINTRAC simulation with prescribed evolution of the plasma current and boundary using a more sophisticated plasma transport model than that in the standard version of the DINA code [5]. Results of the JINTRAC simulations provide, among others, the time evolutions of the following profiles of plasma kinetic parameters: 1) the plasma electric conductivity, 2) the density of non-inductively driven current, 3) the total pressure (produced by the thermal and non-thermal plasma particles), 4) the plasma density (electron and ion) and 5) the plasma temperature (electron and ion). The JINTRAC simulations were carried out up to $t = 200$ s. i.e. beyond the time for which stationary burning plasma conditions are achieved.
- At the *third* step, the JINTRAC produced profiles (with a time step of 1 s and applying linear interpolation between the steps) were used to carry out the scenario simulations using the newly developed DINA code version with prescribed time evolution of the kinetic profiles. The plasma profiles for $t > 200$ s were assumed remain unchanged to the EOB consistent with the JINTRAC modelling findings. The final DINA simulation thus combines the benefits of the DINA plasma magnetic control simulation with accurate calculation of the poloidal magnetic flux and the JINTRAC detailed plasma transport modelling.

**3. Results**

Fig. 1 shows the waveforms of the plasma current and $l_i$ during the first 100 s of the scenario in the simulations with plasma current ramp-ups of 50 s, 70 s and 80 s. The corresponding waveforms of the maximum magnetic field on the PF6 conductor ($max(B_{PF6})$) are shown in Fig. 2. The increase of the current ramp-up duration from 50 s to 70 s is sufficient to reduce $max(B_{PF6})$ from 7.3 T to its design limit of 6.4 T. The increase of the current ramp-up duration to 80 s decreases $max(B_{PF6})$ further down to 6.1 T.

Figs. 3 and 4 show the evolution of the plasma parameters from plasma initiation to the EOB ($t$ = 742 s) in the scenario with 70 s plasma current ramp-up. The duration of burn in this case is about 650 s. The increase of the duration of the plasma current ramp-up to 80 s results in the reduction of the burn duration to about 500 s. It should be noted that the burn duration values depend on the assumptions regarding $Z_{eff}$ during the burn, which in these simulations is $Z_{eff} \approx 1.2$ consistent with high density operation with Be/W plasma facing components at JET. Seeding of extrinsic impurities for the control of power loads to the divertor during the burn could increase the value of $Z_{eff}$ beyond 1.2 (but still lower than $\approx 1.8$ to allow the achievement of $Q \approx 10$ [3]) and correspondingly reduce the burn length inversely proportional to the $Z_{eff}$ value. The reduced burn lengths would still be larger than the minimum of 300 s required for ITER even for the current ramp-up duration of 80 s. In addition, these simulations with 70 and 80 s ramp-up times have all engineering parameters within the design limits.

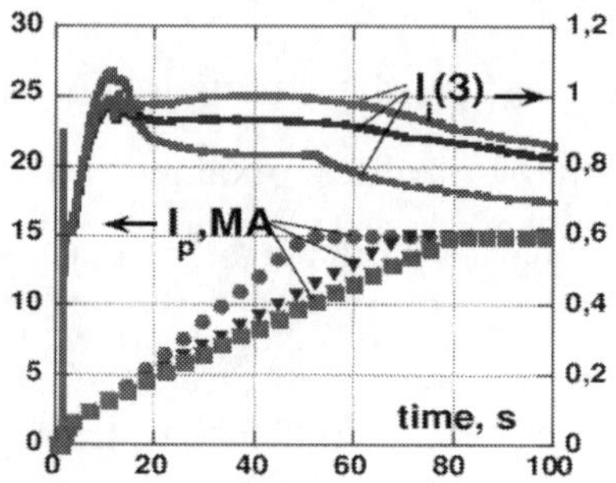

**Fig. 1**: Time traces of $I_p$ and $l_i$ in scenarios with plasma current ramp-up duration of 50, 70 and 80 s.

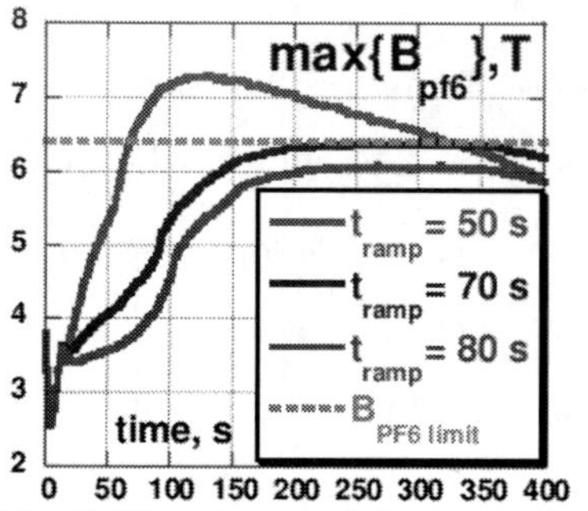

**Fig. 2**: Time traces of the magnetic field maximum on the PF6 conductor at the various rates of the plasma current ramp-up in Fig. 1.

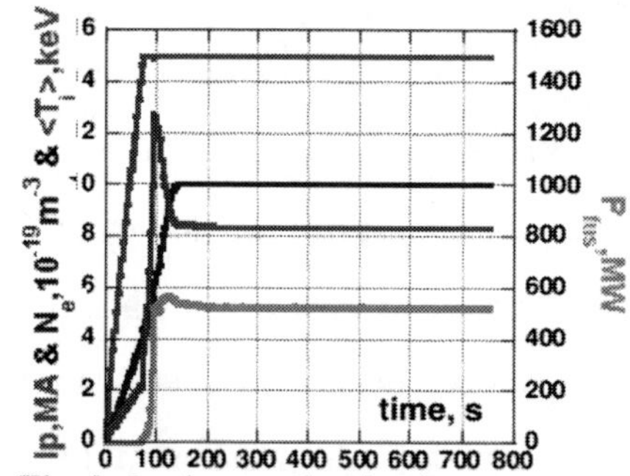

**Fig. 3**: $I_p$, electron density, ion temperature and fusion power in scenario with the plasma current ramp-up duration of 70 s.

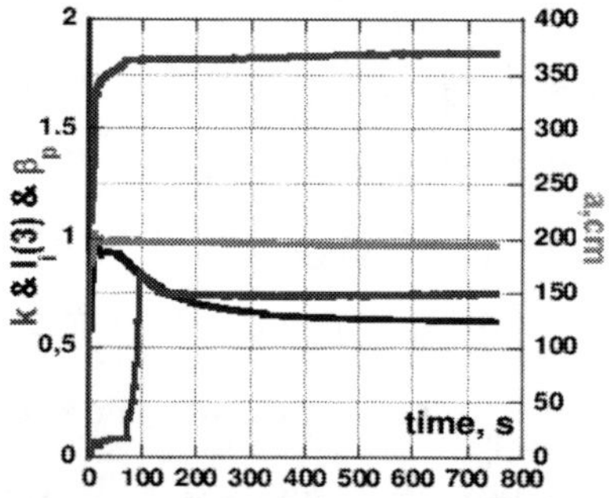

**Fig. 4**: Plasma elongation, $l_i$, $\beta_p$ and minor radius in scenario with the plasma current ramp-up duration of 70 s.

*Disclaimer: The views and opinions expressed herein do not necessarily reflect those of the ITER Organization.*

**References**

[1] V.E. Lukash, et al., 41st EPS Conference on Plasma Physics (2014), P5.010

[2] M. Romanelli, et al., Plasma and Fusion Research, **9** (2014) 3403023

[3] A.R. Polevoi, et al., Nucl. Fusion **55** (2015) 063019

[4] V. Lukash et al., 35th EPS Conf. Plasma Phys., Crete, Greece, (2008), P2.074

[5] V.E. Lukash et al., Plasma Devices and Operations **13** (2005) 143

# Evolution of Fast Particle Driven MHD Instabilities for the ITER 15MA Scenario

M.Yu. Isaev[1], S.V. Konovalov[1] and S.Yu. Medvedev[2]

[1]*National Research Centre Kurchatov Institute, Moscow, Russia*

[2]*Keldysh Institute, Russain Academy of Science, Moscow, Russia*

## 1. Introduction

The present studies are carried out in the frame of the high priority tasks of the ITPA Energetic Particles Topical Group and can be compared with similar studies being performed with the MEGA [1] and the HAGIS-LIGKA [2] codes. Our simulations extend the fast particle driven MHD stability analysis, performed recently for the steady-state ITER case [3] to the ITER Q = 10 15MA scenario.

The spatial structure of toroidal Alfven eigenmodes (TAE) $\eta_{mn}(F_p)$ has been calculated with the KINX code [4], ($F_p$ and $s$ are normalised poloidal and toroidal flux functions). The fast particle nonlinear dynamics, TAE growth rates and wave saturation levels are computed with the VENUS+δf [5] orbit following code. Fourier decomposition in Boozer coordinates ($s$, $\theta$, $\zeta$) of the TAE mode with a poloidal, toroidal indices $m,n$ and mode frequency $\omega$ has a form $\xi = A(t)\ \Sigma\ \eta_{mn}(s)\ \cos(m\theta - n\zeta - \omega t)$. The evolution of the TAE amplitude $A(t)$ is described from the equation for the fast particle-wave interaction

$$dA/dt = -<\int Z\, e\, \delta f\, \boldsymbol{V}\cdot\boldsymbol{E}\, d\tau > / (2K\, \omega^2 A) - A\, \gamma_d, \qquad (1)$$

here $K = \int \rho_b\, \xi^2\, dV$ is the kinetic energy of the plasma perturbation, $\rho_b = m_b\, n_b$ the mass density of the bulk plasma, $dV$ the volume unit, $d\tau = d^3x\, d^3V$ the phase volume unit, $Z\cdot e$ the particle charge, $\delta f$ the perturbed distribution function of fast particles, $\boldsymbol{V}$ the particle velocity vector and $\boldsymbol{E}$ the wave electric field. The ion Landau damping rate $\gamma_d$ is obtained from analytic estimates [6] and included into the VENUS code. The TAE growth rate $\gamma$ is computed from $\gamma = dA/(Adt)$, and the radial component of the TAE perturbation $\delta\boldsymbol{B} = \mathrm{rot}\alpha\boldsymbol{B}$, normalized to the central magnetic field $B_0$, is defined as

$$\delta B_r/B_0 = \delta\boldsymbol{B}\, \nabla s/[|\nabla s|B_0] = -\mu_0(I\partial\alpha/\partial\theta + J\partial\alpha/\partial\zeta)/[|\nabla s|g^{1/2}\, B_0], \qquad (2)$$

where $J$ and $I$ are the toroidal and poloidal current flux functions and $g^{1/2}$ is a Boozer jacobian.

## 2. Simulation results for the ITER 15MA Q =10 scenario

Plasma parameters and profiles predicted by the ASTRA code for the ITER Q = 10 with plasma current of 15MA and monotonic q profile with $q(0) = 1.0$ are described in [1], [2], [6], [7]. At mid-radius, where the drive of alpha-particles and NBI-produced ions is expected to exceed the damping, different TAE modes were explored. Global Alfvenic modes with $n \approx 20$ have found to be amongst the most unstable fast ion driven modes and are the focus of the work in this paper. The plasma cross-section with this perturbation is shown in Fig. 1. Fig. 2(a) shows the profiles of the components of the ballooning TAE mode computed with the KINX code with poloidal indices $m = 20$ (brown), $m = 21$ (blue) with a mode frequency of $\omega = 71$ kHz and the safety factor (black). Fig. 2(b) shows the components of the antiballooning mode with a mode frequency of $\omega = 106$ kHz for the same poloidal indices.

Fig. 3 shows the nonlinear saturation of the ballooning $n = 20$ TAE mode amplitude driven by α-particles with the VENUS code for the ITER Q = 10 scenario without any damping mechanisms. The initial linear normalized growth rate of $\gamma/\omega = 0.4$ (red squares) is high due to the large central α-particle density of $9.26 \times 10^{17}$ $m^{-3}$, which leads to a high amplitude of the perturbation in the range of $\delta B_r/B \approx 10^{-3} - 10^{-2}$ (blue circles) similar to HAGIS results [2].

Different damping mechanisms for the ITER Q = 10 scenario have been considered analytically in [6]. For mid-plasma radius the thermal ion Landau normalized damping is equal to $\gamma_d/\omega = 0.005$, continuum damping is equal to $\gamma_d/\omega = 0.02$ and the radiative and the electron collisional damping are small and can be neglected. From the VENUS simulations, a normalized damping of $\gamma_d/\omega = 0.2$ is required to decrease the ballooning mode saturation level by a factor of 2 with respect to the non-damped case (blue circles on Fig.4) and thus no significant damping effect is expected for this mode in the ITER Q = 10 plasma conditions.

Fig. 5 shows the nonlinear saturation of the antiballooning $n = 20$ TAE mode amplitude (blue circles) computed with the VENUS code without any damping mechanisms. The linear normalized growth rate of $\gamma/\omega = 0.8$ (red squares) for the antiballooning mode is higher by a factor of 2 than the ballooning one due to wider mode structure of the antiballooning mode (Fig. 2(b)) compared to the ballooning one (Fig. 2(a)). The saturated amplitude equals to $\delta B_r/B \approx 10^{-2}$, somewhat larger than the ballooning mode. Similar to the ballooning mode, the damping effects are very small up to $\gamma_d/\omega < 0.1$.

In addition to the α particles drive, we have explored the evolution of TAE $n = 20$ modes driven by the two 16.5MW NBI (1 MeV) beams with different injection geometries: both on axis, both off-axis and one on and the other off-axis. The anisotropic beam particle distributions have been computed with the Monte Carlo NBSOURCE code which includes realistic geometry of the ITER beams [8]. The largest linear growth rates and saturation levels have been obtained for both beams on-axis ($\gamma/\omega = 0.12$ and $\delta B_r/B \approx 10^{-2}$ respectively for the antiballooning mode (Fig.6) without the inclusion of damping effects. The smallest linear growth rates and saturation levels have been obtained for both beams off-axis ($\gamma/\omega = 0.09$ and $\delta B_r/B \approx 3x10^{-3}$, respectively).

## 3. Summary

The structures of the typical and the most driven TAE ballooning and antiballooning eigenmodes with toroidal index $n = 20$ for the ITER Q = 10 plasma have been computed with the KINX code. The largest growth rates on the linear stage of $\gamma/\omega = 0.8$ driven by α-particles and nonlinear mode saturation levels of $\delta B/B \approx 10^{-2}$ have been obtained with the VENUS code for the antiballooning eigenmode with $\omega = 106$ kHz. The damping rates as large as $\gamma_d/\omega = 0.2$ are required to decrease the saturated mode amplitude by a factor of 2. Such a rapid amplitude evolution may violate some of the ordering assumption of our model, i.e. the necessity that $\gamma << \omega$. Further convergence studies for high values of $n$ and simulations with other models should be performed for this case.

NBI ions with realistic distribution functions (including anisotropic pitch angle distributions) with the real ITER NBI beam geometry have been computed with the Monte Carlo NBSOURCE code. The largest linear growth rates and saturation level have been obtained for the two NBI on-axis ($\gamma/\omega = 0.12$ and $\delta B_r/B \approx 10^{-2}$). However, no particle losses have been found from the VENUS orbits simulations for nonlinear saturation levels of $\delta B/B = 10^{-2}$ - $10^{-3}$ for the mode localized at mid-radius. These results are thus positive for the expected fast particle confinement in the ITER Q=10 scenario, along similar findings with the MEGA and HAGIS codes.

**Acknowledgements**

The authors thank Dr. S.P. Hirshman for providing the DESCUR and VMEC codes and Prof. W.A. Cooper for providing us the TERPSICHORE code. The computations were performed on the Kurchatov (Moscow) and CRPP (Lausanne) clusters. The authors thank A. Polevoi, A. Loarte, S. Pinches, ITPA-EP colleagues for their collaboration, support and fruitful discussions.

## References

[1] Y. Todo, A. Bierwage, Plasma and Fus. Research, **9**, (2014) 3403068.

[2] M. Schneller, P. Lauber, S. Briguglio, 25th IAEA FEC; S. Petersburg, Russia; 13-18 Oct 2014; TH/P2-6.

[3] M. Yu. Isaev *et al*, 25th IAEA Fusion Energy Conference; S. Petersburg, Russia; 2014; TH/P3-39.

[4] L. Degtyarev *et al*, Comp. Phys. Comm., 103, 10(1997).

[5] W.A. Cooper *et al*, Plasma Phys. Contr. Fus. 53, 024001(2011).

[6] S.D. Pinches *et al*, Phys. of Plasmas, 22, 021807(2015).

[7] A.R. Polevoi *et al*, J. Plasma Fusion Res, 5, 82 (2002).

[8] P.B. Aleynikov *et al*, 36th EPS Conference on Plasma Physics, P4.154, Sofia, (2009).

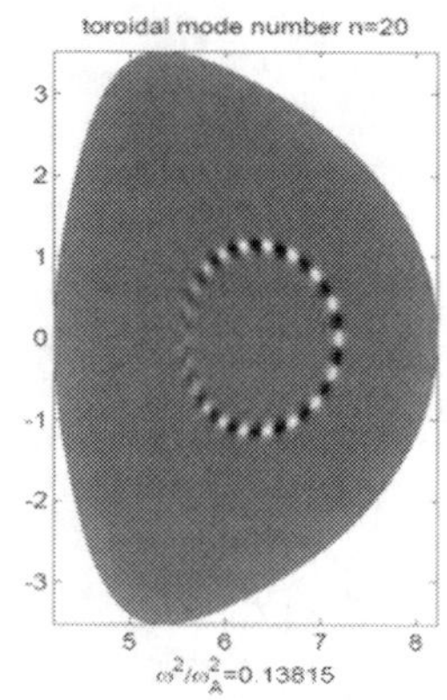


Fig. 1. Plasma cross-section with $n = 20$ TAE mode computed with the KINX code for the ITER Q = 10 plasmas.

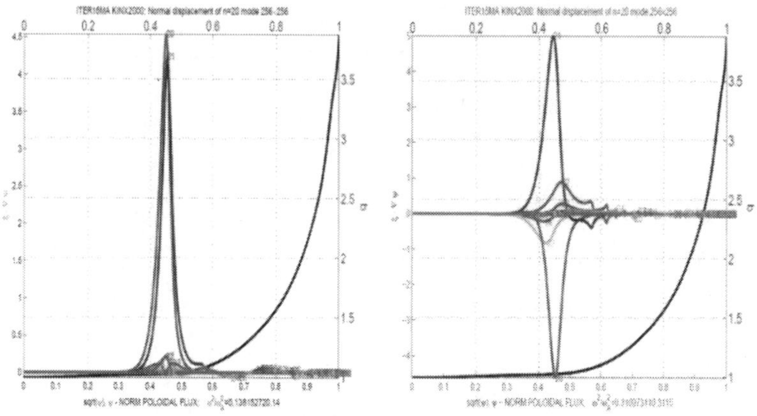

Fig. 2 (a,b). Profiles of the components of the ballooning (a) and antiballooning (b) $n = 20$ TAE mode with poloidal indices $m = 20$ (brown), $m = 21$(blue) and safety factor (black), computed with the KINX code for the ITER Q = 10 plasma.

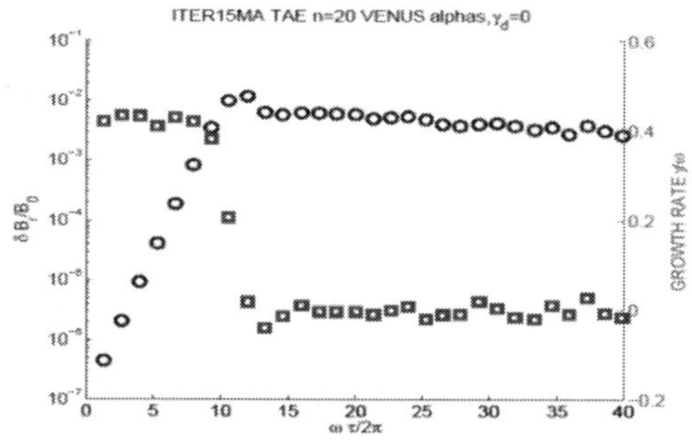


Fig. 3. Nonlinear evolution of the $n = 20$ TAE ballooning mode amplitude (blue circles) and the growth rate (red squares), computed with the VENUS code for the ITER Q = 10 scenario without damping effects.

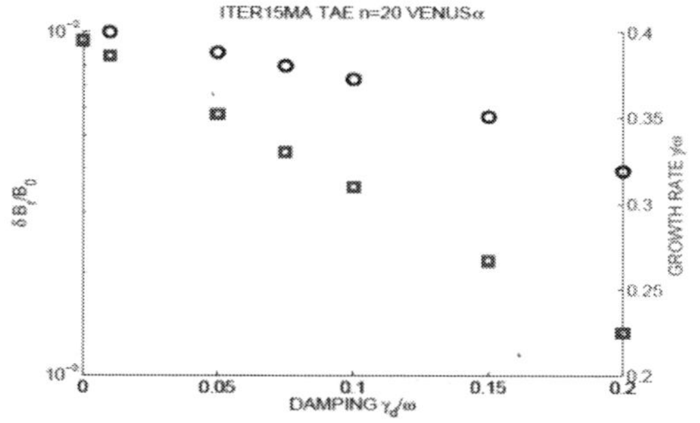


Fig. 4. Maximum amplitude of the ballooning $n = 20$ TAE mode (blue circles) and the total growth rate on the linear stage (red squares) as a function of normalized damping rate $\gamma_d/\omega$, computed with the VENUS code for the ITER Q = 10 scenario.

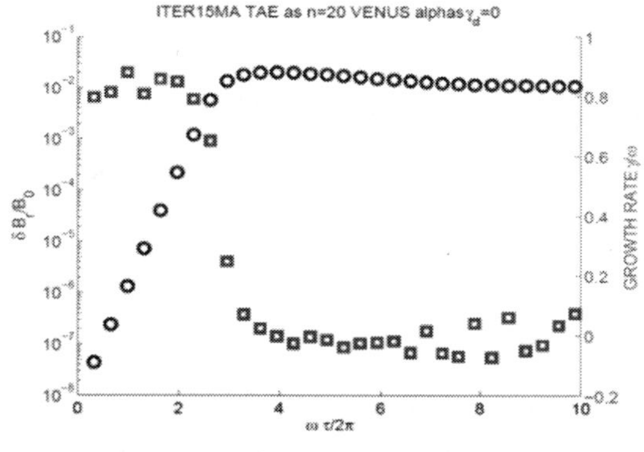


Fig. 5. Evolution of the $n = 20$ TAE antiballooning mode amplitude (blue circles) and of the growth rate (red squares), computed with the VENUS code for the ITER 15MA scenario without damping effects.

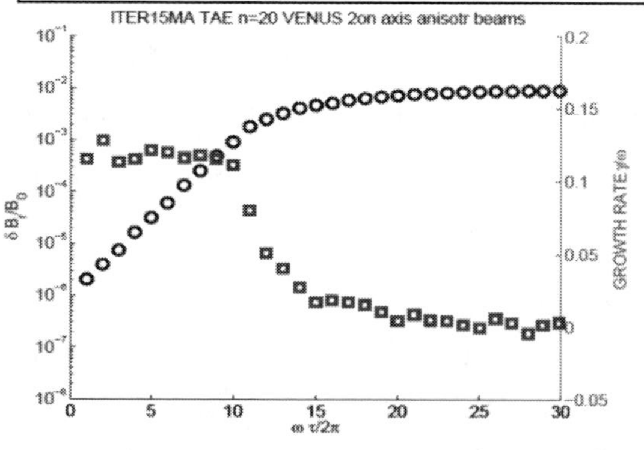


Fig. 6. Evolution of the $n = 20$ TAE antiballooning mode amplitude (blue circles) driven by 2 on-axis NBI ions and of the growth rate (red squares), computed with the VENUS code for the ITER 15MA scenario without damping effects.

# Recovery of relative contributions of SOL emission and divertor stray light in Balmer-alpha spectroscopy in ITER

V.S. Neverov[1], A.B. Kukushkin[1,2], A.G. Alekseev[1], S.W. Lisgo[3]

[1] *National Research Centre "Kurchatov institute", Moscow, Russia*

[2] *National Research Nuclear University "MEPhI", Moscow, Russia*

[3] *ITER Organization, Route de Vinon sur Verdon, St Paul Lez Durance, France*

**1. Introduction.** The verification of a number of essential features of the theoretical model [1] proposed for the H-alpha diagnostics in ITER was carried out in [1], using only spectroscopic data (i.e. without independent knowledge of background plasma parameters) from the recent experiments on tokamak JET with ITER-like wall (JET-ILW). The model [1] is aimed at the recovery of the density of neutral hydrogen, and its isotopic contents, in the SOL by solving a multi-parametric inverse problem with allowance for: (i) strong divertor stray light (DSL) in the signal measured on the lines of sights (LoS) in the main chamber, (ii) significant deviation of the velocity distribution function (VDF) of neutral atoms in the SOL from a Maxwellian, (iii) the data of direct observation of the divertor.

The accuracy of the model should be tested in the frame of a "synthetic" diagnostics which simulates "phantom" experimental spectra using the predictive numerical modelling and allows comparison of "true" values with those recovered from the "phantom" data. Such a test of the recovery of only the VDF parameters in the SOL without the DSL was done in [2] using the modelling by the SOLPS4.3 (B2-EIRENE) code [3] (on an expanded numerical mesh with allowance for the poloidally resolved recycling from the first wall [4]) for background plasma on the flat-top of Q = 10 inductive operation of ITER. Here we extend the test [2] to allow for all three sources of signal on the LoS in the main chamber: inner and outer sections in the SOL on the LoS, and the DSL. We estimate the accuracy of: (i) fitting the Balmer-alpha line shapes of hydrogen isotopes in the SOL with allowance for a strong line shape asymmetry, using the model [5]; (ii) fitting the DSL spectrum using the spectroscopic data for direct observation of divertor from the top on the sixteen LoS; (iii) recovery of partial contributions of the above-mentioned sources in the measured signal. The details of the formalism are reported in [6].

**2. Separation of inner- and outer-SOL light without DSL.** First, we consider the case when the signal measured on the LoS in the main chamber along the major radius from the equatorial port plug contains the contributions from the inner and outer sections of the

SOL plasma and does not contain the DSL. In this case we solve the inverse problem very similar to that formulated and solved for JET-ILW in [1] for the limiter stage of the discharge, but with the "phantom" experimental signal (instead of the measured one) calculated with the local Balmer-alpha emissivity and the VDF values, obtained with SOLPS4.3 (B2-EIRENE) code. Since the "true" value of fraction of the light from the inner/outer section of the SOL is known, we can estimate the error of the inverse problem solution. We use the data calculated for the six ITER operation scenarios (cf. [4,7]): low ("d","e")/middle ("f","g")/high ("h","i") density in the far SOL in the L-mode ("d","f","h")/H-mode ("e","g","i") regime for deuterium plasma. Figure 1 shows the typical results of the inverse problem solution. Without the DSL, the absolute value of the error of recovering the fraction of the inner SOL light in the total signal does not exceed 0.2.

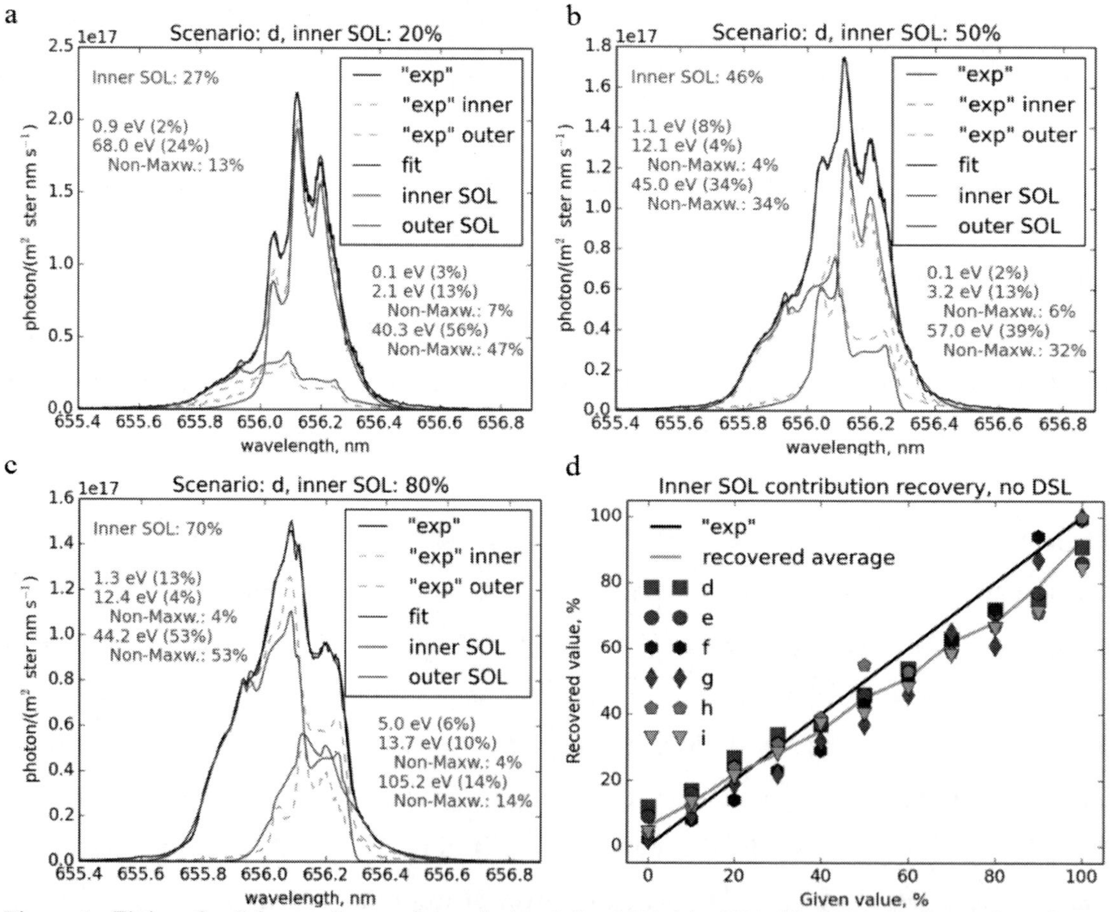


Figure 1. Fitting the "phantom" experimental signal for 20% (a), 50% (b), 80% (c) input (i.e. phantom experimental) fraction of the inner SOL light in the total signal for the scenario "d". The "true" inner and outer SOL light spectra are shown with dashed lines, respectively, while the recovered spectra with solid lines of the same color. Comparison of the recovered values of the fraction of the inner SOL light in the total signal with the "true", phantom experimental values (black curve) for the six ITER operation scenarios (d). The recovered value, averaged over six scenarios, is shown with gray curve.

**3. Separation of inner-and outer-SOL light, and strong DSL.** In this case we need to model the DSL spectrum. In [1] we used the data from direct multiple-LoS observation of the JET-ILW divertor from the top to recover the parameters of the neutral atoms in the divertor and simulate then the DSL spectrum with only one free parameter, the fraction of the pi-component in the Zeeman triplet. Here we perform the same procedure but for the "phantom" experimental data and the 16 observation chords in ITER. Our calculations show that the recovery of the characteristic temperature of the neutral atoms and the isotope ratio in the divertor may be done with a high accuracy. This is not surprising for a single source of the signal and for symmetric lines shapes of the Balmer-alpha emission in the divertor. However the recovered DSL spectrum poorly reproduces the "phantom" one (Figure 2(a)), which is calculated here using the semi-analytical formula (2) from [6]. The difference between these spectra results in a high error of the inverse problem solution in the case of a high fraction of the DSL in the signal (see Figure 2 (b)).

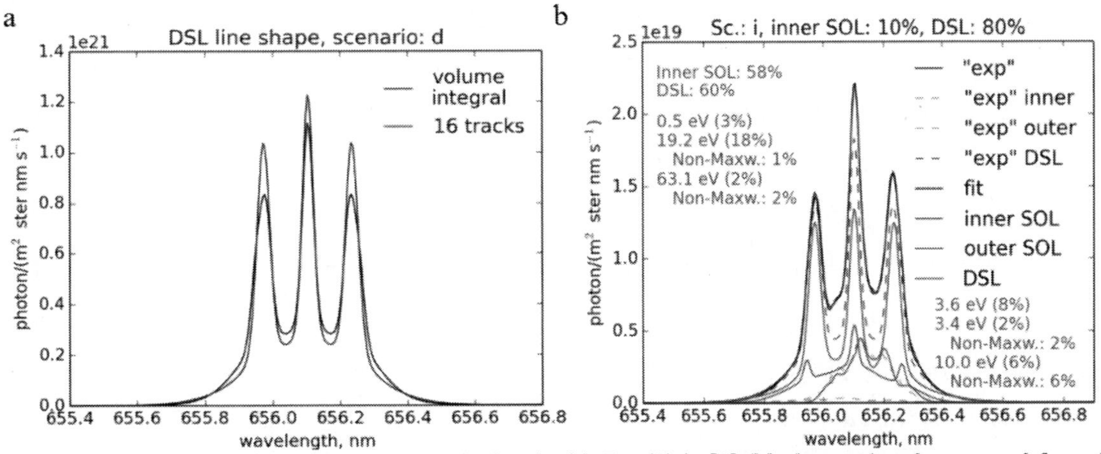

Figure 2. Comparison of the DSL spectra calculated with Eq. (2) in [1] (black curve) and recovered from the direct observation of the divertor on the 16 chords (a). Fitting the "phantom" experimental signal for the 80% fraction of DSL in the measured signal and 10% fraction of the inner SOL light in the total SOL light (b).

It appears that the solver of the inverse problem is confused with distinguishing the contributions of the DSL and the inner-SOL light. Figure 3 shows that already for 40% fraction of the DSL in the total signal the accuracy of identifying the contributions of the inner- and outer-SOL light is unacceptably low and becomes even worse for a higher fraction of the DSL. We have checked that the increase of the accuracy of the DSL modeling, even for an unknown fraction of the Zeeman pi-component, makes it possible to recover the inner and outer SOL contributions with a high accuracy even for the DSL fraction as high as 80%.

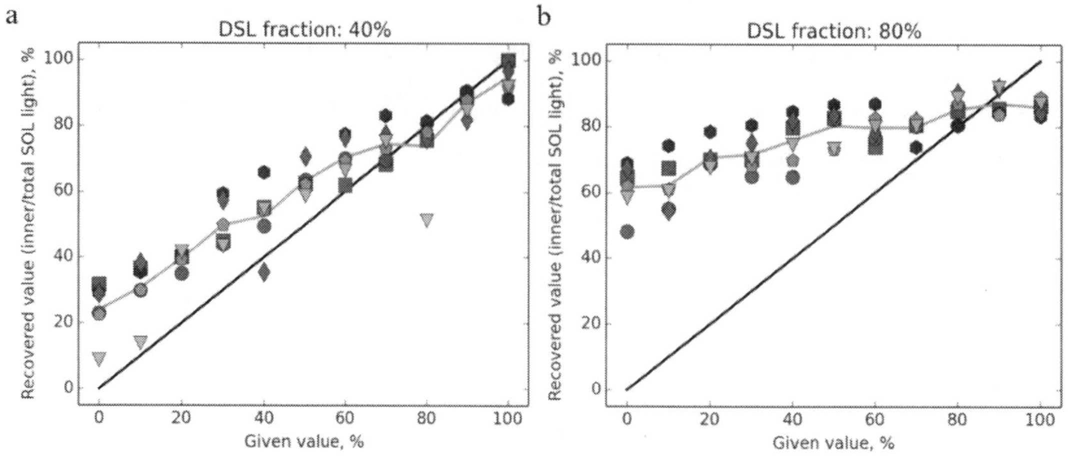


Figure 3. Comparison of the recovered values of the fraction of the inner SOL light in the total SOL light with the "true", phantom experimental values for the six scenarios, for the 40% (a) and 80% (b) DSL fraction in the total measured signal (see the legend in Figure 2(d)).

**3. Conclusions.** We tested the accuracy of the theoretical model [1], proposed for the H-alpha diagnostics in ITER, in the frame of a "synthetic" diagnostics which simulates the "phantom" experimental spectra using the predictive numerical modelling with the SOLPS4.3 (B2-EIRENE) code for the flat-top stage of Q=10 inductive operation of ITER. The error assessment for the line of sight along the major radius from the equatorial port-plug in ITER shows that: (i) without the DSL, the absolute value of the error of recovering the fraction of the inner-SOL light in the total signal does not exceed 0.2; (ii) already for the 40% fraction of the DSL in the total signal, the absolute error of recovering the contributions of the inner- and outer-SOL light is unacceptably high, 0.4, because of poor accuracy of predicting the DSL. Unification of the theoretical models [1] and [7] is needed to increase the accuracy.

**Acknowledgements.** The authors are grateful to all the co-authors [1] for their collaboration in studies on the H-alpha (and Visible Light) Spectroscopy Diagnostics in ITER.

The views and opinions expressed herein do not necessarily reflect those of the ITER Organization.

**References**


[1]. Kukushkin A.B., Neverov V.S., Stamp M.F., A.G. Alekseev, S. Brezinsek, A.V. Gorshkov, M. von Hellermann, M.B. Kadomtsev, V. Kotov, A.S. Kukushkin, M.G. Levashova, S.W. Lisgo, V.S. Lisitsa, V.A. Shurygin, E. Veshchev, D.K. Vukolov, K.Yu. Vukolov, and JET Contributors. Proc. 25th IAEA Fusion Energy Conf., St. Petersburg, 2014, EX/P5-20.

[2]. Neverov V.S. et al. Plasma Phys. Rep., 2015, 41 (2), 103.

[3]. Kukushkin A.S., et al. Nucl. Fusion, 2009, 49, 075008; Braams B.J. PhD thesis. Utrecht: Rijksuniversitet, 1986; Reiter D., Baelmans M., Börner P. Fusion Sci. Tech., 2005, 47, 172.

[4]. Lisgo S.W., Börner P., et al. J. Nucl. Mater. 2011, 415, S965.

[5]. Kukushkin A.B., et al. J. Phys.: Conf. Ser., 2014, 548, 012012.

[6]. Kukushkin A.B., V.S. Neverov, A.G. Alekseev, S.W. Lisgo, A.S. Kukushkin. Proc. 1st IAEA Techn. Meeting "Fusion Data Processing, Validation and Analysis", 1-3 June 2015, Nice, France.

[7]. Kukushkin A.B., et al. Proc. 24th IAEA Fusion Energy Conference, San Diego, CA, 2012, ITR/P5-44.

# Modeling of pre-Thermal Quench stage of disruption in ITER triggered by Massive Gas Injection

V. Leonov, S. Konovalov, V. Zhogolev

*NRC Kurchatov Institute, Moscow, Russian Federation*

Massive Gas Injection (MGI) is considered now as one of the promising methods of Disruption Mitigation (DM). For ITER it is necessary to re-radiate more than 90% of thermal energy to reduce loads on divertor targets. Radiating impurity neutral dynamics was shown [1] to play the dominant role in the pre-TQ stage in MGI mitigated disruption scenarios. In particular, losses of relatively fast secondary CX and recombination neutrals were found to be responsible for low impurity assimilation under MGI. Validation of the simulation model against JET data [1] demonstrated reasonably good agreement between simulation results and experimental data. Application of the model to ITER gave favorable predictions on the capabilities of ITER TM DMS. In-port positioning of the MGI valves (delivery tube length ~1m) was shown to be preferable providing very short, of few ms, response time and effective delivery of the radiating impurity into the plasma. Here we extend analysis of [1] for steady state ITER scenario, examine potential advantages of gas mix (noble gas + deuterium) injection and explore the possibility to use remote gas valve positioning for ITER DMS.

In this work the ASTRA transport code [2] is used for the modeling of the behavior of the bulk plasma parameters and the ZIMPUR code [3] for the simulations of impurity dynamics and radiation. For the description of MGI, model [4] was employed. In the self-consistent simulations standard set of transport equations for $T_i$, $T_e$, $n_d$ was solved with use of the Scaling Based transport Model (SBM): $\chi_e = D_d = D_z = D_0 \cdot F(\rho) \cdot F_{H\text{-}mode} \cdot F_{RS}$, $\chi_i = 2\chi_e$, where radial profile of transport coefficients was taken in the form: $F(\rho) = 1 + 3\rho_N^2$, $\rho_N = \rho/\rho_{max}$. For the radial anomalous pinch velocity we assumed $V_p = 0.5 \cdot D \cdot (\rho_N/a^2)$. Normalization coefficient $D_0$ was adjusted to provide coincidence of energy confinement time with ITER(98y2) scaling [5] with specified H-factor just before the beginning of pre-thermal quench phase and then fixed. In the external transport barrier region ($\rho_N > 0.93$), and in the RS region coefficients $F_{H\text{-}mode}$ and $F_{RS}$ were used to reduce the conductivity $\chi_e$ to the level of the neoclassical ion heat conductivity, and reproduce pedestals and Internal Thermal Barrier on profiles of plasma parameters.

For impurities we consider two groups of neutrals. The primary cold neutrals coming into plasma from the wall and from the MGI system and secondary "warm" neutrals

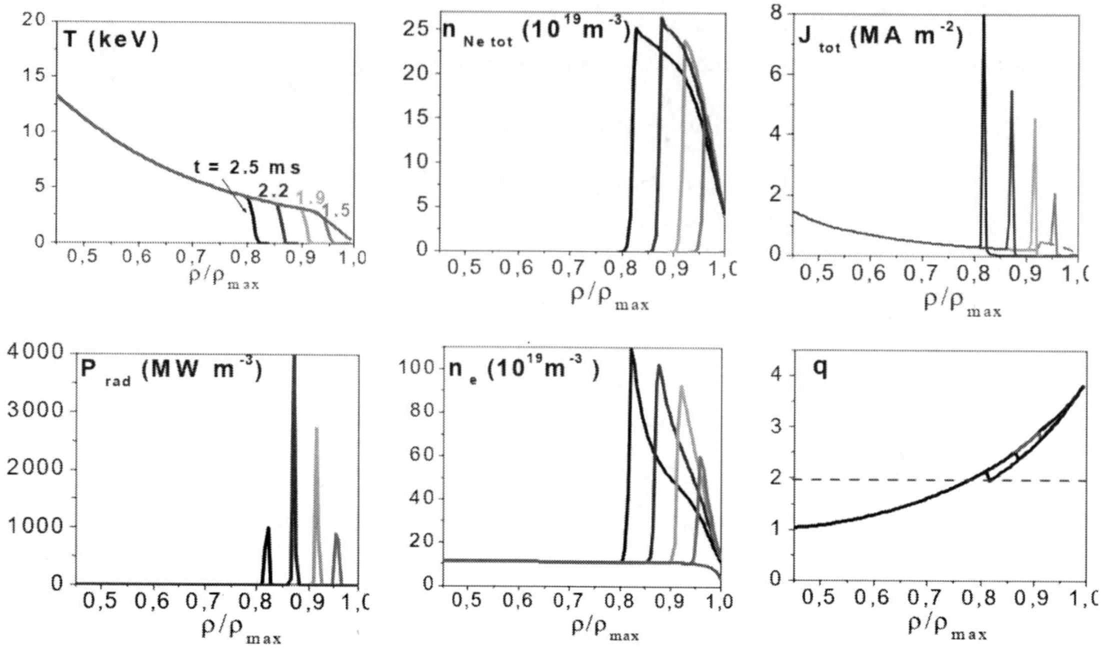


Fig.1 Typical cooling front propagation in Ne MGI forced Pre-TQ for basic inductive 15MA ITER scenario

originated due to the recombination and charge exchange of impurity ions in the plasma. These secondary neutrals can easily escape plasma through the cold periphery and provide appreciable losses of impurities during MGI.

During pre-TQ stage, cooling of the plasma periphery by MGI results in contraction of hot plasma region and current channel. Typical dynamics of radial profiles during Ne MGI for the reference inductive 15 MA ITER scenario is shown in Fig.1. When boundary of current channel approaches q =2 magnetic surface (see t=2.5ms) rise of MHD activity can provoke TQ and plasma mixing. Simulated time dynamics of plasma parameters after Ne and Ar MGI for the basic inductive ITER scenario is shown in Fig.2. In the Ne injection case impurity assimilation coefficient is $f_z \sim 7.2*10^{22}/ 7.4*10^{23}$ ~10% and $\tau_{pre\text{-}TQ}$ ~2.5 ms. In the Ar injection case $f_z \sim 5.5*10^{22}/3.4*10^{23}$ ~16 % and $\tau_{pre\text{-}TQ}$ ~2.7ms but Ar gives

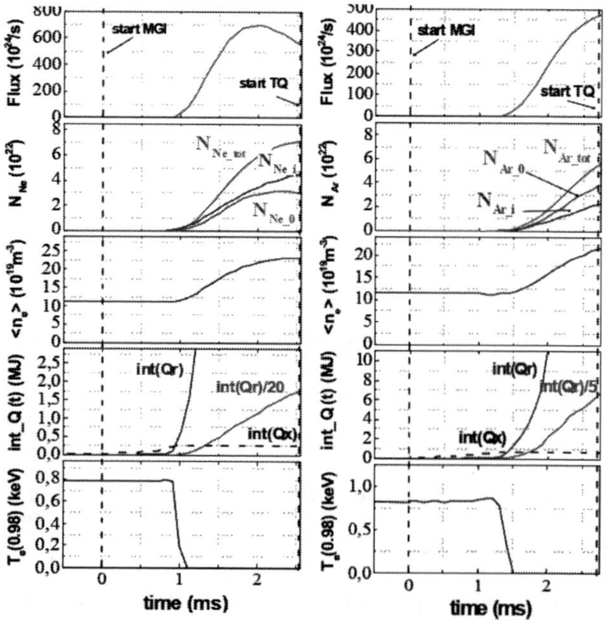


Fig.2 Time traces of various simulated parameters after Ne (left) and Ar (right) MGI for the basic inductive ITER scenario (In-port positioning of TM MGI valves).

smaller $T_e$ in the CQ stage resulting in shortening the CQ duration and, thus, increasing the mechanical loads on conductung structures and danger of RE generation.

Using the mix of a noble gas with deuterium should rise the MGI gas flow velocity. For the first look it can help in achievement of sufficiently short plasma response time with more distant location of the MGI valves. In addition there are experimental evidences that addition of $D_2$ in MGI gas prevents generation of runaway electrons. Fig. 3 presents simulation results for various mixtures of Ne+$D_2$ injection in ITER reference inductive scenario. Fraction of Ne is shown near the curves. Some characteristic values for these cases are given in Table 1. One can see that duration of pre-TQ stage remains sufficiently short and approximately the same in all cases, and cannot serve for selection. Thus, despite the reduction in flow velocity the number of Ne particles in plasma as well as the total number of electrons are higher in the case of pure Ne.

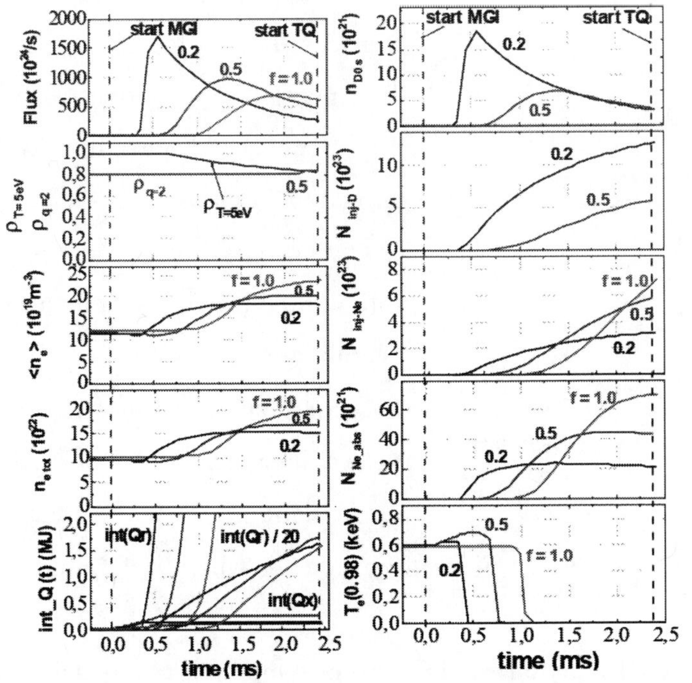

Fig.3 Time dynamics of plasma parameters during Ne+D2 MGI in ITER with different Ne fraction in the injected gas (Scen. #2)

**Table 1**

**($L_{tube}$ = 1m)**

| Ne fraction | $N_{Ne}$ in pl. ($10^{21}$) | $\tau_{pre\text{-}TQ}$ (ms) | TOF (ms) | $\tau_{pre\text{-}TQ}$ – TOF (ms) | $N_{e\,tot}$ ($10^{22}$) | $F_{asimilat}$ (%) |
|---|---|---|---|---|---|---|
| 0.2 | 20.9 | 2.35 | 0.3 | 2.06 | 15.0 | 6.5 |
| 0.5 | 43.5 | 2.3 | 0.46 | 1.84 | 16.7 | 7.5 |
| 1.0 | 69.7 | 2.5 | 0.75 | 1.73 | 19.7 | 9.8 |

Simulations show that the proposed system [in-port positioning of the TM valves] is suitable also for Steady-State 9MA scenario. As it is shown in Fig. 4, calculated Pre-TQ duration time in ITER is more affected by the MGI system parameters than a discharge scenario before disruption.

Simulations of the MGI with second possible length of gas delivery tube of ~8m (Table 2) did not manifested any advantages for the gas mix injection compared to the pure Ne/Ar injection examined in [1].

**Table 2**
**($L_{tube}$ = 8m)**

| Ne fraction | $N_{Ne}$ in pl. ($10^{21}$) | $\tau_{pre\text{-}TQ}$ (ms) | TOF (ms) | $\tau_{pre\text{-}TQ}$ – TOF (ms) | $N_{e\ tot}$ ($10^{22}$) | $F_{asimilat}$ (%) |
|---|---|---|---|---|---|---|
| 0.2 | 19 | 7.1 | 2.49 | 4.6 | 14 | 11.9 |
| 0.5 | 25.5 | 8.3 | 3.61 | 4.7 | 14 | 13.1 |
| 1.0 | 28 | 10.6 | 5.8 | 4.8 | 15 | 14 |

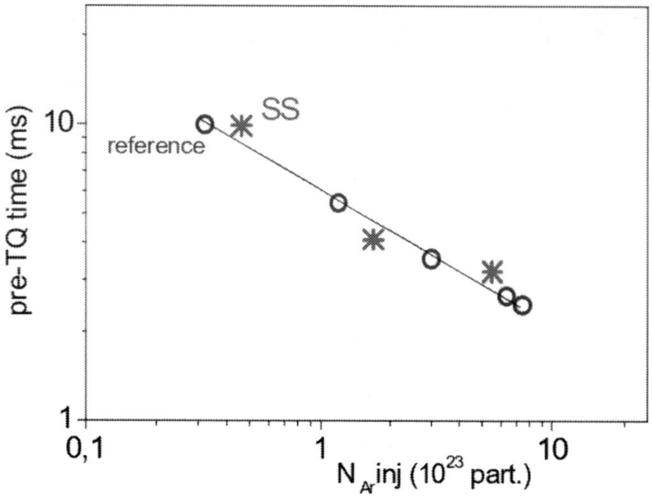


Fig.4 Dependence of pre-TQ time vs total number of assimilated Ne particles for ITER scenarios: blue circles – reference inductive 15 MA scenario, red stars – Steady-State 9 MA scenario.

**Summary :**

- Simulations show that in all considered cases impurity neutral dynamics plays important role in the pre-TQ stage in MGI mitigated disruption scenarios providing large impurity losses and, therefore, low impurity assimilation in plasma.
- Simulations show that pure Ne/Ar injection results in higher accumulation of radiating impurity in the plasma as well as in general rise of electron density which determine the collisional drag for REs.
- Ar MGI could result in too strong reduction of the plasma temperature and, thus, in shortening the CQ time below the tolerable (with respect to EM loads) limit and facilitating the RE generation. Thus Ne injection for TM is preferable.
- In-port positioning of the MGI valves (1m distance from the plasma) is shown to be preferable providing very short, of few ms response time and sufficient (according to estimations of [6]) impurity accumulation for TM for both inductive and steady-state ITER reference scenarios. Increase of delivering tube length to 8m results in 4 times longer pre-TQ time (~10.6 ms) and factor of 2 smaller accumulation of Ne particles in the plasma (~ 3 $10^{22}$). However, even with these reduced characteristics of the MGI system ITER DMS seems to be capable to provide thermal load mitigation.


[1] Leonov V., et al., 25th IAEA Fusion Energy Conf., Snkt.Petersburg, RF, 2014, TH/P3-35

[2] Pereverzev G.V., Yushmanov P.N., Preprint IPP 5/98, Garching. Germany (2002).

[3] Leonov V.M., Zhogolev V.E., Plasma Phys. Control. Fusion, **47** (2005) 903.

[4] Zhogolev V.E., VANT, Nuclear Fusion series, V.37, issue 2, pp. 60-72 (2014) (in Russian).

[5] ITER Physics Basis 1999 Nucl. Fusion 39 2137

[6] Leonov V.M., et al., 38th EPS Conf. on Plasma Phys., Strasbourg, France 2011 P2.108.

# Comparative Analysis of Electron Cyclotron Radiation Power Losses in Tokamaks IGNITOR and ITER

P.V. Minashin[1], A.B. Kukushkin[1,2]

[1]*NRC "Kurchatov Institute", Moscow, Russia*

[2]*Nuclear Research National University MEPhI, Moscow, Russia*

**1. Introduction.** The next generation of fusion experiments will be focused on the research of plasma under ignition conditions (domination of the heating from fusion reactions for times long compared to the most important plasma characteristic time scales). Tokamak ITER, which is currently under construction in France, is the leading project in the magnetically confinement fusion contemporary physics. To demonstrate the feasibility of fusion power the ITER project has to perform two major objectives: reach an energy gain factor of Q = 10 (the ratio of the fusion power produced to auxiliary heating power) and to obtain self-sustained fusion plasma. Another proposed experiment is the IGNITOR project [1]–[4], which aims at demonstrating the possibility of the ignition achieved by the ohmic heating alone (or with minimum use of auxiliary heating). Although ITER and IGNITOR significantly differ from each other in many key parameters, the achievement of plasma ignition conditions require for both projects the increase of the magnetic field as compared to the present devices. This leads to a possible increase of the role of the electron cyclotron (EC) losses in the local electron power balance, which should be considered for optimizing the device operational scenarios.

**2. The EC power losses in tokamaks IGNITOR and ITER.** The first attempt to perform a numerical calculation of total and local net EC power losses in IGNITOR (taking into account the effects of the net absorption of the EC radiation on the periphery of the plasma column [5], [6]) was carried out in [7] for the first version of tokamak IGNITOR project (in [7] the approach [8] is used, which allowed to generalize and improve the method [5]). Figure 1 shows the EC power losses in two basic regimes of tokamak IGNITOR operation [9]. Note that the scenario 2 is close to the second scenario in the recent paper [4] (see Table 1 in [4], the results for the spatial distribution of the main parameters in [4] are not available). The calculations were carried out with numeric codes CYNEQ [7], [10] (using the modified version CYNEQ-B (1D) [11] (account of plasma equilibrium effects)) and CYTRAN [5].

For comparison of the results for tokamak IGNITOR and ITER we show (Fig. 2) the EC power losses for two scenarios of ITER operation. The description of the main characteristics of the "inductive" regime is given in [12], "steady-state" regime — [13], [14].

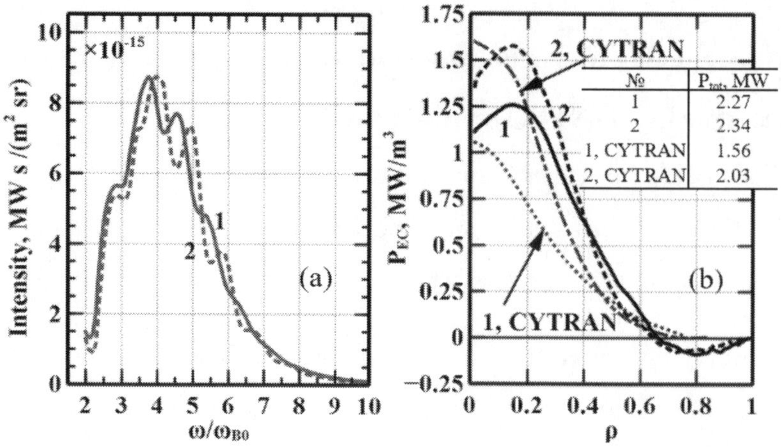


Figure 1. Comparison of the spectral intensity of the outgoing EC radiation intensity ($\omega_{B0}$ is the EC frequency for the vacuum magnetic field on torus axis) (a), spatial profiles of the EC net local losses, $P_{EC}(\rho)$, and total (volume-integrated) losses, $P_{tot}$ (b) in scenarios 1 and 2 of tokamak IGNITOR operation. Calculations are carried out with the CYNEQ-B(1D) code. The wall reflection coefficient for the EC radiation $R_w$=0.6.

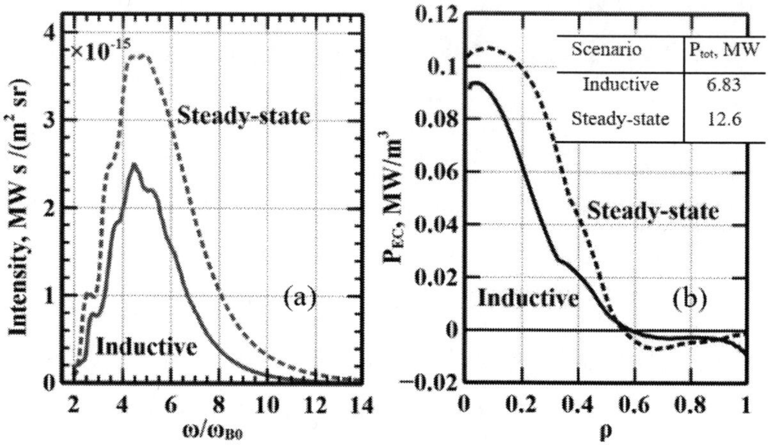


Figure. 2. The same as in figure 1 but for “inductive” and “steady-state” scenarios of tokamak ITER operation.

To assess the role of the EC power losses in IGNITOR one should compare the data of figure 1 with the other components of the electron energy balance (see Fig. 6 a, b and 6, c in [9]). It is seen that the diversification of EC power losses calculations did not affect the conclusion [9] about the role of the EC power losses: the main components of the electron energy balance are the heating by the alpha-particles and the heat losses due to anomalous electron thermal conductivity. For the “inductive” regime of ITER operation, as it follows from the results of [12], the EC power losses do not play a significant role. This conclusion is also valid for the “inductive” scenario of ITER operation according to [9] (see Fig. 4, *c* therein). The situation changes dramatically with increasing electron temperature predicted for the “steady-state” regimes of ITER operation [9]. A detailed analysis of the role of EC power losses in the “steady-state” regime of ITER operation [13] was carried out in [15], [14]: in the central plasma the EC power losses become comparable with the total auxiliary heating power and are the one third of the heating by alpha-particles (see Fig. 3 in [14]).

**3. Influence of the non-Maxwellian VDF on the EC power losses in ITER and IGNITOR.** Influence of non-Maxwellian velocity distribution function (VDF) on the EC power losses is much less analyzed than that in the case of Maxwellian plasma. In [10] the EC radiation power losses was studied in the case of an isotropic bi-Maxwellian VDF with Gaussian distribution of effective temperatures in space coordinates. In [16] the modification of the code RAYTEC [17] is given to take into account an anisotropic non-Maxwellian VDF of electrons.

Here we use the model electron VDF:

$$f_e(\mathbf{p},\mathbf{r}) = n_e(\rho)\left\{\left[1-\delta_{n_e}(\rho)\right]f_{\mathrm{MAXW}}(\mathbf{p},T_e(\rho)) + \delta_{n_e}(\rho)f_{\mathrm{hot}}(\mathbf{p},\mathbf{r})\right\}, \tag{1}$$

where the electrons of the thermal plasma component are described by the relativistic Maxwellian VDF $f_{\mathrm{MAXW}}(\mathbf{p}, T_e)$, and the super-thermal electrons by a model anisotropic VDF:

$$f_{\mathrm{hot}}(\mathbf{p},\ \mathbf{r}) = A\cdot\exp\left[-m_e c^2\left(\sqrt{1+\frac{p_\perp^2}{(m_e c)^2}+\frac{p_\parallel^2}{(m_e c)^2}}-1\right)\left(\frac{\sin^2\theta_p}{T_\perp(\rho)}+\frac{\cos^2\theta_p}{T_\parallel(\rho)}\right)\right], \tag{2}$$

here A is normalization factor, $p_{\perp,\parallel}$ are the components of electron momentum perpendicular and parallel to magnetic field. Spatial distributions of the density and temperature profiles of the super-thermal electrons are given by Gaussian distributions:

$$\delta_{n_e}(\rho) = \delta_{n_e}^{\max}\exp[-(\rho-\rho_0)^2/(\Delta\rho)^2], \tag{3}$$

$$T_{\perp,\parallel}(\rho) = T_{\perp,\parallel}^{\max}\exp[-(\rho-\rho_0)^2/(\Delta\rho)^2], \tag{4}$$

where parameters $\rho_0$, $\Delta\rho$ are taken from [12-13]. The results are given in Figures 3-4.

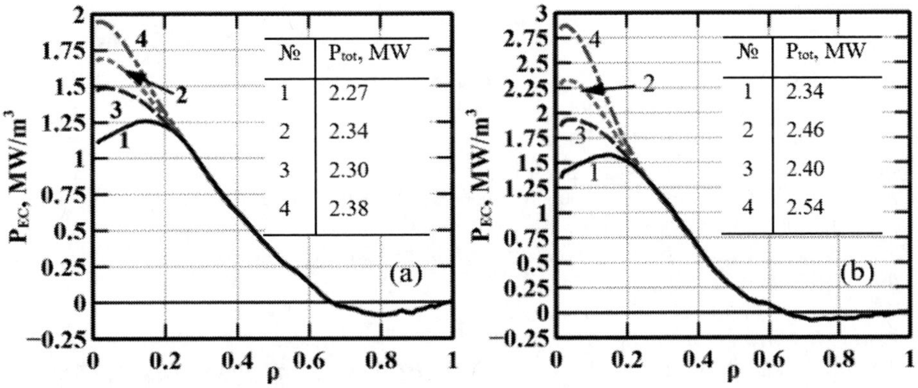


Figure 3. Spatial profile of the EC net local losses, $P_{EC}(\rho)$, and total power losses, $P_{tot}$, for the IGNITOR scenarios 1 ($I_p$ = 7 MA) (a) and 2 ($I_p$ = 11 MA) (b) for the Maxwellian VDF (1), isotropic VDF for super-thermal electrons with $T_\parallel^{\max} = T_\perp^{\max} = 3T_e(0)$ keV (2), anisotropic bi-Maxwellian VDF with $T_\parallel^{\max} = 3T_e(0)+10$ keV, $T_\perp^{\max} = 3T_e(0)-10$ keV (3), anisotropic bi-Maxwellian VDF with $T_\parallel^{\max} = 3T_e(0)-10$ keV, $T_\perp^{\max} = 3T_e(0)+10$ keV (4). Spatial distribution of the density and temperature profiles of the super-thermal electrons are taken the Gaussians with $\rho_0 = 0$, $\Delta\rho = 0.27$, $\delta_{n_e}^{\max} = 0.01$. Calculations are carried out with the CYNEQ-B(1D) code.

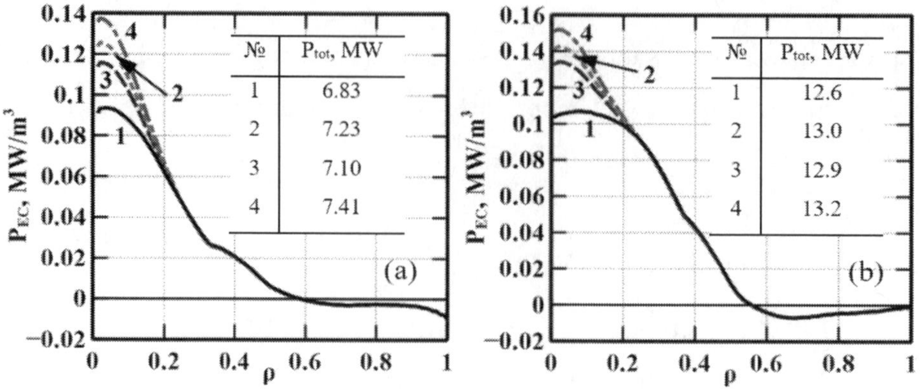


Figure 4. The same as in figure 3 but for the "inductive" (a) and "steady-state" (b) regimes of ITER operation.

**4. Conclusions.** In the present paper we made a comparative analysis of the role of the EC power losses in tokamaks IGNITOR and ITER. The analysis is carried out with the CYNEQ code [7], [10], using the modified version of the CYNEQ-B(1D) code [11]. Although ITER and IGNITOR significantly differ in many key parameters, the relevance of the comparative analysis is related to the use of the high magnetic field (compared to existing devices) to achieve ignition conditions. We demonstrate that despite the strong magnetic field, the EC radiation power losses in tokamak IGNITOR do not play a significant role in the local electron plasma energy balance and do not influence the plasma ignition conditions.

## References


[1] B. Coppi *et al.*, Nuclear Fusion **41**, 1253-7, (2001)
[2] A. Airoldi and G. Cenacchi, Nuclear Fusion **37**, 1117-27, (1997)
[3] A. Airoldi and G. Cenacchi, Nuclear Fusion **41**, 687-93, (2001)
[4] B. Coppi *et al.*, Nuclear Fusion **53**, 104013, (2013)
[5] S. Tamor, Science Applications, Inc. Report SAI-023-81-189LJ/LAPS-72 (1981)
[6] S. Tamor, Nuclear Technology/Fusion **3**, 293, (1983)
[7] A.B. Kukushkin, *Proc.* 14th IAEA Conference on Plasma Physics and Controlled Nuclear Fusion Research (Wuerzburg, Germany) 2 (IAEA), 35-45 (1992)
[8] A.B. Kukushkin, JETP Letters **56**, 487, (1992)
[9] F. Albajar *et al.*, Nuclear Fusion **45**, 642-8, (2005)
[10] K.V. Cherepanov and A.B. Kukushkin, *Proc.* 20th IAEA Fusion Energy Conference (Vilamoura, Portugal) TH/P6-56 (2004)
[11] A.B. Kukushkin and P.V. Minashin, *Proc.* 36th EPS Conference on Plasma Physics (Sofia, Bulgaria) 33E (ECA), P-4.136 (2009)
[12] A.R. Polevoi *et al.*, Journal of Plasma and Fusion Research SERIES **5**, 82-7, (2002)
[13] A.R. Polevoi *et al.*, *Proc.* 37th EPS Conference on Plasma Physics (Dublin, Ireland) 34A (ECA), P2.187 (2010)
[14] A.B. Kukushkin, P.V. Minashin and A.R. Polevoi, Plasma Physics Reports **38**, 211-20, (2012)
[15] A.B. Kukushkin, P.V. Minashin and A.R. Polevoi, *Proc.* 23rd IAEA Fusion Energy Conference (Daejon, South Korea) ITR/P1-34 (2010)
[16] F. Albajar, M. Bornatici and F. Engelmann, *Proc.* 16th Joint Workshop "Electron Cyclotron Emission And Electron Cyclotron Resonance Heating" (Sanya, China) 215-21 (2010)
[17] F. Albajar, M. Bornatici and F. Engelmann, Nuclear Fusion **49**, 115017, (2009)

# Experimental investigation of halo current characteristics in KSTAR

J.G. Bak[1], H.S. Kim[1], S. H. Hahn[1] and J. Kim[1]

[1]*National Fusion Research Institute, Daejeon, Korea*

Small Rogowski coils as halo current monitors (HCMs) are used to measure halo currents during up and downward vertical displacement events (VDEs) under the experimental conditions such as plasma current $I_{p0}$ = 0.2 – 1.0 MA, toroidal field $B_T$ = 1.5 – 3.0 T for elongated plasmas ($\kappa \geq 1.49$) in the KSTAR tokamak. The dependences of halo current $I_h$ on plasma parameters (as plasma current, stored energy, plasma shaping) are investigated in the operational range of $I_{p0}$ and $B_T$. The maximum value of total halo current $I_{h,max}$ and the toroidal peaking factor (TPF) estimated from the measured local halo currents are 40 % of $I_{p0}$ and 5.0, respectively. The value of maximum halo fraction TPF × $I_{h,max}$ / $I_{p0}$ is up to 0.5. In this work, results from the experimental investigations of the halo currents during the campaign of 2014 in the KSTAR are presented.

The vertical displacement events (VDEs), in which plasma generally moves upward or downward, can be usually occurred by the vertical instability due to a failure of vertical control or a result of the gross magnetic perturbation from a minor disruption for elongated plasmas. The VDEs eventually cause major plasma disruptions that generate large electromagnetic forces and high heat loads, which are critical issues in the future magnetic confinement device such as ITER. Some studies on halo currents flowing into the vacuum vessel through in-vessel supporting structures during the VDEs were carried out in tokamaks [1-5]. After the H-mode discharge was successfully produced in the KSTAR, the study on plasma disruption has been carried out because the disruption is critical issue for plasma current of higher than 1.0 MA with higher heating power that is required as one of missions in the KSTAR project. To study on the disruption due to the VDEs, small Rogowski coils, used as halo current monitors (HCMs) [6], were installed on the supporting structures of back-plates for divertors. However we could not get a clear signal from the HCM measurement due to low signal to noise ratio and noise pick-up from the power supply (with a switching frequency of 4kHz) for the in-vessel control coil (IVCC) until the experimental campaign of 2012 in KSTAR. These issues were clearly solved from the campaign of 2014 after the two improvements in the DAQ system for the HCM by changing the *RC* time-constant in the analog integrators for increasing the amplitude of the HCM signal and by adding electronic low pass-filters between the integrator and the digitizer for reducing noise pick-up from the

IVCC. The HCM measurements were carried out during up / downward VDEs for elongated plasmas (κ ≥ 1.49). In this work, the initial experimental investigations from the HCM measurements in the campaign of 2014 were presented.

The KSTAR upper (lower) divertor consists of inboard divertor (ID), central divertor (CD) and outboard divertor (OD), and each divertor has eight-sectored back-plates equally distributed in the toroidal direction. The supporting structures are mounted at the sectored back-plate for each divertor as following; two structures between two sectored back-plates at the ID, two structures between two sectored back-plates and two structures in the middle of each sectored back-plate at the CD, and one structure between two sectored back-plates and one structure in the middle of each sectored back-plate at the OD. These sectored back-plates are electrically connected in the toroidal direction. Small Rogowski coils as HCMs were used to halo currents flowing through the supporting structures at the divertors during VDEs in KSTAR. The details on the HCM measurements were described in Ref. 7.

During the campaign of 2014 in KSTAR, the cases that the VDEs were occurred 1s after plasma start-up were selected to investigate characteristics of the VDEs. Figure 1 shows typical slides of the TV camera during an upward VDE, together with plasma shapes reconstructed by the real-time EFIT in the KSTAR. The plasma parameters for the selected shots were following; $I_{p0}$ = 0.2 – 1.0 MA, $B_T$ = 1.5 – 3.0 T, $q_{95}$ = 3.0 – 8.0, $\kappa$ = 1.49 – 2.1. The maximum plasma current quench rate, $-(dI_p/dt)_{max}$ was up to 200 MA/s during VDEs. There is no clear correlation between $-(dI_p/dt)_{max}$ and $dZ/dt$ as shown in Fig. 2. Here, $dZ/dt$ means the vertical shift velocity of the plasma at time of maximum halo current, and the values are 10 – 110 m/s for the selected shots.

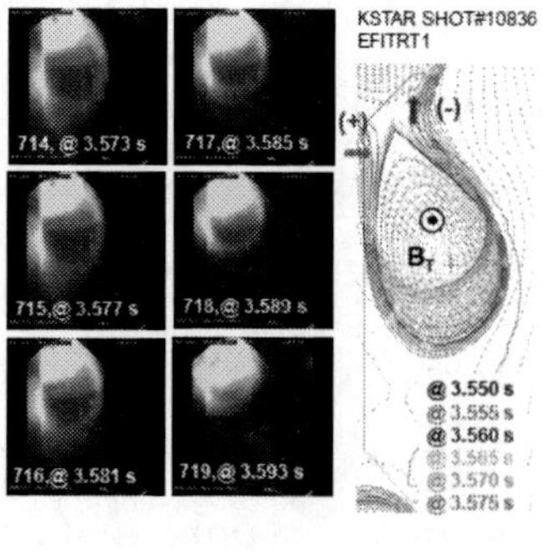


Fig. 1. Upward VDE in KSTAR.

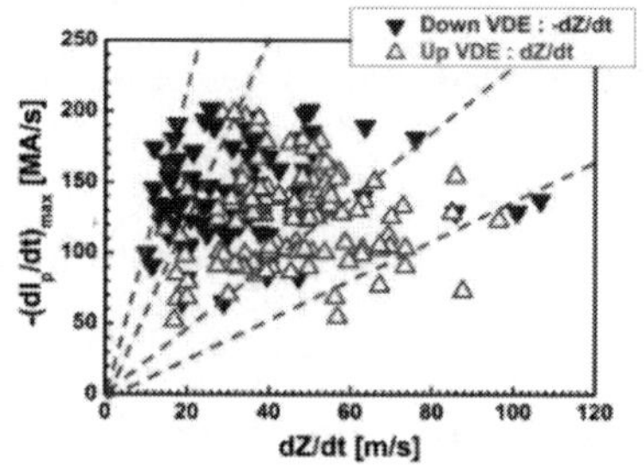


Fig. 2. $-(dI_p/dt)_{max}$ vs. $dZ/dt$ during VDEs.

In addition, there was the linear relationship between the current quench rate and plasma current before disruption as $-(dI_p/dt)_{max}$ [MA/s]= ~ $0.25\times I_{p0}$ [kA]. For a target value of plasma current of 2.0MA in KSTAR tokamak, the maximum plasma current quench rate can be estimated as ~ 500 MA/s from the relationship when the plasma disrupts due to the VDEs.

Since one supporting structure for the outer CD (or the OD), as a halo current channel, covers a toroidal angle of 22.5°, the total halo current $I_h$ ('source current') in the plasma halo region at the outer CD and the OD was estimated from 360/22.5 × ($I_{h,outerCD}$ + $I_{h,OD}$) where $I_{h,outerCD}$ and $I_{h,OD}$ were the average current measured by four HCMs equally distributed in the toroidal direction at the outer CD and OD, respectively. Here, $I_{h,k}$ was obtained from local currents $i_{h,k}(\phi)$measured by the four HCMs. Two different ratios of maximum to the averaged halo current were obtained at outer CD and OD, and the higher one of two values was used as the toroidal peaking factor (TPF). Halo fraction, $f$, was obtained by the ratio of the maximum value $I_{h,max}$ in the time evolution of total halo current and the plasma current in the pre-disruption phase $I_{p0}$.

There are weak corrections between TPF × $f$ and plasma parameters as $I_{p0}$ and stored energy before energy quench $W_{tot0}$ during VDEs: the values become slightly smaller as the parameters increases as shown in Fig. 3, which were similar to the parametric dependency of halo currents reported in JT-60U [3]. In addition, the magnitude of the halo currents tended to increase (decrease) with the increase in elongation $\kappa$ (safety factor $q_{95}$) as shown in Fig. 4, which were also observed in the COMPASS-D [4] and in the JET [5].

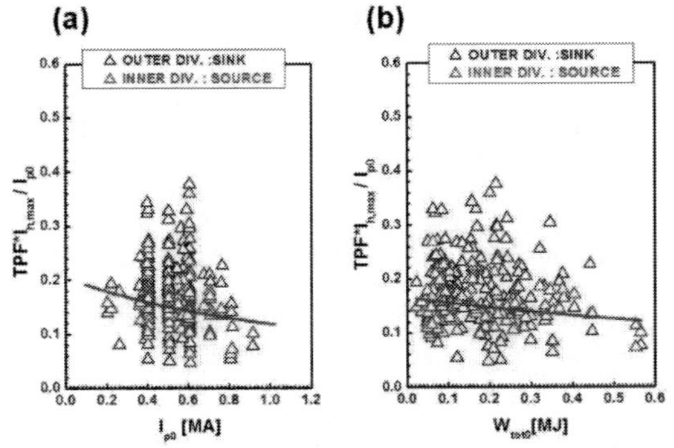


*Fig. 3. Dependence of TPF×f on plasma parameters; (a) $I_{p0}$ and (b) $W_{tot0}$ for upward VDEs. Here, 'source' means halo currents from wall to plasma and 'sink' is the reversed current from plasma to wall.*

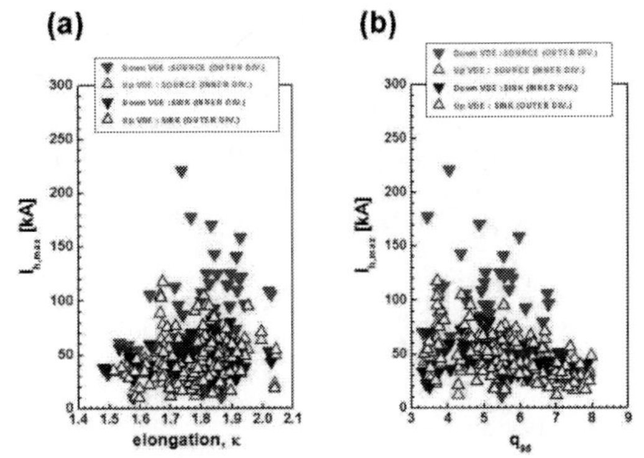


*Fig. 4. Dependence of halo current on (a) elongation and (b) safety factor.*

In the halo current measurements, the value of $f$ is up to 0.4, and the maximum value of TPF is 5.0 as shown in Fig. 5. The value of TPF ×$f$ is 0.5, which is lower than the maximum value required in the ITER database (0.75). Finally, the upper limit of the halo current is investigated as shown in Fig. 6 by using $I_{h,max} \leq \xi^{\xi/(1-\xi)}/(1-\lambda)I_{p0}/q_{95}$ (where $\xi = \tau_p/\tau_{halo}$ is the ratio of the effective $L/R$ resistive time of the plasma and the halo, respectively) [5].

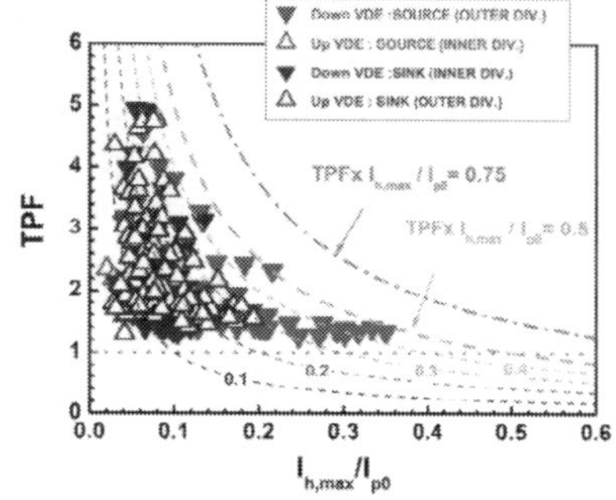


*Fig. 5. TPF versus halo fraction.*

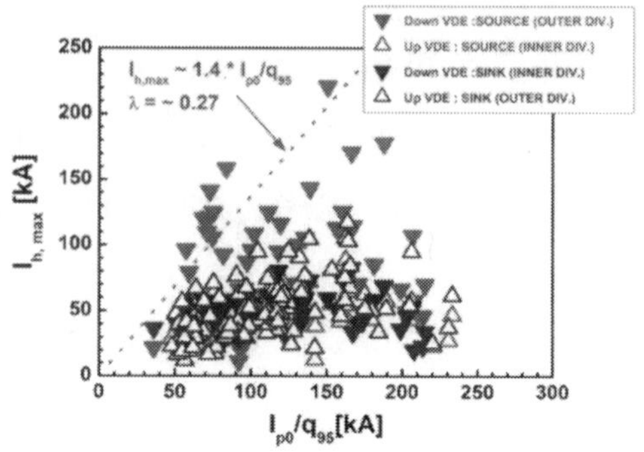


*Fig. 6. Maximum of halo current vs. $I_{p0}/q_{95}$.*

213 VDEs (about 10% of total shots in the campaign of 2014, and 136 / 77 shots for up / down VDEs, respectively) were occurred 1s after plasma start-up, which caused eventually to plasma disruptions. From the initial halo current measurements during VDEs, it was found that there were weak dependencies of halo current upon plasma parameters such that the magnitude slightly decreased as parameters (plasma current, stored energy and $q_{95}$) increased, but the value weakly became higher with as elongation increased. The values of $f$ and TPF $\times f$ were up 0.4 and 0.5, respectively. In addition, the upper limit of the halo current was $I_{h,max} \leq$ ~1.4 $I_{p0}/q_{95}$ ($\lambda$ = ~ 0.27). The further study on halo current characteristics in KSTAR are required as following; investigations to understand the cause of the differences between 'source' and 'sink' currents in the measurements, to interpret plasma decay time during the plasma current quench, to examine the dependence of the halo-current magnitude on the vertical displacement velocity together with behavior of electron temperature in the halo region, and to detect rotating halo current in KSTAR. This research was supported by Ministry of Science, ICT, and Future Planning under KSTAR project contract.

**References**

[1] E.J. Strait, L.L. Lao, J.L. Luxon, and E.E. Reis, *Nuclear Fusion,* **31** (1991) 527.

[2] R.S. Granetz, I.H. Hutchinson, J. Sorci, J.H. Irby, B. Labombard, D. Gwinn, *Nuclear Fusion,* **36** (1996) 559.

[3] Y. Neyatani, R. Yoshino, Y. Nakamura and S. Sakurai, *Nuclear Fusion,* **39** (1999) 559.

[4] P.J. Knight, G.G. Castle, A.W. Morris, A. Caloutsis, C.G. Gimblett, *Nuclear Fusion,* **40** (2000) 325.

[5] V. Riccardo, et al., *Plasma Phys. Control. Fusion,* **46** (2004) 925

[6] J. G. Bak, et al., *Korean Physical Society Meeting,* $21^{st}$ – $23^{th}$ , October, 2009.

[7] J. G. Bak, et al., *Proceeding of $7^{th}$ A3 Foresight Program Workshop* (Chuncheon, Korea, May 19-22, 2015)

# Characterization of edge turbulence in the ELM-crash-suppressed H-mode plasma

J. Lee[1], M.J. Choi[1], G.S. Yun[1], W. Lee[2], H.K. Park[2], N.C. Luhmann, Jr.[3]

[1] *Pohang University of Science and Technology, Pohang, Korea*

[2] *Ulsan National Institute of Science and Technology, Ulsan, Korea*

[3] *University of California, Davis, USA*

## Introduction

The crash of the edge-localized modes (ELMs) is a sudden collapse of the edge pedestal which induce relaxation of energy and particles from the confined plasma into the scrape-off layer [1]. It is a critical issue in magnetic fusion device rely on H-mode plasmas, since the heat load of the ELM crash on plasma components can be damaging. One of most popular methods to control the ELM crash is to perturb the plasma edge by adding small magnetic fields (approximately resonant with the equilibrium magnetic fields) in the edge of plasma, thereby enhancing particle transport and keeping the edge pressure gradient below crash threshold.

Despite successful control of the ELM crash, there are still many open questions concerning the mechanism of ELM crash suppression. In this letter, electron cyclotron emission imaging (ECEI) system on the Korea Superconducting Tokamak Advanced Research (KSTAR) is used to investigate the change of edge transport induced by turbulence during the period of ELM crash suppression by resonant magnetic perturbation (RMP). Especially, 2-D correlation measurement, spectral power density and dispersion relation of edge turbulence have been obtained with high resolution using correlation analysis technique.

## Electron cyclotron emission imaging (ECEI) system in the KSTAR

The ECEI can be regarded as a microwave camera for measuring 2-D electron temperature fluctuations ($T_e$) [2]. The KSTAR ECEI system consists of vertically aligned array of 24 antennas and each antenna is connected to heterodyne detector that resolves the radiation in 8 frequency bands. Therefore the array provides $24 \times 8 = 192$ pixels image of $T_e$ fluctuations ($\tilde{T}_e = \delta T_e / \langle T_e \rangle$, where $\delta T_e = T_e - \langle T_e \rangle$ and $\langle T_e \rangle$ is a time average). Due to flexible local oscillator and large aperture optics system, a view positition of ECEI can be focussed anywhere in poloidal cross-section with variable vertical coverage. In addition, another ECEI system by $1/16^{th}$, to extend the diagnostic capability to quasi 3-D imaging for MHD instabilities as well as correlation ECE diagnostics for turbulence measurement. For the KSTAR ECEI system, the vertical channel spacing $z_0$ of each samle volume is in the range of $1.4-4.0$ cm and the distance

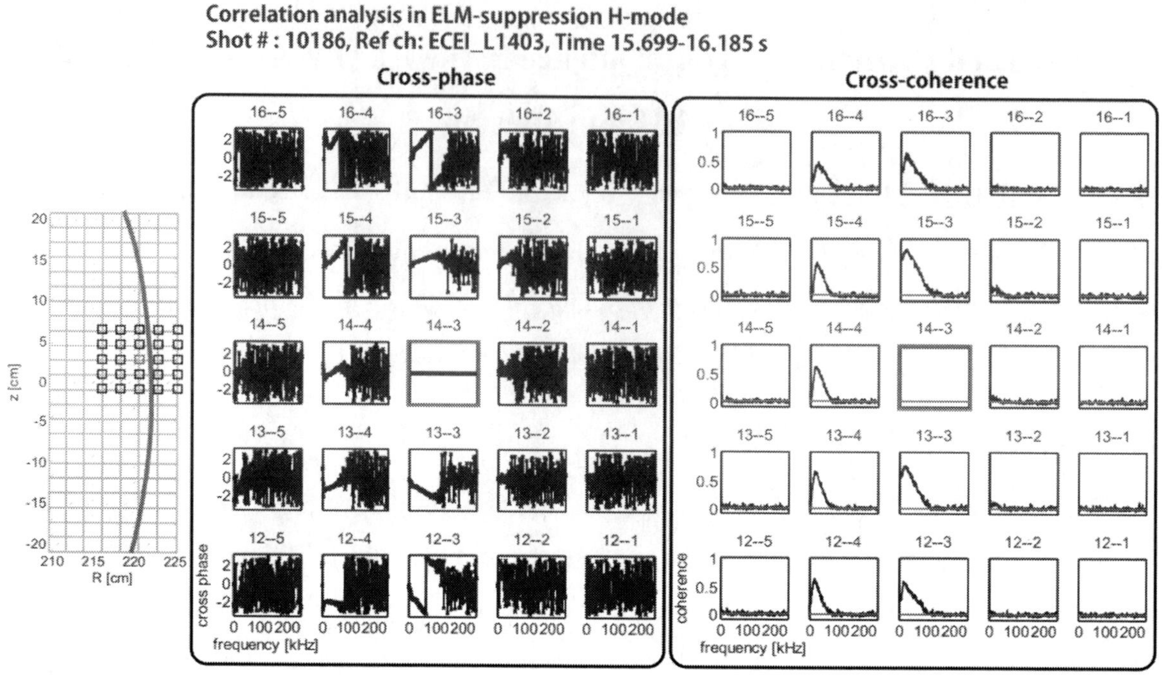


Figure 1: The cross-phase and cross-coherence analysis using 5 (radial) ×5 (poloidal) ECEI channels. Strong poloidal correlation inside the separatrix has been observed in range of $\leq 70$ kHz during ELM-crash-suppressed phase. The red line is the separatrix position.

between adjacent channels in the direction of major radius $R$ is $\sim 2$ cm. Therefore, the upper limit of the system wavenumber resolution is $k_\theta \leq 2.25$ cm$^{-1}$ and $k_r \leq 1.5$ cm$^{-1}$ at the edge.

**Turbulence measurement in ELM-suppression phase**

The correlation analysis has been performed for a steady-state H-mode plasma with ELM-crash-suppressed under $n = 1$ RMP (shot #10186: line-average density $n_{e,l} \sim 2 \times 10^{19}$ m$^{-3}$, and toroidal velocity ($v_{tor} \sim 70$ km/s). Figure 1 shows the cross-phase and cross-coherence for 5 (radial) ×5 (poloidal) channels, which are covering a roughly $10 \times 7$ cm at the plasma midplane. The red box represents the reference channel. The correlationa analysis has been performed in time segment $\sim 0.15$ s which corresponds to $\sim 75000$ data points for 500 kHz sample rate. The result is then resampled by averaging in unit of 10 segments. An interesting observation in Fig. 1 is the existence of broadband and low frequency coherent mode ($f \leq 70$ kHz) along poloidal direction with narrow radial zone. The high frequency components over 100 kHz are sysmatic noise, characterized by random phase.

The approximate size of turbulent eddies can be estimated using cross-correlation coefficients.

Figure 2 is a zero-delay cross-correlation coefficient, $R_{xy}(\tau = 0)$, versus radial and vertical coordinates for turbulent eddies. Gaussian fitting function allows the correlation length to be determined by taking $1/e$ width, $L_{1/e}$ : $f(x) = A_1 exp[-x^2/L_{1/e}^2] + A2$, where $x$ is a spatial coordinate, and $A_1$ and $A_2$ are amplitude and offset terms that were determined by least square fitting. According to the measurement, the poloidal correlation length is $\leq 1.84$ cm and the radial correlation length is $\leq 1.83$ cm which are much smaller than typical size of ELM flux tube. Note that the actual size of turbulent eddy could be smaller than the measurement due to the finite resolution of the ECEI system ($R \sim 2.0$ cm, $z_0 \geq 1.4$ cm).

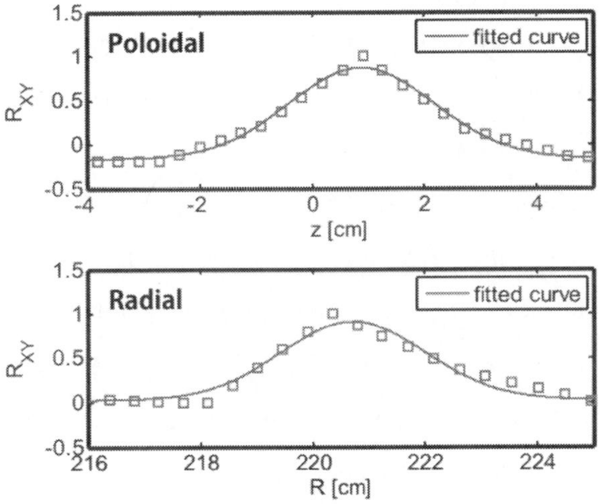


Figure 2: Poloidal and radial size of edge turbulent eddy, estimated from cross-correlation coefficients with zero time lag.

Figure 3 shows the statistical dispersion relation of turbulence in the ELM-crash-suppressed phase represented by local frequency/wavenumber spectrum $S_L(k, f)$ [3]. The $S_L(k, f)$ plot exhibits approximately linear dispersion in a wide range of wavenumbers ($k_\theta \leq 1$ cm$^{-1}$) and frequencies. This implies that the tubulence in the ELM-crash-suppressied phase has long wavelength and average group velocity of $\sim 3$ km/s, propagating in the electron diamagnetic drift direction. The measured wavenumber $k_\theta \leq 1$ cm$^{-1}$ gives the characteristic size of turbulence, which is usually compared to the Larmor radius $\rho_s = \sqrt{2m_iT_e}/eB$, here $m_i$ is ion mass, $e$ is electric charge. In this case, $k_\theta\rho_s \sim$ 0.1 using $\rho_s \sim 1$ mm at the edge. According to the scale size, possible candidates for the observed turbulence are kinetic ballooning mode (KBM), microtearing mode (MTM), resistive

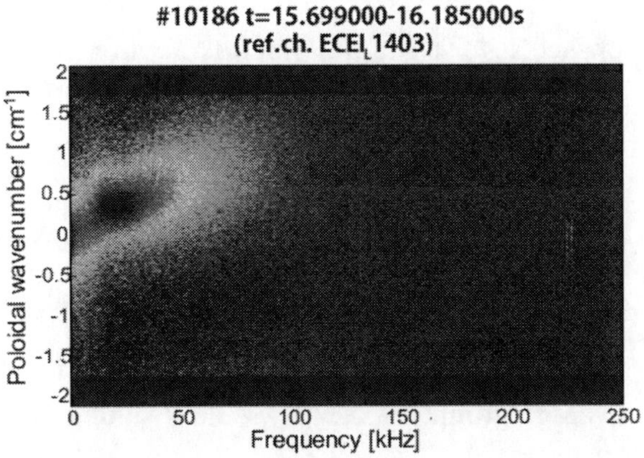


Figure 3: The local frequency/wavenumber spetrum $S_L(k, f)$ during ELM-crash-suppression. The turbulence has long wavelength ($\lambda_{pol} \geq 6$ cm) and average group velocity of $\sim 3$ km/s in electron diamagnetic drift direction.

ballooning mode (RBM) ($k_\theta \rho_s \sim 0.1$ in tokamak) and ion temperature gradient (ITG) mode ($0.1 \leq k_\theta \rho_s \leq 1$) [4]. However, KBM and ITG can be excluded since they propagate in the ion diamagnetic direction. The remaining candidates, RBM and MTM, can be distinguished by the parity analysis with respect to potential fluctuation which will be measured in future experiment.

Since the KSTAR has equipped with two independent ECEI systems, the parallel wavenumber $k_\parallel$ can also be determined by measuring the phase delay $\Delta\phi_{xy}$ between the two systems according to the relation: $k_\parallel = \Delta\phi_{xy}/\Delta l$, where $\Delta l$ is the distance between the two toroidally-separated view positions. It is assumed that turbulent eddies are alinged along magnetic field lines and have parallel wavelength large enough to captured by both ECEI systems simultaneously. Figure 4 is a result of parallel wavenumber measurement during ELM-crash-suppressed phase: $2 \leq \lambda_\parallel \leq 8$ m, which is a wide range in reference to the toroidal circumference $\sim 14$ m of the edge.

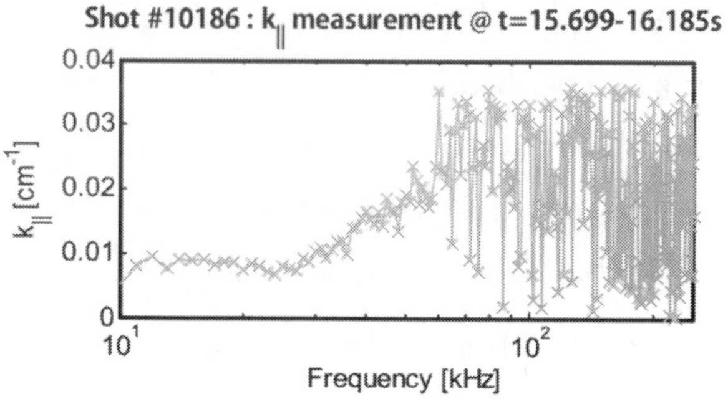


Figure 4: Parallel wavenumber measurement using 3-D ECE imaging technique. The corresponding parallel wavelength range is $2 \leq \lambda_\parallel \leq 8$ m.

## Summary

Edge plasma turbulence during the ELM-crash-suppressed period has been studied by employing correlation analysis technique. The result shows clear cross-phase, cross-coherence and $S_L(k, f)$ spectrum in the plasma edge region: poloidal wavenumbers in the range of $\leq 1$ cm$^{-1}$, average group velocity of $\sim$ 3km/s along electron diamagnetic direction, parallel wavelengths in the range of $2 \leq \lambda_\parallel \leq 8$ m. These are unique features of ELM-crash-suppressed phase and may be critical information for understanding the physics of ELM-crash-suppression.

## References

[1] H.Zohm, Plasma Phys. Control. Fusion **38**, 105 (1996)

[2] GS. Yun, et al., Rev. Sci. Instrum. **81**, 10D930 (2010)

[3] JM. Beall, et al., J. Appl. Phys. **53**, 3933 (1982)

[4] BD. Scott, et al., Phys. Plasmas **12**, 062314 (2005)

# Role of the equilibrium and perturbative central current density in sawtoothing and non-sawtoothing discharges in KSTAR

H. K. Park[1], Y. B. Nam[2], M.J. Choi[2], G. H. Choe[2], W. Lee[1], G. S. Yun[2], S. Jardin[3]

[1]*Ulsan National Institute of Science and Technology, Ulsan, Korea*

[2]*Pohang University of Science and Technology, Pohang, Korea*

[3]*Princeton Plasma Physics Laboratory, Princeton, NJ, United States*

**Abstract**

The recent experiment [1] on excitation of the m/n=3/3 mode with a current blip induced by ECH and successive evolution to the m/n=2/2 and m/n=1/1 mode in the core of the sawtoothing discharge in KSTAR suggests that the central safety factor ($q_0$) may have to change from below ~1 (before crash) to slightly above ~1(after crash). This interpretation is consistent with the q profile condition for MHD simulation necessary for the growth of higher order modes which require $q_0$ slightly above ~1 until the 1/1 mode becomes dominant [2]. Experimental observation of a long lived higher order mode in non-sawtoothing discharge (presumably $q_0>1$) is consistent with the fact that the $q_0$ has to be below ~1 to support the growth of the m/n=1/1 mode and subsequent crash.

**Introduction**

The classical magnetohydrodynamic (MHD) instability, “sawtooth oscillation”, in the core of tokamak plasmas is initiated by the m/n=1/1 kink mode instability driven by an excess core current density [3] in the core. It is believed that the explosive growth of this instability leads to a burst of the m/n=1/1 mode through magnetic reconnection and the core pressure within the q=1 surface is transported to outside of the q=1 layer in a fast time scale (i.e., reconnection time scale) while the growth of the kink instability is in a slow time scale. A decade later the discovery of sawtooth event, experimental measurements of the central safety factor, $q_0$ confirmed that the measured value of $q_0$ remained close to ~0.75 before and after the crash [4, 5]. This suggests that the core current density remains mostly unchanged while the core pressure (density and temperature) is removed in a magnetic reconnection time scale (~hundred μs). Very little change in core current density casted a doubt on theoretical understanding of the sawtooth crash physics and the driving mechanism of the m/n=1/1 kink instability. This experimental confirmation puzzled the community but the controversial issue has been forgotten for long time. Note that there are other experiments that the results are contradicting with these two measurements. The measurement of the higher order modes (m/n=2/2, 3/3, etc.)

induced by the electron cyclotron heating (ECH) in the core of sawtoothing plasmas (presumably $q_0<1$) in KSTAR provided a promising new avenue to revisit this problem, since such higher order modes can have a growth rate only for $q_0$ slight above ~1 and the m/n=1/1 mode can be dominant when $q_0$ is below ~1. This is consistent with the previous theoretical understanding of the sawooth oscillation. Furthermore, a similar experiment in a non-sawtoothing discharge (i.e., presumably $q_0$ is slightly above ~1) sustained a long lived higher order mode for more than ~tens of sawtooth period. This observation further supports that the $q_0$ remains above ~1. In conclusion, experimental excitation of the higher order mode driven by ECH and its time evolution support the required conditions of $q_0$ value; 1) In non-sawtoothing plasmas, excitation of the higher order mode and its sustainment when the $q_0$ is maintained above ~1. 2) In sawtoothing discharges, excitation of the higher order mode after the sawtooth crash supports the hypothesis that $q_0$ is above ~1 and subsequent evolution to the m/n=1/1 mode and following crash suggest that the $q_0$ has to be below ~1 before the crash.

**Generation and evolution of the higher order modes**

KSTAR is a medium size tokamak with major radius $R_0$=1.8m and minor radius $a$=0.5m. The plasma parameters used in this paper are as follow: the average electron density$\langle n_e\rangle$~$3\times10^{19}$m$^{-3}$, the electron temperature$T_{e0}$= 3–3.8 keV at the plasma center, and the elongation factor of the last closed flux surface = 1.5–1.54. In this experiment (#9214), low confinement mode (L-mode) plasma was heated by neutral beam injection (100keV, 1MW) and ECH/CD (170 GHz, 0.7 MW) at the toroidal magnetic field 2.95T. After the plasma current ramp-up ($I_p$=700kA), the ECH/CD vertical position was linearly scanned from $Z$=0 to 16cm at the center in 4s while maintaining the plasma shape and position. The toroidal injection was fixed at 7°in the co-current direction. The observed MHD modes are mainly from the high resolution (space and time) 2-D ECEI system. The ECEI system on KSTAR was developed for visualization of MHD instabilities in 2010. This first ECEI system with the dual arrays provided simultaneous 2-D images of $Te$ fluctuations on the high field side (HFS) and low field side (LFS) at the same toroidal location with spatial resolution ~1–2cm and time resolution ~1–2us. In the 2013, another ECEI system with a single array has been installed at a different toroidal location [6]. The two ECEI views having different fixed toroidal locations were placed radially adjacent to each other so as to provide radially continuous quasi-3D images. Three different mode patterns are m/n= 1/1, 2/2 and 3/3 as the vertical position of the ECH/ECCD is moved from the core to inside edge of the inversion radius [1]. This observation was studied theoretically and confirmed that the required $q_0$ is slightly above ~1[2].

**Study of the equilibrium q-profile condition for higher order mode**

In order to have a systematic comparative study of the time evolution of $q_0$ based on the growth and decay of the higher order mode, an example of time evolution of the m/n=3/3 mode for one sawtooth period (~10 ms) is illustrated in Fig. 1. In Fig.1a, the time trace of the core $T_e$ (black) demonstrates a high frequency oscillation (~17kHz), intermediate oscillation frequency (~11.4kHz) and slower oscillation (~5.7kHz) as the core temperature is increased before the crash. Here, the m/n=3/3 mode is excited after the crash and this mode slowly transforms into the m/n=2/2 mode and then it finally transforms to the m/n=1/1 mode before the crash. The blue time trace in this figure represents the time evolution of the electron temperature at outside of the inversion radius. A sudden rise of the $T_e$ due to the fast crash is followed by an exponential

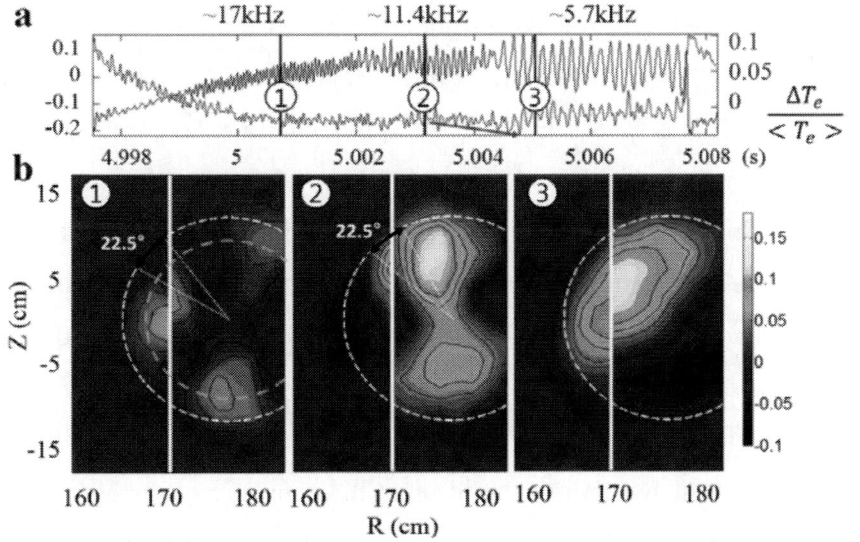


**Figure 1.** Temporal evolution of the m/n=3/3 mode structure. **a)** ECE time traces at the core (black) and outside of the inversion radius (blue). **b)** ECE images of m/n=3/3, 2/2, 1/1 modes corresponding to ~17kHz, ~11.4kHz, and ~5.7kHz oscillation phases. [1]

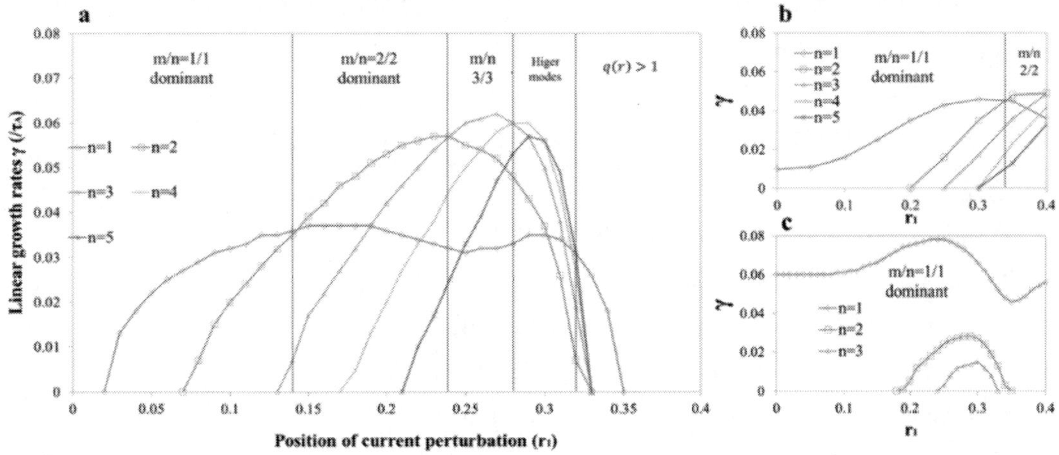


**Figure 2.** Linear growth rate spectra of m≤5 modes with various current perturbation position ($0<r_1<0.4$) and three different $q_0$ values: **a)** $q_0$=1.04 **b)** $q_0$=0.98 **c)** $q_0$=0.80. Higher order modes dominate as $r_1$ move toward the q~1 radius when $q_0$>1.

decay. The corresponding images for each mode are provided in Fig. 1b. Since it is rather difficult to perform a non-linear simulation for whole period, linear simulation with the M3D-C1code [7] has been performed at a fixed time with the equilibrium q profile that supports the highest growth rate for each mode. The result is illustrated in Fig.2. The q profiles with three different central values (0.8, 0.98 and 1.04) within the q~1 surface and various current perturbation positions ($r_1$) with the positive current blip (negative blip in q profile) are used for this simulation. For the m/n=1/1 mode, the growth rate is highest for $q_0$=0.8 and gradually reduced with the increase of the $q_0$ value as shown in this figure. As shown in this figure, the growth of higher order modes (m/n=2/2, 3/3 and higher mode) dominates as the current blip position is moved toward the ~q=1 surface while the $q_0$ is maintained above ~1. The observed time evolution of the m/n=3/3 mode to m/n=2/2, m/n=1/1 and crash, can be interpreted as follows; the observed m/n=3/3 mode right after the crash in Fig.1 supports that the $q_0$ has to be above ~1 and the position of the current blip is between 0.23 and 0.27 to have the growth of the m/n=3/3 mode. This is quite consistent with the experimental condition. As time evolves, the m/n=3/3 mode transforms into the m/n=2/2 mode. This observation may suggest that the current build up due to the local heating and diffusion of the current blip at the same position forms a negative blip slightly toward the center while the $q_0$ is still maintained above ~1. During the period when the dominant mode is the m/n=1/1 mode, the central q value is dropped well below ~1 in a fast time scale due to accumulation of the core current density and current diffusion from the current blip toward the core. In summary, the $q_0$ value has to be above ~1 after the sawtooth crash and initial build-up of the core current is slow (i.e., $q_0$ is above ~1) and a rapid rising phase of the current is followed. In addition, excitation of the higher order mode (m/n=6/6) and sustainment for a long period of time in a discharge with no sawtooth oscillations (presumably $q_0$>1) is further supports the hypothesis of $q_0$ evolution in sawtoothing discharge.

*Work supported by NRF of Korea (grant No. NRF- 2014 M1A 7A1A03029865)

**References**

[1] G. H. Choe *et al*, Nuclear Fusion **55** 013015 (2015)

[2] Andreas Bierwage *et al*, Nuclear Fusion **55** 013016 (2015)

[3] J. Wessen *et al*, "Tokamaks", Oxford University Press (2011)

[4] H. Soltwisch *et al*, Review of Scientific Instrument **59** 1599 (1988)

[5] F. M. Levinton *et al*, Physical Review Letters **63** 2060 (1989)

[6] G.S. Yun *et al*, Review of Scientific Instrument **85** 11D820 (1988)

[7] S.C. Jardin, *et al*. Computational Science & Discovery 5 (2012) 014002

# Studies on RFX-mod discharges with spontaneous rotation

B. Zaniol, T. Bolzonella, R. Cavazzana, P. Innocente, P. Zanca and M. Zuin

*Consorzio RFX (CNR, ENEA, INFN, Universita' di Padova, Acciaierie Venete Spa), Padova (Italy)*

**Introduction** - The Reversed Field Pinch (RFP) has a magnetic configuration characterized by a wide spectrum of m = 0, 1 tearing modes (TM), involved in the dynamo process. These modes have the natural tendency to lock in phase together (slinky-mode) and to the wall, such to create stationary bulges at the plasma surface. In RFX-mod an advance control system of 4 x 48 independently fed saddle coils feedback induces slow rotations in order to avoid a strict wall-locked condition [1]. Recent experiments demonstrated the possibility to recover the tearing modes fast rotation by lowering the plasma current under a certain threshold [2], and opened a new RFX-mod scenario characterized by the "natural" rotation of the dynamo modes. In this new scenario plasma flow rotation changes as well, being affected by the new magnetic perturbation dynamics. Passive toroidal and poloidal impurity flow measurements have been performed during these experiments, despite their low electron densities, $n_e<10^{19}$ $m^{-3}$. The most intense line emission of RFX-mod main impurities could be detected, in particular: O V line at 650 nm, C III line at 464.7 nm and C II line at 464.9 nm. In the following, some of the results are discussed.

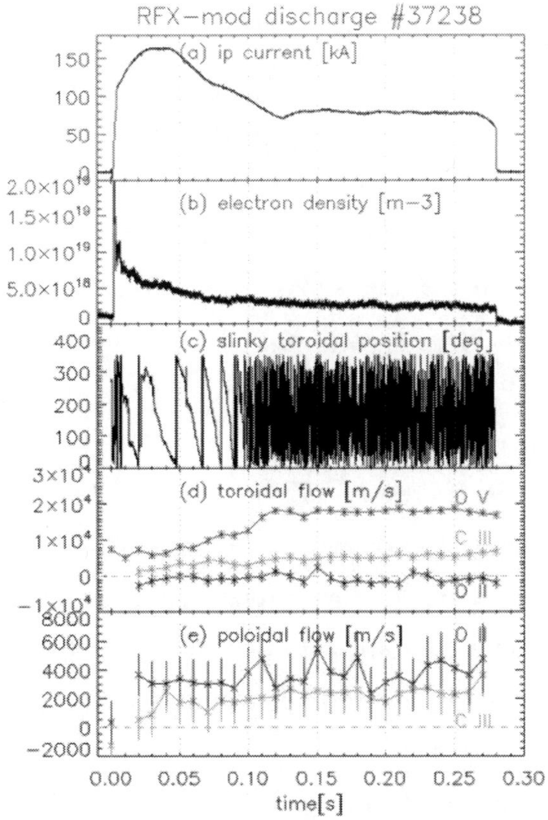


**Figure 1** Time traces of RFX-mod low current experiment. From top panel to bottom one: plasma current, electron density, slinky- mode toroidal position, toroidal impurity flow and poloidal impurity flow.

**Experimental results and discussion** - The time traces of a typical RFX-mod low current experiment are reported in Figure 1. A descending ramp in the plasma current starts soon after the discharge has reached its maximum value at about 150 kA (1a). In this phase the slinky-mode rotation (1c), consequence of the feedback system actively controlling the TM, increases with the lowering of the plasma current, while the toroidal flow of O V grows by a factor three (1d). The TM spontaneous rotation starts at about 100 ms, when the current reaches the threshold value of about 100 kA, and it is clearly seen in the spectra of the m = 0,1 n = -1,-23 modes, Figure 2b. The frequencies of the most internal resonant TM increase of more than one order of magnitude, from few hundreds of Hz to about 15 kHz, Figure 2c. In the following of the discharge the slinky-mode signal is no more reliable because of the screening effect of RFX-mod vacuum vessel (VV) on the fast magnetic modulation.

It is interesting to note that the more external impurities species, C III and O II, are less affected by the mode rotation, both in the poloidal and in the toroidal flow (Figure 1d and 1e).

In RFX-mod the impurity toroidal flow exhibits a strong dependence on electron density [3]. To better study the synergy of mode rotation and electron density the graph reported in Figure 3 has been created. Different colors have been used to represent O V toroidal flow intensity at different plasma currents and electron densities. The dependence on electron density is clear when moving from the right side of the graph to the left, but the highest velocities are reached only by lowering the plasma current under the threshold of 100 kA, when the mode rotation occurs. The C III toroidal flow dependence on plasma current is too

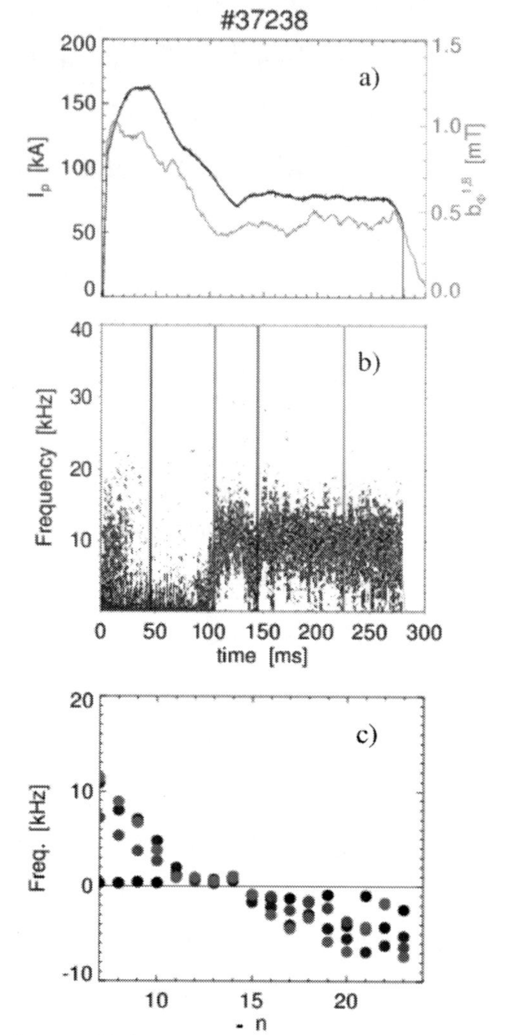


**Figure 2** Magnetic modes frequencies vs discharge time (b), and vs toroidal number n (c). The different colours in (c) correspond to the coloured lines in (b).

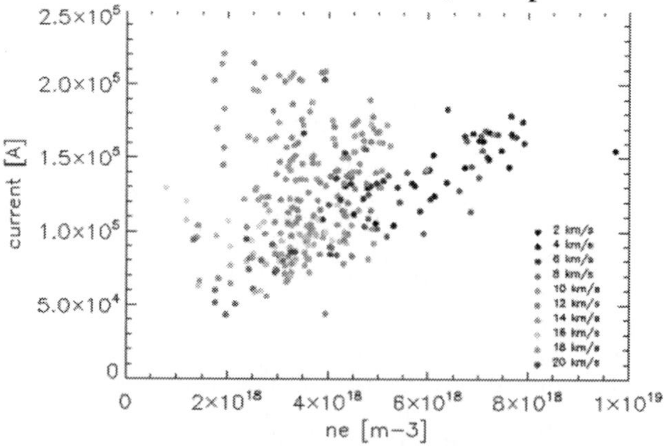


**Figure 3** Dependence of O V toroidal flow on electron density and plasma current.

weak to be seen in Figure 4, where the only clear dependence is on the electron density. A similar graph plotting the poloidal velocity is not significant due to the strong influence of the slinky-mode rotation on the impurity flow. Large oscillations in the poloidal flow up to +/- 20 km/s occur when the slinky-mode passes by the diagnostic line of sight, so affecting the measurements at higher current. Also the poloidal flow measure of O V is not fully reliable due to its low emissivity near the edge, where the poloidal lines of sight are placed. However data seem to suggest an O V poloidal flow of few km/s, consistent with the C III and O II measurements.

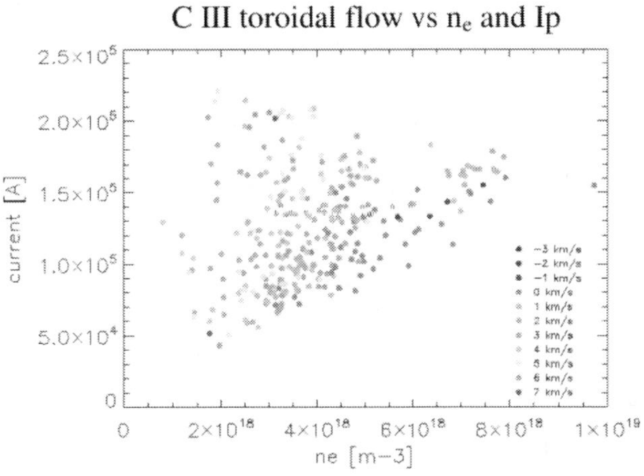


**Figure 4** Dependence of C III toroidal flow on electron density and plasma current.

The averaged radial positions of each ion species have been calculated with the help of a 1-D transport code [4] in order to reconstruct a possible impurity flow profile. The unfavourable signal to noise ratio of the interferometer and of the Thompson scattering measures, not optimised for such low plasma currents, does not allow a detailed reconstruction of electron density, $n_e(r)$, and temperature profiles, $T_e(r)$. Anyhow the observation that $n_e$ and $T_e$ do not change significantly during the current ramp down, neither the impurities influxes from the wall, indicates that the modification of the impurity flow is mainly due to the evolution of the flow radial profile and not to the change of the ions radial positions. An averaged radial normalised position (ρ) of about 0.5 has been found for O V toroidal flow measure, while for C III and O II ρ is 0.78 and 0.81 respectively. Figure 5 shows the modification of the impurity toroidal flow radial profiles when the spontaneous rotation occurs. The measurements cover only a small portion of the radius, but they suggest an increase in the toroidal flow, that affects only marginally the outer region.

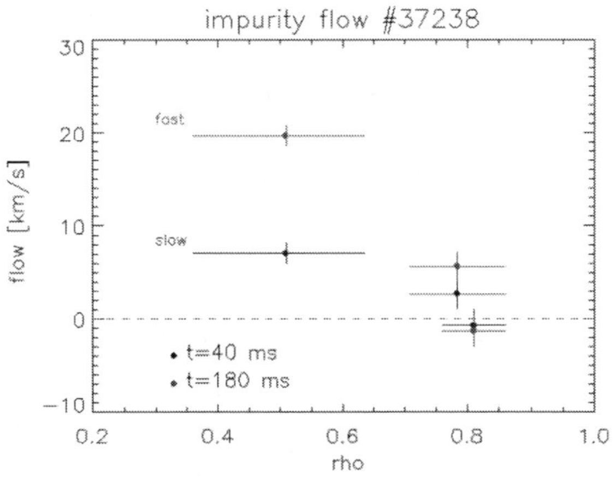


**Figure 5** Radial profile of impurity toroidal flow of the discharge of **Figure 1** at 40 ms, before the transition to fast rotation, and at 180 ms.

To check the consistency of the impurity flow measurements with the magnetic results, according to [2], we considered the following formula that ties together the ion flow angular frequencies Ω (aproximated by the impurity flow measures) and the magnetic modes frequencies $d\varphi^{m,n}/dt$.:

$$\frac{d\varphi^{m,n}}{dt} = -n\Omega_\phi(r_{m,n},t) - m\Omega_\theta(r_{m,n},t) + \frac{nB_\theta}{enR_0B^2}(1+\frac{m^2}{n^2}\frac{R_0^2}{r^2})\frac{d(p_e+p_i)}{dr}\Big|_{r_{m,n}}$$

The last term represent the diamagnetic terms and as stated above can not be calculated with precision due to the uncertainties on $n_e(r)$ and $T_e(r)$ reconstructions. Assuming stationary $n_e$ and $t_e$ profiles, and negligible variations in the poloidal flow, by subtracting the fast rotation condition to the slower one, we obtain:

$$-n\Delta\Omega_\phi(r_{m,n}) = \Delta\left(\frac{d\varphi^{m,n}}{dt}\right) \approx \frac{d\varphi^{m,n}}{dt}\Big|_{fast}$$

Assuming the O Vemission at the position ρ = 0.5 where mode m = 1, n = -9 is resonant, from the toroidal flow increase of about 12 km/s we deduce an angular frequency variation of 0.95 kHz that multiplied by n = -9 matches fairy well with the magnetic mode fast rotation of about 8 kHz. For C III the same calculation fails first of all because of the the fast mode rotation opposite to the current. Supposing to place C III toroidal flow measure at the resonance radius of m = 1 n = -23 mode (the most external measurable mode), a toroidal rotation increase of 3 km/s corresponds to an enancement of the mode frequency of about 5 kHz, that matches the result of the magnetic mode analisys only in absolute value, Fig.2c.

**Conclusion** - The analysis of impurity flow measurements on RFX-mod low current discharges, featuring the unlocking of the magnetic modes and their fast rotation, has started and produced the first results. The impurity toroidal flow increase of more than 10 km/s at middle radius has been consolidated and is consistent with the fast rotation of m = 1, n = -9 TM. Some inconsistencies still persist in the outer radial region where the impurity flow modification is not able to explain the fast counter-current mode rotation. The diamagnetic term may probably have a major role in the outer region; this suggests to better optimize the diagnostics measuring $n_e$ and $t_e$ profiles in future experiments.

**References**

[1] P. Zanca, L. Marrelli, G. Manduchi, G. Marchiori, Nuclear Fusion 47 (2007) 1425.

[2] P. Innocente, P. Zanca, M. Zuin, T. Bolzonella and B. Zaniol, Nucl. Fusion 54, 122001.

[3] B. Zaniol, L. Carraro, E. Gazza, I. Predebon, M.E. Puiatti, P. Scarin, G. Spizzo M. Valisa, 34th EPS Conference 2007.

[4] L. Carraro et al. Nuclear Fusion 36, 1623 1996

# H-mode accessibility in ohmic discharges on RFX-mod

O. Kudlacek[1,2], R. Cavazzana[1], C. Finotti[1], G. Marchiori[1], L. Marrelli[1], P.Piovesan[1], P. Zanca[1], M. Zuin[1],

[1] *Consorzio RFX, Corso Stati Uniti 4, Padova, Italy*
[2] *University of Padova, Via 8 Febbraio, Padova, Italy*

## I. Introduction

RFX-mod is a toroidal device originally designed as a reversed field pinch, presently operated also as a low current tokamak. The major radius of this device is 2 m, minor radius 0.46 m. The on-axis toroidal magnetic field is 0.55 T. The device is equipped by 8 independent shaping coil pairs symmetric with respect to the midplane that allow generation of several types of discharges: limiter discharge with both circular and elongated plasma, double null discharge and as some of the 8 pairs can be disconnected, it is possible to generate single null discharges either with upper or with lower X-point. RFX-mode is also equipped with 192 independent saddle coils for MHD control. Using the saddle coils and sophisticated control algorithms, RFX-mod can operate with the edge safety factor below two or avoid 2/1 mode locking in circular discharges [1]. In future, the same experiments should be performed also for single null discharges, possibly in H-mod. In addition, if the H-mod were reached, RFX-mod saddle coils could be useful tool for ELM mitigation.

Presently, RFX-mod can routinely perform single null discharges. The plasma shape can be computed using the algorithm presented in [2] with precision of several millimetres compared to independent simulations using the Grad-Shafranov solver MAXFEA [3]. The boundary reconstruction method can also provide the value of poloidal magnetic field everywhere in the vacuum. The boundary reconstruction works also in real time and provides the values of the gaps between the plasma and the first wall for 8 lines of sight connecting the centre of the vacuum vessel and the flux loops. Using the real time gap computation, a controller of the plasma shape was designed. It allows following the desired plasma shape reference signals also if the plasma parameters changes [4]. These two tools are useful for the design and performance of the experiments focused on L to H-mod transition. The shape control algorithm can modify the shape during the discharge to modify the L-H threshold power [5] that strongly depends for example on the distance between the X-point and the first wall. The offline boundary reconstruction is useful for evaluation of quantities related to the energy confinement (this property is expected to improve after the LH transition) such as $\beta_p$. Another important parameter that is evaluated is the power through separatrix carried by particles. The computation of these macroscopic plasma parameters is described in the paper.

## II. $\beta_p$ evaluation

In this section, the evaluation of $\beta_p$ for RFX-mod diverted discharges will be presented. The most common method of $\beta_p$ evaluation is based on the diamagnetic measurement. Unfortunately, this method is not reliable for RFX-mod diverted discharges due to hardware limitations. Thus, an independent method to evaluate $\beta_p$ presented in [6] was implemented and used. This method is based on computation of the following integrals along the plasma boundary:

$$s_1 = \frac{2\pi}{V\langle B_p^2\rangle}\int R(s)B_p^2(s)\mathbf{n}(s)\mathbf{p}(s)ds, \qquad (1)$$

where $V$ is the plasma volume, $s$ is the curvilinear coordinate along the plasma boundary, $R$ is the major radius, $B_p$ is the poloidal magnetic field, $\mathbf{n}$ is the normal vector to the plasma boundary and $\mathbf{p}$ is the vector connecting the magnetic axis and the plasma boundary. As we know the plasma boundary and the magnetic field everywhere in the vacuum, the computation is straightforward. The other needed integrals are

$$s_2 = \frac{2\pi R_0}{V\langle B_p^2\rangle}\int R(s)B_p^2(s)\mathbf{n}(s)\mathbf{e}_r ds, \qquad (2)$$

$$s_3 = \frac{2\pi R_0}{V\langle B_p^2\rangle}\int z(s)B_p^2(s)\mathbf{n}(s)\mathbf{e}_z ds, \qquad (3)$$

where $R_0$ is the major radius, $z$ is the vertical coordinate and $\mathbf{e}_{r,z}$ are the unit vectors in radial/vertical direction. Using these integrals, one can estimate the following:

$$\frac{s_1}{2} + s_2 = \beta_p + \frac{l_i}{2} \qquad (4)$$

and

$$l_i = \frac{1}{\alpha - 1}\left(\frac{s_1}{2} + \frac{s_2}{2}(R_t - R_0) - s_3\right), \qquad (5)$$

where $R_t$ is the plasma centroid position and $\alpha = 2k^2/(1+k^2)$, where $k$ is the elongation. The validity of this computation was tested in a few diverted discharges by puffing the gas. After the gas puff, the plasma pressure and thus also $\beta_p$ are expected to increase. A minor increase of $l_i$ is expected as well, since the plasma edge is cooled and thus the peaking of the current increases. The time evolution of $\beta_p$, $l_i$ and $\beta_p + l_i/2$ is shown in Fig. 1. The results of the computation meet the theoretical expectations.

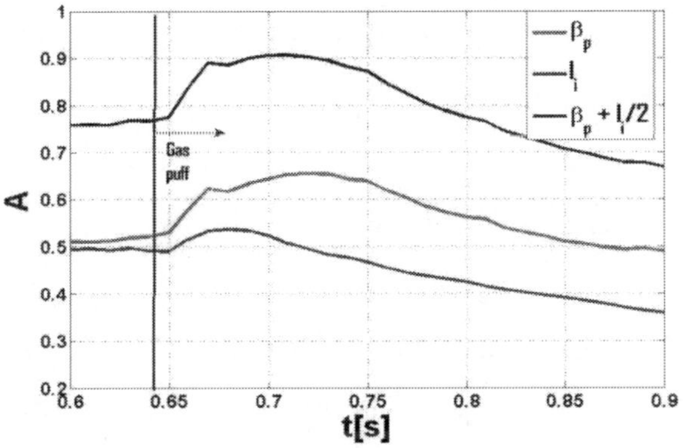


*Fig. 1: The time evolution of $\beta_p$, $l_i$ and $\beta_p + l_i/2$ after gas puff for discharge 37925.*

## III. Estimate of power through separatrix carried by particles

For the L-H transitions, two basic conditions must be met. The electron density and the power through separatrix carried by particles $P_{sep}$ have to be above a certain threshold [7]. The critical density for RFX-mod according to scaling laws presented in [7] is $8 \cdot 10^{18}$ m$^{-3}$ (the real density is usually below this value). The power threshold needed for the LH transition $P_{LH}$ is around 100 kW for the critical density. This value can be exceeded especially in transient phases of the discharge. However, these two values can be different in reality, since the scaling provides just approximate values. The difference between the result of scaling and reality can be easily by a factor of 2. In RFX-mod, the H-mod is not achieved in the flat-top phase of the discharge. Thus, we conclude that the density or $P_{sep}$ are below the threshold: both quantities need to be increased during the experiments focused on LH transition on RFX-mod. The way of increasing the density is out of the scope of this paper.

Let us discuss the possibilities of increasing $P_{sep}$. As there are no additional heating sources on RFX-mod, the only possibility to increase the $P_{sep}$ (besides operation on higher plasma current and reducing radiation) is taking advantage of some transient events such as current ramp-downs when the value of $P_{sep}$ can be temporarily increased. To model this situation, we start from the power balance equation [8]:

$$U \cdot I_p = P_{sep} + I_p\left(L\frac{dI_p}{dt} + \frac{dL}{dt}I_p\right) + \frac{dW}{dt} + P_{rad}, \qquad (6)$$

where $U$ is the applied loop voltage, $L$ is the plasma inductance, $W$ is the plasma energy and $P_{rad}$ is the power radiated by the plasma. All the quantities besides $P_{sep}$ in (6) can be either measured (this is the case of $U$, $I_p$, $P_{rad}$), or computed by the combination of the boundary reconstruction method mentioned in section 1 and the $\beta_p$ and $l_i$ computation described in the previous section. This is the case of $L$ and $W$. The plasma inductance is given by the sum of internal inductance $l_i$ determined in previous section and external inductance that can be computed by elliptic integrals using the knowledge of plasma boundary coordinates. The plasma kinetic energy $W$ can be computed using the value of $\beta_p$: the ratio of the plasma kinetic pressure and the poloidal magnetic field pressure on the plasma boundary. As $\beta_p$ is known and magnetic field at the plasma edge can be provided by the boundary reconstruction code, the computation of $W$ is straightforward.

The time evolution of $P_{sep}$ is in Fig. 2. One can see that it is significantly increased in the plasma current ramp-down phase. This property determines one possible strategy to reach H-mode on RFX-mod: try to obtain the LH transition during the ramp-down phase. As the power required to sustain the H-mode is lower than the power required for the L-H transition, the H-mode might be sustained even after the plasma current value is stabilized. Unfortunately, none of the attempts performed up to now was successful in sustaining H-mode after the current ramp-down, but several cases that could be LH transition were observed transiently during the ramp-down phase.

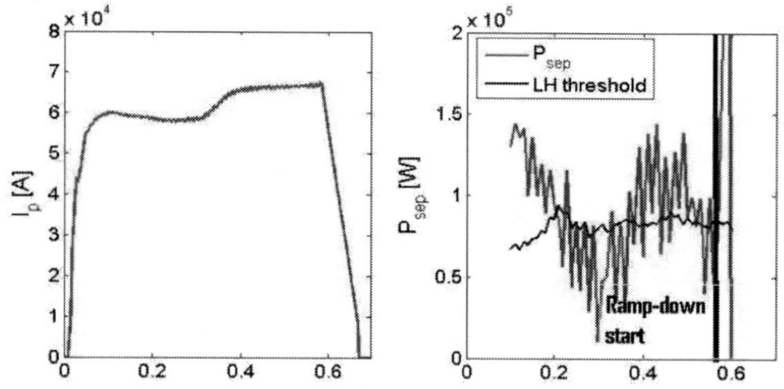


*Fig. 2: The estimate of the $P_{sep}$ power during the plasma current ramp-down phase of discharge number 36914.*

**Acknowledgement**

This project has received funding from the European Union's Horizon 2020 research and innovation programme under grant agreement number 633053 and from Erasmus Mundus International Doctoral College in Fusion Science and Engineering. The views and opinions expressed herein do not necessarily reflect those of the European Commission.

## References

[1] Martin, P.: Fusion Science and Technology, Vol. 59, p. 602-616

[2] Kudlacek O. et al: EPS 2014 Berlin, P5.062

[3] Barabaschi P., The MAXFEA code, Plasma Control Technical Meeting, Naka (Japan) 1993

[4] Marchiori G. et al: EPS 2014 Berlin, P5.040

[5] Ryter F. et al, Plasma Phys Contr Fus 44 (2002) A415.

[6] Lao L.L. et al: *1985* Nucl. Fusion **25** 1421

[7] Ryter F. et al: EPS 2014 Berlin, P2.016

[8] Havlicek J. et al: Global Power Balance in Non-Stationary Discharge Phases in the COMPASS Tokamak, WDS 2014

# Heat flux measurements on graphite samples in RFX-mod

N. Visonà[1,2], A. Canton[2], P. Innocente[2]

[1] *Università degli studi di Padova, Padova, Italy*

*2Consorzio RFX, Associazione EURATOM-ENEA sulla Fusione, Padova, Italy*

mail: nicolo.visona@igi.cnr.it

**Introduction.** The study of plasma-wall interactions (PWI) is interesting in the perspective of a fusion reactor because the behavior of the first wall (FW) can influence directly the plasma core and the achievable fusion power. In particular the FW has to be able to manage the intense, hot plasma exhaust.

Experiments have been carried out in RFX-mod (r/R=0.46/2 m) in order to investigate the plasma edge of the circular shaped reversed field pinch (RFP) configuration from a heat flux point of view analyzing the heat distribution on three graphite samples inserted as limiters up to 12 mm inside the plasma. About half hundred deuterium pulses were executed with plasma current: 0.6÷1 MA, plasma density: $0.5 \div 3*10^{19}$ $m^{-3}$ ($n/n_G$<0.7), reversal parameter F: 0÷-0.12 and peak temperature up to 1500 °C. The samples had two different shapes, one with a corner and one curved, and were also made of two different types of graphite, the Mersen 5890PT which is the same as the present FW, and the Toyo Tanso IG-15 with incremented heat conductivity. An infrared camera (200 Hz resolution) has been set up and used to evaluate the temporal and spatial heat flux distribution from temperature measurement.

**Temperature map on the samples.** The heat print on the samples was different depending on several parameters, first of all the shape of the interacting surface. While the angular sample had the maximum interaction on the corner (incident to the plasma), the curved one had two peaks away from the top (tangent to the plasma) as shown in fig. 1. This is a known result and is explained by the fact that the heat flux is carried to the plasma surface mainly by convection (parallel to the plasma surface) and there is only a small fraction that is diffused (perpendicular). Another thing that is immediately seen is that the electron drift side is always significantly hotter than the other especially at low values of density as $n/n_G$<0.2. This was already been seen in RFX and in other RFP experiments [1] and it is recognized as a signature of the superthermal electrons.

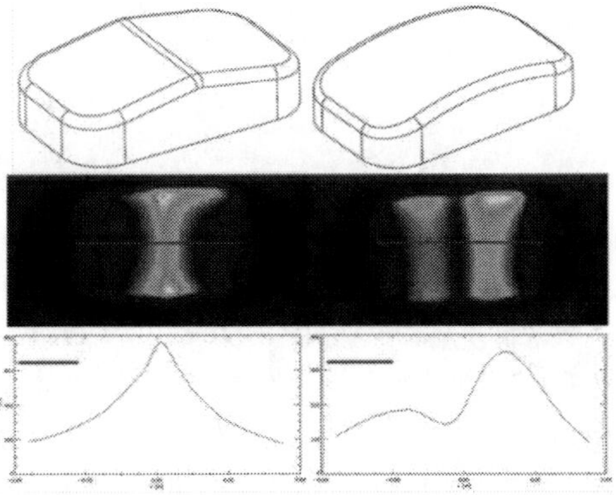

Fig. 1: samples geometries, IR images and temperature profiles along a line.

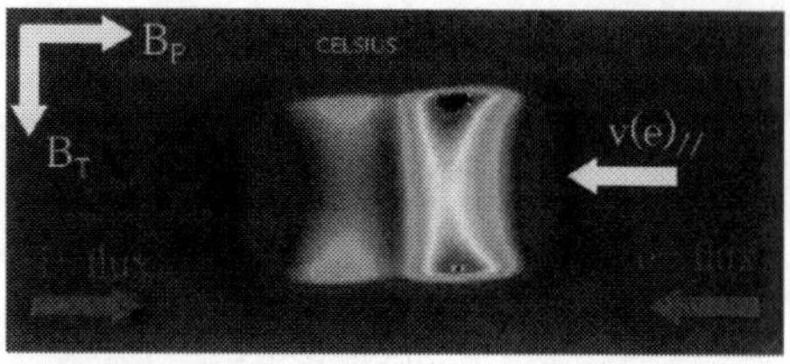


Fig. 2: sample orientation in the plasma

These electrons are accelerated in the plasma core and, since the more energetic are the less interacting, at low density a fraction of them can escape to the edge maintaining the properties of core electrons such as the energy and the parallel component of the velocity. In fact these electrons lay in the high energy tail of the energy distribution at the edge and can contribute to the sustainment of RFP configuration reducing the requirement to the turbulent MHD dynamo.

**Heat decay length.** IR images have been analyzed with the THEODOR code [2] in order to numerically solve the heat conduction inverse problem and to extract heat profiles from temperature maps evolving in time. Given the surface heat power profile ($q_s$) computed along a line, neglecting the perpendicular heat diffusion, the plasma parallel heat flux has been computed considering the local slope ($\alpha$) of the samples by the relation: $q_{//}(x)=q_s(x)/sin(\alpha)$. The approximation of a negligible perpendicular heat flux has been made as a starting point. The heat decay length and the heat peak have been evaluated assuming a heat exponential decay using the usual relation for the parallel heat flux in the SOL $q_{//}(r)=q_{//}(r_0)exp[(r-r_0)/\lambda_q]$ [3] where $\lambda_q$ is the heat decay length and $q_{//}(r)$ is the heat measured at a distance $r-r_0$ the top of the sample outwards; in particular the distance along the sample has been converted in radial distance.

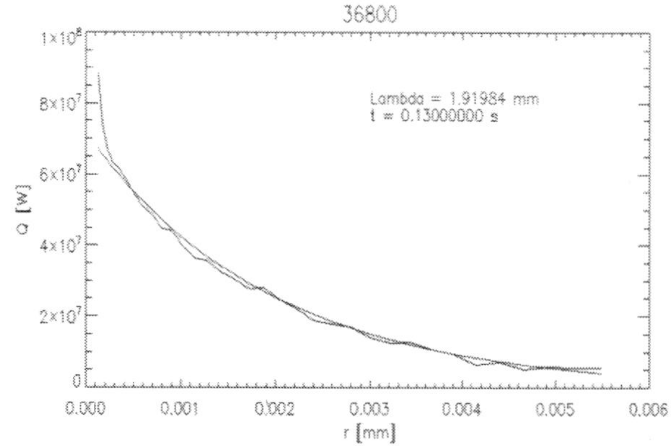


Fig. 3: radial dependence of measured parallel heat flux and computed exponential fit

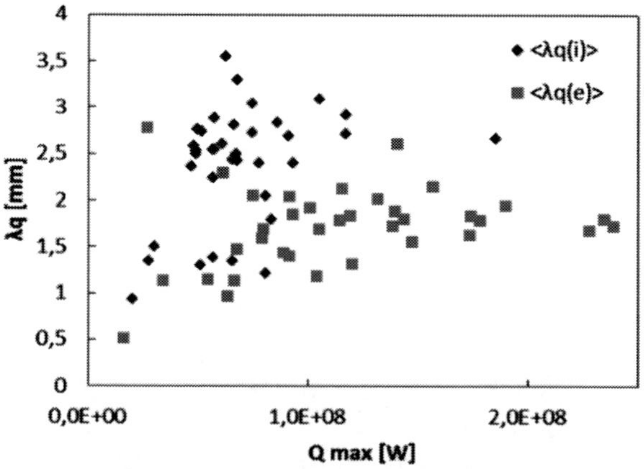


Fig. 4: computed $\lambda_q$ versus $q(r_0)$ for the ion and electron sides (averaged on single shots).

In fig. 3 it can be seen that the exponential decay describes very well the radial dependence of the parallel heat flux in the RFX-mod SOL. This confirms $\lambda_q$ as a good parameter to analyze plasma transport in the SOL, for this reason it has been used to study the dependence of SOL transport on plasma parameters. Fig. 4 shows the dependence of $\lambda_q$ from the heat peak, indicating that larger values of the heat peak correspond to larger heat decay lengths, and it is also dependent on the side; every point in this graph corresponds to a different pulse and is a global average of a side of the sample in time. For the curved sample, inserted at 12 mm closer to the plasma from the surface of the FW, average heat flux have been found around 50 MW, with peaks as high as several times the average while values for $\lambda_q$ are around 2 mm on average for the electron side and around 3 mm for the ion side. Fig. 5 shows that $\lambda_{q.}$ does not depend from $n/n_G$ and fig. 6 shows the correlation between the heat peak and $n/n_G$. While $q(r_0)$ is constant for the ion side, higher values of the heat peak are seen for the lowest values of $n/n_G$ in the electron case, this can be caused again by the superthermal electrons.

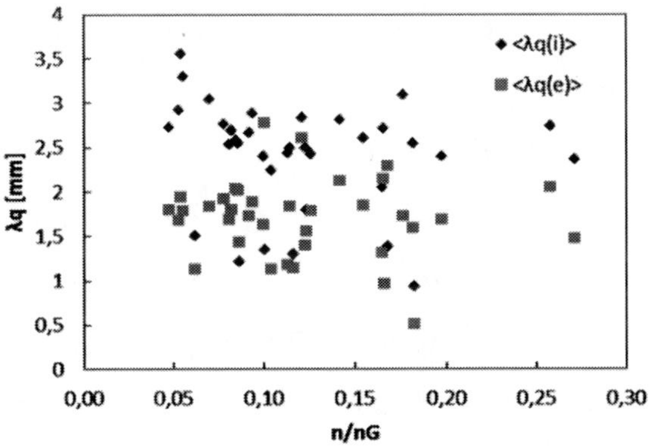


Fig. 5: $\lambda_q$ versus $n/n_G$ for the ion and electron side

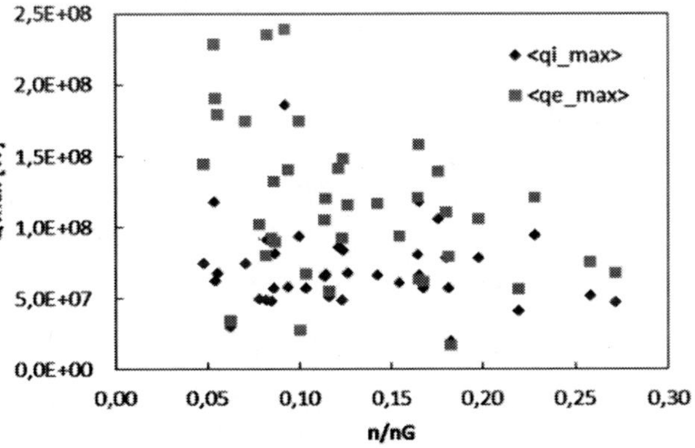


Fig. 6: $q(r_0)$ versus $n/n_G$ for the ion and electron side

One aspect that makes the study of SOL parameters in the RFP devices challenging is the lack of a well-defined interaction point and consequently of the last closed flux surface (LCFS). The problem is mainly related the toroidal local deformations of plasma surface but also to existence of true flux surfaces. Since the toroidal field at the edge is nearly zero and perpendicular transport is not negligible, when the plasma is strongly deformed in one toroidal section, it is not obvious that far from that toroidal section the heat power is 'confined' on the same flux surface. To get an indication on what defines the plasma boundary (and so the SOL region), the heat load and decay length have been compared to the local plasma shift and the local distance of the LCFS due to the deformation at all the toroidal locations

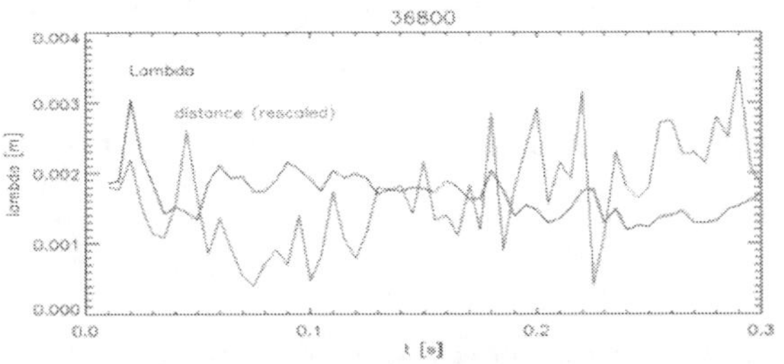


Fig. 7: $\lambda_q(e)$ compared with LCFS distance (rescaled) from sample in a discharge.

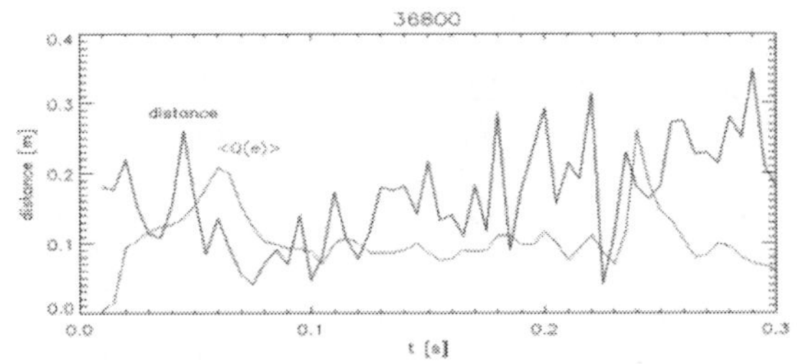


Fig. 8: distance compared with heat averaged on the $e^-$ side surface and rescaled to fit in the graph.

evaluated from magnetic signals [4]. In fig. 7 it is shown the time dependence of the plasma distance and $\lambda_q(e)$, while in fig. 8 the distance is plotted together with the average heat load on the electron drift side. The results of this first analysis do not show a correlation between these parameters, further work is necessary to include also the perpendicular component of the heat flux and better estimates of the position on the plasma with respect to the sample.

**Correlations with plasma parameters.** The dependency of temperature asymmetry on plasma density is shown in fig. 9, where the ratio of the average heat flux between the ion and electron side has been plotted against the ratio between the plasma and the Greenwald density. The data show an asymptotic dependence approaching 1 for values of $n/n_G>0.2$ where the heat flux tends to be equal in both sides, while for $n/n_G<0.2$ the vast majority of points shows a much higher heat flux on the electron side. The few points above 1 refer to low interacting situations.

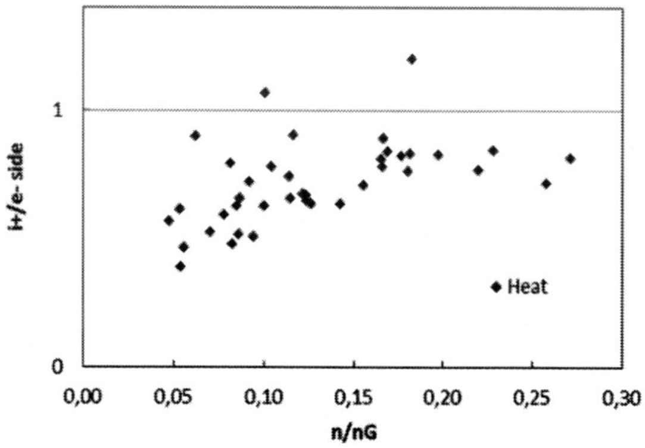

Fig. 9: heat ratio between sides versus normalized density

**Conclusions.** An experimental campaign has been done at RFX-mod in order to measure heat flux parameters of the plasma edge in the reversed field pinch configuration with a circular deuterium plasma during medium power discharges. Three graphite samples have been inserted in a limiter configuration and the measures were obtained with an IR camera.

The heat dynamics has been numerically extrapolated from time and space dependent temperature maps to calculate heat flux and it has been possible to evaluate average heat loads around 50 MW with peaks up to several times the average, and also to calculate the heat decay length, $\lambda_q$, with values around 2 mm for the electron drift side and 3 mm for the ion side. A strong asymmetry in heat fluxes between ion and electron drift side at low density ($n/n_G<0.2$) has been recognized as a distinctive sign of superthermal electrons, which are high energy electrons accelerated in the plasma core and escaped to edge due to low collisionality.

**References.**

[1] Y. Yagi et al. Plasma Phys. Control. Fusion **39** 1925-1927 (1997)

[2] A. Herrmann, 2001 Proc. 28th EPS conference, Funchal, ECA **25A** 2109-2112

[3] G. Arnoux et al. Nucl. Fusion **53** 073016 (2013)

[4] P. Zanca, D. Terranova, M. Valisa, S. Dal Bello, J. Nucl. Mater 363-365 (2007) 733-737

# Exploiting genetic algorithms in transport modelling in RFX-mod

F. Auriemma[1], F. Sattin[1], G. Urso[2], D. F. Escande[3]
[1]*Consorzio RFX (CNR, ENEA, INFN, Università di Padova, Acciaierie Venete Spa), Corso Stati Uniti, 4 Padova, Italy*
[2]*CNR Istituto per le applicazioni del calcolo 'M. Picone', Roma, Italy*
[3]*Aix-Marseille Université, CNRS, PIIM, UMR 7345, 13013 Marseille, France*

The goal of this work is to gain a deeper understanding of the quality of transport reconstruction from modelling of kinetic profiles in the Reversed Field Pinch RFX-mod. The current procedure for main gas transport modelling in stationary discharges in RFX-mod aims at computing the transport coefficients $(D,U)$ (diffusivity and pinch velocity, respectively) by reproducing the experimental data provided by various diagnostics as interferometer, multichannel reflectometer, calibrated edge and core Thomson Scattering. The numerical data are computed via synthetic diagnostics, starting from the radial profile of the generic kinetic quantity $f(r)$ obtained solving the model transport equation:

$$0 = \nabla \cdot (D \nabla f - U\, f) + S \qquad (1)$$

Being an instance of an inverse problem, any conclusion drawn is potentially prone to large uncertainties. In RFX-mod a solidly established procedure involves (optimized) lookup tables from which candidate solutions are sequentially picked up and evaluated against data using the standard $\chi^2$ minimization. This approach proved to be fairly effective on the basis of the experience [1]; however, there is room for alternatives, which may promise improvements regard to, at least, two aspects: (I) minimize the human intervention, with the spin-off of improving the degree of objectivity; (II) provide a more accurate recipe for estimating confidence intervals of estimated quantities.

Genetic Algorithms (GAs) are numerical search tools aiming at finding the global optimum of a given real objective function of one or more real variables, possibly subjected to various linear or non-linear constraints. In this work we apply a Differential Evolution version of the GA (details can be found, e.g., in [2]) to the reconstruction of particle density profiles. The desired output is represented by the particle diffusivity $D(r)$ and convection $U(r)$. In this work the velocity profile $U(r)$ will be constrained in agreement with the theory of transport in a stochastic magnetic field: $U(r) = -0.5 * D(r) * \nabla T_i / T_i$ being $T_i$ the ion temperature profile. Such theory, whose details can be found in [1], has proved to be able to describe the transport regime of RFX-mod discharges. The diffusivity is approximated by a piecewise function:

$$D(r)=\sum_i D_i\,,\quad r_{i-1}<r<r_i \qquad (2)$$

The targets are just the amplitudes $D_i$, but we varied also the number -and accordingly the widths- of the intervals $[r_{i-1}, r_i]$, in order to assess the sensitivity of the results to these parameters. The radial intervals $r_i$ are not equally spaced: they are denser in the edge region, where density gradient and particle source are located.

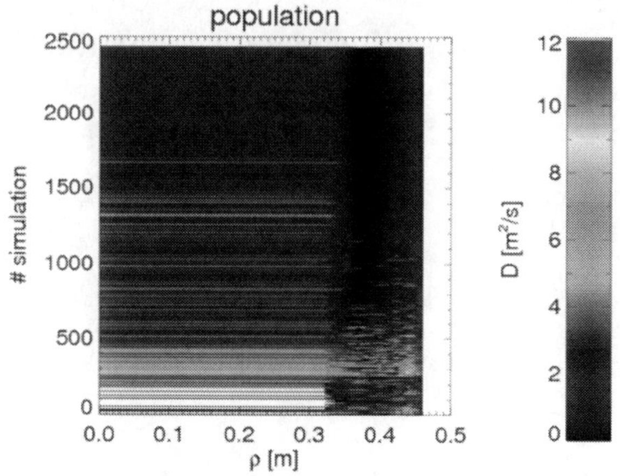


Figure 1: example of evolution of a population: each individual solution is represented by a horizontal line, the colours refer to the $D_i$ magnitude at a specific radius $r_i$. A new generation starts after every 100 individuals. The convergence towards a common target is suggested by the homogeneity of the population after about 2000 simulations (~200 generations).

For a given choice for $D$, we compute the predicted density profile from the transport equation, integrate the solution along the experimental lines of sight and compare it with the interferometer measurements. At each stage within an iterative loop, the GA produces a set ("generation") of potential solutions (2), using as input the previous generations (the starting generation being chosen randomly) and feeds them to the transport code, until an optimum $D$ profile, producing the best data reproduction, is identified. Figure 1 shows an example: each generation consists of 100 different candidate solutions. The first generations contain very different specimens, as highlighted by the wide variations in color between successive generations. Eventually, however, they converge towards a stable, optimal solution. The parameterization (2) is very convenient for a rough modelling of profiles, but has the unfortunate drawback of producing discontinuous profiles for transport coefficients and not penalizing unphysical erratic jumps of $D$ between neighboring intervals (see fig. 2). Thus, we find a possible solution to alleviate this shortcoming. We did not minimize $\chi^2$ itself, but rather the functional:

$$H=\chi^2+\upsilon\, G\,,\ldots G=\sum_i (D_{i+1}-D_i)^2$$

where $G$ is a penalty function set to avoid exceedingly oscillating solutions, with its weight $\upsilon$. Eq. (1) was actually implemented and solved using ASTRA code [3] interfaced with a routine producing $D$ guesses from GA.

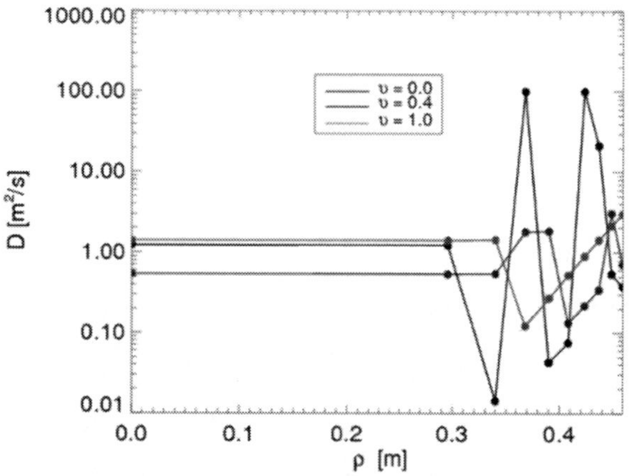


Figure 2: diffusivity profiles for three values of $\upsilon$.

At this stage we performed simulations with different υ, in order to assess its effect and importance. Figure 2 reports three examples of $D$ profiles, being smoother and smoother as υ increases. The final $\chi^2$ is comparable between the three cases. Rigorously, υ should be roughly fixed by the constraint that the optimal solution shall have $\chi^2 \approx 1$. The main outcomes of applying GA's to transport analyses at RFX-mod can be summarized as follows. *(i)* The unsupervised GA method and the supervised traditional look-up table one do yield results of the same quality and provide similar transport level. Figure 3 reports the comparison among the interferometer data of for shot 30056 at 130 ms (black asterisks), the numerical average line integral density for the simulated ASTRA profiles found with the GA (black full diamonds) and the profile obtained with the lookup tables (red full circles): the three sets of data are very close, within the experimental error bars. This is particularly interesting since the two methods do differ also in the spatial structure of the basis functions used to model transport coefficients: the lookup-table method writes ($D,U$) in terms of smooth functions defined globally throughout the whole radius; the GA method employs the piecewise functions (2). *(ii)* Figure 4 shows the $D$ profile for data shown in figure 3: the black line represents the profile with the lowest $\chi^2$ computed with the GA, the grey area shows the local confidence intervals (CI's), defined as the envelopes of $D$ solutions such that the reconstructed measurements lie within an error bar from the measured ones ($\chi^2 \leq 1$). The CI's are larger in those regions where the true level of transport is not effectively probed due to the lack of sources and gradients: i.e. the core inside about $\rho$ = 0.3 m, in the case of particle density. In the same figure, the red smooth line shows the lookup table diffusivity profile, in good agreement with the GA solution. *(iii)* The conclusion from figure. 4 is that three transport zones may be roughly identified: the core region up to $\rho = 0.3$ m, a transport

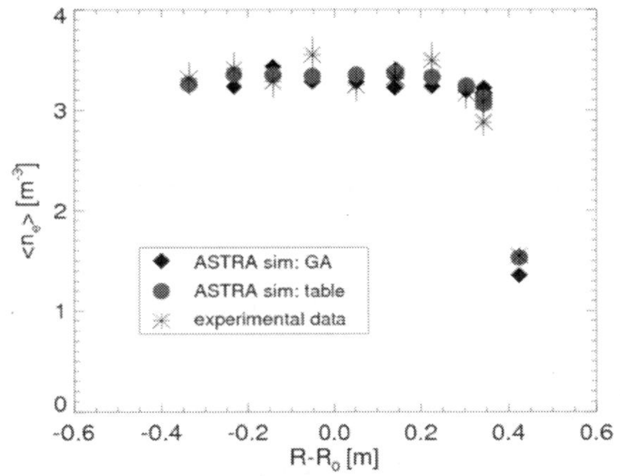


Figure 4: average line integral density: experimental data with corresponding error bars are plotted with asterisks. Black and red full symbols represent respectively the numerical profiles found with the GA and with the lookup-table method

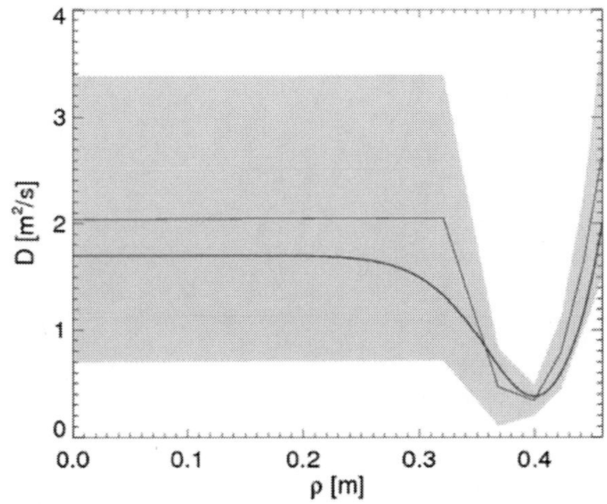


Figure 3: diffusivity profile for shot 30056 at 130 ms computed with υ=0.5, whose experimental data are shown in figure 3. The confidence intervals are shown as grey shaded area. The red smooth line shows the lookup table result.

barrier near to ρ = 0.4 m, a very edge region with high transport. By increasing the number $k$ of intervals $[r_{i-1}, r_i]$ in (2), we refine the spatial resolution of the model, and we may potentially improve the goodness-of-fit. This procedure does not automatically ensure improving the descriptive qualities of the model ,and we use the Akaike Information Criterion (AIC) [4] to identify the optimum number of knots in the solution. The AIC functional is defined as $AIC = 2k + N\chi^2$ ($N$ being the number of interferometer chords). Its minimum value provides the optimal solution as a balance between match of the data (quantified by smaller $\chi^2$) and economy in the description of the model (smaller $k$). Figure 5 shows the AIC functional and the corresponding $\chi^2$ for different k values: (a) the AIC is minimum for $k$ = 3,4, confirming that a three-zones description is an adequate picture, and (b) above $k \geq 3$ no improvement in the matching of the data is actually meaningful, since $\chi^2 \leq 1$, *i.e.*, data are interpolated to within the experimental errors.

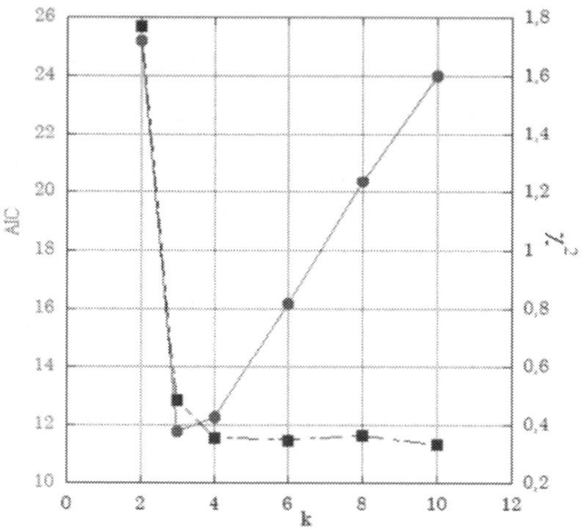

Figure 5: AIC (red symbols) and $\chi^2$ (blue symbols) versus the number of parameters k (Eq. 2).

**Acknowledgments.** This project has received funding from the European Union's Horizon 2020 research and innovation programme under grant agreement number 633053. The views and opinions expressed herein do not necessarily reflect those of the European Commission.

**References**

[1] F. Auriemma *et al*, Nucl. Fusion **55**, 043010 (2015)
[2] F. Sattin *et al*, Plasma Phys. Control. Fusion **56**, 114008 (2014)
[3] G. Pereverzev, P.N. Yushmanov, IPP-report 5/98 (2002)
[4] H. Akaike, IEEE Trans. Autom. Control. **19**, 716 (1974)

# Characterization of Particle Dynamics and Magnetic Reconnection in the RFX-mod Plasmas

M. Zuin[1], L. Stevanato[2], E. Martines[1], F. Auriemma[1], B. Momo[1], R. Cavazzana[1], G. De Masi[1], W. Gonzalez[1], R. Lorenzini[1], P. Scarin[1], S. Spagnolo[1], M. Spolaore[1], N. Vianello[6], W. Schneider[5], D. Cester[2], G. Nebbia[3], L. Sajo-Bohus[4], G. Viesti[2]

*[1]Consorzio RFX, C.so Stati Uniti 4, Padova, Italy*
*[2]Dipartimento di Fisica ed Astronomia dell'Università di Padova, Italy*
*[3]INFN Sezione di Padova, Via Marzolo 8, Padova, Italy*
*[4]Universidad Simón Bolívar, Caracas 1080A, Venezuela*
*[5]Max-Planck –Institut für Plasmaphysik, EURATOM Association, Teilinstitut Greifswald, Wendelsteinstraße 1, D-17491 Greifswald, Germany*
*[6]Centre de Recherches en Physique des Plasmas, Ecole Polytechnique Federale de Lausanne,CH-1015 Lausanne, Switzerland*

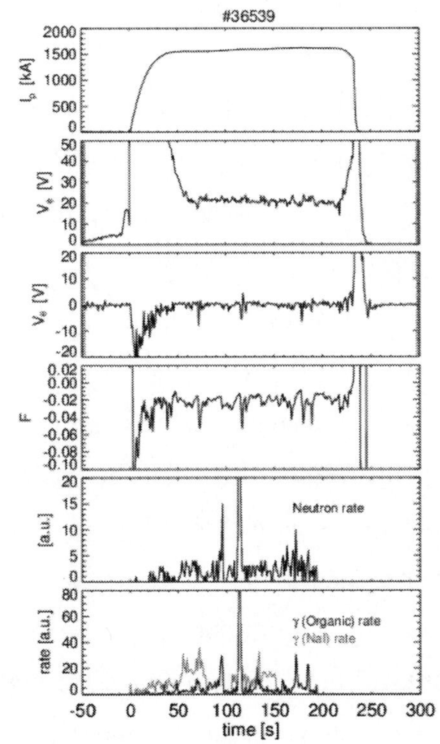


***Fig. 1**: From the top: plasma current as a function of time for a RFP discharge, toroidal and poloidal loop voltages, reversal parameter F, neutron flux and gamma rays from the two scintillators.*

Particle dynamics in magnetically confined plasmas is strongly determined by MHD activity. In this work, an experimental analysis of particle dynamics in D and H plasmas is presented in the RFX-mod device operated in both tokamak and reversed-field pinch (RFP) configurations. Particular attention is devoted to the role of MHD dynamics, in the form of reconnection processes (relaxation processes in the RFP and sawtoothing activity in the tokamak) in determining the energy spectra and the time behaviour of the emission of non-charged particles (neutral atoms, neutrons and gamma's) exiting the plasma.

The diagnostic system makes use of a 51 mm diameter, 51 mm thick EJ-301 liquid scintillator cell and of a calibrated crystal NaI(Tl) larger scintillator coupled to H8500 flat-panel photomultipliers, suitable for operation in a noisy magnetic environment [1]. The liquid scintillator is sensitive to neutron and gamma radiation, signals being classified on the basis of a pulse shape discrimination (PSD) process.

By means of a Neutral Particle Analyzer (NPA) diagnostics it is also possible to follow the time evolution of the distribution function of the neutral atoms produced by charge-exchange processes in the plasma. The NPA has 11 energy channels, being the energy and mass dispersion produced by a E//B (E parallel to B) combination. All diagnostic systems are located on the equatorial plane on the low-field side of the torus with a radial line of sight.

Bursty generation of DD fusion neutrons and gamma rays is observed in RFP plasmas, with some time correlation with the rapid variation of the magnetic field at the edge of the plasma, as shown in Fig. 1, where the time traces of various discharge parameters are given (plasma current, Ip, toroidal and poloidal loop voltages $V_\phi$ and $V_\theta$, respectively, and the reversal parameter $F=B_\phi(a)/\langle B_\phi\rangle$, being $B_\phi$ the toroidal magnetic field).

Indeed, the rapid variation of the F parameter is associated with large impulsive magnetic fluctuations and rearrangement of the magnetic topology, due to magnetic reconnection processes. Bursty, almost cyclic, magnetic reconnection occurs in toroidally localized region of the RFP plasmas, associated with the generation of a poloidal current sheet, as shown in the past [2]. The current sheet is responsible for particle acceleration processes, which are likely to be correlated with the observed γ-ray bursts.

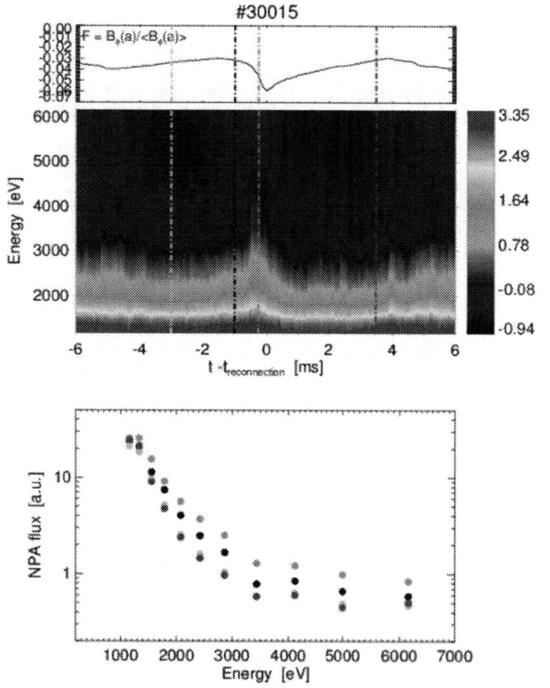


*Fig.2: Neutral atom fluxes collected from the NPA diagnostics as a function of time and energy for a single (conditionally averaged) reconnection event in a RFP plasma. The bottom panel shows the distribution function measured at the indicated time instants (vertical lines in the central panel.*

The "large" reconnections occurring in the plasma are indeed also found to have an impact on the behavior of the distribution functions of neutral atoms collected by the NPA diagnostics, as shown in Fig. 2, where the experimental fluxes are shown as a function of energy and time. Strong variations of both the thermal and the high energy tail (above around 2500eV) are induced by the reconnection process (the collected fluxes have been conditionally averaged for better statistical meaning). In particular, a heating process is recognized on the variation of the exponential slope of the low (Maxwellian) component of the spectrum at the time of maximum magnetic perturbation, which corresponds to an increase of the ion temperature $T_i$ of about 50%.

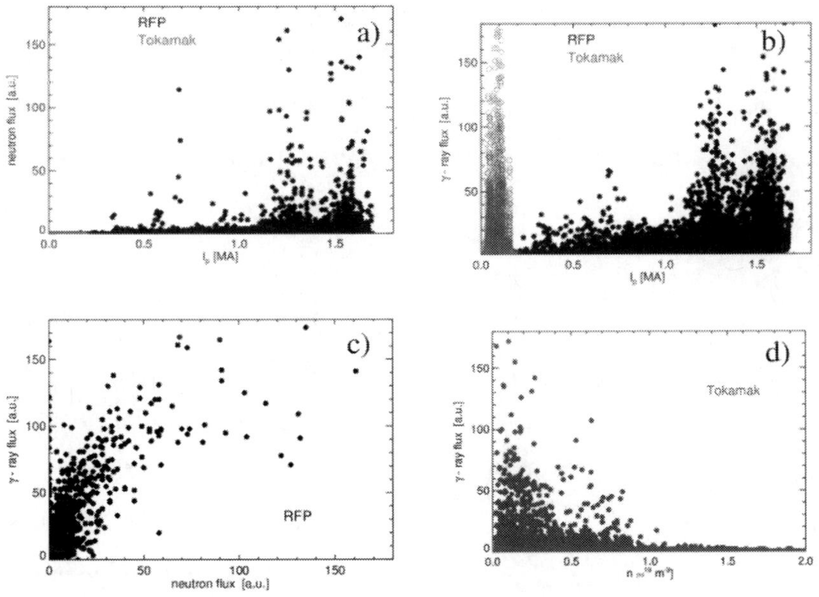


***Fig. 3**: from the top left a) and top-right b) neutron and gamma fluxes as a function of the plasma current $I_p$ in tokamak and RFP discharges, b)top-right gamma fluxes; c) total gamma flux as a function of neutron flux for RFP discharges, d) total gamma flux as a function of the electron density in tokamak plasmas.*

Such non-collisional heating has, indeed, been observed in RFPs [3] and it can be interpreted as responsible for the increased neutron generation from DD reactions as seen in Fig. 1. Neutron production in RFP plasmas is found to depend on the plasma current level, as shown in Fig. 3a, with most of the neutron being collected above 1 MA. In the RFP this is also true for gamma rays (see Fig. 3b). Indeed, gamma and neutron fluxes exhibit good correlation, as is clear in Fig. 3c. Such gamma production would either suggest that along with ion heating also a form of electron heating could occur induced by the reconnection processes or that neutron capture is occurring from the material wall of the RFX-mod device. In Fig. 4, the gamma's spectra collected in various experimental conditions are shown. RFP spectra exhibit a pronounced peak at ~ 500keV, whose origin is still under investigation.

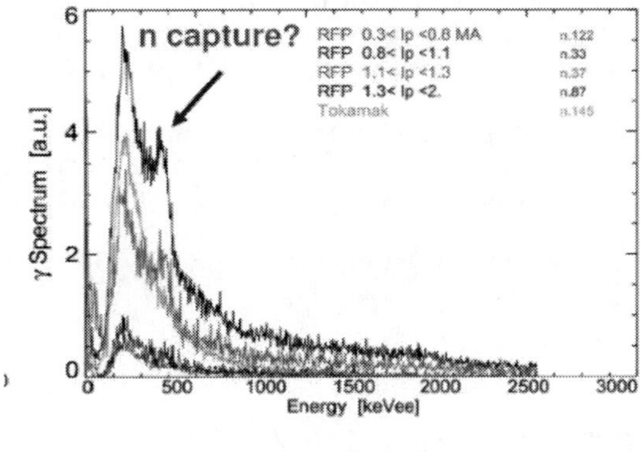


***Fig. 4**: Energy spectra of the gamma rays collected by the organic scintillator in different experimental conditions (tokamak and RFP at various plasma current levels).*

On the other hand, tokamak gamma's spectra appear to be continuous, being most of the gamma's produced in low density discharges (see Fig. 3d), thus likely to be induced by electron runaway population [4]. It is worth saying that almost no neutron have been collected in the low-current (low-toroidal field) tokamak ohmic discharges (Ip< 150kA) produced in the RFX-mod discharges, mainly because of the low ion temperature.

Magnetic reconnection, is found to have an impact on particle dynamics also in tokamak plasmas. In particular, it is found that the sawtoothing activity, mainly observed in low q(a) (<3) ohmic discharges, is related to dramatic variations of both the thermal Maxwellian (usually identified in the energy range between 300 and 800eV) and suprathermal (ion) population, as shown in Fig. 5.

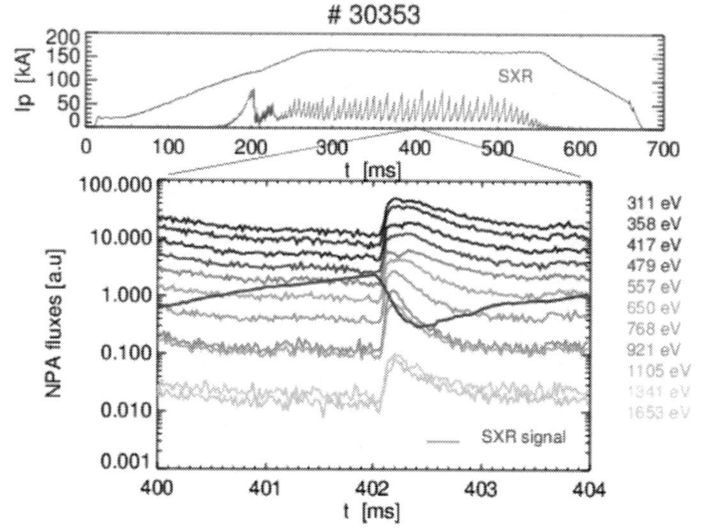


***Fig. 5:*** *Time traces of Ip and SXR signal in a tokamak discharge (top); detail of the time behaviour of the NPA collected fluxes at the various collected energies (from 300 to 1600eV, going from the top trace to the bottom one) along with the detail of the SXR behaviour during sawtoothing activity.*

This observation is in good agreement with what previously observed in spherical tokamaks [5] during internal reconnection events, where the generation of fast tails in the NPA spectra was explained a manifestation of runaway acceleration in the parallel electric field associated with the reconnection process.

In summary, a study of plasma particle dynamics, in particular the phenomenon of impulsive ion heating and the enhancement of high energy tails in the distribution function, is presented. The study is based in the analysis of the signals from passive diagnostics dedicated to neutral particles (atoms, neutrons and gamma rays) exiting the plasma of the RFX-mod device operated both as a tokamak and a RFP.

**References**

[1] *D. Cester et al., Nuclear Instruments and Methods **A 719** (2013) 81–84*
[2] *M. Zuin et al., Plasma Phys. Control. Fusion **51** (2009) 035012*
[3] *R. M. Magee et al., Phys. Rev. Lett. **107**, 065005 (2011)*
[4] *M. Gobbin et al, 4.136, EPS2015*
[5] *P. Helander et al, Phys. Rev. Lett. **23**, (2002) 235002*

# Analysis of Vertical Stability and Resistive Wall Modes in RFX-mod Tokamak Discharges Including 3D Effects

D. Abate[1,2], L. Barbato[1], C. Finotti[2], G. Marchiori[2], N. Marconato[2], S. Mastrostefano[1], F. Villone[1]

[1]*CREATE, DIEI, Università di Cassino e del Lazio Meridionale, 03043, Cassino, Italy*
[2] *Consorzio RFX, Corso Stati Uniti 4 - 35127 Padova, Italy*

## 1. Introduction

The RFX-mod experiment (R/a = 2.0/0.46 m), originally designed to produce high current Reversed Field Pinch plasmas (plasma current up to 2 MA), is currently operated also as a low current tokamak (Bt ~ 0.55 T, Ip ~150 kA @ $q_a \approx 2$). Circular, double-null [1] and single null configurations have been successfully already achieved. Such tokamak configurations may be prone to several MHD instabilities, usually classified resorting to the toroidal mode number $n$: current-driven $n$=1 Resistive Wall Modes [2], due to relatively low safety factor; $n$=0 RWM (vertical instability) [1], due to non-vanishing elongation of the plasma configuration. For the active control of these instabilities, RFX-mod is equipped with a state-of-the-art MHD control system made by 192 (4 poloidally x 48 toroidally) independently fed active coils, with more than 600 magnetic sensors acquired in real time; the feedback system is operated under the MARTe framework.

In particular, we investigate the 3D effects of the conducting structures on different response models of n=0 RWM (vertical instability). Several computational tools will be applied: linearized axisymmetric models (CREATE_L [3]), linearized 3D models (CarMa0 [4]), nonlinear evolutionary equilibrium models including 3D volumetric structures (CarMa0NL [5]). The different assumptions and approximations of the various models allow a clear identification of the key phenomena ruling the evolution of the n=0 vertical instability in RFX-mod tokamak discharges and hence provide fundamental information in the planning and the execution of related experiments and in refining the control system design.

## 2. Mathematical and numerical models

The main assumption is to neglect the plasma mass on the time-scale of interest, which is supposed much longer than the Alfvén time-scale related to plasma inertia. The plasma is assumed to move instantaneously through a sequence of MHD equilibria. This assumption is used for all the computational tools adopted in this study. The linearized axisymmetric plasma response model (CREATE_L [3]) reproduces the features of the plasma behavior relevant for the control of plasma current, position and shape. The plasma is characterized by a small number of global parameters (total plasma current $I_p$, poloidal beta $\beta_p$, internal inductance $l_i$): the model is linearized around a reference configuration in which the state variables are the coil, passive and plasma currents and the profile quantities are seen as disturbances. Assuming an axisymmetric system, Grad-Shafranov equation is solved. The time evolution of the magnetic field is determined by solving the free-boundary equilibrium problem coupled to the circuit equations for the external conductors and similar equation for the total plasma current regarded as an integral form of the Ohm's law. The numerical solution is obtained by using a 2D finite element method in which the unknown is approximated by means of piecewise polynomial functions. The overall plasma response model can be recast in a circuit equation in terms of modified inductance and resistance matrices or equivalently in a state space form. The CarMa0 [4] computational tool self-consistently couples the linearized plasma response model, computed as in CREATE_L, with a 3D time-domain eddy currents integral formulation, which requires only the discretization of the conducting structures. A surface $S$ is chosen in between the plasma region and the conducting structures, through which the interaction can be decoupled as follows. The instantaneous plasma response to a given set of magnetic flux density perturbation on $S$ is computed as a plasma response matrix. The effects of 3D structures on plasma is evaluated by computing the magnetic flux density on $S$ due to 3D eddy currents. The currents induced in the 3D structures by plasma are computed by using an equivalent surface current density on $S$ which produces the same magnetic field as plasma outside the coupling surface. The overall plasma response model can again be recasted in a state space-form. The idea of CarMa0NL [5] is to describe the plasma by solving the non-linear axisymmetric perturbed equilibrium problem, through Newton-Raphson iterations. In this way it is possible to treat self-consistently the non-linear evolution of an axisymmetric plasma surrounded by 3D volumetric conducting structures, providing the means to study situations such as disruptions, ELMs, limiter-diverted transitions, current quenches, etc.

**3. Results**

The equilibrium data of an open loop Single Null (SN) shot ($I_p$=59 kA, $B_t$=0.55 T) have been derived and used to produce the linearized plasma response model by means of CREATE_L code. Table 1 compares the plasma equilibrium parameters obtained with CREATE_L and with the MAXFEA 2D equilibrium code, while Fig. 1 shows the plasma boundary and a comparison of equilibrium values of magnetic fields at sensors.

The dynamical model is characterized by 194 states corresponding to the currents of the active circuits (8 FSC, 4 magnetizing winding (M) sectors, 2 MHD saddle coil null currents), the passive structures (60 Inconel vessel, 59 copper shell, 59 toroidal support structure) and the plasma. The presence of poloidal and inner equatorial cuts in the shell has also been implicitly taken into account by imposing that their total current be null. A vertical instability n=0 is exhibited by the model with a slow growth rate (<10 $s^{-1}$), consistent with the experimental evidence [1]. The model outputs include both the direct estimate of the gaps and the magnetic measurements (poloidal fluxes and poloidal field components). A similar model has been used successfully on double-null discharges [1].

The RFX-mod 3 mm thick shell, clamped over the vacuum vessel, provides the main stabilizing contribution (Table 2) and is characterized by poloidal and inner equatorial gaps, which have been reproduced in the 3D realistic mesh; the toroidal support structure has been neglected. In addition, a fictitious 3D axisymmetric mesh has been generated and used in computations in order to provide a reference and a cross-check with axisymmetric models for the entire procedure. In this case, the modified inductance matrix L* computed over the 3D axisymmetric mesh by CarMa0 has been compared with the CREATE_L results, providing a relative error around 1% and an excellent agreement on growth rates (Table 2).

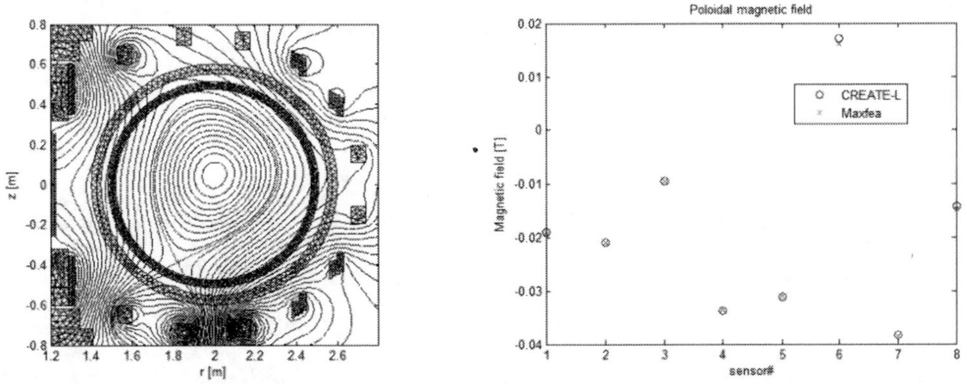


Fig. 1. Plasma boundary and comparison of equilibrium fields.

| | $l_i$ | $\beta_p$ | $R_{X\text{-}point}$ [m] | $Z_{X\text{-}point}$ [m] |
|---|---|---|---|---|
| *CREATE_L* | *1.04* | *0.102* | *1.822* | *-0.387* |
| *MAXFEA* | *1.06* | *0.111* | *1.827* | *-0.384* |

Table 1. Plasma equilibrium values computed by CREATE_L and MAXFEA

| | CREATE_L | CarMa0 (3D axisymm.) | CarMa0 (3D realistic) |
|---|---|---|---|
| $\gamma_{tot}$ [$s^{-1}$] | 7.36 | 7.33 | 7.55 |
| $\gamma_{vessel\ only}$[$s^{-1}$] | 335.4 | 334.2 | N.A. |

Table 2. Comparison of growth rates

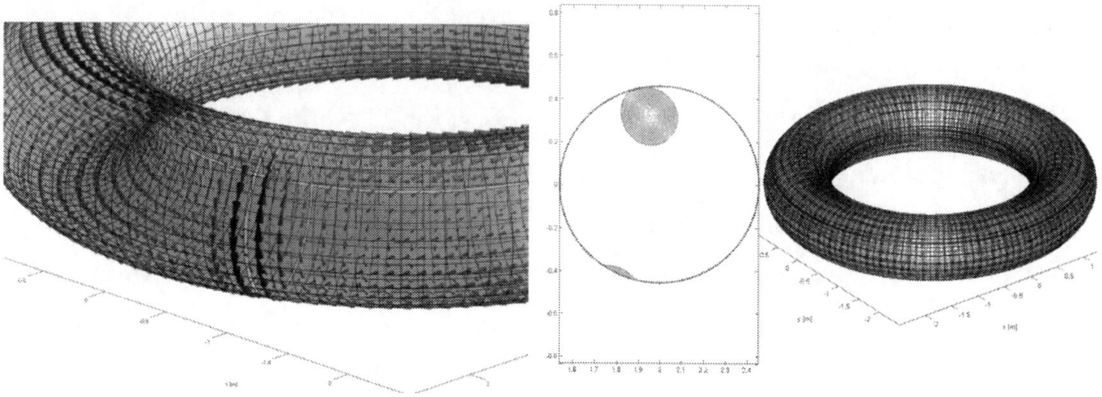

(a) (b)

Fig. 2. Current density patterns and plasma configurations: (a) unstable mode; (b) plasma current quench

From Table 2, we notice that the introduction of the gaps in the shell has a small destabilizing effect. The pattern of the induced current density distribution in the shell is shown in Fig. 2a, where gaps location and their effects on the current distribution are visible. Finally, a fictitious linear plasma current quench event has been considered with CarMa0NL; a typical current density pattern and the corresponding plasma configuration are reported in Fig. 2b.

## 4. Conclusions and perspectives

In this paper we have applied several different computational tools to model the evolution of the n=0 RWM (vertical instability) for SN configurations of RFX-mod. The comparison of different codes and models allows a thorough assessment and cross-check of the results. The next step will be the inclusion of further details in the conducting structures (e.g. the support structure), the comparison with experimental results and the utilization of these models for experimental planning and analysis.

This work was supported in part by Italian MIUR under PRIN grant 2010SPS9B3.

[1] G. Marchiori et al., *41st EPS Conference on Plasma Physics*, Berlin, June 2014, paper P5.040

[2] G. Marchiori et al., *38th EPS Conference on Plasma Physics*, Strasbourg, June 2011, paper P2.110

[3] R. Albanese and F. Villone, *Nucl. Fusion* **38** (1998) 723

[4] F. Villone et al., *Plasma Phys. Control. Fusion* **54** (2012) 085003

[5] F. Villone et al., *Plasma Phys. Control. Fusion* **55** (2013) 09500

# Transport analysis of impurities injected by laser blow-off in ECRH and NBI heated plasmas of TJ-II

B. Zurro[1], J. L. Velasco[1], E.M. Hollmann[2], A. Baciero[1], M.A. Ochando[1], R. Dux[3], K.J. McCarthy[1], F. Medina[1], I. Pastor[1], R.J. Hajjar[2], J.M. García-Regaña[3], D. López-Bruna[1], A. V. Melnikov [4,5], L.G. Eliseev[4], HIBP Team[1,4,6] and TJ-II team[1]

[1]*Laboratorio Nacional de Fusión, CIEMAT, Madrid, Spain*

[2]*University of California-San Diego, La Jolla, CA, USA*

[3]*Max Plank Institute for Plasma Physics, Garching, Germany*

[4]*National Research Centre `Kurchatov Institute', 12382, Moscow, Russia*

[5]*National Research Nuclear University MEPhI, 115409, Moscow, Russia*

[6]*Institute of Plasma Physics, NSC KhIPT, 61108, Kharkov, Ukraine*

**INTRODUCTION.** In previous works in TJ-II, laser blow-off injection (LBI) of BC, LiF, BN and W was used to study impurity confinement in electron cyclotron resonance (ECR) heated plasmas [1, 2]. Herein, the transport analysis of BN impurities injected by LBI into low-density ECR heated plasmas and into higher-density plasmas created during the neutral beam injection (NBI) heating phase of the TJ-II stellarator is studied. These regimes represent two general stellarator regimes: electron-root and ion-root plasmas, respectively. In order to compare these situations, a transport analysis of representative discharges is performed using the impurity transport code STRAHL, from which experimental impurity fluxes are obtained by matching the code results with the temporal behavior of reconstructed local global radiation from bolometer arrays. When compared with neoclassical calculations, poor agreement is found for the low-density plasma while better agreement is found for the latter case. In the higher-density plasma, which is closer to the operational regimes relevant for the stellarator reactor program, impurity accumulation is predicted and observed.

**EXPERIMENTAL.** The experiment herein reported was performed in the TJ-II a four-period, low magnetic shear stellarator with major and average minor radii of 1.5 m and ≤ 0.22 m, respectively [3]. During the NBI heating phase, higher electron densities are achieved, and in contrast to the ECR heated phase, the density profiles (line-average, $n_e \leq 6\times10^{19}$ $m^{-3}$) are rather peaked whilst the electron temperature profiles are flat with core values in the range 0.3 to 0.35 keV. In addition, the majority ion temperature radial profile is flat with a core value of around 80-100 eV in ECRH and 120 eV in NBI discharges.

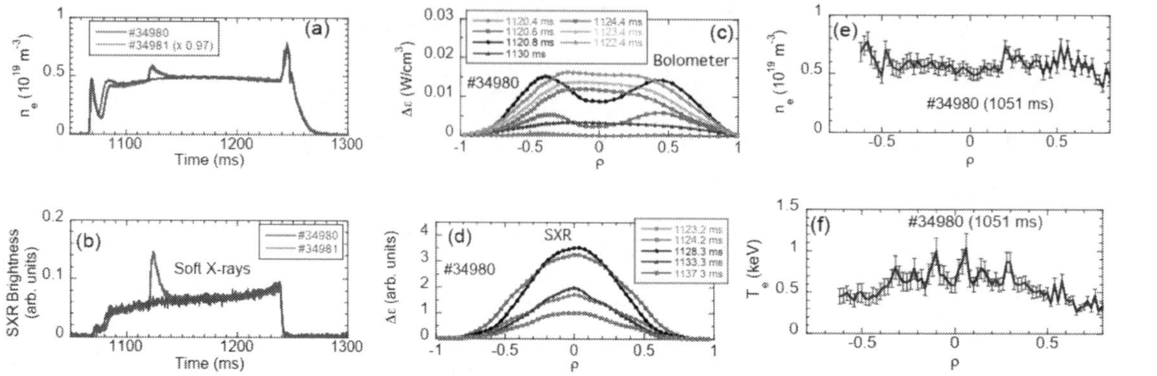


**Fig. 1.** Traces and profiles of interest for the ECRH shots 34980 with BN injection and reference 34981: (a) line average electron density; (b) soft X-ray (fast photodiode with an 8 μm Be filter), close to the injection port; (c) and (d) incremental emissivity profile evolution during the injection as seen by the bolometer and soft X-ray arrays; (e) and (f) electron density and temperature profiles.

In figure 1 the time evolution of selected signals of interest for two ECRH shots, #34980 with BN injection and the non-perturbed #34981, is shown. In boxes a) and b) the brief duration of the perturbation in line-averaged electron density and in the signal from a fast photodiode with an 8 μm thick Be filter (soft X-ray monitor) positioned close to the injection port, can be observed. In figures 1(c) and 1(d) the time evolution of the incremental emissivity due to the impurity injection, from bolometer and soft X-ray arrays, are shown [4, 5]. Finally, the $n_e$ and $T_e$ profiles are depicted in figures 1(e) and 1(f), respectively.

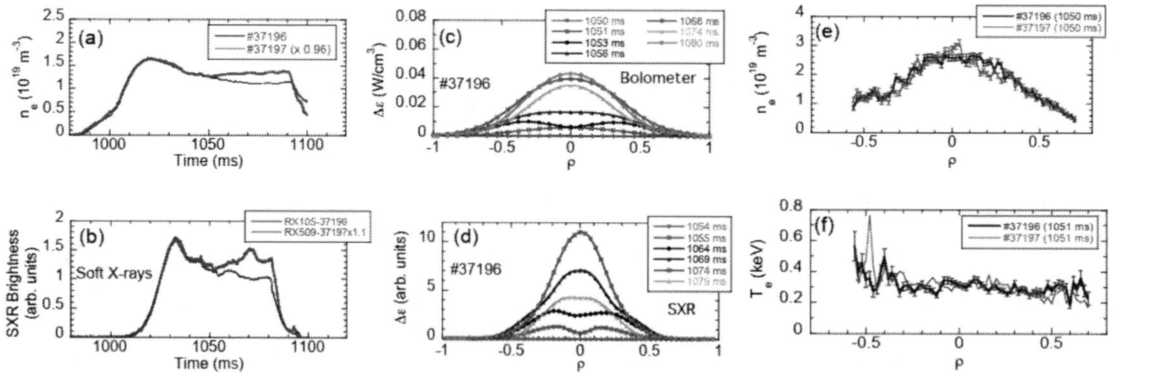


**Fig. 2.** Similar traces and profiles to those of figure 1, but for the NBI discharges #37196 and #37197 (reference).

Figures 2(a) to 2(f) show information equivalent to figure 1 for two NBI heated discharges: #37196 with BN injection and #37197 without perturbation. As in the previous case, impurity injection strongly affects the monitor signals. However, unlike the ECR heated case, the time evolution of radiation signals shows that the perturbation produced by the impurity injection persists for the remainder discharge duration. This would indicate that the plasma does not expel the injected ions, with the exception of the abrupt expulsion that takes place at about 1074 ms, simultaneously with a strong MHD event. The effect of BN injection on the line-averaged electron density, depicted in figure 2(a), shows that electrons delivered by the BN tend to remain in the plasma and that the density increases with time, partly because long impurity confinement times allow higher ionization stages to be reached.

Since the radial electric field is one of the key factors in impurity transport, in figure 3 we show the radial profiles of plasma potential, as determined with a 200 keV $Cs^+$ beam, before BN injection for the two studied discharges [6].

**IMPURITY TRANSPORT RESULTS AND DISCUSSION.** In order to estimate the experimental impurity fluxes we have proceeded to match, using the impurity transport code STRAHL [7], the temporal behavior of reconstructed global radiation signals measured by a bolometer array, with a similar, but improved, method previously reported [8]. The impurity radial flux is parameterized by two radially-dependent coefficients: a diffusive coefficient D and a pinch V. In order to facilitate the simulation, results of neoclassical calculations [9] are used as initial guesses for D and V. In a first step the iteration method consists in scaling by a factor the theoretical D and V profiles given by the neoclassical theory. The simulation involves an iterative process with the possibility of modifying the transport coefficients as well as several experimentally unknown parameters in the code, *e.g.* impurity radial deposition, transport coefficients profiles, etc., in order to achieve a good match between experimental data and simulation results.

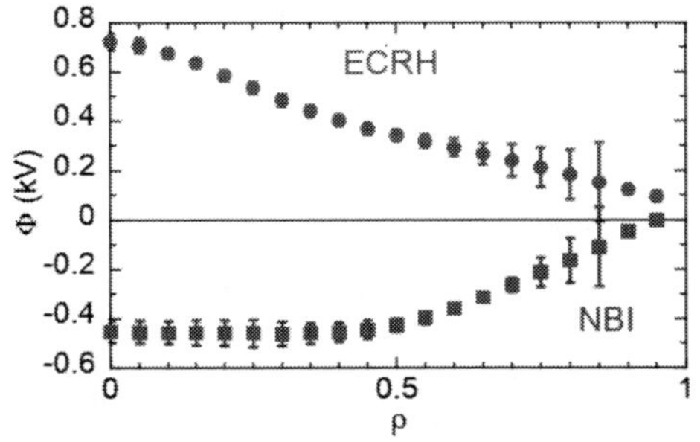


**Fig. 3.** Plasma potential profiles for the two selected discharges chosen for the transport analysis.

The transport analysis results of the ECRH discharge #34980 and NBI shot #37196, where BN was injected by LBI, are depicted in figure 4. In figures 4(a), 4(b) and 4(c), it is presented the comparison between the experimental D&V coefficients and the radial impurity flux determined by the STRAHL code and the results of neoclassical calculation for the ECR heated case. Similar results for the NBI heated shot are shown in plots 4(d), 4(e) and 4(f) of the same figure 4. The discrepancy in two orders of magnitude in the diffusion coefficient for the low density ECR heated discharge must be highlighted. However, a good agreement is found for the NBI discharge at middle radius but differences about two orders of magnitude are seen at the core and about a factor 10 at the periphery. The agreement is better in the case of pinch coefficients for both discharges.

The splitting between D and V coefficients in the experimental flux may be non-unique; in particular, for short discharges in which the impurity density profile does not reach a stationary status, the calculation of D may become inaccurate. With this in mind, we also compare the total radial fluxes: in the low-density case, we observe qualitative agreement, although the absolute value is underestimated in the calculations, especially at outer positions (although it could be that the experimental value is overestimated, e.g. due to the effect of non-maxwellianity of the electron distribution function in the ionization/recombination rate coefficients). In the NBI case, prediction and measurement are probably in agreement within the available precision up to $r/a$=0.6. At outer positions, a radially-localized enhanced diffusion, yet to be understood, is required for explaining the experimental radiation signals.

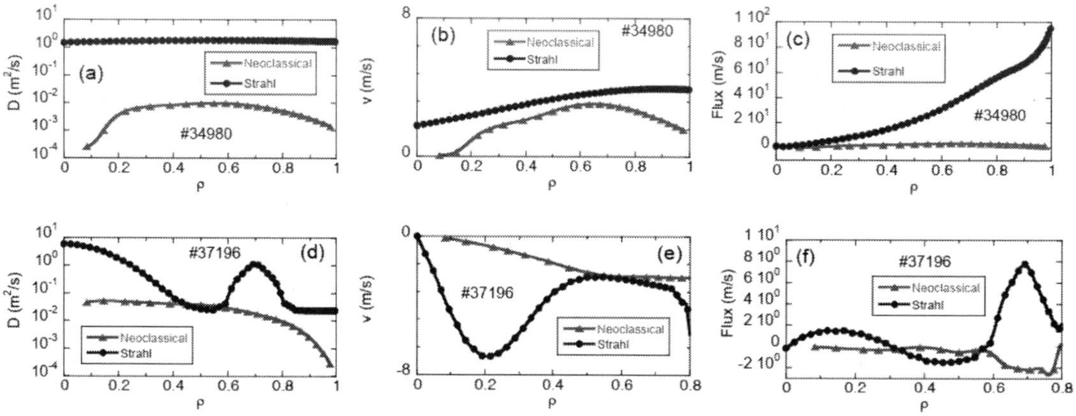


**Fig. 4.** Comparison of transport results deduced from experimental data, using the STRAHL code and calculated with neoclassical theory: (a) D's for ECRH discharge #34980; (b) V's for the same discharge; (c) Fluxes for #34980; (d) D's for NBI discharge #37196; (e) V's for NBI discharge #37196 and (f) Fluxes for #37196.

**Acknowledgements**. This work was partially funded by the Spanish "Ministerio de Economía y Competitividad" under Grants No. ENE2014-56517-R and ENE2012-30832, and by the US Department of Energy under DE-FG02-07ER54917. This work has been carried out within the framework of the EUROfusion Consortium and has received funding from the Euratom research and training programme 2014-2018 under grant agreement No 633053. Plasma potential measurement has been carried out by RSF project 14-22-00193. The views and opinions expressed herein do not necessarily reflect those of the European Commission.

**References**

[1] Zurro B, Hollmann E, Baciero A *et al* 2011 *Nucl. Fusion* **51** 062015

[2] Zurro B. *et al* 2014 *Plasma Phys. Control. Fusion* **56** 124007

[3] Sánchez J., Tabarés F., Tafalla D. *et al* 2009 *J. Nucl. Materials* **390** 852

[4] Ochando MA 2006 *Fusion Sci. & Tech.* **50** 31

[5] Medina F. *et al* 1999 *Rev. Sci. Instrum.* **70** (1) 642-644

[6] Melnikov A.V. *et al.*, 2007 *Fusion Sci. Technol.* **51** (1) 31-37

[7] Dux R, 2014 STRAHL User Manual

[8] Zurro B, Baciero A *et al* 2004 *Rev. Sci. Instrum.* **75** 104231

[9] Velasco JL *et al* 2011 *Plasma Phys. Control. Fusion* **53** 11 115014

# Long-lasting core coherent Modes in TJ-II stellarator plasmas

M. A. Ochando, B. Sun and D. López-Bruna

*Laboratorio Nacional de Fusión, CIEMAT, 28040 Madrid, Spain*

## Introduction

Long lasting internal modes mainly associated with q=1 have been observed in many toroidal devices [see, for instance, 1-5]. In the TJ-II, a low-shear, low-current, low-beta stellarator with $2.2 \geq \iota(0) \geq 1.2$, highly coherent MHD modes with frequencies 4 kHz ≤ f ≤ 15 kHz, are often observed in total radiation signals in the central plasma region ($\rho < 0.5$). They can accompany sawtooth-like oscillations, resembling the precursor modes in tokamak plasmas, get destabilized during internal crashes [6], or appear independently with nearly constant frequency and amplitude. It is rather usual to detect several harmonics (up to 6) that occasionally can have higher intensity than the fundamental mode. They can be found in practically all the magnetic configurations explored to date, independently of plasma volume or shape, magnetic well value or composition. Modes appear mainly in neutral beam heated plasmas although some types can be detected in ECR heated plasmas as well, being their radial extent, intensity and frequencies weakly dependent on rotational transform and plasma pressure profiles. Namely, for a given magnetic configuration, modes are slower and apparently more extended in H-mode (dome-shaped) than L-mode type profiles (bell-shaped), and in strongly He-doped than in pure H2-plasmas.

As core coherent modes (CCM) are almost always invisible to Mirnov coils and the ECE system cannot be used in plasmas with average electron densities higher than $1 \times 10^{19}$ $m^{-3}$, radiation signals from bolometer arrays toroidally distributed around the vacuum vessel [7], have been used to try to explore some of their characteristics. It is worthy to mention that CCMs are hardly detected with the SXR photodiodes.

Since remarkable coincidences with many reported experimental observations of distinct nature in other toroidal devices, say tokamaks, are found in plasmas of a low-shear, moderate-beta, low-current, and high rotational transform stellarator, it is worthy presenting a detailed description of these core modes in order to provide hints to clarify their origin.

## Experimental

The TJ-II is a four-period flexible Heliac with low magnetic shear and major and averaged minor radii of 1.5 m and ≤ 0.22 m, respectively. Plasmas of interest are started either with ECR ($P_{in} \approx 600$ kW, 2 gyrotrons, at 53.2 GHz, 2nd harmonic, X-mode polarization) or with neutral beams of $P_{in} \leq 650$ kW each, tangentially injected co- and counter- the magnetic field direction. The magnetic configurations for which CCMs develop cover almost the whole range of the TJ-II operational space (i. e., $2.2 \geq \iota(0) \geq 1.2$). The diagnostic available to follow CCMs is the bolometer system, consisting in AXUV photodiodes arranged in three 16-channel arrays with identical viewing geometry and three 20-channel arrays for tomography.

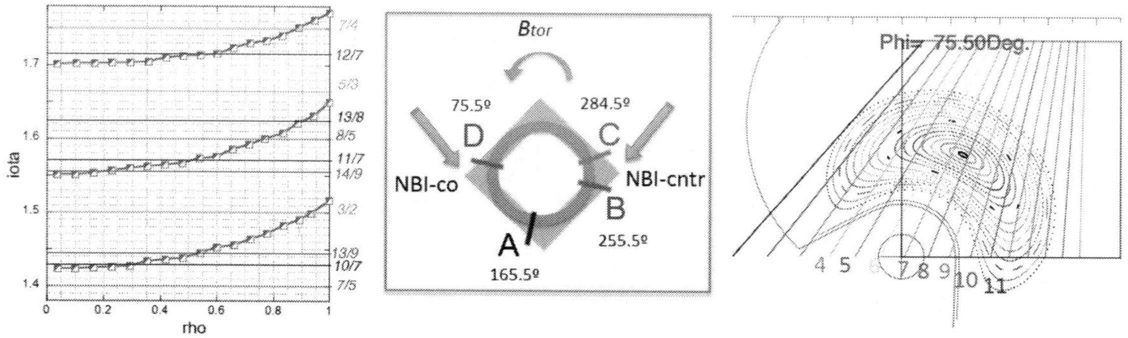


Fig.1 a) iota profiles of some of the studied magnetic configurations b) scheme of the toroidal distribution of the TJ-II bolometer system, c) cross-section and lines of sight of bolometer arrays in A, D and C.

Figure 1 shows: a) the rotational transform profiles in vacuum of the magnetic configurations used in this communication; b) the scheme of the location of the bolometer arrays around the vacuum vessel, the arrows indicate the standard toroidal magnetic field and NBI injection directions; and c) the lines of sight of the 16-channel AXUV arrays over the plasma cross-section corresponding to the centre of the observation ports of modules 7 in sectors labelled A, B and D (the cross-section in sector C-module 2 is their up-dow symmetric.

**Observations**

As above mentioned, CCMs appear in practically all the magnetic configurations and can exhibit different patterns. Some of them are shown in Figure 2: a) a low frequency mode accompanying off-axis sawteeth (OAS), b) a stable and intense ($\Delta$Plin/Plin $\approx$ 6%) mode in a low temperature plasma, and c) a rather modulated frequency mode that survives to a confinement transition. In series of discharges with the same magnetic configuration of the one represented in figure 2a and with pure $H_2$ and strongly He-doped $H_2$ as working gases it could be stated that mode frequency increases as the radiation (and electron density) profiles evolve to more peaked and that for equivalent profiles, modes are faster in plasmas of pure hydrogen. It is worthy to mention that in the shots started directly with NBI, a relatively intense current (-4 to -6 kA) is generated and a non-negligible population of suprathermal electrons remains during the whole discharges duration. Different modes, not modulated but pulsed in intensity, are observed in radiation signals. At present the possible

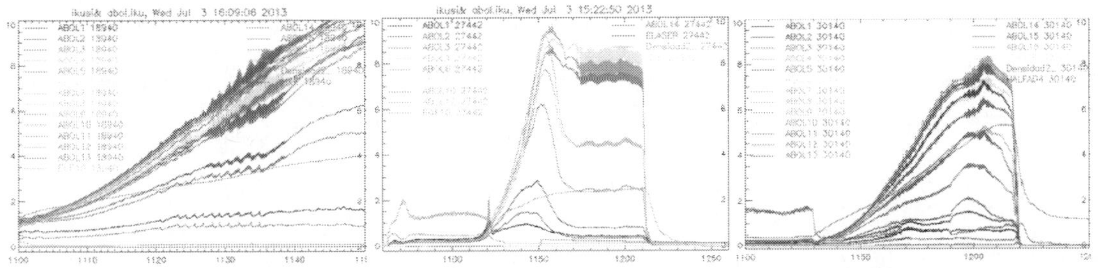

Figure 2. Raw signals of the bolometer array in sector A showing core modes from shots with different plasma conditions (see text) and magnetic configurations: a) $\iota(0) \approx 1.55$, b) $\iota(0) \approx 1.70$ and c) $\iota(0) \approx 1.42$.

role of fast electrons is under investigation [8]. In this communication we will focus on the modes shown in figure 2a.

CCMs associated to OASs resemble the precursor modes in the core of tokamak plasmas, nevertheless their intensity and frequency can be rather insensitive to profile relaxations, as is shown in figure 3a, where three cycles are shown. There it can be seen how the increment in edge radiation due to the outwards radial particle pulses produced by the OASs is detected at the same time in all toroidal locations (compatible with an n=0 mode), whilst the core signal oscillations detected in the different toroidal sectors are shifted. In fact, this is one universal finding for CCMs: for equivalent chords, a toroidal phase-shift $\Delta\phi = \phi/2\pi$ (being $\phi$ the toroidal angle) in the emissivity oscillations appears that is opposite to the momentum input direction (see figure 3b). This would point to a toroidal mode number n = ± 1, with (+) and (-) standing for counter- and co- NB injection respectively (note that this stands for reverse field also), with independence of the iota value. Although not shown, similar modes observed in ECRH plasmas do not exhibit toroidal phase shift. In the laboratory frame, and taking into account that mode frequencies range from 4 kHz to 15 kHz, the deduced phase toroidal velocity is $v_{\text{tor ph}} = 2\pi Lf \approx$ 43-160 km/s. With these sets of bolometers, plus one AXUV array with an UV filter located at $\phi = 104.5°$, we have tried to estimate the fluid toroidal velocity in a shot with a long lasting emissivity perturbation caused by a local impurity source. As is shown in figure 3c, the obtained value is around $v_{\text{tor}}$ = + 10 km/s, with the sign (+) indicating parallel to the momentum input, i.e., opposite to the apparent CCM propagation. This value is in full agreement with the core toroidal velocity, $v_{\text{tor}}$ = 5-10 km/s, determined by CXRS in NB heated plasmas [9].

With respect to the poloidal mode number, it cannot be determined with the three-array tomography system [7] due to the strong shaping of TJ-II plasmas, but according to the inversions in signal oscillations observed in consecutive chords, it seems to depend on the actual magnetic configuration. For a co-injected shot of the same series that the one shown in figure 3a, it was determined that the oscillations propagate in the electron diamagnetic direction at a speed of ≈ 3km/s. In the region $\rho \leq 0.5$ of NBI plasmas, a poloidal rotation velocity of 2 km/s $\leq v_{pol} \leq$ 6 km/s was determined from passive spectroscopy [10].

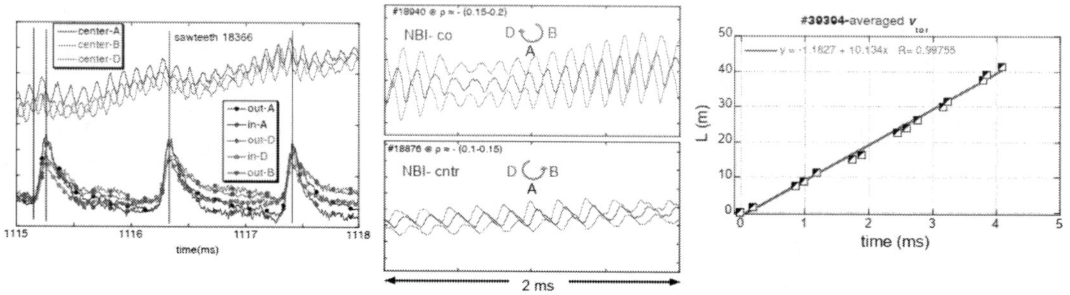


Figure 3. a) Core (upper signals) and edge radiation signals from a OAS discharge, b) toroidal shift of radiation signals from equivalent lines of sight under co- and counter- NB injection for CW toroidal field; the curved arrow indicates the phase propagation direction, and c) fluid velocity deduced from three toroidal transits of impurity-induced radiation increase.

**Summary**

Low frequency modes are observed in the core region ($\rho \leq 0.5$) of plasmas with all magnetic configurations and heating systems. Whilst in ECRH heated plasmas no toroidal phase shift is observed, in NBI heated plasmas a toroidal shift $\Delta\phi = \phi/2\pi$ appears that is not related with the rotational transform. The phase velocity of the oscillations is of the order of the sound velocity and about one order of magnitude higher than the toroidal flow velocity and in reverse direction, i.e., opposite to the NB injection direction. The toroidal mode number is n = ±1. Their poloidal rotation $v_{pol} \approx$ 3 km/s velocity estimated at $\rho \approx 0.3$ is in full agreement with the poloidal rotation velocity measured by spectroscopy in TJ-II for $E_r \approx$ 3 kV/m. The poloidal mode number is uncertain.

**References**

[1] J. H. Harris et al., Phys. Rev. Lett. **53**, 2242 (1984)
[2] J. Manickam et al., Nucl. Fusion **27**, 1461 (1987)
[3] M. P. Gryaznevich et al., Nucl. Fusion **48**, 084003 (2008)
[4] R-B. Zhang et al., Plasma Phys. Control. Fusion **56** 095007 (2014)
[5] L.F. Delgado-Aparicio et al., Nucl. Fusion **53**, 043019 (2013)
[6] D. López-Bruna et al., Nucl. Fusion **53** 073051 (2013)
[7] M. A. Ochando et al., Fusion Sci. Technol. **50**, 316 (2006)
[8] L. F. Delgado-Aparicio et al., Phys. Plasmas **22**, 050701 (2015)
[9] J. Arévalo et al., Nucl. Fusion **53** 023003 (2013)
[10] B. Zurro et al., Fusion Sci. Technol. **50**, 419 (2006)

# Magnetic island rotation in TJ-II plasmas

B. J. Sun, D. López-Bruna, M. A. Ochando, M. A. Pedrosa

*Laboratorio Naciónal de Fusion, Ciemat, 28040-Madrid, Spain*

## Introduction

In order to confine hot plasmas, magnetic-confinement fusion (MCF) devices rely on the establishment of nested flux surfaces. Their breakup to form islands or stochastic regions can degrade plasma confinement and induce disruptions. For these reasons tearing mode studies have been a continued concern, especially in tokamak technology. Over the years, however, there has been accumulating evidence that magnetically resonant regions (prone to open islands) confer also interesting properties to MCF plasmas. It is known, for example, the relation between low order rational values of the safety $q$-profile (typically $\iota = 1/q$ in currentless devices) and the establishment of transport barriers.

The destructive effects of magnetic islands in stellarators are mitigated because the current density shear is much smaller than in tokamaks, thus providing less drive for tearing mode instabilities. Additionally, due to negative magnetic shear, the bootstrap current is believed to be stabilizing [1]. Locked magnetic islands can exist due to error fields and plasma effects, but they do not necessarily grow to threaten confinement. In many cases, the basic dynamics of magnetic islands in stellarators is related to their rotation in the plasma frame. Island rotation is easy to detect, and it may convey important information for plasma control. This work is devoted to magnetic island rotation in plasmas of the TJ-II stellarator and its relation with confinement properties, notably the creation and destruction of transport barriers.

## Phenomenologycal overview

TJ-II is a Heliac-type stellarator with low magnetic shear within a large range $\iota \sim 1$–$2$. ECRH and NBI are used to heat the plasma up to $\sim 1$ keV electron temperatures. Average electron densities, $\bar{n}$, are on the order of $10^{19}$ m$^{-3}$. Low frequency coherent modes are found independently of heating regime, and the mode positions and numbers are often related with some dominant rational surface in the near edge region ($0.4 < \rho < 0.9$, where $\rho$ is a normalized radial coordinate). Typical phenomenology in relation with island rotation is exemplified in figure 1, which corresponds to discharge #33847 in a magnetic configuration such that $\iota = n/m = 8/5$ locates at $\rho \approx 0.76$. During the ECRH phase (Fig. 1(b)) a coherent mode labelled 1 in Fig. 1(a) starts ($t \approx 1110$ ms) with 5 kHz, increases frequency as $\bar{n}$ does, and drops when the gas puffing (not shown) is cut at $t \approx 1140$ ms. The ECRH switches off at $t \approx 1160$ ms provoking about 10 ms of plasma cooling. When the co-NBI is coupled to the plasma the coherent mode re-appears (labelled 2), initially with 21 kHz, slightly increasing frequency with density. After $t = 1220$ ms (dashed vertical line) the frequency increases but also loses coherence and intensity. This process is related with reductions of edge $H_\alpha$ light and magnetic fluctuations, which are typical signatures of plasma preparation for an L-H transition.

The phase indicated with label 3 in figure 1(a) has given rise to much literature because of its rich dynamics. Figures 1 (c) and (d) are just a zoom in this time window for the same discharge. As of $t = 1129$ ms (vertical dashed line), the initially coherent mode undergoes a pulsed activity accompanied by $H_\alpha$ peaks. The $H_\alpha$ becomes larger on average and each pulse separates short phases of rapidly decreasing frequency in the spectrogram. When possible, mode number analyses indicate that modes 1–2–3 share the same poloidal number $m$, always corresponding to a low order rational in the magnetic configuration. Bolometers confirm that the oscillations locate around the expected position of the $\iota = n/m$ resonance. In general, mode 3 can be sustained for long times despite its being the threshold between L- and H-mode confinement regimes [2]. The following sections give more detailed descriptions of steady and bursting modes, and reveal the link to island dynamics.

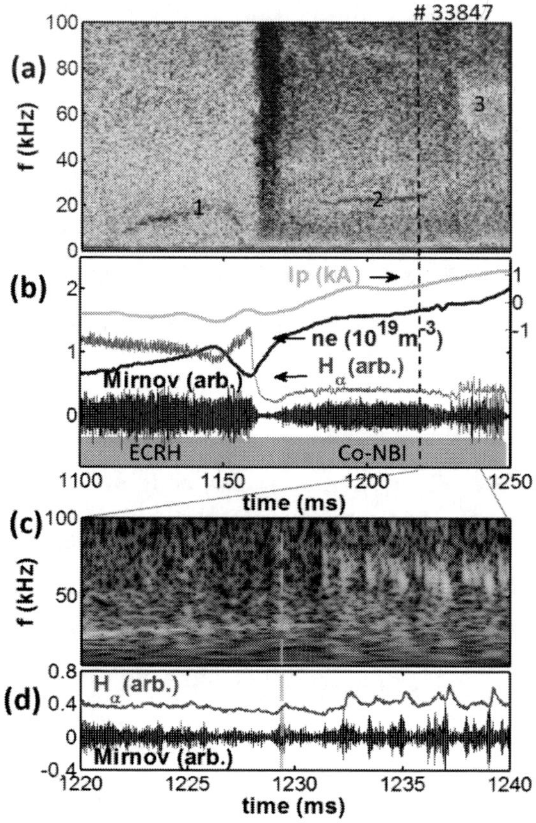


Figure 1: *TJ-II discharge #33847. (a) Power spectrum of Mirnov coil signal; (b) plasma current $I_p$, line average density $\bar{n}$, $H_\alpha$ signal, and raw Mirnov data. Zoom in view: (c) wavelet power spectrum of Mirnov; (d) $H_\alpha$ and Mirnov coil data.*

**Steady island rotation mode**

Steady island rotation, like modes 1 and 2 in Fig. 1(a), is normally seen in low-density ECRH or in NBI plasmas with peaked density profiles, as is typical in L-mode of confinement; and generally with the low order rational surface located around $0.5 < \rho < 0.9$. As expected from rotating islands, the mode frequency scales linearly with $m$ and the slope is compatible with ExB rotation frequency $f_{E\times B} = v_{E\times B}/(2\pi a\rho)$ [3]. Here $a$ is the average minor radius.

Further proof of island rotation is given in figure 2, corresponding to a discharge started and sustained with co-NBI at low density ($\bar{n} = 0.7 \times 10^{19}$). The start-up with NBI causes a negative current $\sim -3$ kA that carries the surface $\iota = 3/2$ to $\rho \approx 0.5$. The 10 kHz coherent mode detected by ECE radiometers shows a clear phase inversion between channels ECE4 ($\rho = 0.45$) and ECE3 ($\rho = 0.53$), indicating a modulation of the average $T_e$ gradient in that region. Steady island rotation can show harmonics, especially for low $m$ rationals. For example, up to the 5th harmonic has been detected for the 3/2 island.

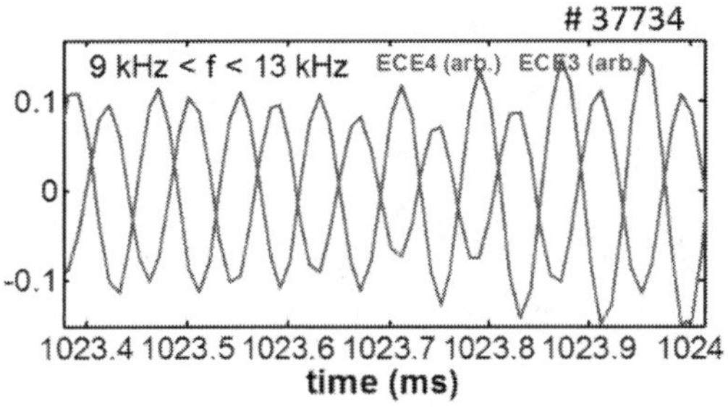


Figure 2: *TJ-II #37734. Electron temperature fluctuation within the 9–13 kHz frequency band for ECE channels #4 ($\rho = 0.45$) and #3 ($\rho = 0.53$).*

**Bursty island rotation mode**

Chirping frequencies (Fig. 1, mode 3) substitute the steady rotation when the plasma approaches the L-H transition. Figure 3 shows time traces of a $\bar{n} = 0.8 \times 10^{19}$ m$^{-3}$ plasma heated and fueled with co-NBI ($I_p \approx 1$ kA) in which $\rho_{8/5} \approx 0.5$. In the shady region, the innermost ECE4 ($\rho = 0.45$) shows a continued increment of electron temperature while the outer ECE3 ($\rho = 0.53$), and more clearly the outermost ECE2 ($\rho = 0.65$), decrease: a transport barrier has developed between the ECE4 and ECE3 locations. Then, a sudden burst on the Mirnov signal coincides with a quick drop of temperature according to ECE4; but ECE3 and ECE2 rise, as expected from the sudden breaking of the barrier. The process is repeated on a ms time-scale.

The correlation between the flattening/steepening of profiles (e.g. electron temperature) around the resonant location and the envelope magnetic fluctuations has been also detected with bolometry and is always associated with a break-up/build-up of transport barriers owing to continuous heating and fueling. This provokes a kind of limit cycle oscillation and flux-surface wide perturbations on many plasma magnitudes.

The amplitude and characteristic times between bursts (e.g. in magnetic or H$_\alpha$ signals) depend mainly on the order of the rational. For higher orders, such as $\iota = 8/5$, the unstable rotation has usually repetition lags about 0.5–1 ms and there are no obvious changes in global plasma parameters, quite like in tokamak grassy ELMs. For the lower order 3/2, each MHD burst can last for several ms causing decrements in $\bar{n}$ and temperatures. Sometimes, clear drops in plasma current and large $H_\alpha$ pulses are found, resembling more closely giant tokamak ELMs.

The frequency of unstable rotation modes also varies linearly with $m$ and can show harmonics for the lowest $m$ values. Frequencies per mode number have been found in the ranges $f/m = 8$–$16$ kHz ($\iota = 8/5$), 7–12 kHz (5/3) and 8–20 kHz (3/2). The lower $f/m$ boundary is quite robust, while the upper limit seems to depend on island size and local plasma parameters.

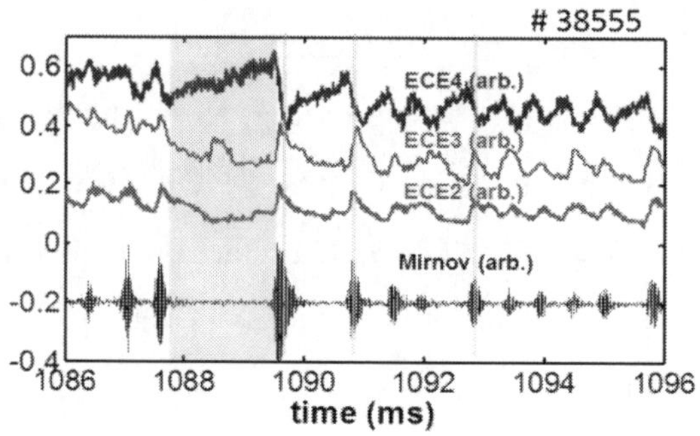


Figure 3: *TJ-II discharge #38555. ECE signals around the* $\iota = 8/5$ *resonance, and Mirnov coil data.*

**Discussion and conclusion**

Low order rational surfaces are singular in MHD equilibrium. They can give birth to magnetic islands that interact non-linearly with the plasma fluid via resistivity, local gradients and plasma flows. These are common elements of L-H transition theories, which makes the rotation of magnetic islands naturally sensitive to the transition threshold, and also viceversa according to TJ-II literature. Such strong link between island dynamics and the formation of transport barriers justifies the existence of a sustainable threshold state, which is the bursty island rotation mode above. The corresponding repetitive creation and breaking of transport barriers provides signatures that are typical of zonal-flow dynamics (e.g. predator-prey cycles, long-range correlations, strong relation with the L-H transition, intermediate modes... ) commonly associated to turbulent self-organization. Whether the latter is relevant or not to interpret the experiments, magnetic island dynamics is a mandatory ingredient in TJ-II plasmas. Similar behaviors can be seen in other experimental devices pointing to common physics, and we suggest that information on the presence of possible magnetic resonances be always given (e.g. the $q$-profile).

In conclusion, magnetic islands can rotate steadily with plasma flow in general TJ-II plasmas. Unstable island rotation, characterized by chirping frequencies and bursts in magnetic activity and plasma transport, is found at the threshold for the L-H transition. The phenomenology resembles typical tokamak activity, suggesting very basic and common physics in MCF plasmas.

**References**

[1] Carrera R et al. 1986 Phys. Fluids **29** 899

[2] López-Bruna D et al. 2013 Nucl. Fusion **53** 073051

[3] Sun B J et al., 41st EPS Conf. on Plasma Physics, Berlin 2014, ECA Vol. 38, P2.090

# Influence of long-scale length radial electric field components on zonal flow-like structures in the TJ-II stellarator

Ulises Losada[1], M. Angeles Pedrosa[1], Arturo Alonso[1], Carlos Silva[2], Boudewijn van Milligen[1], Carlos Hidalgo[1] and the TJ-II team
*Laboratorio Nacional de Fusión, CIEMAT[1]*
*IST, Lisbon[2]*

## 1. Introduction

Mass flows are important in plasma stability and confinement. Experimental and theoretical efforts have been made to improve the understanding of the momentum transport mechanisms that determine the observed plasma rotation profiles. In stellarators, with non-axisymmetric magnetic fields, the neoclassical non-ambipolar fluxes dominate the equilibrium radial electric field [1]. Fluctuating, flux surface-constant zonal flows are thought to be instrumental in turbulent transport control and to be a natural state of turbulent plasmas. Different magnetic configurations can host zonal-flow structures. For stellarators, it has been found that neoclassical optimization could lead to reduced zonal flow (ZF) damping and, as a consequence, to turbulent transport optimization [2].

Previous experiments performed in the TJ-II stellarator have shown that long-range toroidal correlations (LRC) in potential fluctuations are amplified by externally imposed radial electric fields [3], near the L-H confinement edge transition [4] and in plasma regimes with reduced neoclassical viscosity. Gyro-kinetic, linear simulations of collisionless ZF relaxation with the global code EUTERPE [5] have been carried out in the TJ-II geometry. Both the ZF oscillation frequencies and the ZF residual level have been studied under different plasma conditions. In particular, an increase of the residual both with the radial ZF wavelength and the electric field strength is found in the simulations, in agreement with previous works [6]. The residual level increases with the external electric field strength until reaching saturation for intense fields.

Recent results in tokamak plasmas have shown the development of multiple radial scales with spontaneously formed long-lived pattern of $E \times B$ flows together with long-lived pressure corrugations and interspersed with regions of turbulent avalanching [7].

This contribution focuses on the interplay between multi-scale physics mechanisms and long-scale length (neoclassical) radial electric field in the TJ-II stellarator.

## 2. Experimental set-up

Experiments were carried out in the four-period, flexible, low-shear stellarator TJ-II ($B = 1$ T, edge rotational transform $\iota/2\pi(a) = 1.65$, $R = 1.5$ m and $a \leq 0.22$ m). This report is focused on pure (ion root) NBI heated hydrogen plasma discharges ($P_{NBI\,1} \approx 610$ KW, $P_{NBI\,2} \approx 460$ KW). The results reported here were made possible by the use of a unique detection system: two Langmuir probe arrays, named Probe 1 and Probe 2, installed on fast reciprocating drives located at two different toroidal ($\Delta\phi$) / poloidal ($\Delta\theta$) locations, approximately $\Delta\phi = 160^{\circ}$ toroidally apart and poloidally separated $\Delta\theta = 155^{\circ}$. Probe 1 consists of 12 tips in the radial direction spaced 3 mm apart plus 3 tips in inner position spaced 3 mm apart in the poloidal direction. Probe 2 has a similar configuration with 8 tips spaced about 2 mm and 3 mm in the radial and poloidal directions respectively. This set-up permitted the simultaneous investigation of short-range (i.e. plasma radial profiles with spatial resolution in the range of millimeters) and long-range fluctuation scales (i.e. plasma structures at scales of about 10 m) in the plasma boundary region [Fig. 1].

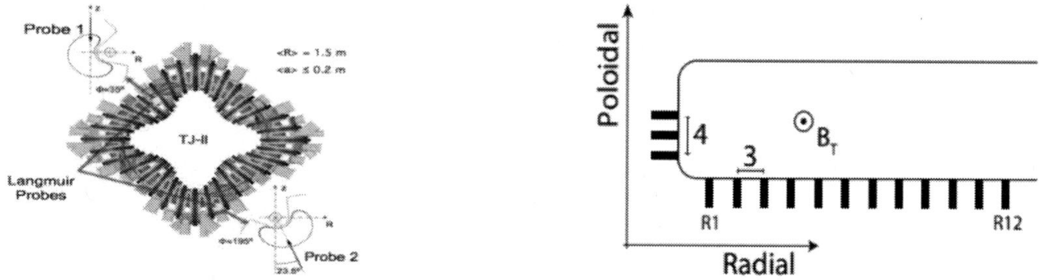


Fig. 1: a) Placement of the two rake probes in the TJ-II stellarator. b) Representation of one of the rake probes (probe 1)

## 3. Experimental results

### 3.1. Long and short scale radial electric fields

The mean velocity of fluctuations perpendicular to $B_T$ and along the magnetic flux surface ($v_\perp$) is computed with the two point correlation technique using floating potential probes poloidally separated 3 mm. In the plasma edge region, just inside the LCFS, $v_\perp$ increases in the electron drift direction up to few km/s. Previous experiments have shown that $v_\perp$ is dominated by the ExB velocity and, in some conditions, by the diamagnetic drift velocity [8, 9]. High correlation was found between the $E_r$ deduced from the gradient in floating potential and the perpendicular phase velocity of fluctuations. Thus, in the present plasma conditions, gradients in floating potential have been used as a proxy for radial electric fields. The long-wave length (~ 20 ion Larmor radius) radial electric fields have been measured from a linear fit of the radial profile of the floating potential; $E_r$ values are strongly dependent on plasma density reaching values up to 3 kV/m. Short scale ( ~ 5 ion Larmor radius) radial electric fields are measured by using Langmuir probes measurements along the radial direction with spatial separation is of 3 mm and time averaging in the range of 1 ms. Its magnitude varies along radial direction, exhibiting an oscillating-like behavior about the mean $E_r$.The amplitude of this oscillation has been determined and is called "corrugation" in $E_r$. This oscillation increases radially inwards and is also bigger for higher values of the global length scale electric field. This result shows coupling between long and short scale length components of $E_r$, with the level of corrugation increasing with the neoclassical radial electric field.

### 3.2. Long Range Correlation

The experimental set-up is ideal for measuring long range correlations. The long-range spectrum is dominated by frequencies below 20 kHz. An increase of the amplitude of Long Range Correlations (LCR) with the long scale radial electric field strength is found reaching saturation for $E_r$values in the order of 1 kV/m [Fig. 2]. These results are qualitatively consistent with GK simulations showing the influence of radial electric fields in the zonal flow residual level. The interplay between neoclassical ExB shear flows and the development of low frequency (zonal flow-like) structures could be explained considering the role of electric fields as a turbulence symmetry-breaking mechanism (i.e. amplifying Reynolds stress driven flows) or/and the influence of radial electric fields on particle orbits (neoclassical damping).

### 3.3. Radial correlation length measurements

The radial correlation ($L_r$) of floating potential fluctuation was characterized using 12 radially spaced (3 mm apart) tips of Probe 1. $L_r$ is defined physically as the distance in which the correlation between two signals decays below 1/e of the maximum of the correlation function. $L_r$ was calculated using two different approaches, cross-correlation function and cross-coherence coefficient showing very similar results. Radial correlation frequency spectra are dominated by small frequencies (<20 KHz). We observe that $L_r$ decreases as $E_r$ increases [Fig. 3.a].

### 3.4. Shear in the electric field

The $E_r$ shearing rate has been calculated at different radial scale lengths. For calculating the short scale shearing rate three Langmuir probes, placed at the middle region of the rake probe, measuring floating potential along the radial direction have been used. In order to obtain the global (long wave length) shear, a kernel function is applied to modulate the number of tips which are used for calculating the second derivative. Experimental results show that the local shear in the electric field is larger than the global one [Fig. 3] reaching values in the order of $10^5 s^{-1}$.

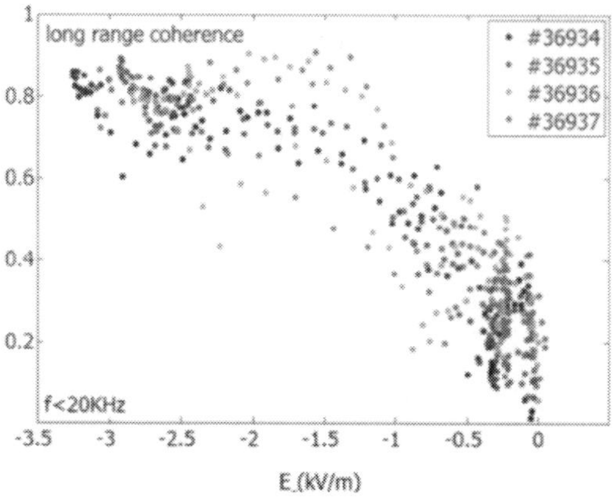


Fig. 2: Amplitude Long Range Correlations versus the long scale radial electric field (global $E_r$) for low frequencies (f < 20 kHz).

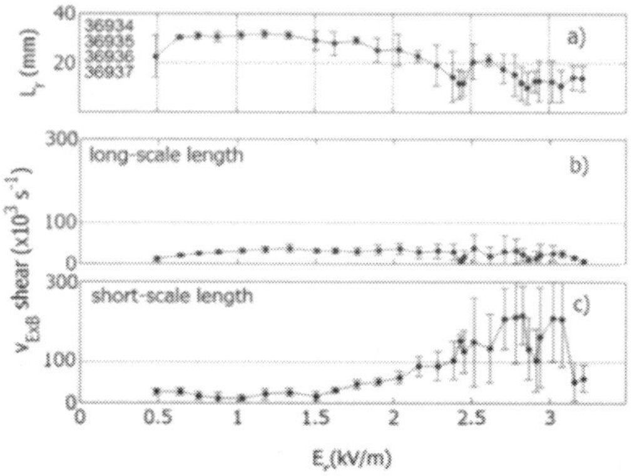


Fig. 3: Radial correlation, global and local $E_r$ shearing rates versus the long scale radial electric field.

## 4. Conclusions

Long-scale length radial electric field (Er) was found to play a role in controlling the amplitude of LRC with zonal flow-like structures, in consistency with gyrokinetic (GK) simulations. At the same time, the radial correlation length of turbulence decreases. Long and short scale length components of the radial electric field are coupled, (level of corrugation increases with the neoclassical radial electric field. Experimental results show that the local shear in the electric field is larger than the global one reaching values in the order of $10^5 s^{-1}$.

**Acknowledgements**
This work has been carried out within the framework of the EUROfusion Consortium and has received funding from the Euratom research under grant agreement No 633053. This work has been partially funded by the Spanish Ministry of Economy and Competitiveness under contract numbers ENE2012-38620-C02-01. The views and opinions expressed herein do not necessarily reflect those of the European Commission.

[1] P. Helander et al., Phys. Rev. Lett. 101 (2008) 145003.
[2] T. H. Watanabe et al., Phys. Rev. Lett. 100 (2008) 195002.
[3] M.A. Pedrosa et al., Phys. Rev. Lett. 100 (2008) 215003.
[4] C. Hidalgo et al., EPL 87 (2009) 55002.
[5] E. Sánchez et al., EPS-2014, Berlin P2.079
[6] A. Mishchenko and R. Kleiber, Phys. of Plasmas 19 (2012) 072316.
[7] Diff Pradalier et al., Phys. Rev. Lett. 114 (2015) 085004
[8] Ch.P. Ritz et al., Phys. Fluids 27 (1984) 2956.
[9] C. Hidalgo et al., Nuclear Fusion 33 (1993) 146.

# Progress on small modular stellarator SCR-1

V.I. Vargas, J. Mora, C. Otárola, E. Zamora, J. Asenjo, A. Mora, E. Villalobos

*Plasma Laboratory for Fusion Energy and Applications, Instituto Tecnológico de Costa Rica, Cartago, Costa Rica*

## Introduction

Since 2009, the Instituto Tecnológico de Costa Rica started a research project on stellarators which aims at designing, constructing and implementing the first device of this technology in Latin America. Small stellarators present clear advantages related to total cost of the project which allows with a relatively low investment train human resources and contribute to the physics and engineering of such devices within the fusion scientific community.

The SCR-1 is a small modular Stellarator for magnetic confinement of plasma developed by the Plasma Laboratory for Fusion Energy and Applications of the Instituto Tecnológico de Costa Rica (ITCR). SCR-1 will be a 2-field period, small modular stellarator with an aspect ratio of $> 4.4$; a low shear configuration with core and edge rotational transform equal to 0.32 and 0.28. It will hold plasma in a 6061-T6 aluminum, torus-shaped, vacuum vessel with an minor plasma radius of 54.11 mm; a volume of 13.76 liters (0.01 $m^3$); and, major radius of R = 238 mm. Plasma will be confined in the volume by an axis magnetic field of 43.8 mT generated by 12 modular coils with 6 turns each, carrying a current of 767.8 A per turn providing a total toroidal field (TF) current of 4.6 kA-turn per coil. The coils will be supplied by a bank of cell batteries of 120 V. Typical length of the plasma pulse will be between 4 s to 10 s. The SCR-1 plasmas will be heated by ECH second harmonic at 2.45 GHz with a plasma density cut-off value of $7.45\times10^{16}$ $m^{-3}$. Two magnetrons with a maximum output power of 2 kW and 3 kW will be used.

## Engineering aspects

The stellarator vacuum vessel along with its coil support, coils and ports were completely constructed in Costa Rica by the Instituto Costarricense de Electricidad (ICE), after hiring their services. All ancillary components and peripherical systems were chosen according to design criterion and they were bought new. Additional stellarator support structures for the peripherical systems (ECH system, Gas Injection System) and the diagnostics were constructed in Costa Rica using 304LN austenitic steel with relative magnetic permeability below 1.01.

The engineering phase, previous to the construction of SCR-1, was divided in the following steps [2]: Predesign, Design, Design review, Detailed design within the manufacturing contracts.

**Vacuum Vessel**

SCR-1 vacuum vessel was manufactured with two 6061-T6 aluminum cylindrical building blocks. Although using austenitic 304L grade stainless steel was analyzed, it was discarded because of the difficulty to manufacture parts according to the device dimensions and because it increased greatly project costs. Internal surface was polished and it was obtained a $R_a$=0.3. It was finally obtained a vacuum vessel with 10 mm thickness and a volume of 0.0418 $m^3$, an external radius of 364.1 mm, an internal radius of the central circle of 112.1 mm, a major radius of R = 238.1 mm, and the internal radius of the vacuum vessel is 94.42 mm.

Each CF port was mechanized according to design dimensions. The vacuum vessel has 24 conflat ports: three of the ports are CF of 6", 10 are CF of 4-5/8", and 9 are CF of 3-3/8". Two rectangular ports, internal dimensions of 90 mm x 40 mm, were added to accommodate future needs. It was decided to use, for all ports, the same CF aluminum design of the vacuum vessel with Cusp Confinement Plasma called MPDX (Madison Plasma Dynamo Experiment) from the University of Wisconsin, Madison. The welding process was certified by manufacturer to AWS welder and it was used a standard procedure as AWS A5.3 and AWS A5.10, which specify aluminum welding procedure.

**Modular field coils**

**a) Coil support system**

In the predesign, design, and design review phases, different ideas on how to place the coils above the vacuum vessel, methods as the one indicated on [1], were considered. Finally in the detailed design phase, it was decided to use coil support or guides that were welded to the vacuum vessel and with the specific shape of each; which should leave a channel in which a conductor wire would be placed. The fabrication of coil supports and guides was done through a cast procedure. This method was chosen because it allows building pieces with the exact shape. Now, to create the casting molds a 3D printer, Dimension DST1200ES, was used. It has a print volume of 254 x 254 x 305 mm and a print precision of 0.254 mm, which is more than acceptable for the creation of the supports because the defined tolerance in the design phase for this process was +/- 2 mm. Once supports were ready with the 3D printer, the casting mold was created and the cast started. The following step was to place the supports in the vacuum vessel, but first it was necessary to mark accurately where the supports were to be welded. It was decided to use a CNC milling machine with a ball mill (to allow to always making contact with the tangency point) to create a channel of 0.2 mm with the coil shape reflected on the vacuum vessel for the welder to know where to place the piece and weld it.

**b) Magnetic coil system**

The SCR-1 magnetic field of confinement will be generated by 12 copper modular coils, whose supports and positioning were done according to the beforehand explanation. Coil geometry was obtained by engineer Vicente Queral [1], from CIEMAT, in Spain. Each modular coil will have 6 turns, made of AWG#4 wire, and a current of 767.8 A per turn providing a total toroidal field (TF) current of 4.6 kA-turn per coil.

Thermal transfer simulations were performed in modular coils [2]. The thermal behavior of copper wire by electrical current passing in the modular coil was simulated using COMSOL Multiphysics software and other methods. Temperature, resistance, voltage and power calculations as a function of time were performed for the electrical circuit under different wire configurations per modular coil to select the power supply taking into account the available budget.

**c) Coils connection and electric current regulator**

Modular coils were connected in series with heterogeneous welding, using tin to create joints in cabling ends. This allows reducing total resistance of 12 coils to 60 mΩ by not using connectors among the coils. Once the coils were connected, they were welded to the end of an AWG #4 cable, AWG #0 copper cable, and it was returned through where the coils are connected in series to cancel magnetic fields generated by peripherical wires to the coils.

To keep a constant current during a period of time less than 10 s, in spite of ohmic heating in coil conductor cable, it was designed an electrical current controller [2].

**Peripherical systems and support structure**

The Peripherical systems and support structure are formed by ECRH system, gas injection system, vacuum pumping system, power supply, control system, support structure. More information about the auxiliary systems can be found at reference [2].

**Status of assembly on site, test program and commissioning tests**

To this date the covers, viewports, vacuum sensor, vacuum pumps, etc, have been placed. The first vacuum testing suggests pressures of $10^{-7}$ Torr in the toroidal vessel. ECRH systems of 2 kW and 3 kW have been tested and operated thanks to the control system indicated above. There have been hydrogen gas injection testing and an injection pattern has been determined for the first discharge. The battery bank is operating and charging; it is ready to be used in the first tests to provide power supply to the coils. Electrical current controller is being assembled and it is expected that by June 2015, it will be operational to be able to do the first plasma discharge. Once all peripherical systems finish the start up, they will undergo magnetic field mapping tests. For the second half of 2015, it is expected to have the first electron density and temperature measures using Langmuir probe and optical spectrometer.

**Conclusions**

A small modular stellarator has been designed and constructed and it is under the implementation phase at Instituto Tecnológico de Costa Rica (See Figure 1b). To this date the aluminum vacuum vessel has surpassed the values of expected low pressures. It was implemented an innovative method to construct modular coils using 3D printer and casting Al. The peripherical systems as ECRH, gas injection, vacuum pumping system, power supply, control system and support structure are ready and operational. It is expected that in June 2015 the start up of the electrical current controller is finished and with that being able to do the first plasma discharge, besides being able to do magnetic field mapping testing.

Figure 1: a) Drawing of all peripherical systems and support structures, b) Status of assembly on site of Stellarator SCR-1.

**References**

[1] V. Queral, "Coil fabrication of the UST_1 modular stellarator and potential enhancements," Fusion Eng. Des., vol. 88, no. 6–8, pp. 683–686, Oct. 2013.

[2] V.I. Vargas, et al. Implementation of Stellarator of Costa Rica 1 (SCR-1), in Proceedings of the 2015 IEEE 26th Symposium on Fusion Engineering (SOFE), May 31-June 4, 2015, Austin, Texas, USA.

# Analysis of the energy distribution of escaping suprathermal ions in neutral-beam injection phase of the TJ-II stellarator

M. Martínez[1], B. Zurro[2], A. Baciero[2], D. Jiménez-Rey[2], V. Tribaldos[3]
[1]*Universidad Carlos III de Madrid, Leganés, Spain.*
[2]*Laboratorio Nacional de Fusión, CIEMAT, Av. Complutense 40, 28040 Madrid, Spain.*
[3]*Departamento de Física, Universidad Carlos III de Madrid, Leganés, Spain*
E-mail contact of main author: *marcos.martinez@externos.ciemat.es*

**INTRODUCTION**. Detailed knowledge of loss mechanisms of fast or suprathermal ions associated to fusion processes or heating methods is of paramount interest in plasma fusion research. In this work, the ion distribution function on suprathermal ions escaping from TJ-II plasmas during the neutral beam injection (NBI) phase of the plasma is being investigated to extend those already reported studies mainly for electron cyclotron resonance heating (ECRH) plasmas [1,2]. These ions are monitored with a flexible luminescent probe (LP) [3], which is operating in a pulse energy discrimination mode [4] and it is located at the edge of the TJ-II stellarator. This work is focused on studying the behaviour of suprathermal ions energy distributions under two different conditions: when the NBI heating starts after an ECRH and when initiating the plasma solely with NBI heating. Also the pellet injection on the second type of discharge (solely with NBI heating), and its effects on suprathermal ions energy distribution are studied. We will discuss its empirical behaviour.

**EXPERIMENTAL**. TJ-II is a four-period, low magnetic shear stellarator with major and averaged minor radii of 1.5 m and ≤0.22 m, respectively. Central electron densities and temperatures up to $1.7\text{x}10^{19}$ $m^{-3}$ and 1 keV respectively are achieved for plasmas created and maintained by ECRH at the second harmonic (f = 53.2 GHz, $P_{ECRH} \leq 500$ kW) [5]. The magnetic confinement of TJ-II plasmas involves four main magnetic field components, two of them (CC and HX) produced by the central coil system [6]. The steady-state values of the CC and HX currents are 10 kA and 4.8 kA, respectively. The associated toroidal electric fields induced are 2.8 and 1 V/turn, respectively, corresponding to a total induced toroidal field of 0.35 V/m, at the lower limit of those used for ohmic break-down in tokamaks [7]. NBI launch times are between 960 and 980 ms, and it is produced for up to 120 ms while CC and HX are on the plateau. A compact multi-barrel pellet injector system developed by the ORNL (Oak Ridge National Laboratory) has being installed in the TJ-II stellarator. It is a four-barrel system destined for use both as an active plasma diagnostic and as a plasma fueling source. In order to achieve both objectives it will be sufficiently flexible to allow frozen hydrogen pellets with diameters from 0.4 to 1 mm to be formed and accelerated to velocities between 100 and 1000 m $s^{-1}$ [8-12]. The used LP has been previously reported [2] and its luminescent screen was made of TG-Green (decay time of 500 ns), with a range of measurement between 1-30 keV.

**EXPERIMENTAL RESULTS AND DISCUSSION.** We have shown in previous works how energy distribution function of escaping suprathermal ions, *f(W)*, behaves in TJ-II plasmas produced solely by ECRH [2,13]. Here, we study how the population of fast ions and their distribution function behave in two types of NBI plasmas, one providing the target plasma by ECRH and the second initiating the plasma by the NBI with the help of a small loop voltage provided by the field rise.

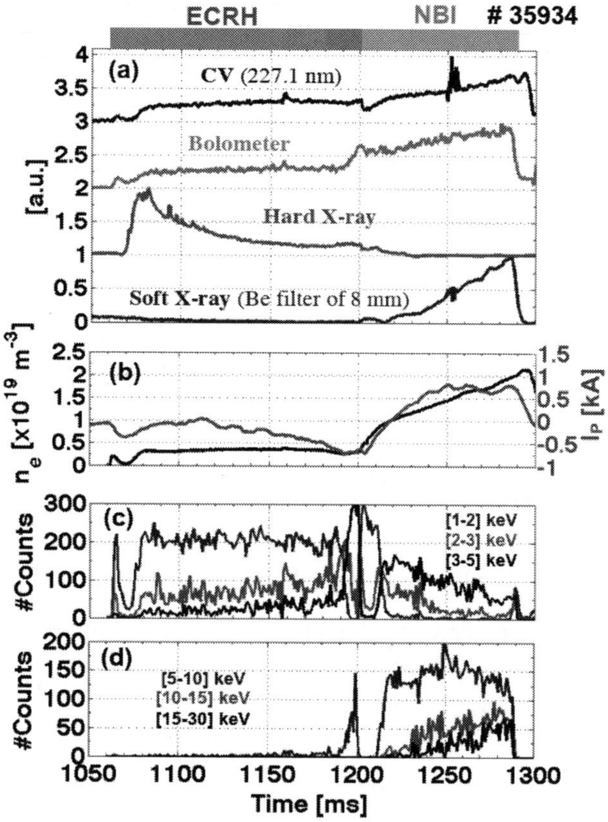


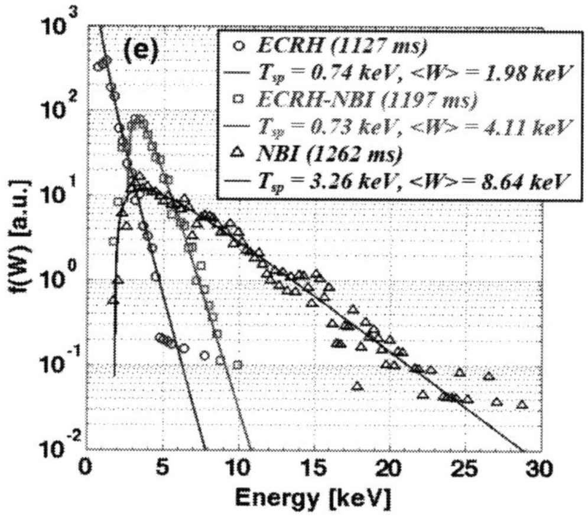


Figure 1. **(a)** Some diagnostics traces for discharge, #35934. **(b)** line-averaged electron density (microwave interferometer, $n_e$, black), plasma current ($I_P$, red). **(c)** and **(d)** Ions counts in different energy ranges. **(e)** *f(W)*, for three different times: in ECRH, 1127 ms; overlap between ECRH and NBI, 1197 ms; NBI heating, 1262 ms.

Figures 1(a) and 1(b) show some diagnostics traces for a discharge with ECRH (1060-1200 ms) and NBI heating (1180-1290 ms). There is an overlap between ECRH and NBI (1180-1200 ms). In figures 1(c) and 1(d), we show the ion counts per millisecond (ms) in different energy windows. It is noted that ions with energy less than 2 keV dominate in ECR heating. In the overlap, with ECRH and NBI heating, the number of ion counts with low (high) energy decreases (increases) in a few milliseconds (< 10 ms). During the NBI heating, the dominant ion counts are those with high energies > 5 keV, that is reflected in the *f(W)*, as shown in figure 1(e), where the *f(W)* for three phases of the discharge is exposed.

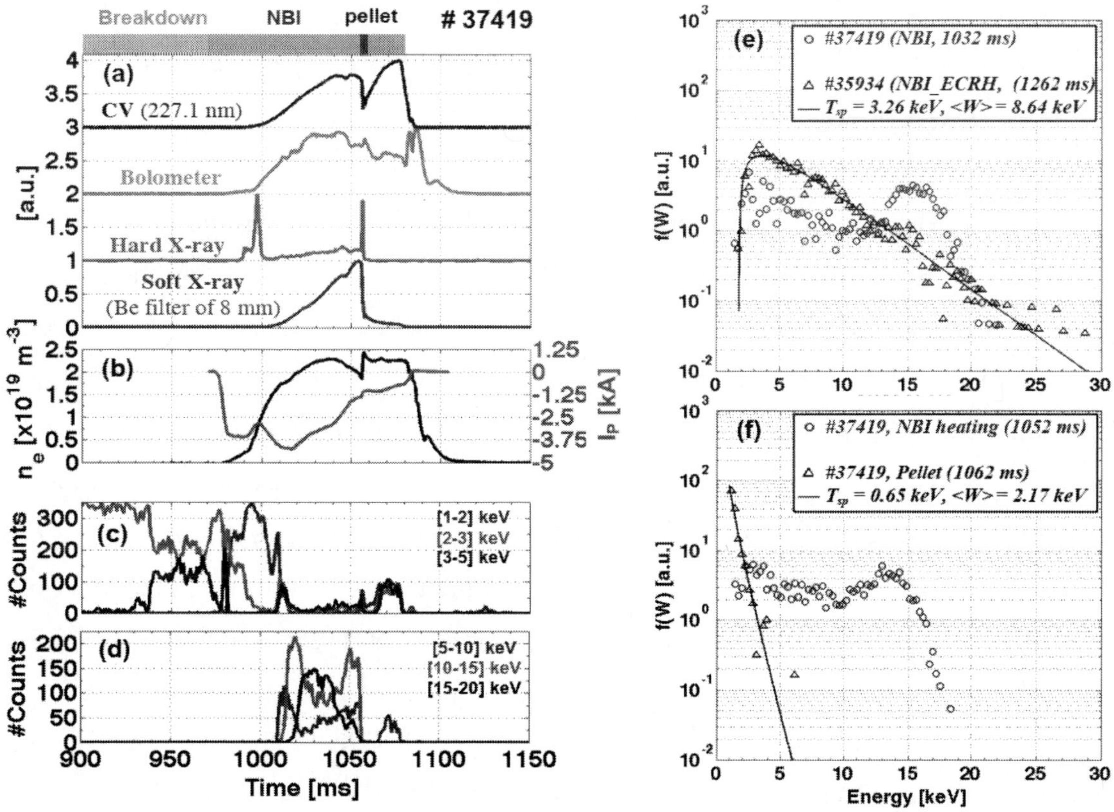


Figure 2. **(a)** Some diagnostics traces for discharge, #37419. **(b)** line-averaged electron density (microwave interferometer, $n_e$, black), plasma current ($I_P$, red). **(c)** and **(d)** shows ions counts in different energy ranges. **(e)** General shapes of *f(W)* for discharges in NBI heating, (blue, #37419) solely initiated by NBI and (black, #35934) started by ECRH. **(f)** *f(W)* for discharge solely with NBI heating. (blue, 1052 ms) *f(W)* before the pellet injection, (black, 1062 ms) *f(W)* during pellet injection.

During ECRH and overlap phase, we see that the *f(W)* are practically the same, including the value of the temperature of suprathermal ions, $T_{sp}$. But expected value of energy, <W>, moves to higher energies. On the other hand, in the NBI heating phase, it is observed high energy tails in *f(W)*. The second experiment (solely NBI heating [5]), needs a breakdown produced by an electric field, see figure 2. This is the best example to show whether it is possible for a loop voltage to generate suprathermal ions. During the ramp-up of current in coils CC and HX, the LP detects fast ions with high energies 2-5 keV (figure 2(c)). The NBI heating begins at 970 ms, then there is a significant change in the energy of ions until 1010 ms, with only ions < 2 keV, what coincides with the rise of the $I_p$ (figure 2(b)), when plasma is warming. During the electron density plateau (> 1010 ms), the ions energy increases up to values higher than 15 keV, see figure 2(d). Now we investigate the effect of the pellet injection (frozen hydrogen of diameter 1mm is introduced at 1057 ms) on suprathermal ions in the discharge # 37419. Figure 2(a) shows how some plasma monitors react at the pellet injection, hard X-ray exhibit a peak in contrast with CV and soft X-rays that both decrease in response to pellet injection as it was to be expected from a sudden cooling of the plasma. Figure 2(b) shows $n_e$ and $I_p$ signals, both exhibit changes in response to pellet injection. Figure 2(d) shows how the suprathermal ion counts are affected, they fell drastically, but after a time ~ 10 ms the LP signal rise again with ions energy < 10 keV, see figure 2(c) and 2(d).

All these changes along discharges affect *f(W)*. In figure 2(e), the first point to note is the shape of the distributions from a discharge with ECRH and NBI heating (#35934, 1262 ms) and another with solely NBI heating (#37419, 1032ms). In the first case, the influence of NBI heating on suprathermal ions gives them more energy and make the tail of the distribution wider. But in the second case (solely NBI heating), it seems that NBI favours that the number of ions increases in a specific range of energies (12-18 keV), and it does not generate tails with higher energies (see figure 2(e)). Finally, the pellet injection causes a thermalization of suprathermal ions distribution (ions < 7 keV, see figure 2(f)). The temperature is very similar to ECRH case (figure 1(e), 1127 ms), where *f(W)* is thermalized. However, this effect only lasts ~ 10 ms, and then the ion energy increases up to < 14 keV, but ions at higher energies (> 14 keV) do not reappear.

**ACKNOWLEDGEMENTS.** This work was partially funded by the Spanish "Ministerio de Economía y Competitividad" under Grant No. ENE2014-56517-R. One of the authors (M. M.) would like to thank CONACYT (Mexico) for his scholarship. This project has received funding from the European Union's Horizon 2020 research and innovation programme under grant agreement number 633053. The views and opinions expressed herein do not necessarily reflect those of the European Commission.

**REFERENCES**

[1] Zurro B, Baciero A, Tribaldos V et al., *Nucl. Fusion* ***53***, 083017 (2013).
[2] Martínez M, Zurro B, Baciero A, Jiménez-Rey D and Tribaldos V, IAEA 25th Fusion Energy Conference (FEC 2014).
[3] Jiménez-Rey D, Zurro B et al, *Rev. Sci. Instrum.* ***79***, 093511 (2008).
[4] Zurro B, Baciero A et al., *Rev. Sci. Instrum.* ***83***, 10, 10D306 (2012).
[5] Tabarés F L, et al. Stellarator News, published by Oak Ridge National Laboratory, Issue 144, August 2014.
[6] Alejaldre C, et al., *Fusion Technol*, **17**, 131 (1990).
[7] Papoular R, *Nucl. Fusion* **16**, 37 (1976).
[8] McCarthy K J et al., Proc. 21st IEEE/NPSS Symposium on Fusion Engineering 2005.
[9] McCarthy K J et al., *Rev. Sci. Instrum.* **79,** 10F321 (2008).
[10] Combs S K et al., Proceedings 2011 IEEE/NPSS 24th Symposium on Fusion Engineering, Chicago (2011).
[11] Combs S K, *Rev. Sci. Instrum.* **64**, 1679 (1993).
[12] Milora S L, Houlberg W A, Lengyel L L, and Mertens V, *Nucl. Fusion* **35**, 657 (1995).
[13] Martínez M, Zurro B, Baciero A, Jiménez-Rey D and Tribaldos V, Proc. 41st EPS conference (2014).

# Electromagnetic turbulence measurements during the L-H transition in the TJ-II stellarator

B. Liu[1,2], M. Spolaore[3], U. Losada[1], A. Alonso[1], M.A. Pedrosa[1], T. Estrada[1], B.Ph. van Milligen[1], N. Vianello[3,4], C. Hidalgo[1] and the TJ-II team[1]

[1]*Laboratorio Nacional de Fusion, CIEMAT, Madrid, Spain*

[2] *Departamento de Física, Universidad Carlos III de Madrid, Madrid, Spain*

[3]*Consorzio RFX, Corso Stati Uniti 4, 35127 Padova, Italy*

[4]*Centre de Recherche en Physique des Plasmas, EPFL, Lausanne, Switzerland*

## I. Introduction

It is well known that sheared flows can suppress turbulence and help to improve plasma confinement. Fluctuating, turbulence-generated zonal flows are a benign non-linear turbulence saturation mechanism, consistently observed in simulations and reported in several experiments. Because of their transport modulation capability zonal flows have been considered natural candidates for triggering the L-H transition. Theoretically, both electrostatic fluctuation-induced Reynolds stress and magnetic fluctuation-induced Maxwell stress can play a role in generating zonal flows [1]. Whereas measurements of Reynolds stress with multi-pin Langmuir probes have been conducted in several devices [2], only a few measurements of Maxwell stress have been reported [3]. In this work we present local measurements of magnetic properties and Maxwell stresses in the proximity of the L-H transition in the TJ-II stellarator.

## II. Experimental setup

Experiments were carried out in the TJ-II stellarator, in the pure NBI heated plasmas (toroidal magnetic field $B_T = 1\,T$, plasma major radius $\langle R \rangle = 1.5\,m$, plasma minor radius $\langle a \rangle \leq 0.22\,m$, $\iota(a)/2\pi \approx 1.6 - 1.9$). The electromagnetic properties were measured using an electromagnetic (EM) probe (also dubbed 'vorticity probe') [4][5] consisting of a Langmuir probe array and three three-axial magnetic coils, which can measure the time derivative of the three components of magnetic field. The three magnetic probes form a right angle in the radial-poloidal plane. They are separated by 3 cm in both radial and poloidal directions. The Langmuir probe array locates in the middle of two magnetic probes in the radial direction. This setup allows for simultaneous measurements of magnetic and electrostatic properties.

### III. Radial profiles of magnetic fluctuation and Maxwell stress

Radial profiles of electromagnetic properties were measured by the EM probe on a shot-to-shot basis in TJ-II. Sixteen shots of similar plasma conditions with probe measured at six different radial positions were taken into account for this analysis. Figure 1a shows the radial profile of magnetic fluctuation (rms). Magnetic fluctuation level increases significantly radially inwards. Maxwell stress was estimated by the time average of the product of radial and poloidal components of magnetic field fluctuation $\langle \tilde{B}_r \tilde{B}_\theta \rangle$. As shown in figure 1b, the Maxwell stress as well as its gradient also increase radially inwards reaching values close to $10^5 m^2/s^2$ at the innermost radial position ($\rho = 0.86$), which were the measurements so far available. When comparing with Reynolds stress previously measured in TJ-II with Langmuir probe, both the level and gradient of Maxwell is relatively low [6]. The frequency resolved Maxwell stress, shown in figure 1c, suggests that the main contribution to Maxwell stress comes from the magnetic fluctuation with frequencies below 50 kHz. This agrees with Maxwell stress measurements from the Extrap-T2R RFP [3].

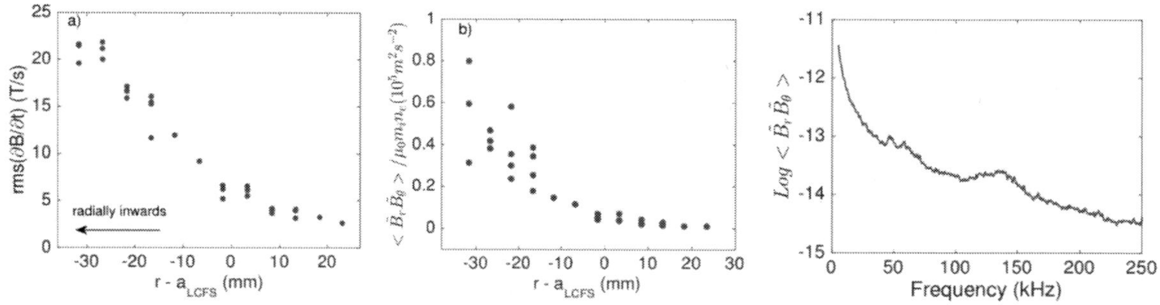


Figure 1. (a) Profile of magnetic fluctuation level, (b) Profile of Maxwell stress, (c) Frequency resolved Maxwell stress.

## IV. Electromagnetic dynamics during L-H transition

The time evolution of plasma parameters and edge fluctuations were monitored by the EM probe in the proximity of the L-H transition for the shot #37638, showing a decrease in $H_\alpha$ and the level of electrostatic and magnetic fluctuations. We also notice that in the middle of the H-mode there is a burst (at around $1027\, ms$) which shows up in almost all measured signals, including $I_s$ and parallel current density ($j_\parallel$). This indicates that the burst may be related to some filamentary electromagnetic structure. Further investigation of this ELM-like event is in progress. The reduction of electrostatic and magnetic fluctuation in a wide frequency range can be seen in figure 3.

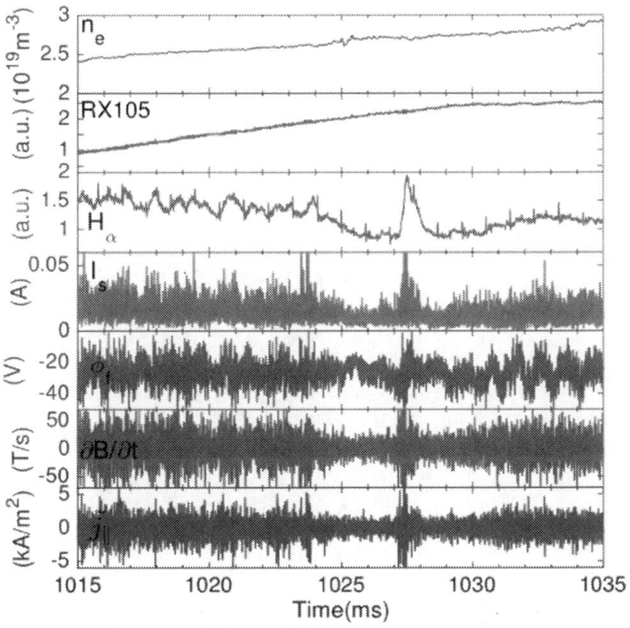

Figure 2. Time evolution of plasma parameters (#37638).

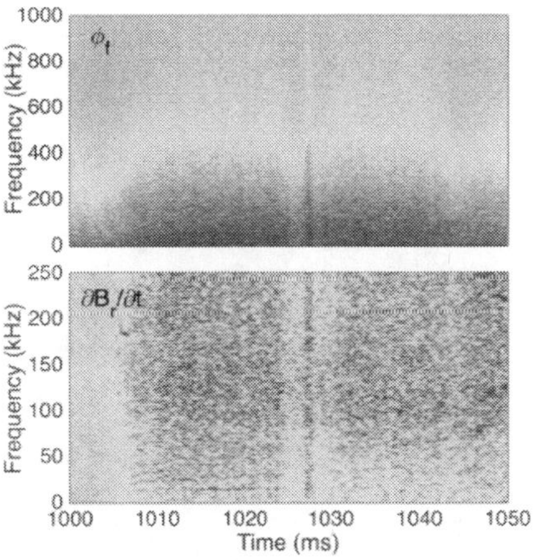

Figure 3. The spectrogram of floating potential and magnetic fluctuation. The sampling rate of floating potential measurement is 2 MHz and magnetic fluctuation measurement is 500 kHz. This results in different frequency scales.

To explore the variation of local electrostatic turbulence structure during L-H transition, we measured the radial correlation length ($L_r$) and poloidal correlation length ($L_\theta$) with two radial and two poloidal Langmuir probe tips using standard two-point correlation technique: the correlation length $L_c$ was estimated from the broadening of fluctuation power spectrum in the wavenumber axis in the wavenumber-frequency plane. Figure 4 shows the time evolution of the local turbulence structure. $L_r$ decreases as approaching the L-H transition. When L-H transition taking place, $L_\theta$ appears to increase.

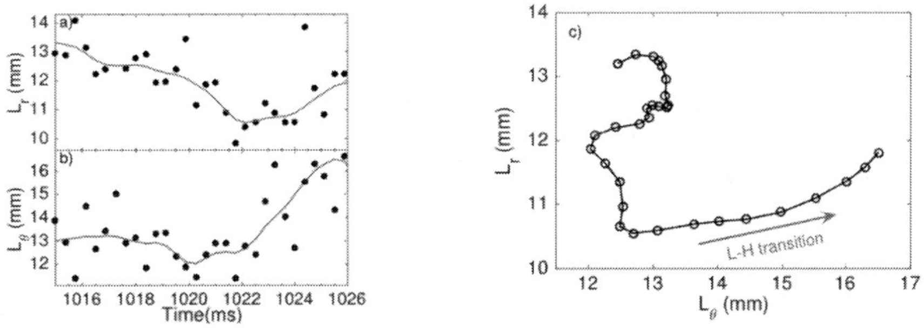

Figure 4. Time evolution of a) radial correlation length ($L_r$) and b) poloidal correlation length ($L_\theta$ ), and c) the portrait $L_r$ and $L_\theta$.

Figure 5 shows the time evolution of Maxwell stress measured by three magnetic probes on the EM probe. Maxwell stress measured by probe 1 is significant larger than by probe 2 and 3. This suggests that a gradient in Maxwell stress is formed at the probe location. Considering the distance between two probe is 3 cm, we estimated the gradient of Maxwell stress can reach the order of $10^7 m/s^2$ at around $\rho = 0.9$. This is about one order of magnitude lower than the gradient of Reynolds stress (~$10^8 m/s^2$) previously measured in TJ-II [6]. We note that this position has not yet reached the transport barrier region which is at $\rho < 0.85$ in TJ-II [7]. It may be interesting to have measurements of Maxwell stress further inside the plasmas in TJ-II in the future. The cross-power spectrogram in the right panel of figure 5 shows the Maxwell stress is mainly due to the fluctuation of magnetic field at low frequencies.

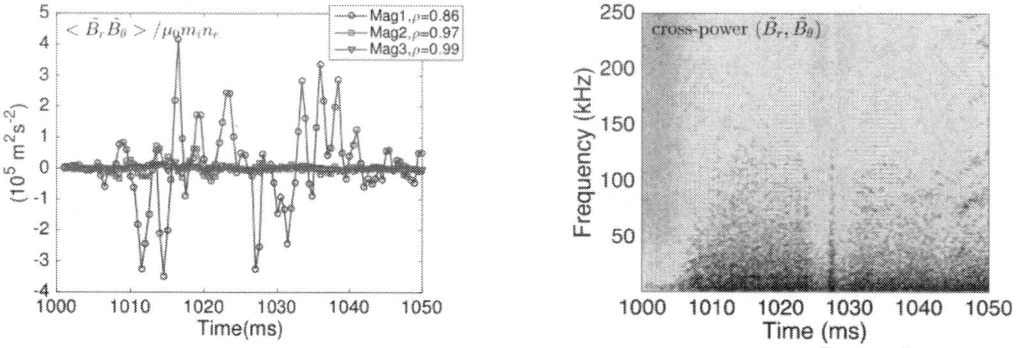


Figure 5. Time evolution of Maxwell stresses (left), and the cross-power spectrogram $\tilde{B}_r$ and $\tilde{B}_\theta$ (right).

## IV. Conclusion

We have conducted electromagnetic measurements of plasma turbulence in the TJ-II stellarator with the electromagnetic probe up to the position of $\rho = 0.86$. We found that magnetic fluctuation and Maxwell stress increase radially inwards; in the proximity of L-H transition, a significant gradient of Maxwell stress was observed at the plasma edge. The value is about one order of magnitude lower than previously measured Reynolds stress [6], however, considering the measurement position was outside the transport barrier [7], it would be interesting to measure the Maxwell stress further inside the plasmas in TJ-II in the future.

## Acknowledgements

This work was supported by the European Commission and carried out within the framework of the Erasmus Mundus International Doctoral College in Fusion Science and Engineering (FUSION-DC). This work has been carried out within the framework of the EUROfusion Consortium and has received funding from the Euratom research and training programme 2014-2018 under grant agreement No 633053. The views and opinions expressed herein do not necessarily reflect those of the European Commission.

## References

[1] P. H. Diamond, S. I. Itoh, K. Itoh, and T. S. Hahm, Plasma Phys. Control. Fusion **47**, R35 (2005).

[2] C. Hidalgo, C. Silva, M. A. Pedrosa et al., Phys. Rev. Lett. **83**, 2203 (1999).

[3] N. Vianello, E. Spada, V. Antoni et al., Phys. Rev. Lett. **94**, (2005).

[4] D. Carralero, Electromagnetic Instability Studies in Fusion Plasmas Edge, PhD thesis (2012).

[5] M. Spolaore, N. Vianello, I. Furno *et al.*, Phys. Plasmas **22**, 012310 (2015).

[6] J. A. Alonso, J. L. Velasco, J. Arévalo *et al.* , Plasma Phys. Control. Fusion **55**, 014001 (2013).

[7] T. Happel, T. Estrada, E. Blanco *et al.*, Phys. Plasmas **18**, 1 (2011).

# The influence of Alfvén waves on fast ion confinement in TJ-II plasmas

J.M. Fontdecaba, F. Castejón, E. Ascasíbar, A. Cappa, T. Estrada, M. Liniers, A. Bustos[1], and TJ-II Team

*Laboratorio Nacional de Fusión, Ciemat, 28040 Madrid, Spain*

*[1]Departamento de Física. Universidad Carlos III, 28911 Leganés (Madrid), Spain*

## 1. Introduction

Fast ion confinement is a key point for fusion devices since alpha particles produced in the fusion reactions must be confined long enough to deposit their energy in the plasma bulk in order to keep the fusion reactions going on[1]. Moreover, the fast ions coming from Neutral Beam Injection (NBI)[2] or Ion Cyclotron Heating must be also well confined in order to increase the heating and current drive efficiencies. Fast ion transport must also be studied in order to find out what are the main escaping points of the ions, since they could damage the vacuum vessel of the device[3]. Fast ions can also develop instabilities[4] that can enhance the transport.

The instabilities caused by fast ions have been previously studied in TJ-II[5], but no direct observation of the dependence of the induced ion transport on the ion energy has been reported so far. In this work, we study the fast ion transport induced by Alfvén wave instabilities, which are themselves driven by fast ions coming from NBI. Several types of Alfvén Eigenmodes (AEs) at different frequencies have been observed in TJ-II and we investigate their effect on the NBI ions confinement properties.

## 2. Experimental set-up

The stellarator TJ-II heliac flexible is a four period medium size device ($R_0$ = 1.5 m; a $\leq$ 0.2 m; $B_0$ = 1 T)[6]. The heating systems are electron cyclotron resonance heating (ECRH) and neutral beam injection (NBI). The ECRH system has two gyrotrons (300 kW each) tuned at the second harmonic of the X-mode (53.2 GHz). The NBI system consists of two injectors of a maximal nominal power of 700 kW each with ion energy of 34 keV in tangential positions (co and counter injectors).

To scan the energy and population of fast ions in the plasma there is a Compact Neutral Particle Analyzer (CNPA)[7]. It can distinguish 16 different energies between 1 and 40 keV with temporal resolution of 1 ms. The line of sight is directed towards the ions coming from the co-injector but it only counts ions coming after a re-neutralization in the plasma because the path between the injector and the CNPA is twisted, so no lost ions from the injector can contaminate the measurements of this diagnostic.

In TJ-II there are three Mirnov coil sets distributed at different poloidal positions of the vacuum vessel. During this work we have used the data of one coil that is located in a set of 12 coils which measures the three components of the magnetic field at four positions located along a vertical line over the plasma column. The sampling frequency of this array of coils is 1 MHz.

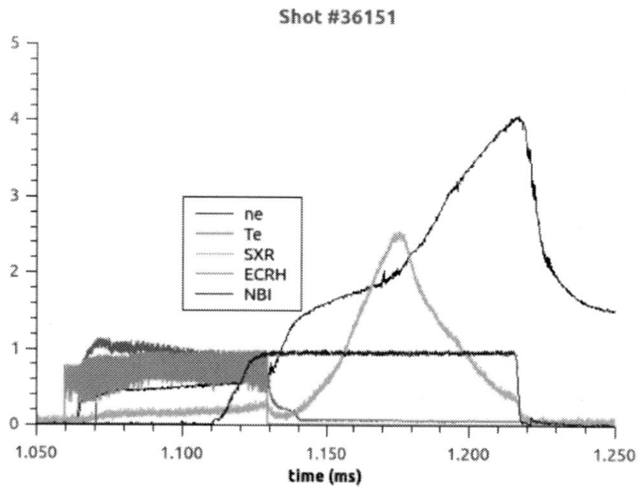


**Figure1.** Typical NBI heated discharge. It shows the line density ($n_e$), electron temperature from a central chord measured by ECE ($T_e$), soft x-rays (SXR) and the heating scheme, ECRH and NBI heating. All the values are arbitrary units except line density in $10^{19}$ $m^{-3}$.

In this case the deuterium plasma is started up with the two gyrotrons with 250 kW each focused slightly off-axis. Once the plasma is created the co-injector is fired and the gyrotrons switched off, so the plasma at the second part of the discharge is heated only by NBI, the bulk plasma being deuterium heated by hydrogen beams. During this series of experiments the voltage and current of the injector were scanned. A design of the heating scheme and main signals of a typical discharge for this experiment are plotted in figure 1.

## 3. Observations

High energy ions are collected by the CNPA and the steady state flux of the neutrals escaping from the plasma after a charge-exchange interaction is measured. We average the spectra in time windows of 10 ms in order to have enough statistics. In the experiments we measure the spectra at two different times of the discharge in order to compare the ion confinement in different conditions. An increase of the steady state flux reflects an increase of the fast ion density and, therefore, an improvement of their confinement.

After the onset of the NBI heating a mode appears at around 150 kHz. At the same time there is a drop of the signal in the high energy channels of the CNPA. This behavior is observed in different configurations and only above a threshold injected NBI power. Figure 2 left shows the signal received by the CNPA at two times of the discharge. Figure 2 at right shows the logarithm of the mean power for a Mirnov coil at the same times. As can be seen there is a clear difference on the signal received by the CNPA, this indicates a difference on

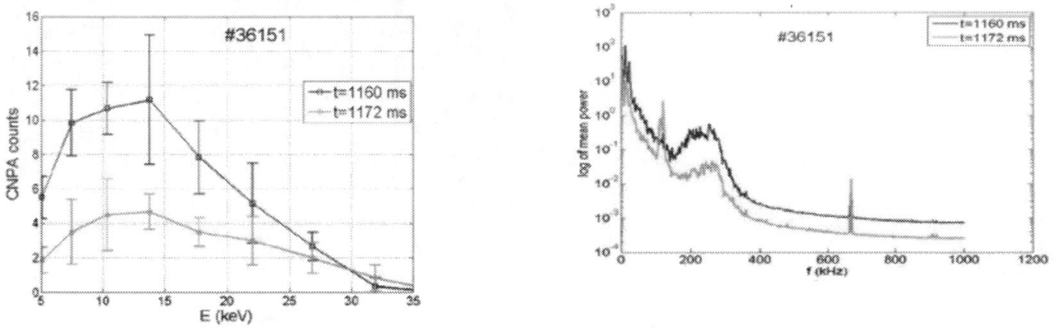


**Figure 2.** CNPA data (left) and log of power spectrum for a Mirnov coil signal in the same discharge at two different times, before and after the trigger of an Alfvén mode.

the fast ion population inside the plasma. The main difference between the two times represented here is in the power spectra of the Mirnov coil, the decrease of the signal in the CNPA coincides with the on-set of an Alfvén mode at about 150 kHz. So the onset of the Alfvén mode leads to a decrease of the fast ion population in the plasma.

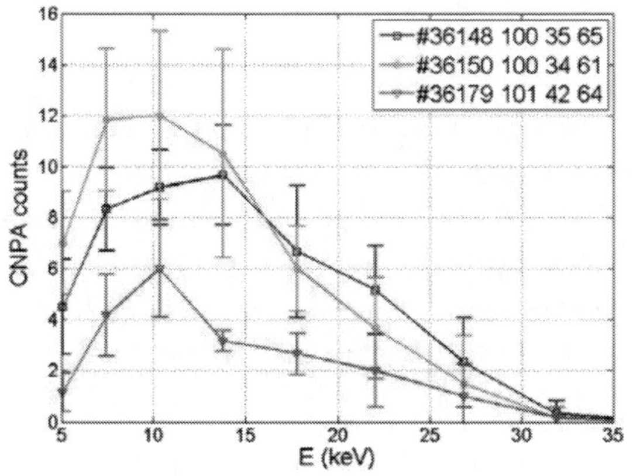


**Figure 3.** Counts in the CNPA for three different configurations with similar plasma parameters.

The change of configuration has indeed effect on the properties of the Alfvén modes. As expected, differences are observed in the CNPA spectra, showing an influence on the energy distribution of fast ions that arrive to the CNPA detectors. Figure 3 shows the results of the measurements for three different magnetic configurations with similar plasma conditions and same heating power. As can be seen on that figure there is a clear difference on the decrease of the counts on the CNPA depending on the magnetic configuration. The main difference between the configurations studied is the iota profile, being the rest of the plasma parameters very similar, so there is an influence of the iota profiles on the fast ion confinement.

On other set of experiments during the NBI phase of the discharge the gyrotrons were switched on again[8]. The heating by microwaves results on the mitigation of the Alfvén modes. Simultaneously the flux of fast ions measured by the CNPA increased with the mitigation of the Alfvén modes as can be seen on figure 4.

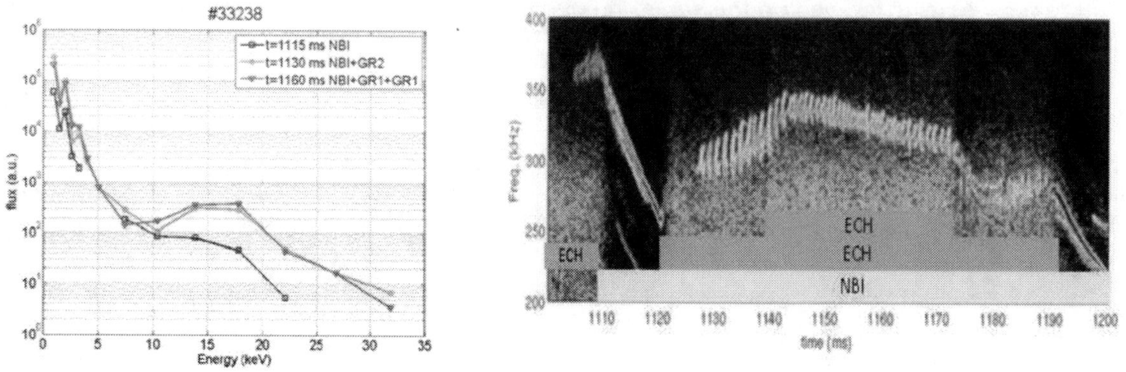


**Figure 4.** Energy spectra measured by the CNPA (left) and log of power spectra of the Mirnov coil for the same discharge. It is clearly seen the difference on the energy on the ions depending on the excited mode.

## 4. Conclusions

Different experiments have shown a decrease of the signal collected by the compact neutral particle analyzer when Alfvén modes are triggered. Also an increase of the flux of the fast ions is observed when the Alfvén modes are mitigated when the ECRH is turned on in a pure NBI plasma. These observations indicate that the Alfvén waves interact with the fast ions in the plasma destabilizing them with the consequence of loss of confinement.

## Acknowledgements

This work has been carried out within the framework of the EUROfusion Consortium and has received funding from the Euratom research and training programme 2014-2018 under grant agreement No 633053. The views and opinions expressed herein do not necessarily reflect those of the European Commission. This work has been partially funded by the Spanish "Ministerio de Economía y Competitividad" under contracts no ENE2013-48109-P and ENE2013-48679-R.

## Bibliography

[1] A. Fasoli et al. Nucl. Fusion **47** (2007) S264-S284
[2] T.H. Stix Plasma Physics **14** (1975) 367-384
[3] R.B. White et al. Phys Plasmas **2** (1995) 2871
[4] C. Z. Cheng and M. S. Chance, Physics of Fluids **29** (1986) 3695
[5] R. Jiménez-Gómez et al. Nucl. Fusion **51** (2011) 033001
[6] J. Sánchez et al. Nucl. Fusion **53** (2013) 104116
[7] F.V. Chernyshev et al. Instr. And Exp. Techniques **47** (2004) 214
[8] K. Nagaoka et al. Nucl. Fusion **53** (2013) 072004

# NBI related gas sources in the TJ-II stellarator

Macarena Liniers, F. Martín-Díaz, J.A. Sebastián, G. Wolfers, R. Carrasco, B. Rojo,
F. Miguel, E. Sánchez, F. Martín†, E. de la Cal, J.M. Fontdecaba, E. Ascasíbar, J. Guasp,
and the TJ-II team

*Laboratorio Nacional de Fusión, Avda. Complutense 40, 28040 Madrid, Spain*

**Introduction** Neutral Beams contribute to plasma fuelling as fast ions confined in the magnetic field, but also through two additional mechanisms: a fraction of the hydrogen gas injected in the NBI plasma source is injected in the machine through the NBI duct alongside the beam. And the fast particle impacts on the vacuum chamber wall give rise to re-emission of the gas adsorbed in the wall between plasma pulses. These two gas sources can be treated as a kind of gas puffing.

In TJ-II, where the NBI system consists of two injectors of 34 keV energy and 1,4 MW total power [1], the contribution of the re-emission is important considering that 11% of the calorimeter power is intercepted in the beam entrance area [2]. In order to characterize these gas sources, fast pressure gauges have been symmetrically located at the four machine periods. It has been observed that the pressure distribution in the machine during the plasma pulse (< 200 ms) presents a high degree of inhomogeneity. The gas conductance between two machine periods has been experimentally determined using pure gas pulses, and the "local" character of the gas sources has been thus established.

The conductance values have been used in the analysis of pressure signals during beam pulses to obtain quantitative information on the re-emitted gas flows.

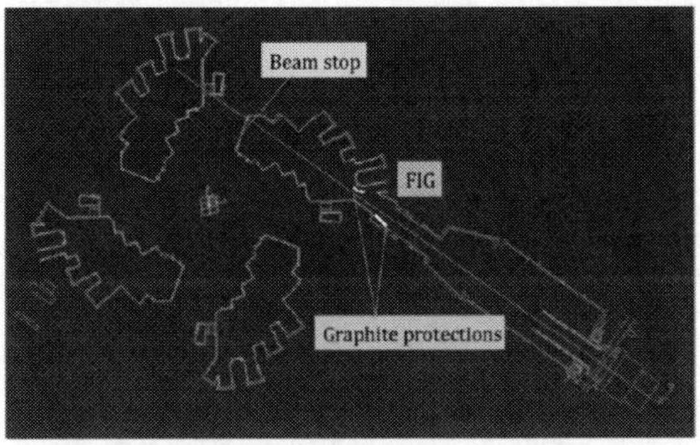


**Figure 1**: CATIA drawing with a section of TJ-II vacuum vessel and NBI#1 through a horizontal plane containing the beam axis

**Gas dynamics in TJ-II: Conductance between periods** TJ-II is a four period stellarator of the heliac type, with major radius R=1.5 m. The high helicity of TJ-II can be best appreciated through the comparison between the plasma minor radius (0.2 m) and the toroidal field coils swing radius (0.28 m). Figure 1 shows a CATIA drawing of the intersection between the vacuum vessel outer wall and a horizontal plane containing the beam axis. In this view, the vacuum vessel has the appearance of four disconnected volumes. Although this discontinuity is only apparent, it does illustrate the effect of the geometry on the propagation of gas particles through the machine. The overall helical geometry, combined with the vacuum vessel design (welded plane wall sectors) determines the gas conductance of the machine.

To obtain a realistic estimate of the conductance between two periods, the pressure traces of four symmetrically located Fast Ion Gauges [3] have been analysed for gas pulses injected in the machine through the NBI ducts. The experimental layout can be seen in Figure 2. Two Penning gauges are installed near the beam ports (D8 and C1) and two Bayard-Alpert gauges are located at sectors D1 and B2. The two Fast Ion Gauges inside the beam boxes (BB1 and BB2) are used in the evaluation of the injected gas flows. The hydrogen gas is injected using the Ion Source piezo-electric valves in both Beam Injectors.

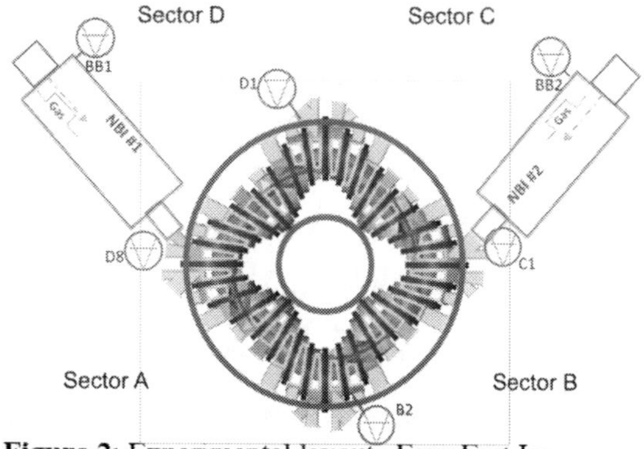

**Figure 2**: Experimental layout. Four Fast Ion Gauges are installed on each of the TJ-II four periods.

When a square-wave gas pulse is injected in the machine through the duct of NBI#1, the gas flows to the neighbouring periods and is pumped away by TJ-II turbomolecular pumps (one pump per period). By adjusting the injected flow, a quasi-steady state can be obtained as shown in Figure 3, where the pressure traces of the five involved Fast Ion gauges are shown.

In Figure 4 a simplified model of the machine from the point of view of gas flow is presented. The machine is modelled as four equal volumes ($V_P$) connected through conductances $C_{ij}$. The flow between volume $i$ and volume $j$ can be written as:

$$Q_{ij} = C_{ij}(P_j - P_i) \qquad (1)$$

Where $P_{i,j}$ take on the pressure readings of the four pressure gauges (D8, D1, C1, B2)

The injected gas flow through any of the NBI ducts is:

$$Q_{0i} = C_{Di}(P_{BBi} - P_i)$$

Each of the four volumes has a turbomolecular pump of pumping speed $S_i$. The gas flow through pump $i$ is written:

$$Q_{Si} = S_i P_i \qquad (2)$$

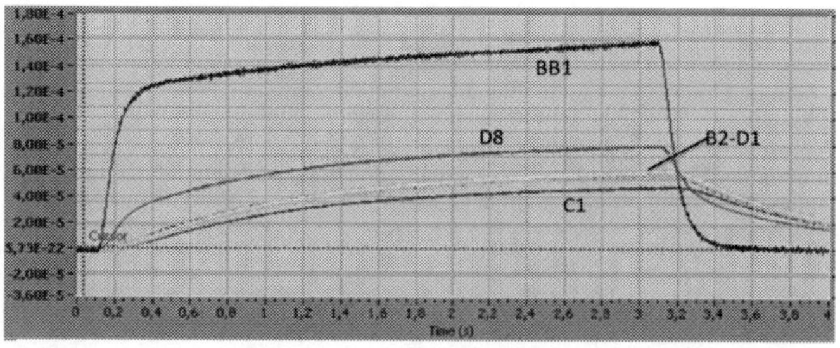

**Figure 3**: Pressure signals from the 5 Fast Ion Gauges for a 3 s square-wave Gas pulse from NBI#1. Pressure unit is mbar.

The equation describing the variation in the amount of gas in volume 1 when a beam is injected through duct#1 is:

$$\frac{d(P_{D8}V_P)}{dt} = Q_D + C_{D1}(P_{BB1} - P_{D8}) - C_{12}(P_{D8} - P_{D1}) - C_{14}(P_{D8} - P_{B2}) - S_1 P_{D8} \qquad (3)$$

Where $Q_D$ is the gas flow re-emitted from the wall.

For a gas pulse, in steady state, $Q_D$ =0 and , $d(P_{D8}V_P)/dt \approx 0$ , so the four flow-balance equations can be written:

$$\begin{aligned} Q_{01} &\equiv C_{D1}(P_{BB1} - P_{D8}) = Q_{12} + Q_{14} + Q_{S1} \\ Q_{12} &= Q_{23} + Q_{S2} \\ Q_{23} &+ Q_{43} - Q_{S3} = 0 \\ Q_{14} &= Q_{43} + Q_{S4} \end{aligned} \tag{4}$$

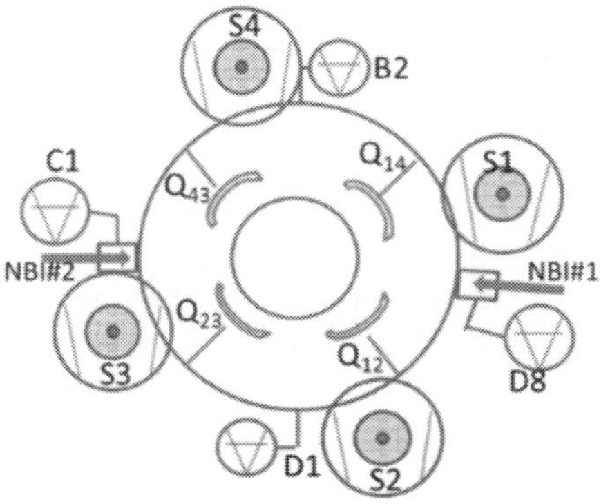

Figure 4: simplified model of TJ-II gas flows as described by eqs. 3 and 4

A fifth equation describes the global flow balance (all the injected gas is pumped away by the turbomolecular pumps):

$$Q_{0i} = Q_{S1} + Q_{S2} + Q_{S3} + Q_{S4} \tag{5}$$

A similar set of equations can be written for the steady state flows established when gas is injected through duct#2. A homogeneous set of linear equations is obtained, with 8 independent equations, and 10 unknown variables: the 4 conductances between periods $C_{ij}$ (to take into account the lack of exact symmetry in the gauges locations), the 2 conductances of the ducts $C_{Di}$ (to allow for small differences in internal pumping inside the beam boxes) and 4 pumping speeds $S_i$ (since some divergence from nominal values is frequently observed). Limits can be readily set for the 10 variables, based on simple estimates (conductances) or nominal values (pumping speeds).

In order to find the "optimal" solution a Monte Carlo approach has been adopted: a stochastic search is carried out in the variables space within their allowed limits, and for each set of 10 random numbers an evaluation of the "unbalance" (UB) of the equations is obtained, comparing the terms containing the $S_i$ with those containing the conductances and establishing the percentual relative differences. Along the optimization procedure, priority is given to the Global Unbalance (GUB) corresponding to the total flow balance (eq. 5). After a great number of repetitions (100 million) an optimal set of values is chosen among those with minimum GUB (<30%) and minimum total UB.

The optimal set of values is presented in table I.

**Estimate of Re-emitted Gas flow** Figure 5 shows the pressure traces for a NBI#1 beam pulse of 100 ms in the absence of magnetic field. The pressure distribution along the torus is highly inhomogeneous for the pulse duration. Analysis of the duct pressure signal (D8) using equation 3 yields the evolution of the re-emitted gas. Table II presents the injected and re- emitted flows as well as the flows between machine periods and to the turbomolecular pump.

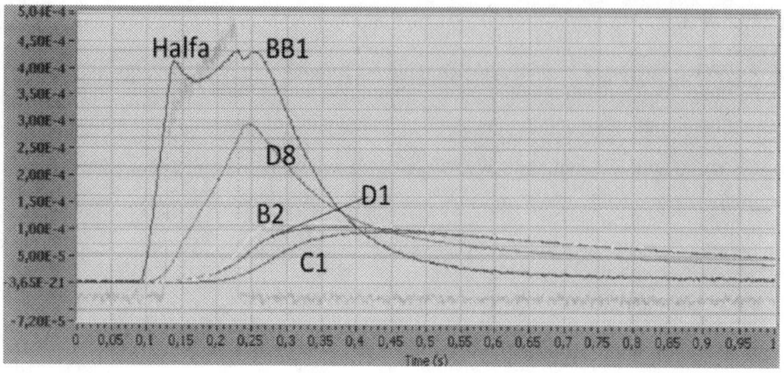


**Figure 5:** Pressure traces of the 5 Fast Ion Gauges during NBI#1 beam pulse (100 ms). Pressure unit is mbar.

**Table I**: optimal set of values for the conductances and pumping speeds (all variables in $l.s^{-1}$)

| | |
|---|---|
| $C_{12}$ | 5026.0 |
| $C_{23}$ | 4900.0 |
| $C_{34}$ | 4901.0 |
| $C_{41}$ | 5128.0 |
| S1 | 1208.0 |
| S2 | 1051.0 |
| S3 | 1500.0 |
| S4 | 1086.0 |
| $C_{D1}$ | 4082.0 |
| $C_{D2}$ | 4071.0 |
| GUB | 28 % |

The magnitude of the re-emitted flow at the end of the pulse is at least a factor 5 that of the injected flow or the propagating flows, therefore establishing the "local" character of the observed pressure rise in the duct region.

**Table II**: The re-emitted flow $Q_D$ as obtained from eq. (3). All quantities in mbar.l.s-1

| t(s) | $D(P_{D8}V_P)/Dt$ | $C_{D1}(P_{BB1}-P_{D8})$ | $C_{12}(P_{D8}-P_{D1})$ | $C_{14}(P_{D8}-P_{B2})$ | $S1*P_{D8}$ | $Q_D$ |
|---|---|---|---|---|---|---|
| 0.13 | 1.95 | 1.32 | 0.12 | 0.12 | 0.02 | **0.89** |
| 0.18 | 2.75 | 0.98 | 0.65 | 0.67 | 0.17 | **3.26** |
| 0.23 | 3.75 | 0.67 | 1.15 | 1.08 | 0.33 | **5.64** |

**Conclusions** An estimate of the gas conductance in TJ-II has been experimentally obtained by studying the gas flows during gas pulses injected from the NBI systems. The obtained value, around 5000 l.s-1, has been used to estimate the gas flow re-emitted from the wall during a beam pulse. The flow of re-emitted gas in the beam entrance area is much larger than the flows to neighbouring periods, which leads to the observed inhomogeneity of the pressure distribution during beam pulses.

**Acknowledgements**

This work has been carried out within the framework of the EUROfusion Consortium and has received funding from the Euratom research and training programme 2014-2018 under grant agreement No 633053. The views and opinions expressed herein do not necessarily reflect those of the European Commission. This work has been partially funded by the Spanish "Ministerio de Economía y Competitividad" under contract no ENE2013-48679-R

**References**

[1] M. Liniers et al., Fusion Technology 1998, **1**, 307-310 (1998)
[2] C. Fuentes et al., Fusion Engineering and Design **82** (2007) 923–932
[3] M. Liniers, et al., Fusion Engineering and Design **88** (2013) 960–963

# Simulating Burning Plasma Operation in ITER

S.H. Kim[1], T.A. Casper[1,2], D.J. Campbell[1], J.A. Snipes[1],
R. Bulmer[3], L.L. LoDestro[3], W.H. Meyer[3], and L.D. Pearlstein[3]

[1]*ITER Organization, Route de Vinon sur Verdon - CS 90046,13067 St Paul-Lez-Durance Cedex, France*
[2]*Now at 1166 Bordeaux St Pleasanton, CA, 94566 USA*
[3]*Lawrence Livermore National Laboratory, P.O. Box 808, Livermore, CA, 94550 USA*

**Introduction** ITER will demonstrate the feasibility of burning plasma operation with a high ratio of fusion power gain Q~10 by operating D-T plasmas in the type-I ELMy H-mode regime. In this work, entry to burn, flat-top burning plasma operation and exit from burn have been investigated using CORSICA [1]. The CORISCA code self-consistently calculates the evolution of 2D free-boundary equilibria with both currents flowing in the surrounding conducting structures and plasma profiles computed with transport and sources, such as 1MeV deuterium neutral beams (NB), 53MHz ion cyclotron (IC), and 170GHz electron cyclotron (EC) heating and current drive. A total of 73 MW of auxiliary heating power is assumed to be available by combining 33MW NB, 20MW EC and 20MW IC. The waveforms of auxiliary heating powers are optimized for the plasma performance and to obtain high fusion power multiplication factor, Q. The effective charge ($Z_{eff}$) initially set to about 4.0 at the beginning of the ramp-up phase is reduced to 1.4~1.7 as the plasma density increases, and then increased to a higher value during the ramp-down as the plasma density decreases. Either Ar or W is used as a high-Z impurity species together with a low-Z impurity, Be. Sawteeth are assumed to be triggered when $q_0$<0.97 and the plasma temperature and current density profiles are modified within the inversion radius.

**Modelling L-H and H-L transitions** Low to high (L-H) and high to low (H-L) confinement transitions are triggered when the net power crossing the plasma separatrix ($P_{tot}$) becomes higher or lower than the L-H transition threshold power ($P_{th}$) obtained from the Martin scaling law [2]. During the L-H/H-L transitions, the plasma density has been increased/reduced linearly with a time constant, tauNe0, and the density profile has been also changed from L-mode (50% $n_{GW}$, parabolic shape) to H-mode (80~90% $n_{GW}$, flat at the core, hyperbolic tangent at the pedestal) with a faster shape transition time constant defined as tauNe0/5 [3]. To model the H-mode pedestal, firstly the hyperbolic tangent density profile at the pedestal has been adjusted with the pedestal top location estimated using the parameterized EPED1 model [4]. Then, the heat diffusivities computed using the Coppi-Tang transport model [5] have been feedback controlled to reproduce the estimated H-mode pedestal pressure. In this way, dithering between L-mode and H-mode is also modelled when the threshold and net powers are frequently crossing over each other.

**Varying the density transition time-scale for L-H and H-L transitions** The influence of the assumed density transition time-scale on L-H and H-L transitions has been examined, assuming Ar and Be as impurity species. When the plasma is in a limited configuration, the ratio of Ar to Be ($n_{Ar}/n_{Be}$) is set to 0.001, and then it is increased to 0.005 when the plasma becomes diverted. When the plasma has completed the L-H transition (t~70s), $n_{Ar}/n_{Be}$ is increased further to 0.075, as if the Ar impurity is injected into the plasma to control the heat load to the diverter. During the flat-top phase, $n_{Be}/n_e$ and $n_{Ar}/n_e$ are set to about 1.8% and 0.14%, consistently with the assumption of $Z_{eff}$(flat-top)~1.7. Three cases with tauNe0=10s, 20s and 30s are compared, and this shows that the increase in the alpha particle heating power becomes slower as the density transition time-scale becomes smaller. However, the alpha particle heating powers become similar after the transient ramps, as the plasma operating conditions for the flat-top phase are not different. It appears that a slow density transition is favorable for reducing the total heating power around t=70s (see figure 1). It is also observed that if the density transition time-scale is reduced to a small value (~5s), $P_{th}$ can be increased faster than the buildup of the alpha particle heating power. This introduces dithering of L-H/H-L transitions and forces the plasma to stay in an L-mode state (figure 1). It appears that the density transition time-scale is an important factor for controlling plasma dynamics for the entry to burn and burn control around the start of flat-top. Another comparison on H-L transitions shows that there is no clear dependency of tauNe0 on the burn exit as the decrease in the alpha particle heating power is much faster than the evolution of $P_{th}$. However, this requires a further study on the potential hysteresis characteristics for H-L transition in ITER. Note that active feedback control of the auxiliary heating power and/or fuel mix, which is useful for reducing the peak total power at the entry to burn, has not yet been attempted in this work.

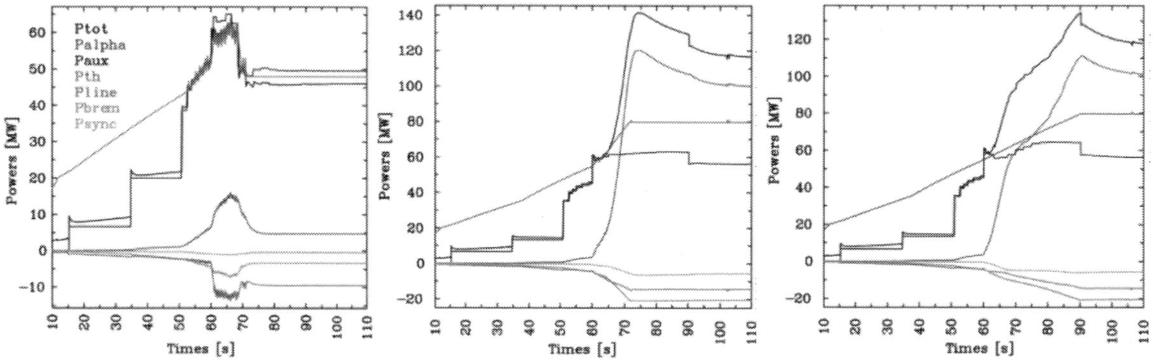


*Figure 1. Evolution of powers during the current ramp-up and at the entry to burn. The density transition time-scale, tauNe0, has been varied from 5s (left) to 10s (centre) and 30s (right).*

**Varying the flat-top plasma density** The flat-top plasma density plays an important role in operating the plasma with high performance, and also in achieving a high fusion power multiplication factor, Q. In this work, several cases with different flat-top densities have been compared. In the cases with Be and Ar impurities, $n_{Be}/n_e$~$1.79\text{x}10^{-2}$ and $n_{Ar}/n_e$~$1.35\text{x}10^{-3}$ were used with an assumption of $Z_{eff}$ (flat-top)~1.7. In other cases with Be and W impurities, $n_{Be}/n_e$~$2.62\text{x}10^{-2}$ and $n_{Ar}/n_e$~$2.62\text{x}10^{-6}$ were used with an assumption of $Z_{eff}$ (flat-top)~1.4. In both cases with different impurity species, similar changes in the plasma parameters have been observed when the volume averaged plasma density during the flat-top phase is varied from $8.82\text{x}10^{19}/\text{m}^3$ to $1.05\text{x}10^{20}/\text{m}^3$. As the flat-top density increases, the radiation powers increases and the pedestal temperature is reduced, whereas the estimated L-H transition threshold power is increased. Therefore, the margin for maintaining the plasma above the threshold power, $|P_{tot}-P_{th}|/P_{th}$, is reduced as the flat-top density increases. Another important observation made in this comparison is that the fusion power multiplication factor is not linearly increased with the flat-top density. The fusion power multiplication factor has increased up to about 10 in the Be/W impurity case at $<n_e>\sim 9.37\times10^{19}/m^3$, however it is slightly reduced down to 9.6 at $<n_e>\sim 1.05\times10^{20}/m^3$. This has happened mainly due to the changes in the pedestal temperature and radiation powers.

**Varying the impurity concentration – Be/W cases** In this study, W is assumed as a high Z impurity while Be is assumed as a low Z impurity. The impurity ion profiles are assumed to be proportional to the ion density profile. The W concentration ($n_W/n_e$) has been varied from $2.62\times10^{-7}$ ($Z_{eff}\sim 1.4$) to $5.24\times10^{-5}$ ($Z_{eff}\sim 1.67$), while the assumed Be concentration ($n_{Be}/n_e\sim 2.62\times10^{-2}$) and the density transition time-scale (tauNe0=20s) are not changed. Three cases are compared as shown in figure 3. The line radiation power is significantly increased when the W concentration is increased. The plasma stayed in L-mode when the assumed W concentration is about $5.24\times10^{-5}$. A similar study performed with a fixed $Z_{eff}$ (flat-top)~1.7 (where $n_{Be}/n_e$ is modified as $n_W/n_e$ varies) has also shown that the marginal W impurity concentration for the entry to burn at 15MA ITER baseline operation is about $3.0\times10^{-5}$.

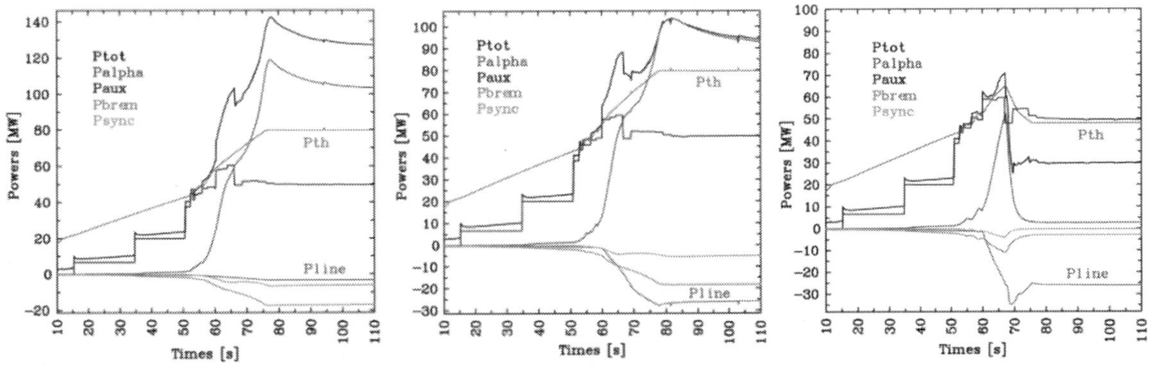


*Figure 2. Evolution of powers during the current ramp-up and entry to burn. The W impurity concentration, $n_W/n_e$, and $Z_{eff}$ have been varied from [$2.62\times10^{-7}$, 1.4] (left) to [$2.62\times10^{-5}$, 1.53] (centre) and [$5.24\times10^{-5}$, 1.67] (right). The Be impurity concentration ($n_{Be}/n_e$) of $2.62\times10^{-2}$ is unchanged.*

**An optimized scenario for 15MA ITER baseline operation** An optimized scenario for the 15MA baseline ITER operation has been developed including all the improved modelling aspects for the entry to burn, flat-top burning plasma operation and exit from burn (see figure 3). The plasma, initially in an inboard limited configuration, is assumed to quickly grow and to be diverted around t=12s and the plasma current is ramped up to 15MA at t=60s. The current flat-top is assumed to end at t=500s, although the flat-top duration can be further extended within the coil current, force and field limits. The plasma current is then ramped down relatively slowly compared to the ramp-up phase, to avoid instabilities associated with the high plasma internal inductance. During the ramp-down phase, the plasma stays in a diverted configuration while it is slowly moving downward. The application of auxiliary heating powers during the ramp-up and ramp-down has been optimized to trigger L-H and H-L transitions within the operation constraints, such as the coil limits, NB shine-through density limit and IC power coupling. The fusion power multiplication factor (Q) of 9.5 and the plasma confinement enhancement factor ($H_{98}(y,2)$) of 1.03 have been achieved with about 50MW of auxiliary heating power during the flat-top. The volume averaged electron density and impurity concentrations ($n_{Be}/n_e$ and $n_{Ar}/n_e$) were $9.37\times10^{19}/m^3$, 1.8% and 0.14%, respectively.

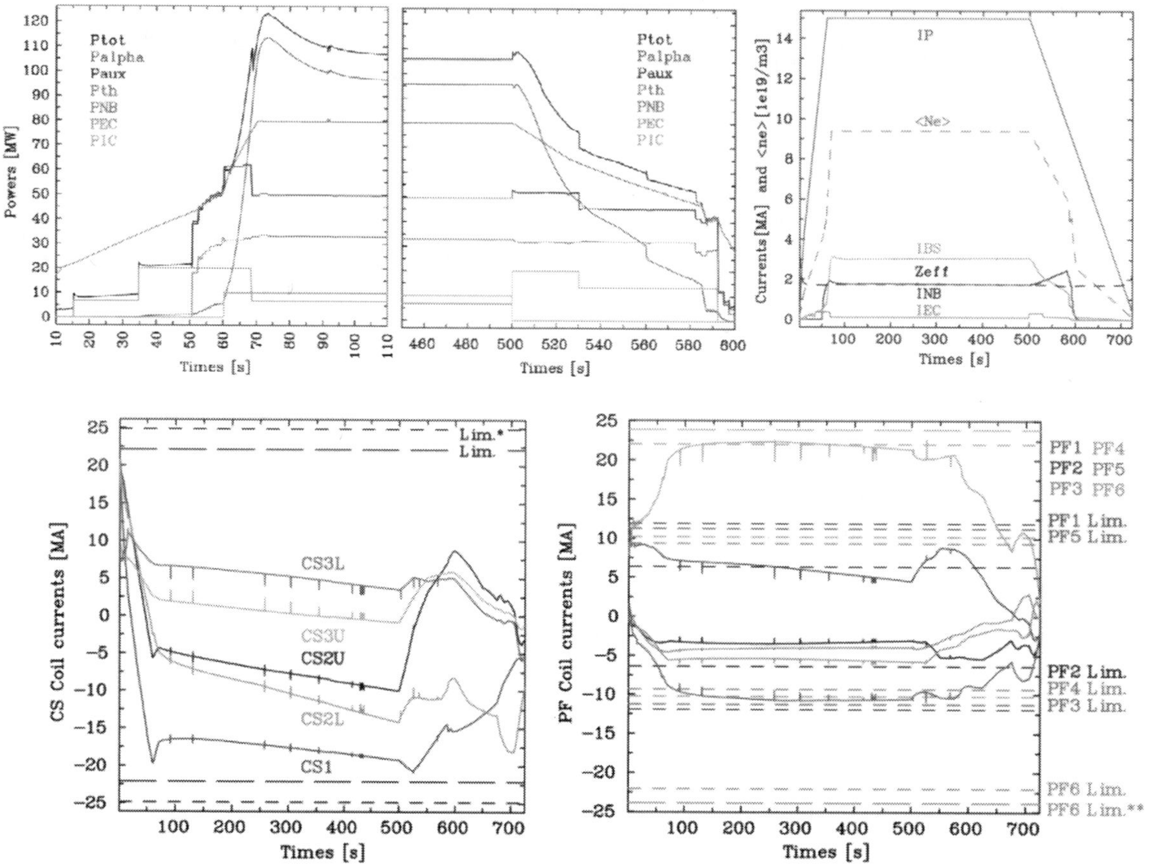


*Figure 3. Evolution of powers (top left), currents, volume-averaged density and $Z_{eff}$ (top right), CS coil currents (bottom left), and PF coil currents (bottom right). The coil current limits are indicated as dashed lines (* and ** represent the extended current limits at reduced applied B-field and/or coil sub-cooling conditions)*

**Summary and Conclusions** In this work, the entry to burn, flat-top burning plasma operation and exit from burn have been investigated with improved modelling capabilities. Feasible high Q burning plasma operation scenarios have been developed for the ITER baseline operation within its operation constraints, such as the coil limits and heat loads. It has been observed that the W concentration needs to be maintained at a low value ($n_W/n_e <$ ~$1.0\times10^{-5}$) to avoid the radiative collapse and early termination of the discharge. This work will be the basis for further optimization study, which will be necessary as the understanding of burning plasma physics improves.

**Disclaimer** ITER is the Nuclear Facility INB no. 174. This paper simulates plasma physics processes, neutron production and fusion performance during ITER operation; nevertheless the nuclear operator is not constrained by the results of this paper. The views and opinions expressed herein do not necessarily reflect those of the ITER Organization.

**References**


[1] Crotinger J A *et al* 1997 *LLNL Report* UCRL-ID-126284; NTIS #PB2005-102154.

[2] Martin Y R *et al.*, 2008 *Journal of Physics; Conference Series* **123**, 012033

[3] Loarte A *et al* 2013 *Nucl Fusion* **53** 083031

[4] Snyder P B *et al* 2004 *Plasma Phys. and Cont. Fusion* **46** A131

[5] Jardin S C *et al* 1993 *Nucl. Fusion* **33** 371

# Reliability study of pellet ELM pace making

G. Kocsis[a], P.T.Lang[b], G. Cseh[a], M. Dunne[b], L. Horváth[d], V. Mertens[b], H. Meyer[c], B. Ploeckl[b], G. I. Pokol[d], P. Zs. Pölöskei[d], B. Sieglin[b], T. Szepesi[a], E. Viezzer[b], EUROfusion MST1 Team[1] and ASDEX Upgrade Team

[a]*Wigner RCP RMI, Konkoly Thege u. 29-33, H-1121 Budapest, Hungary*
[b]*MPI für Plasmaphysik, Boltzmannstr. 2, 85748 Garching, Germany*
[c]*CCFE, Culham Science Centre, Abingdon, Oxon, OX14 3DB, UK*
[d]*BME NTI, Budapest, Hungary*

Injection of cryogenic solid hydrogen isotope pellets turned out to become a potentially powerful tool for ELM pace making in large-scale toroidal machines like ITER. According to scenario requirements, pacing is needed for ELM mitigation and/or to prevent impurity accumulation [1]. Since its discovery at ASDEX Upgrade (AUG) [2], the pellet ELM pacing technique has been confirmed on other tokamaks such as DIII-D [3] and JET [4]. An increase of the spontaneous ELM frequency by a factor of up to 12 was demonstrated at DIII-D, providing the proof that pellet ELM pacing is a reliable control tool[3]. Experiments and simulations showed that ELMs can be initiated by the toroidally localized high pressure edge regions formed by the localized particle deposition by the ablating pellet. However, most of these results had been obtained in devices where at least parts of the plasma facing components had carbon surfaces. In modern, more reactor relevant all-metal wall machines – due to the change in operational conditions imposed by the metallic wall - significant changes of the pedestal conditions and the ELM dynamics were observed. This implies that the reliability of pellet ELM pace making technique should be reconsidered as well.

Accordingly, the EUROfusion MST1 work package and the last campaigns of AUG put special emphasis to investigate the reliability of pellet ELM triggering and to understand the underlying physics of pellet ELM pace making. First experiments on all-metal wall AUG revealed that - in spite of similar pellet caused perturbations - the pellet ELM triggering potential was reduced in certain plasma scenarios. In general it was found that the probability of a pellet triggering an ELM is dependent on the time elapsed since the previous ELM and even "lag times" were observed where this probability drops to zero. It was also revealed that the pellet ELM triggering potential is independent of the pellet parameters (mass, speed, injection frequency), that is, of the pellet imposed perturbation - at least in the technically accessible parameter range. Conversely the triggering potential was clearly dependent on the scenario of the target plasma and it seemed to be that Nitrogen seeding helps to recover some of the pellet triggering potential.

This phenomenon was investigated in detail and the results are presented in this contribution. As target plasma an edge optimised configuration was used with moderate heating resulting in a long stable plateau phase with 300-400kJ plasma energy content and type-I ELMs with an energy drop up to 25kJ. To avoid global plasma perturbation small pellets were injected

[1] See http://www.euro-fusionscipub.org/mst1

into this plateau phase with low frequency (4Hz). In order to provide accurate $T_e$, $T_i$ and $n_e$ profiles in the pedestal region for stability analysis a reference phase was also included into the scenarios before the start of the pellet injection sequence.

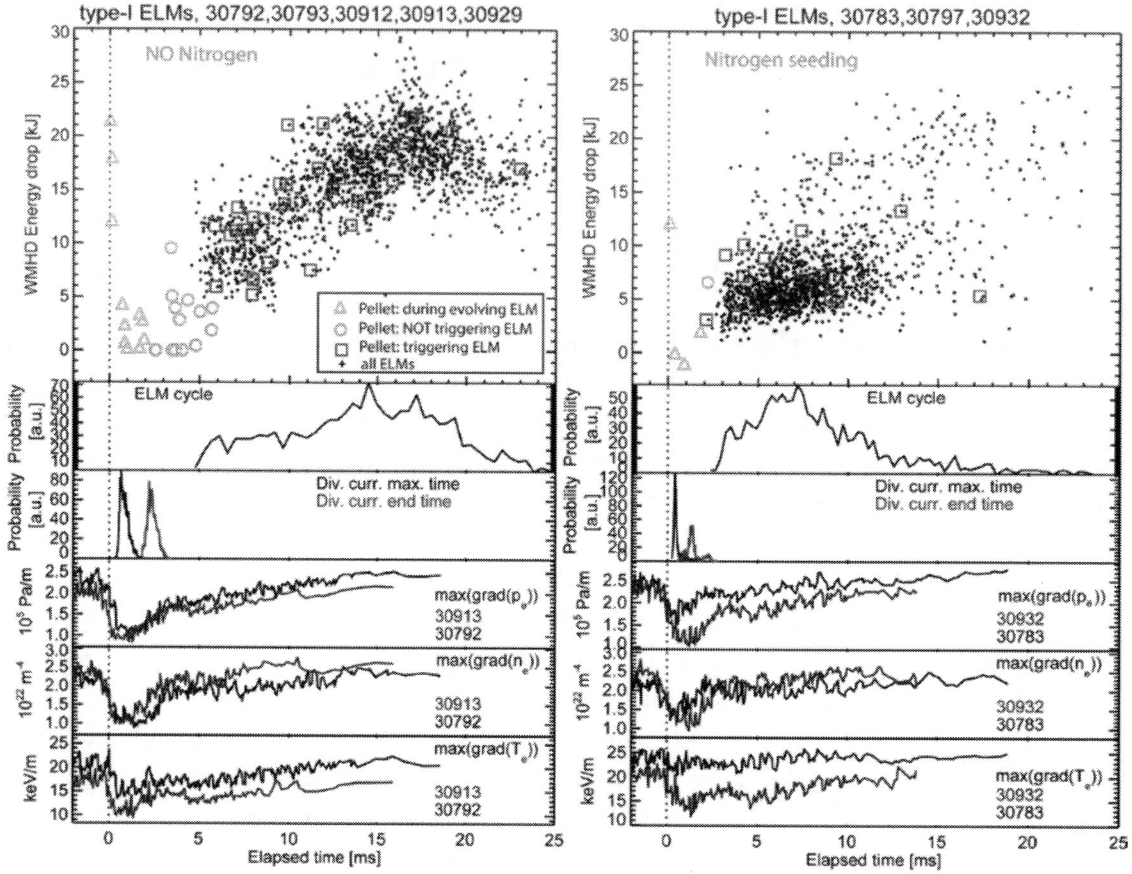


Fig. 1. The ELM provoked plasma energy drop as a function of the time elapsed after the previous ELM both for spontaneous and pellet triggered ELMs. The right column shows the Nitrogen seeded case (puff rate is about $2.5*10^{21}$ $s^{-1}$), while the left column is for similar discharges without Nitrogen puffing. The evolution of the pedestal gradients and statistics about the ELM dynamics (cycle, crash dynamics characterised by the divertor current measurements) are also plotted for both cases.

With this low pellet injection rate the ELM appearance is uncorrelated to the pellets' arrival therefore every pellet hits the plasma at an arbitrary time point in the spontaneous ELM cycle. To collect enough 'pellet events' that is to map the whole ELM cycle the discharge was repeated several times and the experiment was also repeated with Nitrogen seeding applied. The results are summarised in Fig.1 together with the evolution of the pedestal gradients and statistics about the ELM dynamics. Generally it can be observed that pellets can trigger ELMs at elapsed times at which spontaneous ELMs also occur. For the scenario without Nitrogen puffing a lag time of about 5ms (one third of the ELM cycle) is found. Nitrogen seeding (Fig.1., right column) shortens the lag time, but also modifies the target plasma resulting in a better confinement, smaller ELMs, shorter ELM cycle, ELM crash and recovery time. It seems to be that for both seeded and unseeded cases pellets can trigger ELMs after the fast recovery phase of the pedestal (see the evolution of the pedestal gradients). To compensate the confinement improvement caused by the Nitrogen seeding more heating power was also added to the non-seeded scenario (reaching the same plasma energy content as for the seeded case) but this neither changes the ELM cycle and dynamics nor shortens the lag time.

Pellet tracking and the investigation of the ELM onset detected by magnetic pick–up coils revealed that if pellets trigger ELMs the triggering happens shortly after the pellets crossing the separatrix. The time delay of the magnetic ELM onset after the pellet separatrix crossing time lies typically between 0 and 150μs and decreases with increasing elapsed time (similarly to the carbon wall cases). No significant difference was found between the non-seeded and seeded scenario which implies that the pellet imposed perturbation needed to trigger an ELM is independent of the seeding.

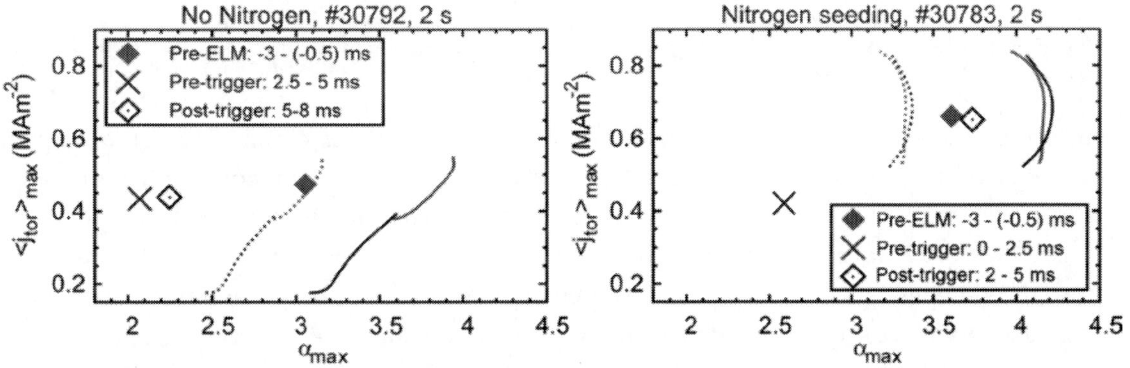


Fig. 2. Stability shown as a function of the stability parameters, the pressure gradient and the current density. $\alpha_{max}$ is the maximum normalized edge pressure gradient and the maximum toroidal edge current density is flux surface averaged. The operational point and the stability boundary are calculated for three phases: just before the ELM (pre-ELM), during the lag time (pre-trigger) and where already **pellets can trigger (post-trigger).** To indicate the accuracy of the calculation the shifted stability boundaries to 80% of the pressure gradient are also over plotted (dotted lines).

The target plasma scenarios were modelled by axial symmetric linear peeling-ballooning (LPB) model which gives indication about the plasma stability. Note that since this analysis assumes toroidal symmetry while the perturbation has a strong 3D character obviously the analysis with full 3D treatment (e.g. by JOREK) would be useful in the future. In the LPB model the kinetic profiles, data from magnetic field and flux difference measurements were used (CLISTE Grad–Shafranov) to calculate a self-consistent plasma equilibrium and full current density profiles and the refined equilibrium was used for stability calculation (MISHKA-fast code)[6]. The results are summarised on Fig.2. both for reference and Nitrogen seeded discharges for three relevant time windows. Since pellets can trigger ELMs at elapsed times at which spontaneous ELMs also uccur, we expect that in these cases the plasma is close to the stability boundary. This is more or less confirmed by the LPB calculations. In the reference case, both the pre- and post-trigger cases are far from the stability boundary, indicating that a natural ELM is unlikely to occur at both of these time points (though they can be observed in the post-trigger phase). In the nitrogen seeded case, the post-trigger case is close to the stability boundary, which is likely to be due to the shorter ELM cycle in this case; a separation of the probabilities of natural and triggered ELMs is only possible in the reference discharge (see also Fig.1 for the probability distribution over the ELM cycle).

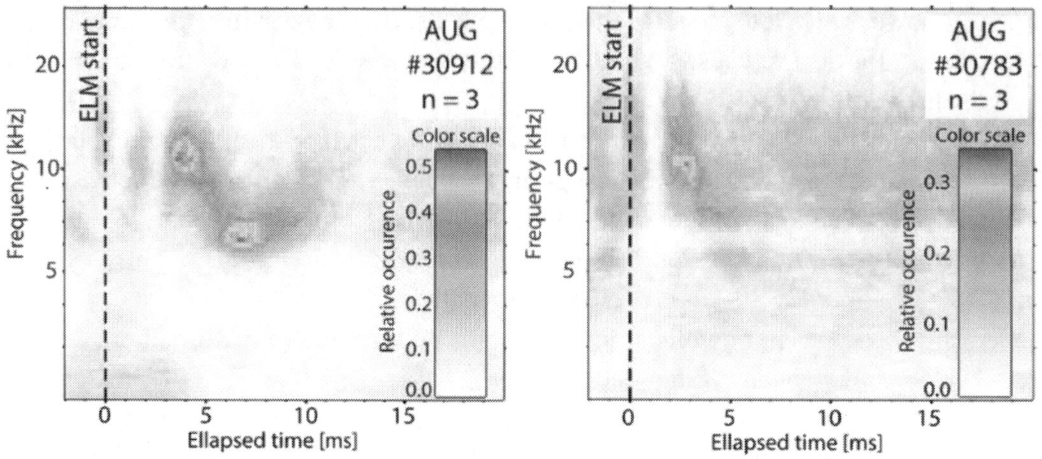


Fig. 3. Time-frequency mode number analysis for toroidal mode number 3 for the non-seeded (left) and seeded scenario (right).

In order to reveal differences for the different stability phases a statistical time-frequency mode number analysis was performed for 283 and 493 ELMs in non-seeded and seeded shots, respectively. The first step of this process was a continuous wavelet decomposition by order 24 Morlet wavelets in the [-2,20 ms] time interval with respect to the ELM start, for the toroidal magnetic probe array. Using the cross-phases of the transforms the most probable mode numbers were determined at each time-frequency point. Fig.3. shows the probability of the occurrence of the n=3 toroidal mode at each time-frequency point aggregated from all ELMs. This mode is the strongest component of the washboard-like modes appearing at equidistant frequencies increasing for increasing mode numbers in the range of n=3-9, propagating in the electron diamagnetic direction. Just after the ELM the washboard-like modes appear at an elevated frequency (10-11kHz), and their frequency gradually decreases to their pre-ELM frequency (6-7kHz) in the post-ELM recovery phase. The characteristic time of this frequency decrease corresponds to the lag-time after the ELMs. This indicates that the washboard mode frequency might be a good indicator for the edge plasma returning into a state susceptible to ELMs.

This work has been carried out within the framework of the EUROfusion Consortium and has received funding from the Euratom research and training programme 2014-2018 under grant agreement No 633053. The views and opinions expressed herein do not necessarily reflect those of the European Commission.

**References**

[1] Loarte A. *et al* 2014 *Nucl. Fusion* **54** 033007
[2] Lang P.T. *et al* 2004 *Nucl. Fusion* **44** 665
[3] Baylor L.R. *et al* 2013 *Phys. Rev. Lett.* **110** 245001
[4] Lang P.T. *et al* 2011 *Nucl. Fusion* **51** 033010
[5] Lang P.T. *et al* 2014 *Nucl. Fusion* **54** 033009
[6] Konz C. *et al* 2009 *36th EPS Conference on Plasma Phys., ECA* **33E** P-1.152

# Relaxation of zonal flows in stellarators: influence of the magnetic configuration

E. Sánchez[1], P. Monreal[1], I. Calvo[1], R. Kleiber [2]

[1] *Laboratorio Nacional de Fusión, CIEMAT, 28040 Madrid, Spain*

[2] *Max Planck Institut für Plasmaphysik, D-17491, Greifswald, Germany*

The importance of zonal flows (ZFs) in the self regulation of turbulent transport is widely accepted nowadays. Although the ZF evolution is a nonlinear problem [1], involving the nonlinear drive and the collisional damping, studying its linear collisionless damping still provides valuable information [2]. In stellarators, in general, the linear collisionless relaxation of a zonal potential perturbation reaches a residual value after an oscillatory phase [3]. Two different characteristic oscillations, a geodesic acoustic mode (GAM) [4] and a low frequency oscillation (LFO), have been identified in stellarators depending on the magnetic configuration [5, 6]. There are indications that both, the ZF oscillations and the residual level, can affect plasma confinement [7, 8]. The residual ZF level as well as the oscillations depend on the radial scale of the perturbation and also on the magnetic configuration.

We take advantage of the flexibility of TJ-II [9] to scan different magnetic configurations and explore the influence of several configuration parameters (magnetic well, magnetic shear, magnetic field curvature and rotational transform) in the linear relaxation of ZFs. The magnetic shear is always very small in TJ-II, except in configurations with ohmic induced current. We have run linear gyrokinetic (GK) collisionless simulations of ZF relaxation with the code EUTERPE [10] for several radial wavelengths and several temperatures in a set of TJ-II configurations covering the ranges of variation in several parameters listed in the Table 1.

| | |
|---|---|
| $a$ | 0.156 - 0.237 m |
| well | 0.013 - 0.068 |
| shear | 0.008 - 0.118 |
| $\varepsilon_t$ | 0.318 - 0.456 |
| $\iota$ | 1.178 - 2.231 |
| $\langle\lvert\kappa_g\rvert\rangle$ | 0.589 - 0.734 m$^{-1}$ |
| $\langle\lvert\kappa_n\rvert\rangle$ | 0.917 - 1.232 m$^{-1}$ |
| $\langle\lvert\nabla s\rvert\rangle$ | 6.927 - 15.153 m$^{-1}$ |
| $k_\Psi$ | $\pi$ - $6\pi$ |
| $T$ | 25 - 800 eV |
| $\langle k\rho\rangle$ | 0.035 - 0.152 |

Table 1: *Ranges of variation of several parameters in the TJ-II configurations studied.*

The simulations are done with adiabatic electrons and are initialized with a perturbation to the ion distribution function of the form $\delta f_i \propto \frac{\langle k_\perp^2 \rho_{ti}^2\rangle}{T}\sin(k_\psi s)F_M$ which produces a zonal potential perturbation, where $F_M$ is a Maxwellian, $\rho_{ti}$ is the thermal ion Larmor radius, $T$ is the temperature, $k_\perp = k_\Psi|\nabla s|$ is the radial wavenumber of the perturbation and $\langle\rangle$ means flux surface average. The evolution of the radial electric field ($\phi'_{00}(t)$) is monitored at several radial positions, where $\phi_{00}$ is the $(0,0)$ Fourier component (in angle variables) of the potential and $'$ means derivative with respect to the radial coordinate $\Psi$. In Fig. 1 a typical time evolution of the radial electric field (normalized to its initial value) is shown together with the Fourier spectrum of the time trace.

Two frequencies are clearly appreciated: a GAM oscillation that is damped very quickly and a LFO that lasts a longer time (see Fig. 1). From the Fourier spectrum the GAM amplitude and frequency is obtained. The LFO amplitude and frequency and the residual level can, in principle, be obtained from the spectrum, but the frequency resolution is poor compared to this frequency and a better result is obtained by fitting the time trace to a model $\frac{\phi'_{00}(t)}{\phi'_{00}(0)} = R + A_{LFO}\sin(\omega_{LFO}t)e^{\gamma t}$. The fit is also shown in Fig. 1. With this fit we obtain the residual ZF level, $R$, the ZF oscillation frequency, $\omega_{LFO}$, and also an amplitude, $A_{LFO}$, and a damping rate, $\gamma$.

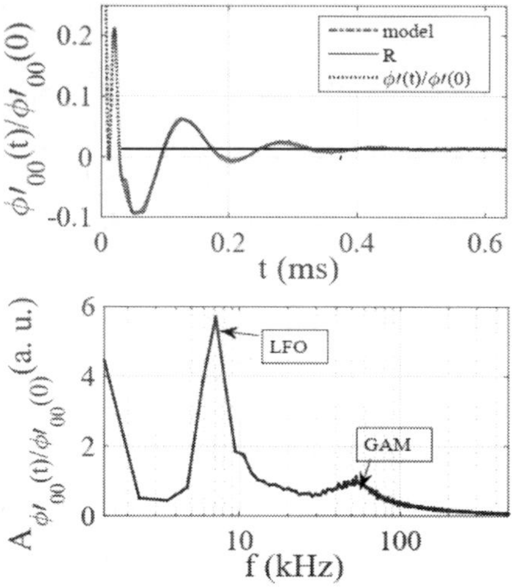

Figure 1: *Time trace (top) and spectra (bootom) of $\phi'_{00}(t)/\phi'_{00}(0)$ for a simulation in the TJ-II standard configuration with $k_\Psi = 3\pi$ and T=100 eV.*

We carried out a multivariate fit of $A_{GAM}$, $\omega_{GAM}$, $R$, $A_{LFO}$, $\omega_{LFO}$ and $\gamma$ with the magnetic configuration parameters. We complement the GK simulations with semi-analytical calculations of the residual ZF level with CAS3D-K [11, 3].

**Residual level**

In [3] the ZF residual level was studied for a wide range of radial wavelengths in the W7-X standard configuration by means of calculations with the code CAS3D-K and the results were compared to simulations with the GENE and EUTERPE GK codes. It was found that in the long-wavelength limit, and using the adiabatic electron approximation, $R = \frac{\phi'_{00}(\infty)}{\phi'_{00}(0)} \approx \frac{(1-\varepsilon_t)}{\langle k_\perp^2 \rho_{ti}\rangle^2 \varepsilon_t}$. That is, the residual level at a specific radial position depends only on three parameters: $\varepsilon_t$, the fraction of trapped particles at the flux surface, and $k_\perp$ and $\rho_{ti}$, defined previously.

It is expected that the details of the magnetic configuration affect the orbits of trapped particles and consequently the ZF residual. In particular, the geodesic curvature enters the radial magnetic drift and could affect it. We carried out a multivariate fit of the CAS3D-K calculations separately for the regions: $k_\perp\rho_{ti} < 0.2$, $0.2 < k_\perp\rho_{ti} < 1$ and $k_\perp\rho_{ti} > 1$. We found that the best fit of the residual in the long wavelength region ($k_\perp\rho_{ti} < 0.2$), including only $k_\perp\rho_{ti}$, $\varepsilon_t$, is $R \propto (k_\perp\rho_{ti})^{1.96\pm0.002}\varepsilon_t^{-1.46\pm0.01}$, close to $\langle k_\perp\rho_{ti}\rangle^2$. In the other regions the best fit is obtained including the averaged geodesic curvature $\kappa_G := \langle|\kappa_g|\rangle$ also. For $0.2 < k_\perp\rho_{ti} < 1$, the best fit is $R \propto (k_\perp\rho_{ti})^{0.88\pm0.01}\varepsilon_t^{-0.44\pm0.04}\kappa_G^{-1.80\pm0.15}$, and $R \propto (k_\perp\rho_{ti})^{-1.60\pm0.004}\varepsilon_t^{0.42\pm0.02}\kappa_G^{-2.45\pm0.07}$ for $k_\perp\rho_{ti} > 1$ .

A multivariate fit of the results from GK simulations in the region with $k_\perp \rho_{ti} < 0.2$ gives as best fit $R \propto (k_\perp \rho_{ti})^{1.57\pm0.01} \varepsilon_t^{-1.69\pm0.04}$ , with a slightly smaller dependency with $k_\perp \rho_{ti}$ as compared to the result obtained with CAS3D-K for the same scales. The long wavelength approximation used in EUTERPE for adiabatic electron simulations does not allow to study the region with $k_\perp \rho_{ti} > 0.5$ with this kind of simulations.

The influence of the trapped particles is much more important at large wavelengths, the corresponding exponent even changing sign for small wavelengths. A negative scaling of the residual with the geodesic curvature is obtained, the influence of $\kappa_G$ on the residual being larger for small wavelengths. This result can be relevant if magnetic configuration optimization is pursued through reduction of the magnetic field curvature. The influence of curvature reduction on the residual level can be expected to be more important for small radial scales ZFs, produced by trapped electron mode turbulence, than for long wavelength ZFs from ion temperature gradient turbulence, typically in the range $k_\perp \rho_{ti} \sim 0.1$.

**GAM oscillation**

The GAM oscillation [4] is produced by the compression of the flux tubes at the high field side and expansion at the low field side. In a tokamak a clear dependency of the GAM frequency with $q^{-1}$ is found. A dependency with $\sqrt{T_e + T_i}$ is also a characteristic of the GAM oscillation. In stellarators a similar dependency was found with $T$ and $q$ [5].

To study the scaling of GAM oscillation characteristics we analyze GK simulations for a set of magnetic configurations and carry out a multivariate fit of the GAM frequency and amplitude obtained from the Fourier spectrum. The best multivariate fit for the GAM frequency gives $\omega_{GAM} \propto T^{0.49\pm0.01} \iota^{0.45\pm0.04} \kappa_G^{0.85\pm0.13}$. Scalings of the frequency with $T$ close to the typical with $\sqrt{T}$ and with $q$ as $1/\sqrt{q}$ are found. An additional positive scaling with the geodesic curvature is found. This scaling is not surprising, as $\kappa_G$ acts as the restoring force in the GAM. The scaling with $\kappa_G$ is positive also in the amplitude.

For the oscillation amplitude the best fit gives $A_{GAM} \propto T^{-0.60\pm0.01} \iota^{0.030\pm0.06} \kappa_G^{0.46\pm0.17}$, with a scaling close to $1/\sqrt{T}$. The amplitude has low dependency with $\iota$, in contrast to the result found for tokamaks and heliotrons, where the GAM damping rate scales with $1/q$.

**Low frequency oscillation**

The low frequency oscillation was studied in [6, 12] and was found to be related to the radial current produced by the radial drift of trapped particles.

From GK simulations in different TJ-II magnetic configurations we made a fit to the damped oscillation model obtaining the frequency, $\omega_{LFO}$, the residual level, $R$, the amplitude, $A_{LFO}$ and the damping rate, $\gamma$, of the LFO. With these data we made a multivariate fit and got the scaling of these quantities with magnetic configuration parameters. It has to be taken into account that the model fitting is a subtle process and the fit is not always done properly so that some data have to be excluded from the analysis. The results of the multivariate fit are as follows:

For the LFO frequency we obtained $\omega_{LFO} \propto T^{0.50\pm0.02} \varepsilon_t^{0.57\pm0.09} \kappa_N^{1.26\pm0.12} \kappa_G^{-0.46\pm0.17} k_\perp^{0.15\pm0.02}$ as best fit. For the amplitude we get $A \propto T^{-0.52\pm0.03} \kappa_N^{1.57\pm0.20} \kappa_G^{-2.69\pm0.33} k_\perp^{-0.53\pm0.05}$ and for the damping rate $\gamma \propto T^{0.63\pm0.03} \varepsilon_t^{-0.33\pm0.13} \kappa_G^{3.11\pm0.34} k_\perp^{0.54\pm0.05}$.

A scaling close to $\sqrt{T}$ for the frequency and to $1/\sqrt{T}$ for the amplitude is found, as it was observed for the GAM oscillation. The damping increases with the temperature as in the case of the GAM oscillation. The frequency has a positive scaling with $\varepsilon_t$, as could be expected because the oscillation is produced by trapped particle's radial drift; consistently the damping rate has a negative scaling with $\varepsilon_t$. The frequency has almost no dependency with the radial scale of the perturbation ($k_\perp$) while it has a clear influence on the damping rate (positive exponent). The averaged normal curvature $\kappa_N := \langle|\kappa_n|\rangle$ has an effect opposite to that of $\kappa_G$ on the oscillation frequency. Interestingly, the scaling of the frequency and amplitude with the geodesic curvature is opposite to that observed for the GAM oscillation and the damping rate is only affected by the geodesic curvature. The amplitude obtained in the fit is affected by the GAM oscillation.

**Aknowledgement**

This work has been partially funded by the Ministerio de Economía y Competitividad of Spain (project ENE2012-30832) and used resources from the Red Española de Supercomputación. We thank A. Könies for providing CAS3D-K. This work has been carried out within the framework of the EUROfusion Consortium and has received funding from the Euratom research and training programme 2014-2018 under grant agreement No 633053. The views and opinions expressed herein do not necessarily reflect those of the European Commission.

**References**

[1] P. H. Diamond, et. al., Plasma Phys. Control. Fusion **47**, R35-R161 (2005).
[2] M. N. Rosenbluth, and F. L. Hinton. Phys. Rev. Lett. **80** 724 (1998).
[3] P. Monreal, et al., submitted to PPCF (2015).
[4] N. Winsor, J. L. Johnson & J. M. Dawson. Phys. Fluids **11**, 11 (1968).
[5] H. Sugama, et al., Phys. Review Lett. **94** 115001 (2005).
[6] A. Mishchenko, et al., Phys. Plasmas **15**, 072309 (2008).
[7] M. Nunami, et al., Phys. Plasmas **20** 092307 (2013).
[8] P. Xanthopoulos, et al., Phys. Rev. Lett. **107** 245002 (2011).
[9] C. Alejaldre, et al., Plasma Phys. Control. Fusion, **41** A539-A548 (1999).
[10] G. Jost, et. al., Phys. Plasmas **7**, 1070664X (2001).
[11] A. Könies, et. al., Phys. Plasmas **7**, 1139 (2000).
[12] P. Helander, A. Mishchenko, R. Kleiber et al. Plasma Phys. Control. Fusion **53**, 5 (2011).

# Inertia and equilibrium impurity flow in 3D magnetic surfaces.

J. A. Alonso, J. L. Velasco and I. Calvo

*Laboratorio Nacional de Fusión, CIEMAT, 28040 Madrid, Spain*

**Abstract**

We derive a compact and exact expression for the parallel inertia of impurity flows in general 3D magnetic surfaces. The form of the flows is constrained to be $E \times B$ in the perpendicular direction but otherwise general in the parallel one. We present computations of parallel inertia for stellarator geometry and several edge conditions and show that, even for the moderate radial electric fields typical of stellarators, inertia-driven impurity density variations cannot in general be neglected for heavy impurities. This is shown to be due to the form of the incompressible streamlines in 3D magnetic surfaces.

## Introduction

Recent works in tokamaks and stellerators have investigated the physical causes of uneven distribution of impurities on flux surfaces [1, 2, 3, 4], and how and to what extent they can alter the radial transport of impurities [5, 6]. Impurities are prone to develop these variations because of their large mass and charge. Inertia, in particular, grows linearly with mass and is known to lead to accumulation in the outer mid-board of rotating tokamaks. In stellarators, their typically much smaller flow over thermal velocity ratio is often invoked to neglect inertial forces in the parallel momentum balance.

A quick order of magnitude estimate of the inertia-driven impurity density variation is obtained by balancing parallel inertia term and pressure gradient terms,

$$\frac{2}{v_{tz}^2}\mathbf{B}\cdot(\mathbf{u}_z\cdot\nabla)\mathbf{u}_z + \mathbf{B}\cdot\nabla\log n_z = 0\ . \tag{1}$$

If one considers similar scale length in the two nabla terms and a characteristic impurity velocity $u_z$ such that $M_z = u_z/v_{thz} \ll 1$, one gets

$$\log n_z \sim 2M_z^2 \rightarrow n_z/\langle n_z\rangle - 1 \sim 2M_z^2\ ,$$

which, for typical $M_z \sim 0.1$, results in a few % relative density variation $\frac{n_z}{\langle n_z\rangle}$. Here $v_{tz} = \sqrt{2T_z/m_z}$, and $\langle A\rangle$ denotes the flux surface average of $A$. Impurities and main ions are considered to be thermally coupled through collisions so that $T_z = T_i$, and the ion temperature is an approximate flux constant $T_i(\psi)$.

In this work we show that, due to the shape of the streamlines of incompressible flows in stellarators, inertia can locally grow to dynamically relevant values in realistic conditions, particularly so for the large $E_r$ values found at the edge transport barrier in H-modes. In particular, the bending of the icompressible streamlines is such that the velocity variation scale-length can become small locally (smaller than or similar to the minor radius) and the velocity large (several times the perpendicular $E \times B$ velocity) due to the parallel Pfircsh-Schlüter components. The return parallel flows resulting from density variations tend to lower the peak values of inertia on a flux surface, smoothing the bending of streamlines and partially compensating the Pfirsch-Schlüeter flow.

**Compact expression for the parallel inertia of impurities**

We choose to work in magnetic coordinates $(\psi, \theta, \phi)$ in terms of which the magnetic field can be written as $\mathbf{B} = \sqrt{g}^{-1}\left(\iota \mathbf{e}_\theta + \mathbf{e}_\phi\right)$. The flux label $\psi$ is the toroidal flux over $2\pi$, $2\pi\psi = \Psi_T$. A impurity velocity field that is $E \times B$ in the perpendicular direction and otherwise arbitrary in the parallel direction can be written as

$$\mathbf{u} = \omega(\psi)\mathbf{e}_\phi + K(\psi, \theta, \phi)\mathbf{B}, \quad \omega(\psi) = -\iota^{-1}\frac{d\Phi}{d\psi}, \tag{2}$$

where $\Phi$ is the electrostatic potential. Using form of the impurity flow the parallel inertia is written in Boozer coordinates as

$$\mathbf{B}\cdot(\mathbf{u}\cdot\nabla)\mathbf{u} = \mathbf{B}\cdot\nabla\left(\frac{K^2B^2}{2} - \frac{\omega^2 g_{\phi\phi}}{2}\right) + B^2\omega\frac{\partial}{\partial\phi}\frac{u_\parallel}{B}. \tag{3}$$

Note that for axis-symmetric systems ($\mathbf{e}_\phi = R\hat{\phi}, g_{\phi\phi} \equiv \mathbf{e}_\phi\cdot\mathbf{e}_\phi = R^2, \partial_\phi = 0$) Equations 2 and 3 reduce to the known tokamak expressions (see e.g. [1]). In order to obtain initial estimates of the impurity density redistribution driven by inertia we start from the constant-density incompressible flow condition:

$$K = \omega(\chi)\left(\frac{\lambda}{B} - \frac{I}{\langle B^2\rangle}\right) + \frac{\langle u_\parallel B\rangle}{\langle B^2\rangle}, \tag{4}$$

where

$$\mathbf{B}\cdot\nabla\frac{\lambda}{B} = -\frac{1}{\sqrt{g}}\frac{\partial\sqrt{g}}{\partial\phi}, \text{ and } \langle\lambda B\rangle = 0\,. \tag{5}$$

The tokamak solution is $\lambda/B = 0$.

Figure 1 shows the numerical evaluation of the inertia and the equivalent density variation for TJ-II at $\rho = 0.75$. For the sake of testing the correctness of the inertia expression 3 we compare the values obtained along a stream line with a finite differences approximation of the inertia term given by $2B^2\mathbf{u}\cdot\nabla u_\parallel/B$. Large values of inertia are found to be due to large variation of the parallel flow component along the streamline combined with large values of $u = |\mathbf{u}|$ caused

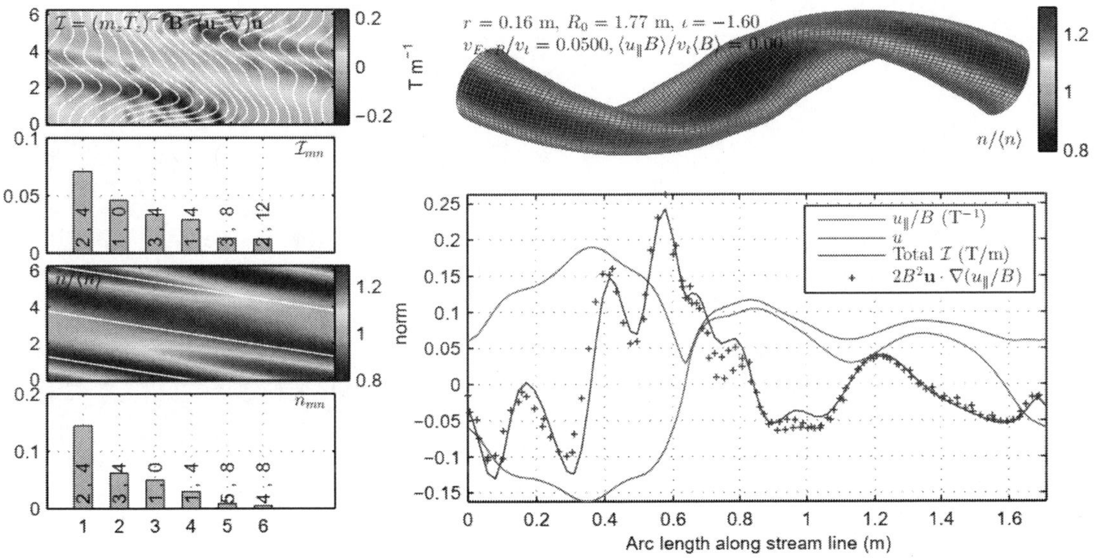


Figure 1: Parallel inertia and equivalent density variation for the $\rho = 0.75$ flux surface of TJ-II. The plot in the lower right corner shows values of $u_\parallel/B, u, \mathscr{I}$ and the finite differences evaluation of approximate inertia expression $2B^2\mathbf{u}\cdot\nabla u_\parallel/B$ along a streamline. Several streamlines of the incompressible flow are shown in the upper-left plot.

by large values of return Pfirsch-Schlüter flows. The equivalent density variation is found by solving the magnetic differential equation 3. To find the actual equilibrium density parallel force balance and particle conservation ($\nabla\cdot n_z\mathbf{u}_z = 0$) need to be solved simultaneously.

Similar calculations are show in figure 2 for the TJ-II, W7-X and LHD configurations in the radial range from $\rho = 0.3$-$0.8$ and for $v_{EtimesB}/v_{tz} = 10^{-2}$. The exact values shown in bars are compared with different estimates of the parallel inertia term. Although substantial differences are observed between the configurations, the order of magnitude estimate $\mathbf{B}\cdot(\mathbf{u}\cdot\nabla)\mathbf{u} \sim < u^2 > /r$ is a fair approximation except for interior positions in LHD.

The relative importance of inertia effects in the impurity dynamics in stellarator configurations will obviously depend on the specific profiles, most critically on the ratio of $E\times B$ over thermal impurity velocity. Values of this ratio for Carbon impurity, $T_z \sim 0.1$-$2$ keV and $E_r/B \sim 1$-$10$ km/s range from $v_{ExB}/v_{tz} \sim 0.005$-$0.25$. Finally, the bulk parallel rotation has not been considered $\frac{\langle u_\parallel B\rangle}{\langle B^2\rangle}$ here. This term, for which the analytic solution of equation 3 can be easily found, is much less effective in driving inertial density variations.

This work has been partially funded by the Spanish Ministry of Economy and Competitiveness under grant number ENE2012-30832 and the EUROfusion work package WP15.S1.A.1. This work has been carried out within the framework of the EUROfusion Consortium and has

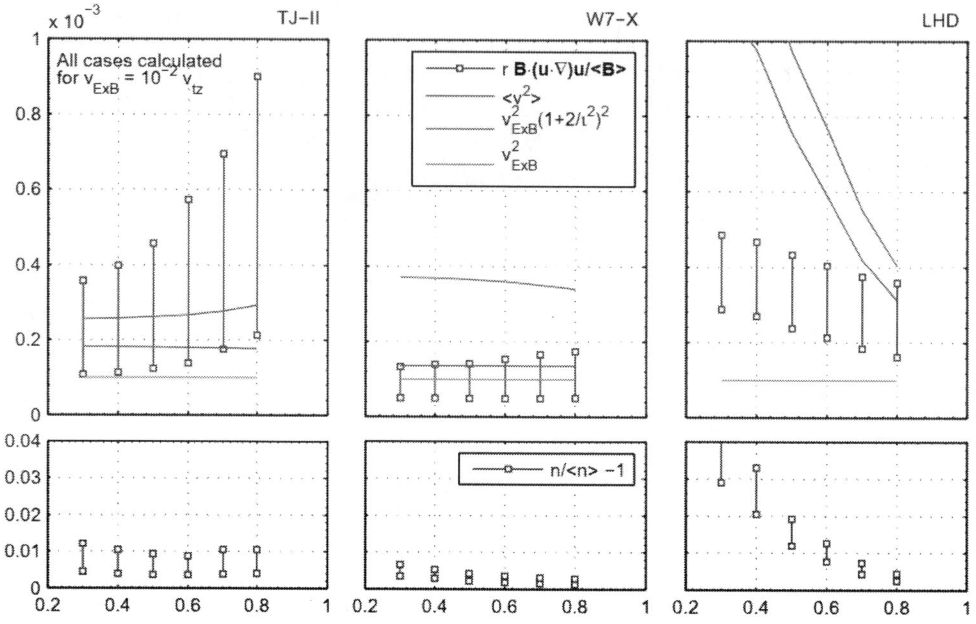


Figure 2: Upper graphs: Typical parallel inertia values for several configurations and radial locations together with different estimates. Lower graphs: equivalent density variations.

received funding from the Euratom research and training programme 2014-2018 under grant agreement No 633053. The views and opinions expressed herein do not necessarily reflect those of the European Commission.

[1] E Viezzer, T Pütterich, E Fable *et al.* 2013 *Plasma Phys. Control. Fusion* **55** 124037.

[2] R M Churchill, B Lipschultz, C Theiler *et al.* 2013 *Nucl. Fusion* **53** 122002.

[3] J Arévalo, J A Alonso, K J McCarthy *et al.* 2014 *Nucl. Fusion* **54** 013008.

[4] F J Casson, C Angioni, E A Belli *et al.* 2015 *Plasma Phys. Control. Fusion* **57** 014031.

[5] J M García-Regaña, R Kleiber *et al.* 2013 *Plasma Phys. Control. Fusion*, **55** 074008.

[6] C Angioni and P Helander 2014 *Plasma Phys. Control. Fusion*, **56** 124001.

# REFMULF: 2D Fullwave FDTD Full Polarization Maxwell Code

F. da Silva[1], S. Heuraux[2],T. Ribeiro[3], D. Aguiam[1]

[1] *Instituto de Plasmas e Fusão Nuclear, Instituto Superior Técnico, Universidade de Lisboa, 1049-001 Lisboa, Portugal*

[2] *Institut Jean Lamour, UMR 7198 CNRS-University of Lorraine, Vandœuvre, France*

[3] *Max-Planck-Institut für Plasmaphysik, 85748 Garching, Germany*

**Abstract**

A novel 2D full-wave FDTD code, REFMULF, includes full polarization waves, allowing the coupling of the transverse-electric mode (TE/X-mode) with the transverse-magnetic mode (TM/O-mode) via a linear vectorial differential equation for **J** with a generic external magnetic field $\mathbf{B_0}$. It extends the capabilities of microwave reflectometry simulation, and being a parallel code, is able to cope with real size problems.

## Introduction

An important tool for the progress of reflectometry is numerical simulation, able to assess the measuring capabilities of existing systems and to predict the performance of future ones in machines such as ITER and DEMO. The newly developed 2D full-wave FDTD code, REFMULF, presented here allows the treatment of full polarization waves, coupling the TE/X-mode with the TM/O-mode via a linear vectorial differential equation for **J** with any kind of external magnetic field $\mathbf{B_0}$.

$$\frac{\mathrm{d}\mathbf{J}}{\mathrm{d}t} = \varepsilon_0\omega_p^2\mathbf{E} - \nu\mathbf{J} + \omega_c \times \mathbf{J}. \qquad (1)$$

This equation further couples wave propagation, described by Maxwell curl equations to the plasma medium. The external magnetic field components of $\mathbf{B_0}$ lying on the propagation plane ($B_x$ and $B_y$) are responsible for linking the TE and TM modes. For a $\mathbf{B_0}$ purely perpendicular to the propagation plane the code describes simultaneously O-mode and X-mode propagation, which enables the simulation of depolarization processes in turbulent plasmas. This offers capabilities unavailable in present day 2D reflectometry codes, closing the gap to the much sought-after usable 3D code. As the code is parallel, is able to cope with real size problems.

## Code description

The code considers propagation in a 2D plane ($x$-$y$ plane) with no gradients along $z$ ($\partial/\partial z = 0$). This assumption avoids a standard linear mode conversion for any arbitrary orientation of the magnetic field. Maxwell equations are solved using the classic Yee schema on two sets of staggered grids, one for the transverse magnetic mode (TMz) with field components ($E_z$, $H_x$ and $H_y$) and another for the transverse electric mode (TEz), with field components ($H_z$, $E_x$ and $E_y$). The TMz and TEz planes are not coupled through Maxwell equations since $\partial/\partial z = 0$ but coupling through the equation for $\mathbf{J}$ is possible through the product $\omega_c \times \mathbf{J}$. Developing equation (1) into components the coupling (emphasized in color) is clear.

$$\frac{dJ_x}{dt} + \nu J_x = \varepsilon_0 \omega_p^2 E_x + \omega_y J_z - \omega_z J_y \quad (2)$$
$$\frac{dJ_y}{dt} + \nu J_y = \varepsilon_0 \omega_p^2 E_y + \omega_z J_x - \omega_x J_z \quad (3)$$
$$\frac{dJ_z}{dt} + \nu J_z = \varepsilon_0 \omega_p^2 E_z + \omega_x J_y - \omega_y J_x. \quad (4)$$

The components $B_x$ and $B_y$ couple the TMz plane to the TEz (in red) and the TEz plane to the TMz (in blue). Equations (2)–(4), with $\nu = 0$, are solved using the Xu-Yuan schema [1] with the modifications proposed in [2] for extended long-run stability. The expression for $J_x$ is given by

$$C_{0x} = 1 + \frac{\omega_x^2 \Delta t^2}{4} \quad (5)$$
$$C_{1x} = 1 + \frac{\omega_y^2 \Delta t^2}{4} - \left( \frac{\omega_x^2 \omega_y^2 \Delta t^4}{16 C_{0x}} - \frac{\Delta t^2 \omega_z^2}{4 C_{0x}} \right)$$
$$\begin{aligned} J_x^{n+1/2}(i, j+1/2) = {} & \frac{1}{C_{1x}} \left[ 1 - \frac{\omega_y^2 \Delta t^2}{4} + \left( \frac{\omega_x^2 \omega_y^2 \Delta t^4}{16 C_{0x}} - \frac{\Delta t^2 \omega_z^2}{4 C_{0x}} \right) \right] J_x^{n-1/2}(i, j+1/2) + \\ & + \frac{\varepsilon_0 \omega_p^2 \Delta t}{C_{1x}} \left[ E_x^n(i, j+1/2) + \left( \frac{\omega_x \omega_y \Delta t^2}{4 C_{0x}} - \frac{\Delta t \omega_z}{2 C_{0x}} \right) E_y^n(i+1/2, j) \right] + \\ & + \frac{\varepsilon_0 \omega_p^2 \Delta t^2}{2 C_{1x}} \left[ \omega_y - \left( \frac{\omega_x^2 \omega_y \Delta t^2}{4 C_{0x}} - \frac{\Delta t \omega_z \omega_x}{2 C_{0x}} \right) \right] E_z^n(i-1/2, j-1/2) + \\ & + \left( \frac{\omega_x \omega_y \Delta t^2}{4 C_{1x}} - \frac{\Delta t \omega_z}{2 C_{1x}} \right) \left[ 1 + \left( \frac{1}{C_{0x}} - \frac{\omega_x^2 \Delta t^2}{4 C_{0x}} \right) \right] J_y^{n-1/2}(i+1/2, j) + \\ & + \frac{1}{C_{1x}} \left[ \omega_y \Delta t - \frac{\omega_x \Delta t}{C_{0x}} \left( \frac{\omega_x \omega_y \Delta t^2}{4} - \frac{\Delta t \omega_z}{2} \right) \right] J_z^{n-1/2}(i-1/2, j-1/2). \end{aligned}$$

Expressions for $J_y$ and $J_z$ can be easily obtained by inspection considering the positive circulation $x \to y \to z$. Note that if $\mathbf{B} = \mathbf{B_z}$, frequencies $\omega_x$ and $\omega_y$ are null and the equations (2)–(3) describe the X-mode propagation as used in REFMULX [2] code while equation (4) describes O-mode propagation as used in REFMUL code [3]. In this case the code can be used within the same scenario to simultaneously run both X- and O-mode cases in order to compare them. To run just a X- or an O-mode case is also possible by keeping the unwanted mode unexcited. Nevertheless that would result in a waste of resources, therefore the code can be compiled not only with the full polarization (default) but also with just X- or O-mode

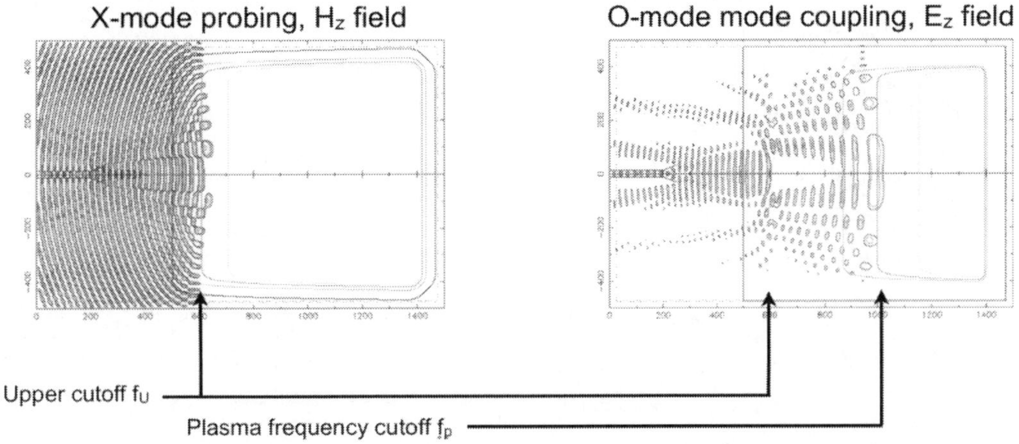


Figure 1: *X-mode is excited on the TEz plane, being reflected at the upper branch cutoff frequency* $f_U$. *The signal couples to O-mode which propagates until the plasma frequency* $f_p$.

executables. With this compile options, REFMULF is intended to replace REFMUL (the O-mode code) and REFMULX (the X-mode code). REFMULF uses Berenger's perfectly matched layer (PML) boundary conditions [4] and an unidirectional transparent source (UTS) [3]. The code is parallelized in the sense that its computational domain is decomposed and distributed among different computer cores using the message-passing interface (MPI) standard [5]. Within each MPI sub-domain, the OpenMP thread-parallelism [6] is further invoked to take advantage of the current trend in modern hierarchical architectures, on which several computer cores are grouped together and locally share memory. Such hybrid parallelization paradigm allows for an efficient scaling up to several hundreds of cores, which is a necessary condition for an extension of the model to three spatial dimensions.

## Full polarization examples

We can observe in Fig. 1 snapshots of the $H_z$ (left) and of the $E_z$ (right) fields for a simulation with a linear density profile and an external magnetic field $B_0 = 0.1\mathbf{e_y} + 0.9\mathbf{e_z}\,$T. A $f_c = 35\,$GHz fixed frequency signal is excited in the TEz plane (*X-mode*, on the left), where it propagates until it is reflected at the upper branch cutoff frequency $f_U$. Due to the $B_y$ component the wave energy is coupled to the TMz plane (*O-mode*, on the right) and can propagate further into the plasma up to the plasma frequency cutoff $f_p$ position, where it is reflected before returning to the antenna. In the complementary situation (Fig. 2), with the same plasma and magnetic field $B_0$, in the TMz plane (*O-mode*), the same frequency $f_c$ is excited and the wave propagates until it reaches the plasma frequency $f_p$ cut-off. There is coupling to the *X-mode* in the TEz plane and reflection at at the upper branch frequency cutoff $f_U$ position. In each of these two examples the wave was only excited on one of the

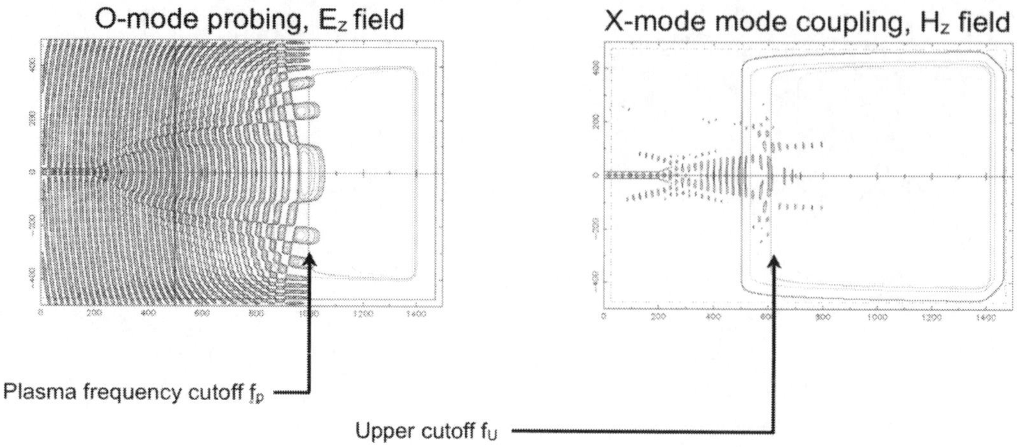


Figure 2: *O-mode is excited on the TMz plane, being reflected at the plasma frequency $f_p$. Signal is coupled to X-mode propagating until reaching the upper branch frequency cutoff $f_U$.*

planes, but more general cases can be treated with excitation on both planes with different frequencies, as well as using different techniques (e.g. pulse, CWFM, etc.).

## Acknowledgements

This work has been carried out within the framework of the EUROfusion Consortium and has received funding from the Euratom research and training programme 2014-2018 under grant agreement No 633053. IST activities also received financial support from *Fundação para a Ciência e Tecnologia* through project UID/FIS/50010/2013 and via the PhD scholarship FCT-SFRH/BD/52414/2013. UL activities received financial support from ANR CHROME ANR-12-BS0006-01. The views and opinions expressed herein do not necessarily reflect those of the European Commission.

## References

[1] Lijun Xu, Naichang Yuan, IEEE Antennas And Wireless Propagation Letters **5**, 335–338 (2006).

[2] Filipe da Silva, Martin Campos Pinto, Bruno Després, Stéphane Heuraux, Journal of Computational Physics **295** (2015) 24–45.

[3] F. da Silva, S. Heuraux, S. Hacquin, M. E. Manso, Journal of Computational Physics **203**, 467–492 (2005).

[4] J-P. Berenger, Journal of Computational Physics **114**, 185–200 (1994).

[5] MPI official website, `http://www.mpi-forum.org/`.

[6] OpenMP official website, `http://openmp.org/wp/`.

# Numerical evaluation of x-mode beam broadening at plasma edge using a FDTD fullwave code

F. da Silva[1], S. Heuraux[2], E. Sysoeva[3], E. Gusakov[3], A. Popov[3]

[1] *Instituto de Plasmas e Fusão Nuclear, Instituto Superior Técnico, Universidade de Lisboa, 1049-001 Lisboa, Portugal*

[2] *Institut Jean Lamour, UMR 7198 CNRS-University of Lorraine, Vandœuvre, France*

[3] *Ioffe Institute, 194021 Politekhnicheskaya st 26, St Petersburg, Russia*

**Abstract**

Microwave beams are commonly used in fusion plasma machines either for diagnose, such as in reflectometry or Collective Thomson Scattering, or for heating purposes being ECRH an example of it. In this work we look into the broadening of the width of probing beam for the case of extraordinary mode propagation in turbulent inhomogeneous plasma for different turbulence k-spectra and plasma conditions using a 2D fullwave finite-difference time-domain code.

## Introduction

The width and positioning of the probing play an important role in reflectometry, namely in wave number selection. It is also important for heating systems for having an efficient deposition of power. Consequently it is an important contemporary topic of research since high levels of edge turbulence and long plasma paths are expected in future fusion machines such ITER, a fact that lead to several works dealing with the enlargement of ordinary mode probing beams, either numerical [3, 1] (reflectometry) or analytical [2] (ECRH). The ECRH studies reach conclusions that are also relevant for reflectometry. We propose to extend these studies on beam behavior to the X-mode using a 2D FDTD full-wave code, REFMULX. The code has two kernels implemented, a classic discretization, the Xu-Yuan kernel (XYK)[4] with modifications (MXYK) and a new innovative energy conservative kernel, proposed by B. Després and M. Campos Pinto, the Després-Pinto kernel (DPK) [5], which changes time centering of **J** and relates **J** and **E** in a Crank-Nicolson way. These kernels' algorithms are different and can be used to crosscheck the results. With plasma movement, a correction is formally needed to the **J** PDE and has been included in the DPK. The weight of the correction will depend on $\partial_t n_e$ and can be either important or negligible. The present results with DPK with and without correction are very similar implying that the correction plays a minute role. It also assures that the use of Xu-Yuan kernel is valid (no correction on **J** is yet implemented on XYK).

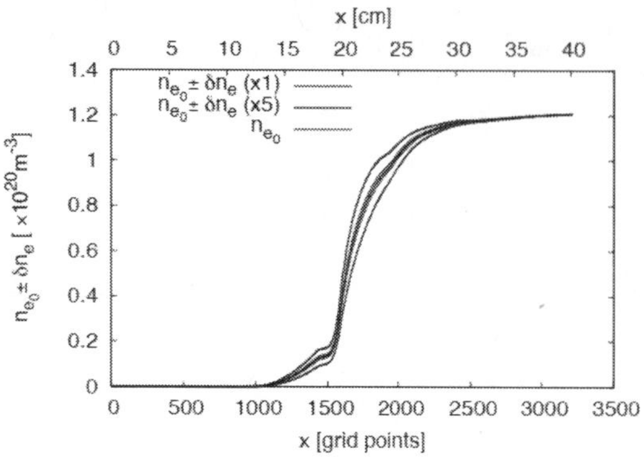


(a) Density profile.

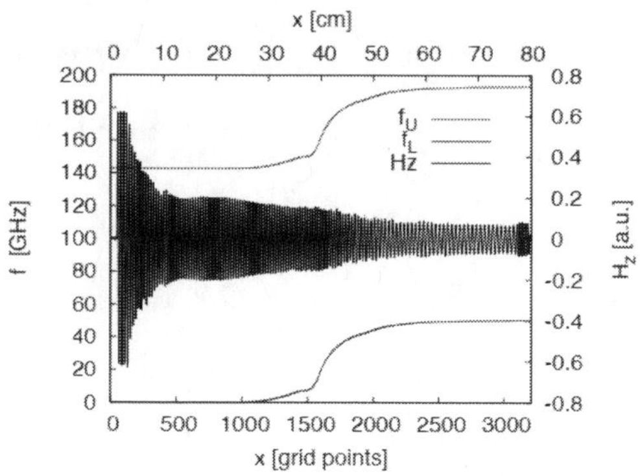


(b) Frequency access and axial cut of $H_z$.

Figure 1: On the left the density profile $n_e$ used is shown in black. Around $n_e$ the RMS envelopes for the added turbulence for the lowest and the highest levels used. On the right the frequency access shows that for a frequency of $f_c = 60\,\text{GHz}$ the wave traverses the plasma *between* the upper and lower frequencies cutoffs.

## Numerical Simulation and results

The approach to this work follows the one used in [2]. We use a plasma slab model with $2400 \times 1900$ grid points ($60 \times 47.5\,\text{cm}$). The size of the grid is $3212 \times 1900$ grid points ($80.3 \times 47.5\,\text{cm}$)—The extra 20.3 cm account for vacuum. A radial density profile, compatible with ITER, was chosen (see Fig. 1a) and probed with a frequency $f_c = 60\,\text{GHz}$. For this value of $f_c$ no cutoff present and the probing beam propagates *above* lower cutoff as observed in Fig. 1b. The number of iterations on each run was 322,240 iterations and for higher statistics averages of several runs were used. For a given amplitude $A(i,j)$ and a random phase $\varphi(i,j)$, a turbulence scenario can be build using Fourier synthesis (calculated with IFFT). Long poloidal turbulence matrices ($L_y = 17{,}640\,\text{gridpoints} = 4.41\,\text{m}$) cycle through the grid with a poloidal velocity $V_{pol}/c = 10\%$. Also as a given input to the problem is the radial density fluctuation profile. The beam is constructed from several poloidal cuts of the field $H_z$ at different radial positions which are integrated on time $\int |H_z|^2 \mathrm{d}t$. This gives a spatial distribution (mapping) of the probing wave density power. Even in the presence of turbulence where the probing beam may be split in several sub-beams at each instant, there will be an average localization of the power deposition correlating with this integrated beam. Running with vacuum the characteristics of the injected beam can be obtained, namely its half-power width. The beam is injected along the antenna axis perpendicular to the plasma. With plasma and a magnetic field $B_0 = +5.1\,\text{T}$ without turbulence ($\delta n_e = 0$) a deviation and enlargement of the beam is noticed. The strong density gradient is at the origin of this deflection [6] and it is put in evidence by experimenting with differ-

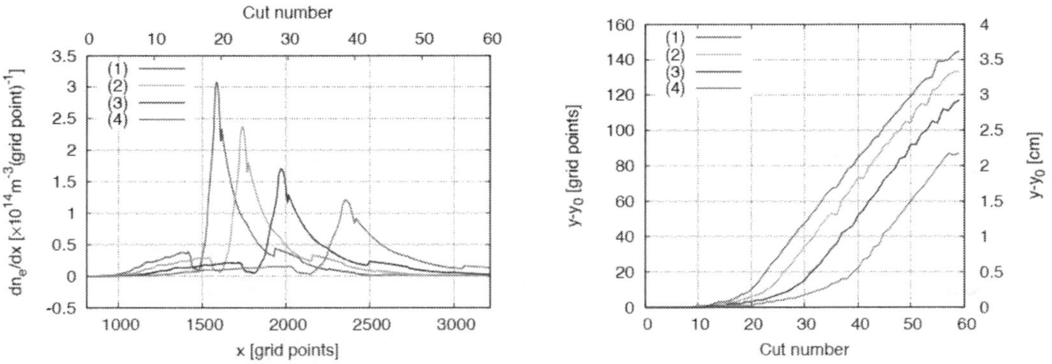


(a) Density gradients $\partial n_e/\partial x$ (b) Position of the maxima of the integrated beam.

Figure 2: The position where density gradients jump up correlate with the augmentation of the deviation suffered by the beam. Density gradient (1) corresponds with the profile displayed in Fig. 1a

ent density gradients $\partial n_e/\partial x$ as shown in Fig. 2a. The maxima of the deflected beams are displayed on Fig. 2b. Reversing the sign of the magnetic field $B_0 = -5.1\,\mathrm{T}$ a change in the direction of the deviation is observed. The value of the maxima is slightly lower though. To notice that both kernels give equivalent consistent results conferring a validation of them. Turbulence adds to the degradation of the beam enlarging it and deviating the position of the maxima. In order to obtain better statistics results were build using an average from 7 runs from different realizations of turbulence matrices making it *equivalent* to $2,255,680$ *iterations* assuming ergodicity.

## Conclusions

Direction of the turbulence velocity plays a role in enlargement of the beam. The results are important to help understand behavior of a probing beam having applications to reflectometry, in selection of wavenumber, probing path and on the effective spot of the beam. Also for uses in power systems for the correct deposition of power. These simulations raise interesting points for follow up work such as the role of the density gradients at the edge (similar to those found in ITER), the impact of turbulence (spectra, radial distribution), the magnetic field bias and the influence of the direction of turbulence flow. Both kernels give consistent results.

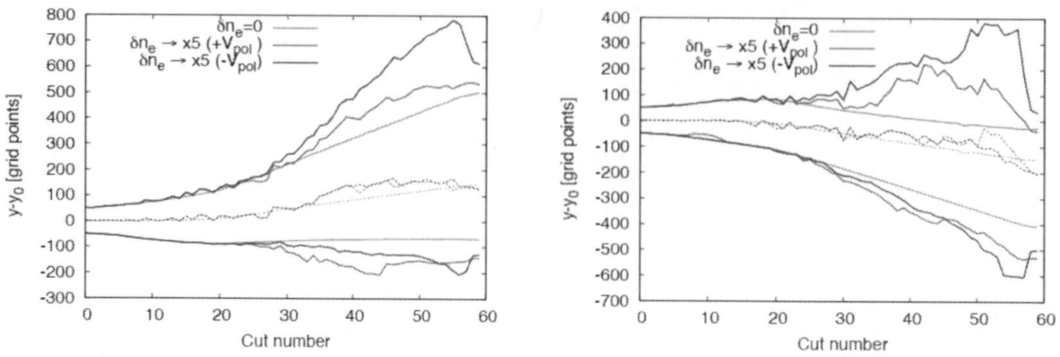


(a) Half width power for $B_0 = +5.1\,\mathrm{T}$.

(b) [Half width power for $B_0 = -5.1\,\mathrm{T}$

Figure 3: Half-power widths of the integrated beams with positive and negative magnetic field $B_0$. The broken lines show the position of the beams' maxima while the continuous lines show the position of the half-power width at each side of the maxima. The *positive* direction of the turbulence flow is displayed in red ($+V_{pol}$) while the *negative* ($-V_{pol}$) in blue. For reference the beam without turbulence appears in grey.

## Acknowledgements

IPFN activities received financial support from *Fundação para a Ciência e Tecnologia* through project UID/FIS/50010/2013. UL activities received support from ANR CHROME ANR-12-BS0006-01.

## References

[1] F. da Silva, S. Heuraux, E. Gusakov, A. Popov, Proc. 11th Intl. Reflectometry Workshop for fusion plasma diagnostics — IRW11 (Padova, May, 2011).

[2] E. Sysoeva, F. da Silva, E. Gusakov, S. Heuraux, A Popov, Nuclear Fusion, **55** (2015).

[3] F. da Silva, S. Heuraux, E. Gusakov, M. Manso and A. Surkov, Proc. 8th Intl. Reflectometry Workshop for fusion plasma diagnostics — IRW8 (St Petersburg, May, 2007).

[4] Lijun Xu, Naichang Yuan, IEEE Antennas And Wireless Propagation Letters **5**, 335–338 (2006).

[5] Filipe da Silva, Martin Campos Pinto, Bruno Després, Stéphane Heuraux, Journal of Computational Physics **295** (2015) 24–45.

[6] R. B. White and F. F. Chen, Plasma Physics **16** (1974) 565.

# Simulation of multiband swept reflectometry for profile evaluation on DEMO using a FDTD Maxwell fullwave code

F. da Silva[1], A. Silva[1], S. Heuraux[2]

[1] *Instituto de Plasmas e Fusão Nuclear, Instituto Superior Técnico, Universidade de Lisboa, 1049-001 Lisboa, Portugal*

[2] *Institut Jean Lamour, UMR 7198 CNRS-University of Lorraine, Vandœuvre, France*

**Abstract**

Numerical simulation constitutes an important tool to understand and access the capabilities of existing or planed reflectometry diagnostics such as ITER and DEMO. Microwave diagnostics, like reflectometry and ECE, with their need for reduced access, front-end robustness, space coverage and spatial resolution are strong candidates to provide DEMO with measurements of electron density and temperature profiles and their associated fluctuations. Frequency modulated reflectometry is a technique widely used in fusion plasmas to measure the electron density and to extract information on density fluctuations. O-mode reflectometry is planed to cover the plasma from edge to core both from high and low field side (LFS) in a range of 18–110GHz. In this work, using finite-difference time-domain full-wave codes for O-mode (REFMUL), we do a first evaluation of a DEMO1 plasma scenario, simulating the beahviour of the Q band (33–50GHz) in the LFS. We conducted a first assessment of possible problems that may be present. These simulations helped us not only to get a first assessment of the DEMO environment from the point of view of microwave diagnostics in general but also of the availability of data and check if the requirements for performing microwave simulations.

## Simulation of the LFS plasma position reflectometer system

A first evaluation of the LFS PPR behavior was done using a 2D full-wave Maxwell FDTD code REFMUL [1]. The main inputs for the simulations come, first from a CAD model of the DEMO machine, which includes the blanket wall. This was used to determine the limits of the wall facing the plasma through which the reflectometer will reach the plasma. At this early state of DEMO development there is not yet a preliminary design of the antennas and how they gain access to the plasma chamber. We have based our model loosely of the same principles planed for ITER. The antennas will be oversized truncated waveguides, one for emission and the second for reception, probing through a gap in the blanket [2, 3], as shown in Fig 1. The second input is a simulation of equilibrium for a DEMO1 scenario from which the data for electronic density $n_e(R,Z)$, needed for this first

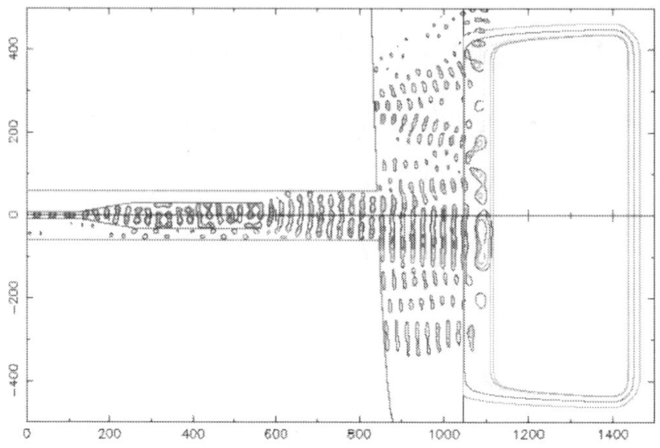


Figure 1: *Snapshot of REFMUL's simulation of FM probing of a LFS DEMO plasma.*

set of simulations is obtained. Electronic temperature $T_e$, and the magnetic field are also available. The first inconsistencies arise when facing the equilibrium with the CAD model. The given position for the separatrix would fall within the blanked (as well as all the SOL). Clearly the equilibrium was calculated without taking the blanket into account (or at least this blanket). This will force us to contract the spatial scale given for the equilibrium in order to fit the density within the machine. Other fixes could be envisaged but we believe that the proper solution, in future simulations, consists in having coherent supplied by the DEMO organization(s). Another problem detected is the quality and completeness of the equilibrium data, and in particular the density. The number of data points available is extremely low. Shown in Tab. 1 are the planned microwave bands, which will range from

Table 1: *Microwave bands planned for DEMO o-mode reflectometers and the availability of DEMO data. In bold, the Q band used in the simulations.*

| Bands | f[GHz] | $n_e[\times 10^{19}\,\mathrm{m}^{-3}]$ | Data |
|---|---|---|---|
| — | 15–22 | 0.028–0.060 | NO |
| — | 22–33 | 0.060–0.135 | NO |
| **Q** | **33–50** | **0.135–0.310** | **NO** |
| V | 50–75 | 0.310–0.698 | Doubtful |
| W | 75–110 | 0.698–1.500 | YES |

15GHz up to 110GHz divided by 5 bands, together with the densities covered and the

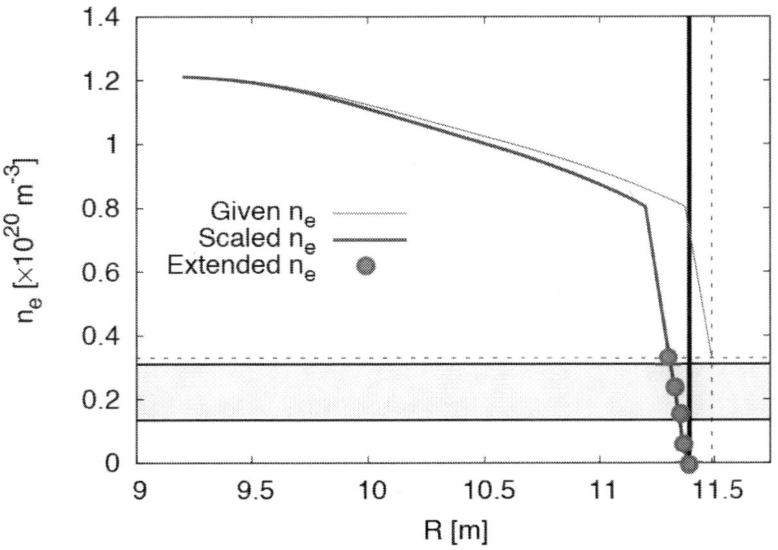


Figure 2: *Axial cut of the 2D density maps, the given data and the modified one, to enable simulations.*

DEMO data supplied for these simulations. One of the goals of reflectometry in DEMO is plasma positioning, which will use frequencies from the first three bands to probe the edge of the plasma within the SOL. Nevertheless the equilibrium provided stops at the separatrix, and for these initial bands no data is available. For the simulations we had to extrapolate from the edge density. In Fig. 2 an axial cut (along $R$) of the density 2D maps summarizes the problems we have encountered and the solutions needed for performing the simulations presented here. The vertical black thick line marks the blanked wall and the vertical broken line the given position of the separatrix. The horizontal broken line marks the limit of the available data. Knowing that the dark band delimits the Q band used in the simulation. The extrapolation and scale contraction needed can be observed comparing the original data supplied in red and the one used in the simulations in blue. The modified 2D $n_e(R,Z)$ map together with the antenna/blanked setup extracted from the CAD model was integrated in REFMUL and a FM frequency swept probing of the plasma simulated. A snapshot of the electric field map appears in Fig. 2. The occurrence of multi-reflections between the steep plasma and the blanket wall and the possibility of excitation of cavity-like modes within the plasma gap can be sources for disturbance of reflectometry signals, being thus important to evaluate its impact. The simulation signals obtained with this synthetic reflectometer can be subjected to the usual signal processing techniques used in the field (or act as a test bed for new ones) to get the relevant information. In Fig. 3 we show the phase derivative obtained using two techniques, an In-phase/Quadrature detection and using a Sliding FFT (SFFT). The I/Q detection being more sensitive to small phase perturbations shows all the impact of

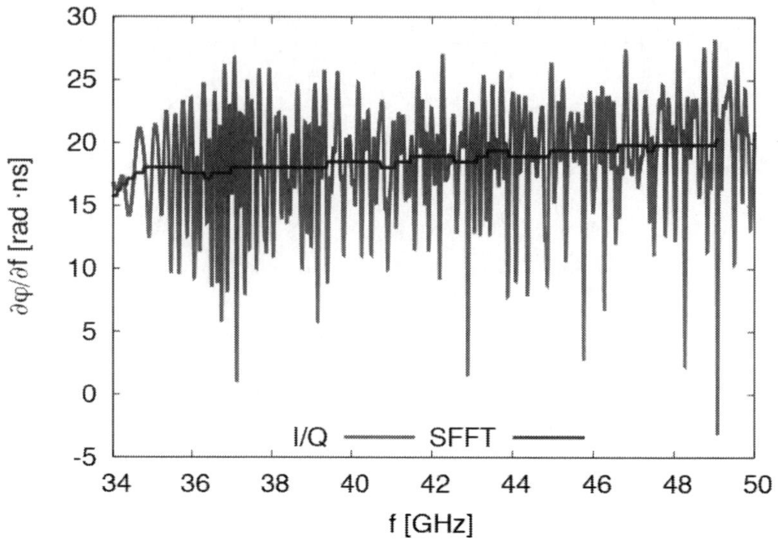


Figure 3: *Phase derivative obtained with a I/Q detection (red) and with a Sliding FFT (blue).*

the multi-reflections and spurious modes occurring during the sweep. The SFFT allows a much cleaner signal but information on the spurious effects is lost. The final use of the data will determine the choice of the technique. These first simulations, using this band and under the limitations forced by the input data, point out reflectometry as a suitable diagnostic to be used in DEMO. This is of course the first step of a work that must be continued and improved. For that is essential better and more accurate edge density profiles.

## Acklowledgements

IPFN activities received financial support from *Fundação para a Ciência e Tecnologia* through project UID/FIS/50010/2013. UL activities received support from ANR CHROME ANR-12-BS0006-01.

## References

[1] F. da Silva, S. Heuraux, S. Hacquin, M. E. Manso, Journal of Computational. Physics **203** 467-492 (2005).

[2] F. da Silva, S. Heuraux and M. Manso, Journal of Plasma Physics **72(6)**, 1205–1208 (2006).

[3] F. da Silva and S. Heuraux, Review of Scientific Instruments **79(10)**, 10F104 (2008).

# Scientific workflows for MHD stability chain analysis of tokamak plasmas

R. Coelho [1], A. Merle [2], G. Vlad [3], D. Yadikin [4], M. Owsiak[5], O. Sauter [2], I. Chapman [6], M. Dunne [7], P. Buratti [3] and JET contributors& and ASDEX Upgrade Team and the EU-IM Team*

*EUROfusion Consortium, JET, Culham Science Centre, Abingdon, OX14 3DB, UK*

[1] Instituto de Plasmas e Fusão Nuclear, Instituto Superior Técnico, Universidade de Lisboa, 1049-001 Lisboa, Portugal

[2] École Polytechnique Fédérale de Lausanne (EPFL), Centre de Recherches en Physique des Plasmas (CRPP), CH-1015 Lausanne, Switzerland

[3] ENEA for EUROfusion, via E. Fermi 45, 00044 Frascati (Roma), Italy

[4] Department of Earth and Space Sciences, Chalmers University of Technology, SE-41296 Göteborg, Sweden

[5] Poznan Supercomputing and Networking Center, IChB PAS, Noskowskiego 12/14, Poznan, Poland

[6] Culham Centre for Fusion Energy, Culham Science Centre, Abingdon, Oxfordshire, OX14 3DB, UK

[7] Max Planck Institute for Plasma Physics, Boltzmannstr. 2, 85748 Garching, Germany

## I – Introduction

The onset of MHD activity in present tokamak plasmas plays a detrimental role in the discharge performance and operational limits. The characterisation of the MHD modes observed is often assisted by predictions from linear MHD codes. Estimation of the MHD stability boundaries in operational plasma scenarios foreseen for ITER and future devices such as DEMO is also crucial. Building confidence on MHD code suite predictions calls for the benchmark of the high-resolution MHD equilibrium and stability solvers involved, a task that is greatly facilitated when all codes share the same physics and machine data ontology and methods for reading and writing the data, as in the European Integrated Modelling (EU-IM) framework [1], presently maintained by the EUROfusion Code Development Work Package. Under its auspices, a Kepler [2] scientific workflow for the analysis of MHD activity in tokamak plasmas was developed (see Figure 1), pluggable to equilibrium reconstruction or discharge simulator workflows. A custom "actor" extracts, from full domain

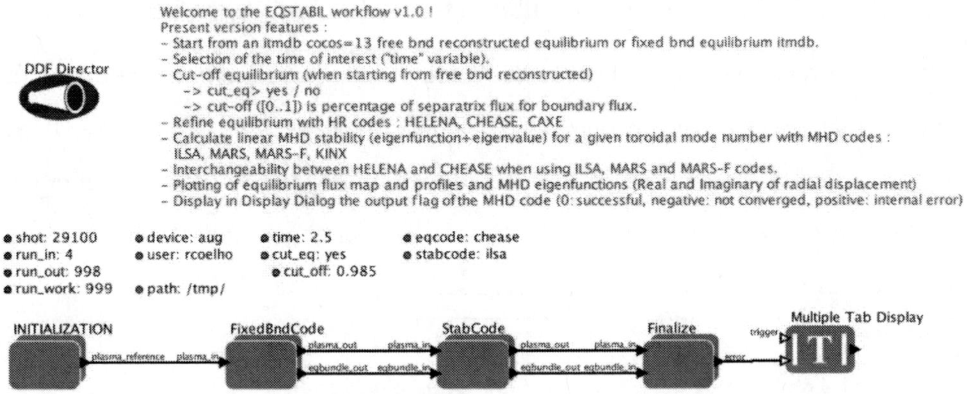


*Figure 1 – Kepler workflow for equilibrium and MHD stability of tokamak discharges*

* See http://www.euro-fusionscipub.org/eu-im

& See the Appendix of F. Romanelli et al., Proceedings of the 25th IAEA Fusion Energy Conference 2014, Saint Petersburg, Russia

equilibrium with a separatrix, a nested closed flux surface equilibrium with plasma boundary flux at selected fraction of the separatrix flux. The workflow presently includes HELENA [3], CHEASE [4] and CAXE [5] equilibrium codes and ILSA [6], MARS [7], MARS-F [8] and KINX [5] linear MHD stability codes. Modular approach allows equilibrium and stability codes to be fully interchangeable for compatible equilibrium metrics e.g. ILSA/MARS/MARS-F, compatible with the jacobian $J \propto R^2$ straight field line coordinate system provided by CHEASE/HELENA. KINX, capable of addressing plasmas with separatrix, is limited to using a custom made quasi-polar quadrangular grid adapted to nested magnetic surfaces provided by CAXE.

## II – Benchmark results

Synthetic equilibria (circular and up-down symmetric elongated with equidistant radial meshes) and experimental (JET #77877 t=46.2s and AUG #29100 t=1.62) equilibria unstable to either n=1 internal or global ideal kink modes were used. The tests focused on the growth rate ($\gamma$) convergence with poloidal harmonic spectra (except for KINX since it is not a spectral code) and with the radial/poloidal mesh size. For the poloidal harmonic (m) convergence, alternate convergence in upper/lower limits was done, starting from a relatively small range $m=[m_1,m_2]$, until a relative error threshold ($\sim 10^{-6}$) in the real part of $\gamma$ was reached. The converged setting of each code was then used for the radial/poloidal grid size scan and verify the expected $1/N_{grid}^2$ scaling ($N_{grid}$ being the number of grid points). ILSA runs were always carried out using the MISHKA-1 code kernel. The circular synthetic equilibria ($N_\psi$=200, $N_\theta$=128) had a perfect conducting wall far away (no-wall limit) from the plasma boundary i.e. $R_{ext}=(R_{wall}-R_{geo})/a=3.5$, while the elongated equilibria ($N_\psi$=200, $N_\theta$=128) had the wall at the plasma boundary. The agreement between all codes in the converged $\gamma$ is very good, with a normalised converged standard deviation in $\gamma$ ($g_N$) of order ~[0.3-1.8]%, as observed from Table I, although poloidal spectra range becomes noticeably larger in ILSA for elongated plasmas. Conversely, m-range in MARS/MARS-F may be reduced when using equal-arc length coordinate system provided by CHEASE (not used in scaling exercises).

| | Circular | | | Elongated | | |
|---|---|---|---|---|---|---|
| Code | m1 | m2 | Growthrate | m1 | m2 | Growthrate |
| ILSA/CHEASE | -3 | 13 | 5.65903380E-02 | -30 | 40 | 6.94658070E-02 |
| ILSA/HELENA | - | - | - | -28 | 44 | 6.94541270E-02 |
| MARS/CHEASE | -2 | 12 | 5.62763000E-02 | -21 | 39 | 6.96735000E-02 |
| MARSF/CHEASE | -6 | 10 | 5.62715461E-02 | -14 | 30 | 6.98718153E-02 |
| CAXE/KINX | -64 | 64 | 5.84452758E-02 | -64 | 64 | 6.98699193E-02 |

*Table I – Poloidal spectra convergence for the synthetic equilibria cases.*

The radial profile and normalised intensity (integral of eigenfunction squared over radial flux coordinate) of the dominant harmonics of the radial velocity eigenfunction show an excellent match among all codes. Agreement of growth rate scaling with radial/poloidal grid size ($1/N^2$) was verified for radial grid size N>200 and poloidal grid sizes covering the poloidal spectra. As anticipated, $g_N$ drops/saturates as grid size is increased. For the equilibria of JET ($B_T$~2.66, $I_p$~1.7MA, $\beta_N$~2.67, $N_\psi$=300, $N_\theta$=256) and AUG ($B_T$~2.4, $I_p$~0.97MA, $\beta_N$~2.55, $N_\psi$=250, $N_\theta$=512), there is a wider scattering in converged $\gamma$ ($g_N$~3.7% for JET and $g_N$~0.9% for AUG) as indicated in Table II. Equilibrium and stability were done with packed radial meshes around rational surfaces and at the plasma boundary and uniform in vacuum. The conducting wall was placed at $R_{ext}$=1.3 for JET (close to marginal stability – $R_{ext}$~1.08) and 1.5 for AUG.

| | JET(#77877) | | | AUG (#29100) | | |
|---|---|---|---|---|---|---|
| Code | m1 | m2 | Growthrate | m1 | m2 | Growthrate |
| ILSA/CHEASE | -21 | 67 | 5.30770720E-03 | -46 | 72 | 3.05979620E-02 |
| ILSA/HELENA | -29 | 67 | 5.49354640E-03 | - | - | - |
| MARS/CHEASE | -19 | 38 | 5.10869000E-03 | -20 | 35 | 3.01958903E-02 |
| MARSF/CHEASE | -9 | 33 | 5.30046053E-03 | -14 | 30 | 3.03251268E-02 |
| CAXE/KINX | -128 | 128 | 5.62557314E-03 | -256 | 256 | 3.06957636E-02 |

*Table II – Poloidal spectra convergence for the JET/AUG equilibria cases.*

The grid scaling tests for the JET and AUG cases are shown in Figure 2 and evidence the different convergence trends of the several codes and the sensitivity to different equilibrium solvers and/or radial meshes (packing is done differently in CHEASE/HELENA/CAXE). Convergence for JET equilibrium is challenged by the proximity to marginal stability and thus to the particular meshing used in the equilibrium and MHD codes. Radial velocity eigenfunctions and normalised intensities are in good agreement and, as anticipated, show slight differences (<5%) depending on the equilibrium code. The rougher grid scaling observed with ILSA when compared to the other codes is also reflected on the eigenfunctions and normalised intensities (closest match between MARS, MARS-F and KINX).

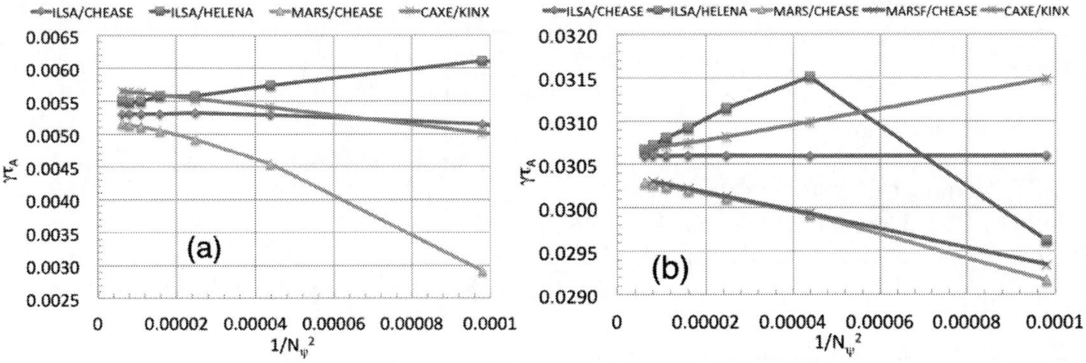


*Figure 2 – Radial grid scaling of n=1 normalised growth rate for JET (a) and AUG (b) equilibrium.*

The dependence of $\gamma$ on $R_{ext}$ was also probed for JET and AUG plasmas (same code parameters as used for the poloidal convergence) and the marginal stability threshold yielded similar results on all codes, with some differences observed closer to the no-wall limit, as indicated in Figure 3, in particular for the JET case. Both plasmas have similar elongation (~1.7) but triangularity is 25% higher in JET and AUG plasma has no upper triangularity thus the JET case results might indicate a higher sensitivity of the codes to the equilibrium used.

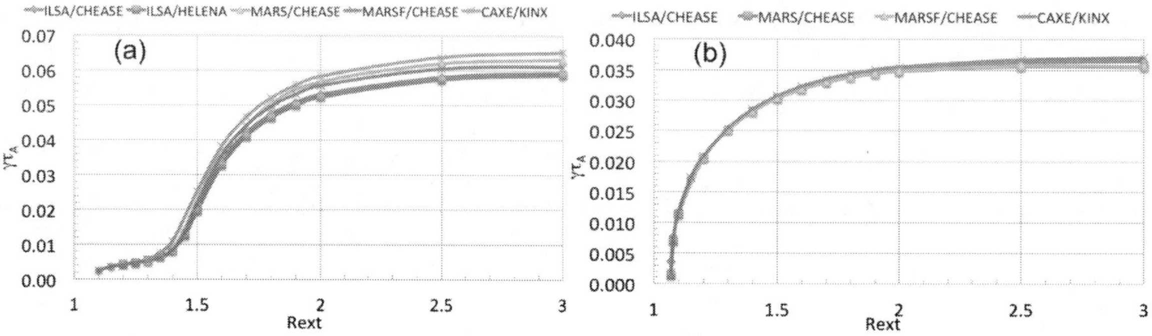


*Figure 3 – Growth rate scaling of n=1 mode with wall position for JET (a) and AUG (b) equilibrium*

## IV – Conclusions

A benchmark of the MHD stability codes installed on the EU-IM modelling platform was performed, addressing code convergence with poloidal harmonics, grid resolution and growthrate dependence on radial location of a conformal conducting wall. Synthetic (circular and elongated) and experimental (JET #77877 and AUG #29100) equilibria were used to study n=1 internal and global kink unstable modes. The agreement on convergence tests, scalings and eigenfunctions/spectra is quite good overall with the experimental JET case showing the largest deviations among codes (though $g_N$ is limited to less than 5%).

**Acknowledgements**
This work has been carried out within the framework of the EUROfusion Consortium and has received funding from the Euratom research and training programme 2014-2018 under grant agreement No 633053. IST activities also received financial support from "Fundação para a Ciência e Tecnologia" through project UID/FIS/50010/2013. The views and opinions expressed herein do not necessarily reflect those of the European Commission.

**References**

[1] F. Imbeaux et al Comp. Phys. Comm. **181**(6):987-998, 2010
[2] https://kepler-project.org/
[3] G.T.A. Huysmans et al., Proceedings of the CP90-Conference on Computational Physics, World Scientific Publishing Co.,1991, p. 371
[4] H. Lütjens et al., Comp. Phys. Comm. **97** (1996) 219
[5] Degtyarev et al., Comp. Phys. Comm. **103** (1997)
[6] C. Konz et al., Scientific Workflows for the Linear MHD Stability Analysis Chain, 37th EPS Conference, Dublin 2010
[7] A. Bondeson et al., Phys. Fluids **B 4** (1992) 1889
[8] Y. Q. Liu et al., Phys. Plasmas **7** (2000) 3681

# Perspective of Negative Triangularity Tokamak as Fusion Energy System

M. Kikuchi[1], S. Medvedev[5], T. Takizuka[6], A. Fasoli[2], Y.Wu[3], P. Diamond[4], X. Duan[7], Y. Kishimoto[8], K. Hanada[9], L. Villard[2], O. Sauter[2], S. Coda[2], B. Duval[2], H. Reimerdes[2], S. Brunner[2], G. Merlo[2], J. Jiang[3], M. Wang[3], M. Ni[3], D. Chen[3], H. Du[3], W. Duan[3], Y Hou[3], A. Ivanov[5], A. Martynov[5], Y. Poshekhonov[5], Y. Ueda[6], L. Yan[7], X. Song[7], G. Zheng[7], J. Liu[7], K. Nagasaki[8], K. Imadera[8], K. Mishra[9], A. Fujisawa[9], K. Nakamura[9], H. Zushi[9], M.J. Pueschel[10], X. Xu[11], P. Zhu[12], D. Told[13], G.Q. Li[14], M. Furukawa[15], T. Ozeki[1], K. Shimizu[1], K. Kawashima[1], H. Urano[1], M. Honda[1], T. Ando[16], M. Kuriyama[16]

[1]*JAEA, Japan,* [2]*CRPP – EPFL, Swiss,* [3]*INEST, China,* [4]*UCSD, US,* [5]*KIAM, Russia,* [6]*OsakaU., Japan,* [7]*SWIP, China,* [8]*KyotoU., Japan,* [9]*KyushuU., Japan,* [10]*U.Wisconsin, USA,* [11]*LLNL, US,* [12]*USTC, China,* [13]*UCLA, US,* [14]*ASIPP, China,* [15]*TottoriU., Japan,* [16]*JAEARetired*

## Introduction

Power and particle control in fusion reactor is quite a challenge and we have studied the negative triangularity tokamak (NTT) as an innovative concept to reduce the transient ELM heat load and the quasi steady-state heat load[1], [2], [3]. A double-null NTT is stable to ideal MHD modes for a reactor relevant $\beta_N > 3$ while it is a magnetic hill configuration[4]. In this paper, we report the configuration study of single-null NTT and its ideal MHD stability.

## Lessons from Fusion Reactor Studies and Experiments

High performance magnetic confinement experiments are limited to short pulse. Fig. 1 a) shows the database of fusion triple product $nT\tau$ v.s duration for existing experiments and the objective regime of ITER and DEMO; the $nT\tau$ is within an order of magnitude while the sustainment time is seven order far from our achievements. The expected steady state heat load is also quite higher compared with existing power sources such as fossil power plants and fission power stations as shown in Figure 1 b). The transient heat load becomes as large as $\sim 1GW/m^2$.

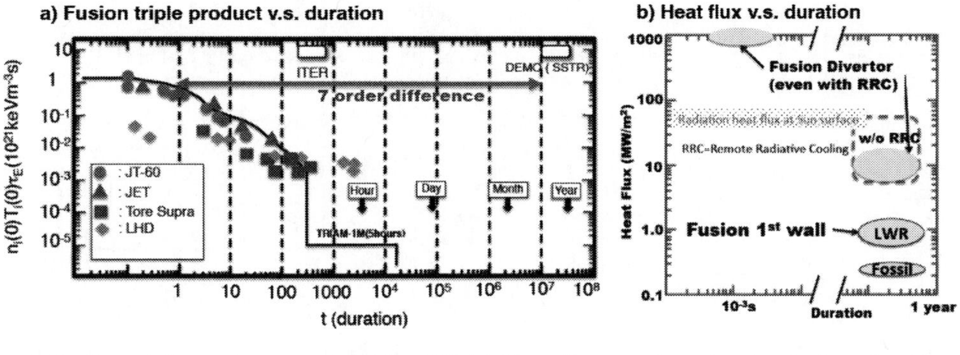


Figure 1: a) Fusion triple product v.s. duration [5]. b) Heat flux v.s. duration for fusion, fossil and fission (LWR) [3]. Fusion has huge transient heat loads by ELM and disruption besides steady heat load.

**Power Handling the First for Fusion Reactor Design**

Present-day standard tokamak operation mode is a H-mode operation with strongly D shaped tokamak to improve core confinement, i.e. "Core the First"- design. ELM strength is enhanced with high edge pedestal in strongly D-shaped plasma associated with high edge bootstrap current and 2nd stability access [6]. Power e-folding length becomes very narrow due to the reduced perpendicular transport in the H-mode edge leading to significantly higher peak heat load at the divertor plate [7]. These issues are associated with this standard tokamak operation mode. If we think the power handling is so critical for the realization of fusion power plant, it may be reasonable to take the design philosophy "Power Handling the First".

**Negative Triangularity Tokamak**

NTT is attractive for the power handling because of its divertor location at the outboard side, which can decrease the heat load without increasing the tokamak major radius, and relevant for the "Power Handling the First" design philosophy. As for the confinement performance, suppression of trapped electron modes (TEM) in the NTT configuration has been observed in TCV[8], which gives a possibility to apply NTT to the future reactor. On the other hand, NTT looses magnetic well property to stabilize interchange instabilities ($D_M - \frac{1}{4} < 0$ is the stability condition and $D_M > 0$ is magnetic hill). Therefore shear stabilization is expected in NTT. We started our assessment on NTT with the double-null (DN) configuration where the strong shear is realized at the edge region. The result shown in Fig.2 was given by Medvedev et al. in FEC 2014 and published in Ref. [4]. Key optimization is to reduce pressure gradient near the core region ($\sqrt{\psi/\psi_s} < 0.5$) to be marginally stable against Mercier mode ($D_M = \frac{1}{4}$) expecting kinetic stabilization[9], while the limiting pressure gradient is determined by ballooning modes and is further lower than marginal stability against Mercier stability at outer half region ($\sqrt{\psi/\psi_s} >$ 0.5). The stability calculations with the KINX code [10] imply the NTT can be stable at the reactor relevant normalized beta $\beta_N > 3$.

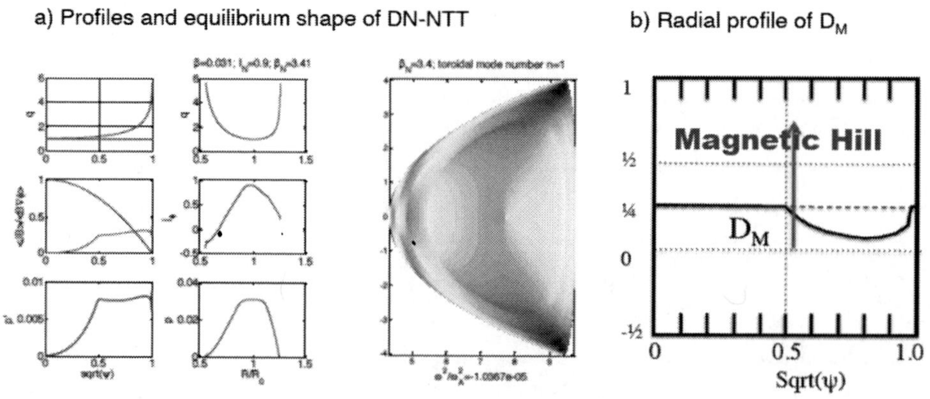


Figure 2: a) $q, j, p' and p$ profiles of DN-NTT [4]. b) Radial profile of $D_M$ [4].

**Single Null Negative Triangularity Tokamak**

While ideal stability of DN-NTT is reasonably good, DN is subject to very careful control of vertical position to equalize heat flux sharing between upper and lower divertors [11]. We have evaluated the stability of single-null (SN) NTT. Figure 3 shows plasma shape for stability calculation and optimized profiles for interchange and kink/ballooning modes. The machine parameters are major radius $R_p = 8.6m$, aspect ratio $A = (R_p/a_p) = 3.5, \kappa = 1.75$, $\delta_u/\delta_x = -0.5/-0.9$, $I_p = 15MA$, $I_N = 1.0(\beta_N = 3.46)$. Stability calculation shows this configuration is stable to $\beta_N = 2.70(n=1), 3.05(n=2), 3.18(n=3), 3.24(n=4), 3.26(n=5)$ and $3.45(n=\infty)$ without wall stabilization. If we place wall at $a_w/a = 1.3$ similar to ITER, we have improved stability $\beta_N(n=1) = 3.11$. The axisymmetric (vertical instability) growth rate is several times larger than for ITER, $\gamma = 24s^{-1}$ but still within the limit $\gamma < 50s^{-1}$ of the active vertical stability control system.

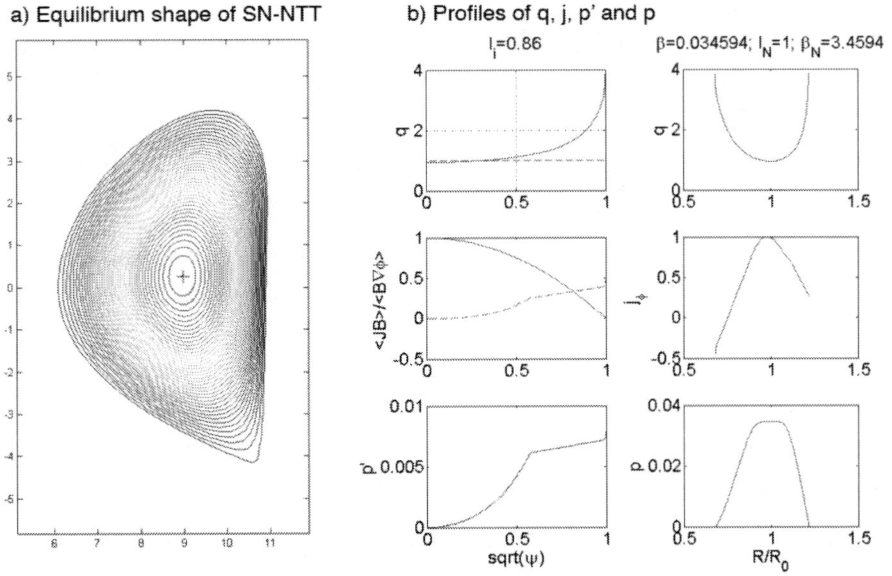


Figure 3: a) Equilibrium shape of SN-NTT. b) $q, j, p'$ and $p$ profiles of SN-NTT.

**Divertor Configuration of SN-NTT**

For the "Power Handling the First" philosophy, $\delta_x = -0.9$ may enable the use of advanced divertor concepts. The Snowflake [12] is one of candidates. Recently Takizuka [13] proposed a new and robust method to reduce heat load to the divertor target called the **Flux-Tube-Expansion** (FTE) by combining long leg and flux tube expansion. The key advantage of FTE divertor is that the divertor leg position can be robust against various plasma perturbations with three sets of divertor control coils. Another more important advantage is that internal coils for flux tube expansion (shown in FTE coils in Fig.4) can be small (each coil carries $\sim 4MAT$) to achieve flux tube expansion of $B_p^0/B_p^{FTE} \sim 3$ which may be realized by using the sector coils.

We may need non-pure tension shape TF coils if we want to eliminate the interlinked PF coils to produce strongly negative upper triangularity $\delta_u$ as shown in Fig. 4 a). Since we expect the stabilization of TEM via Shafranov shift [14], the optimization of upper triangularity depends on improved confinement with Shafranov shift for TEM by operating high $\beta_p$ regime relevant for efficient steady state operation. This is for the future study.

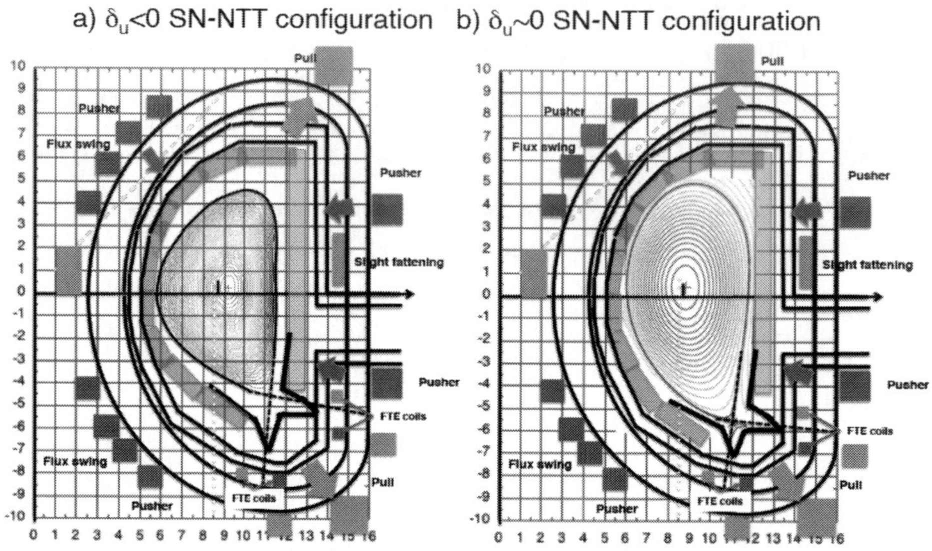


Figure 4: a) $\delta_u = -0.5$, $\delta_x = -0.9$ SN-NTT configuration with flux tube expansion divertor. b) $\delta_u \sim 0$, $\delta_x = -0.9$ SN-NTT configuration with flux tube expansion divertor.

## References

[1] M. Kikuchi, T. Takizuka, Plenary at US-EU TTF 2013 (Santa Rosa, US) 4/9-12, 2013
[2] M. Kikuchi, T. Takizuka, M. Furukawa, JPS Conf. Proc., **1**, 015014 (2014)
[3] M. Kikuchi, A. Fasoli, et al., 1st International E-Conference On Energies 2014, E002
[4] S. Medvedev, M. Kikuchi, et al., Nuclear Fusion **55**, 063013(2015)
[5] M. Kikuchi, M. Azumi, *Frontier in Fusion Research II* , (Springer, 2015) to be published.
[6] A. Loarte et al., Nuclear Fusion **47**, S203(2007)
[7] R.J. Goldston, Nuclear Fusion **52**, 013009(2012)
[8] Y. Camenen et al., Nuclear Fusion **47**, 510(2007)
[9] F. Porcelli, M.N. Rosenbluth, Plasma Phys. Control. Fusion **40**, 481(1998)
[10] L. Degtyarev, A. Martynov, S. Medvedev, et al., Comput. Phys. Commun. **103**(1997)10
[11] T.W. Petri, et al., Proc. 26th EPS, Contr. Fus. Plasma Phys. (Maastricht) **23J**(1999)1237
[12] D.D. Ryutov, Phys. Plasma **14**, 064502(2007)
[13] T. Takizuka, et al., JNM (2014), http://dx.doi.org/10.1016/j.jnucmat.2014.12.065
[14] J.W. Connor, et al., Nuclear Fusion **23**, 1702(1983)

# Development of the multi-fluid transport code on the tokamak flux coordinates

M. Honda[1], A. Fukuyama[2]

[1] *Japan Atomic Energy Agency, Naka, Japan*

[2] *Graduate School of Engineering, Kyoto University, Kyoto, Japan*

## Introduction

We have developed a one-dimensional fluid-type transport code TASK/TX [1]. The code is essentially based on a self-consistent two-fluid model, which consists of two-fluid equations (conservation of mass, momentum and energy) plus Maxwell's equations. It also involves the equations for beam ions [2] and neutrals [3]. It differs from conventional diffusive transport codes mainly in that: the quasi-neutrality condition $n_e = \sum_i Z_i n_i$ and the ambipolar flux condition $\sum_a \Gamma_a = 0$ need not be imposed; the flux-gradient relationship is not used for particle transport; the neoclassical features are self-consistently reproduced. A main drawback is, however, that the governing equations are built on the cylindrical coordinates $(r, \theta, \phi)$, which is equivalent to the large aspect ratio limit of a plasma. In this sense, some physics originating from magnetic geometry such as the Pfirsch-Schlüter flux has been dropped. Furthermore, due to the formulation on the cylindrical coordinates, it is difficult to intuitively understand the models and the results of TASK/TX in comparison with the theory and the models developed on the flux coordinates. Hence we derive the governing equations of TASK/TX on the axisymmetric flux coordinates $(\rho, \theta, \zeta)$ and then numerically implement them.

Hereafter we will expand plasma parameters in terms of a small gyroradius and take into account their non-perturbed (lowest) part. We assume that the drift ordering is appropriate for the momentum equations whereas the transport ordering for the otherwise equations.

## Maxwell's equations

On the axisymmetric flux coordinates, the magnetic field is expressed as $\boldsymbol{B} = \nabla\zeta \times \nabla\psi + I\nabla\zeta$, where $I(\psi) = RB_t$. Maxwell's equations consist of Gauss's law (Poisson's equation), Faraday's law and Ampère's law. From Faraday's law $\boldsymbol{E} = -\nabla\Phi - \dot{\boldsymbol{A}}$, it is found that the covariant components of the electric fields are strongly tied to (the temporal change in) the magnetic fluxes: $E_\theta = \boldsymbol{E} \cdot \sqrt{g}\nabla\zeta \times \nabla\rho = -\dot{\psi}_t, E_\zeta = g_{\zeta\zeta}\boldsymbol{E} \cdot \nabla\zeta = \dot{\psi}(= RE_t)$. Ampère's law relates (the spatial gradient of) the magnetic flux to the current. Taking the scalar product of Ampère's law with $\nabla\zeta$ and the subsequent flux-surface average (FSA) yield (cf. FSA Grad-Shafranov equation)

$$\frac{1}{c^2}\frac{\partial \dot{\psi}}{\partial t} = \frac{1}{V'\langle R^{-2}\rangle}\frac{\partial}{\partial \rho}\left[V'\left\langle\frac{|\nabla\rho|^2}{R^2}\right\rangle\frac{\partial \psi}{\partial \rho}\right] - \mu_0\frac{\langle j^\zeta\rangle}{\langle R^{-2}\rangle}, \quad \frac{\partial \psi}{\partial t} \equiv \dot{\psi}, \tag{1}$$

the latter of which is not only the definition of $\dot{\psi}$ but also one of the governing equations that constitutes Maxwell's equations. This is because it is just alternative form of Faraday's law. Taking the scalar product of Ampère's law with $\boldsymbol{B}$ and then subtracting (1) give the equations for the toroidal flux in the form:

$$\frac{1}{c^2}\frac{\partial \dot{\psi}_t}{\partial t} = V'\left\langle\frac{|\nabla\rho|^2}{R^2}\right\rangle\frac{\partial}{\partial \rho}\left[\frac{1}{V'\langle R^{-2}\rangle}\frac{\partial \psi_t}{\partial \rho}\right] + \mu_0\frac{\langle Bj_\parallel\rangle - I\langle j^\zeta\rangle}{\langle B^\theta\rangle}, \quad \frac{\partial \psi_t}{\partial t} \equiv \dot{\psi}_t. \tag{2}$$

When neglecting the displacement current term that is negligibly small, we readily find that the right-hand sides of (1) and (2) reproduce the expressions of the toroidal and parallel currents. Finally, the Coulomb gauge allows us to simply write FSA Gauss's law as follows:

$$\frac{1}{V'}\frac{\partial}{\partial \rho}\left[V'\langle|\nabla\rho|^2\rangle\frac{\partial \Phi}{\partial \rho}\right] = -\frac{1}{\epsilon_0}\sum_a e_a n_a. \tag{3}$$

**Continuity equations**

A flux-surface-averaged continuity equation is simply given by

$$\frac{1}{V'}\frac{\partial}{\partial t}(V'n_a)\bigg|_\rho + \frac{1}{V'}\frac{\partial}{\partial \rho}[V'n_a\langle(\boldsymbol{u}_a - \boldsymbol{u}_g)\cdot\nabla\rho\rangle] = S_{na}. \tag{4}$$

TASK/TX is distinguished from other transport codes in that $n_a\langle(\boldsymbol{u}_a - \boldsymbol{u}_g)\cdot\nabla\rho\rangle$ is not approximated by the convection-diffusion relationship, but is treated as a dependent variable: The grid velocity $\boldsymbol{u}_g$ and the flux $n_a\langle\boldsymbol{u}_a\cdot\nabla\rho\rangle$ are both self-consistently calculated in the system.

**Momentum equations**

The parallel motion of plasma species regulates neoclassical properties in axisymmetric systems. Using the non-conservative form of the momentum equation, we have up to $O(\delta)$ The lowest order of the viscous term is the neoclassical viscosity. The viscous term and the friction term can be calculated according to the moment approach [4]. Reproducing the driving of poloidal flows, resistivity and the shielding factor of the beam driven current, equations for heat flows have to be simultaneously solved, as is usual for neoclassical transport solvers. Therefore, defining $\langle B\hat{q}_{a\parallel}\rangle = 2\langle Bq_{a\parallel}\rangle/(5p_a)$, the following equations should be solved:

$$m_a n_a\frac{\partial\langle Bu_{a\parallel}\rangle}{\partial t} = -\hat{\mu}_1^a(\langle Bu_{a\parallel}\rangle - BV_{1a}) - \hat{\mu}_2^a(\langle B\hat{q}_{a\parallel}\rangle - BV_{2a}) + \sum_b \ell_{11}^{ab}\langle Bu_{a\parallel}\rangle - \sum_b \ell_{12}^{ab}\langle Bu_{a\parallel}\rangle$$
$$+ e_a n_a\langle BE_\parallel\rangle, \tag{5}$$

$$\frac{5}{2}m_a n_a\frac{\partial\langle B\hat{q}_{a\parallel}\rangle}{\partial t} = -\hat{\mu}_2^a(\langle Bu_{a\parallel}\rangle - BV_{1a}) - \hat{\mu}_3^a(\langle B\hat{q}_{a\parallel}\rangle - BV_{2a}) - \sum_b \ell_{21}^{ab}\langle Bu_{a\parallel}\rangle + \sum_b \ell_{22}^{ab}\langle B\hat{q}_{b\parallel}\rangle. \tag{6}$$

Note that source and sink terms have been omitted here and hereafter.

The toroidal momentum equation is important for TASK/TX because it includes the $\boldsymbol{v}\cdot\nabla\rho$ term that provokes a particle flux as well as a $\boldsymbol{j}\times\boldsymbol{B}$ torque once losses of beam ions produce the non-ambipolar flux. In conventional transport codes the particle transport coefficients evaluated by external modules are directly substituted in the particle transport equation, whereas in TASK/TX coupling of the equations self-consistently brings about particle transport through the continuity equation. While the first-order viscous stress term vanishes due to the CGL form, the second-order term can be expressed as a combination of a convective momentum flux, a.k.a. inward pinch, and a diffusive one plus a residual stress. We have to add a turbulent force $F_a^{\rm QL}$ that drives a turbulence-induced quasilinear particle flux. The toroidal momentum equation is finally given by, where $\langle\mathcal{L}_a\rangle \equiv m_a n_a \langle R u_{a\zeta}\rangle$,

$$\frac{1}{V'}\frac{\partial}{\partial t}(V'\langle\mathcal{L}_a\rangle) = -\frac{1}{V'}\frac{\partial}{\partial\rho}V'\Bigg[\langle|\nabla\rho|\rangle v_{a\zeta}\langle\mathcal{L}_a\rangle + (u_a^\rho - u_g^\rho)\langle\mathcal{L}_a\rangle - \langle|\nabla\rho|^2\rangle\chi_{a\zeta}m_a n_a\frac{\partial\langle R u_{a\zeta}\rangle}{\partial\rho}$$
$$+\langle\Pi_a^{\rm res}\rangle\Bigg] + \sum_b \ell_{11}^{ab}\langle R u_{b\zeta}\rangle - \sum_b \ell_{12}^{ab}\frac{I}{\langle B^2\rangle}\langle B\hat{q}_{b\parallel}\rangle + e_a n_a\langle R E_\zeta\rangle + e_a\frac{\partial\psi}{\partial\rho}n_a u_a^\rho. \quad (7)$$

The radial momentum equation or the radial force balance equation is essentially equivalent to the first-order flow within the flux surface. The leading order is $O(1)$ for pressure and Lorentz force terms, and the other terms are practically ineffective. Thus, we obtain

$$0 = -\frac{1}{m_a}\frac{\partial p_a}{\partial\psi}(\langle B^2\rangle\langle R^2\rangle - I^2) - \frac{e_a}{m_a}n_a\frac{\partial\Phi}{\partial\psi}(\langle B^2\rangle\langle R^2\rangle - I^2) + \frac{e_a}{m_a}n_a I\langle B u_{a\parallel}\rangle - \frac{e_a}{m_a}n_a\langle B^2\rangle\langle R u_{a\zeta}\rangle. \quad (8)$$

**Other equations**

Additionally the heat transport equation and the equation that connects $\langle R u_{a\zeta}\rangle$ and $\langle u_{a\zeta}/R\rangle$ are solved for each species, but their derivation is omitted. As mentioned above, the three equations for neutrals and the two equations for beam ions, if any, are solved as well.

**Numerical results**

We here focus on the neoclassical properties in TASK/TX, which implies that the turbulent particle flux is neglected, i.e., $F_a^{\rm QL} = 0$. For the sake of simplicity, the large aspect ratio plasma with circular cross section is assumed. JT-60U-like plasma parameters are given in the following simulation.

Figure 1 shows the radial profiles of (a) the difference in the charge density and (b) the particle flux for each species, respectively, indicating that automatically the quasi-neutrality is well satisfied and the ambipolar flux condition is satisfied as well. A break of the local charge neutrality in the core region, $(n_{\rm e} - n_{\rm i})/n_{\rm e}$, is of the order of $10^{-7}$, which is always neglected in conventional transport codes, but this tiny discrepancy is indispensable to give rise to the radial electric field

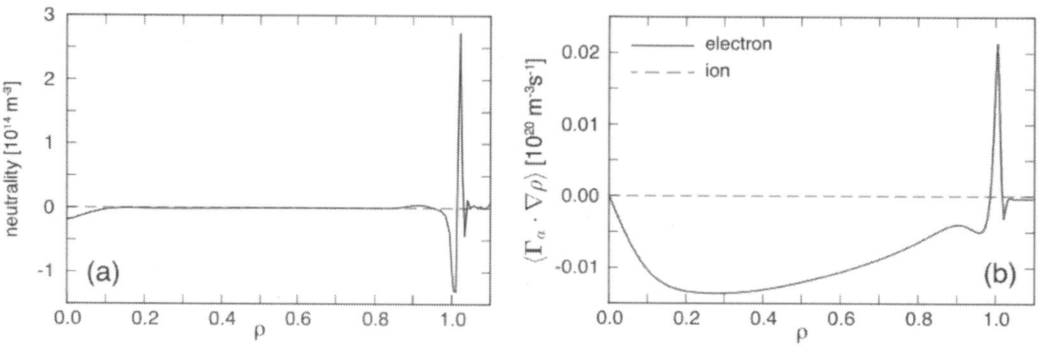


Figure 1: Radial profiles of (a) quasi-neutrality and (b) particle flux for electrons and ions.

$E_r$ through Eq. (3). As obvious in Fig. 1 (b), the cross-field particle flux is automatically ambipolar without imposing ambipolarity. This fact is trivial in the light of neoclassical transport theory because the flux originates from the electric field and friction forces, but the important thing is that it is self-consistently recovered by numerical simulations. The breakdown of the flux cannot be known, but we find that for this case the Ware pinch is dominant over the profile with the aid of NCLASS [5]

Figure 2 displays the comparison of the electron parallel flow $\langle Bu_{e\parallel} \rangle$ calculated in TASK/TX and that by NCLASS: Clearly the good agreement is obtained. In TASK/TX the neoclassical quantities can be internally calculated without external modules such as NCLASS. In this sense, TASK/TX is also a neoclassical transport solver by itself.

This work was supported by JSPS KAKENHI (No. 25820442).

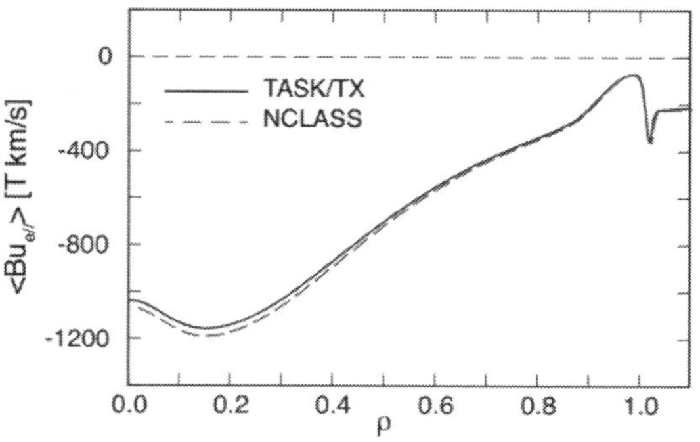


Figure 2: Comparison of $\langle Bu_{e\parallel} \rangle$ calculated in TASK/TX and by NCLASS.

**References**

[1] M. Honda and A. Fukuyama, J. Comput. Phys. **227**, 2808 (2008).

[2] M. Honda et al., Nucl. Fusion **48**, 085003 (2008); *ibid.* **49**, 035009 (2009).

[3] M. Honda et al., J. Plasma Fusion Res. SERIES **9**, 529 (2010).

[4] S. P. Hirshman and D. J. Sigmar, Nucl. Fusion **21**, (1981) 1079.

[5] W. A. Houlberg et al., Phys. Plasmas **4**, 3230 (1997).

# Two-fluid equilibrium analysis of spherical torus plasmas sustained by multi-pulsing coaxial helicity injection

T. Kanki[1], M. Nagata[2]

[1] Japan Coast Guard Academy, Kure, Japan

[2] University of Hyogo, Himeji, Japan

Solenoid-less non-inductive current start-up and steady-state current drive techniques in the spherical torus (ST) have been studied using the coaxial helicity injection (CHI). Recently, a new approach of the CHI operated in multi-pulsing CHI (M-CHI) has been proposed [1]. The M-CHI scenario is that after the plasma current partially decays, a new CHI pulse is applied and the cycle process is repeated to achieve a quasi-steady sustainment and good confinement. To examine the dynamo current drive of ST by the M-CHI, we started the double-pulsing operations in the HIST device, observing that the steep density gradient between the central open flux column (OFC) and closed flux regions causes not only the $\boldsymbol{E}\times\boldsymbol{B}$ drift but also the ion diamagnetic drift to exhibit the two-fluid effect and Hall dynamo [2]. The purpose of this study is to investigate the effect of the M-CHI on the two-fluid equilibrium configurations of ST.

For the two-fluid equilibrium calculations, let us assume axisymmetry about HIST geometric axis in cylindrical coordinates ($r$, $\theta$, $z$). Using the dimensionless variables, an axisymmetric two-fluid flowing equilibrium can be described by a pair of generalized Grad-Shafranov equations for the ion surface variable $Y$ and the electron surface variable $\psi$ [3],

$$\overline{\psi}_i' r^2 \nabla\cdot\left(\frac{\overline{\psi}_i'}{n}\frac{\nabla Y}{r^2}\right)=\frac{r}{\varepsilon}\left(B_\theta\overline{\psi}_i'-nu_\theta\right)+nr^2\left(H_i'-T_iS_i'\right), \tag{1}$$

$$r^2\nabla\cdot\left(\frac{\nabla\psi}{r^2}\right)=\frac{r}{\varepsilon}\left(B_\theta\overline{\psi}_e'-nu_\theta\right)-nr^2\left(H_e'-T_eS_e'\right), \tag{2}$$

and a generalized Bernoulli equations for the number density $n$,

$$\frac{\gamma}{\gamma-1}n^{\gamma-1}\exp[(\gamma-1)S_i]+\frac{u^2}{2}+\phi_E=H_i, \tag{3}$$

$$\frac{\gamma}{\gamma-1}n^{\gamma-1}\exp[(\gamma-1)S_e]-\phi_E=H_e. \tag{4}$$

Here $\boldsymbol{u}$, $\boldsymbol{B}$, $T_\alpha$, $\phi_E$, and $\gamma$ are the ion flow velocity, magnetic field, temperature of species ($\alpha$=$i$, $e$), electrostatic potential, and adiabatic constant, respectively. Also, the two-fluid parameter $\varepsilon$ is defined by the ratio of the ion skin depth $\ell_i=\sqrt{m_i/(\mu_0 ne^2)}$ ($m_i$ and $e$ are the ion mass and electron charge, respectively.) to the system length scale $R$ (the effective radius of the flux conserver (FC) used in the HIST, $R$=0.45 m), and $\varepsilon$=0.0716 is given by using $\ell_i=3.2$ cm through $n = 5.0\times10^{19}$ m$^{-3}$ observed in the HIST experiment. The value of $\gamma$ is assumed as 5/3 in the calculation. Note that $\psi$ corresponds to the familiar poloidal flux function. The poloidal flow stream function $\bar{\psi}_\alpha$, total enthalpy function $H_\alpha$, and entropy function $S_\alpha$ are arbitrary surface functions of their respective surface variables. The three arbitrary functions $\Lambda_\alpha(\equiv\bar{\psi}'_\alpha)$, $H_\alpha$, and $S_\alpha$ for each species can be assumed so as to reflect the experimental results by choosing appropriate function forms,

$$\Lambda_\alpha(x) = \Lambda_{\alpha0} + (\Lambda_{\alpha1} - \Lambda_{\alpha0})\left.\frac{df}{dx}\right|_{x-\Delta x_\alpha;\delta_{\alpha0},\delta_{\alpha1},\delta_{\alpha2}}, \tag{5}$$

$$H_\alpha(x) = H_{\alpha0} + (H_{\alpha1} - H_{\alpha0})\left.\frac{dg}{dx}\right|_{x-\Delta x_{H\alpha};\delta_{H\alpha1},\delta_{H\alpha2}}, \tag{6}$$

$$S_\alpha(x) = S_{\alpha0} + (S_{\alpha1} - S_{\alpha0})\left.\frac{dg}{dx}\right|_{x-\Delta x_{S\alpha};\delta_{S\alpha1},\delta_{S\alpha2}}, \tag{7}$$

$$f(x;\delta_0,\delta_1,\delta_2) = \begin{cases} -\dfrac{1}{\delta_0\delta_1}\left[\dfrac{\delta_0^4}{12}\left(1-\dfrac{2\delta_1}{\delta_0}\right)+x^2\left\{\dfrac{\delta_0\delta_1}{2}+x\left(\dfrac{\delta_0+\delta_1}{3}+\dfrac{x}{4}\right)\right\}\right]; & x<0 \\ -\dfrac{\delta_0^3}{12\delta_1}\left(1-\dfrac{2\delta_1}{\delta_0}\right)-\delta_2\left\{x-\delta_2\log\left(1+\dfrac{x}{\delta_2}\right)\right\}; & x\geq 0 \end{cases}, \tag{8}$$

$$g(x;\delta_1,\delta_2) = \frac{1}{\delta_1+\delta_2}\begin{cases} \delta_1^2\exp[x/\delta_1]; & x<0 \\ \delta_1 x+\delta_2\sqrt{\delta_2^2+x^2}-\delta_2^2+\delta_1^2; & x\geq 0 \end{cases}. \tag{9}$$

Here $\Lambda_{\alpha0}$, $\Lambda_{\alpha1}$, $H_{\alpha0}$, $H_{\alpha1}$, $S_{\alpha0}$, $S_{\alpha1}$, $\Delta x_\alpha$, $\delta_{\alpha0}$, $\delta_{\alpha1}$, $\delta_{\alpha2}$, $\Delta x_{H\alpha}$, $\delta_{H\alpha1}$, $\delta_{H\alpha2}$, $\Delta x_{S\alpha}$, $\delta_{S\alpha1}$, and $\delta_{S\alpha2}$ are constant parameters. To complete two-fluid equilibrium system, we need the auxiliary equations, $Y \equiv \psi + \varepsilon r u_\theta$, $B_\theta = (\bar{\psi}_i - \bar{\psi}_e)/\varepsilon r$, $u_p = |\nabla\bar{\psi}_i|/nr$, and $T_\alpha = n^{\gamma-1}\exp[(\gamma-1)S_\alpha]$. Here $u_p$ is the ion poloidal flow velocity. Equations (1) and (2) have terms of order $1/\varepsilon$ on the right-hand side, and they cause singularities. We employ the nearby-fluids procedure to eliminate the singularities [3].

Next, we consider the boundary conditions for Eqs. (1)-(4). No magnetic flux penetrates the FC, central conductor, and entrance port of the FC. Therefore, $\psi$ is fixed at 0 at the FC wall and surface of the central conductor and entrance port. The bias flux is given by assigning fixed values of $\psi$ to grid points corresponding to the right open end. These values are calculated using the formula,

$$\psi_{\mathrm{bias}}(r) = \frac{4\psi_s}{(R_e - R_c)^2}(r - R_c)(R_e - r), \tag{10}$$

where $R_c$ and $R_e$ are the radii of the central conductor and entrance port, and $\psi_s$ is the maximum value of the bias flux. This Dirichlet boundary condition is used for solving Eq. (2) for $\psi$. The vacuum toroidal magnetic field $B_{t,v}$ is produced by a toroidal field coil current $I_{tf}$ along the geometric axis inside the central conductor. The boundary condition of the flow velocity is required to solve Eq. (1) for $u_\theta$ . We impose the free-slip boundary condition corresponding to no friction at the Dirichlet boundaries. Under the above assumptions and boundary conditions, the equilibrium is numerically determined by using a successive over-relaxation method for updating $\psi$ and a Newton-Raphson method for updating $n$ [4].

On the basis of the numerical results, we investigate the effect of the M-CHI on the two-fluid equilibria. By driving the poloidal current $I_p$ again along the open field lines through applying a new CHI pulse (M-CHI effect), the magnetic helicity is injected again into the plasma through the linkage of the toroidal flux which links the bias poloidal flux. We consider that increasing the poloidal electron flow velocity $u_{ez}$ along the open field lines drives the poloidal current on the open flux region. Therefore, we increase the absolute edge value of the $u_{ez}$ by decreasing the value of $\Lambda_{e0}$ related to the strength of $u_{ez}$. Figure 1 shows the radial profiles of $u_{ez}$ at the midplane. Note that the cases I, II, and III correspond to $\Lambda_{e0}$ =-0.15, -0.16, and -0.17, respectively. In Fig. 1, $u_{ez}$ can be clearly seen to be decreased in the OFC region as the values of $\Lambda_{e0}$ become smaller. The negative values of $u_{ez}$ represent the opposite direction to $I_p$ and $I_{tf}$. As shown in Fig. 2, the poloidal ion flow velocity $u_z$ in the opposite direction to $u_{ez}$, but in the same direction as $I_p$ is increased in the OFC region, enhancing the flow shear around the separatrix in the high magnetic field side, and contributing to the increment of $I_p$. This is because the ion diamagnetic drift velocity $u_{dia}$ is changed in the same direction as the $\boldsymbol{E} \times \boldsymbol{B}$ drift velocity $u_E$ around the separatrix through the negative ion pressure gradient as shown in Fig. 3. Here $u_{dia}$ is comparable to $u_E$, but the ion inertial drift velocity is small. Due to the M-CHI effect, the shear of $u_E$ is enhanced around the separatrix, but the direction of $u_E$ does not change there. As shown in Fig. 4, the density $n$ is decreased in the closed flux region and its negative gradient around the separatrix steepens. The decrease in the density related to the enthalpy, $h_\alpha = (\gamma n^{\gamma-1}/\gamma-1)\exp[(\gamma-1)S_\alpha]$ can be explained using the generalized Bernoulli equations that the increase in the ion flow energy causes the decrease in the enthalpy. The ion temperature profile is almost kept broad one around the separatrix in spite of the M-CHI effect. Therefore, the negative gradient of the ion pressure can be caused depending on that of the density. This steep density gradient can be observed in the HIST experiment when the CHI pulse is applied again during the partial decay of plasma current. As shown in Fig. 5, the toroidal magnetic field $B_\theta$ becomes from a diamagnetic to a paramagnetic profile in the closed flux region due to the increase of $u_z$ in the OFC region, while the diamagnetic $B_\theta$ profile is kept in the OFC region. The toroidal ion flow velocity $u_\theta$ is increased from negative to positive values in the closed flux region as shown in Fig. 5, enhancing the paramagnetic poloidal field. Here the negative $u_\theta$ is the opposite direction to the current. Both the poloidal and toroidal magnetic fields tend to become to the paramagnetic profiles, because both the poloidal and toroidal ion flow velocities tend to contribute to the plasma currents. The M-CHI effect leads the increase of plasma currents and magnetic flux.

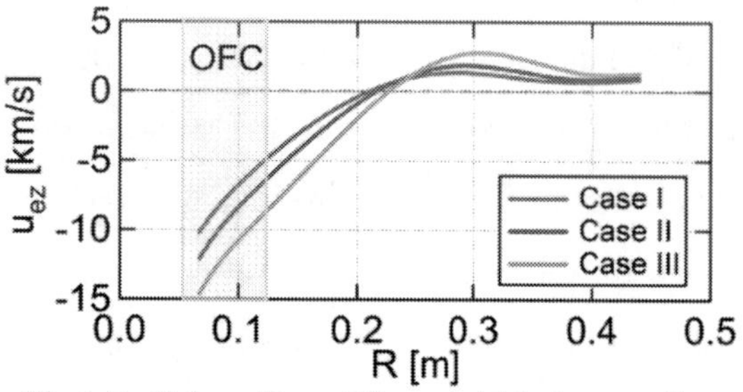


Fig.1 Radial profiles of the poloidal electron flow velocity $u_{ez}$ at the midplane

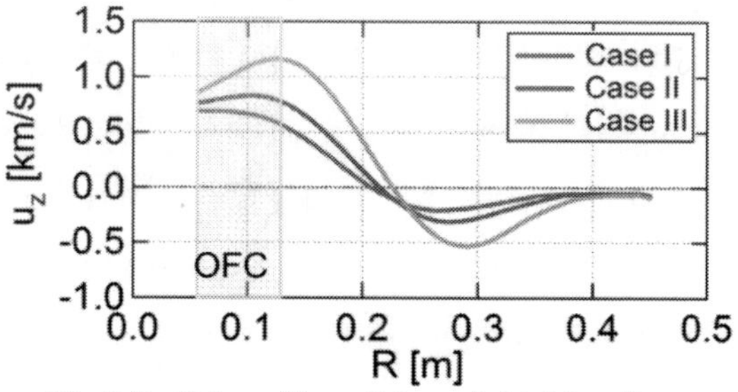


Fig.2 Radial profiles of the poloidal ion flow velocity $u_z$ at the midplane

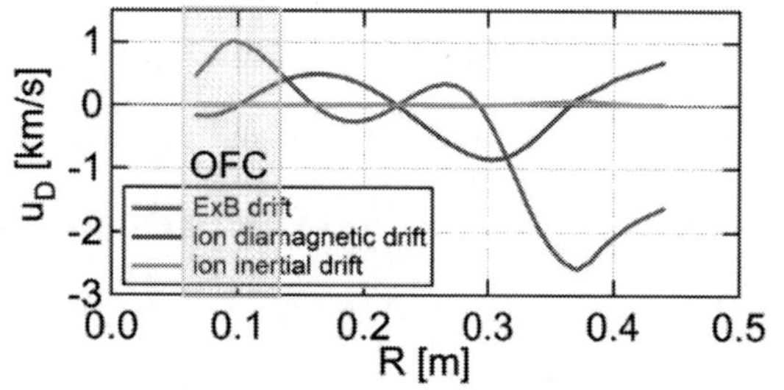


Fig.3 Radial profiles of the ion drift velocity $u_D$ in the poloidal direction at the midplane in the case III

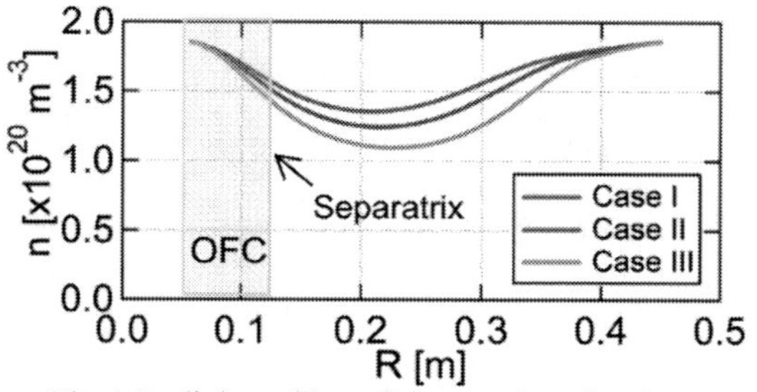


Fig.4 Radial profiles of the number density $n$ at the midplane

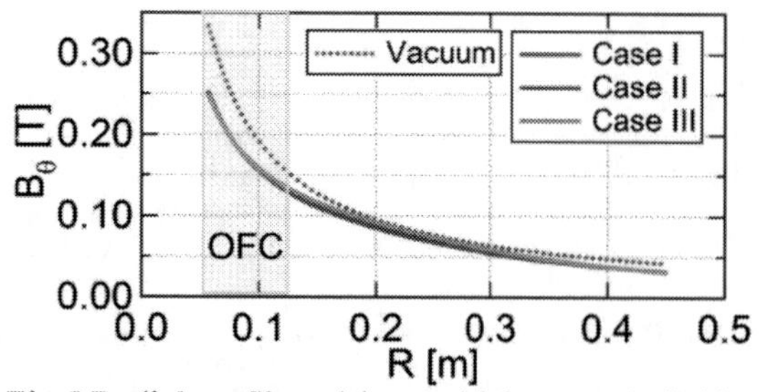


Fig.5 Radial profiles of the toroidal magnetic field $B_\theta$ at the midplane

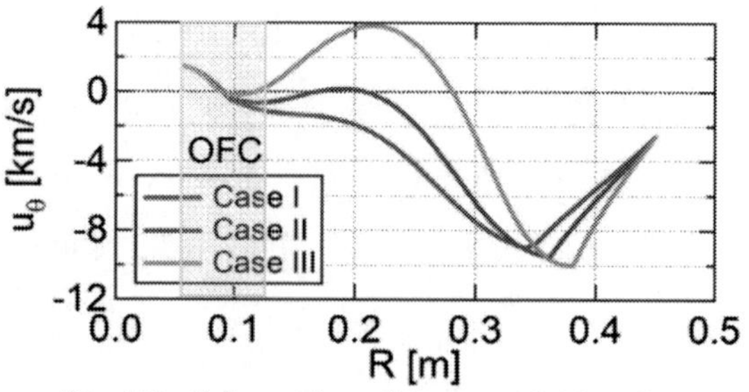


Fig.6 Radial profiles of the toroidal ion flow velocity $u_\theta$ at the midplane

## References

[1] E.B. Hooper, Plasma Phys. Control. Fusion **53**, 085008 (2011).

[2] M. Nagata *et al.*, *24th IAEA Fusion Energy Conference* (San Diego, USA) ICC/1-1Rb (2012).

[3] L.C. Steinhauer and A. Ishida, Phys. Plasmas **13**, 052513 (2006).

[4] T. Kanki *et al.*, Plasma and Fusion Res. **3**, S1066 (2008).

# Impurity transport study based on radial profile analysis of Fe $L_\alpha$ array with n=3-2 transitions of FeXVII to FeXXIV in LHD

X. L. HUANG[1], S. Morita[1,2], T. Oishi[1,2], M. Goto[1,2], H. M. Zhang[1], and I. Murakami[1,2]

[1] *Dept. of Fusion Sci., Graduate University for Advanced Studies, Toki 509-5292, Gifu, Japan*

[2] *National Institute for Fusion Science, Toki 509-5292, Gifu, Japan*

## 1. Introduction

Radial transport of impurity ions still remains several important issues in magnetically confined fusion devices, since the plasma performance is significantly affected by the radiation loss and fuel dilution caused by the impurity. In the Large Helical Device (LHD), the density profile can exhibit a variety of distributions of a peaked, flat or hollow shape, depending on plasma parameters such as $B_t$, $T_e$, and $n_e$. Therefore, it is of great importance to investigate the corresponding impurity transport in the plasma core of LHD [1-3]. As a typical phenomenon in the impurity transport it is reported that the impurity accumulation can be also observed in LHD as well as tokamaks, in which the transport is analyzed based on an assumption of the radial structure of convective velocity [4]. Formation of the hollow or flat density profile in the LHD is quite unique compared to the tokamak. Therefore, LHD plasmas give a good opportunity for studying the impurity transport through analysis of the radial structure in the transport coefficient, in particular, in relation to the density and temperature gradients.

Since iron is an intrinsic impurity in magnetically confined fusion plasmas, it is useful for a comprehensive study of the impurity transport under different plasma parameters or conditions. For the purpose, use of the Fe n=3-2 $L_\alpha$ transition array, emitted in narrow wavelength range of 10 – 18 Å, is of great advantage because the transition array consists of line emissions from iron ions in several charge states of $Fe^{16+}$ through $Fe^{23+}$. In addition, these iron emissions have relatively strong emissivity and are distributed all over the plasma radius of LHD because the ionization energy of such ions ranging between 1 and 2 keV is nearly equal to the electron temperature in core region of LHD plasmas. Radial profiles of the Fe Lα emission have been measured with good quality by a space-resolved extreme-ultraviolet (EUV) spectrometer [5]. It is then possible to determine the profile of transport coefficients, i.e. the diffusion coefficient and the convective velocity, from the emissivity profile with the help of impurity transport simulation code.

## 2. Experimental setup

The EUV system used in the present study working in the wavelength range of 10-130Å mainly consists of an entrance slit, a spatial-resolution slit, a gold-coated concave varied-line-spacing (VLS) laminar-type holographic grating with a groove density of 2400 per mm and a charge-coupled device (CCD) detector with 1024 × 255 pixels (26μm × 26μm/pixel).. The EUV spectrometer is installed on a mid-plane port at the backside of a rectangular vacuum extension chamber. The elevation angle of the central viewing chord is exactly aligned to make possible the vertical profile measurement at upper half of the elliptical plasma, i.e. Z = 0–50 cm.

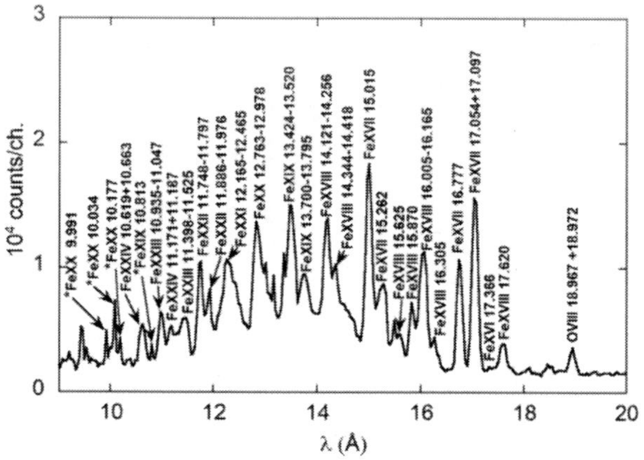


FIG. 1. Fe n = 3–2 $L_\alpha$ spectrum below 20 Å measured after iron impurity pellet injection in NBI discharge with central $T_e$ at 3keV.

The iron spectrum including n = 3-2 $L_\alpha$ lines between 10 and 18 Å has been measured by injecting an iron impurity pellet [6]. A typical result is shown in Fig. 1. A lot of Lα transitions from highly ionized iron are observed in the spectrum. Since many iron lines are blended, the spectrum seems to be a pseudo-continuum. The emissivity profile is derived from the measured vertical profile based on a Fourier Method of Abel inversion [7]. The magnetic surface structure in LHD plasmas is calculated with variation moment equilibrium code (VMEC) [8]. Here, it should be pointed out that the magnetic flux surface calculated with the VMEC is also assumed outside the last close flux surface (LCFS) by extrapolating the magnetic surface contour at LCFS. Although the assumption may cause certain uncertainty, it does not strongly affect the emissivity peak inside the LCFS because the emissivity outside the LCFS is usually weak. The impurity density profile can be then calculated from the emissivity profile based on intensity coefficients obtained with the HULLAC code [9].

## 3. Transport simulation

A one-dimensional impurity transport code is employed to determine the transport coefficient. It is assumed that the impurity transport satisfies the following equation of

$$\Gamma_q = -D_q(r)\frac{\partial n_q}{\partial r} + V_q(r)n_q, \quad (1)$$

where $\Gamma_q$, $n_q$, $D_q$, $V_q$ are the particle flux, the ion density, the diffusion coefficient and the convective velocity of impurity ions in the qth charge state, respectively. Positive and negative velocities stand for outward and inward convections, respectively.

Updated rate coefficients for ionization and recombination have been included in the present transport code [9-11]. With given transport coefficients and radial profiles of plasma parameters, the code yields the impurity ion density profile in all charge states as a function of time. The transport coefficient near the emissivity peak can be determined by comparing the emissivity profile between the experiment and the simulation. In order to determine the transport coefficient at the whole plasma radius, the iron emission is simultaneously analyzed at several charge states. The minimization of deviation between measured and simulated impurity profiles can determine the transport coefficient profile.

**4. Transport analysis of discharges with hollow density profile**

Figure 2 shows the result of transport analysis for a high-density hydrogen discharge with Fe pellet injection. The density profile is clearly hollow. The magnetic axis position is 3.6m and the toroidal field is 2.75T. Here, the radial structure of the transport coefficient is analyzed using measured density profiles of FeXVII through FeXXII. The transport code can well reproduce the profiles, as denoted by solid line in Fig. 2(a). The transport coefficients determined here are shown in Fig. 2(b). The diffusion coefficient gradually increases from the plasma center to the edge, while the convection velocity changes its direction from outward to inward. The transition point is around ρ=0.82 of which the value is equal to the peak position of the hollow density profile. Since the impurity density is less than $10^{-4}$ to the electron density, the proton density should be nearly equal to the electron density. Therefore,

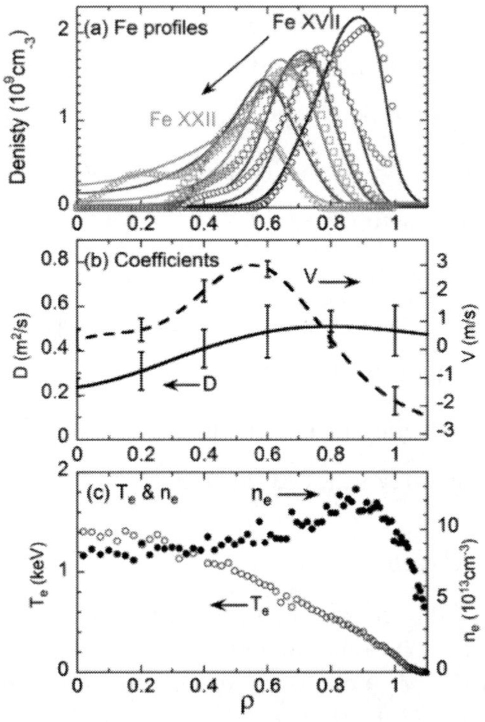


FIG. 2. Experimental (symbols) and simulated (solid lines) profiles of (a) FeXVII through FeXXII, (b) diffusion coefficient (solid line) and convective velocity (dashed line) and (c) $T_e$ (open circles) and $n_e$ (solid circles) .

the result suggests the convection is related to the gradient of ion density profile. This is in accordance with the neoclassical theory of the impurity transport which predicts that the impurity concentrates in the peak position of the ion density. In fact, the impurity further concentrates in the plasma edge when the density profile becomes hollower, as shown in Fig. 3. In this discharge, the outward convection velocity is even larger than that in Fig. 2. The result corresponds to the hollower density profile in spite of the relatively low density discharge.

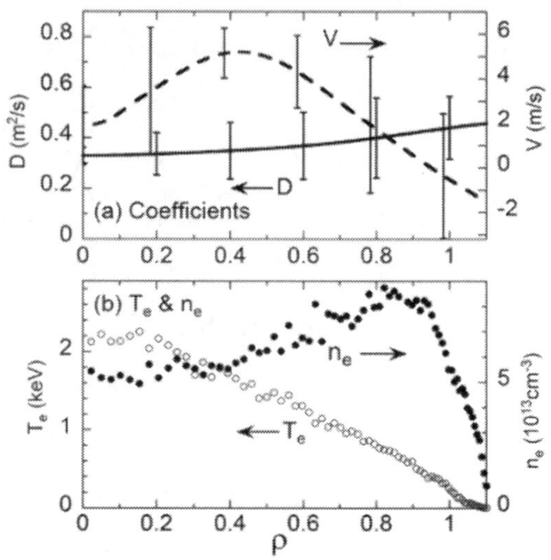


FIG. 3. (a) Diffusion coefficient (solid line) and convective velocity (dashed line) and (b) $T_e$ (open circles) and $n_e$ (solid circles) .

In conclusion, a large outward convection velocity is observed in the plasma center when a hollow density profile occurs. The convection velocity changes the direction at radial location where the density gradient is negative. No significant change in the diffusion coefficient along the plasma radius is observed.

**Acknowledgements**

The authors thank all members of the LHD experimental group for their technical supports. This work was partially carried out under the LHD project financial support (NIFS14ULPP010), the JSPS KAKENHI Grant Number 23340183 and JSPS-NRF-NSFC A3 Foresight Program in the field of Plasma Physics (NSFC: No. 11261140328, NRF: No. 2012K2A2A6000443).

**References**

[1] S. Morita *et al.*, Plasma Sci. Technol. **8**, 55 (2006).
[2] Y. Nakamura *et al.*, Nucl. Fusion **43**, 219 (2003).
[3] M. Yoshinuma *et al.*, Nucl. Fusion **49**, 062002 (2009).
[4] C. F. Dong *et al.*, Plasma Sci. Technol. **15**, 230 (2013).
[5] X. L. Huang *et al.*, Rev. Sci. Instrum. **85**, 043511 (2014).
[6] R. Katai *et al.*, Jpn. J. Appl. Phys. **46**, 3667 (2007).
[7] G. Pretzier *et al.*, Z. Naturforsch. **47a**, 955 (1992).
[8] S. P. Hirshman *et al.*, Comput. Phys. Commun. **43**, 143 (1986).
[9] A. Bar-Shalom *et al.*, J. Quant. Spectrosc. Radiat. Trans., **71**, 169 (2001).
[10] N. R. Badnell, Astrophys. J. Suppl. Ser. **167**, 334 (2006).
[11] N. R. Badnell *et al.*, Astron. Astrophys. **406**, 1151 (2003).
[12] K. P. Dere, Astron. Astrophys **466**, 771 (2007).

# Effect of aspect-ratio on confinement properties of planar-axis and non-planar-axis stellarators

Shoichi Okamura

*National Institute for Fusion Science*

*Oroshi 322-6, Toki, Gifu, Japan*

For the toroidal magnetic confinement fusion, the most basic geometric parameter is the aspect ratio of the torus. It is fundamental both in the physics of the plasma confinement and the technological aspect of the reactor design. In the present stellarator world, we have many different types of the experimental devices which have different values of aspect ratio as well. In tokamak fusion reactor designs, we have made lots of hard discussions in selecting the aspect ratio of the configuration. On the other hand, the reactor designs based on the stellarator configuration have not yet given systematic discussions in selecting the aspect ratio.

Presently we have two large-scale stellarator programs: LHD is running more than 15 years and another Wendelstein 7-X is just starting experiment this year. They have very different magnetic field configurations. As the consequence of these programs, we have two major streams of stellarator reactor design, which adopt almost direct extrapolations of the existing device configurations. Aspect ratio of these devices are relatively large compared to the one of tokamaks. It is because we had been believing that the toroidiciy term in stellarator configuration generally degrade the confinement of neoclassical transport untill we found the idea of the quasi-axisymmetric stellarator. The aspect ratio of LHD is about 6 and the one for W 7-X is 10. These numbers are both more than twice larger than tokamaks but there is a factor of 2 difference between these numbers. In this paper, we will discuss the effect of the modifying the aspect ratio of LHD configurations on the confinement properties considering the origin of the difference between two devices.

Comparison of the confinement properties is made with the effective helical ripples [1] at two minor radii [(r/a) = 1/3 and 2/3] as the measure of the neoclassical confinement and the specific volume of the magnetic flux at the plasma boundary as the measure of the MHD stability. The equilibrium calculation was made for the vacuum field in this paper but the value of the edge specific volume gives a good index of the MHD stability of finite beta equilibria [2]. Because we discuss the general characteristics of stellarator geometry with different aspect ratio, we will select the most basic magnetic configuration of stellarator in the comparison work. In the equilibrium calculation of VMEC [3], we will take only 5 pairs of Fourier elements of r(m,n) and z(m,n) for the boundary shape, which are essential components to construct the stellarator equilibrium field. It was confirmed that these limited components are sufficient for constructing the LHD magnetic field with the magnetic axis position of $R_{ax}$ = 3.6 m, which is the standard configuration of LHD experiments [4].

The confinement properties of LHD magnetic configuration varies very much depending on the magnetic axis position $R_{ax}$. Because the best value of $R_{ax}$ might be different for different aspect ratio values, we compare configurations with different aspect ratios by calculating the dependence of confinement properties on $R_{ax}$ for the configuration of given aspect ratio. Since we calculate the equilibria with the VMEC Fourier components of the boundary shape, we need a good combination of the Fourier elements to controlling the magnetic axis position. It was discussed in the paper [2] that r (0,1) and z (0,1) are the elements that represent the essential change of the magnetic configuration with the shift of the magnetic axis position. Moreover, the complex Fourier element of $\Delta_{11}$ is a good variable as the controlling parameter of the magnetic axis shift effect. The formula to express the boundary shape with such complex Fourier elements is shown below, where $\theta$

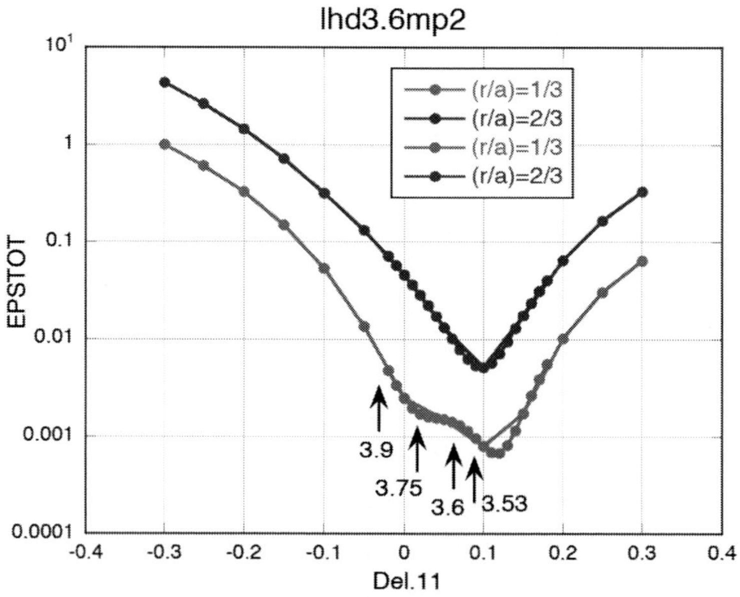


Fig. 1 Effective helical ripples at (r/a) = 1/3 and 2/3 as functions of $\Delta_{11}$. Arrows with numbers show the corresponding values for the LHD configurations with different positions of magnetic axis.

where $\theta$ and $\phi$ are two (poloidal and toroidal) angle parameters mapped on the magnetic surface. $R$ and $Z$ define a point on the magnetic surface in a cylindrical coordinate system.

$$R(\theta,\phi) + iZ(\theta,\phi) = \exp(i\theta)\sum \Delta_{mn} \exp(-im\theta + in\phi)$$

Figure 1 shows the dependence of effective helical ripples on $\Delta_{11}$ for the LHD standard configuration. Arrows point the positions in the figure for the equivalent configurations of LHD experiment (numbers are the magnetic axis positions in meter). In the real operation of LHD experiment, $\Delta_{11}$ is scanned for a small range from -0.05 to 0.1. The purpose of scanning $\Delta_{11}$ in a much wider range in this figure is to compare LHD with W 7-X, which has $\Delta_{11}$ value of 0.3. Such equilibrium calculations were made for different aspect ratio configurations and different number of toroidal period NFP. For example, the equilibrium was calculated for higher NFP value of 12 (LHD has NFP = 10), with the same major radius and minor radius. For this configuration, the geometrical helical torsion of the boundary shape is stronger than LHD. Another equilibrium of NFP = 12 was calculated with a larger major radius keeping the same minor radius. Such configuration has the same geometrical torsion of the boundary but the aspect ratio becomes larger. Finally the equilibrium with a larger aspect ratio but with NFP = 10 was calculated. This configuration has a weaker geometrical torsion than LHD. These calculations were made for NFP = 8, 12, 16, that is, smaller and larger aspect ratio than LHD and for the similar value of the aspect ratio to W 7-X. For the neighboring values of the aspect ratio, configurations with the same geometrical torsion have similar dependence of the effective helical ripples to LHD case. The edge specific volumes are also similar to LHD. The configurations with stronger geometric torsion have the higher effective ripples even for the optimized value of $\Delta_{11}$.

Figure 2 shows the dependence of the effective helical ripples on $\Delta_{11}$ for the very high aspect ratio configuration. The aspect ratio is 9.65 which is similar value to the W 7-X aspect ratio of 10. When we keep the geometric torsion at the same level as LHD for this aspect ratio, NFP should be 16. Such equilibrium has a very sharp peak of small value of the effective helical ripple at $\Delta_{11}$ = 0.05. On the other hand, when we

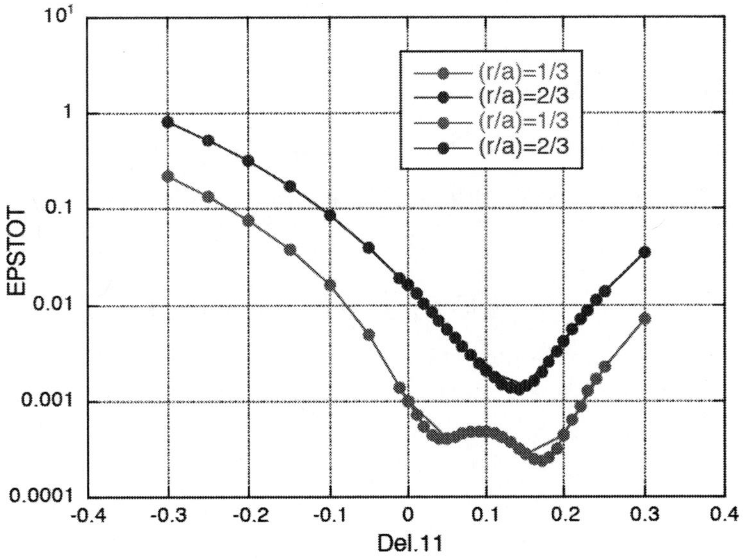


Fig. 2 Effective helical ripples at (r/a) = 1/3 and 2/3 as functions of $\Delta_{11}$ for high aspect ratio LHD-type planar-axis stellarator.

take NFP = 10 same as LHD with a week torsion, a wide range of $\Delta_{11}$ value appears for small effective helical ripples shown in Fig. 2. This means that the confinement does not change much (does not deteriorate) for the modification of the equilibrium due to the plasma pressure increase. This configuration might be good candidate for the high aspect ratio planar-axis stellarator reactor design.

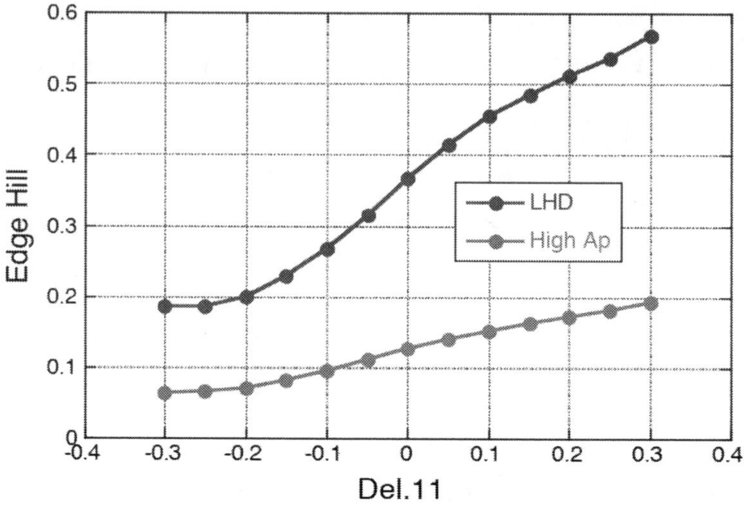


Fig. 3 Specific volume at plasma edge as functions of $\Delta_{11}$ for LHD and high aspect ratio LHD-type planar-axis stellarator.

The comparison of dependence of the edge specific volume on $\Delta_{11}$ for the LHD standard configuration and the high aspect ratio configuration is shown in Fig. 3. The magnetic well for the vacuum field exists in the $\Delta_{11}$ range smaller than -0.05 for LHD configuration, but it does not exist for the high aspect ratio one. However plasmas with high plasma pressure have the magnetic well produced by the Shafranov shift. The small value of edge specific volume for the high aspect ratio configuration would help the magnetic well creation with the plasma pressure.

**References**

[1] V. V. Nemov and S. V. Kasilov, Phys. Plasma **6**, 4622 (1999).

[2] Shoichi Okamura, 40th EPS Conference on Plasma Physics, EPS 2013, **1**, pp. 577-580 (2013).

[3] S. Hirshman, et al., Comput. Phys. Commun. **43**, 143 (1986).

[4] S. Okamura, Contribution to Plasma Physics **50**, 562 (2010).

# Impacts of sheared toroidal rotation on magnetic island in 3D MHD equilibria

Yasuhiro Suzuki[1,2] and Yuji Nakamura[3]

*1) National Institute for Fusion Science, Toki, Japan*

*2) Graduate School for Advanced Studies, SOKENDAI, Toki, Japan*

*3) Graduate School of Energy Science, Kyoto University, Uji, Japan*

## 1. Introduction

In most of situations, tokamak equilibria are analyzed as two-dimensional (2D) systems with the axisymmetry. The nature of this symmetry gives many advantages for its analysis. However, as realistic tokamaks have discreteness of the toroidal field coils, this discreteness yields the toroidal field ripples (TF ripples) and, strictly speaking, realistic tokamaks could not be axisymmetric configurations. In previous work[1)], we pointed out the significance of three-dimensional (3D) effects, which are effects of plasma equilibrium currents along rippled field lines.

On the other hand, in recent tokamak experiments, it is noted that stochastic filed lines reduce strong heat load driven by the edge localize mode (ELM) on the divertor plate. Stochastic field lines are produced by the external helical perturbation and it is called the Dynamic Ergodic Divertor (DED). From the viewpoint of high-β stellarator equilibrium, 3D effects on the stochastic field are very important because finite-β perturbed field produces further stochasticity in the peripheral region. However, in present analysis of DED, 2D MHD equilibrium superimposed vacuum helical perturbed field was still used. In order to consider effects of DED to ELM, considerations of finite-β MHD equilibrium and the impact of 3D effects are critical and urgent issue.

In this study, the fully 3D MHD equilibrium of non-axisymmetric tokamak is solved numerically and the impact of the plasma rotation to the 3D MHD equilibrium is discussed. For this study, we use a 3D MHD equilibrium code HINT[2)], which is widely used to analyze the 3D equilibrium in stellarator researches. Since the HINT uses the real coordinate system, it can treat magnetic island and stochastic field in the computational domain. Thus, as first step, we study the 3D MHD equilibrium including the toroidal rotation. Special attention is the change of the magnetic island due to the toroidal plasma rotation.

## 2. 3D MHD equilibrium calculation including toroidal rotation

At first, we discuss the improvement of the HINT code to include the toroidal rotation. vacuum field in the ITER. The HINT code is a 3D MHD equilibrium calculation code, which is based on the relaxation method. Since the HINT code uses the real coordinate system, which is the cylindrical coordinate, the code can capture the magnetic island and stochastic magnetic field lines in the calculation. The HINT code had been developed for stellarator and heliotron researches and the original version of the code adopted a non orthogonal coordinate system, so-called the rotating helical coordinate system. The HINT code had been updated successfully to the HINT2 code and that code applied to the tokamak calculation with 3D perturbation fields, which are the toroidal field ripple, 3D error field and resonant magnetic perturbation (RMP) fields. However, up to now, the 3D MHD equilibrium is calculated as the magnet static equilibrium. Recently, effects of the plasma rotation to the RMP field, which are shielding and amplification of RMPS, are hot topics in ELM suppression and mitigation experiments. To understand those effects, including the plasma rotation to the 3D MHD equilibrium calculation is urgent issue. In this section, the implementation how to include the plasma rotation is shown.

In this study, only the toroidal rotation is studied for simplicity. The toroidal rotation is prescribed by the function of the toroidal flux and the toroidal flow velocity is defined by the Mach number,

$$M(s) = \frac{v_{\Phi(s)}}{v_{th}},$$

where $v_{\Phi}$is the toroidal flow velocity and $v_{th}$is the ion thermal velocity. The toroidal flow is prescribed as the function of the normalized toroidal flux, $s = \Phi/\Phi_{edge}$. The $\Phi_{edge}$ is the toroidal flux on the last closed flux surface.

The HINT code consists of two parts. First part, step-A, is the relaxation process of the plasma pressure with fixed the magnetic field. Second part, step-B, is the relaxation process of the magnetic field with fixed the plasma pressure. The step-A calculates the pressure distribution satisfying the condition, $\vec{B} \cdot \nabla p = 0$. Instead of calculating that condition, the step-A calculates an averaged plasma pressure along a magnetic field lines, because the condition means no variation of the plasma pressure along the magnetic field lines. Details is shown in Ref. For a case of existing the toroidal flow velocity, the pressure distribution shifts to the outward of the torus by the inertial force. In such a case, from the parallel force balance, the pressure distribution is prescribed by

$$p(s,R) = p\exp(M^2(s)(\frac{R^2}{R_0^2} - 1)).$$

On the other hand, the step-B calculates the time evolution of nonlinear dissipative MHD equations. In these equations, the magnetic field and plasma flow velocity are given by $\vec{v} = \overrightarrow{v_0} + \overrightarrow{v_1}$ and $\vec{B} = \overrightarrow{B_0} + \overrightarrow{B_1}$. Here, $\overrightarrow{B_0}$ is the vacuum magnetic field and $\overrightarrow{B_1}$ is the equilibrium response field. The $\overrightarrow{v_0}$ is a given toroidal flow velocity and $\overrightarrow{v_1}$ is the MHD velocity. Thus, dissipative MHD equations are

$$\frac{\partial \overrightarrow{v_1}}{\partial t} = -\overrightarrow{v_0} \cdot \nabla \overrightarrow{v_0} - \nabla p + \overrightarrow{J_1} \times (\overrightarrow{B_0} + \overrightarrow{B_1}) + \nu \nabla^2 \overrightarrow{v_1},$$

$$\frac{\partial \overrightarrow{B_1}}{\partial t} = \nabla \times ((\overrightarrow{v_0} + \overrightarrow{v_1}) \times (\overrightarrow{B_0} + \overrightarrow{B_1}) - \eta(\overrightarrow{J_1} - \overrightarrow{J_{net}})),$$

$$\overrightarrow{J_1} = \nabla \times \overrightarrow{B_1}.$$

The spatial derivation is approximated by 4th order central finite difference scheme and time marching is calculated by the 4th order Runge-Kutta-Gill scheme. Calculating those two steps iteratively, a steady-state solution is obtained.

**3. Model Calculation of an ITER-like plasma**

In this study, we study a tokamak with a D-shape cross section, which is a comparable to the ITER. In that configuration, the major radius $R_0$ is 3 m and the plasma minor radius $a$ is 1 m.

To study effects of the plasma rotation to the magnetic island, $n$=3 perturbed field is superposed to the plasma. In this configuration, important rationals are $m/n$ = 2/1 and 3/1. Figure 1 (a) shows a puncture map of magnetic field lines for the $M$=0 case, which is zero plasma rotation. $m$ = 3, 4 and 5 island chains appear in the plasma. Special notice is a phase of $m$=3 islands. O-points of $m$=3 islands appear at $\rho = 0.78$.

On the other hand, figure 1 (b) shows a puncture map of magnetic field lines for the $M$=0.3 case. In this study, the toroidal rotation is approximated by a function, $v_\phi(s) \propto (1 - s^7)$. The toroidal rotation stretches the magnetic island. Widths of $m$=4 magnetic islands are almost identical to the $M$=0 case. However, the phase of $m$=3 islands slip poloidally compared with the $M$=0 case. In addition, the width a little bit shrinks. From these results, the shielding of the RMP is very weak. These results are comparable to other results. Similar studies were done by ANIMEC[3)] and NIMROD[4)] code. In that result, the shielding of the plasma rotation is very weak but the phase slip appeared in the NIMROD simulation. These are preliminary results but further results will be discussed in near future.

*(a) M=0* *(b) M=0.3*

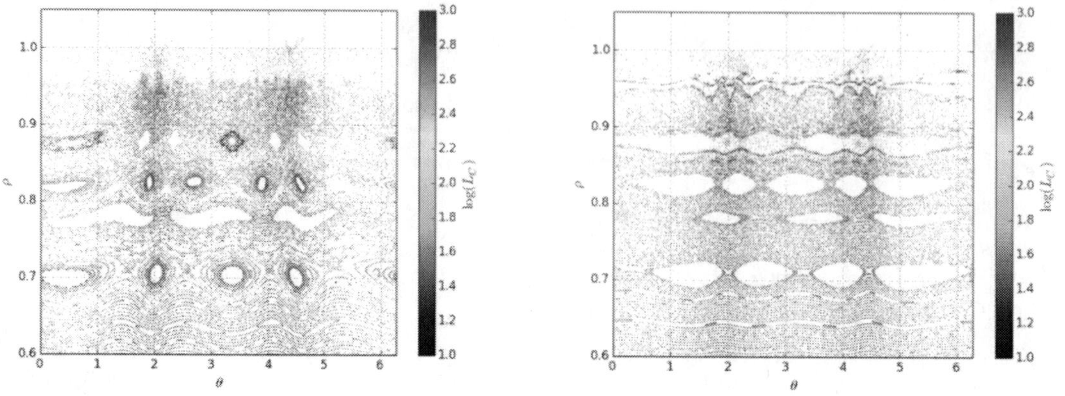


*Figure 1 Puncture maps of magnetic field lines for (a) M=0 and (b) M=0.3 are shown. Pressure and current density profiles are same in both case.*

## 4. Summary

The 3D MHD equilibrium including the plasma rotation is studied for the simple tokamak with circular cross sections. To calculate the 3D MHD equilibrium with the plasma rotation, the 3D MHD equilibrium calculation code, HINT, is improved. In this study, impacts of the toroidal rotation to the magnetic island are studied. If n=3 perturbed field superposes, m=3 and 4 islands appeared. For the M=0.3, the width of magnetic islands is decreased comparing with the zero rotation case. However, m=4 islands slip poloidally but the width is almost identical. Detailed discussions will shown in near future.

1) Y. Suzuki, Y. Nakamura and K. Kondo, Nucl. Fusion **43** (2003) 406

2) Y. Suzuki *et al.*, Nucl. Fusion **46** (2006) L19

3) W. A. Cooper, *et al.*, 40th EPS Conference on Plasma Physics, (2013) P4.124

4) C.Hegna, *et al.*, in private communication.

# Turbulent fluctuations and unstable drift wave in plasma sheared flow

A.Yu. Chirkov, D.N. Karbushev, V.I. Khvesyuk,

*Bauman Moscow State Technical University, Moscow, Russian Federation*

## Introduction

In magnetized plasmas, shear flows reduce the turbulent transport and improve confinement [1–3]. Experimental studies of plasma turbulence in magnetic traps clearly show the relationship of perturbation with drift instabilities [3, 4]: Ion Temperature Gradient (ITG) mode, Trapped Electron Mode (TEM) and, under certain conditions, Electron Temperature Gradient (ETG) mode. Drift-wave turbulence contains spatially coherent structures. A coherent, monochromatic drift wave can drive a band of self-sustaining fluctuating modes. Radially extended convective structures are effective in turbulence spreading. Drift turbulence theories consider decrease of fluctuations as a result of decorrelation of relatively large perturbations under the influence of the stationary or quasi-stationary shear flow [5, 6].

Experimentally established condition for significant reduction in transport is [7]

$$\gamma_s > \gamma, \tag{1}$$

Here $\gamma$ is the typical growth rate of the instability (ITG mode usually), $\gamma_s$ is the shear parameter, $\gamma_s = \partial V / \partial x$ in a slab, $\gamma_s = r\partial(V / r)/\partial r$ in cylindrical geometry, $V$ is the flow velocity.

In the plasma with no sheared flow, diffusivity corresponds the estimation $D_{\perp 0} = k_\perp^{-2}\gamma$, where $k_\perp$ is the typical wave number, $\gamma$ the typical growth rate [8]. The effect of shear flow leads to dependence [9]

$$D_\perp = D_{\perp 0} / [1 + (\gamma_s / \gamma)^2]. \tag{2}$$

Studies of turbulent fluctuations [5, 6] revealed the ratio of the root mean square of the fluctuation level for cases with $\gamma_s \neq 0$ and with $\gamma_s = 0$. In the last case, fluctuation decrease can be explained by the other factors that limit the growth of large perturbations [3].

In the present work, we consider the growth of initial perturbation from the small amplitude harmonic drift wave to the conventional decay. Such decay doesn't mean the disappearance of the wave. It is strong distortion of the wave compared to the original harmonic form, so it causes the non-uniformity of the environment and makes the further existence of the wave impossible. The corresponding amplitude value is typical in terms of limitation of the linear growth of the perturbation and the transition to the nonlinear stage.

The main purpose of the present work is analysis of the decay condition which is adequate to the usually observed turbulence level in magnetized plasmas with the sheared flows.

**The calculation model**

At the initial state amplitude is assumed to be infinitesimal, but during the growth it is considered as finite. For quantitative estimates we use the linear growth rate of ITG mode [10–12]. The influence of flow shear on the fluctuation level is determined by varying the ratio of shear parameter $\gamma_s$ to growth rate $\gamma$. Wave tilting is not considered, as for the drift waves it means a discontinuity of the medium. It is assumed that the state called as the decay corresponds to the close conditions. Turbulent fluctuation level is associated with amplitude value before the decay stage.

Drift wave is considered in slab geometry. The magnetic field is directed along the $z$-axis, the wave propagates in the $y$-direction, the plasma density and temperature decrease in the $x$-direction, the flow velocity $V$ depends on $x$. Waveform is assumed to be independent of $x$ within the layer under consideration. Linear dependence $V(x) = \gamma_s x$ approximates flow velocity in a moving coordinate system. At the initial time waveform $X(y, t)$ is a sine function $X = a_i \sin(k_y y)$, where $a_i$ is the initial amplitude, $k_y = k_\perp$ is wave number. Boundary conditions are the following: $X = 0$ at $y = \pm\pi/k_y, \pm 2\pi/k_y, \pm 3\pi/k_y, \ldots$

The growth of the density perturbations $n^\sim$ is characterized by a constant growth rate as follows: $dn^\sim / dt = \gamma n^\sim$. Deformation field $X(y, t)$ is associated with the perturbed density $n^\sim$ using the relation $n^\sim = -(n_0 / L_n)X$, where $dn_0 / dx = -n_0 / L_n$ is unperturbed density gradient, $n_0$ is unperturbed density, $L_n$ is gradient scale length. For the electron density perturbation the Boltzmann approximation is performed: $(n^\sim / n_0) = (e\varphi / k_B T)$, where $e$ is the electron charge, $k_B$ is the Boltzmann constant, $T$ is electron temperature. Take into account that drift-wave fluctuations are quasi-neutral.

Since the perturbed electron density and wave potential are proportional to the deformation of the medium, then all of the three values $\delta n \sim \varphi \sim X$ are described by the equation

$$\frac{dX}{dt} \equiv \frac{\partial X}{\partial t} + \gamma_s X \frac{\partial X}{\partial y} = \gamma X + D \frac{\partial^2 X}{\partial y^2}. \qquad (3)$$

It is assumed here that $v_y = V(X) = \gamma_s X$. The diffusion operator is introduced in the right-hand side of Eq. (3) to simulate the saturation of the perturbation using effective diffusivity $D$.

Eq. (3) refers to the same type of equations that the Burgers equation [13], which partial solutions are known. Eq. (3) with zero right-hand side matches the cases of the drift motion of particles. In that form it coincides with the equation of nonlinear drift waves [14]. Such an equation also describes the perturbation in the beam of non-interacting particles [13].

**Results of the analysis**

Fig. 1 shows the evolution of the perturbations at $D = D_\perp$, where diffusivity $D_\perp$ is calculated according Eq. (2). To consider the effect of the shear separately, we also examined the case $D \ll D_\perp$. Fig. 2 shows the growth of the amplitude $a = \max(|X|)$ for the two cases: $D = D_\perp$ and $D = 10^{-2}D_\perp$.

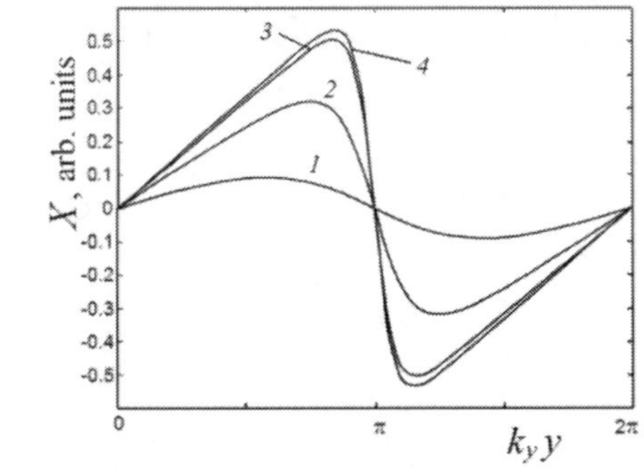


Fig. 1. Temporal evolution of the wave profile at $D = D_\perp$, $\gamma_s/\gamma = 1.5$: $1 - \gamma t = 3$; $2 - 5$; $3 - 7$; $4 - 20$.

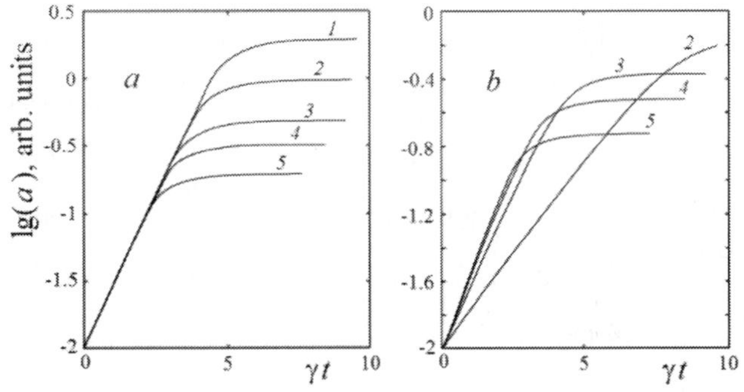


Fig. 2. Amplitude growth at $D = D_\perp$ ($a$) and $D = 10^{-2}D_\perp$ ($b$): $1 - \gamma_s/\gamma = 0.5$; $2 - 1$; $3 - 2$; $4 - 3$; $5 - 5$

At the initial time the amplitude grows linearly. Over time, wave profile tends to a stationary form. It is assumed that the conditions for the existence of the wave no longer hold to the end stage of linear growth. At this time, the amplitude corresponds to the approximate equality of the perturbed and unperturbed gradients [15]:

$$\partial n^{\sim} / \partial y \sim dn_0 / dx. \tag{4}$$

This relationship is considered as a quantitative decay conditions. With no shear flows $k_y\delta n \approx n_0 / L_n$, where $\delta n$ is perturbed density amplitude. It is associated with a wave amplitude $a$ as follows: $\delta n = (n_0 / L_n)a$. An estimate of the amplitude at $\gamma_s = 0$ is

$$a_0 = (\delta n / n_0)L_n = 1/k_y, \tag{5}$$

that corresponds to the experimental data [3, 15].

In our model, when $\gamma_s = 0$ perturbation growth is limited by the diffusion operator. The equality to zero of the right-hand side of Eq. (3) corresponds to the relationship $\gamma \approx D / a^2$. Eq. (5) corresponds to diffusivity $D = D_\perp$ calculated by Eq. (2). Using Eq. (2) for $\gamma_s \neq 0$ obtains

$$a^2 / a_0^2 = 1 / [1 + (\gamma_s / \gamma)^2]. \tag{6}$$

The results of calculations using the formulated decay criteria are shown in Fig. 3. As seen from Fig. 3, calculation results for $D = D_\perp$ practically coincide with the Eq. (6). This coincidence favors the diffusion mechanism of dissipation, at the same time as leading to the plasma transport and to a smoothing of the perturbation profile. We considered the case $D << D_\perp$ also.

**Conclusions**

The analysis showed two regimes with different mechanisms perturbation growth limitation. In the low shear (or no shear) case, dissipative mechanism dominates. It can be presented by the diffusion operator. At strong shear, perturbation growth is limited by deformation of perturbed medium. These mechanisms are in accordance with the Burrell's experimental criterion (Eq. (1)) and result of the theory of Dupree and Itohs (Eq. (2)).

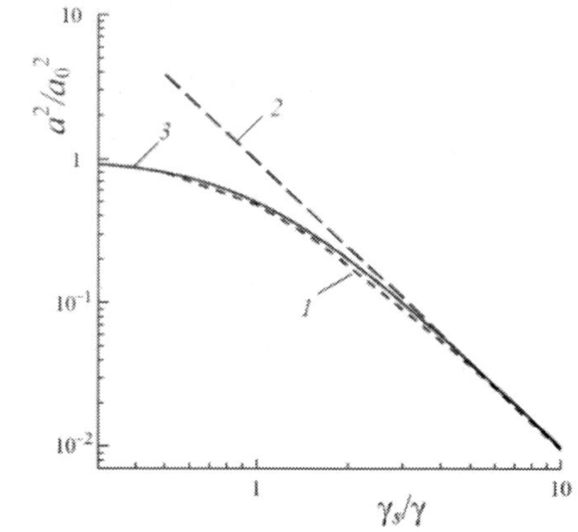

Fig. 3. The relative perturbation amplitude versus the shear parameter: 1 – simulations with $D = D_\perp$; 2 – simulations with $D = 10^{-2} D_\perp$; 3 – Eq. (6)

The analysis is based on the evolution of a single-mode perturbation while really many modes coexist in turbulent medium. The estimates of the perturbation amplitudes (Eqs. (5) and (6)) involve the use of some typical parameters of instability with a certain wave number $k_y$. Such an estimate gives an average result, not a spectrum.

This work was supported by Russian Ministry of Education and Science Contract No. 13.2573.2014/K.

[1] R.C. Wolf, Plasma *Phys. Contol. Fusion*, **45**, R1 (2003).
[2] J.W. Connor, et al., *Nucl. Fusion*, **44**, R1 (2004).
[3] G.D. Conway, *Plasma Phys. Control. Fusion*, **50**, 124026 (2008).
[4] G.R. Tynan, A. Fujisawa, G.R. McKee, *Plasma Phys. Control. Fusion*, **51**, 113001 (2009).
[5] K.C. Shaing, E.C. Crume, Jr., W.A. Houlberg, *Phys. Fluids*, **B2**, 1492 (1990).
[6] Y.Z. Zhang, S.M. Mahajan, *Phys. Fluids*, **B4**, 1385 (1992).
[7] K.H. Burrell, *Phys. Plasmas*, **4**, 1499 (1997).
[8] T.H. Dupree, *Phys. Fluids*, **10**, 1049 (1967).
[9] K. Itoh, S.-I. Itoh, *Plasma Phys. Control Fusion*, **38**, 1 (1996).
[10] A.Yu. Chirkov, V.I. Khvesyuk, *Phys. Plasmas*, **17** (1), 012105 (2010).
[11] A.Yu. Chirkov, V.I. Khvesyuk, *Plasma Phys. Rep.*, **37**, 437 (2011).
[12] A.Yu. Chirkov, J. Fusion Energy, **33** (2), 139 (2014).
[13] B.B. Kadomtsev, V.I. Karpman, *Sov. Phys.-Uspekhi*, **103** (2), 193 (1971).
[14] H. Tasso, *Phys. Lett. A*, **232**, 247 (1997).
[15] W. Horton, Turbulent transport in magnetized plasmas, World Scientific (2012).

# Commissioning of a Non-circular VDE-Free Tokamak with Saddle Coils

T. Kobayashi, S. Hatakeyama, M. Watanabe, H. Tsutsui, S. Tsuji-Iio

*Tokyo Institute of Technology, Tokyo 152-8550, Japan*

## Introduction

In tokamaks, elongated plasmas for high beta and good confinement operation suffer from vertical instabilities. Failure of feedback control of plasma position due to abrupt changes in plasma pressure due to disruptions leads to so-called vertical displacement events (VDEs). The interaction between plasma and first wall due to VDEs can cause the damages of in-vessel components such as blankets by high heat flux and induced electromagnetic forces [1]. On the other hand, it has been shown that the plasma position is robustly stable in a current-carrying stellarator than that in equivalently elongated tokamak plasmas [2]. The aim of our study is to stabilize vertical instabilities of elongated tokamak plasmas by the use of saddle coils as shown in Fig. 1 which generate helically perturbed fields. Although several studies have been made on the effects, they have been conducted in tokamaks with circular cross-sections [3, 4]. For the next step of this study, we constructed a small tokamak device ($R = 0.33$ m, $a = 0.09$ m, $\kappa = 1.8$, $B_t = 0.3$ T) as illustrated in Fig. 2 which has an elongated cross-section to demonstrate the stabilization of VDEs.

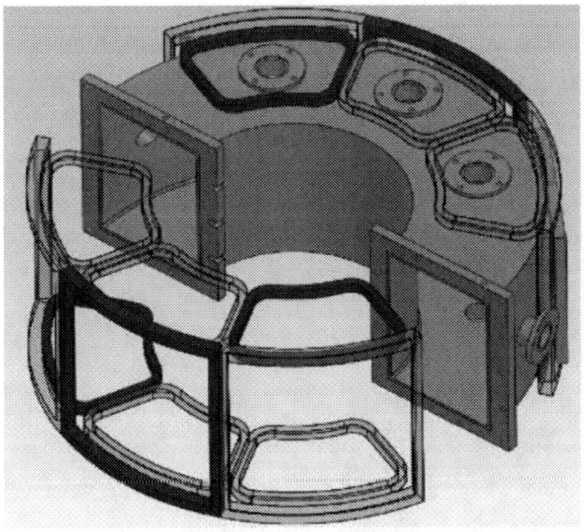

**Fig. 1 Example of saddle coils**

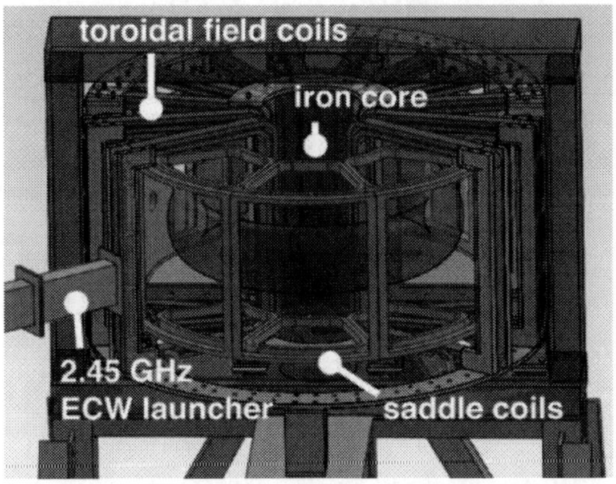


**Fig. 2 Illustration of small tokamak device**

**Effect of saddle coils on VDE**

VDEs occur due to bad curvature of vertical field required to make the plasma elongated. The horizontal magnetic field cannot be changed independently from vertical magnetic field with axially symmetric PF coils alone. In order to stabilize the elongated plasma, it is necessary that the horizontal component of magnetic field ($B_r$) shown in Fig. 3 should be reversed. It is because that the Lorentz force expressed as $I_p \times B$ pulls the elongated plasma vertically when the balance of the force is destroyed with slight vertical plasma shifts. Then if the horizontal magnetic field is reversed near the vacuum vessel, VDEs can be avoidable. Fig. 4 shows the projection of magnetic force lines made by saddle coils on a poloidal cross-section. This indicates that saddle coils can generate horizontal field component which provides a restoring force in combination of the toroidal field.

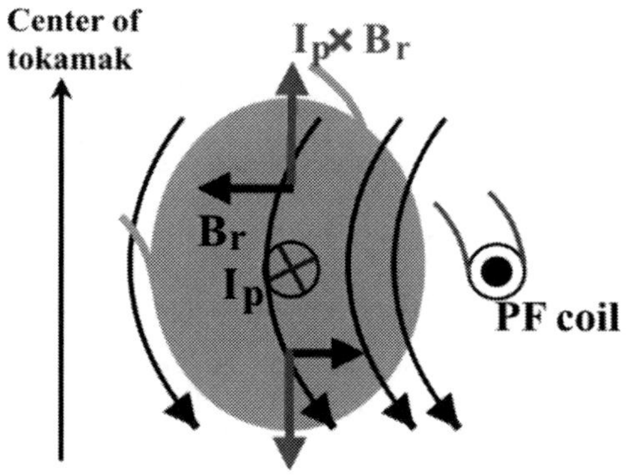


**Fig. 3 Poloidal magnetic fields which elongate the plasma**

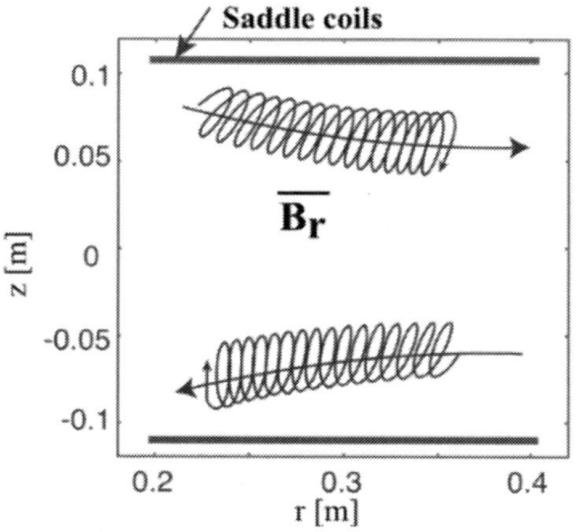


**Fig. 4 Projection of magnetic force lines**

**Commissioning of tokamak plasma**

Our new small tokamak device, shown in Fig. 5, has the coil system comprised of 16 toroidal field (TF) and 2 sets of poloidal field (PF) coils, an ohmic heating (OH) coil with iron core. OH coil was wrapped around 20 turns and driven with capacitors using gate drive as a switch. The TF coils are excited with an inverter-driven induction flywheel generator which is

**Fig. 5 Photo of small tokamak device under commissioning**

controlled to maintain 2.45-GHz electron cyclotron resonance (ECR) at the center of vacuum vessel.

The vertical field generated by the PF coils was constant during discharges by using a DC power supply. PF coils were mounted outside of iron core because this configuration needs no decoupling coil with iron core and is easy for hand-winding. Magnetic sensors were flux loops and a Rogowski coil. Pick up coils will be installed in the shadow of limiters in the poloidal direction. Rogowski coil was wound by hand and achieved current position independence of within 1% error. The signals of flux loops and Rogowski coil were integrated with RC integration circuits in order to measure the plasma current and poloidal magnetic fluxes.

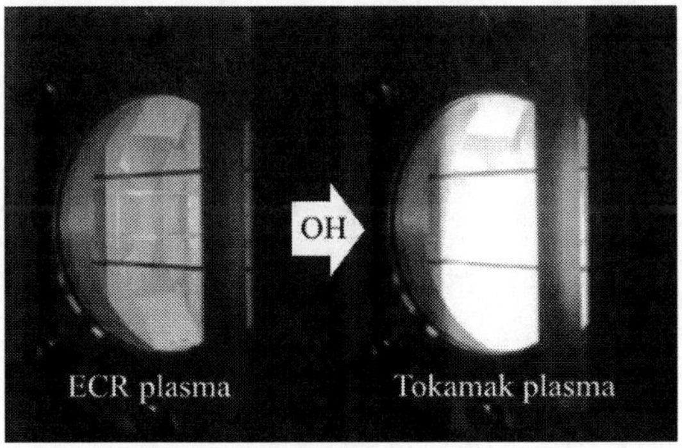


**Fig. 6 Tangential images of plasmas under commissioning**

In this experimental environment, commissioning was conducted. The plasma current was driven by the OH coil in an ECR discharge with 2.45 GHz microwave as shown in Fig. 6. Figure 7 shows the waveforms of flywheel generator employed to excite TF coils which provide resonant magnetic field. By optimization of the vector control, we succeeded in keeping the generator voltage nearly constant. Accordingly the waveform of the coil current was similar to a square in about 1.5 seconds even when the rotator speed dropped to a half.

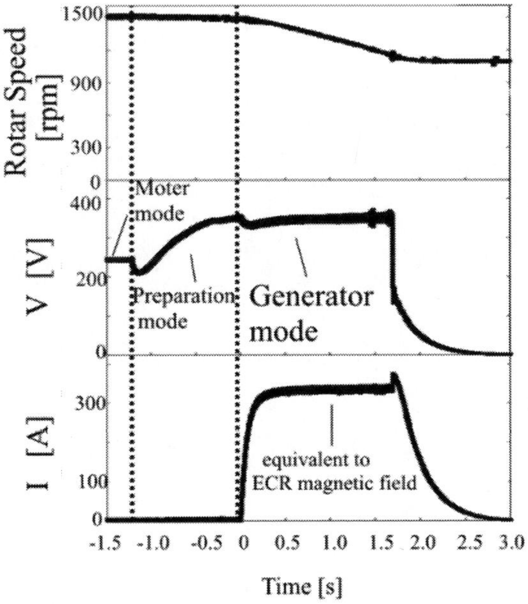


**Fig. 7 Waveforms of induction flywheel generator for ECR discharge**

Figure 8 shows a result of a scan of the vertical magnetic field (Iv = 0 - 2000 A·turns). The maximum value of plasma current was 3.5 kA. With increasing vertical field coil current, the peak of the plasma current was raised except the case of Iv = 2000 A·turns. This can be inferred that too strong vertical magnetic field pushes the plasma inwardly and the enlargement of plasma cross-section was hindered. Additionally, while the plasma current flowed, the one-turn voltage was almost constant but oscillated due to some MHD instabilities. The safety factor when Ip = 3.5 kA is calculated to be about 3.6 by cylindrical approximation. The waveforms of the OH coil current show that the flux swing of the iron core is not fully exploited.

**Summary and future work**

We proposed saddle coils for the position control of elongated plasma and now commissioning a new small tokamak device. We obtained the first tokamak plasma with plasma current of up to 3.5kA driven by the OH coil current and with 2.45 GHz microwave for ECR pre-ionization.

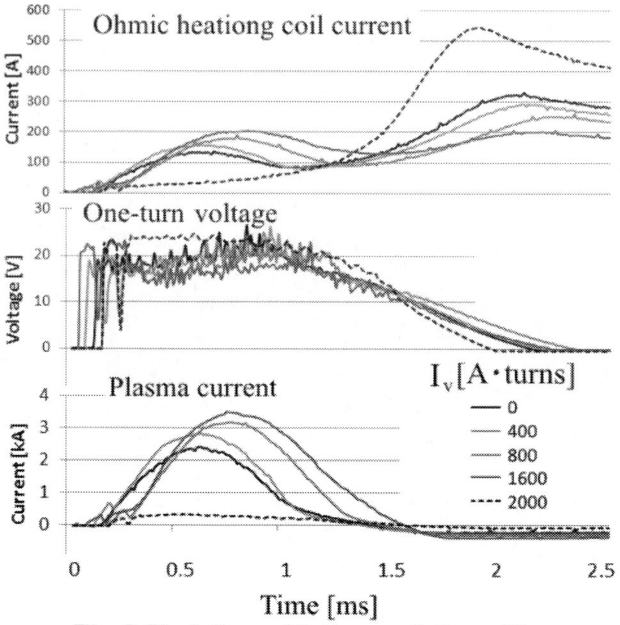

**Fig. 8 Variations of $I_p$ time evolution with different vertical magnetic fields**

Next steps are as follows; firstly, establish tokamak discharges are with toroidal magnetic field up to 0.3 T using a capacitor bank and without ECR pre-ionization. Secondly, elongate the plasma by the feedback control and trigger vertical instabilities by switching off the control. Then the saddle coils will be optimized using a 3-D equilibrium analysis code in order to avoid the VDEs and demonstrate the efficacy of saddle coils.

**References**

[1] T. C. Hender, J. C. Wesley, J. Bialek *et al.*, Nucl. Fusion. 47, S128 (2007).

[2] G. Y. Fu, L. P. Ku, W. A. Cooper, S. H. Hirshman, D. A. Monticello *et al.*, Phys. Plasmas 7, 1809 (2000).

[3] H. Ikezi, K. F. Schwarzenegger, Phys. Fluids 22, 2009 (1979).

[4] S. Hatakeyama, Y. Shibata, S. Kawakami *et al.*, Proc. EPS2012, P1.046 (2012).

# Application of Microwave Frequency Comb for Plasma Reflectometry

S. Inagaki[1,2], T. Tokuzawa[3], T. Mizokami[1], T. Yamada[4], Y. Nagashima[1,2], A. Fujisawa[1,2]

[1] *Institute for Applied Mechanics, Kyushu University, Kasuga, Japan*

[2] *Research Center for Plasma Turbulence, Kyushu University, Kasuga, Japan*

[3] *National Institute for Fusion Science, Toki, Japan*

[4] *Faculty of Arts and Science, Kyushu University, Fukuoka , Japan*

Turbulence transport in magnetized plasmas is considered to be non-diffusive and thus plasma turbulence can generate large-amplitude corrugations in the profile which evolve dynamically [1,2]. In addition, the ballistic propagation of turbulence front is involved in such fine structure formation through an avalanche process related to the self-organized criticality (SOC) models [3, 4]. Observation of such corrugations and associated turbulence dynamics is one of the most challenging problems. Microwave refrectometry is now considered to be very useful tool to observe density profile and also density fluctuations. Thus a multi-channel reflectometer is the most promising to observe the fine structures and their dynamic behaviors simultaneously. We thus applied the microwave frequency comb technique to the reflectometry for core plasma diagnostic [5]. Microwave frequency comb reflectometer is a possible candidate to measure the density profile as a continuous function of radius with high temporal resolution [6]. Here we report an initial experimental result measured with a new microwave frequency comb reflectometer in a linear device.

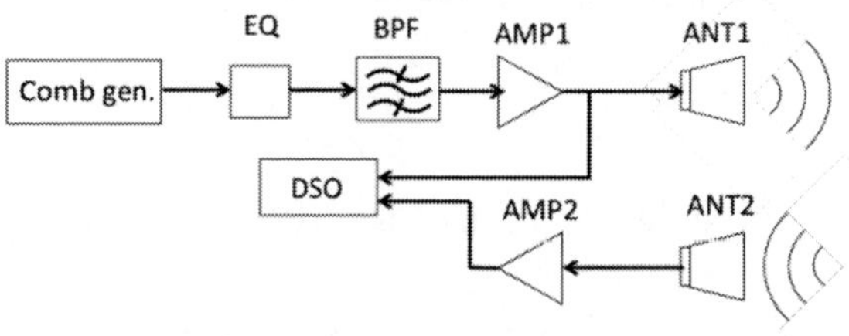


Fig. 1 Block diagram of the system.

Figure 1 shows a block diagram of the frequency comb reflectometer. In our system, the output from comb-generator (the repetition frequency of 0.5 GHz) is filtered by the dual-band (Ku- and K-band, i.e. 12-26 GHz) broad band-pass-filter, where the equalizer is used to obtain the flat level of output power. The band-passed output is linearly amplified and fed to the dual-band rectangular horn antenna by the coaxial cable. Figure 2 shows the typical power spectrum of the incident wave. A lot of frequency components (29 distinct peaks) are shown.

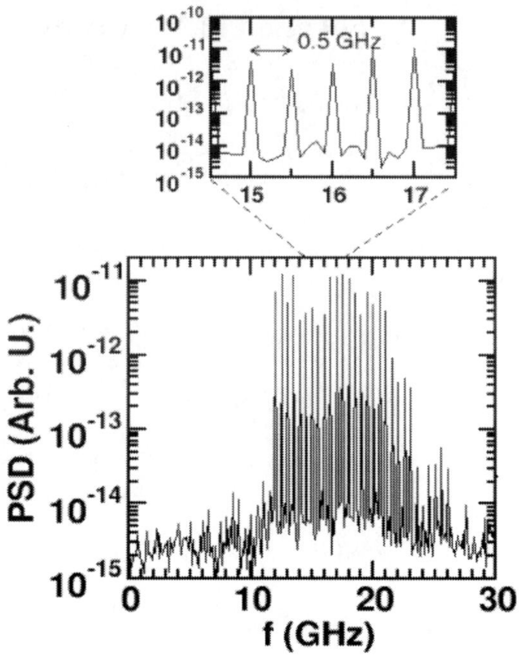

Fig. 2 Power spectrum density of output of comb generator.

In previous work [7], only eight peaks are used because of many micro-wave components, e.g. filter, power divider and mixer are needed. In this study, the incident and reflected wave signals are directly transferred to the digital storage oscilloscope, which has a frequency band of 33/50 GHz (the sampling frequency is 80/160 GHz), so the waveforms of the incident and reflected signals are detected in the form of digital signals with very high temporal resolution. Figure 3 shows the typical incident and reflect microwave waveform.

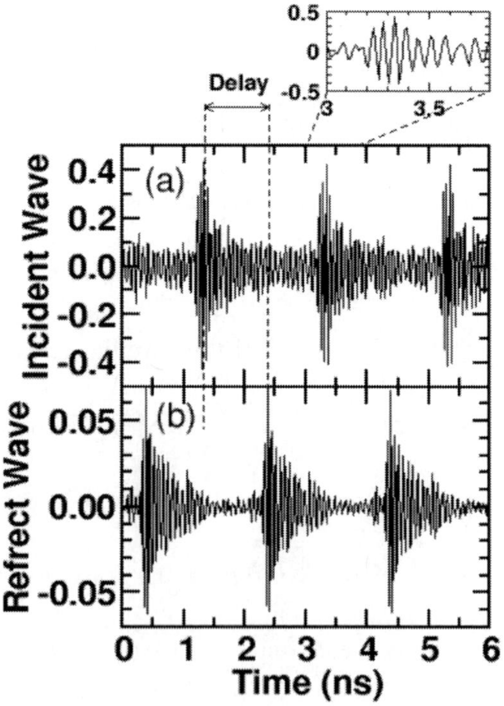

Fig. 3 Typical instantaneous incident- (a) and reflect- microwave (b) by direct signal acquisition

The repetition period is 2 ns (corresponding to a repetition frequency of 0.5 GHz). The delay between the envelope of the incident- and reflect-wave corresponds to the distance between the antenna and reflector. A phase delay $\delta\theta$ due to optical path difference between incident- and reflect-wave is obtained by using of the FFT analysis. Advanced digital signal processing techniques (e.g. Hilbert transform and convolution) allow us to apply ideal filters and improve the signal-to-noise-ratio. For example, the conditional averaging technique over 500 periods (1 period = 1/0.5 GHz = 2 ns) can suppress the 10 % noise in amplitude. The temporal resolution is evaluated to be 1 μs in this case.

The microwave frequency comb reflectometer has been tested. The microwave is launched aiming at the planer metal (SUS) reflector 100-200 mm away from the antenna. The sensitivity of the phase difference is found to be 1.4 rad/mm. The target was switched from metal reflector to a linear magnetized plasma created by radio frequency wave (3 kW/7 MHz) in the PANTA device. The cylindrical vacuum vessel of PANTA has a diameter of 450 mm and a length of 4 m. Operational parameters are as follows; plasma diameter of about 120 mm, axial magnetic field of 0.09 T, and filling argon pressure of 3 mTorr. Typical central density and electron temperature are $1.0 \times 10^{-19}$ $m^{-3}$ and 3 eV, respectively. Probe measurement suggested that the electron density gradient is steep in the radius of $r$ = 30-40 mm, and produces drift wave instability which propagates in the electron diamagnetic direction. The electron density fluctuation was also measured with a 64-channel poloidal probe array and evaluated from the ion saturation current fluctuations. In a single frequency reflectometric measurement the phase of the microwave repeats itself at distance intervals equal to the wavelength. Thus, a phase delay $\delta\theta$ due to optical path difference between incident- and reflect-wave gives the same reflectometric measurement as $(2n\pi + \delta\theta)$, where $n$ is an integer. This "ambiguity problem" makes the distance measurement difficult. One of the techniques for determining $n$ is using two receivers. The simultaneous equation from two receiver can be solved by using integer linear programming. we reconstructed a density profile from the frequency comb reflectometer measurement and compared with the probe measurement. Both are consistent however the error bars of the frequency comb reflectometer measurement are large at this time. One of the candidates of large error bar is reflection from vacuum vessel. Recently, a new data processing technique based on time-frequency tomographic representation is developed [8]. It allows us to separate multiple reflection components. This technique will be applied on our system. Experimental tests are also carried out on toroidal devices (e.g. LHD).

This work is partly supported by a grant-in-aid for scientific research of JSPS, Japan (15H02335, 15H02155, 23244113) and by the collaboration programs of NIFS (NIFS13KOCT001) and of the RIAM of Kyushu University and Asada Science foundation.

[1]. G. Dif-Pradalier et. al., Phys. Rev. E **82** 025401 (2010)

[2] S. Sugita et. al., Plasma Phys. Control. Fusion **54** 125001 (2012).

[3] P. H. Diamond and T. S. Hahm Phys. Plasmas **2** 3640 (1995)

[4] X. Garbet et al., Phys. Plasmas **14** 122305 (2007)

[5] T. Tokuzawa et. al., Plasma Fusion Res. **9** 1402149 (2014).

[6] S. Inagaki et. al., Plasma Fusion Res. **8** 1201171 (2013).

[7] W. A. Peebles et. al., Rev. Sci. Instrum **81** 10D902 (2010).

[8] F. Clairet et. al., Rev. Sci. Instrum. **82** 083502 (2011)

# Parameter Research in Magnetic Field Structure around the outermostsurfaces of L=1 Helical Systems

M. Aizawa and Y. Nagamine

*Institute of Quantum Science, College of Science and Technology,*

*Nihon University, Tokyo, 101-8308, JAPAN*

*E-mail:aizawa@phys.cst.nihon-u.ac.jp*

**Abstract**

The L=1 torsatron systems having a spatial magnetic axis have been studied. The features of magnetic field properties among high and low helical coil aspect ratio systems which are different pitch modulation $\alpha^*$ have been examined. Especially, magnetic field structure around the outermostsurface of each system, have been studied by examining the Lyapunov characteristic exponents. As the result, we propose the desirable properties of magnetic field of helical system.

## 1. Introduction

The trapped particle confinement in the L=1 helical system with a large $N$ is considerable satisfactory by particle orbits tracing and calculating the neoclassical transport particle and heat fluxes [1]. If we consider a compact system, a small field period number of the helical magnetic field $N$ and low coil aspect ratio system is desirable for nuclear fusion device construction. On the contrary, particles transport properties of compact system become worse due to large usual toroidal effects. We have improved particles transport by controlling the effective curvature toroidal term defined as the sum of usual toroidal curvature term and one of the nearest satellite harmonics of helical field term [1]. The improvement of particles confinements evaluated by the Boozer coordinate is observed. But their effects are more limited than that of large aspect ratio cases. And the structures of magnetic field are also studied from viewpoint of the effective curvature term [2].

## 2. Device parameters

We have examined several type devices with different coil aspect ratio $A_C \equiv R_0 / a$, where $R_0$ is a major radius and $a$ is a minor radius of helical system coil. A minor radius $a$ is hold constant $(=0.3[m])$ and a helical coil current is 1000[kA] in each case. The length of one helical field period is also fixed with standard case $N_0 = 17$ device so that new coil aspect ratio will be obtained for an appropriate $N$ by $A_C = NA_{C0} / N_0$. The subscript "0" denotes standard device case. The characteristic parameters are summarized in the reference [2].

### 3. Evaluation of Lyapunov characteristic exponent

The Lyapunov characteristic exponent (LCE) gives the rate of exponential divergence from perturbed initial conditions. The equations of a magnetic field line have three LCEs, and the positive maximum LCE $\lambda$ decides a chaotic property of system. Let the base solution as $\tilde{\boldsymbol{X}}(L)$, where $L$ is the arc length of field line, and perturbed solution as $\hat{\boldsymbol{X}}(L)$. The deviation vector $\boldsymbol{w}(L) = \hat{\boldsymbol{X}}(L) - \tilde{\boldsymbol{X}}(L)$ and its magnitude $\varepsilon(L) = |\boldsymbol{w}(L)|$ are introduced to calculate $\lambda$, where $\lambda$ is defined by

$$\lambda = \lim_{L\to\infty,\varepsilon(0)\to 0} \lambda(L), \quad \lambda(L) = \frac{1}{L}\log\frac{\varepsilon(L)}{\varepsilon(0)}. \tag{1}$$

It is difficult to calculate $\lambda$ by using eq. (1) directly. Because the finite digits of computer treatment cannot process wide range data defined by eq. (1). The practical calculation method considered above problem are rescaling the data and the details are described in the literature [3,4]. We adopt $\lambda(L=1000)$ or extrapolated from $\lambda(L)$ behavior instead of $\lambda = \lambda(L\to\infty)$. The lower graph in the Fig.1 shows $\lambda(L)$ along the magnetic field line which starts from the point inside near the outermostsurface. The value $\lambda(L)$ is converged to 0, which means that this field line construct the nested surface. On the other hand, the upper graph is converged to positive value; this means magnetic field line has no flux surface. These two cases are confirmed by usual Poincare plots method. The value of $\lambda(L=1000)$ is dependent on the starting point, so that we decide $\lambda$ by averaging some neighboring starting points denoted by r0. Figure 2 is that example cases, where five magnetic field lines are followed. It is seen that each $\lambda(L)$ is converged to positive value $\lambda$.

## 4. Magnetic Field Properties

The value of $\lambda$ are shown against the system device period number $N$ in Fig.3. The five different pitch modulation $\alpha^*$ (-0.4, -02, 0.0, +0.2, +0.4) are noted in this figure. These helical systems have a peak to the change of $N$. And, it is possible to say that $N$=12 has a complicated field structure. In Fig.4, the value of $\lambda$ for different $N$ systems are shown against $\alpha^*$. The four or five field lines started from the neighborhood points which are near outside region of the outermostsurface, are followed to evaluate LCE. These four or five data are averaged to obtain the value of $\lambda$. In Fig.4, this average values of $\lambda$ are plotted. In four different $N$ systems, the chaotic properties of magnetic field around the outermostsurface is low in $N = 5$ system and the tendency that the chaotic property is high in $N = 12$ system.

## 5. Conclusion

We have studied the chaotic behavior could be quantitatively evaluated by the LCE. Moreover, it grasped the tendency of the complexity of the field structure of the system in changing parameters of the devices and examining the LCE. The parameter $\alpha^*$ controls the chaotic properties of fields around the outermost surface in large $N$ system. The low $N$ system has low $\lambda$ properties though $\alpha^*$ dependency is weaker than large $N$, namely high $A_c$ cases.

## References

[1] M. Aizawa and S. Shiina; *Phys. Rev. Lett.* Vol. 84, 2638 (2000)

[2] M. Aizawa and Y. Nagamine; *ECA* Vol. **37D** P-4.148 (2013)

[3] M. Aizawa, Y. Nakata and Y. Nagamine; *ECA* Vol. **38F** P-2.082 (2014)

[4] E.Ott; "*Chaos in Dynamical Systems, 2nd ed.*" Cambridge University Press (2002)

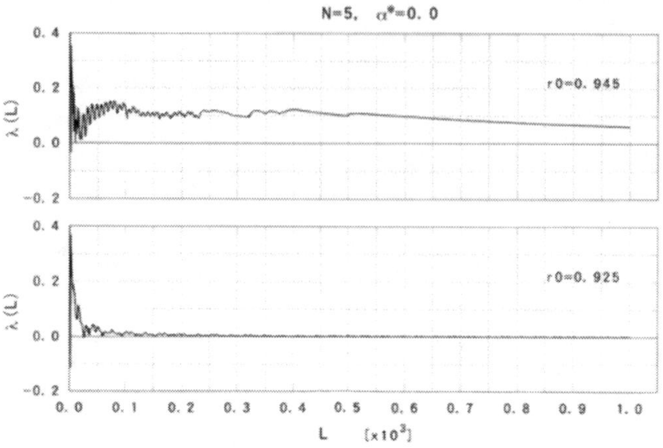


**Fig.1**

Maximum Lyapunov exponents along the magnetic field line are shown. The upper figure shows the field line which starts from outside of the outermostsurface, and the lower figure shows the field line which starts from inside region.

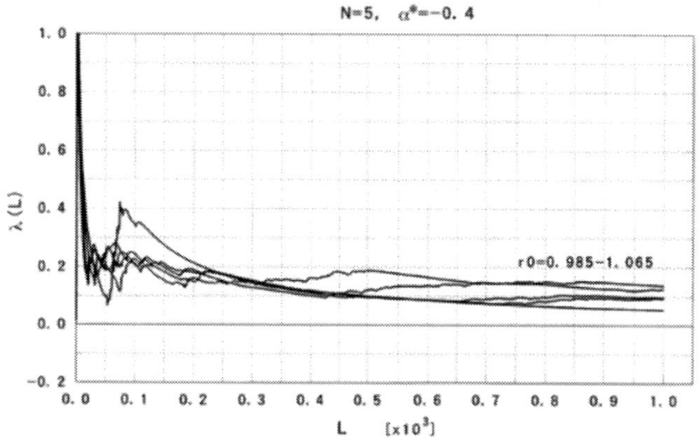


**Fig.2**

Maximum Lyapunov exponents along the magnetic field lines which are outside of the outermostsurface, are converged to positive values, where r0 denotes the starting points.

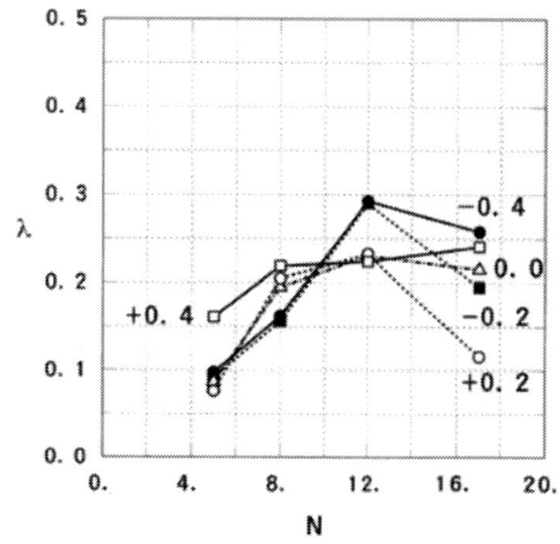


**Fig.3**

The converged maximum Lyapunov exponents $\lambda$ are shown against the system period number $N$, where noted values in the figure describe pitch modulation $\alpha^*$.

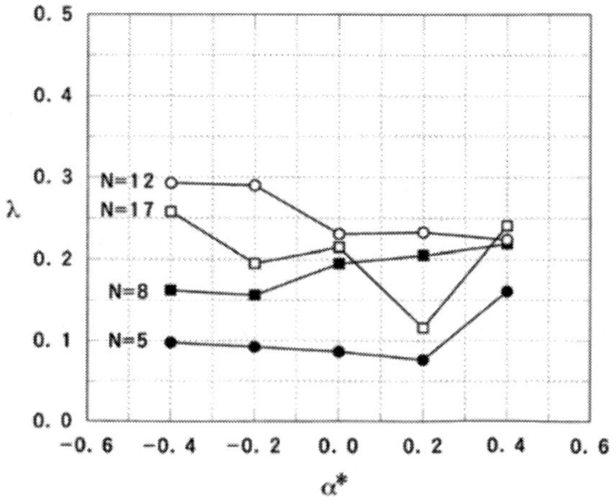


**Fig.4**

For each period number $N$, the averaged maximum Lyapunov exponents $\lambda$ are shown against pitch modulation $\alpha^*$. The compact $N$=5 device has low $\lambda$.

# Control-oriented dynamical model of disruption generated RE beam

D. Carnevale[2], M. Gospodarczyk[2], B. Esposito[1], L. Boncagni[1], M. Sassano[2], S. Galeani[2], W. Bin[4], P. Buratti[1], G. Calabró[1], C. Cianfarani[1], A. Gabrielli[2], G. Ferró[2], G. Granucci[4], D. Marocco[1], J.R. Martin-Solis[4], F. Martinelli[2], L. Panaccione[1], Z. Popovic[3], G. Pucella [1], G. Ramogida[1], M. Riva[1], O. Tudisco[1], and FTU team[1]

[1] *ENEA for EUROfusion, via E. Fermi 45, 00044 Frascati (Roma), Italy*

[2] *Dip. di Ing. Civile e Ing. Informatica, Università di Roma, Tor Vergata, Roma, Italy*

[3] *Universidad Carlos III de Madrid, Avda. Universidad 30, Leganes, 28911-Madrid, Spain*

[4] *IFP-CNR, Via R. Cozzi 53, 20125 Milan, Italy*

We propose a simplified dynamical model of the RE beam current *during the RE plateau* such as

$$\dot{I}_p(t) = -p_1 I_p(t) - p_2 \dot{I}_V(t) - p_3 \dot{I}_T(t) - p_4 FC(t), \tag{1}$$

where the non-negative parameters $p_i$, $i = 1, \dots, 4$ are identified via optimized algorithm based on prediction error minimization, $I_V$ and $I_T$ are the current flowing in the coils V and T shown in Figure 1, respectively, that are both magnetically coupled with the plasma/RE beam. The (real-time) Fission Chamber (FC) signal provides the counts of photoneutrons and photofissions induced by gamma rays with energy higher than 6 MeV (produced by bremsstrahlung of the RE interacting with the metallic plasma facing components). Identification results for a set

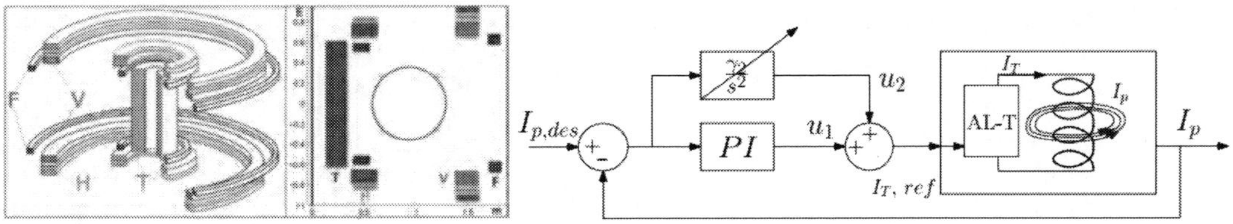


Figure 1: (Left) The active coils at FTU. (Right) The enhanced plasma current control scheme.

of six disruption generated RE beams are shown in Figure 2. The parameters have been optimized for each single shot and their mean value is $[p_1, p_2, p_3, p_4] = [6.7, 7.5, 2.7, 1.3E-8]$ with standard deviation $[1.6, 8.9, 1.8, 1.3E-9]$. The standard deviation of the parameter $p_2$ is very large given that in some experiments where $I_V$ is constant $p_2$ is set to zero. The RE current decay time constant $1/p_1$ clearly depends on cold back-ground plasma density, beam anisotropy and inductance, high-Z impurity particles etc. as in [2, 3]. However we have decided in this preliminary study to neglect these dependences. This is consistent with the aim of designing a robust controller able to cope with parameter uncertainties. Model refinements are the subject of ongoing studies. A standard PI controller plus a pre-programmed term, which resembles a feed-forward action, constitutes the plasma current control loop at FTU. It can be analytically proved that the single integral action of the PI controller is not sufficient to yield zero steady state error for constant desired plasma current. The error can even grow linearly when a ramp-up or ramp-down is required. Since our aim is the RE beam energy reduction by ramping-down its current, the latter issue strongly comes into play. Then, based on the proposed RE beam current approximate model, we have enhanced the PI controller adding a further block that implements a double integrator as shown in the right plot of Figure 1. The double integrator allows to im-

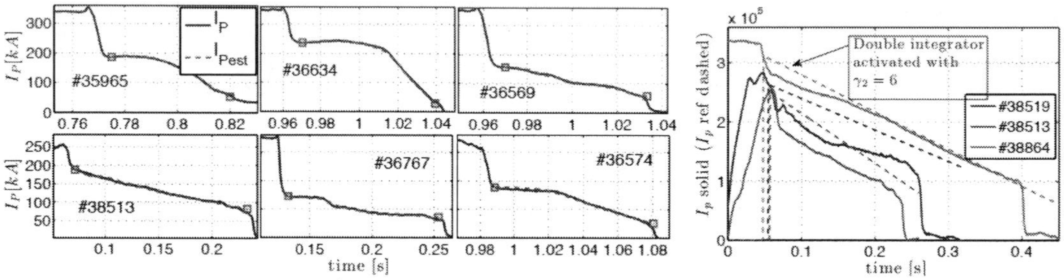


Figure 2: (Left: 6 plots) Comparison between experimental (solid) and estimated (dashed) RE beam current profiles. (Right) In the shot #38864 the double integrator is activated at $0.85s$.

prove performances and the ramp-down tracking error amplitude is (theoretically) proportional to $1/(p_3\gamma_2)$. The value of the gain $\gamma_2$ has been selected based on the identified range of the parameter $p_3$. In [4] it has been shown how the external radius $R_{\text{ext}}$ of the RE beam is a key feature at FTU to avoid RE interaction with the poloidal limiter during RE beam current ramp-down. In order to design a robust controller with improved $R_{\text{ext}}$ tracking performances, we propose the following dynamical model in the *RE beam plateau phase* such as

$$\begin{cases} \dot{x}_1 &= x_2, \\ \dot{x}_2 &= -c_1x_1 - c_2x_2 + ((c_3(3.8I_V + I_F) - c_4I_p) + c_0) + c_5x_3, \\ \dot{x}_3 &= -c_6x_3 + (V_{\text{loop}} - c_7), \end{cases} \tag{2}$$

where $x_1$ is the estimated $R_{\text{ext}}$ and the parameters $c_i$, $i = 1,\dots,7$ have been obtained with the same algorithms adopted for (1). $\ddot{x}_1 = \dot{x}_2$ consists of a term $((c_3(3.8I_V + I_F) - c_4I_p) + c_0)$ that takes into account the vertical magnetic fields produced by the current in the coil F, V, and $I_p$, whereas the term $c_5x_3$ introduces to the Shafranov-shift since $x_3$ is a rough estimate of the RE energy. The identification results are shown in Figure 3 and parameters mean and standard deviation are $[c_1, c_2, c_3, c_4, c_5, c_6, c_7] = [4.3E3, -2.9E6, -0.81, 23, 6.7E5, 6.6E-3, 3.6]$ and $[1.8E3, 8.6E5, 1.2, 21, 1.6E5, 0.3, 2]$, respectively. Model refinements are under investigation to

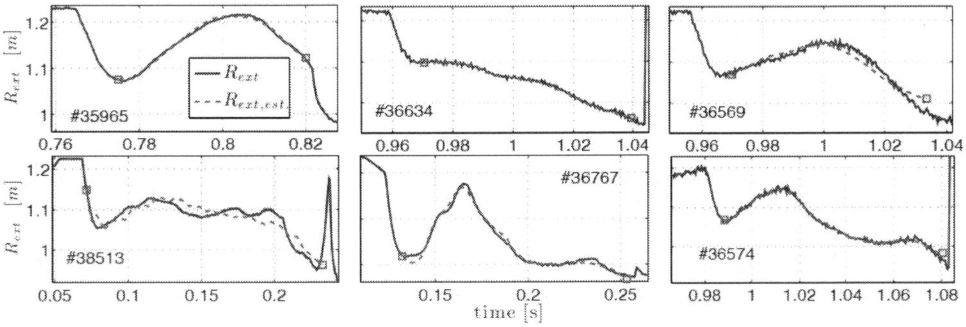


Figure 3: Comparison between experimental (solid) and estimated (dashed) RE beam radial within the time intervals highlighted by red squares. (Right) In the shot #38864 the double integrator is activated improving the tracking performances.

decrease the variation of the parameters among different shots of the model (2).
We have analyzed the differential equation proposed in [2]

$$\dot{W}_e = P_{gain} - P_{coll} - P_{sync} + P_{\text{th}} = e\frac{V_{\text{loop}}}{2\pi R}v\cos(\theta) - \frac{n_e e^4 ln(\Delta)}{4\pi\varepsilon_0^2 m_e v} - \frac{2}{3}r_e m_e c^3\left(\frac{v}{c}\right)^4\gamma^4\left\langle\frac{1}{R^2(t)}\right\rangle + P_{\text{th}}(v), \quad (3)$$

that can be considered to estimate the energy dynamics of an electron as discussed in [2] where we added a further stochastic therm $P_{\text{th}}$ that dominates the other therms when $v$ is low ($W < 1$keV), i.e. when the electron is (thermal) and not runaway. This stochastic term has been designed [6] such that the density distribution of the random variable $W$ satisfies the Maxwell-Boltzmann electron energy distribution. In (3) , $v$ is the electron velocity, $\theta$ its pitch angle, $n_e$ the electron density, $\gamma = 1/\sqrt{1-(v/c))^2}$. RE in FTU are mainly generated by Dreicer effect, so no avalanche effects are taken into account. It is crucial to note that the number of roots of the right-hand side terms in (3), namely equilibrium points of the model, varies depending on the value of $n_e$ and/or $V_{\text{loop}}$. There are values of $n_e$ (high) and $V_{\text{loop}}$ (low) such that there is only one zero corresponding to an asymptotically stable (a.s.) equilibrium (only thermal electrons at stationary conditions). In the other hand with low $n_e$ and high $V_{\text{loop}}$ there exists again only one asymptotically stable equilibrium point at high energy: at steady state all electrons became runaway energy limited by synchrotron radiation loss. There are conditions in between in which three equilibria exist: two of them are a.s. and one, in the middle, is unstable. Then, at steady state, there are two stable (in random sense) population of thermal and runaway electrons depending on the past history of the system (energy history of each electron): this type of differential (stochastic) equation yields an hysteretic behavior of the runaway population. Note that similar results hold also in case of secondary RE generation mechanism as discussed (in a different approach) in recent works [1] and [5]. We show in Figure 4, for the first time to our knowledge, an experimental evidence of the hysteresis in case of RE generation and suppres-

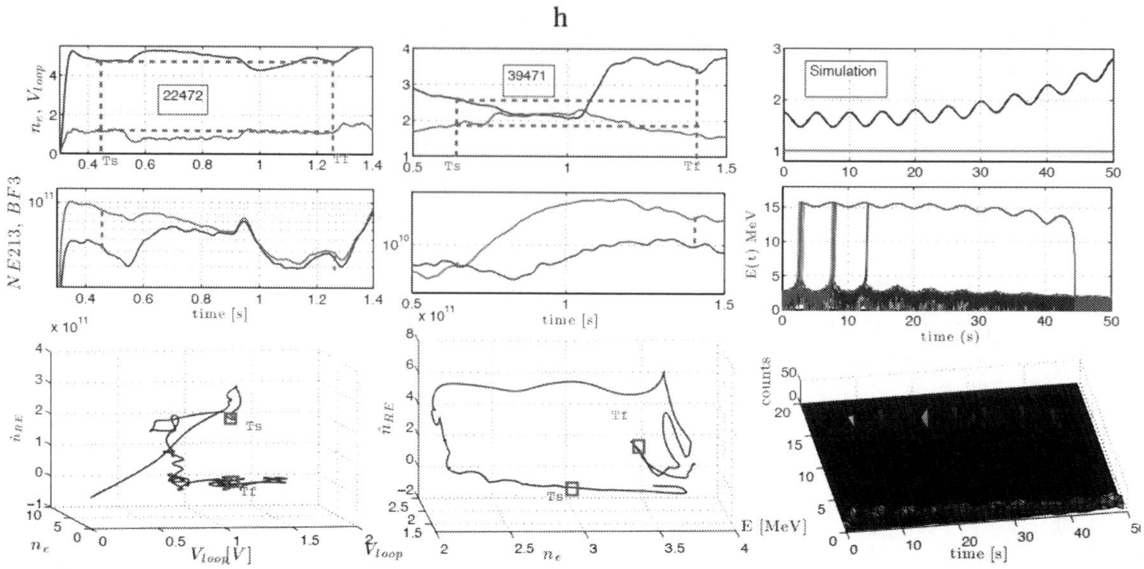

Figure 4: (Left and middle column, top) Density in blue and loop voltage in red (scaled and offset); (middle) Neutrons plus gamma rays counts in red (NE213) and neutrons in blue (BF3); (bottom) Hysteresis graph of the estimated number of RE. (Right column, top) Simulation signal of density (blue) and loop voltage (red); (middle) Energy of 100 $e^-$ obtained integrating numerically (3); (bottom) Number of RE versus energy and time.

sion.

Further studies are necessary to adapt (3) to FTU machine and match experimental results: Figure 4 reports a simulation to show qualitative results obtained by (3) to show the formation and dissipation due to density changes of thermal and RE populations. A vertical displacement RE beam model and studies on magnetic coupling of RE with cold background plasma (magnetically confined by RE) as source of energy loss are currently under investigation to further refine the theoretical thresholds given in [7].

## Acknowledgements

This work has been carried out within the framework of the EUROfusion Consortium and received funding from the Euratom research and training programme 2014-2018 under grant agreement No 633053. The views and opinions expressed herein do not necessarily reflect those of the European Commission.

## References

[1] P. Aleynikov et al., Phys. Rev. Lett., **114**, (2015) 155001.
[2] J. R. Martìn Solìs et al., Phys. Rev. Lett., **105** (2010) 185002.
[3] C. Paz-Soldan et al., Physics of Plasmas **21**, 022514 (2014).
[4] D. Carnevale et al., Proc. 33rd IAEA Conference, St Petersburg (Russian Federation), EX/P2-48 (2014).
[5] J. Decker et al., arXiv:1503.03881 [physics.plasm-ph].
[6] A. Skorokhod and I. I. Gikhman, "Stochastic Differential Equations", Springer Verlag (1972).
[7] B. Esposito et al., Proc. 33rd IAEA Conference, St Petersburg (Russian Federation), EX/P2-50 (2014).

# Theory for neoclassical toroidal plasma viscosity in finite aspect ratio tokamaks with broken symmetry*

K. C. Shaing

*Institute for Space and Plasma Sciences, National Cheng Kung University, Tainan 70101, Taiwan*

## I. Introduction

Real tokamaks do not possess toroidal symmetry. When toroidal symmetry is broken, particle, momentum, and heat transport fluxes are all enhanced [1,2]. The toroidal angular momentum is no longer conserved as expected. A comprehensive transport theory has been developed for large aspect ratio tokamaks with broken symmetry. The theory is summarized in [3,4]. The transport fluxes are calculated on the constant pressure surfaces because of the magnetohydrodynamic (MHD) equilibrium. When magnetic surfaces are not broken, i.e., when KAM surfaces are intact, the constant pressure surfaces are the perturbed magnetic surfaces. Thus, on the perturbed magnetic surfaces, parallel electron heat conductivity $\chi_{\parallel e}$ does not contribute to transport fluxes. The results of the theory are in good agreement with numerical results in the large aspect ratio limit [5-7]. The theory has since been extended to finite aspect ratio tokamaks with broken symmetry [8]. In the superbanana plateau and superbanana regimes, perturbed particle distribution is localized in the phase space. This makes it possible to remove the assumption of the large aspect ratio in these collisionality regimes [8]. Here, we extend the transport fluxes in the collisional boundary layer $\sqrt{\nu}$ regime to include the effects of finite aspect ratio in tokamaks with broken symmetry, where $\nu$ is the typical collision frequency. Again, the extension is made possible by utilizing the fact that the boundary layer solution is localized in the phase space.

## II. Magnetic Coordinates and $|B|$ Spectrum

We adopt Hamada coordinates [9], where magnetic field $\boldsymbol{B}$ can be expressed as

$$\boldsymbol{B} = \psi' \nabla V \times \nabla\theta - \chi \nabla V \times \nabla\zeta, \tag{1}$$

where $V$ is the volume enclosed inside the magnetic surface divided by $4\pi^2$, $\theta$ is the poloidal angle, $\zeta$ is the toroidal angle, $\psi' = \boldsymbol{B} \bullet \nabla\zeta$, $\psi$ is the toroidal flux divided by $2\pi$, $\chi = \boldsymbol{B} \bullet \nabla\theta$, $\chi$

is the poloidal flux divided by $2\pi$, and prime denotes $d/dV$. The $|\boldsymbol{B}|=B$ spectrum on the perturbed magnetic surface can be written as

$$B = B(V,\theta) - B_0 \sum_n \left[A_n(\theta)\cos n\zeta_0 + B_n(\theta)\sin n\zeta_0\right], \tag{2}$$

where $B(V,\theta)$ represents the axisymmetric magnetic field strength on the magnetic surface, $\zeta_0$ = $q\theta$ -$\zeta$ is the field line label, $A_n(\theta)$ = $\sum_m \{b_{mnc}\cos[(m-nq)\theta] + b_{mns}\sin[(m-nq)\theta]\}$, $B_n(\theta)$ = $\sum_m \{-b_{mnc}\sin[(m-nq)\theta] + b_{mns}\cos[(m-nq)\theta]\}$, and $B_0$ is the magnetic field strength on the magnetic axis. The $A_n(\theta)$, and $B_n(\theta)$ terms are the consequences of the broken symmetry in tokamaks. We are interested in the situation that the magnitude of the perturbed magnetic field strength is weak enough so that there are no new classes of trapped particles. Thus, the theory is applicable for rippled tokamaks when ripple trapping is insignificant.

**III. Bounce Averaged Drift Kinetic Equation**

The transport fluxes in the collisional boundary layer regime are calculated form the solution of the bounce averaged drift kinetic equation. The linear version is [1]

$$\left\langle \mathbf{v}_d \bullet \nabla\zeta_0 \right\rangle_b \frac{\partial f_{01}}{\partial \zeta_0} + \left\langle \mathbf{v}_d \bullet \nabla V \right\rangle_b \frac{\partial f_M}{\partial V} = \left\langle C(f_{01}) \right\rangle_b, \tag{3}$$

where $f_{01}$ is the perturbed distribution function, $f_M$ is the equilibrium Maxwellian distribution, $\mathbf{v}_d$ is the drift velocity, the angular brackets denote the bounce average $\langle \bullet \rangle_b$ = $\oint d\theta(\bullet)/\mathrm{v}_\parallel \boldsymbol{n} \bullet \nabla\theta \Big/ \oint d\theta/\mathrm{v}_\parallel \boldsymbol{n} \bullet \nabla\theta$, $\mathrm{v}_\parallel$ is the particle speed that is parallel to $\boldsymbol{B}$, $\boldsymbol{n}$ is the unit vector in the direction of $\boldsymbol{B}$.

The drift velocity $\mathbf{v}_d$ has the conservative form $\mathbf{v}_d$= - $\mathrm{v}_\parallel \boldsymbol{n} \times \nabla(\mathrm{v}_\parallel/\Omega)$, where $\Omega$ is the gyro-frequency. This form for $\mathbf{v}_d$ is valid for low $\beta$ plasmas. Here $\beta$ is defined as the ratio of the thermal pressure to the magnetic field pressure. However, it is trivial to show that it is also valid for finite $\beta$ plasmas in the radial and $\nabla\zeta_0$ directions. It is the $\mathbf{v}_d$ in these two directions that appear in Eq.(3). The expressions for $\langle \mathbf{v}_d \bullet \nabla\zeta_0 \rangle_b$, and $\langle \mathbf{v}_d \bullet \nabla V \rangle_b$ can be found in [8].

**IV. Solution in the Collisional Boundary Layer Regime and Transport Fluxes**

We solve Eq.(3) in the collisional boundary layer regime for tokamaks with broken symmetry. In this regime, the collision frequency is smaller than the toroidal drift frequency. The solution in the region away from the trapped-circulating boundary is

$$f_{01} = \sum_n \frac{McE}{e\chi\langle \mathbf{v}_d \bullet \nabla\zeta_0\rangle_b}\left(\bar{A}_n \cos n\zeta_0 + \bar{B}_n \sin n\zeta_0\right)\frac{\partial f_M}{\partial V}, \quad (4)$$

where $E = \mathrm{v}^2/2$, v is the particle speed,

$$\begin{pmatrix}\bar{A}_n\\ \bar{B}_n\end{pmatrix} = \int_{-\theta_t}^{\theta_t} d\theta \frac{B/B_m}{\left[k^2-(B/B_m-1)/(B_M/B_m-1)\right]^{1/2}}\left[-\lambda\frac{B}{B_m}+2\left(1-\lambda\frac{B}{B_m}\right)\right]\times$$

$$\frac{B_0}{B}\begin{bmatrix}A_n(\theta)\\ B_n(\theta)\end{bmatrix}\left\{\int_{-\theta_t}^{\theta_t} d\theta \frac{B/(B_m\varepsilon')}{\left[k^2-(B/B_m-1)/(B_M/B_m-1)\right]^{1/2}}\right\}^{-1}, \quad (5)$$

$k^2 = (1-\lambda)/\left[\lambda(B_M/B_m-1)\right]$, $\lambda = \mu B_m/E$, and $B_M$ and $B_m$ are respectively the global maximum and minimum of axisymmetric $B(V,\theta)$.

The solution in Eq.(4) does not satisfy the boundary condition and diverges logarithmically at the trapped-circulating boundary where $k^2 = 1$. A boundary layer analysis is needed to make the solution satisfying the boundary condition [10,11]. Because the layer is narrow in the pitch angle space, only pitch angle scattering operator is needed in the boundary layer analysis without the need of the large aspect ratio assumption. The resultant solution that satisfies the boundary condition and matches Eq.(4) when away from the boundary layer is

$$f_{01} = \sum_n \frac{McE}{e\chi}\left(\bar{\alpha}_n \cos n\zeta_0 + \bar{\beta}_n \sin n\zeta_0\right)\frac{\partial f_M}{\partial V}, \quad (6)$$

where

$$\langle \mathbf{v}_d \bullet \nabla\zeta_0\rangle_b^{-1}\begin{pmatrix}\bar{\alpha}_n\\ \bar{\beta}_n\end{pmatrix} = \begin{pmatrix}\bar{A}_n\\ \bar{B}_n\end{pmatrix}\left(1-e^{-\sqrt{|n|y}}\cos\sqrt{|n|y}\right) + \begin{pmatrix}\bar{B}_n\\ -\bar{A}_n\end{pmatrix}\sigma_w e^{-\sqrt{|n|y}}\sin\sqrt{|n|y}, \quad (7)$$

$\sigma_w$ denotes the sign of $\langle \mathbf{v}_d \bullet \nabla\zeta_0\rangle_b$, $y = (1-k^2)\left[\Delta(k^2)\right]^{-1}$ is the layer variable, the layer width is

$$\Delta(k^2) = \left\{\frac{4\nu_D}{\lambda^2(1-B_M/B_m)^2\langle \mathbf{v}_d \bullet \nabla\zeta_0\rangle_b}\frac{\oint d\theta|\mathrm{v}_\parallel/\mathrm{v}|}{\oint d\theta(B/B_M)/(|\mathrm{v}_\parallel/\mathrm{v}|)}\right\}^{1/2}, \quad (8)$$

the pitch angle scattering frequency is $\nu_D = \sum_b \nu_D^{ab}$, $\nu_D^{ab} = \nu_{ab}\left[\Phi(\mathrm{v}/\mathrm{v}_{tb})-G(\mathrm{v}/\mathrm{v}_{tb})\right]/(\mathrm{v}/\mathrm{v}_{ta})^3$, the the basic relaxation frequency $\nu_{ab} = 4\pi N_b(e_a e_b)^2 \ln\Lambda/(\mathrm{v}_{ta}^3 M_a^2)$, $N$ is plasma density, $e$ is the electric charge, $M$ is the mass, $\mathrm{v}_t$ is the thermal speed, the subscripts $a$ and $b$ denote plasma species, $\ln\Lambda$ is the Coulomb logarithm, the Chandrasekhar function is $G(x)$ = $\left[\Phi(x) - x\Phi'(x)\right]/\left(2x^2\right)$, and $\Phi(x)$ is the error function.

Using the solution in Eq.(6), the transport fluxes can be evaluated. The Onsager symmetric flux surface averaged radial particle flux $\Gamma$ and heat flux are

$$\begin{pmatrix} \Gamma \bullet \nabla V \\ q \bullet \nabla V/T \end{pmatrix} = -\frac{1}{4}\left(\frac{Mc}{|e|\chi}\right)^2 \left(\frac{B_M}{B_m} - 1\right)^{-1} \int dv v_D v^6 \frac{\partial f_M}{\partial V} \int_0^1 dk^2 \lambda^{-1} \left(\oint d\theta \left|\frac{v_{\parallel}}{v}\right|\right) \times$$

$$\sum_n \left[\left(\frac{\partial \overline{\alpha}_n}{\partial k^2}\right)^2 + \left(\frac{\partial \overline{\beta}_n}{\partial k^2}\right)^2\right]. \qquad (9)$$

From the flux-force relation [12], these fluxes are related to the toroidal plasma viscosity and can be used to model plasma flow in tokamaks with broken symmetry.

## V. Conclusions

We have extended theory for neoclassical toroidal plasma viscosity for tokamaks with broken symmetry in the superbanana plateau, superbanana, and collisional boundary layer regimes to include effects of finite aspect ratio and finite $\beta$. They can be employed to model plasma flows in finite aspect ratio tokamaks with broken symmetry, invoking flux-force relation.

*This work was supported by Taiwan Ministry of Science and Technology (MOST) under Grant No. 100-2112-M-006-004-MY3.

**References**

[1] K. C. Shaing, Phys. Plasmas **10**, 1443 (2003).

[2] K. C. Shaing, Phys. Rev. Lett. **87**, 245003 (2001).

[3] K. C. Shaing, M. S. Chu, C.T. Hsu, S. A. Sabbagh, J. Seol, and Y. Sun, Plasma Phys. Control. Fusion **54**, 124033 (2012).

[4] K. C. Shaing, S. A. Sabbagh, and M. S. Chu Nucl. Fusion **50**, 025022 (2010).

[5] Y. Sun, Y. Liang, K. C. Shaing, H. R. Kostowski, C. Wiegmann, and T. Zhang, Phys. Rev. Lett. **105**, 145002 (2010).

[6] Y. Sun, Y. Liang, K. C. Shaing, H. R. Kostowski, C. Wiegmann, and T. Zhang, Nucl. Fusion **51**, 053015 (2011).

[7] S. V. Kasilov, W. Kernbichler, A. F. Martitsch, H. Maassberg, Ma. F. Heyn, Phys. Plasmas **21**, 092506 (2014).

[8] K. C. Shaing, J. Plasma Phys. **81**, 905810203 (2015).

[9] S. Hamada, Nucl. Fusion **2**, 23 (1962).

[10] K. C. Shaing, P. Cahyna, M. Becoulet, J.-K. Park, S. A. Sabbagh, and M. S. Chu, Phys. Plamas **15**, 082506 (2008).

[11] K. C. Shaing, J. Seol, Y. Sun, M. S. Chu, and S. A. Sabbagh, Nucl. Fusion **50**, 125008 (2010).

[12] K. C. Shaing, M. S. Chu, and S. A. Sabbagh, Nucl. Fusion **50**, 125012 (2010).

# Improvement of One-Dimensional Fluid Modeling of the SOL-Divertor Plasmas and Neutrals Concerning the Anisotropy of Ion Temperature and the Diffusion Coefficient of Neutrals

S. Togo[1], T. Takizuka[2], M. Nakamura[3], K. Hoshino[3], K. Ibano[2], T. L. Lang[1], Y. Ogawa[1]

[1] *Graduate School of Frontier Sciences, University of Tokyo, Kashiwa, Japan*

[2] *Graduate School of Engineering, Osaka University, Suita, Japan*

[3] *Japan Atomic Energy Agency, Rokkasho, Japan*

## 1. Introduction

In order to design the divertor of future tokamak fusion reactors, simulation code packages of scrape-off layer (SOL) – divertor plasmas play important roles. For the plasma, a fluid model [1] is often adopted instead of a kinetic model. In the plasma fluid model, a second-order-derivative viscosity term is included in the parallel momentum transport equation which requires the boundary condition for Mach number $M$ at the sheath edge, i.e., $M = 1$ based on the Bohm condition. The viscous flux is an approximated form of stress tensor $\pi = 2n(T_{i,//} - T_{i,\perp})/3$ where $n$, $T_{i,//}$ and $T_{i,\perp}$ are plasma density, parallel and perpendicular ion temperature, respectively, and was derived on the assumption that the plasma was collisional enough for $\pi$ to be much smaller than effective isotropic ion pressure $nT_i = n(T_{i,//} + 2T_{i,\perp})/3$ [2]. A kinetic simulation showed, however, that the ion temperature anisotropy of the SOL-divertor plasma was remarkable even in the medium collisional regime [3]. Thus, we developed a one-dimensional (1D) SOL-divertor fluid code which distinguishes between $T_{i,//}$ and $T_{i,\perp}$ [4] by referring to generalized fluid modelling [5]. By introducing anisotropic ion temperatures directly, the parallel momentum transport equation becomes the first-order deferential. Consequently, the boundary condition $M = 1$ at the sheath edge becomes unnecessary. Instead of $M = 1$, in order to model the effects of the divertor plate and sheath region, we developed a virtual divertor (VD) model. This model sets artificial sinks for particle, momentum and energy in the artificial region (VD region) between the two divertor plates according to the image of a "waterfall". It was shown that Bohm condition was automatically satisfied by this VD model [4].

As for the neutral particles, a kinetic Monte-Carlo (MC) model is often used. While kinetic MC model considers every kinetic effect, it has problems of MC noise and long computational time. Fluid model is useful in this point. Thus, we developed a neutral fluid model which can work with VD model. It was pointed out, however, that the simulation results with a fluid model had a significant difference from those with a kinetic MC model [6]. With the fluid model, spread of the neutral particles in the divertor region seemed to be larger as compared to the kinetic MC model. We suspect that this was caused by the overestimation of the diffusion coefficient without the effect of radial loss of neutral particles.

In this paper, we first explain the numerical models in our code in Sec. 2; plasma fluid model which introduces anisotropic ion temperatures, VD model and neutral fluid model which is self-consistent with the plasma via VD model. In Sec. 3, numerical results are shown. Summary and discussion are described in Sec. 4.

**2. Numerical Model**

2.1 *Plasma fluid model*

Basic equations for plasma are given in Ref. [4] in detail and we show here only the equation of parallel momentum transport equation for convenience;

$$\frac{\partial(m_i n V)}{\partial t} + \frac{\partial}{\partial s}\left(m_i n V^2 + nT_{i,//} + nT_e\right) = M_m, \qquad (1)$$

For the definition of notations, refer to Ref. [4]. In Eq. (1), ion pressure gradient term is described only by its parallel component $nT_{i,//}$ and it does not have the viscosity term. Therefore, the boundary condition at the sheath edge, $M = 1$, is no longer necessary.

2.2 *Virtual divertor model*

Instead of boundary condition, $M = 1$, we need to reproduce the effects of the divertor plate and sheath. Thus, we developed a virtual divertor (VD) model [4]. It sets an artificial region between the two divertor plates, as shown in Fig. 1, and sets there artificial sinks for particle, momentum and energy according to the image of a "waterfall". The artificial particle sink, for example, is given as $S^{VD} = -n/\tau^{VD}$. Here, $\tau^{VD}$ is input characteristic time of artificial sinks. For other artificial sinks, refer to Ref. [4]. It was shown that the Bohm condition was automatically satisfied by this VD model [4].

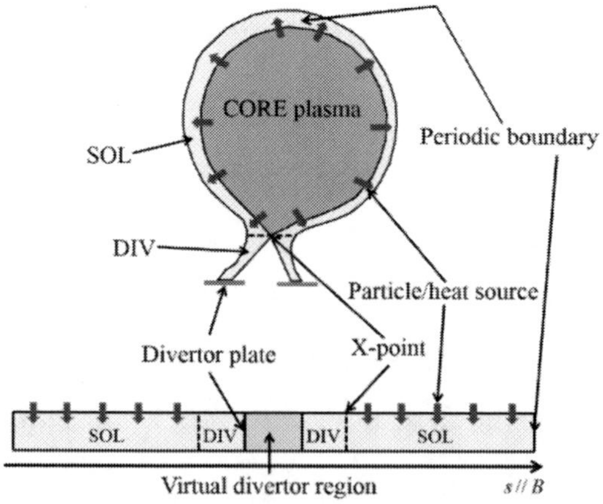

Fig. 1 Schematic picture of virtual divertor model.

2.3 *Neutral model*

Neutral particles are divided into two kinds by referring to the FFCD model [7]; recycling neutrals which are generated at the divertor plate by recycling of ions and diffusion neutrals and transported at a constant velocity, and diffusion neutrals which are generated in the plasma by charge exchange of recycling neutrals and volume recombination of plasma and transported by the charge exchange diffusion. Schematic picture of this neutral model in the plasma region is shown in Fig. 2 (a). The equations for recycling and diffusion neutrals density, $n_{n,recy}$ and $n_{n,diff}$, are given as follows, respectively;

$$\frac{\partial n_{n,recy}}{\partial t}+\frac{\partial n_{n,recy}V_{n,recy}}{\partial x}=-n_{n,recy}\nu_{L,recy}-n_{n,recy}\nu_{iz}-n_{n,recy}\nu_{cx} \quad (2)$$

$$\frac{\partial n_{n,diff}}{\partial t}+\frac{\partial}{\partial x}\left(-D_0\frac{\partial n_{n,diff}}{\partial x}\right)=-n_{n,diff}\nu_{L,diff}-n_{n,diff}\nu_{iz}+n_{n,recy}\nu_{cx}+S_{rc} \quad (3)$$

Here, $x$ is in the poloidal direction. Velocity of recycling neutral is given by $V_{n,recy} = -(2\varepsilon_{FC}/m_i)^{1/2}/2$ with Franck-Condon energy $\varepsilon_{FC} = 3.5$ eV. Radial loss frequencies are defined as $\nu_{L,recy} = 10^{-6}$ s and $\nu_{L,diff} = v_0/d$ where $v_0 = (T_i/m_i)^{1/2}$ is thermal velocity of neutrals and $d$ is the SOL width. The subscripts iz, cx and rc are for ionization, charge exchange and volume recombination, respectively. For the diffusion coefficient, $D_{0,ci} = v_0/(\nu_{cx} + \nu_{iz})$ were used in the preceding studies but, we use $D_{0,ciL} = v_0/(\nu_{cx} + \nu_{iz} + \nu_{L,diff})$, too, which is more strict from the random-walk model point of view. In the VD region, RHSs of Eqs. (2) and (3) are replaced by $S_{n,recy}{}^{VD} = \eta(n/\tau^{VD}) + \eta_0(n_{n,diff}/\tau_0{}^{VD})$ and $S_{n,diff}{}^{VD} = -n_{n,diff}/\tau_0{}^{VD}$, respectively. By this artificial sink and source, this neutral model becomes self-consistent with the plasma via VD model. Recycling rates $\eta$ and $\eta_0$ are for ions and diffusion neutrals, respectively. Schematic picture of this treatment of neutrals in the VD region is shown in Fig. 2 (b).

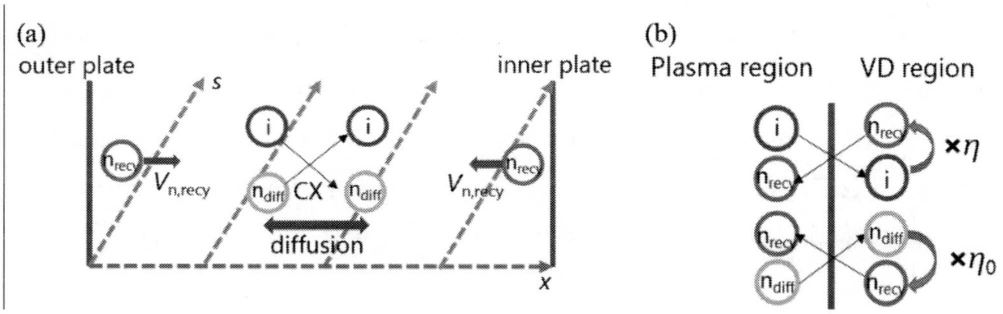


Fig. 2 Schematic picture of the neutral model (a) in plasma region and (b) in VD region.

### 3. Numerical Results

We investigated the effect of neutrals on the ion temperature anisotropy. By the effects of neutrals, the ion temperature anisotropy is relaxed in both SOL and divertor (DIV) regions as shown in Fig. 3. We also investigated the effect of correction of $D_0$ with diffusion-only model and recycling/diffusion model by comparing the plasma and neutral density near the divertor plate between two models for $D_0$ in the case where $\nu_{cx} << \nu_{L,diff}$. With diffusion-only model, in Fig. 4 (a), neutrals spread in wider region in $D_{0,ci}$ case leading to difference in the plasma density profiles. With recycling/diffusion model, in Fig. 4 (b), there was almost no difference. Thus, attention should be paid to the effect of $\nu_{L,diff}$ on $D_0$ particularly when neutrals are described by the diffusion-only model.

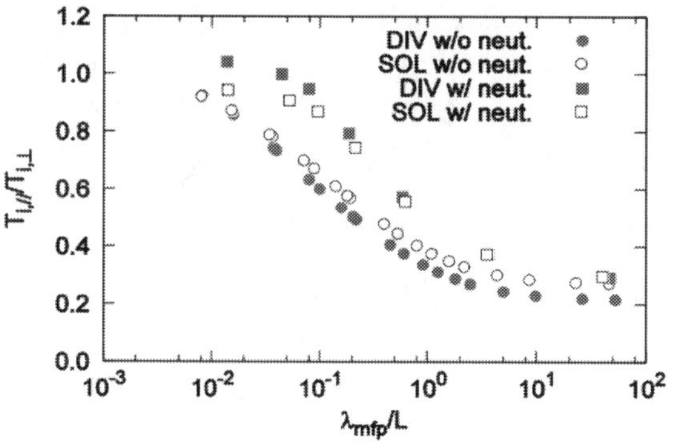


Fig. 3 Anisotropy of ion temperature

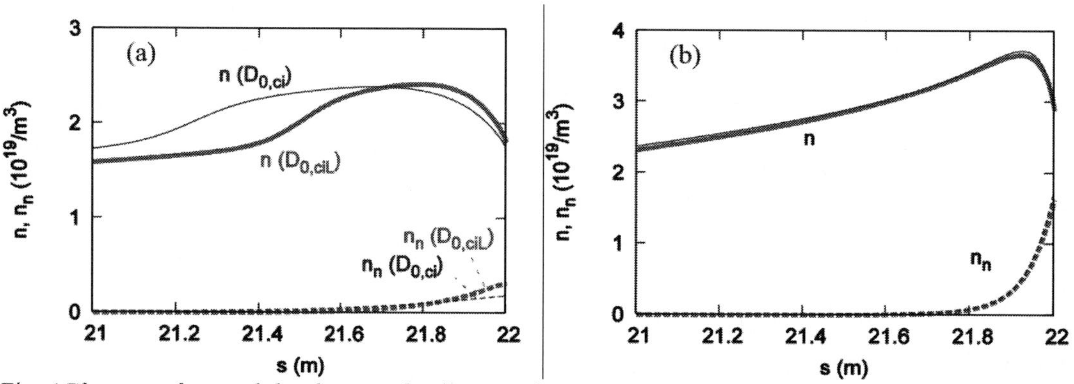


Fig. 4 Plasma and neutral density near the divertor plate at $s$ = 22 m (a) by diffusion neutral model and (b) by recycling and diffusion neutral model. $D_{0,ci} = v_0/(\nu_{cx} + \nu_{iz})$ and $D_{0,ciL} = v_0/(\nu_{cx} + \nu_{iz} + \nu_{L,diff})$ are used.

## 4. Summary and Discussion

In order to reproduce the effects of divertor plates and sheath, instead of the boundary condition for Mach number, $M$ = 1, we use virtual divertor (VD) model. We developed a neutral model which can work with VD model. The effect of neutrals on the ion temperature anisotropy was investigated. The effect of two models for neutral diffusion coefficient $D_0$ was also investigated. Difference of profiles became remarkable when neutrals were described by the diffusion-only model.

## References

[1] S. I. Braginskii, *Reviews of Plasma Physics*, Consultants Bureau (New York, 1965) Vol. 1, p. 205.

[2] P. C. Stangeby, *The Plasma Boundary of Magnetic Fusion Devices*, Institute of Physics Publishing (Bristol and Philadelphia 1999).

[3] T. Takizuka *et al.*, J. Nucl. Mater. **128-129** (1984) 104. [4] S. Togo *et al.* J. Nucl. Mater. **463** (2015) 502.

[5] W. Fundamenski, Plasma Phys. Control. Fusion **47** (2005) R163.

[6] K. Hoshino *et al.*, Contrib. Plasma Phys. **48** (2008) 136. [7] A. K. Prinja *et al.*, J. Nucl. Mater. **196-198** (1992) 340.

# Influence of charge state evolution in the energy deposition of uranium ions traveling through hydrogen plasma

R. Morales[1,2,3,*], M. D. Barriga-Carrasco[2], D. Casas[2], O. Boine-Frankenheim[1], J. Jacoby[3]

[1] *Institut für Theorie Elektromagnetischer Felder (TEMF), Technische Universität Darmstadt, Darmstadt, Germany*

[2] *ETS Ingenieros Industriales, Universidad de Castilla-La Mancha, Ciudad Real, Spain*

[3] *Institut für Angewandte Physik, J. W. Goethe-Universität, Frankfurt am Main, Germany*

*Roberto.Morales1@alu.uclm.es

## 1. Introduction

The nuclear fusion is a promising way to obtain clean energy in the future. The inertial confinement fusion driven by heavy ion beams, is one of the method to obtain energy using fusion reactions. For this reason, understanding the physics of heavy ions traveling through plasmas is an important topic in plasma physics. Heavy ions possess good features to heat small samples of matter and, to reach the necessary temperature and density for the nuclear fusion takes place. On the other hand, conventional stripping techniques are limited in their applicability, e.g. short lifetime in foil stripper and lower efficiency in gas stripper. To reach long lifetime and higher efficiency, the use of plasma as a stripping medium has been studied [1]. In stripper devices, one of the most important thing is the prediction of the final charge state distribution of the ion beam and its total energy loss, which the presented work focuses on.

## 2. Calculation method

According to dielectric formalism, the energy loss of an ion depends on its velocity and on its charge density. Also, it depends on the target through its dielectric function; here the random phase approximation (RPA) model is used because of the excellent results in previous works [2]. The stopping power is defined as the energy loss per unit of length, $dE/dx$. In this work, we use the electronic stopping power, which shows the energy deposition into the plasma by the projectile, taking into account the transitory charge state of the ion,

$$S_p(t) = \frac{2}{\pi\, v^2} \int_0^\infty dk\, \frac{[Z-\rho_e(k,t)]^2}{k} \int_0^{kv} w \,\mathrm{Im}\left[\frac{-1}{\varepsilon(k,w)}\right] dw, \tag{1}$$

where $v$ and $Z$ are respectively the velocity and nuclear charge of the projectile and $\varepsilon(k,w)$ is the dielectric function of the plasma [2]. Here, we use the Brandt-Kitagawa model (BK) [3] to describe the electronic density distribution of the projectile,

$$\rho_e(k,t) = \frac{\bar{N}(t)}{1+(k\Lambda)^2}, \tag{2}$$

where $\bar{N}(t)$ is the transitory mean number of bound electrons in the projectile, and $\Lambda$ is a variational parameter given by,

$$\Lambda(t) = \frac{0.48\,\bar{N}(t)^{2/3}}{Z - \frac{1}{7}\bar{N}(t)}. \quad (3)$$

Finally, the electronic stopping power is given by,

$$S_p = \frac{1}{\tau}\int_0^{\tau} S_p(t)dt, \quad (4)$$

As equation (2) shows, the transitory mean number of bound electrons, $\bar{N}(t)$, must be known. In the present work $\bar{N}(t)$ is calculated by means of two methods.

METHOD 1: SOLVING THE RATE EQUATIONS.

Mean charge state evolution can be calculated by solving the rate equations,

$$\frac{dP_q(t)}{dt} = c_{q+1}P_{q+1}(t) + l_{q-1}P_{q-1}(t) - (c_q + l_q)P_q(t), \quad (5)$$

where $c_q$, $l_q$ are the total capture and ionization cross sections for a projectile with charge state $q$, which must be calculated previously [4], and $P_q$ are the fraction of projectiles in the charge state $q$. Finally, the mean charge state evolution is given by,

$$\bar{Q}(t) = \sum_{q=0}^{Z} qP_q(t), \quad (6)$$

with the normalization $\sum_{q=0}^{Z} P_q = 1$. Then, the transitory number of bound electrons can be calculated as the next equation shows,

$$\bar{N}(t) = Z - \bar{Q}(t). \quad (7)$$

METHOD 2: USING THE EQUILIBRIUM CHARGE STATE.

We can also use the equilibrium charge state of the ion to calculate the transitory number of bound electrons,

$$\bar{N}(t) = N_\infty - (N_\infty - N_0)\,e^{-t/\tau_{ion}}, \quad (8)$$

where $N_0$ is the initial number of bound electrons, and $N_\infty$ is the number of bound electrons in the equilibrium charge state,

$$N_\infty = Z - \langle Q \rangle, \quad (9)$$

where $\langle Q \rangle$ is the equilibrium charge state. In the equation (8), $\tau_{ion}$, is the ionization time,

$$\tau_{ion} = \frac{1}{n_e v \sigma}, \quad (10)$$

where $n_e$ is the free electron density of the plasma, $v$ is the projectile velocity and $\sigma$ is the ionization cross section.

This method requires knowing previously the equilibrium charge state as equation (8) shows. The more complete method makes use of rate equations. In the equilibrium, the fractions of projectiles, $A_q$, are constants, such that, we can obtain the equilibrium charge state supposing that the stationary case in equation (5) is reached,

$$\frac{dP_q(t)}{dt} = 0 \rightarrow l_q P_q - c_{q+1} P_{q+1} = 0. \tag{11}$$

Then we have to solve a ordinary equation system to calculate the $P_q$, and the equilibrium charge state is given by,

$$\langle Q \rangle = \sum_{q=0}^{Z} q P_q \tag{12}$$

However, in the present work, we use the simple formula of Northcliffe with the Kreussler *et al.* stripping criterion [5] adapted to plasma conditions to calculate the charge state equilibrium,

$$\langle Q \rangle = Z \left[ 1 - e^{-\frac{v_r}{Z^{2/3} v_0}} \right], \tag{13}$$

where $v_0$ is the Bohr velocity and $v_r$ is the relative velocity,

$$v_r = \frac{v_e^2}{6v} \left[ \left( \frac{v}{v_e} + 1 \right)^3 - \left| \frac{v}{v_e} - 1 \right|^3 \right]; \quad v_e = \left( \frac{3}{5} v_F^2 + 3V_{the}^2 \right)^{1/2}, \tag{14}$$

where $v_F$ and $v_{the}$ are respectively the Fermi velocity and thermal velocity of plasma electrons.

## 3. Results

Figure 1 shows the mean charge state evolution of an uranium ion beam traveling through fully ionized hydrogen plasma as a function of the initial charge state of the projectile. The smaller charge state of the method using the equilibrium charge state (orange lines) is because of the slight difference in the equilibrium charge state between Kreussler *et al.* model (28.35+) and rate equations (33.6+).

Figure 2 shows the energy loss of the uranium ion beam colliding with hydrogen gas and plasma. The experimental data have been obtained from [6]. It can see that our theoretical calculations agree very good with the experimental data.

## 4. Conclusions

In this work, it has been established two models to calculate the charge state evolution of heavy ions colliding with fully ionized plasma. Furthermore, the initial charge state has been included in the energy loss calculation, concluding that it is very relevant in order to to reproduce the experimental data (Figure 2, left). Finally, it has been demonstrated the necessity of using the Brand-Kitagawa charge state distribution in the energy loss calculation (Figure 2, right).

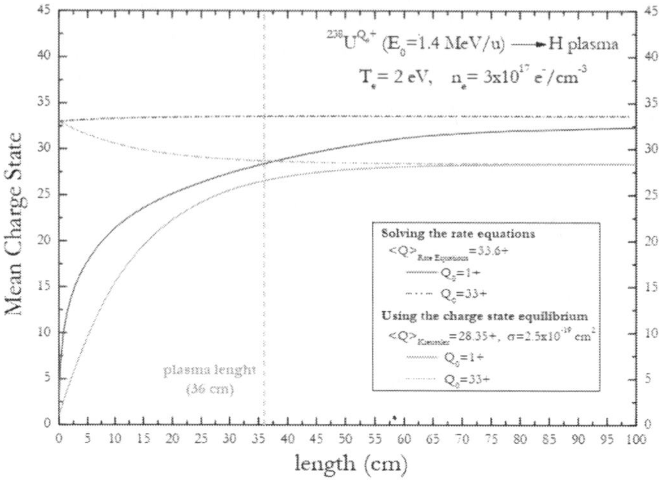


Figure 1: Mean Charge State Evolution

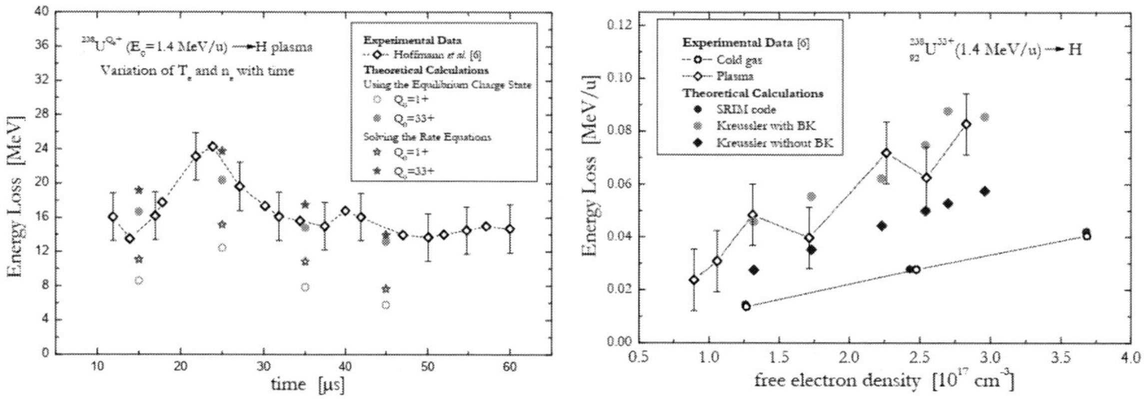


Figure 2: Energy loss as a function of plasma conditions


## Acknowledge

This work has been possible thanks for the funding by the LOEWE program of the HIC for FAIR and for the funding by Erasmus+ program of the Castilla-La Mancha University. This work is also supported by the Spanish National Project ENE2013-45661-C2-1-P.


## References


[1] A. Golubev *et al.*, Nucl. Instrum. Methods Phys. Res. A **464**, 247 (2011)

[2] M. D. Barriga-Carrasco, Phys. Rev. E **82**, 046403 (2010)

[3] W. Brandt and M. Kitagawa, Phys. Rev. B **25**, 9 (1982)

[4] Th. Peter and J. Meyer-ter-Vehn, Phys. Rev. A **43**, 2015 (1991)

[5] S. Kreussler, C. Varelas and W. Brandt, Phys. Rev. B **23**, 82 (1981)

[6] D.H.H. Hoffmann, W. Weyrich, H. Wahl, D. Gardés, R. Bimbot, C. Fleurier, Phys. Rev. B **42**, 4 (1990)

# Generation and transport of γ-rays in ultra-intense laser-matter interactions

T. Nakamura[1], T. Hayakawa[2]

[1] *Fukuoka Institute of Technology, Fukuoka, Japan*

[2] *Japan Atomic Energy Agency, Ibaraki, Japan*

An increase of the focused intensity of the laser pulse has opened new research areas such as high energy density physics. In the regime where the radiation reaction effect plays an important role, the electron motion in the laser field becomes dissipative. This results in the decrease of the maximum energy of electrons [1]. By paying attention to this energy dissipation, it is proposed that high energy photons can be effectively generated when the laser and plasma parameters are properly chosen, which shows the possibility of high power γ-ray source [2, 3]. In this case, a large portion of laser energy is transported inside the target by the photons generated via the radiation reaction effect. In addition, for the further understanding of laser-matter interaction exploring of the energy transport by the high flux of γ -rays is crucial.

We are now developing the Particle-in-Cell (PIC) code with inclusion of the interactions of high energy photons and matters, such as the energy relaxation by Compton scattering, the electron-positron pair creation by Bethe-Heitler process and the nucleon production by the photo-nuclear reactions. These processes are calculated in particle transport code such as PHITS [4] and GEANT4 [5], which are widely used in nuclear physics. In ultra-intense laser-matter interactions, however, the generation processes of γ -rays and collective motions of plasmas as well as the γ -ray transport are supposed to be treated consistently. Then, in this paper, we modelled the above photon transport inside of the material for Particle-in-Cell (PIC) codes which treat the plasma dynamics and their interactions with electromagnetic fields self-consistently.

Here, we consider the laser-plasma interaction in the regime where a quantum effect in the radiation reaction is weak, *i.e.*, $\chi_e = e\hbar\sqrt{(F_{\mu\nu}p^{\nu})^2}/mcE_S \sim \gamma_e E/E_S << 1$, where, e,m,c and $\hbar$ are the electron mass, the electric charge, the speed of light, and the Planck's constant, respectively, $F_{\mu\nu}$ and $p_\mu$ are the field tensor and four-momentum, $E$ is the laser electric field and $E_S \sim 1.3 \times 10^{18}$ V/m is the Schwinger field. The radiation reaction effect is calculated using Landau-Lifshitz equation [6]. The energy spectrum of radiation is evaluated by using classical form of synchrotron radiation formula for a single electron;

$$\frac{dI}{d\omega} \cong 2\sqrt{3}\frac{e^2}{c}\gamma\frac{\omega}{\omega_c}\int_{2\omega/\omega_c}^{\infty} K_{5/3}(x)dx \text{ , (1)}$$

where $\omega_c = 3\gamma^3 c/\rho$ and $K_{5/3}(x)$ is the modifield Bessel function. By summing up the contribution from the electrons which are decelerated by the radiation reaction, we can obtain the energy spectrum of radiated photons via radiation reaction. In Fig. 1, the radiation spectrum via radiaion reaction calculated by the PIC code is plotted as a red dotted line, where the laser pulse having the normalized amplitude of a=150, a power of 10 PW, and duration of 30 fs irradiates on a solid carbon target attached with preplasma having scale length of L=2.5 μm. We then introduce macro-particles of photons in oder to reproduce the above spectrum by the ensemble of the introduced photons. The photon emission of the energy below $E \leq 10\,\text{keV}$ is not included, since we are interested in the physics related to photons in the γ -ray regime. Then the photon energy is determined by randomly samling from the synchrotron radiation profile where the probability is proportional to the spectrum intensity. The weight of the macro-particle is determined from the energy conservation. In this manner, photons are generated in each time step of calculations when the electrons loose their energies by radiation reaction. The energy spectrum calcluated from the photon macro-particle is plotted in Fig.1 as the black line . The spectrum of the photon macro-particle well reproduces the spectrum evaluated by synchrotron formula (more detail in ref. [7]).

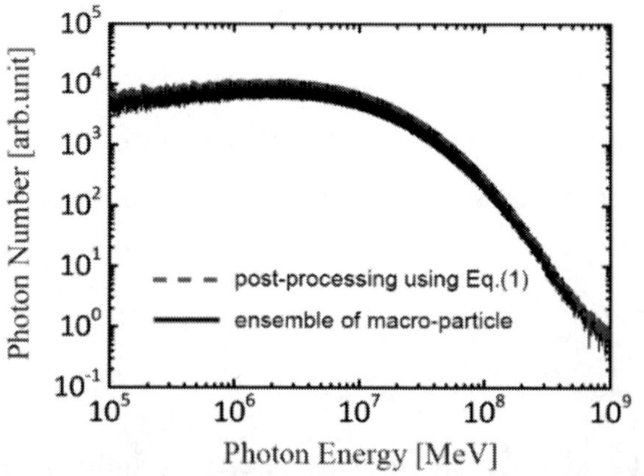


**Figure 1. Comparison of energy spectrum evaluated by synchrotron radiation formula as post-processing ( dotted-line ), and ensemble of the macro-particle of γ -rays introduced in PIC ( solid line ).**

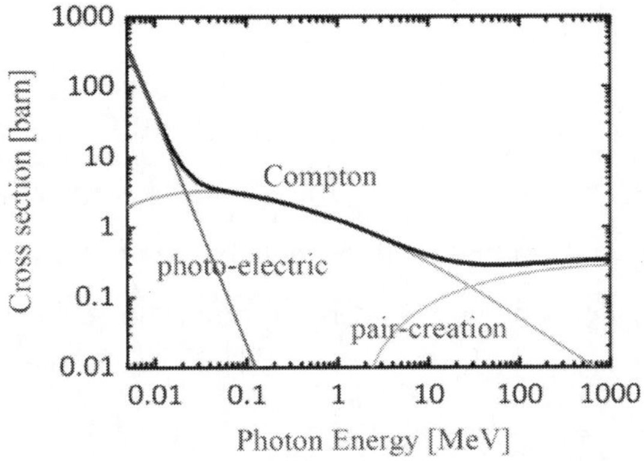


Fig.2 Photo-cross-section for a carbon atom.

When photons propagate through targets, thye interact with atoms and nuclei, which leads to the scattering and abosrption of the photons, and the generations of charged particles and nucleons. Fig. 2 show the photo-cross-section for a carbon target. For photons with energies higher than 10 keV, the cross-section is dominated by that of Compton scattering, where a photon collides with a bound electron and some portion of energy is transferred to the electron becoming a free electron. For photons with energies beyond tens of MeV, electron-positron pairs are created under the electric fields of nuclei. And also photo-nuclear reaction such as (γ,p), (γ,n),( γ,2n), and (γ,α) takes place. These processes are modelled in our PIC simulation code as a Monte Carlo manner. The modelling of photon-matter interactions in our PIC code is verified using a particle transport code PHITS [4]. The detail of the modelling of photo-interactions in our PIC code is explained in ref. [7].

Using the developed code, we can investigate the detail of characteristics generated by the γ -rays transport in targets. Fig. 3 shows the the energy spectrum of the positrons generated by the γ -rays transport, where the simulation conditions are the same as the above.

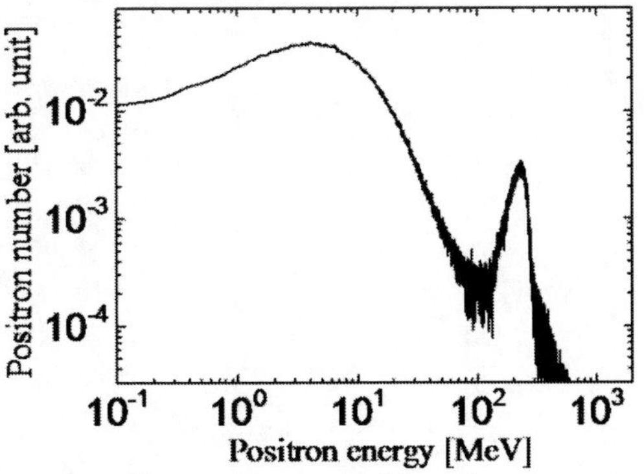


Fig. 3 Energy spectrum of positrons generated by the transport of γ-rays in tergets.

**Acknowledgment**

Authors thank Drs. S.Bulanov, T.Esikepov, M.Kando and J.Koga for discussions. This work has been supported in part by Grants in Aid for Scientific Research (25400540, 15H03665) of Japan. This research used computational resources of the K computer and other computers of the HPCI system provided by the AICS and through the HPIC System Research Project (Project ID : hp120037, hp140122).

**References**

[1] A. Zhidkov, et. al., Phys. Rev. Lett. **88**, 185002 (2002).

[2] T. Nakamura, et. al., Phys. Rev. Lett. **108**, 195001 (2012).

[3] C.P. Rigers, et. al., Phys. Rev. Lett. **108**, 165006 (2012).

[4] T. Sato, et. al., J. Nucl. Sci. Technol. 50, 913 (2013).

[5] J. Allison, et. al., IEEE Trans. Nucl. Sci. 53, 270 (2006).

[6] L. Landau and E.M. Lifshitz, *The classical theory of fields, (Pergamon.1975).*

[7] T. Nakamura and T. Hayakawa, Laser Part. Beams 33, 151 (2015).

# Comprehensive investigations of ablative plasma and crater formation in experiments at PALS related to Shock Ignition concept

T. Pisarczyk[1], S.Yu. Gus'kov[2,3], O. Renner[4], N.N. Demchenko[2], M. Smid[4], R. Dudzak[6,4], T. Chodukowski[1], Kalinowska[1], M. Rosinski[1], P. Parys[1], J. Dostal[6,4] J. Badziak[1], D. Batani[5], S. Borodziuk[1], E. Krousky[6,4], L. Antonelli[5], Y. Maheut[5], G. Cristoforetti[7], L.A. Gizzi[7], P. Koester[7], L. Labate[7], F. Baffigi, J. Ullschmied[6,4], J. Hrebicek[4,6], T. Medrik[4,6], M. Pfeifer[6,4], J. Skala[6,4], P. Pisarczyk[7]

[1] *Institute of Plasma Physics and Laser Microfusion, Warsaw, Poland*

[2] *P.N. Lebedev Physical Institute of RAS, Moscow, Russian Federation*

[3]*National Research Nuclear University (Moscow Eng. Phys. Inst.), Russian Federation*

[4] *Institute of Physics ASCR, Prague, Czech Republic*

[5] *Université Bordeaux, CNRS, CEA, CELIA, Talence, France*

[6] *Institute of Plasma Physics ASCR, Prague, Czech Republic*

[7]*Intense Laser Irradiation Laboratory, INO-CNR, Pisa, Italy*

[8] *Warsaw University of Technology, ICS, Warsaw, Poland*

The shock ignition (SI) is a subject dominating in the last investigations carried out at PALS facility in the frame of international cooperation within the LaserLab projects [1-5]. In the experiments two-layer targets were used, consisting of massive planar Cu coated with a thin polyethylene layer, that in the case of two-beam irradiation geometry simulate conditions related to the SI concept, namely: one of the beams (1ω or 3ω) with a high intensity $>10^{15}$ W/cm$^2$ generates igniting shock wave, while the other one (1ω) with the lower intensity below $10^{14}$ W/cm$^2$ creates a pre-plasma simulating pre-compressed plasma in a real inertial fusion experiment. The experiments concern the influence of the pre-plasma on efficiency of the laser radiation energy transport to a shock wave in which mechanisms of fast electron generation play an important role. This research was based on application of the multi-frame interferometry and a unique set of x-ray and ion diagnostics. In the last studies carried out with the 1ω main beam, femtosecond interferometry turned out to be especially useful [5]. It enabled to determine the changes of the electron density gradients during the laser pulse interaction with a target, which definitively confirmed an occurrence of the resonant absorption in the case of the first harmonic and its influence on the laser radiation energy transport to a shock wave. The investigations presented in this paper are related to the use of 3ω beam to create the igniting shock wave. The desired intensity of the main beam was achieved at constant laser energy on the level of 200 J by changing the focal spot radius of the laser beam on the target in the range of $R_L$ = 50-200 μm. The pre-plasma was created by means of the 1ω laser radiation with the energy of 40 J which was defocused to the focal spot radius of $R_L$ = 300 μm. In the experiments, two-layer targets were used consisting of a planar massive Cu covered by thin layer of polyethylene with different thicknesses. Analogously as in studies [5], the following diagnostics were used: the 3-frame interferometer irradiated by a femtosecond Ti: Sa laser (t = 40 fs and λ = 808 nm), 2D imaging of the Cu $K_\alpha$ emission to determine the distribution of the fast electrons population and their energy, and the grid collectors to measure the angular distributions of the ion emission. As the routine diagnostic, the crater volume measurements were carried to evaluate the efficiency of the laser energy transfer to the shock wave generated in the massive part of the two-layer targets.

**The femtosecond interferometry** was applied to determine the electron density distributions in ablative plasma during the laser beam interaction with the target at different irradiation conditions in the case of absence and presence of the pre-plasma. The obtained electron density distributions prove that the light plasma with the higher pressure generated from the thin plastic layer by an auxiliary beam limits the radial expansion of the central plasma created by the 3ω main laser beam. The radial constrain favours the axial character of the plasma expansion particularly in the case of larger $R_L$. On the basis of the space time density distributions, the information about the maximal density gradients, the scalelength (L) and the total electron number (N) in the ablative plasma were determined. To obtain the information about the density scalelength, an exponential fitting of the axial density profiles has been applied: $n_e(z) = n_0 e^{-z/L}$. The changes of the axial density profiles and the L caused by increasing $R_L$ in the case with and without the pre-plasma are demonstrated in Fig. 1a, while in Fig 1b the dependences of the maximal gradient density are shown.

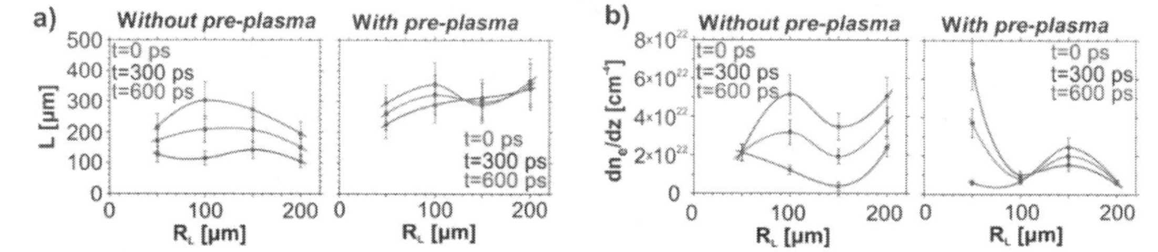


Fig. 1. The changes of L - a) and N-b) as a function of $R_L$ for different expansion times in the case of the absence and the presence of the pre-plasma.

These dependences prove that in the case without the pre-plasma, the non-collisional processes with the L ≅100 μm can develop in the whole range of the varied $R_L$ (such as the resonance absorption). The pre-plasma causes significant extension of L (about 3 times) and creates conditions for the development of the processes with the large-scale gradient that simultaneously require high electron density. This suggests two maxima relating to both the density and its gradient which correspond to the focal spot radii $R_L$ = 50 and 150 μm.

**Crater volume measurements** combined with the interferometric investigations, allowed to obtain the $N/V_{cr}$ parameter (where $V_{cr}$ is the crater volume) which defines a number of thermal electrons participating in a creation of the crater volume unit (1 $cm^3$). The $N/V_{cr}$ parameter was evaluated only for expansion time t= 600 ps, i.e., after the end of the main laser pulse, when both ablation process and processes related to absorption of the laser radiation were completed, Fig. 2.

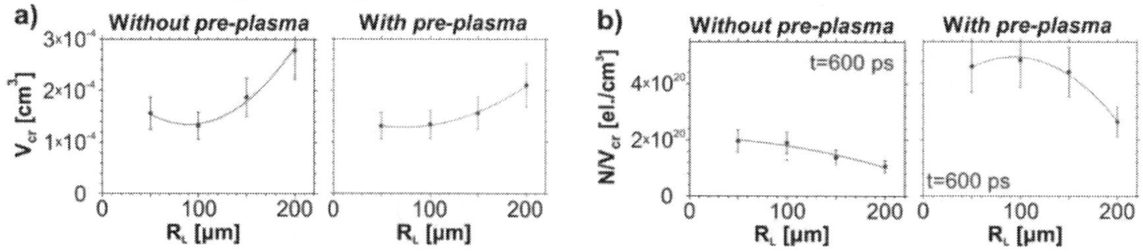


Fig. 2 The changes of $V_{cr}$-a) and $N/V_{cr}$-b) parameter caused by the increasing $R_L$ and by the pre-plasma presence.

In the case without pre-plasma, increase in $V_{cr}$ and decrease in $N/V_{cr}$ with the increasing $R_L$ testifies about dominance of the collision absorption mechanism (inverse bremasstrahlung, IB) in the whole range of $R_L$, when the laser energy transfer to the target is achieved through the electron-thermal conductivity wave (ETC) [1, 3, 5]. For such mechanism and the 3ω laser beam the ablative pressure depends on the 1D geometry of the plasma expansion, which grows with $R_L$. The presence of the pre-plasma does not change the character of $V_{cr}$ and $N/V_{cr}$ dependences, what confirms the presence of the IB absorption and the ETC mechanism. However, the strong decrease in the N/Vcr parameter in the range of $R_L$=150-200 μm due to the more 1D (planar) expansion creates more favourable conditions for the crater creation.

**2D imaging of the Cu Kα line emission** was carried out to investigate the 1D plasma expansion occurring for the 3ω laser radiation due to the fast electron generation under presence and absence of the pre-plasma. The imaging setup was the same as in the previous experiments [5]. To determine the energy of the fast electrons, two-layer targets with different thicknesses of the plastic ($\Delta_{pl}$ = 15, 25, 50 and 100 μm) have been used. Results mapping the Kα emission due to fast electron action are presented in Fig. 3. To determine

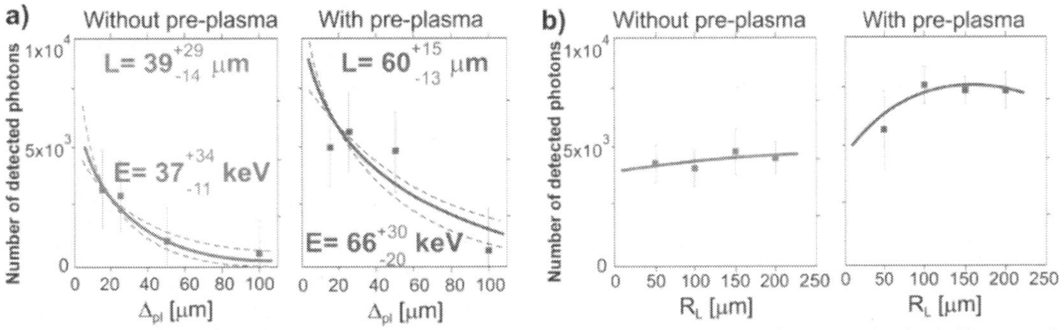


Fig. 3 Measurements of the Cu Kα emission due to generation of fast electrons demonstrate the influence of the pre-plasma on the total population of photons in dependence on: a) $\Delta_{pl}$ and b) $R_L$

the energy of fast electrons, an exponential decay of the photon emission with the increasing thickness of the plastic layer I (x) = A * exp (-x / L) was assumed. It was fitted to the experimental data using a Monte Carlo method assuming Maxwellian distribution of the suprathermal electron energy and electron impact on the double layer target [6]. Calculations showed that the average energy of fast electrons in the case of no pre-plasma), is about 37 keV, i.e., almost two times smaller than in the case of the pre-plasma presence, Fig. 3a. The presence of pre-plasma, Fig. 3b, significantly increases the emission of fast electrons with the increasing $R_L$. Therefore we can conclude that the 1D expansion of the ablative plasma in the presence of the pre-plasma increases both the emission and the energy of the fast electrons, thus suggesting that besides the IB mechanism also non-collisional processes (such as the resonance absorption) associated with the laser energy transport by fast electrons can take place in the process of the laser energy transport and the pressure formation.

**The ion measurements** are in agreement with the interferometry and confirm the limited radial expansion of the central plasma created by means of the 3ω main laser beam due to lower-Z plasma with the higher pressure generated from the thin plastic layer by auxiliary beam.

**2D numerical simulations** were carried out using the 2D hydrodynamic code ATLANT-HE [7] which includes modelling of the laser radiation refraction in the plasma, inverse bremsstrahlung and resonance absorption of the laser radiation, generation of fast electrons due to the resonance absorption, and the fast electron energy transfer with Coulomb collisions. The calculations were made for conditions of the experiment. A satisfactory agreement of calculations with the experiment was obtained in the case of the laser beam interaction with the target in the presence of the pre-plasma, as illustrated by the data in the Table 1 and 2. In the case of the pre-plasma absence, the qualitative character is the same but the observed and simulated L values differ considerably. The large values of experimental L result from the action of the temporally extended beam profile which is characteristic for the PALS laser and contributes to the creation of the low density plasma even in the case of the pre-pulse absence. In Table 2 the numerical simulations of the total ($\delta_{total}$), the resonant ($\delta_{res}$) and the inverse bremsstrahlung $\delta_{ib}$ absorption coefficients are compared for the cases of the pre-plasma absence and presence. It should be noted that in the presence of the pre-plasma, the fast electron energy is close to the experimental values, obtained from Cu Kα line measurements. Table 2 clearly indicates that in the case of the absence of the pre-plasma, the total absorption increases with the increasing focal spot radius from the value 0.593 to 0.876.

In contrast, the fraction of resonant absorption decreases from 0.0883 to 0.0697 and stays small. The average energy of fast electrons $E_{h(max)}/2$ is about 30-35 keV for both with and without pre-plasma cases.. Pre-pulse has a small effect on these quantities and does not alter their dependence on the radius of the main beam. These results differ significantly from the case of the 1ω main pulse which was presented in [5].

Table 1

| L [μm] | | | | |
|---|---|---|---|---|
| | 3ω | | 1ω+3ω | |
| **$R_L$ [μm]** | Exp | 2D | Exp | 2D |
| 50 | 125 | 39 | 220 | 298 |
| 100 | 115 | 34 | 290 | 307 |
| 200 | 100 | 29 | 340 | 283 |

Table 2

| $R_L$ [μm] | 50 | | 100 | | 200 | |
|---|---|---|---|---|---|---|
| | 3ω | 1ω+3ω | 3ω | 1ω+3ω | 3ω | 1ω+3ω |
| $\delta_{total}$ | 0.593 | 0.553 | 0.751 | 0.691 | 0.876 | 0.789 |
| $\delta_{ib}$ | 0.505 | 0.462 | 0.655 | 0.592 | 0.806 | 0.716 |
| $\delta_{res}$ | 0.0883 | 0.091 | 0.0963 | 0.0993 | 0.0697 | 0.0733 |
| $E_{h(max)}$, keV | 69.4 | 73.8 | 62.3 | 69.6 | 58.2 | 54.8 |
| $T_{emax}$, keV | 3.67 | 3.91 | 2.81 | 3.27 | 2.09 | 2.58 |

**Conclusions:** ***Interferometric studies*** of the ablative plasma expansion in combination with the crater volume measurements and the measurements of the Cu $K_\alpha$ emission due to fast electrons confirm the clear influence of the expansion geometry on the efficiency of the laser energy transport to the shock wave and the ablative pressure formation as well as the fast electron generation. The increase of $R_L$ leads to an increase of both the efficiency of the laser energy transport to the target and the fast electron production. Pre-plasma does not significantly change the character of these dependencies, however it strongly influences an increase of the fast electrons energy. ***Numerical simulations*** show that without the pre-plasma, the total absorption is considerably larger for 3ω than that of the 1ω main pulse. This is confirmed by data from ion collector measurements. The fraction of the resonance absorption is about the same. Pre-plasma has a small effect on absorption in the case of the 3ω and strong effect in the case of the 1ω. In the case of the pre-plasma, for the 3ω main pulse the fraction of resonance absorption (and hence the energy contained in the fast electrons) is several times smaller than for the 1ω main pulse. This is confirmed by ion collector measurements, too. The average energy of fast electrons generated by resonant absorption is approximately two times smaller in the case of the 3ω main pulse (~30-35 keV) as compared with the 1ω main pulse (50-70 keV) [5].

**Acknowledgments:** This paper was supported in part by the Access to Research Infrastructure activity in the 7th Framework Program of the EU Contract No.284464, Laserlab Europe III, by the Czech Republic's Ministry of Education, Youth and Sports under PALS RI-LM 2010014 and by the ToIFE Project of the EUROfusion Consortium and by the French-Polish bilateral collaboration program POLONIUM and of the COST Action MP1208 "Developing the physics and the Scientific Community for Inertial Fusion by Russian Foundation for Basic Research (Projects No. 14-02-00010 and No. 13-02-00295). It was also supported in part by National Centre for Science (NCN), Poland, under Grant No. 2012/04/M/ST2/00452.

**Reference:**

1. P. Koester et al, Plasma Physics and Controlled Fusion **55,** 124045 (2013).
2. S.Yu. Gus'kov et al, Laser Particle Beams **32,** 177-915 (2014).
3. T. Pisarczyk et al, Phys. Plasmas **21**, 012708 (2014).
4. D. Batani et al, Phys. Plasmas **21**, 032710 (2014).
5. T. Pisarczyk et al Laser and Particle Beams **33**, 221-236 (2015)
6. F.Salvat, J.M.Fernandez-Varea, and J. Sempau. Penelope-2008: A Code System for Monte Carlo Simulation of Electron and Photon Transport. 2009. ISBN 978-92-64-99066-1.
7. I. G. Lebo et al, Laser Part. Beams **22**, 267–273 (2004).

# X-ray spectroscopic diagnosis of hot electron production in laser-irradiated Cu targets

O. Renner[1], M. Šmíd[1], D. Batani[2], L. Antonelli [2,3,4], F. Barbato[3], F.B. Rosmej[5,6]

[1] *Institute of Physics & ELI-Beamlines, Academy of Sciences CR, Prague, Czech Republic*

[2] *University of Bordeaux, CNRS, CEA, CELIA, UMR 5107, Talence, France*

[3] *Dip. di Ingegneria Industriale, Università degli Studi di Roma "Tor Vergata", Rome, Italy*

[4] *Dipartimento SBAI, Università degli studi di Roma "La Sapienza", Rome, Italy*

[5] *Sorbonne Universités, Pierre et Marie Curie, Paris, France*

[6] *LULI, École Polytechnique, CEA, CNRS, Palaiseau, France*

Generation of suprathermal electrons accompanying interaction of high-intensity lasers with solid targets and their transport through the near-solid density matter [1] have attracted significant interest due to their potential to study the behaviour of matter in extreme conditions. In the same time, hot electrons (hereafter HE) represent one of the key issues for realization of inertial confinement fusion [2]. In the shock ignition approach (SI), a role of hot electrons is ambiguous: they contribute to laser energy transport into pellets and enhancement of ablation pressure but they may also preheat targets. A detailed knowledge of the fast electron generation and transport is therefore of paramount importance for optimization of the SI concept.

A survey of previous HE studies based on x-ray diagnostics can be found, e.g., in paper [3]. Non-Maxwellian electrons affect properties of emitting plasma systems, collisional processes result in creation of holes in inner electronic shells of target atoms and their subsequent filling via radiative transitions. In particular, the $2p \rightarrow 1s$ K-shell fluorescence from relatively cold targets (i.e., K$\alpha$ emission) studied by advanced x-ray methods provides a direct experimental technique for HE investigation. Here we report on a non-conventional approach to HE diagnosis using a combination of monochromatic x-ray imaging and high-resolution K-shell spectroscopy. This novel method has been developed within a series of experiments being performed at PALS laser facility [4] where laser-plasma coupling is studied at intensities up to $5\times10^{16}$ W/cm$^2$, i.e., at parameters of the laser spike envisaged to launch the shock wave igniting the fusion reaction. After a brief description of the experiment, we interpret selected x-ray images and K-shell spectra in terms of HE generation by using a detailed ray-tracing analysis of the implemented instruments and Monte Carlo modelling of HE deposition in targets. Implications of this procedure for routine HE diagnosis are discussed. The reported research contributes to development of advanced diagnostic methods for SI-relevant plasmas.

The experiment was carried out using the fundamental (1ω) and frequency tripled (3ω) radiation of the iodine laser (wavelength $\lambda$ = 1315/438 nm, pulse duration 0.3/0.25 ns, energy 440/170 J focused to intensities $I = 2\times10^{16}/9\times10^{15}$ W/cm$^2$ for 1ω/3ω, respectively) striking massive or thin-foil Cu targets at normal incidence. In several shots, an additional 1ω laser beam (60 J, $1\times10^{14}$ W/cm$^2$) preceding the main pulse by 1.2 ns was used to create a long-scale pre-plasma simulating the compression phase of shock ignition. The diagnostic complex is schematically shown in Fig. 1 and fully described elsewhere [3], here we concentrate on high-resolution x-ray diagnosis with emphasis on high-collection efficiency imaging.

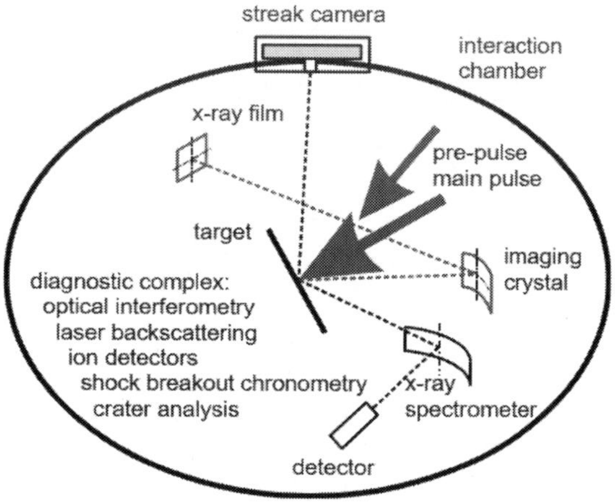


**Fig. 1.** Scheme of the x-ray experimental setup and complementary diagnostics used.

The time-integrated 2D-resolved images of the Cu Kα emission produced by HE interaction with near-surface copper ions were taken at quasi-normal incidence using the crystal of quartz (422) spherically bent to a radius of 380 mm. Images recorded on x-ray film Kodak Industrex AA400 were digitized and converted to incident photon fluxes with respect to the characteristic curve of the film and filter transmission. The range of photon energies $\Delta E_i$ reflected from the crystal surface was rather narrow ($\Delta E_i$=3.86 eV), whereas the range of photon energies relevant for HE diagnosis covers $\Delta E_s$=8–8.265 keV [5]. To correct the recorded signal for the limited spectral coverage, the imaging was complemented by information obtained via high-resolution spectroscopy. The K-shell spectra were observed by using the x-ray spectrometer equipped with the crystal of quartz (223) spherically bent to a radius of 150 mm. The spectra were again recorded on x-ray film, calibrated with respect to dispersion relation of the experimental geometry used and cross-checked via tabulated dominant Cu K-shell transitions. The signal was recalculated to an intensity scale with respect to a variable crystal reflectivity and transmission through filters.

The quantitative interpretation of K-shell spectral emission leans on models correlating the yield of the characteristic x-ray radiation with the action of laser-generated HE [6]. In contrast to our previous estimates of the HE production based on a thin target approximation [7], here the 3D electron trajectories inside the target, generation and transport of x-ray photons were simulated using the Monte Carlo code PENELOPE [8].

The interpretation of K-shell images in terms of the HE generation proceeded in three steps. First the Cu Kα signal recorded above the detector background was recalculated to the source emission. This intermediate result was corrected with respect to the limited spectral range of the imaging by using the information following from the simultaneously recorded x-ray spectrum and finally the resulting K-shell emission was related to the number of suprathermal electrons propagating through the Cu target.

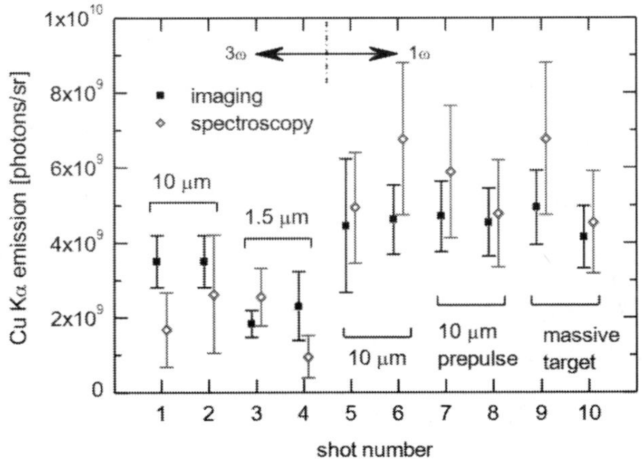


**Fig. 2.** Comparison of photon fluxes recorded at diverse Cu targets by x-ray imaging and spectroscopy in the spectral window of 3.86 eV at the maximum of the Cu K$\alpha_1$ emission.

The performance of the instruments used was analysed via ray-tracing procedure [9] and tested by comparing photon fluxes measured close to the Cu Kα maximum in the spectral window $\Delta E_i$. Comparison of intensities emitted from different-thickness Cu targets under irradiation conditions indicated in Fig. 2 demonstrates that the found fluxes of 1–6×10$^{10}$ photons/4π were identic for imaging and spectroscopy within the error bars (~40%).

The full number of K-shell photons emitted into the energy range of 8–8.265 keV was determined from imaging records by multiplying their integrated signal with a corrective factor relating spectral intensities measured within the above defined ranges of $\Delta E_s$ and $\Delta E_i$. These fluxes were subsequently related to the number of generated HE using the independently measured values of HE temperature $T_{HE}$ and the PENELOPE code predictions. The results recalculated to conversion efficiency $\eta$ of the laser energy in HE are presented in Fig. 3.

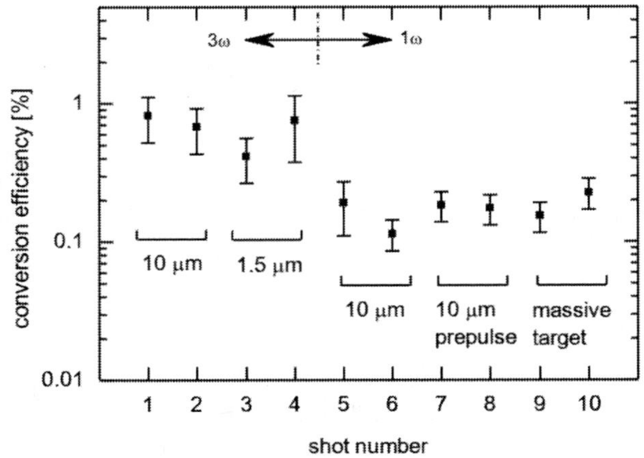


**Fig. 3**. Conversion efficiency of laser energy in hot electrons at two different laser wavelengths, Cu target thicknesses and the plasma scale-length modified by the pre-pulse.

The found values $\eta$ are within 0.1-0.2% and 0.2-0.8% for 1ω and 3ω shots, respectively. Absolute values of these fractions are comparable with the previously published data measured at similar laser intensities, see e.g. [10]. Minimum deviations of $\eta$ observed for different-thickness targets are explained by the low $T_{HE}$≈30 keV found for current experimental conditions, i.e., by limited penetration of suprathermal electrons inside the targets. Indeed, the stopping length of hot electrons with the energy of 30 keV can be estimated using the continuous-slowing-down-approximation to 3.3 μm [11]. This means that most of electrons deposit their energy within the 1.5-μm-thick near-surface layer due to repeated scattering processes. On the other hand, the higher values of conversion efficiency found at 3ω in comparison with 1ω are rather counter-intuitive. The results of our measurements can tentatively be explained by diverse mechanisms of hot electron generation at 1ω and 3ω. The higher values of conversion efficiency found for frequency tripled radiation support a theoretical conjecture of enhanced laser energy absorption by resonance mechanisms and its transport to a flow of fast electrons [12], however the quantitative confirmation of this conclusion is to be done. The detailed interpretation of the experimentally observed data, in particular the role of different plasma scale-lengths and variable laser coupling parameters including their comparison with theoretical models can be found elsewhere [5].

To conclude, relatively easy-to-align and high-collection-efficiency K-shell imaging combined with the high resolution x-ray spectroscopy offers a novel approach to hot electron diagnosis. The correction introduced by spectroscopic information is of paramount importance for quantitative evaluation of recorded quasi-monochromatic images. The described technique is particularly suitable for a routine monitoring of hot electron generation in the partly ionized matter and can be used even at non-plastic-coated targets. The sample results obtained at variable-thickness Cu targets irradiated by laser beams with two different wavelengths demonstrate viability of this method for characterization of suprathermal electron generation in shock-ignition-relevant plasmas.

This work benefitted from the support of the MSMT ELI Project No. CZ.1.05/1.1.00/02.0061, MSMT COST CZ Project LD14089, the COST Action MP1208 "Developing the physics and the Scientific Community for Inertial Fusion", and Academy of Sciences Programme for support of international collaboration, project M100101208.

References

[1] M.H. Key et al, Phys. Plasmas **5**, 1966 (1998)
[2] S.Yu. Guskov, Plasma Phys. Reports **39** (2013) 12013.
[3] P. Köster et al, Plasma Phys. Control. Fusion **55** (2013) 124045.
[4] K. Jungwirth et al, Phys. Plasmas **8** (2001) 2495.
[5] O. Renner et al, New Journal of Physics, submitted.
[6] O.F. Kostenko, N.E. Andreev, Quantum Electronics **43** (2013) 237.
[7] M. Šmíd et al, Phys. Scr. **T145** (2014) 014020.
[8] F. Salvat et al, PENELOPE-2008, http://www.nea.fr/html/science/pubs/2009/nea6416-penelope.pdf.
[9] S.G. Podorov et al, J. Phys. D: Appl. Phys. **34** (2001) 2363.
[10] B. Yaakobi et al., Phys. Plasmas **19** (2012) 012704.
[11] M.J. Berger et al, NIST ESTAR database, http://physics.nist.gov/PhysRefData/Star/Text/ESTAR.html.
[12] S.Yu. Guskov et al, Laser Part. Beams 32 (2014) 177.

# On a highly ionizing shock wave in a gas of high atmospheric pressure

S. P. Sadykova[1], A. A. Rukhadze[2], T. G. Samkharadze[3]

[1] *Forschungszentrum Julich, Jülich Supercomputing Center, Jülich, Germany*

[2] *Prokhorov General Physics Institute, Russian Academy of Sciences, Moscow, Russia*

[3] *Moscow State University of Instrument Engineering and Computer Science Moscow, 107996, Stromynka str. 20, Russia*

The shock wave in a medium is a propagating disturbance. It is characterized by an abrupt, nearly discontinuous change (shock) in thermodynamic parameters of the medium (for simplicity, we consider only the perfect gas). In a case of the plane shock wave the gas parameters (pressure, temperature, density) before and after the shock are constant and homogeneous and for a strong wave they can be interrelated like the following [1] :

$$\frac{P_2}{P_1} = \frac{2\gamma}{\gamma+1}M^2, \; \frac{T_2}{T_1} = \frac{2\gamma(\gamma-1)}{(\gamma+1)^2}M^2, \; \frac{n_2}{n_1} = \frac{\gamma+1}{\gamma-1}, \qquad (1)$$

here the subscripts 1 and 2 are referred respectively to the gas state before and after the shock front, $M$ - Mach number is the ratio of speed of shock wave front and the speed of sound in a gas before the front, $\gamma = \frac{C_p}{C_v}$ is the ratio of the heat capacity at a constant pressure and the heat capacity at a constant volume (for one atomic gas $\gamma = 5/3$). These states are related to the non-isothermal shock. In a case of the isothermal shock when the thermal conductivity (for ex. thermal radiation) in a medium is quite high the relation (1) will transform into the following [1]:

$$\frac{P_2}{P_1} = \frac{n_2}{n_1} = \frac{\gamma+1}{3-\gamma}, \qquad (2)$$

From this it follows that at $\gamma = 1$ the right hand-side will become unity, hence in such a medium the shock wave does not exist. However, it does not mean that nonlinear waves can not exist in such a medium. For example (we will demonstrate it further down), in such a medium the propagation of a soliton is possible whereas the medium state before the wave front side and behind the back side of the front does not change.

Let us return to the shock waves and discuss the structure of the transition layer (wave front width) in a gas. There are a lot of works devoted to such a problem mainly related to the weak shocks ($M \simeq 1$). The waves with the arbitrary intensity were considered by I. Y. Tamm in 1947. However the work was first published in 1965 [2]. In a case of the weak shock wave ($M \simeq 1$) the structure of the transition layer defined in a framework of hydrodynamics is given by the following formula:

$$d = \frac{P_2+P_1}{P_2-P_1}\delta, \;\; \delta = 1.28l, \qquad (3)$$

where $l$ - mean free path of gas particles. From Eq. (3) it follows that the front width is greater than the free path and with an increase of $P_2/P_1$ - $d \to \delta$ limits to the mean free path.

In a case of the strong shock wave ($P_2/P_1 >> 1$) the wave front width obtained in [2] is the following:

$$d = 0.503 l_1 = 2.012 l_2, \tag{4}$$

where $l_1$ and $l_2$ are the particles free paths in the front side and back side of the shock wave respectively. The results obtained in the work [2] in the hydrodynamic approximation were found in a good agreement with those obtained in a frame of the Boltzmann approximation, i.e. in the kinetic approximation. However, a simplified approximation was used: Boltzmann equation with an account of the elastic collisions in a framework of the hard spheres. In gases of high pressure (atmospheric and higher one) the structure of the strong shock waves described in [2] is too simplified. At Mach number greater than 10 due to the gas heating the gas ionization and generation of highly ionized plasma must occur what was not taken into consideration in [2]. As a result, the structure of the strong ionizing wave must have few peculiarities. Namely, right behind the front side of the shock wave at a free path length the gas temperature gets increased to several electron-volts leading to the gas ionization and formation of a strongly ionized non-isothermal plasma. The electron temperature in such a plasma is of order of a gas atom ionization potential whereas the ion temperature is much less than that of electrons. At the same time according to the gas-phase approximation Debye electron radius is less than electron free path. This phase transition occurs at an electron constant temperature of order of an atom ionization potential. It lasts for few orders of an inverse ionization frequencies until the whole gas gets ionized. The width of this domain amounts to few ionization free paths. As a result the fully ionized non-isothermal plasma with the electron temperature of order of few atom ionization potentials gets formed behind the front in which the shock wave should travel. We should ask ourselves: "Is this possible ?". - " No , it is not!" The point is that the acoustic oscillation mode in non-isothermal highly ionized plasma exists only when $T_e >> T_i$ - the ion sound is an isothermal sound with $\gamma = 1$ ($T_e$=const due to the high electron thermal conductivity) [3, 4]. According to Eq. (2) the isothermal shock in a neutral gas at $\gamma = 1$ is not possible. As it will be demonstrated below, in non-isothermal plasma when $T_e >> T_i$ only the propagation of a solitary wave (soliton) with the half-width equal to the Debye electron radius is possible. This wave represents the double electric layer in which the electrons (electron layer) overtake the ions (ion layer) and pull them on. The electric potential of the double layer can be described as a weakly nonideal wave by the following equation [4]:

$$\frac{T_e r_{De}^2}{2e}\frac{d^2\Phi}{d\xi^2} - \frac{T_e}{e}(1 - \frac{V_s}{u})\Phi + \frac{1}{2}\Phi^2 = 0, \tag{5}$$

here $\xi = x - ut$, $r_{De} = \sqrt{T_e/4\pi e^2 n_0}$- Debye electron radius, where $n_e = n_0 \cdot exp(e\Phi/T_e)$, $V_s = \sqrt{T_e/M}$ - ion sound velocity, $u$ - to be determined velocity of a soliton equal to the velocity of the shock wave.

Eq. (5) known as the KortewegâĂŞde Vries equation (KdV Eq. for short) has the following explicit solution:

$$\Phi = \frac{\Phi_{max}}{ch^2(\xi/\Delta)}, \tag{6}$$

where

$$1 - \frac{V_s}{u} = \frac{e\Phi_{max}}{\pi T_e} << 1, \quad \frac{e\Phi_{max}}{6T_e} = \frac{r_{De}^2}{\Delta^2} << 1. \tag{7}$$

As one can see $u \succeq V_S$, i.e. the velocity of a solitary wave is almost equal to that of an ion sound, $\Delta$ - the wave full width at half maximum, $\Delta > r_{De}$ and decreases with increasing wave amplitude.

**Discussion of the results and quantitative estimations**

Eqs. (6) and (7) are exact on implementation domain and describe exactly the solitary wave (6) in a plasma of the highly ionizing shock wave with temperature $T_e \sim 3I_i \sim 3 \cdot 10^5$ K, where $I_i \sim 10$ eV - gas atom ionization potential. At a atmospheric gas pressure - $r_{De} \sim 7 \cdot 10^{-7}$ cm and the mean distance between plasma electrons - $r_a \sim 3 \cdot 10^{-7}$ cm, i.e. the gas-phase approximation condition ($r_a^2 << r_{De}^2$) is satisfied. According to Eq. (7) potential amplitude increases with decreasing $\Delta$ and at $\Delta = 3r_{De}$ (the minimum magnitude at which the equations (5) - (7) are valid) becomes equal to:

$$E_{max} \leq 10^7 V/cm, \tag{8}$$

There exists another constraint for the implementation domain of Eq. (6), namely, the small plasma disturbance by a soliton field when

$$E \leq \sqrt{4\pi n_e T_e} \sim 10^8 V/cm. \tag{9}$$

This field is one order higher than the field (8). The field $E = 10^7$ V/cm is high enough to create the electric discharge in a solid body (for example in a glass). When the condition (8) is violated then the plasma gets exploded, electrons will break apart of ions and get accelerated by the field. However, this stage is out of our consideration and requires a more detailed investigation.

The realization of the above process in atmosphere at normal conditions the Mach number of the shock wave must be of order $M \sim \sqrt{T_e/T_0} \sim 33$. When the Mach number is quite high then the plasma explosion occurs forming the electron gas where the field will exceed the magnitude (8) and the front width will increase with increasing field. This stage is out of our consideration here in this work.

In this way the strong ionizing shock wave in a gas must have a very complex structure: right in the back side of the front at a length of the mean free path of gas particles before the front, the temperature increases to few electron volts and the gas ionization occurs. In the formed plasma behind this temperature-raising shock traveling only of the ion-sound soliton is possible. The soliton half-width is of order of hot electrons Debye radius. In a case of the strong shock wave the ion-sound soliton gets transformed into the extended double layer with the following electric field [5]:

$$E \sim \sqrt{4\pi n_e T_e} = \sqrt{4\pi n_e T_0 T_e/T_0} = M\sqrt{4\pi n_e T_0}, \tag{10}$$

where $T_0$ -gas temperature in the front-side of the front, $M >> 1$ - Mach number, $n_e$ - density of a plasma formed behind the front which is close to the neutral gas density. Relation (1) shows that compared to the normal gas where the pressure behind the front of the strong shock wave is $M^2$ higher than that in front-side, in a plasma, the plasma gets polarized behind the front of the shock wave producing the field with the same pressure. According to the above description, in non-isothermal plasma formation, propagation of the ion-sound soliton (and not of the shock wave) is possible. From Eqs. (5) and (6) it follows that the soliton represents the blob of a electromagnetic field revealing not only the resonance formation of the soliton but also that the soliton is a resonator itself having the electromagnetic field which can lead to the electric discharge at soliton disturbance for ex. at a collision with the hard target like glass. The latter, discharges inside the window glasses, was revealed after the Chelyabinsk meteor explosion that entered Earth's atmosphere over Russia on 15 February 2013 at about 09:20 YEKT. [7]

It is particularly interesting to discuss the concept of the shock waves in plasmas which got increased attention in the literature [8]-[11]. The short overview of these works is given in monography [12]. The authors discuss plasma and the long-range Coulomb interactions of charged plasma particles. However, they consider plasma as an one-atomic gas with the adiabatic constant $\gamma = 5/3$ and the ion temperature $T_i \geq T_e$. At the same time, it is well known that the Coulomb system of charged particles can be treated as a plasma only when the Langmuir plasma frequency is much greater than the electron collisions:

$$\omega_{Le}^2 >> \nu_e^2. \tag{11}$$

We would like to note that the sound in a plasma is isothermal (hence $\gamma = 1$) and can exist in non-isothermal plasma only with $T_e >> T_i$. Namely in such a plasma the density and temperature shocks are impossible, while only solitary waves can propagate therein. If the inverse constraint (11) is satisfied then the peculiar plasma properties cannot be revealed; plasma becomes similar to a gas of neutral particles with remaining charged particles playing no principle role. In our point of view the authors of works [8]-[11] meant such plasma and the words (here citation): " In a plasma an electron is strongly bound to the ion by the charge separation field and they move together as a whole unit" have nothing to do with the plasma.

## References


[1] L.D. Landau, E.M. Lifshitz, *Fluid Mechanics* Vol. 6 (2nd ed.). (ButterworthâĂŞHeinemann, 1987).

[2] I. Y. Tamm, Trudy Ordena Lenina Fizicheskogo Instituta im. P. N. Lebedeva (in Russ.), **29**, P. 239 (1965)

[3] Silin V.P., Rukhadze A. A., *Electromagnetic Properties of Plasma and Plasma-like Media* ( Atomizdat, Moscow 1961).

[4] A.F. Alexandrov, L.S. Bogdankevich, A.A. Rukhadze, *Principles of Plasma Electrodynamics* (Springer, Heidelberg, 1984)

[5] A.A, Rukhadze, Inzhenernaja Phizika (in Russ.) [Engineering Physics], *On the Structure of Strong Ionizing Shock Waves in High-Pressure Gases* , **14**, P. 14, 2014.

[6] M. M. Bashkirov, I. A. Volodin, V. G. Dmitriev, A. A. Konotop, A. A. Rukhadze, V. I. Sergeev, Z. N. Fedorova, A. A. Chaplygin, Nelinejnyj Mir (in Russ.) **8**, N 8, PP. 475-488, 2010.

[7] L. F. Chernogor, Inzhenernaja Phizika (in Russ.) [Engineering Physics], *Plasma, electromagnetic and acoustic effects of meteorite "Chelyabinsk"*, **8**, PP. 23-40, 2013.

[8] Ya. B. Zeldovich, ZHETF (in Russ., in Engl. - JETP) **32**, P. 1126 (1957)

[9] V. D. Shafranov, ZHETF (in Russ., in Engl.- JETP) **32**, P. 1453 (1957)

[10] I.D. Jukes, J. Fluid Mech., **3**, PP. 275 - 285 (1957).

[11] D.A. Tidman, Phys. Rev., **111**, PP. 1439 (1958).

[12] Ya B. Zel'dovich, Yu P. Raizer, *Physics of Shock Waves and High-Temperature Hydrodynamic Phenomena* (Mineola, N.Y.: Dover Publ., 2002)

# Complex space-charge configuration inside and around a grid cathode with hole

C. T. Teodorescu-Soare[1,2], D. G. Dimitriu[1], C. Ionita[2], R. W. Schrittwieser[2]
[1]*Faculty of Physics, Alexandru Ioan Cuza University, Iasi, Romania*
[2]*Institute for Ion Physics and Applied Physics, University of Innsbruck, Innsbruck, Austria*

## 1. Introduction

Complex space charge structures, in form of sheaths, double layers, fireballs, multiple fireballs or inverted fireballs are well-known phenomena appearing in regions of plasma where local constraints are applied, often in form of a localized electric field, pulling the system away from thermodynamic equilibrium. The dynamics of such structures give rise to different plasma instabilities, involving complex oscillations of the plasma parameters or chaotic states under certain experimental conditions [1-3].

We report on investigations of a complex space charge structure in a low-temperature plasma consisting of two fireball-like structures, one of them similar to an inverted fireball [1]. This configuration is obtained by a spherical metallic grid cathode with a small hole.

## 2. Experimental results and discussion

The experiments have been performed in a plasma diode of glass in which the anode is a metallic rectangular plate of 25 cm×20 cm size, while the cathode is a spherical metallic grid with a diameter of 40 mm (with 0.5 mm being the diameter of the metallic wires and 2 mm the mesh width) with a small hole with a diameter of about 6 mm. Fig. 1 shows the static current-voltage characteristics of the cathode, recorded under the following experimental conditions: argon pressure $p = 7.5\times10^{-2}$ mbar, distance between anode and cathode $d = 25$ cm. The complex structure (see Fig. 2) appears suddenly while decreasing the potential on the grid cathode with respect to the anode (i.e. making it more negative) under the value of about –282 V. The static current-voltage characteristic of the grid cathode shows a hysteretic current jump, i.e. increasing the grid potential the fireball structure disappears for a more positive value (–263 V) than the potential where it appeared during decreasing it. The threshold value of the voltage necessary for the appearance of the complex structure decreases (it is more negative) with increasing the gas pressure. Moreover, the width of the hysteresis of the static current-voltage characteristics depends also on the gas pressure.

Electrical probe measurements have shown a positive potential inside the grid cathode with respect to the potential applied on it. This is an indication for the formation of a virtual

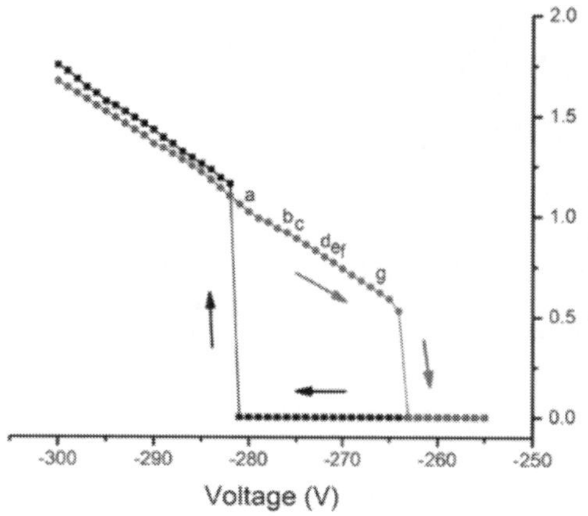


Fig. 1: Static current-voltage characteristic of the grid cathode (the letters indicate the positions where the time series of the discharge current were recorded)

Fig. 2: Complex space charge structure inside and around a grid cathode with a hole

anode inside the grid due to the electrons that perform a pendulum motion similar to the effect inside a hollow cathode. This leads to an amplification of the ionization processes and thereby to a strong increase of the ion density. Thus, we can argue that the inner fireball appears around a virtual anode inside the grid cathode. For more negative potentials, the electrons inside the grid cathode reach sufficient energy to penetrate the inner sheath near the cathode, passing through the hole and giving rise to a second fireball-like structure located outside the cathode (in the vicinity of the hole). This second structure interacts with the negative glow of the discharge (located outside the spherical grid cathode).

The time series of the discharge current reveal strongly nonlinear dynamics of the complex space charge structure. Thus, during re-increasing the grid potential, before disappearing, the structure passes through different dynamic states involving chaos, quasi-periodicity and period-doubling bifurcations, looking like a competition of different routes to chaos, as that one described in [4]. Fig. 3 shows the time series of the discharge current, their FFT's and the reconstructed attractors of the plasma system dynamics (by time delay method), respectively, for different values of the voltage applied on the grid cathode. For low values of the voltage (under –280 V) the behaviour of the plasma system is a chaotic one (intermittent), as it can be observed from the time series of the discharge current in Fig. 3a. Between $V$ = –276 V and $V$ = –272 V the dynamics become quasi-periodic (see Figs. 3b-3d"), and more and more regular as the voltage increases. At $V$ = –271 V a transient phase starts (see Figs. 3e-3e"), driven by a period-doubling bifurcation and ending at $V$ = –270 V with a period-2 dynamics (see Figs. 3f-3f"). The final state of the plasma system (before the complex structure disappears) is a periodic one (see Figs. 3g-3g"), occurring at $V = -266$ V after a reverse (backward) period-doubling bifurcation.

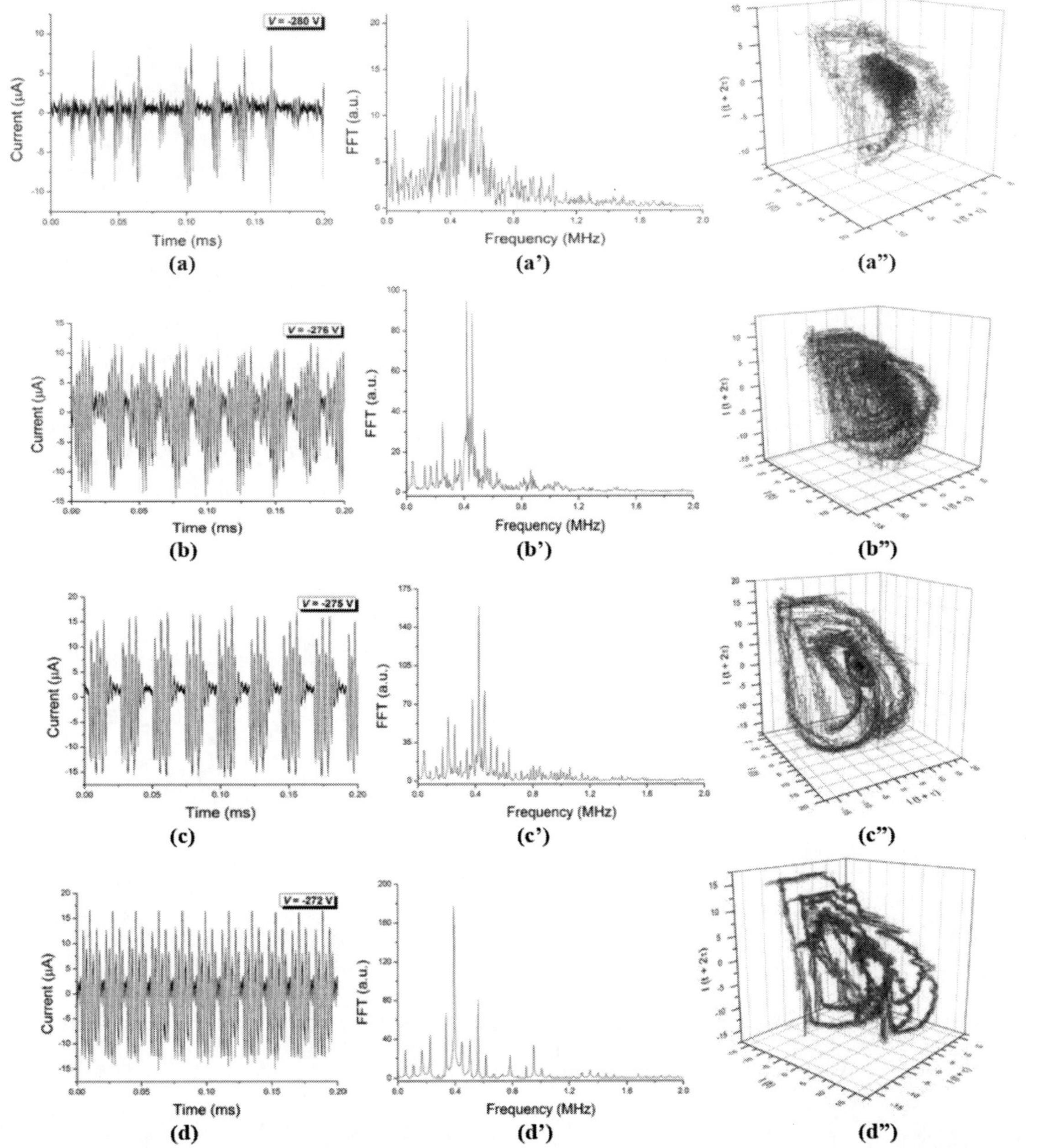


Fig. 3: Time series of the discharge current (left column), their FFT's (central column) and the reconstructed attractors of the plasma systems dynamics by time delay methods (right column), respectively, for different values of the voltage applied on the grid cathode.

## 3. Conclusion

Complex space charge configurations were obtained in low-temperature plasma inside and around a spherical grid cathode with a small hole. The structure appears at potentials below about –260 V. Strong hysteresis is recorded. The dynamics of the complex space charge structure are investigated. The time series of the discharge current reveal that the increase of the applied voltage sets in motion a series of strongly nonlinear phenomena, determining the structure to pass from chaotic/intermittent behaviour through quasi-periodicity, period-doubling to a coherent dynamics towards its disappearance.

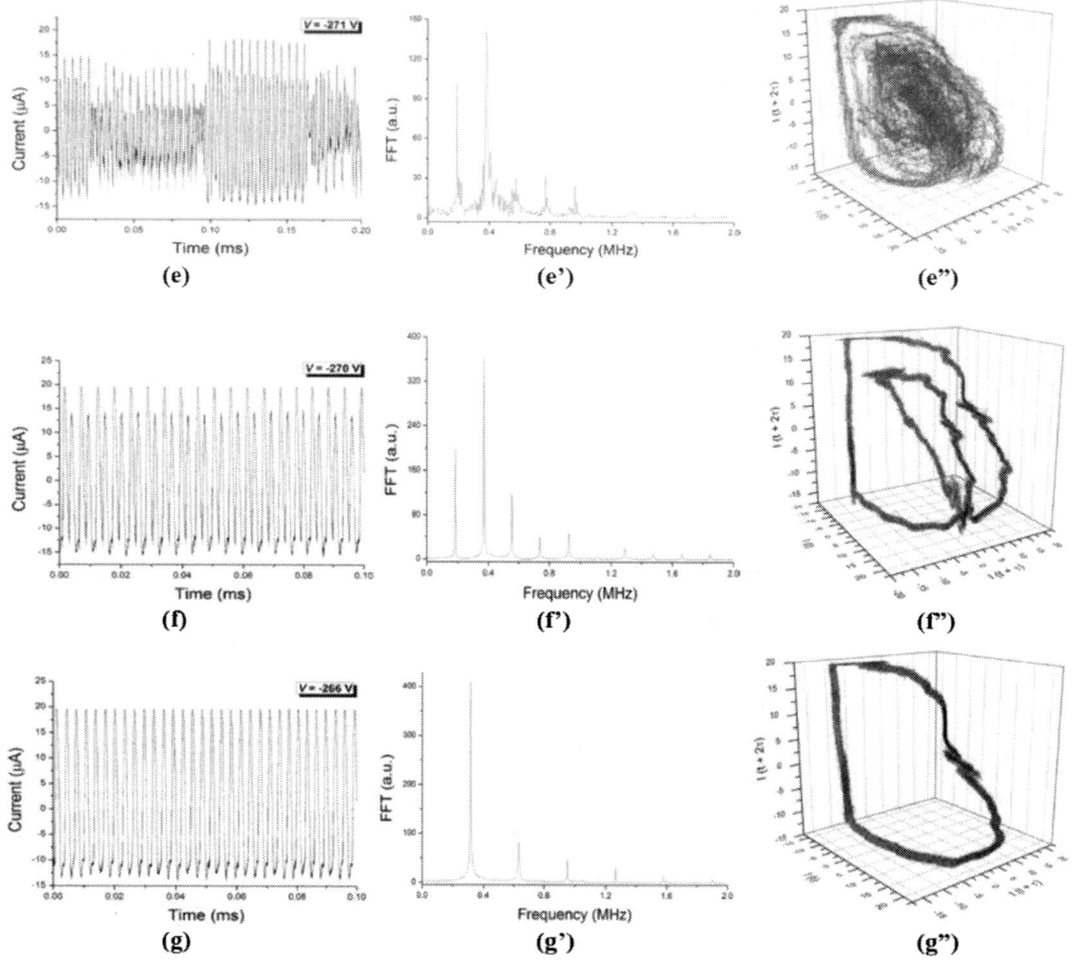


Fig. 3 (continuation)

**Acknowledgement**

This work was supported by the CEEPUS network At-0063 and by the bilateral project RO 07/2104 of the Romanian-Austrian Exchange Services.

**References**

[1] R. L. Stenzel, J. Gruenwald, B. Fonda, C. Ionita, R. Schrittwieser, *Phys. Plasmas* **18** (2011) 012104 (9pp) and **18** (2011) 012105 (9pp).

[2] R. L. Stenzel, J. Gruenwald, C. Ionita, R. Schrittwieser, *Plasma Sources Sci. Technol.* **21** (2012) 015102 (11pp).

[3] D. G. Dimitriu, M. Aflori, L. M. Ivan, V. Radu, E. Poll, M. Agop, *Plasma Sources Sci. Technol.* **22** (2013) 035007 (11pp).

[4] M. Agop, D. G. Dimitriu, L. Vrajitoriu, M. Boicu, *J. Phys. Soc. Japan* **83** (2014) 054501 (11pp).

# Electron kinetics in atmospheric-pressure discharges of helium mixtures with $N_2$ and $O_2$

P. Viegas[1], A. Bourdon[2], L. L. Alves[1], V. Guerra[1]

[1] *Instituto de Plasmas e Fusão Nuclear, Instituto Superior Técnico, Universidade de Lisboa, Lisboa, Portugal*

[2] *Laboratoire de Physique des Plasmas, École Polytechnique, Palaiseau, France*

**1. Introduction.** In last years, helium and helium-based plasmas at atmospheric-pressure and room-temperature have received considerable interest due to their potential for biomedical applications. They can lead to the production of energetic species when interacting with open air, or to the generation of reactive radicals and ions when admixed with a molecular gas. Recently, different experiments have been carried out to study the dynamics of these plasmas in long capillary tubes used for the development of medical devices [1,2]. To optimize these applications one needs to identify and to understand the elementary processes controlling the global behavior of the discharge and the kinetics of the main charged and neutral species in the plasma, which are usually produced using mixtures of helium with small quantities of nitrogen, oxygen or synthesized air (80%$N_2$-20%$O_2$). The modelling of the plasma propagation can also be used for comparison with experiments and to gain physical insight into various phenomena.

**2. Model.** In this work, the tool IST-LoKI (LisbOn KInetics) is adapted and used to study electron-impact reactions in the aforementioned conditions. IST-LoKI is a self-consistent numerical code that solves the two-term electron Boltzmann equation (EBE) together with a system of rate balance equations describing the creation and loss processes of the dominant plasma species. The present study focuses on the electron energy distribution function (EEDF) of helium-containing plasmas, calculated for several values of the stationary reduced electric field ($E/N$), analyzing the effects of small admixtures of $N_2$, $O_2$ and synthesized air, as well as the influence of He metastables involved in stepwise inelastic and superelastic collisions. From the EEDF it is possible to determine the electron kinetic mean energy and all the electron transport parameters and rate coefficients, as a function of $E/N$ [3].

**3. Solution for Helium plasmas.** The electron-neutral scattering cross sections used for the calculations in Helium include the following mechanisms [4]: elastic collisions; direct and stepwise excitation and ionization collisions, with ground-state He($1^1S$) and with metastable excited states He($2^3S$) and He($2^1S$), respectively; de-excitation collisions with metastables He($2^3S$) and He($2^1S$) (these states have long lifetimes and relevant relative densities, between $10^{-9}$ and $10^{-5}$ in a steady-state discharge [4]). For the latter de-excitation mechanisms, the superelastic cross sections are deduced from the corresponding inelastic cross sections, using the Klein-Rosseland relation.

The influence of He metastables on the EEDF is noted mostly at low reduced electric fields. The left figure 1 shows results for 1Td, when the relative density of He($2^3S$) increases from 0 to $10^{-4}$. It is seen that electron-He($2^3S$) superelastic collisions turn the EEDF more energetic, creating a plateau up to the He($2^3S$) threshold at 19.82 eV, corresponding to the energy gained by one electron in a superelastic collision. At very low energies (0-1 eV) the EEDF is very high, presenting a deep fall until $\approx$ 3 eV, the region where the electron-He($2^3S$) inelastic collisions matter the most. After the plateau, the EEDF drops again due to the inelastic collisions with both He($1^1S$) and He($2^3S$), the fall increasing with the metastable density. Similar effects are obtained by introducing He($2^1S$) instead of He($2^3S$), or both metastable states with equal densities. At higher *E/N* values, e.g. between 10 Td and 50 Td, the EEDFs have less energetic bodies and more energetic tails, even in the presence of stepwise-inelastic / superelastic collisions with metastable states that lose importance in this case.

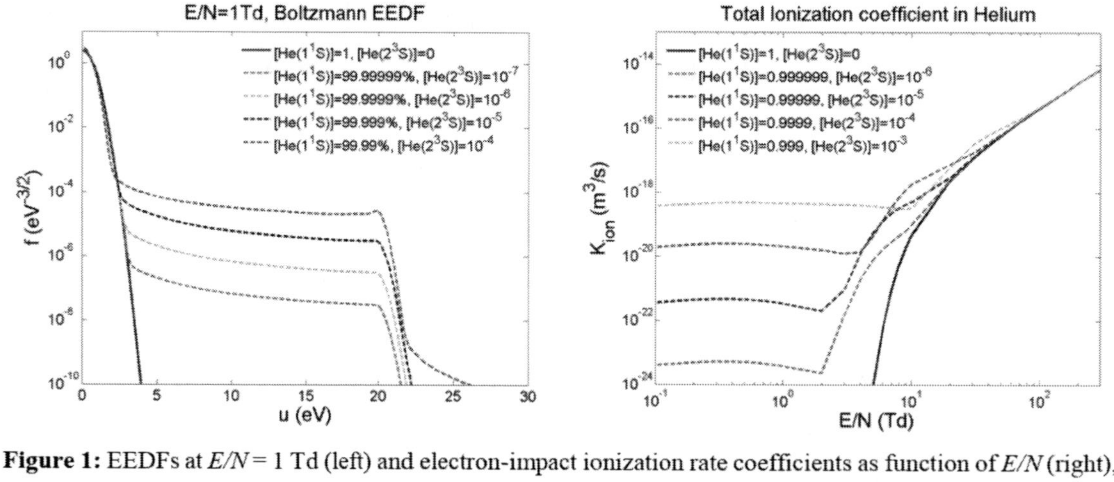


**Figure 1:** EEDFs at *E/N* = 1 Td (left) and electron-impact ionization rate coefficients as function of *E/N* (right), for different relative densities of He($1^1S$) and He($2^3S$).

Ionization is obtained for electron energies above 3.97 eV for collisions with He($2^1S$), 4.77 eV for collisions with He($2^3S$) and 24.59 eV for collisions with He($1^1S$). Therefore, the presence of metastables, with relative densities similar to the ones in figure 1, has a noticeable effect also in the calculated electron rate coefficients, particularly in the ionization coefficient at low *E/N*. The right figure 1 shows this effect for several relative concentrations of He($1^1S$) and

He($2^3S$), plotting the total ionization coefficient over a range of $E/N$ between 0.1 Td and 300 Td. The results demonstrate the influence of stepwise ionization reactions for reduced fields below 20 Td. In particular, for $E/N < 1$ Td the total ionization coefficient clearly increases with the metastable density, due to the very-low energy thresholds associated with stepwise ionization. The EEDF plateau-effect and the influence of stepwise ionization on the total ionization rate coefficient have been previously reported [3,5].

**4. Solution for He-$N_2$-$O_2$ mixtures.** The addition of air constituents to He affects also the EEDF. Helium presents a much more energetic EEDF than molecular gases, such as nitrogen and oxygen. The excited states of He have much higher energy thresholds than the low-energy excited states of $N_2$ and $O_2$, which leads to a depletion of the EEDF tail when they are admixed. The addition of molecular gases to helium, in percentages as small as 1%, is sufficient to produce significant changes in the EEDF, as reported in [6], especially at low fields ($E/N$~1 Td) where low-energy-threshold collisions are dominant. Notice that the description of the electron kinetics in He-$N_2$-$O_2$ plasmas considers the following additional electron-neutral scattering cross sections: elastic, rotational excitation, vibrational excitation (first energy threshold 0.29 eV), electronic excitation (6.17 eV) and ionization (15.5 eV) with $N_2$; and elastic, rotational excitation, vibrational excitation (0.19 eV), electronic excitation (0.98 eV), ionization (12.1 eV) and attachment with $O_2$.

The changes in the EEDF induce differences also in the calculated rate coefficients, particularly in the total electron-impact ionization rate coefficient. The left figure 2 presents this coefficient for several He-$N_2$ mixture compositions, for $E/N$ = 1-100 Td. As expected, the less energetic $N_2$ EEDF leads to an ionization coefficient much lower in pure $N_2$ than in pure He. For the conditions of figure 2, a large decrease in the ionization coefficient is observed above 10% $N_2$, although this quantity presents a maximum for an admixture of only 0.1% $N_2$ at fields below 20 Td (for which the EEDF is not significantly changed), due to the lower $N_2$ ionization-threshold. This means that the rate coefficients for electron-He reactions always decrease by admixing $N_2$ but, on the contrary, the contribution of $N_2$ for direct ionization presents a maximum as a function of its percentage in the mixture. The overall result obtained for He-$N_2$ plasmas suggests that a small admixture of a molecular gas to helium may contribute to create and maintain the discharge.

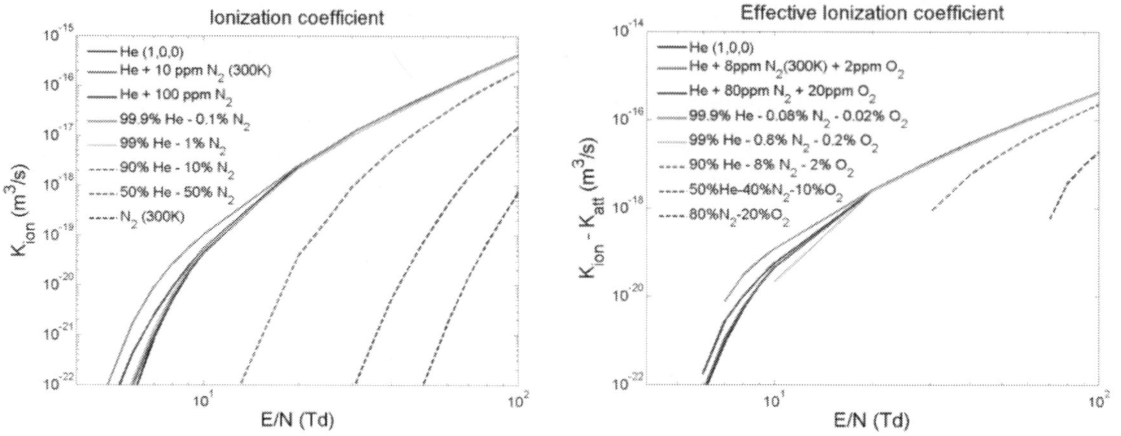


**Figure 2:** Ionization rate coefficients for different relative densities of He($1^1S$) and $N_2$ (left) and effective ionization rate coefficients for different relative densities of He($1^1S$) and dry air (right), as function of *E/N*.

The admixture of $O_2$ to He produces similar changes on the EEDF, but the electronegativity of $O_2$ is shown to have an important additional effect on the plasma behaviour, due to the presence of the electron attachment process, $e^- + O_2 \rightarrow O^- + O$. The right figure 2 presents calculation results of the effective ionization rate coefficient, defined as the difference between the rate coefficients for electron-impact ionization and attachment, obtained for several mixtures of He with dry air, in the range $E/N = 1\text{-}100$ Td. One observes that the addition of $O_2$ is responsible for the appearance of negative values in the effective ionization rate coefficient (due to the attachment mechanism), for low reduced electric fields and low He concentration (see the interrupted curves in figure 2), which suggests enhanced difficulties in the breakdown of oxygen-containing plasmas.

**5. Acknowledgements.** Work partially funded by the Portuguese FCT – Fundação para a Ciência e Tecnologia, under Project UID/FIS/50010/2013.

## 6. References

[1] E. Robert et al., *Plasma Processes and Polymers* 6, 795 (2009).
[2] T. Darny et al., Proceedings of the conference *Gas Discharges* (2014).
[3] G. J. M. Hagelaar and L. C. Pitchford, *Plasma Sources Science and Technologies* 14, 722 (2005).
[4] M. Santos et al., *Journal of Physics D: Applied Physics* 47, 265201 (2014).
[5] C. Gorse et al., *Chemical Physics* 119, 63 (1988).
[6] G. M. Petrov et al., *Plasma Chemistry and Plasma Processing* 20, 2 (2000).

# Vacuum ultraviolet emission from microwave driven hydrogen plasmas

S. Espinho [1], E. Felizardo [1], E. Tatarova [1], and F.M. Dias [1]

[1] *Instituto de Plasmas e Fusão Nuclear, Instituto Superior Técnico, Universidade de Lisboa, Portugal*

## 1. Introduction

Hydrogen discharges, in particular discharges driven by surface waves at microwave frequencies, have the potential to become alternative sources of vacuum ultraviolet (VUV) radiation. VUV radiation falls in the region of the electromagnetic spectrum with wavelengths below 200 nm. Atomic and molecular hydrogen emissions in this spectral range correspond to photon energies high enough to break most organic chemical bonds, initiate surface reactions and modify surface properties [1, 2]. The optimization of these sources requires deep insight of the mechanisms occurring inside the plasma. However, results concerning VUV emissions from discharges operating at microwave frequencies are quite scarce. In this work, we present an experimental investigation of VUV radiation emitted by hydrogen discharges operating at microwave frequencies and low-pressure conditions (0.1 – 0.4 mbar). Emission spectroscopy in the VUV region is used to identify and investigate the excited hydrogen species emitting at these wavelengths as a function of the operational conditions. To complement these results, optical emission spectroscopy in the visible is used to determine the plasma electron density and the temperatures of the gas/excited species.

## 2. Plasma source and experimental conditions

A classical surface-wave-sustained discharge at 2.45 GHz, with a waveguide-surfatron based setup as the field applicator [3] has been used as plasma source [4]. The experimental setup is illustrated in Figure 1. The discharge is produced inside a quartz tube with inner/outer radii equal to 0.15/ 0.25 cm. The gas flow and the pressure is regulated with mass flow controllers and three pressure sensors. The microwave power delivered to the launcher has been varied between 80 and 300 W. The VUV radiation emitted by the plasma is collected by a Horiba Jobin-Yvon Plane Grating Monograph (PGM), which is directly coupled to the discharge tube, thus granting unobstructed line-of-sight over the axis. The PGM is equipped with one of two plane gratings (1800 or 500 gr/mm) allowing the detection of wavelengths from 8 to 125 nm. The VUV photons are then directly collected by a Silicon photodiode detector. To assure a good signal-to-noise response, the light path inside the spectrometer is kept at low pressures $10^{-4}$ - $10^{-5}$ mbar by a turbopump. All the spectra are background corrected. The visible radiation has been collected by an optical fiber positioned perpendicularly to the discharge tube. The optical fiber transmits the visible radiation to the entrance slit of a Jobin-YvonSpex 1250M spectrometer, coupled to a CCD detector. The Jobin-YvonSpex 1250M has a spectral resolution of 0.06 Å. Optical emission spectroscopy of the $H_\beta$ line (486.1 nm) and of the Fulcher-α Q-branch transitions $\left[d^3\Pi_u \rightarrow a^3\Sigma_g^+\right]$ (600 – 616 nm) was performed to determine basic plasma parameters, such as the plasma electron density and the gas/excited species temperatures [5].

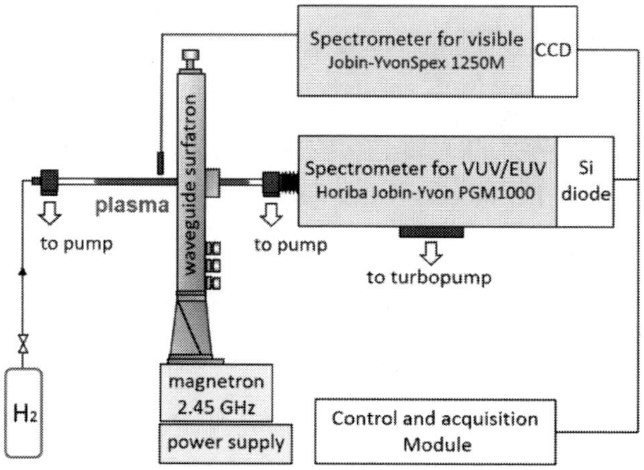


**Figure 1.** Experimental setup for spectroscopy in the visible and VUV spectral ranges.

### 3. Results and discussion

Experimental spectra of hydrogen measured between 8 – 125 nm show the presence of atomic and molecular emissions as seen in Figure 2. The atomic emissions correspond to transitions of the hydrogen Lyman series. The most intense spectral lines correspond to Lyman-α (n=2 → n=1) and Lyman-β (n=3 → n=1) transitions emitting at 121.6 nm and 102.6 nm, with respective energies 10.2 eV and 12.1 eV. Other spectral lines from the Lyman series are also observed at 97.3 nm and 93.8 nm, corresponding to Lyman-γ (n=4 → n=1) and Lyman-ε (n=6 → n=1) transitions. The atomic Lyman series comprises transitions up to level n=12. However, the contribution of the full Lyman series to the spectrum cannot be distinguished from the hydrogen molecular emissions. The observed molecular bands correspond to Lyman $H_2$ ($B^1\Sigma_u^+$ - $X^1\Sigma_g^+$) and Werner $H_2$ ($C^1\Pi_u$ - $X^1\Sigma_g^+$) transitions detected in the 80 - 125 nm spectral range.

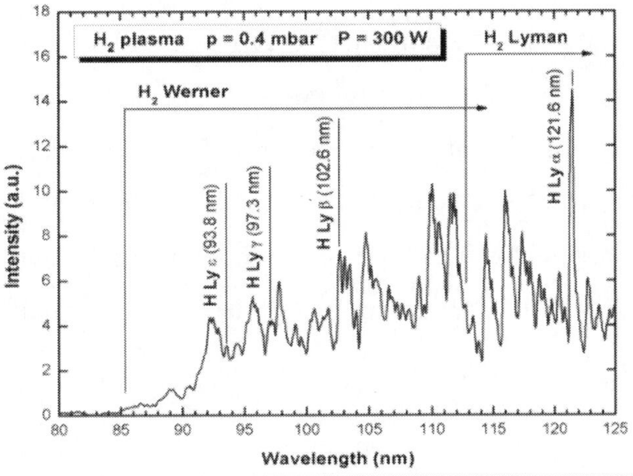


**Figure 2.** Experimental spectrum of hydrogen in the VUV spectral range.

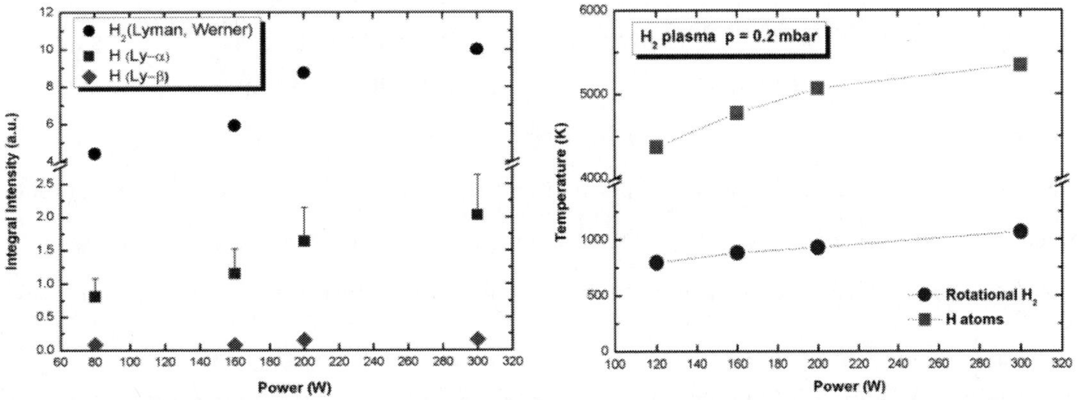


**Figure 3.** (a) Intensity of the Lyman-α, β atomic lines and of the Lyman/Werner molecular bands of hydrogen, as a function of power delivered to the discharge (constant pressure 0.2 mbar). (b) Rotational temperature of hydrogen molecules and kinetic temperature of hydrogen atoms as a function of microwave power.

The relative intensity of the atomic and molecular hydrogen emissions has been investigated as a function of microwave power delivered to the plasma. The integral intensities of the Lyman-α, β lines and of the Werner and Lyman molecular bands are calculated for different powers applied to the discharge at constant pressure 0.2 mbar and the results are shown in Figure 3. The effect of self-absorption on the lines intensity, represented by the bars, was estimated assuming a hydrogen dissociation degree of 50% in the discharge and calculating the axial evolution of the absorbing H atoms density due to axial transport, diffusion, and wall reassociation [6]. The intensity of the molecular bands increases linearly from 100 to 200 W. Both the Lyman-α and Lyman-β line intensities increase by a factor of two when increasing the power from 80 W to 300 W. This is due to the increase of the plasma electron density as the power delivered to the discharge rises, since at low-pressure conditions the main creation mechanism of the excited Lyman-α, β states is by electron impact excitation. Values of the electron density are estimated to be of the order of $10^{12}$ $cm^{-3}$ at the launcher position, decreasing by one order of magnitude towards the end of the plasma column. As shown in figure 3.b, the kinetic temperature of the hydrogen atoms (~5000 K) is found to be much higher than the rotational temperature of the hydrogen molecules (~900 K). This indicates that the hydrogen atoms and the gas are not in thermal equilibrium and the reason for this is connected with the main creation processes of excited atoms as observed in previous works [5].

## 4. Conclusions

Emission spectroscopy was used as a diagnostic, both in the VUV and in the visible regions of the electromagnetic spectrum to study the spectral features of hydrogen microwave driven discharges. It has been shown that hydrogen microwave plasmas have strong and energetic emissions in the 80 – 125 nm range, originating from atomic Lyman-α, β transitions and molecular Werner/Lyman bands. The intensity of these spectral emissions strongly rises as the microwave power is increased. This is due to the strong increase of the plasma electron density with the power delivered to the discharge. As in previous works, it has been found that the excited hydrogen atoms are not in thermal equilibrium with the background gas.

Work partially funded by Portuguese FCT - Fundação para a Ciência e a Tecnologia, under Project UID/FIS/50010/2013 and grant SFRH/BD/52412/2013 (PD-F APPLAuSE).

[1] A. Hollander and M. R. Wertheimer, *Vacuum ultraviolet emission from microwave plasmas of hydrogen and its mixtures with helium and oxygen*, Journal of Vacuum Science & Technology A **12** (3) 879-882 (1994);

[2] E. Tatarova, N. Bundaleska, J. Ph. Sarrette and C. M. Ferreira, *Plasmas for environmental issues: from hydrogen production to 2D materials assembly*, Plasma Sources Sci. Technol. **23**, 063002 (2014);

[3] M. Moisan, M. Chaker, Z. Zakrzewski, & J. Paraszczak, *The waveguide surfatron: a high power surface-wave launcher to sustain large-diameter dense plasma columns*, Journal of Physics E: Scientific Instruments, **20** (11), 1356–1361 (1987);

[4] S. Espinho, E. Felizardo, E. Tatarova, F. M. Dias and C. M. Ferreira, *Vacuum ultraviolet emission from microwave Ar-H2 plasmas*, Applied Physics Letters **102**, 114101 (2013);

[5] E. Tatarova, F. M. Dias, C. M. Ferreira and N. Puač, *Spectroscopic determination of H, He, and H2 temperatures in a large-scale microwave plasma source*, Journal of applied physics **101** (6) 063306, (2007);

[6] B. Gordiets, M. Pinheiro, E. Tatarova, F. M. Dias, C. M. Ferreira and A. Ricard, *A travelling wave sustained hydrogen discharge: modelling and experiment*, Plasma Sources Science and Technology **9** (3) 295 (2000).

# Active Riblets

Ivchenko A.V.

*Samara State Aerospace University, Samara, Russia*

Gas-dynamic processes in the boundary layer can significantly influence on features of near-wall flows and determine parameters for majority of flying machines at subsonic flight regimes [1]. Under these conditions, external disturbances are able to initiate flow instabilities in the boundary layer (Tollmien-Schlichting waves etc.) leading to the laminar-turbulent transition [2]. The turbulent state of the boundary layer is accompanied by a substantial increase in the aerodynamic drag associated with the friction between gas and surface of the moving body [3].

One way to control the laminar-turbulent transition is associated with the formation of spatially-periodic flows by means riblets [4]. Riblets stabilize disturbances of near-wall gas layer and help reduce the body drag (about 6-15%) compared with a smooth surface due to the periodic surface structure [5]. Depending on conditions, the riblets step $\lambda$ (see Fig.1) may be about 0.05-15 mm while a height h overtops the thickness of the shear layer. Thus the riblets effectiveness has limitation and requires optimization of their geometry for specific flow conditions [2].

Interaction between riblets and boundary layer is accompanied by the formation of stable vorticity filaments (into riblets grooves) which called as "vortex cells". Therefore, by influencing the processes in vortex cells you can modify the boundary layer essentially.

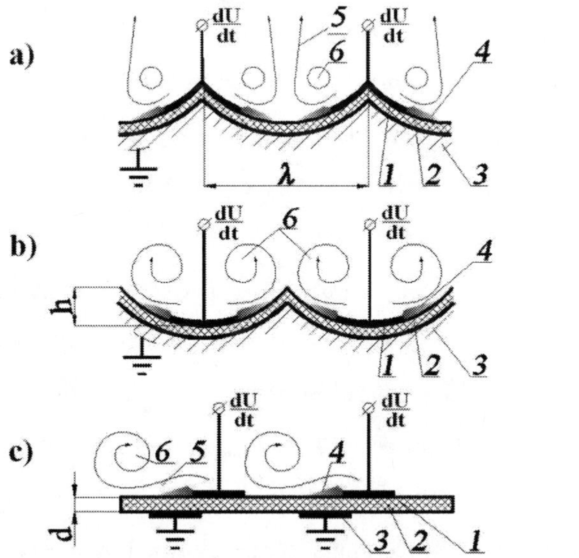


Fig.1. Schemes of active riblets with electrode sets:
a) – electrode system with discharge electrodes disposed on the ribs
b)- electrode system with discharge electrodes disposed in grooves
c)-plasma actuator for the virtual riblets formation
1-discharge electrode,2-dielectric layer, 3-graund electrode, 4-surface discharge, 5-microjet, 6-vortex
$\lambda$-riblets step, h-deep of groove, d-thickness of dielectric layer

The vortex flow modification near the riblets surface can be implemented by using of plasma actuators based on surface discharges [6,7]. According to paper [8], surface discharges can neutralize perturbations in the boundary layer. On the other hand, a short time of plasma excitation ($10^{-7}$-$10^{-5}$s) extends the capabilities of any control system due to the use of reliable (long-time) algorithms. Fig. 1 shows the variants of discharge devices producing vortex-jet modification of the boundary layer near the riblets surface. Here discharge electrodes can be mounted on ribs, into grooves as well as on the smooth surface.

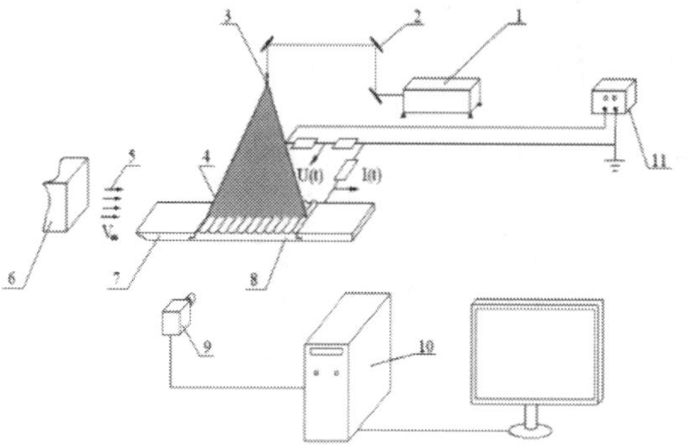


Fig 2. Experimental setup: 1-Laser Solo-120XT-15H2, 2-optical waveguide, 3-lens, 4-light sheet, 5-direction of gas flow, 6-subsonic wind tunnel T-4, 7- air cowl, 8-active riblets, 9- camera FlowSence-2M, 10-computer 11- high-voltage source.

a)

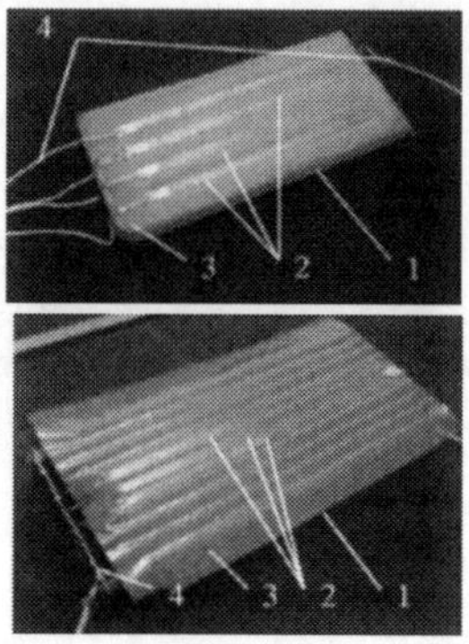

b)

Fig 3 – Multi-electrode sets on flat plate (a) and on the plate with "U"-shaped riblets (b): 1-plate, 2 –discharge electrode, 3-dielectric layer, 4-electrical terminals

Electrode sets in Fig 1 a) and Fig 1 b) are able to generate colliding surface discharges [9-10] wich can be used for changing the flow in vortex cell due to micro-jet formation in vicinity of dielectric surface. On the other hand the asymmetric electrode set (see Fig.1 c) can also create vortices system by acting like virtual riblets.

This study presents the results of the active riblets action for electrode configuration shown in Fig.1 b) and Fig.1 c). For this aim the experimental setup was used (see Fig.2). The smog visualization method was used for registration of flow patterns. The quantitative determination of vector velocity fields (V) and scalar fields vorticity (Γ) was produced by PIV-method [12]. The method of the discharge registration corresponded to paper [9]. If necessary, the external air flow created in one-pass subsonic wind tunnel (V<30 м/c) with test section 200*200*500 mm.

In our case the surface discharge excited by the alternating high voltage with |U|<6.5 kV and f=9 kHz. The plasma formation was performed on the flat plate surface as well as "U"-shape riblets with λ= 10 mm and h = 5 mm (Fig. 3). Both of them had a working surface area about 200 * 100 mm. However dielectric layer for each system was different. For example, in the system (see Fig.3a) fiber-glass plastic (d=0.8 mm) was used while the riblets surface (see Fig.3b) was covered by kapton film (d=0.32 mm,). This electrode set contained three discharge actuators though flat plate had four discharge electrodes.

An investigation of the active riblets interaction with the surrounding space was performed in two stages. In the first stage, the observation of interaction between surface discharges and still air was carried out. It allows you to analyze the flows which were induced by plasma. After that, the registration process of active riblets interaction with transverse external flow (V<2-3 m/s) was executed. It allowed to observe the generated vortices at the sufficient spatial resolution, as well as to investigate the interaction of active riblets with flow separation [2]. PIV-recordings were performed by means of video camera FlowSence-2M with frame frequency $f_{fr}$=4 Hz and laser irradiation impulse energy $W_p$ =45 mJ. The recorded images (200 frames in each series) were processed using a cross-correlative algorithm with sampling window of size 16*16 pixels and 50% overlap [12].

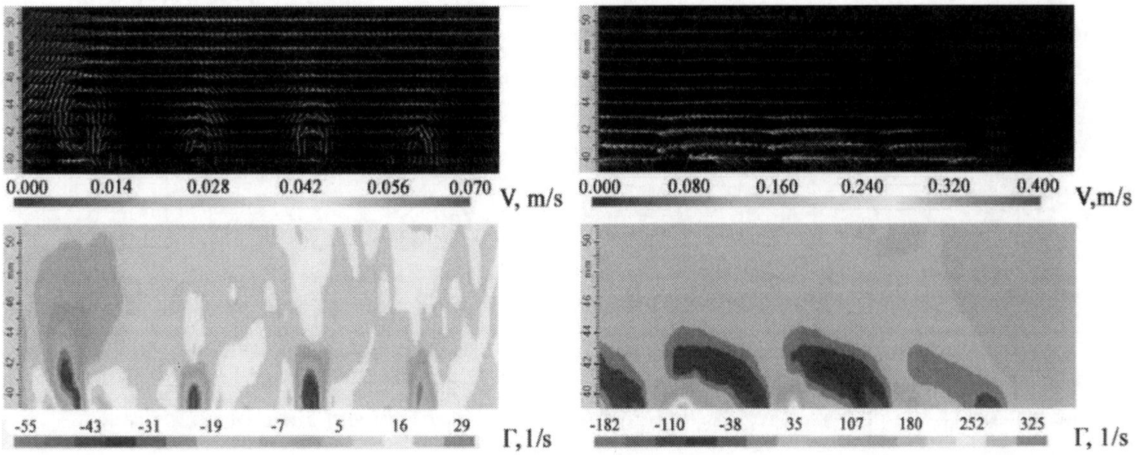


a) b)

Fig.4. Fields of velocity (V) and vorticity (Γ) generated due to the action of virtual riblets in quiescent flow: a)-|U|<2.5 kV; b)-|U|<4 kV

Examples of arising gas flows under the surface discharge action in the quiescent air are presented in Fig.4 and Fig.5. Plasma generation on the flat plate (see Fig.3a) enables the vortex structure formation in vicinity of discharge electrodes (Fig.4) though the operating mode of virtual riblets is not stable. Any external flow can blow the vortices along the smooth surface, by leading to unsteady flow in the boundary layer. Moreover voltage growth causes a transition from the vortex systems to jets which are directed along the streamlined surface (see Fig.4b).

On the other hand vortex cells located above the active riblets surface (Fig. 5) are more stable. Under these conditions, vortex structures are localized in the space (Fig.6) and they are increased in size when voltage increases (Fig. 5b). In the flow, active riblets can improve features of velocity profiles in boundary layer (see Fig.7). It enables to control the flow near the wall efficiently as well as prevent separation flow phenomena.

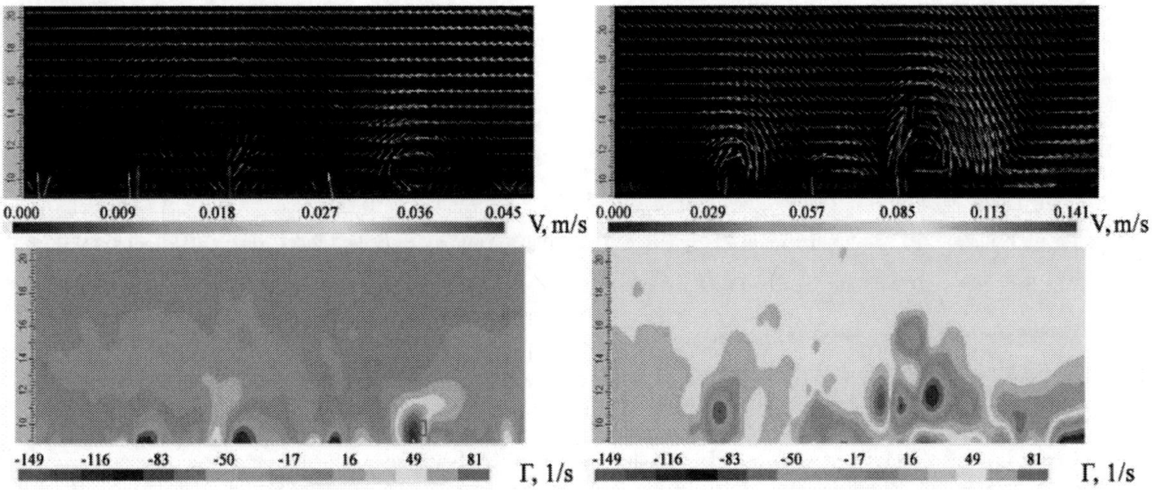


a) b)

Fig.5. Fields of velocity (V) and vorticity (Γ) formed by plasma enhanced riblets in quiescent flow: a)-|U|<2.5 kV; b)-|U|<4 kV

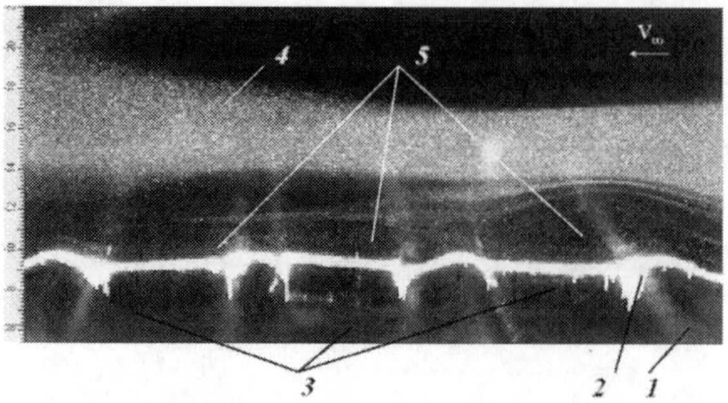


Fig.6. Flow pattern ($V_\infty \cong 1$ м/с) near surface of "U"-shaped active riblets during discharge generation: 1 – riblets vertex; 2 – light sheet; 3-groove containing discharge electrodes; 4-aerosol; 5-vortexes.

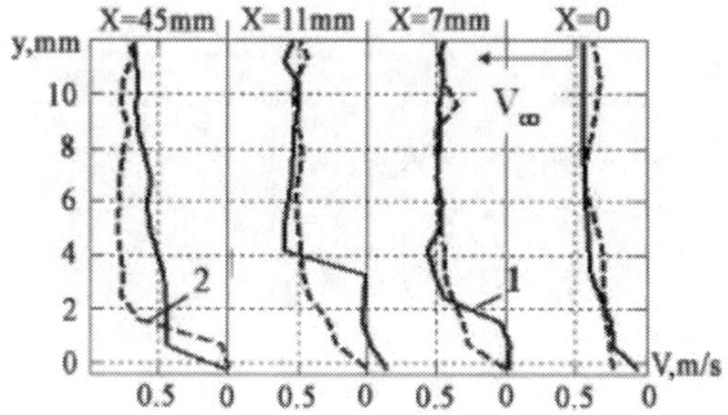


Fig.7 Distribution of tangential velocity component in boundary layer (along flow direction) over the riblets surface: 1-plasma ON; 2-plasma OFF. Coordinate (y=0) correspond to the level of riblets vertexes

In contrast to [8], the presented results demonstrate the ability of effective use of active riblets for flow control in the boundary layer at λ >> 1mm. It extends the capabilities of the presented plasma technology and pays attention to the interaction process research between surface discharges and vortex cells.

The scientific work was realized by partial support of State Program «Scientific and pedagogical staff of innovative Russia» (Contract № P939) and Ministry of Education and Science of the Russian Federation under the Program a competitive recovery of SSAU among the world's scientific and educational centers during 2013-2020.

References

[1] Schlichting, H., Gersten, K. Boundary layer theory - Berlin et al.: Springer. 2003.

[2] Physical basis of the transition to turbulence in non-isolated flows / A.V. Boiko, G. R. Grek et al. - Moskow at al.: Publisher of Computer Science. 2006. [In Russian]

[3] Savill A.M. Drag reduction by passive devices – a review of some recent developments // Structure of Turbulence and Drag Reduction / Ed. by A. Gyr –IUTAM Symposium - Berlin: Springer-Verlag, 1990.-P.429-465.

[4] Grek G. R., Kozlov V.V., Titarenko S. V. An experimental study on the influence of riblets on transition // J. Fluid Mech. – 1996.-Vol. 315. – P. 31-49.

[5] Viswanath P.R. Aircraft viscous drag reduction using riblets // Progress in Aerospace Science. – 2002. – V. 38. – P. 571-600.

[6] Moreau E. Airflow control by non-thermal plasma actuators // J. Phys. D: Appl. Phys. -2007 - V.40. - P.605–636.

[7] Grundmann S., Tropea C. Experimental damping of boundary-layer oscillations using DBD plasma actuators // International Journal of Heat and Fluid Flow – 2009. - V.30. - P.394–402.

[8] Patent US № 8,220,754 International Class: B64C 21/00 Plasma enhanced riblet // McClure et al. Filed: 03 06.2009.

[9] The Properties of Colliding Surface Discharges in Air/ Ivchenko, P.E. Timchenko, V.P. Zakharov, et al // Europhysics Conference Abstracts - 2012.- V. 36F-4p.

[10] Ivchenko A.V. Gas-Dynamic Features of Jets Generated by Colliding Surface Discharges // Europhysics Conference Abstracts - 2013.- V. 37D-4p.

[11] Corke T.C., Post M.L., Orlov D.M. SDBD plasma enhanced aerodynamics: concepts, optimization and applications // Progress in Aerospace since, 2007, V.43. – P.193-217.

[12] Raffel M., Willert C., Kompenhans J. Particle Image Velocimetry: a practical guide – Berlin: Springer, 1998.

# Adaptation of plasma focus type facilities for laboratory simulation of astrophysical jets

V. Krauz[1], V. Myalton[1], V. Vinogradov[1], E. Velikhov[1], S. Ananyev[1], S. Dan'ko[1], Yu. Kalinin[1], A. Kharrasov[1], K. Mitrofanov[2], Yu. Vinogradova[1]

[1] *National Research Centre "Kurchatov Institute", Moscow, Russia*

[2] *Troitsk Institute for Innovation and Fusion Research, Troitsk, Moscow, Russia*

**Introduction.** Laboratory simulation of astrophysical processes is one of the most rapidly progressing research areas [1]. Powerful lasers [2], Z-pinches [3] and other facilities are used for laboratory simulation. Recently, a new series of experiments has been launched in the NRC "Kurchatov Institute" [4] to simulate some of the most striking phenomena in the Universe - astrophysical jets. The facility of plasma focus type, being one of the Z-pinch modifications, is used as an experimental stand. The main goal is to study the mechanisms of the astrophysical jet stabilization, due of which they can propagate along the axis of the accretion disk rotation at distances much greater than their transverse dimensions.

**Experimental setup.** To solve this problem the Filippov-type PF-3 facility was upgraded. A new diagnostic drift chamber was designed to measure the jet and the ambient plasma parameters at the distances up to 100 cm from the place of jet generation (fig. 1). The manifold set of diagnostic tools was used including streak and frame cameras, light collimators, multi-component magnetic probes, laser diagnostics set on the base of nanosecond Nd3+:YAP laser, ballistic pendulum, calorimeter, spectral diagnostics, etc.

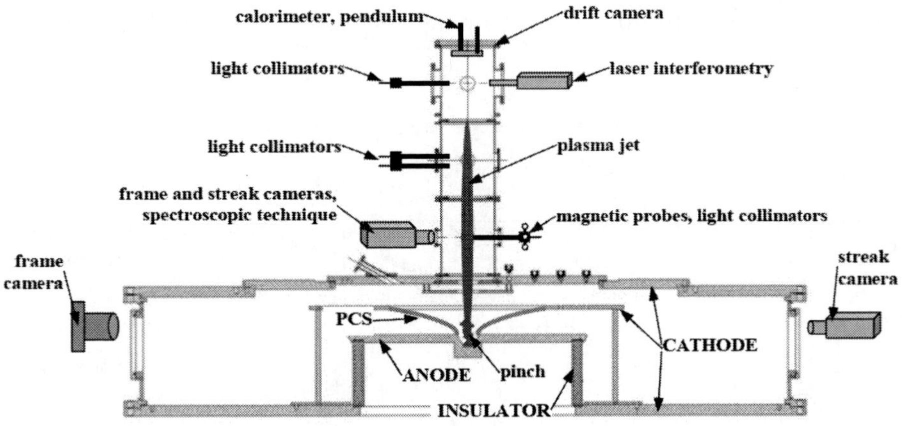


Fig. 1. Experimental setup

Experiments were done at the power supply energy W=290-560 kJ and the discharge current 1 – 2 MA. $H_2$, $D_2$, He, Ne, Ar, Xe and their mixtures at a pressure 1 – 4 Torr were used as a working gases.

**Experimental results.** The stage of the plasma jet formation was studied with the high speed photo recorders. The compact plasma jets moving along the axis occur at the stage of the pinch decay and developing the MHD instabilities (fig. 2a). The initial jet velocity, $V_0 \geq 10^7$ cm/s, exceeds the velocity of the current-carrying plasma sheath in the axial direction and weakly depends on the kind of the working gas.

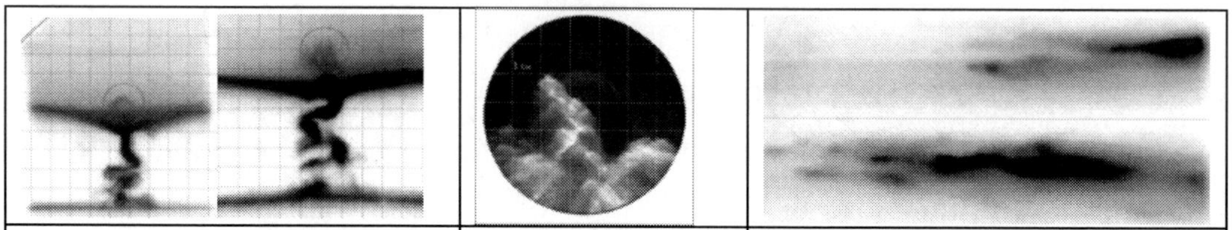

a) b) c)

Fig.2. (a) – Frame camera pictures of the pinch at the stage of jet formation. The scale is 1 cm/div. Frame exposure is 12 ns, time interval between frames is 150 ns. b) - Frame camera picture of the plasma jet at the distance of 35 cm from anode plane. Frame exposure is 12 ns, the scale is 1 cm/div. c) – Streak-camera images of the plasma jet at the slits orientation along the anode plane. The distance between the slits is 4 cm, the sweep duration is 3 μs. The photos were obtained in the discharges in neon.

Frame and streak cameras show preserving the compactness of plasma objects at their propagation over distances much greater than their cross-sectional dimensions. It evidences that the longitudinal velocity of the jet is much greater than the lateral velocity of its expansion. The jet structure consists of separate plasma objects (fig. 2 b, c). Such structures is clearly seen also on the signals of the light collimators, which record the time of arrival of the strongly radiating objects in a certain area of observation (Fig. 3). If use two collimated channels separated by a short distance (16 mm), one can estimate the instantaneous velocity of plasma jet in the point of observation. In some cases, individual peaks on the signal of collimator correlate well with the dips on the current derivative signal.

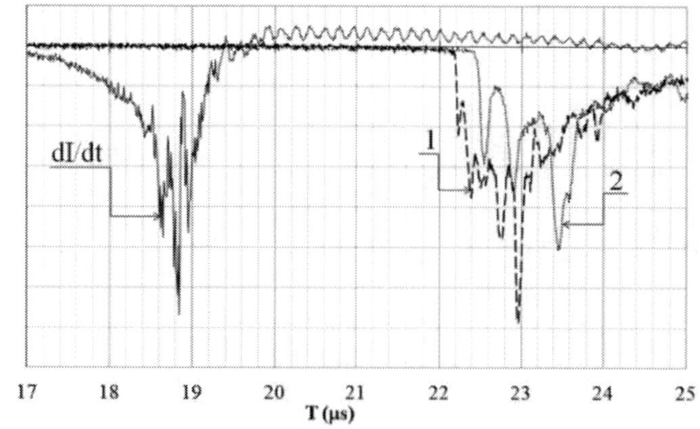


Fig. 3. Signals of discharge current derivative, dI/dt, and optical double collimator (1, 2). The distance between the axis of the two collimator channels is 16 mm. The distance from the anode plane is 35 cm.

In this experimental scheme we used a stationary gas filling. That means that plasma jet propagates in the plasma with finite density arising during ionization of the ambient gas by the emission of the pinch. Therefore information about the parameters of both plasma jet and ambient plasma is key in the study of the jet dynamics. To study the plasma parameters, the diagnostic set including the spectrograph with high resolution in combination with a time-analyzing streak camera was developed. The helium was selected as a working gas. The studies were done at the distance 35 cm from the anode plane. The plasma electron temperature, $T \approx 3 - 8$ eV, was determined from the ratio of the intensities of the two lines, one of which ($\lambda 1 = 5876$ Å) belongs to the neutral helium, and the second one ($\lambda 2 = 4686$ Å) – to the hydrogen-like ion. The plasma concentration estimated from the Stark broadening of the lines of helium due to the electric fields of various nature varies in time and space from $10^{14}$ to $2 \cdot 10^{17}$ cm$^{-3}$.

The presence of the ambient plasma largely determines the evolution of the plasma jet when it moving along the system axis. The plasma jet deceleration was studied with the double optic collimators. It was shown that the jet velocity at different distances from the point of generation is given by the expression $V = V_0 e^{-l/l_0}$, where $V_0$ is the initial jet velocity, $l_0$ is the distance at which the jet velocity decreases in $e$ times. $l_0$ is strongly depend on the variety of working gas, which is associated with different degrees of ionization of background gas by the pinch emission. At the same time, the initial velocity is weakly depend on the kind of gas and corresponds well the velocity determined by the frame cameras at the stage of the jet formation.

The presence of the complex configurations of captured magnetic field in the plasma jet was shown by the multi-channel magnetic probes (Fig. 4). The radial distribution of the azimuthal magnetic field at a distance of 35 cm from the point of generation was built. An example of such distribution for the shot in neon as a working gas in the certain instant of time is shown in Fig. 4b. Such distribution can be explained by the axial current flowing in the region near the axis with the radius 1 – 1.5 cm and closing on the periphery of the plasma jet. Temporal behavior of the signals of various components of the magnetic field, in particular, polarity reversal in some cases (Fig.4a), indicates a possible rotation of the magnetic field induction vector (Fig. 4c). Some evidence of the jet rotation was also obtained with streak cameras. The elucidation of the role of the detected magnetic field configurations in ensuring the jet stability is one of the main directions of our research.

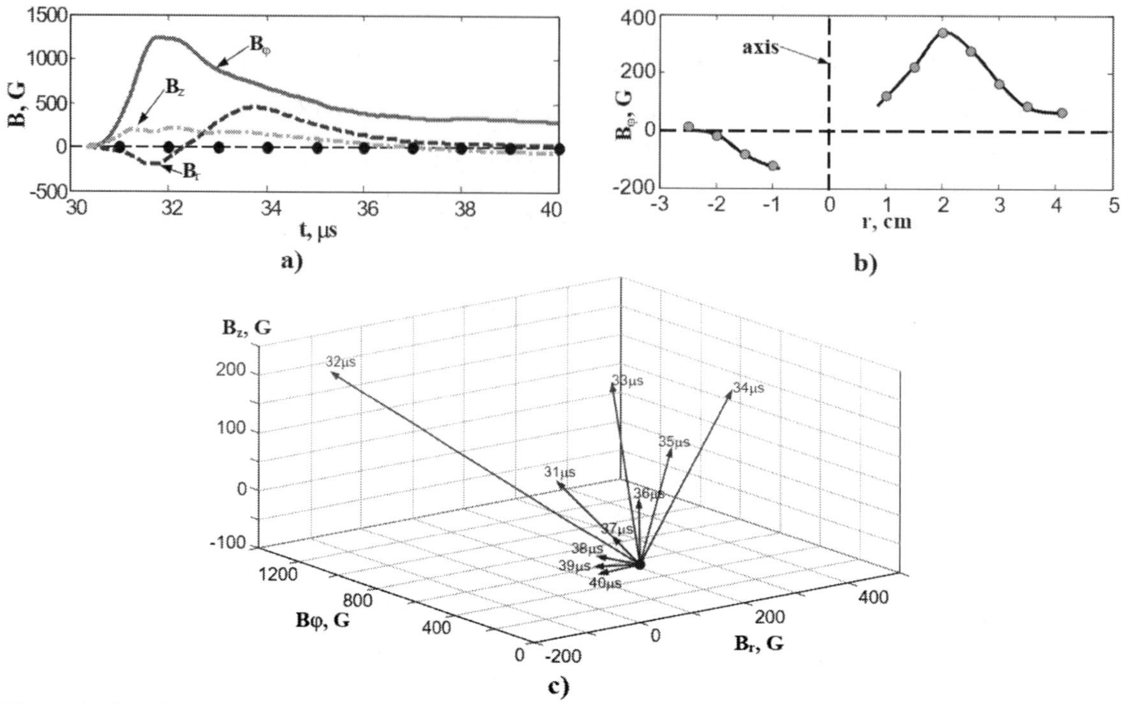


Fig. 4. a) Time dependences: $B_\varphi$, $B_z$, $B_r$ – azimuthal, axial and radial component of the magnetic field induction; b) radial distribution of the azimuthal component of the magnetic field induction, $B_\varphi(r_{i,j},t_i)$; c) direction and value of the sum vector $\vec{B} = \vec{B}_r + \vec{B}_\varphi + \vec{B}_z$, reconstructed for shot on Fig. 4a. The measurements were done at the distance 35 cm from the anode plane. a), c) – shot in $H_2$, b) – shot in neon.

**Conclusion.** The developed experimental stand allows to study the propagation of the plasma jets in the ambient medium to long distances. The experimental scheme and wide possibilities to change the conditions of the experiment allows us to analyze the influence of different factors, such as radiative cooling, the captured magnetic field, hydrodynamic effects, etc. on the collimation and stability of the plasma jet. Achieved parameters of the plasma jet, such as its velocity, $V \geq 10^7$ cm/s, the Mach number, $M \geq 1$, the Reynolds number, $R = 10^4 - 10^5$, the contrast (ratio of the jet density to the density of ambient plasma), K=1-10, the plasma temperature, T= 3 – 8 eV, seems very prospective in the simulation the plasma jets in the young stellar objects known as the objects 'Herbig-Haro'.

The studies were supported in part by the Russian Foundation for Basic Research (projects 14-29-06085, 14-02-01203, №14-02-00179 and 14-02-31473).

References.

1. D. D. Ryutov, et al. Reviews of Modern Physics, 72:167–223, 2000

2. B. Albertazzi et al., Science 17 October 2014: 325-328

3. F Suzuki-Vidal et al., Journal of Physics: Conference Series 511 (2014) 012050

4. V. Krauz et al., Physica Scripta. T161 (2014) 014036

# Laboratory modeling of magnetized accretion dynamics

G. Revet[1,2], D. P. Higginson[1], J. Béard[3], M. Bleicher[4], K. Burdonov[2], M. Borghesi[7], S. N. Chen[1], D. Khagani[5], B. Khiar[6], K. Naughton[7], S. Pikuz[8,10], C. Riconda[1], R. Riquier[1], A. Soloviev[2], T. Vinci[1], O. Willi[4], O. Portugall[3], H. Pépin[9], A. Ciardi[6], J. Fuchs[1,2]
[1]*LULI, École Polytechnique, CNRS, CEA, UPMC, 91128 Palaiseau, France*
[2]*IAP-RAS, Nizhny Novgorod, Russia*
[3]*LNCMI, UPR 3228, CNRS-UFJ-UPS-INSA, 31400 Toulouse, France*
[4]*Institut für Laser-und Plasmaphysik Heinrich Heine Universität Düsseldorf, Germany*
[5]*GSI Helmholtzzentrum für Schwerionenforschung GmbH, 64291 Darmstadt, Germany*
[6]*LERMA, Observatoire de Paris, PSL Research University, CNRS, Sorbonne University, UPMC Univ. Paris 06, F-75005, Paris, France*
[7]*School of Mathematics and Physics, The Queen's University of Belfast, Belfast, UK*
[8]*Joint Institute for High Temperatures RAS, Moscow 125412, Russia*
[9]*INRS-EMT, Varennes, Québec, Canada*
[10]*National Research Nuclear University MEPhI, 115409 Moscow, Russian Federation*

## 1. Introduction

The dynamics of matter accretion in Young Stellar Objects (YSO) is currently a subject of debate and active astrophysical research. Matter ($10^{-11}$–$10^{-13}$ g.cm$^{-3}$) can be extracted from the accretion disk and falls down whilst imbedded in a strong magnetic field (B ~kG) at the free fall velocity (100 – 500 km.s$^{-1}$) onto the stellar surface, in the form of well collimated plasma columns following the magnetic field lines [1,2]. The shock resulting from the impact of these plasma columns onto the stellar surface heat up the infalling flow. Previous experiments have attempted to reproduce accretion shocks in the laboratory using shock-tube configurations. This however created unwanted edge-constraints due the tube walls, and the experiment suffered from a lack of magnetic field [3].

In this paper, we report on an experimental setup which allows us to investigate the accretion dynamics in a magnetized environment. We detail the characteristics of the collimated magnetized plasma jet that is launched onto an obstacle, as well as those of the obstacle before being impacted by the jet. In the light of these parameters, we discuss of the scaling between the astrophysical situation and our laboratory experiment.

## 2. Experimental setup

This experiment was performed at the ELFIE facility, at Ecole Polytechnique (France), using a chirped laser pulse delivering 40 J of energy within a 0.6 ns pulse duration at 1ω (λ = 1057 nm), leading to $I_{max} = 4.5 \times 10^{12}\ W/cm^2$. This laser pulse is used to irradiate a target in the center of a high-current Helmholtz coil, constructed at the Laboratoire National des Champs

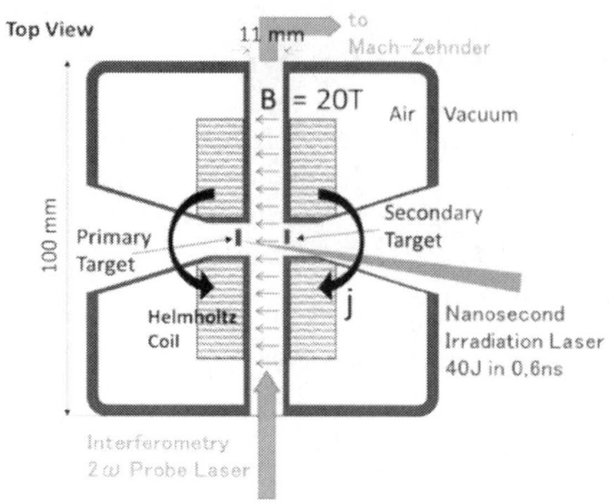


**Fig. 1 :** Top view of the experimental setup. Two targets (primary and secondary targets) look at each other. Their normal is parallel to the magnetic field lines generated by the Helmholtz coil. A longitudinal aperture allows the laser to irradiate the primary target. A radial aperture allows a clear line of sight to the plasma

Magnétiques Intenses (LNCMI, Toulouse, France). The collimated plasma jet originates from the laser interaction with this primary target (see section 3.1) and follows the magnetic field lines to finally hit a secondary Teflon target ($CF_2$). As shown in Fig. 1, two perpendicular apertures allow both targets to be inserted in the coil, and the laser to irradiate the jet source target, as well as a free line of sight for the diagnostics to access the plasma region. The coil is coupled to a 30 kJ/20 kV capacitor bank and delivers a 20 T pulsed (duration ~ 200 μs FWHM) magnetic field in the center of the coil. This magnetic field is homogeneous in the longitudinal and radial direction over, respectively 40 mm and 20 mm [4]. Thus, we can consider the magnetic field as constant across both space and time in comparison with the spatial and temporal scales of the plasma phenomena. Interferometric measurements of the plasma density use a compressed laser pulse (100 mJ in 350 fs, used at 2ω -λ = 528.5 nm-). Three other diagnostics are also used: a FSSR spectrometer yielding a 1D space-resolved measurement of the plasma temperature, a streak camera measuring in both time and 1D space the optical self-emission coming from the heated plasma, and an x-ray radiography diagnostic allowing to probe the accretion dynamics inside dense plasma region (i.e. close to the obstacle surface).

## 3. Accretion protagonists

### *3.1 Collimated jet*

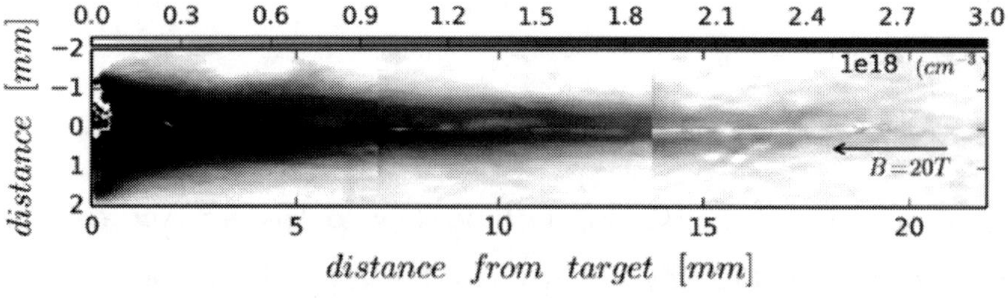


**Fig. 2** : Electron density map of the magnetically collimated jet produced from the laser-irradiated target at 25 ns after laser irradiation, reconstructed from interferometry data via Abel inversion.

The presence of the strong magnetic field induces the collimation of the plasma into a narrow plasma jet with an aspect ratio >10, as shown in Fig. 2. This collimation mechanism is studied in detailed in Ref [1,2]. The wide-angle flow expanding from the heated target, which is non-resistive, carries in its expansion motion the magnetic field lines and compresses them, forming a magnetic wall. Then, the particles are stopped, accumulated and heated at this point, forming a shock that draws the border of a plasma cavity, the center of which presents a lower magnetic pressure as well as a lower plasma density. The flow slides along the wall, forming at the tip of the cavity a conical shock. As a result of this shock and of a new compression of the magnetic field lines at the shock point, the flow is accelerated longitudinally in a very well collimated jet. Fig. 2 displays, at 25ns after laser irradiation, a typical jet resulting from this mechanism in our laser conditions and for a magnetic field of 20 T. To reconstruct this long jet image, the primary target was moved back inside the coil to allow the probing aperture to look at a farther region from the target surface.

The velocity of the jet can be deduced from the time taken by a certain amount of density to reach a certain distance from the source target. The temporal evolution of the density around the obstacle position shows that a density of $1.5 \times 10^{18}\ cm^{-3}$ is reached after 25 ns and stays quite constant over more than hundred nanoseconds. Tracking how much this $1.5 \times 10^{18}\ cm^{-3}$ density point travels along the coil, by scanning different timings of the probe, we find that the speed of the jet ~ 600 $km.s^{-1}$. Preliminary analysis of the FSSR spectrometer data gives a plasma temperature of tens of eV in the jet, at 3-4 mm from the target surface.

*3.2 Obstacle*

We observed that the obstacle is not completely cold when the jet reaches its surface. Indeed, the obstacle is heated by the x-ray emission coming from the interaction between the chirped laser pulse and the primary target. From the density profile of the obstacle measured in shots without magnetic field, and considering a standard self-similar expansion of the plasma, we can deduce a sound speed of 4.2 $km.s^{-1}$. Such a value of the sound speed is of course associated to a low temperature of the plasma (typically around 1 eV), and the gradient scale length of its expansion at the jet arrival time is $C_s \times t = 105\ \mu m$. Below this length, one can consider the density of the obstacle to be close to the solid density.

**4. Scaling with astrophysical objects**

In order to compare our laboratory accretion experiment to astrophysical cases, the table 1 displays some parameters, using the measured characteristics of the jet incident on the obstacle, and the characteristics of accretion in classical T Tauri stars (CTTS) [5, 6]. In this comparison, the astrophysical case is divided in two instances, for two different strengths of the magnetic field, respectively at $5.10^{-3}$ T and $5.10^{-2}$ T, which allows us studying both a case where the magnetic pressure dominates compared to the kinetic pressure (i.e. the magnetic field is able to confine the plasma within the magnetic field lines) and the opposite case.

Although our laboratory case presents a higher temperature compared to the astrophysical case, the higher density of the laboratory jet allows it to reach a Mach number (>1) comparable to the one of the astrophysical configuration. Both cases present a magnetic Reynolds number >1, even if the laboratory presents a regime where $R_m$ is only of a few units. The beta parameters suggests a more accurate correspondence of the laboratory case with the strong magnetic field astrophysical case.

## 5. Conclusion and perspectives

We have detailed a new experimental setup combining laser-produced plasmas and strong external magnetic fields in order to model accretion dynamics in the laboratory. Both the jet and the obstacle characteristics have been detailed; this configuration proves to be scalable to the astrophysical case in some regimes. The pivotal role played by the magnetic field in our experiment leads us to think that this experimental setup will be able to reproduce interesting effects regarding the dynamics of the accretion shock in relation with astrophysical cases.

***Table 1*** : Estimation of some initial plasma parameters for the incoming jet for the laboratory case and the astrophysical CTTS case.

| | Laboratory | CTTS | |
|---|---|---|---|
| *Incident jet* | B-Field = 20T | B-Field = $5\times10^{-3}$T | B-Field = $5\times10^{-2}$T |
| Material | $C_2H_3Cl$ (PVC) | H | H |
| Electronic density [$n.cm^{-3}$] | $1.5\times10^{18}$ | $1.0\times10^{11}$ | $1.0\times10^{11}$ |
| Temperature [eV] | 30 | $2.6\times10^{-1}$ | $2.6\times10^{-1}$ |
| Density [$g.cm^{-3}$] | $8\times10^{-6}$ | $1.7\times10^{-13}$ | $1.7\times10^{-13}$ |
| Speed accretion flow [$km.s^{-1}$] | 600 | 500 | 500 |
| Sound speed [$km.s^{-1}$] | 38 | 6.47 \| 118* | 6.47 \| 118* |
| Mach number | 15.8 | 77.3 \| 4.2* | 77.3 \| 4.2* |
| Alfven speed [$km.s^{-1}$] | $1.96\times10^{2}$ | $3.46\times10^{2}$ | $3.46\times10^{3}$ |
| Alfvenic Mach number | 3.06 | 1.44 | $1.44\times10^{-1}$ |
| Magnetic Reynolds | 2.7 | $1.1\times10^{8}$ | $1.1\times10^{8}$ |
| Magnetic diffusion time [ns] | 71 | $17\times10^{21}$ | $17\times10^{21}$ |
| β | $4.5\times10^{-2}$ | 5 | $6\times10^{-2}$ |
| Mean free path [mm] | $2\times10^{-2}$ | 22.7 | 22.7 |

## References


[1] B. Albertazzi et al., Science 346, 325 (2014)

[2] A. Ciardi et al., Phys. Rev. Lett. 110, 025002 (2013)

[3] U. Chaulagain et al., High Energy Density Physics (2015), http://dx.doi.org/10.1016/j.hedp.2015.01.003

[6] B. Albertazzi et al., Rev. Sci. Instru *84, 043505 (2013)*

[5] S. Orlando et al., A&A 59, A127 (2013)

[6] Argiroffi, C., Maggio, A., & Peres, G. A&A 465, L5 (2007)

# Perturbation Theory for the Collective Interaction of Relativistic Particles with a Localised EM Wave in a Magnetized Plasma

G. Anastassiou[1], Y. Kominis[2], P. A. Zestanakis[1] and K. Hizanidis[1]
[1]School of Electrical and Computer Engineering,
[2]School of Applied Mathematical and Physical Science,
National Technical University of Athens, Greece

## 1. INTRODUCTION

The interaction of a spatially and temporally localized EM wave with charged particles in a magnetized plasma is investigated for the general case of a Gaussian wave-packet and relativistically moving particles, allowing the consideration of applications related to energetic ions and runaway electrons in fusion as well as in space plasmas. The EM field is introduced as a perturbative term, whereas suitable transformations to Guiding Center (GC) variables are employed to represent the kinetic parameters of the relativistically moving particles. The collective characteristics of the interaction are expressed in terms of ensemble averaged variations of the corresponding canonical momenta.

## 2. LIE PERTURBATION SCHEME

The unperturbed Hamiltonian of the interaction expressed in GC canonical variables is given from the expression [1], [2] $\mathcal{H}_0=\sqrt{2mc^2}\sqrt{\Omega P_\phi+P_z^2/2m+mc^2/2}$, where $P_\phi$ and $P_z$ are the canonical momenta conjugate to the gyro-angle $\varphi$ and the GC position $z$ along the magnetic field, respectively, with $P_\varphi=(1/2)m\Omega\rho^2$, $\Omega=qB_0/m$ is the cyclotron frequency, $\rho$ is the Larmor radius and $m$ the rest mass of the particle. The first order perturbative term due to the presence of the EM wave is $\mathcal{H}_1=ec^2(e\mathbf{A}_0-\mathbf{p})\cdot\mathbf{A}_{em}/E_0$ where $\mathbf{B}_0=\nabla\times\mathbf{A}_0$ and $E_0$ is the unperturbed total particle energy.

The EM field is described in terms of the arbitrarily polarized vector potential [3] as $\mathbf{A}_{em}=A_1(\mathbf{r}-\mathbf{V}t,t)[P_{1,\parallel}\hat{\mathbf{z}}+\mathbf{P}_{1,\perp}]e^{i\Psi(\mathbf{r},t)}$ where $\Psi(\mathbf{r},t)=k_\parallel z-\omega t+k_\perp R\sin\theta-k_\perp\rho\sin\varphi=\Psi(\mathbf{R},t)-k_\perp\rho\sin\varphi$ is the phase term of the propagating wave, $\omega$ is the central wave frequency, $\mathbf{V}$ is the group velocity, $\mathbf{R}$ the GC position vector and $k_\parallel$, $k_\perp$ are the wave-vector components with respect to $\mathbf{B}_0$. Applying a Fourier expansion, the first order perturbation term of the Hamiltonian will now become

$$\mathcal{H}_1=\frac{ec^2}{E_0}e^{i\Psi(\mathbf{R},t)}\sum_n e^{-in\varphi}A_1(\mathbf{r}-\mathbf{V}t,t)\left[u_{0\parallel}P_\parallel J_n(k_\perp\rho)+u_{0\perp}\left(P_+J_{n+1}(k_\perp\rho)+P_-J_{n-1}(k_\perp\rho)\right)\right] \quad (1)$$

where $P_{\|}$, $P_+$, $P_-$ are the corresponding parallel, right-hand and left-hand circular polarizations of the vector potential and $J_n$ are Bessel functions. For the vector-potential envelope carrying the spatiotemporal modulation of the wave-packet, a Gaussian-shaped configuration is adopted as being suitable for analytical calculations, that is $A_1(\mathbf{r}-\mathbf{V}t,t)=A_1\,\mathrm{e}^{-\left[\frac{(x-V_x t)^2+(y-V_y t)^2}{a_\perp^2}+\frac{(z-V_z t)^2}{a_\|^2}\right]-\frac{t^2}{a_\tau^2}}$,

where $a_\perp, a_\|$, and $a_\tau$ represent the spatial and the temporal waist of the Gaussian. Applying now a standard Lie-transform canonical perturbation method [4], the first order Lie generating function is

$$w_1=-\int_{t_0}^{t}\mathcal{H}_1(\boldsymbol{\varphi}+\boldsymbol{\omega}_\varphi(\tau-t),\mathbf{J},\tau)d\tau \tag{2}$$

where $\mathbf{J}=(P_z, P_\varphi, P_\theta)$, $\boldsymbol{\varphi}=\boldsymbol{\omega}_\varphi t+\boldsymbol{\varphi}_0$, and $\boldsymbol{\omega}_\varphi=\partial\mathcal{H}_0/\partial\mathbf{J}$ are the corresponding actions, angles and frequencies for the unperturbed system $\mathcal{H}_0$, whereas $P_\theta=(1/2)m\Omega R_\perp^2$ and $\boldsymbol{\varphi}_0$ denotes the initial gyrophase. Moreover, taking the Taylor expansion of the Gaussian and keeping terms up to first order with respect to $\rho/a_\perp$, results in the expression

$$\begin{aligned} w_1=\frac{-ec^2\sqrt{\pi}}{E_0}\tau\sum_n F_n e^{i\mathbf{R}\cdot(\bar{\mathbf{k}}+\tau^2\Delta_n\mathbf{V})}e^{-in\varphi}\Big[G_n-\frac{\rho}{a_\perp}\left(G_{n+1}e^{i\theta}+G_{n-1}e^{-i\theta}\right)\frac{R}{a_\perp}+\\ +\frac{\rho}{a_\perp}\tau^2\left(\mathbf{R}\cdot\mathbf{V}+i\frac{\Delta_n}{2}\right)\frac{V_\perp}{a_\perp}\left(G_{n+1}e^{i\xi}+G_{n-1}e^{-i\xi}\right)\Big]+c.c \end{aligned} \tag{3}$$

where $G_n=u_{0\|}P_\| J_n(k_\perp\rho)+u_{0\perp}P_+J_{n+1}(k_\perp\rho)+u_{0\perp}P_-J_{n-1}(k_\perp\rho)$, $u_{0\|}, u_{0\perp}$ are the components of the mechanical momentum, $\xi=\tan^{-1}(V_y/V_x)$. The function $F_n$ is defined as

$$F_n=e^{-\tau^2|\mathbf{R}\times\mathbf{V}|^2}\,e^{-|\mathbf{R}|^2\frac{\tau^2}{a_\tau^2}}\,e^{-\frac{\tau^2}{4}\Delta_n^2} \tag{4}$$

where $\mathbf{R}=\left[\frac{X}{a_\perp},\frac{Y}{a_\perp},\frac{z}{a_\|}\right]$, $\mathbf{V}=\left[\frac{V_x}{a_\perp},\frac{V_y}{a_\perp},\frac{V_z-P_z/\gamma m}{a_\|}\right]$, $\bar{\mathbf{k}}=\left[0,k_\perp a_\perp,k_\| a_\|\right]$ and $\tau=\frac{1}{\sqrt{\mathbf{V}^2+1/a_\tau^2}}$ is the autocorrelation time of the wave-packet as seen by the particles. Moreover,

$$\Delta_n=k_\| P_z/(\gamma m)-\omega-n\Omega/\gamma \tag{5}$$

with $\gamma=E_0/mc^2$ whereas $\Delta_n=0$ provides the resonance condition [2], [3]. The function $F_n$ describes the most important features of the particle interaction with the localized wave-packet. The first term of (4) describes the dependence of the interaction on the relation between the wave-packet group velocity and the particle GC position as in a scattering-like process, whereas the second term describes the time duration of the interaction. The third term describes the resonance mechanism of the interaction and reveals its localized effect on the action-space of the GC motion, that is centered around $\Delta_n = 0$ and has a characteristic width inversely proportional to $\tau$.

## 3. AVERAGE VARIATIONS OF ACTIONS

The average variation of the actions for an ensemble of particles with uniformly distributed canonical angles is given from [2], [4]

$$\langle \delta P_\ell \rangle = \frac{1}{2}\frac{\partial}{\partial P_\ell}\left\langle \left(\frac{\partial w_1}{\partial \ell}\right)^2 \right\rangle_{z,\varphi,\theta} \tag{6}$$

where $\ell$=($z$, $\varphi$, $\theta$,) and provides the average variation of parallel momentum ($P_z$), Larmor radius ($\rho = (2P_\varphi/m\Omega)^{1/2}$), and transverse GC position ($R_\perp = (2P_\theta/m\Omega)^{1/2}$). In this short paper, we present results only for parallel propagation of the EM wave ($\mathbf{k}_\perp = 0$) for both parallel and circular polarization of the EM wave. In the following figures, we depict the normalized averaged variations of each action $\Delta P_\ell = \langle \delta P_\ell \rangle / (A_1 \tau)^2$ as functions of the parallel momentum $P_z$ for wave-packets with different characteristics, whereas the resonance condition $\Delta_n = 0$, is written in the form

$$P_\varphi = [P_z^2(k_\parallel^2 - \omega^2) - \omega^2 + n^2 - 2nk_\parallel P_z]/2\omega^2 \tag{7}$$

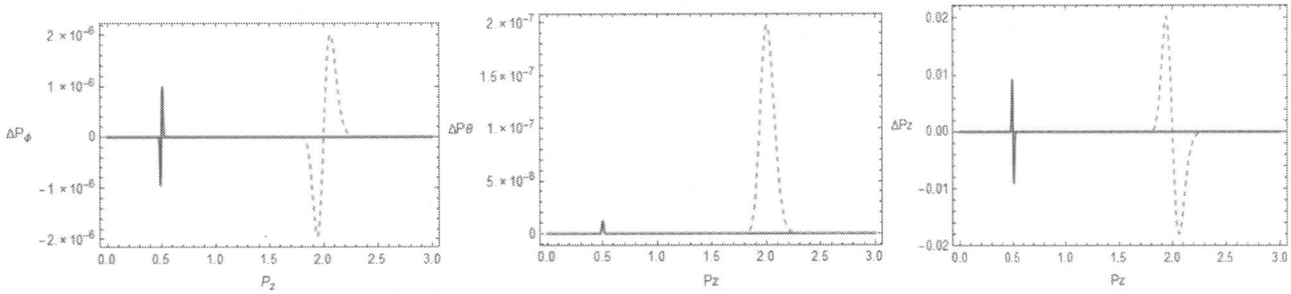


Figure 1:Variations of the averaged actions for $P_\parallel = 1, P_\theta = 1, k_\parallel = 1, \omega = 1, \mathbf{V} = 0, a_\parallel = a_\perp = 10^2, a_\tau = 10^3$, $P_\varphi = 0.5$ (solid line) and $P_\varphi = 2$ (dashed line). In this special case with $k_\parallel = \omega$, the coefficient of the quadratic term in (7) is eliminated, and the resonance condition $\Delta_n=0$ is fulfilled only for $n \neq 0$. Expression (3) suggests that only the terms with $n = \pm 1$ ($n=-1$ for $P_z>0$) are nonzero, a restriction that is imposed by Bessel-functions' properties. This is a physical feature that is due to the finite Larmor-radius effect $\sim\rho/a_\perp$ introduced in our analysis. Moreover, by reducing the transverse spatial extent of the wave packet $a_\perp$ ,we have seen that both $\Delta P_\varphi$ and $\Delta P_\theta$ are rather enhanced, (while the width of the resonances is preserved) whereas $\Delta P_z$ remained unchanged.

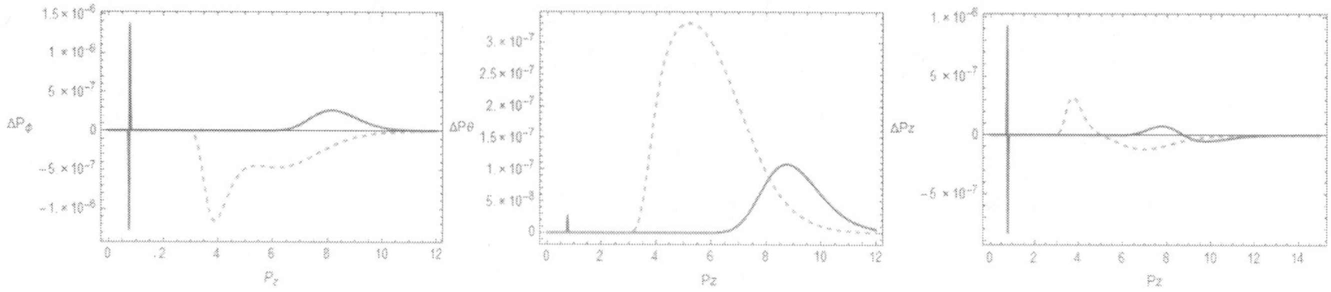


Figure 2: Variations of the averaged actions for $P_{\parallel}=1, P_{\theta}=1, k_{\parallel}=1, \omega=1.1, \mathbf{V}=0, a_{\parallel}=a_{\perp}=10^2, a_{\tau}=10^3$, $P_{\varphi}=0.5$ (solid line) and $P_{\varphi}=2$ (dashed line). In this non-degenerate case with $\omega \neq k_{\parallel}$, the resonance condition between $P_z$ and $P_{\varphi}$ has a parabolic form, which results in two resonant curves, each with different width. Since $k_{\parallel}<\omega$, it is also easy to check that both resonant curves correspond to $n=-1$. As $P_{\varphi}$ is approaching the peak of the parabola, the resonant curves tend to overlap; clearly, beyond that point, the resonance will be infeasible.

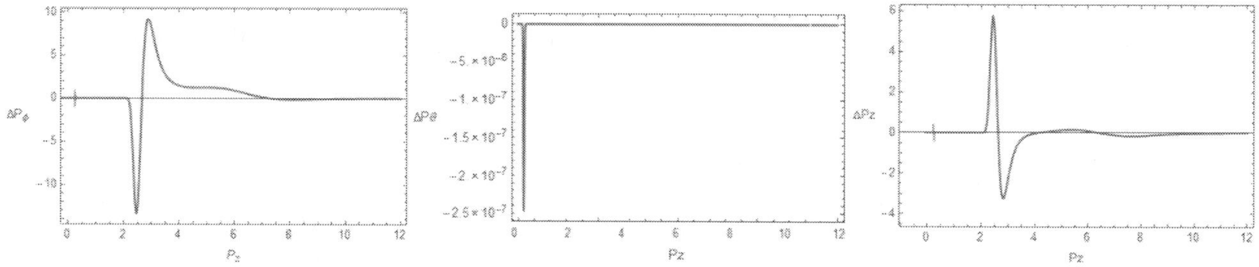


Figure 3:Variations of the averaged actions for $P_{\parallel}=0, P_{+}=1, P_{\theta}=1, k_{\parallel}=1, \omega=1.1, \mathbf{V}=0, a_{\parallel}=a_{\perp}=10^2, a_{\tau}=10^3$ $P_{\varphi}=1.5$. Introducing a circularly (e.g right-hand) polarized EM wave ($P_{\parallel}=P\_=0$) we observe that the existence of higher-order Bessel terms may enrich the spectral content of the interaction. Once again, $n\neq 0$ for $k_{\parallel}<\omega$, and from Bessel function properties the resonances correspond to $n=\{-2,-1\}$. For both $\Delta P_{\varphi}$ and $\Delta P_z$ an overlap between the resonant curves is present, whereas at the same time the amplitude of $\Delta P_{\theta}$ is negligible and only a single resonance is observable, suggesting that the interaction with a circularly polarized wave does not affect considerably the position of the GC.

Work supported in part by the Hellenic National Programme on Controlled Thermonuclear Fusion associated with the EUROfusion Consortium

## 4. REFERENCES

[1] A. J. Lichtenberg, "Regular and Chaotic Dynamics", 2nd ed., 1992

[2] Y. Kominis et al, "Interaction of charged particles with localised electrostatic waves in a magnetised plasma", Phys. Rev. E **85**, 2012

[3] Y. Kominis et al, "Quasilinear theory of electron transport by radio frequency waves and nonaxisymmetric perturbations in toroidal plasmas", Phys. of Plasmas **15**, 2008

[4] J. R. Cary, "Lie Transform Perturbation Theory for Hamiltonian Systems", Phys. Reports 79, No2, 1981.

# Ion-acoustic rogue waves in a magnetized plasma with a q-nonextensive electron velocity distribution: Application to solar wind plasma.

M. Bacha[(1)], M. Shahmansouri[(2)], M. Tribeche[(1)]

[(1)]*Plasma Physics Group Faculty of sciences-Physics, University of Bab-Ezzouar Algiers, Algeria*

[(2)]*Department of Physics, Faculty of Science, Arak University, Arak 38156-88349, Iran*

## 1. Introduction

Ion-acoustic rogue waves (IARW) are addressed in a two-component plasma composed of positively charged fluid ions, as well as electrons obeying the q-nonextensive distribution function in the presence of the external magnetic field **B** with application to the solar wind. The basic set of hydrodynamic equations is presented, and the nonlinear Schrödinger (NLS) equation is derived by using the reductive perturbation method. The electron nonextensivity, the obliqueness and the magnitude of the magnetic field are found to modify the properties of the rogue waves. Our results may aid to explain and interpret the nonlinear oscillations (IARW) that may occur in the solar wind plasma.

## 2. Theoretical model

Let us consider a collisionless, magnetized plasma having ions fluid, nonextensive electrons of densities $n_i$ and $n_e$, respectively. The ion fluid velocity $\mathbf{V_i}=V_{ix}\,\mathbf{e_x}+V_{iy}\,\mathbf{e_y}+V_{iz}\,\mathbf{e_z}$ and the external magnetic field $\mathbf{B}=B_0\cos\theta\,\mathbf{e_x}+B_0\sin\theta\,\mathbf{e_z}$. We assume that the wave length $\lambda \ll$ the s-species particle gyro- radius $\rho_s=Vts/\Omega_s$, (s=e,i) where $V_{ts}=(T_s/m_s)^{1/2}$ and $\Omega_s=eB_0/m_s$ is the thermal velocity and the gyration frequency of s- species particle, respectively. The nonlinear dynamics of the ion- acoustic waves (IAWs), whose phase speed is much smaller (larger) than the electron (ion) thermal velocity, is governed by the normalized basic equations

$$\frac{\partial N_i}{\partial T}+\frac{\partial (N_i V_{ix})}{\partial X}=0 \tag{1}$$

$$\frac{\partial V_{ix}}{\partial T}+V_{ix}\frac{\partial V_{ix}}{\partial X}=-\frac{5\sigma}{3}N_i^{-\frac{1}{3}}\frac{\partial N_i}{\partial X}+\omega_{ci}V_{iy}sin\theta-\frac{\partial \Phi}{\partial X}. \tag{2}$$

$$\frac{\partial V_{iy}}{\partial T}+V_{ix}\frac{\partial V_{iy}}{\partial X}=-\omega_{ci}[V_{ix}sin\theta-V_{iz}cos\theta]. \tag{3}$$

$$\frac{\partial V_{iz}}{\partial T}+V_{ix}\frac{\partial V_{iz}}{\partial X}=-\omega_{ci}V_{iy}cos\theta. \tag{4}$$

$$\frac{\partial^2 \Phi}{\partial X^2}=N_e-N_i. \tag{5}$$

With [1] $$N_e = [1 + (q-1)\Phi]^{\frac{1}{q-1}+\frac{3}{2}}. \quad (6)$$

The parameter q stand for the strength of electron nonextensivity. $\sigma=T_i/T_e$, $T_i$ and $T_e$ are the ion and electron temperatures, respectively. The density and velocity are normalized by the unperturbed ion density $n_{i0}$ and ion-acoustic speed $C_{si}=(k_BT_e/m_i)^{1/2}$, respectively. The electrostatic potential $\Phi$ is normalized by the thermal potential $k_BT_e/e$; e is the magnitude of the electron charge. The time T and the space X variables are normalized by the ion Debye length $\lambda_{Di}=(k_BT_e/4\pi e^2 n_{i0})^{1/2}$ and the ion plasma period $\omega_{pi}^{-1}=(m_i/4\pi e^2 n_{i0})^{1/2}$, respectively. To examine the modulation of the ion-acoustic waves (IAW) propagating in our plasma, we employ the reductive perturbation technique to derive the nonlinear NLS equation. The independent variables [2]- [6] $\xi=\varepsilon(X-V_gT)$ and $\tau=\varepsilon^2 T$ where the small parameter $\varepsilon$ is to be regarded as a measure of the strength of the nonlinearity and $V_g$ is the group velocity of the wave to be determined later. Furthermore, all physical quantities appearing in Eqs. (1- 6) are expanded as $F(X,T) = F_0 + \sum_{m=1}^{\infty} \varepsilon^m \sum_{l=-\infty}^{l=\infty} F_l^{(m)}(\xi,\tau)e^{il\Theta}$, where $\Theta$=kX-ωT,

$F_l^{(m)} = [N_{il}^{(m)}, V_{ixl}^{(m)}, V_{iyl}^{(m)}, V_{izl}^{(m)}, \Phi_l^{(m)}]$ and $F_l^{(0)} = [1,0,0,0,0]^T$. Here k and ω are real variables representing the fundamental (carrier) wave number and frequency, respectively. $F_l^{(m)}$ satisfies the reality condition $F_l^{(m)}= F_l^{(m)*}$ and the asterisk denotes complex conjugate. Substituting these expression along with stretching coordinate into Eqs. (1)- (5) and collecting terms of the same power of ε. The first order approximation (m=1) with the first harmonic (l=1) yields the following dispersion relation

$$D(k,\omega) = 3\omega^4 + \omega^2[-3\omega_{ci}^2 - 5k^2\sigma - 3k^2/(k^2+s)] + \omega_{ci}^2 cos^2\theta[5k^2\sigma + 3k^2/(k^2+s)].$$

Whereas the second order approximation (m=2) with the first harmonic (l=1) gives the expression of the group velocity

$$V_g = \frac{\partial\omega}{\partial k} = \frac{2\alpha^2[5\sigma(k^2+s)^2 + 3s]}{3(k^2+s)^2[2\omega\alpha^2 + \omega\omega_{ci}^4 sin^2\theta cos^2\theta]}$$

Where s=(3q-1)/2 and $\alpha = \omega^2 - \omega_{ci}^2 cos^2\theta$. The third-order (m=3) with (l=1) gives an explicit compatibility condition from which one can derive the following nonlinear Schrödinger (NLS) equation

$$i\frac{\partial\Phi}{\partial\tau} + \frac{P}{2}\frac{\partial^2\Phi}{\partial\xi^2} + Q|\Phi|^2\Phi = 0. \quad (7)$$

We have denoted $\Phi_1^{(1)}=\Phi$. The dispersion and nonlinearity coefficients denote respectively P and Q are given by

$$P = -\frac{2V_gP_1}{\omega k^2[10\sigma(k^2+s)^2+3(k^2+s)-3V_g(k^2-s)]} \quad (8)$$

$$Q = -\frac{3V_g(k^2+s)^2Q_1}{[10\sigma(k^2+s)^2+3(k^2+s)-3V_g(k^2-s)]} \quad (9)$$

The NLS equation (Eq. 7) has a rational solution that is located on a nonzero background and localized in τ and ξ directions [6] as

$$\Phi = \sqrt{\frac{P}{Q}}\left[\frac{4(1+2iP\tau)}{1+4P^2\tau^2+4\xi^2} - 1\right]e^{iP\tau}.$$

## 3. Numerical results and discussion

For the numerical analysis of the present findings, the typical parameters of the solar wind [7] are be considered.

| $r/r_\odot$ | 215 | 10 | 1.5 | 1.03 |
|---|---|---|---|---|
| n($m^{-3}$) | 7 x $10^3$ | 4 x $10^9$ | 1.6 x $10^{13}$ | 2 x $10^{14}$ |
| B(*Tesla*) | 5 x $10^{-9}$ | $10^{-6}$ | 0.5 x $10^{-4}$ | $10^{-4}$ |

Here r and $r_\odot$ represents the distance Sun-Earth and the solar radius, respectively, which is $r_\odot \simeq 7\times10^8$ meters. The variation of P/Q as a function of k for different values of nonextensive parameter q is shown in Figure **1**. This figure shows that the instability sets in when $k<k_c$ and the wave remains stable when $k>k_c$. Besides, it can be seen that the critical value $k_c$, where the instability sets in $P/Q\to\pm\infty$, increases when the nonextensive parameter increases ($q\to1$), i. e., as the electrons evolve toward their thermodynamical equilibrium. The influence of the electrons nonextensivity on the IA rogue waves is displayed in Figures **2** and **3** where we have plotted the wave envelope Φ for nonextensive parameter q=0.9 (Fig. **2**) end q=0.8 (Fig. **3**). Our result depicts that, the amplitude of rogue pulses decreases with the increase of q ($q\to1$), when the electrons evolve toward their thermal equilibrium. The profile of the ion-acoustic rogue waves, which may propagate in solar wind plasma's parameters within the unstable region, for the different physical parameters corresponding to a different distance ($r/r_\odot$) away from the Sun are depicted in the figure **4** for time τ=0. It can be seen that the amplitude of the IA rogue pulses increases with the increase of the distance $r/r_\odot$. As the plasma moves away from the Sun, it would lead to grow the amplitude of the rogue waves.

## 4. Conclusion

To conclude, we have investigated the modulational instability of the envelope ion-acoustic waves in a two-component plasma composed of positively charged fluid ions, as well as nonextensive electrons in the presence of the external magnetic field $B_0$, with application to the solar wind. By using the fluid model and employing the reductive perturbation method, a nonlinear Schrödinger equation has been derived. Our results show that the rogue waves may propagate for plasma parameters within the unstable region. Interestingly, we found that the IA rogue waves may be affected by the electrons nonextensivity depending on whether the parameter q. We also found that as the solar wind moves away from the Sun, a significant concentration amount of energy makes the pulses taller. Our investigation may be of wide relevance to astronomers and space scientists working on the solar wind and interstellar plasmas where nonthermal distributions are turning out to be a very common and characteristic feature.

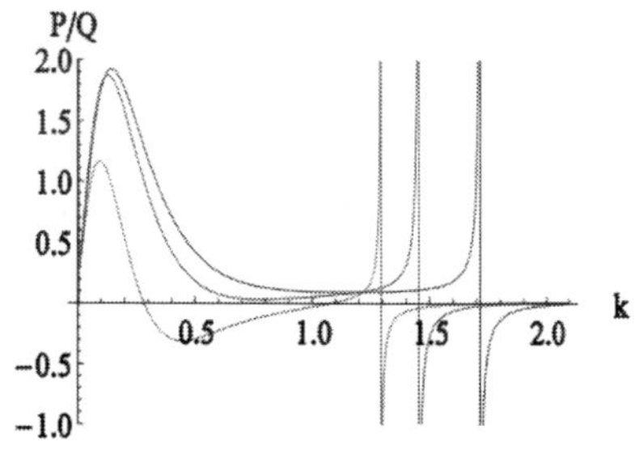


**Figure 1:** Variation of P/Q with k for q=1 (black line), q=0.85 (blue line) and q=0.7 (red line). For the physical parameters of the solar wind to a distance r/ r⊙=215.

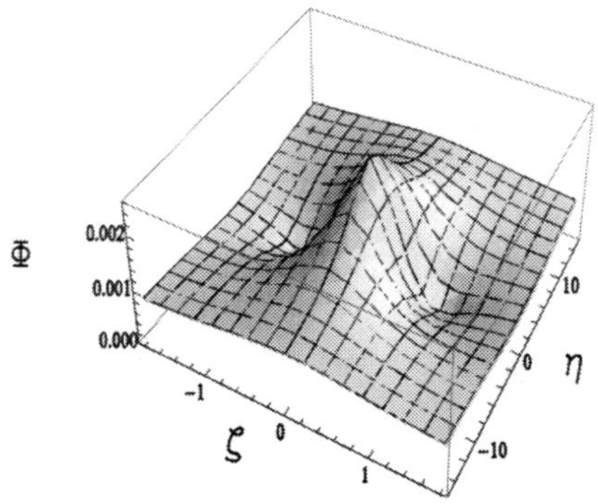


**Figure 2:** The rogue wave profile Φ vs ξ and τ for q=0.9, σ=0.2, k=1.2, θ=20° and r/ r⊙=215.

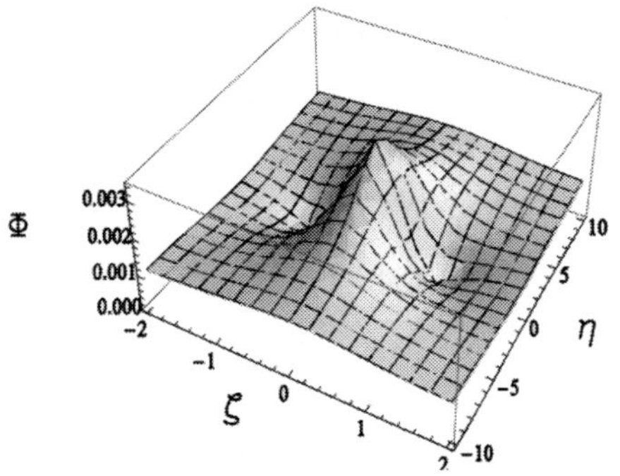


**Figure 3:** The rogue wave profile Φ vs ξ and τ for q=0.9, σ=0.2, k=1.2, θ=20° and r/ r⊙=215.

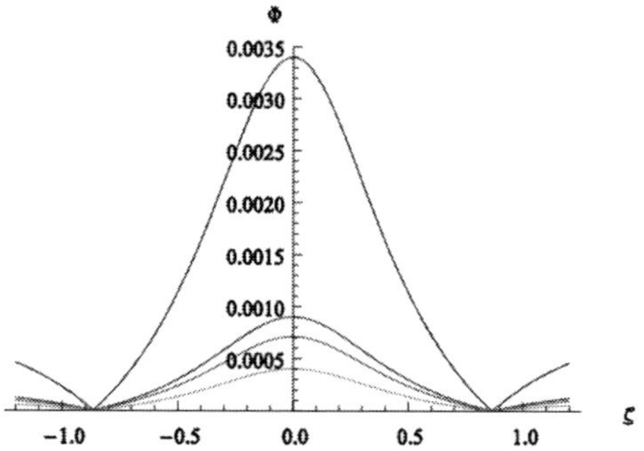


**Figure 4:** The special profile rogue wave Φ vs ξ for τ=0, q=0.8, σ=0.2, k=1.2, θ=20° for different r/ r⊙=215 (black line), r/ r⊙=10 (blue line), r/ r⊙==1.5 (red line) and r/ r⊙=1.03 (green line).

## References


[1] M. Bacha, M. Tribeche, Astrophys. Space Sci. **337**, 253 (2012).

[2] T. Taniuti and N. Yajima, J. Math. Phys. **10**, 1369 (1969).

[3] N. Asano, T. Taniuti and N. Yajima, J. Math. Phys. **10**, 2020 (1969).

[4] T. Kawahara, J. Phys. Soc. Jpn. **35**, 1537 (1973).

[5] R. Sabry, W. M. Moslem, P. K. Shukla, and H. Saleem, Phys. Rev. E **79**, 056402 (2009).

[6] A. Ankievicz, N. Devine, and N. Akhmediev, Phys. Lett. A **373**, 3997 (2009).

[7] G. K. Parks, Physics of Space Plasmas. An Introduction (University of California, Berkley, 1984).

# PIC simulations of collisionless shocks in laboratory-scaled mini magnetospheres

F. Cruz[1], E.P. Alves[1], R.A. Bamford[2], R. Bingham[2,3], R. Fonseca[1,4], L.O. Silva[1]

[1] *GoLP/Instituto de Plasma e Fusão Nuclear, Instituto Superior Técnico, Universidade de Lisboa, 1049-001 Lisboa, Portugal*

[2] *RAL Space, STFC, Rutherford Appleton Laboratory, Harwell Oxford, Didcot OX11 0QX, UK*

[3] *University of Strathclyde, Glasgow, Scotland, UK*

[4] *DCTI/ISCTE-Instituto Universitário de Lisboa, 1649-029 Lisboa, Portugal*

**Abstract**

Available computational resources and laboratory plasma streams allow for an *ab initio* approach to the formation of collisionless shocks on the interaction between plasmas and magnetic obstacles of sizes on the order of the plasma kinetic scales. In this work, we resort to full particle-in-cell simulations to identify the critical obstacle size to generate a compressed plasma region ahead of these objects and show that their effective size depends on the relative orientation between the plasma's and the obstacle's magnetic fields.

## Introduction

Interest has recently risen in the study of mini magnetospheres, mainly motivated by the observation of crustal magnetic anomalies on the lunar surface. The Moon does not possess a global magnetosphere or a bow shock like the Earth [1]. Interestingly, however, it does have localized regions of magnetic field of magnitude $B \sim 10-100$ nT over extensions of $L \sim 100-1000$ km [2]. The typical ion gyration radius around the solar wind magnetic field is of the order of $\rho_i \sim 100-1000$ km at 1 AU, i.e. it is comparable to the lunar magnetic obstacles' spatial scales [3, 4]. A similar scenario has been recently obtained experimentally [5]. Although available laboratory plasma streams have absolute parameters far from those of the solar wind (e.g. plasma density $n \sim 10^{14}-10^{15}$ cm$^{-3}$ and $B \sim 10^{-1}$ T), these flows can share similar Alfvénic Mach numbers $M_A \sim 1-10$ and ratios $\rho_i/L \sim 1$, which allows for the investigation of scaled down space and astrophysical interactions between plasmas and magnetic fields in the laboratory. Correctly modelling the formation of mini magnetospheres and of magnetized, collisionless shocks thus implies understanding the kinetic-scale phenomena of the plasma. Whilst the global MHD dynamics of multi-scale magnetospheric systems is well understood, kinetic-scale physics such as finite Larmor radius effects on magnetized, collisionless bow shocks remain largely unexplored. Particle-in-cell (PIC) simulations play a critical role in this effort since they can capture the important microphysical processes underlying the formation of small-scale magnetospheres [6], as well as of collisionless shocks.

In this work, we perform PIC simulations of a magnetized plasma colliding with a dipolar magnetic field that stops the flow at a distance compared to its kinetic scales. We describe the microphysics of mini magnetospheres and determine the critical obstacle size for the formation of shocks in different 2D simulation planes.

**PIC Simulations**

The numerical experiments presented in this work were performed using OSIRIS [7], a massively-parallel PIC code. OSIRIS operates in normalised plasma units, the independent variable being the plasma density $n_0$. Distances are normalised to the electron skin depth $c/\omega_{pe}$ (where $c$ is the speed of light and $\omega_{pe} = \sqrt{4\pi n_0 e^2/m_e}$ is the plasma frequency, with $e$ and $m_e$ representing the electron charge and mass, respectively) and times are normalised to the inverse of the plasma frequency $1/\omega_{pe}$.

In all the simulations, we use a cold plasma flow with reduced ion-to-electron mass ratio $m_i/m_e = 100$ and speed $v_0 = 100 v_{the}$, where $v_{the}$ is the electron thermal velocity. By using these parameters (the realistic mass ratio is 1836 and the typical solar wind flow velocity is of the order of $10^{-3}c$), we can significantly reduce the computational effort to perform the numerical experiments, although it is still possible to infer about the microphysical properties of these systems. The grid resolution is 10 cells/($c/\omega_{pe}$). The simulation domain is filled with the plasma internal magnetic and electric fields $\mathbf{B}_{\mathrm{IMF}}$ and $\mathbf{E}_{\mathrm{IMF}}$ such that $\mathbf{E}_{\mathrm{IMF}} + \mathbf{v_0} \times \mathbf{B}_{\mathrm{IMF}} = 0$. The magnitude of $\mathbf{B}_{\mathrm{IMF}}$ is chosen such that the flow has an Alfvénic Mach number $M_A = v_0/v_A$, where $v_A = B_{\mathrm{IMF}}/\sqrt{4\pi n_0 m_p}$. The dipole magnetic moment $\mu$ is chosen such that the plasma ram pressure equals the magnetic pressure at a certain distance $L$, comparable to the plasma kinetic scales, given, to first order, by $L \simeq \left(\mu^2/2 n_0 m_i v_0^2\right)^{1/6}$.

Figure 1 shows the results of 2D simulations where the interaction between a plasma of $M_A = 2$ and a dipolar magnetic field can be observed for different dipole intensities. Both the dipolar and the internal plasma magnetic fields point out of the simulation plane. The dipole magnetic moment increases from left to right, such that the density cavities have a size (a) $L < d_i, \rho_i$, (b) $d_i < L < \rho_i$ and (c) $d_i, \rho_i < L$, where $\rho_i$ is the ion Larmor radius defined by $\rho_i = m_i c v_0 / e B_{\mathrm{IMF}} = M_A d_i$. For $L < d_i, \rho_i$ (Fig. 1 a), the dipolar structure can only perturb the plasma creating a faint wake behind it. After the particles are deflected, they are rotated by the internal plasma magnetic field with a Larmor radius $\rho_i > L$, i.e. the plasma is not able to pile up in front of the cavity size and create a compressed (shocked) region of magnetic field. A similar result is observed for $d_i < L < \rho_i$ (Fig. 1 b), even though compressed plasma regions show an

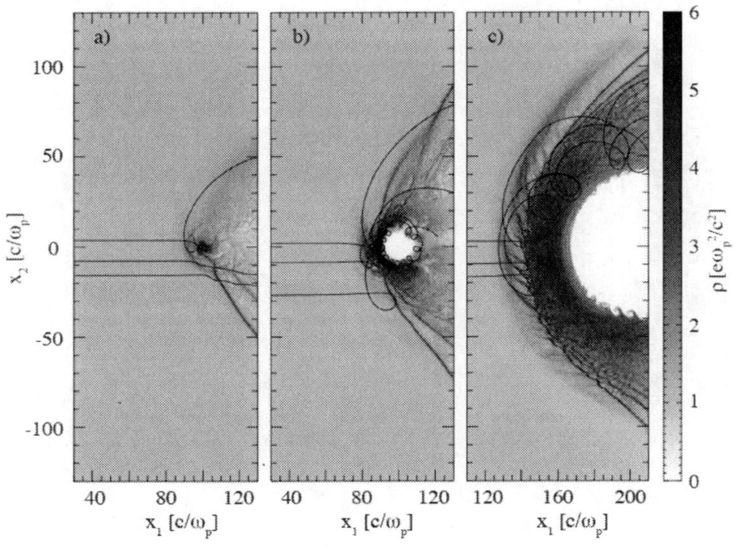


Figure 1: *Critical cavity size for formation of miniature magnetized shocks. A plasma flow with* $v_0 = 0.2c$ *and* $M_A = 2$ *interacts with magnetic dipoles that standoff the plasma at a distance (a)* $L = 0.25\rho_i$, *(b)* $0.75\rho_i$ *and (c)* $2.5\rho_i$. *The black lines represent ion trajectories and illustrate the typical Larmor radius scale after the particles are reflected.*

oscillatory dynamics ahead of the magnetic obstacle. In the case where $L > d_i, \rho_i$ (Fig. 1 c), the plasma ions can be reflected in front of the magnetic obstacle and thus counter-stream with the unperturbed flow, leading to the generation of turbulence via the modified two-stream instability [8]. A curved shock front, clearly identified by the sharp transition between the compressed and unperturbed plasma regions, is formed ahead of the density cavity. These results indicate that $\rho_i$ is the critical kinetic scale that determines the shock formation in mini magnetospheres.

The magnetopause position is controlled by the balance between the plasma ram and the total magnetic pressure. Since the plasma is highly conductive, it compresses the dipolar magnetic field such that its internal magnetic field remains constant. Opposite orientations of $\mathbf{B}_{\mathrm{IMF}}$ can change the total magnetic pressure profile close to the magnetopause and thus inflate or deflate the density cavity. Although these changes may not be relevant in large-scale (e.g. planetary) systems, we find that for mini magnetospheres such inflation/deflation can be on the order of 100% the cavity size for low $M_A$ flows and critically determine the formation of collisionless shocks. We find that the magnetic pressure gradient required to stop the plasma flow occurs farther from the dipole for anti-parallel $\mathbf{B}_{\mathrm{d}}$ and $\mathbf{B}_{\mathrm{IMF}}$ when compared to the opposite relative orientation (see Fig. 2 a, b). This difference is more evident when $\mathbf{B}_{\mathrm{IMF}}$, $\mathbf{B}_{\mathrm{d}}$ point out of the simulation plane. In an in-plane configuration (see Fig. 2 c, d), the field lines can reconnect in front of the magnetopause when $\mathbf{B}_{\mathrm{IMF}}$ and $\mathbf{B}_{\mathrm{d}}$ are anti-parallel. In this case, the simple pressure

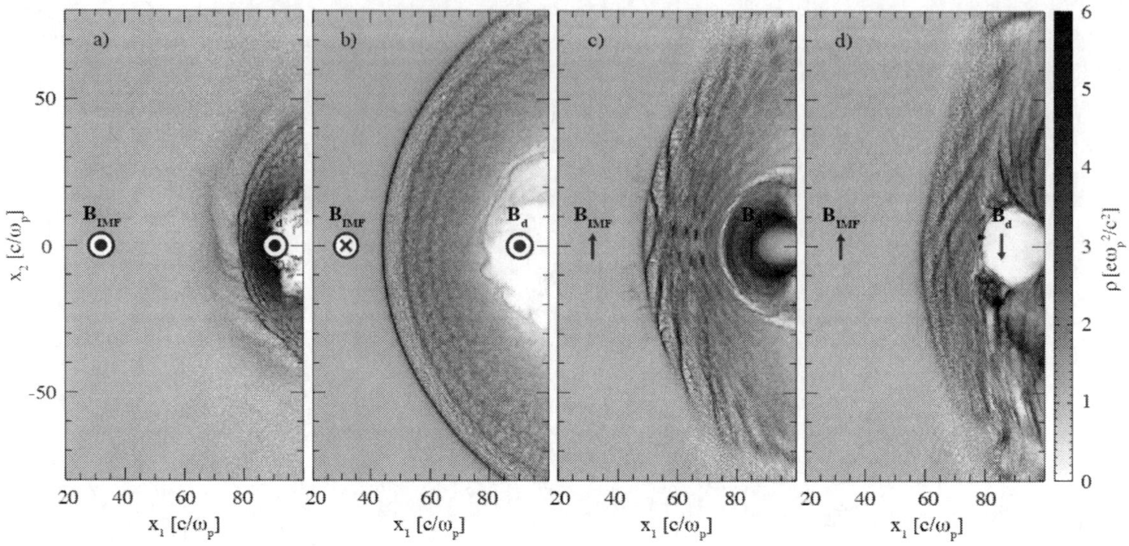


Figure 2: *Effective cavity size is sensitive to* $\mathbf{B}_{IMF}$ *orientation. A plasma flow with* $v_0 = 0.1c$ *and* $M_A = 1.5$ *is collided with a dipolar B-field parallel (a, c)/anti-parallel (b, d) to* $\mathbf{B}_{IMF}$*. The dipolar and plasma internal magnetic fields are set up in out and in-plane configurations in panels (a, b) and (c, d), respectively. The plasma is stopped at a distance* $L = 2d_i$ *according to the macroscopic pressure balance.*

gradient argument presented above is not sufficient to estimate the magnetopause position. In 3D, the interplay between the two planes will be critical to determine the cavity size. This will be investigated in a future publication.

## Conclusions

We show that the critical obstacle size for the formation of shocks in mini magnetospheres is $L > \rho_i$. This result confirms that full PIC simulations are critical to describe the formation of collisionless shocks resulting from the interaction between plasmas and magnetic obstacles. We also observe strong variations of the effective obstacle size depending on the plasma magnetic field orientation. This shows that the global system behaviour can be determined by kinetic-scale physics and suggests that both space and laboratory systems may be highly mutable. Work supported by the European Research Council (Accelerates ERC-2010-AdG 267841).

## References

[1] D.S. Colburn *et al.*, Science **158**, 3804 (1967)
[2] R.P. Lin *et al.*, Science **281**, 5382 (1998)
[3] R.A. Bamford *et al.*, Physical Review Letters **109**, 81101 (2012)
[4] R.A. Bamford *et al.*, arXiv:1505.06304v1 (2015)
[5] P. Brady *et al.*, Physics of Plasmas **16**, 4 (2009)
[6] J. Deca *et al.*, Physical Review Letters **112**, 151102 (2014)
[7] R.A. Fonseca *et al.*, Lecture notes in Computer Science **2331**, pp. 342-351 (2002)
[8] J.B. McBride *et al.*, Physics of Fluids **15**, pp. 2367-2383 (1972)

# POD analysis of heat transport in the solar corona from SDO images

Diana Gamborino[1], Diego del-Castillo-Negrete[2] and Julio J. Martinell[1]

[1] *Instituto de Ciencias Nucleares, UNAM, A. Postal 70-543, México D.F., Mexico*

[2] *Oak Ridge National Laboratory, Oak Ridge, TN, USA*

**I. Introduction.** The solar corona is heated by some mechanism that is still not totally clear but is likely related to upward propagating Alfvén waves carrying energy from the photosphere or to dissipation of magnetic energy by magnetic reconnection. These processes release energy mainly in active regions and then the energy has to propagate to other regions in order to maintain the whole corona at the observed high temperatures. The global heat transport is very much influenced by the complex magnetic field geometry which couples distant regions of the solar corona. This would be expected to produce nonlocal effects in the heat transport leading to non-diffusive processes. Exploring this possibility requires, among other things, a systematic quantitative method to analyze the spatio-temporal evolution of the temperature from observational data. The approach proposed in this paper is based on the use of proper orthogonal decomposition (POD) methods. We analyze images of the solar corona taken by the Solar Dynamics Observatory (SDO) over a period of time following an explosive event focusing on the evolution of a heat pulse. The images are first processed with the SolarSoft package to obtain maps of the Emission Measure and of the temperature over the full solar disk. Then, the region of interest is selected and it is analyzed using two techniques based on POD methods. These are especially well suited for time dependent events.

**II. Images and temperature maps.**

The event analyzed took place on 31/8/2012 at around 20:00 UT. There were seven active regions on the visible solar disk. Region 1562 was the originator of the solar flare having strength C8.4. Using the combination of six filters for the coronal emission as measured by AIA instrument of SDO, the dual maps for the EM and temperature are obtained, following the techniques developed in [1]. The differential emission measure $DEM = d[EM(T,x,y,z)]/dT = n^2 dh/dT$ is used to describe the temperature distribution of plasma emitting along the line of sight $h$ and is calculated using the intensities $F_\lambda(x,y)$ of six EUV lines from different ionization states of iron (Fe VIII, IX, XII, XIV, XVI, XVIII), covering the coronal range from $0.6MK \leq T \leq 16MK$. AIA records a full set of near-simultaneous images in each temperature filter with a fixed cadence of 12 seconds. The DEM distribution can be reconstructed from the six filter fluxes $[F_\lambda(x,y)]$ (in each pixel), by inverting the equation $F_\lambda(x,y) = \int DEM(x,y,T) R_\lambda dT$, where $R_\lambda(T)$ is th filter response. This process is simplified with the method of multi-Gaussian functions in the case of six-filter data, which is a forward-fitting technique with parameterized DEM distribution functions [1]. The temperature maps were computed for a time sequence covering 16.4 minutes with a cadence of 12 seconds, as provided by the AIA data. The EM (left) and temperature (right) maps are shown in Figure 1 for a given time, together with a zoom to our region of interest for a sequence of times.

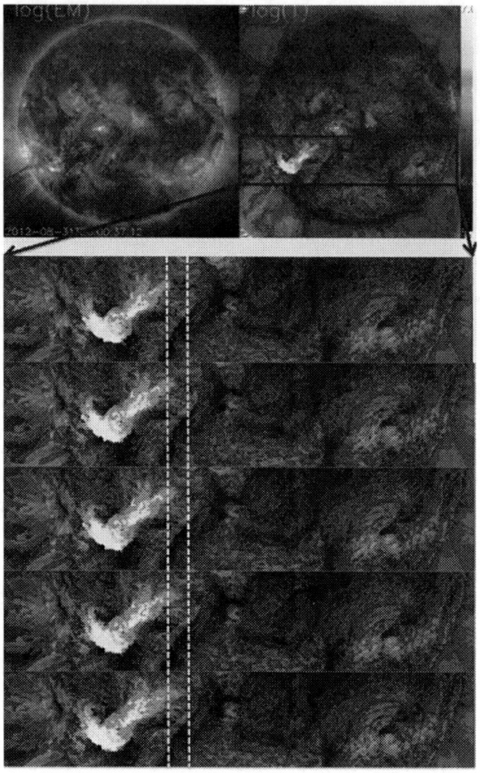

Figure 1: EM and temperature maps with zoom at different times of the region of the heat front.

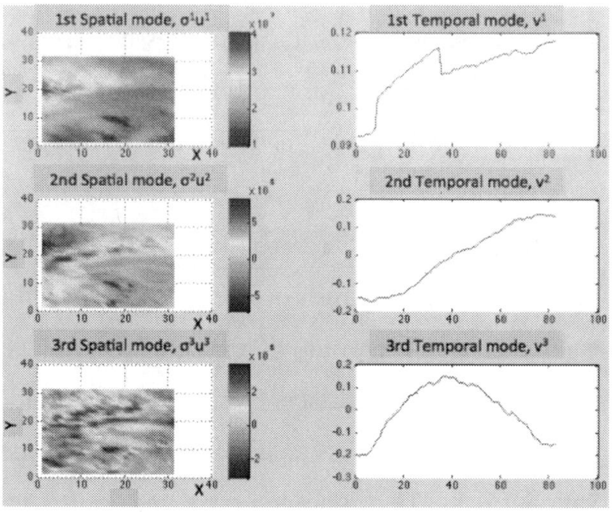


Figure 2: Spatial modes $u_k(r_i)$ (left) and temporal modes $v_k(t_j)$ (right) for $k$ = 1, 2, 3 in topos-chronos analysis. Space modes are "refolded" to a 2D map

## III. POD methods of analysis.

The Singular Value Decomposition (SVD) is, generally speaking, a mathematical method based on matrix algebra that allows us to build a new basis in which the data is optimally represented. It is a powerful tool because it helps us extract dominant features and coherent structures that might be hidden in the data by identifying and sorting the dimensions along which the data exhibits greater variation. In this, a matrix $\mathbf{A}$ of dimension $m \times n$ is decomposed as

$$\mathbf{A} = \mathbf{U}\Sigma\mathbf{V}^T \quad \text{or} \quad A = \sum_{i=1}^{r} \sigma_i u_i v_i^T$$

where **U** and **V** are unitary square matrices and $\Sigma$ is diagonal of dimension $m \times n$. The second expression is the tensor product of two vectors where $r$ =rank($A$). Here $u_i$ and $v_i$ are the $i$-th columns of **U** and **V** respectively. The usefulness of this repserentation is that if a matrix is to be approximated by another of lower rank $k$, the Eckart-Young Theorem assures that

$$A_k = \sum_{i=1}^{k} \sigma_i u_i v_i^T \tag{1}$$

is the optimal (in the L2 norm sense) rank-$k$ approximation of $A$.

SVD is applied to the matrix of the 2D temperature map in the rectangular grid $(x_i, y_j)$, where $i = 1, \ldots, N_x$ and $j = 1, \ldots, N_y$, i.e. $A \to T(x_i, y_j; t)$. The maps change as a function of time. The maximun number of elements in the SVD expansion (eq. (1)) is, therefore, $k = \min\{N_x, N_y\}$. Two different methods based on SVD are applied to extract specific information.

**Topos-Chronos.** The first method separates temporal and spatial dependencies of the temperature to reveal the dominant structures that underlie the process [2]. Since the whole data is a 3D array $T_{ijk} = T(x_i, y_j, t_k)$ $(k = 1, \ldots, N_t)$, it has to be reduced to a 2D matrix by "unfolding" the 2D spatial domain into a 1D vector, i. e. $(x_i, y_j) \to r_n$ with $n = 1, \ldots, N_x \times N_y$. Then, SVD is applied to the space-time 2D matrix $T_{nk} = T(r_n, t_k)$ and the trucation to rank $k$ of the representation is

$$A_{ij}^k = \sum_{m=1}^{k} \sigma^m u^m(r_i) v^m(t_j). \tag{2}$$

where $1 \leq k \leq N^* \equiv \min\{N_x N_y, N_t\}$. The graphic representation of the topos and chronos modes can be seen in Fig. 2 for different ranks.

Using this, it possible to make an estimate of the thermal conductivity associated with a diffusive process in the corona. First, the thermal flux entering the region of interest can be calculated using the time derivative of $v^k(t)$. Then, the spatial gradient along the $x$ axis is computed from $u^k(r)$ and used to determine the heat flux from Fourier's Law ($\vec{q}_T = \kappa \nabla k_B T$) in terms of the *thermal conductivity* ($\kappa$). Equating the fluxes, $\kappa$ can be calculated. In computing $\vec{q}_T$ we have to include the fact that a convective flux $\vec{q}_u$ is also present. Thus, the total flux is $\vec{q}_{Tot} = \vec{q}_u + \vec{q}_T$. From the heat transfer equation and assuming that the heat enters the volume only through the left boundary (the others being negligible), it is possible to show that the flux $q_1$ from this side is

$$\frac{q_1}{k_B}|_{t_j} = \frac{3}{2}\frac{n_e L}{N^2} \sigma^1 \left[ \sum_{i=1}^{N_x N_y} u^1(r_i) \right] \frac{dv^1}{dt}|_{t_j} \quad . \tag{3}$$

including only the lowest mode ($r = 1$, providing an adequate representation), where $N = N_x = N_y$, $L$ is the box size and $n_e = 10^{15} m^{-3}$ the plasma density. This is a function of time but its average value is $\langle q_1 \rangle_t = 2.23 \times 10^3 \mathrm{J\,m^{-2} s^{-1}}$. On the other hand, the convective flux is determined by the pulse speed which is calculated by following an isocontour of temperature in time obtaining $\langle u_x \rangle_t = 7.56 \times 10^3 \mathrm{ms}^{-1}$. For a temperature $T$ =2.5MK, the average convective flux is $\langle q_u \rangle_t = n_e T \langle u_x \rangle = 260 \mathrm{J\,m^{-2} s^{-1}}$. The corresponding value of $\kappa|_t = (q_1|_t - q_u)/|\nabla k_B T|$ averaged over time is $\langle \kappa \rangle_t = 3.02 \times 10^{28} m^{-1} \cdot s^{-1}$. Compared to the collisional value $\kappa_{col} = 3.2 \frac{n k_B T_e \tau_e}{m_e} = 1.315 \times 10^{28} m^{-1} \cdot s^{-1}$ for parallel transport, it is 2.3 times larger, but of the same order.

**GLRAM.** The second method, called Generalized Low Rank Approximation of Matrices (GLRAM) [3, 4], iteratively computes the matrices that minimize the error in the approximation of the matrix $A_i$ by a low range decomposition product, as given by $(min \sum_{i=1}^{n} ||A_i - L M_i R^T||_F^2)$. In each iteration the error is reduced achieving a fast convergence. The crucial difference with SVD is the way to represent the data. Here, the unitary orthogonal bases $L$ and $R$ do not change in time and the time information is stored in the matrices of singular values $M_i$ which is not necessarily diagonal. This allows to find precise contour maps of the temperature as function of time from which we can obtain the widening of the heat pulse, which determines the type of heat transport. Fig. 3 shows the evolution of the area between two temperatures, which is a measure of the dispersion due to transport. The time scaling is of the form $A_{12} \sim \sigma^2 \sim t^{\gamma}$. To exclude advection, only the nose of the front is considered as seen in the upper panel. It is found that the exponent is $\gamma = 0.78 < 1$ indicating that the transport is sub-diffusive.

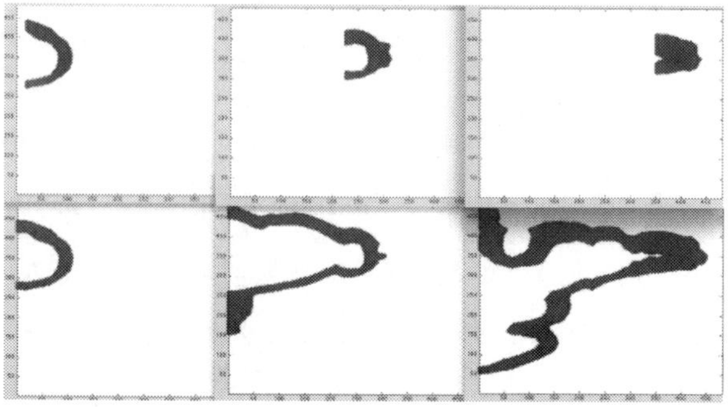

Figure 3: Snapshots of the data interpolation using GLRAM for $n_t = 1, 40, 80$. and for temperature values of 2.5MK. Up: Nose of the thermal pulse. Down: Full structure of thermal pulse.

**Acknowledgements.** This work was partially supported by Projects DGAPA-UNAM IN109115, IN106911 and CONACyT 152905.

**References:**

[1] Aschwanden et al., Solar Phys. **283**, 5 (2011).

[2] S. Futatani, S. Benkadda, and D. del-Castillo-Negrete, Phys. of Plasmas **16**, 042506 (2009).

[3] J. Ye, Mach. Learn. **61**, 167 (2005).

[4] D. del-Castillo-Negrete, S.P. Hirshman, D.A. Spong, and E.F. D'Azevedo, J. Comp. Phys.,**222**, 265 (2007).

# Fluctuation statistics in the scrape-off layer of Alcator C-Mod

R. Kube[1,2], O.E. Garcia[2], B. LaBombard[3], and J.L. Terry[3]

[1] *Institute for Applied Physics and Ion Physics, University of Innsbruck, Austrian*

[2] *Department of Physics and Technology, UiT - The Arctic University of Norway, Tromsø, Norway*

[3] *MIT Plasma Science and Fusion Center, Cambridge, MA, 02139, USA*

## Introduction

The scrape-off layer (SOL) of magnetically confined plasmas is dominated by order unity fluctuations of the particle density and large, concomitant transport events. A large body of research links these phenomena to the radial propagation of plasma filaments which are elongated along the magnetic field and highly localized in the radial poloidal plane, called *blobs* [1]. The observed characteristics of the turbulence in the far SOL plasma is further believed to be universal. For one, the conditionally averaged waveform of large amplitude events in the particle density time series presents a steep rise and a slow decay [2]. On the other hand, feature time series with frequent large amplitude events larger than zero coefficients of sample skewness and excess kurtosis. When sampled at a single radial position, it was shown that the histograms coincide upon normalization [3]. A novel stochastic model for fluctuations in the SOL predicts that, given independent arrival of plasma blobs with exponentially distributed amplitudes, particle density time series are Gamma distributed. This model was successfully used to describe intensity fluctuations, measured by gas-puff imaging in the SOL of Alcator C-Mod [4].

In this contribution, we present analysis of ion saturation current time series, sampled at the outboard mid-plane far SOL (ASP) and at the outer divertor in the Alcator C-Mod tokamak for a series of ohmically heated discharges, where the line-averaged density was between 0.15 and $0.42\overline{n}_\mathrm{e}/n_\mathrm{G}$. All discharges where performed in a lower single-null magnetic geometry with a plasma current of $I_\mathrm{p} = 0.55\mathrm{MA}$ and an on-axis magnetic field of $B_\mathrm{T} = 5.4\mathrm{T}$. The parameters of the discharges, along with the length of the time series and the position of the probe at outboard mid-plane are listed in tab. 1. For all discharges are the two outermost divertor probes at $\rho \approx 8\mathrm{mm}$ and $\approx 10\mathrm{mm}$, where $\rho$ denotes to the distance to the last closed flux surface, as mapped to outboard midplane by the magnetic field.

## Fluctuation statistics

In Fig. 1 we present the time series sampled at outboard midplane in discharge 2. The length of the time series is approx. 0.85s and the histogram spans 4 decades in normalized probability.

| Shot | $\overline{n}_e/n_G$ | $T_e$/eV | ASP position | $t_{start}$/s | $t_{end}$/s | Plot marker |
|---|---|---|---|---|---|---|
| 1 | 0.15 | 35 | near SOL | 0.75(0.75) | 1.10(1.10) | ▼ |
| 2 | 0.28 | 25 | far SOL | 0.65(0.65) | 1.50(1.50) | ◆ |
| 3 | 0.32 | 25 | far SOL | 0.80(–) | 1.10(–) | ■ |
| 4 | 0.31 | 20 | far SOL | 0.80(0.80) | 1.10(1.10) | ● |
| 5 | 0.42 | 20 | far SOL | 0.50(0.50) | 0.70(0.70) | ▲ |

Table 1: List of the plasma parameters and the time interval used for time series analysis. The numbers in parenthesis give the interval on which data from the divertor probe is used. A dash indicates that no data is available.

The best fit of a Gamma distribution on the sample yields a shape parameter of $\gamma = \overline{I}/I_{rms} = 9.77$ and a scale parameter of $\overline{I}/\gamma = 4.14 \times 10^{-3}$A. Here $\overline{\cdot}$ denotes the sample mean and $\cdot_{rms}$ the sample root mean square. This distribution gives a good description of the histogram over the entire range. Coefficients of sample skewness, $S$, and excess kurtosis, $F$, indicate a discernible deviation from normality. Fig. 2 presents histograms of the ion saturation current time series sampled by the two outermost divertor probes in discharge 1. We find that while at $\rho \approx 8$mm the fluctuations rarely exceed twice the mean of the time series, this threshold is exceeded at $\rho \approx 10$mm. From the mean and the variance of the time series we find $\gamma = 8.64(10.1)$ $\rho \approx 8(10)$mm. While a Gamma distribution over estimates the tails of the histogram in the upper panel, it gives a good description of the histogram in the lower panel.

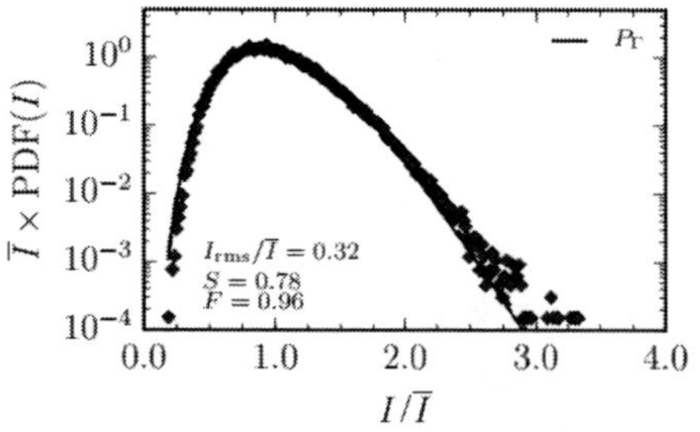


Figure 1: Histogram of the ion saturation current sampled in the far scrape-off layer during discharge 2.

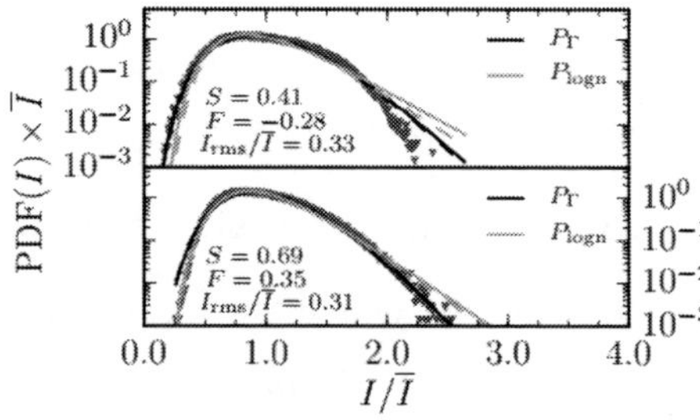


Figure 2: Histogram of the ion saturation current sampled by divertor probes at $\rho \approx 8$mm (upper panel) and at $\rho \approx 10$mm (lower panel) during discharge 1.

We continue by employing conditional averaging [5] to study the ion saturation current time series. In the following, time series are normalized according to $\widetilde{I} = \left(I - \overline{I}\right)/I_{rms}$ and $\widetilde{V} = e\left(V - \overline{V}\right)/T_e$. The waiting times between the arrival of bursts whose amplitude exceeds

2.5 times the root mean square of the time series is shown in Figs. 3 and 4. Compared to the histograms is a maximum likelihood estimate of an exponential distribution, which approximates the histogram well over more than one decade in normalized probability, for all discharges. This implies that large amplitude burst events occur independent. The time scales of the distribution vary between $\sim 120\mu$s and $\sim 260\mu$s in the far scrape-off layer and between $\sim 280\mu$s and $\sim 440\mu$s at the divertor. No systematic variation with the line-averaged plasma density is discernible.

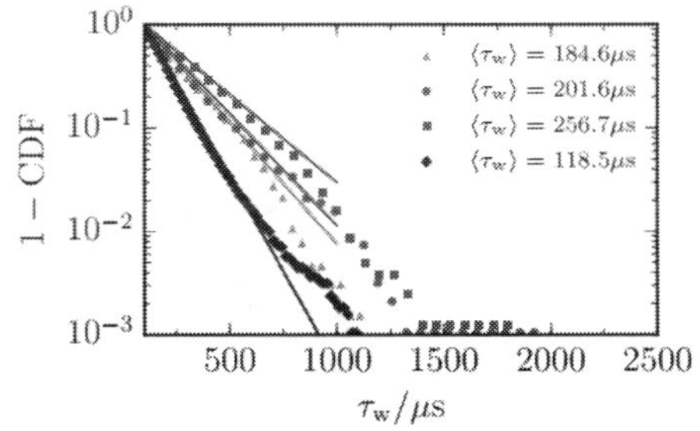


Figure 3: Histogram of waiting times between large amplitude bursts for the time series sampled at outboard mid plane.

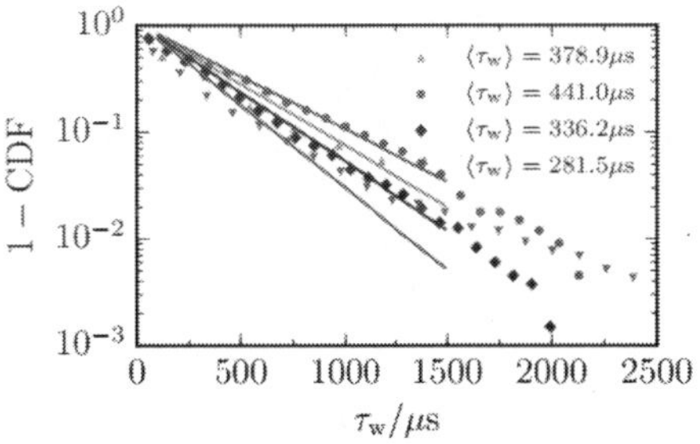


Figure 4: Histogram of waiting times between large amplitude bursts for the time series sampled by the divertor probes.

The dynamics of all time series is governed by the intermittent arrival of large amplitude bursts. This universal dynamics of the time series is reflected in Figs. 5 and 6, which presents histograms of the normalized time series. Within the same normalization, all histogram ms collide. In these figures denotes a triangle up data sampled at outboard mid plane and triangle down data sampled by the divertor probes.

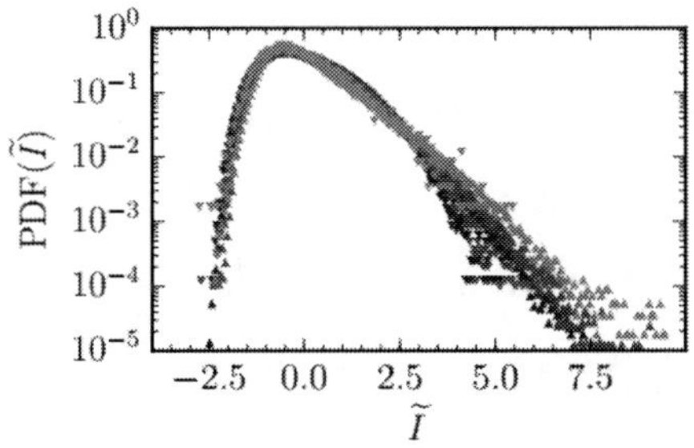


Figure 5: Histogram of all normalized $I$ time series

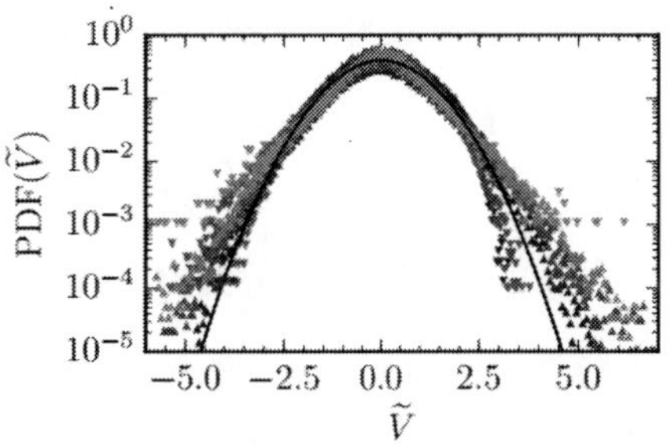


Figure 6: Histogram of all normalized $V$ time series

**Blob detection at the divertor probes**

Fig. 7 shows the conditionally averaged waveform of large amplitude bursts in time series of the ion saturation current as well as the floating potential, as sampled by the Langmuir probes embedded in the divertor. The conditional variance is shown in Fig. 8, where a value of $1-\mathrm{CV} =$

0 denotes a random waveform while $1-\mathrm{CV}=1$ indicates perfect reproducibility. For discharges 1 and 2 we find that the ion saturation current signal shows a steep front and a slow decay. This is accompanied by a dipolar waveform of the floating potential. Discharge 5 features a similar conditionally averaged waveform for $\widetilde{I}$ and a random waveform in the floating potential. This indicates that the plasma blobs are electrically disconnected from the sheaths at this line-averaged plasma density.

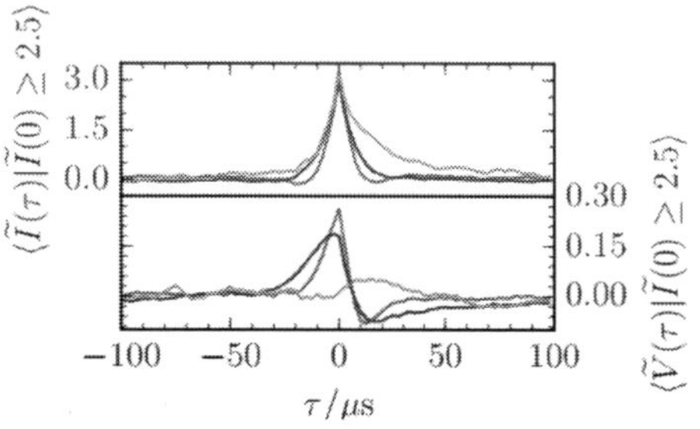


Figure 7: Conditionally averaged waveform of the ion saturation current (upper panel) and floating potential(lower panel), sampled by the outermost divertor probe.

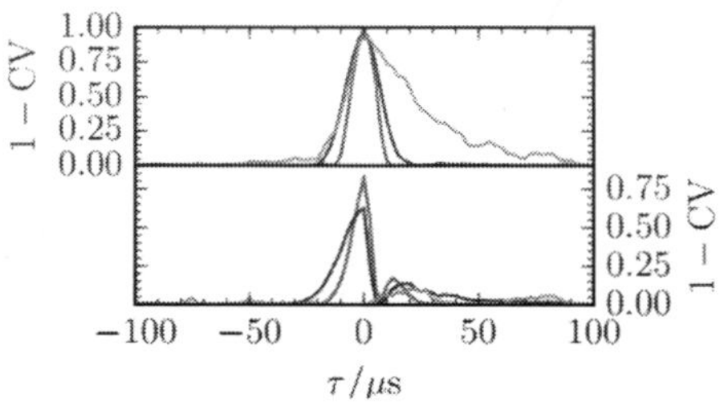


Figure 8: Conditional variance of the wave forms in fig. 7.

**References**

[1] D.A. D'Ippolito, J. R. Myra and S. J. Zweben, Phys. Plasmas **11** 060501(2011)

[2] J.A. Boedo, D.L. Rudakov. R.A. Moyer, et al. Phys. Plasmas **10**, 1670 (2001), D.L. Rudakov, J.A. Boedo, R.A. Moyer, et al. Plasma Phys. Control, Fusion **44** 717 (2002), G.S. Kirnev, V.P. Budaev, S.A. Grashin, et al. Plasma Phys. Control. Fusion **46** (2004) O.E. Garcia, J. Horacek, R.A. Pitts et al. Nucl. Fus. **47** 667 (2007)

[3] J.P. Graves, J. Horacek, R.A. Pitts and K.I. Hopcraft, Plasma Phys. and Control. Fusion **47** L1 (2005), B.P. van Milligen, R. Sanchez, B.A. Carreras, et al. Phys. Plasmas **12**, 052507 (2005), O.E. Garcia, J. Horacek, R.A. Pitts, et al. Nucl. Fus. **47** 667 (2007)

[4] O.E. Garcia, Phys. Rev. Lett. **108** 265001 (2012), O.E. Garcia, S.M Fritzner, R. Kube, et al. Phys. Plasmas **20**, 055901 (2013)

[5] F.J. Oynes, H.L. Pecseli and K. Rypdal Phys. Rev. Lett. **75** 81 (1997)

# Comparison of electron internal transport barrier formation between CHS and Heliotron J

T. Minami[1], N. Kenmochi[2], C. Takahashi[1], K.Nishioka[2],S. Kobayashi[1],
Y. Nakamura[2],H. Okada[1], S. Kado[1], S. Yamamoto[1], S. Ohshima[1], S. Konoshima[1],
G. Weir[1], Y. Ohtani[2], K. Nagasaki[1], T. Mizuuchi[1], F. Sano[1]
[1] *Institute of Advanced Energy, Kyoto University, Kyoto, Japan*
[2] *Graduate School of Energy Science, Kyoto University, Kyoto, Japan*

## Introduction

The electron internal transport barrier (eITB) of helical devices plays an important role on plasma confinement[1, 2]. This barrier is known to be formed due to the radial electric field and the electric field shear is created by the bifurcation of radial electric field ($E_r$) with the electron cyclotron resonance (ECR) heating. The positive radial electric field formation is consistent with the electron-root solution of the ambipolarity condition for $E_r$ of the neoclassical transport. In previous results of Compact Helical System (CHS), the barrier is easily formed in larger effective helical ripple configuration[3]. The barrier formation depends on the magnetic field configuration through the neoclassical transport characteristics.

Recently, the phenomena that have similar characteristics as the eITB by the ECR heating have been observed on Heliotron J, and the steep electron temperature gradient has been observed in the core region[4]. Both the Heliotron J and CHS belong to helical type devices, and both the devices have similar size ( The major and averaged minor radii of Heliotron J are 1.2 and 0.17m and that of CHS are 1m and 0.2m, respectively). On the other hand, both the devices have different magnetic configurations. Heliotron J is helical axis heliotron type, and CHS is heliotron/torsatron type, and the periodicity of Heliotron J is $(l,m) = (1,4)$ and that of CHS is $(l,m) = (2,8)$. Therefore, comparative study of the phenomena is carried out between CHS and Heliotron J to investigate the effect of the magnetic configuration on the eITB formation.

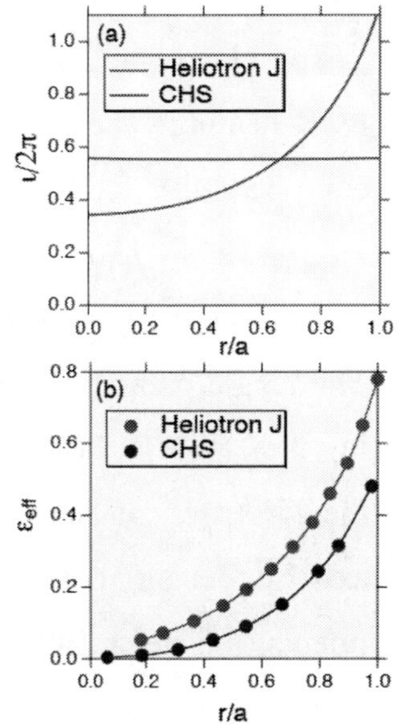


Figure 1: *Rotational transform (a) and effective helical ripple profiles (b) of standard magnetic configuration in Heliotron J and CHS.*

In this paper, the electron temperature and density profiles with eITB are compared between Heliotron J and CHS to clarify the transport characteristics. Differences of the eITB formation dependence on plasma density are presented. And the effect of the magnetic helical ripple on the eITB formation is discussed.

## Magnetic configuration characteristics in Heliotron J and CHS

The experiments have been performed on the standard magnetic configuration of both Heliotron J and CHS, and the magnetic field strength on the magnetic axis of Heliotron J is $B_{ax} = 1.25T$, and that of CHS is $B_{ax} = 0.88T$. The important difference in the magnetic configuration between both the devices is rotational transform profile[5, 6, 7]. Figure 1(a) shows the rotational transform profiles of the standard magnetic configuration of Heliotron J and CHS. The shear of the magnetic field in CHS is positive, while the shear is close to zero in Heliotron J.

The neoclassical transport of the helical plasma is characterized by the effective helical ripple ($\varepsilon_{eff}$), which characterize the helical $1/\nu$ electron transport[8]. The hypothesis of the eITB formation is that the eITB is easily formed in the larger $\varepsilon_{eff}$ magnetic configuration, because the access of the electron-root regime is easy as predicted by the neoclassical transport theory[9]. The difference of the effective helical ripple is shown in Fig.1(b)[7, 10]. The value of the $\varepsilon_{eff}$ of Heliotron J is 2-10 times larger than that of CHS.

## Comparison of eITB formation in Heliotron J and CHS

The plasma with eITB is produced by the ECR heating. The Heliotron J and CHS are equipped with 70GHz ( Injected ECR power: $P_{inj} \sim 120 - 330kW$) and 53GHZ ($P_{inj} \sim 120 - 160kW$) gyrotrons, respectively [5, 6]. The single path absorption of the ECR heating is $\sim 90\%$ in both the experiments. Both the gyrotrons can heat exactly at the magnetic axis by focusing optics. In some CHS experiments, the neutral beam ($P_{inj} \sim 620kW$) is injected (NBI) into the plasma, however, the characteristics of the eITB formation is not different from the ECR heating only plasma, because the deposited power of NBI to the electrons is smaller than the absorbed ECR power due to the low plasma density. The electron temperature and density profiles were measured with Nd:YAG laser Thomson scattering

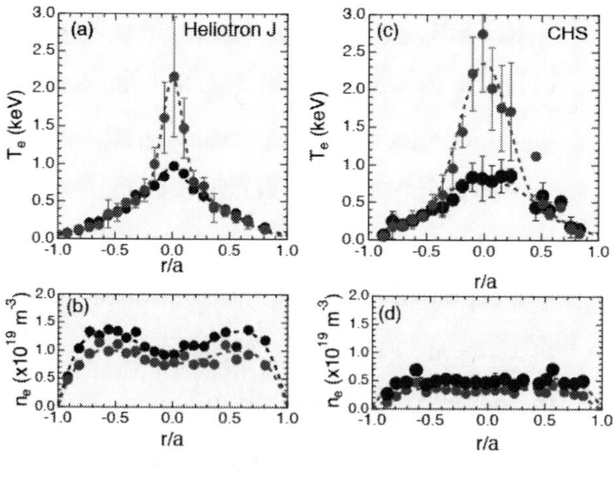


Figure 2: *Typical electron temperature (a)(c) and density profiles (b)(d) with eITB in the Heliotron J (a)(b) and CHS (c)(d). Red points denote the profiles with eITB and blue denote the profiles without eITB.*

system using the same analysis procedure[11, 12].

Figure 2 shows the typical electron and density profiles of Heliotron J and CHS with the eITB formation. Both the profiles have same characteristics. When the eITB is formed, steep electron temperature gradient is created, and peaked temperature profiles are produced in the plasma core, as shown in FIg.2(a)(c). The central electron temperature increases up to similar level of $2-2.5keV$ by the barrier formation. On the other hand, the temperatures on the outside of the peaked profiles with and without eITB are almost equal in both the CHS and Heliotron J. The peaked electron temperature is formed by small reduction of the plasma density, as shown in Fig.2(b) (d). These results show the confinement improvement by the barrier formation in the core region and the confinement degradation due to the profile resilience on the outside of the peaked temperature region[4]. However, both the results have the different electron density when the eITB is formed. The density ($\bar{n}_e \sim 1.2\times10^{19}m^{-3}$) of the Heliotron J is approximately two times larger than that ($\bar{n}_e \sim 0.5\times10^{19}m^{-3}$) of the CHS.

**Density dependence of eITB formation in Heliotron J and CHS**

Figure 3 shows the density dependence on the temperature gradient in the core region and in the outside region of the peaked temperature region. In both plasmas, when the line averaged density is lower than the threshold value, the temperature gradient in the core region increases. However, the threshold electron density in Heliotron J ($1.2\times10^{19}m^{-3}$ at $P_{inj} \sim 330kW$) is two times larger than that in CHS ($0.5\times10^{19}m^{-3}$ at $P_{inj} \sim 130kW$). Consequently, the plasma density regime in which eITB is formed is expanded in Heliotron J.

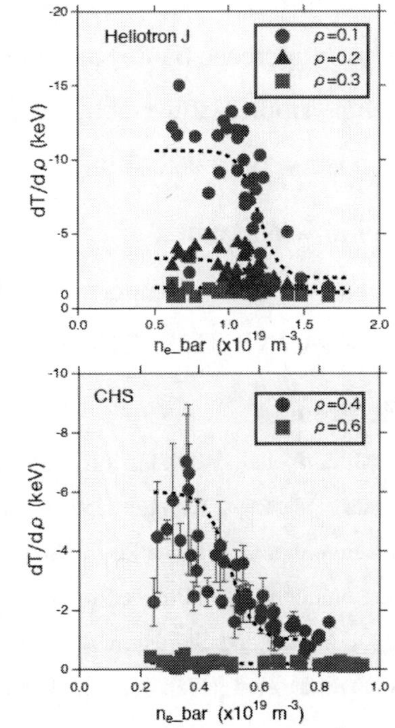


Figure 3: *Density dependence on the temperature gradient in the core region and the outside of the peaked temperature for Heliotron J (a) and CHS (b)*

It is important to take account in the power difference between both the experiments, because the barrier formation depends on the ECR power[4, 9]. Figure 4(a) shows $T_e(0)$ dependence on the injected ECR power ($P_{inj}$) that is normalized by the line averaged density ($\bar{n}_e$). In this figure, the closed and open circles show the Heliotron J and CHS results and the red and blue circles show the profiles with and without the peaked temperature, respectively. Although the threshold value of $P_{inj}/\bar{n}_e$ for the barrier formation is almost equal in both the results, the larger $T_e(0)$ is achieved by the smaller $P_{inj}/\bar{n}_e$ in Heliotron J compared to CHS.

Figure 4(b) shows $T_e(0)$ dependence on the electron collisionality normalized by the bounce frequency at $\rho = 0.2$ ($\nu_h^*$). The $\nu_h^*$ is associated with the ion-root to electron-root transition[2]. The $\nu_h^*$ of the Heliotron J plasma easily reach the collision-less regime compared to CHS due to the larger $\varepsilon_{eff}$. This is because the bounce frequency is higher in Heliotron J due to the higher $\varepsilon_{eff}$. Accordingly, there is a possibility that eITB is easily formed in Heliotron J. However, the eITB formation is realized in higher collisionality in CHS compared to Heliotron J. It shows that the eITB formation is not dominated by the collisionality alone.

## Summary

The comparative study of the eITB formation is carried out between CHS and Heliotron J to investigate the effect of the magnetic configuration on the barrier formation. The threshold electron density for the barrier formation in Heliotron J is two times larger than that in CHS, and the larger $T_e(0)$ is achieved by the smaller $P_{inj}/\bar{n}_e$ in Heliotron J. These results show the possibility of the threshold density increase by the collisionality which related to the effective helical ripple. However, the results also show that the eITB formation is not only determined by the collisionality.

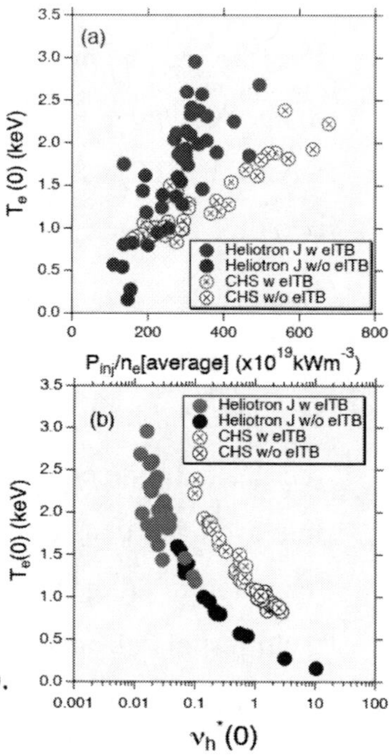


Figure 4: $T_e(0)$ *dependence on the* $P_{inj}/\bar{n}_e$ *(a) and* $\nu_h^*(0)$ *(b). Closed and open circles show Heliotron J and CHS results, red and blue circles show plasma with and without eITB, respectively.*

## Acknowledgments

This work was supported by the NIFS collaborative Research Program (NIFS10KUHK030, NIFS09KUHL, and NIFS10KUHL033).

## References

[1] T. Minami et al., Nucl. Fusion, **44** (2004) 342-349

[2] K.Ida et al., Phys. Rev. Lett., **91**(8) (2003) 085003-1-4

[3] T. Minami et al., Plasma Phys. Control. Fusion, **46** (2004) A285-A290

[4] N. Kenmochi et al., Proceedings of the EPS conference, p.5.131 (2015)

[5] T. Obiki et al., Nucl. Fusion, **41**, No. 7 (2001)

[6] S.Okamura et al.,Nucl. Fusion, **39**, No. 9 (1999)

[7] M.F.Heyn et al., Plasma Phys. Control. Fusion, **43** (2001) 1311-1324

[8] C.D.Beidler and W.N.G.Hitchon, Plasma Phys. Control. Fusion, **36** (1994) 317-353.

[9] M.Yokoyama et al., Nucl. Fusion, **47**, (2007) 1213-1219

[10] F.Sano et al., Plasma Fusion Res., **5** (2010) S2003

[11] T. Minami et al., Rev. Sci. Instrum., **81**, 10D532 (2010)

[12] N.Kenmochi et al., Plasma and Fusion Res., **8** (2013) 2402117.

# Turbulence stabilisation by parallel flows for the DIII-D shortfall case

Y. Camenen[1], F.J. Casson[2], P. Manas[1], A.G. Peeters[3]

[1] *CNRS, Aix-Marseille Univ., PIIM UMR7345, Marseille, France*

[2] *CCFE, Culham Science Centre, Abingdon, OX14 3DB, UK*

[3] *University of Bayreuth, Physics Department, Bayreuth, Germany*

This paper explores the combined impact of toroidal flow $u$, parallel flow shear $u'_{\parallel}$ and perpendicular flow shear $\gamma_E$ on turbulent transport. The study started in the frame of a multi-code benchmark effort based on the DIII-D L-mode shortfall case [1] and was motivated by the unexpectedly strong impact of the toroidal flow on the non-linear heat fluxes observed with the gyrokinetic code GKW [2]. The simulations presented here are performed for the experimental parameters of the DIII-D shortfall case at $r/a = 0.8$ with the electron temperature gradient set to zero, which corresponds to the benchmark case focusing on ion-scale turbulence. Pure toroidal rotation was assumed to compute the $E \times B$ and parallel flow shear from the measured toroidal rotation profile. The corresponding input parameters, given in Table 1, are typical of the edge of an L-mode plasma with moderate NBI heating. In Table 1, the toroidal flow $u = R_0\omega_\varphi/v_{\rm thi}$ and flow shear $u' = -R_0^2/v_{\rm thi}\partial\omega_\varphi/\partial r$ are normalised to the ion thermal velocity $v_{\rm thi} = \sqrt{2T_i/m_i}$ and $R_0$ is a reference major radius. Electromagnetic perturbations ($\phi$ and $A_{\parallel}$), kinetic electrons and

| $R/L_{Ti}$ | $R/L_{Te}$ | $T_e/T_i$ | $R/L_n$ | $u$ | $u'$ | $\gamma_E$ | $\nu_{\rm eff}$ | $Z_{\rm eff}$ | $\beta$ | $\varepsilon$ | $q$ | $\hat{s}$ |
|---|---|---|---|---|---|---|---|---|---|---|---|---|
| 7 | 0 | 0.85 | 3 | -0.15 | -1.25 | -0.2 | 1.45 | 1.33 | $8.10^{-4}$ | 0.28 | 2.8 | 2.05 |

Table 1: *Reference input parameters based on the DIII-D shortfall case at $r/a = 0.8$ and given here in GKW normalised units (see [2] for details on the normalising conventions).*

collisonal pitch-angle scattering are included in the simulations performed with the flux-tube version of GKW. The magnetic equilibrium is specified using the Miller parametrisation.

The destabilising effect of the parallel flow shear $u'_{\parallel}$ and the stabilising effect of the toroidal flow $u$ on the toroidal ITG are well known and have been explored in several studies. Their combined effect, however, has been left mostly unexplored. To tackle this point, a simple dispersion relation emphasising the respective contributions of $u$ and $u'_{\parallel}$ on the linear growth rate is first presented before moving to the numerical results obtained with GKW.

Following Ref. 3, a 1-point (low field side midplane) fluid model is built from the first three moments of the linearised gyrokinetic equation in the local $\delta f$ approximation. The electron response is considered to be adiabatic and finite Larmor radius effects are neglected. In the limit of marginally unstable modes ($\gamma > 0$ and $\gamma \ll \omega_R$ with $\gamma$ the mode growth rate and $\omega_R$ the mode frequency, respectively) and of small parallel symmetry breaking ($u+k_{\parallel N} \ll 1$ with $k_{\parallel N}$ a normalised parallel wave vector), a simple dispersion relation is obtained. This dispersion relation describes the toroidal ITG including flows effects. For $R/L_n = 0$, it leads to the following growth rate:

$$\gamma = k_\theta \rho_i \frac{v_{\text{thi}}}{R_0} \sqrt{2R/L_{Ti} - \frac{49}{3} + 12(u+k_{\parallel N})u'_\parallel - 52(u+k_{\parallel N})^2} \quad (1)$$

This equation is readily obtained starting from Eq. (13) and (16) of Ref. 3 in which the details of the derivation of the fluid model and the normalisations are outlined. Eq. (1) indicates that the mode growth rate depends on the parallel mode structure via the term $k_{\parallel N}$. Owing to the simplicity of the model, a meaningful value of $k_{\parallel N}$ cannot be determined self-consistently: it would at least require to relax the 1-point approximation and the adiabatic electrons assumption. Representative values can nevetheless be assessed from gyrokinetic simulations allowing a qualitative discussion while keeping the model as simple as possible. Toroidal flow and parallel flow shear are known to break the parallel symmetry of the gyrokinetic equation and to generate finite values of $k_{\parallel N}$. This resulting $k_{\parallel N}$ is mostly proportional to $u$ and $u'_\parallel$, with a slope of opposite sign, as shown in Fig. 3 of Ref. 4. Taking this scaling into account, Eq. (1) can be rewritten as:

$$\gamma = k_\theta \rho_i \frac{v_{\text{thi}}}{R_0} \sqrt{2R/L_{Ti} - \frac{49}{3} + a{u'_\parallel}^2 - buu'_\parallel - cu^2} \quad (2)$$

which emphasizes the quadratic dependence of the mode growth rate on $u$ and $u'_\parallel$. The coefficients $a$ and $b$ are usually positive and $c$ is always positive. At $u = 0$, the destabilising effect of the parallel flow shear is recovered, as is the stabilising effect of toroidal rotation at $u'_\parallel = 0$. At finite $u$ and $u'_\parallel$ anything can happen, with $u$ and $u'_\parallel$ having a stabilising or destabilising effect depending on their respective values. This qualitative picture is confirmed in linear gyrokinetic

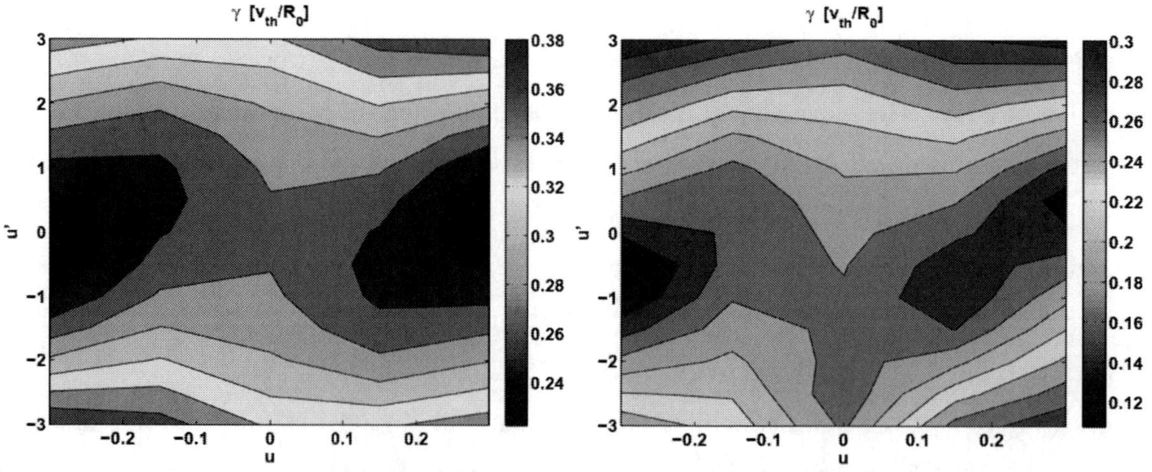


Figure 1: *Linear mode growth rate as a function of the toroidal rotation u and parallel flow shear $u'_\parallel$ at $k_\theta \rho_i = 0.4$ for $k_r \rho_i = 0$ (left plot) and $k_r \rho_i = 0.2$ (right plot).*

simulations for the reference case at $k_\theta \rho_i = 0.4$ (close to the most unstable mode), as shown in Fig. 1a. Running the same simulations for a finite radial wavevector $k_r \rho_i = 0.2$ shows an overall decrease of the mode growth rate, as expected, but also a significant modification of the $u$ and $u'_\parallel$ dependencies, Fig. 1b. This is understood considering that a finite radial wavevector shifts the maximum of the electrostatic potential away from the low field side midplane with a corresponding contribution to $k_{\parallel N}$ independent of $u$ and $u'_\parallel$ and a subsequent modification of Eq. (2).

To investigate how the dependencies of the mode growth rate on $u$ and $u'_\parallel$ affect the non-linear heat fluxes, non-linear simulations are performed for the reference case varying the flow parameters around their nominal values. The perpendicular dynamics is described in Fourier space with 21 poloidal wavevectors ($k_\theta \rho_i = 0$ to $1.36$) and 339 radial wavevectors ($k_r \rho_i = -24.2$ to $24.2$). Finite differences are used in the parallel direction with 32 points along the field line. The velocity space is discretized with 16 $\mu$ points and 64 $v_\parallel$ points. The simulations are run until a converged time average of the heat fluxes is obtained, which depending on the simulations requires between 400 to 900 $R_0/v_{\mathrm{thi}}$ after the non-linear overshoot. In Fig. 2a, the non-linear

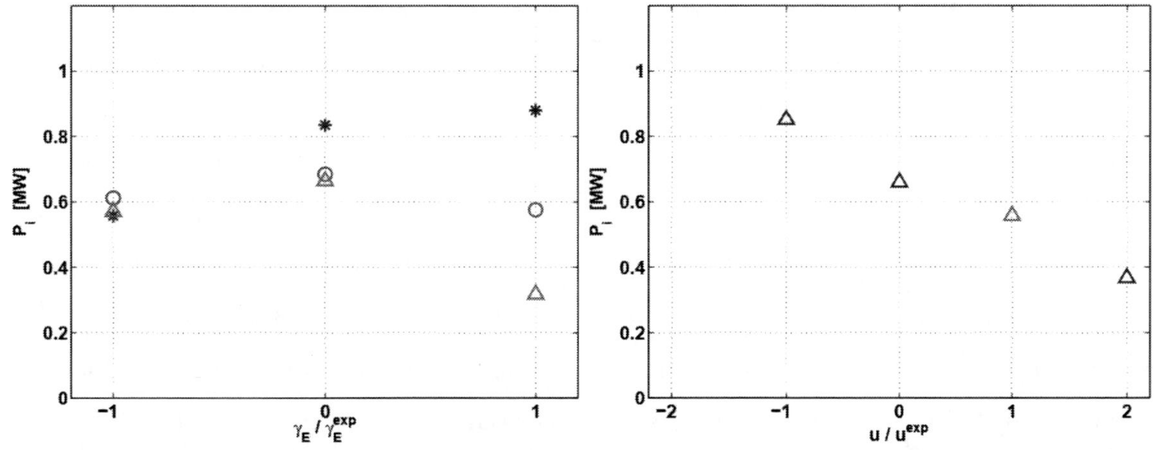


Figure 2: *Left plot: non-linear radial ion heat flux ($P_i = Q_i V'$, with $V$ the flux surface volume) as a function of the $E \times B$ shearing rate $\gamma_E$ for $u'_\parallel = u'_{\parallel exp}$ (blue stars), $u' = 0$ (red circles) and $u' = -u'_{exp}$ (magenta triangles). Right plot: $P_i$ as a function of the toroidal rotation $u$ for the case $\gamma_E = -\gamma_E^{exp}$ and $u'_\parallel = -u'_{\parallel exp}$.*

ion heat flux obtained in these simulations is shown as a function of the $E \times B$ shearing rate for different values of the parallel flow shear. The toroidal rotation is kept at the nominal value: $u = u^{\mathrm{exp}}$. For $u'_\parallel = 0$ (red circles), the ion heat flux is maximum at $\gamma_E = 0$ and decreases at finite $\gamma_E$, irrespectively of its sign, in agreement with the conventional picture. In contrast, at finite parallel flow shear, $u'_\parallel = \pm u'_{\parallel \mathrm{exp}}$, the ion heat flux is no longer symmetric with respect to

$\gamma_E = 0$ and the heat flux reduction depends on the respective signs of $u'_\parallel$ and $\gamma_E$. At $u'_\parallel = u'_{\parallel\exp}$, an increase of the $E \times B$ shearing rate from zero to its nominal value even results in a slight increase of the ion heat flux. Quantitatively, the heat flux reduction obtained when simply changing the sign of the flow and flow shear parameters can be very large: at $\gamma_E = \gamma_E^{\exp}$, changing the sign of $u'_\parallel$ results in a reduction by a factor of almost 3 of the ion heat flux. A similar behavior is observed for the electron heat flux. In Fig. 2a, the different dependency on $u'_\parallel$ obtained at positive or negative $\gamma_E$ is induced by the finite toroidal rotation. To highlight this point, the ion heat flux is shown in Fig. 2b as a function of the toroidal rotation $u$ for the case with $u'_\parallel = -u'_{\parallel\exp}$ and $\gamma_E = -\gamma_E^{\exp}$. An almost linear dependence of the ion heat flux on the toroidal rotation is observed with an increase by about 50% when the direction of the toroidal flow is reversed. These results are qualitatively consistent with the linear picture: the strongest reduction of the linear growth rate and non-linear ion heat flux are obtained when the flow parameters conjugate to generate a large parallel symmetry breaking. The effect obtained in the non-linear simulations is however somewhat stronger than what one would expect from the linear stability analysis.

Turning now to the relevance of this mechanism in the experiments, there are a two main points to be made. The first one is that the DIII-D shortfall case is an L-mode plasma with moderate NBI injection and therefore rather modest flow values. Scaling up the flow parameters, an effect even stronger than what is observed in Fig. 2 would be expected from the linear stability results. The second point concerns the possibility to experimentally decouple the flow parameters. The decoupling of $u'_\parallel$ and $\gamma_E$ depends directly on the poloidal rotation and pressure gradient. In the limit of zero poloidal rotation and pressure gradient, $u'_\parallel$ and $\gamma_E$ are strictly coupled. It is only when the radial electric field is mostly carried by the poloidal rotation or the pressure gradient that $\gamma_E$ can assume values not proportional to $u'_\parallel$. This is typically the case in transport barriers. The decoupling of $u$ from $u'_\parallel$ and $\gamma_E$ is, to some extent, easier to achieve provided the toroidal rotation is driven by NBI and that one has a knob to change the edge toroidal rotation (magnetic perturbations and/or X-point radial position for instance).

## References

[1] C. Holland *et al.*, Phys. Plasmas **16**, 052301 (2009)
`https://fusion.gat.com/theory/Shortfall`

[2] A.G. Peeters *et al.*, Comput. Phys. Commun. **180**, 2650 (2009).

[3] A.G. Peeters *et al.*, Phys. Plasmas **16**, 062311 (2009)

[4] Y. Camenen *et al.*, Phys. Plasmas **16**, 012503 (2009)

# Comparison of Ray-tracing and Full-wave Code Modeling of Helicon Heating and Current Drive in DIII-D*

R.I. Pinsker[1], N. Bertelli[2], D.L. Green[3], E.F. Jaeger[4], C.H. Lau[3], J.M. Park[3], M. Porkolab[5] and R. Prater[1]

[1]*General Atomics, P.O. Box 85608, San Diego, CA 92186-5608, USA*
[2]*Princeton Plasma Physics Laboratory, P.O. Box 451, Princeton, NJ 08543-0451, USA*
[3]*Oak Ridge National Laboratory, P.O. Box 2008, Oak Ridge, TN 37831, USA*
[4]*XCEL Engineering, Inc., Oak Ridge, TN 37830, USA*
[5]*MIT Plasma Science and Fusion Center, Cambridge, MA 02139, USA*

An experimental test of mid-radius current drive using the "helicon" (also known as the "whistler") wave at 476 MHz is being carried out on the DIII-D tokamak to determine the efficiency of coupling to this wave and to test the predicted high current drive efficiency [1]. Up to now, an important open question for this wave has been whether the radial location of the damping and current drive can be controlled by varying the launched parallel index of refraction $n_{\|}$. While ray-tracing calculations [1] predict that the radial location of the helicon damping and current drive is almost independent of the launched parallel wavenumber, very much more computationally expensive full-wave code analyses with STELION [2] have suggested a possible dependence of the radial location of the deposition on the launched parallel wavenumber. In this paper, full-wave solutions of helicon waves launched with a single toroidal mode number obtained with the AORSA code [3] are compared with ray-tracing results from GENRAY [4] in order to investigate this issue.

The reason that the ray-tracing studies show almost no dependence of minor radius of the deposition location on the launched $n_{\|}$ (within the limits imposed by wave accessibility at low $|n_{\|}|$ and practicality of launch at high $|n_{\|}|$) was explained in [5]. The whistler wave ray trajectories make only a small angle with the static magnetic field lines, so to lowest order the rays follow the field lines around the torus, with slow radial penetration. However, unlike the slow wave ("lower hybrid") branch, where that small angle between the rays and the field lines is almost independent of $n_{\|}$, leading to resonance-cone behavior, the fast branch ("helicon" or "whistler") rays make a steeper angle with the field lines as $|n_{\|}|$ is increased. The higher the value of $|n_{\|}|$, the lower the energy of electrons that are Landau resonant with the waves, so if everything were equal, the higher $|n_{\|}|$ waves would damp at lower electron temperatures and hence at larger minor radius. (This is what happens with the lower hybrid branch.) But for the whistler, the more rapid radial penetration of the rays that results from the higher $|n_{\|}|$ largely compensates for the higher damping per unit length along the trajectory, with the result being that for a range of $|n_{\|}|$ the damping occurs at roughly the same minor radius. In the high-performance DIII-D discharge studied in [1], GENRAY predicts that the damping and current drive would occur at normalized minor radius $0.55<\rho<0.6$ as the launched $n_{\|}$ is varied over $2.8<n_{\|}<4.2$. The comparison between a ray starting with $n_{\|}=3$ and

$n_{\|}$=4 is shown in Figs. 1 and 2; the steeper angle at which the ray crosses the flux surfaces at higher initial $n_{\|}$ is visible in the poloidal projection, while the absorption is stronger earlier in the ray trajectory, leading to a broader deposition profile than is obtained with the lower initial $n_{\|}$. However, the slower radial penetration the latter case yields a more rapid absorption and a more radially peaked deposition once $n_{\|}$ and the local electron temperature have risen to a value at which Landau damping sets in. The $n_{\|}$ hardly rises early in the trajectory, especially for the slowly penetrating $n_{\|}$=3 case, because the major radius of the ray decreases only slowly with the above-midplane launch position and the nearly vertical flux surfaces in the outside midplane region. The magnitude of $n_{\|}$ begins to rise only after the field line and hence the ray gets below the midplane and the flux surfaces begin to curve towards smaller major radius. (To lowest order, $|n_{\|}|R\sim$constant.)

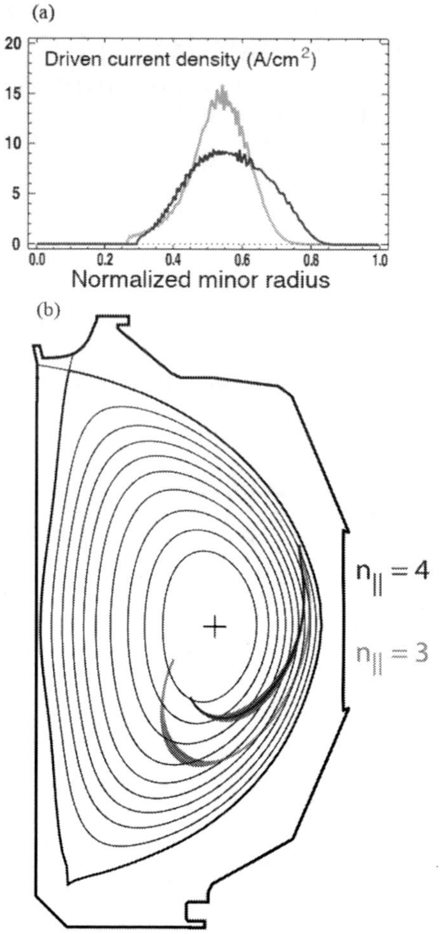


Figure 1. (a) Driven current profile as a function of normalized minor radius for ray with initial $n_{\|}$ of 3 (green) and 4 (blue), in DIII-D equilibrium 122976 calculated by GENRAY. (b) Poloidal projection of ray paths for the two cases, with the thickness of the red shaded regions proportional the rate of absorption along the ray path, (-1/P)(dP/ds).

We expect ray-tracing to be valid in much of the plasma volume, because the helicon's perpendicular wavelength is quite short compared to the scale lengths over which the plasma parameters vary, once the wave is propagating. For example, at 0.5 GHz, density of $4\times10^{19}$ m$^{-3}$, $B_T$=1.4 T, deuterium plasma at $n_{\|}$ = 3, the helicon has a perpendicular wavelength of about 1 cm ($k_{\perp}\sim$630 m$^{-1}$, or $n_{\perp}=k_{\perp}/k_0\sim$60 with $k_0=\omega/c$), while at $n_{\|}$ = 4, the perpendicular wavelength is 1.4 cm ($k_{\perp}\sim$450 m$^{-1}$, $n_{\perp}\sim$43. This is much shorter than the density gradient scale length in DIII-D except at the very edge of the plasma, where the wave has just become propagating and the ray-tracing approximation is invalid in any event. Since the ray must be started within the domain in which the local approximation is valid, the problem of excitation of the waves with a realistic launcher located outside the plasma, where the waves are evanescent, must be addressed with a full-wave calculation like AORSA. In the work presented here, however, we are interested in the wave propagation and absorption in the core, and so do not use a realistic antenna or scrape-off layer model. Calculations of antenna loading in which the antenna is placed in a model of the scrape-off layer are being performed with AORSA and are described elsewhere [6].

In spectral codes such as AORSA, practical limits on computer memory and on computation time determine how fine the computational grid in wavenumber space can be, in turn setting the limit the minimum wavelength perpendicular to the primarily toroidal static magnetic field lines that can be modeled. Since for fast waves the perpendicular wavenumber scales as $k_\perp \sim \omega/v_A \sim f n_e^{0.5} B_0^{-1}$, high frequency, high density and low toroidal field all tend to make the required $k_\perp$ high and a dense grid is needed. For this DIII-D case, the maximum density of the deuterium plasma is nearly $1\times10^{20}$ m$^{-3}$, so the maximum value of $k_\perp$ is in the range 1000-1500 m$^{-1}$ ($n_\perp$ in the range 95-143). The largest perpendicular wavenumber that can be resolved is $2\pi n_{max}/\Delta L$,

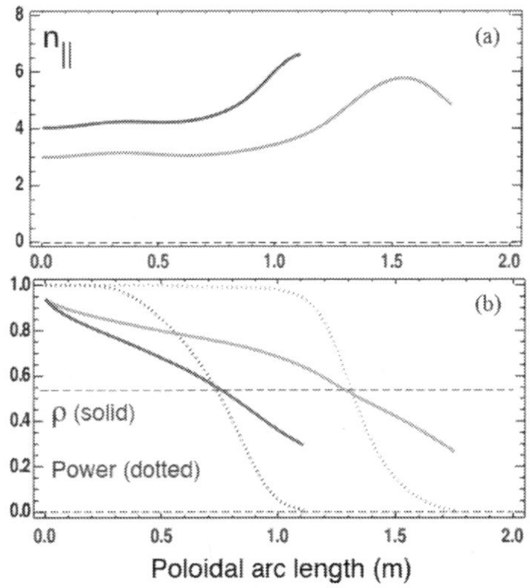

Figure 2. (a) $n_\|$ evolution along the two rays shown in Fig. 1. (b) Normalized minor radius ρ along the rays (solid lines) and the normalized power remaining in the rays (dotted). Color code same as in Fig. 1. The horizontal dashed line drawn at ρ=0.55 shows that the peak in the absorption occurs at almost the same minor radius in the two cases.

where ΔL is the dimension of the computational volume in the relevant direction and the grid in that direction has $n_{max}$ points. The finest grid that has been used so far on this problem is (400,400) (major radius, height), and those dimensions for DIII-D are roughly (ΔR, ΔZ)~(1.4 m, 2.5 m), hence maximum $(k_R,k_Z)$~(1800 m$^{-1}$,1000 m$^{-1}$). We compute the values of the perpendicular index of refraction $n_\perp$ along the GENRAY ray path with the results shown as the heavy line in the top panels of Fig. 3, where it is evident that the maximum $n_\perp$ encountered along the ray path is about 85 for the case with an initial $n_\|$=3, and a less marginal value of about 60 for the case with initial $n_\|$=4. We therefore anticipate a better agreement between AORSA results with a 400 by 400 grid and GENRAY results for the latter case. Contours of the wave electric field amplitude from the two AORSA runs are compared with the GENRAY ray path in the lower panels of Fig. 3, and indeed the maximum wave amplitudes from AORSA coincide with the ray path for the case of higher initial $n_\|$, while in the other case, the AORSA fields appear to penetrate the flux surfaces slightly more rapidly than the ray path. This was further investigated by carrying out a local Fourier transform of the AORSA wave fields along the GENRAY path and comparing the perpendicular wavenumber spectrum with the GENRAY values. This analysis, shown in the top panels of Fig. 3, verifies that the lower values of $n_\perp$ needed for the higher $n_\|$ case are faithfully reproduced by AORSA, while the AORSA spectrum saturates at a level slightly lower than the maximum $n_\perp$ required for the other case.

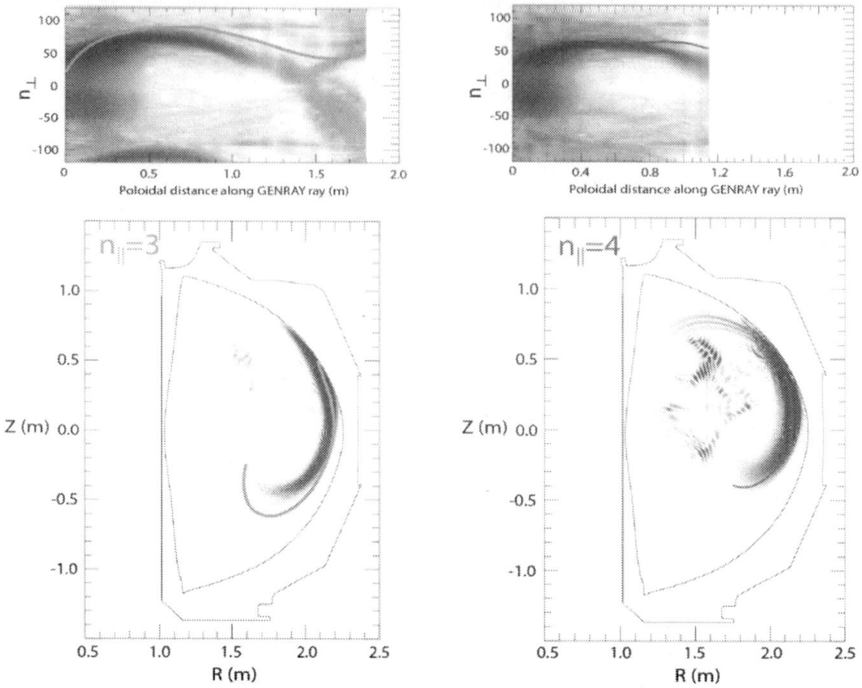


Figure 3. Comparison of ray paths from GENRAY with field ($|E_{rf}|$) contours from AORSA for initial values of $n_{||}$=3 (lower left) and $n_{||}$=4 (lower right). The upper panels show the Fourier spectrum of the AORSA fields for $n_\perp$ at each point along the GENRAY-computed ray path, parameterized by the poloidal distance along the path, compared with GENRAY's computed $n_\perp$ along the ray. Darker colors of the contours correspond to larger field or Fourier amplitudes.

Therefore good agreement between the predicted driven current profiles from the two different computational approaches for the $n_{||}$=4 case reported in [5] is consistent with these results, while to get similarly good agreement with the lower value of $n_{||}$ may require a finer grid and significantly more computer time. We conclude that in this regime, ray-tracing provides a computationally cheap and accurate tool for analysis of current drive with helicon waves, but that it should be spot-checked with the more complete full-wave model to the extent that available computational resources permit.

This work was supported in part by the US Department of Energy under DE-FC02-04ER54698, DE-AC02-09CH11466, and DE-AC05-00OR22725.

[1] R. Prater, C.P. Moeller, R.I. Pinsker, *et al.*, Nucl. Fusion **54**, 083024 (2014)
[2] V. Vdovin, Plasma Phys. Reports **39**, 95 (2013).
[3] E.F. Jaeger, L.A. Berry, E. D'Azevedo, *et al.*, Phys. Plasmas **8**, 1573 (2001)
[4] R.W. Harvey and A.P. Smirnov, *The GENRAY Ray Tracing Code*, CompX Rep. CompX-2000-01 (2001).
[5] R.I. Pinsker, R. Prater, C.P. Moeller, et al., "Off-axis Current Drive With Very High Harmonic Fast Waves for DIII-D", Paper TH/P2-38, IAEA Fusion Energy Conference, St. Petersburg, RF, 2014, to be published in the proceedings.
[6] C. Lau, E.F. Jaeger, N. Bertelli, *et al.*, in Proceedings of 21st Top. Conf. on RF Power in Plasmas, Lake Arrowhead, CA, April 27-29, 2015, to be published.

# Divertor Detachment Studies in DIII-D*

D.N. Hill[1], A.G. McLean[1], A.W. Leonard[2], S.L. Allen[1], A.R. Briesemeister[3], B.M. Covele[4], M. Groth[5], M.A. Makowski[1], T.W. Petrie[2], V.A. Soukhanovskii[1], and H.Y. Guo[2]
[1]*Lawrence Livermore National Laboratory, 7000 East Ave, Livermore, CA 94550, USA*
[2]*General Atomics, PO Box 85608, San Diego, CA 92186-5608, USA*
[3]*Oak Ridge National Laboratory, Oak Ridge, TN 37831, USA*
[4]*University of Texas, Austin, TX 78712-1192, USA*
[5]*Alto University, FI-00076, AALTO, Finland*

ITER and future burning plasma experiments require significant radiative dissipation in the edge and divertor to obtain acceptable surface heat flux (≤10MW/m$^2$) and erosion rates ($T_{e,div}$ < 5eV). The role of scrape-off layer (SOL) transport, divertor geometry, and magnetic configuration in determining the conditions required to achieve such divertor detachment remain uncertain, but has large impact on design and operation of next-step devices. Recent experiments in DIII-D have systematically examined the onset of divertor detachment using a variety of magnetic configurations and new/upgraded diagnostics including divertor Thomson scattering to produce 2D inter-ELM maps of the electron density and temperature.

**Detachment dynamics in DIII-D**

Detachment onset in DIII-D ELMing H-mode discharges is marked by a drop in divertor temperature ($T_e$≤1 eV), heat flux, and pressure as density rises in response to deuterium gas puffing from either the divertor region or the plasma midplane. The divertor Thomson system provides routine measurement of sub-eV electron temperatures between ELMs across both divertor legs by slowly sweeping the radial location of the X-point during a single discharge; the data are then mapped onto a single representative equilibrium, as shown in Fig. 1. It is possible to systematically map the progression of detachment in DIII-D because it is highly reproducible and controllable for fixed core plasma conditions, and the openness of the lower divertor means that conditions are relatively independent of strike point location. Detachment begins first at the inner divertor, followed by the outer divertor at somewhat higher density, as shown in Fig. 1. Reversing $B_{tor}$ ($B \times \nabla B$ away from the divertor) significantly reduces the in/out detachment asymmetry by decreasing the relative density at the inner target as compared to the outer; the data appear qualitatively consistent with expectations based on $E \times B$ particle drifts [1]; quantitative analysis is now underway [2].

While measurements during ELMs are excluded from the data shown in this paper, they undoubtedly affect the overall divertor equilibrium, largely through their impact on global particle balance due to finite recycling times. For fixed shape, ELM frequency increases with power and with gas puffing; typical rates are ~100Hz for the discharges analysed here.

We focus our analysis of the detachment threshold in this and following sections on two essential features beneficial to fusion energy development: reduction of peak heat flux and electron temperature at the divertor targets. The variation of peak divertor heat flux ($q_{div}$), electron temperature, density, and static electron pressure ($n_eT_e$) as a function of line-average core density is shown in Fig. 2, using data from the same discharges as the radial sweeps in Fig. 1 ($I_p$=0.9MA, $B_t$=-1.75T, $P_{beam}$=3.7MW). The ion saturation current as measured by a divertor Langmuir probe near the outer strike point and total radiative loss along the outer leg (determined from 2D reconstruction of bolometer data) are also shown.

The outer leg strike point data in Fig. 2 span a wide range of divertor operating regimes. Below $\bar{n}_e \approx 4.5-4.8\times10^{19} m^{-3}$ , $n_{e,target}$ in both divertors is low and $T_{e,target}$ is high. With

*This work was performed in part under the auspices of the U.S. Department of Energy by LLNL under DE-AC52-07NA27344, General Atomics under DE-FC02-04ER54698, and ORNL under DE-AC05-00OR22725.

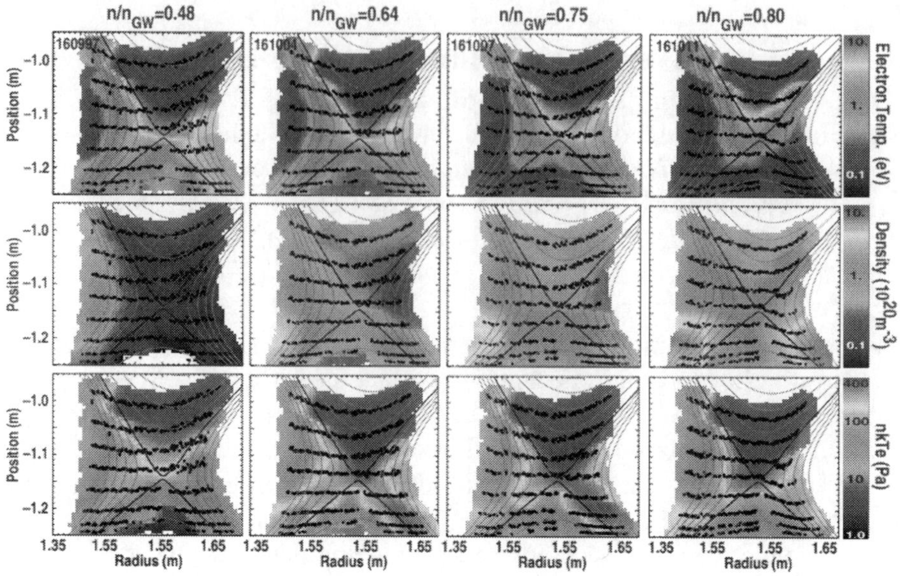


Fig. 1. 2D maps of inter-ELM $T_e$, $n_e$, and $P_e$ at increasing core plasma density (L-R: 3, 4.5, 5.5, 8.5×10$^{19}$m$^{-3}$).

increasing density, high recycling is established at the inner target, increasing $n_e$ and sharply reducing $T_e$ at both divertor targets. At densities between $4.8-5.2\times10^{19}m^{-3}$, high recycling is established at the outer target and static electron pressure increases dramatically. Further gas puffing raises divertor density and radiative loss continue still higher as $T_e$ and divertor heat flux fall even as pressure balance is preserved. Above $\bar{n}_e \approx 6\times10^{19}m^{-3}$ ( $n_{e,div} > 20\times10^{19}m^{-3}$ ), pressure begins to fall and finally, when $T_e$ drops below 3eV near the target, the ion flux drops significantly – the divertor leg is detached from the outer target. At this time, the radiative loss is peaked in the outer SOL near the X-point and there is only a small change (~10-15%) in confinement and pedestal parameters ($n_{e,ped}$, $n_{e,sep}$, $T_{e,sep}$, or $l_{Te,mp}$). Careful analysis of $\nabla T_{e,\parallel}$ and radiative emissivity along the outer divertor leg shows that ion convection dominates parallel heat transport in the detached state, extending the region over which impurity radiation can cool the plasma [3].

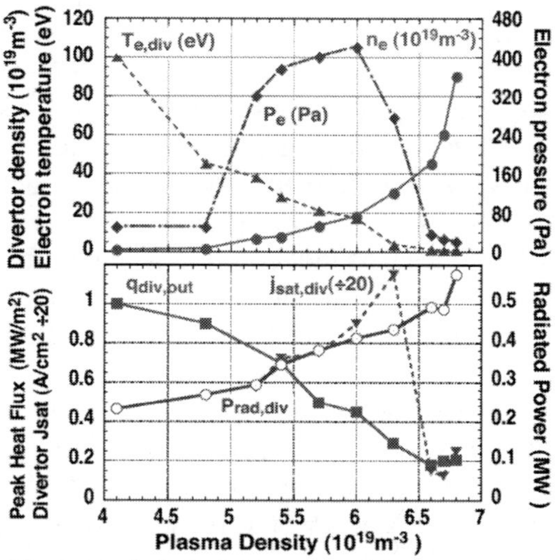


Fig. 2. Variation of plasma conditions at the outer divertor during a multi-shot density scan. Thomson at +1cm above target.

## Detachment in Unit-recycling Helium Discharges without Carbon Radiation

Experiments in helium plasmas in DIII-D provide insight into detached divertor operation in future tokamaks with high temperature metal walls (no hydrocarbon formation and unit recycling). Helium operation in DIII-D eliminates chemical sputtering of carbon, increases ionization mean free paths in the divertor, reduces charge exchange and recombination rates compared to deuterium, and eliminates molecular-hydrogen effects.

Initial "detachment" experiments in L-mode helium plasmas with He-beam heating showed sharp reduction in divertor heat flux and a sharp increase in divertor radiation at $\bar{n}_e \sim 0.6 n_{GW}$, the same as in deuterium. Bolometer measurements showed very similar radiation profiles for both deuterium and helium plasmas, with the bulk of the radiation located in the divertor region. The peak emissivity was about 30% higher helium and the total radiating volume was about twice as large compared to deuterium plasmas. VUV spectroscopy showed that $He^{+}$ at 304Å was responsible for 90% of the radiation loss in the divertor, as compared to deuterium operation, where CIV emission at 1550Å produced more than 50% of the radiative losses.

With helium operation, the surface ion flux measured by Langmuir probes does not drop, even though the peak divertor heat flux drops by ~ a factor of 4, suggesting that volume recombination is not playing a significant role. Recent Thomson measurements confirm that $T_e$ remains above 2eV, where helium recombination is negligible. Further experiments with improved diagnostics will enable extension of previous analysis using 2D edge codes [4].

**Detachment-Onset Threshold Studies**

Our experiments show that the detachment threshold density depends most strongly on the SOL poloidal heat flux, $q_{pol}$, rather than parallel SOL heat flux ($q_{\parallel}$). Whether *local* $q_{pol} = P_{SOL}/\lambda_q$ or $q_{\parallel} = q_{pol}(B_{tot}/B_{pol})$ governs divertor detachment is fundamental to developing approaches for managing peak divertor heat flux and target-plate sputtering in future tokamak fusion reactors because it drives significant design choices about the magnetic topology in the divertor. For example, the only way to reduce $q_{\parallel}$ in the SOL is to increase either the radial transport or the major radius of the divertor leg; poloidal flux expansion in the divertor does not reduce $q_{\parallel}$. Clearly, parallel electron thermal transport dominates energy balance in the main SOL. In the divertor, however, the SOL plasma interacts with material surfaces and with recycled neutrals (fuel and impurities) through atomic physics processes. These interactions are independent of the field line geometry and can be both simultaneously localized in the poloidal dimension and broadly distributed in the parallel dimension along field lines.

To explore the relative importance of $q_{\parallel}$ vs. $q_{pol}$, we varied $q_{pol}$ through a series of power and current scans ($P_{sol} \propto P_{beam}$ and $l_q \propto 1/I_p$), and parallel heat flux $q_{\parallel}$ through a toroidal field scan at constant $P_{sol}$ and $l_q$ ($q_{\parallel} \propto B_{tor}$ at constant $I_p$), results of which appear in Fig. 3. For the current scan, the toroidal field was varied in step with $I_p$ to fix the field line pitch angle and parallel connection length. For the toroidal field scan, the plasma current was held fixed in order to maintain constant midplane heat flux width and $q_{pol}$, which necessarily results in an increase of parallel connection length proportional to the change in $B_{tor}$. The data show that the line-average plasma density required to achieve detached conditions (i.e., $T_{e,div}$<5eV) increases significantly with the increase in $q_{pol}$, while the density at detachment onset actually *decreases* as $q_{\parallel}$ doubles during the toroidal field scan. We speculate that increased field line length in the high neutral density region within a poloidal mean-free path of the divertor target significantly enhances energy dissipation through collisional excitation of fuel and impurity atoms, charge exchange, and pressure loss through volume recombination.

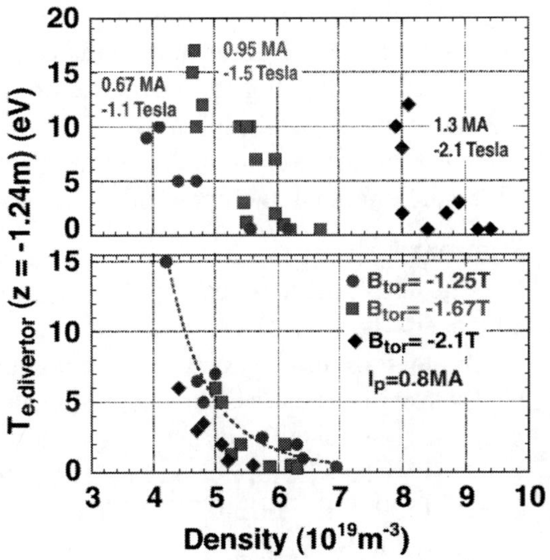


Fig.3. Divertor $T_e$ vs. plasma density. a) current scan at constant q95 ($B_{tor} \propto I_p$). b) toroidal field scan at constant plasma current.

**Divertor Configuration Studies**

The effect of divertor geometry (poloidal length and magnetic flux expansion) was examined by varying the X-point to target-plate distance ($d_{pol}$), and by operating with snowflake and X-divertor configurations. Increasing $d_{pol}$ by about a factor of three, from 0.2m to 0.6m, resulted in a strong reduction in peak target heat flux at high density, even though poloidal flux expansion was reduced by a similar factor. The longer divertor leg at the outer strike point also increased the effectiveness of argon injection in reducing peak heat flux [5].

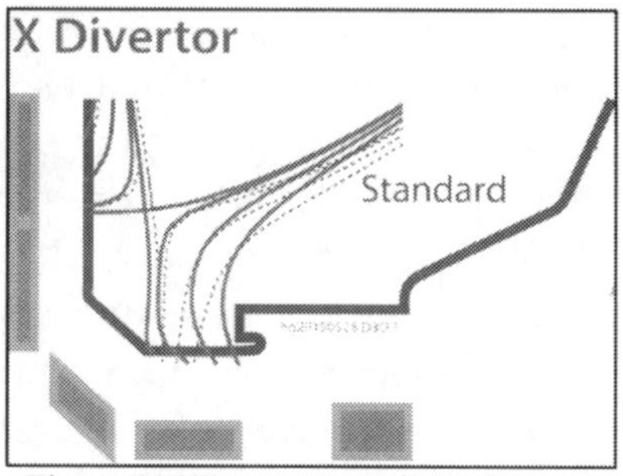

Fig.4. EFIT reconstruction of recent X-divertor configurations in DIII-D.

The effect of increasing poloidal flux expansion near the X-point and near the divertor targets was explored using exact Snowflake [6], Snowflake minus (SF–), and X-divertor configurations [7]. Naturally, the poloidal flux expansion reduced peak divertor heat flux in attached conditions by increasing the contact area by ~4-5 for the SF– configuration and for recent X-divertor discharges (shown in Fig. 4). In the SF– experiments, even the ELM heat pulses were suppressed [8].

Experiments are now being planned to explore the effect of divertor closure (use of shaped divertor structures to enhance neutral trapping near the divetor strike points). In the first phase, we will compare detachment onset density between the open lower divertor with the more closed upper divertor in DIII-D. SOLPS modeling suggests that further enhancement of the degree of divertor closure by small modifications to the shape of the upper divertor tiles may further reduce the detachment threshold density. Fig. 5 shows the concept for the upper divertor modification. The shaped-slot strongly reduces the divertor $T_e$ and target heat flux by redirecting and confining recycling neutrals and impurities near the strike point. This may enable detachment onset at a lower upstream density than for a rectangular slot.

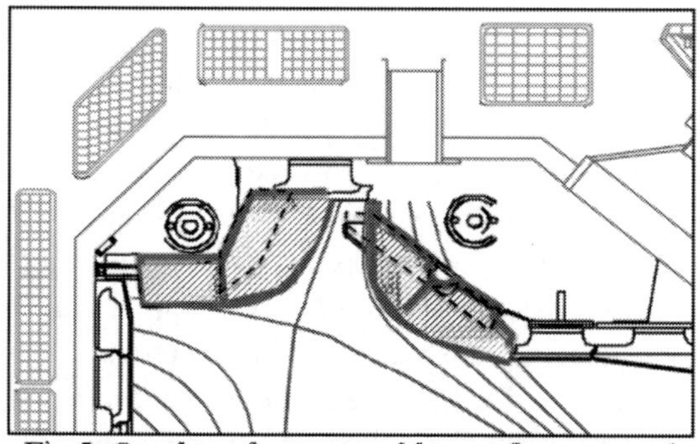
Fig.5. Overlay of conceptual layout for proposed upper-divertor tile modifications (red) in DIII-D.

**Summary**

The experiments described here seek to improve the physics basis for developing an advanced divertor for future burning plasma experiments. Focused variations in plasma parameters, divertor geometry, and magnetic configuration supported by a comprehensive set of divertor diagnostics, show that $q_{pol}$ has a much larger effect on detachment onset than $q_{11}$, providing key data for direct comparison with edge simulation codes.

This work was supported in part by the US Department of Energy under DE-FC02-04ER54609, DE-FG02-04ER54761, DE-AC52-07NA27344, and DE-FG02-08ER54984. (LLNL-CONF-672905)

**References**

1. M. Groth, G.D. Porter, M.E. Rensink, *et al.*, J. Nucl. Mater. **415** (2011) S530.
2. A. McLean, in preparation.
3. A.W. Leonard, M.A. Mahdavi, C.J. Lasnier, *et al.*, Nucl. Fusion **52** (2012) 063015.
4. A. Loarte, A. Kukushkin, G.D. Porter, , et al., Contrib. Plasma Phys. **40** (2000) 508
5. T.W. Petrie, J.M. Canik, C.J. Lasnier, *et al.*, Nucl. Fusion **53** (2013) 113024.
6. D. Ryutov, Phys. Plasmas **14** (2007) 064502
7. P.M. Valanju, M. Kotschenreuther, *et al.*, Phys. Plasmas **16** (2009) 056110
8. D.N. Hill and the DIII-D Team, Nucl. Fusion 53 (2013) 104001.

# Measurements of the Perturbed Rotation in Slowly Rotating Tearing Mode Islands in DIII-D*

N.Z. Taylor[1], R.J. La Haye[2], C.C. Petty[2], N.M. Ferraro[2], C. Bowman[3]

[1] *Oak Ridge Associated Universities, P.O. Box 117, Oak Ridge, Tn 37831, USA*

[2] *General Atomics, P.O. Box 85608, San Diego, CA 92186-5608, USA*

[3] *University of York, Heslington, York, United Kingdom*

## Introduction

Neoclassical tearing modes (NTM) are instabilities which are destabilized and maintained by helical perturbations to the pressure-gradient driven "bootstrap" current. The resulting magnetic islands break up the magnetic surfaces that confine the plasma. The NTM is linearly stable but nonlinearly unstable, and generally requires a "seed" to destabilize a metastable state. In addition, a rotating m/n=2/1 and/or 3/1 island induces eddy currents in the resistive wall that create drag that can bring the island (and plasma) rotation to a stop . This can lead to loss of H-mode and a disruption. A key feature for small island onset is how the ion flow is perturbed by the island. A helically modified ion flow leads to helical ion polarization currents which can either be stabilizing or destabilizing according to both the sign and magnitude of the perturbed flow [1] and [2]. Measuring the local phase shift (if any) and the radial structure at the island of the perturbed flow, with respect to the perturbed magnetic field, is a key measurement to sort out the nature of the ion polarization current [3].

## Measuring the Perturbed Radial Profiles in Freely Rotating Islands

The n=1 island structure is measured with electron cyclotron emission (ECE) radiometry. The ion rotation and temperature are measured by the fastest possible resolution (274 $\mu$s) charge exchange recombination (CER) spectroscopy. A slowly rotating island is needed with a long period of near constant rotation and plasma parameters. This allows the CER diagnostic to resolve the island temporal variation. Coupled m=(1,2,3,4), n=1 modes are slowed down to 1 kHz (max allowed for CER resolution) by near balanced beams. To prevent locking to the wall, the q=2 and q=3 surfaces were placed at smaller normalized radius $\rho$ than usual by setting $q_{95} = 8$; this reduces wall coupling. $\beta_N \simeq 1.8$ ($\beta_\theta \simeq 1.4$) so that the mode was also not large enough for locking. However, the islands were still large enough for the radial resolution of the CER and ECE diagnostics. To get both tangential and vertical rotation data, two different discharges were compared with the 30L neutral beam swapped with the 330L neutral beam to get both the tangential and vertical CER chords. The frequency of the island in the interval of interest is that at which the FFT amplitude of the fluctuating field, measured by Mirnov coils, was at a maximum. Each of the ECE and CER data channels had a linear offset removed and then was fit to a cosine with the frequency fixed to the island frequency and the amplitude and phase as free parameters. After correcting for the expected phase shift in the n=1 signals due to the different toroidal positions of the diagnostics, the complete radial profile of the perturbed quantities with respect to the island can be determined. The resulting perturbed temperature and rotation measurements are shown in Fig. 1. It can be seen that the perturbed $T_i$ profiles agree with the perturbed $T_e$ profiles. The zero crossings and local extrema in the perturbed temperature profiles show the location and size of the islands. The measured perturbed rotation has a local extremum at the island O-points with oppositely oriented extrema located just outside the island.

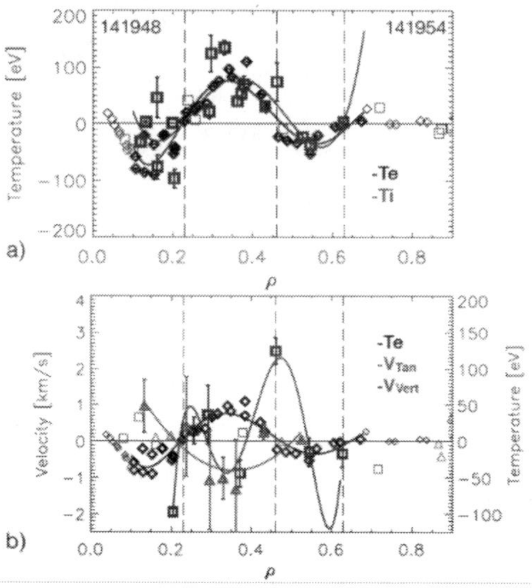


Figure 1: *a) Measured perturbed temperature for 1 kHz island. Dashed lines are q=2,3,4. b) Measured perturbed rotation. Local extrema observed at the location of the island O-point/X-point.*

The perturbed radial profiles of $T_e$, $T_i$ and rotation near the q=3 surface (location where the best resolved perturbed rotation signal was the largest) were fit to polynomials. The order of the polynomial for each profile was determined by minimizing the reduced $\chi^2$. Only data points that had an uncertainty in the phase angle less than 15° from the cosine fitting were included in the polynomial fits. By taking both the phase and amplitude information from the cosine fits, a radial profile for each of the perturbed quantities was determined over an entire island rotation period. The structure of the perturbed rotation profile near a slowly rotating n=1 tearing mode island over two periods is shown in Fig. 2.

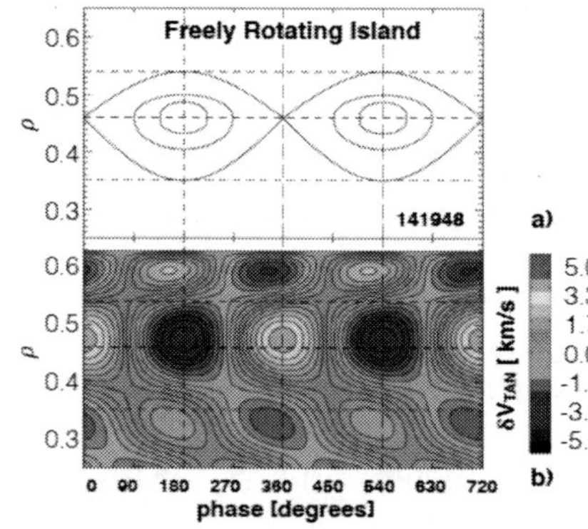


Figure 2: *a) Reconstruction showing the location, size and phase of the m/n=3/1 island, determined by the Te profile. b) The perturbed toroidal rotation near the q=3 surface for two periods. Rotation extrema at the rational surface are in phase with the island.*

The perturbed toroidal rotation is toroidally in phase with the island and has a local minimum at the island rational surface at the O-point. The perturbed rotation also has local maxima located radially just outside the island. The perturbed vertical rotation was relatively small and there was not enough spatial resolution to make any definitive statement on its radial structure. The measured perturbed rotation described above qualitatively agrees with the theoretical picture presented by [1] and [2].

**Measuring the Perturbed Radial Profiles in Entrained Rotating Islands**

An alternate method to producing a slowly rotating n=1 island structure is to allow it to lock and be entrained by a 10 Hz rotating applied field from the "I-coils" [4]. This puts the rotation in the standard 5 ms CER resolution, but the physics of entrained islands may differ from that of freely rotating ones. Here, $q_{95} = 4.8$, $\beta_N = 1.3$ ( as L-mode ) and $\beta_\theta = 0.65$, with the q=2 surface at a more typical (larger) $\rho = 0.57$. As a result, the entrained n=1 island structure includes only m=1 and 2. Figure 3 shows the 10 Hz perturbed temperature and rotation profiles.

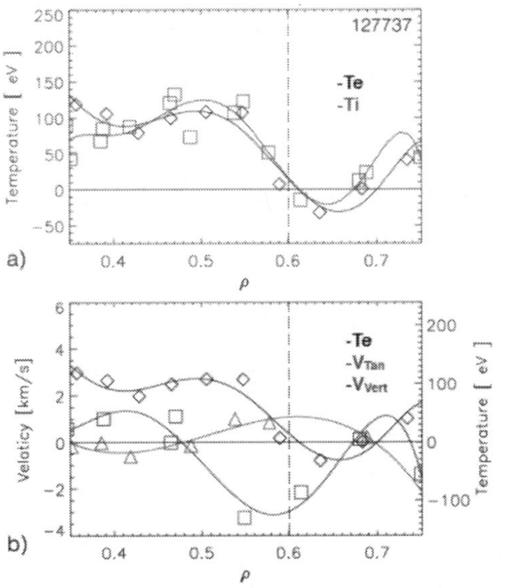


Figure 3: *a) Measured perturbed temperature for 10 Hz island. Dashed line is q=2. b) Measured perturbed rotation. Local extrema observed at the location of the island O-point/X-point.*

As in the freely rotating island case (m/n = 3/1 at 1000 Hz), the perturbed 10 Hz entrained case ion temperature profile agrees with the perturbed $T_e$ profile and shows the location and size of the island. The perturbed rotation profile island has extrema at the island O and X-points, as shown in Fig. 4, with opposite sign extrema in the "wings." The perturbed poloidal rotation is much smaller than the toroidal rotation, but it is also peaked near the q=2 rational surface and in phase with the island O and X-points.

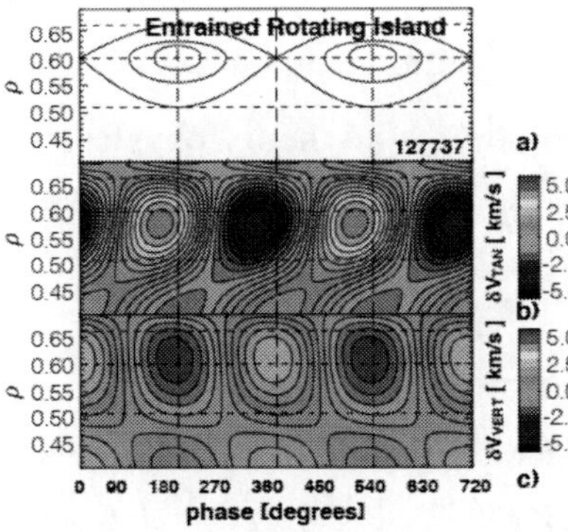


Figure 4: *a) Reconstruction showing the location, size and phase of the m/n=2/1 island, determined by the Te profile. b) The perturbed toroidal rotation near the q=2 surface. c) The perturbed poloidal rotation. the rotation is peaked at the rational surface and in phase with the Island.*

**M3D-C1 Code Comparison for Freely Rotating Island Measurements**

M3D-C1 allows us to compare the measured, saturated, non-linear perturbed rotation in the experiment with the linearly perturbed rotation in an MHD simulation [5]. Given the equilibrium profiles from the DIII-D case with the freely rotating 3/1 island, an M3D-C1 simulation found multiple linearly unstable (n=1, m=2,3,4,5) tearing modes present, the most unstable being a 5/1 mode.

Shown in Fig. 5 are contours of the linearly perturbed poloidal magnetic field and toroidal rotation along the lower-field side midplane. The toroidal rotation appears to be peaked at each of the rational surfaces and 90° out of phase with the tearing modes. The difference in relative phase from the experimental measurements may be due to the non-linear torque of the island on the rotation.

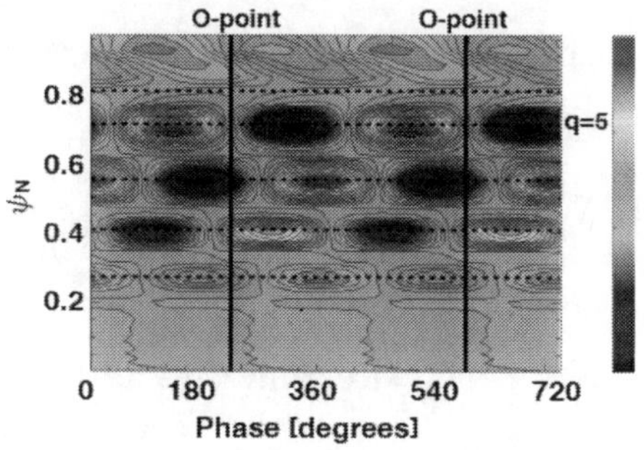


Figure 5: *The n=1 perturbed toroidal rotation from a M3D-C1 simulation. Dotted lines indicate the location of the q=2,3,4,5 and 6 surfaces. The solid lines indicate the phase of the O-point for the 5/1 mode*

## Conclusions

A helically modified ion flow can lead to helical ion polarization currents which can either be stabilizing or destabilizing according to both the sign and magnitude of the perturbed flow. In the freely rotating case presented here the measured perturbed flow was in phase with the island O-point and the inferred ion current ($\delta j = \sum n_i Z_i e V_i$) was about 1/3 the magnitude and in the same direction as the estimated island q=3 helical current density. However, for the entrained case, the inferred ion current, while also peaked on the island rational surface, is both smaller ($\sim$ 1/10) and in the opposite relative sign. These measurements had to be done for large saturated islands in which diagnostic radial resolution was good enough for the island structure. The M3D-C1 code comparison is for the linear "small island" non-saturated evaluation while NTM thresholds from helical ion polarization current theory are only developed for small islands of significance at onset of growth. However, those measurements should be a basis for testing threshold theory physics.

*Work supported in part by the US Department of Energy under DE-FC02-04ER54698 and DE-AC05-06OR23100.

## References

[1] F.L. Waelbroeck, J.W. Connor, and H.R. Wilson, Phys. Rev. Lett. **87**, 215003 (2001)

[2] R.J La Haye, C.C. Petty, E.J. Strait, F.L Waelbroeck, and H.R. Wilson, Phys. Plasmas **10**, 3644 (2003)

[3] R.J La Haye, G.L Jackson, T.C. Luce, K.E.J. Olofsson, W.M. Solomon, F. Turco, Proc. of the 41st EPS Conference on Plasma Physics, Berlin (2014), O5.134

[4] D. Shiraki, R.J. La Haye, N.C. Logan, E.J. Strait, and F.A. Volpe, Nucl. Fusion **54**, 033006 (2014)

[5] N.M. Ferraro, Phys. Plasmas **19**, 056105 (2012)

# Enhanced H-mode pedestals with increased density fluctuations during lithium aerosol injection in DIII-D

Z. Yan[1], G.R. McKee[1], T.H. Osborne[2], G.L. Jackson[2], R. Maingi[3], D.K. Mansfield[3], A.L. Roquemore[3], B.A. Grierson[3], C.P. Chrobak[4], A.G. McLean[5], S.L. Allen[5], D.J. Battaglia[3], A.R. Briesemeister[6], M.Fenstermacher[5], P.B. Snyder[2], and the DIII-D team

[1] *University of Wisconsin-Madison, Madison, WI, USA*
[2] *General Atomics, San Diego, CA, USA*
[3] *Princeton Plasma Physics Laboratory, Princeton, NJ, USA*
[4] *University of California, San Diego, San Diego, CA, USA*
[5] *Lawrence Livermore National Laboratory, Livermore, CA, USA*
[6] *Oak Ridge National Laboratory, Oak Ridge, TN, USA*

Periods of ELM free H-mode (up to 350ms) with increased pedestal pressure and width were observed in the DIII-D tokamak when lithium particles were injected into the plasma. These periods are associated with enhanced density fluctuations (named 'Bursty Chirping Mode' in the later discussion) localized to the pedestal region. The dramatically increased fluctuations are shown to drive outward cross-field particle transport, which reduced the pressure gradient just inside the separatrix. This allowed higher overall pedestal pressure (twice that of normal H-modes) because of the improved peeling-ballooning stability limit.

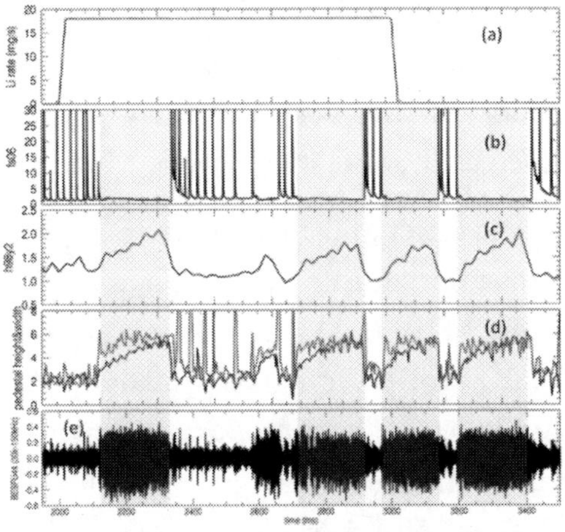


Fig. 1. Experimental parameters for shot 157267 with Li injection, (a) Lithium injection rate; (b) D$\alpha$ light; (c) $H_{98y,2}$; (d) pedestal height (black) and width (red); (e) density fluctuations from the pedestal region from BES measurements. Periods of increased pedestal and density fluctuations are marked with the yellow bands.

The discharges were operated in a lower single null plasma shape with $I_p$=1.2MA, $B_t$=2.0T, $q_{95}$=4.4 and average triangularity ~0.5. Line average electron density was in the range of 3.4-4.7 x$10^{19}$ m$^{-3}$. Li was injected into the plasmas as an aerosol of ~44 μm diameter particles from a Li dropper device located on the top of the DIII-D tokamak. Details of the injector hardware can be found elsewhere [1]. Lithium injection rates of 9-90 mg s$^{-1}$ were used, corresponding to 3.8×$10^5$-3.8×$10^6$ particles per second or 7.8×$10^{20}$-7.8×$10^{21}$ atoms s$^{-1}$. In comparison, a typical DIII-D discharge (assuming $Z_{eff}$=1.5, ne=6×$10^{19}$ m$^{-3}$, plasma volume=21m$^3$) contains ~$10^{21}$ deuterons. As shown in Fig.1, periods of enhanced ELM-free H-mode were observed (marked with the yellow bands) with Li aerosol injection at a rate of 18mg s$^{-1}$ (Fig.1 (a)). $H_{98y,2}$ (Fig.1 (c)), pedestal pressure height and width (Fig.1 (d)) all increase during the ELM free phases.

Density fluctuations were measured with a 2D array of high sensitivity Beam Emission Spectroscopy (BES) [2] located across the pedestal region from $\psi_N$~0.88-1.06. As is shown in Fig. 1(e) density fluctuations from the pedestal region increase dramatically during the enhanced ELM-free phases. Fig. 2 is a spectrogram of such density fluctuations near $\psi_N$~0.995. The frequency of such fluctuations spans from ~40kHz-150kHz. Notably the spectra exhibit a moderately broad frequency width on the time scale of hundreds of ms. However, when inspected on a time scale of a few ms, these fluctuations are bursty in time with each burst lasting a few hundred μs. Within each burst the frequency of the modes changes rapidly both up and down, while the mode itself appears coherent on very short time scales [3]. This is qualitatively different from the broadband turbulence that is typically observed in the pedestal region of H-mode plasmas [4]. Given the rapidly bursting and frequency shifting nature, these modes with Li injection have been named as the 'Bursty Chirping Mode' (BCM).

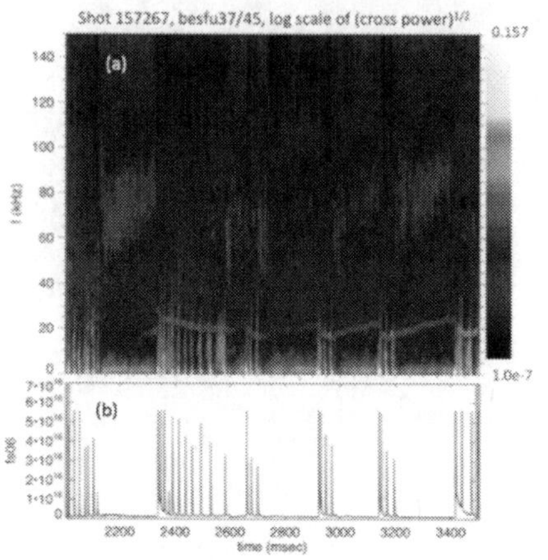


Fig. 2 (a) Spectrogram of density fluctuations from BES at $\psi$~0.995 for shot 157267 with Li injection; (b) edge $D_\alpha$ light.

The onset of the BCM occurs rapidly, typically within 10 ms after an ELM crash. The poloidal wave number of the BCM is $k_\theta$~0.5 $cm^{-1}$ with $k_\theta\rho_s$~0.1-0.2. Fig.3 is a radial profile of the relative density fluctuation amplitudes of these modes as a function of $\psi_N$ from BES measurements integrated over 40-150kHz, where the BCM is observed. The modes are localized to the pedestal region and peak just inside the separatrix. The peak amplitude is nearly ñ/n~8%. This exceeds by ~ tenfold the typical pedestal fluctuations observed in normal H-modes in between ELMs. The modes propagate in the electron diamagnetic direction in the plasma frame [3], consistent with the trapped electron mode (TEM) or micro-tearing mode characteristic. The poloidal correlation length is about 10cm, while the radial correlation length is less than 3cm, showing a strong spatial asymmetry to the mode structure.

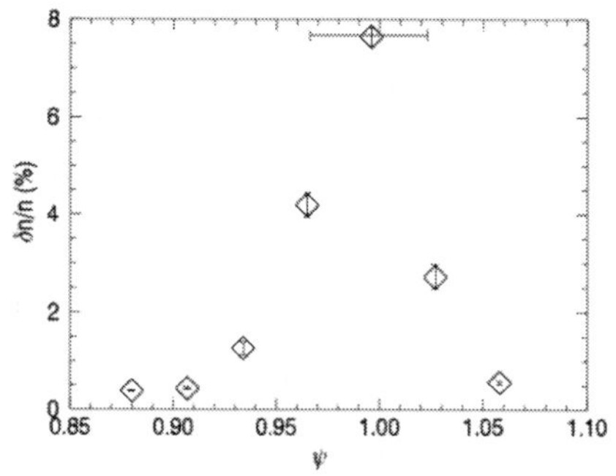


Fig. 3. Radial profiles of relative density fluctuation amplitude integrated over 40kHz to 150kHz from BES measurements.

The time scale of pedestal expansion correlates with the time scale for the growth of the BCM from the BES measurements. During the BCM, each burst correlates well with an increase of the C-III light at both the outer and inner divertor strike point locations, suggesting that the BCM increases plasma loss to the divertor [3]. This can be directly demonstrated from the experimentally inferred particle flux by applying a velocimetry technique to the 2D BES measurements [5]. This technique has the merit of obtaining the velocity field at the sampling time for BES (1 μs) to capture the fast flow dynamics of the turbulent eddy structures. By comparison, the decorrelation time for BCM is of order several μs. Fig. 4 is an example of a set of 2D density fluctuation images with time interval of 1 μs. It shows that the BCM density perturbation propagates in the electron diamagnetic direction (downwards in the image) and outwards. With the measured instantaneous fluctuating radial velocity, the inferred radial particle flux can thus be calculated as $\Gamma = \tilde{n}_e \tilde{V}_r$. Figure 5 are joint probability distribution functions (PDFs) of $\tilde{V}_r$ and $\tilde{n}_e$ at two radial locations,

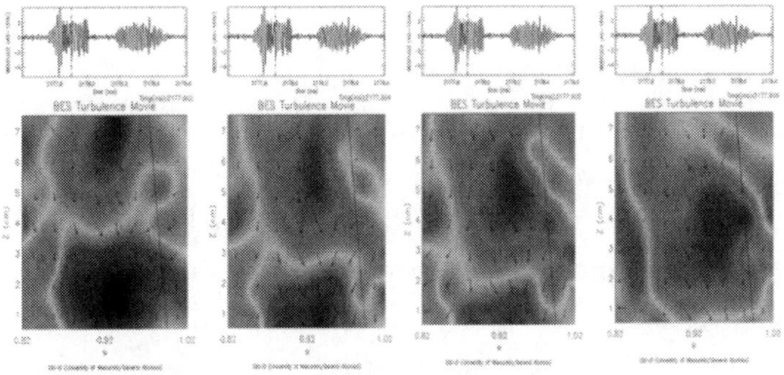

Fig.4 2D images of density fluctuation from BES measurements. Time interval is 1us. Black solid line on each image is the separatrix. Red color indicates positive density perturbations and blue color indicates negative density perturbations.

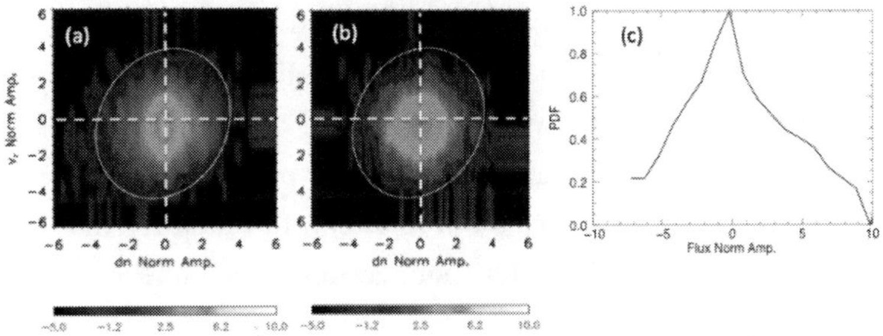


Fig.5 Joint PDF of $\tilde{V}_r$ and $\tilde{n}_e$ at (a) ψ~0.965 and (b) ψ~0.88; (c) PDF of inferred particle flux at ψ~0.965; White dashed lines on (a) and (b) are zero line of $\tilde{V}_r$ and $\tilde{n}_e$. Elliptical lines are exact the same on both (a) and (b) to help comparison.

$\psi_N$~0.965 (Fig.5 (a)) and $\psi_N$~0.88 (Fig.5 (b)). The dashed lines on both figures are zeros of $\tilde{V}_r$ and $\tilde{n}_e$, distributing the figure into four quadrants. It shows that at $\psi_N$~0.965 there are more events in the 1st (upper right section of the figure) and 3rd (lower left section of the figure) quadrant than those at $\psi_N$~0.88, indicating that more positive density perturbations with outward going (positive $\tilde{V}_r$) events and negative density perturbations with inward going (negative $\tilde{V}_r$) events take place, reflecting on average an outward flux. Fig. 5(c) is a PDF of inferred particle flux, Γ, at ψ~0.965. It shows a significant positive skewed PDF indicating intermittent or bursty particle transport events which are characterized by outward-going positive density perturbations. Fig. 6 shows the radial profile of the skewness of the inferred particle flux [i.e., the skewness is the third moment of the probability distribution function measuring the asymmetry of the distribution of the signal, defined as $\Gamma^3 / \langle \Gamma^2 \rangle^{1.5}$ ]. A comparison of Fig. 3 and Fig. 6 shows that the location where the skewness of particle flux is

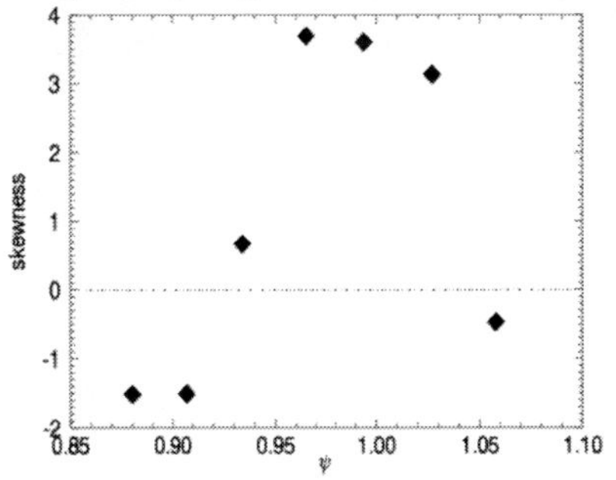


Fig. 6 Skewness of inferred particle flux from BES measurements.

significantly positive is where the amplitude of BCM is maximal. This asymmetry of the skewness across the pedestal and the alignment with the pedestal pressure profile demonstrates that BCM drives outward particle transport that could impact the pedestal profile evolution. This indeed has been shown from the correlation between the initiation and termination of the BCM and the profile evolution during an ELM-free phase [3]. At the BCM onset there is a reduction in electron density and perhaps also temperature in the region near the separatrix where BCM locates. The reduction persists while the BCM is present, and the pedestal continues to expand inboard of this region. When the BCM terminates both temperature and density profiles rebuild near the separatrix. This BCM-induced local flattening of the pressure gradient near the separatrix allows an increased pedestal pressure within the MHD stability constraints. It has been shown that in the Li diluted with BCM present periods the plasmas are operated well below the PBM limit [3].

In summary, significantly enhanced H-mode pedestal pressure and global energy confinement were observed in DIII-D discharges with a small amount of Li injection when large amplitude density fluctuations, the BCM, were present. The modes drive outward particle transport flattening local pressure gradient near separatrix, which allows wider pedestal width and higher pedestal pressure at improved PBM stability. Gyrokinetic simulations will be carried out to attempt to identify the nature and instability of the observed mode, as well as possible connection to a coherent mode observed in EAST during Li injection [6].

This material is based upon work supported by the U.S. DOE, under Awards DE-FG02-89ER53296, DE-FC02-04ER54698, DE-AC02-09CH11466, DESC0012551, DE-FG02-07ER54917, DE-AC52-07NA27344 and DE-AC05-00OR22725.

Reference:

[1] D. Mansfield, et al Fusion Eng. Des. **85**, 890, 2010

[2] G. McKee, et al Rev. Sci. Instrum., **81**, 10D741, 2010

[3] T. Osborne, Nuclear Fusion, **55**, 063018, 2015

[4] Z. Yan, Physics of Plasmas, **18**, 056117, 2011

[5] G. McKee, Rev. Sci. Instrum, **75**, 3490, 2004

[6] J.S. Hu, et. al., Phys. Rev. Letts. **114**, 055001, 2015

# Characterization of Heat Loads from Mitigated Upward or Downward VDEs in DIII-D

E.M. Hollmann[1], N. Commaux[2], N.W. Eidietis[3], C.J. Lasnier[4], R.A. Moyer[1], P.B Parks[3], and D. Shiraki[2]

[1]*University of California San Diego, 9500 Gilman Dr., La Jolla, CA 92093-0417, USA*
[2]*Oak Ridge National Laboratory, PO Box 2008, Oak Ridge, TN 37831, USA*
[3]*General Atomics, PO Box 85608, San Diego, California 92186-5608, USA*
[4]*Lawrence Livermore National Laboratory, Livermore, CA 94550, USA*

Experiments were conducted on the DIII-D tokamak to study vertically unstable plasma (VDE) disruption heat loads when mitigated with massive gas injection (MGI). MGI was performed with neon gas injection from dominantly either above or below the plasma. This was motivated by the present ITER plan to perform disruption thermal quench (TQ) mitigation mostly using valves located above the plasma, causing concern that downward VDEs could be poorly mitigated as a result. However, analysis of localized wall heat loads from IR camera images and radiated power from fast or slow bolometers shows no systematic difference in mitigation effectiveness of upper compared to lower gas valves, regardless of initial impurity injection location. This indicates that TQ mixing of impurities is faster than VDE timescales, resulting in TQ mitigation with similar global effectiveness for upper or lower MGI valve location.

The experiments presented here were performed in the DIII-D tokamak [1]. The target plasmas were an "ITER-like" shape with low triangularity and lower single null. At $t$ = 2000 ms, the vertical stabilization system was turned off and the plasma was given a downward (or upward) kick with the shaping coils. The elongated plasma then goes vertically unstable, drifting into the lower (or upper) divertor. In the absence of mitigation, these plasmas go into a disruption thermal quench (TQ) around time $t$ = 2025 ms. Mitigation is performed here with massive gas injection (MGI) of neon gas. MGI is done from two different locations: an "R+1" port located above the plasma at toroidal angle $\phi = 15°$ and an "R-2" port located below the plasma at toroidal angle $\phi = 135°$. MGI used 300 Torr-L of neon in a pulse about 1.5 ms long (at the valve).

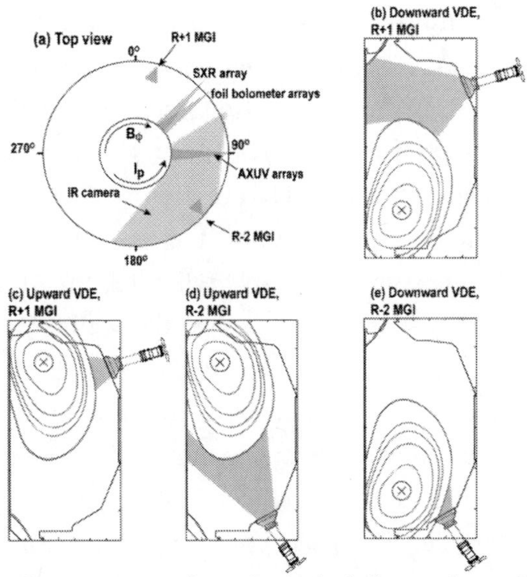


Fig. 1. (a) Overview of principal diagnostics and (b)-(e) schematics of different VDE directions/MGI location combinations.

Figure 1 gives: (a) an overview of principal diagnostics used here and (b) – (e) cartoons of the different mitigation combinations studied here (downward VDE mitigated by R+1 MGI, upward VDE mitigated by R+1 MGI, upward VDE mitigated by R-2 MGI, and downward VDE mitigated by R-2 MGI). The core thermal collapse timing is dominantly diagnosed with soft x-ray arrays. Radiated power is measured with (fast) AXUV photodiode arrays and (slow) foil bolometer arrays. Wall heat loads are measured with a mid-IR (3-5 μm) camera.

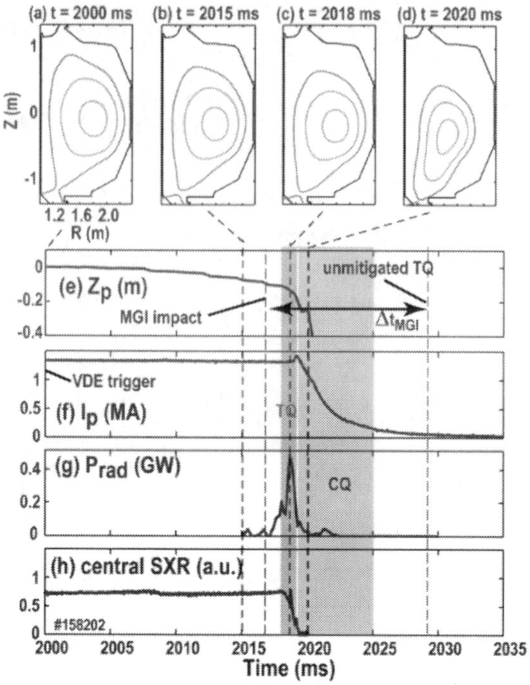


Fig. 2. Overview of mitigated VDE experiment time traces showing (a)-(d) JFIT reconstructions of magnetic flux surfaces, (e) vertical position, (f) plasma current, (g) radiated power, (h) central SXR.

An example of time traces from a VDE mitigated with R-2 MGI are shown in Fig. 2. In this shot, neon MGI is fired into the plasma before the VDE has caused the plasma to limit on the divertor and begin the TQ. The MGI "first light" impact of neon on the plasma edge occurs at $t$ = 2017.2 ms. A rise in radiated power is observed, Fig. 2(g), and 1 ms later the plasma begins the TQ with the core temperature collapsing, Fig. 2(h). The mitigation "timeliness" is characterized by $\Delta t_{MGI}$ , the delay between the MGI impact on the plasma edge and the expected TQ onset time if MGI had not been pre-emptively deployed.

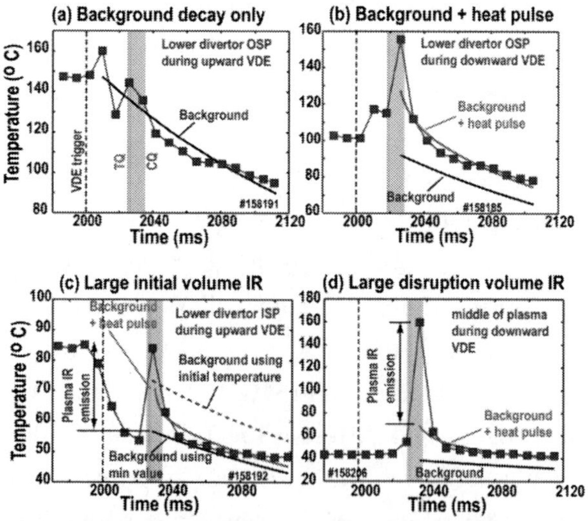


Fig. 3. Examples of fits to post-disruption IR data time sequences at different locations showing: (a) background temperature decay in location with negligible disruption heating, (b) total temperature decay in location with comparable initial and disruption heat pulse heating, (c) temperature decay fits in location with significant initial plasma IR emission, and (d) temperature decay fits in location with significant disruption plasma IR emission.

Wall heat loads are measured with the DIII-D tangential IR camera [2]. The camera is kept to its full field of view during these experiments, thus covering much of the plasma cross section, but limiting the acquisition rate to 8 ms per frame. Because this acquisition rate is slower than the TQ duration (~ 2 ms), the temperature decay of the post-disruption images is used to estimate the time-averaged TQ heat flux to the walls. In the analysis, the temperature decay at each pixel is fit as a function of time after the disruption is over, assuming a separable contribution from the initial temperature of the surface (due to steady-state plasma heating) plus the contribution from the disruption heat pulse. Heat diffusion into a semi-infinite plane is assumed [3]. Blackbody radiation, lateral heat diffusion, and variation of wall properties with temperature are ignored. The background temperature cooling time constant $\tau_\infty$ is fit experimentally from lower divertor pixels after upward VDEs, where initial heating dominates. Typically, $\tau_\infty \approx 170$ ms is found, as shown in Fig. 3(a). A sample fit where both initial wall temperature and disruption heat pulse are significant is shown in Fig. 3(b). In some pixels, volume plasma IR emission prior to the disruption is significant; an example of this is shown in Fig. 3(c). This is corrected for by lowering the initial temperature $T_0$ until the background decay curve lies below all data points. To avoid volume IR emission during the disruption, fits are performed only on data points past the end of the CQ. Frequently, the data point(s) falling during the CQ time window are observed to lie well above the fit; this is interpreted as being at least partially due to volume IR emission during the CQ. An example of this is shown in Fig. 3(d), where a very large deviation from the post-disruption decay fit is seen during the CQ.

Figure 4 shows an example of IR images and resulting reconstructed disruption heat flux for an unmitigated upward VDE. Figure 4(a) gives an example pre-disruption IR image, Fig. 4(b) shows a during-disruption IR image, and Fig. 4(c) shows an image of the resulting calculated disruption heat flux. Figure 4(d) shows a CAD model of the IR camera view.

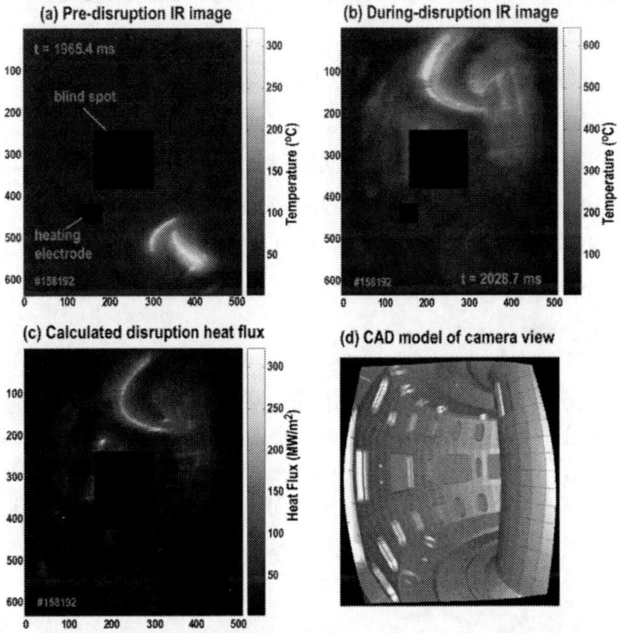


Fig. 4. Example IR images for an unmitigated upward VDE showing: (a) pre-disruption image, (b) during-disruption image, (c) calculated disruption heat flux, and (d) CAD model of IR camera view.

Figure 5 gives an overview of 0D trends in effectiveness of MGI mitigation of VDEs. The data are plotted as a function of the MGI impact delay $\Delta t_{MGI}$. Figure 5(a) shows the total radiated energy estimated by tomography of the slow bolometers. Figure 5(b) shows the energy going into localized wall heating estimated from IR images and assuming toroidal symmetry (this includes both conducted and convected contributions). Figure 5(c) shows vertical vessel motion, giving a qualitative picture of vessel forces. Figure 5(d) gives the magnitude of the plasma current decay rate. This gives a rough picture of the amount of impurities in the CQ plasma, with larger impurity levels giving a lower temperature and a faster current decay rate. The data with $\Delta t_{MGI} = 5$ ms actually corresponds to unmitigated VDE disruptions with no MGI.

Figure 5 indicates that there is no significant global improvement in mitigation effectiveness of VDEs when using a MGI port closer to the VDE direction. In contrast, MGI timing delay $\Delta t_{MGI}$ clearly has a large effect, with earlier MGI resulting in higher radiated energies, lower localized heat fluxes, and lower vessel motion. Even injecting during the TQ ($\Delta t_{MGI} = 0$) appears to have some small mitigation benefit over not using MGI at all. This data thus suggests that TQ mixing of heat and impurities is very effective even when the plasma is limited on the divertor. This effective TQ mixing tends to smooth out the effect of injection location on global disruption mitigation indicators. One trend in the data which is not understood at present, is the large difference in heat loads seen between upward and downward VDEs, Fig. 5(b). This may be due to upper and lower divertor surfaces having different material properties, such as IR emissivity.

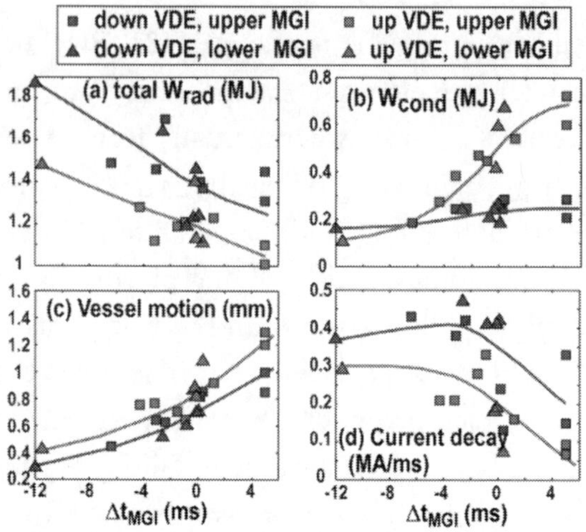


Fig. 5. Global trends in VDE mitigation effectiveness for different VDE direction/MGI location combinations as a function of MGI trigger delay $\Delta t_{MGI}$ showing (a) total radiated energy, (b) total conducted/convected energy, (c) vessel vertical motion, and (d) CQ current decay rate.

This material is based upon work supported by the U.S. Department of Energy, Office of Science, Office of Fusion Energy Sciences, using the DIII-D National Fusion Facility, a DOE Office of Science user facility, under Awards DE-FG02-07ER54917, DE-FC02-04ER54698, DE-AC05-00OR22725, DE-AC52-07NA27344, and DE-AC05-06OR23100. DIII-D data shown in this paper can be obtained in digital format by following the links at https://fusion.gat.com/global/D3D_DMP.

[1] J. L. Luxon, Nucl. Fusion **42** (2002) 614.
[2] C.J. Lasnier, et al., Rev. Sci. Instrum. **85**, 11D855 (2014).
[3] H.S. Carslow and J.C. Jaeger, Conduction of Heat in Solids, 2nd Ed. (Oxford, Claredon, 1995).

# Inter-ELM Pedestal Evolution and the Role of Edge Fluctuations in the C-Mod and DIII-D Tokamaks

A. Diallo[1], R.J. Groebner[2], J.W. Hughes[3], T.L. Rhodes[4], S.-G. Baek[3], B. LaBombard[3], J.L. Terry[3] , I. Cziegler[5], J. Walk[3], A.E. Hubbard[3], D. Smith[6], T.H. Osborne[2], J.M. Canik[7], W. Guttenfelder[1], P.B. Snyder[2], and the Alcator C-Mod and DIII-D Teams

[1] *Princeton Plasma Physics Laboratory, P.O. 451, Princeton, NJ, USA*

[2] *General Atomics, P.O. Box 85608, San Diego, CA, USA*

[3] *MIT Plasma Science and Fusion Center, Cambridge, MA, USA*

[4] *Physics Department., University of California Los Angeles, Los Angeles, CA, USA*

[5] *University of California San Diego, 9500 Gilman Dr., La Jolla, CA, USA*

[6] *University of Wisconsin-Madison, 1500 Engineering Dr., Madison, WI, USA*

[7] *Oak Ridge National Laboratory, P.O. Box 2008, Oak Ridge, TN, USA*

Understanding the physics leading to the recovery of the edge transport barrier following edge-localized-modes (ELM) crashes and buildup to the next ELM crash is an active and critical area of edge physics research for ITER projections. In one of the hypotheses of the EPED model [1], the pedestal pressure profile in H-mode discharges is predicted to be limited by micro-instabilities. One such micro-instability is the kinetic ballooning mode (KBM), which is hypothesized to provide a "soft" limit that regulates the edge transport, thereby restricting the local pressure gradient. Experiments were recently performed on the C-Mod and DIII-D devices to search for instabilities correlated with the pedestal evolution between ELMs [2, 4, 3].

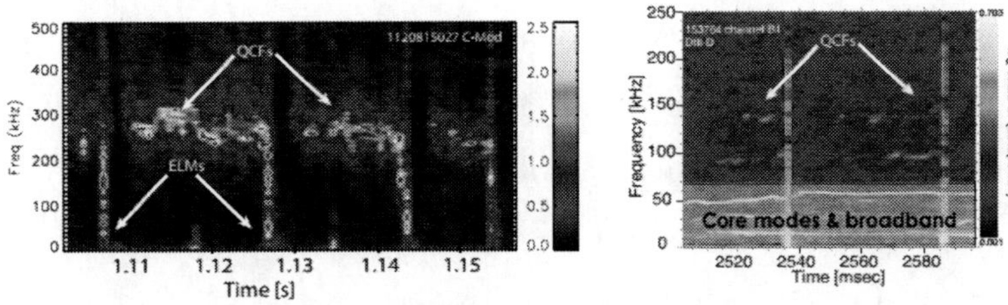


Figure 1: Magnetic spectrograms. Left panel: C-Mod (see Ref. [2]).Right panel DIII-D (see Ref. [3]).

Dedicated ELMy discharges were obtained on C-Mod with $B_T = 5.4$ T and the plasma current set at 0.9 MA in LSN magnetic configuration and high triangularity ($\delta \sim 0.8$). On DIII-D the ELMy discharges were obtained at $B_T = 1.9$T for three sets of plasma currents (0.75, 1.0, 1.6 MA). These experiments were targeted at capturing the pedestal recovery after an ELM crash using the recently upgraded Thomson scattering system on DIII-D [5]. To accurately resolve the inter-ELM dynamics, the lasers were fired in a bunch mode, which enabled temporal resolution of up to 100 $\mu$s. This increase in temporal

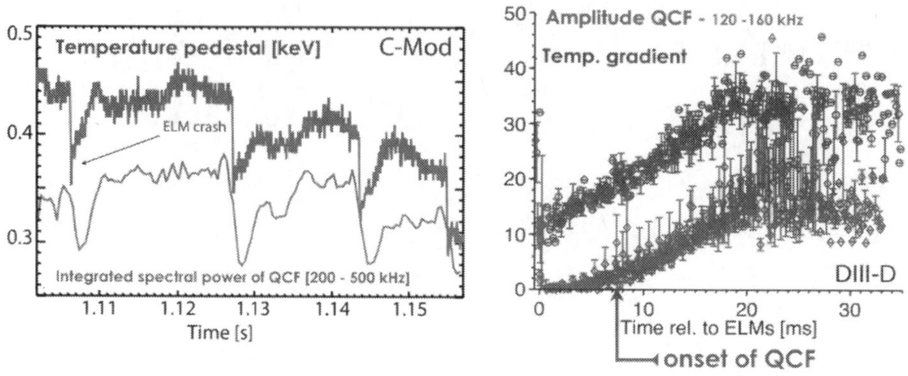


Figure 2: Left panel - C-Mod: the pedestal temperature is plotted in blue, and the integrated spectral power of the QCF is displayed in green [2, 4]. Right panel - DIII-D: the red symbols represent the amplitude of the QCF as a function of time relative to an ELM. The blue symbols display the corresponding temperature gradient evolution [3].

resolution is achieved by using multiple lasers in the same path with pulses interleaved closely in time.

Measurements of the pedestal recovery following ELMs on C-Mod and DIII-D reveal new evidence that fluctuations play an important role in the saturation of the edge gradient in the inter-ELM period. Identification of edge quasi-coherent fluctuations (QCFs) on multiple diagnostics is observed when the pressure gradient in the edge pedestal region of the plasma saturates following the ELM. Figure 1 displays the spectrograms of the magnetic fluctuations on both machines. The left panel represents the spectrogram as measured on C-Mod. The right panel of this figure displays the spectrogram on DIII-D, which indicates multiple QCFs. The lower frequencies (below 50 kHz) show core modes and broadband fluctuations uncorrelated with the ELM events. On C-Mod and DIII-D, the QCFs are observed to precede the ELM onset. The QCFs disappear after the ELM crash and later appear leading up to the next ELM.

Correlations between the QCF's onset and the pedestal parameters are shown in figure 2. On the left panel of this figure, the C-Mod results are shown, which display the temperature pedestal evolution and the integrated spectral power from the magnetic fluctuations to highlight the correlation between the onset of the QCFs and the pedestal temperature. Such correlation is clearly seen on the DIII-D results displayed on the right panel of figure 2. This panel captures the pedestal gradient evolution until saturation. At near 10 ms after the ELM crash, the QCF is observed and grows until saturation, which appears to occur nearly at the same time as the temperature gradient saturation. While causality is difficult to prove at this stage, it is suggested that the QCFs onset at a given temperature gradient and the amplitude of the QCF grows until sufficient transport is provided to clamp the gradient evolution. Coincidentally, this saturation corresponds to the saturation of the QCF's amplitude.

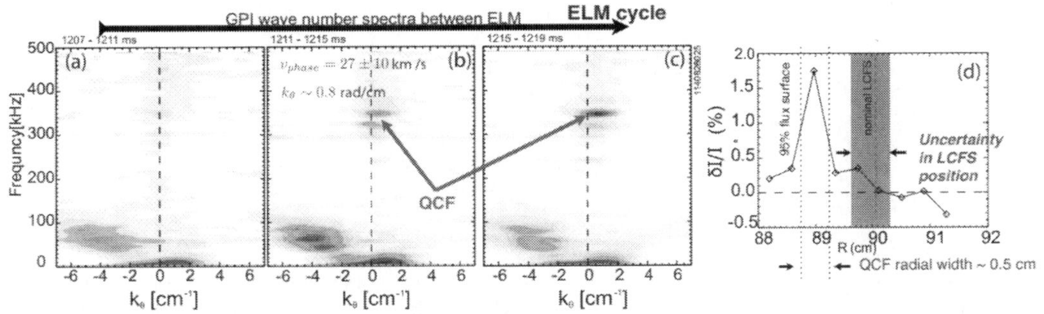


Figure 3: QCF spectral power evolution and localization using GPI on C-Mod. (a) - (c) represent the wavenumber - frequency spectrum during the inter-ELM phase on C-Mod. (d) The QCF radial profile is determined using GPI and the peak appears to be in the pedestal (see Ref [4] for details.)

The localization of the QCFs has been determined on C-Mod using the density fluctuations from the reflectometer system (see ref. [2]) as well as the GPI system. Figure 3(a)-(c) displays the GPI wavenumber frequency spectra during the the ELM cycle. Early in the ELM cycle, no QCF can be observed. Late during the ELM cycle, however, QCFs with poloidal wavenumber $k_\theta \sim 0.8$ rad/cm propagating at $v_{phase} \sim 27 \pm 10$ km/s emerge. The localization of the QCFs is indicated in figure 3(d). In this figure, the QCF is shown to be localized in the pedestal region. On DIII-D, figure 4 displays three spectrograms of the BES signals spanning the pedestal region. The location where the BES spectrograms were performed are labelled on figure 4(d). The spectrograms of the density fluctuations from BES indicate that the QCFs fluctuations appear to be localized in a narrow region of the pedestal.

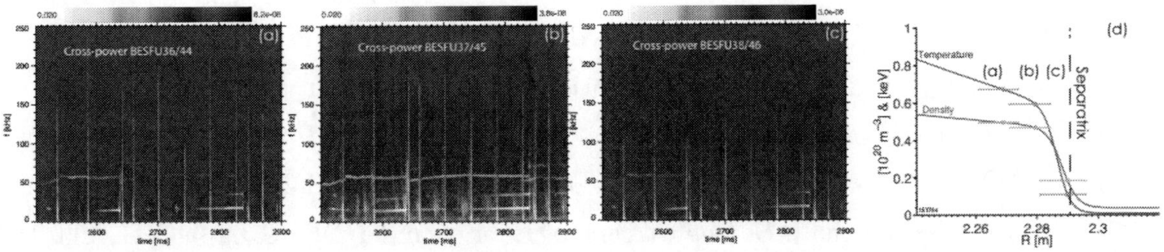


Figure 4: QCF spectral power evolution and localization using BES on DIII-D. (a) - (c) display the cross-power spectrograms at three radii, which is keyed in (d) on the temperature and density profiles [3].

Initial linear microstability properties of these edge plasmas have been analyzed for the high current case using the initial value gyrokinetic code GS2 [6]. To get an idea what kind of microinstabilities are likely to be responsible for driving the turbulence for our nominal parameters in various $\beta$ regimes, linear gyrokinetics calculations in the pedestal region in C-Mod and DIII-D are performed, as displayed in figures 5 (a) and (b). Figure 5 (a) shows that the most unstable mode at low k peaks at values of $k_\theta \rho_s \sim 0.04$, which is in agreement with experimental observations on C-Mod. This low-k mode is predicted to

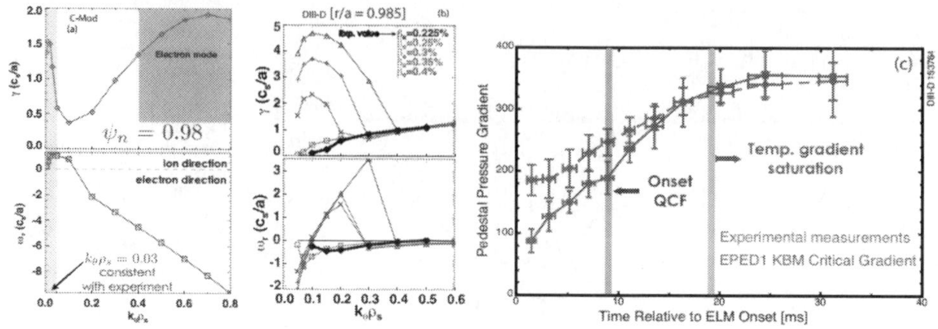


Figure 5: Left and middle panels display the results of GS2 calculations in the pedestal on C-Mod [4] and DIII-D [3]. The top of these panels are the growth rates and the bottom plots are the real frequencies. The right most panel shows the dynamic of the pedestal using EPED1 calculations. Overlaid on this plot are the time of the onset of the QCF and time when the temperature gradient saturates.

propagate in the ion direction. On DIII-D, the linear stability analysis predict an electron directed mode as shown in figure 5(b) at $\beta_e \sim 0.25\%$ slightly larger than $\beta_e^{exp}$. A $\beta$ scan further indicates that at the pedestal top the most unstable mode is near KBM criticality. The propagation direction, on the other hand is predicted to be in the electron direction.

Figure 5(c) displays the gradient evolution against the time relative to an ELM. In this figure, the red squares indicate the binned averages of experimental data. From each of the data points, the EPED1 model requires eight input parameters, ($B_T$, $I_p$, R, a, $\kappa$, $\delta$, global $\beta$, and $n_e^{ped}$) from which it outputs the pedestal pressure height and width (see Ref. [7] for further details). Here, given that experimental observations give fast measurements of $n_e^{ped}$ and $T_e^{ped}$, the pressure pedestal is approximated by $2n_e^{ped}T_e^{ped}$ to track the dynamics. In figure 5(c), the closed blue symbols represent the EPED1 KBM constraint. EPED1 model indicates that pedestal pressure gradient reaches approximately the KBM critical gradient when the fluctuations are observed (as indicated by the green vertical line). In addition the EPED1 model predicts saturation of the pressure gradient, i.e., the observed gradient does not exceed the prediction after the onset of the instability. Overall, the QCFs on DIII-D and C-Mod appear to be consistent with the characteristics of KBM.

Work supported by the U.S. Department of Energy under DE-AC02-09CH11466, DE-FC02-04ER54698, DE-FC02-99ER54512, DE-FG02-08ER54984, DE-FG02-07ER54917, DE-FG02-89ER53296, DE-FG02-08ER54999 and DE-AC05-00OR22725.

## References

[1] P.B. Snyder et al. *Nuclear Fusion*, 51(10):103016, 2011.
[2] A. Diallo, et al. *Phys. Rev. Lett.*, 112:115001, Mar 2014.
[3] A. Diallo et al. *Physics of Plasmas*, 22(5) 05611 , 2015.
[4] A. Diallo et al. *Nuclear Fusion*, 55(5):053003, 2015.
[5] D. Eldon et al. *Review of Scientific Instruments*, 83(10):10E343, 2012.
[6] M. Kotschenreuther et al. *Computer Physics Communications*, 88:128–140, 1995.
[7] P. B. Snyder et al. *Phys. Plasmas*, 16:056118, 2009.

# Avoidance of m=2, n=1 tearing mode wall-locking by torque-balance control with magnetic feedback in DIII-D and RFX-mod

P. Zanca[1], M. Okabayashi[2], R. Paccagnella[1], E.J. Strait[3], A.M. Garofalo[3], J.M. Hanson[4], Y. In[5], R.J. La Haye[3], L. Marrelli[1], P. Martin[1], P. Piovesan[1], C. Piron[1], L. Piron[6], D. Shiraki[4*], F.A. Volpe[4], and the DIII-D and RFX-mod Teams

*[1]Consorzio RFX (CNR, ENEA, INFN, Università di Padova, Acciaierie Venete Spa), Corso Stati Uniti 4, 35127 Padova, Italy; [2]Princeton Plasma Physics Laboratory, PO Box 451, Princeton, NJ 08543-0451, USA; [3]General Atomics, PO Box 85608, San Diego, CA 92186-5608, USA; [4]Columbia University, 2960 Broadway, New York, NY 10027-6900, USA; [5]FAR-TECH, Inc., San Diego, CA 92121-1136, USA; [6]CCFE, Culham Science Centre, Abingdon, Oxon OX14 3DB, UK*

**Permanent Address: Oak Ridge National Laboratory, PO Box 2008, Oak Ridge, TN, USA*

Recent experiments performed in DIII-D [1] and RFX-mod operated as a tokamak [2] have demonstrated that magnetic feedback realized by active saddle coils is able to push a wall locked $m$=2, $n$=1 tearing mode (TM) into slow rotation with frequency of the order of several tens of Hertz. The result is not obvious, given the many differences of the two experiments both in terms of layout and plasma condition. In DIII-D ($R_0$=1.6m, $a\approx$0.6m) diverted, D-shaped, high-$\beta$ plasmas, controlled by coils placed inside the 5ms time-constant vacuum vessel have been performed. Instead, in RFX-mod ($R_0$=2m, $a$=0.459m) we realized circular cross-section, limiter, ohmic discharges with low edge safety factor, $2<q(a)<3$, feedback controlled by active coils placed outside the 100ms time-constant copper shell. As shown in [3] the feedback-induced frequencies can be interpreted as an equilibrium condition in which the viscous torque due to the plasma flow balances the electromagnetic torque developed by the interaction between TM and the external passive and active conductive structures. For mode amplitudes above the wall-locking threshold, electromagnetic torque $T_{EM}$ is the dominant effect (see figure A.1 of [3]), and the feedback-induced frequencies are approximately the solutions of the equation $T_{EM}(\omega)$=0, satisfying the stability requirement $\partial T_{EM}/\partial\omega$<0. A cubic equation for $\omega$ is obtained, as shown in [1] on the basis of a model which describes the coupling between TM and the conductive structures in terms of mutual inductances [4]. A similar equation can be derived using the cylindrical Newcomb's model described in the appendix of [3], considering one single wall, and radial field sensors placed, for the sake of simplicity, on it. A wall-locked solution $\omega$=0 is selected for feedback proportional gain $K$ ($K<0$) such that $K/K_{crit}<1$. Here $K_{crit}<0$ is a critical value, which assumes different expressions depending on the active coils location with respect to the wall (out-shell for RFX-mod, in-vessel for DIII-D). It depends on the ratio $\tau_w/\Delta t$ between the wall time-constant and the global delay of the feedback chain. For $K/K_{crit}>1$ two opposite feedback frequencies $\omega=\pm\sqrt{K/K_{crit}-1}/\Delta t$ are obtained, which amount to some tens of Hertz for both RFX-mod and DIII-D, as experimentally observed. In DIII-D a control variant with $K$ complex, corresponding to a small phase shift, is routinely used, since it increases the stability of the rotations [1]. With complex gains the previous formula assumes a slightly different form [1].

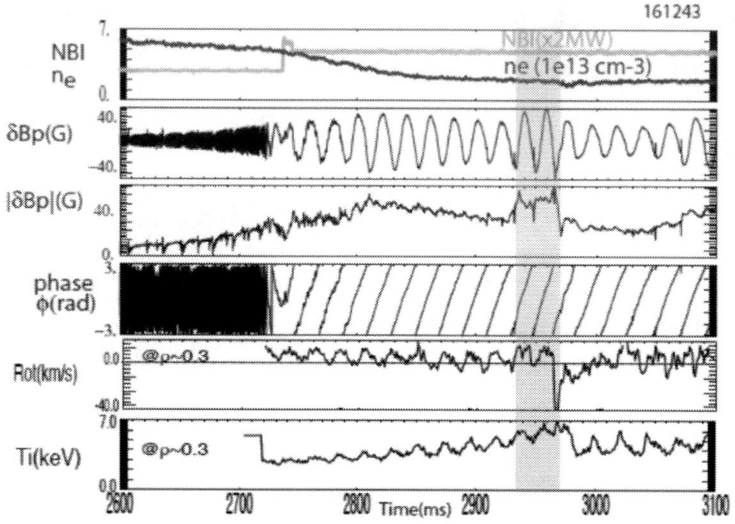


**Fig. 1**. Example of $m=2$, $n=1$ TM control in DIII-D. The green vertical band indicates the occurrence of an internal reconnection event.

Since such a simple and fundamental model unifies the interpretation in two very different configurations, it can be concluded that the phenomenon does not depend on many details of plasma parameters and is robust and reproducible. An example of the resiliency of this process against unexpected events such as a major internal reconnection is presented in Figure 1 for DIII-D. Here, the feedback induced rotation at about 50Hz is established (fourth panel) as soon as the TM amplitude increases (second and third panel, where the $n=1$ harmonic is plotted; a dominant $m=2$ component is identified) to a level that would lead to wall-locking without control. This rotation produces modulations in the ion velocity signal (fifth panel) and temperature (sixth panel). The density reduction from $4\times10^{13}cm^{-3}$ to $1.5\times10^{13}cm^{-3}$ (first panel) induces an increase of the central ion temperature from 4keV to 7keV (sixth panel). The consequent rise of the core pressure gradient produces an increase of the MHD activity, highlighted by the green band, culminating in a minor disruption at t=2965ms. Internal MHD events such this can cause significant modification of localized radial electric field with important impact on plasma flow velocity. In fact, a significant reversal of the velocity, concomitant with the temperature drop, follows the reconnection event (the time delay of the measured $T_i$ drop with respect to the rotation negative spike is presumably related to the toroidal asymmetry of the fast collapse process). In spite of this, the feedback induced frequency remains almost unchanged.

There is a crucial difference between TM rotations induced by feedback and those produced by locking the mode to a feed-forward perturbation of given amplitude and frequency. Generally speaking, in the former case the phase difference with respect to the external perturbation stabilizes the amplitude, so the mode saturation level is reduced. Instead, in the latter case the phase difference is destabilizing [5]. However, even the feed-forward induced rotation can mitigate the mode amplitude owing to the screening effect of the passive conductive structure [5] (at least a vacuum vessel is present). Though an experimental comparison of the two techniques for control of wall-locked TM in the same experiment is lacking (in [6] both methods are examined, but they are applied to fast rotating $m$=2, $n$=1), simulations with the RFXlocking code [2] indicate a superiority of the feedback, at least with the external active coils of the RFX-mod layout so far considered. An example is shown in figure 2 for an equilibrium where a growing amplitude $m$=2, $n$=1 TM tends to wall-lock. The feedback is active up to 0.2s, and afterwards it is turned off imposing a feed-forward $m$=2, $n$=1 current on the active coils with amplitude and frequency respectively factors 4 and 0.5 of the values reached during the previous feedback phase. Adjustment of this kind is necessary to sustain feed-forward rotations. Rutherford's equation for the island width is considered only after 0.06s, whereas the width is kept constant to very low level during the first part of the simulation, to allow the mode frequency to settle on the natural value (some kHz). The asymptotic island width of the feed-forward phase, met after the damping of the oscillations, is slightly smaller than the saturation value 0.146m that would be reached in the absence of any control (not shown in the plot), but significantly larger than the value obtained with feedback. One may object that using feed-forward the TM frequency can be forced to a desired value, whereas under feedback this is not possible. However, relaxing the

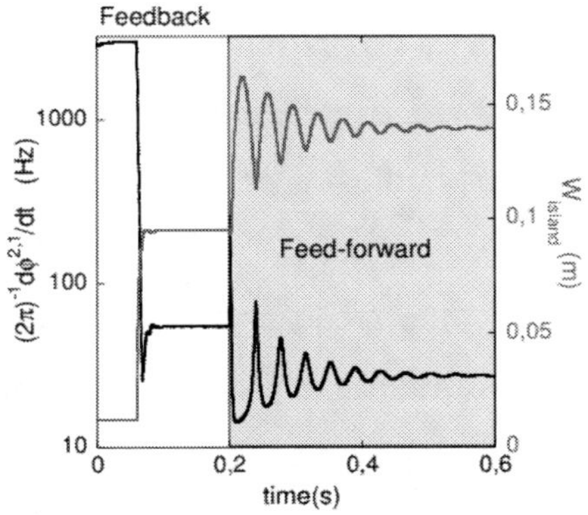


**Fig.2**. RFXlocking simulation of $m$=2, $n$=1 control, showing mode frequency (black; log-scale) and island width (blue; a=0.459).

constraint of zeroing the feedback variable with the addition of a small-amplitude rotating reference it is possible to select the desired TM frequency, in a range of reasonable requirements. This variant of the feedback scheme is shown in Figure 3 for a RFX-mod discharge. Schematically, the feedback law is $I_{coil}^{2,1} = K\left[b_r^{2,1}(r_{sens}) - b_{ref}\exp(i\omega_{ref}t)\right]$. In this example the TM slows down from the fast natural frequency (some kHz) such as to become visible on the feedback sensors, which are located inside the shell, but are screened by the innermost 3ms time-constant vacuum-vessel, only in the time window between 0.2s and 0.45s, where a significant amplitude is detected (plot a). As shown by plot b, during this interval feedback rotation is established with $\omega_{ref}/2\pi$=25Hz imposed by the small non-zero reference ($b_{ref}$=1G). Outside this interval no plasma mode is seen, and sensors detect only the rotating reference: the radial field extrapolated at the plasma radius becomes noisy as the toroidal field signal, which is used to compute it. The transitions from the fast to the feedback rotation at t≈0.2s and its inverse at t≈0.45s are arguably due to equilibrium changes not measurable with the present RFX-mod diagnostics.

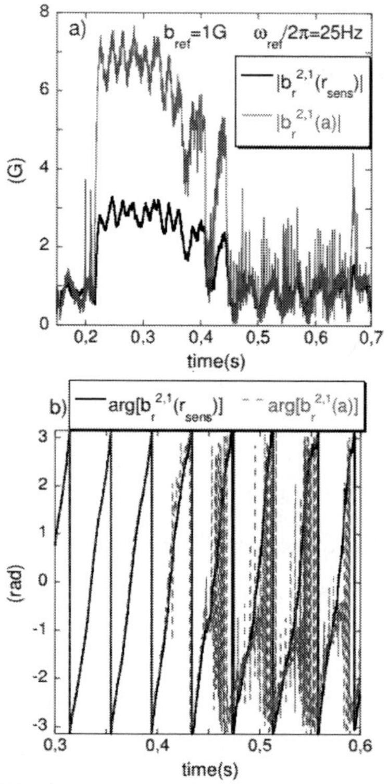


**Fig.3** Example of $m$=2, $n$=1 TM feedback control with non-zero reference at 25Hz in RFX-mod. We show the radial field harmonic at the sensor radius (black), and the extrapolation at the plasma surface (red), computed with the inclusion of the toroidal field signal. Plot a) refers to the amplitude, plot b) to the phase.

In conclusion these experiments performed at DIII-D and RFX-mod have demonstrated that magnetic feedback is a robust technique to avoid wall-locking of the $m$=2, $n$=1 tearing mode, that can be implemented, in principle, in any tokamak device.

[1] M. Okabayashi *et al*, EX/P2-42, 25th IAEA Fusion Energy Conference, Saint Petersburg, Russia (2014); [2] P. Zanca et al, *Nucl. Fusion* **55** (2015) 043020; [3] P. Zanca, L. Marrelli, G. Manduchi, G. Marchiori, *Nucl. Fusion* **47** (2007) 1425; [4] E. J. Strait, *Phys. Plasmas* **22** (2015) 021803; [5] R. Fitzpatrick, *Nucl. Fusion* **33** (1993) 1049; [6] G. A. Navratil *et al*, *Phys. Plasmas* **5** (1998) 1857

*Work supported in part by the U.S. Department of Energy under DE-AC02-09CH11466, DE-FC02-04ER54698, DE-FG02-04ER54761, DE-SC0008520, and DE-FG02-08ER85195.

# Understanding the Physics of EHO in QH-mode on DIII-D Including the Role of Rotation Shear

Xi Chen[1], K.H. Burrell[1], N.M. Ferraro[1], T.H. Osborne[1], C.M. Muscatello[1], C.C. Petty[1], A.M. Garofalo[1], P.B. Snyder[1], R.J. Groebner[1], R. Nazikian[2], G.J. Kramer[2], B.J. Tobias[2], W.M. Solomon[2], Z. Yan[3], G.R. McKee[3], X. Ren[4], N.C. Luhmann Jr.[4], and M.E. Austin[5]

[1]*General Atomics, P.O. Box 85608, San Diego, CA 92186-5608, USA*
[2]*Princeton Plasma Physics Laboratory, P.O. Box 451, Princeton, NJ 08543-0451, USA*
[3]*University of Wisconsin-Madison, 1500 Engineering Dr., Madison, WI 53706, USA*
[4]*University of California Davis, 347 Memorial Un, Davis, CA 53706 USA*
[5]*Institute for Fusion Studies, University of Texas-Austin, Austin, TX 78712, USA*

Quiescent H-mode (QH-mode)[1] is a stationary operation mode with sustained H-mode edge pedestal and confinement but without detrimental edge-localized modes (ELMs). QH-mode operation was first discovered on DIII-D, and was subsequently seen on ASDEX Upgrade, JET, and JT-60U [2]. One of the key components of successful QH-mode operation is the edge harmonic oscillation (EHO), which opens an additional transport channel that maintains the plasma edge at conditions just below the ELM stability boundary [3]. Previous experimental and theoretical studies suggest that the EHO is the saturated state of a kink-peeling mode destabilized by edge rotation [3,4]. This paper presents linear simulations of the EHOs using the M3D-C1 MHD code [5]. The modeled mode structure agrees well with experiments and the theory of EHO being a kink-peeling mode. The rotation and rotational shear effects on the mode growth rate found in the modeling is consistent with theory and experiments. Measurements showing the EHO and the sometimes co-existing broadband MHD might be different modes are also presented.

A typical ELM-free QH-mode discharge (#157102) that is sustained until t~3500ms and accompanied by a strong steady coherent EHO on the DIII-D tokamak is modeled. The plasma is highly shaped, and slightly biased to lower single null (κ=1.9, δ=0.58, dRsep= -6mm). Starting from t=1500ms, the plasma major radius at the outer mid-plane is scanned by ±2cm at 5Hz [Fig. 1(d)] while other plasma parameters are maintained constant to improve the spatial resolution of the diagnostics that have fixed measurement locations. A suite of

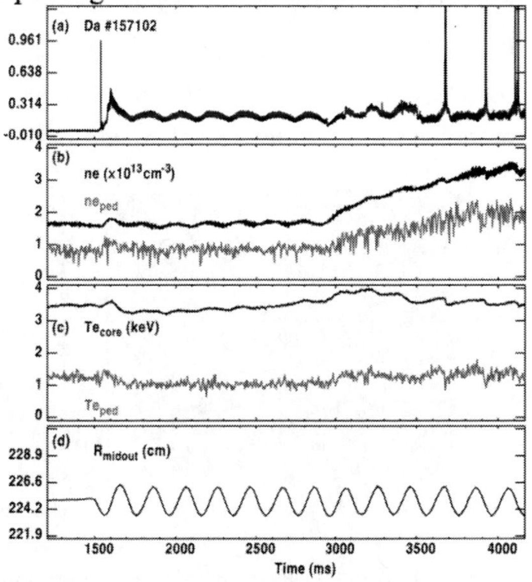


Figure 1: Time evolution of (a) Da light, (b) line-averaged and pedestal (blue) (the density rise after 3000ms is associated with the decreased NBI torque) (c) core and edge (blue) electron temperature, (d) the major radius at outer mid-plane. $|B_T| \sim 2.06$T, $|I_p| \sim 1.1$MA, $q_{95} \sim 5.4$.

pedestal diagnostics provides important data for resolving edge modes with very fine structure such as EHO. The high-resolution edge Thomson scattering system measures electron density and temperature. The edge charge exchange recombination (CER) spectroscopy system provides ion density, temperature and plasma rotation measurements by measuring carbon ions (the dominant impurity on DIII-D). A coherent EHO with an n=1 fundamental and harmonics up to n=7 is detected on edge magnetics sensors (n is the toroidal mode number). Besides the magnetics, the coherent EHO is also detected by electron cyclotron emission (ECE) radiometer, ECE-Imaging (ECE-I), beam emission spectrometer (BES) and microwave imaging reflectometer (MIR). The profile fitting is obtained by mapping data from t=2420±17.5ms onto flux surfaces using EFIT [6] equilibria reconstructed using magnetics data. Then a kinetic EFIT is generated where the bootstrap current is calculated using Sauter model [7] and neutral beam driven current is derived from a ONETWO [8] calculation.

Those profiles and equilibrium are then input into the M3D-C1 code in order to model the EHO. M3D-C1 solves the fluid equations in real 3D toroidal geometry. Linear single fluid M3D-C1 modeling with a resistive wall of n=1 to 7 modes has been conducted for the discharge in Fig. 1 at t~2420ms. Because these are linear calculations, we only compare the modeled mode structure (not the absolute amplitude) to the theory and experiments. Qualitative agreement is found in the comparison between the modeling results and various diagnostics measurements, as seen in Fig. 2. Figure 2(a) shows the modeled 2D mode structure of n=5 mode that agrees well with the ideal ELITE code [9] prediction. As shown in Fig. 2(b), the temperature fluctuation due to the n=5 mode from the modeling and from the ECE measurement both peak near ~227cm with a width of ~2.5cm. The strong density fluctuations (FWHM ~2cm, peaked at 2.5 cm inside the separatrix) associated with the modeled n=5 mode are also comparable to the BES measurements. The modeled 2D temperature and density fluctuations due to the edge mode qualitatively agree with those measured by ECE-I and MIR, respectively. The poloidal wavenumbers of n=1 to 7 derived from the M3D-C1 are consistent with those inferred from measurements at poloidally separated channels of various diagnostics, for example, for n=2, $k_\theta$ ($cm^{-1}$) = 0.04(M3D-C1), 0.04(Magnetics), 0.03(BES), 0.03(MIR) and 0.05(ECE-I). More

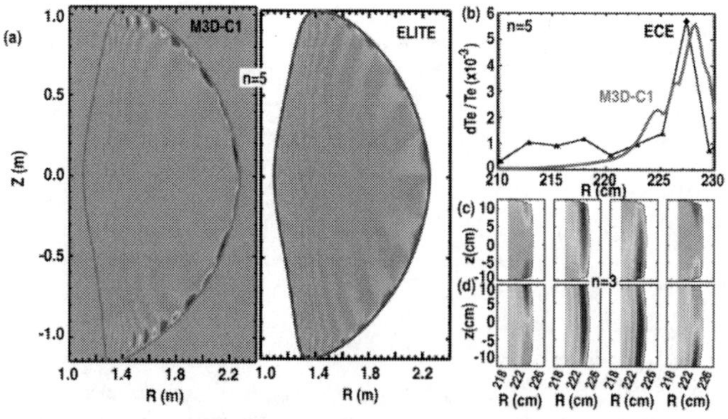


Figure 2: (a) 2D M3D-C1 pressure fluctuations (left) and ELITE predicted displacement (right) of n=5 EHO (both are without rotation); (b) Radial profile of temperature fluctuations of n=5 EHO from ECE (black) and M3DC1 (red); (c) Synthetic MIR signal with (d) M3D-C1 input of 2D density fluctuation of n=3 EHO at four different toroidal locations.

detailed comparisons with MIR measurements will be made by comparing the measurements to the synthetic MIR signal. The synthetic signal is obtained by interposing M3D-C1 modeled density fluctuations due to the EHO into a synthetic MIR diagnostic (simulating the diagnostic response), an example of which is shown in Fig. 2(c), (d) for an n=3 mode.

For a given discharge, different rotation profiles are input into the M3D-C1 code and the corresponding linear growth rates of modes in a range of toroidal mode numbers are studied. In these linear single fluid calculations, the magnitude of both the electron and ion drift velocity perpendicular to the magnetic field is the *E*x*B* drift, which is calculated from CER measurements. Here we modeled a QH plasma (#153440) which has a coherent EHO with dominant n=2 component near t=1725ms. In the series of modeling where the input rotation profile was scaled by 0, 25%, 50%, 75% and 100%, the linear growth rate of n=1 to n=10 modes was calculated. The mode stabilization by rotation and rotational shear occurs for all modes with n≥4. Rotation and rotational shear destabilizes the n=2 mode while barely affecting n=1 and 3 modes. The growth rates of these modes with 0% and 100% of the experimental rotation value are compared in Fig. 3 as an example. This trend is in line with previous ELITE stability calculations showing the rotational shear destabilizing low-n modes while stabilizing high-n modes [4]. The linear growth rate of n=2 increases roughly linearly with increasing rotation and rotational shear, and becomes the highest at the experimental rotation level. This is consistent with the experiment (saturated state) where n=2 EHO is dominant. Peeling-ballooning theory predicts that the EHO can exist with both signs of rotation and previous experiments have confirmed that [10]. We found consistent results in M3D-C1 modeling, that is, the linear growth rate only changes <10% when the rotation direction is reversed.

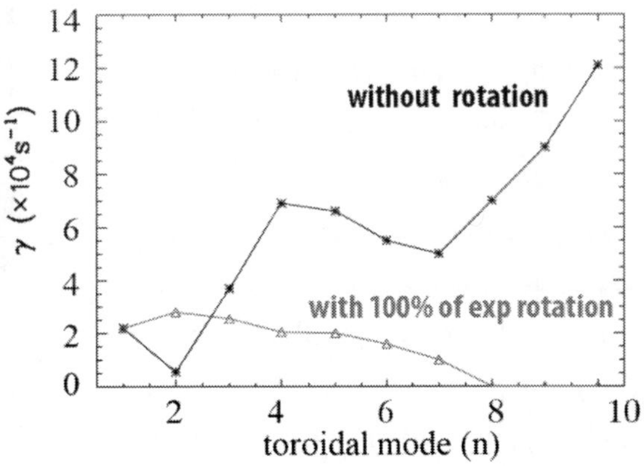

Figure 3: Linear growth rates of n=1 to 10 modes from simulations without (black) and with (red) experimental rotation.

It is often observed that some broadband MHD co-exists with the coherent EHO [Fig. 4(a), t<2700ms]. The coherent EHO rotates toroidally in the direction of plasma rotation while the broadband MHD rotates in the plasma current (Ip) direction; their poloidal phase velocities have opposite signs during counter-Ip neutral beam injection [Fig. 4(b), (c)]. The different rotation characteristics suggest they are different modes. In some highly shaped plasmas at low rotation, the coherent EHO goes away and only the broadband MHD remains [e.g., Fig. 4(a) 2700-3800ms]. These plasmas can develop a higher but wider pedestal that operates below the peeling-ballooning stability limit with good confinement. The increased broadband fluctuations detected by BES suggest the plasma might move to a turbulence-transport dominant regime and further investigation is on going.

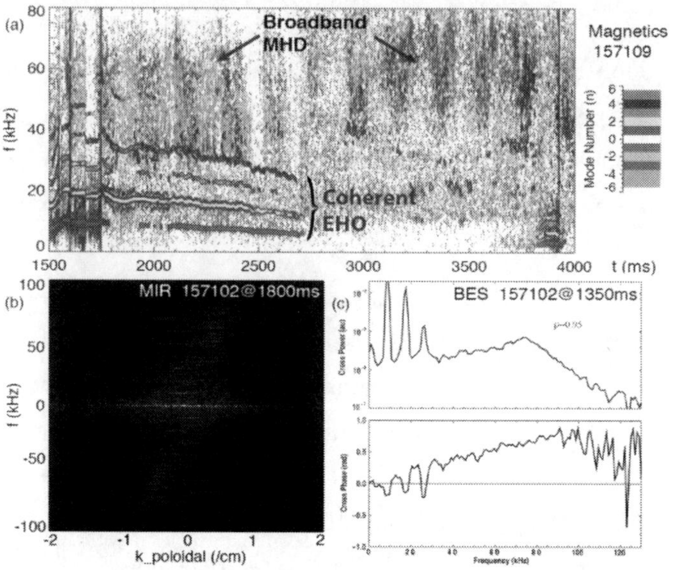


Figure 4: (a) Fourier analysis of magnetics array signals, (b) spectrogram of MIR signal (c) crosspower and phase between two poloidal BES signals of coherent EHO and broadband MHD.

In summary, M3D-C1 is being applied to model the EHO in QH plasmas on DIII-D. The calculated linear eigenmode structure (~2 cm wide radially, located at the edge steep gradient region, $k_\theta$ of 0.02~0.2 $cm^{-1}$ increasing with n) exhibits similar features as the measured EHO by magnetics, ECE, BES, ECE-I and MIR. Consistent with experiments and the initial theory, M3D-C1 modeling shows rotation and rotational shear destabilizes low-n modes while stabilizing high-n modes. Measurements show that the coherent EHO and broadband MHD have different rotation characteristics, suggesting different identities. These findings advance the physics basis for developing stationary ELM-free plasmas for ITER and beyond. However, nonlinear simulations are needed to understand the saturation mechanism for the EHO.

This work was supported by the U.S. Department of Energy under DE-FC02-04ER5468, DE-AC02-09CH11466, DE-FG02-89ER53296, DE-FG02-99ER54531, and DE-FG03-97ER54415.

[1] K.H. Burrell, et al., Bull. Am. Phys. Soc. 44, 127 (1999)
[2] W. Suttrop, et al., Plasma Phys. Controlled Fusion 45, 1399 (2003); W. Suttrop, et al., Nucl. Fusion 45, 721 (2005); Y. Sakamoto, et al., Plasma Phys. Controlled Fusion 46, A299 (2004); N. Oyama, et al., Nucl. Fusion 45, 871 (2005)
[3] T.H. Osborne, et al., J. of Phys. 123, 012014(2008); K.H. Burrell, et al., Nucl. Fusion 49, 085024 (2009)
[4] P.B. Snyder, et al., Nucl. Fusion 49, 961 (2007); A.M. Garofalo, et al., Phys. Plasmas 22, 056116 (2015)
[5] N.M. Ferraro, S.C. Jardin, J. Comp. Phys. 228, 7742 (2009)
[6] L.L. Lao, et al., Nucl. Fusion 25, 1611 (1985)
[7] O. Sauter, C. Angioni and Y.R. Lin-Liu, Phys. Plasmas 9, 5140 (2002)
[8] R.J. Goldston, et al., J. Comp. Phys. 43, 61 (1981)
[9] P.B. Snyder, H.R. Wilson, et al., Phys. Plasmas 9, 2037 (2002).
[10] K.H. Burrell, et al., Phys. Rev. Lett. 102, 155003 (2009)

# Experimental observation of axial ion loss during the Alfven ion-cyclotron instability in the GDT mirror

*A.V.Anikeev[1,2], P.A.Bagryansky[1,2], O.A.Korobeinikova[2], S.V.Murakhtin[1,2], K.V.Zaytsev[1], D.V.Yurov[1].*

*[1]Budker Institute of Nuclear Physics, Novosibirsk, 630090, Russia,*

*[2]Novosibirsk State University, Novosibirsk, 630090, Russia S.V.Murakhtin@inp.nsk.su*

## Introduction

The gas dynamic trap (GDT) is an axisymmetric open system with a high mirror ratio for confinement of collision plasma and anisotropic hot ion population produced by an oblique deuterium neutral beam injection [1]. The hot ion component has the mean energy $E_h \approx 9$ keV and the maximum density $n_h \approx 3 \cdot 10^{13} cm^{-3}$. The collision warm plasma has the electron temperature up to $T_w \approx 250$ eV and density $n_w \approx 5 \cdot 10^{13} cm^{-3}$. The local relative plasma pressure β approaching 0.6 was measured [2]. These parameters allow to propose a project of the 2 MW / m$^2$ fusion neutron source based on the GDT with a simple geometry and relatively low construction and operation costs [3]. The high β coupled with the anisotropic distribution function of the hot ions potentially leads to the instability excitation and, consequently, plasma confinement degradation. The instability relevant to the plasma activity at frequencies of the order of the ion-cyclotron frequency was previously observed and investigated in the GDT experiment [4]. As a result, the instability was preliminarily identified as an Alfvén ion-cyclotron (AIC) instability, and as it has been obtained from the experiment the longitudinal velocity range of particles responsible for the AIC instability excitation is on the order of the initial longitudinal speed of injected ions. This range is consistent with the theory [5]. This paper presents an investigation of the hot ion losses in the GDT experiment during the AIC instability and comparison with the appropriate theoretical model.

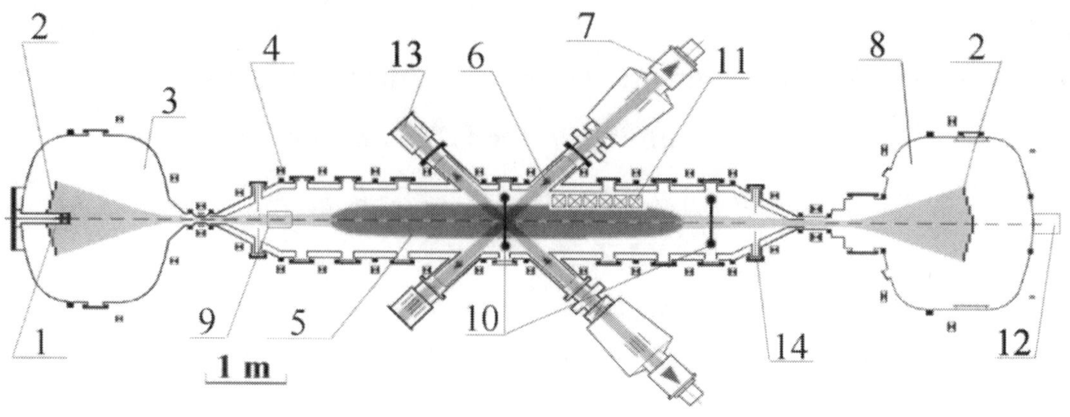


**Fig. 1.** Layout of the GDT experiment. 1 – plasma gun, 2 – plasma absorber, 3 – gun tank (eastern expander), 4 – coil of the magnetic system, 5 – fast ions motion region, 6 – deuterium beams, 7 – atomic injection system, 8 – western expander, 9 – gas box, 10 – diamagnetic loops, 11 – magnetic probes, 12 – energy analyzer, 13 – heating beams receivers tanks, 14 - limiter.

## Experimental setup

A layout of the GDT experimental setup is presented on Fig.1. The longitudinal ion velocity is an important parameter responsible for AIC developing. Moreover it is required to measure magnetic field value, excited wave frequency and its longitudinal length in instability generation region. We have used the magnetic probes (11 on Fig. 1) for identifying wave parameters listed above.

The experimental investigation of longitudinal loss spectra were made by the modernized 5-channel energy analyzer [6] (see 12 on Fig.1), installed in at the end wall in the GDT expander cell. The magnetic analyzer was the main diagnostic for measurement of an axial fast ion loss value during AIC developing. Particles leaving the central cell continue driving along the force lines and enter through the vacuum valve to the chamber with the magnetic separator, which divides particles depending on its energy into different registration channels. As a result, the hot ion loss spectra in the range of 5 – 25 keV were obtained.

## Numerical simulation

The energy distribution functions of hot ions leaving the trap through magnetic mirrors were numerically simulated by a Monte-Carlo transport code MCFIT [7]. In addition to this method a fast one-dimensional code DOL [8] was employed for hot ion spectra modeling. The experimentally measured energy distribution of ions leaving the GDT along magnetic lines is demonstrated on Fig. 2. Distributions obtained in numerical simulations using MCFIT and DOL codes are shown there for comparison with the experimental data.

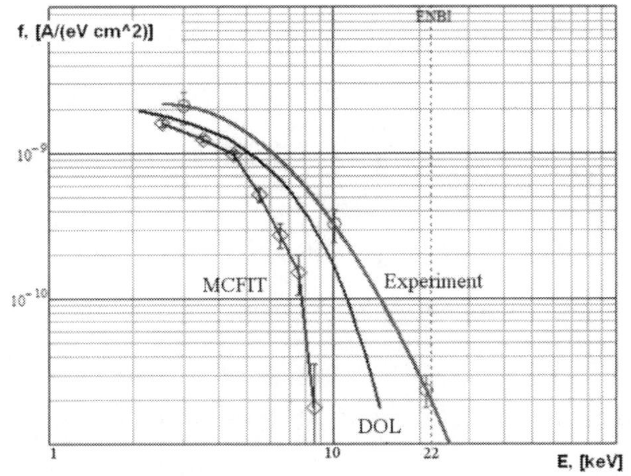


**Fig.2.** Experimentally measured ion energy distribution on the GDT end wall and results of simulations. The energy of injected atoms $E_{NBI}$=22 keV is pointed by a vertical line.

The tolerable agreement of the experimental spectrum with the simulation results points first, to the Coulomb character of hot ions interaction with warm target plasma and, second, to the validity of measurement method submitted. Insignificant distinction of absolute values in spectra shown on Fig. 2 is explained by the difference between input data in the models and real experiment parameters.

## Results and discussion

The measured energy spectra of hot ions leaving the central cell facility during AIC developing are shown on Fig. 3.

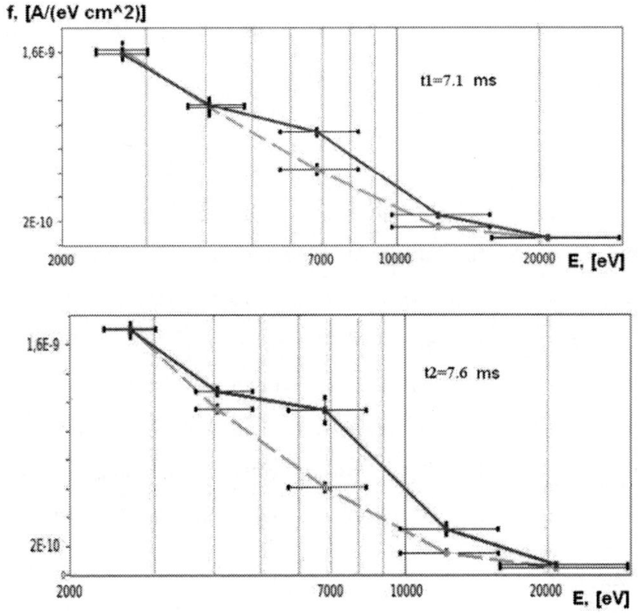


**Fig. 3.** Energy distributions of fast ions during AIC growth (solid line) and without it (dotted line) for different time points of the GDT experiment.

As it is seen on Fig.3 the ion flow increases during AIC growth in energy range 4 - 12 keV. These values are certainly less than energy of injected deuterium atoms $E_{inj}$= 22 keV. Obtained results confirm a hypothesis that the only minor group of hot ions is found to be in resonant conditions with the Alfven wave. This group has energies closed to ultimate value of $E_{inj}$ and angle dispersion closed to injected atoms dispersion. The particles leave the resonance region during their relaxation without achieving the loss cone boundary. In that way the ions efficiently give their energy to electrons and heat it up.

The relative AIC power loss estimated as a $(P_{||}^{AIC} - P_{||}^{noAIC})/P_{tr}$ , were $P_{tr}$ is a power deposited to the plasmas by powerful atomic beams, gives us the values of 0.2% and 0.5% for different time points presented on Fig.3. This experimental fact shows that energy losses corresponded to the AIC are negligible.

## Conclusion

The energy spectrum of hot ions leaving the trap through magnetic mirrors in the energy range of 2.5 keV – 22 keV was obtained for regime with AIC developing and without one. A fact that the ion loss flow has increased during the AIC growth in middle energy range indicates effective participation of hot ions in plasma heating, so that the drag time of hot ions less than the time of scattering to the loss cone.

The energy losses corresponded to the AIC are less than 1%. The result obtained confirms the theoretical prediction that only small group of hot ions actively interacts with the wave for the GDT experimental conditions. Those ions are located in the narrow region of phase space near the injection point. A small drag or scattering is sufficient for taking out from the resonance area.

## References

[1] V.V.Mirnov and D.D.Ryutov, Sov. Tech. Phys, Lett. **5**, 678 (1979).

[2] P.A.Bagryansky et al. Fusion Science and Technology, Volume **59**, Number 1t, Pages 31-35 (2011).

[3] P. A.Bagryansky, et al. Fusion Engineering and Design **70**, (2004).

[4] K V Zaytsev et al 2014 Physica Scripta. 2014 014004.

[5] I. S.Chernoshtanov and Ya. A.Tsidulko Fusion Sci. Technol. **63** 319.

[6] V.V.Afrosimov et al. JTP, 1975. Volume **45.** Number 1. Pages 56-63.

[7] A.V.Anikeev, A.N.Karpushov et al. // Transaction of Fusion Technology. 2001. V.39, No.1T, PP. 183-186.

[8] D.V.Yurov et al. // Fusion Science and Technology. 2013. V.63, No. 1T, pp. 313-315.

# Study of the NBCD in the spherical tokamak Globus-M

V.B. Minaev [1], P.B. Shchegolev [1], N.N. Bakharev [1], F.V. Chernyshev [1], V.K. Gusev [1], G.S. Kurskiev [1], I.V. Miroshnikov [2], M.I. Patrov [1], Yu.V. Petrov [1], N.V. Sakharov [1], I.Yu. Senichenkov [2], A.Yu. Telnova [2], S.Yu. Tolstyakov [1], E.G. Zhilin [3]

[1] *Ioffe Institute, St. Petersburg, Russia*

[2] *Peter the Great St. Petersburg Polytechnic University, St. Petersburg, Russia*

[3] *Ioffe Fusion Technology Ltd., St. Petersburg, Russia*

Non-inductive current drive becomes a pressing need in application of the spherical tokamak for the reactor usage. Therefore, one of the main operational aims of the Globus-M machine is to investigate neutral beam current drive (NBCD). Noticeable NBCD in Globus-M has been observed tentatively in a number of experiments. The latest experimental results are reported in the paper together with the outcome of performed numerical simulations.

**Globus-M experimental set-up and main techniques**

The Globus-M spherical tokamak [1] is equipped with the mid plane neutral beam injector (NBI), which provides hydrogen or deuterium beams with the energy up to 30 keV [2]. It is capable of delivering up to 1.0 MW beam power. The impact parameter is 0.32 m (see fig. 1). Electromagnetic system of the Globus-M tokamak allows wide possibilities for plasma shaping. Typical configurations used for studying of NBCD and heating during the last experimental campaign are shown in fig. 2. The target plasmas were hydrogen and deuterium.

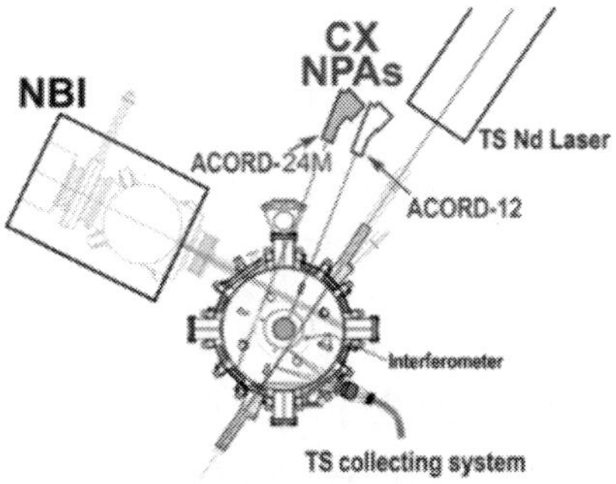


Fig. 1 Globus-M experimental set-up

The experiments were performed in the plasma current range of 0.13-0.20 MA and with the toroidal magnetic field 0.4 T. The co-injected beam was also either hydrogen or deuterium. In the experiment we used IPM-2 ion source with lower in comparison with

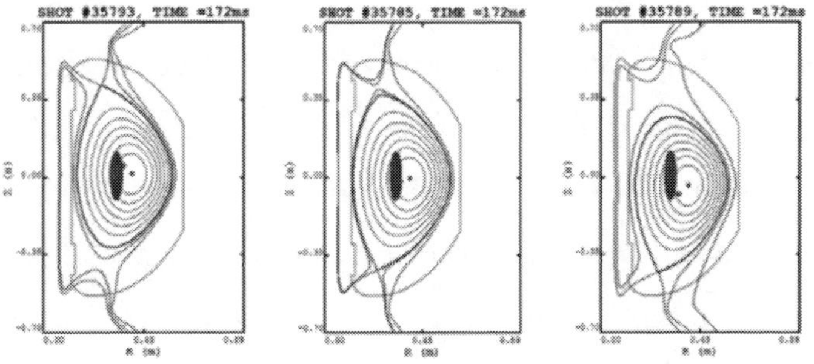

Fig. 2 Magnetic configurations with different vertical displacement (from left to right: +5cm, -2cm, -7cm correspondently). Brown ellipse – neutral beam footprint; red dot– magnetic axis; greyish-blue point – geometric center

IPM-1 output power (up to 0.5 MW), but with smaller beam cross-section of about 20×4 cm. Plasma line averaged density was monitored in vertical direction by means of the microwave interferometer. Electron temperature and density profiles were measured in mid plane with the help of Thomson scattering diagnostics. Ion component behavior was investigated with two neutral particle analyzers, which aimed in transversal and parallel to the beam line directions. Also the ion temperature was measured by means of CXRS. Magnetic configuration was reconstructed using the EFIT code [3]. The target plasma parameters were simulated with the help of ASTRA transport code [4,5]. Light impurity (carbon) spatial distribution fits to loop voltage signal. A full 3D fast ion tracking algorithm combined with the solution of the Boltzmann kinetic equation [6] was applied to calculate the beam power deposition for each beam energy fraction.

**Experimental results**

Plasma shot with moderate current (170 kA) was chosen as a basic target. As it is seen from fig. 3, the effect of the NBCD appeared in a short-term rise of the plasma current and prolonged drop of the loop voltage. Such behavior is defined by response of the plasma current feedback loop to additionally driven current. Initially we chose radial plasma column position, where loop voltage drop was most noticeable. A two centimeter shift of plasma column towards the tokamak axis relative to the vacuum vessel center seems to be optimal. Under these conditions we investigated a dependence of non-inductive fraction of the current on plasma density. Hydrogen beam (0.5 MW, 28 keV) was injected into deuterium plasmas during stationary phase of a discharge. The averaged density varied in the range of 1.5 -5.0 × $10^{19}$ $m^{-3}$ from shot to shot (see fig.4). The values of non-inductive current calculated using ASTRA code are shown in fig.5. One can see that beam-driven current reduces with plasma density rise. Alternatively, the boot-strap current increases approximately by the same value. The total

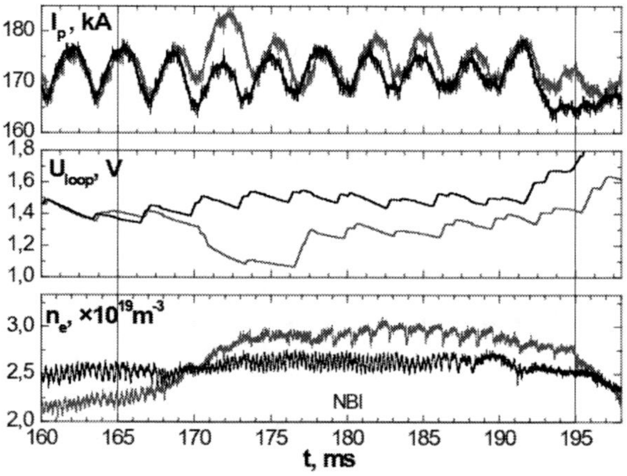


Fig. 3 Evolution of the plasma parameters during NB injection in #35783 (red) and complementary OH #35784 (black) shots. H-beam (28 keV, 0.5 MW) was injected into D-plasmas

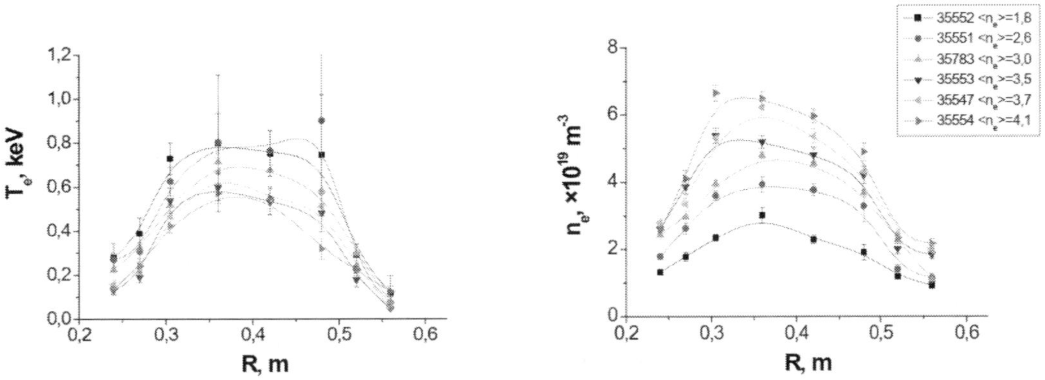


Fig. 4 Electron temperature and density spatial distribution in shots during NBI

fraction of the non-inductive current was about 17-25 % in reported experiment. The beam power, which was absorbed by plasma, did not exceed 30%.

The comparison of hydrogen and deuterium NBI into hydrogen and deuterium plasmas was performed in passed campaign. The boot-strap current is independent of isotopic combinations and increases with density definitely. Efficiency of the beam driven current is higher for hydrogen injection in comparison with deuterium one. This fact can be explained by significantly higher level of the direct (shine-through and first-orbit) losses for deuterium beam in the Globus-M tokamak because of the relatively low toroidal magnetic field. As it was reported in [6,7] direct losses can exceed 70% for low density plasmas. Experimental conditions have to change appreciably in the Globus-M2 machine [8] with significantly higher magnetic field (up to 1 T).

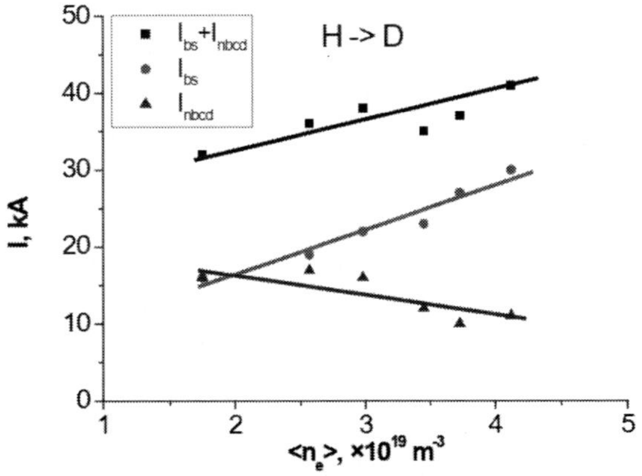


Fig. 5 Non inductive current dependence on plasma density. H-beam (28 keV, 0.5 MW) was injected into D-plasmas

The attempt of off-axis injection was made recently by displacing the Globus-M plasma by up to ±0.08 m vertically. Typical plasma column positions were shown in fig.2. We had to reduce plasma current down to 130 kA maintaining identical plasma target parameters in this experiment (see fig. 6). No significant difference in a value of NBCD was found. Simulations by means of the ASTRA code showed that beam driven current is about 13-14 kA for all shots. Nevertheless, the poloidal magnetic flux consumption $\Psi_{pol}$ was minimal for non-shifted column position. Such behavior could be explained by some reduction of electron temperature and increase of impurity influx in the case of plasma column displacement. At the same time plasma column displacement range was at least twice smaller than the dimension of the neutral beam in vertical direction and effect was weak.

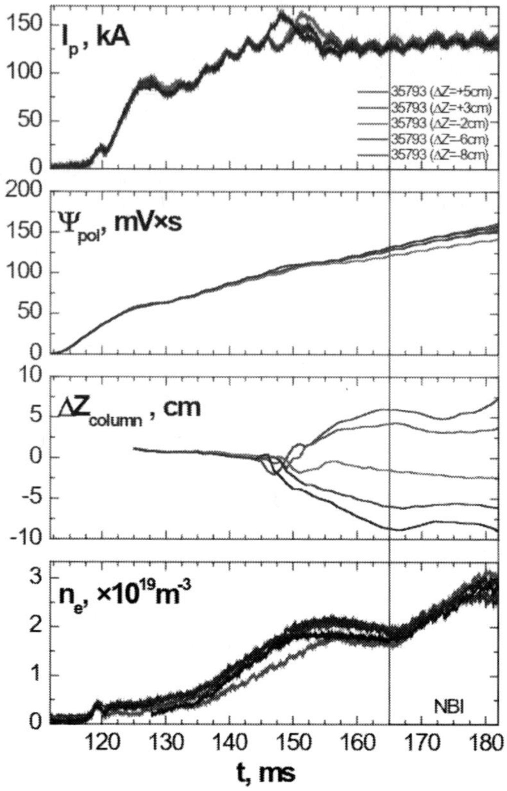

Fig. 6 Evolution of plasma parameters in shots with different plasma column vertical displacement. H-beam (28 keV, 0.5 MW) was injected into D-plasmas

## Conclusions and future plans

The performed simulations of the NBI in Globus-M indicated that around 25% of the plasma current is driven non-inductively, where the beam driven fraction is up to one half. The efficiency of NBCD reduces as the plasma density rises. Alternatively the boot-strap current increases following concentration. The off-axis NBI experiment showed negligible dependence of beam driven current on plasma column vertical displacement in Globus-M conditions.

Further experiments using higher NBI power are planned for the near future in order to increase non-inductive current fraction. However we expect significant progress in NB heating and current drive experiments in upgraded Globus-M2 machine, where the toroidal magnetic field attains 1 T and neutral beam direct losses fall manifold. By then a new neutral beam injector (1 MW, 50 keV, 1 s) will be installed in addition to existing one.

## Acknowledgments

The work has been carried out in the Joint Research Center “Materials science and characterization in advanced technology”. V.B. Minaev, V.K. Gusev and N.V. Sakharov are grateful to the Ministry of Education and Science of the Russian Federation for the financial support (Agreement 14.621.21.0007, id RFMEFI62114X0007).

**References:**

[1] Gusev V.K., Golant V.E., Gusakov E.Z., et al., Technical Physics, 44 (1999) No. 9, 1054-1057

[2] Gusev V.K., Dech A.V., Esipov L.A., et al., Technical Physics, 52 (2007) No. 9, 1127-1143

[3] Lao L.L., John H. St., Stambaugh R.D., et al., Nucl. Fusion 25 (1985) 1611

[4] Pereverzev G.V., Yushmanov P.N., Dnestrovskii A.Yu., et al.,, Repor t IPP 5/42, 1991

[5] Polevoi A., Shirai H. and Takizuka T., JAERI -Data/Code 97-014, 1997

[6] Bakharev N.N., Chernyshev F.V., Goncharov P.R., et al., Nucl. Fusion, 55 (2015) 043023

[7] Gusev V.K., Azizov E.A., Alekseev A.B., et al., Nucl. Fusion, 53 (2013) 9, #093013

[8] Minaev V., Gusev V., Patrov M., et al., Proc. of 25th FEC IAEA conf., Saint-Petersburg, 2014, FIP/P8-25

# Modeling of the fast ion behavior in the Globus-M spherical tokamak

N.N. Bakharev[1)], F.V. Chernyshev[1)], P. R. Goncharov[2)], V.K. Gusev[1)], A.D. Iblyaminova[1)], V.A. Kornev[1)], G.S. Kurskiev[1)], A.D. Melnik[1)], V.B. Minaev[1)], M.I. Mironov[1)], M.I. Patrov[1)], Yu.V. Petrov[1)], N.V. Sakharov[1)], P.B. Shchegolev[1)], S.Yu. Tolstyakov[1)], G.V. Zadvitskiy[2,3)]

[1)] *Ioffe Institute, St. Petersburg, Russia*

[2)] *Peter the Great Polytechnic University, St. Petersburg, Russia*

[3)] *IJL UMR 7198 CNRS - Université de Lorraine, BP 70239, F-54506 Vandoeuvre-lès-Nancy, Cedex France*

**Introduction**

Fast ions will play a key role in the future compact fusion neutron sources (CFNS) and two-energy component fission-fusion hybrid reactors based on the concept of a spherical tokamak [1]. Understanding of their behavior is essential for the design of future machines and choice of their operation modes. However it is not studied in details in contrast to the fast ion behavior in conventional tokamaks due to the difficulties in interpretation and a low number of spherical tokamaks with significant auxiliary heating, which is the source of fast ions in present day laboratory-scale experiments. Globus-M spherical tokamak (R/a = 0.36/0.24 = 1.5, $B_{tor}$ = 0.4 T, $I_p$ = 0.2 MA, neutral beam injection (NBI) of 0.3-1 MW, 18-30 keV ) offers opportunity to study energetic particle behavior and to compare modeling results with the experiments. Globus-M2 (R/a = 0.36/0.24 = 1.5, $B_{tor}$ = 1 T, $I_p$ = 0.5 MA, two NBIs of 1 MW, 18-30 and 20-50 keV) [2], which is under construction now, will expand the scope of the available experimental conditions. The current report continues the series of work aiming 1) to create and test a model that adequately describes the fast ion losses, plasma heating, current generation, yields and spectra of generated neutrons; 2) to verify this model in experiments on Globus-M and Globus-M2; 3) to compare it with results obtained by the other computer codes such as NUBEAM [3].

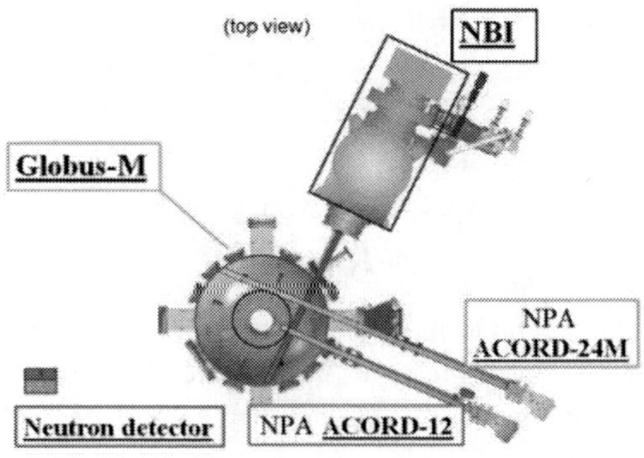


Figure 1. Experimental set-up.

**Experimental set-up and main techniques**

Experimental set-up is shown in Figure 1. The main used diagnostic tools are two charge exchange (CX) neutral particle analyzers (NPA) and neutron detector. ACORD-12 NPA is used for studying the evaluation of the ion temperature, isotope composition and energy spectrum of the trapped ions. ACORD-24M NPA, which observes plasma with the same

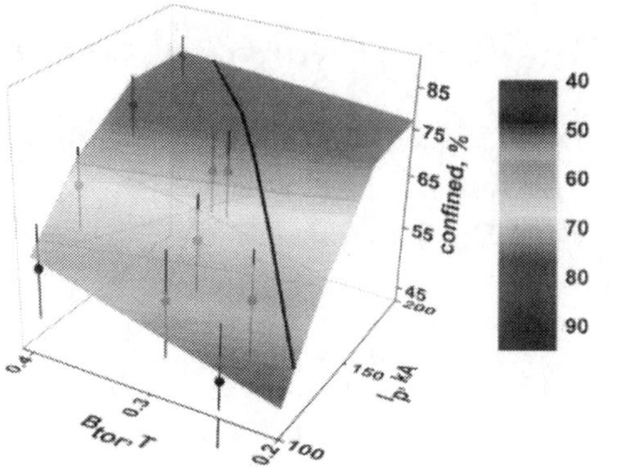


Figure 2. Simulated dependence of the confined fast ion fraction on plasma current and toroidal magnetic field for the 18 keV H NBI (3D surface) and experimental estimates of the percentage of the confined particles (dots).

impact parameter as the NBI, is used for recording of the energy spectrum of the passing particles. Electron temperature and density profiles were measured by means of the Thomson scattering diagnostics. Plasma shape was reconstructed by means of the EFIT code [4] and neutral density distribution was calculated using DOUBLE code. Two codes utilizing different principles were applied for the calculation of the fast ion distribution: the NUBEAM module and the full 3D fast ion tracking algorithm combined with the solution of the Boltzmann kinetic equation [Goncharov P.R. et al. Phys. Plasmas 17 (2010) 112313]. Fast ion tracking algorithm is used to calculate the source function for the Boltzmann equation taking into account direct losses. The Boltzmann kinetic equation describes the ion slowing down. It contains the Landau collision term, pitch angle scattering term, slowing-down losses term and velocity diffusion term.

**The influence of the magnetic field and plasma current on fast ion confinement**

Parametric dependence of the confinement of fast particles produced by the 18 keV H beam on magnetic field and plasma current was studied in Globus-M. The modeling result obtained by the tracking algorithm is shown as a 3D surface in Figure 2. Experimental estimates are indicated by points with error bars. The experimental border separating the MHD stable region for the set of the performed experiments from the unstable one is drawn as a solid line of constant Ip/B ~ 500 kA/T. The strong variation of direct losses with plasma current is seen. The reason of this lies in the fact that the change in plasma current completely changes orbit shape while the increased magnetic field only makes the orbits more narrow, which is illustrated in figure 3 where orbits calculated by the 3D tracking algorithm are shown.

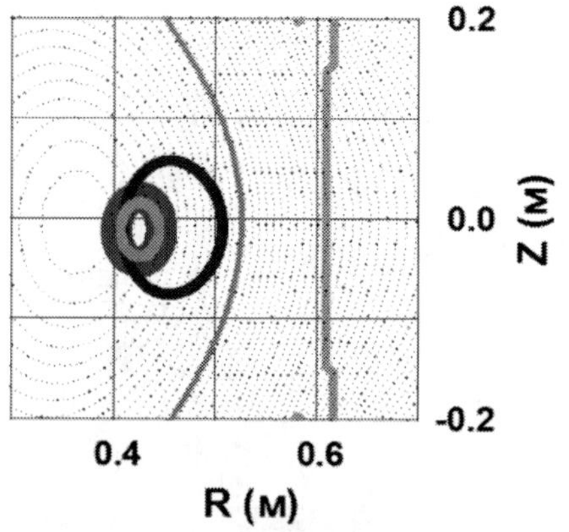


Figure 3. Orbits of the 18 keV H ions ionized in the same point for the case of 200 kA and 0.4 T (green), 105 kA and 0.4 T (blue) and 200 kA and 0.2 T (red).

## Counter-NBI

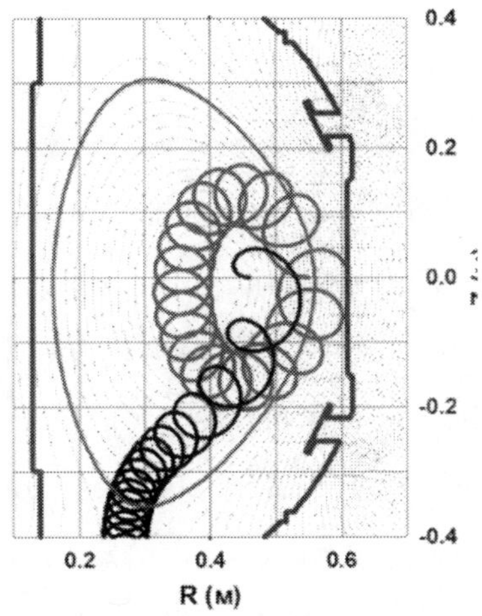


Figure 4. Typical orbit of the lost ion during co- (green) and counter- (black) NBI. Plasma border is shown by the red line.

First experiments on counter-NBI were carried out in the Globus-M spherical tokamak. ELM-free H-mode was obtained. The main feature of these experiments is extremely high first orbit (FO) losses of the fast ions. Orbit modeling shows that even for the case of the 18 keV H NBI they are around 85-95% depending on the experimental conditions. For comparison in the case of the co-NBI FO losses are around 15%. NUBEAM simulations show more optimistic result – FO losses are less than 50%, while total losses are ~ 30%. This significant difference is obviously due to the guiding center approximation used in NUBEAM. Despite the fact that the "generalized" finite Larmor radius correction, adapted for the spherical tokamaks, is incorporated in NUBEAM, nonnegligible difference from the result obtained without approximation by tracking algorithm exists for these specific conditions. It is interesting to note, that orbits of the lost ions during counter-NBI differ from the orbits of co-injected lost ions. While during co-NBI energetic ions which undergone FO losses hit the outer wall, during counter-NBI they get into the divertor region. Typical orbits of the lost ions in co- and counter-NBI cases are shown in Figure 4.

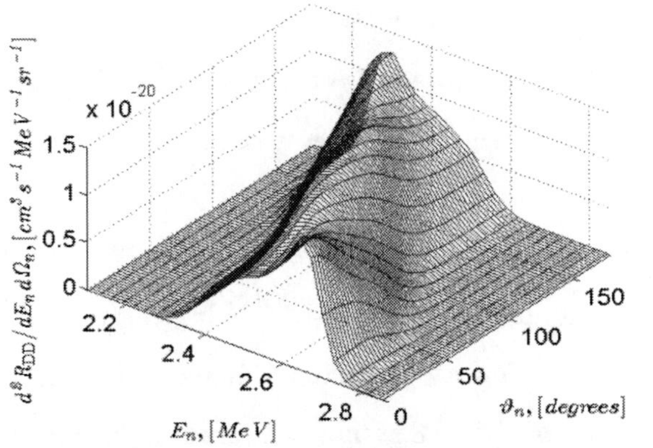


Figure 5. Calculated double differential local rate coefficient of deuterium-deuterium fusion reaction depending on neutron energy and laboratory angle in the center of Globus-M2.

High fast particle losses result in an insignificant ion heating as compared to the case of co-NBI. Only 10-20% ion temperature growth is observed, while during co-injection threefold increase is easily obtained [5]. This fact is consistent with the simulation results. Modeling and experiment did not reveal strong dependence of FO losses on plasma current and inward plasma shift in contrast to the case of the co-NBI [6]. This leads to the fact that in Globus-M2 FO losses during counter-NBI will be around 70% both for the 30 and 50 keV beams. Further 10% FO losses reduction is possible by decreasing beam impact parameter by 4 cm. Despite the fact that fast ion losses during counter-NBI still significant, the full absorbed NBI power will be higher by order of magnitude. It will be up to 0.5 MW for the 2-injector case, which is greater than absorbed NBI power in Globus-M during co-NBI.

**Neutron spectra calculations**

Since one of the main Globus-M/M2 tasks is the support of the compact neutron sources line, it is important not only to be able to calculate neutron yield, which is routinely done by many computer codes, but also to calculate beam-plasma neutron spectra. Figure 5 shows the energy and angle distribution of the local DD fusion neutron source calculated for the parameters in the center of Globus-M2 plasma column using the formula derived in [7] and a simple analytical model for the anisotropic velocity distribution of fuel nuclei described therein. At the next step, the calculations of expected distributions for the particular observation geometry of a neutron spectrometer on Globus-M2 are planned on the basis of a more profound model.

**Conclusions**

Investigation of fast ion confinement in Globus-M tokamak was continued. New modeling techniques were developed and successfully applied. Dependence of the direct losses on plasma current and magnetic field was studied. Peculiarities of the energetic particle behaviour during counter-NBI were investigated. Neutron spectrum for the upcoming Globus-M2 was calculated.

**Acknowledgements**

The work has been carried out on unique scientific device “Globus-M”. N.N. Bakharev, V.K. Gusev, V.B. Minaev, N.V. Sakharov and S. Yu Tolstyakov thank for financial support Ministry of Education and Science of the Russian Federation (Agreement 14.619.21.0001, id RFMEFI61914X0001).

**References**

[1] Masayuki Ono and Robert Kaita, Physics of plasma 22 (2015) 040501

[2] Gusev V.K. et al. Nucl. Fusion 53, 093013 (2013)

[3] Pankin A. et al. Comp. Phys. Comm. 2004 159 157

[4] Lao L.L. et al. 1985 Nucl. Fusion 25 1611

[5] Bakharev N.N. et al. 2013 TPL 39 1085

[6] Bakharev N.N. et al. Nucl. Fusion 55 (2015) 043023

[7] Goncharov P.R. Nucl. Fusion, 55 (2015) 063012

# Plasma investigation using SPD XUV silicon photodiodes on the Globus-M tokamak

A.D. Iblyaminova[1], A.G. Alekseyev[2], P.N. Aruev[1], N.N. Bakharev[1], V.K. Gusev[1], N.A.Khromov[1], G.S. Kurskiev[1], V.B. Minaev[1], E.E. Mukhin[1], A.I. Panov[2], M.I. Patrov[1], Yu.V. Petrov[1], N.V. Sakharov[1], P.B Shchegolev[1], S.Yu. Tolstyakov[1], A.V. Voronin[1], V.V. Zabrodsky[1] and the Globus-M Research Team[1]

[1] *Ioffe Institute, Saint-Petersburg, Russia*

[2] *SRC RF TRINITI, Troitsk, Moscow, Russia*

## Introduction

One of the effective ways for high temperature plasma investigation is measurement of radiation losses. Diagnostic methods based on appliance of such silicon photodiodes as AXUV (Absolute eXtreme UltraViolet) [1] or SPD (Silicon Precision Detector) [2] are connected with developing comprised system. Diagnostic complex should take into account drop of sensitivity of the photodiodes to electromagnetic radiation in <30 eV region [3-5]. The contribution of this band to the total radiation power could be significant as was obtained by methods of filtered AXUV with VUV spectrometer [5] and by comparing with thermal bolometers as was made in a wide range of experiments on tokamaks. Nevertheless high sensitivity, temporal resolution and relative simplicity of AXUV or SPD based diagnostics make them attractive instruments for plasma radiation exploring.

## Experiment

Experiment was performed on the compact spherical tokamak Globus-M ($B_T$=0.4 T, $I_p$<0.25 MA, R/a=1.5) [6]. The diagnostics for radiated power measurements based on SPD was supplemented with four-channel filtered SPDs. Detectors are sensitive to radiation in the following spectral ranges: channel 1 – 1.12–7.7 eV, channel 2 – 1.12–10.9 eV, channel 3 – 330 eV–15 keV, channel 4 – 1.12 eV–15 keV (10% cut-off transmittance). Thus low energy range is covered by first two channels using 0.95 mm magnesium fluoride and 0.7 mm quartz filters respectively. Soft X-ray range is measured by channel 3 using silicon nitride film with aluminum coating [7] instead of traditional beryllium foil. Channel 4 remained unfiltered. Spectral sensitivities of the channels are presented in

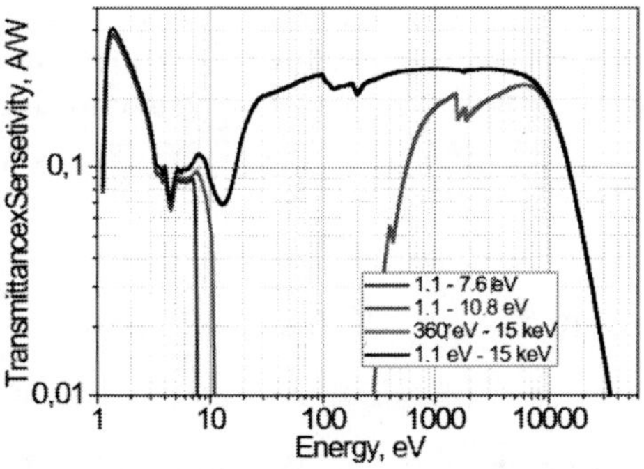


Figure 1. Spectral curves of 4-channel spectroscopic system based on SPD.

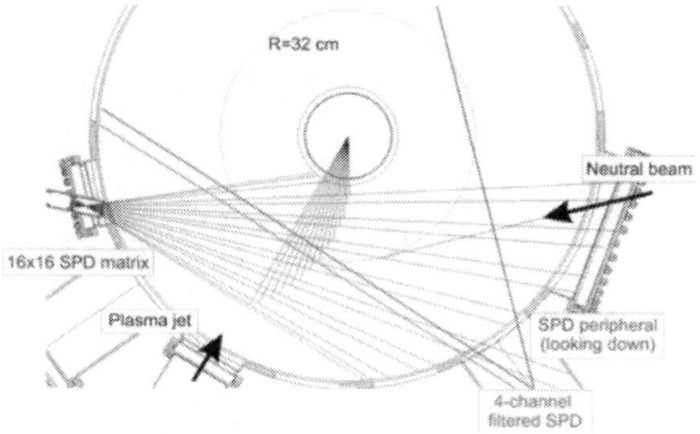


Figure 2. FOV layout of SPDs in the equatorial midplane of Globus-M (cones are not shown). Lines of view of 2 remained channels in filtered system are duplicating.

the figure 1. Detected radiation spectrum was split in 4 bands: A (1–7.7) eV, B (7.7–10.9) eV, C (10.9–330) eV, D (330–$1.5\cdot10^4$) eV. Radiation in A and D bands was measured directly; radiation in B band was obtained as difference between two first channels, radiation in C band was derived by subtraction of radiated power measured by channels 2 and 3 from 4th. Each channel is collimated by two 1 mm apertures and has tangential field of view (FOV) through plasma core. SPD FOV layout on the tokamak is shown in the figure 2. Experiments were performed with deuterium plasmas, plasma current was 170 kA. Deuterium beam with $P_{NBI}$=450 kW, $E_{beam}$=27 keV was used as auxiliary heating method.

**Results**

Radiated power in various spectral ranges and its contribution to the total spectral range (1.12eV – 15 keV) – $P_i/P_{tot}$ – was investigated. It was found nearly linear dependence of radiated power in the whole and SXR ranges on electron density for both ohmic heated and NBI discharges (figure 3a, b). As seen from the figures radiated power detected by filtered SPD based system for NB heated discharges was higher than for OH for the whole spectrum and D band.

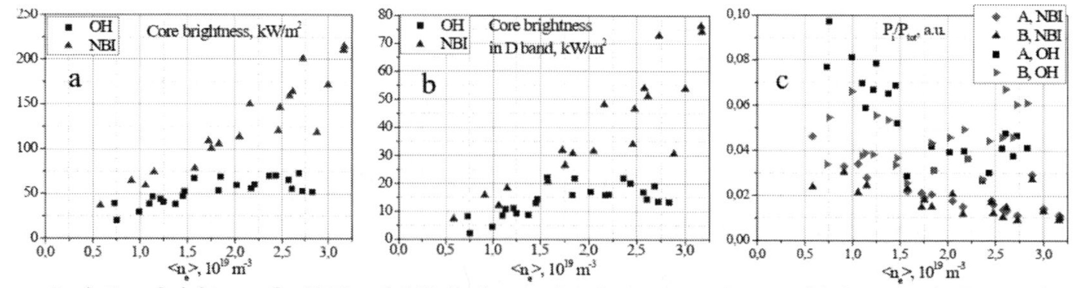


Figure 3. a) Core brightness for NBI and OH discharges in whole spectral range, b) the same in D SXR band. c) Dependence of $P_i/P_{tot}$ from chord average electron density at R=42 cm for NBI and OH shots in A, B spectral ranges; $P_i$ – radiated power in i-th spectral range (i=A, B), $P_{tot}$ – power radiated in the whole spectral range.

Slight increase of radiation in B range, where Ly-alpha and CIV (8 eV) are apparently dominated, for OH discharges was also observed. Besides lower electron temperature in OH, gas puff might cause more radiation by cold neutrals injection to the plasma periphery. Radiated power in A band and CIII (465 nm) line intensity have shown no significant difference between two heating regimes. So contribution $P_i/P_{tot}$ for low energy ranges up to

11 eV (A and B ranges), for ohmic discharges was higher than for NBI heated plasmas (fig.3c).

Possible reason of increased radiation in NB heated regimes could be higher impurity content due to high energy particle interactions with plasma facing materials. However further spectroscopic measurements and precise calculation of ionization balance are needed.

For the discharges contribution to radiation losses was: in C band ~ 60-70%, in D band ~ 30-20%, A and B bands up to ~ 15%. Introducing <*S*> as the ratio of photocurrent in 4th channel and sum of radiated power in all spectral ranges led to the values of ~0.21-0.22 A/W mean spectral responsivity in various regimes.

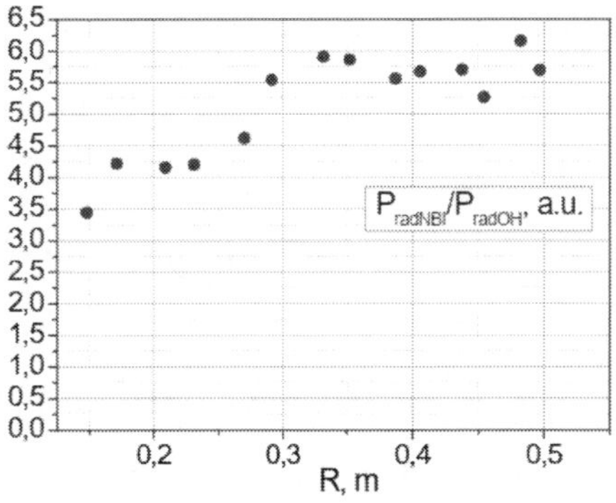


Figure 4. Ratio of radiation losses in NB and OH discharges from equatorial linear array of 16x16 SPD for $<n_e>=3*10^{19}$ m$^{-3}$.

Brightness distributions measured by 16x16 SPD hybrid array were compared for two heating regimes. Typical ratio of brightness in NBI and OH discharges for $<n_e>=3\cdot10^{19}$ m$^{-3}$ from equatorial arrays is presented in the figure 4. It is seen that for plasma core and outer border the ratio is increased compare to values measured by SPD from filtered system (fig. 3a). Also values obtained by single peripheral SPD with chord of view passing down tangentially to the outer border of plasma showed radiation augmentation in NBI shots only up to 30%. Even though there is no accurate spectral responsivity to high energy particles for SPDs their contribution to array signal due to 16x16 matrix viewing geometry is obvious.

Injection of helium and deuterium jet using plasma gun [8] into the tokamak with deuterium discharges for diagnostic and fuel delivery purposes correspondingly was carried out. Plasma jet parameters were: density ~$10^{22}$ m$^{-3}$, kinetic energy ~200 - 300 eV, vacuum plasma velocity ~ 150 km/s. After He injection electron density rose in accordance with diffusion time along whole

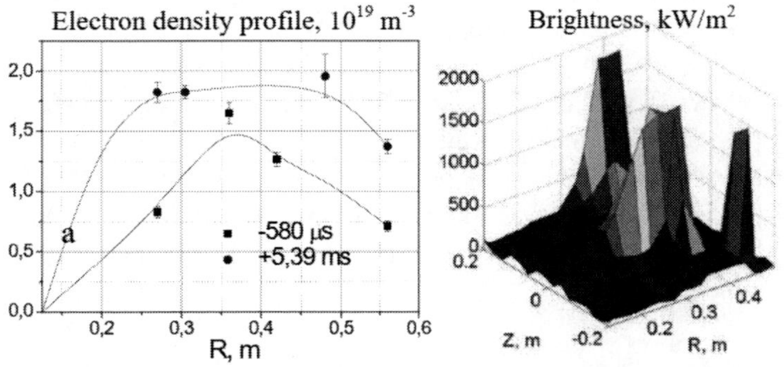


Figure 5. He jet penetration in tokamak plasma. a) Electron density profile before and after jet injection. b) Brightness distribution at t=46 μs after He jet injection obtained by 16x16 SPD matrix array; R-major radius, Z – vertical coordinate.

major radius of plasma (fig. 5a). Slightly decrease of electron temperature was also observed. Brightness distribution corresponding to maximum He jet penetration is presented in the figure 5b. Permeation up to the plasma core was seen. In the deuterium injection case poor penetration was achieved. Core density rose insignificantly, although electron temperature fell down almost twice. Evolution of the images registered by matrix array showed increase in radiation spreading toward to plasma core with high velocity.

**Conclusions**

For the goals of more accurate radiation losses measurements using silicon photodiodes diagnostic system containing 4 channels based on SPD with filters was installed on the Globus-M tokamak. Filtered channels provide different energy cut-offs in SPD spectral sensitivity and allow to take into account reduced sensitivity of SPDs in low energy range. The values of effective spectral sensitivity that could be used for direct calculation was achieved, although additional spectroscopic data in UV range will be necessary for more precise calculation. Radiation losses measurements for NB heated plasmas have shown up to 4-fold higher level compare to OH regimes. This could be probably because of additional impurity inflow. No increase in radiation for NB plasmas in 1-10.9 energy band was observed. Remarkable contribution of high energy particles to 16x16 SPD array response is revealed for NBI shots. Penetration of He plasma jet up to plasma core was obtained with high reproducibility.

**Acknowledgments**

The work has been carried out on unique scientific device “Globus-M”. A.D. Iblyaminova, N.N. Bakharev, G.S. Kurskiev, M.I.Patrov and P.B Shchegolev thank for financial support RFBR (grant 14-02-31152-mol-a).

**References**

[1] Korde R. et al., Proc. SPIE, 1140, **126** (1989);
[2] Scholze F. et al. Metrologia **43**, S6 – S10 (2006);
[3] Alekseyev A.G. et al., Plasma & Fusion Research, 2007, **2**, S1061;
[4] Watanabe F. et al., Plasma & Fusion Research, 2007, **2**, S1066;
[5] D. S. Gray, S et al. Rev. of Sci. Instr., **75**, No. 2 (2004);
[6] Gusev V.K. et al., Technical Physics, **44**, No. 9, 1054 (1999);
[7] Pekka T. Torma et. al. IEEE Trans. on Nuclear Science, **61**, No. 1, 695 (2014);
[8] Voronin A.V. et al., Nucl. Fusion, **45**, 1039–1045 (2005).

# GOL-NB Project: New Multiple-Mirror Trap with NBI-Heated Plasma

V.V. Postupaev[1,2], V. I. Batkin[1,2], A. V. Burdakov[1,3], I. A. Ivanov[1,2], K. N. Kuklin[1], K. I. Mekler[1], S. V. Polosatkin[1,3], A. F. Rovenskikh[1], A. V. Sudnikov[1,2]

[1] *Budker Institute of Nuclear Physics, 11 Lavrentjev Avenue, 630090 Novosibirsk, Russia*

[2] *Novosibirsk State University, 2 Pirogova st., 630090 Novosibirsk, Russia*

[3] *Novosibirsk State Technical University, 20 Karl Marx Avenue, 630092 Novosibirsk, Russia*

## I. INTRODUCTION

Existing modern concepts of fusion-grade open traps usually incorporate special end sections to improve axial confinement of plasma. These sections utilize either effect of ambipolar plasma potential (in projects of tandem-mirror-based reactors) or an effective friction force between a corrugated magnetic field and plasma stream (in projects of multiple-mirror reactors). In a multiple-mirror confinement system [1], plasma expansion along a corrugated magnetic field is slowed down due to an effective friction force between populations of locally-trapped and transiting particles. Under optimum conditions (i.e. at the free path length of ions $\lambda_i$ is close to the corrugation period $l$), the particle confinement time $\tau$ scales as the square of the device length $L$: $\tau \approx R^2(L^2/\lambda_i v_{Ti})$, where $v_{Ti}$ is the ion thermal speed and $R = B_{max}/B_{min}$ is the mirror ratio.

The multiple-mirror approach to fusion was experimentally studied in Novosibirsk in the experiments with sub-fusion plasma on GOL-3 device [2]. The plasma was heated by a high-power electron beam; it was highly turbulent therefore. Confinement of turbulent plasma in GOL-3 is governed by several specific collective mechanisms that strongly differ the existing device from a classical trap that was considered by theory. This prevents us from the explicit demonstration of confinement scalings in a sufficient parameters range.

GOL-3 experiments have demonstrated an efficient multiple-mirror confinement at sub-fusion plasma parameters. Further advance to a fusion-grade device can be either along the existing paradigm of a turbulent plasma confinement or towards a new configuration featuring a central section for confinement of NBI-heated plasma and end multiple-mirror solenoids that will reduce plasma losses along the magnetic field.

In this paper, we discuss plans for a major reconfiguration and upgrade of the existing GOL-3 device. The main physical objective of new facility will be the direct demonstration of capacity of multiple-mirror end sections to reduce the end losses.

II. NEW PHYSICAL PROGRAM FOR GOL-3

Modern understanding of linear confinement systems was implemented in the GDMT project [3] that will integrate features from both the gasdynamic (GDT) and multiple-mirror traps. NBI-heated two-component plasma will be confined in GDMT in a GDT-like central magnetic section. Two multiple-mirror end sections will slow down the axial plasma flow. In the pure GDT concept [4], an ambipolar potential forms that retards electrons and changes particle losses from the trap. The mean energy $\varepsilon$ of an e-i pair that escapes the trap is $\varepsilon \approx 8T_e$, where $T_e$ is the electron temperature of the warm plasma [4]. The lifetime of fast ions and fusion efficiency are determined by the electron temperature that in turn is maintained by the power balance between the drag for the fast ions and power loss through the mirrors. It can be increased if some method of reducing the power losses is implemented.

A fast-track verification of the main multiple-mirror scalings can be achieved at the existing GOL-3 facility. We plan to install a new low-field central trap in the middle of the existing solenoid. Plasma will be heated there by a 1.5 MW NBI system. Then the plasma will flow through the solenoids to end expander tanks. The solenoids can be reconfigured from a simple solenoidal configuration with 4.8 T field to a multiple-mirror one with 22 cm period and 1.4 mirror ratio - see Fig. 1.

In this project, we assume that all physics known from GDT experiments [4] will work, including classical slowing down of fast ions, absence of turbulence-related additional significant power losses, scalings for the rate of gasdynamic losses, and methods of MHD stabilization. Figure 2 demonstrates results from a simple energy balance model that includes main physical processes in the system. Here, the shaded area restricts the parameter space for the multiple-mirror confinement. Lower dashed line corresponds to a pure GDT configuration. The optimal density for start experiments is in the region of $(3-5)\times10^{19}$ m$^{-3}$. For this density, the temperature in the pure GDT regime will be already suitable for the multiple-mirror confinement. The temperature should grow with the activation of the multiple-mirror configuration. Two steps of research are proposed. During the first step, a classical multiple-mirror confinement will work in the solenoids. The expected plasma

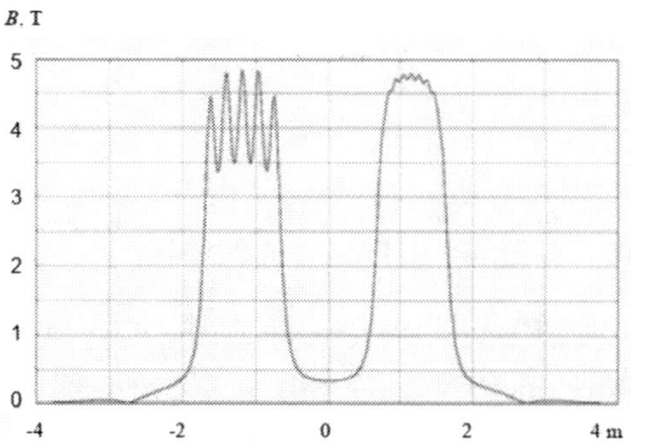


Fig. 1. The magnetic field profile at the axis for a configuration with the central trap for NBI heating, two solenoids and two expanders with cusps. The left solenoid is in the standard multiple-mirror mode, the right one operates as the simple solenoid.

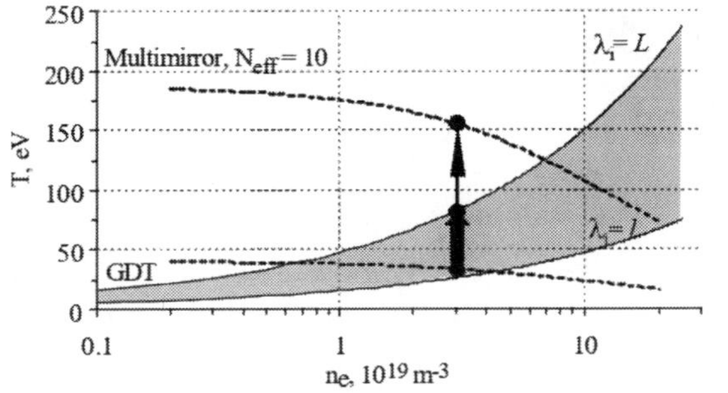


Fig. 2. Parameter space for the proposed experiment (calculated with a simple model for 1.5 MW NBI, plasma Ø 20 cm, $B_{center}$ = 0.3 T, $B_{mirror}$ = 4.5 T). Solid lines constrain the domain of multiple-mirror confinement. Lower dashed line indicates a pure GDT regime, upper dashed line corresponds to a maximal possible 10-fold reduction of axial losses.

parameters will correspond to the upper solid line in Fig. 2, i.e. we expect about twofold temperature growth. Advance to higher temperatures is possible if we will be able to implement a technology of a non-binary particle scattering in the solenoids.

Summarizing this Section, the baseline physical program includes the following [5]. Task 1: transform GOL-3 to the configuration with the central trap and two simple solenoids and demonstrate the GDT-like confinement. Task 2: demonstrate the temperature growth in the multiple-mirror configuration. Task 3: extend the achievable parameter space to higher temperatures with control of the particle free path length. Task 4: improve the plasma parameters with other plasma heating methods.

## III. UPGRADE PLANS FOR GOL-3

The existing GOL-3 is a multipurpose plasma physics facility; current research programs cover the areas listed in the Introduction. The proposed upgrade of GOL-3 includes separation of the existing facility into two partly-independent devices that are more dedicated to specific physics. The first device GOL-3T will continue beam-plasma-related research. The rest of the existing 11-m-long solenoid will be adapted to an autonomous operation. For some time, this longer part of the solenoid will be used as a testbed for several new technologies for the next-step experiments in the multiple-mirror confinement, including new plasma creation system and two new 25 keV, 0.7 MW, 1 ms neutral beam injectors [6]. After that, new vacuum chamber, new pumping system, and new plasma creation system will replace the existing hardware in that section. During the next modernization step, the central trap and two end expander tanks will be connected to the solenoid. The neutral beam injectors will be mounted at the central trap. At this step, MHD stabilization will be provided by either the magnetic configuration or the plasma biasing. The plasma sustainment will be as long as NBI injection, i.e. 1 ms.

The further physical program supposes application of additional plasma heating methods and turbulence control in the multiple-mirror sections. This should improve plasma parameters and expand the available parameter space. The detailed hardware specifications for these steps will be completed later.

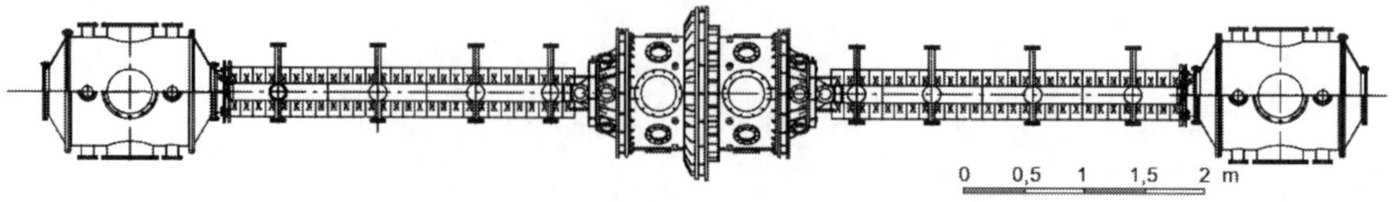


Fig. 3. Layout of GOL-NB after the proposed reconfiguration of GOL-3 (front view). The device consists of the central trap with 2 × 0.75 MW NBI, two multiple-mirror solenoids, two expander tanks and a plasma gun.

## VI. SUMMARY

For more than four decades since the introduction of the idea of multiple-mirror plasma confinement, GOL-3 was the only physical facility that provided experimental data on confinement of hot plasma with sub-fusion parameters. Achievements in the GOL-3 program have demonstrated that the multiple-mirror approach works even better than it was initially expected for the case of turbulent beam-heated plasma. However, the use of the high-power electron beam for plasma heating challenges scalability of the system to truly reactor-grade steady-state parameters, though visions of such systems exist [7].

In this paper, we introduce plans for the GOL-NB device that will be optimized for the multiple-mirror confinement studies. Two main ideas behind the proposed upgrade are the separation of plasma heating and confinement processes and the achievement of a quasi-stationary plasma state. In the proposed configuration, a new low-field central section with 1.5 MW NBI heating will be embedded in the center of the existing multiple-mirror solenoid. The central section will work as a miniaturized GDT trap with the same well-established physics. Depending on the magnetic configuration of the adjacent multiple-mirror section, the baseline plasma losses through mirrors will change. Preliminary calculations predict growth of the electron temperature from 30 eV up to 75 eV for $3\times10^{19}$ $m^{-3}$ density at a transition from a GDT-like confinement to a supposed multiple-mirror regime. Currently, design and modeling activities are carried out.

## ACKNOWLEDGMENTS

The authors are grateful to GOL-3 team, GDT team, and BINP plasma theory group for their valuable discussions and comments.

## REFERENCES

1. G.I. Budker, et al., *JETP Letters*, **14**, 212 (1971).
2. A. Burdakov, et al., *Fusion Sci. Technol.*, **55** (No. 2T), 63 (2009).
3. A. Beklemishev, et al., *Fusion Sci. Technol.*, **63** (No. 1T), 46 (2013).
4. A.A. Ivanov, V.V. Prikhodko, *Plasma Phys. Control. Fusion*, **55**, 063001 (2013).
5. V.V. Postupaev, et al., *Fusion Sci. Technol.*, **68** (2015). doi: 10.13182/FST14-846
6. V.I. Batkin, et al., *Fusion Sci. Technol.*, **59** (No. 1T), 262 (2011).
7. A.V. Burdakov, et al., *Fusion Sci. Technol.*, **59** (No. 1T), 9 (2011).

# First observation of an electron internal transport barrier in Heliotron J

N. Kenmochi[1,2], T. Minami[3], C. Takahashi[3], S. Tei[1], K. Nishioka[1], T. Mizuuchi[3], S. Kobayashi[3], K. Nagasaki[3], Y. Nakamura[1], H. Okada[3], S. Kado[3], S. Yamamoto[3], S. Ohshima[3], S. Konoshima[3], G. M. Weir[3], Y. Otani[1], F. Sano[3].

[1] *Graduate School of Energy Science, Kyoto University, Gokasho, Uji, Japan*

[2] *Research Fellow of Japan Society for the Promotion of Science*

[3] *Institute of Advanced Energy, Kyoto University, Gokasho, Uji, Japan*

## Introduction

Electron Internal Transport Barrier (ITB) formation is reported in many helical devices from the viewpoint of Core Electron-Root Confinement (CERC), which is characterized by highly peaked electron temperature profiles together with large positive ambipolar radial electric fields (neoclassically defined as the electron-root solution) [1]. In Heliotron J, an ITB with a peaked electron temperature profile caused by improved thermal transport has recently been observed with on-axis Electron Cyclotron Heating (ECH).

This paper is the first report on the characteristics of the observed ITB plasma in Heliotron J. The experimental transport analysis and a comparison with neoclassical calculations are carried out, and the results are discussed.

## Experimental Setup

Heliotron J is a medium-sized helical-axis heliotron device (the averaged plasma major radius is $R$ = 1.2 m, the averaged minor radius is $a$ = 0.17 m, the averaged magnetic field strength at the magnetic axis is $B_{ax}$ = 1.35 T) with an $L/M$ = 1/4 helical coil, where $L$ is the pole number and $M$ is the pitch number of the helical coil [2, 3]. Figure 1(a) shows the time evolution of plasma parameters, the line averaged electron density, $\overline{n_e}$, and the plasma stored energy, $W_p^{dia}$, with ECH injection power, $P_{ECH}$. The second-harmonic 70 GHz ECH beam with the extraordinary mode is perpendicularly injected and focused on the magnetic axis. The field strength is adjusted to the resonance condition of the microwaves at the focused magnetic axis. In this experiment, $P_{ECH}$ is controlled to be gradually reduced from 330 kW to 120 kW after $t$ = 210 ms. The line averaged electron density is kept approximately constant and equal to $\overline{n_e} \sim 1.0 \times 10^{19}$ m$^{-3}$. The stored energy slightly decreases with reduced $P_{ECH}$.

## Experimental Results

Figures 1(b) and 1(c) show typical electron temperature ($T_e$) and density ($n_e$) profiles respectively for two heating powers, $P_{ECH}$ ~ 240 kW and 175 kW. Here, the $T_e$ and $n_e$ profiles are measured using a Nd:YAG Thomson scattering system (YAG-TS)[4, 5]. Reduction of $P_{ECH}$ causes decreased $T_e$ gradient in the core region ($r/a$ < 0.3), while $T_e(r/a)$ in the outer region ($r/a$ > 0.3) is almost the same between the two heating powers. The maximum value of $T_e(0)$ reaches 1.5 keV with a steep gradient of 30 keVm$^{-1}$ at $r/a$ = 0.2.

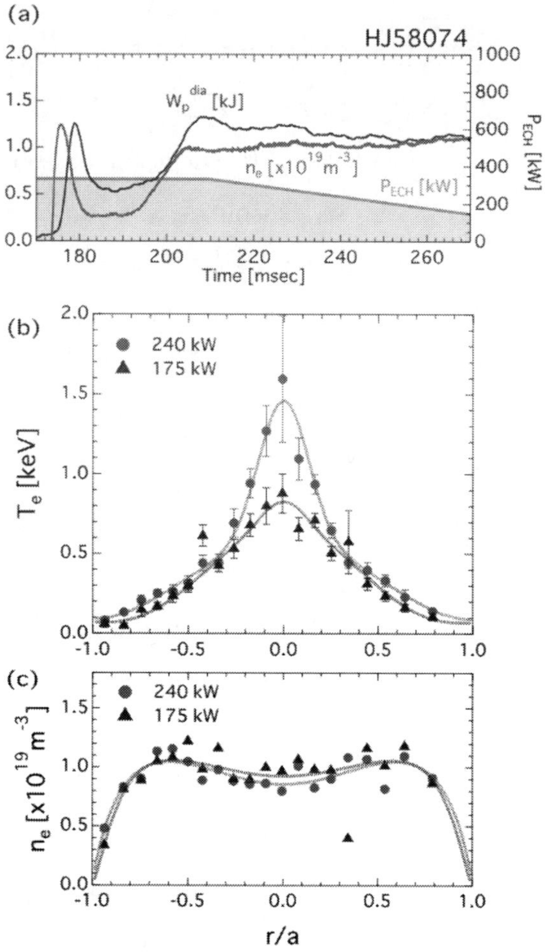


Figure 1. (a) Time evolution of the plasma stored energy, line averaged electron density and injection power of ECH. Profiles of (b)$T_e$ and (c)$n_e$ with ECH (injection power: 175 kW and 240 kW).

## Heat Transport Analysis using Experiential Results

The effective electron thermal diffusivity profiles ($\chi_e^{eff} = (Q_{ECH}/n_e)/\nabla T_e$) are evaluated by using $T_e(r/a)$, $n_e(r/a)$ and the single-pass ECH deposition profiles ($Q_{ECH}$) for the cases of $P_{ECH}$ ~ 240 kW and 175 kW (Fig. 2). Here, $Q_{ECH}$ is calculated using the TRAVIS ray tracing code [6, 7]. In this estimation of $\chi_e^{eff}$, we neglect the electron-ion energy transfer and impurity radiation losses, whose contributions are negligible especially in the core region. The uncertainty in $\chi_e^{eff}$ comes from that in the $T_e$ and $n_e$ profile measurements. For the low injection power of 175 kW, the $\chi_e^{eff}$ is approximately constant in the whole confinement region, which reflects the constant $T_e$ gradient of entire region. For the 240 kW case, the $\chi_e^{eff}$ decreases significantly from 10 m$^2$/s to 4 m$^2$/s between $r/a$ = 0.4 and $r/a$ = 0.2; however, the $\chi_e^{eff}$ is higher in the outer region of the plasma ($r/a$ > 0.3) than at the lower power. This is come from that, the shape of $T_e$ profile does not change relating to profile stiffness, despite the large difference of $P_{ECH}$ (240 kW and 175 kW). In the core region ($r/a$ < 0.3), the large increase of $T_e$ gradient indicates a reduction of $\chi_e^{eff}$ to the lower level than that for the case of $P_{ECH}$ ~ 175 kW.

The transport analysis shows the clear reduction of $\chi_e^{eff}$ for $r/a < 0.3$ in the highly peaked $T_e$-profile case, which suggests improved confinement in the core region.

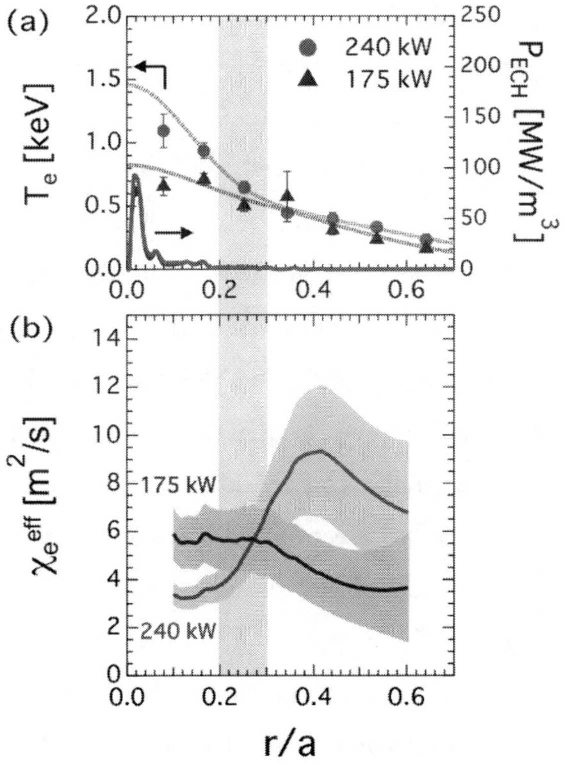


Figure 2. Radial profiles of (a) the $T_e$, ECH deposition and (b) the electron thermal diffusivity, $\chi_e^{eff}$, for the plasmas with and without electron ITB.

## Comparison with Neoclassical Theory

A neoclassical (NC) calculation using the DKES code [8, 9] with the measured $n_e$ and $T_e$ profiles is carried out for the plasma shown in Fig. 1 to compare the experimental results to NC theory. Figures 3(b) and 3(c) show the theoretically expected ambipolar radial electric field, $E_r$, and the neoclassical prediction of the electron thermal diffusivity ($\chi_e^{NC}$), respectively. For both ITB and non-ITB plasmas within $r/a \sim 0.3$, only the electron root (large positive $E_r$) is predicted, while electron and ion roots are predicted to coexist in the outer region ($0.3 < r/a < 0.6$). This discussion focuses on the core region where only the electron root is expected, while experimental measurements of $E_r$ is required to determine which root is realized in the outer region. At $r/a \sim 0.2$, a strong positive $E_r$ (12 kV/m) is expected with the ITB plasma while the value of $E_r$ reduces to almost half its value for the non-ITB plasma as shown in Fig. 3(b). The $E_r$ shear also increases around the plasma center ($0.2 < r/a < 0.3$) for the plasma with ITB, which could suppress fluctuation leading to the reduction of anomalous transport.

Within $r/a < 0.3$, the predicted values of $\chi_e^{NC}$ for the ITB plasma and the non-ITB plasma are almost the same and much smaller than the values of $\chi_e^{eff}$. In addition, the difference of the $\chi_e^{eff}$ between the ITB and non-ITB plasma is much larger than that of $\chi_e^{NC}$. Since the experimentally obtained $\chi_e^{eff}$ contains both neoclassical and anomalous transport effects, the difference in $\chi_e^{eff}$ is attributed to a reduction of anomalous transport in the core region with ITB formation.

## Summary

In Heliotron J, an ITB caused by improved thermal transport in the core region was observed with centrally focused ECH microwaves injected into plasma for the first time. The $T_e$ exceeds 1.5 keV in the core region, and the heat transport analysis shows significant improvement of the effective electron thermal diffusivity $\chi_e^{eff}$ in plasma with an ITB over that without the ITB. The DKES calculation predicts that a large positive $E_r$ is formed in the core region, although a transition from ion- to electron- root has not yet been experimentally observed and verification will be required. The expected $E_r$ profile should have a strong $E_r$ shear around $r/a \sim 0.3$, which can lead to suppressed anomalous thermal transport and improved confinement.

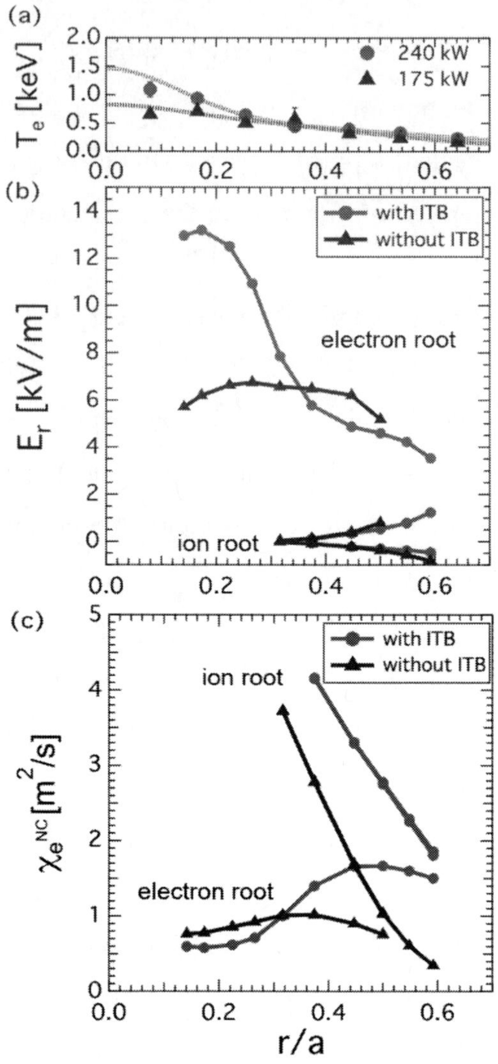


Figure 3. Results of the neoclassical calculation using the DKES code for the plasmas with and without ITB formation. (a) Radial profiles of (a) the $T_e$, (b) electric-field, and (c) electron thermal-diffusivity profiles.

## Acknowledgments

This work was partially supported by the NIFS collaborative Research Program (NIFS10KUHK030, NIFS09KUHL, and NIFS10KUHL033) and JSPS KAKENHI Grant Number 15J09177.

### Reference

[1] M. Yokoyama *et al.*, Nucl. Fusion **47** (2007) 1213

[2] F. Sano *et al.*, J. Plasma Fusion Res. Ser. **3** (2000) 26

[3] T. Obiki *et al.*, Nucl. Fusion **41** (2001) 833

[4] N. Kenmochi *et al.*, Plasma and Fusion Res. **8** (2013) 2402117

[5] N. Kenmochi *et al.*, Rev. Sci. Instrum. **85** (2014) 11D819

[6] Marushchenko N. B. *et al.*, Plasma Fusion Res. **2** (2007) S1129

[7] K. Nagasaki *et al.*, Nucl. Fusion **51** (2011) 103035

[8] Hirshman. S. P. *et al.*, Phys. Fluids **29** (1986) 2951-9

[9] Van Rij *et al.*, Phys. Fluids B **1** (1989) 563

# Temporal evolution of high-density plasmas produced by advanced fuelling techniques in Heliotron J

Y. Ohtani[1], S. Ohshima[2], N. Asavathavornvanit[1], T. Akiyama[3], T. Minami[2], K. Tanaka[3], F. Sano[2], K. Nagasaki[2], Y. Nakamura[2] , H. Okada[2], S. Kado[2], S. Kobayashi[2], S. Yamamoto[2], S. Konoshima[2], G.M. Weir[2], N. Kenmochi[1], X. Lu[1] and T. Mizuuchi[2]
[1] *Graduate School of Energy Science, Kyoto University, Gokasho, Uji 611-0011, Japan*
[2] *Institute of Advanced Energy, Kyoto University, Gokasho, Uji 611-0011, Japan*
[3] *National Institute for Fusion Science, 322-6 Oroshi-cho, Toki 509-5292, Japan*

## 1. Introduction

Advanced fuelling control methods have recently been developed for the production of high density plasma and control of density profile and are utilized in many devices. In Heliotron J, such high performance plasmas have successfully been realized using a supersonic molecular beam injection (SMBI) or a short pulsed high intensity gas puff (HIGP) techniques [1,2,3].

Recently, a new far infrared (FIR) laser interferometer with 1 μs time resolution has been developed to measure such high density plasmas, and the first results were successfully obtained [4]. The temporal evolution of density profile shape obtained with FIR- and microwave- interferometers and the comparison between SMBI and the HIGP plasmas are discussed in this article.

## 2. Experimental setup and results

The Heliotron J device is a medium sized ($\langle R_0\rangle/\langle a_p\rangle$ = 1.2m/0.17m, $\langle B_0\rangle \leq$ 1.5T) helical axis Heliotron device with L/M = 1/4 helical coil, where L and M are the pole number of the helical coil and helical pitch number, respectively. Figure 1 shows the top view of the Heliotron J device, and the arrangement of heating and fuelling systems and major diagnostic equipment. For density control, a conventional gas-puffing system is usually used, and HIGP is realized using the gas-puffing system shown "GAS" in Fig.1. A SMBI system is installed on 11.5 port shown "SMBI" in Fig.1 and injects SMB at 1~2 km/s.

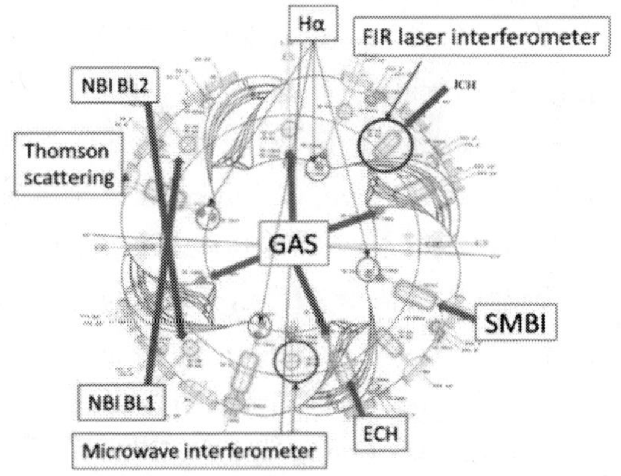

Fig. 1 Top view of Heliotron J device and the arrangement of heating and fuelling system and diagnostics.

A FIR laser interferometer and a microwave interferometer are utilized to measure line averaged density in this study, are combined to estimate time evolution of density profile shape. As shown in Fig. 2 the microwave- and FIR- interferometers are located at #8.5 port and # 14.5 port with on-axis and off-axis, vertical chords, respectively.

In this experiment, the plasma is produced with 70 GHz ECH of 340 kW, and is sustained with additionally injected tangential NBI of 600 kW (NBI BL2), as shown in Fig. 3. Additional fuelling are injected at 235 ms with the time duration of 10 ms for HIGP, and at 233 ms with 1 ms short pulse for SMBI. In both cases, the maximum line averaged density are adjusted to have almost the same value with $\bar{n}_e \sim 2 \times 10^{19} m^{-3}$.

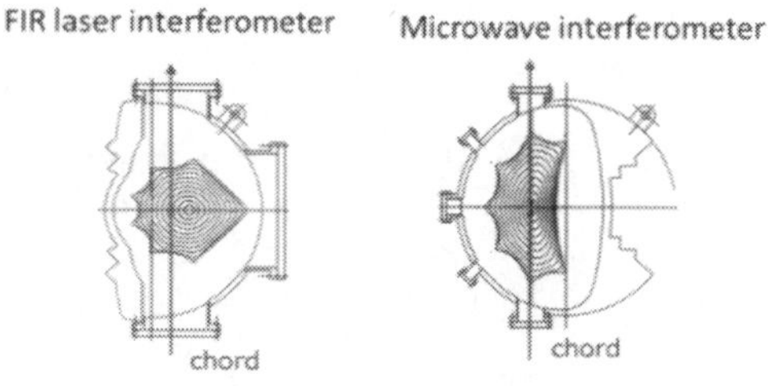


Fig. 2. Chord of (left) FIR laser interferometer and (right) microwave interferometer

Figure 3 and 4 shows waveforms for the line averaged density measured with (a) microwave interferometer ($\bar{n}_e^{MICRO}$) and FIR interferometer ($\bar{n}_e^{FIR}$), (b) stored energy $W_p$, and (c) Hα signals in HIGP and SMBI discharges, respectively. In both cases, increase in density and decrease in $W_p$ are observed after the additional fuelling. The stored energy is kept constant at low level from ~ 245 to ~ 270 ms, and then $W_p$ starts to recover spontaneously and take a maximum value of ~ 1.8 kJ at ~ 280 ms. Several differences for the temporal responses are observed although basic features are similar. For instance, quick and sharp increases of $\bar{n}_e$ ($d\bar{n}_e/dt \sim 0.29 \times 10^{19} m^{-3}$ / ms) and the Hα signal are observed after the SMB injection, compared with the response of HIGP ($d\bar{n}_e/dt \sim 0.15 \times 10^{19} m^{-3}$ / ms). In addition, $\bar{n}_e$ and the Hα signals continue to increase for 5-7 ms after 1.8 ms short pulse of SMBI, which is different from the case of HIGP plasma where the such increase immediately stops just after the injection of HIGP. These observation should relate to the differences on fuelling process and efficiency between SMBI and HIGP, and in practice, strong toroidal asymmetry of Hα signal is observed in SMB injection in previous study [5], which is characteristics of the local, strong fuelling on the SMBI technique.

## 4. Time evolution of profile shape in HIGP and SMB injection.

The ratio of the line averaged density obtained with off-axis FIR- to that of on-axis microwave interferometer, $\bar{n}_e^{FIR}/\bar{n}_e^{MICRO}$, is a parameter to relate to the profile shape. For example, a higher value of $\bar{n}_e^{FIR}/\bar{n}_e^{MICRO}$ indicates that the contribution of edge electron density in the region of $\rho > 0.35$ is higher, which means the profile is broader. On the other hand, a lower value of the ratio means more peaked density profile. Using the ratio, temporal change of profile shape can be evaluated.

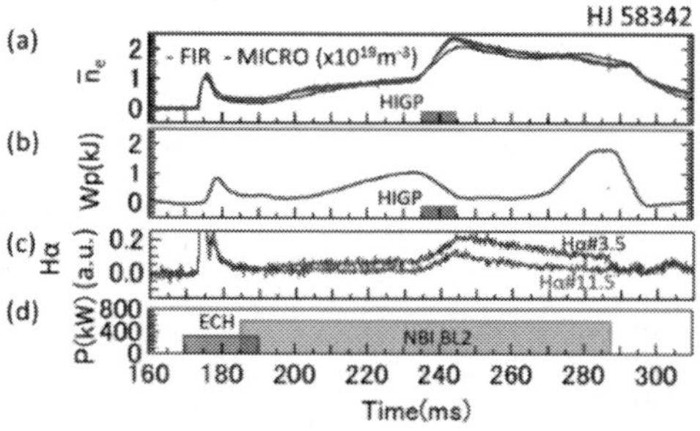


Fig. 3. Waveform of HIGP plasma experiment

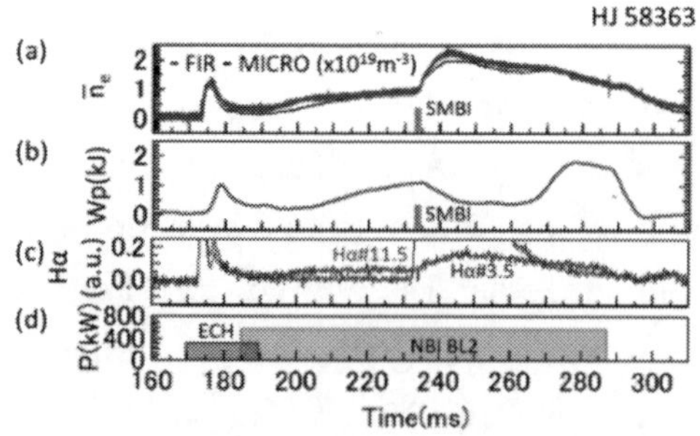


Fig. 4. Waveform of SMBI plasma experiment. Hα radiation indicates the timing of SMBI.

Figure 5 (a) and (b) show the time development of $\bar{n}_e^{FIR}$ and $\bar{n}_e^{MICRO}$, and the ratio, $\bar{n}_e^{FIR}/\bar{n}_e^{MICRO}$, in HIGP plasma. The density and temperature profile obtained with the results of Thomson scattering measurement are shown in Fig. 5 (c) [6].

A broadening of density profile is observed during HIGP injection. In Fig.5 (c), $\bar{n}_e^{FIR}/\bar{n}_e^{MICRO}$ increases during HIGP injection, which suggests that the HIGP strongly fuels to outside region of ρ > 0.35 and modifies density profile to broader shape. Electron temperature is cooled down during the density increase phase in entire region of plasma, as shown in the temperature profile at 240 ms in Fig. 5 (c). After the termination of HIGP injection, the density profile shape start to change to be peaked moderately, and then keeps almost constant value from 250 to 270 ms.

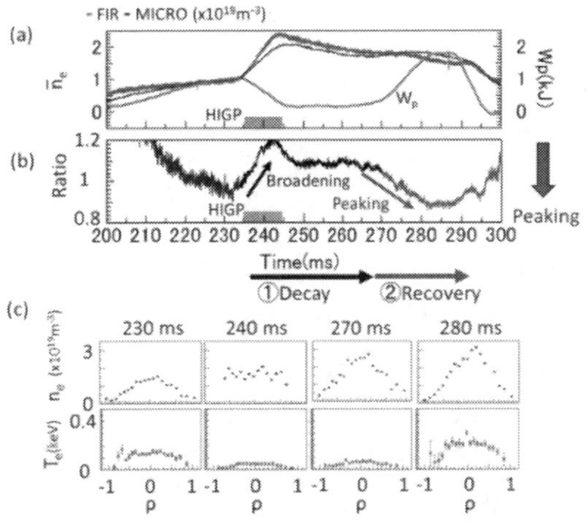


Fig. 5. (a) Time development of $\bar{n}_e^{FIR}$ and $\bar{n}_e^{MICRO}$, (b) $\bar{n}_e^{FIR}/\bar{n}_e^{MICRO}$, (c) $n_e$ and $T_e$ profile obtained with Thomson scattering measurement.

In the recovery phase of stored energy, a peaking of density profile shape was found in the HIGP plasma, which can be seen in the decrease in $\bar{n}_e^{FIR}/\bar{n}_e^{MICRO}$ ratio. The line averaged density slightly decreases but does not change so much during the recovery phase, however, electron temperature significantly rises as seen in the temperature profile at 270 and 280 ms of Fig. 5 (c). The recovery of $W_P$ is, therefore, mainly attributed to the increase of temperature.

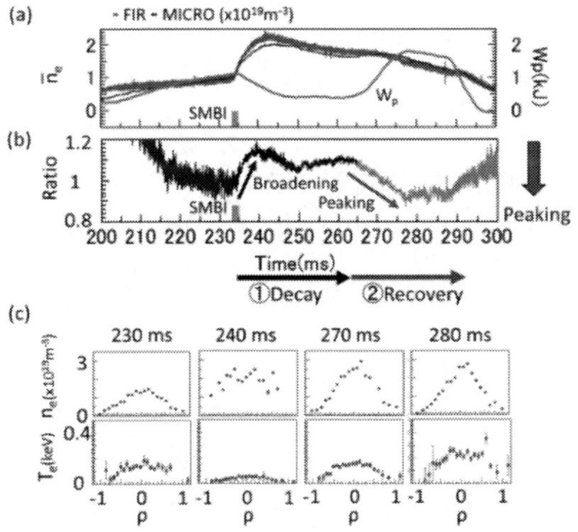


Fig. 6. (a) Time development of $\bar{n}_e^{FIR}$ and $\bar{n}_e^{MICRO}$, (b) $\bar{n}_e^{FIR}/\bar{n}_e^{MICRO}$, (c) $n_e$ and $T_e$ profile obtained with Thomson scattering measurement.

The result of SMBI plasma also shows similar time evolution, as shown in Fig. 6. (a-c), however, density profile broadening after SMBI is more moderate, compared to the case of HIGP plasma. The increment of $\bar{n}_e^{FIR}/\bar{n}_e^{MICRO}$ is from ~ 1.0 to ~ 1.15, and is smaller than that from ~0.95 to ~1.20 in HIGP case, which indicate the profile change of SMB injected plasma is smaller than that of HIGP plasma.

The dependences of profile shape parameter $\bar{n}_e^{FIR}/\bar{n}_e^{MICRO}$ on stored energy $W_P$ are plotted in Figure 7 (a) and (b) for HIGP and SMBI plasmas. These figures clearly show that, the density profile is broaden with the decay of stored energy, and the profile is peaked with the recovery of the $W_P$ in both cases. In addition, the variation of profile shape and stored energy are a little larger in HIGP plasma although the maximum stored energy are almost the same values (~1.8 kJ), which suggests a deeper penetration to core plasma in SMB plasma.

## 4. Summary

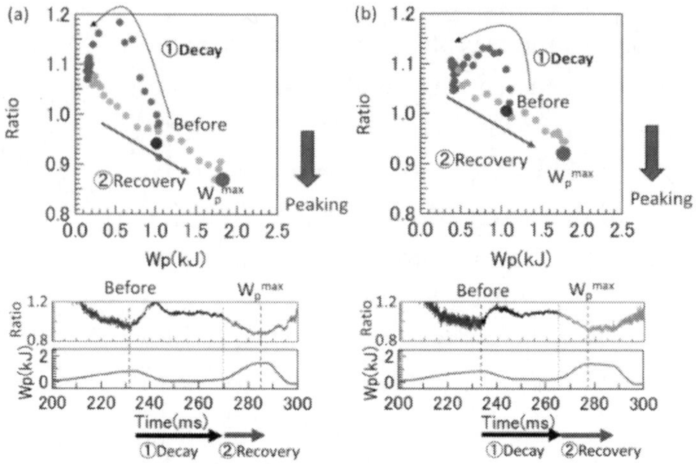


Fig. 7. The dependence of $\bar{n}_e^{FIR}/\bar{n}_e^{MICRO}$ on the stored energy and time development of $\bar{n}_e^{FIR}/\bar{n}_e^{MICRO}$ and stored energy for (a) HIGP plasma, and (b) SMBI plasma

The line averaged density of the high density plasma with HIGP and SMBI in Heliotron J are measured using FIR laser and microwave interferometers, and the time evolution of density profile shapes are evaluated. In both HIGP and SMBI cases, profile broadening is observed after additional fuelling with the decay of stored energy. After HIGP/SMBI injection, the line averaged density is kept high value and the stored energy remains low value, continuously. After that, the density starts to become peaked profile spontaneously with the recovery of stored energy. The increase of electron temperature is the main contribution to the recovery of stored energy. The temporal response of density profile shape and its relationship with stored energy are similar, however, smaller disturbance to the plasma parameters are observed in SMBI plasma, compared to that of HIGP, which might be caused by SMB penetrating into plasma more deeply than HIGP.

**Acknowledgements**

This work is performed with the support and under the auspices of the Collaboration Program of the Laboratory for Complex Energy Processes, Institute of Advanced Energy, Kyoto University and the NIFS Collaborative Research Program (NIFS10KUHL030, NIFS09KUHL028 and NIFS13KUHL056).

References

[1] T. Mizuuchi, et al., J. Nucl. Mater., 415 (2011) 443

[2] S. Kobayashi et al., Proc. 40th EPS Conf., 1-5 July (2013)P1.148

[3] T. Mizuuchi et al., Proc. IAEA2014, 13-18 Oct. (2014)EX/P4-29

[4] Y. Ohtani et al., submitted to Plasma Fusion Res.

[5] T. Mizuuchi, et al., Contrib. Plasma Phys. 50, No. 6-7, 639-645 (2010)

[6] N. Kenmochi et al. Rev. Sci. Instrum., vol.8, 2402117 (2013)

# Plasma start-up by using transient coaxial helicity injection on HIST

M. Nagata, T. Hanao, N. Oki, Y. Uesaka, T. Kawai, Y. Kikuchi, N. Fukumoto and T. Kanki*

*Graduate School of Engineering, University of Hyogo, Himeji, Hyogo 671-2280, Japan*

**Japan Coast Guard Academy, Kure, Hiroshima 737-8512, Japan*

## 1. Introduction

An advantage of the Spherical Torus (ST) is the low aspect ratio and so elimination of a central solenoid coil is required for attractive high-beta fusion reactors based on the ST concept. The steady-state Coaxial Helicity Injection (CHI) is one of useful current drive methods for ST and spheromak plasmas, where dynamo activities are required to transport inwardly injection currents toward core regions, leadings to a degradation of confinements. In the transient-CHI (T-CHI), only axisymmetric reconnection during plasmoid ejection process from an injector is thought to be necessary for generating a high quality closed flux. The T-CHI without requiring for dynamo is a promising candidate for the non-inductive plasma start-up in a reactor-size ST device. So far, the T-CHI method has been successfully applied to NSTX for the start-up followed by inductive ramp-up. This coupled discharge has achieved plasma currents larger than 1 MA [1]. However, understanding the physics of the flux closure during T-CHI still remains as a key issue [2,3]. Internal magnetic field measurements in a smaller machine make it possible to confirm the generation of closed flux surfaces. Thus, we have recently tried the internal magnetic probe measurement to investigate the flux closure in the T-CHI experiments on HIST ($R$=0.30 m, $a$=0.24 m, $A$=1.25) [4]. In the recent experiment, T-CHI has generated toroidal currents up to $I_t$=40-60 kA under the condition of the presence of closed flux. A result from the measurement has provided us a verification of the generation of an X-point followed by magnetic reconnection due to the $\boldsymbol{J}\times\boldsymbol{B}$ plasma flow from the gun. The closed poloidal flux increases proportionally with the toroidal plasma current as increasing the injection voltage across the inner and outer gun electrodes. We also has found that MHD stabilities of the formed magnetic configurations depends on an amplitude of the injector (gun) bias flux. The lower bias flux increases the edge current density larger after the flux closure, causing the kink instability of the open field line which was already observed in the steady-state CHI current derive.

## 2. Experimental set-up

The HIST device can form and sustain the high and low-q ST plasmas by utilizing the variation of the external toroidal field (TF) coil current $I_{tf}$=0~125 kAturns. Figure 1 shows the schematic diagram of the HIST device and main diagnostics. The detail explanation of the HIST is represented in the reference [5]. The start-up discharge bank ($V_{max}$=10 kV, $C$=2.6 mF) are used to breakdown the gas injected between the coaxial electrodes.

The HIST device has surface poloidal pick-up coils to calculate plasma currents, electrostatic double probes, Ion Doppler Spectrometer (IDS) and so on. A six-channel $\lambda$ probe incorporating small size Rogowski and flux loops is used to measure toroidal current density $J_t$ and toroidal flux profiles on the flux conserver (FC) midplane. To measure time evolutions of poloidal flux $\Psi_p$ contours, we inserted 2D magnetic pick-up coil arrays on the poloidal cross section as shown by Fig.1. This 2D flux plots data provide us the maximum value of the poloidal flux. We can measure the radial profile of the electron temperature and the density by using the electrostatic probe with voltage sweeping.

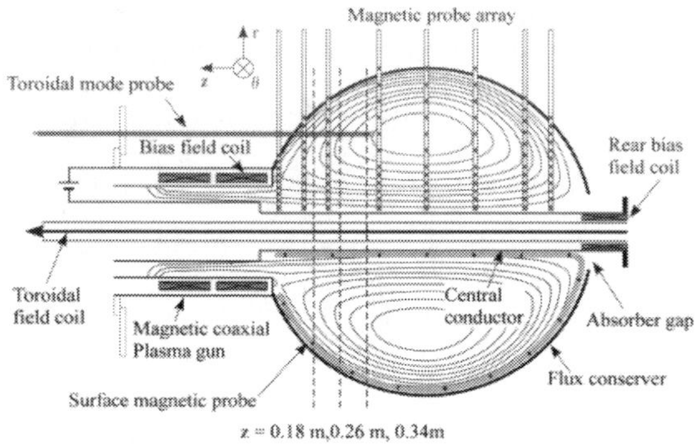


*FIG.1. Schematic diagram of the HIST device and diagnostics. 2D magnetic pick up coils location for measurements of poloidal magnetic flux $\Psi_p$ contours.*

## 3. Experimental results and summary

The gun capacitor bank was sized so as to be able to provide the energy needed for gas breakdown of all the injected gas and heating to meet and exceed the inductive stored energy in the resulting plasma discharge and capable of providing adequate current to exceed the bubble burst current requirement. Figure 2 shows time evolutions of the toroidal current $I_t$, the closed poloidal flux $\Psi_p$, the gun current $I_{gun}$ and the gun voltage $V_g$ in the high and low gun bias operations. The $\Psi_{p.closed}$ is calculated by the flux contours measurement. The bias flux is about 1.6 mWb for the low

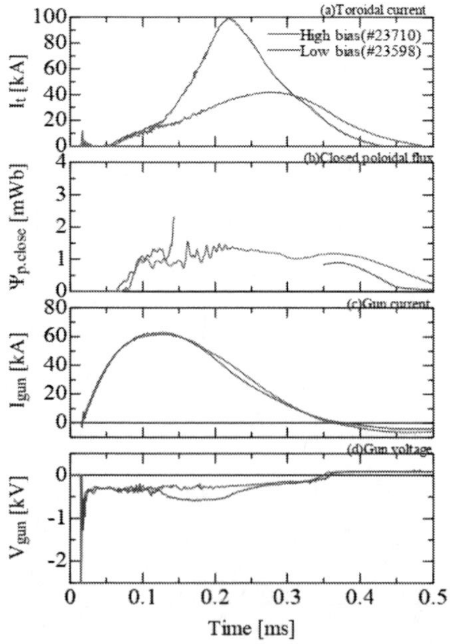


*FIG.2. Time evolutions of the toroidal current $I_t$, the closed poloidal flux $\Psi_{p.closed}$, the gun current $I_{gun}$ and the gun voltage $V_{gun}$.*

bias flux. The toroidal current peaks at $t$=0.25-0.3 ms and then decays due to the resistivity. The total poloidal flux is amplified over the bias flux. The gun current $I_{inj}$ injected effectively into the FC region is less than about 10% of the measured gun current $I_{gun}$, because the gun current dissipates almost in the gun electrode region. The toroidal current in the low bias case is two times larger than that in the high bias case, but for the former case, the plasma becomes unstable and ordered closed flux surfaces disappear after $t$=0.14 ms.

Figure 3 (a) (b) shows the time evolution of poloidal flux contours in the both high and low bias cases during the current start-up phase. The plasmoid is ejected from the gun muzzle, after then the magnetic reconnection occurs to create the X-point. We can see in the figure that the closed flux region is surrounded by the open field lines intersecting with electrodes. The plasmoid ejected from the gun has not reached to the bottom of the FC until $t$=0.1 ms. The ratio of the closed flux to the total flux is about 40 %. After reaching the bottom of the FC, the plot of poloidal flux contours at $t$=0.165 ms are distorted in the low bias case as shown by Fig.3 (b). This means that the assumption of axi-symmetry in calculating the poloidal flux from a measured $B_z$ component is not valid at this time. However, this unstable configuration interestingly recovers to the original axi-symmetric state when the plasma starts to decay. During this plasmoid injection phase, the toroidal current density $j_t$ profile is peaked around the magnetic axis ($R$=0.3 m). After the plasma reaches the bottom of the FC, the $j_t$ profile becomes hollow rapidly because the gun current injected continuously tends to flow along the inner and outer edge open field lines. The higher toroidal current seen in the low bias case is attributed to the injection current flowing in the edge region. During the decay phase, the inner edge current diffuses toward the magnetic axis.

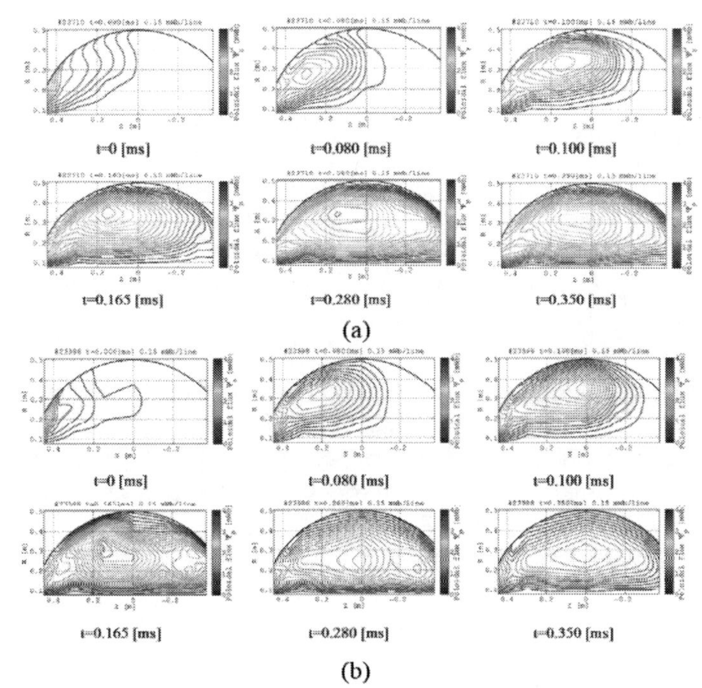


*FIG. 3(a)(b) Comparison of poloidal flux contour plots between high bias (a) and low bias (b) operations*

We have measured the toroidal $n$-mode using 8 pick-up coils located in the

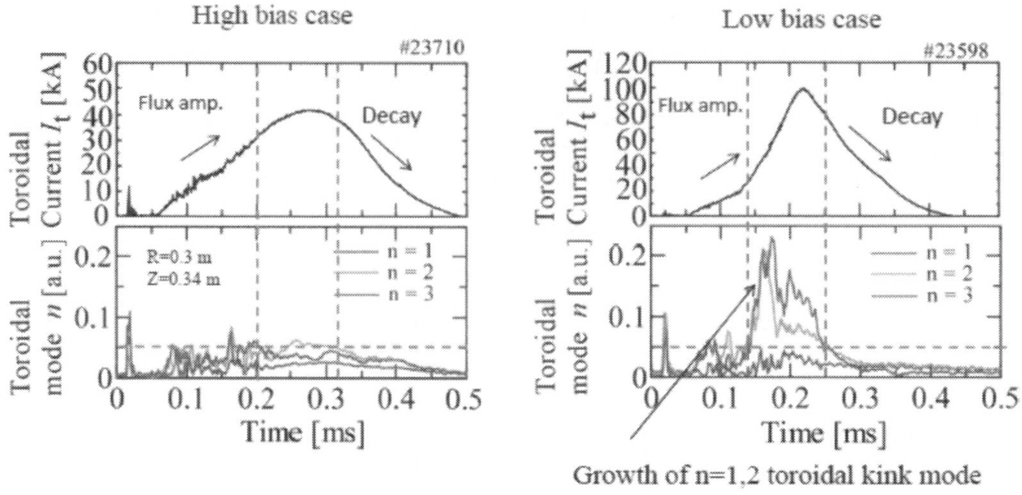


*FIG. 4. Comparison of toroidal mode n between high and low bias cases*

toroidal direction. Figure 4 shows the comparison of the $n$-mode between high and low bias cases. The time evolution of the $n$-mode for the case of lower bias shows the rapid growth of the $n$=1 and 2 modes at $t$=0.14 ms corresponding to the distortion time of the flux contours. The result indicates an occurrence of the kink instability due to an excess injection current at the edge. When the configuration relaxes to the original state around $t$=0.25 ms, correspondingly the amplitude of these modes decreases below the critical level of triggering the kink mode.

We have verified the formation of the closed flux (flux closure) in the T- CHI start-up by measuring internal magnetic fields. The excess current injected at the edge after the formation causes the kink instability leading to destroy of the axisymmetric configuration. To avoid this phenomenon, it is important to shorten a pulse length of the injector current and drop to zero level just after the magnetic configuration is formed.

References

[1] R. Raman, et al., Phys. Rev. Lett. **104**, 095003 (2010).
[2] E. B. Hooper, et al., Phys. Plasmas **20**, 095003 (2013).
[3] F. Ebrahimi and R. Raman, Phys. Rev. Lett. **114**, 205003 (2015)
[4] M. Nagata, et al., 25th IAEA Fusion Energy Conference, Saint Petersburg, EX/P4-30 (2014).
[5] M. Nagata, et al., Phys. Plasmas **10** (2003) 2932.

# Fishbone Instability Triggered Nonlocal Transient Transport in HL-2A NBI Plasmas

W. Chen[1], Y. Xu[1], X.T. Ding[1], Z.B. Shi[1], M. Jiang[1], W.L. Zhong[1], X.Q. Ji[1], D.L. Yu[1], Z. C. Yang[1], L.M. Yu[1], J.X. Li[1], Y. G. Li[1], J.Y. Cao[1], X.M. Song[1], M. Xu[1], L.W. Yan[1], Yi. Liu[1], Q.W. Yang[1], and X.R. Duan[1]

[1] *Southwestern Institute of Physics, P.O. Box 432 Chengdu 610041, China*

A new-type nonlocal transient transport[1][2] induced by the fishbone instability has been observed in HL-2A ($R/a = 1.65m/0.4m$) low-density ($ne < 1.5 \times 10^{19} m^{-3}$) NBI plasmas with the lower single-null divertor configuration. This phenomenon is well reproducible. Figure 1 shows typical experimental results with the fishbone instability and nonlocal transport during NBI. The sawteeth are gradually suppressed by beam ions after the NBI is switched on at $t = 410ms$. The fishbones with bursting-amplitude and chirping-frequency are driven by energetic ions during $t = 550-650ms$. The repetitive fishbones produce the similar effect like the modulated heating/cooling pulses. The lifetime of each fishbone is $t_f = 3-5ms$, and the interval-time of two fishbones is $\Delta t = 3-7ms$. The deuterium-alpha radiation ($D_\alpha$) in the divertor drops significantly while the fishbone bursts, and the line-averaged density does hardly change.

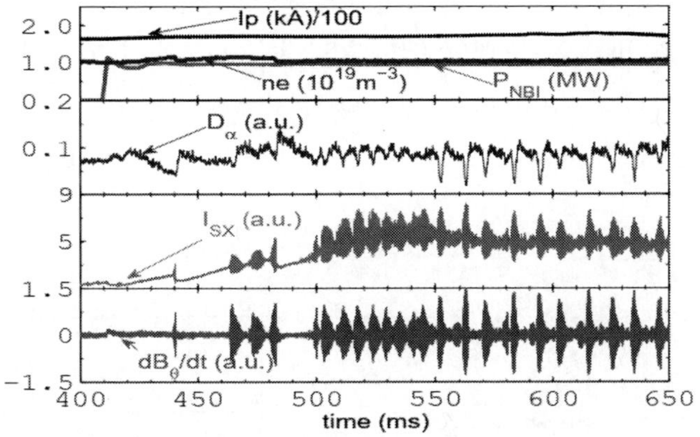


Figure 1: Typical discharge with fishbone (fb) instability and nonlocal phenomenon on HL-2A. Plasma current, $Ip$; electron density, $ne$; NBI power, $P_{NBI}$; deuterium-alpha radiation in the divertor, $D_\alpha$; soft X-ray signal in the core, $I_{SX}$; Mirnov signal in the midplane, $dB_\theta/dt$.

The radial localization and fluctuation amplitudes of the m/n=1/1 fishbone are measured by the multi-channel ECE radiometer. Figures 2(a-d) show spectrograms of autopower of ECE signals at radial positions $\rho = 0.07$, $\rho = 0.15$, $\rho = 0.23$ and $\rho = 0.30$. Furthermore, the fishbones do not occur at spectrograms of autopower of ECE signals in the region of $\rho \geq 0.30$, but appear at spectrograms of crosspower of Mirnov data and ECE signals. Therefore it is inferred that

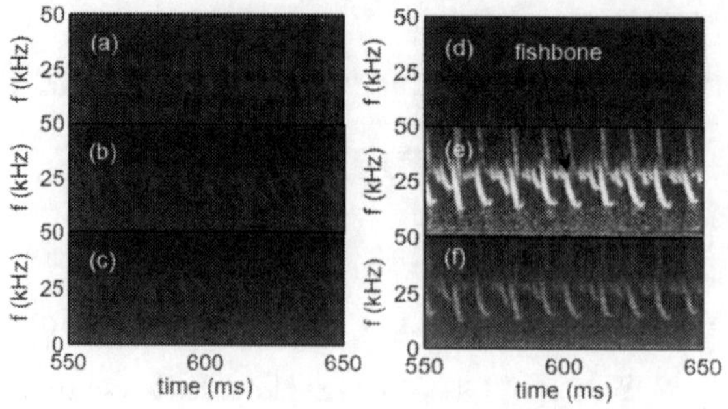


Figure 2: Spectrograms (a-d) of autopower of ECE signals at different radial positions, (a) $\rho = 0.07$, $\rho = 0.15$, $\rho = 0.23$ and $\rho = 0.30$; spectrogram (e) of autopower of the Mivnov signal; spectrogram (f) of crosspower of Mirnov data and ECE signal at $\rho = 0.30$. Spectrograms of autopower of ECE signals and spectrograms of crosspower of Mirnov data and ECE signals are similar and not shown here while $\rho > 0.3$

the fishbones localize in the core, $\rho < 0.30$, and the fishbone-induced temperature perturbation spreads to the plasma edge. The traces of electron temperatures are shown in Figure 3. With the fishbone bursting it is found that the core (localization of fishbone, $\rho < 0.30$) heating leads to a simultaneous decrease (but not slow increase) in temperature at the plasma edge. The reversal radius of temperatures is $\rho \simeq 0.55$, which is near q=3/2 rational surface. The effect reveals fast anomalous transport of core heat pulses to plasma edge, not compatible with diffusive time scales. It is a new-type nonlocal response which is a counterpart to that induced by the edge heating/cooling pulse. The perturbation source of the new-type nonlocal is magnetic, and it is produced by the plasma itself.

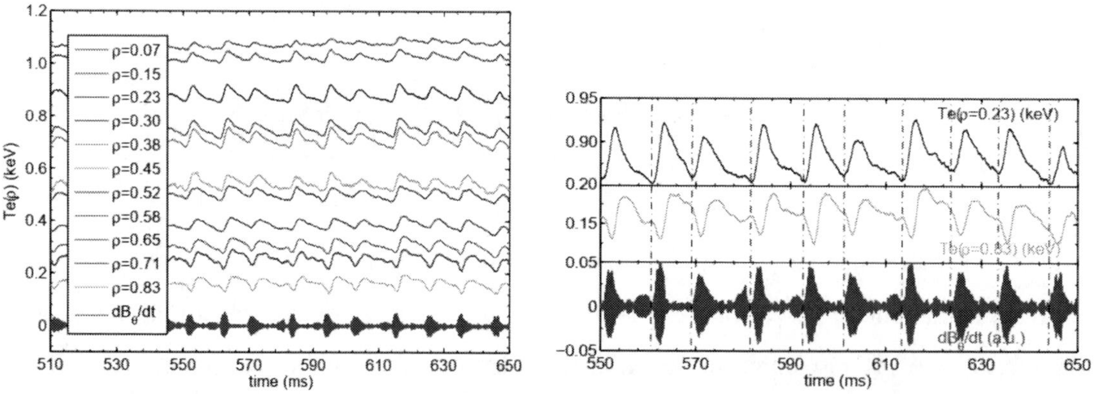


Figure 3: Time evolution of the Mirnov signal ($dB_\theta/dt$) with strong ion fishbone instabilities and electron temperatures ($Te$) at different radial positions. Panoramic view (left) and local view (right).

A novel result, the nonlinear coupling between the fishbone and background turbulence during the nonlocal, has been analyzed by the bispectrum technique. The squared bicoherence is given by $\hat{b}^2(f_1,f_2) = \mid \hat{B}_{Te\tilde{B}_\theta\tilde{B}_\theta}(f_1,f_2) \mid / < \mid Te(f_1)\tilde{B}_\theta(f_2) \mid^2 >< \mid \tilde{B}_\theta(f_3) \mid^2 >$ with the Fourier bispectral $\hat{B}_{Te\tilde{B}_\theta\tilde{B}_\theta}(f_1,f_2) =< Te(f_1)\tilde{B}_\theta(f_2)\tilde{B}^*_\theta(f_3) >$, $f_3 = f_1 \pm f_2$ and $0 < \hat{b}^2(f_1,f_2) < 1$. The squared-bicoherences of magnetic and temperature fluctuations are shown in Figure 4. It is found that $\mid f_2 \mid = k * f_{fb}$ and $\mid f_2 \pm f_1 \mid = k * f_{fb}$ ($k$ is a positive integer) are strong coupling lines. It indicates that the nonlinear interaction between the fishbone and background turbulence and there exists energy transfer between fishbone and temperature fluctuations. This may explain why the core ($\rho < 0.3$) electron temperature increases abnormally while the fishbone bursts. It also suggests that the multi-scale (macro-transport, meso-EP/MHD and micro-turbulence) physics is responsible for the global nonlocal response.

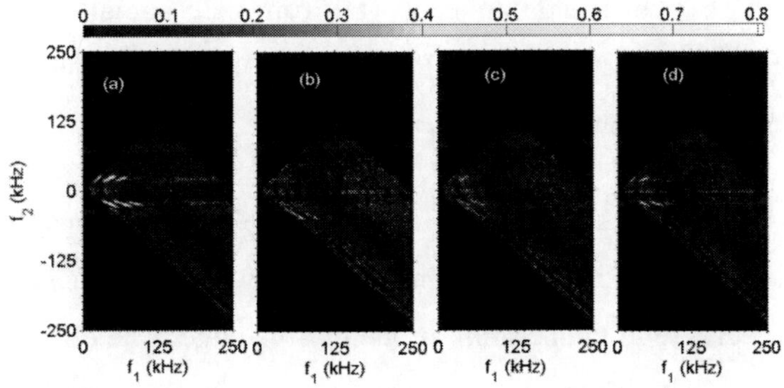


Figure 4: Squared bicoherences (a-d) $\hat{b}^2_{Te\tilde{B}_\theta\tilde{B}_\theta}(f_1,f_2)$ at different radial positions. (a) $\rho = 0.23$, (b) $\rho = 0.38$, (c) $\rho = 0.52$, and (d) $\rho = 0.65$.

Figure 5 shows the auto-correlation function (ACF) coefficients (a-b) of ECE signals at two radial positions ($\rho = 0.23 < \rho_{rev}$ and $\rho = 0.65 > \rho_{rev}$) and spatial profiles (c) of Hurst exponents (H, obtained by R/S method) from ECE signals. It is found that ACFs and Hurst exponent both enhances during the fishbone and nonlocal transport. And so the new-type nonlocal transport is potentially linked to self-organized critical (SOC) dynamics. Figure 5(d) gives that 2D pattern (d) of cross-correlation coefficients from ECE signals. It shows that the fluctuations induced by the fishbone is a long-range correlation, and the local micro-fluctuations communicate with those at distant radius via the long-range correlation, and the heat energy with $f = 20-22kHz$, which is the power-spectral peak of the fishbone, rapidly propagates outward. Meanwhile, it is observed that there are two phase jumps near the q=1 ($0.2 < \rho < 0.3$) and q=3/2 ($0.5 < \rho < 0.6$) rational surfaces. The phase jump is likely due to the presence of the shear flow which alters the mode propagation. It is found from figure 5(d) that the response time of the nonlocal transport is around $\tau \simeq 50\mu s$ in the radial distance, $0 < \rho < 0.8$, and the velocity of the radial propagation is $v_r \simeq 0.8a/\tau \simeq 6km/s$.

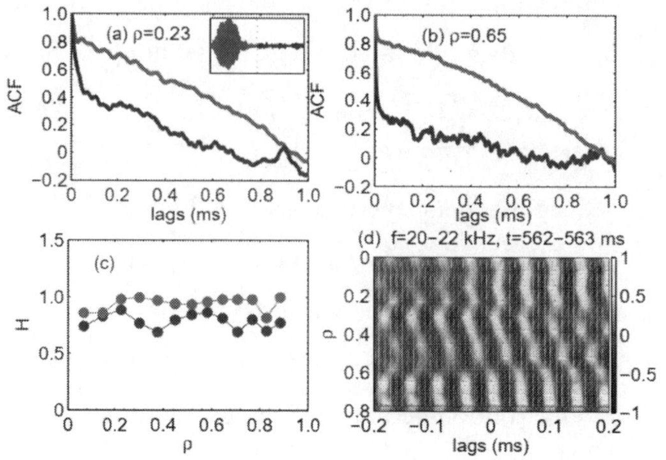


Figure 5: Auto-correlation function (ACF) coefficients (a-b) of ECE signals at two radial positions; Spatial profiles (c) of Hurst exponents (H) from ECE signals; 2D pattern (d) of cross-correlation coefficients from ECE signals. Red line with fishbone, blue line without fishbone. ACFs are similar and not shown here at other positions.

In summary, an experimental observation of a new-type nonlocal transport, which is triggered by the ion fishbone instability, is presented in a laboratory plasma. The rapid core heating leads to a simultaneous decrease in temperature in the plasma edge. The effect reveals anomalously fast transport of core heat pulses to plasma edge, not compatible with diffusive time scales. These results prove clearly that the magnetic perturbation, long-range correlation, meso-scale structure as well as $E \times B$ flow play crucial roles in the nonlocal response, and demonstrate explicitly that the fast response ($\sim 100\mu s$) of the nonlocal is electromagnetic but not electrostatic. The experimental results suggest there is a complexity coupling between the strong fishbone excitation and fluctuation-induced transport. The nonlinear coupling induces the energy transfer between the fishbone and background turbulence, and it is one of mechanisms of the fishbone-induced temperature increase in the core regions. The magnetic perturbation along with shear flow should dominate the nonlocal transient transport. The results also indicate that the nonlocal response is potentially linked to SOC dynamics.

This work is supported in part by the ITER-CN under Grants No. 2013GB104000 and 2013GB106004, and by the NNSF of China under Grant No. 11261140326, 11475058 and 11405049.

## References

[1] Callen, J.D. and Kissick, M.W. *Plasma Phys. Control. Fusion* **39**, B173-B188 (1997).

[2] Ida, K. et al. *Nucl. Fusion* **55**, 013022 (2015).

# Investigation of non-local heat transport and its interplay with neoclassical tearing modes (NTMs) in the HL-2A tokamak

Y. Xu[1], X. Q. Ji[1], O. Pan[1], Yi Liu[1], W. L. Zhong[1], Z. B. Shi[1], M. Jiang[1], B. B. Feng[1], D. L. Yu[1], Y. Zhou[1], J. Cheng[1], M. Xu[1], W. Chen[1], Yuan Xu[1], Y. B. Dong[1], L. W. Yan[1], X. T. Ding[1], Q. W. Yang[1], X. R. Duan[1], Yong Liu[1] and the HL-2A team

[1] *Southwestern Institute of Physics, P. O. Box 432, Chengdu 610041, People's Republic of China*

C. Hidalgo[2]

[2] *Laboratorio Nacional de Fusion, CIEMAT, 28040 Madrid, Spain*

In fusion plasmas, a number of experiments for the study of transient transport events have revealed a striking phenomenon, i. e., a fast "non-local" response of the core temperature rising to an edge cooling executed at the plasma periphery [1-6]. A few common features are observed during this non-local transport (NLT): (i) the core electron temperature ($T_e$) quickly increases before the edge cooling pulse diffuses in; (ii) the NLT usually happens below a critical density or collisionality; (iii) the reversal radius of the opposite $T_e$ variation on two sides is often located in the region of $1<q<2$. This transient and non-diffusive response with reversed polarity calls seriously into question of the standard "local" transport model, and hence, has puzzled scientists for more than twenty years. For understanding the physical processes of the NLT, various theoretical models have been proposed to explain the provocative non-local effects [3]. Among them the self-organized criticality (SOC) paradigm attempts to unravel the transport enigma via the interrelation between individual turbulent eddies and large-scale transport events like the "sand-pile" avalanche [7-10].

In this paper, we present the experimental investigation on the properties of SOC dynamics during the NLT at the HL-2A tokamak. The experiments were carried out in ohmically or ECRH heated plasmas. Typical discharge parameters were $R$=1.65 m, $a$=0.38-0.40 m, $B_T$=1.3-1.45 T, $I_p$=150-165 kA, $P_{ECRH}$=1.2-2.0 MW and $\bar{n}_e$ = (0.8-2.0) $\times 10^{19}$ m$^{-3}$. The non-local effects were induced in low density discharges by employing a SMBI system which made cold pulses at the plasma boundary [4]. For measuring the equilibrium and fluctuating $T_e$, a 16 channel ECE diagnostic was used and digitized at a sampling rate of 1.25 MHz. The Doppler reflectometer and Langmuir probes detected edge turbulence and poloidal plasma rotations, and Mirnov coils measured magnetic fluctuations.

Figure 1 plots discharge waveforms and multi-channel ECE signals in a typical NLT shot. Figure 1(a) shows that the $I_p$ is constant and the SMB was injected at $t$=1084 ms during the stationary phase of the discharge. After the SMBI, the $\bar{n}_e$ slightly increases due to fueling effects. Figure 1(b) displays the evolution of $T_e$ at different radii. At the outer region, the SMBI cooled the plasma and caused reduction of $T_e$. However, in the core area the red curves show, surprisingly, increases in $T_e$. The reverse position, $r/a$=0.47, is around $q$=3/2 surface. It has been found that the density and temperature fluctuation levels are nearly unchanged before and during the NLT [11], implying that the turbulent transport is not directly contributive to the core

$T_e$ rising via confinement improvement.

To investigate the role of SOC in the non-local transport, we have analyzed the Hurst parameter ($H$) in temperature fluctuations ($\tilde{T}_e$) before and during the NLT. The $H$ is calculated by structure function (SF) [12] and R/S [13] methods in $\tilde{T}_e$ data. Figure 2(a) shows an increase of $H$ (with NLT) from 0.78 to 0.85 at $r/a$=0.77. The radial profile of Hurst exponents further indicates that the H values are generally enhanced in all radii in the nonlocal phase (see Fig. 2(b)). Besides, the cross-correlation function (CCF) of $\tilde{T}_e$ between ECE channels has also been computed. The results show an increase of the radial correlation length in the propagation of avalanche events and stronger inward transport during the NLT [11]. These results explicitly point out an enhancement of SOC paradigm during the non-local period. This enhanced SOC behavior with NLT might be related to the reduced flow shear in the plasma edge, which is caused by increased damping (or cooling effects) of SMBI, as depicted in Fig. 2(c).

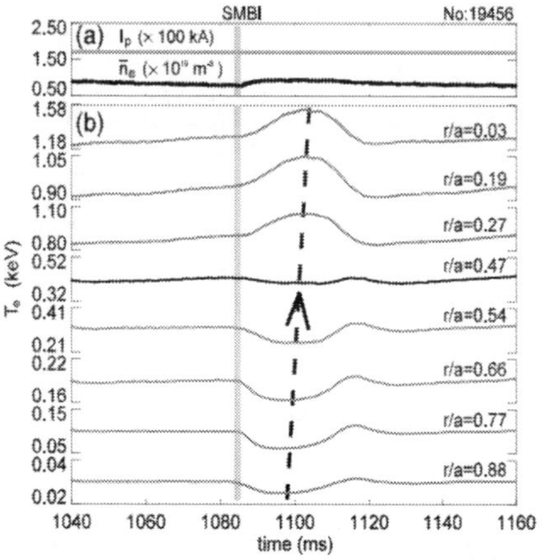


FIG. 1: Typical discharge waveforms of non-local transport induced by SMBI (vertical yellow bar) at HL-2A. (a) Plasma current and line-averaged density; (b) Multi-channel ECE signals showing decrease of $T_e$ in the plasma edge (blue) and increase of $T_e$ in the core (red).

In this study, during the NLT at HL-2A, an intimate interaction between the neoclassical tearing modes (NTMs) and the nonlocality has been observed in ERCH heated plasmas [6]. As shown in Figs. 3(a) and (b), after injection of SMB at 838 ms the NLT is induced, and then an NTM is excited at t≈844 ms. The mode numbers have been identified as $m/n$=3/2 and frequency $f$≈7 kHz, as shown in the frequency spectrum of the SXR signal in Fig. 3(c). The neoclassical nature of the tearing modes is illustrated in Fig. 3(d), where the variation in the fluctuation amplitude of the radial magnetic field (square root of $\tilde{B}_r \propto$ island width) around the mode frequency ($f$=6-9

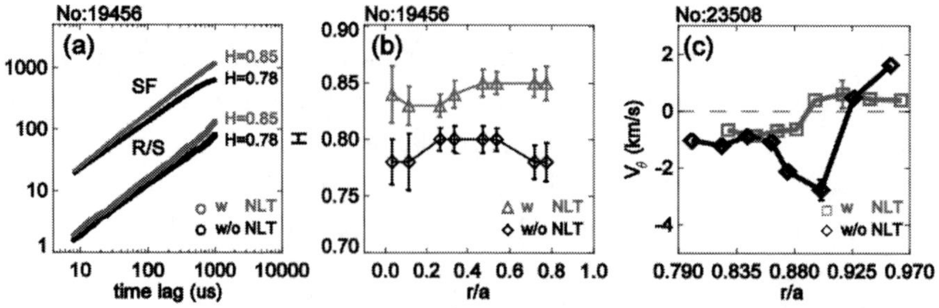


FIG. 2: Comparison of SOC features without NLT (black) and with NLT (red). (a) Hurst exponents calculated by SF and R/S methods at $r/a$=0.77; (b) radial dependence of Hurst parameters and (c) radial profile of the poloidal flow at plasma boundary.

kHz) exhibits approximately a linear relation with $\beta_N$ in five shots having the 3/2 islands. It is found that the reversion surface of the NLT is usually located at q=3/2 or 2/1surface. Nearby the reversion surface, the increase of local temperature (or pressure) gradient results in increase of bootstrap currents, and hence, onset of NTMs at q=3/2 or 2/1 surface. As shown in Fig 3, no seeding island was seen prior to NTM onset, implying that the local pressure gradient grows strong enough to linearly drive the NTMs. Because of the non-local effect, the NTM can be triggered at lower $\beta_N$ [14].

In the present study, significant influence of NTMs on the magnitude of nonlocal responses has been observed. Plotted in Figs. 4(a)-(e) are time evolutions of plasma density, multi-channel ECE signals and contour-plot of frequency spectrum of the SXR signal in a discharge with two nonlocal transport induced by edge gas-puffing. After the first gas-puffing at $t$=570 ms, the nonlocal effect appears and then the 3/2 NTM is excited with $f$=6-8 kHz, similar to that depicted in figure 3. After the NTM occurrence, the second gas-puffing is injected at $t$=770 ms, which also triggers a nonlocal transport. However, for the second nonlocality the changes of $T_e$ in both edge and core regions are much smaller than the first one. To understand this phenomenon, we calculated the Hurst exponents and CCF in $\tilde{T}_e$ during the first (without NTM) and second (with NTM) nonlocal periods. With NTM, the H values drop abruptly in the inner region, whereas in the outer area they are nearly unvaried (see Fig. 4(f)). The reduction of avalanche-like transport by the NTM can be more clearly seen in Figs. 4(g) and (h), where the contour-plot of CCF between ECE signals in the first and second NLT period are plotted. The figures show that with NTM the radial correlation in $\tilde{T}_e$ remains large in the outer zone ( $\rho\approx$0.6-0.8), consistent with high Hurst parameters shown in Fig. 4(f). However, in the inner region ($\rho$<0.6), the radial correlation of turbulent events is broken by the NTM. In HL-2A, we have measured the toroidal plasma rotation with the presence of a magnetic island [14] and found that the local flow shear inside the island is strongly enhanced when the island is wide enough, similar to that observed in LHD [15]. According to theories and

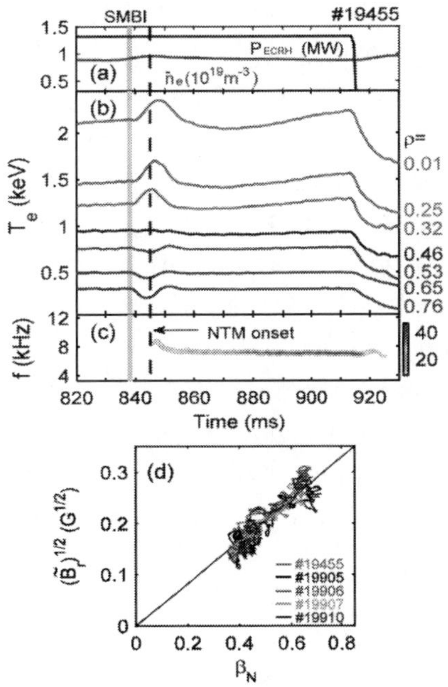


FIG. 3: Typical discharge waveforms with an onset of a 3/2 NTM during the nonlocal transport induced by SMBI in ECRH-heated plasmas at HL-2A. (a) the line-averaged density and ECRH power; (b) multi-channel ECE signals showing decrease of $T_e$ in the plasma edge and increase of $T_e$ in the core; (c) contour-plot of frequency spectrum of the SXR signal showing the onset of a 3/2 NTM ($f\approx 7$ kHz) during the nonlocal phase; (d) variation of the square root of $\tilde{B}_r$ of the 3/2 island versus $\beta_N$ in five discharges.

modeling [8, 10], such sheared flows should restrain the SOC or avalanche behaviors. Hence, the mechanism governing the nonlocal transport is weakened. This might explain the damping effect of an NTM island on the nonlocal transport.

In conclusion, in the HL-2A tokamak, we have investigated the properties of SOC dynamics in the nonlocal transport. Experimental evidence shows that, along with reduction of flow shear at the plasma edge, the SOC or avalanche behaviors are strongly enhanced during the prompt nonlocal phase, suggesting important role of the SOC paradigm in governing the long-distance nonlocal heat transport.

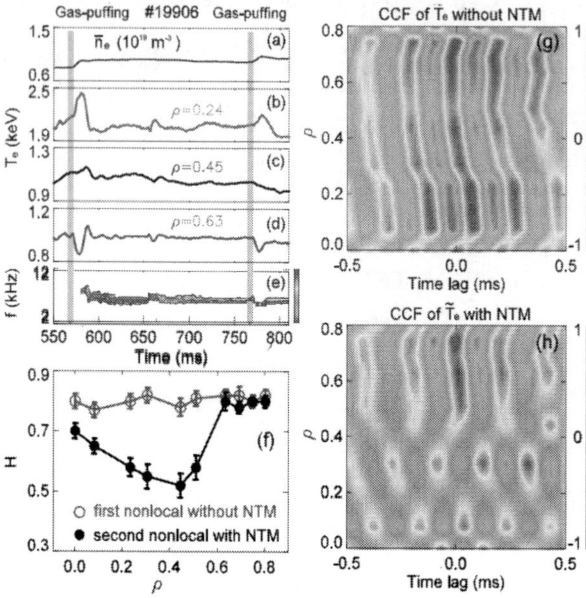


FIG. 4: Impact of NTMs on the magnitude of nonlocal transport. (a) to (e) are time evolutions of plasma density, multi-channel ECE signals and contour-plot of frequency spectrum in the SXR signal showing reduction of nonlocal effects with the presence of the 3/2 NTM; (f) radial profiles of Hurst parameters during the nonlocal phase without (red) and with (black) the NTM; (g) and (h) are contour-plot of CCF calculated in filtered ECE signals (4-7 kHz) during the nonlocality without and with the NTM, respectively.

In addition, during the nonlocal experiments at HL-2A, the NTMs can be triggered by transient increase of the local pressure gradient nearby the reversion surface of the nonlocality. Meanwhile, it is found that the NTM plays a damping role on the nonlocal effects due to locally sheared flow generated at the magnetic island of the NTM.

[1] M. W. Kissick *et al.*, Nucl. Fusion 34, 349 (1994).
[2] K. W. Gentle *et al.*, Phys. Rev. Lett. 74, 3620 (1995).
[3] J. D. Callen and M. W. Kissick, Plasma Phys. Controlled Fusion 39, B173 (1997).
[4] H. J. Sun *et al.*, Plasma Phys. Controlled Fusion 52, 045003 (2010).
[5] S. Inagaki *et al.*, Nucl. Fusion 52, 023022 (2012).
[6] X. Ji *et al.*, *IAEA Conference* (St. Petersburg, Russia, 2014), EX/6-4.
[7] T. Hwa and M. Kardar, Phys. Rev. A 45, 7002 (1992).
[8] P. H. Diamond and T. S. Hahm, Phys. Plasmas 2, 3640 (1995).
[9] B. A. Carreas *et al.*, Phys. Plasmas 3, 2903 (1996).
[10] D. E. Newman *et al.*, Phys. Plasmas 3, 1858 (1996).
[11] O. Pan *et al.*, submitted to Nucl. Fusion.
[12] A. Davis *et al.*, J. Geophys. Res. 99, 8055 (1994).
[13] H. E. Hurst, Trans. Am. Soc. Civ. Eng. 116, 770 (1951).
[14] X. Ji *et al.*, to be submitted to Phys. Rev. Lett.
[15] K. Ida et al., Phys. Rev. Lett. 88, 015002 (2002).

# Coupling optimization experiment on HL-2A based on Passive-Active Multijunction antenna for 3.7GHz Lower Hybrid system

Xingyu Bai, Hao Zeng, Bo Lu, Xiaolan Zou[1], Roland Magne[1], Annika Ekedahl[1], Julien Hillairet[1], Chao Wang, Jun Liang, He Wang, Yali Chen, Junwei He, Jieqiong Wang, Kun Feng and Jun Rao

*Southwestern Institute of Physics, Chengdu, China*

*[1] CEA, IRFM, 13108 Saint Paul-lez-Durance, France*

**System Introduction.** A new Lower Hybrid Wave (LHW) system (3.7GHz/2MW/2s) has been built on HL-2A tokamak. A Passive-Active-Multijunction (PAM) concept antenna [1], [2] was designed and built, fed by 500kW × 4 TH2103A Klystrons, delivering ITER-relevant power density, i.e. 25MW/m$^2$ at f = 3.7GHz [3],[4]. The antenna is designed to launch a peak parallel refractive index of $N_{\parallel}$=2.75 with a low theoretical Reflection Coefficient (RC), i.e. less than 1% [5], [6]. A specific gas puffing system for improving the antenna coupling and a set of Langmuir probes for antenna mouth density measurements are separately located on one side of the antenna each. An auxiliary vacuum system is installed at the rear of the antenna to improve the pumping efficiency. Vacuum leak tests and low power microwave scattering parameter measurements were done before the PAM antenna was installed on HL-2A, allowing the antenna to be conditioned at RF power of 100kW/100ms before starting plasma operation.

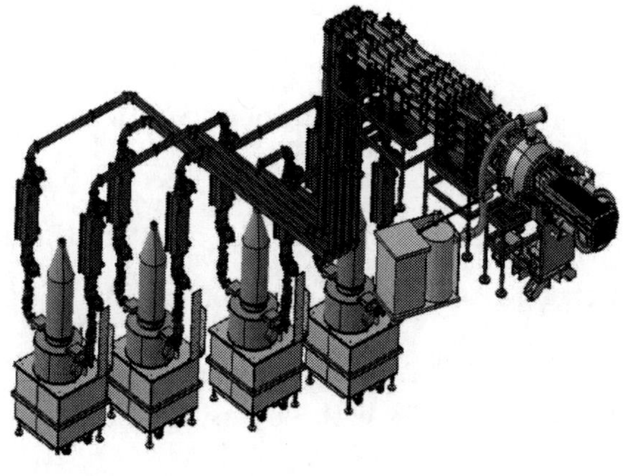

Fig. 1: New LHW system (3.7GHz/2MW/2s) of HL-2A. 500kW × 4 TH2103A Klystrons are located on the ground floor and the antenna is on the first floor, with an auxiliary vacuum system beside it.

Fig. 2: PAM launcher was developed and installed, facing to the plasma of HL-2A. A set of eight Langmuir probes is located on the right side, while gas puffing system is on the left side.

**Commissioning.** Commissioning work for the transmission line and the PAM antenna was carried out after the klystrons had been operated on test bed at full output power and duration.

Initially, the antenna was set about 3cm behind the limiter to guarantee safe operation, since the antenna had not been baked before installation. The commissioning work lasted about one month at RF power of 100kW/100ms. After the commissioning, the coupling was found to be improved, i.e. power reflection coefficient (RC) of ~10-30%, compared to almost 100% before the commissioning. The coupling optimization was then carried out.

**Coupling optimization experiments.** Coupling optimization experiments were carried out in both limiter and divertor configuration on HL-2A with the PAM antenna. From the calculated results by the ALOHA code (Fig. 3) it is found that optimum coupling, i.e. lowest power reflection coefficient (RC), is obtained for $n_e \sim 5\times10^{17}m^{-3}$. The RC revolution with density is shown in Fig. 3, which also shows that RC increases strongly when approaching the cut-off density ($n_{co} = 1.7\times10^{17}m^{-3}$ for f = 3.7 GHz). The edge density at the antenna mouth affects LH coupling much according to the calculation results, which was also obtained during the optimization experiments, as shown in Fig. 4.

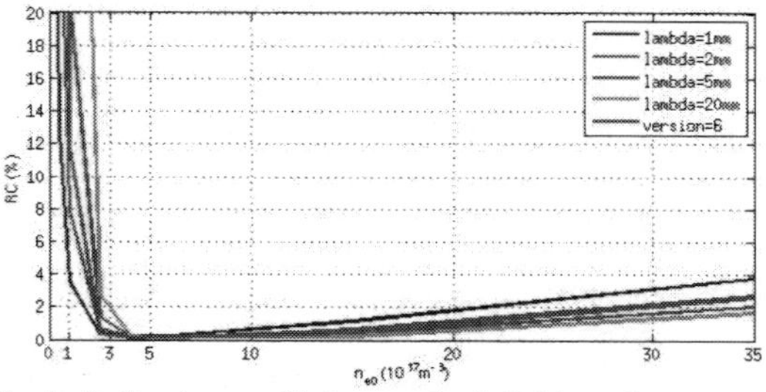

Fig. 3: Reflection coefficient on the PAM launcher versus electron density at the launcher mouth from ALOHA code results.

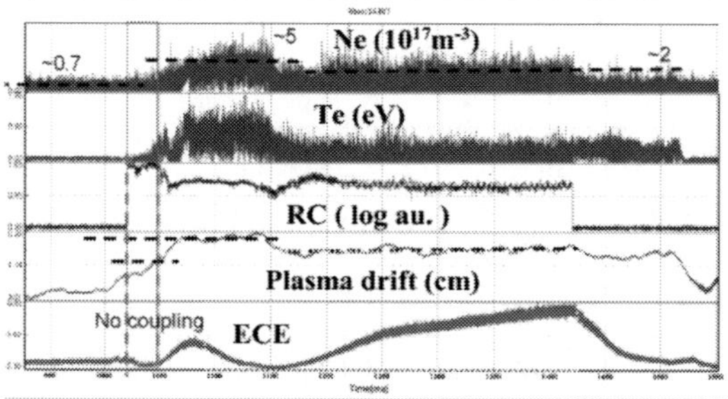


Fig. 4: RC revolution during electron density changes at the launcher mouth. The density is measured by Langmuir probes. (Note that RC is displayed in log scale).

The experimental results show that if the density at the launcher mouth is less than $1\times10^{17}m^{-3}$, i.e. below the cut-off density, the wave is not coupled. If the density is around $5\times10^{17}m^{-3}$, the RC is low, in accordance with ALOHA calculations. Good coupling was obtained both in limiter and divertor configuration in the optimization experiments with the help of different methods, such as gas puffing, launcher position change, plasma position control and plasma density change. The injected LH power reached 500kW. The experiments were performed in the optimized parameter range of average RC lower than 10% with 400ms duration, which makes it possible to carry out further LH physics experiments on HL-2A with the PAM.

When the antenna is not very close to the plasma, the density at the antenna mouth is often lower than the cut-off density $n_{co}$, and the coupling is very weak. In these cases a gas puffing in front of the antenna mouth is always needed for improving the LH coupling. Due to the non-uniformity of the gas puffing at the antenna mouth, the RC distribution in different grills was not uniform either. The coupling is worse at the top grills than that at the bottom ones.

This asymmetry was more obvious in the divertor configuration, due to the single null with lower X-point of HL-2A. Generally it has been found that more the gas puffing, better the coupling. Parameters controlling the gap between the antenna mouth and the plasma include the antenna position, the limiter position and the plasma drift in horizontal direction in the divertor configuration. The gap was selected between 1cm and 10cm during the optimization experiments. The coupling is better with smaller gap as expected shown in Fig. 5. But sometimes smaller gap caused plasma disruptions, since the plasma touched the wall in the divertor. The plasma shape control was therefore also very important for the optimization of the coupling. Another interesting parameter affecting coupling was plasma drift in vertical direction, shown in Fig. 6. When the plasma is slightly moved upwards, the value of RC at the bottom grill of the antenna is reduced from 19% to 12%, while that at the top grill is reduced from 99% to 58%.

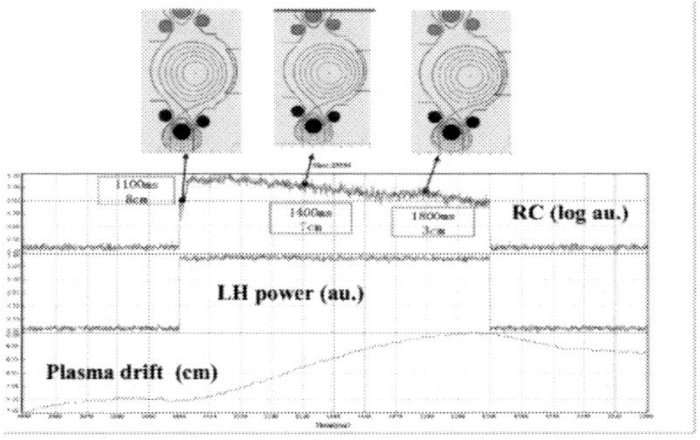

Fig. 5: Gap affects coupling in divertor configuration.

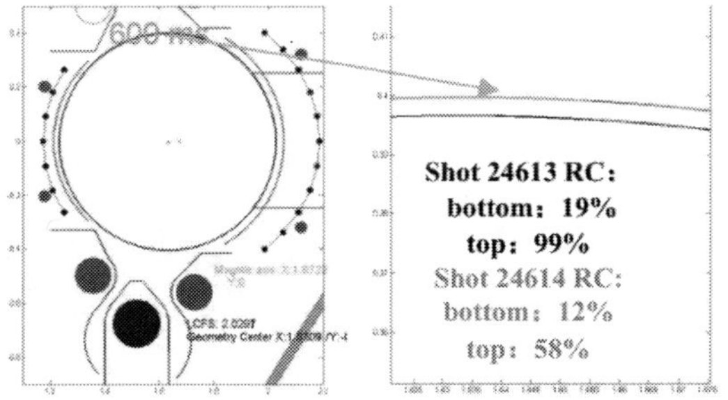

Fig. 6: Plasma drift in vertical direction affects coupling in limiter configuration.

**Plasma behaviour.** Lower Hybrid Current Drive（LHCD） and Heating (LHH) effects have been observed in the coupling experiments as shown in Fig.7. Indeed, during the LHW phase, the loop voltage is clearly decreased due to the non-inductive current driven by suprathermal electrons generated by LHCD, while the stored energy is increased showing the LHW heating effect. However the LHW power is not high enough to give fully non-inductive current drive in the present LHCD experiments. Fig.8 shows the temperature profile measured by ECE diagnostic at different time. The ECE signals at t=1300ms show clearly that the temperature profile is strongly affected by suprathermal electrons generated by LHCD, especially the edge channels. ECE signals, extremely sensitive to the suprathermal electron population in the plasma, can be considered as good indications for the suprathermal electrons.

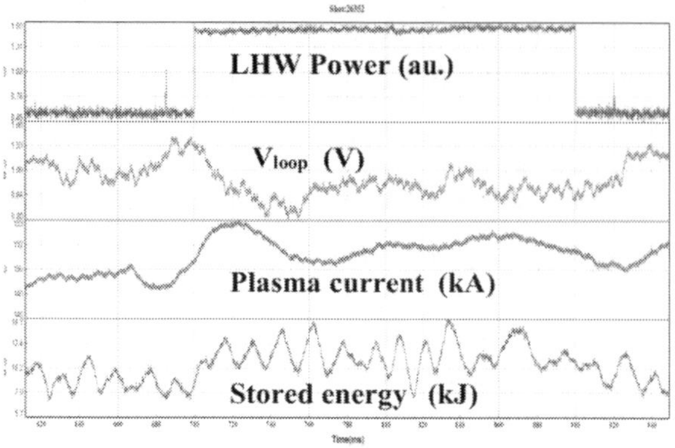


Fig. 7: LHCD & LHH effects observed during LHW injection in shot No. 26352. The loop voltage dropped due to LHCD and the stored energy increased due to LHH.

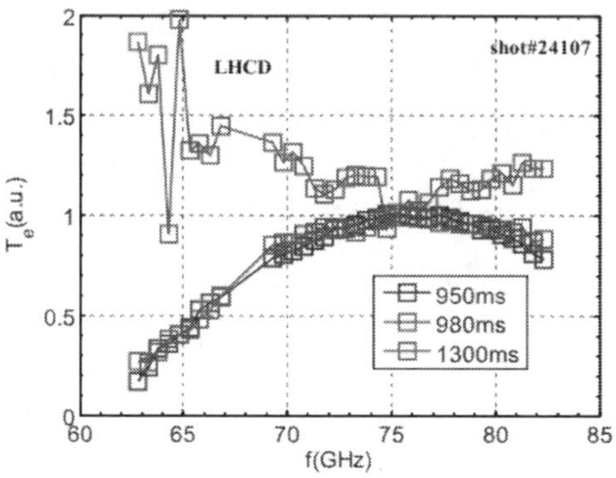


Fig. 8: Temperature profile measured by ECE diagnostic at different time: 950ms, 980ms in the phase without LHCD, and 1300ms with LHCD, The profile illustrates the effect on the temperature of suprathermal electrons generated by LHCD.

**Summary and Perspective.** A new 3.7GHz LHW system was built and exploited on the HL-2A tokamak. The PAM concept antenna was installed and commissioned. Coupling optimization experiments of the LHW system were carried out, in which the coupled LH power reached 500kW/400ms. A power reflection coefficient (RC) less than 10% was obtained in both limiter and divertor configuration. The key parameters affecting LH coupling were also studied. Effects on the plasma, such as LHCD, LHH and suprathermal electrons were observed in those experiments. These first promising experimental results will allow performing further physics experiments with the PAM antenna on HL-2A. In the near future, the transmission line efficiency would be increased by improving the flexible waveguides. And the gas puffing system would also be modified to make the gas quantity more controllable. With those hardware improvements, the LHW system will be able to deliver higher power in longer duration (2MW/2s), which would be high enough to give fully non-inductive current drive, or to trigger H-mode with LH alone.

**Acknowledgements.** The authors gratefully acknowledge the support of the HL-2A Team, in particular the control & operation group, the vacuum & gas puffing group, the diagnostic group and also the ECRH group. This work, supported by the collaboration between SWIP and CEA, was carried out with the help of the LHCD group of Tore Supra (CEA).

**References.**

[1] D. Guilhem et al., Fusion Eng. Des. **86** (2011) 279

[2] A. Ekedahl et al., Nucl. Fusion **50** (2010) 112002

[3] G.T. Hoang et al., Nucl. Fusion **49** (2009) 075001

[4] Ph. Bibet et al., Fusion Eng. Des. **74** (2005) 419

[5] J. Hillairet et al., Fusion Eng. Des. **84** (2009) 953

[6] X. Bai et al., Nuclear Fusion and Plasma Physics Vol. **29** (2009) P. 224-228

# First observation of thermal ion internal transport barrier on the HL-2A tokamak

D.L. Yu, Y. L. Wei, L. Liu, J. Y. Cao, A. P. Sun, Z. B. Shi, B.S. Yuan, W. J. Chen, Q. Ma, X. X. He, K.J. Zhao, X.Q. Ji, Y. Zhou, M. Jiang, W. L. Zhong, W. Deng, Y.G. Li, J.M. Gao, Yi Liu, Y. Xu, L. W. Yan, Q. W. Yang, X. T. Ding, J.Q. Dong, X. R. Duan, Yong Liu and HL-2A team

*Southwestern Institute of Physics, Chengdu, China*

The peaked density and temperature plasma is always pursued as it ensures the possibility of an operating merits characterized by high beta, high energy confinement time, and a large fraction of bootstrap current. The discharge with the internal transport barriers (ITBs), which origins from the pellet injection experiments [1], is a promising regime to satisfy the demand. The previous investigation reveals that the reversed (weak) magnetic shear is very important to develop the ITBs [2]. On ASDEX-U and JET tokamaks, the q = 1 ITB characteristics are compared [3]. Both of them are accompanied the fishbone oscillations, the q profiles are rather flat in the plasma core. The fishbone and the long lasting mode (LLM) are observed during the NBI heating on HL-2A tokamak. They are driven by the energetic particles in tokamaks; the former is a typical internal kink mode [4], and the latter is internal kink-like mode [5]. The appearance of these modes indicates the q profiles in the core are weakly or reversely sheared [5]. Recently, a 32/64-channel charge exchange recombination spectroscopy (CXRS) diagnostic system is developed on the HL-2A tokamak (R = 1.65 m, a = 0.4 m), monitoring plasma ion temperature and toroidal rotation velocity simultaneously with spatial and temporal resolutions up to ~1 cm and 5 ms [6, 7]. And the ion ITBs are observed during the NBI heating. Besides, a 32-channel fast electron cyclotron emission (ECE) system can provide the electron temperature ($T_e$) with temporal and spatial resolutions up to 0.8 μs and 1 cm[8].

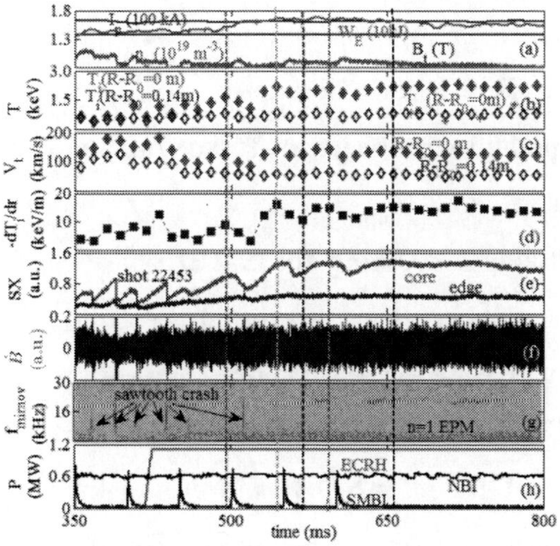

Figure 1. Typical ITB discharge on HL-2A

In shot 22453, the powers of neutral

beam injection (NBI) and ECRH are about 0.6 MW and 1.1 MW, respectively; the line-averaged plasma density is 0.8~0.9 $\times 10^{19}$ $m^{-3}$, as shown in figure 1(a). The ion temperature in the core $T_i(0)$ is slightly higher than $T_i(r/a = 0.37)$ before 490 ms, indicating that the gradient in the core is small; however, the gradient becomes steeper at around 500 ms, and $T_i(0)$ is three times as high as that of $T_i(r/a = 0.37)$, whereas the core electron temperature $T_e(0)$ is the same as the $T_i$ at $r/a = 0.37$, as shown in figure 1(b); the sawtooth oscillation disappeared after 512 ms. The energetic particle mode (n = 1) can be observed after 570 ms, as shown in figure 1(g). Meanwhile, the stored plasma energy is increased when the ion temperature gradient

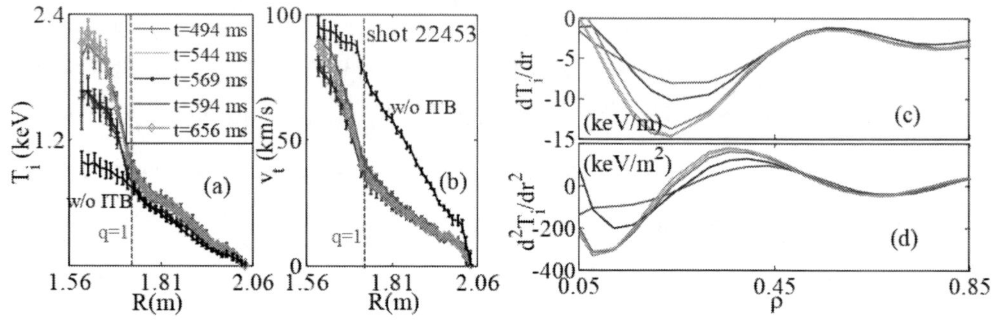


Figure 2. $T_i$ (a) and $v_t$ (b) profiles and the first (c) and second (d) derivative of Ti.

becomes steeper, as shown in figures 1(a) and (d). It is observed that the $T_i$ and $v_t$ are peaked after the increase of stored energy, as shown in figure 2 (a) and (b). The $T_i$ during the stored energy increase (t = 494 ms) is similar to that after the SMBI pulse (t = 569 ms), as shown in figure 2(a); and it is clearly lower than those of other time, as the former is not well developed and the latter is influenced by the SMBI.

Furthermore, the first and second derivative of ion temperature is analyzed, as shown in figure 2(c) and (d). The first derivative of ion temperature can be up to -15 keV/m, and it is about -10 ~ -7 keV/m for the sawtooth and during the SMBI phases, respectively. The maximum second derivative of ion temperature can be up to 175 keV/m$^2$, this value decreases to 2/3 and 1/2 after SMBI and during the sawtooth phases, respectively. For the

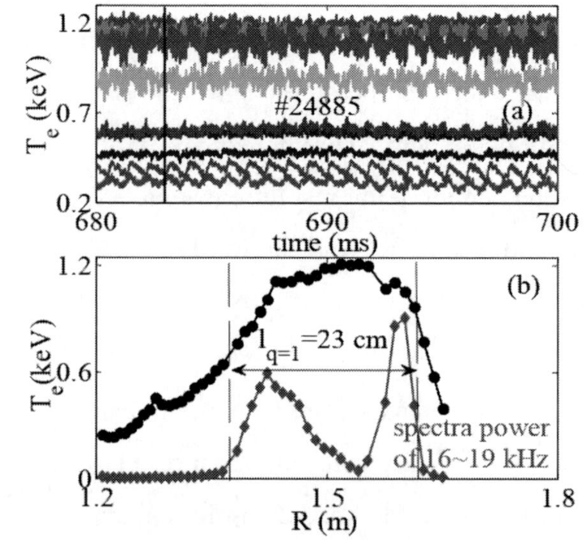


Figure 3. The position of ITB foot. Time traces of the $T_e$ (a) and the radial profiles (b).

second derivative of ion temperature, there exist two extrema, one locates at the foot of the peaked ion temperature and the other is close to the plasma core.

According to the ion temperature profiles, the ITB foot is around ~ 12~14 cm away from the plasma core. The location of the ITB foot is a key parameter to understand the mechanisms and the formation of the ITB. Therefore, it is necessary to obtain the location from other plasma parameters. The 32-channel ECE system has the spatial resolution of 1~2 cm. The time evolution of the $T_e$ and the radial profiles are shown in figure 3(a) and (b), respectively. The n = 1 LLM can be observed in the soft x-ray arrays, ECE system and Mirnov coils. To determine the q = 1 surface, the ECE signals are filtered and the radial profile of the spectrum power of 16~19 kHz can be obtained, as shown in figure 3(b). As the LLM is an internal-kink-like mode, they decrease sharply close to q=1 surface, therefore, the q = 1 surfaces are determined by the e-fold decay of the maximum intensity. And the averaged r = 11.5 cm of q=1 surface can be derived. It is about 2~3 cm smaller than the CXRS due to the spatial resolution is around ~3 cm in the plasma core.

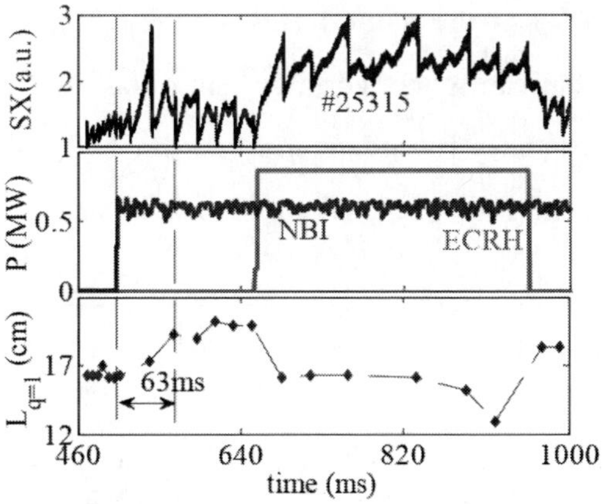

Figure 4. Time traces of soft X-ray(a), power of NBI and ECRH(b), the width of q=1(c).

The reason of ITB tending to be formed at the beginning of the NBI is due to the current drive effect of the NBI. As the motional Stark effect polarimeter is not available, the current drive effect of NBI is verified by the width of q = 1 measured by the ECE system during the sawtooth crash. The $L_{q=1}$ is defined as the width between the q=1 surfaces at high field and low field sides. For shot 25315, the $L_{q=1}$ is around 16 cm during the Ohmic phase, and is increased to ~20 cm during the NBI heating; the width is decreased to the Ohmic level during the ECRH and NBI phase; and when the ECRH is switched off, the width is again increased to ~20 cm. This experiment indicates that the NBI can increase the $L_{q=1}$ and hence the weak shear of the q profiles are available. Another reason for the ITB formation may relay on the plasma rotation shear at the beginning of the NBI heating phase. The plasma rotation is anti-current direction at first; and the direction changes due to the input torque from the NBI. On the other hand, the NBI torque deposit more in the core than that at the edge, therefore, the increase in the core is much evident and hence the velocity shear can be observed at the vicinity of q=1 surface.

The normalized ion temperature gradient ($R/L_{Ti}$) is compared for 17 shots, as shown in figure 5. For the ITB shots 22453, 24885, 24887, 24889, 24893, 25092 and 25099 etc., the $R/L_{Ti}$ is clearly higher; among these shots, the long lasting modes (fishbone oscillations) are observed during the peaked ion temperature and velocity. For shots 22865 and 22870, sawtooth and short duration fishbone oscillations are observed during the peaked phases. Another interesting phenomenon is that the $T_i(0.5)$ of the discharges without the ITBs seems not lower than that with the ITBs. Further analysis indicates that they are decreased with the maximum value of normalized ion temperature gradient inside the ITB. The $R/L_{Ti}$ at the $r/a = 0.5$ has the same trend with $T_i(0.5)$, indicating that better confinement inside the ITB will result in worse confinement outside it. However, the mechanism is unknown. In other words, it is easy to understand that the $T_i(0)$ is higher when the confinement becomes better, but the trend for $T_i(0.5)$ is not the case.

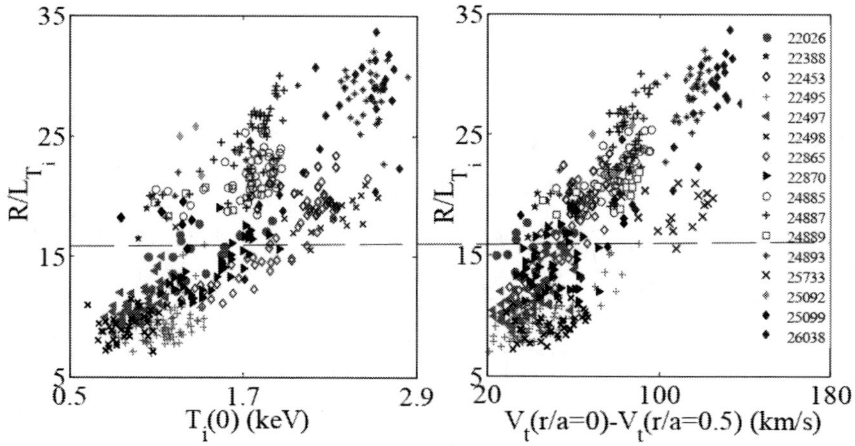


Figure 5. The $R/L_{Ti}$ versus $T_i(0)$ (a) and $v_T(0)$ (b).

The thermal ion internal transport barriers are firstly observed during the neutral beam injection (NBI) on the HL-2A tokamak. The ITBs are accompanied with the long-lasting mode, and the position of the barrier foot is at the $q = 1$ surface. The ITB can be more frequently observed at the beginning of the NBI heating due to the current drive effect.

## References

[1] A. Weller *et al.*, Physics Review Letters **59** 2303(1987).
[2] M. Hugon *et al.*, Nuclear Fusion **32** 33 (1992).
[3] E. Joffrin *et al.*, Plasma Physic and Controlled Fusion **44** 1203 (2002).
[4] G.Y. Fu *et al.*, Physics of Plasmas **13** 052517 (2006).
[5] R. B. Zhang *et al.*, Plasma Physic and Controlled Fusion **56** 095007 (2014).
[6] D.L. Yu *et al.*, Review Scientific Instruments **85** 11E402 (2014).
[7] Y. L. Wei *et al.*, Review Scientific Instruments **85** 103503 (2014).
[8] Z.B. Shi *et al.*, Review Scientific Instruments **85** 023510 (2014).

# VDE Simulation of the Snowflake Divertor Configuration in HL-2M

L. Xue[1], X.R. Duan[1], G.Y. Zheng[1], Y.D. Pan[1], S.L. Yan[1],
V.V. Dokuka[2], V.E. Lukash[3,4], R.R. Khayrutdinov[2,4]
[1]*Southwestern Institute of Physics, Chengdu, China*
[2]*TRINITI, Troitsk, Russia*
[3]*NRC Kurchatov Institute, Moscow, Russia*
[4]*National Research Nuclear University MEPhL (Moscow Engineering Physics Institute), Moscow, Russia*

1. ***Introduction.*** The vertical displacement event (VDE) is one of essential issues for vertically-elongated tokamaks. Thermal and electromagnetic loads during VDE are severe damage to in-vessel components of tokamaks. Estimating the magnitude of these loads is necessary for an engineering design. HL-2M is a vertically-elongated tokamak under construction, which will achieves advanced divertor configurations, such as the snowflake divertor. In this paper, the VDE under snowflake divertor configuration in HL-2M is investigated by DINA code [1-3].

2. ***Model.*** In HL-2M [4] model, sixteen PF coils, one CS coil as well as eighty-eight magnetic probes and fifty-two loops are installed. The vacuum vessel (VV) is divided into two hundred and ninety filaments. Divertor target plates are created by the limiter module. The computation domain covering 0.9~2.6m on the major radius direction R and -1.5~1.5m on the vertical direction Z is divided into 8385 grid elements. The geometry of HL-2M model for DINA code is shown in Fig. 1.

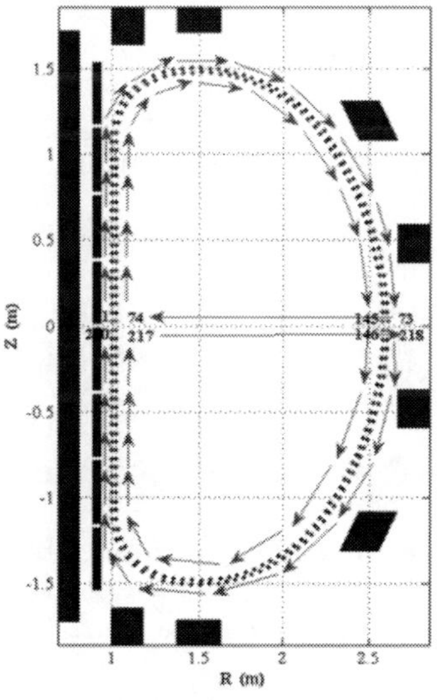

Figure 1. The geometry of HL-2M model for DINA

3. ***VDE away from null-points.*** The snowflake divertor configuration [5] has a larger poloidal weak field around the two null points than the standard one. In order to investigate the effect from the poloidal weak field when plasma moves away from null-points, a lower single-null snowflake configuration with upward VDE is simulated. Then it is compared to the

case of a lower single-null standard configuration with same main plasma parameters shown in Fig. 2. During the simulation, in the case of the feedback system fails, the plasma uncontrollably move upward duo to small perturbations of plasma parameters, evolving into an upper single-null standard divertor configuration. When the plasma column contacts divertor baffles, a limiter configuration forms. Then the thermal quench (TQ) and current quench (CQ) occur almost at the same time. The duration of current quench is 13.5 milliseconds and the peak of normalized poloidal halo current reaches 33.7 percent. By comparing with VDE of the standard divertor configuration, we found no obvious difference between the two VDE cases. This can also be reflected by evolution of these main plasma parameters shown in Fig. 3. Thus it seems that the lower poloidal weak field of the snowflake configuration plays an unimportant role when plasma moves upward.

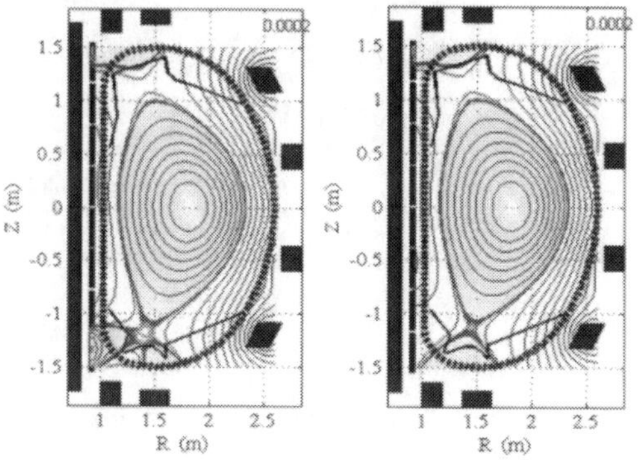


Fig. 2. The lower single-mull snowflake divertor configuration (left) and standard divertor configuration (right) in the flat phase of discharge.

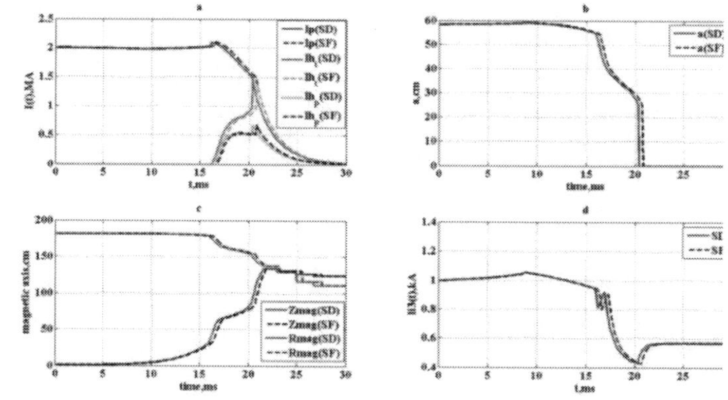

Fig. 3. Evolution of main plasma parameters: plasma current (a), halo current (a), the minor radius (b), the magnetic axis (c), the internal inductance (d).

4. ***VDE towards null-points.*** HL-2M has symmetrical poloidal field coils system and has the ability to generate double-null configurations. A double-null snowflake divertor configuration with upward VDE is simulated to investigate VDE towards null- points. Based on the analyzed result of part 3, effect from the poloidal weak field around lower null-points can be excluded. Here we still use the double-null standard

configuration with same main plasma parameters to compare as shown in Fig. 4. The change of the last closed flux surface (LCFS) under the snowflake VDE per 0.1 millisecond during the VDE is shown in Fig. 5. During the simulation of snowflake VDE, although the vertical instability still develops before the TQ, the upper poloidal

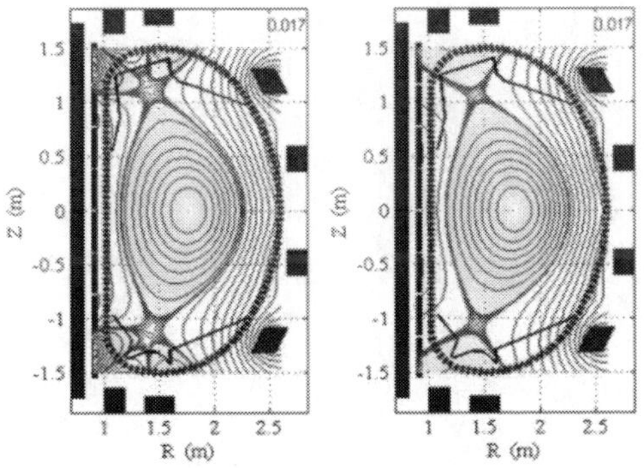


Fig. 4. The double-null snowflake divertor configuration (left) and standard divertor configuration in the flat phase of discharge.

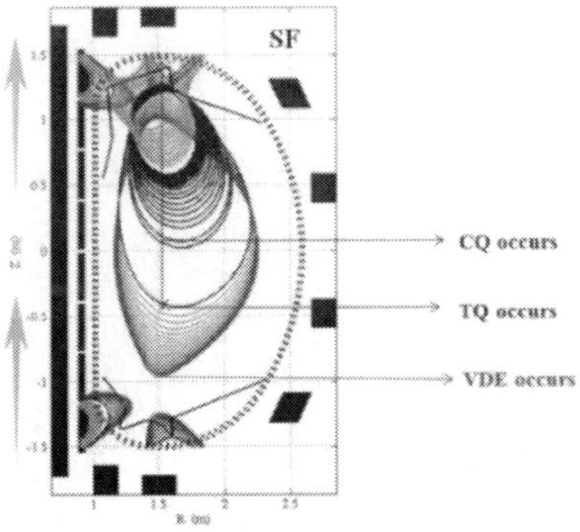


Fig. 5. Change of the LCFS of snowflake divertor configuration during VDE.

weak field may partly slow down the development. Under this effect, the vertical movement velocity of the plasma column before the TQ is slower than the case of standard VDE. During the TQ, sudden large perturbations of plasma parameters cause the velocity reaches a peak. After that the vertical instability is restrained duo to the energy release and the effect from upper divertor baffles as shown in Fig. 6. As the safety factor decreases during the CQ, the core plasma becomes more and more unstable, and finally the current decay rate suddenly reaches a peak. This will give rise to the spike of halo current. By the comparison, it is found value of the normalized poloidal halo current peak under snowflake VDE reaches 51.3 percent, which is 30 percent larger than the case of standard. Meanwhile a shorter duration of

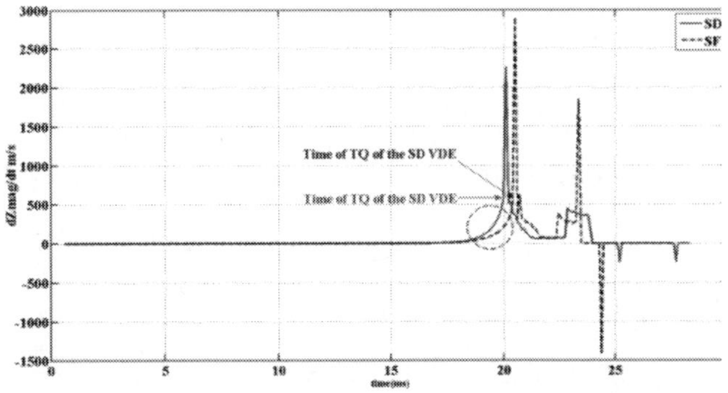


Fig. 6. Time trace of the velocity: standard (red) and snowflake (blue).

the CQ, about 7.4 milliseconds, is also obtained under snowflake VDE shown in Fig. 7.

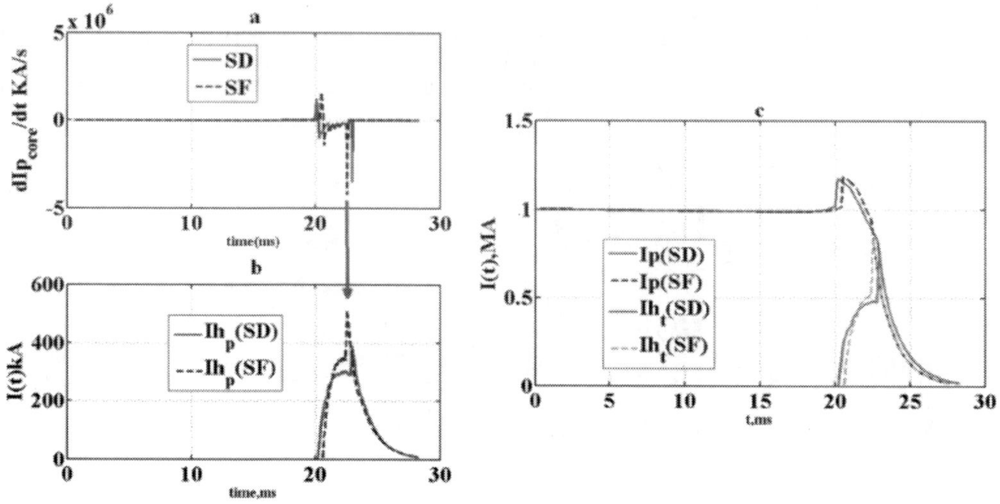


Fig. 7. (a) Evolution of the core plasma decay rate, (b) evolution of the poloidal halo current, (c) evolution of the total plasma current and toroidal current.

5. ***Summary.*** Simulation results show that the magnetic configuration on the plasma movement direction actually influences the process of VDE. For the case towards null-points, comparing with the standard, vertical velocity of the snowflake VDE is slower duo to the effect from the poloidal weak field around two null-points. This may be beneficial for control. But the higher peak of normalized poloidal halo current and shorter duration of the CQ may give rise to larger electromagnetic force on in-vessel components. This should be focused on the tokamak engineering design.

***Acknowledgments.*** This work was supported by Chinese ITER Plan Project Foundation under Grant Nos. 2013GB113001 and 2014GB110004.

***References***

[1] R.R. Khayrutdinov, *et al., Plasma Phys. Control. Fusion.* **43**, 321 (2001).

[2] V.E. Lukash et al. Evolution of the DINA-CH tokamak full discharge simulator 2004 31th EPS Conf. on Plasma Phys. (UK, London, 2004) p 2.151 http://epsppd.epfl.ch/London/pdf/P2_151.pdf.

[3] Masayoshi Sugihara, et al., *Plasma Phys. Control. Fusion.* **46**,1581 (2004).

[4] G.Y. Zheng, *et al., Fusion Engineering and Design.* **89**, 2621 (2014).

[5] D.D. Ryutov, *et al., Plasma Phys. Control. Fusion.* **54**, 124050 (2012).

# Current ramp-up scenario with reduced central solenoid magnetic flux consumption in JT-60SA

T. Wakatsuki[1], T. Suzuki[1], N. Hayashi[1], J. Shiraishi[1], S. Ide[1], Y. Takase[2]

[1] *Japan Atomic Energy Agency, Naka, Japan*

[2] *The University of Tokyo, Kashiwa, Japan*

## 1. Introduction

The amount of magnetic flux swing capability of central solenoid (CS) is a fundamental parameter of a tokamak reactor, since it affects the overall size and the output power of the reactor. If it is required that the plasma current should be ramped-up solely by the CS induction, the minimum size of the CS imposes a strong constraint on the reactor designs. In this study, we have investigated reduction of the CS flux required in the current ramp-up phase in JT-60SA using an integrated modeling code suite (TOPICS). In the previous study on current ramp-up with reduced CS flux consumption in JT-60SA, we developed a scenario in which the plasma current is ramped-up from 0.6 MA to 2.1 MA without additional CS flux consumption by overdriving the plasma current using neutral beams (NB) and electron cyclotron (EC) waves[1]. The investigation of the ramp-up scenarios with several prescribed density profiles revealed that a pressure profile with an H-mode pedestal and a wide internal transport barrier (ITB) whose foot location is at a large minor radius is required in order to obtain a large bootstrap current within the MHD stability limit. However, the pressure profiles were strongly dependent on the prescribed density profiles in the previous study. If the particle transport is solved, the width and the location of the density ITB might be different from the prescribed density profiles. Therefore, in this study we investigate the possibility of modification of the pressure profiles using the heating and current drive (H&CD) actuators in JT-60SA by solving both the particle transport and the thermal transport.

## 2. Modeling tools and assumed experimental conditions

TOPICS is an integrated modeling code suite and its main part solves the 1-D transport equations in accordance with the 2-D free boundary equilibrium. Several turbulent models can be used for the integration of the anomalous heat transport in TOPICS. Among them, we use CDBM model which demonstrated its ability to reproduce plasma profiles with ITB in JT-60U[2,3]. As for the particle transport calculation, we assume that the anomalous particle diffusivity is proportional to the thermal diffusivity according to experimental results of JT-60U. In the reversed shear plasmas on JT-60U, the effective particle diffusivity in the ITB region was estimated to be 0.04-0.2 times the ion thermal diffusivity when only the diffusion term was considered[4]. Thus, we assume an effective anomalous particle diffusivity $D_{ano} = 0.2 \times \chi_{CDBM,i}$ and calculate the particle transport assuming that the particle diffusivity is a sum of neoclassical and anomalous diffusivities with zero particle pinch velocity. The neoclassical diffusivity is needed for including a strong neoclassical diffusion inside the reversed shear region. Particle sources are NBI and the edge gas puff. The volume averaged density is feedback controlled by the edge gas puff.

JT-60SA will be equipped with two tangential negative ion based neutral beams (NNB), 24 positive ion based neutral beams (PNB) and a steerable EC wave launcher. The beam energy of the NNB will be 500 keV while that of the PNB will be 85 keV. One of the NNB will be injected on-axis and the other will be off-axis to the plasma magnetic axis, which can be used to modify the current profile, and each beam power will be 5 MW. There will be three groups in the PNB, which will be co-tangential beams, counter-tangential beams and perpendicular beams to the plasma current. Co- and counter-tangential beams consist of four beams with 1 MW power each, respectively, and perpendicular beams consist 16 beams with 1 MW power each. The maximum power of the EC wave will be 7 MW.

## 3. Controllability of density and temperature profile during current ramp-up

Figure 1 is one of the results of the ramp-up scenario simulation. At the low current (0.6 MA) phase, only co-tangential PNB and EC can be used because shine through losses of on- and off-axis NNB and perpendicular PNB are large. Then, the plasma current is overdriven using co-tangential PNB and EC from 2 s. The toroidal injection angle of EC wave is 10 degrees in the co-direction and the EC is locally absorbed at $\rho = 0.45$. On the other

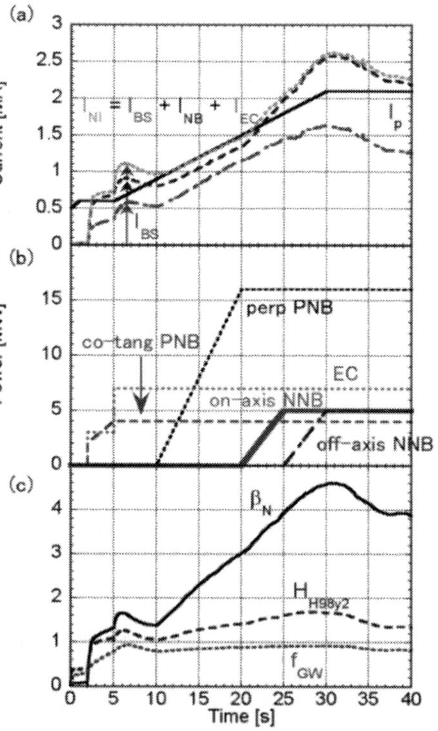


Figure 1. The TOPICS simulation of the current ramp-up scenario from 0.6 MA to 2.1 MA without resistive flux consumption. Time evolutions of (a) the plasma current, the bootstrap current and non-inductively driven current, (b) input powers used for the auxiliary heating and current drive and (c) the confinement enhancement factor ($H_{H98y2}$), the ratio of the electron density to the Greenwald density limit ($f_{GW}$), the normalized beta ($\beta_N$) .

hand, once the plasma current exceeds 1 MA and the electron density becomes higher than 2 x $10^{19}$ $m^{-3}$, shine through losses of on- and off-axis NNB and perpendicular PNB become less than 5%. The electron density in the scenario is relatively high and the fraction of the electron density to the Greenwald density limit ($f_{GW}$) is kept approximately 0.8 throughout the current ramp-up in order to obtain a high bootstrap current fraction. The confinement enhancement factor from ITER-98(y,2) scaling ($H_{H98y2}$) is less than 1.7 and the normalized beta ($\beta_N$) is less than 4.6.

The response of the density and the temperature profiles to the change in H & CD input is investigated at the low current phase (0.6 MA at 5 s) by changing a fraction of the input power of co-tangential PNB and EC while the total power is kept 7 MW. As shown in Fig. 2, a broader ITB in the density and the temperature profiles are formed when 4MW co-tangential PNB and 3MW EC are injected. With only EC, the width of ITB becomes narrower and the ITB foot moves slightly inwards. As a result, the bootstrap current reduces. If the location of EC is moved from $\rho$ = 0.45 to $\rho$ = 0.55 by changing the injection angle of the EC wave, the modification of the $q$-profile becomes small because the current drive efficiency decreases. As a result, the ITB becomes weak and the bootstrap current reduces more than 25 % although the ITB foot is moved outward. The current drive efficiency of co-tangential PNB is greater than EC. Therefore it is better to use co-tangential PNB with assist of EC for obtaining not only a large NB driven current but also a large bootstrap current by forming a broader pressure profile.

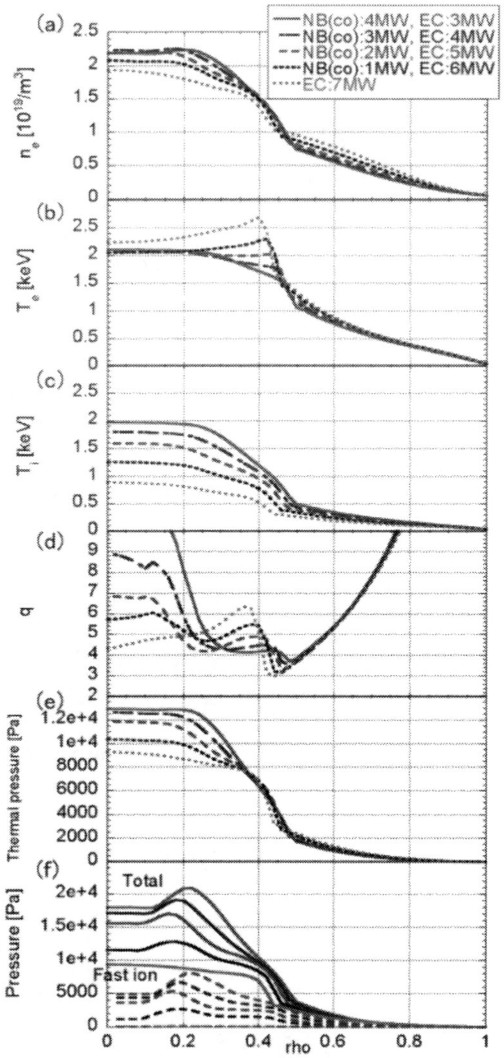


Figure 2. The change of (a) the density profile, (b) the electron temperature profile, (c) the ion temperature profile, (d) the safety factor profile, (e) the thermal pressure profile and (f) the total pressure profiles and the fast ion pressure profiles at the low current phase (0.6 MA at 5 s) in the scenario shown in figure 1 when input power fraction of co-tangential PNB and EC is changed.

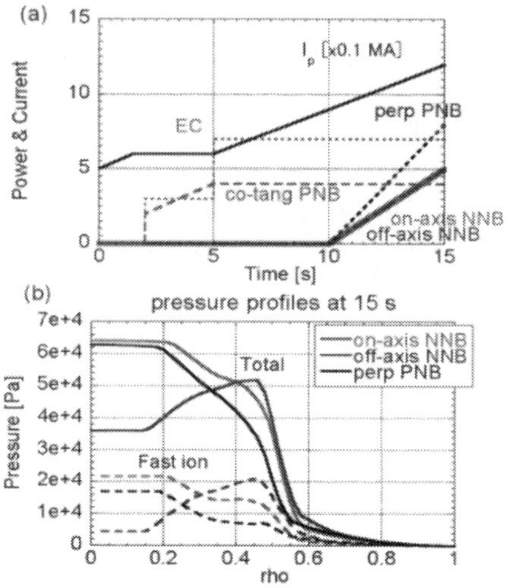


Figure 3. (a) Time evolutions of the plasma current and input powers. Note that on-axis NNB, off-axis NNB and perpendicular PNB are not injected simultaneously but alternatively. (b) The difference of the pressure profile at 15 s in the cases when the additional input power is on-axis NNB, off-axis NNB or perpendicular PNB.

As for the investigation of the response of the density and the temperature profiles to the change in H & CD input at the middle current phase (1.2 MA at 15 s), three scenarios are calculated. As shown in Fig. 3(a), on- and off-axis NNBs or perpendicular PNB are started to be injected in addition to 4 MW co-tangential PNB and 7 MW EC from 10 s and ramped-up to 5 MW or 8 MW at 15 s, respectively. ITB foots can be moved outward from the position at 10 s ($\rho = 0.52$) in all the three cases, as shown in Fig. 3(b). Among them, the largest ITB foot radius can be obtained by off-axis NNB. Therefore, this case might be suitable for further H & CD. However, the bootstrap current in the off-axis NNB case is 0.74 MA and slightly smaller than that in the on-axis NNB case (0.82 MA) and the perpendicular PNB case (0.84 MA) because the ITB width becomes narrower due to a strong neoclassical diffusion inside ITB which is caused by a sharp rise of the safety factor. The currents driven by on-axis NNB and off-axis NNB are large and the overdriven currents are more than 0.4 MA. Therefore, the total input power required for overdriving the plasma current can be reduced and a gentler pressure gradient which is preferable for the MHD stability could be obtained in the on-axis case and the off-axis NNB case by reducing the input power of EC, for example.

## 4. Summary

The possibility of control of the density and the temperature profiles in the reduced CS flux current ramp-up scenario is investigated using TOPICS. Both in the low current (0.6 MA) phase and the middle current phase (1.2 MA) the pressure profiles can be modified by changing the H & CD method. The optimization of the pressure profile considering the ideal MHD stability is remained as a future work.

This work is supported by JSPS Grant-in-Aid for Research Activity Start-up 26889069.

[1] Wakatsuki, T., et al., Plasma Phys. Control. Fusion **57**, 065005 (2015)

[2] Honda, M., et al., Nucl. Fusion **46**, 580 (2006)

[3] Fukuyama, A., et al., Plasma Phys. Control. Fusion **37**, 611 (1995)

[4] Takenaga, H., et al., Nucl. Fusion **43**, 1235 (2003)

# Advance in integrated modelling towards prediction and control of JT-60SA plasmas

N. Hayashi, M. Honda, J. Shiraishi, Y. Miyata, T. Wakatsuki, K. Hoshino, M. Toma, T. Suzuki, H. Urano, K. Shimizu, K. Hamamatsu, S. Ide and JT-60SA Team

*Japan Atomic Energy Agency, Japan*

## 1. Introduction

The project mission of JT-60SA [1] is to contribute to early realization of fusion energy by supporting the exploitation of ITER and by complementing ITER in resolving key physics and engineering issues for DEMO reactors. The device aims to achieve high-beta and high-bootstrap-fraction plasmas, in which the linkage between many physics factors is strong, resulting in complicated behavior of self-regulating plasmas. This self-regulation is expected to be strong in ITER and DEMO burning plasmas. In order to predict, control such plasmas, and to develop operation scenarios, we are developing codes/models which can describe each of both physics and engineering factors, and integrating them to one code suite TOPICS, as shown in Fig.1. It had been formally referred to as TOPICS-IB [2]. Now it is simply referred to as TOPICS, since the core transport module which was called TOPICS has now evolved integrating the other functions and modules so as to represent the suite.

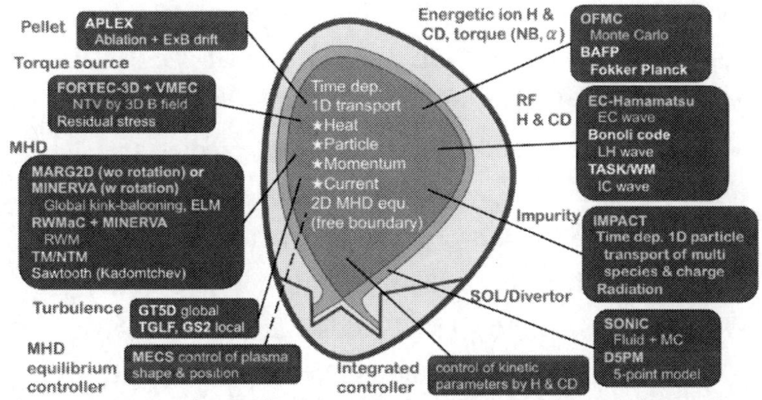

Fig.1 Schematic view of integrated code TOPICS and its modules where yellow denotes codes/models used in this paper.

## 2. Physics modelling

To study RWM in high-beta plasmas, TOPICS is coupled with a stability code MINERVA/RWMaC, in which the kinetic-MHD model is extended to include the toroidal rotation effect self-consistently. The energy exchange between MHD modes and particle's motion is generalized to yield three additional terms, representing Coriolis, centrifugal and rotation shear effects, compared with the conventional model. Figure 2 shows a stability diagram in a plane of rotation shear and rotation for a JT-60U experiment analyzed by MINERVA / RWMaC. For the rotation Ω=30-40 krad/s, increased rotation shear leads to the RWM

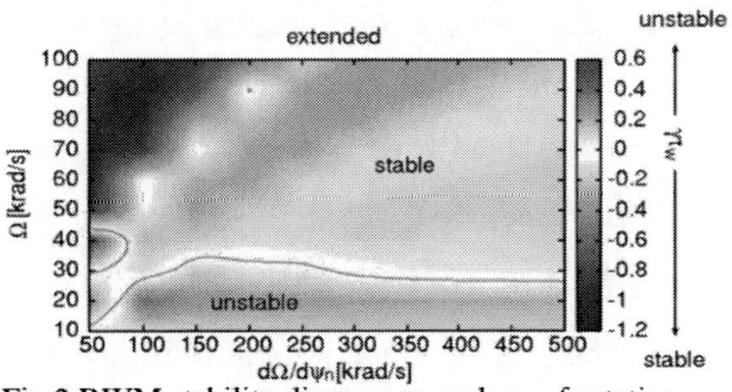

Fig.2 RWM stability diagram on a plane of rotation shear $d\Omega/d\psi_n$ and rotation $\Omega$, where $\psi_n$ is normalized poloidal flux, for a JT-60U experiment.

stabilization and this tendency is consistent with experiment. While the conventional model shows that the rotation shear destabilizes RWM and unstable regions appear for faster rotation $\Omega$>60 krad/s, the extended model removes the unstable regions and enlarge the stable region [3]. Simulations of MINERVA/RWMaC and TOPICS with the MHD equilibrium variation by the centrifugal force, which largely affects the RWM stability, realizes self-consistent analyses of RWM stability on the plasma evolution.

To study the toroidal rotation driven by various torques, TOPICS is coupled with an orbit-following code OFMC for NB induced collisional and jxB torques, with a 3D MHD equilibrium code VMEC and a $\delta f$ drift-kinetic code FORTEC-3D for a NTV torque, and with a toroidal momentum boundary model based on a zero gradient of radial electric field at the separatrix observed in JT-60U experiments. Analyses for JT-60U experiments show including NTV improves the reproducibility of toroidal rotation, and the boundary model in conjunction with a SOL/divertor model D5PM predicts well the toroidal rotation at the boundary. This integrated model is applied to predict the toroidal rotation in a hydrogen L-mode plasma in ITER. Figure 3 shows profiles of torques and toroidal rotation. Since the minimum toroidal field (TF) ripple created by 18 TF coils and ferritic steel tiles is considered here, the NTV torque is small compared to NB ones and the core rotation of ~2% of Alfven speed is predicted. However, NB ports and the tritium blanket module can enhance the TF ripple and thus NTV [4].

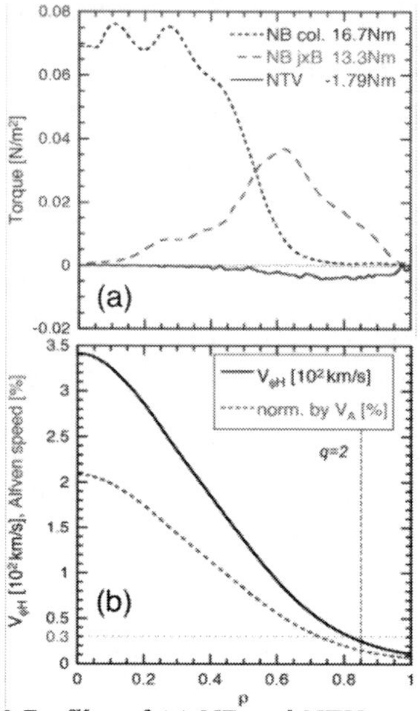


Fig.3 Profiles of (a) NB and NTV torques, and (b) toroidal rotation for a ITER hydrogen L-mode plasma.

We develop a core impurity transport code IMPACT, integrate it with TOPICS and study the core accumulation of impurity seeded to reduce the divertor heat load below a preferable level. For a JT-60SA steady-state (SS) scenario, an integrated divertor code SONIC showed that the low divertor heat load (< 10 MW/m$^2$) with low SOL density (< 1.5 × 10$^{19}$ m$^{-3}$), which is required for the scenario, was achieved by Ar gas puffing of 0.86 Pa m$^3$/s to the divertor region [5]. The Ar inflow to the core is evaluated from the SONIC result

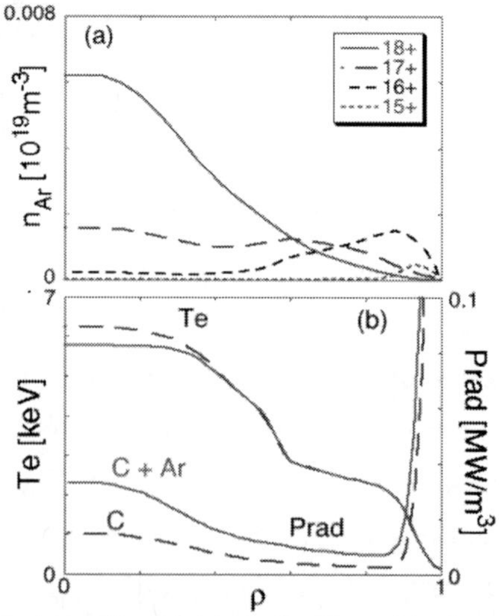


Fig.4 Profiles of (a) Ar density with charge states 15-18+, (b) electron temperature and radiation without/with Ar (intrinsic C exists in both cases) for a JT-60SA SS scenario.

and is assumed to be injected as MC neutrals in IMPACT, the neoclassical transport is calculated by NCLASS, and anomalous diffusivities are set to the neoclassical level to consider the maximum accumulation. Figure 4 shows simulation results of IMPACT and TOPICS where the same amount of intrinsic C exists in both cases without and with Ar. Ar with charge states 15-18+ are accumulated in the core and the radiation increases slightly (+ 1.5 MW). The Ar accumulation is so mild that the plasma performance can be recovered by additional heating within the machine capability.

TOPICS is applied to study the reduction of flux consumption of central solenoid (CS) during the current ramp up, which can relax a constrain on DEMO. JT-60SA simulations coupled with an ideal MHD stability code MARG2D show that the plasma current can be ramped-up to reach $\beta_N \geq 3$ with MHD modes stabilized by an ideal wall, and with high bootstrap and NB driven currents, resulting in no additional CS flux consumption [6].

**3. Control engineering modelling**

In the SS operation, the real-time control of plasma parameters is essential in avoiding MHD activities and sustaining the self-regulating plasma at the high performance within the machine capability. To study the controllability of kinetic parameters and their profiles via external actuators, an integrated real-time controller is developed on the basis of PID control scheme in a matrix form for multi-input and multi-output control, and coupled with TOPICS. Figure 5 shows waveforms of real-time control of $\beta_N$ and $V_{loop}$ for a JT-60SA SS scenario, where positive-ion-based NB and ECH is used for the $\beta_N$ control and negative-ion-based NBCD for $V_{loop}$ control. The integrated controller commands NB and EC powers, which are used as inputs of a simple 1D Fokker Planck code for NBH/CD and a EC-hamamatsu code for ECH. By optimizing control gains, $\beta_N$ follows well its reference raised by 3 steps and $V_{loop}$~0 is achieved during $\beta_N$=4 as its zero reference value. Simultaneous control of $\beta_N$ and $V_{loop}$ is possible at $\beta_N \geq 4$ in either high ($f_{GW}$=0.88, Fig.5) or low density ($f_{GW}$=0.52) [7].

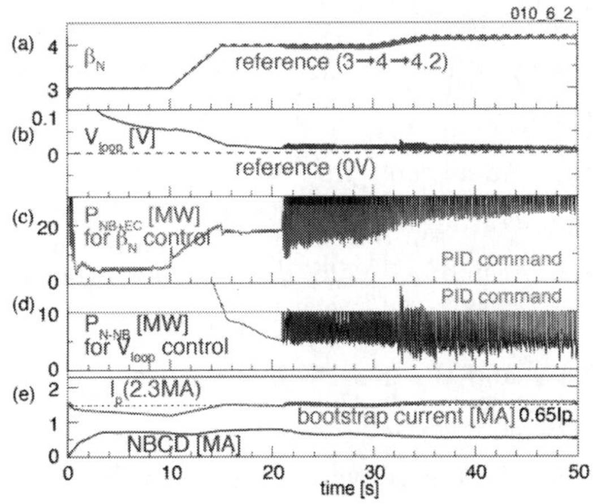


Fig.5 Real-time control of $\beta_N$ and $V_{loop}$ for a JT-60SA SS scenario. (a) $\beta_N$ and reference, (b) $V_{loop}$ and reference, (c) P-NB and EC heating power controlling $\beta_N$, (d) N-NBCD controlling $V_{loop}$, (e) plasma current, bootstrap current and NBCD current.

A MHD equilibrium control simulator (MECS) has been developed to study control techniques with the power supply capability of PF coils and the reconstructed plasma boundary by Cauchy condition surface (CCS) method. MECS consists of modules as shown in Fig.6. The isoflux controller modifies coil currents to reduce residuals between the poloidal magnetic flux at LCFS and that at the control points which specify the plasma position and shape, and also between actual and reference values of plasma current. The MHD equilibrium solver predicts the equilibrium for given $\beta_p$ and $l_i$ at present and this part will be coupled with TOPICS. Figure 7 shows a MECS simulation with and without CCS for a modelled heating phase in JT-60SA. Waveforms with CCS agree well with an ideal case without CCS where the shape is assumed known without diagnostics. This indicates the MHD equilibrium can be feedback controlled on the basis of combination of isoflux scheme and CCS method [8].

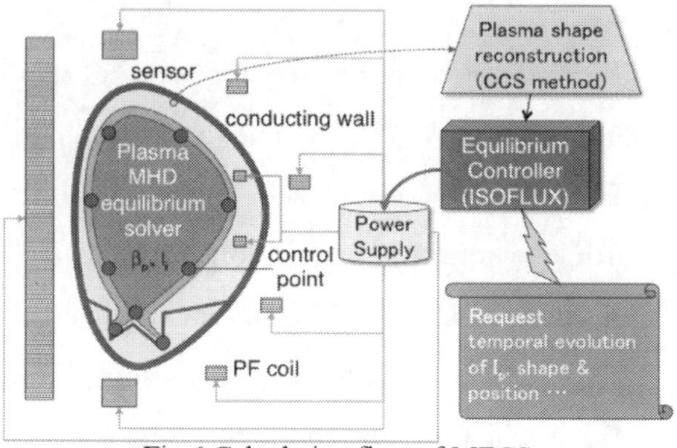


Fig.6 Calculation flow of MECS

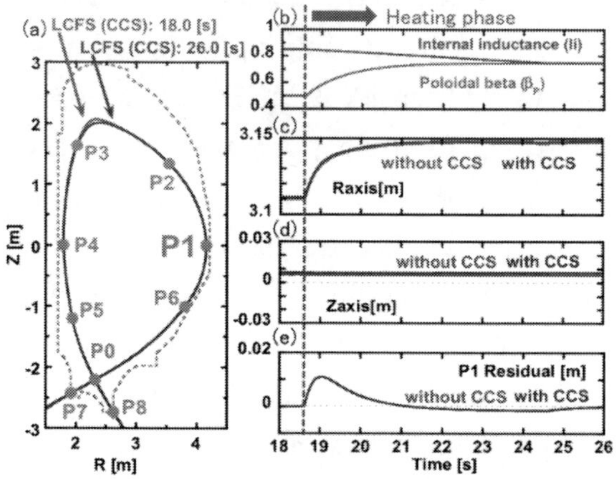


Fig.7 MECS simulation for JT-60SA. (a) Locations of control points and LCFS by CCS. Waveforms of (b) $\beta_p$ and $l_i$, (c) $R_{axis}$, (d) $Z_{axis}$ and (e) P1 residual without and with CCS.

## 4. Summary

Towards the prediction and control of JT-60SA plasmas, integrated modelling has advanced in both physics and engineering aspects; physics modelling: RWM stability with the toroidal rotation and kinetic effects, toroidal rotation with NTV and BC model, impurity transport in core and SOL / divertor plasmas, plasma current ramp up with reduced CS flux consumption, control engineering modelling: integrated real-time control for kinetic parameters, MHD equilibrium control. To study the burning plasma control in JT-60SA, a module to simulate $\alpha$ heating by a part of NBs will be developed and the integrated control will be improved to control it. For more self-consistent modelling, couplings of IMPACT+SONIC and MECS+TOPICS will be considered. These developed models will be further applied to the development of JT-60SA scenario.

This work was partly supported by JSPS KAKENHI Grant Number 26420862 and was partly carried out using HELIOS supercomputer system at International Fusion Energy Research Centre, Aomori, Japan, under Broader Approach collaboration between Euratom and Japan, implemented by Fusion for Energy and JAEA.

**Reference:**


[1] JT-60SA Research Plan v3.1, http://www.jt60sa.org/pdfs/JT-60SA_Res_Plan.pdf

[2] N. Hayashi et al. 2010 Phys. Plasmas **17**, 056112

[3] J. Shiraishi et al. 2014 Proc. IAEA FEC, TH/P6-12

[4] M. Honda et al. 2014 Proc. IAEA FEC, TH/5-1

[5] K. Hoshino et al. 2014 Contrib. Plasma Physics **54**, 404

[6] T. Wakatsuki et al. this conference, P5.144

[7] T. Suzuki et al. 2013 Proc. 40th EPS Conf. on Plasma Physics, P2.136

[8] Y. Miyata et al. 2014 Plasma Fusion Res. **9**, 3403045

# Dependence of pedestal structure on collisionality in JT-60U

H. Urano[1], N. Aiba[2], K. Kamiya[1], Y. Kamada[1] and the JT-60 Team[1]
[1] *Japan Atomic Energy Agency, Naka, Ibaraki 311-0193 Japan*
[2] *Japan Atomic Energy Agency, Rokkasho, Aomori 039-3212, Japan*

## 1. Introduction

The characterisation of pedestal structure has intensively been addressed from the perspective of edge dimensionless parameters, aiming at the extrapolation towards ITER. It has been recognised that the spatial width in H-mode pedestal region depends strongly on the poloidal beta value as [1–4]:

$$\Delta_{\Psi_N} = \beta^{1/2} \cdot f\left(\nu^*, \kappa, \epsilon, \cdots\right) \tag{1}$$

where $\Delta_{\Psi_N}$ denotes the pedestal width in the normalized poloidal flux space. There is a strong correlation existing between $\beta$ and $\rho^*$ because of the constraint of ELMs. Thus, the dependence of pedestal width on $\beta$ and $\rho^*$ has mainly been studied at a given magnetic geometry in many tokamaks. In other words, the dependence of pedestal width on the other dimensionless parameters, such as $\nu^*$, $\kappa$, $\epsilon$, etc, has not been fully understood.

This paper reports the experimental result on the dependence of pedestal width on $\nu^*$ in JT-60U. In a metallic wall, high gas puff rate is necessary to have the screening effect to avoid high $Z$ impurity influxes where the pedestal $\nu^*$ may be high. The characteristics of pedestal structure in high $\nu^*$ regime is also important. Thus, there are two $\nu^*$ scans conducted in ITER-relevant low $\nu^*$ regime ($0.03 \le \nu^* \le 0.2$) and high $\nu^*$ regime ($0.2 \le \nu^* < 0.8$).

## 2. Experiments in ITER-relevant low $\nu^*$ regime

In ITER-relevant low $\nu^*$ regime, a dimensionless collisionality scan experiment has been conducted. The pure collisionality scan requires the experimental setup which satisfies the following conditions:

$$n \propto I_p^0, \quad T \propto I_p^2, \quad I_p \propto B_t \tag{2}$$

where $n$, $T$, $I_p$ and $B_t$ denote the density, temperature, plasma current and magnetic field, respectively. The other dimensionless parameters relevant to the magnetic geometry such as $q$, $\kappa$, $\epsilon$, etc are fixed. These conditions lead to the experiment in which the plasma density remains constant with sufficient heating for satisfying $T \sim I_p^2$. Figs. 1(a) and (b) show the temporal evolution of plasma parameters at low and high $\nu^*$, respectively. The $I_p$ and $B_t$ are 1.79 MA and 4.0 T in the low $\nu^*$ discharge whereas those are 1.03 MA and 2.3 T in the high $\nu^*$ discharge, so that $q_{95}$ is $\sim$ 3.9 for both cases. The plasma configuration is fixed at $R = 3.3$m, $a = 0.8$m, $\delta = 0.34$, $\kappa = 1.5$. The line-averaged electron density is also controlled at $\bar{n}_e \sim 2.8 \times 10^{19}\text{m}^{-3}$. The scan ranges of $I_p$ and $B_t$ were chosen base on the experimental condition with practically controllable values of $\beta_p$ and $\bar{n}_e/n_{GW}$.

Figs. 2(a) and (b) show the spatial profiles of $n_e$ and $T_i$ at two time slices indicated by a broken line in figure 1, respectively. The $\nu^*$ was varied by the factor of 5 from 0.04 to 0.19 whereas the other dimensionless parameters were almost fixed. The $n_e$ profiles are approximately similar. Edge $T_i$ profiles are self-similar and differ by the factor of $\sim$ 4 with nearly the same pedestal width. However, the difference in the pedestal $T_i$ is larger than

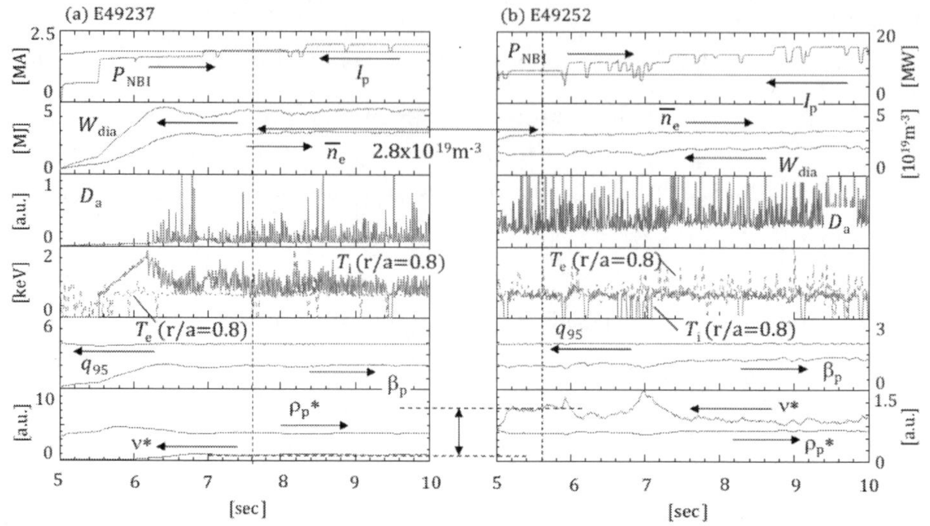


FIG. 1: Temporal evolutions of plasma parameters in H-mode discharges at (a) low and (b) high $\nu^*$. The $\rho_p^*$ and $\nu^*$ are measures of the volume-averaged values in arbitrary unit.

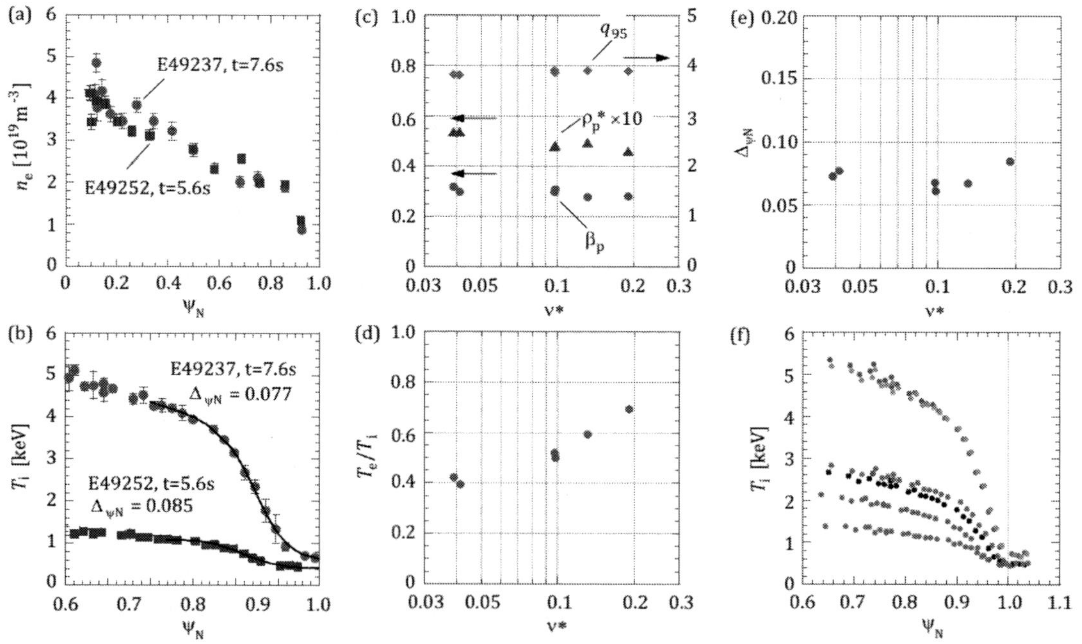


FIG. 2: Spatial profiles of (a) $n_e$ and (b) $T_i$ for low $\nu^*$ (= 0.04) and high $\nu^*$ (= 0.19) cases. The variations of (c) $q_{95}$, $\rho_p^*$, $\beta_p$ and (d) $T_e/T_i$ at the pedestal as a function of $\nu^*$. (e) Dependence of the pedestal width $\Delta_{\psi N}$ on $\nu^*$. (f) The pedestal $T_i$ profiles in the variation of $\nu^*$.

the expected value of the factor of $(1.79/1.03)^2 \simeq 3$ because $T$ should scale as $I_p^2$. Fig. 2(c) shows the variation of $q_{95}$, $\rho_p^*$ and $\beta_p$ at the pedestal as a function of $\nu^*$. The $\nu^*$ scan with the variation of $I_p = 1.03 - 1.79$MA ($B_t = 2.3 - 4.0$T) enabled us to fix $q_{95}$, $\rho_p^*$ and $\beta_p$ at the pedestal. Fig. 2(d) shows the variation of $T_e/T_i$ at the pedestal as a function of $\nu^*$. The electron-ion decoupling is enhanced at lower $\nu^*$ because ions are predominantly heated by the NBIs with the acceleration energy of $\sim$ 85keV. This decoupling causes the difference in $T_i$ which is larger than predicted.

Fig. 2(e) shows the dependence of the pedestal width $\Delta_{\psi N}$ on $\nu^*$. The pedestal widths

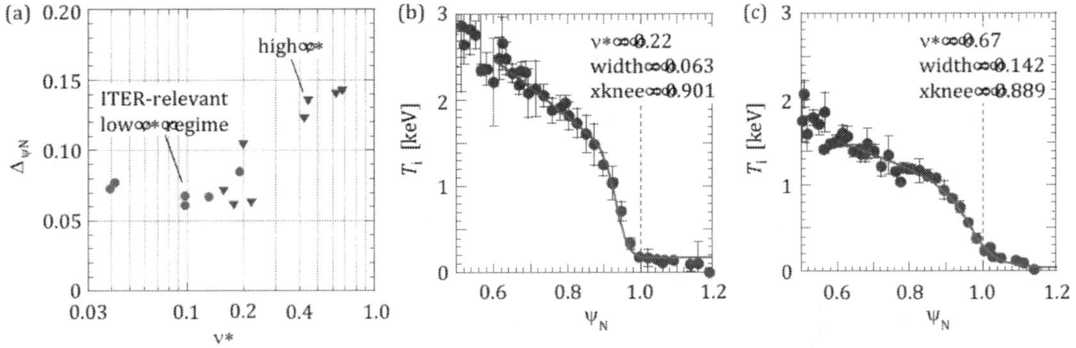


FIG. 3: (a) Dependence of the pedestal width $\Delta_{\psi N}$ on $\nu^*$. The pedestal $T_i$ profiles at $\nu^*$ of (b) 0.22 and (c) 0.67 in high $\nu^*$ regime.

have been evaluated by a fit with `mtanh` function. In the range of $\nu^*$ from 0.04 to 0.19, the pedestal width does not change against the variation in $\nu^*$. Fig. 2(f) shows the pedestal $T_i$ profiles in this $\nu^*$ scan. The experimental result indicates that the pedestal width is independent of $\nu^*$ in the ITER-relevant low $\nu^*$ regime.

## 3. Experiments in high $\nu^*$ regime

In high $\nu^*$ regime, $\nu^*$ was varied from 0.16 to 0.67 by deuterium gas puff rate. In this sense, this experiment is not a pure $\nu^*$ scan like the one performed in the low $\nu^*$ regime. The main aim of this study is to identify whether the pedestal width is determined only by $\beta_p^{0.5}$. As long as the pedestal $\beta_p$ and magnetic geometry are fixed, the deuterium gas puff scan can be a good dataset with the variation in $\nu^*$ to see the pedestal structure other than the effect of $\beta_p$. The experiment was conducted at 1.2 MA and 2.5 T ($q_{95} \sim 3.4$) [5]. The plasma configuration is fixed at $R = 3.4$m, $a = 0.8$m, $\delta = 0.36$, $\kappa = 1.4$. The $\bar{n}_e$ is varied by gas puff from 2.4 to $3.4 \times 10^{19}$m$^{-3}$.

Fig. 3(a) shows the dependence of the pedestal width $\Delta_{\psi N}$ on $\nu^*$ for both scans in the low and high $\nu^*$ regime. Note that the $\Delta_{\psi N}$ values cannot simply be compared between two scans because the experimental conditions are different. In contrast to the low $\nu^*$ regime, the pedestal width becomes greater with increased $\nu^*$ in the high $\nu^*$ regime. Figs. 3(b) and (c) show the pedestal $T_i$ profiles at low $\nu^*(= 0.22)$ and high $\nu^*(= 0.67)$, respectively. The pedestal broadening in the high $\nu^*$ regime indicates that the pedestal width cannot simply be explained by $\beta_p^{0.5}$ at the pedestal.

## 4. Pedestal stability analysis

Figs. 4(a) and (b) show the pedestal $j - \alpha$ diagrams in the ITER-relevant low $\nu^*$ regime. In the low $\nu^*$ regime, the pedestal is close to the peeling-ballooning mode boundary at intermediate $n$ toroidal mode number. The most unstable mode numbers are $n = 10$ and 16 at $\nu^*$ of 0.04 and 0.19, respectively. A larger $j_{bs}$ of $\sim 0.7$MA/m$^2$ is obtained in the low $\nu^*$ case whereas $j_{bs}$ is $\sim 0.3$MA/m$^2$ in the high $\nu^*$ case. The pressure gradient in the steep gradient region is not significantly changed with the variation of the edge current when the pedestal stays along the peeling-ballooning boundary with the intermediate $n$ toroidal mode number.

Figs. 4(c) and (d) show the pedestal $j - \alpha$ diagrams in the high $\nu^*$ regime. In the high $\nu^*$ regime, the pedestal is close to high $n$ ballooning mode boundary. The most unstable

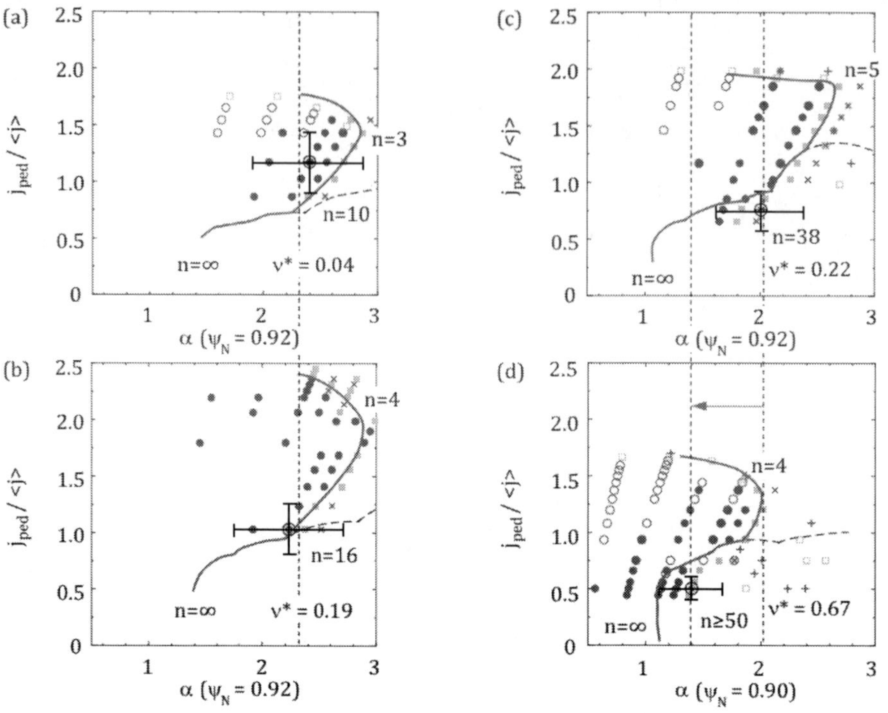


FIG. 4: The pedestal $j - \alpha$ diagrams for the ITER-relevant low $\nu^*$ scan at (a) $\nu^*$ of 0.04 and (b) 0.19 and for the high $\nu^*$ scan at (c) $\nu^*$ of 0.22 and (d) 0.67.

mode numbers are $n = 38$ and $\geq 50$ at $\nu^*$ of 0.22 and 0.67, respectively. When the pedestal is destabilised by high $n$ ballooning mode, the $\alpha$ decreases significantly from 2.0 to 1.3 at the steepest gradient position (or from 2.0 to 1.2 at $\psi_N = 0.92$) with the reduction in the edge current. The reduction in the pressure gradient at fixed pedestal pressure is consistent with the observation of the broadening of the pedestal width.

## 5. Conclusions

Dependence of pedestal width on collisionality has been investigated in JT-60U. In the ITER-relevant low $\nu^*$ regime of $0.04 < \nu^* < 0.2$, the pedestal width does not change in the variation of $\nu^*$. In the high $\nu^*$ regime of $0.2 < \nu^* < 0.7$, the pedestal width broadens with increased $\nu^*$. The pedestal pressure gradient is not significantly changed with the variation of the edge current at low $\nu^*$ whereas the pressure gradient decreases with the reduction in the edge current. The pedestal broadening is observed when the pedestal is unstable at high $n$ ballooning mode. The experimental observation brings us a hypothesis that at fixed $\mathrm{d}p/\mathrm{d}\psi$ the pedestal can be destabilised by broadening the pedestal width to increase the edge current large enough to destabilise the peeling-ballooning mode. The analysis during the inter-ELM phase will help proving this hypothesis in the future issue.

## References

[1] Urano, H., et al., Nucl. Fusion **48** (2008) 045008.
[2] Beurskens, M., et al., Phys. Plasmas **18** (2011) 056120.
[3] Kirk, A., et al., Plasma Phys. Control. Fusion **51** (2009) 065016.
[4] Walk, J.R., et al., Nucl. Fusion **52** (2012) 063011.
[5] Urano, H., et al., Nucl. Fusion **55** (2015) 033010.

# Plasma density fluctuations and turbulent thermal fluxes in the transient process at ECH in the L-2M stellarator

G.M. Batanov, V.D. Borzosekov, N.K. Kharchev, Yu.V. Kholnov, E.M. Konchekov, D.V. Malakhov, K.A. Sarksyan and D.G. Vasilkov.

*A.M. Prokhorov General Physics Institute RAS, Moscow, Russia*

Studying of anomalous heat transport and plasma turbulence role in it is still of interest to magnetic confinement devices. In the paper the measurements of short-wavelength ($k_\perp = 30\ \mathrm{cm}^{-1}$) and long-wavelength density fluctuations ($k_\perp = 1\ \mathrm{cm}^{-1}$) at a transient process in the L-2M stellarator are presented. Such process in the L-2M stellarator is usually attributed to electron cyclotron heating (ECH) switch on/off and it is accompanied by density pump-out effect [1,2] and missing power effect [3,4].

The major radius of the L-2M stellarator $R$ = 100 cm, average radius of the last closed flux surface $a_0$ = 11.5 cm, rotational transform angle at the magnetic axis $\iota(0) = 0.17$ and at the last closed flux surface (LCFS) $\iota(a_0) = 0.8$. The ECH system consists of two gyrotrons with frequencies $f$ = 75 GHz. The implemented heating scenario is X2 (heating with extraordinary wave at the second harmonic of electron gyrofrequency) with microwave power injected in two adjacent standard external ports of toroidal vacuum vessel. By varying gyrotrons switch on/off time it is possible to achieve pedestal-like form of ECH power time evolution. It should be noted that one gyrotron is used both for heating and microwave scattering diagnostics. Thus gyrotrons switching sequence does matter for density fluctuations measurements.

Short-wavelength density fluctuations were measured using backscattering of gyrotron radiation [5]. Long-wavelength density fluctuations were studied using small-angle scattering of gyrotron radiation [6]. Both diagnostics are based on homodyne detection. Core electron electron temperature $T_e$ was measured by 2nd harmonic ECE (electron cyclotron emission) diagnostics $f$= 76 GHz. Average electron density $n$ was measured by a central chord microwave interferometer $f$ = 141 GHz. Discharges with different ECH power, electron density time evolution and ECH absorption region location (central, non-central heating) were investigated.

Plasma density fluctuations are routinely characterized by their mean square value $\langle \tilde{n}^2 \rangle$ (angle brackets mean averaging herein). However, turbulent heat and particle fluxes also depend on particle temperature. Let us assume that the turbulence has drift-wave nature. Then electron heat flux will be:

$$q_e = c k_\perp T_e^2 \langle \tilde{n}^2 \rangle (\langle n \rangle e B_0)^{-1} \cdot \xi(t),$$

here $c$ — speed of light, e — charge of an electron, $B_0$ — magnetic field, $T_e$ — electron temperature, $\xi(t)$ — function of phase shift between density and potential fluctuations. The fact that plasma oscillations speed in wave depends on $T_e$ $\tilde{v} = c\tilde{E}/B_0 = ck_\perp T_e(\langle\tilde{n}\rangle/\langle n\rangle)(B_0 e)^{-1}$ was taken into account in the expression of heat flux $q_e$. Since focus of this paper is to investigate dependence of turbulent heat flux on ECH power then further in the paper time evolution of $\langle\tilde{n}^2\rangle$ and $T_e^2\langle\tilde{n}^2\rangle/\langle n\rangle$ will be used to characterize change of the flux.

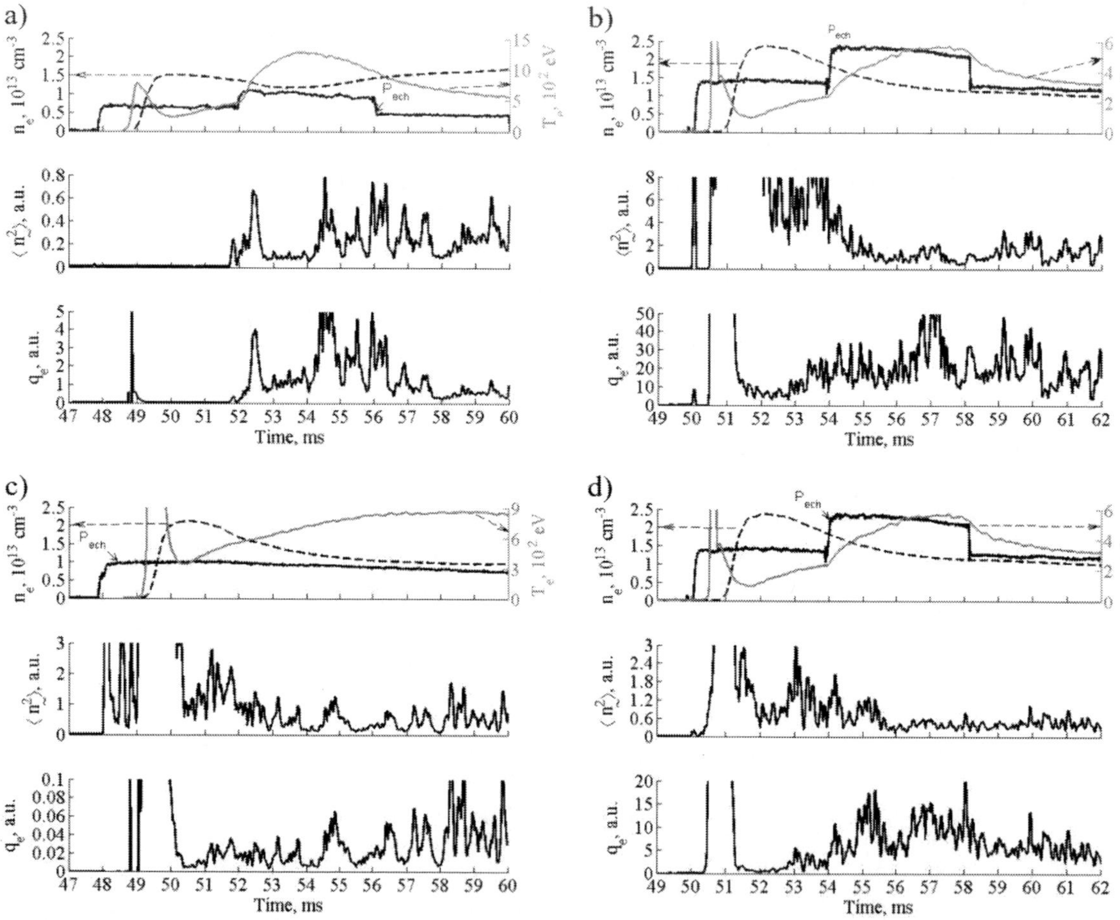


Fig.1. For each case: upper graph — electron density $n_e$, electron temperature $T_e$ and microwave heating power $P_{ech}$ time evolution; middle graph —density fluctuations energy $\langle n_\sim^2\rangle$; lower graph — turbulent heat flux $q_e$. a) Central ECH, short-wavelength density fluctuations, b) non-central ECH, short-wavelength density fluctuations . c) reference case, short-wavelength density fluctuations, d) Non-central ECH, long-wavelength density fluctuations.

Measurements results for central ECH case are presented in fig.1(a). Here each gyrotron has power of 130—150 kW. Short-wavelength density fluctuations level $\langle\tilde{n}^2\rangle$ was measured by backscattering diagnostics. At second (first and second is related to switching on time sequence herein) gyrotron switch on (52 ms) there is already high level of density fluctuations in plasma that was created and heated by first gyrotron. Fluctuations level increase appears with a time delay 2—2.5 ms, which is comparable with energy confinement time, after second gyrotron switch on at 52 ms and further the level remains increased even after first gyrotron switch off (56 ms). Turbulent heat flux (related to short-wavelength fluctuations) changes synchronously with electron temperature change. Thus turbulent heat flux after doubling the ECH power is increased up to 4 times at a central heating.

Measurements results for non-central ECH case (at magnetic field $B_0$ = 1.29 T absorption region is shifted towards inner wall on $a_0/4$) are presented in fig.1(b). In this case each gyrotron has microwave power 400 kW. The gyrotron which also used for giagnostics worked during all discharge length, thus density fluctuations measurements are available for full ECH length. Second gyrotron had only short pulse from 54 ms till 58 ms. As can be seen from fig.1(b) minimum of ECE signal is at 52 ms, when electron density is maximal. This minimum correspond to $T_e$ = 100—200 eV. At this time short-wavelength density fluctuations level is maximal. Then the level is decreased almost by an order to 55 ms, when second gyrotron already heats the plasma for 1 ms. After second gyrotron switch off at 58 ms fluctuations level is increased up to 1.5—2 times. Thus, time delay of density fluctuations level increase, after second gyrotron switch on, is 4 ms at non-central ECH. The level remains increased till the end of heating (62 ms), despite of second gyrotron switch off at 58 ms. Electron temperature increase determines electron heat flux increase in this case too. It appears that turbulent heat flux is maximal at the end of the second gyrotron pulse.

A discharge with constant ECH power is presented in fig.1(c) as a reference case. Here ECH power $P_{ECH}$ = 400 kW and electron density time evolution close to that of previous case. It can be seen that time evolution of the fluctuations level is similar to the non-central heating case with one exception: after decrease in the beginning of the discharge the level is not increased, as it was with some delay after second gyrotron switch on, but remains constant till the ECH pulse end. The turbulent heat flux evolves accordingly to electron temperature evolution.

Time evolution of long-wavelength density fluctuations level was investigated for the case of non-central heating at pedestal-like form of ECH power time evolution. These measurements, which are presented in the fig.1(d), were carried out simultaneously with the short-wavelength fluctuations measurements from fig.1(b). Therefore density and temperature time evolutions are the same as in fig.1(b). Time evolution of the long-wavelength density fluctuation level resembles time evolution of short-wavelength density fluctuations level in the reference case: highest fluctuations level is observed at the time period of plasma breakdown and sharp density increase (50.5—51.5 ms); the level is decreased till the time of minimal electron temperature and further it remains approximately constant. However, turbulent heat flux is increased after second gyrotron switch on at 54 ms and is decreased after second gyrotron switch off at 58 ms till the ECH end at 62 ms. Thus the heat flux increase after doubling ECH and decrease after it ends is typical both for long-wavelength density fluctuations and for short-wavelength density fluctuations

Another important issue to mention is a high value of the fluctuations amplitude $\tilde{a} = \tilde{v}/\omega$ that means non-linearity of the observed phenomena. This fact was pointed out in [7] for density fluctuations with $k_\perp = 6\ \mathrm{cm}^{-1}$ and it was shown that Reynolds number $R_E = k_\perp r_n (e\tilde{\varphi}/T_e) > 1$ ($r_n$— density gradient scale length, $\tilde{\varphi}$ — potential fluctuations level). $R_E > 1$ is a condition for excitation of drift wave vortices [8]. For the case of non-central ECH presented in this paper estimation of relative density fluctuations level yields $\langle\tilde{n}\rangle/\langle n\rangle = 10^{-3}$. At $T_e = 600$ eV it gives $\tilde{v} \approx 10^5$ cm/s. Taken into account that approximately 50% of $\langle\tilde{n}^2\rangle$ is in the 25 kHz frequency range [9] then, given that $\omega = 10^5\ \mathrm{s}^{-1}$, fluctuations amplitude $\tilde{a} = \tilde{v}/\omega = 1$ cm, i.e. amplitude higher than wavelength more than an order magnitude. Unlikely that in such case linear or quasilinear approach will be reliable. For our case dispersive spreading of drift waves is damped by nonlinearity at $r_n > 33$ cm (for $R_E > 1$ condition). Such density profile flattening is typical for L-2M stellarator currentless plasmas created and sustained by ECH.

To conclude this short paper we should emphasize factors responsible for enhancing of anomalous transport: electron temperature increase itself, fluctuation speed increase (caused by temperature increase) and density fluctuations level increase. All that is enhanced by a strong turbulence case which suppose drift wave vortices formation.

This work was supported by the RFBR grant 14-02-00589.

[1] V. Erkmann et al. Plasma Phys. Control Fusion. 1986. V.28. P.1277.
[2] H. Renner et al. Plasma Phys. Control Fusion. 1989. V.31. P1579.
[3] V. Stroth et al. Plasma Phys. Control Fusion. 1996. V.38. P.611.
[4] V.F. Andreev et al. Plasma Phys. Control Fusion. 2004. V.46. P.319.
[5] G.M. Batanov et al. Plasma Phys. Rep. 2013. V.39. N.6. P.444.
[6] D.K. Akulina et al. Plasma Phys. Rep. 2008. V.34. N.12. P.979.
[7] G.M. Batanov et al. Plasma Phys. Rep. 1993. V.19, P.628.
[8] W. Horton. Phys. Fluids B. 1989. V.1. P.524.
[9] G.M. Batanov et al. Plasma Phys Rep. 2014. V.40 P.265

# Formation of Non-Maxwellian SXR Spectra at Various Plasma Parameters and ECRH Powers at the L-2M Stellarator

A.I. Meshcheryakov, D.G. Vasilkov, A.A. Vasileva, I.Yu. Vaphin, E.V. Voronova, I.A. Grishina, V.P. Logvinenko and V.D. Stepakhin

*Prokhorov General Physics Institute of the Russian Academy of Sciences, Moscow, Russia, e-mail: meshch@fpl.gpi.ru*

1. EXPERIMENTAL SETUP

Experiments on plasma creation and heating by microwave radiation of high power density of $P_{ECRH}/V_P$ = (2.0 -4.0) MW/m$^{-3}$ were carried out at the L-2M stellarator [1]. In these experiments, nonmaxwellian SXR spectra were measured.

The L-2M device is a classical two-pole stellarator ($l$ = 2, $N$ = 7) with a major radius of $R$ = 1 m, minor plasma radius of $a$ = 0.115 m, and toroidal magnetic field of $B_0$ = 1.34 T. The rotational transform varies from $\iota$ = 0.18 at the magnetic axis to $\iota$ = 0.78 at the plasma edge. A gyrotron with a power of up to 600 kW is used to create and heat plasma at the frequency 75 GHz (the second harmonic of the electron gyrofrequency). The resonance region was in the center of plasma. Microwaves were delivered to the plasma by the waveguide ($\mathbf{k}\perp\mathbf{B}_0$). The section in which heating was performed was spaced at a distance of about 1 m from the section in which the SXR spectrometer was arranged.

Spectral measurements were carried out by the SXR spectrometer which is capable of measuring spectra in the energy range from 1 to 80 keV and has the count rate of $V = 1.7\cdot10^5$ photons per second. Such a high counting rate was achieved at the cost of deterioration of the spectral resolution of the instrument which is $\Delta E$ = 320 eV at energy of about 6 keV.

SXR radiation was received in the direction of the central chord in the equatorial plane of the stellarator. To obtain one spectrum, spectral data was summed during several pulses of the facility in the quasi-stationary stage of the discharge when the main plasma parameters become steady-state. In all ECRH operation modes, nonmaxwellian SXR spectra with two different slopes in the low and high energy ranges were observed. Typical SXR spectrum is shown in Fig. 1. This spectrum consists of the thermal part and epithermal "tail". The breaking point of the spectra corresponds to energy of $E$ = 5 keV. To quantitatively estimate the deviation of the measured SXR spectrum from the Maxwellian one, we will use two parameters. The first one is the slope of the spectrum plotted in semilogarithmic coordinate axes. For thermal electrons, the slope of the spectrum is the electron temperature which we denote as $T_e^{bulk}$. We will conventionally characterize the slope of the epithermal part of the

SXR spectrum by the "tail temperature" $T_e^{tail}$. The second parameter is electron density. Let us denote as $n_e^{bulk}$ and $n_e^{tail}$ densities of thermal and epithermal electrons, respectively. These parameters can be determined from the measured SXR spectra because, in our case, these spectra plotted in semilogarithmic coordinate axes consist of two straight segments.

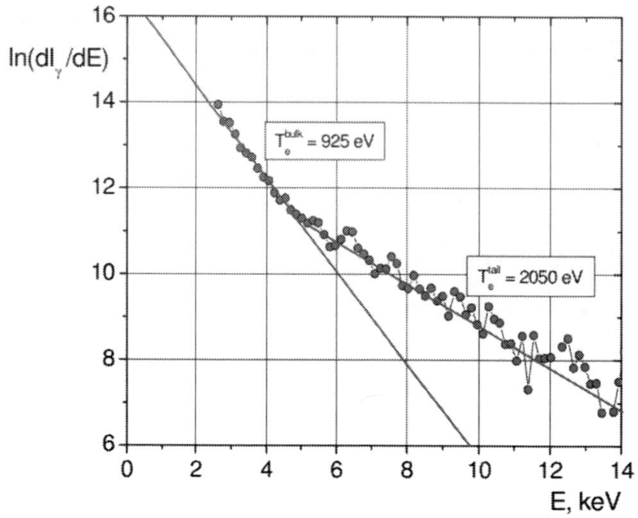


Fig.1. Typical SXR spectrum measured in the ECR heating mode. $P_{ECRH}$ = 430 kW, and $n_e$ = $1.75*10^{19}m^{-3}$.

## 2. EXPERIMENTAL RESULTS

To clarify the reasons for occurrence of epithermal tails in SXR spectra, we carried out the SXR spectra measurements at various plasma parameters and ECRH powers. Plasma electron density and ECRH power were varied in the ranges of $n_e = 0.7\cdot10^{19}M^{-3}$ - $2.5\cdot10^{19}M^{-3}$ and $P_{ECRH}$ = 200 - 600 kW, respectively.

Figure 2 shows the temperatures of the bulk of thermal electrons $T_e^{bulk}$ (triangles) and the "tail temperatures" $T_e^{tail}$ (black dots) depending on the heating power normalized to the electron density of $n_e = 10^{19}$ m$^{-3}$. The temperature of the thermal part of the spectrum (lower curve) considerably increases with growing normalized power and that is quite reasonable. At the same time, the epithermal tail "temperature" $T_e^{tail}$ (upper curve) remains practically constant within the measurement errors.

The dependences of the $T_e^{bulk}$ and $T_e^{tail}$ temperatures on the electron density of plasma measured at the constant heating power are shown in Fig. 3. Curves 1 and 2 are the dependences of the thermal temperature $T_e^{bulk}$ on the electron density corresponding to heating powers of 200 and 400 kW, respectively. Curve 3 shows the dependence of the "temperature" of epithermal electrons on the electron density for both heating powers.

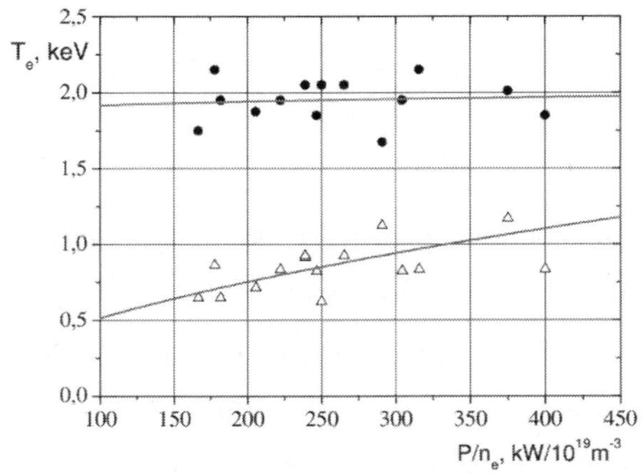


Fig.2. Dependence of the two SXR spectrum slopes $T_e^{bulk}$ and $T_e^{tail}$ on the normalized ECRH power $P_{ECRH} / n_e$.

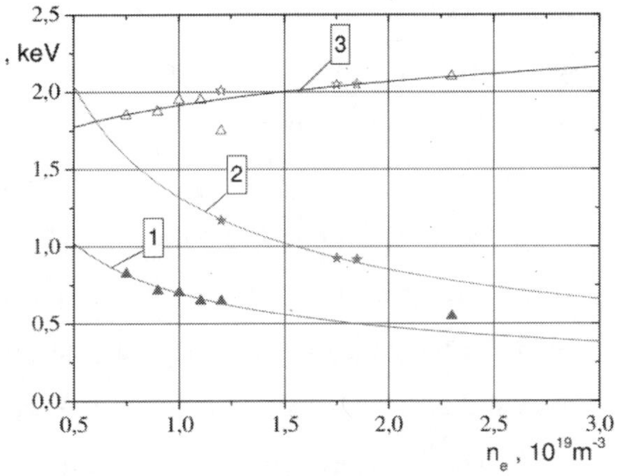


Fig.3. Dependences of the $T_e^{bulk}$ and $T_e^{tail}$ temperatures on the electron density of plasma at two ECRH powers 200 kW (curve 1) and 400 kW (curve 2).

As seen from the figure, the temperature of thermal electrons $T_e^{bulk}$ decreases with growing electron density since the heating power per one electron becomes lower. In contrast, the "tail temperature" $T_e^{tail}$ slightly increases with growing electron density and it does not depend on the heating power.

Figure 4 shows the dependence of density of epithermal electrons $n_e^{tail}$ on the total electron density of plasma at two ECRH powers 200 and 400 kW. The density of epithermal electrons decreases with growing electron density of plasma and, for a higher power of 400 kW, it decreases sharper.

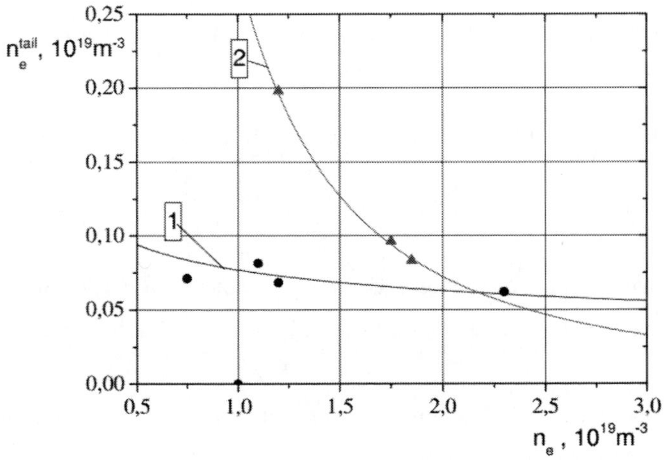


Fig. 4. Dependence of density of epithermal electrons on electron density of plasma at two ECRH powers 200 kW (curve 1) and 400 kW (curve 2).

Explanation of the experimental dependences (Figs. 2 – 4) requires understanding of the mechanism of generation of epithermal electrons in these experiments which is not yet clear.

3. DISCUSSION

SXR spectra with two slopes were previously observed at the T-10 [2] and TCV [3] tokamaks in experiments on current drive. In [2], the occurrence of epithermal electrons was associated with the generation of drag currents caused by changes in the distribution function of electrons over energies. In both papers, the authors were unable to give satisfactory explanation of the "two-temperature" shape of the measured SXR spectra.

We also can not suggest the well-studied explanation of the shape of the measured SXR spectra consisting of two straight sections. We can only hypothesize that the observed shape of the SXR spectra occurs due to the geometry of microwaves' absorption. In [4], it was shown that, in our experiments, radial width of the area of microwaves' absorption is about 0.5-0.7 cm. Apparently, the temperature profile is very sharp within the absorption area. The SXR spectrometer received radiation coming along the central chord of plasma section. Radiation from both the area of plasma heating and the colder areas of plasma fall into its field of view. Tentative simulations of X-ray flux from plasma carried out at various given profiles of plasma density and temperature showed that, in some cases, "two-temperature" spectra can occur in plasma. The studies in this field are in progress.

Thus, in the ECRH experiments at the L-2M stellarator, the SXR spectra with two "temperatures" were observed. It was found that, with growing heating power and with decreasing density, the temperature of the thermal part of the spectrum increases and the epithermal tail "temperature" remains practically constant within the measurement errors. The mechanism of generation of the epithermal part of the electron distribution function over energies is not yet clear.

The work was supported by the Russian Fund for Basic Research (project no. 15-02-09192).

**References**

[1]. G.M. Batanov, V.I. Belousov, Yu.F. Bondar et al, A New MIG-3 Gyrotron Complex for Creation and Heating of Plasma in the L-2M Stellarator and the First Experimental Results, Plasma Physics Reports, V.39, No 13, p. 1088 (2013)

[2]. Yu. V. Esipchuk, N.A. Kirneva, A.A. Martynov, V.M. Trukhin, Suprathermal SXR emission in ECCD experiments at the T-10 tokamak, Plasma Physics Reports, (1995), V. 21, No 7,

[3]. P.Blanchard, S.Alberti, S.Coda, H. Weisen, P. Nikkola, and I. Klimanov, Hight field side measuraments of non-thermal electron cyclotron emission on TCV plasmas with ECH and ECCD, Plasma Phys. and Control. Fussion 44 (2002) pp.2231-2249.

[4]. A. S. Sakharov and M. A. Tereshchenko, Kinetic Simulation of EC Plasma Heating and Current Drive in the L-2M Stellarator, Plasma Physics Report, (2002), Vol.28, No. 7, pp. 539-548.

# Modelling of non-inductive current generation upon superposition of vertical field on EC heated plasma in the LATE device

T. Maekawa, K. Kuroda, M. Wada, M. Uchida, H. Tanaka

*Graduate School of Energy Science, Kyoto University, Kyoto, Japan*

A toroidal current was generated in a number of experiments when a weak vertical field (Bv) was superposed in EC-heated toroidal plasmas [1-4]. Furthermore closed flux surfaces were often formed via a rapid current increase in small aspect ratio devices [3, 4]. These results are important since realization of compact tokamak reactors depends on reduction or elimination of a central solenoid from the reactors.

In order to understand the current generation mechanism and the equilibrium characteristics of plasmas in the external field composed of Bv and the toroidal field Bt, non-inductive current generation experiments have been carried out in the Low Aspect ratio Torus Experiment (LATE) device using 2.45GHz, 2s microwaves of 1-2 kW [5 (P5.150 in this conference)]. After the microwave of 1.5 kW is injected into the steady external field of Bv=6.8G and Bt=480G (both at R=25cm), a plasma current is generated and quickly increases up to 920A (typically in ~0.5s), and then is kept steady for 1.5 s until the end of the microwave pulse. Figure 1 shows profiles of electron density ($n_e$), temperature ($T_e$), pressure ($p_e$) and space potential ($V_s$) measured with Langmuir probe at the steady phase. In this experiment a top panel (Z=30cm), a bottom panel (Z=-30cm), and an ion energy analyser just behind an arced fine mesh slit on the top panel were installed. The radial profiles of vertical current onto the bottom and top panels, and the ion collector are shown in Fig.1(c).

In this paper the experimental results are examined using the plasma fluid model and also particle models. Radial and vertical force balances are primal to the force balance in the toroidal direction. Therefore, the radial and vertical components of the force balance

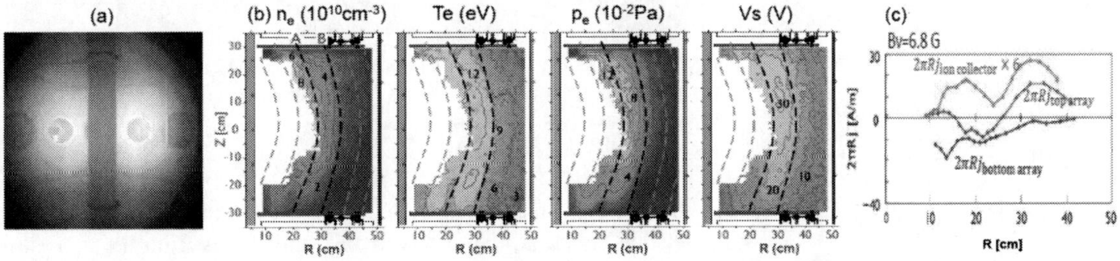


Fig.1 (a) Plasma image, (b) profiles of electron density, temperature, pressure and space potential, (c) radial current profiles onto the bottom and top panels and the ion collector during steady phase of discharge. $R_{ECR}$=13.8cm. The white area shows the probe tip locations where the interferometer line density along the chord Rt=12cm deteriorates by more than 10% compared with the values without probe insertion.

equation ($\mathbf{j} \times \mathbf{B} = \nabla p$) and the condition of steady current circulation ($\nabla \mathbf{j} = 0$) constitute the basic set of equations to analyse experimental results. Using R and Z components of $\mathbf{j} \times \mathbf{B} = \nabla p$, and $\nabla \mathbf{j} = 0$,

$$\frac{\partial p}{\partial Z} = -j_\varphi B_R - \frac{R}{2}\left[\frac{\partial(j_\varphi B_R)}{\partial R} + \frac{\partial(j_\varphi B_Z)}{\partial Z}\right]$$

The solution reads $p(R,Z) = p_{VC}(R) + p_{TC}(R)$, where

$$p_{TC}(R,Z) = \frac{R j_\varphi B_Z}{2} - \int^{Z}\left[j_\varphi B_R + \frac{R}{2}\frac{\partial(j_\varphi B_R)}{\partial R}\right] dZ'$$

is related to the toroidal current $j_\varphi$, and $p_{VC}(R)$ is responsible to the vertical current that circulates via the conducting vessel [6]. We may define $p_{VC}$ as

$$p_{VC}(R) = \frac{R j_{VC} B_\varphi}{2},$$

where $j_{VC}$ is the vertical current density.

Toroidal current profile is estimated from the magnetic analysis using 13 flux loops. First we employ a model of a single D-shaped area of toroidal current in which every local current has the same $\varphi$ direction. The current smoothly becomes zero at the boundary and there is no current beyond the boundary. The model has 10 fitting parameters for ellipticity, triangularity, broadness or peaking factor for profile, shift of current peak position, location of D shape, etc. We search for a set of fitting parameters so that the flux values match to the measured flux values. Note that all the power supplies for toroidal and poloidal field coils are regulated with transistors and the noise levels on the flux loop signals are quite low. The best fit current profile for the plasma in Fig. 1 is shown in Fig. 2(a) with the fitting results to the flux loop signals in Fig. 2(b). No current exists near the bottom panel and this is

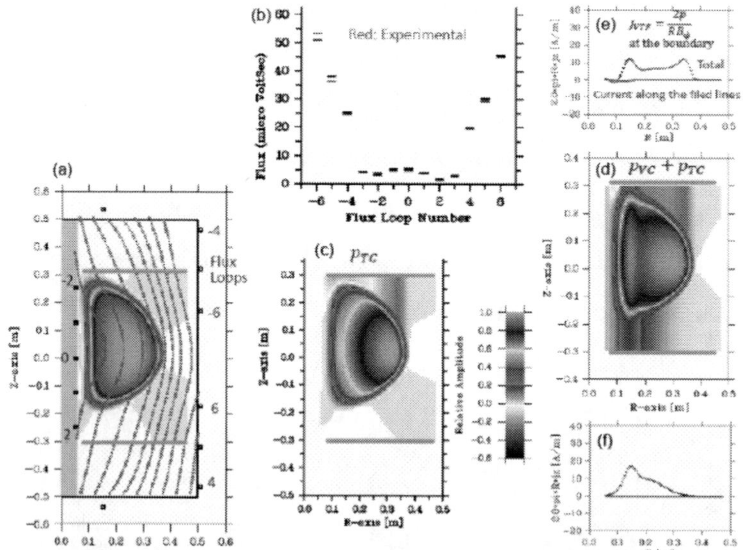


Fig. 2. Analysis with a single D-shape area of toroidal current. (a) $j_\varphi$ and poloidal flux contours, (b) observed flux and fluxes from the model, (c) $p_{TC}$ profile, (d) $p_{TC} + p_{VC}$ profile, (e)and(f) radial profiles of current flowing onto the top panel and flowing out from the bottom panel, respectively.

the salient feature when we use this model. The pressure profile ascribed to the toroidal current is shown in Fig. 2(c). In order to match the vertical current through the bottom and top panels as well as its profile on the bottom panel to the experimental ones, we add an appropriate pressure profile ($p_{VC}$) responsible to the vertical current. Sum of both pressures ($p_{VC} + p_{TC}$) is shown in Fig. 2(d) with radial profiles of current flowing onto the top panel (Fig. 2(e)) and flowing out from the bottom panel (Fig. 2(f)). In this model parallel current along the field line is zero at the panel surfaces and therefore there is no contribution from the parallel current to the current onto the panels.

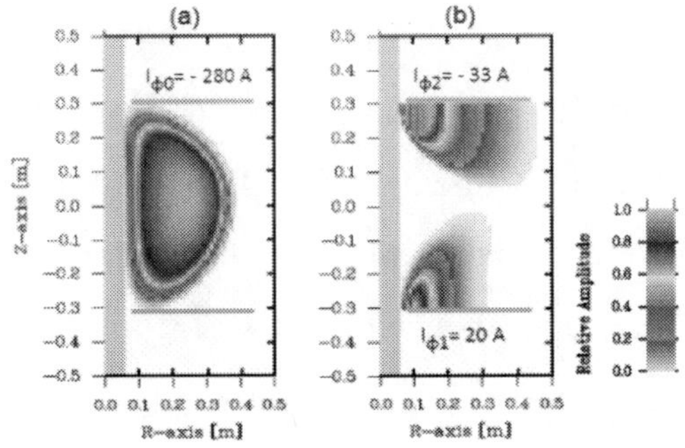


Fig.3. Profiles of (a) main toroidal current (b) toroidal components of the currents flowing onto the top and bottom panels along the field lines.

Next we take into account the contribution to $j_\varphi$ from the parallel currents that flow onto the top and bottom panels along the field lines as shown in Fig. 3(b). The total current profile and the fitting results to the flux loop signals are shown in Figs. 4(a) and 4(b), respectively. In order to match the vertical current through the bottom and top panels as well as its profile on the bottom panel to the experimental ones we add an appropriate pressure profile responsible to the vertical current. Figs.4(e) and 4(f) show radial profiles of current flowing onto the top panel and flowing out from the bottom panel, respectively. Here the current onto the panels is contributed from the parallel current flowing along the field lines in addition to the vertical VTF drift current which is proportional to the edge pressure in front of the panels.

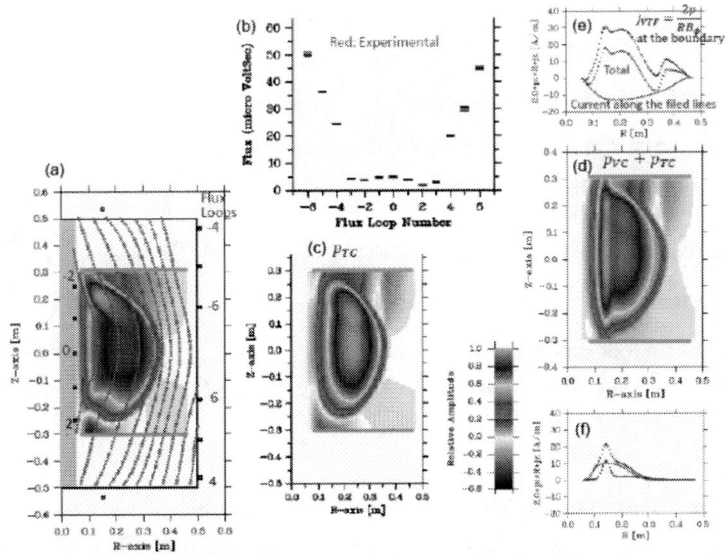


Fig. 4. Analysis by taking into account the contribution from the currents flowing onto the top and bottom panels along the field lines as those shown in Fig. 3. (a) $j_\varphi$ and poloidal flux contours, (b) observed flux and fluxes from the model, (c) $p_{TC}$ profile, (d) $p_{TC} + p_{VC}$ profile, (e)and(f) radial profiles of current flowing onto the top panel and flowing out from the bottom panel, respectively.

Comparison with the experimental results shows that the former model using a single current profile is not appropriate since the pressure profile near the bottom panel does not match the experimental one (Fig. 1(c)), while the latter model fairly matches to the experimental profile. However, the absolute value of pressure is significantly larger than the experimental one, suggesting that contribution to pressure from the tail electrons is significant.

Even in the latter model the radial current profile onto the top panel does not match to the experimental one. This may show that fluid description for ion behavior fails near the top panel, where a steep potential slope onto the top panel exists as shown in Fig.1(b). Fig.5 shows typical ion orbits starting from the line from R=30cm to 48cm at Z=20cm. The full orbits deviate from the guiding center orbits just before the top panel at inside of R<25cm corresponding to the fluid result in Fig.4(e). Note that the guiding center drift description is equivalent to the fluid description.

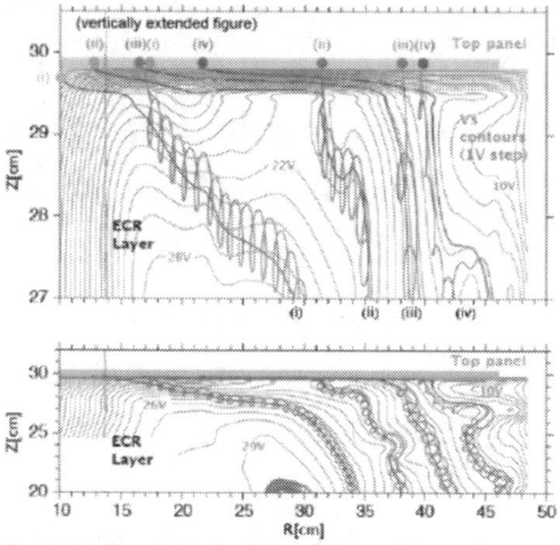

Fig.5. Full orbits (red) and the orbits using guiding center drift approximation (blue) starting from Z=20cm line with the initial kinetic energy of 2eV and the pitch angle of 63 degrees. The upper figure is vertically extended to show the detail.

Using lots of full orbits of protons starting from the line R=30cm to 48cm at Z=20cm, where the ions are assumed to have a Maxwell distribution with a temperature of Ti=2eV and the same density profile as the measured electron density profile, the ion behavior near the top panel is simulated. The simulated current profile onto the top panel ($j_{\mathrm{simulation}}$) matches to the current profile measured with the ion analyzer as shown in Fig.6.

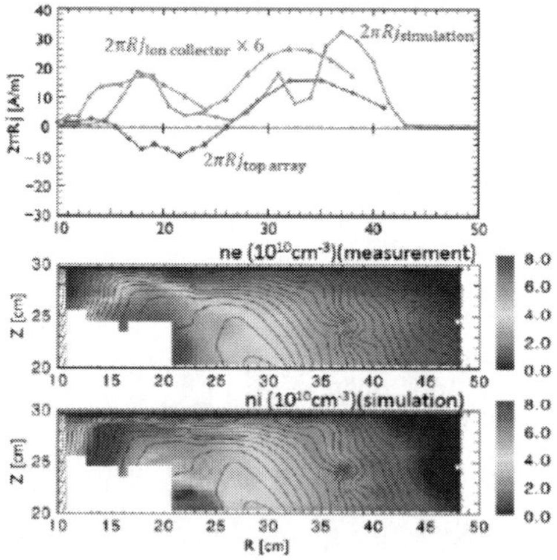

Fig.6. Comparison of experimental results and the results from full orbit analysis for ions.

**Acknowledgements**

The present work was supported by KAKENHI (Grant Number 23360411 and 26289357).

**References**

[1] C. B. Forest et al., Phys. Rev. Lett 68 (1992) 3559.
[2] V. Shevchenko et al., Proc. 25th Fusion Energy Conf., EX/P6-22 (2014)
[3] T. Yoshinaga et al, Phys. Rev. Lett 96 (2006) 125005
[4] M. Uchida et al, Nucl. Fusion 51 (2011) 063031
[5] K. Kuroda, et al., 42nd EPS Conference on Plasma Phys. Lisbon, P5.150 (2015)
[6] T. Maekawa, et al., Nucl. Fusion 52 (2012) 083008

# Observation of toroidal and vertical currents in EC heated plasmas under vertical field in the LATE device

K. Kuroda, M. Wada, M. Uchida, H. Tanaka and T. Maekawa

*Graduate school of Energy Science, Kyoto University, Kyoto, Japan*

Non-inductive current drive and formation of a closed flux surface by EC-heating have been studied in the low aspect ratio torus experiment (LATE) device [1-4], which are important for stable start-up and reducing the size of the reactor [5,6]. A model for the non-inductive formation of a closed flux surface in EC-heated toroidal plasmas has been proposed [7]. In a toroidal ECR plasma a vertical charge separation current is generated by the vertical drifts of ions and electrons due to the toroidal field gradient and curvature. Hereafter the vertical drift is referred to as vacuum toroidal field (VTF) drift. The current density is written in terms of local electron and ion pressure $p_e$ and $p_i$ as

$$j_{VTF} = \frac{2(p_e + p_i)}{RB_\varphi},$$

where $B_\varphi$ is the toroidal field at the radial coordinate R in the cylindrical coordinate (R, φ, Z). Here we consider the case of $B_\varphi>0$. Since $p_e$ is much higher than $p_i$ in the main part of ECR plasmas, the current density is almost proportional to $p_e$. When no vertical field is superposed, the vertical charge separation current flows from the bottom to the top and circulates through the vacuum vessel. When a vertical field is superposed, an internal another return path along helical field lines is provided in addition to the external path through the vessel. The vertical component of the internal return current along helical field lines is cancelled and the toroidal component becomes a toroidal current. As the toroidal current increases and its self field is developed, the cross field passing (CFP) orbit electrons which are EC-heated and asymmetrically confined tail electrons increase. These electrons provide an additional toroidal current and a closed flux surface is formed eventually.

We have observed the equilibrium characteristics that maintain discharges in the various toroidal ECR plasmas superposed a steady vertical field in the range of Bv=0G to 10.5G at which a closed flux surface forms. The vacuum vessel of the LATE device is a cylinder with an inner size of R=5.8~50cm and Z=-50~50cm. The top and bottom panels which almost cover the ceiling and the flooring of the vacuum vessel are installed at Z=30cm and Z=-30cm, respectively. A radial array of small electrodes is embedded on the each panel and an ion energy analyzer is attached behind the top panel. Plasmas are generated between the panels using 2.45GHz microwave pulse of 2s. The toroidal field is $B_\varphi$=480G at R=25cm in all discharges.

Figure 1 shows plasma images captured by a CCD camera, plasma profiles measured by Langmuir probes, the toroidal current profiles obtained by signals of thirteen flux loops surrounding the vessel (refer to P5 149 in this conference [8]) and radial profiles of the vertical circulation current flowing to the top and bottom panels measured by the radial arrays and ion current flowing to the top panel measured by the ion analyzer in the discharges with Pinj=1.5kW and Bv=0G, 2.7G, 6.8G and with Pinj=2.0kW and Bv=10.5G, respectively. Here Bv denotes the Z-component of external field at R=22cm. The toridal current flowing in the negative φ direction and vertical circulation current through the external path are generated. A closed flux surface is observed in the discharge with Pinj=2.0kW and Bv=10.5G.

In the case of Bv=0G, a vertically uniform electron pressure ridge drives a vertically uniform current by the electron VTF drift and a space potential hill which crosses the field and shifts upward regulates the drift flows of electrons and ions without accumulating charges. [9]

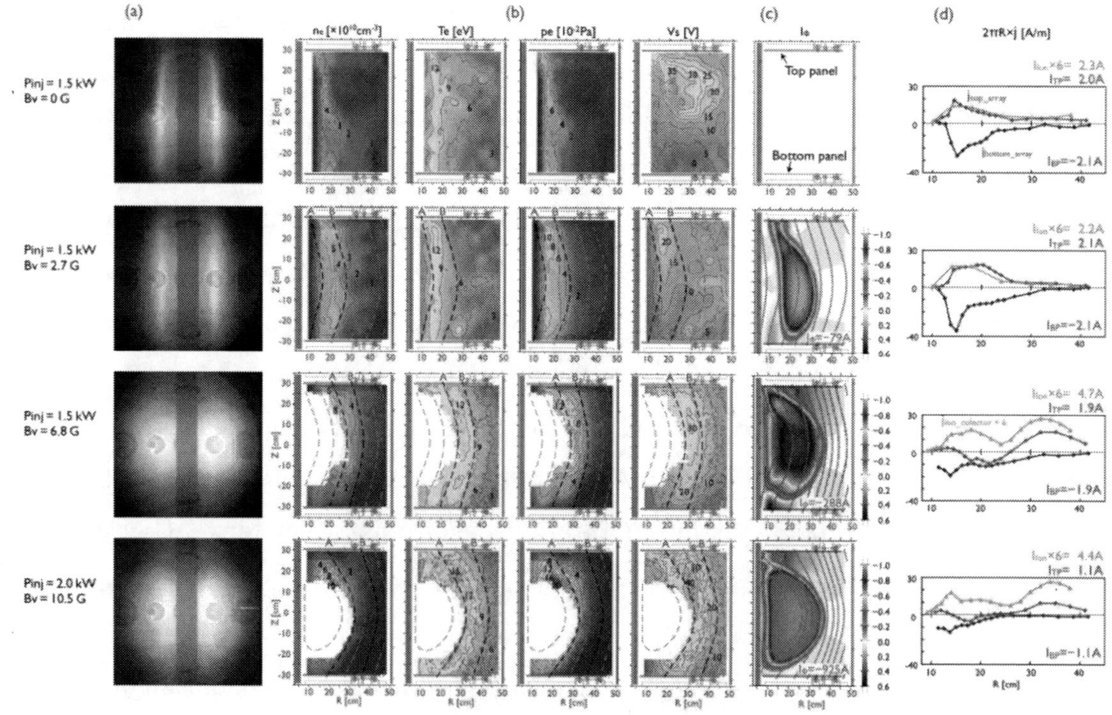


*Figure 1 (a) CCD plasma images. (b) Plasma profiles of electron density (ne), temperature (Te) and pressure (pe) and space potential (Vs). The data are not shown when the disturbance of a interferometer line density of the chord Rt=12cm becomes more than 10% by probe insertion. (c) Toroidal current profiles. (d) Radial profiles of vertical circulation current densities on the top and bottom, $j_{top\ array}$ and $j_{bottom\ array}$, and ion current density on the top, $j_{ion\ collector}$. The radial profiles are weighted by $2\pi R$. $I_{TP}$ and $I_{BP}$ are total amounts of current on the top and bottom. $I_{ion}$ is a estimated total amount of ion current on the top.*

In the case of Bv=6.8G and 10.5G, as shown in Fig. 1(d), the negative current flowing at R~18cm on the top panel shows there are more electrons flowing along field lines to the panel than the ions flowing by the drifts. The ion's flow by the drift increase at R~32cm on the top panel, where a steep potential slope forms, and $I_{TP}$ and $I_{BP}$ are balanced. The electron density decreases near the bottom panel and the electron VTF drift current is low relative to the current flowing to the panel. The result shows the electron's flow along field lines to the bottom panel decreases the electron density near the panel and the flow contributes to the toroidal current which is opposite direction of the toroidal current in the main part.

The main electron's flow changes from the flow by the drifts across the field to the flow along helical field lines by superposing a week vertical field and the plasma profiles change drastically. The space potential hill across field at Bv=0G is relaxed significantly at Bv=2.7G as shown in Fig. 1(d). Along helical field lines, the electron temperature becomes uniform and the density obeys the Bolzman's law "$n_e \propto \exp\{V_S/T_e\}$" at Bv=6.8G and 10.5G as shown in Fig. 2 . On the other hand ion's flow along helical field lines is much slower than the flow by E×B drift, which is simulated by ion particle orbits (refer to P5 149 in this conference [8]).

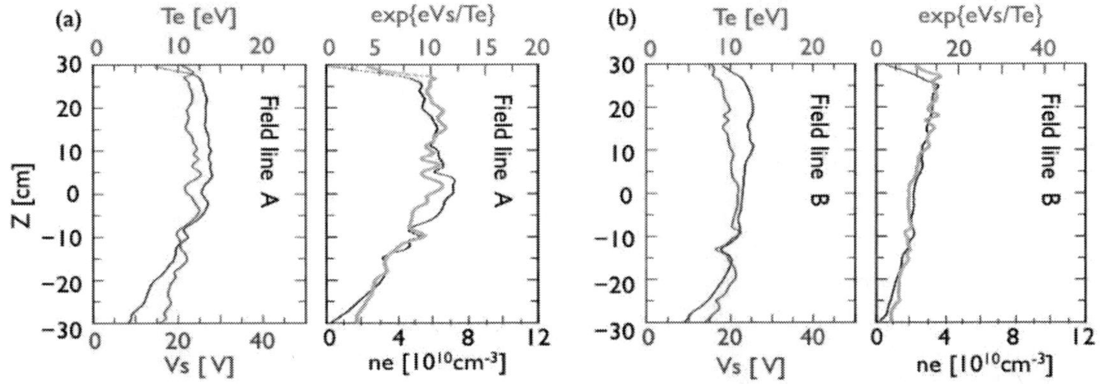


*Figure 2 Profiles of electron density and temperature, space potential and the value of Boltzmann's law "exp{eVs/Te}" along field lines (a) A and (b) B in Fig 1(b) in the discharge with Pinj=1.5W and Bv=6.8G.*

Figure 3(a) shows that the toroidal current increases linearly with Bv up to Bv=6.8G and increases furthermore at Bv=10.5G where a closed flux surface is formed, while the vertical circulation current is almost constant at about 2A at Bv=0~6.8G and decreases to about 1A after formation of a closed flux surface. The ballooning force of plasma pressure is balanced with the counter forces by the interaction of the toroidal current and the vertical circulation current with the field. Figure 3(b) shows the counter forces roughly estimated as follows (for detailed analyses, refer to P5 149 in this conference [8]).

$$F_{CF}[I_\varphi \times B_Z] = I_\varphi \times B_V \times 2\pi \times 0.25m$$
$$F_{CF}[I_Z \times B_\varphi] = I_{BP} \times 0.048T \times 0.6m$$

The former force due to the toroidal current is estimated using the current of $I_\varphi$, the field strength of Bv and the current path length of $2\pi \times 0.25m$ and the later force due to the vertical circulation current is estimated using the current of $I_{BP}$, the field strength of 480G that is $B_\varphi$ at R=25cm and the current path length of 60cm that is the plasma height. While the ballooning force of plasma pressure is balanced by only $F_{CF}$ $[I_Z{\times}B_\varphi]$ due to the vertical current at Bv=0G, $F_{CF}$ $[I_\varphi{\times}B_Z]$ increases much larger than $F_{CF}$ $[I_Z{\times}B_\varphi]$ at Bv>5G. The plasma becomes mainly maintained by the toroidal current and the pressure profile should change depending on the toroidal current. Figure 1 shows that as Bv is increased, the electron pressure and the plasma image change from cylindrical profiles to spherical profiles and spread outward in corresponding to the increasing toroidal current and the spreading area of the current profile. The line integrated density along the chord of Rt=12cm increases as shown in Fig. 3(a) and the peaks of electron density, temperature and pressure also rise with increasing the toroidal current as shown in Fig. 1.

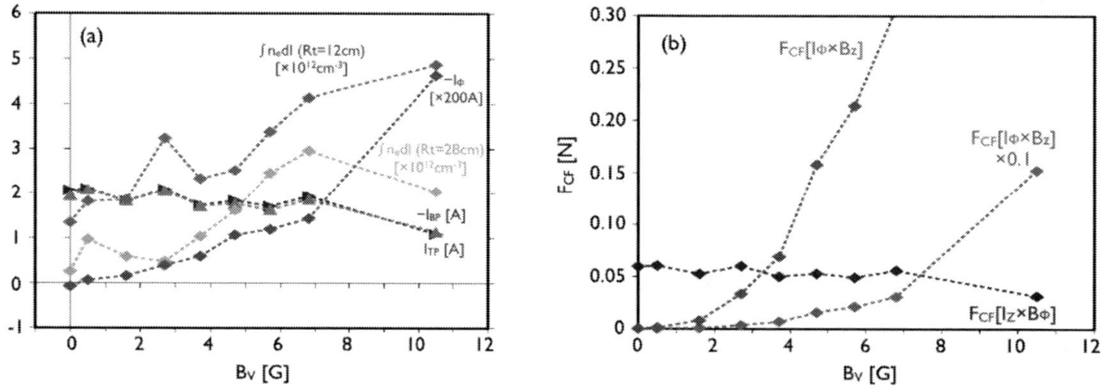


*Figure 3 Changes with Bv of (a) the line integrated densities along the mid-plannes chord of Rt=12cm and Rt=28cm, toroidal current, $I_\varphi$ and vertical circulation current on the top and bottom panels, $I_{TP}$ and $I_{BP}$, and (b) roughly estimated counter forces by "$I_\varphi{\times}B_Z$" and "$I_Z{\times}B_\varphi$".*

**Acknowledgements**

The present work was supported by KAKENHI (Grant Number 23360411 and 26289357).

**References**

[1] M. Uchida, T. Maekawa, H. Tanaka, S. Ide, Y. Takase et al., Nucl. Fusion. **51** (2011) 063031

[2] T. Yoshinaga, M. Uchida, H. Tanaka and T. Maekawa, Phys. Rev. Lett. **96** (2006) 125005

[3] T. Maekawa, T. Terumichi, H. Tanaka et al., Nucl. Fusion. **45** (2005)1439

[4] S. Nishi, T Sakabe, M. Uchida, H. Tanaka and T. Maekawa, Plasma Phys. Control. Fusion **96** (2010) 125004.

[5] S. Nishio et al., Proc. 20th Int. conf. on Fusion Energy 2004 (Vilamoura Portugal) (Vienna: IAEA) CD-ROM file FT/P7-35 and http://www-naweb.iaea.org/napc/physics/fec/fec2004/datasets/index.html.

[6] K. Tobita et al., Fusion Eng. Des. **81** (2006) 1151.

[7] T. Maekawa, T. Yoshinaga, M. Uchida, F. Watanabe and H. Tanaka Nucl. Fusion. **52** (2012) 083008.

[8] T. Maekawa et al, 42nd EPS Conference on Plasma Phys. Lisbon, P5.149 (2015).

[9] K. kuroda, M. Wada, M. Uchida, H. Tanaka and T. Maekawa, Plasma Phys. Control. Fusion **57** (2015) 075010.

# Correlation study between two-dimensional impurity distribution and edge magnetic structure in Large Helical Device

H. M. Zhang[1], S. Morita[1,2], T. Ohishi[1,2], X. L. Huang[1], Y. Narushima[1,2], M. Goto[1,2] and M. Kobayashi[1,2]

[1] *Dept. of Fusion Sci., Graduate University for Advanced Studies, Toki 509-5292, Gifu, Japan*

[2] *National Institute for Fusion Science, Toki 509-5292, Gifu, Japan*

## 1. Introduction

The transport in edge plasmas can be controlled in both the tokamak and helical devices, when the resonant magnetic perturbation (RMP) is applied. In tokamaks, the RMP is mainly used to mitigate the edge-localized mode (ELM) [1]. In the Large Helical Device (LHD), on the other hand, edge magnetic island (m/n=1/1) formed by the RMP is used to sustain a steady state detached plasma without impurity gas puff [2]. It means that the edge impurity transport can be changed by the RMP. In a previous study [3], it is found that the RMP changes the impurity screening effect in the ergodic layer as a result of data analysis on the impurity emission in extreme ultraviolet (EUV) and vacuum ultraviolet (VUV) ranges. However, any impurity profile analysis has not been done until now. Recently, space-resolved EUV spectrometers have been developed in LHD to measure a vertical profile and 2-D distribution of the impurity line emission [4,5]. In the present study, the vertical profile and 2-D distribution of the impurity line emission in the EUV wavelength range are analyzed to investigate the effect of the m/n=1/1 magnetic island formed by the RMP on the edge impurity transport.

## 2. Experimental setup

In LHD, the m/n=1/1 magnetic island formed by the RMP coil system was originally used in the local island divertor experiment to realize an efficient particle exhaust. The O-point of the m/n=1/1 magnetic island is formed at the outboard side of horizontally elongated plasma cross section in #6-O ($\phi$=180°) or #7-O ($\phi$=216°) toroidal location, of which the configuration is called 6-O or 7-O island, respectively. Since the magnetic axis ($R_{ax}$) is set to 3.75 m in the present study, the radial location of the m/n=1/1 island is located just inside of the last closed flux surface (LCFS). The edge magnetic field structure with RMP is temporally kept constant during the discharge in LHD.

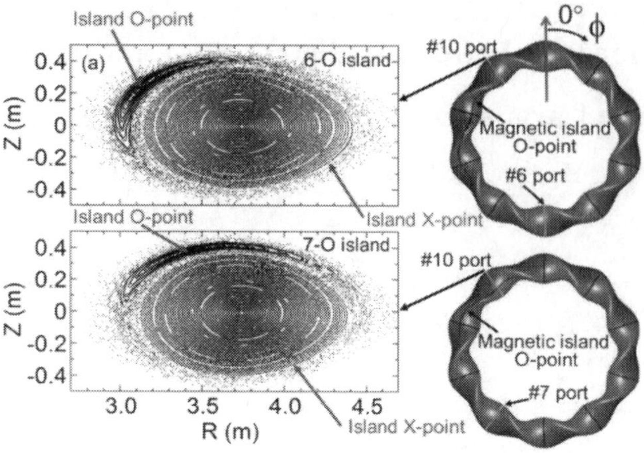


Fig.1 Poincare plot of flux surface in vacuum condition at #10-O toroidal location for (a) 6-O and (b) 7-O islands.

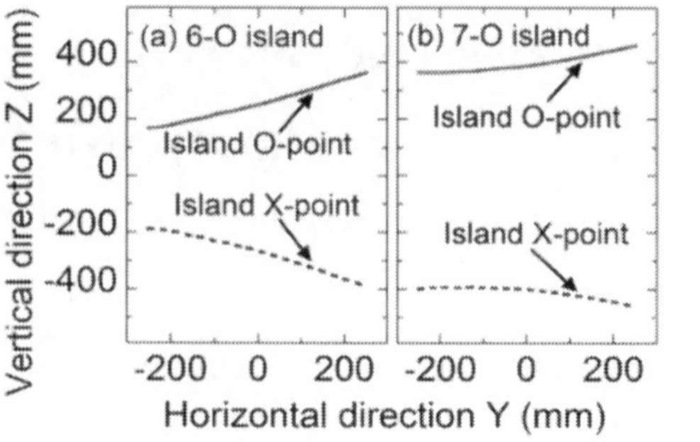


Fig.2 Vertical position of island O-point (solid line) and X-point (dash line) of the m/n=1/1 magnetic island within the observation area of EUV_Long2 spectrometer installed at #10-O LHD port for (a) 6-O and (b) 7-O islands, respectively.

The EUV spectroscopies are installed at the #10-O LHD port ($\phi$=324°). Two EUV spectrometers named EUV_Short and EUV_Long are always used to monitor the impurity behavior of LHD discharges by measuring the impurity line emission in wide wavelength ranges of 10–130 and 30–650 Å, respectively [6]. Another two space-resolved EUV spectrometers named EUV_Short2 and EUV_Long2 measure the vertical profile and 2-D distribution of impurity line emissions, respectively [4,5]. As shown in Fig. 1, the local O-point of m/n=1/1 island at #10-O LHD port is localized at the inboard side for both the configurations of 6-O and 7-O islands. The 2-D observation area of the EUV_Long2 spectrometer (500≤Z≤500mm and -250≤Y≤250mm) and the position of island O- and X-points are shown in Fig.2. The vertical position of the magnetic island O-point at 7-O island moves a larger distance (Y=165-365mm) compared to 6-O island (Y=365-460mm).

## 3. Experimental results and discussion

Radial profiles of electron density and temperature are shown in Figs.3 (a), (b) and (c) for normal discharges without RMP, 6-O and 7-O islands, respectively. A flattening of the $T_e$ profile is clearly seen at $\iota/2\pi=1$ reflecting the presence of magnetic island for both 6-O and 7-O island cases. The width of the local flattening shape of the $T_e$ profile reflects the size of the magnetic island.

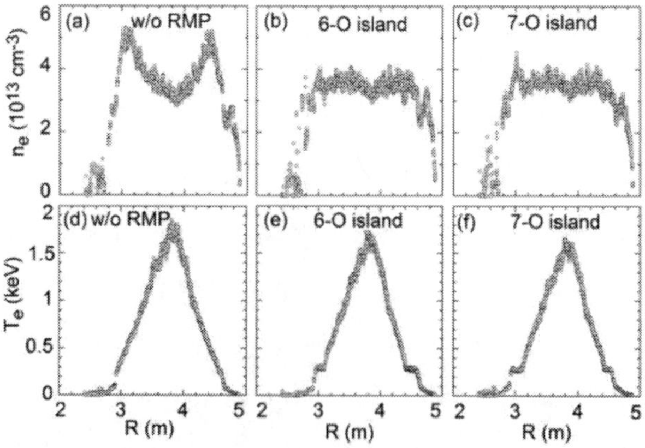


Fig.3 Radial profiles of $n_e$ and $T_e$ in discharges without RMP, with 6-O island and with 7-O island.

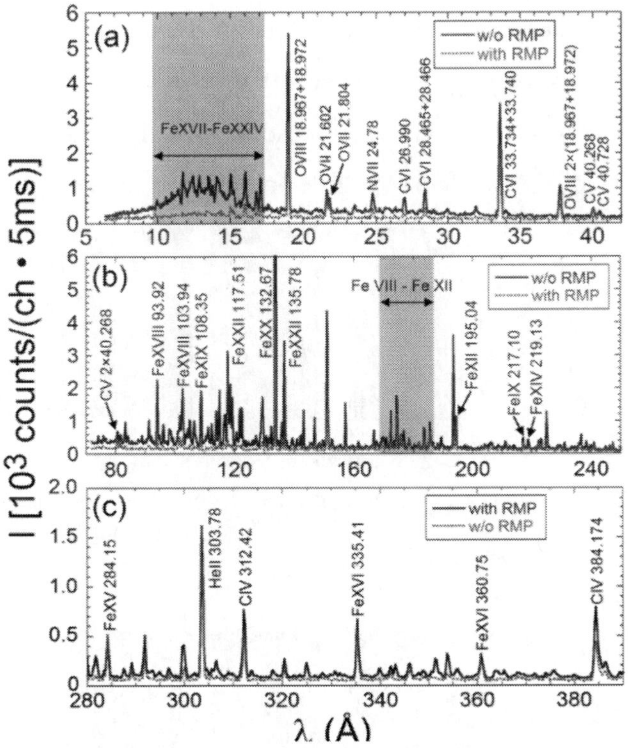

Fig.4 EUV spectra of impurity emissions from discharges without RMP (solid line) and with RMP (dotted line).

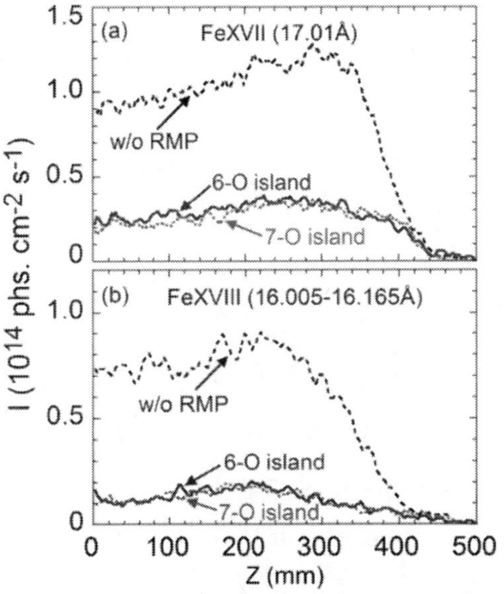

Fig.5 Vertical profiles of (a) FeXVII and (b) FeXVIII in discharges without RMP (dashed line), with 6-O island (solid line) and with 7-O island (dotted line).

Impurity spectra are compared between discharges with and without island in wider wavelength range, as shown in Figs. 4 (a), (b) and (c). It is very clear that emissions from highly ionized iron at plasma core, e.g. FeXVII–FeXXII at 10-17Å, significantly decrease when the RMP is applied. Both the 6-O and 7-O island cases indicate the same result. On the other hand, emissions from low-ionized iron at plasma edge, e.g. FeVIII–FeXII at 170-177Å show very little change. The same result is also observed in in the vertical profile of FeXVII and FeXVIII, as shown in Figs. 5 (a) and (b), respectively. This large emission change in the plasma core cannot be explained by a small change in $T_e$ and $n_e$ profiles. Therefore, the result strongly suggests that the presence of the m/n=1/1 magnetic island prevents the transport of iron ions from the ergodic layer to the plasma core.

The effect of RMP is also clearly seen in the 2-D distribution of FeXVI emission. The result is shown in Fig.6. In the case of 6-O island in Fig.6 (b), the FeXVI emission indicates a up/down asymmetry, while the emission from the case without island in Fig.6 (a) shows almost uniform intensity. The up/down asymmetry becomes more remarkable in the 7-O island case. In addition, the emission is stronger at the right hand side than that at the left side. When the 2-D FeXVI distribution is compared between Figs. 6 (b) and (c), the vertical position of island O-point shown in Figs. 2 (a) and (b) has to be considered. It is then understood that the FeXVI emission can be enhanced near the island O-point. In contrast to this, the 2-D distribution of CIV emission has no significant change among the three cases, as shown in Figs. 7(a), (b) and (c). The reason is very clear because the CIV is located in the ergodic layer outside LCFS where the effect of RMP is basically absent. We then conclude the m/n=1/1 magnetic island formed just inside LCFS does not give any obvious effect to the impurity transport in the ergodic layer.

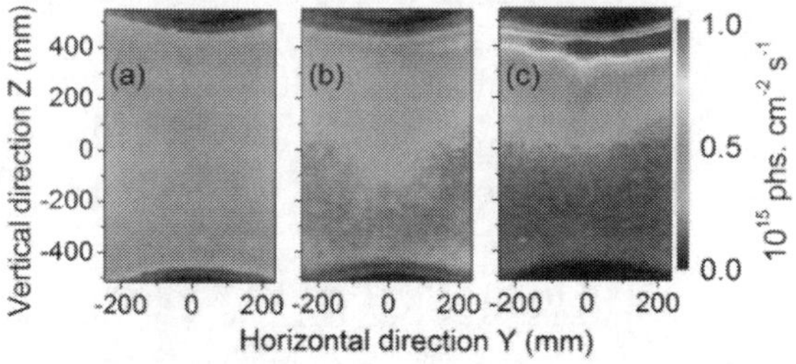


Fig.6, 2-D distributions of FeXVI in discharges (a) without RMP, (b) with 6-O island and (c) with 7-O island.

## 4. Summary

When the RMP is applied, iron emissions from plasma core significantly decrease, whereas iron emissions from plasma edge do not change. It indicates that the m/n=1/1 magnetic island can prevent the impurity transport from the plasma edge to plasma core. It is also found that the FeXVI emission is selectively enhanced near the O-point of the island showing an extremely large up/down asymmetry.

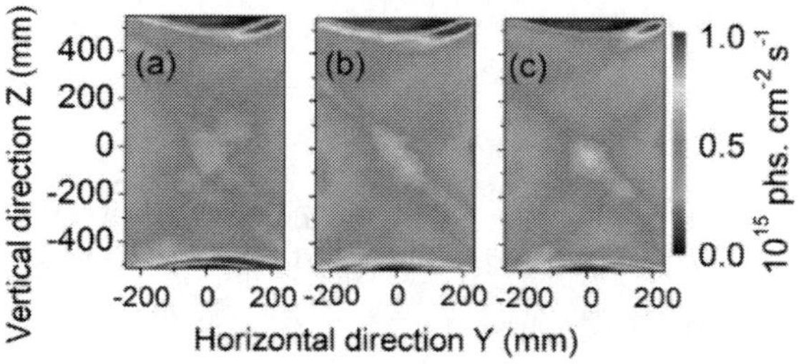


Fig.7, 2-D distributions of CIV in discharges (a) without RMP, (b) with 6-O island and (c) with 7-O island.


## Acknowledgements

The authors thank all members of the LHD experimental group for their technical supports. This work was partially carried out under the LHD project financial support (NIFS14ULPP010), the JSPS KAKENHI Grant Number 23340183 and JSPS-NRF-NSFC A3 Foresight Program in the field of Plasma Physics (NSFC: No. 11261140328, NRF: No. 2012K2A2A6000443).


## Reference


[1] T.E. Evans et al, Nucl. Fusion **48,** 024002 (2008).

[2] M. Kobayashi, et al., Phys. Plasmas **17**, 056111 (2010).

[3] M. B. Chowdhuri, et al., Phys. Plasmas **16**, 062502 (2009).

[4] E. H. Wang, et al., Rev. Sci. Instrum. **83**, 043503 (2012).

[5] X. L. Huang, et al., Rev. Sci. Instrum. **85**, 043511 (2014).

[6] M. B. Chowdhuri, et al., Rev. Sci. Instrum. **78**, 023501 (2007).

# Observation of asymmetric radial profile of tungsten ions in low ionization stages located in edge plasmas of Large Helical Device

T. Oishi[1,2], S. Morita[1,2], X. L. Huang[2], H. M. Zhang[2], M. Goto[1,2], the LHD Experiment Group[1]

[1] *National Institute for Fusion Science, Toki, Japan*

[2] *Department of Fusion Science, Graduate University for Advanced Studies, Toki, Japan*

Tungsten is regarded as a leading candidate material for the plasma facing components in ITER and future fusion reactors. Considering transport processes of tungsten impurity in ITER, firstly neutral tungsten atoms are released from the divertor plates, then tungsten ions at lower ionization stages are transported in the edge plasmas, and finally tungsten ions at higher ionization stages are distributed in the confinement region. To understand behaviors of the tungsten impurity ions, diagnostics for line emissions from tungsten ions have been intensively conducted, such as visible spectroscopy for neutral tungsten atoms around 4000 Å and extreme ultraviolet (EUV) spectroscopy for tungsten ions at higher ionization stages around 15-70 Å [1,2]. However, tungsten ions at lower ionization stages have not been measured except for several cases of vacuum ultraviolet (VUV) spectroscopy in basic plasma experiments even though it is necessary for accurate evaluation of tungsten influx and comprehensive understanding of the tungsten impurity transport in high temperature plasmas [3,4]. In the present study, VUV spectra of emissions released from tungsten ions are measured in the Large Helical Device (LHD) for exploration of tungsten lines at low ionization stages which will contribute spectroscopic studies of edge plasmas in ITER and other tungsten-wall machines.

LHD is a superconducting heliotron device with the major/minor radii of 3.6/0.64 m, the plasma volume of 30 $m^3$, and the toroidal magnetic field of 3 T in the standard configuration [5]. Tungsten ions are introduced in the LHD plasma by injecting a tungsten impurity pellet which consists of a small piece of tungsten wire covered by a polyethylene or carbon tube. The length and diameter of tungsten wire is 0.6 mm and 0.15 mm, respectively. Owing to the advantages of steady-state current-free helical systems, the LHD plasma can be sustained even though substantial amount of tungsten atoms are injected [6]. Figure 1(a) illustrates the top view of the instruments together with the optical axis of the VUV spectrometer and the incident orbit of the impurity pellet. A 3m normal incidence VUV spectrometer (McPherson model 2253) is installed on an outboard midplane diagnostic port which is the same as the impurity pellet injector [7,8]. A back-illuminated CCD detector (Andor model DO435-BN: 1024 × 1024 pixels) is placed at the position of the exit slit of the spectrometer. A high wavelength dispersion of 0.037 Å/pixel enables measurements of the Doppler profiles of the impurity lines in the working wavelength range of 300-3200 Å. Time resolutions are 50 ms for the "full-binning" measurement in which all 1024 vertical pixels are replaced by single channel and the spatial resolution is entirely eliminated and 100 ms for "space-resolved" measurement in which the wavelength-dispersed image of the VUV emission is projected on a corresponding vertical position on CCD through a space-resolved slit mounted between the entrance slit and the grating. Figure 1(b) illustrates the vertical view angle of the VUV spectroscopy and the pellet orbit on the horizontally-elongated cross section of the magnetic field in LHD.

Figure 2 shows VUV spectra in the wavelength range between 600 to 705 Å in the time frame just after the tungsten pellet injection in hydrogen discharge in LHD. The plasma was initiated by the electron cyclotron heating, and three neutral hydrogen beams based on negative ion sources with total port-through power of 8 MW were injected. Central electron density, $n_e(0)$, and temperature, $T_e(0)$, just before the pellet injection was $2 \times 10^{13}$ cm$^{-3}$ and 3 keV, respectively. Bright WVI lines with 5d-6p transitions located at the wavelengths of 605.926 Å, 639.683 Å, and 677.722 Å have been observed. They are the first observation of line spectra from tungsten ions in low ionization stages in fusion plasma experiments.

We also succeeded a measurement of the vertical profile of the emission intensity of the WVI line. Figure 3(a) shows a typical temporal evolution of $n_e(0)$ and $T_e(0)$. At the pellet injection, $n_e(0)$ increases and $T_e(0)$ decreases. Figure 3(b) shows a temporal evolution of the intensity of the second order of WVI 639.683 Å line measured using a 20 cm normal incident VUV spectrometer [9]. A large emission is observed at the same time of the pellet injection, which is an emission from the pellet ablation cloud. Figure 3(c) shows a vertical profiles of WVI 639.683 Å line in time frame just after the pellet injection. The emission

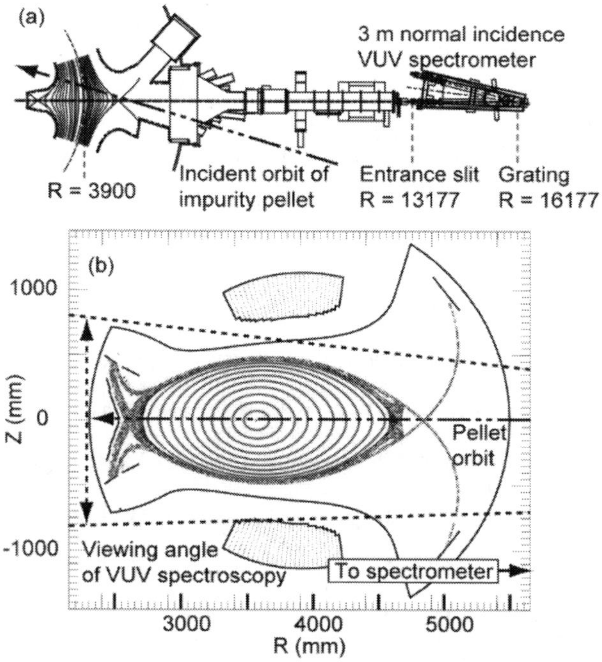


Figure 1. Schematic drawing of VUV spectroscopy in impurity pellet injection experiments. (a) Top view of the instruments together with the optical axis of the VUV spectrometer and the incident orbit of the impurity pellet. (b) Vertical viewing angle and the pellet orbit on the poloidal cross section of the magnetic field in LHD.

profile consists of two components; one is emissions from a pellet ablation cloud located at the vertical center of the plasma and the other is emissions from the plasma confinement region. The enlarged figure for this time frame around the signal level of the emission from plasma confinement region is shown in Fig. 3(d). The positions of the last closed flux surface (LCFS) in the vacuum magnetic field and an edge boundary denoted as $a_{99}$ which is the effective minor radius in which 99% of the plasma kinetic energy is confined are shown together. The profile has a vertical asymmetry that the signal intensity in the lower half is significantly larger than that in the upper half. This asymmetry with an intensity peak close to the lower edge of the plasma is more obvious in the time frame 100-200 ms after the pellet injection as shown in Fig. 3(e). Mechanisms forming the asymmetrical profile of WVI line should be investigated in

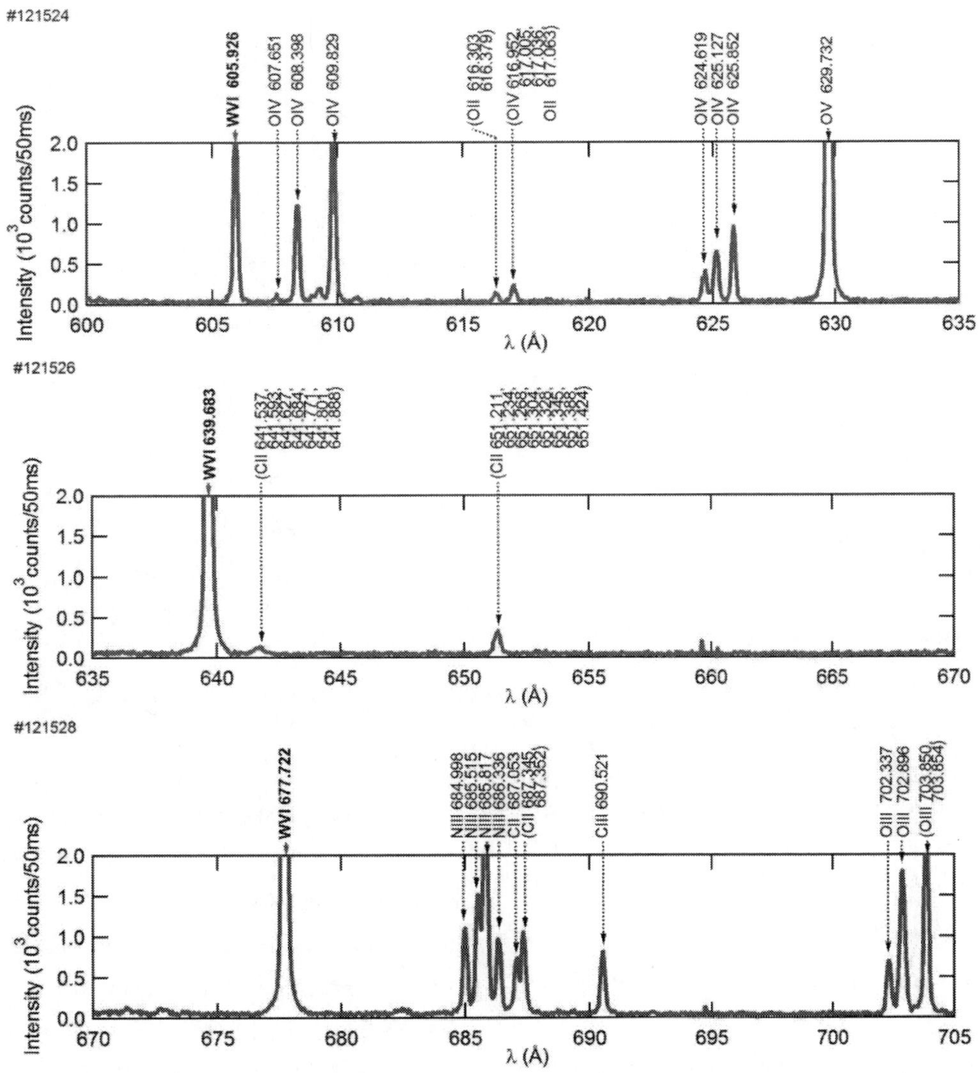


Figure 2. VUV spectra in the wavelength range between 600 to 705 Å measured using the 3 m normal incidence VUV spectrometer in the time frame just after the tungsten pellet injection.

future studies. Candidate factors responsible for the asymmetry include a thick stochastic magnetic field layer called "ergodic layer" located outside the core plasma of LHD, because it has been observed that intrinsic impurities located in the ergodic layer have asymmetrical profiles [10].

The authors thank all the members of the LHD team for their cooperation with the LHD operation. This work is partially conducted under the LHD project financial support (NIFS14ULPP010). This work was also supported by Grant-in-Aid for Young Scientists (B) 26800282 and partially supported by the JSPS-NRF-NSFC A3 Foresight Program in the field of Plasma Physics (NSFC: No.11261140328, NRF: No.2012K2A2A6000443).

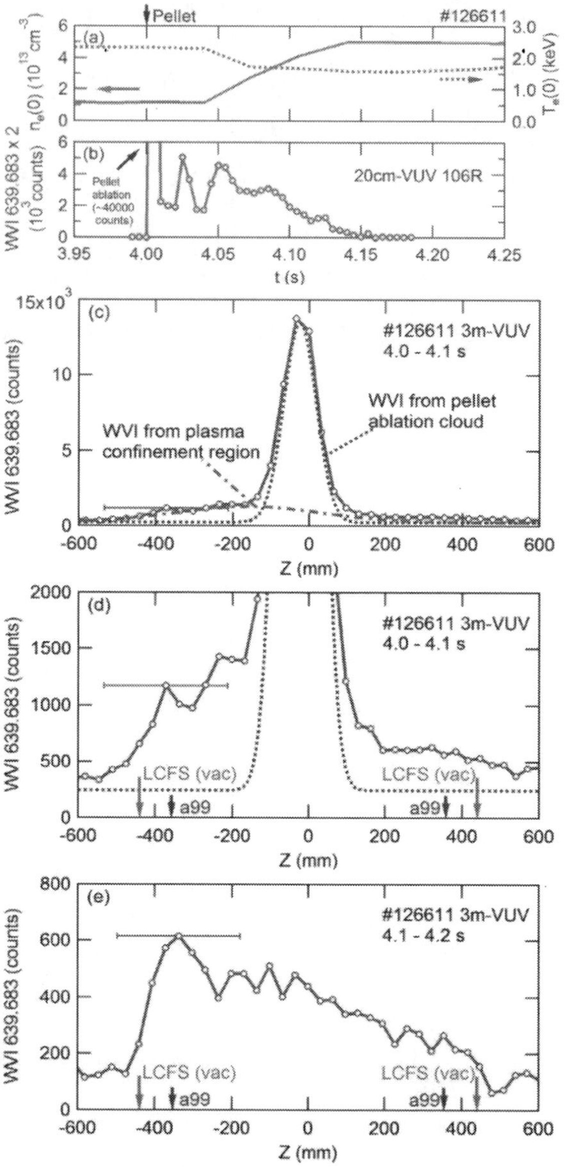


Figure 3. Temporal evolutions of (a) central electron density and temperature with tungsten pellet injection and (b) intensity of WVI 639.683 × 2 Å line. Vertical profiles of WVI 639.683 Å line in time frames (c) 0-100 ms after the pellet injection, (d) the enlarged figure for the same time frame around the signal level of the emission from plasma confinement region, and (e) 100-200 ms after the pellet injection.

**REFERENCES**


[1] T. Pütterich *et al.*, Plasma Phys. Control. Fusion **50** (2008) 085016.

[2] S. Morita *et al.*, AIP Conf. Proc. **1545**, 143 (2013); *Proceedings of ICAMDATA-2012*, Gaithersburg, 30 September–4 October 2012.

[3] R. Radtke *et al.*, *Atomic and Plasma-Material Interaction Data for Fusion* (IAEA, 2007), Vol. 13, p. 45.

[4] J. Clementson *et al.*, J. Phys. B: At. Mol. Opt. Phys. **43**, 144009 (2010).

[5] H. Yamada for the LHD Experiment Group, Nucl. Fusion **51**, 094021 (2011).

[6] X. L. Huang *et al.*, Rev. Sci. Instrum. **85**, 11E818 (2014).

[7] T. Oishi *et al.*, Appl. Opt. **53**, 6900 (2014).

[8] T. Oishi *et al.*, Rev. Sci. Instrum. **85**, 11E415 (2014).

[9] T. Oishi *et al.*, Plasma Fusion Res. **10**, 3402031 (2015).

[10] R. Katai *et al.*, Rev. Sci. Instrum. **77**, 10F307 (2006).

# Electron temperature measurements using the LHD Thomson scattering system having four scattering configurations

I. Yamada[1], R. Yasuhara[1], H. Funaba[1], K. Narihara[1], H. Hayashi[1],
E. Yatsuka[2], T. Hatae[2], H. Tojo[2], M. Yoshikawa[3], and T. Minami[4]
[1] *National Institute for Fusion Science, Toki, Japan 509-5292*
[2] *Japan Atomic Energy Agency, Naka, Japan, 311-0193*
[3]*Plasma Research Center, University of Tsukuba, Tsukuba, 305-8577, Japan*
[4]*Institute of Advanced Energy, Kyoto University, Uji, Kyoto 611-0011, Japan*

## Introduction

The Large Helical Device (LHD) Thomson scattering system has routinely measured electron temperature ($T_e$) and density ($n_e$) profiles of LHD plasmas. [1][2] The LHD Thomson scattering system has two scattering configurations: backward and forward Thomson scatterings, as reported in the previous EPS conferences [3][4]. There are two objectives of the upgrade; extension of the measurable electron temperature ($T_e$) range and the search for the electron temperature anisotropy in fusion plasmas. We installed new light collection optics to observe $T_e$ component parallel to the LHD magnetic field line around the plasma center. We describe the LHD Thomson scattering system having four scattering configurations and the first result obtained by using the new light collection optics.

## Scattering configuration of the LHD Thomson scattering system

Figure 1 shows the schematic diagram of the LHD Thomson scattering system. In the original design of the LHD Thomson scattering system, the backward Thomson scattering light is observed. By installing a beam returning system, the LHD Thomson scattering system now can observe both the backward and forward Thomson scattering lights. The typical scattering angles are 167° and 13° for the backward and forward scattering configurations respectively. In the original design, a rectangular window, whose dimensions are 600 mm x 350 mm, was installed, and 144 optical fibers and

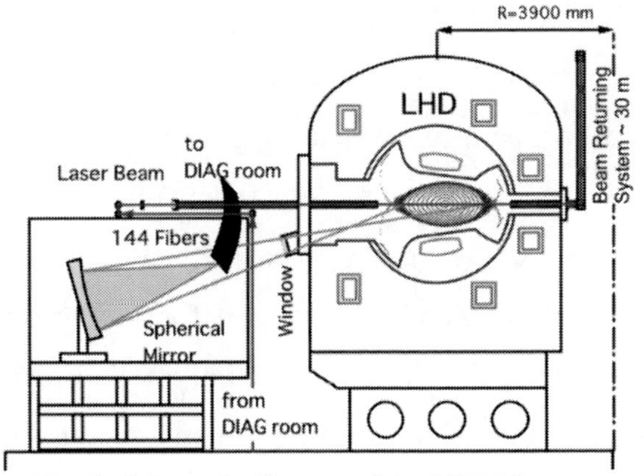

Fig. 1. Schematic diagram of the LHD Thomson scattering system.

polychromators have been used to measure $T_e s$ and $n_e s$ at 144 spatial points along the LHD major radius. In addition to the main light collection system, we now installed the second window, whose diameter is 340 mm. The center of the second window is located on the LHD mid-plane as shown in Fig. 2. A schematic diagram of the geometrical configuration of the two windows is show in Fig. 3. The scattering angles in the new system are 171.2 ° and 8.8 ° for the backward and forward scattering configurations, respectively. At the present, the second light collection optics observes $T_e$ and $n_e$ at a spatial point (plasma center). The number of measurement points will be increased by adding optical fibers and polychromators.

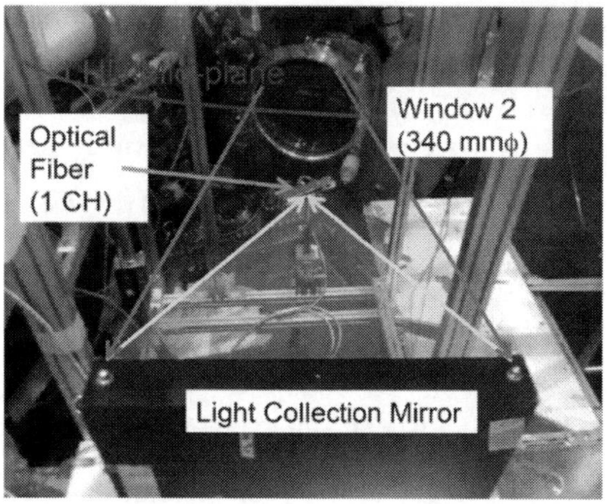


Fig. 2. Newly installed second light collection optics.

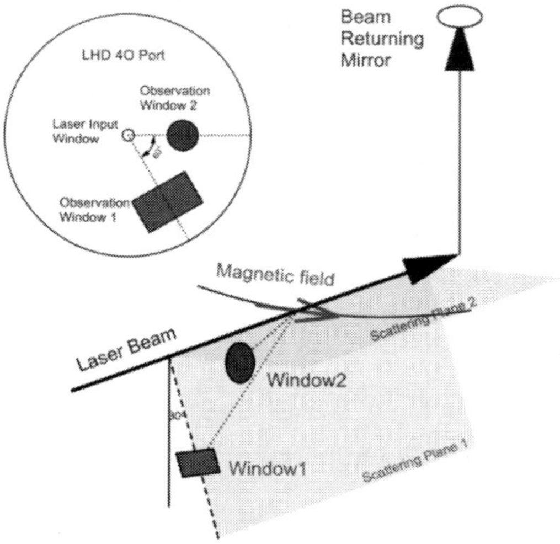


Fig. 3. Schematic diagram of the two view windows.

By using the two windows and the two scattering configurations, total four directional components of $T_e s$ are obtained. Figure 4 shows the angles between the magnetic field line and the four $T_e$ components. The magnetic axis is located at $R$ = 3.65 m, and the positions of the last closed magnetic surface is $R$ = 2.5 m and 4.9 m. In the backward scattering measurements, the angles between the magnetic field line and measured temperature direction are about 90° for both the Windows 1 and 2 in the whole observation region. On the other hand, the angles are about 0° at $R$ = 5 m and 3.7 m for the forward scattering measurements using the Windows 1 and 2, respectively. Therefore, forward scattering measurements using the Windows 1 and 2 are suitable for the observation of the $T_e$ component parallel to the magnetic field line in the plasma edge and center regions, respectively.

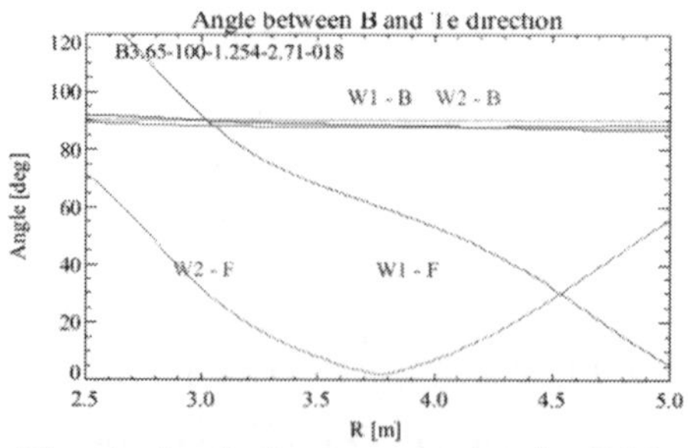


Fig. 4. Angle between magnetic field line and the observed direction of electron temperature.

## Experimental

In the experiment, we used a newly constructed 7-wavelength channel polychromator. It is specially optimized to both the backward and forward scattering measurements. The spectral response is shown in Fig. 5. In the forward scattering measurement, the three longer wavelength channels, CH.1-3, play an important role because the width of forward Thomson scattering spectrum is much narrower than that for backward scattering. On the other hand, all the seven channels are used to determine backward scattering $T_e$s. In the experiment, we used a fast digitizer system based on four fast oscilloscopes to observe both the backward and forward scattering signals simultaneously. Figure 6 shows an example of Thomson scattering signals measured by two polychromators that see around the plasma center through Windows 2 and 1. The upper two figures show the backward and forward scattering signals obtained by

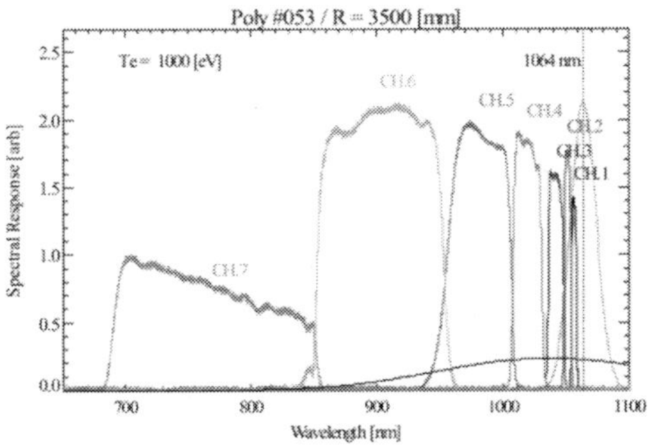


Fig. 5. Spectral response of the 7-channel polychromator. Thomson scattering spectrum for the backward and forward scatteings at 1000 eV are also plotted.

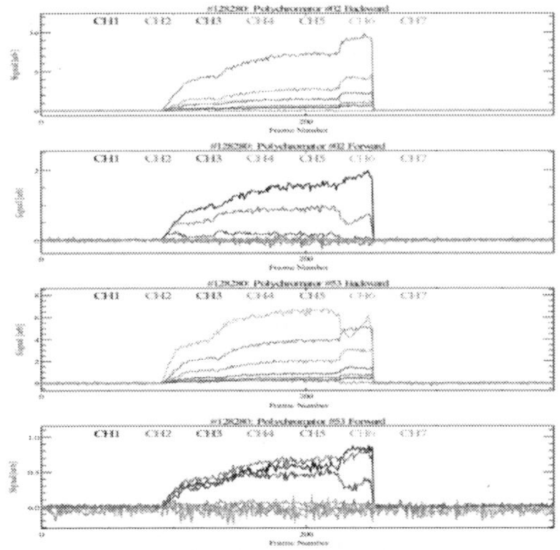


Fig. 6. Thom scattering signals obtained by two polychromators.

a) b)

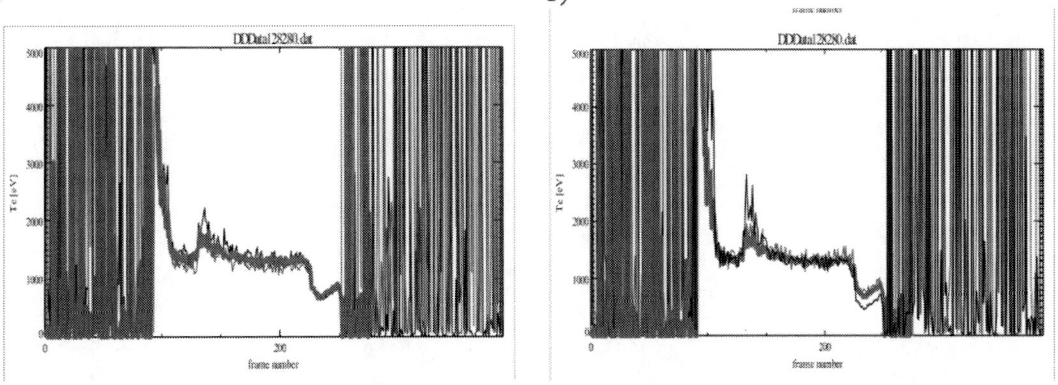


Figs. 7. Figures a) and b) show comparisons of the backward and forward scattering electron temperatures obtained by using Window 2 and Window 1 respectively. Thick curves in the figures show Te obtained by the main LHD Thomson scattering system. They show good agreements within the experimental errors of ~10-20 %.

a polychromator for the Window 2. The lower two figures show those for a polychromator for the Window 1. A $T_e$ component is obtained from each set of the Thomson scattering signals

Figures 7 a) and b) show backward and forward scattering $T_e$s obtained by the new and main light collection optics respectively. The $T_e$ determined from the main LHD Thomson scattering system is also plotted (thick curves). The $T_e$s show good agreements within the experimental errors of about 10-20%, suggesting the system for measuring four $T_e$s components works well.

## Summary and Future Plan

We have installed new experimental apparatus on the LHD Thomson scattering system to observe four electron components by using backward and forward scattering configurations, and two light collection optics. The first result shows that the system works well. However, further optimization of polychromators and development of data analysis procedure will be needed to reduce experimental errors.

## Acknowledgements

This work was supported by the NIFS LHD project budgets (NIFS13ULHH005 and NIFS14ULHH005) and JSPS KAKENHI Grant Number 25289341.

### References

[1] I. Yamada *et al.*, Fusion Sci. Tech., 58, 345 (2010).

[2] I. Yamada *et al.*, Rev Sci. Instrum., 83, 10E340 (2012).

[3] I. Yamada *et al.*, Europhysics conference abstracts, 37D, O2.112 (2013).

[4] I. Yamada *et al.*, Europhysics conference abstracts, 38F, P4.009 (2014).

# Study of ICRF Heating Characteristics of Various Antenna in the Large Helical Device

S. Kamio[1], T. Seki[1], K. Saito[1], H. Kasahara[1], R. Seki[1], G. Nomura[1], M. Goto[1], M. Osakabe[1], T. Mutoh[1], and LHD Experiment Group[1]

[1] *National Institute for Fusion Science, Toki 509-5292, Japan*

Ion cyclotron range of frequencies (ICRF) heating is studied using three types of antenna in the Large Helical Device (LHD) shown in Fig. 1. As shown in Fig. 1, One toroidal array antenna (HAS antenna (a)) which can change the parallel wave number $k_{//}$ and two poloidal array antennas (FAIT antenna (b) and PA antenna (c)). FAIT antenna has large power density with the compact head, and high loading resistance with attached the field aligned impedance transformer. PA antenna does not have faraday shield (FS) because recently removed to investigate the effect of the removing the FS. With these antennas, minority ion heating is adopted in helium majority and hydrogen minority plasma in the LHD experiment. In the ICRF minority ion heating, the ICRF heating efficiency strongly depends on the minority ion ratio. The heating efficiency also depends upon the resonance layer, plasma loading, plasma parameter, and antenna shape such as with/without faraday shield (FS) because these parameters change the fast ion tail, the absorption layer and/or heating power to peripheral region. In the recent experimental campaign, the FAIT antenna was installed and the FS on PA antenna was removed, and by the upgrade of the heating devices and the control system, high power ICRF (more than 4 MW) can be injected. Therefore the evaluation of ICRF heating characteristics of various antenna is important in the recent experimental condition.

In order to verify these physical properties, we measured the heating efficiency of ICRF minority ion heating in terms of minority ion ratio, electron density and temperature, magnetic field strength, ICRF phase, and plasma-antenna distance in the short pulse (less than 10 sec) and in the long pulse (approximately 40 sec) discharges. In the experiments, we investigated using various diagnostics such as silicon-diode-based fast neutral analyser (Si-FNA)[1] for measuring the fast ion tail, diamagnetic loop, charge exchange spectroscopy with NBI, and Tomson scattering for measuring the plasma stored energy $w_p$, $w_{pi}$, and $w_{pe}$. As a results of these experiments, we summarized the ICRF heating characteristics each three types of antenna. The heating efficiency strongly depends on the plasma density, minority ion ratio, and antenna current phase. They have strong dependence, and results suggest their mutual dependence, however, therefore the conditions are too complex and difficult to make full database in the limited experimental campaign. Therefore, we summarized the dependences on minority ratio and antenna current phase separately. Figure 2 shows the ICRF heating efficiency dependence on (a) minority ratio and (b) antenna current phase. The heating efficiency were calculated by ICRF power modulation experiments[2]. Here, we used the simple model for calculating the heating efficiency, defined by the ratio of the absorption power $P_{abs}$ and the ICRF $P_{ant,}$

$$\eta \equiv \frac{P_{\text{abs}}}{P_{\text{abs}}} = \frac{\omega}{\sin\delta}\frac{|\delta W_{\text{p}}|}{|\delta P_{\text{abs}}|},$$

We also have to consider the influence of the high-energy ion. The power balance equations are written as

$$\frac{dW_{\text{tail}}}{dt} = \eta P_{\text{ant}} - \frac{W_{\text{tail}}}{\tau_{\text{se}}/2} - \frac{W_{\text{tail}}}{\tau_{\text{sHe}}/2} - \frac{W_{\text{tail}}}{\tau_{\text{tail}}},$$

$$\frac{dW_{\text{bulk}}}{dt} = \frac{W_{\text{tail}}}{\tau_{\text{se}}/2} + \frac{W_{\text{tail}}}{\tau_{\text{sHe}}/2} - \frac{W_{\text{bulk}}}{\tau_{\text{tail}}}.$$

In this paper, we used simple model for evaluate the heating efficiency, therefore some results may have over estimation.

Toroidal array antenna (HAS) and poloidal array antennas (FAIT and PA) show different dependence upon the minority ratio and antenna current phase. The antenna current phases of good efficiency are shown in Fig. 1 red arrow. Especially, the PA antenna (FS-less) has good efficiency at monopole which cancel the co-axial mode. Because FS-less antenna has large $E_{\perp}$ in front of the antenna head, therefore the cancelling $E_{\perp}$ is important for heating plasma.

High power ICRF injection experiments are demonstrated with various diagnostics, its power and electron density are shown in Fig. 3 (a). Figure 3 (b) and (c) shows the ion and electron temperature, and stored energy. Ion and electron temperature were increased during the increased of the ICRF power, and stored energy of high energy tail calculated by

$$W_{\text{p}} = W_{\text{pi}} + W_{\text{pe}} + W_{\text{pi\_tail}},$$

the stored energy of the high energy ion also increased by increasing the ICRF power. The amount of the high energy particle measured by Si-FNA was increase shown in Fig. 3 (d) however, the effective temperature was almost same. Figure 4 shows the density and temperature profiles. The central ion temperature was increased, and density profile became hollow. We are preparing the ICRF heating calculation, and comparison and evaluation with these experimental results are future works.

[1] M. Osakabe *et al.*, Rev. Sci. Instrum. 72, 788-791 (2001)

[2] Y. Torii *et al.*, Plasma Phys. Control. Fusion, 43, 1191-1210 (2001)

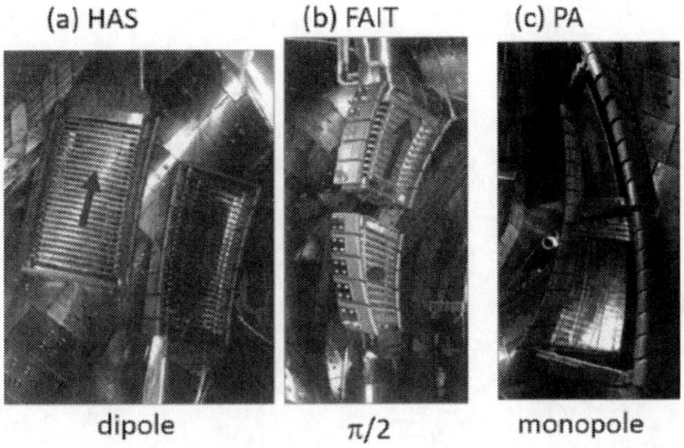


Fig. 1. Pictures of three types of the ICRF antennas, HAS antenna, FAIT antenna, and PA antenna installed in the LHD. Red arrows show the antenna current phase for good heating efficiency.

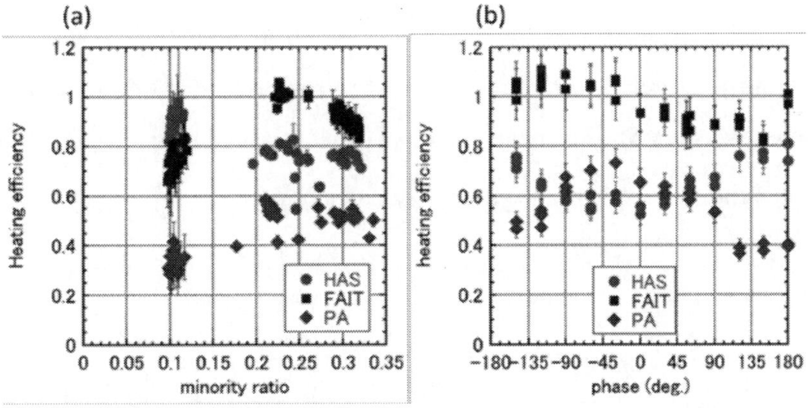


Fig. 2. Heating efficiency of various (a) minority ratio and (b) antenna current phase. As experimental conditions, (a) $n_e = 1 - 1.3 \times 10^{19}\mathrm{m}^{-3}$, and current phases of (HAS, FAIT, PA antenna) are (180, 57, -120 deg.) respectively, and (b) $n_e = 1 - 1.3 \times 10^{19}\mathrm{m}^{-3}$ and minority ratio is 0.2-0.25.

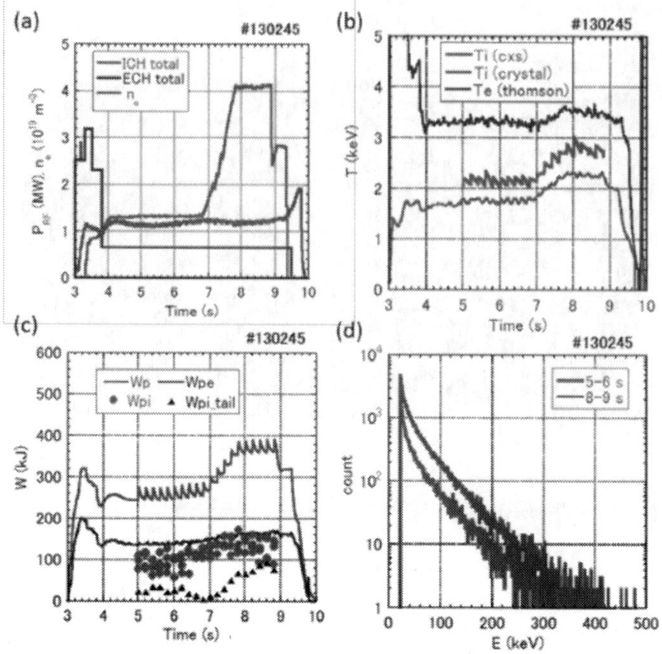


Fig. 3. High power ICRF injection experiments are demonstrated with various diagnostics. (a) The injected RF powers and line averaged electron density, (b) the ion and electron temperature, (c) the stored energy $w_p$, $w_{pe}$, and $w_{pi}$ measured by diamagnetic loop, charge exchange spectroscopy, Thomson scattering respectively the $w_{pi\_tail}$ is obtained by $w_p - (w_{pi} + w_{pe})$, (d) the results of the high energy particle measurement.

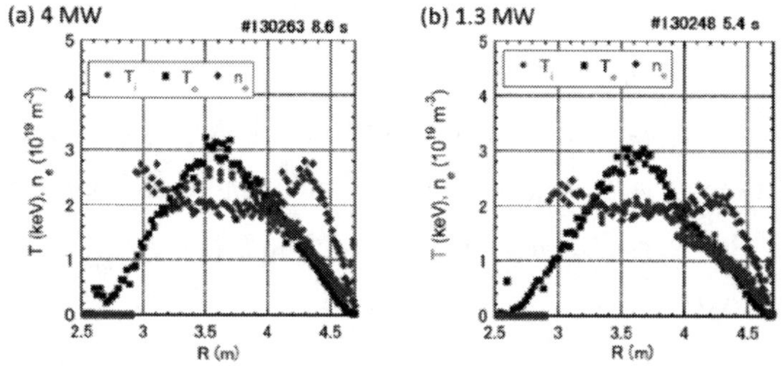


Fig. 4. The results of the electron density profile and the ion and electron temperature profiles measurement in the ICRF (a) high and (b) low power injection.

# Study of the large-scale MHD mode and its effect on GAM in the T-10 tokamak

L.G. Eliseev[1], N.V. Ivanov[1], A.M. Kakurin[1], S.E. Lysenko[1], S.G. Maltsev[1],
A.V. Melnikov[1, 2], S.V. Perfilov[1], and V.N. Zenin[1]
[1] *National Research Centre 'Kurchatov Institute', Moscow, Russian Federation*
[2] *National Research Nuclear University MEPhI, Moscow, Russian Federation*
*e-mail*: reonid@yahoo.com

Geodesic acoustic modes (GAMs) are considered as a possible mechanism, regulating turbulence and affecting the transition to the improved confinement [1]. To study GAMs and MHD modes we used two main diagnostics: Heavy Ion Beam Probe (HIBP) [2] and poloidal Magnetic Probe Array (MPA) [3]. HIBP measures the change of probing beam energy after ionisation in plasma and obtains the mean value of plasma potential $\varphi$ and its fluctuation $\Delta\varphi$; measures the change of beam current $I_{tot}$ and finds the relative fluctuation of density $\Delta n/n$; measures the toroidal shift of the beam and determines the relative fluctuation of poloidal magnetic field $\Delta B_p/B_p$ in the radial range 6 cm < $r$ <31 cm. MPA can define the amplitude $\Delta B_p$ and poloidal number $m$ of MHD mode. Also data from bolometer, multichannel interferometer, and diamagnetic loops were used.

The large-scale mode (LSM) under consideration with low $m$=2 and/or $m$=3 was recently studied [3] in Ohmic plasmas by a combination of MPA and HIBP in the T-10 tokamak ($a$=0.3 m, $R$=1.5 m, $B_t$=1.6-2.3 T, $I_p$=145-220 kA, $q(a)$~2.8-3.9). Owing to low frequency ($f$=1.5 - 5 kHz) the mode has an audible sound. LSM is seen by various diagnostics. In the example presented in Fig. 1, the amplitude of the fluctuation is as large as: for density $\Delta n_e/n_e$<8%, for potential $\Delta\varphi$~25-50V and for poloidal field $\Delta B$<1 mT. According to the MPA data, LSM has the poloidal mode number $m$~2-3 ($m$=2 in Fig. 2) and its amplitude varies from 0.1 to 1 mT. The LSM mode development leads to some degradation of confinement, indicated by the reduction in the stored energy $W_{dia}$, influx of impurities. When the mode amplitude exceeds some critical level of about 1 mT, the plasma confinement falls, feedback control system failures and major disruption occurs. This agrees with earlier data [4]. LSM is typically excited in a narrow window of $q(a)$ or plasma currents, e.g. $\Delta I_p$ ~5 kA, in the range 138-143 kA, Fig.3.

In the whole variety of the T-10 discharges GAM are typically observed with an amplitude 10-100 V [5]. Contrasting to this observation, the onset of LSM is accompanied by suppression of GAM. After spontaneous decay of LSM, the quiescent discharge is restored, i.e., the values of the plasma stored energy, potential and GAM amplitude recover.

LSM mainly observed at low densities $1< \bar{n}_e< 2.5\times10^{19}$ m$^{-3}$, Fig. 3, lower box. Similar to the typical T-10 tearing modes ($m$=2, $f$=7 kHz), LSM has a pronounced component in the plasma density, observed by HIBP. In contrast to $f$=7 kHz tearing mode LSM has strong electric potential component $\Delta\varphi \sim 40$ V.

The radial extent of potential perturbation associated with LSM is found to be as large as 4-6 cm and typically located at outer $^1/_3$ of the plasma column, as measured with HIBP radial scan, Fig.4. This data is supported by another experiment with the alternation of the HIBP sample volume position from 26 to 29 cm. Fig. 5 shows that at $r$=26 cm the LSM amplitude is small, while at $r$=29 cm the amplitude is large in agreement with scanning data.

Figure 6 shows discharge with periodical vanishing of LSM. When LSM disappears, the density and stored energy increase. In quiescent discharges without LSM the potential decreases with density [6]. In the discharges with strong LSM the deviation from this

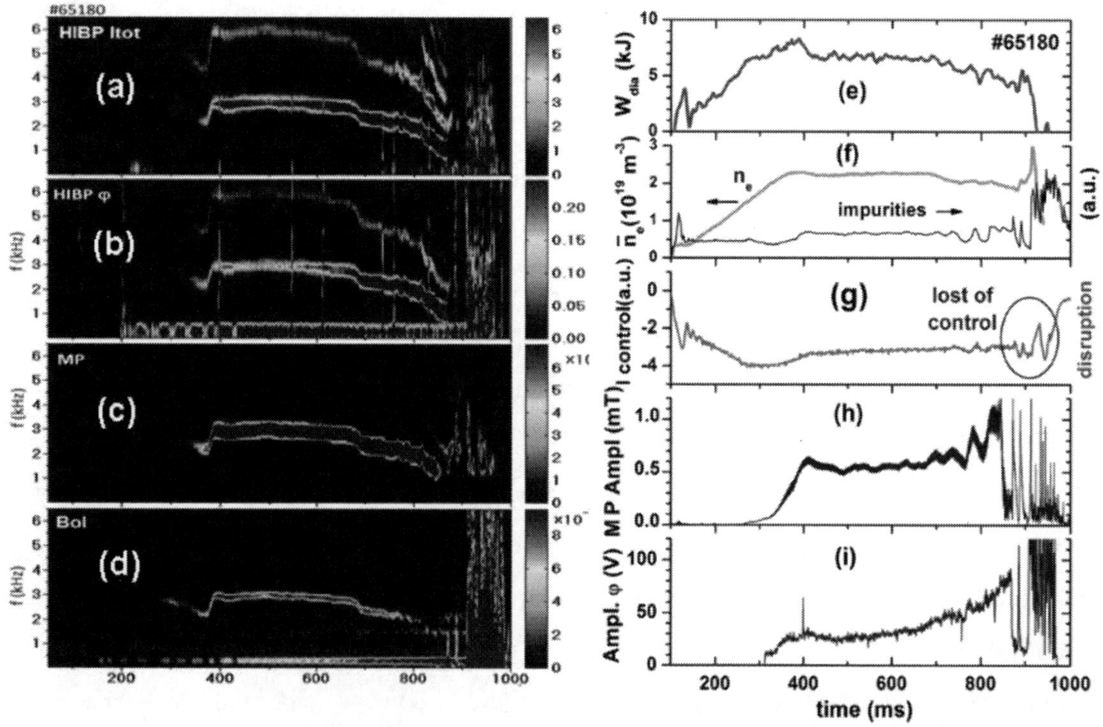


Fig.1. Low-frequency large-scale mode is seen on many parameters: on density (a), on potential (b) and (i), on magnetic probes (c) and (h), (d) on bolometer. Grows of mode amplitude leads to degradation of energy (e) and particle (f) confinement, influx of impurities (f), lost of feedback control and disruption (g).

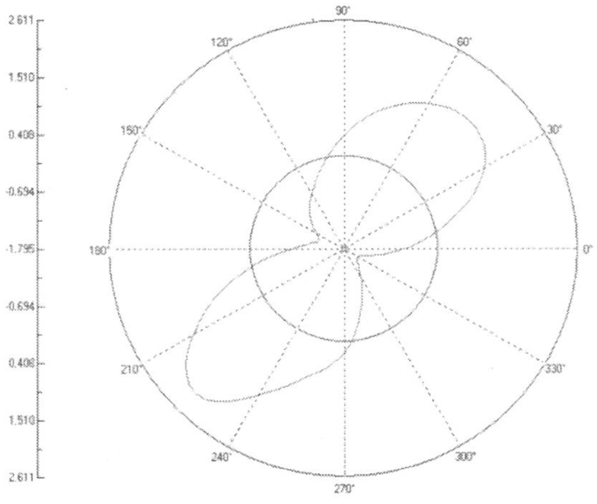

Fig. 2. Poloidal number of LSM $m$=2, obtained by MPA

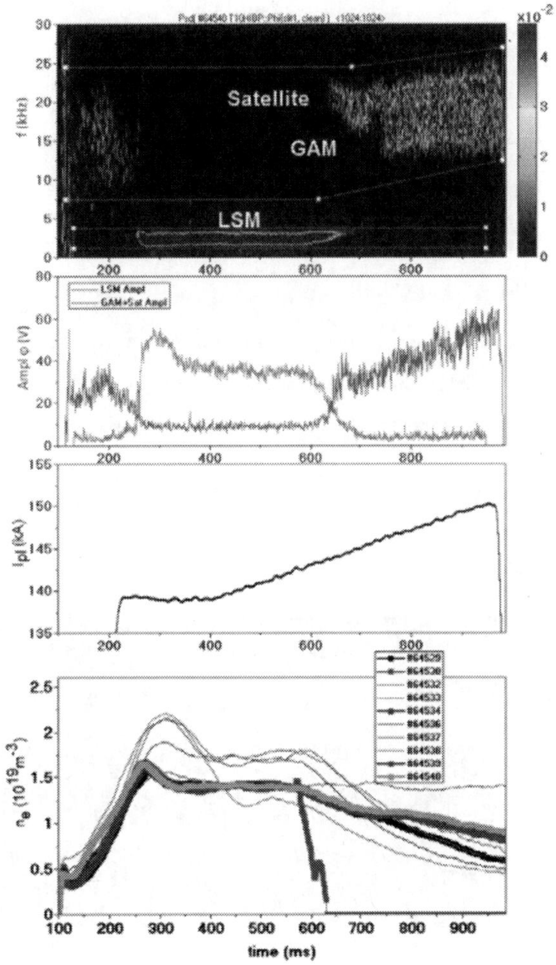


Fig.3. LSM exists in the narrow window of currents ~ 5 kA around the resonant value and in for specific densities (fat lines).

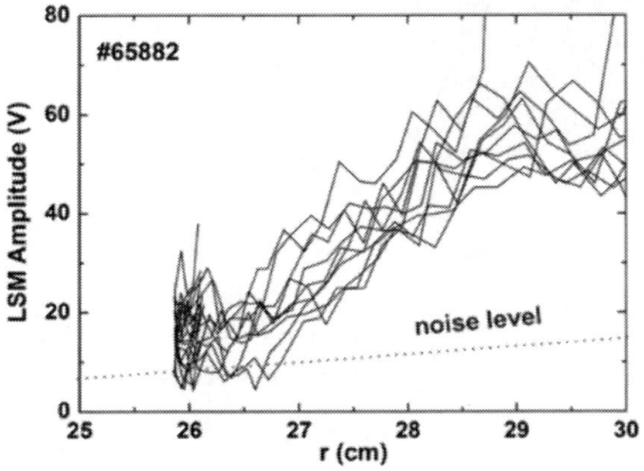


Fig. 4. Radial localization of the potential perturbation, associated to LSM.

Fig. 5. Evolution of LSM measured in two alternating points at $r$=26 cm and $r$=29 cm. ⇨

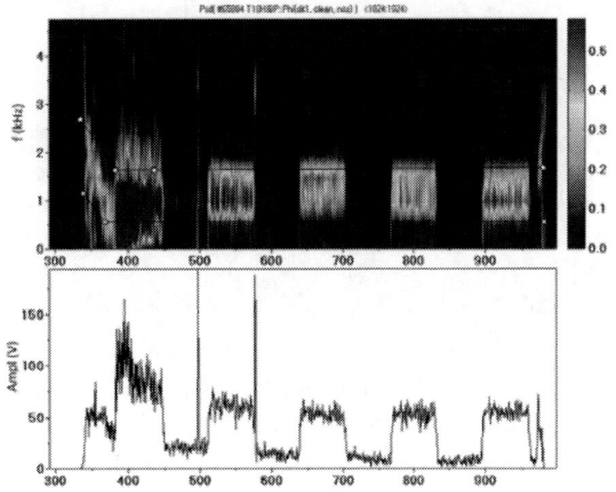

dependence is observed and the potential increases, evolves towards the positive direction, Fig. 7.

**Acknowledgement**

The work was carried out by Russian Science Foundation, project 14-22-00193.

**References**

1. Wagner F. 2007 Plasma Phys. Control. Fusion **49** B1
2. Melnikov A. V. et al 1995 Rev. Sci. Instrum. **66** 317
3. Eliseev L.G. et al. 2015 Phys. Plasmas **22** 052504
4. Mirnov S.V. Physical processes in tokamak plasmas, 1983, Energoatomizdat, (In Russian)
5. Melnikov A.V. et al. 2015 Nucl. Fusion **55** 063001
6. Melnikov A.V. et al., 2011 Nucl. Fusion **51** 083043

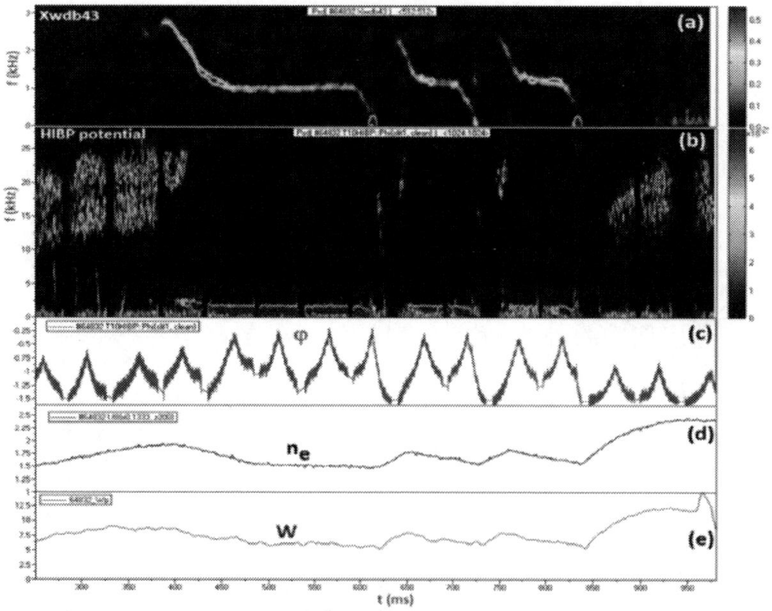


Fig. 6. When LSM periodically vanished, the stored energy and plasma density increase.
(a) soft X-ray signal,
(b) HIBP potential spectrogram,
(c) HIBP potential signal (scanning),
(d) line-averaged density,
(e) the stored energy.
$B_z = 2.3$ T, $I_{pl} = 220$ kA

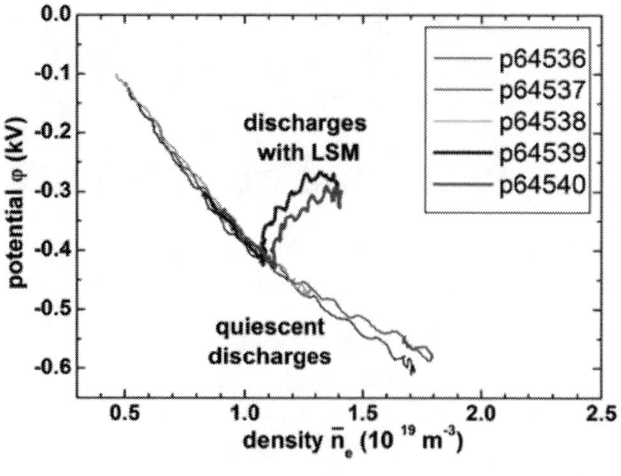


Fig. 7. Effect of LSM on the plasma potential, measured for different densities at $r$=0.14 m for the series of the similar discharges. Each curve is labeled by discharge number.
$B_z = 1.6$ T, $I_{pl} = 144$ kA

# Characteristics of turbulence and electric fields in the peripheral plasma of the T-10 tokamak

R.Y. Solomatin, V.P. Budaev, S.A. Grashin, E.V. Dzyatko, L.G. Eliseev, A.V. Melnikov*, S.V. Perfilov, V.N. Zenin

*National Research Center "Kurchatov Institute", Moscow, Russia*
**National Research Nuclear University MEPhI, Moscow, Russia*

Measurements made in the Scrape-Off Layer (SOL) and in the vicinity of the Last Closed Flux Surface (LCFS) indicate that the particle transport across the magnetic field is highly anomalous [1, 2], leading to an essential increase of particle and energy flux reaching the wall of the vacuum chamber. This transport is determined by a strong intermittent turbulence associated with formation and radial propagation of the high density plasma structures ("blobs") [3, 5]. Investigations of the turbulence characteristics and the electric fields on the plasma periphery and in the SOL give the opportunity for understanding the mechanisms of particle transport and plasma confinement in tokamak. Investigation of the spectral characteristics of density and plasma potential turbulence clarifies the role of different types of oscillations like Geodesic Acoustic Mode (GAM) or MHD-modes in the particle and energy transport.

Turbulence and electric field characteristics were studied on T-10 tokamak by Movable Langmuir Probes (MLP) [3] and by Heavy Ion Bean Probe (HIBP) [4]. The average values and fluctuations of density, plasma potential and electric fields were measured in the core (by HIBP) and in the edge (by HIBP and MLP) plasmas. T-10 tokamak has a major radius of $R_0 = 1.5$ m, a circular poloidal graphite limiter with radius of 0.33 m and a rail graphite limiter installed at a radius 0.30 m. Experiments were performed in deuterium plasma with ohmic heating. In these experiments the average electron density $N_e$ was in the range $1{\cdot}10^{19} \div 4{\cdot}10^{19}$ m$^{-3}$, the toroidal magnetic field $B_t = 2.3$ T, the plasma current $I_p = 0.2 \div 0.3$ MA.

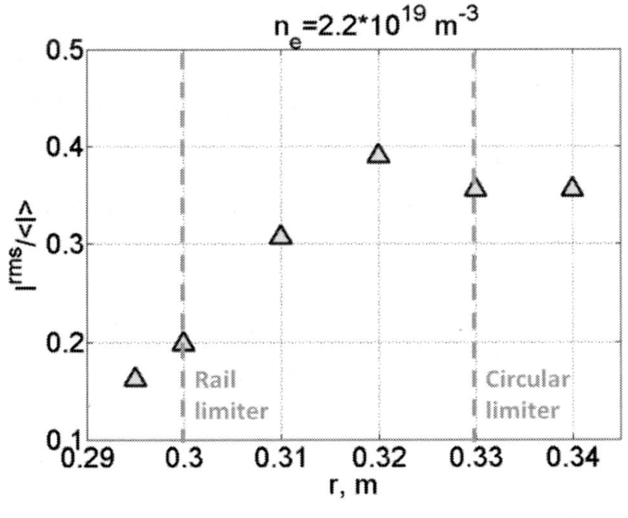


Fig. 1. Radial variation of the relative level of ion saturation current fluctuations. LP ion saturation current is proportional to electron density $N_e$.

Fluctuations of the plasma parameters (density, electric fields, turbulent particle flux) are highly intermittent in the SOL of

T-10 [3, 6]. Plasma structures with high density were observed in the SOL. The relative level of the fluctuations rises with the radius beyond the LCFS (Fig. 1), time interval between blobs and their duration increases (from 50 to 400 μs).

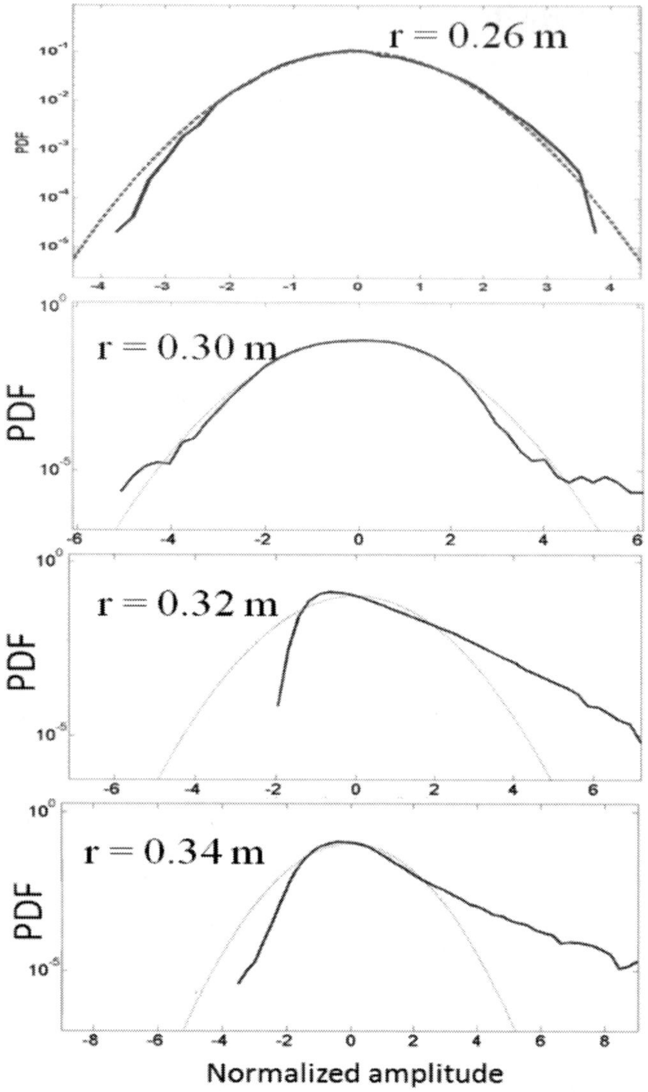


Fig. 2. Probability Density Function (PDF) of the density fluctuations measured by HIBP (r = 0.26 m) and LP (r = 0.30 - 0.34 m).

Measurements made by the MLP and HIBP indicate that in the periphery plasma (r = 0.25-0.3 m), where the level of intermittent turbulence is low, the statistics of plasma turbulence are close to the Gaussian one. In the SOL, the deviation of PDF from the Gaussian statistics rises (Fig. 2).

Radial electric field is positive in the SOL (up to 4 kV/m) and changes its sign near the LCFS (Fig. 3). Also the poloidal rotation velocity and the $E\times B$ drift velocity change their sign near the LCFS (Fig. 4). That can be connected with the change of particle transport behavior. For estimation of poloidal rotation velocity of the plasma structures were used a technique based on the time-delay of LP signals from the two poloidally separated probes. The measured poloidal plasma rotation velocity and the $E\times B$ drift velocity in the SOL are close to each other in value (-1.0-1.5 km/s) and have the same direction. In the SOL they directed towards the ion diamagnetic drift.

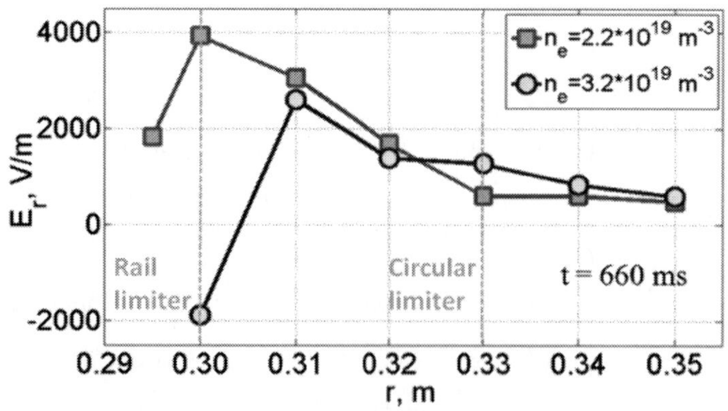


Fig. 3. Radial variation of the radial electric field.

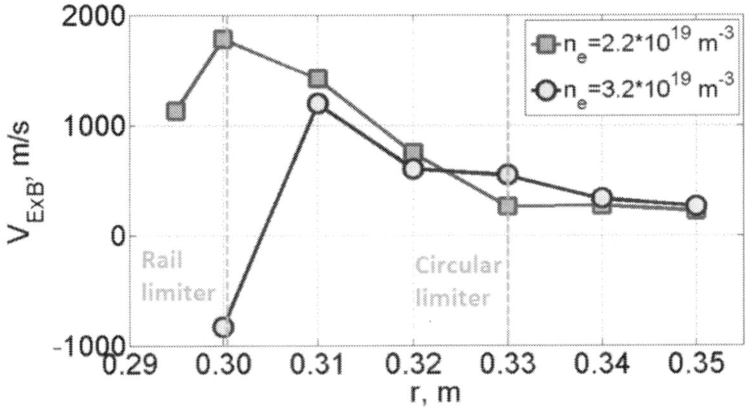


Fig. 4. Radial variation of the $E\times B$ drift velocity.

Measurements of the poloidal electric fields allows us to estimate the cross-field turbulent particle flux $\Gamma_\perp$.

$$\Gamma_\perp = \langle n_e V_r \rangle = \frac{c}{B}\langle \tilde{n}_e \tilde{E}_\theta \rangle = \frac{c}{B}\gamma_{nE}\,\tilde{n}_e \tilde{E}_\theta$$

where $n_e$, $V_r$ and $E_\theta$ are the root-mean-square values of the density, the radial velocity and the poloidal electric field; $\gamma_{nE}$ is a coefficient of the cross-correlation between $n_e$ and $E_\theta$ fluctuations. Radial turbulent flux reaches the value of $5\cdot10^{21}$ $m^{-2}s^{-1}$ near the LCFS and represents the significant part of the total particle flux across the plasma boundary.

In the vicinity of the LCFS Geodesic-Acoustic-Mode-like (GAM) oscillations were observed at 12-18 kHz on the floating potential and ion saturation current LP signals. GAM oscillations have the intermittent behavior. The only low-frequency (1-9 kHz) fluctuations observed at larger radii (closer to the wall of the tokamak) (Fig. 5). Measurements made by the LP and HIBP at the plasma periphery indicate also a quasi-coherent peak of oscillations with frequencies in the range 50-100 kHz.

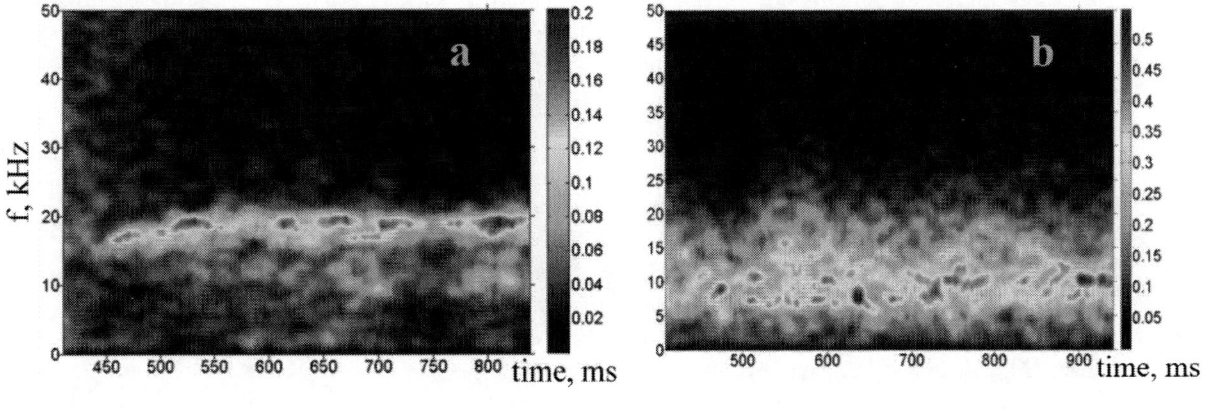


Fig. 5. Power spectral density of the floating potential oscillations in the SOL of T-10. $B_t$ = 2.3 T, $I_p$ = 220 kA, $N_e = 2\cdot10^{19}$ m$^{-3}$. **a.** $r$ = 0.3 m. **b.** $r$ = 0.32 m.

The frequency of the GAM oscillation rises with the electron temperature $T_e$ (Fig. 6), according to the theoretical predictions for GAM [7].

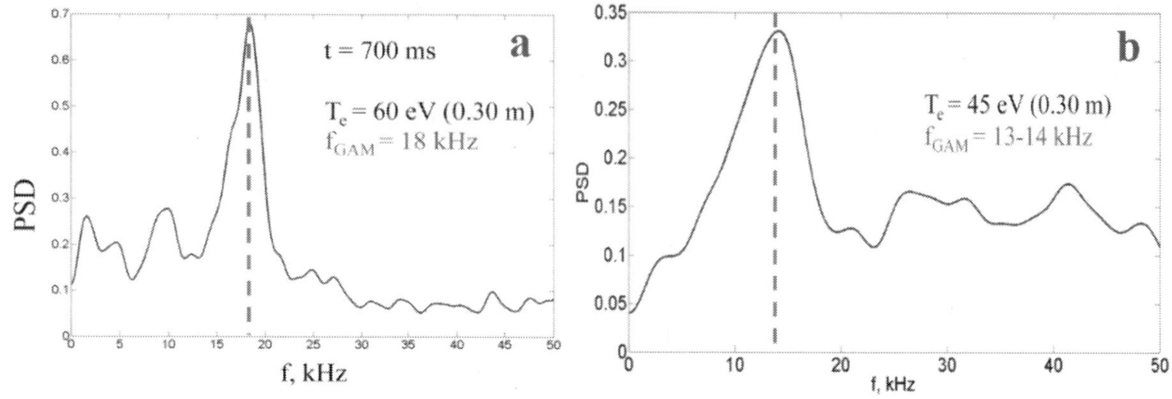


Fig. 6. Power spectral density of the floating potential turbulence near the LCFS of T-10. $B_t$ = 2.3 T, $I_p$ = 220 kA, $r$ = 0.3 m. **a.** $N_e = 2\cdot10^{19}$ m$^{-3}$. **b.** $N_e = 3.2\cdot10^{19}$ m$^{-3}$.

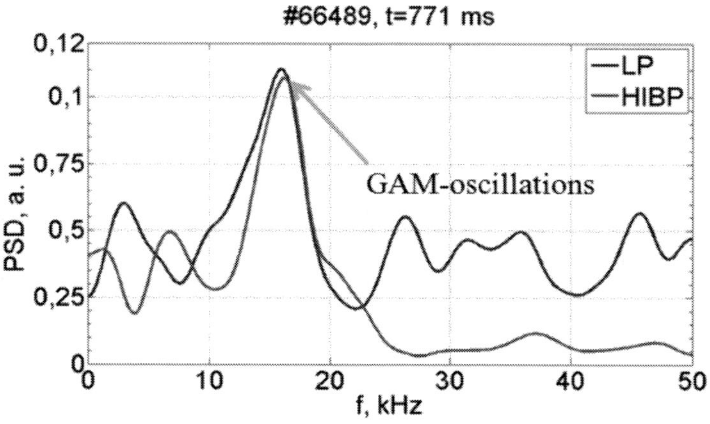


Fig. 7. Power spectral density of the MLP (r = 0.30 m) and the HIBP (r = 0.25-0.26 m) potential signals.

The GAM frequency measured by MLP corresponds to the HIBP measurements in the peripheral and core plasma (Fig. 7) [8]. In some cases the correlation between the MLP and HIBP potentials were found at the GAM frequency.

The work was carried out by Russian Science Foundation project 14-22-00193.

**References.**

[1]. Zweben S. J. et al. Plasma Phys. Control. Fusion **49** S1-23 (2007).
[2]. Kirnev G. S., Budaev V.P., et al. Journal of Nuclear Materials v.337-339, p.352-356 (2005).
[3]. R.Y. Solomatin et al. 41st EPS Conf. on Plasma Phys. and Nucl. Fusion, Berlin 2014, P4.038.
[4]. Melnikov A.V. et al. Nucl. Fusion **53** (2013) 092002.
[5]. Budaev V.P., Savin S.P., Zelenyi L.M. Physics-Uspekhi **54** (2011) 875–918.
[6]. Kirnev G. S., Budaev V.P., et al. Nuclear Fusion **45** (2005) 459-467.
[7]. P.M. Diamond et al Plasma Phys. Control Fusion **47** (2005) R35-R161.
[8]. A.V. Melnikov et al Nucl. Fusion **55** (2015) 063001.

# Measurements of impurities concentrations using modernized CXRS diagnostics at T-10

L. Klyuchnikov, V. Krupin, K. Korobov, A. Nemets, M. Nurgaliev, A. Dnestrovskij, A. Barsukov, N. Naumenko*, S.Tugarinov, S.Serov, V.Troinov, A.Gorbunov, D.Denshchikov

*NRC «Kurchatov institute», Moscow, Russia, lklyuchnikov@list.ru*

** IPh NAS RB, Minsk, the Republic of Belarus*

Active spectroscopic diagnostics (CXRS) is developed at T-10 tokamak for measurements of the plasma ion temperature and impurities concentrations [1,2] . Modernized scheme of the diagnostics includes 3 High Etendue Spectrometers (HES) as shown in figure 1. Upgraded design allows providing simultaneous measurements of beam $H_\alpha$ line and 2 impurities CXRS spectra. Light from two collecting systems is transmitted by optical fibers to spectroscopy control room. Each system collects light from 9 lines of sight, uniformly distributed from -0,3$a_L$ to $a_L$ ($a_L$=30 *cm,* rail limiter radius). Registration system includes 3 spectrometers and 3 scientific cameras: two EMCCD PhotonMAX 512B and PIXIS 512B. Spectrometers based on Echelle gratings are marked as HES1 and HES2, spectrometer with holographic grating for 529 nm is marked as HES3. HES1 and HES3 spectrometers are coupled in one box as shown on figure 2. Light of 529±10 nm diffracts on the first grating with ~50% efficiency, while the rest of light outside its operational wavelengths range passes through the holographic grating with ~0.8-0.9 transmission coefficient, allowing to measure spectra of another CXRS line or $H_\alpha$ beam line, depending on experiment requirements. Combined spectrometer allows performing precise concentration profiles measurements in case of simultaneous impurity CXRS line and beam $H_\alpha$ line spectra, because in that case the most of uncertainties and possible measurements errors are canceled [3].

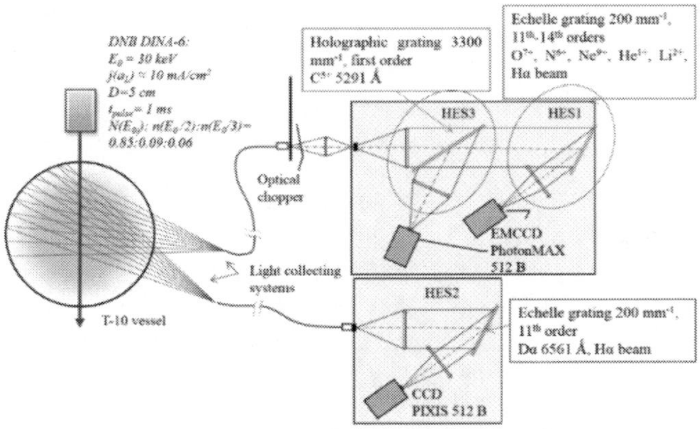


Fig 1. Scheme of CXRS measurements at T-10

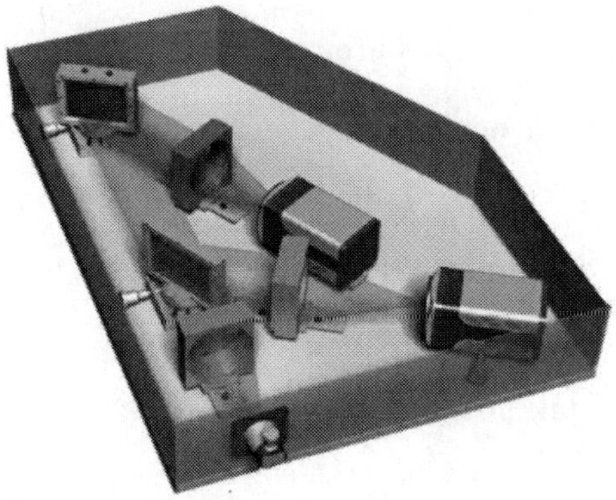

Fig 2. Combined spectrometer HES13

Spectrometer with 3 holographic gratings for 468, 529 and 656 nm is a prototype developed for CXRS diagnostics of ITER edge plasma ($\rho \geq 0.5$). Chosen scheme and used elements have shown very promising results during tests at T-10 machine.

Profile measurements of plasma intrinsic impurities (C, N, O) and injected impurities (He, Li, Ne) are provided using the modernized diagnostic scheme in various OH and ECRH discharges. It is found that in T-10 wall conditions in operational regimes there are similar high values of carbon and oxygen concentrations (1.5–3% of $n_e$) without any significant nitrogen concentration. Therefore, in 2014 T-10 experimental campaign the combined spectrometer was used for measurements of two main light impurities, oxygen and carbon, while spectrometer HES2 was used for measurements of beam $H_\alpha$ line. This approach provides reliable concentrations measurements, but it requires accurate relative calibrations using beam in gas emission and absolute calibrations of all spectrometers.

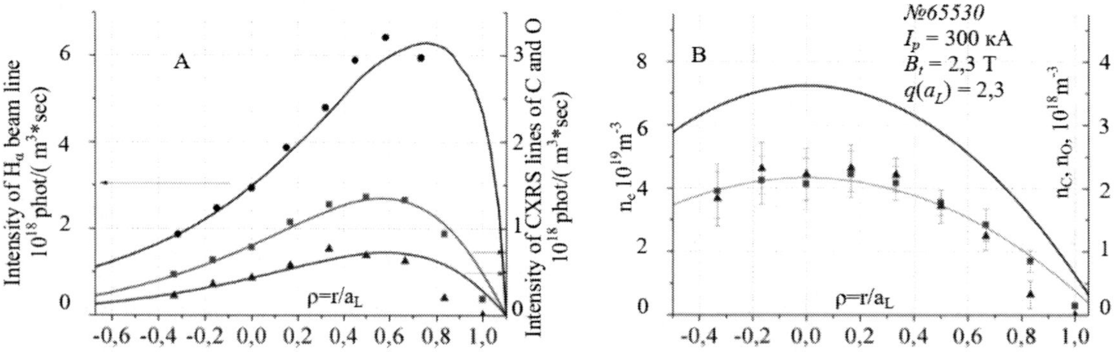


Fig 3. A - distributions of intensities of main energy $H_\alpha$ beam line and CXRS lines of carbon and oxygen. ● –$H_\alpha$ beam, ■ – CXRS $C^{5+}$, ▲ – CXRS $O^{7+}$, dots – measured data, curves – numerical calculation. B – measured carbon and oxygen profiles, black line – $n_e(r)$, red line – normalized $n_e(r)$

For verification of CXRS data, measurements in high plasma current OH discharge are shown. This kind of discharges at T-10 is characterized by measured $Z_{eff}$=const(r) profile, thus impurities distributions are similar to $n_e$ profile. Figure 3A shows intensities distributions of main energy $H_\alpha$ beam line and CXRS lines of C and O in ohmic discharge with $I_p$=300 kA and $\bar{n}_e = 4{,}8 \cdot 10^{19}\ m^{-3}$. Dots correspond to experimental data, curves are results of calculations with ADAS [4] data. Measured concentration profiles of carbon and oxygen nuclei (Fig. 3B) produce effective ion charge profile that is in good agreement with independent measurements of $Z_{eff}(r)$ via bremsstrahlung. The deviation of the C and O nuclei concentrations from the calculated line at the plasma edge ($\rho$>0.8) can be caused by the charge exchange with main gas atoms which was not taken into account in the numerical model.

**Impurities behavior in OH plasma**

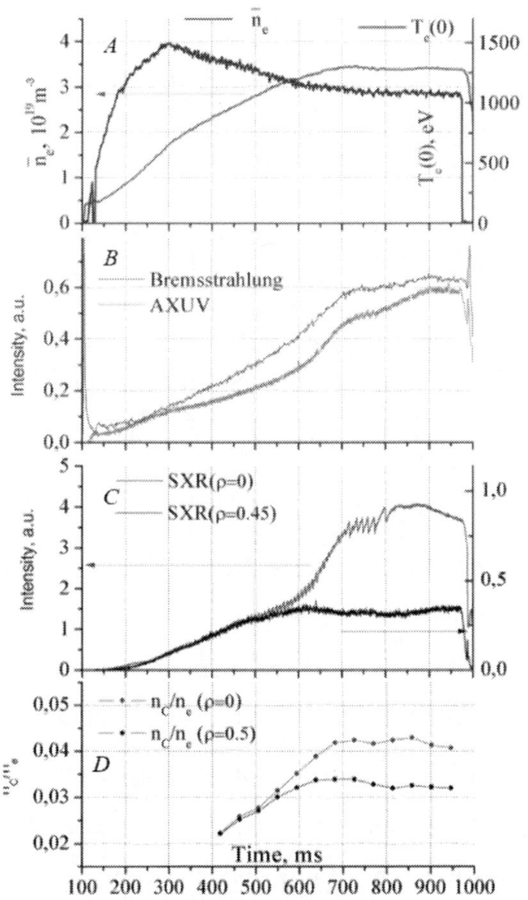


Fig 4. Plasma parameters in OH discharge at plasma current 180 kA with termination of saw-tooth activity.

Measurements with the modernized CXRS system allow revealing a strong dependence of light impurities behavior at OH discharges on plasma current and electron density. In high plasma current case measured impurity profiles is similar to $n_e(r)$, indicating the dominance of the anomalous transport, but at lower plasma current and higher electron density the impurity accumulation is observed that can be caused by the increase of neoclassical processes against the anomalous transport. This accumulation can lead to a termination of the sawtooth activity without any external action on plasma (see fig.4C). The accumulation of impurities is confirmed by bremsstrahlung measurements of $Z_{eff}$, and also by AXUV and SXR radiation dynamics. At the first stage of the discharge the total impurity concentration increases (fig.4, t<600ms), but impurity profiles stay similar to $n_e(r)$. Then the process of impurities peaking is observed that can lead to the current profile flattening and the saw-tooth stabilization. Peaking of impurities in contaminated plasma indicates strong influence of $Z_{eff}$ value on ratio of neoclassical and anomalous transport.

**Impurities behavior in ECRH plasma**

Light impurities removal from plasma center with ECRH is investigated on T-10 using the CXRS diagnostics. Figure 5 shows $n_e$ and impurities profiles in OH and ECRH discharges with low plasma current ($I_p$=180 kA). At OH stage carbon and oxygen are more peaked than electron density $n_e$. After central ECRH switch-on, concentrations of C and O drop by a factor of ~2–2.5 and radial profiles of impurities become similar to $n_e(r)$ within the error limits. Observed effect at ECRH stage is related to the initial neoclassical accumulation of impurities in the OH plasma that is disturbed by the ECRH impact. Similar effect was observed in other machines and the closest result was obtained at TCV [5]. The impurity transport study in T-10 in OH and ECRH regimes is described in [2].

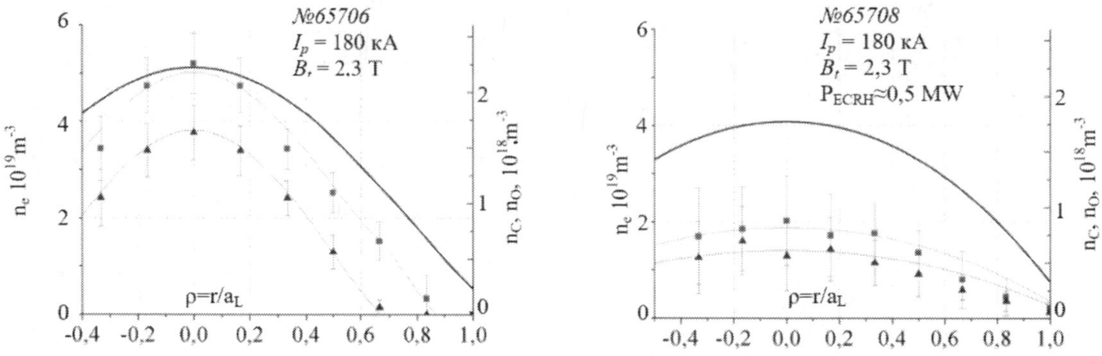


Fig 5. Electron (——), C(■) and O(▲) density profiles in OH (left) and ECRH (right) discharges at Ip=180 kA.

Efficiency of impurities removal increases with higher plasma density, $Z_{eff}$ value and with lower plasma current in OH stage and with higher heating power in ECRH stage. In discharges with the highest $n_e \cdot Z_{eff}/I_p$ ratio the nuclei impuritiy concentration can decrease up to 3 times in ECRH regime comparing to OH values. In conditions of the sufficient change of the transport from the neoclassical one in OH stage to the strong anomalous in ECRH stage, there is a possibility for the determination of impurities anomalous transport coefficients D and V in ECRH stage using light impurities nuclei dynamics measured by CXRS and dynamics of impurity sources. Despite the fact that confinement of impurities in central zone with ECRH significantly decreases indicating increase of anomalous impurity transport, measured density turbulence at radii ρ>0.4 decreases with ECRH relatively to OH stage [6]. Study of relation of plasma turbulence and impurities transport is the subject of further investigation at T-10.

**Acknowledgements**

Work was carried out by the funding of Russian Science Foundation Project 14-22-00193.

**References**

1. V.A. Krupin et al. – Plasma Physics Reports, 2013, Volume 39, 8, pp 632-643
2. L.A. Klyuchnikov et al. – IAEA FEC 2014, EX/P1-44
3. Bespamyatnov I.O. et al. – Review of Scientific Instruments 2010, vol. 81, 10D709.
4. The Atomic Data and Analysis Structure (ADAS) http://www.adas.ac.uk/
5. Emiliano Fable, – "Experimental and Theoretical Study of Particle Transport in the TCV Tokamak", 2009
6. V. Vershkov et al, – Nucl. Fusion 55 (2015) 063014.

# Hydrogen pellet initiated H-mode in TUMAN-3M tokamak

L.G. Askinazi, A.A. Belokurov, V.A. Kornev, S.V. Lebedev, A.I. Smirnov,
A.S. Tukachinsky, N.A. Zhubr
*Ioffe Institute, Saint-Petersburg, Russia*

H-mode is necessary operational regime for ITER, thus methods of plasma fuelling and LH-transition initiation applicable in plasma with high temperature and density should be developed. Classic technology of plasma fuelling via peripheral gas puffing proves itself ineffective when it comes to large-scale tokamaks. That is also unfortunate because in smaller devices gas puffing may also lead to LH-transition. One method that could solve both problems – fuelling and LH-transition initiation – is pellet injection. In present paper we focus on H-mode initiation aspect.

Pellet injection creates steep density gradient and also causes plasma cooling in the ablation region. These two effects are responsible for creation of strong and inhomogeneous radial electric field [1]:

$$E_r = \frac{T_i}{e}\left[\frac{\partial \ln n}{\partial r} + k_T \frac{\partial \ln T_i}{\partial r}\right] \tag{1}$$

If the pellet is sufficiently large and the penetration depth is small, the region of pellet-induced gradients occurs close the LCFS, and shear of such perturbed radial electric field could be large enough to create a self-sustaining transport barrier. LH-transition triggering by a pellet was observed for the first time on TUMAN-3 tokamak in 1991 [2] with LiD pellets. On DIII-D hydrogen pellet injection was found to lower substantially the LH-transition power threshold [3].

TUMAN-3M tokamak is now equipped with pneumatic injector, capable of shooting up to four frozen hydrogen or deuterium pellets with diameter of 0.7-1.0 mm and velocity 150-1000 m/s in tangential direction. Tangential injection provides prolonged pellet trajectory in the peripheral plasma region and localization of the evaporation region outside $\rho = (r/a) = 0.4$. For the means of ablation measurements there are used Hα/Dα line emission detectors with aperture directed at pellet impact area. Also a triple electrostatic probe is used for peripheral electron temperature measurements.

In the recent experiments hydrogen pellets were injected into deuterium plasma. For most shots pellet diameters were chosen 0.9 and 1.0 mm, velocity 300-500 m/s. Typical average plasma density before the injection was about $(1\div1.5)\cdot10^{19}$ m$^{-3}$. In the experiments different types of temporal evolution of Hα/Dα emission in the evaporation region were observed, see fig. 1.

One is typical ablation curve with gradual growth during 0.2-0.5 ms and much steeper decay. The ablation curve is traditionally observed when a solid-state pellet is evaporated in tokamak plasma. Density increase with characteristic time of ablation is observed. The Hα/Dα emission evolution of another kind has smooth shape with relatively short growth period and long (up to 5 ms) decay. This type of evolution probably is a result of injection of hydrogen gas due to pellet disintegration and evaporation in the pellet-guide. Difference in density growth rate depends on pellet evaporation localization – solid pellet, evaporated inside the plasma, yields fast increase (fig. 1a), while gas, only spreading on periphery, leads to slow density increase (fig. 2b).

However, both scenarios of fuelling may result in LH-transition (fig. 2). In several shots the evolution of plasma parameters typical for LH-transition was observed after the injection: average density growth, continuing after the pellet ablation, and peripheral Hα/Dα emission decrease. Probe measurements near LCFS have also shown that local density and electron temperature decrease after the injection, which corresponds with predictions for evolution of $n_e$ and $T_e$ outside the transport barrier.

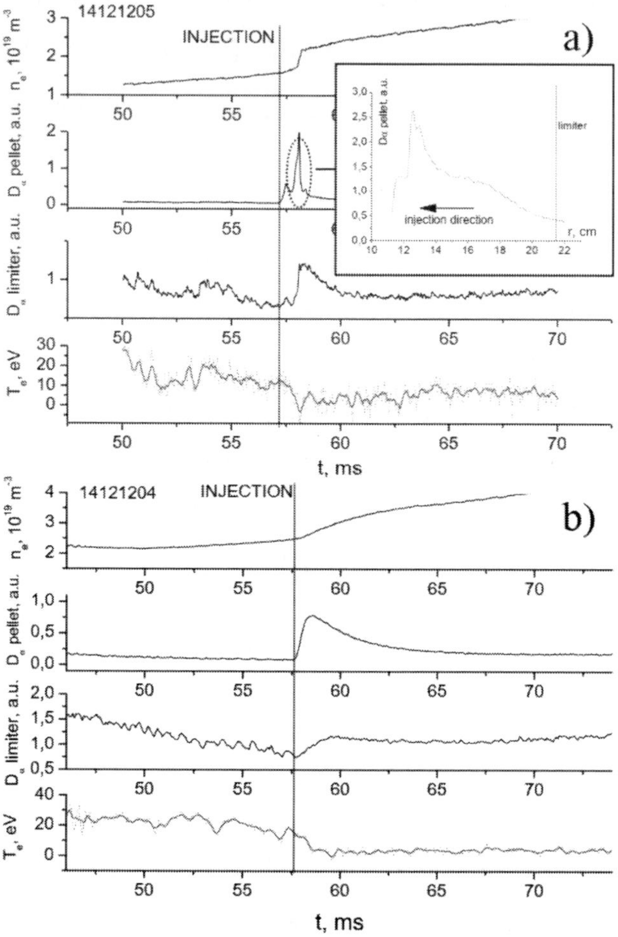


Fig.1 Evolution of plasma density, Hα/Dα emission and peripheral electron temperature for two types of experimental ablation scenarios: a) solid pellet evaporation with slow growth and steep decay; b) snow-like disintegrated pellet evaporation with steep growth and long decay. «Dα pellet» directly observes evaporation region, «Dα limiter» observes plasma periphery far from pellet and used for LH-transition indication. Fragment in a) shows the spatial distribution of emission, calculated on the base of ablation curve.

To understand possibility and criteria of pellet-induced transition, a modelling of plasma density profile evolution with pellet-modified particle source was made. There was modelled an evolution of plasma density:

$$\frac{\partial n(r,t)}{\partial t}-\frac{1}{r}\frac{\partial}{\partial r}r\cdot\left(D(r,t)\cdot\frac{\partial n(r,t)}{\partial r}-v(r,t)\cdot n(r,t)\right)=S(r)+S_{PELLET}(r,t) \quad (2)$$

Here diffusion coefficient $D$ and convective velocity $v$ are conjectured to depend on shear of radial electric field $\omega_{ExB}$ (calculated in accordance with formula (1)):

$D(r,t)=k(\omega_{ExB})\cdot D_0(r)$; $v(r,t)=k(\omega_{ExB})\cdot v_0(r)$

where $k(\omega_{ExB})$ in the form, described in [4], varies from 1 for L-mode to 0.1 for the case of strongly suppressed anomalous transport.

Confinement improvement mechanism, conjectured in the model, is the following: pellet ablation creates a perturbation of particle source $S_{PELLET}$, which is introduced into diffusion equation in the form derived from experimental pellet ablation curve (with its spatial and temporal characteristics). $S_{PELLET}$ is normalized to full particle content in the pellet.

Localized pellet-induced source and thus density perturbation creates additional density gradient in the ablation area; based on the gradient, profiles of radial electric field and its shear are calculated. Strong shear leads to transport coefficients damping, thus modifying density profile and creating a transport barrier – if shear is strong enough, transport barrier becomes self-sustaining and system remains in state with steep peripheral gradients, which is considered as H-mode. According to modeling results, injection of pellets with low velocity, resulting in short penetration depth

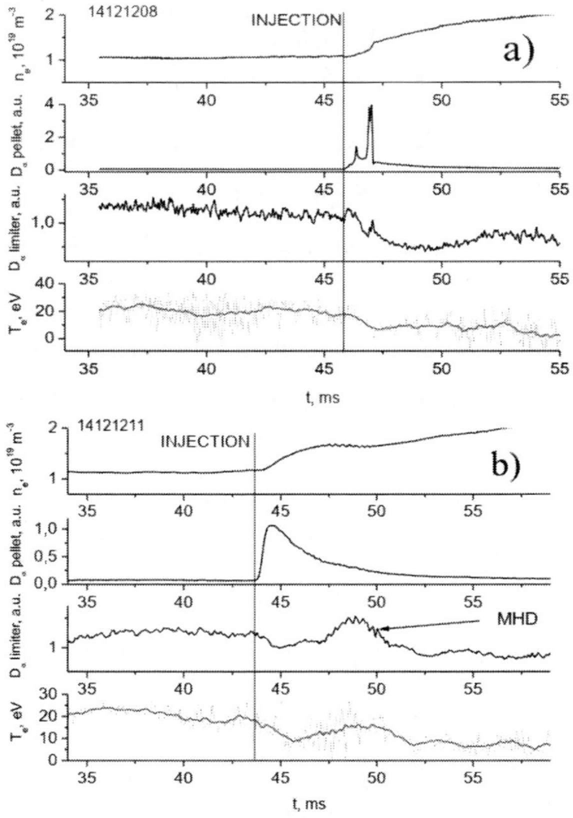


Fig.2 Evolution of plasma density, Hα/Dα emission and peripheral electron temperature in different scenarios of fueling, which result in LH-transition initiation: a) solid pellet, b) disintegrated pellet. Average density growth, peripheral Dα and Te decrease indicate the transition.

($\rho = r/a > 0.8$) and ablation in the area of "essential" gradients, is the most beneficial for LH-transition initiation.

Scenarios with disintegrated pellet (long evolution) were not modeled, because it is impossible to obtain information about spatial distribution of particle source in that scenario. Also, these scenarios are similar to initiation of H-mode with gas puffing, suitable only for small-scale tokamaks, and thus are of little interest to fusion goals. Ion temperature in peripheral area is a crucial parameter, as radial electric field value depends on it. In fig. 3 there are presented several peripheral density gradient evolution curves for different ion temperature values. If this value exceeds some threshold value, system remains if state with increased gradient.

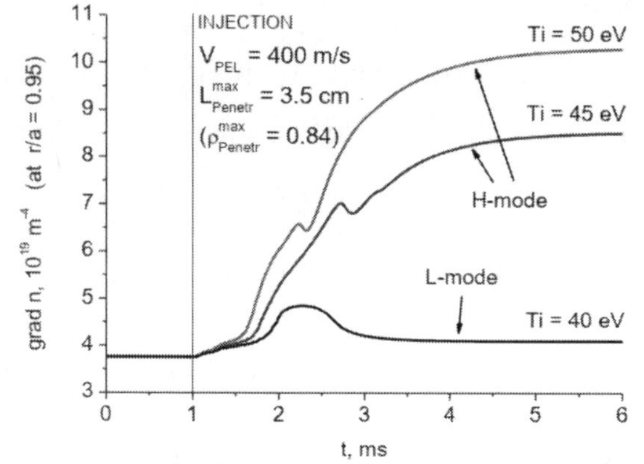


Fig.3 Modeling of peripheral density gradient during pellet-injection. If ion temperature exceeds certain value, system remains in improved confinement state with high peripheral gradient value.

This model has several drawbacks, e.g. it does not calculate temperature profile evolution due to poor knowledge about temperature (especially $T_i$) profiles in TUMAN-3M, thus $T_i$ profile was taken the same as density profile.

There are more sophisticated models, simulating LH-transition [5], which conclude that gradient-driven mean flow plays major role in particle injection induced LH-transition. This conclusion makes the results of our simple modeling somewhat reliable for rough estimates of pellet-induced LH-transition possibility and criteria.

The work has been carried out with cooperation and equipment of unique scientific device "Globus-M" team. Authors thank for partial financial support Ministry of Education and Science of the Russian Federation (Agreement 14.619.21.0001, id RFMEFI61914X0001).

**References:**

[1] S.P. Hirshman, D.J. Sigmar, Neoclassical transport of impurities in tokamak plasmas. 1981 Nucl. Fusion.21 1079.

[2] Lebedev S.V. et al., Ohmic H-mode studies on TUMAN-3, 1994 PPCF 36 B289

[3] Gohil P. et al., Investigations of H-Mode Plasmas Triggered Directly by Pellet Injection in the DIII-D Tokamak, 2001 PRL 86 644

[4] G.M. Staebler, Theory of internal and edge transport barriers, 1998 PPCF 40 569–580

[5] K. Miki, P.H. Diamond et al., Physics of Stimulated L→H Transitions, 2013 PRL 110, 195002

# Connecting Microscopic L-H Transition Physics to the Power Threshold*

L. Schmitz,[1] B.A. Grierson,[2] Z. Yan,[3] P. Gohil,[4] G.R. McKee,[3] L. Zeng,[1] T.L. Rhodes,[1] J.A. Boedo,[5] D. Eldon,[2] L. Bardoczi,[1] C. Chrystal,[6] J.R. Groebner,[4] K.H. Burrell,[4] E.J. Doyle,[1] G. Wang[1]

[1]*University of California-Los Angeles, Los Angeles, CA 90095-7799, USA*
[2]*Princeton Plasma Physics Laboratory, Princeton, NJ 08543-0451, USA*
[3]*University of Wisconsin-Madison, Madison, WI 53706, USA*
[4]*General Atomics, PO Box 85608, San Diego, CA 92186-5608, USA*
[5]*University of California San Diego, La Jolla, CA 92093, USA*
[6]*Oak Ridge Associated Universities, Oak Ridge, TN 37830, USA*

A physics-based model of the L-H transition power threshold is needed to confidently extrapolate auxiliary heating requirements for burning plasmas. Two important scalings that need a physics basis are the isotope scaling, and the non-monotonic scaling of the power threshold with plasma density. The goal of this work is to link differences in the microscopic turbulence-flow interaction to the resulting macroscopic power threshold density and isotope scaling.

Evidence from several recent experiments has pointed towards a synergistic role of turbulence-driven flows [Zonal Flows (ZFs)] and pressure-gradient-driven flows in the trigger and evolution of the L-H transition [1,3-7]. L-H transitions preceded by limit cycle oscillations (LCO) between the turbulence amplitude and the $\boldsymbol{E}\times\boldsymbol{B}$ flow have been obtained in DIII-D [8,9] and other devices near the L-H power threshold $P_{\mathrm{th}}$, across a large range of plasma density, current, and safety factor ($1\mathrm{x}10^{19}\ \mathrm{m}^{-3} \leq \langle \mathrm{n} \rangle \leq 5.5\mathrm{x}10^{19}\ \mathrm{m}^{-3}$, $0.6 \leq I_{\mathrm{p}} \leq 1.5$ MA, $3.8 \leq q_{95} \leq 8$). Detailed measurements of the $\boldsymbol{E}\times\boldsymbol{B}$ velocity and turbulence time evolution across the L-LCO-H-mode transition have been reported previously [8,9]. Here we present direct evidence that the $\boldsymbol{E}\times\boldsymbol{B}$ velocity modulation at the start of the LCO phase ($t_{\mathrm{LCO}}$) is dominated by the ion poloidal velocity [Fig.1(a)] that is driven via the perpendicular Reynolds stress [Fig.1(b)]. We use a phase-lock analysis technique to improve the signal-to-noise ratio of the main ion CER data and extract an estimate of the time-resolved poloidal ion flow velocity modulation. The $\boldsymbol{E}\times\boldsymbol{B}$ velocity is measured via Doppler Backscattering [10,11]. Fig.1(a) shows that the measured poloidal ion flow modulation accounts within error margins for the measured flow modulation (the offset is due to L-mode diamagnetic flow and toroidal flow related to NBI co-injection). No modulation of the ion pressure gradient $\nabla P_i$ is found initially during the LCO phase [1,9] or during the initial trigger phase of regular fast L-H transitions [8,12], but it has been shown previously that the pressure gradient dominates the $\boldsymbol{E}\times\boldsymbol{B}$ shear ultimately later in the LCO phase, locking in the H-mode transport barrier. The modulation of the toroidal flow velocity (not shown here) is a small fraction of the mean parallel velocity resulting from the co-NBI beam torque, and is out-of-phase with the poloidal velocity.

The poloidal ion flow acceleration via the Reynolds stress $<\tilde{v}_r\tilde{v}_\theta>$ can be expressed as $(1+2q^2)\partial< v_{i\theta} >/\partial t = -\partial< \tilde{v}_r\tilde{v}_\theta >/\partial r - \mu < v_{i\theta} >$ [13], where $\mu$ is the flow damping rate towards the neoclassical flow.

Fig. 1(b) shows a comparison of the measured flow acceleration with the Reynolds stress gradient evaluated via BES (Beam Emission Spectroscopy) velocimetry. The poloidal flow damping (in the plateau regime) is expected to be small here compared to the first two terms in the above equation, and is not evaluated. It is clear from Fig.1(b) that the Reynolds stress radial gradient is sufficiently large across the edge layer to account for the observed (dipolar) poloidal flow acceleration during the early LCO phase.

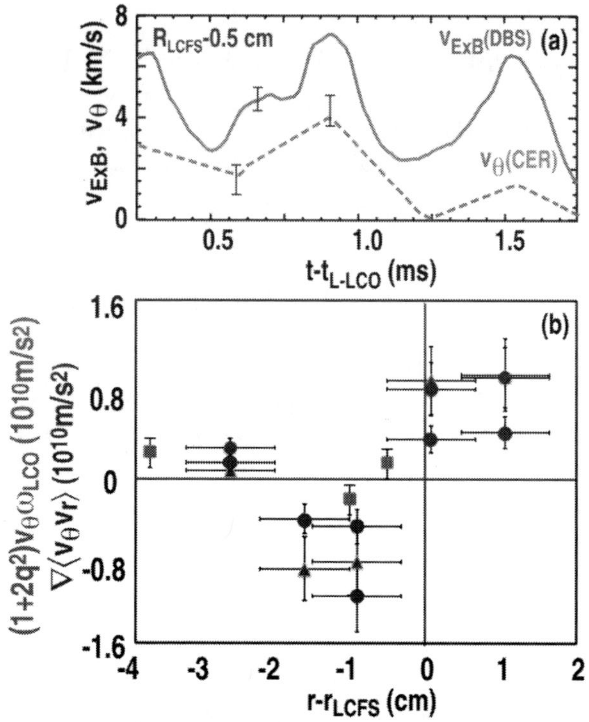


*Fig.1: (a) Time evolution ExB velocity and poloidal main ion velocity during the initial LCO cycles; (b) Measured poloidal flow acceleration compared to Reynolds stress radial gradient ($r_{LCFS}$ is the separatrix radius).*

The energy transfer from the ambient L-mode turbulence into directional meso-scale flows is an important quantity that couples the L-H trigger dynamics to the power threshold scaling. This energy transfer has been described by a rate equation for the (low frequency) flow energy $W$ (where $<\tilde{v}_\theta\tilde{v}_r>$ is the Reynolds stress and $v_\theta$ is the poloidal ion flow [14]):

$$\frac{\partial W}{\partial t} = <\tilde{v}_\theta\tilde{v}_r> \frac{\partial v_\theta}{\partial r} - \gamma_{ZF} W - \frac{\partial}{\partial r} <\tilde{v}_\theta\tilde{v}_r> v_\theta \qquad (1)$$

Neglecting the Zonal flow damping rate $\gamma_{ZF}$ and the fourth term describing radial energy spreading, the energy transfer rate is proportional to the (L-mode) seed flow shear $\partial v_\theta / \partial r$ and to the Reynolds stress. The measured maximum $\boldsymbol{E}\times\boldsymbol{B}$ and $\mathbf{v}\times\boldsymbol{B}$ flow shearing rates ~0.5 cm inside the LCFS (the latter quantity evaluated from $\omega_{\mathbf{E}\times\mathbf{B}}$ via subtracting the measured diamagnetic shear) are shown in Figure 2(a), for deuterium (D) and hydrogen (H) plasmas, in lower single null ITER-similar (ISS) shape. Both the (total) $\boldsymbol{E}\times\boldsymbol{B}$ shearing rate and the $\mathbf{v}\times\boldsymbol{B}$ flow shearing rate in equation (1) increase towards low and high plasma density, and are smaller near the power threshold minimum [Fig.2(c), $<n>_{min}$~2-3x$10^{19}$ $m^{-3}$]. The diamagnetic shear [shown in Fig. 2(b)] increases also at low density, and is larger in deuterium plasmas. Figure 2(c) shows the power threshold (loss power across the separatrix) for D and H "regular" fast L-H transitions in the ISS shape as well as LCO transitions (D) with a lower x-point height and correspondingly reduced power threshold [15].

From Fig. 2 we can conclude that (i) the L-mode $\boldsymbol{E}\times\boldsymbol{B}$ shear depends on density and peaks at low density; (ii) the L-mode seed flow shear is small near $<n>_{min}$ but is significant at low and high density, and therefore scales similarly with density as the L-H transition power threshold; (iii) the L-mode diamagnetic flow shear, and the seed flow shear just before the transition is higher in D than in H, particularly in the low density branch of the power threshold, and may facilitate the L-H transition in D as observed in the power threshold isotope scaling.

Fig.3(a) shows that the increased diamagnetic shear in the D-plasma is primarily due to a lower L-mode edge density gradient. Very little difference was found in the ion temperature gradient (not shown here). Fig. 3(b) shows that the $\boldsymbol{E}\times\boldsymbol{B}$ velocity inside the last closed flux surface has a less pronounced well structure in H-plasmas compared to D-plasmas, and that the velocity shear in the outer and in the inner flow shear layer is reduced. Fig. 3(c) shows that density fluctuations levels (at $k_\theta\rho_s$~0.5) are higher in the H-plasma, consistent with either higher instability growth rates or lower $\boldsymbol{E}\times\boldsymbol{B}$ shear (as observed here), However, larger scale density fluctuations measured by BES show the opposite behavior in H and D.

We have presented evidence that turbulence-driven poloidal ion flow is the dominant contribution to the $\boldsymbol{E}\times\boldsymbol{B}$ velocity modulation early during the LCO phase. As such it is responsible for the initial turbulence suppression during the LCO

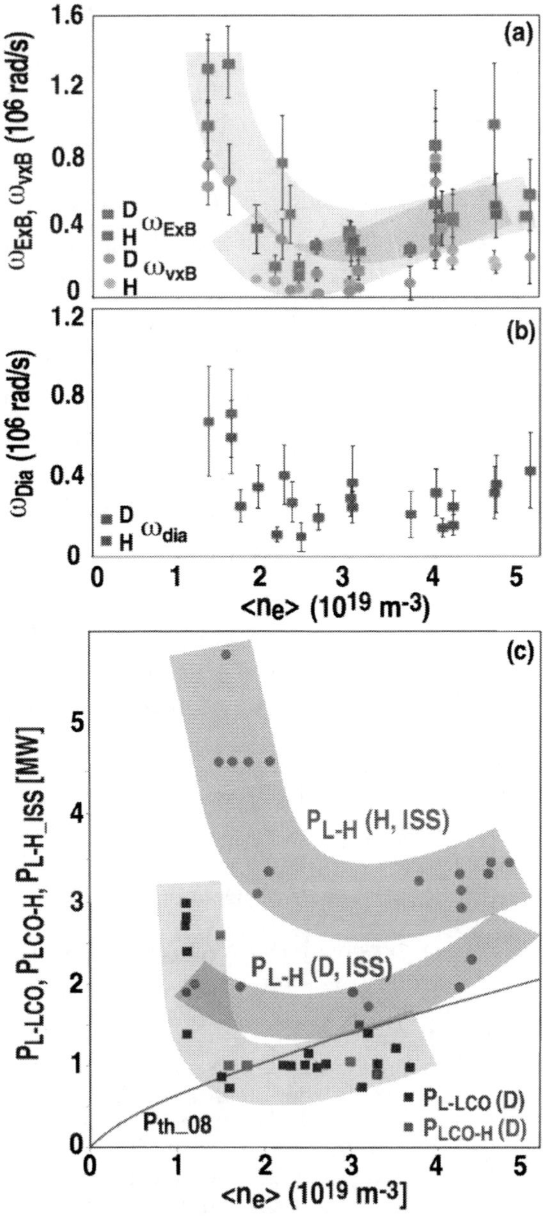


*Fig.2: (a) L-mode* $\boldsymbol{E}\times\boldsymbol{B}$ *flow shearing rate and* $\boldsymbol{v}\times\boldsymbol{B}$ *component vs density; (b) L-mode diamagnetic flow shear in D and H ISS plasmas; (c) measured power threshold for L-mode-LCO and LCO-H-mode transitions (D) and L-H transitions in ISS-shape (D and H) compared to the 2008 empirical (Martin) scaling* ($P_{th\_08}$ *[3]).*

phase, as demonstrated earlier [9]. It is also shown here that the L-mode seed flow shear increases in particular in the low density branch of the power threshold [Fig.2(a)], and that the L-mode ***E***×***B*** shear and diamagnetic shear increase with decreasing line averaged density below $n_{\mathrm{min}}$, the density where the power threshold is lowest, and depend on ion species [Fig.2(a,b)]. This result is in contrast to previous reports where no density dependence of the L-mode ***E***×***B*** flow shear was observed [16,17]. Accordingly, the turbulence-flow energy transfer rate may increase with the threshold power in the low- and high-density branches of the power threshold scaling. Further work is needed to clarify whether this effect contributes to, or explains, the non-monotonic dependence of the L-H transition power threshold on plasma density.

*This work was supported by the U.S. DOE under DE-FG03-01ER54615, DE-AC02-09CH11466, DE-FG02-89ER53296, DE-FG02-08ER54999, DE-FG02-07ER54917 and DE-FG02-08ER54984.

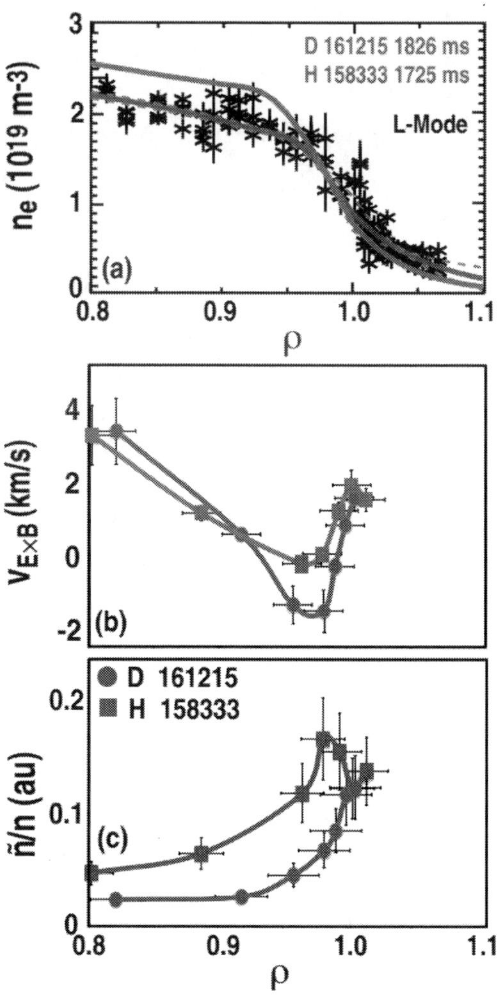


*Fig.3: (a) Plasma density, (b)* ***E***×***B*** *velocity profile, and (c) normalized DBS density fluctuation level in D- and H-plasmas (<n>=3x10$^{19}$ m$^{-3}$). Loss power past LCFS is within 1 MW above $P_{th}$.*

[1] L. Schmitz, L. Zeng, T.L. Rhodes, et al., Phys. Rev. Lett. **108**, 155002-5 (2012).
[2] K. Miki and P.H. Diamond, Phys. Plasmas **19**, 092306 (2012).
[3] G.R. McKee et al., Nucl. Fusion **49**, 115016 (2009).
[4] S.J. Zweben, R.J.Maqueda, R. Hager, et al., Phys. Plasmas **17**, 102502 (2010).
[5] G.D. Conway, C. Angioni, F. Ryter, et al., Phys. Rev. Lett. **106**, 065001 (2011).
[6] G.S. Xu et al., Phys. Rev. Lett. **107**, 125001 (2011).
[7] T. Estrada, C. Hidalgo, T. Happel, et al., Phys. Rev. Lett. **107**, 245004 (2011).
[8] L. Schmitz et al., Nucl. Fusion **54** 073012 (2014).
[9] L. Schmitz et al., Paper EX/11-4, 25th IAEA Fusion Energy Conference, October 13-18, 2014, St. Petersburg, Russian Federation
[10] W.A.Peebles, T.L. Rhodes, J.C. Hillesheim, et al., Rev. Sci. Instrum. **81**, 10D902 (2010).
[11] J.C. Hillesheim et al., Rev. Sci. Instrum. **81**, 10D907 (2010).
[12] I. Cziegler, G.R. Tynan, P.H. Diamond, et al., Plasma Phys. Control. Fusion **56** 075013 (2014).
[13] P.H. Diamond, and Y.-B. Kim, Phys. Fluids **B3**, 1626 (1991)
[14] G. Tynan, et al. Nucl. Fusion **53,** 073053 (2012).
[15] P. Gohil, et al., Nucl. Fusion **51**,103020 (2011).
[16] P. Sauter, C. Puetterich, F. Ryter, et al., Nucl. Fusion **52**, 012001 (2012).
[17] C. Maggi, E. Dalabie, T.M. Biewer, et al. Nucl. Fusion **54** 023007 (2014)

# Ion velocity distribution function in the plasma of its own gas

V. Soukhomlinov[1], A. Mustafaev[2], A. Grabovskiy[2] and M. Ainov[2]

[1]*Saint-Petersburg State University, Russia*

[2]*National Mineral Resources University (Mining University), Saint-Petersburg, Russia*

The paper proposed a new probe method for anisotropic ion velocity distribution function (IVDF) diagnostics. Potentialities of this method have been demonstrated under conditions, when on a mean free path an ion acquires velocity, comparable with average thermal velocity of atoms. Energy and angular dependencies of the first seven coefficients of IVDF expansion in Legendre series for $Hg^+$ in Hg, $He^+$ in He and $Ar^+$ in Ar have been measured, and the polar diagrams of the ion motion have been plotted. In order to verify the reliability and accuracy of the method the analytic solution of the kinetic Boltzmann equation for the ions in plasma of its own gas has been found. The conditions under which resonant charge exchange is a dominant process has been considered, whereas the electrical field magnitude is arbitrary. For the ambipolar field relation between the resonant charge exchange cross-section and the relative velocity has been taken into account. The results of theoretical and experimental data, with provision for the instrumental function of the probe method, correlate very accurately.

IVDF research is of special interest for modern applications: plasma technologies, ion surface treatment, technology of selective etching and creation of relief by ion bombardment, new generation of nanoelectronics (single-electron transistors, spintronics, etc). [1, 2].

In this context, experimental methods of IVDF measurements in different discharges, in particular, in DC self-sustained discharge plasmas are of special interest. The authors do not know papers, where the IVDF measurement data for such a discharge have been published. In reference to theoretical research, despite many existing papers it's difficult to select the ones where IVDF for the DC self-sustained discharge plasmas has been calculated.

The ion drift was theoretically considered in [3-6]. In [3] the drift velocity for ions of inert gases in plasma of its own gas in proximity of the strong field was analyzed, the distribution of atoms according to energies being given by delta-function. The results of drift velocity calculations for inert gas ions in its own gas were presented in [4], however, the author did not publish the analytical expression for IVDF.

Neglecting all processes, except charge exchange, in [5] a time interval of the ion movement in its own gas has been calculated, during which an ion has a velocity component along the electric field in the range of $v_{iz}$ and $v_{iz} + dv_{iz}$. These data have been calculated for interpretation of the experimental results of [7]. However it is difficult to obtain IVDF for the total velocity using these results.

The treatise considers [6] IVDF problem solution in its own gas neglecting the creation of ions if they have Maxwellian distribution by velocities as a result of a charge exchange. Thus, for the uniform cross-section of a charge exchange a Maxwellian type of IVDF has been obtained with the temperature, determined by an electrical field in plasma.

In [8, 9] a new method of calculation of matrix elements of collision integral is represented which is intended for solving the non-steady Boltzmann equation by a method of moments for ions in conditions, when the main process is resonant charge exchange. However, in strong fields, when thermal energy of an atom and ion, acquired on the mean free path, is less than 0.1, this approach is difficult for calculation of a steady IVDF.

In [10] an analytical solution of Boltzmann equation for ions in BGK-model [11] and a numerical solution for uniform cross-section of resonant charge exchange are compared. It should also be emphasized that the analytical solution gives incorrect asymptotics for the drift velocity of an ion in case of strong fields.

The paper [13] deals with analysis of IVDF in unbounded uniform plasma in the presence of steady (in time and space) electric fields and under condition that the main process that has an effect on the drift velocity of ions is resonant charge exchange. The problem is solved both numerically (by Monte-Carlo method) and analytically. When solving analytically a charge exchange cross-section is assumed to be independent on relative motion of colliding an ion and atom.

The paper [14] represents the problem of determining the IVDF in plasma in its own gas in the presence of strong E-field to have been analytically solved. Calculation results correspond well with the known experimental data for drift velocities in strong fields. It has been found, that IVDF differs significantly from Maxwellian and depends on two parameters. In addition, in [15] the authors state the IVDF for $Hg^+$ in $Hg$ vapors to have been measured for the first time. Experimental data correlate very accurately with calculation results.

In this work a new experimental probe method for full IVDF reconstruction in conditions of arbitrary electric field has been proposed. The main limiting factor for the method is a small thickness of Debye near-probe layer in comparison with the probe size.

In order to verify the reliability and accuracy of the developed method the analytical solution of Boltzmann kinetic equation for ions in plasma in its own gas has been found in case of resonant charge exchange being the dominant process. The E-field strength in plasma may be arbitrary and a charge exchange cross-section depends on relative velocities of an ion and atom.

Feasibilities of the developed method are demonstrated in this work. The energy and angular dependencies of seven Legendre coefficients have been measured and full IVDF for $Hg^+$ in Hg, $He^+$ in He and $Ar^+$ in Ar has been reconstructed for the first time (fig. 1a, b). A comparison of measured and calculated of Legendre coefficients and corresponding IVDF angular dependencies for $Hg^+$ в $Hg$, $He^+$ in $He$ and $Ar^+$ in $Ar$ has been made (Fig 1 and 2, respectively). The results of calculation and experimental data give good fit.

Figure 2a shows that when ion velocities are less than the average thermal velocity of atoms, IVDF is isotropic, despite the presence of a strong field. When ion velocities increase the distribution becomes more elongated in the direction of the electric field (fig. 2b), indicating its increasing anisotropy.

This work was supported by the Ministry of Education of the Russian Federation.

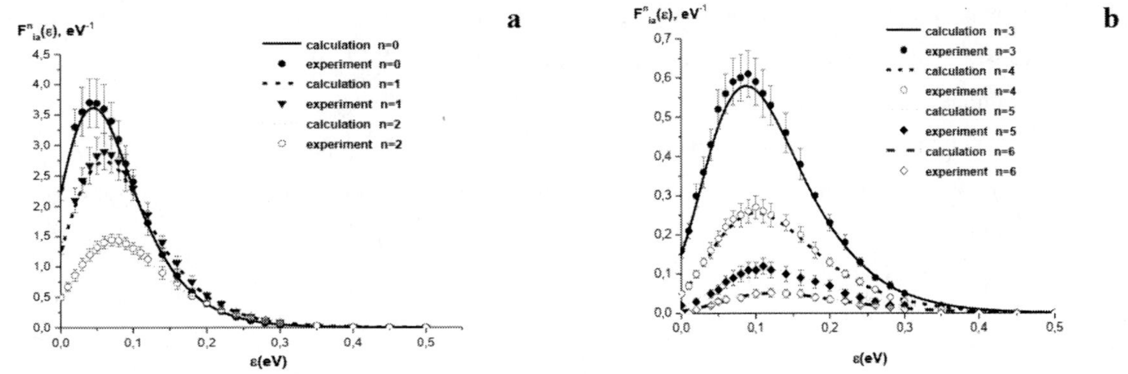


**Fig. 1a.** The energy dependencies of the first three expansion coefficients (n=1-3) of IVDF $Ar^+$ in $Ar$ by the Legendre polynomials for value of differentiating signal $\Delta\varepsilon = 0.05\ V$. **Fig. 1b.** The same conditions but n=3-6.

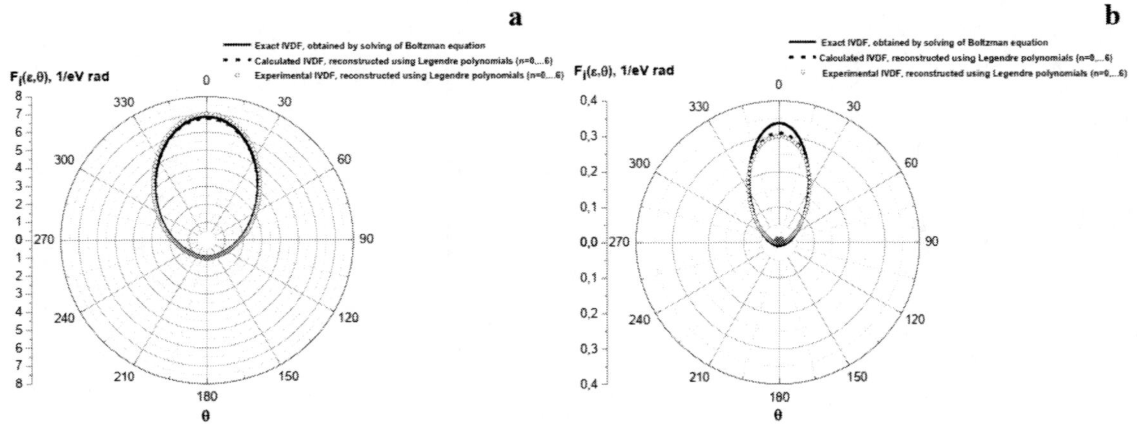


**Fig. 2a.** Angular dependency of the IVDF for the same conditions as indicated in fig. 1, a. Ions energy $\varepsilon = 0.1\ eV$. **Fig. 2b.** The same, but for ions with energy $\varepsilon = 0.3\ eV$.

## Reference list


1. H.Abe, M.Yoneda and N.Fujiwara. Developments of Plasma Etching Technology for Fabricating Semiconductor Devices // Jpn. J. Appl. Phys. 2008. Vol.47. P. 1435.
2. Michael A. Lieberman. Plasma Processing for Nanoelectronics – History and Prospects // Bull. of the APS. 2010. Vol. 55. N 7. P. 105.
3. Smirnov B.M. Mobility of ions in own gas // JTP. 1966. Vol. 36. № 10. P. 1864.
4. Perel' V.I. Calculation of drift velocity of ions in own gas // JTP. 1957. Vol. 32. P. 526.
5. Fok V.A. On motion of ions in plasma // JTP. 1948. Vol. 18. P. 1048.
6. Golant V.E., Zhilinskii A.P., Sakharov S.A. Plasma Physics basics. Moscow, Atomizdat. 1977. 150 p.
7. Frisch S.E., Kagan Y.M. Spectral study of ion motion in plasma // JTP. 1947. Vol. 17. P. 577.
8. A.Y. Ender, I.A. Ender // JTP. 2010. Vol. 80. № 2. P. 8.
9. A.Y. Ender, I.A. Ender // JTP. 2010. Vol. 80. № 2. P. 18.
10. D. Else, R. Kompaneets and S.V. Vladimirov // Physics of plasmas. 2009. Vol. 16. P 062106.
11. P.L. Bhatnagar, E.P. Gross and M. Krook // Phys. Rev. 1954. Vol. 94. P. 511.
12. M. Lampe, T.B. Röcker, G.Joyce, S.K. Zhdanov, A.V. Ivlev, and G.E. Morfill. Ion distribution function in a plasma with uniform electric field // Physics of Plasmas. 2012. Vol.19. P. 113703.
13. Mustafaev A.S., Soukhomlinov V.S., Ainov M.A. Experimental and theoretical study of strongly anisotropic IDF by velocities in own gas plasma for strong fields//JTP. 2014. (in print).
14. Lapshin V.F., Mustafaev A.S. Method of flat single-sided probe for diagnostics of anisotropic plasma // JTP. 1989. Vol. 59. P. 35.
15. Mustafaev A.S. Dynamics of E-beams in plasmas // JTP. 2001. Vol. 71. P. 111.

# Application limits of runaway electron modeling based on analytical formulas of generation and loss rates

G.I. Pokol[1], R. Lohner[1], G. Papp[2], B. Erdos[1] and the EU-IM Team*

[1] *Institute of Nuclear Techniques, Budapest University of Technology and Economics, Hungary*

[2] *Max-Planck/Princeton Center for Plasma Physics*

**Introduction** Runaway electrons pose one of the most critical problems of reactor-size tokamaks and a serious challenge for modeling at the same time. Description of runaway electrons is fundamentally a kinetic problem, however, self-consistent plasma simulations demand less computationally expensive models than the full kinetic description. This has brought to life an approach of modeling the evolution of the first moments of the runaway electron distribution function (namely density and current) by using analytical estimates of runaway generation and loss rates derived as approximative solutions of the kinetic problem. This allows for a self-consistent simulation of tokamak disruptions including energy balance affected by impurity penetration and electric field diffusion along with the dynamics of runaway electron current [1].

Recently a three-level modeling approach was adopted to runaway electron simulation within the European Integrated Modeling (EU-IM) framework [2]. The first level of modeling is limited to the indication if runaway electron generation is possible or likely. The second level adopts a similar approach as described above, using analytical formulas to estimate changes in the runaway electron density and current. The third level is foreseen to be based on the solution of the full electron kinetics using the LUKE code [3].

This paper analyses the limitations of the first two levels of modeling in view of recent theoretical and numerical results in runaway electron theory.

**Models for indication of runaway electrons** Within the EU-IM framework [2], a "Runaway Indicator" module ("*actor*" in a Kepler workflow) was developed to provide an indication for when to expect run-away tail formation. This *actor* is integrated into the "Instantaneous events" *composite actor* of the European Transport Simulator (ETS) [4] to provide warning messages of possible runaway electron generation during simulation-time. Another possibility is to do the check as post-processing of simulations in a separate workflow to provide confirmation that no runaway electrons were generated when their generation was unexpected.

The most obvious parameter to check is if the toroidal electric field exceeds the critical electric field ($E_c$) needed for the existence of runaway electrons by definition [5]:

*See http://www.euro-fusionscipub.org/eu-im.

$$E_c = n_e \frac{e^3 \ln \Lambda}{4\pi \varepsilon_0^2 m_0 c^2}, \tag{1}$$

which only depends on the electron density $n_e$ and constants - there is only a weak dependence on temperature through the Coulomb logarithm $\ln \Lambda$.

This criterion can be used as a conservative check, however we do not tend to see runaway electrons in experiments just above this threshold, which is mostly explained by the runaway electron generation rate being limited to very low values for near-critical electric field [6]. For this reason a second check was implemented to verify if the primary Dreicer generation rate is above a pre-defined threshold. A convenient choice for the formula for the Dreicer generation rate is the final (67) formula in the classical paper by Connor and Hastie [5], which is also used in the GO code [1]. However, one should be aware that this formula is the non-relativistic approximation of formula (66) of the same paper, which is in turn a high electric field approximation of formula (63). It is evident from Figure 1 that there is a region of the parameter space where these formulas give significantly different results. This high temperature ($T > 500$ eV and low electric field ($E/E_c < 10$) region is not particularly relevant in disruptions, that is the typical application of GO [1], however, it is just the region that might occur in the present ETS simulations [4]. In this case, the applied formula (67) is a conservative approximation for distribution functions close to Maxwellian.

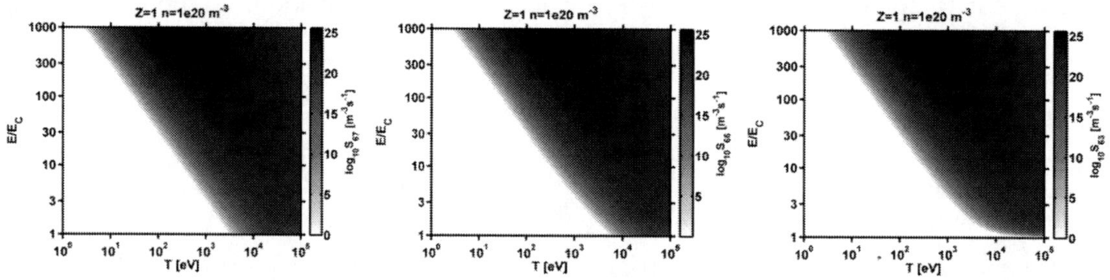


Figure 1: *Comparison of Dreicer-generation growth rates for the different approximations of formulas (67), (66) and (63) of the classical paper by Connor and Hastie [5].*

It was also shown that the critical electric field of equation 1 and the Dreicer growth rate can be significantly modified by the momentum loss due to synchrotron emission [6, 7]. Taking this into account is not necessary for the runaway electron warning, but would be desired for the actual modeling of runaway electron current.

**Runaway electron current estimation by analytical formulas** A second *actor* was developed to provide an estimation of the non-inductive current due to runaway electrons using computationally cheap analytical estimates of runaway electron growth rates and transport. This *actor* is to be integrated into the "Heating & Current Drive Workflow" which will allow simple benchmarking with more advanced kinetic models.

In order to provide a reasonable estimate for the current carried by runaway electrons, a minimum of three mechanisms are to be considered: Dreicer generation, avalanche generation and runaway electron radial transport. Dreicer generation has been discussed in the previous section, but here the high T necessitates full relativistic description that is formula (63) of Connor and Hastie [5]. This agrees well with numerical calculations by Kulsrud [8] for moderately high electric field ($E/E_c \approx 100$), but at low electric field the correction due to synchrotron radiation [6, 7] will need to be handled. This could be achieved by applying the correction factor to the growth rate as given in Figure 3 of the paper by Stahl et.al. [6]. At low temperature, where the plasma is only partially ionized, we encounter a different problem as the stopping power becomes affected by the bounded electrons. For this effect no simple corrections exist at the moment [9], however, currently this is outside the typical application range of EU-IM workflows. For low aspect ratio tokamaks a correction factor for the effect of toroidicity is to be applied as suggested by Nilsson et.al. [10, 11].

The classical way of describing the avalanche was put forward by Rosenbluth and Putvinski [12]. Their formula for the growth rate is, however, only valid for high electric fields, when the momentum loss of runaway electrons in knock-on collisions is negligible compared to the strong momentum input from the electric field. At low electric field a momentum-conserving approach to the knock-on collisions is to be applied [7]. Figure 2 (based on Figure 4 of [7]) shows the comparison of the avalanche growth rate at low electric field calculated by the Rosenbluth and Putvinski model [12], and the model put forward by Aleynikov and Breizmann [7]. The new estimate of the avalanche growth rate features a threshold ($E_a$), and then quickly converges to the classical expression for higher electric field values. Here we propose an approximation of estimating the threshold electric field by formula (8) by the paper by Aleynikov et.al. [7], and using the Rosenbluth and Putvinski model for higher electric field values.

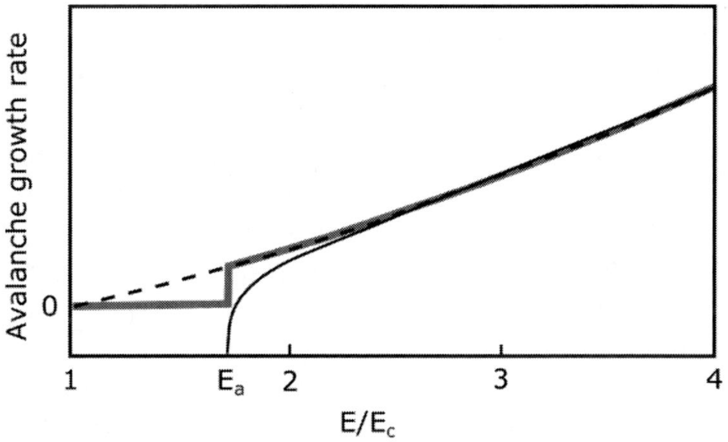


Figure 2: *Comparison of avalanche growth rate by Rosenbluth and Putvinski (dashed black line) [12] and the reduced kinetic model by Aleynikov and Breizmann (black line) as presented in [7]. Here proposed simple model is overplotted by red.*

Recent studies with LUKE showed that the avalanche growth rate can be significantly reduced at toroidal magnetic surfaces with high mirror ratio due to the trapping of the high energy electrons generated in the knock-on collisions. Formula (A.4) of the paper by Nilsson et.al. [11] (also published in [10] could be applied to have a flux-surface dependent correction factor.

Regarding the radial transport of runaway electrons the classical treatment is to use the diffusion approximation; either collisional or as derived by Rechester and Rosenbluth [13]. Recently the shortcomings of this approach were shown by a comparison with a result obtained by following the orbits of a large set of runaway electrons in a perturbed magnetic field [14]. A simple and correct treatment is still to be proposed. Meanwhile, the diffusive transport model is to be used, as it can reproduce one key aspect and that is that the radial transport is the dominating loss mechanism.

**Conclusions** In the present paper a modeling approach of runaway electron was put forward based on analytical kinetic formulas, which would be extremely useful in EU-IM [2]. Since at present the ETS does not address disruption conditions, a number of correction factors in the runaway model are needed, to account for high temperature and low electric field, as described in this paper. Following the implementation of the proposed model in the EU-IM framework, a benchmark within it is foreseen to full kinetic calculations, like LUKE [3].

**Acknowledgements** The authors are grateful to A. Stahl and P. Aleynikov for fruitful discussions. Authors at BME NTI acknowledge the support of the Hungarian State grant NTP-TDK-14-0022. This work has been carried out within the framework of the EUROfusion Consortium and has received funding from the Euratom research and training programme 2014-2018 under grant agreement No 633053. The views and opinions expressed herein do not necessarily reflect those of the European Commission.

## References

[1] G. Papp *et al.*, Nuclear Fusion **53** 123017 (2014)
[2] G.L. Falchetto, et. al., Nuclear Fusion **54**, 043018 (2014)
[3] Y. Peysson and J. Decker, Fus. Sci.& Tech. **65**, 22 (2014)
[4] D. Kalupin et al., Nucl. Fusion **53**, 123007 (2013)
[5] J.W. Connor and R.J. Hastie, Nucl. Fusion **15**, 415 (1975)
[6] A. Stahl, et.al., Phys. Rev. Lett. **114**, 115002 (2015)
[7] P. Aleynikov and B.N. Breizmann, Phys. Rev. Lett. **114**, 155001 (2015)
[8] R.M. Kulsrud, et.al., Phys. Rev. Lett. **31**, 690 (1973)
[9] P. Aleynikov, et.al., Proceedings of IAEA FEC 2014, TH/P3-38 (2014)
[10] E. Nilsson, et.al., 41st EPS Conference on Plasma physics, 2.303 (2014)
[11] E. Nilsson, et.al., Plasma Phys. Contr. Fusion *submitted for publication* (2015) arXiv:1504.00175
[12] M.N. Rosenbluth and S.V. Putvinski, Nucl. Fusion **37**, 1355 (1997)
[13] A.B. Rechester, M.N. Rosenbluth, Phys. Rev. Lett. **40**, 38 (1978)
[14] G. Papp *et.al.*, J. Plasma Phys. *accepted for publication* (2015) arXiv:1504.05074

# Dynamic equilibria and MHD instabilities in toroidal plasmas with non-uniform transport coefficients

Shimpei Futatani[1,2], Jorge A. Morales[3], Wouter J. T. Bos[1]

[1] *LMFA-CNRS UMR 5509, Ecole Centrale de Lyon, Université de Lyon, 69134 Ecully, France*

[2] *Department of Computer Applications in Science and Engineering (CASE), Barcelona Supercomputing Center, Barcelona, Spain*

[3] *CEA, IRFM, F-13108 Saint-Paul-Lez-Durance, France*

## Introduction

Plasmas confined by toroidal magnetic and electric fields are generally considered the best candidates to achieve sustainable nuclear fusion. Tokamaks and Reversed Field Pinches (RFPs) are two of such configurations which are currently investigated intensively. In the ideal case the plasmas in such reactors would remain quietly confined within the magnetic field in order to allow their core to reach the temperature needed for thermonuclear fusion.

To study fusion plasmas, magnetohydrodynamics (MHD) has been widely applied as an accessible description considering a force-free static equilibrium. Such an equilibrium only exists if the pressure forces are balanced by the Lorentz-forces that originate from the imposed magnetic and electric fields. In straight cylinder configurations such force-free states can be easily defined, considering for instance z-pinch and $\theta$-pinch devices [1]. Even though such geometries can be subject to MHD instabilities, a force-free state can be defined. This changes in toroidal geometry.

In the simplest case in which the toroidal electric field is generated by a central solenoid, without external current drive, and where the toroidal magnetic field is induced by the poloidally orientated coils, ignoring ripples and other details, the imposed electro-magnetic fields have a simple form. It was shown in previous studies [2, 3, 4, 5] that for this case, assuming uniform electric resistivity, such an equilibrium is not possible. These studies showed, following an increasing level of complexity, that the velocity can never be zero if the current density is linked to the electric field by Ohm's law. The present work builds upon the results of these studies, increasing by one step the complexity, considering spatially non-uniform electric resistivity and viscosity profiles. Indeed, in practice, strong pressure, density and temperature gradients will influence the local values of the viscosity and resistivity in the plasma. The present approach takes this into account in the coarsest way, by defining profiles as a function of the minor radius. In particular we will show how different types of dynamic equilibrium and MHD instabilities appear as a function of the strength and the spatial distribution of the transport coefficients.

## MHD equations

In the magnetohydrodynamics approximation, plasma is modeled as a charge-neutral electromagnetic conducting fluid. In the incompressible description, the dynamics are governed by

$$\rho \frac{D\boldsymbol{u}}{Dt} = -\nabla p + \nabla \cdot \boldsymbol{\sigma} + \boldsymbol{j} \times \boldsymbol{B}, \qquad \frac{\partial \boldsymbol{B}}{\partial t} = -\nabla \times \boldsymbol{E}, \tag{1}$$

$$\boldsymbol{E} = \eta \boldsymbol{j} - \boldsymbol{u} \times \boldsymbol{B}, \qquad \nabla \cdot \boldsymbol{u} = 0, \qquad \nabla \cdot \boldsymbol{B} = 0, \tag{2}$$

with the current density $\boldsymbol{j} = \nabla \times \boldsymbol{B}$ and the pressure $p$. These equations consist of dimensionless values, using the toroidal Alfven speed $C_A = B_0\sqrt{\rho\mu_0}$ as a reference velocity, with $\rho$ the mass density and $\mu_0$ the magnetic permeability constant. The stress tensor $\sigma_{ij}$ is given by

$$\sigma_{ij} = \mu \left( \frac{\partial u_i}{\partial x_j} + \frac{\partial u_j}{\partial x_i} \right). \tag{3}$$

In the case of a spatially uniform viscosity and density and resisitivity, equation (1) is simplified to

$$\frac{D\boldsymbol{u}}{Dt} = -\frac{1}{\rho}\nabla p + \nu \Delta \boldsymbol{u} + \boldsymbol{j} \times \boldsymbol{B}, \qquad \frac{D\boldsymbol{B}}{Dt} = \boldsymbol{B} \cdot \nabla \boldsymbol{u} + \eta \Delta \boldsymbol{B}. \tag{4}$$

The viscosity and resistivity profiles we consider in the present work mimic the profiles induced by the strong drop of temperature newar the edge of experimental fusion reactors and are given by $\eta(r) = \eta_0 f(r)$ and $\nu(r) = \nu_0 f(r)$ with $f(r) = \left(1 + (r/a)^8\right)^{5/4}$ where $r/a$ is the normalized minor radius. In the following both uniform and non-uniform transport coefficients will be considered:

I. Constant magnetic-resistivity $\eta_0$ and kinetic-viscosity $\nu_0$,

II. Space dependent magnetic-resistivity $\eta(r)$ and kinetic-viscosity $\nu(r)$.

The profile of the magnetic-resistivity induces the profile of the current density through Ohm's law (Eq.3) if $\boldsymbol{u} = 0$. We investigate the influence of the profile of the magnetic resistivity $\eta$ on the dynamics, for a given fixed toroidal current, i.e. $\int j_{\eta(r)} dV = \int j_{\eta_0} dV$.

## Simulation results

The calculations are performed for several toroidal magnetic field values $B_{tor}$, ranging from 0.6 to 0.05 for Case I (uniform profile of coefficients) while keeping the poloidal current constant. The other, Case II, the toroidal magnetic field is ranging from 1.2 to 0.1. If the toroidal magnetic field $B_{tor}$ is large enough, corresponding to the tokamak regime ($B_{tor} = 0.6$ for Case I and $B_{tor} = 1.2$ for Case II), the evolution of the magnetic field fluctuations is small and the

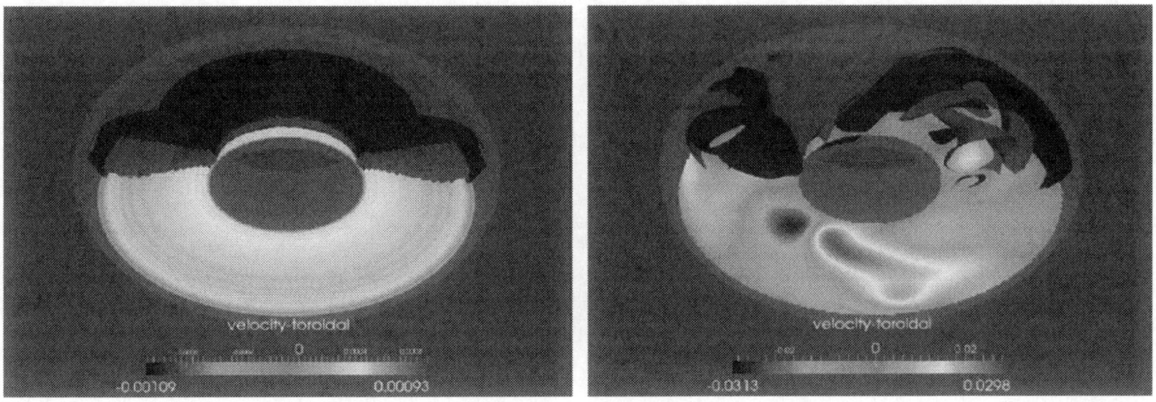


Figure 1: Isosurfaces for a value of 20% of the maximum toroidal velocity for case I,R (left) and Case I,T (right) (red contours show positive velocity and blue contours negative) at steady state for Lundquist number $M = 2000$. Also, a cross-section at a fixed value of the vertical coordinate is shown, quantifying the toroidal velocity in the plane [6].

safety factor $q$ does not change much either. Concerning the shape of the q-profile in non-uniform resistivity case (Case II), it is more parabolic than in the uniform resistivity case (Case I), which is more realistic behavior for fusion plasmas [6]. By decreasing the toroidal magnetic field $B_{tor}$, to $B_{tor} = 0.05$ for Case I and to $B_{tor} = 0.1$ for Case II, MHD instabilities generate an important poloidal flow, and the dynamo process increases the toroidal magnetic field $B_{tor}$ at the core of the plasma. As a consequence, the plasma at the steady state shows the dynamics characteristic of RFP devices as shown in Fig. 1. Here, $M$ is the Lundquiest number which is the dimensionless ratio of an Alfven wave crossing timescale to a resistive diffusion timescale. For both cases, the highest pinch-ratio case, where $B_{tor} = 0.05$ (Case I), $B_{tor} = 0.1$ (Case II), will be called the R-Regime (for RFP) and the opposite, $B_{tor} = 0.6$ (Case I), $B_{tor} = 1.2$ (Case II) the T-Regime (for Tokamak).

Figure 2 shows the visualizations of the velocity magnitude of Case I and II in the RFP-regime. The velocity gradients are more important in Case II. This is related to the fact that the comparison of the two cases is done for a given equal current. In the case of a non-uniform resistivity profile, the current is more concentrated in the center of the domain, so that the local pinch ratio will be higher. This generates a wider variety of unstable modes, as is reflected in the visualizations in Fig. 2.

## Conclusions

The work presents a comprehensive numerical study of the dynamics of incompressible, visco-resistive MHD in a toroidal geometry. Both uniform and spatially variable viscosity and

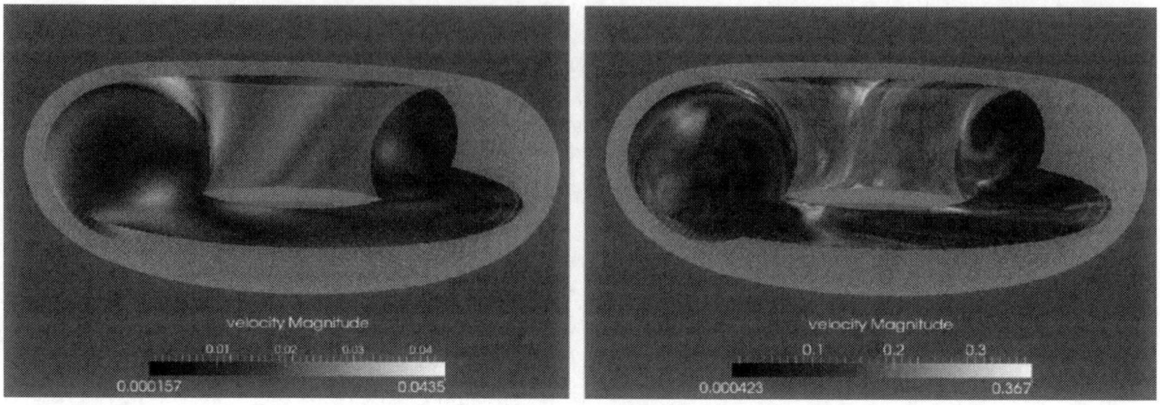


Figure 2: Visualizations of velocity fluctuations at the steady state of the Lundquist number 2000 ($M$=2000) simulations of the RFP-regime, $B_{tor}$ = 0.05 for Case I (left), and $B_{tor}$ = 0.1 for Case II (right) [6].

resistivity are considered. The resulting dynamics included dynamic equilibria for the quiescent, tokamak like regime, and the chaotic RFP-like regime.

In the present work it is shown how a spatial inhomogeneity of the viscosity and resisitivity coefficients influences a toroidal equilibrium. Both parameters in a stable, tokamak-like regime and an unstable, RFP-like regime are considered. For a given total current, the velocities and magnetic fluctuations are more important by over an order of magnitude for the case where the transport coefficients are varying as a function of radius. Furthermore, the reversal parameter drops more importantly as a fuction of the pinch-ratio for the case of space-dependent coefficients. We have studied the parametric dependence of the Lundquist number ($M$) on the toroidal velocity and the poloidal velocity in detail in detail in [6].

**References**

[1] J. Freidberg, Rev. Mod. Phys. 54, 801 (1982).

[2] D. Montgomery and X. Shan, Comments on Plasma Physics and Controlled Fusion 15, 315 (1994)

[3] J. Bates and D. Montgomery, Phys. Plasmas 5, 2649 (1998).

[4] L.P. Kamp and D.C. Montgomery, Journal of Plasma Physics 70, 113 (2004)

[5] Jorge A. Morales, Wouter J. T. Bos, Kai Schneider, and David C. Montgomery, Phys. Rev. Lett. 109, 175002 (2012).

[6] Shimpei Futatani, Jorge A. Morales, Wouter J. T. Bos, Phys Plasmas 22, 052503 (2015).

# Resonances of Energetic Particle Instabilities During Frequency Chirping*

D.C. Pace[1], W.W. Heidbrink[2], M.A. Van Zeeland[1]

[1] *General Atomics, P.O. Box 85608, San Diego, CA 92186-5608, USA*

[2] *University of California Irvine, University Dr., Irvine, CA 92697, USA*

Convective energetic ion transport results from Alfvén eigenmodes (AEs) and energetic particle modes (EPMs) due to resonances between these modes and the orbit topology of the ions. Experiments at the DIII-D tokamak [1] show that ion/EPM resonance space can be selectively probed as limited resonances produce energetic ion losses that are coherent at the mode frequency. A particular set of modes are the focus of the present work: these modes exhibit fast frequency chirping and represent a highly nonlinear interaction with beam ions. This phenomena is especially sensitive to details of the energetic ion phase space, therefore making it an incredibly useful scenario for validating models of energetic ion transport due to MHD. A simplified analysis readily identifies these resonances using experimental quantities, and these results are applied to the development of future experiments.

Neutral beam injection in the DIII-D tokamak ($R = 1.7\,\text{m}$, $a = 0.6\,\text{m}$) occasionally results ($\approx 3\%$ of discharges) in the excitation of energetic ion modes characterized by a fast downward frequency chirp as shown in the density spectrogram of Fig. 1. These modes evolve in frequency much faster than the equilibrium (i.e., much faster than the more standard frequency evolution of a reversed shear Alfvén eigenmode), often chirping through $\Delta f = 20$ kHz in approximately 1 ms.

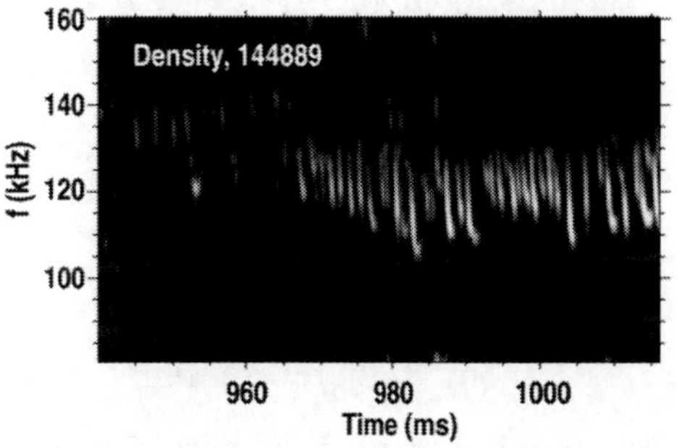


Figure 1: *Cross-power spectrum from two channels of the interferometer diagnostic indicating density fluctuations that exhibit frequency chirping.*

Resonances between AEs and energetic ions (beam ions in these DIII-D experiments) are identified when the following resonance condition is met,

$$\omega + (m+l)\omega_\theta - n\omega_\zeta = 0 \qquad (1)$$

where $\omega$ is the mode frequency, $m$ is the poloidal mode number, $l$ is an integer quantum number, $\omega_\theta$ is the ion poloidal transit frequency, $n$ is the toroidal mode number, and $\omega_\zeta$ is the ion toroidal transit frequency. The ion orbit frequencies are calculated by a constants of motion code that calculates gyrocenter trajectories [2]. The code identifies values of $p = m + l$ that satisfy the resonance condition through $p = (\omega - n\omega_\zeta)/\omega_\theta$. In this application, the value of the polooidal mode number is set as $m = nq_{\text{mode}}$, where $q_{\text{mode}}$ is the value of $q$ at the mode location. The actual functional dependence, $m = m(n, q_{\text{mode}})$, depends on the mode frequency, e.g., for a TAE with $\omega = v_A/(2qR)$ we have $m = nq_{\text{mode}} - 1/2$. The approximation $m = nq_{\text{mode}}$ is sufficient because a range of $p$-values are searched for resonances. Even if the value of $m$ is incorrect for a particular mode, the resonances will still be identified based on the wide range of $p$ that is searched (a range determined by the user-provided range of $l$-values). This allows for an efficient and relatively quick calculation of the resonance space that can help to explain experimental observations.

The efficiency of this analysis is the motivation for applying it to experiments. All of the quantities used in this consideration are readily measured in a tokamak experiment. Codes that identify resonances based on solving for complete mode eigenfunctions (e.g., HAGIS [3]), require considerably more input effort and computational resources (they will provide a more complete physics output, however). In balance, the simplified techniques discussed here provide immediate guidance concerning the ion-wave interactions, and ways to modify them, while not precluding the application of more comprehensive physics models in future analysis.

An example of the beam ion - AE resonance phase space is shown in Fig. 2 for a chirping observation. The color contours represent the initial energetic ion distribution resulting from beam deposition. Beam ion velocities for injection at $E_{\text{beam}} = 81$ keV are sub-Alfvénic for this plasma with $B_T = 1.9$ T and $\bar{n}_e = 2.1 \times 10^{19}$ m$^{-3}$, therefore the beam ion distribution is shown only for the full energy (the interaction between beam ions and AEs is largest for the ions with the greatest velocity). The phase space is plotted as a function of toroidal canonical momentum ($P_\phi$) normalized to the poloidal magnetic flux at the separatrix ($\Psi_{\text{wall}}$), and the magnetic moment ($\mu$) normalized to the central value ($E/B$). Concentrations of beam ion density result from the individual beam geometries. Overlaid on the beam ion distribution is a set of AE resonances calculated based on the measured properties of one of the chirping modes. The central frequency of the chirp range, 70 kHz, is used along with the toroidal mode number of $n = 3$. Coherent beam ion losses at 70 kHz are measured by the fast ion loss detector (FILD) [4] in this plasma, and the location of that obser-

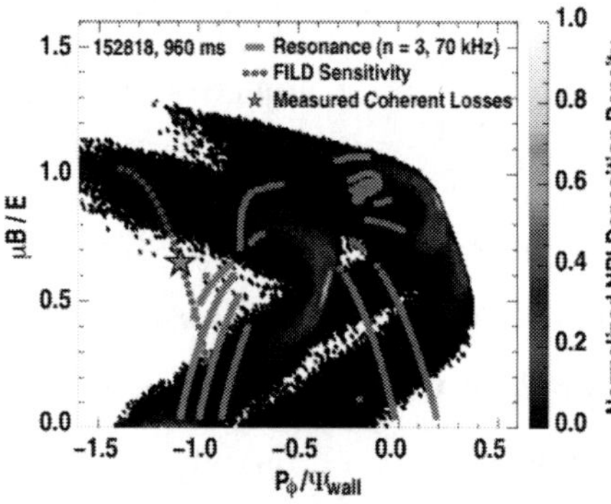


Figure 2: *Beam ion deposition (color contours) along with resonances for the observed $n = 3$, 70 kHz Alfvénic chirping mode. The sensitivity (at $E_{ion} = 81$ keV) of the FILD is indicated by the dotted line, and the $\star$-symbol marks the observed coherent beam ion losses (i.e., occurring at 70 kHz).*

vation is indicated by the $\star$-symbol along the dotted line representing the entire sensitivity of the FILD system. Since the FILD observations also involve ions at $E = 81$ keV, this analysis effectively maps the phase space transport of beam ions: the ions originate from the resonance-deposition overlap areas that are closest to the FILD measurement marker. A second FILD [5] does not measure any coherent losses from the chirping mode. This lack of observed losses is sensible considering that the sensitivity of the second FILD exists at values of $P_\phi/\Psi_{\text{wall}}$ that are more negative than the detector shown in Fig. 2, and therefore it exists too far away from the resonances (in phase space) to detect the interaction.

Considering the full energy beam ions as the primary source of interaction with the Alfvénic modes also allows for an assessment of the relative strength of the resonances. This is described by Fig. 3. Figure 3(a) shows a beam ion deposition distribution and resonance curves from a shot in which chirping modes were observed with $n = 1$ at a central frequency of 54.4 kHz. While a complete assessment of this situation requires acknowledging the actual beam ion slowing down distribution that occurs over time, it is reasonable to consider the full beam energy component as the complete energetic ion distribution for time periods shortly following the initial beam injection. The resonances from this shot are plotted in Fig. 3(b) according to the total number of beam ions with which they overlap.

The source of free energy for driving the modes is actually caused by gradients in phase space, and those traces are plotted in Fig. 3(c). The strongest resonant interaction is the one that passes nearly perfectly through the beam deposition curves resulting from the combined injection of tangential and perpendicular beam lines. This serves as an example of an area in which a more comprehensive code is ultimately needed; the energy transfer between the AE and ion depends on $\int_{\text{orbit}} \vec{v} \cdot \vec{E}$, where $\vec{v}$ is the ion velocity, $\vec{E}$ is the electric field of the AE, and the integral is taken over the trajectory of the ion [6]. The beam ion density gradient analysis of Fig. 3(c) does not account for the $\vec{v} \cdot \vec{E}$ term over the actual trajectory of the ion, therefore, this is not an accurate measure of the energy exchange.

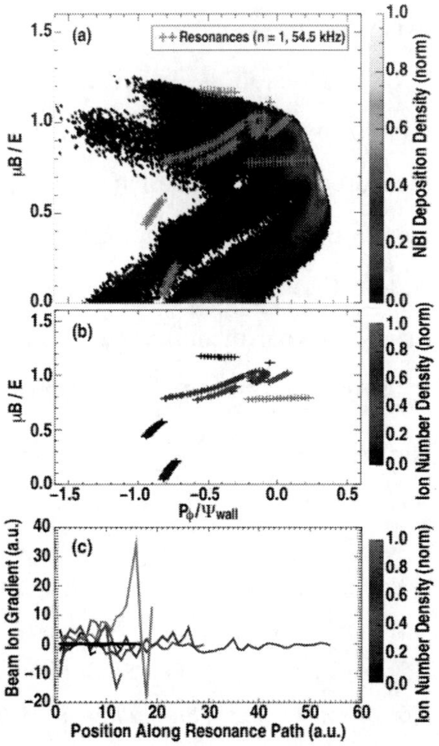


Figure 3: *Shot 152818: (a) Beam ion deposition (color contour) along with resonances from a chirping Alfvén eigenmode (+-symbols). (b) Resonances paths from panel (a) plotted according to the number of beam ions with which they overlap. (c) Beam ion phase space gradient along the resonance paths.*

Considering its range of applicability, a host of experimental ideas follow from this type of phase space analysis. The beam ion gradients corresponding to strongest resonant interaction identified in Fig. 3(c) can be manipulated through appropriate changes in plasma or beam parameters. For example, replacing the perpendicular beams with all tangential beams would allow for the same plasma parameters (i.e., the auxiliary heating power can be matched), while changing (likely reducing) the beam ion gradients in that area of phase space. Developing experiments that systematically probe beam ion distributions in this way is important for the mode validation effort since the applicability and confidence of such models, as they are applied to predictions for future fusion devices, is determined based on their ability to reproduce existing experiments.

In conclusion, a simplified analysis of resonances between Alfvén eigenmodes and beam ions is applied to DIII-D experiments featuring fast frequency chirping modes. Combining the measured AE characteristics with beam ion orbit frequencies allows for an efficient method of generating a toroidal momentum - magnetic moment phase space that readily identifies interactions that likely lead to convective beam ion loss. This information is used to identify adjustments in experimental parameters (e.g., beam duty cycles) that provide for improved systematic investigation of beam ion transport due to AEs and EPMs.

## References

[1] J.L. Luxon, *Nucl. Fusion* **42**, 614 (2002)

[2] M.A. Van Zeeland, et al., *Phys. Plasmas* **18**, 056114 (2011)

[3] S.D. Pinches, et al., *Comp. Phys. Comm.* **111**, 133 (1998)

[4] X. Chen, et al., *Rev. Sci. Instrum.* **83**, 10D707 (2012)

[5] R.K. Fisher, et al., *Rev. Sci. Instrum.* **81**, 10D307 (2010)

[6] W.W. Heidbrink, *Phys. Plasmas* **15**, 055501 (2008)

*This work was supported by the US Department of Energy under DE-FC02-04ER54698 and SC-G903402.

# Helium fuelling for edge plasma parameters control in the thermonuclear experiments

S.I. Lashkul, A.B. Altukhov, A.D. Gurchenko, E.Z. Gusakov, V.V. Dyachenko, L.A. Esipov, M.Yu. Kantor, D.V. Kouprienko

*Ioffe Institute, Politekhnicheskaya 26, St. Petersburg, Russia*

In deuterium–tritium plasmas of ITER helium produced in fusion reactions could be applicable as diagnostic test instrument for plasma parameters control in the divertor SOL. Also the He additionally injected fraction of plasma could be useful for monitoring of the ITER full-W divertor in stationary or slow transient cycles during non-active H/He phases of DT operations [1, 2].

**I.** This paper describes FT-2 ($R = 0.55$ m, $a = 0.08$ m, $B_T \leq 3$T, $I_{pl} = 19 \div 40$ kA) tokamak experiments where additional helium puffing into the hydrogen/deuterium plasma is considered as diagnostic tool [3]. The feature of sensitivity of the helium line emission to electron temperature and density can be applied as effective diagnostics for these plasma parameters.

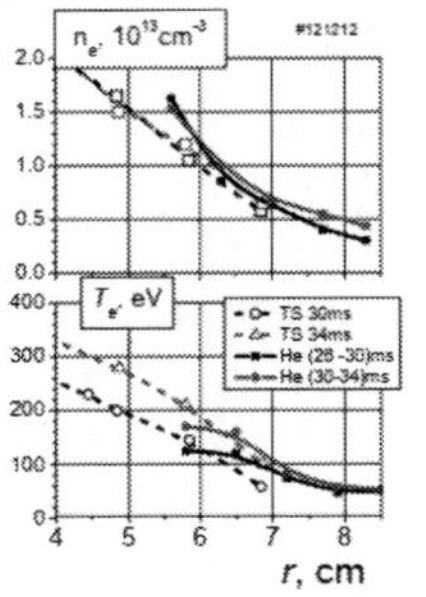


*Fig. 1 Comparing of $n_e(r)$ and $T_e(r)$ measured by spectroscopic and TS diagnostics methods in CRU experiment.*

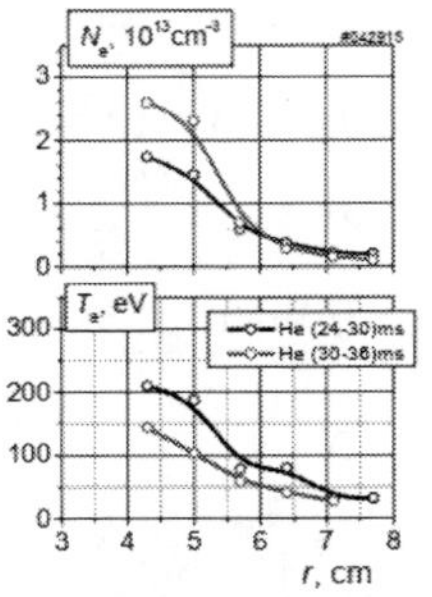


*Fig. 2 $n_e(r)$ and $T_e(r)$ measured by He lines ratio method in LHCD experiment*

The region of pulsed gas puffing was observed through an upper port of tokamak vessel. The AvaSpec double-channel absolutely calibrated spectrometer (553 – 765 nm and 410 – 495 nm) was used for simultaneous measurements of the chosen spectral lines in the cloud of the injected gas localized near the bottom of the diagnostic cross section. The plasma cross section was scanned in the horizontal plane from the high field side (HFS) of the torus. Using series of well reproducible discharges, the radiation profiles were measured from shot to shot with a spatial resolution of 0.5 cm. Variations of $n_e(r)$ and $T_e(r)$ at the plasma edge were monitored by the method based on the proportionality of the ratio between the singlet helium lines $R_1(n,T_e)$ = HeI(668nm)/HeI(728nm) ($1s3d^1D$ - $1s2p^1P/1s3s^1S$ - $1s2p^1P$) and the singlet triplet lines $R_2(T_e,n_e)$ = HeI(728nm)/HeI(706nm) ($1s3s^1S$-$1s2p^1P/1s3s^3S$-$1s2p^3P$), correspondingly. The factors $R_1$, $R_2$ interconnected the intensity ratios of the above lines to the plasma parameters and Calculated by ADAS code are given in different papers [3 and 4]. One has to notice that application of this spectroscopy method for $T_e(r)$ and $n_e(r)$ measurements involves specific difficulties [4]. Some of them are the effects of the radiation trapping of the resonance line ($I_{501}$) [5] and dependence of intensities of triplet lines ($I_{706}$) from the fraction of metastable atoms, which varies as helium penetrates into the plasma. That is why it is necessary to take into account the percentage of atoms in the $2^3S$ state for experimental correction of $R_2$.

For experimental verification of the calculated factors $R_1$ and $R_2$ the pure helium ohmic heating (OH) plasma experiment with passive spectroscopy was performed. The relationship between the line intensity ratios and plasma parameters is resulted by $n_e(r)$ and $T_e(r)$ progressive approximation approach. Characteristic features of that calibration method are based on comparing of spectroscopic data versus $T_e(r)$ and $n_e(r)$ measurements, provided by multi-pass intracavity Thomson scattering (TS) [6] and microwave interferometer. Results of spectral measurements of $n_e(r)$ and $T_e(r)$ on the base of the corrected line intensity ratio $R_1$ and $R_2$ of the HeI emission in the dynamical diverse tokamak experiments are presented here.

We arranged H/He experiment with fast plasma current ramp up (CRU) from 20kA up to 32kA started at 30ms with 0.5ms rise time. As the result the rise of $T_e$ with changeless $n_e(r)$ profile observed by He lines ratio at out radii (r > 6cm) of the plasma cross section are well satisfied with the TS data shown in Fig. 1. Integral time of spectral measurements is 4 ms. For lower hybrid heating (LHH) experiment in D/He plasma an increase of the density and central $T_e(r = 0\text{cm})$ followed by cooling of the plasma periphery during RF pulse applied at 30ms (F = 920 MHz, $\Delta t_{RF} = 6\text{ms}$) are feature of the plasma parameters change discussed in [3, 6 and 7]. The similar characteristic variation of the $n_e(r)$ and $T_e(r)$ data has been obtained with the HeI line intensity ratio method measured at exterior radii ($r > 4\text{cm}$) as it is shown in Fig. 2.

**II.** The paper also focuses on the influence of He fuelling of H/D plasma on the

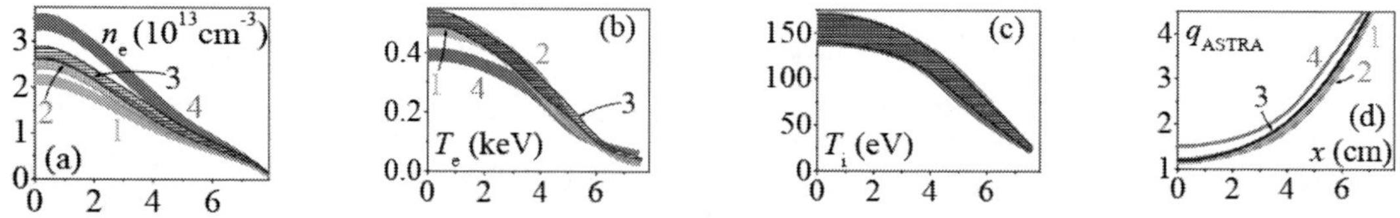


***Fig. 3.*** *Regime with intensive He puffing. Experimental radial profiles of (a) the density; (b) electron and (c) ion temperature. (d) The safety factor profile. (1 - 25 ms, 2 - 27 ms, 3 - 29 ms and 4 - 38 ms)*

anomalous heat transport development and suppression controlled by interaction of multi-scale turbulence, characterized by microwave Doppler backscattering diagnostics: correlative enhanced scattering and reflectometry [8]. The medium-scale geodesic acoustic mode (GAM) turbulence impacted by He puffing is compared with dynamics of the turbulence level profiles. The preliminary set of measurements characterizing GAMs was carried out in hydrogen discharge with a low plasma current 20 kA and toroidal magnetic field 2.2 T. Similar initial parameters and profiles (shown in Fig. 3 for different time moments, together with ASTRA code reconstruction for the safety factor $q$) were chosen in hydrogen regime with helium fuelling. During the time interval 25-38 ms in H/He regime the plasma current decreases from 21.5 till 20.5 kA and $Z_{eff}$ increases from 2 till 3.2. The ratio of the densities for helium $He^{2+}$ and hydrogen $H^+$ at $x \sim 5$ cm was roughly estimated as $n_{He2+}/n_{H+} \sim 6$ [3].

The correlative Doppler enhanced scattering technique [8, 9] was implemented for local measurements of the turbulence radial wave number spectra, turbulence level $|n|_0^2$ and Doppler shift ($f_D$) of the frequency spectrum: $f_D = k_\theta V_\theta/(2\pi)$, where $V_\theta$ is the poloidal phase velocity and $k_\theta$ is the fluctuation poloidal wave number. The frequency shift is composed of the drift-wave frequency and the Doppler frequency shift associated with plasma poloidal rotation. The spectrum of the $f_D(t)$ dependence in H regime averaged over 8 FFT realizations taken one by one within 1.6 ms interval between dashed lines (shown in Fig. 4) is presented in Fig. 5(a) by pink curve 1. An intensive spectral line is clearly seen in this spectrum at $F \sim 54$ kHz. The black solid line 2 corresponds to a mean noise level, which can be attributed to perturbations of a radial electric field $E_r$ and density, as well as distortions of the upper hybrid resonance and magnetic surfaces [8]. The ratio of the line's amplitude to a noise level at peak's frequency is a contrast $C = 3.5\pm1.1$. The radial profiles of the spectral line's contrast $C$ and frequency $F$ at $t \sim 29$ ms are shown in Fig. 5(b,c) by circles. The mean value of the $f_D(F)$ noise in the terms of contrast corresponds to the level $C = 1$. The grey curve in Fig. 5(b) is the standard deviation of this noise. It should be explained that the line's frequency $F$ was determined only for cases when the contrast error bars were not crossing this grey curve in Fig. 5(b). The theoretical estimations of GAMs frequency ($F_G$) are also plotted in Fig. 5(c).

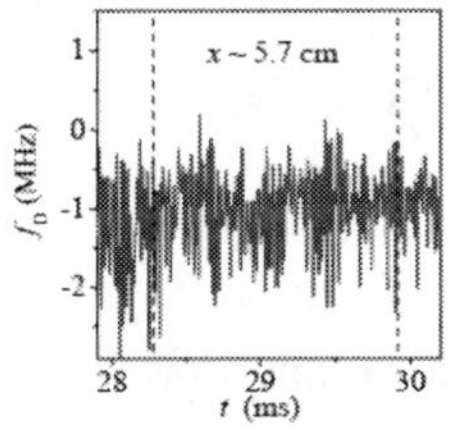


**Fig. 4.** Time trace of the Doppler frequency shift $f_D$.

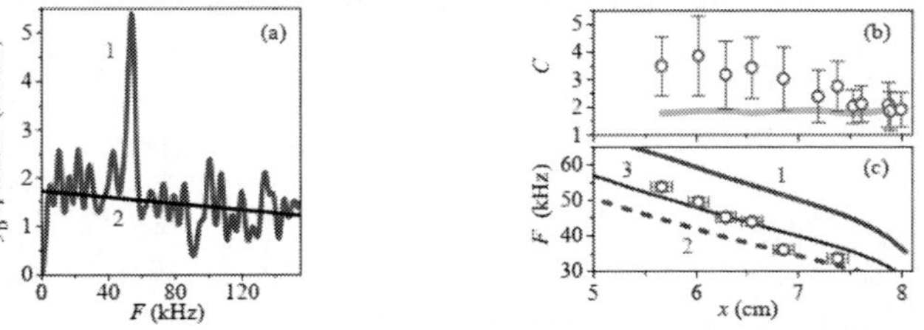


**Fig. 5.** H regime. (*a*) The spectrum of $f_D$ time trace. Radial profiles of the dominating line's (*b*) contrast and (*c*) frequency.

The olive curve 1 was obtained with formula [8]: $F_G^2 \approx (7T_i/4 + Z_i T_e)(2\pi^2 R^2 m_i)^{-1}$, where $m_i = m_H$ (the ion hydrogen mass) and $Z_i = 1$ (for $H^+$). An application of this formula with $T_i = 0$ gives the dashed olive curve 2 (as well as dashed curves in Fig. 6(b)). The blue curve 3, which is the most appropriate for experimental points, is based on the analytical prediction for GAMs frequency, obtained in [10] for plasma with two ion species, when the $O^{8+}$ impurity component was taken into account. This result demonstrates one of arguments [8] that the dominating line in the $f_D(F)$ spectrum corresponds to the GAM. The influence of the isotope effect on GAM frequency was checked in H/He regime. Olive curves 1 (solid) and 2 (dashed) in Fig. 6(b) were calculated as $F_G$ estimates for pure hydrogen $H^+$ case with $m_i = m_H$, $Z_i = 1$ in the same manner as it was done in Fig. 5(c). Two underlying blue curves 3 (solid) and 4 (dashed) in Fig. 6(b) were calculated as $F_G$ estimates for pure helium $He^{2+}$ case with $m_i = 4m_H$, $Z_i = 2$. As it is clearly seen experimental $F$-values have shifted from the region between two olive curves to an area between blue ones in accordance with the isotope dependence of the GAM frequency.

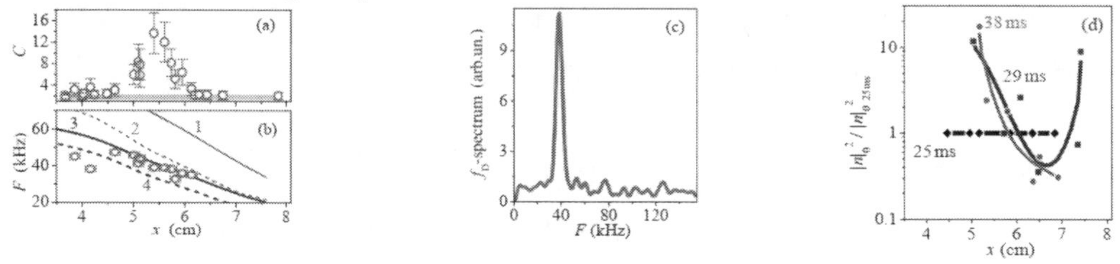


**Fig. 6.** H/He regime. Profiles of the dominating line's (a) contrast and (b) frequency at $t \sim 27$ ms. (c) The $f_D$ spectrum at $x \sim 5.4$ cm, $t \sim 27$ ms. (d) The turbulence level profiles normalized by its value at $t \sim 25$ ms.

An example of the $f_D(F)$ spectrum in H/He regime at $x \sim 5.4$ cm is shown in Fig. 6(c) for the same time point as in Fig. 5(a). Exactly in this time the contrast of the main peak reaches the highest value $C \sim 13.7 \pm 3.8$. As it seen from the $C(x)$ profile in Fig. 6(a) GAMs oscillations in H/He regime are more intensive and they are observed in a narrower zone, compared to the H case shown in Fig. 5(b). This effect is explained by growth of the Landau damping in the gradient zone and by suppression of the low frequency turbulence [11] at $x > 6$ cm as demonstrated by its typical level $|n|_0^2$ in the logarithmic scale in Fig. 6(d). It should be stressed that in H/He regime the turbulence level significantly increased at $x = (5$-$6)$ cm for $t > 25$ ms that leads to the local enhancement of the GAM amplitude ($C$) relative to pure H case or the initial stage of H/He regime ($t = 25$ms) without visible improvement in the energy confinement in investigated case. However an evidence of diffusion modulation by GAMs revealed in other experiments and numerical simulations [12] supports the possibility of the density profile steepening, finally leading to the L-H transition.

This work was supported in part by the Russian Foundation for Basic Research project nos. 13-02-00614, 13-08-00411 and 14-08-00476.

**References**

1. S. Carpentier-Chouchana, T. Hirai, F. Escourbiac et al. *Phys. Scr.* **T159** (2014) 014002
2. A.S. Kukushkin, H.D. Pacher,V. Kotov et al. *Nucl. Fusion* **53** (2013) 123025
3. S.I. Lashkul, A.B. Altukhov, V.V. D'yachenko et al. *Plasma Phys. Rep.* **38** (2012) 851
4. E de la Cal. *Plasma Phys. Control. Fusion* 43 (2001) 813–823
5. Shin Kajita et al. Phys. Plas. 13, 013301 2006
6. S.I. Lashkul et al Proc. 7th Int. Workshop, Nizhny Novgorod (2009), vol. 2, pp. 312 – 317
7. S.I. Lashkul, A.B. Altukhov, A.D. Gurchenko et al. *Nucl. Fusion* 55 (2015) to be published
8. A.D. Gurchenko, E.Z. Gusakov. A.B. Altukhov et al. *Plasma Phys. Control. Fusion* **55** (2013) 085017
9. A.D. Gurchenko, E.Z. Gusakov, S.I. Lashkul et al. *Plasma Phys. Rep.* **39** (2013) 337
10. Guo W, Wang S and Li J 2010 *Phys. Plasmas* **17** 112510
11. Gurchenko A D, Gusakov E Z, Kouprienko D V, Leerink et al 2010 *Plas. Phys. Control. Fusion* **52** 035010
12. L.G. Askinazi, V.I. Afanasyev, A.B. Altukhov et al., *Nucl. Fusion* 55 (2015) to be published

# Experimental studies of cyclical plasma effects on tungsten

A.V. Voronin[1], S.E. Alexandrov[1], G.F Avdeeva[2], B.Ya.Ber[1], P.N. Brunkov[1], A.A. Bormatov[2], V.K. Gusev[1], E.V. Demina[3], D.Yu. Kazantsev[1], S.G. Konnikov[1], I.V. Miroshnikov[2], E.E. Mukhin[1], A.N. Novokhatsky[1], S.I. Pavlov[1], Yu.V. Petrov[1], M.D. Prusakova[3], N.V Sakharov[1], G.Yu. Sotnikova[1]

[1] *Ioffe Institute, St. Petersburg, Russia*

[2] *St. Petersburg State Polytechnical University, St. Petersburg, Russia*

[3] *A. A. Baikov Institute of Metallurgy and Material Sciences (IMET), Moscow, Russia*

Plasma-surface interaction is an important issue of creating protective materials for ITER divertor. Its main function is absorption of heat flow out of the edge plasma and extraction of helium which is a product of burning in the reaction of thermonuclear fusion. It is assumed that the walls of the divertor will be irradiated with a flow of helium ions $10^{22}$-$10^{24}$ He m$^{-2}$ s$^{-1}$ (<500 eV), as well thermal stresses as a result of ELM-events will be occurred. Existing tokamaks can not generate heat flow needed for the reactor. Currently the most adequate interaction of the plasma with the wall of ITER can be experimentally studied and modeled in real tokamaks equipped with samples previously irradiated (damaged) using alternative high power heat sources. Such experiments focus on study of the properties of materials and the behavior of the parameters of the tokamak plasma. This work aims at an experimental study of cyclic heat flow on tungsten produced by plasmas of a coaxial gun [1] and the tokamak Globus-M [2]. Experimental results on the interaction of tungsten tiles with hydrogen, deuterium and helium plasma are presented.

***Investigation of tungsten irradiated by the plasma of the coaxial gun***. We have already presented some results of cyclic irradiation of tungsten with a hydrogen plasma jet simulating transient heat loads (ELM-events) [3,4]. Significant changes in the morphology of the surface layer were found with the help of scanning electron microscopy (SEM) (JEOL JSM-7001F) (Fig. 1a). After 100 cycles of irradiation columnar structure

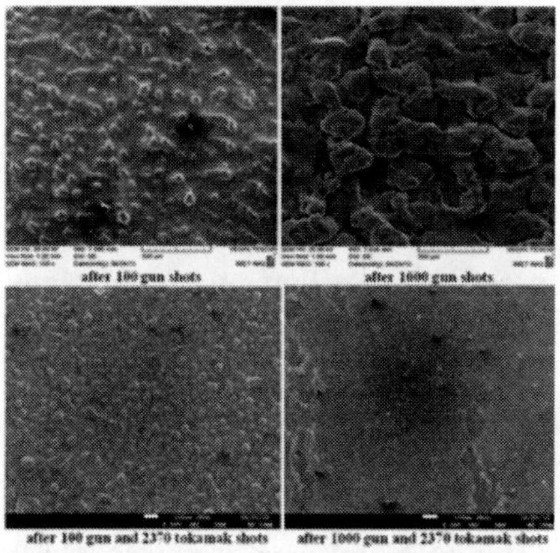


*Fig. 1: SEM images of surface of Plansee tungsten samples after cyclic irradiation by a hydrogen plasma jet and deuterium plasma of Globus-M. The plasma jet parameters: pulse duration 15 $\mu$s; power density $\leq$ 100 GW/m$^2$; heat flux factor $\leq$ 230 MJ m$^{-2}$c$^{-1/2}$. The Globus-M plasma parameters: pulse duration~70 ms; power density $\leq$ 2.5 MW/m$^2$; heat flux factor $\leq$ 1 MJ m$^{-2}$ c$^{-1/2}$*

perpendicular to the irradiated surface was formed. After 1000-fold cycles the failure mechanism was changed - the columns melted to form a droplet-shape structure. The surface became rough forming recrystallized layer with depth of approximately 100 μm and a friable layer under the irradiated surface. Two-color high-speed pyrometer developed by IoffeLED Ltd allowed measurements of the surface temperature of tungsten during irradiation by the plasma jet. (Fig. 2a) [5]. One can see that the surface temperature was heated up to 4000 $^0$C and quickly cooled. The whole process lasted ~ 10 μs.

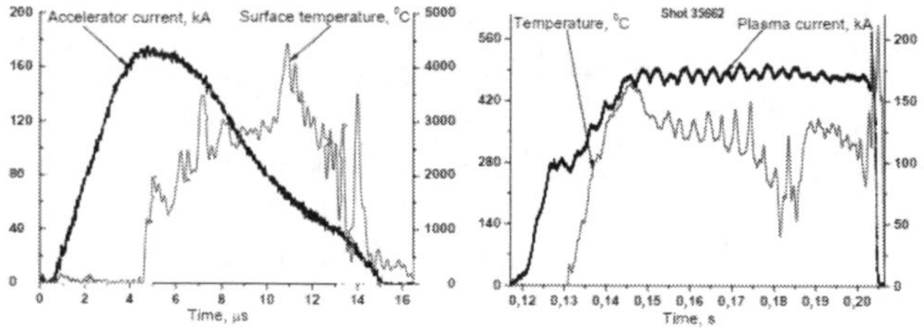


*a* *b*

*Fig. 2: Dependences of the plasma current and the surface temperature on time. Tungsten irradiated by: a) jet of plasma gun (pulse duration 15 μs; power density ≤ 100 GW/m$^2$; heat flux factor ≤ 230 MJm$^{-2}$c$^{-1/2}$); b) plasma of Globus-M (pulse duration ~70 ms; power density ≤ 2.5 MW/m$^2$; heat flux factor ≤ 1 MJm$^{-2}$c$^{-1/2}$)*

Multiple irradiation of tungsten with a helium plasma jet simulating helium ash divertor condition was conducted. In this case the temperature did not reach the melting point of tungsten and was about 1000 $^0$C (Fig. 3). It is evident that multiple plasma gun irradiations created bubbles and pores. In large areas of blocks a structural transmutation associated with the accumulation of energy was occurred. Recrystallized processes at submicron level of tungsten were happened. The helium damaged samples were installed in the Globus –M tokamak for further investigation. The studies have shown that the jet of the plasma gun allows us to simulate the impact of ELM-events and helium ions on the material in conditions similar to the ITER divertor ones.

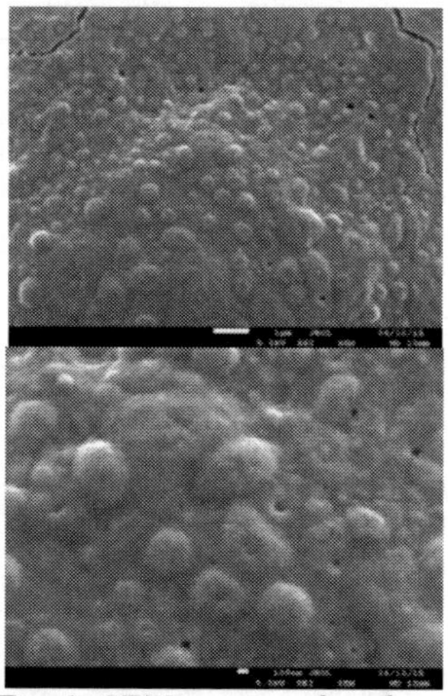

*Fig. 3: SEM images of surface of JSC POLEMA tungsten samples after cyclic irradiation by helium plasma jet (number of pulses 1000; pulse duration 15 μs; distance between the gun and the sample 250 mm; fluence 10$^{23}$ – 10$^{24}$ He/m$^2$; power density ≤ 20 GW/m$^2$; heat flux factor ≤ 50 MJm$^{-2}$c$^{1/2}$)*

***Investigation of tungsten irradiated by tokamak plasma of Globus-M.*** The tungsten samples a priory irradiated by the plasma gun were installed near the outer strike point region of the Globus-M lower divertor. Specific power load to the divertor tiles reached ~ 2.5 MW/m$^2$. The first results of the interaction of deuterium tokamak plasma with the surface of the tiles damaged by ELM-events have been reported in [2, 6]. IR camera measuring the temperature field of the tungsten surface showed that shortly after discharge termination the temperature was much higher on the tiles damaged by the hydrogen plasma jet. The heat removal from the bulk of the tiles damaged with cyclic loading equivalent to 1000 ITER ELM events by the plasma jet was obstructed by formation of a loose layer. Two color pyrometer with spatial resolution about 1 cm registered surface temperature of low divertor region (Fig.2b). One can see that the temperature of the tile could reach the value ≥ $400^0$C.

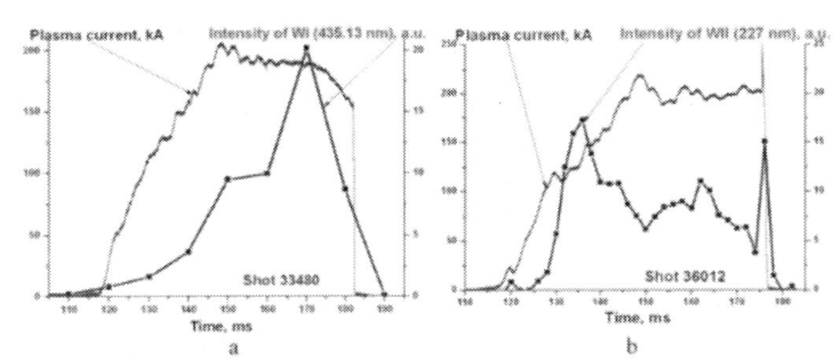


*Fig. 4: Dependences of plasma current and line radiation of tungsten on time in Globus-M*

Radial displacement of the outer branch of the separartrix could cause a variation of the temperature on time. Tungsten shielded divertor did not change total radiation losses noticeable. But spectroscopy detected some neutral and ionized tungsten in plasma of the Globus-M (Fig.4). Figure 4a shows that amount of the neutral tungsten increased by the end of the discharge. The total plasma radiation losses had similar behavior. Ionized tungsten partially burned out at the current rise and recombined at discharge termination (Fig.4b).

After 2370 shots (total duration ~ 200 s) the Plansee double forged tungsten samples were taken out for analysis. Structure of tungsten surface layers irradiated by the hydrogen plasma jet and their subsequent exposure in deuterium plasma of the tokamak Globus-M was modified (Fig.1b). Investigation of morphology showed that the columnar and droplet-shape structures were smoothed. An in-depth distribution of D, H and other elements in tungsten was determined by SIMS CAMECA IMS 7f (Fig.5). One can see that characteristic depth of the layer accumulating impurities was more than 0.5 μm. The sample irradiated with 1000 gun and 2370 tokamak pulses had the greatest depth and quantity of the impurity (c). Mainly the layer accumulated impurity of boron (boronization of the tokamak chamber was conducted regularly in order to improve the discharge reproducibility). The amount of other elements in the layer was not more than a few percents.

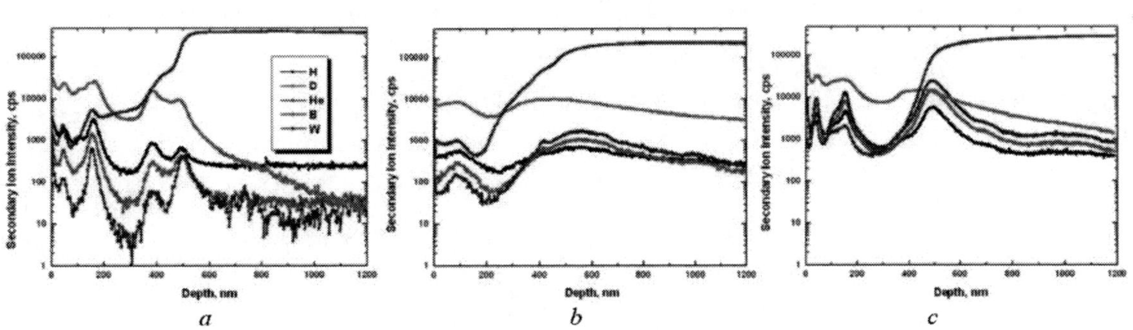


*Fig. 5: SIMS in-depth profiles of elements in Plansee double forged tungsten after irradiation by: a) 2370 shots of the Globus-M; b) 100 shots of plasma gun and (a); c) 1000 shots of plasma gun and (a). Hydrogen plasma jet parameters: pulse duration 15 μs; power density ≤ 100 GW/m$^2$; heat flux factor ≤ 230 MJ m$^{-2}$ c$^{-1/2}$. Deuterium Globus-M plasma parameters: pulse duration~70 ms; power density ≤ 2.5 MW/m$^2$; heat flux factor ≤ 1 MJ m$^{-2}$ c$^{-1/2}$*

**Conclusions**

Multiple irradiations of tungsten by pulse flows of hydrogen, deuterium and helium plasma were conducted. The surface temperature during irradiation by the plasma jet (up to 4000 $^0$C) and the tokamak plasma (~400 $^0$C) with spatial and time resolution was measured. Investigations of morphology of the surface layer of tungsten after multiple plasma irradiations under various conditions were performed. The helium jet of the plasma gun allows simulation of the impact of helium ions in the ITER divertor materials - creating bubbles. It was shown that the structure of tungsten surface layers irradiated by the hydrogen plasma jet and their subsequent exposure in deuterium plasma of the tokamak Globus-M was modified - structure gets smoothed. The tiles did not increase total radiation losses appreciable, but spectroscopy detected some neutral and ionized tungsten in Globus-M. The characteristic depth of the layer accumulating impurities reached more than 0.5 μm. The samples irradiated with the gun and the tokamak had the greatest damaged depth and quantity of the impurity. Mainly the layer accumulated the boron impurity. The amount of other elements in the layer was not exceeded a few percents. The results of successively plasma irradiation of tungsten with the help of the plasma gun and the tokamak confirmed a possibility of modeling of plasma-wall interaction in the ITER divertor.

This work was supported in part by the IAEA research contracts 16939, 16960 and MINOBRNAUKA, agreement 14.621.21.0007 (unique identifier RFMEFI62114X0007).

**References**

[1] A.V. Voronin, et al. Parameter measurement of the plasma jet in material irradiation. Technical Physics 2013, Issue. 8, p. 36.

[2] V.K. Gusev, et al., Review of Globus-M Spherical Tokamak Results. Proceedings of 25th IAEA FEC, Saint Petersburg, Russia, 13-18 October 2014, OV/P-03.

[3] A.V. Voronin, et al. Study of tungsten at cyclic interaction with jet of plasma gun. Technical Physics Letters, 2014, Vol. 40, No. 24.

[4] A.V. Voronin, et al. Study of tungsten at multiple ELM-power loads. 41th European Physical Society Conference on Plasma Physics, Berlin, Germany, 22nd—27th July 2014, P4-035.

[5] http://www.ioffeled.com/index.php?option=com_content&view=article&id=31&Itemid=80

[6] A.N. Novokhatsky, et al, Testing of Mock-ups for a Full Tungsten Divertor on Globus-M Tokamak. Proceedings of 25th Fusion Energy Conference St. Petersburg, Russia, 13-18 October 2014, MPT/P8-15.

# Modeling of impurity distribution and radial transport in the H-mode edge plasma

E. Kaveeva, V. Rozhansky, I. Senichenkov , I. Veselova, S. Voskoboynikov

*Peter the Great St.Petersburg Polytechnic University, St.Petersburg, Russia*

Possible impurity accumulation is one of the crucial problems for the H-mode operation of future fusion plant. The impurity flows at the plasma edge determine the pedestal impurity density and therefore determine impurity level in the core. Here an analytical model describing the poloidal distribution of impurities and their radial fluxes is presented for high density H-mode pedestal. The discrepancies with the neoclassical predictions are emphasized. The results of the B2SOLPS5.2 [1] modeling in the ASDEX-Upgrade and Globus-M geometries with account of impurities are compared with the analytical model.

The proof of principle modeling of the impurity flows in the case of non-zero electron conductivity associated with resonant magnetic perturbations (RMPs) is also discussed.

## 1. Analytical model

To understand qualitatively the behavior of the impurity let us discuss the test ($Z_{eff}-1<<1$) highly ionized impurity $Z_a>>1$ in the tokamak with $\varepsilon = r/R << 1$, circular cross section and without Shafranov shift. The toroidal coordinate system $(r,\theta,\phi)$ is used. First, let us consider the main contributions to parallel momentum balance for impurities

$$-b_\theta\ \partial(n_a T_i)/r\partial\theta - b_\theta n_a e Z_a\ \partial\varphi/r\partial\theta + \nu_a n_a m_a (V_\parallel - V_{\parallel a}) + \alpha b_\theta n_a Z_a^2\ \partial T_i/r\partial\theta = 0 \qquad (1)$$

Pressure gradient, electrostatic force, main ion-impurity friction and thermal force with coefficient $\alpha \geq 1$ are taken into account while inertia is neglected, $b_\theta = B_\theta / B$. The term $\partial\varphi/\partial\theta$ can be found using the parallel momentum balance for electrons and the condition $\partial n_i(T_e+T_i)/\partial\theta \approx 0$. Then $V_{\parallel a}$ can be expressed using Eq. (1). Substituting this velocity into the continuity equation for impurities, where the radial fluxes are neglected, we have:

$$\frac{\partial}{\partial\theta} R\left[ b_\theta n_a \left( V_\parallel - \frac{b_\theta D}{n_a}\frac{\partial n_a}{r\partial\theta} + \frac{b_\theta D^T}{T_i}(1+\beta)\frac{\partial T_i}{r\partial\theta}\right) + n_a V_\theta^E + n_a V_{a\theta}^{dia} \right] = 0 \qquad (2)$$

with $R = R_0 + r\cos\theta$, $D = T_i/(\nu_a m_a)$, $D^T = \alpha Z_a^2 T_i/(\nu_a m_a)$, $\beta = [Z_a T_e/(T_e+T_i) - 1]/(\alpha Z_a^2) \leq 1$. The parallel velocity $V_\parallel$ of the main ions is determined by neoclassical effects [2]: $b_\theta V_\parallel + V_\theta^E = -V_\theta^{dia} - (k^T - 1)R_0/(eB_0 R)\partial T_i/\partial r$, the subscript 0 denotes average values at the flux surface. This expression can be substituted into Eq.(2), and the impurity density can be sought as $n_a = n_0(r) + n_1(r)\cos\theta + n_2(r)\sin\theta$ assuming $n_0 >> n_1, n_2$, $T_i = T_{i0} - \delta T_i \sin\theta$,

$\delta T_i << T_{i0}$. The linearization of Eq.(2) gives:

$$n_1 = n_0\left[\alpha Z_a^2(1+\beta)\delta T_i / T_i - 2(A+C)\varepsilon\right]A/(1+A^2),\ n_2 = -n_1 / A \tag{3}$$

with the parameters

$$A = \frac{r}{b_\theta^2 D}(V_{a\theta 0}^{dia} - V_{\theta 0}^{dia} - (k^T-1)\frac{1}{eB_0}\frac{\partial T_i}{\partial r}),\qquad C = \frac{r}{b_\theta^2 D}(k^T-1)\frac{1}{eB_0}\frac{\partial T_i}{\partial r}. \tag{4}$$

If $A >> 1$ then the density maximum is situated at the inner or outer midplane [3], if $A << 1$ – at the top or the bottom of the flux surface. The radial flow of the impurity can be obtained from the toroidal momentum balance. The other way is to use the diamagnetic and $\vec{E}\times\vec{B}$ drifts defined from the poloidal force balance, and it gives the same result:

$\Gamma_r = -T_i /(Z_a eBR)\left[n_2 + \alpha\beta Z_a^2(n_0 + 0.5 n_1 / \varepsilon)\delta T_i / T_i\right]$.

Let us discuss two limiting cases: i) $A << 1$ and therefore $n_2 \sim n_0 A\varepsilon,\ n_1 << n_2$. Then

$$\Gamma_r \approx \left(\alpha Z_a^2 \delta T_i / T_i - 2(A+C)\varepsilon\right) n_0 T_i /(Z_a eBR) \tag{5}$$

This expression can be transformed to the neoclassical one [4]. By the order of magnitude $\Gamma_r \sim \Gamma^{\nabla B}\cdot A\cdot\varepsilon$ (here $\Gamma^{\nabla B} = n_0 T_i /(Z_a eBR)$ is impurity $\nabla B$ drift flow). The contribution to $\Gamma_r$ of the $\vec{E}\times\vec{B}$ drift is small comparing to the $\nabla B$ drift.

ii) $A >> 1$, then $n_1 \sim \varepsilon n_0$, $n_1 >> n_2$ and

$$\Gamma_r \approx \frac{n_0 T_i}{Z_a eBR}\left(\frac{\delta T_i}{T_i}\frac{\alpha(1+\beta)Z_a^2}{A^2} - 2\frac{A+C}{A^2}\varepsilon - \alpha^2\beta(1+\beta)\left(\frac{Z_a^2}{\varepsilon}\frac{\delta T_i}{T_i}\right)^2\frac{\varepsilon}{2A} + \alpha\beta Z_a^2\frac{\delta T_i}{T_i}\frac{C}{A}\right) \tag{6}$$

The first two terms in the sum in the parentheses are mostly responsible for the transport of the impurity by the $\nabla B$ drift, while the electric drift corresponds to the third and fourth terms. The first two terms are of the order of $\Gamma^{\nabla B}\cdot\varepsilon / A$ while in case of the Pfirsch-Schlueter regime $\delta T_i / T_i \sim A\varepsilon / Z_a^2$ and the third and fourth terms are of the order of $\Gamma^{\nabla B}\cdot\varepsilon A / Z_a$. Increasing the value $A > \sqrt{Z_a}$ we come to the situation when the contribution of the $\vec{E}\times\vec{B}$ drift into the radial transport of impurities is bigger than that of the $\nabla B$ drift, which is not typical for standard neoclassical solution.

In the case of radial conductivity caused by RMPs two corrections are necessary:

i) the non-zero sin-like pressure perturbation for the main plasma component (ions plus electrons) $p = n_i(T_e + T_i) = p_0 - \delta p\sin\theta$ should be taken into account in the calculation of the potential poloidal perturbation. It gives rise to modification of the parameter $\beta$. The new coefficient is $\beta_1 = 1/(\alpha Z_a^2)\left[(1-(\delta p / p)/(\delta T_i / T_i + T_e))Z_a T_e /(T_e + T_i) - 1\right]$. For A >> 1 the $\vec{E}\times\vec{B}$ drift in the radial flows should decrease due to RMPs, Eq. (6). The $\delta p$ estimate through the RMP pump-out flow [5] is $\delta p \approx eBR\Gamma^{pump-out} \leq \delta T_i \; n_i T_i B_\theta^2 /(m_i r^2 v_i^2)$ in the Pfirsch-Schlueter regime. ii) The poloidal rotation of the main ions deviates from the neoclassical value in the presence of RMPs. As a result the term $-(k^T - 1)/(eB_0)(\partial T_i / \partial r)$ representing the neoclassical poloidal velocity in Eq.(4) changes to the equilibrium poloidal velocity for RMPs [6]. For $A >> 1$, $A$ decreases due to RMPs and the radial flows associated with the electric drift decrease as well.

Since the sum $A+C$ does not change with the onset of RMPs, for $A<<1$ both the density perturbation and the impurity radial flow do not change. The changes in this case can be only associated with the pump-out and consequent density gradient decrease of the main ions, which enters the $A+C$ term.

**Modeling**

The modeling was done for AUG with carbon impurity in the geometry of shot #17151.The parameter $A$ was increased by rising the main ion density and corresponding decreasing of temperatures and toroidal velocities. The density, ion temperature, and $C^{6+}$ profiles at the outer midplane are shown in Fig.1a-c. The Mach number at the inner boundary of the computational domain is 0.2 for all calculations. The values of parameter $A$ for $C^{6+}$ in the pedestal 2 *cm* from the separatrix at the outer midplane are 0.2, 1.5, 17 for the small, intermediate and big density of deuterium. The radial profiles of the flow components of $C^{6+}$ ions through the flux surface are shown in Fig.2a-c. For low deuterium density the dominant in the pedestal is $\nabla B$-drift induced flow directed outwards, Eq.(5), while for big density the $\vec{E} \times \vec{B}$ drift induced flow directed inwards gives the main contribution, Eq.(6). Radial profiles of the fluxes for the calculation with radial conductivity due to RMPs are presented in Fig.2d. The conductivity level is chosen big, so that the radial electric field changes the direction. The radial $\vec{E} \times \vec{B}$ drift induced flow in the region with RMPs is significantly decreased, in agreement with the analytical model. Still, the detailed analysis of the $C^{6+}$ particle balance in modeling shows that the divergence of the radial diffusive flow is significant in the barrier region and therefore the analytical model can be inaccurate. The other effect which can improve the model is the inclusion of centrifugal force, leading to accumulation of both main ions and impurities at the outer midplane[7]. In the modeling the parallel momentum balance shows that this effect is not big. Still, for big toroidal velocities it can change the radial impurity flows. The modeling was also performed for spherical tokamak Globus-M shot #34439 with the central density $3.7 \cdot 10^{19}$ $m^{-3}$ and temperature 490 eV, plasma current 114 kA. For such parameters the convective radial transport deviates from neoclassical one, Fig. 2e.

**Conclusions.** The modeling and analytical consideration show the significant deviations of the impurity transport from the neoclassical model in the H-mode and pedestal of high density discharges. The $\vec{E} \times \vec{B}$ radial drift in combination with inner-outer midplane asymmetry of the impurity gives significant impurity inward radial flow. At the same time the neoclassical contribution associated with $\nabla B$ flow is decreased due to the smoothening of the impurity top-bottom density asymmetry by the poloidal rotation.

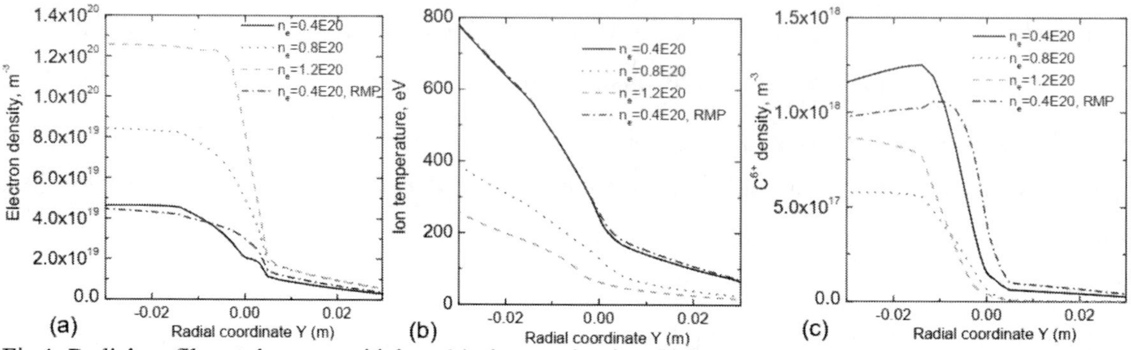


Fig.1. Radial profiles at the outer midplane (a) electron density, (b) ion temperature, (c) carbon ion density

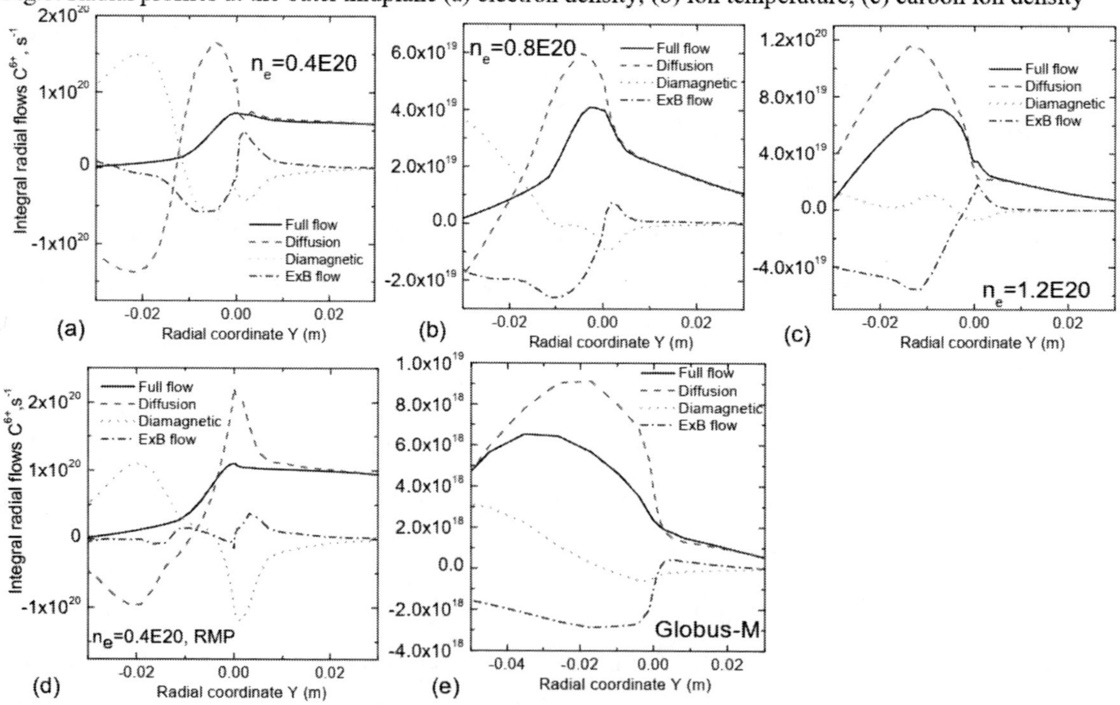


Fig. 2. Radial flows of carbon ions through the flux surface

**Acknowledgements**

The work was supported by grant of Russian Ministry of Education and Science "Filaments and stochastization". Authors are also grateful for financial support from the Russian Ministry of Education and Science (Agreement No. 14.619.21.0001, 15.08.2014, id RFMEFI61914X0001).

**References**

[1] V. Rozhansky *et al.*, Nucl. Fusion **49** (2009) 025007
[2] F. L. Hinton and R. D. Hazeltine, Rew. of Modern Physics **48** (1976) 239
[3] V. Rozhansky and M. Tendler in Rev. Plasma Phys. Ed. by B.B. Kadomtsev, Cons. Bureau N.Y. **v. 19** 1996
[4] P. H. Rutherford, Phys. Fluids **17** (1974) 1782
[5] V. Rozhansky *et al.*, Nucl. Fusion **50** (2010) 034005
[6] E. Kaveeva and V. Rozhansky, Plasma Phys. Control. Fusion **56** (2014) 125015
[7] E. A. Belli, J. Candy and C. Angioni, Plasma Phys. Control. Fusion **56** (2014) 124002

# 3D probe diagnostics of plasmas

A. Mustafaev, A. Grabovskiy, A. Strakhova

*National Mineral Resources University (Mining University), Saint-Petersburg, Russia*

This paper deals with the further development of the probe technique for the investigation of nonequilibrium anisotropic plasmas. The flat probe method, which allows to determine full electron velocity distribution function (EVDF) in plasmas with any degree of anisotropy, but only with axial symmetry, has been improved in direction of measurements in plasma without symmetry at all. The theoretical basis of the method has been developed. For probes of different geometries the analytical expressions, connecting the second derivative of probe current with respect to the probe potential with the Legendre components of the EVDF has been obtained. The method has been experimentally tested in positive column of glow discharge. It has been demonstrated, that proposed method provide a number of new possibilities, such as investigations near the plasma boundaries and obtaining of non-traditional information about the processes of particle escape to the walls, which could be useful for investigations of Langmuir paradox in plasma.

One of the current fundamental problems of plasma physics is the so-called "Langmuir paradox", which has been long known and widely discussed in the literature [1-5].

In the paper [6] L.D. Tsendin attempted to explain the Langmuir paradox in plasma of low pressure. As a result it has been concluded, that in this case the main mechanism of EVDF formation is escape of the electrons to the walls, determined by their elastic scattering in the so-called "loss cone".

The first systematic experimental investigations of EVDF in conditions, when Langmuir paradox is exist, have been undertaken by Y.M. Kagan with collaborators [7]. The result was the assumption that EVDF Maxwellization occurs due to the unknown “wall” mechanism.

Paper [8] demonstrates the results of experiments, where spherical and cylindrical probes, oriented perpendicular to the discharge axis have been used. At this orientation, the cylindrical probe is unable to "feel" the loss cone, and Maxwellian EVDF have been registered in a wide range of energies. Most clearly the depletion of EVDF, associated with the escape of the electrons to the walls, was observed in experiment [8], where flat wall probe has been used. In [9] also noted the differences in the second derivatives, recorded at the two perpendicular orientations of the cylindrical probe - along and across the discharge axis.

The first focused investigations of electron loss cone have been conducted by the authors of [10] in argon high-frequency discharge. The EVDF has been measured by flat probe, located at a distance of 1 cm from the wall. It has been established, that increasing of electrons energy leads to the emergence of EVDF characteristic features, associated with the escape of electrons from plasma volume.

In order to register the loss cone and move forward in research of Langmuir paradox, it is vital to develop new reliable method of plasma diagnostics, providing the possibilities of EVDF investigations with angular resolution near the plasma boundaries, measurements of radial electric field profiles, near-wall potential jump, electron concentration, etc.

To solve this problem we propose new probe method for diagnostics of plasmas with arbitrary symmetry. This method has been experimentally tested and experimental results, testifying in favor of the conclusions [6, 10] have been obtained.

**Theory**

The known method of flat single-sided probe consist in the EVDF expansion in a series of Legendre polynomials and registration of the second derivative at different angles of the probe orientation with respect to the discharge axis [11]. It allow us to obtain the most complete information about the strongly nonequilibrium anisotropic plasmas. The basis of the method is given by following set of equations:

$$j_u''(\vec{r}; eu; \phi_0) = \frac{2\pi e^3}{m^2} \sum_{j=0}^{\infty} F_j(\vec{r}; eu) P_j(\cos\phi_0); \tag{1}$$

$$F_j(\vec{r}; eu) = f_j(\vec{r}; eu) - \int_{eu}^{\infty} f_j(\vec{r}; \varepsilon) \frac{\partial}{\partial(eu)} P_j\left(\sqrt{\frac{eu}{\varepsilon}}\right) d\varepsilon. \tag{2}$$

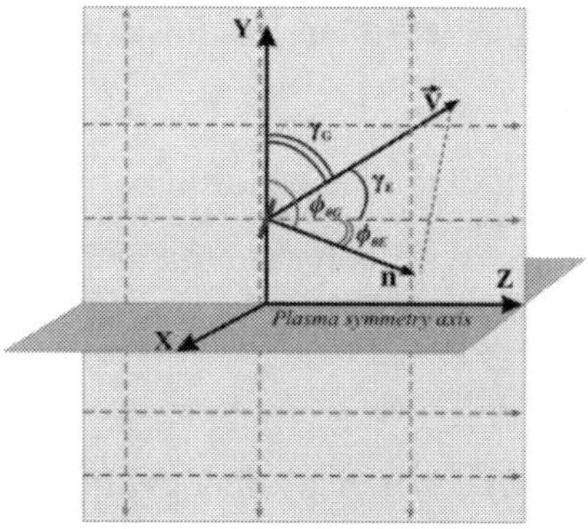


Fig. 1. The geometry of the problem in the case of probe, located in radial area of the interelectrode gap

However, the scope of this method is limited by the requirement of plasma axial symmetry. Fig. 1 demonstrates a case, where flat probe is located near the boundary of the plasma volume (radial area) and it is necessary to take into account the asymmetry, caused by presence of a radial electric field.

In this paper we propose new probe method for diagnostics of the plasma with arbitrary symmetry. In this case the equations of the flat probe method (1, 2) are transformed into the following equations:

$$j_u''(\vec{r}; eu) = \frac{2\pi e^3}{m^2} \sum_{j=0}^{\infty} \{F_{jE}(\vec{r}; eu) P_j(\cos\phi_{0E}) + F_{jG}(\vec{r}; eu) P_j(\cos\phi_{0G})\}; \tag{3}$$

$$F_{jE}(\vec{r}; eu) = f_{jE}(\vec{r}; eu) - \int_{eu}^{\infty} f_{je}(\vec{r}; eu) \frac{\partial}{\partial(eu)} P_j\left(\sqrt{\frac{eu}{\varepsilon}}\right) d\varepsilon; \tag{4}$$

$$F_{jG}(\vec{r}; eu) = f_{jG}(\vec{r}; eu) - \int_{eu}^{\infty} f_{jG}(\vec{r}; eu) \frac{\partial}{\partial(eu)} P_j\left(\sqrt{\frac{eu}{\varepsilon}}\right) d\varepsilon, \tag{5}$$

where $\phi_{0E}$; $\phi_{0G}$ - the angles between the normal to the flat probe surface and directions of the charged particles fluxes, caused by electric field and diffusion, respectively.

Similarly to the method of the flat probe, reconstruction of the components $f_{jE}$ $f_{jG}$ is connected with the solution of corresponding integral Volterra equations of second kind, binding Legendre components of the EVDF and of the second derivative.

**Experiment**

Probe measurements have been carried out in the range of helium pressure 0.1-1 torr and discharge current 0.1 A to 0.5 A at a distance of 1 cm from the plasma wall for two orientations of the probe with respect to the discharge axis (0 and $180^0$). Corresponding second derivatives have been registered, angular structure of the EVDF in the loss cone has been reconstructed and polar diagrams of the electron motion have been plotted (Fig. 2).

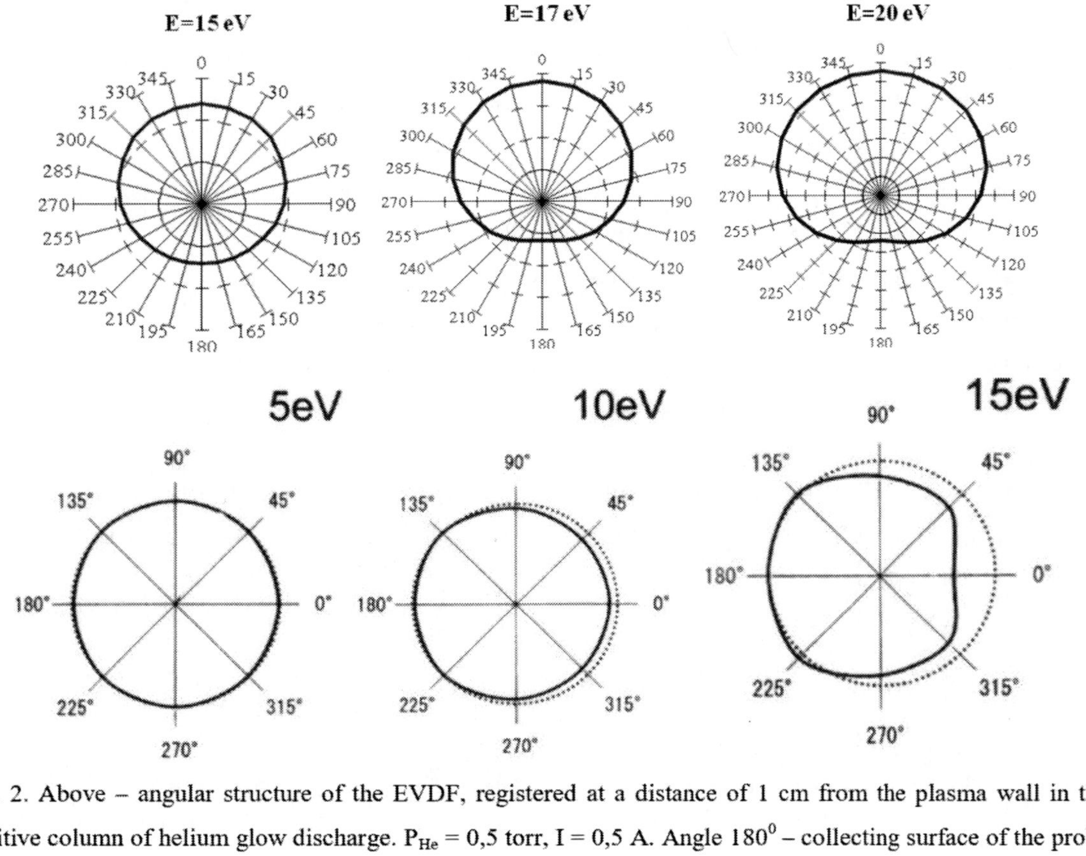


Fig. 2. Above – angular structure of the EVDF, registered at a distance of 1 cm from the plasma wall in the positive column of helium glow discharge. $P_{He}$ = 0,5 torr, I = 0,5 A. Angle $180^0$ – collecting surface of the probe is turned to the wall.

Below - angular structure of the EVDF, registered in plasma of high-frequency discharge in argon by authors [10]. Angle $0^0$ – collecting surface of the probe is turned to the wall

According to the paper [10], when energy of the electrons increase, on the diagrams of electron motion is clearly visible the depletion of EVDF, associated with the escape of electrons to the walls. If the wall potential exceeds the potential of the probe, the probe "feels" EVDF in the loss cone. Increasing of current and pressure does not affect on EVDF shape. Rotation of the probe by $180^0$ let us to measure EVDF in the main part of the plasma volume. In this case the polar diagrams demonstrate us isotropic EVDF, typical for positive column of glow discharge.

The obtained data correlate quite accurately with theoretical and experimental results of [6, 10]. This is evident about reliability of the developed method and could be a confirmation of the hypothesis, that Langmuir paradox in plasma is associated with physical characteristics of the EVDF formation as a combination of known mechanisms, not with a hypothetical mechanism of EVDF Maxwellization.

This work was supported by the Ministry of Education of the Russian Federation.

**References**

1. Langmuir I. // Phys. Rev. 1925. Vol. 26. P. 585–613.
2. Gabor D., Ash E.A., Dracott D. // Nature. 1955. Vol. 176. N 11. P. 916–919.
3. Granovskiy V.L. Electrical current in gas. Sustained current. M.: Nauka, 1971. 272 P.
4. Chen F. Introduction to plasma physics. Plenum press. 1974. 421 p.
5. Kadomtsev B.V. // Problems of plasma theory / Ed. by. M.A. Leontovich. Vol. 4. M.: Atomizdat, 1964. 325 P.
6. L D Tsendin 2003 Plasma Sources Sci. Technol. 12 S51
7. Kagan Y.M. // Spectroscopy of gas discharge plasmas. L.: Nauka, 1970. P. 201-223.
8. Rayment S.W., Twiddy N.D. // Proc. Soc. A. 1968. Vol. 340. P. 87–98.
9. Godyak V.A., Kuzovnikov A.A., Khadir M.A. // Mess. Of the MSU. Phys and astron. 1971. № 3. P. 336-338.
10. Ishijima T., Uenuma M., Tsendin L.D. and Sugai H. Cont. pap. of ESCAMPING 2002. Grenoble, France. Vol 1, p. 221.
11. Mustafaev A.S. Probe Method for Investigation of Anisotropic EVDF. Electron Kinetics and Applications of Glow Discharges // NATO Int. Sci. Session / Ed. By Kortshagen U., Tsendin L. N.Y. - London: Plenum Press, 1998. V. 367. P. 531.

# Analytical evaluation of the forces during fast current quenches in tokamaks

V.D. Pustovitov

*National Research Centre Kurchatov Institute and National Research Nuclear University MEPhI (Moscow Engineering Physics Institute), Moscow, Russia*

**1. Introduction.** Recently a problem of the sideways forces during and because of the plasma disruptions in tokamaks became a hot topic of theoretical debates [1–3]. Discussions of the related aspects and comparison of some predictions with experimental data from JET can be found in [4–6]. The problem belongs to the research priorities of ITER studies because significant uncertainties remain in quantitative modeling of the electromagnetic loads associated with unmitigated disruptions in ITER.

Several mechanisms are discussed [1–6] that can contribute to the force, but the approaches are still fragmentary and incomplete. For example, the opinion that "the repelling force between the plasma and the vessel is of little significance in asymmetric events" [5] is completely opposite to the reasoning in [2]. Also, there is no yet agreement on the key elements of the process and boundary conditions in the task [1–3, 6].

Here the model of electromagnetic interaction between the current carrying systems [5] is re-examined and the estimates of the forces during fast current quenches in tokamaks are presented. In [5], the eddy current forces have been neglected, but here we take them into account and show that their contribution into the integral force can be large.

**2. Formulation of the problem.** Our analysis is essentially based on classical results of the plasma equilibrium theory [7, 8]. In those papers (see also [9]), the force balance has been calculated for a tokamak treated as a system of three interacting elements: plasma, toroidal field coils and poloidal field coils. Here we add the fourth element, the wall with a current either dynamically induced or resulting from the halo currents. Another difference from [7–9], where a static equilibrium has been examined, is that we consider the current quench stage of the tokamak disruptions. This means a fast evolution of the plasma with its inevitable motion. Though this is far from equilibrium, the plasma inertia plays a minor role and can be neglected compared to the electromagnetic forces described by $\mathbf{j}\times\mathbf{B}$, where $\mathbf{B}$ is the magnetic field and $\mathbf{j}=\nabla\times\mathbf{B}/\mu_0$ is the current density. In ITER, the current quench duration $\tau_{tq}$ is expected to be about 36 ms [10], while the resistive wall time is estimated as $\tau_w \approx 0.34$ s [11]. With $\tau_{tq} << \tau_w$, the fast change in $\mathbf{B}$ inside the tokamak chamber will be shielded by the currents induced in the chamber wall. Here we assume finite $\tau_{tq}\partial\mathbf{B}/\partial t$ inside, but with $\partial\mathbf{B}/\partial t = 0$ in the outer region.

**3. Calculation of the force.** Here we calculate the ballooning force on the wall,

$$F_w \equiv \int_{wall} (\mathbf{j}\times\mathbf{B})\cdot\mathbf{e}_R dV, \tag{1}$$

where $\mathbf{e}_R = \nabla R$ is the unit vector along the major radius. This requires $\mathbf{j}$ in the wall, but the integral nature of (1) and careful account of the interaction between all 4 elements of the system allows estimation of $F_w$ without knowledge of $\mathbf{j}$.

The radial force balance in this toroidal system is

$$F_p + F_w = F_{\Sigma\Sigma} + F_{\Sigma c} + F_{\Sigma t}, \tag{2}$$

where

$$F_p \equiv \int_{plasma} (\mathbf{j}\times\mathbf{B} - \nabla p)\cdot\mathbf{e}_R dV \tag{3}$$

is the force on the plasma (its pressure $p$ can be neglected at a current quench), and

$$F_{ik} \equiv \int_i (\mathbf{j}_i \times\mathbf{B}_k)\cdot\mathbf{e}_R dV_i \tag{4}$$

is the integral force acting on the element $i$ (with current density $\mathbf{j}_i$) from the element $k$ (creating the magnetic field $\mathbf{B}_k$). Accordingly, $\Sigma$ means wall ($w$) plus plasma ($p$), $c$ the poloidal coils, and $t$ the toroidal coils. A force $F_{ii}$ describes the interaction of the current $\mathbf{j}_i$ with the field $\mathbf{B}_i$ produced by the same current.

It is clear that (with or without plasma-wall contacts)

$$F_{\Sigma\Sigma} = F_{pp} + F_{ww} \tag{5}$$

because always $F_{pw} + F_{wp} = 0$. If $\partial\mathbf{B}/\partial t = 0$ in the outer region, the forces on the coils must remain constant during the quench. Therefore

$$F_{\Sigma c} + F_{\Sigma t} = F_{pc}^0 + F_{pt}^0, \tag{6}$$

where $F_{pc}^0$ and $F_{pt}^0$ are the values of $F_{pi}$ before the disruption. In the initial state we have

$$F_p^0 = F_{pp}^0 + F_{pc}^0 + F_{pt}^0 = 0. \tag{7}$$

Combination of (2), (5), (6) and (7) yields

$$F_w = F_{pp} + F_{ww} - F_{pp}^0 - F_p. \tag{8}$$

According to [7–9]

$$F_{pp} = \frac{\mu_0 J^2}{2}\left(\ln\frac{8R_0}{a} - 1 + \frac{\ell_i}{2} + \frac{\beta_p}{2}\right), \tag{9}$$

where $J$ is the net plasma current, $a$ is the plasma minor radius, $R_0$ is its major radius,

$$\ell_i \equiv \overline{B_\theta^2} / B_J^2 \qquad (10)$$

is the internal inductance per unit length of the plasma column ( $\overline{f}$ stands for the averaged over the plasma transverse cross section),

$$B_J = \frac{\mu_0 J}{2\pi a} \qquad (11)$$

is the poloidal field $B_\theta$ at the plasma boundary,

$$\beta_p \equiv 2\mu_0 \overline{p} / B_J^2 \qquad (12)$$

is the 'poloidal beta' with $p$ the plasma pressure and $\mu_0 = 4\pi\times10^{-7}$ H/m the vacuum magnetic permeability (SI units are used here).

The same logic and approach as described in [7–9] can be also applied to a system Σ (plasma plus wall). The result is a natural generalization of (9):

$$F_{pp} + F_{ww} = \frac{\mu_0 (J+J_w)^2}{2}\left( \ln\frac{8R_0}{r_w} - 1 + \frac{\ell_i^\Sigma}{2} + \frac{\beta_p^\Sigma}{2} \right), \qquad (13)$$

where $\ell_i^\Sigma$ and $\beta_p^\Sigma$ are defined similarly to $\ell_i$ and $\beta_p$, but now with averaging $< f >$ over the volume inside the vessel ( $r < r_w$ ) instead of $\overline{f}$, $r_w$ is the minor radius of the wall, and $J_w$ is the net current in the wall. At the current quench, the plasma pressure contribution is negligible. Then $\beta_p^\Sigma = 0$, and we need only $\ell_i^\Sigma$ here.

For a circular large-aspect-ratio plasma, we have in the plasma-wall vacuum gap

$$B_\theta \approx B_J a / r \qquad (14)$$

with $B_J$ defined by (11) and $r$ the radius counted from the plasma centre in the perpendicular cross-section. The integration over the volume inside the vessel yields

$$\int_{p+g} B_\theta^2 dV = V_p B_J^2 \left( \ell_i + 2\ln\frac{r_w}{a} \right), \qquad (15)$$

where $V_p = 2\pi^2 a^2 R_0$ is the plasma volume. In the geometrically thin wall, the magnetic energy must be small. Disregarding it, but allowing for $B_\theta$ variation across the wall, we obtain

$$\ell_i^\Sigma \equiv < B_\theta^2 > / B_\theta^2(r_{w+}) = \frac{J^2}{(J+J_w)^2}\left( \ell_i + 2\ln\frac{r_w}{a} \right), \qquad (16)$$

where $r_{w+}$ means the outer side of the wall.

The ideal wall assumption implies that

$$J + J_w = J_0, \qquad (17)$$

where $J_0$ is the plasma current before the quench. Under this constraint, combining (13) with $\beta_p^\Sigma = 0$ and (16), we obtain from Eqs. (8) and (9)

$$F_w = \frac{\mu_0 J_0^2}{2}\left(\frac{\ell_i - \ell_i^0}{2} - \frac{\beta_p^0}{2}\right) + \frac{\mu_0(J^2 - J_0^2)}{2}\left(\ln\frac{r_w}{a} + \frac{\ell_i}{2}\right). \qquad (18)$$

The first term describes the force that appears due to the thermal quench when $J_w = 0$ and $\beta_p$ drops from $\beta_p^0$ to zero. It is constant at the current quench when an additional force $\sim (J^2 - J_0^2)$ appears that is related to the magnetic energy redistribution in the inner region.

**4. Discussion.** The latter formula is obtained at $\partial\mathbf{B}/\partial t = 0$ in the outer region. In reality the disruption duration may be longer than the wall time [5]. In such cases equation (18) can serve as an upper estimate for JET. However, the ideal-wall limit seems to be pertinent at the mentioned ITER expectations [10, 11]: $\tau_{tq} \approx 36$ ms and $\tau_w \approx 340$ ms.

Being essentially different from the Noll's formula [2, 4–6], equation (18) represents a mechanism that was not accounted for in [2, 4–6]. Accordingly, the force (8) is an addition to the Noll's and other disruption forces discussed in [1–6, 10]. For tokamaks with a large current, formula (18) predicts quite a large value of $F_w$. Thus, this force must be an important contribution of the force balance in ITER.

The presented approach is based on the integral force balance required for a plasma equilibrium in a tokamak. The algorithm outlined by equations (1)–(8) is quite general and can also be used in other tasks where the wall becomes a current carrier. Here the outward penetration of the plasma-produced perturbation is assumed weak because of the skin effect in the wall. For slower events, the wall resistivity has to be properly incorporated.

[1] H. R. Strauss, R. Paccagnella and J. Breslau, Phys. Plasmas **17**, 082505 (2010).
[2] L. E. Zakharov, S. A. Galkin and S. N. Gerasimov, Phys. Plasmas **19**, 055703 (2012).
[3] H. R. Strauss, et al., Nucl. Fusion **53**, 073018 (2013).
[4] V. Riccardo, S. Walker, and P. Noll, Fusion Eng. Des. **47**, 389 (2000).
[5] V. Riccardo, P. Noll, and S. P. Walker, Nucl. Fusion **40**, 1805 (2000).
[6] S. N. Gerasimov, et al., Nucl. Fusion **54**, 073009 (2014).
[7] V. D. Shafranov, J. Nucl. Energy, Part C, **5**, 251 (1963).
[8] V. D. Shafranov, in *Reviews of Plasma Physics*, Ed. by M. A. Leontovich (Consultants Bureau, New York, 1966), Vol. 2, p. 103.
[9] K. Miyamoto, *Fundamentals of Plasma Physics and Controlled Fusion*. NIFS-PROC-48 (Toki, 2001).
[10] T. C. Hender, et al., Nucl. Fusion **47**, S128 (2007).
[11] Y. Gribov and V. D. Pustovitov, 19th IAEA Fusion Energy Conf., Lyon, (2002) IAEA-CN-94/CT/P-12; http://www.iaea.org/programmes/ripc/physics/fec2002/pdf/ctp_12.pdf

# Simulation of Resonant Magnetic Field Penetration Dynamics with the TEAR Code

N.V. Ivanov, A.M. Kakurin

*National Research Centre «Kurchatov Institute», Moscow, Russia*

It is well known that externally applied Resonant Magnetic Perturbation (RMP) can generate magnetic islands and consequently penetrate through the resonant magnetic surface [1-5]. This effect attracts considerable attention in the experimental and theoretical magnetic fusion research because RMP loops are often used in the present tokamak experiments and are designed for future tokamaks including ITER.

The process of RMP penetration is usually analyzed under assumption that the mode rotation is coupled with the rotation of resonant plasma layer. This well known no-slip constraint has been introduced in [2]. Besides that the poloidal rotation of plasma is omitted in some papers on the RMP penetration. However, there are some evidences [6] that the no-slip approximation is not valid for sufficiently small magnetic islands because the island velocity can deviate from plasma velocity due to finite plasma resistivity and consequent magnetic reconnections. The plasma poloidal rotation can also play some role in the field penetration process despite the high value of poloidal viscosity. This is expected because the ratio between viscous torque and plasma viscosity is much less for poloidal direction than for toroidal one for similar rotation velocities.

Results of TEAR code simulation and analysis of the RMP penetration are presented. The effects of the magnetic island and plasma velocity decoupling and of the plasma poloidal rotation on the RMP penetration are the main subjects of this paper. The TEAR code [6-8] is based on two-fluid MHD approximation that gives coupled diffusion-type equations for the magnetic flux perturbation and for the plasma rotation velocities in toroidal and poloidal directions. The radial distribution of the magnetic flux perturbation is calculated with account of an externally applied RMP and RMP component generated by eddy current in the resistive vacuum vessel. Variations of magnetic island width and rotation velocity under RMP due to finite plasma resistivity are taken into account [6]. An externally applied magnetic flux can diffuse across the plasma in the so-called magnetic-reconnection layer where the ideal-plasma assumption breaks down. The flux diffusion results in rearrangement of magnetic field topology that depends on the angular position of RMP with respect to the mode.

Integration of the equation of motion over the resonant layer gives equations for the toroidal and poloidal rotation velocities of the resonant layer depending on toroidal and poloidal moments of inertia of the plasma resonant layer, electromagnetic and viscous torques. Viscous friction torques are calculated under the assumption that poloidal viscosity is two orders of magnitude higher than toroidal one.

Simulation results of the RMP penetration dynamics are presented with the view of TEAR code validation against COMPASS-C tokamak experimental data [1]. We consider typical Ohmic COMPASS-C discharge in which application of helical (2, 1) RMP causes mode penetration. The experimental dynamics of this process is shown in *Figure 2* in Ref. [1]. In this figure, one can see that locked mode signal indicates the RMP penetration as a sudden growth of a large stationary mode. Available COMPASS-C plasma characteristics are used for the modelling. We use values of resistive time $\tau_R$ = 45 ms, toroidal viscous time $\tau_v$ = 22 ms, intrinsic value of the mode natural frequency $\Omega_{\mathrm{nat}\,0}/2\pi = \left(mV_{\theta 0}/r_S - nV_{\varphi 0}/R - \omega^*\right)\Big|_{r_S}/2\pi = -14\ \mathrm{kHz}$, initial toroidal, $\omega_{\varphi 0}/2\pi = 3\ \mathrm{kHz}$, and poloidal, $\omega_{\theta 0} = 0$, rotation frequencies and frequency of electron diamagnetic drift $\omega^*/2\pi = 11\ \mathrm{kHz}$.

The results of COMPASS-C experiment simulations are shown in FIG.1 and FIG.2. The simulations taking into account the effect of magnetic island and plasma velocity deviations are shown in the left columns (*a*) of these figures. The results of simulations fulfilled under the no-slip approximation are shown for comparison in the right columns (*b*). The RMP penetration process appearing as a change of natural frequency to $\Omega_{\mathrm{nat}} = 0$ value and a quick rise of the non-rotating island is in good agreement with the experimental data. In FIG.1(*a*) and FIG.2(*a*), the penetration is preceded by formation of stationary non-rotating small magnetic island in rotating plasma (see also [9]). In addition, the simulation shows a spin-up event similar to the experiment after the RMP is switched off. The induced island first unlocks from the RMP field and then decays away on a resistive time-scale.

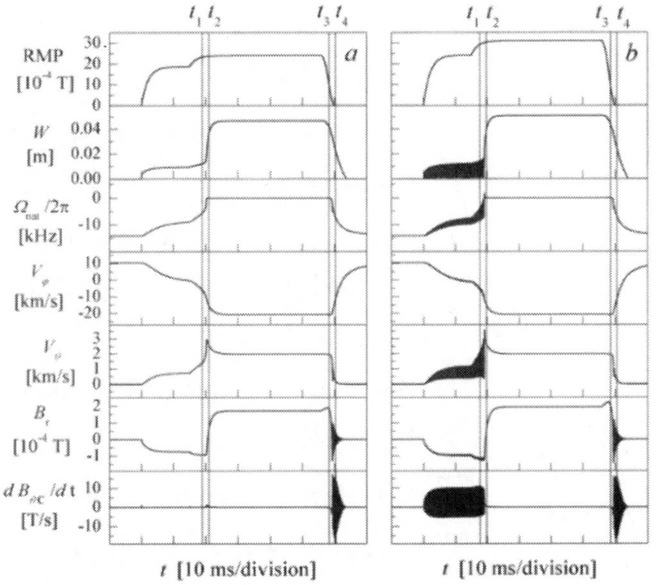


FIG.1. Simulation results of the RMP penetration dynamics in the COMPASS-C tokamak: radial component of RMP, magnetic island width ($W$), instantaneous value of mode natural frequency ($\Omega_{nat}$), toroidal ($V_\varphi$) and poloidal ($V_\theta$) velocities of resonant plasma layer, radial component of the mode magnetic field ($B_r$) at plasma boundary, and poloidal field time-derivative ($dB_{\theta C}/dt$)

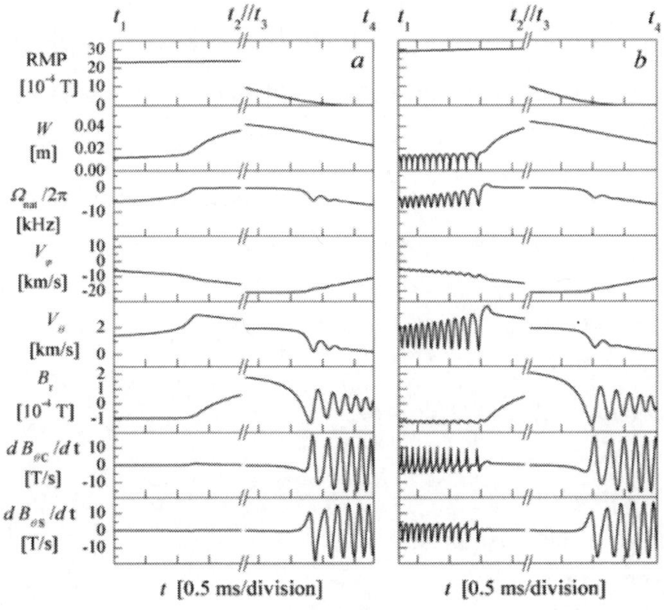


FIG.2. Simulation results of the RMP penetration dynamics in the COMPASS-C tokamak in the time intervals shown with vertical straight lines in FIG.1

Like in the experiment (see *Figure 2* in Ref. [1]), calculations made under the assumption that magnetic island and plasma velocity deviations are not forbidden (see left panels in FIG.1 and FIG.2) do not give oscillations of the $dB_{\theta C}/dt$ waveform before start of the RMP penetration process. On the contrary, the calculations under the no-slip approximation shown in the right ($b$) panels in FIG.1 and FIG.2 give frequent oscillations of the magnetic island width, $W$, before start of the field penetration process. These $W$ oscillations are followed by a significant rise of the $dB_{\theta C}/dt$ oscillation amplitude. One should note that these $dB_{\theta C}/dt$ oscillations are not observed in the experiment.

Similar to the experiment, one can see in FIG.1 and FIG.2 a change of the plasma toroidal velocity to $V_\varphi = -21$ km/s value in the process of RMP penetration. According to calculations with no plasma poloidal rotation, the toroidal velocity turns to $V_\varphi = -38.4$ km/s after the RMP penetration. This value differs from the experimental one.

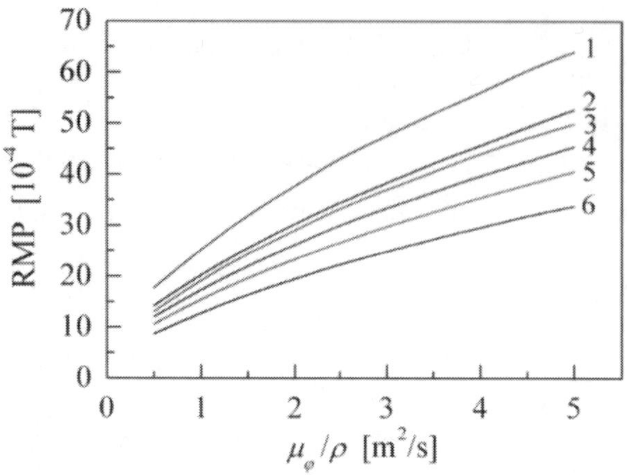


FIG.3. Dependences of RMP penetration threshold values on toroidal plasma viscosity in conditions: no-slip approximation and no poloidal rotation (1); account of island and plasma velocity deviations, no poloidal rotation (2); no-slip approximation, account of poloidal rotation (3); full model (island and plasma velocity deviations, account of poloidal rotation) $\mu_\theta/\mu_\varphi = 200$ (4), $\mu_\theta/\mu_\varphi = 100$ (5), $\mu_\theta/\mu_\varphi = 50$ (6)

The comparison of the left (a) and right (b) panels in Fig.1 and FIG.2 shows that the RMP penetration threshold is lower if the no-slip constraint is omitted and deviations of magnetic island velocity with respect to plasma are taken into account. Besides that the account of the plasma poloidal rotation also decreases the value of RMP penetration threshold. These results are also presented in FIG.3. In addition, one can see that the threshold value of the RMP penetration increases with the rise of the toroidal and poloidal plasma viscosity. The curves in FIG.3 are proportional to $(\mu_\varphi/\rho)^{7/12}$ that is similar to the prediction in [10].

**Summary**

In this paper the main attention was paid to dynamics of RMP penetration. The results of simulations with account of mode and plasma decoupling and plasma poloidal rotation are consistent with experimental data. The deviations between rotation velocities of magnetic island and plasma due to final plasma resistivity as well as plasma poloidal rotation reduce the RMP penetration threshold.

[1] Hender T.C., et al. Nucl. Fusion **32** (1992) 2091.

[2] Fitzpatrick R. Nucl. Fusion **33** (1993) 1049.

[3] Kikuchi Y., et al. Plasma Phys. Control. Fusion **48** (2006) 169.

[4] Yu Q., et al. Nucl. Fusion **48** (2008) 024007.

[5] Wang N., et al. Nucl. Fusion **54** (2014) 064014.

[6] Eliseev L.G. et al. Physics of Plasmas **22** (2015) 052504.

[7] Ivanov N., et al. 24th IAEA FEC (2012) TH/P3-22, PAS&T/TF 36, v. 2, p. 55 (2013), see http://vant.iterru.ru/vant_2013_2/6.pdf.

[8] N.V. Ivanov and A.M. Kakurin. Physics of Plasmas **21** (2014) 102502.

[9] N.V. Ivanov and A.M. Kakurin, in 38th EPS Conference on Plasma Physics (2011), P2.080, see http://ocs.ciemat.es/EPS2011PAP/pdf/P2.080.pdf; PAS&T/TF (2012) 1, p. 64, see http://vant.iterru.ru/engvant_2012_1/6.pdf.

[10] R.J. Buttery, et al. Nucl. Fusion **39** (1999) 1827.

# Complex Diagnostics of Electron Velocity Distribution with Thomson Scattering and Electron Cyclotron Emission in Tokamaks

P.A. Sdvizhenskii[1], P.V. Minashin[1], A.B. Kukushkin[1,2], G.S. Kurskiev[3], E.E. Mukhin[3]

[1]*National Research Center "Kurchatov Institute", Moscow, Russian Federation*
[2]*National Research Nuclear University MEPhI, Moscow, Russian Federation*
[3]*Ioffe Physical Technical Institute, St. Petersburg, Russian Federation*

**1. Introduction.** Under conditions of strong auxiliary plasma heating in tokamaks a difference between temperature values obtained by Thomson scattering (TS) diagnostics and ones obtained by electron cyclotron (EC) emission spectra (from thermal EC emission on the first or second harmonics of fundamental EC frequency) is observed. The cause of this difference is often treated as a deviation of the electron velocity distribution function (eVDF) from Maxwellian one. Here we suggest a new algorithm for diagnostics of main parameters of the eVDF under condition of a substantial deviation from the Maxwellian VDF. The algorithm combines two formerly developed algorithms of the eVDF assessment: (i) from TS spectra [1] and (ii) from EC emission in the spectral range where plasma is optically thin (i.e. for moderate- or high-number harmonics of fundamental EC frequency) [2]. The work is stimulated by the search [3] for the improvement of the TS diagnostics in tokamak-reactors like DEMO (multicolor laser model, cooperation with other diagnostics of the eVDF). The important example of interpretation beyond the frame of assumption about Maxwellian VDF in present tokamak experiments with a strong auxiliary heating is given in [4]. Here we give preliminary analysis of the complex diagnostic opportunities suggested for the case of a strong deviation from the Maxwellian eVDF. We consider the case of a hypothetic experiment, where we use the current technical parameters of the core plasma TS diagnostic system in ITER and extend the method [5] of its accuracy assessment. Plasma is described by the predictive modelling data for the so-called steady-state scenario of discharge in ITER [6].

**2. Complex eVDF diagnostics algorithm.**

In the thermal energy range, the eVDF in each space point is the Maxwellian with the temperature $T_e$ as a free parameter. In the weakly/moderate superthermal energy range the eVDF is assumed to be an anisotropic quasi-Maxwellian with only a couple of unknowns, $T_{e,\parallel}$ and $T_{e,\perp}$:

$$f(\mathbf{p}) = (1-\delta_{\text{Hot}})\cdot f_{\text{Maxw}}(\mathbf{p}) + \delta_{\text{Hot}}\cdot f_{\text{Hot}}(\mathbf{p}_\parallel,\mathbf{p}_\perp), \tag{1}$$

$$f_{\text{Hot}}(\mathbf{p}_\parallel,\mathbf{p}_\perp) = C_{\text{Hot}}\exp\left[-mc^2\left(\frac{\sqrt{1+(p/mc)^2}-1}{p^2}\right)\left(\frac{p_\parallel^2}{T_{e,\parallel}}+\frac{p_\perp^2}{T_{e,\perp}}\right)\right], \tag{2}$$

where $f_{Maxw}$ is the relativistic Maxwellian distribution; $\mathbf{p}_{\|}$ and $\mathbf{p}_{\perp}$ are the components of the momentum $\mathbf{p}$, respectively, parallel and perpendicular to the local magnetic field; $C_{\mathrm{Hot}}$ is the normalization factor (in the momentum space), and $\delta_{\mathrm{Hot}}$ the fraction of superthermal electrons in the total electron density. At higher energies, where the most significant deviations from Maxwellian are expected and the TS diagnostics does not work, the eVDF is an arbitrary function of $\mathbf{p}_{\|}$ and $\mathbf{p}_{\perp}$ (see [2] for more details).

The TS-based eVDF diagnostic algorithm [1] includes the calculation of the number of photoelectrons, $\left[N_{\text{ph-el}}^{(j)}\right]_{\text{Laser}}$, in a given spectral channel of the detector (where $j$ numerates spectral channels) in terms of the normalized cross-section σ of the Thomson scattering, which is averaged over the assumed model eVDF (1), (2). For each spectral channel, the difference between total $\left[N_{\text{ph-el}}^{(j)}\right]_{\text{Total}}$ and background $\left[N_{\text{ph-el}}^{(j)}\right]_{\text{Backgr}}$ signals is calculated. Randomization of input parameters is applied to estimate the diagnostic's accuracy. The values of the sought-for parameters are recovered via solving the inverse problem, which relies on the minimization of the difference between the "phantom" experimental laser-scattering signal and the respective variable calculated signal:

$$\sum_{j}\left|\left\{\left[N_{ph-el}^{(j)}(\xi_{assum})\right]_{Total}-\left[N_{ph-el}^{(j)}\right]_{Backgr}\right\}_{Random}-\left[N_{ph-el}^{(j)}(\xi)\right]_{Laser}\right|\xrightarrow[\xi]{}\min, \tag{3}$$

where the summation goes over all lasers and spectral channels. Further development of the eVDF recovery algorithm [1] with the correct error assessment was continued in [5]. It was shown that the large error of recovering the parameter $\delta_{\mathrm{Hot}}$ does not influence the accuracy of recovering the mean electron energy for the total non-Maxwellian VDF. Thus, the inverse problem solution is stable with respect to the recovery of the mean energy regardless of the particular form of the deviation from a Maxwellian in the thermal and weakly superthermal range of the electron energy. Nevertheless, for higher mean energies this statement becomes incorrect, that's why using other diagnostics data is necessary.

Recovery of the eVDF, based on the EC emission for moderate- or high-number harmonics of fundamental EC frequency, does not need the knowledge about the functional type of the eVDF. So the algorithm [2] does not recover the values of any parameters characterizing it (as it is in case of algorithm [1]), but directly recover the eVDF's values instead. Momentum space is divided into sectors, and their contributions to the EC spectrum are calculated. The spectrum is considered as a sum of two terms – contributions from Maxwellian $I_{\mathrm{esc}}^{\mathrm{Maxw}}(\omega)$ and hot $I_{\mathrm{esc}}^{\mathrm{Hot}}(\omega)$ plasma components, $I_{\mathrm{esc}}(\omega)=I_{\mathrm{esc}}^{\mathrm{Maxw}}(\omega)+I_{\mathrm{esc}}^{\mathrm{Hot}}(\omega)$. Here

$I_{esc}^{Maxw}(\omega)$ is assumed to be known (e.g., from the TS data). To recover the eVDF, an inverse problem is solved via minimization of the sum of differences between the "phantom" experimental EC emission spectrum and the respective calculated signal.

The first stage of the present complex diagnostic is the preliminary estimation of the sought-for parameters. Their values are recovered for each space point, where a substantial contribution of superthermal electrons to the EC emission is expected. To do this the inverse problem (3) -- with the known TS spectrum and the assumed eVDF (1), (2) -- is solved. After that, the combined optimization is performed. The solution is searched closely to the preliminary estimation. Thus, the recovery from the TS data is improved due to taking into account the additional information about hot electrons from the EC emission data for moderate- or high-number harmonics of fundamental EC frequency.

Figure 1 shows the results of the estimation exclusively from the TS data.

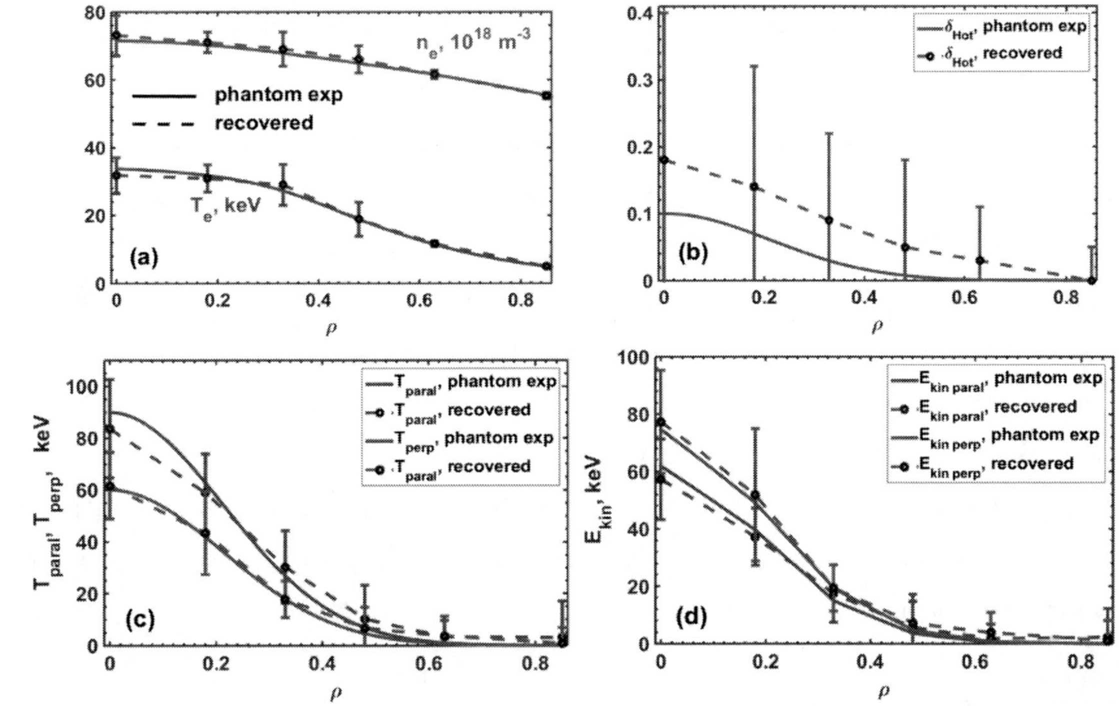


Fig. 1. The profiles as a function of effective normalized minor radius coordinate: (a) electron Maxwellian bulk's temperature and density; (b) hot electrons' fraction in the total density; (c) parallel and perpendicular effective temperatures for hot electrons; (d) their mean kinetic energies for parallel and perpendicular to magnetic field motion. Solid curves on "a" correspond to ITER steady-state scenario [6], while on "b", "c" and "d" they are taken the Gaussians (see [2]). Dashed curves show the inverse problem solution with the error bars (2.5 standard deviation) estimated with the Monte-Carlo simulations for random input parameters (see [1, 5] for more details).

The parameters of the Maxwellian plasma component are recovered rather well (Fig. 1(a)), whereas for non-Maxwellian component the accuracy is essentially worse (Fig. 1(b)-(d)). The latter may be improved via using additionally the EC diagnostic data (Fig. 2).

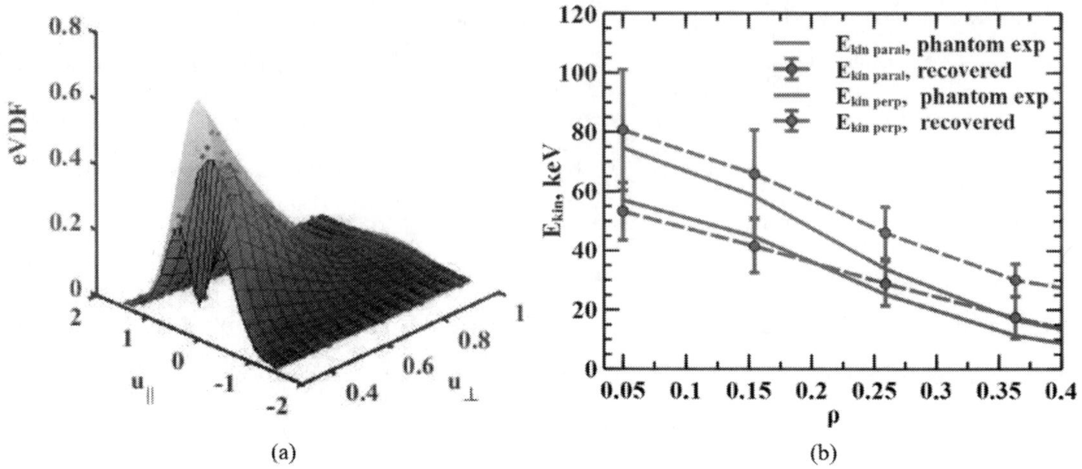


Fig. 2. The results of the eVDF recovering from the EC+TS data: (a) the eVDF for hot electrons as a function of momentum components normalized to $m_ec$; (b) mean energies of hot electrons (error bars correspond to 2.5 standard deviation).

**3. Conclusions.** A new algorithm is suggested for complex diagnostics of the electron velocity distribution function (eVDF) under condition of a substantial deviation from the Maxwellian eVDF. An inverse problem is formulated, which is based on the TS spectrum diagnostic and the EC emission (for moderate- or high-number harmonics of fundamental EC frequency) spectrum diagnostic. It is shown that there is a possibility to improve the recovery of the hot electrons eVDF parameters from the TS diagnostics via using the combined TS+EC diagnostics.

**Acknowledgements.** We thank V.I. Poznyak for helpful discussion of the EC diagnostics experimental issues. The part of work related to inverse problems solutions is supported by the Russian Foundation for Basic Research (grant RFBR-15-07-07850-a).

## References

[1] G.S. Kurskiev, P.A. Sdvizhenskii, A.B. Kukushkin, E.E. Mukhin, M. Bassan. Proc. 41st EPS Conf. Plasma Physics, Berlin, 2014, P5.009.

[2] P.V. Minashin, A.B. Kukushkin. International Review of Atomic and Molecular Physics (IRAMP), **4** (2), 93-104 (2013).

[3] E.E. Mukhin, et al. AIP Conf. Proceedings **1612**, 69-72 (2014).

[4] G. Zhuang, et al. Plasma Physics Control. Fusion, 47, 1539–1558 (2005).

[5] G.S. Kurskiev, P.A. Sdvizhenskii, M. Bassan, P. Andrew, A.N. Bazhenov, I.M. Bukreev, P.V. Chernakov, M.M. Kochergin, A.B. Kukushkin, A.S. Kukushkin, E.E. Mukhin, A.G. Razdobarin, D.S. Samsonov, V.V. Semenov, S.Yu. Tolstyakov, S. Kajita and S.V. Masyukevich, Nucl. Fusion 55 (2015).

[6]. A.R. Polevoi, et al. Proc. 37th EPS Conf. Plasma Physics, Dublin, 2010, P2.187.

# A Model of Recovering the Fast Nonlocal Transport Parameters in Magnetic Fusion Plasmas

A.B. Kukushkin[1,2], P.A. Sdvizhenskii[1], V.V. Voloshinov[3], A.A. Kulichenko[1,2]

[1]*National Research Center "Kurchatov Institute", Moscow, Russian Federation*
[2]*National Research Nuclear University MEPhI, Moscow, Russian Federation*
[3]*Institute for Information Transmission Problems (Kharkevich Institute) of Russian Academy of Science, Moscow, Russian Federation*

**1. Introduction.** Superdiffusion formalisms (an integral equation with nonlocal, longer than diffusive, spatial correlations and respective dominance of the long mean-free-path energy carriers) were suggested to describe the anomalous heat transport in magnetized fusion plasmas. These included the steady-state heat transport by the electron/ion Bernstein waves [1] and the electron cyclotron waves [2]; perturbative heat transport by the electron Bernstein waves [3], which appeared to be described by the Biberman-Holstein (B-H) equation known from the theory of excitation transport in atomic/ionic spectral lines; fractional diffusion model [4] for perturbative transport (see more references in the survey [5,6] of fast nonlocal transport interpretations). The increasing evidences for the fast nonlocal transport in magnetic fusion devices (e.g., "cold pulse" experiments in tokamaks and stellarators) suggest further diversification of perturbative models to identify possible physics mechanisms of nonlocality.

Here we suggest a model, which modifies and extends the approach [7]. We formulate an inverse problem for the recovery of main features of the kernel of the heat transport equation from the space-time dependence of the observed temperature perturbations. The developed model is applied to interpreting the data from the "cold pulse" experiments in tokamak TFTR [5, 8] and stellarator LHD [9].

**2. A model of fast nonlocal heat transport (FNHT).** The model aims at recovering the main parameters of unknown fast carriers of the heat via solving an inverse problem. The model is based on the following assumptions. (i) We assume that at the initial stage of the FNHT events, like those caused by the "cold pulse" or similar perturbations (see [5]), the observed unexpected dependence of electron temperature $T_e$ (namely, the immediate, in the diffusion time scale, rise of $T_e$ in the central plasma in the case of fast cooling of the edge plasma) is fully determined by the unknown fast carriers (presumably, of electromagnetic origin), i.e. their intensity (heat flux) exceeds that of any other heat transport mechanism at this stage. Also, their group velocity is assumed to be high enough to neglect the retardation effects in the course of the carrier's travel between the spatial points of emission ("birth") and absorption ("death") of the carrier. (ii) The energy density of the carriers is small as compared to that of plasma. (iii) The carriers are (almost) fully reflected from the plasma boundary or the vacuum chamber wall. This makes their intensity spatially homogeneous and isotropic in the group velocity angles and determined by the current parameters of the plasma at each time instant. (iv) To exclude possible contribution of the diffusion processes to the observed evolution of $T_e$, the inverse problem is solved only in the central plasma, to satisfy the limitations of the model (see the first item) both in the space and time. Correspondingly, we treat the impact of the fast cooling of the edge upon the temperature dynamics in the central plasma (namely, approximately linear growth of Te at initial stage of the event, see [5, 8]), whereas the source function in the central plasma is assumed to be not disturbed by the $T_e$ perturbation in the edge. This enables us to express the time derivatives of $T_e$ in the core in terms of the abrupt change ("jump") of the $T_e$ profile in the edge plasma. (v) The source function, $S(\omega,T_e(\mathbf{r}))$, which gives the distribution of the carriers' source in their energy (described by the wave frequency $\omega$), is assumed to be determined by the local value of $T_e$, and taken a Gaussian with unknown, sought-for dependence of three parameters, namely, the total power density Q(T), the mean energy/frequency $\overline{\omega}(T)$ , and the energy dispersion (the width of the "line shape"), $\Delta$(T), on the temperature. The dependence of these parameters on the other parameters, e.g., plasma density, is neglected.

The above model allows the derivation of an equation in the frame of the B-H approach (namely, assuming the complete redistribution over energy/frequency within the spectral line shape of the carriers' source during the process of the absorption and subsequent emission of the carrier in a given spatial point). However, preliminary results of solving such an inverse problem revealed too much freedom for the final solutions. At this stage we decided to restrict ourselves to evaluating the possibility to find the spectrum-integrated source function, Q(T), that formally corresponds to a monochromatic ("one-energy-group") transport. This finally gives the following equation:

$$\kappa\big(T_0(\rho)\big)\int_0^1 S\big(T_0(\rho_1)\big)\rho_1 d\rho_1\left[\frac{1}{\int_0^1 \kappa\big(T_1(\rho_2)\big)\rho_2 d\rho_2}-\frac{1}{\int_0^1 \kappa\big(T_0(\rho_2)\big)\rho_2 d\rho_2}\right]=\Pi(\rho),$$

$$0<\rho<\rho_{max}, \tag{1}$$

where $\Pi(\rho)$ is the time derivative of plasma energy density, which is taken from experimental data, ρ is the normalized minor radius of plasma in the 1D transport model, the source function $S$ is related to the absorption coefficient $\kappa$ with the Kirchhoff law (in the Rayleigh-Jeans limit) at a certain frequency,

$$S(T)\equiv\frac{Q(T)}{4\pi}=const\ T\ \kappa(T)\,, \tag{2}$$

the profiles of temperature before the “cold pulse”, $T_0(\rho)$, and after that, $T_1(\rho)$, differ only in the edge plasma (the profile $T_1(\rho)$ enters Eq. (1) only as an integrand so that its evolution after abrupt jump may be neglected within time of approximate constancy of $\Pi(\rho)$). The inverse problem is formulated as a minimization of the deviation from the equality in Eq. (1) with respect to sought-for function Q(T). This function is assumed to be the sum of several terms of the power-law expansion, including the half-integer power.

The inverse problem solutions give quite different results for stellarator LHD (Figs. 1-2) and tokamak TFTR.

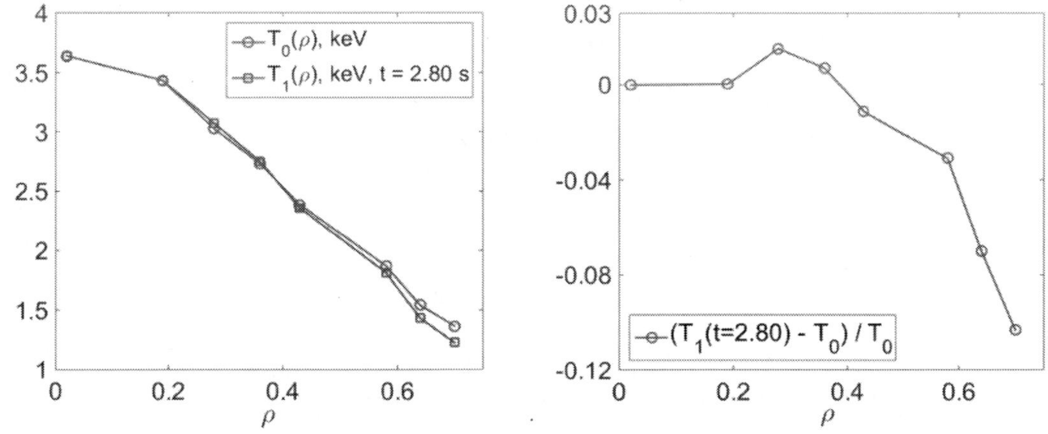


Figure 1. Left: the profiles of electron temperature before, $T_0$, and after, $T_1$, pellet injection when $\Pi(\rho)$, the time derivative of Te, is approximately constant in time in the central plasma (see Fig. 1 in [9]). Right: the relative difference of the above-mentioned temperature profiles.

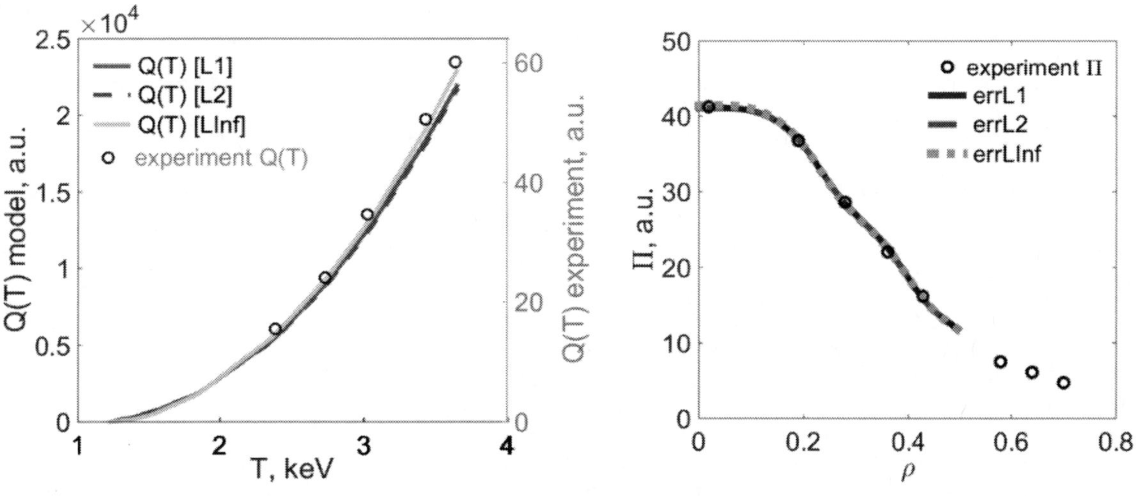


Figure 2. Left: the results of recovering the temperature dependence of the source function for energy carriers from Eq. (1) for the data [9] with $\rho_{max}=0.5$, for minimization of: the least squares error (L2), the sum of error's absolute values (L1), the global absolute error (Linf). The circles point on the values of $T_e$, which correspond to experimental data for $\Pi(\rho)$. The right axis corresponds to the recovery of only the functional form of Q(T) directly from proportionality of $\Pi(\rho)$ and $k(T(\rho))$ in Eq. (1). Right: fitting of the experimental data for $\Pi(\rho)$ for various minimization criteria.

The growth of the absorption coefficient with increasing temperature in the present model explains the effect of the inward heat flux in the cold pulse experiments: the edge plasma after fast cooling absorbs less energy from the circulating flux of energy carriers so that their increasing intensity produced higher absorption in the central plasma. Interestingly, such a mechanism is symmetric with respect to the sign of the fast temperature change in the periphery (the observed effect of "inverse polarity" of the temperature change in the center and the periphery, produced by the fast heating of the periphery, is discussed in [5]).

For tokamak TFTR data [5,8], already the first step of solving the inverse problem, namely the recovery of only the functional form of $Q(T)$ directly from proportionality of $\Pi(\rho)$ and $k(T(\rho))$ in Eq. (1), shows incapability of Eq. (1) to provide reasonable solutions. Our analysis suggests that the present model (i)-(v) should be modified for tokamaks to allow for strong internal transport barriers at magnetic surfaces with rational value of the safety factor q (for reliable identification of these barriers in tokamak RTP see [10]; the models of the MHD equilibria in a sectioned plasma were suggested in [11, 12]). E.g., for the shot in Fig. 1 in [5,8] the peak of $\Pi(\rho)$ is located at q=2 magnetic surface.

**3. Conclusions.** We suggest a model to interpret the observed fast nonlocal transport events where heat flux is directed opposite to conventional diffusion flux and the response is almost immediate within diffusion time scale. We suggest the following explanation with a hypothetical energy carriers (presumably, of electromagnetic wave origin): the abrupt cooling of the edge plasma leads to an abrupt decrease of the absorption in the edge that, in turn, allows more transparent circulation of the energy carriers and, under condition of almost full reflection from the edge, an increase of the intensity/flux of the carriers. The latter leads to a linear (in time) growth of temperature in the central plasma at initial stage of the event.

**Acknowledgements.** The part of work related to the inverse problems solutions is supported by the Russian Foundation for Basic Research (grant RFBR-15-07-07850-a).

## References


[1]. M.N. Rosenbluth, C.S. Liu. Phys. Fluids **19**, 815 (1976).
[2]. S. Tamor. Nucl. Instr. and Meth. Phys. Res. **A271**, 37 (1988).
[3]. A.B. Kukushkin, V.S. Lisitsa, Yu.A. Saveliev. JETP Lett. **46**, 448 (1987).
[4]. D. del-Castillo-Negrete, P. Mantica, V. Naulin, J.J. Rasmussen. Nucl. Fusion **48**(7), 075009 (2008).
[5]. J.D. Callen, M.W. Kissick, Plasma Phys. Contr. Fusion, **39**, Suppl. 12B, 173 (1997).
[6]. A.B. Kukushkin, K.V. Cherepanov. AIP Conf. Proceedings, **1154**, 83-94 (2009).
[7]. A.B. Kukushkin. Proc. 24th EPS Conf. on Plasma Phys. and Contr. Fusion, Berchtesgaden, Germany, 1997, Europhys. Conf. Abs., vol. 21A, Part II, pp. 849-852.
[8]. M.W. Kissick, J.D. Callen, E.D. Fredrickson, A.C. Janos, G. Taylor. Nucl. Fusion, **36**, 1691 (1996).
[9]. N. Tamura, *et al.*, Nucl. Fusion **47**, 449 (2007).
[10]. N.J. Lopez Cardoso, et al., Plasma Phys. and Contr. Fusion, **39** (Suppl. 12B), 303 (1997).
[11]. A.B. Kukushkin, V.A. Rantsev-Kartinov Proc. 26th EPS Conf. on Contr. Fusion Plasma Phys. (Maastricht, The Netherlands), 1999, vol. 23J, 1737–40 http://epsppd.epfl.ch/Maas/web/pdf/p4096.pdf .
[12]. M.J. Hole, R. Mills, S.R. Hudson and R.L. Dewar. Nucl. Fusion **49** (2009) 065019 (7pp).

# Analytical model of wall force produced by kink modes combined with plasma vertical displacement

D.V. Mironov[1] and V.D. Pustovitov[1,2]

[1]*Moscow Institute of Physics and Technology, Dolgoprudny, Moscow Region, Russia,*

[2]*National Research Centre "Kurchatov Institute", Moscow, Russia*

**1. Introduction.** This task is related to evaluation of the electromagnetic forces on the conducting structures during plasma disruptions [1–7]. Such forces are observed on Joint European Torus (JET) [2] accompanying with the vertical displacement events (VDEs). In [6], an expression for the force produced by kink modes together with VDEs was derived using Eq. (43) from [4]. The latter is shown to be incorrect in [7]. This is a sufficient reason for revision of the predictions made in [6].

Here we calculate the non-axisymmetric (sideways) wall force produced by kink modes combined with VDEs. We use a model described in [7], but extend it here by considering three modes and their nonlinear beating.

**2. Formulation of the problem.** As in [4, 6, 7], we consider a cylindrical plasma with minor radius $r_{pl}$ surrounded by a coaxial resistive wall of uniform conductivity $\sigma$, radius $r_w$ and thickness $d_w$. The plasma-wall gap and space behind the wall are treated as a vacuum. The magnetic field is described as $\mathbf{B}=\mathbf{B}_0+\mathbf{b}$, where $\mathbf{B}_0$ is the axisymmetric equilibrium field ($\partial\mathbf{B}_0/\partial t=0$) and $\mathbf{b}$ is the perturbation. The latter induces the eddy current $\mathbf{j}=\sigma\mathbf{E}$ in the wall, where $\mathbf{E}$ is the electric field governed by $\nabla\times\mathbf{E}=-\partial\mathbf{b}/\partial t$. Then, a force with a volume density $\mathbf{f}=\mathbf{j}\times\mathbf{B}$ will act on the wall. A lateral (sideways) force is defined by

$$F_X \equiv \int_{wall} \mathbf{f}\cdot\mathbf{e}_X dV = R_0 r_w d_w \oint f_r \cos\theta\cos\zeta\, d\theta d\zeta\,, \qquad (1)$$

where $\mathbf{e}_X$ is the unit vector along a fixed horizontal direction, $R_0$ is the major radius, $\theta$ and $\zeta$ are the poloidal and toroidal angles, respectively, and

$$f_r \equiv \frac{1}{d_w}\int_{in}^{out} (\mathbf{j}\times\mathbf{B}_0+\mathbf{j}\times\mathbf{b})\cdot\mathbf{e}_r dr \qquad \text{so that} \qquad d_w f_r = (B_{0\theta}b_\theta + B_{0\zeta}b_\zeta)\Big|_{out}^{in} + \frac{1}{2}b_\theta^2\Big|_{out}^{in}. \qquad (2)$$

In the quadratic term here, we disregarded a small contribution from $b_\zeta$. Equation (1) shows that only $f_r \propto \cos\theta\cos\zeta$ can give a non-zero sideways force. Therefore, $m/n=1/1$ mode can contribute into linear term of Eq. (2). This case was discussed in [7]. In quadratic term, we need $b_\theta^\perp \propto \cos\theta\cos\zeta$. This can be produced by the beating of $m/n$ and $m\pm 1/n\pm 1$ modes. Here we consider the perturbations in form

$$b_\theta = |b_{\theta,11}|\cos(\theta-\zeta+\delta_{11}) + |b_{\theta,VDE}|\sin\theta + |b_{\theta,21}|\sin(2\theta-\zeta+\delta_{21}) \qquad (3)$$

with $\delta_{11}$ and $\delta_{21}$ that are toroidal phase shifts of the modes. VDE denotes $m/n = 1/0$ mode.

With such $b_\theta$ for ideal conducting wall, Eqs. (1)–(2) give a sideways force

$$F_X = F_1 + F_2 = F_l \frac{|b_{\theta,11}|}{B_{0\theta}} + F_q \frac{|b_{\theta,21} b_{\theta,VDE}|}{B_{0\theta}^2}, \qquad (4)$$

where $F_l = -F_{Xm}\left(1 - \frac{r_{pl}^2}{q_{pl} r_w^2}\right)\cos\delta_{11}$ was calculated in [7], $F_q = F_{Xm}\frac{1}{2}\frac{r_{pl}^2}{q_{pl} r_w^2}\cos\delta_{21}$,

$F_{Xm} \equiv \frac{\pi r_w}{2} J B_{0\zeta}$, $b_\theta$ and $B_{0\theta}$ are taken at the wall, $q_{pl}$ is the safety factor $q \equiv r B_{0\zeta} / (R_0 B_{0\theta})$ at the plasma boundary, $J$ is the total current in the plasma. With ITER parameters [8]: $J = 15$ MA, $B_{0\zeta} = 5.3$ T, $r_{pl} = 2$ m, $r_w / r_{pl} = 1.3$, we have $F_{Xm}^{ITER} = 325$ MN. A large plasma vertical displacement was observed on JET at $q_{pl} = 1.1$ [9]. Using the latter and ITER data the term $F_1 \geq 37.5$ MN requires $|b_{\theta,11}| / B_{0\theta} > 0.25$, while the ITER is designed for 48 MN [9]. However, this limit can be exceeded by the contribution from quadratic term of Eq. (4), which gives $F_2 \geq 22$ MN at $|b_{\theta,21} b_{\theta,VDE}| / B_{0\theta}^2 > 0.25$. On the other hand, $F_X = 0$ at

$$\cos\delta_{21} = -2\cos\delta_{11}\left|\frac{B_{0\theta} b_{\theta,11}}{b_{\theta,21} b_{\theta,VDE}}\right|\left(1 - q_{pl}\frac{r_w^2}{r_{pl}^2}\right). \qquad (5)$$

A result similar to Eq. (4) is described by Eqs. (11)–(15) in [6]. These equations predict a force as a monotone decreasing function of $\gamma\tau_w$ with a maximum at $\gamma = 0$ instead of $\gamma\tau_w \approx 1$ as claimed in [6] for parameters presented in [4], where $\gamma$ is the kink growth rate and $\tau_w$ is the wall penetration time. On the contrary, analysis in [7] shows that $F_1$ force must be maximal in the ideal wall limit at $\gamma\tau_w \to \infty$. Here we obtain the same result for $F_2$ and compare it with that in [6].

**3. Calculation of the force.** We calculate quadratic term of Eq. (2). At $nr << mR_0$ equation $\nabla^2\varphi = 0$ with $\mathbf{b} = \nabla\varphi$ and $\varphi = \sum_{mn} \mathrm{Re}\left[\varphi_{mn} \exp(im\theta - in\zeta)\right]$ gives us

$$\varphi_{mn} = -\frac{r_w}{m} B_m \left[x^{-m} + \frac{\Gamma_m}{2m}(x^{-m} + x^m)\right] \qquad (6)$$

in the plasma-wall vacuum gap with $x \equiv r/r_w$ and $\Gamma_m = -2m B_m^{out}/B_m$ (see also [7,10, 11]). Then,

$$b_{\theta,mn}\Big|_{in} = -i(1+\Gamma_m/m) b_{r,mn}(r_w) \qquad \text{and} \qquad b_{\theta,mn}\Big|_{out} = -i b_{r,mn}(r_{w+}), \tag{7}$$

where $b_{r,mn} \equiv \partial\varphi_{mn}/\partial r$ is the amplitude of $\mathbf{b}\cdot\nabla r$ and $r_{w+} = r_w + d_w$. Using these expressions we introduce $d_w f_{r2}^{side} \equiv b_{\theta,21} b_{\theta,VDE}\Big|_{out}^{in}$ :

$$d_w f_{r2}^{side} \equiv \frac{1}{4}\mathrm{Re}\left[\Gamma_\Sigma (1+\alpha_2) b^*_{r,VDE}(r_w) b_{r,21}(r_w) \exp(i\theta - i\zeta)\right] \tag{8}$$

that is a force volume density only contributes to sideways force in quadratic term with

$$\alpha_2 = \frac{2}{\Gamma_\Sigma}\left[1 - \frac{b^*_{r,VDE}(r_{w+}) b_{r,21}(r_{w+})}{b^*_{r,VDE}(r_w) b_{r,21}(r_w)}\right] \qquad \text{and} \qquad \Gamma_\Sigma = 2\Gamma_1^* + \Gamma_2 + \Gamma_2\Gamma_1^*, \tag{9}$$

where superscript * denotes a complex conjugate variables.

In [6], the force is finally expressed through the radial component $\xi_r$ of the plasma displacement $\boldsymbol{\xi}$. To move in this direction we use the consequence of (6)

$$\frac{b_{r,mn}(r)}{b_{r,mn}(r_w)} = \left[1 + \frac{\Gamma_m}{2m}(1 - x^{2m})\right] \cdot x^{-m-1} \tag{10}$$

valid in the plasma-wall gap. If the plasma is ideal as in Ref. [4, 6, 7], we have $\mathbf{b} = \nabla\times(\boldsymbol{\xi}\times\mathbf{B}_0)$, $b_r = \mathbf{B}_0\cdot\nabla\xi_r$ and

$$r_{pl} b_{r,mn}(r_{pl}) = i B_J (m - n q_{pl}) \xi_{r,mn} \tag{11}$$

with $B_J \equiv B_{0\theta}(r_{pl})$ and $\xi_{r,mn}$ taken at the plasma boundary. Then, Eq. (8) takes a form of

$$d_w f_{r2}^{side} \equiv 2\,\mathrm{Re}\left[\frac{\Gamma_\Sigma}{\Gamma_1^*\Gamma_2} \frac{B_J^2(2-q_{pl})(1+\alpha_2)\kappa_1^5}{(2/\Gamma_1^* + 1 - \kappa_1^2)(4/\Gamma_2 + 1 - \kappa_2^2)} \frac{\xi_{r,21}\xi^*_{r,VDE}}{r_{pl}^2} \exp(i\theta - i\zeta)\right], \tag{12}$$

where $\kappa_m \equiv (r_{pl}/r_w)^m$ and free parameters: $\alpha_2$ and $\Gamma_\Sigma$ are defined by Eq. (9).

**4. Comparisons, estimates, and discussion.** In [4, 6], a thin shell approximation was used that requires continuity of the normal component of **b** (here, $\mathbf{b}\cdot\nabla r$) at the wall. Hence $\alpha_2 = 0$ and $\Gamma_m$ is related to $\gamma_{mn}$ by (see [10, 11] and the references therein)

$$\Gamma_m = \gamma_{mn}\tau_w \qquad \text{with} \qquad \tau_w \equiv \sigma r_w d_w. \tag{13}$$

In contrast to [4, 6], expression (12) gives the wall force increases monotonically with $\gamma$, starting from $f_{r2}^{side} = 0$ at $\gamma = 0$ and maximal at $\gamma = \infty$. Using $\xi_{r,21}\xi^*_{r,VDE} = |\xi_{r,21}\xi_{r,VDE}| e^{i\delta_{21}}$ and $\gamma = \infty$, Eq. (12) after integration across the wall volume gives us the sideways force:

$$F_2^{id} = 4F_q \frac{\text{[illegible]}}{r_w^3} \frac{\text{[illegible]}}{(1-\kappa_1^2)(1-\kappa_2^2)} \frac{\text{[illegible]}}{r_{pl}^2}. \qquad (14)$$

The result $F_x^{21}$ described by Eqs. (11)–(12) in [6] at $\gamma = 0$ and at $\gamma = \infty$ differs from one another in 1.33 times for chosen parameters in [6]. Comparing them with Eq. (14):

$$F_x^{21} = 2F_2^{id}(r_w^2/r_{pl}^2 - 1) \qquad (15)$$

at $\gamma = \infty$. With $r_w/r_{pl} = 2$ as in [6] we have $F_x^{21} = 6F_2^{id}$. This shows us that the result described by Eqs. (11)–(12) in [6] overestimates the wall force 6 times. Let us notice that comparison of $F_x^{11}$ in Eqs. (13)–(14) in [6] with $F_{(43)}^S$ obtained after integration across the wall volume of Eq. (43) in [4] at $\gamma = \infty$ gives us

$$F_x^{11} = 2F_{(43)}^S r_w^3/r_{pl}^3 . \qquad (16)$$

At $r_w/r_{pl} = 3$ as in [4], $F_x^{11} = 54F_{(43)}^S$ despite the statement made right before Eq. (13) in [6] that $F_x^{11}$ and $F_{(43)}^S$ are the same.

In [2], equation $F_X^{Noll} \equiv 2F_{Xm}\xi_z/r_w$ was obtained. It gives a sideways force on the rigid ring with current $J$ in the toroidal field $B_{0\zeta}$ when this ring is tilted about the axis $X$ by a small angle $\alpha = \xi_z/R_0$ so that $\xi_z$ is the amplitude of displacement. With ITER [8] parameters, mentioned in section 2, $F_X^{Noll} \geq 135$ MN requires $\xi_z/r_{pl} \geq 0.27$. At $q_{pl} = 1.1$, $|\xi_{r,11}|/r_{pl} \geq 0.27$ and $|\xi_{r,21}\xi_{VDE}|/r_{pl}^2 \geq 0.073$ the sideways wall force $F_2^{ITER} \geq 40$ MN, while $F_1^{ITER} \geq 16$ MN calculated in [7]. Thus at $q_{pl}$ near 1 the quadratic term $F_2$ of Eq. (4) dominates over linear, which $F_1 \to 0$ in this case.

**5. Conclusion.** In contrast to [4, 6], we have shown that the force is monotonically increasing with $\gamma\tau_w$ and maximal at $\gamma = \infty$. With the latter the sideways force $F_2$ produced by $m/n = 2/1$ kink mode combined with a plasma vertical displacement can be larger than 48 MN that ITER is designed for [9] at $q_{pl} \to 1$ and $\xi_{r,21}\xi_{VDE}/r_{pl}^2 \geq 0.073$, while $F_1 \to 0$ [7].

[1] V. Riccardo, P. Noll, and S. P. Walker, Nucl. Fusion **40**, 1805 (2000).
[2] V. Riccardo, S. Walker, and P. Noll, Fusion Eng. Des. **47**, 389 (2000).
[3] V. Riccardo and S. P. Walker, Plasma Phys. Controlled Fusion **42**, 29 (2000).
[4] H. R. Strauss, R. Paccagnella and J. Breslau, Phys. Plasmas **17**, 082505 (2010).
[5] L. E. Zakharov, S. A. Galkin and S. N. Gerasimov, Phys. Plasmas **19**, 055703
[6] H. R. Strauss, et. al, Nucl. Fusion **53**, 073018 (2013).
[7] D.V. Mironov and V.D. Pustovitov, Phys. Plasmas **22**, 052502 (2015).
[8] T. C. Hender *et al.*, Nucl. Fusion **47**, S128 (2007).
[9] V. Riccardo, *et al.*, Plasma Phys. Controlled Fusion **52**, 124018 (2010)
[10] V. D. Pustovitov, Phys. Plasmas **19**, 062503 (2012).
[11] V. D. Pustovitov, Plasma Phys. Rep. **38**, 697 (2012).

# Improvement of energetic particle confinement through stellarator optimization

S.A. Lazerson[1], P. Bolgert[1], D. Gates[1], M. McMillan[2]

[1] *Princeton Plasma Physics Laboratory, Princeton, U.S.A.*

[2] *Christ's College, Cambridge, U.K.*

Stellarator designs with good predicted energetic particle confinement are key to forwarding the stellarator concept as a reactor design. In a nuclear fusion reactor, alpha particles produced by the deuterium-tritium reaction must stay inside the plasma volume long enough to thermalize with the background plasma. Numerical optimisation of stellarator equilibria has already been used to demonstrate improvements in neoclassical confinement [1], the possibility of reduced turbulent transport [2], and inherent MHD stability [3] (no disruptions). In this paper, the capability to optimise stellarator designs for energetic particle confinement using the STELLOPT [5] and BEAMS3D [4] codes is documented.

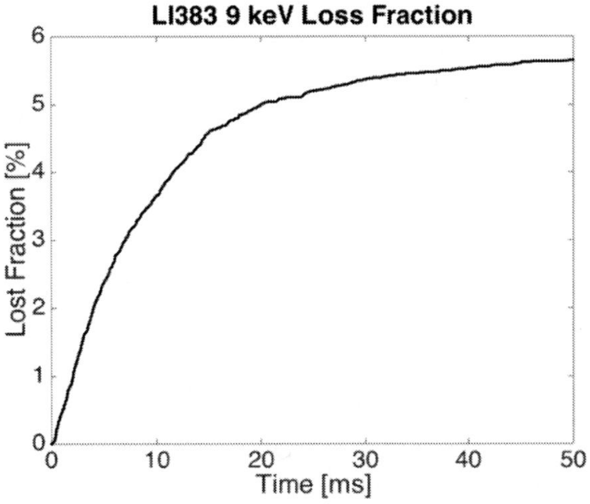


Figure 1: *9 keV proton loss rate in the LI383 configuration.*

Optimisation of stellarator equilibria has been carried out using the STELLOPT optimiser coupled to the BEAMS3D guiding center code. The STELLOPT code is designed to optimise the input parameters of an ideal MHD equilibrium (VMEC [6]) to various target physics parameters using Levenberg-Marquardt, Differential Evolution, and Particle Swarm optimisation techniques. In this work, the boundary harmonics (in Hirshman-Breslau form) of the NCSX fixed boundary baseline (LI383) are optimised with regards to Ballooning stability (provided by COBRA) and energetic particle targets. Sigmas are chosen so as to weight the chi-squared function toward energetic particle confinement. The BEAMS3D code provides a means to calculate the energetic particle orbits. Originally designed for neutral beam injection simulations, the code has been interfaced to STELLOPT allowing a user defined particle birth profile to be chosen. In this work, two energetic particle starting grids were used. The initial and 'optimised' equilibrium were evaluated using

12,000 particles, with particles born on a 20 by 20 grid in poloidal and toroidal space. In these runs, 30 pitch-angles were considered. In order to make the optimisation tractable on the Hydra cluster, a reduced set of particles was considered focusing on larger pitch angles particles (which were driving the majority of the losses). In all cases the particles were launched from the $s = 0.05$ (normalised toroidal flux) surface and had a total energy of 9 keV. Protons of this energy in NCSX have the same gyroradius to minor radius ratio as 3.5 $MeV$ alpha particles in the ARIES-CS device. Particles were followed for 50 ms of simulated time as the loss fraction appears to asymptote by this point (Figure 1). The total time for conducting such a simulation is reduced through the parallel computation of the BEAMS3D code and that of STELLOPT.

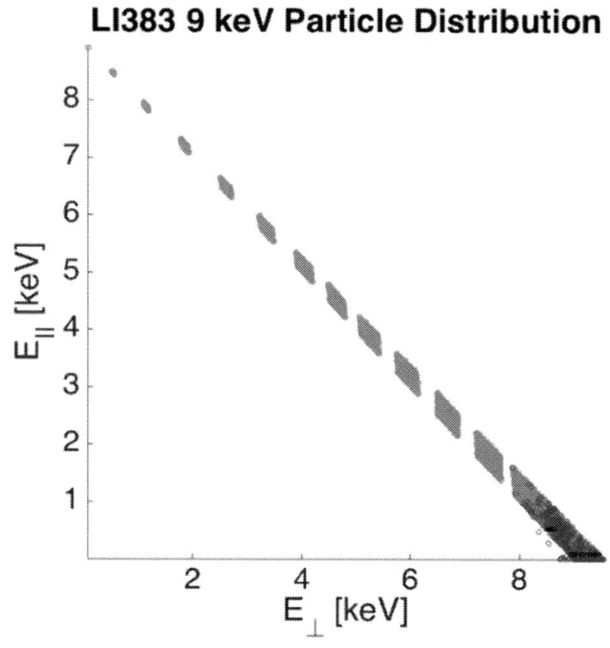


Figure 2: *Initial (sqaure) and final (circle) 9 keV particle velocity space distribution. Lost particles (red) are localised to large pitch angle, while confined particles (green) span make up a significant fraction of the population.*

In order to determine if particles are lost, a wall collision model has been integrated into the BEAMS3D code. The wall is modelled using triangular tessellation, allowing both simple and complex geometries. Calculation of collisions between a particle trajectory and the wall is done using ray tracing. Every time-step the code determines if the ray formed by the particle position at the previous time-step and the current time-step intersects one the the triangles defining the wall. If the trajectory strikes the wall, the strike location is saved and trajectory integration for that particle ends. For the fixed boundary equilibrium used in this work, the VMEC boundary for a given equilibrium is used to generate a tessellated wall composed of 120 points in the poloidal and toroidal direction.

The optimisation begins with an initial evaluation of particle confinement in the LI383 equilibrium. In this configuration, particles are uniformly lost from the flux surface. No one starting location for the particles appears to be more ill-conditioned to confine particles than another. Examining the velocity space distribution it becomes clear that losses are coming from particles with fairly large pitch angles (Figure 2). While in experimental devices this can be solved by tangential neutral beam injection, a reactor suffers from a more isotropic alpha particle distribution. It should also be noted that the total percentage of lost particles is in line for predictions made for the NCSX stellarator at 40 keV neutral beam injection [7]. Note that in these calculations no slowing down or scattering effects were included.

Figure 3 depicts the successive reduction in particle losses as the optimiser found new minima. In each case changes to the equilibria were small, highlighting the high sensitivity of the equilibrium to boundary changes. Examination of the parameter space Jacobian shows that the $n = 1$ modes tend to dominate the descent. These modes show the steepest gradients. The Levenberg trajectory, calculated by STELLOPT, found minima which were not of lower value than those found during the Jacobian evaluation. Thus the STEPOPT routine turns on which evaluates combinations of the finite difference calculation yielding the greatest reduction in Chi-squared ($\chi^2 = \sum \left(f_i^{target} - f_i^{equil.}\right)^2 / \sigma_i^2$). This contributed to the relatively small changes in the equilibrium.

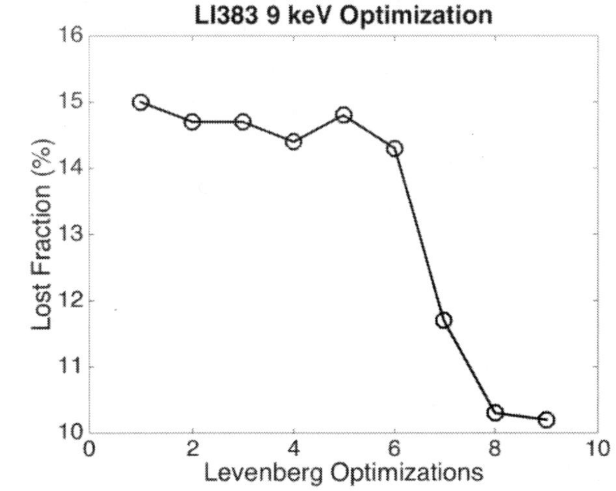

Figure 3: *Optimisation of particle losses using reduced particle set. The bump at iteration 5 is due to the equilibrium being restarted from iteration 3.*

The final optimised configuration shows a reduction in lost particle fraction from 5.4% to 5.2% overall. This comes after the sub-population, with pitch angles in the range of $v_{\|}/v_{\perp} = [0.1, 1.0]$, saw a reduction of lost particles from 15% down to just above 10%. Work is underway to develop runs with larger particle populations in order to drive down losses. The current runs are limited by the available processors and 24 hours simulation time.

The capability to optimise stellarator equilibrium for improved energetic particle confinement is being developed using the NCSX baseline configuration. In this work, 9 keV protons were utilised as a proxy for 3.5 MeV alpha particles in ARIES-CS. Given the current computational resources (10,000 processors for 24 hrs), runs with over 10,000 particles proved problematic for optimisations. This could be alleviated by longer runs and a larger number of processors. Given the scaling documented on the Hydra machine, a single Levenberg optimisation using 10,000 particles in 24 hours time would require around 90,000 processors. It may also be possible to improve the results using a reduced set of particles by increasing the number of particles modestly.

## References

[1] J. M. Canik *et al.*, Phys. Rev. Lett. **98**, 085002 (2007)
[2] H. Mynick *et al.*, Plasma Phys. Control. Fusion **56**, 094001 (2014)
[3] G. Y. Fu *et al.*, Fusion Sci. and Tech. **51**, 218 (2007)
[4] M. McMillan and S. A. Lazerson, Plasma Phys. and Control. Fusion **56**, 095019 (2014)
[5] D. A. Spong *et al.*, Nucl. Fusion **41**, 711-6 (2001)
[6] S. .P. Hirshman and J. C. Whitson, Phys. Fluids **26**, 3553 (1983)
[7] M. C. Zarnstorff *et al.*, Plasma Phys. and Control. Fusion **43**, A237 (2001)

# Mysteries of PEDESTAL poloidal ASYMMETRIES revealed

Silvia Espinosa and Peter J. Catto

*Plasma Science and Fusion Center (PSFC-MIT), Cambridge, USA*

## Introduction and motivation

Since high confinement operation was discovered, understanding tokamak pedestal physics has become one of the most crucial challenges facing the development of magnetic fusion energy. Gaining insight into flows and transport in this region may lead to the achievement of even higher energy confinement times and, consequently, fusion performance.

It has been suggested that the sudden transition between states of low and high confinement, the L-H transition, involves the suppression of turbulence by sheared radial electric fields [1]. For H-mode pedestals, the amount of turbulence may be only large enough to affect high order phenomena, such as heat transport. Neoclassical collisional theory may than be expected to properly treat low order phenomena, such as flows.

Nevertheless, high confinement mode edge pedestals on Alcator C-Mod [2, 3] (Fig. 1 and 2) exhibit significantly stronger poloidal asymmetry in both boron temperature and density than predicted by the most comprehensive neoclassical theoretical models developed to date [5]. First, the boron temperature is larger on the outboard (LFS) side compared to the inboard side (HFS) by up to a 70%. Hence, it seems not to be a flux surface as assumed by the previous models. Second, there is accumulation of boron density on the high field side up to six fold, which almost doubles the theoretical predictions. Thus, the magnetic field in-out asymmetry seems not to be its only drive. Importantly, the largest in-out asymmetry in density and temperature are not reached at the same radial location (Fig. 2). As a result, there is concern [3] that turbulence might affect even the lowest order phenom-

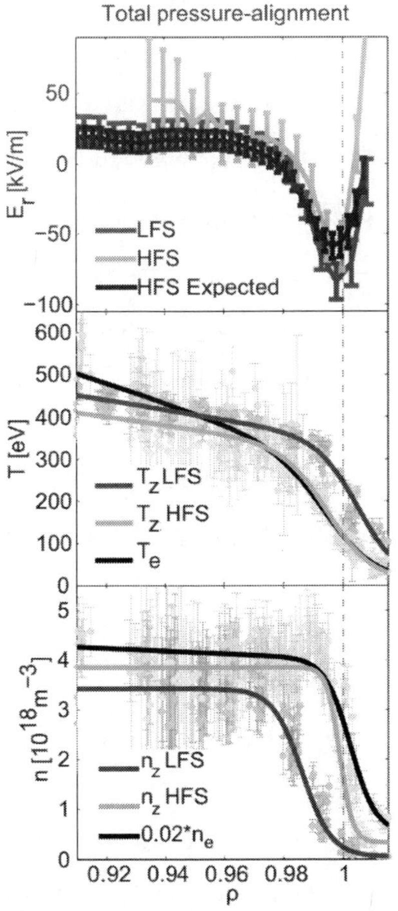


**Figure 1:** *High and low field side radial profiles of electric field, boron temperature and density [3].*

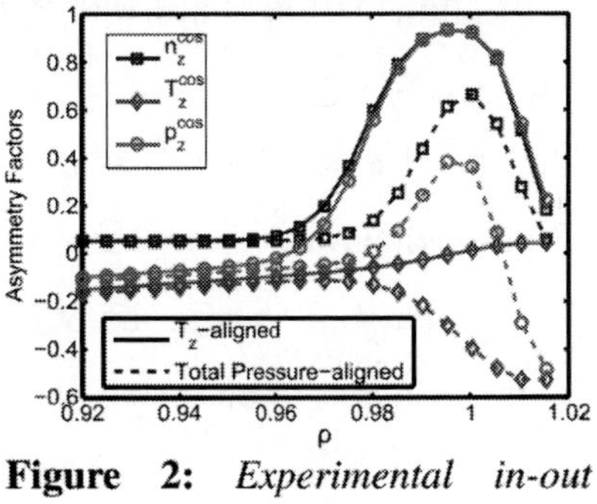


**Figure 2:** *Experimental in-out asymmetries,* $\frac{HFS-LFS}{HFS+LFS}$ *[3].*

ena, via for instance ballooning modes.

We propose a novel self-consistent neoclassical theoretical model that allows us to explain these poloidal asymmetries, without the need of invoking anomalous transport. Its main novel features are listed on Table 1. First, impurity temperature asymmetries can be driven by inertial effects, which are significant when impurities are allowed to reach sonic speeds. Second, a much stronger impurity density in-out asymmetry than given by just the magnetic field can be introduced by the poloidally varying impurity diamagnetic drift.

| **Physics captured** | [4] | [5] | Us |
|---|---|---|---|
| Non-trace impurities | | ✓ | ✓ |
| Strong radial electric field | ✓ | | ✓ |
| Diamagnetic flow effects | | | ✓ |
| Strong temperature gradient | | | ✓ |
| Sonic background ions | | | ✓ |
| Poloidal density variation | | Weak | ✓ |
| Poloidal temperature variation | | | ✓ |

**Table 1:** *Physics captured and comparison with previous pedestal models.*

**Self-consistent predictive neoclassical model for pedestal poloidal asymmetries**

In this section, the novel physics included in the improved neoclassical theory developed and their implications are introduced. The mathematical details will be found in a publication to follow.

*Orderings:* The theoretical model consists of highly charged non-trace collisional impurities, bulk ions and almost collisionless electrons. Strong impurity temperature and density poloidal variation are allowed. A small aspect ratio limit is assumed, so the poloidal background ion gyroradius is much larger than the ion gyroradius. The pedestal width of electrostatic potential can be comparable to the poloidal background ion gyroradius, so that the impurity and background ion flows can reach the ion thermal speed. In addition, the pedestal width of the impurity density is such that the impurity diamagnetic drift contribution is allowed to be comparable to the impurity poloidal and toroidal flow terms when measuring the radial electric field, as experimentally observed [2]. Sonic behavior is assumed to happen for the background ion diamagnetic term. The bulk ion temperature pedestal width is taken to be of the same order as that of its density. Finally, it is assumed that the impurity and background ion temperatures are of the same order at all pedestal radial locations, and hence its

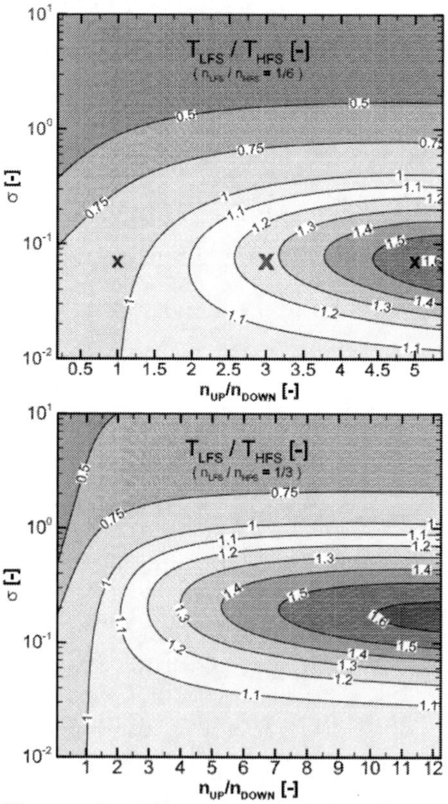


**Figure 3:** *Effect on the ratio of outboard to inboard impurity temperature of the impurity density up-down asymmetry and ratio between poloidal and collisional impurity flow, σ. Both strong (up) and weak (down) impurity density in-out asymmetry are presented.*

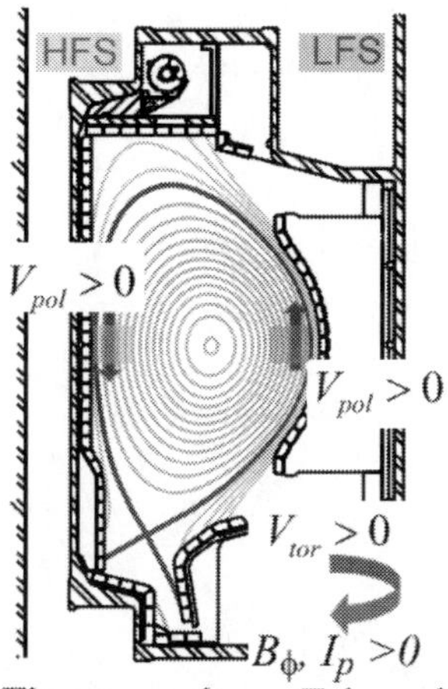


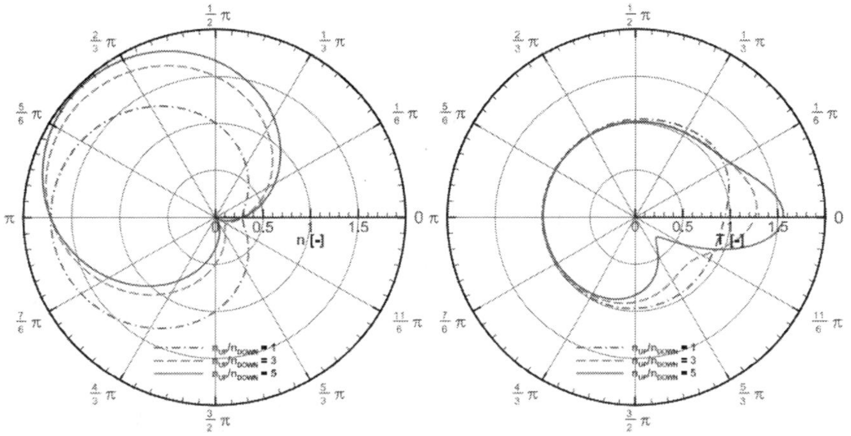


**Figure 4:** *Tokamak poloidal cross section [2].*

**Figure 5:** *Dimensionless impurity density (left) and temperature (right) polar profiles, for points marked on Fig.3.*

pedestals widths are of the same order. As a result, the impurity density pedestal width is smaller than that of its temperature by approximately the ratio of impurity to background atomic numbers, which is also in agreement with the experiments (Fig. 1).

*Conservation equations for impurities:* In order to analyse the parallel dynamics on each flux surface separately, the parallel equilibrium is assumed to be reached much faster than the cross field equilibrium. The perpendicular momentum conservation is assumed to be given by

$$\mathbf{V}_{z\perp} = \frac{c}{B^2}\mathbf{B} \times \left(\nabla\Phi + \frac{1}{z_z e n_z}\nabla p_z\right) \tag{1}$$

where $\mathbf{V}_z$ is the impurity flow, $\mathbf{B}$ is the magnetic field, $\Phi$ the electrostatic potential and $z_z$, $n_z$ and $p_z$ the impurity atomic number, density and pressure, respectively. Using conservation of particles on the flux surface,

$$\mathbf{V}_z = -cR^2\nabla\zeta\left(\frac{\partial\Phi}{\partial\psi} + \frac{1}{z_z e n_z}\frac{\partial p_z}{\partial\psi}\right) + \frac{K_z}{n_z}\mathbf{B} \tag{2}$$

where $R$ is the tokamak major radius, $\zeta$ and $\psi$ the poloidal angle and flux and $K_z$ a flux function.

The parallel momentum and energy conservation equations are assumed to be the following:

$$m_z n_z \mathbf{V}_z \cdot \nabla\mathbf{V}_z \cdot \frac{\mathbf{B}}{B} = R_{zi\|}, \qquad R_{zi\|} = \frac{1}{2} m_z n_z \nu_{zi}\left(V_{i\|} - V_{z\|}\right) + O\left(\sqrt{\varepsilon}\right) \tag{3}$$

$$\frac{n_z^2}{T_z^{\frac{1}{2}}}\mathbf{V}_z \cdot \nabla\left(\frac{T_z^{\frac{3}{2}}}{n_z}\right) = Q_{zi}, \qquad Q_{zi} = \frac{3}{2} n_z \nu_{zi}\left(T_i - T_z\right) + O\left(\sqrt{\varepsilon}\right) \tag{4}$$

where $m_z$ is the impurity mass and $\nu_{zi}$ the collision frequency of bulk ions and impurities, $V_{i\|}$ the bulk ion parallel flow, $\varepsilon$ the aspect ratio and $T_z$ and $T_i$ the impurity and bulk ion temperatures.

## Conclusions

Given an impurity density profile, such us $n \approx 1 + C\cos\theta + S\sin\theta$, the energy equation can be solved numerically for the impurity temperature (Fig. 3); where $n \equiv \frac{n_z}{\langle n_z\rangle}$, $T \equiv \frac{T_z}{T_i}$, $\sigma \equiv \frac{K_z \mathbf{B}\cdot\nabla\theta}{\nu_{zi}\langle n_z\rangle} \sim$

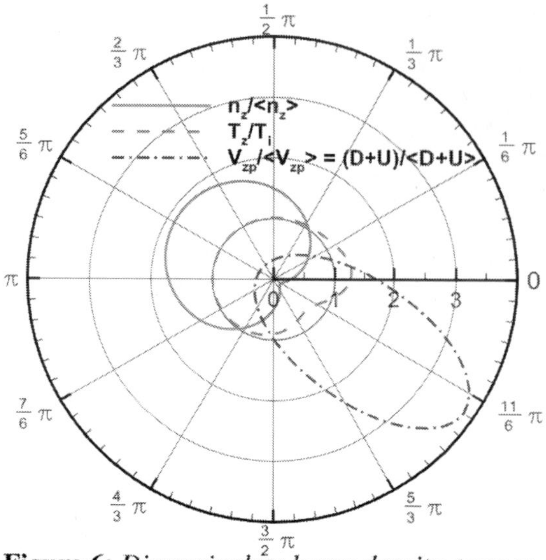


**Figure 6:** *Dimensionless boron density, temperature and poloidal velocity polar profiles, for the green point marked on Fig.3.*

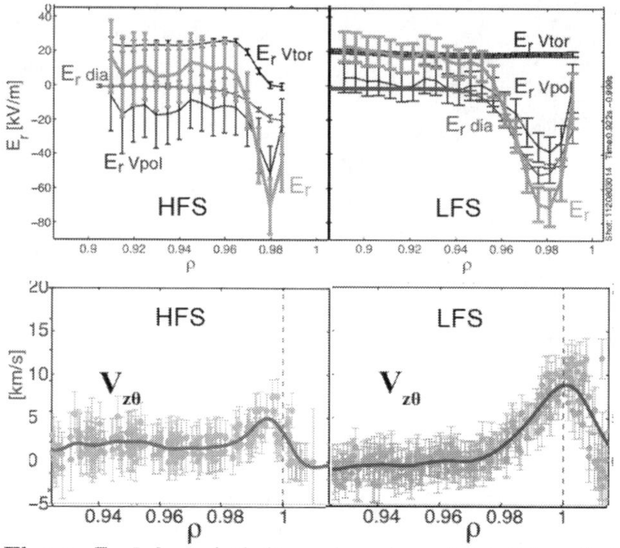


**Figure 7:** *Inboard (left) and outboard (right) radial profiles of the radial electric components [2] (top) and boron poloidal flow (down) [3].*

$\frac{V_{zp}}{V_{zi}qR}$, $\theta$ the poloidal angle and $\langle\rangle$ denotes flux surface average. It can be seen that, in the small aspect ratio limit, the model developed can capture the experimental temperature by selecting appropriately the up-down asymmetry of the density (Fig. 3 and 5).

When the bulk ion density and temperature vary slowly poloidally, the parallel momentum equation has an analytical solution given by $n = \frac{1}{D+U} + O\left(\sqrt{\varepsilon}\right)$; where $U \equiv \frac{BV_{\parallel i}\langle n_z\rangle}{K_z\langle B^2\rangle} \sim \frac{V_{\parallel i}}{V_{z\theta}}$ and $D \equiv \frac{c\langle n_z\rangle\sqrt{\langle R^2\rangle}}{K_z\sqrt{\langle B^2\rangle}}\left(\frac{\partial\Phi}{\partial\psi} + \frac{1}{z_z e n_z}\frac{\partial p_z}{\partial\psi}\right) \sim \frac{V_{z\zeta}}{V_{z\theta}}$. It can be solved for the poloidal variation of the impurity diamagnetic flow, which is similar to that of the impurity poloidal flux (Fig. 6). These predictions successfully agree with the experiments (Fig. 7). Hence, this theory may be used in the future to get information, for instance, about all poloidal angles or the background.

We conclude that these extensions in neoclassical theory successfully capture the strong experimental poloidal impurity temperature and density variation, providing a more realistic predictive model for pedestal observations without the need of invoking anomalous transport.

## References


[1] J.W. Connor and H.R. Wilson, Plasma Phys Contr Fusion **42**:R1 (2000)

[2] C. Theiler et al., Nucl Fusion **54**, 083017 (2014)

[3] R.M. Churchill et al., Phys.Plasmas, **22**, 056104 (2015)

[4] G. Kagan and P.J. Catto, Plasma Phys Contr Fusion **50**, 085010 (2008)

[5] P. Helander, Phys Plasmas **5**, 3999 (1998)


The authors are indebted to Prof. Ian Hutchinson and Jeffrey Freidberg for stimulating discussions.Work supported by DOE Grant DE-FG0291ER54109 and La Caixa Fellowship.

# Propagation of radio frequency waves through fluctuations in plasmas

A.K. Ram[1], K. Hizanidis[2], S. Valvis[2]

[1] *Plasma Science and Fusion Center, M.I.T., Cambridge, MA., USA*

[2] *National Technical University of Athens, Association EURATOM, Athens, Greece*

## Introduction

The scrape-off layer in a tokamak fusion plasma is replete with coherent fluctuations in the form of blobs or filaments [1, 2, 3, 4]. Radio frequency (RF) waves, excited by an external antenna structure, have to propagate through this turbulent region on their way toward the core of the plasma. Since the plasma permittivity inside the fluctuations is different compared to the background plasma, the characteristic properties of the incident RF waves can change during their transit through the fluctuations. It is common to study these changes using the geometric optics approximation (see [5] and references therein). In this approximation, rays, representing plane waves, are refracted due to changes in the plasma permittivity. The domain of validity of this approximation is quite limited; it requires that $\delta n \equiv |n_f - n_b|/n_b \ll 1$ ($n_f$ is the density of electrons inside the fluctuation and $n_b$ is the background density). However, from experimental observations, typically, $\delta n \gtrsim 1$ in the scrape-off layer. Consequently, a study of the scattering process requires a more general approach that is not limited by the bounds of the geometric optics approximation.

In this paper, we present a theoretical model for the scattering of RF waves by density filaments using the full-wave Maxwell's equations. The model is based on the conventional Mie theory that is used for scattering of vacuum electromagnetic waves by spherical dielectrics [6]. Our model applies to magnetized plasmas for which the permittivity is given by a tensor, rather than a scalar, and is derived from our previous study on the scattering of RF waves by spherical blobs [7, 8]. The present model pertains to scattering off a cylindrical filament with its axis aligned along the toroidal magnetic field line. It assumes that the background plasma, as well as the plasma in the filament, is cold and has uniform density. However, it is not restricted to small density fluctuations – $\delta n$ is completely arbitrary. The theoretical framework applies to plasma waves of any frequency; it is valid for ion cyclotron, lower hybrid, and electron cyclotron waves.

Beyond their respective domain of validity, there are three primary differences between the full-wave theory and the geometric optics approximation. First, whereas geometric optics is used for describing refractive changes in the ray propagation, the full-wave model also includes reflection and diffraction. Second, in geometric optics, the character of the wave does not change as the RF ray propagates through fluctuations. For example, an incident ordinary wave in the electron cyclotron range of frequencies will remain an ordinary wave during its encounter with fluctuations. In the full-wave model, the fluctuations can couple power to other plasma waves. Thus, for the example considered, the ordinary wave couples power to the extraordinary wave. This is not nonlinear parametric coupling; it is linear coupling facilitated by the fluctuations. Third, due to diffraction of waves by the filament, the scattered waves propagate in all radial directions relative to the magnetic field line. Consequently, fluctuations can scatter some of the incident wave power to surface waves which do not propagate into the core plasma. This effect cannot be described within the geometric optics approximation.

**Basic equations and boundary conditions**

We assume that the axis of the cylindrical filament is aligned along the magnetic field. The center of the coordinate system is taken to be the center of the filament. The filament is assumed to be stationary and of infinite extent in the axial direction. For a cold fluid plasma, described by a linearized set of continuity and momentum equations for electrons and ions, Faraday's and Ampere's equations in Maxwell's system of equations can be combined [9] to yield the following equation for the spatial variation of the electric field,

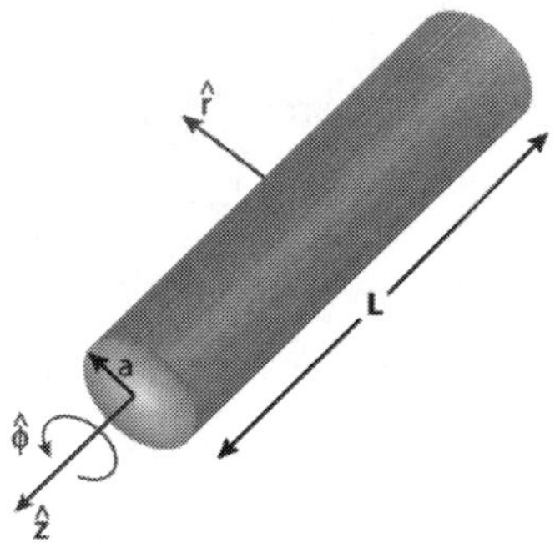

Figure 1: *Coordinate system with respect to the cylindrical filament*

$$\nabla \times \nabla \times \mathbf{E}(\mathbf{r}) - \frac{\omega^2}{c^2} \overleftrightarrow{\mathbf{K}}(\mathbf{r}) . \mathbf{E}(\mathbf{r}) = 0, \tag{1}$$

where $\omega$ is the angular frequency, $c$ is the speed of light, and $\overleftrightarrow{\mathbf{K}}(\mathbf{r})$ is the plasma permittivity tensor. We have assumed that the plasma equilibrium is time independent, while the linearized perturbed electromagnetic fields and the plasma density have a time dependence of the form $e^{-i\omega t}$. In the cylindrical coordinate system, in which the ambient magnetic field $\mathbf{B}_0 = B_0 \hat{\mathbf{z}}$ is aligned along the $z$-axis [9],

$$\overleftrightarrow{\mathbf{K}} = \begin{pmatrix} K_\rho & -iK_\phi & 0 \\ iK_\phi & K_\rho & 0 \\ 0 & 0 & K_z \end{pmatrix}, \tag{2}$$

with

$$
\begin{aligned}
K_\rho &= 1 - \frac{\omega_{pe}^2}{\omega^2 - \omega_{ce}^2} - \sum_\iota \frac{\omega_{pi}^2}{\omega^2 - \omega_{ci}^2}, \\
K_\phi &= -\frac{\omega_{ce}}{\omega}\frac{\omega_{pe}^2}{\omega^2 - \omega_{ce}^2} + \sum_\iota \frac{\omega_{ci}}{\omega}\frac{\omega_{pi}^2}{\omega^2 - \omega_{ci}^2}, \\
K_z &= 1 - \frac{\omega_{pe}^2}{\omega^2} - \sum_\iota \frac{\omega_{pi}^2}{\omega^2},
\end{aligned} \tag{3}
$$

where $\omega_{pe}$ ($\omega_{pi}$) and $\omega_{ce}$ ($\omega_{ci}$) are the angular electron (ion) plasma frequency and cyclotron frequency, respectively, and the index $\iota$ represents all the ion species in the plasma. The plasma and cyclotron frequencies can, in general, be functions of space. The permittivity tensor of the background plasma and of the filament are expressed in terms of their respective densities and ion compositions.

A critical component in the description of full-wave scattering is that the electromagnetic fields satisfy boundary conditions at the interface between the filament and the background plasma. The boundary conditions which follow from Maxwell's equations, with the requirement that there be no free surface charges or currents at the interface, are

$$
\Delta(\hat{\mathbf{r}}.\mathbf{B})_{r=a} = 0, \qquad \Delta(\hat{\mathbf{r}}.\mathbf{D})_{r=a} \equiv \Delta\left(\hat{\mathbf{r}}.\overleftrightarrow{\mathbf{K}}(\mathbf{r}).\mathbf{E}\right)_{r=a} = 0, \tag{4}
$$

$$
\Delta(\hat{\mathbf{r}}\times\mathbf{E})_{r=a} = 0, \qquad \Delta(\hat{\mathbf{r}}\times\mathbf{B})_{r=a} = 0. \tag{5}
$$

Here $\mathbf{B}(\mathbf{r})$ is the wave magnetic field, $\mathbf{D}(\mathbf{r})$ is the displacement electric field, $a$ is the radius of the filament, and $\Delta$ is the difference, at $r = a$, of the enclosed quantity evaluated inside the filament and in the background plasma. Of these six scalar conditions, it can be shown that only four are independent.

**Results**

In order to demonstrate the capability of the theoretical model, we consider the scattering of the slow lower hybrid wave [9] by a filament. The background plasma and the plasma inside the filament has deuterium ions with electron densities $2\times10^{19}$ cm$^{-3}$ and $2.5\times10^{19}$ cm$^{-3}$, respectively. The ambient magnetic field is 4.5 T, and the incident plane wave has a frequency of 4.6 GHz with a parallel refractive index of 2. Figure 2 shows the magnitude of the Poynting flux (normalized to the Poynting flux of the incident wave) in the $x-y$ plane. The incoming plane wave is incident from

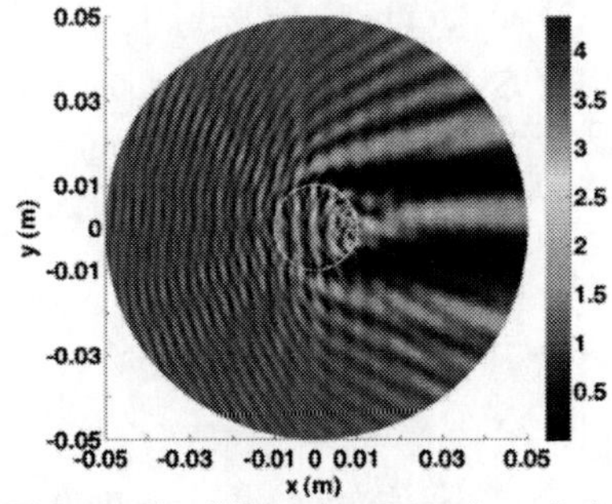


Figure 2: *The magnitude of the total Poynting flux, normalized to the Poynting flux of the incident wave, in the $x-y$ plane.*

the left hand side and the cylindrical filament, outlined in white, has a radius of 0.01 m. The pattern on the left hand side indicates a beating between the incident and the back-scattered wave. The enhanced flux in the forward direction, near the boundary of the filament, is a consequence of large electric and magnetic fields of the wave packet – a result of Gauss' law. The large amplitude fields can be the seeds for nonlinear parametric processes. The scattering in the forward direction includes diffraction and shadowing due to the filament. The detailed wave pattern emphasizes the point that the wave power propagating into the core of the plasma has a more complicated structure than a plane wave. Indeed, a spectral analysis of the electric fields in the forward direction points to a broadening of the wave vector spectrum along the magnetic field line. The spectrum is centered around the wave number of the incident wave. The side scattering of the lower hybrid wave is evident as there is power propagating in the $y$-direction. The incident Poynting vector is in the $x-z$ plane. For the parameters used to obtain Fig. 2, the fast branch of the lower hybrid wave is evanescent in the background plasma. The slow branch is the only propagating wave outside the filament. Inside the filament, both the slow and fast branches of the lower hybrid wave are propagating. This affects the properties of the scattered waves due to coupling, through the boundary conditions, inside the region separating the filament from the background plasma.

The research reported in this paper has been supported in part by the US Department of Energy Grant numbers DE-FG02-91ER-54109 and DE-FG02-99ER-54525-NSTX, and by National Programme on Controlled Thermonuclear Fusion associated with the EUROfusion Consortium.

## References

[1] S. I. Krasheninnikov, *Phys. Lett. A* **283**, 368–370 (2001).

[2] J. R. Myra, D. A. D'Ippolito, D. P. Stotler, S. J. Zweben, B. P. LeBlanc, J. E. Menard, R. J. Maqueda, and J. Bodeo, *Phys. Plasmas* **13**, 092509-1–092509-10 (2006).

[3] O. Grulke, J. L. Terry, B. LaBombard, S. J. Zweben, *Phys. Plasmas* **13**, 012306-1–012306-7 (2006).

[4] S. J. Zweben, R. J. Maqueda, D. P. Stotler, A. Keesee, J. Boedo, C. E. Bush, S. M. Kaye, B. LeBlanc, J. L. Lowrance, V. J. Mastrocola, R. Maingi, N. Nishino, G. Renda, D. W. Swain, J. B. Wilgen, and the NSTX Team, *Nucl. Fusion* **44**, 134-153 (2004).

[5] K. Hizanidis, A. K. Ram, Y. Kominis, C. Tsironis, *Phys. Plasmas* **17**, 022505-1–022505-10 (2010).

[6] H. C. van de Hulst, *Light Scattering by Small Particles*, Dover, New York, 1957, Chapter 9.

[7] A. K. Ram, K. Hizanidis, and Y. Kominis, *Phys. Plasmas* **20**, 056110-1–056110-10 (2013).

[8] A. K. Ram and K. Hizanidis, *Radiation Effects and Defects in Solids: Incorporating Plasma Science and Plasma Technology* **168**, 759–775 (2013).

[9] T. H. Stix, *Waves in Plasmas*, Springer, New York, 1992, pp. 3–12.

# Role of Electric Field in the Formation of Detached Regime

I.Senichenkov[1], E. Kaveeva[1], V. Rozhansky[1], E. Sytova[1], I. Veselova[1],
S. Voskoboynikov[1], D. Coster[2]

*[1]Peter the Great St.Petersburg Polytechnic University, Saint Petersburg, Russia*

*[2]Max-Planck Institut für Plasmaphysik, EURATOM Association, Garching, Germany*

The detached regime of divertors is considered now as the main working regime of ITER, and the physics of its formation is actively studied experimentally [1-3] and theoretically [4].

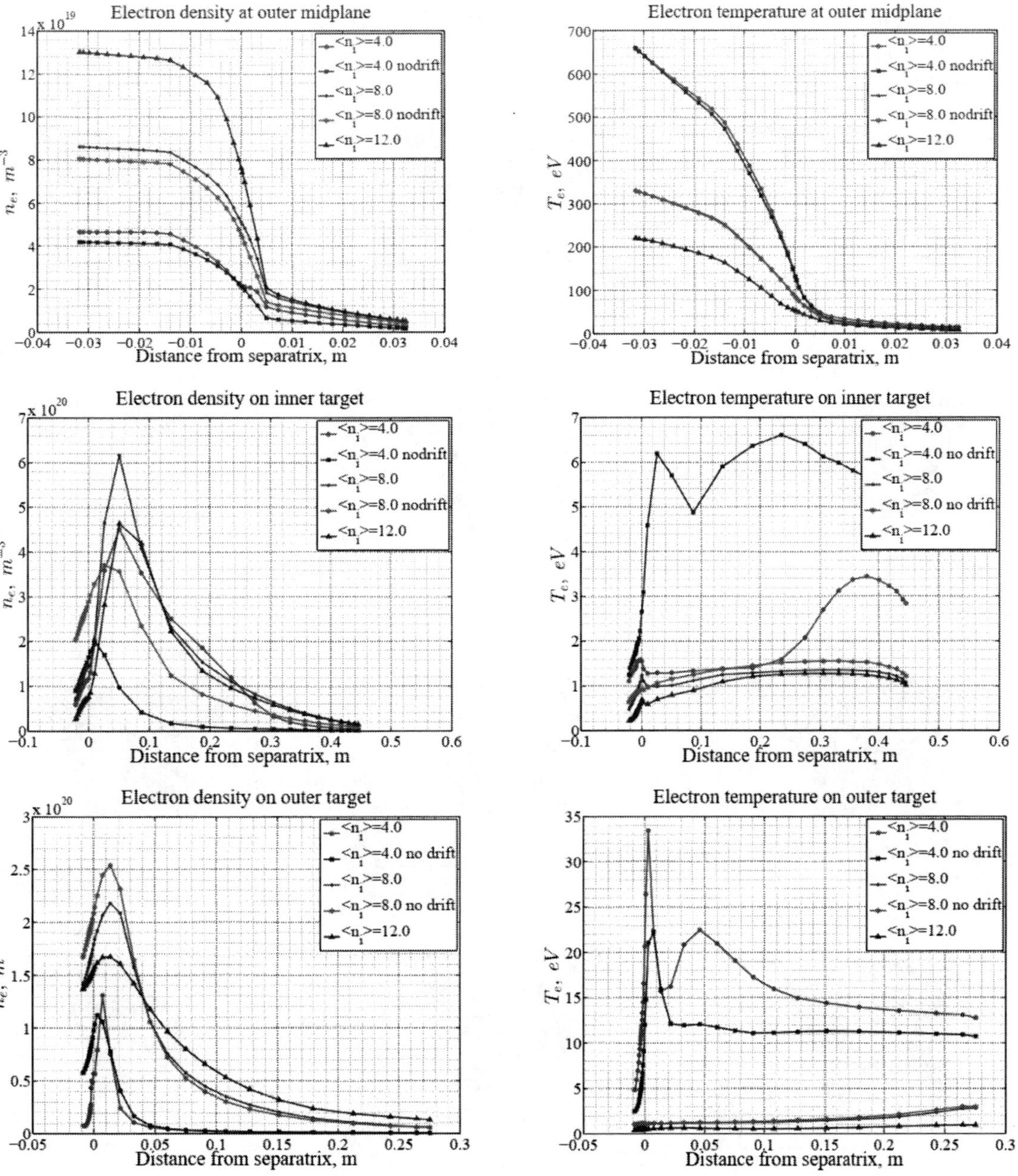


*Figure 1: Outer midplane and divertor profiles of electron density and temperature for all the runs considered. Here and later on runs are labeled by core boundary density of deuterium.*

In the present report the role of electric field and corresponding $\vec{E}\times\vec{B}$ drift and currents in the detached regime formation is studied using modeling with the SOLPS-ITER code [5]. As a base modeling scenario the ASDEX Upgrade shot #17151 (with carbon impurity) was chosen, then the transition to detachment was modelled by increasing the density at the core boundary of computational domain and a corresponding decreasing of temperatures keeping the product $n_e(T_e+T_i)$ constant. Total power through separatrix P ≈ 3 MW (without radiation) was the same for all variants. Additionally runs were made with drifts turned off as reference cases, see Figs 1 and 2. For $<n_i>=4.0\cdot10^{19}$ m$^{-3}$ the inner divertor is detached and outer is in high recycling regime. For $<n_i>=8.0\cdot10^{19}$ m$^{-3}$ and $<n_i>=12.0\cdot10^{19}$ m$^{-3}$ the outer divertor also becomes detached. As the core density increases, the particle flux first increases too (being proportional to saturation current), and then falls down, i.e. the rollover occurs.

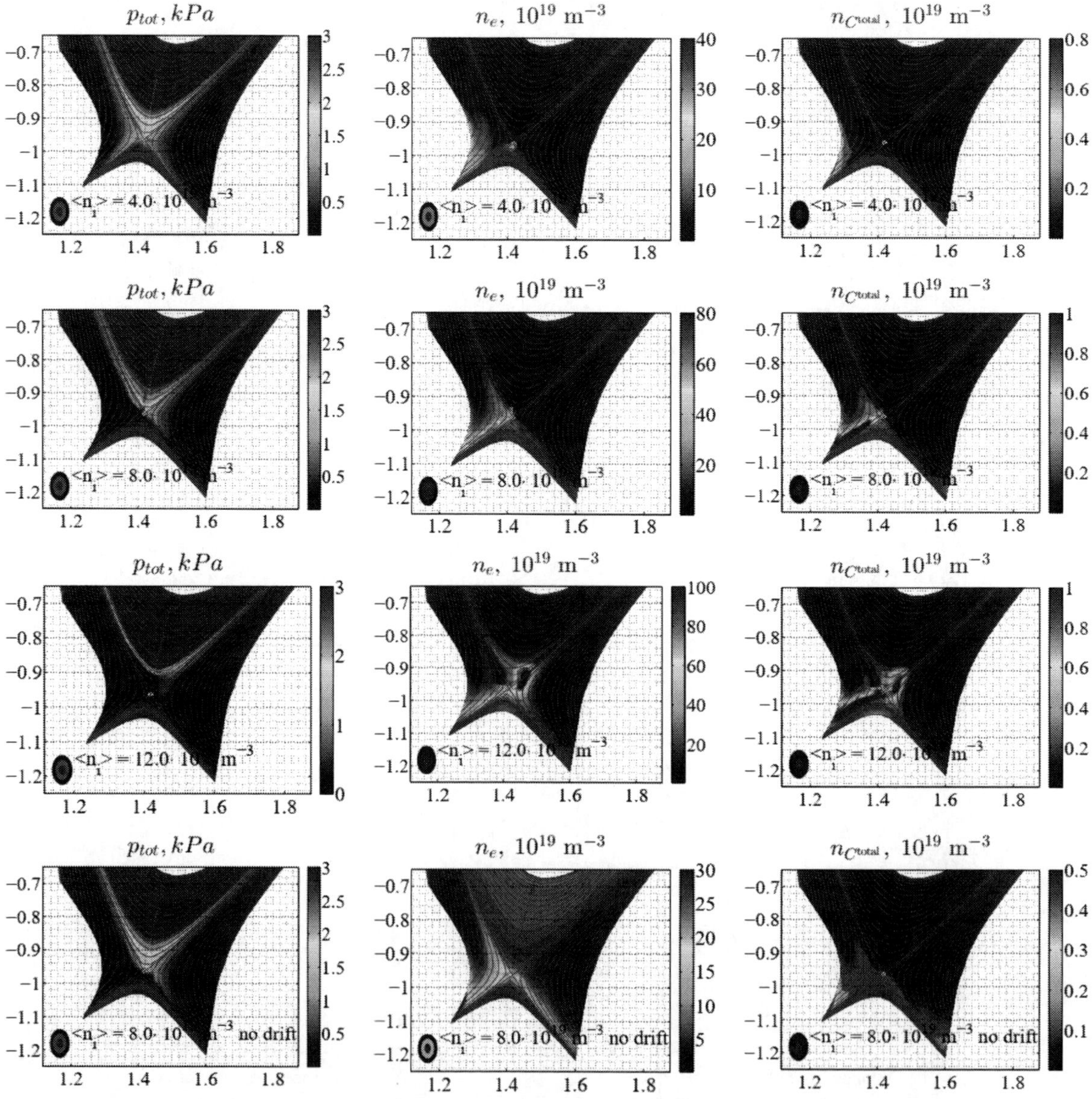


*Figure 2. Total pressure, electron and carbon density in several runs*

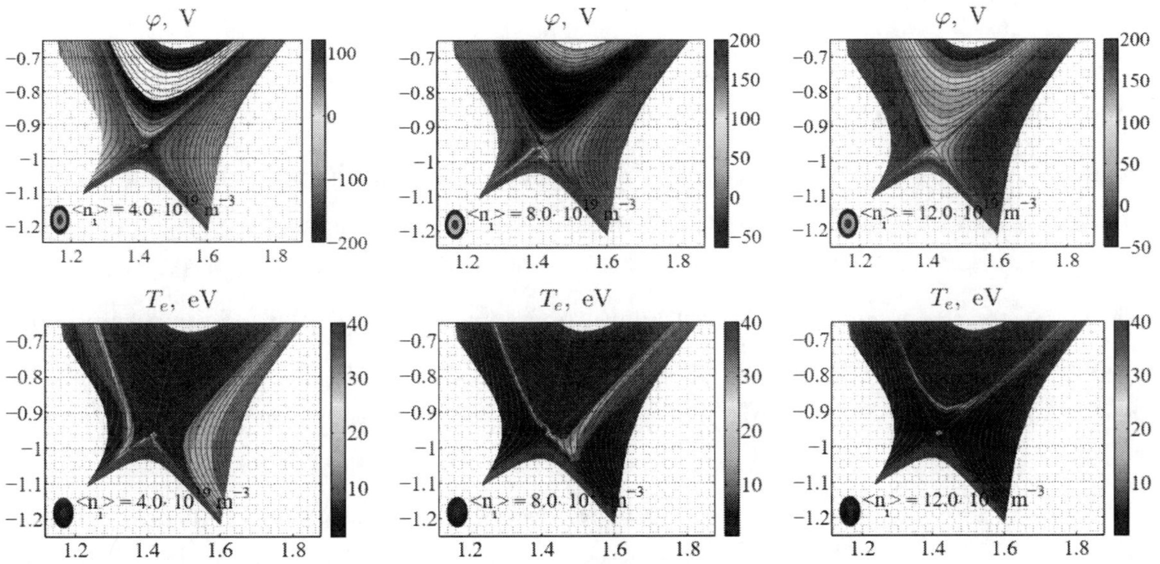


*Figure 3. Electrostatic potential and electron temperature for runs with* $<n_i>$=4.0, 8.0, 12.0 × $10^{19}$ m$^{-3}$

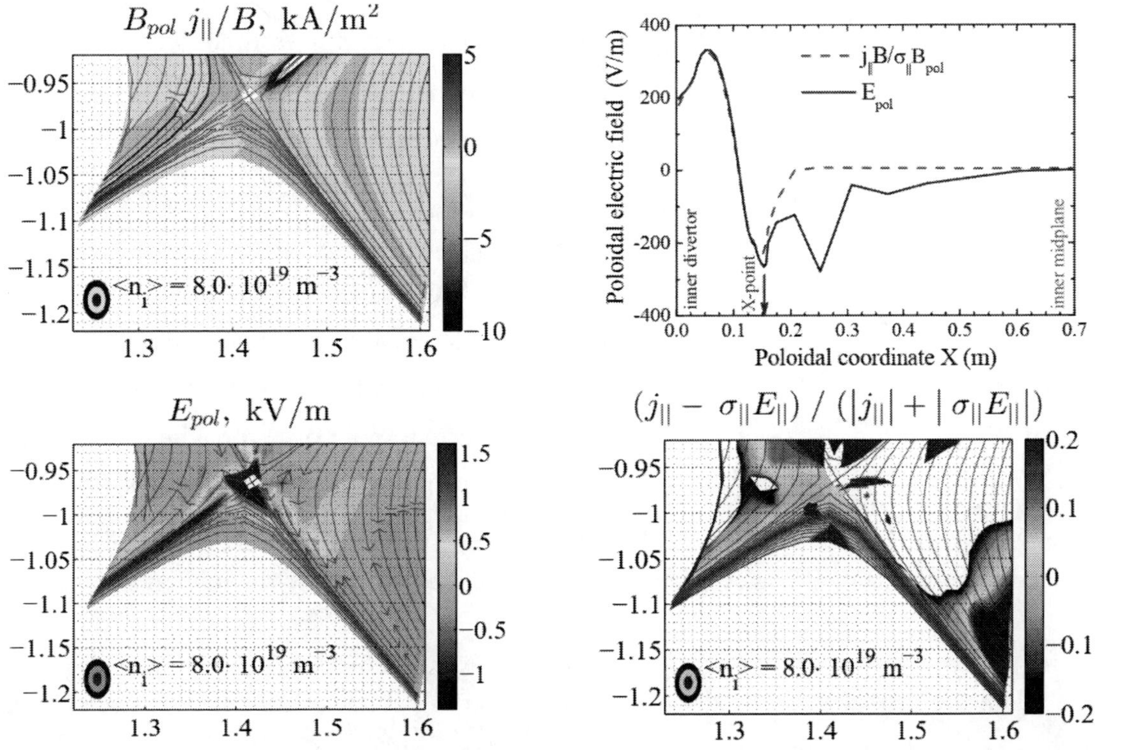


*Figure 4. Left top: Poloidal projection of parallel current density, colour arrows represent the direction. Left bottom: Poloidal electric field. Right top: Poloidal electric field along flux surface lying between two bold curves on left top graph. Right bottom: Zones where the relation* $j_{\|} \approx \sigma_{\|} E_{\|}$ *is satisfied.*

It is demonstrated that the typical feature of a detached regime of a divertor is the formation of rather strong potential in the divertor region, up to 100 V, which can be order of magnitude larger than $T_e/e$, see Figure 3. Strong poloidal electric fields in the cold regions of low Spitzer conductivity are required to provide parallel current $j_{\|} = \sigma_{\|}\left(E_{\|} + \nabla_{\|}(n_e T_e)/en_e + 0.71\nabla_{\|} T_e/e\right)$. Thus, in contrast to attached divertor, where electric field balances electron pressure gradient and thermal force, so that $E_{\|}+\nabla_{\|}(n_eT_e)/en_e+0.71\nabla_{\|}T_e/e\approx 0$ and $E_{\|}$ is of the order of $\nabla_{\|}T_e/e$, in the detached regime the relation $j_{\|}\approx\sigma_{\|}E_{\|}$ is satisfied in the cold region, and $|E_{\|}|$ becomes much larger than $|\nabla_{\|}T_e|/e$, see Fig. 4. The parallel current is the sum of parallel Pfirsch-Schlüter current which closes vertical $\nabla B$ driven current and thermal current flowing from the hotter to the colder divertor. Note that when both inner and outer divertors fall into detached regime, the thermal current is reduced significantly compared with the case when at least one divertor is attached. This results in the significant reduction of the net current on divertor plates down to values comparable with total radial currents through the outer SOL and PFR boundaries of computational domain, see Figure 5. Strong radial electric field also arises due to electrostatic character of the potential. As a result vortex flow of plasma in the divertor region is observed, which is shown schematically in Figure 6. This vortex flow is responsible for the convective transport of cold particles from inner target into the confined region. At $<n_i>=12.0\cdot 10^{19}$ m$^{-3}$ cold and dense highly radiating region in the X-point vicinity arises, which affects the core plasma (Fig 2). Another result of the $\vec{E}\times\vec{B}$ drifts is additional particle flows along the plates. These flows are comparable with the diffusive flows and could lead to the spreading of the density profiles along the plates towards the SOL. Effect is more pronounced at inner divertor, Fig. 1. However the structure of the poloidal electric field is complicated (Fig. 4), since it is determined by the parallel current, required to compensate PS and thermal currents, and therefore E-field could change sign.

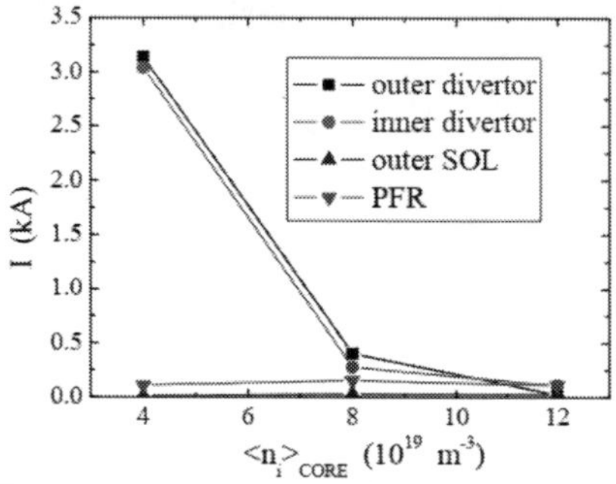


*Figure 5. Net current through boundaries of computational domain.*

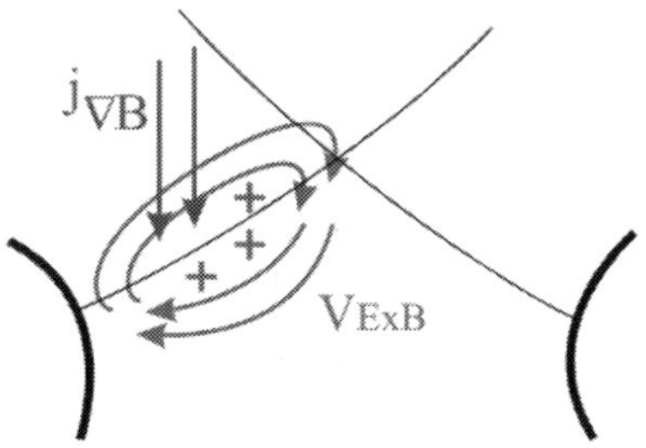


*Figure 6. Scheme of $\vec{E}\times\vec{B}$ vortex flow.*

## Conclusions

It is demonstrated that in the detached regimes strong electric field and $\vec{E}\times\vec{B}$ drift arise, which are completely different from those in high recycling regimes. The radial $\vec{E}\times\vec{B}$ drifts are comparable with the diffusive fluxes and responsible for density redistribution along the plates and near the X-point.

## References

[1] S. Potzel et al., Nucl. Fusion **54** (2014) 013001
[2] F. Reimold et al., Nucl. Fusion **55** (2015) 033004
[3] A. Kallenbach et al, Nucl. Fusion **55** (2015) 053026
[4] F. Reimold et al., Journal of Nuclear Materials 463 (2015) 128–134
[5] S. Wiesen et al., Journal of Nuclear Materials 463 (2015) 480–484

# A VLASOV CODE SIMULATION OF THE AMPLIFICATION OF SEED PULSES BY BRILLOUIN BACKSCATTERING IN PLASMAS

M. Shoucri[1], J-P. Matte[2], F. Vidal[2]

*1-Institut de recherche d'Hydro-Québec (IREQ), Varennes, Québec, Canada, J3X-1S1*

*2-INRS-Energie, Matériaux et Télécommunications, Université du Québec, Varennes, Québec, Canada J3X-1S2*

We use an Eulerian Vlasov code to study the problem of the amplification of seed pulses by Brillouin backscattering in the strong-coupling regime. In this process, there is an energy transfer, mediated by a resonant ion wave, from a long, high energy pump electromagnetic wave, to an initially counter-propagating ultra-short seed pulse. The code solves the one-dimensional Vlasov-Maxwell set of equations, and has been previously used to study the problem of the plasma-based backward Raman amplification of seed pulses in underdense plasmas [1], and the problem of the Brillouin backscattering amplification of seed pulses in overdense plasma in the strong coupling regime [2]. In the example we present here, we use the parameters which have been used in [2], but we increase the plasma amplifier length from the value of $1000c/\omega_{pe}$ to $1600c/\omega_{pe}$, which is the length used in the PIC simulations reported in [3], where essentially the same parameters were used for the same problem. For such a length of interaction, in the PIC simulations [3] the pump becomes depleted due to scattering from numerical noise. Here, the noiseless Vlasov code allows to follow the amplification of the seed pulses by Brillouin backscattering until the formation of steep gradients of the amplified pulses. We present simulation results in the plateau of a high density ($n/n_{cr}=0.3$) uniform plasma. This corresponds to a ratio of the pump frequency to the plasma frequency $\omega_{0p}/\omega_{pe}=1.826$. The normalized vector potential or quiver momentum $a_0$ verify $a_0^2=I\lambda_0^2/1.368\text{x}10^{18}$, where $I$ is the laser intensity in W/cm$^2$ and $\lambda_0$ the wavelength in microns. For the chosen pump beam intensity $I_p=10^{16}$ W/cm$^2$, and wavelength $\lambda_{0p}=1.0\mu m$, the pump normalized vector potential is $a_{0p}=0.0855$, assumed constant during the simulation. The normalized time $\omega_{pe}^{-1}$ and length $c\omega_{pe}^{-1}$ units correspond to 0.97 fs and 0.29 $\mu$m, respectively. Hydrogen plasma is assumed, with electron and ion temperatures $T_e$=500eV and $T_i$=50eV. The Debye length is $\lambda_D=0.0313$. With $N$=50000 grid points in space, we have for the mesh-size $\Delta x = 0.032$, which is of the order of the Debye length. We use for the seed pulse an intensity of $I_s = 10^{15}$ W/cm$^2$ at the same wavelength as the pump pulse, which leads to $a_{0s} = 0.027$. The seed frequency $\omega_{0s}$ is equal to the pump frequency $\omega_{0s} = \omega_{0p} = \omega_0$. From the dispersion relation $\omega_{0p}^2 = 1 + k_{0p}^2$ we have for the corresponding wavenumber $k_{0p} = 1.528$. In a similar simulation presented in [4], the length of the initially uniform plasma system was restricted to $273c/\omega_{pe}$ because of the noise level in the PIC simulation. In the simulation results presented here, the length is $L_p = 1600c/\omega_{pe}$, about 6 times the length used in [4].

The forward propagating linearly polarized pump laser beam penetrates the plasma at $x$=0, with a value $E^+ = 2E_{0p}\cos(\omega_0 t)$. Figure (1) show the time evolution of the pump (full curve), and of the seed pulse (broken curve). In the right of Fig.(1a), at $t$=1600, the precursor of the pump, which propagates at the speed of light, has just reached the boundary at $x$=1600, while the seed pulse has just penetrated (see the small bump at the right, in the Fig.(1a)). At time $t$=1600-2$\tau_s$,where $\tau_s = 100$ is the duration of the seed pulse, the backward seed wave $E^- = -2E_{0s}P_r(t)\cos(\omega_0\tau)$ is injected at the right boundary at $x$=1600, where $\tau = t - 1400$. The seed pulse has a temporal shape: $P_r(t) = \sin^2(\pi\tau/2\tau_s)$ , 1400<$t$<1600 and zero otherwise. In our normalized units $E_{0p,s} = \omega_0 a_{0p,s}$. We present in Figs.(1b-1d) a sequence showing the growth of the seed pulse which is propagating towards the left (broken curve). Behind the seed peak, the pump (full curve) is slowly depleted, while in front of the growing seed pulse, the incident pump has a constant amplitude. The simulation is stopped at $t$=2816 in Fig.(1d), due to the formation of a very steep gradient behind the seed front, which is followed by a numerical instability. At that time, the seed amplitude reached a factor about 6 times higher than that of the pump, or about 20 times the initial seed peak amplitude. So we have been able to push the calculations a little bit further in time compared to the results presented in [2]. Comparing Fig.(1b) and Fig.(1d), we see that the front edge of the seed pulse has compressed, and detached from the tail. Figs.(2) and (3) present the electron density and ion density profiles at three different times, while Fig.(4) presents the longitudinal electric field. Figs.(5) and (6) present the phase-space contour plots at different positions for the electrons and the ions. More details on the phase-space plots can be found in [2].

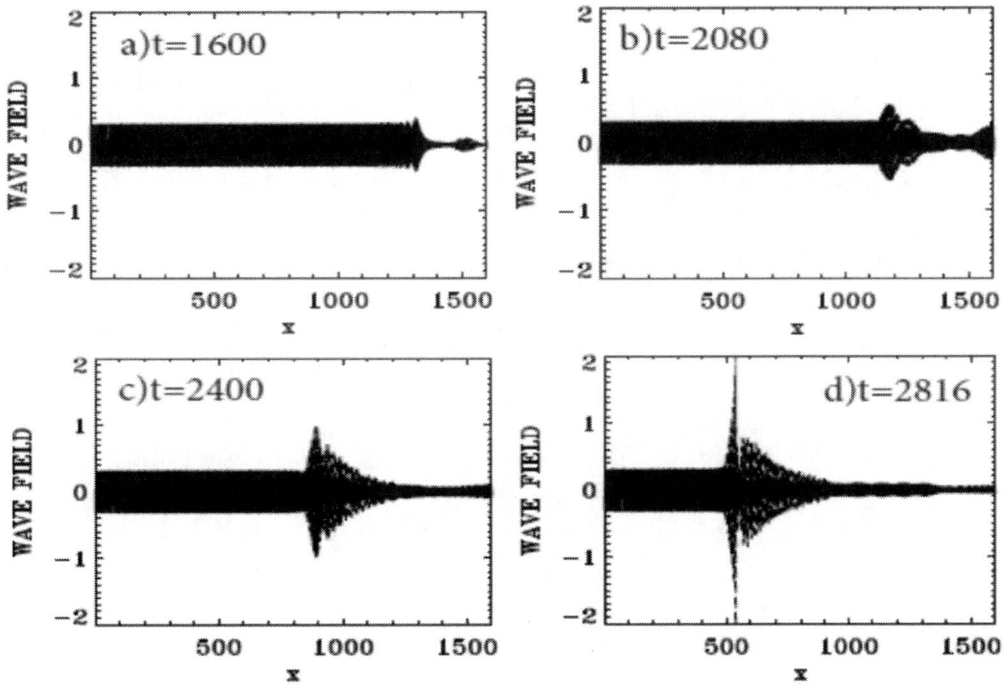


Figure 1. Incident pump wave $E^{+}$ (full curve) and backward seed wave $E^{-}$ at different times

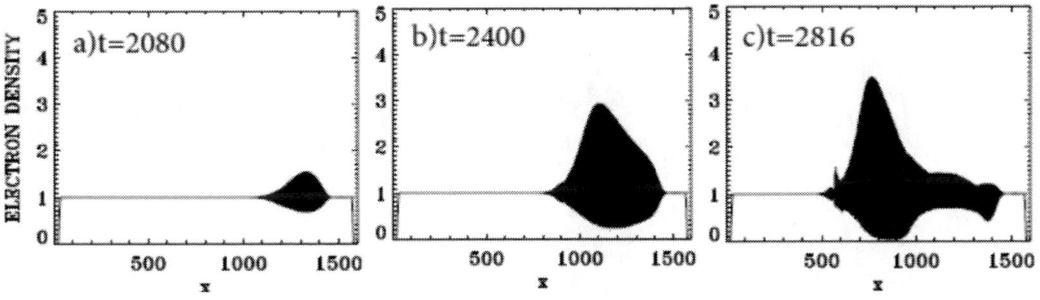


Figure 2. Electron density profiles at different times.

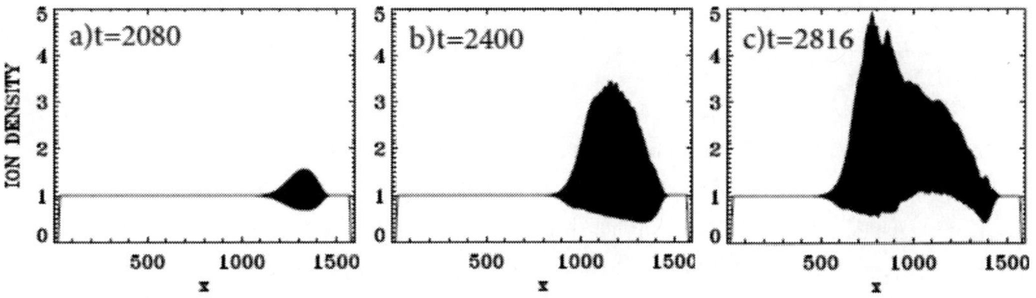


Figure3. Ion density profiles at different times

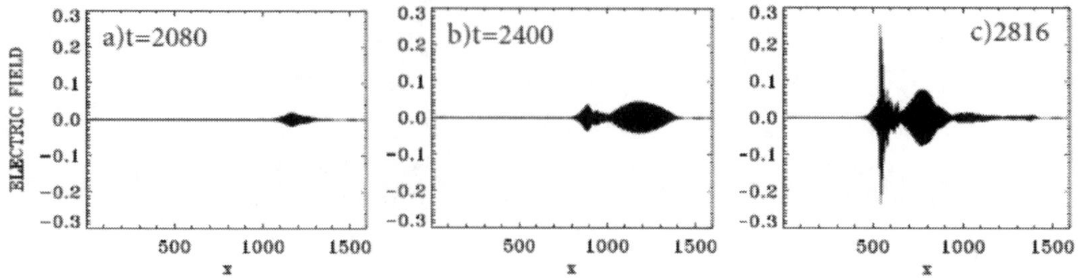


Figure 4. Longitudinal electric field at different times.

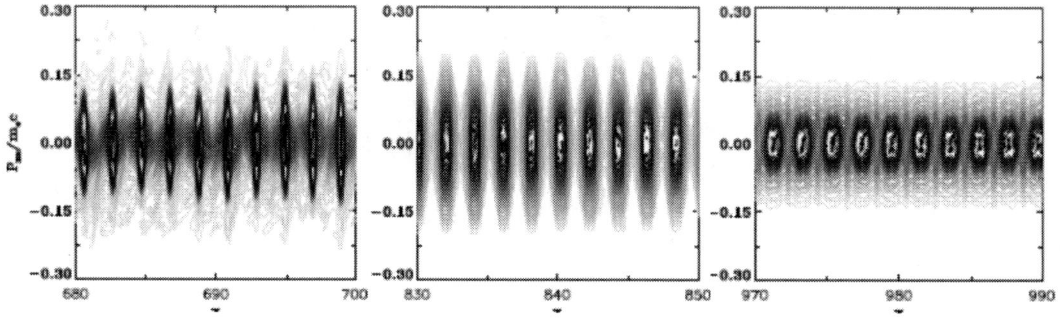

Figure 5. Phase-space contour plots of the electron distribution function at $t$=2816.

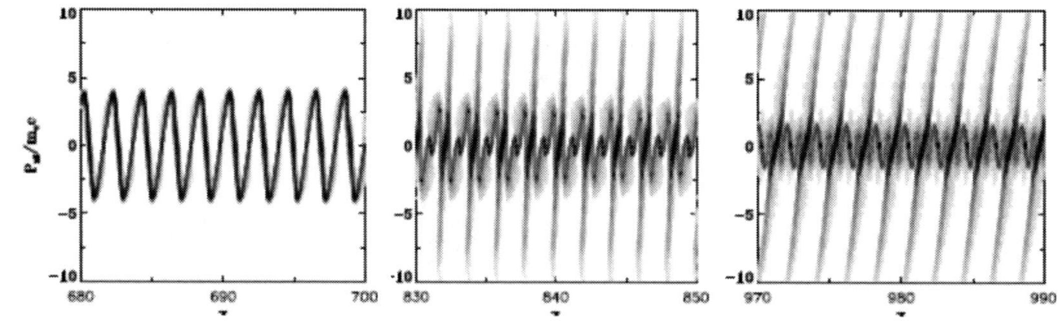

Figure 6. Phase-space contour plots of the ion distribution function at $t$=2816.

The authors are grateful to the Centre de calcul scientifique de l'IREQ (CASIR) for the computer time used for these simulations.

References

[1] M. Shoucri, J.-P. Matte, F. Vidal Proc. 41rst EPS Conf., Berlin (2014)

[2] M. Shoucri, J.-P. Matte, F. Vidal Phys. Plasmas 22, 053101 (2015),

[3] K.A. Humphrey, R.M.G.M. Trines, F. Fiuza, D.C. Speirs et al. , Phys. Plasmas 20, 102114, (2013)

[4] A. Andreev, C. Riconda, V. Tichonchuk, S. Weber Phys. Plasmas 13, 053110 (2006)

# Backward Raman amplification in extended regimes of pump pulse parameters

Z. Toroker[1], V.M. Malkin[2], N. J. Fisch[2]

[1] *Technion - Israel Institute of Technology, Haifa, Israel*

[2] *Princeton University, Princeton, USA*

## Introduction

The maximum output in intensities reachable through backward Raman amplification of laser pulses in plasma can, in principle, exceed by orders of magnitude methods based on material gratings and amplifiers. The backward Raman amplification (BRA) in plasma is a resonant interaction in which a long duration laser pulse (pump) transfers most of its energy to a counter propagating short duration laser pulse (seed) via the mediation of the Langmuir wave.

In cold plasma the BRA in plasma, taking into account the lowest order relativistic electron nonlinearity and group velocity dispersion effects for the amplified pulse, can be described by one-dimensional fluid model in the form of coupled 3-wave equations [1]:

$$a_t + c_a a_z = V_3 bf\ ,\ \ f_t = -V_3 ab^*,$$
$$b_t - b_z c_b = -V_3 af^* - i\kappa b_{tt} + iR|b|^2 b. \qquad (1)$$

Here $a$, $b$, and $f$ are envelopes of the pump, seed and Langmuir wave. The envelopes normalized to electron quiver velocities in fields of the pump, seed, and Langmuir wave, respectively. The subscripts $t$ and $z$ signify time and space derivatives; $c_a$ and $c_b$ are group velocities of the pump and amplified pulses; $V_3$ is the 3-wave coupling constant ; $R$ is the coeffcient of nonlinear frequency shift due to the relativistic electron nonlinearity (REN); $\kappa$ is the group velocity dispersion (GVD) coefficient; The relativistic electron nonlinearity is taken into account only for the seed pulse, because the seed is ultimately amplified to intensities much higher than that of the pump. The group velocity dispersion is also taken into account only for the seed pulse, because the seed is of much shorter duration than the pump, and therefore more sensitive to the dispersion effect.

## Relativistic electron nonlinear regime

For pump intensity below the wavebreaking it was assumed that the maximum output intensity can be obtained when the leading spike of the pi-pulse reaches saturation. The saturation of the leading spike is the result of the relativistic electron nonlinearity which induces a phase shift between the three waves. At the saturation point the accumulated phase shift is large and as a

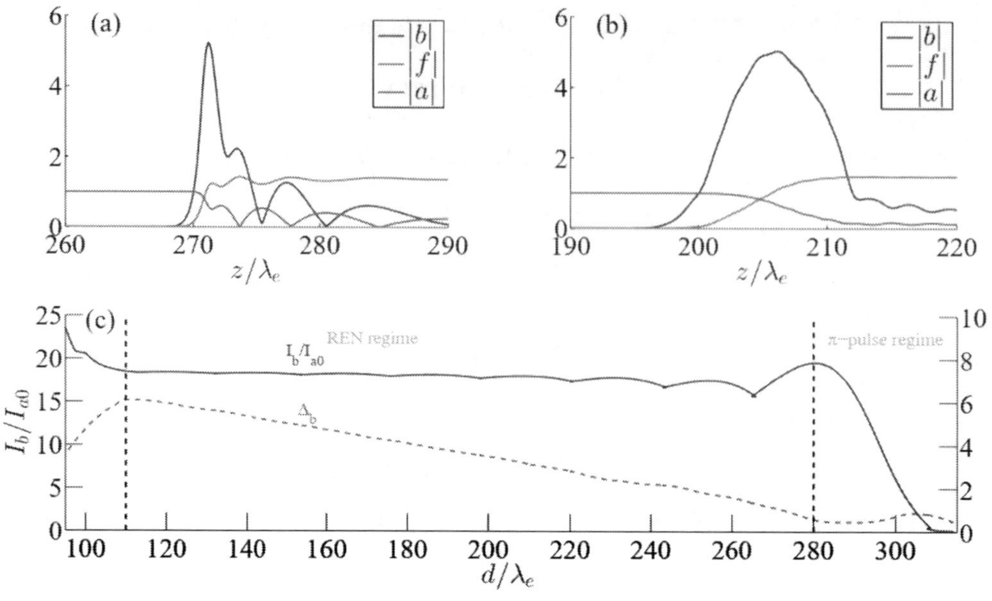


Figure 1: The three envelopes when the has traversed a length of (a) 64 $\mu$m ($\pi$-pulse regime) and (b) 137 $\mu$m (REN regime). (c) The seed amplitude and width as function of the seed peak location.

result prevents from efficient energy transfer from the pump to the amplified seed. The concern was that immediately this point a modulational instability or pulse breakup could be developed. Thus, to avoid from reaching the modulational instability, the pump width was limited (and also the plasma length) till the saturation point.

However, we present an illustrative example of seed amplification beyond the saturation point (a general formulation can be found in Refs. [1] and [2]). Consider an injected pump intensity of 55 PW/cm$^2$ with a wavelength of 0.351 $\mu$m, injected into a 354 $\mu$m long plasma. The initial counter-propagating seed has intensity of 4 PW/cm$^2$ with duration of 13.2 fsec. The plasma density is $n_e = 8 \times 10^{20}$ cm$^{-3}$. Fig. 1c shows in the pi-pulse regime, that the amplified pulse intensity is 1.045 EW/cm$^2$ with full width at half maximum (FWHM) duration of 4.7 fsec after the seed has traversed a length of 64 $\mu$m. In this example the dispersion parameter is $Q = 0.23 > 0.03$ and as a result it is expected the GVD will be a dominant effect. Since the strong GVD tends to stretch the width of the seed, the overall effect with the Raman amplification is that the amplified pulse appears as a single spike. In this example when the amplified seed travels a distance of 137 $\mu$m (Fig. 1d), the intensity is 0.99 EW/cm$^2$, with a FWHM duration of about 52.8 fsec. Hence, while the ratio between the seed intensities in the REN regime to the pi-pulse regime is nearly the same the fluence ratio is about ten times larger.

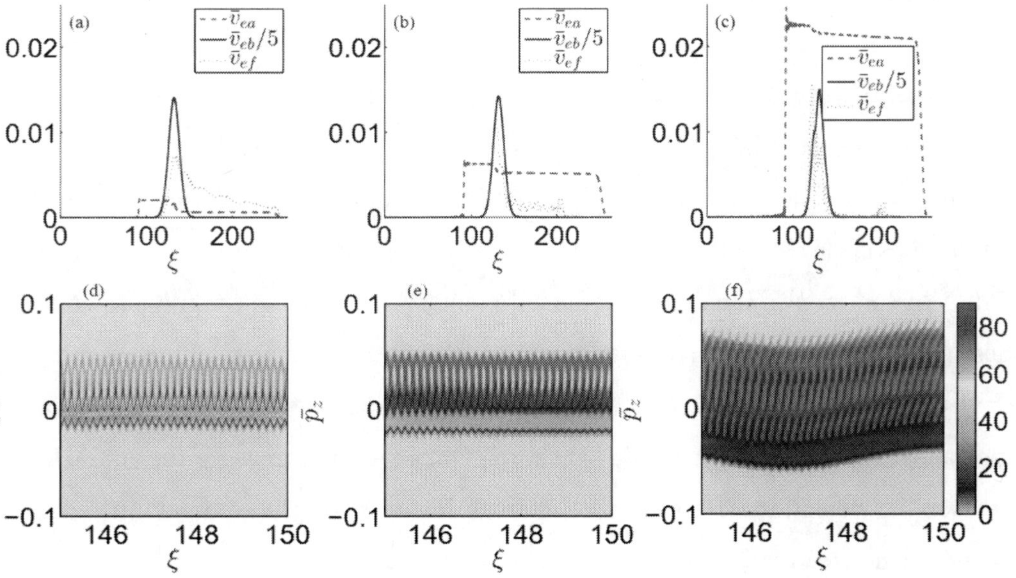


Figure 2: (a) Envelopes of the pump (dashed curve), the seed (solid curve), and the Langmuir wave (dotted curve) for pump amplitude $v_{ea}/v_{br} = 0.5$, which is below the wavebreaking threshold; (b) pump amplitude $v_{ea}/v_{br} = 1.5$, which is in the mild wavebreaking regime; (c) pump amplitude $v_{ea}/v_{br} = 5.5$, which is in the strong wavebreaking regime. (d)-(f) The corresponding electron phase space distributions in the region behind the seed pulse maximum spike.

**Wavebreaking regime**

The BRA can be used to compress soft X-ray of nano meter wavelength laser pulses, where the laser to plasma frequency ratio is significantly larger on the order of 20 or more than in the optical regime. In this soft X-ray regime the dominant degrading effect is Langmuir wave-breaking assuming the plasma temperature low enough to avoid Landau damping. Langmuir wave-breaking occurs when the longitudinal quiver velocity of electrons is approximately equal to the phase velocity of the Langmuir wave. The electrons can then no longer perform simple oscillatory option in support of the wave. Based on the Manley-Rowe relations the threshold for the Langmuir wavebreaking can be expressed in terms of the initial pump amplitude. More specifically, the wavebreaking threshold is $v_{br} = c(2\omega_a/\omega_e)^{-3/2}$, where $v_{br}$ is the electron quiver velocity in the pump pulse.

To explore the Langmuir wavebreaking, we consider the case where $\omega_a/\omega_e = 20$. In addition, to avoid Landau damping, the initial electron temperature is set to 10 eV. Since the wavebreaking is a kinetic effect, the BRA was calculated by a Vlasov-Maxwell (VM) code [3]. In the regime where the initial pump amplitude is below the wavebreaking threshold ($v_{ea}/v_{br} = 0.5$) the pump depletion is 90% (Fig. 2a). For initial pump amplitude that is 1.5 larger than the wavebreaking threshold (mild wavebreaking regime) the pump depletion is reduced to about 30% (Fig. 2b). For pump amplitude significantly larger than the wavebreaking threshold ($v_{ea}/v_{br} = 5.5$) the pump depletion is reduced to about 10% (Fig. 2c). While in the regime below the wavebreaking (Fig. 2d) and in the mild wavebreaking regime (Fig. 2e) most of the electrons are distributed according to the initial Maxwellian distribution, in the strong wavebreaking regime, a large fraction of particles are accelerated (Fig. 2f).

The VM code confirmed that moderate efficient BRA can occur for pump intensity up to a few times larger than the wavebreaking threshold. However, for initial pump intensity larger than more by ten times the wavebreaking threshold, the amplification efficiency decreases. Another important result is that in the mild wavebreaking regime increasing the initial seed intensity increases the depletion of the pump, whereas in the strong wavebreaking regime the pump depletion is not seen affected by an intense seed pulse.

**Conclusions**

To summarize, below the Langmuir wavebreaking and at plasma temperature in the region where the Landau damping and inverse bremsstrahlung can be neglected, the amplification efficiency is mainly limited by the relativistic electron nonlinearity and seed group velocity dispersion. Our numerical simulations indicate that the seed can be amplified beyond the saturation of its leading spike, in other words, beyond the so-called $\pi$-pulse regime. More specifically, for large dispersion paramter or dense plasma the amplitude of the seed is saturated and its width can be increased substantially by a factor of ten due to the strong GVD which leads to high output fluence.

In addition to the possible extension of the pump width, it turns out that the pump intensity can exceed the Langmuir wavebreaking threshold. More specifically, for pump intensity up to a few times the wavebreaking threshold, it is still possible to have moderately efficient Raman amplification. However, for pump intensities larger than tenfold the wavebreaking threshold, the efficiency significantly decreases.

**References**

[1] V. M. Malkin, Z. Toroker, and N. J. Fisch, Phys. Plasmas **21**, 093112 (2014).

[2] V. M. Malkin, Z. Toroker, and N. J. Fisch, Phys. Rev. E **90**, 063110 (2014).

[3] Z. Toroker, V. M. Malkin, and N. J. Fisch, Phys. Plasmas **21**, 113110 (2014).

# A transport beamline solution to control optically accelerated proton beams

F. Schillaci[1], A. Amato[1], L. Ando'[1], L. Attili[2], M. Borghesi[3], G. Candiano[1], G. A. P. Cirrone[1], L. Cosentino[1], M. Costa[1], G. Cuttone[1], G. De Luca[1], D. Doria[3], G. Gallo[1], G. Larosa[1], R. Leanza[1,4], G. Korn[5], R. Manna[1], M. Maggiore[6], V. Marchese[1], D. Margarone[5], A. Maugeri[1], G. Milluzzo[1,4], A. Musumarra[1], L. Pandola[1], G. Petringa[1], S. Pulvirenti[1], D. Rifuggiato[1], D. Rizzo[1], F. Romano[1], S. Salamone[1], M. Sedita[1], A. Seminara[1], V. Scuderi[1,5], A. Tramontana[1,4], B. Trovato[1], C. Viglianisi[1]

[1] *INFN-LNS, Via S. Sofia 62 - 95125 Catania, Italy*

[2] *INFN-Sezione di Torino, Via P. Giuria 1, Torino, Italy*

[3] *Centre for Plasma Physics School of Mathematics and Physics The Queen's University of Belfast, United Kingdom of Great Britain*

[4] *University of Catania,Via S. Sofia 64 - 95125 Catania, Italy*

[5] *Institute of Physics of the ASCR, ELI-Beamlines Project, Harfa Office Park Ceskomoravská - 2420/15a CZ 190 93 Prague, Czech Republic*

[6] *INFN-LNL,Viale dell'Universitá 2 - 35020 Legnaro (PD), Italy*

**Abstact**

Laser-target interaction represents a very promising field for several potential applications, from the nuclear physics to the radiobiology. However optically accelerated particle beams are characterized by some extreme features, not suitable for many applications. Therefore, beyond the improvements at the laser-target interaction level, many researchers are spending their efforts for the development of specific beam transport devices in order to obtain controlled and reproducible output beams.In this background, the ELIMED (ELI-Beamlines MEDical applications) project was born. Within 2017, a dedicated transport beam-line coupled with dosimetric systems, named ELIMED, will be installed at the Extreme Light Infrastructure Beamlines (ELI-Beamlines) facility in Prague (CZ),as a part of the ELIMAIA (ELI Multidisciplinary Applications of laserâĂŞIon Acceleration) beamline.

**Introduction**

Many experimental as well as theoretical studies demonstrate that proton/ion beams acceleration (up to hundreds of MeVs) based on the interaction of high power (order of PW) lasers with matter can represent a concrete future alternative in the field of particle acceleration [1]. However the energy and the angular spread of laser-driven beams represent the main issues for a muldisciplinary application and different solutions for dedicated beam-transport lines as well as diagnostics and dosimetric systems have been proposed [2, 3, 4]. Bearing in mind these purposes the ELIMED (ELI-Beamlines MEDical applications) collaboration was defined. In detail,at the end of 2012, the ELIMED initiative was presented at the ISAC (International Scientific Advisory Committee) of ELI-Beamlines and finally accepted at the end of 2013. In 2013 the ELI-Beamlines Institute officially started the realization of the ELIMAIA (ELI Multidisciplinary Applications of laser-Ion Acceleration) beamline, specifically dedicated to the ion acceleration and to their applications. Finally, as a result of a public tender, launched in 2014 by the FZU (ELI-Beamlines),the INFN-LNS has been officially appointed to realize the beam transport, the dosimetric and irradiation section in the ELIMAIA room.

**The transport beam line @ ELIMAIA**

Ions accelerated by laser-matter interaction are characterized by high intensities, multiple-species and charge states, a wide energy spectrum and a large energy-dependent angular distribution. Therefore,in order to make these beams suitable for multidisciplinary applications, the design of specific beam line transport elements, coupled to appropriate diagnostic devices, seems to be mandatory. The main aims of the transport elements will be the control of the optically accelerated ions final energy and angular distributions, the beam-line reliability as well as the flexibility and reproducibility of the final beam spot size and dose distribution. The beam transport solution that will be proposed for the ELIMAIA beamline will consist of two main elements: a collecting-focusing system and an energy selector.

The aim of the collection system is to collect the particle within a certain energy range, correct the angular divergence of the beam and inject the particles in the selection system. The beams coming out from this first part of the beam-line will have characteristics similar to the conventional beams and, hence, easy to transport and shape with conventional magnetic lenses, such as resistive quadrupoles and steerers, which will be placed in the last part of the beam-line. The above description of the beam-line, makes it clear that the ESS is the core element around whom all the other magnetic devices have to be designed and realized.

In order to understand the different mechanical and experimental issues, crucial for the design and the realization of the ELIMAIA beamline, a prototype has been already realized at INFN-LNS. It consists of a collecting-focusing sector and of an energy selector system (ESS), as well. As regarding the quadrupole system prototype, it is composed of four remotely controlled permanent quadrupoles with magnetic field gradients of 110T/m and 114T/m, 20mm bore and with lengths of 40 and 80mm, respectively [9]. This system allows to cover a wide energy operational range, from 0 up to 30 MeV. The final beam energy refinement is then obtained by means of the second transport device, i. e. the energy selector, elesewhere detailed shown [5, 6, 7, 8]. The ESS prototype has been already tested with conventional protons up to 12 MeV, at INFN-LNS and INFN-LNL, as well as with non-conventional beams at the Queens' University of Belfast. Recently also the focusing and collecting system prototype has been tested with 10 MeV conventional proton beams, delivered by the Tandem at INFN-LNS. Figure 1 reports the comparison between the 10 MeV proton beam spot distribution predicted by the optics simulations and the experimental data taken using Gafchromic film, EBT3 type. As shown, a quite good agreement between the simulation output and the measured spot size has been obtained. Simulations have been performed using the TraceWin analytical code.

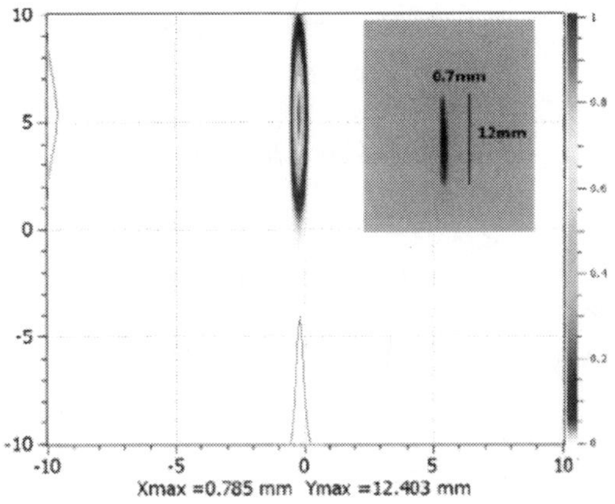


Figure 1: *Comparison between the simulated optics results and the experimental data for the 10 MeV proton beam spot along the PMQs system.*

## ACKNOWLEDGMENTS

This work has been performed within the ELIMED activities supported by the INFN (Italian Institute for Nuclear Physics), the MIUR (Italian Ministry of Education, Research and University) and by the ELI-Beamlines Contract no S 14-187, under Laser Gen (CZ.1.07/2.3.00/30.0057), under the Ministry of Education, Youth and Sports of the Czech Republic (ELI-Beamlines reg. No. CZ.1.05/ 1.1.00/02.0061), the institution Fyzikalni ustav, AV CR, v.v.i and under the project, co-financed by the European Social Fund and the state budget of the Czech Republic.

## References

[1] G. Mourou,T.Tajima,S.V.Bulanov,Reviews of Modern Physics 78 (2006) 309.

[2] S. Busold et al., Phys. Rev. ST Accel. Beams 17 (2014) 031302

[3] S. Sinigardi et al., Phys. Rev. ST Accel. Beams 14 (2011) 121301.

[4] C. Richter et al., Phys. Med. Biol. 56 (2011) 1529 - 1543

[5] M.Maggiore et al.,in:II ELIMED Workshop and Panel, AIP Conference Proceedings 1546 (2013) 34

[6] F.Romano et al.,in: II ELIMED Workshop and Panel, AIP Conference Proceedings 1546 (2013) 63

[7] V. Scuderi et al., Nuclear Instruments and Methods in Physics Research Section A 740 (2014) 87

[8] A.Tramontana etal., JNIST (2014) C05065

[9] F. Schillaci et al., JINST 10 (2015) T05001

# Adjustable temporal contrast for laser plasma experiments at the PHELIX laser facility

F. Wagner[1,2], S. Bedacht[3], C. P. João[4], A. Ortner[3], M. Roth[3], T. Stöhlker[1,2], A. Tauschwitz[5] and V. Bagnoud[1,2]

[1] *GSI Helmholtzzentrum für Schwerionenforschung GmbH, Darmstadt, Germany*

[2] *Helmholtz Institute Jena, Jena, Germany*

[3] *Technische Universität Darmstadt, Darmstadt, Germany*

[4] *GoLP/IPFN Instituto Superior Técnico, Lisbon, Portugal*

[5] *Frankfurt University, Frankfurt, Germany*

## Temporal contrast in laser plasma experiments

The temporal contrast of short laser pulses, which is defined by the ratio of the peak intensity to the intensity at a given time before this maximum, is an important parameter for experiments using ultra-intense lasers. The intrinsic noise in laser amplifiers results in degradation of the temporal contrast, leading to pre-ionization of the target material already a few nanoseconds before the impact of the laser pulse maximum. This early interaction creates a pre-plasma that strongly influences laser-matter-interaction and even compromises several physical phenomena. Some applications such as high harmonic generation from the surface of solid targets [1] or experiments which are based on the interaction with ultrathin targets, e.g. exploring new laser ion acceleration mechanisms [2, 3], cannot tolerate pre-ionization up to a given time before the impact of the pulse maximum. On the other hand in some cases a certain pre-plasma can even improve the outcome of the experiment [4], e.g. due to enhanced absorption and self-focusing effects. Consequently, the optimum pre-plasma depends on the application and an adjustable contrast level is very beneficial for user laser facilities which undertake a wide spectrum of experiments. Different contrast improvement techniques such as plasma mirrors (PM) [5] or cross-polarized wave generation (XPW) [6] are currently applied at several high-power laser systems around the world. However, all these methods either allow for switching between only two configurations, enabling high- and low contrast pulses, respectively, or they even enable just one option that features the highest contrast possible.

We developed a novel technique which allows for continuous tuning of the temporal contrast in petawatt-class lasers [7]. Our method was applied to the PHELIX (Petawatt High Energy Laser for heavy Ion eXperiments) [8] user laser facility at GSI Helmholtzzentrum für Schwerionenforschung GmbH and has been used there as a standard option for the last two years. An arbitrary level of amplified spontaneous emission (ASE) from the standard contrast ratio of $10^6$ to an optimum of $10^{11}$ can be applied and this range can be further extended to an ultrahigh contrast ratio of approximately $10^{13}$ by combining our technique with an additional plasma mirror. This unique feature of the PHELIX laser enables performing several new experiments of which a few have been carried out within the last 24 months.

**Tunable contrast at the PHELIX laser facility**

In Fig. 1 a schematic of the PHELIX short pulse beamline is shown. The short (100 fs) pulses from the oscillator are directly amplified in an ultrafast optical parametric amplifier (uOPA). Using a dedicated laser-diode-pumped pump laser [9] which provides 5 mJ pulses at 520 nm with a pulse duration of about 1 ps the oscillator pulses can be amplified up to 100 μJ without any degradation of the nanosecond and picosecond temporal-contrast. These pulses are then amplified in the PHELIX chirped pulse amplification (CPA) system [8] up to maximum energies of 250 J with minimum pulse durations of 500 fs. By focussing these pulses peak intensities between $10^{20}\,W/cm^2$ and $10^{21}\,W/cm^2$ are achieved depending on the used focussing parabola.

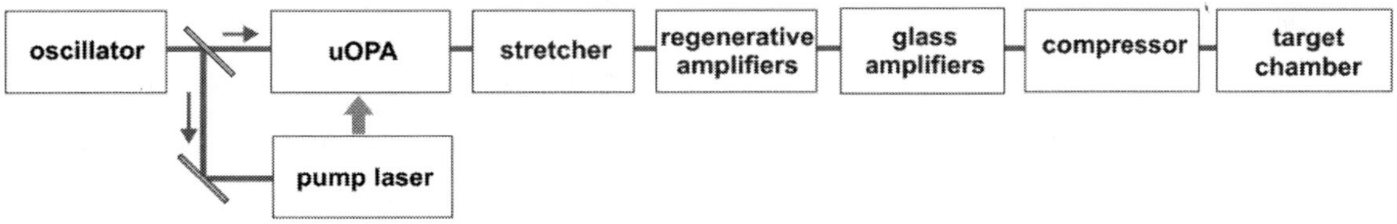


Figure 1: *Schematic of the PHELIX short pulse beamline*

The adjustable ASE level is accomplished by tuning the gain between the ASE-free uOPA and the first of two regenerative laser amplifiers which generates ASE, while the total gain of the system is kept constant. Fig. 2 (left) shows two pulse profiles recorded with a scanning third order cross-correlator (Sequoia, Amplitude Technologies) at an auxiliary compressor after the regenerative amplifiers. When the total gain is provided by the laser amplifiers the pulse profile (red curve) is similar to the profiles measured with standard CPA systems. The ASE level is about six orders of magnitude below the peak corresponding to an intensity of approximately $10^{14}\,W/cm^2$ clearly above the ionization threshold of any target material (indicated as a coloured bar in the graph). However, when a gain of $10^4$ is provided by the uOPA (black curve) the ASE is reduced by four orders of magnitude reaching a contrast level of $10^{10}$, which equals the detection limit of the used cross-correlator. Fig. 2 (right) shows the ASE levels that are accessible for different gain values of the regenerative amplifier. A higher gain in the regenerative amplifier corresponds to less gain in the uOPA, and vice versa. The graph shows a nearly linear lowering of the ASE level when the gain of the regenerative amplifier and the uOPA are decreased and increased, respectively. The minimum ASE level which can be achieved with the maximum parametric gain value of $10^5$ can be determined by extrapolating the experimental values. Whith this maximum contrast pre-ionization is precluded up to an instant of 100 ps before the peak for most of the used target materials. This measurement shows that an arbitrary ASE level between six and eleven orders of magnitude below the peak intensity can be applied with our particular design of the uOPA. Furthermore it demonstrates that our concept is scalable to even higher contrast values proving its suitability for future laser systems which are aimed at intensities beyond $10^{21}\,\mathrm{W/cm^2}$ and ultrahigh temporal contrast.

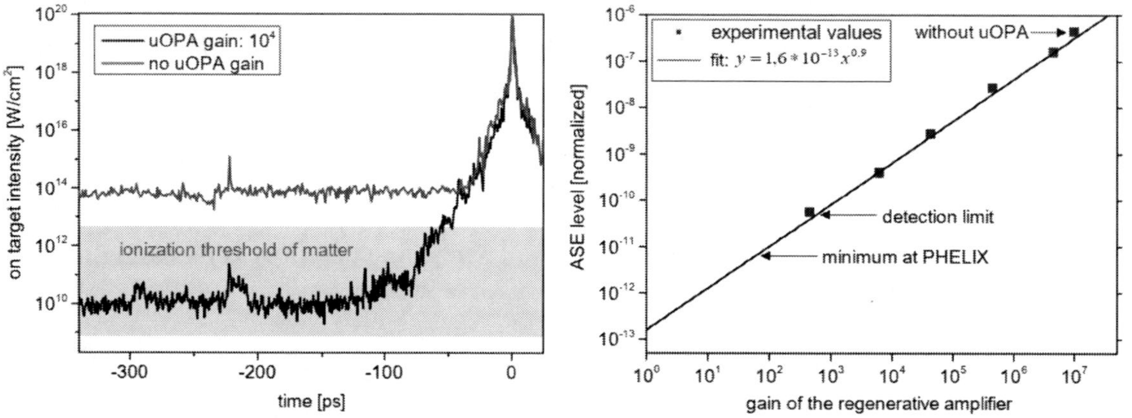


Figure 2: *Left: Temporal profile of the PHELIX pulse scaled to a peak intensity of* $10^{20}\,\mathrm{W/cm^2}$ *for an uOPA gain of* $10^4$ *(black) and no uOPA gain (red). Right: ASE levels normalized to the pulse maximum for different gain values of the regenerative amplifier. The combined gain of the uOPA and the regenerative amplifier was kept constant at* $10^7$ *for this measurement. Blue squares are experimental values taken from measured pulse profiles. The black line represents a fit to the experimental values.*

**Effect on laser plasma experiments**

The effect of different ASE levels on the produced pre-plasma during the interaction of an ultra-intense laser pulse with solid targets has been studied with 2-dimensional hydrodynamic simulations which were affirmed by time-resolved shadographic measurements [10]. The results for the pre-plasma size, which we define as the distance from the initial target surface to the position where the electron density goes down to a value of $10^{20}\,\mathrm{cm^{-3}}$, are summarized in table1. The used targets were flat copper foils and the duration of the ASE was 2.5 ns. These results identify the initial conditions for experiments exploring the interaction of high-power lasers with matter and can help to analyze experimental results. Furthermore they allow for applying the optimum pre-plasma for each experiment by choosing appropriate gain values for the uOPA. A more detailed description of this pre-plasma characterization is given in [10].

| ASE intensity [W/cm$^2$] | $2.5 \cdot 10^{10}$ | $10^{12}$ | $5 \cdot 10^{13}$ |
|---|---|---|---|
| preplasma size[µm] | $6 \pm 1$ | $13 \pm 3$ | $34 \pm 5$ |

Table 1: *Pre-Plasma size (defined as the distance from the initial target surface to the position where the electron density goes down to* $10^{20}\,\mathrm{cm}^{-3}$*) from flat copper targets for different ASE intensities.*

This unique tunable contrast feature facilitates various new experiments at the PHELIX facility. In particular the ultra-high contrast enables the investigation of new effects that require ultrathin targets. Several experiments with sub-µm targets have been successfully performed within the last two years. For instance, using plastic targets with thicknesses between 100 nm and 1200 nm shot under an angle of incidence of 10° we could observe two angularly separated laser-driven proton beams which can be attributed to two different acceleration mechanisms acting simultaneously [11]. For an optimum match of target and laser conditions we observed maximum proton energies beyond 60 MeV. In a similar experimental configuration and by the use of deuterated plastic targets deuterons have been accelerated to energies greater than 85 MeV which were used to generate intense bursts of neutrons [13]. Even thinner targets ($< 100$ nm) have been successfully used for laser ion acceleration by combining the ultra-high contrast frontend with an additional plasma mirror [14].

## Acknoledgments

This work was supported by the BMBF project 05P12RDFA1 and the Helmholtz Institue Jena and it has received funding from the European Unions's Horizon 2020 research program under grant agreement number 633053.

## References

[1] B. Dromey, et. al., Nature Physics **2**, 456 (2006)
[2] A. Henig, et. al., Physical Review Letters **103**, 245003 (2009)
[3] L. Yin, et. al., Physics of Plasmas **14**, 056706 (2007)
[4] P. McKenna, et. al., Laser and Particle Beams **26**, 591 (2008)
[5] B. Dromey, et. al., Review of Scientific Instruments **75**, 645 (2004)
[6] A. Jullien,. et. al., Optics Letters **30**, 920 (2005)
[7] F. Wagner, et. al., Applied Physics B **116**, 429 (2014)
[8] V. Bagnoud, et. al., Applied Physics B **100**, 137 (2010)
[9] C. P. João, et. al., Applied Physics B **118** (3), 401 (2015)
[10] F. Wagner, et. al., Optics Express **22**, 29505 (2014)
[11] F. Wagner, et. al., Physics of Plasmas **22**, 063110 (2015)
[12] M. Roth, et. al., Physical Review Letters **110**, 044802 (2013)
[13] G.A. Wurden, et. al., EPS 2015 conference contribution P5.214
[14] L. Senje, et. al., Review of Scientific Instruments **85** (1), 113302 (2014)

# Slowing-down of proton beams in plasmas and gas-jet targets with non-uniform features

D. Casas[1], M.D. Barriga-Carrasco[1], A.A. Andreev[2], M. Schnürer[2]

[1] *E.T.S.I. Industriales, Universidad de Castilla-La Mancha, Ciudad Real 13071, Spain*

[2] *Max Born Institute, Max Born Str. 2a, D-12489, Berlin, Germany*

## 1. Introduction

Laser-accelerated proton beams are used to analyse plasmas and gas-jet targets. These beams have a continuous of proton kinetic energies of a few MeV that allow tracing the temporal evolution of the target using the Thomson parabola technique [1]. The proton energy can be selected by means of magnetic fields and slits when the time of flight method is used to resolve the energy loss [2].

The estimation of the energy loss of a proton bunch that traverses a plasma target is made by means of electron stopping power, which is divided into two contributions: free electron stopping power and bound electron stopping power. The first one, due to plasma electrons, is calculated using a kind of Random Phase Approximation (RPA) dielectric function. The latter one, estimated for electrons bound to plasma ions, is calculated by means of an interpolation between low and high projectile velocities [3].

## 2. Calculation methods

The stopping power of a free electron gas can be calculated by means of RPA dielectric function using this expression:

$$Sp_f(v) = \frac{2Z^2}{\pi v^2}\int_0^{\infty}\frac{dk}{k}\int_0^{kv}\omega d\omega \operatorname{Im}\left[\frac{-1}{\varepsilon_{RPA}(k,\omega)}\right] \tag{1}$$

However, this calculation could be computationally hard in some cases. An alternative way to obtain free electron stopping power using RPA calculations consists to interpolate from a complete database [4]. The stopping power depends mostly on temperature and density, for this reason a bilinear interpolation of these two variables is used in a 2D grid.

Bound electron stopping power is calculated by means of this expression:

$$Sp_b(v) = \frac{4\pi Z^2 n_{at}}{v^2} L_b(v), \tag{2}$$

where $L_b(v)$ is interpolated between a high and low projectile velocities:

$$L_b(v)=\begin{cases} L_H(v)=\ln\frac{2v^2}{I}-\frac{2K}{v^2} & \text{for } v>v_{int} \\ L_B(v)=\frac{\alpha v^3}{1+Gv^2} & \text{for } v\le v_{int} \end{cases} \tag{3}$$

$$v_{int}=\sqrt{3K+1.5I} \tag{4}$$

and $G$ is given by $L_H(v_{\text{int}})=L_B(v_{\text{int}})$, $K$ is the electron kinetic energy, $I$ is the mean excitation energy and $\alpha$ is the friction coefficient for low velocities [5].

Mean excitation energy can be estimated using oscillator strengths method [2, 5] where the sum rules referred to energy momenta are defined as [6]:

$$S(\mu)=\sum_n f_{0n}E_{0n}^{\mu}+\int_{IP}^{\infty}\left(\frac{\mathrm{d}f}{\mathrm{d}E}\right)E^{\mu}\mathrm{d}E \tag{5}$$

Where $S(-1)$ is related to quadratic mean radius, and $S(1)$ is proportional to electron kinetic energy:

$$\langle r^2\rangle=\frac{3}{2}S(-1) \qquad K=\frac{3}{4}S(1) \tag{6}$$

Then, the mean excitation energy, $I$, is obtained,

$$I=\sqrt{\frac{2K}{\langle r^2\rangle}}=\sqrt{\frac{S(1)}{S(-1)}} \tag{7}$$

The quantities $S(1)$ and $S(-1)$ are calculated using the Flexible Atomic Code (FAC) and tabulated data [7-9].

The energy loss of a proton beam in a material, like plasma or gas-jet target, is a dynamic process. When it impacts with an initial energy, $E_{p0}$, it starts losing energy with a rate that is given by the free and bound electron stopping power estimated before. Using an iterative scheme, this energy loss could be calculated. The method is to divide the plasma length in segments and to evaluate the energy loss in the ith step by means of:

$$E_{L_i}=\frac{Sp_i}{\Delta x} \tag{8}$$

where $Sp_i$ is the stopping power in the ith segment and $\Delta x$ its length. The target is divided in many thin layers where temperature, density, and ionization are chosen constants. The layer thickness is calculated so that gradients of plasma features inside are small and, they can be supposed uniforms along the layer length.

## 3. Results

Using the previous equations, it is possible to evaluate the target density profile effects for different density distributions. For instance, a rectangular shape with a constant density and a piecewise approximation of a trapezium shape with a density profile given by [10]:

$$n_i(z) = \frac{2n_{i\,\max}}{1+\exp\left[\frac{2\varsigma\theta(\varsigma)}{l_r} - \frac{2\varsigma\theta(-\varsigma)}{l_{fr}}\right]} \tag{9}$$

Both cases conserve the particle quantity. The density profiles and energy losses are given in the two graphs of Figure 1.

Stopping power of helium gas-jet was also analysed [2]. The density of this jet was modelled using a Gaussian profile and a mean atomic density of $2.675\times10^{20}\ at/cm^3$ was obtained. At this density, our theoretical calculations, using the excitation energy calculated by means of Eqs. (5), (6), and (7) (see Table 1), are close to experimental data, see Figure 2.

| | |
|---|---|
| *I* | 2.292 |
| *S*(1) | 2.612 |
| *S*(-1) | 0.497 |

Table 1. Mean excitation energy and energy momenta obtained using oscillator strength method for atomic helium in atomic units.

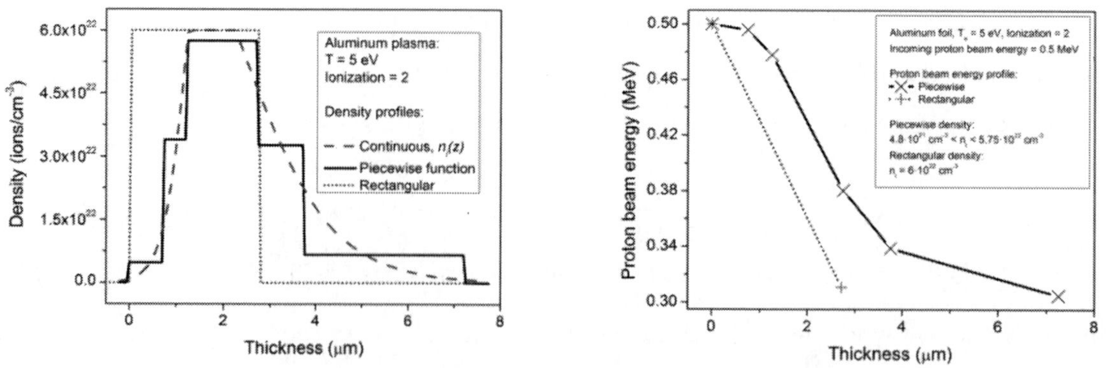


Fig. 1. The target density profiles (left) and its corresponding energy loss functions (right).

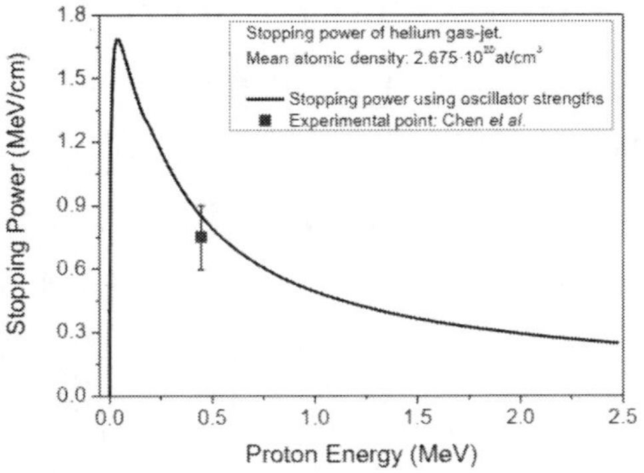


Fig. 2. Stopping power of a helium gas-jet as function of proton beam energy.

## 4. Conclusions

The combination of interpolation formulas with an iterative scheme for the stopping power calculation has been a useful tool to analyse the proton beam energy loss in laser-created plasma and gas-jet target. The influence of the target density profile in the shape of energy loss function has been shown. It supposes a slight difference in the final energy of the proton beam due to the quasi-linearity dependence on density in the stopping power expressions. The stopping power of helium gas-jet has been calculated from its average atomic density, obtained from its Gaussian profile, and its mean excitation energy, estimated by means of oscillator strength method. This theoretical estimation has been close to the experimental result.

## Acknowledgements

This work is supported by the German Project Transregio 18 (DFG) and the Spanish ENE2013-45661-C2-1-P.

## References

[1] F. Abicht *et al.*, Applied Physics Letters **105**, 034101 (2014)

[2] S. N. Chen *et al.*, Nuclear Instruments and Methods in Physics Research A **740**, 105–106 (2014)

[3] D. Casas, M. D. Barriga-Carrasco, and J. Rubio, Physical Review E **88**, 033102 (2013).

[4] http://www.uclm.es/area/amf/manuel/programas.htm

[5] M. D. Barriga-Carrasco and David Casas, Laser and Particle Beams **31**(1), 105-111 (2013)

[6] J. Berkowitz, Journal of Physics B: Atomic, Molecular and Optical Physics, **30** 881–892 (1997)

[7] http://sprg.ssl.berkeley.edu/~mfgu/fac/

[8] http://physics.nist.gov/PhysRefData/ASD/lines_form.html

[9] R. Rodriguez *et al.*, Laser and Particle Beams **26**, 433–448 (2008)

[10] A. A. Andreev *et al.*, Physics of Plasmas **16**, 013103 (2009)

# Study of Hot Electron production in SI-relevant regime

M.Afshari[1], G.Peres[1], O.Renner[2], M.Smid[2, 3], E.Krousky[4]

[1]*Dip.F.C, University of Palermo, Palermo, Italy*

[2]*Institute of Physics of the ASCR, v.v.i., Prague, Czech Republic*

[3]*Czech Technical University in Prague, FNSPE, Prague, Czech Republic*

[4]*Institute of Plasma Physics CAS, v.v.i., Prague, Czech Republic*

**Abstract:**

In this paper we study the generation and propagation of hot electrons produced by high intensity laser pulses through multiple targets with different thickness in an interaction regime relevant to shock ignition (SI) approach for ICF schemes. The main diagnostic was Cu K-α fluorescence imaging by spherically bent quartz (422) crystal. The average values of hot electron temperature and conversion efficiency are 26.2 ± 1.8 keV and 0.7 ± 0.4 %, respectively. The experiment is in a series of experiments to shock ignition feasibility which performed at PALS lab (Prague Asterix Laser System) in Prague, Czech Republic.

## Introduction

At laser intensities up to $10^{16}$ W/cm$^2$, envisaged in the SI scheme, laser interaction with the long scale plasma may lead to the generation of hot electrons (**HE**). The experiment aims at investigating of generation of such HE and their role in the generation of the shock wave [3]. We used two types of targets. The thin multilayer targets, CHCl+Ti+Cu, CHCl+Cu+Al, and thick targets, CHCl+Massive Cu with different thickness of CHCl layer. Hot electrons were detected by K-α emission of Cu and Ti layers. The images were recorded with the crystal of quartz (422) spherically bent to a radius of 380 mm. the interaction of hot electrons with Bound atomic structures of tracer layers (Cu or Ti) produce K-α emission which reflected by spherically bent crystal to film plates and produce K-α spot (black circle) with a Gaussian distribution as shown in Fig.1.

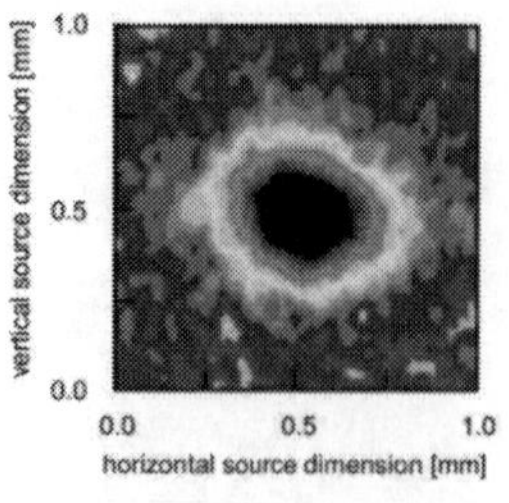


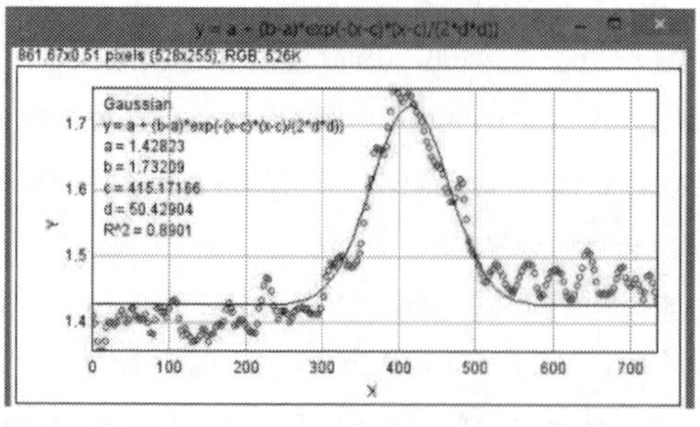


Fig.1: Left: A typical 2D-resolved image of the Cu K-α spot as imprinted on film slice. Right: Gaussian distribution of hot electrons.

**Experimental results:**

From the obtained images, we can get two types of quantitative information: K-α spot and Total K-α Number of Photons **(NPhs).** As soon as K-α spot and NPhs determined we can calculate some important parameters as come in following:

**1- Spreading angle of hot electrons**

The spot size, measured as a function of target thickness, can be used to evaluate the angle of divergence of the fast electron beam. The importance of spreading angle of HE can be seen especially in fast igniting scheme which strongly depends on the generation of HE, their collimation, transport, and energy deposition in the over-dense region of the plasma. Hence the estimation of divergence of HE is, therefore extremely crucial.

For this purpose K-α spot of tracer layers of CHCl+Ti+Cu (Cu and Ti) and CHCl+Cu+Al (Cu) plotted versus total thickness of their targets as presented in Fig.2.

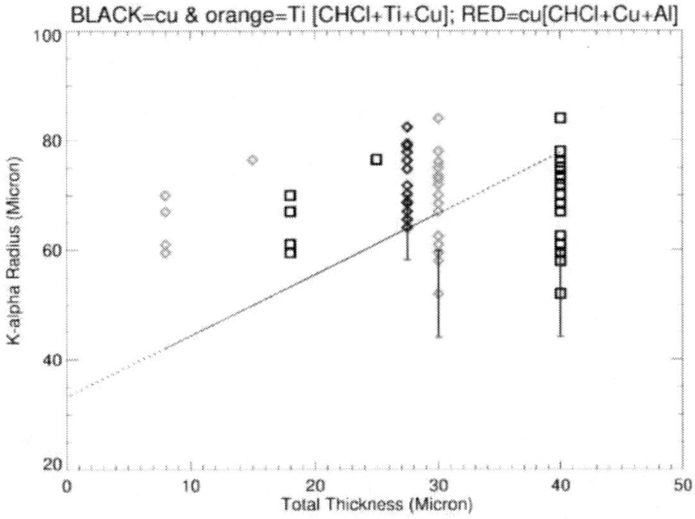


Fig.2: K-α spot of tracer layers of CHCl+Ti+Cu (Cu and Ti) and CHCl+Cu+Al (Cu) plotted versus total thickness of their targets

By interpolation data (Radius R vs. Thickness d) with a straight line R(d) = A +Bd, the tangent of slope (parameters B) will give the spreading angle of HE. In our case the spreading angle estimated to be **$\theta \approx 47.5^{\circ}$.** Typical values reported in literature are between 40–50°.

**2- HE penetration depth**

The penetration depth of HE can be inferred from the measurements of the K-α photon flux versus different thicknesses of the plastic layer as presented in Fig.3. For penetration depth measurement thick targets (CHCl+Massive Cu) must be used to prevent electron refluxing.

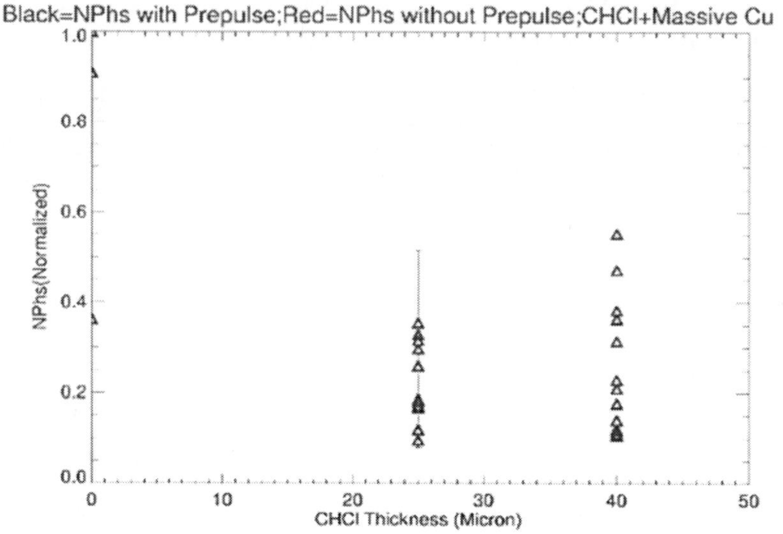


Fig.3: K-α photon flux versus different thicknesses of the plastic layer for penetration depth measurement.

The measured flux as a function of plastic layer thickness fitted with an exponential function, N(x) = a exp (-x/L), where N, x and L indicate the measured photon number, the plastic layer thickness and the penetration depth and a are free fitting parameter. We fitted the function for two data set, first for all data and then for data without prepulse. In first case the penetration depth was found to be **L ≈ 27.4 μm** while for data without prepulse it was **L ≈ 26.9 μm.** As we can see in the presence of prepulse, penetration depth of HE was increased that could be considered as a positive effect of preplasma effect.

## 3-Hot electron beam temperature

One of key issue in HE experiments is HE Temperature since by determination of $T_{HE}$ we can calculate conversion efficiency. But instead of a more realistic electron energy distribution we assume a mono energetic electron beam. Temperature estimated with different methods:

### 3-1 Beg's scaling law:

By using Beg's scaling law **$T_e$ (KeV) = 215($I_{18}$ $\lambda^2_{\mu m}$)** $^{1/3}$ (ONLY for 3w mode, thin target approximation). According to variation of laser intensity between 2.7 e+15 - 1.07 e+16 (w/cm$^2$) and $\lambda_{3\omega}$=438nm, $T_e$ ranged **17.3-27.3 (keV)** with the average **Te ≈ 26.2 ±1.8 (KeV).**

### 3-2 slope of penetration depth diagram

The Experimental slope of penetration depth [N(x) = a exp (-x/L)] enables to infer first the penetration depth of HE (section 2) and second to deduce the corresponding HE temperature.

In the case of CHCl + Massive Cu the penetration depth was found **L ≈ 27.4 μm** that by comparison with the electron stopping range in Mylar the HE energy estimated to be ~50±10 keV. In the case of CHCl+Ti+Cu target, by increasing CHCl layer, the Kα signal was found rather flat so we couldn't fit exponential function to deduce HE temperature.

### 3-3 Ratio of Kα signals of Ti and Cu versus mean HE energy

Another way to measure $T_{hot}$ is based on the ratio of K-α signals coming from the Ti and Cu layers. In one hand, experimental data are averaged, and error bars are deduced from the standard deviation σ (around 20 % in relative). Then the experimental values was compared with Monte-Carlo GEANT4 simulations. The hot electron temperatures were inferred by fitting experimental data with simulations, as illustrated in Fig. 4 .Depending on the CHCl thickness, the $T_{hot}$ ranges from **18-31.5 keV** which is comparable to ones by Beg's law. [4]

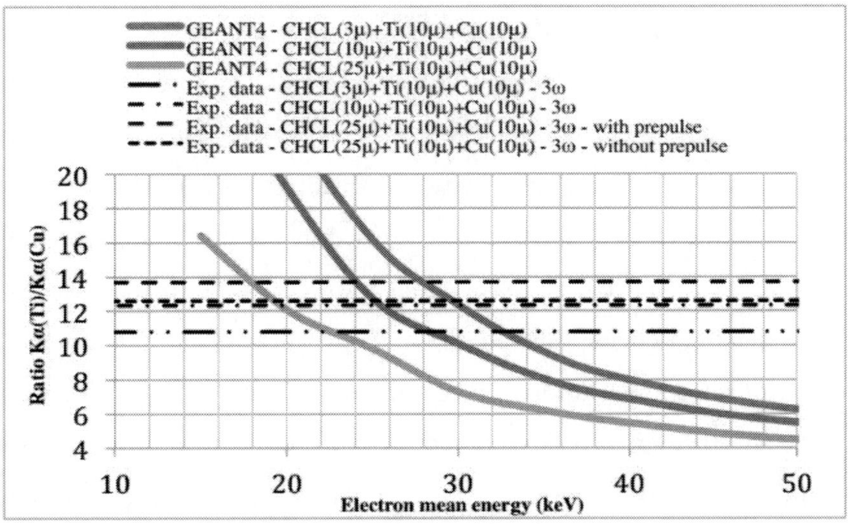


Fig.4: Comparison between GEANT4 simulations and experimental data (for simplicity, error bars are not represented here)

### 4- Conversion efficiency

By assuming a single-temperature electron distribution function and knowing the energy of the laser $E_{laser}$, we can infer the laser-to-electrons conversion efficiency (η) by estimating, with GEANT4, the electron total energy $E_e$ needed to reproduce the measured absolute Kα signal. Depending on the CHCl thickness, η ranged from 0.45-1.1 %.

### References

[1] R. Betti, et al., Phys. Rev. Lett. 98, 155001 (2007)

[2] J. Lindl, Phys. Plasmas 2, 3933 (1995)

[3] D.Batani et al., Nucl. Fusion54 (2014) 054009 (29pp)

[4] Data Interpretation related to hot electrons generation (3ω) L. Antonelli et al.

# Effect of electron temperature on magnetized plasma sheaths contaminated by multi-sized impurities

S. Chekour[1], A. Tahraoui[1], D. Benlemdjaldi[1,2]
[1] *Quantum Electronics Laboratory, Faculty of Physics, U.S.T.H.B.*
*BP 32 El-Alia Bab Ezzouar, Algiers 16111, Algieria*
[2] *Département de Physique, Ecole Normale Supérieure*
*BP 92 Vieux Kouba, Algiers 16050, Algeria*
*Corresponding author: che-samir@hotmail.com*

## Abstract

In the present work, the effect of electron temperature on magnetized electrostatic sheaths structure is investigated. A three dimensional and stationary theoretical model is established. Electrons are considered in thermodynamic equilibrium whereas ions are described by fluid equations. The impurities are spherical dust grains with different sizes, which are also modeled by fluid equations. The grain size distribution is described by a Gaussian law [1,2] and the charge variation is described by the orbit motion limed model [3,4]. Numerical results show that the increase of electron temperature induces the increase of the electrostatic sheath thickness. Indeed, the increase of the electron temperature increases their probability of attachment to dust grains and so the charge separation. Consequently, the electrostatic sheath thickness is enlarged. The presence of the magnetic field reduces the electrostatic sheath thickness. The effect of the other parameters are examined and discussed.

## Introduction

When plasma is in contact with a solid wall, such as an electrode in discharge plasmas, it acquires a negative potential with respect to the bulk plasma, due to the high mobility of the electrons. An electrostatic sheath that is a boundary layer where the plasma departs from quasi neutrality gets formed.

In the present work, we focus on the effect of electron temperature on magnetized plasma sheaths contaminated by multi-sized impurities.

## Theoretical Model

The electrons *(e)* are assumed to be in thermal equilibrium thus, their number densities $n_e$ satisfy the Boltzmann relation.

$$n_e = n_{e0}\exp(e\phi / T_e), \tag{1}$$

where $T_e$ is the electron temperature and $\phi$ is the electrostatic potential.

The positive ions and dust grains are described by fluid equations [5],

$$\frac{\partial n_i}{\partial t} + \nabla.(n_i \vec{v}_i) = k_i\, n_i\, n_e, \tag{2}$$

$$\frac{\partial \vec{v}_i}{\partial t} + \vec{v}_i.\nabla\vec{v}_i = \frac{\vec{F}_c}{m_i} + \frac{e}{m_i}\left(\vec{E} + \vec{v}_i \times \vec{B}\right) - \frac{\vec{\nabla} p_i}{n_i\, m_i}, \tag{3}$$

$$\frac{\partial n_{dk}}{\partial t} + \nabla.(n_{dk}\,\vec{v}_{dk}) = 0, \tag{4}$$

$$m_{dk}\left(\frac{\partial \vec{v}_{dk}}{\partial t} + \vec{v}_{dk}.\nabla\vec{v}_{dk}\right) = \vec{f}_{ek} + \vec{f}_{gk} + \vec{f}_{idk} + \vec{f}_{ndk}. \tag{5}$$

where $B$ is magnetic field making an angle $\theta$ with the vertical axis $Oz$, $n_i$, $v_i$ and $m_i$ are the density, the fluid velocity and the mass of the ions respectively, $q_d$, $m_d$ and $v_d$ are the charge, the mass and the fluid velocity of dust grains respectively, $g$ being the acceleration of gravity. The forces actives on the dust grains are the electromagnetic force $f_{ek}$, the gravitational force $f_{gk}$, the ion drag force $f_{idk}$ and the neutral drag force $f_{ndk}$ respectively. Their expressions are given by:

$$\vec{f}_{ek} = q_{dk}\left(\vec{E} + \vec{v}_{dk} \times \vec{B}\right), \tag{6}$$

$$\vec{f}_{gk} = m_{dk}\vec{g} = \frac{4}{3}\pi r_{dk}^3 \rho_d \vec{g}, \tag{7}$$

$$\vec{F}_{idk} = \pi\, m_{ik}\, n_{ik}\, r_{dk}^2\, v_{sk}\, \vec{v}_{ik}\left(1 - 2 b_{\pi/2k}^2 / r_{dk} + 4\Gamma_k\, b_{\pi/2k}^2 / r_{dk}^2\right), \tag{8}$$

where $b_{\pi/2k} = e q_{dk} / m_{ik} \, v_{sk}^2$ is the orbital impact parameter, $\Gamma$ is the Coulomb logarithm and $v_{sk} = \left(v_{ik}^2 + 8T_i / \pi m_i\right)^{1/2}$ is total ion velocity.

In order to relate the self-consistent potential to electron, negative and positive ions as well as dust densities in the sheath, we use Poisson's equation

$$\Delta\phi = -\frac{1}{\varepsilon_0}\left( n_i e - n_e e + \sum_k n_{dk} q_{dk} \right). \quad (9)$$

The dust grains charge is described by the orbit motion limited model (OML) [3-4]

$$\vec{v}_{dk} . \nabla q_{dk} = I_{ek} + I_{ik} \quad (10)$$

Finally, the dust size distribution is modeled by a Gaussian Law [1-2],

$$f(r_d) = D\exp\left(-\mu\left(r_d - r_{dm}\right)^2\right), \quad \text{for } r_{d\min} \le r_d \le r_{d\max}, \quad (11)$$

otherwise, the distribution has zero value, where $r_{dm}$ is average dust radius, $D$ is the normalization constant and the constant $\mu$ is given by

$$f(r_{d\min}) = f(r_{d\max}) = 0.01 f(r_{dm}), \quad (12)$$

$r_{d\min}$ and $r_{d\max}$ being the lower and the upper limits of the dust grains radius.

## Numerical results

In this section, we present the numerical results of our theoretical model. We have used a non reactive gas flow, i.e., argon as background gas and neglected any gas flow $(v_n = 0)$.

The physical parameters used are $T_i = 0.01eV$, $T_n = 0.01eV$, $n_{i0} = 10^9 cm^{-3}$, $P_n = 10 mTorr$, $B = 1T$, $\theta = 20°$, $\delta_d = 10^{-4}$, $\rho_d = 2 g cm^{-3}$, $r_{d\min} = 0.01 \mu m$, $r_{d\max} = 5 \mu m$, $u_{i0} = 1.5$ and $u_{d0} = 2.5$.

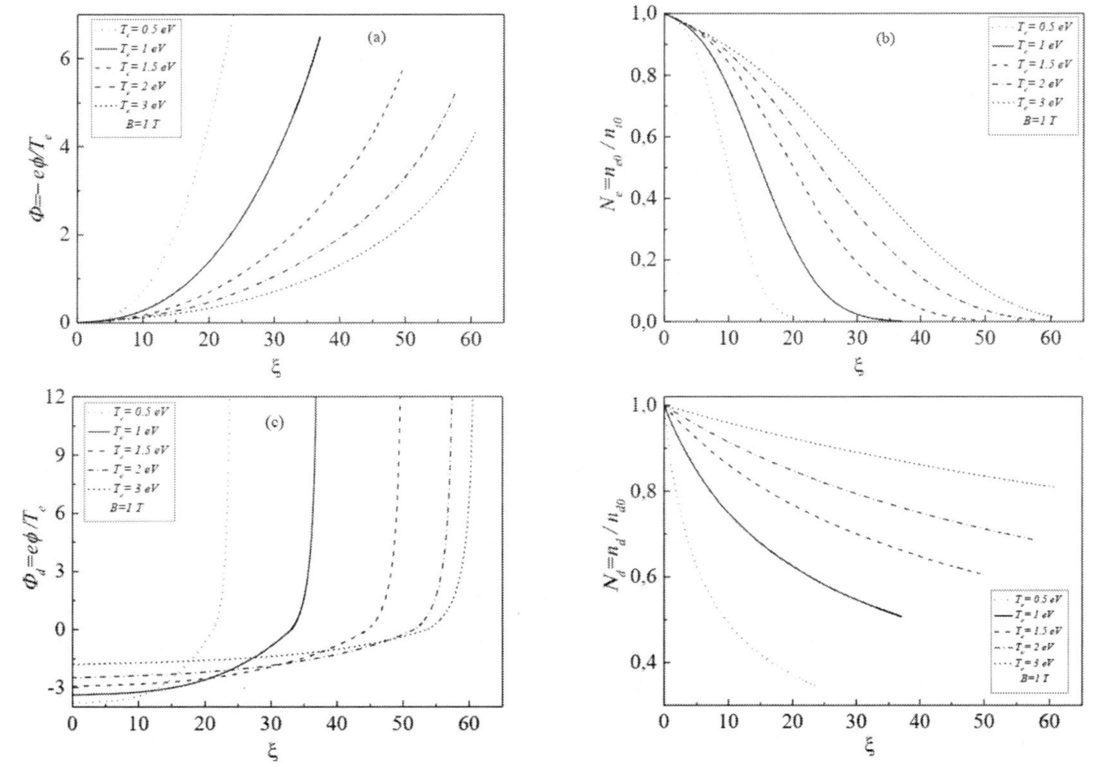


**Figure1:** Normalizes electrostatic potential (a), normalized electron density (b), normalized dust grains surface potential (c) and normalized dust grains density (d) as function $\xi = z / \lambda_{Di}$.

The numerical results show that the increase of electrons temperature induces the increase of the electrostatic sheath thickness. Indeed, the increase of the electron temperature increases their probability of attachment to dust grains and so the charge separation. Consequently, the electrostatic sheath thickness is enlarged. On the other hand figure (c) shows that the dust charge remains negative even for distances too close to wall and then changes sign abruptly just before the wall is reached. This is due to the fact that the density of electrons falls faster than that of ions in the sheath.

## References

[1] W. S. Duan, Phys. Plasmas 8, 3583 (2001).

[2] B. Zham, A. Tahraoui, D. Benlemdjaldi and S. Chekour, Phys. Plasmas **21**, 123702 (2014).

[3] P. K. Shukla and A. A. Mamun, *Introduction to Dusty Plasma Physics* (Institute of Physics, Bristol, 2001).

[4] J. E. Allen, Phys. Scr. 45, 497 (1992).

[5] A. Tahraoui, B. Zaham, R. Annou, J. Bougdira and R. Hugon, Proceedings of the First International Conférence on Laser Plasma Applications in Materials Science, Algeria 23-26 June 2008, AIP Conference Proceedings 1047, p. 268.

# Conversion of Methane to $C_2$ Hydrocarbons and Hydrogen Using Microwave ′tornado′-type Plasma

D. Tsyganov, N. Bundaleska, E. Tatarova

*Instituto de Plasmas e Fusão Nuclear, Instituto Superior Técnico,*

*Universidade de Lisboa, Lisboa, Portugal*

A microwave (2.45 GHz) "tornado" - type plasma with a high-speed tangential gas injection (swirl) working at atmospheric pressure conditions has been applied for reforming of methane. Theoretical model based on a set of nonlinearly coupled, spatially dependent differential equations describing plasma thermodynamics, gas heating and flow has been developed.

**Keywords:** microwave plasma, methane, hydrogen.

## Introduction.

The interest in reforming technologies for hydrogen production is motivated by the ongoing need for energy and environmental pollution. As well hydrogen has the potential to effectively reduce our dependence on fossil fuels, since it is present in compounds such as water, hydrocarbons, biomass, etc. Among different conversion technologies - steam reforming, partial oxidation and autothermal reforming - steam methane reforming is the most common commercially used method for hydrogen production [1, 2]. Plasma systems offer an adequate solution to this problem since plasma comprises thermal and chemical reactor functions, as well as, catalytic properties.

Microwave plasma operating at atmospheric pressure demonstrates important advantages compared to other plasma sources as a conversion tool, since it can achieve extremely high densities of active species of interest at relatively low gas temperatures [3].

In this study, a surface wave driven microwave plasma operating at atmospheric pressure conditions has been applied for reforming of methane. The optical emission spectrum in the range 250-700 nm has been investigated to detect the presence of the species of interest in the plasma. Kinetic model describing the methane decomposition in the argon plasma environment have been proposed.

## Experimental setup.

Surface-wave driven plasma is a special type of microwave plasma where the microwave power is transferred to the plasma by the electric field of a guided wave [3]. The scheme of the experimental setup is shown in Fig. 1 along with the swirl injector and a photo of the plasma torch. Microwaves at 2.45 GHz are provided by a generator with maximum power of 2 kW. The microwaves travel through a waveguide system, which includes a water-cooled circulator, directional couplers, a 3-stub tuner, a moveable short-circuit and a waveguide surfatron as a field applicator.

The discharge ignites in a quartz tube, placed vertically and perpendicularly to the waveguide wider wall. The injected into the discharge gas mixture consists of premixed argon and methane. The total gas flow introduced into the discharge was $Q^{tot} = Q_{Ar} + Q_{CH4}$. The microwave power was kept P = 500 W, argon flow $Q_{Ar}$ = 250 sccm and methane flow $Q_{CH4}$ = 0.6 sccm. The discharge is sustained under vortex-type flow conditions, which are provided by a swirl system (Fig. 1b). Relative concentration of $C_2$, $H_2$ and H were monitored using OES.

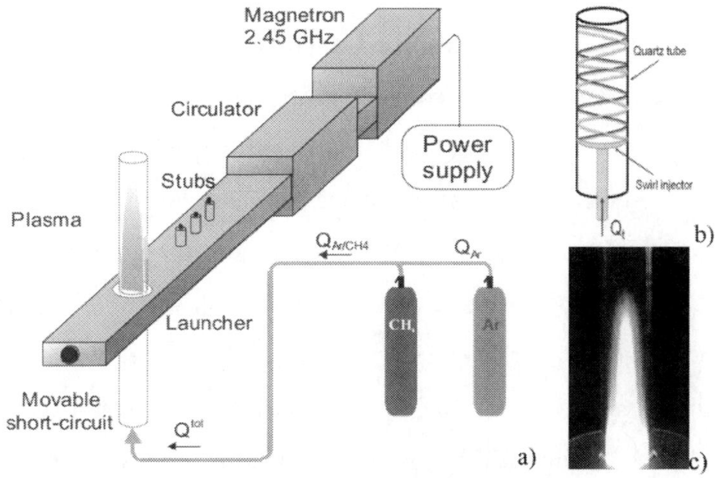

Fig 1. Scheme of the experimental setup (a), sketch of the swirl system (b) and a photo of the plasma torch (c)

**Theoretical model.**

The "active" microwave plasma region of the reactor considered is composed of two zones. The first one is the surface wave sustained discharge zone, including the zone inside the launcher and the extended "hot" plasma zone outside the launcher. Here, the surface wave power is absorbed by plasma electrons, which transfer the power to heavy particles via elastic and inelastic collisions, resulting in high gas temperatures. The gas temperature maintains nearly constant in the discharge zone when moving away from the launcher (up to about 10 cm) and then drops sharply in the "near" (10 - 13 cm) afterglow plasma zone. The second zone in axial direction is plasma afterglow (~20 cm) zone. The model input parameters are externally controlled, i.e. gas flows, total power delivered to the launcher, pressure and background gas mixture composition, i.e. Ar/$CH_4$ injected into the discharge.

A kinetic scheme similar to that presented in [4] is used. The scheme includes 57 components and about 390 chemical reactions. The model was applied for the heterogeneous case, i.e. considering that both gas and solid phase carbon nuclei are formed in the plasma. The gas thermal balance equation can be written as: $\frac{p_0}{k_{Bol}} V_0 \frac{1}{T_0} C_p \frac{dT_a}{dz} = -\frac{4\chi(T)}{R^2}(T_a - T_w) + \frac{\delta}{S}\frac{dP}{dz}$ here $V_0$ - initial gas velocity [m/s], $T_0$ - initial gas temperature, $p_0$ - gas pressure, $k_{Bol}$ - Boltzmann's constant, $C_p = 5/2k_{Bol}$, $S$ - plasma cross-section and $P$ - absorbed microwave power, $\delta$ - coefficient expressing the fraction of absorbed power. (LLA: Some quantities are not defined)

**Results and discussion.**

The equilibrium values of the stable substances and intermediate complexes along the "hot" plasma zone are useful to understand the general workings of methane decomposition and formation of carbon precursors. The calculated equilibrium values, i.e., the calculated concentrations of the main substances are shown in Fig. 2a and 2b as a function of the gas temperature. The equilibrium dissociation of methane begins at around 500 K and is complete at 1000 K. If the equilibrium heterogeneous process is considered, solid carbon and molecular hydrogen are the products formed in the range 1000 – 2500 K. Further, hydrogen dissociates and acetylene formation begins, reaching maximum concentration at ~3000 K.

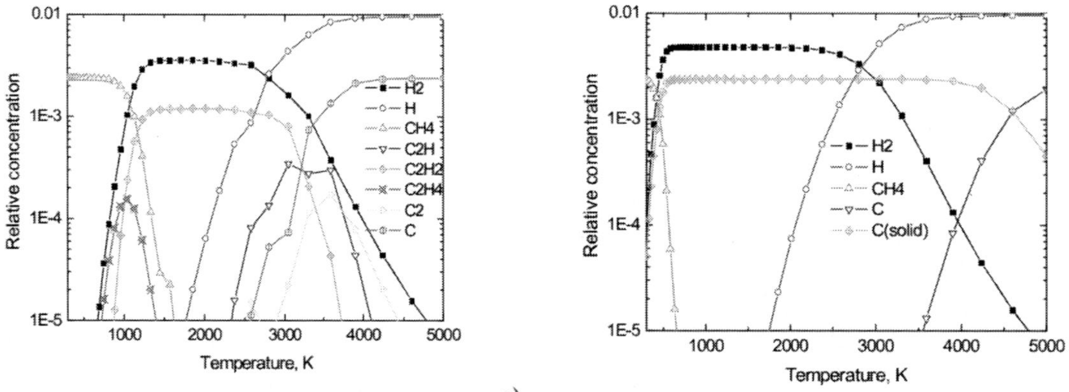


a) b)

Fig 2. Simplified equilibrium homogenous (a) and equilibrium heterogeneous (b) diagrams for the main methane decomposition products

The non-equilibrium results presenting the axial distribution of the decomposition products for heterogeneous case are shown in Fig. 3. Methane dissociation begins in "hot" plasma zone (~ 0.5 cm). The major initial reaction is C-H bond breaking with simultaneous formation of $CH_3$, $CH_2$, CH, H, $C_2$ and C at "hot" plasma zone. These radicals are involved in the formation of ethane and ethylene. The recombination of radicals can lead to the formation of acetylene through the following reactions:

$$CH_3 + CH_3 \rightarrow C_2H_6 \rightarrow C_2H_4 + H_2 \rightarrow C_2H_2 + 2H_2, \quad CH_2 + CH_2 \rightarrow C_2H_4 \rightarrow C_2H_2 + H_2,$$

$$CH + CH \rightarrow C_2H_2, \; CH_3 + CH \rightarrow C_2H_2 + H_2, \; CH_2 + CH \rightarrow C_2H_2 + H.$$

The acetylene is a stable molecule and can be broken down to the simplest C and H atoms at sufficiently high temperatures (above 4000 K). Diffusion of gas-phase carbon takes place from the "hot" plasma zone towards the tube wall, where the temperature is lower and nucleation of carbon species takes place. The main zone where nucleation occurs is located near the tube walls. (LLA: repetition) Part of the solid carbon nuclei is deposited on the wall; however the major part is gradually removed with the outlet gas stream.

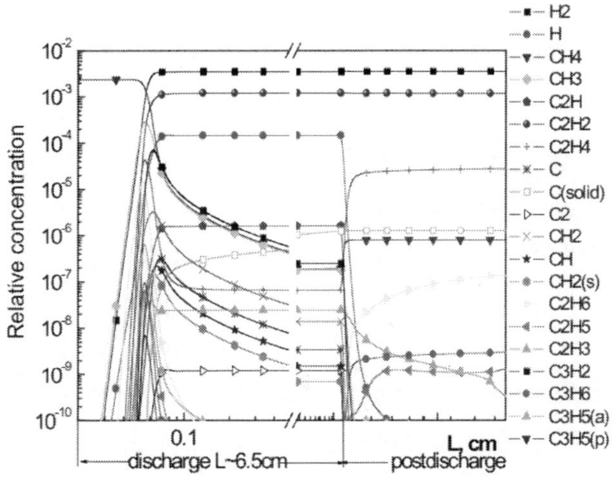


Fig 3. Simplified nonequilibrium heterogeneous diagram for the main methane decomposition products.

## Conclusion.

A microwave, "tornado"-type plasma, with swirl injection of the background gas, operating at 2.45 GHz and atmospheric pressure has been used to decompose methane in argon and argon-water vapor plasma environments. The theoretical model based on a set of non-linear rate balance equations describing the chemical kinetics and the gas thermal balance has been developed. Hydrogen, solid carbon and acetylene are the main outlet products. Production of acetylene can be promoted by preventing the formation of carbon. The results of the present investigation clearly show that surface wave sustained microwave plasma is a promising tool for methane reforming. Further investigation is in progress to analyze in detail the carbon balance, including measurements of the carbon deposited on the walls as well as the solid carbon carried away by the outlet gas flow.

## Acknowledgments.

This work was supported by the Fundação para a Ciência e Tecnologia (FCT) under projects PTDC/FIS/108411/2008 and Pest-OE/SADG/LA/0010/2011.

## References.

1. *Lange J.P., Jong K.P., Ansorge J., Tijm P.J.A.* Studies in Surface Science and Catalysis 1997. V.107 doi:10.1016/S0167-2991(97)80320-5
2. *Dias F.M., Bundaleska N., Henriques J., Tatarova E., Ferreira C.M.* Journal of Physics Conference Series Issue 06. 2014. doi:10.1088/1742-6596/516/1/012002
3. *Tatarova E., Bundaleska N., Dias F. M., Tsyganov D., Saavedra R. and Ferreira C. M.* Plasma Source Sciences and Technology V22. 2013. doi:10.1088/0963-0252/22/6/065001
4. *Bundaleska N., Tsyganov D., Tatarova E., Dias F.M., Ferreira C.M.* International Journal of Hydrogen Energy Vol.11 2014. Issue 39. doi:10.1016/j.ijhydene.2014.01.194

# Numerical simulations of astrophysical cyclotron-maser emission

D.C. Speirs[1], R. Bingham[1,2], K. Ronald[1], A.D.R. Phelps[1], S.L. McConville[1], K.M. Gillespie[1], R.A. Cairns[3], B.J. Kellett[1], I. Vorgul[3], A.W. Cross[1]

[1] *Department of Physics, SUPA, University of Strathclyde, Glasgow, Scotland, G4 0NG, U.K*

[2] *Department of Physics, SUPA, University of Strathclyde, Glasgow, Scotland, G4 0NG, U.K.*

[3] *School of Mathematics and Statistics, University of St. Andrews, Fife, Scotland, KY16 9SS, U.K.*

The electron cyclotron-maser instability plays a significant role in the generation of highly non-thermal, polarised radio emission from the auroral region of planets and stars having a dipole-like magnetic field [1, 2]. More recently, this radiation model has been extended to a variety of astrophysical environments including Blazar jets, collisionless shocks and brown dwarfs [3–7]. Planetary auroral radio emission has been a topic of interest for many years with several mechanisms having been proposed [8–13]. As the dispersion relation describing the electromagnetic wave growth only depends on the factor by which the magnetic field increases and on the ratio of plasma and cyclotron frequencies, the mechanism can scale effectively over many orders of magnitude both in wavelength and spatial scale. In the Earth's auroral region, observations by the Viking spacecraft [11] and the FAST satellite [12] led to the suggestion that the free energy source for AKR is a population of downward accelerated electrons having a large perpendicular velocity produced by a combination of parallel accelerating electric fields and converging magnetic field lines. The electron distribution takes the form of a horseshoe or crescent shape due to the first adiabatic invariance as the electrons move into the stronger magnetic field. Scaled laboratory experiments also demonstrate the potential for horseshoe distributions to generate narrowband cyclotron-maser emission at GHz frequencies over distances of a few meters [15–18].

Although the generation mechanism is well established, there remain important questions with regard to how the radiation may propagate and ultimately escape. Models and observations [19, 20] assume near field-aligned R-mode polarisation prior to escape, however it is well known that the radiation is generated near perpendicular to the magnetic field in the X-mode. The manner by which the radiation may effectively be redirected along the magnetic field and ultimately escape is a question recently addressed by Mutel [20] and Meniett [21], with a model of upward refraction due to altitude variation in the background plasma density and an initial radiation propagation vector that is mildly against the direction of the electron flux (negative axial wavenumber). Current theoretical models for X-mode generation via the cyclotron-maser instability do not predict growth with a negative axial wavenumber, however we have developed a new theoretical model that clearly demonstrates an enhanced growth rate for finite negative axial wavenumbers and moderate electron energy spread. This model is verified by the results of full 3D PiC (Particle-in-cell) code simulations, demonstrating for the first time that auroral cyclotron-maser emission is indeed characterised by a finite negative axial wavenumber.

The theoretical model [22] assumes an analytic form for the electron distribution function with different pitch angles, energy spread and density ratios between the hot component making up the horseshoe element and a background Maxwellian component. This is then used in a dispersion relation for the R-X mode which is obtainable from the susceptibility tensor given by Stix [23]. The derived growth rate is approximately given by $Im(\omega) = \frac{Im(D)}{\frac{\partial}{\partial\omega}(Re(D))}$, where $D = 0$ is the dispersion relation. The imaginary part $Im(D)$ does not vary greatly for small $\pm n_{\parallel}$ (where $n_{\parallel} = k_{\parallel}c/\omega$) however the value of $\partial D/\partial\omega$ rises dramatically as a function of frequency, due to approaching the upper hybrid resonance, where $D$ has a singularity. This means that the growth rate for positive $n_{\parallel}$ (higher frequency and forward-wave) is reduced and maximum growth occurs at lower frequency (negative $n_{\parallel}$ and backward-wave). Beam spreading in velocity space tends to reduce growth rate for negative $n_{\parallel}$, therefore enhanced backward wave coupling should break down for increased energy or velocity spread.

We compare the theoretical analysis with the results of PiC simulations, undertaken using the particle-in-cell code VORPAL [24]. The simulation geometry comprised an axial length of $144\lambda_{ce}$ (where $\lambda_{ce}$ is the vacuum wavelength of radiation at the peak electron cyclotron frequency within the simulation) and transverse perfectly-matched-layer boundaries to mitigate reflection. A 20keV electron beam was injected parallel to an axial magnetic field increasing by a factor of 15 over $45\lambda_{ce}$. The peak magnetic flux density was plateaued over the remaining $99\lambda_{ce}$ of beam propagation path to maintain cyclotron resonance for a significant number of Larmor steps with $\omega_{ce}/\omega_{pe} = 20$.

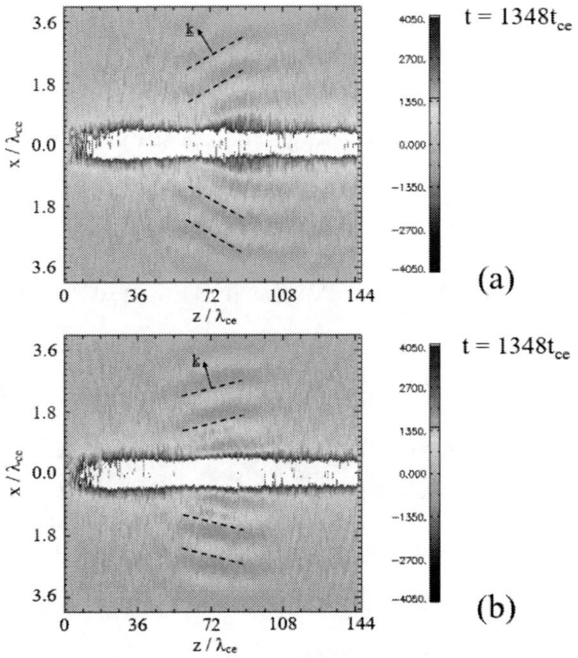


Figure 1: *Contour plots of Ey in an x-z plane at* $t/t_{ce} =$ 1348 *for (a)* 2% *and (b)* 6% *Gaussian energy spreads*

An analysis of the transverse electric field profile $E_y$ in an x-z plane (Fig. 1b,c) for 2% and 6% energy spread reveals a well-defined and temporally modulated electromagnetic emission at an axial coordinate of $80\lambda_{ce}$ and polarised in the X-mode with a significant negative axial wavenumber (backward-wave component). The backward wave character of emission is evidenced by an oblique wave front propagation angle with respect to the axis of the system. The effective backward-wave angle for 2% energy spread is $\sim 3°$ from perpendicular - consistent with requirements for the aforementioned model of upward refraction of the generated radiation [21]. It is evident in figure 1 that as the energy spread increases to 6%, there is well-defined change in the wavevector with a reduction in backward propagation angle to $\sim 2°$. This emission profile not only indicates a reduction in backward wave coupling with increased energy spread, but a distinct curvature can also be observed in the wavefront for 6% energy spread (figure 1c). This curvature indicates a progressive shift towards perpendicular propagation with increasing axial coordinate, showing that velocity spreading due to cyclotron-wave coupling reduces efficiency for -ve $k_{\|}$. This result is consistent with predictions of the analytical theory.

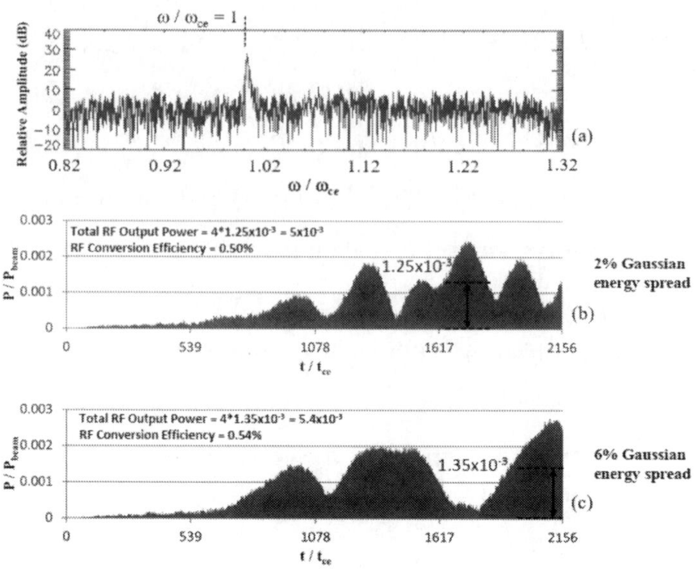


Figure 2: *Contour plots of Ey in an x-z plane at* $t/t_{ce} =$ 1348 *for (a) 2% and (b) 6% Gaussian energy spreads*

Figure 2a contains a Fourier transform of $E_y$ at time $2156t_{ce}$ and an axial position of $86\lambda_{ce}$. A single spectral component is present at $\omega/\omega_{ce} = 1$, corresponding to narrowband emission at the relativistic electron cyclotron frequency. Looking at the corresponding transverse Poynting flux over a y-z plane in figure 2b, a temporal modulation to the emission is evident, with a $810t_{ce}$ lead time for significant growth of the radio frequency (RF) Poynting flux. A peak saturated output power of $P/P_{beam} = 1.25 \times 10^{-3}$ is apparent (as the space-time average is $1/2$ of the plotted peak amplitude), which integrated over the transverse dimensions of the system (factoring in the 4 enclosing Poynting flux planes) equates to an RF conversion efficiency of 0.50%. Looking at 6% energy spread (figure 2c) a more stable output is observed, with longer period temporal fluctuations and a marginally increased peak beam-wave conversion efficiency of 0.54%. It would therefore appear that as the electron energy spread is increased, not only does the wavevector of emission tend from backward-wave to near perpendicular, but a more temporaly stable cyclotron-maser emission process is observed.

In conclusion, we have shown that the electron cyclotron-maser instability driven by a horseshoe-shaped velocity distribution has an inherently enhanced growth rate for wave propagation with finite negative $k_{||}$ corresponding to a backward-wave propagation angle a few degrees from perpendicular. This natural backward-wave character to the emission serves as a suitable precursor to the recently proposed model of upward refraction and field-aligned escape of terrestrial auroral kilometric radiation [20, 21].

This work was supported by EPSRC grant no. EP/G04239X/1.

## References

[1] P. Zarka, Adv. Space Res. **12**, 99 (1992).
[2] R. Bingham, R. A. Cairns, and B. J. Kellett, Astron. Astrophys. **370**, 1000 (2001).
[3] M. C. Begelman, R. E. Ergun, and M. J. Rees, Astrophys. J. **625**, 51 (2005).
[4] R. Bingham, et al., Space Sci. Rev. *10.1007/s11214-013-9963-z* (2013).
[5] R. A. Treumann, Astron. Astrophys. Rev. **13**, 229 (2006).
[6] R. Bingham et al., Astrophys. J. **595**, 279 (2003).
[7] G. Hallinan et al., Astrophys. J. **684**, 644 (2008).
[8] C. S. Wu and L. C. Lee, Astrophys. J. **230**, 621 (1979).
[9] R. M. Winglee and P. L. Pritchett, J. Geophys. Res. **91**, 13531 (1986).
[10] D. B. Melrose, *Instabilities in Space and Laboratory Plasmas* (Cambridge University Press, 1986).
[11] P. Louarn et al., J. Geophys. Res. **95**, 5983 (1990).
[12] R. E. Ergun et al., Astrophys. J. **538**, 456 (2000).
[13] R. Bingham and R. A. Cairns, Phys. Plasmas **7**, 3089 (2000).
[14] A. Roux et al., J. Geophys. Res. **98**, 11657 (1993).
[15] K. Ronald et al., Phys. Plasmas **15**, 056503 (2008).
[16] S. L. McConville et al., Plasma Phys. Controlled Fusion **50**, 074010 (2008).
[17] D. C. Speirs et al., Plasma Phys. Controlled Fusion **50**, 074011 (2008).
[18] A. G. Shalashov et al, Plasma Phys. Controlled Fusion **54**, 085023 (2012).
[19] R. A. Cairns, I. Vorgul, and R. Bingham, Phys. Rev. Lett. **101**, 215003 (2008).
[20] R. L. Mutel, I. W. Christopher, and J. S. Pickett, Geophys. Res. Letters **35**, L07104 (2008).
[21] J. D. Menietti et al., J. Geophys. Res. **116**, A12219 (2011).
[22] D.C. Speirs et al., Physical Review Letters, 113, 155002, (2014).
[23] T. H. Stix, *Waves in Plasma* (AIP, New York, 1992).
[24] C. Nieter and J. R. Cary, J. Comput. Phys. **196**, 448 (2004).

# A TWT-Upgrade to study wave-particle interactions

M. Caetano de Sousa[1], D. Guyomarc'h[2], and F. Doveil[2]

[1] *Instituto de Física, Universidade de São Paulo, São Paulo, Brasil*

[2] *Laboratoire PIIM, UMR 7345 CNRS-Aix Marseille Université, France*

Widely used as HF wave amplifiers for industrial applications, Traveling Wave Tubes (TWT) are also used to mimic and carefully study wave-particle interaction in a plasma [1-3]. We describe a recently built upgrade of a TWT whose slow wave structure consists of a 4 m long helix, with 3.4 cm diameter and 1 mm pitch, made of a 0.6 mm diameter Be-Cu wire wrapped in an insulating tape. The helix is inserted in a glass tube evacuated by two ion pumps on both ends. At one end, an electron gun made of a triode produces an electron beam propagating along the axis of the helix and radially confined by a constant axial magnetic field. Four movable probes capacitively coupled to the helix through the glass tube are used to launch and monitor a spectrum of waves generated by an arbitrary waveform generator at a few tens of Mhz. At the other end of the helix, a trochoidal analyzer allows to reconstruct the electron distribution functions of the beam after its self-consistent interaction with the waves. Linear properties of this new device will be reported.

## Principle of the experiment

The 1D beam-plasma system is a paradigm to study self-consistent wave-particle interaction in plasma turbulence. The bulk plasma acts as a linear dielectric to support the wave propagation and can be advantageously replaced by a slow wave structure, such as the helix in a Traveling Wave Tube (TWT). One is left with studying the essentially 1D interaction of the beam, propagating close to the axis of the helix since confined by a strong magnetic field (500G), with the waves, propagating along the structure and which, close to the axis, mainly consist in a longitudinal electric field.

## Up-grade of the experimental set-up

We had to face a forced up-grade of our TWT since, as shown in Fig.1 (left), the glass tube was broken. In the previous design of the TWT the helix was held inside the evacuated glass tube by three alumina rods visible on the figure. This resulted in a rather fragile system. In the TWT-upgrade, we opted for a much more mechanically robust design: the 4 m long helix, with 3.4 cm diameter and 1 mm pitch, made of a 0.6 mm diameter Be-Cu wire is wrapped in an insulating tape and then inserted into the glass tube, as shown in Fig.1 (right). As in the previous TWT, absorbing tapes are used at both ends of the helix to reduce electromagnetic wave reflection.

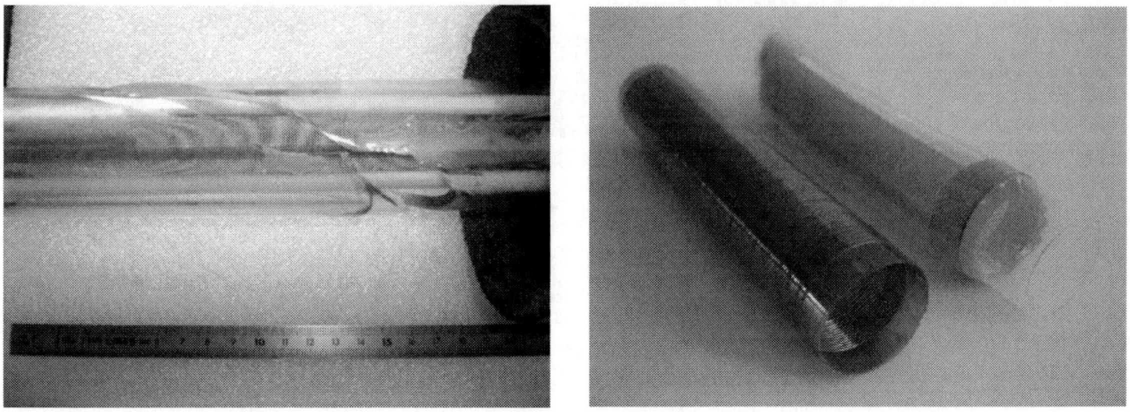

Fig. 1 : left) view of the broken helix, supported by three alumina rods in the glass tube, right) helix glued inside tape and inserted inside the glass tube.

The evacuated glass tube is set on the axis of a slanted wave-guide to shield electromagnetically the TWT. Four probes are capacitively coupled to the helix through the glass tube. In the TWT-upgrade, great care has been given to improve the precision and repeatability of the probes position as shown in Fig.2. Probes can be used to launch or receive a spectrum of waves generated by an arbitrary waveform generator, which allows to control the frequency (between 10 and 80 MHz), amplitude and phase of each wave.
As in the previous TWT, at one end, a triod with a pierced anode (3mm diameter) is used as an electron gun (10-100 eV) and, at the other end, a trochoidal analyzer allows to measure the electron energy distribution function after the beam has interacted with the waves.

**Experimental results**

Since, in the TWT-upgrade, the helix is much closer to the probes than in the previous version, the measured signal is larger but much more sensitive to the distance between probe and helix. Great care needs to be given to measure the coupling coefficient between each probe and the helix at any given axial position.

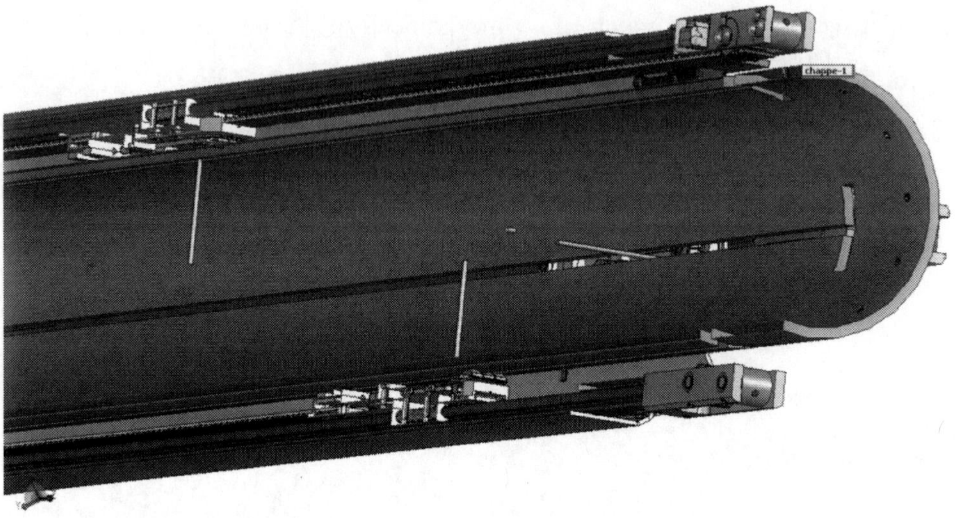


Fig. 2 : Movable probes on their trails (inside precision rail)
positioned by PU timing belts inside slanted wave-guide

The points in Fig.3 show the dispersion relation as measured using an interferometer method, with one probe as an emitter at a given frequency and another probe as a receiver. An excellent agreement is obtained with the the theoretical prediction of a sheet model.

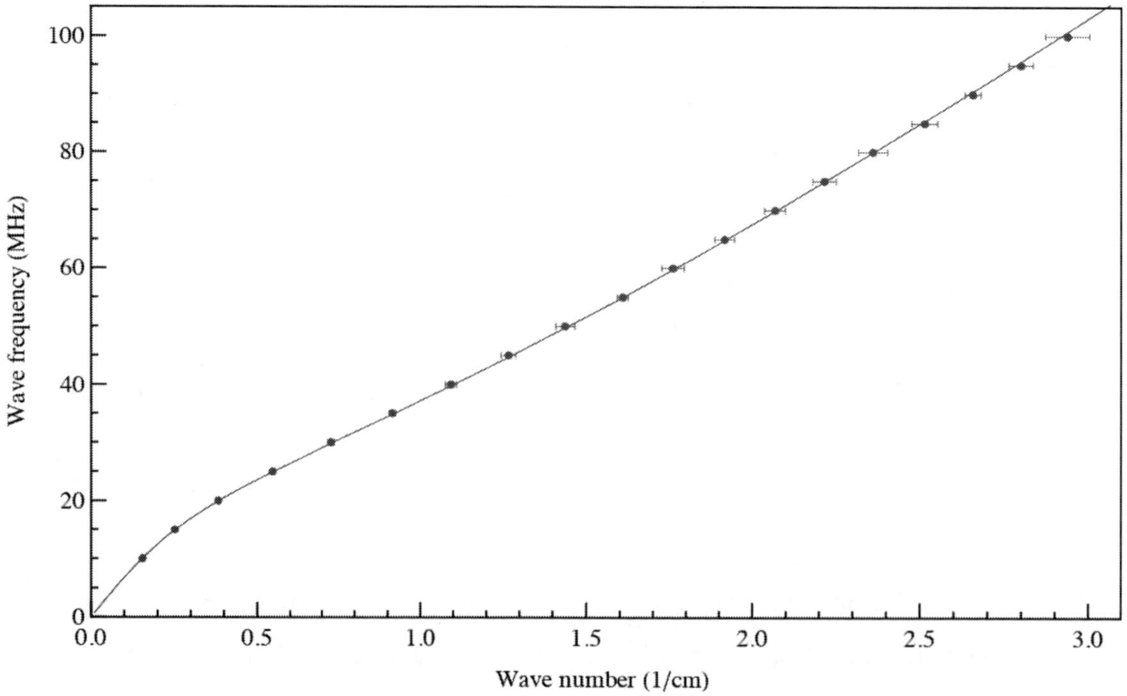


Fig. 3 : Measured and theoretically predicted dispersion relation of the TWT-upgrade.

**Perspectives**

In TWT-upgrade, we plan to study the synergy between the chaotic behavior observed with a low intensity test beam [3] and the self-consistent effects leading to wave amplification [1] when increasing the cold beam intensity.

TWT-upgrade will serve to bench mark industry oriented codes [4] .

We also plan to launch packets of electrons [5] with a well defined phase with respect to a wave to further explore the electrons phase space.

The authors thank J-B. Faure and V. Long for their efficient help in designing and using a new way to build the helix.

References

1] G. Dimonte, and J.H. Malmberg, Phys. Fluids 21, 1188 (1978).
[2] S.I. Tsunoda, F. Doveil, and J.H. Malmberg, Phys. Rev. Lett. 58, 1112 (1987).
[3] F. Doveil, A. Macor, and A. Aïssi, Celest. Mech. Dyn. Astr. 102, 255 (2008).
[4] P. Bernardi, F. André, J-F. David, A. Leclair, and F. Doveil, IEEE Trans. El. Dev. 58, 1761 (2011).
[5] A. Macor, F. Doveil, and E. Garabedian, Nonlinear Phen. Compl. Syst., 10, 180 (2007).

# Freak Waves and Modulational Dynamics in Plasmas with Negative Ions

I. S. El-Kamash[1,2], B. Reville[1] and I. Kourakis[1]

[1] *Centre for Plasma Physics, Queen's University Belfast, BT7 1NN Northern Ireland, UK*

[2] *Physics Department, Faculty of Science, Mansoura University, 35516 Mansoura, Egypt*

**Abstract**

The nonlinear dynamics of modulated electrostatic wavepackets propagating in negative-ion plasmas is investigated from first principles. A nonlinear Schrödinger equation is derived by adopting a multiscale technique. The stability of breather- like (bright envelope soliton) structures, considered as a precursor to *freak wave* (*rogue wave*) formation, is investigated and then tested via numerical simulations.

**Introduction.** Plasmas incorporating negative ions, in addition to positive ions and electrons exist in the laboratory [1, 2] and also in Space environments [3, 4, 5]. Such multicomponent plasmas support a variety of nonlinear modes, related with various instabilities and relevant phenomena [6, 7, 8]. The propagation of ion-acoustic (IA) waves in negative-ion plasmas (NIP) was studied theoretically within a nonlinear Schrödinger equation (NLSE) framework by Saito *et al* [9], who discussed modulational instability as a precursor for envelope modes in such a system. That work was based on the modified Korteweg-de Vries (mKdV) formalism (applicable in a "special" plasma configuration [10], where the nonlinearity coefficient in the ordinary KdV equation vanishes and higher-order nonlinearity dominates), and was thus limited to the weak-amplitude superacoustic region. The problem was later investigated experimentally by Bailung and coworkers [11, 12], who reported the observation of *freak-wave* (*rogue wave*) type structures, and adopted the interpretation provided by Saito *et al* [9] for their interpretation.

In this study, we adopt a multiscale perturbation technique to derive a nonlinear Schrodinger equation for modulated ion-acoustic wavepackets in negative ion plasma. Explicit conditions and instability thresholds for envelope soliton formation and modulational instability are determined, in terms of the negative-to-positive density and mass ratio(s). The stability of the solutions obtained has been investigated numerically via a Crank-Nicolson method. The main aim of our study is to extend the criteria (existence region) for freak waves in negative-ion plasmas, beyond the simplifying assumptions (weak-amplitude, weakly superacoustic regime) implicitly adopted in [11, 12]. In particular, we have shown, as anticipated, that the occurrence of freak wave structures is not necessarily due to the "special" plasma configuration that leads to the mKdV description [9], but may be observed in an extended region of parameter values.

**Theoretical model.** We consider a three component collisionless unmagnetized plasma consisting of electrons (mass $m_e$,charge $e$), positive ions (mass $m_+$,charge $q_+ = Z_+ e$) and negative ions (mass $m_-$, charge $q_- = -Z_- e$), and electrons (mass $m_e$, charge $e$). A reduced (dimensionless) fluid model for this configuration, in one-dimensional (1D) geometry, reads

$$\partial_t n_+ + \partial_x (n_+ u_+) = 0, \tag{1}$$

$$\partial_t u_+ + u_+ \partial_x u_+ = -\nabla\phi, \tag{2}$$

$$\partial_t n_n + \partial_x (n_- u_-) = 0, \tag{3}$$

$$\partial_t u_- + u_- \partial_x u_- = \delta\nabla\phi, \tag{4}$$

$$\partial_{xx}\phi = -n_+ + \beta n_- + (1-\beta)\exp(\phi), \tag{5}$$

The fluid velocity ($u_\pm$), density ($n_\pm$) and electrostatic potential ($\phi$) variables are scaled by $V_0 = (Z_+ k_B T_e/m_+)^{1/2}$, $n_{0,+}$ and $k_B T_e/e$, respectively. Space and time are respectively scaled by $\omega_{p,+}^{-1} = (4\pi e^2 n_{+,0} Z_+^2/m_+)^{-1/2}$ and $\lambda_{D,+} = (k_B T_e/4\pi Z_+ e^2 n_{+,0})^{1/2}$. We have defined the quantities $\beta = n_{-,0} Z_- /(n_{+,0} Z_+)$ and $\delta = (Z_-/m_-)/(Z_+/m_+)$. Here, $\omega_{p,+}$ is the positive-ion plasma frequency, $T_e$ is the electron temperature and $k_B$ is the Boltzmann constant.

**Amplitude modulation modelling.** Let us define the state (column) vector $\mathbf{S} = (n_+, u_+, n_-, u_-, \phi)^T$ and the corresponding equilibrium state $\mathbf{S}^{(0)} = (1,0,1,0,0)^T$. We shall consider small deviations from equilibrium by taking $\mathbf{S} = \mathbf{S}^{(0)} + \varepsilon\mathbf{S}^{(1)} + \varepsilon^2\mathbf{S}^{(2)} + ... = \mathbf{S}^{(0)} + \sum_{n=1}^{\infty}\varepsilon^n\mathbf{S}^{(n)} = \mathbf{S}^{(0)} + \sum_{n=1}^{\infty}\varepsilon^n\sum_{l=-n}^{n}\mathbf{S_{j,1}}^{(n)}\exp[il(kx-\omega t)]$, where $\varepsilon \ll 1$ is a small real parameter. We define the stretched (slow) space and time variables $T_r = \varepsilon^r t$ and $X_r = \varepsilon^r x$ (for $r = 0,1,2,3,...$). The fast scales ($r = 0$) affect the dynamics via the phase $\theta = kX - \omega T$ (only), while the slow scales ($r \geq 1$) enter the argument of the $l-$th harmonic amplitude $\mathbf{S}_l^{(n)}$, allowed to vary along $x$.

*Linear dynamics*: The first-order (linear) expressions lead: $n_1 = \frac{\phi_1}{v_{ph}^2}$, $u_1 = \frac{\phi_1}{v_{ph}}$, along with the *(*dispersion relation): $\omega^2 = \frac{(1+\delta\beta)k^2}{k^2+(1-\beta)}$. This provides (in the long-wavelength limit) the expression $v_{ph} = \omega/k \simeq \left(\frac{1+\delta\beta}{1-\beta}\right)^{1/2}$ for the true sound speed (affected by the negative ions ia the parameters $\beta$ and $\delta$). Note that an e-i plasma is recovered upon setting $\beta = 0$.

The 2nd-order equations provide a set of expressions (here omitted for brevity) for the amplitude(s) of the 2nd, 1st and 0th harmonic corrections at this order. Annihilation of secular terms prescribes the group velocity as $v_g = \frac{d\omega}{dk} = \frac{(1-\beta)(1+\delta\beta)^{1/2}}{[k^2+(1-\beta)]^{3/2}}$, viz. $S_{j,1}^{(1)} = S_{j,1}^{(1)}(X_1 - v_g T_1)$.

Proceeding to order $\sim\varepsilon^3$, one obtains the Nonlinear Schrödinger Equation (NLSE)

$$i\partial_\tau\Psi + P\partial_{\xi\xi}\Psi + Q|\Psi|^2\Psi = 0 \tag{6}$$

where $\tau = T_2 = \varepsilon^2 t$ and $\xi = X_1 - v_g T_1 = \varepsilon(x - v_g t)$. Note that the *dispersion coefficient* $P =$

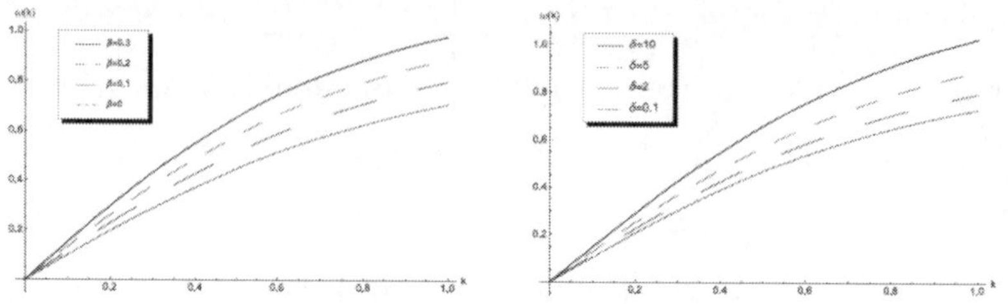


Figure 1: (Color online) The angular frequency $\omega$ is depicted versus the wavenumber $k$, for different values of $\beta$ and $\delta$.

$\omega''(k)/2$ is negative, while the *nonlinearity coefficient* $Q$, due to carrier wave self-interaction, is given as a perplex function $Q = Q(\{\omega;\beta,\delta,...\})$ (here omitted).

**Freak waves as breather-type solutions of the NLSE.** Various solutions of the NLSE (6) have been proposed as prototypical forms of freak (rogue) waves. These are summarized, e.g., in [13] and thus need not be presented here.

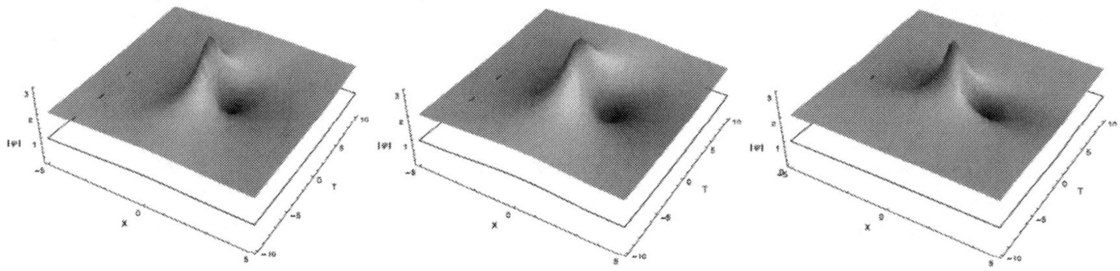

Figure 2: (Color online) The *Peregrine soliton* is depicted for different values of $\beta$ ($= 0.2, 0.4, 0.6$), with $k = 0.1$ and $\delta = 1$ (e.g. for $H^+/H^-$ plasma).

**Numerical analysis.** In order to test our predictions for the nature and stability of bright envelope solitons, we have integrated the NLSE via an implicit integration method, in different cases. The known bright soliton solution was adopted as initial condition (IC) in the numerical integration of the NLSE (6). In some runs, the plasma parameters were taken to differ between IC and numerical integrator, with the aim to assess the impact of the variation of e.g., $\beta$ on a given energy lump injected in the system as initial condition. In brief, a bright pulse changing shape (but retaining its stability) in the bright ($PQ > 0$) region, while it decays and spreads in the dark ($PQ < 0$) region.

Our results will be reported in full detail elsewhere.

**Acknowledgments.** This work was supported from Queen's University Belfast (Centre for Plasma Physics) funding. One of us (I.S. El-Kamash) acknowledges the financial support by an Egyptian Government fellowship.

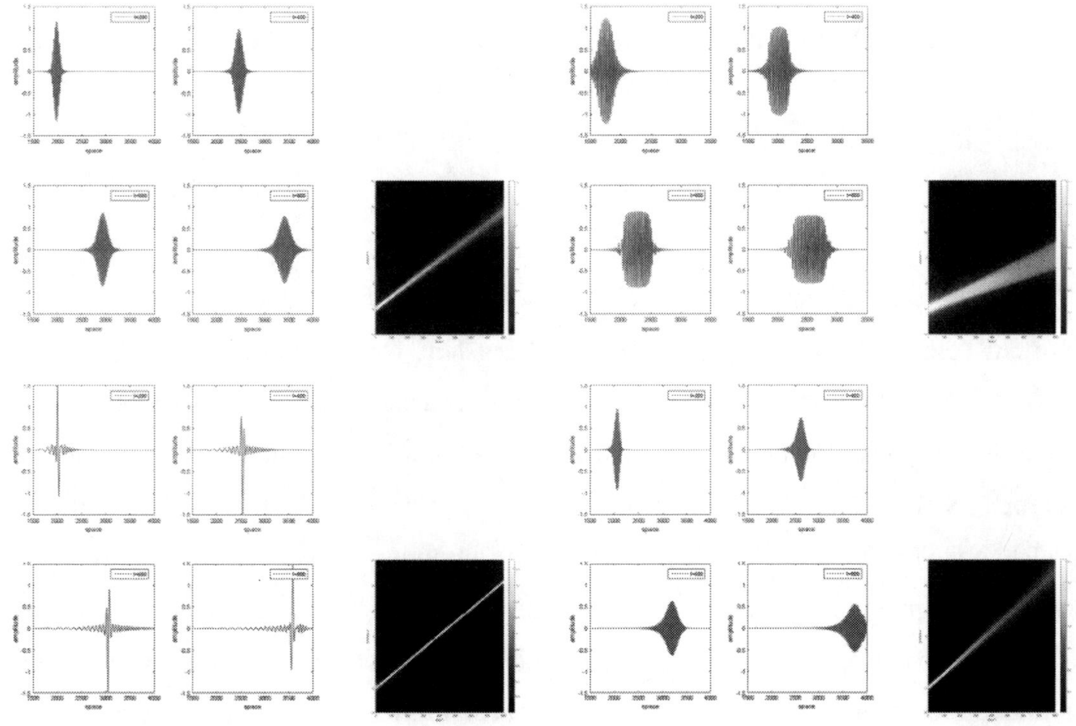

Figure 3: (Color online) Time evolution and propagation of a breather (bright soliton) on the space-time plane. Top left panel: IC ($\beta = 0.1, \delta = 1, k = 0.1$, $PQ > 0$), NLSE ($\beta = 0.2, \delta = 1, k = 0.1$, $PQ > 0$). Top right panel: IC ($\beta = 0.2, \delta = 1, k = 0.1$, $PQ > 0$), NLSE ($\beta = 0, \delta = 1, k = 0.1$, $PQ < 0$). Bottom left panel: IC ($\beta = 0.2, \delta = 1, k = 0.1$, $PQ > 0$), NLSE ($\beta = 0.4, \delta = 2.1, k = 0.7$, $PQ > 0$). Bottom right panel: IC ($\beta = 0, \delta = 1, k = 0.1$, $PQ < 0$), NLSE ($\beta = 0.2, \delta = 1, k = 0.1$, $PQ > 0$).

## References


[1] Cooney, J. L. *et al* (1993), Plasma Sources Sci. Technol. **2**, 73.

[2] Sharma, S. K., Devi, K., Adhikary, N. C. and Bailung, H. (2008), Phys. Plasmas, **15**, 082111.

[3] Swider, W. (1988), Pure Appl. Geophys., **127**, 403.

[4] Chaizy, P. H., et al. (1991), Nature, **34**, 393.

[5] Coates, A. J., et al. (2007), Geophys. Res. Lett., **34**, L22103 1.

[6] A. Y. Wong, D. L. Mamas, and D. Arnush(1975), Phys. Fluids **18**, 1489.

[7] T. Intrator and N. Hershkowitz(1983), Phys. Fluids **26**, 1942.

[8] M. S. Ruderman, T. Talipova and E. Pelinovsky (2008), J. Plasma Physics **74**, 639.

[9] Saito, M., Watanabe, S., and Tanaka, H. (1984), J. Phys. Soc. Jpn., **53**, 2304.

[10] Frank Verheest (2001), *Waves in Dusty Space Plasmas* (Kluwer Publishing).

[11] Bailung, H., Sharma, S. K., and Nakamura, Y. (2011), Phys. Rev. Lett. **107**, 255005.

[12] Sharma, S. K., Bailung, H. (2013), J. Geophys. Res. **118**, 919.

[13] Veldes, G.P. *et al* (2013), J. Opt. **15**, 064003; McKerr, M. *et al* (2014), Plasma Phys. Cont. Fus. **56**, 035007.

# Theory of the ordinary and extraordinary mode coupling in fluctuating plasmas

E.D. Gospodchikov, A.G. Shalashov, T.A. Khusainov

*Institute of Applied Physics of Russian Academy of Sciences, Nizhny Novgorod, Russia*

Linear conversion of the ordinary (O) wave to the extraordinary (X) wave plays an important role in excitation of the electron Bernstein waves, which in turn provide an effective way for high-frequency (electron cyclotron) heating and diagnostics of overdense plasma in spherical tokamaks and optimized stellarators [1]. In spite of rather clear theoretical background behind this process, exiting theory tends to overestimate efficiency the O-X coupling in all available experiments. Following Laqua significant discrepancy between the predicted and observed efficiencies is usually explained by turbulent fluctuations of the cut-off layer where the mode conversion occurs [2]. However, only a qualitative model based on essentially one-dimensional theory has been proposed, in which poloidal fluctuations are taken into account by effective broadening of the incident wave spectrum over the poloidal wave numbers. In the present paper we propose a more rigorous approach based on two-dimensional theory of the mode conversion that can adequacy describe fluctuations without any qualitative speculations.

## 1. Reference wave equations

Following [3], Maxwell equations in a region of the O-X wave coupling can be reduced to

$$\begin{cases} N_\parallel (i\partial/\partial x - \partial/\partial y) F_\parallel = \sqrt{2}k_0 (\varepsilon_+ - N_\parallel^2) F_+ \\ N_\parallel (i\partial/\partial x + \partial/\partial y) F_+ = \sqrt{2}k_0\, \varepsilon_\parallel F_\parallel \end{cases} . \quad (1)$$

Here $k_0 = \omega/c$, $N_\parallel = k_\parallel / k_0$ is the refractive index parallel to the magnetic field, $F_{+,\parallel} = E_{+,\parallel} \exp(-ik_\parallel z + i\omega t)$ are slow field amplitudes and $\varepsilon_{+,\parallel}$ are the dielectric tensor components in the Stix frame: $\varepsilon_\parallel = 1 - \omega_{pe}^2/\omega^2$, $\varepsilon_+ = 1 - \omega_{pe}^2/\omega(\omega + \omega_{ce})$. Effective coupling occurs when r.h.s. of both equations goes to zero, i.e. $\varepsilon_\parallel$, $\varepsilon_+ - N_\parallel^2 \to 0$ (plasma cut-offs). There are three spatial scales in the above equations: the density variation $L_n = n_e/\nabla n_e$, the magnetic field variation $L_B = B/\nabla B$, and the wave coupling length $L_\nabla \sim \sqrt{L_n / k_0}$ .

Equations (1) have been obtained for a weakly inhomogeneous plasma, $k_0 L_n$, $k_0 L_B >> 1$, so in the tokamak conditions the coupling is highly localised, $L_\nabla << L_n < L_B$.

To study the effect of fluctuations we consider the simplest case in which

$$n_e = (x / L_n)\, n_{cr} + \delta n_e(x, y), \quad \mathbf{B} = (1 + x / L_B)\, B_0\, \mathbf{z}_0, \tag{2}$$

i.e. only density is perturbed, and without this perturbation the problem is reduced to one-dimensional case. Note that $x$ and $y$ represent here the radial and poloidal directions correspondingly, $n_{cr}$ is the plasma cut-off density at which $\varepsilon_\parallel = 0$. Having in mind that O-X coupling occurs on scale $L_\nabla$ shorter then all plasma inhomogeneity scales, we retain only linear dependences over $x$ in the magnetic field and the non-perturbed density. Then, the r.h.s. of equations (1) may be simplified noting that

$$\varepsilon_\parallel \propto x' + \delta_s + \delta_a, \quad \varepsilon_+ - N_\parallel^2 \propto x' + \delta_s - \delta_a, \tag{3}$$

where $x' = x / L_\nabla$ is the normalized radial coordinate, and

$$\delta_s = \frac{\delta n_e}{n_{cr}} \frac{L_n}{L_\nabla} - \delta_a, \; \delta_a = \frac{\delta n_e}{2n_{cr}} \frac{L_n}{L_\nabla} \frac{L_n}{L_B(1 + \omega / \omega_{ce}) - L_n}, \tag{4}$$

represent the "symmetrical" and "asymmetrical" contributions of the density fluctuations. Note that asymmetrical part appears due the regular variation of a magnetic field, and therefore $\delta_a << \delta_s$ for large aspect ratio tokamaks while $\delta_a \sim \delta_s$ for spherical tokamaks.

Study of wave equations (1) can be done after transformation to new field variables $F_+ \propto A^+ + A^-$, $F_\parallel \propto A^+ - A^-$ introduced in such a way that in the WKB limit $A^+$ and $A^-$ are amplitudes of waves that propagate, respectively, in the positive and negative directions along the radial coordinate:

$$\begin{cases} (i\,\partial/\partial x' + x' + \delta_s)A^+ = (-\partial/\partial y' + \delta_a)A^- \\ (i\,\partial/\partial x' - x' - \delta_s)A^- = (-\partial/\partial y' - \delta_a)A^+ \end{cases}. \tag{5}$$

One can see that the r.h.s. of these equations describes the coupling between counter propagating waves which is affected only by the asymmetrical part of the fluctuations. Opposite to it, the symmetrical part does not disturb the wave coupling however can modify propagation of the separate waves outside the coupling region. So even from a structure of wave equations we can argue that symmetric and asymmetric contributions of density fluctuations act essentially different. Physically it may be understood as $\delta_s$ describes the shift of the evanescent region, while $\delta_a$ describes the fluctuation of its width, see figure 1.

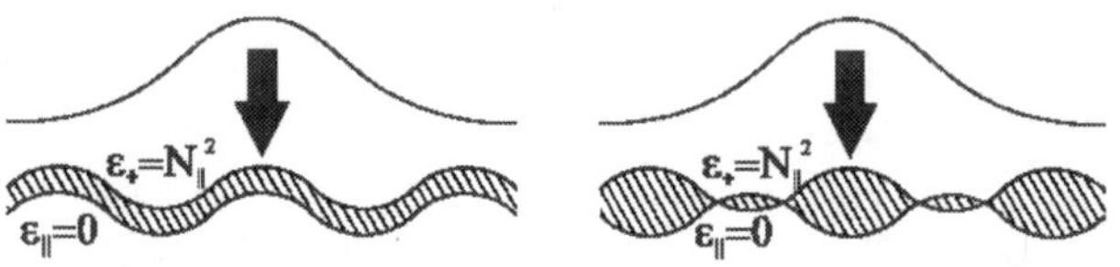


Fig. 1. Schematic of the symmetric (left) and asymmetric (right) contributions of density fluctuations. Dashing shows the region where studied electromagnetic modes are evanescent.

## 2. Coupling efficiency for plasma with harmonic fluctuations

Equations (5) were investigated in details in paper [3]; the main results of these studies may be summarized as follows.

For radial harmonic fluctuation, $\delta_{s,a} = a_{s,a}\cos(\kappa x' + \varphi)$, the problem is reduced to the essentially one-dimensional case, $\partial/\partial y' \to ik_y$, that may be treated in a standard Born approximation. In for case of effective mode coupling, $k_y << 1$, one can calculate the transmission coefficient of incident plane wave averaged over a random phase $\varphi$,

$$T_{1D} = 1 - \int_0^{2\pi} \frac{|A^-(-\infty)|^2}{|A^+(-\infty)|^2}\frac{d\phi}{2\pi} = \pi k_y^2 + \tfrac{1}{2}\pi a_a^2 - \pi^2 k_y^2 (a_a^2 + 4a_s^2 k_y^2/\kappa^2)\sin^2(\kappa^2/4) + \ldots \quad (6)$$

For poloidal harmonic fluctuation, $\delta_{s,a} = a_{s,a}\cos(\kappa y' + \varphi)$, the symmetric contribution ($a_s$) results in coupling of co-propagating harmonics, so one can neglect the reflection (take the r.h.s. of Eq.(5) equal to zero). In this case an exact analytic solution may be obtained, that describes the slow modification of the poloidal spectrum of an incident beam as it approaches the plasma cut-off. This is actually well-known diffusion in k-space induced by fluctuations [4]. It is important to note, that this effect is collected on a much longer path than the localized reflection region where the incident wave couples with a counter-propagating wave. Inside this zone, the symmetric contribution is of minor importance for all meaningful parameters being investigated. The asymmetric contribution ($a_a$) couples counter-propagating harmonics close to the plasma cut-off. The phase-averaged transmission coefficient for the plane wave with $k_y = 0$ may be estimated as

$$T_{2D} \approx \exp[-\tfrac{1}{2}\pi(1+\pi\kappa^2)^{-1}a_a^2]. \quad (7)$$

## 3. Discussion

Basing on the developed theory we revise the role of fluctuations in the OXB experiment. Note that our definitions of fluctuation parameters (4) contain product of small and large terms. As a result, in a fusion experiment the symmetric contribution $\delta_s$ may be of order of unity even though the relative density perturbations are small. Nevertheless, study of the typical parameter space shows that the impact of symmetric contribution of both radial and poloidal fluctuations on the O-X coupling remains small even for unrealistically large fluctuation level ($\delta T \sim 1\%$ for $\delta n_e / n_e \sim 20\%$). However, poloidal fluctuations can modify the amplitude and phase distributions in the incident quasi-optical beam far outside the coupling region located near the cut-off layer, what affect the consequent mode-conversion process ($\delta T \sim 10\%$ for the path 10 cm and $\delta n_e / n_e \sim 2\%$). The asymmetric contribution resulted from a magnetic field inhomogeneity typically can be neglected in large aspect ratio devices but not in compact tori since $\delta_a / \delta_s \sim L_n / L_B \sim a / R$. Therefore, in a conventional tokamak fluctuations result only in distortion of a wave beam on its way towards the mode coupling region but have no influence on the efficiency of the mode interaction inside the coupling region. In a spherical tokamak with $\delta_a \sim \delta_s$ the asymmetrical part of density perturbations may be significant inside the coupling region resulting in degradation the transmission efficiency by up to 25% for realistic conditions.

Summarizing, we find that plasma fluctuations can not be fully responsible for the low efficiency of OXB heating of overdense plasma observed in present-day experiments. We guess that the main factor impeding the tunneling of the electromagnetic waves through the plasma cut-off in these experiments may be associated with a curvature of toroidal magnetic surfaces [5].

Acknowledgements. The research was supported by RFBR (pr. 14-02-31024, 15-32-20770) and the President Council for Grants (pr. MD-1736.2014.2).

References

[1] H. P. Laqua, PPCF, 2007, 49 R1–42.
[2] H. P. Laqua et al., Phys. Rev. Lett., 1997, 78, 3467.
[3] A. G. Shalashov, E. D. Gospodchikov, PPCF, 2014, 56, 125011
[5] E. V. Sysoeva et al., PPCF, 2013, 55, 115001.
[6] E. D. Gospodchikov et al., PPCF, 2012, 54, 045009.

# Investigation of the O-mode anomalous absorption in the UH resonance

V.I. Arkhipenko[1], E.Z. Gusakov[2], L.V. Simonchik[1], E.V. Sysoeva[2], M.S. Usachonak[1]

[1] *Institute of Physics NAS of Belarus, Minsk, Belarus*

[2] *Ioffe Institute, St-Petersburg, Russia*

High power ECRH is widely used nowadays for heating of electron component, current drive, control of neoclassical tearing modes and is considered for application in ITER. Since 80th nonlinear effects such as parametric decay instabilities (PDIs) were believed to be deeply suppressed in the second harmonic extraordinary mode and first harmonic ordinary mode ECRH experiments in toroidal magnetic fusion devices [1]. Nevertheless, during the last decade a number of anomalous phenomena observations have been reported in second harmonic ECRH experiments [2-4] such as fast ion generation [2, 3] and anomalous backscattering [4]. An explanation of these observations proposed recently is based on possibility of low-threshold parametric excitation of decay waves trapped in plasma due to non-monotonic behaviour of plasma density in radial direction [5, 6] associated with density pump out effect, magnetic islands, drift-wave eddies or blobs and density maximum at the discharge axis. In principle, this mechanism is not specific for the second harmonic ECRH and can occur in the case of first harmonic O-mode heating in contemporary devices and in ITER. In this report we analyse a possibility of low-threshold parametric excitation by the O-mode pump of the upper hybrid (UH) waves localised in the axially symmetric plasma filament elongated in the magnetic field direction. The theory predictions are compared to the results of model experiments on interaction of ECR frequency range microwave O-mode pump with inhomogeneous plasma performed at the linear plasma device “Granit” [7].

**Theoretical approach and results.**

We consider the decay of the ordinary pump wave propagating perpendicular to the magnetic field and possessing the electric field given by expression $E_{0z} = \sqrt{\dfrac{8P_0}{\pi ZYc}} \exp\left(-\dfrac{z^2}{2Z^2} - \dfrac{y^2}{2Y^2} - i\omega_0 t\right)$, where $z$, $y$ are directions along and transverse to the magnetic field respectively and $P_0$ is the pump wave power. The wave number of the pump wave is supposed to be negligible compared to wave numbers of the decay waves. We use cylindrical coordinate system below with the axial direction along the magnetic field. The basic equations describing generation of UH wave $\varphi_{uh} = \tilde{\varphi}_{uh}(r)\exp(ik_z z + im\phi - i\omega t)$ and low frequency (LF) daughter wave $\varphi_{lf} = \tilde{\varphi}_{lf}(r)\exp(-ik_z z - im\phi - i\Omega t)$ are as follows:

$$\left[\left(\varepsilon+\beta\,\Delta_\perp\right)\Delta_\perp+\eta\frac{\partial^2}{\partial z^2}\right]\varphi_{uh}=-4\pi\rho_{uh} \tag{1}$$

$$\hat{L}\varphi_{lf}=-4\pi\rho_{lf} \tag{2}$$

Here the permittivity tensor components are given by $\varepsilon=1+\delta\varepsilon_e(\omega)$ ; $\delta\varepsilon_e(\omega)=-\frac{\omega_{pe}{}^2}{\omega^2-\omega_{ce}{}^2}$;

$\eta_{uh}=1-\frac{\omega_{pe}^2}{\omega^2}$ ; $\beta=\frac{\omega_{pe}{}^2 v_{Te}{}^2\left(\omega^2+3\omega_{ce}^2\right)}{\left(\omega^2-\omega_{ce}^2\right)^3}$ ; $v_{Te}=\sqrt{2T_e/m_e}$ ; $\Delta_\perp=\Delta-\frac{\partial^2}{\partial z^2}$ ; $\hat{L}$ is the LF wave operator and the nonlinear charge densities are given by $\rho_{uh}=\frac{i}{4\pi}\delta\varepsilon_i(\Omega)\delta\varepsilon_e(\omega)\frac{E_0k_z}{\omega_0^2}\frac{e}{m_e}\Delta\varphi_{lf}^*;\rho_{lf}=-\frac{i}{4\pi}\delta\varepsilon_i(\Omega)\delta\varepsilon_e(\omega)\frac{E_0k_z}{\omega_0^2}\frac{e}{m_e}\Delta\varphi_{uh}^*$ ; $\delta\varepsilon_i(\Omega)=-\frac{\omega_{pi}^2}{\Omega^2}$.

In the absence of the pump wave (1) the WKB representation $\tilde{\varphi}_{uh}=\varphi_0 e^{i\int k_r dr}$ for the UH wave potential results in the following dispersion relation

$$D\left(k_r,r,\omega\right)=\left(\varepsilon(r)+\beta\left(k_r{}^2+\frac{m^2}{r^2}\right)\right)\left(k_r{}^2+\frac{m^2}{r^2}\right)+\eta k_z{}^2=0 \tag{3}$$

In the case of standard plasma density profile possessing maximal value higher than the UHR density at the filament axis this dispersion relation describes UH waves trapped in the plasma. In particular, for the azimuthally propagating waves ($k_r{}^2 \ll \frac{m^2}{r^2}$) the UH wave localization takes place in the vicinity of the point where the following set of equations is satisfied for $k_{rl}=0$: $\left\{D(\omega_l,k_{rl},r_l)=0;\ \left.\frac{\partial D}{\partial r}\right|_{\omega_l,k_{rl},r_l}=0\right\}$. In the vicinity of this point equation (1) for the potential takes the form

$$\left[\left(\omega-\omega_l\right)D_\omega+\delta k_z D_{k_z}+\frac{\left(r-r_l\right)^2}{2}D_{rr}-\frac{1}{2}D_{k_rk_r}\frac{\partial^2}{\partial r^2}-i\delta D\right]\tilde{\varphi}_{uh}(r)=-4\pi\rho_{uh}e^{-im\phi}e^{-ik_zz} \tag{4}$$

where $\delta D=\frac{\omega_{pe}{}^2}{2\omega_{ce}{}^2}\sqrt{\pi}\frac{m^2}{r_0^2}\exp\left(-\frac{\left(\omega-\omega_{ce}\right)^2}{k_z{}^2v_e{}^2}\right)$ stands for electron cyclotron damping. In the absence of the pump and damping we put $\delta k_z=0$ and for $\omega_n=\omega_l-\left(2n+1\right)\frac{\sqrt{D_{k_rk_r}D_{rr}}}{D_\omega}$ obtain the localized solution of equation (4) in the form

$$\tilde{\varphi}_{uh}^{(0)}\left(r,n\right)=\exp\left[-\left(r-r_l\right)^2\sqrt[4]{\frac{D_{rr}}{D_{k_rk_r}}}\right]H_n\left(\left(r-r_l\right)^2\sqrt[4]{\frac{D_{rr}}{D_{k_rk_r}}}\right) \tag{5}$$

where $H_n\left(\zeta\right)$ stands for Hermit polynomial. Assuming the nonlinear interaction and UH wave damping small and accounting for them using the perturbation method we obtain the LF wave amplitude at frequency $\Omega = \omega_0 - \omega$ in the form $\tilde{\varphi}_{lf}(r) = -4\pi \int G(r, r')\rho_{lf}(r')dr'$ where $G$ is the Green function corresponding to the $L$ operator in equation (2). Finally the perturbation of the UH wave's axial wavenumber associated with nonlinear interaction and damping is given in the perturbation theory by

$$\delta k_z^n = \frac{-4\pi \int \tilde{\varphi}_{uh}(r,n)\rho_{uh}(r)dr + i\int \tilde{\varphi}_{uh}(r,n)\delta D(r)dr}{D_{k_z}\int [\tilde{\varphi}_{uh}(r,n)]^2 dr} \qquad (6)$$

where $-4\pi\rho_{uh} = \delta\varepsilon_i(\Omega)\delta\varepsilon_e(\omega)\frac{E_i k_z}{\omega_i^2}\frac{e}{m_e}\Delta\left[\int G(r,r')\delta\varepsilon_i(\Omega)\delta\varepsilon_e(\omega)\frac{E_i k_z}{\omega_i^2}\frac{e}{m_e}\Delta\varphi_{uh}(r,n)dr'\right]$. The threshold of convective PDI in this case is determined by condition $\delta k_z Z > 1$. In the case of ion acoustic LF wave the operator $L$ takes a form $\hat{L} = \frac{\omega_{pi}^2}{c_s^2} - \frac{\omega_{pi}^2}{\Omega^2}\Delta$, where $c_s = \sqrt{T_e/m_i}$ allowing explicit calculation of the Green function and determination of the PDI threshold using (6). Theoretically predicted dependence of the threshold power on the UH wave axial wavenumber is shown in Fig.1 for experimental parameters ( $H$=500Oe, $n_e$=$8\times10^{10}cm^{-3}$, $T_e$=$1eV$) . This dependence possesses a clear minimum of 10W determined by competition of nonlinear drive and cyclotron damping both growing with increasing wavenumber.

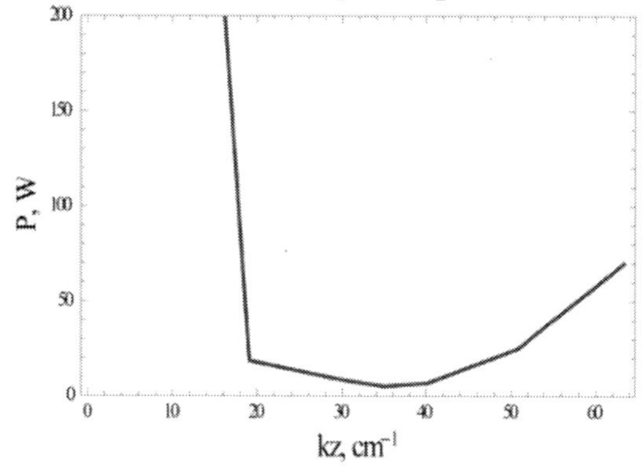


**Fig.1.** Dependence of the PDI threshold on axial wavenumber of UH wave.

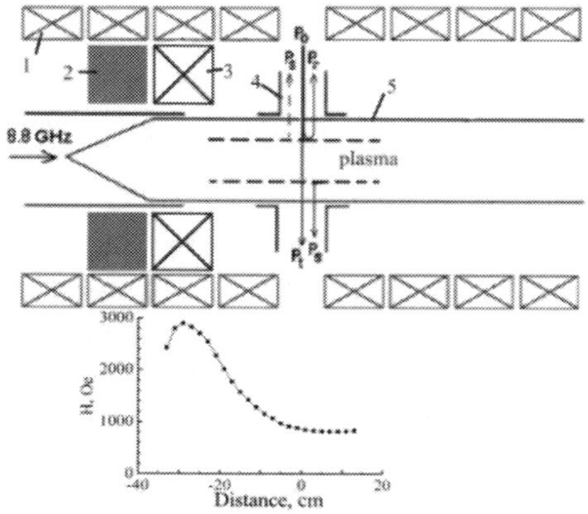


**Fig.2.** The experimental scheme and the magnetic field distribution. 1 – coils, 2 – iron insertion, 3 – additional coil, 4 – waveguide 72×34 mm$^2$, 5 – glass tube. $P_0$, $P_t$, $P_r$– incident, transmitted and reflected RF powers.

**The experimental observations.**

The experiments were carried out on the linear plasma device "Granit" [7] possessing two section with different magnetic field shown in Fig.2. The region at the magnetic system end with high field (~0.3 T) is used for background plasma generation by ECR discharge at frequency 8.8 GHz. In low magnetic field region in the center of magnetic system the plasma volume (2.5 cm in diameter) filled by argon at the pressure of 1 Pa is crossing the waveguide 72 x 34 mm$^2$ using which the microwave at frequency $f_0$ = 2.35 GHz and power up to 200W is launched. The microwave electric field is parallel to the external magnetic field. Magnetic field in the waveguide region is homogeneous and variable in the range 450-850Oe. The incident RF absorption in the waveguide was characterized using plasma light radiation intensity as well as transmitted and reflected RF power variation. The absorption dynamics for the microwave pulses (8 μs) at frequency $f_0$=2.35 GHz was investigated at power up to 200 W for magnetic field variation in the range of 450-850Oe. It is shown, that at small electron densities ($n_e \sim 10^9$ cm$^{-3}$) absorption takes place only in plasma in narrow range of magnetic field – 750-830 Oe corresponding to ECR (see Fig.3). The additional absorption range steadily grows at higher electron densities. At the volume averaged density $n_e \sim 5\times10^{10}$ cm$^{-3}$ the maximal absorption is observed at much lower magnetic field (~550Oe), as it is seen in Fig.4. This effect could be related both with the PDI excitation and the two-dimensional hybrid resonance phenomena. To exclude the latter possibility the radiation intensity dependences on the pump power at different magnetic fields was investigated. As it is seen in Fig.5, in the high magnetic field case the dependence is linear crossing the origin, which is typical for linear ECR absorption. On contrary, in the low-field case a threshold-like dependence is observed, crossing the power axis at $P$= 20 – 30W and indicating the possibility of anomalous absorption. The experimental threshold value is close to the theory prediction for localized UH wave parametric excitation.

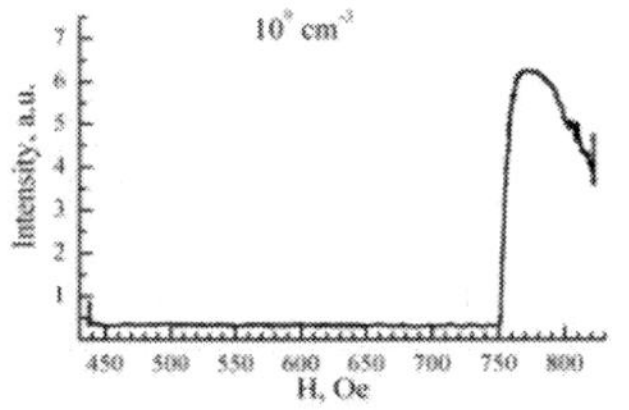


**Fig.3.** Dependence of additional radiation on magnetic field at low plasma density.

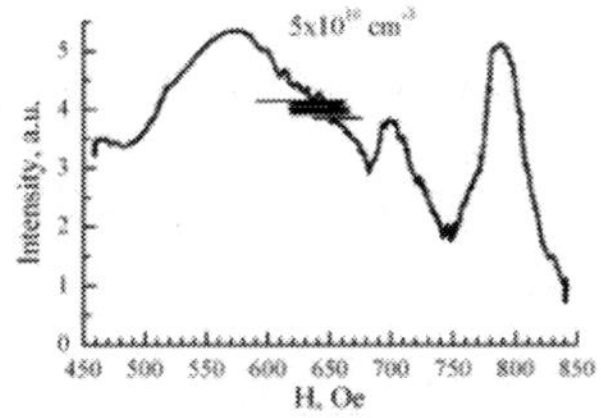


**Fig.4.** Dependence of additional radiation on magnetic field at high plasma density.

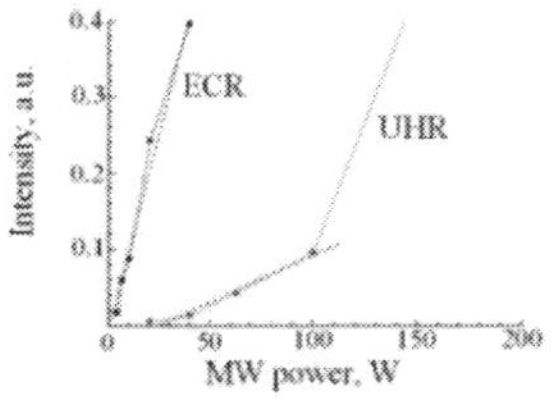


**Fig.5.** Dependence of additional radiation on the pump power for different magnetic fields.

**Acknowledgements**

The computations were performed at facilities of USD "Globus-M". Partial financial support of BRFBR-RFBR grants (F14R-047, 14-02-90003) and RF MES (Agreement 14.619.21.0001, id RFMEFI61914X0001) is acknowledged.

**References**

[1] M. Porkolab, B.I. Cohen, 1988 Nucl. Fusion 28, 239

[2] A.N.Karpushov et al. in Proc. 30th EPS Conf. on Plasma Physics (2003), 27A, P–3.123.

[3] Rapisarda D. et al 2007 Plasma Phys. Control. Fusion 49 309

[4] Westerhof E. et al 2009 Phys. Rev. Lett. 103 125001

[5] Gusakov E. et al 2010 Phys. Rev. Lett. 105 115003

[6] Gusakov E. et al 2014 Plasma Phys. Control. Fusion 56 015010

[7] Arkhipenko, V.I., Budnikov, V.N. et al. Plasma Phys. Reports, 2000, 26, 314.

# Spontaneous excitation of convective cells by kinetic Alfvén waves*

Fulvio Zonca[1,2], Yu Lin[3], and Liu Chen[2,4]

[1] *C.R. ENEA Frascati, Via E. Fermi 45 – C.P. 65, 00044 Frascati, Italy*

[2] *Institute for Fusion Theory and Simulation and Department of Physics, Zhejiang University, Hangzhou, 310027 People's Republic of China*

[3] *Physics Department, Auburn University, Auburn, Alabama 36849-5311, USA*

[4] *Department of Physics and Astronomy, University of California, Irvine CA 92697-4575, USA*

## Introduction

Zonal structures such as zonal flows are known to play crucial roles in dynamically regulating plasma transport in tokamak plasmas. The analogues in uniform plasma are the convective cells, which have been extensively studied in the 1970's [1, 2] in the context of cross-field transport [3]. In particular, it is worthwhile mentioning the extensive studies of convective cells excitation by kinetic Alfvén waves (KAW), with applications to flows in the upper ionosphere [4, 5].

It has been recently demonstrated that, generally, electrostatic (ESCC) and the magnetostatic (MSCC) convective cells are excited simultaneously by KAW [6]. Proper theoretical treatment of spontaneous excitation of ESCC and MSCC by KAW must take into account: (i) the finite ion Larmor radius (FILR) corrections to the Reynolds stress; (ii) the finite coupling between ESCC and MSCC. In this work, theoretical predictions are compared with numerical simulation results of a kinetic ion–fluid electron hybrid model [7]. Excellent agreement is obtained for mode structure and generation rate of convective cells (CCs) by KAW, demonstrating that ESCC and MSCC are indeed coupled, and that spontaneous CC excitation is suppressed at long wavelength, showing the crucial role of FILR effects.

As KAW are efficiently excited by mode conversion with the shear Alfvén continuous spectrum, it can be assumed that their perpendicular $\boldsymbol{k}$ spectrum (to the ambient $\boldsymbol{B}$ field) is anisotropic and predominantly directed across magnetic flux surfaces, with negligible impact on cross field transport. Present results imply that CC are not only efficiently excited by KAW via modulational instability at short wavelengths, but also that they isotropize the KAW spectrum as consequence of scattering of the initial KAW spectrum off the CCs. Thus, this effect has potentially great impact on cross field transport in both space as well as fusion plasmas [6, 7].

*Work supported by US DoE, NSF, ITER-CN, and NSFC grants.

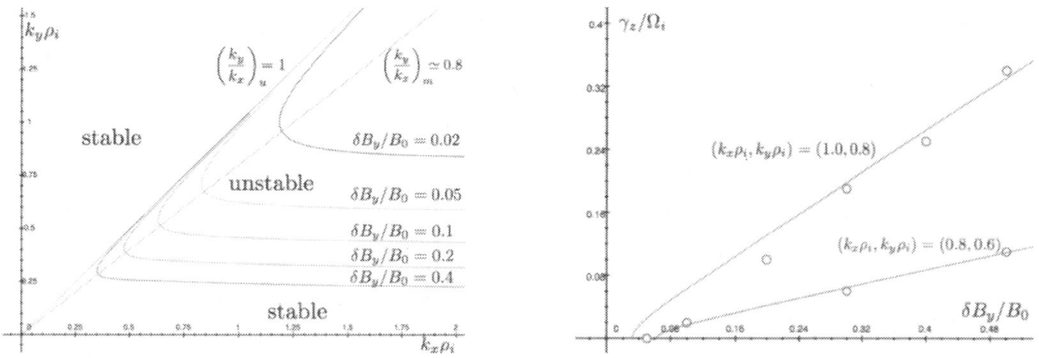


Figure 1: (left) Marginal stability curves in the $(k_x\rho_i, k_y\rho_i)$ plane for fixed $k_{\parallel 0}\rho_i = 0.02$, $\tau = 1$ and $\beta_e = \beta_i = 0.2$ and different values of $\delta B_y/B_0$. (right) Modulational instability growth rate from Eq. (3) (continuous line) vs. $\delta B_y/B_0$ is compared with hybrid simulation results (open circles) for $(k_x\rho_i, k_y\rho_i) = (0.8, 0.6)$ (blue) and $(k_x\rho_i, k_y\rho_i) = (1.0, 0.8)$ (red).

**Modulational instability dispersion relation**

Consider an infinite, uniform, low-$\beta$ plasma with $\boldsymbol{B}_0 = B_0\hat{\boldsymbol{z}}$. We adopt $\delta\phi$ and $\delta A_\parallel$ as the field variables, with $|k_\perp\rho_i|$ formally of $\mathscr{O}(1)$ and $\rho_i$ the ion Larmor radius. Consider also 4-wave interactions among the pump KAW $\Omega_0 = (\omega_0, \boldsymbol{k}_0)$, the convective cell (CC), $\Omega_z = (\omega_z, \boldsymbol{k}_z)$ with $\boldsymbol{k}_z \cdot \hat{\boldsymbol{z}} = 0$; and the KAW sidebands, $\Omega_\pm = (\omega_\pm, \boldsymbol{k}_\pm)$, where $\omega_+ = \omega_z + \omega_0$, $\omega_- = \omega_z - \omega_0^*$ and $\boldsymbol{k}_\pm = \boldsymbol{k}_z \pm \boldsymbol{k}_0$. The pump KAW satisfies the dispersion relation and polarization condition

$$\varepsilon_{A0} = (1-\Gamma_0)/b_0 - \sigma_0 k_{\parallel 0}^2 v_A^2/\omega_0^2 = 0 \quad \text{and} \quad \delta\psi_0 \equiv \omega_0 \delta A_{\parallel 0}/(k_{\parallel 0}c) = \sigma_0 \delta\phi_0 \ ; \qquad (1)$$

where $\Gamma_0 = I_0(b_0)e^{-b_0}$, $I_0$ is the modified Bessel function, $b_0 = k_{\perp 0}^2\rho_i^2$, $\rho_i = \Omega_i^{-1}(T_i/m_i)^{1/2}$, $\sigma_0 = 1 + \tau(1-\Gamma_0)$ with $\tau = T_e/T_i$, and $v_A$ is the Alfvén speed in the considered uniform equilibrium. Due to the interaction of the KAW pump $\Omega_0$ with the CC $\Omega_z$, up and down-shifted (in frequency) KAW sidebands are generated. It is possible to demonstrate that maximum nonlinear coupling occurs for $\boldsymbol{k}_z \perp \boldsymbol{k}_0$ [6], so that $b_\pm = (k_{\perp\pm}^2/k_{\perp 0}^2)b_0 = b_z + b_0$, $\Gamma_+ = \Gamma_-$, $\varepsilon_{A+} = \varepsilon_{A-}$, and

$$b_\pm\varepsilon_{A\pm} = 2\frac{(1-\Gamma_\pm)/\omega_0}{(1\pm\omega_z/\omega_0)^2}(\pm\omega_z - \Delta_\pm + \omega_z^2/2\omega_0) \ , \ \frac{\Delta_\pm}{\omega_0} = \frac{b_\pm\sigma_\pm(1-\Gamma_0) - b_0\sigma_0(1-\Gamma_\pm)}{2b_0\sigma_0(1-\Gamma_\pm)}; \qquad (2)$$

with $\Delta_\pm = \Delta$ the frequency mismatch of KAW sidebands from the resonance condition $\varepsilon_{A\pm} = 0$. Here, for simplicity, we consider $\boldsymbol{k}_{\perp 0} = \hat{\boldsymbol{x}}k_{\perp 0}$ and, thus, $\boldsymbol{k}_z = \hat{\boldsymbol{y}}k_z$. Then, field equations for KAW sidebands can be solved for $\delta\phi_\pm$ and $\delta\psi_\pm \equiv (\omega_0 \pm \omega_z)\delta A_{\parallel\pm}/(k_{\parallel 0}c)$ as a function of $\delta\phi_z$, $\delta\psi_z \equiv \omega_0\delta A_{\parallel z}/(k_{\parallel 0}c)$, and the KAW pump amplitude. These expressions are then substituted into the evolution equations for $\delta\phi_z$ and $\delta\psi_z$; which, letting $\omega_z = i\gamma_z$, can be cast as [6]

$$\begin{aligned}(\gamma_z^2 + \Delta^2)\,\delta\phi_z &= -\alpha_\phi(\delta\phi_z - \delta\psi_z) + \beta_\phi\delta\psi_z \ , \\ (\gamma_z^2 + \Delta^2)\,\delta\psi_z &= -\alpha_\psi(\delta\phi_z - \delta\psi_z) + \beta_\psi\delta\psi_z \ ,\end{aligned} \qquad (3)$$

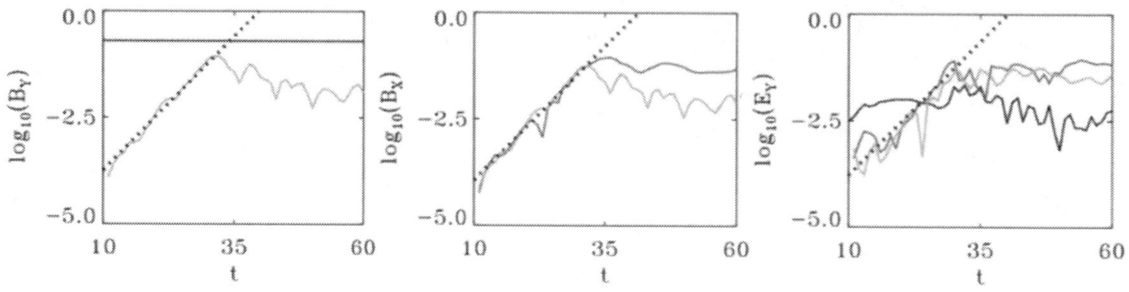


Figure 2: Time evolution of $B_y$, $B_x$, and $E_y$ in the pump KAW (black), daughter convective cell (red), and the matching KAW (green) modes.

which yield CC mode structure and modulational instability dispersion relation. Here, $\alpha_\phi$, $\beta_\phi$, $\alpha_\psi$ and $\beta_\psi$ are known functions of $|\delta B_y/B_0| = |(\sigma_0 c/\omega_0 B_0)k_{\perp 0}k_{\parallel 0}\delta\phi_0|$; and $\omega_0$, $k_{\perp 0}$ and $k_z$ [6]. In the long wavelength limit, $b_0 \ll 1$, we generally have $\delta\psi_z/\delta\phi_z \sim \mathcal{O}(1)$. However, Eq. (3) reduces to $\gamma_z^2 + \Delta^2 \simeq (\alpha_\psi - \alpha_\phi) < 0$. Thus, KAW can not spontaneously excite CC in the $b_0 \ll 1$ limit. Meanwhile, assuming that $\delta\phi_z$ (ESCC) is suppressed, delivers the erroneous claim that MSCC can be spontaneously excited by KAW for $b_0 \ll 1$.

In the short wavelength limit, $b_0 \gg 1$, we have again that $\delta\psi_z/\delta\phi_z \sim \mathcal{O}(1)$ and Eq. (3) reduces to $\gamma_z^2 + \Delta^2 \simeq \alpha_\phi[\tau - (1+\tau)(b_z/2b_0)]$, with $\alpha_\phi \propto b_z^2$. Thus, $b_z < 2b_0\tau/(1+\tau)$ is requested for modulational instability, and maximum growth rate is obtained for $(b_z/b_0)_m \simeq (4/3)\tau/(1+\tau)$. Meanwhile, $\alpha_\phi\tau > \Delta^2$ is also needed for modulational instability, which poses a lower bound on $b_z$ for effective CC excitation.

**Numerical solution of the modulational instability dispersion relation**

From Eq. (3), it is generally expected that $\delta\psi_z/\delta\phi_z \sim \mathcal{O}(1)$; i.e., ESCC and MSCC should be excited by KAW simultaneously for $b_0 \sim \mathcal{O}(1)$. This is shown by numerical solution of Eq. (3) in Fig. 1(left), reporting the marginal stability curves in the $(k_x\rho_i, k_y\rho_i) = (k_{\perp 0}\rho_i, k_z\rho_i)$ plane for different values of the KAW pump, $\delta B_y/B_0 = 0.02, 0.05, 0.1, 0.2, 0.4$. Here, fixed parameters are $k_{\parallel 0}\rho_i = 0.02$, $\tau = 1$ and $\beta_e = \beta_i = 0.2$. The amplitude dependent lower threshold in $k_y\rho_i = k_z\rho_i$, and the upper threshold at $k_y = k_x$ for $\tau = 1$ predicted theoretically are clearly visible. The maximum CC growth rate condition $(k_y/k_x)_m \simeq (2/3)^{1/2} \simeq 0.8$ for $\tau = 1$ is also well verified. For the same fixed parameters, Fig. 1(right), meanwhile, shows the modulational instability growth rate vs. the KAW pump amplitude $\delta B_y/B_0$ for $(k_x\rho_i, k_y\rho_i) = (0.8, 0.6)$ and $(k_x\rho_i, k_y\rho_i) = (1.0, 0.8)$, compared with numerical results from hybrid simulations.

## Hybrid Simulation

The hybrid simulation scheme is similar to that of *Lin, Johnson, and Wang* [7, 8], in which ions are treated as fully kinetic particles moving in a self-consistent electromagnetic field, and electrons are treated as a massless fluid. The computation scheme is fully nonlinear, but our analysis of the modulational instability focuses only on the early stage of its exponential growth. Fixed parameters are those of Fig. 1, lengths are normalized to $\rho_i$, and the time to $\Omega_i^{-1}$.

The pump mode is imposed everywhere as a steady driver at each time step, with magnetic field $\delta\mathbf{B}_0 = (0, \delta B_y, 0)\sin(\mathbf{k}_0 \cdot \mathbf{x} - \omega_0 t)$ and specified wave number and frequency, which, in the hybrid simulation, are consistent with Eq. (1). From $t = 0$-$10$, the system is filtered to keep only the Fourier mode $\mathbf{k_0}$ of initial pump, which allows a self-consistently development of the pump mode structure. For $t > 10$, more Fourier modes are released in order to examine the excitation of the convective cell modes. Figure 2 shows the time evolution of $B_y$, $B_x$, and $E_y$ in the resulting three wave interaction for a case with constant pump amplitude $\delta B_y = 0.4$ at $\mathbf{k}_0 = (1.25, 0, 0.02)$. The black solid curves show the pump mode, for which $B_y$ is constant in time. The $E_y$ component of the pump is two orders of magnitude smaller than $E_x$ (not shown). The excitation of the CC mode, with wave number $\mathbf{k}_z = (0, 0.8, 0)$, is shown with the red curve. Both $B_x$ and $E_y$ increase nearly exponentially with time from $t = 10$, as fitted by the straight dotted line, reaching the saturation levels at $t \simeq 31$. The growth rate is measured to be $\gamma/\Omega_i = 0.29$. No power is excited in the $B_x$ component, consistent with $k_x = 0$ in the CC mode. The green curve depicts the matching KAW mode with $\mathbf{k}_+ = (1.25, 0.8, 0.02)$, in which $B_x$, $B_y$, and $E_y$ are all seen to also grow exponentially with $\gamma/\Omega_i = 0.29$.

## References

[1] J. B. Taylor and B. McNamara, Phys. Fluids **14**, 1492 (1971).

[2] H. Okuda and J. M. Dawson, Phys. Fluids **16**, 408 (1973).

[3] P. K. Shukla, M. Y. Yu, H. U. Rahman and K. H. Spatschek, Phys. Rep. **105**, 227 (1984).

[4] R. Z. Sagdeev, V. D. Shapiro and V. I. Shevchenko, Zh. Eksp. Teor. Fiz. Pis'ma Red. **27**, 361 (1978).

[5] R. Z. Sagdeev and V. D. Shapiro and V. I. Shevchenko, Sov. Phys. JETP **27**, 340 (1978).

[6] F. Zonca and L. Chen, "Spontaneous excitation of convective cells by kinetic Alfvén waves". Presented at the 2014 International Sherwood Fusion Theory Conference, San Diego, California, March 24 - 26 (2014).

[7] Y. Lin, J. R. Johnson and X. Wang, Phys. Rev. Lett. **109**, 125003 (2012).

[8] Y. Lin, J. R. Johnson, and X. Y. Wang, J. Geophys. Res. **115**, A04208 (2010).

## AUTHOR INDEX

# AUTHOR INDEX

# AUTHOR INDEX

# AUTHOR INDEX

# AUTHOR INDEX

# AUTHOR INDEX

# AUTHOR INDEX

# AUTHOR INDEX

# AUTHOR INDEX

# AUTHOR INDEX

# AUTHOR INDEX

# AUTHOR INDEX

# AUTHOR INDEX

# AUTHOR INDEX

# AUTHOR INDEX

# AUTHOR INDEX

# AUTHOR INDEX